Study Guide and Solutions Manual

to accompany

ORGANIC CHEMISTRY

FOURTH EDITION

G. Marc Loudon

Purdue University

Joseph G. Stowell

Purdue University

New York Oxford

OXFORD UNIVERSITY PRESS

2003

Oxford University Press

Oxford New York
Auckland Bangkok Buenos Aires Cape Town Chennai
Dar es Salaam Delhi Hong Kong Istanbul Karachi Kolkata
Kuala Lumpur Madrid Melbourne Mexico City Mumbai Nairobi
São Paulo Shanghai Singapore Taipei Tokyo Toronto

Published by Oxford University Press, Inc.
198 Madison Avenue, New York, New York, 10016
http://www.oup-usa.org

ISBN-13 978-0-19-512000-4
ISBN 0-19-512000-0

Printing number: 9 8 7 6 5 4

Printed in the United States of America
on acid-free paper

Preface

The purpose of this preface is to give you an overview of the organization and features of this Study Guide and Solutions Manual so that you might use it more effectively.

Organization

Each chapter corresponds to the chapter of the text with the same number and title. The following sections are found within each chapter, each headed by its own icon.

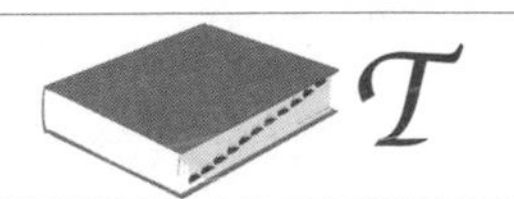

TERMS

This is a list of new or important terms encountered within the corresponding chapter of the text along with the location of each term. The lists of terms in this manual comprise the vocabulary of organic chemistry that you will need to master the subject. These lists will probably be most useful if you can *write* a definition of each term *in your own words* and give an example, if appropriate. Don't forget that if you need a more general index of terms, the text itself has a very detailed index that can be used to locate the definition of any term of interest.

CONCEPTS

This is a summary of the concepts within the chapter in outline form. In some cases, these are presented from a somewhat different perspective, and are organized somewhat differently, than they are in the text. These outlines can be used in various ways. For example, they can provide a quick review after you have read the chapter; they can be used for ready reference while you are working problems; or they can serve as a refresher prior to class periods. Some of the authors' students use them as their first encounter with the material and refer to the text for additional clarification when necessary.

REACTIONS

This summary consolidates the reactions of the chapter in one place and presents not only the reactions themselves but also the essential features of their mechanisms and stereochemistry for cases in which these issues are discussed in the text. Studying and learning reactions is a skill that should be cultivated, and Study Guide Link 5.4 on page 122 of this manual will help you to use these reaction summaries to best advantage.

SCIENTISTS

This is a list of the scientists whose work is cited by name in the text, along with their birth and death dates. Usually some anecdotal information is provided about them in the text.

STUDY GUIDE LINKS

These are short extensions of text material that are called out by margin icons within the text. Study Guide Links are of two types. A Study Guide Link of the first type, flagged with a checked (✓) icon, provides additional explanations of topics that typically cause difficulty. For example, Study Guide Link 5.4, "How to Study Organic Reactions," (mentioned in item 3 above) is flagged with a check at a point in the text (p. 172) at which some students begin to have difficulty learning reactions. This Study Guide Link provides detailed hints on how to study and learn organic reactions efficiently. A Study Guide Link of the second type, called out in the text by an unchecked icon, provides more in-depth information about the particular topic or a different way of looking at the same topic. An example is Study Guide Link 5.3, "Mechanism of Organoborane Oxidation," which is called out on text p. 171. Although it may at first seem that the material in the Study Guide Links should have been included in the text, doing so would have made the text unacceptably long, and, in some cases, would have interrupted the logical flow of material.

SOLUTIONS TO PROBLEMS

Solutions to the problems within the text come first, and these are followed by solutions to the additional problems. We have endeavored not only to provide answers to problems but also, in many cases, to show how these answers are deduced. In the previous edition, we provided answers to only selected problems. As the result of numerous student requests, in the fourth edition we have provided solutions for all of the problems.

The best use of solutions is to try to work problems with the solutions in this manual closed or covered, and consult a solution only when you have answered the problem or have given it a reasonable effort. In many cases, you'll find that many of the problems are paired; that is, there are two problems of a specific type. In such cases, it will sometimes prove effective to study the solution to the first problem of the pair, and then use it as a guide to work the second problem of the pair.

Features

This manual contains certain features that are designed to help you use it more effectively.

1. **Typesetting Conventions**

   The text and typesetting conventions are as close as possible to those used in the text.

2. **Index Tabs**

   You will find on the first right-hand page of each Solutions section a black rectangle at the edge of the page. (For example, the first of these appears on page 13.) The edges of these rectangles can be seen when the manual is viewed end-on and the pages are bent back slightly. These serve as index tabs that can be used to locate the solutions sections rapidly. They are arranged in three rows of nine. Thus, to locate the solutions for Chapter 12, bend back the pages and open to the third index mark in the second row.

3. Icon Comments

Within the solutions to the problems, you will occasionally find comments marked with icons of two different types.

This icon indicates a comment that provides additional information about the topic covered in the problem, an alternate correct answer, or another way of looking at the problem.

This icon indicates a special caution about some aspect of the problem or the material on which the problem is based—something you should be careful about to avoid confusion.

Acknowledgments

This manual has undergone three separate accuracy checks by the authors and a critical reading by two accuracy checkers. We have tried hard to make this manual as free of errors as possible. Nevertheless, if errors are found, we would like to know about them so that we can correct them on reprint if possible. You can send these or any comments to us by electronic mail at *marc.loudon.1@purdue.edu,* by FAX at 765-494-7880, or by mail at RHPH Building, Purdue University, West Lafayette IN 47907-1330. A list of text errors is posted on the Oxford University Press web site (*http://www.loudon4e.org*), and we anticipate that a list of errors for this Guide, if necessary, will be posted there as well.

We would like to gratefully acknowledge the many comments of the users of the previous edition of this manual, especially Professor Ron Magid, Professor Mark Cushman, and the many students who sent me comments and corrections. I am particularly indebted to Harvard student Keith Santiago, who worked every problem in the third edition and caught many errors, and to the many students at the University of Tennessee who also sent me corrections. We particularly appreciate the conscientious work of the accuracy checkers, graduate students Brian Fox and Michelle Martin.

We sincerely hope that you find this manual useful in your study of organic chemistry.

G. Marc Loudon and Joseph G. Stowell

About the Chapter-Opening Art

The banner art that opens each chapter (as well as the Preface and Table of Contents) symbolizes the purpose of this book: to help you "crystallize" the concepts of organic chemistry. This is a photomicrograph of crystals of pentamidine isethionate, a compound used in the treatment of pneumonia, taken under polarized light.

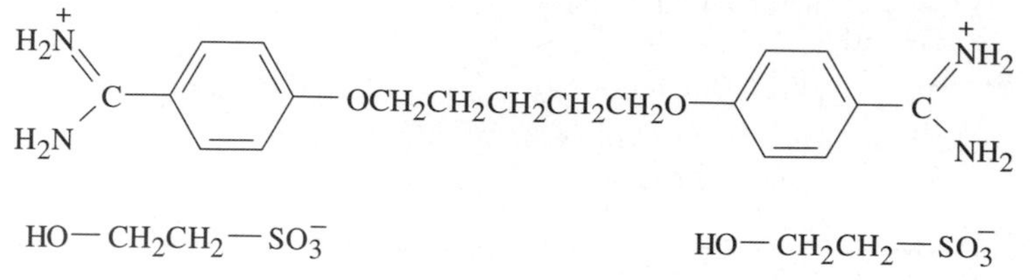

$H_2\overset{+}{N}$ $\qquad$ $\overset{+}{N}H_2$

C — $OCH_2CH_2CH_2CH_2CH_2O$ — C

H_2N $\qquad$ NH_2

$HO{-}CH_2CH_2{-}SO_3^-$ $\qquad$ $HO{-}CH_2CH_2{-}SO_3^-$

pentamidine isethionate

The photograph shows three crystal forms: one predominates on the right of the figure, a second predominates on the left of the figure, and the "stars" are crystallites of the hydrated form, that is, a form that incorporates water into the crystal.

This photograph was taken by Ulrich J. Griesser and is used by permission of SSCI, Inc., of West Lafayette, Indiana, a company that provides comprehensive research and analytical services to the pharmaceutical, specialty, industrial, and fine chemical industries in the characterization of the chemistry of solid materials.

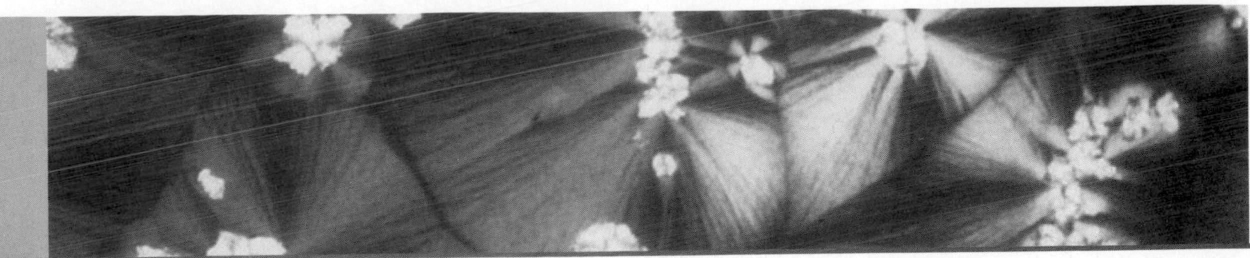

Table of Contents

1

Chemical Bonding and Chemical Structure

TERMS

Term	Section
allowed state	1.6A
angular momentum quantum number ℓ	1.6A
antibonding molecular orbital (MO)	1.8A
atomic orbital	1.6A
Aufbau principle	1.7
bond angle	1.3B
bond dipole	1.2D
bond length	1.3B
bond order	1.3B
bonding molecular orbital (MO)	1.8A
chemical bond	1.2
covalent bond	1.2C
cylindrical symmetry	1.8A
debye D	1.2D
dihedral angle	1.3B
dipole moment μ	1.2D
double bond	1.2C
electron diffraction	1.3A
electronegativity	1.2D
electronic configuration	1.7
electrostatic attraction	1.2B
formal charge	1.2C
Heisenberg uncertainty principle	1.5
Hund's rules	1.7
hybrid orbital	1.9A
ionic bond	1.2B
ionic compound	1.2B
isoelectronic	1.2B
Lewis structure	1.2C
linear molecule	1.3B
magnetic quantum number *m*	1.6A
microwave spectroscopy	1.3A
molecular connectivity	1.3
molecular geometry	1.3
molecular models	1.3B
molecular orbital	1.8A
node	1.6B
octet rule	1.2A
organic chemisty	1.1
p orbital	1.6A
Pauli exclusion principle	1.7
polar bond	1.2D
polar molecule	1.2D
principal quantum number *n*	1.6A
quantum mechanics	1.5
quantum number	1.6A
regular tetrahedron	1.3B
resonance hybrid	1.4
resonance structure	1.4
s orbital	1.6A
sigma bond (σ bond)	1.8A
sp^3 orbital	1.9A
spin quantum number *s*	1.6A
structure	1.3
tetrahedron	1.3B
torsion angle	1.3B
trigonal planar geometry	1.3B
trigonal pyramidal geometry	1.3B
triple bond	1.2C
unshared electron pair	1.2C
valence electrons	1.2A
valence orbital	1.7
valence shell	1.2A
wave function	1.6A
wave peak	1.6B
wave trough	1.6B
wave-particle duality	1.5
X-ray crystallography	1.3A

The glossary in this and subsequent chapters is a list of the key terms and concepts contained in the chapter. These terms and concepts will be used throughout the text. It probably will not help you to memorize the exact definitions given in the text; rather, define each of these terms and concepts in your own words and give an example if appropriate.

CONCEPTS

I. ATOMS

A. Basics

1. A neutral atom of each element contains a number of both protons and electrons equal to its atomic number.
2. The ease with which neutral atoms lose electrons to form positive ions increases to the left and towards the bottom of the periodic table.
3. The ease with which neutral atoms gain electrons to form negative ions increases to the right and towards the top of the periodic table.
4. Two different species that have the same number of electrons are said to be isoelectronic.
5. Chemistry happens because of the behavior of electrons in atoms and molecules.
   a. Organic chemistry is the branch of science that deals generally with compounds of carbon. Initially, the term organic implied a connection with living things.
   b. In the modern concept, organic chemistry deals with compounds of carbon regardless of their origin.
   c. As a science, organic chemistry lies at the interface of the physical and biological sciences.

B. Valence Electrons

1. The electrons in the outermost shell (the valence shell) of an atom are called valence electrons.
2. The number of valence electrons for any neutral atom in an A group of the periodic table (except He) equals its group number.
3. Stable ions are formed when atoms gain or lose valence electrons in order to have the same number of electrons as the noble gas of closest atomic number.

II. CHEMICAL BONDS

A. Ionic Bonds

1. The ionic bond—a force that hold atoms together within molecules—is an electrostatic attraction between ions of opposite charge.
2. The ionic bond is most likely to form between atoms at opposite ends of the periodic table.
3. The formation of ions tends to follow the octet rule—each atom is surrounded by eight valence electrons (two electrons for hydrogen).
4. The ionic bond is the same in all directions; that is, it has no preferred orientation in space.
5. When an ionic compound dissolves in water:
   a. ionic bonds are broken;
   b. the ionic compound dissociates into free ions.

$$Na^+Cl^- \xrightarrow{H_2O} Na^+ + Cl^-$$

B. Covalent Bonds

1. Covalent bonds are formed when the orbitals of different atoms overlap and share electrons.
2. Both electrons in a covalent bond are shared between the bonding atoms.
3. Covalent bonding can be understood to arise from the filling of bonding molecular orbitals by electron pairs.
4. The bonding in covalent compounds tends to follow the octet rule—each atom is surrounded by eight valence electrons (two electrons for hydrogen).
5. The covalent bond has a definite direction in space.
6. All organic molecules contain covalent bonds.

C. Polar Covalent Bonds

1. In polar covalent compounds, electrons are shared unequally between bonded atoms, and a bond dipole results.
   a. The tendency of an atom to attract electrons to itself in a covalent bond is indicated by its electronegativity.
   b. A polar bond is a bond between atoms of significantly different electronegativities.
2. Partial charge in a polar covalent bond is indicated by a delta (δ), which is read "partially" or "somewhat."

$$\overset{\delta+}{\mathrm{H}}-\overset{\delta-}{\mathrm{Cl}}$$

3. The uneven electron distribution in a compound containing covalent bonds is measured by the dipole moment μ in units of debyes D:

$$\mu = q\mathbf{r}$$

   where q is the magnitude of the separated charge and $\mathbf{r}$ is a vector from the site of positive charge to the site of negative charge.
   a. Dipole moments of typical polar organic molecules are in the 1–3 D range.
   b. Each polar bond has associated with it a dipole moment contribution, called a bond dipole.
4. A bond dipole is a vector quantity and μ and $\mathbf{r}$ have the same direction; the dipole moment is oriented from the partial positive charge to the partial negative charge.
   a. The vector sum of all bond dipoles in a molecule is its dipole moment.
   b. Two vectors of equal magnitude oriented in opposite directions always cancel.

$$\underset{\delta-}{\mathrm{O}}=\underset{\delta+}{\mathrm{C}}=\underset{\delta-}{\mathrm{O}}$$

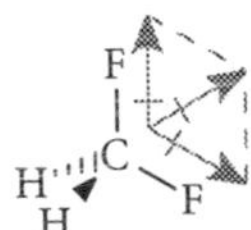

5. Molecules that have permanent dipole moments are called polar molecules.
6. The polarity of a molecule can significantly affect its chemical and physical properties.

D. Octet Rule

1. An atomic species tends to be especially stable when its valence shell contains eight electrons.
2. The tendency of atoms to gain or lose valence electrons to form ions with the noble-gas configuration is known as the octet rule.
   a. The octet rule is often obeyed in covalent bonding.
   b. The sum of all bonding electrons and unshared pairs for atoms in the second period of the periodic table will under no circumstances be greater than eight (octet rule). These atoms may, however, have fewer than eight electrons.
   c. In some cases, atoms below the second period of the periodic table may have more than eight electrons. However, these cases occur so infrequently that rule b should be followed until exceptions are discussed.
3. To determine if an atom has a complete octet, count all unshared valence electrons and all bonding electrons.

E. Lewis Structures

1. Lewis structures are molecular structures that use lines and/or dot symbols between atoms to denote bonds and dot symbols on atoms to denote unshared electrons.
   a. Nonvalence electrons are not shown in Lewis structures.
   b. A hydrogen can share no more than two electrons.
2. A bond consisting of one electron pair is called a single bond (bond order = 1).
3. A bond consisting of two electron pairs is called a double bond (bond order = 2).

4. A bond consisting of three electron pairs is called a triple bond (bond order = 3).

F. Formal Charge

1. Formal charge is an electronic bookkeeping system that assigns a charge to specific atoms.
2. The sum of the formal charges on the individual atoms must equal the total charge of the ion.
   a. Assign a valence electron count to the atom by adding all unshared valence electrons on the atom and one electron from every covalent bond to the atom.
   b. Subtract this electron count from the group number of the atom in the periodic table (equal to the number of valence electrons in the neutral species).
   c. The resulting difference is the formal charge.

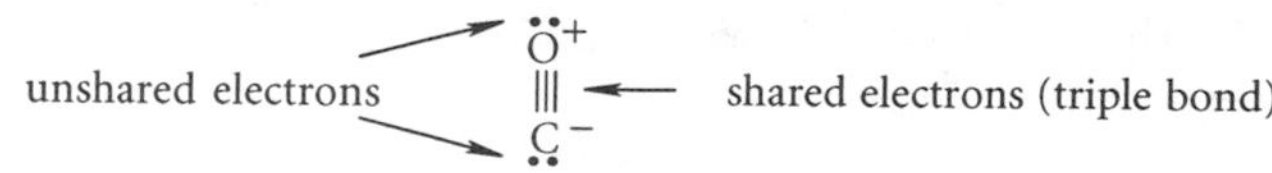

III. MOLECULAR STRUCTURE

A. General

1. The structure of a molecule is determined by its molecular connectivity (how atoms are connected) and molecular geometry (how they are situated in space).
2. Sources of molecular structure are
   a. X-Ray crystallography
   b. electron diffraction
   c. microwave spectroscopy
3. How a molecule reacts chemically is closely linked to its structure.

B. Bond Lengths

1. The bond length is the distance between the centers of the bonded nuclei (in Å; Å = 10^{-10} m).
2. Bond lengths are governed, in order of importance, by:
   a. The row of the periodic table from which the bonded atoms are derived: longer bonds are associated with higher periods (lower rows).
   b. Bond order (the number of covalent bonds shared by two atoms): bonds of higher order are shorter.
   c. The column of the periodic table from which the bonded atoms are derived: longer bonds are associated with higher atomic number.

C. Bond Angles

1. The bond angle is the angle between each pair of bonds to the same atom.
2. The bond angles within a molecule determine its shape.
3. Approximate bond angles can be predicted by assuming that the groups bound to a central atom are as far apart as possible.
   a. An atom bonded to four groups has tetrahedral geometry (bond angle = 109.5°).
   b. An atom bonded to three groups has trigonal planar geometry (bond angle = 120°).
   c. An atom bonded to two groups has linear geometry (bond angle = 180°).
   d. An unshared electron pair can be considered as a bond without a nucleus at one end.

D. Molecular Geometry

1. The geometry of a molecule is the spatial relationship of the bonds on adjacent atoms and is important for molecules that contain many atoms.
2. The angular relationship of bonds on adjacent atoms when projected onto the same plane is called the dihedral (torsional) angle.
3. The geometries of a molecule are described by specifying the bond lengths, bond angles, and dihedral angles between bonds on adjacent atoms.

E. Resonance Structures

1. A molecule that is not adequately described by a single Lewis structure is represented as a resonance hybrid.
   a. A resonance hybrid is a weighted average of two or more fictitious Lewis structures.
   b. When resonance structures are identical, they are equally important descriptions of the molecule.
   c. Resonance between two Lewis structures is shown by a double-headed arrow.
   d. Resonance structures are not in equilibrium; a molecule has a single structure that is a hybrid of its fictitious Lewis structures.

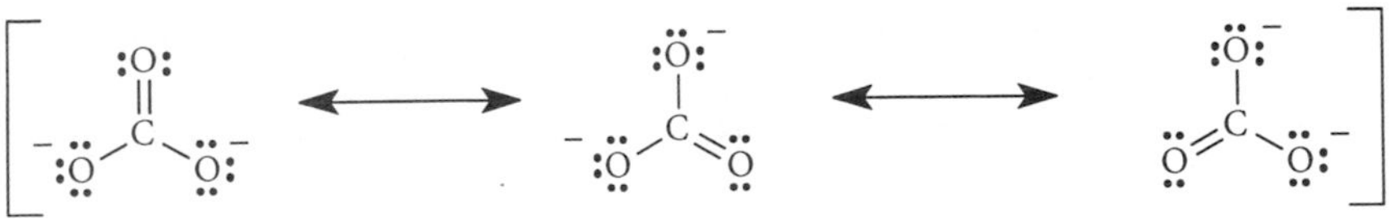

2. A molecule represented by resonance structures is more stable than its fictional resonance contributors and is said to be resonance-stabilized.

IV. NATURE OF THE ELECTRON

A. Wavelike Properties

1. In the submicroscopic world of the electron and other small particles, there is no real distinction between particles and waves.
   a. The behavior of very small objects simultaneously as waves and particles is called the wave-particle duality.
   b. Orbitals are descriptions of the wave properties of electrons in atoms and molecules, including their spatial distribution.
   c. As a consequence of their wave properties, electrons in atoms and molecules can exist only in certain allowed energy states (orbitals).
   d. Waves have peaks and troughs, which are regions where the waves are, respectively, at their maximum and minimum heights.
   e. A node is a point, or in a three-dimensional wave, a surface, at which the wave is zero.
   f. An atomic orbital in physics is described by a mathematical function called a wavefunction.
      i. The sign of the wavefunction for an electron is not the same as the charge on the electron.
      ii. A wave peak is a region in which the wavefuntion has a positive (+) sign.
      iii. A wave trough is a region in which the wavefuntion has a negative (–) sign.
      iv. The wavefunction is zero at a node.
2. The theory that describes the behavior of electrons in molecules is called quantum mechanics.
   a. Heisenberg uncertainty principle: the exact position and velocity of an electron cannot be specified; only the probability that it occupies a certain region of space can be specified.
   b. Pauli exclusion principle: no two electrons may have all four quantum numbers the same. A maximum of two electrons may be placed in any one atomic orbital, and these electrons must have opposite spins.
   c. Hund's rules: electrons are distributed among identical orbitals of equal energy, single electrons are placed into separate orbitals before the orbitals are filled, and the spins of these unpaired electrons are the same.
   d. Aufbau principle: to determine the electronic configuration of an atom, electrons are placed one by one into orbitals of the lowest possible energy in a manner consistent with the Pauli exclusion principle and Hund's rules.
3. Electron density within an orbital is a matter of probability (Heisenberg uncertainty principle); think of an orbital as a "smear" of electron density.

B. Quantum Numbers

1. Electrons in orbitals are characterized by quantum numbers which, for atoms, are designated n, ℓ, m, and s.

a. The principal quantum number n can assume any integral value greater than zero ($n = 1, 2, 3, \ldots$).
b. The angular momentum quantum number ℓ depends on the value of n and can assume any integral value from zero through $n - 1$ ($\ell = 0, 1, 2, \ldots, n - 1$) where $\ell = 0 \Rightarrow s$; $\ell = 1 \Rightarrow p$, $\ell = 2 \Rightarrow d$, and $\ell = 3 \Rightarrow f$.
c. The magnetic quantum number m depends on the value of ℓ and can assume integral values between $-l$ and $+l$.
d. The spin quantum number s can assume a value of $+\frac{1}{2}$ or $-\frac{1}{2}$.

2. Each orbital is described by a series of three quantum numbers:
   a. The principal quantum number n governs the energy and size of an orbital; orbitals of higher n have higher energy.
   b. The angular momentum quantum number ℓ governs the shape of an orbital: $\ell = 0$ (s orbitals) are spheres; orbitals with $\ell = 1$ (p orbitals) are dumbbells.
   c. The magnetic quantum number m governs the orientation (directionality) of an orbital.

C. Energies

1. The energy of an electron is quantized, that is, limited to certain values.
   a. The higher the principal quantum number n of an electron, the higher the energy.
      i. In atoms other than hydrogen, the energy is also a function of the ℓ quantum number.
      ii. Electrons with the same principal quantum number n but different values of ℓ have different energies.
   b. Gaps between energy levels become progressively smaller as the principal quantum number increases.
   c. The energy gap between orbitals that differ in principal quantum number is greater than the gap between two orbitals within the same principal quantum number.

V. ORBITALS

A. Atomic Orbitals

1. Some orbitals contain nodes which separate the wave-peak parts of the orbitals from the wave-trough parts.
   a. An orbital with principal quantum number n has $n - 1$ nodes.
   b. The number of nodes in an atomic orbital increases with its principal quantum number.
   c. The greater number of nodes in orbitals with higher n is a reflection of their higher energies.
2. Each atomic orbital is characterized by a three-dimensional region of space in which the electron is most likely to exist.
   a. The size of an atomic orbital is governed by its principal quantum number n; that is, the larger is n, the larger the region of space occupied by the orbital.
   b. The angular momentum quantum number ℓ governs the shape of an atomic orbital.
   c. The magnetic quantum number m governs the directionality of an atomic orbital.
3. Atomic orbitals are populated with electrons according to the Aufbau principle.
4. The distribution of electron density in a given type of orbital has a characteristic arrangement in space:
   a. An electron in an s atomic orbital is a smear of electron density in the shape of a sphere.
   b. An electron in a p orbital has a smear of electron density in the shape of a dumbbell consisting of two lobes that are part of the same orbital and are directed in space.
   c. The three p orbitals are mutually perpendicular.

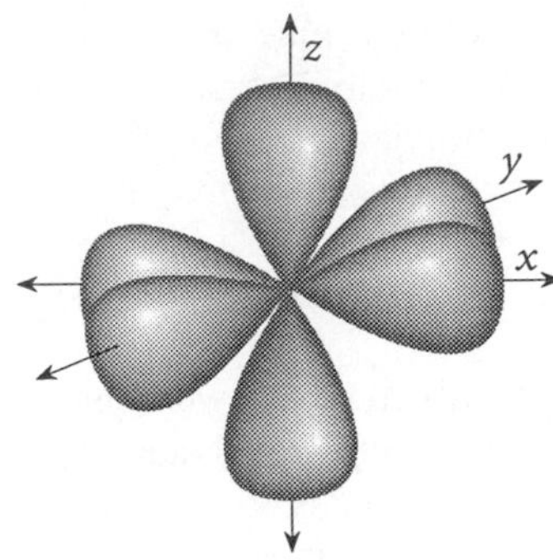

5. The electronic configuration of an atom or ion describes the distribution of electrons among the atomic orbitals.

chlorine atom $1s^2 2s^2 2p^6 3s^2 3p^5$

6. Within the A group elements, the valence electrons of an atom are the electrons that occupy the orbitals with the highest principal quantum number.
   a. The valence orbitals of an atom are the orbitals that contain the valence electrons.
   b. Chemical interactions between atoms involves their valence electrons and valence orbitals.

B. Hybrid Orbitals

1. Hybrid orbitals arise from the mixing of atomic orbitals on the same atom with different ℓ quantum numbers.
2. The hybridization of an atom and the geometry of the attached atoms are closely related.
   a. Hybridization affords bonds that are as far apart as possible.
   b. Hybridization of an atom provides orbitals that have the bulk of their electron density directed toward the nuclei of the bonded atoms.
   c. Hybridization gives stronger, more stable bonds.
3. One 2*s* orbital and three 2*p* orbitals can be mixed to give four equivalent orbitals which are one part *s* character and three parts *p* character, called sp^3 orbitals.
   a. The electron density of an sp^3 orbital is highly directed in space.
   b. One of the lobes is very large and is directed towards the bonded atom; the other lobe is very small.
   c. All sp^3-hybridized atoms have tetrahedral geometry.

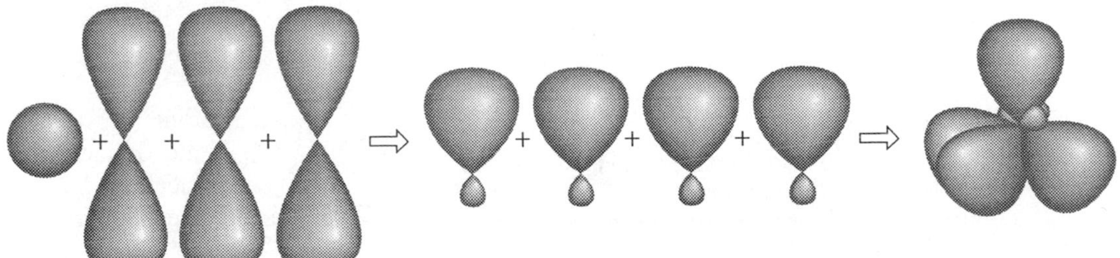

4. Unshared electrons are in orbitals with more *s* character.
   a. Orbitals with greater *s* character occupy more space.
   b. Unshared pairs have a greater spatial requirement than bonds.

C. Molecular Orbitals

1. Chemical bonding is an energetically favorable process.
2. When atoms combine into a molecule, the electrons of each atom are no longer localized on individual atoms but "belong" to the entire molecule.

3. Combination of x atomic orbitals gives x molecular orbitals.
   a. One molecular orbital of the hydrogen molecule is formed by the addition (additive overlap) of the individual atomic orbitals (a bonding molecular orbital).
      i. Bonding molecular orbitals arise through wave peak-wave peak or wave trough-wave trough interactions.
      ii. The energy of an electron in the bonding molecular orbital is lower than that of an electron in an isolated atom.
   b. The second molecular orbital of the hydrogen molecule is generated by the subtraction (subtractive overlap) of the individual atomic orbitals (an antibonding molecular orbital).
      i. Antibonding molecular orbitals arise through wave peak-wave trough interactions.
      ii. The energy of an electron in an antibonding molecular orbital is higher than that of an electron in an isolated atom.
4. Molecular orbitals of greater energy have more nodes.
5. Molecular orbitals have different energies and are arranged in order of increasing energy, then populated with electrons according to the Pauli principle, Hund's rules, and the Aufbau principle.
   a. The electrons occupy the bonding molecular orbitals before occupying the antibonding orbitals.
   b. The most common bonding situations occur when bonding molecular orbitals contain electron pairs and antibonding molecular orbitals are empty.
6. The Lewis view of the electron-pair bond is approximately equivalent to the quantum-mechanical idea of a bonding molecular orbital occupied by a pair of electrons.
   a. Bonds that have cylindrical symmetry about the internuclear axis are called σ bonds (sigma bonds).
   b. The carbon in methane has an arrangement of orbitals that is different from the orbitals in atomic carbon.

SCIENTISTS

Marcellin Berthelot (1827–1907) 1.1B
Jöns Jacob Berzelius (1779–1848) 1.1B
Petrus (Peter) Josephus Wilhelmus Debye (1884–1966) [Chemistry 1936] 1.2D
Friedrich Hermann Hund (1896–1997) 1.7
Alfred Wilhelm Hermann Kolbe (1818–1884) 1.1B
Walter Kossel (1888–1956) 1.2A
Antoine Laurent Lavoisier (1743–1794) 1.1B
Gilbert Newton Lewis (1875–1946) 1.2C
Wolfgang Ernst Pauli (1900–1958) [Physics 1945] 1.7
Linus Carl Pauling (1901–1994) [Chemistry 1954, Peace 1962] 1.9A
Theophrastus Bombastus von Hohenheim (alias Paracelsus) (ca. 1493–1541) 1.1B
Friedrich Wöhler (1800–1882) 1.1B

Dates in brackets refer to Nobel Prize awards.

STUDY GUIDE LINKS

✓1.1 Formal Charge

After you have a small amount of practice assigning formal charges, you'll begin to notice certain patterns that emerge. For example, an oxygen with three single bonds and one unshared electron pair *always* has a +1 formal charge:

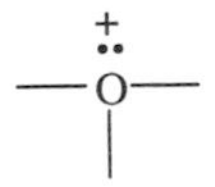

Some other common patterns are shown below.

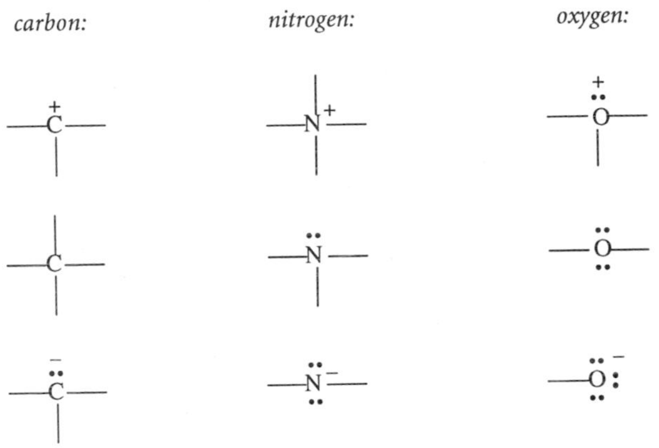

carbon: *nitrogen:* *oxygen:*

(Verify these patterns by calculating the formal charge in each one!) Recognition of these patterns can be a great time-saver in applying formal-charge rules. Conversely, when you calculate a formal charge, always double-check your calculation if you appear to be violating one of these common patterns.

1.2 Dipole Moments

As noted in the text, the mathematical definition of the dipole moment μ is as follows:

$$\mu = q\mathbf{r} \qquad \text{(SG1.1)}$$

where q is the magnitude of the separated charge and **r** is a vector from the site of positive charge to the site of negative charge. When μ is a bond dipole, as in H—Cl, the magnitude of **r** is simply the bond length; in the case of H—Cl, it is the length of the H—Cl bond. In the HCl molecule the dipole moment vector μ lies along the H—Cl bond, and points from the H (the site of positive charge) to the Cl (the site of negative charge), as shown in the text. That is, μ and **r** *have the same direction.*

The dimensions of the dipole moment, as suggested by Eq. SG1.1 above, correspond to charge × length. As the text indicates, the units of the dipole moment are called *debyes.* A dipole moment of one debye results when opposite charges, each with a magnitude of 1×10^{-10} electrostatic unit (esu), are separated by one Ångstrom. That is,

$$1 \text{ debye} = 1 \times 10^{-10} \text{ esu Å}$$

These units were established prior to the current trend toward the use of standard international units. To put these units on more familiar ground, the charge on an electron is 4.8×10^{-10} esu, and one Ångstrom = 10^{-10} meter = 100 picometers. (In standard international (SI) units, in which the unit of length is the meter and the unit of charge is the coulomb (C), 1 D = 3.34×10^{-30} C-m, or coulomb-meter.)

An actual calculation of the charge separation in H—Cl can provide a more quantitative sense of the meaning of the dipole moment. A table of experimentally measured dipole moments gives $\boldsymbol{\mu}$ =1.08 D for HCl. A table of bond lengths gives the length of the H—Cl bond as 1.274 Å; this is **r** in Eq. SG1.1. The only unknown remaining in Eq. SG1.1 is q. Because $\boldsymbol{\mu}$ and **r** have the same direction, q can be calculated by dividing the magnitude of $\boldsymbol{\mu}$ by that of **r.**

$$q = |\boldsymbol{\mu}| \div |\mathbf{r}| = 1.08 \times 10^{-10} \text{ esu Å} \div 1.274 \text{ Å} = 0.848 \times 10^{-10} \text{ esu}$$

As noted above, the charge on an electron is 4.8×10^{-10} esu. Consequently, the charge separation in HCl in electronic charge units is

$$\begin{aligned} q \text{ (in electrons)} &= (0.848 \times 10^{-10} \text{ esu}) \div (4.8 \times 10^{-10} \text{ esu per electron}) \\ &= 0.18 \text{ electron} \end{aligned}$$

In other words, the dipole moment of HCl, 1.08 D, means that a partial negative charge of 0.18 units ("18% of an electron") is on the chlorine of HCl, and a partial positive charge of 0.18 units is on the hydrogen.

Notice that the definition of dipole moment in Eq. SG1.1 contains two elements: the *amount* of charge separated (q) and the *distance* between the separated charges (**r**). A smaller amount of charge separated by a long bond can result in as significant a dipole moment as a larger amount of charge separated by a shorter bond. (See Problem 1.31 in Chapter 1.)

✓1.3 Structure Drawing Conventions

Don't be confused by typesetting conventions that seem to ignore what you've just learned about molecular geometry. For example, you might see methane (CH_4) written as follows:

```
     H
     |
  H—C—H
     |
     H
```

You now know that methane is tetrahedral. But there is a good reason for using simpler but less accurate structures anyway. When geometrically accurate structures are not needed, there is no point in going to the additional trouble (and in the case of typesetting, the expense) of using them. In other words, we use the simplest structures that accomplish the purpose at hand, and so should you! When the shape of a molecule is an important issue, use line-and-wedge formulas or other types of structures that convey spatial information. Otherwise, use the simpler structures.

1.4 Electron Density Distribution in Orbitals

The orbitals in the text (Fig. 1.7, Figs. 1.8–1.11) are drawn as closed geometrical figures designed to encompass about 90% of the electron density within the orbital. These figures do not show the distribution of the density within the orbital. That is the purpose of this study guide link.

In the text you learned that the wavemotion of an electron in an orbital is described by a function ψ called a *wavefunction.* It turns out that the electron density at any point in space of an electron in an orbital ψ is proportional to the *square* of the wavefunction, ψ^2. Consequently, by plotting the square of the wavefunction as a function of spatial coordinates we can obtain a description of the electron density at any point in space. Because we are dealing with a function of three spatial dimensions, such a plot would require four dimensions, or at least a contour plot in three dimensions. A simpler plot that has all the information we need is obtained by plotting the electron density ψ^2 as a function of the radial distance r along any straight line originating at the nucleus. Consider, for example, the 1s orbital. Because the orbital is a sphere, we would get the same plot in any direction. The result for a 1s orbital is the plot in Fig. SG1.1a (dark line). From this plot you can see that the electron density is greatest at the nucleus and then falls off exponentially. There are no nodes—no regions of zero electron density—in this plot.

The plot just described gives the electron density at a *point*. A different view of electron density is obtained if we plot the probability of finding the electron in a *spherical shell* at some distance r from the nucleus. The appropriate function for this probability is $r^2\psi^2$. The reason this makes sense is that the further away we are from the nucleus, the more surface area there is in a spherical shell. Another way of thinking about this is to imagine the dimples on two basketballs, one large and one small. The dimples on both are the same size. Then ask yourself, which basketball has more dimples? The large one, of course. Then think of each dimple as one probability unit for finding an electron. The larger the sphere (that is, the greater is r), the larger is the probability of finding an electron. Since the surface area of a sphere increases as the square of the radius, that is, r^2, then multiplying the probability at one point ψ^2 by r^2 gives the probability of finding the electron in a spherical shell of radius r. The gray curve in Fig. SG1.1a is a plot of this probability. Notice that this product has a maximum at 0.529 Å. This is called the *Bohr radius*. It is exactly the same as the radial dimension of the electron orbit calculated in an earlier, planetary model of the atom. This model said that the electron was in a fixed orbit with a radius of 0.529 Å; the quantum model says that the *maximum likelihood* of finding an electron is on a spherical shell 0.529 Å from the nucleus; but there is a finite probability at *any* distance from the nucleus.

A similar plot of electron density for the 2*s* orbital is given in Fig. SG1.1b. In the plot of ψ^2 you can clearly see the node—the region of zero electron density—at 1.1 Å. There is a small maximum in the spherical-shell electron probability at the Bohr radius, but the largest maximum is in the "outer ball" of electron density is at about 3 Å. If you contrast the value of r at which this maximum occurs with that for the 1*s* orbital you can get a very graphic sense of the different sizes of the two orbitals.

Plots of electron density for the 2*p* and 3*p* orbitals are given in Fig. SG1.2a and SG1.2b, respectively. These plots can be imagined to be taken on a line running through the nucleus and along the axis of the orbital. Both positive and negative values of r are shown to stress the lobe character of the orbitals. Both orbitals have planar nodes at the nucleus that separate the orbital into two lobes. The 3*p* orbital has, in addition, a spherical node separates the inner and outer regions of each lobe. This node is shown in Fig. SG1.2b as two dashed lines, which represent the $+r$ and $-r$ values, *i.e.*, the radius, of the spherical node. The maximum in the spherical-shell electron

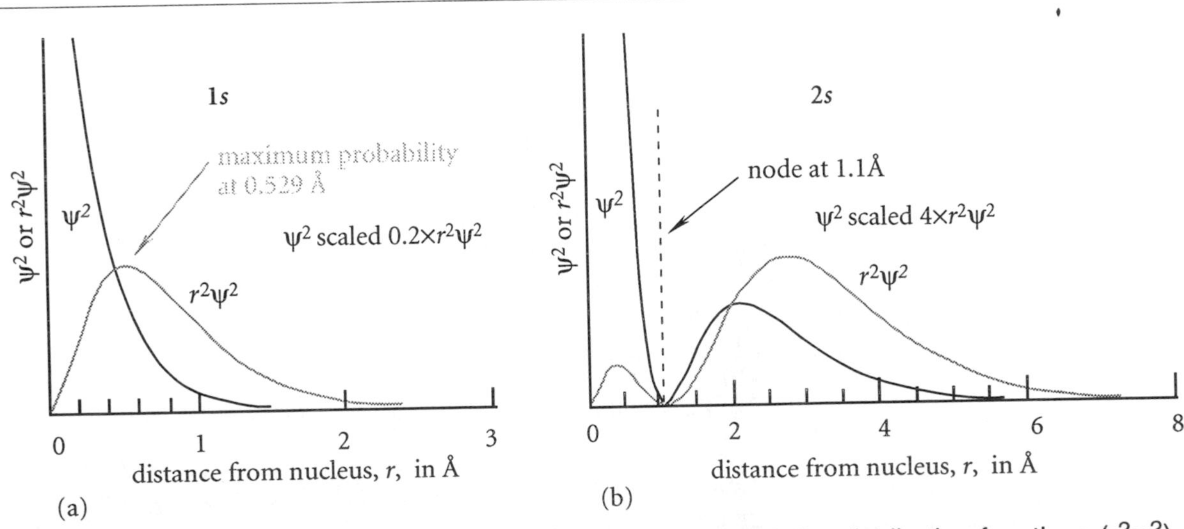

Figure SG1.1. Plots of wavefunctions (ψ^2) and their probability distribution functions ($r^2\psi^2$) for (a) a 1*s* orbital; (b) a 2*s* orbital. The "scaling" of ψ^2 is simply so that this quantity can be shown on the same plot. The "sizes" of the orbitals can be approximated by the distances at which $r^2\psi^2$ is a maximum. Thus, the 1*s* orbital has a maximum at 0.529 Å; the 2*s* orbital has a maximum around 3 Å.

probability $r^2\psi^2$ for the $2p$ orbital (grey line, Fig. SG1.2a) is at about 4 Bohr radii (2.1 Å), whereas the largest maximum in the $3p$ orbital occurs at about 12 Bohr radii (6.3 Å). Again, notice the larger size of the orbital with the larger quantum number.

These orbital pictures are derived from the *hydrogen atom.* Orbitals of atoms with higher atomic number have basically the same shape and nodal characteristics, but their sizes differ.

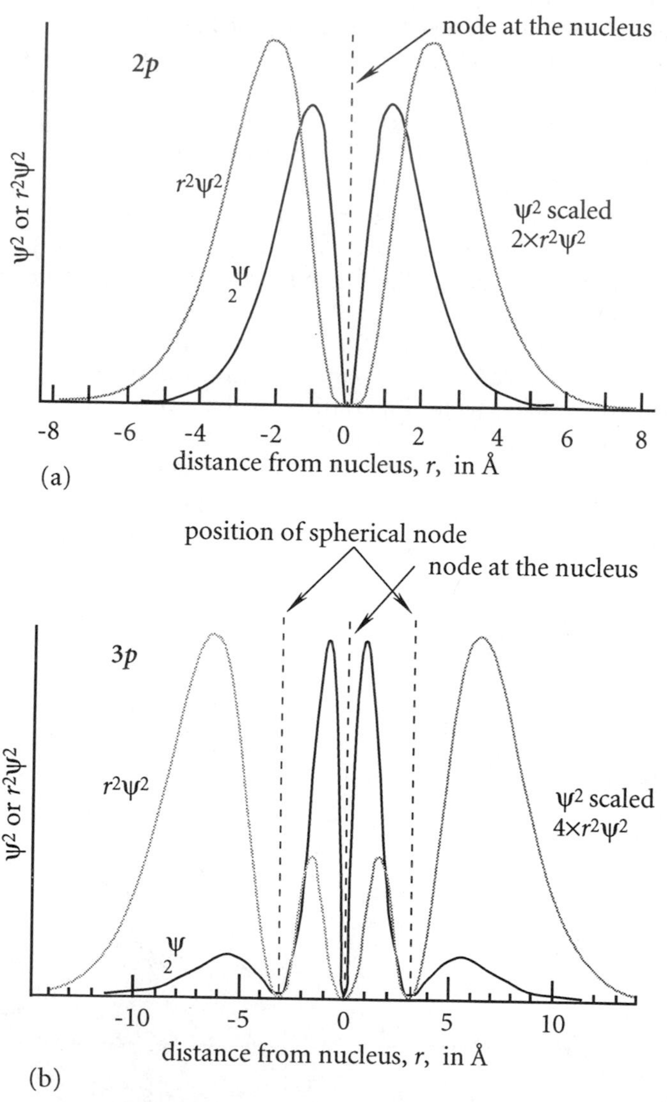

Figure SG1.2 Plots of wavefunctions (ψ^2) and their probability distribution functions ($r^2\psi^2$) for (a) a $2p$ orbital; (b) a $3p$ orbital.

SOLUTIONS TO PROBLEMS

Solutions to In-Text Problems

1.1 (a) Because sodium (Na) is in Group 1A, it has one valence electron.
(b) The neutral calcium atom has a number of valence electrons equal to its group number, that is, 2.
(c) Oxygen has six valence electrons; hence, O^{2-} has eight.
(d) Neutral Br, being in Group 7A, has seven valence electrons; therefore, Br^{+} has six.

1.2 (a) Because neon has ten electrons, the negative ion requested in the problem also has ten. The singly charged negative ion with ten electrons is F^{-}.
(b) The dipositive ion isoelectronic with Ar is Ca^{+2}.
(c) The neon species isoelectronic with neutral fluorine has nine valence electrons, and is therefore Ne^{+}.

1.3 (a) (b) (c) (d)

```
      ••
     :Cl:            H                               +
      |         ••   |+            ••                ••
  H — C — Cl:    H — N — H     H — N — H       H — O — H
      |         ••   |             |               |
     :Cl:            H             H               H
      ••
```

chloroform — ammonium ion — ammonia — hydronium ion

1.4 One structure is that of ethanol, and the other is that of dimethyl ether.

```
    H   H                        H         H
    |   |   ••                   |    ••   |
H — C — C — O — H            H — C — O — C — H
    |   |   ••                   |    ••   |
    H   H                        H         H
```

ethanol — dimethyl ether

In all but the simplest cases there are many structures that have the same atomic composition. For molecules of moderate size hundreds or even thousands of structures are possible. Compounds that have the same atomic composition, but different atomic connectivities, are called **constitutional isomers.** Thus, dimethyl ether and ethanol are constitutional isomers. You'll learn about isomers in Chapter 2.

1.5

```
    H
    |
H — C — C≡N:      acetonitrile
    |
    H
```

1.6 The formal charge on boron is –1; the formal charge on nitrogen is +1; the formal charge on the hydrogens is zero. The net charge on the entire structure is zero.

1.7 Of the atoms in the molecule, carbon and hydrogen differ least in electronegativity (Table 1.1). Consequently, the C—H bonds are the least polar bonds. The carbon with the most partial positive character is the one bound to the greatest number of electronegative atoms:

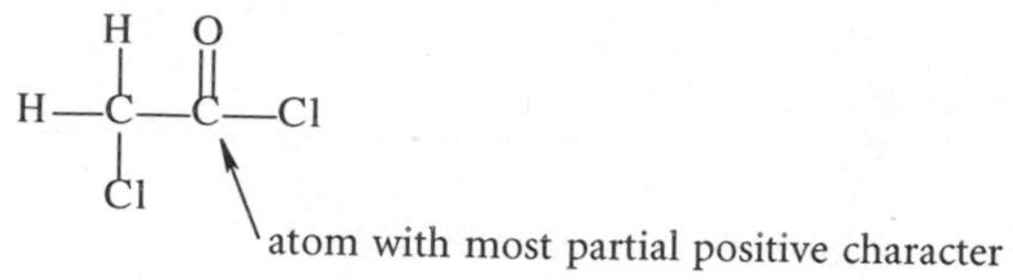

atom with most partial positive character

1.8 (a) Water has bent geometry; that is, the H—O—H bond angle is approximately tetrahedral. Repulsion between the lone pairs and the bonds reduces this bond angle somewhat. (The actual bond angle is about 105°.)

(b) The tetrafluoroborate anion is tetrahedral, that is, it has F—B—F bond angles of 109.5°.

(c) The formaldehyde molecule has trigonal planar geometry. Thus, both the H—C—H bond angle and the H—C═O bond angle are about 120°.

(d) In acetonitrile, each of the H—C—H bond angles, as well as the H—C—C bond angle, is tetrahedral (about 109.5°). The C—C≡N bond angle is linear (180°).

1.9 Bond angles: *aa, ab, bc, bd, cd, de, df,* and *ef* are all about 120°, because all are centered on atoms with trigonal-planar geometry; *fg* is predicted to have the tetrahedral value of 109.5°, but actually, is closer to 105° because of electron-pair repulsion (see Solution 1.8a).

Bond lengths: $g < a < e < b < f < d < c$. Bonds g and a are shorter than the others because they involve the smallest atom (H); bond g is shorter than bond a because oxygen is smaller than carbon. Bonds e and b are the next shortest because they are double bonds; of these, bond e is shortest because oxygen is smaller than carbon. Bond f is shorter than bond d for the same reason. Bond c is longest because it involves the largest atom (Br). Notice that the decisions in this problem require an appreciation of the relative importance of the trends in bond length given in the text.

1.10 In the following planar projections, the near carbon is shown as a circle and the far carbon is hidden from view.

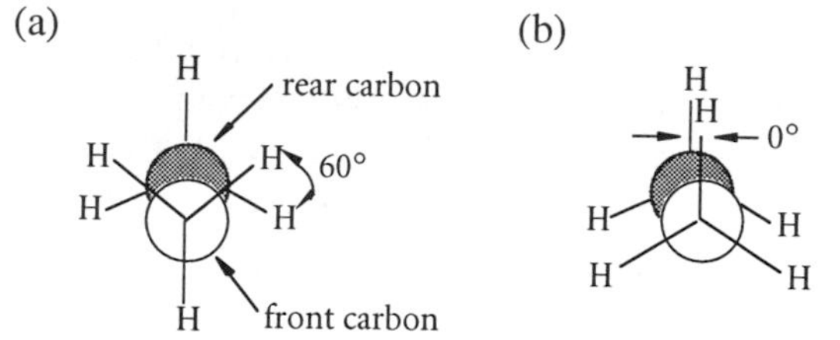

(a) rear carbon; 60°; front carbon

(b) 0°

1.11 The resonance structures of benzene:

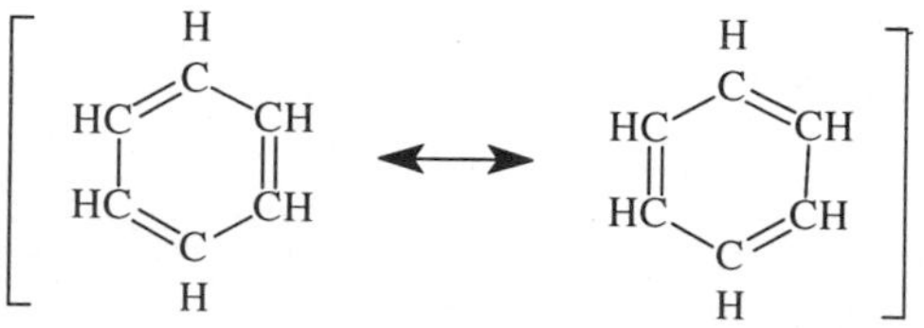

Notice that each bond is a single bond in one structure and a double bond in the other. Hence, on average, each bond is 1.5 bonds.

1.12 Fig. SG1.3 shows a plot of $\psi = \sin nx$ for the different values of n. Evidently, the number of nodes is one less than the "quantum number" n; that is, $nodes = n - 1$.

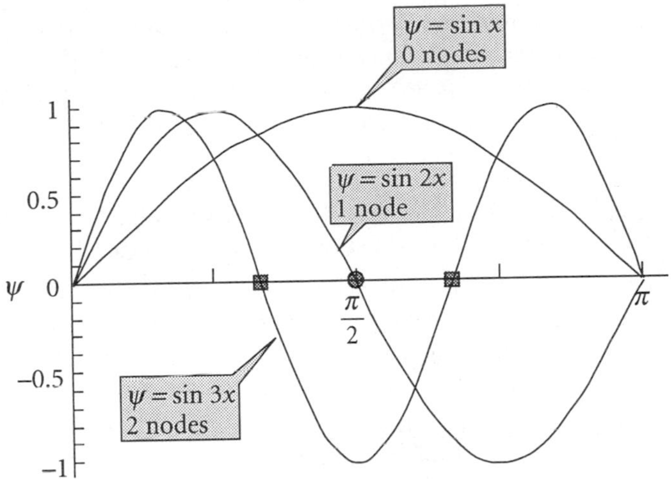

Figure SG1.3. A plot of the function $\psi = \sin nx$ for $n = 1$, 2, and 3 for the solution to Problem 1.12. The nodes are given by filled squares (for $\psi = \sin 3x$) and a filled circle (for $\psi = \sin 2x$).

Orbitals are described by mathematical functions of three dimensions, whereas $\psi = \sin nx$ is a function of one dimension. However, the quantum numbers of orbitals have the same effect as n in this simple equation: they control the number of nodes.

1.13 (a) A 3*s* orbital is three concentric spheres of electron density, each separated by a node, as shown in the "cutaway" diagram (a) in Fig. SG1.4.

(b) A 4*s* orbital is four concentric spheres of electron density, each separated by a node, as shown in "cutaway" diagram (b) in Fig. SG1.4.

1.14 (a) Oxygen atom (atomic number = 8, therefore 8 electrons): $(1s)^2(2s)^2(2p_x)^2(2p_y)(2p_z)$. Notice that two of the 2*p* electrons are unpaired; each of these is in a different 2*p* orbital. The valence orbitals are the 2*s* and 2*p* orbitals, and the valence electrons are the six electrons that occupy these orbitals.

(b) Chloride ion, Cl^- (atomic number = 17 and one negative charge, therefore 18 electrons): This ion has the same electronic configuration as argon: $(1s)^2(2s)^2(2p)^6(3s)^2(3p)^6$. The valence orbitals are the 3*s* and 3*p* orbitals, and the valence electrons are the seven electrons that occupy these orbitals.

(c) Sodium atom (atomic number = 11, therefore 11 electrons): $(1s)^2(2s)^2(2p)^6(3s)^1$. The valence orbital is the 3*s* orbital, and the valence electron is the one electron that occupies this orbital.

(solution continues)

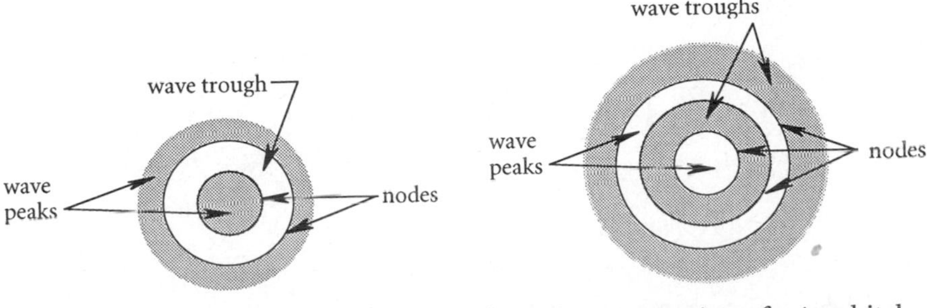

Figure SG1.4. Cross sections of (a) 3*s* and (b) 4*s* orbitals in the solution to Problem 1.13.

(d) Potassium ion, K^+ (atomic number = 19 and one positive charge, therefore 18 electrons): This ion, like the chloride ion has the same electronic configuration as argon: $(1s)^2(2s)^2(2p)^6(3s)^2(3p)^6$. The valence orbital of potassium ion is the same as the valence orbital of the potassium *atom*, which is the 4*s* orbital. The one electron that occupies this orbital in the K *atom* is lost to give the potassium *ion*, K^+. Thus, the potassium ion, K^+, has *no* valence electrons.

1.15 The pictures and energy levels of the molecular orbitals are exactly the same as they are for the dihydrogen molecular orbitals in Fig. 1.14 of the text. However, the various species differ in their electron occupancies.

(a) The He_2^+ ion contains three electrons. By the Aufbau principle, two electrons occupy the bonding molecular orbital, and one occupies the antibonding molecular orbital. This is shown in the electron-occupancy diagram (a) of Fig. SG1.5. the bonding molecular orbital contains a greater number of electrons than the antibonding molecular orbital, this species is stable.

(b) The H_2^- ion contains three electrons. By the Aufbau principle, two occupy the bonding molecular orbital, and one occupies the antibonding molecular orbital. This is also shown in the electron-occupancy diagram (a) of Figure SG1.5. Because the bonding molecular orbital contains a greater number of electrons than the antibonding molecular orbital, this species is stable. Notice that in terms of electron occupancy, H_2^- and He_2^+ are identical.

(c) The H_2^{2-} ion contains four electrons; two occupy the bonding molecular orbital, and two occupy the antibonding molecular orbital. This is shown in the electron-occupancy diagram (b) of Fig. SG1.5. In this species, the energetic advantage of the electrons in the bonding molecular orbital is cancelled by the energetic disadvantage of the same number of electrons in the antibonding molecular orbital. This species consequently has no energetic advantage over two dissociated hydride ($H:^-$) ions, and therefore it readily dissociates.

(d) The H_2^+ ion contains one electron in a bonding molecular orbital. This is shown in the electron-occupancy diagram (c) of Figure SG1.5. Because the bonding molecular orbital contains a greater number of electrons than the antibonding molecular orbital, this species is stable.

1.16 Because the H_2^+ ion has one electron in a bonding molecular orbital, it should have about half the stability of H_2 itself relative to dissociated fragments. Thus, 217 kJ/mol (52 kcal/mol) is an estimate of the bond dissociation energy of this species.

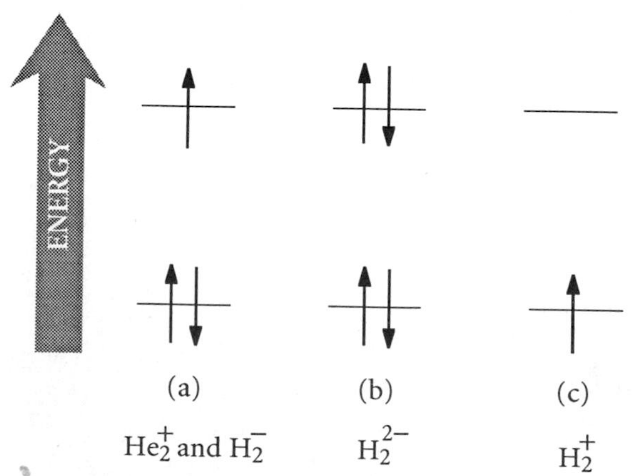

Figure SG1.5 Electron-occupancy diagram for the solution to Problem 1.15.

1.17 (a) If the oxygen of water is sp^3-hybridized, then two of the sp^3 hybrid orbitals contain unshared electron pairs. The other two sp^3 hybrid orbitals contain one electron each; each of these overlaps with the $1s$ orbital of a hydrogen atom (which contains one electron) to give the two sp^3-$1s$ σ bonds (the O—H bonds) of water.

(b) Because unshared electron pairs prefer s orbitals, the orbitals containing the unshared pairs have more $2s$ character, and therefore the orbitals used to form the bonds have more $2p$ character. By the reasoning discussed in the text for ammonia, bonds involving orbitals with more $2p$ character are separated by smaller bond angles. Hence, the H—O—H angle in water is less than 109.5°. (In fact, it is 104.8°.)

Solutions to Additional Problems

1.18 Compound (2) would be most likely to exist as an ionic species, because Cs and F come from opposite corners of the periodic table.

1.19 Species (b) and (c) have complete octets on every atom (duets for hydrogens); in (a), (d), and (e), carbon, boron, and iodine, respectively, have electronic sextets. The formal charges are:

(a) +1 on carbon
(b) 0 on all atoms
(c) –1 on carbon
(d) 0 on all atoms
(e) +1 on iodine

1.20 (a) C_3H_8 propane (b) C_2H_3Cl chloroethylene (c) C_2H_2O ketene

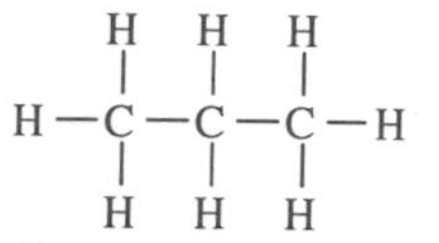

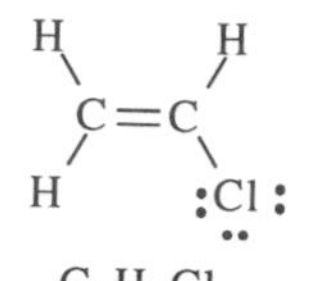

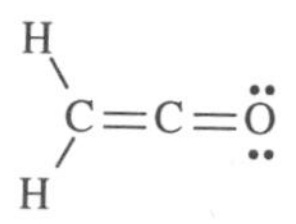

1.21 (a) perchlorate: –1 on each of the oxygens, +3 on the Cl, –1 overall.
(b) trimethylene oxide: +1 on the nitrogen, –1 on the oxygen, 0 on all others, 0 overall.
(c) methylene: 0 charge on everything, and of course 0 charge overall.
(d) ozone:

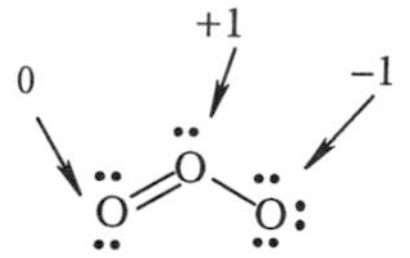

0   +1   –1

0 charge overall

(e) hypochlorite: 0 on the chlorine, –1 on the oxygen, –1 overall.
(f) ethyl radical: 0 charge on everything, 0 charge overall.

1.22 (a) chlorine atom (atomic number = 17; therefore 17 electrons): $(1s)^2(2s)^2(2p)^6(3s)^2(3p_x)^2(3p_y)^2(3p_z)$
(b) fluoride ion (atomic number = 9 and one negative charge;therefore 10 electrons): $(1s)^2(2s)^2(2p)^6$
(c) argon atom (atomic number = 18; therefore 18 electrons): $(1s)^2(2s)^2(2p)^6(3s)^2(3p)^6$
(d) magnesium atom (atomic number = 12; therefroe 12 electrons): $(1s)^2(2s)^2(2p)^6(3s)^2$

1.23 The 2*d* orbital is not permitted. The reason is that *d* means that the *l* quantum number = 2; the value of the *l* quantum number cannot exceed *one less than* the value of the principle quantum number *n*, which in this case is 2. Thus, *l* cannot exceed 1 in principal quantum level 2.

1.24 (a) The carbon is bound to three "groups," the two Hs and an electron pair. Hence, this carbon is trigonal planar (if you count the electron pair) and the molecule is "bent" with an H—C—H angle of about 120° (or perhaps somewhat less because of the tendency of unshared pairs to reduce bond angles).
(b) The beryllium (Be) bears two groups, and therefore has linear geometry. The H—Be—H bond angle is 180°.
(c) The carbon is bound to three groups; hence, this ion is trigonal-planar, and the H—C—H bond angles are 120°.
(d) The silicon is bound to four chlorines, and is therefore tetrahedral. The Cl—Si—Cl bond angle is 109.5°.
(e) The central oxygen of ozone is bound to three "groups," two oxygens and an electron pair. Hence, this oxygen is trigonal planar (if you count the electron pair), and ozone is "bent" with an O—O—O bond angle of about 120°.
(f) The terminal carbons are bound to three groups, and are therefore trigonal planar. The central carbon is bound to two groups, and is therefore linear. Hence, the H—C—H and H—C═C bond angles are 120°;

the C=C=C bond angle is 180°.

(g) The nitrogen is trigonal planar; the O=N—O bond angle is 120°, as are both C—N—O bond angles. The carbon is bound to four groups, and is tetrahedral; all bond angles centered on carbon are about 109.5°.

1.25 (a) The allyl cation is a hybrid of the two structures. Because the structures are identical, each structure contributes equally to the structure of the hybrid. Because each carbon-carbon bond is a single bond (bond order = 1) in one resonance structure and a double bond (bond order = 2) in the other, the *average* bond order is (1 + 2)/2 = 1.5. In other words, each carbon-carbon bond is equivalent to a single bond plus "half" of a double bond.

(b) By analogous reasoning, each of the terminal carbons bears +1 charge in one structure and 0 charge in the other. Hence, the charge on the ion is the average of these two, or (+1 + 0)/2 = +0.5. The charge on the central carbon is zero in all resonance structures and therefore zero overall.

(c) The resonance structure on the right violates the octet rule for nitrogen!

Notice from this example that a positive charge does *not* necessarily mean that an atom lacks an octet. This point will be considered further in Chapter 3.

1.26 We use reasoning analogous to that used in the previous problem. This ion is a hybrid of three identical structures. Hence, each property will be the sum of that property in each structure divided by 3.

(a) Negative charge on each oxygen: (–1 + 0 + –1)/3 = –2/3.

(b) Bond order of each carbon-oxygen bond: (1 + 2 + 1)/3 = 4/3 or 1.33.

1.27 (a) Figure 1.8 shows that a 2*s* orbital has a single spherical node; and Figure 1.10 shows that the 2*p* orbital has one planar node. Figure 1.11 shows that the 3*p* orbital has two nodes—one planar node and one spherical node. In all cases, the total number of nodes is one less than the principal quantum number. In the 2*s* orbital, the value of *l* is 0, and the number of planar nodes is zero. In the 2*p* and 3*p* orbitals, the value of *l* is 1, and each orbital has one planar node.

(b) The total number of nodes is *n* – 1, or 4. The value of *l* (0, which implies an *s* orbital) gives the number of planar nodes, which is 0. Hence, the 5*s* orbital has four spherical nodes. By analogous reasoning, a 3*d* orbital has two nodes, both of which are planar; hence, it has no spherical nodes. (Notice this property of the 3*d* orbital in the solution to the following problem.)

1.28 There are two planar nodes, one along the *x*-axis and one along the *y*-axis.

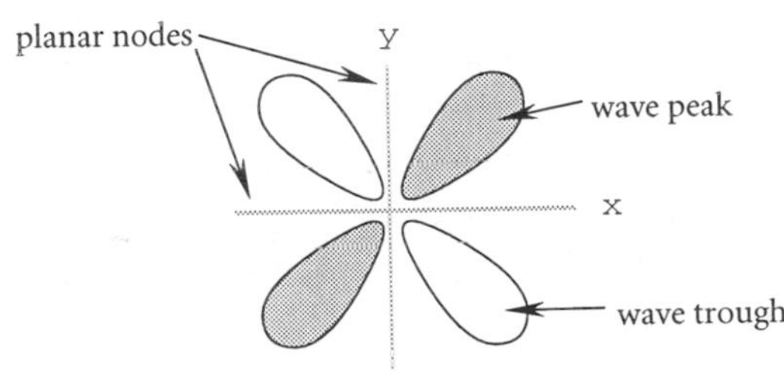

1.29 (a) The number of equivalent *f* orbitals equals the number of possible values of the quantum number *m*, which is 7 (±3, ±2, ±1, 0).That is, there are *l* + 1 = 7 equivalent *f* orbitals.

(b) The orbital designation *f* means *l* = 3. Orbitals with *l* = 3 (that is, *f* orbitals) first appear in quantum level 4.

(c) There are *n* – 1 = 5 – 1 = 4 nodes. (Three of them are planar, and one is spherical; see Problem 1.27. Can you sketch a 4*f* orbital?)

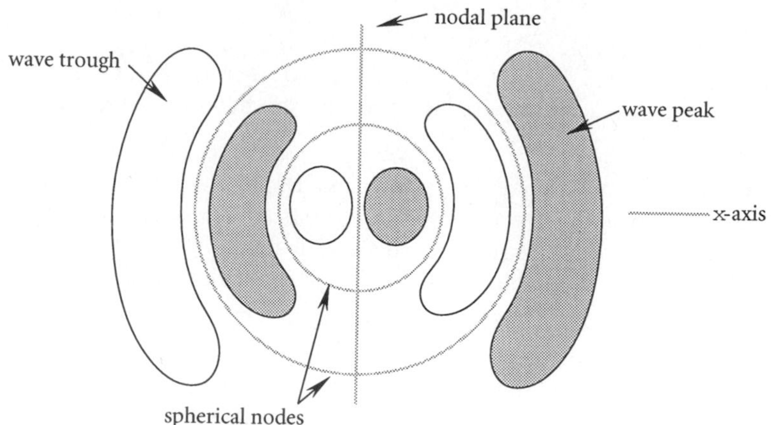

Figure SG1.6. A cutaway diagram of a 4*p* orbital showing the nodes is the solution to Problem 1.30. The actual shape of the orbital is the volume of revolution about the *x*-axis (that is, the volume "swept out" when the figure is rotated around the *x*-axis).

1.30 A 4*p* orbital is somewhat like a 3*p* orbital, except that it has one additional spherical node and two additional regions of electron density. A "cross section" is shown in Fig. SG1.6.)

1.31 According to Eq. 1.4 in the text, the dipole moment is determined by *two* factors. The first is the separated charge, which should be greater for the compound containing the more electronegative atom, that is, for $H_3C—F$. However, the second factor is the bond length **r**, which is greater for the compound containing the *larger* atom, that is, for $H_3C—Cl$. Evidently, these two opposing factors just about compensate in their effects on the dipole moments of the two compounds.

1.32 No matter how any CH_2 group is turned, the resultant bond dipole is the same:

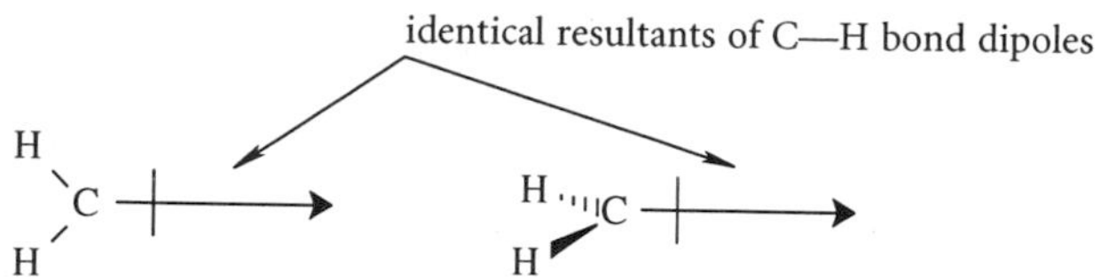

Hence, both conformations of ethylene should have zero dipole moment because the resultants of the two CH_2 groups cancel each other in either conformation. Thus, the observation of zero dipole moment does not permit a choice between these two conformations.

1.33 A linear water molecule would have zero dipole moment, because the O—H bond dipoles would be oriented in opposite directions and would cancel. Hence, the nonzero dipole moment for water shows that it is a bent molecule. Out with Professor Szents!

1.34 To solve this problem, we first have to know the geometry of CCl_4. This should be easy: it's tetrahedral, because the carbon is bonded to four identical groups. Now, as suggested in the hint, let's do a vector addition for the bond dipoles of each Cl—C—Cl pair. (Use a dipole arrow of arbitrary length as long as it is the same for both. We'll use an arrow with the length of the C—Cl bonds.) Here's the result of the vector addition, which we've done graphically, for the first pair. Notice that, because the C—Cl bonds are identical, the resultant must *bisect* the angle between the two bonds.

Cl Cl C Cl Cl

Now let's do the resultant for the second pair. Because the C—Cl bonds are identical to those of the first pair, it follows that the resultant is the same length, and it again bisects the Cl—C—Cl bond angle:

Cl Cl C Cl Cl

resultant from first vector addition

resultant from second vector addition

As the diagram shows, and as you should be able to see from your models, the two resultants are oriented at an agle of 180°; that is, ther are pointed in opposite directions. Since they are identical in magnitude, they cancel. Hence the dipole moment of CCl_4 is zero.

Here's a brief refresher on graphical vector addition. (1) The relative lengths of vectors are proportional to their magnitudes. (2) Any vector may be moved at will so long as its angular orientation is not changed. (3) Two parallel vectors can be added by adding their lengths, and the resultant points in the same direction as the original pair. (4) Two antiparallel vectors (that is, two vectors that are parallel but pointing in opposite directions) can be added by subtracting their lengths, and the resultant points in the direction of the longer vector of the pair. (5) Two vectors may be added by constructing a parallelogram in which two sides consist of the vectors themselves and the remaining two sides are parallel to these vectors (dashed lines in the diagram above). The vector addition product, or resultant, is the diagonal of the resulting parallelogram. (6) The resultant of many vectors can be determined as the resultants of successive pairs.

1.35 (a) The three conformations of hydrogen peroxide are as follows: In the corresponding planar projections, the rear oxygen is hidden from view by the front oxygen:

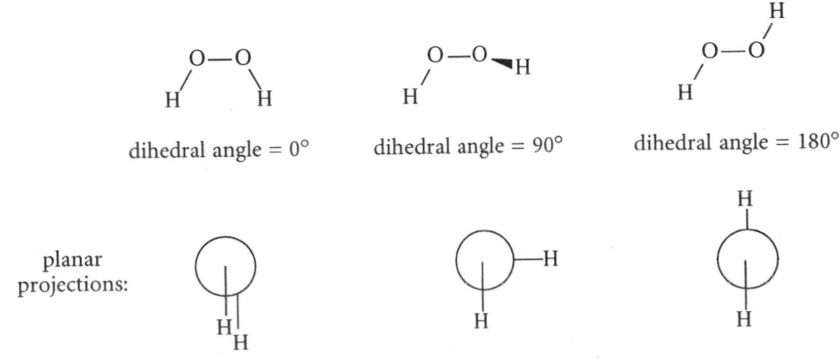

Interconversion of these conformations requires internal rotation about the O—O bond.

(b) When the O—H bonds lie at a dihedral angle of 180° their bond dipoles cancel; consequently, a hydrogen peroxide molecule in this conformation would have a dipole moment of zero. This conformation is *ruled out* by the observation of a significant dipole moment.

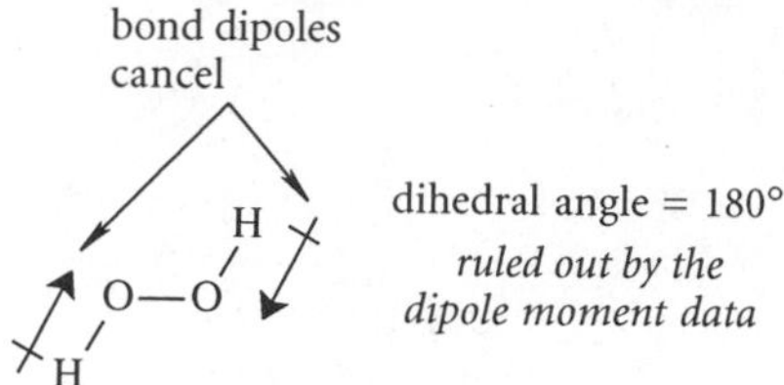

(c) Here's a useful construction:

y
H
y-component of μ_{OH}
μ_{OH} = 1.52 D
O
O
x
96.5°
x-component of μ_{OH}

The x-components of the two O—H bond dipoles point in opposite directions regardless of the bond angle or the dihedral angle. Hence, the x-components cancel. Therefore, we can deal with the y-component of the O—H bond dipole exclusively. First, we have to calculate the y-component. This is just 1.52 cos (6.5°) = 1.52 (0.9935) = 1.51 D.

Note that 96.5° is not the dihedral angle, but the bond angle. The idea is to calculate the dihedral angle. From the analysis in the previous paragraph, we know that the resultant dipole moment of the molecule (given as 2.13 D) is derived from vector addition of the y-components of the two O—H bond dipoles, which we now know are 1.51 D. All we have to do is determine the angle θ that will give 2.13 D as the resultant from vector addition of two 1.51 D dipoles.

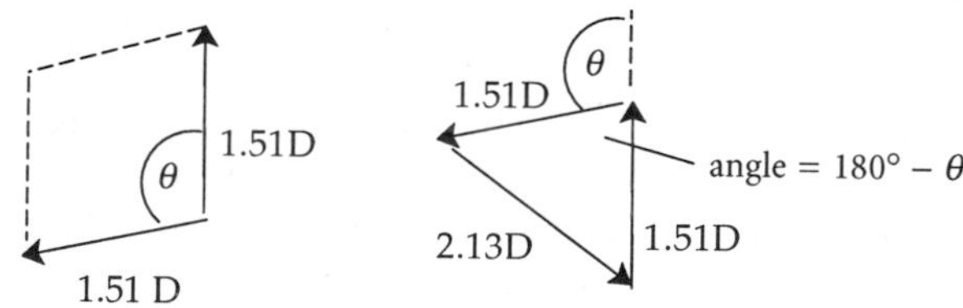

The proper construction is the arrangement on the left; because we can move a vector at will so long as its magnitude and direction remain unchanged, we derive the construction on the right. The resultant, 2.13D, is the vector that completes a head-to-tail connection of the two bond dipoles. From the law of cosines, this vector is

$$(2.13\ \text{D})^2 = (1.51\ \text{D})^2 + (1.51\ \text{D})^2 - 2(1.51\ \text{D})(1.51\ \text{D})\cos(180° - \theta)$$

or

$$4.54\ \text{D}^2 = 2.28\ \text{D}^2 + 2.28\ \text{D}^2 - (4.56\ \text{D}^2)\cos(180° - \theta)$$

from which we have

$$\frac{-0.02\ \text{D}^2}{-4.56\ \text{D}^2} = \cos(180° - \theta) = 0.00439$$

Using the "arccos" function on a calculator, we find that $180° - \theta = 89.8°$. Thus, the desired dihedral angle is 180° – 89.8° = 90.2°. Therefore, it appears that the dihedral angle in hydrogen peroxide is about 90°. (The actual structure of hydrogen peroxide determined by electron diffraction shows a dihedral angle of 93.8°!) Thus, the conformation of hydrogen peroxide is best represented as follows:

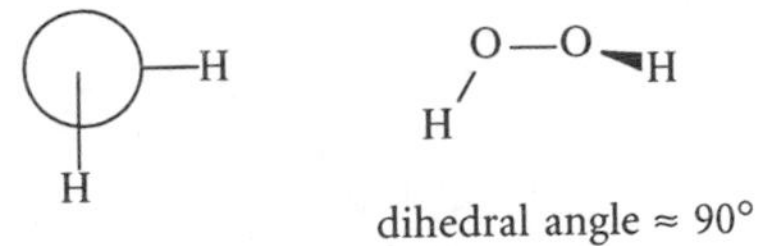

1.36 (a) The bonding molecular orbital is derived by the constructive overlap of wave peaks. The antibonding molecular orbital is derived by the destructive overlap of a peak with a trough. The resulting molecular orbitals are shown in Fig. SG1.7 on p. 12.

(b) The nodes are shown in Fig. SG1.7. Both of the MOs contain the nodes of the original $2p$ orbitals. The antibonding MO has in addition a node that results from the destructive overlap of a peak and a trough.

(c) The resulting bond *is* a σ bond because it is cylindrically symmetrical.

1.37 (a) The bonding molecular orbital is derived by the constructive "side-to-side" overlap of peaks with peaks and troughs with troughs. The antibonding MO is derived by interchanging peaks and troughs of one of the $2p$ orbitals and then allowing destructive overlap of troughs with peaks and peaks with troughs. These molecular orbitals are shown in Fig. SG1.8 on p. 12.

(b) The nodes are also shown in Fig. SG1.8. Both of the MOs have the nodes of the original $2p$ orbitals, which merge into a single node because the nodal plane is common to both orbitals. The antibonding MO has in addition a node between the original orbitals that results from the destructive overlap of a peak and a trough.

(c) When two electrons occupy the bonding molecular orbital, the resulting bond is *not* a σ bond because the molecular orbital is not cylindrically symmetrical.

This type of bond, called a pi (π) bond, is important in the carbon-carbon double bond. Pi bonds are discussed in Chapter 4.

1.38 When a hydrogen molecule absorbs light, an electron jumps from a bonding molecular orbital into an antibonding molecular orbital:

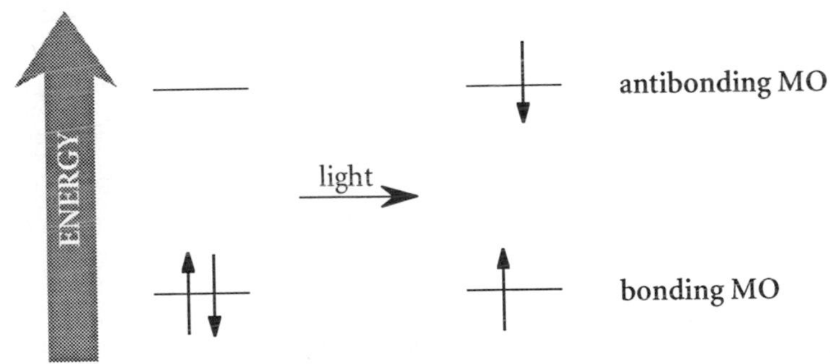

After light absorption, the hydrogen molecule has no excess of bonding electrons, and there is no energetic advantage to bonding. Hence, the molecule can readily dissociate into hydrogen atoms.

1.39 (a) Because there are three values of spin, each orbital can hold three electrons.

(b) The configuration of the atom with atomic number = 8: $(1s)^3(2s)^3(2p)^2$

(c) The second noble gas has a filled $2p$ orbital; but there are only two $2p$ orbitals ($m = 0, +1$); so the $2p$ level holds six electrons (three in each of the $2p$ orbitals). With filled $1s$, $2s$, and $2p$ levels, the second noble gas has $(3 + 3 + 6) = 12$ electrons; hence, its atomic number is 12.

(d) Because a filled shell contains nine electrons (three electrons each in one s and two p orbitals), the corresponding rule might be called a "nonet rule."

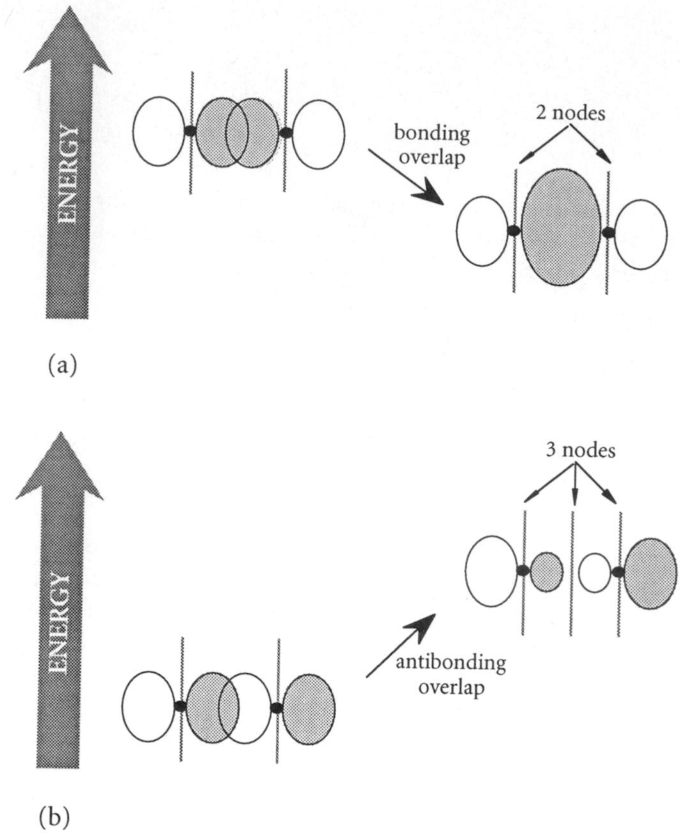

Figure SG1.7. The molecular orbitals resulting from a head-to-head overlap of 2*p* orbitals (Problem 1.36). (a) Formation of the bonding MO. (b) Formation of the antibonding MO.

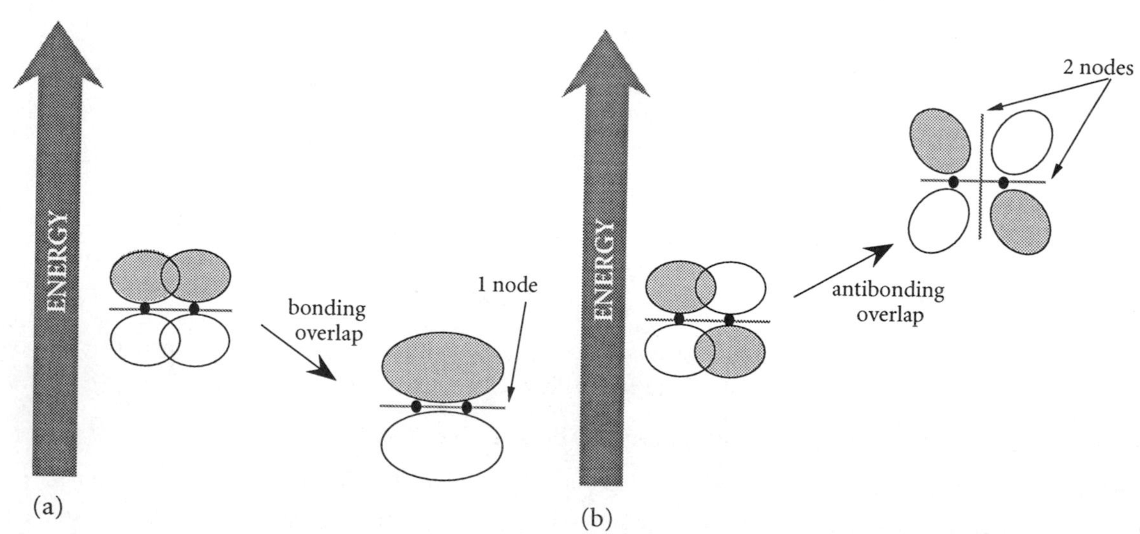

Figure SG1.8 The molecular orbitals resulting from a side-to-side overlap of 2*p* orbitals (Problem 1.37). (a) Formation of the bonding MO. (b) Formation of the antibonding MO.

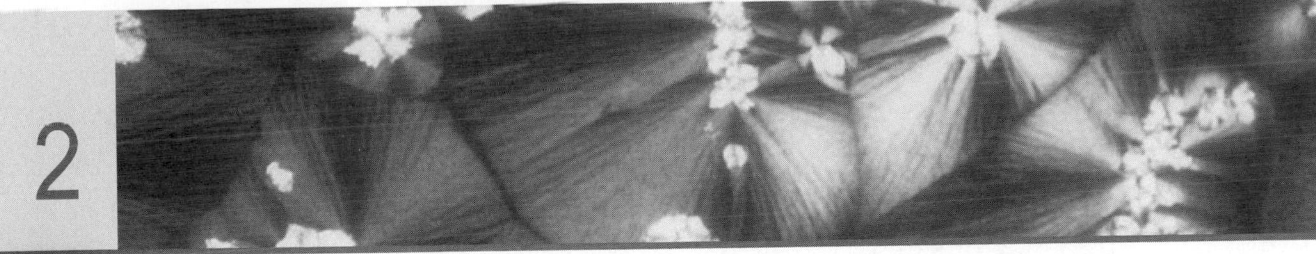

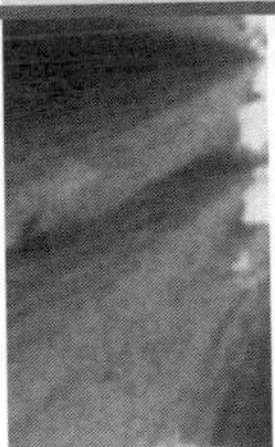

2 Alkanes

TERMS

CONCEPTS

I. HYDROCARBONS

A. General

1. Hydrocarbons are compounds that contain only the elements carbon and hydrogen.
2. Hydrocarbons are derived from petroleum and are important raw materials for the preparation of other organic compounds.

a. Alkanes and aromatic hydrocarbons are separated by a technique called fractional distillation.
b. Catalytic cracking is used industrially to convert alkanes of higher molecular mass into alkanes and alkenes of lower molecular mass.
c. An industrial process called reforming converts unbranched alkanes into branched-chain alkanes, which have superior ignition properties as motor fuels.

B. Aliphatic hydrocarbons

1. Alkanes (paraffins) contain only single bonds.
   a. Normal alkanes (*n*-alkanes) contain unbranched carbon chains.
   b. Alkanes have sp^3-hybridized carbon atoms with tetrahedral geometry.
   c. Alkanes may contain unbranched chains, branched chains, or rings (cycloalkanes).

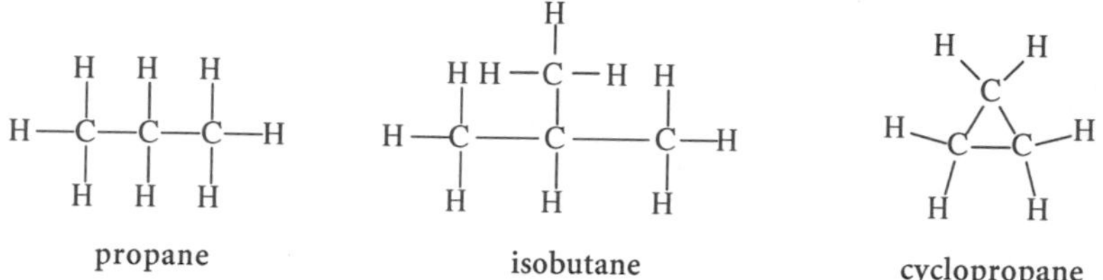

propane isobutane cyclopropane

   d. Carbon atoms can be classified by their branching patterns.
      i. A primary carbon is bonded to one other carbon.
      ii. A secondary carbon is bonded to two other carbons.
      iii. A tertiary carbon is bonded to three other carbons.
      iv. A quaternary carbon is bonded to four other carbons.

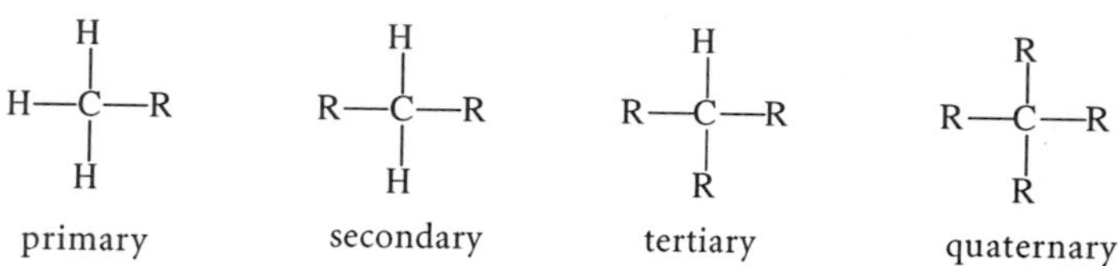

primary secondary tertiary quaternary

2. Alkenes (olefins) contain carbon-carbon double bonds.
3. Alkynes (acetylenes) contain carbon-carbon triple bonds.

C. Aromatic Hydrocarbons

1. Aromatic hydrocarbons consist of benzene and its substituted derivatives.
2. Aromatic hydrocarbons are also called arenes.

benzene toluene

II. MOLECULAR STRUCTURE

A. Empirical Formula

1. The empirical formula (obtained from elemental analysis) of a compound gives the smallest whole-number molar proportions of the elements present in the compound.
   a. Convert the relative masses of the elements into molar proportions by dividing the mass percent of each element by its atomic mass.

b. Divide the molar proportion of each element by that of the element present in the smallest proportion.
c. Multiply the resulting proportions by successive integers (2, 3, 4, …) until whole-number proportions for all elements are obtained.

butane: 82.66 % C; 17.34 % H ⇨ $C_{\frac{82.66}{12.011}}H_{\frac{17.34}{1.008}}$ ⇨ $C_{6.88}H_{17.2}$ ⇨ $C_1H_{2.5}$ ⇨ C_2H_5

B. Molecular Formula

1. The molecular formula of a compound gives its atomic composition.
   a. All noncyclic alkanes have the general formula C_nH_{2n+2} in which n is the number of carbon atoms in the alkanes.

   decane $C_{10}H_{22}$ ($n = 10$, $H = 2n + 2 = 22$)

   b. The family of unbranched alkanes form a series in which successive members differ from one other by each —CH_2— (methylene) group.
   c. A series of compounds that differ by the addition of methylene groups is called a homologous series.
2. The molecular formula may or may not differ from the empirical formula.

   butane C_2H_5 (empirical formula) C_4H_{10} (molecular formula)

C. Isomers

1. Isomers are different compounds with the same molecular formula.
2. Isomers that differ in the connectivity of their atoms are termed constitutional isomers (structural isomers).

C_4H_{10} constitutional isomers:

```
                                          H
    H   H   H   H                         |
    |   |   |   |                     H—C—H
H—C—C—C—C—H   butane           H      |       H          isobutane (2-methylpropane)
    |   |   |   |                  |      |       |
    H   H   H   H               H—C———C———C—H
                                   |      |       |
                                   H      H       H
```

D. Structural Formula

1. The structural formula of a molecule is its Lewis structure, which shows the connectivity of its atoms.
   a. Connectivity is the order in which the atoms are connected.
   b. A condensed structural formula conveys the same information as that of the structural formula.

```
    H   H   H   H
    |   |   |   |
H—C—C—C—C—H   butane ⇨   CH3CH2CH2CH3
    |   |   |   |                  condensed formula
    H   H   H   H
```

E. Skeletal structures

1. Skeletal structures show only the carbon-carbon bonds.
2. Carbons are located at each vertex and at the ends of the structures.

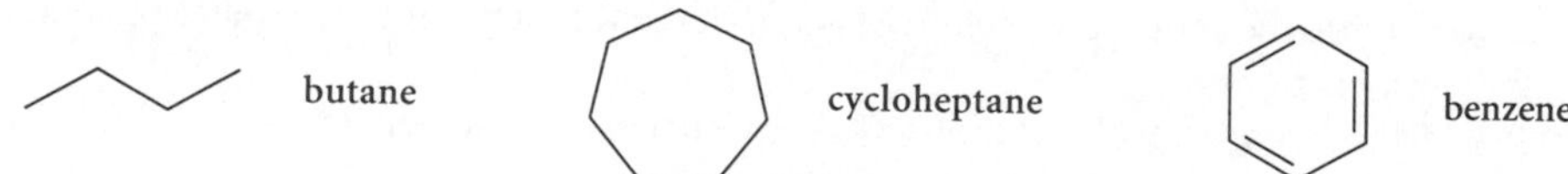

F. Functional Group

1. A functional group is a characteristically bonded group of atoms that has about the same chemical reactivity whenever it occurs in a variety of compounds.
2. Alkyl groups are substituent groups derived from alkanes; R is the abbreviation of an alkyl group.
3. Aryl groups are substituent groups derived from benzene and its derivatives; the phenyl group is the simplest aryl group.
   a. Ph is the abbreviation for a phenyl group.
   b. Ar is the abbreviation for a general aryl (substituted phenyl) group.

III. MOLECULAR CONFORMATION

A. General

1. The conformation of a molecule is the spatial arrangement of its atoms when all of the dihedral angles are specified.
2. Alkanes exist in various staggered conformations that are rapidly interconverted at room temperature by internal rotation.
   a. Internal rotation is the rotation about a single bond.
   b. Staggered conformations about single bonds are favored.
3. The conformation that minimizes van der Waals repulsions has the lowest energy and is the predominant one.
   a. The van der Waals radius is the distance from the nucleus of one atom that the nucleus of a second *nonbonded* atom can approach without causing a repulsive interaction.
   b. The van der Waals radius is a measure of an atom's effective size.
   c. Van der Waals repulsion is the extra energy required to force two nonbonded atoms within the sum of their van der Waals radii.
4. Conformations in which larger groups are brought closer together are less stable than conformations in which these groups are farther apart.
5. A Newman projection is a type of planar projection along *one* bond, called the projected bond.
   a. To draw a Newman projection, start with a circle.
   b. The bonds to the nearer atom of the projected bond are drawn from the center of the circle.
   c. The bonds to the farther atom of the projected bond are drawn from the edge of the circle.

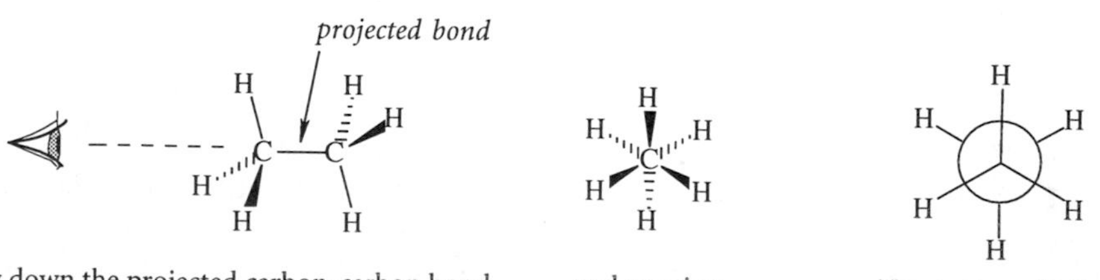

6. The most stable conformation—the conformation of lowest energy—is present in the greatest amount.

B. Ethane Conformations

1. The staggered conformation of ethane has a dihedral angle of 60°.
2. The eclipsed conformation of ethane has a dihedral angle of 0°.

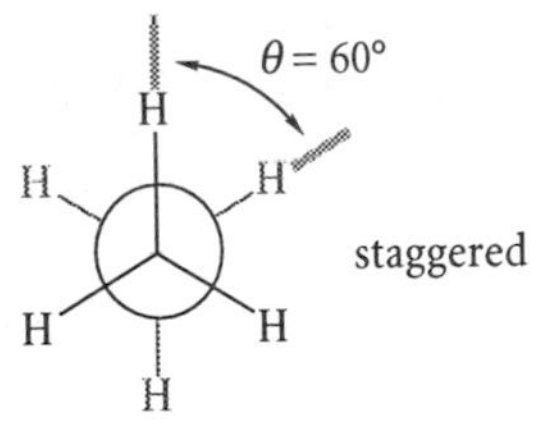

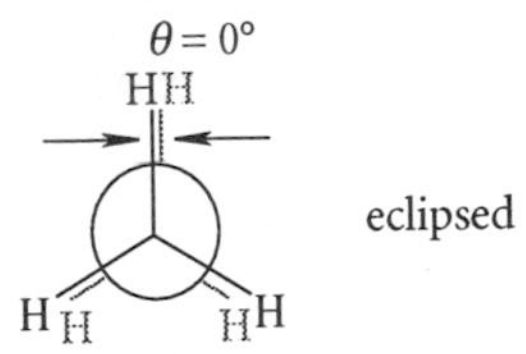

$\theta = 60°$ staggered

$\theta = 0°$ eclipsed

C. Butane Conformations

1. The anti conformation of butane has a dihedral angle of 180° between C—CH_3 bonds and is the major conformation.
2. The gauche conformations of butane have dihedral angles of ±60° between C—CH_3 bonds and exist in lower concentration than the anti conformation.

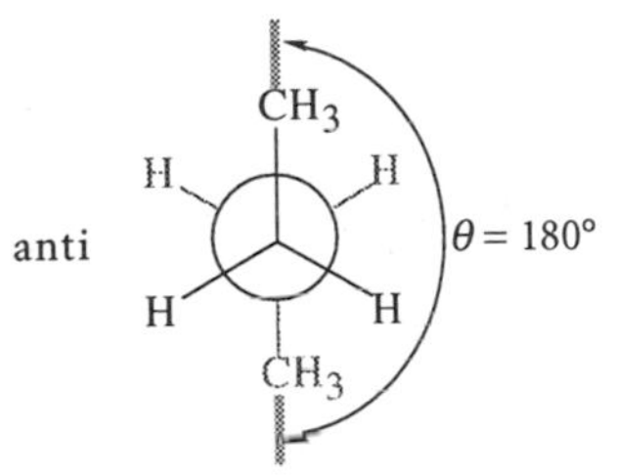

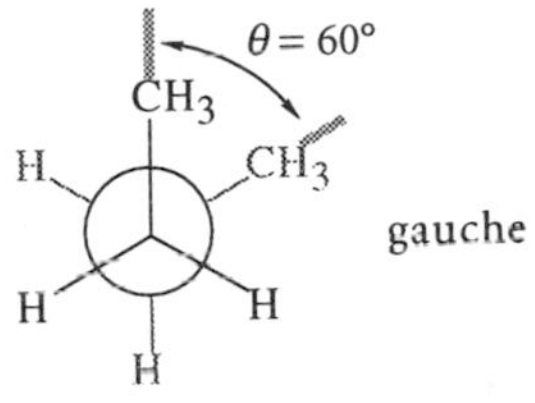

anti $\theta = 180°$

$\theta = 60°$ gauche

D. Energetics

1. Identical conformations have identical energies.
2. For ethane:
   a. The eclipsed conformation of ethane is characterized by an energy maximum.
   b. The staggered conformation of ethane is characterized by an energy minimum.
   c. The staggered conformation is the more stable conformation of ethane.
   d. The higher energy associated with the eclipsing of bonds is called torsional energy or torsional strain.
3. For butane:
   a. The anti conformation of butane is characterized by an energy minimum.
   b. The gauche conformations of butane are characterized by two identical energy minima.
   c. The anti conformation is the more stable conformation of butane; its energy is less than that of the two gauche conformations.

IV. NOMENCLATURE

A. General

1. Alkanes are named systematically according to the substitutive nomenclature rules of the IUPAC.
2. The name of a compound is based on its principal chain, which for an alkane is the longest continuous carbon chain in the molecule.
   a. Branching groups are in general termed substituents; substituents derived from alkanes are called alkyl groups.
   b. An alkyl group is named by substituting the *ane* suffix in the name of the parent alkane with a *yl* suffix: alkyl group name = alkane name – *ane* + *yl*.

B. Unbranched Alkanes

1. The names of the unbranched alkanes are derived from the number of carbons.
2. The names of the unbranched alkanes are listed in Table 2.1 on page 45 of the text.

8 carbons H—C—C—C—C—C—C—C—C—H (each carbon bearing two H) octane

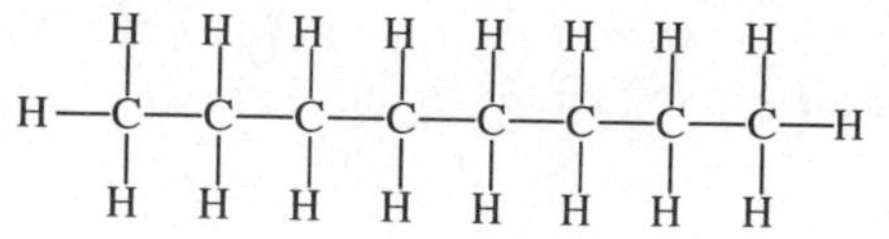

C. Branched Alkanes

1. Determine the principal chain: alkanes containing branched carbon chains are named according to the principal chain.
   a. The principal chain is the longest continuous chain in the molecule no matter how the molecule is drawn.
   b. In structures having two or more chains with the same length, the principal chain is the one containing the greater number of branches.

⇨

2. Number the carbons of the principal chain consecutively from one end to the other in the direction that gives the branches the lower numbers.
   a. When there are multiple substituent groups on the principal chain, each substituent receives its own number.
   b. When there are substituent groups at more than one carbon of the principal chain, alternative numbering schemes are compared number by number, and the one is chosen that gives the lower number at the first point of difference.
   c. When the numbering of different groups is not resolved by the other rules, the first-cited group receives the lowest number.

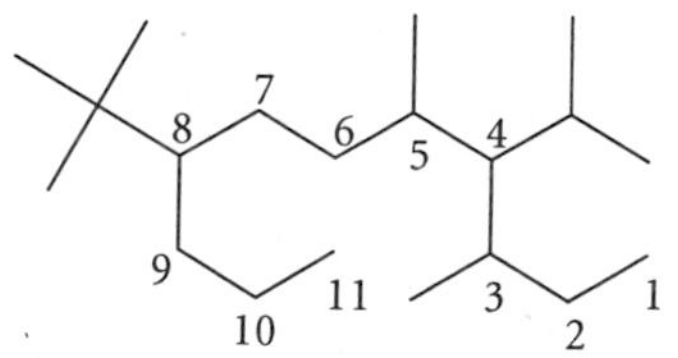

3. Name each branch.
   a. Identify the carbon number of the principal chain at which it occurs.
   b. The prefixes di-, tri-, tetra-, etc., are used to indicate the number of identical substituents.

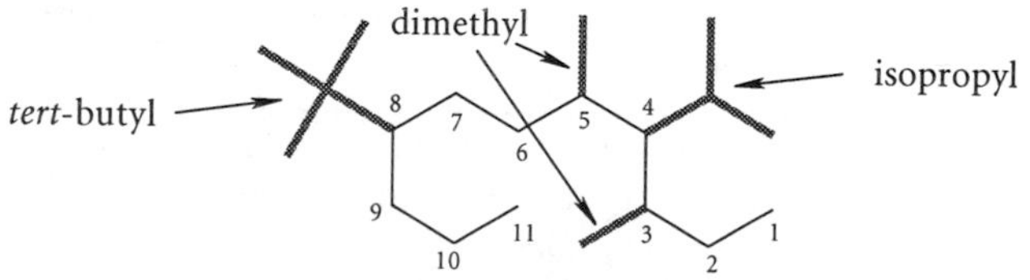

dimethyl

tert-butyl

isopropyl

4. Alphabetize the substituent groups regardless of their location on the principal chain.
   a. The prefixes iso-, neo-, and cyclo- are considered in alphabetizing.
   b. Numerical prefixes such as di- and tri- are ignored in alphabetizing.
   c. The prefixes *tert*- and *sec*- are ignored in alphabetizing.

*Alphabetical order of citation: tert-*butyl, isopropyl, dimethyl

5. Construct the name by writing the carbon number of the principal chain at which the substituent occurs, a hyphen, the name of the branch, and the name of the alkane corresponding to the principal chain.

8-*tert*-butyl-4-isopropyl-3,5-dimethylundecane

D. Cycloalkanes

1. Cycloalkanes are alkanes that contain carbon chains in closed loops, or rings.
2. The ring is treated as the principal chain if it has the greatest number of carbons.
   a. The prefix cyclo- is attached to the alkane name corresponding to the number of carbons in the ring.
   b. The numerical prefix 1- is not necessary for cycloalkanes having only one substituent.
3. When a noncyclic carbon chain contains more carbons than an attached ring, the ring is treated as a substituent.

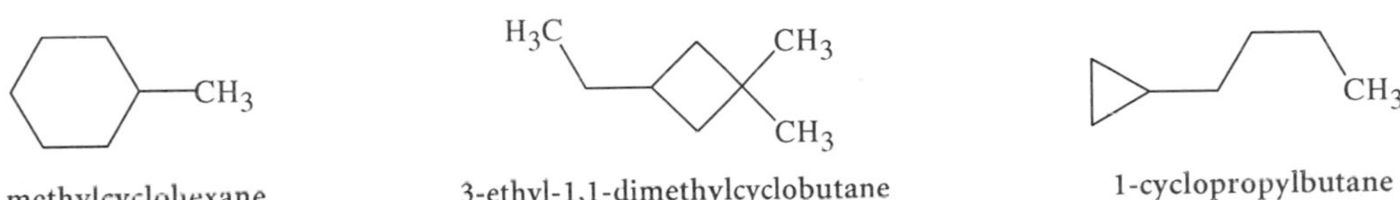

methylcyclohexane 3-ethyl-1,1-dimethylcyclobutane 1-cyclopropylbutane

V. PHYSICAL PROPERTIES

A. General

1. The physical properties of an organic compound determine the conditions under which the compound is handled and used, and are also important in characterizing the compound.
2. When two molecules approach each other closely, as in a liquid, the electron clouds of one molecule interact with the electron clouds of the other.
   a. As a result, both molecules temporarily acquire small localized separations of charge called induced dipoles.
   b. The attraction between the dipoles of different molecules is an example of an attractive van der Waals force (dispersion force).
3. Generally, physical properties within a homologous series vary in a regular way, and members of a homologous series undergo essentially the same reactions.

B. Boiling Point

1. The boiling point is the temperature at which the vapor pressure of a substance is equal to atmospheric pressure.
   a. The boiling point is a crude measure of the attractive (van der Waals) forces among molecules in the liquid state compared to those in the gaseous state.
   b. Boiling points increase with increasing molecular weight within a homologous series—typically 20–30 °C per carbon atom.
2. The shape of a molecule is also important in determining its boiling point.
   a. Large molecules have relatively high boiling points.
   b. Highly branched molecules have relatively low boiling points because they have less molecular surface available for van der Waals attractions.

C. Melting Point

1. The melting point of a substance is the temperature above which it is transformed spontaneously from the solid to the liquid state.
   a. Melting points tend to increase with increasing molecular weight within a series.
   b. Highly symmetrical molecules have particularly high melting points.
2. A sawtooth pattern of melting point behavior is observed within some series.

D. Other Physical Properties

1. A molecule's dipole moment (see Sect. 1.2D) determines its polarity, which in turn, affect its physical properties.
   a. Alkanes have negligible dipole moments.
   b. Alkanes are nonpolar molecules.
2. Solubilities are important in determining which solvents can be used to form solutions; most reactions are carried out in solution.
   a. Water solubility is a crucial factor in the activity of drugs and other biologically important compounds.
   b. Alkanes are, for all practical purposes, insoluble in water.
3. The density of a compound is the weight of a compound per unit volume at a given temperature.
   a. Alkanes have considerably lower densities than water.
   b. A mixture of an alkane and water will separate into two distinct layers with the less dense alkane layer on top.

REACTIONS

I. REACTION OF HYDROCARBONS

A. Combustion

1. Combustion is the most important reaction of alkanes.
2. Complete combustion affords carbon dioxide and water as the only combustion products.

$$C_nH_{2n+2} + \frac{3n+1}{2} O_2 \longrightarrow n\, CO_2 + (n+1)\, H_2O$$

3. Combustion can be used in elemental analysis, which is the quantitative determination of elemental compositions.
   a. The mass percents of the elements obtained from a combustion analysis in a compound can be used to determine directly the empirical formula.

SCIENTISTS

STUDY GUIDE LINKS

✓2.1 Newman Projections

The discussion of butane in the text utilizes Newman projections about the central carbon-carbon bond. You should realize that a Newman projection can be drawn for *any* bond in a molecule. For example, a Newman projection about the bond between carbons 1 and 2 in butane is as follows:

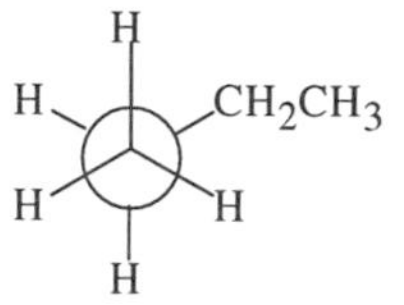

or

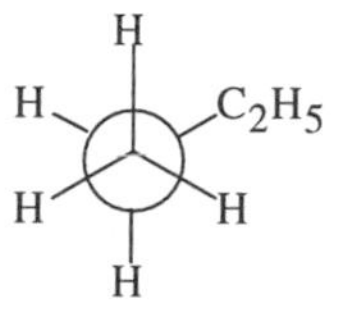

Of course, this projection doesn't show the gauche or anti relationship between the CH_3 groups at each end of the molecule; this relationship is shown only by a Newman projection about the *central* carbon-carbon bond. When we use Newman projections, we project the bond that shows the relationships of interest.

2.2 Atomic Radii

You may be familiar with the *atomic radii* as a measure of atomic size. An atom's atomic radius is its effective size when it is *covalently bonded to another atom*. The van der Waals radius, in contrast, measures an atom's effective size when it is *not covalently bonded* to another atom.

We would expect the covalent radius of an atom to be much smaller than its van der Waals radius. Why? Because electrons are shared in a chemical bond, nuclei are drawn closer together than they would be in the absence of such sharing. *Nonbonded* atoms do not share their electrons, by definition.

Let's look at a simple example. The bond length of H_2 (H—H, or dihydrogen) is 0.746Å. If we define the covalent radius as half of the separation of the two hydrogen nuclei, then the covalent radius of hydrogen is 0.746Å/2 = 0.378Å. As the text states, the *van der Waals radius* of hydrogen is 1.2Å. Clearly, a covalently bound hydrogen is separated much less from its bonded partner than it is if it approaches another atom.

If nonbonded atoms are pushed together so that their electron clouds overlap, they resist. This resistance is caused by the van der Waals repulsions. Pushing atoms together can result from applying pressure. We know if we apply pressure to a balloon, the instant we release the pressure, the balloon springs back to its original shape. This is because the molecules of gas inside the balloon resist being pushed together. In some structures, atoms which are not bonded to each other are forced to violate each other's van der Waals radii; *gauche*-butane is such a structure. The resulting van der Waals repulsions cause the molecule to have a greater internal energy.

In summary: When two nonbonded atoms interact, it is their *van der Waals radii* that define their spatial requirement. Forcing two nonbonded atoms to violate their van der Waals radii results in *van der Waals repulsions.*

✓2.3 Nomenclature of Simple Branched Compounds

The IUPAC recognizes as valid the following older names for branched alkane isomers of four, five, and six carbon atoms because of their common historical usage.

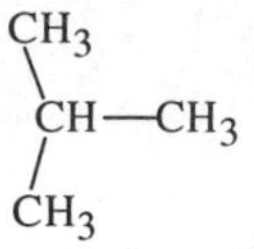

$$\begin{array}{l} CH_3 \\ \ \ \backslash \\ \ CH{-}CH_3 \\ \ / \\ CH_3 \end{array} \quad \text{or} \quad (CH_3)_3CH$$

isobutane

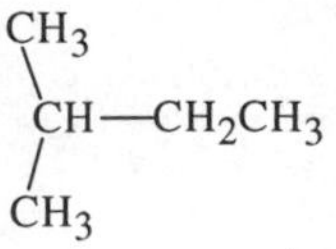

$$\begin{array}{l} CH_3 \\ \ \ \backslash \\ \ CH{-}CH_2CH_3 \\ \ / \\ CH_3 \end{array} \quad \text{or} \quad (CH_3)_2CHCH_2CH_3$$

isopentane

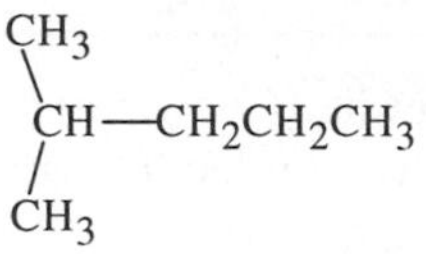

$$\begin{array}{l} CH_3 \\ \ \ \backslash \\ \ CH{-}CH_2CH_2CH_3 \\ \ / \\ CH_3 \end{array} \quad \text{or} \quad (CH_3)_2CHCH_2CH_2CH_3$$

isohexane

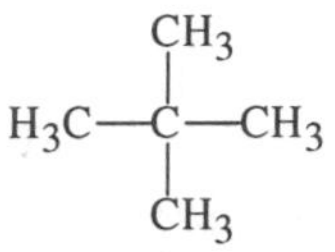

$$\begin{array}{c} CH_3 \\ | \\ H_3C{-}C{-}CH_3 \\ | \\ CH_3 \end{array} \quad \text{or} \quad (CH_3)_4C$$

neopentane

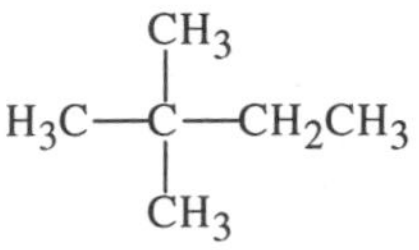

$$\begin{array}{c} CH_3 \\ | \\ H_3C{-}C{-}CH_2CH_3 \\ | \\ CH_3 \end{array} \quad \text{or} \quad (CH_3)_3CCH_2CH_3$$

neohexane

The distinguishing feature of an alkane named with the "iso" prefix is a pattern of two methyl branches at the end of a carbon chain. The distinguishing feature of an alkane with a "neo" prefix is a pattern of three methyl branches at the end of a carbon chain.

If you refer to Table 2.2 in the text, you'll see that the names of the branched alkyl groups in that table are derived from the names shown above for the corresponding alkanes.

SOLUTIONS TO PROBLEMS

Solutions to In-Text Problems

2.1 (a) The number of hydrogens in an alkane (branched or unbranched) is $2n + 2$, where n is the number of carbons. Consequently, there are $2(18) + 2 = 38$ hydrogens in the alkane with eighteen carbons.

(b) No; all alkanes—indeed, all hydrocarbons—must have an even number of hydrogens. Thus, if $2n + 2 = 23$, then n would have to be 10.5—an impoosible number of carbons.

2.2 The difference in the boiling points of nonane, decane, and undecane in Table 2.1 is about 20° per carbon. Consequently, the boiling point of tridecane is estimated by adding 20° to the boiling point of dodecane to obtain 236°. (The actual value is 235°.) Structural formulas of tridecane are as follows:

$CH_3CH_2CH_2CH_2CH_2CH_2CH_2CH_2CH_2CH_2CH_2CH_2CH_3$

structural formula of tridecane

$CH_3(CH_2)_{11}CH_3$

condensed structural formula of tridecane

2.3 (a) The staggered conformations of isopentane are *A*, *C*, and *E*; the eclipsed conformations are *B*, *D*, and *F*.

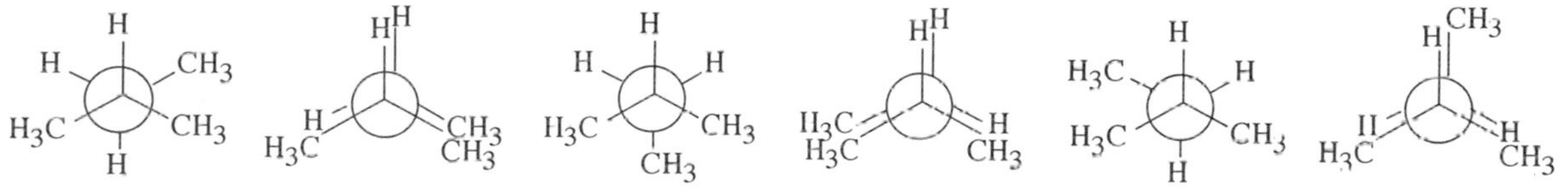

(b) The curve of potential energy versus angle of rotation is shown in Fig. SG2.1.

(c) Conformations *A* and *E* have the lowest energy because they have one fewer gauche interaction than conformation *C*, hence, either of these conformations is present in greater concentration than conformation *C*.

2.4 We might number the chain from the left, in which case the two branches receive the numbers (2,4), or from the right, in which case the two branches receive the numbers (3,5). Because 2 is lower than 3, we number from the left, and the correct name is 2,4-dimethylhexane.

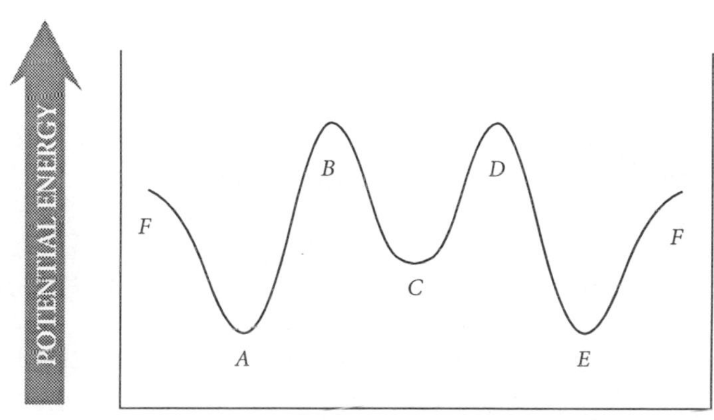

Figure SG2.1. A curve of potential energy versus angle of internal rotation to accompany the solution to Problem 2.3. The conformations are shown in the problem.

2.5 This type of problem requires a systematic approach. First, draw the structure of heptane itself.

$$CH_3CH_2CH_2CH_2CH_2CH_2CH_3 \quad \textbf{heptane}$$

Next, draw the structures with a principal chain of six carbons and one methyl branch:

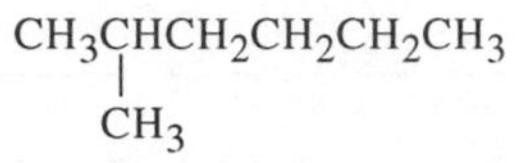

$$\begin{array}{c} CH_3\underset{\displaystyle |\atop\displaystyle CH_3}{C}HCH_2CH_2CH_2CH_3 \\ \textbf{2-methylhexane} \end{array} \qquad \begin{array}{c} CH_3CH_2\underset{\displaystyle |\atop\displaystyle CH_3}{C}HCH_2CH_2CH_3 \\ \textbf{3-methylhexane} \end{array}$$

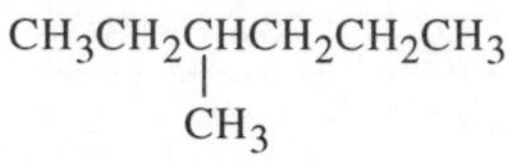

Notice that moving the methyl branch one or more carbons to the right gives identical structures. Next, draw the structures with five carbons in their principal chains and two methyl branches:

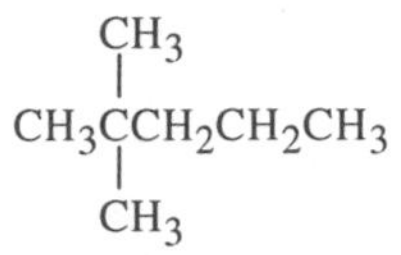

$$\begin{array}{c} CH_3\overset{\displaystyle CH_3\atop\displaystyle |}{\underset{\displaystyle |\atop\displaystyle CH_3}{C}}CH_2CH_2CH_3 \\ \textbf{2,2-dimethylpentane} \end{array} \quad \begin{array}{c} CH_3CH_2\overset{\displaystyle CH_3\atop\displaystyle |}{\underset{\displaystyle |\atop\displaystyle CH_3}{C}}CH_2CH_3 \\ \textbf{3,3-dimethylpentane} \end{array} \quad \begin{array}{c} CH_3\overset{\displaystyle CH_3\atop\displaystyle |}{C}H\underset{\displaystyle |\atop\displaystyle CH_3}{C}HCH_2CH_3 \\ \textbf{2,3-dimethylpentane} \end{array} \quad \begin{array}{c} CH_3\overset{\displaystyle CH_3\atop\displaystyle |}{C}HCH_2\underset{\displaystyle |\atop\displaystyle CH_3}{C}HCH_3 \\ \textbf{2,4-dimethylpentane} \end{array}$$

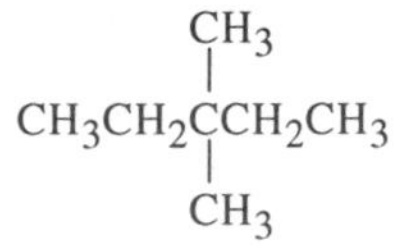

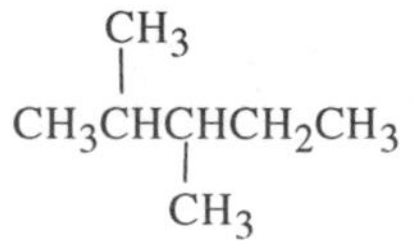

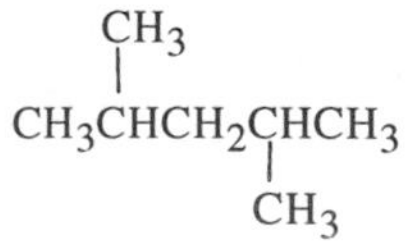

Then there is the structure with one ethyl branch:

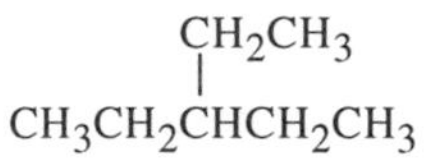

$$\begin{array}{c} CH_3CH_2\overset{\displaystyle CH_2CH_3\atop\displaystyle |}{C}HCH_2CH_3 \\ \textbf{3-ethylpentane} \end{array}$$

Finally, there is one structure with a four-carbon principal chain:

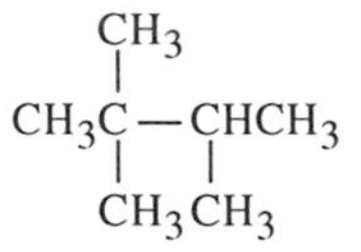

$$\begin{array}{c} CH_3\overset{\displaystyle CH_3\atop\displaystyle |}{\underset{\displaystyle |\atop\displaystyle CH_3}{C}}-\underset{\displaystyle |\atop\displaystyle CH_3}{C}HCH_3 \\ \textbf{2,2,3-trimethylbutane} \end{array}$$

Notice that any other structures with substituents on a four-carbon chain are identical to structures already considered.

2.6 (a) The designation and numbering of the principal chain are shown below.

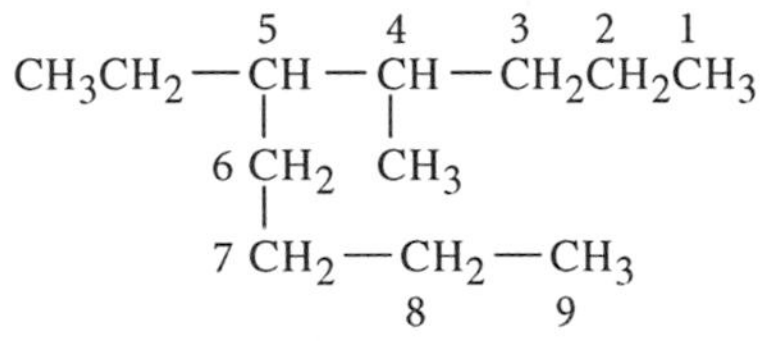

$$\begin{array}{l} \qquad\qquad\quad\ 5 \qquad\ \ 4 \qquad\ \ 3 \quad 2 \quad 1 \\ CH_3CH_2-\underset{\displaystyle |\atop\displaystyle 6\ CH_2}{C}H-\underset{\displaystyle |\atop\displaystyle CH_3}{C}H-CH_2CH_2CH_3 \\ \qquad\qquad\quad\ \ | \\ \qquad\qquad\ 7\ CH_2-\underset{8}{CH_2}-\underset{9}{CH_3} \end{array}$$

Numbering is in the direction shown because it gives the lower number at first point of difference. The compound is a substituted nonane; it has branches at carbon-4 and carbon-5. The name is 5-ethyl-4-methylnonane. (Notice the alphabetical citation of substituents.)

(b) Numbering from either direction gives the branch numbers (4,5,5,6); for that reason, alphabetical citation (Rule 10) provides the basis for numbering. The name is 4-ethyl-5,5,6-trimethylnonane.

2.7 The compound is 3-ethyl-4-propylheptane, and the expanded structure is as follows:

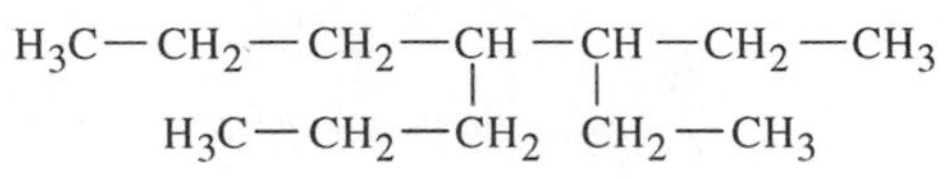

$$\begin{array}{l} H_3C-CH_2-CH_2-CH-CH-CH_2-CH_3 \\ \qquad\qquad\qquad\qquad\ \ |\qquad\ \ | \\ \qquad\ \ H_3C-CH_2-CH_2\ \ CH_2-CH_3 \end{array}$$

Your structure might look different from this one. Any structure with the same connectivity as this one is correct.

2.8 The structure of 4-isopropyl-2,4,5-trimethylheptane:

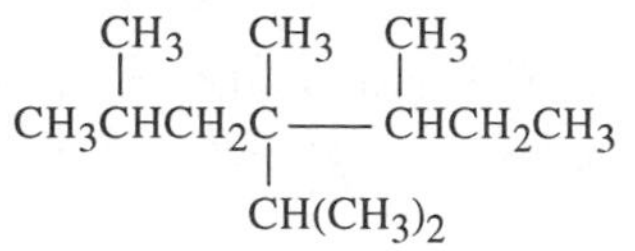

```
        CH3    CH3    CH3
        |      |      |
CH3CHCH2C ——— CHCH2CH3     4-isopropyl-2,4,5-trimethylheptane
        |
        CH(CH3)2
```

2.9 (a) In the following structure, p = primary, s = secondary, t = tertiary, and q = quaternary.

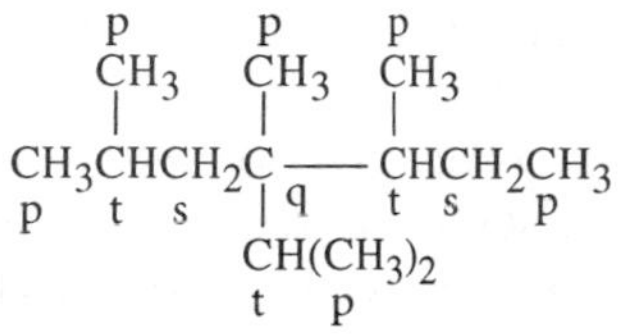

```
        p      p      p
        CH3    CH3    CH3
        |      |      |
CH3CHCH2C ——— CHCH2CH3
 p  t  s |q    t  s   p
        CH(CH3)2
        t    p
```

(b) The primary hydrogens are those in the above structure at the carbons labeled "p"; the secondary hydrogens are those at the carbons labeled "s"; and the tertiary hydrogens are those at the carbons labeled "t."

(c) The methyl groups are the CH_3 groups in the structure of part (a) with carbons that are labeled with a "p". The other groups are circled in the structures below.

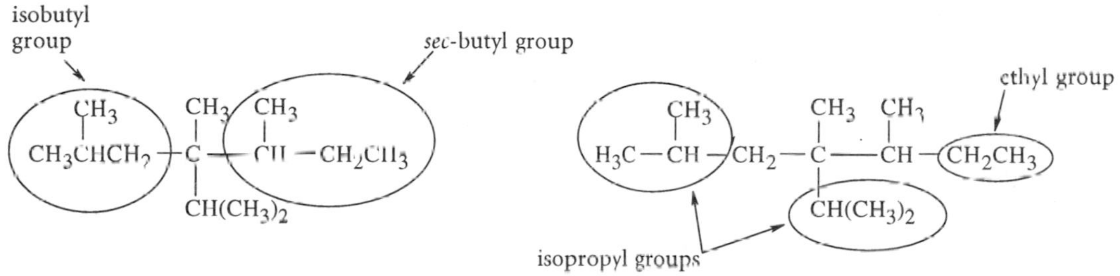

```
 CH3        CH3  CH3
 |          |    |
CH3CHCH2 —  C —  CH — CH2CH3
            |
            CH(CH3)2

      CH3            CH3  CH3
      |              |    |
H3C — CH — CH2 — C ——— CH — CH2CH3
                 |
                 CH(CH3)2
```

Note that a compound can contain alkyl groups that are not explicitly cited in the name. For example, the ethyl group in this compound is part of the principal chain and therefore is not cited in the name. Some of the methyl groups are part of larger groups.

2.10 (a) (b)

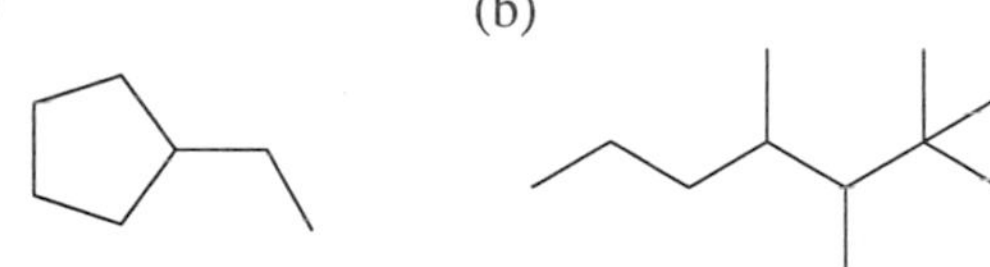

Notice that other correct structures may not look exactly like these. Don't forget that there are many equivalent ways to draw a given structure. The important point is that the connectivities of your structures and the ones above should be the same.

2.11 (a) The numbering scheme is 1,2,4-. The ethyl group receives the number 1 by Rule 10, text p. 59. The name is 1-ethyl-2,4-dimethylcyclopentane.

(b) The name is 1,1,2,4-tetraethyl-3-methylcyclobutane.

2.12 The more branched and compact molecule 2,2,3,3-tetramethylbutane should have the lower boiling point; therefore, its boiling point is 106.5°. Because it is very symmetrical, it should also have the higher melting point; its melting point is 100.7°.

Note that this compound is a liquid over only about a six-degree range; as a solid, it has a very high vapor pressure, especially at temperatures approaching its melting point. For this reason, this compound readily *sublimes* (is converted directly from a solid into a gas and vice-versa). If we gently heat the solid, it will evaporate and condense as a solid on a cooled condenser. This process, called *sublimation,* is used to purify solids that have a relatively high vapor pressure.

2.13 Water is not very effective in distinguishing a gasoline fire because the gasoline is not "smothered" by the water. Because it is less dense than water, gasoline always rises to the surface of water where the air supports further combustion.

2.14 (a) Recall that water has a density of 1.00 g/mL. Because water is less dense than bromoethane, water is the upper layer and bromoethane is the lower layer.

(b) Clearly, two of the three substances are mutually soluble. From part (a), we know that water and bromoethane are insoluble—they form two layers. From the reading, we know that hexane, a hydrocarbon, is insoluble in water. We can only conclude that one layer is water and the other is a solution of hexane and bromoethane. Of course, if volumes are nearly additive, the hexane-containing layer will be the one with larger volume. But we are not given which layer has the larger volume, so let's decide the answer in another way.

Because we use equal volumes of hexane and bromoethane, a reasonable conclusion is that the density of the mixture is the average of the two densities, that is, (1.46 g/mL + 0.66 g/mL)/2 = 1.06 g/mL. The mixture is somewhat more dense than water, and is therefore the lower layer.

2.15 C, 48.64 mass %; H, 8.16 mass %; O, 43.20 mass %.

2.16 The empirical formula of a compound with molecular formula $C_3H_6O_2$ is the same as the molecular formula, because this formula is the smallest one that contains whole-number ratios for the elements.

2.17 (a) Follow the procedure used in Study Problem 2.6. First, the number of mg of carbon in the compound is the fraction of CO_2 that is carbon (= 12.01/44.01) times the number of mg of CO_2 produced.

$$\text{mg C} = (12.01\text{ mg C}/44.01\text{ mg CO}_2)(31.95\text{ mg CO}_2) = 8.72$$

The number of mg of hydrogen must be 10.00 – 8.72 = 1.28. Let's verify this:

$$\text{mg H} = (2 \times 1.008\text{ mg H}/18.02\text{ mg H}_2\text{O})(11.44\text{ mg H}_2\text{O}) = 1.28$$

Thus, the mass percent carbon in the sample is (8.72/10.00) × 100 = 87.2; the mass percent hydrogen is 12.8. Notice that the entire mass of *X* is accounted for by carbon and hydrogen, as it must be if compound *X* is an alkane. The empirical formula is then $C_{(87.2/12.01)}H_{(12.8/1.008)}$ or $C_{7.26}H_{12.7}$. Dividing through by 7.26, the formula becomes $C_1H_{1.75}$ or $C_{4/4}H_{7/4}$. Multiplication by 4 gives the empirical formula of C_4H_7. This cannot be the molecular formula, because a hydrocarbon must have an even number of hydrogens. The molecular mass contribution of C_4H_7 is 55.1. Since the molecular mass of *X* is 110, then the molecular formula is twice C_4H_7, or C_8H_{14}.

(b) Combustion of a hydrocarbon C_8H_{14} follows the equation

$$C_8H_{14} + \tfrac{23}{2}O_2 \longrightarrow 8\,CO_2 + 7H_2O$$

According to this equation, 11.5 moles of oxygen are consumed per mole of hydrocarbon. Because 10.00 mg (0.000091 mole) of hydrocarbon is used in the analysis, then (11.5 mole O_2/mole *X*)(0.000091 mole *X*) mole = 0.00105 mole O_2. This amount of oxygen is equal to (0.00105 mole O_2) × (16 g O_2/mole O_2) 0.0336 g O_2 = 33.6 mg of O_2, which is the amount consumed.

2.18 (a) Any structure with five carbons, an —OH group, and *either* a ring or a double bond is correct. Two of many examples are the following:

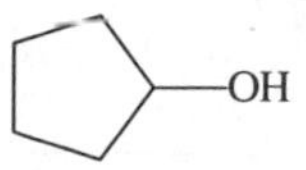

(cyclopentyl)—OH $H_2C{=}CH{-}CH_2CH_2CH_2{-}OH$

(b) A carboxylic acid has the form RCO_2H. The mass percents result in an empirical formula $C_{3.33}H_{6.66}O_{3.33}$, or CH_2O. Because a carboxylic acid must have at least two oxygens, the smallest possible molecular formula is $C_2H_4O_2$. A structure corresponding to this formula is that of acetic acid:

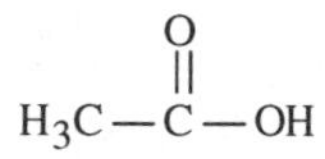

$$H_3C{-}\overset{\overset{\large O}{\|}}{C}{-}OH \quad \text{acetic acid}$$

2.19 Any structures containing nitrogen are ruled out; thus, an amide is excluded. Any structures that have double or triple bonds are ruled out, because a compound containing only carbon, hydrogen, and oxygen with five carbons and twelve hydrogens can have no double bonds. Thus, a carboxylic acid, a phenol, and an ester are also excluded. The compound could be either an ether or an alcohol or both. There are many examples of structures having the molecular formula $C_5H_{12}O_2$ that contain two ether groups, two alcohol groups, or one ether group and one alcohol group. Three examples are the following:

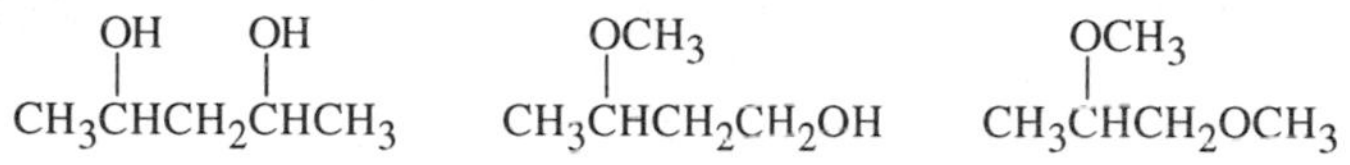

$$CH_3\overset{\overset{OH}{|}}{C}HCH_2\overset{\overset{OH}{|}}{C}HCH_3 \qquad CH_3\overset{\overset{OCH_3}{|}}{C}HCH_2CH_2OH \qquad CH_3\overset{\overset{OCH_3}{|}}{C}HCH_2OCH_3$$

Solutions to Additional Problems

2.20 (a) Use the rule of thumb that an additional carbon adds about 20–30° to the boiling point. Because the second compound has one more carbon than the first, its boiling point should be about 20–30° more. The estimated boiling point is 175–185 °C.

(b) Because the two compounds have the same functional groups and the same number of carbons, their boiling points should be very similar. An estimate is 152 °C.

2.21 The isomers of octane with five carbons in their principal chains:

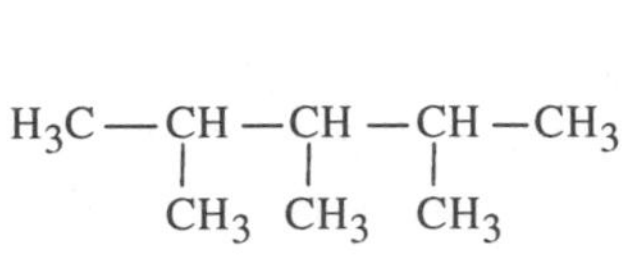

$H_3C-CH(CH_3)-CH(CH_3)-CH(CH_3)-CH_3$
2,3,4-trimethylpentane

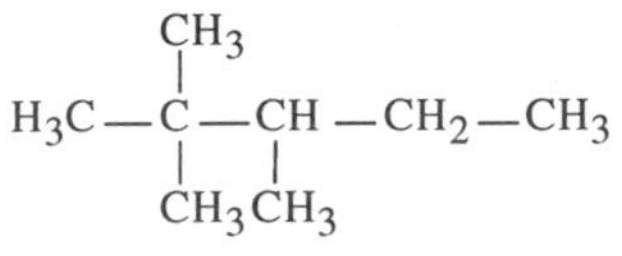

$H_3C-C(CH_3)_2-CH(CH_3)-CH_2-CH_3$
2,2,3-trimethylpentane

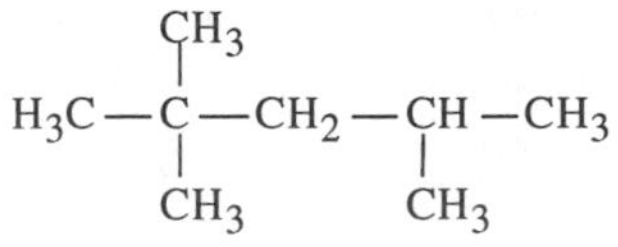

$H_3C-C(CH_3)_2-CH_2-CH(CH_3)-CH_3$
2,2,4-trimethylpentane

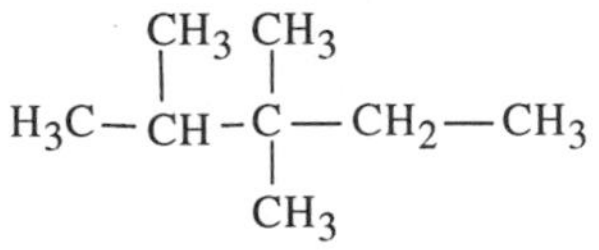

$H_3C-CH(CH_3)-C(CH_3)_2-CH_2-CH_3$
2,3,3-trimethylpentane

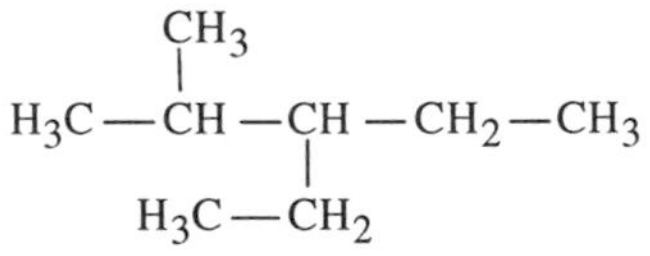

$H_3C-CH(CH_3)-CH(CH_2CH_3)-CH_2-CH_3$
3-ethyl-2-methylpentane

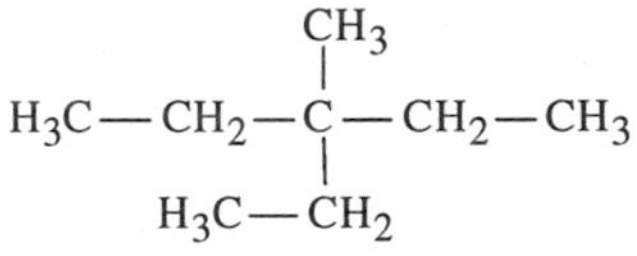

$H_3C-CH_2-C(CH_3)(CH_2CH_3)-CH_2-CH_3$
3-ethyl-3-methylpentane

2.22 (a) In the following structures, p = primary, s = secondary, t = tertiary, and q = quaternary.

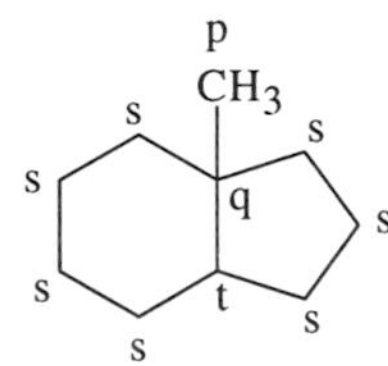

Don't forget that in skeletal structures enough hydrogens are *understood* to be present at each vertex to give four bonds to each carbon.

(b) The abbreviations are given in part (a).

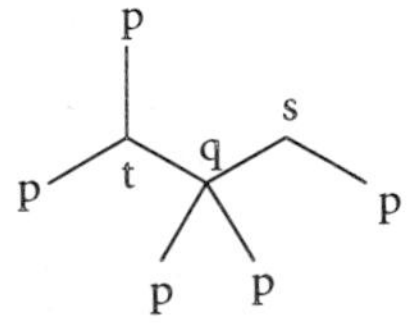

p p t q s p p p

2.23 (a) Neopentane has five carbons, and all hydrogens are primary:

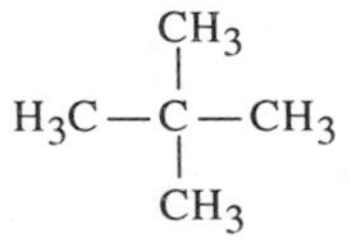

$H_3C-C(CH_3)_2-CH_3$ neopentane

(b) Cyclopentane has five carbons and only secondary hydrogens:

cyclopentane

(c) This one is tricky. The structures that fit this description have several rings fused together. Among these are the following, along with their trivial names.

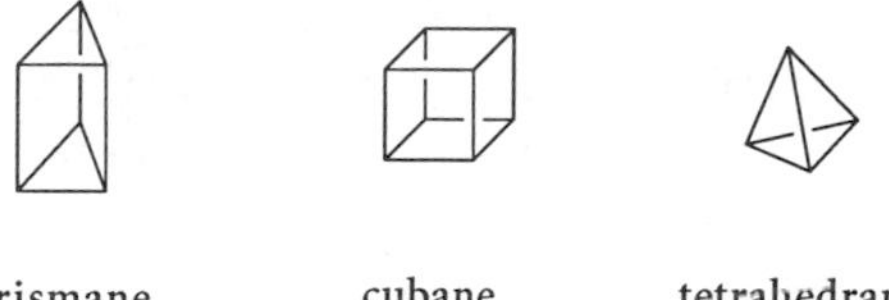

prismane cubane tetrahedrane

2.24 (a) 5-ethyl-4,4-dimethyloctane. Note that the numbering is governed by the "first point of difference" rule (Rule 8, text page 58).

(b) 2,3-dimethylhexane. Note that the principal chain need not be drawn in a linear manner.

(c) 4-isopropyloctane

(d) In this example, alternative numbering schemes are compared: (1,2,2,3), (1,1,2,5), and (1,4,5,5); notice that the first point of difference among these schemes occurs at the second number, and that the *second* of these schemes thus has the lowest number at the first point of difference. The name, then, is 2,5-diethyl-1,1-dimethylcyclopentane.

2.25 (a)

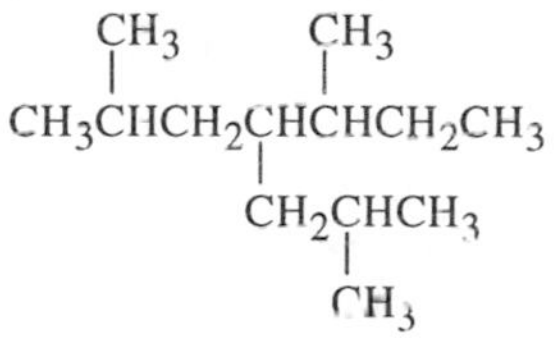

CH_3 CH_3
$CH_3CHCH_2CHCHCH_2CH_3$
CH_2CHCH_3
CH_3

4-isobutyl-2,5-dimethylheptane

(b)

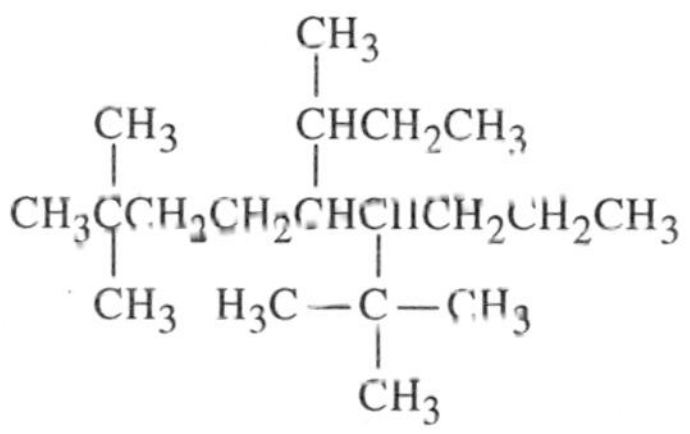

CH_3
CH_3 $CHCH_2CH_3$
$CH_3CCH_2CH_2CHCHCH_2CH_2CH_3$
CH_3 $H_3C-C-CH_3$
CH_3

5-*sec*-butyl-6-*tert*-butyl-2,2-dimethylnonane

2.26 (a) This compound should be named 2,4,6,6-tetramethyloctane:

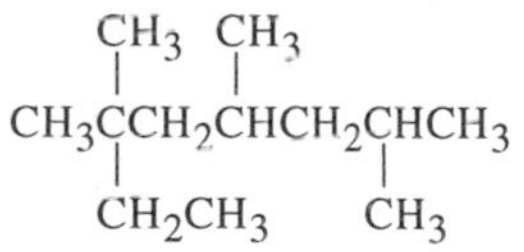

CH_3 CH_3
$CH_3CCH_2CHCH_2CHCH_3$
CH_2CH_3 CH_3

2,4,6,6-tetramethyloctane

(b) 5-neopentyldecane is a correct name.

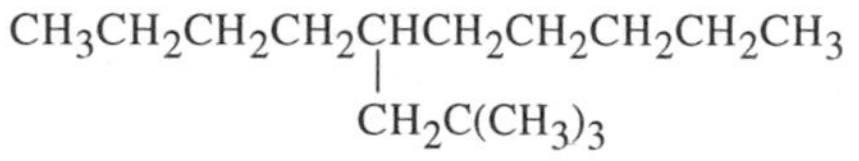

$CH_3CH_2CH_2CH_2CHCH_2CH_2CH_2CH_2CH_3$
$CH_2C(CH_3)_3$

5-neopentyldecane

(c) Ima forgot about the "first point of difference" rule.

CH_3
CH_3

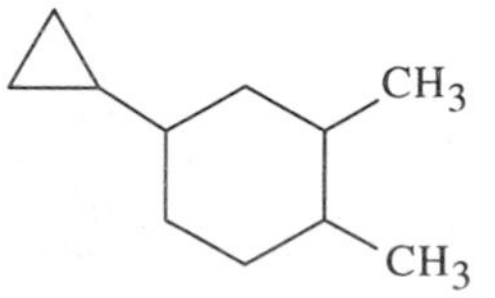

4-cyclopropyl-1,2-dimethylcyclohexane

(d) This compound should be named 4-*tert*-butyloctane:

```
    CH3 CH2CH2CH2CH3
     |   |
H3C—C —CHCH2CH2CH3        4-tert-butyloctane
     |
    CH3
```

2.27 (a) The connectivity of *A*, starting near the observer, is [CH_3, C connected to C_2H_5 and to CH_3, $CH(CH_3)_2$]. The connectivity of *B* is [$(CH_3)_2CH$, CH connected to C_2H_5, CH_2, CH_3], which is the same as [$(CH_3)_2CH$, $CH(C_2H_5)_2$], which is the same as *C*. Hence, structures *B* and *C* are Newman projections of the same compound, 3-ethyl-2-methylpentane. Compound *A* is 2,3,3-trimethylpentane. You should have discovered that naming the compounds is a lot easier if you draw out the structures rather than working directly from the Newman projections.

A common mistake in interpreting Newman projections is to forget the carbon that is "hidden." Be sure to remember that the Newman projection represents a bond to *two* carbons: the nearer one for which bonds are drawn to the center of the circle; and the farther one, for which bonds are drawn to the periphery of the circle.

(b) Structures *A* and *C* both have the connectivity [$(CH_3)_2CH$, CH_2, $CH(CH_3)$, CH_2, CH_2, CH_3]. The connectivity of *B* is different. Consequently, structures *A* and *C* represent the same compound, 2,4-dimethylheptane. Compound *B* is 3-ethyl-2-methylheptane.

2.28 All staggered conformations (*A*) of chloroethane are identical and thus have identical energies; all eclipsed conformations (*B*) are identical, and also have identical energies. The energy barrier is the energy difference between conformations *B* and *A*. The diagram is shown in Fig. SG2.2 on the following page.

Note that setting the eclipsed conformation *B* at 0° is arbitrary; we could just as easily place the staggered conformation *A* at 0°, that is, offset the curve 60° to the right or left. The important point about your diagram is that all the minima should be at an identical energy, and all the maxima should be at an identical, higher energy, and minima and maxima should be separated by 60°. Because a dihedral angle of 0° represents an eclipsed conformation, and a dihedral angle of 180° represents a staggered conformation, it is conventional to let the angles of rotation correspond to the dihedral angles, that is, to place an eclipsed conformation at 0° and a staggered conformation at 180°.

2.29 Both curves would have the same general appearance, that is, three barriers of equal height. (See Fig. SG2.3, following page.) However, the curve for 2,2,3,3-tetramethylbutane should have greater potential-energy barriers because each eclipsed conformation has three severe methyl-methyl interactions.

2.30 (a) The conformations of 1,2-dichloroethane:

D A E B F C

The graph of energy versus angle of rotation is shown in Fig. SG2.4 on p. 5. Notice that the *C-F* barrier is identical to the *A-E* barrier; the *C-D* barrier is identical to the *A-D* barrier.

(b) Conformation *B* is present in greatest amount, because it has the lowest energy.

(Solutions text continues on p. 44)

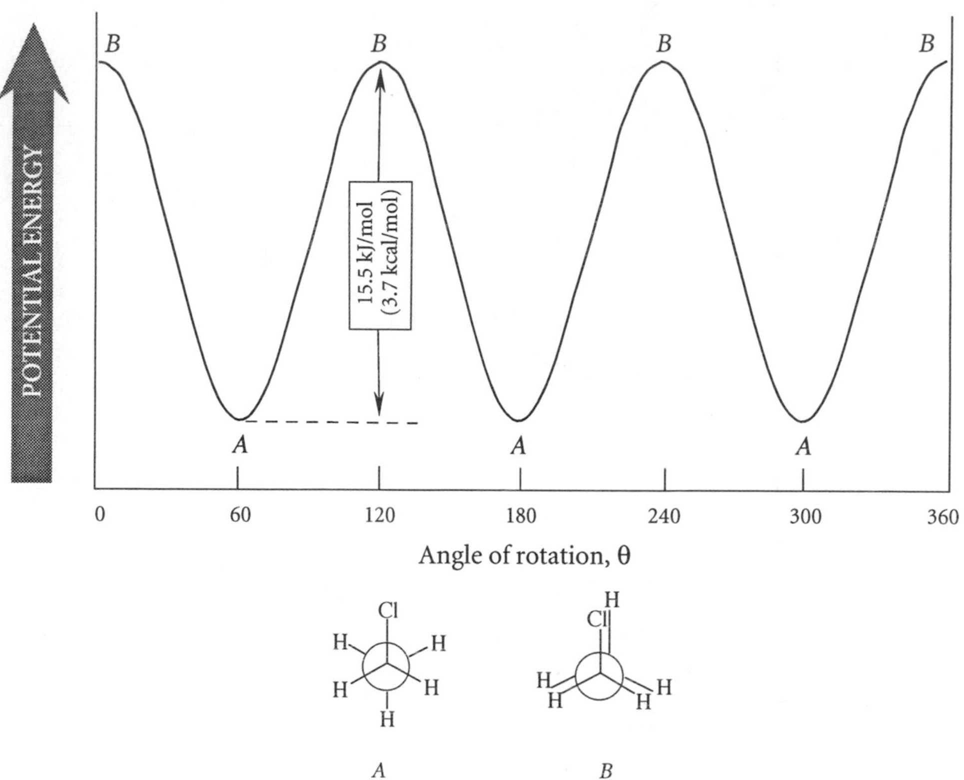

Figure SG2.2. A curve of potential energy versus angle of internal rotation for chloroethane (Problem 2.28). The conformations are shown in the problem.

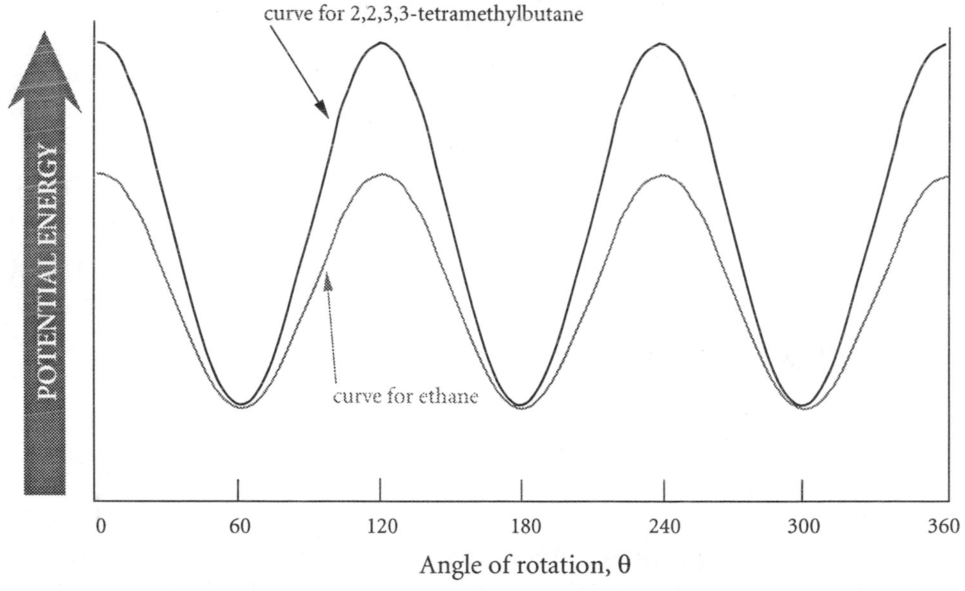

Figure SG2.3 Comparison of the curves for potential energy versus angle of internal rotation for 2,2,3,3-tetramethylbutane and ethane (Problem 2.29). The minima of the two curves have been arbitrarily placed at the same energy for comparison purposes.

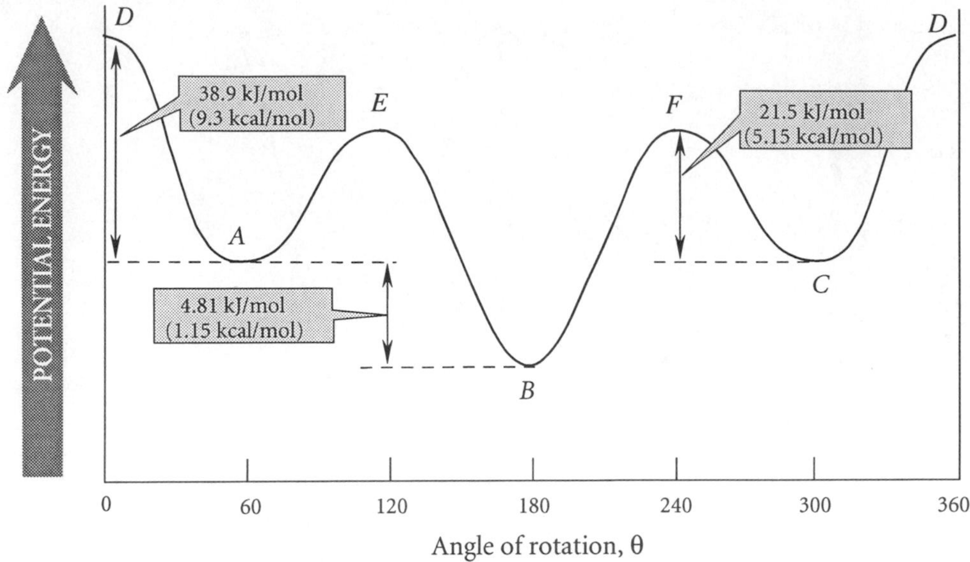

Figure SG2.4 Potential energy versus angle of rotation for 1,2-dichloroethane (Problem 2.30). Notice that the curve is identical to that for butane, except that the energies are somewhat different. The conformations are identified by letter and drawn in the solution.

2.31 Because severe van der Waals repulsions occur between the *tert*-butyl groups in compound *A* at normal bond lengths and angles, the only way for the molecule to relieve these repulsions is to stretch the appropriate bonds and flatten itself by widening the C—C—C angles. This also costs energy, but presumably not as much as retaining the smaller bond lengths and angles.

2.32 The conformational analysis of butane shows that the anti conformation is favored at each internal carbon-carbon bond. Anti conformations at each carbon-carbon bond of hexane lead to a zig-zag shape for the carbon backbone. The model and a line-wedge formula are as follows:

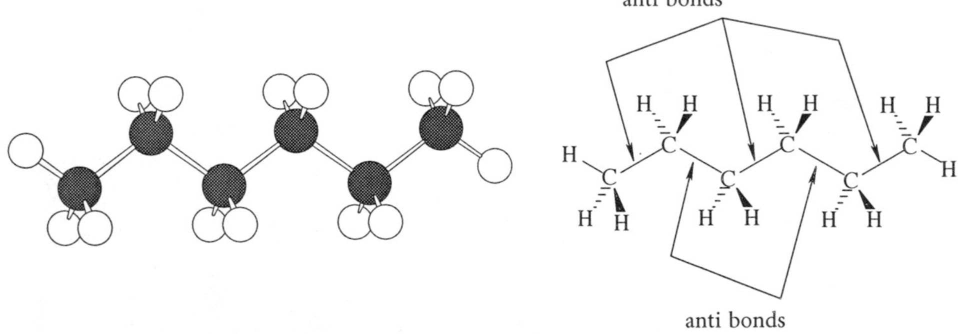

2.33 A major contributor to the barrier to internal rotation is the van der Waals repulsions between the methyl groups. In compound *B* these repulsions are smaller than they are in compound *A* because the distance between methyl groups is greater in *B*. This distance is greater because the Si—Si bond is longer than the C—C bond.

2.34 The C—O bond is somewhat shorter than the C—C bond. Consequently, when the methyl groups are gauche, the distance between them is smaller in the ether (compound A) than in butane (compound B). Consequently, van der Waals repulsions in the gauche conformation of the ether should be larger than those of the gauche conformation in butane, and the energy of the gauche conformation of the ether should be higher relative to the anti conformation, in which van der Waals repulsions are absent. In other words, the ether should have the greater energy difference between the gauche and anti conformations.

2.35 (a) The expected dipole moment for the anti conformation is zero, because the polar C—Br bonds are oriented in opposite directions; hence their bond dipoles cancel. Likewise, corresponding to every C—H bond is another C—H bond pointing in the opposite direction.

(b) It follows from part (a) that the observed dipole moment is due solely to the relatively small amount of gauche conformation present. Hence,

$$\mu = \mu_{anti}N_{anti} + \mu_{gauche}N_{gauche}$$

or $$1.0 = (0)(0.82) + \mu_{gauche}(0.18) = \mu_{gauche}(0.18)$$

Solving for μ_{gauche} gives $\mu_{gauche} = 5.56$ D. This is a very large dipole moment. But this is what would be expected for a compound in which two C—Br bond dipoles (as well as the smaller C—H bond dipoles) are oriented in nearly the same direction.

2.36 (a) A balanced equation for combustion of a cycloalkane C_nH_{2n}:

$$C_nH_{2n} + (3n/2)O_2 \longrightarrow nCO_2 + nH_2O$$

(b) Intuitively, because there is more hydrogen in a noncyclic alkane, more water is formed on combustion, and therefore more oxygen should be required. Let's see if that is correct. A balanced equation for conbustion of an alkane was given in Eq. 2.2 in the text as

$$C_nH_{2n+2} + \frac{3n+1}{2} \longrightarrow (n+1)\,H_2O + n\,CO_2$$

Comparing the coefficients of O_2 in the two equations, it's clear that $(3n + 1)/2$ is larger than $3n/2$. Hence, our intuition is correct: combustion of a noncyclic alkane requires more oxygen than combustion of a cyclic alkane.

2.37 We count the carbons and then fill in the hydrogens: two on each carbon except for carbons at the ring juctions, which have one hydrogen. The molecular formula of decalin is thus $C_{10}H_{18}$. The molecular mass of decalin is $(10)(12.01) + (18)(1.0079) = 138.24$ g mol^{-1}. The mass of carbon in decalin is 120.1 g mol^{-1}. Hence, the mass percent carbon in decalin is

$$\frac{120.1}{138.24} \times 100 = 86.88\%$$

2.38 (a) Follow the procedure in Study Problem 2.7 to obtain $C_{7.26}H_{12.72}$, or $CH_{1.75}$, or C_4H_7. The molecular formula of a hydrocarbon must have an even number of hydrogens; consequently, the smallest possible molecular formula is C_8H_{14}.

(b) Two of several structures consistent with the analysis in part (a) that have two tertiary carbons and all others secondary:

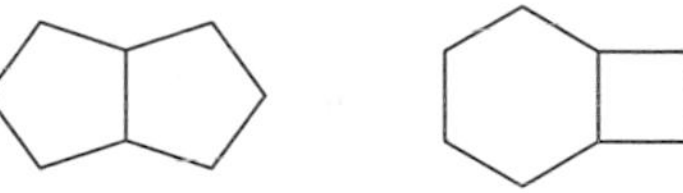

(c) An alkane consistent with this formula that has one quaternary carbon and all other carbons secondary (that is, no primary hydrogens and no tertiary carbons):

2.39 (a) First, determine the fraction of the mass of CO_2 that is oxygen (O):

$$\text{fraction O in } CO_2 = (32.00 \text{ mg of O})/(44.01 \text{ mg of } CO_2) = 0.7271 \text{ mg of O per mg of } CO_2$$

The mass of oxygen in the CO_2 formed is then

$$\text{mass of O} = (0.7271 \text{ mg of O per mg of } CO_2)(21.96 \text{ mg of } CO_2) = 15.97 \text{ mg}$$

Likewise, the mass of oxygen in the H_2O formed is

$$\text{mass of O} = (16.00 \text{ mg of O}/18.02 \text{ mg of } H_2O)(8.99 \text{ mg of } H_2O) = 7.98 \text{ mg}$$

By conservation of mass, the total oxygen consumed is equal to the total oxygen in the products, and is therefore

$$15.97 \text{ mg} + 7.98 \text{ mg} = 23.95 \text{ mg}.$$

This weight of oxygen consumed corresponds to

$$(23.95 \text{ mg of O})/(32.00 \text{ mg of O per mmol of } O_2) = 0.748 \text{ mmol of } O_2$$

At 0 °C (273 K), this corresponds to

$$(0.748 \text{ mmol})(22.4 \text{ mL/mmol}) = 16.76 \text{ mL}$$

Applying a gas-law correction to this number, the volume at 25 °C (298 K) is

$$(16.76 \text{ mL})(298 \text{ K})/(273 \text{ K}) = 18.29 \text{ mL}$$

(b) To calculate the molecular formula, apply the techniques described in Study Problem 2.6. The mass of carbon in the sample is derived from the amount of carbon dioxide, and the mass of hydrogen is derived from the amount of water:

$$\text{Mass of C} = (21.96 \text{ mg of } CO_2)(12.01 \text{ mg of C}/44.01 \text{ mg of } CO_2) = 5.99 \text{ mg}$$
$$\text{Mass of H} = (8.99 \text{ mg of } H_2O)(2.016 \text{ mg of H}/18.02 \text{ mg of } H_2O) = 1.006 \text{ mg}$$

The empirical formula is $C_{(5.99/12.01)}H_{(1.006/1.008)} = C_{0.5}H_1 = CH_2$. Because the group mass of CH_2 is 14 mass units, the molecular formula of the hydrocarbon is $(CH_2)_{10}$, or $C_{10}H_{20}$.

2.40 (a) ketone
(b) nitrile
(c) alcohol
(d) ether

2.41 Consult the inside front cover of the text for a list of common functional groups. The functional groups in acebutolol are identified and the compound classes are labeled in the following structure.

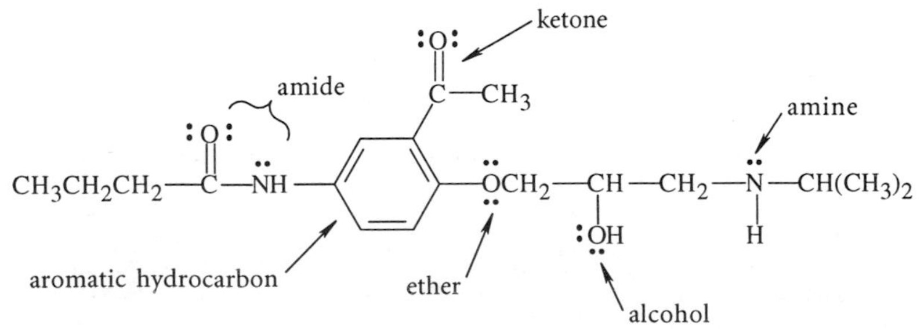

3 Acids and Bases; The Curved-Arrow Notation

TERMS

amphoteric compund 3.4A
bond dissociation energy 3.6A
Brønsted acid 3.4A
Brønsted base 3.4A
Brønsted-Lowry acid-base concept 3.4A
carboxylate ion 3.6B
conjugate acid 3.4A
conjugate acid-base pair 3.4A
conjugate base 3.4A
curved-arrow notation 3.1C
dissociation constant, K_a 3.4B
electron affinity 3.6A
electron attack 3.1C
electron-deficient compound 3.1A
electron-donating polar effect 3.6B
electron-pair displacement reaction 3.2A
electron-withdrawing polar effect 3.6B
electrophile 3.1B
electrostatic law 3.6B
element effect 3.6A
inductive effect 3.6B
ionization potential 3.6A
Lewis acid 3.1B
Lewis acid-base association reaction 3.1B
Lewis acid base dissociation reaction 3.1C
Lewis base 3.1B
nucleophile 3.1B
pK_a 3.4B
polar effect 3.6B
standard free energy of dissociation, ΔG°_a ... 3.5

CONCEPTS

I. ACIDS AND BASES

A. Electron-Deficient Compounds

1. Some compounds contain atoms that are short of a valence-electron octet by one or more electron pairs; these species are termed electron-deficient compounds.
   a. The octet rule stipulates the maximum number of electrons.
   b. Electron-deficient compounds have a tendency to undergo chemical reactions that complete their valence-shell octets.

B. Lewis Acids and Bases

1. Lewis acids
   a. A Lewis acid is a species that accepts an electron pair to form a new bond in a chemical reaction.
   b. Electron-deficient compounds constitute an important type of Lewis acid.
   c. A Lewis acid is an electrophile (an “electron-loving” species).
2. Lewis bases
   a. A Lewis base is a species that donates an electron pair to form a new bond in a chemical reaction.

b. A Lewis base is a nucleophile (a "nucleus-loving" species).

3. All Lewis acid-base reactions involve either the reactions of Lewis bases with electron-deficient compounds, or electron-pair displacements.
   a. A Lewis acid-base association reaction occurs when a Lewis acid and a Lewis base combine to give a single product.

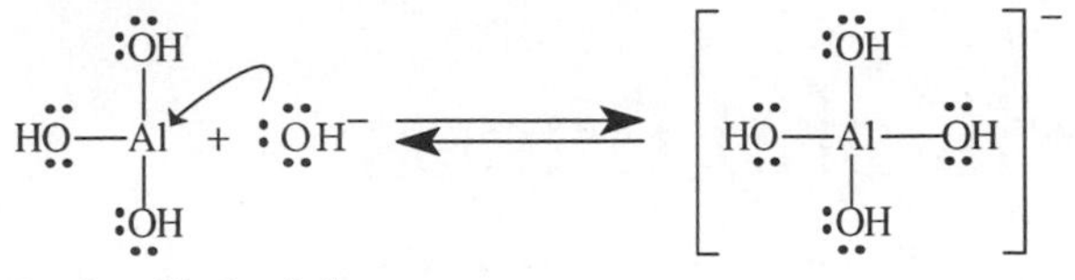

$$Al(OH)_3 + {:}\ddot{O}H^- \rightleftharpoons [Al(OH)_4]^-$$

Lewis acid Lewis base

   b. A Lewis acid-base dissociation reaction is the reverse of a Lewis acid-base association reaction.
   c. An electron-pair displacement reaction occurs when a Lewis base attacks an atom that is not electron-deficient.

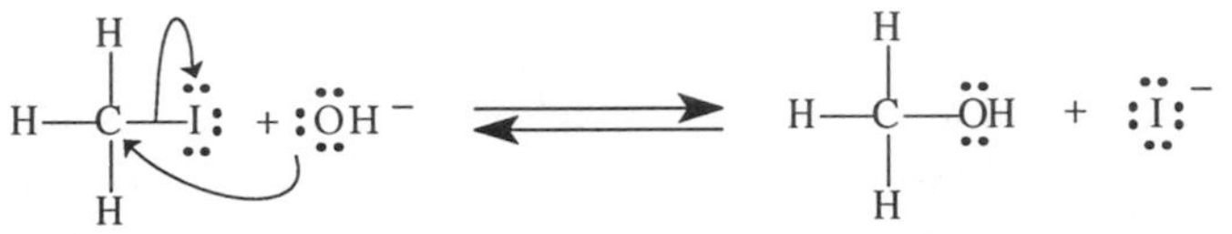

$$H_3C{-}\ddot{I}{:} + {:}\ddot{O}H^- \rightleftharpoons H_3C{-}\ddot{O}H + {:}\ddot{I}{:}^-$$

Lewis acid Lewis base

B. Brønsted-Lowry Acids and Bases

1. A Brønsted acid is a species that reacts by donating a proton.
   a. When a Brønsted acid loses a proton its conjugate base is formed. For example, the conjugate base of a carboxylic acid is called a carboxylate ion.
   b. A Brønsted acid and its conjugate base constitute a conjugate acid-base pair.

$H_3C{-}C(=O){-}OH$ — acetic acid (a carboxylic acid) Brønsted acid

$H_3C{-}C(=O){-}O^-$ — acetate ion (a carboxylate ion) conjugate base

2. A Brønsted base is a species that reacts by accepting a proton.
   a. When a Brønsted base accepts a proton, its conjugate acid is formed.
   b. A Brønsted base and its conjugate acid comprise a conjugate acid-base pair.

$CH_3\ddot{O}H$ — Brønsted base

$CH_3\overset{+}{\ddot{O}}H_2$ — conjugate acid

3. All Brønsted acid-base reactions involve two conjugate acid-base pairs; the conjugate acid-base relationship is across the equilibrium arrows.

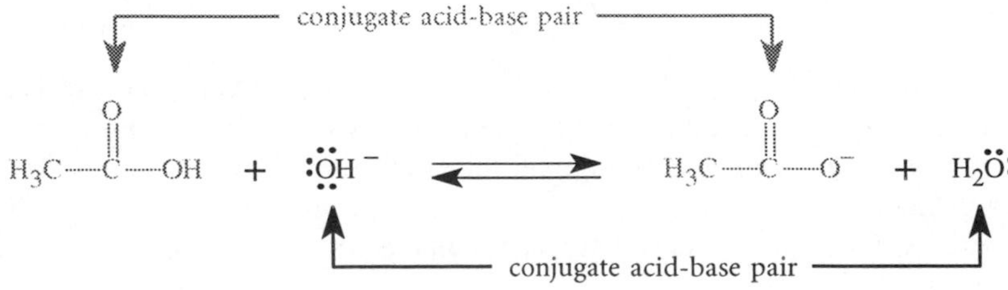

conjugate acid-base pair

$$H_3C{-}C(=O){-}OH + {:}\ddot{O}H^- \rightleftharpoons H_3C{-}C(=O){-}O^- + H_2\ddot{O}{:}$$

conjugate acid-base pair

4. An amphoteric compound can act as either an acid or a base. For example, water is an amphoteric compound.

$H_3\overset{+}{\ddot{O}}$:	$H_2\ddot{O}$:	$:\ddot{O}H^-$	$H_2\ddot{O}$:
Brønsted acid	conjugate base	Brønsted base	conjugate acid

5. Brønsted acid-base reactions are electron-pair displacement reactions in which a proton is transferred from one atom to another.
   a. The equilibrium in a Brønsted acid-base reaction always favors the side with the weaker acid and weaker base.
   b. Two structural effects on Brønsted acidity are the element effect and the polar effect.

C. Dissociation Constant

1. The dissociation constant K_a is a measure of the strength of an acid, specifically, how well a Brønsted acid donates a proton to water, the "standard" Brønsted base.
   a. The dissociation constant K_a for each Brønsted acid is unique.
   b. The strength of an acid is measured by the magnitude of its dissociation constant.
2. A logarithmic pK_a scale is used, in which pK_a = –log K_a; a stronger acid has a smaller pK_a value.
3. The strength of a Brønsted base is conveniently expressed in terms of the pK_a of its conjugate acid.
   a. If the base is strong, the conjugate acid is weak.
   b. A stronger base has the conjugate acid with the greater (more positive or less negative) pK_a.

II. FREE ENERGIES OF REACTIONS AND STRUCTURAL EFFECTS ON ACIDITY

A. Standard Free Energy $\Delta G°$

1. The standard free energy of dissociation $\Delta G_a°$ is equal to the standard free energies of the ionization products minus the standard free energy of the un-ionized acid.
   a. When $\Delta G°$ is positive, the concentration of ionization products is less than the concentration of un-ionized acid—the acid is a weak acid.
   b. When $\Delta G°$ is negative, the concentration of ionization products is greater than the concentration of the un-ionized acid—the acid is a strong acid.
2. The equilibrium constant K_{eq} for a reaction is related to the standard free-energy difference $\Delta G°$ between products and reactants by the relationship $\Delta G° = -2.3RT \log K_{eq}$.
   a. Reactions with $K_{eq} < 1$ have positive $\Delta G°$ values and favor reactants at equilibrium.
   b. Reactions with $K_{eq} > 1$ have negative $\Delta G°$ values and favor products at equilibrium.
3. The equilibrium constant for a reaction indicates which species in a chemical equilibrium are present in highest concentration.
   a. A chemical equilibrium favors the species of lower free energy.
   b. The more two compounds differ in free energy, the greater is the difference in their concentrations at equilibrium.
   c. When the free energy of a conjugate base is lower (when a conjugate base is more stable), the pK_a of the acid is smaller (the acid is more acidic).

B. Element Effect

1. For an acid H—A, a smaller bond dissociation energy increases acidity.
2. For an acid H—A, a larger electron affinity of A• increases acidity.
3. Brønsted acidity increases as the atom to which the acidic proton is attached has a greater atomic number within a column (group) of the periodic table.
   a. This trend is dominated by the H—A bond dissociation energies of the acids.
   b. Stronger acids have weaker H—A bonds.
4. Brønsted acidity increases as the atom to which the acidic proton is attached is farther to the right within a row (period) of the periodic table.
   a. This trend is dominated by electron-attracting abilities (electron affinities) of the elements A in the acids H—A.

b. Stronger acids have A groups that are more electron-attracting.

C. Polar Effect

1. The polar effect (inductive effect) on acidity is caused largely by the interaction of charges formed in an acid-base reaction with the bond dipoles of polar bonds or with other charged groups in an acid or base molecule.
   a. The electrostatic law (Eq 3.33, text p. 103) governs the interaction of charges and/or dipoles.
   b. Electronegative groups exert an electron-withdrawing effect by pulling electrons towards themselves and away from the carbon to which they are attached.
   c. Some atoms exhibit an electron-donating polar effect by pushing electrons away from themselves and towards the carbon to which they are attached.
2. Electronegative substituent groups such as halogens increase the acidities of carboxylic acids by stabilizing their conjugate-base carboxylate ions.

III. THE CURVED-ARROW NOTATION

1. The curved-arrow notation is a symbolic device for depicting the flow of electron pairs in chemical reactions.
   a. The electrons flow from the electron donor (Lewis base) to the electron acceptor (Lewis acid).
   b. The arrow is drawn form the electron source to the electron acceptor.
      i. Both the reaction of a Lewis base with an electron-deficient compound (a Lewis acid-base association reaction) and its reverse (a Lewis acid-base dissociation reaction) require one curved arrow.

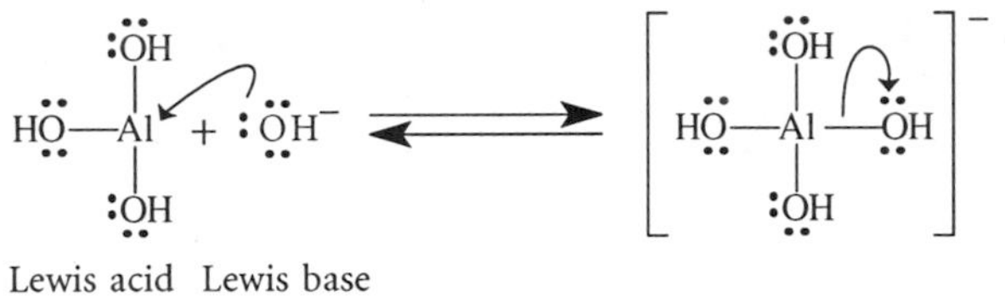

$Al(OH)_3 + {}^-OH \rightleftharpoons [Al(OH)_4]^-$

Lewis acid Lewis base

      ii. An electron-pair displacement reaction requires two or more curved arrows.

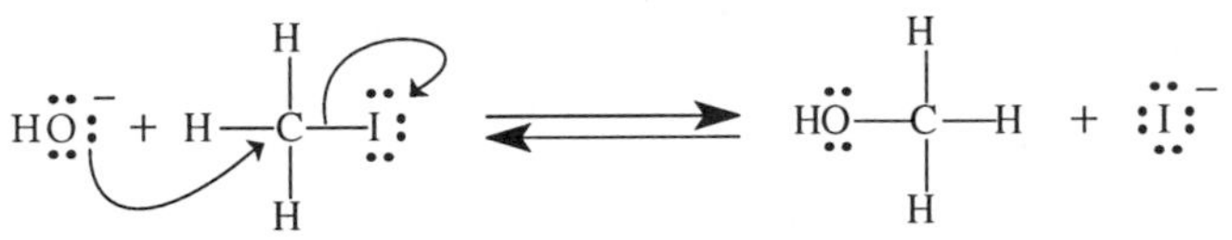

$HO^- + H_3C{-}I \rightleftharpoons HO{-}CH_3 + I^-$

      iii. A Brønsted acid-base reaction is an example of an electron-pair displacement reaction in which a proton is the atom that is attacked by the base.

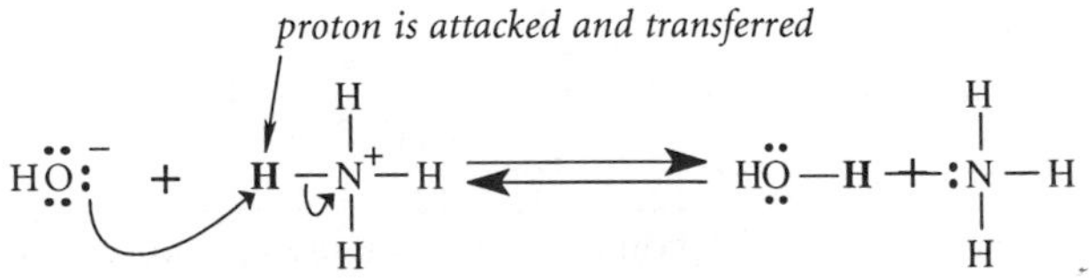

proton is attacked and transferred

$HO^- + H{-}NH_3^+ \rightleftharpoons HO{-}H + :NH_3$

2. The curved arrow originates at the source of the electrons and terminates at the destination of the electron pair.
   a. Redraw all atoms just as they were in the reactants.

b. Draw the bonds and electron pairs that do not change.

c. Draw the new bonds or electron pairs indicated by the curved-arrow notation.

d. Complete the formal charges to give the product. Remember, the algebraic sum of the charges on the reactants and products must be equal.

3. The curved-arrow notation can be used to derive resonance structures that are related by the movement of one or more electron pairs.

a. The interconversion of resonance structures is *not* a reaction.
b. The resonance structures are, taken together, a representation of a *single molecule*.

SCIENTISTS

STUDY GUIDE LINKS

✓3.1 The Curved-Arrow Notation

The curved-arrow notation is a symbolic device that must be used precisely. However, some things matter and some things don't. In drawing curved arrows, the only things that matter are *the starting point and the destination of the arrows.* It doesn't matter whether an arrow curves up or down, or whether it goes to the top or bottom of the atom. Your curved arrows might appear different from the ones in the text or solutions manual; what matters is where they start and where they end. For example, in each of the following examples the curved arrows mean the same thing: breaking of a B—F bond with the electron pair moving onto the fluorine.

:F—B—F: means the same thing as :F—B—F:

It also doesn't matter where the arrowhead points on an atom. For example, in the above example, the arrow can point directly to the F, or to an electron pair on the F. In each case, the meaning is that fluorine is taking on another electron pair.

The following is an example of the curved-arrow notation written in three different ways for an electron-pair displacement reaction. All of these are correct. They all refer to the same process: removal of a proton from the ammonium ion by hydroxide ion to give water and ammonia.

All describe the same process that gives the following as products:

HO—H + :N—H

✓3.2 Rules for Use of the Curved-Arrow Notation

Now that you've seen several examples of the curved-arrow notation, let's summarize its use with six simple rules. If you follow these you will always use the curved-arrow notation correctly. The first two rules reiterate points that have been made repeatedly in the text. Examples of the last four rules are provided below.

Rules for use of the curved-arrow notation:

The first two rules are very important and general:

1. Curved arrows indicate the flow or "movement" of electron pairs, not atoms or nuclei.
2. A curved arrow is always drawn with its tail at the source of electrons (a bond or unshared pair) and its head at the destination.

The second two rules deal with the manipulation of unshared electron pairs:

3. When an atom donates an unshared electron pair, a curved arrow must indicate that the electron pair is used to form an additional bond to the same atom.
4. When an atom accepts an unshared electron pair, a curved arrow must show that the electron pair originates from a bond to the same atom.

The following example illustrates both of these rules:

The oxygen that donates the electron pair. . .

. . .is involved in the newly-formed bond

$$CH_3\ddot{O}H + H{-}\ddot{Cl}: \longrightarrow CH_3\overset{+}{O}H(H) + :\ddot{Cl}:^-$$

The new electron pair on Cl came from a bond to the same Cl.

The last two rules deal with the manipulation of bonds:

5. When a bond is lost, a curved arrow must show its conversion either into an unshared electron pair on one of the two bonded atoms, or into a new bond involving one of the two original bonded atoms.
6. When a bond is formed, a curved arrow must show that it originates from either an electron pair on one of the two bonded atoms, or from a bond involving one of the two bonded atoms.

The following example illustrates these rules:

Bond lost from Br becomes an electron pair on the same Br

$$H\ddot{O}:^- \; H{-}\ddot{O}{-}CH_2{-}\ddot{Br}: \longrightarrow H\ddot{O}{-}H + \ddot{O}{=}CH_2 + :\ddot{Br}:^-$$

Bond between O and H becomes a new bond to one of the two atoms (O)

When confronted with a series of rules like these for use of the curved-arrow notation, some students are tempted to memorize them. *Don't do it!* What should you do with these rules? First, go back in the text and look at about three or four equations that show the curved-arrow notation and demonstrate to yourself that these rules are applied correctly. Then, the next few times you use the notation (as in Problem 3.3, for example) use your intuition to answer the problem, but then check your result against these rules before you look at the answer. If you do this three or four times, you're not likely to misuse the notation.

✓3.3 Identification of Acids and Bases

This section stresses an important subtlety in dealing conceptually with acids and bases. An acid or a base—whether of the Lewis type or the Brønsted type—is defined by a *reaction,* and not by a *structure.* For example, some compounds react by donating a proton in certain reactions, in which case they are Brønsted acids; in other reactions, they react in other ways, in which case, they are *not* Brønsted acids.

You can sometimes tell from a structure whether a compound *might* react as an acid or a base. When someone looks at a structure and says, "That compound is an acid," what that person really means is "That compound is *capable of acting as an acid.*" It *is* an acid only in reactions in which it is acting like one. Thus, a compound with an unshared electron pair *might* react as a Lewis base;

but if the same compound has an acidic hydrogen, it might also react as a Brønsted acid. The water molecule, with its amphoteric character, is an example of exactly this situation. Whether water is an acid or a base depends on the reaction. An electron-deficient compound certainly has the capability to react as a Lewis acid, and we might say, "That compound is a Lewis acid," but it is not a Lewis acid until it reacts like one. It might undergo other types of reactions in which it does not act as a Lewis acid. You have to see what happens to a compound in a reaction before you can classify the compound as an acid or base; and then the acid or base character is defined for that particular reaction.

Consider the following analogy. Suppose you have a friend who has just purchased a sleek, red convertible. He or she asks you, "Is that a fast car, or what?" Logically, you can't answer the question until you see the car perform, even though you suspect that it may have the capability of running at great speed. In the Indianapolis 500, your friend's car may *not* be fast at all. On the back roads of rural Indiana, it may be the fastest car anyone has ever seen. When someone shows you a structure and asks, "Is this a strong Brønsted acid or what?" your logical answer should be, "Show me the reaction and I'll tell you."

Another point is worth noting about the identification of acids and bases. *All* Brønsted acid-base reactions are Lewis acid-base reactions, but the converse is not true: some Lewis acid-base reactions are not Brønsted acid-base reactions. An alternative definition of Brønsted acid-base reactions that stresses this point is as follows. A **Brønsted acid-base reaction** is a type of Lewis acid-base reaction in which an electron-pair displacement occurs on a proton.

✓3.4 The Difference between pK_a and pH

Some students are confused by the difference between pK_a and pH. The pK_a is a *property of an individual acid.* We can't alter the pK_a without altering the structure of the acid. The pH is a *property of a solution.* It is $-\log\,[H_3O^+]$, that is, the negative log of the hydrogen-ion concentration. The pH of a solution can be experimentally altered.

Suppose we have an aqueous solution containing 0.1 *M* of an acid HA. The pK_a of HA determines the pH of the solution; the stronger the acid, the lower the pH. However, *we can alter the* pH *of the solution experimentally by adding hydrogen ions or hydroxide ions;* this has *no effect* on the pK_a of the acid. What such an addition *does* affect is the relative amounts of the acid and its conjugate base that are present. We can see this from text Eq. 3.14:

$$K_a = \frac{[A^-][H_3O^+]}{[HA]}$$

Dividing through by $[H_3O^+]$, we have

$$\frac{K_a}{[H_3O^+]} = \frac{[A^-]}{[HA]}$$

Taking logs, and identifying pH as $-\log\,[H_3O^+]$, and pK_a as $-\log K_a$,

$$\text{pH} - \text{p}K_a = \log\left(\frac{[A^-]}{[HA]}\right) \tag{SG3.1}$$

Notice that the relative amounts of HA and its conjugate base A^- are determined by how we adjust the pH of the solution, *and the relationship of the pH to the* pK_a *of the acid.* If we don't adjust the pH—if we just add the acid to water—then the pH (as well as the ratio of HA and A^-) is determined only by the strength of the acid in accordance with the laws of chemical equilibrium. Alternatively, if we fix the ratio $[A^-]/[HA]$ by adding a certain ratio of an acid HA and its conjugate base to solution, the pH is automatically determined by the log of this ratio and the pK_a value of the acid.

The experimentalist can control independently *either* the pH *or* the ratio $[A^-]/[HA]$, but not both at the same time, as they depend on each other. The pK_a is never adjustable by the experimentalist; it is an intrinsic property of the acid/base system.

To summarize:

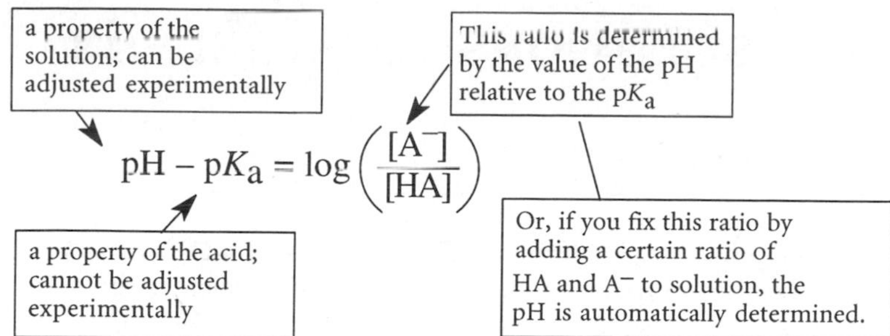

✓3.5 Basicity Constants

In the older literature, a quantity called the **basicity constant**, K_b, was used as a measure of base strength. The basicity constant is essentially a measure of the ability of a base to accept a proton from water rather than from H_3O^+. For example, the reaction in which ammonia accepts a proton from water is as follows:

$$NH_3 + H_2O \rightleftharpoons {}^+NH_4 + {}^-OH$$

The equilibrium constant for this reaction is

$$K_{eq} = \frac{[{}^+NH_4][{}^-OH]}{[NH_3][H_2O]} \quad \text{(SG3.2)}$$

Since $[H_2O]$ is effectively a constant, it is convenient to incorporate it into the equilibrium constant. Multiplying through this equation by $[H_2O]$ gives the *basicity constant*, K_b:

$$K_b = K_{eq}[H_2O] = \frac{[{}^+NH_4][{}^-OH]}{[NH_3]} \quad \text{(SG3.3)}$$

It is easy to show that the K_b for a base is simply related to the dissociation constant K_a of its conjugate acid. By multiplying the numerator and denominator of this equation by $[H_3O^+]$, and recognizing that the ion-product constant of water K_w equals $[H_3O^+][{}^-OH]$, we find that

$$K_b = \frac{K_w}{K_a} \quad \text{(SG3.4)}$$

You should verify this equation. The value of K_w is $10^{-14} \cdot mol^2 \cdot L^{-2}$, and therefore p$K_w$ = 14. Taking negative logarithms, and recognizing that pK_w = 14, we have the relationship between the pK_a of an acid and the pK_b of its conjugate base:

$$pK_b = 14 - pK_a \quad \text{(SG3.5)}$$

The pK_b is falling into disuse, because most people realize that the pK_a is perfectly useful to characterize both the strength of an acid and the strength of its conjugate base. Why have a second number that in effect does the same thing? Acids with higher pK_a are weaker, and their conjugate bases are stronger, than acids with lower pK_a.

3.6 Factors That Affect Acidity

The bond dissociation energies and electron affinities used to explain the element effect are *gas phase* data. Yet the pK_a values presented in the text are solution data. Without the gas-phase data we could not be sure whether the trends in solution pK_a values come from some effect of the solvent (water) or from the molecules themselves. It is the very nature of the gas phase that the properties of gaseous molecules at low pressure are essentially unaffected by surrounding molecules. Consequently, because the element effect is observed in the gas phase, we know that this same effect in solution data is caused by the structures of the molecules themselves.

SOLUTIONS TO PROBLEMS

Solutions to In-Text Problems

3.1 (a) In (1), the product results from attack of the oxygen of H_2O, which has unshared pairs, on the electron-deficient carbon of the cation. In (2), the electron-deficient boron of BF_3 is attacked by the nitrogen of ammonia, which has unshared electron pairs. (The electron pairs on the fluorine are not shown because they have no direct part in the reaction.)

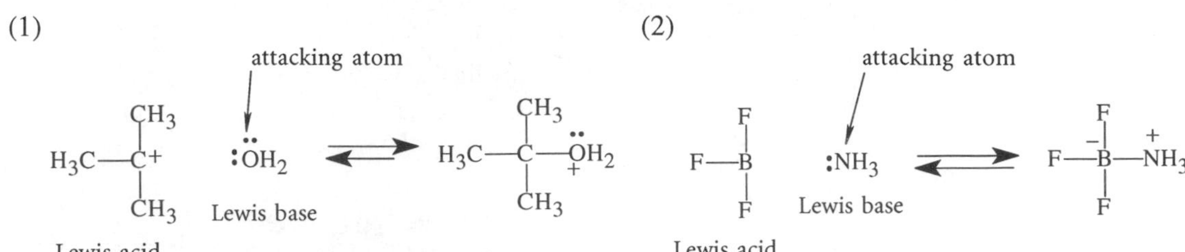

(b) The curved-arrow notation is as follows:

(1) $(CH_3)_3C^+ \;\; :OH_2 \rightleftharpoons (CH_3)_3C{-}\overset{+}{O}H_2$

(2) $F_3B \;\; :NH_3 \rightleftharpoons F_3\overset{-}{B}{-}\overset{+}{N}H_3$

3.2 (a) $HO{-}CH_2{-}CH_3$ $\;\; :Cl:^-$ (b) $(CH_3)_2\overset{+}{C}{-}CH_2{-}H$ $\;\; :Br:^-$ (c) H_3Al $\;\; H{-}CH_3$ $\;\; :Br:^-$ (d) $H_2C{-}O{-}O{-}O{-}CH_2$ (ring)

3.3 (a) $H_3N: \;\; H_3C{-}Br:$ (b) $CH_3\ddot{O}:^- \;\; H{-}CH_2{-}CH(CH_3){-}Br:$

3.4 (a) Deriving the resonance structures of the allyl cation requires a single arrow that indicates the delocalization of one bond of the double bond toward the electron-deficient carbon:

$$\left[H_2\overset{+}{C}{-}CH{=}CH_2 \longleftrightarrow H_2C{=}CH{-}\overset{+}{C}H_2 \right]$$

Note that the number of bonds to each carbon is the average of the bonds in the two structures: (1 + 2)/2 = 1.5, as required.

(b) Deriving the resonance structure that shows sharing of the negative charge and double-bond character requires use of the curved-arrow notation for electron-pair displacement reactions:

$$\left[H_2\overset{..}{\underset{-}{C}}{-}CH{=}CH_2 \longleftrightarrow H_2C{=}CH{-}\overset{..}{\underset{-}{C}}H_2 \right]$$

(c) The resonance structures for benzene are obtained by a series of electron-pair displacements in a cyclic fashion:

3.5 (a) On the left side of this reaction the hydroxide ion acts as the base and the ammonia as the acid; on the right, the amide anion (H_2N^-) acts as the base and the water as the acid.

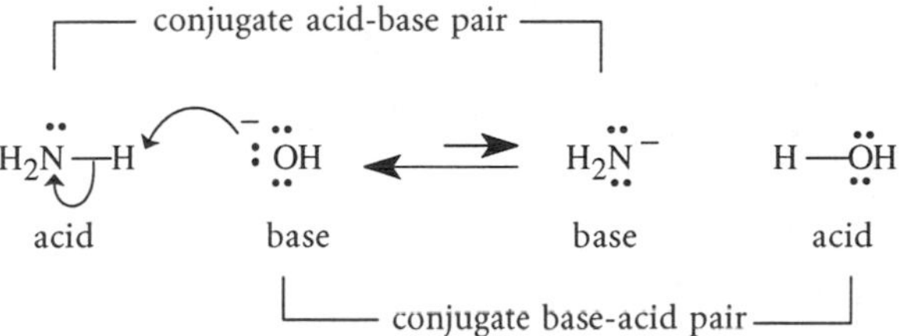

(b) On the left side of this reaction, one ammonia molecule acts as an acid and the other acts as a base. On the right, the conjugate acid of ammonia, the ammonium ion, acts as the acid, and the conjugate base of ammonia, the amide ion, acts as the base.

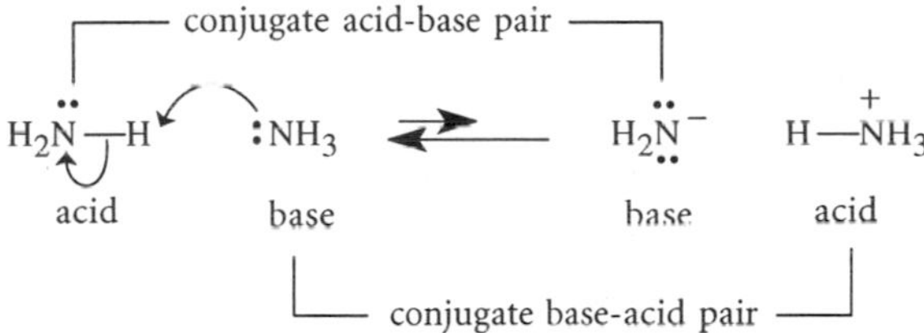

Don't hesitate to re-draw structures or to move electron pairs to different locations on the same atom if it helps you to accomplish the purpose at hand. Notice that the structure of one ammonia molecule was re-drawn in the structure on the left so that the curved-arrow formalism could be drawn clearly.

3.6 In the pair $H_2O/^-OH$, H_2O is the acid, because it has one more proton than ^-OH; in the other pair, CH_3OH is the acid for the same reason. Thus, the reaction required is one in which the acid H_2O donates a proton to the base $^-OCH_3$. (Of course, it would be equally correct to write the reaction in the reverse direction.)

conjugate base-acid pair

$$HO—H \quad :\bar{O}—CH_3 \rightleftharpoons HO:^- + H—O—CH_3$$

conjugate acid-base pair

3.7 (a) $pK_a = 3$ (b) $pK_a = 5.2$ (c) $pK_a = -1.7$

If you need a refresher on determining logarithms with your calculator, here are the keystrokes for the solution to part (b) on an RPN-style (Hewlett-Packard) calculator:

5.8 [E] 6 [+/–] [ENTER] [LOG] [+/–]

The result is your answer. On other types of calculators, the keystrokes for the same operation are

5.8 [EXP] 6 [+/–] [LOG] [+/–]

3.8 (a) $K_a = 10^{-4}$ (b) $K_a = 1.6 \times 10^{-8}$ (c) $K_a = 10^2 = 100$

To determine a K_a value from a pK_a value with your calculator, first enter the pK_a, for example, 7.8, and change the sign ([+/–]) to give –7.8. Then press the 10^x key to get the answer. If the result is a very small number, you will have to display it in exponential notation to avoid an answer of zero on your calculator.

3.9 (a) The strongest acid in Problem 3.7 is the one with the largest dissociation constant, that is, acid (c).
(b) The strongest acid in Problem 3.8 is the one with the smallest (that is, least positive or most negative) pK_a, that is, acid (c).

3.10 (a) The strongest base in the second column of Table 3.1 is the conjugate base of the weakest acid. Because the weakest acid in the table is ammonia, the strongest base is the amide ion $^-NH_2$.
(b) The weakest base in Table 3.1 is the conjugate base of the strongest acid. Because the strongest acid is perchloric acid, the weakest base is the perchlorate ion, ClO_4^-.

3.11 (a) From Table 3.1, the pK_a of HCN is 9.40, and the pK_a of the ammonium ion is 9.25. In the reaction, ammonia reacts with HCN to give ammonium ion, $^+NH_4$, and cyanide ion, ^-CN. From the procedure in Study Problem 3.5, the log of equilibrium constant K_{eq} for the reaction of ammonia as a base with the acid HCN is

$$\log K_{eq} = pK_a(^+NH_4) - pK_a(HCN)$$

or

$$\log K_{eq} = 9.25 - 9.40 = -0.15$$

from which K_{eq} is calculated by taking the antilogarithm:

$$K_{eq} = 10^{(-0.15)} = 7.08 \times 10^{-1} = 0.708$$

(b) Use the procedure in part (a). Fluoride ion reacts with HCN to give HF and cyanide ion, ^-CN. The relevant pK_a values are those of HCN (9.4) and HF (3.2). If the equilibrium constant for the reaction is K_{eq}, then

$$\log K_{eq} = pK_a(HF) - pK_a(HCN)$$
$$\log K_{eq} = 3.2 - 9.4 = -6.2$$

or

$$K_{eq} = 10^{-6.2} = 6.3 \times 10^{-7}$$

3.12 (a) When ammonia acts as a base toward water, the acidic species on the left is water, and the relevant pK_a from Table 3.1 is 15.7. The other relevant pK_a is that of the ammonium ion, the acid on the right side, which is 9.25.

$$H_2O + NH_3 \rightleftharpoons HO^- + \overset{+}{N}H_4$$

acidic species on this side of the equation

(b) When ammonia acts as an acid toward the base water, the acidic species on the left side is ammonia, and the relevant pK_a is 35 for ammonia. The other relevant pK_a is that of H_3O^+, which is –1.7.

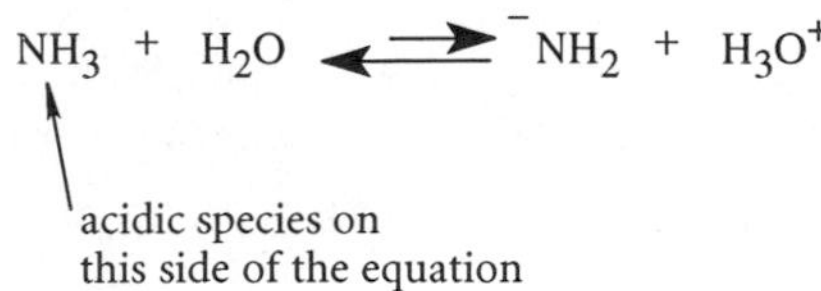

The log K_{eq} for reaction (a) is 9.25 – 15.7 = –6.45, and $K_{eq} = 3.6 \times 10^{-7}$. The log K_{eq} for reaction (b) is –36.7, and $K_{eq} = 2 \times 10^{-37}$. Although both equilibrium constants are small, K_{eq} of (b) is *much* smaller than that of (a). Clearly, reaction (a) is much more important than reaction (b) in an aqueous solution of ammonia.

3.13 (a) Use Eq. 3.24b in the text, with $2.303RT = 5.706$ kJ/mol at 298 K:

$$K_{eq} = 10^{-(-14.6)/5.706} = 362$$

(b) For this calculation use text Eq. 3.23.

$$\Delta G° = -2.303RT \log K_{eq} = -5.706 \log(305) \text{ kJ/mol} = -5.706(2.48) \text{ kJ/mol}$$

or

$$\Delta G° = -14.2 \text{ kJ/mol}$$

In this Solutions Manual, kJ/mol is used as the unit of energy. If you wish to convert kJ/mol to kcal/mol, simply divide by 4.184 kJ/kcal.

3.14 (a) To complete this calculation, first calculate the equilibrium constant, and then use it to do a mass-action calculation.

$$K_{eq} = 10^{-\Delta G°/2.303RT} = 10^{-(-2.93)/5.706} = 10^{0.513} = 3.26$$

Make a table of concentrations for the three species in the equilibrium:

	A	*B*	*C*
Initially:	0.1	0.2	0
Change:	$-x$	$-x$	x
At equilibrium:	$0.1 - x$	$0.2 - x$	x

The equilibrium constant can then be written as

$$K_{eq} = \frac{[C]}{[A][B]} = \frac{x}{(0.1 - x)(0.2 - x)} = 3.26$$

from which is obtained the quadratic expression

$$x^2 - 0.607x + 0.02 = 0$$

The quadratic formula gives

$$x = \frac{0.607}{2} \pm \frac{\sqrt{0.368 - 0.080}}{2} = 0.303 \pm 0.268$$

Choose the physically meaningful value of x to obtain

$$[C] = 0.035\ M,\ [B] = 0.2 - x = 0.165\ M, \text{ and } [A] = 0.1 - x = 0.065\ M.$$

(b) If the $\Delta G°$ is positive, the equilibrium constant is less than unity and the reaction is less favorable; consequently, there should be much less *C* and more *A* and *B* at equilibrium.

An actual calculation shows that $[C] = 0.0061$. Then $[A] = 0.1 - x = 0.0939\ M$, and $[B\} = 0.2 - x = 0.1939$. The concentration of *C* is 10% of that in part (a), a result consistent with the statement in part (b).

3.15 (a) Thiols are more acidic than alcohols (element effect), and a chlorine-substituted alcohol is more acidic than the unsubstituted alcohol (polar effect). Hence the order of decreasing pK_a (increasing acidity) is

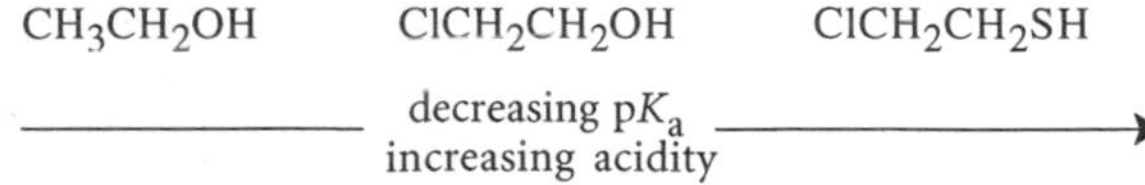

Remember that the element effect is usually *much* more important than the polar effect in determining acidity.

(b) Because oxygen is an electronegative atom, increasing the number of $—OCH_3$ groups increases the acid-strengthening polar effect. Hence, the order of decreasing pK_a (increasing acidity) is

$$H_3C—CO_2H \qquad CH_3O—CH_2—CO_2H \qquad \underset{OCH_3}{\underset{|}{CH_2}}—\underset{OCH_3}{\underset{|}{CH}}—CO_2H$$

decreasing pK_a
increasing acidity →

3.16 (a) Use Eq. 3.22 in the text, with $2.3RT = 5.706$ kJ/mol at 298 K.

$$\Delta G° = 2.3RT(\mathrm{p}K_a) = (5.706)(2.66)\text{ kJ/mol} = 15.2\text{ kJ/mol}$$

This means that it takes 15.2 kJ/mol to convert one mole of fluoroacetic acid into one mole of hydronium ion and one mole of fluoroacetate anion.

(b) Use the same procedure as in part (a):

$$\Delta G° = 2.3RT(\mathrm{p}K_a) = (5.706)(4.76)\text{ kJ/mol} = 27.2\text{ kJ/mol}$$

3.17 To say that acetic acid is weaker than fluoroacetic acid means that the conjugate base of acetic acid is less stable relative to the un-ionized acid than the conjugate base of fluoroacetic acid is to its un-ionized acid (Fig. 3.2). "Less stable" means that more free energy is required to convert the weaker acid into its conjugate base, and this reduced stability is reflected in the larger standard free energy of ionization for acetic acid. The reason for this difference is that the negative charge in the conjugate-base anion of fluoroacetic acid is stabilized by its interaction with the bond dipole of the fluorine substituent—in other words, the charge-stabilizing polar effect of the fluorine substituent.

Solutions to Additional Problems

3.18 Compound (a) is *not* electron-deficient because each atom has an octet of electrons. Compound (b) is electron-deficient, because the carbon has a sextet of electrons. Compound (c) is electron-deficient because the nitrogen has a sextet of electrons.

It might seem strange that a compound such as (c) with an unshared electron pair should be electron-deficient. It is the valence-electron count, not the presence or absence of unshared pairs, that determines electron deficiency. Any species is electron-deficient when an atom is surrounded by fewer than four pairs of electrons (or one pair for hydrogen).

3.19 The general approach to all parts is to identify an electron-deficient compound, and use it as a Lewis acid by attacking the electron-deficient atom with an electron pair on the other molecule or ion.

(a)

$$H_3C-\overset{CH_3}{\underset{CH_3}{C^+}} \quad :\ddot{\underset{..}{Cl}}:^- \longrightarrow H_3C-\overset{CH_3}{\underset{CH_3}{C}}-\ddot{\underset{..}{Cl}}:$$

(b)

$$(H_3C)_2\ddot{O}: \quad BF_3 \longrightarrow (H_3C)_2\overset{+}{\ddot{O}}-\overset{-}{B}F_3$$

Don't forget that you can redraw any structure to suit the purposes of the problem at hand.

(c)

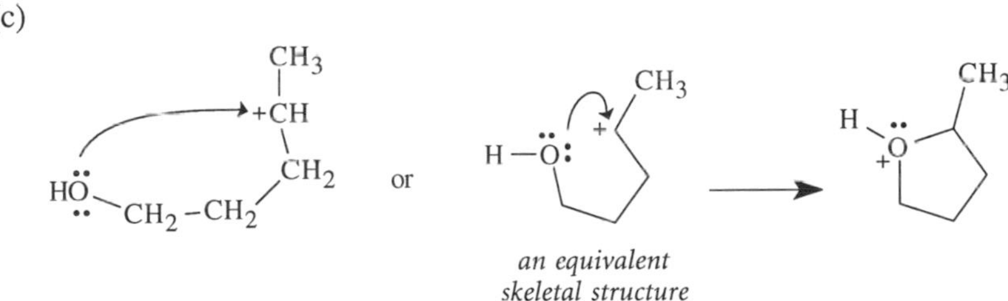

an equivalent skeletal structure

Don't forget that carbon chains are flexible and can undergo internal rotations. Groups that appear far apart in structures that are drawn linearly can in many cases be brought into proximity by internal rotations. If this is not clear, construct a model for this case, and convince yourself that the —OH group can easily be brought into bonding distance with the positively charged carbon.

(d) We use the electron pair on the carbon to attack the electron-deficient boron. If we were to use the electron pair on the oxygen, oxygen would end up with *two* plus charges. Oxygen is too electronegative to bear two positive charges in its "ordinary" chemistry. In contrast, attack of the carbon gives a product in which the corresponding carbon is neutral.

$$(CH_3)_3B \quad :\overset{-}{C}\equiv\overset{+}{O}: \longrightarrow (CH_3)_3\overset{-}{B}-C\equiv\overset{+}{O}:$$

(e) Notice here that we use the electron pair on the atom that has the complete octet. Even though the carbon of the :CH_2 contains an electron pair, this carbon is electron-deficient.

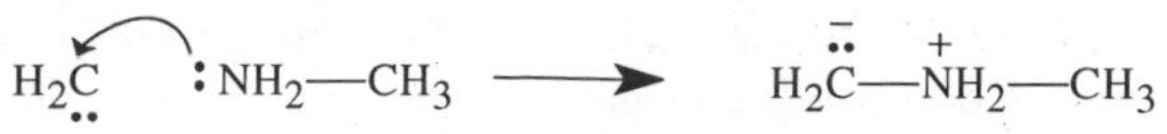

$$H_2\ddot{C}\;\; :NH_2—CH_3 \longrightarrow H_2\ddot{C}^{-}—\overset{+}{N}H_2—CH_3$$

3.20 (a)

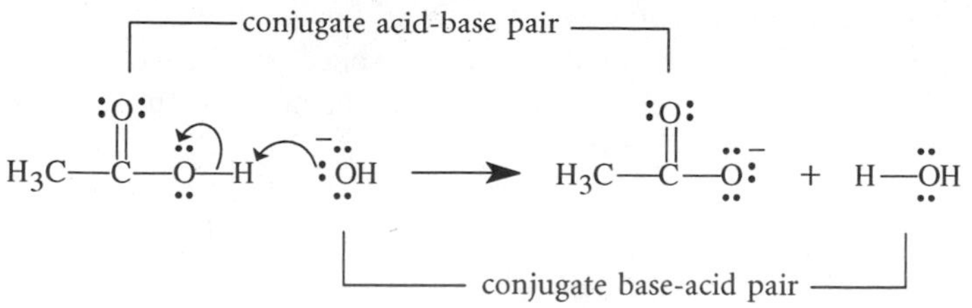

conjugate acid-base pair

$$H_3C—C(=O)—\ddot{O}—H \;\; :\ddot{O}H^- \longrightarrow H_3C—C(=O)—\ddot{O}:^- + H—\ddot{O}H$$

conjugate base-acid pair

(b)

conjugate acid-base pair

$$H_3C—C(=O)—\ddot{O}—H \;\; :\ddot{O}H_2 \longrightarrow H_3C—C(=O)—\ddot{O}:^- + H—\overset{+}{O}H_2$$

conjugate base-acid pair

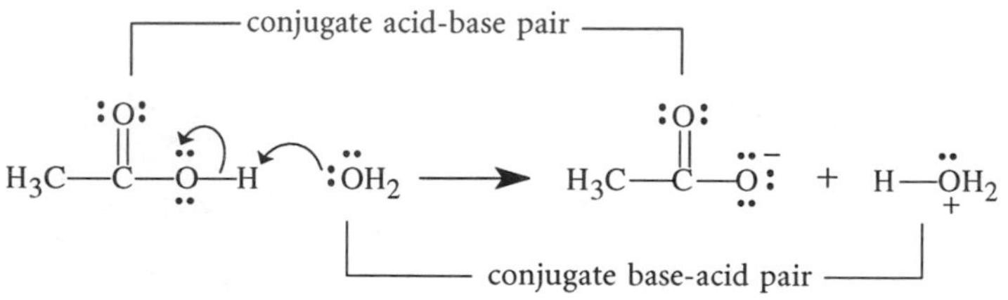

(c) In this case, the acids and bases are groups within the same molecule.

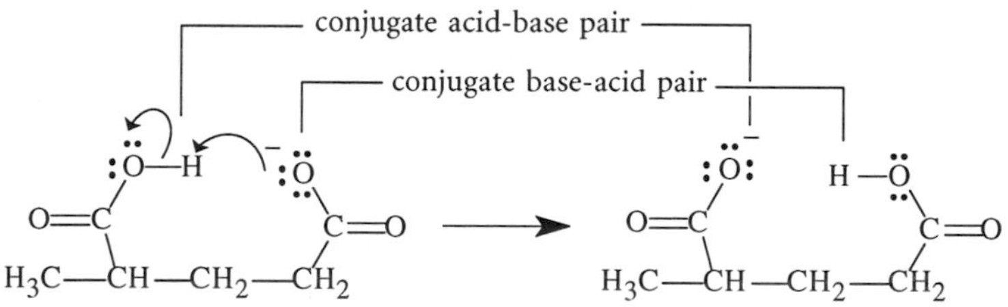

conjugate acid-base pair

conjugate base-acid pair

$$O{=}C(\ddot{O}—H)—CH(CH_3)—CH_2—CH_2—C(=O)\ddot{O}:^- \longrightarrow O{=}C(\ddot{O}:^-)—CH(CH_3)—CH_2—CH_2—C(=O)\ddot{O}—H$$

(d)

conjugate acid-base pair

$$(CH_3)_2C{=}CH_2 \;\; H—\overset{+}{O}H_2 \longrightarrow (CH_3)_2\overset{+}{C}—CH_2H + :\ddot{O}H_2$$

conjugate base-acid pair

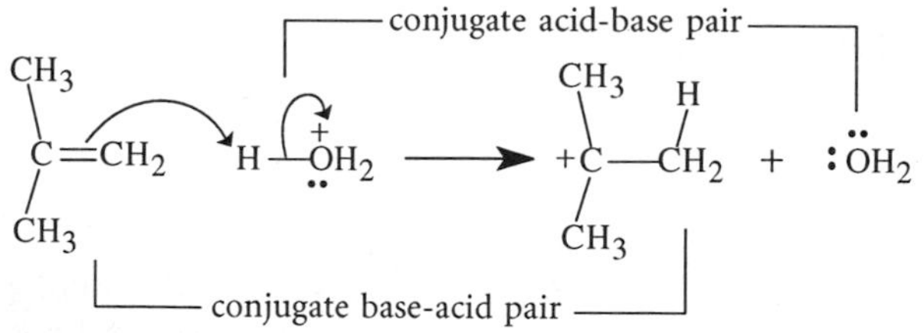

3.21 ***Step 1:***

(a) (1), (4), and (5). The reaction is Lewis acid-base reaction; the proton that is transferred is the Lewis acid, and the —OH of the alcohol is the Lewis base. (Note that all Brønsted acid-base reactions are Lewis acid-base reactions.)

(b) and (c) The conjugate pairs and the curved-arrow notation are as follows:

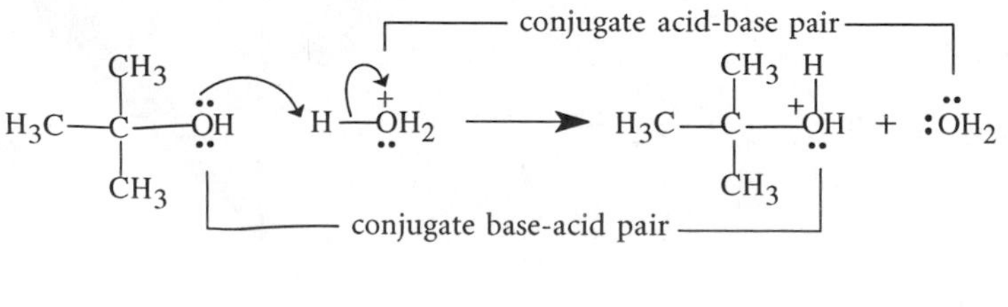

conjugate acid-base pair

$$H_3C—C(CH_3)_2—\ddot{O}H \;\; H—\overset{+}{O}H_2 \longrightarrow H_3C—C(CH_3)_2—\overset{+}{O}H(H) + :\ddot{O}H_2$$

conjugate base-acid pair

Step 2:

(a) (1), (3) The reaction is a Lewis acid-base dissociation, which is a type of Lewis acid-base reaction.

(b) The reaction is not a Brønsted acid-base reaction.

(c) The curved-arrow notation is as follows:

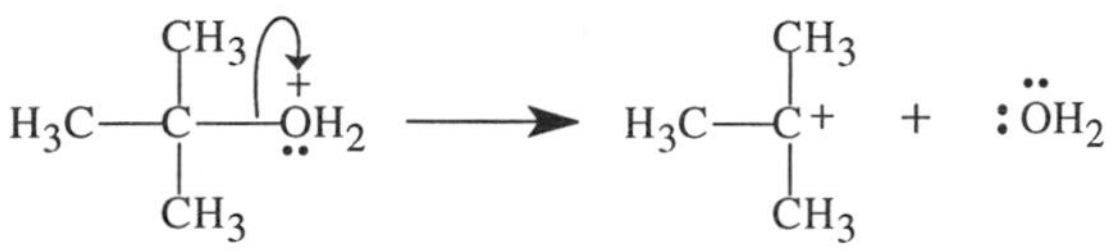

$$H_3C-C(CH_3)_2-\overset{+}{O}H_2 \longrightarrow H_3C-\overset{+}{C}(CH_3)_2 + :\ddot{O}H_2$$

Step 3:

(a) (1), (4), (5). The reaction is a Brønsted acid-base reaction.

(b) and (c) The conjugate pairs and the curved-arrow notation are as follows:

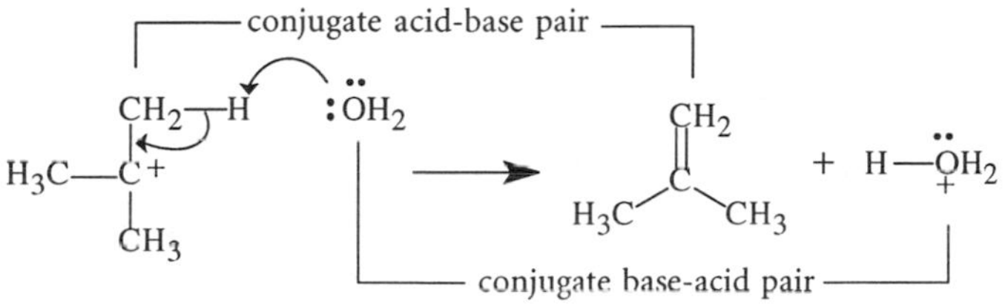

conjugate acid-base pair

$$H_3C-\overset{+}{C}(CH_3)-CH_2-H \quad :\ddot{O}H_2 \longrightarrow (H_3C)(CH_3)C{=}CH_2 + H-\overset{+}{\ddot{O}}H_2$$

conjugate base-acid pair

3.22 (a) We use the electron-pair displacement notation to form a double bond with one of the oxygen unshared pairs, and to convert the existing double bond to a single bond:

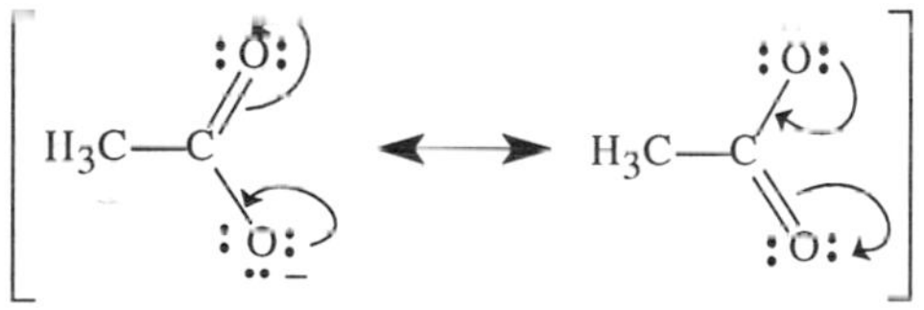

$$\left[H_3C-C(=\ddot{O}:)-\ddot{O}:^- \longleftrightarrow H_3C-C(-\ddot{O}:^-)=\ddot{O}: \right]$$

(b)

$$\left[:C{=}\ddot{O}: \longleftrightarrow :\overset{-}{C}{\equiv}\overset{+}{O}: \right]$$

3.23 (a)

$$\left[:\ddot{O}{=}\overset{+}{\ddot{O}}-\ddot{O}:^- \longleftrightarrow {}^-:\ddot{O}-\overset{+}{\ddot{O}}{=}\ddot{O}: \right]$$

(b)

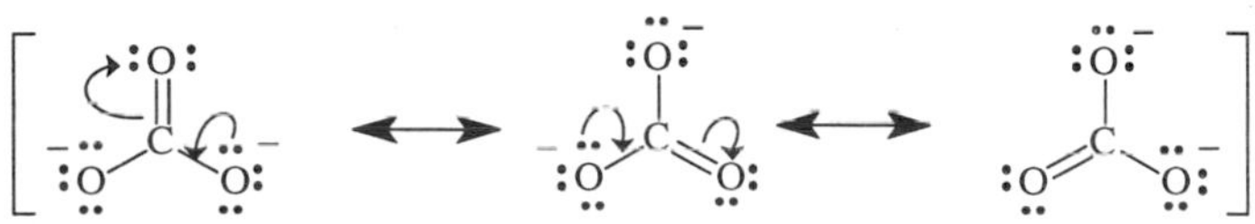

$$\left[{}^-:\ddot{O}-C(=O:)-\ddot{O}:^- \longleftrightarrow {}^-:\ddot{O}-C(-\ddot{O}:^-)=\ddot{O}: \longleftrightarrow :\ddot{O}{=}C(-\ddot{O}:^-)-\ddot{O}:^- \right]$$

(c)

$$\left[H_2C{=}\overset{+}{\ddot{O}}-H \longleftrightarrow H_2\overset{+}{C}-\ddot{O}-H \right]$$

3.24 (a) $(CH_3)_2(H_3C)C^+ + :N{\equiv}N:$

(b) $H_3C{-}C(H)(\ddot{O}{-}MgBr){-}CH_3$

(c) $H_3C{-}\ddot{C}H^-{-}CH{=}\ddot{O} + (CH_3)_3Si{-}\ddot{F}:$

(d) $H_3C{-}C(=\ddot{O}){-}\ddot{N}H_2 + {}^-:\ddot{O}C_2H_5$

3.25 (a) (b) (c) (d)

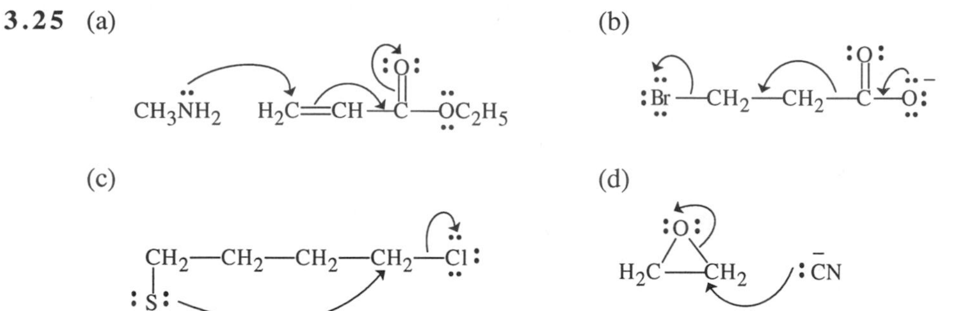

For part (c), see the note following the solution to Problem 3.19(c).

(e) To show the curved-arrow notation you have to draw out a C—H bond that is not explicitly shown in the problem.

$:\ddot{B}r{-}CH_2{-}CH(H){-}Ph \quad {}^-:\ddot{O}CH_3$

3.26 (a) The curved arrow is going in the wrong direction; it should go from the electrons on the ^-OH to the acidic hydrogen:

$H_3C{-}C(=O){-}\ddot{O}{-}H \quad {}^-:\ddot{O}H$

(b) The one curved arrow is shown correctly, but a second one is needed:

$CH_3\ddot{O}:^- \quad H_3C{-}\ddot{B}r:$

3.27 The resonance structures of naphthalene:

[naphthalene ⟷ naphthalene ⟷ naphthalene]

3.28 (a) For purposes of this solution let 2,2-dimethylpropane = D and pentane = P. Then for the hypothetical reaction $P \rightleftharpoons D$, we define the equilibrium constant $K_{eq} = [D]/[P]$, and the standard free energy change

$\Delta G°$ is therefore $G°(D) - G°(P) = -6.86$ kJ/mol. Using Eq. 3.24b in the text,

$$K_{eq} = 10^{-(-6.86)/5.706} = 10^{1.20} = 15.9 = [D]/[P]$$

Taking the total hydrocarbon as 1.000, and noting from above that $[D] = 15.9[P]$,

$$1.000 = [D] + [P] = 15.9[P] + [P] = 16.9[P]$$

or $$[P] = 1.000/16.9 = 0.059$$

and $$[D] = 1.000 - 0.059 = 0.941$$

Consequently, at equilibrium the mixture contains 5.9% *P* and 94.1% *D*. Notice that *D* is present in greater amount because it is more stable.

(b) Let K_{eq} be the ratio [anti]/[gauche]; then $\Delta G°$ for gauche Q anti is –3.8 kJ/mol. (Because gauche has higher energy than anti, $\Delta G°$ is negative.) Then use Eq. 3.24b in the text.

$$K_{eq} = 10^{-\Delta G°/2.3RT} = 10^{-(-3.8)/5.706} = 10^{0.6667} = 4.63$$

This is the ratio of the anti conformation to any *one* gauche conformation. Taking the total fraction of butane as 1.0, and noting from above that [anti] = 4.63[gauche],

$$1.0 = [\text{anti}] + 2[\text{gauche}] = 4.63[\text{gauche}] + 2[\text{gauche}] = 6.63[\text{gauche}]$$

or $$[\text{gauche}] = 1.0/6.63 = 0.151$$

Consequently, the total fraction gauche conformation is $2 \times 0.151 = 0.302$ (that is, about 30% gauche conformation); and the total fraction anti conformation is $1.0 - 0.302 = 0.698$ (that is, about 70% anti conformation) in a sample of butane.

3.29 To arrange in order of decreasing pK_a means to arrange in order of *increasing acidity*.

(a) CH_3CH_2OH is less acidic than $ClCH_2CH_2OH$, which is less acidic than Cl_2CHCH_2OH. Chlorine substituents increase acidity because of their electron-withdrawing polar effect; the more chlorines there are, the greater the effect.

(b) Because of the element effect, both thiols (the compounds with —SH groups) are more acidic than ethanol (CH_3CH_2OH). The compound $ClCH_2CH_2SH$ is the more acidic of the two thiols because of the polar effect of the chlorine substituent.

(c) The element effect down a column of the periodic table predicts that $HN(CH_3)_2$ should be less acidic than $HP(CH_3)_2$, which should be less acidic than $HAs(CH_3)_2$. Notice that these acidities refer to the reaction in which the *neutral* molecules lose a proton to give the *anionic* conjugate bases (M = N, P, or As):

$$\text{H—}\ddot{M}(CH_3)_2 + \text{base}:^- \rightleftharpoons {}^-:\ddot{M}(CH_3)_2 + \text{base—H}$$

The acidity of $H_2\overset{+}{As}(CH_3)_2$ refers to the reaction in which this positively charged species loses a proton to give the neutral conjugate base $H\ddot{As}(CH_3)_2$. The cationic acid $H_2\overset{+}{As}(CH_3)_2$ is *much* more acidic than the neutral acid $H\ddot{As}(CH_3)_2$ because a positively charged atom is considerably more electronegative than the same neutral atom, and is therefore a better electron acceptor. (This is the same reason that H_3O^+ is more acidic than H_2O.) The better the electron acceptor, the stronger the acid.

(d) CH_3CH_2OH is less acidic than $(CH_3)_2NCH_2CH_2OH$ because of the polar effect of the electronegative nitrogen substituent. In the last compound, not only is the polar effect of the nitrogen present, but also a positive charge, which can stabilize the conjugate-base anion. Furthermore, the positively-charged nitrogen is closer to the —OH group. Consequently, $(CH_3)_3\overset{+}{N}$—OH is the most acidic of the three compounds.

3.30 In each part, we apply Eq. 3.18b and the procedure used in Study Problem 3.5.

(a) The information needed is the pK_a of H—CN, the acid on the left side of the equation. From Table 3.1, the pK_a of H—CN is 9.40. The equilibrium favors the side with the weaker acid, namely the right side, but by very little, because the two acids have about the same pK_a. Applying Eq. 3.18b,

$$K_{eq} = 10^{9.76-9.40} = 10^{+0.36} = 2.3$$

(b) Again we apply text Eq. 3.18b with pK_a = 15.7 for H_2O:

$$K_{eq} = 10^{15.7-10.5} = 10^{5.2} = 1.58 \times 10^5$$

3.31 (a) Apply Eq. 3.23 using the K_{eq} from the solution to problem 3.30(a).

$$\Delta G° = -2.3RT \log K_{eq} = -5.706 \log (2.3) = -2.1 \text{ kJ/mol}$$

(b) If the equilibrium concentrations of $(CH_3)_3N$ and HCN are both x, then the equilibrium concentrations of $(CH_3)_3\overset{+}{N}H$ and ^{-}CN are both $(0.1 - x)$. Substituting in the equilibrium-constant expression,

$$K_{eq} = 2.3 = \frac{[(CH_3)_3\overset{+}{N}H][^{-}CN]}{[(CH_3)_3N][HCN]} = \frac{(0.1-x)^2}{x^2}$$

This leads to the quadratic expression $x^2 + 0.154x - 0.00769 = 0$, which is solved to give $x = 0.0397$ *M*. Thus, at equilibrium, the concentrations of the species on the left side of the equation are both 0.0397 *M*, and the concentrations of the species on the right side are both 0.1000 – 0.0397 = 0.0603 *M*.

3.32 (a) By Eq. 3.22, text p. 96, the compound with the smaller pK_a has the smaller standard free energy of dissociation ΔG_a°. Consequently, phenylacetic acid has the smaller ΔG_a°.

(b) The problem is asking for the standard free energy of dissociation ΔG_a°. Use Eq. 3.22 in the text. (Assume 25 °C or 298 K.)

$$\Delta G_a^\circ = 2.3RT(pK_a) = 5.706(4.31) = 24.6 \text{ kJ/mol}$$

(c) Phenylacetic acid has a lower pK_a than acetic acid. Hence, the phenyl substituent is acid-strengthening; it stabilizes the conjugate-base anion in the same sense that a chlorine does. Consequently, its polar effect, like that of a chlorine, is evidently electron-withdrawing.

3.33 (a) For the first ionization (pK_a = 2.86), either of the two equivalent carboxylic acid groups can ionize:

$$H-O-\overset{O}{\overset{\|}{C}}-CH_2-\overset{O}{\overset{\|}{C}}-O-H + H_2O \rightleftharpoons H-O-\overset{O}{\overset{\|}{C}}-CH_2-\overset{O}{\overset{\|}{C}}-O^- + H_3O^+$$

For the second ionization (pK_a = 5.70), the remaining carboxylic acid group ionizes:

$$H-O-\overset{O}{\overset{\|}{C}}-CH_2-\overset{O}{\overset{\|}{C}}-O^- + H_2O \rightleftharpoons {}^{-}O-\overset{O}{\overset{\|}{C}}-CH_2-\overset{O}{\overset{\|}{C}}-O^- + H_3O^+$$

(b) The first pK_a is lower than that of acetic acid because, in the product of the first ionization, the electron-withdrawing polar effect of the un-ionized carboxylic acid group stabilizes the conjugate-base anion on the other. (The two electronegative oxygens are responsible for the polar effect of the carboxylic acid group.) The second pK_a is greater than that of acetic acid because one negative charge in the dianionic product of the second ionization interacts unfavorably (repulsively) with the second negative charge, and the energy of the dianion is thereby raised. (See Fig. 3.2, text p. 103.)

(c) As *n* becomes larger, the polar groups become more remote. Hence, *both* polar effects discussed in part (b) become less significant, and *both* pK_a values approach that of acetic acid; that is, the pK_a values of the two ionizations become more similar. (In terms of the hint, as *r* in the denominator of the electrostatic law becomes larger, the energy of interaction of one charge with another becomes smaller.)

Even if n is very large, the two pK_a values are not exactly equal. Theoretically, the first pK_a should be 0.6 smaller than the second pK_a (that is, the first K_a should be 4 times larger than the second K_a). The reason is that either of *two* equivalent acidic hydrogens can undergo the first ionization, but only one can undergo the second; this means that the first ionization is twice as probable statistically. Furthermore, either of *two* equivalent carboxylate ions can be protonated in the conjugate base of the monoacid, but only one is available in the conjugate base of the diacid. Thus, protonation of the dianion is twice as probable as protonation of the monoanion, which means that the second ionization is half as probable as the first—or again, that the first ionization is twice as probable as the second. Both of these factors multiply; thus, the first ionization is four times as likely as the second on purely statistical grounds; log (4) = 0.6.

3.34 The primary effect of a polar substituent is on the stability of a charged species. The charged species in these equilibria are the cations. The question then becomes, how does a fluorine substituent affect the stability of a cation? As Eq. 3.32 in the text shows, the polar effect of the fluorine *stabilizes negative charge;* however, it *destabilizes positive charge* because the positive ends of the C—F dipoles interact repulsively with the positive charge of the cation, as shown by the following diagram:

C-F bond dipole

repul;sive interaction

CH_3

F—C—$\overset{+}{C}$

CH_3

Consequently, the cation product of the second equilibrium is destabilized relative to the neutral starting material. This is shown in Fig. SG3.1. Because this cation has a higher energy, the energy required to form it is larger, and the equilibrium constant for its formation is smaller. In other words, Eq. (1) has the equilibrium constant most favorable to the right.

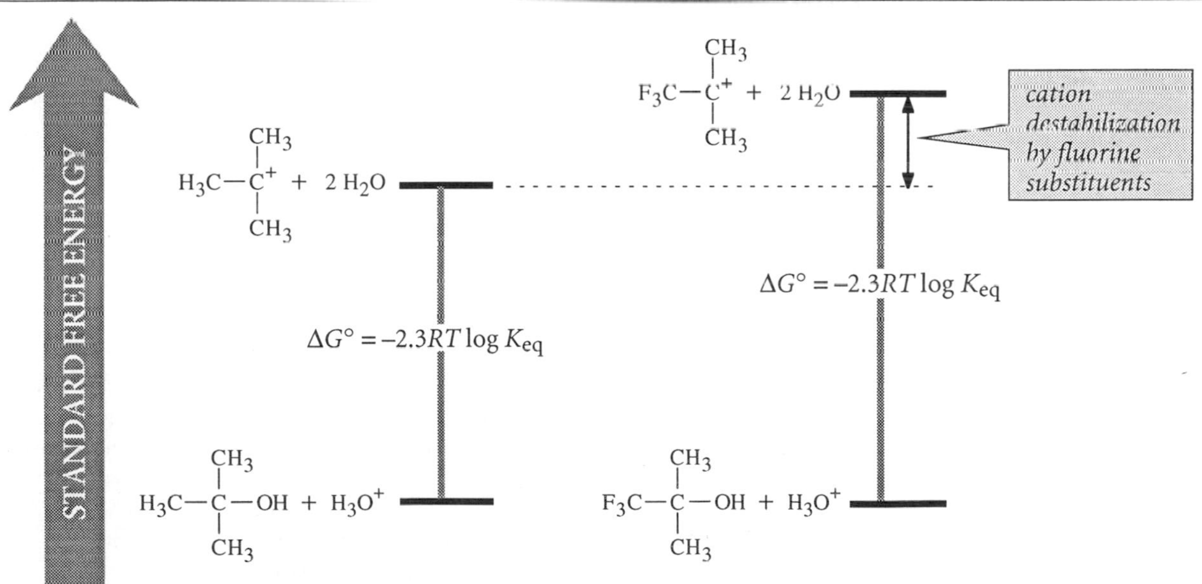

Figure SG3.1 Free-energy diagram to accompany the solution to Problem 3.34. The polar effect of the fluorine substituent raises the energy of the cation, and therefore reduces the equilibrium constant for its formation. (The two alcohols, which are uncharged, are arbitrarily placed at the same energy for comparison purposes.)

Any time you are asked to assess how a change in structure will affect a compound's chemical properties, you should turn this into a question of relative energies, construct an energy diagram, and then reason from it. Problem 3.36 below provides another opportunity for this sort of approach.

3.35 Given the similar sizes of Cl and CH_3, van der Waals repulsions in the corresponding conformations of 1,2-dichloroethane and butane should be similar. Hence, the larger energy barrier for 1,2-dichloroethane is due to some factor other than van der Waals repulsions. When the two chlorines are eclipsed, their bond dipoles are aligned. The proximity of like charges generates a repulsion that adds to the energy of the eclipsed conformation.

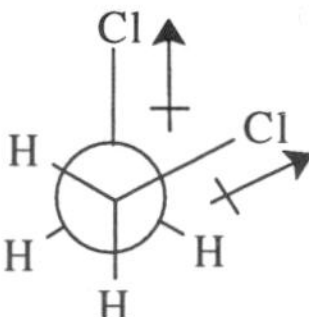

gauche conformation of 1,2-dichloroethane: C—Cl dipoles are not fully aligned

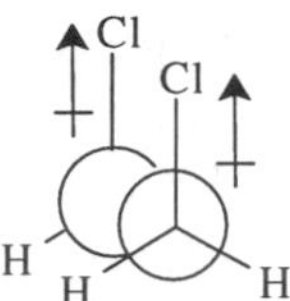

eclipsed conformation of 1,2-dichloroethane: C—Cl dipoles are aligned

In the gauche conformation, this dipole alignment is not present. Because the height of the energy barrier between gauche conformations is the difference in energy between the eclipsed conformation and either gauche conformation, the increased energy of the eclipsed conformation in 1,2-dichloroethane raises this barrier. This dipole effect is not present in butane.

3.36 According to text Fig. 3.2, selective destabilization of the conjugate acid form of an acid-base pair *raises* the energy of the acidic form relative to its conjugate base and *reduces* its pK_a.

Now let's apply this reasoning to the specific example in the problem. This analysis is not unlike the one used in the solution to Problem 3.34. First assume that the chlorine affects the stability of the *charged species,* which in this case is the conjugate *acid*:

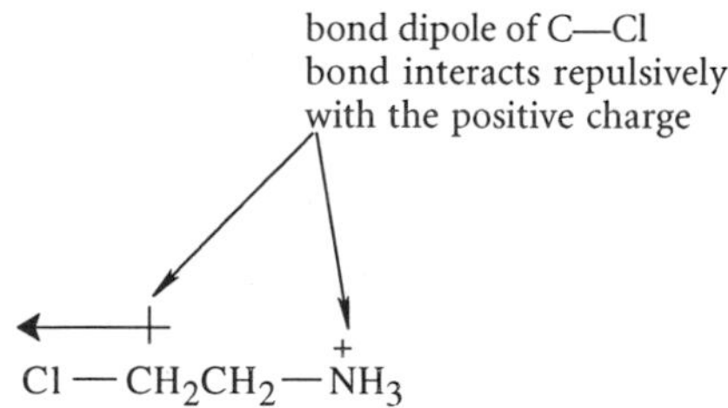

The repulsive interaction between the positive charge and the positive end of the C—Cl bond dipole selectively raises the energy of the conjugate acid. This should reduce the pK_a of the *conjugate acid* of *A* relative to the pK_a of the *conjugate acid* of *B*. This is shown diagrammatically in Fig. SG3.2 on the following page. This means that *B* is the more basic of the two conjugate bases. The actual data bear out this prediction: chlorine substitution reduces the pK_a of ethylammonium ion by almost two units:

conjugate acid of *A*:	conjugate acid of *B*:
$Cl-CH_2CH_2-\overset{+}{N}H_3$	$CH_3CH_2-\overset{+}{N}H_3$
2-chloroethylammonium ion	ethylammonium ion
$pK_a \approx 8.8$	$pK_a \approx 10.6$

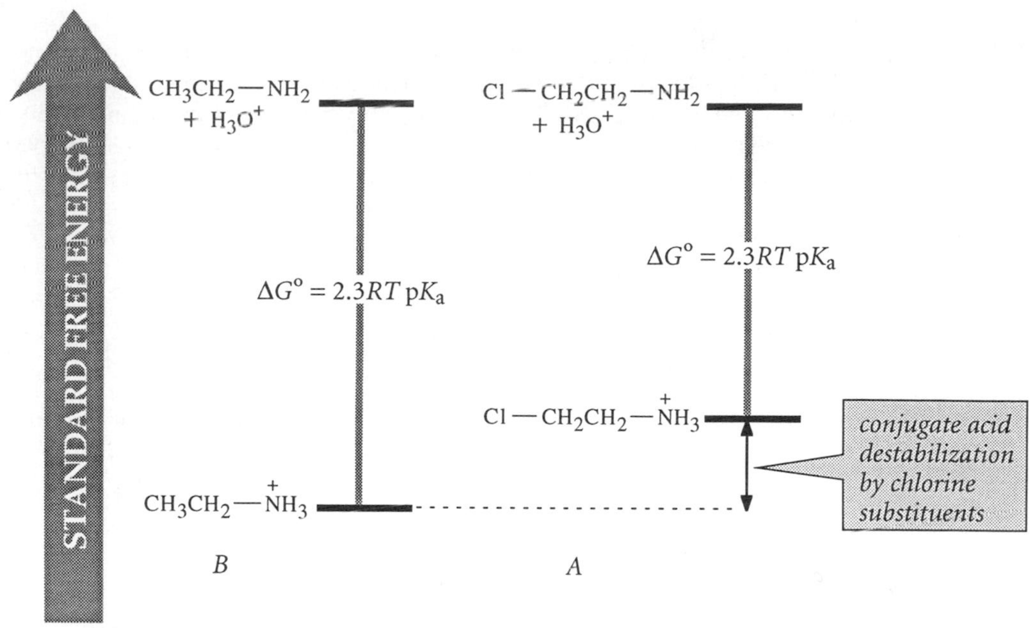

Figure SG3.2 The effect of chlorine substitution on the energy and pK_a of the conjugate acid of ethylamine to accompany the solution to Problem 3.36. Note that by raising the energy of the conjugate acid, chlorine substitution reduces the pK_a. (The two conjugate bases, which are uncharged, are arbitrarily placed at the same energy for comparison purposes.)

3.37 The equilibrium constant K_{eq} for the reaction of text Eq. 3.17 is given by

$$K_{eq} = \frac{[BH][A^-]}{[AH][B^-]}$$

Following the hint, first show that this equals K_{AH}/K_{BH}. We do that by applying the definition of K_{AH} and K_{BH}.

$$K_{AH}/K_{BH} = \frac{[A^-][H_3O^+]}{[AH]} \times \frac{[BH]}{[H_3O^+][B^-]} = \frac{[BH][A^-]}{[AH][B^-]} = K_{eq}$$

or

$$K_{eq} = K_{AH}/K_{BH}$$

Then take logarithms of this equation and apply the definition of pK_a:

$$\log K_{eq} = \log K_{AH} - \log K_{BH} = pK_{BH} - pK_{AH}$$

Taking antilogs of both sides proves the assertion:

$$10^{(\log K_{eq})} = K_{eq} = 10^{(pK_{BH} - pK_{AH})}$$

3.38 Both HI and HCl are much stronger acids than H_3O^+. Consequently, the following equilibrium lies well to the right for both X = Cl and X = I:

$$HX + H_2O \rightleftharpoons H_3O^+ + X^-$$

Since the major acidic species in solution in each case is H_3O^+, and 10^{-3} *M* of this species is present in each case, both solutions have the same pH value of 3.

Note that the most acidic species that can exist in a solvent is the conjugate acid of the solvent; in water, this is H_3O^+. Likewise, the most basic species that can exist in a solvent is the conjugate base of the solvent; in water this is ^{-}OH. Because acids or bases stronger than the solvent react to give the conjugate acid or base of the solvent, respectively, their greater acidity or basicity is not reflected in the pH of the resulting solution. This effect is sometimes termed the *leveling effect* of solvent.

3.39 The first step in the reaction is shown in Eq. 3.6 of the text:

$$H_3\overset{-}{B}—H \quad H—\ddot{\underset{..}{O}}H \longrightarrow H_3B + H_2 + \overset{-}{:}\ddot{\underset{..}{O}}H$$

The second step involves attack of the Lewis base ^{-}OH on the electron-deficient Lewis acid BH_3:

$$H_3B \quad \overset{-}{:}\ddot{\underset{..}{O}}H \longrightarrow H_3\overset{-}{B}—\ddot{\underset{..}{O}}H$$

3.40 To solve the various parts of this problem, recognize that At is a halogen that is below iodine on the periodic table, and apply trends in the periodic table.

(a) Because bond dissociation energy decreases down a column of the periodic table, the H—At bond should be weaker than the H—I bond. The bond dissociation energy of H—At is smaller.

(b) Because electron affinities decrease down a column of the periodic table, the electron affinity of At should be smaller than that of I.

(c) Because dissociation energies dominate the dissociation constant within a *column* of the periodic table, H—At should be a stronger acid than H—I.

4

Introduction to Alkenes; Structure and Reactivity

TERMS

- addition reaction 4.6
- alkene .... Introduction
- alkene hydration 4.9B
- alkyl halide 4.7
- Cahn-Ingold-Prelog system 4.2B
- carbocation 4.7B
- carbocation rearrangement 4.7D
- carbonium ion 4.7B
- catalyst 4.9
- catalyst poison 4.9
- catalytic hydrogenation 4.6, 4.9A
- cis-trans isomerism 4.1D
- degree of unsaturation U 4.3
- E 4.2B
- E,Z system 4.2B
- endothermic reaction 4.5A
- energy barrier 4.8A
- enzyme 4.9C
- exothermic reaction 4.5A
- Hammond's postulate 4.8C
- heat of formation ΔH_f° 4.5A
- Hess's law of constant heat summation 4.5A
- heterogeneous catalyst 4.9
- homogeneous catalyst 4.9
- hydration 4.9B
- hydride shift 4.7D
- hyperconjugation 4.7C
- Markownikoff's rule 4.7A
- molecular configuration 4.1B
- multistep reaction 4.8B
- olefin .... Introduction
- π bond 4.1A
- π molecular orbital 4.1A
- π^* molecular orbital 4.1A
- primary carbocation 4.7C
- protonation 4.7B
- rate-determining step 4.8B
- rate-limiting step 4.8B
- reaction coordinate 4.8A
- reaction free-energy diagram 4.8A
- reaction mechanism 4.7B
- reaction rate 4.8A
- reactive intermediate 4.7B
- rearrangement 4.7D
- regioselective reaction 4.7A
- sp^2-hybridized carbon 4.1A
- sp^2 orbital 4.1A
- saturated hydrocarbon .... Introduction
- secondary carbocation 4.7C
- standard free energy of activation $\Delta G^{\circ\ddagger}$ 4.8A
- stereocenter 4.1B
- stereogenic atom 4.1B
- stereoisomer 4.1B
- tertiary carbocation 4.7C
- transition state $\ddagger$ 4.8A
- unsaturated hydrocarbon .... Introduction
- unsaturation number U 4.3
- unstable intermediate 4.7B
- Z 4.2B

CONCEPTS

I. STRUCTURE, BONDING, AND HEAT OF FORMATION IN ALKENES

A. General

1. Alkenes (sometimes called olefins) are hydrocarbons that contain carbon-carbon double bonds.
   a. The carbon-carbon double bond consists of a σ bond and a π bond.
   b. The π electrons of a double bond are weakly basic and can attack Brønsted or Lewis acids.
   c. For alkenes in general, the carbons of the double bond and the atoms attached to them all lie in the same plane.
2. Because alkenes have fewer hydrogens than the corresponding alkanes, they are classified as unsaturated hydrocarbons.

B. Hybridization

1. The carbon atoms of a double bond, as well as other trigonal-planar atoms, are sp^2-hybridized.
   a. The valence orbitals on each carbon of a double bond are hybridized so that two of the three $2p$ orbitals mix with the $2s$ orbital to form three sp^2 orbitals.
      i. An sp^2orbital has 33% *s* character
      ii. Electron density within an sp^2orbital is concentrated slightly closer to the nucleus than an sp^3orbital.
   b. One $2p$ orbital remains unhybridized.

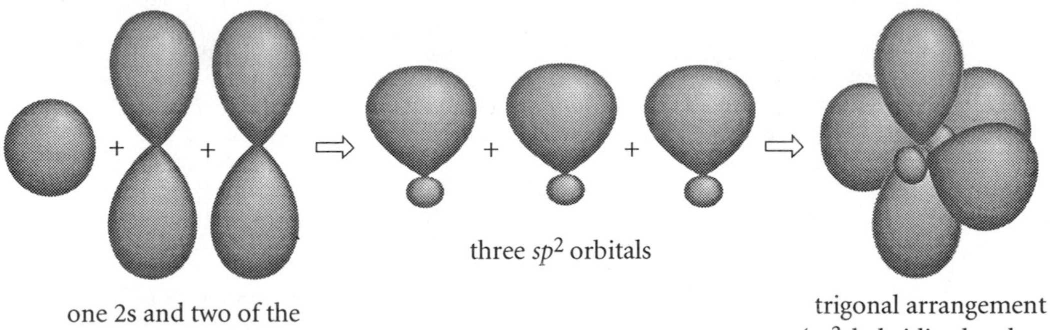

2. Hybridization and geometry are related.
   a. Whenever a main-group atom has trigonal-planar geometry, its hybridization is sp^2.
   b. Whenever a main-group atom has tetrahedral geometry, its hybridization is sp^3.
3. Trigonal carbon withdraws (attracts) electrons from alkyl groups. That is, sp^2-hybridized carbon is electronegative relative to sp^3-hybridized carbon.

C. The π Bond

1. The second bond of a carbon-carbon double bond, called a π bond, arises from the overlap of a $2p$ orbital on one carbon atom with a $2p$ orbital on the other.
   a. Bonding (π) molecular orbitals, resulting from additive overlap of the two carbon $2p$ orbitals, are at lower energy than the isolated p orbitals.
   b. Antibonding molecular orbitals (π^*), resulting from subtractive overlap of the two carbon $2p$ orbitals, are at higher energy than the isolated p orbitals.
   c. The π molecular orbital has a nodal plane; the π^* molecular orbital has two nodal planes.

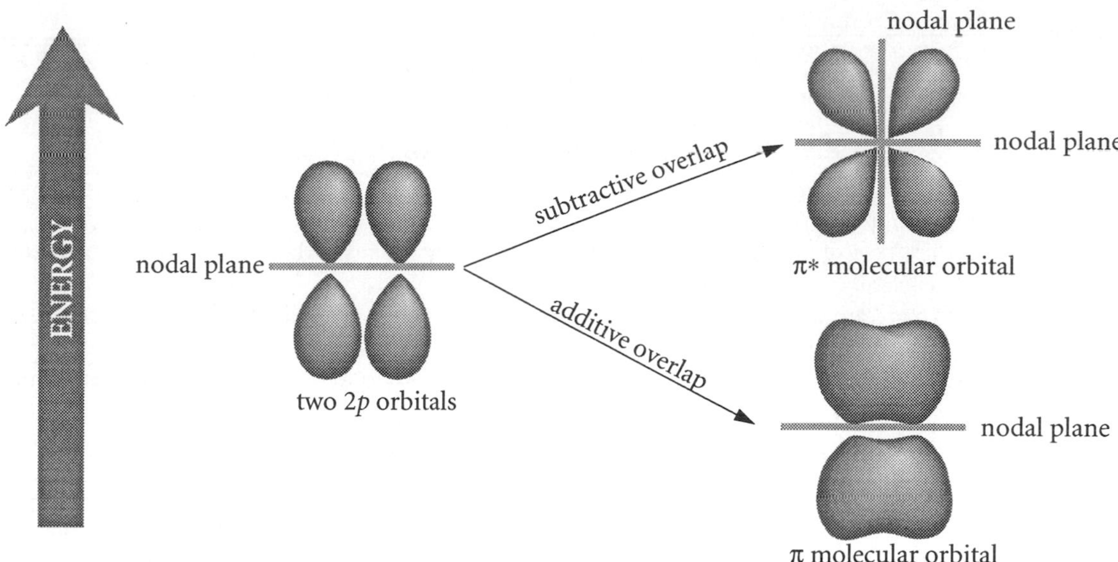

2. The π bond is one bond with two lobes; the π bond thus has electron density both above and below the plane defined by the carbons of the double bond and the directly attached atoms.

D. Stereoisomerism

1. Compounds with identical connectivities that differ in the spatial arrangement of their atoms are called stereoisomers.
   a. Rotation about the alkene double bond does not occur under normal conditions. In order for such an internal rotation to occur, the π bond would have to break; thus, some alkenes can exist as cis and trans isomers.
   b. Cis and trans isomers are named systematically with the *E,Z* priority system.

H_3C, H / C=C / H, CH_3 ← stereoisomers → H, H / C=C / H_3C, CH_3

E-2-butene (or *trans*-2-butene) — *Z*-2-butene (or *cis*-2-butene)

2. A stereocenter (stereogenic atom) is an atom at which the interchange of two groups gives a stereoisomer.

H, CH_3 / C=C / H_3C, H (stereocenter) ← stereoisomers → H, H / C=C (← stereocenter) / H_3C, CH_3

3. Isomeric alkenes that differ in the positions of their double bonds are examples of constitutional isomers.

H_3C, H / C=C / H, CH_3 ← constitutional isomers → H_3C, H / H_2C—C / CH_2

E-2-butene (or *trans*-2-butene) — 1-butene

E. Heats of Formation

1. The standard heat of formation ΔH_f° of a compound is the heat of the reaction in which the compound is formed from its elements in their natural states at 1 atm pressure and 25 °C.
   a. A reaction in which heat is liberated is said to be an exothermic reaction.
   b. A reaction in which heat is absorbed is said to be an endothermic reaction.
2. Chemical reactions and their associated energies can be added algebraically (Hess's law of constant heat summation).
   a. The heat of any reaction is equal to the enthalpy of the products minus the enthalpy of the reactants.
   b. The relative enthalpies of many organic compounds are available in standard tables as heats of formation.
3. Heats of formation (enthalpies of formation) can be used to determine the relative stabilities of various bonding arrangements.

E. Relative Stabilities of Alkene Isomers

1. An alkene is stabilized by alkyl substituents on the double bond. Of two alkene isomers, the one with the greater number of alkyl branches on the double bond is usually the more stable.
2. For noncyclic alkenes, *trans*-alkenes are in most cases more stabile than their cis stereoisomers. The higher energies of a *cis*-alkene is due to van der Waals repulsions between cis groups on its double bond.

II. NOMENCLATURE

A. Unbranched Alkenes

1. An unbranched alkene is named by replacing the *ane* suffix in the name of the corresponding alkane with the ending *ene*.
2. The carbons are numbered from one end to the other so that the double bond receives the lower number.

$$\overset{6}{H_3C}-\overset{5}{CH_2}-\overset{4}{CH_2}-\overset{3}{CH}=\overset{2}{CH}-\overset{1}{CH_3}$$ 2-hexene

B. Branched Alkenes

1. The names of alkenes with branched chains are derived from their principal chains.
   a. The principal chain is defined as the carbon chain containing the greatest number of double bonds, even if this is not the longest chain

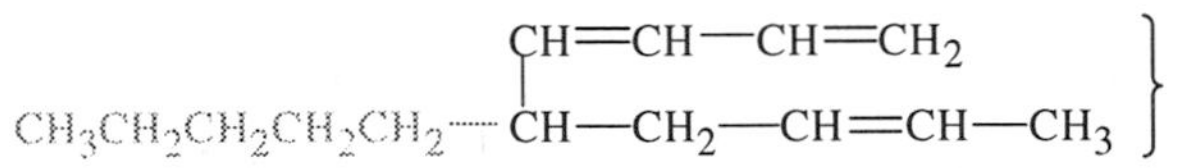

$$\left.\begin{matrix} CH=CH-CH=CH_2 \\ | \\ CH_3CH_2CH_2CH_2CH_2-CH-CH_2-CH=CH-CH_3 \end{matrix}\right\}$$ principal chain

   b. If more than one candidate for the principal chain have equal numbers of double bonds, the principal chain is the longest of these.

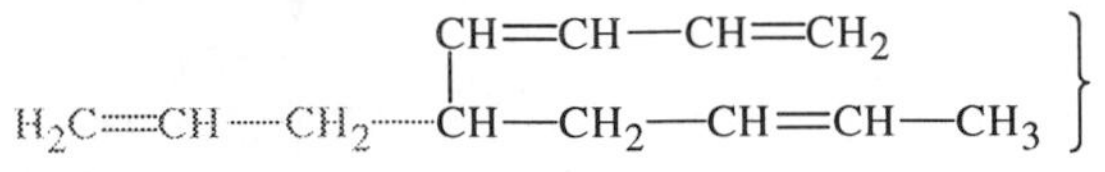

$$\left.\begin{matrix} CH=CH-CH=CH_2 \\ | \\ H_2C=CH-CH_2-CH-CH_2-CH=CH-CH_3 \end{matrix}\right\}$$ principal chain

2. The principal chain is numbered from the end so as to give the lowest numbers to the double bonds at the first point of difference.
   a. If the name remains ambiguous after determining the correct number for the double bonds, then the principal chain is numbered so that the lowest numbers are given to the branches at the first point of difference.
   b. The position of the double bond is cited in the name after the names of the substituents.

c. If a compound contains more than one double bond, the *ane* ending of the corresponding alkane is replaced by *adiene* (if there are two double bonds), *atriene* (if there are three double bonds), etc.

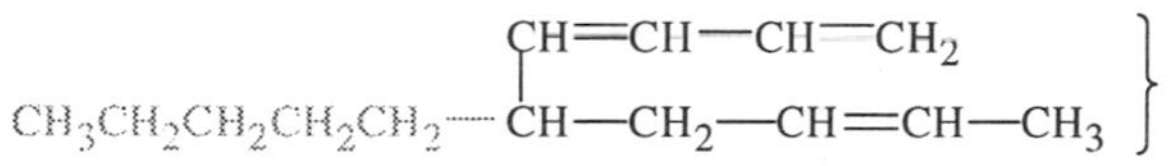

$$\begin{array}{l} \quad\quad\quad\quad\quad\quad\quad\quad\quad CH{=}CH{-}CH{=}CH_2 \\ CH_3CH_2CH_2CH_2CH_2{-}CH{-}CH_2{-}CH{=}CH{-}CH_3 \end{array}$$

a triene: 5-pentyl-1,3,7-nonatriene

3. Substituent groups may also contain double bonds.
   a. Some widely occurring groups of this type have special names that must be learned.
   b. The names of other groups are constructed from the name of the parent hydrocarbon by dropping the final *e* and replacing it with *yl*, and they are numbered from the point of attachment to the principal chain.

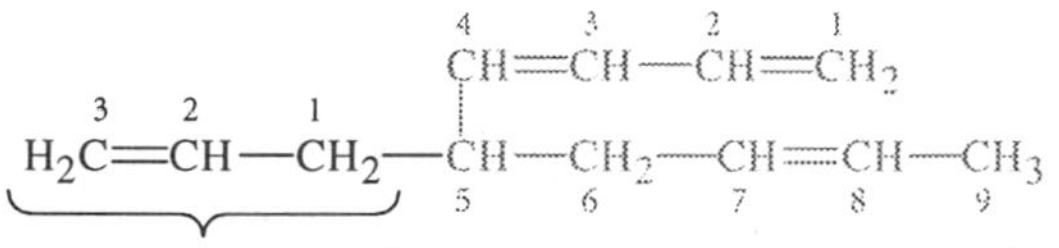

$$\begin{array}{l} \quad\quad\quad\quad\quad\quad\quad\quad \overset{4}{C}H{=}\overset{3}{C}H{-}\overset{2}{C}H{=}\overset{1}{C}H_2 \\ H_2\overset{3}{C}{=}\overset{2}{C}H{-}\overset{1}{C}H_2{-}\underset{5}{C}H{-}\underset{6}{C}H_2{-}\underset{7}{C}H{=}\underset{8}{C}H{-}\underset{9}{C}H_3 \end{array}$$

allyl group or 2-propenyl group

5-(2-propenyl)-1,3,7-nonatriene
or
5-allyl-1,3,7-nonatriene

4. Some alkenes have nonsystematic traditional names that are recognized by the IUPAC.

$Ph{-}CH{=}CH_2$ styrene

C. The Cahn-Ingold-Prelog (*E,Z*) System

1. The Cahn Ingold-Prelog system is applied to alkenes by assigning relative priorities to the two groups on each carbon of the double bond.
   a. The designation *Z* (*zusammen*; together) is used when the groups of highest priority are on the same side of the double bond.
   b. The designation *E* (*entgegen*; across) is used when the groups of highest priority are on different sides of the double bond.

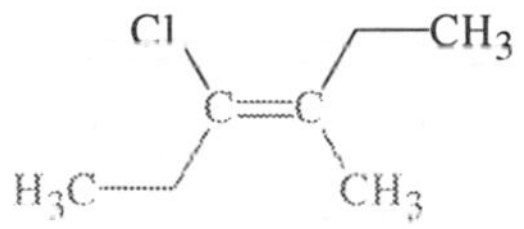

Cl, H_3C / C=C / CH_3, CH_3

(*Z*)-3-chloro 4 methyl-3-hexene

groups of higher priority at each carbon are shown in black

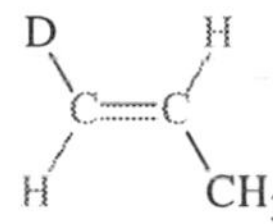

D, H / C=C / H, CH_3

(*E*)-1-deuteriopropene

2. To assign relative priorities of groups at a given carbon of a double bond, examine the atoms directly attached to that carbon.
   a. Assign higher priority to the group containing the atom of higher atomic number (or atomic mass in the case of isotopes).
   b. If the two atoms directly attached to the carbon of the double bond are the same, compare the sets of atoms attached to these carbons.
      i. Arrange the attached atoms within each set in descending priority order, and make a pairwise comparison of the atoms in the two sets.
      ii. The higher priority is assigned to the atom of higher atomic number (or atomic mass in the case of isotopes) at the first point of difference.
      iii. Double bonds are treated by a special convention, in which the double bond is rewritten as a single bond and the atoms at each end of the double bond are duplicated.
      iv. The treatment of triple bonds requires triplicating the atoms involved.
   c. If the sets of attached atoms are identical, move away from the double bond within each group to the next atom following the path of highest priority and repeat part b above.

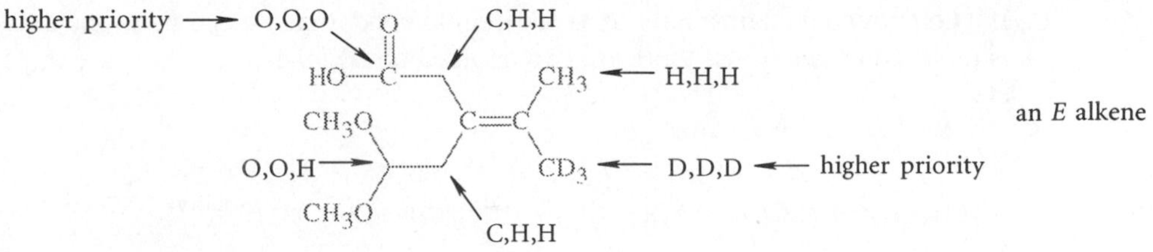

higher priority → O,O,O
C,H,H
HO—C(=O)
CH_3 ← H,H,H
CH_3O
C=C
an *E* alkene
O,O,H →
CD_3 ← D,D,D ← higher priority
CH_3O
C,H,H

D. Unsaturation Number

1. The unsaturation number *U* of a molecule is equal to the total number of rings and multiple bonds in the molecule.
2. The unsaturation number can be calculated from the molecular formula:

$$U = \frac{2C + 2 + N - X - H}{2} = \text{number of rings} + \text{multiple bonds}$$

where C = the number of carbons;
N = the number of nitrogens;
X = the number of halogens;
and H = the number of hydrogens.

3. Interpretation of the formula:
   a. The number of hydrogens in a fully saturated molecule is equal to $2C + 2 + N - X$.
   b. Every ring or double bond reduces the number of hydrogens from this maximum by two (2).
   c. Each halogen atom in an organic compound reduces the maximum possible number of hydrogens by one (1).
   c. Each nitrogen atom present increases the maximum possible number of hydrogens by one (1).
   d. Each oxygen atom present has no effect on the maximum possible number of hydrogens.

$C_8H_8ClNO_2$

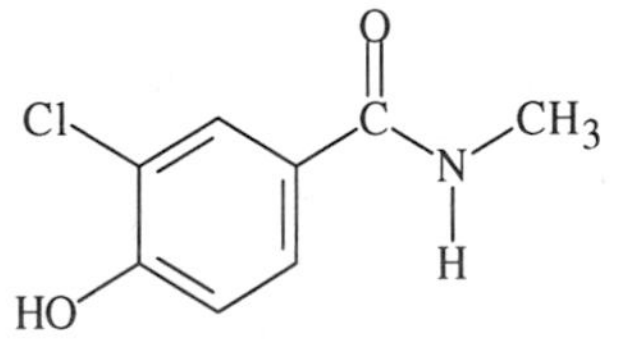

Cl, HO, O, C, N, CH_3, H

$$U = \frac{2(8) + 2 + (1) - (1) - (8)}{2} = 5 = \text{1 ring + 4 double bonds}$$

III. PRINCIPLES OF REACTIONS

A. Regioselective Reactions

1. A reaction that gives only one of several possible constitutional isomers is said to be a regioselective reaction.
2. Many alkene addition reactions are regioselective when they involve alkenes with different numbers of substituents on the two carbons of their double bonds.
3. The regioselectivity observed in the addition reactions of hydrogen halides or water to alkenes is a consequence of several factors:
   a. The rate-limiting transition state of each reaction resembles a carbocation.
   b. The relative stability of carbocations is tertiary > secondary > primary.
   c. The structures and energies of transition states for reactions involving unstable intermediates (such as carbocations) resemble the structures and energies of the unstable intermediates themselves (Hammond's postulate).
4. Reactions involving carbocation intermediates show rearrangements in some cases.

B. Reaction Rates

1. The rate of a chemical reaction can be defined for our purposes as the number of reactant molecules converted into products in a given time.

2. The relative free energies of transition state and reactants determine the reaction rate.
3. Two factors that govern the intrinsic reaction rate are:
   a. the size of the energy barrier (standard free energy of activation $\Delta G^{\circ\ddagger}$): reactions with larger $\Delta G^{\circ\ddagger}$ are slower.
   b. the temperature: reactions are faster at higher temperatures.
4. The equilibrium constant for a reaction tells us absolutely nothing about its rate.

C. Catalysts

1. A substance that increases the rate of a reaction without being consumed is called a catalyst.
   a. A catalyst cannot affect the position of an equilibrium but it does affect the rate at which a reaction comes to equilibrium.
   b. A catalyst accelerates both forward and reverse reactions of an equilibrium equally.
2. Catalysts are of two types: heterogeneous and homogeneous.
   a. A catalyst that is soluble in a reaction solution is called a homogeneous catalyst. Acid-catalyzed hydration of alkenes is a reaction involving homogeneous catalysis.
   b. A catalyst that exists in a separate phase from the reactants is called a heterogeneous catalyst. Catalytic hydrogenation of alkenes is a reaction involving heterogeneous catalysis.
3. Enzymes are biological catalysts.

D. Multistep Reactions, the Rate-Limiting Step, and Mechanisms

1. Each step of a multistep reaction has its own characteristic rate and therefore its own transition state.
2. The slowest step in a multistep chemical reaction is called the rate-limiting step (rate-determining step) of the reaction.
   a. The rate-limiting step is the step with the transition state of highest free energy.
   b. The overall rate of a reaction is equal to the rate of the rate-limiting step.
   c. Anything that increases the rate of this step increases the overall reaction rate.
3. The complete description of a reaction pathway, including any reactive intermediates such as carbocations, is called the mechanism of the reaction.

E. Transition-State Theory

1. As reactants change into products, they pass through an unstable state of maximum free energy called the transition state.
   a. The transition state has a higher energy than either the reactants or products.
   b. The transition state represents an energy barrier to the interconversion of reactants and products.
2. The energy barrier $\Delta G^{\circ\ddagger}$, called the standard free energy of activation, is equal to the difference between the standard free energies of the transition state and reactants.
   a. The size of the energy barrier $\Delta G^{\circ\ddagger}$ determines the rate of a reaction: the higher the barrier, the lower the rate.
   b. This energy barrier is shown graphically in a reaction free-energy diagram in which the pathway of the reaction from reactants to products is called the reaction coordinate.

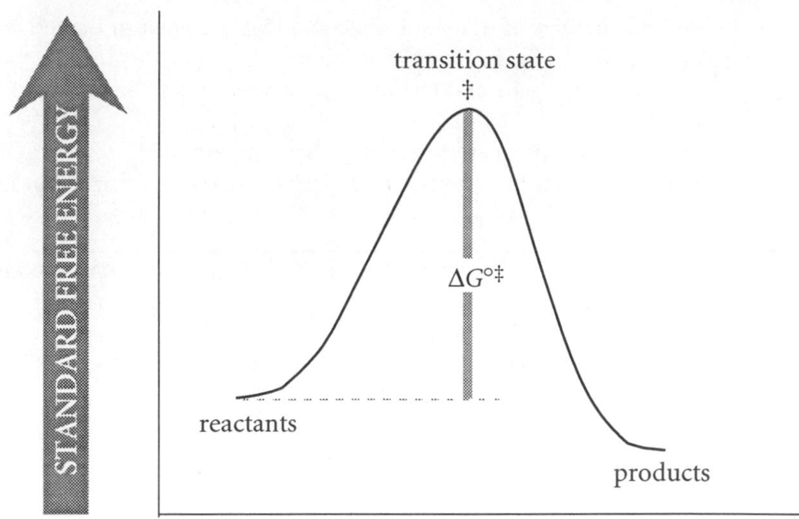

3. In general, molecules obtain the energy required for a reaction from their thermal motions.
4. The average energy of molecules in a chemical system can be increased by raising the temperature. Thus, reaction rates increase with increasing temperature because a greater fraction of molecules has sufficient energy to react.

F. Hammond's Postulate

1. The assumption that the structure and energy of a transition state closely resembles the structure and energy of a corresponding unstable intermediate is known as Hammond's postulate.
2. It is not the relative stabilities of the reactive intermediates themselves that determine which of two competing reactions is faster; rather, it is the relative free energies of the transition states for the formation of these intermediates that determine the relative rates of reactions.
3. Hammond's postulate draws the connection between transition states and the corresponding reactive intermediates.

G. Structure and Stability of Carbocations

1. A species with a positively charged, electron-deficient carbon is called a carbocation (carbonium ion in older literature).
   a. Carbocations are examples of reactive (unstable) intermediates—species that react so rapidly that they never accumulate in more than very low concentrations.
   b. Carbocations are powerful electron-deficient Lewis acids and thus are potent electrophiles.
2. The relative stability of carbocations is in the order:

tertiary > secondary > primary

   a. A primary carbocation has one alkyl group bound to the electron-deficient carbon.
   b. A secondary carbocation has two alkyl groups bound to the electron-deficient carbon.
   c. A tertiary carbocation has three alkyl groups bound to the electron-deficient carbon.

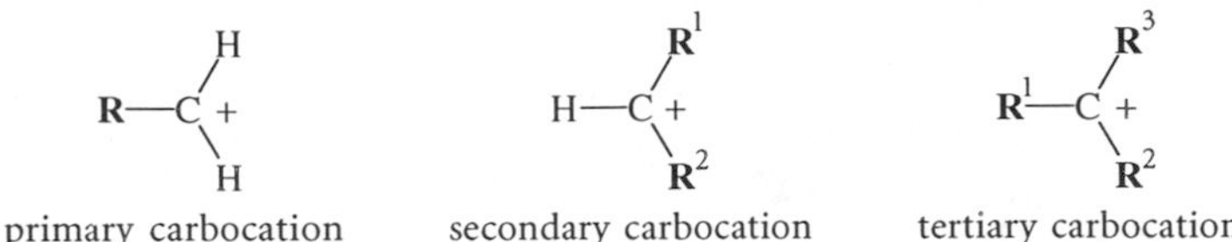

primary carbocation   secondary carbocation   tertiary carbocation

3. The explanation for the stabilization of carbocations by alkyl branching is a phenomenon called hyperconjugation, which is the overlap of bonding electrons from the adjacent σ bonds with the unoccupied $2p$ orbital of the carbocation.

a. The energetic advantage of hyperconjugation comes from the additional bonding symbolized by the double bond.

$$\left[H_2C(H){-}C^{+} \longleftrightarrow H^{+}\; H_2C{=}C \right]$$

H. Carbocation Rearrangements

1. Carbocations can react with nucleophiles or rearrange to more stable carbocations.
2. In a rearrangement, a group or atom from the starting material has moved to a different position in the product.

$$(CH_3)_3C{-}\overset{+}{C}H{-}CH_3 \xrightarrow{\text{rearrangement}} (H_3C)_2\overset{+}{C}{-}C(CH_3)_2{-}H$$

secondary carbocation — tertiary carbocation

a. Rearrangement of a carbocation is favored when the rearranged ion is more stable.
b. Some rearrangements involve a hydride shift—the migration of a hydrogen with its two bonding electrons.

$$H_3C{-}CH(CH_3){-}\overset{+}{C}H{-}CH_3 \xrightarrow{\text{hydride shift}} (H_3C)_2\overset{+}{C}{-}CH(CH_3){-}H$$

secondary carbocation — tertiary carbocation

REACTIONS

I. ADDITION REACTIONS OF ALKENES

A. Addition of Hydrogen Halides to Alkenes

1. Hydrogen halides add to alkenes in a regioselective manner so that the hydrogen adds to the less branched carbon and the halide group to the more branched carbon of the double bond (Markownikoff's rule).
2. The products of these reactions are called alkyl halides.

1-methylcyclopentene $\xrightarrow{HBr}$ 1-bromo-1-methylcyclopentane

an alkyl halide

3. The addition of a hydrogen halide to an alkene is a regioselective reaction that occurs in two successive steps.

a. In the first step, protonation of the alkene double bond occurs at the carbon with the fewer alkyl branches so that the more stable carbocation is formed—the one with the greater number of alkyl branches at the electron-deficient carbon.
b. In the second step, the halide ion attacks the electron-deficient carbon.

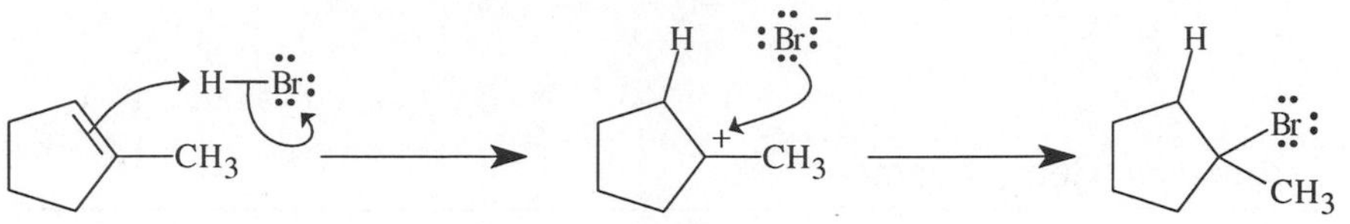

4. Hydrogen halide addition to an alkene is a regioselective reaction because addition involves the transition state that resembles the more stable of the two possible carbocation intermediates.
5. The formation of a carbocation from an alkene is an electron-pair displacement reaction in which the π bond acts as a Brønsted base towards the Brønsted acid H—X.

B. Catalytic Hydrogenation of Alkenes

1. Catalytic hydrogenation is an addition of hydrogen in the presence of a catalyst and is one of the best ways to convert alkenes into alkanes.
   a. Because hydrogenation catalysts are insoluble in the reaction solution, they are examples of heterogeneous catalysts.
   b. Benzene rings are inert to conditions under which normal double bonds react readily, but under conditions of high temperature and pressure they can be hydrogenated.

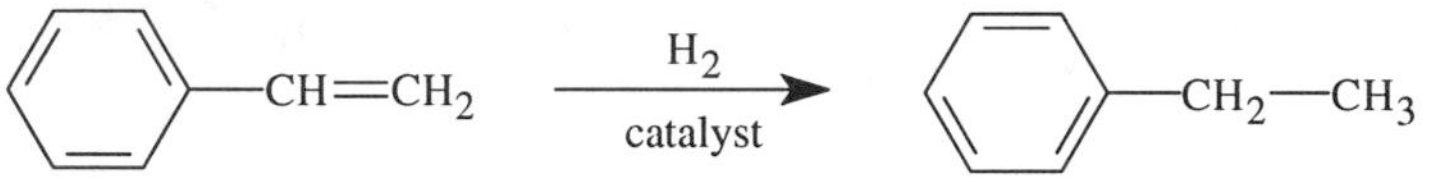

C. Hydration of Alkenes

1. The addition of water to the alkene double bond, called alkene hydration, is an acid-catalyzed reaction.
   a. Because the catalyzing acid is soluble in the reaction solution, it is a homogeneous catalyst.
   b. Hydration is a regioselective reaction.

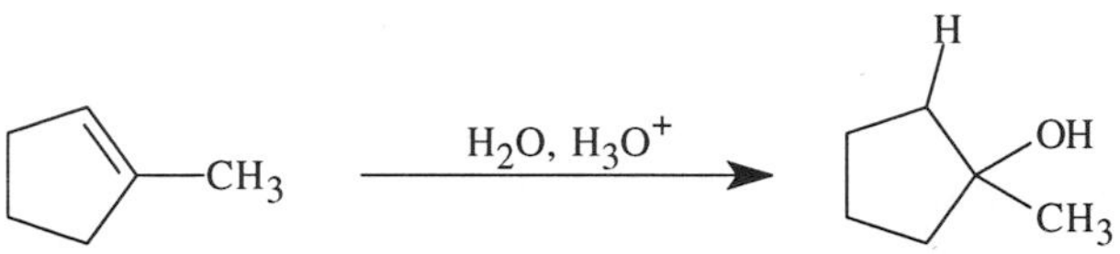

2. Alkene hydration is a multistep reaction.
   a. In the first step of the reaction (the rate-limiting step), the double bond is protonated to give a carbocation.
   b. In the next step of the hydration reaction, the carbocation is attacked by the Lewis base water in a Lewis acid-base association reaction.
   c. Finally, a proton is lost to solvent in another Brønsted acid-base reaction to give an alcohol product and regenerate the catalyzing acid.

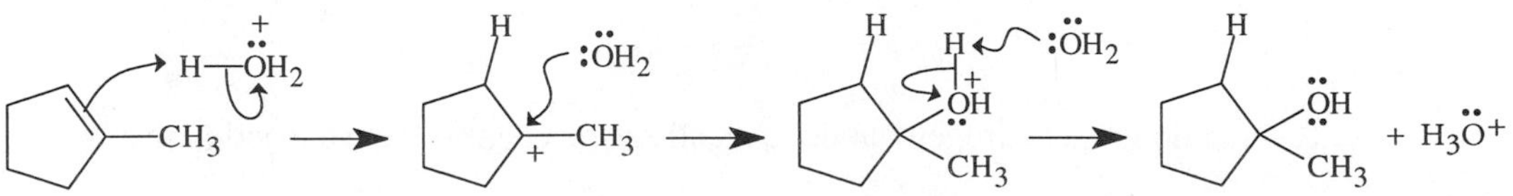

3. Because the hydration reaction involves carbocation intermediates, some alkenes give rearranged hydration products.

SCIENTISTS

Robert Sidney Cahn (1899–1981)	4.2B
George Simms Hammond (born 1921)	4.8C
Germain Henri Hess (1802–1850)	4.5A
Christopher Kelk Ingold (1893–1970)	4.2B
Vladimir Vasilevich Markownikoff (1838–1904)	4.7A
Vladimir Prelog (1906–1998) [Chemistry 1975]	4.2B
Frank Clifford Whitmore (1887–1947)	4.7D

Dates in brackets refer to Nobel Prize awards.

STUDY GUIDE LINKS

4.1 Different Bonding Models for Ethylene

It may seem strange that we are using one bonding model—the hybrid-orbital model—to describe the σ bonds of ethylene, and another model—the molecular-orbital model—to describe the π bond. We use the hybrid-orbital model for the σ bonds because it gives us a picture of individual bonds. The molecular-orbital picture, in contrast, shows the σ framework as an "electron pudding" with imbedded nuclei. Individual σ bonds cannot be associated with individual molecular orbitals because the molecular orbitals occupied by the σ electrons encompass the entire molecule. (A similar problem was discussed for the molecular orbitals of methane; text Sec. 1.9A.) For the π bond, however, there is a direct correspondence between the bond and the bonding molecular orbital.

We could avoid a molecular orbital description of the π bond by simply describing it as the result of overlap of two $2p$ orbitals. But when we do that, we lose sight of the antibonding molecular orbital, which will turn out to be very important in understanding both ultraviolet spectroscopy (Chapter 15) and certain types of chemical reactions (Chapter 25).

A complete molecular-orbital analysis of ethylene shows that the π molecular orbitals are quite separate from the molecular orbitals associated with the σ electrons. Hence, a separate treatment of the π orbitals is not only conceptually useful but also theoretically justified.

✓4.2 Different Ways to Draw the Same Structure

The discussion in the text requires you to realize that there are several different ways to draw the same alkene stereoisomer. For example, two equivalent ways to represent *cis*-2-butene are the following:

H_3C, CH_3 / C=C / H, H and H, H / C=C / H_3C, CH_3

cis-2-butene, or (*Z*)-2-butene

If this is not clear, think of the structure as an object. Rotate the structure 180° about an axis through the C=C bond, as follows:

H_3C, CH_3 / C=C / H, H —180°→ H, H / C=C / H_3C, CH_3

(Of course, when we perform such a rotation, the atom labels, for example, CH_3, are maintained "right side up"; that is, they are not inverted.)

Similarly, two equivalent ways of drawing *trans*-2-butene are as follows:

H_3C, H / C=C / H, CH_3 and H, CH_3 / C=C / H_3C, H

trans-2-butene, or (*E*)-2-butene

The two structures in text Eq. 4.3 are examples of identical structures drawn differently. Some students think that structures such as these are stereoisomers, but in fact they represent the same molecule. As these structures demonstrate, the ultimate test of identity of two structures is their congruence—the superimposability of each atom of one structure and an identical atom of the other.

✓4.3 Drawing Structures from Names

When asked to draw a complicated structure from a name, as in Problem 4.5b, you should review again the procedure used in Study Problem 2.3. This procedure requires you to be *systematic*. The natural tendency is to try to write the finished structure immediately. Instead, you should take it one step at a time:

1. Write the carbon skeleton of the principal chain; do not be concerned with stereochemistry.
2. Add the substituents to the principal chain.
3. Add the hydrogens.
4. Decide on relative group priorities at the double bonds to which an *E* or *Z* configuration must be assigned. Jot them down on your structure.
5. Redraw your structure with proper stereochemistry. It helps to maintain proper 120° bond angles at the double bonds.

Let's illustrate with an example. Suppose you are asked to draw the structure of (2*E*,4*Z*)-3-isobutyl-2,4-hexadiene. Follow the above steps in order.

1. The principal chain contains six carbons with double bonds at carbons 2 and 4:

   C—C=C—C=C—C

2. There is an isobutyl group at carbon 3. Add the hydrogens within this group, since you won't have to manipulate this group further.

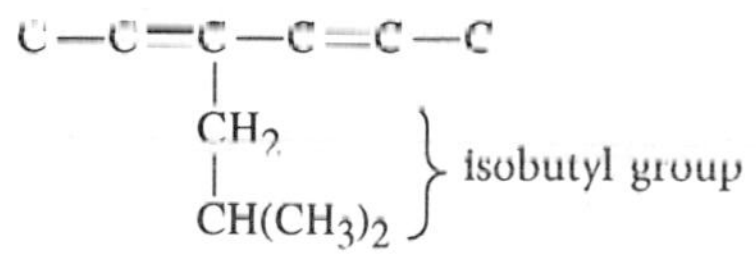

   C—C=C—C=C—C (with CH_2—$CH(CH_3)_2$ on carbon 3: isobutyl group)

3. Add the missing hydrogens.

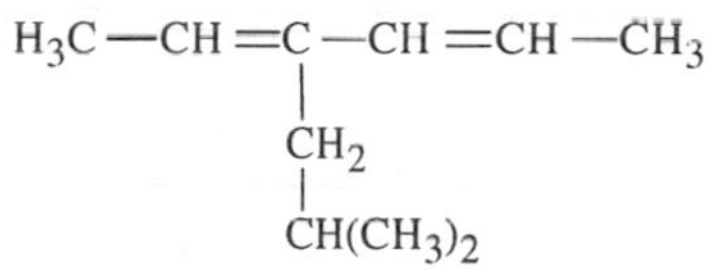

   $H_3C—CH=C—CH=CH—CH_3$ (with CH_2—$CH(CH_3)_2$ on carbon 3)

4. Indicate the relative priorities of the groups at each double bond. For example, focusing on the leftmost double bond (carbons in boldface below), the relative priorities of the attached groups are indicated by numbers:

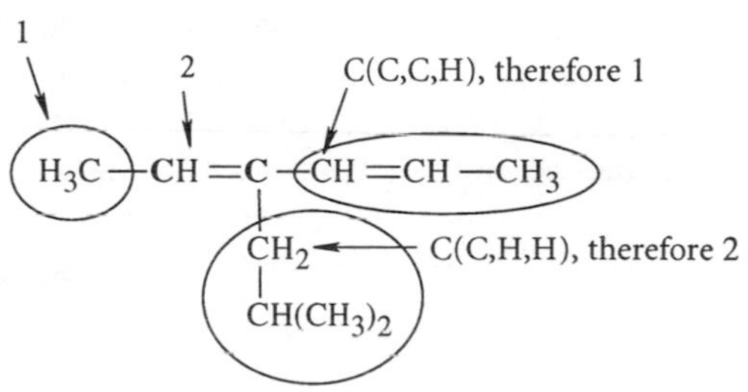

   H_3C (1) — CH (2) =C— CH=CH—CH_3: C(C,C,H), therefore 1
   CH_2—$CH(CH_3)_2$: C(C,H,H), therefore 2

   (You should complete the priorities for the rightmost double bond.)

(list continues)

5. Redraw the structure with proper stereochemistry.

(2*E*,4*Z*)-3-isobutyl-2,4-hexadiene

You should work Problem 4.5b using this step-by-step procedure.

In many cases, it is easier to work a complicated problem if it can be broken into smaller "chunks." When you learn to bypass the anxiety created by a complex problem and adopt this approach, you have made significant progress toward becoming a good problem solver.

4.4 Sources of Heats of Formation

Heats of formation are not obtained by direct measurement. Rather, they are calculated from more readily available data by applying Hess's law. Two types of data that are commonly used are heats of combustion and heats of hydrogenation. To illustrate, suppose that we want to calculate the heat of formation of *trans*-2-butene from the following known heats of combustion: hydrogen, –241.8 kJ/mol; carbon, –393.5 kJ/mol; and *trans*-2-butene, –2530.0 kJ/mol. The combustion reaction (Sec. 2.7) is the reaction of each species with oxygen to give water (in the combustion of hydrogen), carbon dioxide (in the combustion of carbon), and both water and carbon dioxide (in the combustion of hydrocarbons).

Hess's law allows us to express the formation of *trans*-2-butene from its elements as the sum of three combustion reactions. Identical species on opposite sides of the equations cancel, and the resulting equations and their enthalpies are added algebraically:

Equations:	$\Delta H°$ (kJ/mol)
$4H_2 + 2O_2 \longrightarrow 4H_2O$	$4(-241.8) = -967.2$
$4C + 4O_2 \longrightarrow 4CO_2$	$4(-393.5) = -1574.0$
$4CO_2 + 4H_2O \longrightarrow 6O_2$ + *trans*-2-butene	+2529.6
Sum: $4C + 4H_2 \longrightarrow$ *trans*-2-butene	$\Delta H_f^\circ = -11.6$ kJ/mol

Note that because four moles of both H_2 and C are required, their respective enthalpies of combustion must be multiplied by 4. Note also that the combustion of *trans*-2-butene must be written in reverse so that the formation equation comes out with *trans*-2-butene on the right. Consequently, the sign of the enthalpy of this combustion is also reversed.

As this example illustrates, heats of formation derived from combustion are in many cases small differences between large numbers and are therefore subject to considerable uncertainty. Another reaction that has been used for obtaining heats of formation is *catalytic hydrogenation.* In this reaction, hydrogen is added to an alkene double bond in the presence of a catalyst (Sec. 4.9A). For *trans*-2-butene, this reaction is as follows:

$$\text{trans-2-butene} + H_2 \xrightarrow{\text{catalyst}} H_3C-CH_2-CH_2-CH_3 \text{ (butane)}$$

The overall $\Delta H°$ of this reaction can be measured with excellent precision, and is typically much smaller than a heat of combustion. (For example, the $\Delta H°$ of hydrogenation of *trans*-2-butene is –115.5 kJ/mol.) The heats of formation of simple alkanes are known with good precision. For example, the heat of formation of butane is –127.1 kJ/mol. The heat of formation of H_2 is zero by definition. Hence, it is possible to calculate the heat of formation of the alkene (try it!). See Problems 4.14 and 4.16 for calculations of heats of formation from combution and hydrogenation data, respectively.

Fortunately, calculations like these are not necessary every time a heat of formation is needed, because such calculations have already been done to provide the data available in standard tables of heats of formation.

4.5 Relationship between Free Energy and Enthalpy

To review: If you want to calculate an equilibrium constant, the *standard free energies* (ΔG° values) of the reactants and products are the relevant quantities. If you want to ask which of two molecules has a less energetic (or more stable) arrangement of bonds, then you need the *standard enthalpies of formation* (ΔH° values) of the two compounds.

This section discusses why two different types of energy are needed for different purposes. The discrepancy between the enthalpy change for any process (such as a reaction) and the free-energy change lies in something called the *entropy change.* The relationship between the free-energy change, the enthalpy change, and the entropy change is deceptively simple:

$$\Delta G^\circ = \Delta H^\circ - T\Delta S^\circ \qquad \text{(SG4.1)}$$

In this equation, ΔS° is the entropy change and T is the absolute temperature in kelvins. What is the meaning of the entropy change?

Entropy measures molecular randomness, or *freedom of motion.* For example, consider the following reaction:

$$\text{cyclobutane} \longrightarrow H_2C{=}CHCH_2CH_3 \ \text{(1-butene)}$$

cyclobutane 1-butene

The experimentally determined ΔH° for the reaction at 25 °C is –26.53 kJ/mol (–6.34 kcal/mol), and the ΔG° for the reaction at 25 °C is –40.13 kJ/mol (–9.59 kcal/mol). Solving Eq. SG4.1, $T\Delta S^\circ$ at T = 298 K is 13.60 kJ/mol (3.25 kcal/mol); that is, ΔS° is positive. A positive ΔS° means that there is more *randomness* or *freedom of motion* in the products than the reactants.

What is it about 1-butene that is more "random" than cyclobutane? Notice that 1-butene has two carbon-carbon single bonds. Internal rotation can occur readily about these bonds, just as in butane. However, in cyclobutane, *internal rotations cannot occur* because the various carbon atoms are constrained into a small ring. (If this is not clear to you, make a model of cyclobutane and convince yourself that internal rotation about its carbon-carbon bonds cannot occur.) Thus, there is *greater freedom of motion* in 1-butene than in cyclobutane. Because of the internal rotations in 1-butene, the hydrogens on the carbons connected by single bonds can move through a greater volume of space than the same hydrogens in cyclobutane. Thus, *their positions are more random.* Hence, the randomness—or *entropy*—in 1-butene is greater than that in cyclobutane.

Because of entropy—the "randomness factor"—the energy that controls the position of a chemical equilibrium (ΔG°) is different from the energy stored in chemical bonds (ΔH°). Eq. SG4.1 shows that an increase of randomness in a reaction—a positive ΔS°—gives the product an additional advantage in a chemical equilibrium *over and above* that which results from the formation of more stable bonds.

4.6 Molecular-Orbital Description of Hyperconjugation

A molecular-orbital treatment shows why hyperconjugation is a stabilizing effect. If you understand the molecular-orbital treatment of π bonds (text Fig. 4.6), you should have no difficulty with the treatment that follows.

We'll use as our starting orbitals the empty $2p$ orbital of the carbocation and the sp^2-$1s$ σ-bond orbital of the adjacent C—H bond. Notice that both orbitals contain a node. When the axes of the two orbitals are coplanar, they can overlap. This overlap *is* hyperconjugation.

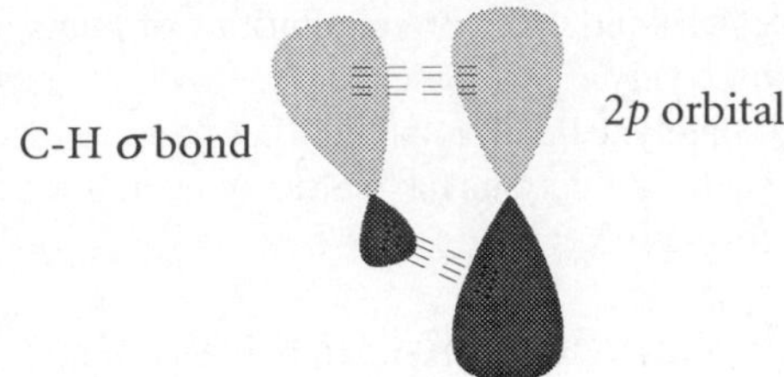

To construct molecular orbitals from these two orbitals, we imagine both bonding overlap and antibonding overlap:

The energy diagram for molecular-orbital formation is as follows: (The sp^2-$1s$ σ-bond orbital has lower energy than the $2p$ orbital; why?)

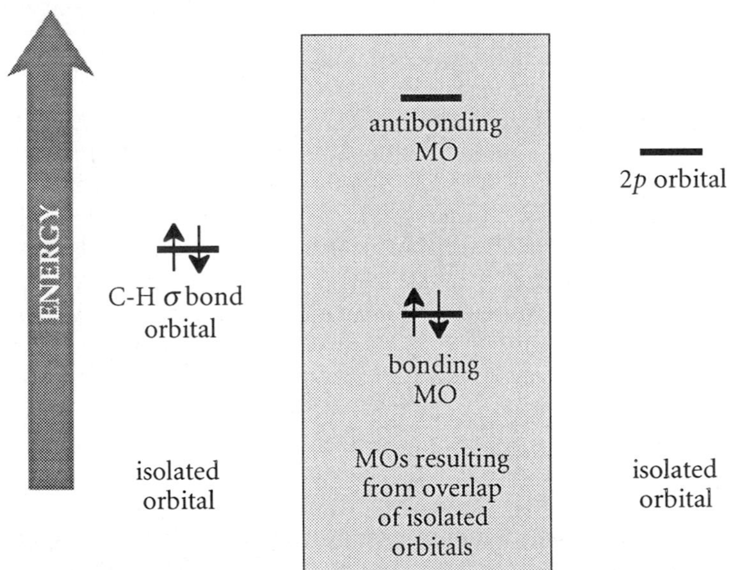

Two electrons must be distributed between the two molecular orbitals. As usual, these both go with opposite spin into the bonding molecular orbital, as shown above. Because this orbital has a lower energy than the original σ-bond orbital, molecular-orbital formation (that is, hyperconjugation) lowers the energy of the system.

It is not strictly correct to use a bond orbital as a starting orbital for this treatment, since a bond orbital is not a proper molecular orbital of the carbocation. In fact, a full-fledged molecular-orbital treatment of the *tert*-butyl cation would start with all the carbons and hydrogens in the proper geometry and allow all their valence orbitals to interact. Adopting such an approach gives a more rigorous and complete molecular-orbital picture, but (as noted in Study Guide Link 4.1) discrete bonds disappear and the descriptive value of the treatment is diminished. However, such a full-fledged molecular-orbital approach *does* show the importance of coplanar $2p$-orbital and

C—H axes for reducing the energy of the cation. That is, the molecular orbitals that result when these axes are *not* coplanar give a cation of higher energy. Moreover, in the cation with the correct geometry for hyperconjugation, the overlap described above can be discerned as a component of the complete molecular orbitals.

✓4.7 Solving Reaction Problems When the Product Is Known

You'll encounter problems such as Problem 4.18 and 4.20 throughout the text. They are designed to test your understanding of the reactions you'll study. In many cases, students who make a reasonable effort to understand each reaction find that supplying the products of a reaction when the starting materials are given, as in Problem 4.18, is a relatively straightforward exercise. However, problems such as Problem 4.20, in which you are given a product and asked to supply an appropriate reactant (or *two* appropriate reactants!) sometimes presents difficulties. This Study Guide Link discusses a systematic approach to such problems.

Suppose you are asked to give the structure of an alkene that would give the following compound as the major product of H—Br addition:

$$CH_3CH_2CH_2\underset{\displaystyle Br}{\underset{|}{C}}HCH_3$$

It is not hard to understand that the bromine of the product must have come from the H—Br. However, there are many hydrogens in the product! Which ones were there to start with, and which one came from the H—Br?

First, recognize that the carbon bearing the bromine must have originally been one carbon of the double bond. It then follows that the other carbon of the double bond must be an *adjacent* carbon (because two carbons involved in the same double bond must be adjacent.). Use this fact to construct *all possible alkenes* that *might* be starting materials. Do this by removing the bromine and a hydrogen from *each adjacent carbon in turn.*

$$CH_3\overset{3}{C}H_2\overset{2}{\underset{\displaystyle Br}{\underset{|}{C}}}H\overset{1}{C}H_3$$

remove Br from carbon-2 and H from carbon-1 ⇨ $CH_3CH_2CH_2CH{=}CH_2$
1-pentene

remove Br from carbon-2 and H from carbon-3 ⇨ $CH_3CH_2CH{=}CHCH_3$
2-pentene

(The symbol ⇨ means, "Implies as starting materials.")

Which of these is correct? Or are they both correct? You haven't finished the problem until you've mentally carried out the addition of HBr to *each compound.* Doing this and applying the known regioselectivity of HBr addition leads to the conclusion that the desired alkyl halide could be prepared as the major product from 1-pentene. However, both carbons of the double bond of 2-pentene bear the same number of alkyl groups. Eq. 4.17, text p. 135, indicates that from this starting material we should expect not only the desired product, but also a second product:

$$CH_3CH_2\underset{\displaystyle Br}{\underset{|}{C}}HCH_2CH_3$$

Furthermore, the two products should be formed in nearly equal amounts. Consequently, 1-pentene is the only alkene that will give the desired alkyl halide as the *major* product (that is, the one formed almost exclusively).

Solving this type of problem is something like taking apart an engine or a household appliance to make a repair. Taking it apart is only half the battle. Once it is apart, you then have to put it back together. Suppose you find that the parts seem to fit together in several different ways. You must then decide which of the ways will allow the engine or the appliance to run properly and which will not. Analogously, once you have identified potential starting materials, you must determine whether they really will work, given the known characteristics of the reaction.

✓4.8 Solving Structure Problems

A number of problems ask you to deduce structures that are initially unknown by piecing together chemical data. Problems 4.46 and 4.47 are of this type. This study guide link illustrates a systematic approach to this type of problem by beginning the solution to Problem 4.46.

1. If the formulas of any of the unknowns are given, deduce all the information you can from the formulas. Begin with the unsaturation numbers (Sec. 4.3).

   In Problem 4.46, compound *X* has an unsaturation number $U = 2$, and compound *Y* has $U = 1$. This means that *X* has one ring and one double bond, or two double bonds. (It has to have at least one double bond, because it undergoes addition.) Compound *Y* results from addition of HBr, because the formula of *Y* is equal to that of *X* plus the elements of HBr.

2. Write all the information in the problem in equation form. This process gives you the entire problem at a glance.

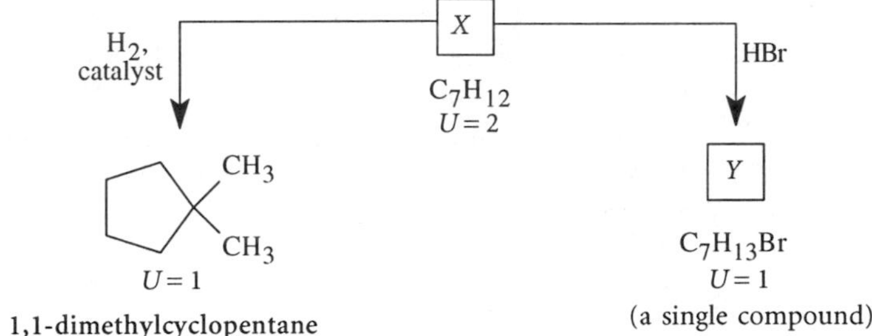

3. If there is a structure given explicitly anywhere in the problem, *even at the end of the problem*, deduce what you can from this structure.

   Notice that 1,1-dimethylcyclopentane is given explicitly in the problem as the catalytic hydrogenation product of *X*. Barring rearrangements, the structure provides the real key to solving the problem: it provides *the carbon skeleton of compound X*. It follows that one of the degrees of unsaturation of *X* must be a ring, and the other, as previously deduced, is a double bond. Thus, compound *X* must be a five-membered ring containing a double bond with two methyl groups on one carbon. The carbon bearing the methyl groups cannot be part of the double bond (why?).

4. Write out *all possibilities* and choose between (or among) them from the evidence presented, if possible.

   This step is one that meets great resistance from many students for some reason. But it is essential. Suppose a detective arrests the first person he or she meets that fits the description of a burglar; chances are, the detective will make a lot of false arrests. Instead, he or she ideally should list *all* the suspects and decide which among them fit all the data. The only "suspects" for *X* are the following:

   CH_3 / CH_3 and CH_3 / CH_3

   X1 *X2*

   But what about the following structure?

CH_3
CH_3

This is the same as *X1*! Don't forget that most structures can be drawn in different ways.

To finish solving the problem, notice that compound *Y* is a *single* compound. Only one of the "suspects" above would give a *single compound* on addition of HBr; the other would give a mixture. What is the solution? Which compound is *X*?

✓4.9 Solving Mechanistic Problems

Chapter 4 has stressed the importance of mechanisms in understanding two alkene reactions: hydrogen halide addition and hydration. The principles learned in these sections, along with the curved-arrow notation and acid-base reactions covered in Sections 3.1–3.4, can be used to understand reactions you've never seen before. Problem 4.58 asks that you propose a rational series of steps that can account for such a reaction. This reaction has not been presented in this text. It contains several strange elements that might discourage you initially from trying to understand it. First, a ring is formed. Second, the product is an ether; none of the reactions covered so far give ethers as products. Nevertheless, a rational approach can be used to understand this reaction and its mechanism.

Break the problem-solving strategy into steps.

1. Examine the reactants and products and *label corresponding atoms.*

The result of such an analysis is as follows:

$(H_3C)_2C{=}CHCH_2CH_2\ddot{O}H$ (atoms numbered 1, 1, 2, 3, 4, 5, 6) — reactant

H_3C (1), H_3C (1), C (2), CH_2 (3), CH_2 (4), CH_2 (5), $\ddot{O}$ (6) — product

reactant product

Notice that the atoms at the vertices of the skeletal structure in the product have been explicitly written out. You should always do this so that you are reminded how many hydrogens are located at each carbon. Both CH_3 groups are labeled "1" since they are indistinguishable. Because these CH_3 groups are attached to carbon-2 in the reactant, postulate the same relationship in the product. The identity of the oxygen is clear, since it is the only oxygen in the molecule. Because it is attached to carbon-5 in the starting material, postulate the same for the product. The identities of carbons 3 and 4 then follow. Note that the correspondences proposed are hypotheses—but they are *reasonable* hypotheses. You are asked to *propose* a rational mechanism, not prove that it is correct.

2. Describe what has happened to the various functional groups in the molecule. Does the transformation fit the pattern of any reactions you have seen before?

Observe that the double bond in the starting material is missing in the product. In fact, carbon-3 of the double bond, the less branched carbon, has added a hydrogen, and carbon-2, the more branched carbon of the double bond, has added oxygen-6 from within the same molecule. This is, therefore, an *addition* reaction. It is also a *regioselective* addition in the same sense as hydration or HBr addition: hydrogen has added to the carbon with fewer alkyl branches, and the more electronegative atom (oxygen) has added to the carbon with more alkyl branches. Furthermore, the acid used is not consumed, and is therefore a catalyst. *What reactions have you studied that are acid-catalyzed regioselective additions?* Recall from Sec. 4.9B that hydration is such an addition. The reaction in this problem is indeed very similar to hydration, except that oxygen-6 of the —OH group, rather than water, has taken part in the addition.

3. Write a mechanism using the curved-arrow notation and principles that you know from studying closely related reactions. Use separate structures for each mechanistic step; that is, do not attempt to represent several mechanistic steps on the same structure.

In hydration, the π electrons attack the acidic hydrogen of the catalyzing acid (see Eq. 4.38a, text p. 152) so that this hydrogen is transferred to the carbon of the double bond that has the smaller number of alkyl branches, and a carbocation is left at the carbon with the greater number of alkyl branches. Such a step would provide the extra hydrogen required at carbon-3. Using the curved-arrow notation, write this *Brønsted acid-base reaction*:

$$(H_3C)_2C{=}CH{-}CH_2CH_2{-}OH + H{-}\overset{+}{O}H_2 \longrightarrow (H_3C)_2\overset{+}{C}{-}CH_2{-}CH_2CH_2{-}OH + \ddot{O}H_2$$

(Because the H_2SO_4 is present in dilute aqueous solution, the catalyzing acid is H_3O^+.) The analysis in Step 1 showed that, in order to form the ring, oxygen-6 must attack carbon-2. Since the electron-deficient carbon is a Lewis acid, and the oxygen is a Lewis base, the situation is ripe for this *Lewis acid-base reaction*.

$$(H_3C)_2\overset{+}{C}{-}CH_2{-}CH_2{-}CH_2{-}\ddot{O}H \longrightarrow \text{ring: } (H_3C)_2C{-}CH_2{-}CH_2{-}CH_2{-}\overset{+}{O}(H){-}$$

(This step is analogous to the attack of water on the carbocation in hydration; see Eq. 4.38b, text p. 152.) The carbocation is redrawn above so that the oxygen is poised in the correct place to form the ring. (Remember that rotations about carbon-carbon single bonds occur readily; so redrawing the molecule in this way is reasonable.)

Completion of the mechanism requires only loss of the proton from the oxygen. A water molecule, which was generated in the first step as the conjugate base of the acid catalyst H_3O^+, can remove this proton to re-form H_3O^+ in a *Brønsted acid-base reaction*:

$$:\ddot{O}H_2 + \text{ring } H{-}\overset{+}{O} \longrightarrow \underset{\text{product}}{\text{ring } :\ddot{O}{-}} + \underset{\text{catalyst regenerated}}{H{-}\overset{+}{O}H_2}$$

The final product is drawn differently from the one in the problem, but it is the same molecule. (You can prove this to yourself, if necessary, by analyzing the connectivities of the two.) Evidently, this reaction is *virtually identical* to hydration. The only difference is that the nucleophile (Lewis base) is derived from oxygen-6—the oxygen from within the same molecule—rather than water.

Notice that the mechanism consists of three simple acid-base reactions. Acid-base reactions, or closely analogous reactions, play a key role in the mechanisms of many organic reactions.

Writing mechanisms for known reactions, as in this problem, should help you to *understand* reactions rather than *memorize* them. This is the first step toward the development of a skill that is powerful indeed: the ability to *predict* the course of unknown reactions before they are ever run in the laboratory!

One final point: You should practice Step 1 above *every time* you see a new reaction in this text. The ability to postulate a correspondence between atoms of the reactant and product will help you to read and understand the discussions of organic reactions much more efficiently.

Problems 4.59 and 4.60 in the text offer you additional practice in solving mechanistic puzzles. Try your hand at them. But remember: be systematic, and use the curved-arrow notation correctly.

SOLUTIONS TO PROBLEMS

Solutions to In-Text Problems

4.1 (a) 1-Pentene cannot exist as cis, trans isomers.
(b) 3-Hexene can exist as both cis and trans isomers; both carbons of the double bond (carbons 3 and 4) are stereocenters.
(c) 1,3-Pentadiene can exist as cis, trans isomers at the bond between carbons 3 and 4; both of these carbons are stereocenters. (Cis-trans isomerism is *not* possible at the double bond between carbon-1 and carbon-2.)
(d) 3-Methyl-2-pentene cannot exist as cis, trans isomers because two of the groups at one carbon of the double bond are the same (the methyl groups). This molecule contains no stereocenters.
(e) In principle, cyclobutene, which is a *cis*-alkene, might also exist as a *trans*-isomer. However, the model-building exercise should have convinced you that such an isomer is far too strained to exist. The carbons of the double bond are stereocenters.

4.2 (a)

$$H_2C{=}\underset{\displaystyle CH_3}{\underset{|}{C}}{-}CH_3$$

2-methylpropene

(b)

$$H_2C{=}CH{-}CH{=}\underset{\displaystyle CH_3}{\underset{|}{C}}{-}CH_2CH_3$$

4-methyl-1,3-hexadiene

(c)

(cyclopentene ring)–$C(=CH_2)CH_3$

1-isopropenylcyclopentene

Many chemists were educated in nomenclature to understand that if the position of a substituent is unambiguous it need not be cited by number. Thus, compound (a) could be called methylpropene, because the methyl substituent *must* be on carbon-2. That is, if it is on carbon-1, the compound would not be named as a propene; it would be named as a butene. None of the formal "rule books" on nomenclature support this idea. Hence, we number *all* substituents. The 1993 nomenclature recommendations of the IUPAC went a step further and recommended numbering not only substituents, but also all double bonds, whether ambiguous or not; furthermore, the number is *always* cited before the suffix "ene." Thus, compound (a) in the 1993 system is called 2-methylprop-1-ene. (The 1993 recommendations are discussed in Appendix I of the text.) A positive aspect of this recommendation is that it creates consistency in the way that all alkene names are constructed. However, this recommendation has not yet been generally accepted, and for that reason it has not been used in this text. However, students interested in learning the new system can do so with an excellent brief tutorial, *Organic Nomenclature, A Programmed Introduction,* Fifth Edition, by James. G. Traynham, published by Prentice-Hall.

(d)

$$H_2C{=}CH{-}CH{=}CH{-}\underset{\displaystyle CH_2CH_2CH{=}CHCH_3}{\underset{|}{CH}}{-}CH{=}CH{-}CH{=}CH{-}CH_3$$

5-(3-pentenyl)-1,3,6,8-decatetraene

4.3 (a) 3-heptene
(b) 1,3-dimethylcyclopentene
(c) 5-allyl-2,6-nonadiene

Remember, the 2-propenyl group is called the *allyl group;* see text page 121.

4.4 (a) (*Z*)-3-isopropyl-2-hexene

(b) First, name the compound without considering stereochemistry: 4-isobutyl-1,4-hexadiene. Next, determine the stereochemistry at the double bond between carbons 4 and 5. (*E,Z* isomerism is not possible at carbon-1). Start by writing the structure with the double bond between carbons 1 and 2 replicated, and decide on the path of highest priority within each group (shown in boldface in the structure below):

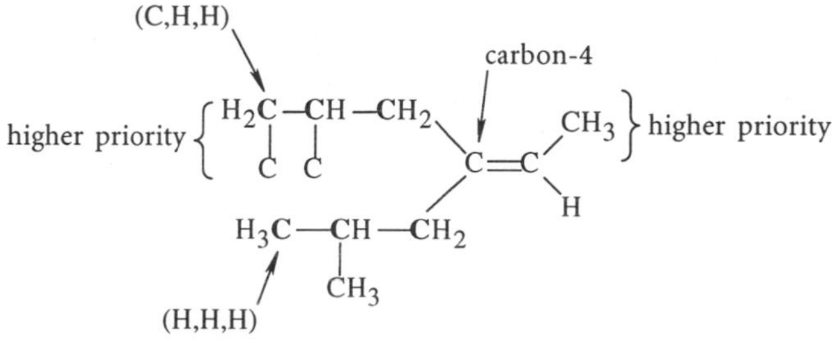

(C,H,H)

carbon-4

higher priority { $H_2C-CH-CH_2$ (with C, C duplicated below) } $C=C$ { CH_3 } higher priority; H

$H_3C-CH-CH_2$; CH_3

(H,H,H)

As the diagram above shows, the first point of difference occurs at the third carbon out; thus, the allyl group has higher priority than the isobutyl group. Because the groups of higher priority are on the same side of the double bond, the stereochemistry at this double bond is *Z*, and the full name is therefore (*Z*)-4-isobutyl-1,4-hexadiene. No number before the *Z* is necessary when only one double bond has a stereochemical designator.

4.5 (a)

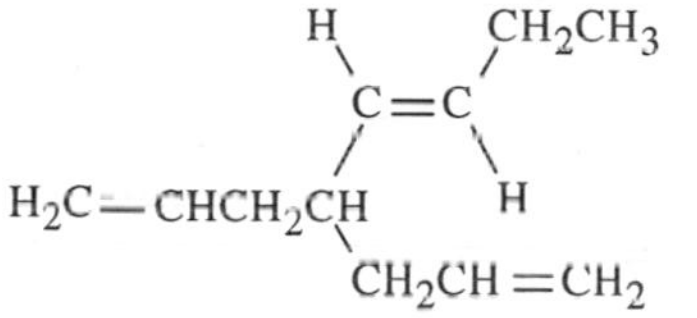

(*E*)-4-allyl-1,5-octadiene

(b)

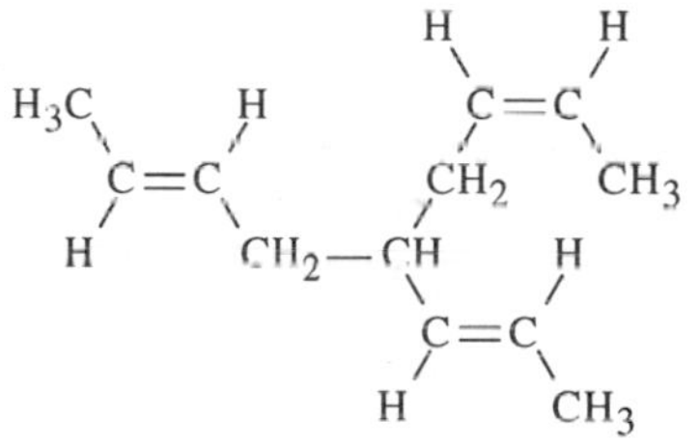

(2*E*,7*Z*)-5-[(*E*)-1-propenyl]-2,7-nonadiene

4.6 (a) CH_3CH_2–CH(—)–CH_3O ; O(C) (higher priority) *vs.* Cl_2CH–CH(—)–HO ; O(H)

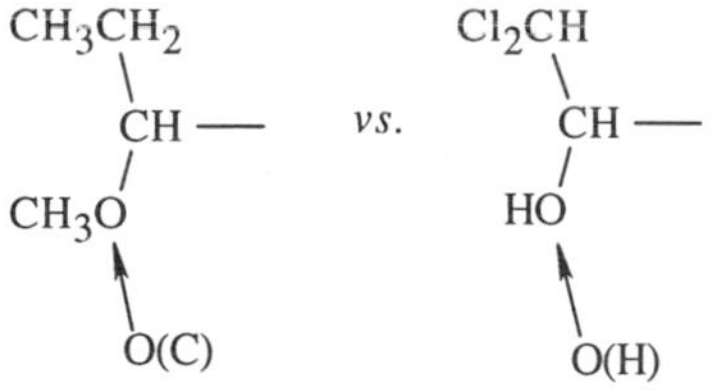

(b) C(C,H,H) (cyclobutyl) (higher priority) *vs.* C(*,*,*) ; H_2C-CH- with C, C duplicated

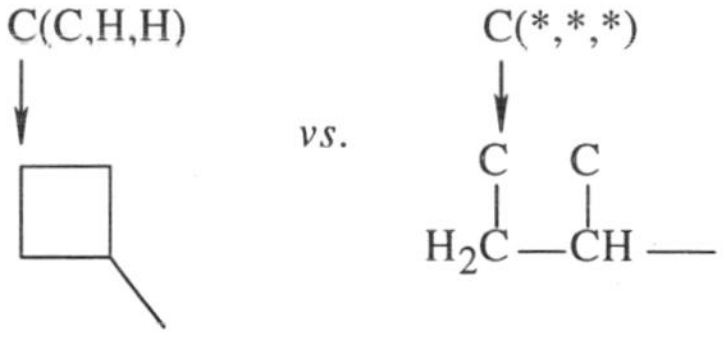

(c) $(CH_3)_3C$— ; C(C,C,C) versus CH_3–CH(—)–OCH_3 ; C(O,C,H) (higher priority)

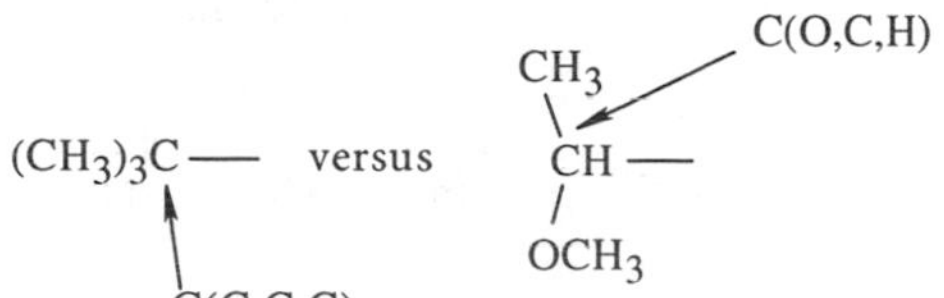

(d) $(CH_3)_3C$— ; C(H,H,H) versus HC(C)(C)—C(C)(C)— ; C(C,C,H) (higher priority)

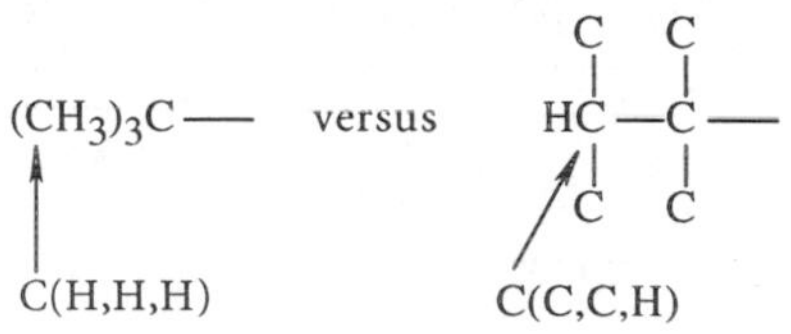

4.7 (a) Use Eq. 4.6 to obtain $U = 0$.
(b) Use Eq. 4.7 to obtain $U = 3$.

4.8 (a) The prefix "cyclo" in the name indicates that the compound has one ring; consequently, $U = 1$.
(b) The suffix "triene" in the name indicates that the carbon has three double bonds; consequently, the

unsaturation number is 3.

4.9 The empirical formula is $C_{(85.60/12.01)}H_{(14.40/1.008)}$ or $C_{7.127}H_{14.29}$ or CH_2. Any compound with this empirical formula can have only one degree of unsaturation. Thus, $U = 1$; it can have *either* one ring *or* one double bond.

4.10 A triple bond contributes two units to the unsaturation number.

4.11 (a) For simplicity, ignore the H—C bond dipoles. *Cis*-2-butene has the greater dipole moment because the bond dipoles of the H_3C—C bonds tend to add. In contrast, the dipole moment of *trans*-2-butene is zero because corresponding bonds are oriented in opposite directions, and their bond dipoles consequently cancel.

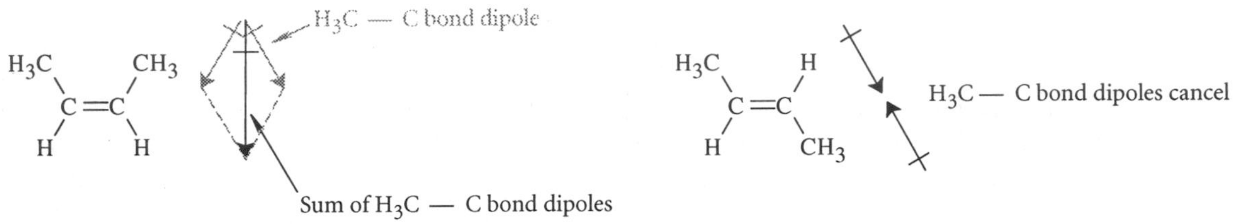

H_3C — C bond dipole

Sum of H_3C — C bond dipoles

H_3C— C bond dipoles cancel

(Although the C—H bond dipoles are not considered explicitly, each one is oriented at 180° to a H_3C—C bond dipole, and is smaller than the H_3C—C bond dipole; hence, the resultant for *cis*-2-butene is not so large as that shown above, but it is nevertheless different from zero; and the dipole moment of *trans*-2-butene is zero.)

(b) 2-Methylpropene should have the greater dipole moment because the additional H_3C—C bond makes a dipole contribution that is not present in propene itself. (In the following analysis, the H—C bond dipoles of the CH_2 groups are ignored because they cancel in the comparison of the two molecules.)

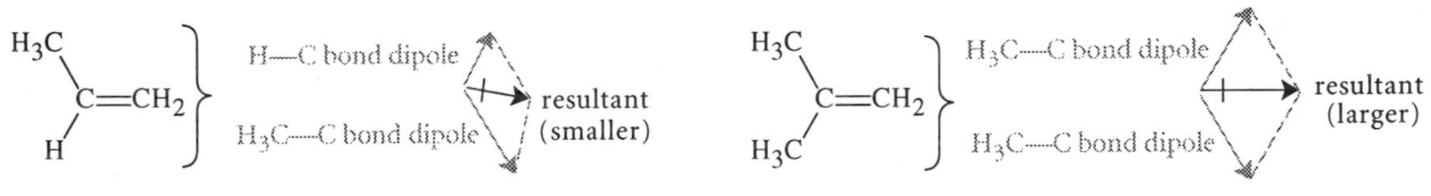

H—C bond dipole
H_3C—C bond dipole
resultant (smaller)

methylpropene

H_3C—C bond dipole
H_3C—C bond dipole
resultant (larger)

2-methylpropene

See the brief refresher on graphical vector addition on p. 21 of this Study Guide.

4.12 (a) The enthalpy of reaction is obtained by subtracting the ΔH_f° of the reactant, 1-butene, from that of the product, 2-methylpropene. Therefore, $\Delta H^\circ = -17.30 - (-0.30) = -17.0$ kJ/mol (4.06 kal/mol).

(b) Because the calculation is formulated in such a way that 2-methylpropene is the product, and the ΔH° of the process is negative, then the product has lower energy (is more stable) than the reactant; that is, 2-methylpropene is more stable.

4.13 (a) The overall ΔH° of the process, +15.3 kJ/mol, must equal the difference in the ΔH_f° values of product and reactant. Hence,

$$\begin{aligned} +15.3 &= \Delta H_f^\circ(\text{1-hexene}) - \Delta H_f^\circ(\text{2-ethyl-1-butene}) \\ &= -40.5 - x \end{aligned}$$

Solving, we have that $x = \Delta H_f^\circ$(2-ethyl-1-butene) = –55.8 kJ/mol

(b) Because the enthalpy change for the process is positive, the product is more stable than the reactant. Hence, 2-ethyl-1-butene is the more stable of the two compounds.

4.14 The combustion reaction for 1-heptene is

$$H_2C{=}CHCH_2CH_2CH_2CH_2CH_3 + 10.5\ O_2 \longrightarrow 7\ CO_2 + 7\ H_2O$$

The given $\Delta H°$ for this reaction is –4385.1 kJ/mol. This is equal to the heats of formation of products minus those of the reactants. The heats of formation of CO_2 and H_2O are the same as the heats of combustion of C and H_2, respectively, because these combustion reactions form CO_2 and H_2O from their elements.

$$-4385.1\text{ kJ/mol} = 7\Delta H_f^\circ(CO_2) + 7\Delta H_f^\circ(H_2O) - \Delta H_f^\circ(\text{1-hexene}) - 10.5\Delta H_f^\circ(O_2)$$

Letting ΔH_f°(1-hexene) = x,

$$-4385.1\text{ kJ/mol} = 7(-393.51) + 7(-241.83) - x - 10.5(0)$$

or $x = \Delta H_f^\circ$(1-hexene) = –62.28 kJ/mol (–14.89 kcal/mol).

This calculation illustrates the point made in Study Guide Link 4.4 that heats of formation derived from heats of combustion are small difference between large numbers and are therefore subject to some uncertainty.

4.15 (a) Compound *A* is more stable than compound *B* because the double bond in *A* has one more alkyl substituent.

(b) Compound *B* is more stable than compound *A*, because compound *A*, which is a *cis*-alkene, is destabilized by significant van der Waals repulsions between the *tert*-butyl group and the methyl group. Because compound *B* is a *trans*-alkene, these destabilizing repulsions are absent.

4.16 Hydrogenation of either (*E*)- or (*Z*)-3-hexene gives the same product, hexane. This allows us to use the heats of hydrogenation in a Hess's law calculation:

$CH_3CH_2(H)C{=}C(H)CH_2CH_3$ + H_2 ⟶ $CH_3CH_2CH_2CH_2CH_2CH_3$ $\Delta H° = -121.6 = -167.2 - x$

(*Z*)-3-hexene, $\Delta H_f^\circ = x$; hexane, $\Delta H_f^\circ = -167.2$

From this we obtain the ΔH_f° of (*Z*)-3-hexene as –45.68 kJ/mol (–10.90 kcal/mol). A similar calculation for (*E*)-3-hexene gives the ΔH_f° as –49.30 kJ/mol (–11.78 kcal/mol). Notice that the *E* isomer is more stable, as we expect. (Why?)

4.17 (a) $H_2C(Br)—CH_2(Br)$ (b) $H_3C—CH_2—I$ (c) $CH_3CH_2—B(CH_2CH_3)—CH_2CH_3$

4.18 In (a), the chlorine goes to the carbon of the alkene with the methyl branch. In (b), the bromine goes to the carbon with the methyl branch.

(a) $(H_3C)_2CH—Cl$ (b) 1-bromo-1-methylcyclohexane (CH_3, Br on the same ring carbon)

4.19 We follow the patterns in text Eqs. 4.19 and 4.20. First the double bond is protonated on the *less-branched carbon*by HBr to give a carbocation on the more branched carbon and Br^-. Note that even through a proton is transferred, we use the curved-arrow notation to show *the flow of electrons.* In the second step, the Br^- attacks the carbocation in a Lewis acid-base association reaction to give the product.

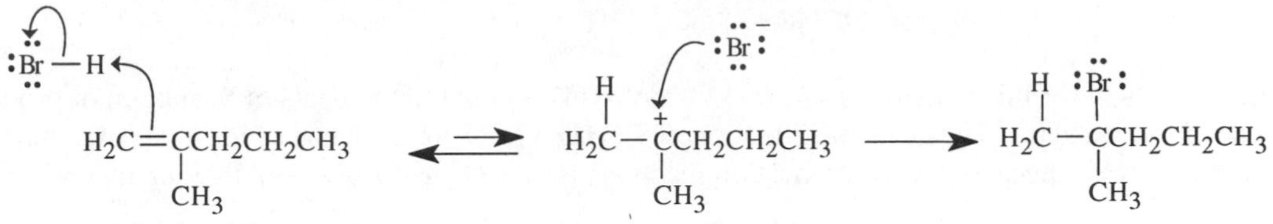

$:\ddot{Br}-H$ + $H_2C{=}C(CH_3)CH_2CH_2CH_3$ ⇌ $H_2C(H)-\overset{+}{C}(CH_3)CH_2CH_2CH_3$ + $:\ddot{Br}:^-$ ⟶ $H_2C(H)-C(Br)(CH_3)CH_2CH_2CH_3$

2-methyl-1-pentene; a tertiary carboocation

4.20 (a)

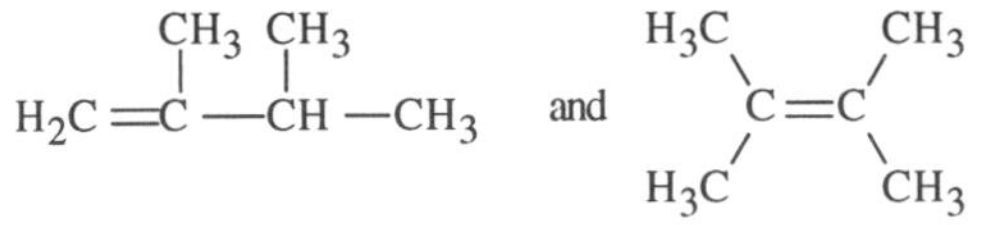

$H_2C{=}C(CH_3)-CH(CH_3)-CH_3$ and $(H_3C)_2C{=}C(CH_3)_2$

(b)

1-methylcyclohexene (CH_3) and methylenecyclohexane (CH_2)

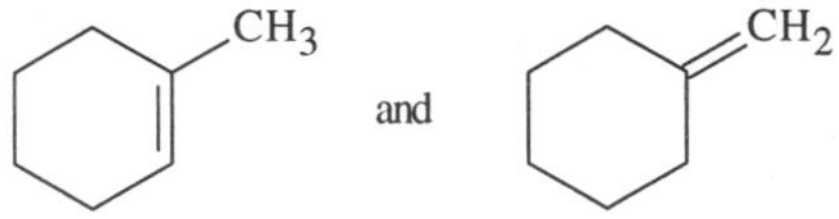

4.21 Protonation of the double bond gives a carbocation intermediate *A* which can be attacked by bromide ion to give the minor product *B*.

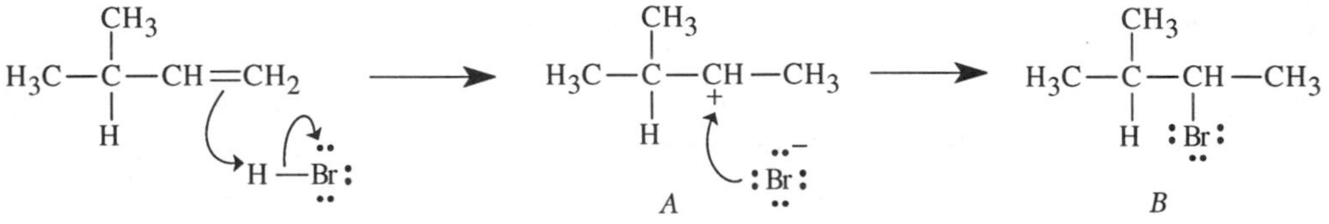

$H_3C-C(CH_3)(H)-CH{=}CH_2$ + $H-\ddot{Br}:$ ⟶ $H_3C-C(CH_3)(H)-\overset{+}{C}H-CH_3$ + $:\ddot{Br}:^-$ ⟶ $H_3C-C(CH_3)(H)-CH(Br)-CH_3$

A *B*

Carbocation *A* can also rearrange by a hydride shift to give carbocation *C*, which is attacked by bromide ion to give the major product *D*:

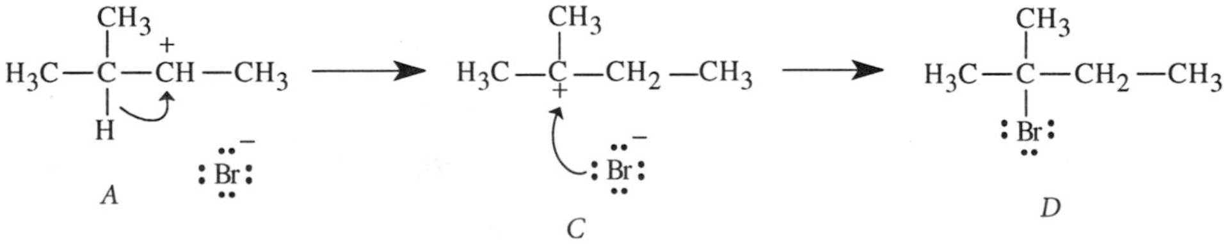

$H_3C-C(CH_3)(H)-\overset{+}{C}H-CH_3$ ⟶ $H_3C-\overset{+}{C}(CH_3)-CH_2-CH_3$ + $:\ddot{Br}:^-$ ⟶ $H_3C-C(CH_3)(Br)-CH_2-CH_3$

A *C* *D*

4.22 If these alkyl halides could be prepared from alkenes, the alkene starting materials would have to be *A1*, *B1*, and *C1* for *A*, *B*, and *C*, respectively:

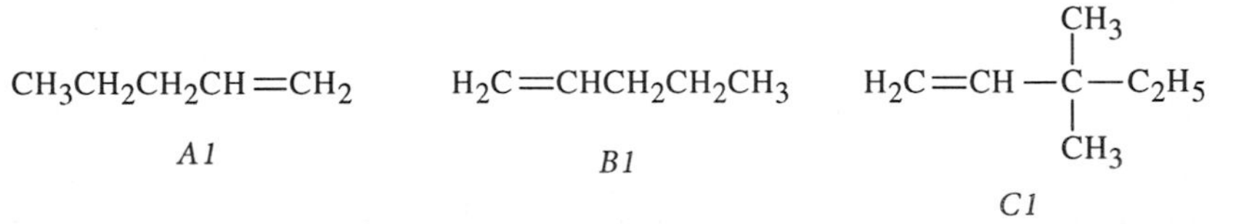

$CH_3CH_2CH_2CH{=}CH_2$ (*A1*) $H_2C{=}CHCH_2CH_2CH_3$ (*B1*) $H_2C{=}CH-C(CH_3)_2-C_2H_5$ (*C1*)

(Note that *B1* is the same as *A1*; it is simply drawn differently.) Reaction of *A1* (or *B1*) with HBr would give *B* but not *A* because the bromine goes to the carbon of the double bond with the alkyl branch. Hence, compound *B*, but not compound *A*, could be prepared by HBr addition to alkene *B1*. Addition of HBr to alkene *C1* would not give solely compound *C*, but would also give one or more rearrangement products.

4.23 (a) Apply Eq. 4.29b with $\Delta G_B^{\circ\ddagger} - \Delta G_A^{\circ\ddagger} = -15$ kJ/mol. Then

$$\log\left(\frac{\text{rate}_A}{\text{rate}_B}\right) = \frac{\Delta G_B^{\circ\ddagger} - \Delta G_A^{\circ\ddagger}}{2.3RT} = \frac{-15}{5.706} = -2.63$$

or

$$\frac{\text{rate}_A}{\text{rate}_B} = 10^{-2.63} = 2.3 \times 10^{-3}$$

Thus, reaction *A* is 0.0023 as fast as reaction *B*, or reaction *B* is (1/0.0023) or 427 as fast as reaction *A*.

(b) The text states that a reaction rate approximately doubles for every 10° rise in temperature. The question is, then how many "doublings" in rate correspond to the rate factor 427 above?

$$
\begin{aligned}
427 &= 2^n \\
\log 427 &= n \log 2 \\
2.63 &= 0.301n \\
n &= 8.74
\end{aligned}
$$

Evidently, a rate factor of 427 corresponds to almost 9 doublings of the rate—8.74, to be exact. The temperature at which reaction *A* is run would have to be raised an estimated 87° to make it as fast as reaction *B*.

4.24 (a) We again apply Eq. 4.29b.

$$\log\left(\frac{\text{rate}_A}{\text{rate}_B}\right) = \log 10^{-6} = -6 = \frac{\Delta G^{\circ\ddagger}_B - \Delta G^{\circ\ddagger}_A}{2.3RT}$$

$$\Delta G^{\circ\ddagger}_B - \Delta G^{\circ\ddagger}_A = (-6)(5.706) = -34.24 \text{ kJ/mol}$$

With $\Delta G^{\circ\ddagger}_A = 90$ kJ/mol, we solve for $\Delta G^{\circ\ddagger}_A$ to obtain

$$\Delta G^{\circ\ddagger}_B = 90 - 34.2 = 55.8 \text{ kJ/mol (13.3 kcal/mol)}$$

(b) The two free energy-reaction coordinate diagrams are shown in Fig. SG4.1 on the following page.

4.25 The reaction-free energy diagram shown in Fig. SG4.2 on the following page meets the criteria in the problem.

4.26 (a) The curved arrows show where bonds are being formed and broken. Thus, there is a partial bond between the HO^- and the carbon, and a partial bond between the CH_3 and the Br. (Notice that the locations of the different groups are not defined by the curved-arrow formalism; the arrangement shown below is arbitrary. The important points are the partial bonds and the partial charges.)

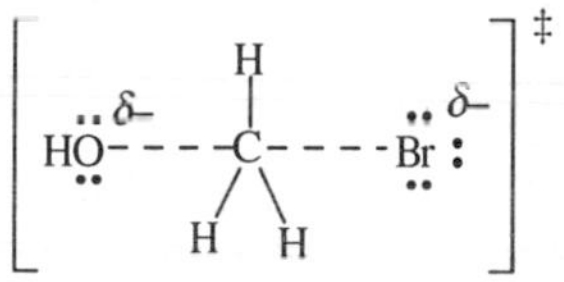

$$\left[\overset{\delta-}{\ddot{\text{H}\ddot{\text{O}}}} \text{- - -} \underset{\text{H}\ \ \text{H}}{\overset{\text{H}}{\text{C}}} \text{- - -} \overset{\delta-}{:\ddot{\text{Br}}:} \right]^\ddagger$$

(b) The bond between bromide ion and the electron-deficient carbon of the carbocation is partially formed:

$$:\ddot{\text{Br}}:^- \curvearrowright \overset{+}{\text{C}}(CH_3)_3 \qquad \left[:\overset{\delta-}{\ddot{\text{Br}}} \text{- - - - - - -} \overset{\delta+}{\text{C}}(CH_3)_3 \right]^\ddagger$$

4.27 Because the transition states of the two reactions should resemble the respective carbocation intermediates (Hammond's postulate), examine the structures of the two carbocations and determine which should be more stable.

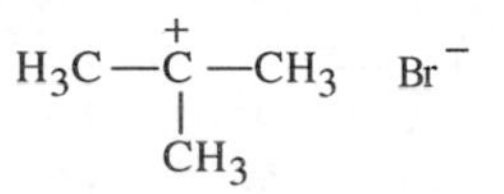

$$H_3C-\overset{+}{\underset{CH_3}{C}}-CH_3 \quad Br^- \qquad\qquad H_3C-\overset{+}{C}H-CH_2-CH_3 \quad Br^-$$

carbocation intermediate from HBr addition to 2-methylpropene

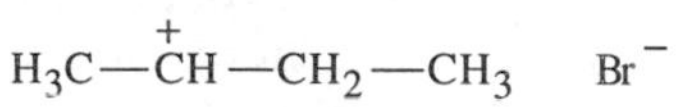

carbocation intermediate from HBr addition to *trans*-2-butene

The carbocation intermediate in the addition of HBr to 2-methylpropene is tertiary, and is thus more stable than the carbocation intermediate in the addition of HBr to *trans*-2-butene, which is secondary. If the transition states for the two addition reactions resemble the carbocation intermediates, the transition state for the reaction of 2-methylpropene is also more stable. The relative rates of the two reactions are governed by the *differences*

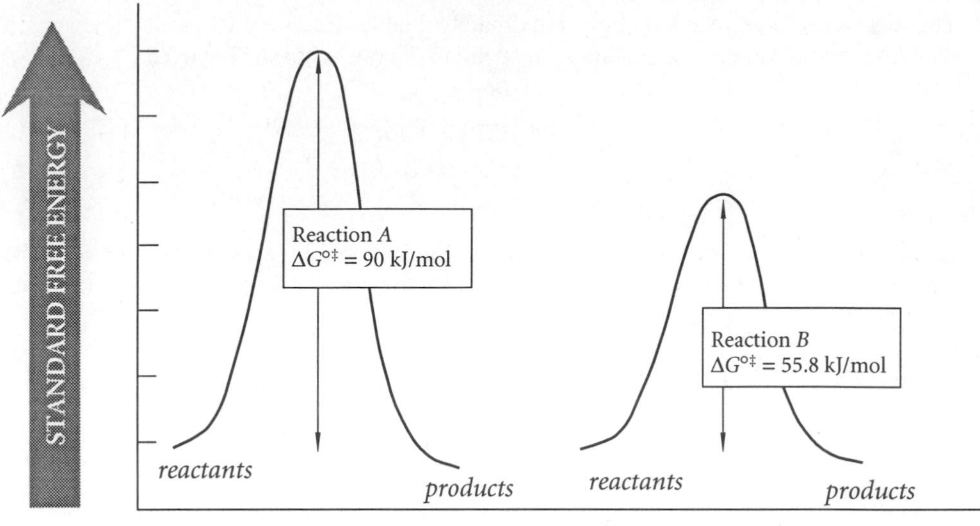

Figure SG4.1. A reaction free-energy diagram to accompany the solution to Problem 4.24b. The two curves are drawn to scale.

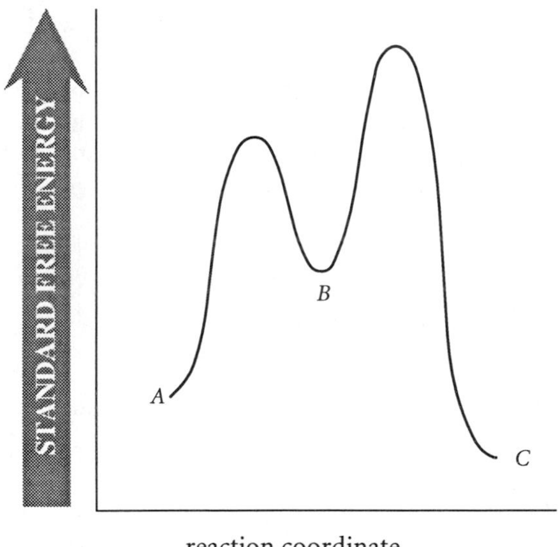

Figure SG4.2 A reaction free-energy diagram to accompany the solution to Problem 4.25.

between the standard free energies of the transition states and their respective starting materials. Therefore, we have to consider the free energies of *both* the alkenes and the transition states. *If the two alkenes do not differ appreciably in energy, then the relative rates of the two reactions are governed only by the relative free energies of the two transition states.* (This is a fairly good assumption, because both alkenes have two alkyl branches.) Under this assumption, addition of HBr to 2-methylpropene is faster, because the transition state for this reaction has lower free energy.

4.28 (a) The product is pentane, $CH_3CH_2CH_2CH_2CH_3$.
(b) The product is hexane, $CH_3CH_2CH_2CH_2CH_2CH_3$. (Hydrogen adds to both double bonds.)

4.29 Five alkenes with the formula C_6H_{12} that give hexane as the product of catalytic hydrogenation:

$H_2C{=}CHCH_2CH_2CH_2CH_3$	$CH_3CH{=}CHCH_2CH_2CH_3$	$CH_3CH_2CH{=}CHCH_2CH_3$
1-hexene	(*E*)- and (*Z*)-2-hexene	(*E*)- and (*Z*)-3-hexene

4.30 The mechanism of the reaction in Eq. 4.39:

$$(CH_3)_2CH{-}CH{=}CH_2 \;+\; H{-}\overset{+}{O}H_2 \longrightarrow (CH_3)_2CH{-}\overset{+}{C}H{-}CH_3 \;(:OH_2) \longrightarrow$$

$$H_3C{-}\overset{+}{C}(CH_3){-}CH_2{-}CH_3 \;(:OH_2) \longrightarrow H_3C{-}C(CH_3)(\overset{+}{O}H{-}H){-}CH_2{-}CH_3 \;(:OH_2) \longrightarrow H_3C{-}C(CH_3)(HO){-}CH_2{-}CH_3 + H_3\overset{+}{O}:$$

4.31 The mechanism of this reaction is essentially identical to the mechanism of HCl addition to the same alkene given in Eqs. 4.25–4.26 on text page 140–1, except that the Brønsted acid is H_3O^+ and the nucleophile is H_2O.

$$(CH_3)_3C{-}CH{=}CH_2 \;+\; H{-}\overset{+}{O}H_2 \longrightarrow (CH_3)_3C{-}\overset{+}{C}H{-}CH_3 \;(:OH_2) \longrightarrow$$

$$H_2O: \;+\; H_3C{-}\overset{+}{C}(CH_3){-}CH(CH_3){-}CH_3 \xrightarrow{H_2O:} H_3C{-}C(CH_3)(\overset{+}{O}H{-}H){-}CH(CH_3)_2 \;(:OH_2) \longrightarrow \boxed{H_3C{-}C(CH_3)(HO){-}CH(CH_3)_2} + H_3\overset{+}{O}$$

predicted product

4.32 The hydration of ethylene involves formation of a primary carbocation. Hence, the transition state resembles such a cation. As you know, primary carbocations are *much* less stable than secondary and tertiary carbocations; hence, transition states that resemble primary carbocations are also unstable. An unstable transition state means that the transition state has a very high energy. When a reaction has a transition state of very high energy, the reaction is slow.

4.33 The hydration of propene involves the usual steps of alkene protonation to form a secondary carbocation followed by attack of water on the carbocation. The structure of isopropyl alcohol follows from the structure of the carbocation. Notice that a secondary carbocation rather than a primary carbocation is involved (Why?).

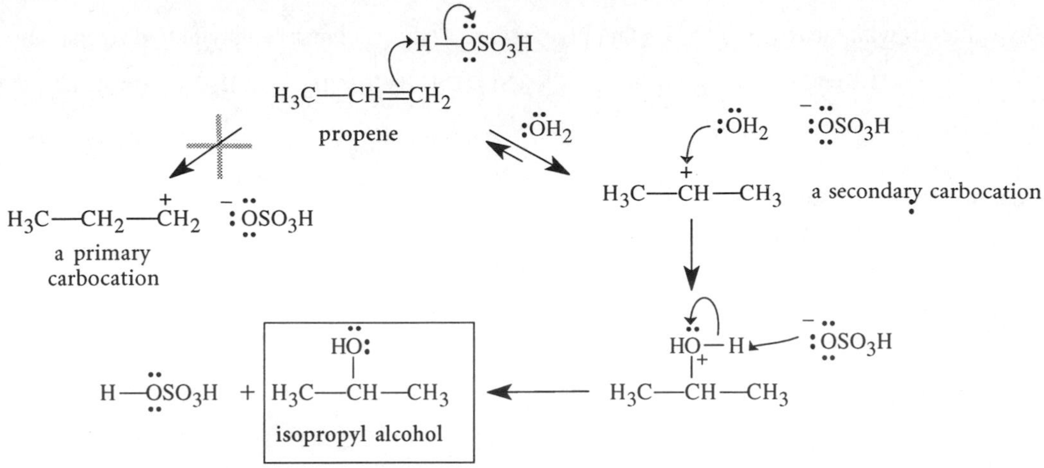

$H-\ddot{O}SO_3H$

$H_3C-CH=CH_2$

propene

$:\ddot{O}H_2$

$H_3C-CH_2-\overset{+}{C}H_2 \quad :\ddot{O}SO_3H^-$

a primary carbocation

$:\ddot{O}H_2 \quad :\ddot{O}SO_3H^-$

$H_3C-\overset{+}{C}H-CH_3$ a secondary carbocation

$H\overset{+}{O}-H \quad :\ddot{O}SO_3H^-$

$H_3C-CH-CH_3$

$H-\ddot{O}SO_3H + $ [$HO:$ / $H_3C-CH-CH_3$] isopropyl alcohol

4.34 The alcohol formed in the hydration of methylenecyclobutane:

OH

CH₃

1-methylcyclobutanol

Solutions to Additional Problems

4.35 The structures and substitutive names of the methylpentene isomers:

$H_2C{=}C(CH_3)CH_2CH_2CH_3$ — 2-methyl-1-pentene

$H_2C{=}CHCH(CH_3)CH_2CH_3$ — 3-methyl-1-pentene

$H_2C{=}CHCH_2CH(CH_3)CH_3$ — 4-methyl-1-pentene

$CH_3C(CH_3){=}CHCH_2CH_3$ — 2-methyl-2-pentene

$CH_3CH{=}C(CH_3)CH_2CH_3$ — 3-methyl-2-pentene (*E*) and (*Z*) isomers

$CH_3CH{=}CHCH(CH_3)CH_3$ — 4-methyl-2-pentene (*E*) and (*Z*) isomers

4.36 Each of the following methylpentene isomers should give mostly one product when it reacts with HBr: 2-methyl-1-pentene; 4-methyl-1-pentene; 2-methyl-2-pentene; and both stereoisomers of 3-methyl-2-pentene. In contrast, 3-methyl-1-pentene should give a rearrangement product as well as the product of normal addition; and each of the stereoisomers of 4-methyl-2-pentene can give two products of normal addition as well as a rearrangement product.

4.37 Arranging the alkenes in order of increasing heats of formation is the same as arranging them in order of decreasing stability. The major factors that govern alkene stability are (1) the number of alkyl branches on the double bond; and whether the alkene is cis or trans. The order of increasing heats of formation:

(*E*)-3-methyl-2-pentene $\lesssim$ 2-methyl-2-pentene $\approx$ (*Z*)-3-methyl-2-pentene $<$ (*E*)-4-methyl-2-pentene $\approx$

2-methyl-1-pentene $<$ (*Z*)-4-methyl-2-pentene $<$ 3-methyl-1-pentene $\approx$ 4-methyl-1-pentene

4.38

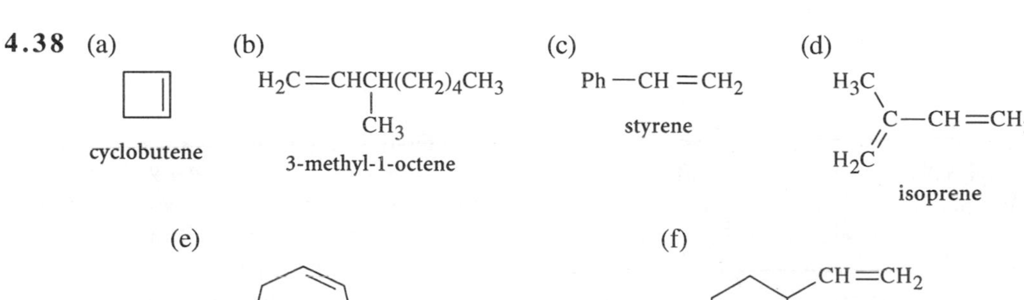

(a) cyclobutene

(b) $H_2C{=}CHCH(CH_3)(CH_2)_4CH_3$ 3-methyl-1-octene

(c) $Ph{-}CH{=}CH_2$ styrene

(d) $H_2C{=}C(CH_3){-}CH{=}CH_2$ isoprene

(e) 5,5-dimethyl-1,3-cycloheptadiene

(f) 1-vinylcyclohexene

4.39 (a) 6-methyl-1-heptene
(b) 3-ethylcyclopentene
(c) 2-isopropyl-1-methyl-1,3-cyclohexadiene
(d) 1-(1-cyclopentenyl)cyclopentene
(e) (*E*)-2-butyl-1,3-pentadiene
(f) (*E*)-5-[(*Z*)-1-propenyl]-1,6-nonadiene

4.40 First draw the structure corresponding to the name, and then rename it if necessary.

(a) $CH_3CH_2CH{=}CH_2$

1-butene

(b) $(H_3C)_3C$—CH=CH—CH_3 (H and H on opposite sides)

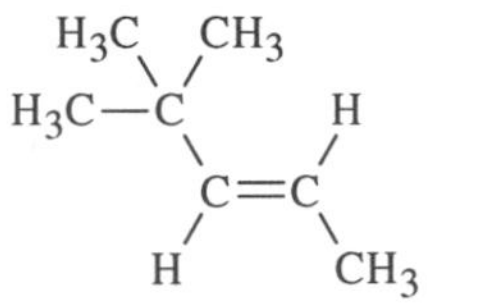

trans-4,4-dimethyl-2-pentene
or (*E*)-4,4-dimethyl-2-pentene

(c) H_3C—CH=CH—$CH_2CH_2CH_3$ (H and H on the same side)

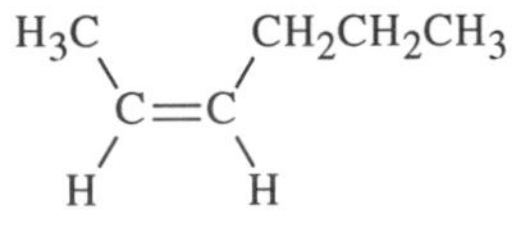

(*Z*)-2-hexene
(The name is correct)

(d) 4-methylcycloheptene

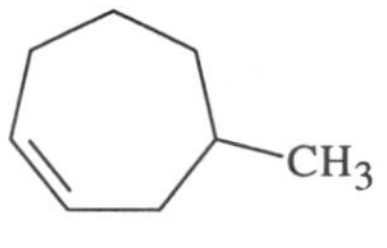

4.41 (a) *Z*; the decision is on the basis of the preference for a heavy isotope over a lighter one in both cases.(b)
(b) *E*
(c) *E*; note in this case that the two branches of the ring are treated as separate substituents; the fact that the two branches are tied into a ring has no effect on their relative priorities.
(d) The structure is drawn below with the double bond of the cyclobutenyl group in replicated form. Because groups of higher priority are on opposite sides of the double bond, this compound has the *E* configuration.

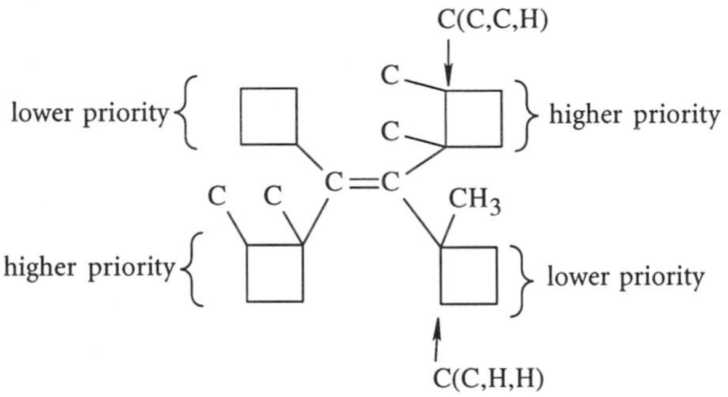

C(C,C,H)

lower priority | higher priority

C=C

higher priority | lower priority

C(C,H,H)

4.42 (a) C (b) N (c) I (d) C (e) S

4.43 BF_3 is trigonal planar with F—B—F bond angles of 120°. Because hybridization and geometry are related, and we know that other trigonal-planar atoms (*e.g.*, carbon of alkenes or carbocations) are sp^2-hybridized, it is reasonable to suppose that boron of BF_3 is also sp^2-hybridized. An orbital diagram (following page) is much like the one in Fig. 4.3 of the text, except that boron has three rather than four valence electrons. In this diagram, the valence electrons are shown in gray. To form BF_3, each of the sp^2 orbitals in boron, which contain a single electron, overlap with an orbital of the fluorine atom that contains a single electron. This makes three B—F electron-pair bonds. Because the boron orbitals used to form the bonds are oriented at 120° in a common plane, then the three B—F bonds are also oriented at 120° in a common plane.

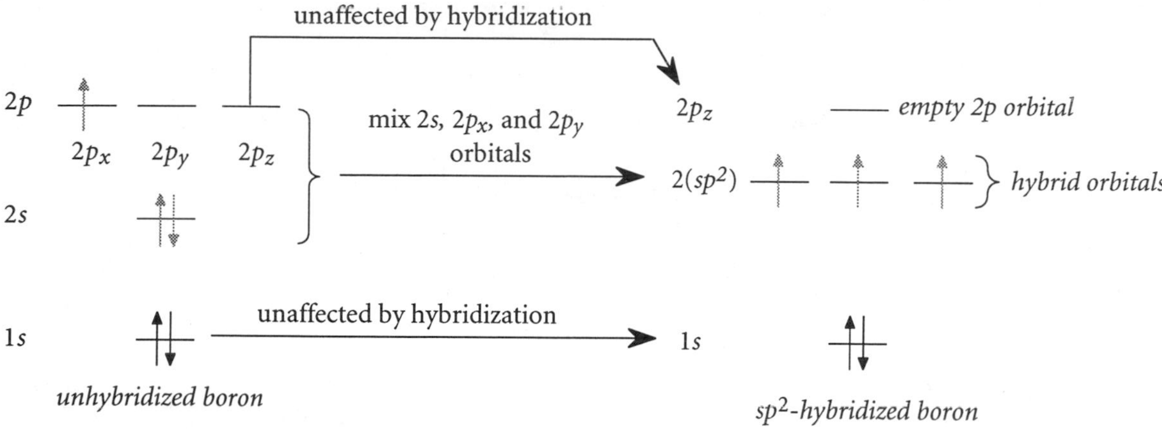

4.44 (a) an sp^3(C)–1s(H) σ bond
(b) an sp^3(C)–sp^2(C) σ bond
(c) an sp^2(C)–sp^2(C) σ bond and a carbon–carbon π bond
(d) an sp^2(C)–sp^2(C) σ bond
(e) an sp^2(C)–1s(H) σ bond

4.45 (a) Either of the following two alkenes would react with HBr to give the alkyl halide product shown in the problem:

CH_3 and CH_2

1-methylcyclopentene methylenecyclopentane

(b) Although the bromine is on the same carbon in the products from the respective alkenes, the hydrogen from HBr (and hence the deuterium from DBr) is not. Consequently, different products are obtained from DBr addition to the two alkenes.

CH_3 —DBr→ H_3C Br D CH_2 —DBr→ H_2CD Br

4.46 The process to be used in working this problem is thoroughly explained in Study Guide Link 4.8. The structures of *X* and *Y* are as follows:

CH_3 CH_3 (*X*)   Br CH_3 CH_3 (*Y*)

4.47 Both *cis*- and *trans*-3-hexene should give the same *single* alkyl halide product when they react with HI, and both would undergo catalytic hydrogenation to give hexane:

$$\underset{\textit{cis}\text{-3-hexene}}{(C_2H_5)HC{=}CH(C_2H_5)} \quad \text{or} \quad \underset{\textit{trans}\text{-3-hexene}}{(C_2H_5)HC{=}CH(C_2H_5)} \xrightarrow{HI} \underset{\text{3-iodohexane}}{C_2H_5CH(I)CH_2C_2H_5}$$

$$\downarrow H_2\text{, catalyst}$$

$$\underset{\text{hexane}}{CH_3CH_2CH_2CH_2CH_2CH_3}$$

4.48 Hydration, because of its mechanism, yields an alcohol in which the —OH group is positioned at the *more branched* carbon of the double bond. Al Keyhall's proposal would require a different regioselectivity, and therefore would not work.

$$(CH_3)_2CHCH{=}CH_2 \xrightarrow[\times]{\textit{Al Keyhall proposal}\; H_2O,\, H_3O^+} (CH_3)_2CHCH_2CH_2OH$$

$$\downarrow H_2O,\, H_3O^+$$

$$\underset{\textit{observed products}}{(CH_3)_2CHCH(OH)CH_3 + (CH_3)_2C(OH)CH_2CH_3}$$

(For the origin of the rearrangement product, see the solution to Problem 4.30.)

4.49 (a) Before drawing the reaction-free energy diagram, convert K_{eq} into a free energy. (Assume 25 °C or 298 K.)

$$\Delta G° = -2.3RT \log K_{eq} = -5.706 \log(150) = -12.4 \text{ kJ/mol } (-2.96 \text{ kcal/mol})$$

This number is the energy difference between A and B. Clearly, B has the lower energy, since it is favored in the equilibrium. The reaction free-energy diagram is shown in Fig. SG4.3.

(b) The standard free energy of activation for the reaction in the B to A direction is 12.4 kJ/mol + 96 kJ/mol = 108 kJ/mol (25.9 kcal/mol). This follows geometrically from the diagram in part (a); it is the difference between the standard free energy of B and the standard free energy of the transition state.

4.50 (a) Compound D is present in greatest amount when the reaction comes to equilibrium, because it has the smallest standard free energy. Compound C is present in least amount because it has the largest standard free energy.

(b) The rate-limiting step in this reaction is the $B \rightleftharpoons C$ step, because it has the transition state of highest standard free energy.

(c) The standard free energy of activation is the energy difference between the transition state of highest free energy and the reactant. This is labeled in part (c) of the diagram in Fig. SG4.4 for the case when the reactant is A. (What is $\Delta G^{°\ddagger}$ when the reactant is D, that is, in the case of the reverse reaction?)

(d) The reaction of $C \longrightarrow D$ is faster than the reaction $C \longrightarrow B$ because the energy barrier for the $C \longrightarrow D$ reaction is smaller. (See part (d) of the diagram in Fig. SG4.4.)

4.51 The first step of hydration, protonation of the double bond, is rate-limiting (text p. 152), and the reactive intermediate that results from this process is a carbocation.According to Hammond's postulate, the transition state closely resembles this intermediate.

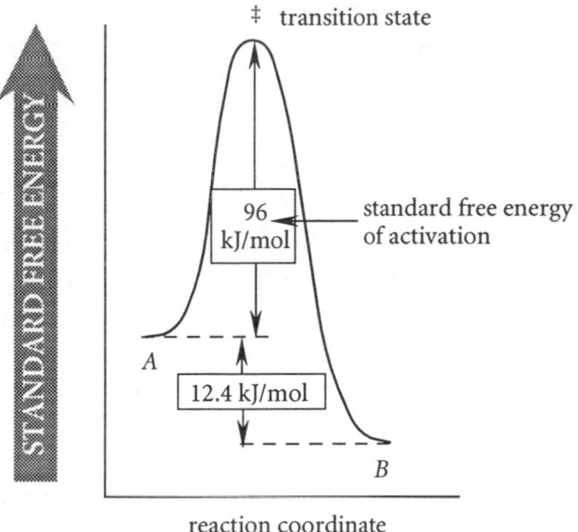

Figure SG4.3 Reaction free-energy diagram to accompany the solution to Problem 4.49.

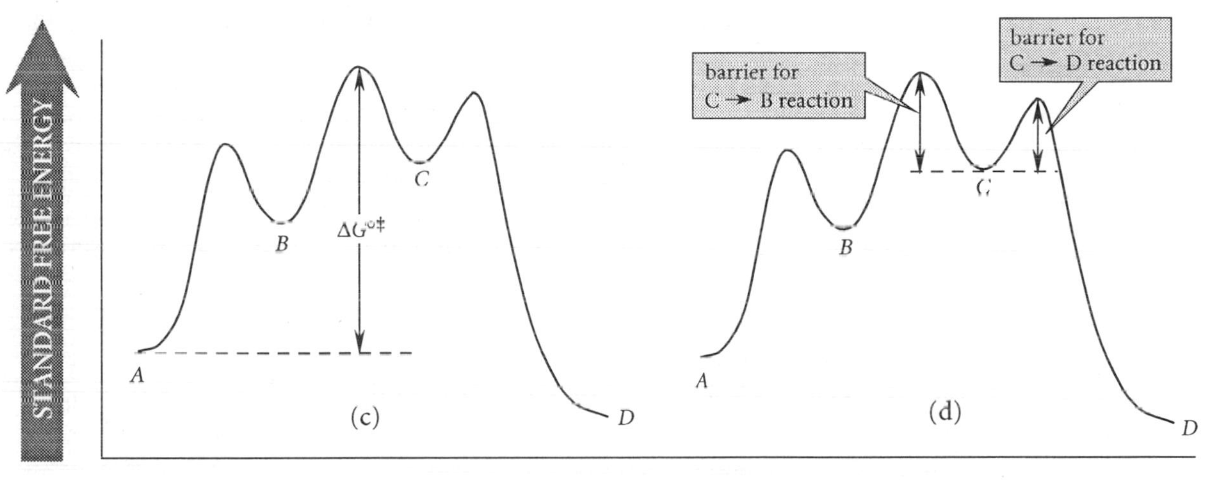

reaction coordinates

Figure SG4.4 Reaction free-energy diagrams to accompany the solutions to Problem 4.50c and 4.50d.

CH_3 OH_2

carbocation intermediate from protonation of 1-methylcyclohexane

δ+ CH_3 δ+ H---$\ddot{O}H_2$ ‡

transition state for protonation of 1-methylcyclohexene

4.52 (a) The bonding arrangement in 1,3-pentadiene is more stable because this diene has the smaller heat of formation.

When two double bonds are separated by one single bond, the double bonds are said to be *conjugated.* The conjugated arrangement of double bonds is a particularly stable arrangement. The reason for this is discussed in Chapter 15.

(b) The relevant combustion equation is

	1,3-pentadiene (C_5H_8) +	$7O_2$	⟶	$5CO_2$ +	$4H_2O$
heats of formation (kJ/mol)	75.8	0		5(–393.5)	4(–241.8)

(Note that the heats of formation of CO_2 and H_2O are by definition identical to the heats of combustion of carbon and hydrogen, respectively.) The $\Delta H°$ for this reaction is 4(–241.8) + 5(–393.5) – 75.8 = –2859.0 kJ/mol. The heat of combustion (always expressed as a positive number for heat liberated) is therefore 2858.9 kJ/mol (683.3 kcal/mol).

4.53 Following the hint, an energy diagram for the relationships involved is shown in Fig. SG4.5. As this diagram emphasizes, the enthalpy "distance" from 5C + $5H_2$ to $5CO_2$ + $5H_2O$ is the same in both cases. (The enthalpy of O_2 required for combustion is ignored, and in any case is zero by convention.) Because this total $\Delta H°$ is known from the 3-methyl-1-butene case, the required enthalpy x can be calculated. According to the diagram, then,

$$-27.4 - 3149.3 = -35.1 + x$$

Solving, $x = -3141.6$ kJ/mol. Thus, 3141.6 kJ/mol of heat is liberated when 2-methyl-1-butene is burned. Less energy is liberated than when 3-methyl-1-butene is burned because 2-methyl-1-butene has lower energy than 3-methyl-1-butene. (Why?)

4.54 If you made the models as requested you found that it is very difficult to close the ring containing the double bond with the E (or trans) configuration, whereas the ring containing the Z (or cis) double bond can be closed without difficulty. (If this conclusion is not clear, be sure you built a model of the *cyclic* alkenes and not a pair of the acyclic heptenes.) Thus, (E)-cycloheptene is less stable than (Z)-cycloheptene. Because the relative heats of formation of two isomers are the same as their relative stabilities, it follows that (Z)-cycloheptene has the smaller (less positive or more negative) heat of formation.

4.55 In each *trans*-isomer, the large groups have modest if any van der Waals repulsions with the neighboring hydrogens. However, the *tert*-butyl groups in the *cis*-isomer of (1) suffer more severe van der Waals repulsions than the methyl groups in the *cis*-isomer of (2) because *tert*-butyl groups are much larger than methyl groups. Consequently, relative to the *trans*-isomer, the *cis*-isomer of (1) has much higher energy than the *cis*-isomer of (2). Consequently, more energy is released in the conversion shown in (1) than in the conversion shown in (2). This is summarized in the Fig. SG4.6 on the following page, in which the two *trans*-isomers have been placed at the same energy for comparison.

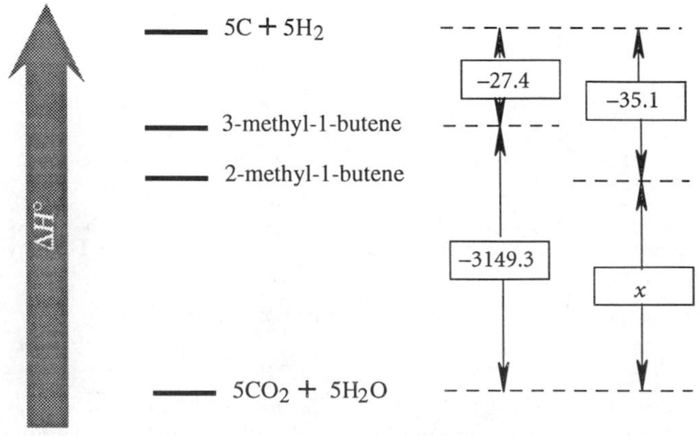

Figure SG4.5 Enthalpy relationships for the solution to Problem 4.53.

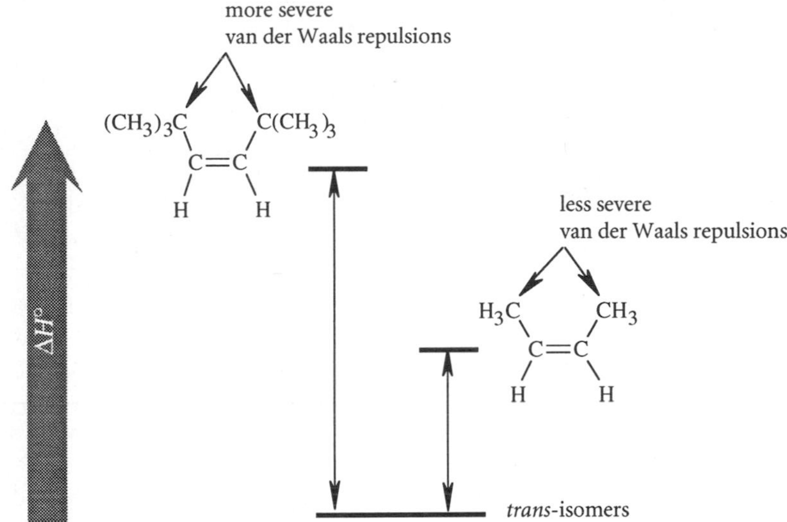

Figure SG4.6 Enthalpy relationships for the solution to Problem 4.55. The two *trans*-isomers have been arbitrarily placed at the same enthalpy for comparison.

4.56 (a) The data show that methyl groups are more electron-donating than hydrogens. That is, the H_3C—C= bond dipole is larger than the H—C= bond dipole. This follows because each H_3C—C bond dipole in the first structure, and each H—C bond dipole in the second structure, is parallel to a C—Cl bond dipole within the same structure. As shown in the following vector diagram, in which R = CH_3 or R = H, the parallel dipoles reinforce to give a larger resultant *A*; and two such resultants add to give a resultant *B* for the entire molecule.

(See the note following the solution to Problem 1.34, p. 21 of this manual, on graphical vector addition, if necessary.) It follows from this diagram that the larger is the R—C bond dipole, the larger is the molecular dipole moment. Because the compound containing the methyl groups has the larger dipole moment, it follows that the H_3C—C bond dipole is larger than the H—C bond dipole. In other words, a methyl group is more electron donating than a hydrogen towards a double bond.

(b) Assume that the major contributors to the molecular dipole moment are the H_3C—C and the C—Cl bond dipoles as indicated in part (a). These bond dipoles reinforce in the *E* isomer, and are nearly opposed in the *Z* isomer. Consequently, the *E* isomer has the greater dipole moment.

4.57 The curved-arrow notation is as follows:

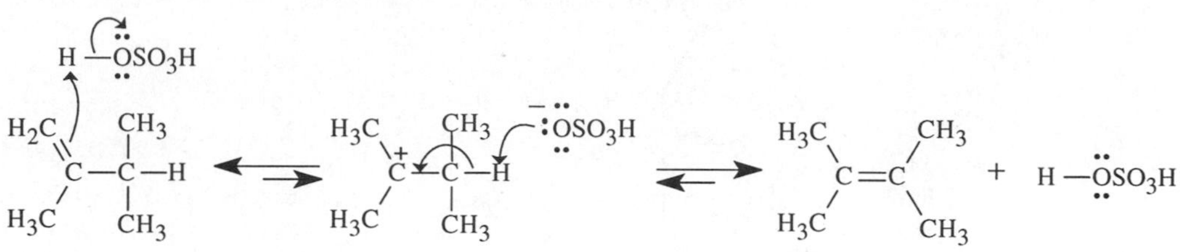

$$(H_3C)_2C{=}\overset{CH_3}{C}(CH_3)H \;\; H{-}\ddot{O}SO_3H \rightleftharpoons (H_3C)_2\overset{+}{C}{-}C(CH_3)_2H \;\; {}^{-}:\ddot{O}SO_3H \rightleftharpoons (H_3C)_2C{=}C(CH_3)_2 + H{-}\ddot{O}SO_3H$$

4.58 The solution to this problem is discussed thoroughly in Study Guide Link 4.9.

4.59 This is an acid-catalyzed addition of methanol (CH_3OH) to the double bond, and is completely analogous to hydration. The proton that begins the addition comes from the acid catalyst; notice that the ionized form of H_2SO_4 in methanol is protonated methanol ($CH_3\overset{+}{O}H_2$).

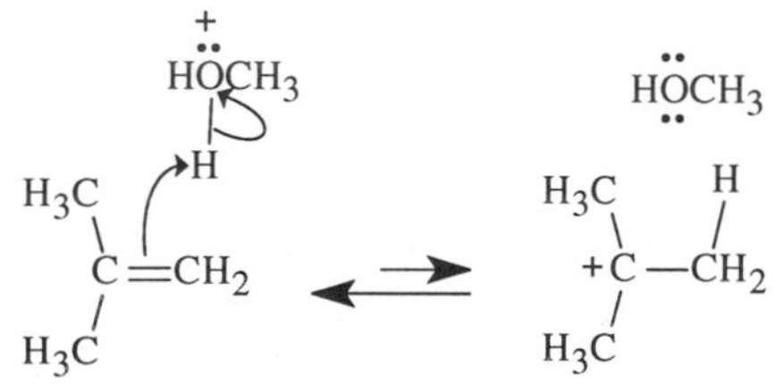

$$(H_3C)_2C{=}CH_2 + H\overset{+}{\ddot{O}}(H)CH_3 \rightleftharpoons (H_3C)_2\overset{+}{C}{-}CH_2H + H\ddot{O}CH_3$$

The nucleophile methanol attacks the carbocation:

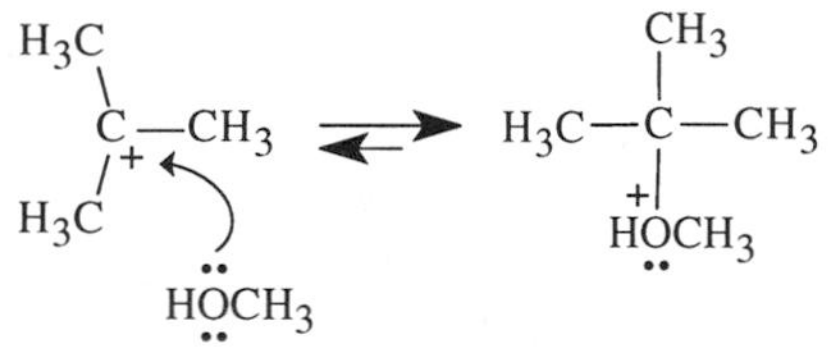

$$(H_3C)_2\overset{+}{C}{-}CH_3 + H\ddot{O}CH_3 \rightleftharpoons H_3C{-}C(CH_3)_2{-}\overset{+}{O}(H)CH_3$$

Finally, a proton is transferred to solvent to regenerate the catalyst and complete the mechanism.

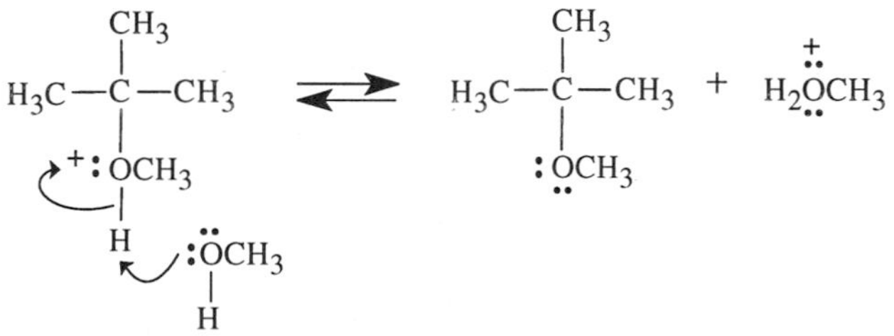

$$H_3C{-}C(CH_3)_2{-}\overset{+}{O}(H)CH_3 + :\ddot{O}(H)CH_3 \rightleftharpoons H_3C{-}C(CH_3)_2{-}\ddot{O}CH_3 + H_2\overset{+}{\ddot{O}}CH_3$$

4.60 Protonation occurs at carbon-5 (see numbered structure below) to form a tertiary carbocation. (Protonation at carbon-1 would give a secondary carbocation.) Because the tertiary carbocation is more stable, it is formed more rapidly (Hammond's postulate). The tertiary carbocation resulting from protonation then adds to the other double bond (the only choice) to give a secondary carbocation, which is attacked by water.

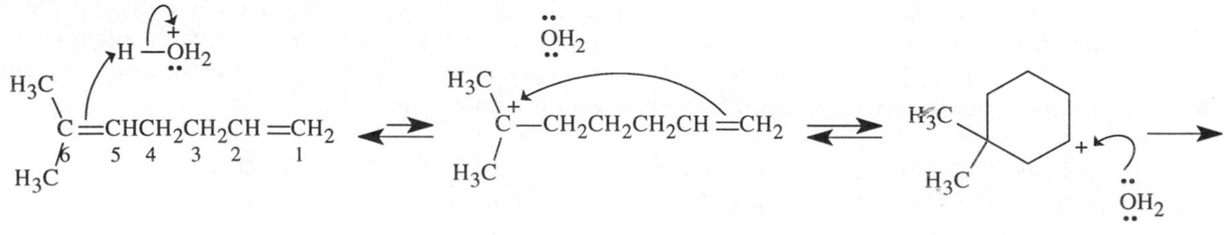

$$(H_3C)_2\underset{6}{C}{=}\underset{5}{C}H\underset{4}{C}H_2\underset{3}{C}H_2\underset{2}{C}H{=}\underset{1}{C}H_2 + H{-}\overset{+}{\ddot{O}}H_2 \rightleftharpoons (H_3C)_2\overset{+}{C}{-}CH_2CH_2CH_2CH{=}CH_2 \;\; (\ddot{O}H_2) \rightleftharpoons \text{1,1-dimethylcyclohexyl cation (H}_3\text{C, H}_3\text{C)} + \ddot{O}H_2 \longrightarrow$$

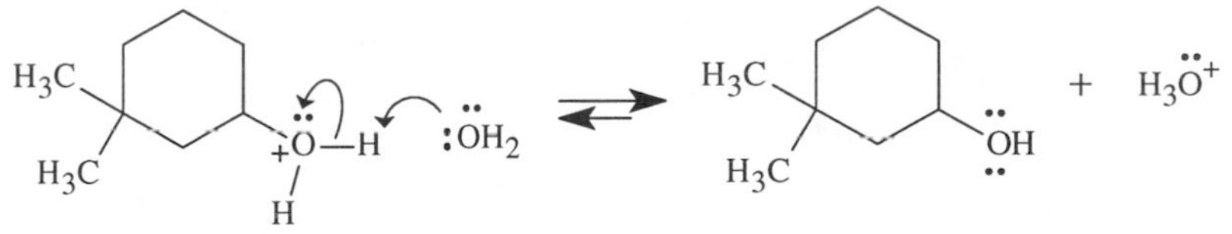

Why. in the second step, would a tertiary carbocation be converted into a secondary one? Because in this particular process a π bond is also replaced by a σ bond. A σ bond is *much* stronger than a π bond—strong enough to compensate for the formation of a secondary carbocation. In other words, the energy loss from σ-bond formation is enough to offset the energy gain for formation of a secondary carbocation.

4.61 (a) We refer to the alkene on the left, 2,3-dimethyl-1-butene, as A, and the alkene on the right, 2,3-dimethyl-2-butene, as B. Compound B is favored at equilibrium because it has the smaller (less positive) ΔG_f°. (Notice that this alkene has more alkyl substituents on its double bond.) The equilibrium constant is obtained from Eq. 3.24b. Define K_{eq} as $[B]/[A]$:

$$K_{eq} = 10^{-\Delta G^\circ/2.3RT} = 10^{-(-3.18/5.706)} = 10^{0.557} = 3.61$$

(b) The equilibrium constant tells us *nothing* about how rapidly this interconversion takes place. For example, in the absence of a catalyst, the rate of this interconversion is essentially zero. However, this reaction does occur at a convenient rate in the presence of an acid catalyst, as suggested by Problem 4.57.

4.62 (a) First draw the structures of the two compounds:

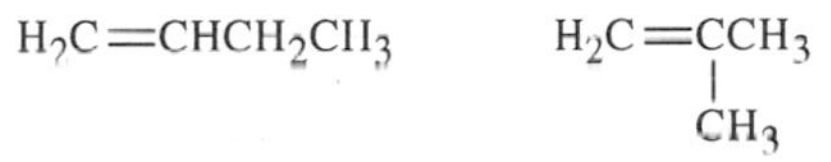

$H_2C{=}CHCH_2CH_3$ (1-butene) $\qquad$ $H_2C{=}C(CH_3)CH_3$ (2-methylpropene)

2-Methylpropene has two alkyl branches on its double bond, whereas 1-butene has only one. Because alkyl branches on the double bond stabilize an alkene, 2-methylpropene is the more stable of the two alkenes.

(b) The reaction with the smaller standard free energy of activation ($\Delta G^{\circ\ddagger}$) is faster. Consequently, hydration of 2-methylpropene is faster.

(c) Let 1-butene = A, and let 2-methylpropene = B. The reaction-free energy diagrams for hydration are shown in Fig. SG4.7 on the following page. (Only the first, rate-limiting, step of the mechanism is shown in each diagram.)

(d) The diagrams in Fig. SG4.7 require the following equality:

$$13.39 + \Delta G^{\circ\ddagger}(A) = x + \Delta G^{\circ\ddagger}(B)$$

Rearranging,

$$\Delta G^{\circ\ddagger}(A) - \Delta G^{\circ\ddagger}(B) = x - 13.39$$

The difference between the standard free energies of activation is given in the problem as 22.84 kJ/mol. Hence,

$$x = 22.84 + 13.39 = 36.23 \text{ kJ/mol}$$

Thus, the transition state for hydration of 1-butene is 36.23 kJ/mol *less stable* than the transition state for hydration of 2-methylpropene. From Sec. 4.9B, the mechanism of hydration involves a carbocation intermediate, and the first step, protonation of the double bond to form this intermediate, is rate-limiting. In the case of 2-methylpropene, the carbocation intermediate is tertiary; in the hydration of 1-butene, it is secondary. Tertiary carbocations are more stable than secondary carbocations. Because the transition states resemble the respective carbocations in structure and energy (Hammond's postulate), the transition state for formation of the more stable carbocation is the more stable transition state, as observed.

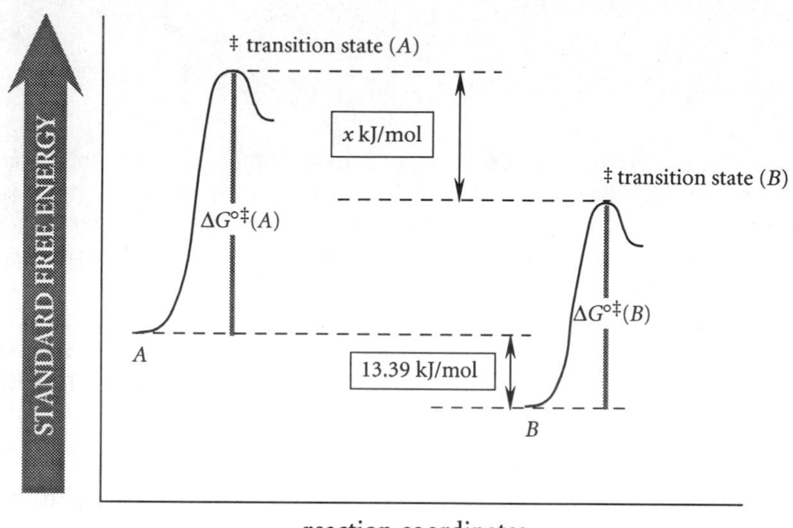

Figure SG4.7 Reaction free-energy diagrams to accompany the solution to Problem 4.62.

5

Addition Reactions of Alkenes

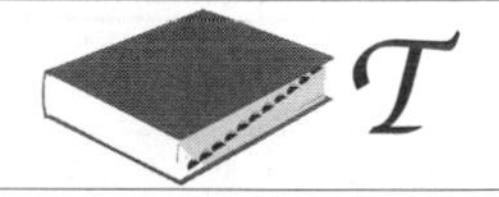

TERMS

CONCEPTS

I. REACTIONS INVOLVING ALKENES (See also Section R)

A. Addition Reactions

1. The most common reaction of alkenes is addition to the double bond, which occurs by:
   a. mechanisms involving carbocation intermediates (addition of hydrogen halides, hydration)
   b. mechanisms involving cyclic ion intermediates (oxymercuration, halogenation)
   c. mechanisms involving free-radical intermediates (free-radical addition of HBr, polymerization)
   d. concerted mechanisms (hydroboration, ozonolysis)
2. Many addition reactions are regioselective because of their mechanisms.

B. Other Reactions

1. Some useful transformations of alkenes involve additions followed by other transformations.
   a. Oxymercuration followed by reduction affords alcohols.
   b. Hydroboration followed by oxidation affords alcohols.
   c. Ozonolysis followed by:
      i. treatment with $(CH_3)_2S$ affords aldehydes and ketones by cleavage of the double bond.
      ii. treatment with H_2O_2 affords ketones and carboxylic acids by cleavage of the double bond.
2. Alkenes such as ethylene and propene are produced by cracking alkanes at high temperature.

II. WRITING ORGANIC REACTIONS

A. Reaction Equations

1. The most thorough way to write a reaction is to use a complete, balanced equation.
2. In many cases it is convenient to abbreviate reactions by showing only the organic starting materials and the major organic products; reaction conditions (solvents, catalysts, etc.) are written over and/or under the arrow.

B. Product Yields

1. The theoretical yield is the maximum amount of product that can be formed in a reaction if there are no losses or side-reactions.
2. The actual yield is the amount of product isolated from a reaction.
3. The percentage yield is the actual yield divided by the theoretical yield times 100%.

III. BOND DISSOCIATIONS AND FREE RADICALS

A. Types of Bond Dissociations

1. An electron-pair bond may break so that one of the bonding atoms receives both electrons and the other becomes electron-deficient; this type of bond breaking is called heterolytic cleavage or heterolysis.

—C—I: —(heterolytic bond dissociation)→ —C⁺ + :I:⁻

a carbocation   an anion

2. An electron-pair bond may break so that each bonding atom retains one electron of the chemical bond; this type of bond breaking is called homolytic cleavage or homolysis.
   a. Any species with at least one unpaired electron is called a free radical.

b. A different curved-arrow notation, called the fishhook notation, is used for homolysis. In this notation, electrons are moved individually rather than in pairs.

$$\text{—C—I:} \xrightarrow{\text{homolytic bond dissociation}} \text{—C}\cdot \;+\; \cdot\text{I:}$$

a free radical   a free radical

B. Bond Dissociation Energies

1. The bond dissociation energy of a covalent bond measures the energy required to break the bond homolytically to form two free radicals.
   a. For a bond between two atoms A—B, the bond dissociation energy is defined as the enthalpy $\Delta H°$ of the homolytic reaction.
   b. The lower a bond dissociation energy is, the lower is the temperature required to rupture the bond in question and form free radicals at a reasonable rate.
   c. A bond dissociation energy measures the intrinsic strength of a chemical bond.
2. The $\Delta H°$ of reactions can be calculated or estimated by subtracting the bond dissociation energies of the bonds formed from those of the bonds broken.
3. Although bond dissociation energies apply to reactions in the gas phase, they can be used to compare the enthalpies of two reaction in solution provided that the effect of solvent is either negligible or is the same for both of the reactions being compared.

C. Free Radicals and Reactions

1. Free radicals behave as reactive intermediates; that is, they react before they can accumulate in significant amounts.
2. Free-radical reactions typically involve three types of steps: initiation steps, propagation steps, and termination steps.
3. Three types of reactions that free radicals can undergo are:
   a. addition to a double bond, in which a free radical reacts with a π bond to give a new free radical.

$$\text{R}\cdot + \text{C=C} \longrightarrow \text{R—C—C}\cdot$$

   b. atom abstraction, in which a free radical removes an atom from another molecule, and a new free radical is formed

$$\text{R}\cdot + \text{H—C—C—} \longrightarrow \text{R—H} + \cdot\text{C—C—}$$

   c. recombination (the reverse of bond rupture), in which free radicals react to form a covalent bond

$$\text{R}\cdot + \text{R}\cdot \longrightarrow \text{R—R}$$

4. If a reaction occurs in the presence of heat, light, or a known free-radical initiator (peroxides, AIBN, etc.), but does not occur in its absence, it is fairly certain that the reaction involves free-radical intermediates.
5. Free-radical reactions that involve propagation steps and their associated initiation and termination steps are said to be free-radical chain reactions.

a. For each free radical consumed in the propagation steps one is produced.
b. The free radicals involved in the propagation steps of a chain reaction are said to propagate the chain.

D. Free-Radical Chain Reaction Mechanisms

1. In initiation steps, the free radicals that take part in subsequent steps of the reaction are formed from a molecule that readily undergoes homolysis, called a free-radical initiator.
   a. The initiator is in effect the source of free radicals.

$$\mathrm{RO{-}OR} \longrightarrow 2\,\mathrm{RO}\cdot$$

   b. Sometimes heat or light initiates a free-radical reaction in which the additional energy promotes homolysis of the free-radical initiator.
   c. Some free-radical initiation reactions take place in more than one step. For example, in the free-radical addition of HBr to alkenes, a hydrogen atom from HBr is removed by the free radical that was formed in the first initiation step.

$$\mathrm{RO}\cdot + \mathrm{H{-}Br} \longrightarrow \mathrm{RO{-}H} + \mathrm{Br}\cdot$$

2. In propagation steps, even though products are formed there is no net consumption or destruction of free radicals. The free-radical by-product of one propagation step serves as the starting material for another propagation step.
3. In termination steps, free radicals are destroyed by recombination reactions.
   a. In a recombination reaction, which is the reverse of a homolytic dissociation reaction, two free radicalscombine to form a covalent bond.
   b. The recombination reactions of free radicals are in general highly exothermic; they have very favorable (negative) $\Delta H°$ values.

E. The Peroxide Effect on HBr Addition to Alkenes

1. The peroxide effect: A small amount of a peroxide added to the reaction mixture reverses the regioselectivity of HBr addition to the double bond of an unsymmetrically branched alkene. The Br adds to the less branched carbon of the double bond, and the H adds to the more branched carbon.

$$\mathrm{R_2C{=}CH_2} + \mathrm{HBr} \xrightarrow{\text{no peroxides}} \mathrm{R{-}CBr(R){-}CH_2H}$$

$$\mathrm{R_2C{=}CH_2} + \mathrm{HBr} \xrightarrow{\text{+ peroxides}} \mathrm{R{-}CH(R){-}CH_2Br}$$

2. The reaction involves free-radical intermediates.
3. The large bromine atom reacts at the least branched carbon of the double bond because otherwise it would experience van der Waals repulsions with the hydrogens in the branches.
   a. Any effect on a reaction that can be attributed to van der Waals repulsions is termed a steric effect.
   b. Because the reaction with the transition state of lower energy is the faster reaction, reaction of the Br atom at the less branched carbon of the alkene double bond (giving the more branched free radical) is faster than reaction at the more branched carbon.
4. Addition of the bromine atom to the less branched carbon of the double bond gives the more stable free radical.
   a. The stability of free radicals is in the order tertiary > secondary > primary.
   b. The effect of branching on free-radical stability is considerably smaller than the effect of branching on carbocation stability.

c. The stability order implies that free radicals, like alkenes and carbocations, are stabilized by alkyl group substitution at sp^2-hybridized carbons.

5. The "peroxide effect" is not limited to peroxides; any good free-radical initiator will bring about the same effect.

REACTIONS

I. REACTIONS OF ALKENES WITH HALOGENS

A. Addition of Chlorine and Bromine

1. Bromine and chlorine undergo addition to alkenes to give compounds with halogens on adjacent carbons called vicinal dihalides. (X = Br or Cl)

$$\text{C=C} \xrightarrow{X_2} \text{X—C—C—X}$$

a vicinal dihalide

2. The rapid disappearance of the red bromine color during bromine addition is a useful qualitative test for alkenes.
3. The equilibrium for addition of I_2 favors the alkene. Addition of F_2 occurs, but the reaction is violent and is accompanied by other reactions.
4. Bromine addition to alkenes in many cases involves a reactive intermediate called a bromonium ion.
   a. In a bromonium ion, a positively charged bromine with an octet of electrons is bonded to two other atoms.

a bromonium ion

   b. Attack of the bromide ion at either of the carbons bound to bromine in the bromonium ion completes the addition of bromine.

a vicinal dibromide

B. Formation of Halohydrins

1. When an alkene reacts with Cl_2, Br_2, or I_2 in a solvent containing a large amount of water, water attacks the cyclic ion. (X = Cl, Br, or I)

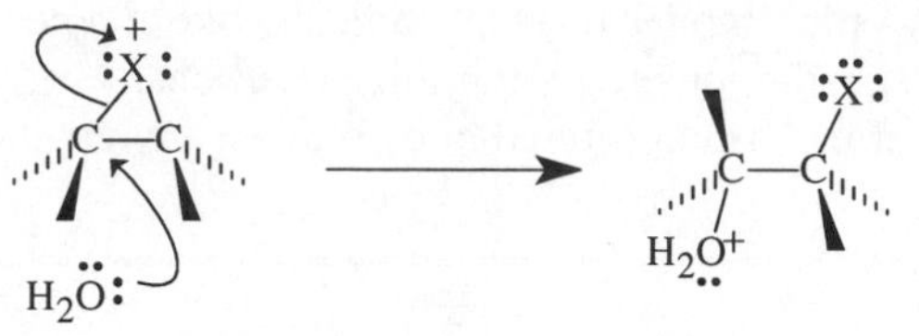

2. Loss of a proton from the oxygen gives a compound containing both an —OH and a halogen (—X group) called a halohydrin.

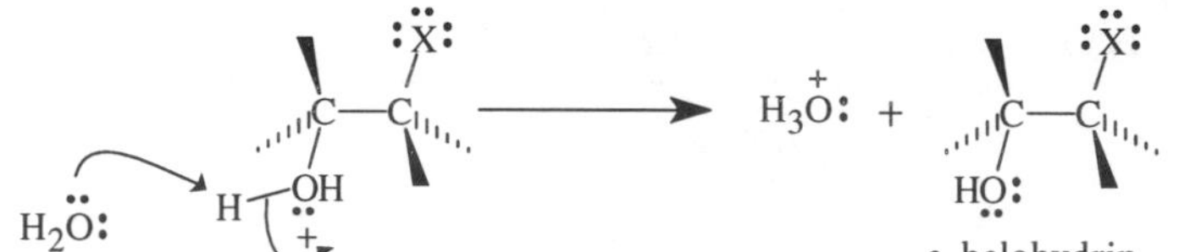

a halohydrin

   a. Halohydrins are compounds that contain both a halogen and an —OH group.
   b. The reaction is completely regioselective when one carbon of the alkene is bonded to two alkyl substituents and the other is bonded to two hydrogens.

C. Free-Radical Addition of Hydrogen Bromide to Alkenes

1. Alkenes react with hydrogen bromide in the presence of peroxides so that the bromine adds to the less branched carbon of the double bond.

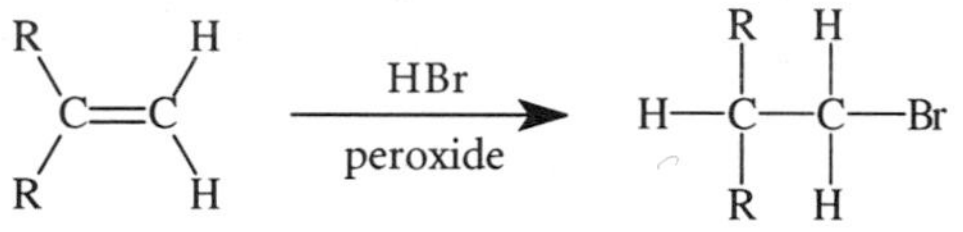

$$R_2C{=}CH_2 \xrightarrow[\text{peroxide}]{\text{HBr}} H{-}CR_2{-}CH_2{-}Br$$

   a. This reversal of regioselectivity in HBr addition in the presence of a peroxide is termed the peroxide effect.
   b. This mechanism involves reactive intermediates called free radicals.

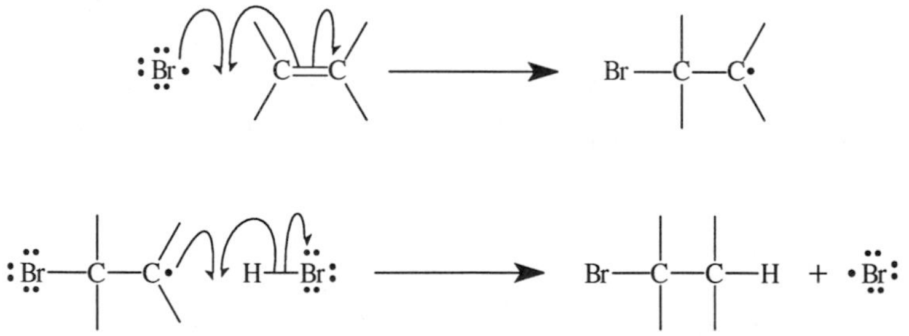

2. The regioselectivity of HI or HCl addition to alkenes is not affected by the presence of peroxides.

II. CONVERSION OF ALKENES INTO ALCOHOLS

A. Oxymercuration-Reduction of Alkenes

1. Alkenes react with mercuric acetate in aqueous solution to give addition products in which an —HgOAc (acetoxymercuri) group and an —OH (hydroxy) group derived from water have added to the double bond; this reaction is called oxymercuration.
   a. The first step of the reaction mechanism involves the formation of a cyclic ion called a mercurinium ion.

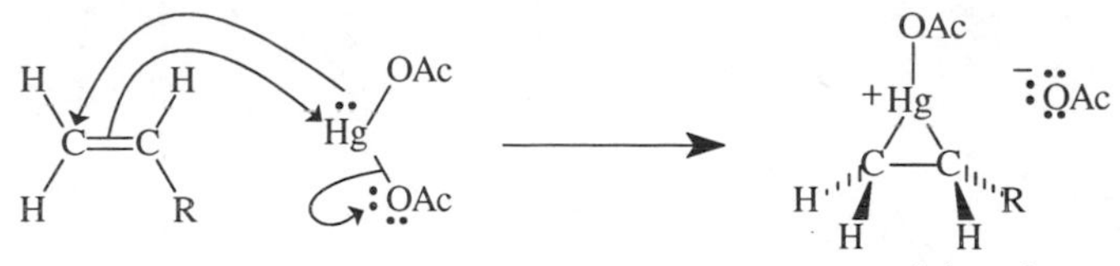

a mercurinium ion

b. The mercurinium ion is attacked by the solvent water exclusively at the more branched carbon.

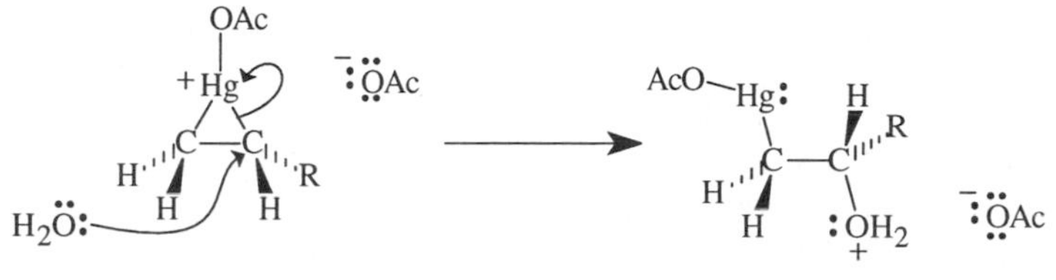

c. The addition is completed by transfer of a proton to the acetate ion that is formed.

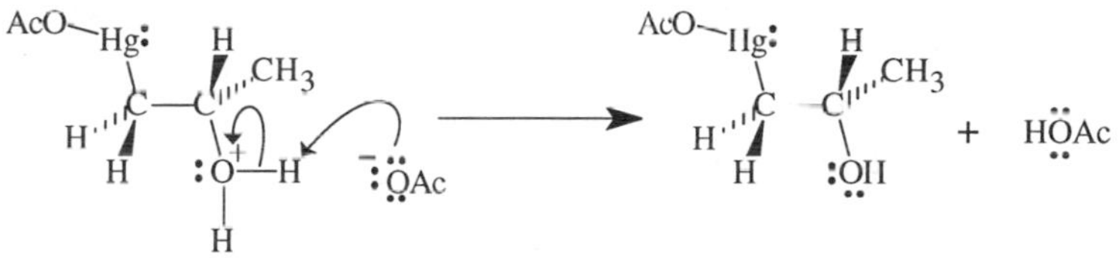

2. Oxymercuration adducts are easily converted into alcohols by treatment with the reducing agent sodium borohydride ($NaBH_4$) in base. In this reaction, the carbon-mercury bond is replaced by a carbon-hydrogen bond.

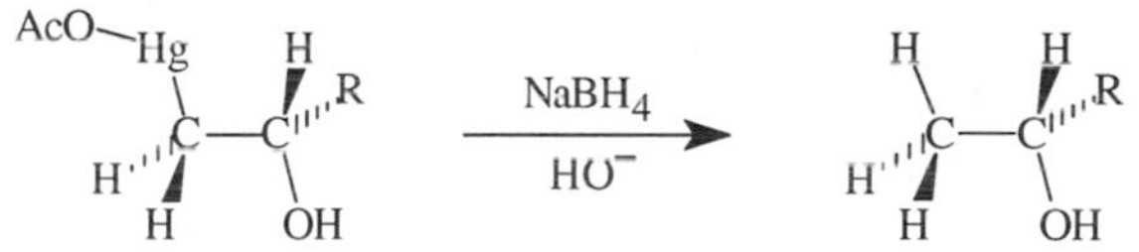

3. The overall reaction sequence, called an oxymercuration-reduction, results in the net addition of the elements of water (H and OH) to an alkene double bond in a regioselective manner.
   a. The —OH group is added to the more branched carbon of the double bond.
   b. The product is free of rearrangements and other side reactions that are encountered in hydration.

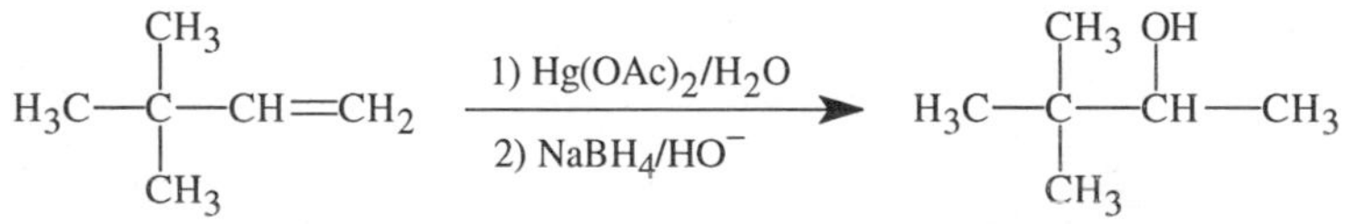

$$H_3C-C(CH_3)_2-CH{=}CH_2 \xrightarrow[\text{2) } NaBH_4/HO^-]{\text{1) } Hg(OAc)_2/H_2O} H_3C-C(CH_3)_2-CH(OH)-CH_3$$

B. Hydroboration-Oxidation of Alkenes

1. Borane BH_3 adds regioselectively to alkenes so that the boron becomes bonded to the less branched carbon of the double bond, and the hydrogen becomes bonded to the more branched carbon.
   a. The addition of BH_3 to alkenes, called hydroboration, occurs in a single step without intermediates, that is, by a concerted mechanism.
   b. One borane molecule can add to three alkene molecules; the product is a trialkylborane.

$$\text{C=C} \xrightarrow{BH_3} (\text{H—C—C})_3\text{B}$$

2. Trialkylboranes are oxidized to alcohols using basic hydrogen peroxide, H_2O_2. The net result of this transformation is replacement of the boron by an —OH in each alkyl group.

$$(\text{H—C—C})_3\text{B} \xrightarrow[HO^-]{H_2O_2} 3\ \text{H—C—C—OH} + \text{borate salts}$$

3. The overall reaction sequence, called a hydroboration-oxidation, results in the net addition of the elements of water (H and OH) to an alkene in a regioselective manner.
   a. The —OH group is added to the least branched carbon atom of the double bond.
   b. The product is free of rearrangements and other side reactions that are encountered in hydration.

$$H_3C\text{—}C(CH_3)_2\text{—}CH{=}CH_2 \xrightarrow[2)\ H_2O_2/HO^-]{1)\ BH_3} H_3C\text{—}C(CH_3)_2\text{—}CH_2\text{—}CH_2OH$$

C. Comparison of Methods for the Synthesis of Alcohols from Alkenes

1. Hydration of alkenes is a useful industrial method for preparing a few alcohols, but it is not a good laboratory method.
2. Hydroboration-oxidation is a good laboratory method that gives an alcohol in which the —OH group has been added to the least branched carbon of the double bond.
3. Oxymercuration-reduction is a good laboratory method that gives an alcohol in which the —OH group has been added to the more branched carbon of the double bond.

III. OTHER REACTIONS

A. Ozonolysis of Alkenes

1. The reaction of an alkene with ozone to yield products of double-bond cleavage is called ozonolysis.
2. Ozone, O_3, adds to alkenes at low temperature in a concerted cycloaddition reaction to yield unstable cyclic compounds called molozonides.

$$\text{C=C} \xrightarrow{O_3} \text{a molozonide}$$

3. The molozonide cycloaddition product is unstable and spontaneously forms the ozonide.

$$\text{molozonide} \longrightarrow \text{an ozonide}$$

4. If the ozonide is treated with dimethyl sulfide, $(CH_3)_2S$, the ozonide is split into products that contain a C=O group.

a. If a carbon of the double bond in the starting material bears a hydrogen, an aldehyde is formed.
b. If a carbon of the double bond bears no hydrogens, a ketone is formed.

$(CH_3)_2S$

an aldehyde a ketone

5. If the ozonide is simply treated with water, hydrogen peroxide (H_2O_2) is formed as a by-product and carboxylic acids are formed instead of aldehydes.

H_2O_2, H_2O

a carboxylic acid a ketone

B. Free-Radical Polymerization of Alkenes

1. In the presence of free-radical initiators such as peroxides, many alkenes (monomers) react to form polymers in a reaction called polymerization.

$n\ H_2C{=}CHCl$ —polymerization→ $\left(-CH_2-CHCl-\right)_n$

vinyl chloride poly(vinyl chloride)

a. Polymers are very large molecules composed of repeating units.
b. Polyethylene is an example of an addition polymer—a polymer in which no atoms of the monomer unit have been lost as a result of the polymerization reaction.
c. Polyethylene formation occurs by a free-radical mechanism and is thus a free-radical polymerization.

C. Thermal Cracking of Alkenes

1. Simple alkenes are produced industrially from alkanes in a process called thermal cracking, which breaks larger alkanes into a mixture of smaller hydrocarbons, some of which are alkenes.
2. Ethylene, the alkene of greatest commercial importance, is produced by thermal cracking of ethane.

SCIENTISTS

Herbert Charles Brown (born 1912) [Chemistry 1979] 5.3B
Morris Selig Kharasch (1895–1957) 5.5A
Roy Joseph Plunkett (1910–1994) 5.6
Karl Waldemar Ziegler (1898–1973) [Chemistry 1963] 5.6

Dates in brackets refer to Nobel Prize awards.

STUDY GUIDE LINKS

✓5.1 Transition Elements and the Electron-Counting Rules

Equation 5.12a in the text involves the element mercury, which is a transition element. If you have been paying careful attention to formal charges and electron counts, you may be puzzled about how to deal with an element such as Hg. First, don't try to apply the usual rules of electron counting to transition elements, because these elements contain more than eight valence electrons, and the octet rule does not apply. (We'll discuss electron counting in transition elements in Sec. 18.5.) It's also hard to calculate formal charge by the rules given in Chapter 1 because transition elements have more than eight valence electrons, and we can't show them all without making a hopelessly complex-looking structure. In Eq. 5.12a, we've shown as an unshared pair only two of the *ten* valence electrons on mercury in $Hg(OAc)_2$. However, we can calculate the *change* in the formal charge on an element in a reaction by applying the following rules of formal-charge calculation:

1. The electrons in unshared pairs on an atom are both assigned to the atom.
2. One electron from each electron-pair bond is assigned to an element.
3. The change in formal charge on an element in any reaction is equal to the number of electrons assigned to the element in the starting material minus the number in the product.

(These rules are completely consistent with the formal-charge rules you've already learned, and work on *any* atom.)

Let's apply these rules to calculate the formal charge on Hg in the last step of text Eq. 5.12a. We are *given* that Hg in the structure on the left has a formal charge of zero.

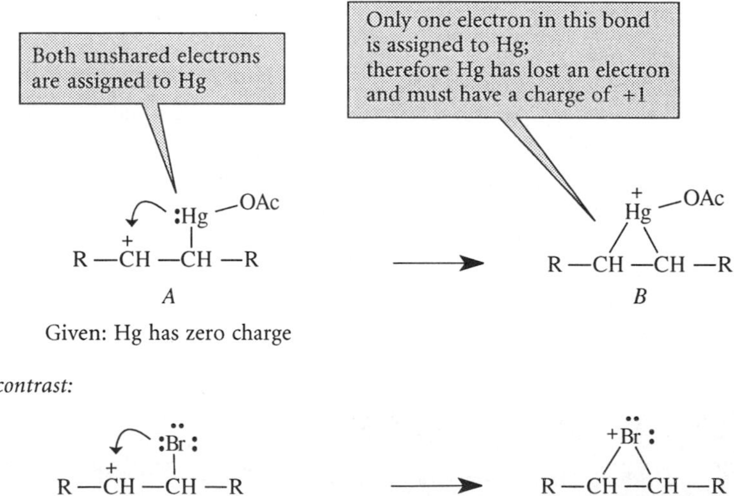

In species *A*, both unshared electrons and one electron in each covalent bond are assigned to Hg, for a total of 4 electrons. In species *B*, 3 electrons (one from each covalent bond) are assigned to Hg. The change in charge is $4 - 3 = +1$. Given that Hg has zero charge in *A*, it must have a +1 charge in *B*.

✓5.2 Mechanism of Hydroboration

Although the mechanism of hydroboration does not involve carbocation intermediates, let's *think* of it as a two-step mechanism involving carbocations, because such an approach will help us to see the relationship between hydroboration and the other addition reactions we've studied.

In the first step, BH_3 reacts with the π electrons of the double bond in the starting material *S* because BH_3 is electron deficient:

$$\underset{S}{(H_3C)_2C{=}CH_2 + BH_3} \longrightarrow \underset{A}{(H_3C)_2\overset{+}{C}{-}CH_2{-}\overset{-}{B}H_3}$$

The boron is bonded to the carbon of the double bond with fewer branches because, as a result, the carbocation is formed in which the electron-deficient carbon has the greater number of branches. In the second step, one hydrogen on the boron *with its bonding electrons* then moves to the carbocation, leaving boron electron-deficient and ready for a second addition:

$$\underset{A}{(H_3C)_2\overset{+}{C}{-}CH_2{-}\overset{-}{B}H_3} \longrightarrow \underset{B}{H_3C{-}\underset{CH_3}{\overset{H}{C}}{-}CH_2{-}BH_2}$$

Now, we imagine both steps occurring at the same time. That's the concerted mechanism. The *transition state* of the actual reaction is something that incorporates features of the transition states of both of these steps at the same time:

$$\left[(H_3C)_2\overset{\delta+}{C}\cdots H\cdots\overset{\delta-}{B}H_2\cdots CH_2\right]^{\ddagger}$$

The bond between H and C represented with dotted lines is very weak, and that is why there is partial positive charge on carbon.

There's nothing wrong with thinking about concerted mechanisms in a stepwise fashion. This mode of thinking helps us to see the connection with other mechanisms that actually are two-step mechanisms, such as HBr addition. Notice that we took the same approach in the text with bromine addition (Eq 5.4b, text p. 161) and oxymercuration (Eq. 5.12b, text p. 166). We simply have to keep in the back of our minds that other phenomena associated with carbocations (such as rearrangements) are not observed in such concerted reactions. There's other evidence that carbocations are not involved as well; we'll discuss this in Sec. 7.9D.

5.3 Mechanism of Organoborane Oxidation

Although the oxidation step of hydroboration-oxidation may appear mysterious, it is nothing more than a series of Brønsted and Lewis acid-base reactions.

First, the hydroperoxide anion is produced by ionization of H_2O_2. As you can see from the pK_a values below, this ionization is quite favorable when ^{-}OH is used as the base:

$$HO^{-} + \underset{pK_a = 11.6}{H{-}O{-}OH} \rightleftharpoons \underset{pK_a = 15.7}{HO{-}H} + \underset{\text{hydroperoxide anion}}{^{-}O{-}OH}$$

Next, this anion attacks the boron of the organoborane in a Lewis acid-base reaction. (The alkyl groups on boron are abbreviated as R groups.)

$$R_3B \leftarrow \bar{:}\ddot{O}-\ddot{O}H \longrightarrow R_3\bar{B}-\ddot{O}-\ddot{O}H$$

An alkyl group on boron moves from boron to oxygen with breaking of the (very weak) bond between the two oxygens and loss of the $^-$OH group:

$$R_2\bar{B}(R)-\ddot{O}-\ddot{O}H \longrightarrow R_2B-\ddot{O}-R + {}^-:\ddot{O}H$$

This is the step that replaces a carbon-boron bond with a carbon-oxygen bond. Why would such a reaction take place?

Again, let's use the device of imagining that this process occurs in a stepwise manner. The O—O bond is very weak, because two electronegative atoms are "pulling" at the bonding electrons. Imagine that $^-$OH is expelled, leaving an *electron-deficient oxygen:*

$$R_2\bar{B}(R)-\ddot{O}-\ddot{O}H \longrightarrow R_2\bar{B}(R)-\ddot{O}^+ + {}^-:\ddot{O}H$$

The electron-deficient oxygen is like a carbocation, except with *much* higher energy. (Notice: this is not the same situation that occurs in $H_3\overset{+}{O}$, which is *not* electron deficient!) This very unstable cation can rearrange by taking electrons from the adjacent C—B bond:

$$R_2\bar{B}(R)-\ddot{O}^+ \longrightarrow R_2B-\ddot{O}:-R$$

Thus, the driving force for this rearrangement is the flow of electrons from the relatively electropositive boron to the very electronegative oxygen of $^-$OH.

In subsequent steps of the reaction, the bonds between the other R groups are converted in a similar manner to R—O bonds:

$$R_2B-OR \xrightarrow[\text{(same mechanism)}]{H_2O_2/^-OH} RO-B(OR)-OR$$

a trialkyl borate

Finally, the resulting trialkyl borate undergoes hydrolysis to the alcohol product. The final hydrolysis breaks the boron-oxygen bonds but leaves the oxygen-carbon bonds unaffected.

$$4\ {}^-OH + RO-B(OR)_2 \xrightarrow{H_2O} 3\ H-OR + \bar{B}(OH)_4$$

(You should be able to write the multistep mechanism of this hydrolysis. Remember, $^-$OH is a Lewis base and $B(OR)_3$ is a Lewis acid.)

✓5.4 How to Study Organic Reactions

You have now studied several organic reactions, and perhaps it is apparent that you are going to study many more. How can you keep all these reactions straight? How can you recall just the right reaction to use in solving a particular problem?

There are four keys to learning reactions; each of these points is discussed further below.

1. Be active when you read the text.
2. Be organized.
3. Review frequently.
4. Study in small chunks.

Be active when you read the text. To see what this means, let's go through one section of the text right now. Open your text to Section 5.3A, "Oxymercuration-Reduction of Alkenes." This section is fairly typical of those in the text that describe important reactions. First comes a statement of the reaction. When you read this, don't just stare at the page. Instead, *be active!* Try to make a hypothesis about where the different atoms come from. (The use of color in the equations is designed to help you with this.) For example, where does the —OH group in the product of text Eq. 5.11 originate? Where do the two —OAc groups end up? If you think you have the right idea, write the same reaction using a different alkene starting material.

Next, the text provides a few practical facts about the reaction, for example, some information about the solvent or the reaction conditions. These details can be very important, even crucial (in which case the text will say so), or they might be added for completeness or for your later use as reference material. In most cases, you should read them but not bother to memorize them, because the object of our first pass through the chapter is to understand the reaction itself.

Next follows a discussion of the *mechanism* of the reaction. What should you do with the mechanism? What you *should not* do is memorize it. What you *should do* is to follow each step of the mechanism to see whether your hypothesis about where the various atoms come from in the product is correct, and how they get there.

Here's an important point. Each step in a curved-arrow mechanism *must* be one of three types.

1. Lewis acid-base association (text p. 83)
2. Lewis acid-base dissociation (text p. 84)
3. Electron-pair displacement reaction (text p. 86)
   a. Brønsted acid-base reaction (text p. 91)
   b. Other electron-pair displacement
   c. Combination of (a) and (b)

Reactions of type (1) and (2) require *single* curved arrows. Reactions of type (3) require *two or more* curved arrows. Reactions of type (3a) involve by definition electron-pair displacements on a *proton.* As you examine a mechanism, *classify each mechanistic step* using the list above. Then *look for points in common with other mechanisms.* For example, *all* of the alkene reactions you have studied so far start with the π electrons of an alkene attacking a Lewis acid. Then consider the reactive intermediates, if any (for example, carbocations). Ask where you have seen such intermediates before.

One purpose of mechanisms is to help you see certain unifying ideas that seemingly different reactions have in common. (You may have noticed, for example, that carbocation theory keeps creeping into the discussion of many of the addition reactions you have studied.) As you go further along in the text, you might even try to write a mechanism on your own before looking at the one in the text. (Use Study Guide Link 4.9 to help you.) Comparing your mechanism to the one in the text will help you to refine your mechanism-writing skills. Finally, after studying the mechanism, *be active!* Start with a different alkene starting material and write the mechanism on your own.

Sometimes there are further applications of the reactions or additional facts—sort of a "wrap-up." For example, Eq. 5.15 illustrates the point that no rearrangements are observed in oxymercuration.

Finally come the problems. The answers to all of the problems in the text are provided in this manual. This means that you might be tempted to read the answers before you work the problems. Instead, work problems immediately when you come to them, and work them *with the answer book closed!* These provide a valuable test over your understanding of the chemistry involved. Put a mark by the ones you find difficult; re-work these problems before an exam. Another strategy is to work alternate problems initially; then use the others as a "pre-test" for exam study when you think you understand the material.

As you read the section, you may come across certain unfamiliar terms. For example, in the very first sentence of Section 5.3A is the term "addition." Do you know what this means? *The great temptation is to skip over unfamiliar terms, hoping that they will become clear in context.* This strategy often works when reading a novel, but you *can't* do this in scientific reading! You *must* understand each term before proceeding to the next. Ask yourself, "What is an addition reaction?" If you don't know, don't feel that you're intellectually deficient—just find out the answer. *Everyone* requires a number of repetitions in order to learn new ideas. (The author can't keep repeating the definition, because it would make the book even larger than it is!) How do you find out? *Use the index!* Or, in some cases, a cross-reference to an earlier section will be given. Look it up! If you get in the habit of taking this approach, you will find yourself retaining more, and you ultimately will have to do less work. If an unfamiliar term is one you have had to look up before, make yourself a glossary: Write the term and its definition *in your own words*—*not* necessarily the words of the text—and review it periodically. One prominent theory of learning says that students construct their own knowledge. What this means is that until you have processed something in *your own mind* you don't know it. Hence, when you see a new term, *read* the definition; *process it* (that is, frame it terms that *you* can understand, no matter how unsophisticated you might think your description is), and then *write it* (that is, be active). In summary, here's what you should be doing:

1. *Read*
2. *Process*
3. *Write*

You have probably noticed that each chapter in this Study Guide opens with a list of terms. A good use of this list is to see how many of the terms you can *write* without looking them up. We *purposefully* did *not* provide definitions because it is important for you to construct your own definitions.

If you do what has been suggested, you'll notice that you have been very active. You have a piece of scratch paper on your desk, and it should be filled with notes and structures.

Two other suggestions: First, *do not* underline (or highlight) large sections of the text! This represents an effort to be active, when in fact it represents in many cases *physical* activity coupled with *mental* passivity. If there is something you really want to emphasize, then by all means, highlight away! Otherwise, don't bother. It's a waste of time and effort. Second, if you outline the chapter, don't go into an "autopilot" mode; think about what you write, and don't passively copy sections of the text. Pretend you are a teacher and that you're writing the outline for someone who can't understand the text at all. This Study Guide has also provided chapter outlines for you. In most cases, topics have been reorganized and regrouped in these outlines so that you think about topics in a different way. For example, part of the outline might be called "Reactions;" another part, "Rules;" etc. This should be useful to you if you want to overview the chapter quickly, but is not meant to substitute for summaries that you write for yourself. Some students use these outlines for a "first pass" over a chapter before reading the chapter in detail.

Be organized. After you have studied a reaction, write it down in a general form for later review. One good way is to fold a sheet of paper in half. Write the reactants on the left and the products on the right. Write two arrows, one on each side. Use "R" groups or "dangling bonds" to make the reaction as general as possible. Let's illustrate. A review sheet for oxymercuration–reduction might look as shown in Fig. SG5.1 on the following page. (Some people prefer to use "flash cards," with one side of a reaction written on the front of the card, and the other on the back.) Each time you learn a new reaction, enter it on your sheet. Notice above that not only the individual reactions of oxymercuration and reduction have been entered, but also a summary reaction. The choice of three "R" groups is arbitrary; you might instead use an alkene starting material R—CH=CH_2 The point here is to use enough "R" groups that the regioselectivity

fold

$$R_2C{=}CH{-}R + Hg(OAc)_2 + H_2O \xrightarrow{THF/H_2O} \;\Big|\; \xrightarrow{THF/H_2O} R{-}C(R)(OH){-}CH(HgOAc){-}R + HOAc$$

$$R{-}C(R)(OH){-}CH(HgOAc){-}R \xrightarrow{NaBH_4/^{-}OH} \;\Big|\; \xrightarrow{NaBH_4/^{-}OH} R{-}C(R)(OH){-}CH(H){-}R$$

$$R_2C{=}CH{-}R \xrightarrow[2)\ NaBH_4/^{-}OH]{1)\ Hg(OAc)_2/H_2O} \;\Big|\; \xrightarrow[2)\ NaBH_4/^{-}OH]{1)\ Hg(OAc)_2/H_2O} R{-}C(R)(OH){-}CH_2{-}R$$

Figure SG5.1 Part of a typical reaction review sheet for Chapter 5.

of the reaction is apparent. Notice also that not all reactions are balanced; the goal is to focus on what happens to the organic compound. Each time you finish a study session, fold that page in half. With the left side in view, complete the right on a piece of scratch paper. Then, with the right side in view, complete the starting materials. (Notice that the arrow bearing the reaction conditions is repeated on both sides of the fold for this purpose.) You should find that this process helps you learn reactions both ways.This knowledge will become particularly valuable when you start combining reactions in multistep syntheses. Next, ask yourself why the reaction is reasonable. For the oxymercuration step, you should recognize that the reaction is an addition. Go through the mechanism. What is the electrophile? Why does it react with the alkene π electrons? Why is the reaction regioselective? (Note that the text does not cover the mechanism of the reduction step.) Then write a specific example of the reaction using R groups of your choice.

If you do this *every time you read,* and *if you understand what you read,* then what you are doing here is *not* memorizing the reactions, at least not in a rote fashion, because you will have already learned them by careful study. Rather, you are cataloging them for use as a future study tool. If you can't complete one side or the other of a reaction, look at the missing side. If it looks unfamiliar, return to the text and focus exclusively on the part that is unfamiliar.

Review frequently. Review your reaction sheet *every time you study!* Go back about three assignments for each review, dropping older material from your review. Then review the older material again once every week or two, or right before an exam or quiz. (Notice that this gives you something *active* to do in preparing for an examination.) Do the same thing with the list of terms. You will find that relatively few reactions and terms are covered in each assignment, and that learning each day's assignment takes relatively little time. If you get behind, however, the number of reactions and terms that must be learned will grow rapidly!

Many students and professors suffer from adherence to a theory that one of the author's colleagues calls the "immunization theory of learning." The student's version of this theory is: "I've studied this once, and therefore I should never have to see it again . . . *ever.*" The professor's version of the theory is, "Students studied this once, and therefore they should automatically know it. There is no need for me to review, and I should be annoyed if a student does not have total recall of everything he/she has ever been taught!" Those who adhere to these ideas deny a fundamental caveat about learning: *Continued reinforcement is one of the best ways to learn.* The author is frequently amazed how often students are frustrated because they didn't learn something on the first pass. You should *expect* to forget material, and you should expect to re-learn it, probably more than once. Each relearning, however, takes less time and brings with it deeper understanding. What would happen to the college basketball team if the coach taught them a play *once* and they never reviewed it before a game? What would happen to a musician who practiced a piece once before a concert? Why should studying chemistry (or anything else, for that matter) be any different?

Study in small chunks. Set realistic goals for each study session and pursue them regularly. If you can only afford six hours per week, it is far better to study six times per week for one hour than one time for six hours. If you can allow twelve hours per week, then break it into six two-hour sessions. Notice how this strategy is built into the discussion of the reaction sheet above: you should study your reaction sheet over and over again, but confine your concentration to relatively small parts of it.

If you have ever played an organized sport, or studied a musical instrument seriously, you should understand the "small chunk" strategy well. Does the coach of a winning team hold one ten-hour practice per week, or five two-hour practices? Why? Because the mental and physical learning capacities of most individuals erode significantly as fatigue sets in. Does a skilled performer on a musical instrument practice once per week for twelve hours, or six times per week for two hours?

What is remarkable is that when you *stay organized* and *study in small chunks*, your capacity for further study increases! Just as an athlete's endurance increases with regular workouts, your ability to concentrate for longer periods of time will increase. Just as an athlete's speed and skill increase with each workout, you will also find that you will become more efficient: you will accomplish much more in less time!

Obviously, the suggestions given here for learning reactions are not the only ones that will work. But if you are at a loss about how to study organic chemistry on your own, try these suggestions, and then refine them to suit your own needs. If you have some other great ideas, send them to the author! Perhaps they will appear in the *next* edition of this Study Guide!

✓5.5 Mechanism of Ozonolysis

You may have noticed that several mechanisms, such as bromonium ion formation, mercurinium ion formation, and (in Study Guide Link 5.2) hydroboration have been presented as stepwise processes to show how these mechanisms relate to mechanisms involving carbocations. Let us take this approach with ozonolysis.

First, let's look at the structure of ozone. The central oxygen, which is a positively charged electronegative atom, wants to accept electrons:

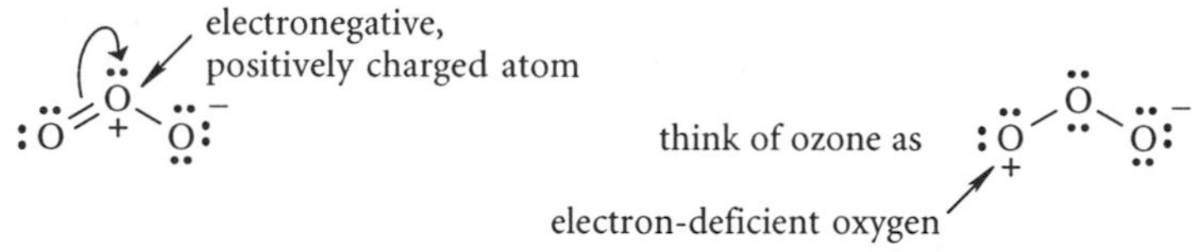

:Ö=Ȯ⁺—Ö:⁻ ← electronegative, positively charged atom

think of ozone as :Ö⁺—Ö—Ö:⁻ ← electron-deficient oxygen

As a result, one of the terminal oxygens is electron-deficient. This electron-deficient oxygen is a powerful Lewis acid, because there is a strong drive for an electron-deficient electronegative atom to complete its octet. If this oxygen accepts the π electrons of the alkene, a carbocation is generated:

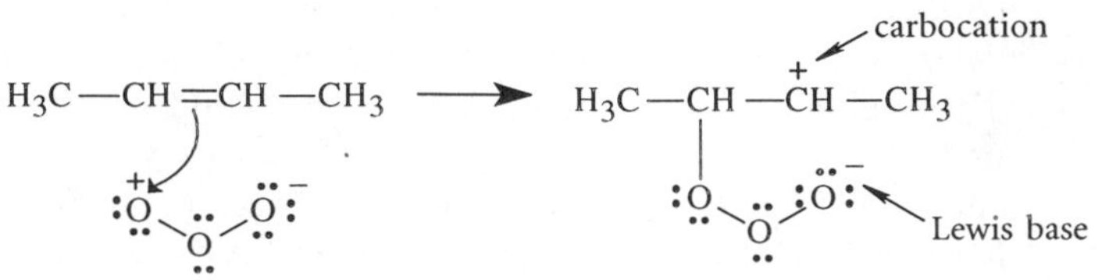

$$H_3C—CH{=}CH—CH_3 \longrightarrow H_3C—CH(—O—O—O^-)—\overset{+}{C}H—CH_3 \qquad \text{(SG5.3a)}$$

carbocation

Lewis base

The carbocation, a Lewis acid, is attacked by the negatively charged oxygen, which is a Lewis base. This completes the addition.

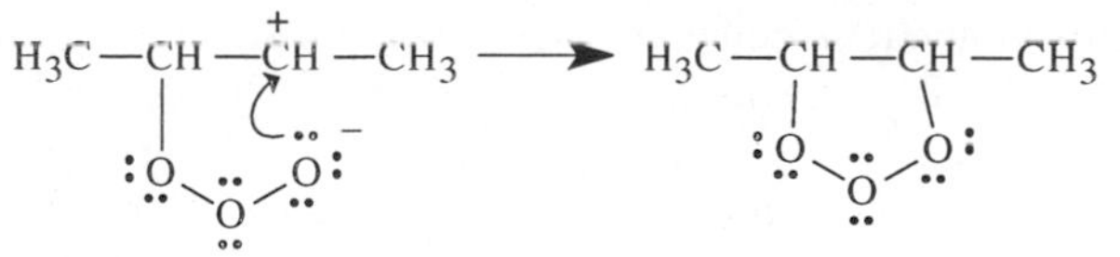

H$_3$C—CH—$\overset{+}{C}$H—CH$_3$ ⟶ H$_3$C—CH—CH—CH$_3$ (SG5.3b)

Even though the mechanism is concerted, thinking of it as a two-step process, as shown above, helps to understand why the reaction takes place.

The following discussion gives more detail on the conversion of the molozonide to the ozonide. The molozonide decomposes in a "reverse cycloaddition."

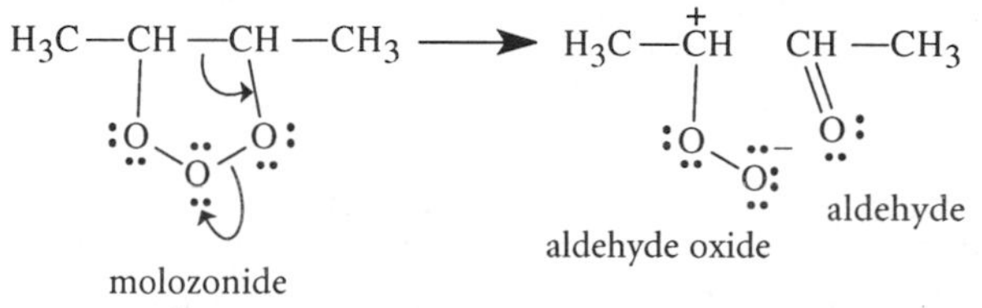

H$_3$C—CH—CH—CH$_3$ ⟶ H$_3$C—$\overset{+}{C}$H CH—CH$_3$ (SG5.4a)

molozonide aldehyde oxide aldehyde

This occurs because the O—O bond is a very weak bond, and oxygen, an electronegative atom, readily accepts an electron pair. Notice that the "aldehyde oxide" is a molecule a lot like ozone: it has an electron-deficient (Lewis acid) end, that is, the carbocation, and an electron-rich (Lewis base) end, that is, the —O$^-$. Thus, it should not be surprising that it *reacts* a lot like ozone. It simply adds to the double bond of the aldehyde. Now, there are two ways that this addition can occur. One way is simply the reverse of Eq. SG5.4a. In the other way, the aldehyde adds in the opposite sense:

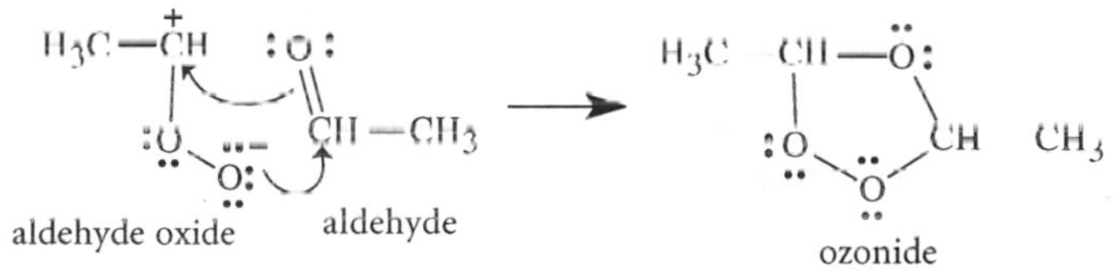

H$_3$C—$\overset{+}{C}$H CH—CH$_3$ ⟶ H$_3$C CH—O CH CH$_3$ (SG5.4b)

aldehyde oxide aldehyde ozonide

In summary, the molozonide-to-ozonide conversion involves a fragmentation-cycloaddition sequence of reactions.

✓5.6 Why Are Free-Radical Reactions Usually Chain Reactions?

Some students have problems proposing mechanisms for free-radical reactions because they do not understand why free-radical reactions are almost always chain reactions. For example, it is not uncommon for a student to propose the following as a mechanism for free-radical addition of HBr to an alkene:

R—CH=CH—R + H• + •Br ⟶ R—CH(H)—CH(Br)—R

There is certainly nothing wrong with the fishhook formalism, but this mechanism cannot possibly be right. Why not?

First, where do the H• and the Br• come from? Although the source of Br• is shown in the text, there is no readily apparent source of •H. Hence, only one of the two radicals is produced on initiation.

Second, this reaction destroys free radicals. If this were the mechanism, there would have to be a separate initiation step for every molecule of product produced. This, in turn, would require a stoichiometric amount of initiator rather than the 1% or so that is commonly used.

Finally, remember that free radicals are present in very low concentration. The mechanism above requires a collision between *three* molecules, two of which are present in *minuscule* concentrations. The probability of such a collision is infinitely small. It is much more likely that a single radical (such as Br•) will collide with a reactant (such as an alkene molecule), as in the propagation steps shown in text Eqs. 5.43a and 5.43b on text p. 181. Recall from the text that termination reactions also require the collision of two free radicals, and these reactions do not compete with propagation steps until all the reactants are exhausted. Hence, the mechanism above should be unlikely for a similar reason.

The message is this: When you think about the mechanism of a free-radical reaction, you should always think about it in terms of a chain process, specifically, its propagation steps. Problem 5.32 (text p. 197) allows you to test whether you understand these points.

✓5.7 Bond Dissociation Energies and Heats of Reaction

Study Problem 5.3 in the text shows how bond dissociation energies can be used to calculate heats of reactions: subtract the bond dissociation energies of the bonds formed from those of the bonds broken. This study guide link shows why this process works.

Essentially, this is a Hess's law calculation. This can be illustrated by using bond dissociation energies to calculate the $\Delta H°$ of the following gas-phase reaction:

$$\cdot CH_3 + Cl_2 \longrightarrow Cl{-}CH_3 + Cl\cdot \qquad \text{(SG5.5)}$$

The term "bond dissociation energy" will be abbreviated with the acronym "BDE" for convenience. The bond formed is a H_3C—Cl bond, with BDE = 356 kJ/mol (85 kcal/mol). The bond broken is a Cl—Cl bond, with BDE = 247 kJ/mol (59 kcal/mol). The $\Delta H°$ for the reaction, then, is 247 – 356 = –109 kJ/mol, or 59 – 85 = –26 kcal/mol.

The reaction in Eq. SG5.5 can be viewed as the sum of the following two reactions:

$$Cl_2 \longrightarrow \cancel{2}Cl\cdot \qquad \text{(SG5.6a)}$$

$$\cdot CH_3 + \cancel{Cl\cdot} \longrightarrow Cl{-}CH_3 \qquad \text{(SG5.6b)}$$

$$\textit{Sum:}\quad \cdot CH_3 + Cl_2 \longrightarrow Cl{-}CH_3 + Cl\cdot \qquad \text{(SG5.5)}$$

The $\Delta H°$ of Eq. SG5.5 is then the sum of the $\Delta H°$ values of the two reactions, SG5.6a and SG5.6b:

	Reaction	$\Delta H°$
	$Cl_2 \longrightarrow \cancel{2}Cl\cdot$	BDE of Cl_2
	$\cdot CH_3 + \cancel{Cl\cdot} \longrightarrow Cl{-}CH_3$	–BDE of Cl—CH_3
Sum:	$\cdot CH_3 + Cl_2 \longrightarrow Cl{-}CH_3 + Cl\cdot$	(BDE of Cl_2) – (BDE of Cl—CH_3)

As shown above, the $\Delta H°$ of the first reaction, by definition, is equal to the BDE of Cl_2. Because the second reaction is simply the bond dissociation of H_3C—Cl written in reverse, its $\Delta H°$ is the negative BDE of H_3C—Cl. Thus, the total $\Delta H°$ is then the BDE of the bond broken less the BDE of the bond formed.

✓5.8 Solving Structure Problems

Before working Problem 5.38 in the text, be sure you have read Study Guide Link 4.8 (also called "Solving Structure Problems").

Problem 5.38 contains a lot of information. The temptation is to solve this problem from the top down. It bears repeating that this is not usually the best way to deal with this type of problem. What is known about *A*? Its empirical formula can be determined from the elemental analysis. (Do this now.) The reaction with Br_2, the fact that it undergoes hydrogenation, and the fact that it reacts with ozone all suggest that *A* is an alkene. Immediately you should look for a *structure*. There are two structures in the problem: one given in words, the other explicitly. The hydrogenation product of *A* is 1-isopropyl-4-methylcyclohexane:

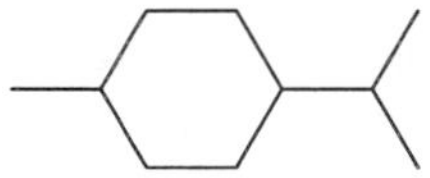

1-isopropyl-4-methylhexane

This tells us *immediately* the carbon skeleton of *A*. Only the double bonds are missing. This structure further implies that *A* contains a ring. Since the empirical formula of *A* is known, the number of carbons (10) in this hydrogenation product also reveals the molecular formula of *A*. (What is the formula?) Since the molecular formula of *A* is known, its unsaturation number can be calculated. Remember that the ring accounts for one degree of unsaturation. How many double bonds does compound *A* have?

The second structure given in the problem is the ozonolysis product. Remember that ozonolysis produces *two* C=O groups for every double bond in the molecule. However, in this ozonolysis product, there are *three* C=O groups. But there are also *nine carbons*. Evidently, one carbon and one C=O group are missing. This means that the missing carbon is part of the missing C=O group, and that a one-carbon fragment was liberated by ozonolysis as a second product which was evidently not recovered. This conclusion is consistent with the unsaturation number, which indicates that there are two double bonds. Two double bonds, after ozonolysis, gives four C=O groups.

This information establishes a partial structure for *A:* compound *A* must contain a $=CH_2$ group. Such a group can come from the known carbon skeleton in only two ways:

or

1 *2*

Since there are two double bonds in *A*, one additional double bond must be provided to convert either *1* or *2* into possible structures for *A*. From the ozonolysis product, this double bond must be *within* the ring, because ozonolysis opens the ring. Now it is time to *write all the possibilities.* (Recall that this point was stressed in Study Guide Link 4.8.) From compound *1* the only possibilities for *A* are the following:

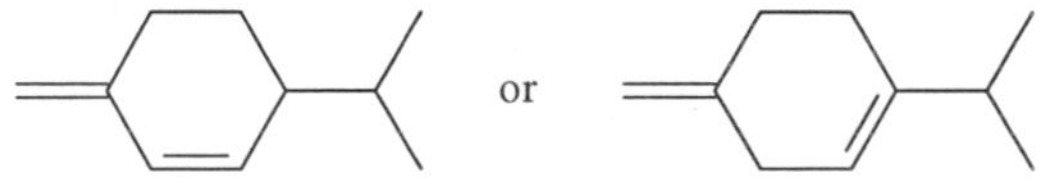

or

1-1 *1-2*

From compound *2* the only possibilities are the following:

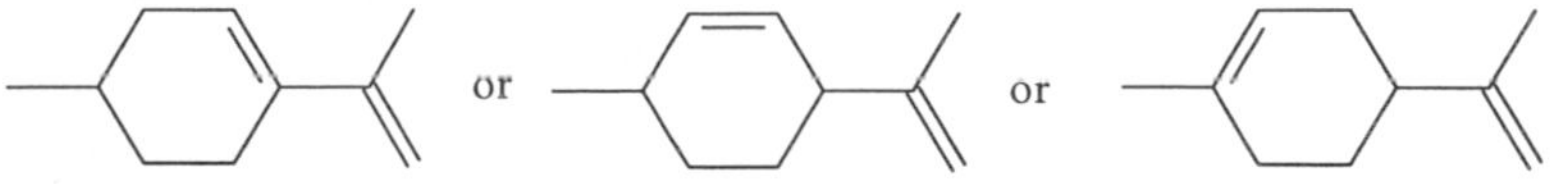

or or

2-1 *2-2* *2-3*

There are two ways to finish the problem. One is to apply ozonolysis to each of the five structures and see which one gives the observed product. Notice that the observed product contains two groups of the following type:

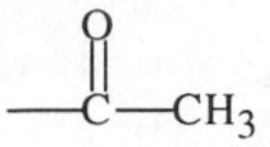

$$-\overset{\overset{\displaystyle O}{\|}}{C}-CH_3$$

Which structures would give *two* of these groups after ozonolysis? You should finish the problem by deciding between these.

The second way to finish the problem is to reverse ozonolysis mentally in all possible ways. The carbons involved in double bonds have to be the ones "labeled" with oxygens. These are carbons 1, 2, and 3 in the following structure of the ozonolysis product:

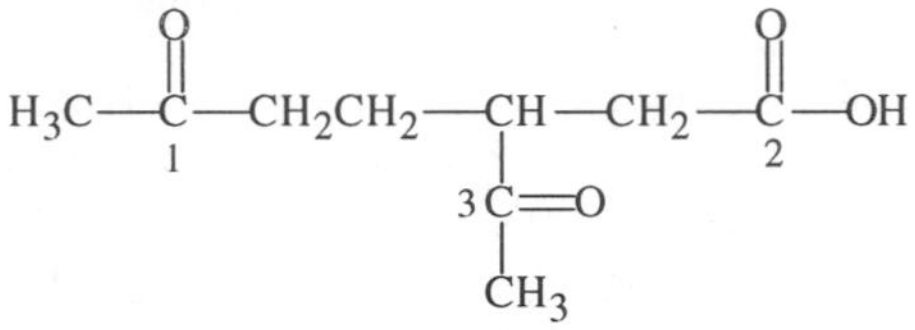

$$H_3C-\underset{1}{\overset{\overset{\displaystyle O}{\|}}{C}}-CH_2CH_2-\underset{\underset{\displaystyle CH_3}{|}}{\underset{3\,C=O}{\underset{|}{CH}}}-CH_2-\underset{2}{\overset{\overset{\displaystyle O}{\|}}{C}}-OH$$

(And remember, one carbon is missing, which will be referred to as carbon-4.) Once again, *write out all possibilities.* All possible connections are:

1. C1 with C2, C3 with C4
2. C1 with C3, C2 with C4
3. C1 with C4, C2 with C3

And all but one can be ruled out by recalling that compound *A* has to contain a six-membered ring. The only structure for *A* that has a six-membered ring is obtained by connecting C1 to C2. Hence, C3 and C4 are also connected. You should finish reconstructing compound *A*. (The final solution is in the Additional Solutions part of this chapter.)

You can see that working a problem such as this is like solving a big puzzle. You need to understand the reactions involved, but you need something else: the ability to reason analytically. If you keep working problems of this sort, you can learn to develop this reasoning ability. This ability can help you not only in chemistry; it will help you in your career (whatever it is)—and it can even help you solve ordinary day-to-day problems.

SOLUTIONS TO PROBLEMS

Solutions to In-Text Problems

5.1 (a) The mechanism involves a bromonium ion intermediate, as shown in Eqs. 5.4

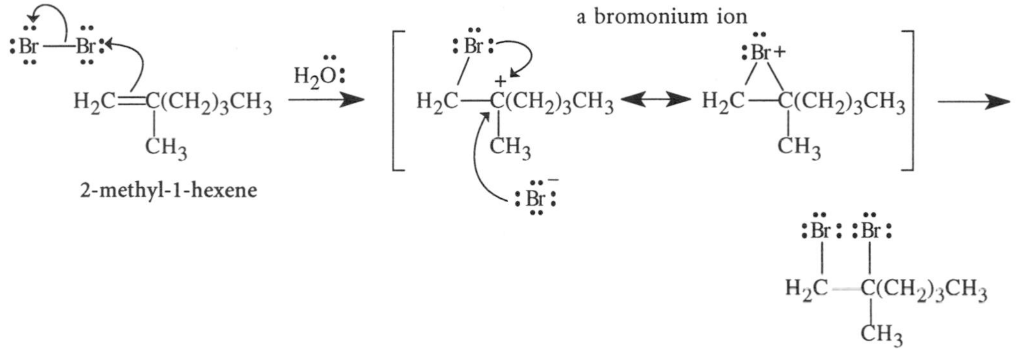

a bromonium ion

$H_2\ddot{O}$:

2-methyl-1-hexene

(b) The bromonium ion formed in part (a) underges attack by water to give the product.

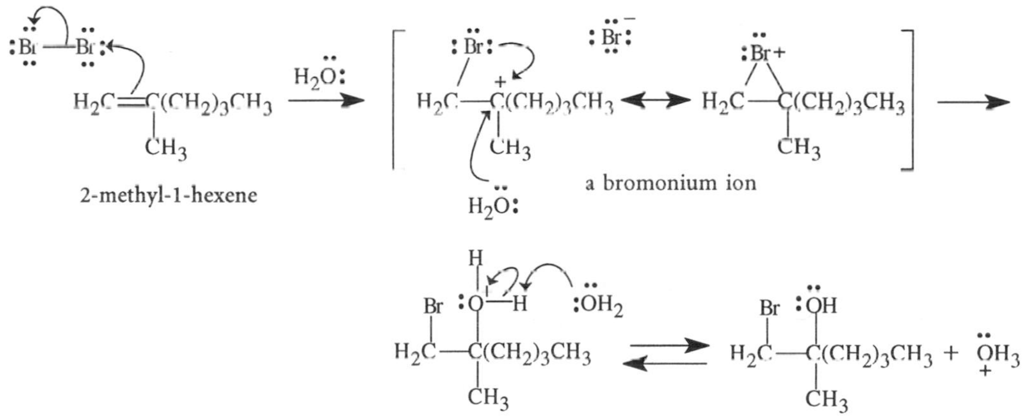

$H_2\ddot{O}$:

2-methyl-1-hexene

a bromonium ion

(c) An iodonium ion analogous to the bromonium ion in parts (a) and (b) undergoes attack by azide ion.

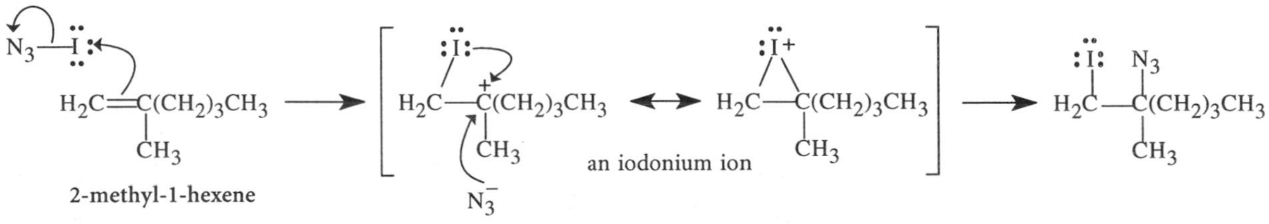

2-methyl-1-hexene

an iodonium ion

5.2 The alkene that would react with Cl_2 in H_2O to give the chlorohydrin *B* in Study Problem 5.1:

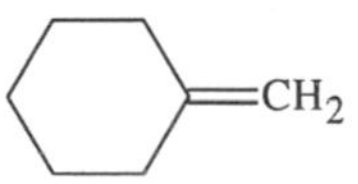

$=CH_2$ methylenecyclohexane

5.3 (a)

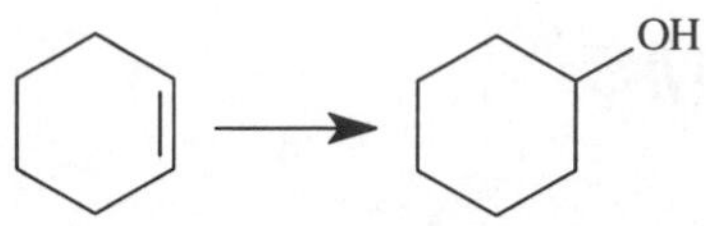

cyclohexene → cyclohexanol (OH)

cyclohexene

(b)

$(CH_3)_2C{=}CHCH_2CH_3 \longrightarrow (CH_3)_2C(OH)CH_2CH_2CH_3$

2-methyl-2-pentene

(c) In this case two products are formed because the carbons of the alkene double bond have the same number of alkyl substituents (one); consequently, there is no basis for significant regioselectivity.

$H_3C(H)C{=}C(H)CH(CH_3)_2 \longrightarrow CH_3CH(OH)CH_2CH(CH_3)_2 + CH_3CH_2CH(OH)CH(CH_3)_2$

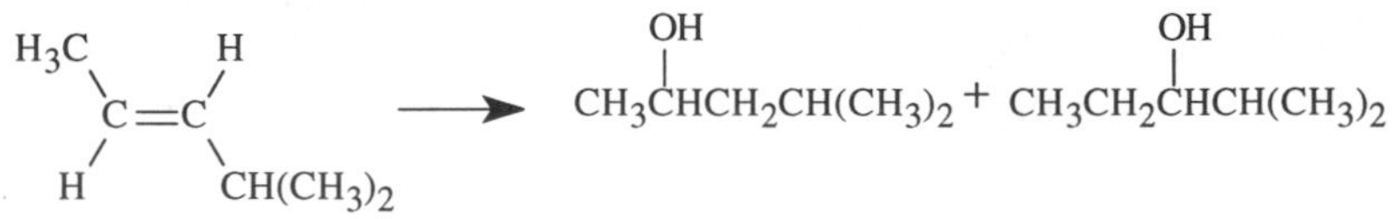

trans-4-methyl-2-pentene

5.4 (a) Either of the following alkenes would give the alcohol in part (a) as the major product:

$CH_3CH_2C({=}CH_2)CH_2CH_3$ or $CH_3CH_2C(CH_3){=}CHCH_3$

2-ethyl-1-butene

3-methyl-2-pentene (either (*E*) or (*Z*) stereoisomer)

(b)

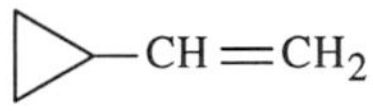

cyclopropyl$-CH{=}CH_2$ cyclopropylethylene

5.5 The products of hydroboration-oxidation are the same as the ones in parts (a) and (c) of Problem 5.3. In (a), the alkene is symmetrical; consequently, regioselectivity has no meaning for this alkene. In part (c), the carbons of the double bond have the same number of alkyl substituents; consequently, there is no reason to expect significant regioselectivity.

(b)

$(CH_3)_2C{=}CHCH_2CH_3 \longrightarrow (CH_3)_2CHCH(OH)CH_2CH_3$

2-methyl-2-pentene

5.6 The products are the same for part (a) of both problems because both alkenes are symmetrical; consequently, regioselectivity has no meaning. For part (c) of both problems, the carbons of the double bond have the same number of alkyl substituents; consequently, there is no basis for expecting substantial regioselectivity in either reaction, although there might be small differences in the relative amounts of corresponding products. In part (b), the products are different because the two reactions give opposite regioselectivities when the number of branches on each carbon of the double bond is different.

5.7 (a) The alkene required is methylenecyclohexane. (The structure is in the solution to Problem 5.2 above.)

(b)

$CH_3CH_2C(CH_3){=}CHCH_3$

3-methyl-2-pentene (either *E* or *Z* stereoisomer)

5.8 (a) Because *cis*-2-butene is a symmetrical alkene, it reacts to give the same product in either oxymercuration-reduction or hydroboration-oxidation.

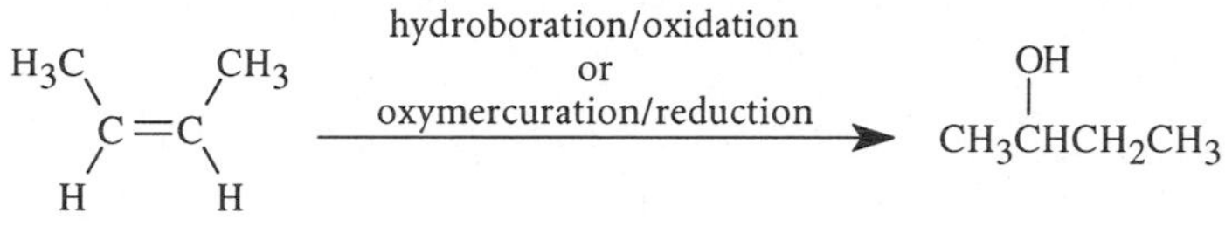

$$\underset{\textit{cis}\text{-2-butene}}{H_3C(H)C{=}C(H)CH_3} \xrightarrow[\text{oxymercuration/reduction}]{\text{hydroboration/oxidation or}} CH_3\overset{OH}{\overset{|}{C}}HCH_2CH_3$$

(b) Because 1-methylcyclohexene has different numbers of alkyl branches at the carbons of its double bond, it gives different products in the two reactions:

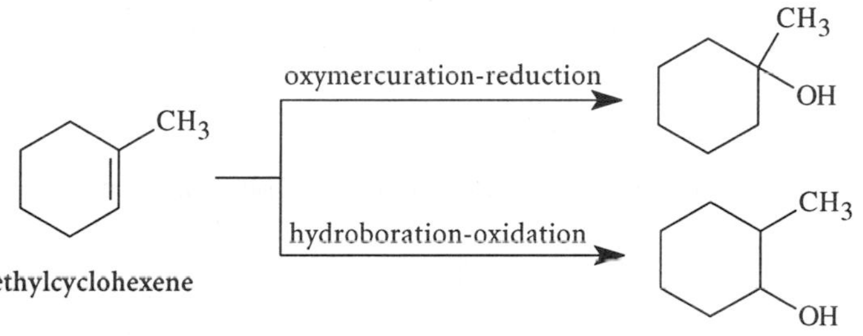

1-methylcyclohexene —oxymercuration-reduction→ 1-methylcyclohexanol (CH_3 and OH on the same carbon)

1-methylcyclohexene —hydroboration-oxidation→ 2-methylcyclohexanol (CH_3 and OH on adjacent carbons)

5.9 (a)

$$\underset{\text{3-ethyl-2-pentene}}{CH_3CH_2-\underset{C_2H_5}{\underset{|}{C}}{=}CHCH_3} \xrightarrow[\text{THF/H}_2\text{O}]{\text{Hg(OAc)}_2} \xrightarrow{\text{NaBH}_4} (C_2H_5)_3C{-}OH$$

(b)

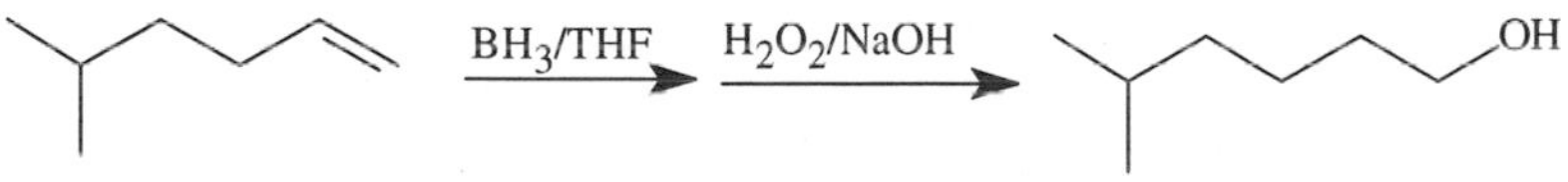

5-methyl-1-hexene $\xrightarrow{\text{BH}_3\text{/THF}}\xrightarrow{\text{H}_2\text{O}_2\text{/NaOH}}$ 5-methyl-1-hexanol (OH)

(c) Although hydroboration-oxidation is used in the reaction below, oxymercuration-reduction would in principle work equally well because the alkene is symmetrical, and regioselectivity is therefore not an issue.

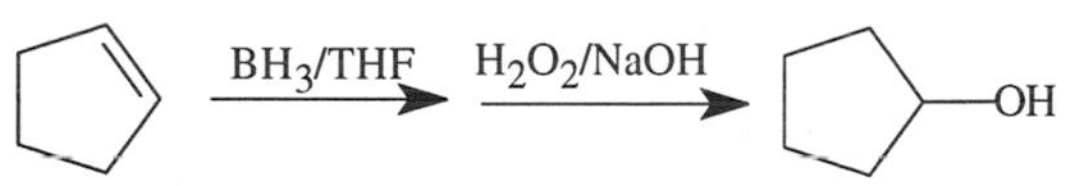

cyclopentene $\xrightarrow{\text{BH}_3\text{/THF}}\xrightarrow{\text{H}_2\text{O}_2\text{/NaOH}}$ cyclopentanol (—OH)

5.10 (a)

$$\underset{\text{3-methyl-2-pentene}}{CH_3CH{=}\underset{CH_3}{\underset{|}{C}}CH_2CH_3} \xrightarrow{O_3}\xrightarrow{(CH_3)_2S} CH_3CH{=}O + CH_3\overset{O}{\overset{\|}{C}}CH_2CH_3$$

(b)

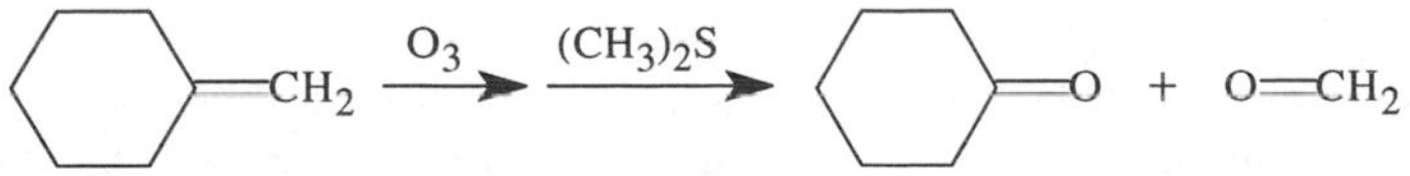

methylenecyclohexane (=CH_2) $\xrightarrow{O_3}\xrightarrow{(CH_3)_2S}$ cyclohexanone (=O) + $O{=}CH_2$

(c)

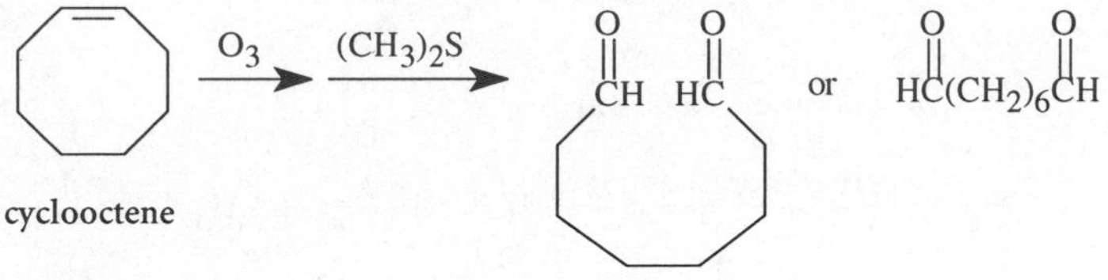

cyclooctene $\xrightarrow{O_3}$ $\xrightarrow{(CH_3)_2S}$ (cyclic dialdehyde: O=CH HC=O bridged by ring) or $HC(=O)(CH_2)_6C(=O)H$

(d) 2-Methylpentane, $(CH_3)_2CHCH_2CH_2CH_3$, is an alkane; because it has no double bond, it does not undergo ozonolysis.

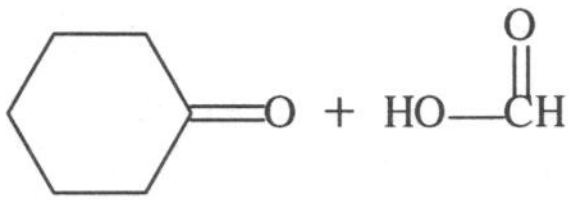

5.11 (a) CH_3COH (with C=O) + $CH_3CCH_2CH_3$ (with C=O)

(b) cyclohexanone + HO—CH (with C=O)

(c) $HOC(CH_2)_6COH$ (each C with C=O)

(d) 2-Methylpentane is an alkane; because it has no double bond, it does not undergo ozonolysis.

5.12 (a) $CH_3CH_2CH_2CH{=}CHCH_2CH_2CH_3$

4-octene
(either *E* or *Z* stereoisomer)

(b) CH_3

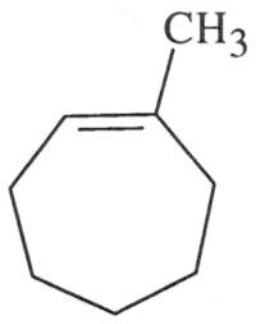

1-methylcycloheptene

(c)

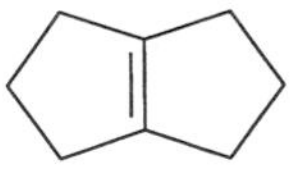

5.13 As the solution to Problem 5.12(a) illustrates, *E* and *Z* isomers of an alkene give the same ozonolysis products. Therefore, ozonolysis cannot be used to determine the stereochemistry at the alkene double bond.

5.14 (a) $2\ (CH_3)_3C\cdot$

(b) $(CH_3)_2\dot{C}—CH_2$ with :Br: on CH_2

(c) $H_2\dot{C}—CH{=}CH—CH{=}CH_2$

(d) $R\cdot\ +\ H_2C{=}CH_2$

Part (c) illustrates that the fishhook notation can be used to derive resonance structures of free radicals.

5.15 (a) This reaction is heterolytic, because the reaction requires movement of an electron *pair.*

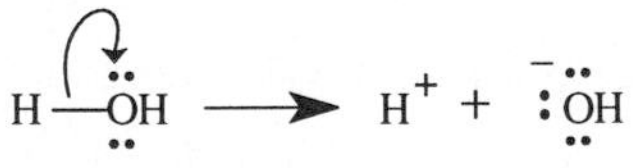

$$H—\ddot{O}H \longrightarrow H^+ + {:}\ddot{O}H^-$$

(b) This reaction is homolytic, because the reaction involves movement of single electrons.

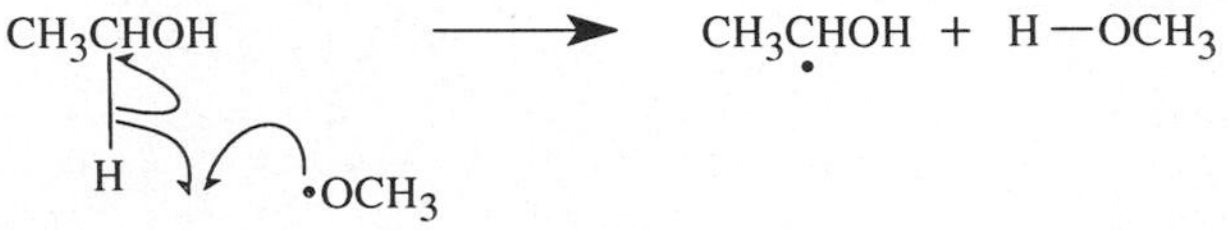

$$CH_3CHOH\ (\text{with H}) + \cdot OCH_3 \longrightarrow CH_3\dot{C}HOH + H—OCH_3$$

Notice that when dealing with radical species it is necessary only to write the unpaired electron; unshared pairs (for example, those on oxygen in the previous solution) need not be written explicitly when they are not involved in the reaction.

5.16 (a) The initiation step is the formation of bromine atoms by abstraction of a hydrogen atom from HBr by the initiating radical. (The formation of this radical is shown in Eq. 5.41, text p. 180.)

$$NC-\dot{C}(CH_3)_2 + H-Br \longrightarrow NC-C(CH_3)_2H + \cdot Br$$

The propagation steps are as follows:

$$\text{cyclohexene} + \cdot Br \longrightarrow \text{2-bromocyclohexyl radical} \xrightarrow{HBr} \text{bromocyclohexane} + \cdot Br$$

(b) Recombination (termination) products can result from the recombination of any of the radical intermediates involved in the propagation steps. These are

Br Br (bi(2-bromocyclohexyl)) Br—Br (1,2-dibromocyclohexane) Br Br

5.17 In propagation step (1), the initating free radical adds to ethylene:

initiating free radical

$$H_2C{=}CH_2 + \cdot Br \longrightarrow H_2\dot{C}-CH_2Br$$

A

In propagation step (2), radical *A* reacts with Br_2 to generate the addition product and another Br·, which propagates the chain.

$$Br-Br + H_2\dot{C}-CH_2Br \longrightarrow BrH_2C-CH_2Br + Br\cdot$$

A addition product

Radical *A* could form the product by reacting with another initiating radical Br·. But this does not happen! The reason is that the initiating radical is present in trace amounts, whereas the reactant Br_2 is present in large amounts. Be sure to read Study Guide Link 5.6 if you missed this point.

5.18 (a)

$$H_2C{=}CHCH_2CH_2CH_3 + HBr \xrightarrow{\text{peroxides}} BrCH_2CH_2CH_2CH_2CH_3$$

1-pentene

(b) In this case, both of the two possible reaction pathways involve secondary free radicals. Recall, however, from the text that the regioselectivity of free-radical HBr addition is controlled in part by steric factors. Hence, the Br goes to the less congested side of the double bond.

$$(H_3C)HC{=}CH(C(CH_3)_3) + HBr \xrightarrow{\text{peroxides}} CH_3CH(Br)CH_2C(CH_3)_3 + CH_3CH_2CH(Br)C(CH_3)_3$$

(*E*)-4,4-dimethyl-2-pentene (major product)

5.19 (a) One H—Cl bond (431 kJ/mol) and one H_3C—Cl bond (356 kJ/mol) are formed; and one H_3C—H bond (439 kJ/mol) and one Cl—Cl bond (247 kJ/mol) are broken. As shown by Eq. 5.57 on text p. 188, subtract the bond dissociation energies of the bonds formed from those of the bonds broken to obtain the desired estimate for $\Delta H°$.

$$\Delta H° = 247 + 439 - (431 + 356) = -101 \text{ kJ/mol}$$

(b) Two Cl—CH_2 bonds are formed (335 kJ/mol each); a Cl—Cl bond (247 kJ/mol) and a carbon-carbon π bond ($\approx$ 270 kJ/mol) are broken.

$$\Delta H° = 247 + 270 - 2(335) = -153 \text{ kJ/mol}$$

The value used in this solution for the bond dissociation energy of the π bond (270 kJ/mol) is in the middle of the range of values given in Table 5.3. If you used a different value, your answer will, of course, be correspondingly different.

5.20 (a) Calculation of the $\Delta H°$ values involves subtracting the bond dissociation energies of the bonds formed from those of the bonds broken, as in the previous problem. In the second propagation step, a secondary C—H bond is formed, and an H—Br bond is broken.

$$\Delta H° = 368 - 410 = -42 \text{ kJ/mol}$$

(b) The corresponding calculation of $\Delta H°$ for HCl is

$$\Delta H° = 431 - 410 = +21 \text{ kJ/mol}$$

(c) As noted in the text, endothermic propagation steps are much slower than exothermic ones. Because the second propagation step of the free-radical addition of HCl is endothermic, it is to slow to compete with the very exothermic recombination reactions that effectively shut down the chain mechanism.

5.21 The second propagation step in which Br is abstracted rather than H involves formation of a secondary C—Br bond rather than a secondary C—H bond; the bond broken (H—Br) is the same. Hence, we use the calculation in the solution of Problem 5.20(a), except that the bond dissociation energy of a C—Br bond is substituted for that of a C—H bond. The closest value in Table 5.3 is the bond dissociation energy of CH_3CH_2—Br, 284 kJ/mol. The required $\Delta H°$ is therefore

$$\Delta H° = 368 - 284 = +84 \text{ kJ/mol}$$

If anything, the bond dissociation energy of a secondary C—Br bond will be even lower than 284 kJ/mol (why?). Hence, abstraction of Br is a very *unfavorable* process energetically ($\Delta H° >> 0$), whereas the abstraction of H is a very favorable process ($\Delta H° < 0$). As is usually the case in free-radical reactions, the relative enthalpies of the processes governs their relative rates. Hence, abstraction of H is *much* more favorable, and therefore is much faster, than abstraction of Br.

5.22 The polymer structure is

$$\left(\underset{\substack{|\\ Cl}}{CH} - CH_2 \right)_n$$ poly(vinyl chloride)

Solutions to Additional Problems

5.23 (a) Br, C_2H_5, Br (b) C_2H_5, O, O, O (c) O, C—C_2H_5, C—H, O (d) O, C—C_2H_5, C—OH, O (e) CO_2 + H_2O (f) Br, C_2H_5

(g) OH, C_2H_5, I (h) C_2H_5 (i) C_2H_5, Br (j) C_2H_5, 3, B (k) C_2H_5, OH

(l) OH, C_2H_5, HgOAc (m) OH, C_2H_5 (n) I, C_2H_5 (o) I, C_2H_5

Notice in part (o) that there is no "peroxide effect" on HI addition; that is, free radical initiators such as peroxides or AIBN have no effect on the regioselectivity of HI addition.

5.24 (a) Any symmetrical alkene with six carbons is an acceptable answer. Three examples are the following:

—CH_3 $CH_3CH_2CH{=}CHCH_2CH_3$

(b) To answer this, you have to imagine double bonds at all possible places in the same carbon skeleton.

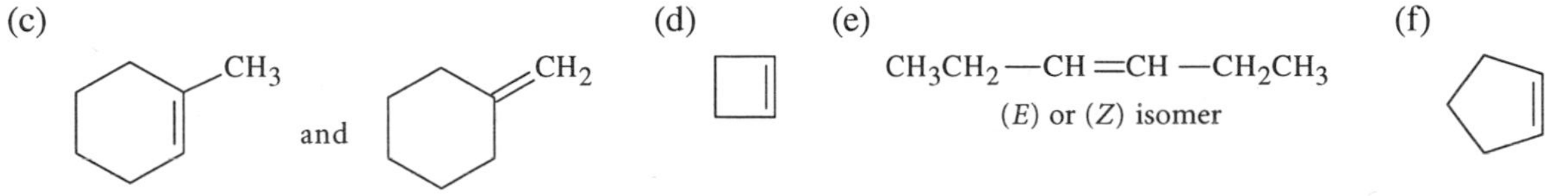

(A double bond at any other position gives a structure identical to one of these.)

(c) CH_3 and CH_2 (d) (e) $CH_3CH_2—CH{=}CH—CH_2CH_3$ (*E*) or (*Z*) isomer (f)

5.25 (a) CO_2H, CO_2H (b) O, $CH{=}O$ (c) $=CH_2$

5.26 (a)

$$CH_3CH_2{-}\underset{\displaystyle CH_2CH_3}{\underset{|}{C}}{=}CHCH_3 \xrightarrow[\text{2) NaBH}_4]{\text{1) Hg(OAc)}_2/\text{H}_2\text{O}} CH_3CH_2{-}\overset{\displaystyle OH}{\overset{|}{\underset{\displaystyle CH_2CH_3}{\underset{|}{C}}}}{-}CH_2CH_3$$

Remember that reaction schemes in which the reagents are numbered over the arrow mean two *separate* reactions. Thus, in the reaction above, the starting alkene reacts with $Hg(OAc)_2$ and water; then, in a *separate reaction*, $NaBH_4$ is added. Because there are two reactions, the intermediate compound that reacts with $NaBH_4$ is not shown; what is it? When you write synthetic schemes this way, be sure to number the reagents. If you were to write the following:

$$CH_3CH_2{-}\underset{\displaystyle CH_2CH_3}{\underset{|}{C}}{=}CHCH_3 \xrightarrow[\text{NaBH}_4]{\text{Hg(OAc)}_2/\text{H}_2\text{O}} CH_3CH_2{-}\overset{\displaystyle OH}{\overset{|}{\underset{\displaystyle CH_2CH_3}{\underset{|}{C}}}}{-}CH_2CH_3$$

it would mean that the $Hg(OAc)_2/H_2O$ and $NaBH_4$ are added to the reaction mixture at the same time. This would *not* give the desired product.

(b)

$$H_2C{=}CHCH_2CH_2CH_2CH_3 \xrightarrow[\text{2) NaBH}_4]{\text{1) Hg(OAc)}_2/\text{H}_2\text{O}} CH_3\overset{\displaystyle OH}{\overset{|}{C}}HCH_2CH_2CH_2CH_3$$

(c)

$$H_2C{=}CHCH_2CH_2CH_2CH_3 \xrightarrow[\text{2) H}_2\text{O}_2/\text{NaOH}]{\text{1) BH}_3/\text{THF}} HOCH_2CH_2CH_2CH_2CH_2CH_3$$

(d)

$$H_2C{=}CHCH_2CH_2CH_3 \xrightarrow{\text{HBr, peroxides}} BrCH_2CH_2CH_2CH_2CH_3$$

(e)

$$CH_3CH{=}\underset{\displaystyle CH_3}{\underset{|}{C}}CH_2CH_3 \xrightarrow{Br_2} CH_3\overset{\displaystyle Br}{\overset{|}{C}}H{-}\overset{\displaystyle Br}{\overset{|}{\underset{\displaystyle CH_3}{\underset{|}{C}}}}CH_2CH_3$$

(f)

$$CH_3CH_2\underset{\displaystyle CH_2}{\underset{\|}{C}}CH_2CH_3 \xrightarrow{Br_2/H_2O} CH_3CH_2\overset{\displaystyle OH}{\overset{|}{\underset{\displaystyle CH_2Br}{\underset{|}{C}}}}CH_2CH_3$$

(g)

$$\text{1-methylcyclohexene} \xrightarrow[\text{2) (CH}_3)_2\text{S}]{\text{1) O}_3} H{-}\overset{\displaystyle O}{\overset{\|}{C}}(CH_2)_4\overset{\displaystyle O}{\overset{\|}{C}}{-}CH_3$$

(h)

$$\text{cyclopentene} \xrightarrow[\text{2) H}_2\text{O}_2/\text{H}_2\text{O}]{\text{1) O}_3} HO{-}\overset{\displaystyle O}{\overset{\|}{C}}(CH_2)_3\overset{\displaystyle O}{\overset{\|}{C}}{-}OH$$

(i)

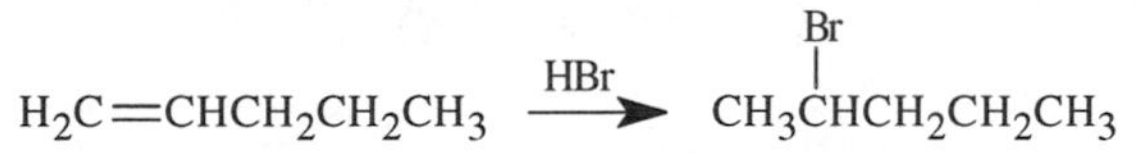

$$H_2C{=}CHCH_2CH_2CH_3 \xrightarrow{HBr} CH_3\underset{}{\overset{Br}{\overset{|}{C}}}HCH_2CH_2CH_3$$

(j)

$$CH_3CH_2{-}\underset{\underset{CH_2CH_3}{|}}{C}{=}CHCH_3 \xrightarrow{H_2,\ Pt/C} CH_3CH_2\underset{\underset{CH_2CH_3}{|}}{C}HCH_2CH_3$$

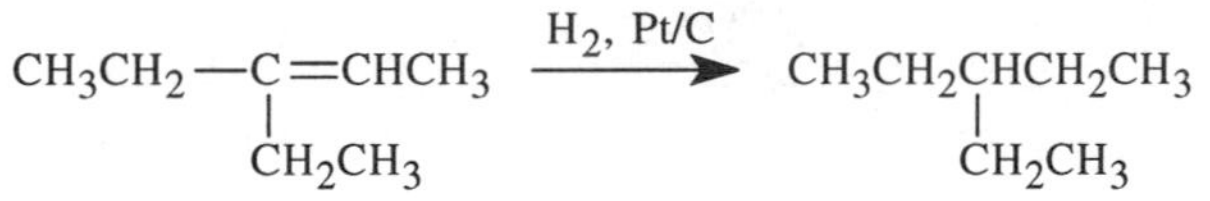

(The starting material for this transformation could be any alkene with the same carbon skeleton.)

5.27 (a) Given that this is hydroboration-oxidation, the source of deuterium must be BD_3.

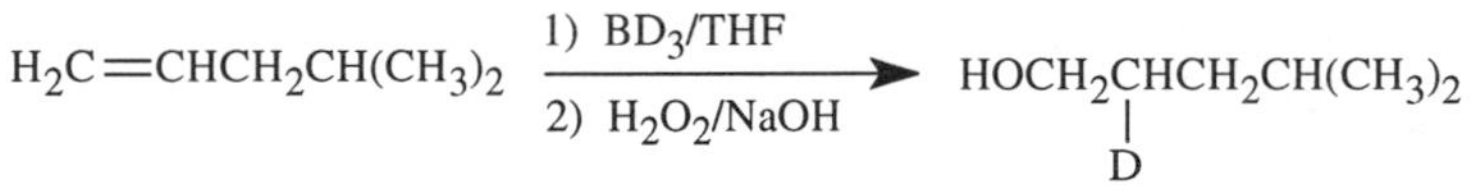

$$H_2C{=}CHCH_2CH(CH_3)_2 \xrightarrow[\text{2) } H_2O_2/NaOH]{\text{1) } BD_3/THF} HOCH_2\underset{\underset{D}{|}}{C}HCH_2CH(CH_3)_2$$

(b) Since $NaBH_4$ provides the hydrogen that is changed to deuterium, then $NaBD_4$ is the required reagent.

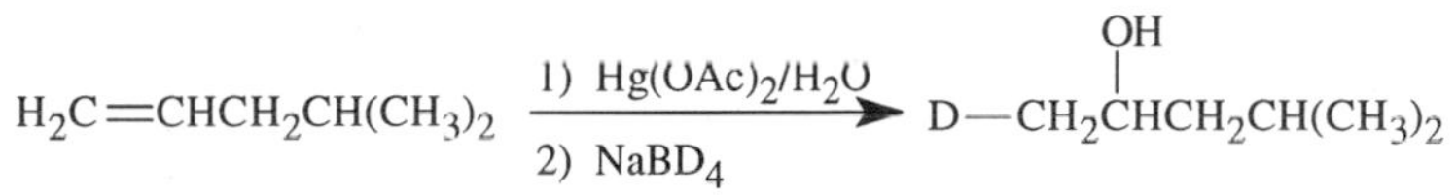

$$H_2C{=}CHCH_2CH(CH_3)_2 \xrightarrow[\text{2) } NaBD_4]{\text{1) } Hg(OAc)_2/H_2O} D{-}CH_2\overset{\overset{OH}{|}}{C}HCH_2CH(CH_3)_2$$

5.28 (a) In an inert solvent, the only nucleophile available to attack the bromonium ion intermediate is bromide ion. Hence, the product is the dibromoalkane:

$$H_2C{=}\underset{\underset{CH_3}{|}}{C}CH_2CH_3 \xrightarrow[CH_2Cl_2]{Br_2} BrCH_2\overset{\overset{Br}{|}}{\underset{\underset{CH_3}{|}}{C}}CH_2CH_3$$

2-methyl-1-butene → 1,2-dibromo-2-methylbutane

(b) When water is present, H_2O is available as a nucleophile to attack the bromonium ion as shown in Eq. 5.5a, text p. 162. The product is the following bromohydrin:

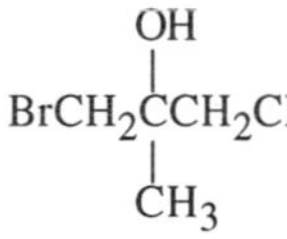

$$BrCH_2\overset{\overset{OH}{|}}{\underset{\underset{CH_3}{|}}{C}}CH_2CH_3 \quad \text{1-bromo-2-methyl-2-butanol}$$

(c) When the bromination is carried out in methanol, then methanol, like water in part (b), is the nucleophile present in greatest concentration; consequently, it attacks the bromonium ion to give an ether:

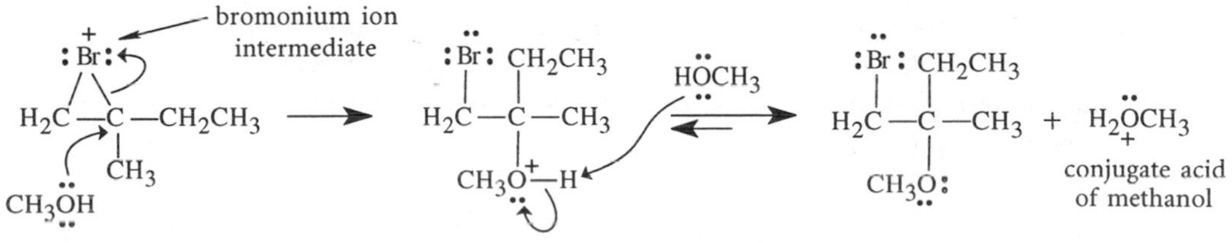

bromonium ion intermediate

$$\text{(bromonium ion of } H_2C{-}C(CH_3){-}CH_2CH_3\text{)} + CH_3\ddot{O}H \longrightarrow \underset{\underset{CH_3\overset{+}{O}{-}H}{|}}{\overset{:Br:}{\overset{|}{H_2C}}{-}\overset{CH_2CH_3}{\overset{|}{C}}}{-}CH_3 \xrightarrow{H\ddot{O}CH_3} \underset{\underset{CH_3\ddot{O}:}{|}}{\overset{:Br:}{\overset{|}{H_2C}}{-}\overset{CH_2CH_3}{\overset{|}{C}}}{-}CH_3 + H_2\overset{+}{\ddot{O}}CH_3$$

conjugate acid of methanol

(d) When the solution contains both methanol and bromide ion, both nucleophiles compete for the bromide ion. Which one "wins" the competition depends on their intrinsic effectiveness as nucleophiles and on their relative concentrations. If the bromide concentration is high enough, the product will be largely the dibromoalkane shown in the solution to part (a). At lower bromide concentrations the product will be a mixture of this dibromoalkane and the ether shown in the solution to part (c).

5.29 The logic is very similar to that used in the solution to Problem 5.28. The reactive intermediate involved is a mercurinium ion (see text p. 166). The only difference is the nucleophile that attacks this ion. A general mechanism for this process is

$$AcO-\overset{+}{Hg}\ (\text{bridging } H_2C-CH-(CH_2)_3CH_3;\ \text{mercurinium ion intermediate}) + R-\ddot{O}H \longrightarrow AcO-Hg-H_2C-CH(\overset{+}{O}(R)H)-(CH_2)_3CH_3 \xrightarrow{:\ddot{O}Ac^-} \rightleftharpoons$$

$$H\ddot{O}Ac + AcO-Hg-H_2C-CH(\ddot{O}R)-(CH_2)_3CH_3 \xrightarrow{NaBH_4} \underset{P}{H_3C-CH(\ddot{O}R)-(CH_2)_3CH_3}$$

The product *P* is the same in all parts (a)–(c) except for the identity of the R group. In (a), R = H; in (b), R = isopropyl; and in (c), R = acetoxy.

(a) $H_3C-CH(H\ddot{O}:)-(CH_2)_3CH_3$ — an alcohol

(b) $H_3C-CH(\ddot{O}-CH(CH_3)_2)-(CH_2)_3CH_3$ — an ether

(c) $H_3C-CH(\ddot{O}-C(=\ddot{O})-CH_3)-(CH_2)_3CH_3$ — an ester

5.30 (a) Only the first two products are formed in the absence of peroxides, and only the third is formed in the presence of peroxides. Different products are formed because different mechanisms and reactive intermediates are involved.

(b) The mechanism for the formation of the first two products is identical to that shown for reaction of the same alkene with HCl in Eqs. 4.25 and 4.26a–b, text pp. 140–1, except, of course, that HBr is used instead of HCl. The first product results from a carbocation rearrangement, and the second from normal regioselective ("Markownikoff") addition. The third product is the consequence of a free-radical addition mechanism, the propagation steps of which are as follows:

$$(CH_3)_3C-CH{=}CH_2 + \cdot Br \longrightarrow (CH_3)_3C-\dot{C}H-CH_2Br \xrightarrow{H-Br} (CH_3)_3C-CH_2-CH_2Br + \cdot Br$$

(The initiation steps are shown in Eqs. 5.40 and 5.42 on pp. 180–1 of the text.)

(c) Peroxide-promoted addition is in competition with normal addition and rearrangement. The normal processes occur at the same rate at which they occur in the absence of peroxides. The fact that *only* the product of peroxide-promoted addition is observed, then, means that this process is much faster than the other, competing, processes.

5.31 (a) Reactive intermediate: $CH_3CH_2(CH_3)C-CH_2$ bridged by $\ddot{Br}^+$ — a bromonium ion

Product: $CH_3CH_2C(Br)(CH_3)CH_2Br$

(b) Reactive intermediate: $CH_3CH_2-\overset{+}{C}(CH_3)-CH_3$ — a carbocation

Product: $CH_3CH_2C(Br)(CH_3)CH_3$

(c) Reactive intermediate:

Product:

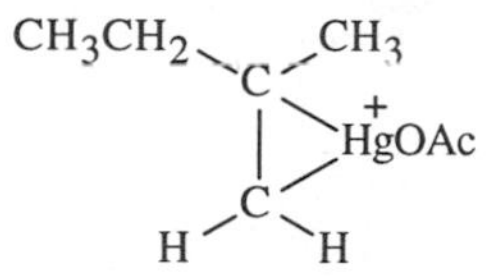

CH_3CH_2, CH_3, C, $\overset{+}{H}gOAc$, C, H, H

a mercurinium ion

OH
$CH_3CH_2CCH_3$
CH_2—HgOAc

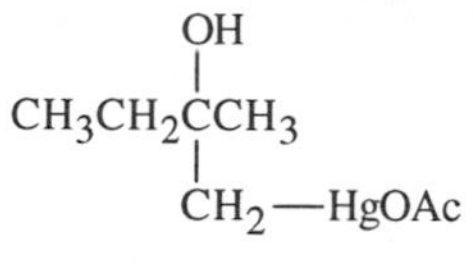

5.32 (a) (Be sure to read Study Guide Link 5.6.) Because the CF_3 group has added to the carbon of the double bond with no alkyl branches, the radical $\cdot CF_3$ must be the first one that adds to the alkene. The radical that results from this addition abstracts an iodine atom from CF_3I to give the product.

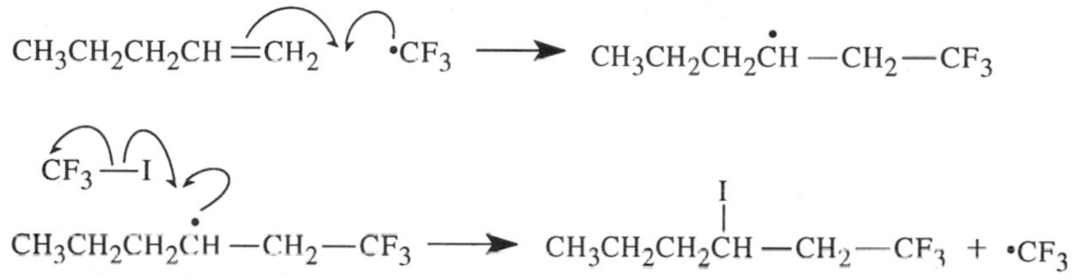

$$CH_3CH_2CH_2CH{=}CH_2 \;\; \cdot CF_3 \longrightarrow CH_3CH_2CH_2\dot{C}H{-}CH_2{-}CF_3$$

$$CF_3{-}I \quad CH_3CH_2CH_2\dot{C}H{-}CH_2{-}CF_3 \longrightarrow CH_3CH_2CH_2CH(I){-}CH_2{-}CF_3 + \cdot CF_3$$

(b) Addition of $\cdot CF_3$ to 2-methyl-1-pentene occurs at an *unbranched* carbon to give a *tertiary* free radical; addition of $\cdot CF_3$ to (*E*)-4-methyl-2-pentene must occur at a *branched* carbon to give a *secondary* free radical. Because addition to 2-methyl-1-pentene gives the more stable free radical, and because there should be less severe van der Waals repulsions in the transition state, this addition should be the faster of the two.

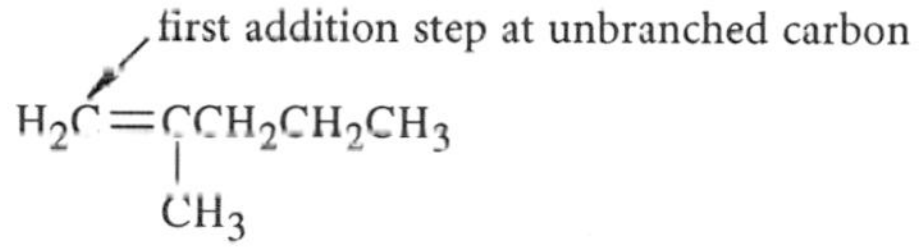

first addition step at unbranched carbon

$H_2C{=}C(CH_3)CH_2CH_2CH_3$

2-methyl-1-pentene

first addition step must occur at a branched carbon

H_3C, H, C=C, H, $CH(CH_3)_2$

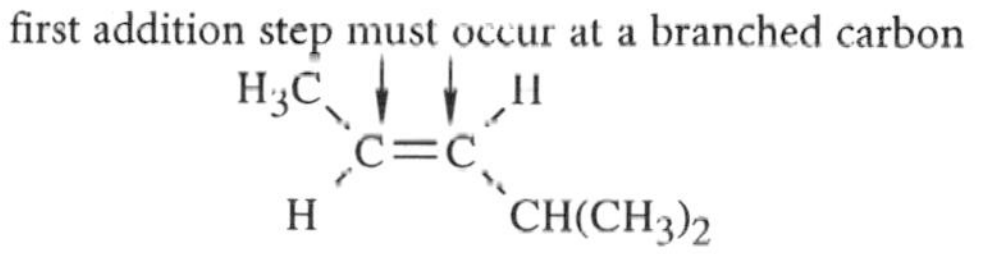

(*E*)-4-methyl-2-pentene

5.33 (a) Because carbon-carbon bonds are weaker than carbon-hydrogen bonds (see Table 5.3), a carbon-carbon bond should break more easily. Breaking of a bond to any of the methyl groups (pathway (1) below) gives a methyl radical and a tertiary free radical. Breaking of the central carbon-carbon bond (pathway (2) below) gives two tertiary free radicals. Because the latter process gives the more stable radicals, it is the one that should occur most readily.

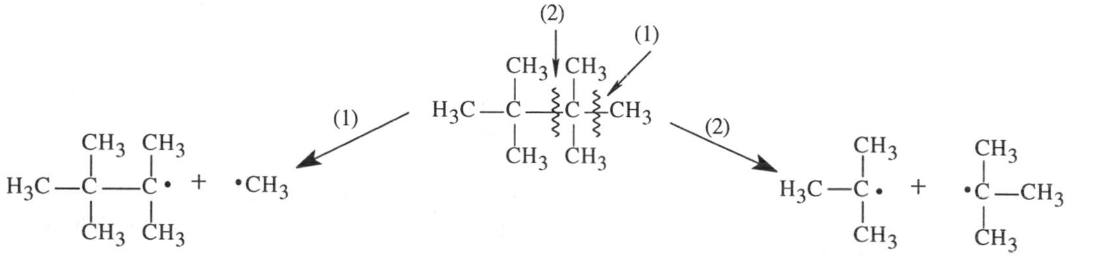

$$(CH_3)_3C{-}C(CH_3)_2\dot{} + \cdot CH_3 \xleftarrow{(1)} H_3C{-}C(CH_3)_2{-}C(CH_3)_2{-}CH_3 \xrightarrow{(2)} H_3C{-}\dot{C}(CH_3)_2 + \cdot\dot{C}(CH_3)_2{-}CH_3$$

more stable combination of radicals

(b) The faster a reaction, the lower the temperature at which it occurs at a given rate. The cracking of 2,2,3,3-tetramethylbutane is the faster reaction, and therefore occurs at lower temperature. The cracking of 2,2,3,3-tetramethylbutane (as shown in part (a)) is faster because it gives two tertiary radicals; the cracking of ethane, in contrast, gives two methyl radicals. The bond that breaks to give the radical with the greater number of alkyl branches is the weaker bond, and is therefore the bond that is more easily broken.

(c) The $\Delta H°$ for the reaction in part (a) is the difference between the heats of formation of products and reactants. Thus,

$$\Delta H° = 2\Delta H_f°(\bullet C(CH_3)_3) - \Delta H_f°(\text{2,2,3,3-tetramethylbutane})$$
$$= 2(48.5) - (-225.9) = 322.9 \text{ kJ/mol}$$

The $\Delta H°$ for the cracking of ethane is

$$\Delta H° = 2\Delta H_f°(\bullet CH_3) - \Delta H_f°(\text{ethane})$$
$$= 2(145.7) - (-84.7) = 376.1 \text{ kJ/mol}$$

This calculation shows that the cracking of ethane is 53 kJ/mol less favorable than the cracking of 2,2,3,3-tetramethylbutane. This demonstrates quantitatively the assertion in part (b), namely, that 2,2,3,3-tetramethylbutane undergoes cracking more readily. The underlying assumption is that the free energies of activation and enthalpies of the cracking reaction are similar.

5.34 (a) Ozonolysis breaks the polymer at its double bonds:

```
—CH2      CH2—CH2      CH2—                                CH2—CH2
   \      /       \     /          1) O3                    /       \
    C=C            C=C           ——————————>       O=C            C=O
   /     \        /     \        2) aqueous H2O2       \            /
CH3       H     CH3       H                             OH        CH3
```

(b) Because E and Z isomers of alkenes give the same ozonolysis products, gutta-percha is evidently either the all-E stereoisomer of natural rubber or a polymer containing both E and Z alkene units. (In fact, it is the all-E stereoisomer.)

```
 —(CH2      H   )
     \     /
      C=C
     /    \
  (CH3      CH2)n—
```

gutta-percha

5.35 (a) Because the H—CN bond is stronger than the O—H bond, the abstraction of a hydrogen atom from HCN by the *tert*-butoxy radical is endothermic by 519 – 435 = 84 kJ/mol. (These numbers are from Table 5.3 with the bond energy of CH_3O—H as an approximation for that of $(CH_3)_3CO$—H.) Because both initiation reactions are highly endothermic, the reaction is not likely to generate a high enough concentration of radicals to initiate a chain reaction.

(b) In this reaction, formation of a C—H bond releases 410 kJ/mol of energy; breaking the H—CN bond requires 519 kJ/mol of energy. Consequently, this step is endothermic by 119 kJ/mol. Free-radical chain reactions with highly endothermic propagation steps generally do not occur. (See text p. 189.)

5.36 (a) In the general reaction, the C—H bond of methane and the X—X bond of the halogen are broken; the C—X bond of the alkyl halide and the H—X bond of the hydrogen halide are formed. The overall energetics of the reactions are as follows, in kJ/mol:

	Breaking of C—H bond	Breaking of X—X bond	Formation of H_3C—X bond	Formation of H—X bond	Overall energy change ($\Delta H°$)
X = Cl	439	247	–356	–431	–101
X = Br	439	192	–293	–368	–30
X = I	439	151	–238	–297	+55

These calculations show that chlorination and bromination are exothermic, whereas iodination is endothermic. Consequently, iodination does not occur because it is energetically unfavorable.

(b) Since the iodination reaction is energetically unfavorable, its reverse is energetically favorable. Consequently, H—I (from the trace of acid) reacts with H_3C—I as follows:

$$H\text{—}I + H_3C\text{—}I \longrightarrow I\text{—}I + CH_4$$

Clearly, iodine is formed in this reaction.

5.37 (a) The structure of polystyrene:

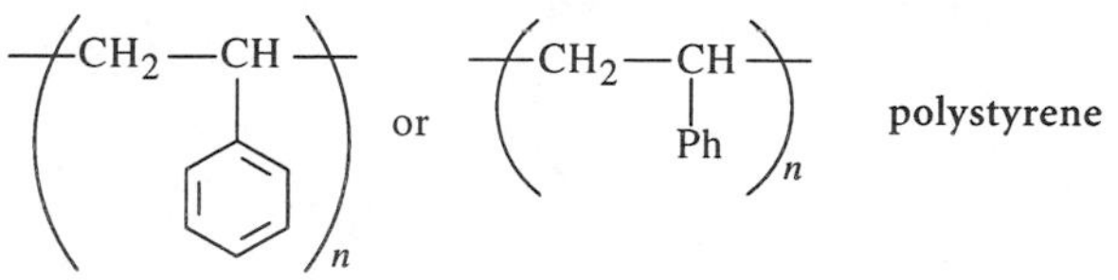

$$\left(CH_2-CH(C_6H_5) \right)_n \quad \text{or} \quad \left(CH_2-CH(Ph) \right)_n \quad \textbf{polystyrene}$$

(b) Because both "ends" of 1,4-divinylbenzene can be involved in polymer formation, addition of 1,4-divinylbenzene serves to connect, or *crosslink,* polymer chains. Such a crosslink is shown with bolded bonds in the following structure:

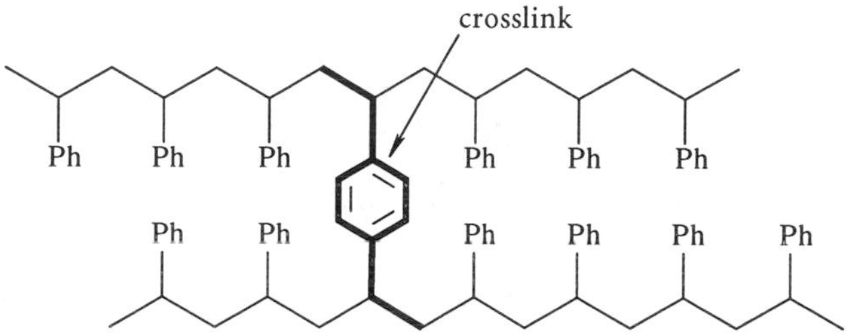

crosslink

Ph Ph Ph Ph Ph Ph

Ph Ph Ph Ph Ph Ph

Notice that because only a small amount of 1,4-divinylbenzene is used, divinylbenzene does not polymerize with itself.

Crosslinks are introduced into polymers to increase their strength and rigidity.

5.38 A detailed discussion of the approach to solving this problem is found in Study Guide Link 5.8. Compound *A* is limonene, a natural product obtained from the oils of lemons and oranges.

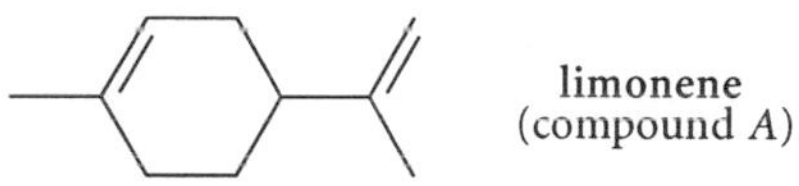

limonene
(compound *A*)

5.39 (a) Protonation of the double bond gives a tertiary carbocation which is attacked by one of the oxygens within the same molecule.

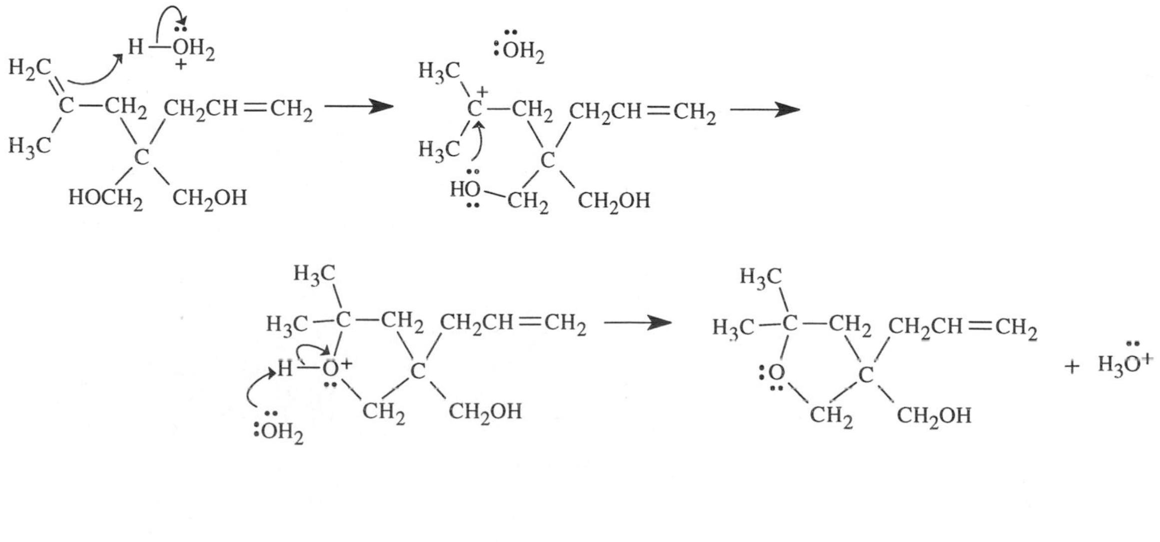

$$(H_3C)(H_2C{=})C-CH_2C(CH_2OH)(HOCH_2)CH_2CH{=}CH_2 \xrightarrow{H-\overset{+}{O}H_2} (H_3C)_2\overset{+}{C}-CH_2C(CH_2OH)(CH_2OH)CH_2CH{=}CH_2 \longrightarrow$$

$$\text{oxonium ion} \xrightarrow{:OH_2} \text{cyclic ether} + H_3O^+$$

(b) The nitrogen within the same molecule attacks a bromonium ion intermediate.

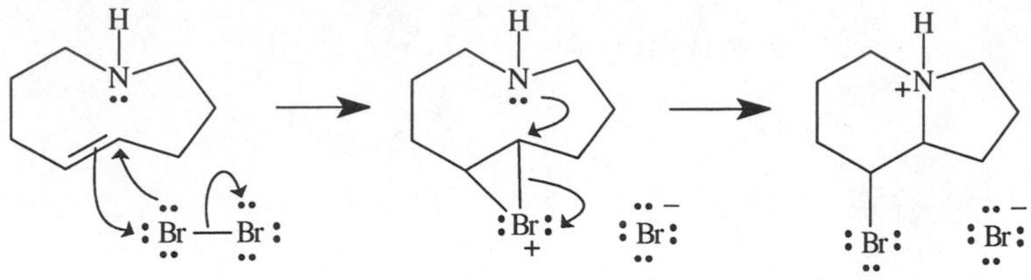

(c) The oxygen of a hydroxy group introduced in one oxymercuration reaction attacks a mercurinium ion within the same molecule. Two products are formed because the mercurinium ion can be attacked at either of two carbons.

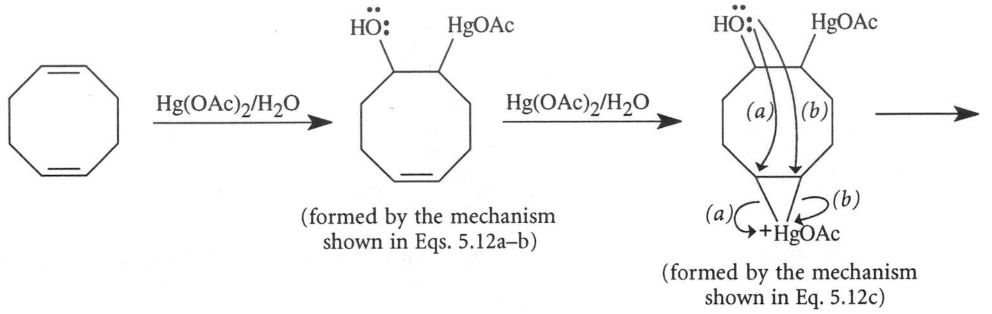
HO:
HgOAc
$Hg(OAc)_2/H_2O$
$Hg(OAc)_2/H_2O$
(a)
(b)
+HgOAc
(formed by the mechanism shown in Eqs. 5.12a–b)
(formed by the mechanism shown in Eq. 5.12c)

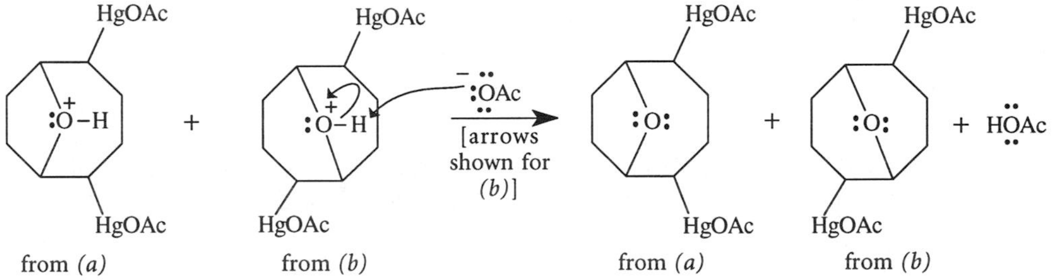
HgOAc
:OAc
[arrows shown for (b)]
+ HOAc
from (a)
from (b)
from (a)
from (b)

Reaction with $NaBH_4$ results in the formation of the products shown in the problem by replacement of the —HgOAc groups with —H.

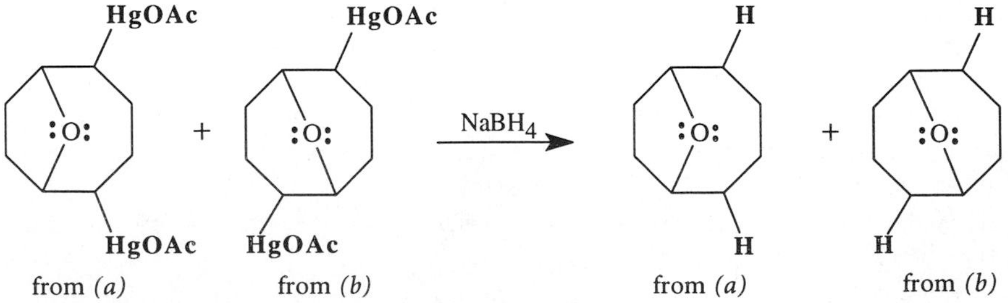
HgOAc
$NaBH_4$
H
from (a)
from (b)
from (a)
from (b)

(d) In the initiation step, an ethylthio radical is formed by reaction of C_2H_5SH with an alkoxy radical derived from the homolytic cleavage of peroxides:

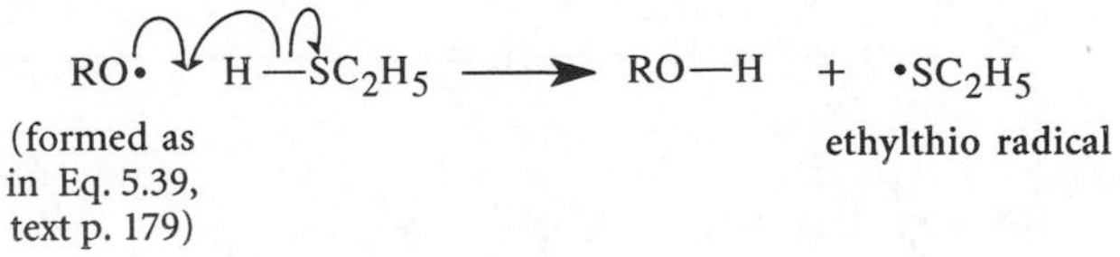
RO•
H—SC_2H_5
RO—H + •SC_2H_5
(formed as in Eq. 5.39, text p. 179)
ethylthio radical

The propagation steps of the chain reaction are as follows:

$$\text{1-methylcyclohexene} + \cdot SC_2H_5 \xrightarrow{H-SC_2H_5} \text{(2-ethylthio-1-methylcyclohexyl radical)} + H-SC_2H_5 \longrightarrow \text{1-methyl-2-(ethylthio)cyclohexane (H added)} + \cdot SC_2H_5$$

(e) A •CBr3 radical formed in the initiation steps adds to the double bond; the resulting radical then abstracts a bromine atom from Br3C—Br to propagate the chain.

$$Br_3C-Br \;\; \cdot OR \longrightarrow Br_3C\cdot + Br-OR$$

$$Br_3C\cdot \;\; H_2C=CH(CH_2)_5CH_3 \xrightarrow{Br-CBr_3}$$

$$Br_3C-CH_2-\dot{C}H(CH_2)_5CH_3 \;\; Br-CBr_3 \longrightarrow Br_3C-CH_2-CH(Br)(CH_2)_5CH_3$$

5.40 (a) We would expect the methyl radical to react at the less-branched carbon of the double bond because this gives the more branched radical, and because the less-branched carbon is the least sterically hindered site of attack.

$$A \qquad H_2C=CH_2 + \cdot CH_3 \longrightarrow H_2\dot{C}-CH_2-CH_3$$

$$B \qquad (CH_3)_2C=CH_2 + \cdot CH_3 \longrightarrow (CH_3)_2\dot{C}-CH_2-CH_3$$

$$C \qquad (CH_3)_2C=CCH_3 + \cdot CH_3 \longrightarrow (CH_3)_2\dot{C}-CH(CH_3)-CH_3$$

(b) When we compare reactions *B* and *A*, we see that two methyl groups at the site of the radical accelerate the reaction modestly. This is probably due to the stability of the free radical (tertiary *vs.* primary) and the reflection of this stability in the transition state for its formation (Hammond's postulate). It might be surprising that the effect on rate is so modest. However, if you consider the effect of branching on alkene stability (Table 4.1) and the effect of branching on free-radical stability (Table 5.2), you can see that these effects are not very different. Remember that relative rates reflect the *difference* in standard free energy of transition states and reactants. Because branching affects both starting material and transition state in the same way, and because the effect is of a similar magnitude in both alkenes and free radicals, the residual effect on the transition state is rather modest.

When we compare reactions *B* and *C*, we start with a *more* branched alkene in *B* and go to an identically branched radical. This should lower the energy of the alkene starting material more than that of the transition state. If so, reaction *C* should be slower—again, not because the transition state is *higher* in energy, but because the starting material is *lower* in energy. (When you climb two energy mountains, the top of the mountains may be at the same elevation above sea level, but if the base camp is lower for one mountain, then that mountain requires a greater climb.) In addition, a steric effect operates on reaction *C* that does not operate on reaction *B*, because the site of attack in reaction *C* is more branched. This would tend to raise the energy of the transition state for reaction *C* and lower the rate, as observed.

5.41 These are both examples of steric effects. Placing three highly branched groups around a central boron results in van der Waals repulsions. These repulsions are severe enough that only two groups can be bound to boron in the case of disiamyl borane, and only one group in the case of thexylborane. (Note the greater branching in thexylborane.) Now, alkyl branches stabilize sp^2-hybridized boron (see the solution to Problem 4.43) just as they stabilize sp^2-hybridized carbon; but if the alkyl branches are *themselves* branched, they form a thicket of methyl groups that interact repulsively as more of these branches are accumulated. These repulsive interactions reduce the relative stability of the trialkylboranes, evidently so much so that they cannot form.

5.42 You can see from the product that the *branched* carbon of one alkene molecule adds to the *unbranched* carbon of another (or vice-versa). Because we know that Lewis acids react with double bonds, it would seem that if one alkene molecule could be converted into a Lewis acid such as a carbocation that this could get the process started. The acidic conditions are just right for carbocation formation:

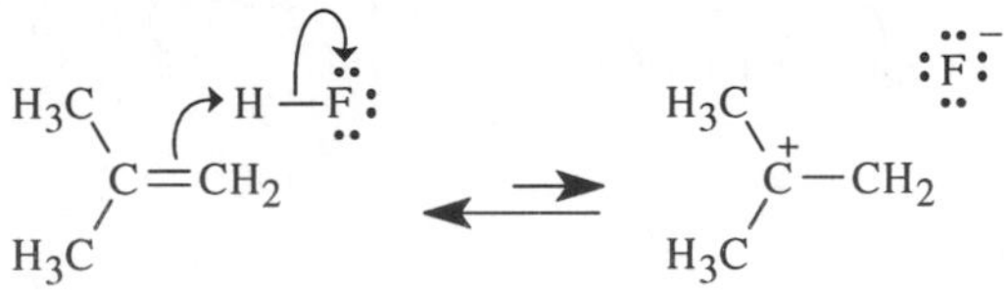

$$(H_3C)_2C{=}CH_2 + H{-}F \rightleftharpoons (H_3C)_2\overset{+}{C}{-}CH_3 \quad F^-$$

A
a carbocation

Carbocation *A* serves as a Lewis acid and is attacked by the π electrons of another alkene molecule:

$$(H_3C)_2C{=}CH_2 + (H_3C)_2\overset{+}{C}{-}CH_3 \;\; F^- \longrightarrow (H_3C)_2\overset{+}{C}{-}CH_2{-}C(CH_3)_2{-}CH_3 \;\; F^-$$

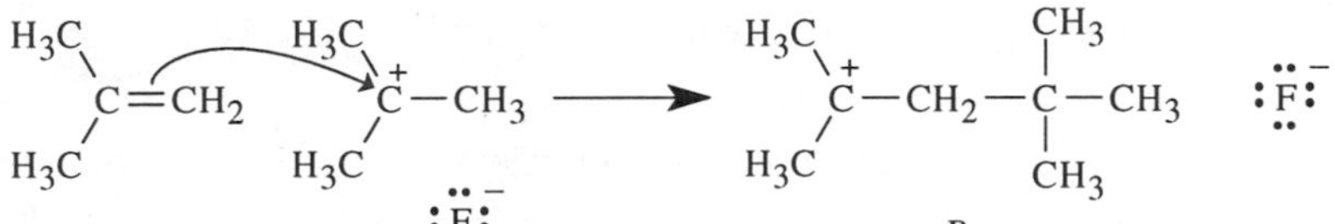

B

Carbocation *B* is attacked by the π electrons of yet another alkene molecule:

$$(H_3C)_2C{=}CH_2 + (H_3C)_2\overset{+}{C}{-}CH_2{-}C(CH_3)_2{-}CH_3 \;\; F^- \longrightarrow (H_3C)_2\overset{+}{C}{-}CH_2{-}C(CH_3)_2{-}CH_2{-}C(CH_3)_2{-}CH_3 \;\; F^-$$

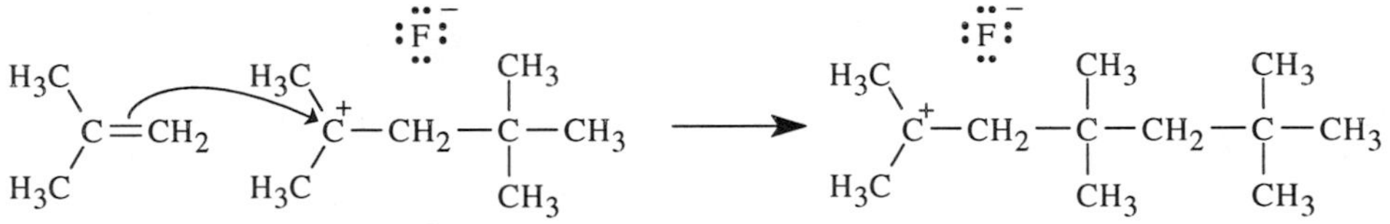

B → *C*

This process continues indefinitely:

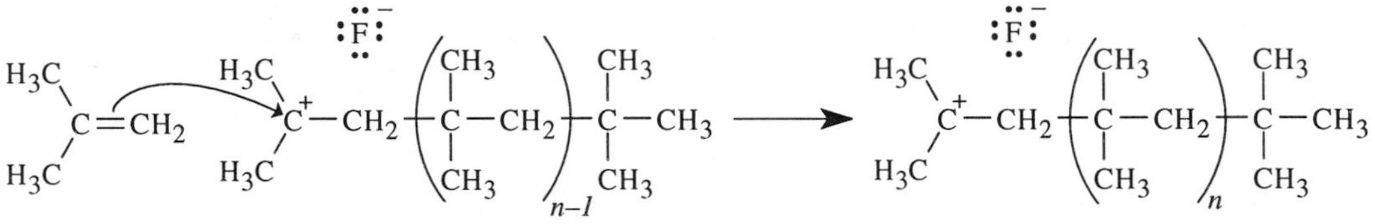

$$(H_3C)_2C{=}CH_2 + (H_3C)_2\overset{+}{C}{-}CH_2{+}C(CH_3)_2{-}CH_2{+}_{n-1}C(CH_3)_2{-}CH_3 \;\; F^- \longrightarrow (H_3C)_2\overset{+}{C}{-}CH_2{+}C(CH_3)_2{-}CH_2{+}_{n}C(CH_3)_2{-}CH_3 \;\; F^-$$

D

Eventually, the alkene concentration drops to such a low value that fluoride ion competes for attack on the carbocations:

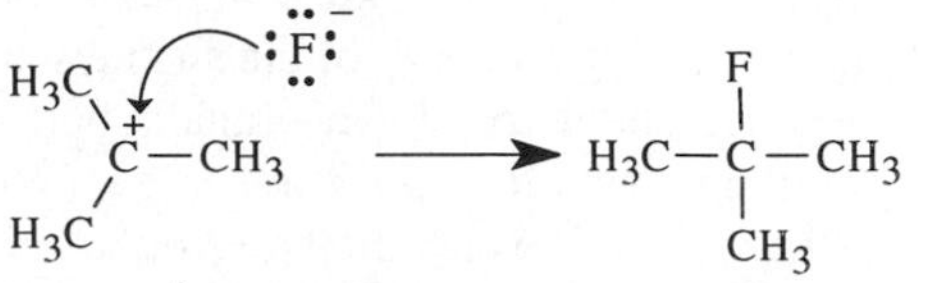

$$(H_3C)_2\overset{+}{C}{-}CH_3 + F^- \longrightarrow H_3C{-}CF(CH_3){-}CH_3$$

A → *tert*-butyl fluoride

The polymer chain can terminate either by attack of fluoride on the carbocation or by loss of a proton to give an alkene. In either case, *n* is so large that the terminal groups are ignored in the final structure.

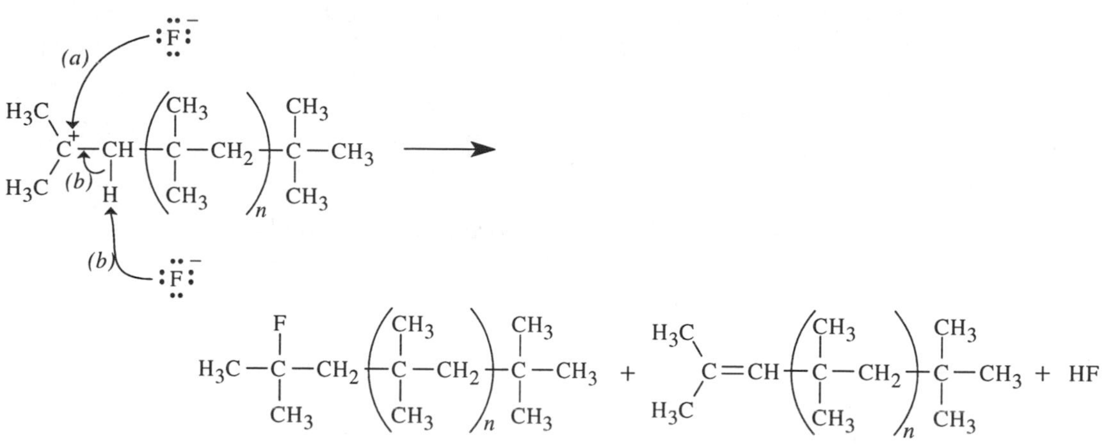

$$(H_3C)_2\overset{+}{C}-CH(H)-\left(C(CH_3)_2-CH_2\right)_n-C(CH_3)_2-CH_3 \xrightarrow{\;:\ddot{F}:^-\ (a)\text{ or }(b)\;}$$

$$H_3C-CF(CH_3)-CH_2-\left(C(CH_3)_2-CH_2\right)_n-C(CH_3)_3 \;+\; (H_3C)_2C{=}CH-\left(C(CH_3)_2-CH_2\right)_n-C(CH_3)_3 + HF$$

from *(a)* from *(b)*

5.43 The starting alkene contains six carbons; acetone, the ozonolysis product of *B*, contains three carbons. Because acetone is the only product formed, it is reasonable to suppose that *two equivalents* of it are produced. The elemental analysis of *B* shows that it has the empirical formula CH_2; thus, *B* is evidently an alkene that contains one double bond. From the ozonolysis data, the structure of *B* is

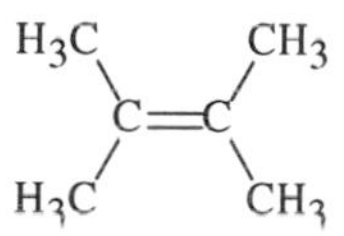

$$(H_3C)_2C{=}C(CH_3)_2$$

2,3 dimethyl 2 butene
compound *B*

Comparison of the carbon connectivities of compound *B* and the alkene starting material shows that a rearrangement has occurred during the transformation of the alkene starting material to *B*. A source of the rearrangement is a reaction that involves carbocations; the HBr addition is such a reaction. Indeed, the reactions shown in Eqs. 4.25–4.26, text pp. 140–1, with HBr instead of HCl, shows that the structure of compound *A* is likely to be

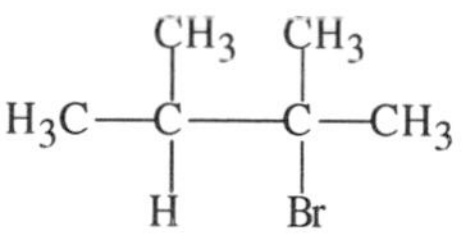

$$H_3C-CH(CH_3)-CBr(CH_3)-CH_3$$

2-bromo-2,3-dimethylbutane
compound *A*

This structure has the same carbon connectivity as compound *B*. In the reaction that converts *A* to *B*, the strong base in the problem removes a proton and a bromide ion is expelled as NaBr. Given the structure of compound *B* shown above, the following analysis is reasonable:

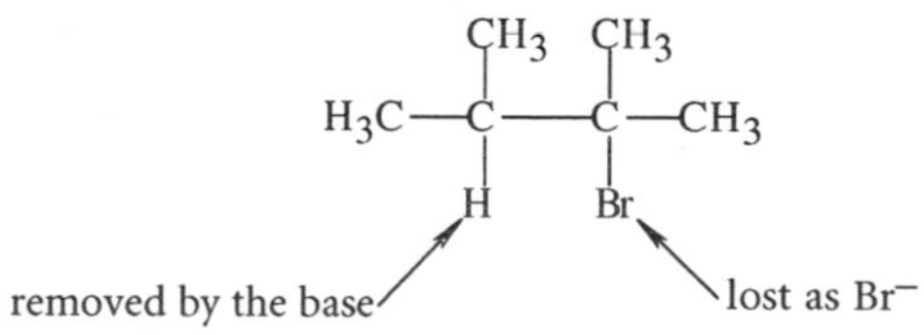

$$H_3C-CH(CH_3)-CBr(CH_3)-CH_3$$

removed by the base (H) lost as Br^- (Br)

A curved-arrow mechanism consistent with this analysis is the following:

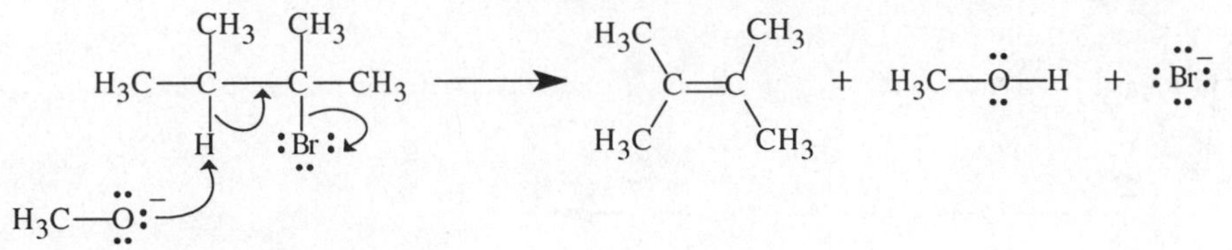

You may wonder why only the tertiary hydrogen is removed by the base. In fact, a competing reaction occurs in which one of the primary hydrogens is removed to give a second product:

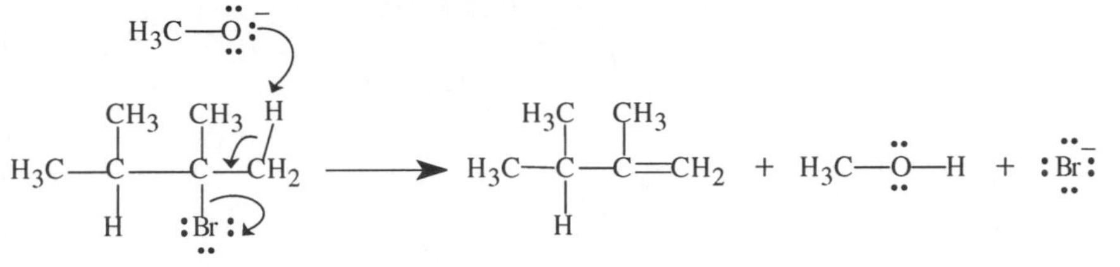

The problem, however, focused on only one of the products for simplicity.

6

Principles of Stereochemistry

TERMS

absolute configuration 6.5
absolute stereochemistry 6.5
achiral 6.1A
amine 6.10B
amine inversion 6.10B
asymmetric carbon atom 6.1B
Biot's law 6.12
Cahn-Ingold-Prelog rules 6.2
chiral 6.1A
chirality 6.1A
conformational diastereomer 6.10A
conformational enantiomer 6.10A
congruent 6.1A
dextrorotatory 6.3B
diastereoisomer 6.6
diastereomer 6.6
diastereomeric relationship 6.6
diastereomeric salt formation 6.8
enantiomer 6.1A
enantiomeric relationship 6.1A
enantiomeric resolution 6.4, 6.8
enantiomerically pure 6.3C
Fischer projection 6.11
internal mirror plane 6.1C
levorotatory 6.3B
meso compound 6.7
noncongruent 6.1A
optical rotation, α 6.3B
optically active 6.3B
plane of symmetry 6.1C
plane-polarized light 6.3A
polarimeter 6.3B
polarized light 6.3A
racemate 6.4
racemic mixture 6.4
racemization 6.4
R 6.2
resolving agent 6.8
S 6.2
specific rotation, $[\alpha]$ 6.3B
stereocenter 6.1B
stereochemical configuration 6.2
stereochemical correlation 6.5
stereochemistry .... Introduction
stereoisomer .... Introduction
symmetry element 6.1C

CONCEPTS

I. CHIRALITY AND SYMMETRY

A. Symmetry and Mirror Images

1. The symmetry of any object can be described by certain symmetry elements, which are lines, points, or planes that relate equivalent parts of an object.
   a. A plane of symmetry is sometimes called an internal mirror plane.

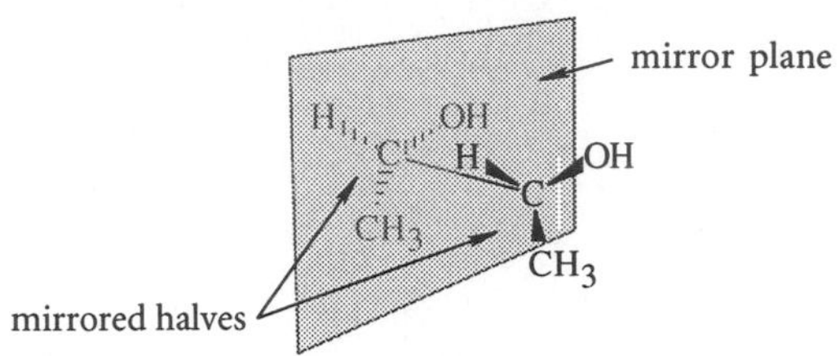

   b. If a molecule has a plane of symmetry, it is not chiral.
2. Some molecules are congruent to their mirror images.
   a. The congruence of a molecule and its mirror image shows that they are identical.
   b. If a molecule is congruent to its mirror image, it is achiral; if not, it is chiral.

B. Asymmetric Carbon

1. An asymmetric carbon is a carbon bonded to four different groups.
   a. An asymmetric atom is another type of stereocenter, or stereogenic atom.

asymmetric atom — stereocenter

   b. All asymmetric carbons are stereocenters, but not all stereocenters are asymmetric carbons.
      i. A stereocenter is an atom at which the interchange of two groups gives a stereoisomer.
      ii. The carbon atoms involved in double bonds of *E,Z* isomers are also stereocenters, but they are not asymmetric carbons.
2. A molecule that contains only one asymmetric carbon is chiral.
   a. Chiral molecules lack certain symmetry elements such as a plane of symmetry.
   b. Molecules that are not chiral are said to be achiral.
3. An asymmetric carbon is not a necessary condition for chirality; some chiral molecules contain no asymmetric atoms.

mirror — enantiomers of 2,3-pentadiene

4. Some molecules with more than one asymmetric carbon are achiral; these are called meso compounds.

H_3C—C(H)(Br)—C(H)(H)—C(H)(Br)—CH_3 (asymmetric carbon ← → asymmetric carbon)

meso-2,4-dibromopentane

a. If a molecule contains n asymmetric carbons, then it has 2^n stereoisomers unless there are meso compounds.
b. If there are meso compounds, then there are fewer than 2^n stereoisomers.

II. ISOMERIC RELATIONSHIPS AND NOMENCLATURE

A. Isomers

1. Isomers are different compounds that have the same molecular formula.
   a. Constitutional isomers have different atomic connectivities.
   b. Stereoisomers are molecules that have the same atomic connectivity but differ in the arrangement of their atoms in space.
      i. The study of stereoisomers and the chemical effects of stereoisomerism is called stereochemistry.
      ii. The *cis-trans* (or *E*,*Z*) isomers of alkenes are examples of stereoisomers.

$(H)_2C{=}C(CH_2CH_3)_2$ ← constitutional isomers → $H_3C(H)C{=}C(CH_2CH_3)(CH_3)$ ← stereoisomers → $H_3C(H)C{=}C(CH_3)(CH_2CH_3)$

2-ethyl-1-butene (*Z*)-3-methyl-2-pentene (*E*)-3-methyl-2-pentene

2. Two types of stereoisomers are
   a. enantiomers, which are molecules that are related as object and noncongruent mirror image.
   b. diastereomers, which are stereoisomers that are not enantiomers and differ in all of their physical properties.

Two nonidentical molecules
↓
Same molecular formula? —No→ The molecules are *not* isomers.
↓ Yes
Same connectivity? —No→ The molecules are *constitutional isomers.*
↓ Yes
Mirror images? —No→ The molecules are *diastereomers.*
↓ Yes
The molecules are *enantiomers.*

B. Enantiomers

1. Molecules that are noncongruent mirror images are called enantiomers.
   a. Enantiomers must not only be mirror images; they must also be noncongruent mirror images.
   b. Molecules that can exist as enantiomers are said to be chiral.

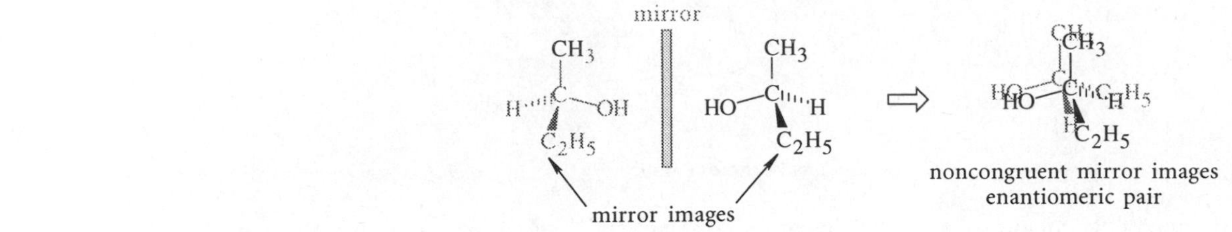

mirror

mirror images

noncongruent mirror images
enantiomeric pair

2. For a pair of chiral molecules with more than one asymmetric carbon to be enantiomers, they must have different configurations at every asymmetric carbon.

C. Racemates

1. A mixture containing equimolar amounts of two enantiomers is called a racemate or racemic mixture.
   a. Racemates typically have physical properties that are different from those of the pure enantiomers.
   b. The optical rotation of any racemate is zero; a racemate is optically inactive.
2. The process of forming a racemate from a pure enantiomer is called racemization.
3. The separation of a pair of enantiomers is called an enantiomeric resolution.
   a. In this process, a pair of enantiomers are allowed to react with a chiral resolving agent, a process that forms diastereomers, which, unlike enantiomers, have different physical properties.
   b. After separating the resulting diastereomers, the pure enantiomers are separated from the resolving agent.
4. Salt formation is such a convenient reaction that it is often used for the enantiomeric resolution of amines and carboxylic acids.

D. Conformational Stereoisomers

1. Enantiomers that are interconverted by a conformational change are called conformational enantiomers.
   a. The two gauche conformations of butane are examples of conformational enantiomers.

← conformational enantiomers →

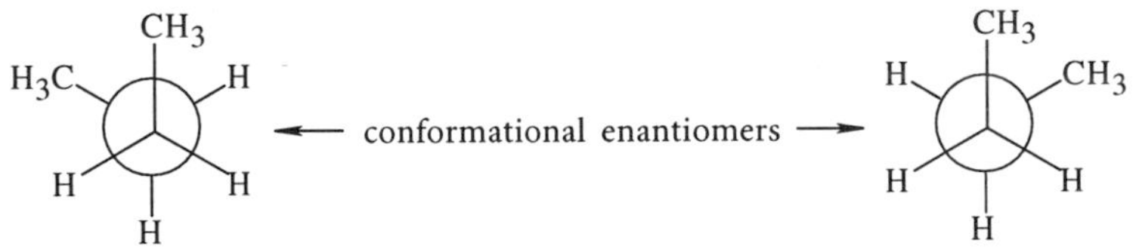

2. Diastereomers that are interconverted by a conformational change are called conformational diastereomers.
   a. *Anti-* and *gauche*-butane are conformational diastereomers.

← conformational diastereomers →

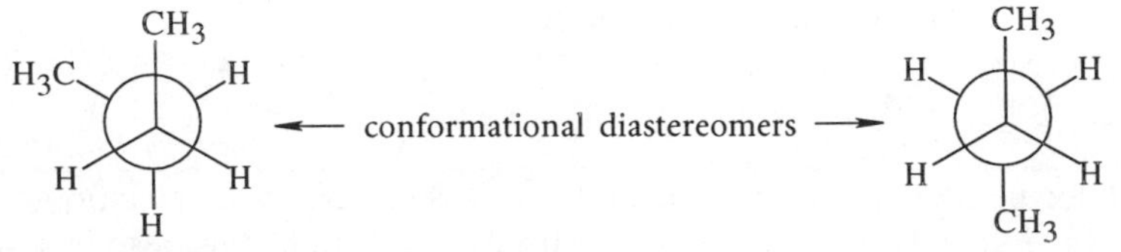

3. A compound which, under ordinary conditions, consists of rapidly interconverting enantiomers is considered to be achiral.
   a. Even though some butane molecules exist in chiral gauche conformations, butane is considered to be achiral because the enantiomeric gauche conformations rapidly interconvert.
   b. Some amines undergo rapid interconversion of stereoisomers in a process called amine inversion.

CH_3 — H···N(lone pair) — C_2H_5 ⇌ (amine inversion) — (lone pair)N···H — CH_3, C_2H_5

c. The interconversion of enantiomeric *gauche*-butanes and amine inversion are two examples of racemization.

E. Nomenclature of Enantiomers: The *R*,*S* System

1. Cahn-Ingold-Prelog priority rules (the same rules used to assign *E* and *Z* conformations to alkene stereoisomers) are used to assign the stereochemical configuration, or arrangement of atoms, at each asymmetric carbon.
2. The following steps are used to assign the configuration of an asymmetric carbon.
   a. Identify an asymmetric carbon and the four different groups bound to it.

D, H, C, CH_2CH_3, $CH_2CH_2CH_3$ — four different groups: D, H, CH_2CH_3, $CH_2CH_2CH_3$

   b. Assign priorities to the four different groups according to the rules given below. The convention used in the text is that the highest priority = 1.
      i. Assign higher priority to the atom of higher atomic number (or atomic mass in the case of isotopes) at the first point of difference.
      ii. If two or more of the attached atoms are identical, move away from the chiral atom within each group to the next atom following the path of highest priority and repeat part *i* above.

asymmetric carbon; priority (3) → D; lowest priority (4) → H; CH_2CH_3 ← priority (2) (HHH); $CH_2CH_2CH_3$ ← highest priority (1); CHH ← first point of difference

      iii. Multiple bonds are treated by a special convention, in which the multiple bond is rewritten as a single bond and the atoms at each end of the multiple bond are replicated.
   c. Arrange the attached atoms in descending priority order.
   d. View the molecule along the bond from the asymmetric carbon to the group of lowest priority, that is, with the asymmetric carbon nearer and the lowest priority group farther away.
   e. Consider the clockwise or counterclockwise order of the remaining group priorities.
      i. If the priorities of these groups decrease in the clockwise direction, the asymmetric carbon is said to have the *R* configuration.
      ii. If the priorities of these groups decrease in the counterclockwise direction, the asymmetric carbon is said to have the *S* configuration.

priority 4 → H; $CH_2CH_2CH_3$ ← priority 1; priority 3 → D; CH_2CH_3 ← priority 2; clockwise order of priorities, thus *R* configuration

   f. A stereoisomer is named by indicating the configuration of each asymmetric carbon before the systematic name of the compound.

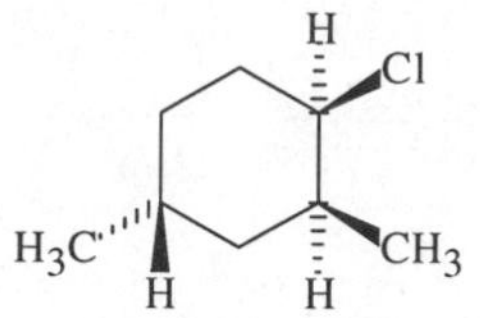

(1*R*,2*S*,4*R*)-1-chloro-2,4-dimethylcyclohexane

3. Assigning an *R* or *S* configuration to every asymmetric carbon in a molecule specifies its absolute stereochemical configuration, or absolute stereochemistry.

III. PHYSICAL PROPERTIES AND REPRESENTATIONS OF STEREOISOMERS

A. Polarized Light and Optical Activity

1. A pair of enantiomers have identical physical properties except for their optical activities. The optical rotations of a pair of enantiomers have equal magnitudes but opposite signs.
   a. Plane-polarized light, or simply polarized light, is light with an electric field that oscillates in only one plane; polarized light is obtained by passing ordinary light through a polarizer.
   b. If plane-polarized light is passed through one enantiomer of a chiral substance, the plane of polarization of the emergent light is rotated.
2. A substance that rotates the plane of polarized light is said to be optically active.
   a. Individual enantiomers of chiral substances are optically active.
   b. Optical activity is measured in a device called a polarimeter.
      i. If the sample rotates the plane of polarized light in the clockwise direction, the optical rotation is given a plus sign (+) and the sample is said to be dextrorotatory.
      ii. If the sample rotates the plane of polarized light in the counterclockwise direction, the optical rotation is given a minus sign (–) and the sample is said to be levorotatory.
   c. The number of degrees α that the analyzer must be turned to measure the angle of rotation is the optical rotation of the sample.
      i. The observed optical rotation α is proportional to both the concentration c (in grams per milliliter) of the sample and the length l (in decimeters) of the sample container; the dependence of optical activity on concentration is sometimes called Biot's law.

$$\alpha = [\alpha]cl \qquad \text{Biot's law}$$

      ii. The specific rotation $[\alpha]$ is used as the standard measure of optical activity and is a constant for a particular compound; it is also a function of temperature and wavelength.

specific rotation → $[\alpha]_D^{20}$ ← temperature (°C); ← D-line of sodium lamp

3. There is no simple relationship between the sign of optical rotation and absolute configuration.
4. A sample of a pure chiral compound uncontaminated by its enantiomer is said to be enantiomerically pure.
5. Optical activity and chirality formed the logical foundation for the postulate of tetrahedral bonding geometry at carbon.

B. Stereochemical Correlation

1. The absolute configurations of most organic compounds are determined experimentally by using chemical reactions to correlate them with other compounds of known absolute configurations in a process called stereochemical correlation.
2. The most secure way of relating absolute configurations is to use reactions that do not break the bonds at the asymmetric carbon.

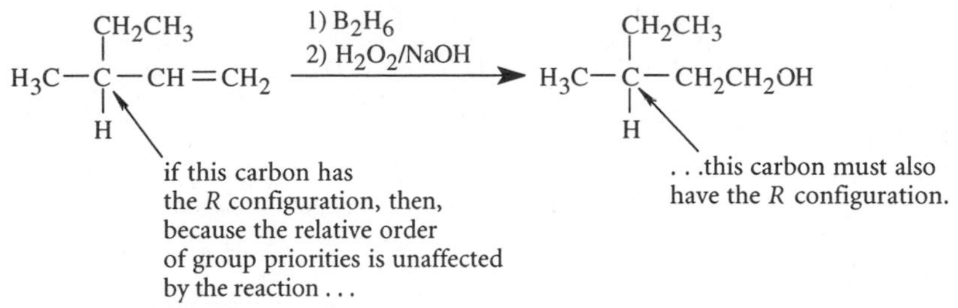

$$H_3C-\overset{CH_2CH_3}{\underset{H}{C}}-CH{=}CH_2 \xrightarrow[2)\ H_2O_2/NaOH]{1)\ B_2H_6} H_3C-\overset{CH_2CH_3}{\underset{H}{C}}-CH_2CH_2OH$$

if this carbon has the *R* configuration, then, because the relative order of group priorities is unaffected by the reaction . . .

. . .this carbon must also have the *R* configuration.

C. Fischer Projections

1. The structures of chiral compounds can be drawn in planar representations called Fischer projections.
   a. All vertical bonds are assumed to be oriented away from the observer.
   b. All horizontal bonds are assumed to be oriented toward the observer.
   c. The asymmetric carbons themselves are not drawn, but are assumed to be located at the intersections of vertical and horizontal bonds.

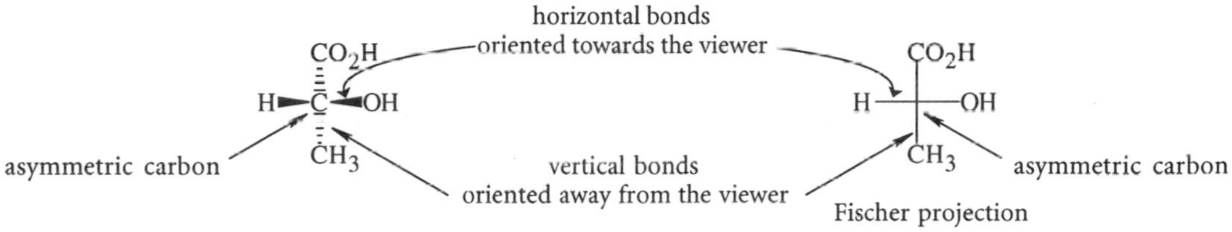

horizontal bonds oriented towards the viewer

CO_2H, H, C, OH, CH_3 — asymmetric carbon

vertical bonds oriented away from the viewer

CO_2H, H, OH, CH_3 — asymmetric carbon

Fischer projection

2. Several valid Fischer projections can be drawn for most chiral molecules.
3. Different Fischer projections for the same molecule can be derived by the following rules:
   a. A Fischer projection may be turned 180° in the plane of the paper

CO_2H / H—OH / CH_3 —180°→ CH_3 / HO—H / CO_2H

equivalent Fischer projections

   b. A Fischer projection may not be turned 90°

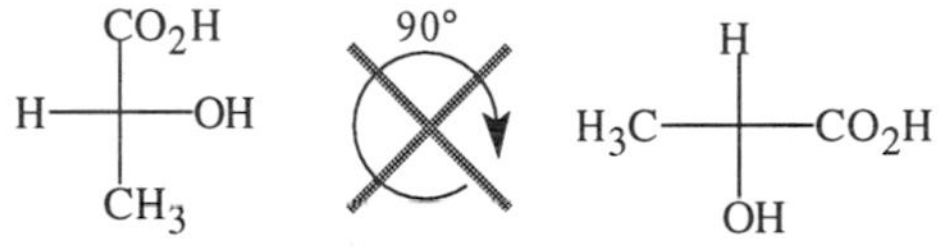

CO_2H / H—OH / CH_3 —90° (crossed out)→ H / H_3C—CO_2H / OH

Fischer projections of enantiomers, therefore not equivalent Fischer projections.

   c. A Fischer projection may not be lifted from the plane of the paper and turned over

CO_2H / H—OH / CH_3 —180° (crossed out)→ CO_2H / HO—H / CH_3

Fischer projections of enantiomers, therefore not equivalent Fischer projections.

   d. The three groups at either end of a Fischer projection may be interchanged in a cyclic permutation; that is, all three groups can be moved at the same time in a closed loop so that each occupies an adjacent position

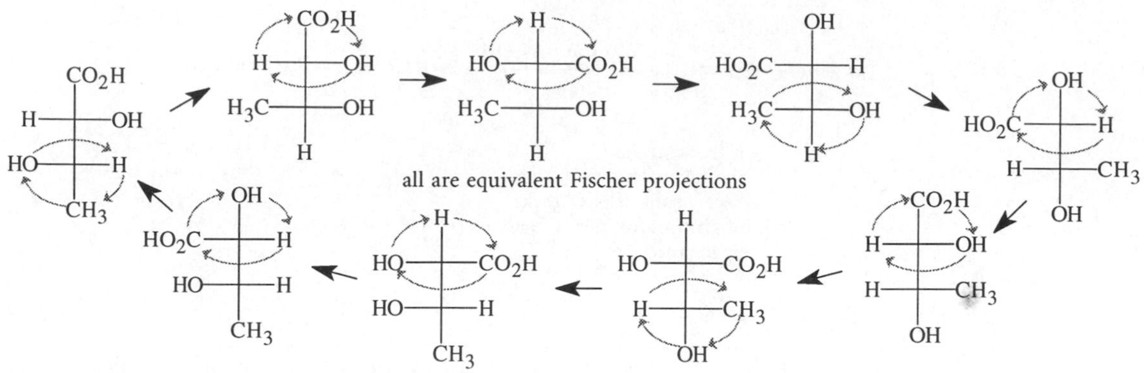

all are equivalent Fischer projections

3. It is particularly easy to recognize enantiomers and meso compounds from the appropriate Fischer projections, because planes of symmetry in the actual molecules reduce to lines of symmetry in their projections.

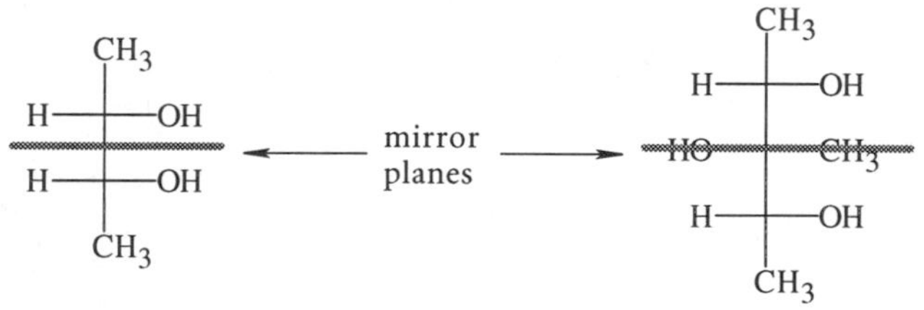

mirror planes

both compounds are meso

SCIENTISTS

Jean-Baptiste Biot (1774–1862) 6.12
Robert Sidney Cahn (1899–1981) 6.2
(Herman) Emil Fischer (1852–1919) [Chemistry 1902] 6.11
Abbé René-Just Haüy (1743–1822) 6.12
John Frederick William Herschel (1792–1871) 6.12
Christopher Kelk Ingold (1893–1970) 6.2
Adolph Wilhelm Herman Kolbe (1818-1884) 6.12
Joseph Achille Le Bel (1847–1930) 6.12
Louis Pasteur (1822–1895) 6.12
Vladimir Prelog (1906–1998) [Chemistry 1975].......... 6.2
Jacobus Hendricus van't Hoff (1852–1911) [Chemistry 1901] 6.12

Dates in brackets refer to Nobel Prize awards.

STUDY GUIDE LINKS

✓6.1 Finding Asymmetric Carbons in Rings

Finding asymmetric carbons in rings requires that each arm of a ring be treated as a separate group. To illustrate, consider the carbon of the structure below that is indicated with an asterisk.

CH_3 * CH_3 CH_3

This carbon bears a CH_3 group and a hydrogen. (The hydrogen is not shown explicitly, because this is a skeletal structure). The other two groups are the two "arms" of the ring. *Even though the two groups are tied together into a ring, they can be considered as separate groups.* Proceeding counterclockwise from the asterisked carbon, the connectivity sequence is —CH_2—CH_2—; proceeding clockwise, the connectivity sequence is —CH_2—$C(CH_3)_2$—; a difference occurs at the second carbon out. Hence, all the groups on the asterisked carbon are different, and thus this carbon is an asymmetric carbon. Notice that we continue as far as necessary along each branch to find a difference.

The following structure, in contrast, has no asymmetric carbon:

CH_3 1 2 3 4 5 6

For example, carbon-1 bears a methyl, a hydrogen, and the two "arms" of the ring. Proceeding around the ring from carbon-1, at carbon-2 and carbon-6 the arms are the same (—CH_2— groups); at carbon-3 and carbon-5 the arms are the same (again, —CH_2— groups); and the two arms join at carbon-4. If no difference is found when proceeding from the carbon in question to the point at which the arms of the ring join, the two arms of the ring are identical.

✓6.2 Stereocenters and Asymmetric Carbons

Stereochemical terminology changes as chemists acquire a more rigorous understanding of the geometrical aspects of stereochemistry. The purpose of this section is to alert you to a problem that still exists with the term *stereocenter.* The definition used in this text—"an atom at which interchange of two groups gives a stereoisomer"—is the one accepted by all experts in the field of stereochemistry. A number of texts, however, have defined *stereocenter* to be an atom with four different groups attached. This is the same definition that is used in this text for *asymmetric carbon.* Thus, some texts do not distinguish between the terms *asymmetric carbon* and *stereocenter.* As shown in the text, however, all asymmetric carbons are stereocenters, but the converse is not true: not all stereocenters are asymmetric carbons. The cases discussed in Sec. 4.1B of the text demonstrate this point: the carbons of the double bond in *E* and *Z* isomers are stereocenters but are not asymmetric carbons. Because so many texts misdefine this term, you should be alert to the fact that you may see it misused.

A number of chemists have objected to the use of the term *asymmetric carbon* because a carbon atom, of course, is not "asymmetric." In fact, many chiral *molecules* themselves are not "asymmetric;" they do have certain types of symmetry. Perhaps a better term for asymmetric carbon would be *asymmetrically substituted carbon.* Other terms you might hear are "chiral center" and "chiral carbon," but objections have been raised to these also. The term *asymmetric carbon* has been so widely used that everyone knows what it means, and this text has continued to use it. The important point is that, whatever we call it, it is *not* the same as a stereocenter.

✓6.3 Using Perspective Structures

The most foolproof way to assign absolute configurations from a perspective structure (that is, a structure in which the bonds are represented as lines, wedges, and dashed wedges) is to build a model and use it to assign configurations. However, once you gain experience in doing this, you will probably be able to use perspective structures themselves to assign configurations.

Before trying to use perspective structures, be sure you understand how to interpret them. This was discussed in Chapter 1, p. 14, of the text. To review:

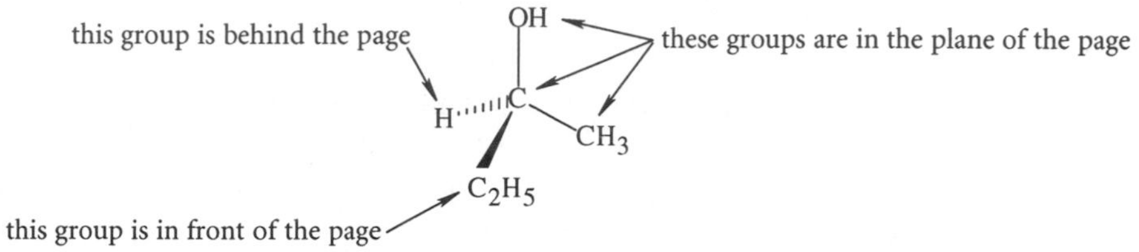

this group is behind the page — OH — these groups are in the plane of the page — H····C — CH_3 — C_2H_5 — this group is in front of the page

The groups with ordinary bonds, along with the central carbon, are in the plane of the page. The group connected to the central carbon by the solid wedge is in front of the page. The group connected to the central carbon by the dashed wedge is behind the page.

If you use journals, monographs, or other texts, you will find that there are variations on the conventions used for representing perspective. In some cases, solid bonds rather than solid wedges are used; dashed bonds rather than dashed wedges are used. Thus, all of the following structures mean the same thing:

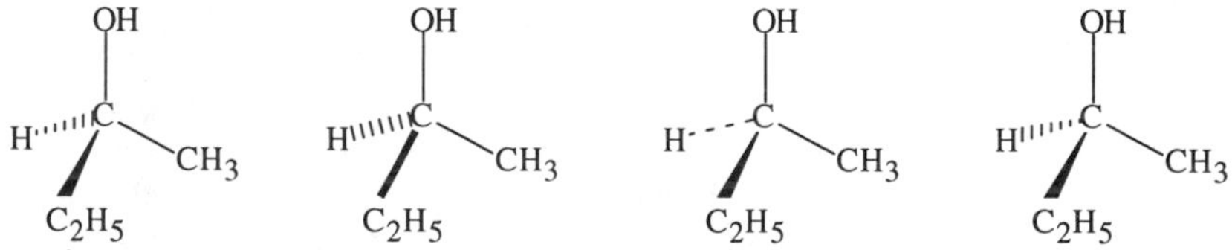

OH, H····C, CH_3, C_2H_5 OH, H····C, CH_3, C_2H_5 OH, H- - -C, CH_3, C_2H_5 OH, H····C, CH_3, C_2H_5

Notice that the last structure is the same as the first except for the orientation of the dashed wedge. Although this last convention is widely used, this text uses a convention in which the thick end of the dashed wedge is nearest the observer, because it more accurately conveys perspective.

If you have a choice in the matter, it is probably easiest to assign the configuration of an asymmetric carbon using perspective formulas if you place the group of lowest priority behind the page—that is, on a dashed wedge. The structure above has been drawn in this manner. When you do this, you are automatically viewing the molecule in the proper manner for assigning configuration:

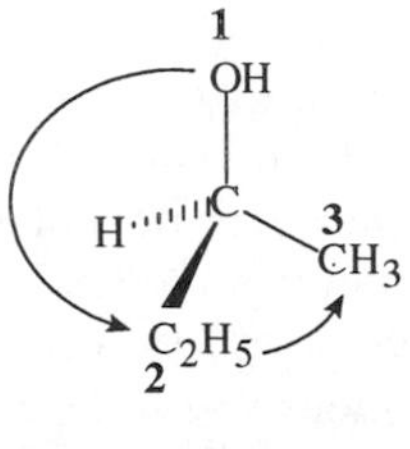

1 OH, H····C, 3 CH_3, C_2H_5 2

The H is farthest from the observer, and we are almost looking down the C—H bond. Of course, it doesn't matter whether the dashed bond is slanted to the right (as in the text, Study Problem 6.2) or to the left, as above.

If you are confronted with a structure that is *not* drawn in this standard manner, and if you need to assign a configuration, then either you have to build a model, or you have to learn to manipulate the perspective structure mentally. Let's consider these two situations in turn.

If you build a model, keep your model as simple as possible. *Never build more of a model than necessary to solve the problem at hand.* (For example, if a molecule contains a methyl group and you don't have to deal explicitly with the hydrogens, then simply use a carbon.) When you are asked to assign configuration, a handy device is to dedicate four atoms of your model set for this purpose. Take an atom and paint a "1" on it with "white-out" (the stuff used for covering typing errors). Paint a "2" on another atom; a "3" on another; and a "4" on another. Then, when you see a perspective formula you can't interpret directly, first assign relative priorities to the groups in the structure. Then build a model using a carbon and only your four labeled atoms. Just make sure they go in the same positions of relative priority as the groups of the structure you are trying to interpret. Then manipulate your model so that group "4" is farthest away from you, and assign the configuration.

The ability to manipulate structures mentally generally comes by working problems in stereochemistry with models. However, the following short exercise might help you develop this ability. Start with a tetrahedral model, and imagine turning it 120° in the plane of the page as shown by the arrows below:

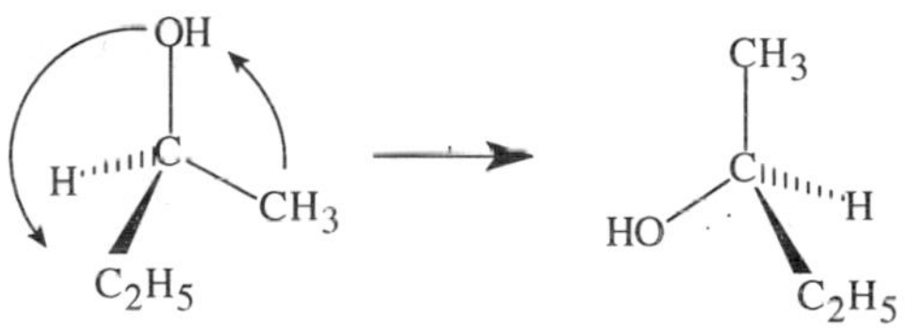

Now imagine spinning it about each bond in turn so that each group moves into the adjacent position, as follows:

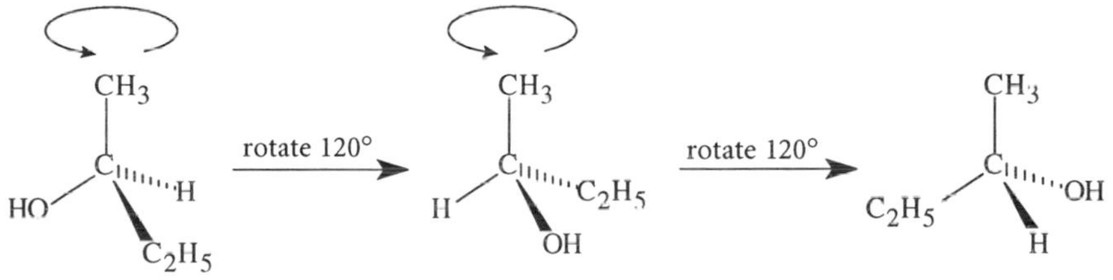

Then practice turning the structure about the other bonds and draw the result. Use a model to check yourself. If you spend a few minutes carrying out these or similar exercises, you should quickly become more proficient in visualizing perspective structures in three dimensions.

6.4 Terminology of Racemates

Nowadays the terms *racemic mixture* and *racemate* are often used interchangeably to indicate *any* equimolar mixture of enantiomers. However, the term *racemic mixture* also has a more precise use. This term is sometimes used to indicate a *solid state* of a racemate that is a mixture of two crystal types, one for each enantiomer. In other words, a racemic mixture is a conglomerate of two enantiomeric crystal forms.

The term *racemic compound* is sometimes used, and it, too, has a more precise definition. This term indicates a *single* crystal form of a racemate that contains equal numbers of molecules of each enantiomer. It is not a conglomerate, but rather a single crystal that contains equal amounts of each enantiomer within its crystal structure.

Racemic mixtures and racemic compounds can be distinguished by an analysis of their melting behavior as a function of composition.

6.5 Center of Symmetry

The manipulation shown at the top of page 222 in the text demonstrates that *meso*-2,3-butanediol is congruent to its mirror image, and is therefore achiral. You may have noticed that the conformation of 2,3-butanediol used in this equation has *no internal mirror plane.*

How can a molecule be achiral if it has no internal mirror plane? This molecule has one of the other chirality-excluding symmetry elements that are less frequently encountered. The symmetry element in this molecule is called a **center of symmetry.** A center of symmetry is a point in the exact center of an object. More precisely, a center of symmetry is a point such that *all* straight lines drawn through the point touch equivalent and indistinguishable parts of the objects at identical distances. Thus, the two oxygens lie on a line (heavy line) through the center of symmetry (heavy dot) and are equidistant from the center.

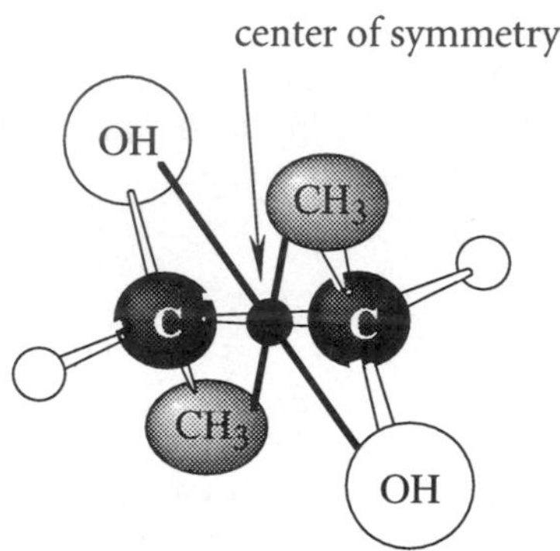

The two methyl groups are related in the same way, as are the hydrogens, the two asymmetric carbons, or any other equivalent elements of the molecule. The anti conformation of butane is another molecule that owes its absence of chirality to a center of symmetry.

A structure has a center of symmetry when reflection in a mirror followed by a 180° rotation reproduces the same structure.

reflect → rotate 180° →

identical structures

Except for planes of symmetry, centers of symmetry are the most frequently occurring symmetry elements in achiral molecules.

6.6 Stereochemical Nomenclature of Meso Compounds

How do we designate a meso compound in the *R,S* system? To take the case of 2,3-butanediol discussed in Section 6.7 of the text, should it be called (2*S*,3*R*)-2,3-butanediol, (2*R*,3*S*)-2,3-butanediol, or something else?

Because numbering in IUPAC substitutive nomenclature is based on *connectivity,* and not on stereochemistry, one cannot arbitrarily assign the lower number to one of the two asymmetric carbons. Thus, *Chemical Abstracts,* the index to the world's chemical literature, names *meso*-2,3-butanediol as (*R**,*S**)-2,3-butanediol. The asterisks on *R** and *S** signify that only the *relative configurations* of the two asymmetric carbons are being specified. In other words, *R** and *S** do not indicate *absolute configuration,* but rather indicate that the configurations of the two carbons are simply *different.* Because of the symmetry of the molecule, the knowledge that the two carbons have different configurations essentially specifies the compound as a meso compound.

You'll find more on designating relative configurations in Study Guide Link 7.1.

6.7 Isolation of Conformational Enantiomers

Although butane undergoes internal rotation too rapidly for its enantiomeric gauche forms to be isolated, chemists have succeeded in preparing other compounds in which internal rotation about a single bond is so slow that conformational enantiomers can be isolated and purified. The following compound is an example:

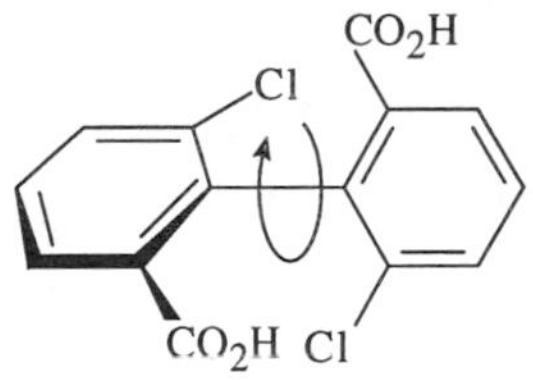

conformational enantiomers can be isolated at room temperature.

(How might you separate the enantiomers of this compound?) The two benzene rings are forced to occupy perpendicular planes; if they were to occupy the same plane, substituents on one ring would "bump into" those on the other, and very severe van der Waals repulsions would result. The conformation of the compound shown above is chiral. (You can verify its chirality by drawing or constructing its mirror image and showing that it is noncongruent.) Interconversion of the two enantiomeric conformations requires a rotation about the central bond, indicated by the circular arrow, which takes the rings into the same plane. Because of the van der Waals repulsions just noted, this process requires such high energy that it cannot occur at room temperature.

In contrast, the compound in which the —CO_2H groups are replaced by hydrogens and the chlorine substituents are replaced by fluorines *cannot* be separated into its enantiomers at room temperature. In this case, racemization requires a planar conformation in which two small atoms (F and H) are close. Because van der Waals repulsions are much smaller here, this process requires less energy, and racemization occurs readily at room temperature.

✓6.8 Additional Manipulations with Fischer Projections

Two other "tricks" can prove useful in handling Fischer projections. The first has to do with assignment of configuration. If the lowest-priority group is in one of the vertical positions, the configuration can be assigned directly, as shown in the text. But what if you are dealing with a Fischer projection in which the lowest-priority group is in a horizontal position? Of course, you could re-draw the molecule in a new Fischer projection, or you could simply build a model and assign configuration from the model. But these procedures are too time-consuming; there is an easier way:

If the lowest-priority group is in a horizontal position, simply assign the configuration using the other three groups—but, since the assignment will be wrong, reverse it!

Let's illustrate with the following Fischer projection by assigning the configuration at the carbon marked with an asterisk:

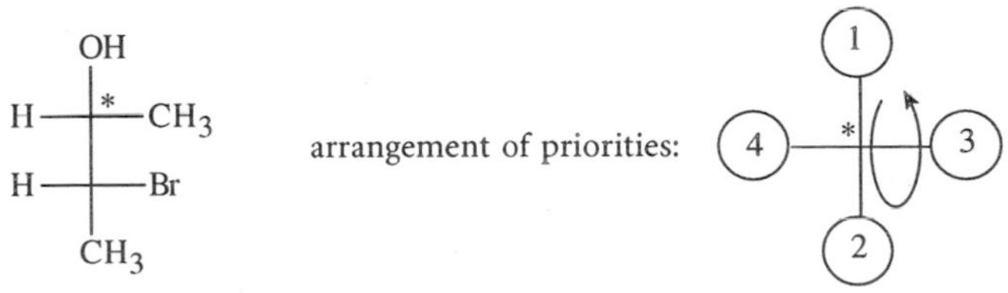

Groups 1, 2, and 3 lie in a counterclockwise pattern. Normally, this would indicate an *S* configuration. But because the viewing convention has been violated, this is not correct. Hence, the configuration is *R*.

Since there are only two possible outcomes, knowing the wrong answer is as good as knowing the correct one. (The author tells his students that this is one of the few situations in which two wrongs make a right.)

The second "trick" stems from the definition of a stereocenter. By definition, the structure of a stereoisomer can be generated by interchanging two groups at a stereocenter. Thus, *interchanging the positions of any two groups at a stereocenter in the Fischer projection changes the configuration of the stereocenter.* This works as well in Fischer projections as it does in three-dimensional models. To illustrate, let's generate the enantiomer of the compound shown above. In an enantiomer, the configuration of *every* asymmetric carbon must be reversed. To do this, simply interchange any two groups at *each* asymmetric carbon:

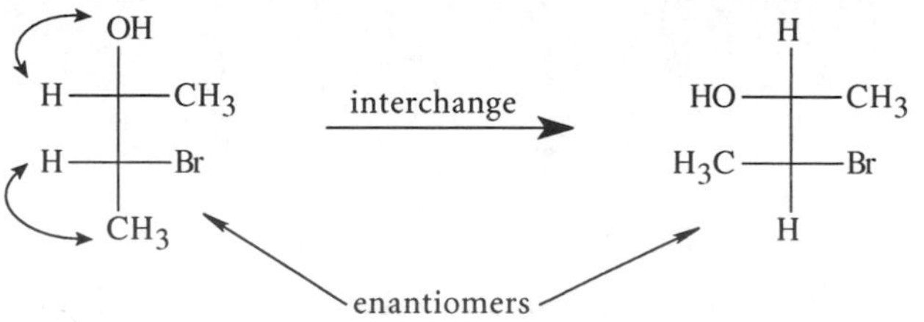

OH; H—C—CH_3; H—C—Br; CH_3 —interchange→ H; HO—C—CH_3; H_3C—C—Br; H (enantiomers)

These two Fischer projections certainly do not appear to be mirror images. Nevertheless, they are Fischer projections of enantiomers. (Try to perform allowed manipulations of the second Fischer projection so that it is a mirror image of the first.)

It follows that if interchanging the positions of two groups at a stereocenter changes the configuration of the stereocenter, then doing it again makes the configuration what it was originally. That is, *an even number of group interchanges at a stereocenter leave the configuration of the stereocenter unchanged.* Perhaps you can convince yourself that the "cyclic permutation" at a stereocenter (Rule 4 and Eqs. 6.12–6.13 on text p. 234), which leaves the configuration of the stereocenter unchanged, is completely equivalent to two group interchanges at the same stereocenter!

SOLUTIONS TO PROBLEMS

Solutions to In-Text Problems

6.1 (a) Chiral
(b) Achiral
(c) Chiral. Note that nitrogen is the asymmetric atom in this case.

6.2 (a) Chiral
(b) Achiral (neglecting the writing)
(c) Achiral, assuming that it is sharpened with a symmetrical point and that there is no writing on the side.
(d) Achiral (neglecting the internal organs and any superficial characteristics)
(e) Achiral
(f) Chiral. Actually, the author, a right-hander, didn't know this until a reviewer, a southpaw, commented on how badly the scissor industry discriminates against left-handers.

6.3 (a) A plane of symmetry in methane is a plane containing any two C—H bonds and bisecting the other two. (See illustration below.)
(b) A plane of symmetry is any plane containing the tip of the cone and a diameter of the base. (See illustration below)
(c) A plane of symmetry is any plane containing the tip and bisecting opposite sides of the base; or, equivalently, any plane containing the tip and the center of the base.
(d) A plane of symmetry is the plane containing the two C—Cl bonds and bisecting the angle between the two C—H bonds (see illustration below), or any plane containing the two C—H bonds and bisecting the angle between the two C—Cl bonds.

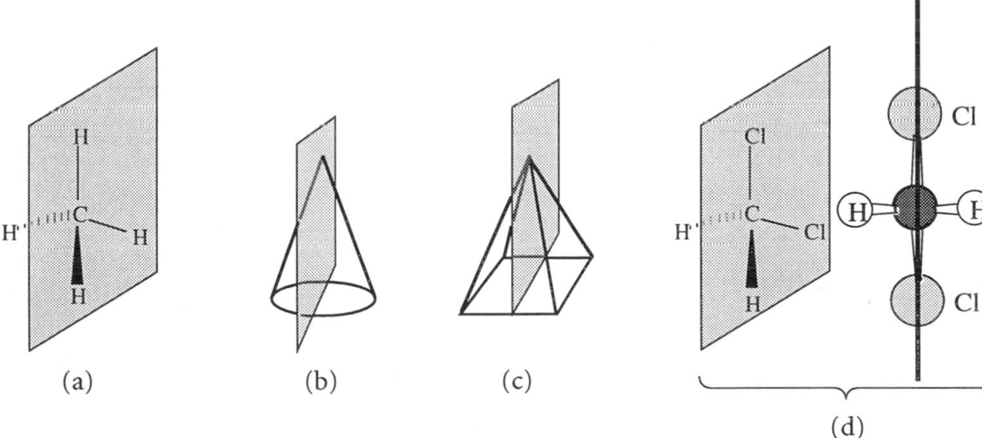

6.4 (a) (b)

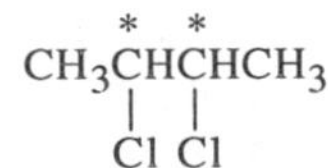

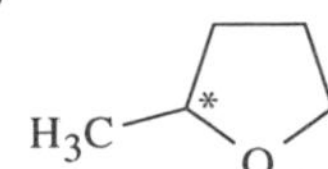

Be sure to read Study Guide Link 6.1 if part (b) was troublesome for you.

6.5 (a) (b)

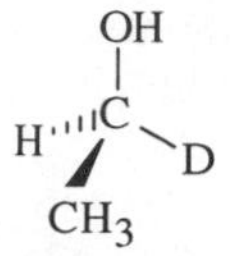

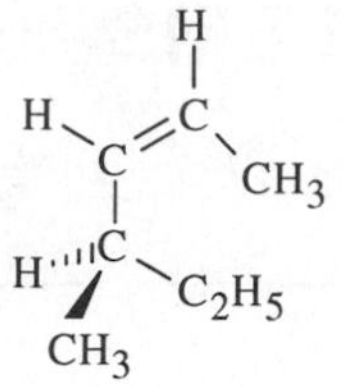

6.6 (a) The asymmetric carbon in the given stereoisomer of alanine has the *S* configuration.
(b) The asymmetric carbon in the given stereoisomer of malic acid has the *S* configuration.
(c) The asymmetric atom is a nitrogen, and it has the *S* configuration. [See the solution to Problem 6.1(c)].

6.7 Halve the concentration. If the rotation at the first concentration is +10°, the rotation at the lower concentration should read +5°; if the rotation at the first concentration is –350°, the rotation at the lower concentration should be –175°, which is the same as +185°; if the rotation at the higher concentration is +370°, the rotation at the lower concentration should be +185°. To decide between the last two possibilities, halve the concentration again.

6.8 (a) The asymmetric carbon of thalidomide is marked with an asterisk in the structure below:

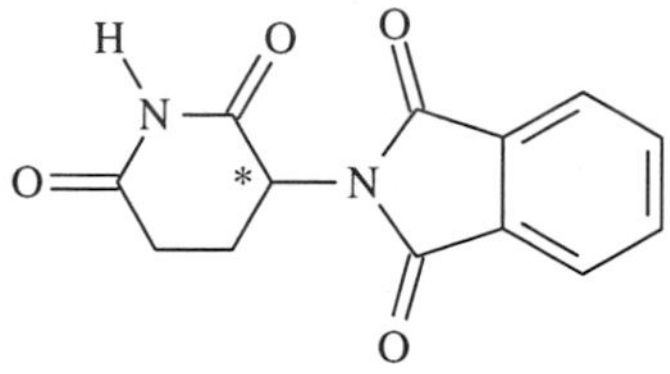

thalidomide

(b) One of several correct perspective structures of (*S*)-thalidomide:

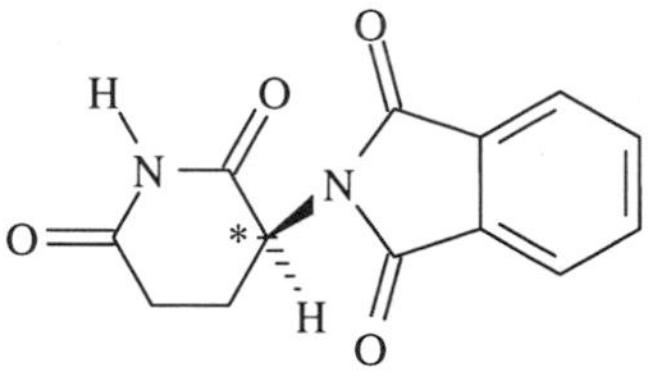

(*S*)-thalidomide

6.9 (a) A concentration of 0.1 *M* is the same as a concentration of (0.1 mol/L)(150 g/mol)(0.001 L/mL) = 0.015 g/mL. Using this number for *c* in Eq. 6.1,

$$[\alpha] = \frac{\alpha}{c \cdot l} = \frac{+0.20 \text{ deg}}{(0.015 \text{ g}\cdot\text{mL}^{-1})(1 \text{ dm})} = +13.3 \text{ deg}\cdot\text{mL}^{-1}\cdot\text{g}^{-1}\cdot\text{dm}^{-1}$$

(b) Because the two enantiomers are present in equal amounts, their rotations cancel, and the observed rotation is zero.
(c) Because the concentration is halved, the observed rotation is also halved. Thus, it is 0.10°.
(d) The specific rotation is the same, because by definition this parameter is independent of concentration.
(e) The specific rotation of *L* is –13.3 deg·mL^{-1}·g^{-1}·dm^{-1}; enantiomers have specific rotations of equal magnitudes and opposite signs. Again, the specific rotation is the same regardless of concentration.
(f) In this solution, the 0.005 mole of *L* cancels the rotation of 0.005 mole of *D*. The resulting rotation is as if only 0.005 mole, or a concentration of 0.05 mol/L, of *D* were present. This amount of *D* corresponds to a concentration of (0.05 mol/L)(150 g/mol)(0.001 L/mL) = 0.0075 g/mL. Since this concentration is half of the original concentration, the observed rotation is also half of the original, that is, 0.10°.

6.10 The racemate has no effect on the observed rotation other than to dilute the sample. Hence, after addition of the racemate, the concentration of the excess (*R*)-2-butanol is 0.75 *M*. This corresponds to (0.75 mol/L)(74.12 g/mol)(0.001 L/mL) = 0.05559 g/mL. Use this as the value of *c* in Eq. 6.1 with $[\alpha] = -13.9$ deg·mL·g^{-1}·dm^{-1}:

$$\alpha = (-13.9 \text{ deg·mL·g}^{-1}\text{·dm}^{-1})(0.05559 \text{ g·mL}^{-1})(1 \text{ dm}) = -0.773°$$

6.11 Begin with the perspective structure of the reactant; in the product, corresponding groups are in corresponding positions, because the reaction breaks none of the bonds to the asymmetric carbon.

1 Ph; 4 H; 3 $CH_2CH_2CH_3$; $CH{=}CH_2$ 2 $\xrightarrow[\text{Pd/C}]{H_2}$ 1 Ph; 4 H; 2 $CH_2CH_2CH_3$; CH_2CH_3 3

(*S*)-(+)-enantiomer — (−)-enantiomer; has the *R* configuration

This analysis shows that the (−)-enantiomer of the product has the *R* configuration. Notice that the "change of configuration" caused by this reaction is merely an artifact of the relative group priorities, shown by the numbers in the preceding structures.

6.12 Proceed in the manner suggested by the solution to Problem 6.10. The absolute configuration of the alkene in Eq. 6.2 is known. Carry out the following catalytic hydrogenation:

Ph; H; $CH{=}CH_2$; CH_3 $\xrightarrow[\text{Pd/C}]{H_2}$ Ph; H; CH_2CH_3; CH_3

(*R*) enantiomer — (*R*) enantiomer

If we assume that hydrogenation proceeds in the normal manner, then the product must have the *R* configuration. Determine the sign of its specific rotation. If positive, then the product shown is the (*R*)-(+)-enantiomer, and it is the dextrorotatory enantiomer; if negative, then the product shown is the (*R*)-(−)-enantiomer, which means that the (*S*)-(+)-enantiomer is the dextrorotatory enantiomer.

6.13 (a) In order for a molecule to have a meso stereoisomer, it must have more than one asymmetric atom, and it must be divisible into constitutionally identical halves (that is, halves that have the same connectivities relative to the dividing line). By these criteria, compound (a) does possess a meso stereoisomer.

$CH_3CH(Cl)—CH_2—CH(Cl)CH_3$ (Cl, H, Cl above; H below; dividing line through the central C)

Notice that the dividing line may pass through one or more atoms.

(b) This compound does not have a meso stereoisomer. A meso compound must have at least two asymmetric carbons.

(c) *Trans*-2-hexene cannot exist as a meso stereoisomer because it has no asymmetric atoms.

6.14 As suggested by the hint, the internal mirror plane passes through the central carbon atom, the hydrogen, and the OH group. This carbon can have either of two configurations. The only requirement for a meso compound (in addition to the identical connectivities of the two half-structures relative to a dividing line) is that the asymmetric carbons at the end of the structure must have opposite configurations.

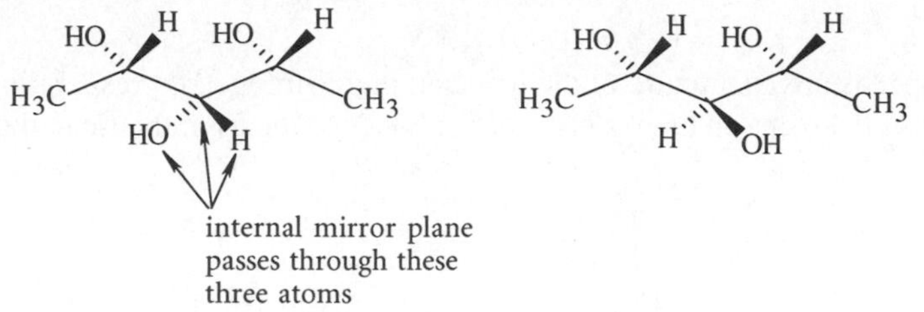

internal mirror plane passes through these three atoms

Stereocenters in the center of meso compounds, such as the central carbons of the structures above, are sometimes termed *pseudoasymmetric centers* because the only difference between two of the groups attached to such a carbon is their stereochemical configurations. Although there is a configurational notation for such carbons, it is not discussed in the text.

6.15 A resolving agent must be *chiral* and it must be *enantiomerically pure.* Only *A* meets both of these criteria.

6.16 Compounds (a), (b), and (d) are chiral and have stereocenters as shown in the following structures. The stereocenter in (a) is an asymmetric carbon. One of the three stereocenters in (b) is an asymmetric carbon. Compound (d) has three stereocenters, but none are asymmetric carbons, even though the molecule is chiral. Compound (c) is not chiral but it has two stereocenters.

(a) asymmetric carbon and a stereocenter

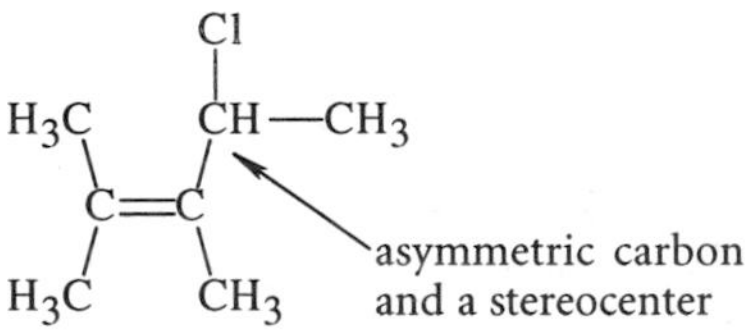

(b) stereocenter and asymmetric carbon; stereocenters

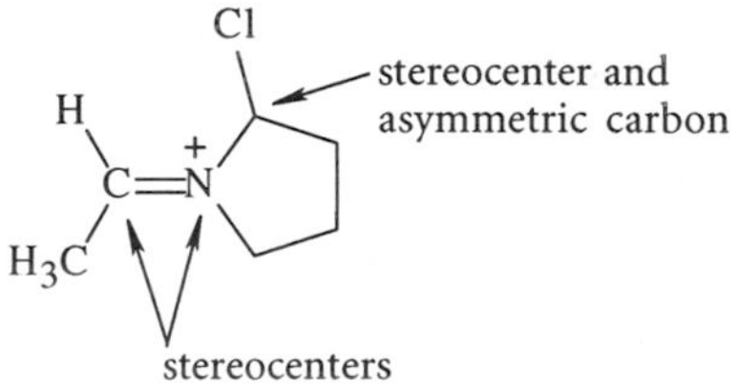

(c) stereocenters

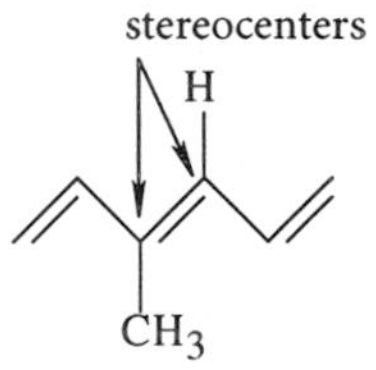

(d) stereocenters

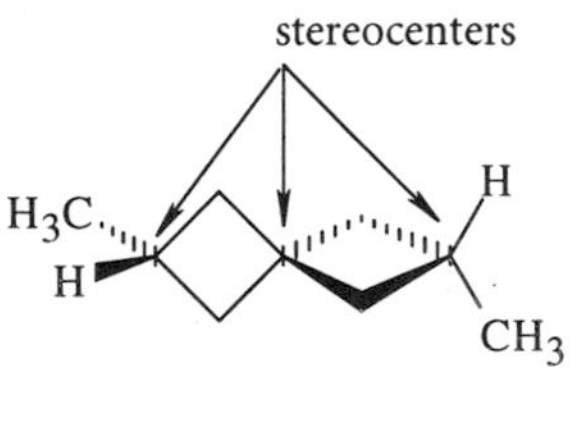

The example in part (d) shows that a tetrahedral stereocenter need *not* be the same thing as an asymmetric carbon.

6.17 In *anti*-butane, the two —CH_2— carbons are stereocenters, because interchange of a CH_3 group and an H gives a stereoisomer (*gauche*-butane).

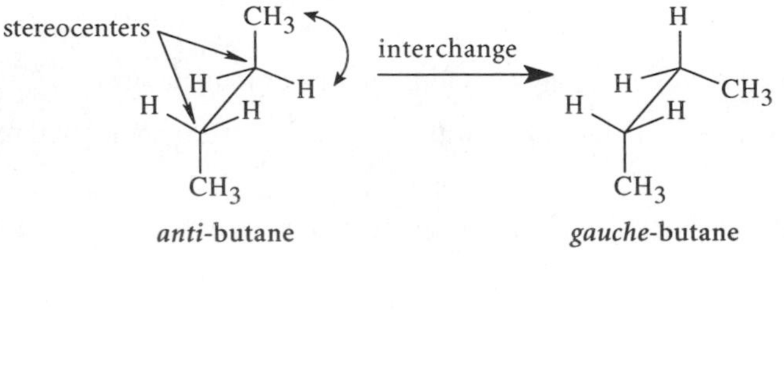

stereocenters — interchange

anti-butane

gauche-butane

As the term *stereocenter* is normally used, it is not applied to individual, rapidly interconverting conformations. Consequently, butane has no stereocenters. However, under conditions of very low temperature at which the conformations of butane are stable, or (more practically) for molecules that consist of conformations that do not interconvert rapidly, the term can be applied.

6.18 (a) The three conformations of *meso*-2,3-butanediol:

A: OH; H_3C, H; H, CH_3; OH

B: OH; H_3C, H; HO, H; CH_3

C: OH; H_3C, H; H_3C, OH; H

Conformation *A* is achiral; it is congruent to its mirror image. (This was demonstrated on text page 222.) Conformations *B* and *C* are enantiomers. This is not obvious from the way they are drawn above, but if you will make a model of each, you should be able to see that they are noncongruent mirror images.

(b) Because its conformations interconvert rapidly, *meso*-2,3-butanediol cannot be optically active. As the text indicates, molecules that consist of rapidly interconverting enantiomers are said to be achiral. However, at very low temperatures, conformations *B* and *C* could in principle be isolated; each would be optically active, and the two conformations would have rotations of equal magnitudes and opposite signs.

6.19 (a) The carbon stereocenter is fixed in the *S* configuration, but the nitrogen stereocenter undergoes inversion between *R* and *S* configurations. Consequently, the molecule is a mixture of C(*S*),N(*S*) and C(*S*),N(*R*) conformations; that is, it is a rapidly equilibrating mixture of diastereomers.

(b) Each of the rapidly interconverting species in part (a) has an enantiomer: C(*S*),N(*S*) has an enantiomer C(*R*),N(*R*), and C(*S*),N(*R*) has an enantiomer C(*R*),N(*S*). Because inversion of the nitrogen stereocenter does not affect the configuration of the carbon stereocenter, it would be possible to resolve the racemate of this compound into enantiomeric sets of rapidly interconverting *diastereomers*.

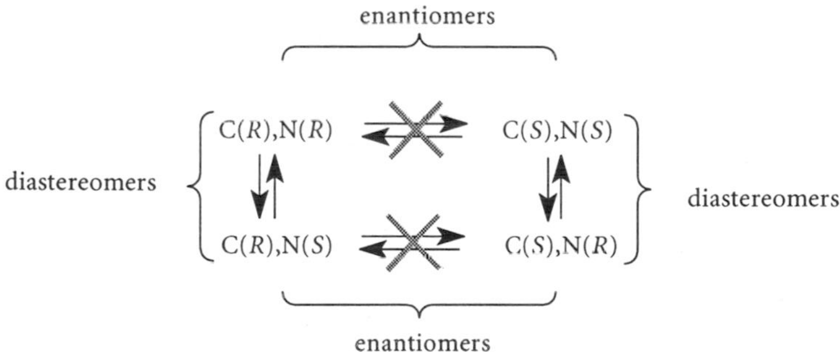

6.20 (a) Many Fischer projections are possible for a compound with three asymmetric carbons. The two below are related by a 180° rotation in the plane of the page. If yours don't look like these, try to use the allowed manipulations of Fischer projections to see whether yours are equivalent. (There are eighteen possible correct Fischer projections!)

H; HO_2C—Cl; H—Cl; $HOCH_2$—Br; H

H; Br—CH_2OH; Cl—H; Cl—CO_2H; H

(b) The following two of the nine possible Fischer projections of (*S*)-2-butanol are related by a cyclic permutation of the ethyl, —OH, and the methyl group.

OH
C_2H_5—+—CH_3
H

C_2H_5
H_3C—+—OH
H

6.21 (a) The way to solve a problem of this sort is to manipulate one of the Fischer projections so that two groups in one projection are in the same positions as the corresponding groups in the other; then the relationship between the two molecules should be clear. Thus, manipulate the projection on the right so that the OH groups are in the vertical positions. It then becomes apparent that this structure and the one on the left in the problem are enantiomers.

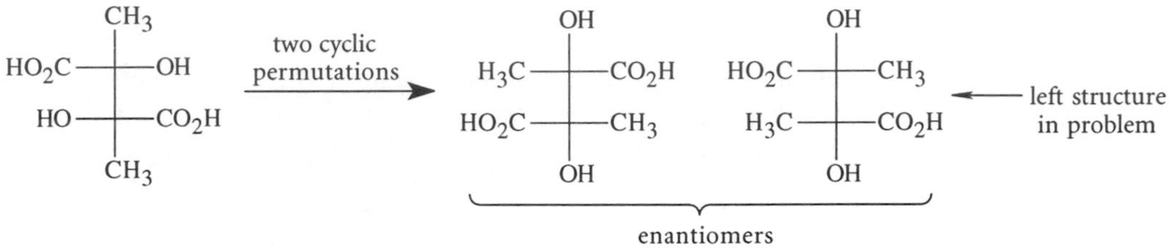

CH_3
HO_2C—+—OH
HO—+—CO_2H
CH_3

two cyclic permutations →

OH
H_3C—+—CO_2H
HO_2C—+—CH_3
OH

OH
HO_2C—+—CH_3
H_3C—+—CO_2H
OH

← left structure in problem

enantiomers

(b) Turning either projection 180° in the plane of the page establishes that it is identical to the other projection.

6.22 (a) *R*
(b) *S*
(c) The upper asymmetric carbon has the *R* configuration; the lower one has the *S* configuration. (Notice that this is a meso compound; hence, once you determine the configuration of one asymmetric carbon, the other must have the opposite configuration.)
(d) Both asymmetric carbons have the *R* configuration.

6.23 Pasteur isolated the chiral stereoisomers, that is, the (2*R*,3*R*) and the (2*S*,3*S*) stereoisomers. (Because they are optically active, they must be chiral.) Yet to be isolated was the meso stereoisomer. Like all meso compounds, *meso*-tartaric acid is achiral and therefore is optically inactive.

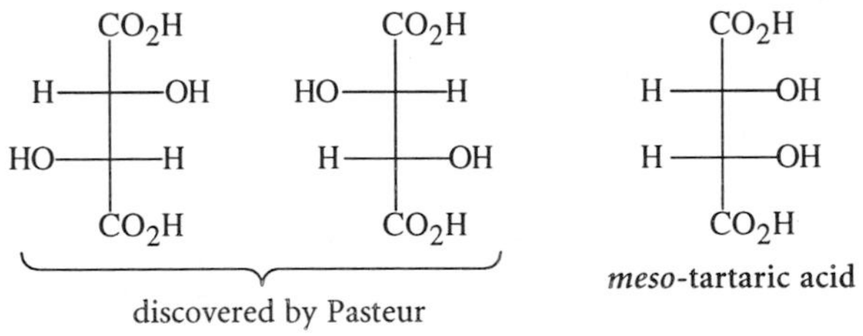

CO_2H
H—+—OH
HO—+—H
CO_2H

CO_2H
HO—+—H
H—+—OH
CO_2H

discovered by Pasteur

CO_2H
H—+—OH
H—+—OH
CO_2H

meso-tartaric acid

6.24 The "chiral reference" was the first crystal that he separated. The handedness of each subsequent crystal was either "like" that of the first one or "opposite" to it.

Solutions to Additional Problems

6.25 (a) $H_2C{=}CHCH_2CH(CH_3)_2$

4-methyl-1-pentene

has no stereocenters or asymmetric carbons

(b) stereocenters; asymmetric carbon and stereocenter

$H_3C(H)C{=}C(H)CHCH_2CH_3$ with CH_3 on CH

(*E*)-4-methyl-2-hexene

(c) CH_3 — asymmetric carbon and stereocenter; stereocenters

3-methylcyclohexene

In part (c), the two carbons of the double bond are stereocenters, but *trans*-cyclohexene is too unstable to exist. That these carbons are stereocenters, however, has practical significance when cyclohexene undergoes chemical reactions. This point is discussed in Chapter 7 on text p. 283.

6.26 (a) First, draw the structure of 3,4,5,6-tetramethyl-4-octene.

$$CH_3CH_2\overset{3}{C}H(CH_3)\overset{4}{C}(CH_3){=}\overset{5}{C}(CH_3)\overset{6}{C}H(CH_3)CH_2CH_3$$ 3,4,5,6-tetramethyl-4-octene

This compound exists as six stereoisomers: (3*S*,4*Z*,6*S*), (3*R*,4*Z*,6*R*), (4*Z*,*meso*), (3*S*,4*E*,6*S*), (3*R*,4*E*,6*R*), and (4*E*,*meso*).

(b) Carbons 3, 4, 5, and 6 are stereocenters.

(c) Carbons 3 and 6 are asymmetric carbons.

6.27 (a) $CH_3CH_2\overset{*}{C}H(CH_3)$–cyclohexyl

(b) $H_3C{-}\overset{*}{C}H(NH_2){-}CH_2OH$

(c) CH_3 (no asymmetric carbons)

(d) CH_3, OH (bicyclic structure with asymmetric carbons marked *)

(e) HO (steroid structure with asymmetric carbons marked *)

6.28 (a) *R*; CH_3, O, OCH_3, H_2C, CH_2, C, H_3C, H_3C

(b) CH_3, *S*; H, OCH_3; H, OH; CH_3, *R*

(c) The asymmetric atom is a phosphorus, and it has the *R* configuration. (Notice that the —$N(CH_3)_2$ group is the group of lowest priority.)

(d) In a meso compound, asymmetric carbons at corresponding distances from the line of symmetry have opposite configurations.

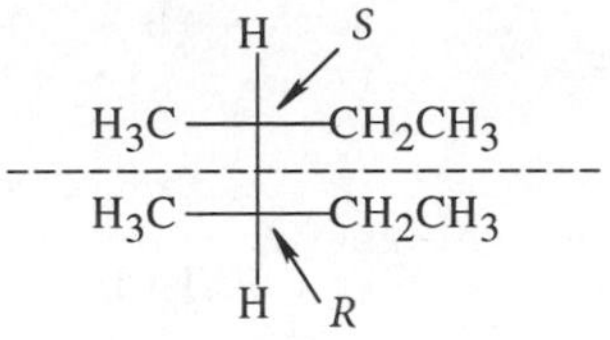

meso-3,4-dimethylhexane

6.29 (a) The two stereoisomers of ibuprofen:

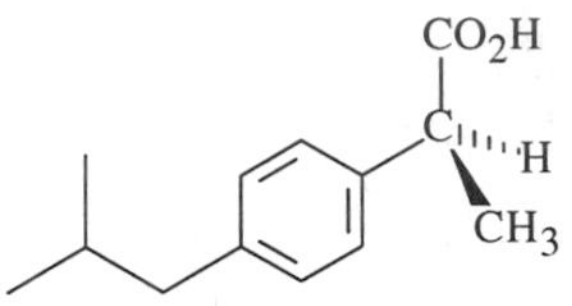

(*S*)-ibuprofen 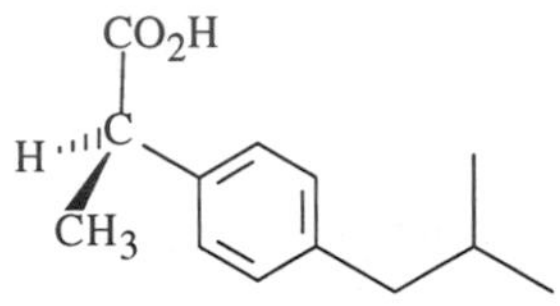 (*R*)-ibuprofen

(b) The *S* enantiomer is drawn on the left.

Don't forget that there are many correct ways to draw a given perspective structure. The structures above were drawn with the group of lowest priority in a "rear" (dashed wedge) position, because it is easiest to assign configuration in such a situation. If you wonder whether one of your structures is the same as one of those above, build a model of each and test them for congruence. If they are congruent, they are the same.

6.30 Start with an arbitrary Fischer projection of the proper configuration and rotate it 180° in the plane of the paper:

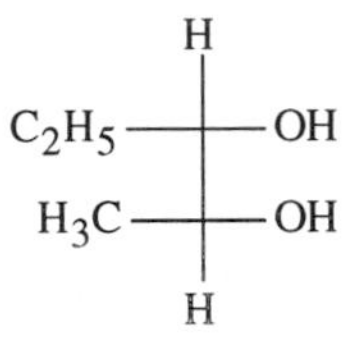

A 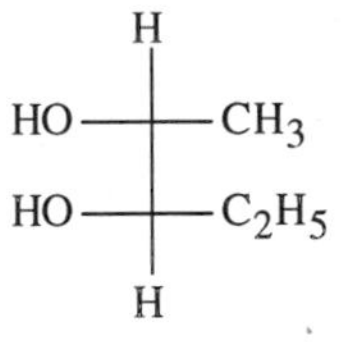 *B*

Perform cyclic permutations of *A* at the bottom carbon to generate two new projections:

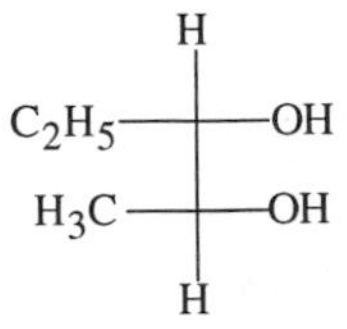

A

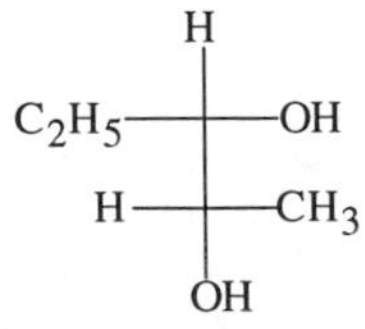

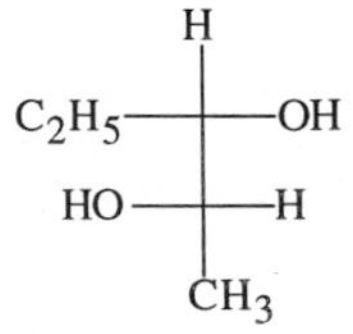

Next, perform a cyclic permutation of *A* at the top carbon to generate a new projection *A*′ and perform two cyclic permutations on the lower carbon of *A*′ to generate a total of three new projections:

C_2H_5 / HO—H / H_3C—OH / H — *A′*

C_2H_5 / HO—H / H—CH_3 / OH

C_2H_5 / HO—H / HO—H / CH_3

Then do a cyclic permutation on the top carbon of *A′* to give *A″*, and perform two cyclic permutations on the lower carbon of *A″* to get three more projections:

OH / H—C_2H_5 / H_3C—OH / H — *A″*

OH / H—C_2H_5 / H—CH_3 / OH

OH / H—C_2H_5 / HO—H / CH_3

Now repeat the same routine on structure *B*:

H / HO—CH_3 / HO—C_2H_5 / H — *B*

H / HO—CH_3 / H—OH / C_2H_5

H / HO—CH_3 / C_2H_5—H / OH

OH / H_3C—H / HO—C_2H_5 / H — *B′*

OH / H_3C—H / H—OH / C_2H_5

OH / H_3C—H / C_2H_5—H / OH

CH_3 / H—OH / HO—C_2H_5 / H — *B″*

CH_3 / H—OH / H—OH / C_2H_5

CH_3 / H—OH / C_2H_5—H / OH

6.31 Either of the following isomeric heptanes is a correct answer:

C_2H_5—CH(CH_3)—$CH_2CH_2CH_3$
3-methylhexane

C_2H_5—CH(CH_3)—$CH(CH_3)_2$
2,3-dimethylpentane

6.32 The following stereoisomer of 1,2-dimethylcyclopropane is chiral. If you're not convinced, build a model of it and another model of its mirror image and test them for congruence.

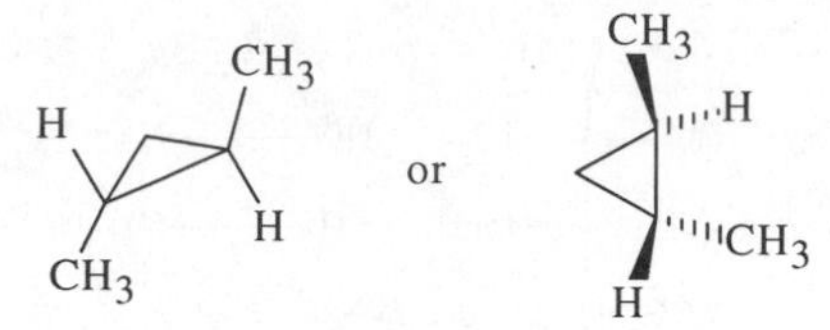

H CH_3 H CH_3 or CH_3 H CH_3 H

6.33 (a) True. 2-Chlorohexane and 3-chlorohexane are examples of chiral constitutional isomers.

(b) True by definition.

(c) False. An achiral molecule is congruent to its mirror image. (See Fig. 6.1 for an example.)

(d) True by definition

(e) False. For example, 2-chlorohexane has no diastereomer. In order for a chiral molecule to have a diastereomer it must have two or more stereocenters.

(f) False. Some *E,Z* isomers are not chiral (*e.g.*, (*E*)- and (*Z*)-2-butene). Likewise, any meso compound is an achiral diastereomer of a compound containing asymmetric carbons.

(g) False. Meso compounds have asymmetric carbons and are not chiral.

(h) False, because some stereoisomerism is not associated with chirality, for example, *E,Z* stereoisomerism.

(i) True. By definition, interchanging two groups at a stereocenter generates a stereoisomer.

(j) False. Molecules with a mirror-image relationship must be either enantiomeric or identical.

(k) False. Every chiral compound is optically active, although it is possible in some cases for the optical activity to be too small to be detected experimentally.

(l) False. Optical rotation has no general relationship to *R* and *S* configuration.

(m) False, because certain achiral molecules have symmetry elements other than planes of symmetry. (See text pp. 205 and 222 and Study Guide Link 6.5.)

(n) True, because the presence of a plane of symmetry is sufficient (although not necessary) to eliminate chirality.

(o) True. An asymmetric carbon is one type of stereocenter.

6.34 The isomeric alkyl chlorides $C_6H_{13}Cl$ with the carbon skeleton of 3-methylpentane:

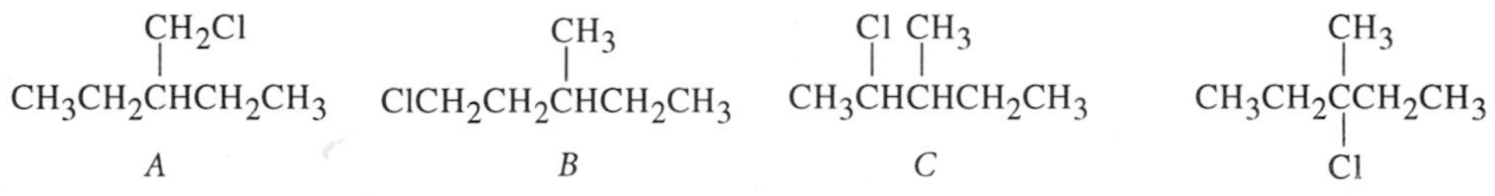

CH_2Cl $CH_3CH_2CHCH_2CH_3$	CH_3 $ClCH_2CH_2CHCH_2CH_3$	$Cl\ CH_3$ $CH_3CHCHCH_2CH_3$	CH_3 $CH_3CH_2CCH_2CH_3$ Cl
A	*B*	*C*	*D*

Compounds *B* and *C* are chiral. The four compounds *A–D* are constitutional isomers. Compound *B* can exist as a pair of enantiomers; compound *C* can exist as two diastereomeric pairs of enantiomers (four stereoisomers total).

6.35 Meso compounds must be achiral compounds with at least two asymmetric carbons and symmetrical branching patterns.

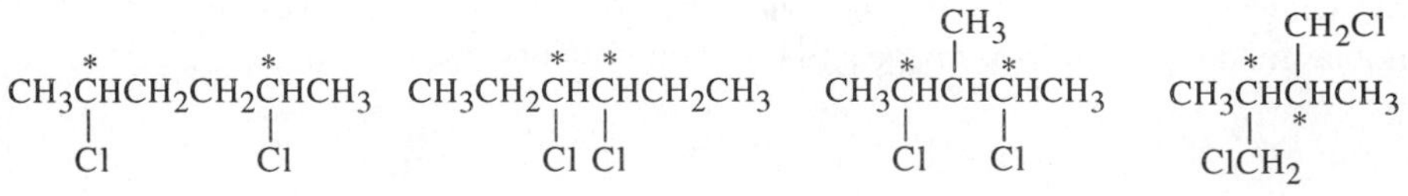

$\overset{*}{}\ \ \overset{*}{}$ $CH_3CHCH_2CH_2CHCH_3$ $Cl\qquad Cl$	$*\ *$ $CH_3CH_2CHCHCH_2CH_3$ $Cl\ Cl$	CH_3 $*\ \ *$ $CH_3CHCHCHCH_3$ $Cl\quad Cl$	CH_2Cl $*$ $CH_3CHCHCH_3$ $*$ $ClCH_2$
A	*B*	*C*	*D*

Compound *C* can exist as two different meso compounds. Here are their Fischer projections:

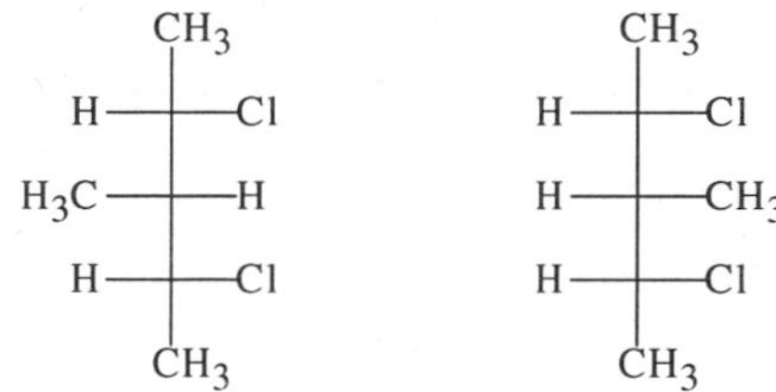

In a case like this, the central carbon is called a *pseudoasymmetric carbon.* Like an asymmetric carbon, a pseudoasymmetric carbon is bound to four different groups. However, two of the groups have identical connectivities and themselves contain one or more asymmetric carbons at which they differ in their *R* or *S* configuration(s).

6.36 The conformations of isopentane (generated by clockwise rotations of the rear carbon):

CH_3, H, CH_3, H, H, CH_3 — *A*; CH_3, H_3C, H, H, H, CH_3 — *B*; CH_3, H_3C, CH_3, H, H, H — *C*

(a) Conformations *A* and *B* are chiral; they are conformational enantiomers.

(b) These two conformations interconvert so rapidly that they cannot be isolated. As the text states on page 229, a molecule that consists of rapidly interconverting enantiomers is considered to be achiral.

(c) The heats of formation of *A* and *B* are identical, and are lower than the heat of formation of *C*, because *C* contains two *gauche*-methyl interactions, whereas *A* and *B* each have only one. The heats of formation of *A* and *B* are equal because enantiomers have identical physical properties.

6.37 Compound *B* can undergo amine inversion, which rapidly interconverts its enantiomeric forms. Compound *A* cannot undergo inversion because it does not have an unshared electron pair; hence, its two enantiomeric forms cannot interconvert and can therefore be isolated.

6.38 A 0.5 *M* solution of this alkene contains (0.5 mol/L)(146.2 g/mol)(0.001 L/mL) = 0.0731 g/mL. Using Eq. 6.1, with l = 0.5 dm, the observed rotation is

$$(+76 \text{ deg}\cdot\text{mL}\cdot\text{g}^{-1}\cdot\text{dm}^{-1})(0.0731 \text{ g}\cdot\text{mL}^{-1})(0.5 \text{ dm}) = +2.78 \text{ deg}$$

6.39 (a) To be sure that every possibility is covered, draw all $2^3 = 8$ possibilities:

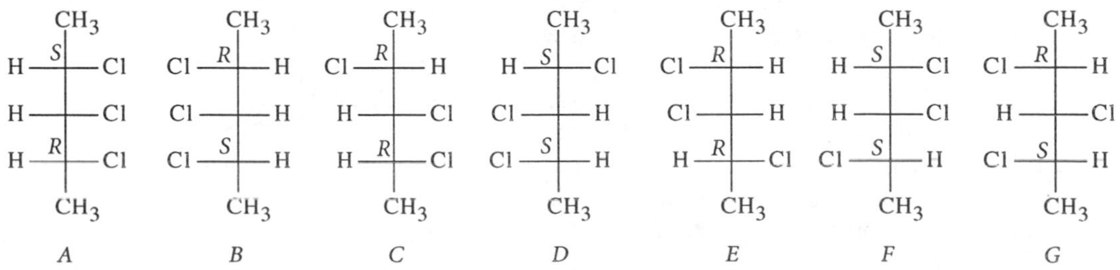

By turning projections 180° in the plane of the page you can show that *A* = *B*, *C* = *E*, *D* = *F*, and *G* = *H*. Consequently, there are four stereoisomers of this compound. (Which structures are identical is also clear from the configurations assigned in part (b).)

(b) See structures in part (a) for the asymmetric carbon configurations.

(c) The stereoisomers that do *not* have internal mirror planes are chiral, that is, structures *C* (=*E*) and *D* (=*F*). Their mirror images are noncongruent. Structures *A* (=*B*) and *G* (=*H*) are meso and therefore achiral.

(d) Carbon-3 is a stereocenter in *A* (=*B*) and *G* (=*H*) because interchange of the H and the Cl in each case gives a diastereomer. Carbon-3 is *not* a stereocenter in structures *C* (=*E*) and *D* (=*F*). (See the note following the solution to Problem 6.14 on p. 166 of this manual.)

6.40 Because hydrogenation causes two different groups to become identical, the carbon that was asymmetric before the reaction is not asymmetric after the reaction. Because the product is achiral, it cannot be optically active.

$$H_2C{=}CH{-}CH(CH_3){-}CH_2CH_3 \xrightarrow{H_2,\ \text{catalyst}} CH_3CH_2{-}CH(CH_3){-}CH_2CH_3$$

chiral → achiral

6.41 Salts that are either identical or enantiomers should have identical solubilities. The solution to the problem then hinges on determining the relationship between each pair of salts. If they have the same configurations at corresponding asymmetric carbons, they are identical and have identical solubilities. If they have different configurations at both of the corresponding asymmetric carbons, they are enantiomers and have identical solubilities, because enantiomers have identical properties. If they have different configurations at one asymmetric carbon and the same configuration at the other, they are diastereomers and have different solubilities, because diastereomers have different properties. For *A* and *B*:

A: cation (Fischer projection: top Ph, left H, right CH_3, bottom $\overset{+}{N}H_3$) — *S*; anion (top H, left ^-O_2C, right CH_3, bottom Ph) — *S*

B: cation (top $\overset{+}{N}H_3$, left H, right Ph, bottom CH_3) — *S*; anion (top H, left H_3C, right CO_2^-, bottom Ph) — *R*

(Notice that we have turned the anionic component of *A* 180° in the plane of the paper for ease of compariuson to *B*.) The two salts are therefore diastereomers and should have different solubilities in methanol. Similar analyses give the following conclusions:

A and *C* are diastereomers and should have different solubilites.
A and *D* are enantiomers and have identical solubilities.
B and *C* are enantiomers and have identical solubilities.
B and *D* are diastereomers and should have different solubilities.
C and *D* are diastereomers and should have different solubilities.

6.42

A and *B*:	diastereomers	*A* and *C*:	identical (turn 180° in plane)
A and *D*:	diastereomers	*A* and *E*:	diastereomers
B and *C*:	diastereomers	*B* and *D*:	enantiomers (do a cyclic permutation on top carbon of *D*)
B and *E*:	diastereomers	*C* and *D*:	diastereomers
C and *E*:	diastereomers	*D* and *E*:	diastereomers

6.43 (a) If the geometry is tetrahedral, there are two enantiomeric stereoisomers (drawn in Fischer projection):

(b) For square-planar geometry, there are three stereoisomers. These are found by making all possible pairwise switches between adjacent groups and ruling out identities. Note that switching nonadjacent groups gives the same structure, because these are *truly planar* structures (not Fischer projections of nonplanar structures), and can therefore lifted out of the plane of the page and turned over.

different ways of drawing the same compound

(c) For pyramidal geometry, there are three diastereomeric sets of enantiomers (six stereoisomers total):

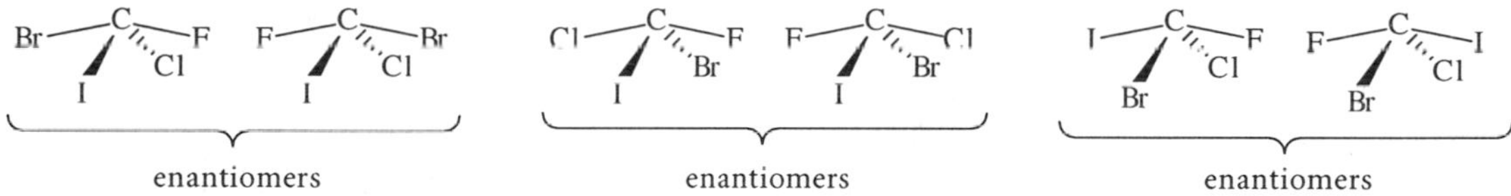

enantiomers enantiomers enantiomers

6.44 In a compound of the form X_2ZY_2 with square-planar geometry, there are two ways to arrange groups X and Y about atom Z: with like groups in adjacent corners, or with like groups in opposite corners:

diastereomers

Because these are stereoisomers, and they are not enantiomers, they must be diastereomers. Tetrahedral compounds of the form X_2ZY_2 (for example, H_2CCl_2) *cannot* exist as stereoisomers. Hence, the fact that $Cl_2Pt(NH_3)_2$ exists as stereoisomers with different properties shows that these stereoisomers are diastereomers, and hence that their geometry about platinum is square-planar.

6.45 The ultimate test for chirality is to make the mirror image and test it for congruence:

mirror images

turn 180° in the page

not congruent

Because the structure is not congruent to its mirror image, the compound is chiral.

6.46 Stereocenters are indicated with an asterisk (*) in the following structures.

(a) (b) (c) (d)

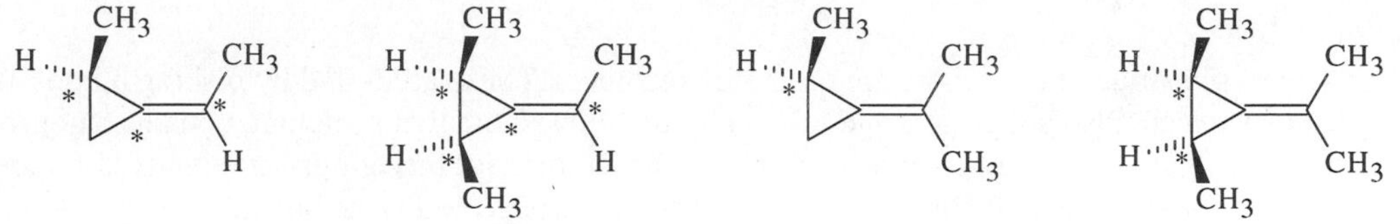

In part (b), identification of the stereocenter at the junction of the ring and the double bond is tricky. Breaking the two bonds of the ring and interchanging them gives a stereoisomer:

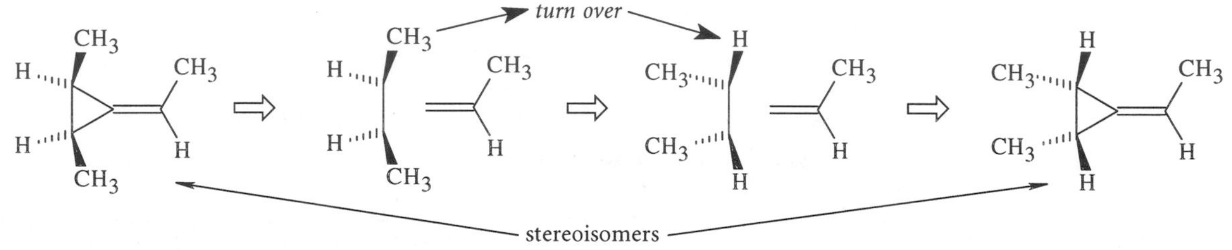

turn over

stereoisomers

6.47 The optical rotations and melting points show that Fischer obtained enantiomers as a result of this transformation. Consider what each of the two possible geometries predicts for an interchange of the two groups. For tetrahedral carbon, the interchange of any two groups at an asymmetric carbon gives enantiomers. Consequently, the experimental result is consistent with tetrahedral geometry.

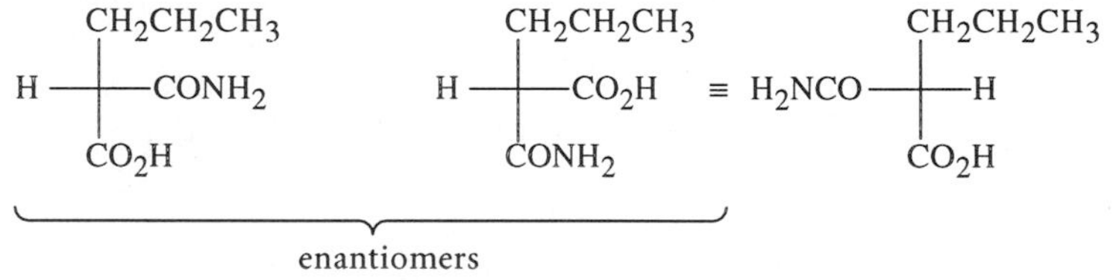

enantiomers

If the molecule were pyramidal, and if Fischer *happened* to be dealing with a stereoisomer in which the —CO_2H and —$CONH_2$ groups were at opposite (rather than adjacent) corners of the pyramid, he would also have obtained enantiomers from his experiment:

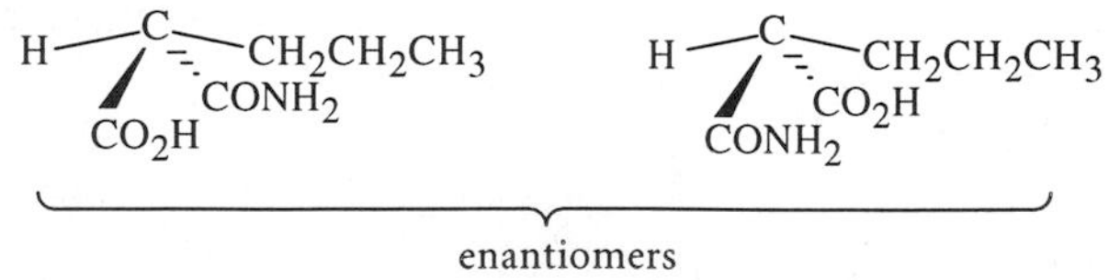

enantiomers

But if Fischer *happened* to be dealing with a stereoisomer in which the —CO_2H and —$CONH_2$ groups were at adjacent corners, the exchange should have given diastereomers. Thus, the conversion of one stereoisomer into its enantiomer could have been the result of two things occurring simultaneously: (1) pyramidal geometry, and (2) the fortuitous choice of a particular stereoisomer. Consequently, Fischer's result was consistent with either tetrahedral or pyramidal geometry.

6.48 Neither result taken alone rules out pyramidal geometry. However, for an atom with two identical groups and pyramidal geometry to be achiral, the two identical groups must be at the opposite corners of the pyramid.

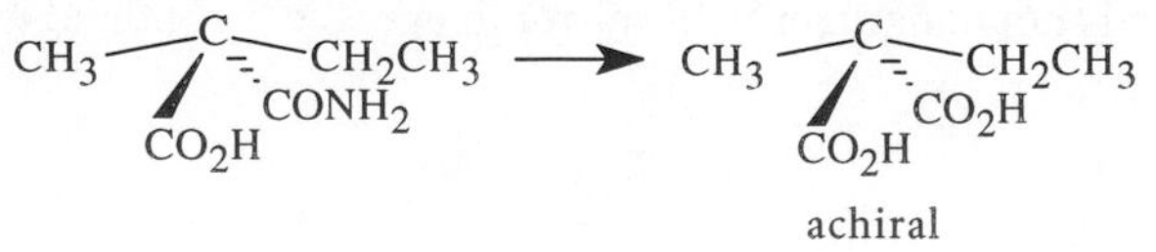

achiral

If the two groups had been at opposite corners of the pyramid in the *first* experiment, then in the *second experiment with the same starting material,* the two groups must have been at adjacent corners, and this transformation would have yielded a chiral compound:

CH_3—C(—CH_2CH_3)(—$CONH_2$)(—CO_2H) ⟶ CH_3—C(—CH_2CH_3)(—$CONH_2$)(—CH_3)

a chiral compound

(Be sure to convince yourself that the molecule on the right is chiral. Do this by drawing (or making a model of) its mirror image and show that the two are noncongruent.) On the other hand, if the starting compound had tetrahedral geometry, then making *any* two groups identical would give an achiral compound. Since this was the experimental result, pyramidal geometry was ruled out, and tetrahedral geometry thus remained the only (and, as it turned out, correct) possibility.

7 Cyclic Compounds; Stereochemistry of Reactions

TERMS

addition reaction 7.9A
angle strain 7.2A, 7.5B
angular methyl group 7.6D
anti-addition 7.9A
axial hydrogen 7.2A
backside attack 7.9C
"banana" bond 7.5B
"bent" bond 7.5B
bicyclic compound 7.6A
boat conformation 7.2B
Bredt's rule 7.6C
bridged bicyclic compound 7.6A
bridgehead carbon 7.6A
chair conformation 7.2A
chair flip 7.2B
chair-to-chair interconversion 7.2B
cis-disubstituted cycloalkane 7.4A
cis ring fusion 7.6B
conformational analysis 7.3
decalin 7.6B
1,3-diaxial interaction 7.3
down carbon 7.2A
envelope conformation 7.5A
equatorial hydrogen 7.2A
flagpole hydrogen 7.2C
fused bicyclic compound 7.6A
inversion of configuration 7.9B
monocyclic compound 7.1
polycyclic compound 7.6A
puckered cyclobutane conformation 7.5B
retention of configuration 7.9B
spirocyclic compound 7.6A
stereoselective reaction 7.9B
stereospecific reaction 7.9C
steroid 7.6D
substitution reaction 7.9B
syn-addition 7.9A
trans-disubstituted cycloalkane 7.4A
trans ring fusion 7.6B
twist-boat conformation 7.2C
up carbon 7.2A

CONCEPTS

I. CYCLOALKANES

A. Monocyclic Compounds

1. A compound that contains a single ring is called a monocyclic compound.
2. The relative stabilities of the monocyclic alkanes can be determined from their heats of formation.
3. Of the cycloalkanes with ten or fewer carbons, cyclohexane has the smallest heat of formation per CH_2; thus, cyclohexane is the most stable of these cycloalkanes.
4. Cyclohexane has the same stability as a typical unbranched alkane (–20.7 kJ/mol or –4.95 kcal/mol per CH_2 group).
5. Cyclopentane has somewhat higher energy than cyclohexane.
6. Cyclobutane and cyclopropane are the least stable of the monocyclic alkanes, cyclopropane being the least stable cycloalkane.
7. Rings larger than cyclopropane are puckered; that is, their carbon skeletons are not planar.

B. Bicyclic and Polycyclic Compounds

1. Bicyclic compounds share two or more common atoms (called bridgehead atoms) between two rings.
   a. A bicyclic compound is classified as a fused bicyclic compound when the bridgehead atoms are adjacent.

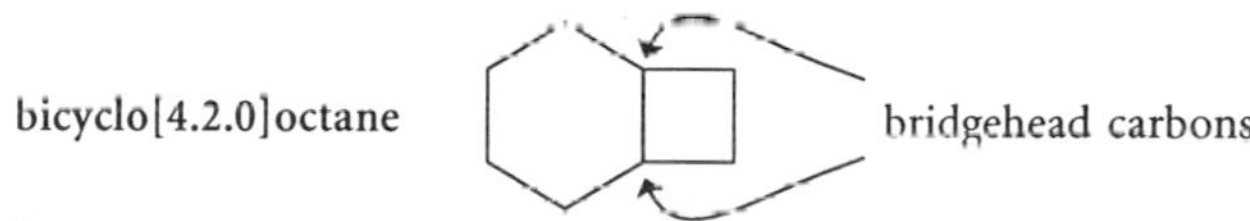

   b. A bicyclic compound is classified as a bridged bicyclic compound when the bridgehead atoms are not adjacent.

2. Spirocyclic compounds share a single common atom between two rings.

spiro[4.3]octane — common carbon

3. The name of a polycyclic structure consists of the following components in the given order:
   a. The substituents and their positions are indicated as in monocyclic compounds; the bridges are numbered beginning at a bridgehead and proceeding around the rings (in decreasing size).
   b. The appropriate prefix *bicyclo*, *tricyclo*, *tetracyclo*, etc., indicates the number of rings present.
   c. Brackets contain, in decreasing size, the number of carbon atoms in each bridge.
   d. The hydrocarbon name is that of the open-chain hydrocarbon containing the same total number of carbon atoms.

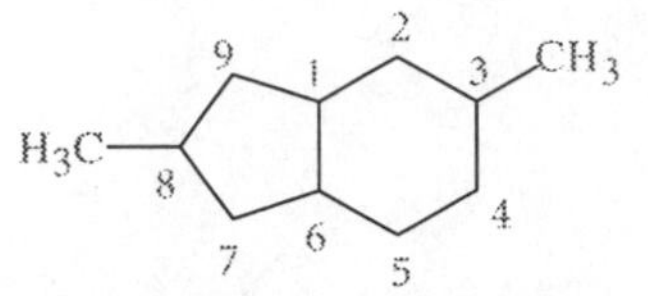

3,8-dimethylbicyclo[4.3.0]nonane

4. Polycyclic compounds contain many rings joined at common atoms.

C. *Trans*-cycloalkenes

1. A double bond at a bridgehead in small rings introduces a great amount of strain, and requires twisting the molecule about the double bond, thus weakening the overlap of $2p$ orbitals involved in the π bond.
2. In a bicyclic compound, a bridgehead atom contained solely within small rings (seven or fewer atoms) cannot be part of a double bond (Bredt's rule). Compounds that violate Bredt's rule are very unstable.
3. Bicyclic compounds that have bridgehead double bonds in rings of more than seven atoms are more stable and can be isolated.

II. CONFORMATIONS OF CYCLOHEXANE

A. Chair Conformation

1. The most stable conformation of cyclohexane is called the chair conformation because of its resemblance to a lawn chair.
   a. Bonds on opposite sides of the ring are parallel.
   b. Cyclohexane contains two types of hydrogen.
      i. If a model of a chair cyclohexane rests on a table, six C—H bonds are perpendicular to the plane of the table; these hydrogens are called axial hydrogens.
      ii. Six C—H bonds point outward along the periphery of the ring; these hydrogens are called equatorial hydrogens.

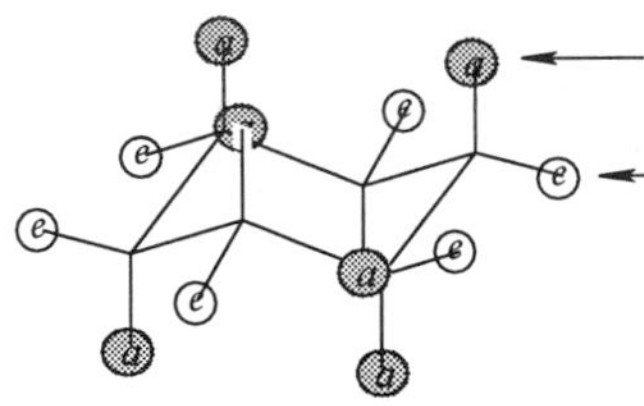

axial hydrogen
equatorial hydrogen

      iii. If an axial hydrogen is up on one carbon, the two neighboring axial hydrogens are down, and vice versa. The same is true of the equatorial hydrogens.
   c. In the chair conformation, all bonds are staggered.
2. A cyclohexane ring is usually drawn in a slightly tilted and rotated perspective so that all of its bonds are visible.
   a. Three alternating carbons (referred to as up carbons) define a plane that is above the plane defined by the other three carbons (referred to as down carbons).
   b. Two perspectives are commonly used for cyclohexane rings; a rotation of either perspective by an odd multiple of 60° about an axis through the center of the ring gives the other perspective.

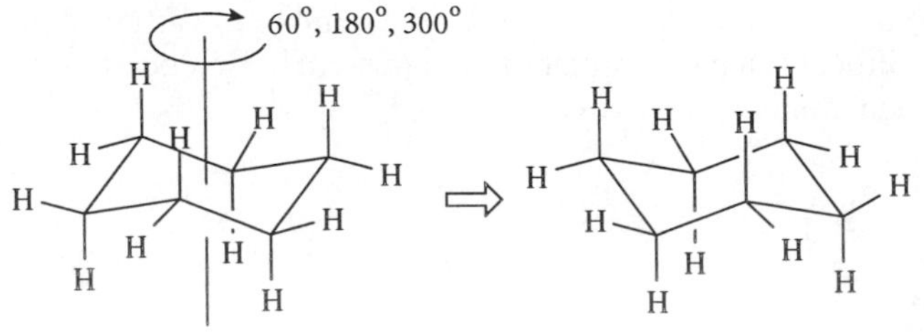

60°, 180°, 300°

c. Drawing cyclohexane carbon skeletons

i. Draw two parallel bonds slanted to the left for one perspective, or slanted to the right for the other perspective.

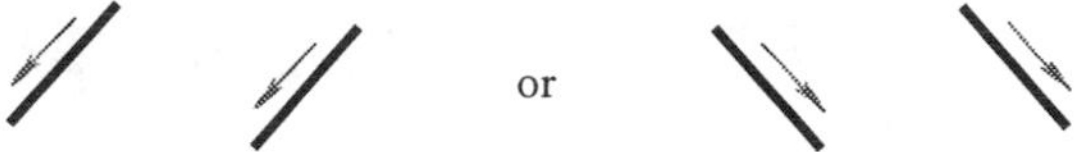

ii. Connect the tops of the slanted bonds with two more bonds in a uneven "V" arrangement.

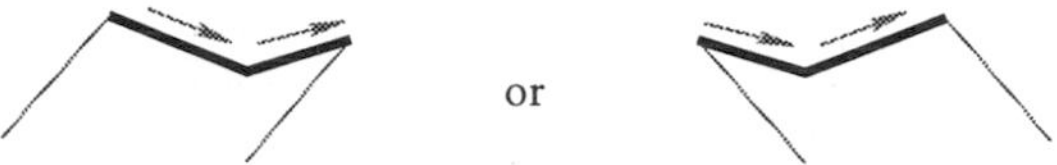

iii. Connect the bottoms of the slanted bonds with the remaining two bonds in an uneven "inverted V" arrangement.

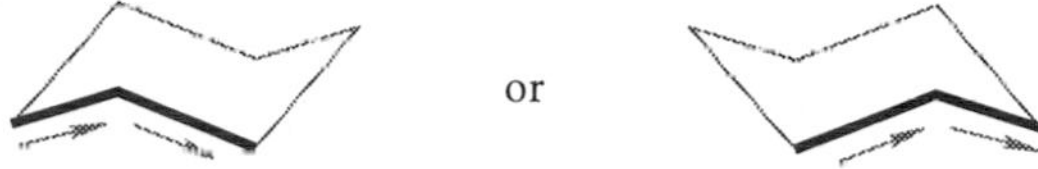

3. The stability of cyclohexane is due to the absence of angle strain and to the fact that all of its bonds can be staggered without compromising the tetrahedral carbon geometry.
4. When a cyclohexane molecule undergoes internal rotations, a change in the conformation of the ring occurs.
5. Movement of each down carbon up and up-carbon down changes one chair conformation into another completely equivalent chair conformation.

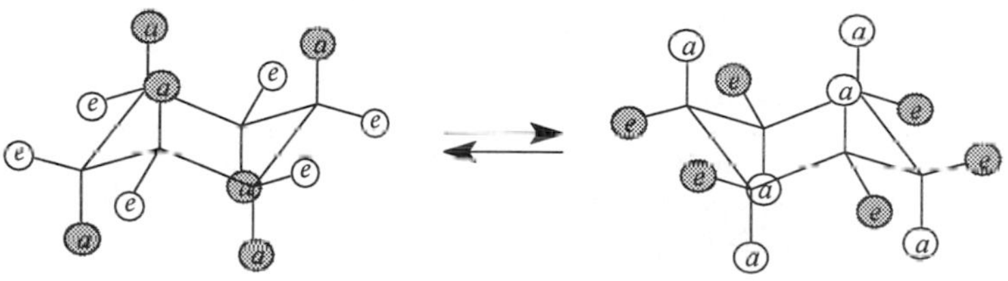

   a. The equatorial hydrogens become axial, and the axial hydrogens become equatorial.
   b. The up carbons become down carbons, and the down carbons become up carbons.
6. This interconversion of two chair forms of cyclohexane is called the chair interconversion or the chair flip.
7. The chair flip causes axial and equatorial hydrogens to change positions rapidly; thus, averaged over time, these hydrogens are equivalent and indistinguishable.

B. Boat Conformation

1. The boat conformation is the result of simultaneous internal rotations about all carbon-carbon bonds except those to one carbon.
2. The boat conformation is not a stable conformation of cyclohexane; it contains two sources of instability:
   a. Certain hydrogens are eclipsed.
   b. The "flagpole" hydrogens experience van der Waals repulsion.

"flagpole" hydrogens

carbon flips up

chair conformation

boat conformation

eclipsed hydrogens

C. Twist-boat Conformation

1. The boat conformation undergoes very slight internal rotation that results in another stable conformation of cyclohexane called a twist-boat conformation.
2. The twist-boat conformation is less stable than the chair conformation.
3. A boat conformation itself can be thought of as the transition state for interconversion of two twist-boat conformations.

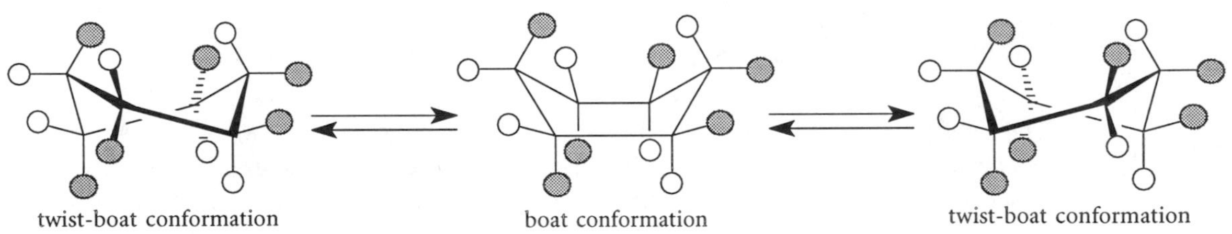

III. SUBSTITUTED CYCLOHEXANES: PLANAR REPRESENTATION AND CONFORMATIONAL ANALYSIS

A. Monosubstituted Cyclohexanes

1. Methylcyclohexane is a mixture of two conformational diastereomers.
   a. Diastereomers have different energies, one form is more stable than the other, and the more stable form is present in greater amount.
   b. The equatorial conformation of a substituted cyclohexane is more stable than the axial conformation.
   c. When the methyl group is in an axial position, van der Waals repulsions occur between one of the methyl hydrogens and the two axial hydrogens on the same face of the ring.
      i. Such unfavorable interactions between axial groups are called 1,3-diaxial interactions.
      ii. Each methyl-hydrogen 1,3-diaxial interaction "costs" the same amount of energy (3.7 kJ/mol) as a *gauche*-butane interaction.

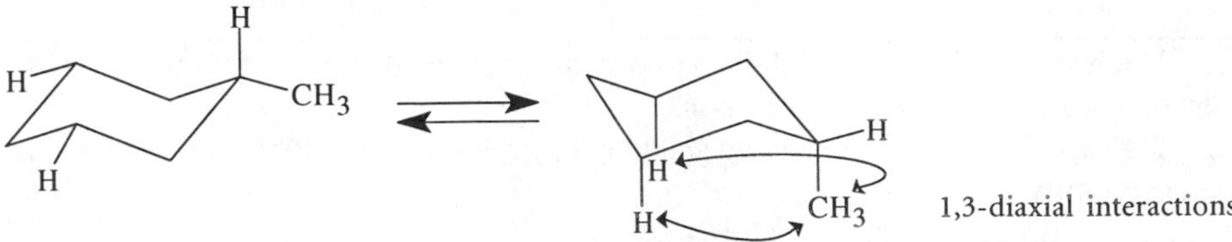

      iii. Because it contains two methyl-hydrogen 1,3-diaxial interactions, axial methylcyclohexane is $(2 \times 3.7) = 7.4$ kJ/mol less stable than equatorial methylcyclohexane.
2. The larger the substituent group on a cyclohexane ring, the more the equatorial conformation is favored.
3. The investigation of molecular conformations and their relative energies is called conformational analysis.

B. Disubstituted Cyclohexanes; Cis-Trans Isomerism in Cyclic Compounds

1. The *trans* designation is used with cyclic compounds when two substituents have an up-down relationship.

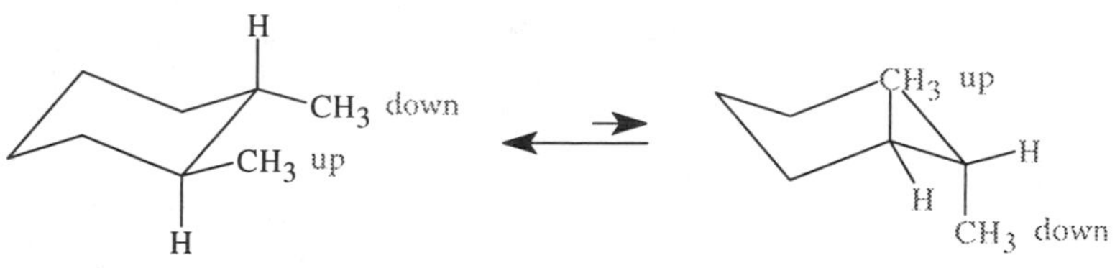

2. The *cis* designation is used with cyclic compounds when two substituents have an up-up or down-down relationship.

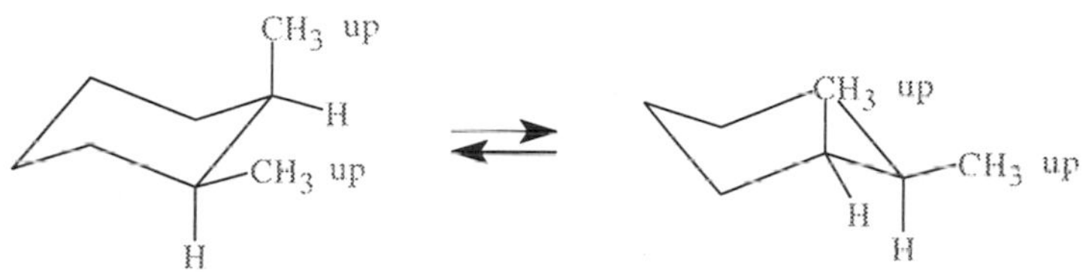

3. The cis and trans designations specify the relative stereochemical configurations of two asymmetric carbons in the ring. These configurations are not affected by the conformation equilibrium.

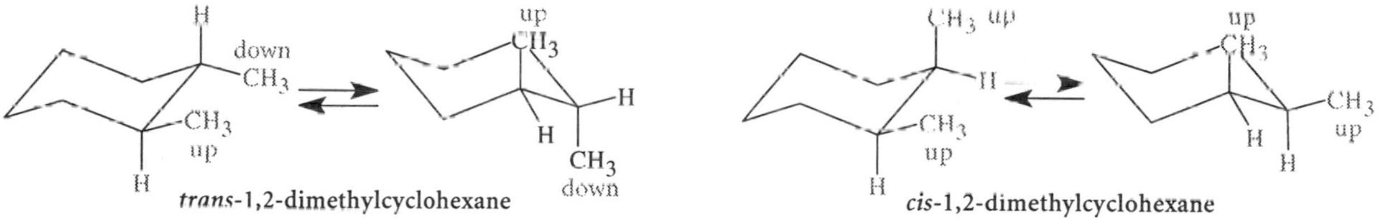

trans-1,2-dimethylcyclohexane

cis-1,2-dimethylcyclohexane

4. When a ring contains more than two substituents, the *R*,*S* system must be used to indicate configuration.
5. When two groups on a substituted cyclohexane conflict in their preference for the equatorial position, the preferred conformation can usually be predicted from the relative conformational preferences of the two groups.

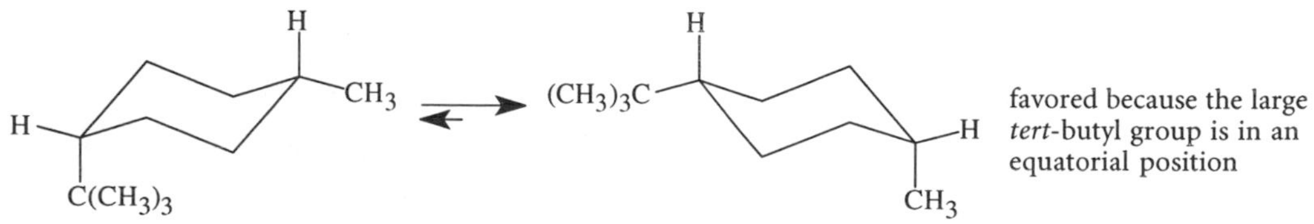

favored because the large *tert*-butyl group is in an equatorial position

C. Planar Representation of Cyclic Compounds

1. Planar structures of cyclic compounds can be used for situations in which conformational details are not important.
   a. The bond of an up substituent (towards the observer) is represented by a solid wedge.
   b. The bond of a down substituent (away from the observer) is represented by a dashed wedge.

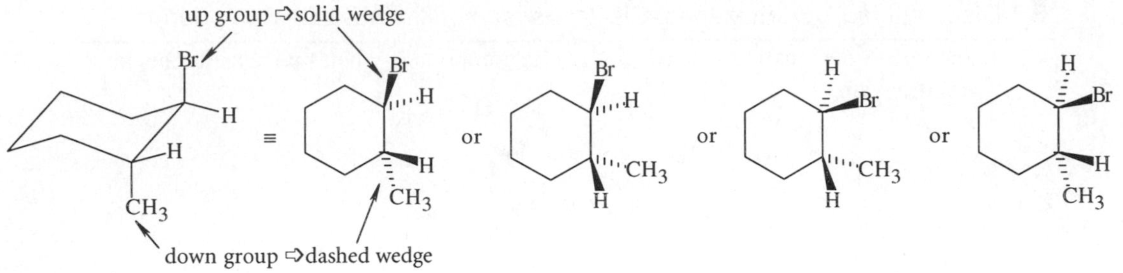

up group ⇨solid wedge

down group ⇨dashed wedge

2. Planar structures are particularly useful for assessing the chirality of cyclic compounds.

D. Stereochemical Consequences of the Chair Flip

1. Some molecules are rapidly interconverting mixtures of diastereomers.

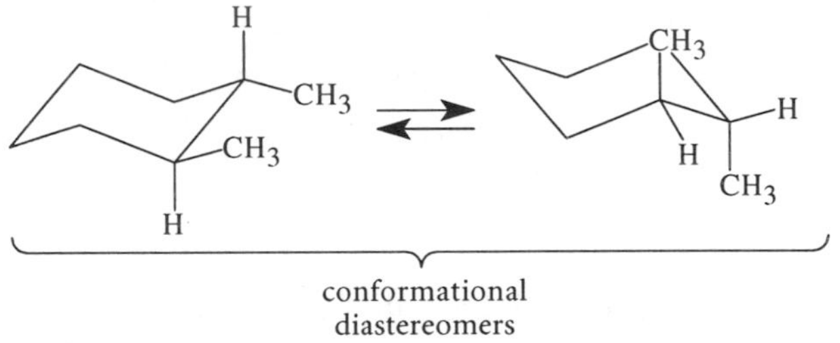

conformational diastereomers

a. Such molecules can be resolved into enantiomers; that is, they can be isolated in optically active form.
b. The chirality of such molecules is evident from their planar structures.

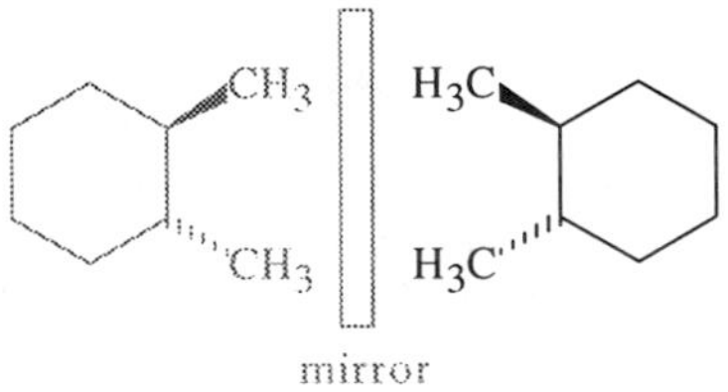

mirror

noncongruent mirror images

2. Some molecules are rapidly interconverting mixtures of enantiomers.

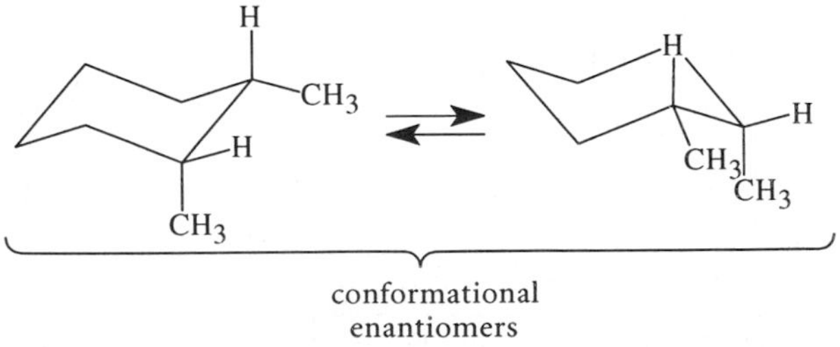

conformational enantiomers

a. Such molecules behave as if they were achiral.
b. The planar structures of such compounds are meso. However, the compounds are actually racemates that consist of rapidly interconverting enantiomers.

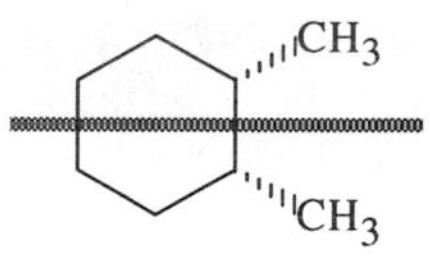

CH_3

mirror plane

CH_3

3. Some molecules are mixtures of rapidly interconverting diastereomeric meso conformations.

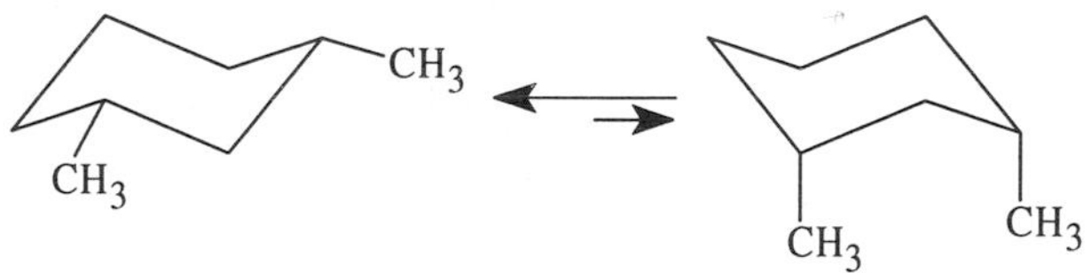

CH_3 CH_3 CH_3 CH_3

IV. CYCLOPENTANE, CYCLOBUTANE, AND CYCLOPROPANE

A. Cyclopentane

1. The carbon skeleton of cyclopentane exists in an envelope conformation.

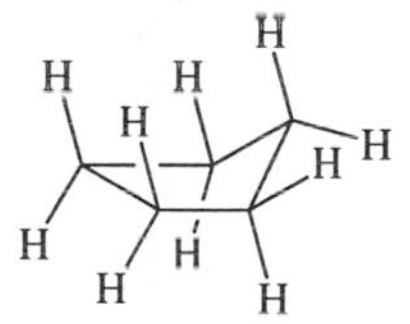

2. The envelope conformation undergoes rapid conformational changes in which each carbon alternates as the "point" of the envelope.
3. Substituted cyclopentanes also exist in envelope conformations, and the substituents adopt positions that minimize van der Waals repulsions with neighboring groups.

the conformation with an equatorial methyl is more stable

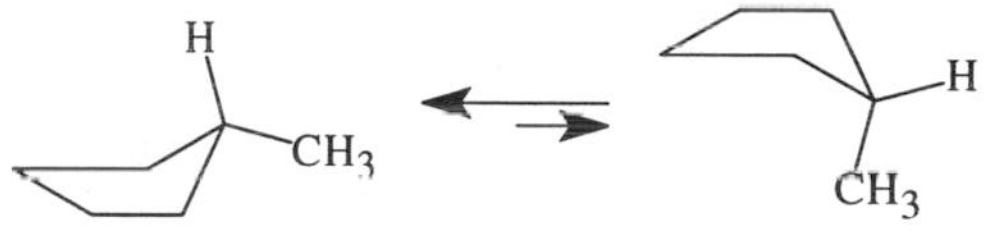

H CH_3 H CH_3

4. When a cyclopentane ring has two or more substituent groups, cis and trans relationships between the groups are possible, just as in cyclohexane.

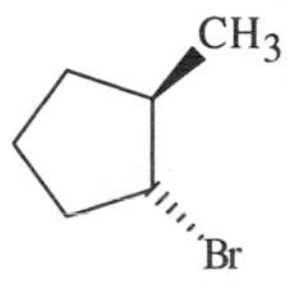

CH_3 Br

trans-1-bromo-2-methylcyclopentane

B. Cyclobutane

1. Cyclobutane consists of two puckered conformations in rapid equilibrium.

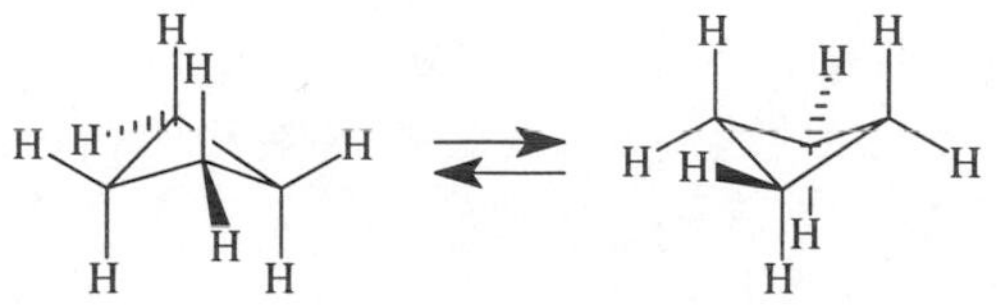

2. Cyclobutane and cyclopropane contain angle strain, or ring strain, which is the increase in energy that results when the angles between carbon-carbon bonds are constrained by the size of the ring to be much smaller than the optimum tetrahedral angle.

C. Cyclopropane

1. Cyclopropane has a planar carbon skeleton.

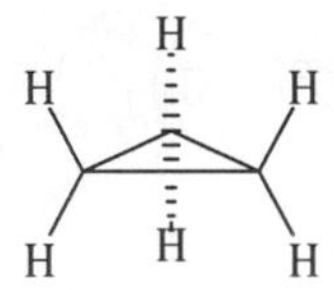

2. Neither angle strain nor eclipsing interactions between hydrogens in cyclopropane can be relieved by puckering.
3. Each carbon-carbon bond of cyclopropane is bent in a "banana" shape between the two carbons.
4. Bent bonds reduce angle strain but do so at a cost of less effective overlap between the carbon orbitals.

V. CONFORMATIONS OF BICYCLIC AND POLYCYCLIC COMPOUNDS

A. Cis and Trans Ring Fusion

1. Two rings in a fused bicyclic compound can be joined in a cis or trans arrangement.
2. Bicyclo[4.4.0]decane, commonly called decalin, can exist as two stereoisomers.
   a. In *cis*-decalin, two —CH_2— groups of ring *B* (shaded circles) are cis substituents on ring *A*; likewise, two —CH_2— groups of ring *A* (shaded squares) are cis substituents on ring *B*.

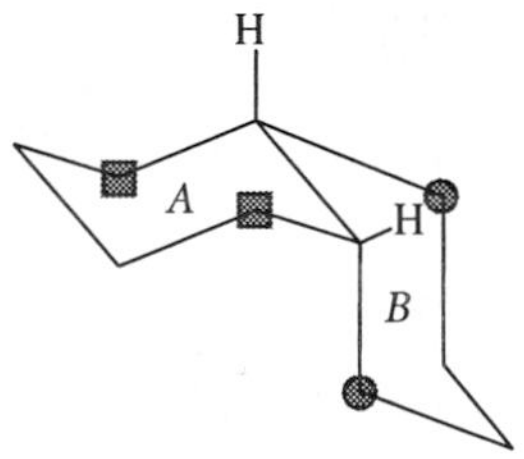

cis-decalin

   b. In *trans*-decalin, the shaded —CH_2— groups in one ring are *trans*-diequatorial substituents on the other. The bridgehead hydrogens are *trans*-diaxial.

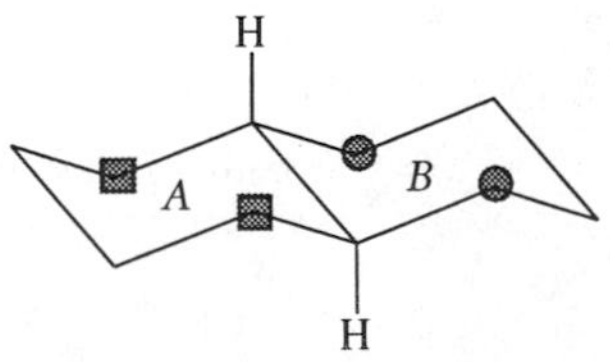

trans-decalin

3. Each cyclohexane ring in *cis*-decalin can undergo the chair flip.

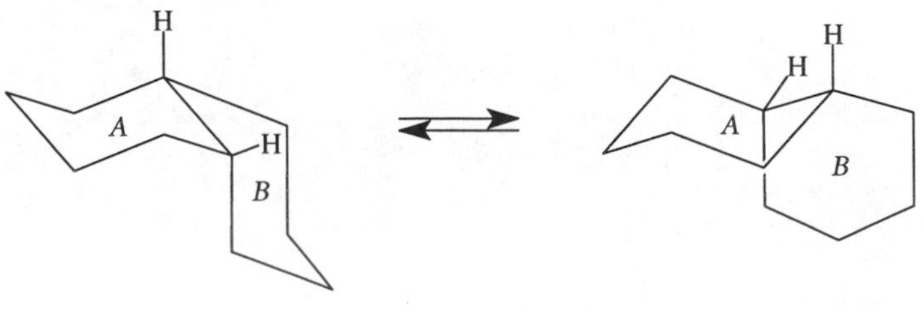

4. In *trans*-decalin, the cyclohexane rings can assume twist-boat conformations but cannot flip into their alternate chair conformations.
5. *Trans*-decalin is more stable than *cis*-decalin because it has fewer 1,3-diaxial interactions.
6. In rings of six atoms or greater, both *cis*- and *trans*-fused isomers are well known, but the trans-fused ones are more stable because 1,3-diaxial interactions are minimized.
7. Trans fusion introduces ring strain in rings of fewer than six atoms.

B. Steroids

1. The steroids include many naturally occurring compounds that have fused rings.
2. In many cases all ring fusions are trans; all-trans ring fusion causes a steroid to be conformationally rigid and relatively flat.
3. Many steroids have axial methyl groups, called angular methyl groups, at carbons 10 and 13.
4. The face of the steroid occupied by the angular methyl groups is called the β-face, whereas the other face is called the α-face.

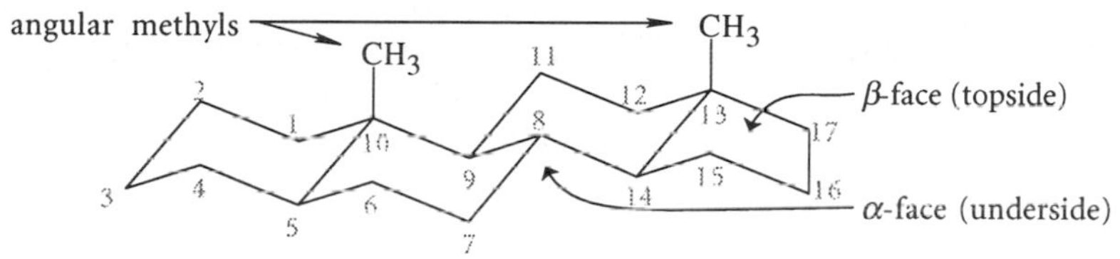

angular methyls CH_3 CH_3 β-face (topside) α-face (underside)

VI. STEREOCHEMISTRY AND CHEMICAL REACTIONS

A. General

1. When chiral products are formed from achiral starting materials, both enantiomers of a pair are always formed at identical rates; the product is always a racemate.

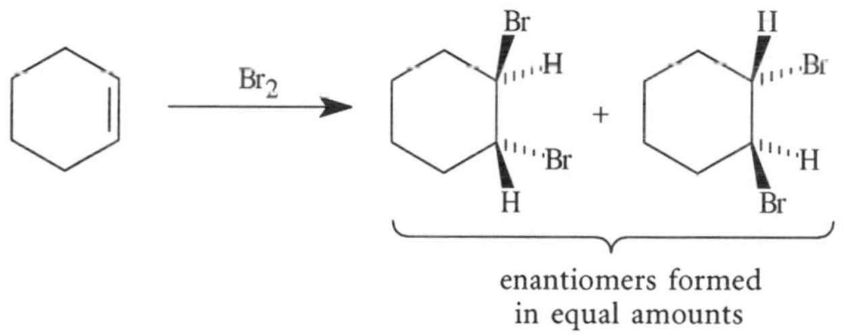

Br_2

enantiomers formed in equal amounts

   a. The enantiomeric transition states have identical free energies, as do the enantiomeric starting materials.
   b. Optical activity never arises spontaneously in the reactions of achiral compounds.
2. Enantiomers have identical reactivities with achiral reagents because enantiomers have the same physical properties.
3. A pair of enantiomers differ in their chemical or physical behavior only when they interact with other chiral objects or forces.
   a. The transition state for the reaction of one enantiomer is the diastereomer of the transition state for the reaction of the other.
   b. Enantiomers have different reactivities with a chiral reagent because the resulting diastereomeric transition states have different energies.
   c. Most enzymes accept only one enantiomer of a chiral substrate.
4. When diastereomeric products can be formed in a reaction, they are in principle formed at different rates and different amounts because they are formed through diastereomeric transition states; one transition state has lower standard free energy than its diastereomer.

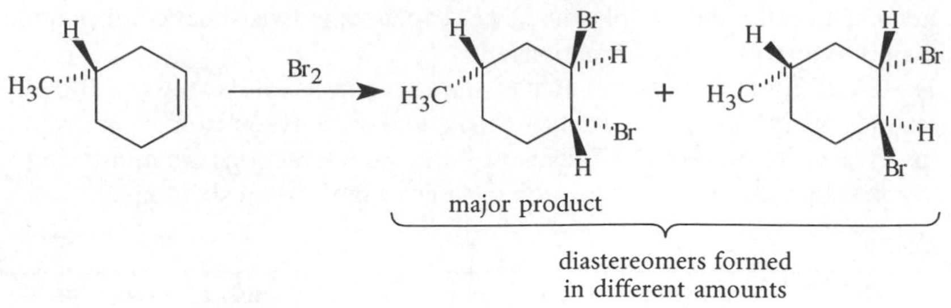

5. Diastereomers in general have different reactivities towards any reagent, whether the reagent is chiral or achiral because diastereomers have different free energies.
6. Each carbon at which a chemical change occurs must be a stereocenter in the product in order for the stereochemistry of the reaction to be determined.

B. *Syn*- and *Anti*-Addition to Alkenes

1. The stereochemistry of an addition can be determined only when the stereochemically different modes of addition give rise to stereochemically different products, that is, when both carbons of the double bond become carbon stereocenters in the product.
2. The *syn* and *anti* modes of addition can be distinguished by analyzing the stereochemistry of the products.
   a. In a *syn*-addition, two groups add to a double bond from the same side or face.

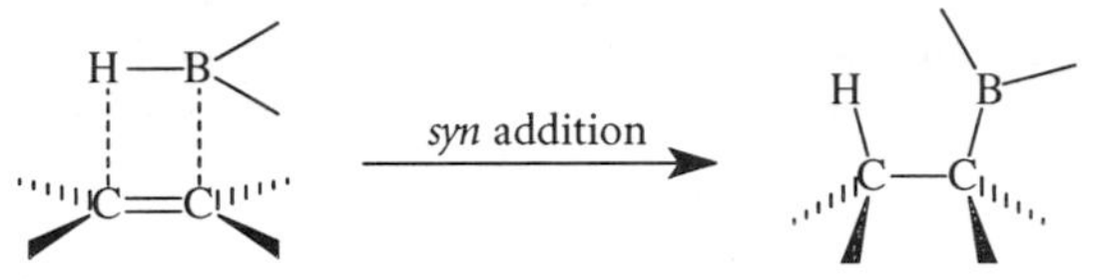

   b. In an *anti*-addition, two groups add to a double bond from opposite sides or faces.

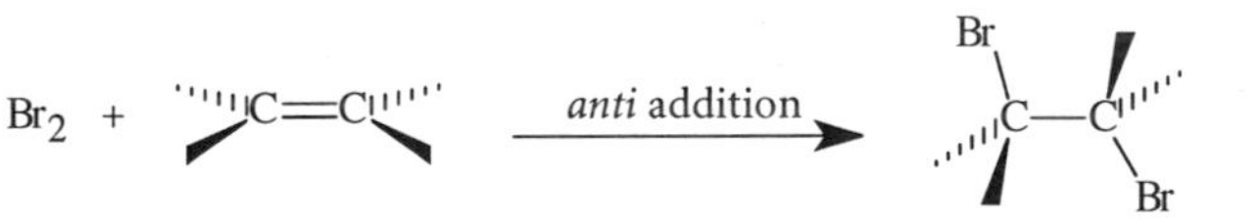

C. Reactions at Asymmetric Carbons

1. A substitution reaction, a reaction in which one group is replaced by another, can occur in two stereochemically different ways.
   a. Retention of configuration occurs when substitution results in the substituting group and the leaving group having the same relative stereochemical positions.

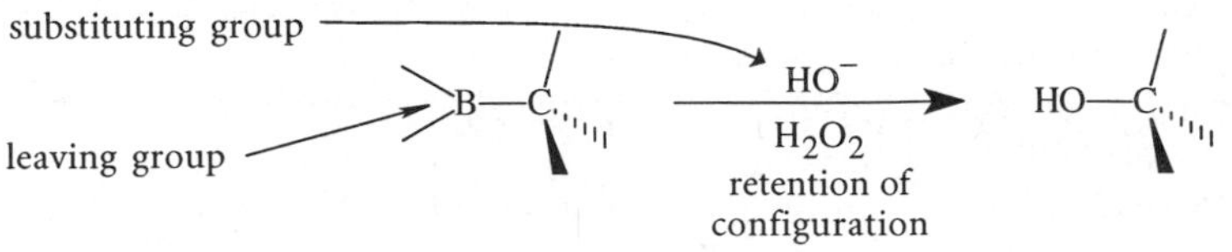

   b. Inversion of configuration occurs when substitution results in the substituting group and the leaving group having different relative stereochemical positions.

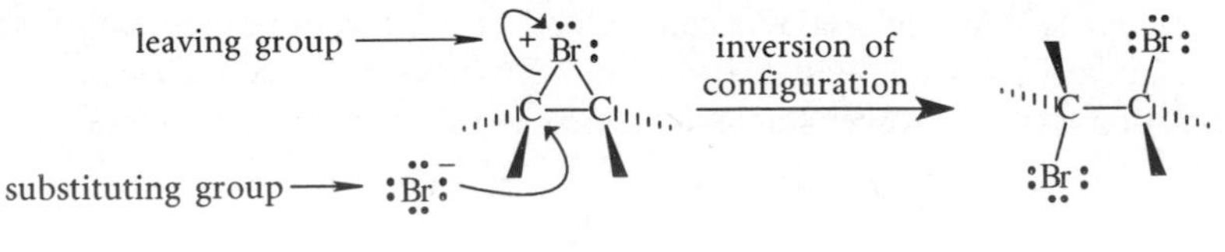

c. Loss of configuration occurs when there is a mixture of both retention and inversion of configuration.

C—HgOAc —($NaBH_4$, loss of configuration)→ C—H + H—C

2. A reaction in which particular stereoisomer(s) of the product are formed to the exclusion of other(s) is said to be a stereoselective reaction.

VII. ALKENE ADDITION REACTIONS

A. Bromine Addition to Alkenes

1. Bromine addition to alkenes is in many cases a stereoselective *anti*-addition.
   a. A bromonium ion forms at either face of the alkene in a *syn*-addition.
   b. Backside attack of bromide on a carbon of the bromonium ion results in inversion of configuration and accounts for the observed *anti*-addition of Br_2 to alkenes.

H, H, + :Br—Br: —(*syn*-addition)→ H, Br:+, H, :Br:− —(substitution with inversion of configuration)→ :Br:, H, Br:, H

overall *anti*-addition

B. Hydroboration–Oxidation of Alkenes

1. Hydroboration is a stereoselective *syn* addition and is a direct consequence of the concerted mechanism of the reaction.
2. The oxidation of organoboranes is a substitution reaction that occurs with retention of configuration.
3. The hydroboration-oxidation of an alkene brings about the net *syn*-addition of the elements of H—OH to the double bond.

CH_3 —(BH_3/THF, *syn* addition)→ H, CH_3, H, B —(H_2O_2/OH^-, substitution with retention of configuration)→ H, CH_3, H, OH

C. Catalytic Hydrogenation of Alkenes

1. Catalytic hydrogenation of most alkenes is a stereoselective *syn*-addition.

CH_3, CH_3 —(H_2, catalyst)→ H, CH_3, CH_3, H

D. Oxymercuration–Reduction of Alkenes

1. Oxymercuration-reduction is not a stereoselective reaction.

a. Oxymercuration of alkenes is typically a stereoselective *anti*-addition.
b. The mercury is replaced by hydrogen with loss of stereochemical configuration.

CH_3 (1-methylcyclohexene) $\xrightarrow{Hg(OAc)_2}$ (mercurinium ion: CH_3, $\overset{+}{Hg}OAc$, H) $\xrightarrow{H_2O}$ (CH_3, OH, HgOAc, H) $\xrightarrow{NaBH_4}$ (CH_3, OH, H, H)

2. Oxymercuration-reduction, although not stereoselective, is a regioselective reaction and is very useful in situations in which stereoselectivity is not an issue.

VIII. STEREOCHEMISTRY AND MECHANISM

1. Stereochemistry provides important details about the mechanism of a reaction.
2. Mechanisms are postulated that are consistent with all known evidence.
3. Mechanisms are revised when necessary to accommodate new experimental evidence.
4. No mechanism is ever proved.

SCIENTISTS

Konrad Julius Bredt (1855–1937) 7.6C
Charles Hackett Bushweller (born 1939) 7.3
Philip E. Eaton (born 1936) 7.6A
Frederick Richard Jensen (1925–1987) 7.3
Russel E. Marker (1902–1995) 7.6D
Leo Armand Paquette (born 1934) 7.6A
Thomas Winston Cole, Jr. (born 1941) 7.6A

Dates in brackets refer to Nobel Prize awards.

STUDY GUIDE LINKS

7.1 Designating Relative Configuration

The use of *cis* and *trans* for designating relative stereochemistry in disubstituted rings has been adopted by *Chemical Abstracts,* the index to the world's chemical literature.

Chemical Abstracts adopted a somewhat different system for designating absolute configuration of chiral cyclic compounds than the one used in the text, which is based on the original IUPAC rules. The *Chemical Abstracts* system involves the use of relative configurations. Consider the following compound:

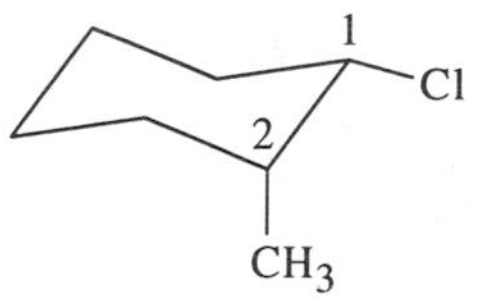

The IUPAC name for this compound is (1*S*,2*R*)-1-chloro-2-methylcyclohexane. The *Chemical Abstracts* name is (1*S*-*cis*)-1-chloro-2-methylcyclohexane. In this system, the absolute configuration is given for the stereocenter of lower number (in this case the carbon bearing the chlorine), and then the configuration of the other stereocenter is given with the relative descriptor *cis* or *trans*.

The 1,4-disubstituted cyclohexanes are particularly interesting.

A 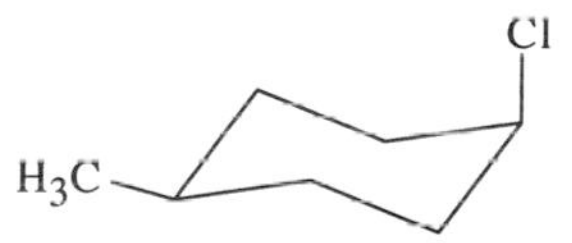 *B*

Notice that these compounds are *achiral*, but each one contains two carbon stereocenters. There are *no* asymmetric carbons. (This is another illustration of the point that not all carbon stereocenters are asymmetric carbons.) Hence, the *R* and *S* system cannot be applied. The stereochemistry of these compounds is fully conveyed by the designation *trans* or *cis*. Thus, Compound *A* is *trans*-1-chloro-2-methylcyclohexane; compound *B* is *cis*-1-chloro-2-methylcyclohexane.

As the text states, the cis-trans system is cumbersome when a ring contains more than two substituents. Although various extensions of this system have been proposed, *Chemical Abstracts* uses the terms α and β to describe relative stereochemistry. The term α means "on the same side of the ring in a planar projection" and β means "on the opposite side of the ring in a planar projection." The positions of all substituents are referred to the position of the substituent on a reference carbon, which, in the simplest cases, is the lowest-numbered carbon. The substituent at the reference carbon by definition is in the α position. Thus we have the following example.

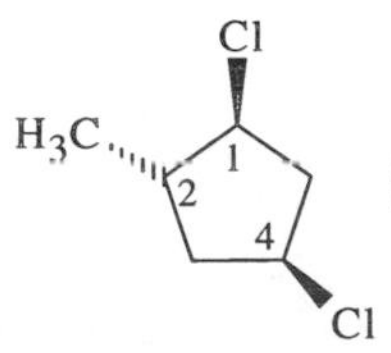

[1*S*-(1α,2β,4α)]-1,4-dichloro-2-methylcyclopentane

The chlorine at carbon-1 is α by definition. Since it is "up", the other "up" substituent (the other chlorine) is also α, and the "down" substituent (the methyl group) is β. If we did not know the absolute configuration, but only the relative configuration, we would leave off the initial "1*S*" and name the compound as (1α,2β,4α)-1,4-dichloro-2-methylcyclopentane.

The following two cases are examples of achiral compounds with three stereocenters:

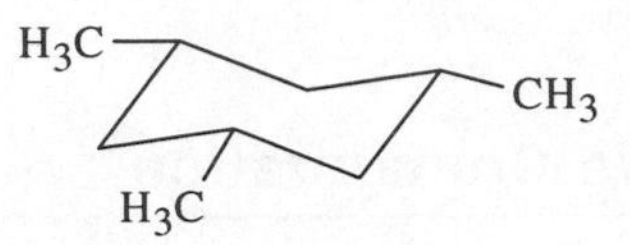

$(1\alpha,3\alpha,5\alpha)$-1,3,5-trimethylcyclohexane

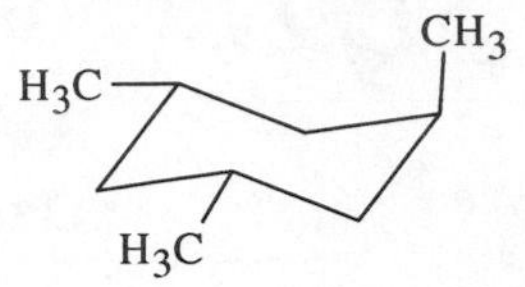

$(1\alpha,3\alpha,5\beta)$-1,3,5-trimethylcyclohexane

In the compound on the left, the three methyl groups are all designated as α because they are on the same "side" of the ring, that is, all "down."

The *Chemical Abstracts* system has additional layers of complexity for more complicated cases, but the above introduction covers most of the cases that one would ordinarily encounter.

✓7.2 Relating Cyclohexane Conformations

Although you can often resort to models to visualize the stereochemical relationships between chair conformations of substituted cyclohexanes, it is useful to be able to carry out on paper manipulations like the ones used in the text. To gain greater facility with such manipulations, try the following exercises, using *cis*-1,2-dimethylcyclohexane as an example. Draw the result of each manipulation, and then check your drawing against the results shown. *Be sure to follow each manipulation with a model if there is any question about it!*

First, turn the structure 180° about a horizontal ("*x*") axis:

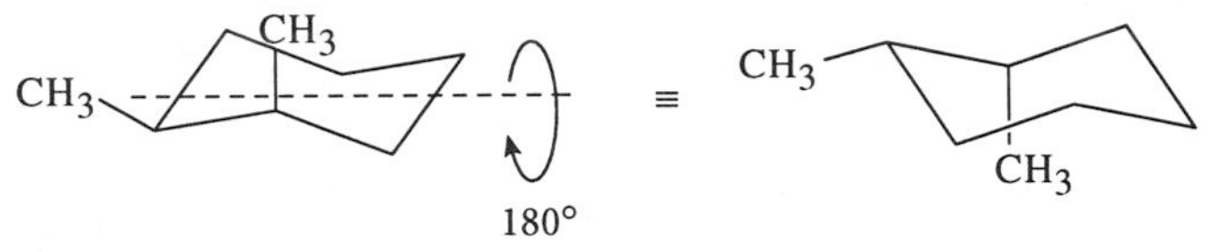

180°

Now turn the original structure 180° about a "*y*" axis:

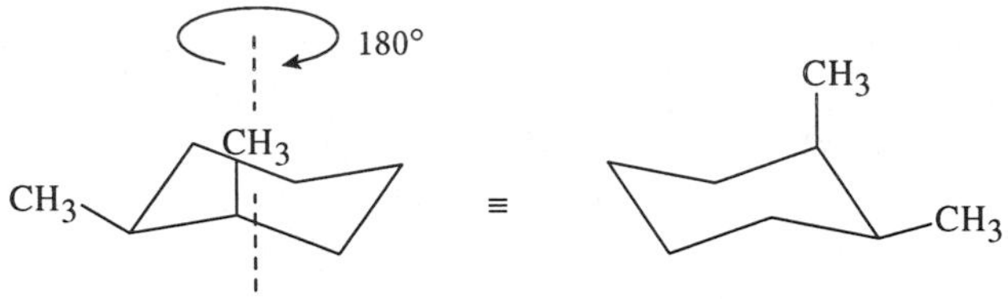

180°

Practice turning the structure in 60° increments about the same axis:

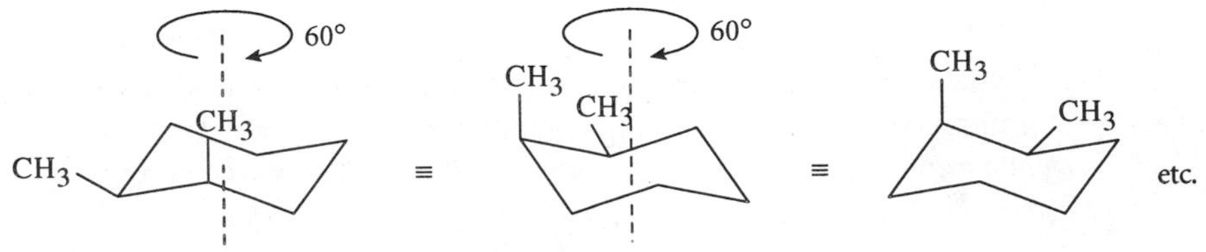

60° 60° etc.

Finally, turn the original structure 180° about an axis perpendicular to the page ("*z*" axis):

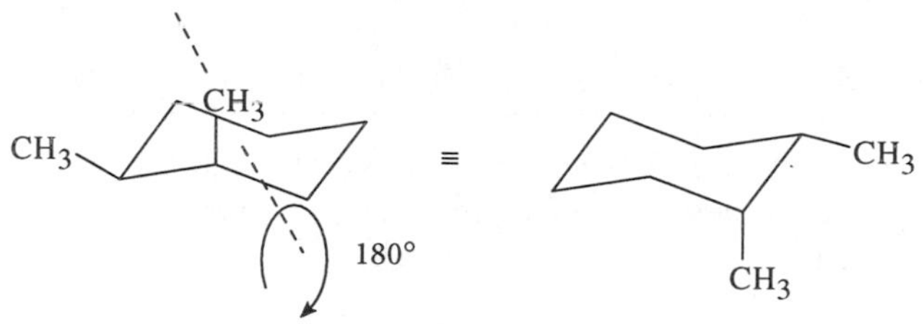

If you can readily perform manipulations like these without recourse to models, you can solve virtually any problem involving cyclohexane stereochemistry.

Let's use these skills to solve the following problem. What is the stereochemical relationship between the two chair conformations of *trans*-1,3-dimethylcyclohexane?

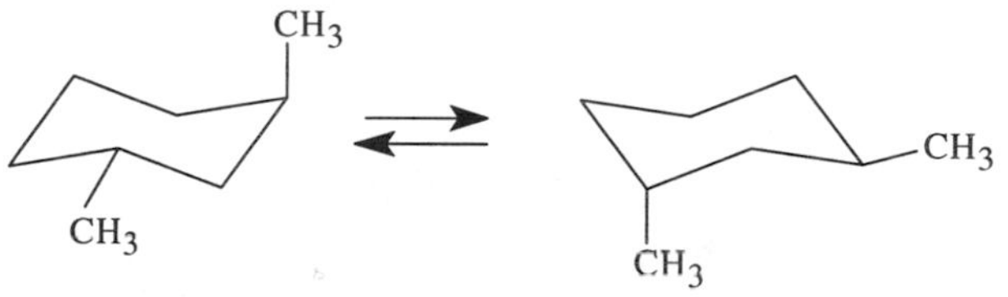

(SG7.1)

There are three possible relationships between these conformations: they could be conformational diastereomers; they could be conformational enantiomers; or they could be identical. The strategy here is to orient one of the methyl groups in one structure in the same way as a methyl group in the other structure. Then compare the relative positions of the remaining methyl groups. To do this, rotate the second structure about an axis perpendicular to the page, as follows:

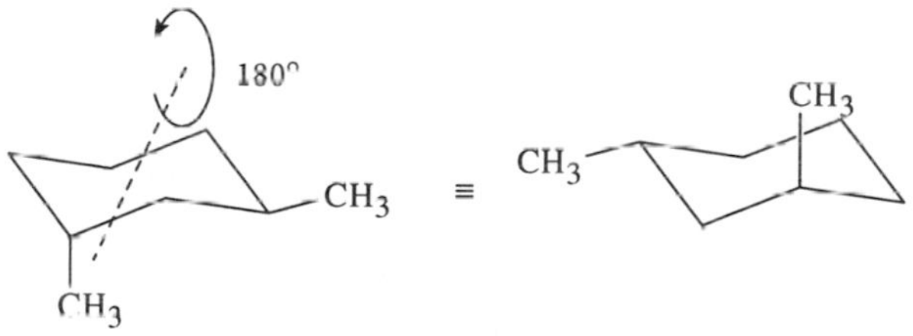

Finally, rotate the resulting structure 60° about an axis in the page, as follows:

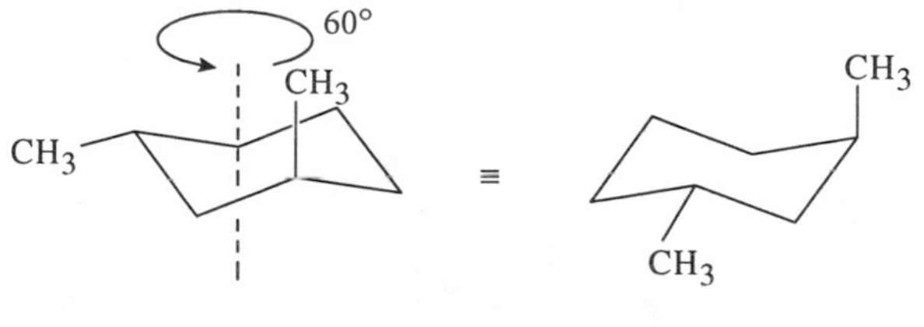

This structure is identical to the first conformation in Eq. SG7.1. Consequently, the chair flip in this case interconverts *identical species.*

It is worth noting that some people, even after practice, have difficulty with manipulations like these because they have inherent difficulty with spatial perception. (Don't assume you are such a person until you have practiced!) This has nothing to do with your intelligence! If you are such a person, don't hesitate to use models. The answer you get may take a little more time, but it is just as correct.

7.3 Alkenelike Behavior of Cyclopropanes

The large amount of strain in cyclopropanes causes them to be more reactive than ordinary cycloalkanes. For example, cyclopropanes can be hydrogenated, although not so readily as alkenes.

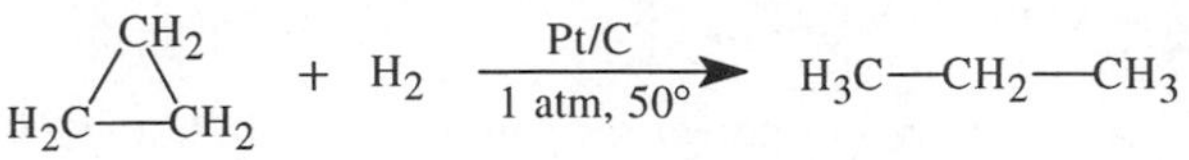

$$\text{cyclopropane } (CH_2)_3 + H_2 \xrightarrow[\text{1 atm, 50°}]{\text{Pt/C}} H_3C{-}CH_2{-}CH_3$$

Although reactions like this suggest that cyclopropanes react like alkenes, an alternative view is that alkenes are cycloalkanes with "two-membered rings."

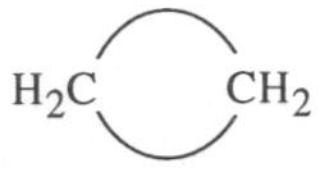

H_2C ⌒ CH_2 ethylene viewed as a cyclic alkane

In fact, it can be argued from hybrid-orbital considerations that a view of the alkene double bond as two "bent bonds" (similar to those in cyclopropane) is just as valid as the picture adopted by this text in which the two bonds are different, one a σ bond and one a π bond. (Many model sets represent double bonds as "bent bonds.")

If ethylene indeed is visualized as a "two-membered ring"—that is, *cycloethane*—it would be expected to be very strained, and that this strain would be reflected in a very large heat of formation. Indeed, the heat of formation of ethylene per CH_2 is even greater than that of cyclopropane.

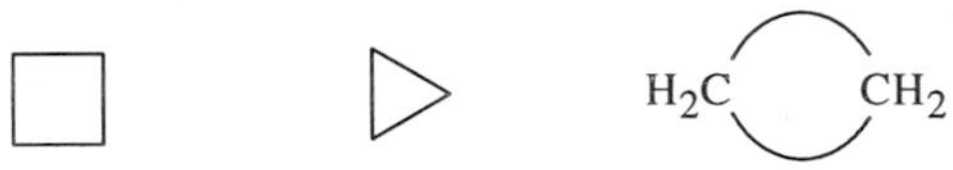

	cyclobutane	cyclopropane	"cycloethane" (ethylene)
ΔH_f° per CH_2 (kJ/mol):	6.90	17.8	26.2
(kcal/mol):	1.65	4.25	6.25

In the conventional π model of the alkene double bond, the high energy of the bonds in ethylene resides completely in the relatively weak π bond. In the bent-bond model, the two bonds are equivalent. They are stronger than a π bond, but weaker than a σ bond.

7.4 Optical Activity

Optical activity is an important example of the principles discussed in the text. Plane-polarized light is actually the *vector sum of right and left circularly polarized light,* two forms of light in which the electric field vector propagates as a right- and left-handed helix, respectively. (See Fig. SG7.1.) These two forms of light are noncongruent mirror images—that is, they are enantiomers. In other words, plane-polarized light is "enantiomeric light." Let's call the right-handed form of light *R,* and the left-handed form *S.* Suppose that plane polarized light—consisting of equal amounts of *R* and *S* forms—is passed through an enantiomerically pure sample that has the *S* configuration. When the light interacts with the sample, two types of interaction occur: *S* sample with *S* light, and *S* sample with *R* light. *These two interactions are diastereomeric,* because an *S,S* combination is the diastereomer of an *S,R* combination. Because these two interactions have different energies, one is stronger than the other. The form of light with the stronger interaction is retarded in its passage through the sample. This in turn causes the plane of the polarized light to rotate—that is, it causes optical activity. *The enantiomeric forms of light differ when they interact with a chiral object.*

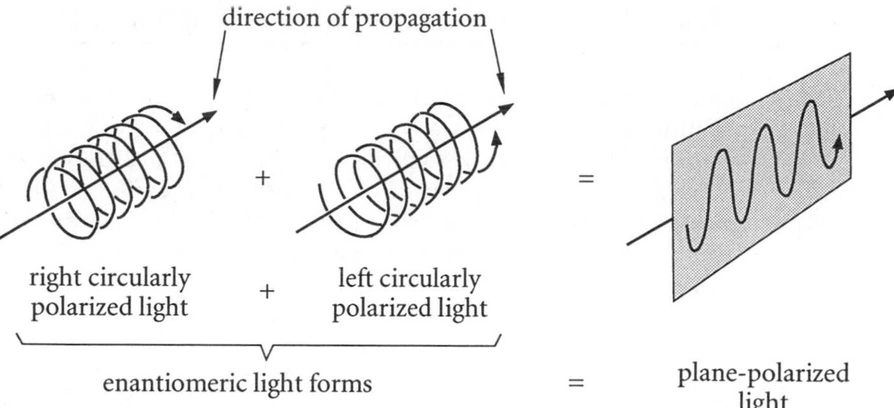

Figure SG7.1 The electric fields of right and left circularly polarized light propagate through space as helices, and the two add vectorially to give plane-polarized light. Notice that the two helices are chiral; consequently, plane-polarized light, their vector-addition product, is a racemate.

✓7.5 Reactions of Chiral Molecules

Notice carefully that Sec. 7.8A of the text deals only with reactions of *achiral molecules* that give enantiomeric pairs of products. What happens when *chiral* molecules react to give stereoisomeric products? No general principles cover this situation. In some cases, stereoisomers of the product may be formed in equal amounts; in other cases, one stereoisomer may predominate. The result can be determined by the reaction mechanism, that is, the stereochemical nature of chemical changes that occur at stereocenters. (A situation of this sort is discussed in Sec. 7.9B.) In other cases reactants or intermediates are racemized in the reaction, in which case the principles of Sec. 7.7A apply to each enantiomer of the racemate.

It is even possible for chiral molecules to react in some cases to give achiral products even when a stereocenter does not undergo a chemical change. Thus, the following hydrogenation reaction involving a chiral alkene gives an achiral alkane because the asymmetric carbon in the alkene is no longer an asymmetric carbon in the alkane product:

$$H_2C{=}CH{-}C(H)(CH_3)CH_2CH_3 + H_2 \xrightarrow{\text{catalyst}} CH_3CH_2{-}C(H)(CH_3)CH_2CH_3$$

chiral and optically active — achiral and optically inactive

Optical activity (and chirality) is lost in this reaction because the product is not a chiral molecule. In *this particular case,* the reaction causes a symmetry to occur within the structure of the product that does not exist in the reactant. Optical activity would *not* be lost in the same reaction of a different alkene in which the product is chiral:

$$H_2C{=}CHCH_2{-}C(H)(CH_3)CH_2CH_3 + H_2 \xrightarrow{\text{catalyst}} CH_3CH_2CH_2{-}C(H)(CH_3)CH_2CH_3$$

chiral and optically active — chiral and optically active

(See also Problem 6.40, text p. 241.)

✓7.6 Analysis of Reaction Stereochemistry

Notice carefully that a given type of reaction might yield diastereomers with one starting material, enantiomers with another, and no stereoisomers at all with a third. For example, the bromine addition to cyclohexene in Study Problem 7.5 on text p. 282gives a pair of diastereomers, and one of these is formed as a pair of enantiomers. In bromine addition to propene, only a pair of enantiomers is formed. In bromine addition to ethylene, no stereoisomers at all are formed. For a given reaction, no single stereochemical result applies to all compounds. Each reaction must be analyzed individually.

✓7.7 Stereoselective and Stereospecific Reactions

Some texts define the term *stereospecific reaction* to mean "100% stereoselective within limits of detection." This idea can be conveyed by saying simply that a reaction is "highly stereoselective." This definition of *stereospecific* is not generally accepted; be careful not to confuse the terms *stereospecific* and *stereoselective* in this way.

7.8 Stereochemistry of Organoborane Oxidation

The mechanism of organoborane oxidation was discussed in Study Guide Link 5.3.

Retention of stereochemistry in this reaction can be understood in terms of this mechanism. An alkyl group on boron moves from boron to oxygen in such a way that the configuration of the migrating carbon is unaffected:

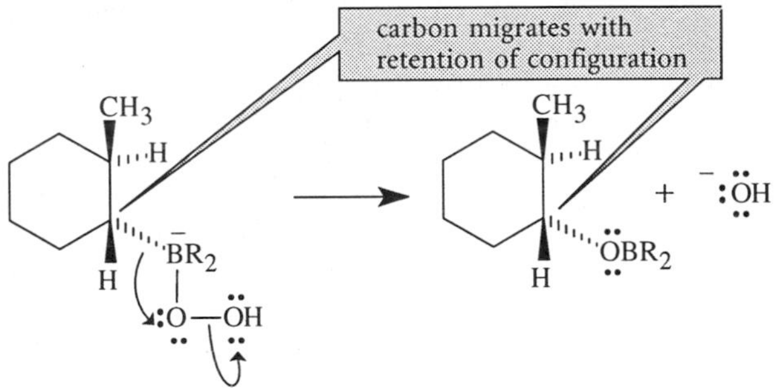

carbon migrates with retention of configuration

CH_3, H, $\overset{-}{B}R_2$, H, :O—OH → CH_3, H, OBR_2, H + $^{-}$:OH

Thus, a carbon-boron bond is replaced by a carbon-oxygen bond without changing the stereochemical configuration of the carbon.

In subsequent steps of the reaction, the bonds between the other R groups are converted in a similar manner into R—O bonds:

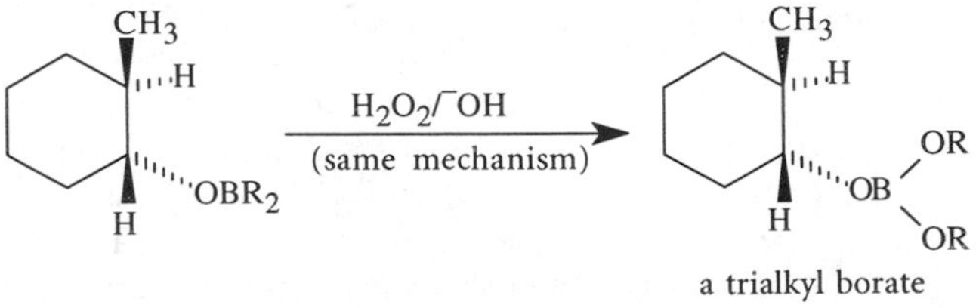

CH_3, H, OBR_2, H —H_2O_2/$^{-}$OH (same mechanism)→ CH_3, H, OB(OR)(OR), H

a trialkyl borate

Finally, the resulting trialkyl borate undergoes hydrolysis to the alcohol product. The final hydrolysis breaks the boron-oxygen bonds but leaves the oxygen-carbon bonds unaffected. Because these bonds are not affected, the configuration at carbon does not change.

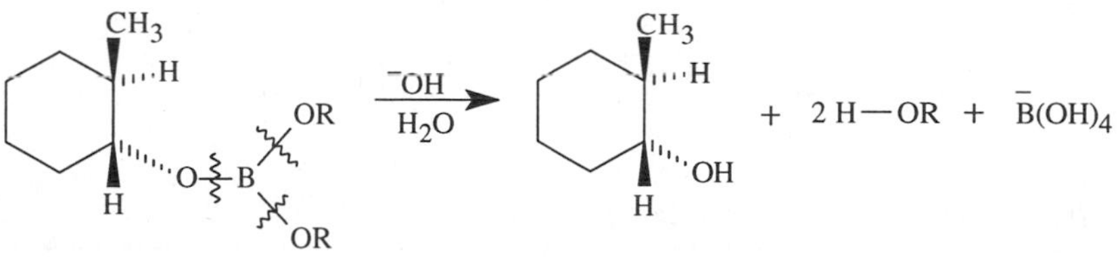

$$\text{(trialkylborate of 2-methylcyclohexanol)} \xrightarrow[H_2O]{^-OH} \text{trans-2-methylcyclohexanol} + 2\,H{-}OR + \bar{B}(OH)_4$$

✓7.9 When Stereoselectivity Matters

The stereoselectivity of a reaction (or lack of it) is of no concern if stereochemically different modes of a reaction give the same products. Let's see how this statement would apply in the case of bromine addition to alkenes.

The addition of bromine to ethylene does not give stereoisomers; hence, it makes absolutely no difference in the outcome of the reaction whether the reaction is a *syn*- or *anti*-addition or a mixture of the two.

$$H_2C{=}CH_2 \xrightarrow{Br_2} Br{-}CH_2{-}CH_2{-}Br$$

Furthermore, we cannot even tell whether the addition is syn or anti. It is often assumed (reasonably) that additions which occur in one particular way on certain compounds occur in the same way on similar compounds for which the stereoselectivity cannot be determined. Thus, because addition of bromine is an *anti*-addition with other alkenes, we assume that it is an *anti*-addition for ethylene as well.

In fact, this assumption has been tested: the stereochemistry of bromine addition to isotopically substituted ethylenes was investigated in 1998.

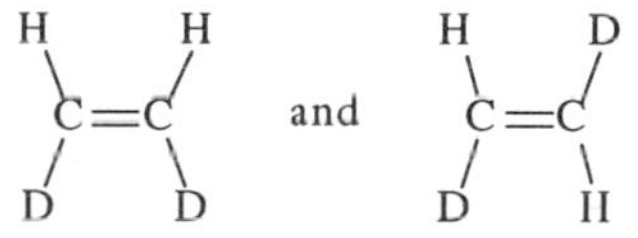

$$(H)(D)C{=}C(H)(D) \quad \text{and} \quad (H)(D)C{=}C(D)(H)$$

This experiment established that *anti* addition does in fact occur. Most people would accept that *anti*-addition to the deuterium-substituted ethylenes is proof that ethylene itself undergoes *anti*-addition.

In the addition of bromine to propene, stereoisomers are formed:

$$H_3C{-}CH{=}CH_2 \xrightarrow{Br_2} H_3C{-}\underset{Br}{\underset{|}{C}H}{-}CH_2Br$$

Two *enantiomers* are possible, but they *must* be formed in the same amounts, because the reagents are achiral. Because carbon-1 of the product is not a stereocenter, the question of a *syn*- or *anti*-addition is irrelevant to the composition of the product mixture in the addition of bromine to propene.

Remember the point made in Sec. 7.9A of the text: the stereochemistry of an addition is relevant only when *both* carbons of the double bond become stereocenters as a result of the addition. Thus, one can study the mode of bromine addition to *cis*- and *trans*-2-butene, as in Sec. 7.9C, because the two carbons of the double bond in the addition product are stereocenters. In a practical sense, bromine addition would not be as useful when applied to the 2-butenes or to cyclohexene if it were not stereoselective, because a mixture of diastereomers would be formed.

The text shows that oxymercuration-reduction is not a stereoselective reaction. The lack of stereoselectivity matters only when it is applied to an alkene that would give a mixture of diastereomeric products. Many reactions that are not stereoselective are useful as long as their use is confined to situations in which stereoselectivity is not an issue.

SOLUTIONS TO PROBLEMS

Solutions to In-Text Problems

7.1 (a) It is given that the half-chair conformation is the transition state for the interconversion of the chair and twist-boat. This means that it is the conformation at the energy maximum labeled as 44.8 kJ/mol (the greatest maximum in text Fig. 7.5). Just as there are two chairs, there are also two half-chairs. (Where is the other one on the energy diagram of text Fig. 7.5?)

(b) First, the half-chair conformation contains significant ring strain. (Your model may pop back from a half-chair to a boat or chair if you try to place it on a table without holding it.) Second, there is significant eclipsing between some of the hydrogens. As we learned when we studied the conformations of ethane, eclipsing interactions are destabilizing.

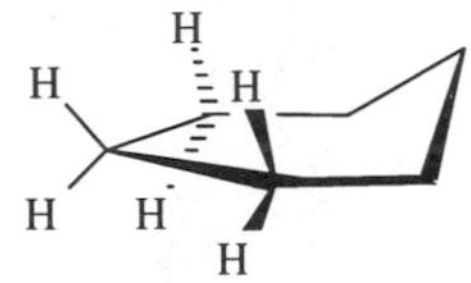

half-chair conformation
(eclipsed hydrogens are shown)

7.2 We apply text Eq. 3.24b, with $\Delta G° = 7.3$ kJ/mol.

$$K_{eq} = 10^{-\Delta G°/2.3RT} = 10^{-7.3/5.706} = 10^{-1.28} = 5.25 \times 10^{-2}$$

Letting the concentration of axial conformation = $[A]$ and that of equatorial = $[E]$, then this equilibrium constant means that

$$\frac{[A]}{[E]} = 0.0525$$

In one mole of cyclohexane, $[A] + [E] = 1.00$. Using the previous equation, then $[A] = 0.0525[E]$. Substituting,

$$[A] + [E] = 0.0525[E] + [E] = 1.00$$

Solving, $[E] = 0.950$; therefore $[A] = 0.050$. The equatorial conformation, then, comprises 95% of methyl cyclohexane, and the axial conformation 5%.

7.3 Because the axial conformation of *tert*-butyl cyclohexane is even more unstable (relative to the equatorial conformation), there is even less of it present. Going through a similar calculation, with $\Delta G° = 20$ kJ/mol, the $K_{eq} = [A]/[E]$ is 3.13×10^{-4}. This means, in turn, that a sample of *tert*-butylcyclohexane contains about 99.97% of the equatorial conformation.

7.4 (a) Because the axial conformation of fluorocyclohexane has *two* fluorine-hydrogen 1,3-diaxial interactions (just as the axial conformation of methylcyclohexane has two methyl-hydrogen 1,3-diaxial interactions), the energetic cost of a single 1,3-diaxial interaction in fluorocyclohexane is 0.50 kJ/mol (0.13 kcal/mol).

(b) The gauche conformation of 1-fluoropropane has a single gauche fluorine-methyl interaction.

gauche fluorine and methyl

F, H, CH_3, H, H, H

This has about the same energy cost as a *single* fluorine-hydrogen 1,3-diaxial interaction in the axial conformation of fluorocyclohexane. From part (a), the energy cost of this is 0.50 kJ/mol (0.13 kcal/mol) relative to the anti conformation.

7.5 A rapid rotation can occur about the bond between the cyclohexane ring and the ethyl group so that the van der Waals radii of the methyl group and the axial hydrogens do not overlap.

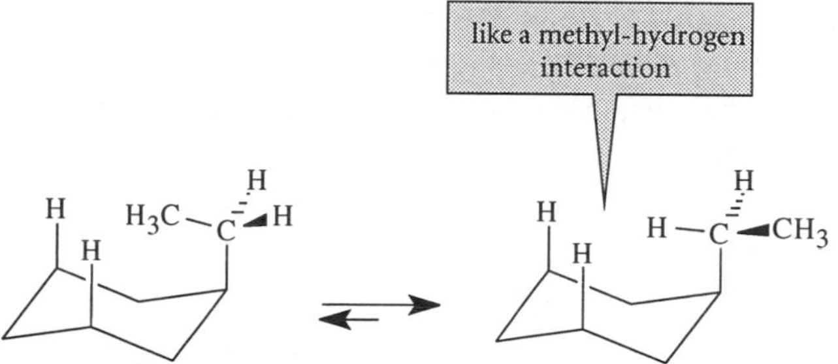

In the conformation on the right, the interactions of the axial hydrogens with the hydrogens of the ethyl group are about the same as the corresponding interactions in a methyl-hydrogen 1,3-diaxial interaction. As a result, the ethyl-hydrogen interaction has about the same energy cost as a methyl-hydrogen interaction.

7.6 (a) The two chair conformations of *cis*-1,3-dimethylcyclohexane:

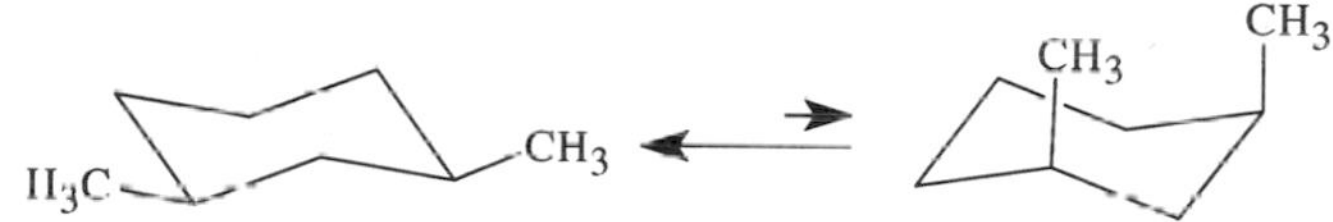

Don't forget that a *cis*-1,3-disubstituted cyclohexane can be drawn in many valid ways. If the structures you drew look different from those above, try to manipulate your structures mentally to see whether they can be made to look identical to the ones above. (Study Guide Link 7.1 provides some exercises that should be helpful.)

(b) The conformations of *trans*-1-ethyl-4-isopropylcyclohexane:

$(CH_3)_2CH$ CH_2CH_3 ⇄ $(CH_3)_2CH$ CH_2CH_3

7.7 (a) Two boat conformations of *cis*-1,3-dimethylcyclohexane:

H_3C CH_3 CH_3 CH_3

(b) Two boat conformations of *trans*-1-ethyl-4-isopropylcyclohexane:

$(CH_3)_2CH$ CH_2CH_3 $(CH_3)_2CH$ CH_2CH_3

7.8 First, draw the two chair conformations of *trans*-1,4-dimethylcyclohexane:

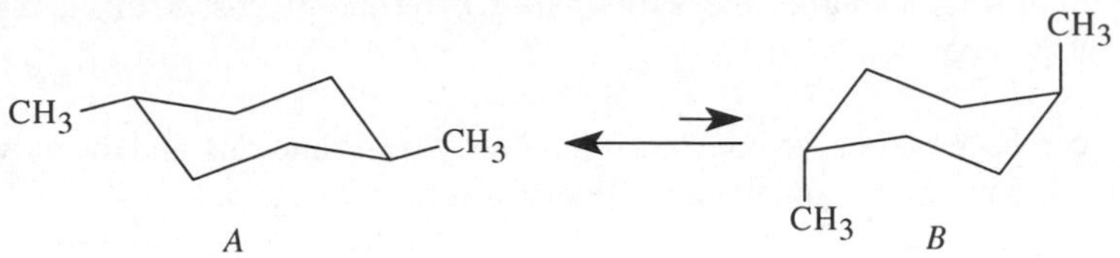

CH_3 CH_3 *A* ⇌ CH_3 CH_3 *B*

Conformation *A* has no 1,3-diaxial methyl-hydrogen interactions. Conformation *B* has four—two for each methyl group. At 3.7 kJ/mol each, these interactions make conformation *B* 14.8 kJ/mol less stable than conformation *A*.

7.9 (a) CH_3 CH_3 or CH_3 CH_3

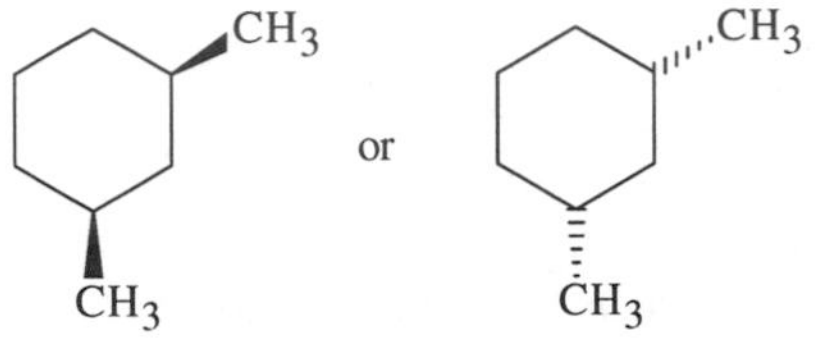

cis-1,3-dimethylcyclohexane

(b) CH_3 CH_3 or CH_3 CH_3

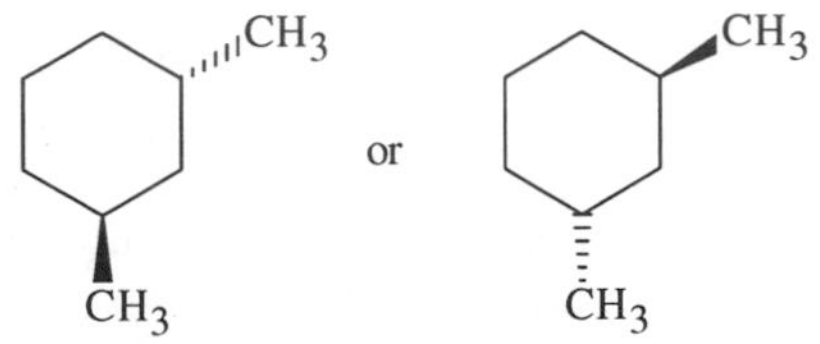

trans-1,3-dimethylcyclohexane

7.10 (a) The larger *tert*-butyl group assumes the equatorial position:

$(CH_3)_3C$ CH_3

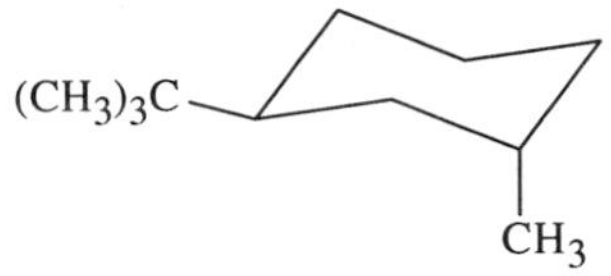

(b) The most stable conformation is the one that has the greater number of groups in the equatorial position:

Cl Cl Cl

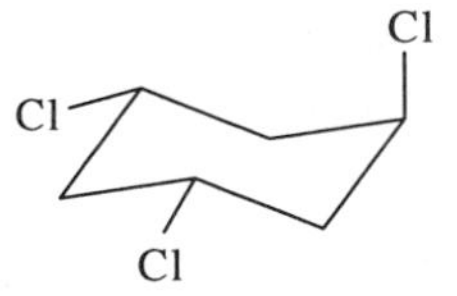

(c) The larger isopropyl group assumes the equatorial position. The presence of oxygen does not alter the fact that the ring exists in a chair conformation.

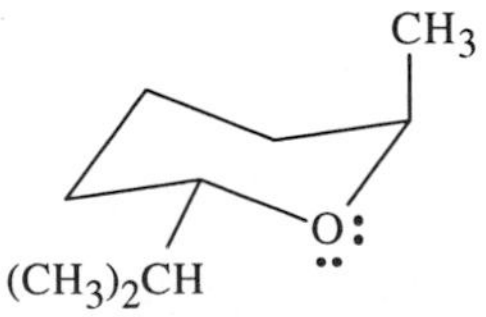

CH_3 O: $(CH_3)_2CH$

7.11 As the discussion in this section of the text has shown, a molecule is chiral (that is, can be isolated in optically active form) when its planar structure is noncongruent to its mirror image.

(a) *Trans*-1,3-dimethylcyclohexane can be isolated in optically active form. (See structures below.)
(b) 1,1-Dimethylcyclohexane is achiral, and therefore cannot be optically active.
(c) *Cis*-1,4-dimethylcyclohexane is achiral, and therefore cannot be optically active.
(d) *Cis*-1-ethyl-3-methylcyclohexane can be isolated in optically active form. (See structures below.)

for part (a):

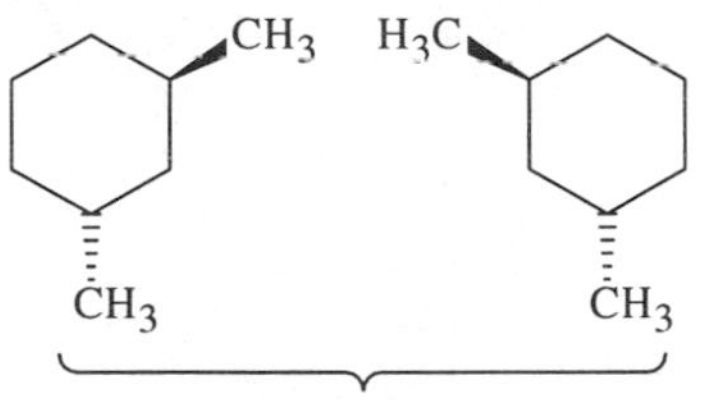

noncongruent mirror images
trans-1,3-dimethylcyclohexane

for part (d):

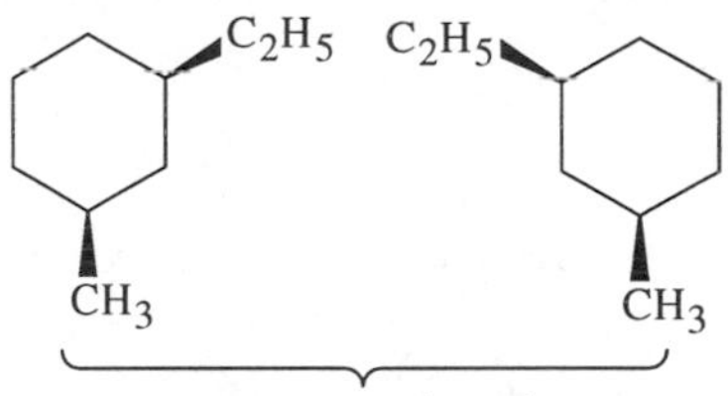

noncongruent mirror images
cis-1-ethyl-3-methylcyclohexane

7.12 (a) *Trans*-1,4-dimethylcyclohexane does *not* have asymmetric carbons.

(b) *Trans*-1,4-dimethylcyclohexane has two stereocenters, because interchange of the methyl and hydrogen groups at either one gives *cis*-1,4-dimethylcyclohexane, a stereoisomer.

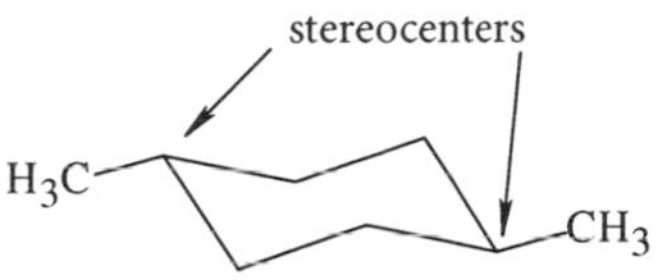

(c) The two chair conformations of *trans*-1,4-dimethylcyclohexane are conformational diastereomers:

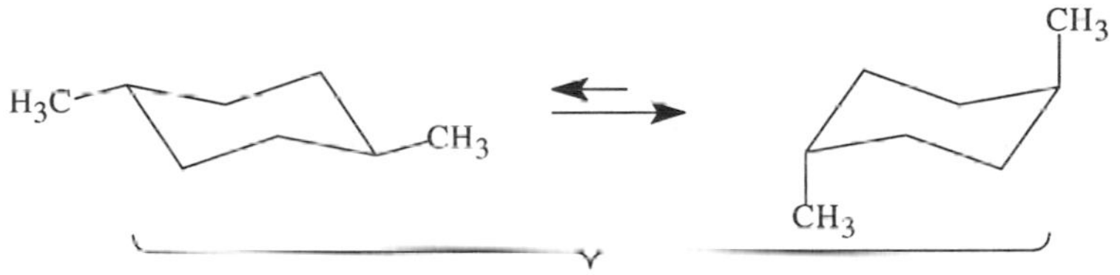

conformational diastereomers

(d) *Trans*-1,4-dimethylcyclohexane is not chiral because it has an internal mirror plane.

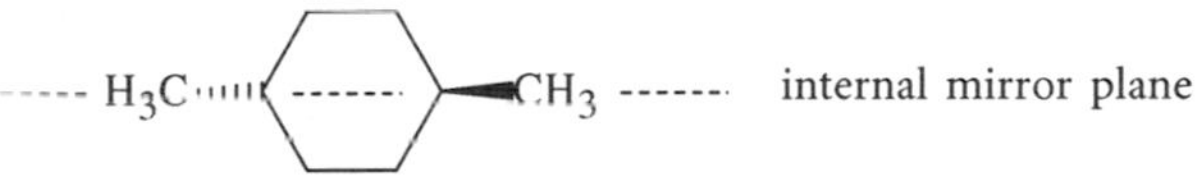

trans-1,4-dimethylcyclohexane

7.13 (a) The two structures are diastereomers. (They are stereoisomers but not enantiomers.)

(b) The two structures differ in configuration at both asymmetric carbons. They are enantiomers.

(c) Turning the molecule over by rotation about the indicated axis shows that one structure can be converted into the other; thus, the structures are identical.

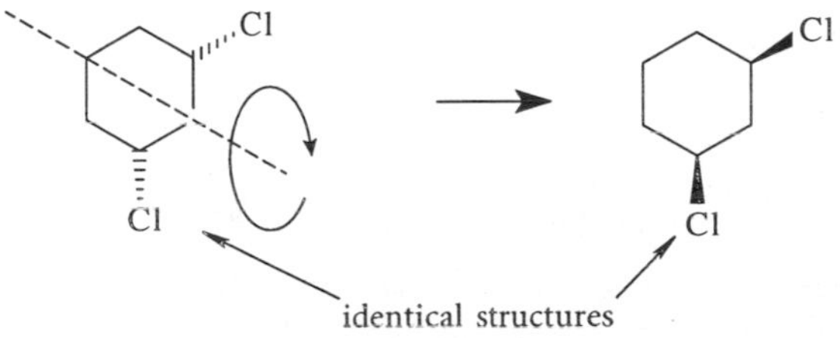

identical structures

7.14 In planar *trans*-1,3-dibromocyclobutane, the carbon-bromine bonds are directed in opposite directions; hence, their bond dipoles would cancel, and the net dipole moment of the molecule would be zero. In puckered *trans*-1,3-dibromocyclobutane, the bond dipoles do not cancel.

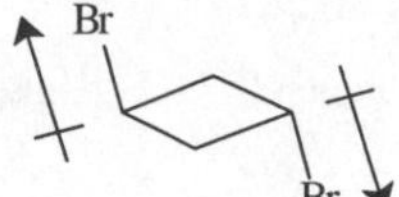

planar *trans*-1,3-dibromocyclobutane:
bond dipoles cancel

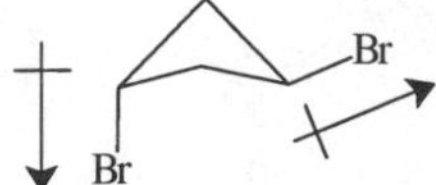

puckered *trans*-1,3-dibromocyclobutane:
bond dipoles do not cancel

7.15 (a) *Cis*-1,2-dimethylcyclopropane is achiral; it is a meso compound.

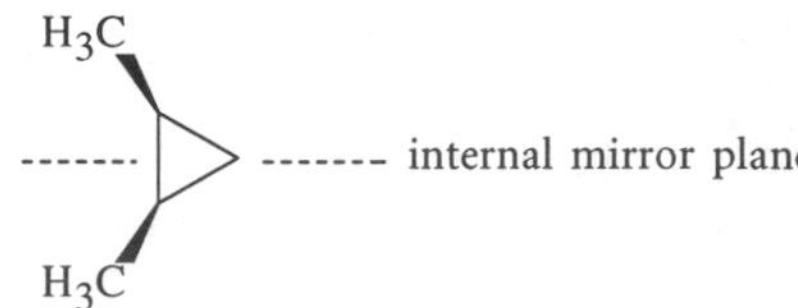

cis-1,2-dimethylcyclopropane

(b) *Trans*-1,2-dimethylcyclopropane is chiral.

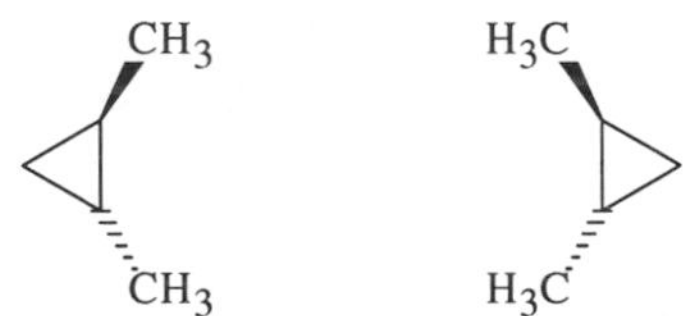

enantiomeric *trans*-1,2-dimethylcyclopropanes

7.16 (a) Bicyclo[2.2.2]octane
(b) Bicyclo[3.2.0]heptane

7.17 A fused bicyclic compound, by definition, has a "zero-carbon bridge," and will have a zero in the number associated with the name. Consequently, compound *B* is a fused bicyclic compound; compound *A* is a bridged bicyclic compound.

7.18 The two labeled carbons in the diagram of *cis*-decalin below are axial substituents in the other ring; consequently, the hydrogens on these carbons are involved in 1,3-diaxial interactions (dashed lines in the following structure.). There are, as in other axially-substituted alkylcyclohexanes, two 1,3-diaxial interactions per axial carbon, except that one is common to both carbons; hence, there are a total of three 1,3-diaxial interactions. Because there are no 1,3-diaxial interactions in *trans*-decalin, it is more stable than *cis*-decalin by $3 \times 3.7 = 11.1$ kJ/mol (2.65 kcal/mol).

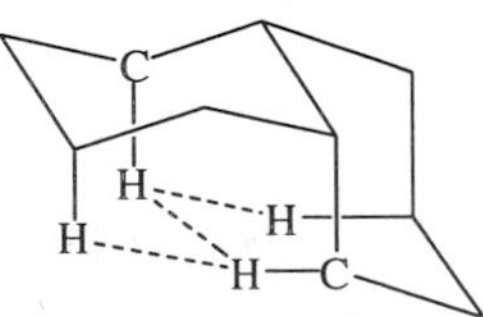

7.19 (a) The model of *cis*-bicyclo[3.1.0]hexane is much more easily built. The smaller is a ring, the closer is the dihedral angle of the trans bonds to 180°. Forcing this dihedral angle to much smaller values introduces strain within the ring. Such a large dihedral angle causes the ends of trans bonds to be too far apart to be readily bridged by only one carbon.

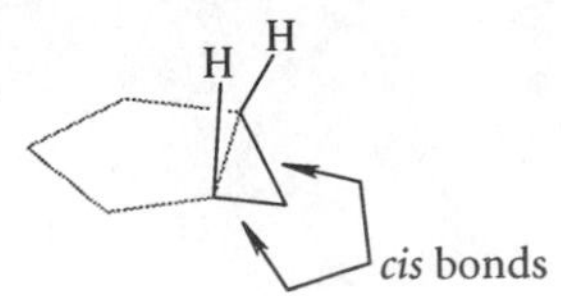

cis-bicyclo[3.1.0]hexane

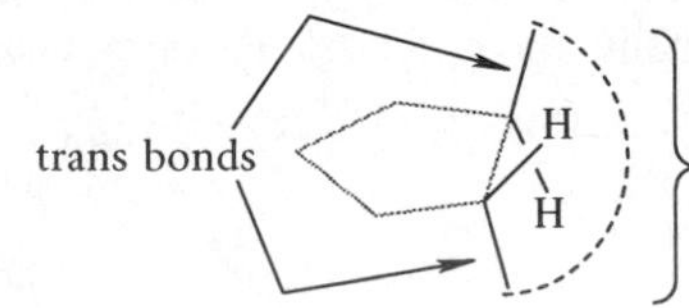

Be sure that you build the models called for in this and the following problem. Models are much more revealing, and provide you with much greater insight, than simply reading verbal descriptions.

(b) The model of *trans*-bicyclo[5.3.0]decane is easier to build. The larger is a ring, the easier it is to compress the dihedral angle of the trans bonds at the ring junction without introducing significant strain in the ring. This angle must be very close to 0° in order to accommodate a fused cyclopropane, that is, to bridge the ends of trans bonds with only one carbon.

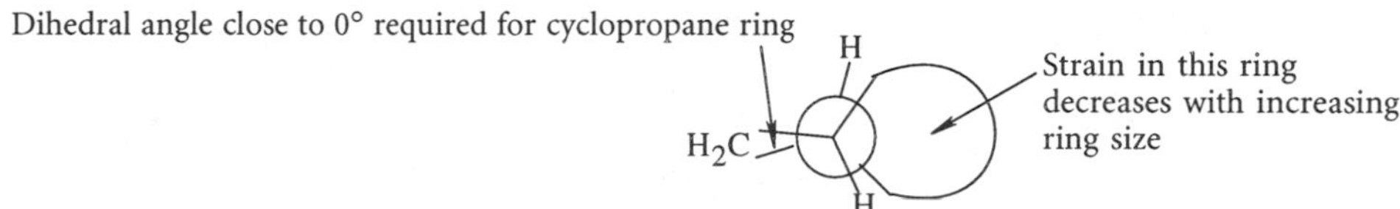

7.20 (a) Compound *B* is less stable because it has a bridgehead double bond within a small ring. Consequently, it would have the greater (more positive or less negative) heat of formation.

(b) Although both molecules have bridgehead double bonds, the double bond in compound *B* is more twisted, and a model of this molecule is more difficult to build. Consequently, compound *B* is less stable and therefore would have the greater (more positive or less negative) heat of formation.

7.21 (a) Because a nail and a hammer are achiral objects, all interactions of these objects with any chiral object (such as a hand) and its enantiomer are identical. Hence, assuming equal strength, a right and a left hand should be equally adept at driving a nail. (This assumes, of course, that the person driving the nail doesn't accidentally bend it into some chiral shape.) In contrast, when it drives a screw, a chiral hand interacts with the helical (and chiral) threads of the screw. For this reason, the right and left hands of the same person should differ in their ability to drive a screw.

(b) Premeds could take out their stethoscopes and determine which side of each person the heart is on. Or, you could offer to shake hands. Mr. L would extend what he calls his right hand, but it would to us be his left. Or, you could ask them to write their names. What would be the anticipated result? In any case, the point is that the two mirror images are distinguished by comparing a chiral reference element (our bodies, our right hands, or the way we write our names) with the corresponding elements of the two people.

7.22 The following stereoisomers of 2,3-dibromobutane (shown in Fischer projections) could form in bromine addition to *cis*-2-butene:

CH_3	CH_3	CH_3
Br—+—H	H—+—Br	H—+—Br
H—+—Br	Br—+—H	H—+—Br
CH_3	CH_3	CH_3
(2*R*,3*R*)	(2*S*,3*S*)	meso

(2*R*,3*R*) and (2*S*,3*S*): enantiomers

enantiomers and meso: diastereomers

The enantiomers must be formed in identical amounts because the starting materials are achiral. The diastereomers are formed in different amounts.

7.23 Hydroboration-oxidation of *trans*-2-butene gives racemic 2-butanol, that is, two enantiomers formed in equal amounts.

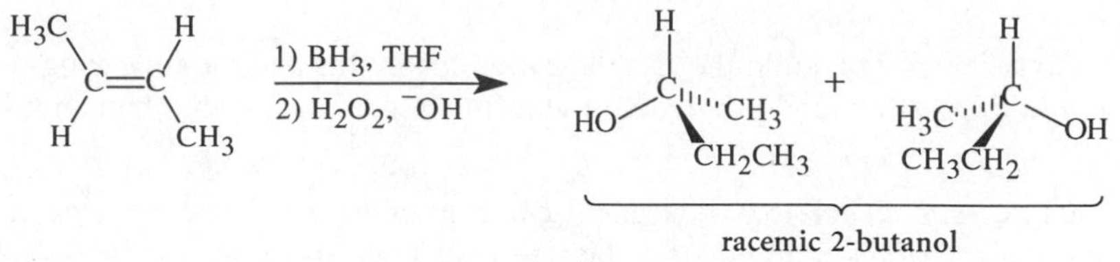

$$\text{(}E\text{)-2-butene} \xrightarrow[\text{2) } H_2O_2,\ ^-OH]{\text{1) } BH_3,\ THF} \text{racemic 2-butanol}$$

racemic 2-butanol

7.24 Write all possible diastereomers resulting from addition to one enantiomer of the starting material. These are in principle formed in different amounts. (Problem 7.26 considers this case further.)

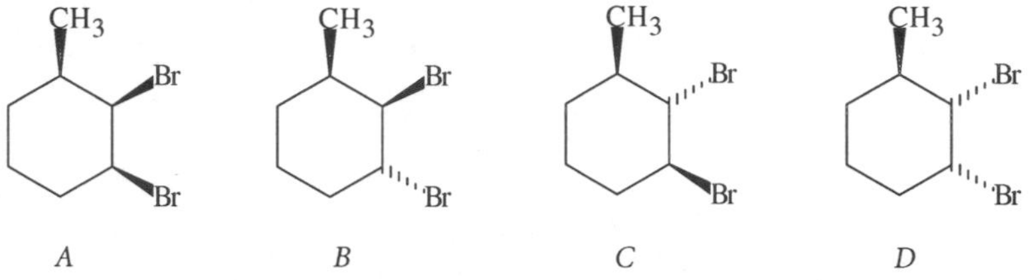

A B C D

Each of these structures has an enantiomer; drawing these gives four additional diastereomers, which are also formed in different amounts. (Diastereomers are *always* formed in different amounts in principle.)

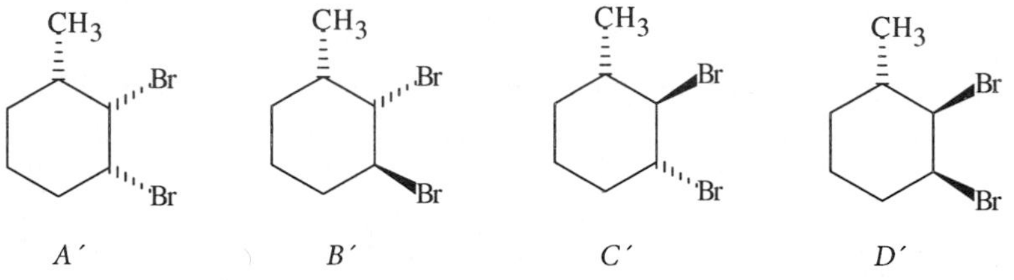

A′ B′ C′ D′

Each point in the process that gives *A* (that is, each reactive intermediate, each transition state, etc.) is enantiomeric to the corresponding point in the process that gives *A*′; therefore the two processes must occur with identical rates. Because the starting alkene is racemic, the same amounts of the enantiomeric alkenes are involved in each process; therefore, the enantiomers of each pair are formed in identical amounts. Thus, *A* and *A*′ are formed in identical amounts; *B* and *B*′ are formed in identical amounts; and so on.

7.25 Because the bromonium-ion mechanism involves *anti*-addition, bromine addition to cyclohexene must give *trans*-1,2-dibromocyclohexane as a product.

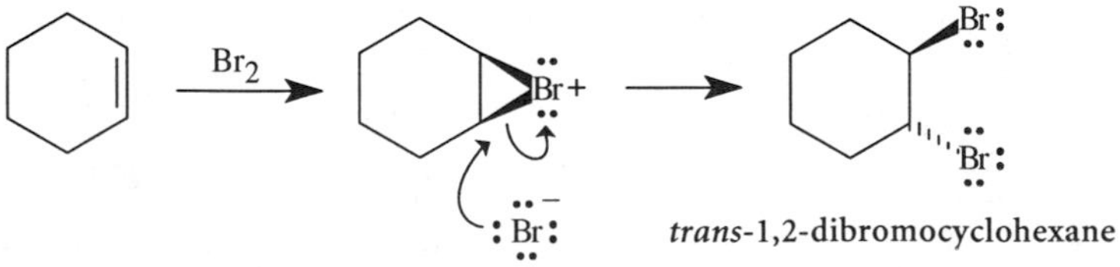

$$\text{cyclohexene} \xrightarrow{Br_2} \text{bromonium ion} + :\ddot{Br}:^- \longrightarrow \textit{trans}\text{-1,2-dibromocyclohexane}$$

trans-1,2-dibromocyclohexane

7.26 (Refer to the structures in the solution to Problem 7.24.) Because the bromines of the product are trans, products *B* and *C*, and their respective enantiomers *B*′ and *C*′, are the major products.

7.27 (a) Because hydroboration-oxidation is a net *syn*-addition, the H and OH are added to the same face of the alkene. This addition can occur in two different ways to give enantiomeric products, which are formed in identical amounts.

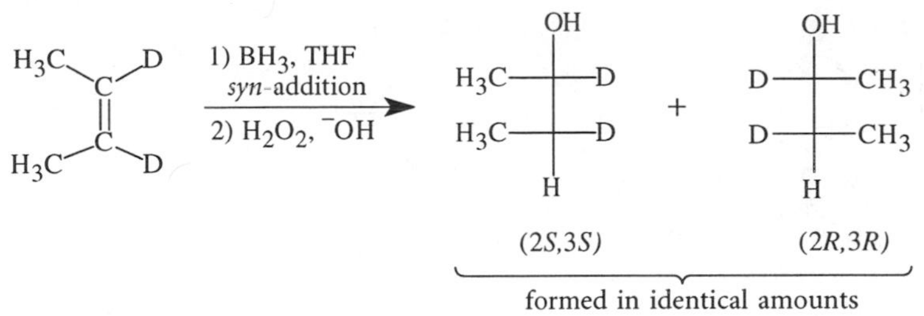

1) BH_3, THF *syn*-addition
2) H_2O_2, ^-OH

(2*S*,3*S*) (2*R*,3*R*)

formed in identical amounts

(b) As we expect for a stereospecific reaction, use of a diastereomeric starting material gives the diastereomeric set of products.

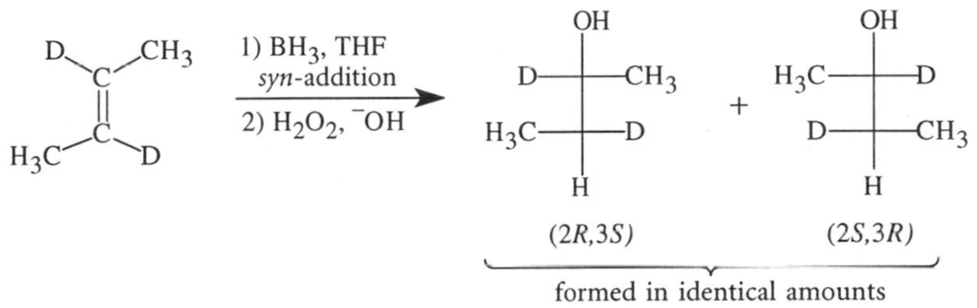

1) BH_3, THF *syn*-addition
2) H_2O_2, ^-OH

(2*R*,3*S*) (2*S*,3*R*)

formed in identical amounts

7.28 The presence of deuteriums in Problem 7.27 gives rise to diastereomeric products. If the starting material is not isotopically substituted, then the two products are enantiomeric, and the same pair of enantiomers is formed whether the *cis-* or *trans-*alkene is used as the starting material.

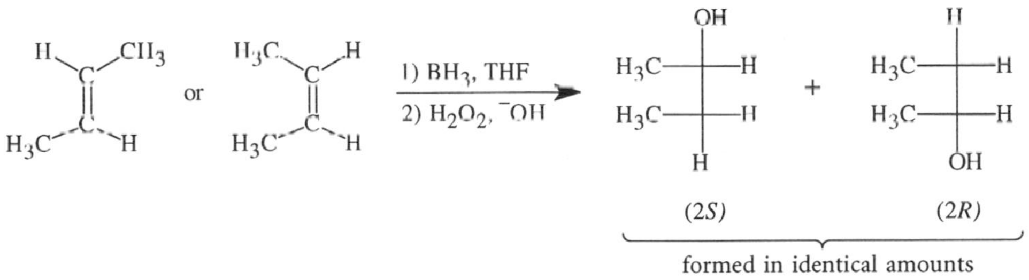

1) BH_3, THF
2) H_2O_2, ^-OH

(2*S*) (2*R*)

formed in identical amounts

7.29 (a) We know that oxymercuration is an *anti*-addition.

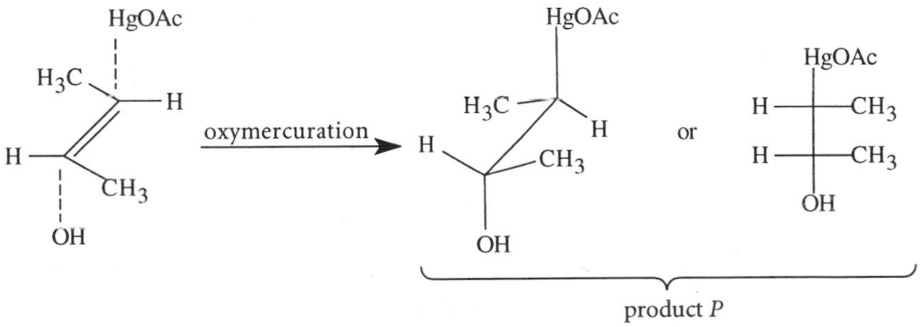

oxymercuration

product *P*

Of course, along with *P* must be formed an equal amount of its enantiomer *P*´ (why?). You can develop the structure of *P*´ by having —H and —OH add to carbons at opposite ends of the double bond, or by having each attack same carbon as shown above from the opposite face.

(b) The enantiomeric products of the reaction in part (a) (*P* and its enantiomer, *P*´) are shown below. Because the $NaBD_4$/NaOH reaction occurs with *loss of stereochemistry,* two diastereomers are formed from each of the enantiomers *P* and *P*´. As a result, all four possible stereoisomers are formed.

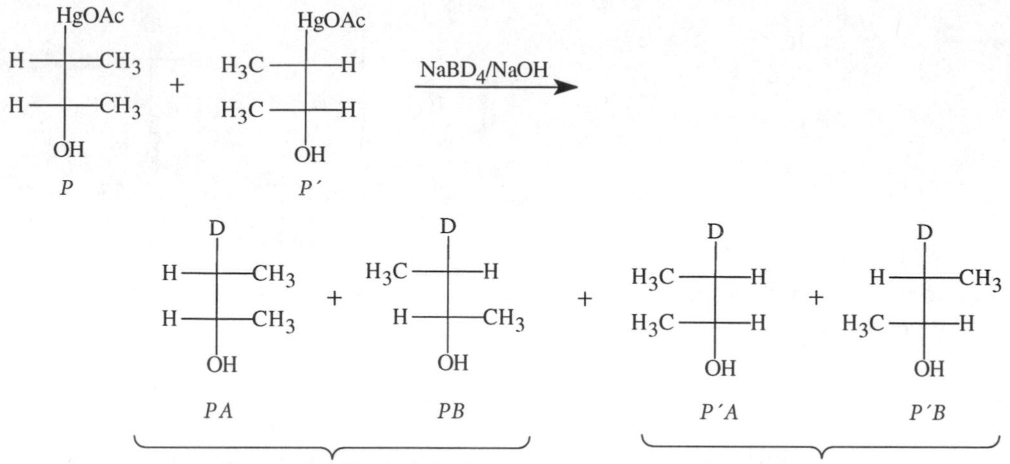

HgOAc; H—CH$_3$; H—CH$_3$; OH (*P*) + HgOAc; H$_3$C—H; H$_3$C—H; OH (*P*′) $\xrightarrow{NaBD_4/NaOH}$

D; H—CH$_3$; H—CH$_3$; OH (*PA*) + D; H$_3$C—H; H—CH$_3$; OH (*PB*) + D; H$_3$C—H; H$_3$C—H; OH (*P*′*A*) + D; H—CH$_3$; H$_3$C—H; OH (*P*′*B*)

diastereomeric pair from *P* — diastereomeric pair from *P*′

(*PA* and *P*′*A* are enantiomers; *PB* and *P*′*B* are also enantiomers.)

In Eq. 7.55, two products are shown; but, if the starting material in that equation is racemic, as it must be if the reagents from which it is formed are achiral, then the same four products are formed. Here are the structures of those products as shown in the text. You should show the identity of each one to the corresponding structure above by using allowed manipulations of its Fischer projection.

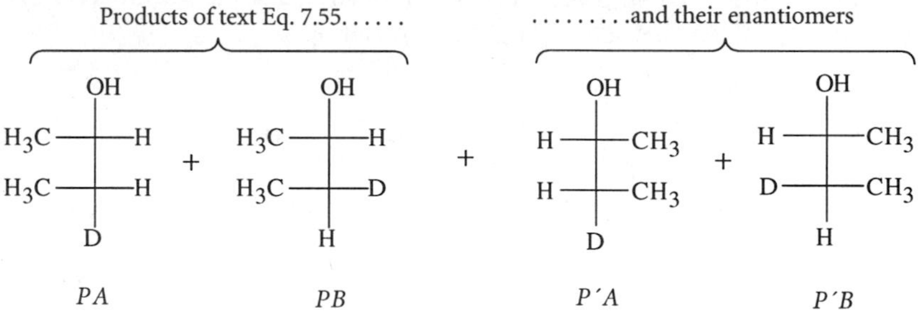

For example, to show the identity of the two *PB* structures:

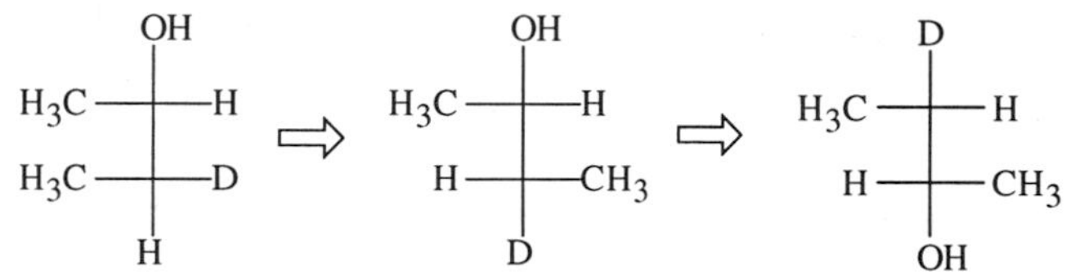

OH; H$_3$C—H; H$_3$C—D; H ⇨ OH; H$_3$C—H; H—CH$_3$; D ⇨ D; H$_3$C—H; H—CH$_3$; OH

Another way to show this identity, of course, is to relate the *R* or *S* configurations of corresponding carbons.

Solutions to Additional Problems

7.30 (a) Two of four that could be drawn are the following: (b)

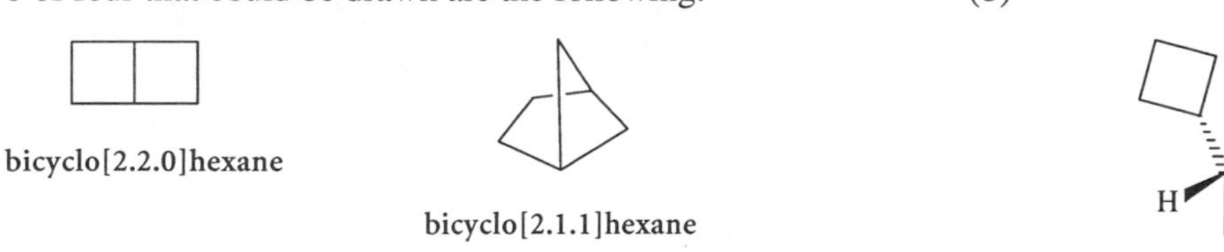

bicyclo[2.2.0]hexane

bicyclo[2.1.1]hexane

H

(*S*)-4-cyclobutylcyclohexene

7.31 Methylcyclohexane and (*E*)-4-methyl-2-hexene are *constitutional isomers.* Hence, they have the same molecular mass but different properties.

(a) A molecular mass determination would not distinguish between the two compounds because their molecular masses are identical.

(b) Uptake of H_2 over a catalyst would distinguish between the two compounds because the alkene would take up hydrogen in the presence of a catalyst, whereas the cycloalkane would not.

(c) The reaction with Br_2 would distinguish between them because the alkene would decolorize Br_2 solution (because alkenes react with Br_2), whereas the alkane would not.

(d) Determination of the empirical formula would not distinguish between the two compounds because both compounds are isomers; they have the same formula, and hence, the same empirical formula.

(e) Because constitutional isomers have different physical properties, determination of the heat of formation would distinguish between the two compounds. (Which would have the greater heat of formation?)

(f) Enantiomeric resolution would distinguish between the two because (*E*)-4-methyl-2-hexene is chiral, and can be resolved into enantiomers, whereas methylcyclohexane, an achiral compound, cannot be.

7.32 Distinguishing between enantiomers requires a technique that has a chiral attribute.

(a) Boiling point does not distinguish between two enantiomers.

(b) Optical rotation does have a chiral aspect, and therefore can distinguish between two enantiomers. (See Study Guide Link 7.4.)

(c) Because hexane is an achiral solvent, enantiomers have identical solubilities in hexane.

(d) Density is not a chiral property, and therefore the two enantiomers have identical densities.

(e) The solubilities of two enantiomers in principle differ in an enantiomerically pure chiral solvent.

(f) Dipole moment is not a chiral property, and therefore the two enantiomers have identical dipole moments.

(g) Because taste buds are chiral, they differentiate between two enantiomers. One common example of this is the two enantiomers of carvone:

CH_3 O H_3C CH_2

carvone

The (+)-enantiomer is largely responsible for the odor and flavor of carraway (the seeds on rye bread), and the (–)-enantiomer is the major component of spearmint oil.

7.33

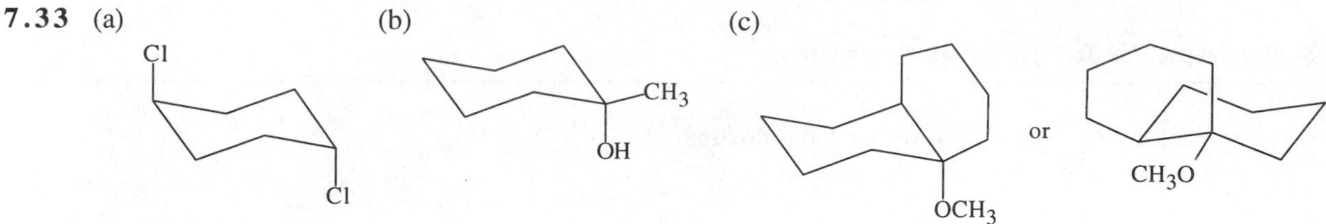

7.34 (a) In this compound the large *tert*-butyl group must assume the equatorial position. The positions of the other groups follow from their positions relative to the *tert*-butyl group.

(b) In choosing which conformation to draw, the goal is to put the maximum number of methyl groups possible in equatorial positions.

7.35 (a) The equilibrium constant is given as 2.07 for the ratio [equatorial]/[axial]. The standard free-energy difference is calculated by applying Eq. 3.23, text p. 97:

$$\Delta G° = -2.3RT \log (2.07) = -(5.706 \text{ kJ/mol})(0.3160) = -1.80 \text{ kJ/mol}$$

In other words, for the equilibrium axial $\rightleftarrows$ equatorial, the equatorial form has the lower standard free energy, and is thus more stable.

(b) The standard free energy change for the reaction equatorial $\rightleftarrows$ axial is +9.2 kJ/mol; the axial conformation has the higher standard free energy. The equilibrium constant is calculated by applying Eq. 3.24b, text p. 97.

$$K_{eq} = 10^{-\Delta G°/2.3RT} = 10^{-9.2/5.706} = 10^{-1.61} = 2.4 \times 10^{-2}$$

That is, the ratio [axial]/[equatorial] is 0.024. This means that there is 1/0.024 = 42 times as much equatorial conformation as there is axial conformation at equilibrium.

7.36 Compound *A* can be prepared by hydroboration-oxidation from either *cis*- or *trans*-3-hexene because the alkene is symmetrical and regioselectivity is not an issue.

$$\underset{\text{3-hexene}}{CH_3CH_2CH{=}CHCH_2CH_3} \xrightarrow[\text{2) } H_2O_2,\ ^-OH]{\text{1) } BH_3/THF} \underset{\substack{\text{3-hexanol} \\ \text{(compound } A\text{)}}}{CH_3CH_2\overset{OH}{\overset{|}{C}}HCH_2CH_2CH_3}$$

Hydroboration-oxidation of 1-methylcyclopentene would give compound *D*, not compound *B*, because this reaction sequence results in a *syn*-addition.

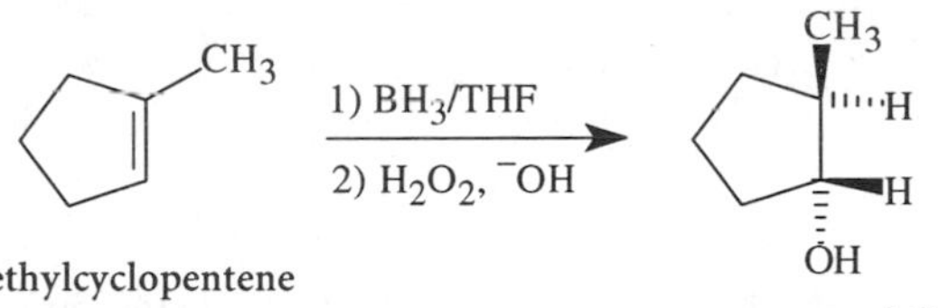

1-methylcyclopentene $\xrightarrow[\text{2) } H_2O_2,\ ^-OH]{\text{1) } BH_3/THF}$ compound *D*

(Of course, compound *D* is racemic, and is therefore a 50:50 mixture of the compound shown above and its enantiomer.)

Compound *C* cannot be produced by hydroboration-oxidation of an alkene because the OH group would have to end up at the more branched carbon of any possible alkene starting material.

7.37 *For reaction (1):*

(a) The products:

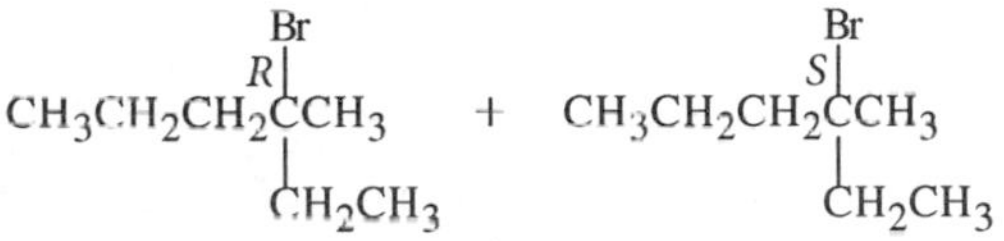

$CH_3CH_2CH_2C(Br)(CH_2CH_3)CH_3$ (*R*) + $CH_3CH_2CH_2C(Br)(CH_2CH_3)CH_3$ (*S*)

(b) The two products are enantiomers.
(c) The two products are formed in identical amounts, because the starting materials are achiral. In other words, the product is a racemate.
(d) The two products would have identical boiling points and melting points, and would thus not be separable by fractional distillation or conventional fractional crystallization.

For reaction (2):

(a) The products:

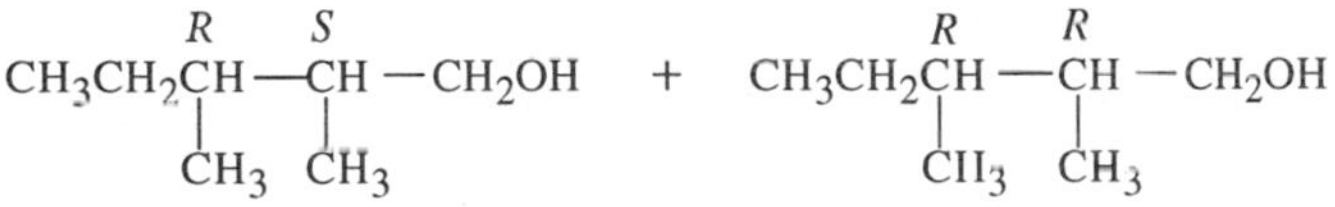

$CH_3CH_2CH(CH_3)—CH(CH_3)—CH_2OH$ (*R*, *S*) + $CH_3CH_2CH(CH_3)—CH(CH_3)—CH_2OH$ (*R*, *R*)

(b) The two products are diastereomers.
(c) The two products are formed in different amounts.
(d) The two products have different boiling points and melting points, and could therefore be separated (in principle) by fractional distillation or fractional crystallization.

For reaction (3):

(a) The products:

$CH_3CH_2CH(Br)CH_2Br$ (*S*) + $CH_3CH_2CH(Br)CH_2Br$ (*R*)

(b) The two products are enantiomers.
(c) The two products are formed in identical amounts, because the starting materials are achiral. In other words, the product is a racemate.
(d) The two products would have identical boiling points and melting points, and would thus not be separable by fractional distillation or conventional fractional crystallization.

For reaction (4):

(a) The starting alkene is racemic, and is therefore an equimolar mixture of two enantiomers. To predict the products, make the prediction for each enantiomer separately and then combine the results. Starting arbitrarily with the *S* enantiomer, the products are the following two diastereomers:

$$\underset{A}{\overset{S\ \ S}{CH_3\underset{Ph}{C}H\underset{Br}{C}HCH_2Br}} \quad + \quad \underset{B}{\overset{S\ \ R}{CH_3\underset{Ph}{C}H\underset{Br}{C}HCH_2Br}}$$

The *R* enantiomer of the alkene starting material gives a corresponding pair of diastereomers:

$$\underset{C}{\overset{R\ \ R}{CH_3\underset{Ph}{C}H\underset{Br}{C}HCH_2Br}} \quad + \quad \underset{D}{\overset{R\ \ S}{CH_3\underset{Ph}{C}H\underset{Br}{C}HCH_2Br}}$$

(b) Compounds *A* and *B* are diastereomers, as are compounds *C* and *D,* compounds *A* and *D,* and compounds *B* and *C*; compounds *A* and *C* are enantiomers, as are compounds *B* and *D.*
(c) The diastereomers are formed in different amounts; the enantiomers are formed in identical amounts.
(d) Any pair of diastereomers have different melting points and boiling points; any pair of enantiomers have identical melting points and boiling points.

For reaction (5):

(a) The product results from *syn*-addition of H_2 across the double bond; the product is called *cis*-decalin.

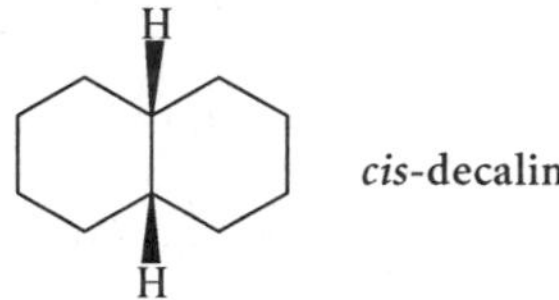

cis-decalin

(b) This is the only product formed. Because the product is achiral, it has no enantiomer.
(c) Because only one product is formed, this part does not apply.
(d) Because only one product is formed, this part does not apply.

7.38 (a) 1,1-Dimethylcyclohexane and *cis*-1,4-dimethylcyclohexane have identical chair conformations.

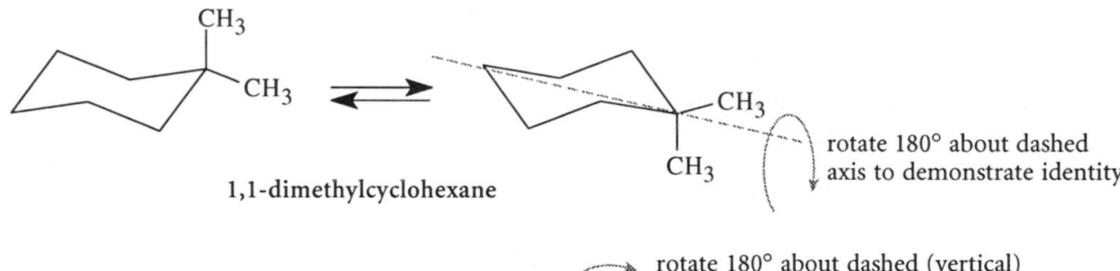

1,1-dimethylcyclohexane

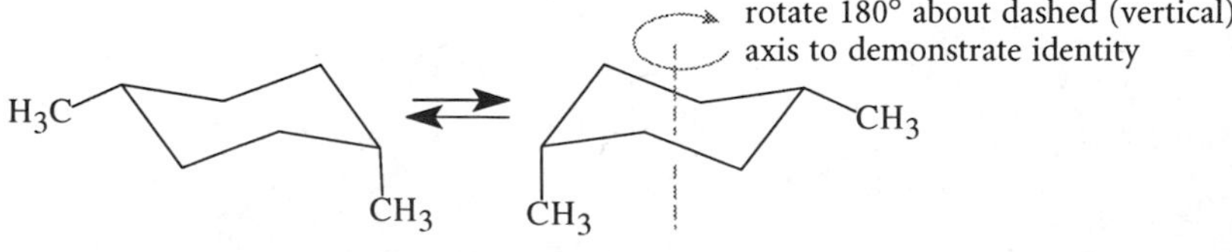

cis-1,4-dimethylcyclohexane

(b) Both *cis*-1,3-dimethylcyclohexane and *trans*-1,4-dimethylcyclohexane are achiral and exist as conformation diastereomers.

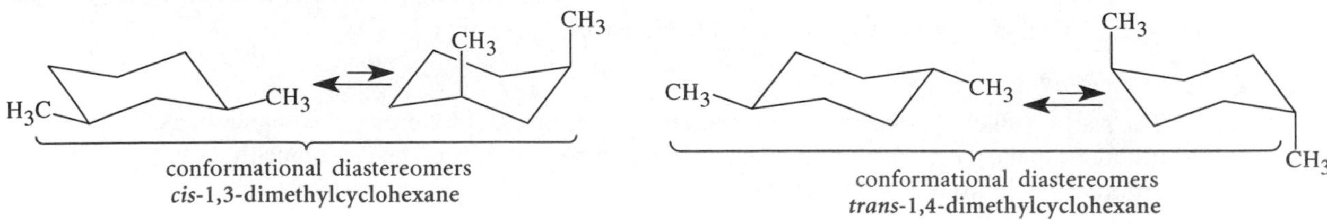

conformational diastereomers
cis-1,3-dimethylcyclohexane

conformational diastereomers
trans-1,4-dimethylcyclohexane

(c) *Trans*-1,2-dimethylcyclohexane is chiral and exists as rapidly equilibrating conformational diastereomers. (The analysis of this case is on text p. 264.)

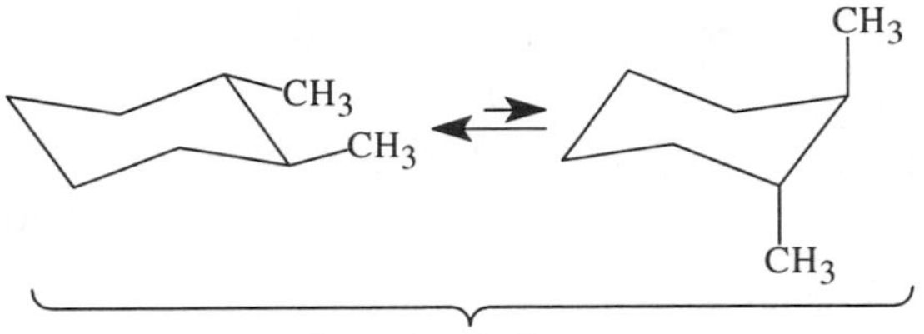

CH_3 CH_3 ⇄ CH_3 CH_3

conformational diastereomers

trans-1,2-dimethylcyclohexane

(d) *Cis*-1,2-dimethylcyclohexane exists as two chair conformations that are conformational enantiomers. (The analysis of this case is on text pp. 262–3.)

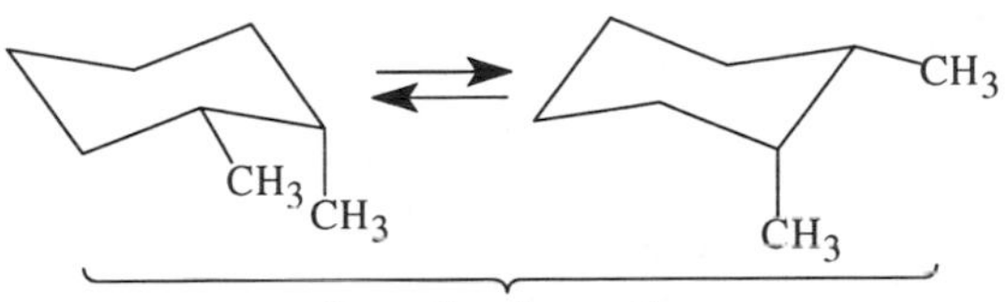

CH_3 CH_3 ⇄ CH_3 CH_3

conformational enantiomers

cis-1,2-dimethylcyclohexane

7.39 A conformational representation of the steroid: note that all ring junctions are trans.

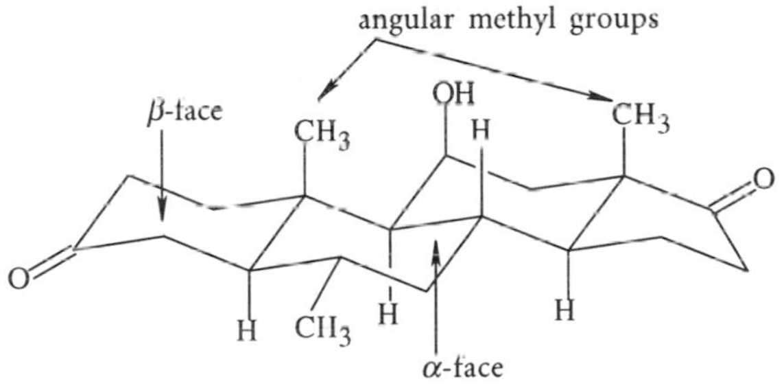

angular methyl groups

β-face CH_3 OH H CH_3 O O H CH_3 H H α-face

7.40 The two chair conformations of α-(+)-glucopyranose are as follows; conformation *A* is favored at equilibrium. (Why?)

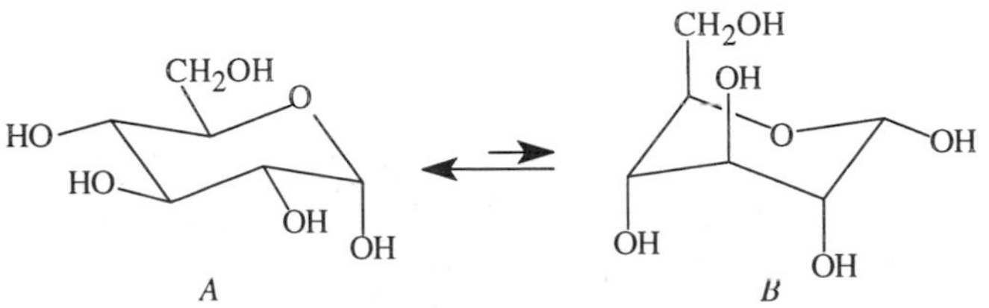

CH_2OH O HO HO OH OH *A* ⇄ CH_2OH OH O OH OH OH *B*

7.41 (a) *Anti*-addition can occur to give two chiral diastereomers. Each is a single enantiomer because the starting material is a single enantiomer.

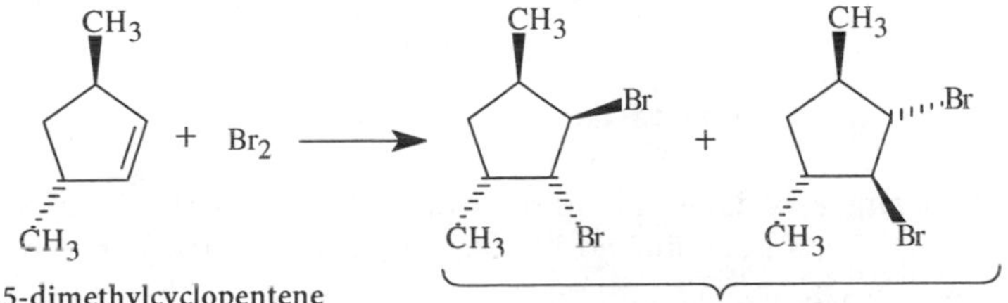

CH_3 CH_3 + Br_2 ⟶ CH_3 Br CH_3 Br + CH_3 Br CH_3 Br

(3*R*,5*R*)-3,5-dimethylcyclopentene

chiral diastereomers

(b) The bromonium ion derived from attack of cyclopentene on bromine undergoes backside attack by water to give the *trans*-bromohydrin, which is chiral. Because the starting materials are achiral, the chiral product is obtained as a racemate.

$$\text{cyclopentene} + Br_2 + H_2O \longrightarrow \textit{trans}\text{-2-bromocyclopentanol} + HBr$$

(racemate)

7.42 The two dibromides are the two diastereomeric *trans*-disubstituted *trans*-decalin derivatives:

The different compounds result from attack of bromide on the bromonium ion at the two possible carbons. (Be sure to work through this mechanism.) Because each compound is chiral and the starting materials are not, each diastereomer is obtained as a racemate.

7.43 Given that addition of bromine at each double bond is *anti,* two diastereomers can be formed:

meso racemate

Because these are diastereomers, they have different physical properties; evidently, one melts at 255°, the other at 188°, although the data do not determine which is which.

Assuming you had samples of the two compounds but didn't know which was which, what experiment could you do that might identify the two compounds? (*Answer:* carry out an enantiomeric resolution. The compound that can be separated into optically active components—its two enantiomers—is identified as the racemate.) It turns out that the meso compound is the higher-melting stereoisomer.

7.44 (a) This compound is *not* consistent with the data because it is meso and cannot be optically active.

(b) This compound is consistent with the data because it is chiral but would hydrogenate to give *trans*-1,4-dimethylcyclohexane, an achiral product.

(c) This compound is consistent with the data because it is chiral and would hydrogenate to give a mixture of *cis*- and *trans*-1,4-dimethylcyclohexane, which are both achiral products.

(d) This compound is *not* consistent with the data, because, although it is chiral and thus optically active, it would hydrogenate to give a chiral and optically active product, (1*S*,3*S*)-1,3-dimethylcyclohexane.

(e) This compound is *not* consistent with the data because it is not chiral and therefore could not be optically active.

In summary, only alkenes (b) and (c) are consistent with *all* the data.

7.45 Because this is an amine, both the chair flip and amine inversion can occur. Note that the amine inversion interchanges the axial and equatorial positions of the electron pair and the hydrogen on the nitrogen *without* flipping the ring. Thus, four chair conformations are in rapid equilibrium:

7.46 (a) The only stereocenter in the molecule is the nitrogen, which rapidly undergoes inversion, a process that rapidly interconverts enantiomers. Consequently, this compound cannot be resolved into enantiomers at room temperature.

(b) Because nitrogen inversion interconverts enantiomers, the compound cannot be resolved into enantiomers.

(c) The bicyclic structure prevents the nitrogens from inverting (see the solution to Problem 7.62), and both nitrogens are stereocenters. Consequently, the molecule is chiral and can be resolved into enantiomers.

(d) The asymmetric ring carbon causes the molecule to be chiral; inversion of the nitrogen interconverts diastereomers. Hence, the compound can be resolved into two enantiomeric sets of rapidly interconverting diastereomers.

7.47 In the transition state for inversion of both amines, the nitrogen is *sp*^{2}-hybridized. The C—N—C angle required by this hybridization is 120°. This angle, which has an optimal C—N—C angle of 60°, is much more difficult to achieve within a three-membered ring than it is in within a five-membered ring, which has an optimal C—N—C angle of 109°. Hence, there is much more angle strain in the transition state for inversion of 1-methylaziridine than there is in the transition state for inversion of 1-methylpyrrolidine. For this reason, the transition state for inversion of the aziridine derivative has a very high energy, and nitrogen inversion in this derivative is very slow. (This is a very unusual situation, since inversion of most amines is very rapid; see Sec. 6.10B.)

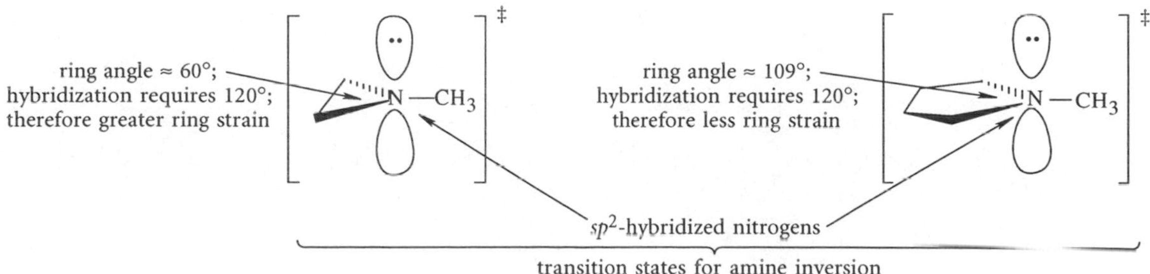

7.48 The *cis* isomer of the starting alkene was used. This is shown by starting with the *cis* isomer and imagining the *syn* addition of the two —OH groups to the same face of the double bond.

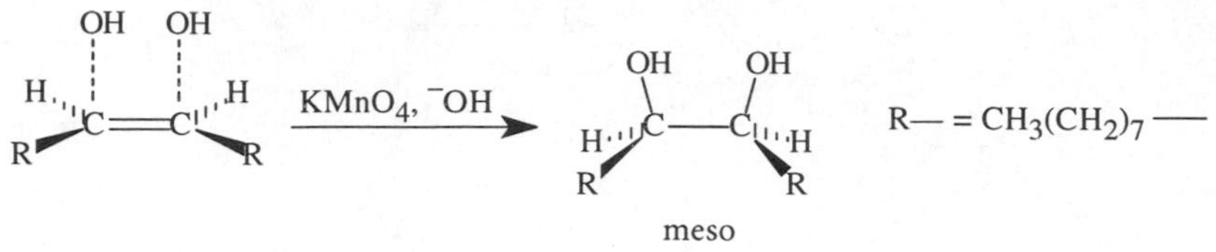

$KMnO_4$, ^-OH → meso

R— = $CH_3(CH_2)_7$—

By a similar analysis, *syn* addition to the *trans* isomer of the alkene gives the racemate. (You should demonstrate this to yourself.)

As this example and Sec. 7.9C show, there is a relationship between the *stereochemistry of the alkene,* the *stereochemistry of the reaction,* and the *stereochemistry of the product.* If we specify any two of these, the other is determined.

7.49 (a) First, convert the Fischer projection of the product into a three-dimensional perspective. (Build a model if necessary.) Then internally rotate about the central bond until the two —CO_2^- groups have a dihedral angle of 180°, as they do in the starting material. Finally, examine the relationship of the —D and —OD groups. This reveals that the product results from *anti*-addition of the —D and the —OD to the alkene:

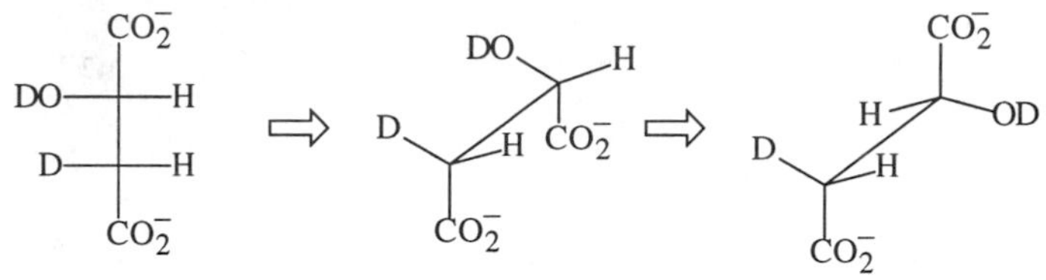

—D and —OD are on opposite sides of the molecule; therefore, addition is anti.

(b) The use of D_2O allows us to differentiate the protons of the solvent from those of the starting material. The stereochemistry of the addition cannot be determined without this distinction.

7.50 (a) Following the hint in the problem, interpret each chair structure as a planar "line-and-wedge" structure. If these two structures are congruent, then the two chair structures are *either* identical *or* are conformational stereoisomers. Decide between these by using the chair structures themselves. If the two chair structures are not congruent, then use their planar representations to decide on their relationship.

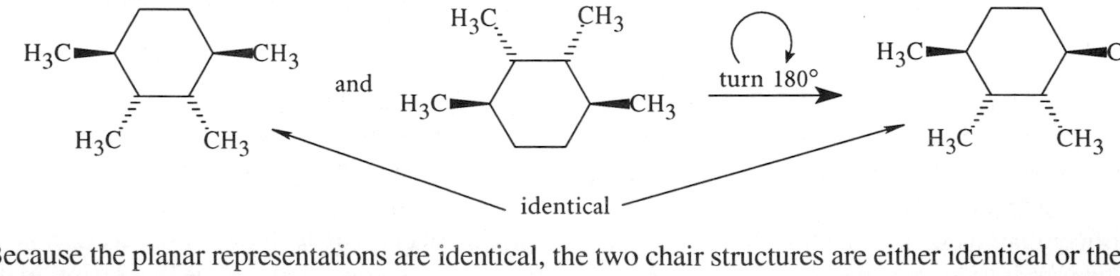

and — turn 180° — identical

Because the planar representations are identical, the two chair structures are either identical or they are conformational stereoisomers. Examination of the chair structures is necessary to reveal which alternative is correct. Let's rotate the second structure 180° about the axis shows to place all substituents in the same relative positions.

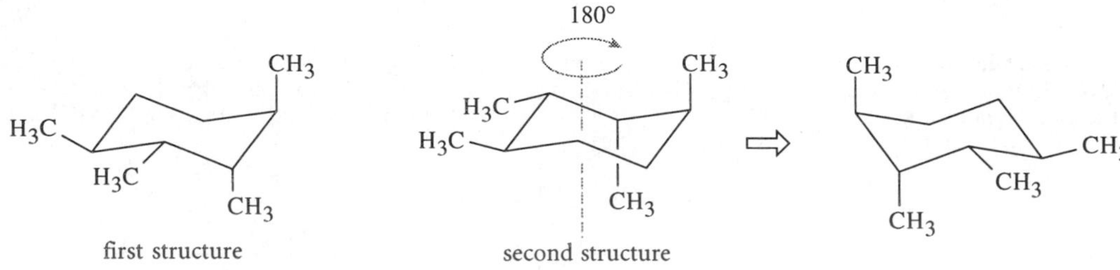

first structure — second structure

These two chair structures are clearly mirror images, and they are noncongruent mirror images. A chair flip on the first structure converts it into the second; hence, the two are chair structures are conformational enantiomers.

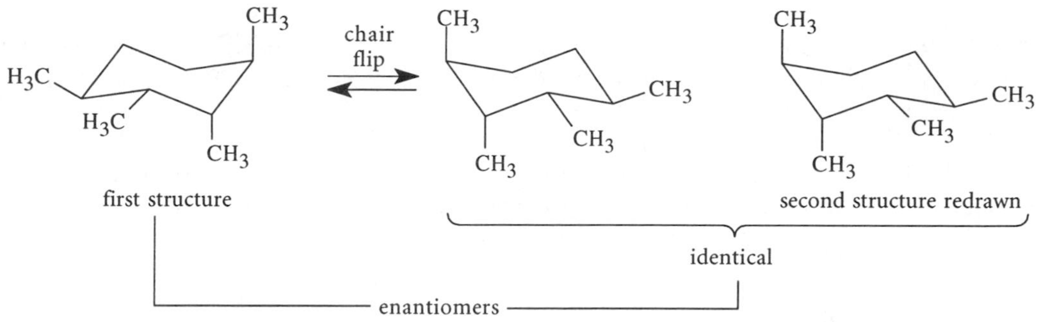

first structure — chair flip — second structure redrawn — identical — enantiomers

(b)

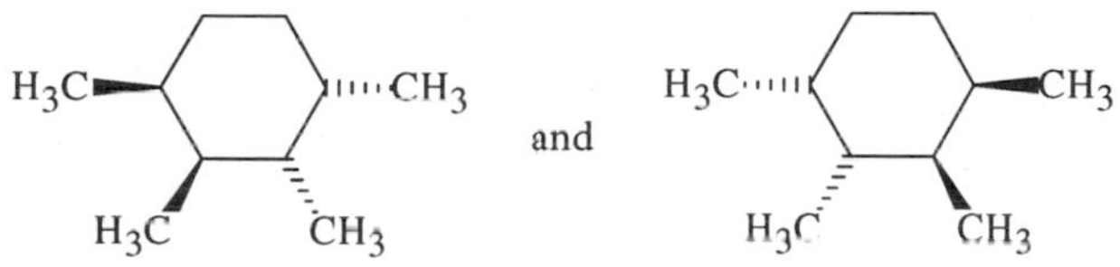

and

These two planar structures are noncongruent mirror images and are therefore enantiomers. Notice that their enantiomeric relationship might not be obvious from the two chair structures without subjecting one of them to the chair flip!

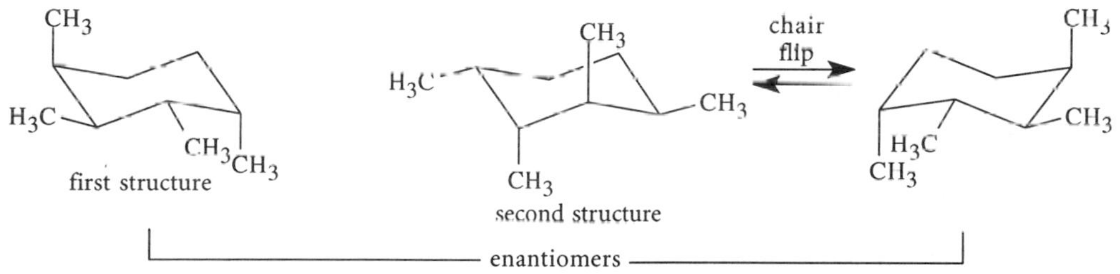

first structure — second structure — chair flip — enantiomers

Like *trans*-1,2-dimethylcyclohexane (text p. 264), each molecule is a mixture of two conformational diastereomers.

(c)

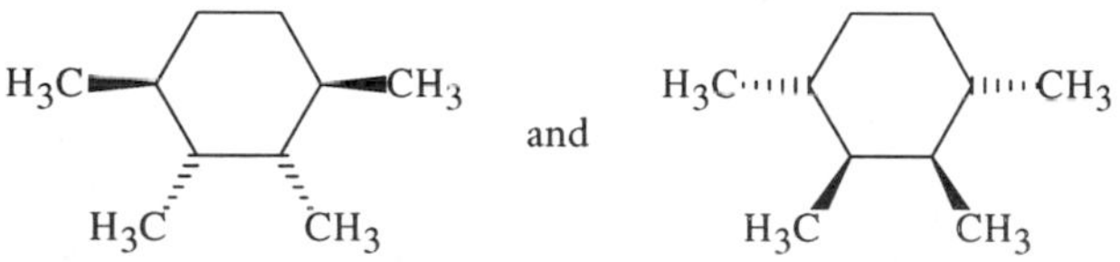

and

That the two planar structures are congruent can be determined by lifting one out of the page and turning it over onto the other (so that "up" bonds become "down" bonds and vice-versa). Consequently, the two chair conformations are either identical or are conformational stereoisomers. Rotating the second structure 180° about the axis shown demonstrates that the two chair structures are in fact identical.

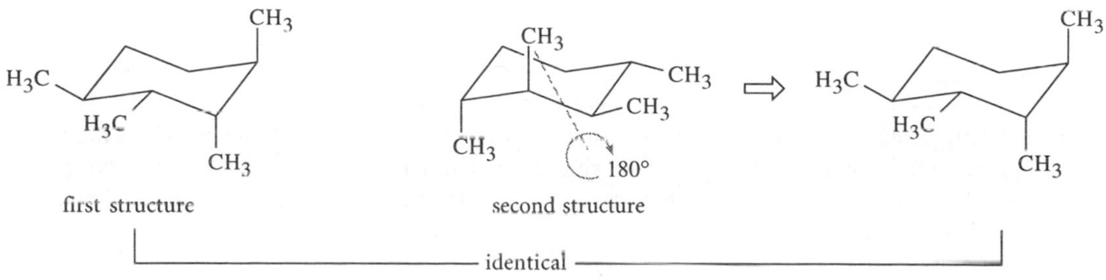

first structure — second structure — 180° — identical

Remember that the use of models to solve problems in which two structures are compared requires *two models,* one of the original structure and another of the structure with which it is compared.

Problem 7.50 is a particularly good test of your ability to manipulate cyclohexane rings. If you were able to solve it without a great deal of difficulty, you have in fact mastered the art of handling chair conformations!

7.51 (a) Hydration involves a carbocation intermediate. Because a carbocation can be attacked by nucleophiles at either face of its vacant $2p$ orbital, a mixture of stereoisomeric hydration products is expected.

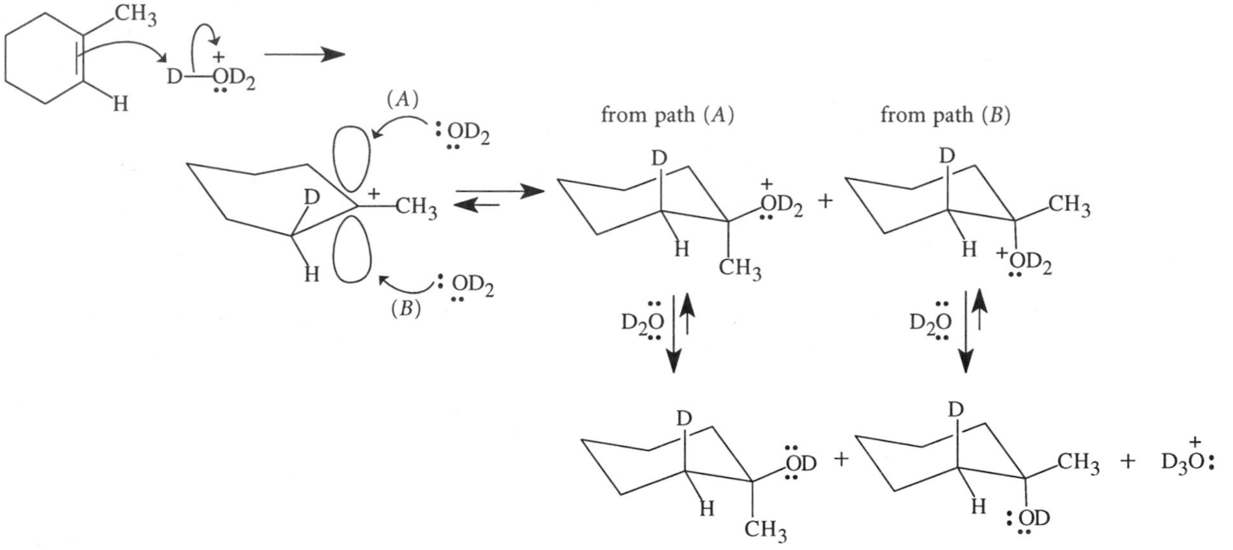

from path (*A*) from path (*B*)

A *B*

(b) As in the solution to Problem 7.49(b), the use of D_2O allows us to differentiate the protons of the solvent from those of the starting material.

(c) The two hydrogens referred to in part (b) could also be distinguished if the one in the starting alkene is replaced by a deuterium, and H_3O^+ is used as the acid, that is, if an alkene of the following structure is used:

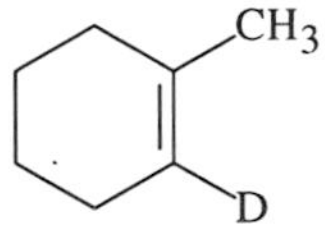

7.52 (a) The two chiral stereoisomers are the following enantiomeric pair:

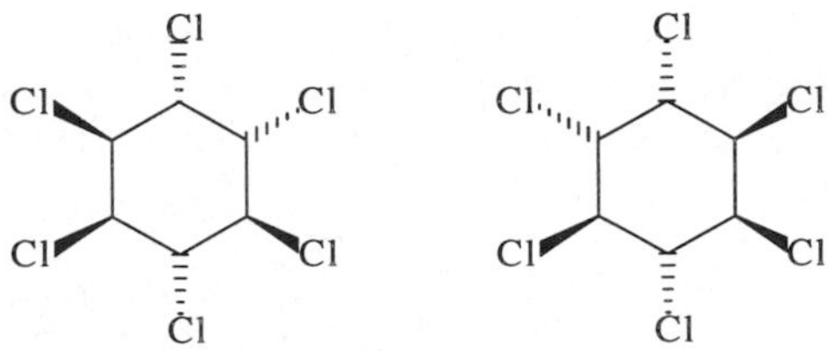

(b) The only candidate structures are those in which the number of axial chlorines equals the number of equatorial chlorines. In addition, the two chlorines within each of the three pairs that have a 1,4-relationship must differ in whether their positions are axial or equatorial. Thus, each of the following stereoisomers has two identical chair conformations.

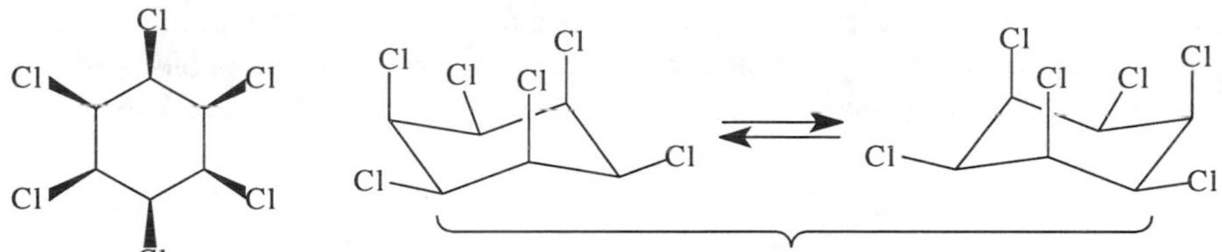

identical

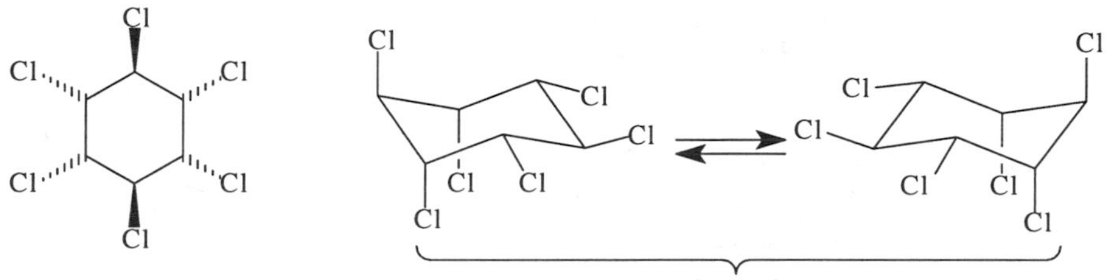

identical

The identity within each pair can be seen by rotating one structure 180° about a vertical axis through the center of the molecule to give the other, for example:

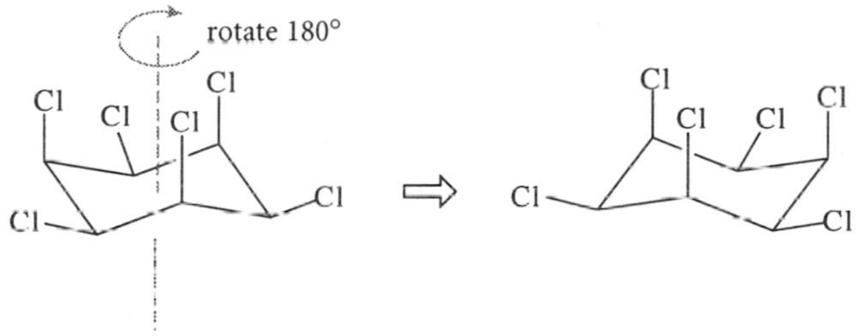

rotate 180°

7.53 The true statements about the relationship of *cis*- and *trans*-decalin (see text p. 270) are (c) and (d).

7.54 In order for a compound to exist in a relatively high-energy twist-boat conformation, there must be a good reason for it to avoid the chair. Indeed, compounds *B* and *D* have very severe 1,3-diaxial interactions between a *tert*-butyl group and a methyl group in the chair conformations shown. Compound *D*, a *cis*-decalin derivative, can avoid its 1,3-diaxial interaction by undergoing a chair flip. However, compound *B*, a *trans*-decalin derivative, cannot undergo the chair flip (see text p. 270–1). Hence, the only way it can avoid the 1,3-diaxial interaction is for the ring containing the *tert*-butyl group to assume a twist-boat conformation. The correct answer, then, is *B*.

7.55 *Cis*-1,3-di-*tert*-butylcyclohexane can exist in a chair form (*A*) in which both *tert*-butyl groups are equatorial. However, in either chair form of *trans*-1,3-di-*tert*-butylcyclohexane (*B*), there is an axial *tert*-butyl group. The axial *tert*-butyl group can be avoided if compound *B* exists in a twist-boat form *C*.

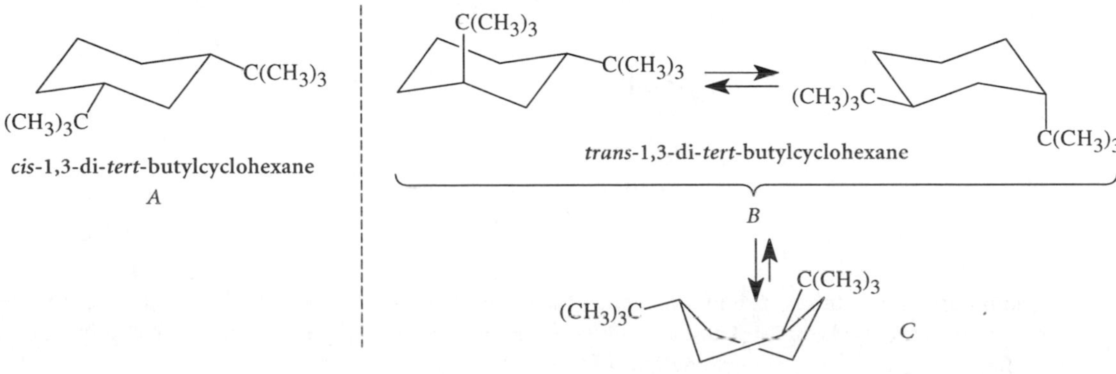

7.56 Within each structure, count the 1,3-diaxial interactions between the methyl group and axial ring hydrogens four carbons away (counting the methyl as one carbon). These are shown with double-headed arrows in the structures of *A* and *B* below. (There are no methyl-hydrogen 1,3-diaxial interactions in compound *C*.)

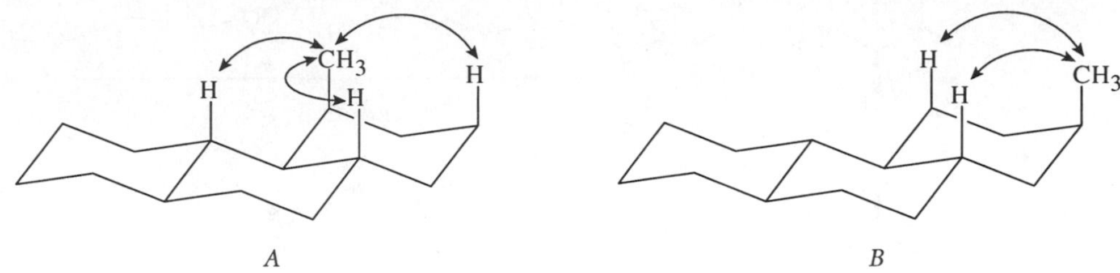

Each interaction shown adds 3.7 kJ/mol to the heat of formation relative to that of *C*. Consequently, *C* has the lowest heat of formation; the heat of formation of *B* is $(2 \times 3.7) = 7.4$ kJ/mol greater; and the heat of formation of *A* is $(3 \times 3.7) = 11.1$ kJ/mol greater than that of *C*.

7.57 (a)

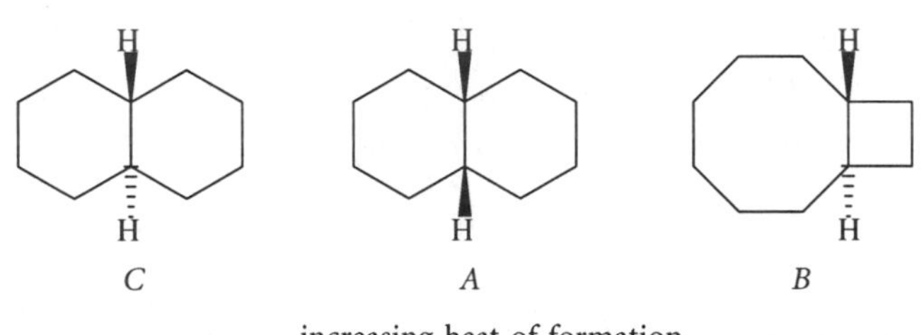

Trans-decalin, *C*, is more stable than *cis*-decalin, *A* (see Problem 7.18, text p. 271). However, neither compound has angle strain. Compound *B* is least stable because of the strain in its four-membered ring. You can see from Table 7.1, text p. 245, that the strain in a cyclobutane ring ($4 \times 6.9 = 27.6$ kJ/mol) is far more destabilizing that the three 1,3-diaxial interactions ($3 \times 3.7 = 11.1$ kJ/mol) in *cis*-decalin.

(b)

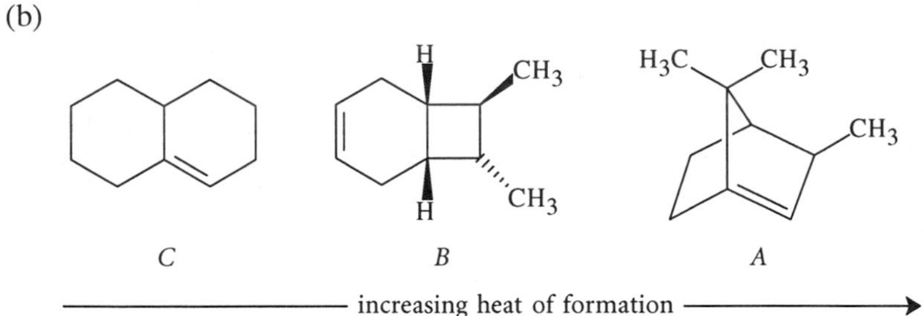

The ring strain in *B* makes it less stable than *C*; and the twisted double bond in *A* (violation of Bredt's rule) makes it so unstable that it cannot be isolated.

7.58 (a) Assuming an overall *syn* addition, the two diastereomeric products are the following:

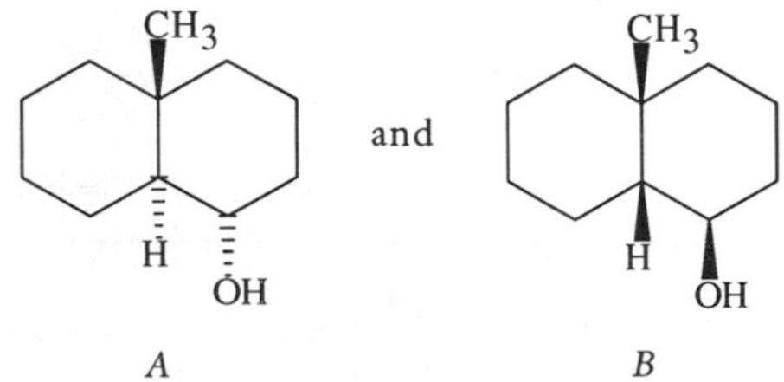

(b) The borane-THF complex should approach the alkene from the less hindered face, that is, from the face opposite to the methyl group, to give, after oxidation, product *A*. In effect, the methyl group acts as an "umbrella" to block reaction at one face of the double bond:

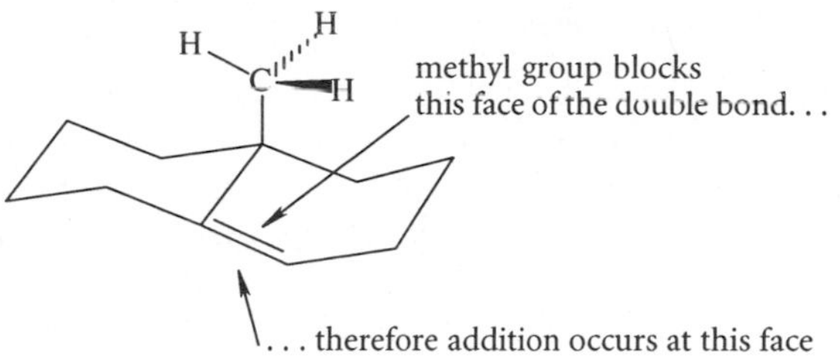

methyl group blocks this face of the double bond. . .

. . . therefore addition occurs at this face

This problem illustrates a very common trick in performing chemistry on steroids that contain axial methyl groups. The methyl group is used to block access to one face of the steroid, thus allowing the formation of single stereoisomers.

7.59 The organoborane can approach the π bond of a *cis*-2-butene molecule in two distinguishable ways. Approach from one face gives a transition state in which there is severe van der Waals repulsions between a methyl group of the alkene and a methyl group on the organoborane (structure *A*). Approach from the opposite face gives a transition state in which the van der Waals repulsion is between a *hydrogen* of the alkene and a methyl group of the borane (structure *B*).

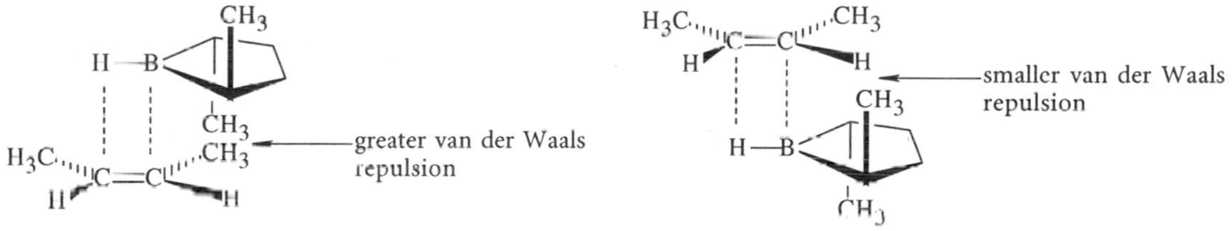

greater van der Waals repulsion

smaller van der Waals repulsion

A *B*

Given that transition state *B* has fewer van der Waals repulsions, the product borane is derived from that transition state. When oxidized, it gives the corresponding alcohol, which has the *S* configuration.

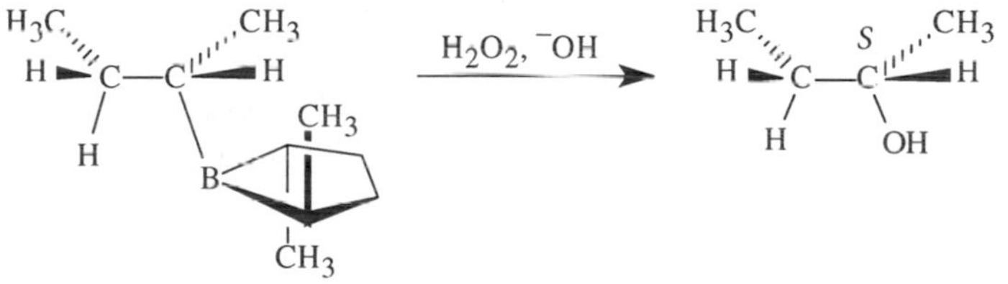

H_2O_2, ^{-}OH

7.60 (a) Conformation *A* contains four 1,3-methyl-hydrogen diaxial interactions; the energy cost of these is (4 × 3.7) = 14.8 kJ/mol. In addition, conformation *A* contains a *gauche*-butane interaction for an additional 3.7 kJ/mol, for a total energy cost of 3.7 + 14.8 = 18.5 kJ/mol. Conformation *B* contains two methyl-hydrogen 1,3-diaxial interactions, at an energy cost of (2 × 3.7) = 7.4 kJ/mol; a *gauche*-butane interaction at 3.7 kJ/mol; and the methyl-methyl 1,3-diaxial interaction, at an energy cost to be determined. Let the energy cost of a methyl-hydrogen 1,3-diaxial interaction or a *gauche*-butane interaction be $\Delta G°$(Me-H) = 3.7 kJ/mol, and let the energy cost of a methyl-methyl 1,3-diaxial interaction be $\Delta G°$(Me-Me). It is given that the $\Delta G°$ of the overall equilibrium, $\Delta G°$(eq), equals 8.4 kJ/mol and that conformation *A* has lower energy. Assuming that all energies except for the 1,3-diaxial interactions are about the same, then

$$\Delta G°(\text{eq}) = \Delta G°(B) - \Delta G°(A) = 8.4 \text{ kJ/mol} = 3\Delta G°(\text{Me-H}) + \Delta G°(\text{Me-Me}) - 5\Delta G°(\text{Me-H})$$

or

$$8.4 \text{ kJ/mol} = \Delta G°(\text{Me-Me}) - 2\Delta G°(\text{Me-H}) = \Delta G°(\text{Me-Me}) - 2(3.7 \text{ kJ/mol})$$

Solving,

$$\Delta G°(\text{Me-Me}) = 8.4 \text{ kJ/mol} + 7.4 \text{ kJ/mol} = 15.8 \text{ kJ/mol } (3.78 \text{ kcal/mol})$$

This calculation shows that a methyl-methyl 1,3-diaxial interaction is more than four times as costly in energy terms as a methyl-hydrogen 1,3-diaxial interaction. This should make sense, because a methyl group is much larger than a hydrogen.

(b) Conformation *C* has no 1,3-diaxial interactions, whereas conformation *D* has one methyl-methyl 1,3-diaxial interaction, which, from part (a), costs 15.8 kJ/mol, and two methyl-hydrogen 1,3-diaxial interactions, which cost $(2 \times 3.7) = 7.4$ kJ/mol. Thus, conformation *D* is $(15.8 + 7.4) = 23.2$ kJ/mol less stable than conformation *C*. The standard free energy change for the equilibrium is therefore +23.2 kJ/mol (5.54 kcal/mol).

7.61 (a) These compounds are noncongruent mirror images and are therefore enantiomers.

(b) We know that in most common cases (for example, *cis*- and *trans*-2-butene), *E,Z* isomers are diastereomers. However, all that is needed to apply the *E,Z* nomenclature is a difference in priorities between the two groups attached to *each* carbon of a double bond. Using compound *A* as an example, the issue, then, is whether the two groups beginning with the carbons *(a)* and *(b)* have different priorities.

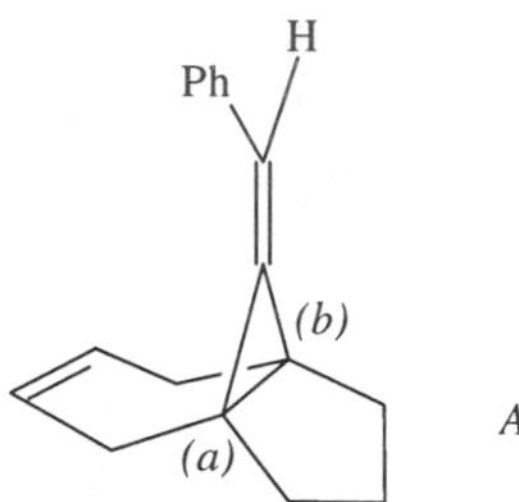

Certainly, we can discern no differences in *connectivity* if we begin at carbons (*a*) and (*b*) and move outwards along the path of highest priority within each group. However, there is a difference at the carbons themselves, and that difference is in their *stereochemical configurations.* Because we did not cover this possibility in Chapter 4, you would have been justified in making up your own priority rule; but Cahn, Ingold and Prelog decided arbitrarily that an *R* configuration receives the higher priority. Because carbon (*b*) has the *R* configuration, and it is on the side of the double bond opposite to the higher-priority phenyl (Ph) group, then alkene *A* is the *E* isomer. It follows that alkene *B* is the *Z* isomer. This, then, is an amusing and unusual situation in which *E,Z* isomers are actually enantiomers.

7.62 (a) Two rings in a bridged bicyclic compound can be joined in three stereochemically different ways which might be termed (*out,out*), (*in,out*), and (*in,in*). Assuming hydrogens at both bridgeheads, these ways can be represented schematically as follows.

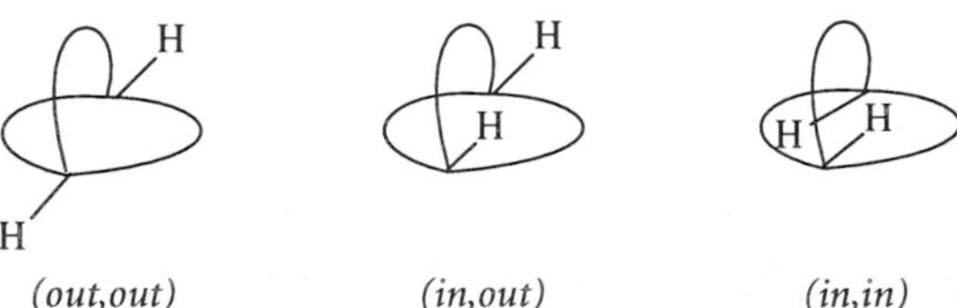

(b) As these simple diagrams suggest, the (*in,out*) and (*in,in*) patterns require that one and two hydrogens, respectively, occupy the region of space *within* a ring. This is sterically impossible for small rings. For small rings, this would actually require such a distortion of the tetrahedral structure that all four bonds to the bridgehead carbon with the *in* configuration would lie on one side of a plane through this carbon. (If the impossibility of achieving this sort of structure is not clear, try to build a model of (*in,out*)-bicyclo[2.2.2]octane.) The larger the rings, the larger the space within the ring, and for very large rings, the "inner" hydrogens can be accommodated easily within the ring and the tetrahedral bonding geometry of the bridgehead carbons can be achieved without significant strain. Hence, the (*in,out*) and (*in,in*) stereoisomers of the 77-carbon bicyclic alkane are stable.

8

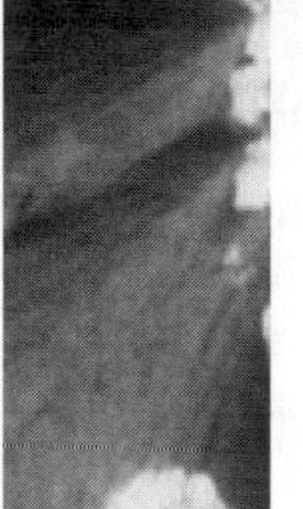

Introduction to Alkyl Halides, Alcohols, Ethers, Thiols, and Sulfides

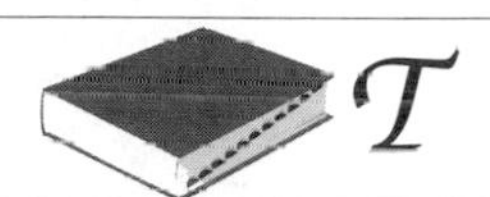

TERMS

CONCEPTS

I. NOMENCLATURE OF ALKYL HALIDES, ALCOHOLS, ETHERS, THIOLS, AND SULFIDES

A. General

1. Organic compounds are named by both common and substitutive nomenclature.
2. Common nomenclature, or radicofunctional nomenclature, is generally used only for the simplest and most common compounds.
   a. Do not confuse the allyl group with the vinyl group.

$H_2C{=}CH{-}CH_2{-}$ allyl group $\qquad$ $H_2C{=}CH{-}$ vinyl group

   b. Do not confuse the benzyl group with the phenyl group.

$C_6H_5{-}CH_2{-}$ benzyl group $\qquad$ $C_6H_5{-}$ phenyl group

3. In substitutive nomenclature, the name is based on the principal group and principal chain; other groups are cited as substituents.
   a. Identify the principal group: the chemical group on which the name is based and cited as a suffix in the name.
      i. Hydroxy (—OH) and thiol (—SH) groups can be cited as principal groups.
      ii. A compete list of principal groups and their relative priorities are summarized in Appendix I of the text.
      iii. If a compound does not contain a principal group, it is named as a substituted hydrocarbon.
   b. Identify the principal carbon chain on which the name is based by applying the following criteria in the order given below until a decision can be made (an extensive list is given in Appendix I of the text):
      i. greatest number of principal groups;
      ii. greatest number of double and triple bonds;
      iii. greatest length;
      iv. greatest number of other substituents.
   c. Number the principal chain consecutively from one end by applying the following criteria in the order given below until there is no ambiguity:
      i. lowest number for the principal group;
      ii. lowest numbers for multiple bonds, with double bonds having priority over triple bonds;
      iii. lowest numbers for other substituents;
      iv. lowest number for the substituent cited first in the name.
   d. Begin construction of the name with the name of the hydrocarbon corresponding to the principal chain.
      i. Cite the principal group by its suffix and number; its number is the last one cited in the name.
      ii. Cite the names and numbers of the other substituents in alphabetical order at the beginning of the name.
4. Common and substitutive nomenclature should not be mixed.

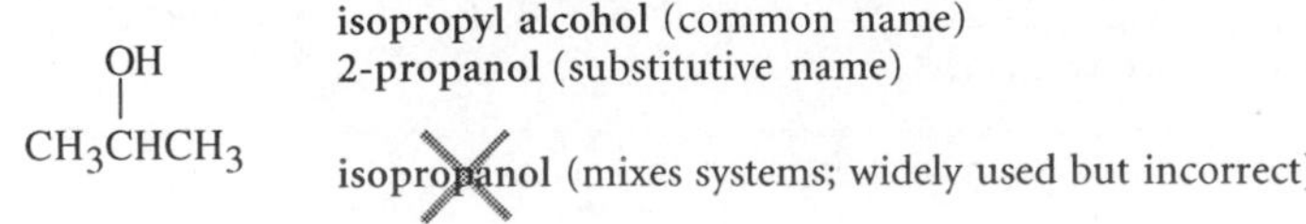

5. Cyclic compounds with rings that contain more than one type of atom are called heterocyclic compounds.

B. Alkyl Halides

1. In an alkyl halide, a halogen atom is bonded to the carbon of an alkyl group.
2. Alkyl halides are classified as methyl, primary, secondary, or tertiary, depending of the number of alkyl groups attached to the carbon bearing the halogen. (Arrows show bonds to alkyl groups.)

	$H_3C—F$	$(H_3C)_2CH—CH_2—Br$	cyclohexyl—Cl	$H_3C—C(CH_3)_2—I$
type of alkyl halide:	methyl	primary	secondary	tertiary
number of alkyl groups:	none	1	2	3

3. The common name of an alkyl halide is constructed from the name of the alkyl group followed by the halide as a separate word; the methyl trihalides are called haloforms.

$H_3C—F$	$(H_3C)_2CH—CH_2—Br$	cyclohexyl—Cl	$H_3C—C(CH_3)_2—I$	$CHCl_3$
methyl fluoride	isobutyl bromide	cyclohexyl chloride	*tert*-butyl iodide	chloroform

4. The IUPAC substitutive name of an alkyl halide is constructed by applying the rules of alkane and alkene nomenclature.
   a. Halogens are always treated as substituents.
   b. The halogens substituents are named fluoro, chloro, bromo, or iodo.

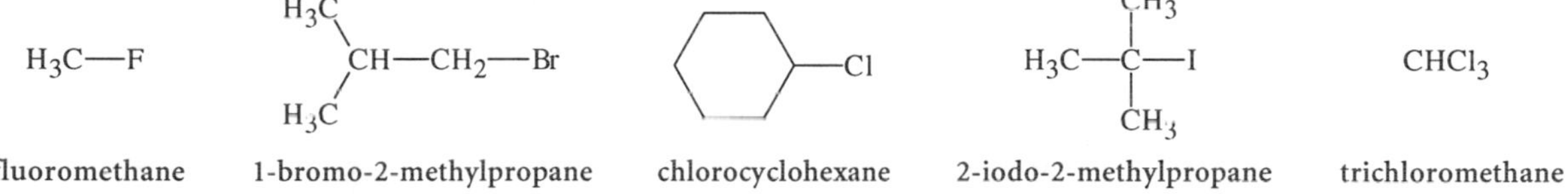

fluoromethane 1-bromo-2-methylpropane chlorocyclohexane 2-iodo-2-methylpropane trichloromethane

C. Alcohols and Thiols

1. In an alcohol, a hydroxy group, —OH, is bonded to the carbon of an alkyl group.
2. In a thiol (also called a mercaptan), a sulfhydryl group (also called a mercapto or hydrosulfide group), —SH, is bonded to an alkyl group; thiols are the sulfur analogs of alcohols.
3. Alcohols are classified as methyl, primary, secondary, or tertiary, depending of the number of alkyl groups attached to the carbon bearing the hydroxy group. (Arrows show bonds to alkyl groups.)

	$H_3C—OH$	$(H_3C)_2CH—CH_2—OH$	cyclohexyl—OH	$H_3C—C(CH_3)_2—OH$
type of alcohol:	methyl	primary	secondary	tertiary
number of alkyl groups:	none	1	2	3

4. Compounds that contain two or more hydroxy groups on adjacent carbons are called glycols.

$HO—CH_2—CH_2—OH$

ethylene glycol

5. In common nomenclature:
   a. The common name of an alcohol is derived by specifying the alkyl group to which the —OH group is attached, followed by the separate word alcohol.
   b. Thiols are named in the common system as mercaptans or hydrosulfides.

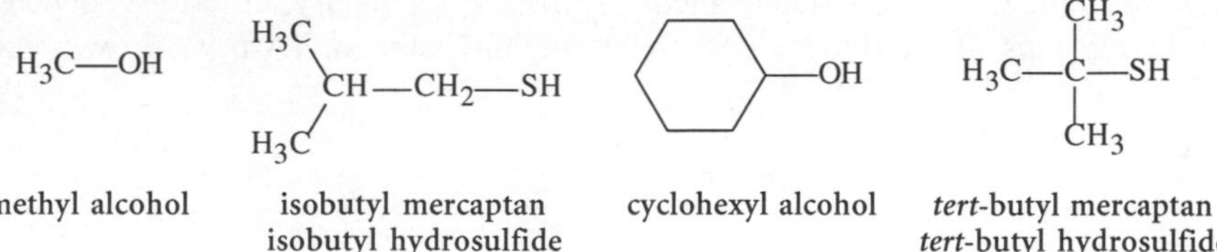

methyl alcohol | isobutyl mercaptan, isobutyl hydrosulfide | cyclohexyl alcohol | *tert*-butyl mercaptan, *tert*-butyl hydrosulfide

6. In substitutive nomenclature:
   a. The name of an alcohol is constructed by dropping the final *e* from the name of the parent alkane and adding an *ol* suffix.
   b. To name an alcohol containing more than one —OH group, the suffixes *diol*, *triol*, etc., are added to the name of the appropriate alkane without dropping the final *e*.
   c. For simple thiols, the —SH group is the principal group; the name is constructed by adding a *thiol* suffix to the name of the parent alkane (the final *e* is retained).

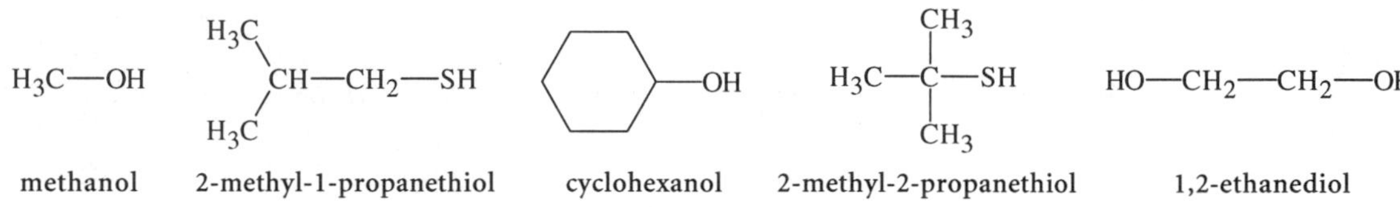

methanol | 2-methyl-1-propanethiol | cyclohexanol | 2-methyl-2-propanethiol | 1,2-ethanediol

D. Ethers and Sulfides

1. In an ether, an oxygen is bonded to two carbon groups, which may or may not be identical.
2. A thioether, or sulfide, is the sulfur analog of an ether.
3. In common nomenclature:
   a. The common name of an ether is constructed by citing as separate words the two groups attached to the ether oxygen in alphabetical order, followed by the word *ether*.
   b. The common name of a sulfide is constructed by citing as separate words the two groups attached to the sulfur in alphabetical order, followed by the word *sulfide* (or *thioether* as in older literature).

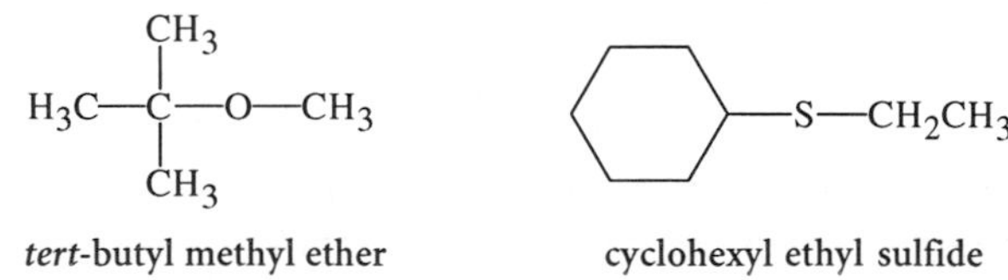

tert-butyl methyl ether | cyclohexyl ethyl sulfide

   c. The names of the following common heterocyclic ethers and sulfides should be learned:

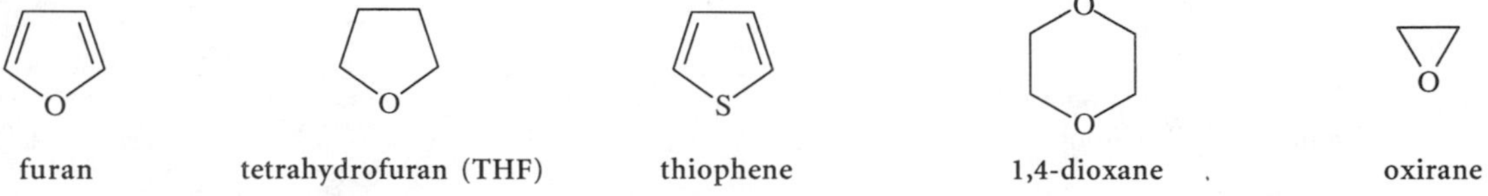

furan | tetrahydrofuran (THF) | thiophene | 1,4-dioxane | oxirane

4. In substitutive nomenclature:
   a. Ethers and sulfides are never cited as principal groups; alkoxy groups (—OR) and alkylthio groups (—SR) are always cited as substituents.

b. An —SR group is named by adding the suffix *thio* to the name of the R group; the final *yl* is not dropped.

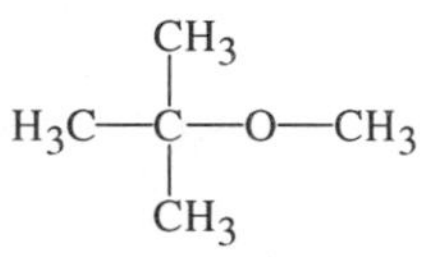

2-methoxy-2-methylpropane 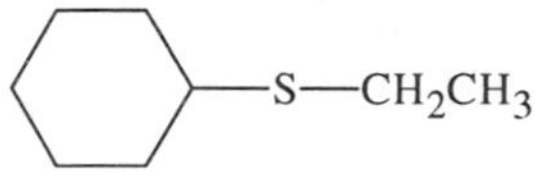 (ethylthio)cyclohexane

5. Oxirane is the parent compound of a special class of heterocyclic ethers, called epoxides, that contain three-membered rings.
   a. Most epoxides are named substitutively as derivatives of oxirane.
   b. The atoms of the epoxide ring are numbered consecutively with the oxygen receiving the number 1 regardless of the substituents present.

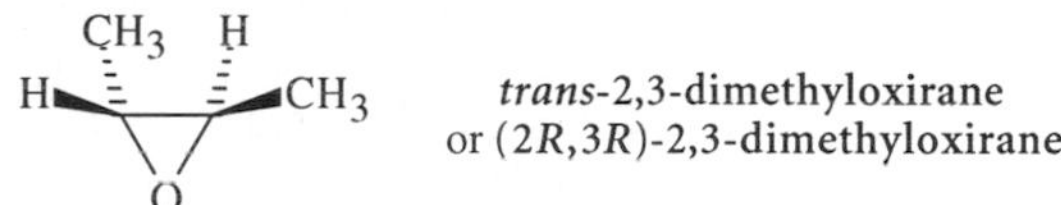

trans-2,3-dimethyloxirane
or (2*R*,3*R*)-2,3-dimethyloxirane

II. PHYSICAL PROPERTIES OF ALKYL HALIDES, ALCOHOLS, ETHERS, THIOLS, AND SULFIDES

A. Structure

1. Alkyl halides, alcohols, thiols, ethers, and sulfides have bond angles at carbon which are very close to the tetrahedral value of 109.5°.
2. Oxygen and sulfur have four "groups": two electron pairs and two alkyl groups or hydrogens.
   a. The unshared electron pairs on sulfur occupy orbitals derived from quantum level 3 that take up more space than those on oxygen, which are derived from quantum level 2.
   b. The angle at sulfur is generally found to be closer to 90° than the angle at oxygen.
3. Within a column of the periodic table, bonds to atoms of higher atomic number are longer.
4. Within a row of the periodic table, bond lengths decrease towards higher atomic number.
5. For a given molecular mass, alkyl halide molecules have smaller volumes than alkane molecules.

B. Hydrogen Bonding

1. The hydrogen bond results from the combination of two factors: a weak covalent interaction between a hydrogen on the donor atom and unshared electron pairs on the acceptor atom, and an electrostatic attraction between oppositely charged ends of two dipoles.
   a. In the liquid state, hydrogen bonding is a force of attraction between molecules.
   b. In the case of alcohols, hydrogen bonding is a weak association of the O—H proton of one molecule with the oxygen of another.
   c. In the gas phase, hydrogen bonding is much less important and, at low pressures, does not exist.
2. Formation of a hydrogen bond requires two partners:
   a. the hydrogen-bond donor (the atom to which the hydrogen is fully bonded).
      i. The best hydrogen-bond donor atoms in neutral molecules are oxygens, nitrogens, and halogens.
      ii. A hydrogen-bond donor is analogous to a Brønsted acid; thus, all strong proton acids are also good hydrogen-bond donors.
   b. the hydrogen-bond acceptor (the atom to which the hydrogen is partially bonded).
      i. The best hydrogen-bond acceptors in neutral molecules are the electronegative first-row atoms oxygen, nitrogen, and fluorine.
      ii. The hydrogen-bond acceptor is analogous to a Brønsted base; thus, all strong Brønsted bases are also good hydrogen-bond acceptors.
   c. Some atoms are donors but not acceptors.

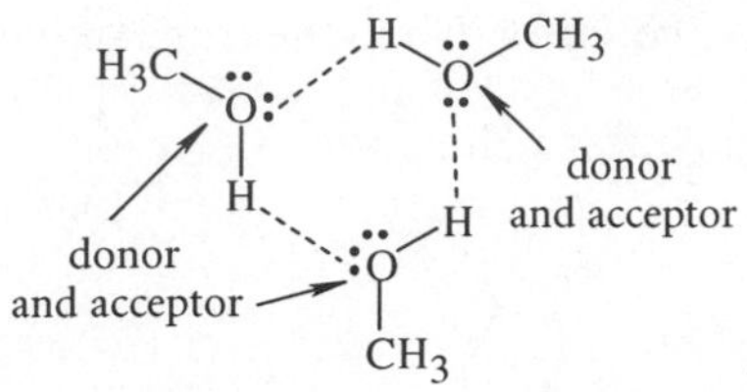

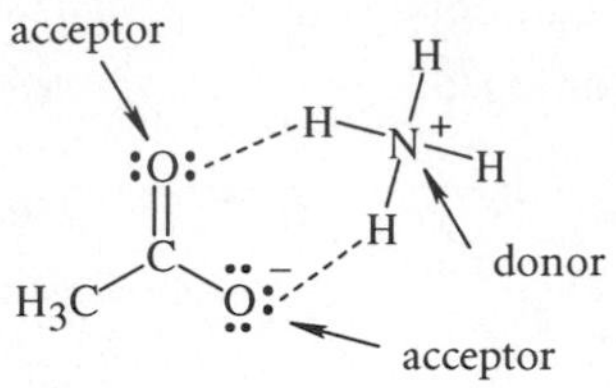

3. The hydrogen bond between two molecules resembles the same two molecules poised to undergo a Brønsted acid-base reaction.

C. Boiling Points

1. Noncovalent association of molecules in the liquid state increases their boiling points; the most important forces involved in intermolecular associations are
   a. hydrogen bonding ⇨ hydrogen-bonded molecules have greater boiling points;
   b. attractive van der Waals forces (greatest for larger, more extended molecules), which are influenced by:
      i. molecular size ⇨ larger molecules have greater boiling points;
      ii. molecular shape ⇨ more extended, less spherical molecules have greater boiling points;
   c. attractive interactions between permanent dipoles ⇨ molecules with permanent dipole moments have higher boiling points.
2. The polarity of a compound affects its boiling point.
   a. Most alkyl halides, alcohols, and ethers are polar molecules; that is, they have permanent dipole moments.
   b. When molecules with the same shape and molecular mass are compared, in many cases the more polar molecule has the higher boiling point.
3. A higher boiling point results from
   a. greater attraction between molecules in the liquid state.
   b. larger intermolecular attractions.
4. In the case of alkyl halides, the effects of molecular volumes and polarity oppose each other.
   a. They nearly cancel in the case of alkyl chlorides.
   b. Alkyl bromides and iodides have lower boiling points than the alkanes of about the same molecular mass.
5. The boiling points of alcohols, especially alcohols of lower molecular mass, are unusually high in comparison to those of other organic compounds.
   a. The unusual trend in the boiling points of alcohols is the result of hydrogen bonding.
   b. In order to vaporize a hydrogen-bonded liquid, energy is required to break the hydrogen bonds between molecules. Thus, hydrogen bonding raises the boiling point of a liquid.

D. Safety Hazards of Ethers

1. Samples of ethers can accumulate dangerous quantities of explosive peroxides and hydroperoxides by autoxidation, a spontaneous oxidation by oxygen in air.
2. Ethers have low flash points. A flash point is the minimum temperature at which a compound is ignited by a small flame under standard conditions.

III. SOLVENTS IN ORGANIC CHEMISTRY

A. Classification of Solvents

1. Solvents, liquids used to dissolve one or more compounds, can be classified in three ways, which are not mutually exclusive:
   a. A solvent is classified as protic or aprotic, depending on its ability to donate hydrogen bonds.
      i. A protic solvent consists of molecules that can act as hydrogen-bond donors.
      ii. Solvents that cannot act as hydrogen-bond donors are termed aprotic solvents.
   b. A solvent is classified as polar or apolar, depending on the size of its dielectric constant.
      i. A polar solvent has a high dielectric constant; an apolar solvent has a low dielectric constant ϵ.

  ii. A polar solvent separates, or shields, ions effectively from each other.
  iii. The electrostatic law shows that when the dielectric constant ϵ is large, the magnitude of the energy of interaction between the ions in solution is small.
 c. A solvent is classified as donor or nondonor, depending on its ability to act as a Lewis base.
  i. Donor solvents consist of molecules that can donate unshared electron pairs, that is, molecules that can act as Lewis bases.
  ii. Nondonor solvents cannot act as Lewis bases.
2. Molecular polarity and solvent polarity are different concepts.
 a. Molecular polarity, or dipole moment, is a property of individual molecules.
 b. Solvent polarity, or dielectric constant, is a property of many molecules acting together.
3. Polar molecules are attracted to each other because they can align in such a way that the negative end of one dipole is attracted to the positive end of the other.
4. When a polar molecule contains a hydrocarbon portion of even moderate size its polarity has little effect on its physical properties.

B. Solubility

1. Three solvent properties contribute to the solubility of ionic compounds:
 a. polarity (high dielectric constant), by which solvent molecules separate ions of opposite charge;
 b. proticity (hydrogen-bond donor capacity), by which solvent molecules solvate anions;
 c. electron-donor ability, by which solvent molecules solvate cations through Lewis-base and ion-dipole interactions; unshared electron pairs enable a solvent molecule to act as a Lewis base (electron donor) toward an electron-deficient cation.
2. Solubility is the factor that determines whether a given compound will dissolve in a particular solvent.
 a. For a covalent compound, the best rule of thumb for selecting a solvent is "like dissolves like."
 b. For ionic compounds, the best solvents are polar, protic, donor solvents.
3. Ionic compounds in solution can exist in several forms, two of which are ion pairs and associated ions.
 a. In an ion pair, each ion is closely associated with an ion of opposite charge.
 b. Dissociated ions move more or less independently in solution and are surrounded by several solvent molecules called collectively the solvent shell, or solvent cage, of the ion.
4. The ability of a solvent to shield ions from one another is measured by its dielectric constant ϵ.
5. Three solvent properties enhance the solubility of ionic compounds:
 a. highly polar solvents separate ions from each other;
 b. protic solvents stabilize anions by hydrogen bonding;
 c. donor solvents stabilize cations by donor interactions.
6. Solvation is a term used to describe the favorable interaction of a dissolved molecule with solvent; when solvent molecules interact favorably with an ion, they are said to solvate the ion. Hydrogen bonding and donor interactions are two solvation mechanisms for ions.
7. The effects of solvents on chemical reactions are closely tied to the principles of solubility.

C. Cell Membranes

1. Cell membranes consist of mostly phospholipids, cholesterol (a steroid), and some imbedded proteins.
 a. A typical phospholipid contains a polar head group (a group containing one or more charged atoms) and a long hydrocarbon tail that consists of 15–17 carbons atoms in an unbranched chain.
 b. The phospholipids are arranged in a phospholipid bilayer in which the polar head groups are oriented towards water and the hydrocarbon tails are on the interior.
 c. The transport of ions through cell membranes requires special carriers or pores, which are proteins imbedded in the membrane.
2. The lipid bilayer is generally impermeable to ions; a number of uncharged molecules diffuse readily through the cell membrane.
 a. Drugs that are completely insoluble in hydrocarbons cannot pass through membranes.
 b. Drugs with moderate solubility in hydrocarbons can get through the membrane.
 c. Drugs that are highly soluble in hydrocarbons can get trapped within the membrane.

3. Ion channels, or "ion gates," provide passageways for ions into and out of cells.

D. Crown Ethers and Ionophore Antibiotics

1. Crown ethers are heterocyclic ethers containing a number of regularly spaced oxygen atoms that interact with cations through the same mechanism used by donor solvents.
   a. Because the metal ion must fit within the cavity, the crown ethers have some selectivity for metal ions according to size.

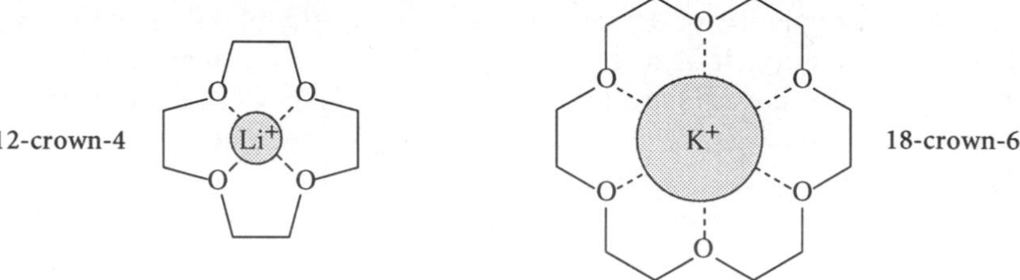

   b. Because crown ethers have significant solubilities in hydrocarbon solvents, they can be used to dissolve salts in hydrocarbon solvents.
2. Ionophores are molecules that form strong complexes with specific ions; crown ethers are one type of ionophore.
3. Crown ethers and other ionophores form complexes with cations by creating artificial solvation shells for them.
4. The ionophore antibiotics form strong complexes with metal ions in much the same way as crown ethers.
   a. An antibiotic is a compound that interferes with the growth or survival of one or more microorganisms.
   b. Ionophore antibiotics act by disrupting the ion balance within the cell.

IV. ACIDITY AND BASICITY

A. Acidity of Alcohols and Thiols

1. Alcohols and thiols are weak acids.
   a. Typical primary alcohols have pK_a values near 15–16 in aqueous solution.
   b. Thiols, with pK_a values near 10–11, are substantially more acidic than alcohols.
2. The conjugate bases of alcohols are called alkoxides, or alcoholates.
   a. The common name of an alkoxide is constructed by deleting the final *yl* from the name of the alkyl group and adding the suffix *oxide*.
   b. In substitutive nomenclature, the suffix *ate* is simply added to the name of the alcohol.

$$CH_3CH_2\ddot{\underset{..}{O}}:^- \quad Na^+$$

sodium ethoxide (common)
sodium ethanolate (systematic)

3. The conjugate bases of thiols (mercaptides or sulfides) are called sulfides in common nomenclature and thiolates in substitutive nomenclature

$$CH_3CH_2\ddot{\underset{..}{S}}:^- \quad Na^+$$

sodium ethyl sulfide (common)
sodium ethanethiololate (substitutive)

4. The relative acidities of alcohols and thiols are a reflection of the element effect.

B. Structural Effects on Alcohol Acidity

1. The acidities of alcohols in solution are in the order

methyl > primary > secondary > tertiary

   a. The acidity order in solution is due primarily to the effectiveness with which alcohol molecules solvate their conjugate-base anions.
   b. The relative acidity order of different types of alcohols is reversed in the gas phase (that is, in the absence of solvent) compared to the relative acidity order in solution.
2. The acidity of alcohols in solution is reduced by branching near the —OH group and increased by electron-withdrawing substituents.
   a. Substituted alcohols and thiols show the same type of polar effects on acidity as do substituted carboxylic acids; electronegative substituents increase acidity.

CH_3CH_2OH	CF_3CH_2OH
$pK_a = 15.9$	$pK_a = 12.4$

   b. The polar effects of electronegative groups are more important when these groups are closer to the —OH group.
3. Branched alcohols are more acidic than unbranched ones in the gas phase because α-alkyl substituents stabilize alkoxide ions.

C. Basicity of Alcohols, Thiols, and Ethers

1. Alcohols, thiols, and ethers are weak Brønsted bases, and react with strong acids to form positively charged conjugate-acid cations that have negative pK_a values.
   a. Alcohols and ethers do not differ greatly from water in their basicities; their conjugate acids have pK_a values of –2 to –3.
   b. Thiols and sulfides, however, are much less basic; their conjugate acids have pK_a values of –5 to –7.
   c. Alcohols and thiols, like water, are amphoteric substances; that is, they can both gain and lose a proton.
2. Tertiary alkoxides are more basic than primary alkoxides; that is, tertiary alcohols have higher pK_a values than primary alcohols.
3. Water, alcohols, and ethers are good Lewis bases.

REACTIONS

I. REACTIONS OF ALKYL HALIDES AND ALKANES

A. Formation of Grignard and Organolithium Reagents

1. Compounds that contain carbon-metal bonds are called organometallic compounds.
   a. A Grignard reagent is an organometallic compound of the form R—MgX, where X = Br, Cl, or I.
   b. Organolithium reagents are organometallic compounds of the form R—Li.
2. Both Grignard and organolithium reagents are formed by adding the corresponding alkyl or aryl halides to rapidly stirred suspensions of the appropriate metal.

$$\text{R—X} + \text{Mg} \xrightarrow[\text{ether}]{} \underset{\text{Grignard reagent}}{\text{R—Mg—X}} \qquad (\text{X} = \text{Cl, Br, or I})$$

$$\text{R—X} + 2\,\text{Li} \xrightarrow[\text{hydrocarbon}]{} \underset{\text{organolithium reagent}}{\text{R—Li}} + \text{LiX} \qquad (\text{X} = \text{Cl, Br, or I})$$

a. Reaction of alkyl halides with magnesium metal yields Grignard reagents.
   i. Ether solvents must be used for the formation of Grignard reagents; the ether solvent plays a crucial role in their formation because the ether associates with the metal in a Lewis acid-base interaction.

$$(C_2H_5)_2\ddot{O}\!: \rightarrow \text{Mg(Br)(R)} \leftarrow :\!\ddot{O}(C_2H_5)_2$$

   ii. Grignard reagents are formed on the surface of the magnesium metal.
   iii. The formation of Grignard reagents is believed to involve radical intermediates.
b. Reaction of alkyl halides with lithium in hydrocarbon solvents yields organolithium reagents.
c. Because they react vigorously with water and oxygen, Grignard and organolithium reagents must be prepared under rigorously moisture-free and oxygen-free conditions.

3. Grignard and organolithium reagents behave as strong Brønsted bases and react violently with acids, including water and alcohols, to give alkanes. (See Part II.B of this section.)
4. All reactions of Grignard and organolithium reagents can be understood in terms of the polarity of the carbon-metal bond.
   a. Because carbon is more electronegative than either magnesium or lithium, the negative end of the carbon-metal bond is the carbon atom.
      i. A carbon bearing three bonds, an unshared electron pair, and a negative formal charge is termed a carbon anion, or carbanion.
      ii. Carbanions are powerful Brønsted bases because their conjugate acids, the corresponding alkanes, are extremely weak acids.
   b. Grignard and organolithium reagents react as if they were carbanions; however, they are not true carbanions because they have covalent carbon-metal bonds.

$$\overset{\delta-}{\text{—C—}}\overset{\delta+}{\text{MgBr}} \;\; \text{(polarized bond)} \qquad \text{reacts as if it were} \qquad \text{—C:}^- \;\text{(carbanion)} + {}^+\text{MgBr}$$

B. Free-Radical Halogenation of Alkanes

1. Alkanes react with bromine and chlorine in the presence of heat or light in free-radical substitution reactions to give alkyl halides.
2. The mechanism of this reaction in fact follows the typical pattern of other free-radical chain reactions; it has initiation, propagation, and termination steps.
   a. The reaction is initiated when a halogen molecule absorbs energy from heat or light and dissociates homolytically into halogen atoms.

$$\text{X—X} \xrightarrow[\text{or heat}]{\text{light}} \text{X}\cdot + \text{X}\cdot \qquad (\text{X} = \text{Cl or Br}) \qquad \text{initiation step}$$

   b. The ensuing chain reaction has the following propagation steps.

$$\text{R—H} + \cdot\text{X} \longrightarrow \text{R}\cdot + \text{H—X}$$

$$\text{R}\cdot + \text{X—X} \longrightarrow \text{R—X} + \text{X}\cdot$$

propagation steps

c. Termination steps result from the recombination of radical species.

$$\text{R}\cdot + \text{X}\cdot \longrightarrow \text{R—X}$$

one of several possible termination steps

3. Free-radical halogenations with chlorine and bromine proceed smoothly, halogenation with fluorine is violent, and halogenation with iodine does not occur.

II. ACID-BASE REACTIONS OF ALCOHOL, THIOLS, GRIGNARD REAGENTS, AND ORGANOLITHIUM REAGENTS

A. Formation of Alkoxides and Mercaptides

1. Alkoxides can be formed from alcohols with stronger bases such as sodium hydride, NaH, which is a source of the hydride ion, H:^-.

$$\text{R—}\ddot{\text{O}}\text{H} + \text{NaH} \longrightarrow \text{R—}\ddot{\text{O}}\text{:}^- \ \text{Na}^+ + \text{H}_2$$

2. Sodium metal reacts with an alcohol to afford a solution of the corresponding sodium alkoxide; the rate of this reaction depends strongly on the alcohol.
3. Because thiols are much more acidic than water or alcohols, they, unlike alcohols, can be converted completely into their conjugate-base mercaptide anions by reaction with one equivalent of hydroxide or alkoxide.
4. Thiols form insoluble mercaptides (sulfides) with many heavy-metal ions, such as Hg^{2+}, Cu^{2+}, and Pb^{2+}.

$$\text{R—}\ddot{\text{S}}\text{H} + \text{NaOH} \xrightarrow[\text{H}_2\text{O}]{} \text{R—}\ddot{\text{S}}\text{:}^- \ \text{Na}^+ + \text{H}_2\text{O}$$

$$2\ \text{R—}\ddot{\text{S}}\text{H} + \text{Hg}^{2+} \xrightarrow[\text{H}_2\text{O}]{} \text{R}\ddot{\text{S}}\text{—Hg—}\ddot{\text{S}}\text{R} + 2\ \text{H}_3\text{O}^+$$

B. Protonolysis of Organometallic Reagents

1. All Grignard or organolithium reagents react vigorously with even relatively weak acids, such as water and alcohols, to give the conjugate-base hydroxide or alkoxides and the conjugate-acid hydrocarbon of the carbanion; this reaction is an example of protonolysis.

$$\text{R—MgX} + \text{R'—OH} \longrightarrow \text{R—H} + \text{Mg}^{2+} + \text{X}^- + \text{R'O}^-$$

2. A protonolysis is a reaction with the proton of an acid that breaks chemical bonds.
   a. Protonolysis of Grignard reagents is also useful, because it provides a method for the preparation of hydrocarbons from alkyl halides.
   b. A particularly useful variation of this reaction is the preparation of hydrocarbons labeled with the hydrogen isotopes deuterium (D, or ^{2}H) or tritium (T, or ^{3}H) by reaction of a Grignard reagent with the corresponding isotopically labeled water.

$$\text{1-bromo-1-methylcyclopentane (Br, CH}_3) \xrightarrow[\text{ether}]{\text{Mg}} \text{(MgBr, CH}_3) \xrightarrow{\text{D}_2\text{O}} \text{(D, CH}_3) + \text{Mg}^{2+} + \text{Br}^- + \text{DO}^-$$

SCIENTISTS

Philippe Antoine Barbier (1848–1922) 8.8
Donald James Cram (b. 1919) [Chemistry 1987] 8.5B
Paul Jozef Crutzen (b. 1933) [Chemistry 1995] 8.9B
Jean-Baptiste André Dumas (1800–1884) 8.9B
François Auguste Victor Grignard (1871–1935) [Chemistry 1912] 8.8
Jean-Marie Pierre Lehn (born 1939) [Chemistry 1987] 8.5B
Mario Jose Molina (b. 1943) [Chemistry 1995] 8.9B
Paul Hermann Müller (1899–1965) 8.9B
Charles John Pedersen (1904–1989) [Chemistry 1987] 8.5B
Frank Sherwood Rowland (b. 1927) [Chemistry 1995] 8.9B

Dates in brackets refer to Nobel Prize awards.

STUDY GUIDE LINKS

✓8.1 Common Nomenclature

Sometimes you may see an *n*- prefix used in the common nomenclature of some organic compounds, as in the following example:

$CH_3CH_2CH_2CH_2CH_3$

pentane
(sometimes called ***n*-pentane**)

$CH_3CH_2CH_2CH_2$—Br

1-bromobutane (substitutive nomenclature)
butyl bromide (common nomenclature)
(sometimes called ***n*-butyl bromide**)

The *n*- prefix stands for "normal." At one time, the prefix *n*- and the word normal were used to indicate an isomer containing a functional group at the end of an unbranched carbon chain. However, this prefix is superfluous and unnecessary. For example, the name butyl bromide itself means the structure shown on the right above; an additional prefix is unnecessary. Branched-chain isomers have other names, such as isobutyl bromide, *sec*-butyl bromide, or *tert*-butyl bromide. Because common names are unambiguous without the prefix *n*-, the IUPAC recommended abandoning it. Despite this recommendation, the prefix continues to be used.

✓8.2 Boiling Points and Solubilities

Some students confuse the prediction of boiling points with the prediction of solubilities, probably because both phenomena involve interactions *between* molecules.

Solubility is governed by the interaction between *different* molecules: the solute and the solvent. Thus, ethyl methyl ether (CH_3CH_2—O—CH_3) has a significant solubility in water because it accepts hydrogen bonds from water.

The boiling point is governed by the interaction between *like* molecules. For example, the boiling point of ethyl methyl ether is governed by the interactions between only ethyl methyl ether molecules. The fact that this ether is a hydrogen bond acceptor is *irrelevant* to the question of its boiling point, because this ether cannot donate hydrogen bonds to itself.

When thinking about intermolecular interactions, be sure you are analyzing the proper interactions for the situation at hand.

8.3 Solubility of Covalent Compounds: A Deeper Look

The "like-dissolves-like" rule is a convenient and useful way of predicting solubilities, but it does not deal with the underlying basis of solubility.

To deal with solubility in quantitative terms, we look at the formation of a solution as a chemical process and investigate its standard free energy. Consider the solution of a substance *A* in a solvent *S*:

$$A + S \rightleftharpoons \text{Solution of } A \text{ in } S$$

The standard free energy of this process is ΔG°_{soln}. If ΔG°_{soln} is negative, this process is favorable, and *A* will be soluble; if it is positive, this process is unfavorable, and *A* will not be soluble.

Any free-energy change is the result of contributions from the enthalpy change (ΔH°) and the entropy change (ΔS°). Because $\Delta G^\circ = \Delta H^\circ - T\Delta S^\circ$, then

$$\Delta G^\circ_{soln} = \Delta H^\circ_{soln} - T\Delta S^\circ_{soln} \qquad \text{(SG8.1)}$$

Let's consider the enthalpy and entropy contributions separately.

We can interpret ΔH°_{soln} in terms of *how interactions between molecules change* during the solution process. Let's say we are trying to dissolve pentane in water, a process we know is unfavorable. (Recall that alkanes are insoluble in water.) The ΔH°_{soln} for this process is the difference between the enthalpies of the molecular interactions in the solution state and the enthalpies of molecular interactions in the pure liquids:

$$\Delta H^\circ_{soln} = \Delta H^\circ(\text{interactions between molecules in the solution state}) - \Delta H^\circ(\text{interactions between molecules in the pure liquids})$$

The interactions in each state can be "itemized" to give the following five terms:

$$\begin{aligned}\Delta H^\circ_{soln} = {} & \Delta H^\circ(\text{pentane-water interactions in solution}) + \\ & \Delta H^\circ(\text{pentane-pentane interactions in solution}) + \\ & \Delta H^\circ(\text{water-water interactions in solution}) - \\ & \Delta H^\circ(\text{pentane-pentane interactions in pure pentane}) - \\ & \Delta H^\circ(\text{water-water interactions in pure water})\end{aligned}$$

Let's think about this dissection term-by-term. (Remember that negative ΔH° values are favorable.) We might think that the first term [ΔH°(pentane-water interactions in solution)] should be positive because pentane and water cannot interact by hydrogen bonding. This term, however, is actually slightly negative (*i.e.*, favorable), because of an attractive interaction between the permanent dipoles of water molecules and induced dipoles of pentane molecules. (These are the van der Waals attractions shown in Fig. 2.8, Sec. 2.6A, of the text, except that one of the interacting molecules has a permanent dipole moment.) The fourth term [ΔH°(pentane-pentane interactions in pure pentane)] should be negative, because pentane molecules interact favorably with each other by van der Waals attractions. Because the first and fourth terms are subtracted, and because they have the same sign, they *tend* to cancel.

The second term [ΔH°(pentane-pentane interactions in solution)] should be zero, because, by definition of a solution, pentane molecules are completely dispersed in the solvent water and thus do not interact with each other.

To complete our analysis of ΔH°_{soln} consider the third and fifth terms. At first, we might think that these terms should also cancel, because it might seem reasonable that the interactions between water molecules should not change when we dissolve pentane. However, this is not the case! The water molecules in the vicinity of pentane molecules cannot form hydrogen bonds with pentane. Some of the hydrogen-bonding sites on these water molecules are left unsatisfied. This situation would increase the ΔH° were it not for the fact that these water molecules find another way to satisfy the missing hydrogen bonds: they form *more* hydrogen bonds with nearby water molecules. Hence, the water molecules in the vicinity of pentane molecules become *more* hydrogen-bonded than the water molecules in bulk water. Some scientists believe that this water resembles ice in the sense that these water molecules are more rigid and less mobile. (The same phenomenon is responsible for surface tension at an air-water interface. Water molecules at the surface are more strongly hydrogen-bonded—so much so that the surface becomes less penetrable. You may recall that it is possible to float a coin on the surface of water.) The net result is that the balance of the third and fifth terms is negative. It turns out that the difference between the third and fifth terms dominates the overall ΔH°_{soln}: the ΔH°_{soln} for dissolving pentane in water is slightly negative (favorable)! The measured value for ΔH°_{soln} is –8.79 kJ/mol (–2.1 kcal/mol). In other words, the fifth term—the interactions between water molecules in a solution of pentane—tips the balance to make the ΔH°_{soln} favorable for dissolving pentane in water.

If ΔH°_{soln} were the only factor controlling solubility, then pentane should dissolve in water! Because we know that dissolving pentane in water is unfavorable, then it must be true that the ΔG°_{soln} for dissolving pentane in water is positive. The only way this can happen is for $-T\Delta S^\circ_{soln}$ to be positive, that is, for ΔS°_{soln} to be negative. We can understand this by viewing entropy as a measure of *randomness*, an interpretation that was first discussed in Study Guide Link 4.5. Another way to think of entropy is that *entropy decreases when molecular motion decreases, or when molecular order increases.* When we dissolve one substance in another, randomness increases, because we go from two pure states (pure solvent and pure solute) to a mixed state. It's as if we took a bag of red balls and a bag of blue balls and threw them together to give a bag of balls in which red and blue balls were randomly mixed. We call the associated entropy change the *entropy of mixing,* $\Delta S^{\circ}_{\text{mixing}}$. Hence, $\Delta S^{\circ}_{\text{mixing}}$ is *always* positive, and thus the entropy of mixing *always* favors a solution over two pure substances. Consequently, the entropy of mixing cannot be the negative entropy we are looking for.

However, we found in our discussion of $\Delta H^{\circ}_{\text{soln}}$ above that changes in the interactions among solvent (water) molecules can accompany solution formation. Associated with these changes is an entropy change, which we'll call $\Delta S^{\circ}_{\text{solvent}}$. Recall that water molecules in the vicinity of dissolved pentane molecules become more hydrogen-bonded, and thus more rigid and less mobile. This means that these water molecules become more ordered; thus their motion becomes *less random*. Hence, $\Delta S^{\circ}_{\text{solvent}}$ is negative. The $\Delta S^{\circ}_{\text{soln}}$ is the sum of $\Delta S^{\circ}_{\text{mixing}}$ and $\Delta S^{\circ}_{\text{solvent}}$; that is

$$\Delta S^{\circ}_{\text{soln}} = \Delta S^{\circ}_{\text{mixing}} + \Delta S^{\circ}_{\text{solvent}}$$

Multiplying each ΔS° value by $-T$ gives its contribution to the standard free energy (Eq. SG8.1). Specifically, $-T\Delta S^{\circ}_{\text{mixing}}$ for dissolving anything in one liter of water at 298 K (25 °C) is about –12.5 kJ/mol (–3 kcal/mol). The value of $-T\Delta S^{\circ}_{\text{solvent}}$ for dissolving one mole of pentane is +143 kJ/mol (+34.2 kcal/mol). Adding these two values, the $-T\Delta S^{\circ}_{\text{soln}}$ for dissolving pentane in water at 298 K (25 °C) is therefore +130.5 kJ/mol, or 31.2 kcal/mol. Recall from the previous paragraph that $\Delta H^{\circ}_{\text{soln}}$ is –8.79 kJ/mol (2.1 kcal/mol). Hence, substituting these values in Eq. SG8.1 gives

$$\begin{aligned}\Delta G^{\circ}_{\text{soln}} &= \Delta H^{\circ}_{\text{soln}} - T\Delta S^{\circ}_{\text{soln}}\\ &= -8.79 + 130.5 = +121.7 \text{ kJ/mol}\\ &= -2.1 + 31.2 = +29.1 \text{ kcal/mol}\end{aligned}$$

Thus, $\Delta G^{\circ}_{\text{soln}}$ is positive for dissolving pentane in water (that is, pentane is insoluble) because of the changes that take place in water—the increased order, or decreased entropy—that result. Notice that the $\Delta H^{\circ}_{\text{soln}}$ of pentane in water is actually negative, but the inability of pentane molecules to satisfy the hydrogen-bonding requirements of nearby water molecules forces these water molecules into low-entropy interactions with each other.

We've considered in detail the solution process for two "unlike" substances. Now let's consider the solution process for two "like" substances. First, consider the solution of methanol in water. Both substances are protic. When we dissolve methanol in water, we're "trading in" methanol-methanol interactions and water-water interactions for methanol-water interactions. Because hydrogen bonding is the dominant interaction in all three cases, the trade is almost enthalpy-neutral. Because methanol molecules can form hydrogen bonds with water molecules, water molecules in the solution are not forced into low-entropy water-water interactions; the hydrogen-bonding capability of water is satisfied by methanol. Hence, the entropy of solution is dominated by the entropy of mixing, which is positive. Hence, $\Delta G^{\circ}_{\text{soln}}$ is negative.

When we dissolve an apolar aprotic substance in another apolar aprotic substance, say, pentane in hexane, we're "trading in" pentane-pentane and hexane-hexane interactions for pentane-hexane interactions. Because the same sorts of intermolecular interactions are involved in the three cases—van der Waals attractions, or dispersion forces—once again the $\Delta H^{\circ}_{\text{soln}}$ is nearly zero. Distributing pentane molecules in hexane does not cause any significant change in the solvent hexane because pentane is very much like hexane. Hence, the solution process is once again dominated by the $\Delta S^{\circ}_{\text{mixing}}$, and, as a result, $\Delta G^{\circ}_{\text{soln}}$ is negative: pentane is miscible in hexane.

As you can see from this discussion, the mixing of two pure phases always results in a positive entropy change (the entropy of mixing) that favors the formation of solutions. When two substances are very similar, the $\Delta H^{\circ}_{\text{soln}}$ is nearly zero. When two substances are very different, at least when one is water, formation of the solution causes changes in the interactions between solvent molecules that result in a negative entropy change. This change opposes the entropy of mixing and resists formation of the solution. The basis of the "like-dissolves-like" rule, at least in the case of water as a solvent, is that solvent molecules can interact with "like" solute molecules as well as they can with each other, thus avoiding the necessity of additional low-entropy interactions within the solvent itself.

8.4 Solvation of Tertiary Alkoxides

The discussion in the text shows that the relative basicities of alkoxides in solution is clearly governed by solvation, because the basicity order in solution (tertiary > secondary > primary) is different from that in the gas phase (tertiary < secondary < primary). How can we visualize this solvation effect?

At first, it was thought that this might be a steric effect; that is, the alkyl branches in a tertiary alkoxide "get in the way" of the solvent that would normally be solvating the alkoxide oxygen. However, this was shown not to be the case in the following way. There is some indication that the $\Delta H°$ of solvation for tertiary alkoxides is actually negative. Steric effects normally are reflected by *increases* in $\Delta H°$. (For example, this is why *cis*-2-butene has a greater $\Delta H_f°$ than *trans*-2-butene.) The unfavorable $\Delta G°$ for solvation, then, resides in a negative $\Delta S°$. (See Study Guide Link 8.3; read about the solubility of hydrocarbons in water.) The negative $\Delta S°$ means that the solvent is becoming more ordered, less chaotic, when the tertiary alkoxide dissolves in it.

When the tertiary alkoxide dissolves, the solvent surrounding the hydrocarbon groups cannot form hydrogen bonds to the hydrocarbon groups, so they form stronger hydrogen bonds to each other. This releases energy; therefore, the $\Delta H°$ decreases. However, the tighter hydrogen bonds reduce the random motions of the solvent molecules; because randomness decreases, $\Delta S°$ decreases. In effect, the solvent around the alkoxide becomes more like a solid (ice, if the solvent were water). Because of the tightened hydrogen-bond structure of the solvent, the solvent resists disruption by interaction with the alkoxide oxygen, and the alkoxide oxygen is left "naked," that is, unsolvated. This is what we mean when we say that the hydrocarbon groups adversely affect the solvation of the alkoxide ion.

Think of the following analogy. Imagine a lot of little children running around wildly on a playground, more or less at random. A new child (let's call him Josh) enters the playground carrying candy. A few of the children see this and run up to Josh and interact "favorably" with him by accepting some of his candy.

In a second scenario, let's imagine that a teacher is directing a game in which the students have to hold hands tightly and skip around her in a very ordered circle. The teacher has a bag of candy, but the students can't interact with her "candy" because they are holding hands. The "candy" is analogous to the alkoxide; the teacher is analogous to the hydrocarbon groups; and the children are analogous to solvent molecules. Because the students are forced to interact (hold hands) with each other, they can't reach out for the candy.

SOLUTIONS TO PROBLEMS

Solutions to In-Text Problems

8.1 (a) Isobutyl fluoride is a primary alkyl halide.
(b) Hexyl iodide is a primary alkyl halide.
(c) Cyclopentyl bromide is a secondary alkyl halide.
(d) *Tert*-butyl chloride is a tertiary alkyl halide.

8.2 (a) chlorocyclopropane

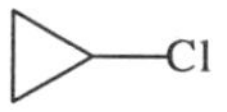

(b) $CH_3C(Cl)_2CH_2CH_2CH(CH_3)CH_3$

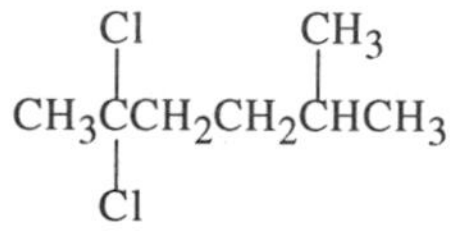

2,2-dichloro-5-methylhexane

(c) 6-bromo-1-chloro-3-methylcy-clohexene

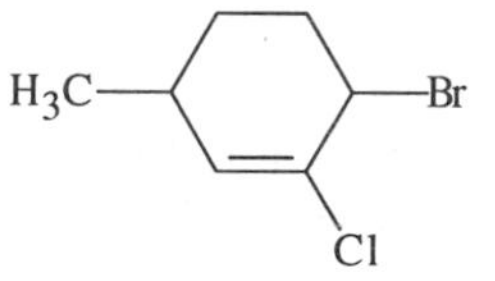

(d) $I—CH_2—I$

methylene iodide

8.3 (a) (*Z*)-3-Chloro-2-pentene
(b) 3-Bromo-3-chloro-1-methylcyclopropene
(c) 3-Bromo-1,1,1-trichloro-2-fluorobutane
(d) Chloroform ($HCCl_3$) is the traditional name for trichloromethane.
(e) Neopentyl bromide [$(CH_3)_3C—CH_2—Br$] is the common name for 1-bromo-2,2-dimethylpropane.
(f) 1-Chloro-3-isopropyl-1-methylcyclohexane
(g) 1,3-Dibromocyclobutane

8.4 (a)

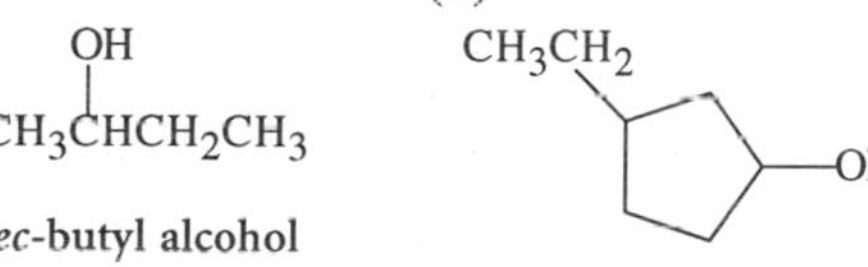

$CH_3CH(OH)CH_2CH_3$

sec-butyl alcohol

(b) 3-ethylcyclopentanol

(c) $CH_3CH(OH)CH_2$—CH=CH—$CH(Cl)CH_3$

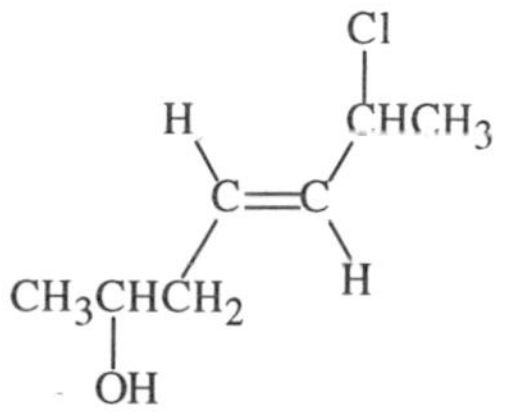

(*E*)-6-chloro-4-hepten-2-ol

(d) 2-cyclohexenol

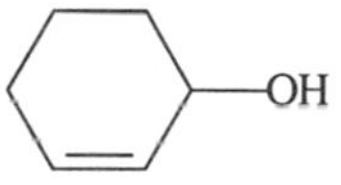

8.5 (a) 1-Butanol
(b) 3-Bromo-1-butanol
(c) (*E*)-1-Chloro-3-methyl-3-penten-2-ol
(d) 2-Chloro-5-methyl-2-cyclopentenol
(e) 3-Butyl-2,4-pentanediol (See rule 2a on text p. 305.)
(f) 2,5-Cyclohexadienol
(g) 2-Methyl-2-propanethiol
(h) 1-Mercapto-2-pentanol

8.6 (a) $CH_3CH_2OCH_2CH_2CH_3$

ethyl propyl ether

(b)

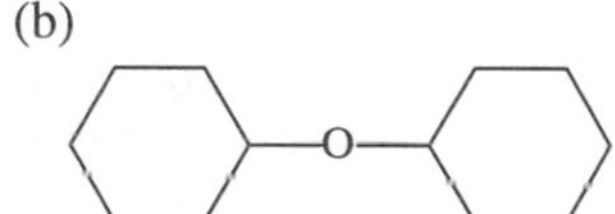

dicyclohexyl ether

(c)

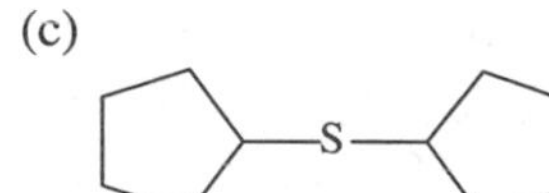

dicyclopentyl sulfide

(d) $(CH_3)_3C—S—CH(CH_3)_2$

tert-butyl isopropyl sulfide

(e) $H_2C{=}CHCH_2OCH_2Ph$

allyl benzyl ether

(f) C_6H_5—O—CH$=$$CH_2$

phenyl vinyl ether

(g) (2*R*,3*R*)-2,3-dimethyloxirane

(h) $CH_3CH(CH_3)CH_2CH_2CH(SCH_2CH_3)CH_2CH_3$

5-(ethylthio)-2-methylheptane

8.7 (a) 2-methoxy-2-methylpropane
(b) 2-ethoxyethanol
(c) 1-(isobutylthio)-2-methylpropane
(d) (*E*)-5-methoxy-3-penten-1-ol

8.8 Notice from Table 8.1 that the bond lengths in a given period are about 0.04 Å greater in the column containing oxygen and sulfur than in the column containing the halogens. Assuming that this trend holds true for CH_3SeH, then the predicted carbon-selenium bond length in this compound is about 1.98 Å. The prediction for $(CH_3)_2Se$ can be refined by noting in Fig. 8.1, text p. 310, that the carbon-sulfur bond length in dimethyl sulfide is about 0.02 Å smaller than that in methanethiol. Assuming a similar trend in the corresponding selenium derivatives, the predicted bond length for $(CH_3)_2Se$ is 1.96 Å.

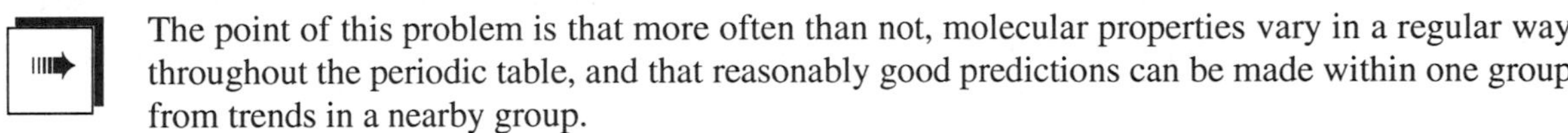

The point of this problem is that more often than not, molecular properties vary in a regular way throughout the periodic table, and that reasonably good predictions can be made within one group from trends in a nearby group.

8.9 If bonds to an atom involved *only* its *p* orbitals, then the bond angle would be 90°. It follows, then, that the closer the bond angle is to 90°, the more *p* character in the bond. From Fig. 8.1 on text p.310, it's clear that the C—S—H bond angle is much closer to 90° than is the C—O—H bond angle of methanol. Hence, bonds to sulfur (that is, the sulfur orbitals used to make these bonds) have more *p* character than bonds to oxygen. A corollary is that the unshared pairs on sulfur have more *s* character than those on oxygen.

8.10 Because of the molecular geometry of *trans*-1,2-dichloroethylene, the C—Cl bonds are oriented in opposite directions, as are the C—H bonds. Consequently, their bond dipoles cancel, and the molecular dipole moment of *trans*-1,2-dichloroethylene is zero. (The cancellation of the C—Cl bond dipoles is shown in the diagram below; the much weaker C—H bond dipoles also cancel for the same reason.) The corresponding bond dipoles of *cis*-1,2-dichloroethylene do not cancel; consequently it has a significant dipole moment. The interaction of the molecular dipole moments of *cis*-1,2-dichloroethylene molecules provides a cohesive force in the liquid state that is not present in the trans stereoisomer. (See top of text p. 312.) Because such cohesive forces enhance boiling point, *cis*-1,2-dichloroethylene has the higher boiling point of 60.3°.

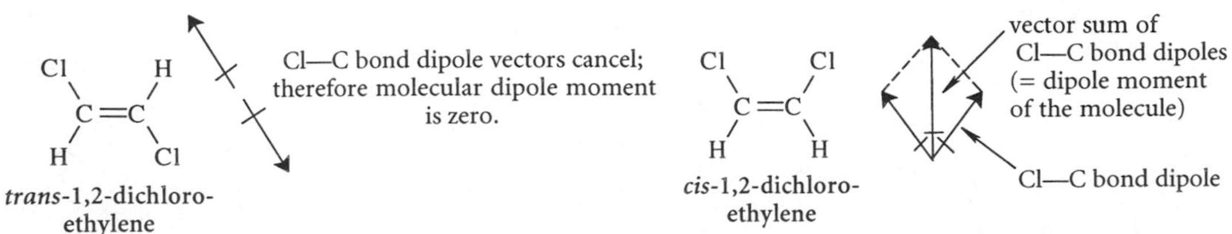

8.11 (a) Recall that a bromine contributes about the same molecular mass (80 units) as a pentyl group (73 units), and that alkyl bromides have boiling points *lower* than alkanes of about the same molecular mass; then 2-bromopropane should have a boiling point that is lower than that of an eight-carbon alkane. Because the two alkanes listed have more than eight carbons, their boiling points are correspondingly higher. Because the two alkanes have the same type of branching, their boiling points should be in the same order as their molecular masses. These considerations suggest that the order of increasing boiling points is 2-bromo-propane < 4-ethylheptane < 4-ethyloctane. (The actual boiling points are 59°, 141°, and 168°, respectively.)

(b) A chlorine contributes about the same molecular mass (35 units) as an ethyl group (29 units), and alkyl chlorides have about the same boiling points as alkanes of the same molecular mass. Hence, chloromethane has about the same boiling point as propane, which has a lower boiling point than the five-carbon alkene 1-pentene. The alcohol has the highest boiling point because it has about the same molecular mass as 1-pentene, but can donate and accept hydrogen bonds. Consequently, it has the highest boiling point of all. The order of increasing boiling points is, therefore, chloromethane (–42°) < 1-pentene (30°) < 1-butanol (118°).

8.12 (a) HBr is primarily a hydrogen-bond donor because it is a strong acid; —Br is a poor acceptor.

(b) Hydrogen fluoride is an excellent hydrogen-bond donor, and the fluorine is an excellent hydrogen-bond acceptor.

(c) Acetone is a hydrogen-bond acceptor; the oxygen can accept hydrogen bonds at its unshared electron pairs. (See structure below.)

(d) *N*-methylacetamide can serve as both a hydrogen-bond donor and a hydrogen-bond acceptor. (See structure below.)

(e) Phenol is both a hydrogen-bond donor and a hydrogen-bond acceptor. The O—H can participate in hydrogen-bond donation, and the oxygen can accept hydrogen bonds at its unshared electron pairs.

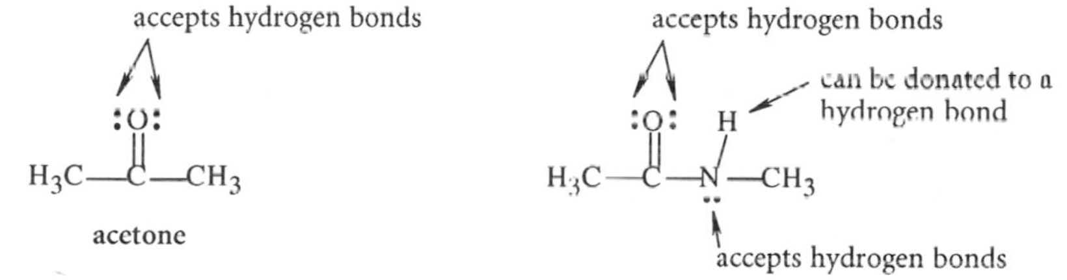

accepts hydrogen bonds

$H_3C—C(=O)—CH_3$

acetone

accepts hydrogen bonds

can be donated to a hydrogen bond

$H_3C—C(=O)—N(H)—CH_3$

accepts hydrogen bonds

N-methylacetamide

(f) The ethylammonium ion can donate its N—H hydrogens to hydrogen bonds, but it cannot accept hydrogen bonds because it has no unshared electron pairs.

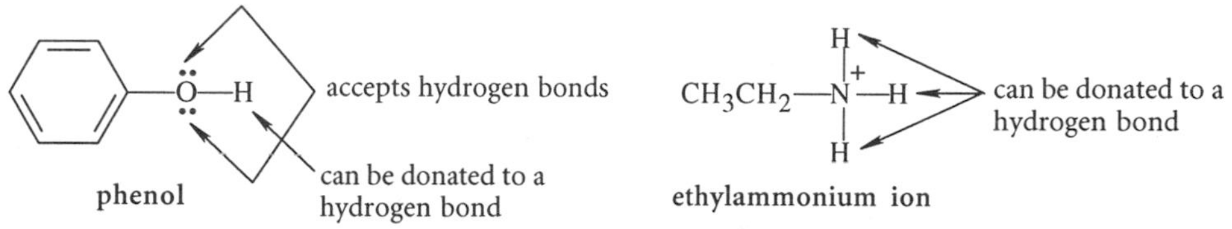

accepts hydrogen bonds

can be donated to a hydrogen bond

phenol

$CH_3CH_2—\overset{+}{N}H_3$

can be donated to a hydrogen bond

ethylammonium ion

8.13 (a) 2-Methoxyethanol is a polar, protic, donor solvent. (The dielectric constant of 17 is borderline for a polar solvent.)

(b) 2,2,2-Trifluoroethanol is a polar, protic, donor solvent.

(c) This compound, 2-butanone, is a polar, aprotic, donor solvent.

(d) 2,2,4-Trimethylpentane (a major component of gasoline) is an apolar, aprotic, nondonor solvent.

8.14 Use the "like-dissolves-like" criterion. Hexane should be least soluble in ethanol because hexane is an aprotic, apolar, nondonor substance, whereas ethanol is a polar, protic, donor solvent. 1-Octanol is also protic, but its structure has a rather substantial hydrocarbon portion that would interact more favorably than ethanol with hexane.

8.15 (a) The lower layer is methylene chloride because it has a higher density than water. Methanol should be more soluble in water because of its ability to form hydrogen bonds. Hence, methanol is mostly in the upper (water) layer.

(b) Certainly the 1-nonanol is less soluble in water than methanol is, and its long hydrocarbon chain would both reduce its solubility in water and promote its solubility in methylene chloride. The data on text p. 319 show that even 1-hexanol has a very low solubility in water. The solubility of 1-nonanol would be much lower still. The 1-nonanol would be found primarily in the methylene chloride layer.

8.16 Water can accept a hydrogen bond from the hydrogen of the N—H bond, and can donate a hydrogen bond to the oxygen of the C═O bond as well as to the nitrogen. This hydrogen bonding tends to solubilize acetanilide in water. In contrast, the CH_3 group and the phenyl ring cannot hydrogen-bond with water; consequently, these apolar ("greasy") groups tend to make acetanilide insoluble.

Groups such as alkyl and phenyl groups that reduce water solubility are sometimes termed *hydrophobic groups.*

8.17 The crown ether is ideally constructed to accommodate hydrogen bonding to the methylammonium ion. (Although this ion is actually puckered, we treat is as planar for simplicity.) The important hydrogen-bonding and donor interactions are shown below.

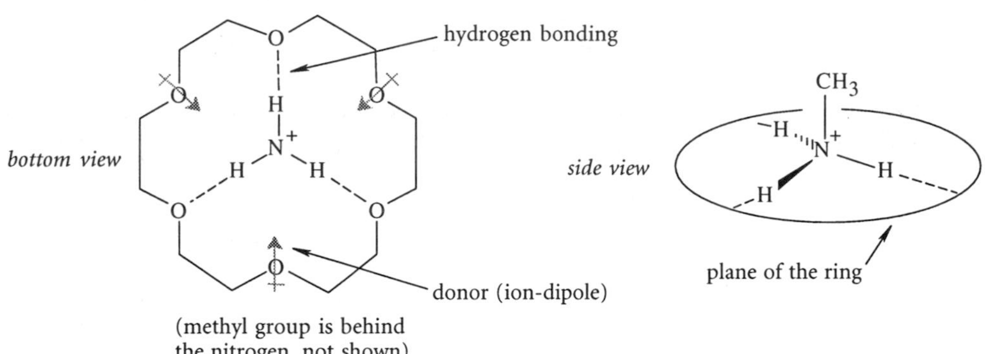

8.18 (a) $Na^+ (CH_3)_2CH—O^-$ sodium isopropoxide

(b) $K^+ (CH_3)_3C—O^-$

potassium *tert*-butoxide

(c) $\left[CH_3CH_2C(CH_3)_2CH_2O^- \right]_2 Mg^{2+}$

magnesium 2,2-dimethyl-1-butanolate

8.19 (a) Calcium methoxide or calcium methanolate.

(b) Cuprous ethanethiolate [or copper(I) ethanethiolate]

8.20 (a)

$$\underset{C}{Cl(CH_2)_3OH} < \underset{A}{ClCH_2CH_2OH} < \underset{B}{Cl_2CHCH_2OH}$$

Compound *B* is more acidic than compound *A* because it has more chlorines at the same separation from the oxygen; and compound *A* is more acidic than compound *C* because the separation of the electronegative chlorine from the oxygen is greater in *C*. (Recall that polar effects decrease with increasing separation of the two interacting groups; see text p. 103.)

(b)

$$\underset{C}{CH_3CH_2OH} < \underset{B}{ClCH_2CH_2OH} < \underset{A}{ClCH_2CH_2SH}$$

Compound *A* is more acidic than compound *B* because of the element effect (Sec. 3.6A). Compound *B* is more acidic than compound *C* because of the polar effect of the chlorine, which is acid-strengthening.

(c)

$$\underset{A}{CH_3CH_2CH_2CH_2OH} < \underset{B}{CH_3OCH_2CH_2OH}$$

Compound *B* is more acidic than compound *A* because of the electron-withdrawing polar effect of the oxygen in the methoxy group.

8.21 (a)

$$(CH_3)_2CH—Br + Mg \xrightarrow{\text{ether}} (CH_3)_2CH—MgBr$$

(b)

$$Ph—Br + 2\,Li \xrightarrow{\text{hexane}} Ph—Li + LiBr$$

(c) Organoboranes are prepared by hydroboration of alkenes (see text Sec. 5.3B). (Don't forget what you've already learned!)

$$3\,(CH_3)_2C{=}CH_2 + BH_3 \xrightarrow{\text{THF}} [(CH_3)_2CH—CH_2]_3B$$

8.22 (a)

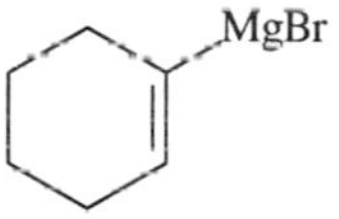

MgBr

1-cyclohexenylmagnesium bromide

(b) $(CH_3)_3C—Li$ *tert*-butyllithium Lithium chloride, Li^+Cl^-, is a by-product.

8.23 (a) Whenever a Grignard or organolithium reagent reacts with a protic substance such as water, an alcohol, or an acid, a Brønsted acid-base reaction occurs, as illustrated in Eqs. 8.23–8.25 on text p. 337.

$$CH_3\ddot{O}—H \quad H_3C—Li \longrightarrow CH_3\ddot{O}{:}^- \; Li^+ + \underset{\text{methane}}{CH_4}$$

(b)

$$H\ddot{O}—H \quad (CH_3)_2CHCH_2—MgCl \longrightarrow H\ddot{O}{:}^- \; Cl\overset{+}{M}g + \underset{\text{isobutane}}{(CH_3)_3CH}$$

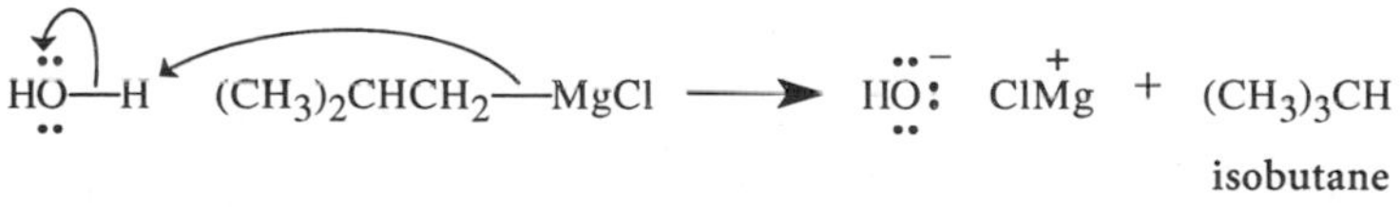

8.24 (a) $(CH_3)_2CH—MgBr$ and $CH_3CH_2CH_2—MgBr$ would both react with water to give propane.

(b) In this case the products would be different because of the isotope: $(CH_3)_2CH—D$ and $CH_3CH_2CH_2—D$

8.25 The free-radical chain mechanism for bromination of ethane:

Initiation step:

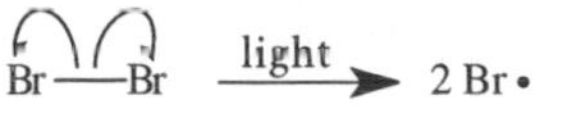

$$Br—Br \xrightarrow{\text{light}} 2\,Br\bullet$$

Propagation steps:

$$Br\cdot \; H{-}CH_2CH_3 \longrightarrow Br{-}H + \cdot CH_2CH_3$$
ethyl radical

$$Br{-}Br \; \cdot CH_2CH_3 \longrightarrow Br\cdot + Br{-}CH_2CH_3$$
ethyl bromide

Remember that when dealing with radical species it is necessary only to write the unpaired electron; unshared pairs (for example, those on bromine in the previous solution) need not be written explicitly when they are not involved in the reaction.

8.26 Small amounts of butane can arise from dimerization (that is, the reaction of two molecules) of ethyl radical, which is formed during the first propagation step in the previous solution.

$$CH_3\dot{C}H_2 \quad \cdot CH_2CH_3 \longrightarrow CH_3CH_2{-}CH_2CH_3$$
ethyl radical ethyl radical butane

Solutions to Additional Problems

8.27 The answers to all four parts are as follows:

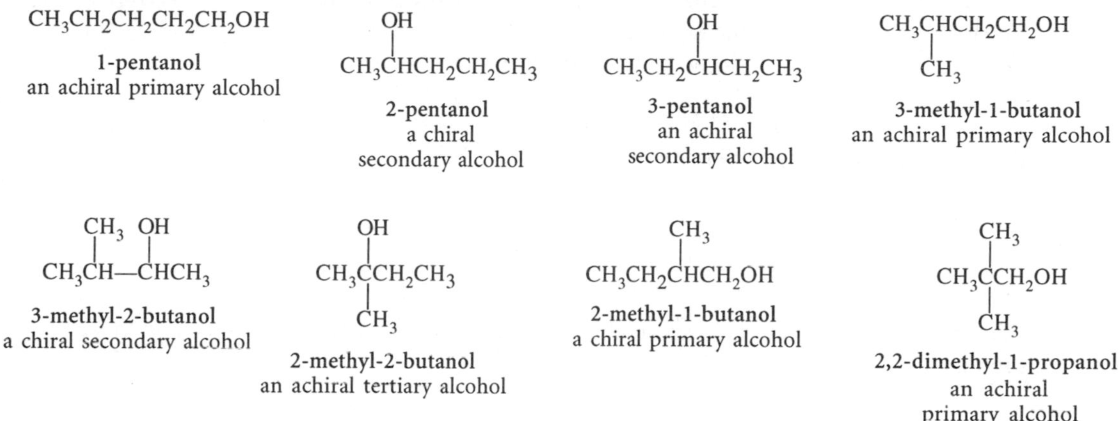

8.28 (a) The systematic name of halothane is 2-bromo-2-chloro-1,1,1-trifluoroethane.
(b) The systematic name of methoxyflurane is 2,2-dichloro-1,1-difluoro-1-methoxyethane.

8.29 (a) 3-Methyl-1-butanethiol
(b) 2-Butene-1-thiol

8.30 (a) The order of boiling points is 2-methyl-1-pentanol < 1-pentanol. These compounds are constitutional isomers and their molecular masses are the same. The issue here is molecular shape: increased branching lowers boiling point. (See Study Problem 8.4 on text p. 315; the actual boiling points are 129° and 137°, respectively.)

(b) The order of boiling points is *tert* butyl alcohol < 2-pentanol < 1-hexanol. (The actual boiling points are 82°, 119°, 158°.) This follows the order of molecular masses. The increased branching of 2-pentanol relative to 1-hexanol, and *tert*-butyl alcohol relative to 2-pentanol, makes the differences between these boiling points even greater than they would be for unbranched alcohols of the same molecular masses.

(c) The order of boiling points is 1-hexene < 1-chloropentane < 1-hexanol. (The actual boiling points are 63°, 108°, 158°.) Use the fact that the mass of a chlorine (35 units) is about the same as that of an ethyl group (29 units), and that alkyl chlorides have about the same boiling points as alkanes (and alkenes) of the same molecular mass. Consequently, 1-chloropentane has about the same boiling point as 1-heptene, which, in turn, has a higher boiling point than 1-hexene. 1-Hexanol has about the same molecular mass as heptane, and its boiling point is considerably higher because of hydrogen bonding.

(d) The order of boiling points is propane < diethyl ether < 1,2-propanediol. (The actual boiling points are –42°, 37°, 189°.) Diethyl ether has a higher boiling point than propane because diethyl ether is more polar and because it has a greater molecular mass. 1,2-Propanediol has the highest boiling point because of hydrogen bonding.

(e) The order of boiling points is cyclobutane < chlorocyclobutane < cyclooctane. Use the facts that the chlorine has about the same mass as an ethyl group, and that alkyl chlorides and alkanes of the same molecular mass have about the same boiling points. Then order the boiling points by molecular mass. (The actual boiling points are 12°, 83°, and 151°, respectively.)

8.31 (a) The first compound, acetic acid, has a higher boiling point than the second, ethyl acetate, because hydrogen bonding between acetic acid molecules can take place; the O—H hydrogen on one molecule can be donated to form a hydrogen bond to either of the oxygens on another molecule. Because ethyl acetate has no O—H, molecules of ethyl acetate cannot associate by hydrogen bonding. The intermolecular hydrogen bonds must be broken in order to vaporize acetic acid, and the energy required to break these hydrogen bonds is reflected in a higher boiling point.

(b) Use the same reasoning as in part (a). The N—H hydrogens of the first compound, acetamide, can be involved in hydrogen bonds between molecules that involve the oxygen or the nitrogen as an acceptor. The second compound, *N,N*-dimethylacetamide, has no hydrogens that can be involved in hydrogen bonding. The hydrogen bonding in the liquid state of acetamide is reflected in a higher boiling point, despite the higher molecular mass of the second compound.

8.32 (a) The unsaturation number of the ether is 1; because it has no double bonds, it must contain a ring. Two of the many acceptable possibilities are

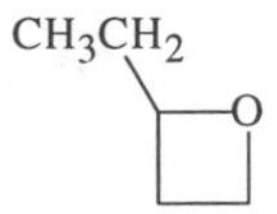

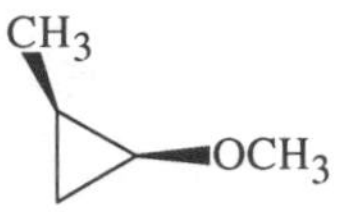

CH_3CH_2 O CH_3 OCH_3

(b) The unsaturation number is 2, and both rings and/or multiple bonds are allowed in this case. Two of several possibilities are

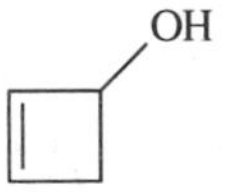

OH

$HC{\equiv}C{-}CH(OH){-}CH_3$

(c) The following glycol, although it has asymmetric carbons, is achiral because it is meso.

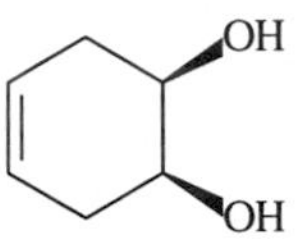

OH OH

(d) 2,3-Butanediol can exist as a pair of enantiomers and a meso stereoisomer.

$CH_3CH(OH){-}CH(OH)CH_3$
2,3-butanediol

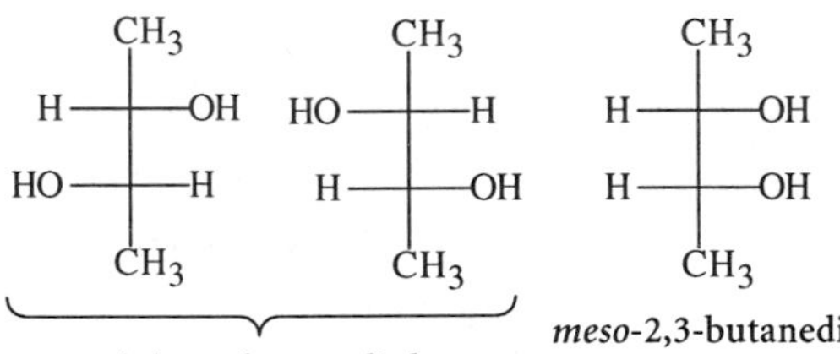

(±)-2,3-butanediol enantiomers; *meso*-2,3-butanediol

(e) The stereoisomeric forms of 1,2-butanediol consist of a pair of enantiomers:

$HOCH_2{-}CH(OH){-}CH_2CH_3$ 1,2-butanediol

(f)

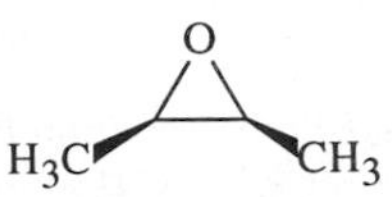

cis-2,3-dimethyloxirane (a meso compound)

trans-2,3-dimethyloxirane (exists as two enantiomers)

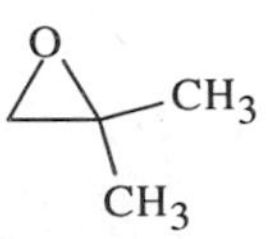

2,2-dimethyloxirane

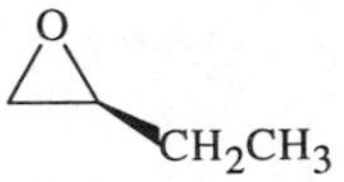

2-ethyloxirane (exists as two enantiomers)

8.33 (a) Dihydrogen, H_2
(b) Dideuterium, D_2
(c) Ethane, CH_3CH_3

(d) Dihydrogen, H_2
(e) Propane, $CH_3CH_2CH_3$

Remember that when a Grignard reagent reacts with a proton source, whether it is water or an alcohol, the *conjugate acid* of the "carbon anion" part of the Grignard reagent is formed. (See Eqs. 8.23 and 8.25, text p. 337.)

8.34 (a) The water-miscible compound is propyl alcohol; alcohols of low molecular mass are miscible with water because of their hydrogen-bonding interactions with water. Isomeric ethers are considerably less soluble, because they cannot donate hydrogen bonds.

(b) Allyl methyl ether, $H_2C{=}CH{-}CH_2{-}OCH_3$, decolorizes a solution of bromine in CCl_4 because the bromine adds to the double bond; the alcohol has no double bond to react with bromine.

(c) The compounds must be cyclic because they contain no double bonds; and they must be alcohols because they give off a gas (methane) when treated with the Grignard reagent methylmagnesium iodide. The compounds can only be the four stereoisomers of 2-methylcyclopropanol: the enantiomeric pair of cis isomers, and the enantiomeric pair of trans isomers.

(d) Only the alcohol, 2-methylcyclohexanol, will react with sodium hydride to give a gaseous product (H_2).

8.35 (a) Compound *A* is soluble in hydrocarbon solvents by the "like-dissolves-like" criterion. The alkyl groups, particularly the large cetyl ($C_{16}H_{33}$) group, resemble the hydrocarbon chains of the solvent. In contrast, there is no hydrogen-bond acceptor available in a hydrocarbon solvent to stabilize ammonium chloride.

(b) Although compound *A* is soluble in hydrocarbon solvents, it exists in such solvents as ion pairs and higher aggregates rather than as free ions. The reason is that hydrocarbon solvents have a very low dielectric constant ($\epsilon \cong 2$), whereas a high dielectric constant is required to separate ionic aggregates into free ions. Furthermore, a hydrocarbon offers no solvation by hydrogen bonding to the bromide counter-ion. Hence, this anion remains in proximity to its positive partner.

8.36 (a) Neopentane, $(CH_3)_4C$, gives only one monochlorination product, $(CH_3)_3C{-}CH_2Cl$.

(b) Isobutane, $(CH_3)_3CH$, gives two achiral monochlorination products: isobutyl chloride, $(CH_3)_2CHCH_2Cl$, and *tert*-butyl chloride, $(CH_3)_3C{-}Cl$.

8.37 (a) *tert*-butyl alcohol < isopropyl alcohol < propyl alcohol < 1-propanethiol. Alcohols with a greater degree of branching are less acidic because their conjugate bases are more poorly solvated; and thiols are substantially more acidic than alcohols (element effect).

(b) 2-chloroethanol < 3-chloro-1-propanethiol < 2-chloro-1-propanethiol. Thiols are more acidic than alcohols (element effect). 2-Chloro-1-propanethiol is the more acidic thiol because the electronegative chlorine is closer to the site of negative charge in the conjugate-base thiolate anion (polar effect).

(c) $CH_3NHCH_2CH_2CH_2OH < CH_3NHCH_2CH_2OH < (CH_3)_3\overset{+}{N}CH_2CH_2OH$. The last compound is most acidic because it has a full-fledged positive charge that stabilizes a negative charge in the conjugate-base anion. In the other two compounds, the nitrogen stabilizes the conjugate-base anion by its electron-withdrawing polar effect; in the more acidic of these compounds, the nitrogen is closer to the oxygen, which is the site of negative charge in the conjugate-base alkoxide anion (polar effect).

(d) $^{-}O{-}CH_2CH_2{-}OH < CH_3CH_2CH_2{-}OH < CH_3O{-}CH_2CH_2{-}OH$. 2-Methoxyethanol is most acidic because of the electron-withdrawing polar effect of the oxygen. The anion is least acidic because the negative charge on the oxygen interacts repulsively with a second negative charge formed on ionization of the O—H group:

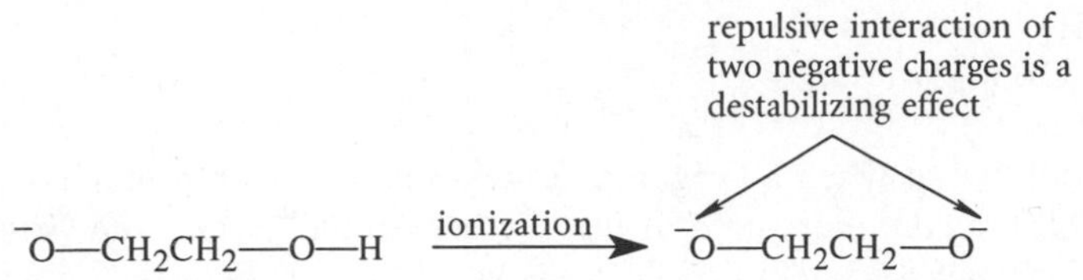

repulsive interaction of two negative charges is a destabilizing effect

$$^{-}O—CH_2CH_2—O—H \xrightarrow{\text{ionization}} {}^{-}O—CH_2CH_2—O^{-}$$

(e) The order of increasing acidity is

$$CH_3CH_2\ddot{O}H < CH_3CH_2\overset{+}{\underset{|}{O}}(H)CH_2CH_3 < CH_3\overset{+}{S}(H)CH_3$$

The protonated ether is more acidic than the alcohol for the same reason that H_3O^+ is more acidic than H_2O: a positively charged oxygen is much more electronegative than a neutral oxygen. The sulfonium ion is most acidic because of the element effect.

8.38 Water forms a Lewis acid-Lewis base complex with Cu^{2+} ion:

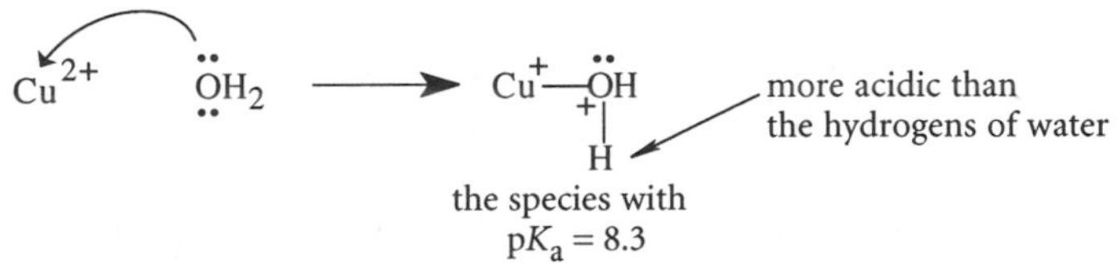

$$Cu^{2+} \quad \ddot{O}H_2 \longrightarrow Cu^{+}—\overset{+}{O}H(H)$$

the species with $pK_a = 8.3$

more acidic than the hydrogens of water

Ionization of a hydrogen of the complexed water removes a positive charge from the oxygen and eliminates the charge-charge repulsion in the complex. Ionization of water itself does not have this driving force. Hence, water that is complexed to copper is more acidic.

8.39 In the presence of concentrated acid, dibutyl ether is protonated. The protonated ether is an ionic compound, and ionic compounds are soluble in water:

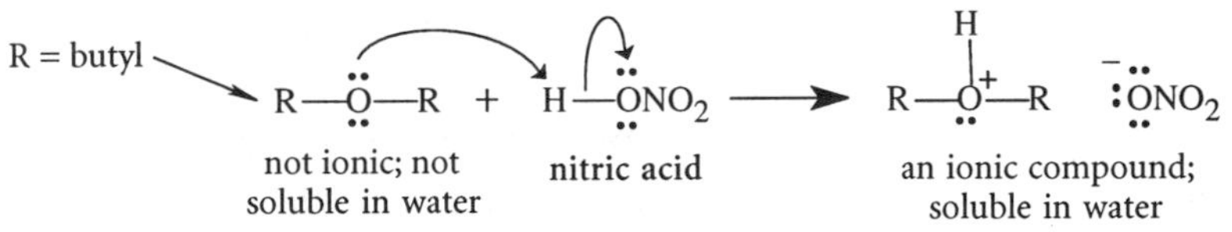

R = butyl

$$R—\ddot{O}—R + H—\ddot{O}NO_2 \longrightarrow R—\overset{+}{O}(H)—R \quad {}^{-}:\ddot{O}NO_2$$

not ionic; not soluble in water — nitric acid — an ionic compound; soluble in water

8.40 Think of the dissociation reaction of the crown ether-K^+ complex as you would any other equilibrium:

$$\text{crown ether–}K^+ + \text{solvent} \rightleftharpoons \text{crown ether} + K^+\text{(solvated)}$$

Anything that stabilizes the species on the right side of the equation will increase the dissociation constant of the complex. Because potassium ion, a typical positive metal ion, is solvated more effectively by water than by ether, the dissociation constant for the complex is greater in water.

8.41 Hexethal should be (and is) the more potent sedative, because it is more soluble in membranes, and therefore can readily pass through them. It is more soluble because the longer alkyl chain is more like the interior of a membrane than is the shorter ethyl chain. Furthermore, the long alkyl chain makes hexethal *less* soluble in water, because long alkyl groups are not effectively solvated by water.

8.42 As Eq. 8.19, text p. 335, shows, ethers solubilize Grignard reagents by forming Lewis acid-base complexes with them. Because a tertiary amine is also a Lewis base, it can solubilize Grignard reagents by the same type of interaction:

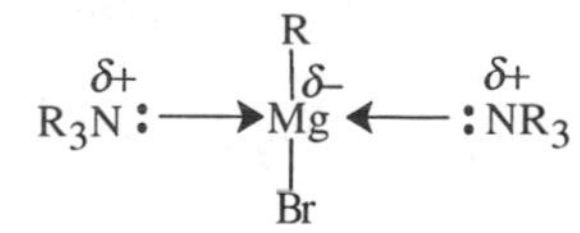

$$R_3N\overset{\delta+}{:}\longrightarrow \overset{\overset{R}{|}\,\delta-}{\underset{\underset{Br}{|}}{Mg}} \longleftarrow \overset{\delta+}{:}NR_3$$

8.43 (a) By LeChatelier's principle, when the concentration of ethanol increases, the concentration of dimer increases.

(b) Use the usual relationship between $\Delta G°$ and K_{eq}; recall that $2.3RT$ at 298 K is 5.71 kJ/mol.

$$\Delta G° = -2.303RT(\log K_{eq}) = -5.71(\log 11) = -5.94 \text{ kJ/mol}$$

(c) Neglect the volume of ethanol and assume that 1 mole of ethanol is dissolved in 1 L of solution so that the initial concentration of ethanol is 1 *M*. (This introduces an error of only 2%.) Assume that x *M* of ethanol react to form dimer. The concentration of ethanol remaining is then $(1.0 - x)$ *M*, and the concentration of dimer is $x/2$ *M*. (Remember, it takes two ethanol molecules to form one molecule of dimer.) The equilibrium-constant expression for dimerization becomes

$$K_{eq} = 11 = \frac{x/2}{(1-x)^2}$$

From the quadratic formula, $x = 0.808$ *M*. Thus, the concentration of dimer is $x/2 = 0.404$ *M*; that of free ethanol is $(1 - x)$ or 0.192 *M*. Thus, about 80% of the ethanol is dimerized in a 1 *M* solution.

(d) Because dimerization occurs through hydrogen-bond formation, then the greater equilibrium constant for dimerization of ethanol shows that ethanol has the stronger hydrogen bond.

(e) The alcohol $CH_3SCH_2CH_2OH$ would have the greater solubility, because it can both donate and accept hydrogen bonds to water. An —SH group is a much poorer hydrogen bond acceptor and hydrogen bond donor than —OH.

8.44 Flick's fragrant flailings were fundamentally futile because his ether is also an alcohol. The Grignard reagent was destroyed by the —OH group of the alcohol in a protonolysis reaction:

$$CH_3CH_2\text{—}MgBr + H\text{—}OCH_2CH_2OC_2H_5 \longrightarrow CH_3CH_3 + BrMg^+ \ {}^-OCH_2CH_2OC_2H_5$$

8.45 (a) The free-radical bromination of 2-methylpropane (isobutane) gives the following two alkyl bromides:

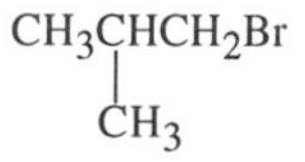

$CH_3CH(CH_3)CH_2Br$ and $CH_3C(Br)(CH_3)CH_3$

isobutyl bromide (1-bromo-2-methylpropane) — *tert*-butyl bromide (2-bromo-2-methylpropane)

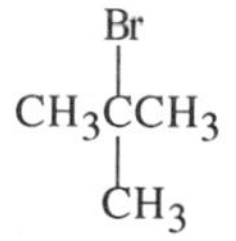

(b) The propagation steps in the formation of alkyl halides from alkanes are as follows, with X = Br and R = alkyl. (See Eqs. 8.30–8.32, text p. 338.)

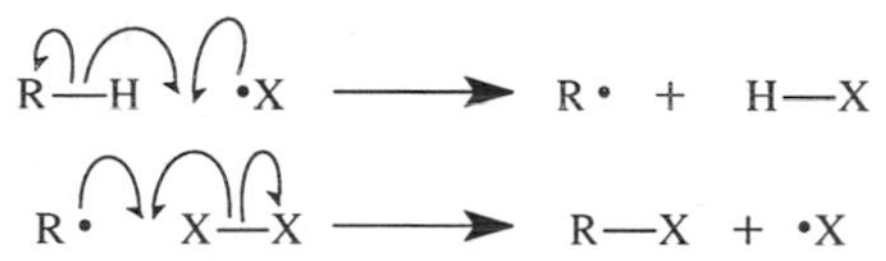

$$R\text{—}H + \cdot X \longrightarrow R\cdot + H\text{—}X$$

$$R\cdot + X\text{—}X \longrightarrow R\text{—}X + \cdot X$$

When the primary hydrogens of isobutane react, the free-radical intermediate R• is the *isobutyl radical,* $(CH_3)_2CH\dot{C}H_2$, a *primary* free radical. When the tertiary hydrogen of isobutane reacts, the free-radical intermediate R• is the *tert-butyl radical,* $(CH_3)_3\dot{C}$, a *tertiary* free radical. Table 5.3, text p. 188, shows that formation of a primary free radical by dissociation of a C—H bond requires about 16 kJ/mol more energy than formation of a tertiary free radical. Because the *tert*-butyl radical is more stable than the isobutyl radical, its formation should be faster; consequently, the product resulting from its reaction with

Br_2, namely, *tert*-butyl bromide, should be (and is) the major product formed.

Notice that isobutane has *nine* primary hydrogens and only *one* tertiary hydrogen. Thus, if the reaction were random, nine times as much isobutyl bromide would be formed as *tert*-butyl bromide. The fact that *tert*-butyl bromide is the major product, despite its statistical disadvantage, shows the importance of the relative stabilities of the free-radical intermediates in determining the outcome of the reaction.

8.46 The outcome of the Grignard protonolysis shows that all of the alkyl halides have the same carbon skeleton, that of 2,4-dimethylpentane. The protonolysis in D_2O confirms the fact that the bromines are bound at different places on the carbon skeleton. The only three possibilities for the alkyl halides are

$CH_3C(CH_3)(Br)CH_2CH(CH_3)CH_3$ (*A*) $BrCH_2CH(CH_3)CH_2CH(CH_3)CH_3$ (*B*) $CH_3CH(CH_3)CH(Br)CH(CH_3)CH_3$ (*C*)

Compound *B* is the chiral alkyl halide, and compounds *A* and *C* are the other two. (Of course, the absolute configuration of *B* (that is, whether it is *R* or *S*) is not known.

8.47 We know from Chapter 7 that hydroboration is a stereospecific *syn*-addition. Hence, the product of hydroboration is compound *A*, the *syn*-addition product. We know that protonolysis of Grignard reagents and related organometallic replaces the metal with a hydrogen. On the assumption that the same thing happens with organoboranes (and it does), the product is the meso compound. Notice that this product is the same one that would be produced by catalytic hydrogenation of the same alkene.

[Structure: organoborane with H, D, CH_2CH_3 / D, CH_2CH_3] or [Fischer projection: H top; D—CH_2CH_3; D—CH_2CH_3; B bottom]
A (as the racemate)
$\xrightarrow[\text{protonolysis}]{H_3C-C(=O)-OH}$ [Fischer projection: H top; D—CH_2CH_3; D—CH_2CH_3; **H** bottom]
meso

8.48 Figure 5.2, text p. 186, shows that free radicals are sp^2-hybridized, and that the unpaired electron resides in a 2*p* orbital. Combining this fact with the mechanism shown in the solution to Problem 8.45(b) shows that the free-radical intermediate involved in the bromination of *sec*-butylbenzene has the structure shown on the left of the following equation. Because the radical is planar, and therefore achiral, the chirality of the starting *sec*-butylbenzene is irrelevant; the radical can react with Br_2 equally well at either the top lobe *(a)* or bottom lobe *(b)* of the 2*p* orbital. Because the two reaction pathways are enantiomeric, they occur with equal rates. Consequently, the product of the reaction is the *racemate* of (1-bromo-1-methylpropyl)benzene, and therefore this product should be optically inactive, even if the *sec*-butylbenzene starting material is optically active.

[Scheme: planar radical H_3C, Ph, C—CH_2CH_3 with 2*p* orbital; Br—Br attacks top lobe *(a)* → H_3C, Ph, C(Br)—CH_2CH_3 + •Br; attacks bottom lobe *(b)* → H_3C, Ph, C(Br)—CH_2CH_3 + •Br; the two products: enantiomers]

8.49 (a) In compound *A*, the equatorial position of the —OH group is preferred for the usual reason: it avoids 1,3-diaxial interactions with hydrogens on the ring. However, compound *B* prefers to have an axial —OH group because it affords the opportunity for intramolecular hydrogen bonding; this additional bonding is a stabilizing effect, and lowers the energy of the axial conformation below that of the equatorial conformation.

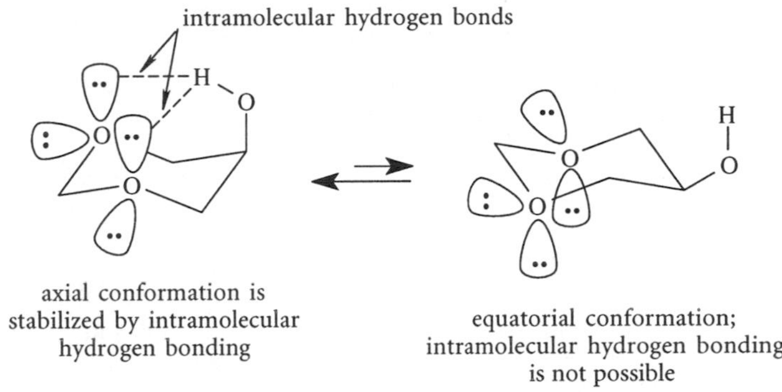

axial conformation is stabilized by intramolecular hydrogen bonding

equatorial conformation; intramolecular hydrogen bonding is not possible

(b) The result in part (a) shows that intramolecular hydrogen bonding can stabilize conformations that otherwise might be less stable. In this case, intramolecular hydrogen bonding can stabilize the *gauche* conformations of both stereoisomers. Indeed, in either enantiomer of the racemate, such hydrogen bonding can occur in one conformation in which the large *tert*-butyl groups are anti to each other. However, in the meso stereoisomer, the necessity that the hydroxy groups be gauche in order for hydrogen bonding to occur also means that the large *tert*-butyl groups must also be gauche. The stabilizing effect of intramolecular hydrogen bonding cannot compensate for the magnitude of the resulting van der Waals repulsions between the *tert*-butyl groups.

one hydrogen-bonding conformation of (3*R*,4*R*)-2,2,5,5-tetramethyl-3,4-hexanediol; *tert*-butyl groups are anti

gauche conformation of *meso*-2,2,5,5-tetramethyl-3,4-hexanediol

severe van der Waals repulsions

8.50 (a) In the chair conformation, for every C—O bond dipole in a given direction, there is another C—O bond dipole of the same magnitude pointing in the opposite direction. Thus, in the following diagram, the gray dipoles cancel each other, and the black dipoles cancel each other.

chair conformation of 1,4-dioxane

Because all bond dipoles cancel, the net dipole moment of this conformation is zero.

(b) To the extent that the twist-boat conformation is present, it will contribute a nonzero dipole moment, because the C—O bond dipoles do not cancel in this conformation. (We use the simpler boat conformation to illustrate this idea.)

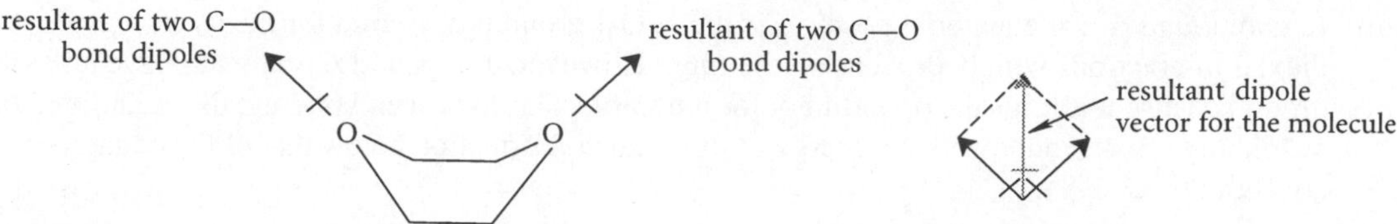

boat conformation of 1,4-dioxane

The dipole moment of the twist-boat conformation is actually rather large, but it is present in very small concentration. The dipole moment of any molecule is the weighted average of the dipole moments of individual conformations. In other words, the nonzero dipole moment of 1,4-dioxane results from the presence of a very small amount of a conformation that has a large dipole moment.

Taking the bond dipole of a C—O bond as 1.52 D, you should be able to calculate the dipole moment of the twist-boat conformation from the data in this problem and the assumption of some reasonable C—O—C bond angles, which you can find in tables in the text. Then, knowing that the net dipole moment of 1,4-dioxane is 0.38 D, you should be able to calculate the fraction of twist-boat conformation in a sample of dioxane!

8.51 (a) Equilibrium (2) contains more of the conformation with the methyl group in an axial position than equilibrium (1) does; that is, equilibrium (2) lies farther to the right. The reason is that, because C—O bonds are shorter than C—C bonds, the axial methyl group in equilibrium (1) is closer to the axial hydrogens than the axial methyl group in equilibrium (2). Therefore, the 1,3-diaxial interactions in the axial-methyl conformation of equilibrium (1) are somewhat more severe than those in the axial-methyl conformation of equilibrium (2).

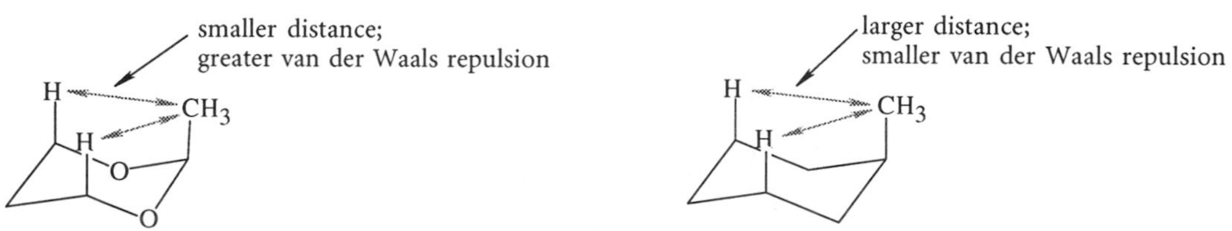

(b) The reasoning is much the same as that in part (a). Because the C—O bonds are shorter in the ether A than the corresponding C—C bonds in butane B, the methyl groups are brought closer together in the gauche conformation of the ether than they are in the gauche conformation of butane. Consequently, van der Waals repulsions in the gauche conformation of the ether are somewhat greater than they are in the gauche conformation of butane. In contrast, the shorter bonds of the ether should have little effect on the ebergy of the anti conformation, in which the methyl groups are far apart. Hence, the greater energy of the gauche conformation of the ether (relative to the anti conformation) causes less of the gauche conformation to be present at equilibrium. Thus, butane contains more gauche conformation at equilibrium.

8.52 Because BF_3 is electron-deficient, it readily reacts as a Lewis acid with the unshared pairs of the Lewis base ethanol:

$$C_2H_5\ddot{\underset{..}{O}}H \quad BF_3 \longrightarrow C_2H_5\overset{\displaystyle H}{\overset{|}{\underset{..}{O}}}{}^{+}\!\!-\!\overset{-}{B}F_3$$

Fluoride ion dissociates from the boron in a Lewis acid-base dissociation, and then removes the proton from the oxygen in a Brønsted acid-base reaction:

$$C_2H_5\overset{\displaystyle H}{\overset{|}{\underset{..}{O}}}{}^{+}\!\!-\!\overset{\displaystyle :\ddot{F}:}{\overset{|}{\underset{-}{B}}}F_2 \longrightarrow C_2H_5\overset{\displaystyle H}{\overset{|}{\underset{..}{O}}}{}^{+}\!\!-\!BF_2 \quad :\ddot{\underset{..}{F}}:^{-} \longrightarrow C_2H_5\ddot{\underset{..}{O}}\!-\!BF_2 + H\!-\!\ddot{\underset{..}{F}}:$$

The resulting species contains an electron-deficient boron, which is attacked by ethanol, and a similar sequence of reactions is repeated:

$$C_2H_5O{-}BF_2 + C_2H_5\ddot{O}H \longrightarrow C_2H_5O{-}\overset{\overset{:\ddot{F}:}{|}}{\bar{B}}F{-}\overset{+}{O}(H)C_2H_5 \longrightarrow C_2H_5O{-}BF{-}\overset{+}{O}(H)C_2H_5 + :\ddot{F}:^- \longrightarrow C_2H_5O{-}BF{-}\ddot{O}C_2H_5 + H{-}\ddot{F}:$$

A third, exactly analogous, sequence, which we leave for you to write, finishes the mechanism.

9

Chemistry of Alkyl Halides

TERMS

anti-elimination 9.5D
bimolecular reaction 9.4A
carbene 9.8A
carbenoid 9.8B
α-carbon 9.1B
β-carbon 9.1B
dichloromethylene 9.8A
E1 mechanism 9.6A
E1 reaction 9.6A
E2 mechanism 9.5A
E2 reaction 9.5A
elimination reaction 9.1B
α-elimination 9.8A
β-elimination 9.1B
first-order reaction 9.3B
intramolecular substitution reaction 9.1A
kinetic order 9.3B
leaving group 9.1A
Le Chatelier's principle 9.2
methylene 9.8B
nucleophile 9.1A
nucleophilic displacement reaction 9.1A
nucleophilic substitution reaction 9.1A
nucleophilicity 9.4D
primary deuterium isotope effect 9.5C
product-determining step 9.6B
rate constant 9.3B
rate law 9.3B
reaction rate 9.3
second-order reaction 9.3B
Simmons-Smith reaction 9.8B
S_N1 mechanism 9.6A
S_N1 reaction 9.6A
S_N2 mechanism 9.4A
S_N2 reaction 9.4A
solvolysis 9.6
steric effect 9.4C
syn-elimination 9.5D
unimolecular reaction 9.6A
Zaitsev elimination 9.5E
Zaitsev's rule 9.5E

CONCEPTS

I. GENERAL REACTION ISSUES

A. Equilibria

1. Knowledge of the equilibrium constant for a reaction provides no information about the rate at which the reaction takes place.
   a. Whether the equilibrium in a nucleophilic substitution reaction is favorable can be predicted from an analysis of the corresponding Brønsted acid-base reaction.
   b. If the Brønsted acid-base reaction strongly favors one side of the equation, then the analogous substitution reaction likewise favors the same side of the equation.
2. The equilibrium in any nucleophilic substitution reaction, as in an acid-base reaction, favors release of the weaker base.

$$\underset{}{I\text{—}CH_3} + \underset{\text{stronger base}}{Na^+\ {}^-C{\equiv}N} \rightleftharpoons H_3C\text{—}C{\equiv}N + \underset{\text{weaker base}}{Na^+\ I^-}$$

a. LeChatelier's principle states that if an equilibrium is disturbed, the components of the equilibrium will react so as to offset the effect of the disturbance.
b. Some equilibria that are not too unfavorable can be driven to completion by applying LeChatelier's principle, *e.g.*, by precipitation of a by-product.

B. Rate Law

1. The mathematical statement of how a reaction rate depends on concentration is called the rate law.
   a. Each reaction has its own characteristic rate law.
   b. The concentration terms of the rate law indicate what atoms are involved in the rate-limiting transition state.
2. In the rate law, the constant of proportionality, k, between the rate and the concentrations of reactants is called the rate constant.
   a. In general, the rate constant is different for every reaction, and it is a fundamental physical constant for a given reaction under particular conditions of temperature, pressure, solvent, etc.
   b. The rate constant is the rate under standard conditions (1 M concentration of each reactant). The rates of two reactions are compared by comparing their rate constants.
   c. The rate constant is related to the standard free energy of activation $\Delta G^{\circ\ddagger}$.
      i. The standard free energy of activation, or energy barrier, determines the rate of a reaction under standard conditions.
      ii. If the rates of two reactions A and B are compared, the relationship between their rate constants and their standard free energies of activation is

$$-2.3RT\log(k_A/k_B) = \Delta G_A^{\circ\ddagger} - \Delta G_B^{\circ\ddagger}$$

   d. The dimensions of the rate constant depend on the kinetic order of the reaction.
      i. The overall kinetic order for a reaction is the sum of the powers of all the concentrations in the rate law.
      ii. The kinetic order in each reactant is the power to which its concentration is raised in the rate law.
3. A rate law is determined experimentally by varying the concentration of each reactant (including any catalysts) independently and measuring the resulting effect on the rate.
4. The rate law of a reaction provides fundamental information about the mechanism of a reaction.
   a. Mechanisms not consistent with the rate law are ruled out.
   b. Of the chemically reasonable mechanisms consistent with the rate law, the simplest one is provisionally adopted.
   c. The mechanism of a reaction is modified or refined if required by subsequent experiments.
   d. The concentration terms of the rate law indicate which atoms are present in the transition state of the rate-limiting step.
   e. Although the rate law indicates what atoms are present in the transition state, it provides no information about how they are arranged.

C. Nucleophilicity and Solvents

1. Nucleophilic reactivity is termed nucleophilicity.
   a. The nucleophilicity of a base is determined by the rate of its reaction with a standard Lewis acid.
   b. When the attacking atom is the same, the relative nucleophilicities of different nucleophiles correlate roughly with their relative Brønsted basicities.
   c. Nucleophilicity, not Brønsted basicity, governs the rate of an S_N2 reaction.
2. The relationship between nucleophilicity and basicity depends strongly on the solvent.
   a. In protic solvents:
      i. Nucleophilicity increases towards greater atomic number within a group (column) of the periodic table.

ii. Hydrogen bonding occurs between protic solvent molecules and nucleophilic anions and thus decreases the availability of unshared electron pairs for nucleophilic reactions.
iii. Within a group (column) of the periodic table, the most strongly basic anions (those of lower periods) are most strongly hydrogen-bonded, and are therefore least nucleophilic.

b. In aprotic solvents:
i. There is no hydrogen bonding to reduce the nucleophilicity of dissolved anions; consequently, anions dissolved in such solvents are considerably more nucleophilic than they are in protic solvents.
ii. Within a group of the periodic table, the most basic anions are the best nucleophiles.
iii. The nucleophilicity of an anion is much greater in a polar, aprotic solvent than in a protic solvent.

3. Solvents can affect the reactivity of nucleophiles.
   a. Polar, protic solvents best solubilize charged ions, but anionic nucleophiles are least reactive in these solvents.
   b. Polar, aprotic solvents in many cases provide the best combination of solubility and reactivity for anionic nucleophiles.
   c. The reactivity of uncharged nucleophiles shows less dependence on solvent.

D. Nucleophiles

1. A nucleophile generally means a Lewis base that attacks an atom other than hydrogen.
2. Most strong Brønsted bases are good nucleophiles, but some excellent nucleophiles are relatively weak Brønsted bases.
   a. Many nucleophiles are anions, others are uncharged, or in a few cases, even positively charged.
   b. Uncharged nucleophiles show a much weaker dependence of nucleophilicity on solvent.
3. The strongest bases are generally the most reactive nucleophiles; however, the reverse is true in protic solvents for anions in which the attacking atoms come from within a group (column) of the periodic table.
4. The nucleophilicities of anionic nucleophiles are much greater in a polar, aprotic solvent than in a protic solvent.

E. Primary Deuterium Isotope Effect

1. When a hydrogen is transferred in the rate-limiting step of a reaction, a compound in which that hydrogen is replaced by its isotope deuterium will react more slowly in the same reaction.
   a. This effect of deuterium substitution on reaction rates is called a primary deuterium isotope effect.
   b. The theoretical basis for the primary deuterium isotope effect lies in the comparative strengths of C—H and C—D bonds.
      i. The bond to the heavier isotope D is stronger (and thus requires more energy to break) than the bond to the lighter isotope H.
      ii. The energy barrier, or free energy of activation, for the compound with the C—D bond is greater; as a result, its rate of reaction is smaller.
2. A primary deuterium isotope effect is observed only when the hydrogen that is transferred in the rate-determining step is substituted by deuterium.

II. REACTIONS OF ALKYL HALIDES

A. Nucleophilic Substitution Reactions

1. A nucleophilic substitution reaction, or nucleophilic displacement reaction, is a very general type of reaction:
   a. It is a *substitution* because one group is substituted for (or displaces) another group, which departs.
      i. The group that is displaced in a nucleophilic substitution reaction is termed the leaving group.
      ii. The leaving group in a nucleophilic displacement reaction is the nucleophile in the reverse reaction, and vice-versa.

iii. The best leaving groups give the weakest bases as products.

b. It is a *nucleophilic substitution* because the substituting group acts as a nucleophile, or Lewis base.

i. Many nucleophiles are anions, others are uncharged, or in a few cases, even positively charged.

ii. The roles of nucleophile and leaving group are reversed if the reaction can be run in the reverse direction.

c. Nucleophilic substitution reactions occur by two mechanisms:

i. The S_N2 mechanism occurs in a single step with inversion of stereochemical configuration and is characterized by a second-order rate law.

Nuc:⁻ + C—X ⟶ Nuc—C + X:⁻

$$\text{rate} = k[\text{Nuc:}^-][\text{C—X}]$$

ii. The S_N1 mechanism is characterized by a first-order rate law that contains only a term in alkyl halide concentration and by significant loss of stereochemical configuration.

—C—X ⟶ (slower) —C+ + X:⁻ ⟶ (Nuc:⁻, faster) —C—Nuc

$$\text{rate} = k[\text{—C—X}]$$

2. The reaction of an alkyl halide with a solvent in which no base or nucleophile has been added is termed a solvolysis.
   a. The solvolysis reactions of tertiary alkyl halides are fastest in polar, protic donor solvents.
   b. Tertiary and secondary alkyl halides undergo solvolysis reactions by the S_N1 and E1 mechanisms; tertiary alkyl halides are more reactive.
   c. Rearrangements are observed in S_N1 solvolysis reactions that involve rearrangement-prone carbocations as intermediates.
3. An intramolecular substitution reaction is a reaction in which the nucleophile and the leaving group are part of the same molecule.
4. Nucleophilic substitution reactions can be used to transform alkyl halides into a wide variety of other functional groups. (See Table 9.1 on text p. 353.)
5. For the most part, polar protic solvents, polar aprotic solvents, or mixtures of these are used in substitution reactions.
6. Each nucleophilic substitution reaction is conceptually similar to a Brønsted acid-base reaction.

B. Elimination Reactions

1. An elimination reaction is a reaction in which two or more groups are lost from within the same molecule.
   a. In an alkyl halide, the carbon bearing the halogen is often referred to as the α-carbon, and the adjacent carbons are referred to as the β-carbons.

β-hydrogen ⟶ H H ⟵ α-hydrogen
—C—C—X
β-carbon, α-carbon

   b. An elimination that involves loss of two groups from adjacent carbons is termed a β-elimination.
   c. A β-elimination reaction is conceptually the reverse of an addition to an alkene.
2. An α-elimination, as in the formation of dichloromethylene from chloroform, is an elimination of two groups (in this case the elements of HCl) from the same atom.

$$\text{Base:}^- + \text{H—CCl}_2\text{—Cl} \longrightarrow \text{Cl}_2\text{C:} + \text{Base—H} + \text{:Cl:}^-$$

chloroform — dichloromethylene

3. β-Elimination reactions occur by two mechanisms:
   a. The E2 mechanism, which competes with the S_N2 mechanism, has a second-order rate law and occurs with anti stereochemistry.

$$\text{Base:}^- + \text{H—C—C—X} \longrightarrow \text{C=C} + \text{X:}^- + \text{Base—H} \qquad \text{rate} = k[\text{Base:}^-][\text{H—C—C—X}]$$

anti-elimination

   b. The E1 mechanism is an alternative product-determining step of the S_N1 mechanism in which a carbocation intermediate loses a β-hydrogen to form an alkene.

$$\text{H—C—C—X} \xrightarrow{\text{rate-limiting}} \text{H—C—C}^+ + \text{X:}^- \longrightarrow \text{C=C} + \text{HX} \qquad \text{rate} = k[\text{H—C—C—X}]$$

4. Base-promoted β-elimination reactions in many cases occur by an E2 mechanism; they typically follow a rate law that is second order overall and first order in each reactant.
   a. A Lewis base acting as a Brønsted base attacks a β-hydrogen of the alkyl halide, not a carbon atom as in a nucleophilic substitution reaction.
   b. If the reacting alkyl halide has more than one type of β-hydrogen atom, then more than one β-elimination reaction is possible.
   c. Strong bases, such as sodium ethoxide [$Na^+ \, {}^-OCH_2CH_3$] and potassium *tert*-butoxide [$K^+ \, {}^-OC(CH_3)_3$], promote the β-elimination reactions of alkyl halides.
   d. Often the conjugate-acid alcohols of these bases are used as solvents, for example, sodium ethoxide in ethanol or potassium *tert*-butoxide in *tert*-butyl alcohol.

C. Carbenes and Carbenoids

1. A haloform reacts with base in an α-elimination reaction to give dihalomethylene, a carbene.
2. A carbene is an unstable and highly reactive species having a divalent carbon atom.
   a. The divalent carbon of a carbene can act as a nucleophile and an electrophile at the same time.
      i. An atom with an unshared electron pair reacts as a Lewis base, or nucleophile.
      ii. An atom that lacks an electronic octet is an electron-deficient Lewis acid that can act as an electrophile by accepting an electron pair without losing one at the same time.
   b. The carbon atom of dichloromethylene bears three groups (two chlorines and an unshared pair of electrons) and therefore has approximately trigonal-planar geometry.

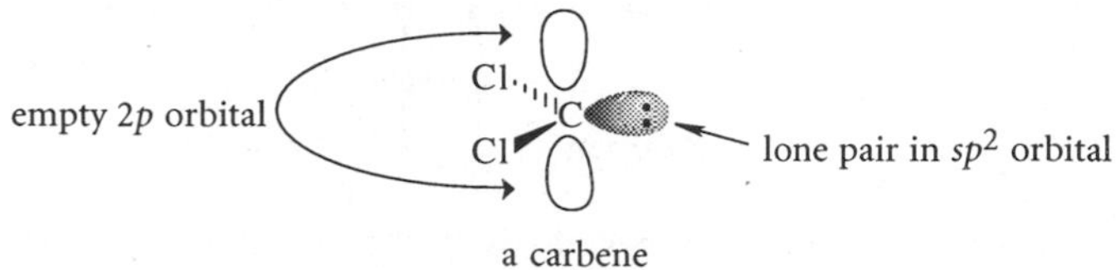

a carbene

3. The *syn*-addition of carbenes or carbenoids to alkenes yields cyclopropanes in a reaction that forms new carbon-carbon bonds.
   a. Dihalomethylene adds to alkenes to give dihalocyclopropanes.
   b. Methylene iodide ($CH_2 I_2$) reacts with a zinc-copper couple to give a carbenoid organometallic reagent which adds to alkenes to give cyclopropanes.

III. BIMOLECULAR REACTIONS

A. S_N2 Reaction

1. A mechanism in which attack of a nucleophile on an atom (usually carbon) displaces a leaving group from the same atom in a concerted manner is called an S_N2 mechanism.
   a. The meaning of the "nickname" S_N2 is **S** for substitution, **N** for nucleophilic, and **2** for bimolecular (meaning that the transition state of the reaction involves two species).
   b. Reactions that occur by S_N2 mechanisms are called S_N2 reactions.
   c. An S_N2 mechanism, because it is concerted, involves no reactive intermediates.
2. In general, secondary alkyl halides undergo S_N2 reactions much more slowly than typical primary alkyl halides, and tertiary alkyl halides are even less reactive.
   a. The reaction rate is second order overall: first order in the nucleophile and first order in the alkyl halide.
   b. The mechanism involves backside attack of the nucleophile on the alkyl halide and inversion of stereochemical configuration.

Nuc:⁻ + C—X ⟶ Nuc—C + X:⁻

   c. The reaction is retarded by branching (a steric effect) at both the α- and β-carbon atoms; alkyl halides with three β-branches are unreactive.
   d. The best leaving groups in the S_N2 reaction are those that react to give the weakest bases as products.
   e. The S_N2 mechanism is especially rapid in polar, aprotic solvents.
   f. The strongest bases are generally the most reactive nucleophiles; however, the reverse is true in protic solvents for anions with attacking atoms from within a group (column) of the periodic table. For example, RS^- is a better nucleophile than RO^- in a protic solvent.
3. In the transition state for an S_N2 reaction on carbon, the nucleophile and the leaving group are partially bonded to opposite lobes of the carbon 2*p* orbital.
   a. The central atoms is turned "inside out," and is approximately sp^2-hybridized in the transition state.

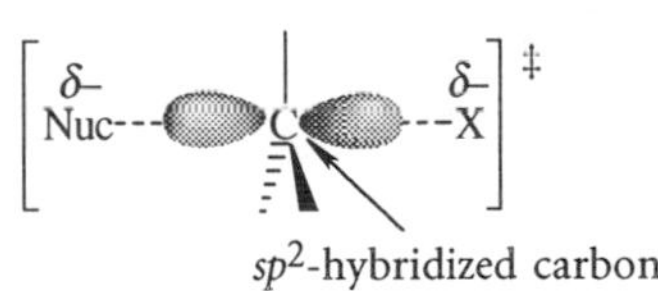

sp^2-hybridized carbon

   b. Van der Waals repulsions (steric effects) raise the energy of the transition state and therefore reduce the reaction rate. Thus, alkyl halides that are highly branched at the α- and β-carbons react slowly or not at all by the S_N2 mechanism.

B. E2 Mechanism

1. A mechanism involving concerted removal of a β-proton by a base and loss of a halide ion is called an E2 mechanism.
   a. Reactions that occur by the E2 mechanism are called E2 reactions.
   b. The meaning of the "nickname" E2 is **E** for elimination and **2** for bimolecular (meaning that the transition state of the reaction involves two species).
2. The E2 reaction is a β-elimination reaction of alkyl halides that is promoted by strong bases.
   a. The rates of E2 reactions are second order overall: first order in base and first order in the alkyl halide.
   b. E2 reactions are normally *anti*-eliminations. (An *anti*-elimination is conceptually the reverse of an *anti*-addition.)

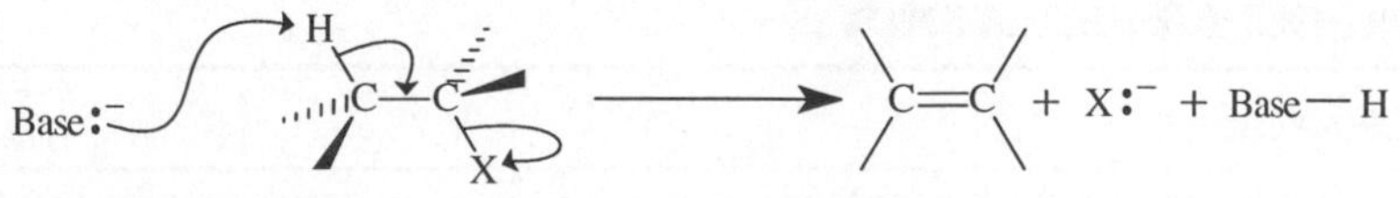

anti-elimination

c. *Anti*-elimination is preferred for three reasons:
   i. *Anti*-elimination occurs through a transition state that has a staggered conformation, whereas *syn*-elimination occurs through a transition state that has a higher-energy eclipsed conformation.
   ii. *Anti*-elimination requires the base and leaving group to be on opposite sides of the molecule, out of each other's way, whereas *syn*-elimination requires the base and leaving group to be on the same side of the molecule, where they can interfere sterically with each other.
   iii. *Anti*-elimination is more favorable because it involves all-backside electronic displacements, whereas *syn*-elimination requires a frontside electronic displacement on the carbon-halogen bond.

d. The best leaving groups give the weakest bases as products.

3. The rates of E2 reactions show substantial primary deuterium isotope effects at the β-hydrogen atoms because a β-hydrogen is removed in the transition state.
4. When an alkyl halide has more than one type of β-hydrogen, more than one alkene product can be formed.
   a. The most stable alkene (the alkene with the greatest numbers of alkyl substituents at their double bonds) are formed in greatest amounts.
   b. The standard free energy of the E2 transition state, like that of an alkene, is lowered by branching.
5. The greatest use of the E2 elimination for the preparation of alkenes is when the alkyl halide has only one type of β-hydrogen, and only one alkene product is possible.

C. S_N2 versus E2 Reaction

1. The S_N2 and E2 reactions are competing processes; the reaction pathway that occurs most rapidly is the one that predominates.
   a. A greater fraction of S_N2 reaction is observed when the attacking atom is a good nucleophile yet is a relatively weak base, for example, RS^- or I^-.
   b. Elimination is favored by:
      i. alkyl branches in the alkyl halide at the α- or β carbon atoms,
      ii. alkyl branches in the base; and
      iii. stronger bases.
2. When simple alkoxide bases such as methoxide and ethoxide are used, the predominant product of an E2 reaction is usually the most stable alkene isomer.
   a. The reaction that leads to the alkene product with more branching at the developing double bond is the faster reaction.
   b. The alkene products are stable under the conditions of the reaction.
   c. Because the product mixture, once formed, does not change, the distribution of products reflects the relative energies of the competing transition states.
3. The effect of alkyl halide structure of the alkyl halide:
   a. alkyl halides with greater amounts of branching at the α-carbon give greater amounts of elimination.
   b. Tertiary alkyl halides give more elimination than secondary, which give more than primary.
   c. Alkyl halides with greater amounts of branching at the β-carbon give greater amounts of elimination as long as there is at least one β-hydrogen.
4. The effect of base structure:
   a. More highly branched bases give a greater fraction of elimination than unbranched ones.
   b. More highly branched bases give a greater fraction of the alkene product with less branching at the double bond.
   c. Weaker bases that are good nucleophiles give a greater fraction of substitution.

IV. UNIMOLECULAR REACTIONS

A. The S_N1 Reaction

1. A substitution mechanism that involves a carbocation intermediate is called an S_N1 mechanism.
   a. Substitution reactions that take place by the S_N1 mechanism are called S_N1 reactions.
   b. The meaning of the S_N1 "nickname" is **S** for substitution, **N** for nucleophilic, and **1** for unimolecular (meaning that the transition state of the reaction involves one species other than solvent).
2. The S_N1 mechanism is observed mostly with tertiary and secondary alkyl halides because the reaction involves a carbocation intermediate, the formation of which is promoted by branching at the α-carbon.
3. The S_N1 mechanism occurs mostly in polar, protic solvents.

B. The E1 Reaction

1. An elimination mechanism that involves carbocation intermediates is termed an E1 mechanism.
   a. Reactions that occur by E1 mechanisms are called E1 reactions.
   b. The meaning of the E1 "nickname" is **E** for elimination and **1** for unimolecular (meaning that the transition state of the reaction involves one species other than solvent).
2. A strong base is not required for the E1 reaction as it is for the E2 reaction.
3. When an alkyl halide contains more than one type of β-hydrogen, more than one type of elimination product can be formed.
4. The alkene with the greatest number of alkyl substituents on the double bond predominates in the product.

C. S_N1 versus E1 Reactions

1. S_N1-E1 reactions are most rapid with tertiary alkyl halides, they occur more slowly with secondary alkyl halides, and they are almost never observed with primary alkyl halides.
   a. The reactivity order of the alkyl halides is fluorides <<< chlorides < bromides < iodides.
   b. If an alkyl halide has β-hydrogens, elimination products formed by the E1 reaction accompany substitution products formed by the S_N1 reaction.
2. The S_N1 and E1 reactions of an alkyl halide share a common intermediate, the carbocation.
   a. The first step, ionization of the alkyl halide to the carbocation, is the rate-limiting step and thus has the transition state of highest free energy.
      i. This step, a Lewis acid-base dissociation, is the rate-limiting step of both the substitution and elimination reactions.

$$\text{H}-\overset{|}{\underset{|}{\text{C}}}-\overset{|}{\underset{|}{\text{C}}}-\text{X} \longrightarrow \text{H}-\overset{|}{\underset{|}{\text{C}}}-\text{C}^{+} + \text{X:}^{-}$$

      ii. Substitution and elimination products arise from competing reactions of the carbocation.

elimination: $\text{Base:}^{-} + \text{H}-\overset{|}{\underset{|}{\text{C}}}-\text{C}^{+} \longrightarrow \text{C}=\text{C} + \text{Base}-\text{H}$

substitution: $\text{H}-\overset{|}{\underset{|}{\text{C}}}-\text{C}^{+} \;\; \text{Nuc:}^{-} \longrightarrow \text{H}-\overset{|}{\underset{|}{\text{C}}}-\overset{|}{\underset{|}{\text{C}}}-\text{Nuc}$

   b. The different products that can be formed in S_N1-E1 reactions reflect three reactions of carbocation intermediates:
      i. reaction with a nucleophile (the S_N1 reaction);
      ii. loss of a β-proton (the E1 reaction); and

   iii. rearrangement of the initially formed carbocation intermediate to a more stable carbocation followed by (*i*) or (*ii*).

c. Because the relative rates of these steps determine the ratio of products, they are said to be the product-determining steps.

d. The rates of the product-determining steps have no effect on the rate at which the alkyl halide reacts.

3. The S_N1 and E1 reactions are accelerated by polar, protic donor solvents; ionic dissociation is favored by solvents that separate ions and by solvents that solvate ions.

REACTIONS

(See also Concepts, Secs. II.A. and II.B.)

I. FORMATION OF CYCLOPROPANES

A. Dihalocarbenes

1. Reaction of a haloform with base in the presence of an alkene yields a 1,1-dihalocyclopropane.
2. The addition of a dihalomethylene to an alkene is a concerted *syn*-addition reaction.

$$CHCl_3 + (H)(CH_3)C{=}C(H)(CH_3) \xrightarrow[(CH_3)_3COH]{K^+\ ^-OC(CH_3)_3} \text{1,1-dichloro-2,3-dimethylcyclopropane (cis)}$$

B. Carbenoids: The Simmons-Smith Reaction

1. Cyclopropane without halogen atoms can be prepared by allowing alkenes to react with methylene iodide (CH_2I_2) in the presence of a zinc-copper couple (a copper-activated zinc preparation) in a reaction called the Simmons-Smith reaction.

$$CH_2I_2 \xrightarrow{\text{Zn–Cu couple}} I{-}CH_2{-}ZnI \xrightarrow{(H_3C)(H)C{=}C(H)(CH_3)} \text{cis-1,2-dimethylcyclopropane}$$

   a. The active reagent in the Simmons-Smith reaction is believed to be an α-halo organometallic compound, a compound with halogen and a metal on the same carbon.
   b. The formation of this species is analogous to formation of a Grignard reagent.
   c. The Simmons-Smith reagent can be conceptualized as methylene (**:**CH_2, the simplest carbene) coordinated to the zinc atom.
   d. Because they show carbenelike reactivity, α-halo organometallic compounds are sometimes termed carbenoids, that is, a reagent that is not a free carbene but has carbenelike reactivity.

2. Addition reactions of methylene from the Simmons-Smith reagent to alkenes, like the reactions of dichloromethylene, are *syn*-additions.

SCIENTISTS

STUDY GUIDE LINKS

9.1 Reaction Rates

Because rates are usually functions of concentration, and because concentrations change as a reaction proceed, you can perhaps appreciate that the rate of a reaction changes as the reaction proceeds. If the rate of a reaction is proportional to the concentrations of one or more reactants, the rate diminishes during a reaction because the concentration of reactants diminishes. (See Problem 9.7 in the text.)

Reaction rates, like other rates of change, are more precisely expressed in the language of calculus. Suppose, for example, we have a reaction $S \rightarrow P$ that follows a rate law that is first order in A. Then the rate can be written

$$\text{rate} = \frac{d[P]}{dt} = -\frac{d[S]}{dt} = k[S] \tag{SG9.1}$$

This is a first-order differential equation that can be solved for $[S]$ as a function of time. If S_0 is the concentration of S at the start of the reaction ($t = 0$), then the solution to this equation is

$$[S] = S_0 e^{-kt} \tag{SG9.2}$$

(If you are not familiar with differential equations, you can easily verify this solution by differentiating Eq. SG9.2, as follows.

$$d[S]/dt = -kS_0 e^{-kt} \tag{SG9.3}$$

Substituting Eqs. SG9.2 into Eq. SG9.1 yields Eq. SG9.3 and thus verifies the solution.)

Equation SG9.2 gives the expression for the actual concentration of $[S]$ as a function of time, and shows that it is a decaying exponential. Each rate law has its own characteristic solution for the dependence of concentration on time. Computer-fitting actual concentration versus time data for a reaction of known kinetic order to the appropriate rate law allows the experimentalist to determine the value of the rate constant k. This is how rate constants are usually derived in practice.

✓9.2 Deducing Mechanisms from Rate Laws

Suppose you are an expert in traffic safety, and you are asked to determine as much as you can about traffic accidents on a bridge. In fact, your job is to come up with a description for a "typical" accident and formulate a strategy to eliminate the accidents. Of course, you would interview eyewitnesses and read police reports to gather your data. But suppose eyewitness accounts are not available to you for some reason. This is the situation in deducing reaction mechanisms. We cannot directly observe transition states (that is, molecular collisions).

Imagine that the traffic data you have available are the number of accidents as a function of the traffic density on the bridge in the two directions. Suppose you find that the number of accidents is proportional to the number of cars traveling east *and* to the number of cars traveling west. From this information you can construct a mental picture—a mechanism—for a typical collision. Because the number of accidents in a given time is proportional to the traffic flow in *both* directions, you could reasonably conclude that a typical accident is a head-on collision between a car in one lane and a car in the other. An *unreasonable* picture would be two cars in the same lane having a rear-end collision, because, in this case, the number of accidents in a given time would be proportional to some function of the traffic density in only that lane. The head-on collision is as good a picture of an accident as any other until other data force you to refine it. In fact, it suggests an experiment for reducing the number of accidents: build a wall between the two lanes of traffic. If this "experiment" reduces the number of accidents, it would support your picture of a typical accident. Notice that it would *not prove your picture to be correct;* it would only *show that your picture is not incorrect.* Nothing short of eyewitnesses can prove you correct.

Similarly, a starting point for deducing a chemical mechanism is the rate law. The rate law tells us about "molecular collisions" that lead to a reaction, but does not tell us whether these collisions occur by a head-on mechanism, a rear-end mechanism, or some less direct pathway. The simplest mechanism possible is adopted, and a modification is considered only when required by new data.

9.3 Molecular-Orbital Basis of Backside Attack

Molecular orbital theory shows very clearly why backside attack takes place. Recall that the molecular orbitals of a molecule consist of equal numbers of *bonding* and *antibonding* molecular orbitals. It turns out that the antibonding molecular orbitals play a crucial role in some chemical reactions, and the S_N2 reaction is one such reaction.

When a nucleophile attacks an alkyl halide, it donates a pair of electrons to the alkyl halide. Where in the alkyl halide molecule does the electron pair go? In the curved-arrow mechanism using Lewis structures, we don't worry about it; we simply say that the electron pair on the nucleophile displaces the leaving group. However, if you think about it, you'll realize that the electrons from the nucleophile have to interact with a *molecular orbital* of the alkyl halide molecule; otherwise, these electrons have nowhere to go. Because all bonding molecular orbitals are filled, the only empty molecular orbitals in an alkyl halide molecule are the *antibonding* molecular orbitals.

The electron pair of the nucleophile can in principle interact with any of the antibonding molecular orbitals of the alkyl halide. In order to pay the lowest energy price, it stands to reason that this electron pair would interact with the *antibonding molecular orbital of lowest energy.* This molecular orbital has a nickname: the **LUMO:** "lowest unoccupied molecular orbital." Let's have a look at the LUMO of a typical alkyl halide and see what insight it offers us about the S_N2 reaction.

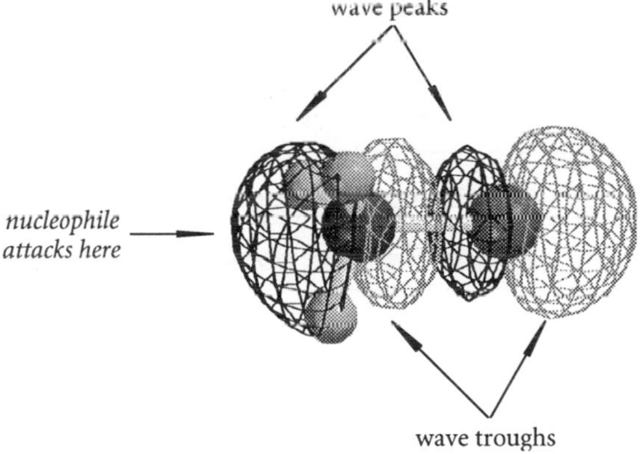

LUMO of methyl chloride

This picture was derived from an actual quantum-mechanics calculation of the molecular orbitals of the CH_3Cl (methyl chloride) molecule. This shows the peaks and troughs of the LUMO as "nets"—as if the lobes are formed out of "chicken wire." These are superimposed on a ball-and-stick model of the molecule. The wave peaks are shown in black, and the wave troughs are in gray. (Notice that this lobe has three nodes; we expect nodes in an antibonding orbital.) Notice that *the largest wave peak is located at the back side of the C—Cl bond.* Very simply, the unshared pair of the nucleophile attacks this orbital at the position that would give the most probable electron density, and this is backside to the C—Cl bond.

Filling of an antibonding orbital weakens the overall bonding. (Use the dihydrogen molecular orbitals as your mental model. Recall that when two electrons occupy its antibonding molecular orbital, dihydrogen will dissociate into two $H:^-$ ions; see Problem 1.15(c) on text p. 34, and the discussion of why He is monatomic on text p. 33.) Because the LUMO happens to be uniquely associated with the C—Cl bond, it stands to reason that the C—Cl bond will be weakened when the nucleophile attacks the LUMO and "injects" two electrons into it. In fact, it is weakened so much that it breaks. This, then, is the molecular-orbital description of the process in which a nucleophile uses an unshared electron pair to "kick out" a leaving group from an alkyl halide.

When the chloride ion departs, the LUMO shown previously is transformed into a filled (bonding) molecular orbital of the product that involves the nucleophile, and the the leaving group (Cl^-) departs with its unshared electron pair.

✓9.4 Reaction Stereochemistry and Fischer Projections

It may be tempting to attempt to deduce reaction stereochemistry directly from a Fischer projection. For example, a student might look at the starting material in Eq. 9.37a on p. 374 of the text and say, "Obviously, the phenyls are on the same side of the molecule, so that's the way they have to be in the product." The problem with this reasoning is that it ignores the fact that *Fischer projections do not indicate conformation; they are only meant to indicate stereochemical configuration.* Suppose that the same starting material is redrawn in a different Fischer projection:

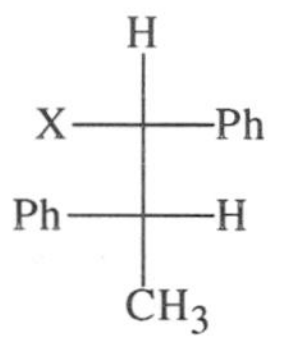

H
X—Ph
Ph—H
CH_3

This is exactly the same molecule, but drawn in a Fischer projection in which the phenyls happen to be on opposite sides of the projection. You can see that the reasoning above would not work, because it predicts that the phenyl groups would end up trans. An experimental result can hardly depend on how we choose to draw a structure!

The Fischer projection stands for a particular stereoisomer that might exist in many conformations. In order to examine a particular conformation—which is what must be done to understand the elimination stereochemistry—the Fischer projection must be translated into the conformation of interest. This translation process was explained in Sec. 6.11 of the text, but let's review it again in the present context.

First, interpret the Fischer projection directly as an eclipsed conformation:

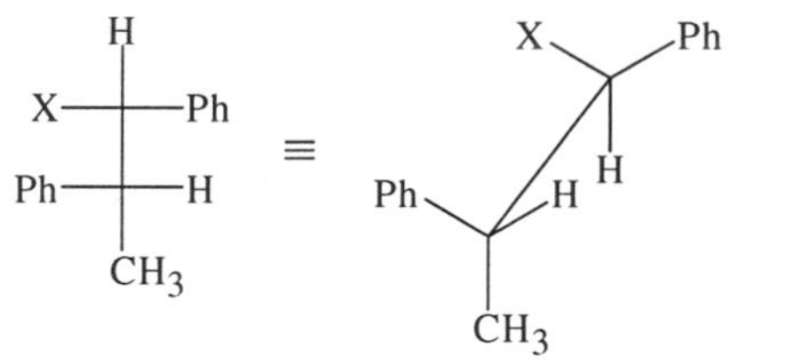

H
X—Ph
Ph—H
CH_3

≡

X Ph
H
Ph H
CH_3

Then change the conformation so that the X on the α-carbon and the H on the β-carbon are anti. (We do this because we want to see the stereochemical consequences of *anti*-elimination.)

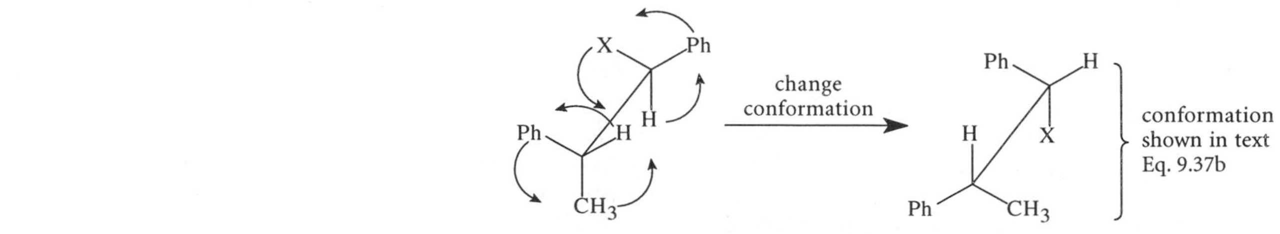

X Ph
H
Ph H
CH_3

change conformation →

Ph H
H X
Ph CH_3

conformation shown in text Eq. 9.37b

Notice that when the halogen on one carbon and the β-hydrogen on the other are anti, the phenyl groups are on the same side of the molecule; that is why they are also on the same side of the molecule in the product—that is, they are cis.

The message here is that if you are looking at the stereochemistry of a reaction, and any of the molecules involved are drawn in a Fischer projection, the first thing you should do is interpret the Fischer projections in terms of three-dimensional perspective structures. *Don't try to see the reaction stereochemistry directly from the Fischer projection.*

You might ask why we just don't draw the appropriate conformation in the text, and forget about Fischer projections. The reason is that the starting material exists in a variety of conformations. The Fischer projection system has nothing to say about conformation. To draw a single conformation as a starting material for Eq. 9.37a might suggest that the molecule exists only in that conformation, and such an implication would be wrong.

✓9.5 Branching in Cyclic Compounds

Study Problem 9.1 on text p. 380 refers to β-branches in the starting alkyl halide *B* and branching at the double bond of alkene *C*. Some students have difficulty in seeing branches within a cyclic compound, even though they have no problem with acyclic compounds. The reason for this difficulty is that the branches in a cyclic compound are "tied together" into a ring. For example, "branching at the double bond" in compound *C* should be viewed as follows:

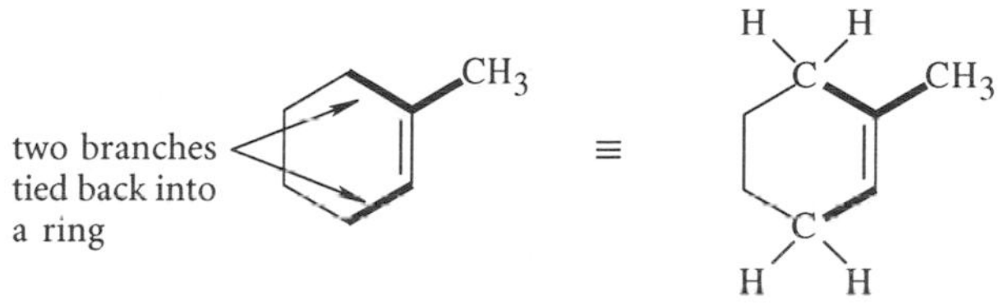

two branches tied back into a ring

CH_3 ≡ H H C CH_3 C H H

(The methyl group, of course, is a third branch on the double bond, and it is not tied back.) A similar approach is taken to the analysis of β-branching in alkyl halide *B*:

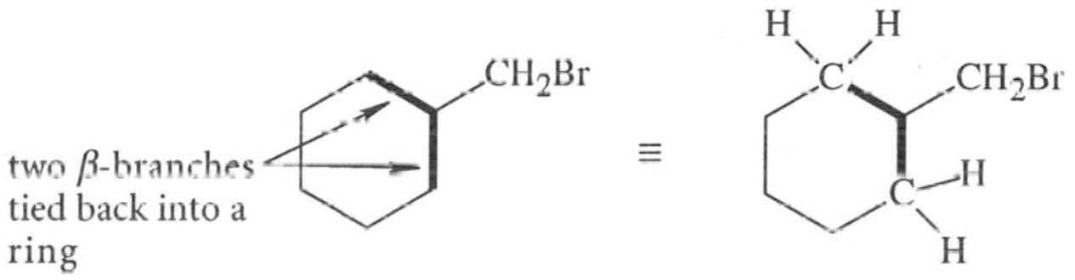

two β-branches tied back into a ring

CH_2Br ≡ H H C CH_2Br C H H

The fact that the branching carbons are linked in a ring doesn't change the fact that they are branches. After all, tying your hands together wouldn't change the fact that you still have two hands, would it?

✓9.6 Diagnosing Reactivity Patterns in Substitution and Elimination Reactions

The last case study in Study Problem 9.2 (text p. 392) points out a danger in using a guide such as Table 9.6 on text p. 392: *you can sometimes expect special cases not covered by the table or borderline cases that are hard to predict.* You can't diagnose every possible reaction with a summary table any more than a physician can diagnose a patient with a table of symptoms and diseases! When you encounter borderline cases that are difficult to predict, you should be able to narrow the possibilities of what can happen. The important thing is for you to be able to predict trends and to understand the reasoning behind them. Table 9.6 will prove most useful if you *make the effort to understand each entry in terms of the principles that underlie it,* and then *think* about each problem you are asked to solve in terms of these principles rather than attack it by rote.

A number of professors of organic chemistry, the author included, find that the section of our organic chemistry courses that deals with substitution and elimination reactions—the section covered by this chapter—is a point at which a number of students have particular difficulty. We have discerned at least two reasons for this problem. First, this is a point at which the number of reactions and ideas have accumulated to the stage that the student who is not carefully organized is overwhelmed. If you are in this situation, you should read (or re-read) Study Guide Link 5.4, "How to Study Organic Reactions," on p. 122 of this manual, and follow the suggestions there. The second reason for difficulty is the fact that all four mechanisms of substitution and elimination—S_N1, S_N2, E1, and E2—can potentially happen simultaneously, and it can be difficult to keep straight exactly when each process can take place. Again, to use an analogy from the medical world, when a physician diagnoses a problem with a patient, he/she can't focus simply on one system, such as the heart; the physician must understand that the many systems in the body interact, and that there can be many root causes contributing to a given symptom. In some cases, the problems are so complex that they can't be readily solved! Table 9.6 is designed to help you keep all the possibilities in substitution and elimination reactions organized.

A broader philosophical issue is relevant to this discussion. The temptation is for students to believe that there are "recipes" or set formulas for solving every problem. This section on substitutions and eliminations shows that there are *principles* that can guide you into the right "ballpark," but inevitably, uncertainties will arise in some situations. Again: Your goal should be not to memorize every conceivable case, but rather to bring to bear your knowledge of the principles so as to narrow the possibilities.

If you were a laboratory scientist trying to predict the outcome of a reaction, you would make your best prediction using the principles that you know, and then you would go into the laboratory and run an experiment to see whether you are correct. A physician facing such uncertainty runs tests to gather as many facts as possible, adopts a hypothesis, and checks it by administering appropriate medication or carrying out other tests. Two people with difficulties in a relationship adopt a hypothesis as to the nature of the problem and, if they are behaving rationally, change their behavior and see whether the problem is solved. For most problems there aren't magic recipes and tables that allow us to solve problems by rote, nor are there simple formulas into which we "plug" data for an answer. Uncertainty is in the very nature of science, and indeed, of life itself. If you think about it, life would be awfully dull without the challenge of such uncertainty. *One reason that organic chemistry is valued as a prerequisite course of study by various disciplines is that a student must, in order to succeed, develop a body of principles that can be applied to situations in which there is inherent uncertainty.* The good news is that if you master this skill in organic chemistry, you'll find it easier to do so in your other endeavors as well.

✓9.7 Mechanism of Carbene Addition to Alkenes

In Study Guide Links 5.2 and 5.5, we treated the mechanisms of hydroboration and ozonolysis, respectively, as stepwise reactions involving carbocation intermediates so we could see their connection to other addition mechanisms that do involve carbocations. Bearing in mind that carbene additions are concerted reactions, let us take the same approach here.

Because the carbene is electron-deficient, it can act as a Lewis acid towards the π electrons of the alkene:

$$Cl_2\ddot{C} \leftarrow (CH_3-CH{=}CH-CH_3) \longrightarrow Cl_2\bar{\ddot{C}}-CH(CH_3)-\overset{+}{C}H-CH_3$$

The carbon derived from the carbene in the resulting species now has a complete octet and an unshared pair; consequently, it can act as a nucleophile. One carbon derived from the alkene is now electron deficient and can react as an electrophile. The logical consequence is for an *intramolecular* Lewis acid-base association to take place; when nature thus takes its course the result is a cyclopropane.

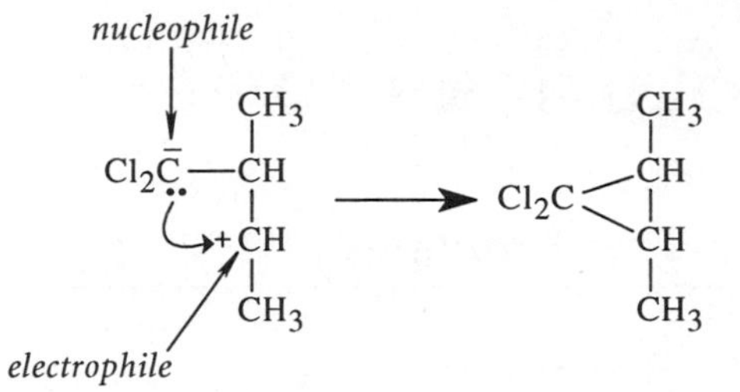

nucleophile

$$Cl_2\bar{\ddot{C}}-CH(CH_3)-\overset{+}{C}H-CH_3 \longrightarrow Cl_2C\langle CH(CH_3)-CH(CH_3)\rangle$$

electrophile

Can you see how this reaction really is conceptually a lot like HBr addition?

SOLUTIONS TO PROBLEMS

Solutions to In-Text Problems

9.1 (a) $CH_3CH_2CH_2CH_2S{-}CH_3 + Na^+ I^-$

butyl methyl sulfide

(b) $CH_3CH_2\overset{+}{N}H_3\ I^-$

ethylammonium iodide

9.2 (a) Two elimination products are expected because there are two types of β-hydrogen in the starting alkyl halide.

2-bromo-2,3-dimethylbutane, $H_3C{-}C(CH_3)(Br){-}CH(CH_3){-}CH_3$ (β-hydrogens labelled *(a)* on the methyl groups of C-2 and *(b)* on C-3) $\xrightarrow{Na^+\ CH_3CH_2O^-}$ $H_2C{=}C(CH_3){-}CH(CH_3)_2$ + $(CH_3)_2C{=}C(CH_3)_2$

from elimination of a hydrogen *(a)* and bromine; from elimination of hydrogen *(b)* and bromine

(b) As in part (a), because there are two types of β-hydrogen, two alkenes can be formed:

1-chloro-1-methylcyclohexane (ring hydrogens labelled *(a)*, methyl hydrogens labelled *(b)*) $\xrightarrow{Na^+\ C_2H_5O^-}$ 1-methylcyclohexene ($-CH_3$) + methylenecyclohexane ($=CH_2$)

from elimination of a hydrogen *(a)* and chlorine; from elimination of a hydrogen *(b)* and chlorine

9.3 (a) Notice that there are two types of β-hydrogen, and thus correspondingly two possible β-elimination products (three, if we count stereoisomers).

$CH_3CH(Br)CH_2CH_3$ (hydrogens *(a)* on CH_3, *(b)* on CH_2) $\xrightarrow[CH_3OH]{Na^+\ CH_3O^-}$ $CH_3CH(OCH_3)CH_2CH_3$ + $H_2C{=}CHCH_2CH_3$ + $CH_3CH{=}CHCH_3$ (cis and trans)

substitution product; from elimination of a hydrogen *(a)* and bromine; from elimination of a hydrogen *(b)* and bromine

(b) Methyl iodide can only undergo substitution to give dimethyl ether, dimethyl ether, $H_3C{-}O{-}CH_3$. It of course cannot undergo β-elimination because it has no β-hydrogens.

(c) The substitution product can in principle be obtained as two stereoisomers. (Sec. 9.4B of the text shows which of these is formed and why.)

1-bromo-3-methylcyclohexane (CH_3 wedge, Br hashed) $\xrightarrow[CH_3OH]{Na^+\ CH_3O^-}$ OCH_3 (hashed) isomer + OCH_3 (wedge) isomer + 4-methylcyclohexene (CH_3) + 3-methylcyclohexene (CH_3)

substitution products; elimination products

9.4 The principle to be applied is that each reaction favors the side with the weaker base.

(a) Because iodide ion is a weaker base than chloride ion, the equilibrium lies to the left.
(b) Because fluoride ion is a stronger base than chloride ion, the equilibrium lies to the right.
(c) Because azide ion is a stronger base than chloride ion, the equilibrium lies to the right.
(d) Because methoxide ion is a much stronger base than chloride ion, the equilibrium lies to the right.

9.5 (a) The reaction is third order overall; the reaction is first order in alkene and second order in bromine. The dimensions of the rate constant are such that the overall rate has the dimensions $\text{mol}\cdot\text{L}^{-1}\cdot\text{sec}^{-1}$; thus, the rate constant has the dimensions $\text{L}^2\cdot\text{mol}^{-2}\cdot\text{sec}^{-1}$, or $M^{-2}\cdot\text{sec}^{-1}$.

(b) The reaction is first order overall, and first order in alkyl halide. The rate constant has the dimensions of sec^{-1}.

9.6 (a) Use Eq. 9.23a:

$$(k_A/k_B) = 10^{(\Delta G^{\circ\ddagger}_B - \Delta G^{\circ\ddagger}_A)/2.3RT} = 10^{(14\ \text{kJ/mol}/5.706\ \text{kJ/mol})} = 10^{2.45} = 282$$

(b) We transform Eq. 9.23b to get the difference between the standard free energies of activation.

$$2.3RT \log (k_A/k_B) = (5.706)(\log 1/450) = (5.706)(-2.65) = -15.1 = \Delta G^{\circ\ddagger}_B - \Delta G^{\circ\ddagger}_A$$

In other words, the $\Delta G^{\circ\ddagger}$ of reaction A is greater than that of reaction B by 15.1 kJ/mol.

In problems like this, always check your answers for reasonableness to be sure you haven't committed a sign error. For example, in part (a), reaction A has the lower standard free energy of activation and must be faster than reaction B; hence, the desired ratio must be >1. In part (b), reaction B is faster and therefore must have the smaller $\Delta G^{\circ\ddagger}$; hence, the difference $\Delta G^{\circ\ddagger}_B - \Delta G^{\circ\ddagger}_A$ must be negative.

9.7 Because *D* and *E* are converted into *F*, their concentrations decrease with time; hence, the rate also decreases with time. Since the rate is the *slope* of the curves of [reactants] or [products] with time, then the slopes of these curves continually decrease with time, and approach zero at long times. A plot of these concentrations with time is shown in Fig. SG9.1.

Note that the *rate* of a reaction is not the same as a *rate constant.* The rate changes with time because concentrations change with time; the rate constant does not change. The rate law for any reaction shows that the rate constant is the rate under a *defined set of conditions*, that is, 1 *M* concentration of each reactant. You will often see the term *relative rate*, which is used when we compare the rates of two reactions. Because rates change with time, such a comparison is meaningful only when it is made under a defined set of conditions. Hence, the term *relative rate* is used to mean *the relative value of the rate constants.*

9.8 A mechanism consistent with the rate law is a bimolecular process in which ammonia displaces acetate ion from the proton:

$$CH_3C(=O)-\ddot{O}-H \quad :NH_3 \longrightarrow \left[CH_3C(=O)-\overset{\delta-}{\ddot{O}}\text{---}H\text{----}\overset{\delta+}{:NH_3} \right]^{\ddagger} \longrightarrow CH_3C(=O)-\ddot{O}:^- + \overset{+}{N}H_4$$

acetate ion

9.9 Because the S_N2 mechanism involves a molecule of alkyl halide and a molecule of nucleophile in a bimolecular reaction, the expected rate law is second order, first order in alkyl halide and first order in cyanide:

$$\text{rate} = k[C_2H_5Br][^-CN]$$

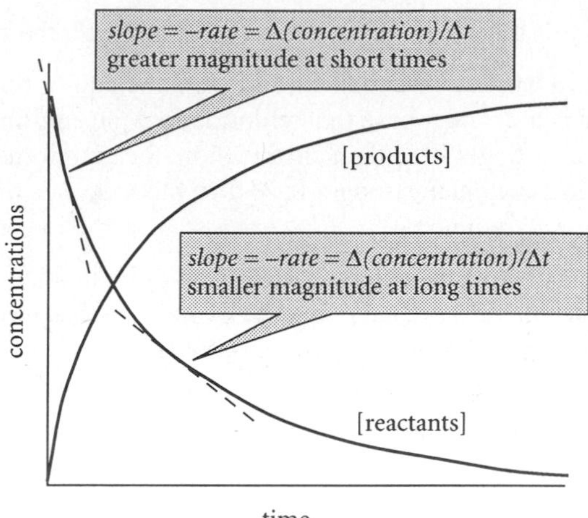

Figure SG9.1 A plot of reactant and product concentrations to accompany the solution to Problem 9.7. Notice that the rate is the *slope* of the plot at a given time. Because the rate is proportional to concentration, the rate decreases as the concentrations decrease.

9.10 If we assume that the S_N2 mechanism is operating, the reaction occurs with inversion of configuration to give (*S*)-1-iodobutane.

$$\text{(R)-chloroethane-1-}d \;[\text{H, D, Cl, }CH_2CH_2CH_3\text{ on C}] + K^+ I^- \longrightarrow K^+ Cl^- + \text{(S)-iodoethane-1-}d \;[\text{I, D, H, }CH_2CH_2CH_3\text{ on C}]$$

(*R*)-chloroethane-1-*d* (*S*)-iodoethane-1-*d*

9.11 In an S_N2 reaction, the stronger base is the better nucleophile when the same attacking atom is involved. Because ethoxide is a much stronger base than ethanol, it is the better nucleophile.

9.12 The discussion of this section shows that elements in higher periods within a group are better nucleophiles in protic solvents than elements in lower periods of the same group. Thus, triethylphosphine, $:P(C_2H_5)_3$, is a better nucleophile than triethylamine, $:N(C_2H_5)_3$. The products are as follows:

$$H_3C{-}I + :X(C_2H_5)_3 \longrightarrow H_3C{-}\overset{+}{P}(C_2H_5)_3\ I^- \quad \text{or} \quad H_3C{-}\overset{+}{N}(C_2H_5)_3\ I^-$$

X = P X = N

9.13 (a) The products of the S_N2 reaction between potassium acetate and ethyl iodide:

$$H_3C{-}C({=}O){-}O{-}CH_2CH_3 \quad + \quad K^+ I^-$$

ethyl acetate potassium iodide

(b) Potassium acetate is a better nucleophile in acetone because ethanol is a protic solvent and reduces the nucleophilicity of potassium acetate by hydrogen-bond donation. Consequently, potassium acetate in acetone reacts more rapidly with ethyl iodide than a solution of the same nucleophile in ethanol.

9.14 (a) The discussion of leaving-group effects in Sec. 9.5B indicates that leaving-group effects are about the same in S_N2 and E2 reactions; and Eq. 9.34 on text p. 372 indicates that bromide is a better leaving group than chloride; alkyl bromides are about fifty times more reactive than alkyl chlorides. Therefore, compound *A* reacts most rapidly. Because of the primary deuterium isotope effect, compound *C* reacts about 3–8 times more rapidly than compound *B*. (Note that leaving-group effects are more important than isotope effects.) Thus, the order of increasing reactivity is $B < C < A$.

(b) Because iodide is a much better leaving group than fluoride (see Eq. 9.34, text p. 372), compound *B* is much more reactive than compound *A*. The reactivity order is $A << B$.

9.15 (a) The hydration of styrene should be (and is) slower in D_2O, because the transferred proton is a deuterium, and a rate-retarding primary deuterium isotope effect operates when the proton transferred in the rate-limiting step is isotopically substituted.

The products differ by isotopic substitution. In H_2O/H_3O^+, the hydration product is compound *A* below; in D_2O/D_3O^+, the hydration product is compound *B*.

OH
PhCHCH$_3$ *A*

OD
PhCHCH$_2$D *B*

(b) The hydration rate of the deuterium-substituted styrene should differ very little, if at all, from that of styrene itself, because the deuteriums are not transferred in the rate-limiting step.

A small effect of isotopic substitution occurs in this case because of the differential effect of deuterium and hydrogen on the rehybridization of carbon in the transition state. (The carbon bearing the deuteriums rehybridizes from sp^2 to sp^3 in the rate-limiting step.) However, this effect on rate amounts to only a few percent. Effects of this sort are called *secondary deuterium isotope effects.*

9.16 (a) Analyze one enantiomer of (±)-stilbene dibromide; the result is the same for the other. First, draw the Fischer projection; then translate it into a conformational projection (*Step 1* below). Next, internally rotate about the central carbon-carbon bond until the H and Br that are eliminated are anti (*Step 2* below). Finally, draw the alkene that is formed and note its stereochemistry (*Step 3* below). The process starts with the Fischer projection because it is easy to distinguish between the chiral and meso stereoisomers of the starting material with a Fischer projection. Of course, you can start at whatever point in the process is easiest.

(1) (2) (3) $Na^+ C_2H_5O^-$ (E2 reaction) → + C_2H_5OH + $Na^+ Br^-$

anti H and Br

(±)-stilbene dibromide

alkene product has the *Z* configuration

(b) Reasoning identical to that used in part (a) shows that the alkene formed has the *E* configuration.

(1) (2) (3) $Na^+ C_2H_5O^-$ (E2 reaction) → + C_2H_5OH + $Na^+ Br^-$

meso-stilbene dibromide

alkene product has the *E* configuration

9.17 The diastereomer of the starting alkyl halide in Eq. 9.37a gives the diastereomer of the product. This can be verified by using a process similar to that used in the solution to Problem 9.16.

Fischer projection: X (top), Ph—C—H, H—C—Ph, CH_3 (bottom) $\xrightarrow{Na^+\ C_2H_5O^-,\ C_2H_5OH}$ Ph(H)C=C(H$_3$C)(Ph)

(*E*)-α-methylstilbene

9.18 (a) Let isobutyl bromide react with dimethyl sulfide, preferably in a polar aprotic solvent (although the reaction of a third-period nucleophile would occur at a useful rate in a polar solvent such as ethanol).

$$(CH_3)_2CHCH_2Br + (CH_3)_2\ddot{S}: \xrightarrow{\text{acetone}} (CH_3)_2CHCH_2\overset{+}{\underset{\cdot\cdot}{S}}(CH_3)_2\ Br^-$$

isobutyl bromide; dimethyl sulfide; isobutyldimethylsulfonium bromide

(b) In this case, the nucleophile is the conjugate base of ethanethiol, sodium ethanethiolate. Although a polar aprotic solvent could be used, ethanol would probably be the most convenient solvent (for solubility reasons). Because the nucleophile is from the third period, the reaction would occur at a convenient rate in a protic solvent; hence, experimental convenience determines the choice of solvent. The nucleophile would be easily formed from ethanethiol with one equivalent of sodium ethoxide.

$$(CH_3)_2CHCH_2Br + Na^+\ C_2H_5\ddot{\underset{\cdot\cdot}{S}}:^- \xrightarrow{C_2H_5OH} (CH_3)_2CHCH_2\ddot{\underset{\cdot\cdot}{S}}CH_2CH_3 + Na^+\ Br^-$$

isobutyl bromide; sodium ethanethiolate; isobutyl ethyl sulfide

(c) In contrast, this reaction is an elimination that could be brought about by the reaction of isobutyl bromide with any strong base, such as $Na^+\ CH_3CH_2O^-$ (sodium ethoxide) or $K^+\ (CH_3)_3C—O^-$ (potassium *tert*-butoxide).

$$(CH_3)_2CHCH_2Br \xrightarrow[(CH_3)_3COH]{K^+\ (CH_3)_3CO^-} (CH_3)_2C{=}CH_2$$

isobutyl bromide; 2-methylpropene (isobutylene)

9.19 Analyze the branching pattern in each alkyl halide: branching at both the α- and β-positions increases the ratio of elimination to substitution. Compound *D* has both an α-branch and two β-branches; compound *B* has two β-branches; compound *C* has one β-branch; and compound *A*, of course, has no β hydrogens and therefore cannot undergo β-elimination. Therefore, the ratio of E2 elimination to S_N2 substitution decreases in the order $D > B > C > A$.

9.20 Because α-branching in the base promotes a greater proportion of elimination, the order is $C > A > B$.

9.21 (a) The tertiary halide will undergo the S_N1-E1 process to give substitution products that result from attack of both water and ethanol, respectively, on the carbocation intermediate; and this carbocation can lose a β-proton to solvent (water or ethanol, abbreviated ROH below) to form two alkenes. Of the two alkenes, compound *D* will be formed in greater amount because it has a greater number of alkyl branches at the double bond.

$CH_3C(CH_3)(Br)CH_2CH_3 \xrightarrow{\text{ionization}} CH_3\overset{+}{C}(CH_3)CH_2CH_3 \; Br^-$

H_2O → $CH_3C(CH_3)(OH)CH_2CH_3$ (A); C_2H_5OH → $CH_3C(CH_3)(OC_2H_5)CH_2CH_3$ (B); ROH, alkene formation → $H_2C{=}C(CH_3)CH_2CH_3$ (C) and $(CH_3)_2C{=}CHCH_3 + R\overset{+}{O}H_2\,Br^-$ (D)

(b) Products *A* and *B* result from attack of ethanol and water, respectively, on the carbocation intermediate that is formed by ionization of the alkyl halide starting material. Products *C* and *D* result from attack of ethanol and water, respectively, on the rearranged carbocation intermediate. (Compound *C* is the product shown in Eq. 9.57.)

$H_3C{-}C(CH_3)_2{-}CH(CH_3)Cl \xrightarrow[-Cl^-]{\text{ionization}} H_3C{-}C(CH_3)_2{-}\overset{+}{C}H(CH_3)$ (unrearranged carbocation) $\xrightarrow{\text{rearrangement}} H_3C{-}\overset{+}{C}(CH_3){-}CH(CH_3){-}CH_3$ (rearranged carbocation)

Unrearranged carbocation: C_2H_5OH → $H_3C{-}C(CH_3)_2{-}CH(CH_3){-}OC_2H_5$ (A); H_2O → $H_3C{-}C(CH_3)_2{-}CH(CH_3){-}OH$ (B)

Rearranged carbocation: C_2H_5OH → $H_3C{-}C(CH_3)(OC_2H_5){-}CH(CH_3){-}CH_3$ (C); H_2O → $H_3C{-}C(CH_3)(OH){-}CH(CH_3){-}CH_3$ (D)

Alkenes can form by loss of a β-proton from either of the two carbocation intermediates. (The solvent, indicated below by ROH, acts as the base that removes the proton.)

$H_3C{-}C(CH_3)_2{-}\overset{+}{C}H(CH_3)$ (unrearranged carbocation) $\xrightarrow{\text{ROH}} H_3C{-}C(CH_3)_2{-}CH{=}CH_2 + R\overset{+}{O}H_2$ (E)

$H_3C{-}\overset{+}{C}(CH_3){-}CH(CH_3){-}CH_3$ (rearranged carbocation) $\xrightarrow{\text{ROH}} (H_3C)_2C{=}C(CH_3)_2$ (F) $+ H_2C{=}C(CH_3){-}CH(CH_3){-}CH_3$ (G) $+ R\overset{+}{O}H_2$

Alkene *F* would be formed in greatest amount because it has the greatest amount of branching at the double bond. Alkene *G* would be formed in smaller amount; alkene *E* would be formed in least amount for two reasons: it comes from a carbocation that can rearrange, and it has only one branch at the double bond.

9.22 The mechanism for the formation of the rearrangement product in Eq. 9.57:

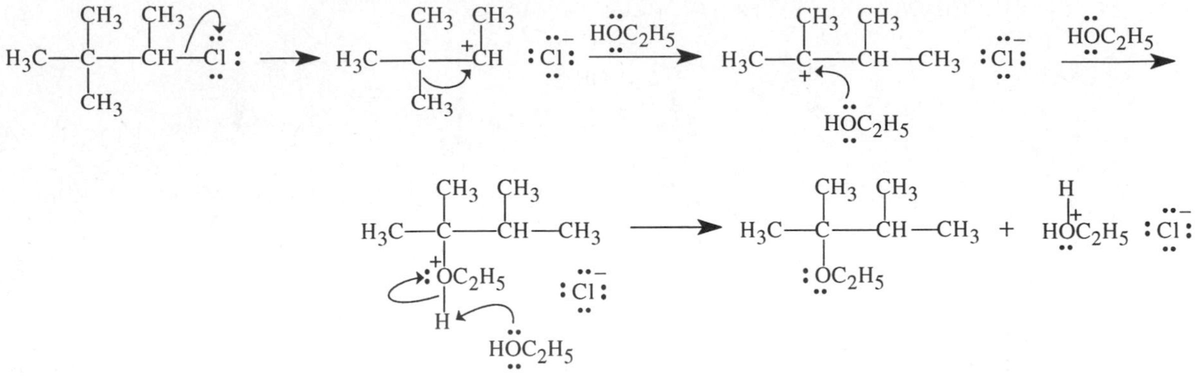

$$H_3C{-}C(CH_3)_2{-}CH(CH_3){-}\ddot{Cl}: \longrightarrow H_3C{-}C(CH_3)_2{-}\overset{+}{C}H(CH_3)\ \ :\ddot{Cl}:^- \xrightarrow{H\ddot{O}C_2H_5} H_3C{-}\overset{+}{C}(CH_3){-}CH(CH_3){-}CH_3\ \ :\ddot{Cl}:^-\ \ H\ddot{O}C_2H_5 \xrightarrow{H\ddot{O}C_2H_5}$$

$$H_3C{-}C(CH_3)(\overset{+}{\ddot{O}}(H)C_2H_5){-}CH(CH_3){-}CH_3\ \ :\ddot{Cl}:^-\ \ H\ddot{O}C_2H_5 \longrightarrow H_3C{-}C(CH_3)(:\ddot{O}C_2H_5){-}CH(CH_3){-}CH_3 + H\overset{+}{\ddot{O}}(H)C_2H_5\ :\ddot{Cl}:^-$$

9.23 In terms of Fig. 9.10 on text p. 390, if the carbocation is considerably more stable, then it should last long enough to form a fully solvated ion. Such an ion is truly achiral; hence, attack of solvent (water in Fig. 9.10) from either face is equally likely, and complete racemization should be observed.

9.24 (a) Entry 3 of Table 9.6 on text p. 392 covers this case; the major product is the substitution product 1-methoxybutane , which is formed by an S_N2 mechanism:

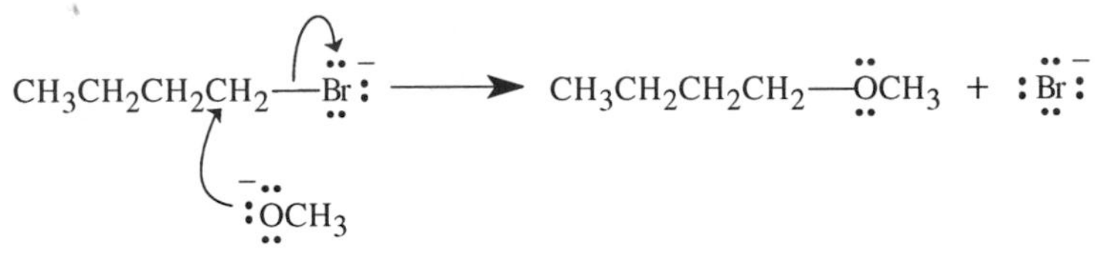

$$CH_3CH_2CH_2CH_2{-}\ddot{Br}: \ \ \ :\ddot{O}CH_3^- \longrightarrow CH_3CH_2CH_2CH_2{-}\ddot{O}CH_3 + :\ddot{Br}:^-$$

A small amount of the alkene 1-butene will also be formed:

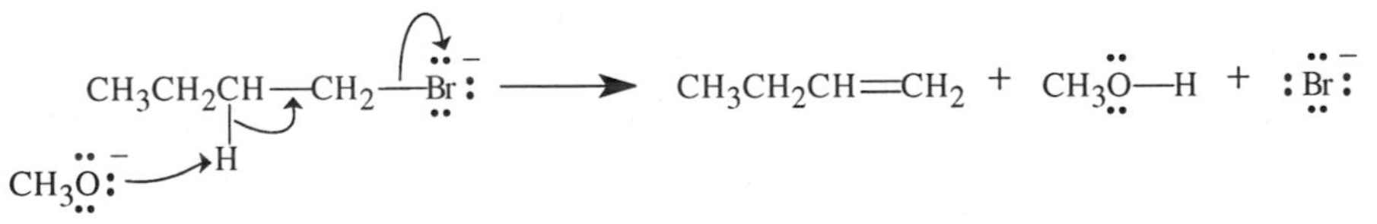

$$CH_3CH_2CH(H){-}CH_2{-}\ddot{Br}:\ \ \ CH_3\ddot{O}:^- \longrightarrow CH_3CH_2CH{=}CH_2 + CH_3\ddot{O}{-}H + :\ddot{Br}:^-$$

(b) 2-Bromobutane is a secondary alkyl halide, and potassium *tert*-butoxide is a strong, highly branched base. Entry 7 of Table 9.6 covers this case. Thus, the E2 reaction is the major process that occurs. Two possible alkenes, 1-butene and 2-butene, can form. Either or both can be considered as correct answers. However, it happens that attack of a *highly branched* base such as *tert*-butoxide occurs on a proton at the *least branched* carbon because (a) attack at a hydrogen on a more branched carbon is retarded by steric hindrance, and (b) a carbon with fewer branches bears more hydrogens that can react. Hence, 1-butene is the major product formed.

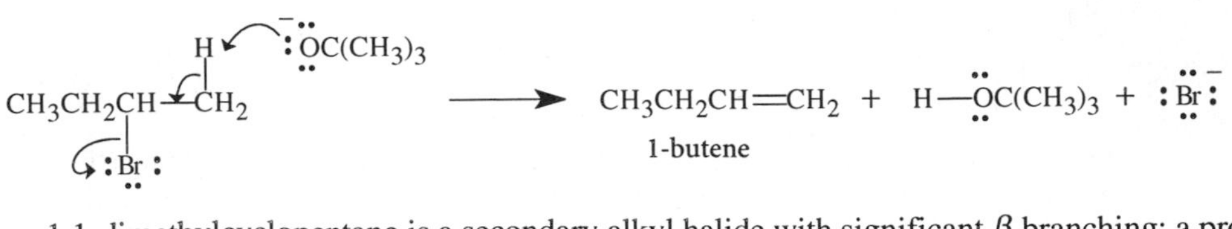

$$CH_3CH_2CH(:\ddot{Br}:){-}CH_2(H)\ \ \ :\ddot{O}C(CH_3)_3^- \longrightarrow \underset{\text{1-butene}}{CH_3CH_2CH{=}CH_2} + H{-}\ddot{O}C(CH_3)_3 + :\ddot{Br}:^-$$

(c) 2-Bromo-1,1-dimethylcyclopentane is a secondary alkyl halide with significant β-branching; a protic solvent is used without a strong base. Entry 9 in Table 9.6 covers this case. Rearrangement from the initially formed secondary carbocation *a* to the tertiary carbocation *b* is likely to occur.

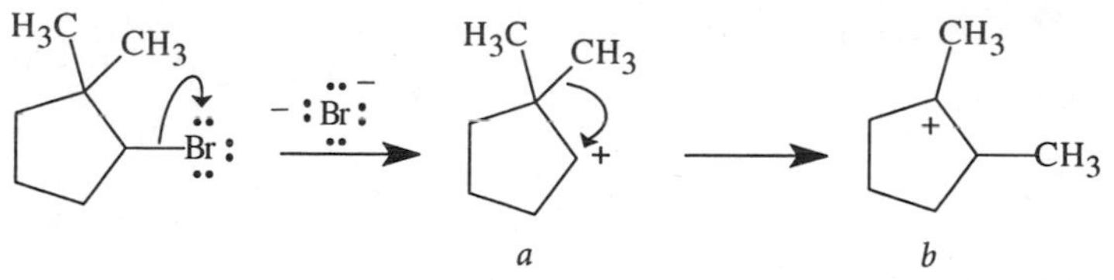

(Note that rearrangement of the carbocation *a* to another tertiary carbocation *c* is also possible:

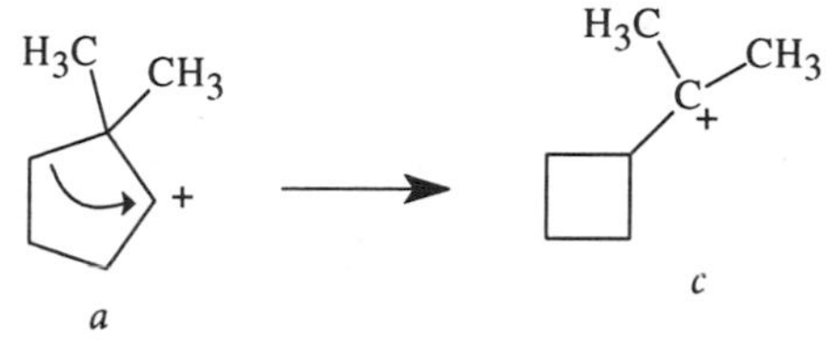

This rearrangement will not occur because it produces a carbocation with significantly greater ring strain.)

Each carbocation can react to give S_N1 substitution and E1 elimination products. Carbocation *a* gives the substitution product *A* and the elimination product *B*.

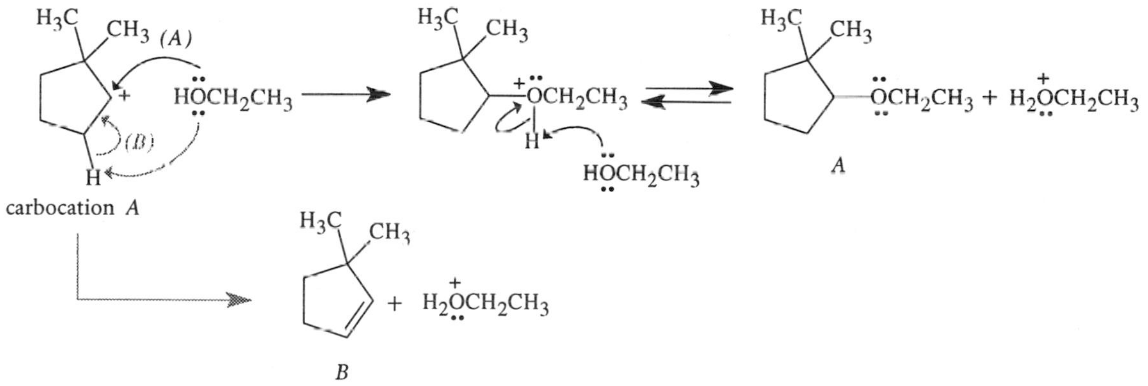

Carbocation *b* gives the substitution product *C* and the elimination products *D–F*. (The curved-arrow mechanism for formation of *C* is analogous to the mechanism shown above for the formation of *A*; and the curved-arrow mechanisms for the formation of alkenes *D–F* (by the E1 pathway) are analogous to the mechanism shown above for the formation of *B*.)

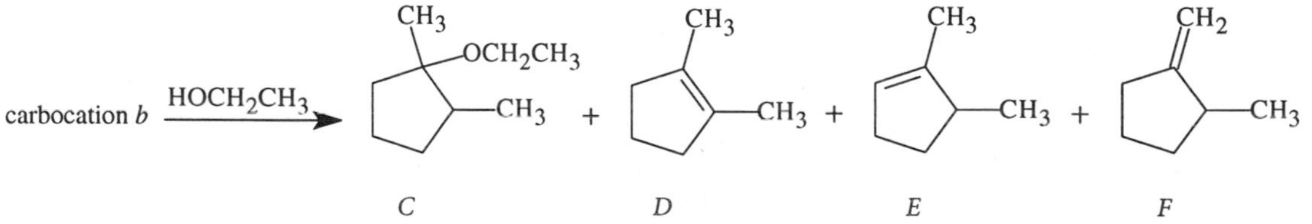

Compounds *C*, *D*, and *E* are expected to be the major products of the reaction.

(d) Bromocyclohexane is a secondary alkyl halide; methanol is a polar, protic solvent; and there is no strong base present. Entry 9 of Table 9.6 covers this situation; both S_N1 product *A* and E1 product *B* are formed.

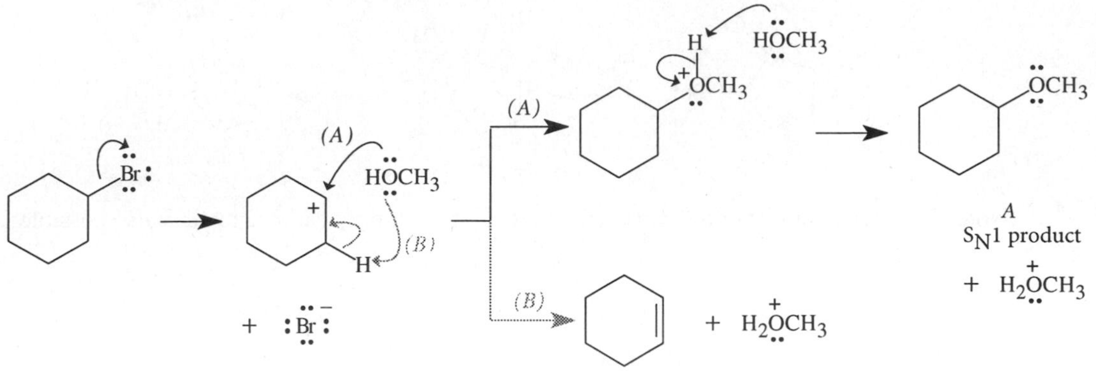

(A) (A) $HOCH_3$ (B) H OCH_3 $HOCH_3$ OCH_3

+ $:Br:^-$ (B) + $H_2\overset{+}{O}CH_3$

A
S_N1 product
+ $H_2\overset{+}{O}CH_3$

B
E1 product

9.25 (a)

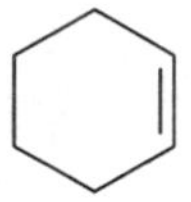

+ $HCBr_3$
bromoform

cyclohexene

(b)

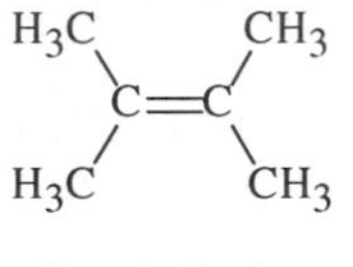

H_3C CH_3 C=C H_3C CH_3 + $PhCH_2Br$
benzyl bromide

2,3-dimethyl-2-butene

9.26 (a)

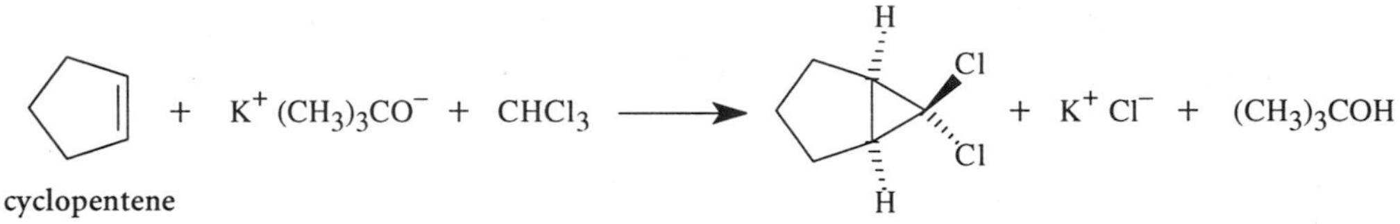

+ $K^+ (CH_3)_3CO^-$ + $CHCl_3$ ⟶ (H, Cl, Cl, H) + $K^+ Cl^-$ + $(CH_3)_3COH$

cyclopentene

This product is the only one formed, given that the reaction is a *syn*-addition.

(b)

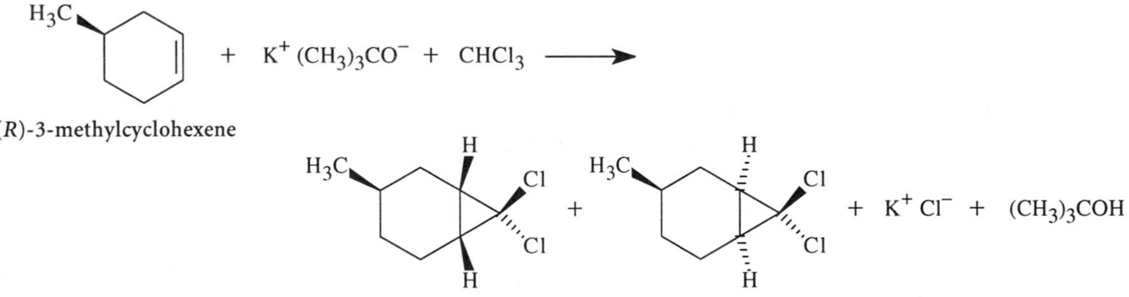

H_3C + $K^+ (CH_3)_3CO^-$ + $CHCl_3$ ⟶

(*R*)-3-methylcyclohexene

H_3C (H, Cl, Cl, H) + H_3C (H, Cl, Cl, H) + $K^+ Cl^-$ + $(CH_3)_3COH$

A *B*

Compounds *A* and *B* are diastereomers and are formed in different amounts. (See Sec. 7.8B.) If the attack on the carbene occurs at the face of the ring opposite to the methyl group, we might expect compound *A* to be the major one.

9.27 (a)

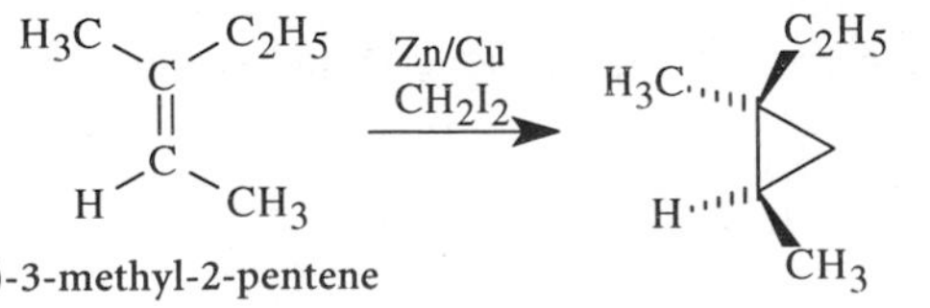

H_3C C_2H_5 C=C H CH_3 $\xrightarrow{Zn/Cu,\ CH_2I_2}$ H_3C C_2H_5 H CH_3

(*Z*)-3-methyl-2-pentene (racemate)

(b)

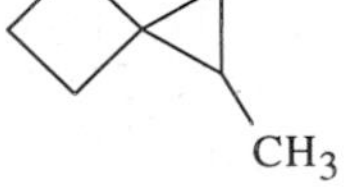

CH_3

(racemate)

9.28 (a)

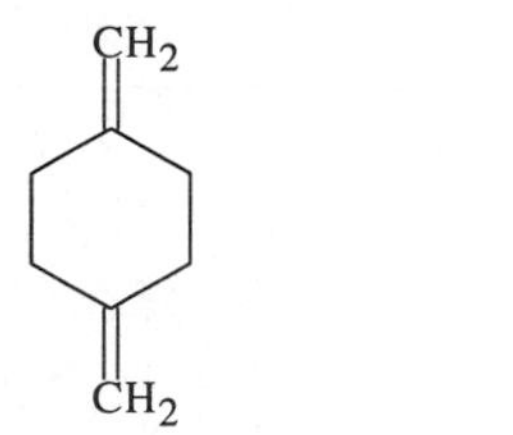

(b)

CH3

In the case of (b), attack of the reagent from the face of the double bond *opposite* to the methyl group gives the stereoisomer shown in the problem. In principle, a second stereoisomer could be formed by attack from the same side as the methyl group, but this is less likely because of the steric effect.

Solutions to Additional Problems

9.29 The first step in any problem that requires structures is to draw the structures:

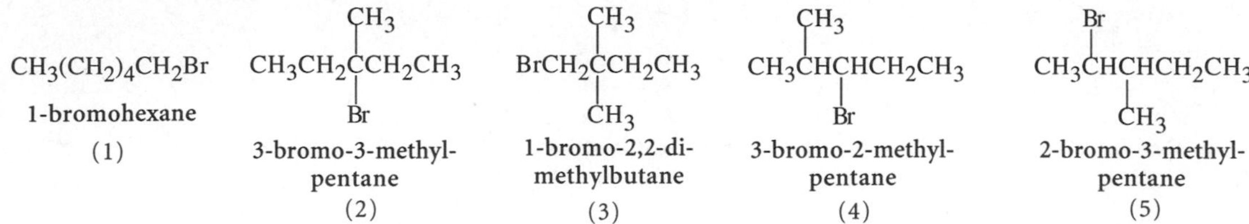

(a) Compounds (4) and (5) can exist as enantiomers.
(b) Compound (5) can exist as diastereomers because it has two asymmetric carbons.
(c) Compound (1) gives the fastest S_N2 reaction with sodium methoxide because it is the only primary alkyl halide.
(d) Compound (3) is least reactive to sodium methoxide in methanol because it cannot undergo a β-elimination—it has no β-hydrogens—and the three β-branches make it virtually unreactive in the S_N2 reaction, much like neopentyl bromide.
(e) Compound (1) can give only one alkene in the E2 reaction.
(f) Compound (2) will give an E2 but no S_N2 reaction with sodium methoxide in methanol.
(g) Compounds (4) and (5) undergo the S_N1 reaction with rearrangement. If compound (3) is forced to react, it will also undergo the S_N1 reaction with rearrangement, but under ordinary conditions it will not react.
(h) Compound (2) will give the fastest S_N1 reaction because it is the only tertiary alkyl halide.

9.30 The products of the reactions of isopentyl bromide, $(CH_3)_2CHCH_2CH_2Br$: (When the solvent is a by-product it is not shown.)

(a) $(CH_3)_2CHCH_2CH_2I$ + $K^+ Br^-$

(b) $(CH_3)_2CHCH_2CH_2OH$ + $(CH_3)_2CHCH_2CH_2OC_2H_5$ + $(CH_3)_2CHCH{=}CH_2$ (small amount) + $K^+ Br^-$

(c) $(CH_3)_2CHCH{=}CH_2$ + $K^+ Br^-$

(d) $(CH_3)_2CHCHCH_3$ (Br on C-3) + $(CH_3)_2CCH_2CH_3$ (Br on C-2) (from rearrangement)

(e) $(CH_3)_2CHCH_2CH_2F$ + $Cs^+ Br^-$

(f) $(CH_3)_2CH$–cyclopropane with Cl, Cl + $(CH_3)_3COH$

(g) $(CH_3)_2CH$–cyclopropane + ZnI_2

(h) $(CH_3)_2CHCH_2CH_3$ + $Li^+ Br^-$ + $Li^+ \, ^-OCH_2CH_3$

(i) $(CH_3)_2CHCH_2CH_2OCH_3$ + $(CH_3)_2CHCH{=}CH_2$ (small amount) + $Na^+ Br^-$

(j) $(CH_3)_2CHCH_2CH_2{-}D$ + Mg^{2+} + ^-OD + Br^-

9.31 The products of the reactions of 2-bromo-2-methylhexane:

$CH_3C(CH_2)_3CH_3$ (CH3 above, Br below) 2-bromo-2-methylhexane

(a)

$CH_3C(CH_3)(OH)(CH_2)_3CH_3$ + $CH_3C(CH_3)(OC_2H_5)(CH_2)_3CH_3$ + $CH_3C(CH_3){=}CHCH_2CH_2CH_3$ (predominant alkene formed) + $H_2C{=}C(CH_3)(CH_2)_3CH_3$ + H_3O^+ + $C_2H_5\overset{+}{O}H_2$ + Br^-

(b)

$CH_3C(CH_3){=}CHCH_2CH_2CH_3$ + $H_2C{=}C(CH_3)(CH_2)_3CH_3$ + $Na^+ Br^-$

(c)

$CH_3C(CH_3)(OH)(CH_2)_3CH_3$ + $CH_3C(CH_3)(I)(CH_2)_3CH_3$ + $K^+ Br^-$ + the alkenes in part (b) + $H_3O^+ Br^-$

In part (c), the relative amounts of alkyl iodide depend on the concentration of iodide ion in the solution. The iodide will solvolyze to the alcohol (assuming it remains soluble in the aqueous acetone).

(d)

$CH_3CH(CH_3)CH(Br)CH_2CH_2CH_3$ + $BrCH_2CH(CH_3)(CH_2)_3CH_3$

(e)

$CH_3C(CH_3)(OH)CH_2CH_2CH_2CH_3$

+ the by-products of oxymercuration-reduction shown in Eq. 5.13, text p. 167

(f)

$CH_3CH(CH_3)CH(OH)CH_2CH_2CH_3$ + $HOCH_2CH(CH_3)(CH_2)_3CH_3$

+ the by-products of hydroboration-oxidation shown in Eq. 5.21, text p. 170

9.32 The first thing to do, of course, is to draw out the structures.

CH_3Br	$CH_3CH(Br)CH_2CH_3$	$CH_3CH_2C(CH_2Br)(CH_3)CH_2CH_3$	$BrCH_2CH_2CH_2CH_2CH_3$	$BrCH_2CH(CH_3)CH_2CH_3$
A	*B*	*C*	*D*	*E*

The order of increasing S_N2 reaction with KI in acetone is $A < B < C < E < D$. Tertiary alkyl halides such as *A* react slowly, if at all, in S_N2 reactions; secondary alkyl halides such as *B* react more rapidly; primary alkyl halides with two β-branches (*C*) react even more rapidly; and unbranched primary alkyl halides (*D* and *E*) react most rapidly. Of the two primary alkyl halides shown, the one with the better leaving group (*D*, with the bromide leaving group) reacts more rapidly.

9.33 The order of increasing S_N2 reaction rates is $C < B < E < D < A$. Alkyl halides with three β-branches (*C*) are virtually unreactive in S_N2 reactions. Secondary alkyl halides with no β-branches (*B*) react more slowly than primary alkyl halides with two β-branches (*E*), and the latter react more slowly than unbranched primary alkyl halides (*D*). Methyl halides (*A*) react most rapidly.

9.34 To deduce the structure of the nucleophile, disconnect the appropriate atom from the ethyl group, add a pair of electrons and calculate the formal charge, if any. Of course, an appropriate counterion (such as Na^+ or K^+) is also present.

(a) $:\overset{-}{C}N$ (b) $:\ddot{\overset{-}{O}}CH_2CH_2CH_3$ (c) $:\ddot{\overset{-}{O}}CH_2CH_2CH_2OCH_3$ (d) 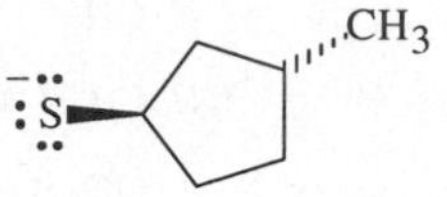 $:\ddot{\overset{-}{S}}$–cyclopentyl–CH_3 (e) $(CH_3)_3N:$

9.35 (a) (*R*)-2-bromopentane reacts to give an S_N2 product with inversion along with some alkenes formed by the E2 mechanism.

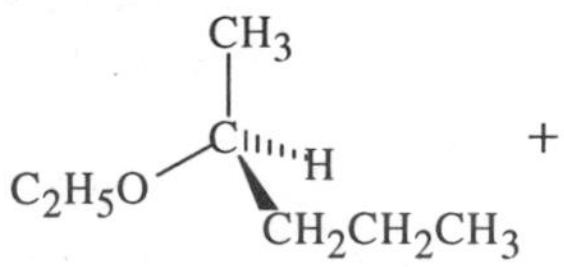

$$C_2H_5O\text{–}C(CH_3)(H)(CH_2CH_2CH_3) + CH_3CH{=}CHCH_2CH_3 + H_2C{=}CHCH_2CH_2CH_3 + Na^+\,Br^-$$

(*S*)-2-ethoxypentane; cis and trans

(b) The substitution reaction occurs by an S_N2 mechanism with inversion of configuration. (Little or no alkene is formed because the alkyl halide is primary.)

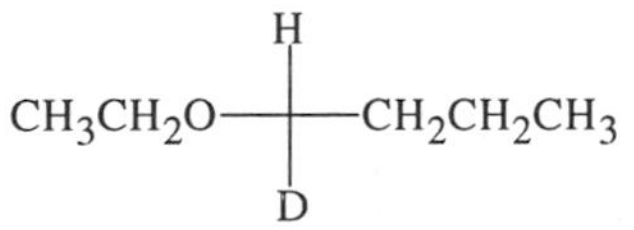

$$CH_3CH_2O\text{–}C(H)(D)\text{–}CH_2CH_2CH_3 + Na^+\,Br^-$$

9.36 (a) This compound can react to give a mixture of alkenes, because it contains more than one type of β-hydrogen:

$$CH_3CH{=}C(CH_3)_2 + CH_3CH_2C(CH_3){=}CH_2$$

(b) This compound can react to give only one alkene, cyclohexene.

(c) This compound can react to give a mixture of alkenes:

$$C_6H_{11}\text{–}CH{=}C(CH_3)_2 + C_6H_{11}\text{–}CH_2C(CH_3){=}CH_2$$

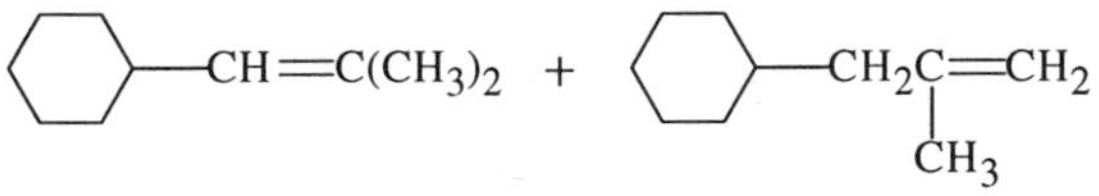

(d) Because there is only one β-hydrogen, this compound can react to give only one alkene:

$$CH_3CH_2C(CH_3){=}CH_2$$

9.37 Notso has attempted the S_N2 reaction of a *tertiary* alkyl halide with a strong base. As you no doubt realize, and Notso would have realized, had he consulted a superior organic chemistry text such as yours, that these are precisely the conditions that promote the E2 reaction. Consequently, Notso formed an alkene. Ima, on the other hand, allowed a strong base to react with methyl iodide, which cannot undergo elimination; the only alternative is the desired S_N2 reaction.

9.38 In base, some of the chlorines are eliminated as chloride ion in E2 reactions. One such reaction is as follows; can you find another? (S_N2 reactions probably do not occur; why?)

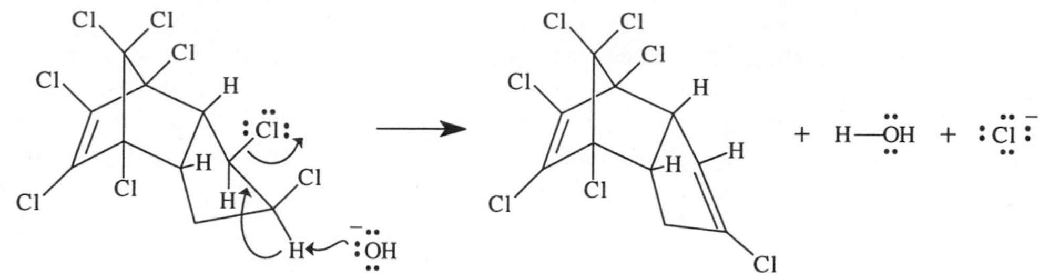

Cl Cl Cl Cl H :Cl: H Cl Cl H Cl H :OH → Cl Cl Cl Cl H H H Cl Cl Cl + H—OH + :Cl:$^-$

9.39 (a) Draw either enantiomer of the molecule in two staggered conformations in which a β-hydrogen or a β-deuterium, respectively, is anti to the bromine. These conformations reveal that, when the deuterium is anti to the bromine, the methyl groups are on opposite sides of the molecule; hence, elimination of D and Br gives a *trans*-alkene that does not contain deuterium. When the hydrogen is anti to the bromine, the methyl groups are on the same side of the molecule; hence, elimination of H and Br gives a (*Z*)-alkene that contains deuterium.

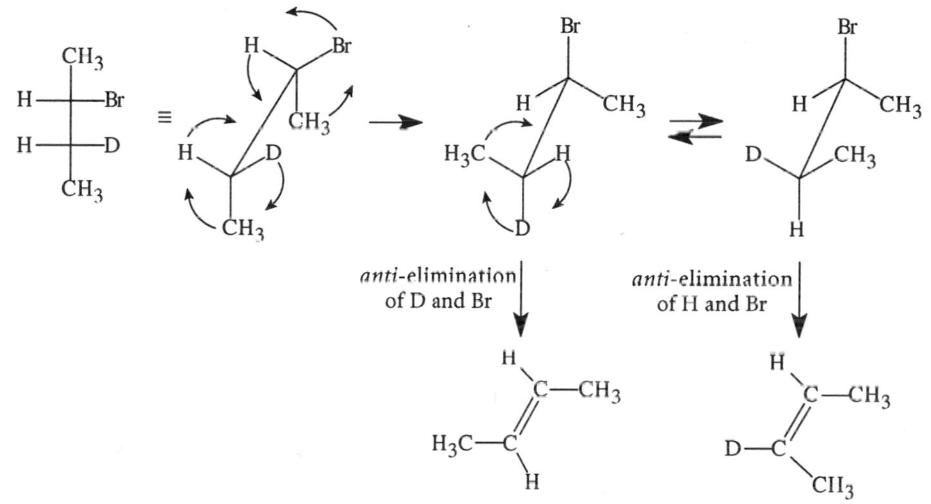

anti-elimination of D and Br

anti-elimination of H and Br

(b) The reasoning is much the same as in part (a). An *anti*-elimination of H and Br requires the methyl groups to be on opposite sides of the molecule; and an *anti*-elimination of D and Br requires the methyl groups to be on the same side of the molecule.

anti-elimination of D and Br

anti-elimination of H and Br

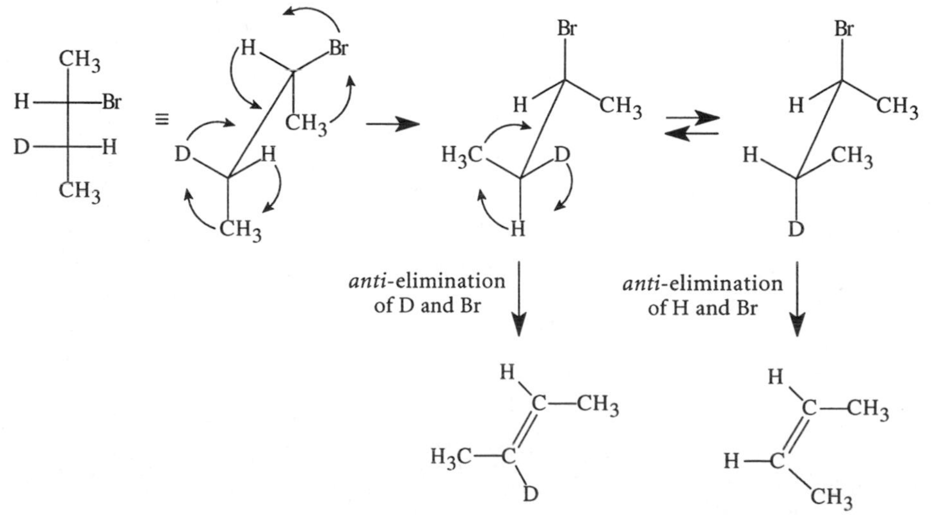

9.40 In the transition state of the S_N2 reaction, the carbon at which substitution occurs is sp^2-hybridized; consequently, the ideal bond angles at this carbon are 120°. (See Fig. 9.2 on text p. 364.) This requirement for a 120° bond angle means that as cyclopropyl iodide is converted into the S_N2 transition state, the bond angle within the cyclopropane ring should increase. The increasing angle causes additional angle strain in the cyclopropane ring that raises the energy of the transition state. Consequently, S_N2 reactions of cyclopropyl halides are unusually slow.

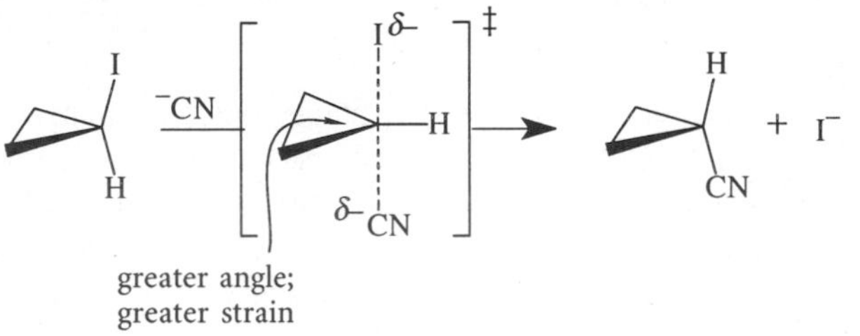

greater angle;
greater strain

Such rate-retarding angle strain is present in neither compound (b), isopropyl iodide, nor its S_N2 transition state. Consequently, the S_N2 reactions of isopropyl iodide are faster than those of cyclopropyl iodide.

9.41 If this compound were to undergo solvolysis, it would have to form carbocation *A*:

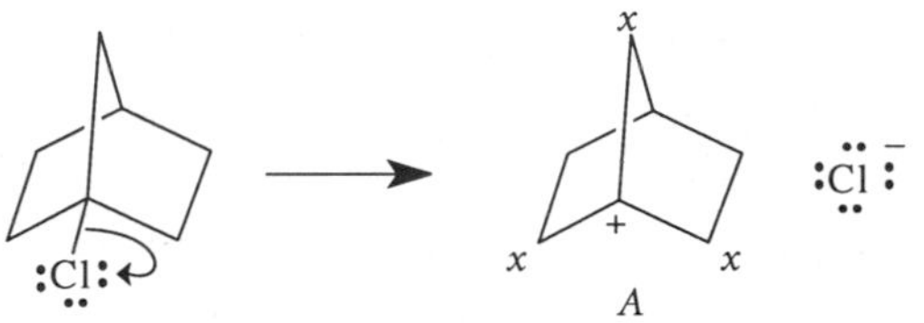

A

As a model will verify, the three carbons marked "*x*" cannot become coplanar because of the constraints of the bicyclic ring system. Yet sp^2 hybridization requires trigonal-planar geometry. Remember: *hybridization and geometry are connected.* If a molecule can't achieve the geometry for a given hybridization, then that hybridization will not occur. Because sp^2 hybridization is the lowest-energy hybridization for a carbocation, the inability to achieve this hybridization raises the energy of the carbocation.

But that's not all. Solvation of the carbocation can only occur from one side, because the back side of the electron-deficient carbon is blocked by part of the ring system. The absence of effective solvation, then, also raises the energy of this carbocation. Any solvolysis reaction involving such an unstable carbocation is slow.

9.42 Potassium fluoride, an ionic compound, is not soluble in hydrocarbons such as benzene. Consequently, it will not react in such solvents because its dissolved concentration is essentially zero. When the crown ether is added, the crown ether forms a benzene-soluble complex with the potassium ion, which, by electrical neutrality, includes an accompanying "naked" (that is, essentially unsolvated) fluoride ion. This soluble but unsolvated fluoride ion is a very good nucleophile, and reacts with benzyl bromide to yield benzyl fluoride, $PhCH_2F$, plus "naked" bromide ion, which then exchanges with insoluble potassium fluoride to give more dissolved fluoride and solid KBr. The newly dissolved fluoride can react with more benzyl bromide. The cycle is repeated until one of the reactants is exhausted. To summarize:

crown ether (soluble) + $K^+ F^-$ (insoluble) (excess) ⇌ crown ether • K^+ F^- (soluble) (soluble) + $K^+ F^-$ (insoluble) $\xrightarrow{PhCH_2Br}$

$PhCH_2F$ + crown ether • K^+ Br^- (soluble) (soluble) + $K^+ F^-$ (insoluble) ⇌ crown ether • K^+ F^- (soluble) (soluble) + $K^+ Br^-$ (insoluble)

This scheme shows that complexation of the cation by the crown ether is essential for the reaction to occur. Because [18]-crown-6 does not bind the smaller lithium cation—it is selective for the larger potassium cation—lithium fluoride is not solubilized by the crown ether, and is therefore unreactive whether the crown ether is present or not.

9.43 (a) The S_N1 solvolysis products result from attack of the respective solvent molecules on the carbocation intermediate:

$$(CH_3)_3C{-}Cl \rightleftharpoons (CH_3)_3C^+ \; Cl^-$$

$$(CH_3)_3C^+ + H_3C{-}C(=O){-}OH \rightarrow H_3C{-}C(=O){-}O{-}C(CH_3)_3 + HCl \quad \textit{tert-butyl acetate}$$

$$(CH_3)_3C^+ + H{-}C(=O){-}OH \rightarrow H{-}C(=O){-}O{-}C(CH_3)_3 + HCl \quad \textit{tert-butyl formate}$$

(b) The reaction is faster in formic acid, because its higher dielectric constant promotes the separation of the chloride ion leaving group from the carbocation. (This effect is discussed in Sec. 9.6C on text p. 387.)

9.44 (a) In a hydrogen-bonding solvent such as ethanol, reactions of alkyl halides with strong-base nucleophiles are retarded by hydrogen bonding of the solvent to the nucleophile. Hence, within a column of the periodic table, the element effect on nucleophilicity is that weaker bases make the best nucleophiles. Hence, the thiolate ion (which is a weaker base than the alkoxide ion) will out-compete the alkoxide ion for a limited amount of alkyl halide. As a result, the product of the reaction is $CH_3CH_2S{-}CH_3$.

(b) In polar aprotic solvents, which cannot form hydrogen bonds to strong bases, stronger bases within a column of the periodic table are the best nucleophiles. In this case, the alkoxide ion out-competes the thiolate for the alkyl halide, and the product is thus $CH_3CH_2O{-}CH_3$.

9.45 (a) The thiosulfate dianion contains two types of nucleophilic atom: the anionic oxygen and the anionic sulfur. Hence, alkylation could occur at the oxygen or the sulfur:

$$Na^+ \; {}^-S{-}S^{+2}(O^-)_2{-}O{-}CH_3 \quad + \quad H_3C{-}S{-}S^{+2}(O^-)_2{-}O^- \; Na^+$$

A
product of *O*-alkylation

B
product of *S*-alkylation

(b) In a hydrogen-bonding solvent such as methanol, the more weakly basic atom is the better nucleophile. (The principle is the same as in the previous problem.) Hence, product *B* is the major one observed. (What solvent change could you make to obtain more of product *A*?)

9.46 (a) The principle to apply in this case is that the right side of the equilibrium is favored by the solvent that better separates and solvates ions. Because ethanol is the more polar solvent—it has the higher dielectric constant—it better separates ions. Because it is a protic, donor solvent, it better solvates ions; consequently, the equilibrium lies farther to the right in ethanol. Furthermore, ethanol's ability to donate hydrogen bonds allows it to provide better solvation for the bromide anion than does diethyl ether.

(b) In this case, the dielectric constants of the two solvents are the same. Consequently, the ability of the solvent to solvate the ions determines the relative position of the equilibrium. Because dimethylacetamide is *aprotic,* it does not solvate anions as well as the aqueous methanol solvent, which has a hydrogen-bonding solvation mechanism available. Consequently, the equilibrium lies farther to the right in aqueous methanol.

9.47 The reaction between methyl iodide and sodium ethoxide is an S_N2 process that has the following rate law:

$$\text{rate} = k[CH_3I][CH_3CH_2O^-]$$

This means that the rate of the reaction depends on the concentration of the nucleophile. In the first case, the nucleophile concentration changes during the reaction from 0.1 *M* to 0 *M*; in the second case, the nucleophile concentration changes from 0.5 *M* to 0.4 *M*. Clearly, in the second case, the reaction is faster because the concentration of the nucleophile is higher at all times during the reaction. However, in either case, there is sufficient nucleophile present to react completely with the alkyl halide. If we wait long enough, the yield of the reaction will be the same in either case.

9.48 In methanol, the iodide ion is an excellent nucleophile but, because of its weak basicity, also a good leaving group. Hence, the iodide ion reacts rapidly with methyl bromide to give methyl iodide. The methyl iodide then reacts more slowly with the methanol solvent in a slow S_N2 solvolysis reaction.

$$:\ddot{I}:^- \quad H_3C{-}Br \xrightarrow{\text{faster}} :\ddot{I}{-}CH_3 + :\ddot{Br}:^- \xrightarrow[\text{slower}]{H{-}\ddot{O}CH_3} :\ddot{I}:^- + H_3C{-}\overset{+}{O}(H)CH_3 \rightleftarrows H_3C{-}\ddot{O}{-}CH_3 + H_2\overset{+}{O}CH_3$$

(with $H\ddot{O}CH_3$ removing the proton)

iodide ion disappears rapidly

iodide ion reappears slowly

9.49 Because it is a tertiary alkyl halide without β-hydrogens, trityl chloride can react only by the S_N1 mechanism. A characteristic feature of this mechanism is that the first step is rate-limiting, whereas attack of the nucleophile(s) occurs on the carbocation intermediate in the second step. Hence, the rate of reaction of trityl chloride is *independent of the nucleophile concentration,* and therefore independent of the nucleophile that reacts with the carbocation.

$$\underset{\text{trityl chloride}}{Ph_3C{-}Cl} \xrightarrow{\text{rate-limiting step}} \underset{\text{trityl cation}}{Ph_3C^+ \ Cl^-}$$

However, the product-determining steps for the various reactions are different. In the product-determining step, a nucleophile attacks the carbocation. The principle to apply is that the better nucleophile reacts more rapidly with the carbocation intermediate; hence, the major product is derived from the better nucleophile.

Thus, in reaction (1), the only nucleophile available is water. After attack of water on the carbocation to give $Ph_3C{-}\overset{+}{O}H_2$, a proton is lost to solvent to give trityl alcohol as the solvolysis product. In reaction (2), both water and azide ion are in competition for the trityl cation. The basicities of water and azide are determined by the pK_a values of their respective conjugate acids. The conjugate acid of water, H_3O^+, has a pK_a value of –1.7. The conjugate acid of azide ion, HN_3, has a pK_a value of 4.7. Consequently, azide is the more basic nucleophile, and is evidently the better nucleophile. When, in reaction (2), sodium azide is added to the reaction mixture, trityl chloride ionizes at the same rate, but the carbocation intermediate reacts preferentially with azide ion.

When hydroxide ion is added in reaction (3), the rate of ionization of trityl chloride again remains unchanged; however, because both hydroxide ion and azide are present, both nucleophiles react with the carbocation to give a mixture of trityl alcohol and trityl azide.

9.50 (a) If retention is the stereochemical scenario, each act of substitution has no affect on the optical activity, and $k°/k_S = 0$. If inversion is the stereochemical scenario, each act of substitution inverts the stereochemical configuration of one alkyl halide molecule, which gives it the opposite sign but same magnitude of optical rotation. In this case, *two* molecules' worth of optical activity become zero. Therefore, $k°/k_S = 2$. This situation can be envisioned with the following "cartoon:"

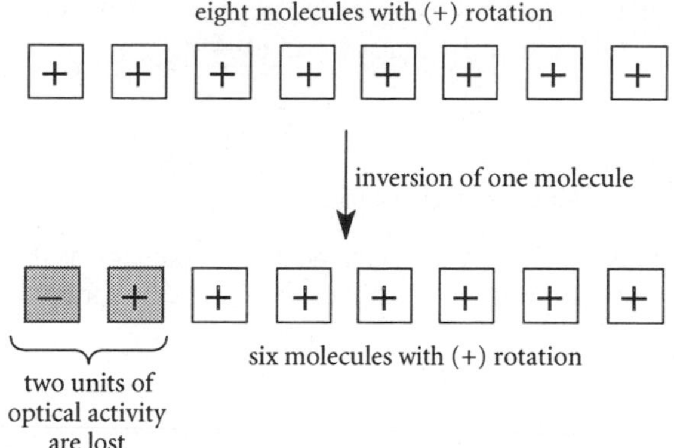

If equal amounts of retention and inversion occur, then for every event that results in the loss of two units of optical activity, we have another event that results in the loss of no optical activity; on average, $k°/k_S = 1$.

(b) The results show clearly that $k°/k_S = 2$, thus establishing inversion as the stereochemical outcome.

9.51 Remember that the key to working problems of this type is to find a known structure and work from it. (See point 3 in Study Guide Link 4.8, p. 88 of this manual.) Compound *B* is an alkene, because it decolorizes Br_2 in an inert solvent. The structure of the ozonolysis product *C* leads to the identity of *B*.

HO_2C — CO_2H ⇒ ≡

C *B*

The elemental analysis of *A* leads to the molecular formulay $C_8H_{13}Br$, which is equivalent to that of the alkene *B* plus the elements of HBr. Because the carbon skeleton must be bicyclic, there can be no additional unsaturation in *A*, a point confirmed by the fact that it does not add bromine. Evidently, compound *A* is an alkyl halide that undergoes a reaction with a strong base to give the alkene *B;* this reaction is then an E2 reaction. The alkyl halide *A,* and its E2 reaction, are as follows:

Br $\xrightarrow{K^+\ (CH_3)_3C—O^-}$

A *B*

9.52 The fact that protonolysis reactions of the corresponding Grignard reagents give the same hydrocarbon indicates that the two compounds have the same carbon skeleton. The conditions of ethanol and no added base are S_N1 conditions. Since compound *A* reacts rapidly to give a solution containing bromide ion, it must be an alkyl halide that readily undergoes an S_N1 reaction, and therefore it is probably a tertiary alkyl bromide. Because the two alkyl halides give the same ether, the product from compound *B* must be formed in a rearrangement. The only tertiary alkyl halide with the formula $C_5H_{11}Br$ is 2-bromo-2-methylbutane, and this is therefore compound *A:*

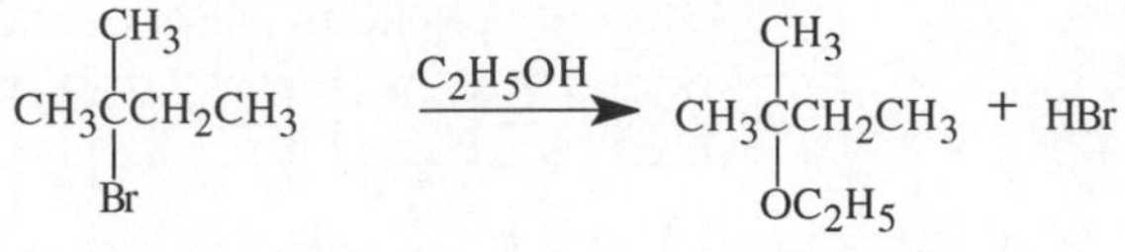

$$\underset{\substack{A\\ \text{2-bromo-2-methylbutane}}}{CH_3\overset{CH_3}{\underset{Br}{C}}CH_2CH_3} \xrightarrow{C_2H_5OH} \underset{C}{CH_3\overset{CH_3}{\underset{OC_2H_5}{C}}CH_2CH_3} + HBr$$

There are two possible alkyl halides with the same carbon skeleton as *A* that could rearrange in respective S_N1 reactions to give the same carbocation, and hence the same ether product, as *A*; these are labeled *B1* and *B2* below:

$$\underset{B1}{BrCH_2\underset{CH_3}{CH}CH_2CH_3} \qquad \underset{B2}{CH_3\overset{CH_3}{CH}\underset{Br}{CH}CH_3}$$

However, only *B2* can react in an E2 reaction with sodium ethoxide to give an alkene that furnishes acetone as one of its ozonolysis products:

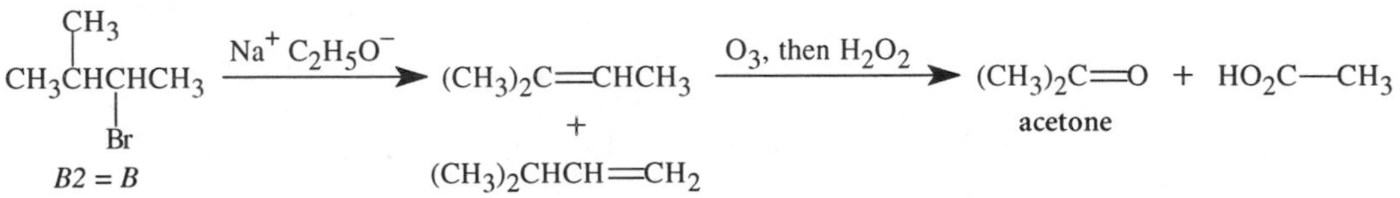

$$\underset{\substack{B2 = B\\ \text{2-bromo-3-methylbutane}}}{CH_3\overset{CH_3}{CH}\underset{Br}{CH}CH_3} \xrightarrow{Na^+\ C_2H_5O^-} (CH_3)_2C{=}CHCH_3 \xrightarrow{O_3,\text{ then } H_2O_2} \underset{\text{acetone}}{(CH_3)_2C{=}O} + HO_2C{-}CH_3$$

$$+\ (CH_3)_2CHCH{=}CH_2 \text{ (does not form acetone on ozonolysis)}$$

Consequently, compound *B2* is compound *B,* 2-bromo-3-methylbutane.

9.53 Compound *A* is an alkyl halide with two degrees of unsaturation; because one unsaturation is accounted for by the cyclohexane ring found in subsequent products, the other must be a double bond. Subsequent elimination reactions of *A* yield two dienes *B* and *C*. The catalytic hydrogenation of these two compounds establishes that compounds *A–C* and 1-isopropyl-4-methylcyclohexane have the same carbon skeleton. Because ozonolysis leaves the cyclohexane ring intact, the double bond must be external to the ring; and the identity of product *F* shows that both double bonds are in fact attached to the ring. There is only one possibility for compound *B*:

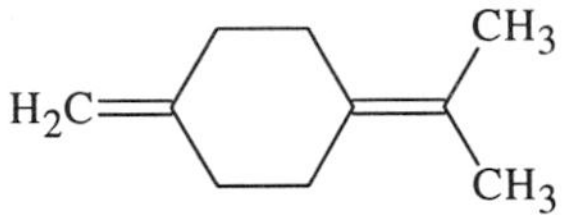

$H_2C{=}$(cyclohexane)${=}C(CH_3)_2$ compound *B*

The following possibilities for *A* can be enumerated:

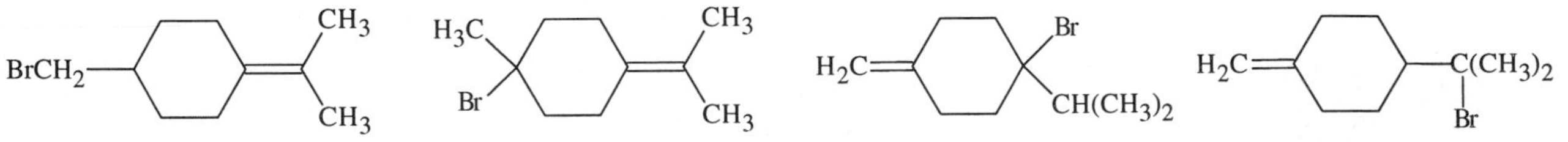

A1: $BrCH_2$–(cyclohexane)$=C(CH_3)_2$

A2: H_3C, Br–(cyclohexane)$=C(CH_3)_2$

A3: $H_2C=$(cyclohexane)–Br, $CH(CH_3)_2$

A4: $H_2C=$(cyclohexane)–$\underset{Br}{C}(CH_3)_2$

Notice that any other possibility for *A* is ruled out both by the structure of *B* and by the fact that compound *A* is achiral. Compound *A1* is ruled out by the fact that it can only give one diene in the E2 reaction; yet compound *A* gives two dienes, compounds *B* and *C*. That compound *A* undergoes S_N1 solvolysis rapidly also rules out compound *A1,* because it is a primary alkyl halide that would not react under S_N1 conditions. Because compound *A* undergoes ozonolysis to give acetone—and because structure *A1* was ruled out—*only structure A2 fits all the data.* The structure of alkene *B* is given above; the structure of alkene *C*, the other E2 reaction product of compound *A* = *A2*, is as follows:

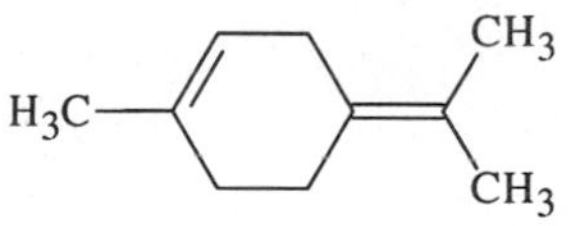

compound *C*

Now that we know compound *A*, we are in a position to derive the structures of compounds *D* and *E*. These are the bromine-addition products of compound *A*. The two products are the two possible diastereomers:

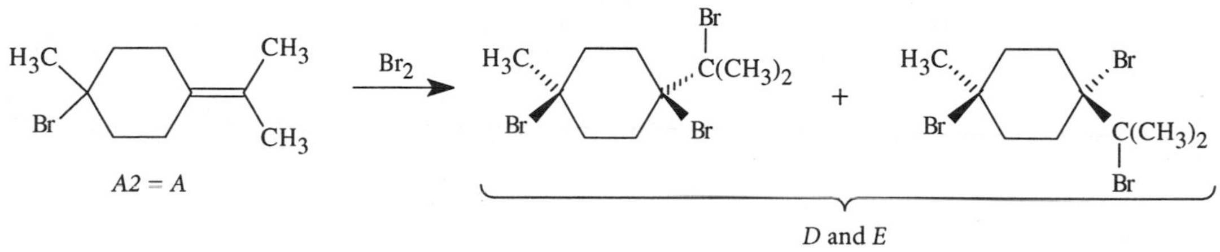

A2 = *A*

D and *E*

9.54 The important keys to solving this problem are (1) that the E2 reaction is fastest when the hydrogen and chlorine are anti; and (2) the substituted cyclohexanes can undergo the chair flip rapidly. In order for menthyl chloride to undergo *anti*-elimination, it must undergo the chair flip so that the chlorine is axial:

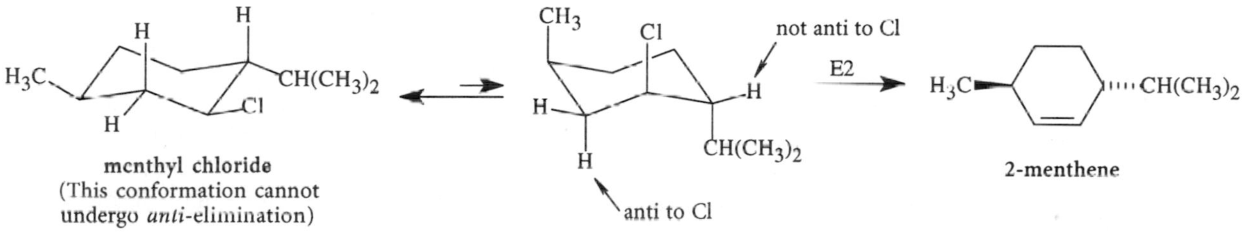

menthyl chloride
(This conformation cannot undergo *anti*-elimination)

not anti to Cl

anti to Cl

2-menthene

The only alkene that can be formed by *anti*-elimination is 2-menthene. In contrast, neomenthyl chloride has two β-hydrogens, H^a and H^b, that can be lost in *anti*-eliminations along with the chlorine to give 2-menthene and 3-menthene, respectively:

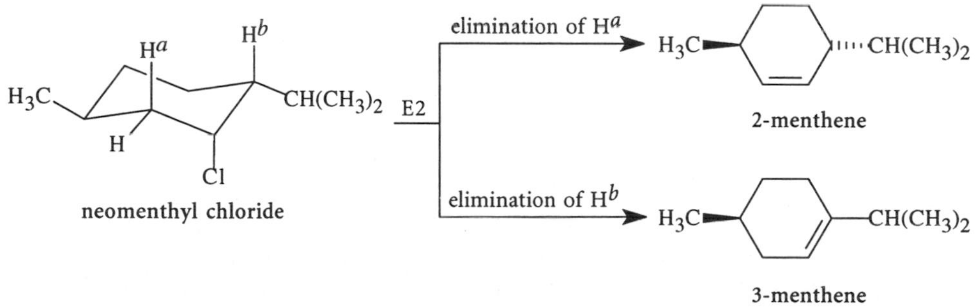

neomenthyl chloride

elimination of H^a

2-menthene

elimination of H^b

3-menthene

More 3-menthene is formed because it is the alkene with the greater number of alkyl branches at its double bond.

9.55 (a) Because the alkene product is cis, the transition state for the elimination must have the two methyl groups on the same side of the molecule. Drawing the molecule in a conformation with the methyl groups on the same side shows that the bromines are eliminated from opposite sides of the molecule; that is, the elimination is anti.

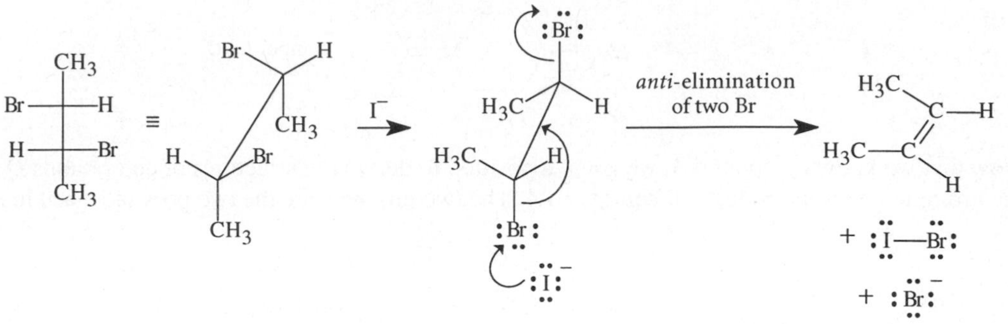

(b) Because the phenyl groups are on opposite sides of the product alkene, the transition state for elimination must also have the phenyl groups on opposite sides. Drawing a conformation that meets this condition reveals that the hydrogen and the acetoxy group are eliminated from the same side of the molecule; that is, the elimination is syn.

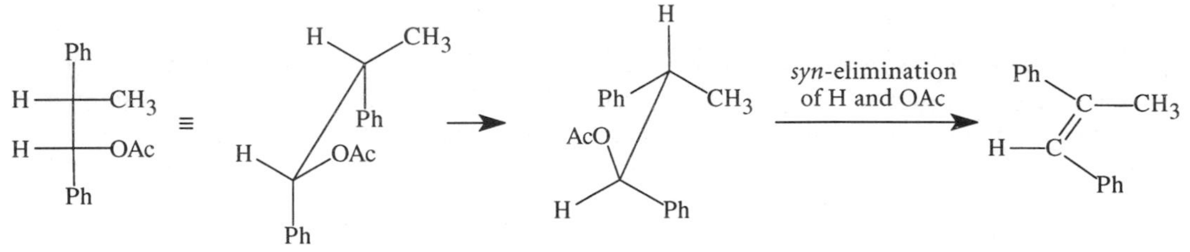

Notice that the elimination requires a transition state that has an eclipsed, or nearly eclipsed, conformation; otherwise, the alkene would be formed with a twisted double bond. The reason for the syn stereochemistry is that an oxygen of the acetoxy group removes a proton intramolecularly:

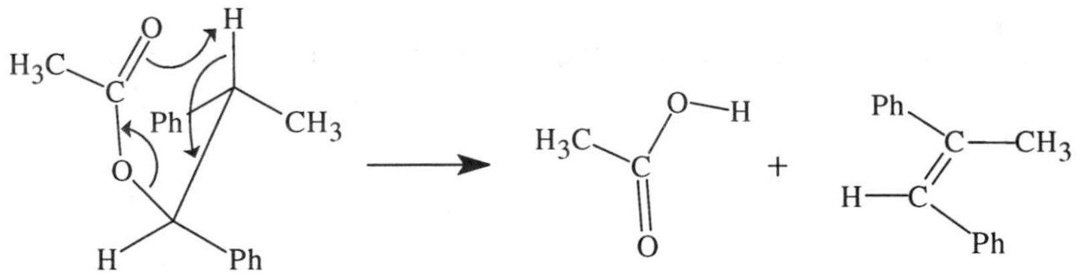

(An intramolecular *anti*-elimination involving a similar proton transfer is stereochemically impossible.)

9.56 As in Problem 9.55, the key to the solution is to assume that E2 reactions are faster when they can occur with anti stereochemistry. In the first compound, two β-hydrogens can be eliminated with the bromine: the hydrogen at the ring junction, and a hydrogen of the methyl group. The hydrogen at the ring junction is eliminated because it gives the alkene with the greater number of alkyl branches on the double bond.

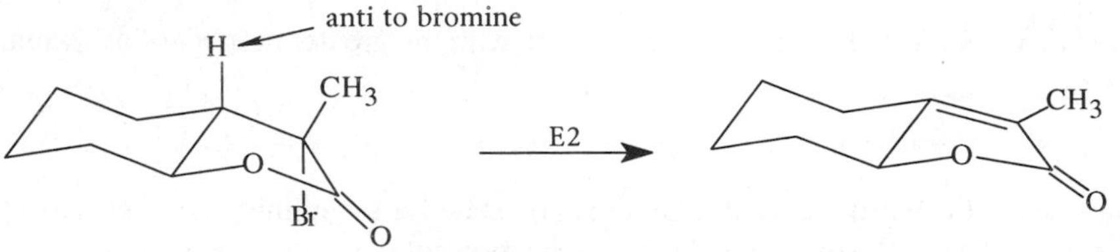

In the second compound, the hydrogen at the ring junction cannot be anti to the bromine; hence, the only anti hydrogen available for elimination is a hydrogen of the methyl group:

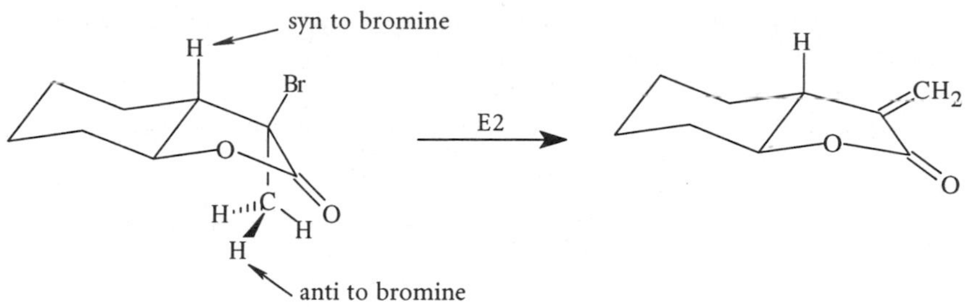

syn to bromine

E2

anti to bromine

9.57 (a) Initiation of the reaction by AIBN (see text p. 180) suggests a free-radical chain mechanism. The first initiation step is shown in Eq. 5.41, text p. 180. Let the free-radical product of this reaction be abbreviated as R•, and let the butyl group be abbreviated as Bu. The second initiation step is

$$R\bullet \quad H{-}Sn(Bu)_3 \longrightarrow R{-}H + \bullet Sn(Bu)_3$$

The propagation steps are as follows:

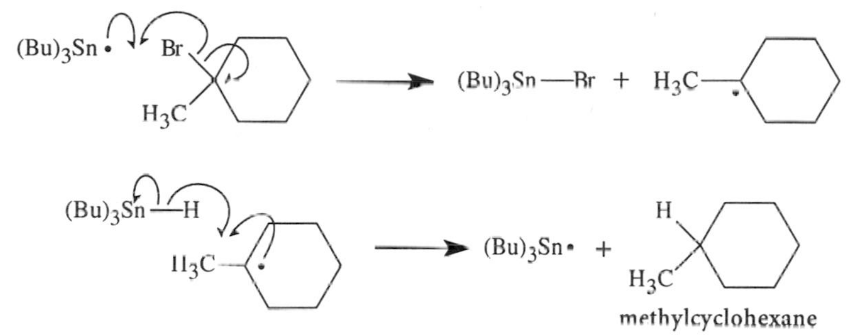

$(Bu)_3Sn\bullet$ + 1-bromo-1-methylcyclohexane ⟶ $(Bu)_3Sn{-}Br$ + 1-methylcyclohexyl radical

$(Bu)_3Sn{-}H$ + 1-methylcyclohexyl radical ⟶ $(Bu)_3Sn\bullet$ + methylcyclohexane

methylcyclohexane

(b) One sequence of reactions is to convert the alkyl halide into a Grignard reagent, which is then treated with water:

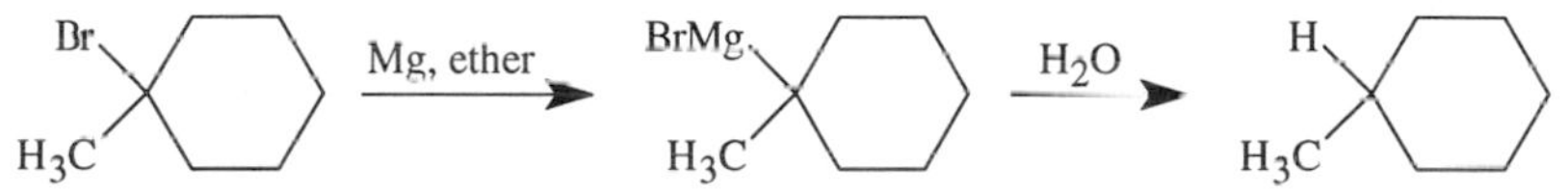

1-bromo-1-methylcyclohexane $\xrightarrow{\text{Mg, ether}}$ (BrMg, H_3C)cyclohexane $\xrightarrow{H_2O}$ methylcyclohexane

Another is to carry out an E2 reaction to obtain an alkene (or a mixture of alkenes), which is then subjected to catalytic hydrogenation to give methylcyclohexane.

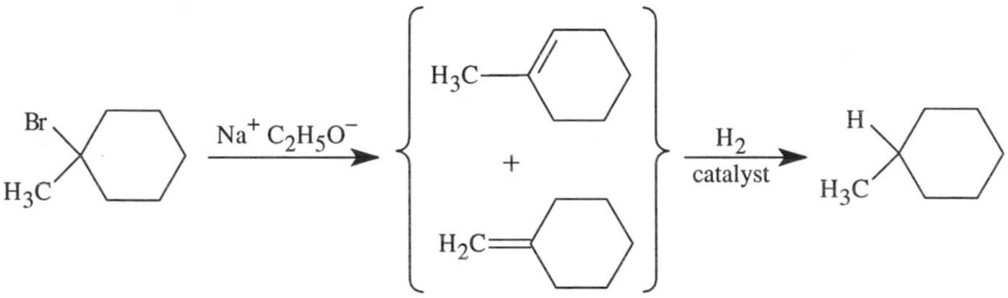

1-bromo-1-methylcyclohexane $\xrightarrow{Na^+\ C_2H_5O^-}$ { 1-methylcyclohexene (H_3C—) + methylenecyclohexane (H_2C=) } $\xrightarrow[\text{catalyst}]{H_2}$ methylcyclohexane

9.58 The mechanism given would not fit Fact #1. The mechanism as written is a concerted substitution by an S_N2 mechanism, the nucleophile being a hydride (a hydrogen with its two bonding electrons) from $^-BH_4$. Because S_N2 reactions are so slow that they are not useful on tertiary carbons, the fact that the reaction actually "works well" (which means it proceeds at a convenient rate in high yield) on a tertiary organomercury compound is inconsistent with an S_N2 mechanism.

The mechanism given would also not fit Fact #2. An S_N2 mechanism would predict inversion of stereochemical configuration.

A reactive intermediate that would fit these facts would be either a carbocation or a radical. Both intermediates are sp^2-hybridized and therefore have trigonal planar geometry. Either species is achiral (barring asymmetric carbons in the three attached groups) and would therefore have to give both enantiomers of the product in equal amounts.

Although the problem did not ask for it, a curved-arrow carbocation mechanism could be the following:

$$R_3C-Hg-\ddot{O}Ac \longrightarrow R_3C^+ + \ddot{H}g + {:}\ddot{O}Ac^-$$

$$R_3C^+ \quad H-\bar{B}H_3 \longrightarrow R_3C-H + BH_3$$

The problem with this mechanism is that it predicts that rearrangements should be observed with certain secondary organomercury compounds; yet rearrangements are not observed in oxymercuration-reduction.

A generally accepted mechanism is the following. In the first step, borohydride acts as a nucleophile in a substitution at *mercury*. (You can think of this as an S_N2 reaction on mercury, although the mechanism may not be concerted.)

$$R_3C-Hg-\ddot{O}Ac + H-\bar{B}H_3 \longrightarrow R_3C-Hg-H + {:}\ddot{O}Ac^- + BH_3$$

Fragmentation of the carbon-mercury bond (which is rather weak) produces the radical $R_3C\cdot$ that propagates the chain:

$$R_3C-Hg-H \longrightarrow R_3C\cdot \quad \cdot Hg-H$$

$$R_3C\cdot \quad H-Hg-CR_3 \longrightarrow R_3C-H + \ddot{H}g + \cdot CR_3$$

9.59 The reaction of butylamine with 1-bromobutane is a typical S_N2 reaction:

$$CH_3CH_2CH_2CH_2-\ddot{N}H_2 \quad {:}\ddot{Br}-CH_2CH_2CH_2CH_3 \longrightarrow CH_3CH_2CH_2CH_2-\overset{+}{N}H_2-CH_2CH_2CH_2CH_3 \quad {:}\ddot{Br}{:}^-$$

This mechanism is consistent with the second-order rate law, because the rate law requires one molecule of amine and one molecule of alkyl halide in the transition state.

The second reaction is also a nucleophilic substitution reaction, but, because it is intramolecular (that is, the nucleophile and carbon that it attacks are part of the same molecule), the reaction is first-order.

$$Br(CH_2)_4\ddot{N}H_2 \longrightarrow \text{(cyclic)}\ \overset{+}{N}H_2 \quad {:}\ddot{Br}{:}^-$$

9.60 (a) Product *A* results from a conventional E2 reaction:

$$HO-C_6H_{10}-Cl \underset{}{\overset{^-OH}{\rightleftharpoons}} \text{(ring flip, axial Cl, axial H)} \xrightarrow{{:}\ddot{O}H^-} HO-C_6H_9 + H-\ddot{O}H + {:}\ddot{Cl}{:}^-$$

3-cyclohexenol
A

(The same sort of elimination reaction gives compound *A* from *cis*-4-chlorocyclohexanol.) Product *B* is formed by an intramolecular nucleophilic substitution reaction of the alkoxide that occurs in the twist-boat conformation:

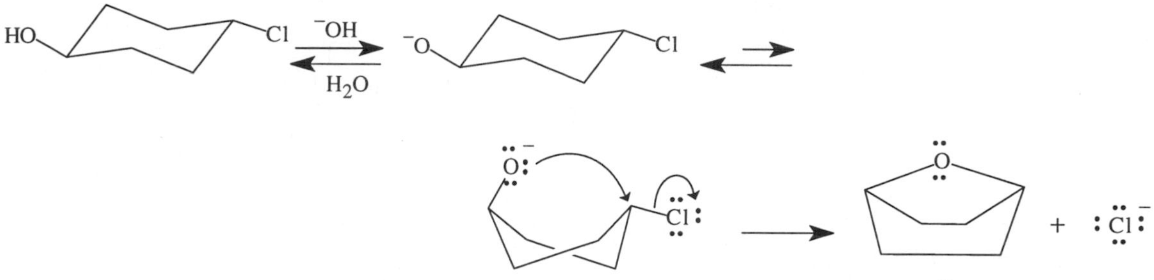

Product *C* results from an ordinary S_N2 reaction:

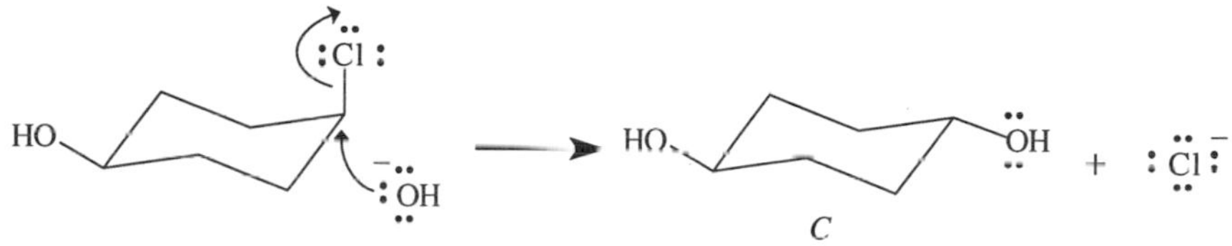

(b) The intramolecular substitution that yields the bicyclic compound *B* requires, like all nucleophilic substitutions, approach of the nucleophile to the carbon from the backside. This is only possible in the twist-boat form of the trans stereoisomer. Because a conformation that permits backside attack of the intramolecular alkoxide is not available to the cis stereoisomer, it reacts instead by an ordinary S_N2 reaction with the external nucleophile ^-OH.

Notice that the formation of product *B* instead of *cis*-1,4-cyclohexanediol from the trans stereoisomer of the starting material means that the *intramolecular* nucleophilic substitution reaction is much faster than the S_N2 reaction of the same compound with hydroxide. Intramolecular reactions are in many cases faster than analogous intermolecular reactions. This point is explored in more detail in Sec. 11.7 of the text.

9.61 (a) The reaction occurs largely with retention of configuration and only a small amount of inversion. This is of course inconsistent with stereochemical inversion observed with other S_N2 reactions, and caused a stir for that reason. (By the time this study was done, stereochemical inversion in the S_N2 reaction had become dogma.)

(b) The initial formation of *trans-P* occurs by the usual S_N2 mechanism with inversion. The subsequent formation of *cis-P* occurs when iodide ion in solution displaces iodide, again with inversion, from *trans-P* by a second S_N2 reaction with inversion.

(c) *Cis-P* is favored because both substituents can exist in equatorial arrangements. (See Sec. 7.11 and Problem 7.14 for a discussion of the puckering of cyclobutane rings.) As in substituted cyclohexanes, substituted cyclobutanes are destabilized by 1,3-diaxial interactions.

cis-P
both substituents equatorial
(more stable)

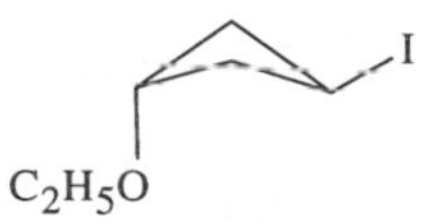

trans-P
one substituent axial
(less stable)

9.62 (a) Bromine addition to alkenes is anti; consequently, the stereochemistry of compound *B* is as follows:

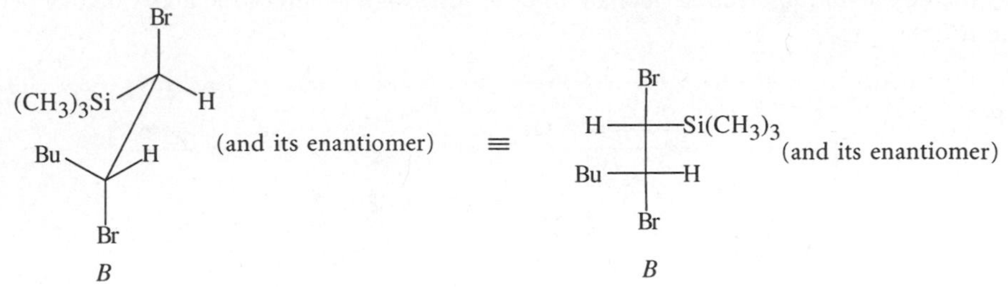

Br, $(CH_3)_3Si$, H, Bu, H, Br (and its enantiomer) ≡ Br, H—$Si(CH_3)_3$, Bu—H, Br (and its enantiomer)

B *B*

(b) Draw the structure of compound *B* in a conformation in which the butyl (Bu) group and the Br that remains after the elimination are on opposite sides of the molecule, because this is the way they are in the alkene product. This shows that the trimethylsilyl group and the bromine are anti; consequently, the elimination shown is an *anti*-elimination.

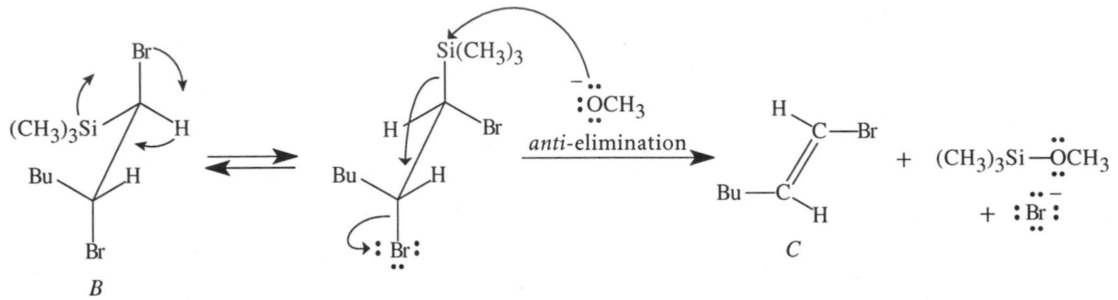

B ⇌ [conformer with $Si(CH_3)_3$, H, Br, Bu, H, :Br:] + $^{-}$:OCH_3 —*anti*-elimination→ H(Br)C=C(H)Bu (*C*) + $(CH_3)_3Si—OCH_3$ + :Br:$^-$

(c) The *E* stereoisomer of compound *A* would give the *Z* stereoisomer of compound *C*. This can be shown by the analysis used in parts (a) and (b) with the positions of the trimethylsilyl group and the hydrogen interchanged.

9.63 (a) Sodium hydroxide acts as a base to form trichloromethyl anion, which then forms dichlorocarbene. (See Eq. 9.59a–b on text p. 395.) Dichlorocarbene reacts with iodide ion to form a new anion, which is protonated by water. The reaction is not observed in the absence of NaOH because a base is required to form trichloromethyl anion. Also, the fact that the reaction is not observed in the absence of NaOH rules out a simple S_N2 reaction of iodide ion with chloroform.

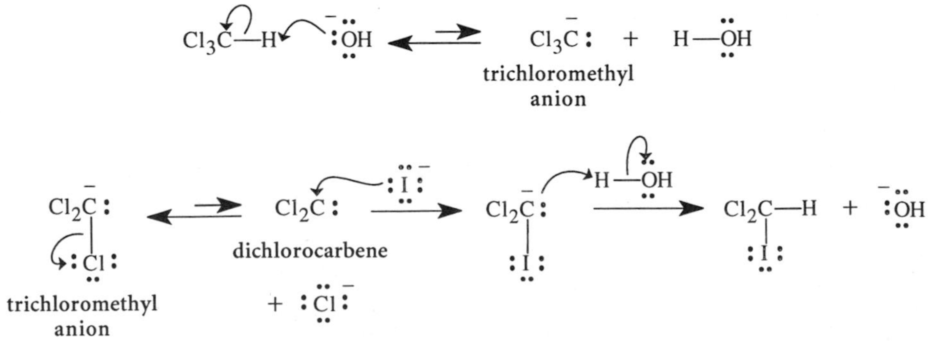

$$Cl_3C—H + {}^{-}:\ddot{O}H \rightleftharpoons Cl_3\bar{C}: + H—\ddot{O}H$$

trichloromethyl anion

$Cl_2\bar{C}$:—:Cl: (trichloromethyl anion) ⇌ Cl_2C: (dichlorocarbene) + :Cl:$^-$; Cl_2C: + :I:$^-$ → $Cl_2\bar{C}$:—:I: ; + H—OH → Cl_2C(I)—H + $^-$:OH

(b) Butyllithium acts as a base to form an anion that decomposes to phenylcarbene, which then reacts with the alkene to form the cyclopropane.

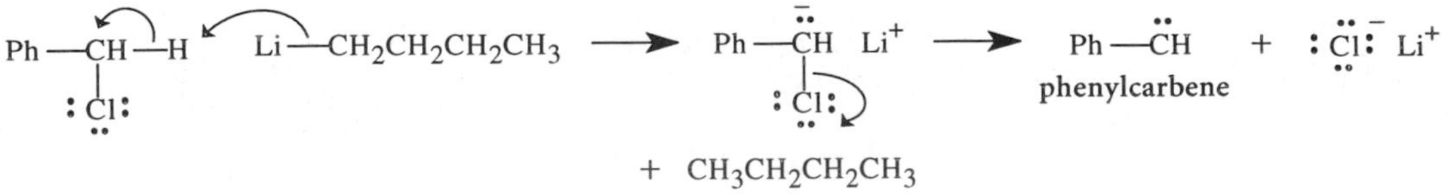

Ph—CH(Cl)—H + Li—$CH_2CH_2CH_2CH_3$ → Ph—$\ddot{C}H^-$(Cl) Li^+ + $CH_3CH_2CH_2CH_3$ → Ph—$\ddot{C}H$ (phenylcarbene) + :Cl:$^-$ Li^+

Ph—$\ddot{C}$H → Ph—

10

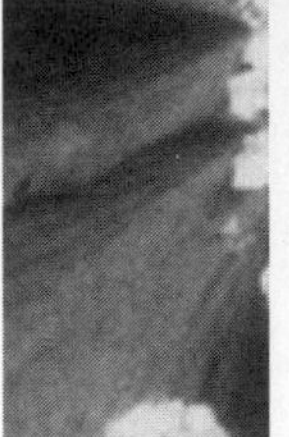

Chemistry of Alcohols and Thiols

TERMS

Term	Section
alcohol dehydrogenase	10.7, 10.8B
alkylating agent	10.3B
alkylated	10.3B
anaerobic fermentation	10.7
carbonyl compound	10.6A
chemically equivalent	10.8A
chemically nonequivalent	10.8A
chlorosulfite ester	10.3D
chromate ester	10.6A
coenzyme	10.7
Collins reagent	10.6A
constitutionally equivalent	10.8A
constitutionally nonequivalent	10.8A
dehydration	10.1
diastereotopic	10.8A
disulfide	10.9
enantiotopic	10.8A
half-reaction	10.5A
homotopic	10.8A
mesylate	10.3A
multistep synthesis	10.11
NADH	10.7
NAD^+	10.7
niacin	10.7
organic synthesis	10.10
oxidation	10.5A
oxidation level	10.5A
oxidation number	10.5A
oxidation state	10.5A
oxidizing agent	10.5B
principle of microscopic reversibility	10.1
reducing agent	10.5B
reduction	10.5A
substitution test	10.8A
sulfenic acid	10.9
sulfinic acid	10.9
sulfonate anion	10.3A
sulfonate ester	10.3A
sulfonic acid	10.3A, 10.9
target molecule	10.11
thionyl chloride	10.3D
p-toluenesulfonyl chloride	10.3A
tosyl chloride	10.3A
tosylate	10.3A

CONCEPTS

I. OXIDATION AND REDUCTION IN ORGANIC CHEMISTRY

A. General

1. In organic chemistry, whether a transformation is an oxidation or a reduction is determined by the oxidation state of the reactants and products.
   a. An oxidation is a transformation in which electrons are lost in the corresponding half-reaction.
      i. The product has a greater (more positive, less negative) oxidation number than the reactant.
      ii. The reagent that brings about the transformation is called an oxidizing agent.
   b. A reduction is a transformation in which electrons are gained in the corresponding half-reaction.
      i. The product has a smaller (less positive, more negative) oxidation number than the reactant.
      ii. The reagent that effects the transformation is called a reducing agent.

c. Oxidations and reductions always occur in pairs; whenever something is oxidized, something else is reduced.

B. Oxidation Numbers

1. Assign an oxidation level to each carbon that undergoes a change between reactant and product by the following method:
   a. Assign a −1 for:
      i. every bond from the carbon to a less electronegative element (including hydrogen)
      ii. every negative charge on the carbon
   b. Assign a 0 (zero) for:
      i. every bond from the carbon to another carbon atom
      ii. every unpaired electron on the carbon
   c. Assign a +1 for:
      i. every bond from the carbon to a more electronegative element
      ii. every positive charge on the carbon
   d. Add the numbers assigned under (a), (b), and (c) to obtain the oxidation level of the carbon under consideration
2. Determine the oxidation numbers N_{ox} of both the reactant and product by adding, within each compound, the oxidation levels of all the carbons computed in step 1.
3. Compute the difference N_{ox}(product) − N_{ox}(reactant) to determine whether the transformation is an oxidation, reduction, or neither.
   a. If the difference is a positive number, the transformation is an oxidation.
   b. If the difference is a negative number, the transformation is a reduction.
   c. If the difference is zero, the transformation is neither an oxidation nor a reduction.

$H_3C-CH(OH)-CH(OH)-CH_3 \longrightarrow H_3C-C(=O)-C(=O)-CH_3$

	Reactant	Product
oxidation levels of carbons	0 0	+2 +2
oxidation numbers of compounds	$N_{ox} = 0$	$N_{ox} = +4$

N_{ox}(products) − N_{ox}(reactants) = +4 − 0 = +4
Therefore the reaction is an oxidation.

C. Oxidations and Reductions

1. Whether a reaction is an oxidation or a reduction does not necessarily depend on the introduction or loss of oxygen.
2. A balanced half-reaction shows the loss or gain of electrons.
   a. Use H_2O to balance missing oxygens.
   b. Use H^+ to balance missing hydrogens.
   c. Use e^- to balance charges.
   d. The change in oxidation number is the same as the number of electrons lost or gained in the corresponding half-reaction.
3. The change in oxidation number for a transformation indicates whether an oxidizing or reducing agent is required to bring about the reaction.
4. The oxidation-number concept can be used to organize organic compounds into functional groups with the same oxidation level; carbons with larger numbers of hydrogens have a greater number of possible oxidation states.

D. Oxidations in Biological Systems

1. An understanding of the fundamental types of organic reactions and their mechanisms is useful in the study of biochemical processes.

2. A naturally occurring oxidation catalyzed by the enzyme alcohol dehydrogenase is the conversion of ethanol into acetaldehyde by NAD^+.
3. Nicotinamide adenine dinucleotide (NAD^+) is a biological oxidizing agent.
   a. NAD^+ is an example of a coenzyme: a molecule required, along with an enzyme, for certain biological oxidation reactions to occur.
   b. NADH is the reduced form of NAD^+ and is a reducing agent.
   c. The coenzymes NAD^+ and NADH are derived from the vitamin niacin.

II. CHEMICAL AND STEREOCHEMICAL EQUIVALENCE

A. Chemical Equivalence and Nonequivalence

1. It is sometimes important to know when two groups in a molecule are chemically equivalent, that is, when they behave in exactly the same way towards a chemical reagent.
2. Two groups within a molecule can be classified as constitutionally equivalent or constitutionally nonequivalent according to their connectivity relationship.
   a. Groups within a molecule are constitutionally equivalent when they have the same connectivity relationship to all other atoms in the molecule.
3. Whether two constitutionally equivalent groups are chemically equivalent depends on their stereochemical relationship.
   a. Constitutional nonequivalence is a sufficient but not a necessary condition for chemical nonequivalence.
   b. Constitutionally nonequivalent groups are chemically nonequivalent.
4. Constitutionally equivalent groups are of three types:
   a. Homotopic groups are chemically equivalent and indistinguishable under all circumstances.
   b. Enantiotopic groups are chemically equivalent towards achiral reagents, but are chemically nonequivalent towards chiral reagents.
   c. Diastereotopic groups are chemically nonequivalent under all conditions.

B. Determination of Constitutional Equivalence

1. The stereochemical relationship between constitutionally equivalent groups is revealed by a substitution test.
   a. Substitute each constitutionally equivalent group in turn with a fictitious circled group and compare the resulting molecules.
   b. Their stereochemical relationship determines the relationship of the circled groups.
2. When the substitution test give identical molecules, as in this example, the constitutionally equivalent groups are said to be homotopic.

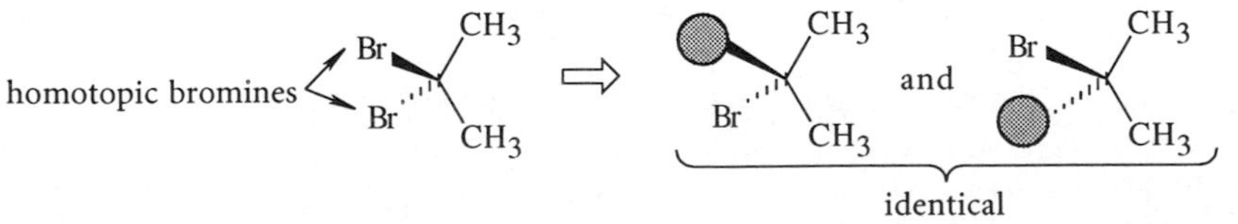

homotopic bromines — identical

3. When the substitution test gives enantiomers, the constitutionally equivalent groups are said to be enantiotopic.

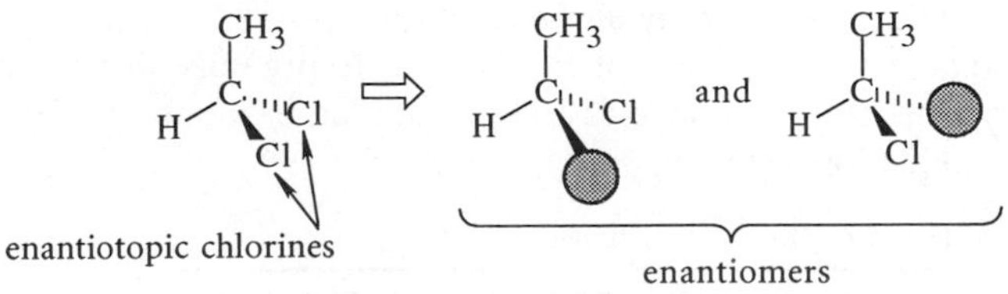

enantiotopic chlorines — enantiomers

4. When a substitution test gives diastereomers, the constitutionally equivalent groups are said to be diastereotopic.

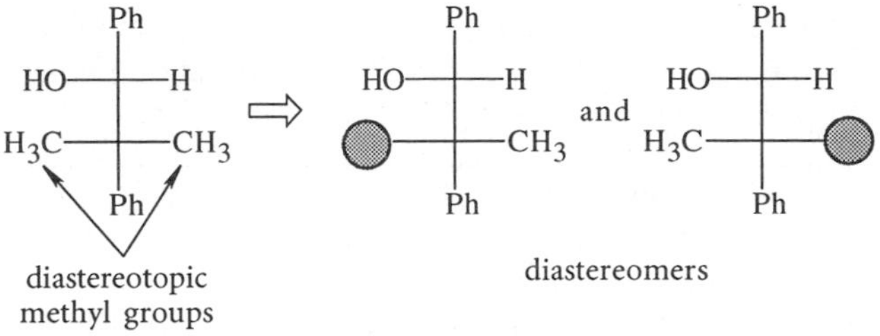

Ph; HO—H; H_3C—CH_3; Ph ⇒ Ph; HO—H; ●—CH_3; Ph and Ph; HO—H; H_3C—●; Ph

diastereotopic methyl groups

diastereomers

a. Diastereotopic groups are easily recognized when two constitutionally equivalent groups are present in a molecule that contains an asymmetric carbon (as in the example above).
b. Diastereotopic groups are easily recognized when two groups on one carbon of a double bond are the same and the two groups on the other carbon are different.

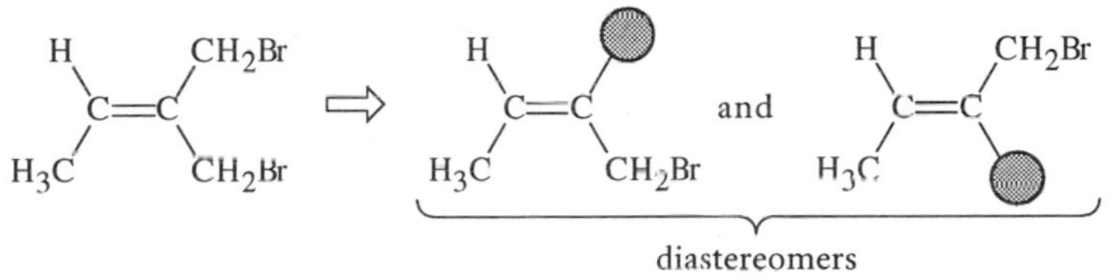

H, H_3C C=C CH_2Br, CH_2Br ⇒ H, H_3C C=C ●, CH_2Br and H, H_3C C=C CH_2Br, ●

diastereomers

III. ORGANIC SYNTHESIS AND REACTIONS

A. Organic Synthesis

1. The preparation of organic compounds from other organic compounds by the use of one or more reactions is called organic synthesis.
2. A synthesis involving a sequence of several reactions is called a multistep synthesis.
3. The molecule to be synthesized is called the target molecule.
4. In order to assess the best route to the target molecule from the starting material, work backward from the target molecule towards the starting material.

B. Principle of Microscopic Reversibility

1. Any reaction and its reverse proceed by the forward and reverse of the same mechanism.
   a. This statement is known as the principle of microscopic reversibility.
   b. It follows from this principle that forward and reverse reactions must have the same intermediates and the same rate-limiting transition states.
   c. The rate-limiting step of a reaction is the same as the rate-limiting step of its reverse.
2. Any reaction catalyzed in one direction is also catalyzed in the other.

C. Dehydration Reactions

1. A reaction in which the elements of water are lost from the starting material is called a dehydration reaction.
   a. Strong acids such as H_2SO_4 and H_3PO_4 catalyze a β-elimination reaction in which water is lost from a secondary or tertiary alcohol to give an alkene.

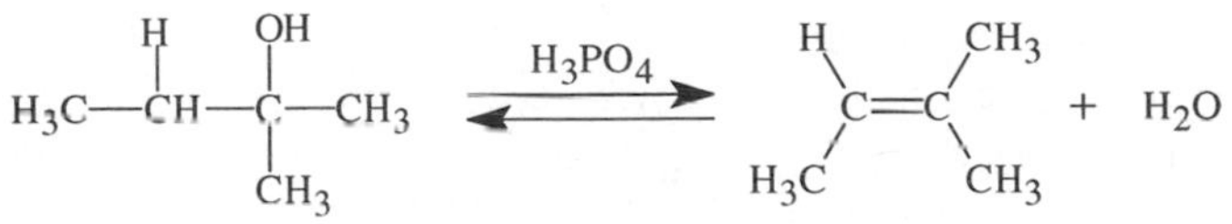

$$H_3C\text{—}CH(H)\text{—}C(OH)(CH_3)\text{—}CH_3 \xrightleftharpoons{H_3PO_4} (H)(H_3C)C{=}C(CH_3)(CH_3) + H_2O$$

b. Lewis acids such as alumina (aluminum oxide, Al_2O_3) and/or heat can also be used to catalyze or promote dehydration reactions.

2. Most acid-catalyzed dehydrations of alcohols are reversible reactions and can easily be driven towards the alkene products by applying LeChatelier's principle.
3. The dehydration of alcohols is the reverse of the hydration of alkenes.

D. Alkylating Agents

1. In a nucleophilic substitution, an alkyl group is transferred from a leaving group to a nucleophile.
   a. A nucleophile is said to be alkylated by an alkyl halide or a sulfonate ester in the same sense that a Brønsted base is protonated by a strong acid.
   b. Alkyl halides, sulfonate esters, and related compounds are sometimes referred to as alkylating agents; that is, they react rapidly with nucleophiles in S_N2 or S_N1 reaction to transfer an alkyl group.
2. Alkyl esters of strong inorganic acids are typically very potent alkylating agents because they contain leaving groups that are very weak bases and thus react very rapidly.

IV. REACTIONS INVOLVING ALCOHOLS AND THIOLS

A. Substitution and Elimination Reactions of Alcohols

1. Several reactions of alcohols involve breaking the C—O bond.
   a. In acid-catalyzed dehydration, the —OH group of an alcohol is converted by protonation into a good leaving group and is eliminated as water to give an alkene.
   b. In the reaction with hydrogen halides, the —OH group of an alcohol is converted by protonation into a good leaving group and is displaced by halide (in reactions with hydrogen halides) to give an alkyl halide.
   c. In the reaction of an alcohol with $SOCl_2$, the reagent itself converts the —OH into a good leaving group, which in a subsequent reaction is displaced by halide ion.
   d. In the reaction with a sulfonyl chloride, the alcohol is converted into a sulfonate ester, which is an excellent leaving group in substitution or elimination reactions.
2. Alkyl halide formation and dehydration to alkenes are alternative branches of a common mechanism.
3. Hydration of alkenes and dehydration of alcohols are the forward and reverse of the same reaction; both hydration and alkenes to alcohols and dehydration of alcohols to alkenes are catalyzed by acids.

B. Considerations for the Conversion of Alcohols into Alkyl Halides

1. A variety of reactions that can be used to convert alcohols into alkyl halides are:
   a. reaction with hydrogen halides;
   b. formation of sulfonate esters and displacement with halide ions; and
   c. reaction with $SOCl_2$ or PBr_3.
2. The method of choice depends on the structure of the alcohol and on the type of alkyl halide to be prepared; in some cases two or more methods will work equally well.
3. In order to break the carbon-oxygen bond, the —OH group must first be converted into a good leaving group by
   a. protonation: protonated alcohols are intermediates in both dehydration to alkenes and substitution to give alkyl halides.
   b. conversion into sulfonate esters, inorganic esters, or related leaving groups: these esters, to a useful approximation, react like alkyl halides.
   c. reaction with thionyl chloride
4. Reactions of alcohols with hydrogen halides:
   a. When a primary alcohol is the starting material, the reaction occurs as a concerted displacement of water from the protonated alcohol by halide ion.
      i. The reaction is an S_N2 reaction in which water is the leaving group.

$$RCH_2{-}\ddot{O}H \xrightleftharpoons{H{-}\ddot{Br}:} RCH_2{-}\overset{+}{O}H_2 \;\; :\ddot{Br}:^- \longrightarrow RCH_2{-}\ddot{Br}: + \ddot{O}H_2$$

ii. Primary alcohols require heating for several hours.

b. The reactions of secondary alcohols with hydrogen halides tend to occur by the S_N1 mechanism.

i. When carbocation intermediates are involved in the reactions of alcohols with hydrogen halides, rearrangements occur in certain cases.

ii. Rearrangements are best avoided in the preparation of secondary alkyl halides by the reaction of a halide ion with a sulfonate ester in a polar aprotic solvent.

c. In the reactions of tertiary alcohols, protonation of the alcohol oxygen is followed by carbocation formation.

i. Once the alcohol is protonated, the reaction is essentially an S_N1 reaction with H_2O as the leaving group.

ii. The carbocation reacts with the halide ion, which is present in great excess.

$$R_3C{-}\ddot{O}H \xrightleftharpoons[\text{(excess)}]{H_3\overset{+}{O}\;:\ddot{Br}:^-} R_3C{-}\overset{+}{O}H_2\;\; :\ddot{Br}:^- \rightleftharpoons R_3C^+ + \ddot{O}H_2 \xrightarrow{:\ddot{Br}:^-} R_3C{-}\ddot{Br}:$$

carbocation

iii. Typically, tertiary alcohols react with hydrogen halides within a few minutes.

5. Reactions of alcohols with thionyl chloride:

a. Thionyl chloride effects the conversion of alcohols into chlorosulfite esters, which are converted within the reaction mixture into alkyl chlorides.

i. When an alcohol reacts with thionyl chloride, a chlorosulfite ester intermediate is formed.

ii. The chlorosulfite ester reacts readily with nucleophiles because the chlorosulfite group, —O—SO—Cl, is a very good leaving group.

iii. The chlorosulfite ester is usually not isolated, but reacts with the chloride ion formed to give the alkyl chloride.

iv. The displaced $^-$O—SO—Cl ion is unstable and decomposes to SO_2 and Cl^-.

$$CH_3(CH_2)_3CH_2{-}\ddot{O}H + Cl_2S{=}O + C_5H_5N: \longrightarrow CH_3(CH_2)_3CH_2{-}\ddot{O}{-}S({=}O)Cl + :\ddot{Cl}:^- + C_5H_5\overset{+}{N}{-}H$$

pentanol, thionyl chloride, pyridine, pentyl chlorosulfite, pyridinium ion

$$:\ddot{Cl}:^- + CH_3(CH_2)_3CH_2{-}\ddot{O}{-}S({=}O)Cl \longrightarrow CH_3(CH_2)_3CH_2{-}\ddot{Cl}: + {}^-:\ddot{O}{-}S({=}O)Cl$$

chloropentane

$$^-:\ddot{O}{-}S({=}\ddot{O}:){-}\ddot{Cl}: \longrightarrow :O{=}\ddot{S}{=}O: + :\ddot{Cl}:^-$$

chlorosulfite ion, sulfur dioxide

b. The thionyl chloride method is most useful with primary alcohols; it can also be used with secondary alcohols, although rearrangements in such cases have been known to occur.

6. Reaction with phosphorous tribromide:
   a. Phosphorus tribromide is similar to thionyl chloride in the sense that it converts the —OH group into a good leaving group and at the same time provides a nucleophile (bromide) to effect the substitution reaction.
   b. All three bromines can be utilized.
   c. The risk of rearrangement, although not totally absent, is less than with HBr.
   d. PBr_3 can be used with alcohols that contain acid-sensitive functional groups which would not survive treatment with HBr.

C. Considerations for the Dehydration of Alcohols to Alkenes

1. Acid-catalyzed alcohol dehydration occurs by a three-step mechanism involving a carbocation intermediate (E1 mechanism).
   a. In the first step, the —OH group of the alcohol accepts a proton from the catalyzing acid in a Brønsted acid-base reaction to become a very good leaving group (water). Thus, the basicity of alcohols is important to the success of the dehydration reaction.
   b. In the second step, the carbon-oxygen bond of the alcohol breaks in a Lewis acid-base dissociation to give water and a carbocation.
   c. In the third step, water removes a β-proton from the carbocation in another Brønsted acid-base reaction to generate the alkene product and regenerate the catalyzing acid H_3O^+.

$$\text{H—C—C—}\ddot{\text{O}}\text{H} + \text{H—}\overset{+}{\text{O}}\text{H}_2 \longrightarrow \text{H}_2\ddot{\text{O}}\text{:} + \text{H—C—C—}\overset{+}{\text{O}}\text{H}_2$$

$$\text{H—C—C—}\overset{+}{\text{O}}\text{H}_2 \longrightarrow \text{H—C—C}^{+} + \text{H}_2\ddot{\text{O}}\text{:}$$

$$\text{H}_2\ddot{\text{O}}\text{:} + \text{H—C—C—}\overset{+}{\text{O}}\text{H}_2 \longrightarrow \text{H}_3\overset{+}{\text{O}} + \text{C=C} + \text{H}_2\ddot{\text{O}}\text{:}$$

2. If the alcohol has more than one type of β-hydrogen, then a mixture of alkene products can be expected.
3. Alcohols that react to give rearrangement-prone carbocation intermediates yield rearranged alkenes.
4. The most stable alkene (the one with the greatest number of branches at the double bond) is the alkene formed in greatest amount.
5. The relative rates of alcohol dehydration are in the order tertiary > secondary >> primary.
6. Dehydration of primary alcohols is generally not a useful laboratory procedure for the preparation of alkenes.

D. Oxidation of Alcohols and Thiols

1. Primary and secondary alcohols can be oxidized to carbonyl compounds with Cr(VI).
   a. Chromate (CrO_4^{2-}) and dichromate ($Cr_2O_7^{2-}$) reagents are customarily used under strongly acidic conditions.
   b. CrO_3 is often used in pyridine.
   c. A complex between CrO_3 and two molecules of pyridine in methylene chloride solvent is commonly known as Collins reagent.

2. The oxidation of alcohols with Cr(VI) reagents involves an acid-catalyzed displacement of water from chromic acid by the alcohol to form a chromate ester.
   a. After protonation of the chromate ester, it decomposes in a β-elimination reaction.
   b. The last step is much like an E2 reaction, except that it does not involve a strong base.

$$\underset{\text{ethanol}}{\mathrm{CH_3CH_2OH}} + \underset{\text{chromic acid}}{\mathrm{HO{-}CrO_2{-}OH}} \xrightarrow{H_3O^+} \underset{\text{a chromate ester}}{\mathrm{CH_3CH_2{-}O{-}CrO_2{-}OH}} + H_2O$$

$$\mathrm{CH_3CH_2{-}O{-}Cr(=O)(=\ddot{O}{:}){-}OH} + \mathrm{H{-}\overset{+}{O}H_2} \rightleftarrows \mathrm{CH_3CH_2{-}O{-}Cr(=O)(={:}\overset{+}{O}H){-}OH} + H_2\ddot{O}{:}$$

$$H_2\ddot{O}{:} + \mathrm{CH_3CH_2{-}O{-}Cr(=O)(={:}\overset{+}{O}H){-}OH} \longrightarrow H_3\overset{+}{O}{:} + \underset{\text{acetaldehyde}}{\mathrm{CH_3CH{=}O}} + \underset{\text{Cr(IV)}}{\left[\mathrm{HO{-}\overset{+}{}\ddot{Cr}^{-}(=O){-}OH} \longleftrightarrow \mathrm{H\ddot{O}{-}\ddot{Cr}(=O){-}OH}\right]}$$

   c. Primary alcohols are oxidized to aldehydes (in the absence of water) or carboxylic acids (in the presence of water).
   d. Secondary alcohols are oxidized to ketones.
   e. Tertiary alcohols are not oxidized.
3. Primary alcohols can be oxidized to carboxylic acids with $KMnO_4$.
4. Thiols are oxidized at sulfur rather than at the α-carbon.
5. Disulfides and sulfonic acids are two common oxidation products of thiols.

REACTIONS

I. SUBSTITUTION AND ELIMINATION REACTIONS OF ALCOHOLS AND THEIR DERIVATIVES

A. Sulfonate Ester and Inorganic Esters of Alcohols

1. Sulfonate esters are derivatives of sulfonic acids, which are compounds of the form R—SO_3H.
   a. A sulfonate ester is a compound in which the acidic hydrogen of a sulfonic acid is replaced by an alkyl or aryl group.

$$\underset{p\text{-toluenesulfonic acid}}{\mathrm{H_3C{-}C_6H_4{-}SO_2{-}OH}} \qquad \underset{\substack{\text{ethyl } p\text{-toluenesulfonate (ethyl tosylate)}\\ \text{a sulfonate ester}}}{\mathrm{H_3C{-}C_6H_4{-}SO_2{-}OCH_2CH_3}}$$

   b. Sulfonate esters are prepared from alcohols and other sulfonic acid derivatives called sulfonyl chlorides.

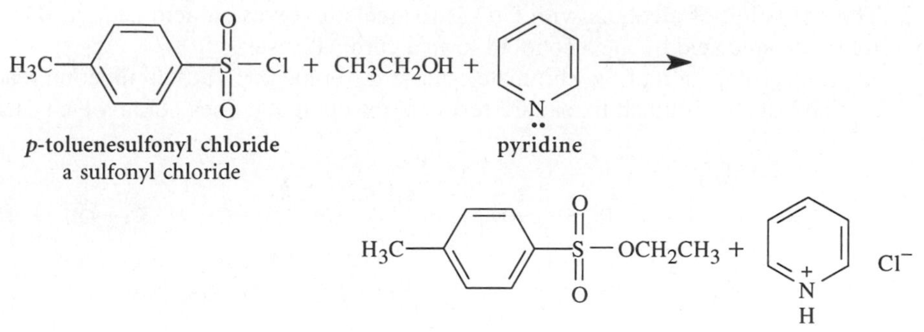

$H_3C-C_6H_4-SO_2-Cl + CH_3CH_2OH +$ pyridine $\longrightarrow$ $H_3C-C_6H_4-SO_2-OCH_2CH_3$ + pyridinium ($C_5H_5NH^+$) Cl^-

p-toluenesulfonyl chloride
a sulfonyl chloride

pyridine

i. This is a nucleophilic substitution reaction in which the oxygen of the alcohol displaces chloride ion from the tosyl chloride.

ii. The pyridine used as the solvent is a base; besides catalyzing the reaction, it also prevents HCl from forming in the reaction.

2. Sulfonate esters have approximately the same reactivities as the corresponding alkyl bromides in substitution and elimination reactions.
   a. Sulfonate anions, like bromide ions, are good leaving groups.
   b. Sulfonate anions are weak bases; they are the conjugate bases of sulfonic acids, which are strong acids.

3. Sulfonate esters prepared from primary and secondary alcohols, like the corresponding alkyl halides, undergo S_N2 reactions in which a sulfonate ion serves as the leaving group.

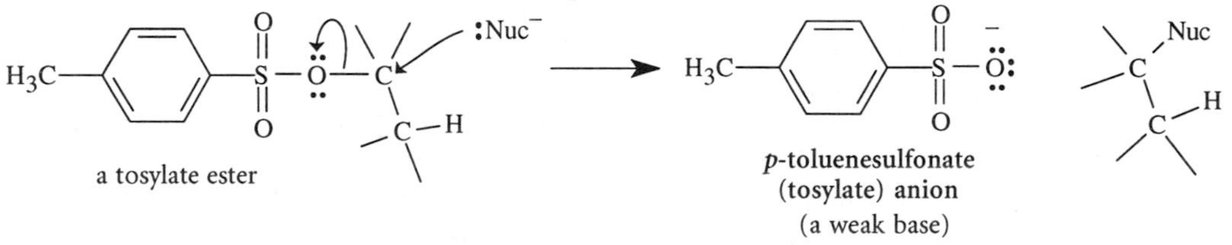

a nucleophile

:Nuc⁻

a tosylate ester

p-toluenesulfonate
(tosylate) anion
(a weak base)

4. Secondary and tertiary sulfonate esters, like the corresponding alkyl halides, also undergo E2 reactions with strong bases.

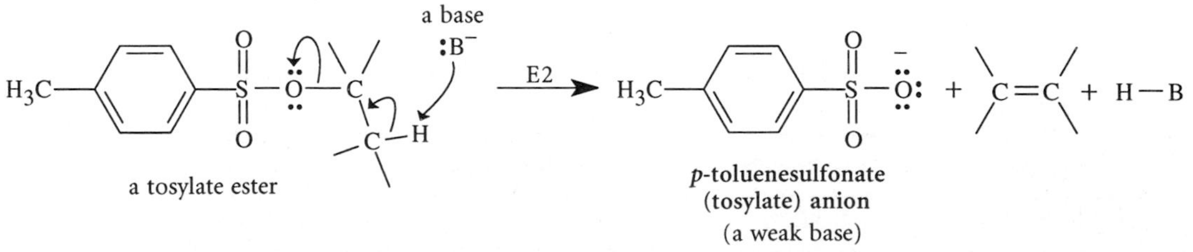

a base

:B⁻

E2

$+ \; C{=}C \; + \; H-B$

a tosylate ester

p-toluenesulfonate
(tosylate) anion
(a weak base)

5. Secondary and tertiary sulfonate esters, like the corresponding alkyl halides, also undergo S_N1-E1 solvolysis reactions in polar protic solvents.

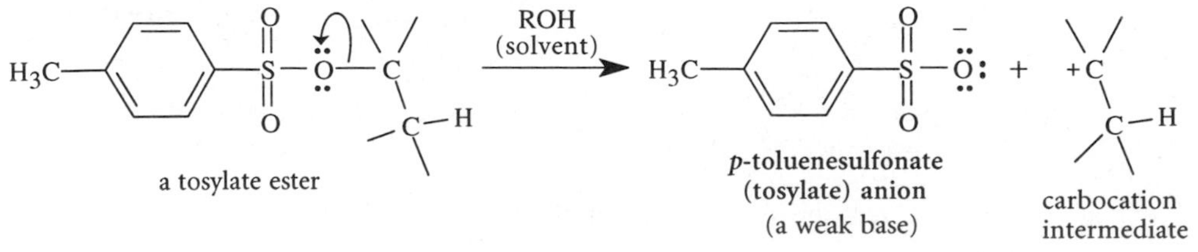

ROH
(solvent)

a tosylate ester

p-toluenesulfonate
(tosylate) anion
(a weak base)

carbocation
intermediate

$$\text{carbocation intermediate } \xrightarrow{\text{ROH}} \text{ solvolysis } (S_N1) \text{ product} + \text{elimination (E1) product } (C{=}C) + R\overset{+}{O}H_2$$

carbocation intermediate — solvolysis (S_N1) product — elimination (E1) product

B. Preparation of Primary Alkyl Halides

1. Primary alkyl chlorides are best prepared from primary alcohols by the thionyl chloride method.

$$CH_3(CH_2)_2CH_2OH \xrightarrow[CH_2Cl_2]{SOCl_2} CH_3(CH_2)_2CH_2Cl$$

2. Primary alkyl bromides are prepared from primary alcohols by the reaction of the alcohol with concentrated HBr or with PBr_3.

$$CH_3(CH_2)_2CH_2OH \xrightarrow[\text{or } PBr_3]{HBr} CH_3(CH_2)_2CH_2Br$$

3. Primary alkyl iodides are prepared from primary alcohols by the reaction of the alcohol with concentrated HI, which is often provided by mixing KI and H_3PO_4.

$$CH_3(CH_2)_2CH_2OH \xrightarrow{KI/H_3PO_4} CH_3(CH_2)_2CH_2I$$

4. The sulfonate ester method works well for primary alkyl halides but requires two separate reactions:
   a. formation of the sulfonate ester
   b. reaction of the ester with halide ion.

$$CH_3(CH_2)_2CH_2OH \xrightarrow[\text{pyridine}]{RSO_2Cl} \underset{\text{a sulfonate ester}}{CH_3(CH_2)_2CH_2OSO_2R} \xrightarrow[\text{acetone}]{X^-} CH_3(CH_2)_2CH_2X$$

5. Because all these methods have an S_N2 mechanism as their basis, alcohols with a significant amount of β-branching do not react under the usual conditions.

C. Preparation of Secondary Alkyl Halides

1. Secondary alkyl chlorides can be prepared from secondary alcohols by the thionyl chloride method if the secondary alcohol has relatively little β-branching.

$$(CH_3)_2CH{-}OH \xrightarrow{SOCl_2} (CH_3)_2CH{-}Cl$$

2. Secondary alkyl bromides can be prepared from alcohols by the PBr_3 method if the desired alkyl halide product is derived from a carbocation that does not rearrange.

$$H_3C{-}CH_2{-}\underset{}{CH(OH)}{-}CH_3 \xrightarrow{PBr_3} H_3C{-}CH_2{-}CH(Br){-}CH_3$$

3. Secondary alkyl halides can be prepared from secondary alcohols via the sulfonate ester, which should be treated with the appropriate halide ion in a polar aprotic solvent to avoid rearrangements.

$$\text{HO}\!-\!\text{C}_6\text{H}_{10}\!-\!\text{CH}_3 \xrightarrow[\text{pyridine}]{\text{CH}_3\text{SO}_2\text{Cl}} \text{H}_3\text{C}-\text{S}(=\text{O})_2-\text{O}\!-\!\text{C}_6\text{H}_{10}\!-\!\text{CH}_3 \xrightarrow[\text{acetone or DMSO}]{\text{X}^-} \text{X}\!-\!\text{C}_6\text{H}_{10}\!-\!\text{CH}_3$$

4. Specialized methods (not discussed in the text) are required for primary and secondary alcohols that have significant β-branching.

D. Preparation of Tertiary Alkyl Halides

1. Tertiary alkyl chlorides can be prepared from the corresponding tertiary alcohols by reaction with HCl under mild conditions.

$$\text{HO}-\text{C}(\text{CH}_3)_2-\text{CH}_3 \xrightarrow[\text{H}_2\text{O}]{\text{HCl}} \text{Cl}-\text{C}(\text{CH}_3)_2-\text{CH}_3$$

2. Tertiary alkyl bromides can be prepared from the corresponding tertiary alcohols by reaction with HBr under mild conditions.

$$\text{HO}-\text{C}(\text{CH}_3)_2-\text{CH}_3 \xrightarrow[\text{H}_2\text{O}]{\text{HBr}} \text{Br}-\text{C}(\text{CH}_3)_2-\text{CH}_3$$

3. Tertiary sulfonates do not undergo S_N2 reactions.

E. Preparation of Alkenes

1. Alkenes can be prepared by the acid-catalyzed dehydration of alcohols.

$$\text{H}_3\text{C}-\text{CH(H)}-\text{C(OH)(CH}_3)-\text{CH}_3 \xrightleftharpoons{\text{H}_3\text{PO}_4} (\text{H})(\text{H}_3\text{C})\text{C}=\text{C}(\text{CH}_3)(\text{CH}_3) + \text{H}_2\text{O}$$

2. The equilibrium in this reaction is driven towards the alkene by removal of the volatile alkene, removal of water, or both.

3. The relative rates of alcohol dehydration are in the order tertiary > secondary >> primary.

4. Sulfonate esters readily undergo the E2 reaction with alkoxide bases instead of undergoing S_N2 reactions.

$$\text{C}_6\text{H}_{11}-\text{CH}_2\text{OH} \xrightarrow[\text{pyridine}]{\text{PhSO}_2\text{Cl}} \text{C}_6\text{H}_{11}-\text{CH}_2-\text{O}-\text{S}(=\text{O})_2-\text{Ph} \xrightarrow[(\text{CH}_3)_3\text{COH}]{\text{K}^+\ (\text{CH}_3)_3\text{CO}^-} \text{C}_6\text{H}_{10}=\text{CH}_2$$

4. This reaction is especially useful
   a. when the acidic conditions of alcohol dehydration lead to rearrangements or other side reactions.
   b. for primary alcohols in which dehydration is not an option.

II. OXIDATION OF ALCOHOLS AND THIOLS

A. Oxidation of Alcohols to Aldehydes and Ketones

1. Primary and secondary alcohols are oxidized by reagents containing Cr(VI), that is, chromium in the +6 oxidation state, to give carbonyl compounds (compounds containing the carbonyl group, $>C=O$).
   a. The α-carbon atom of an alcohol must bear a hydrogen atom for oxidation to an aldehyde or ketone to occur.
   b. Primary alcohols react with Cr(VI) reagents to give aldehydes, but if water is present, aldehydes are further oxidized to carboxylic acids.

$$CH_3(CH_2)_4CH_2OH \xrightarrow[CH_2Cl_2]{CrO_3(\text{pyridine})_2} CH_3(CH_2)_4-\overset{O}{\overset{\|}{C}}-H$$

hexanol → hexanal

$$CH_3(CH_2)_4CH_2OH \xrightarrow[H_2SO_4/H_2O]{K_2Cr_2O_7} CH_3(CH_2)_4-\overset{O}{\overset{\|}{C}}-OH$$

hexanol → hexanoic acid

B. Oxidation of Alcohols to Carboxylic Acids

1. Primary alcohols can be oxidized to carboxylic acids using aqueous solutions of Cr(VI) such as aqueous potassium dichromate ($K_2Cr_2O_7$) in acid.

$$\text{(cyclohexenyl)}-CH_2OH \xrightarrow[H_2SO_4/H_2O]{K_2CrO_7} \text{(cyclohexenyl)}-CO_2H$$

2. Primary alcohols can be oxidized to carboxylic acids with potassium permanganate ($KMnO_4$) in basic solution, followed by acidification.

$$CH_3(CH_2)_5CH_2OH \xrightarrow{KMnO_4,\ ^-OH} \xrightarrow{H_3O^+} CH_3(CH_2)_5\overset{O}{\overset{\|}{C}}-OH$$

1-nonanol, a primary alcohol → nonanoic acid, a carboxylic acid

   a. Because $KMnO_4$ reacts with alkene double bonds, Cr(VI) is preferred for the oxidation of alcohols that contain double or triple bonds.
   b. Potassium permanganate is not used for the oxidation of secondary alcohols to ketones because many ketones react further with the alkaline permanganate reagent.

C. Oxidation of Thiols

1. Oxidation of a thiol takes place not at the carbon, but at the sulfur.
   a. The most commonly occurring oxidation products of thiols are disulfides and sulfonic acids.

$$R-S-S-R$$

general structure of a disulfide

$$R-\overset{O}{\overset{\|}{\underset{O}{\underset{\|}{S}}}}-OH$$

general structure of a sulfonic acid

   b. The Lewis structures of these derivatives require either violation of the octet rule or separation of formal charge.

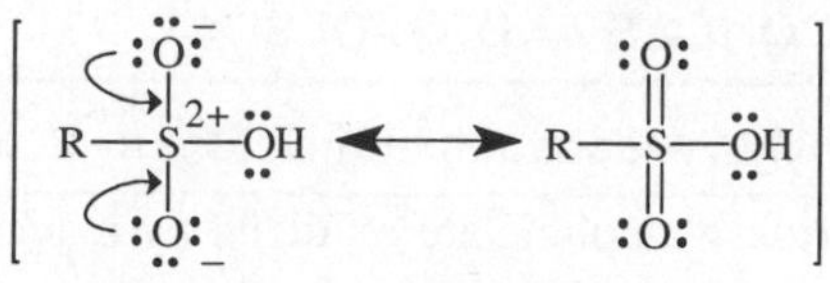

$$\left[R{-}\overset{:\ddot{O}:^{-}}{\underset{:\ddot{O}:^{-}}{S^{2+}}}{-}\ddot{O}H \longleftrightarrow R{-}\overset{:O:}{\underset{:O:}{\overset{\|}{\underset{\|}{S}}}}{-}\ddot{O}H \right]$$

c. Sulfur can accommodate more than eight valence electrons because, in addition to its 3*s* and 3*p* orbitals, it has unfilled 3*d* orbitals of relatively low energy.

d. The same oxidation-number formalism used for carbon can be applied to oxidation at sulfur.

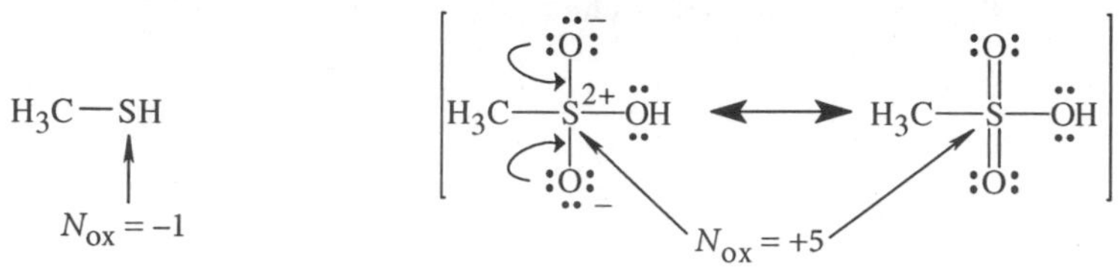

$$H_3C{-}SH \quad (N_{ox} = -1) \qquad \left[H_3C{-}\overset{:\ddot{O}:^{-}}{\underset{:\ddot{O}:^{-}}{S^{2+}}}{-}\ddot{O}H \longleftrightarrow H_3C{-}\overset{:O:}{\underset{:O:}{\overset{\|}{\underset{\|}{S}}}}{-}\ddot{O}H \right] \quad (N_{ox} = +5)$$

2. Sulfonic acids are formed by vigorous oxidation of thiols or disulfides with $KMnO_4$ or nitric acid (HNO_3).

$$\underset{\text{methanethiol}}{CH_3SH} \xrightarrow{HNO_3} \underset{\text{methanesulfonic acid}}{CH_3SO_3H}$$

3. Many thiols spontaneously oxidize to disulfides merely on standing in air (O_2).
4. Thiols can also be converted into disulfides by mild oxidants such as I_2 in base or Br_2 in CCl_4.
   a. These reactions can be viewed as a series of S_N2 reactions in which halogen and sulfur are attacked by thiolate anion.

$$\underset{\text{methanethiol}}{H_3C{-}S{-}H} \xrightarrow[-H_2O]{^{-}OH} H_3C{-}S^{-} \xrightarrow[-I^{-}]{I_2} H_3C{-}S{-}I \xrightarrow[-I^{-}]{RS^{-}} \underset{\text{methyl disulfide}}{H_3C{-}S{-}S{-}CH_3}$$

   b. When thiols and disulfides are present together in the same solution, an equilibrium among them is rapidly established.

$$\underset{\text{methanethiol}}{CH_3SH} + \underset{\text{diethyl disulfide}}{CH_3CH_2S{-}SCH_2CH_3} \rightleftharpoons \underset{\text{ethanethiol}}{CH_3CH_2SH} + \underset{\text{ethyl methyl disulfide}}{CH_3S{-}SCH_2CH_3}$$

STUDY GUIDE LINKS

✓10.1 Dehydration of Alcohols

The text stresses the close analogy between the E1 reaction and alcohol dehydration. Could the analogy between the chemistry of alcohols and that of alkyl halides be pursued even further? Could an alcohol be dehydrated by an E2 reaction that uses a strong base? Let's consider whether this would be reasonable. First, in order for an elimination to take place, the —OH group must depart as *hydroxide ion,* $^{-}$OH. Chapter 9 showed that —OH does not act as a leaving group in S_N2 reactions—it is too strong a base. For the same reason, it also cannot act as a leaving group in E2 reactions. (There are a few exceptions that involve cases in which the β-hydrogen is unusually acidic.)

Notice that in the dehydration mechanism shown in Eq. 10.3, text p. 410, *water* is the base that removes the β-hydrogen, not hydroxide. When H_3O^+ acts as the acid, in most cases its conjugate base H_2O acts as the base. *Solutions that contain substantial amounts of hydronium ion cannot also contain enough hydroxide ion to act as a base in base-catalyzed reactions.* Likewise, in solutions that contain substantial hydroxide ion, water, not hydronium ion, serves as the acid, because hydronium ion is not present to a significant extent in solutions of hydroxide.

✓10.2 Rearrangements Involving Cyclic Carbon Skeletons

The rearrangement shown in Eq. 10.6 of the text occurs within a ring. Following such rearrangements sometimes proves to be tricky even for students who have a good understanding of carbocation rearrangements. The key to following such rearrangements, at least at first, is to *draw out the individual carbon atoms implied by the skeletal formula,* as demonstrated below. As this equation shows, the rearrangement is nothing more than the migration of carbon-5 from carbon-2 to carbon-1.

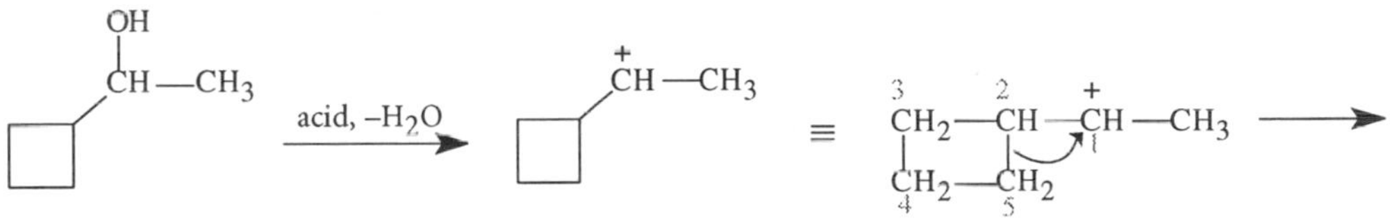

acid, $-H_2O$

carbocation intermediate

carbocation intermediate with all carbons shown

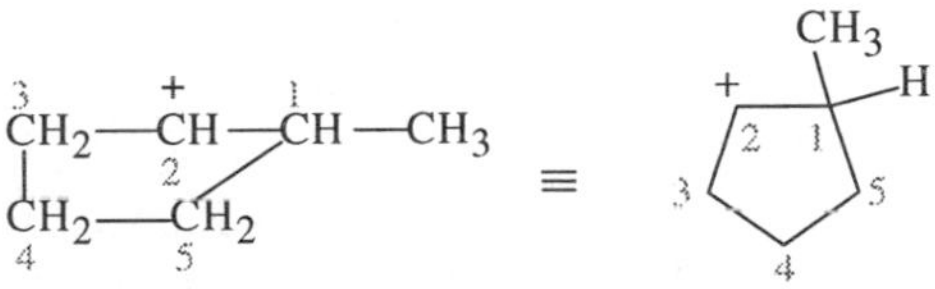

rearranged carbocation

rearranged carbocation, skeletal structure

Notice that the rearranged carbocation is drawn first with all the atoms as they were before rearrangement; then the structure is drawn more conventionally.

As a result of this rearrangement, carbon-2 becomes electron-deficient, and carbon-1 becomes part of a ring that is now one carbon larger. Conceptually, this ring expansion is analogous to the expansion of a noose in a rope:

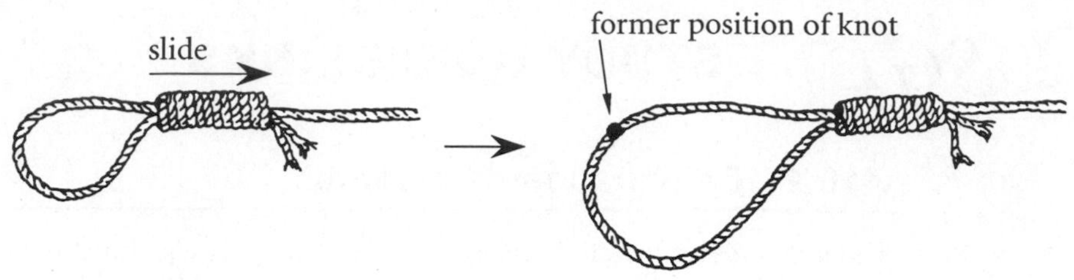

The product is formed by loss of a β-proton from the rearranged carbocation:

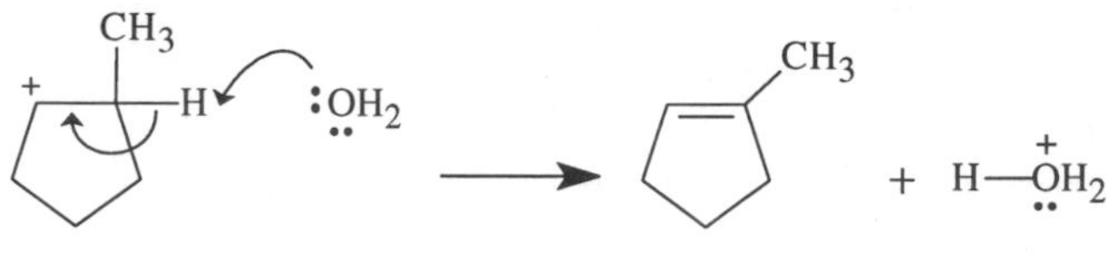

10.3 Mechanism of Sulfonate Ester Formation

As the text suggests, the formation of sulfonate esters is another nucleophilic substitution reaction. The role of the pyridine is to form a small amount of the conjugate base alkoxide ion of the alcohol:

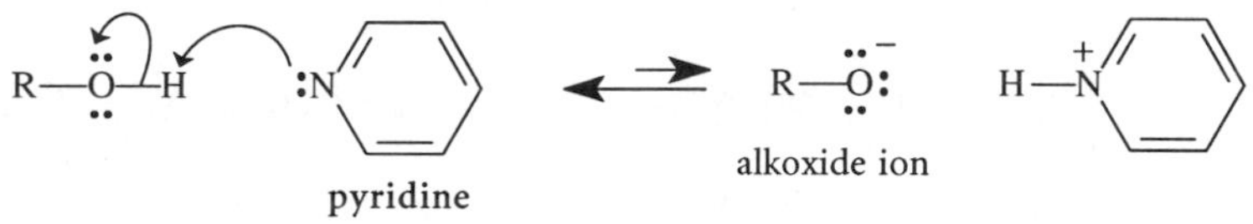

This alkoxide, acting as a nucleophile, displaces chloride ion from the sulfonyl chloride in a nucleophilic substitution reaction to form the sulfonate ester.

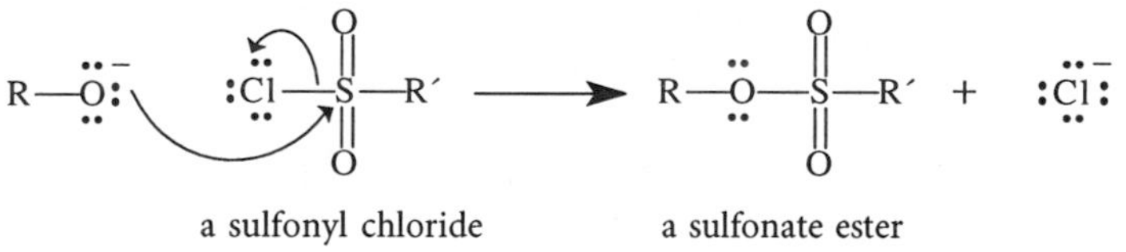

Notice that there is really nothing very new in this mechanism; the only difference is that a sulfonyl chloride instead of an alkyl halide undergoes substitution.

✓10.4 Oxidations and Reductions

You may have learned more extensive rules for assigning oxidation numbers in general chemistry. In organic chemistry, the major concern is the oxidation states of carbon atoms, because the vast majority of oxidations in organic chemistry involve oxidation at carbon. Occasionally, the same ideas are applied to other atoms, such as sulfur, but such cases follow essentially the same rules.

✓10.5 More on Half-Reactions

The technique used for balancing the half-reaction in Eq. 10.28a–d of the text, shown in Study Problem 10.4, text p. 428, is relatively straightforward when oxygens are involved. How can half-reactions be written when oxygens are not involved? For example, what is the half-reaction in the following oxidation?

$$H_2C{=}CH_2 \xrightarrow{Br_2} Br{-}CH_2{-}CH_2{-}Br$$

Since oxygens are balanced with H_2O, the bromines can be balanced with HBr. Taking this approach results in the following half-reaction:

$$2\,\text{HBr} + \text{H}_2\text{C}{=}\text{CH}_2 \longrightarrow \text{Br—CH}_2\text{—CH}_2\text{—Br} + 2e^- + 2\text{H}^+$$

The corresponding inorganic half-reaction is

$$2e^- + 2\text{H}^+ + \text{Br}_2 \longrightarrow 2\,\text{HBr}$$

If the two half-reactions are added, and if the "dummy electrons" and the HBr molecules on each side of the equation are canceled, the sum is the overall bromine addition reaction:

$$2\,\cancel{\text{HBr}} + \text{H}_2\text{C}{=}\text{CH}_2 \longrightarrow \text{Br—CH}_2\text{—CH}_2\text{—Br} + \cancel{2e^-} + \cancel{2\text{H}^+}$$

$$\cancel{2e^-} + \cancel{2\text{H}^+} + \text{Br}_2 \longrightarrow 2\,\cancel{\text{HBr}}$$

Sum: $$\text{H}_2\text{C}{=}\text{CH}_2 + \text{Br}_2 \longrightarrow \text{Br—CH}_2\text{—CH}_2\text{—Br}$$

Although the notion of a half-reaction can thus be applied to bromine addition, it isn't very useful because not only "dummy electrons" but also "dummy HBr" molecules have to be used to make it work; HBr, as you know, is not involved in the bromine addition reaction of alkenes. On the other hand, water and protons *really are* involved in the oxidation of ethanol to acetic acid, so that the idea of a half-reaction is somewhat more realistic.

When you want to determine the number of "electrons lost" and "electrons gained" in an oxidation or reduction reaction, you'll find it much simpler as a rule to calculate the change in oxidation number than to balance a half-reaction. The only time half-reactions are worth the effort in organic chemistry is when you have to balance a complicated oxidation reduction equation, as demonstrated in Study Problem 10.5 (text p. 428). (One certainly doesn't need half-reactions to balance the bromine-addition reaction above.) Nevertheless, it is important for you to understand the concept of a half-reaction because biochemists frequently discuss redox processes in terms of half-reactions. If you take biochemistry, you'll probably be glad you learned (or re-learned) these concepts here.

10.6 Symmetry Relationships among Constitutionally Equivalent Groups

In Sec. 6.1A, you learned that enantiomers are molecules that are noncongruent mirror images. You also learned in Sec. 6.1C that enantiomers lack certain *symmetry elements,* such as internal mirror planes. The purpose of this study guide link is to show that the different types of constitutionally equivalent groups—homotopic, enantiotopic, and diastereotopic groups—can also be classified by their symmetry relationships.

Homotopic groups are interchanged by a rotation of the molecule that gives an indistinguishable structure. For example, rotation of the methylene chloride molecule 180° about an axis bisecting the two C—Cl (or the two C—H) bonds (dashed line) exchanges the hydrogens and the chlorines, yet leaves the molecule looking exactly as it did before the rotations:

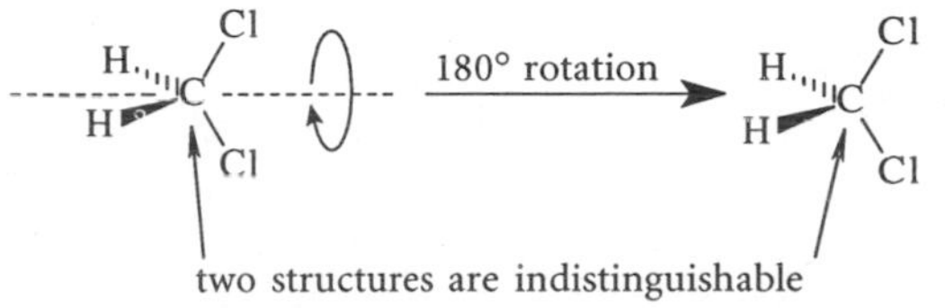

Cl, H, C, H, Cl — 180° rotation → Cl, H, C, H, Cl

two structures are indistinguishable

Consequently, the hydrogens are homotopic, and the chlorines are also homotopic. (The hydrogens of CH_3Cl are also homotopic, as shown in the text; perhaps you might want to demonstrate to yourself the rotation of this molecule that interchanges the three hydrogens while leaving the molecule invariant.)

Enantiotopic groups cannot be interchanged by such a rotation. For example, a rotation that interchanges the two enantiotopic α-protons of ethanol gives a molecule that looks different:

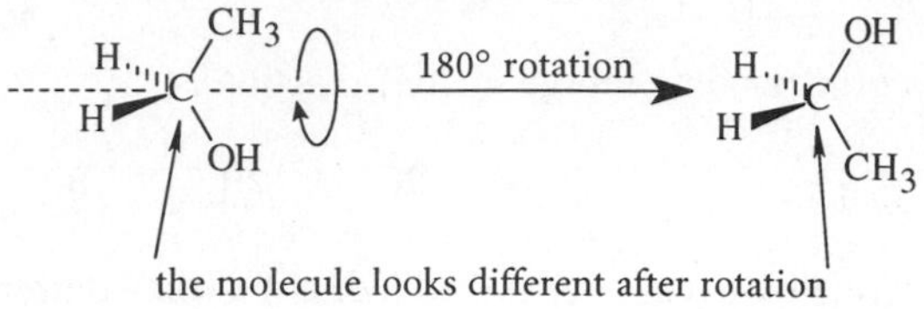

That is, the CH_3 group is "up" before the rotation and the OH group is "down." However, after the rotation; these groups have changed positions. The second structure is not congruent to the first without turning it over again.

Enantiotopic groups are related as object and mirror image by an internal mirror plane of the molecule. The internal mirror plane of the ethanol molecule (dashed line in the following diagram) contains the CH_3, the OH, and the central carbon, and it bisects the angle between the two C—H bonds at the α-carbon.

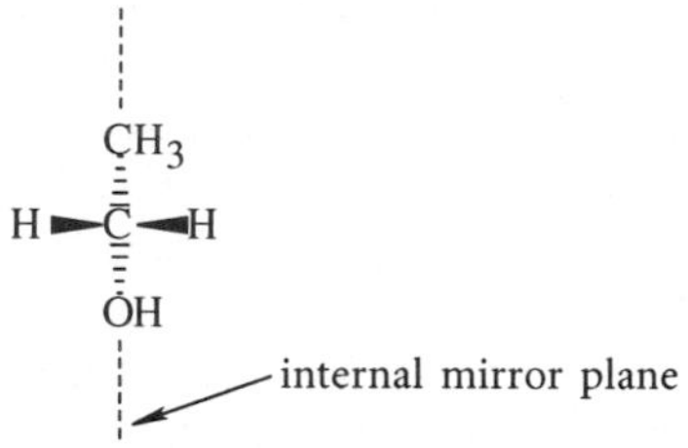

The two hydrogens of ethanol are related as object and mirror image by this plane; hence, these two hydrogens are enantiotopic. Notice that two homotopic groups might also be related by an internal mirror plane, but in order to be homotopic, they must *in addition* be related by a rotation of the molecule as described above. Since a chiral molecule by definition cannot contain an internal mirror plane, it follows that a chiral structure *cannot* contain enantiotopic groups (but it can contain homotopic and diastereotopic groups!).

Finally, diastereotopic protons have no symmetry relationship whatsoever.

SOLUTIONS TO PROBLEMS

Solutions to In-Text Problems

10.1 (a) (b)

$$\begin{matrix} Ph & & H \\ & C{=}C & \\ H & & Ph \end{matrix}$$

$CH_3CH{=}C(CH_3)CH_2CH_2CH_2CH_3$ (*E* and *Z*) *A* + $CH_3CH_2C(CH_3){=}CHCH_2CH_2CH_3$ (*E* and *Z*) *B* + $CH_3CH_2C({=}CH_2)CH_2CH_2CH_2CH_3$ *C*

10.2 In part (b), the *E* isomers of *A* and *B* will be formed because the *E* isomers are considerably more stable. Because alkenes *A* and *B* have more branching at the double bond than *C*, the *E* isomers of *A* and *B* will be formed in greater amount than alkene *C*. (Recall that the E1 reaction, of which dehydration is an example, gives the most stable alkene isomers.)

10.3 (a) 1-Methylcyclohexanol and either *cis*- or *trans*-2-methylcyclohexanol should give 1-methylcyclohexene as the major product of dehydration, because this is the most stable alkene that could be formed in each case. The tertiary alcohol 1-methylcyclohexanol should react most rapidly because it involves a tertiary carbocation intermediate; dehydrations of the other alcohols involve a less stable secondary carbocation intermediate.

1-methylcyclohexanol, 2-methylcyclohexanol $\xrightarrow{H_3O^+}$ 1-methylcyclohexene

Note that the carbocation that is initially formed from 2-methylcyclohexanol would undoubtedly rearrange to the same tertiary carbocation involved in the dehydration of 1-methylcyclohexanol; the same product must result from the same intermediate. 1-Methylcyclohexene should also be the major product formed from the unrearranged carbocation because it is the more branched of the two possible alkenes.

(b) Both 3-methyl-3-pentanol and 3-methyl-2-pentanol should give 3-methyl-2-pentene as the major product. The tertiary alcohol 3-methyl-3-pentanol should dehydrate more rapidly.

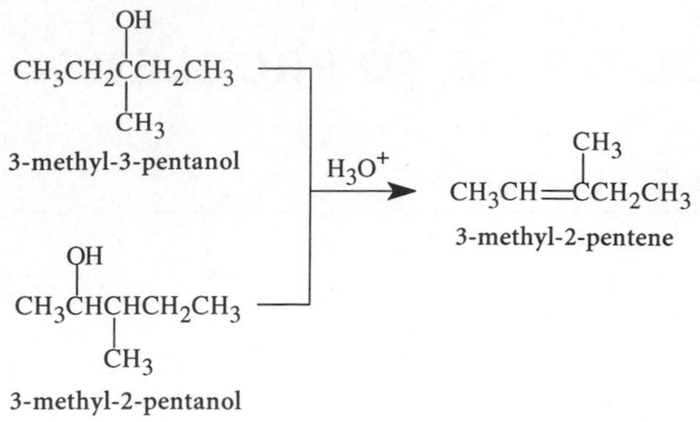

$$CH_3CH_2C(OH)(CH_3)CH_2CH_3 \text{ (3-methyl-3-pentanol)} \xrightarrow{H_3O^+} CH_3CH{=}C(CH_3)CH_2CH_3 \text{ (3-methyl-2-pentene)}$$

$$CH_3CH(OH)CH(CH_3)CH_2CH_3 \text{ (3-methyl-2-pentanol)} \xrightarrow{H_3O^+} CH_3CH{=}C(CH_3)CH_2CH_3 \text{ (3-methyl-2-pentene)}$$

10.4 (a) As the text suggests, the mechanism involves a rearrangement of the initially formed carbocation to a more stable carbocation.

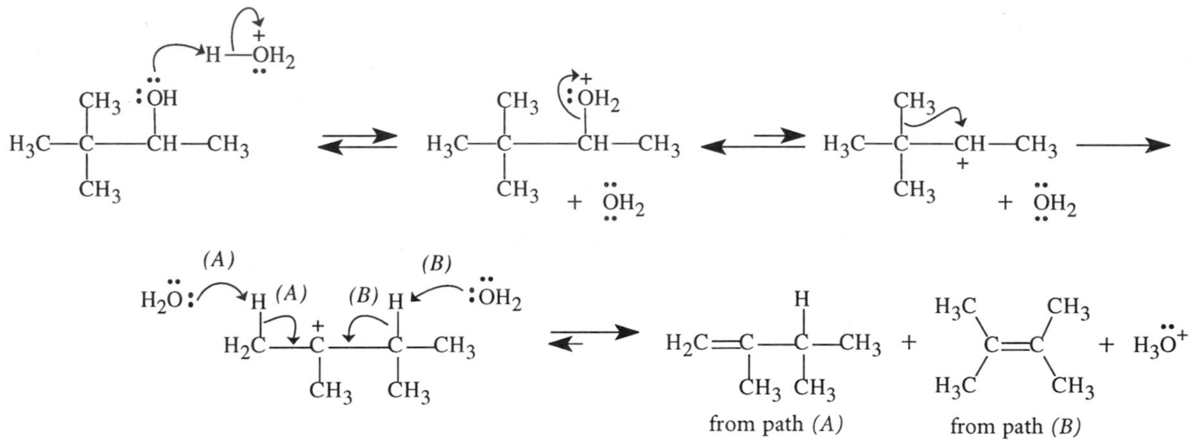

$$(CH_3)_3C{-}CH(OH){-}CH_3 + H{-}\overset{+}{O}H_2 \rightleftharpoons (CH_3)_3C{-}CH(\overset{+}{O}H_2){-}CH_3 \rightleftharpoons (CH_3)_3C{-}\overset{+}{C}H{-}CH_3 + OH_2 \longrightarrow$$

$$H_2C{-}\overset{+}{C}(CH_3){-}CH(CH_3){-}CH_3 \text{ (paths (A) and (B), with } H_2O\text{)} \rightleftharpoons H_2C{=}C(CH_3){-}CH(CH_3){-}CH_3 + (H_3C)_2C{=}C(CH_3)_2 + H_3O^+$$

from path (A) — from path (B)

(b) Rearrangement occurs because a more strained secondary carbocation is converted into a less strained, and therefore more stable, secondary carbocation.

10.5 (a)

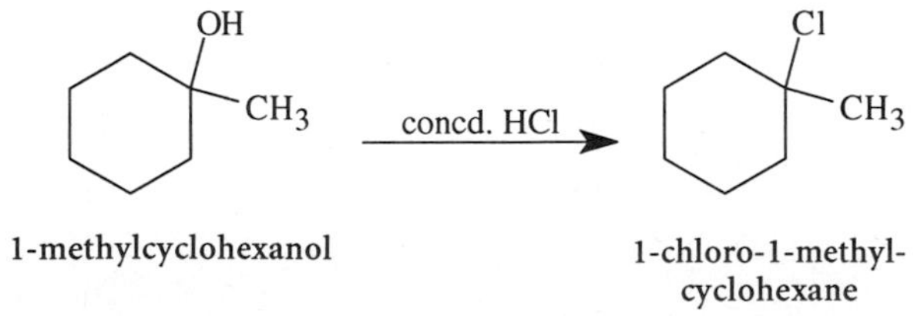

$$\text{1-methylcyclohexanol} \xrightarrow{\text{concd. HCl}} \text{1-chloro-1-methylcyclohexane}$$

(b)

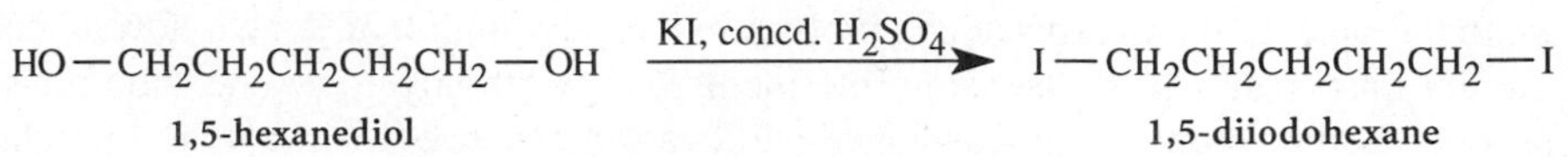

$$HO{-}CH_2CH_2CH_2CH_2CH_2{-}OH \xrightarrow{\text{KI, concd. } H_2SO_4} I{-}CH_2CH_2CH_2CH_2CH_2{-}I$$

1,5-hexanediol — 1,5-diiodohexane

In part (b), why not use concentrated HI? In fact, this is a perfectly acceptable answer. However, let's consider some practical issues. Concentrated HI itself is a really noxious material that is difficult to store without significant decomposition to I_2. As you recall from general chemistry, an iodide salt plus strong acid provides a solution of HI. Because KI and H_2SO_4 are readily stored and relatively stable compounds, mixing them in the reaction vessel to provide HI *in situ* (that is, "on the spot," or "in the reaction vessel") is a much more convenient way to supply HI.

10.6 Aha! Another carbocation rearrangement:

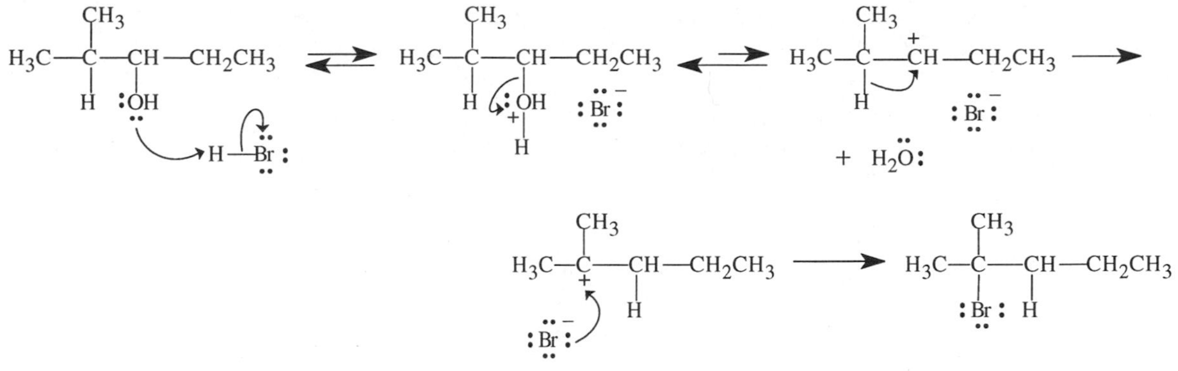

$$H_3C{-}C(CH_3)(H){-}CH(\ddot{O}H){-}CH_2CH_3 + H{-}\ddot{Br}: \rightleftharpoons H_3C{-}C(CH_3)(H){-}CH(\overset{+}{O}H_2){-}CH_2CH_3 \; :\ddot{Br}:^- \rightleftharpoons H_3C{-}C(CH_3)(H){-}\overset{+}{C}H{-}CH_2CH_3 \; :\ddot{Br}:^- + H_2\ddot{O}: \longrightarrow$$

$$H_3C{-}\overset{+}{C}(CH_3){-}CH(H){-}CH_2CH_3 \; :\ddot{Br}:^- \longrightarrow H_3C{-}C(CH_3)(:\ddot{Br}:){-}CH(H){-}CH_2CH_3$$

10.7 (a)

$$CH_3CH_2CH_2OH \xrightarrow[H_2SO_4]{\text{concd. HBr}} CH_3CH_2CH_2Br$$

1-propanol → 1-bromopropane

(b) $I{-}CH_2CH_2CH_2{-}I$

1,3-diiodopropane

(c)

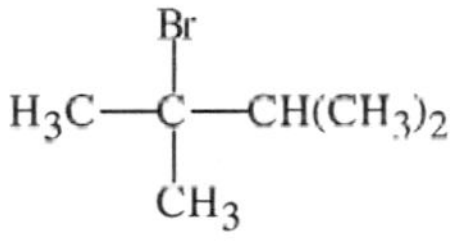

$$H_3C{-}C(Br)(CH_3){-}CH(CH_3)_2$$

(a rearrangement product)

(d) This compound, neopentyl alcohol, will not react, because it is primary, and therefore cannot undergo the S_N1 reaction; and it has too much β-branching to undergo the S_N2 reaction.

10.8 (a)

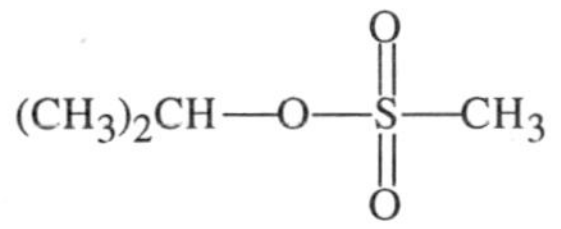

$$(CH_3)_2CH{-}O{-}S(=O)_2{-}CH_3$$

isopropyl methanesulfonate
(isopropyl mesylate)

(b)

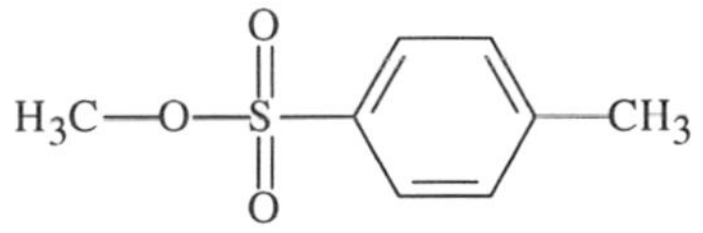

$$H_3C{-}O{-}S(=O)_2{-}C_6H_4{-}CH_3$$

methyl *p*-toluenesulfonate
(methyl tosylate)

(c)

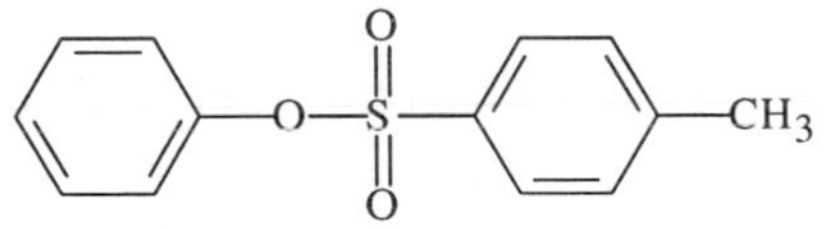

$$C_6H_5{-}O{-}S(=O)_2{-}C_6H_4{-}CH_3$$

phenyl tosylate
(phenyl *p*-toluenesulfonate)

(d)

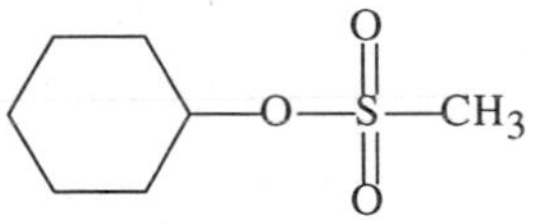

$$C_6H_{11}{-}O{-}S(=O)_2{-}CH_3$$

cyclohexyl mesylate
(cyclohexyl methanesulfonate)

10.9 (a)

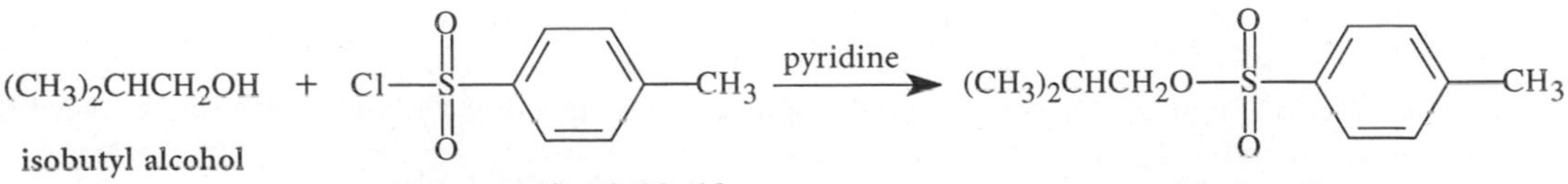

$$(CH_3)_2CHCH_2OH + Cl{-}S(=O)_2{-}C_6H_4{-}CH_3 \xrightarrow{\text{pyridine}} (CH_3)_2CHCH_2O{-}S(=O)_2{-}C_6H_4{-}CH_3$$

isobutyl alcohol + *p*-toluenesulfonyl chloride (tosyl chloride) → isobutyl tosylate

(b)

cyclohexanol-OH + $Cl-SO_2-CH_3$ $\xrightarrow{\text{pyridine}}$ cyclohexyl-$O-SO_2-CH_3$

cyclohexanol; mesyl chloride (methanesulfonyl chloride); cyclohexyl mesylate

Remember that when we outline a synthesis we provide key starting materials and products, but do not necessarily show by-products and do not necessarily balance the reaction. Notice that by-products are missing in the previous solutions. This is acceptable as long as the request is for a synthesis rather than for a balanced reaction.

10.10 (a)

$$(CH_3)_2CHCH_2CH_2CH_2OH \xrightarrow{\text{TsCl, pyridine}} (CH_3)_2CHCH_2CH_2CH_2OTs \xrightarrow[\text{polar aprotic solvent}]{Na^+\ I^-} (CH_3)_2CHCH_2CH_2CH_2I$$

(b) There are two possible syntheses. In the first, the alkoxide 3-cyclopentyl-1-propanolate is allowed to react with methyl tosylate:

cyclopentyl-$CH_2CH_2CH_2OH$ $\xrightarrow[-H_2]{\text{NaH}}$ cyclopentyl-$CH_2CH_2CH_2O^-\ Na^+$ $\xrightarrow{H_3CO-SO_2-C_6H_4-CH_3\ \text{(methyl tosylate)}}$ cyclopentyl-$CH_2CH_2CH_2OCH_3$

3-cyclopentyl-1-propanol; sodium 3-cyclopentyl-1-propanolate; (3-methoxypropyl)cyclopentane

In the second, the tosylate of 3-cyclopentyl-1-propanol is allowed to react with sodium methoxide:

cyclopentyl-$CH_2CH_2CH_2OH$ $\xrightarrow[\text{pyridine}]{Cl-SO_2-C_6H_4-CH_3\ (p\text{-toluenesulfonyl chloride})}$

3-cyclopentyl-1-propanol

cyclopentyl-$CH_2CH_2CH_2O-SO_2-C_6H_4-CH_3$ $\xrightarrow{Na^+\ ^-OCH_3}$ cyclopentyl-$CH_2CH_2CH_2OCH_3$

3-cyclopentylpropyl tosylate; (3-methoxypropyl)cyclopentane

Because methyl tosylate cannot undergo the competing E2 reaction, the first synthesis is probably the better one, although both are reasonable.

10.11 (a) The major product is the nitrile that results from the S_N2 reaction of cyanide ion with the tosylate:

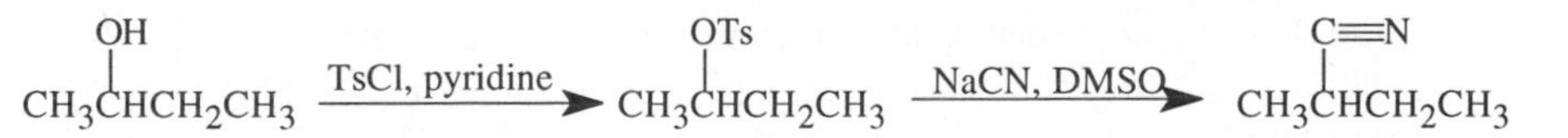

$$\underset{\text{2-butanol}}{CH_3\overset{OH}{\overset{|}{C}}HCH_2CH_3} \xrightarrow{\text{TsCl, pyridine}} \underset{\textit{sec}\text{-butyl tosylate}}{CH_3\overset{OTs}{\overset{|}{C}}HCH_2CH_3} \xrightarrow{\text{NaCN, DMSO}} \underset{\text{2-methylbutanenitrile}}{CH_3\overset{C\equiv N}{\overset{|}{C}}HCH_2CH_3}$$

(b) The first reaction forms the methanesulfonate (mesylate) ester. Because a strong, highly branched alkoxide base is used, the major product of the second reaction is the alkene 4-methyl-1-pentene.

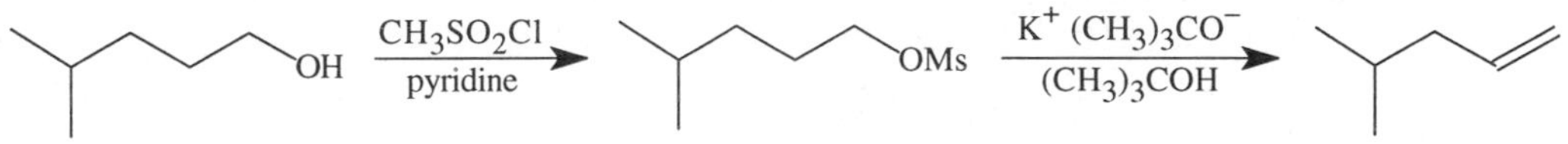

4-methyl-1-pentanol $\xrightarrow[\text{pyridine}]{CH_3SO_2Cl}$ 4-methyl-1-pentyl mesylate (OMs) $\xrightarrow[(CH_3)_3COH]{K^+\ (CH_3)_3CO^-}$ 4-methyl-1-pentene

10.12 (a) $CH_3O-P(=O)(OCH_3)-OCH_3$ trimethyl phosphate

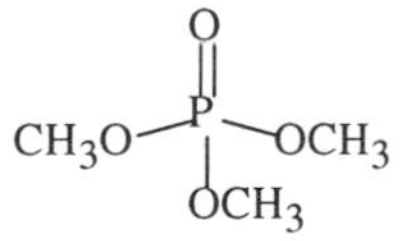

(b) $CH_3CH_2O-P(=O)(OH)-OH$ ethyl phosphate

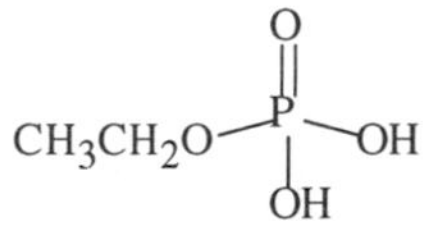

10.13 (a) Reaction with water:

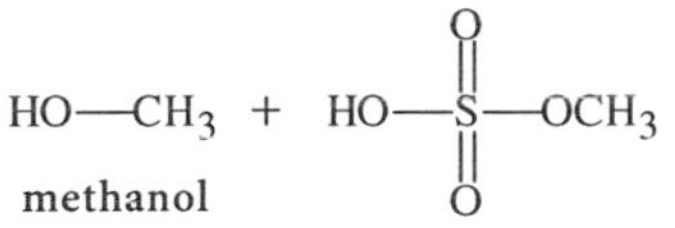

$HO—CH_3 + HO—S(=O)_2—OCH_3$

methanol

(b) Reaction with methylamine:

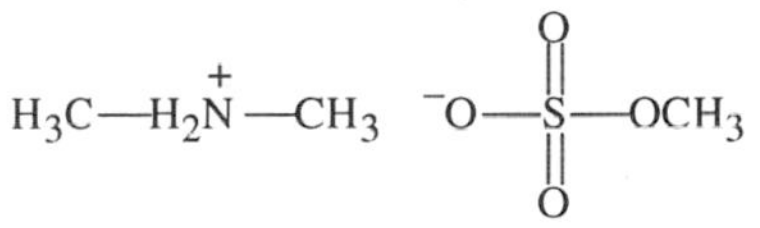

$H_3C—\overset{+}{H_2N}—CH_3 \quad {}^-O—S(=O)_2—OCH_3$

dimethylammonium methylsulfate

(c) Reaction with sodium ethoxide:

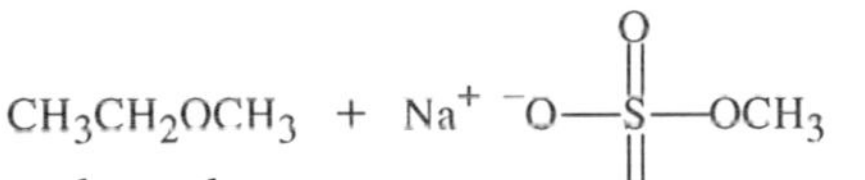

$CH_3CH_2OCH_3 + Na^+ \; {}^-O—S(=O)_2—OCH_3$

methoxyethane

(d) Reaction with sodium 1-propanethiolate:

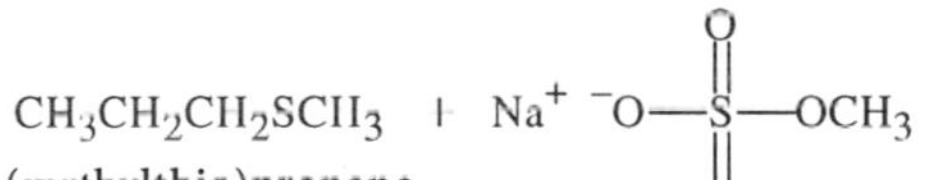

$CH_3CH_2CH_2SCH_3 + Na^+ \; {}^-O—S(=O)_2—OCH_3$

1-(methylthio)propane
(methyl propyl sulfide)

10.14 1-Butanol can be converted into 1-bromobutane by treating it with phosphorus tribromide (PBr_3). (See Eq. 10.24 on text p. 421.) 1-Butanol can also be converted into 1-chlorobutane by first converting it into a sulfonate ester such as butyl tosylate, then treating that ester with sodium bromide in a polar aprotic solvent such as DMSO. Finally, and perhaps most simply, 1-butanol can be converted into 1-bromobutane by treatment with concentrated HBr and H_2SO_4.

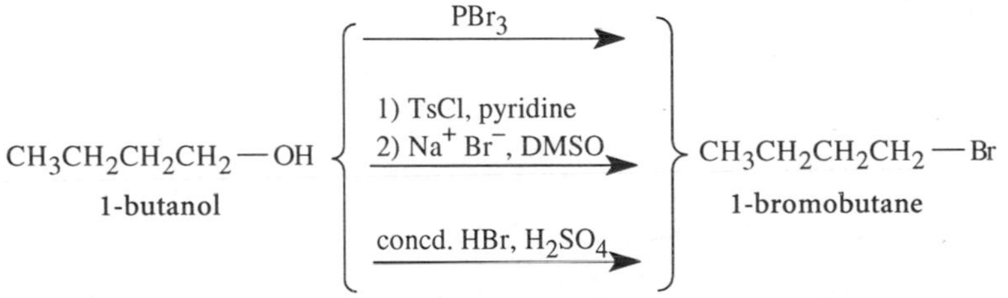

$CH_3CH_2CH_2CH_2—OH$ (1-butanol) → $CH_3CH_2CH_2CH_2—Br$ (1-bromobutane), by any of:
- PBr_3
- 1) TsCl, pyridine 2) $Na^+ Br^-$, DMSO
- concd. HBr, H_2SO_4

10.15 (a) Eq. 10.23 of the text shows that the attack of the chloride ion on the chlorosulfite ester is an S_N2 reaction; consequently, it should take place with inversion. It follows that the product should be the *R* enantiomer of $CH_3CH_2CH_2CHD—Cl$.

(b) Eq. 10.25 of the text shows that the attack of bromide ion on the reactive intermediate occurs by a concerted (S_N2) substitution reaction. Therefore, the reaction should occur with inversion of stereochemistry, and the product would then be (*S*)-2-bromopentane. The S_N2 reaction occurs at an acceptable rate on a secondary carbon in the absence of β-branching. In addition, the reaction is fast because the leaving group is a *very* weak base. However, it is possible that some S_N1 mechanism could occur; this is hard to predict. To the extent that the S_N1 reaction occurs, some racemization also takes place.

10.16 (a) Because the alcohol groups are primary, concentrated HBr/H_2SO_4 will bring about the desired reaction. Treatment with PBr_3 will also work.

(b) The simplest method for effecting the conversion shown is to treat the alcohol with thionyl chloride and pyridine. Conversion of the alcohol to a sulfonate ester and treatment of the ester with sodium chloride in a polar aprotic solvent would also work, but involves more steps.

(c) The alcohol reacts with HBr to give a carbocation intermediate that rearranges. However, this is precisely the reaction desired; reaction of the rearranged carbocation with bromide ion gives the desired product. (See Study Guide Link 10.2 on p. 307 of this manual.)

(d) Because a carbocation intermediate and hence rearrangements are a distinct possibility if this alcohol is exposed to acidic reagents such as HBr, the sulfonate ester method should be used. Thus, treat the alcohol with tosyl chloride and pyridine, and treat the resulting tosylate with sodium bromide in a polar aprotic solvent. This type of solvent suppresses carbocation formation.

10.17 Cyclopentanol and cyclohexanol are two secondary alcohols (among others) that could be converted into the corresponding alkyl bromides without rearrangements. If hydride shifts occur within a carbocation intermediate derived from these alcohols, the same carbocation would be formed.

OH OH

cyclopentanol cyclohexanol

10.18 (a) The bromination of methane to give methyl bromide is a two-electron oxidation.

(b) This transformation (the conversion of toluene into benzoic acid) is a six-electron oxidation.

(c) This transformation (the conversion of an alkyl iodide into an alkane) is a two-electron reduction.

(d) This transformation (the dihydroxylation of an alkene by $KMnO_4$) is a two-electron oxidation.

(e) The ozonolysis of 2-methyl-2-butene is a six-electron oxidation. Notice that both carbons of the double bond must be considered, even though the molecule is "split in two" as a result of the reaction.

(f) The addition of HBr to an alkene is neither an oxidation nor a reduction; although one carbon of the double bond undergoes an oxidation, the other undergoes a compensating reduction.

(g) This is a two-electron reduction; note that the negative charge contributes –1 to the oxidation number of the product.

10.19 (b) $2H_2O + PhCH_3 \longrightarrow PhCO_2H + 6e^- + 6H^+$

(d) Notice that two electrons are "lost" in this half-reaction, and that this is consistent with its classification in the solution to the previous problem (10.18d) as a two-electron oxidation:

$$CH_3CH{=}CHPh + 2H_2O \longrightarrow CH_3\underset{|}{\overset{OH}{C}}H{-}\overset{OH}{\underset{|}{C}}HPh + 2e^- + 2H^+$$

10.20 (a) This is an oxidation-reduction reaction; the alkene is reduced, and the H_2 is oxidized.

(b) This is an oxidation-reduction reaction; the organic compound is reduced, and the $^-AlH_4$ is oxidized.

(c) This is an oxidation-reduction reaction; the alkene is oxidized, and the Br_2 is reduced.

(d) This is an oxidation-reduction reaction; the organic compound is oxidized, and oxygen is reduced.

10.21 First balance the equation; then, from the balanced equation, determine how many moles of permanganate per mole of toluene are required.

$$2H_2O + PhCH_3 \longrightarrow PhCO_2H + 6e^- + 6H^+$$

$$\overset{8}{\cancel{4}}H^+ + \overset{2}{}MnO_4^- + \overset{6}{\cancel{3}}e^- \longrightarrow \overset{2}{}MnO_2 + \overset{4}{\cancel{2}}H_2O$$

$$PhCH_3 + 2MnO_4^- + 2H^+ \longrightarrow PhCO_2H + 2H_2O + 2MnO_2$$

This balanced equation shows that two moles of permanganate are required to oxidize one mole of toluene.

A shorter way to solve a problem of this type is to count the number of electrons lost and gained in each half reaction, and then use enough of each reagent to make these balance. To illustrate, the oxidation of toluene is a six-electron oxidation, and the reduction of permanganate to MnO_2 is a three-electron reduction. Thus two moles of permanganate are required to "accept" the six electrons necessary to oxidize one mole of toluene.

10.22 (a) The primary alcohol is oxidized, but the tertiary alcohol is not. Because water is absent, oxidation by Cr(VI) stops at the aldehyde.

[Structure: cyclobutane bearing $CH_2CH{=}O$, and H_3C and HO on the same ring carbon]

(b) Each primary alcohol is oxidized; the product is the dialdehyde $O{=}CHCH_2CH{=}O$ (malonaldehyde).

10.23 (a) The carboxylic acid could be prepared from $(CH_3)_2CHCH_2CH_2CH_2CH_2OH$ (5-methyl-1-hexanol) by oxidation either with aqueous chromic acid [or any other form of aqueous Cr(VI)], or with aqueous potassium permanganate followed by acid.

(b) This compound (3-pentanone) can be prepared by a Collins oxidation of the corresponding alcohol, 3-pentanol. (Aqueous dichromate, illustrated below, could also be used.)

$$\underset{\text{3-pentanol}}{CH_3CH_2\overset{OH}{\overset{|}{C}}HCH_2CH_3} \xrightarrow{Na_2Cr_2O_7,\ H_2SO_4,\ H_2O} \underset{\text{3-pentanone}}{CH_3CH_2\overset{O}{\overset{||}{C}}CH_2CH_3}$$

(c) The aldehyde can be prepared from 5-methyl-1,5-hexanediol by the Collins oxidation or other variation of nonaqueous oxidation with Cr(VI).

$$\underset{\text{5-methyl-1,5-hexanediol}}{CH_3\overset{OH}{\overset{|}{\underset{CH_3}{\underset{|}{C}}}}CH_2CH_2CH_2CH_2OH} \xrightarrow{CrO_3(pyridine)_2} CH_3\overset{OH}{\overset{|}{\underset{CH_3}{\underset{|}{C}}}}CH_2CH_2CH_2CH{=}O$$

(d) This aldehyde can be prepared by a Collins oxidation of the corresponding alcohol.

[Structure: cyclopentene ring–CH_2OH] $\xrightarrow{CrO_3(pyridine)_2}$ [Structure: cyclopentene ring–$CH{=}O$]

(2-cyclopentenyl)methanol

10.24 This oxidation occurs in much the same way that NAD^+ oxidations occur: by transfer of a hydride from the alcohol to the carbocation. (Notice that it is a *hydride* (a hydrogen with two electrons), not a proton, that is transferred.)

$$\underset{\underset{\text{Ph}_3\text{C}^+}{\text{H}}}{\overset{:\ddot{\text{O}}\text{H}}{\text{CH}_3\text{CCH}_2\text{CH}_2\text{CH}_2\text{CH}_2\text{CH}_3}} \xrightarrow{\text{hydride transfer}} \left[\overset{:\ddot{\text{O}}-\text{H}}{\text{CH}_3\overset{+}{\text{C}}\text{CH}_2\text{CH}_2\text{CH}_2\text{CH}_2\text{CH}_3} \longleftrightarrow \overset{+\ddot{\text{O}}-\text{H}}{\text{CH}_3\overset{\|}{\text{C}}\text{CH}_2\text{CH}_2\text{CH}_2\text{CH}_2\text{CH}_3}\right] + \text{Ph}_3\text{CH} \xrightarrow{\text{F}-\bar{\text{B}}\text{F}_3} \text{H}-\text{F} + \text{BF}_3 + \overset{:\text{O}:}{\text{CH}_3\overset{\|}{\text{C}}\text{CH}_2\text{CH}_2\text{CH}_2\text{CH}_2\text{CH}_3}$$

10.25 (a) The two methyl groups *a* and *b* are constitutionally equivalent and diastereotopic. The pair of methyl groups *a* and *c* are constitutionally nonequivalent, as are the pair *b* and *c*.

Notice that when investigating the stereochemical relationships among several groups, the appropriate technique is to consider *each pair.* Thus, three relationships must be categorized within three groups (although it is possible that some relationships might be the same.) Here's an amusing little puzzle that uses the same sort of reasoning. Man *A* looks at a picture and comments about the person *B* in the picture, saying, "Brothers and sisters have I none; this man's father is my father's son." What is the relationship between *A* and *B*?

(b) Hydrogens *a* and *b* are constitutionally equivalent and enantiotopic. (The analysis of this case is essentially identical to the analysis of the α-hydrogens of ethanol; see Eq. 10.48, text p. 438.)

(c) Hydrogens *a* and *b* within a particular chair conformation are constitutionally equivalent and diastereotopic. However, the chair flip interchanges the positions of these two hydrogens and makes them completely equivalent, that is, homotopic, *over time.*

(d) The two fluorines are constitutionally equivalent and diastereotopic.

(e) By turning the Fischer projection 180°, you can verify that hydrogen *a* is indistinguishable from hydrogen *d,* and that hydrogen *b* is indistinguishable from hydrogen *c.* Consequently H^a and H^d are constitutionally equivalent and homotopic, as are H^b and H^c.

CH3
(H)—+—H
H—+—H
CH3

180° rotation in the page shows that the two molecules are congruent; hence, the two circled hydrogens are homotopic. →

CH3
H—+—H
H—+—(H)
CH3

Replacing H^a and H^b in turn with a "circled H" shows that these hydrogens are constitutionally equivalent and enantiotopic, as are H^c and H^d.

CH3
(H)—+—H
H—+—H
CH3

The two molecules are enantiomers; hence, the two circled hydrogens are enantiotopic.

CH3
H—+—(H)
H—+—H
CH3

The replacement test along with the allowed rotation of Fischer projections 180° in the plane of the page shows that H^a is constitutionally equivalent and enantiotopic to H^c, and that H^b is constitutionally equivalent and enantiotopic to H^d.

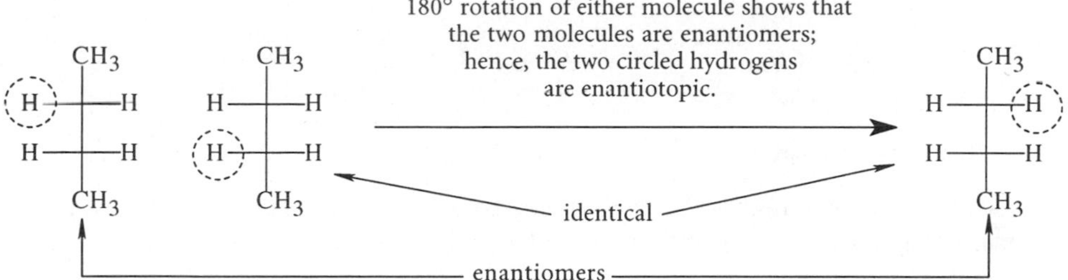

It is particularly easy to see the enantiotopic relationships discussed in the last two cases by applying the symmetry criterion discussed in Study Guide Link 10.6, beginning on p. 309 of this manual. Recall that a molecule containing enantiotopic groups must contain at least one internal mirror plane, and that enantiotopic groups are related as object and mirror image by such an internal mirror plane.

(f) Carbon-2 and carbon-4 are constitutionally equivalent and enantiotopic. Hydrogens *a* and *d* are constitutionally equivalent and diastereotopic, as are hydrogens *b* and *c*. Hydrogens *a* and *c* are constitutionally equivalent and enantiotopic, as are hydrogens *b* and *d*. Finally, hydrogens *a* and *b* are constitutionally equivalent and diastereotopic, as are hydrogens *c* and *d*.

10.26 In both, the hydrogen (or deuterium) is delivered from the pro-*(R)* position on NADH (the "up" position when NADH is drawn as it is in the problem) so that it ends up in the pro-*(R)* position of deuterated ethanol. In part (a), the resulting isotopically substituted ethanol is chiral. In part (b), because deuterium is delivered, the α-carbon of the resulting ethanol bears two deuteriums and it therefore has no asymmetric carbon; hence, the molecule is achiral.

(a) H (top), H_3C — — OH, D (bottom)

(*S*)-ethanol-1-*d*

(b) D (top), H_3C — — OH, D (bottom)

ethanol-1,1-d_2
(achiral)

10.27 In a thiol, the oxidation number of sulfur is –1; in a sulfonic acid, it is +5; hence, the change in oxidation number is +6. The oxidation of a thiol to a sulfonic acid is a six-electron oxidation. (This can be verified by writing a balanced half-reaction.)

10.28 The reaction is a nucleophilic substitution in which one sulfur attacks another. Because a thiolate is a much better nucleophile than a thiol—thiolates are *much* more basic than neutral thiols—a thiolate serves as the nucleophile in this reaction. A base is required to form the thiolate anion. The reaction is faster in the presence of base because the reaction rate depends on the nucleophile concentration, and base increases the concentration of the nucleophile.

$$CH_3CH_2\ddot{S}—H \quad :\ddot{O}^-CH_2CH_3 \rightleftharpoons CH_3CH_2\ddot{S}:^- + H—\ddot{O}CH_2CH_3$$

$$CH_3CH_2\ddot{S}:^- \quad \underset{\displaystyle CH_2CH_2CH_3}{S}—\ddot{S}CH_2CH_2CH_3 \rightleftharpoons CH_3CH_2\ddot{S}—SCH_2CH_2CH_3 + :\ddot{S}^-CH_2CH_2CH_3$$

10.29 (a) 2-Methyl-3-pentanol can be prepared by hydroboration-oxidation of 2-methyl-2-pentene, which can be prepared by dehydration of 2-methyl-2-pentanol:

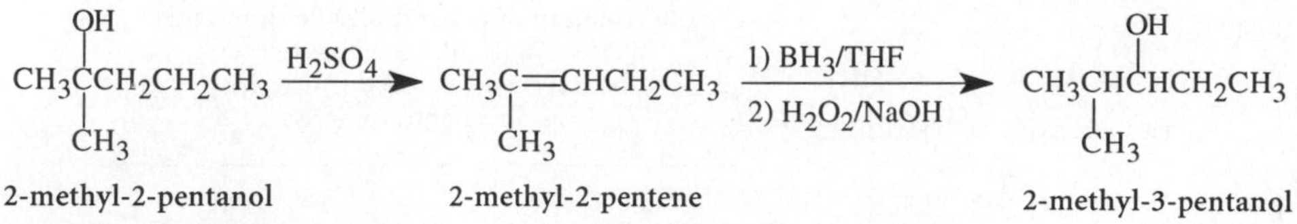

$$\underset{\text{2-methyl-2-pentanol}}{CH_3\overset{OH}{\underset{CH_3}{C}}CH_2CH_2CH_3} \xrightarrow{H_2SO_4} \underset{\text{2-methyl-2-pentene}}{CH_3\underset{CH_3}{C}{=}CHCH_2CH_3} \xrightarrow[\text{2) } H_2O_2/NaOH]{\text{1) } BH_3/THF} \underset{\text{2-methyl-3-pentanol}}{CH_3\underset{CH_3}{C}H\overset{OH}{C}HCH_2CH_3}$$

(b) The deuterium-containing alkane can be prepared by protonolysis of a Grignard reagent in D_2O; the Grignard reagent can be prepared from an alkyl halide; and the alkyl halide can be prepared from an alcohol.

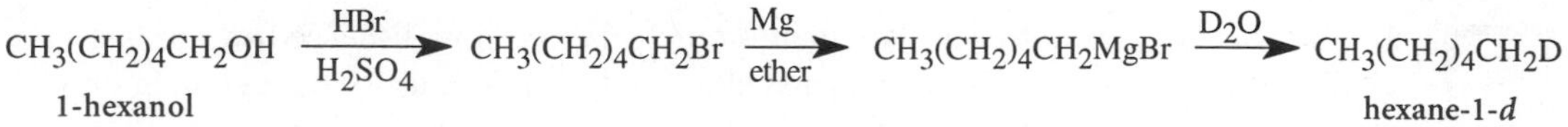

$$\underset{\text{1-hexanol}}{CH_3(CH_2)_4CH_2OH} \xrightarrow[H_2SO_4]{HBr} CH_3(CH_2)_4CH_2Br \xrightarrow[\text{ether}]{Mg} CH_3(CH_2)_4CH_2MgBr \xrightarrow{D_2O} \underset{\text{hexane-1-}d}{CH_3(CH_2)_4CH_2D}$$

(c) The carboxylic acid can be prepared by oxidation of a primary alcohol, which, in turn, can be prepared from the alkene starting material by hydroboration-oxidation.

$$\underset{\text{methylenecyclohexane}}{C_6H_{10}{=}CH_2} \xrightarrow[\text{2) } H_2O_2/NaOH]{\text{1) } BH_3/THF} \underset{\text{cyclohexylmethanol}}{C_6H_{11}{-}CH_2OH} \xrightarrow[\text{2) } H_3O^+]{\text{1) } KMnO_4/NaOH} \underset{\text{cyclohexanecarboxylic acid}}{C_6H_{11}{-}CO_2H}$$

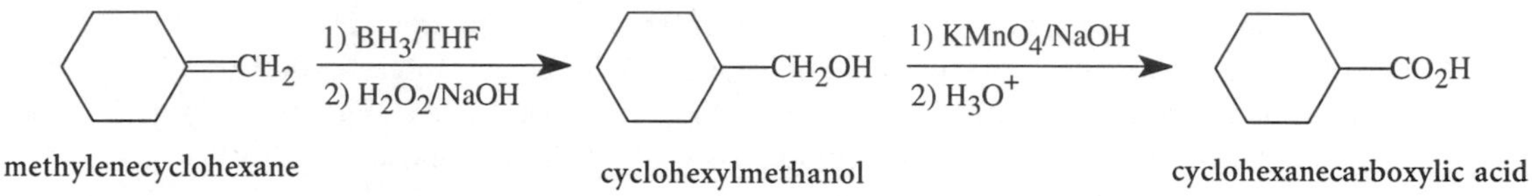

(d) The aldehyde can be prepared by oxidation of a primary alcohol; the required primary alcohol can be prepared by hydroboration-oxidation of an alkene; and the required alkene can be prepared by an E2 reaction of a primary alkyl halide using a branched base.

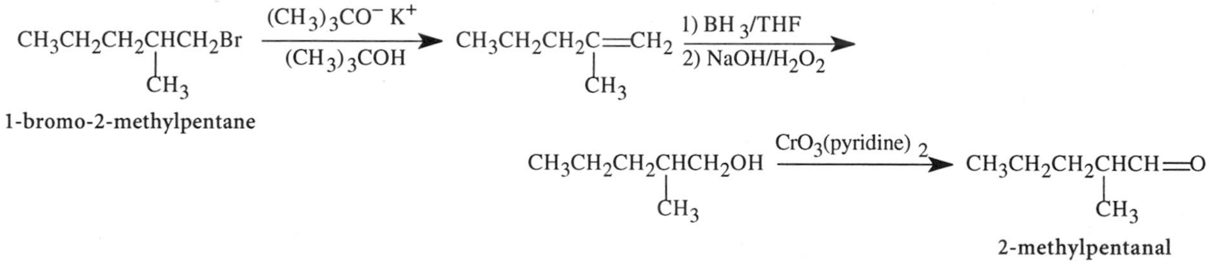

$$\underset{\text{1-bromo-2-methylpentane}}{CH_3CH_2CH_2\underset{CH_3}{C}HCH_2Br} \xrightarrow[(CH_3)_3COH]{(CH_3)_3CO^-\ K^+} CH_3CH_2CH_2\underset{CH_3}{C}{=}CH_2 \xrightarrow[\text{2) } NaOH/H_2O_2]{\text{1) } BH_3/THF}$$

$$CH_3CH_2CH_2\underset{CH_3}{C}HCH_2OH \xrightarrow{CrO_3(\text{pyridine})_2} \underset{\text{2-methylpentanal}}{CH_3CH_2CH_2\underset{CH_3}{C}HCH{=}O}$$

10.30 This reaction is an S_N2 reaction of iodide ion with the alkyl chloride. It occurs with inversion of stereochemistry at the α-carbon because the nucleophile displaces the leaving group by a backside-attack mechanism. (See Fig. 9.2, text p. 364.) An S_N2 reaction can be carried out successfully only on primary and secondary alkyl halides; even in these cases there must not be too much β-branching. The reason is that branches suffer van der Waals repulsions with both the nucleophile and the leaving group; these repulsions raise the energy of the transition state and thus retard the reaction. (See Fig. 9.3, text p. 365.)

Solutions to Additional Problems

10.31 (a) $CH_3CH_2CH_2CH_2Br$

(b) $CH_3CH_2CH_2CH_2\overset{+}{O}H_2\ HSO_4^-$

(c) $CH_3CH_2CH_2CH{=}O + Cr^{3+}$

(d) $CH_3CH_2CH_2CH_2O^-\ Na^+ + H_2$

(e) $CH_3CH_2CH_2CH_2OCH_3 + Na^+\ I^-$

(f) $CH_3CH_2CH_2CH_2OTs$ + (pyridinium) $\overset{+}{N}H\ Cl^-$

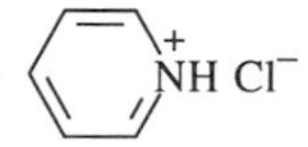

(g) $CH_3CH_2CH_2CH_2O^-\ {}^+MgBr + CH_3CH_2CH_3$

(h) $CH_3CH_2CH_2CH_2Cl + SO_2$ + (pyridinium) $\overset{+}{N}H\ Cl^-$

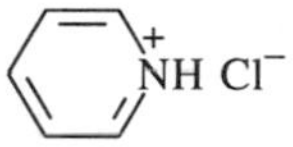

(i) $CH_3CH_2CH_2CH_2Br$

(j) $CH_3CH_2CH_2CH_2MgBr$

(k) $CH_3CH_2CH{=}CH_2 + K^+\ TsO^- + (CH_3)_3COH$

10.32 (a) $H_3C{-}C(CH_3)_2{-}Cl$

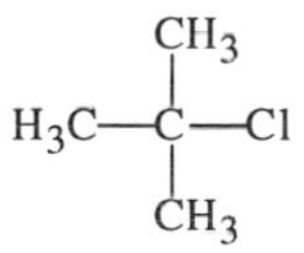

(b) no reaction

(c) $(CH_3)_2C{=}CH_2$

(d) no reaction

(e) $H_3C{-}C(CH_3)_2{-}O^-\ K^+ + H_2$

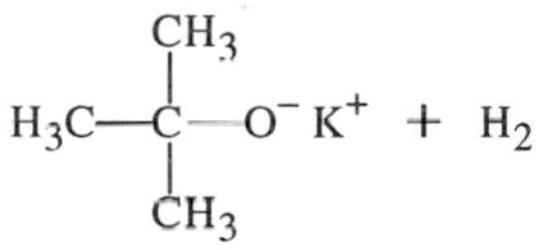

(f) $H_3C{-}C(CH_3)_2{-}OMs$ + (pyridinium) $\overset{+}{N}H\ Cl^-$

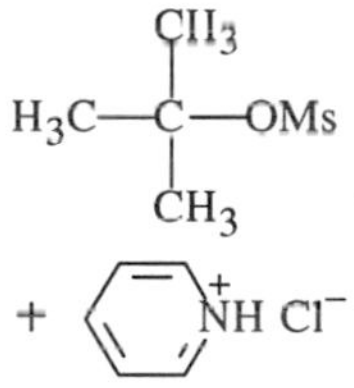

(g) $(CH_3)_2C{=}CH_2 + Na^+\ {}^-OMs + H_2O$

(h) $(CH_3)_2C{=}CH_2 + K^+\ Cl^- + (CH_3)_3COH$

10.33 (a)

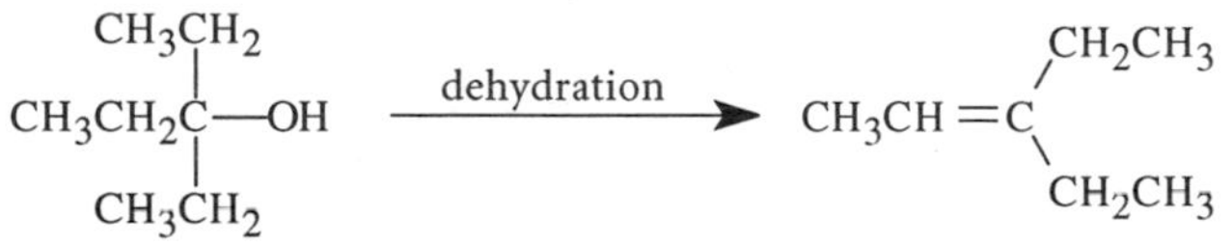

$$CH_3CH_2C(CH_2CH_3)_2{-}OH \xrightarrow{\text{dehydration}} CH_3CH{=}C(CH_2CH_3)_2$$

3-ethyl-3-pentanol

(b)

$$PhCH(OH)CH_2Ph \xrightarrow{\text{dehydration}} PhCH{=}CHPh \xrightarrow{O_3,\ \text{then}\ (CH_3)_2S} 2\ PhCH{=}O$$

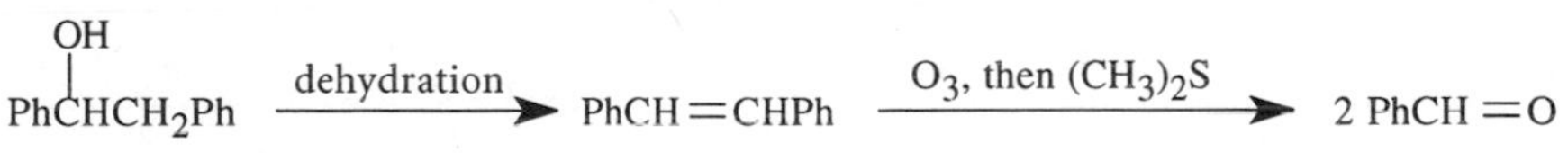

1,2-diphenylethanol

(c) Oxidation of 2-methyl-1-butanol gives the same product as ozonolysis of *trans*-3,6-dimethyl-4-octene:

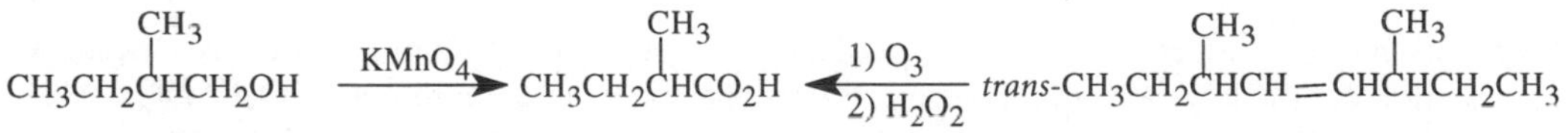

$$CH_3CH_2CH(CH_3)CH_2OH \xrightarrow{KMnO_4} CH_3CH_2CH(CH_3)CO_2H \xleftarrow[2)\ H_2O_2]{1)\ O_3} trans\text{-}CH_3CH_2CH(CH_3)CH{=}CHCH(CH_3)CH_2CH_3$$

2-methyl-1-butanol — 3,6-dimethyl-4-octene

10.34 This is a triester of 1,2,3-propanetriol (glycerol) and nitric acid, known by its traditional name of nitroglycerin. It is derived from glycerol (a triol) and nitric acid:

$$\underset{\text{1,2,3-propanetriol (glycerol, or glycerin)}}{HOCH_2-CH(OH)-CH_2OH} + \underset{\text{nitric acid}}{HONO_2} \longrightarrow \underset{\text{glyceryl trinitrate (nitroglycerin)}}{O_2NOCH_2-CH(ONO_2)-CH_2ONO_2}$$

10.35 In the following answers, chemically equivalent items have the same letter.

(a) In this compound, fluorines *a* are constitutionally equivalent and homotopic; fluorines *b* are constitutionally equivalent and enantiotopic; fluorines *a* are constitutionally nonequivalent to fluorines *b*.

$$F^a_3C-CF^b_2-H$$

(b) Fluorines *a* are constitutionally equivalent and diastereotopic; fluorines *a* are constitutionally nonequivalent to fluorine *b*.

$$(F^a)_2C=C(F^b)CH_3$$

(c) Fluorines *a* are constitutionally equivalent and diastereotopic; fluorines *a* are constitutionally nonequivalent to fluorine *b*.

$$H-C(F^a)_2-C(F^b)(CH_3)-H$$

10.36 (a)

$$\overset{a}{CH_3}\overset{b}{CH_2}(\overset{c}{H})C=C(\overset{c}{H})\overset{b}{CH_2}\overset{a}{CH_3}$$

three chemically nonequivalent sets

(b)

$$\overset{a}{CH_3}O-C(H^b)(H^c)-C(Br)(Cl)-CH^d$$

four chemically nonequivalent sets

(c) five chemically nonequivalent sets

(d)

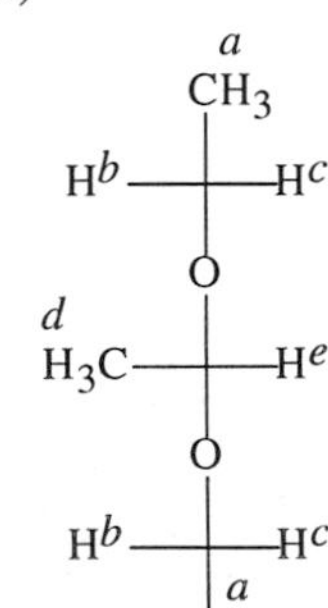

five chemically nonequivalent sets

10.37 Protonation of the —OH group, formation of a carbocation, and attack of isotopically labeled water on this carbocation are the key steps in this mechanism. (The character ○ represents ^{18}O.)

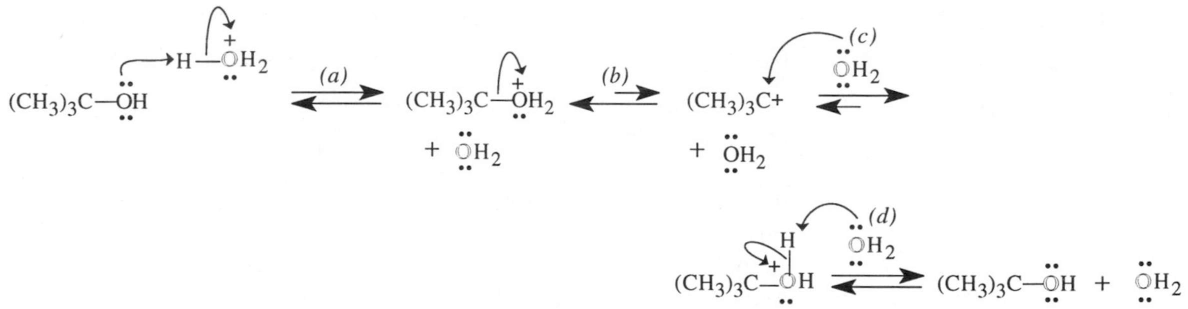

$$(CH_3)_3C{-}\ddot{O}H + H{-}\overset{+}{O}H_2 \xrightleftharpoons{(a)} (CH_3)_3C{-}\overset{+}{O}H_2 + \ddot{O}H_2 \xrightleftharpoons{(b)} (CH_3)_3C^+ + \ddot{O}H_2 \xrightleftharpoons{(c)\ \ddot{O}H_2}$$

$$(CH_3)_3C{-}\overset{+}{O}H(H) + \ddot{O}H_2 \xrightleftharpoons{(d)} (CH_3)_3C{-}\ddot{O}H + \overset{+}{O}H_3$$

Steps (*a*) and (*d*) are Brønsted acid-base reactions. (Be sure you can identify the acid and the base.) Step (*b*) is a Lewis acid-base dissociation. Step (*c*) is a Lewis acids-base association.

10.38 (a) Neither an oxidation nor a reduction. An internal oxidation-reduction has occurred, but there is no *net* oxidation or reduction.
(b) A two-electron reduction.
(c) A two-electron oxidation.
(d) A two-electron oxidation.
(e) Neither an oxidation nor a reduction.

10.39 (a) We have to use a method that involves one inversion of configuration. Thus, we convert the alcohol to a tosylate; this reaction does not affect the C—O bond. Then we displace the tosylate with inversion using isotopic $^-$OH (that is, $^{18}\bar{O}H$):

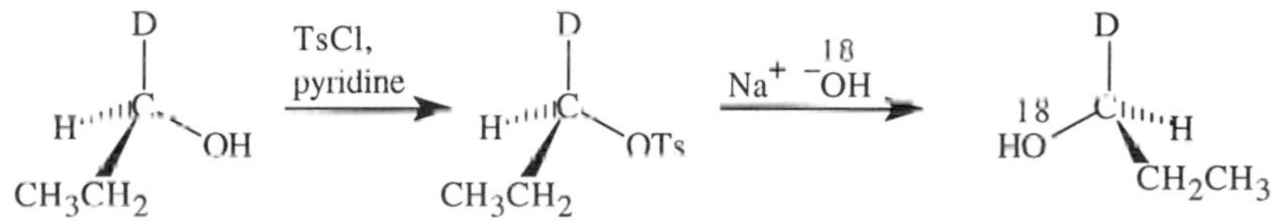

$$\text{H(D)(CH}_3\text{CH}_2\text{)C–OH} \xrightarrow{\text{TsCl, pyridine}} \text{H(D)(CH}_3\text{CH}_2\text{)C–OTs} \xrightarrow{Na^+\ ^{-18}OH} \text{H}^{18}\text{O–C(D)(H)CH}_2\text{CH}_3$$

(b) In this case, convert the alcohol into a bromide using a method that involves an inversion of configuration. (Either PBr_3 or the two-step alcohol → tosylate → alkyl bromide sequence shown below will work.) Then, in a second inversion step, displace the bromide with $^{18}\bar{O}H$ to provide the alcohol with the desired configuration.

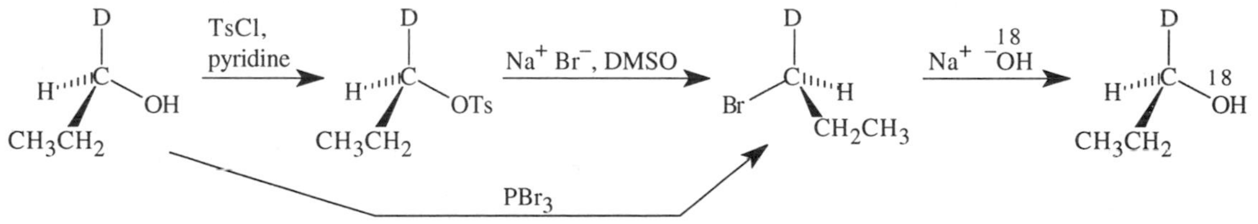

$$\text{H(D)(CH}_3\text{CH}_2\text{)C–OH} \xrightarrow{\text{TsCl, pyridine}} \text{H(D)(CH}_3\text{CH}_2\text{)C–OTs} \xrightarrow{Na^+\ Br^-,\ \text{DMSO}} \text{Br–C(D)(H)CH}_2\text{CH}_3 \xrightarrow{Na^+\ ^{-18}OH} \text{H(D)(CH}_3\text{CH}_2\text{)C–}^{18}\text{OH}$$

(alcohol $\xrightarrow{PBr_3}$ bromide directly)

10.40 (a) To introduce a deuterium we need to start with a Grignard reagent, which, in turn, requires an alkyl bromide. Alkyl bromides can be prepared from alcohols, but only after we move the —OH group from carbon-1 to carbon-2 via the alkene.

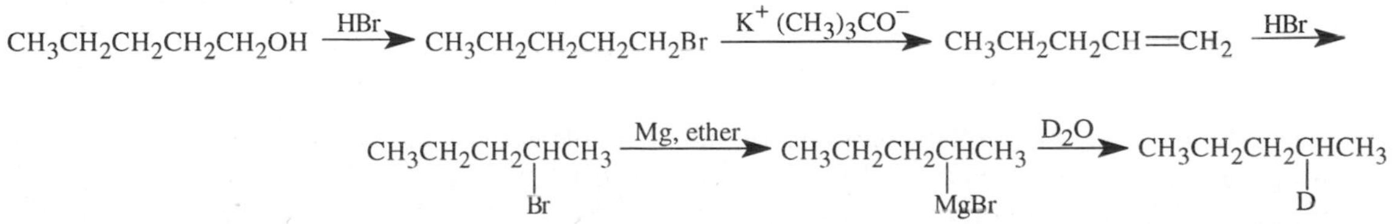

$$CH_3CH_2CH_2CH_2CH_2OH \xrightarrow{HBr} CH_3CH_2CH_2CH_2CH_2Br \xrightarrow{K^+\ (CH_3)_3CO^-} CH_3CH_2CH_2CH{=}CH_2 \xrightarrow{HBr}$$

$$CH_3CH_2CH_2CH(Br)CH_3 \xrightarrow{\text{Mg, ether}} CH_3CH_2CH_2CH(MgBr)CH_3 \xrightarrow{D_2O} CH_3CH_2CH_2CH(D)CH_3$$

(b)

$$\text{C}_5\text{H}_9\text{—CH=CH}_2 \xrightarrow[\text{2) H}_2\text{O}_2,\ ^-\text{OH}]{\text{1) BH}_3\text{/THF}} \text{C}_5\text{H}_9\text{—CH}_2\text{CH}_2\text{OH} \xrightarrow[\text{2) H}_3\text{O}^+]{\text{1) KMnO}_4,\ ^-\text{OH}} \text{C}_5\text{H}_9\text{—CH}_2\text{CO}_2\text{H}$$

In many cases, there is more than one acceptable synthesis for a given target, or, within a synthesis, there may be more than one acceptable method for accomplishing a particular conversion. For example, in part (a) above, you might have elected to convert the starting alcohol into a sulfonate ester such as a tosylate and carry out the elimination on the tosylate. In part (b), you might have elected to oxidize the primary alcohol with a different oxidizing agent, such as aqueous dichromate. Be sure to ask your teaching assistant or professor if you have a different synthesis and you are not sure whether it is acceptable.

(c)

$$\text{C}_5\text{H}_9\text{—CH}_2\text{CH}_2\text{OH} \xrightarrow{\text{CrO}_3\text{(pyridine)}_2} \text{C}_5\text{H}_9\text{—CH}_2\text{CH=O}$$

[prepared in part (b)]

(d) The key point here is that a carbon-carbon bond is broken.

$$\text{C}_5\text{H}_9\text{—CH=CH}_2 \xrightarrow[\text{2) H}_2\text{O}_2]{\text{1) O}_3} \text{C}_5\text{H}_9\text{—CO}_2\text{H}$$

(e)

$$\text{C}_5\text{H}_9\text{—CH=CH}_2 \xrightarrow{\text{H}_2\text{, catalyst}} \text{C}_5\text{H}_9\text{—CH}_2\text{CH}_3$$

(f) Pursuing the hint alerts you that cyanide ion can be used as a nucleophile in S_N2 reactions.

$$\underset{\text{1-butene}}{\text{CH}_3\text{CH}_2\text{CH=CH}_2} \xrightarrow[\text{2) H}_2\text{O}_2,\ ^-\text{OH}]{\text{1) BH}_3\text{/THF}} \text{CH}_3\text{CH}_2\text{CH}_2\text{CH}_2\text{OH} \xrightarrow{\text{TsCl, pyridine}}$$

$$\text{CH}_3\text{CH}_2\text{CH}_2\text{CH}_2\text{OTs} \xrightarrow{\text{NaCN, DMSO}} \text{CH}_3\text{CH}_2\text{CH}_2\text{CH}_2\text{CN}$$

One could also convert the alcohol into the alkyl bromide with HBr/H_2SO_4 and allow the bromide to react with cyanide. Another alternative is that 1-butene could be allowed to react with HBr in the presence of peroxides and then the resulting 1-bromobutane could be treated with NaCN in DMSO.

$$\underset{\text{1-butene}}{\text{CH}_3\text{CH}_2\text{CH=CH}_2} \xrightarrow{\text{HBr, peroxides}} \underset{\text{1-bromobutane}}{\text{CH}_3\text{CH}_2\text{CH}_2\text{CH}_2\text{Br}} \xrightarrow{\text{NaCN, DMSO}} \text{CH}_3\text{CH}_2\text{CH}_2\text{CH}_2\text{CN}$$

10.41 The principle to apply is that the best leaving groups are the weakest bases. The question becomes, then, which of the two leaving groups is the weaker base. The triflate anion is a weaker base than the mesylate anion because the *polar effect* of the fluorines stabilizes the negative charge in the triflate anion. (See Sec. 3.6B of the text.) Consequently, the triflate group is the better leaving group.

$^-O-S(=O)_2-CF_3$ is a weaker base than $^-O-S(=O)_2-CH_3$

triflate anion — mesylate anion

$R-O-S(=O)_2-CF_3$ is a better leaving group than $R-O-S(=O)_2-CH_3$

10.42 Use the balanced equation to determine the stoichiometry and convert the required molar equivalents to grams. The conversion of CrO_3 into Cr^{+3} is a three-electron reduction, and the conversion of 2-heptanol into the corresponding ketone, 2-heptanone, is a two-electron oxidation. For the number of electrons lost to be equal to the number gained, two moles of CrO_3 are required for every three of the alcohol that are oxidized, that is, 2/3 mol of CrO_3 per mole of alcohol. (If you don't follow this argument, balance the full equation.) Ten grams of 2-heptanol (molecular mass 116.2 g/mol) equals 0.08606 mol. The amount of CrO_3 (molecular mass = 99.99 g/mol) required is

$$\left(\frac{2 \text{ mol } CrO_3}{3 \text{ mol alcohol}}\right)(0.08606 \text{ mol alcohol})\left(\frac{99.99 \text{ g } CrO_3}{1 \text{ mol } CrO_3}\right) = 5.74 \text{ g}$$

10.43 First, write a balanced equation:

$$\overset{3}{H_2O} + \overset{3}{CH_3OCH_2CH_2OH} \longrightarrow \overset{3}{CH_3OCH_2CO_2H} + \overset{12}{\not{4}e^-} + \overset{12}{\not{4}H^+}$$

$$\overset{12}{\not{3}e^-} + \overset{12}{\not{3}H^+} + \overset{4}{HNO_3} \longrightarrow \overset{4}{NO} + \overset{8}{\not{2}H_2O}$$

Sum:

$$3CH_3OCH_2CH_2OH + 4HNO_3 \longrightarrow 3CH_3OCH_2CO_2H + 4NO + 5H_2O$$

Because the alcohol oxidation is a four-electron oxidation, and the reduction of nitric acid to NO is a three-electron reduction, it takes four moles of nitric acid to oxidize three moles of the alcohol, that is, 4/3 mole of nitric acid per mole of alcohol. Therefore, it takes (4/3)(0.100) = 0.133 moles of nitric acid to oxidize 0.1 mole of the alcohol.

10.44 The essence of this solution is a balanced equation for the oxidation of ethanol by dichromate and conversion of the stoichiometry into a value for percent blood alcohol. Because the dichromate is aqueous, assume that ethanol is oxidized to acetic acid (a four-electron oxidation). A balanced equation for this process is as follows (verify this):

$$16H^+ + 3CH_3CH_2OH + 2Cr_2O_7^{2-} \longrightarrow 3CH_3CO_2H + 11H_2O + 4Cr^{3+}$$

This equation shows that 3/2 mole of ethanol is oxidized for every mole of $Cr_2O_7^{2-}$ ion consumed. Required to solve the problem are the grams of ethanol oxidized and the mL of blood in which that ethanol is contained. The process can be diagrammed as follows:

moles of dichromate ⇨ moles of ethanol ⇨ g of ethanol

mL of air ⇨ mL of blood

} divide and multiply the result by 100

The amount of ethanol oxidized is (3/2 mol ethanol per mol of dichromate)(0.507×10^{-6} mol of dichromate) = 0.761×10^{-6} mol of ethanol. The molecular mass of ethanol is 46.07 g/mol; hence, the g of ethanol in the sample of blood is (46.07 g/mol)(0.761×10^{-6} mol) = 3.50×10^{-5}.
The 52.5 mL of air collected is equivalent to 0.0250 mL of blood:

$$\text{mL blood} = \frac{52.5 \text{ mL of air}}{2100 \text{ mL of air per mL of blood}} = 0.0250$$

Finally, the grams of ethanol per mL of blood is (3.50×10^{-5} g of ethanol) ÷ (0.0250 mL of blood) = 1.40×10^{-3}, and, using the formula in the problem, the percent blood alcohol content is this number times 100, or 0.14%. Consequently, Bobbin is legally intoxicated and Officer Order should make the arrest.

10.45 (a) The glycol is oxidized. This is clear from the fact that a bond to carbon is replaced by a bond to oxygen at each carbon of the glycol. The other participant in the reaction, periodate, must therefore be reduced. Indeed, the ionic product iodate (IO_3^-) contains one fewer oxygen bound to the iodine.

(b) The number of electrons involved in the oxidation half-reaction is determined from the oxidation numbers of the carbons that change:

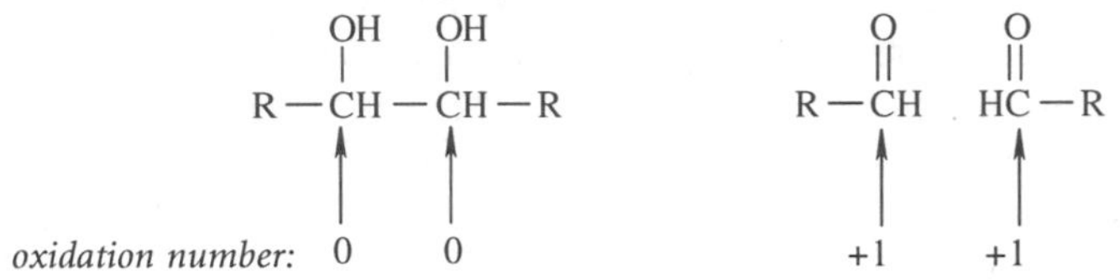

R—CH(OH)—CH(OH)—R → R—CH(=O) HC(=O)—R

oxidation number: 0 0 +1 +1

The number of electrons lost is (+1 + +1) – (0 + 0) = +2. (Of course, this could equally well be determined from a balanced half-reaction.)

The iodine can be assigned an oxidation number of +7 in periodate. and +5 in iodate. How do we know this? Assign +2 to every oxygen because oxygen is divalent and presumably has two bonds to the iodine. (See the top of text p. 428 for a similar case.) Assign a –1 for every negative charge. Hence, the reduction of periodate is a two-electron reduction. However, we can verify this with a balanced half-reaction:

$$2e^- + IO_4^- + 2H^+ \longrightarrow IO_3^- + H_2O$$

(c) You could, of course, balance the entire reaction, but it is easier just to use the fact that the electrons lost (2) equals the electrons gained (2) in each half-reaction. Thus, to oxidize a mole of glycol requires a mole of periodate; to oxidize 0.1 mole of glycol requires 0.1 mole of periodate.

10.46 Because thiols and disulfides equilibrate to give a mixture of all possible thiols and disulfides (see Eq. 10.58a–b, text p. 446), Stench has evidently obtained a complex mixture of di(2-octyl) disulfide, dibutyl disulfide, and butyl 2-octyl disulfide. Although the desired product is *statistically* favored (it is twice as likely to be formed as the other disulfides), it is not likely to be more than 50% of the reaction mixture.

10.47 The reactivity data and the molecular formula of *A* indicate that compound *A* is an alkene with one double bond. The identity of compound *D* follows from the oxidation of 3-hexanol; it can only be 3-hexanone (see following equation). It is given that 3-hexanone is an ozonolysis product of alkene *A* (along with H—CO_2H (formic acid), not shown in the following equation). Since alkene *A* has seven carbons and one double bond, and 3-hexanone has six carbons, the carbon of alkene *A* not accounted for by 3-hexanone must be part of a $=CH_2$ group. Therefore, the identity of alkene *A* is established as 2-ethyl-1-pentene. The identities of compounds *B* and *C* follow from the reactions of *A*.

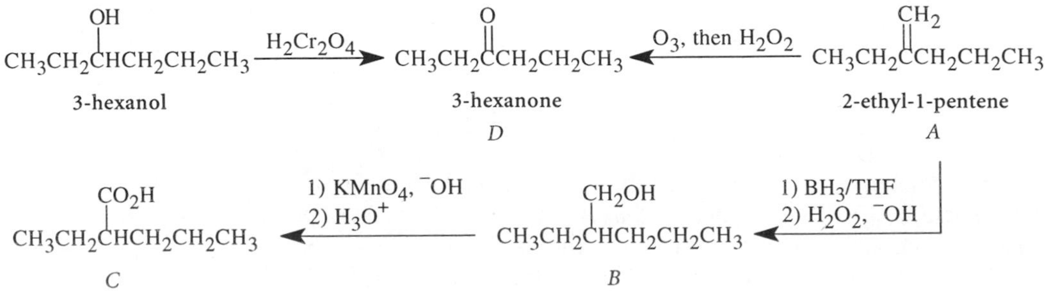

$CH_3CH_2CH(OH)CH_2CH_2CH_3$ (3-hexanol) $\xrightarrow{H_2Cr_2O_4}$ $CH_3CH_2C(=O)CH_2CH_2CH_3$ (3-hexanone, *D*) $\xleftarrow{O_3,\ \text{then}\ H_2O_2}$ $CH_3CH_2C(=CH_2)CH_2CH_2CH_3$ (2-ethyl-1-pentene, *A*)

A $\xrightarrow{1)\ BH_3/THF;\ 2)\ H_2O_2,\ ^-OH}$ $CH_3CH_2CH(CH_2OH)CH_2CH_2CH_3$ (*B*) $\xrightarrow{1)\ KMnO_4,\ ^-OH;\ 2)\ H_3O^+}$ $CH_3CH_2CH(CO_2H)CH_2CH_2CH_3$ (*C*)

10.48 Their reactions with NaH, their formulas, and the fact that they can be resolved into enantiomers show that compounds *A* and *B* are chiral isomeric alcohols. Their eventual conversion into methylcyclohexane shows that they have the same carbon skeleton as methylcyclohexane. When *optically active A* and *B* are converted into tosylates and subjected to the E2 reaction with potassium *tert*-butoxide they give *optically active,* and therefore *chiral,* alkenes. The only possible chiral alkenes with the carbon skeleton of methylcyclohexane are 4-methylcyclohexene and 3-methylcyclohexene, which are compounds *C* and *D;* the data do not determine which is which. To summarize:

CH_3 CH_3 or H_2, catalyst CH_3

4-methylcyclohexene 3-methylcyclohexene methylcyclohexane

(compounds *C* and *D*; the data do not determine which is which)

Because compounds *A* and *B* are chiral, their tosylates must also be chiral. The two chiral tosylates that would both give a mixture of compounds *C* and *D* are the tosylates of the stereoisomeric 3-methylcyclohexanols. Because alcohols *A* and *B* have different melting points, they must be the *diastereomeric* 3-methylcyclohexanols. (Notice that a 4-methylcyclohexanol is ruled out because neither stereoisomer of 4-methylcyclohexanol can be chiral.) The individual identities of these alcohols follow from an analysis of the E2 reactions of their tosylates. Note that *anti*-elimination can take place only when the tosylate group is in an axial position. In the tosylate of *cis*-3-methylcyclohexanol, the tosylate group can only assume the axial position required for E2 elimination when the molecule is in a very unstable conformation in which the methyl group is also axial:

CH_3 OTs tosylate of *A* OTs CH_3 severe 1,3-diaxial interaction K^+ $(CH_3)_3CO^-$ (slower) CH_3 + CH_3

In the tosylate of *trans*-3-methylcyclohexanol, the tosylate group can assume the axial position in a conformation in which the methyl group is equatorial.

CH_3 OTs tosylate of *B* OTs CH_3 K^+ $(CH_3)_3CO^-$ (faster) CH_3 + CH_3

The compound containing the greater amount of the conformation in which the tosylate group is in the axial position—the tosylate of *B*—undergoes elimination more rapidly. Consequently, compound *A* is *cis*-3-methylcyclohexanol, and compound *B* is *trans*-3-methylcyclohexanol.

CH_3 OH CH_3 OH

cis-3-methylcyclohexanol (compound *A*) *trans*-3-methylcyclohexanol (compound *B*)

10.49 (a)

$$\text{PhCH}_2\text{OH} \xrightarrow{\text{tosyl chloride, pyridine}} \text{PhCH}_2\text{OTs} \xrightarrow{\text{NaBr, DMSO}} \text{PhCH}_2\text{Br}$$

(b)

$$\text{Ph—CH}_2\text{CH}_2\text{CH}_2\text{I}$$

(c)

$$(CH_3)_2CH\text{—}SH \xrightarrow[CH_3OH]{CH_3O^-} (CH_3)_2CH\text{—}S^- \xrightarrow[\text{(see Eq. 10.20, text p. 419)}]{\text{dimethyl sulfate}} (CH_3)_2CH\text{—}SCH_3$$

(d)

$$CH_3\underset{\displaystyle Cl}{C}HCH_2CH_2Ph$$

(e)

$$H_3C\text{—}\overset{Br}{\underset{CH_3}{C}}\text{—}CH_2\text{—}\overset{Br}{CH}\text{—}CH_3$$

(f)

CH_3 (bicyclic alkene, angular methyl)

(racemate)

Note in part (f) that, because the starting material is achiral, the chiral product must be formed as the racemate; Sec. 7.8A.

(g)

$$\text{cyclohexanol (OH)} \xrightarrow{\text{conc. HBr}} \text{bromocyclohexane (Br)} \xrightarrow[(CH_3)_3COH]{K^+\ (CH_3)_3CO^-} \text{cyclohexene}$$

(h) After the reaction comes to equilibrium, all possible thiols and all possible disulfides are present. (See the solution to Problem 10.28 for the rationale.)

$$(CH_3)_2CHCH_2CH_2\text{—}S\text{—}S\text{—}C_2H_5 \ + \ (CH_3)_2CHCH_2CH_2\text{—}S\text{—}S\text{—}CH_2CH_2CH(CH_3)_2$$

$$+ \ C_2H_5\text{—}S\text{—}S\text{—}C_2H_5 \ + \ (CH_3)_2CHCH_2CH_2SH \ + \ C_2H_5SH$$

10.50 (a) The primary deuterium kinetic isotope effect of 6.6 suggests that the step of the mechanism shown in Eq. 10.39c, text p. 432, is rate-limiting, because the isotopically substituted hydrogen is transferred in this step.

(b) Because of the primary isotope effect in part (a), the deuterium is removed more slowly than the hydrogen (about seven times more slowly). Thus, more of the aldehyde containing deuterium is formed. Because $CrO_3(\text{pyridine})_2$ is an achiral reagent, it cannot differentiate between the two enantiotopic α-hydrogens of ethanol; each is removed with equal frequency. Consequently, the isotope effect causes isotopic discrimination *regardless* of the stereochemical positions of the hydrogen and deuterium.

(c) The enzyme alcohol dehydrogenase, which is a chiral catalyst, distinguishes between the enantiotopic α-hydrogens of ethanol so completely that it removes *only* the pro-(*R*) hydrogen whether this hydrogen is isotopically substituted or not. Thus, when this hydrogen is substituted by deuterium, the deuterium is removed, and, as a result, the aldehyde contains no deuterium. Although there may be an isotope effect associated with removal of this hydrogen, it is not reflected in the relative amounts of H and D removed from a given enantiomer because the stereochemical discrimination (which may be a factor of 1000 or more) is a much larger effect than the primary deuterium isotope effect (which could be a factor only as large as 7 or so).

10.51 The two carbon branches of citric acid are *enantiotopic*. Hence, they are chemically distinguishable to a chiral catalyst such as an enzyme. Evidently, the difference is such that the dehydration occurs into the unlabeled branch.

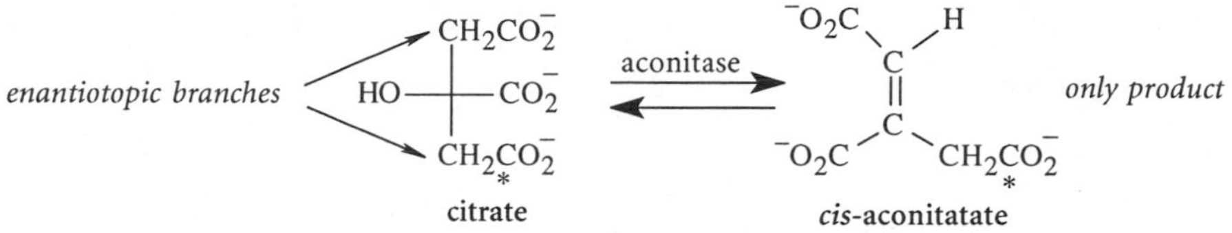

enantiotopic branches HO—C($CH_2CO_2^-$)(CO_2^-)($\overset{*}{C}H_2CO_2^-$) (citrate) $\xrightarrow{\text{aconitase}}$ ^-O_2C(H)C=C(^-O_2C)($\overset{*}{C}H_2CO_2^-$) (*cis*-aconitatate) *only product*

One difference is that in H_2SO_4 solution the carboxylate groups are not ionized, but this is not the key difference. The point of the problem is that an achiral laboratory reagent will not make the distinction between enantiotopic groups. Hence, equal amounts of dehydration should occur into each branch.

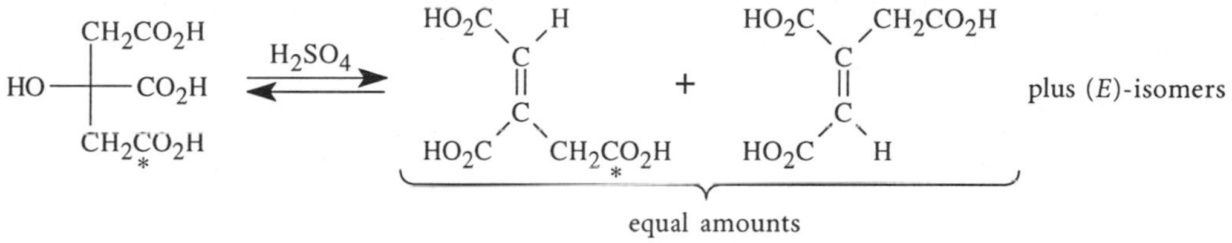

HO—C(CH_2CO_2H)(CO_2H)($\overset{*}{C}H_2CO_2H$) $\overset{H_2SO_4}{\rightleftharpoons}$ HO_2C(H)C=C(HO_2C)($\overset{*}{C}H_2CO_2H$) + HO_2C(CH_2CO_2H)C=C(HO_2C)(H) plus (*E*)-isomers

equal amounts

Note that the alkene product in the enzyme-catalyzed reaction is the (*Z*)-isomer (which biochemists call "*cis*-aconitate"). It would not be surprising to find in the product mixture from the H_2SO_4-catalyzed reaction some of the corresponding (*E*)-isomers as well.

10.52 The stereochemistry of the addition reaction as shown indicates that the pro-(*R*) hydrogen at carbon-3 in the product is the one added from solvent. Thus, by the principle of microscopic reversibility (text p. 410), it is the one eliminated in the reverse reaction. When this hydrogen is substituted by deuterium, then deuterium is eliminated. When the pro-(*S*) hydrogen is substituted by deuterium, it is retained.

(a) The product retains deuterium because the hydrogen that is eliminated comes from carbon-3, whereas the deuterium in the starting material is at carbon-2.

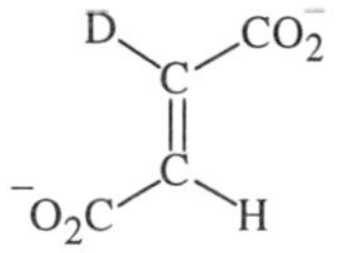

D(CO_2^-)C=C(^-O_2C)(H)

(b) The product is the same as (a), but for a different reason. In this case, the deuterium at carbon-3 is in the pro-(*S*) position; since the pro-(*R*) hydrogen is eliminated, the fumarate retains the deuterium.

(c) In this case, dehydration cannot occur because the enzyme is stereoselective and stereospecific. Only the (2*S*) enantiomer reacts; the enantiomer shown is the (2*R*) enantiomer. The position of the deuterium is irrelevant.

10.53 Unlike HBr, HCN is a rather weak acid ($pK_a = 9.4$), and it provides very little cyanide ion because it is essentially un-ionized. In addition, it is not acidic enough to protonate the alcohol to any reasonable extent, because the pK_a of a protonated alcohol is –2. Because the —OH group of the alcohol is not protonated, it is not a good leaving group. With a nucleophile concentration of nearly zero and a poor leaving group, a substitution reaction cannot take place.

Addition of H_2SO_4 to the reaction mixture would indeed solve the leaving-group problem, because the H_2SO_4 would protonate the —OH group. However, added protons suppress the ionization of HCN even more than it is suppressed normally. Without a reasonable concentration of nucleophile, the reaction is doomed. Bluelip is doomed, too, because addition of H_2SO_4 to a solution of cyanide ion produces lethal HCN gas.

10.54 Ethanol is partially protonated under the acidic conditions. Water is displaced from a protonated ethanol molecule by another ethanol molecule:

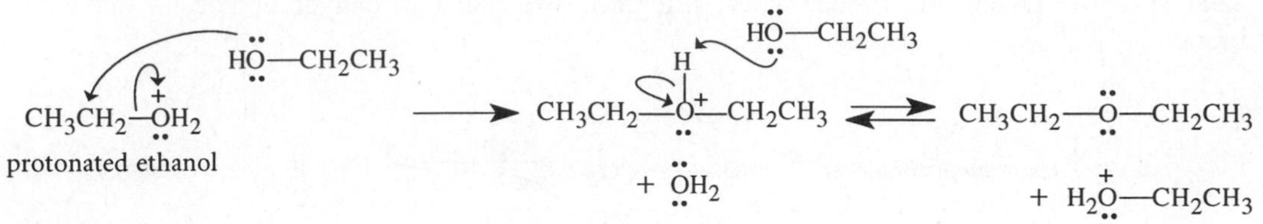

$$CH_3CH_2-\overset{+}{O}H_2 \text{ (protonated ethanol)} + HO-CH_2CH_3 \longrightarrow CH_3CH_2-\overset{+}{O}(H)-CH_2CH_3 + OH_2 \rightleftharpoons CH_3CH_2-O-CH_2CH_3 + H_2\overset{+}{O}-CH_2CH_3$$

Ethanol is a rather weak base; recall (Eq. 8.15b, text p. 333) that the conjugate acid of ethanol has a pK_a of –2 to –3. Because ethanol is a weak base, it is a poor nucleophile. This is why the reaction requires heat.

10.55 (a) The carbocation rearrangement in this solution, in which a tertiary carbocation is converted into a secondary carbocation, is driven by relief of ring strain.

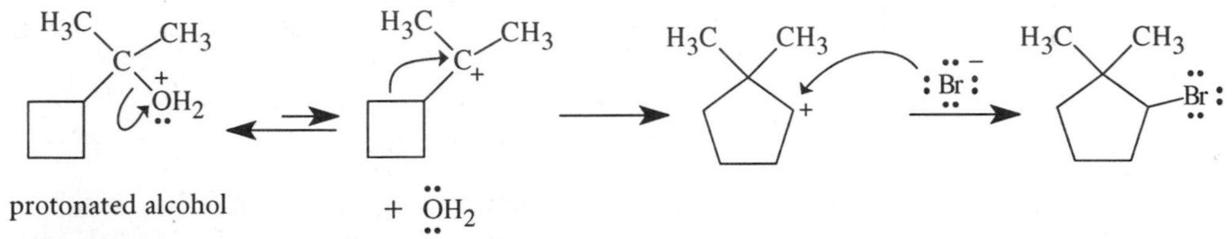

protonated alcohol ⇌ + OH_2 → → (Br⁻)

(b) The carbocation rearrangement in this mechanism also involves a ring expansion. In this case, a tertiary carbocation is converted into another tertiary carbocation.

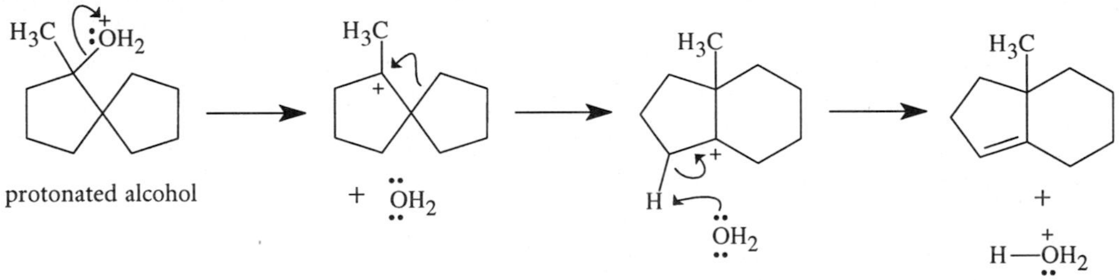

protonated alcohol → + OH_2 → (H, OH_2) → + $H-\overset{+}{O}H_2$

(c) The product contains eight carbons; hence, it involves the reaction of two 2-methylpropene carbon skeletons, joined as follows:

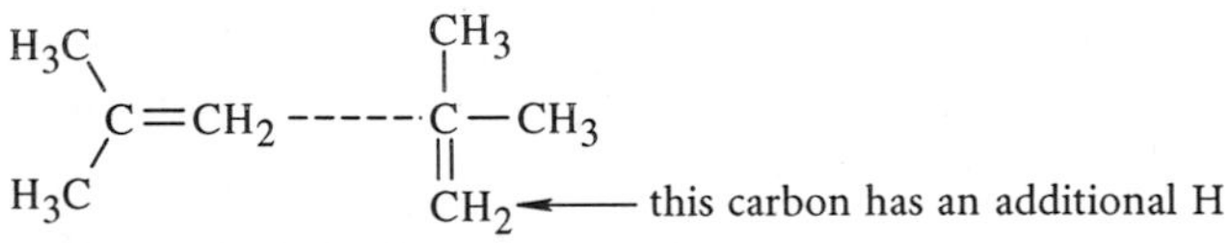

$$(H_3C)_2C=CH_2 \text{-----} C(CH_3)_2=CH_2$$

this carbon has an additional H

The first step in this process is the reaction of one alkene molecule with HF to give a *tert*-butyl cation *A*:

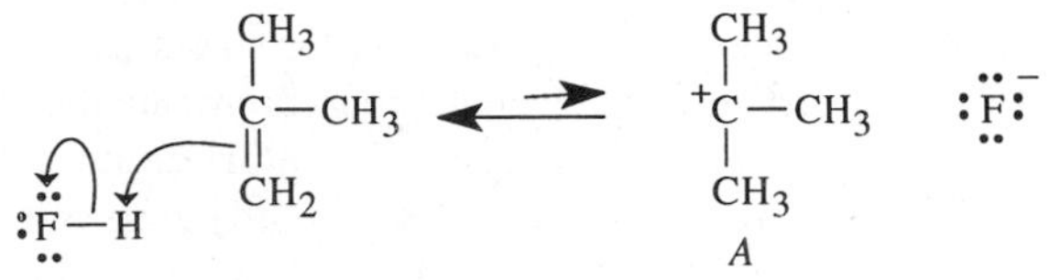

$$:\ddot{F}-H + CH_2=C(CH_3)_2 \rightleftharpoons {}^{+}C(CH_3)_3 \ (A) \quad :\ddot{F}:^-$$

(We won't show the fluoride counterion in subsequent steps.) Carbocation *A* then reacts as a Lewis acid with a second molecule of alkene to give carbocation *B*:

$$(CH_3)_2C{=}CH_2 + {}^{+}C(CH_3)_3 \;(A) \longrightarrow (CH_3)_2\overset{+}{C}{-}CH_2{-}C(CH_3)_3 \;(B)$$

Finally, carbocation *B* reacts with isobutane by removing a *hydride* (a hydrogen with two electrons). This is exactly like a carbocation rearrangement by hydride shift, except that the reaction occurs across two different molecules:

$$\underset{\text{isobutane}}{(CH_3)_3C{-}H} + \underset{B}{{}^{+}C(CH_3)_2{-}CH_2{-}C(CH_3)_3} \longrightarrow \underset{\substack{tert\text{-butyl cation}\\ A}}{(CH_3)_3C^{+}} + H{-}C(CH_3)_2{-}CH_2{-}C(CH_3)_3$$

The product is accompanied by a *tert*-butyl cation *A*, which is the cation produced in the initial protonation step. This cation can thus react with another molecule of alkene to give another carbocation *B*, and so on. Notice that if a *tert*-butyl cation *A* reacts with a molecule of isobutane by a hydride shift, another *tert*-butyl cation *A* and another isobutane are formed, that is, such a "side reaction" (which undoubtedly occurs) has no net effect. Finally, the hydride-shift mechanism accounts for the results with deuterium.

Notice that this is actually a *chain reaction*, in which the *tert*-butyl cation propagates the chain. Although most chain reactions are *free-radical* chain reactions, this is an interesting example of a *cationic* chain reaction.

You might compare this problem with Problem 5.42 (text p. 200) and see if you can tell what it is about the conditions of the two reactions, both of which involve 2-methylpropene and HF, that causes the outcome to be different.

11

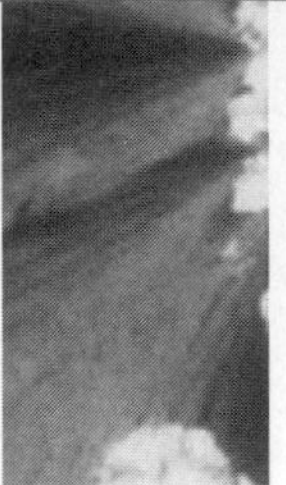

Chemistry of Ethers, Epoxides, Glycols, and Sulfides

TERMS

alkoxymercuration-reduction 11.1B
amine oxide.. 11.5B
anchimeric assistance 11.7
episulfonium ion 11.7
epoxide.. Introduction
ether cleavage 11.3
glycol 11.5
intermolecular reaction 11.7
intramolecular reaction 11.7
neighboring-group participation 11.7
osmate ester 11.5B
oxonium salt 11.6A
peracid 11.2A
periodic acid 11.5B
peroxyacid 11.2A
peroxycarboxylic acid 11.2A
sulfide .... Introduction
sulfone 11.8
sulfonium salt 11.6A
sulfoxide 11.8
Williamson ether synthesis 11.1A

CONCEPTS

I. GENERAL SYNTHETIC AND MECHANISTIC CONCEPTS

A. Ethers, Epoxides, Glycols, and Sulfides

1. The chemistry of ethers is closely related to the chemistry of alkyl halides, alcohols, and alkenes; however, ethers are considerably less reactive.
2. Epoxides are very reactive heterocycles in which the ether linkage is part of a three-membered ring.
   a. Epoxides, like their carbon analogs, the cyclopropanes, possess significant ring strain.
   b. Because of this strain, the bonds of an epoxide are weaker than those of an ordinary ether, and thus they are more easily broken.
3. Glycols are compounds that contain hydroxy groups on adjacent carbon atoms; they undergo some reactions of alcohols but also have unique chemistry that is related to that of epoxides.
4. Sulfides (thioethers) are sulfur analogs of ethers but differ from ethers in the way that they react in oxidation reactions.

B. Neighboring-Group Participation

1. Reactions involving the covalent participation of neighboring groups are said to occur with anchimeric assistance, or neighboring-group participation.
   a. Neighboring-group mechanisms involve intramolecular reactions that are in competition with ordinary intermolecular reactions.
   b. Reactions involving neighboring-group participation are generally faster than analogous reactions that do not involve neighboring-group participation.

i. An intramolecular reaction has a greater probability of occurring.
ii. Reactions that occur with greater probability have larger rates.
iii. The formation of a three-membered ring occurs despite the strain in the intermediate.

2. Neighboring-group participation in nucleophilic substitution reactions is common for cases in which three-, five-, and six-membered rings are formed.

C. The Three Fundamental Operations of Organic Synthesis

1. Most reactions used in organic synthesis involve one or more of three fundamental operations:
   a. Functional-group transformation: conversion of one functional group into another (the most common type of synthetic operation).
   b. Control of stereochemistry (accomplished with stereoselective reactions).
   c. Formation of carbon-carbon bonds (a particularly important synthetic operation). Two examples:
      i. Cyclopropane formation from carbenes or carbenoids and alkenes.
      ii. Reaction of Grignard reagents with ethylene oxide.

REACTIONS

I. SYNTHESIS OF ETHERS, EPOXIDES, GLYCOLS, AND SULFIDES

A. Williamson Ether Synthesis

1. The Williamson ether synthesis is an S_N2 reaction in which the conjugate base of an alcohol (or thiol) reacts as a nucleophile with an unhindered alkyl halide or sulfonate ester to form an ether (or sulfide).

$$(CH_3)_3C{-}O^-\,Na^+ + H_3C{-}OTs \longrightarrow (CH_3)_3C{-}O{-}CH_3 + TsO^-\,Na^+$$

sodium 2-methyl-2-propanolate + methyl tosylate → 2-methoxy-2-methylpropane (or *tert*-butyl methyl ether) + sodium tosylate

$$Ph{-}S^-\,Na^+ + CH_3CH_2Br \longrightarrow Ph{-}S{-}CH_2CH_3 + Na^+\,Br^-$$

sodium thiophenolate + bromoethane → (ethylthio)benzene (or ethyl phenyl sulfide) + sodium bromide

   a. Methyl halides, primary alkyl halides, or the corresponding sulfonate esters can be used in a Williamson synthesis.
   b. Tertiary and many secondary alkyl halides cannot be used in a Williamson synthesis because the E2 (elimination) reaction occurs instead.

$$H_3C{-}O^-\,Na^+ + (CH_3)_3C{-}I \longrightarrow H_3C{-}O{-}H + Na^+\,I^- + (CH_3)_2C{=}CH_2$$

sodium methanolate + *tert*-butyl iodide (or 2-iodo-2-methylpropane) → methanol + sodium tosylate + 2-methylpropene

2. In principle, two different Williamson syntheses are possible for any ether with two different alkyl groups.
   a. The preferred synthesis is usually the one that involves the alkyl halide with the greater S_N2 reactivity.
   b. Tertiary and secondary alkyl groups should be derived from the alkoxide.

B. Alkoxymercuration-Reduction of Alkenes

1. An alkoxymercuration-reduction reaction occurs when oxymercuration-reduction is carried out in an alcohol solvent instead of water.

$$(CH_3)_2C{=}CH_2 \xrightarrow[(CH_3)_2CHOH]{Hg(OAc)_2} (CH_3)_2C(OCH(CH_3)_2){-}CH_2HgOAc \xrightarrow{NaBH_4} (CH_3)_2C(OCH(CH_3)_2){-}CH_2H$$

2. The mechanism of the alkoxymercuration reaction is completely analogous to the mechanism of oxymercuration, except that an alcohol instead of water is the nucleophile that attacks the mercurinium ion intermediate.
3. The alkoxymercuration-reduction reaction can be used to prepare ethers that cannot be prepared by a Williamson ether synthesis.

C. Ethers from Alcohol Dehydration and Alkene Addition

1. Two molecules of an alcohol can undergo dehydration to give an ether (a reaction used primarily in industry).

$$\underset{\text{ethanol}}{2\,CH_3CH_2OH} \xrightarrow[140°]{H_2SO_4} \underset{\text{diethyl ether}}{CH_3CH_2{-}O{-}CH_2CH_3}$$

   a. This is generally restricted to the preparation of symmetrical ethers derived from primary alcohols.
   b. Secondary and tertiary alcohols cannot be used because they undergo dehydration to alkenes.
   c. The formation of ethers from primary alcohols is an S_N2 reaction in which one alcohol displaces water from another molecule of protonated alcohol.

$$\underset{\text{protonated alcohol}}{RCH_2{-}\overset{+}{O}H_2} \quad HO{-}CH_2R \xrightarrow{S_N2} RCH_2{-}\overset{+}{O}(H){-}CH_2R \;\; (:OH_2) \longrightarrow RCH_2{-}O{-}CH_2R + H_3O^+$$

2. Tertiary alcohols can be converted into unsymmetrical ethers by treating them with dilute solutions of strong acids in an alcohol solvent.

$$\underset{\textit{tert}\text{-butyl alcohol}}{(CH_3)_3C{-}OH} + \underset{\substack{\text{methanol}\\\text{(solvent; in excess)}}}{CH_3OH} \xrightarrow{H_2SO_4\text{ (trace)}} \underset{\textit{tert}\text{-butyl methyl ether}}{(CH_3)_3C{-}O{-}CH_3} + H_2O$$

   a. The formation of an ether from a tertiary alcohol and a primary alcohol solvent is an S_N1 solvolysis reaction in which the protonated alcohol loses water to give a relatively stable carbocation that is attacked by solvent to give the product.

$$\underset{\substack{\text{conjugate acid of the}\\\text{tertiary alcohol}}}{(CH_3)_3C{-}\overset{+}{O}H(H)} \rightleftharpoons (CH_3)_3C^+ + H_2O \xrightarrow[\text{solvent}]{HO{-}CH_3} \underset{\substack{\text{conjugate acid}\\\text{of the product}}}{(CH_3)_3C{-}\overset{+}{O}(H){-}CH_3} \underset{}{\overset{HO{-}CH_3}{\rightleftharpoons}} (CH_3)_3C{-}O{-}CH_3 + H_2\overset{+}{O}{-}CH_3$$

   b. One of the alcohol starting materials must readily lose water after protonation to form a relatively stable carbocation; typically this is a tertiary alcohol.

c. The alcohol that is used in excess must be one that either cannot lose water after protonation to give a carbocation or should form a carbocation much less readily than the tertiary alcohol. Typically this is a primary alcohol.
d. Any alkene that does form (by an E1 pathway) is not removed but is reprotonated to give back a carbocation, which eventually reacts with the alcohol solvent.
e. Ether formation from tertiary alcohols and the dehydration of tertiary alcohols are alternative branches of a common mechanism. (See Study Guide Link 11.2 and text pp. 450–60.)

3. Treatment of an alkene with a large excess of alcohol in the presence of an acid catalyst gives an ether provided that a relatively stable carbocation intermediate is involved.

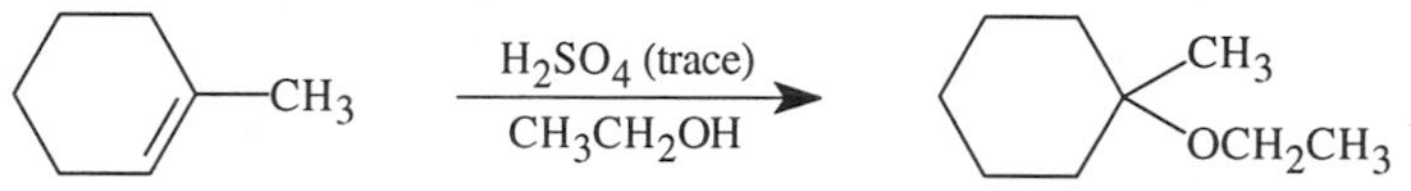

CH_3 $\xrightarrow[CH_3CH_2OH]{H_2SO_4\text{ (trace)}}$ CH_3 OCH_2CH_3

1-methylcyclohexene 1-ethoxy-1-methylcyclohexane

D. Oxidation of Alkenes with Peroxycarboxylic Acids

1. One of the best laboratory preparations of epoxides involves the direct oxidation of alkenes with peroxycarboxylic acids.

+ R—C(=O)—O—O—H $\xrightarrow{CH_2Cl_2}$ (H, O, H) + R—C(=O)—OH

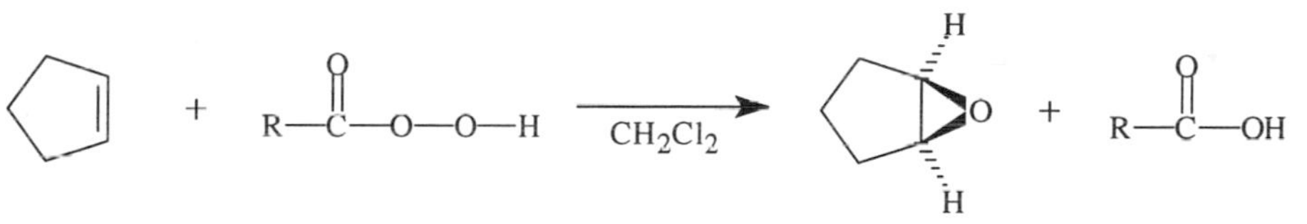

cyclopentene a peroxycarboxylic acid cyclopentene oxide a carboxylic acid

a. The oxidizing agent, a peroxycarboxylic acid, is a carboxylic acid that contains an —O—O—H (hydroperoxy) group rather than an —OH (hydroxy) group.
b. The general terms peroxyacid or peracid, denoting any acid containing a hydroperoxy group, are sometimes used instead of peroxycarboxylic acid.
c. Many peroxycarboxylic acids are unstable and are formed just prior to use by mixing a carboxylic acid with hydrogen peroxide.

2. The formation of an epoxide from an alkene and a peroxycarboxylic acid is a concerted addition reaction. (But see Study Guide Link 11.3 for a further discussion of this mechanism.)

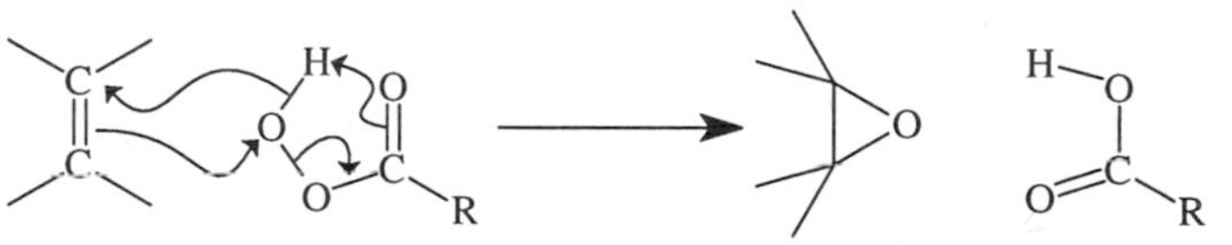

C H O O C O R → O H—O O=C R

alkene peroxycarboxylic acid epoxide carboxylic acid

a. This mechanism is very similar to that for the formation of a bromonium ion in bromine addition to alkenes.
b. The formation of epoxides with peroxycarboxylic acids is a stereoselective reaction; it takes place with complete retention of the alkene stereochemistry. (See example under point 1.)

E. Cyclization of Halohydrins

1. Epoxides can be synthesized by the treatment of halohydrins with base.

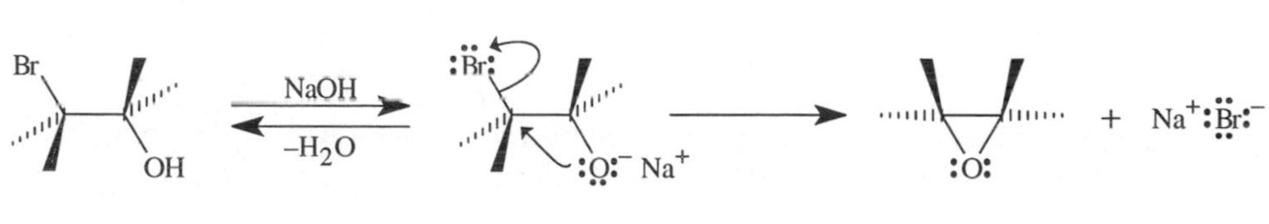

Br OH $\underset{-H_2O}{\xrightarrow{NaOH}}$:Br: :O:⁻ Na^+ → :O: + Na^+:Br:⁻

a. This reaction is an intramolecular variation of the Williamson ether synthesis in which the alcohol and the alkyl halide are part of the same molecule.
b. The alkoxide anion, formed reversibly by reaction of the alcohol with NaOH, displaces halide ion from the neighboring carbon.

2. Like other S_N2 reactions, this reaction takes place by backside attack of the nucleophilic oxygen anion at the halide-bearing carbon.
   a. The attacking oxygen and the leaving halide assume an anti relationship in the transition state of the reaction.
   b. Halohydrins derived from cyclic compounds must be able to assume the required anti relationship through a conformational change if epoxide formation is to succeed.

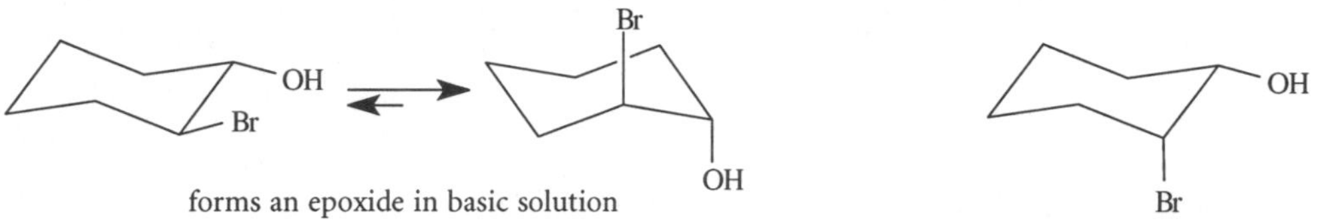

forms an epoxide in basic solution

does not form an epoxide in basic solution

F. Preparation of Glycols

1. Some glycols can be prepared by the acid-catalyzed reaction of water with epoxides.

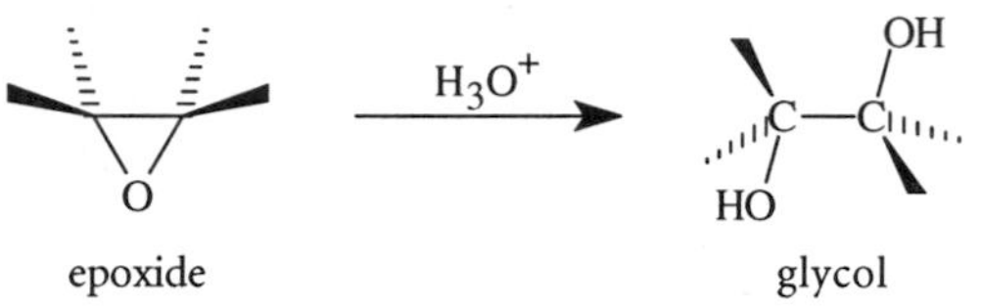

epoxide $\xrightarrow{H_3O^+}$ glycol

2. Alkenes can be oxidized to glycols with OsO_4.
   a. The osmium in OsO_4 is in a high oxidation state, Os(VIII), and thus attracts electrons.
      i. This electron-attracting ability results in a concerted cycloaddition reaction between OsO_4 and an alkene that reduces the Os(VIII) to Os(VI) and gives an intermediate called an osmate ester..

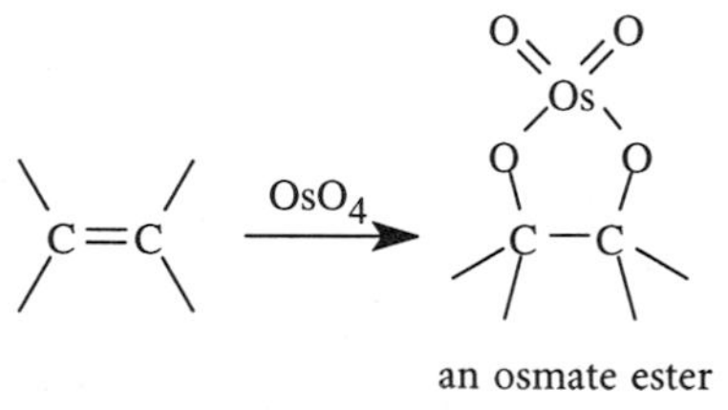

$\xrightarrow{OsO_4}$ an osmate ester

      ii. A glycol is formed when the cyclic osmate ester is treated with water.

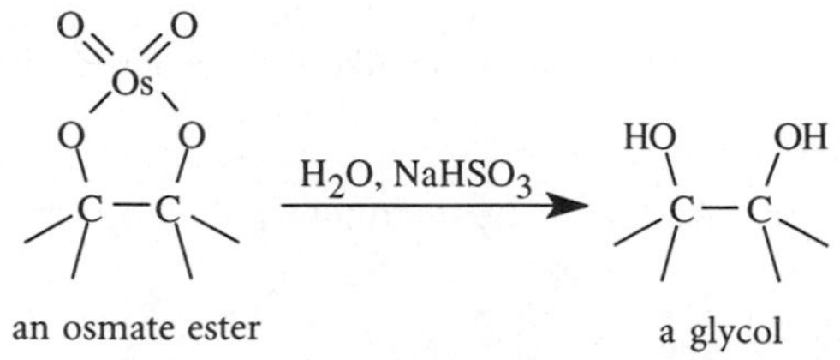

an osmate ester $\xrightarrow{H_2O,\ NaHSO_3}$ a glycol

   b. A mild reducing agent such as sodium bisulfite ($NaHSO_3$) is added in many cases to convert the osmium-containing by-products into reduced forms of osmium that are easily removed by filtration.

c. Osmium and its compounds are very toxic and are quite expensive. By including an amine oxide (another oxidizing agent) in the reaction mixture, only catalytic amounts of OsO_4 are required because the amine oxide converts the osmium by-products back into OsO_4.

an alkene + OsO_4 (catalytic amount) —water/*tert*-butyl alcohol (solvent) pyridine→ osmate ester —2 H_2O→ a glycol + Os(VI)

Os(VI) + $R_3\overset{|}{N}-O^-$ (an amine oxide) → OsO_4 (catalyst regenerated) + $R_3\ddot{N}$ (an amine) + H_2O

d. The formation of glycols from alkenes is a stereospecific *syn*-addition. The five-membered osmate ester ring is easily formed when two oxygens of OsO_4 are added to the same face of the double bond by a concerted mechanism.

cyclohexene —OsO_4→ osmate ester —H_2O, $NaHSO_3$→ *cis*-1,2-cyclohexanediol

e. Hydrolysis of epoxides and the OsO_4 oxidation of alkenes are complementary reactions because they provide glycols with different stereochemistry.

3. Glycol formation from alkenes can also be carried out with potassium permanganate ($KMnO_4$), usually under aqueous alkaline conditions.

cyclohexene —$KMnO_4$, ^-OH, H_2O / acetone→ *cis*-1,2-cyclohexanediol

a. This reaction is also a stereospecific *syn*-addition, and its mechanism is probably similar to that of OsO_4 addition.
   i. Visually, when oxidation occurs, the brilliant purple color of the permanganate ion is replaced by a murky brown precipitate of magnesium dioxide (MnO_2).
   ii. This color change can be used as a test (called the Baeyer test) for functional groups that can be oxidized by $KMnO_4$.

b. A problem with the use of $KMnO_4$ is that yields are low in many cases because overoxidation occurs.

II. REACTIONS OF ETHERS, EPOXIDES, GLYCOLS, AND SULFIDES

A. Cleavage of Ethers

1. Ethers are relatively unreactive compounds; however, the C—O bonds of ethers react under acidic conditions (HI or HBr) to give alcohols and alkyl halides in a reaction called ether cleavage.

$$R{-}O{-}R' + H{-}I \longrightarrow R{-}OH + R'{-}I$$

$$R{-}OH \xrightarrow[\text{(in some cases)}]{H{-}I} R{-}I + H_2O$$

   a. Ethers containing only methyl and/or primary alkyl groups require strong acid and relatively harsh conditions to bring about ether cleavage.
   b. Secondary and tertiary ethers cleave more readily than primary or methyl ethers because tertiary carbocation intermediates can be formed.
   c. The alcohol formed in the cleavage of an ether can go on to react with HI or HBr to give a second molecule of alkyl halide.
2. The cleavage mechanism of ethers containing only methyl and/or primary alkyl groups involves
   a. protonation of the ether oxygen, followed by
   b. attack of the protonated ether by the iodide or bromide ion (a good nucleophile) in an S_N2 reaction to form an alkyl halide and liberate an alcohol as a leaving group.

$$\underset{\text{1-methoxyhexane}}{CH_3(CH_2)_4CH_2{-}\ddot{O}{-}CH_3} \underset{\text{heat}}{\overset{\text{HBr}}{\rightleftharpoons}} CH_3(CH_2)_4CH_2{-}\overset{H}{\overset{|}{O^+}}{-}CH_3 \;\; :\ddot{Br}:^- \longrightarrow \underset{\text{1-hexanol}}{CH_3(CH_2)_4CH_2{-}\ddot{O}H} + \underset{\text{bromomethane}}{H_3C{-}\ddot{Br}:}$$

$$\underset{\text{1-hexanol}}{CH_3(CH_2)_4CH_2{-}OH} \xrightarrow[\text{heat}]{\text{HBr}} \underset{\text{1-bromohexane}}{CH_3(CH_2)_4CH_2{-}Br}$$

3. The mechanism of cleavage in ethers containing tertiary alkyl groups (which occurs under milder conditions) involves
   a. protonation of the ether oxygen, followed by
   b. formation of a carbocation by loss of the alcohol leaving group, and finally
   c. attack of the carbocation by halide ion.
4. Because the S_N1 reaction is faster than competing S_N2 processes, none of the primary alkyl halide is formed.

$$\underset{\textit{tert}\text{-butyl methyl ether}}{H_3C{-}\ddot{O}{-}C(CH_3)_3} \overset{\text{HI}}{\rightleftharpoons} H_3C{-}\overset{H}{\overset{|}{O^+}}{-}C(CH_3)_3 \;\; I^- \longrightarrow \underset{\text{methyl alcohol}}{H_3C{-}\ddot{O}H} + \underset{\text{a tertiary carbocation}}{(CH_3)_3C^+ \; :\ddot{I}:^-} \longrightarrow \underset{\textit{tert}\text{-butyl iodide}}{(CH_3)_3C{-}\ddot{I}:}$$

B. Reactions of Epoxides with Acid or Base

1. Because of their ring strain, epoxides undergo ring-opening reactions with ease; the opening of an epoxide relieves the strain of the three-membered ring.
2. The ring opening of epoxides by bases involves backside attack of the nucleophile on the epoxide carbon.
   a. A reaction of this type is essentially an S_N2 reaction in which the epoxide oxygen serves as the leaving group.

b. Under basic or neutral conditions, nucleophiles typically attack unsymmetrical epoxides at the carbon with fewer branches.

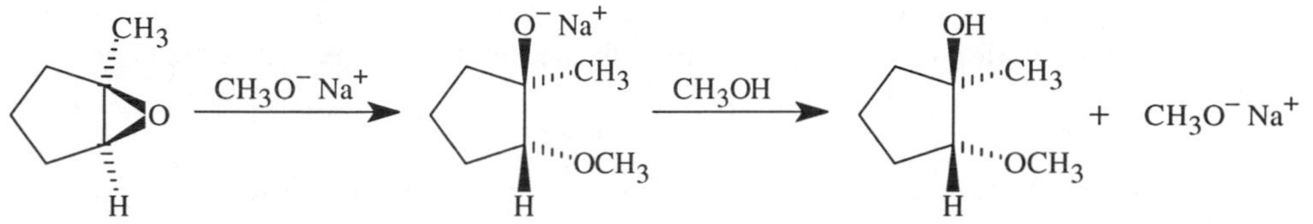

i. This regioselectivity is expected from the effect of branching on the rates of S_N2 reactions.
ii. Branching retards the rate of attack; hence, attack at the unbranched carbon is faster and leads to the observed product.

c. Inversion of configuration is observed if attack occurs at a stereocenter.

3. The regioselectivity of the ring-opening reaction is different under acidic conditions.
   a. The nucleophile reacts at the more branched carbon of the epoxide.
   b. If neither carbon is tertiary, a mixture of products is formed in most cases.

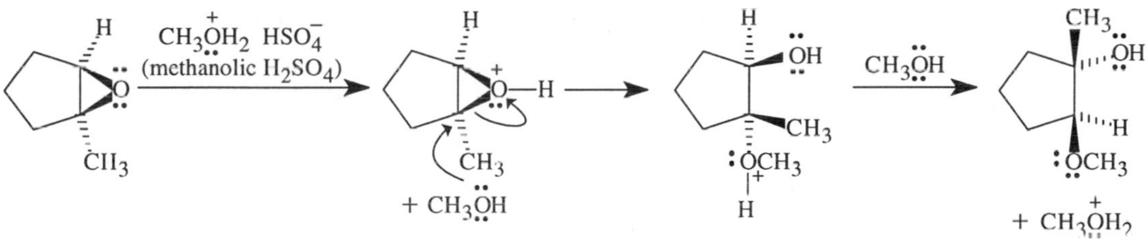

i. Bonds to tertiary carbon atoms are weaker than bonds to primary carbon atoms and the protonated oxygen is a good leaving group.
ii. The regioselectivity of acid-catalyzed epoxide ring opening is very similar to attack on bromonium ions.

4. Acid-catalyzed epoxide hydrolysis is generally a useful way to prepare glycols.

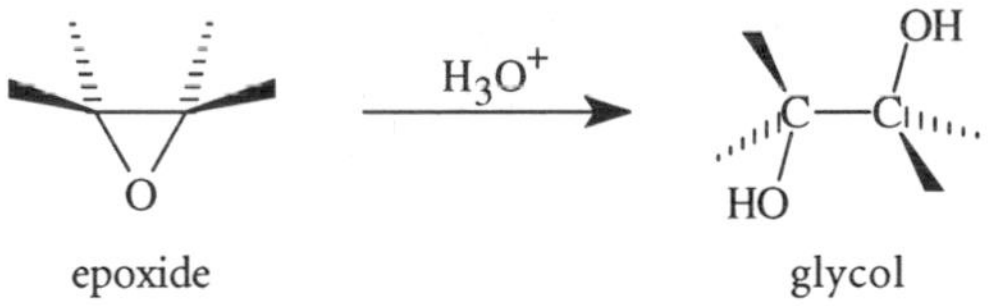

epoxide glycol

   a. This reaction occurs by backside attack of water on the protonated epoxide.
   b. This reaction, when combined with epoxide formation from an alkene, gives glycols with different stereochemistry from that obtained from the OsO_4 reaction with alkenes. (Compare with the reaction in Sec. I.F.2.d.)

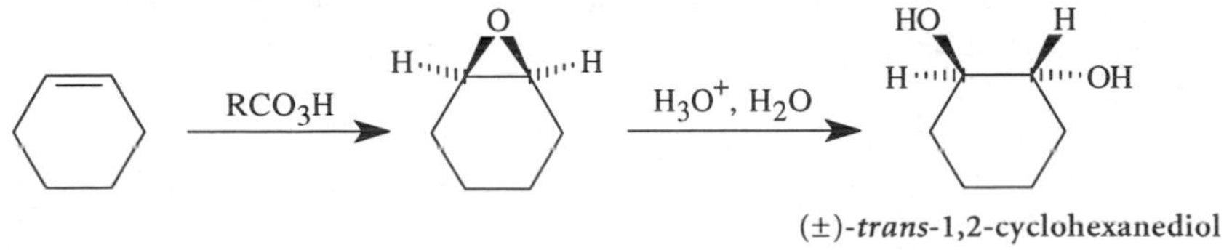

(±)-*trans*-1,2-cyclohexanediol

5. Base-catalyzed hydrolysis of epoxides also gives glycols, although, in some cases, polymerization occurs as a side reaction under the basic conditions.

C. Reaction of Ethylene Oxide with Grignard Reagents

1. Grignard reagents react with ethylene oxide to give, after a protonation step, primary alcohols.

a. The carbon of the C—Mg bond of the Grignard reagent has carbanion character and attacks the epoxide as a nucleophile.
b. The magnesium of the Grignard reagent is a Lewis acid and coordinates with the epoxide oxygen.
c. Coordination of an oxygen to a Lewis acid makes the oxygen a better leaving group.

Br—Mg—Ph + H_2C—H_2C epoxide :Ö:--Mg(Ph)(Br) ⟶ Ph—CH_2CH_2—Ö:⁻--Mg(Ph)(Br) $^{+}$MgBr ⇌ Ph—CH_2CH_2—Ö:⁻ $^{+}$MgBr + Ph—Mg—Br

d. After the Grignard reagent has reacted, the alkoxide is converted into the alcohol product in a separate step by the addition of water or dilute acid.

$$H_3O^+ + Ph—CH_2CH_2—\ddot{O}:^- \ {}^{+}MgBr \longrightarrow Ph—CH_2CH_2—\ddot{O}H + H_2O + Mg^{2+} + Br^-$$

2. The reaction of a Grignard reagent with ethylene oxide provides a method for the formation of carbon-carbon bonds.
3. Grignard reagents react with unsymmetrical epoxides to give mixtures of products caused by rearrangements and other side reactions.

D. Oxidative Cleavage of Glycols

1. The carbon-carbon bond between the —OH groups of a vicinal glycol can be cleaved with periodic acid to give two carbonyl compounds.
   a. The cleavage of glycols with periodic acid takes place through a cyclic periodate ester intermediate that forms when the glycol displaces two —OH groups from H_5IO_6.
   b. The cyclic ester spontaneously breaks down by a cyclic flow of electrons in which the iodine accepts an electron pair.

OH, OH, CH_3 —H_5IO_6 periodic acid⟶ (O, OH, OH, O, O, OH, CH_3) a cyclic periodate ester + 2 H_2O ⟶ H—C(=O)—⋯—C(=O)—CH_3 + :I(OH)(OH)(=O)(=O) iodine (V)

2. A glycol that cannot form a cyclic ester intermediate is not cleaved by periodic acid.
3. The periodate cleavage reaction has been used as a test for glycols as well as for synthesis.

E. Oxidation of Ethers and Sulfides

1. Except for peroxide formation (which occurs when ethers are allowed to stand in air) and combustion, ethers are relatively inert towards oxidizing conditions.
2. Sulfides oxidize at sulfur rather than at carbon when they react with many common oxidizing agents; sulfides are readily oxidized to sulfoxides and sulfones.
   a. Sulfoxides and sulfones can be prepared by the direct oxidation of sulfides with one and two equivalents, respectively, of hydrogen peroxide, H_2O_2.

R—S—R′ (a sulfide) —H_2O_2⟶ R—S(=O)—R′ (a sulfoxide) —H_2O_2⟶ R—S(=O)(=O)—R′ (a sulfone)

b. Other common oxidizing agents such as $KMnO_4$, HNO_3, and peroxyacids readily oxidize sulfides.

3. Nonionic Lewis structures for sulfoxides and sulfones cannot be written without violating the octet rule.

$$\left[R-\overset{:O:}{\overset{\|}{\underset{..}{S}}}-R \longleftrightarrow R-\overset{:\ddot{O}:^-}{\overset{|}{\underset{..}{S^+}}}-R \right]$$

sulfoxide

$$\left[R-\overset{:O:}{\overset{\|}{\underset{\underset{:O:}{\|}}{S}}}-R \longleftrightarrow R-\overset{:\ddot{O}:^-}{\overset{|}{\underset{\underset{:\ddot{O}:^-}{|}}{S^{2+}}}}-R \right]$$

sulfone

F. Reactions of Oxonium and Sulfonium Salts

1. If the acidic hydrogen of a protonated ether is replaced with an alkyl group, the resulting compound is called an oxonium salt; the sulfur analog of an oxonium salt is a sulfonium salt.

$(CH_3)_3O^+\ ^-BF_4$ — trimethyloxonium tetrafluoroborate

$(C_2H_5)_3S^+\ ^-BF_4$ — triethylsulfonium tetrafluoroborate

2. Oxonium and sulfonium salts react with nucleophiles in S_N2 reactions.

$$R-OH + (CH_3)_3O^+\ ^-BF_4 \longrightarrow R-O-CH_3 + CH_3-O-CH_3 + HBF_4$$

an alcohol; trimethyloxonium tetrafluoroborate (an oxonium salt); a methyl ether; dimethyl ether; tetrafluoroboric acid

a. Oxonium salts are among the most reactive alkylating agents known (they react very rapidly with most nucleophiles). They are usually isolated as $^-BF_4$ salts because the $^-BF_4$ counterion is not nucleophilic. These salts must be stored in the absence of moisture.

b. Sulfonium salts are considerably less reactive than the corresponding alkyl chlorides in S_N2 reactions and therefore are handled more easily than oxonium salts.

c. Sulfonium salts are somewhat less reactive than the corresponding alkyl chlorides in S_N2 reactions.

3. The important biological sulfonium salt *S*-adenosylmethionine (SAM) is a methylating agent for biological nucleophiles.

SCIENTISTS

STUDY GUIDE LINKS

✓11.1 Learning New Reactions from Earlier Reactions

Quite often the text points out the close relationship between two reactions—a new one, and one you have already studied. Such connections will help you to understand new reactions in terms of what you already know and will substantially reduce the amount of *really new* material that you must master. Alkoxymercuration-reduction is such a case; this reaction is closely related to oxymercuration-reduction. In order to make the connection between the two reactions, it almost goes without saying that you have to know what oxymercuration-reduction is! If you did not know the meaning of this term when you read this section, did you go back and find out?

The question at the end of Study Problem 11.1 on text p. 456—why *tert*-butyl bromide and sodium methoxide will not work in a Williamson ether synthesis—is another effort to get you to think about earlier material, in this case the reactivities of alkyl halides in the presence of a strong base. Did you answer this question? (If you can't answer it, review Sec 9.5F as suggested in the problem.)

Remember that *continued review* is one of the keys to successful study in organic chemistry. You must try constantly to remain active rather than passive when you study. Answering questions such as the ones posed in the text will assist you. If you can't answer one of these questions, write it down and get help at the next available opportunity!

✓11.2 Common Intermediates from Different Starting Materials

An important thing to notice about this section is that the same reactive intermediate—in this case, a carbocation—in some cases can be generated in several different ways. For example, consider the many different ways that a *tert*-butyl cation can be formed as a reactive intermediate:

$$(CH_3)_3C{-}OH \xrightarrow{\text{"}H^+\text{"},\ -H_2O} (H_3C)_2\overset{+}{C}{-}CH_3$$

$$(H_3C)_2C{=}CH_2 \xrightarrow{\text{"}H^+\text{"}} (H_3C)_2\overset{+}{C}{-}CH_3 \xleftarrow[\text{(solvolysis)}]{-Br^-} (CH_3)_3C{-}Br$$

tert-butyl cation

Whether a carbocation is formed by protonation of an alkene, by dehydration of an alcohol, or by sovolysis of an alkyl halide, the same product is obtained if a large excess of the same nucleophile is present in each case. Thus, if methanol is used as the solvent, *tert*-butyl methyl ether is the product regardless of the origin of the carbocation. On the other hand, if concentrated HBr is used, *tert*-butyl bromide is the product.

It is natural at first to think of hydrogen halide addition to alkenes, alcohol dehydration, alkene hydration, and solvolysis of tertiary alkyl halides as different reactions. What you now should be able to see is that these reactions are all linked through a common intermediate, the carbocation. Which product is obtained depends on the conditions used.

✓11.3 Mechanism of Epoxide Formation

The curved-arrow mechanism of epoxide formation incorporates elements of several steps which, when viewed separately, may make the mechanism easier to follow. First, imagine that the alkene π-electrons attack the oxygen of a peroxycarboxylic acid and that a carbocation is formed:

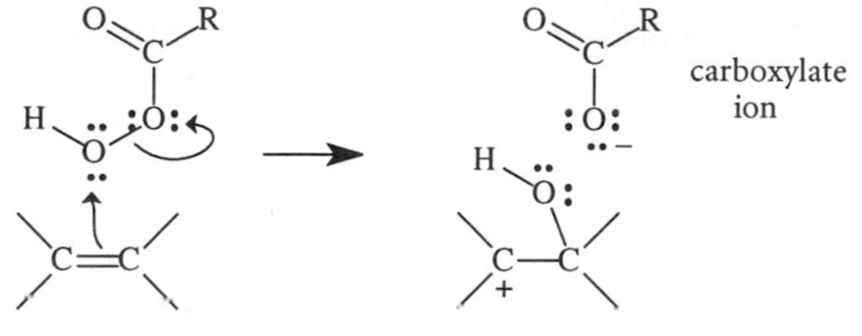

carboxylate ion

carbocation

This step breaks a weak O—O bond and liberates a good leaving group, the carboxylate ion, which is a weak base. Next, the oxygen of the —OH group captures the carbocation with an unshared electron pair in an internal Lewis acid-base association reaction:

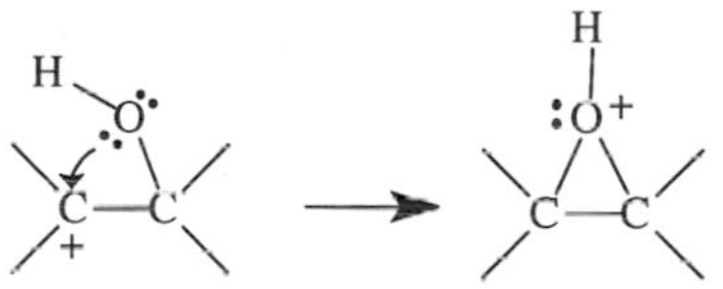

protonated epoxide

Finally, the carboxylate ion expelled in the first step removes the proton from the protonated epoxide:

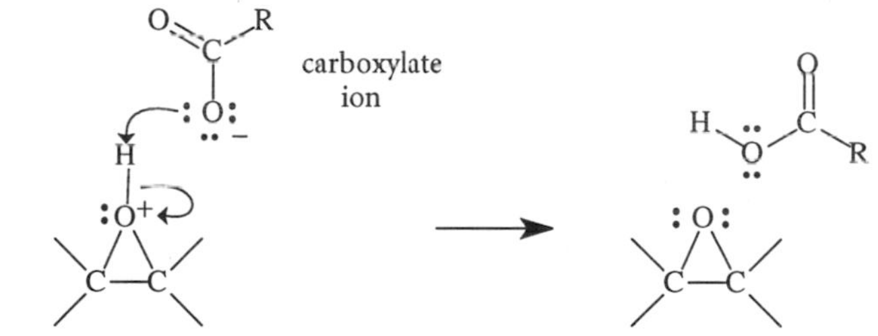

carboxylate ion

protonated epoxide

epoxide

Although the actual mechanism of epoxide formation is *not* stepwise, and does *not* involve carbocations, thinking of it this way helps to see how it is related to other mechanisms that do involve carbocations. A similar approach has been used to help you understand bromine addition (Eq. 5.4b, text p. 161), hydroboration (Study Guide Link 5.2, p. 120 of this manual), and ozonolysis (Study Guide Link 5.5, p. 126 of this manual).

✓11.4 Mechanism of OsO_4 Addition

Even though the OsO_4 addition is a concerted reaction—a reaction that occurs in a single step—let's break the reaction into individual steps so that we can see the relationship of this reaction to other electrophilic additions we have studied.

First, let's think about the electronic character of OsO_4. Remember that osmium in a +8 oxidation state is very electronegative. Therefore, we can think of the reagent in the following way:

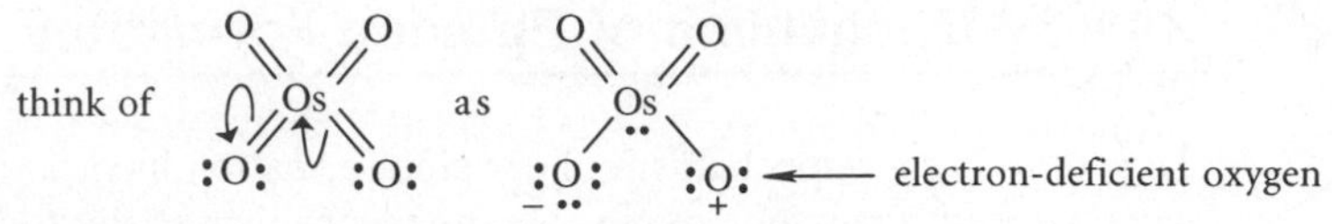

think of as electron-deficient oxygen

In this picture, osmium has satisfied its "hunger" for electrons, but as a result, one oxygen has become electron-deficient. This is why the π electrons of the alkene can attack this oxygen. Compare with the corresponding structure of ozone:

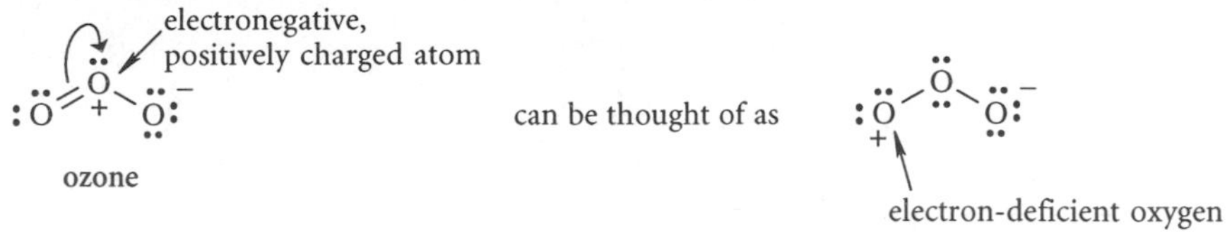

electronegative, positively charged atom

ozone

can be thought of as

electron-deficient oxygen

The mechanism of OsO_4 addition can now be viewed as a close analogy to the mechanism of ozone addition (Study Guide Link 5.5). The electrophilic oxygen accepts the pair of π electrons from the alkene, and a carbocation is generated:

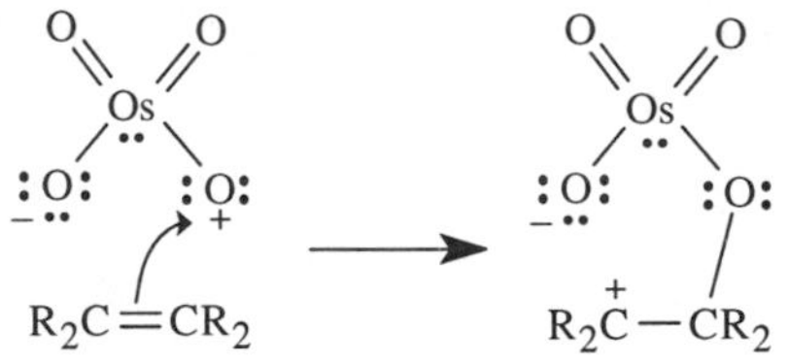

$R_2C{=}CR_2$ ⟶ $R_2\overset{+}{C}{-}CR_2$

The negatively charged oxygen can act as a Lewis base to complete the addition:

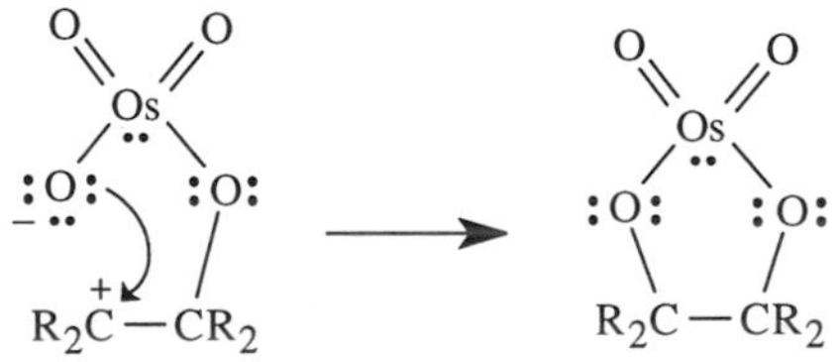

$R_2\overset{+}{C}{-}CR_2$ ⟶ $R_2C{-}CR_2$

osmate ester

Even though the mechanism is concerted, thinking of it as a two-step process, as shown above, helps to understand why the reaction takes place, and helps to see the connection to other reactions that involve carbocations explicitly.

11.5 Reaction Probability and Entropy

Recall that a standard free-energy change consists of a contribution from both an enthalpy change and an entropy change (Study Guide Link 4.5 on p. 85 of this manual):

$$\Delta G° = \Delta H° - T\Delta S°$$

Likewise, the standard free energy of activation $\Delta G^{°\ddagger}$ consists of contributions from the *enthalpy of activation* $\Delta H^{°\ddagger}$ and the *entropy of activation* $\Delta S^{°\ddagger}$ according to the equation

$$\Delta G^{°\ddagger} = \Delta H^{°\ddagger} - T\Delta S^{°\ddagger} \qquad \text{(SG11.1)}$$

The "reaction probability" discussed in Sec. 11.7 is roughly equivalent to the entropy of activation; the ring strain contributes to the enthalpy of activation. Study Guide Link 4.5 discusses the relationship of entropy and probability in chemical terms. A loss of *randomness* corresponds to a negative entropy change, which, according to Eq. SG11.1, raises the free-energy barrier for a reaction, and therefore decreases the rate. When two molecules collide to form one molecule, the translational motion associated with two molecules moving about randomly is converted into the translational motion of a single molecule moving about randomly; hence, an *intermolecular* reaction causes a loss of randomness, or a decrease in *translational entropy.* When a molecule reacts *intramolecularly,* there is no change in the number of molecules; to a useful approximation, the translational entropy change is zero. Hence, the "reaction probability," or entropy of activation, is less negative (or more positive) for an intramolecular reaction. Counterbalancing this is the loss of internal rotations associated with ring formation; consequently, an intramolecular reaction that results in ring formation also suffers a loss of randomness as well, but the entropy (randomness) associated with internal rotations is *much* less than that associated with translation.

Forming a three-membered ring introduces strain, which is reflected in an increased $\Delta H^{\circ\ddagger}$. However, as suggested in the text, when intramolecular and intermolecular reactions are compared, the increase in $\Delta H^{\circ\ddagger}$ resulting from an intramolecular reaction that forms a three-membered ring is more than offset by the decrease in $\Delta S^{\circ\ddagger}$ associated with an intermolecular reaction. The balance of $\Delta H^{\circ\ddagger}$ and $\Delta S^{\circ\ddagger}$ is such that three, five, and six-membered rings seem to be most often involved in cases of neighboring-group participation.

SOLUTIONS TO PROBLEMS

Solutions to In-Text Problems

11.1 (a) In this reaction, an alkoxide, formed by reaction of the alcohol with sodium, is alkylated by methyl iodide to give isopropyl methyl ether:

$$(CH_3)_2CHOH \xrightarrow{Na} (CH_3)_2CHO^- \, Na^+ + H_2 \xrightarrow{CH_3I} (CH_3)_2CHO{-}CH_3 + Na^+ \, I^-$$

isopropyl methyl ether

(b) A thiolate ion, formed by reaction of NaOH with the thiol, is alkylated by allyl chloride to give $H_2C{=}CH{-}CH_2{-}S{-}CH_3$ (allyl methyl sulfide) + $Na^+ \, Cl^-$. Notice that NaOH is a strong enough base to form the thiolate anion, but would not be strong enough to form the conjugate base of an alcohol.

(c) Alkoxide bases react with tertiary alkyl halides to give elimination products rather than substitution products. (See Sec. 9.5F and Problem 9.37 on p. 401 of the text.) Consequently, the products are $(CH_3)_2C{=}CH_2$ (2-methylpropene), $Na^+ \, Br^-$ (sodium bromide), and CH_3OH (methanol).

(d) Neopentyl halides and sulfonate esters do not undergo S_N2 reactions at room temperature; furthermore, they do not undergo β-elimination because there are no β-hydrogens. Thus, no reaction occurs.

11.2 (a)

$$C_6H_{11}{-}CH_2CH_2OH \xrightarrow{NaH} C_6H_{11}{-}CH_2CH_2O^- \, Na^+ + H_2 \xrightarrow{I{-}CH_2CH_3} C_6H_{11}{-}CH_2CH_2OCH_2CH_3 + Na^+ \, I^-$$

A second Williamson synthesis, the reaction of sodium ethoxide with (bromoethyl)cyclohexane, is in principle possible; however, because the alkyl halide has a β-branch, this reaction would give, in addition to the desired product, a significant amount of alkene resulting from elimination.

(b)

$$(CH_3)_2CHSH \xrightarrow[C_2H_5OH]{Na^+ \, C_2H_5O^-} (CH_3)_2CHS^- \, Na^+ + C_2H_5OH \xrightarrow{I{-}CH_3} (CH_3)_2CHSCH_3 + Na^+ \, I^-$$

The alternative synthesis, reaction of the methanethiolate ion with isopropyl bromide, is less desirable because secondary alkyl halides react more slowly than methyl halides and give some elimination products.

(c) Di-*tert*-butyl ether cannot be prepared by a Williamson synthesis, because its preparation would require an S_N2 reaction of *tert*-butoxide with a *tert*-butyl halide. With this combination of reagents, an E2 reaction would occur instead, and 2-methylpropene and *tert*-butyl alcohol would be the only organic products formed. (See part (c) of Problem 11.1.)

11.3 (a) Let $-R = -CH_2CH_2CH_2CH_3$. The mechanism is essentially identical to that shown in Eq. 5.12a–b on text p. 166, except that the nucleophile is isopropyl alcohol rather than water.

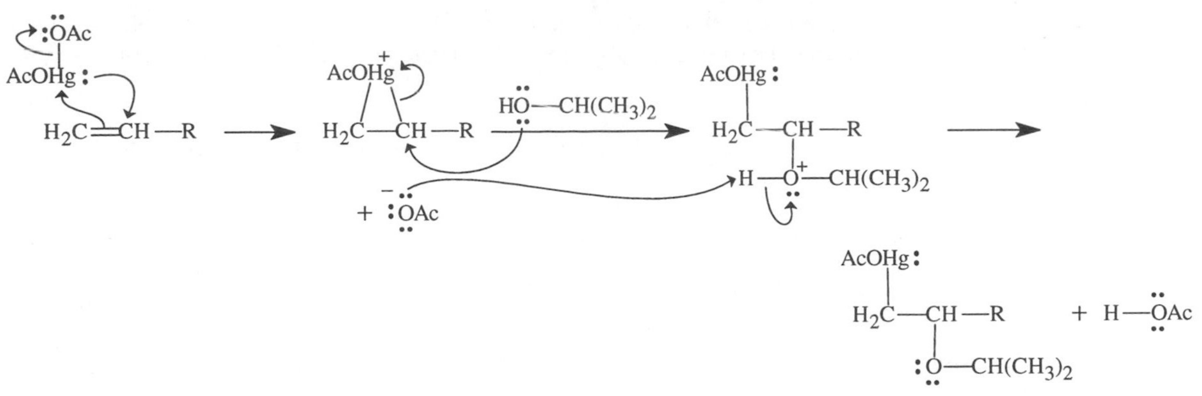

:ÖAc

AcOHg: $H_2C{=}CH{-}R$ ⟶ $AcO\overset{+}{H}g$ $H_2C{-}CH{-}R$ + :ÖAc$^-$ $\xrightarrow{HÖ{-}CH(CH_3)_2}$ AcOHg: $H_2C{-}CH{-}R$, $H{-}\overset{+}{O}{-}CH(CH_3)_2$ ⟶

AcOHg: $H_2C{-}CH{-}R$, :Ö$-CH(CH_3)_2$ + H—ÖAc

Attack of the alcohol on the mercurinium ion intermediate occurs at the more branched carbon because this carbon-mercury bond is weaker (that is, easier to break) than the bond to primary carbon. Another way to state the same point is to say that the mercurinium ion is described by three resonance structures, of which structure *B* is more important than structure *C*, because in structure *B* electron deficiency resides on the more branched carbon:

[(C) $AcO\overset{+}{H}g$ (B) $H_2C{-}CH{-}R$ ⟷ AcOHg: $H_2C{-}\overset{+}{C}H{-}R$ ⟷ AcÖHg $H_2\overset{+}{C}{-}CH{-}R$]

A *B* *C*

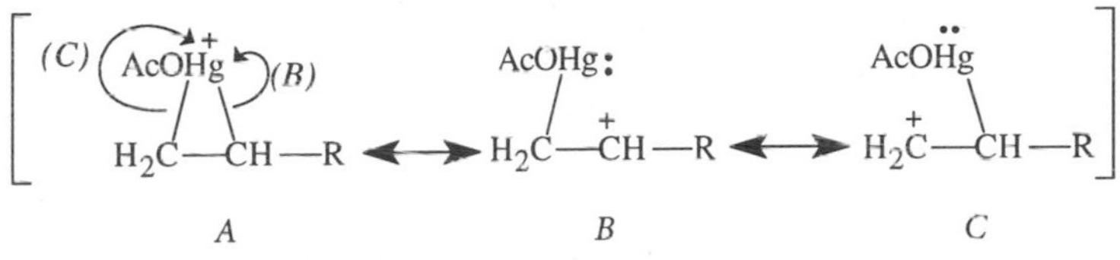

(b) Synthesis of the ether in Eq. 11.6b would require a secondary alkyl halide and an alkoxide with α-branching. A significant, if not predominant, amount of alkene by-product would be obtained by the E2 mechanism.

11.4 The product results from oxymercuration of the double bond followed by reduction (by $NaBH_4$) to remove the mercury. The net effect is overall addition of the elements of ethanol to the double bond with the ethoxy group on the more branched carbon.

$(CH_3)_2CH\underset{|}{\overset{OC_2H_5}{C}}HCH_3$ **2-ethoxy-3-methylbutane**

11.5 (a) Cyclohexene is subjected to oxymercuration in cyclohexanol, and the product is then reduced with $NaBH_4$:

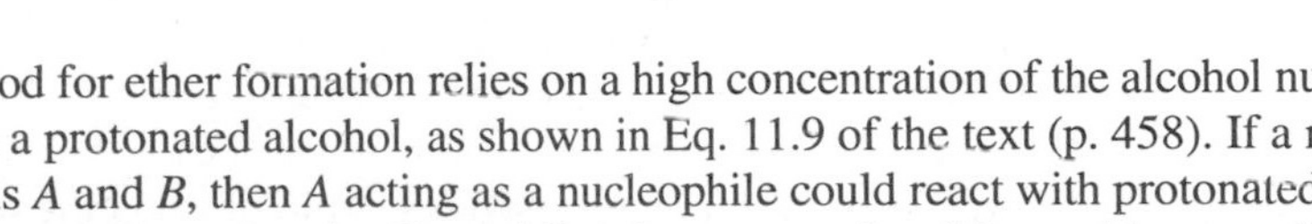

cyclohexene + HO–cyclohexane $\xrightarrow[\text{2) } NaBH_4]{\text{1) } Hg(OAc)_2}$ cyclohexyl–O–cyclohexyl

dicyclohexyl ether

(b)

$(CH_3)_2C{=}CH_2$ + $HOCH_2CH(CH_3)_2$ $\xrightarrow[\text{2) } NaBH_4]{\text{1) } Hg(OAc)_2}$ $(CH_3)_3C{-}OCH_2CH(CH_3)_2$

2-methylpropene **isobutyl alcohol** ***tert*-butyl isobutyl ether**

11.6 The dehydration method for ether formation relies on a high concentration of the alcohol nucleophile (as the solvent) reacting with a protonated alcohol, as shown in Eq. 11.9 of the text (p. 458). If a reaction mixture contained two alcohols *A* and *B*, then *A* acting as a nucleophile could react with protonated *A* (to give ether *A*—*A*) or with protonated *B* (to give ether *A*—*B*); and *B* acting as a nucleophile could react with protonated *A* (to give ether *A*—*B*), or protonated *B* (to give ether *B*—*B*). Hence, a mixture of three different ethers would be obtained. This would have to be laboriously separated, and the yield of no one component would be very high, barring unusually high reactivity.

The reason we can let a *tertiary* alcohol react with a primary alcohol to give an unsymmetrical ether is that the tertiary alcohol forms a carbocation in acidic solution *much* faster than either it or the primary ether react by the S_N2 mechanism; and, once the carbocation is formed, it is snatched up by the large excess of primary alcohol that is present.

11.7 A carbocation from the tertiary alcohol is attacked by methanol to give the following ether:

$$\mathrm{Ph{-}C(CH_3)_2{-}OCH_3}$$

11.8 Use the discussion in Study Guide Link 11.2: What alkene would give the same carbocation as the starting alcohol in Problem 11.7?

$$\left.\begin{array}{l}\mathrm{Ph{-}C(CH_3){=}CH_2} \xrightarrow{H_2SO_4} \\ \mathrm{Ph{-}C(OH)(CH_3){-}CH_3} \xrightarrow[-H_2O]{H_2SO_4}\end{array}\right\} \mathrm{Ph{-}\overset{+}{C}(CH_3){-}CH_3}\ \ \mathrm{HSO_4^-} \xrightarrow{CH_3OH} \mathrm{Ph{-}C(OCH_3)(CH_3){-}CH_3}$$

11.9 (a) Because this ether is symmetrical, and because both alkyl groups are primary, alcohol dehydration is the appropriate method. This ether can be prepared by treating $ClCH_2CH_2OH$ (2-chloroethanol) with H_2SO_4 at high temperature.

(b) Because this is an ether with one tertiary alkyl group and one methyl group, it can be prepared by dehydration of the tertiary alcohol in the presence of methanol, or by acid-catalyzed addition of methanol to the appropriate alkene.

$$\left\{\begin{array}{c}\mathrm{CH_3C(OH)(CH_3)CH_2CH_3} \\ \text{2-methyl-2-butanol}\end{array}\ \text{or}\ \begin{array}{c}\mathrm{H_2C{=}C(CH_3)CH_2CH_3} \\ \text{2-methyl-1-butene} \\ \text{or} \\ \mathrm{(CH_3)_2C{=}CHCH_3} \\ \text{2-methyl-2-butene}\end{array}\right\} + \begin{array}{c}\mathrm{CH_3OH} \\ \text{(solvent)}\end{array} \xrightarrow{H_2SO_4} \begin{array}{c}\mathrm{CH_3C(OCH_3)(CH_3)CH_2CH_3} \\ \text{2-methoxy-2-methylbutane}\end{array}$$

(c) Because this ether has a tertiary alkyl group, it can be derived by acid-catalyzed addition of 2-propanol to an alkene, or by dehydration of *tert*-butyl alcohol in the presence of 2-propanol as solvent.

$$\left.\begin{array}{c}\mathrm{(CH_3)_2C{=}CH_2} \\ \text{2-methylpropene} \\ \text{or} \\ \mathrm{(CH_3)_3C{-}OH} \\ \textit{tert}\text{-butyl alcohol}\end{array}\right\} + \begin{array}{c}\mathrm{HOCH(CH_3)_2} \\ \text{2-propanol} \\ \text{(solvent)}\end{array} \xrightarrow{H_2SO_4} \begin{array}{c}\mathrm{(CH_3)_3C{-}O{-}CH(CH_3)_2} \\ \textit{tert}\text{-butyl isopropyl ether}\end{array}$$

Another good way to prepare a laboratory-sized sample of this ether would be by alkoxymercuration-reduction (Sec. 11.1B) of 2-methylpropene in 2-propanol solvent.

11.10 (a) H_3C, H_3C C=CH_2 (b) 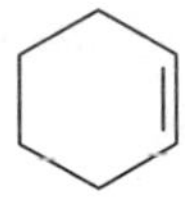 (c) 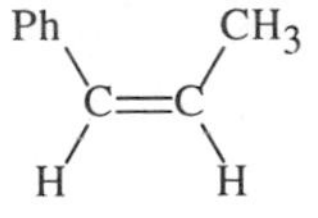 Ph, CH_3, C=C, H, H (d) Ph, H, C=C, H, CH_3

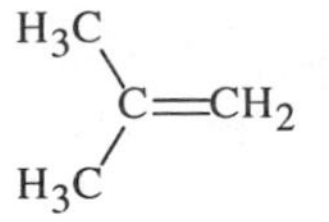

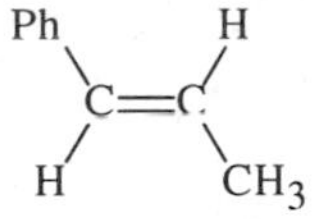

Notice in parts (c) and (d) that the stereochemistry of the product and the net *syn*-addition dictate the stereochemistry of the alkene starting material.

11.11 (a)

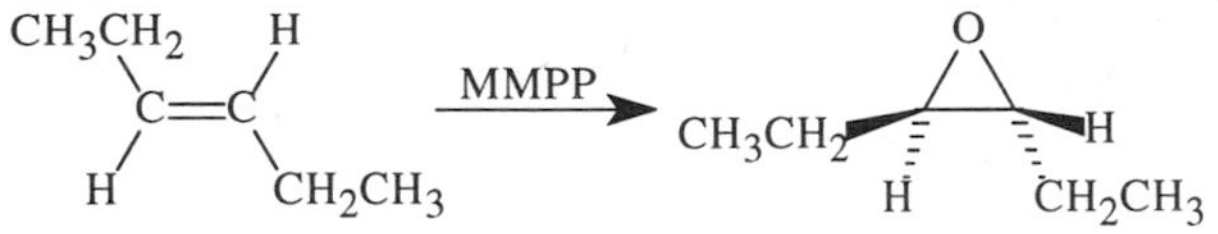

CH_3CH_2, H, C=C, H, CH_2CH_3 —MMPP→ CH_3CH_2, O, H, H, CH_2CH_3

trans-3-hexene *trans*-2,3-diethyloxirane

(b)

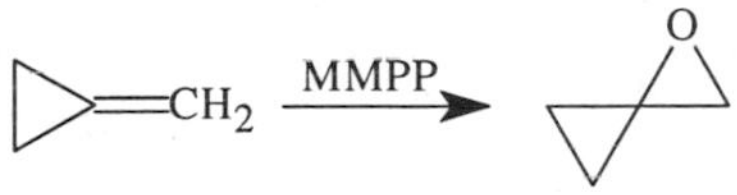

=CH_2 —MMPP→ O

11.12 The key to solving this problem is to realize that, in order for epoxide formation to occur, the oxygen of the conjugate-base alkoxide must be anti to the bromine in the transition state. Draw each of the stereoisomeric alkoxides in such a conformation and evaluate the relative energies of the resulting transition states.

In the transition state for the reaction of the alkoxide conjugate base of *A*, the methyl groups are anti:

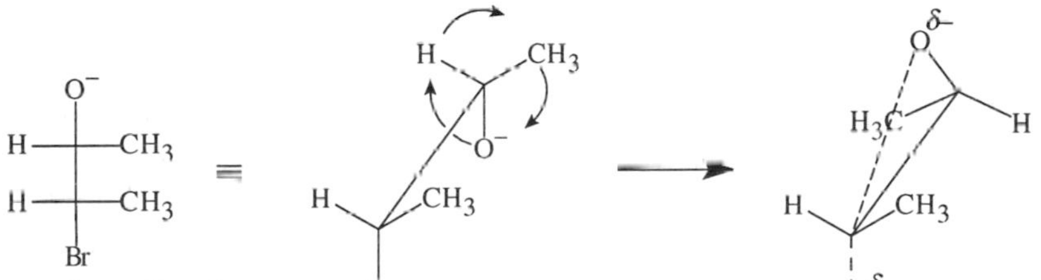

conjugate-base alkoxide of compound *A*

transition-state conformation for reaction of *A*: methyl groups are anti.

In contrast, the methyl groups are gauche in the transition state for the reaction of the alkoxide derived from *B:*

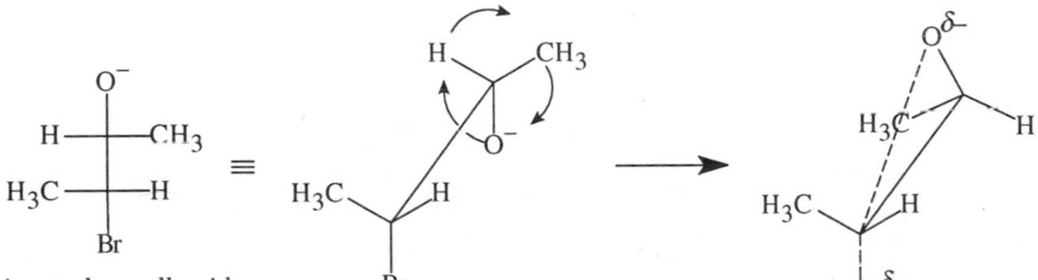

conjugate-base alkoxide of compound *B*

transition-state conformation for reaction of *B*: methyl groups are gauche.

The unfavorable gauche interaction between the two methyl groups makes the transition state derived from *B* more energetic and therefore more difficult to achieve. Consequently, the reaction of *A* is faster.

11.13 As in the solution to Problem 11.12, the alkoxide oxygen and the halogen leaving group must be anti in the transition state. This anti relationship is possible in a chair conformation only in the reaction of the trans stereoisomer. In the cis stereoisomer, such an anti relationship is not possible. Because the cis stereoisomer cannot achieve the appropriate transition-state conformation for epoxide formation, it is unreactive.

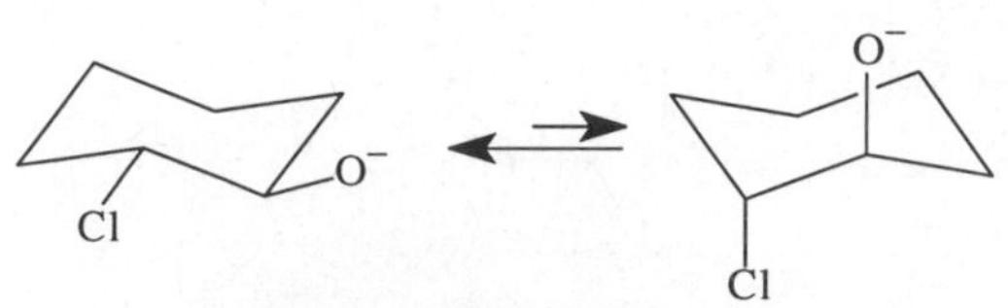

trans-2-chlorocyclohexanol
Anti relationship of O and Cl
is possible.

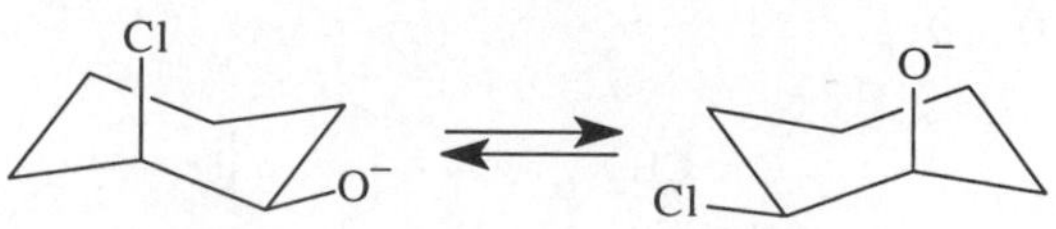

cis-2-chlorocyclohexanol
Anti relationship of O and Cl
is not possible.

11.14 (a) In the protonated ether (see Eq. 11.25a, text p. 465, for a typical protonation reaction) the iodide ion can attack either the carbon of the methyl group or the α-carbon of the butyl group. Because this is an S_N2 reaction, and because S_N2 reactions of methyl compounds are faster than those of butyl compounds (why?), attack occurs at the methyl carbon to give methyl iodide and 1-butanol.

(b) The discussion on text p. 465 indicates that the cleavage reactions of tertiary ethers, which are S_N1 reactions, occur under milder conditions than the S_N2 cleavage reactions of methyl and primary ethers. Hence, cleavage of the protonated ether occurs so as to give the *tert*-butyl cation and methanol. (See Eq. 11.26a–b on text p. 465 for the mechanism of a similar reaction.) Attack of iodide on the *tert*-butyl cation gives *tert*-butyl iodide.

(c) As indicated in part (b), acidic cleavage of tertiary ethers occurs by the S_N1 mechanism. As in part (b), the protonated ether reacts to give the *tert*-butyl cation and methanol. Loss of a β-proton from the cation gives 2-methylpropene; although this can protonate to regenerate the *tert*-butyl cation, the conditions of the reaction (distillation of low-boiling compounds) drives the volatile alkene from the reaction mixture as it is formed.

(d) Tertiary ethers and sulfides cleave by an S_N1 mechanism. The rate-limiting step in S_N1 reactions is formation of the carbocation by dissociation. The rate of this dissociation reaction is proportional to the concentration of the reacting species—in this case, the conjugate acid of the ether or sulfide. Because sulfides are much less basic than ethers (Sec. 8.7), the concentration of the reacting species is much smaller in the case of sulfides. Consequently, sulfides are less reactive than ethers toward acid-promoted cleavage.

(e) This case speaks to the balance between S_N1 or S_N2 cleavage of an ether at a secondary alkyl group vs. S_N2 cleavage of the same ether at a methyl group. Notice that the asymmetric carbon (*) is the α-carbon of the ether in the secondary alkyl group:

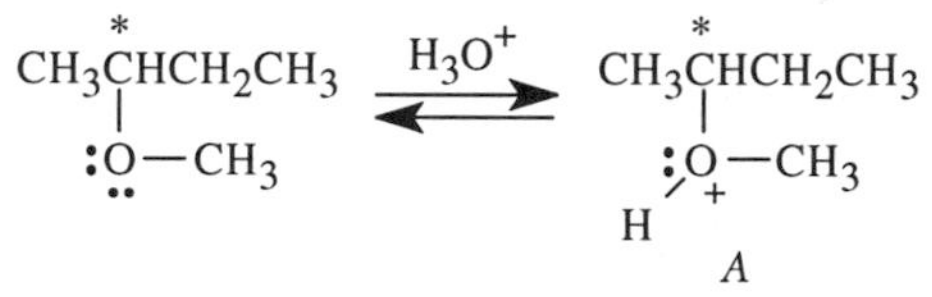

$$CH_3\overset{*}{C}HCH_2CH_3 \;(\text{with } {:}\ddot{O}{-}CH_3) \underset{}{\overset{H_3O^+}{\rightleftharpoons}} CH_3\overset{*}{C}HCH_2CH_3 \;(\text{with } H{-}\overset{+}{\ddot{O}}{-}CH_3)$$

A

Thus, optical activity serves as a barometer of what is going on mechanistically at this carbon. If cleavage of conjugate acid *A* occured by attack of Br^- at the *sec*-butyl group by an S_N2 mechanism, we would get *sec*-butyl bromide and methanol, and the product would be optically active. The observation of optical activity, *but not the identity of the products,* fits the data. Hence, this cannot be what is happening.

If cleavage of conjugate acid *A* occurred by an S_N1 mechanism at the asymmetric carbon of the alkyl group, a carbocation would be formed that would be in principle racemic. Hence, optical activity (or some optical activity) would be lost, and the product would be *sec*-butyl bromide and methanol. *Neither the optical activity data nor the identity of the products is consistent with the observations.* Thus, this mechanism is ruled out.

Finally, if bromide ion attacked the methyl carbon in conjugate-acid *A*, then the products would be methyl bromide and optically pure 2-butanol. The observations are consistent with this prediction.

Of course, for completeness, we should consider S_N1 cleavage of *A* at the methyl group; but this would require the formation of $^+CH_3$, a very unstable carbocation. This can be rejected on the basis of what we know about carbocations.

11.15 (a) The cleavage occurs by the S_N1 mechanism at either alkyl group. The products are therefore isopropyl iodide (2-iodopropane, $(CH_3)_2CH{-}I$) and isopropyl alcohol (2-propanol, $(CH_3)_2CH{-}OH$). (Under the reaction conditions, 2-propanol can also react to give 2-iodopropane.)

$$\underset{\text{diisopropyl ether}}{H_3C{-}CH(CH_3){-}O{-}CH(CH_3){-}CH_3} \xrightarrow{HI,\ H_2O} \underset{\text{2-iodopropane}}{H_3C{-}CH(CH_3){-}I} + \underset{\text{2-propanol}}{HO{-}CH(CH_3){-}CH_3}$$

(b) The cleavage occurs at the tertiary alkyl group by the S_N1 mechanism to give the tertiary iodide 2-iodo-2,3-dimethylbutane along with ethanol.

$$\underset{\text{2-ethoxy-2,3-dimethylbutane}}{CH_3C(OCH_2CH_3)(CH_3)CH(CH_3)_2} \xrightarrow{HI,\ H_2O} \underset{\text{2-iodo-2,3-dimethylbutane}}{CH_3C(I)(CH_3)CH(CH_3)_2} + \underset{\text{ethanol}}{HO{-}CH_2CH_3}$$

11.16 (a) Ammonia attacks the epoxide at the carbon with fewer branches:

$$(CH_3)_2CH(H_3C)\overset{O}{C{-}CH_2} + NH_3 \longrightarrow H_3C{-}C(O^-)((CH_3)_2CH){-}CH_2{-}\overset{+}{N}H_3 \underset{}{\overset{\text{proton transfers}}{\rightleftharpoons}} H_3C{-}C(OH)((CH_3)_2CH){-}CH_2{-}NH_2$$

(b) Azide attacks the epoxide at the carbon with fewer branches:

$$CH_3CH_2(H_3C)\overset{O}{C{-}CH_2} + Na^+\ N_3^- \longrightarrow H_3C{-}C(O^-\ Na^+)(CH_3CH_2){-}CH_2{-}N_3 \underset{R\,=\,C_2H_5\text{ or }R\,=\,H}{\overset{\text{ROH (solvent)}}{\rightleftharpoons}} H_3C{-}C(OH)(CH_3CH_2){-}CH_2{-}N_3 + RO^-\ Na^+$$

11.17 The strategy in this problem is to let the —OH group originate from the epoxide oxygen. In the starting material, this oxygen must be attached to the same carbon as the —OH group in the product as well as to an adjacent carbon; the nucleophile becomes attached to the adjacent carbon. To summarize:

$$R{-}CH(OH){-}\overset{|}{\underset{|}{C}}{-}X \Rightarrow R{-}\overset{O}{CH{-}C}\!\!<\ +\ X{:}^-$$

(a) The strategy outlined above suggests that the desired alcohol could be obtained by the reaction of cyanide ion with 2-pentyloxirane:

$$\underset{\text{2-pentyloxirane}}{CH_3(CH_2)_4\overset{O}{CH{-}CH_2}} + {}^-CN \longrightarrow CH_3(CH_2)_4CH(O^-)CH_2CN \xrightarrow{H_3O^+} CH_3(CH_2)_4CH(OH)CH_2CN$$

The strategy outlined above suggests another possibility:

$$CH_3(CH_2)_4\overset{O}{CH{-}CHCN} + \underset{\text{hydride ion}}{H{:}^-}$$

If you came up with this idea, you are reasoning correctly. However, this strategy will not work in this case because the most common source of nucleophilic hydride, $LiAlH_4$ (lithium aluminum hydride), also reacts with the —CN group.

(b) Apply the strategy outlined in the previous part of this problem. Notice that the ring-opening reaction occurs with inversion of configuration.

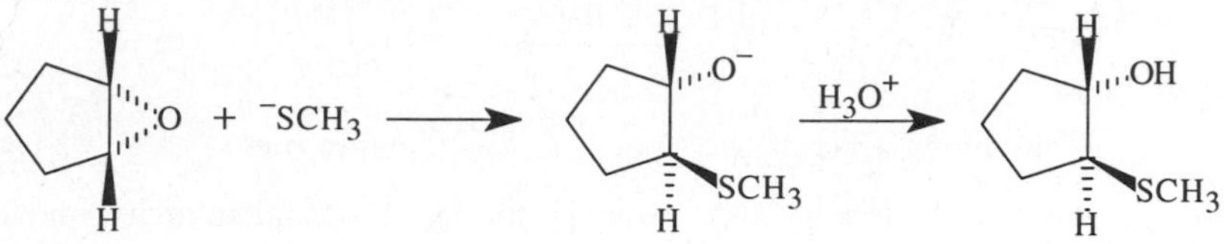

$$\text{cyclopentene oxide} + {}^{-}SCH_3 \longrightarrow \text{alkoxide (}O^-\text{, }SCH_3\text{, trans)} \xrightarrow{H_3O^+} \text{trans-2-(methylthio)cyclopentanol (OH, }SCH_3\text{)}$$

11.18 (a) Attack of methanol on the protonated epoxide occurs with inversion to give (3*S*,4*R*)-4-methoxy-3-hexanol:

$$\text{protonated epoxide} + HO\!-\!CH_3 \longrightarrow \text{intermediate} \rightleftharpoons \text{product} + H_3C\!-\!\overset{+}{O}H_2$$

(3*S*,4*R*)-4-methoxy-3-hexanol

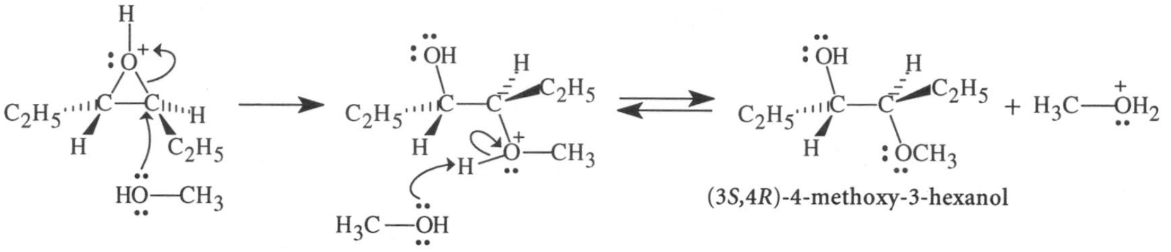

(Verify that attack of methanol at the other carbon gives the same result.)

(b) The enantiomer of the epoxide in part (a) gives the enantiomer of the product formed in part (a), (3*R*,4*S*)-4-methoxy-3-hexanol.

11.19 Under acidic conditions, attack of water occurs at the more branched carbon of the epoxide ring; consequently, the —OH group at that carbon is enriched in ^{18}O. In basic solution, attack of hydroxide occurs at the less branched carbon (the CH_2 carbon) of the epoxide, and the —OH group at that carbon is enriched in ^{18}O. ($^*O = {}^{18}O$.)

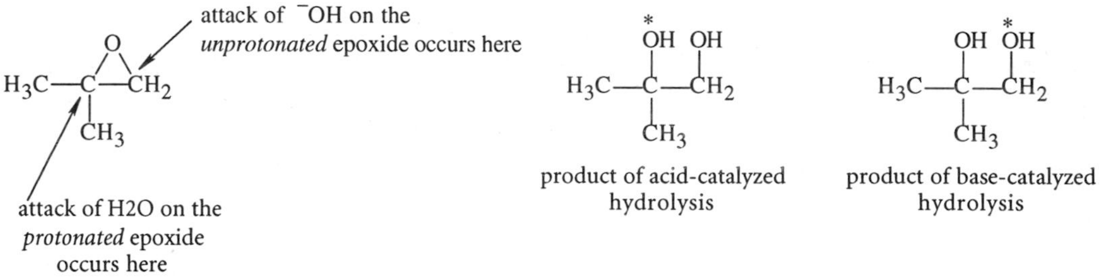

attack of ^{-}OH on the *unprotonated* epoxide occurs here

attack of H2O on the *protonated* epoxide occurs here

$$H_3C\!-\!\overset{*OH}{\underset{CH_3}{C}}\!-\!\overset{OH}{C}H_2 \qquad\qquad H_3C\!-\!\overset{OH}{\underset{CH_3}{C}}\!-\!\overset{*OH}{C}H_2$$

product of acid-catalyzed hydrolysis — product of base-catalyzed hydrolysis

11.20 Subtract the group added to the Grignard reagent, namely, the $—CH_2CH_2OH$ group, from the product, and add back the —MgBr to what remains. (See color coding in Eq. 11.40, text p. 473.)

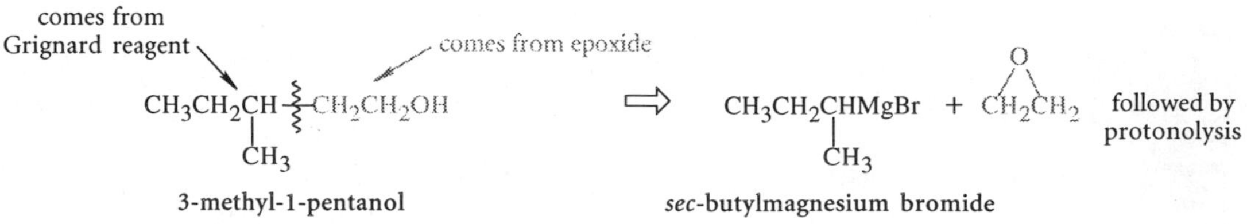

comes from Grignard reagent — comes from epoxide

$$CH_3CH_2\underset{CH_3}{CH}\!-\!CH_2CH_2OH \Longrightarrow CH_3CH_2\underset{CH_3}{CH}MgBr + \overset{O}{CH_2CH_2}$$ followed by protonolysis

3-methyl-1-pentanol — *sec*-butylmagnesium bromide

11.21 The product is 2-cyclopentylethanol:

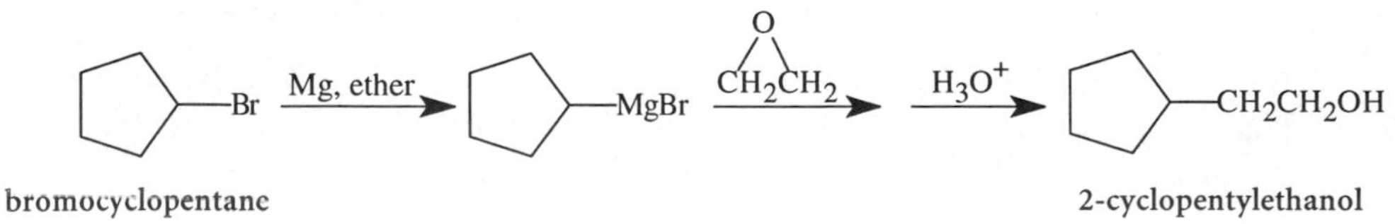

$$\text{C}_5\text{H}_9\!-\!Br \xrightarrow{\text{Mg, ether}} \text{C}_5\text{H}_9\!-\!MgBr \xrightarrow{\overset{O}{CH_2CH_2}} \xrightarrow{H_3O^+} \text{C}_5\text{H}_9\!-\!CH_2CH_2OH$$

bromocyclopentane — 2-cyclopentylethanol

11.22 The products are the corresponding glycols. Note the *syn*-addition in both cases.

(a)

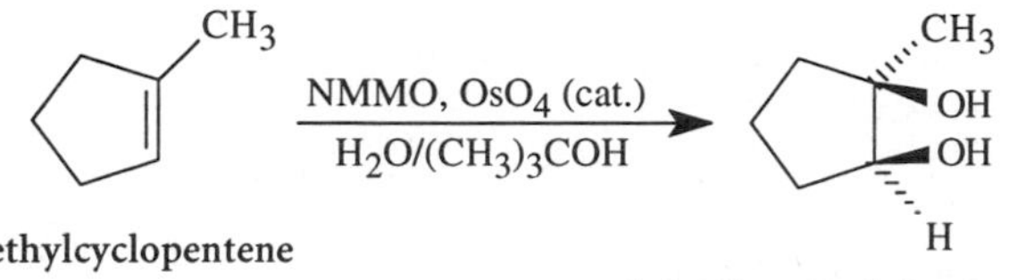

1-methylcyclopentene $\xrightarrow[H_2O/(CH_3)_3COH]{NMMO,\ OsO_4\ (cat.)}$ (1*R*,2*S*)-1-methyl-1,2-cyclopentanediol (and its enantiomer; the racemate)

(b)

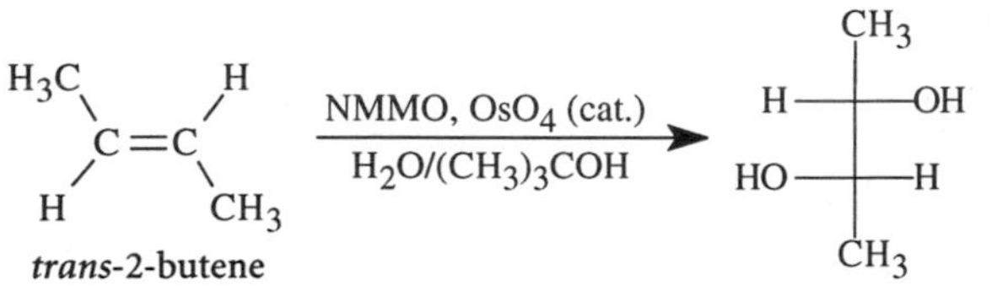

trans-2-butene $\xrightarrow[H_2O/(CH_3)_3COH]{NMMO,\ OsO_4\ (cat.)}$ (±)-2,3-butanediol (racemate)

11.23 The double bond in the starting material is between the glycol carbons of the product.

(a) $CH_3CH_2OCH_2CH_2CH{=}CH_2$

4-ethoxy-1-butene

(b) methylenecyclobutane

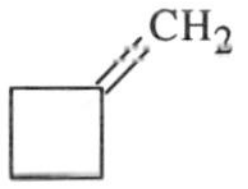

(c) Here we must be concerned with the stereochemistry of the product. Given a *syn*-addition, formation of a meso product demands a *cis*-alkene starting material. (Be sure to go through the complete analysis if you do not see why this is so.)

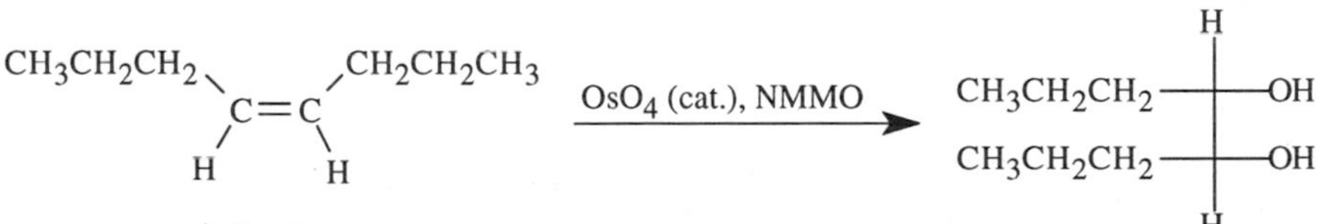

cis-2-octene $\xrightarrow{OsO_4\ (cat.),\ NMMO}$ *meso*-4,5-octanediol

11.24 A process occurs that is analogous to the one shown in Eq. 11.46 of the text for OsO_4 addition:

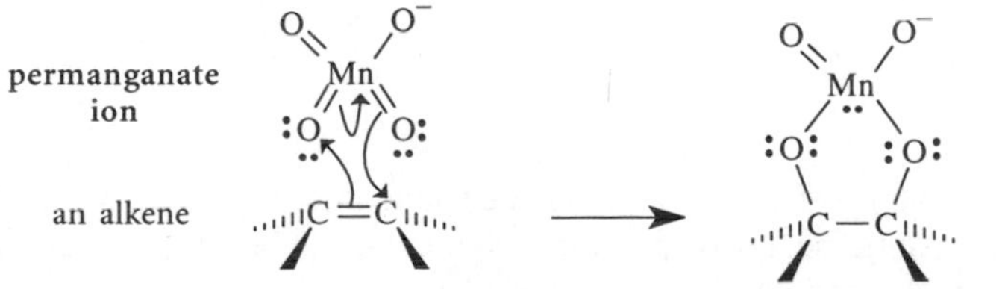

permanganate ion + an alkene ⟶ a permanganate ester

11.25 In each case the carbon-carbon bond between the OH groups is cleaved and an aldehyde or ketone is formed at each carbon. Note in part (c) that when the glycol is cyclic ring opening occurs.

(a) $PhCH_2CH{=}O + CH_2{=}O$

(b) cyclobutyl–CH=O + $HC(=O)CH_3$

(c) ring with CH=O and CH=O

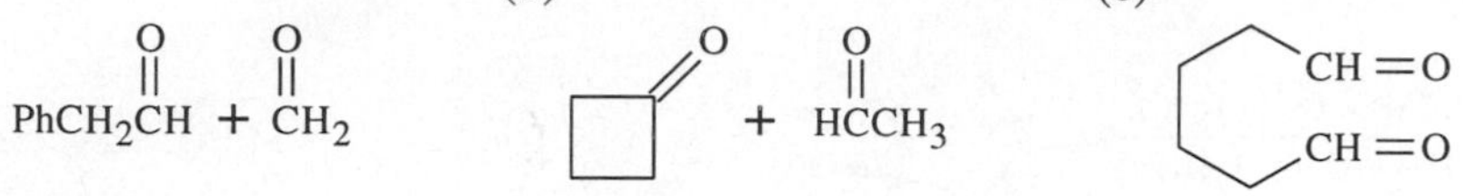

11.26 To solve this problem, mentally reverse the glycol cleavage reaction:

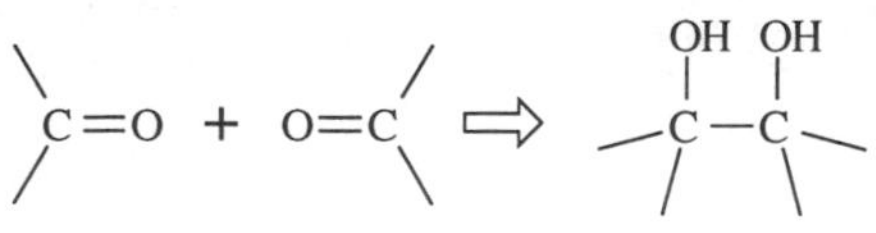

$$\rangle C{=}O + O{=}C\langle \;\Rightarrow\; -\overset{OH}{\underset{|}{C}}-\overset{OH}{\underset{|}{C}}-$$

(a) $H_3C-C(OH)(CH_3)-C(OH)$(cyclopentyl)

(b) bicyclic decalin diol with OH at both ring-fusion carbons

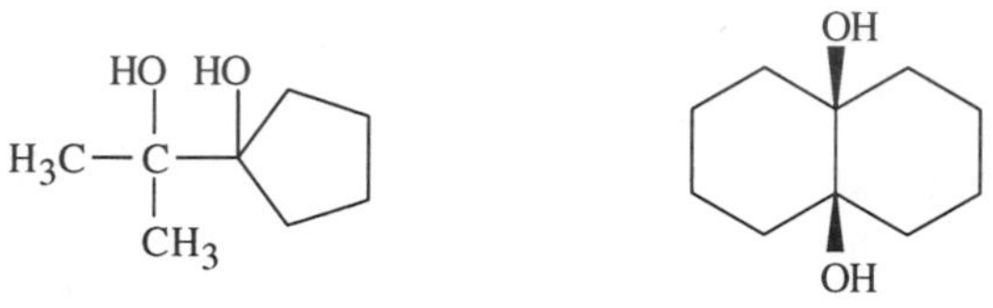

11.27 Because iodide ion is a good nucleophile and the trimethyloxonium ion is an excellent alkylating agent, alkylation of the iodide ion occurs to give dimethyl ether and methyl iodide.

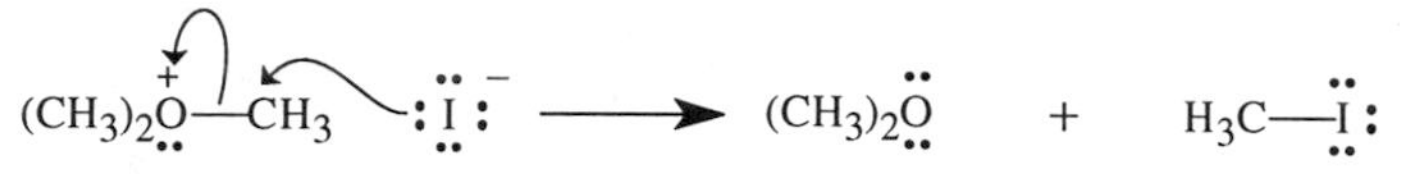

$$(CH_3)_2\overset{+}{O}{-}CH_3 \;\; :\ddot{I}:^- \longrightarrow (CH_3)_2\ddot{O} + H_3C{-}\ddot{I}:$$

trimethyloxonium iodide — dimethyl ether — methyl iodide

11.28 In each case a nucleophile (Lewis base) is alkylated by a methyl group of the oxonium ion in an S_N2 reaction:

(a) $(CH_3)_2O$ + N-methylpyridinium $^-BF_4$

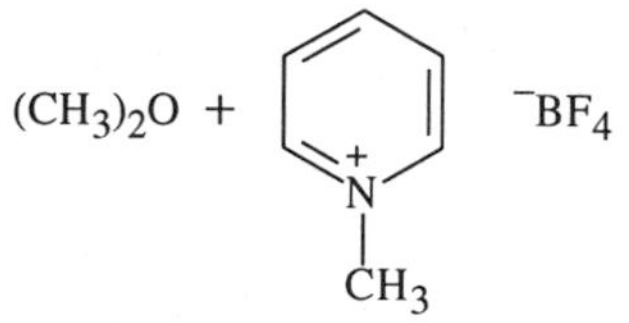

(b) $(CH_3)_3S^+ \; ^-BF_4 + (CH_3)_2O$

11.29 (a) If neighboring-group participation does not occur, only the alcohol $C_2H_5S{-}CH_2\overset{*}{C}H_2{-}OH$ (compound *C* in the diagram below) will form (by direct attack of water on the carbon α to the chlorine).

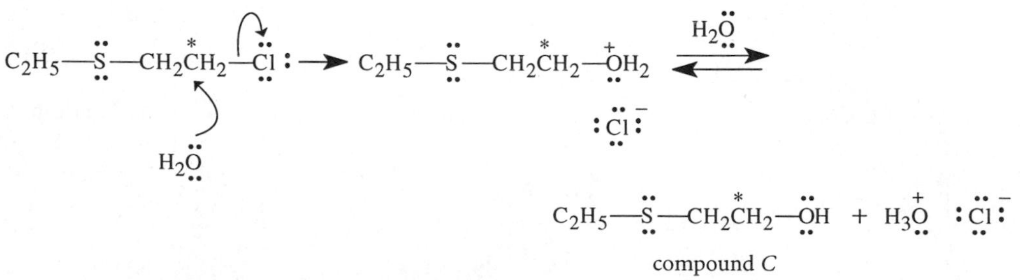

$$C_2H_5{-}\ddot{S}{-}CH_2\overset{*}{C}H_2{-}\ddot{Cl}: \xrightarrow{H_2\ddot{O}} C_2H_5{-}\ddot{S}{-}CH_2\overset{*}{C}H_2{-}\overset{+}{O}H_2 \;\; :\ddot{Cl}:^- \xrightleftharpoons{H_2\ddot{O}} C_2H_5{-}\ddot{S}{-}CH_2\overset{*}{C}H_2{-}\ddot{O}H + H_3\overset{+}{O} \;\; :\ddot{Cl}:^-$$

compound *C*

(b) If neighboring-group participation occurs, the cyclic intermediate *A* will be formed. Because this intermediate is symmetrical (except for the isotope), it can be attacked at either carbon to give a mixture consisting of equal amounts of the two labeled alcohols *B* and *C*.

intermediate *A*

$$C_2H_5\text{—}\ddot{S}\text{—}CH_2\overset{*}{C}H_2\text{—}\ddot{Cl}: \longrightarrow :\ddot{Cl}:^- + \underset{\ddot{O}H_2}{\overset{C_2H_5\text{—}S^+ \ (B)\ (C)}{CH_2\overset{*}{C}H_2}} \longrightarrow \underset{\text{from } (B)}{H_2\overset{+}{\ddot{O}}\text{—}CH_2\overset{*}{C}H_2\text{—}\ddot{S}\text{—}C_2H_5} + \underset{\text{from } (C)}{C_2H_5\text{—}\ddot{S}\text{—}CH_2\overset{*}{C}H_2\text{—}\overset{+}{\ddot{O}}H_2}$$

$$\updownarrow H_2O \qquad \updownarrow H_2O$$

$$H_3O^+ + \underset{B}{H\ddot{O}\text{—}CH_2\overset{*}{C}H_2\text{—}\ddot{S}\text{—}C_2H_5} + \underset{C}{C_2H_5\text{—}\ddot{S}\text{—}CH_2\overset{*}{C}H_2\text{—}\ddot{O}H}$$

intermediate *A*

$$C_2H_5\text{—}\ddot{S}\text{—}CH_2\overset{*}{C}H_2\text{—}\ddot{Cl}: \longrightarrow :\ddot{Cl}:^- + \underset{\ddot{O}H_2}{\overset{C_2H_5\text{—}S^+ \ (B)\ (C)}{CH_2\overset{*}{C}H_2}} \longrightarrow \underset{\text{from } (B)}{H_2\overset{+}{\ddot{O}}\text{—}CH_2\overset{*}{C}H_2\text{—}\ddot{S}\text{—}C_2H_5} + \underset{\text{from } (C)}{C_2H_5\text{—}\ddot{S}\text{—}CH_2\overset{*}{C}H_2\text{—}\overset{+}{\ddot{O}}H_2}$$

$$\updownarrow H_2O \qquad \updownarrow H_2O$$

$$H_3O^+ + \underset{B}{H\ddot{O}\text{—}CH_2\overset{*}{C}H_2\text{—}\ddot{S}\text{—}C_2H_5} + \underset{C}{C_2H_5\text{—}\ddot{S}\text{—}CH_2\overset{*}{C}H_2\text{—}\ddot{O}H}$$

11.30 In this example, the methyl branch serves essentially as a "label" for one of the two carbons of the alcohol. The formation of a single chloride is reasonable if a cyclic intermediate much like the one in Problem 11.29 is formed, and if this intermediate is attacked only at the carbon bearing the methyl group. Because the same intermediate is formed in both reactions, the same product should also be formed.

cyclic intermediate

$$\underset{\text{protonated alcohol}}{C_2H_5\ddot{S}\text{—}CH_2\underset{CH_3}{C}H\text{—}\overset{+}{\ddot{O}}H_2} \longrightarrow \ddot{O}H_2 + \overset{C_2H_5\text{—}:S^+}{CH_2\text{—}CH\text{—}CH_3} \xrightarrow{:\ddot{Cl}:^-} C_2H_5\ddot{S}\text{—}CH_2\text{—}\overset{:\ddot{Cl}:}{C}H\text{—}CH_3$$

(The formation of the cyclic intermediate from one of the alcohols is shown; be sure to show that the other alcohol forms the same intermediate and hence, the same product.)

A separate question is why the chloride ion should attack only at the more branched carbon. The cyclic intermediate is much like a protonated epoxide; hence, attack occurs at the more branched carbon for the same reason it does in a protonated epoxide. (See discussion of Eq. 11.33b on text p. 470.)

11.31 The intramolecular product tetrahydrofuran results from an internal nucleophilic substitution reaction of the alkoxide on the alkyl halide. The intermolecular product 1,4-butanediol results from the S_N2 reaction of hydroxide ion with the alkyl halide.

$$\underset{\text{4-bromo-1-butanol}}{Br\text{—}CH_2CH_2CH_2CH_2\text{—}OH} \xrightarrow{NaOH} \underset{\substack{\text{tetrahydrofuran} \\ \textit{intramolecular} \\ \textit{substitution product}}}{\text{(tetrahydrofuran ring, O)}} \quad \text{or} \quad \underset{\substack{\text{1,4-butanediol} \\ \textit{intermolecular} \\ \textit{substitution product}}}{HO\text{—}CH_2CH_2CH_2CH_2\text{—}OH}$$

To form the cyclic product tetrahydrofuran, ^{-}OH must react with the alcohol to ionize it; the conjugate base anion of the alcohol then cyclizes in an *internal* substitution reaction. In contrast, to form 1,4-butanediol, ^{-}OH must react with the alkyl halide in an S_N2 reaction. Ionization reations are much faster than S_N2 reactions; and once the ionization has occurred, the cyclization, because it is intermolecular, is *much* faster than the S_N2 reaction between hydroxide and the alkyl halide. Hence, tetrahydrofuran would be the major product.

Another reasonable *intermolecular* possibility you may have considered is following:

$$Br—CH_2CH_2CH_2CH_2—OH \xrightarrow{NaOH} Br—CH_2CH_2CH_2CH_2—O—CH_2CH_2CH_2CH_2OH$$

4-bromo-1-butanol → 4-(4-bromobutoxy)-1-butanol

To form this product, ionized 4-bromobutanol must react with the alkyl halide "end" of another molecule of 4-bromobutanol in an S_N2 reaction. Because such a reaction is bimolecular, the intramolecular reaction of the same ion to give tetrahydrofuran is *much* faster.

11.32 (a) Lengthening a carbon chain by two carbons is required; what better way than the reaction of a Grignard reagent with ethylene oxide?

$$(CH_3)_2C{=}CH_2 \xrightarrow[2)\ H_2O_2,\ ^{-}OH]{1)\ BH_3/THF} (CH_3)_2CHCH_2OH \xrightarrow[H_2SO_4]{conc.\ HBr,} (CH_3)_2CHCH_2Br \xrightarrow{Mg,\ ether}$$

2-methylpropene; isobutyl alcohol; isobutyl bromide

$$(CH_3)_2CHCH_2MgBr \xrightarrow[2)\ H_3O^+]{1)\ \text{ethylene oxide}} (CH_3)_2CHCH_2CH_2CH_2OH \xrightarrow[2)\ H_3O^+]{1)\ KMnO_4,\ ^{-}OH} (CH_3)_2CHCH_2CH_2CO_2H$$

An alternate synthesis of isobutyl bromide is to treat 2-methylpropene with HBr in the presence of peroxides.

(b) Oxidize the isobutyl alcohol, which is obtained in the hydroboration-oxidation reaction in part (a).

$$(CH_3)_2CHCH_2OH \xrightarrow[2)\ H_3O^+]{1)\ KMnO_4,\ ^{-}OH} (CH_3)_2CHCO_2H$$

prepared in part (a)

(c) The sulfone must be obtained by oxidation of dibutyl sulfide; and the sulfide is obtained from the S_N2 reaction of the thiolate conjugate base of 1-butanethiol with 1-bromobutane.

$$CH_3CH_2CH_2CH_2SH \xrightarrow[C_2H_5OH]{Na^+\ C_2H_5O^-} \xrightarrow[\text{1-bromobutane}]{CH_3CH_2CH_2CH_2Br} CH_3CH_2CH_2CH_2SCH_2CH_2CH_2CH_3 \xrightarrow{2\ equiv.\ H_2O_2} CH_3CH_2CH_2CH_2S(=O)_2CH_2CH_2CH_2CH_3$$

1-butanethiol; dibutylsulfone

(d) The approach is similar to that used in Study Problem 11.6.

cyclopentene $\xrightarrow{mCPBA}$ cyclopentene oxide $\xrightarrow[C_2H_5OH]{dil.\ H_2SO_4,}$ trans-2-ethoxycyclopentanol (OH, OC_2H_5) $\xrightarrow[2)\ CH_3I]{1)\ NaH}$ (OCH_3, OC_2H_5)

(±)-*trans*-1-ethoxy-2-methoxycyclopentane

(All chiral compounds are formed as the racemates.) Epoxide opening with acidic methanol and alkylation with ethyl iodide would also work; however, methyl halides are better alkylating agents than ethyl halides (why?).

Solutions to Additional Problems

11.33 (a) Two of the many nine-carbon ethers that cannot be prepared by the Williamson ether synthesis:

$H_3C-C(CH_3)_2-O-C(CH_3)_2-CH_2CH_3$ — 2-methyl-2-*tert*-butoxybutane

$H_3C-C(CH_3)_2-O-CH_2-C(CH_3)_2-CH_3$ — *tert*-butyl neopentyl ether

The Williamson ether synthesis involves an S_N2 reaction between an alkoxide ion and an alkyl halide or sulfonate ester. Thus, one of the alkyl groups of an ether to be prepared by the Williamson synthesis must be derived from an alkyl halide that will undergo the S_N2 reaction. The foregoing ethers contain only tertiary and neopentyl alkyl groups, and neither tertiary nor neopentyl alkyl halides undergo the S_N2 reaction.

(b) Any ether containing only methyl or primary alkyl groups is a correct answer. Two of several possible examples are $CH_3O(CH_2)_7CH_3$ (methyl octyl ether, or 1-methoxyoctane) and $CH_3(CH_2)_3O(CH_2)_4CH_3$ (butyl pentyl ether, or 1-butoxypentane).

(c) Because the alkyl halides formed are in the same molecule, the ether is cyclic:

tetrahydrofuran

(d) Dipropyl ether $(CH_3CH_2CH_2)_2O$, would give propyl bromide as the only alkyl halide.

(e) In each of the following cases the different glycols would be diastereomers.

$H_3C(H)C{=}C(H)CH_3$ (cis) or $H_3C(H)C{=}C(H)CH_3$ (trans) or cyclobutene

(f) 1-Butene $H_2C{=}CHCH_2CH_3$ gives racemic 1,2-butanediol with either reagent.

(g) Since the alkene has an unsaturation number of 3 and forms diepoxides, it must also have a ring.

1,4-cyclohexadiene gives monoepoxide and two stereoisomeric diepoxides (cis + trans)

Notice that 1,3-cyclohexadiene is not a correct answer because it gives two stereoisomeric mono-epoxides and three stereoisomeric di-epoxides.

(h) A correct answer must be an alkene that cannot give a glycol that could exist as diastereomers. The alkene 1-hexene $(H_2C{=}CHCH_2CH_2CH_2CH_3)$ is one such compound.

11.34 First, draw the starting material, and then apply the reaction conditions:

$(H_3C)(CH_3CH_2)C(-O-)CH_2$ — 2-ethyl-2-methyloxirane

(a) and (b) (c) 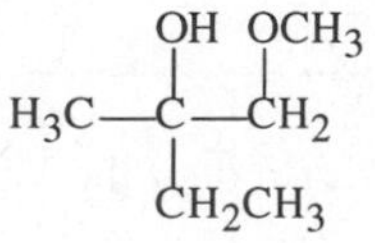 (d) 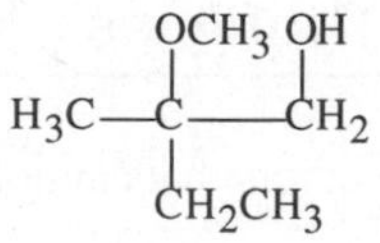 (e)

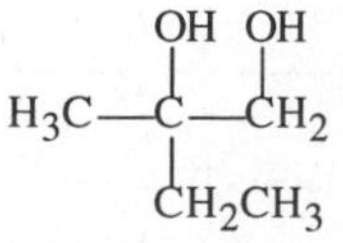

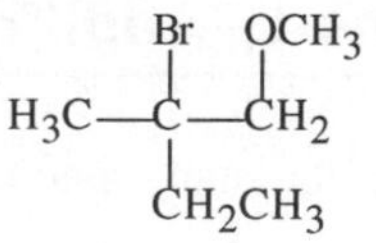

In (e), why would reaction of the alcohol with HBr take place faster than formation of the primary alkyl halide from the ether? Because ether cleavage involves a tertiary carbocation, and is very fast for that reason. The product (f) below has a similar explanation: in this case, the ether is cleaved because a tertiary carbocation is involved.

(f) (g) 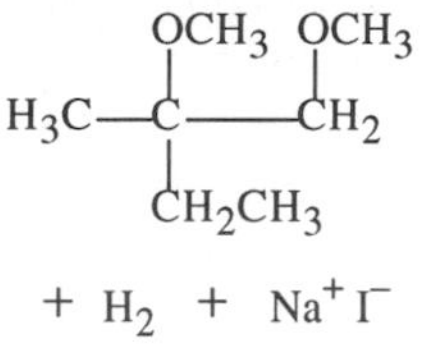 (h)

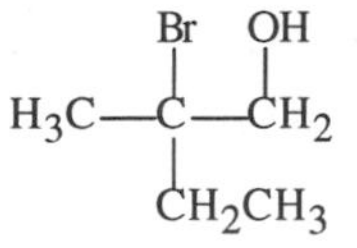

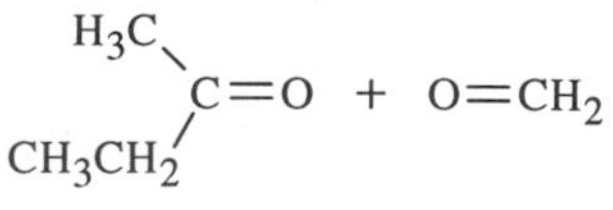

$+ H_2 + Na^+I^-$

H_3C, CH_3CH_2 $C{=}O + O{=}CH_2$

(i) 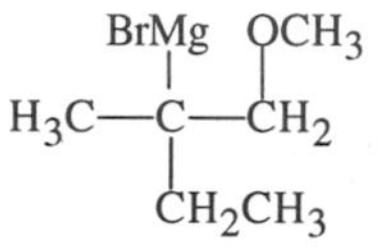 (j)

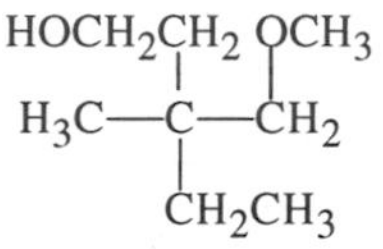

11.35 Because the C—S bond is weaker than the C—O and C—C bonds, and because a thiolate ion is less basic than an alkoxide ion or carbanion, the thiirane (the sulfur compound) reacts most rapidly.

11.36 The first reaction, opening of the epoxide, should occur most rapidly, because the epoxide is considerably more strained than tetrahydrofuran. It is the ring strain that causes epoxide opening to be so fast.

11.37 (a) 3-Ethoxypropene undergoes bromine addition, and therefore decolorizes a solution of Br_2 in an inert solvent; 1-ethoxypropane does not.

(b) 1-Pentanol, the alcohol, is fairly soluble in water, whereas the ether is not. Alternatively, the alcohol evolves H_2 when treated with NaH or Na, or evolves methane (CH_4) when treated with the Grignard reagent CH_3MgBr; the ether, which has no O—H group, does not.

(c) 1-Methoxy-2-chloro-2-methylpropane, a tertiary alkyl halide, undergoes the S_N1 reaction on gentle heating in a protic solvent such as aqueous acetone. This results in an acidic solution of aqueous HCl, the formation of which can be detected with litmus paper. Alternatively, the chloride ion produced can be precipitated with silver ion.

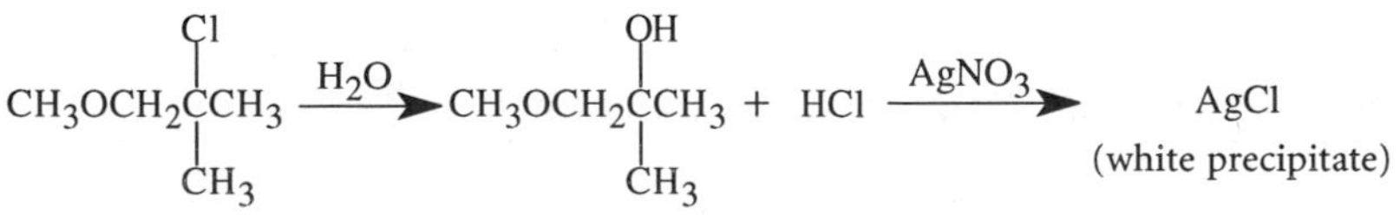

$$CH_3OCH_2C(Cl)(CH_3)CH_3 \xrightarrow{H_2O} CH_3OCH_2C(OH)(CH_3)CH_3 + HCl \xrightarrow{AgNO_3} AgCl$$

1-methoxy-2-chloro-2-methylpropane; + E1 products; AgCl (white precipitate)

The first compound, 1-methoxy-2-methylpropane, is not an alkyl halide.

11.38 Epoxides add HCl to give neutral chloro alcohols. This reaction is so fast and so favorable that HCl can be quantitatively (that is, completely) removed by adding 2-methyloxirane to a reaction mixture.

$$\text{2-methyloxirane} + HCl \longrightarrow H_3C{-}\underset{}{\overset{Cl}{\overset{|}{C}}}H{-}CH_2OH$$

2-methyloxirane

11.39 2-Pentanol cannot be prepared by the reaction of a Grignard reagent with ethylene oxide, because this method can only be used for the preparation of certain *primary* alcohols. However, 1-pentanol, a primary alcohol, can be prepared by the reaction of propylmagnesium bromide with ethylene oxide followed by protonolysis.

$$CH_3CH_2CH_2{-}MgBr + H_2C{-}CH_2\,(\text{O bridge}) \xrightarrow{\text{ether}} \xrightarrow{H_3O^+,\ H_2O} CH_3CH_2CH_2CH_2CH_2OH$$

propylmagnesium bromide; ethylene oxide; 1-pentanol

11.40 (a) Sodium ethoxide reacts with water to give ethanol and sodium hydroxide. Although the pK_a values of water and ethanol are similar, water is present in excess because it is the solvent, and the equilibrium therefore favors sodium hydroxide. Consequently, the alcohol $(CH_3)_2CHCH_2CH_2OH$ rather than the ether will be formed as the major substitution product.

$$Na^{+}\ {}^{-}OC_2H_5 + H_2O \rightleftharpoons HOC_2H_5 + Na^{+}\ {}^{-}OH$$

H_2O (solvent); $Na^{+}\ {}^{-}OH$ reacts with the alkyl halide

Another problem is that the alkyl halide is insoluble in water. This insolubility makes any reaction that does occur very slow because the concentration of alkyl halide is limited to the very small amount that will dissolve. Changing the solvent to ethanol would solve all of these problems.

(b) The alkyl halide from which the Grignard reagent is formed is also an alcohol. Because alcohols react rapidly with Grignard reagents in Brønsted acid-base reactions, the Grignard reagent will be destroyed. (See Eq. 8.26, text p. 337.)

11.41 (a) Reaction of ethylene ($H_2C{=}CH_2$) with OsO_4 in the presence of water gives ethylene glycol ($HO{-}CH_2CH_2{-}OH$). This compound is achiral and thus cannot be resolved into enantiomers.

(b) The reaction of OsO_4 with any alkene is a *syn*-addition. The reaction with *cis*-2-butene gives *meso*-2,3-butanediol. Because meso compounds are achiral, this compound cannot be resolved into enantiomers.

$$\textit{cis}\text{-2-butene} \xrightarrow{OsO_4} \xrightarrow{NaHSO_3/H_2O} \textit{meso}\text{-2,3-butanediol}$$

cis-2-butene; *meso*-2,3-butanediol

(c) In contrast, the *syn*-addition of two —OH groups to *trans*-2-butene gives (±)-2,3-butanediol, that is, the racemate. Any racemate can in principle be resolved into enantiomers.

$$\textit{trans}\text{-2-butene} \xrightarrow{OsO_4} \xrightarrow{NaHSO_3/H_2O} (\pm)\text{-2,3-butanediol}$$

trans-2-butene; (±)-2,3-butanediol

(d) The *syn*-addition of two —OH groups to *cis*-2-pentene gives (±)-2,3-pentanediol. This compound is a racemate and can be resolved into enantiomers.

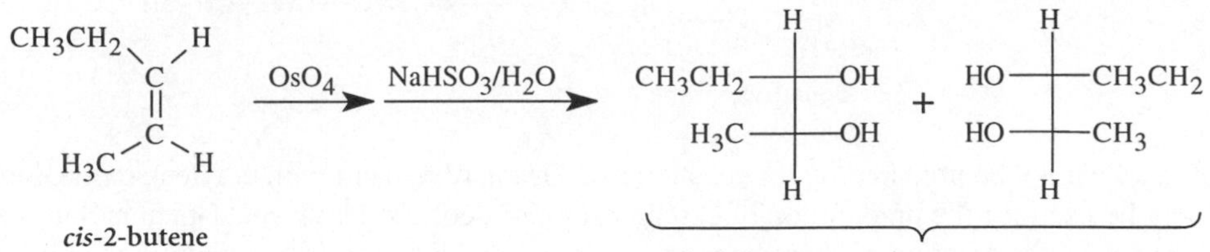

cis-2-butene

(±)-2,3-pentanediol (the racemate)

This outcome is different from that of the reaction in (b) because *cis*-2-butene and 2,3-butanediol have symmetry not present in *cis*-2-pentene and 2,3-pentanediol.

11.42 Because the hydroxide ion reacts with an epoxide at the less branched carbon (the CH_2 group), the configuration at the asymmetric carbon is unaffected. Consequently, (+)-2-methyloxirane has the *R* configuration.

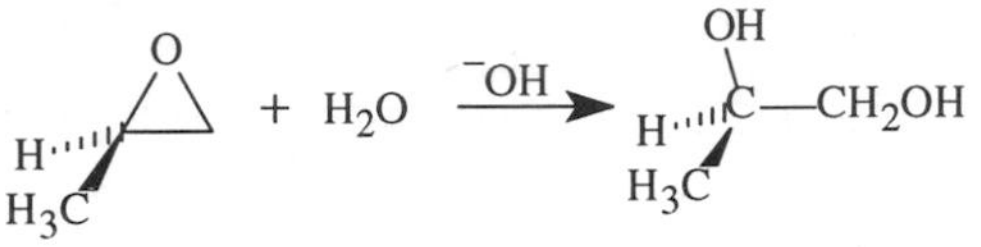

(*R*)-(+)-2-methyloxirane (*R*)-(−)-1,2-propanediol

11.43 In this reaction the nucleophile (water) attacks the protonated epoxide at the more branched carbon with inversion of stereochemical configuration.

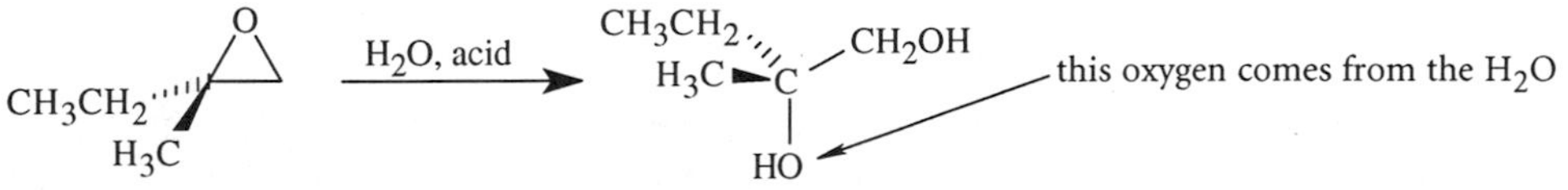

this oxygen comes from the H_2O

(*S*)-2-ethyl-2-methyloxirane (*R*)-2-methyl-1,2-butanediol

11.44 The formula of the product indicates one degree of unsaturation. Because the double bond undergoes addition with mercuric acetate, this unsaturation cannot be due to a double bond and therefore is accounted for by a ring. As the hint and the molecular formula of the product suggest, this is an alkoxymercuration reaction in which the —OH group from within the same molecule serves as the nucleophile:

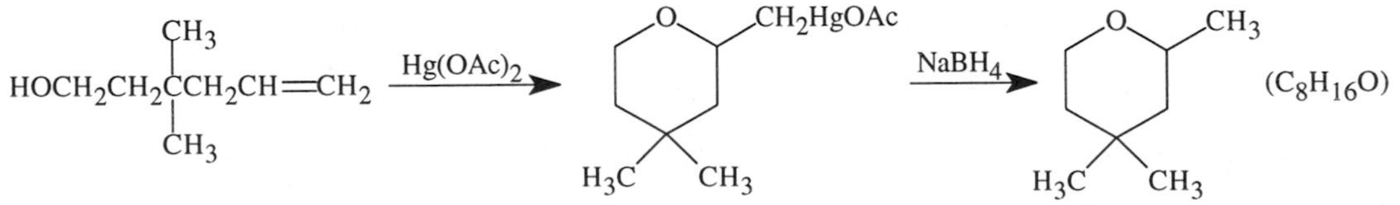

$$HOCH_2CH_2C(CH_3)_2CH_2CH{=}CH_2 \xrightarrow{Hg(OAc)_2} \text{(ring)}-CH_2HgOAc \xrightarrow{NaBH_4} \text{(ring)}-CH_3 \quad (C_8H_{16}O)$$

11.45 As a result of this reaction, carbon-2 becomes asymmetric, and diastereomers are formed corresponding to the two possible configurations of this carbon. The product with the *S* configuration at carbon-2 (compound *A*) is optically active; however, the other product (compound *B*) is a meso compound, and hence, is optically inactive.

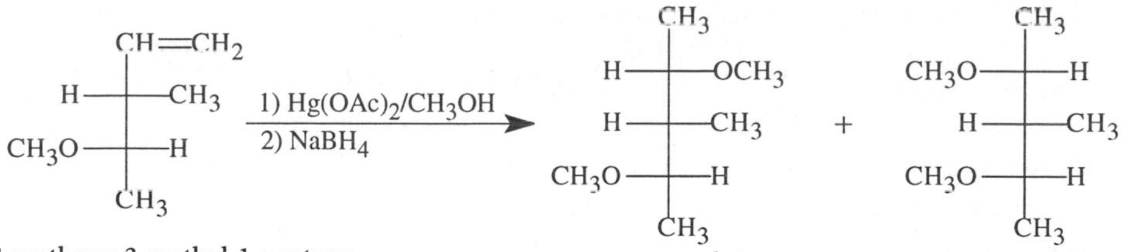

(3*S*,4*S*)-4-methoxy-3-methyl-1-pentene — compound *A* (optically active) — compound *B* (a meso compound)

11.46 Synthesis (1) will not accomplish the desired objective, because the —OH group of the alcohol would be lost as water. (See the mechanism in Eqs. 11.11–11.12, text p. 459.) In contrast, synthesis (2) would result in complete incorporation of the isotopic oxygen into the ether. Because this is a conventional Williamson synthesis, the carbon-oxygen bond to the ring is not broken. (See Eq. 11.3, text p. 456.) Consequently, you should advise Flaskclamper to use the second synthesis.

11.47 Structure (2) is meso and therefore a compound with this structure cannot be optically active; hence, *D* has structure (2). A compound with structure (3) would give the same products of periodic acid oxidation as a compound with structure (2); hence, *C* has structure (3). Because a compound with the structure (1) is not a vicinal glycol, it would not react with periodic acid; hence, *A* has structure (1). Compound *B* therefore has structure (4); and, indeed, it would react to give periodate cleavage products that are different from those obtained from compound *C* (structure 3) or compound *D* (structure 2).

11.48 (a) This is a Williamson ether synthesis by the S_N2 mechanism that gives $CH_3CH_2CH_2$—O—CH_2CH_3 (ethyl propyl ether, or 1-ethoxypropane) + Na^+ Br^-.

(b) A strong base with a tertiary halide gives elimination by the E2 mechanism.

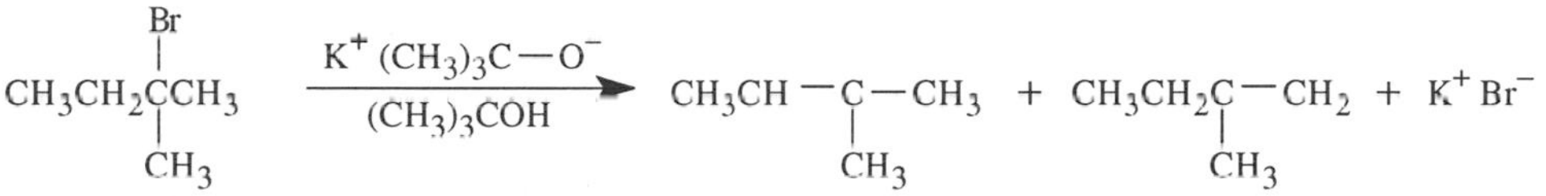

2-bromo-2-methylbutane — 2-methyl-2-butene — 2-methyl-1-butene

Although 2-methyl-2-butene has more branches at the double bond and is the more stable alkene, 2-methyl-1-butene is the predominant product because of the use of a highly branched base. (See text p. 379.)

(c) This is an alkoxymercuration-reduction that gives the following ether:

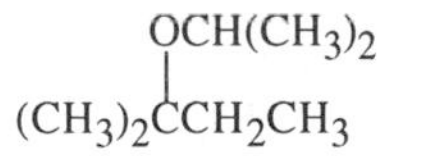

2-methyl-2-isopropoxybutane

(d) This is an epoxidation with retention of alkene configuration:

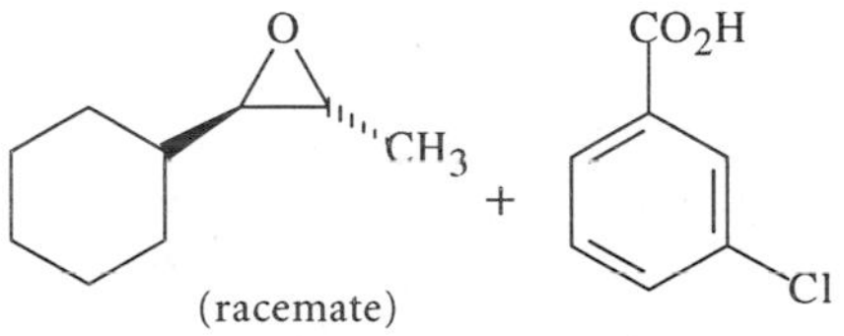

(racemate)

(e) A bromohydrin is formed in the first step, and this is oxidized to an α-bromo ketone by the CrO_3.

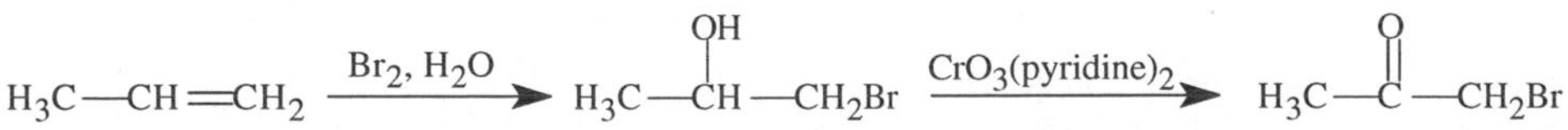

(f) The epoxide is opened to a glycol by the acidic aqueous reaction conditions, and the glycol is cleaved as it is formed by the periodic acid.

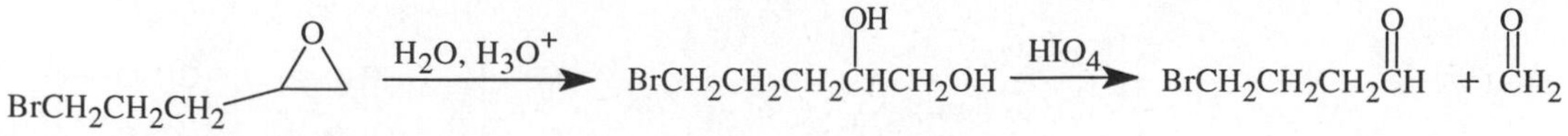

$$BrCH_2CH_2CH_2\text{-epoxide} \xrightarrow{H_2O,\ H_3O^+} BrCH_2CH_2CH_2CH(OH)CH_2OH \xrightarrow{HIO_4} BrCH_2CH_2CH_2CH{=}O + CH_2{=}O$$

(g) This is glycol formation with $KMnO_4$. The syn stereochemistry of the addition and the cis stereochemistry of the starting material determine the stereochemistry of the product:

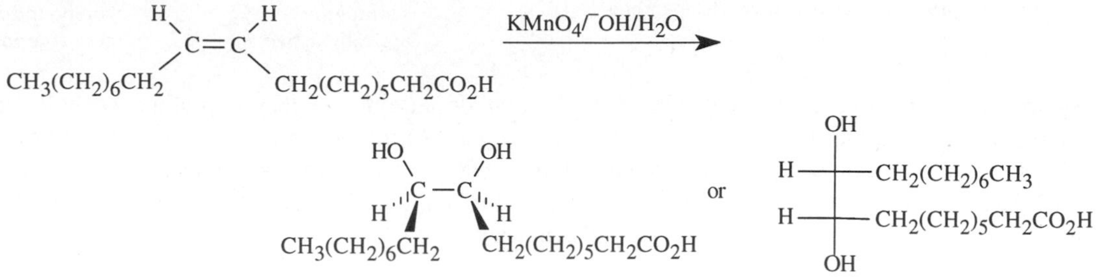

$$\text{cis-}CH_3(CH_2)_6CH_2CH{=}CHCH_2(CH_2)_5CH_2CO_2H \xrightarrow{KMnO_4/^-OH/H_2O}$$

HO, OH, H, H; $CH_3(CH_2)_6CH_2$ $CH_2(CH_2)_5CH_2CO_2H$ (racemate) or OH, H—$CH_2(CH_2)_6CH_3$, H—$CH_2(CH_2)_5CH_2CO_2H$, OH (racemate)

(h) The nucleophile azide ion opens the epoxide with inversion of configuration:

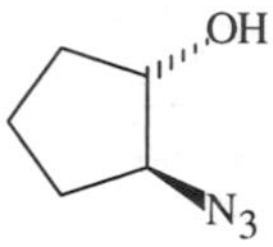

OH, N_3 (racemate)

(i) Sulfide ion displaces one chloride to give a thiolate ion, which then reacts in an internal nucleophilic substitution to give the cyclic sulfide tetrahydrothiophene.

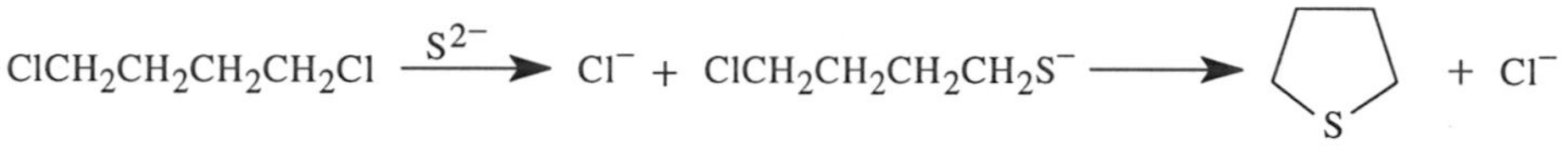

$$ClCH_2CH_2CH_2CH_2Cl \xrightarrow{S^{2-}} Cl^- + ClCH_2CH_2CH_2CH_2S^- \longrightarrow \text{(S ring)} + Cl^-$$

tetrahydrothiophene

(j) A tosylate of one —OH group is formed. (Introduction of a second tosylate is much slower because of the resulting van der Waals repulsions caused by the introduction of a second large tosylate group.) The resulting hydroxy tosylate reacts like a bromohydrin. (Remember, you can think of the tosylate group as a "fat halide.") Thus, in base, the —OH group ionizes to an alkoxide, which undergoes an internal nucleophilic substitution reaction to give an epoxide. This is like epoxide formation from a bromohydrin.

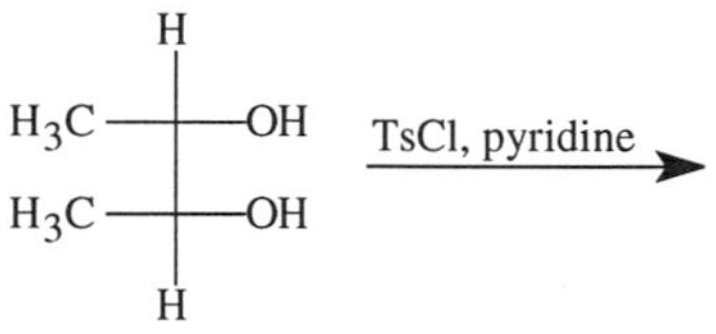

H; H_3C—OH; H_3C—OH; H $\xrightarrow{\text{TsCl, pyridine}}$

meso-2,3-butanediol

H; H_3C—OTs; H_3C—OH; H ⇨ OTs; H—CH_3; H—CH_3; OH ⇨ HO, H; H_3C—C—C—CH_3; H, OTs $\xrightarrow{NaOH}$ H_3C (epoxide) CH_3 + $Na^+\ ^-OTs$

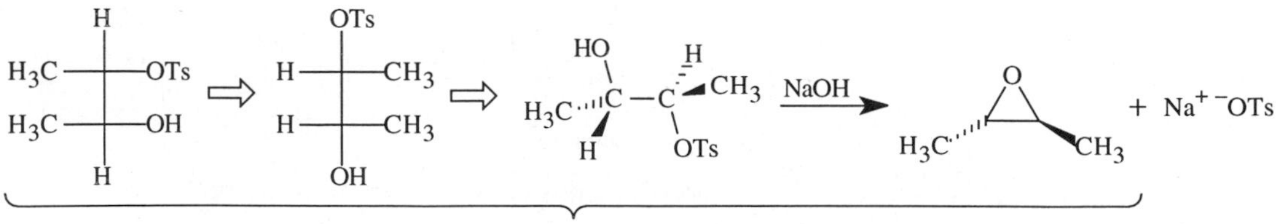

all are racemates

Because the starting diol is achiral (it is meso) and the reagents are all achiral, all chiral compounds in this reaction scheme are racemates.

11.49 (a)

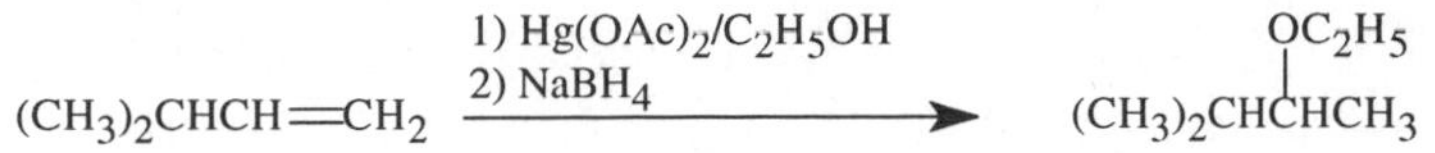

$$(CH_3)_2CHCH{=}CH_2 \xrightarrow[\text{2) } NaBH_4]{\text{1) } Hg(OAc)_2/C_2H_5OH} (CH_3)_2CHCH(OC_2H_5)CH_3$$

3-methyl-1-butene → 2-ethoxy-3-methylbutane

(b)

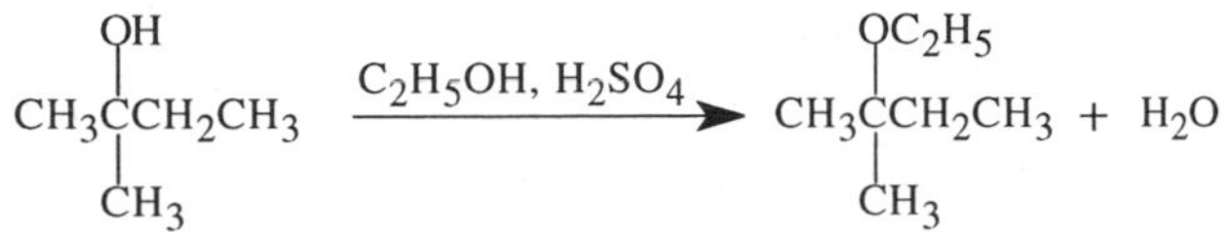

$$CH_3C(OH)(CH_3)CH_2CH_3 \xrightarrow{C_2H_5OH,\ H_2SO_4} CH_3C(OC_2H_5)(CH_3)CH_2CH_3 + H_2O$$

2-methyl-2-butanol → 2-ethoxy-2-methylbutane

(c)

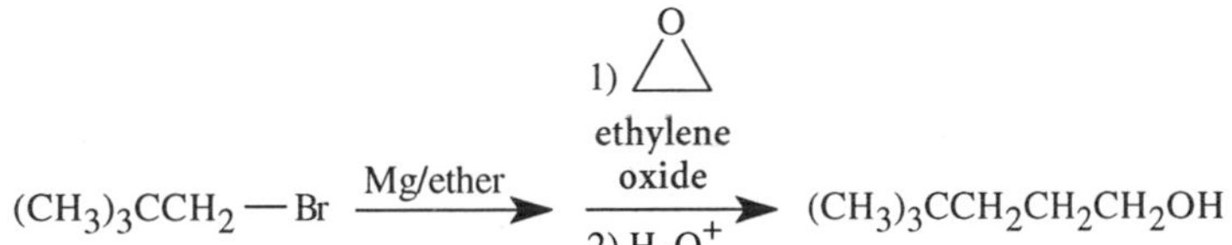

$$(CH_3)_3CCH_2{-}Br \xrightarrow{\text{Mg/ether}} \xrightarrow[\text{2) } H_3O^+]{\text{1) ethylene oxide}} (CH_3)_3CCH_2CH_2CH_2OH$$

4,4-dimethyl-1-pentanol

(d)

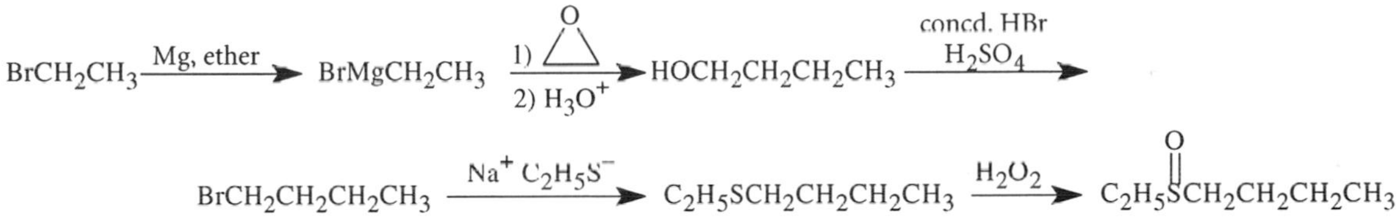

$$BrCH_2CH_3 \xrightarrow{\text{Mg, ether}} BrMgCH_2CH_3 \xrightarrow[\text{2) } H_3O^+]{\text{1) ethylene oxide}} HOCH_2CH_2CH_2CH_3 \xrightarrow[H_2SO_4]{\text{concd. HBr}}$$

$$BrCH_2CH_2CH_2CH_3 \xrightarrow{Na^+\ C_2H_5S^-} C_2H_5SCH_2CH_2CH_2CH_3 \xrightarrow{H_2O_2} C_2H_5S(=O)CH_2CH_2CH_2CH_3$$

(e)

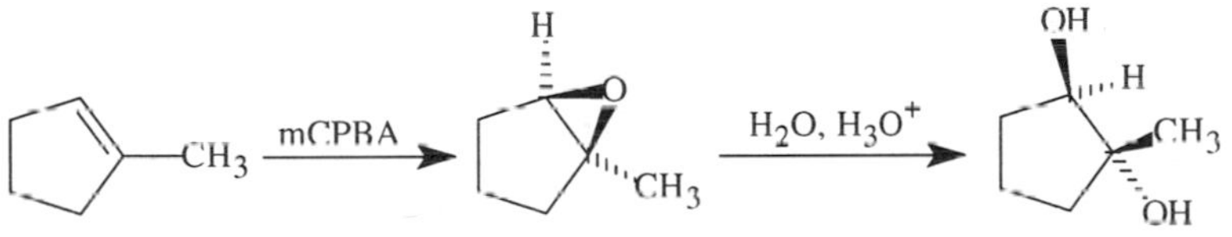

1-methylcyclopentene $\xrightarrow{\text{mCPBA}}$ 1-methyl-1,2-epoxycyclopentane (racemate) $\xrightarrow{H_2O,\ H_3O^+}$ trans-1-methylcyclopentane-1,2-diol (racemate)

The chiral compounds are, of course, racemates; why?

(f)

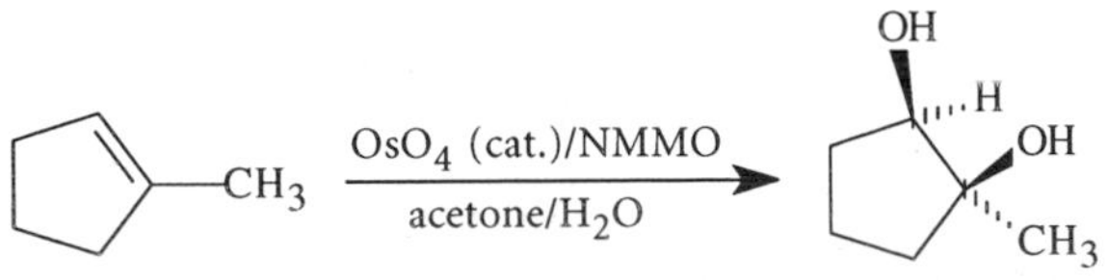

1-methylcyclopentene $\xrightarrow[\text{acetone}/H_2O]{OsO_4\text{ (cat.)/NMMO}}$ cis-1-methylcyclopentane-1,2-diol (racemate)

(g)

cyclohexene $\xrightarrow[\text{2) } NaBH_4]{\text{1) } Hg(OAc)_2/(CH_3)_2CHOH}$ $C_6H_{11}OCH(CH_3)_2$

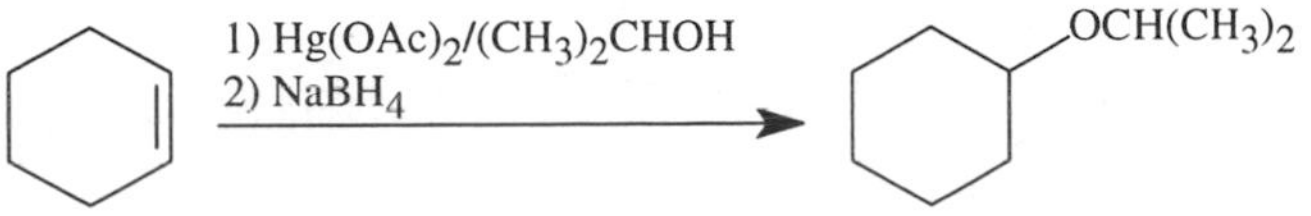

cyclohexene → cyclohexyl isopropyl ether

(h)

$$(CH_3)_2CHCH{=}CH_2 \xrightarrow[2)\ H_2O_2,\ ^-OH]{1)\ BH_3/THF} (CH_3)_2CHCH_2CH_2OH \xrightarrow[H_2SO_4]{\text{concd. HBr}}$$

3-methyl-1-butene

$$(CH_3)_2CHCH_2CH_2Br \xrightarrow{\text{Mg, ether}} (CH_3)_2CHCH_2CH_2MgBr \xrightarrow[2)\ H_3O^+]{1)\ \text{ethylene oxide}}$$

$$(CH_3)_2CHCH_2CH_2CH_2CH_2OH \xrightarrow{CrO_3(\text{pyridine})_2} (CH_3)_2CHCH_2CH_2CH_2CH{=}O$$

The alkyl halide in this synthesis could be prepared by treating the starting alkene, 3-methyl-1-butene, with HBr in the presence of peroxides.

(i)

$$CH_3CH_2CH(CH_3)CH{=}CH_2 \xrightarrow{\text{mCPBA}} CH_3CH_2CH(CH_3)\overset{O}{CH{-}CH_2} \xrightarrow[CH_3OH]{Na^+\ CH_3O^-} CH_3CH_2CH(CH_3)CH(OH)CH_2OCH_3 \xrightarrow{H_2Cr_2O_7} CH_3CH_2CH(CH_3)\overset{O}{\overset{\|}{C}}CH_2OCH_3$$

3-methyl-1-pentene

(j)

$$(CH_3)_2C{=}CH_2 \xrightarrow{\text{mCPBA}} (CH_3)_2\overset{O}{C{-}CH_2} \xrightarrow{CH_3CH_2OH,\ H_2SO_4} (CH_3)_2C(OCH_2CH_3){-}CH_2OH \xrightarrow{CrO_3(\text{pyridine})_2} (CH_3)_2C(OCH_2CH_3){-}CH{=}O$$

2-methylpropene

11.50 (a) The desired compound can be prepared by a ring-opening reaction of (2*R*,3*R*)-2,3-dimethyloxirane with acidic methanol, because epoxide ring-opening occurs with inversion of configuration:

$$\text{(2}R\text{,3}R\text{)-2,3-dimethyloxirane} \xrightarrow{CH_3OH,\ H_2SO_4} \text{(2}R\text{,3}S\text{)-3-methoxy-2-butanol}$$

(2*R*,3*R*)-2,3-dimethyloxirane (2*R*,3*S*)-3-methoxy-2-butanol

(Be sure to convince yourself that attack of methanol at either carbon of the epoxide ring gives the same stereoisomer of the product.)

(b) Carbon-3, the asymmetric carbon, has the *S* configuration, which is the same configuration that carbon-3 has in the product of part (a). Consequently, oxidation of the alcohol product of part (a) gives the desired compound:

$$\text{(2}R\text{,3}S\text{)-3-methoxy-2-butanol} \xrightarrow{H_2Cr_2O_7} H_3C{-}\overset{O}{\overset{\|}{C}}{-}CH(CH_3)(OCH_3) \equiv \text{Fischer projection: } \overset{O}{\overset{\|}{C}}{-}CH_3 \text{ (top)},\ CH_3O{-}\text{(left)},\ {-}H \text{ (right)},\ CH_3 \text{ (bottom)}$$

(2*R*,3*S*)-3-methoxy-2-butanol

(c) Analyze the configurations of the desired compound and the alcohol prepared in part (a):

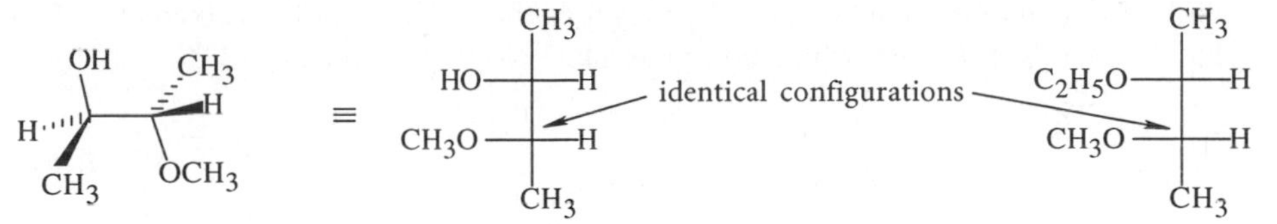

(2R,3S)-3-methoxy-2-butanol ≡ identical configurations — target compound

This analysis shows that if the —OH group can be converted into an ethyl ether without changing the configuration of either carbon, the problem will be solved. The Williamson synthesis is a good way to accomplish this objective, because it does not break a carbon-oxygen bond. (The reasoning is similar to that used in solving Problem 11.46.)

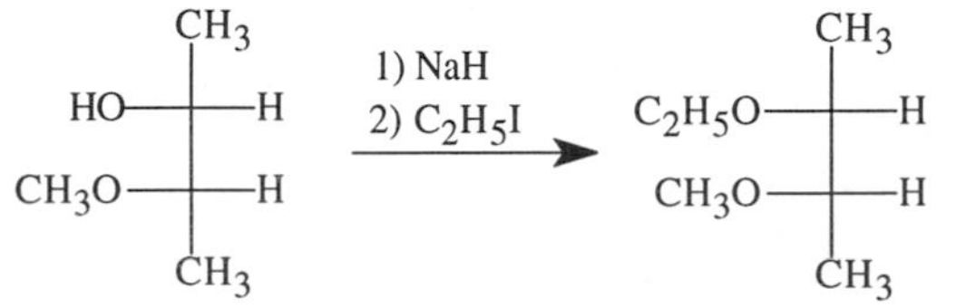

1) NaH
2) C_2H_5I

(d) Use the reasoning employed in solving parts (a) and (c). Because the carbon bearing the ethoxy group is the one that must be inverted, use the ring-opening reaction in part (a), except substitute ethanol for methanol. Then carry out the Williamson synthesis used in part (c) with methyl iodide instead of ethyl iodide.

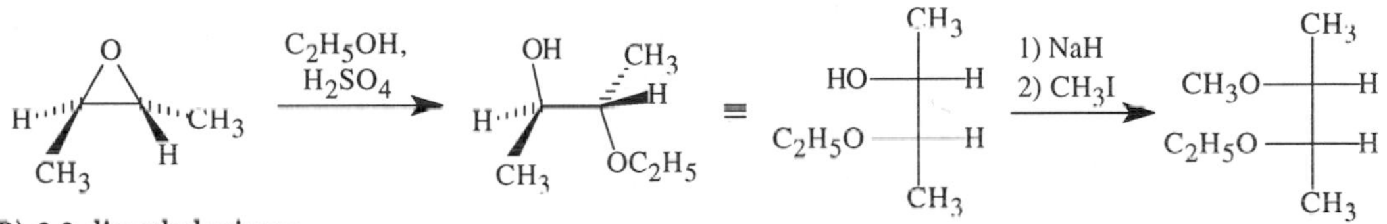

(2R,3R)-2,3-dimethyloxirane — C_2H_5OH, H_2SO_4 → (2R,3S)-3-ethoxy-2-butanol — 1) NaH 2) CH_3I →

11.51 Compound *A* is an octene because it undergoes typical alkene reactions and it gives octane on catalytic hydrogenation. Compounds *C* and *D* are stereoisomeric glycols, and glycol *D* is achiral. The only octene isomer that would give an achiral glycol is one of the stereoisomers of 4-octene. *Cis*-4-octene reacts with mCPBA to give an *achiral* (meso) epoxide, which, upon ring opening in aqueous acid, gives *chiral* 4,5-octanediol. The same alkene reacts in a *syn*-addition with OsO_4 to give *meso*-4,5-octanediol, an achiral compound. Therefore, compound *A* is *cis*-4-octene. To summarize:

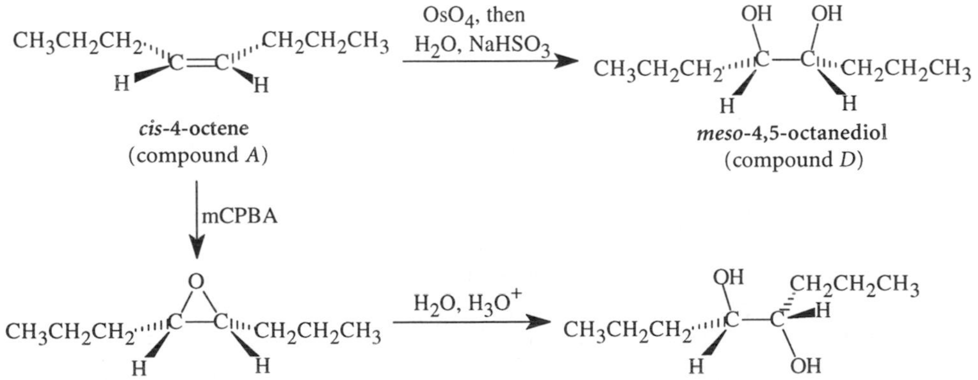

cis-4-octene (compound *A*) — OsO_4, then H_2O, $NaHSO_3$ → *meso*-4,5-octanediol (compound *D*)

mCPBA ↓

meso-2,3-dipropyloxirane (compound *B*) — H_2O, H_3O^+ → (±)-4,5-octanediol (compound *D*)

11.52 (a) The OsO_4 (in the presence of water), of course, effects glycol formation from the alkene starting material. This produced Os(VI), which is re-oxidized to Os(VIII) by periodic acid. One equivalent of the periodic acid effects this oxidation, continually recycling the osmium until the alkene is fully utilized. The second equivalent of the periodic acid then brings about cleavage of the glycol. To summarize:

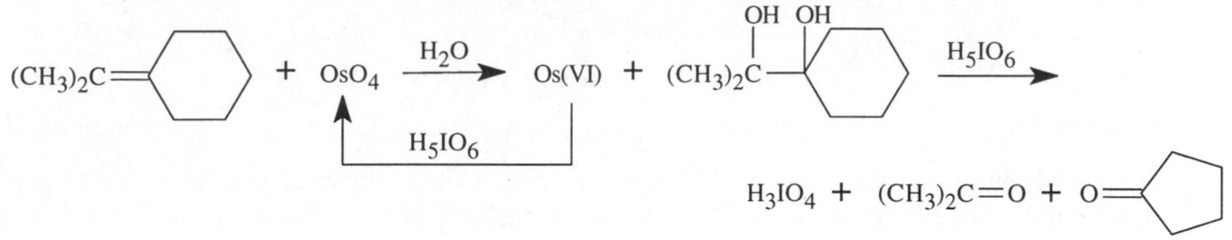

$(CH_3)_2C$=cyclohexylidene + OsO_4 $\xrightarrow{H_2O}$ Os(VI) + $(CH_3)_2C(OH)$–C(OH)(cyclohexane) $\xrightarrow{H_5IO_6}$ H_3IO_4 + $(CH_3)_2C{=}O$ + O=cyclopentane...

(Os(VI) is reoxidized to OsO_4 by H_5IO_6.)

(b)

bicyclic alkene + OsO_4 $\xrightarrow{H_2O}$ Os(VI) + diol (OH, OH) $\xrightarrow{H_5IO_6}$ H_3IO_4 + dialdehyde ≡ cyclohexane with two CH=O groups

(Os(VI) is reoxidized to OsO_4 by H_5IO_6.)

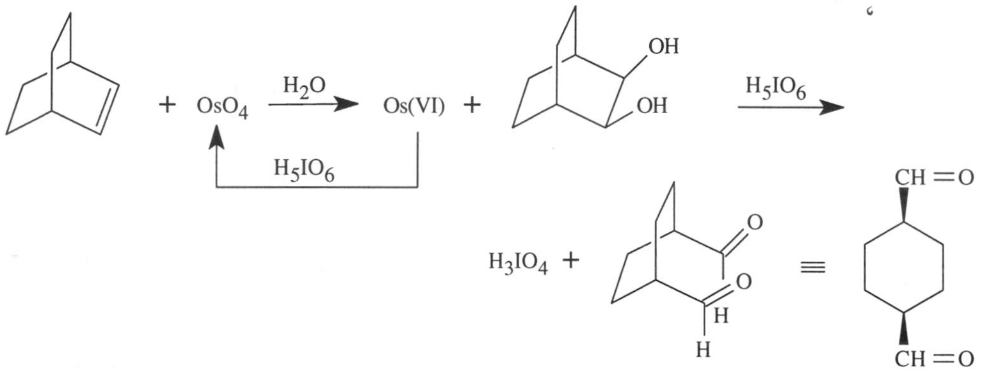

11.53 In both parts the mCPBA reacts at the side of the alkene π bond that involves the less severe van der Waals repulsions.

(a) One of the methyl groups must be axial, and therefore it creates a 1,3-diaxial interaction with the incoming reagent on one face of the ring. Therefore, attack occurs at the other face.

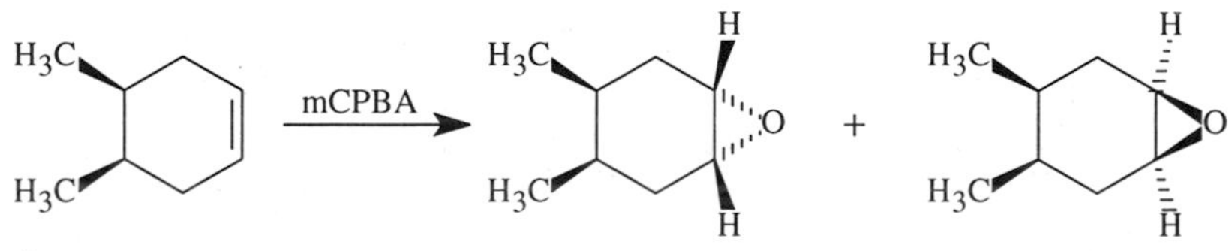

H_3C, H_3C cyclohexene $\xrightarrow{\text{mCPBA}}$ epoxide *A* (major product) + epoxide *B*

A
(major product)

B

(b) The face of the ring on the *same side* of the methyl group is blocked; hence, reaction occurs at the opposite face.

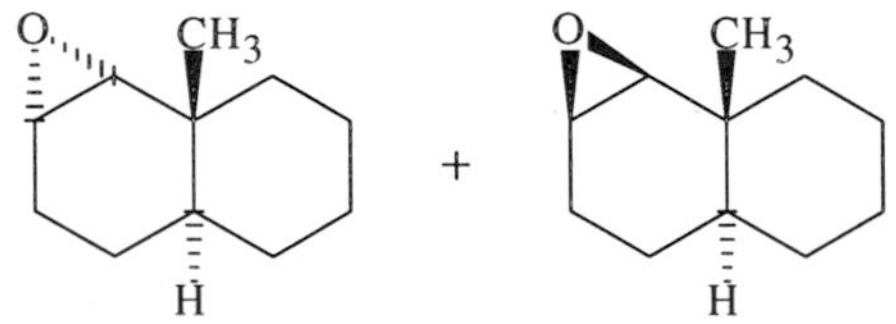

O, CH_3, H + O, CH_3, H

A
(major product)

B

11.54 (a) The curved-arrow mechanism for salt formation:

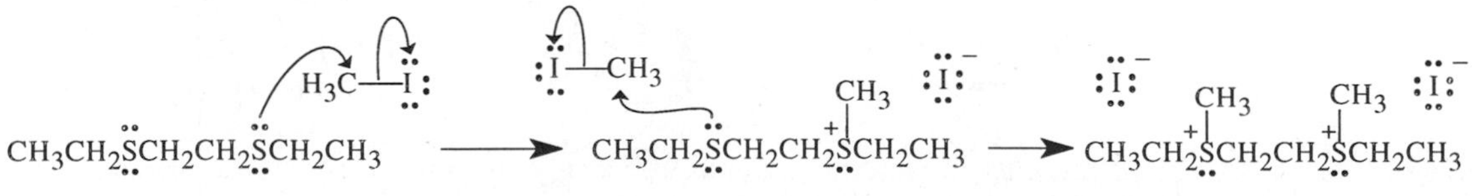

$CH_3CH_2\ddot{S}CH_2CH_2\ddot{S}CH_2CH_3$ + H_3C–$\ddot{I}$: $\longrightarrow$ $CH_3CH_2\ddot{S}CH_2CH_2\overset{+}{S}(CH_3)CH_2CH_3$:$\ddot{I}$:$^-$ + CH_3–I $\longrightarrow$:$\ddot{I}$:$^-$ $CH_3CH_2\overset{+}{S}(CH_3)CH_2CH_2\overset{+}{S}(CH_3)CH_2CH_3$:$\ddot{I}$:$^-$

(b) Counting the electron pair, each sulfur in the product bears four different groups. Consequently, the sulfurs are stereocenters. Because the sulfurs do not undergo rapid inversion, stereoisomeric sulfonium salts that differ in configuration at *sulfur* can be isolated. Because each salt contains two sulfur stereocenters, the two salts are diastereomers . One diastereomer is the meso, or (*R*,*S*), stereoisomer; the other is the racemate, a 1:1 mixture of the (*R*,*R*) and (*S*,*S*) stereoisomers.

11.55 The anion that results from the ring-opening reaction with hydroxide reacts as a nucleophile in another epoxide ring-opening reaction; and so on.

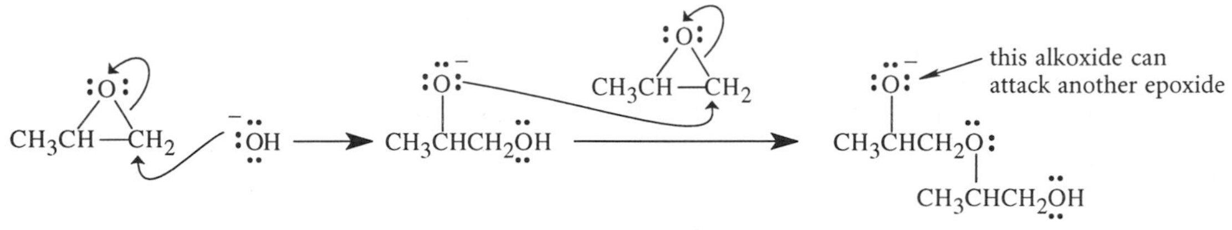

CH_3CH—CH_2 (epoxide, :O:) + $^-$:OH ⟶ CH_3CHCH_2OH (with :O:$^-$) + CH_3CH—CH_2 (epoxide) ⟶ CH_3CHCH_2O: / CH_3CHCH_2OH (with :O:$^-$ — this alkoxide can attack another epoxide)

11.56 (a) The rate acceleration suggests a mechanism involving neighboring-group participation. The product of this reaction results from the net substitution of the chlorines by —OH groups to give a diol. The mechanism for the first substitution is shown below in detail; you should write the mechanism for the second.

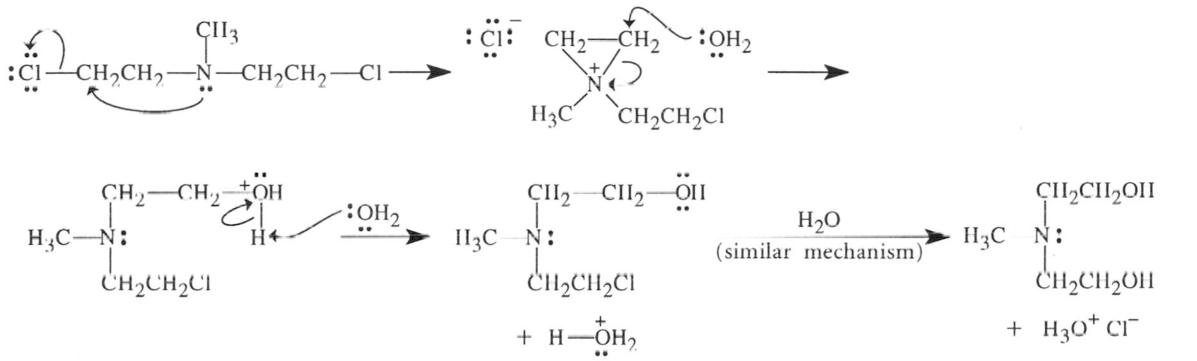

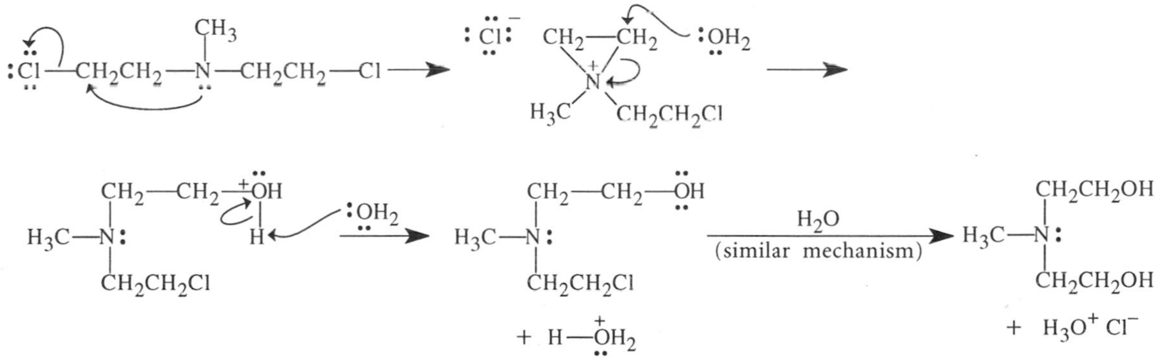

:Cl—CH_2CH_2—N(CH_3)—CH_2CH_2—Cl ⟶ :Cl:$^-$ + CH_2—CH_2 (aziridinium, $\overset{+}{N}$, H_3C, CH_2CH_2Cl) + :OH_2 ⟶

H_3C—N:(CH_2—CH_2—$\overset{+}{O}H$—H)(CH_2CH_2Cl) + :OH_2 ⟶ H_3C—N:(CH_2—CH_2—OH)(CH_2CH_2Cl) + H—$\overset{+}{O}H_2$ $\xrightarrow[\text{(similar mechanism)}]{H_2O}$ H_3C—N:(CH_2CH_2OH)(CH_2CH_2OH) + H_3O^+ Cl^-

:Cl—CH_2CH_2—N(CH_3)—CH_2CH_2—Cl ⟶ :Cl:$^-$ + CH_2—CH_2 (aziridinium, $\overset{+}{N}$, H_3C, CH_2CH_2Cl) + :OH_2 ⟶

H_3C—N:(CH_2—CH_2—$\overset{+}{O}H$—H)(CH_2CH_2Cl) + :OH_2 ⟶ H_3C—N:(CH_2—CH_2—OH)(CH_2CH_2Cl) + H—$\overset{+}{O}H_2$ $\xrightarrow[\text{(similar mechanism)}]{H_2O}$ H_3C—N:(CH_2CH_2OH)(CH_2CH_2OH) + H_3O^+ Cl^-

(b) The reaction of mechlorethamine with an amine R_3N: would follow a similar pattern in which the amine rather than water serves as the ultimate nucleophile. The final product is

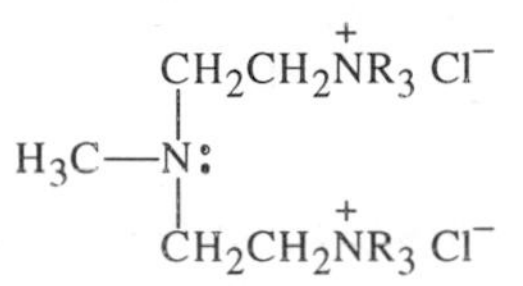

H_3C—N:($CH_2CH_2\overset{+}{N}R_3$ Cl^-)($CH_2CH_2\overset{+}{N}R_3$ Cl^-)

11.57 (a) Because the large *tert*-butyl group virtually locks the molecule into the chair conformation in which this group is equatorial, the two —OH groups in the compound *A* are locked into a trans-diaxial arrangement.

The cyclic periodate ester intermediate cannot be formed from two hydroxy groups in this arrangement without introducing significant ring strain. Because the required intermediate cannot form, the oxidation of compound *A* cannot occur. In contrast, compound *B* can be oxidized with periodic acid because the two —OH groups are trans-diequatorial.

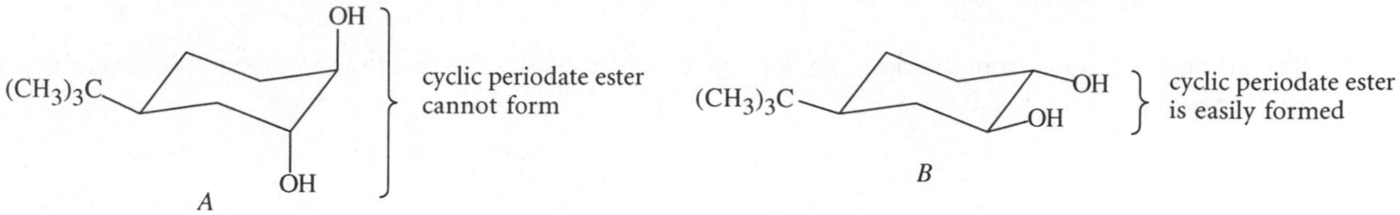

$(CH_3)_3C$ OH OH *A* — cyclic periodate ester cannot form

$(CH_3)_3C$ OH OH *B* — cyclic periodate ester is easily formed

(b) Compound *B* is locked into a conformation in which the two hydroxy groups are trans-diaxial; consequently, this compound cannot be oxidized for the same reason that compound *A* cannot be oxidized in part (a).

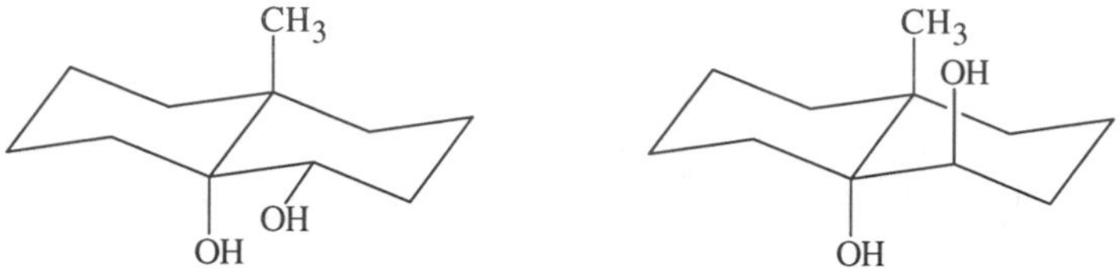

CH_3 OH OH *A* (can be oxidized)

CH_3 OH OH *B* (cannot be oxidized)

11.58 Notice that the transformation of compound *A* to compound *B* occurs with *retention of configuration.* A retention of configuration can be the result of *two successive inversions of configuration.* Fact (3) suggests that neighboring-group participation is occurring. Indeed, a neighboring-group mechanism involves a meso and therefore *achiral* cyclic sulfonium-ion intermediate *D*, formed by an intramolecular substitution with inversion of configuration:

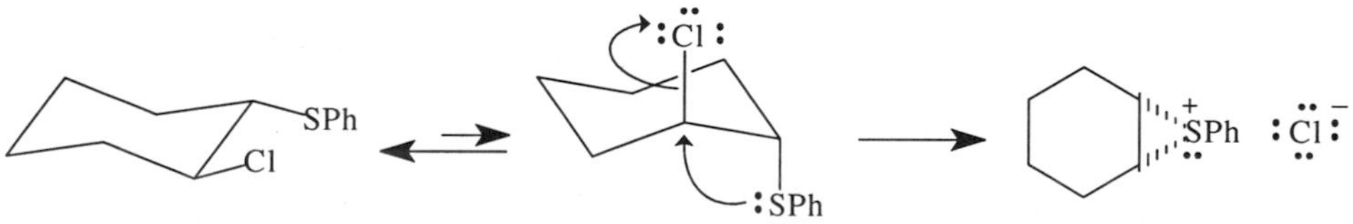

SPh Cl *A* ⇌ :Cl: :SPh → SPh :Cl: *D*

Because intermediate *D* is achiral, any chiral products that result from it must be formed as racemates (Sec. 7.8A); this accounts for fact (2). Reactions of intermediate *D* with ethanol at the two carbons of the three-membered ring give respectively the two enantiomers of product *B*. (The formation of one enantiomer is shown below; you should show formation of the other.) This substitution reaction occurs with inversion of configuration; coupled with the first inversion that takes place in the formation of *D*, this accounts for the overall retention noted in fact (1).

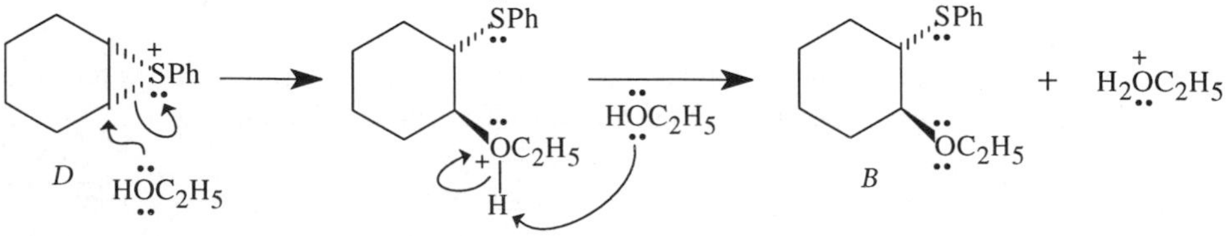

D SPh HOC_2H_5 → SPh OC_2H_5 H → (HOC_2H_5) SPh OC_2H_5 *B* + $H_2\overset{+}{O}C_2H_5$

11.59 Reaction (1) is an S_N2 reaction in which oppositely charged species react. This reaction should be promoted by conditions that increase the interaction between charges, that is, by a smaller dielectric constant. Therefore, reaction (1) is faster in ethanol. Reaction (2) is an S_N1 reaction in which the rate-limiting step involves dissociation of an alkyl halide to a carbocation and a chloride ion; this reaction is promoted by conditions that foster separation of charged groups, that is, by a greater dielectric constant. Consequently, reaction (2) is faster in water. Reaction (3) is an S_N1 reaction, but, because the starting sulfonium ion is positively charged, formation of the carbocation intermediate does not involve any change in the number of charges. Consequently, this reaction is not affected very much by a change of solvent.

11.60 (a) In this reaction, the initially formed bromonium ion is attacked intramolecularly by the oxygen of the —OH group. (The mechanism for the formation of a bromonium ion is shown in Eq. 5.4a, text p. 161.)

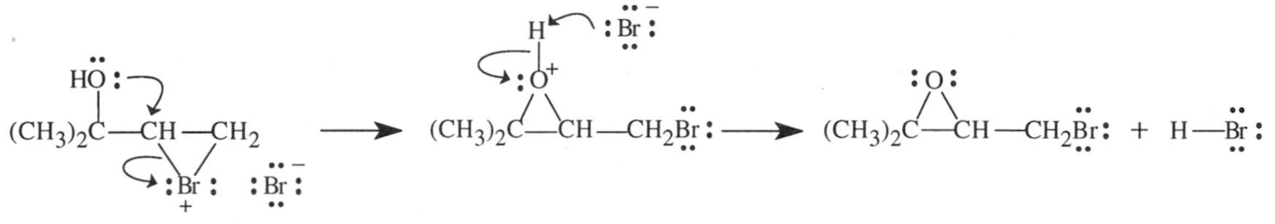

$$(CH_3)_2C(OH)—CH—CH_2\ (\text{bromonium ion}) + :Br:^- \longrightarrow (CH_3)_2C—CH—CH_2Br\ (\text{protonated epoxide}) \longrightarrow (CH_3)_2C—CH—CH_2Br\ (\text{epoxide}) + H—Br$$

bromonium ion

(b) The alkoxide, formed by ionization of the alcohol —OH group, opens the epoxide intramolecularly. The hydroxide catalyst is regenerated by protonation of the resulting alkoxide ion.

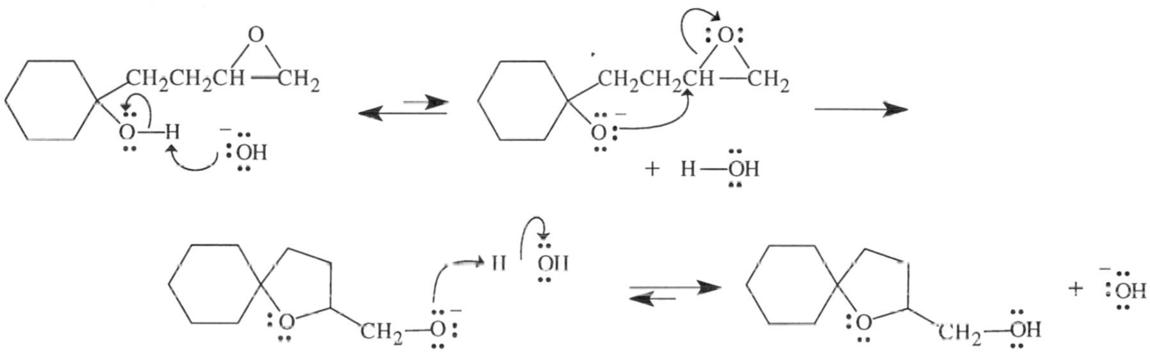

$$\text{(cyclohexyl)}(O—H)CH_2CH_2CH—CH_2\ (\text{epoxide}) + :OH^- \rightleftharpoons \text{(cyclohexyl)}(O^-)CH_2CH_2CH—CH_2\ (\text{epoxide}) + H—OH \longrightarrow$$

$$\text{spiro-tetrahydrofuran}—CH_2—O^- + H—OH \rightleftharpoons \text{spiro-tetrahydrofuran}—CH_2—OH + :OH^-$$

(c) This mechanism is very similar to the mechanism for the intramolecular substitution involved in Problem 9.60(a); see pp. 290–291 of this manual. The thiolate anion, formed by ionization of the thiol, displaces the tosylate group within a twist-boat conformation.

$$^-S\text{(chair)}OTs \rightleftharpoons ^-S\text{(twist-boat)}OTs \longrightarrow \text{(bicyclic sulfide)} + :OTs^-$$

thiolate ion formed by ionization of thiol starting material

twist-boat conformation places thiolate ion in position for backside attack

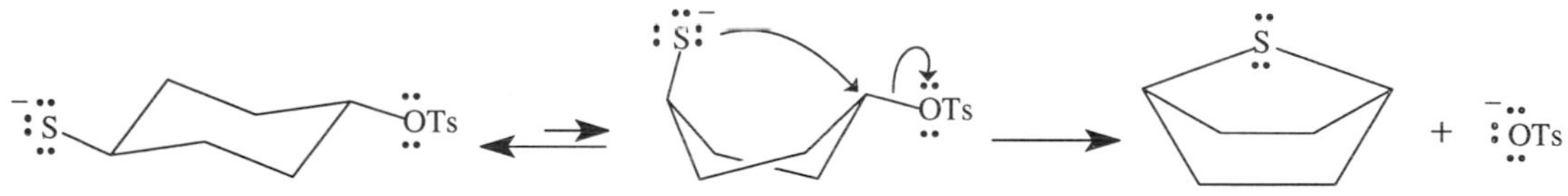

(d) The epoxide chemistry in this chapter has focused on ring-opening substitution reactions; however, you have learned that elimination competes with substitution. Thus, this reaction is an E2-like elimination reaction in which the two possible β-hydrogens are removed and the epoxide oxygen serves as a leaving group. The mechanism for the formation of the major alkene product is shown below; the minor product is formed by an essentially identical mechanism involving the other β-hydrogens. Substitution (that is, ring-opening) evidently does not occur because of the van der Waals repulsions that would result in the transition state of such a reaction between the alkyl branches of the base and those of the epoxide.

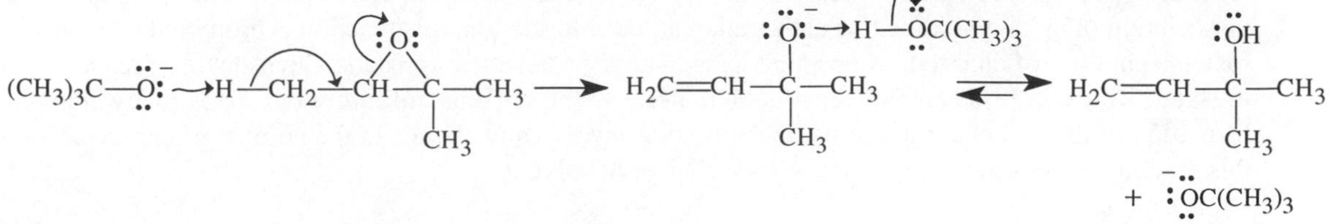

(e) The carbocation intermediate, which is formed by protonation of the double bond, is attacked intramolecularly by the sulfur nucleophile:

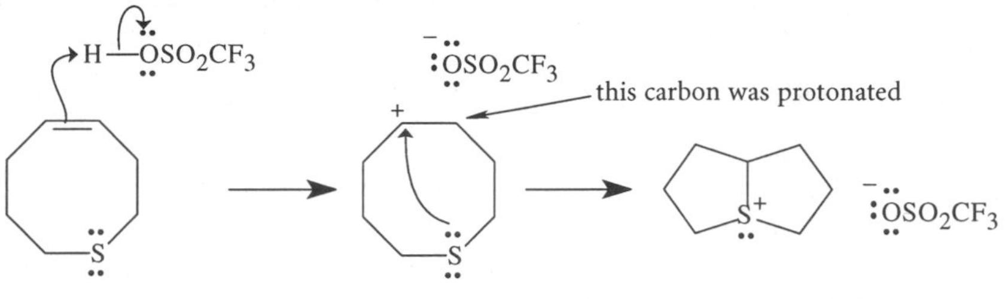

this carbon was protonated

(f) This appears to be a relatively rare instance of neighboring-group participation involving a four-membered ring.

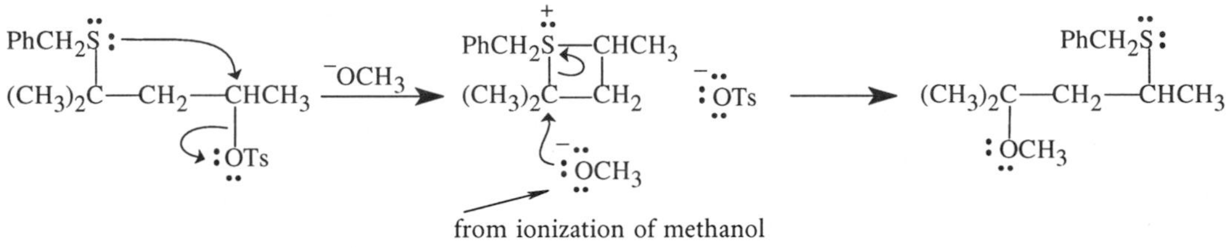

from ionization of methanol

11.61 The principles to use in solving this problem are that (*a*) the *tert*-butyl group, because it is so large, must assume the equatorial position; and (*b*) the —OH and —Cl groups must be able to achieve an anti relationship in order for backside attack (and thus epoxide formation) to occur. The required anti relationship is impossible for compound (3) because the —OH and —Cl groups are cis and are thus fixed in either chair conformation at a dihedral angle of about 60°. Therefore, compound (3) is *A*. In the chair conformation in which the *tert*-butyl group is equatorial, the —OH and —Cl groups in compound (1) are trans-diaxial, and are ideally set up for the formation of an epoxide. Consequently, compound (1) is *B*, and the epoxide formed from it is *D*.

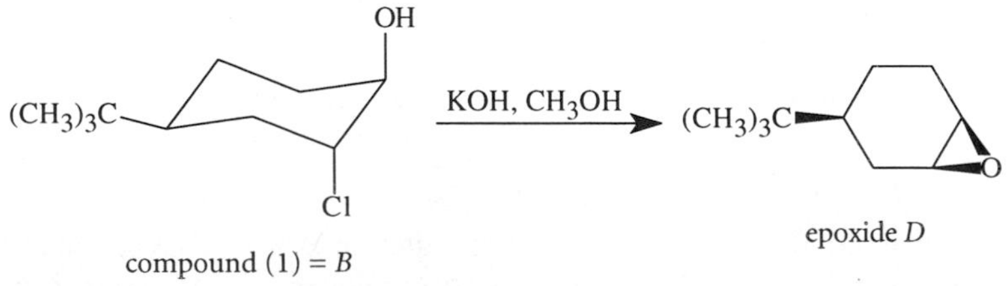

compound (1) = *B*

epoxide *D*

Finally, consider compound (2): the —OH and —Cl groups are trans, but they are diequatorial. The only way that they can achieve an anti relationship is either for the ring to undergo a chair flip, or for it to assume a twist-boat conformation. In either case, the conformation required for epoxide formation has a very high energy, and very little of it will be present. Because the transition-state conformation has very high energy, epoxide formation from compound (2), although possible, is very slow. Therefore, compound (2) = *C* and the epoxide formed from it is *E*. Note that epoxides *D* and *E* are stereoisomers, as required by the problem.

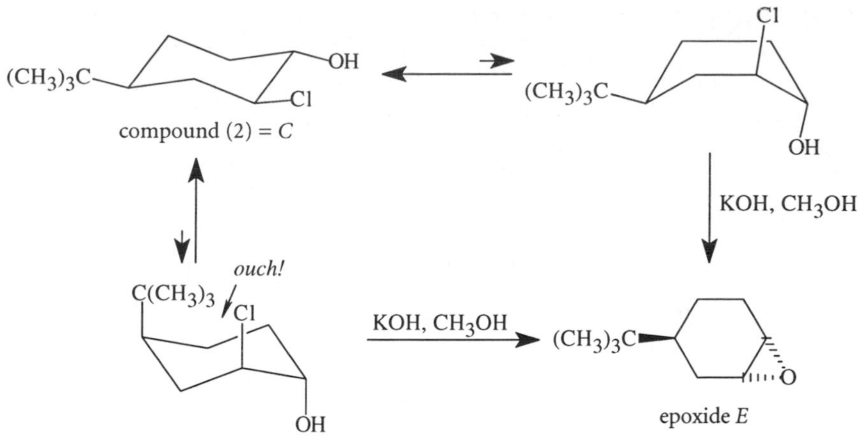

(CH$_3$)$_3$C— —OH —Cl ⟷ (CH$_3$)$_3$C— Cl OH

compound (2) = C

KOH, CH$_3$OH

ouch! C(CH$_3$)$_3$ Cl OH

KOH, CH$_3$OH → (CH$_3$)$_3$C— O

epoxide E

11.62 As illustrated by Eq. 11.22 on text p. 464, and as discussed in the solution to the previous problem, the —O$^-$ and —Br groups must be able to assume a trans-diaxial arrangement in the transition state for backside attack to occur.

Immediately we rule out compound *D*, as the —OH and —Br groups are cis.

The —OH and —Br groups in compounds *A–C* are trans. Because all of the compounds shown are trans-decalin derivatives, they cannot undergo the chair flip. Hence, the —OH and —Br groups could be trans-diequatorial or trans-diaxial. Examine each one in turn:

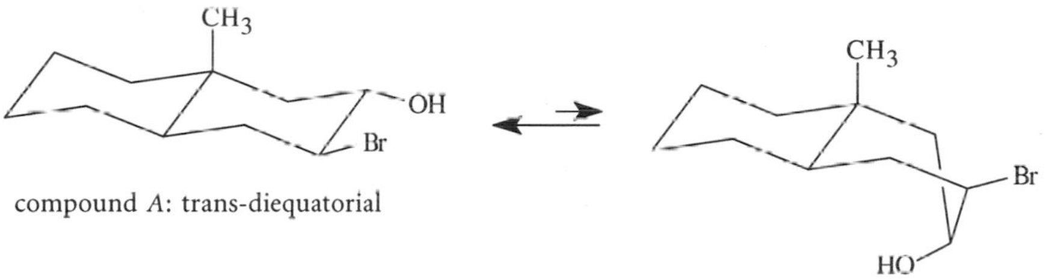

CH$_3$ OH Br ⟷ CH$_3$ Br HO

compound *A*: trans-diequatorial

must adopt a twist-boat conformation to achieve the anti arrangement required for the reaction

Compound *A* must assume a twist-boat conformation in order to undergo epoxide formation by backside attack. Because this conformation, and hence the transition state of the reaction, has a very high energy, epoxide formation is likely to be very slow.

The remaining two bromohydrins *B* and *C* have trans-diaxial arrangements and thus react readily:

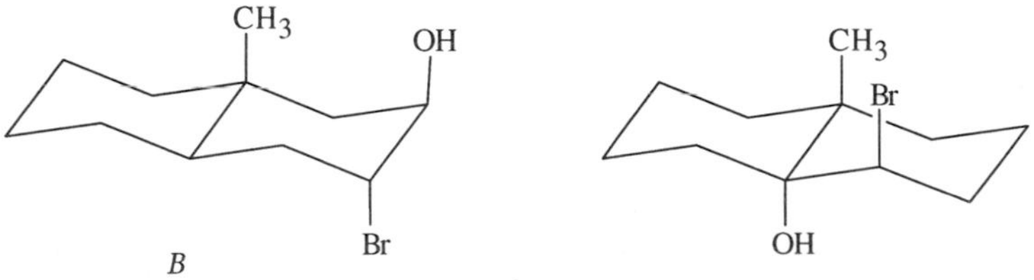

CH$_3$ OH Br

B

CH$_3$ Br OH

C

11.63 In this reaction, the bromine assists the departure of the protonated —OH group by a backside-attack mechanism, forming a bromonium ion in the process with inversion of configuration. The bromonium ion is opened with a second inversion of configuration by bromide ion from the ionization of HBr.

In this example, the bromonium ion is meso; opening of this ion by bromide ion gives chiral 2,3-dibromobutane; but, because the bromonium ion is achiral, the 2,3-dibromobutane *must* be formed as the racemate. The formation of one enantiomer of (±)-2,3-dibromobutane is shown below; you should show formation of the other.

bromonium ion
(a meso compound)

$-H_2\ddot{O}:$

protonated alcohol

one enantiomer of
(±)-2,3-dibromobutane

What stereoisomer(s) of the product would be formed from *(a)* the enantiomer of the starting material? *(b)* From a diastereomer of the starting material? *Answers: (a)* The same racemate shown above. *(b) Meso*-2,3-dibromobutane. Can you justify this answer?

12

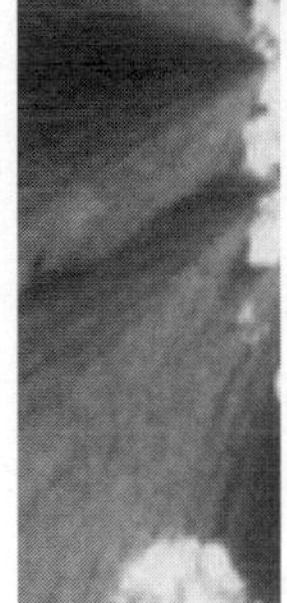

Introduction to Spectroscopy: Infrared Spectroscopy and Mass Spectrometry

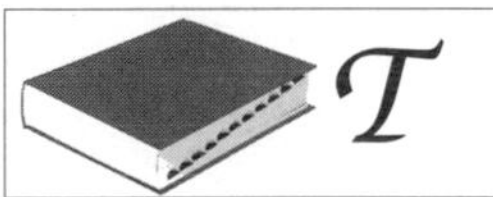

TERMS

Term	Section
absorption spectroscopy	12.1B
base peak	12.6A
bending vibration	12.3A
bond vibration	12.2B
chemical ionization (CI)	12.6E
CI mass spectrometry	12.6E
cycles per second, cps	12.1A
α-cleavage	12.6C
EI mass spectrometry	12.6E
electromagnetic radiation	12.1A
electromagnetic spectrum	12.1A
electrospray mass spectrometry	12.6F
even-electron ion	12.6D
exact mass	12.6F
fast-atom bombardment (FAB)	12.6F
FAB mass spectrometry	12.6F
FTIR spectroscopy	12.5
fragment ion	12.6A, 12.6C
fragmentation reaction	12.6A
frequency, ν	12.1A
hertz, Hz	12.1A
high-resolution mass spectrometer	12.6F
inductive cleavage	12.6C
infrared-active absorption	12.3B
infrared-inactive absorption	12.3B
infrared (IR) spectrometer	12.2A, 12.5
infrared (IR) spectroscopy	12.1B, 12.2
infrared (IR) spectrum	12.2A
in-plane bend	12.3A
inverse centimeters, cm^{-1}	12.2A
isotopic peak	12.6B
mass spectrometer	12.6A
mass spectrometry	12.1B, 12.6
mass spectrum	12.6A
mass-to-charge ratio, m/z	12.6A
micrometers, μm	12.2A
molecular ion	12.6A
nominal mass	12.6F
normal vibrational mode	12.3A
odd-electron ion	12.6D
out-of-plane bend	12.3A
peak position	12.2A
percent transmittance, $\%T$	12.2A
photon	12.1A
Planck's constant, h	12.1A
radical cation	12.6A
Raman spectroscopy	12.3B
reciprocal centimeters, cm^{-1}	12.2A
relative abundance	12.6A
rocking vibration	12.3B
scissoring vibration	12.3A
spectrometer	12.1B
spectrophotometer	12.1B
spectroscopy	12.1
spectrum	12.1B
stretching vibration	12.3A
symmetrical vibration	12.3A
twisting vibration	12.3B
unsymmetrical vibration	12.3A
wagging vibration	12.3A
wavelength, λ	12.1A
wavenumber, $\tilde{\nu}$	12.2A

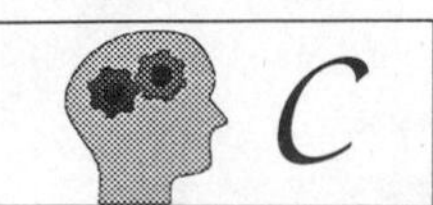

CONCEPTS

I. INTRODUCTION TO SPECTROSCOPY

A. Electromagnetic Radiation

1. The total range of electromagnetic radiation is called the electromagnetic spectrum.
   a. Visible light as well as X-rays, ultraviolet radiation, infrared radiation, microwaves, and radio waves are various forms of electromagnetic radiation.
   b. Electromagnetic radiation is characterized by its energy, wavelength, and frequency, which are interrelated.
2. Electromagnetic radiation has wave characteristics and can be characterized by its wavelength λ.
   a. The wavelength is the distance between successive peaks or successive troughs in the wave.
   b. The frequency ν of a wave is the number of wavelengths that pass a point per unit time when the wave is propagated through space.
      i. The frequency of a wave is given by the equation

$$\nu = \frac{c}{\lambda}$$

      in which c = the velocity of light (3×10^8 m/s).
      ii. The frequency has the dimensions of s^{-1} (cycles per second, cps) or hertz, Hz.
3. Light can also show particlelike behavior.
   a. The light particle is called a photon.
   b. The relationship between the energy of a photon and the wavelength or frequency of light is a fundamental law of physics:

$$E = h\nu = \frac{hc}{\lambda}$$

   where h is Planck's constant, a universal constant (6.625×10^{-27} erg · s or 6.625×10^{-34} J · s).
   c. The energy and frequency of light increase as the wavelength decreases.

B. Absorption Spectroscopy

1. Spectroscopy is the study of the interactions of matter and light (or other electromagnetic radiation).
   a. Spectroscopy can be used to determine unknown molecular structures.
   b. The most common type of spectroscopy used for structure determination is absorption spectroscopy.
      i. Matter can absorb energy from certain wavelengths of electromagnetic radiation.
      ii. This absorption is determined as a function of wavelength, frequency, or energy in an instrument called a spectrophotometer or spectrometer.
2. An absorption spectroscopy experiment requires
   a. a source of electromagnetic radiation
   b. the sample to be examined
   c. a detector that measures the intensity of the unabsorbed radiation that passes through the sample
   d. a recorder to graph the result as radiation transmitted (or radiation absorbed) versus wavelength (or frequency).
      i. The graph is commonly called a spectrum of the sample.
      ii. The spectrum of a compound is a function of its structure.
3. The three types of spectroscopy of greatest use (which differ conceptually only in the frequency of radiation used), and the general type of information each provides, are
   a. infrared (IR) spectroscopy—the functional groups present

b. nuclear magnetic resonance (NMR) spectroscopy—the number, connectivity, and functional-group environment of carbons and hydrogens
c. ultraviolet-visible (UV-VIS) or simply ultraviolet (UV) spectroscopy—the types of π-electron systems present.

II. INFRARED SPECTROSCOPY

A. The Infrared Spectrum

1. Infrared spectroscopy deals with the absorption of infrared radiation by molecular vibrations.
2. The infrared (IR) spectrum is measured in an instrument called an infrared spectrophotometer.
3. An IR spectrum is a plot of the infrared radiation transmitted through a sample as a function of the wavenumber or wavelength of the radiation.
   a. The quantity plotted on the lower horizontal axis is the wavenumber $\tilde{\nu}$ of the light.
      i. The wavenumber, in units of reciprocal centimeters or inverse centimeters (cm^{-1}), is simply another way to express the wavelength or frequency of the radiation.
      ii. Wavenumber is inversely proportional to the wavelength λ.

$$\left.\begin{aligned} \tilde{\nu} &= \frac{10^4}{\lambda} \\ \lambda &= \frac{10^4}{\tilde{\nu}} \end{aligned}\right\} \quad (\lambda \text{ in micrometers, } \tilde{\nu} \text{ in cm}^{-1})$$

      iii. The frequency ν and wavenumber $\tilde{\nu}$ are proportional.
   b. The quantity plotted on the vertical axis is percent transmittance (the percent of the irradiation falling on the sample that is transmitted to the detector).
   c. Sometimes an IR spectrum is not presented in graphical form, but is summarized completely or in part using descriptions of peak positions, intensities, and shapes.
      i. Intensities are often expressed qualitatively using the designations vs (very strong), s (strong), m (moderate), or w (weak).
      ii. Some peaks are narrow (or sharp, abbreviated sh), whereas others are wide (or broad, abbreviated br).
4. The absorptions observed in an IR spectrum are the result of vibrations within a molecule.
   a. Absorption of energy from infrared radiation can occur only when there is an exact match between the frequency of the radiation and the frequency of the bond vibration.
   b. Energy is absorbed and the intensity of the bond vibration increases when radiation of the matched frequency interacts with a vibrating bond.

B. Infrared Absorption and Chemical Structure

1. Each peak in the IR spectrum of a molecule corresponds to absorption of energy by the vibration of a particular bond or group of bonds.
   a. Certain absorptions are diagnostic; that is, they indicate with reasonable certainty that a particular functional group is present.
      i. The =C—H stretching and bending absorptions and the C=C stretching absorptions are very useful for the identification of alkenes.
      ii. The O—H stretching absorption is diagnostic for alcohols.
   b. Other peaks are confirmatory; that is, similar peaks can be found in other types of molecules, but their presence confirms a structural diagnosis made in other ways.
      i. The absorptions in the 1000 cm^{-1} to 1600 cm^{-1} region of the spectrum, known as the "fingerprint region," are generally not interpreted in detail, but they serve as a valuable "molecular fingerprint."
   c. The absence of absorptions in an IR spectrum restricts the possible structures under consideration.

2. Two aspects of IR absorption peaks are particularly important:
   a. the position of the peak (the wavenumber or wavelength at which it occurs); and
   b. the intensity of the peak.

C. Factors That Determine IR Absorption Position

1. The IR absorptions of stronger bonds—bonds with greater bond dissociation energy—occur at higher wavenumber.
2. The vibration frequency for a bond between two atoms of different mass is governed primarily by the mass of the lighter atom; vibrations of lighter atoms occur at higher wavenumbers than vibrations of heavier atoms.
3. The type of vibration being observed affects the position of the absorption in the IR spectrum.
   a. The allowed vibrations of a molecule are termed its normal vibrational modes.
   b. A stretching vibration occurs along the line of the chemical bond.
   c. A bending vibration is any vibration that does not occur along the line of the chemical bond.
      i. Bending vibrations occur at lower frequencies (higher wavelengths) than stretching vibrations of the same groups.
      ii. Bending vibrations can be such that the hydrogens move in the plane of the group or out of the plane of the group.
   d. Stretching and bending vibrations can be symmetrical or unsymmetrical with respect to a plane between the vibrating group.

D. Factors That Determine IR Absorption Intensity

1. A greater number of molecules in the sample (that is, a more concentrated sample or a greater optical path) and more absorbing groups within a molecule give a more intense spectrum.
2. Absorptions can be expected only for vibrations that cause a change in the molecular dipole moment.
   a. Any vibration that gives rise to an IR absorption is said to be infrared active.
   b. Certain symmetrical (or nearly symmetrical) molecules lack IR absorptions that otherwise might be expected to be observed.
      i. Absorptions that result in no dipole moment change are infrared inactive.
      ii. The infrared-inactive vibrations occur, but they are simply not observed in the IR spectrum.
      iii. IR-inactive vibrations can be observed by a less common type of spectroscopy called Raman spectroscopy.
   c. Because the intensity of an IR absorption depends on the size of the dipole moment change that accompanies the corresponding vibration, IR absorptions differ widely in intensity.

E. The Infrared Spectrometer

1. In a conventional IR spectrometer, the spectrum is obtained by slowly scanning wavelength through the range of interest; this technique takes several minutes.
2. In a newer type of IR spectrometer, called a Fourier-transform infrared spectrometer (FTIR spectrometer), the IR spectrum can be obtained in just a few seconds.

III. MASS SPECTROMETRY

A. General

1. The utility of mass spectrometry:
   a. It can be used to determine the molecular mass of an unknown compound, and
   b. It can be used to determine the structure (or a partial structure) of an unknown compound by an analysis of the fragment ions in the spectrum.
2. A mass spectrum is a graph of the relative amount of each ion (called the relative abundance) as a function of the ionic mass (or mass-to-charge ratio m/z).
   a. The mass spectrum can be determined for any molecule that can be vaporized in a high vacuum, and this includes most organic compounds.
   b. Only ions are detected by the mass spectrometer; neutral molecules and radicals do not appear as peaks in the mass spectrum.

3. The instrument used to obtain a mass spectrum is called a mass spectrometer.
   a. In one type of mass spectrometer, a compound is vaporized in a vacuum and bombarded with an electron beam of high energy in which an electron is ejected from the molecule.
      i. A species formed in this process is both a radical and a cation—a radical cation.
      ii. Radical cations decompose in a series of reactions called fragmentation reactions to give positively charged fragment ions of differing mass.
   b. In the mass spectrometer, the fragment ions are separated according to their mass-to-charge ratio, m/z (m = mass, z = the charge of the fragment).
4. The ion derived from electron ejection before any fragmentation takes place is known as the molecular ion, and abbreviated M.
   a. The molecular ion occurs at an m/z value equal to the molecular mass of the sample molecule.
   b. Except for peaks due to isotopes, the molecular ion peak is the peak of highest m/z in any ordinary mass spectrum.
5. The base peak is the ion of greatest relative abundance and is arbitrarily assigned a relative abundance of 100%. Other peaks in the mass spectrum are scaled to the base peak.
   a. The relative abundances of the fragment ions are recorded as a function of their mass-to-charge ratios m/z, which, for most ions, equal their masses.
   b. Both molecular masses and partial structures can be derived from the masses of these ionic fragments.

B. Fragmentation Mechanisms

1. The molecular ion is formed by loss of an electron.
   a. A stable molecular ion decomposes slowly and is detected by the mass spectrometer as a peak of large relative abundance.
   b. A less stable molecular ion decomposes, in some cases completely, into smaller pieces detected as smaller ions, called fragment ions.
      i. A type of cleavage called inductive cleavage, involving the heterolytic cleavage of a bond to a positively charged electronegative atom, is very common in ethers, alkyl halides, and other compounds containing a very electronegative element.

$$C_2H_5{-}CH(CH_3){-}\ddot{O}{-}CH(CH_3){-}CH_3 \xrightarrow{-e^-} C_2H_5{-}CH(CH_3){-}\dot{\overset{+}{O}}{-}CH(CH_3){-}CH_3 \longrightarrow C_2H_5{-}\overset{+}{C}H(CH_3) + {:}\dot{\ddot{O}}{-}CH(CH_3){-}CH_3$$

inductive cleavage

      ii. A type of cleavage called α-cleavage, involving the homolytic cleavage of a bond β to a positively charged electronegative atom, is important in secondary and tertiary ethers, alcohols, alkenes, and several other types of compounds. (This type of cleavage is also known as β-scission.)

$$CH_3CH_2{-}CH(CH_3){-}\ddot{O}{-}CH(CH_3)_2 \xrightarrow{-e^-} CH_3CH_2{-}CH(CH_3){-}\dot{\overset{+}{O}}{-}CH(CH_3)_2 \longrightarrow CH_3\dot{C}H_2 + \left[(H)(H_3C)C{=}\overset{+}{\ddot{O}}{-}CH(CH_3)_2 \longleftrightarrow (H)(H_3C)\overset{+}{C}{-}\ddot{O}{-}CH(CH_3)_2\right]$$

α-cleavage (β-scission)

2. Molecular ions derived from compounds of even mass, as well as some fragment ions produced in other ways, have even masses.
   a. When such a compound undergoes fragmentation to give separate radical and cation species, each of these must have an odd mass.

b. When a compound contains only C, H, and O, its fragment ions of odd mass must be even-electron ions and its fragment ions of even mass must be odd-electron ions.
   i. Most of the fragment ions observed in the mass spectra of molecules containing only C, H, and O have odd molecular masses.
   ii. Since the unpaired electron is carried off in the radical fragment, the carbocation fragment detected by the mass spectrometer must have no unpaired electrons.

3. Analyze a mass spectrum by thinking of fragment ions as coherent pieces that result from the breaking of chemical bonds in the molecular ion.
   a. Every possible fragment may not be seen in the mass spectrum of a compound.
   b. The relative abundances of the various fragments in a mass spectrum depend on their relative lifetimes, that is, the relative rates at which they break apart into smaller fragments.
   c. Typically the most stable ions have the longest lifetimes and are detected as the largest peaks.
4. The masses of fragments lost as well as the masses of the fragments observed can be used to postulate the structures of fragment ions and thus convey information about their electronic structures.
   a. Ions containing no unpaired electrons are called even-electron ions and are formed by such processes as α-cleavage, inductive cleavage, and direct fragmentation at a σ bond.
   b. Ions containing an unpaired electron (a radical cation) are called odd-electron ions.
      i. Hydrogen-atom transfer followed by loss of a stable neutral molecule is a very common mechanism for the formation of odd-electron ions in primary alcohols, alkyl halides, and other compounds.

$$\mathrm{R{-}CH(H){-}CH_2{-}\overset{+}{\ddot{O}}H} \longrightarrow \mathrm{R{-}\dot{C}H{-}CH_2{-}\overset{+}{\ddot{O}}H{-}H} \longrightarrow \mathrm{R{-}\dot{C}H{-}\overset{+}{C}H_2} + \mathrm{H{-}\ddot{O}H}$$

molecular ion of an alcohol — odd-electron ion

C. Identifying the Molecular Ion

1. The molecular ion is the most important peak in the mass spectrum because it provides the molecular mass of the molecule under study and thus is the basis for calculating the losses involved in fragmentation.
2. Whether the ion of highest mass is the molecular ion can be determined by:
   a. knowing that all compounds containing only the elements C, H, and O have even molecular masses and therefore must have a molecular ion of even mass;
   b. calculating the losses of fragments observed from the peak that is assumed to be the molecular ion and determining whether these losses correspond to reasonable combinations of atoms;
   c. noting the mass change that results from converting the compound into a derivative by known reactions; or
   d. using a different ionization technique. (See E.1.B.)

D. Isotopic Peaks

1. Associated with each peak in a mass spectrum are other peaks at higher mass that arise from the presence of isotopes at their natural abundance.
   a. A peak that occurs one mass unit higher than the molecular ion (M) is termed an M + 1 peak.
   b. Every peak in a mass spectrum has isotopic peaks.
2. The relative abundance of each isotopic peak is directly related to the natural abundance of the isotope involved.
   a. Isotopic peaks are particularly useful for diagnosing the presence of elements that consist of more than one isotope of high natural abundance such as chlorine and bromine.
   b. The naturally occurring isotopes of chlorine (^{35}Cl and ^{37}Cl) exist in a ratio of 3:1.
   c. The naturally occurring isotopes of bromine (^{79}Br and ^{81}Br) exist in about equal amounts.
3. Isotopes are especially useful because they provide specific labels at particular atoms without changing their chemical properties.

a. The fate of specifically labeled atoms in the products of a reaction, determined by mass spectrometry, can be used to decide between two reaction mechanisms.
b. Nonradioactive isotopes can be used for biological metabolic studies (studies that deal with the fates of chemical compounds when they react in biological systems).

E. The Mass Spectrometer

1. Various techniques are available to obtain the ions necessary for observation in a mass spectrometer:
   a. Electron-impact (EI) mass spectrometry involves bombardment of the vaporized sample with a beam of electrons.
   b. Chemical ionization (CI) mass spectrometry involves treating the vaporized sample with a proton source.
   c. Fast-atom bombardment (FAB) involves compounds in solution that are converted directly into gas-phase ions by subjecting them to a beam of heavy atoms (such as xenon, argon, or cesium) that have been accelerated to high velocities.
   d. Electrospray mass spectrometry involves atomizing a solution of the sample within highly charged droplets.
2. More complex mass spectrometers, called high-resolution mass spectrometers, can resolve ions that are separated in mass by only a few thousandths of a mass unit.
   a. Nominal mass is the mass of the ion to the nearest whole number.
   b. Exact mass is the mass of the ion to four or more decimal places.
   c. The exact mass provides an elemental analysis of the molecular ion as well as an elemental analysis of each fragment in the mass spectrum.

STUDY GUIDE LINKS

12.1 FTIR Spectroscopy

This section will give you a better idea how an FTIR spectrometer works. In an FTIR spectrometer, radiation to be passed through the sample is split and sent over two paths simultaneously by a system consisting of two mirrors, one moving and one fixed (a Michelson interferometer). The difference between the lengths of the two light paths at various times is precisely calibrated with a laser. The superimposed beams of light from the two paths is passed through the sample. The interference patterns, or interferograms, that result can be thought of as superimposed plots of absorption *vs.* time from both light beams in which all wavelengths in the spectrum contribute simultaneously. The interferograms are stored in a small computer and analyzed by *Fourier transformation.* Fourier transformation is a mathematical technique for decomposing time-based wave motions into their contributing frequency components. (For example, the unique sound of a bell as it dies out over time might be analyzed in terms of the different contributing pitches by the same technique.) Fourier analysis converts the interferograms into a plot of absorption versus wavenumber—the IR spectrum. The entire process takes just a few seconds.

Besides saving time, FTIR has much better wavenumber resolution than conventional IR spectroscopy. (This means that two closely-spaced peaks would appear as separate peaks rather than as a single broad peak.) In addition, multiple spectra from the same sample can be digitally added; because "noise" is random, it decreases when multiple spectra are summed, and the signal due to the IR spectrum grows. Consequently, many spectra can be accumulated from the same sample in a very short time, and thus strong spectra can be obtained from very dilute samples. The principles of FTIR have been known for many years, but only in relatively recent times has the technique become practical because of the availability of inexpensive, high-powered computers. The digital revolution has come to spectroscopy just as it has to sound and video technology!

12.2 The Mass Spectrometer

This section will give you a more quantitative understanding of how a mass spectrometer operates. When an ion in a mass spectrometer is accelerated to a voltage V its kinetic energy is $mv^2/2$, where m is its mass, and v is its velocity. Since all ions have, to a good approximation, the same velocity as a result of the accelerating voltage, then it is the *ionic mass* that gives one ion a different kinetic energy from another. The basic equation describing the bending of an ion by a magnetic field $\mathbf{B}$ is

$$m/z = k\frac{\mathbf{B}^2r^2}{V}$$

where r is the radius of the circular path over which the ion is bent, z is the charge on the ion, and k is a proportionality constant. This equation shows that the paths of ions of a larger mass are bent over a circle of greater radius. Similarly, for an ion of a given mass, increasing the magnetic field $\mathbf{B}$ decreases the radius of its path.

In the mass spectrometer, a collector is positioned to sense only the ions traversing a path of a particular radius r. By increasing the magnetic field $\mathbf{B}$ in a regular way, ions of progressively increasing mass are focused onto the collector. The relative intensity of the ion beam of a given mass is recorded as an ion current. Thus is obtained a plot of ion current versus m/z—the mass spectrum.

SOLUTIONS TO PROBLEMS

Solutions to In-Text Problems

12.1 (a) Apply Eq. 12.1.

$$\nu = \frac{(3.00 \times 10^8 \text{ m/sec})}{(9.00 \times 10^{-6} \text{ m})} = 3.33 \times 10^{13} \text{ sec}^{-1}$$

(b) Apply Eq. 12.1 and include the conversion factor 10^{-10} m/Å.

$$\nu = \frac{(3.00 \times 10^8 \text{ m/sec})}{(4800 \text{ Å})(10^{-10} \text{ m/Å})} = 6.25 \times 10^{14} \text{ sec}^{-1}$$

12.2 (a) Multiply the frequency obtained in the solution to Problem 12.1(a) times Planck's constant:

$$E = h\nu = (3.99 \times 10^{-13} \text{ kJ sec mol}^{-1})(3.33 \times 10^{13} \text{ sec}^{-1}) = 13.3 \text{ kJ/mol}$$

(b) Use the same procedure as in part (a) of this problem:

$$E = h\nu = (3.99 \times 10^{-13} \text{ kJ sec mol}^{-1})(6.25 \times 10^{14} \text{ sec}^{-1}) = 249 \text{ kJ/mol}$$

12.3 (a) The energy of X-rays is greater than that of any visible light, including blue light. (In fact, the energy is so much greater that prolonged exposure to X-rays is harmful.)

(b) The energy of radar (that is, microwaves) is less than that of any visible light, including red light.

If the energy of microwaves is much less than that of red light, why is it that microwave energy rapidly cooks food, whereas red light does not? This is so because microwaves are absorbed by molecules in food, and the absorbed energy is rapidly converted into heat. In contrast, red light is reflected or transmitted, but it is mostly not absorbed, and is therefore not converted into heat.

12.4 (a) Apply Eq. 12.7a:

$$\tilde{\nu} = \text{wavenumber} = (1 \times 10^4 \text{ }\mu\text{m/cm}) \div 6.0 \text{ }\mu\text{m} = 1667 \text{ cm}^{-1}$$

(b) Apply Eq. 12.7b:

$$\lambda = \text{wavelength} = (1 \times 10^4 \text{ }\mu\text{m/cm}) \div 1720 \text{ cm}^{-1} = 5.81 \text{ }\mu\text{m}$$

12.5 From Eq. 12.8 on text p. 502 we know that

$$\tilde{\nu} = c\nu$$

Convert the frequency into wavenumber and then consult Fig. 12.4 on text p. 499.

$$\tilde{\nu} = \nu/c = (9 \times 10^{13} \text{ sec}^{-1})/(3 \times 10^{10} \text{ cm sec}^{-1}) = 3000 \text{ cm}^{-1}$$

The group of peaks around 3000 cm^{-1} are due to the C—H stretching vibrations in nonane.

12.6 Using Eq. 12.8, convert the wavenumber to a frequency, which is the "times per second" equivalent of wavelength or wavenumber.

$$\nu = c\tilde{\nu} = (3 \times 10^{10} \text{ cm sec}^{-1})(2143 \text{ cm}^{-1}) = 6.43 \times 10^{13} \text{ sec}^{-1}$$

12.7 According to the text, bond strength is the major determining factor of IR absorption frequency as we go across a row of the periodic table. Table 5.3 is essentially a table of bond strengths. From this table, we find that bond strengths are in the order

$$\text{H—F} > \text{H—OH} > \text{H—NH}_2 > \text{H—CH}_3$$

Because the strongest bonds vibrate with the highest frequencies, the order of stretching frequencies is the same; that is, H—F has the highest frequency, and H—CH_3 the lowest.

12.8 Take the ratio of two equations like 12.13, one for the C—H bond, and the other for the C—D bond. Notice that everything cancels except the frequencies and the square roots of the masses.

$$\frac{\tilde{\nu}_D}{\tilde{\nu}_H} = \sqrt{\frac{m_H}{m_D}}$$

or $$\tilde{\nu}_D = (\tilde{\nu}_H)\sqrt{\frac{m_H}{m_D}} = (3090 \text{ cm}^{-1})\sqrt{\frac{1}{2}} = 2185 \text{ cm}^{-1}$$

In fact, C—D vibrations appear in the IR at lower energy than the corresponding C—H vibrations (to the right in conventional IR spectra).

12.9 (a) Active. The triple bond is located asymmetrically; hence, even a small dipole moment will be affected by the stretching vibration.
(b) Inactive. The zero dipole moment of this alkyne is not changed by stretching the triple bond.
(c) Active. The C═O dipole is increased by the stretch because its length changes. Recall (Eq. Eq. 1.4, text p. 10) that dipole moment is proportional to length.
(d) Inactive. The zero dipole moment of this alkene is not changed by stretching the double bond.
(e) Inactive. The zero dipole moment of cyclohexane is unaffected by this symmetrical vibration.

12.10 The moderately strong C═C stretching absorption at about 1650 cm^{-1} and the C—H bending vibration at 890 cm^{-1} suggest that (a) is the spectrum of 2-methyl-1-hexene. The C—H bending absorption at about 970 cm^{-1} and the virtual absence of a C═C stretching absorption in spectrum (b) confirms that this is the spectrum of *trans*-2-heptene.

12.11 The absence of a C═C stretching absorption near 1650 cm^{-1} and an O—H stretching absorption near 3300 cm^{-1}, and the presence of a strong absorption near 1100 cm^{-1} (C—O stretching absorption), show that this is the spectrum of the ether.

12.12 The two C—O bonds of an ether can undergo both symmetrical and unsymmetrical stretching vibrations; each of these normal vibrational modes has an associated infrared absorption. (These vibrations are described on the first two lines of Fig. 12.7 on text p. 510.)

12.13 Think of molecular association by hydrogen bonding as a reaction like any other and apply LeChatelier's principle.

$$2\ \text{RO—H} \rightleftharpoons \text{RO—H}\cdots\text{OR(—H)}$$

unassociated alcohols → hydrogen-bonded alcohols

Increasing the concentration of alcohol drives this equilibrium to the right, and decreasing the concentration of the alcohol drives this equilibrium to the left. At sufficiently low concentrations of alcohol, the spectrum is that of the unassociated alcohol. The O—H bond in a hydrogen-bonded O—H absorbs at a different frequency than the unassociated O—H.

12.14 Table 12.3 shows that Si has the following isotopic abundances: ^{28}Si, 92.21; ^{29}Si, 4.67; and ^{30}Si, 3.10. These abundances result in the following relative intensities for the m/z = 74 and 75 peaks:

m/z = 74: relative intensity (4.67)/(92.21) = 0.0506, or 5.06%
m/z = 75: relative intensity (3.10)/(92.21) = 0.0336, or 3.36%

To this must be added the contribution due to ^{13}C. The molecular mass of $(CH_3)_4Si$ is 88; therefore, the base peak at $m/z = 73$ results from loss of 15 mass units (CH_3). Hence, the $m/z = 74$ peak will have a contribution from three isotopic carbons equal to $3(1.11)/(92.21) = 0.0361$ (3.61%). Add this contribution to that calculated above to get the final value for the $m/z = 74$ peak:

$$m/z = 74\text{: relative intensity} = 5.06\% + 3.61\% = 8.67\%$$

The contribution of ^{13}C to the $m/z = 75$ peak is negligible. Hence, the $m/z = 74$ peak has a relative intensity of 8.67%, and the $m/z = 75$ peak has a relative intensity of 3.36%.

12.15 Assume the molecular ion (base) peak at $m/z = 50$ has an abundance of 100%. This peak is due to molecules that contain ^{12}C, ^{1}H, and ^{35}Cl. The M + 1 peak is due to molecules that contain either ^{13}C, ^{1}H, and ^{35}Cl or ^{12}C, ^{2}H, and ^{35}Cl. The intensity of the M + 1 peak due to ^{13}C *relative to the base peak* is $(0.0111/0.989) = 1.12\%$, and the relative intensity due to ^{2}H is $3(0.00015)/(0.99985) = 0.0004$, or 0.04%. Hence, the M + 1 peak at $m/z = 51$ is due almost entirely to molecules that contain ^{13}C, ^{1}H, and ^{35}Cl, but we'll include the contribution of ^{2}H for completeness. Because there is one carbon, the intensity of the $m/z = 51$ peak is 1.16% (1.12% + 0.04%). The M + 2 peak at $m/z = 52$ is due almost entirely to ^{12}C, ^{1}H, and ^{37}Cl; according to Table 12.3, the ratio of this peak to the base peak should be $24.23/75.77 = 0.320$, or 32%. (This peak has a contribution from molecules that contain ^{13}C, ^{2}H, and ^{35}C equal to $3 \times (0.0111/0.989)(0.00015/0.99985) = 0.00017\%$ that can be ignored). Finally, there is a peak at M + 3, or $m/z = 53$, which is due to molecules that contain either ^{13}C, ^{1}H, and ^{37}Cl or to ^{12}C, ^{2}H, and ^{37}Cl. The contribution of ^{2}H is only 0.0004. Thus, the relative abundance from the *simultaneous* presence of ^{13}C and ^{37}Cl is $(0.0116)(0.320) = 0.00368$, or 0.36%. This peak is almost negligible.

12.16 (a) 2-Methyl-2-pentanol has molecular mass = 102. The fragment at $m/z = 59$ corresponds to the loss of 43 mass units (a propyl group). This loss can occur by an α-cleavage mechanism:

$$(CH_3)_2C(\ddot{O}H)\text{—}CH_2CH_2CH_3 \xrightarrow{-e^-} (CH_3)_2C(\overset{+}{\dot{O}}H)\text{—}CH_2CH_2CH_3 \longrightarrow (CH_3)_2C{=}\overset{+}{\ddot{O}}H + \cdot CH_2CH_2CH_3$$

2-methyl-2-pentanol | molecular ion M | $m/z = 59$

(b) The peak at $m/z = 87$ corresponds to loss of 15 mass units (a methyl group). Begin with the same molecular ion as in part (a).

$$H_3C\text{—}C(\overset{+}{\dot{O}}H)(CH_3)\text{—}CH_2CH_2CH_3 \longrightarrow H_3C\cdot + H_3C\text{—}C({=}\overset{+}{\ddot{O}}H)\text{—}CH_2CH_2CH_3$$

M redrawn | $m/z = 87$

12.17 One fragmentation of *sec*-butyl isopropyl ether by inductive cleavage is shown in Eq. 12.23(b) of the text, and this is followed by the comment that a similar mechanism accounts for the formation of the $m/z = 43$ fragment. That inductive cleavage mechanism is as follows:

$$CH_3CH(C_2H_5)\text{—}\overset{+}{\dot{O}}\text{—}CH(CH_3)_2 \longrightarrow CH_3CH(C_2H_5)\text{—}\dot{\ddot{O}}\colon + {}^{+}CH(CH_3)_2$$

molecular ion of *sec*-butyl isopropyl ether | $m/z = 43$

12.18 (a) Because even-electron ions of compounds containing only C, H, and Cl have odd mass, the ion with $m/z = 57$ is the even-electron ion and the ion with $m/z = 56$ is the odd-electron ion.

(b) HCl can be lost to give the odd-electron ion.

(c) The molecular mass of 2-chlorobutane is 92 (for the ^{35}Cl isotope). The $m/z = 57$ peak is due to loss of a chlorine atom from the molecular ion by an inductive-cleavage mechanism to give the *sec*-butyl cation:

$$CH_3\underset{\overset{|}{:\ddot{Cl}\cdot^{+}}}{C}HCH_2CH_3 \longrightarrow CH_3\overset{+}{C}HCH_2CH_3 + :\ddot{Cl}\cdot$$

molecular ion of 2-chlorobutane — sec-butyl cation, $m/z = 57$

The m/z = 56 peak is due to hydrogen-atom transfer within the molecular ion followed by loss of HCl in a process analogous to that shown for loss of water in Eq. 12.27 of the text:

$$H_3C—\overset{:\ddot{Cl}\cdot^{+}}{CH}—\overset{H}{C}HCH_3 \longrightarrow H_3C—\overset{\overset{+}{:\ddot{Cl}}—H}{CH}—\dot{C}HCH_3 \longrightarrow :\ddot{Cl}—H + H_3C—\overset{+}{C}H—\dot{C}HCH_3$$

molecular ion of 2-chlorobutane — $m/z = 56$

12.19 Formation of a methyl ether adds 14 mass units to the molecular mass of an alcohol. Recall from the text that the CI technique (because the molecular ion is protonated) gives a molecular ion one mass unit higher than the actual molecular ion. Hence, the molecular mass of the methyl ether is 116. Postulate, then, that the molecular mass of alcohol *A* is (116 – 14) = 102. The prominent peaks at m/z = 87 and m/z = 73 in the spectrum of the alcohol, then, represent mass losses of 15 units (a methyl group) and 29 units (an ethyl group), respectively. Now, subtract the masses of the known fragments in the alcohol from its molecular mass, and thus determine what fragment is missing. A C—OH unit contributes 29 mass units; the ethyl and methyl groups noted above contribute a total of 44 mass units. A group or groups totaling 29 mass units is (are) required to complete the structure; another ethyl group is consistent with this requirement. A tertiary alcohol containing one methyl and two ethyl groups at the α-carbon, that is, 3-methyl-3-pentanol, fits the data. Notice that an ethyl group could be lost by α-cleavage in two ways; consequently, a peak corresponding to a mass loss of 29 units, that is, a peak at m/z = 73, should be particularly strong, as observed.

$$H_3C—\underset{CH_2CH_3}{\overset{OH}{C}}—CH_2CH_3 \quad \text{3-methyl-3-pentanol}$$

Solutions to Additional Problems

12.20 The three factors that determine the wavenumber, or position, of an infrared absorption are: strength of the bonds involved, masses of the atoms involved, and the type of vibration.

12.21 The intensity of an infrared absorption is affected by (1) the number of absorbing groups and (2) the size of the dipole moment change when the molecule undergoes the bond vibration. The number of absorbing groups, in turn, depends on (1) the number of groups of interest within a given molecule and (2) the concentration of molecules in the sample.

12.22 (a) Catalytically hydrogenate the alkene, and confirm the reaction by observing loss of the C═C stretching absorption at 1660–1675 cm^{-1}.

CH_3 — $\xrightarrow{H_2,\ Pt/C\ catalyst}$ — CH_3

1-methylcyclohexene methylcyclohexane

(b) Carry out a Williamson ether synthesis:

$$CH_3(CH_2)_4CH_2OH \xrightarrow{NaH} \xrightarrow{CH_3I} CH_3(CH_2)_4CH_2OCH_3$$

1-hexanol hexyl methyl ether (1-methoxyhexane)

Confirm the reaction by observing loss of the O—H stretch of the alcohol in the 3200–3400 cm^{-1} region of the IR spectrum.

12.23 The principle is that compounds which have different physical properties have different IR spectra.

(a) Because 3-pentanol and racemic 2-pentanol are constitutional isomers, they have different IR spectra.

(b) Because enantiomers have identical physical properties, the IR spectra of (*R*)- and (*S*)-2-pentanol are identical.

(c) This one is a little tricky! The two chair conformations of cyclohexanol are diastereomers and, because diastereomers have different physical properties, the two conformations also have different IR spectra! The problem is *observing* the two conformations independently. Any sample of cyclohexanol at room temperature is a mixture of these diastereomeric conformations, and the IR spectrum of cyclohexanol is therefore a spectrum of the mixture, with each conformation's spectrum present in proportion to its concentration.

12.24 Spectrum (1) has no peaks characteristic of alkenes or alcohols, but it does have a strong peak near 1100 cm^{-1} that is a C—O stretching absorption. This is the spectrum of compound (d), dipropyl ether.

Spectrum (2) shows the typical absorptions of a —CH═CH_2 group: the double C—H bending absorption at 910 and 990 cm^{-1}, and the strong C═C stretching absorption at 1640 cm^{-1}. This is the spectrum of compound (a), 1,5-hexadiene.

Spectrum (3) is virtually featureless; the only strong absorptions are the C—H stretching absorptions in the 2800–3000 cm^{-1} region. This is the spectrum of compound (f), cyclohexane.

Spectrum (4) shows both O—H stretching and C═C stretching absorptions, and is the spectrum of compound (c), 1-hexen-3-ol. The double peaks at 910 and 990 cm^{-1} confirm the presence of a —CH═CH_2 group. Compound (g), 3-hexanol, is ruled out because it should not have alkene absorptions in its IR spectrum.

Spectrum (5) has C═C stretching absorption at 1675 cm^{-1} and shows the typical high-wavenumber shoulder for the alkene C—H stretching absorption. The position of the C═C stretching absorption along with the C—H bending absorption near 800 cm^{-1} indicates an alkene with three alkyl substituents on its double bond. This is the spectrum of compound (b), 1-methylcyclopentene. The definite absence of a C—H bending absorption near 970 cm^{-1} rules out compound (e), *trans*-4-octene.

12.25 Only spectrum (2) has the C═C stretching absorption near 1640 cm^{-1} and the high-wavenumber shoulder for the alkene C—H stretching absorption; this is therefore the spectrum of compound *A*, and spectrum (1) is the spectrum of compound *B*.

12.26 (a) Because stronger bonds absorb at higher wavenumber (text p. 507), the order of increasing C—H bond strength is alkane < alkene < alkyne. This deduction can be verified by consulting the relevant C—H bond dissociation energies in Table 5.3, text p. 188; these bond energies are alkane (418 kJ/mol), alkene (460 kJ/mol), and alkyne (548 kJ/mol).

(b) From Eq. 12.11 (text p. 507) and the associated arguments, the stretching frequencies are proportional to the square roots of the force constants κ. The force constant, in turn, are roughly proportional to the bond strengths. We then have

$$\frac{\tilde{\nu}_{\text{alkyne C—H}}}{\tilde{\nu}_{\text{alkane C—H}}} = \frac{3300}{2850} = 1.157 = \sqrt{\frac{548\ \text{kJ·mol}^{-1}}{x\ \text{kJ·mol}^{-1}}}$$

Solving,

$$x = (548\ \text{kJ·mol}^{-1})/(1.157)^2 = 409\ \text{kJ·mol}^{-1}$$

As you can see from Table 5.3, the actual value is 418 kJ·mol^{-1}. Thus, this calculation gives an answer that is within 2% of the accepted value.

12.27 The solution to the previous problem noted the correlation between bond dissociation energies and IR absorption frequencies. Triple bonds are clearly stronger than double bonds, which are stronger than single bonds. The question is which of the two double bonds is stronger. Table 5.3 on text p. 188 shows that a C═C double bond has a bond dissociation energy of about 715 kJ/mol, and a C═O double bond has a bond dissociation energy of 732 kJ/mol. On this basis, the C═O bond is predicted to absorb at higher frequency. (This is realized in practice; typical C═O absorptions occur in the 1710 cm^{-1} region, whereas C═C absorptions occur in the 1650 cm^{-1} region.) In summary, then, the order of increasing bond strengths and increasing IR absorption frequencies is:

C—C < C═C < C═O < C≡C

12.28 (a) The bond dipole of the S—H bond is much less than that of the O—H bond because the electronegativities of sulfur and hydrogen differ less than do the electronegativities of oxygen and hydrogen (Table 1.1, text p. 9). As a result, the S—H vibration is less active in the infrared, and the corresponding absorption is less intense. The S—H absorption occurs at lower frequency mostly because the S—H bond is much weaker than the O—H bond; compare the bond dissociation energies for H—SH and H—OH in Table 5.3, text p. 188. (The difference between the masses of sulfur and oxygen has little effect on the vibration frequency because the mass of the smaller atom, hydrogen, is the major determinant of the vibration frequency, as shown by Eq. 12.13 on text p. 508.)

(b) Compound *B* has a typical strong, broad O—H stretching absorption in the 3200–3400 cm^{-1} region, and is therefore the alcohol. Compound *A*, then, is the thiol. Note the considerably weaker S—H stretching absorption in the spectrum of compound *A* at about 2530 cm^{-1}, as suggested by part (a) of the problem.

12.29 (a) Loss of a propyl radical from the molecular ion of 3-methyl-3-hexanol by α-cleavage gives a fragment with m/z = 73.

CH_3CH_2—C($\ddot{O}H^{+\bullet}$)(CH_3)—$CH_2CH_2CH_3$ ⟶ CH_3CH_2—C(═$\ddot{O}H^+$)(CH_3) + $\bullet CH_2CH_2CH_3$

molecular ion of 3-methyl-3-hexanol

m/z = 73

(b) The parent compound has molecular mass = 115. The fragment with $m/z = 72$ corresponds to a loss of 43 mass units, which is accounted for by loss of a propyl radical by an α-cleavage mechanism that is very much like the mechanism shown in part (a):

$$\underset{\text{molecular ion}}{CH_3CH_2-\overset{H_2\ddot{N}^{\bullet+}}{\underset{CH_3}{C}}-CH_2CH_2CH_3} \longrightarrow \underset{m/z = 72}{CH_3CH_2-\overset{H_2\ddot{N}+}{\underset{CH_3}{\overset{\|}{C}}}} + \bullet CH_2CH_2CH_3$$

(c) 1-Pentanol has molecular mass = 88; therefore, a fragment with $m/z = 70$, which must be an odd-electron ion, could result from the loss of H_2O (18 mass units.) A hydrogen-transfer mechanism followed by loss of H_2O, exactly as shown for 1-heptanol in Eq. 12.27 on text p. 527, accounts for this ion.

$$\underset{\text{molecular ion of 1-pentanol}}{CH_3(CH_2)_2\overset{H}{C}H-\overset{\bullet\ddot{O}H}{C}H_2} \longrightarrow CH_3(CH_2)_2\overset{\bullet}{C}H-\overset{H-\overset{+}{\ddot{O}}H}{C}H_2 \longrightarrow \underset{m/z = 70}{CH_3(CH_2)_2\overset{\bullet}{C}H-\overset{+}{C}H_2} + H-\ddot{O}H$$

(d) Fragmentation of neopentane at any one of its four carbon-carbon bonds gives a methyl radical and the *tert*-butyl cation, which has the correct mass:

$$\underset{\text{neopentane}}{H_3C-C(CH_3)_3} \xrightarrow{e^-} H_3C \overset{+}{\vdots} C(CH_3)_3 \longrightarrow H_3C\bullet + \underset{\substack{tert\text{-butyl cation} \\ m/z = 57}}{\overset{+}{C}(CH_3)_3}$$

12.30 (a) Both compounds have the same absorptions, except that the absorptions of compound C in Fig. 12.23 are displaced to lower frequency, an observation that implies a higher mass for the absorbing group; see the discussion of the mass effect associated with Eq. 12.13 on text p. 508. In particular, a peak at 3000 cm^{-1} in compound D, undoubtedly a C—H stretching absorption, is displaced to 2240 cm^{-1} in compound C. Hence, compound C is $CDCl_3$.

Eq. 12.13 on text p. 508 gives the quantitative basis of the mass effect. If the force constants of C—H and C—D bonds are nearly the same (and they are), then the ratio of stretching frequencies of these two bonds should be equal to the square root of the ratios of the mass terms. Eq. 12.13 predicts a ratio of $\sqrt{2} = 1.414$; the more exact equation, Eq. 12.10 (text p. 507), predicts a ratio of 1.36. The actual ratio, (3000/2240) = 1.34, is very close to this prediction.

(b) The two compounds could be distinguished by mass spectrometry by the masses of their parent ions. Both $CHCl_3$ and $CDCl_3$ have four molecular ions (why four?). Each molecular ion peak of $CDCl_3$ lies at one unit higher mass than the corresponding molecular ion peak of $CHCl_3$.

12.31 (a) A mass of 28 units could be accounted for by a molecule of ethylene, $H_2C{=}CH_2$.

(b) Loss of 18 mass units could indicate loss of H_2O. (See the solution to Problem 12.29(c) for an example of such a loss.)

(c) The data indicate the presence of chlorine, which has a mass of 35; loss of 36 mass units, then, could correspond to loss of HCl. (See the solution to Problem 12.18(c) for an example of such a process.)

12.32 First ask what products are expected from the reaction. Two constitutional isomers, both with molecular mass = 88, are anticipated:

$$\underset{\textit{trans}\text{-2-pentene}}{\overset{H_3C\quad\quad H}{C{=}C}\atop{H\quad\quad CH_2CH_3}} \xrightarrow[2)\ H_2O_2/NaOH]{1)\ BH_3/THF} \underset{\text{2-pentanol}}{CH_3\overset{OH}{C}HCH_2CH_2CH_3} + \underset{\text{3-pentanol}}{CH_3CH_2\overset{OH}{C}HCH_2CH_3}$$

Spectrum (b) is consistent with 3-pentanol; the base peak at $m/z = 59$ corresponds to loss of CH_3CH_2—, and there are two ethyl groups in this compound that could be lost by α-cleavage. Spectrum (a) has a base peak at $m/z = 45$ that corresponds to loss of 43 units (a propyl group). 2-Pentanol has a propyl branch that could be lost as a radical by α-cleavage.

When unknown compounds come from a chemical reaction, use what you know about the reaction as a starting point for postulating structures.

12.33 2-Methoxybutane, molecular mass = 88, can lose either a methyl group (15 mass units) or an ethyl group (29 mass units) by α-cleavage. Such losses would give rise to peaks at $m/z = 73$ and $m/z = 59$, respectively. Spectrum (2) contains peaks at these masses, and is therefore the spectrum of 2-methoxybutane. 1-Methoxybutane, in contrast, can lose a propyl group (43 mass units) by α-cleavage to give a peak at $m/z = 45$, as observed in spectrum (1).

$$H_3C\text{-}\wr\text{-}\overset{OCH_3}{C}H\text{-}\wr\text{-}CH_2CH_3 \qquad CH_3CH_2CH_2\text{-}\wr\text{-}\overset{OCH_3}{C}H_2$$

α-cleavage here gives a peak at $m/z = 59$ — α-cleavage here gives a peak at $m/z = 73$

2-methoxybutane

α-cleavage here gives a peak at $m/z = 45$

1-methoxybutane

The odd-electron fragment in the spectrum of 1-methoxybutane at $m/z = 56$ comes from a hydrogen transfer and loss of methanol; the mechanism is similar to that shown in Eq. 12.27, text p. 527, or in the solution to Problem 12.29(c).

12.34 The three peaks correspond to dibromomethane molecules that have, respectively, two atoms of ^{79}Br, one atom each of ^{79}Br and ^{81}Br, and two atoms of ^{81}Br. The probability of any combination of two isotopes is the product of their separate probabilities. Make a table of the possibilities:

	Br #1	Br #2		
$m/z = 172$	79	79	relative probability = $(0.5069)^2 = 0.257$	
$m/z = 174\}$	79	81	relative probability = (0.5069)(0.4931) =	0.250
	81	79	relative probability = (0.5069)(0.4931) =	0.250
			total relative probability	0.500
$m/z = 176$	81	81	relative probability = $(0.4931)^2 = 0.243$	

The probabilities, and therefore the relative abundances, of the molecular ions in dibromomethane are in the ratio 0.257:0.500:0.243, or 1.00:1.95:0.95, or about 1:2:1, as asserted in the problem.

The middle entry of the table above shows that two *different* isotopes can occur within the same compound in two ways. An analogy is the combinations that can be rolled with a pair of six-sided dice. There is only one way to roll a "2" (a "1" on each of the dice), but there are two ways to roll a "3" (a "1" on one of the dice and a "2" on the other, and vice-versa).

12.35 Make a table much like that used in the solution to Problem 12.34.

	Cl #1	Cl #2		
$m/z = 84$	35	35	relative probability = $(0.7577)^2 = 0.574$	
$m/z = 86\}$	35	37	relative probability = (0.7577)(0.2423) =	0.184
	37	35	relative probability = (0.7577)(0.2423) =	0.184
			total relative probability	0.368
$m/z = 88$	37	37	relative probability = $(0.2423)^2 = 0.059$	

Taking the peak at $m/z = 84$ as 100%, the ratios of the peaks are 100%, 64.1%, and 10.3%, respectively.

12.36 (a) The ratio of the M + 1 to the M peak is 1.06/19 = 0.056. If each carbon atom contributes 0.011 to this ratio, then this ratio indicates a compound containing five carbon atoms. Five carbon atoms account for 60 mass units. One oxygen accounts for 16 mass units. (A compound with molecular mass = 86 and five carbons can have no more than one oxygen.) Ten hydrogens are required to fulfill the molecular mass of the compound. The formula is $C_5H_{10}O$.

(b) The abundance of the M + 1 peak relative to the M peak is 2.5/37, or 0.068. At 0.011 per carbon, this accounts for six carbon atoms, or 72 mass units. It is given that the remaining mass is due to hydrogen. Therefore the formula is C_6H_{10}.

The use of isotopic peaks was once important in determining formulas. However, this has been supplanted by the use of *exact masses,* discussed on text p. 529. Typically, a formula of each peak is provided to the investigator as part of the output of a typical mass spectrometer. The use of exact masses was made possible by the advent of high resolution mass spectrometers.

12.37 (a) The molecular mass of ethyl bromide (CH_3CH_2Br) is 108 (for ^{79}Br) and 110 (for ^{81}Br). Consequently, the ion with $m/z = 110$ is the molecular ion of the molecule containing the heavier isotope. It is formed by ejection of an electron from one of the bromine unshared pairs:

$$H_3C{-}CH_2{-}\ddot{\underset{..}{Br}}: \xrightarrow{-e^-} H_3C{-}CH_2{-}\overset{+\cdot}{\underset{..}{Br}}:$$

$m/z = 108, 110$

(b) The ion with $m/z = 108$ is the molecular ion containing the lighter bromine isotope, and it is formed by the process shown in part (a).

(c) The fragment at $m/z = 81$ corresponds to the formation of a bromine cation and an ethyl radical from the compound containing the heavier bromine isotope:

$$CH_3{-}CH_2{-}\overset{+\cdot}{\underset{..}{Br}}: \longrightarrow CH_3{-}\dot{C}H_2 + \overset{..+}{\underset{..}{Br}}:$$

$m/z = 79, 81$

(d) The ion with $m/z = 79$ is the bromine cation formed from the molecular ion containing the lighter bromine isotope, and it is formed by the mechanism shown in part (c).

(e) The fragment at $m/z = 29$ is the ethyl cation formed by inductive cleavage at the carbon-bromine bond:

$$CH_3{-}CH_2{-}\overset{+\cdot}{\underset{..}{Br}}: \longrightarrow CH_3{-}\overset{+}{C}H_2 + :\overset{\cdot}{\underset{..}{Br}}:$$

$m/z = 29$

(f) The ion with $m/z = 28$ is an odd-electron ion formed from the molecular ion by hydrogen transfer followed by loss of HBr:

$$\overset{\cdot}{C}H_2—CH_2 \text{ (with } H \text{ and } \overset{\cdot\cdot +}{\cdot Br:}) \longrightarrow \overset{\bullet}{C}H_2—CH_2(H—\overset{\cdot\cdot +}{Br:}) \longrightarrow \overset{\bullet}{C}H_2—\overset{+}{C}H_2 + H—\ddot{\underset{\cdot\cdot}{Br}}:$$

$m/z = 28$

(g) The ion at $m/z = 27$ could arise by elimination of a hydrogen atom by α-cleavage from the ion with $m/z = 28$; see part (f).

$$H_2\overset{\bullet}{C}—\overset{+}{C}H(H) \longrightarrow H_2C{=}\overset{+}{C}H + \bullet H$$

$m/z = 28$ $m/z = 27$

12.38 The presence of an odd number of nitrogens in a molecule containing, as the other atoms, any combination of C, H, O, and halogen reverses the odd-electron/even-electron mass correlations in Sec. 12.6D on text p. 527 because such a molecule must have an odd molecular mass. (Molecules containing no nitrogen or an even number of nitrogens have even molecular masses.)

(a) The molecular ion of any molecule, because it is formed by ejection of a single electron, must be an odd-electron ion.

(b) A fragment of even mass containing one nitrogen is an even-electron ion. (The ion in the solution to Problem 12.29(b) is an example.)

(c) A fragment ion of odd mass containing a single nitrogen must be an odd-electron ion.

12.39 (a) Sigma (σ) electrons are held closer to the nucleus than π electrons; by the electrostatic law, attraction between particles of opposite charge (such as an electron and a nucleus) is greater when the particles are closer. Hence, there is a greater energy of attraction of the nucleus for σ electrons. This means that more energy must be used to remove a σ electron; hence, a π electron is removed more easily. An equivalent explanation is that the bond dissociation energy of a typical π bond is considerably less than that of a σ bond (see Table 5.3, text p. 188). It should take less energy to eject an electron from a weaker bond—a π bond—than it does from a stronger bond—a σ bond.

(b) The fragment with $m/z = 41$ is the allyl cation, which is formed by α-cleavage of the molecular ion.

$$H_2C{=}CH—CH_2—CH_2CH_2CH_2CH_3 \xrightarrow{-e^-}$$

1-heptene

$$\left[H_2\overset{\bullet}{C}—\overset{+}{C}H—CH_2—CH_2CH_2CH_2CH_3 \longleftrightarrow H_2\overset{+}{C}—\overset{\bullet}{C}H—CH_2—CH_2CH_2CH_2CH_3\right] \longrightarrow$$

molecular ion

$$H_2\overset{+}{C}—CH{=}CH_2 + \bullet CH_2CH_2CH_2CH_3$$

allyl cation
$m/z = 41$

The resonance structures of the molecular ion may seem strange because they are interconverted by movement of a single electron between adjacent carbon atoms. That this is reasonable follows from a consideration of the bonding π molecular orbital of the alkene. The molecular ion is formed by removal of a single electron from that molecular orbital, but the molecular orbital remains with its one electron in the product radical-cation. The single remaining electron is shared between the two carbons involved in the π orbital just as two electrons are shared between the same two carbons in the starting alkene.

You will learn in Chapter 15 that an allyl cation is considerably more stable than an ordinary primary carbocation.

(c) The ion product of Eq. 12.27 (text p. 527) is identical to the molecular ion in part (b) above. Because identical ionic intermediates are involved, mass spectra containing the same peaks resulting from the fragmentation of this ion are observed.

Nuclear Magnetic Resonance Spectroscopy

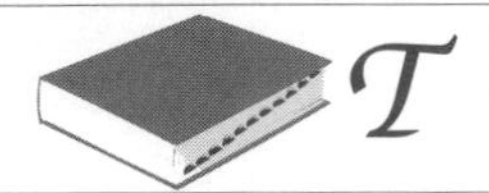

TERMS

absorption (in NMR) 13.1
allylic proton 13.7A
applied magnetic field, $\mathbf{B_0}$ 13.2
carbon NMR (CMR) spectroscopy 13.9
chemical exchange 13.7D
chemical shift 13.1, 13.3B
chloroform-*d*
(deuterochloroform, $CDCl_3$) 13.6
coupled proton 13.4A
coupling constant, *J* 13.4A
D_2O shake 13.7D
delta (δ), ppm 13.1
DEPT technique 13.9
diastereotopic proton 13.3D
doublet (d) 13.4C
doublet of doublets 13.5A
enantiotopic proton 13.3D
first-order spectrum 13.5B
Fourier transform NMR (FT-NMR) 13.11
fundamental equation of NMR 13.2
gauss 13.2
geminal splitting 13.7A
^{1}H NMR 13.1
homotopic proton 13.3D
integral 13.3E
leaning 13.4A
lines (in NMR) 13.1
magnetic field (applied or external) $\mathbf{B_0}$ 13.2
magnetic field (on a proton) $\mathbf{B_p}$.................... 13.2
magnetic resonance imaging (MRI) 13.12
magnetogyric ratio γ 13.2
multiplet (m) 13.4C
multiplicative splitting 13.5
$n + 1$ rule 13.4A
NMR spectrometer 13.2
NMR spectroscopy 13.1, 13.2
NMR tomography 13.12
nuclear magnetic resonance (NMR) 13.2
nuclear spin 13.2
operating frequency, ν_0 .... 13.1, 13.3B
parts per million (ppm), δ 13.1
proton NMR (PMR) spectroscopy 13.1
proton spin decoupling 13.9
proton-decoupled CMR spectrum 13.9
pulse-FT-NMR 13.11
quartet (q) 13.4C
radiofrequency, rf 13.2
resonance (in NMR) 13.1
shielding parameter 13.3A
singlet (s) 13.4C
solid-state NMR 13.12
spin 13.1, 13.2
spin flipping 13.2
spin state 13.2
spin state energy difference $\Delta\epsilon$ 13.2
splitting 13.4
splitting diagram 13.5A
tetramethylsilane (TMS) 13.1
triplet (t) 13.4C
vinylic proton 13.7A

CONCEPTS

I. NMR SPECTROSCOPY

A. Introduction

1. Some nuclei have the property of spin, which is analogous to the spin of electrons.
   a. The hydrogen nucleus ^1H, that is, the proton, has a nuclear spin that can assume either of two values, designated by quantum numbers $+\frac{1}{2}$ and $-\frac{1}{2}$.
   b. A nucleus with spin acts like a tiny magnet; hydrogen nuclei, whether alone or in a compound, respond to a magnetic field.
      i. In the absence of a magnetic field, the nuclear magnetic poles are oriented randomly.
      ii. After a magnetic field is applied, the magnetic poles of some nuclei are oriented parallel to the magnetic field. These nuclei have a spin of $+\frac{1}{2}$.
      iii. After a magnetic field is applied, the magnetic poles of some nuclei are oriented antiparallel to the magnetic field. These nuclei have a spin of $-\frac{1}{2}$, and have somewhat higher energy than the spins with $+\frac{1}{2}$ spin.
      iv. There is a slight excess of nuclei with spin of $+\frac{1}{2}$ because these spins have lower energy.
2. Any nucleus with a spin can be detected by NMR spectroscopy.
   a. Proton NMR (PMR or ^1H NMR) is used to detect protons.
   b. Carbon NMR (CMR or ^{13}C NMR) is used to detect carbon atoms.
   c. Phosphorous NMR (or ^{31}P NMR) is used to detect phosphorous atoms.

B. Physical Basis of NMR Spectroscopy

1. NMR spectroscopy is based on the magnetic properties of nuclei that result from a property called nuclear spin.
2. The presence of a magnetic field causes the two spin states in a given proton p to have different energies; the energy difference $\Delta\epsilon_p$ between the two spin states for the proton is given by the fundamental equation of NMR:

$$\Delta\epsilon_p = \frac{h\gamma_H}{2\pi}\mathbf{B}_p$$

   a. h is Plank's constant (3.99×10^{-23} kJ · s · mol^{-1})
   b. $\mathbf{B}_p$ is the magnitude of the magnetic field at the proton (in gauss)
   c. γ_H is the fundamental constant of the proton, called the magnetogyric ratio.
3. When the nuclei in a magnetic field are subjected to electromagnetic radiation of energy E_p exactly equal to $\Delta\epsilon_p$, this energy is absorbed by some of the protons in the $+\frac{1}{2}$ state.
   a. The absorbed energy causes these nuclei to invert or "flip" their spins to the more energetic state with $-\frac{1}{2}$ spin.
   b. The electromagnetic radiation with this energy consists of radio waves; that is, the radiation is in the radiofrequency, or "rf," region of the electromagnetic spectrum.

$$\text{radiofrequency required for absorption} = \nu_p = \frac{E_p}{h} = \frac{\Delta\epsilon_p}{h} = \frac{\gamma_H}{2\pi}\mathbf{B}_p$$

4. This energy absorption by nuclei in a magnetic field is termed nuclear magnetic resonance (NMR).
   a. NMR can be detected in a type of absorption spectrophotometer called a nuclear magnetic resonance spectrometer (NMR spectrometer).
   b. NMR spectroscopy requires a magnetic field because energy absorption by nuclei requires spin states of different energy. This situation, in turn, exists only in the presence of a magnetic field.
   c. The study of NMR absorption is called NMR spectroscopy.

5. Because almost all organic compounds contain hydrogens, proton NMR spectroscopy is especially useful to the organic chemist.

C.The NMR Spectrometer

1. An NMR spectrometer requires a strong magnetic field to establish the tiny energy differences between the nuclear spin states.
2. The radiofrequency (rf) required for radiation and detection of the absorption is managed in a small wire coil surrounding the sample compartment.
3. The sample is contained in a glass tube that is rapidly spun about its longitudinal axis.
4. In pulse-Fourier transform NMR techniques
   a. all of the proton spins are excited instantaneously with the rf pulse containing a broad band of frequencies, and
   b. the spectrum is obtained by analyzing the emission of rf energy as the spins return to equilibrium.

II. THE NMR SPECTRUM

A. Chemical Shift

1. The local magnetic field $\mathbf{B_p}$ experienced by a proton is different from the external magnetic field $\mathbf{B_0}$ provided by the NMR spectrometer.
   a. In general, **B** is less than $\mathbf{B_0}$. The difference between $\mathbf{B_p}$ and $\mathbf{B_0}$ for a given proton *p* is depends on its shielding parameter σ_p, which has a positive value less than 1.

$$\mathbf{B_p} = (1 - \sigma_p)\mathbf{B_0}$$

   b. Electrons circulating in the vicinity of a proton exert their own magnetic fields that oppose the external field and are said to "shield" the proton.
      i. Electronegative groups near a proton decrease the surrounding electron density and thus decrease the shielding of the proton (give a smaller σ_p).
      ii. Electropositive groups near a proton increase the surrounding electron density and thus increase the shielding of the proton (give a larger σ_p).
2. The chemical shift, in Hz, of any proton is defined as the difference in the frequencies at which the proton and the reference compound TMS absorb electromagnetic radiation.

$$\mathrm{H_3C{-}Si(CH_3)_2{-}CH_3}$$

tetramethylsilane
(TMS)
δ 0.0 by definition

   a. The chemical shift of TMS is set to zero by definition.
   b. TMS is used as a standard not only because it is volatile and inert, but also because its NMR absorption occurs at a lower frequency than the absorption of most organic compounds.

B. Chemical Shift Scales

1. The absorption frequency of any proton *p* relative to that of TMS ($\Delta\nu_p = \nu_p - \nu_{TMS}$) is plotted across the top of the spectrum.
2. Absorption is plotted on the vertical axis in arbitrary units, increasing from bottom to top; that is, absorption peaks are registered as upward deflections in the spectrum.
   a. Absorption peaks are termed absorptions, lines, or resonances.
   b. Absorption positions are always cited by their chemical shifts.
3. The chemical shift $\Delta\nu_p$ for a proton *p* is proportional to the applied field $\mathbf{B_0}$:

$$\Delta\nu_p = \nu_p - \nu_{TMS} = \frac{\gamma_H}{2\pi}\mathbf{B_p} - \frac{\gamma_H}{2\pi}\mathbf{B_{TMS}} = \frac{\gamma_H}{2\pi}(\sigma_{TMS} - \sigma_p)\mathbf{B_0}$$

where σ_{TMS} and σ_p are the shielding parameters for TMS protons and p protons, respectively.

4. The operating frequency ν_0 of the NMR spectrometer is also proportional to the applied field $\mathbf{B_0}$:

$$\nu_0 = \frac{\gamma_H}{2\pi}\mathbf{B_0}$$

5. The chemical shift in parts per million (ppm or δ) for a proton p is defined as the ratio of the chemical shift in Hz ($\Delta\nu_p$) and the operating frequency ν_0. Because both of these quantities are proportional to the applied field, their ratio is independent of the applied field:

$$\delta_p \text{ (ppm)} = \frac{\Delta\nu_p}{\nu_0} = (\sigma_{TMS} - \sigma_p)$$

   a. Each nucleus has a characteristic δ value.
   b. This δ value is independent of the operating frequency (and applied field).
   c. The ppm scale is shown on the lower horizontal axis of the spectrum.
   d. The ppm scale increases from right to left.
   e. NMR spectra taken at different operating frequencies have the same chemical shift scale in ppm but a different chemical shift scale in Hz.
   f. The chemical shifts of most organic compounds are in the range δ 0 to δ 10.

C. Relationship of Chemical Shift to Structure

1. Each chemically nonequivalent set of protons in a molecule gives (in principle) a different resonance in the NMR spectrum.
   a. The chemical shift of a proton provides information about nearby groups.
   b. Factors that increase the proton chemical shift are:
      i. Increasing electronegativity of nearby groups.
      ii. Increasing number of nearby electronegative groups.
      iii. Decreasing distance between the proton and nearby electronegative groups.
2. Neighboring groups affect the chemical shift of a proton through their effects on the magnetic fields caused by the nearby electrons.
   a. Circulation of electrons in the vicinity of a proton causes a magnetic field that opposes the external field.
   b. The more electron density there is near a proton, the greater will be the shielding; thus, the greater will be the shielding parameter σ_p.
   c. Electronegative groups near a proton pull electrons away and decrease the shielding of a proton; thus, such groups decrease σ_p. Thus, protons near electronegative groups have a large chemical shift.
   d. Electropositive groups near a proton increase the surrounding electron density and increase the shielding of the proton; thus, such groups increase σ_p. Hence, protons near electropositive groups have relatively large chemical shifts.
3. The chemical-shift contributions of different groups are approximately additive.
4. The chemical-shift contributions of various groups can be found in Figure 13.4, text page 550, or in Appendix III, text page A-5. (Other, more quantitative, approaches are discussed in Study Guide Link 13.1.)

D. The Number of Absorptions in an NMR Spectrum

1. Chemically equivalent protons have identical chemical shifts.
   a. For two protons to be chemically equivalent, it is necessary but not sufficient that they have the same connectivity relationship to the rest of the molecule, that is, they must be constitutionally equivalent.
      i. It is possible for chemical-shift differences to be so small that they are not detectable.
      ii. It is possible for the resonances of chemically different groups to accidentally overlap.

   b. Homotopic protons are chemically equivalent and have identical chemical shifts under all circumstances.
   c. Enantiotopic protons are chemically equivalent as long as they are in an achiral environment, such as an achiral solvent.
2. Chemically nonequivalent protons in principle have different chemical shifts.
   a. Constitutionally nonequivalent groups are chemically nonequivalent.
   b. Diastereotopic groups are constitutionally equivalent but are chemically nonequivalent; diastereotopic protons in principle have different chemical shifts.
   c. In an optically active chiral solvent, enantiotopic protons in principle have different chemical shifts.
3. The minimum number of chemically nonequivalent sets of protons in a compound of unknown structure can be determined by counting the number of different groups of resonances in its NMR spectrum.

E. Counting Protons with the Integral

1. The size of an NMR absorption is governed essentially by the number of protons contributing to it.
2. The intensity of an NMR absorption is determined by the total area under the peak.
   a. NMR instruments are equipped with an integrating device that can be used to display the integral on the spectrum.
   b. The relative height of the integral (in any convenient units, such as chart spaces or millimeters) is proportional to the number of protons contributing to the peak.
      i. The integrals are calculated by the computer that controls the NMR spectrometer.
      ii. The integrals are displayed in arbitrary units.
      iii. It is the ratios of the integrals of the different protons in a spectrum, not their absolute values, that are important.
   c. The number of absorptions in principle indicates the number of chemically nonequivalent sets of protons.
   d. If the total number of protons is known, the absolute number of protons in each set can be calculated.

F. Splitting and the $n + 1$ Rule

1. Splitting occurs because the magnetic field due to the spin of a neighboring proton adds to or subtracts from the applied magnetic field and affects the total field experienced by an observed proton.
   a. The splitting pattern gives the number of protons adjacent to the protons being observed.
   b. No splitting is observed between chemically equivalent protons.
   c. Splitting is caused by the interaction of proton spins, which is transmitted through the electrons in the intervening chemical bonds.
2. The relationship between the number of lines in the splitting pattern for an observed proton and the number of adjacent protons is given by the $n + 1$ rule.
   a. If there are n adjacent nonequivalent protons, the resonance of an observed proton is split into $n + 1$ lines.
   b. When two sets of protons split each other they are said to be coupled.
   c. With saturated carbon atoms, splitting is observed only between protons on adjacent carbon atoms.
      i. The spacing between adjacent peaks of a splitting pattern (in Hz) is called the coupling constant, J.
      ii. Two coupled protons must have the same J value.
      iii. The coupling constant does not vary with the operating frequency or the applied magnetic field strength.
3. In most cases, a splitting pattern can be discerned by the relative intensities of its component lines which have well-defined ratios (see Table 13.2, text page 560).
   a. The chemical shift of a split resonance in most cases occurs at or near the midpoint of the splitting pattern.
   b. The departure from the normal intensity ratios is called leaning.

   i. Leaning is most severe when the chemical-shift difference between two absorptions that split each other is small.
   ii. Leaning is less pronounced when the chemical-shift difference between the two signals is large.
4. A set of protons can be split by protons on more than one adjacent carbon.
5. NMR spectra are often recorded in abbreviated form:
   a. the chemical shift of each resonance
   b. the integral
   c. the splitting (and coupling constant if known) using the following abbreviations:
      i. s—singlet.
      ii. d—doublet.
      iii. t—triplet.
      iv. q—quartet.
      v. m—multiplet.

G. Complex NMR Spectra

1. The NMR spectra of some compounds contain splitting patterns that do not appear to be the simple ones predicted by the $n + 1$ rule. This behavior can be due to
   a. multiplicative splitting—the splitting of one set of protons by more than one other set of protons. (When all the coupling constants are different, the splitting is in general multiplicative.)
   b. a breakdown of the $n + 1$ rule in certain cases.
      i. Spectra that conform to the $n + 1$ rule are called first-order spectra; some compounds contain splitting patterns that are more complex than predicted by the $n + 1$ rule.
      ii. First-order NMR spectra are generally observed when the chemical shift difference (in Hz) between coupled protons is greater than their coupling constants by a factor of about five or more. (When the chemical shift difference is less than this, non-first-order splitting patterns usually can be anticipated.)
2. When multiplicative splitting occurs and the coupling constants between each pair of coupled protons are equal, the splitting is as predicted by the $n + 1$ rule.
3. The conditions for first-order behavior are more likely to be met using an instrument of higher operating field.
   a. Chemical shifts (in Hz) increase with increasing operating frequency ν_0 (and thus with increasing magnetic field $\mathbf{B}_0$ as well).
   b. Coupling constants do not vary with the operating frequency.
   c. Intensity ratios within splitting patterns more closely approach the idealized values.
4. A splitting diagram is used to visualize multiplicative splitting.
   a. The resonance for each nonequivalent set of protons is drawn as a single, vertical unsplit line at the appropriate chemical shit with a height proportional to its integral.
   b. The different splittings are then applied successively to each line using known values of the coupling constants to obtain the actual spectrum.
   c. The splitting diagram is independent of the order in which the splittings are applied.

III. FUNCTIONAL-GROUP NMR ABSORPTIONS

A. NMR Spectra of Alkanes and Cycloalkanes

1. The protons in a typical alkane are in very similar chemical environments.
   a. The NMR spectra of alkanes and cycloalkanes cover a very narrow range of chemical shifts, typically δ 0.7–1.7.
   b. The splitting in such spectra shows extensive non-first-order behavior.
2. The chemical shifts of protons on a cyclopropane ring are unusual for alkanes; they absorb at unusually low chemical shifts, typically δ 0–0.5.
   a. The unusual chemical shifts are caused by a magnetic field of an electron current present in the cyclopropane ring that is induced by the applied field.

b. This induced field is oriented so that the chemical shifts of the cyclopropane protons are decreased.

B. NMR Spectra of Alkenes

1. A characteristic proton NMR absorption for alkenes is the absorption for the protons on the double bond; such protons are called vinylic protons.
   a. The chemical shifts of vinylic protons (δ 4.5–6.0) are much greater than would be predicted from the electronegativity of the alkene functional group.
   b. The applied field $\mathbf{B_0}$ induces a circulation of the π electrons in closed loops above and below the plane of the alkene. (See Fig. 13.15, text p. 575.)
   c. A magnetic field is induced that opposes the applied field $\mathbf{B_0}$ at the center of the loop.
      i. The induced field augments the applied field at the vinylic protons.
      ii. The vinylic protons are thus deshielded and thus have greater chemical shifts.
   d. The absorptions of internal vinylic protons occur at somewhat greater chemical shifts than absorptions of terminal vinylic protons.
2. Splitting between vinylic protons in alkenes depends strongly on the geometrical relationship of the coupled protons.
   a. Vinylic protons of a *cis*-alkene have smaller coupling constants than those of their *trans*-isomers.
   b. The very weak splitting (called geminal splitting) between vinylic protons on the same carbon stands in contrast to the much larger cis and trans splitting.
3. Another characteristic proton NMR absorption for alkenes is the absorption for the protons on carbons adjacent to the double bond, called allylic protons.
   a. Allylic protons have greater chemical shifts (δ 1.5–2.5) than ordinary alkyl protons.
   b. Allylic protons have considerably smaller chemical shifts than vinylic protons.
4. Small splitting in alkenes is sometimes observed between protons separated by more than three bonds; these long-distance interactions between protons are transmitted by the π electrons.

C. NMR Spectra of Alkyl Halides and Ethers

1. The chemical shifts caused by the halogens are usually in proportion to their electronegativities; chloro groups and ether oxygens have about the same chemical-shift effect on neighboring protons.
2. Epoxides have considerably smaller chemical shifts than acyclic ethers, just as cyclopropanes have smaller chemical shifts than acyclic alkanes.
3. Proton resonances are split by neighboring fluorine (which, like the proton, has a nuclear spin of $\pm\frac{1}{2}$) in the same general way that they are split by neighboring protons; the same $n + 1$ splitting rule applies.
   a. Values of H-F coupling constants are larger than H-H coupling constants.
   b. Coupling between protons and fluorines can sometimes be observed over as many as four single bonds.
4. The common isotopes of chlorine, bromine, and iodine do not cause detectable proton splittings.

D. NMR Spectra of Alcohols

1. Protons on the α-carbons of primary and secondary alcohols generally have chemical shifts in the same range as ethers (δ 3.2–4.2).
2. Since tertiary alcohols have no α-protons, the observation of an OH stretching absorption in the IR spectrum accompanied by the absence of the CH—O absorption in the NMR is good evidence for a tertiary alcohol (or a phenol).
3. The chemical shift of the OH proton in an alcohol depends on the degree to which the alcohol is involved in hydrogen bonding under the conditions that the spectrum is determined.
   a. The presence of water, acid, or base causes collapse of the OH resonance to a single line and obliterates all splitting associated with this proton.
      i. This effect is caused by a phenomenon called chemical exchange—an equilibrium involving chemical reactions that take place very rapidly as the NMR spectrum is being determined.
      ii. This type of behavior is quite general for alcohols, amines, and other compounds with a proton bonded to an electronegative atom.

b. Rapidly exchanging protons do not show spin-spin splitting with neighboring protons.
   i. Acid and base catalyze this exchange reaction, accelerating it enough that splitting is no longer observed.
   ii. In the absence of acid or base, this exchange is much slower, and splitting of the OH protons and neighboring protons is observed.

4. The assignment of the OH proton can be confirmed by what is called "the D_2O shake." (When a drop of D_2O is added to the NMR sample tube and the tube is shaken, the OH protons rapidly exchange with the protons of D_2O to form OD groups on the alcohol and thus become invisible to NMR.)

IV. OTHER ASPECTS OF NMR SPECTROSCOPY

A. Use of Deuterium in Proton NMR

1. Although deuterium has a nuclear spin, deuterium NMR and proton NMR require different operating frequencies.
   a. Deuterium NMR absorptions are not detected under the conditions used for proton NMR.
   b. Deuterium is effectively "silent" in proton NMR.
   c. The coupling constants for proton-deuterium splitting are very small.
   d. Deuterium substitution can be used to simplify NMR spectra and assign resonances.

2. Solvents used in NMR spectroscopy must either be devoid of protons, or their protons must not have NMR absorptions that obscure the sample absorptions.
   a. $CDCl_3$ (chloroform-*d*, or "deuterochloroform") has no proton-NMR absorption, but it has all the desirable solvent properties of chloroform.
   b. Carbon tetrachloride (CCl_4) is a useful solvent because it has no protons, and therefore has no NMR absorption.

B. Carbon-13 NMR

1. Any nucleus with a nuclear spin can be studied by NMR spectroscopy; for a given magnetic field strength, different nuclei absorb energy in different frequency ranges.

2. The NMR spectroscopy of carbon is called carbon NMR or CMR; the only isotope of carbon that has a nuclear spin is ^{13}C.
   a. The relative abundance of ^{13}C suggests that carbon NMR spectra should be about 1.1% as intense as proton NMR spectra.
   b. The resonance of a ^{13}C nucleus is also intrinsically weaker than that of a proton because of the magnetic properties of the carbon nucleus.
   c. Because of these two factors, ^{13}C spectra are about 0.0002 times as intense as ^{1}H spectra.
   d. Special instrumental techniques have been devised for obtaining such weak spectra.

3. Coupling (splitting) between carbons generally is not observed.
   a. The reason is the low natural abundance of carbon-13.
   b. Two adjacent ^{13}C atoms almost never occur together within the same molecule.

4. The range of chemical shifts is large (~200 ppm) compared with that in proton NMR; trends in carbon chemical shifts parallel those for proton chemical shifts. (See Fig. 13.21 on text p. 586.)
   a. Chemical shifts in CMR are more sensitive to small changes in chemical environment.
   b. It is in many cases possible to observe distinct resonances for two carbons in very similar chemical environments.

5. Splitting of ^{13}C resonances by protons (^{13}C-^{1}H splitting) is so large that it adds unwanted complexity to CMR spectra.
   a. Splitting is eliminated by a special technique called proton spin decoupling; a single unsplit line is observed for each chemically nonequivalent set of carbon atoms.
   b. Spectra in which proton coupling has been eliminated are called proton-decoupled carbon NMR spectra.

c. The DEPT technique yields separate carbon-NMR spectra for methyl, methylene, and methine carbons, and each line in these spectra corresponds to a line in the complete CMR spectrum; lines in the complete CMR spectrum that do not appear in the DEPT spectra are assumed to arise from quaternary carbons.

6. CMR spectra are generally not integrated because the instrumental technique used for taking the spectra gives relative peak integrals that are governed by factors other than the number of carbons.
   a. The decoupling technique enhances the peaks of carbons that bear hydrogens.
   b. Peaks for carbons that bear no hydrogens are usually considerably smaller than those for other carbons.

C. NMR Spectroscopy of Dynamic Systems

1. The spectrum of a compound involved in a rapid equilibrium is a single spectrum that is the time-average of all species involved in the equilibrium.
2. The time-averaging effect of NMR is not limited to simple conformational equilibria; the spectra of molecules undergoing any rapid process, such as a chemical reaction, are also averaged by NMR spectroscopy.
3. As the temperature of an equilibrium mixture is lowered, the interconversion rate of the species in the equilibrium is reduced. If the rate is reduced sufficiently, the time-average spectrum of an equilibrium mixture observed at higher temperature is in some cases changed at lower temperature to the superimposed spectra of the individual species in the equilibrium.

D. Other Uses of NMR

1. In pulse-Fourier transform NMR (FT-NMR), a large number of spectra are recorded and stored in a computer.
   a. The random background noise is reduced by adding together large numbers of spectra.
   b. The signals from the individual spectra are mathematically added to give a much stronger spectrum than could be obtained from a single spectrum.
2. Solid-state NMR is used to study the properties of important solid substances.
3. Phosphorus-31 NMR is used to study biological processes.
4. NMR tomography, or magnetic-resonance imaging (MRI), monitors the proton magnetic resonances in signals from water in various parts of the body; using this technique, clinical scientists can achieve organ imaging without using X-rays or other potentially harmful types of radiation.

E. Solving Structure Problems with Spectroscopy

1. Write down everything about the molecular structure that is known.
   a. From the mass spectrum determine, if possible, the molecular mass.
   b. If an elemental analysis is given, calculate the molecular formula and determine the unsaturation number.
   c. Look for evidence in both the IR and NMR spectra for any functional groups that are consistent with the molecular formula; write down any structural fragments indicated by the spectra.
2. Use the proton NMR spectrum and, if possible, the CMR spectrum, to determine the number of nonequivalent sets of carbons and/or protons or to set some limits.
   a. Use the total integral of the entire spectrum and the molecular formula to determine the number of integral units per proton.
      i. Use the integral of each absorption and the result from (a) above to determine the number of protons in each set.
      ii. Use the splitting pattern to determine the number of protons on atoms adjacent to the observed proton.
   b. Determine from the chemical shift of each set which set must be closest to each of the functional groups that are present.
   c. Write down partial structures that are consistent with each piece of evidence, and then write down all possible structures that are consistent with all the evidence.
   d. Estimate the chemical shifts of the protons in each structure, and, if possible, choose the structure that best reconciles the predicted and observed chemical shifts.

3. Rationalize all spectra for consistency with the proposed structure.
4. Most spectra contain redundant structural information.

SCIENTISTS

Dates in brackets refer to Nobel Prize awards.

STUDY GUIDE LINKS

✓13.1 Quantitative Estimates of Chemical Shifts

Figure 13.4 on text p. 550 can be used as a simple method to estimate chemical shifts. However, other relatively simple methods can be used to give more accurate estimates of chemical shift. The basic principle of these methods is that the effect of groups on nearby protons is essentially additive.

The chemical shifts of compounds $H_3C—G$, where G is a substituent group, can be read directly from the second column of Table SG13.1. For example, the chemical shift of CH_3I is δ 2.2.

Table SG13.1 also provides the data that can be used for calculating the chemical shifts of methylene groups in a compound $G^1—CH_2—G^2$, where G^1 and G^2 are attached groups. To estimate the chemical shift, we simply add the *shift contributions* of the two groups to 0.2. That is,

$$\delta(G^1—\mathbf{CH_2}—G^2) = 0.2 + \sigma(G^1) + \sigma(G^2) \qquad \text{(SG13.1)}$$

where $\sigma(G^1)$ and $\sigma(G^2)$ are the shift contributions of the two groups. These shift contributions are given in the last column of Table SG13.1.

For example, suppose you want to calculate the chemical shift of the methylene protons of methylene chloride (CH_2Cl_2). In this example, $G^1 = G^2 = Cl$. Applying Eq. SG13.1, with $G^1 = G^2 = 2.5$, we have δ (Cl—CH_2—Cl) = 0.2 + 2.5 + 2.5 = 5.2. The actual chemical shift of methylene chloride is δ 5.3. (This method gives results that are generally accurate to within a few tenths of a ppm.)

Here are some other examples:

$H_3CO\mathbf{CH_2}C{\equiv}N$: 0.2 + 2.4 + 1.7 = δ 4.3; observed is δ 4.25.
$H_3CO\mathbf{CH_2}OCH_3$: 0.2 + 2.4 + 2.4 = δ 5.0. (Figure 13.1 shows that the actual shift is δ 4.6.)
$H_3CO\mathbf{CH_2}Cl$: 0.2 + 2.5 + 2.4 = δ 5.1; observed is δ 5.5.

When using this method, the effect of a β-halogen, β-ether, or β-OH group can be taken into account by adding another 0.5 ppm. For example, let's estimate the chemical shift of the boldfaced protons in the following compound:

Br—$\mathbf{CH_2}$—CH_2—Cl

In this example, the chlorine is β to the protons of interest. Using Eq. SG13.1, we have

$$\delta = 0.2 + 0.6 + 2.3 + 0.5 = 3.6$$

where the last 0.5 is the contribution for the β-halogen. (The observed shift is δ 3.3.)

A review of this approach can be found in two articles in the *Journal of Chemical Education,* which you should be able to find in your college library:

E. C. Friedrich and K. G. Runkle, *J. Chem. Educ.,* **1984,** *61,* 830–832; *ibid.,* **1986,** *63,* 127–129.

The second article extends the method discussed in the text to the chemical shifts of methine protons.

Another method for estimating chemical shifts makes use of a very useful tabulation of chemical-shift data that was assembled by Tom Curphey (now a professor in the Dartmouth School of Medicine) while a graduate student at Harvard in 1961, and subsequently modified by Harry Morrison (now Professor Emeritus of Chemistry and Dean Emeritus of Science at Purdue). The Curphey-Morrison data are given in Table SG13.2 on pp. 405–406.

(Text continues on p. 404)

Table SG13.1 Group Contributions for Estimating Chemical Shifts of CH_2 Protons Using Eq. SG13.1

Functional group G	δ for CH_3—G	σ(G) for —CH_2—G
—H	0.2	0.0
—CR_3 or —CR_2— (R = H, alkyl)	0.9	0.6
—F	4.3	3.6
—Cl	3.0	2.5
—Br	2.7	2.3
—I	2.2	1.8
—CR=CR_2 (R = H, alkyl)	1.8	1.3
—C≡CR (R = H, alkyl)	2.0	1.4
—OH	3.5	2.6
—OR (R = alkyl)	3.3	2.4
—OR (R = aryl)	3.7	2.9
—SH, —SR	2.4	1.6
—C_6H_5 (Ph, phenyl)	2.3	1.8
—C(=O)—R	2.1 (R = alkyl) 2.6 (R = H)	1.5
—C(=O)—OR (R = alkyl, H)	2.1	1.5
—O—C(=O)—OR	3.6 (R = alkyl) 3.8 (R = aryl)	3.0
—C(=O)—NR_2 (R = alkyl, H)	2.0	1.5

(table continues)

Table SG13.1, continued

Functional group G	δ for CH_3—G	σ(G) for —CH_2—G
—NR—C(=O)—R (R = alkyl, H)	2.8	—
—NR_2 (R = alkyl, H)	2.2	1.6
—N(—)—R (R = aryl)	2.9	—
—C≡N	2.0	1.7

The use of this table can be illustrated by a calculation of the chemical shifts of the various protons of the following compound.

$$\overset{a}{H_3C}—\underset{b}{\overset{Cl}{CH}}—\overset{c}{CH_2}—\overset{d}{CH_2}—Cl$$

Protons *a* are methyl protons; therefore begin the calculation with a base shift of δ 0.9. To this add δ 0.63 for a chlorine β to a methyl group (first line, last column of the chlorine entry). The predicted chemical shift of protons *a* is δ 1.53. (The observed shift is δ 1.60.)

Proton *b* is a methine proton; begin with a base shift of δ 1.55. To this add δ 2.55 for an α-chlorine contribution (third line for chlorine under "Alpha shift") for a predicted chemical shift of δ 4.10. (The observed shift is δ 4.27.)

Protons *c* are methylene protons with *two* β-chlorines. The predicted shift is δ 1.20 + δ 0.53 + δ 0.53 = δ 2.26. (The observed shift is δ 2.15.)

You should try to predict the chemical shift of the protons *d*. (The observed shift is δ 3.72.)

You should notice from the comparisons given above between calculated and observed shifts that the chemical shifts calculated by this (or any other) method are estimates; do not rely on them for accuracy to more than a few tenths of a part per million.

The Curphey-Morrison table gives explicitly the contributions for β shifts as well as the shifts for methine protons. Its advantage over the table in the text is that it can be used to calculate the shifts of methine protons. However, it does not predict very well the shifts of CH_2 or CH protons bound to two or more electronegative groups. (For example, the predicted shift of CH_2Cl_2 is δ 5.8; the method above using Eq. SG13.1 predicts δ 5.2; the observed shift is δ 5.3.)

✓13.2 Approaches to Problem Solving

The approach given on pp. 555–6 the text for solving NMR structure problems is not necessarily the only approach. For example, you might take step 5 first. The point is that *you should be systematic and extract all the information you can from each piece of evidence in the spectrum before starting to write complete structures.*

(Text continues on p. 406)

Table SG13.2 The Curphey-Morrison Table for Estimating Chemical Shifts[a]

Effect of a Substituent G on the Chemical Shifts of Hα and Hβ

$$\text{G—}\underset{\text{H}\alpha}{\overset{|}{\underset{|}{\text{C}}}}\text{—}\underset{\text{H}\beta}{\overset{|}{\underset{|}{\text{C}}}}\text{—}$$

Base Values of Shift Positions: $—CH_3$ δ 0.90; $—CH_2—$ δ 1.20; $—\overset{|}{C}H—$ δ 1.55

Functional group G	Type of hydrogen	Alpha shift	Beta shift
—Cl	CH_3	2.43	0.63
	CH_2	2.30	0.53
	CH	2.55	0.03
—Br	CH_3	1.80	0.83
	CH_2	2.18	0.60
	CH	2.68	0.25
—I	CH_3	1.28	1.23
	CH_2	1.95	0.58
	CH	2.75	0.00
—Aryl	CH_3	1.40	0.35
	CH_2	1.45	0.53
	CH	1.33	0.22
—CH=O and —CR=O	CH_3	1.23	0.25
	CH_2	1.10	0.30
	CH	0.95	–
$—CO_2H$ and $—CO_2R$	CH_3	1.23	0.25
	CH_2	1.05	0.30
	CH	1.05	–
—C=C—	CH_3	0.78	0.05
	CH_2	0.75	0.10
	CH	1.25	0.00
—OH and —OR	CH_3	2.45	0.35
	CH_2	2.30	0.15
	CH	2.20	–
—O—Aryl	CH_3	2.95	0.40
	CH_2	3.00	0.45
	CH	3.30	–

(table continues)

Table SG13.2, continued

Functional group G	Type of hydrogen	Alpha shift	Beta shift
$—OCO_2R$	CH_3	2.88	0.38
	CH_2	2.98	0.43
	CH	3.45	0.36
$—NR_3$	CH_3	1.25	0.13
	CH_2	1.40	0.13
	CH	1.35	0.00
—C≡N	CH_3	1.10	0.45
	CH_2	1.10	0.40
	CH	1.05	0.47
$—NO_2$	CH_3	3.50	0.65
	CH_2	3.15	0.85
	CH	3.05	–
—C≡C—	CH_3	0.90	0.15
	CH_2	0.80	0.05
	CH	0.35	–

[a]Reprinted by permission from the Ph. D. dissertation of T. J. Curphey, Harvard University, 1961, as modified by T. J. Curphey and H. A. Morrison

A good analogy (suggested to the author by a reviewer) is the various approaches to working a jigsaw puzzle. Some people start by working around the borders, and others by looking for unique patterns or colors in the picture on the box in which the puzzle is packaged. The important point is that fitting together pieces of the puzzle at random is not likely to yield a solution in a reasonable amount of time; *your approach must be systematic.*

The dangers of using "recipes" for problem solving were discussed in Study Guide Link 9.6 on p. 265 of this manual. Remember that exceptions to any recipe will occur, and you will be best equipped to deal with such situations by developing your own approach and understanding the basis of each step. Do not let any "recipe" inhibit you from trying your own style of problem solving. Ultimately the style that works for you is the right one.

Needless to say, learning to work any type of problem requires practice. The text provides plenty of practice problems for you; you have to use them to become proficient.

✓13.3 More NMR Problem-Solving Hints

When you solve structure problems with NMR (or any sort of spectroscopy), it is very important that you be *intellectually honest* with yourself. Do *not* force a structure to fit the data. That is, write down each observation from the spectrum and then write down what it tells you. This point can be illustrated with the jigsaw-puzzle analogy in Study Guide Link 13.2. Anyone who has worked a jigsaw puzzle has encountered pieces that look as if they should go together, but do not quite fit; the temptation to force-fit the pieces is often great—but force-fitting never solves the puzzle. When you propose a structure, be able to defend it on the basis of *evidence that you can cite.* If you must guess a structure, then you haven't really solved the problem. If you are stymied by a structural problem, write down *all possible structures* that seem to fit the data. Ask yourself carefully whether each of these structures fits all the information available: chemical shift, integration, and splitting. Sometimes the act of writing down and analyzing the information in this way will lead to a solution. If you must consult the solution without having completely solved the problem, then at least you have defined the points of uncertainty. If, after looking at the solution, you still don't see why your structures are ruled out, seek assistance! You can sometimes learn more from a situation like this, as frustrating as it is, as you can from a correctly worked problem!

Now to some specific tips. The author has observed a tendency among some students to assume that protons which are physically adjacent in a structure must also have absorptions that are adjacent in their NMR spectra. Don't be caught in this trap. Although this may be true in specific cases, in general it is not something that you can rely on.

The last hint is to remind you about common splitting patterns in NMR spectra that appear over and over again. These patterns are so easy to recognize that they *immediately* suggest the presence of certain groups. Let's consider a few of these.

A three-proton triplet at high field along with a two-proton quartet at lower field [Fig. SG13.1(a)] is the earmark of an ethyl group in a structure CH_3CH_2—X, in which the group X has no coupled protons. Ethyl esters and ethyl ethers show this type of pattern. The chemical shift of the quartet provides information about the nature of the group X.

A one-proton septet or multiplet at low field along with a six-proton doublet at high field [Fig. SG13.1(b)] is characteristic of an isopropyl group. Sometimes the one-proton pattern, because of its complexity, is buried among other absorptions, or is otherwise difficult to distinguish. In such cases, the six-proton doublet alone can be used to diagnose an isopropyl group. Likewise, a three-proton doublet at relatively high field is characteristic of the methyl group in the following partial structure:

three-proton doublet

$$\rangle CH—CH_3$$

A three-proton singlet almost always suggests a methyl group. For example, a three-proton singlet in the δ 3–3.5 region in many cases corresponds to a CH_3O— group. A nine-proton singlet a *tert*-butyl group. Be on the lookout for these common patterns when you solve unknowns with NMR.

Several collections of known spectra are available in many libraries. Two excellent collections are the *Aldrich Library of NMR Spectra,* and a newer volume, the *Aldrich Library of Carbon-13 and 300 MHz Proton NMR Spectra.* Although these volumes contain a limited number of spectra, they do encompass a large number of simpler compounds. A more extensive collection is the *Sadtler Standard NMR Spectra,* which includes literally thousands of spectra, most of which are 60 MHz spectra. Because these spectra are taken at lower field, they are in some cases more complex than spectra of the same compounds taken at higher field. (This point is discussed in Sec. 13.5B.) However, this collection nevertheless contains a wealth of relatively simple spectra. (Sadtler has also published a more limited catalog of 300 MHz spectra.) A corresponding collection of carbon-13 spectra also exists. (Carbon NMR is discussed in Sec. 13.9 of the text.) All of these collections contain good indexes, including formula indexes, that allow you to find spectra quickly. A good way to improve your ability to interpret NMR spectra is to look through the formula indexes of these collections for compounds with relatively few carbons—say, eight carbons or fewer for aliphatic compounds—and try to write the structure from the name. Then predict the spectrum, and, finally, check your prediction against the actual spectrum. If your prediction doesn't match the actual spectrum, try to explain why.

13.4 Fourier-Transform NMR

In FT-NMR, the sample is pulsed with a very intense, short pulse of rf radiation that causes all the nuclear spins to absorb energy simultaneously. The decay of absorption is then monitored for a second or two. During this time, all the nuclear spins in the sample return to their normal equilibrium distribution of +1/2 and –1/2 spins and, in doing so, emit radiofrequency energy, each

(Text continues on p. 409)

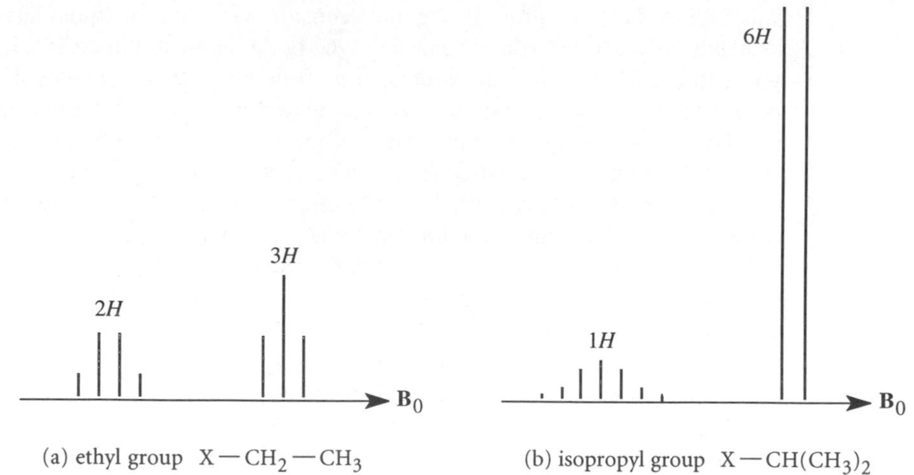

Figure SG13.1 Some splitting patterns commonly observed in proton NMR spectra. In this diagram, absorption peaks are represented as lines.

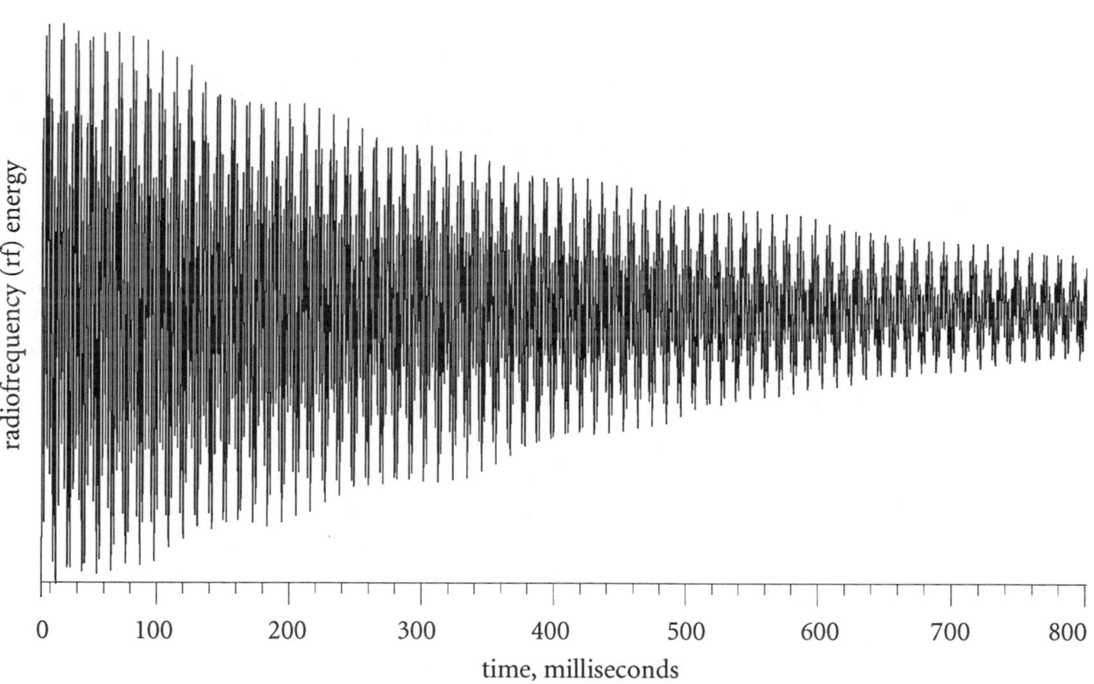

Figure SG13.2 An NMR "ringdown." This is the emission of all frequencies as a plot of radiofrequency (rf) energy versus time following the pulse of dimethoxymethane with a short, powerful burst of rf energy. Fourier transformation of these data give the spectrum in Fig. 13.1 of the text.

at its characteristic frequency. An analogy is the ringing of a bell. When we "pulse" (strike) the bell, all frequencies (pitches) in the bell are activated simultaneously. As the bell rings, the pitches add together to give the characteristic sound of the bell, which dies out over time. An actual graph of this rf energy decay for dimethoxymethane (the compound whose spectrum is shown in Fig. 13.1 of the text) is shown in Fig. SG13.2 on p. 407. Remember, this is the total rf emitted from all the different proton frequencies simultaneously. You can see the die-out over time, as well as the periodic pulsing of the rf, in the same sense that you can hear pulsations in the sound of a bell when it rings. (In fact, this trace is sometimes called a "ringdown.") A mathematical technique called *Fourier transformation* can be used to translate the ringing of the bell over time into a plot of sound intensity *vs.* pitch (frequency). Such a plot would be a spectrum showing the different pitches of which the sound is composed and their relative intensities. The same technique can be applied to the decaying NMR signal over time to convert it into a plot of absorption intensity versus frequency (or field), which is the conventional NMR spectrum. Thus, Fourier transformation of the data in Fig. SG13.2 on p. 408 provides the actual NMR spectrum of dimethoxymethane shown in Fig. 13.1 of the text. This mathematical process, once very laborious, was performed in a fraction of a second on the author's desktop computer.

SOLUTIONS TO PROBLEMS

Solutions to In-Text Problems

13.1 (a) Use Eq. 13.4 on text p. 544:

$$\nu = \frac{\gamma_H}{2\pi}\mathbf{B} = \frac{26{,}753 \text{ rad·gauss}^{-1}\text{·sec}^{-1}}{2\pi \text{ rad}}(117{,}430 \text{ gauss}) = 500 \times 10^6 \text{ sec}^{-1} = 500 \text{ MHz}$$

A 500 megahertz NMR spectrometer would be required.

Whenever π enters into a formula, its dimension are "radians." Remember that the hertz (Hz) is the name of the dimension represented by the unit sec^{-1}.

(b) Use the same equation and solve for **B**:

$$\nu = 900 \times 10^6 \text{ Hz} = \frac{\gamma_H}{2\pi}\mathbf{B} = \frac{26{,}753 \text{ rad·gauss}^{-1}\text{·sec}^{-1}}{2\pi \text{ rad}}\mathbf{B}$$

Solving,

$$\mathbf{B} = 211{,}373 \text{ gauss}$$

Because ν and **B** are proportional, we could also have used the result of the last part and multiplied the value of **B** in the last part by the ratio of frequencies (= 900 MHz/500 MHz).

13.2 (a) Apply Eq. 13.15 on text p. 548 with $\delta_p = 3.35$:

$$\Delta\nu = (\delta_p)(\nu_0) = (3.35)(90) = 301.5 \text{ Hz}$$

Alternatively, take the value of $\Delta\nu$ in Fig. 13.1 (1005 Hz) and multiply it by the ratio of ν_0 values (90/300).

(b) This is the same problem with different numbers, and the answer is $\Delta\nu = 2010$ Hz.

13.3 (a) Subtract two equations like Eq. 13.12 (text p. 547) for two different frequencies, $\nu(1)$ and $\nu(2)$. The difference $\delta(1) - \delta(2)$ is the desired chemical-shift difference. Notice that the chemical shift $\Delta\nu$ is defined as $\nu_H - \nu_{TMS}$. When we take the difference of chemical shifts, the ν_{TMS} subtracts out. Thus,

$$\delta(1) - \delta(2) = \frac{\nu(1) - \nu(\text{TMS})}{\nu_0} - \frac{\nu(2) - \nu(\text{TMS})}{\nu_0} = \frac{\nu(1) - \nu(2)}{\nu_0} = \frac{45}{60} = 0.75 \text{ ppm}$$

(b) At 300 MHz the chemical-shift difference is 45 Hz/300 MHz = 0.15 ppm.

13.4 According to Eq. 13.5, as the shielding goes up, the local field at the proton decreases. This means that the energy separation between the proton spin states decreases. This, in turn (Eq. 13.8a), means that the frequency required for resonance decreases. Of course, δ is proportional to the frequency. Hence, the lowest chemical shift corresponds to the most shielded proton. The protons at δ 1.3 are therefore the most shielded, and those at δ 5.5 are the least shielded.

13.5 Parts (a)–(c) are answered in the text discussion that follows the problem.

(d) Because Si is more electropositive (less electronegative) than any of the other atoms in the table, hydrogens near the Si are *more shielded*. Because chemical shift decreases with increased shielding, this means that $(CH_3)_4Si$ has a smaller chemical shift. A derivative $(CH_3)_xM$, in which M is an element more electropositive than Si, should have a negative chemical shift. $(CH_3)_2Mg$ and $(CH_3)_4Sn$ are two of several possible correct answers.

13.6 (a) The order is $C < B < A$. The protons of dichloromethane (CH_2Cl_2) have the greatest chemical shift, because chlorine is more electronegative than iodine, and because CH_2Cl_2 has the largest number of chlorines. The chemical shift of methylene iodide is greater than that of methyl iodide because two iodines have a greater chemical shift contribution than one.

(b) The order is $B < A < D < C$. The proton in compound *C* (1,1,2,2-tetrachloroethane) has the greatest chemical shift of compounds *A–C* because it is adjacent to two chlorines. In comparing *C* and *D*, compound *C* has the greater chemical shift because of the β chlorines in *C*, as well as the fact that the proton in *C* is a methine proton, whereas the protons in *D* are methylene protons. In comparing *A* to *B*, the protons in *A* are methylene protons that are α to a chlorine, whereas in *B* the protons are methyl protons that are β to chlorine. Finally, in comparing *A* with *D*, the protons in *D* are α to two chlorines, whereas the protons in *A* are α to only one chlorine. The effect of an α-chlorine is much greater than that of a β-chlorine, because chemical shift falls off with distance of the electronegative group from the proton.

(c) The order is $B < C < A$. This follows the electronegativity order of the central atom; the protons in the compound containing the least electronegative element have the smallest chemical shift. (See the solution to Problem 13.5(d).)

13.7 (a) Because protons H^a and H^b are diastereotopic, their chemical shifts are different.

(b) Because protons H^a and H^b are constitutionally nonequivalent, their chemical shifts are different.

(c) Because protons H^a and H^b are diastereotopic, their chemical shifts are different. Because the chair flip interconverts the positions of these two protons, their chemical shifts are the same when averaged out over time. (Sec. 13.8 addresses this point.)

(d) Because protons H^a and H^b are enantiotopic, their chemical shifts are identical.

13.8 The question asks essentially how many chemically nonequivalent sets of protons there are in each case.

(a) This compound has two chemically nonequivalent sets of protons; hence, its NMR spectrum consists of two resonances.

(b) Four resonances are observed.

(c) Three resonances are observed. Notice that the two methyl groups are diastereotopic and hence magnetically nonequivalent.

13.9 (a) This is the structure given in Problem 13.8(a). As the solution to that problem postulated, there are indeed two absorptions.

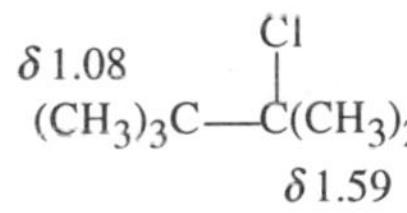

δ 1.08 → $(CH_3)_3C$—C(Cl)$(CH_3)_2$ ← δ 1.59

Another structure consistent with the data is the following:

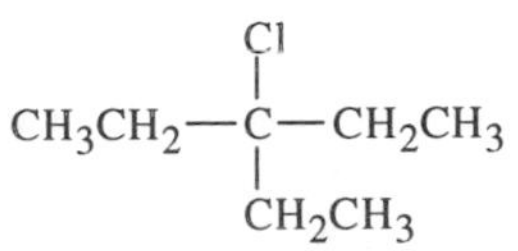

CH_3CH_2—C(Cl)(CH_2CH_3)—CH_2CH_3

This is a satisfactory answer at this point in the chapter. However, as you will learn in Section 13.4 of the text, the spectrum of this compound will consist of a number of lines that result from a phenomenon called splitting, which occurs when protons are on *adjacent carbons*.

(b)

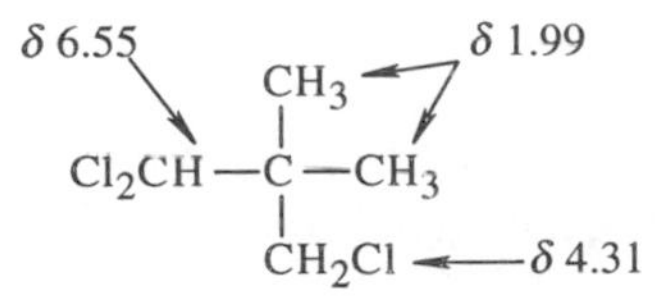

Cl_2CH (δ 6.55)—C(CH_3)(CH_2Cl)—CH_3; CH_3 groups δ 1.99; CH_2Cl δ 4.31

1,1,3-dichloro-2,2-dimethylpropane

13.10 (a) This is a quantitative analysis problem that can be solved by assigning the resonances and using their integrals. The resonance at δ 1.8 is that of *tert*-butyl bromide, and the resonance at δ 2.2 is that of methyl iodide. (These assignments follow from Table 13.1 and Fig. 13.4 on text pp. 549 and 550, respectively.) The ratio of methyl iodide to *tert*-butyl bromide is 15:1. This follows from the fact that there are three times as many protons in the *tert*-butyl bromide molecule as in the methyl iodide molecule; thus, for a given concentration, the *tert*-butyl bromide gives a resonance that is three times as intense as that of methyl iodide.

$$\frac{5 \text{ protons } CH_3I \times (1 \text{ molecule}/3 \text{ protons})}{1 \text{ proton } (CH_3)_3CBr \times (1 \text{ molecule}/9 \text{ protons})} = \frac{15 \text{ molecules of } CH_3I}{1 \text{ molecule of } (CH_3)_3CBr}$$

The mole percent of CH_3I is the number of moles of CH_3I (15) divided by the total moles of both compounds (16) expressed as a percent. Therefore, the mole percent of CH_3I is (15/16) × 100, or 94 mole percent; the mole percent of $(CH_3)_3CBr$ is (1/16) × 100, or 6 mole percent.

(b) The $(CH_3)_3CBr$ impurity in CH_3I is more easily detected, because a given mole fraction of $(CH_3)_3CBr$ gives a resonance that is three times as strong as the resonance for the same amount of CH_3I, as the solution to part (a) of this problem demonstrated.

13.11 In the following discussions, the integral is not mentioned. Of course, it corresponds to the number of protons under observation in each case.

(a) The resonance of the CH_3 protons should be a doublet (one neighboring proton) in the alkyl region, which is δ 0.7–δ 1.7. Because of the β-chlorines this doublet will probably be at the higher end of this range. The —$CHCl_2$ proton is a quartet at considerably greater chemical shift. Fig. 13.4 puts δ 4.2 as the limit on the chemical shift of a CHCl proton; because there are two such protons, the resonance would be at higher shift. (The shift of CH_2Cl_2 in Table 13.1 is δ 5.3; a methine proton would be shifted even more.) Leaning would make the rightmost lines of the quartet and the leftmost line of the doublet, that is, the inner lines of both patterns, somewhat larger than the outer lines. That is, the intensities of the patterns would deviate slightly from those predicted by the idealized ratios in Table 13.2 (p. 560) of the text.

(b) The protons of the $ClCH_2$— group would be a doublet at δ 4.2 or higher. The protons of the —$CHCl_2$ group would be a triplet at considerably greater chemical shift. This chemical shift would be similar to, and probably somewhat greater than, the chemical shift of the —$CHCl_2$ protons in the previous part. (The greater shift would be caused by the β chlorine.) Leaning would make the inner lines of both patterns—the rightmost lines of the triplet and the leftmost line of the doublet—taller than the outer lines.

(c) The methyl groups would be a large doublet in the δ 0.7–1.7 range, probably near the middle. (The β halogen would increase the shift, but methyl protons are usually near the low end of the ranges given in Fig. 13.4, p. 550.) The methine proton would be a septet near δ 3.0.

(d) This compound, 1,3-dimethoxypropane, should have three resonances. The methylene resonances for the boldfaced protons $CH_3O\mathbf{CH_2}CH_2\mathbf{CH_2}OCH_3$ should be a triplet in the δ 3.5 region, because there are two neighboring protons. The central —CH_2— should be a quintet (a five-line pattern), because there are four neighboring protons. This would normally be in the δ 1.0 region, but should appear at somewhat greater chemical shift, perhaps around δ 1.8, because of the β-oxygens. The methyl groups should be one large singlet near δ 3.2.

(e) This compound (oxetane) is really a "tied-down" version of the compound in (d) without the methyl groups. The methylenes closest to the oxygen are a triplet near δ 3.5, and the central —CH_2— should be a quintet (a five-line pattern) near δ 1.8.

(f) The spectrum of this compound consists of two singlets. The —CH_2— resonance occurs at a chemical shift greater than δ 4.2 (because of the β-chlorine), and the methyl groups should appear in the δ 0.7–1.7 region, probably towards the higher end because of the chlorines.

More precise estimates of chemical shift can be made using the methods discussed in Study Guide Link 13.1 on p. 402 of this manual.

↑↑↑ all spins $+^1/_2$
(one way; relative probability 1)

↑↑↓ ↑↓↑ ↓↑↑ two spins $+^1/_2$, one spin $-^1/_2$
(three ways; relative probability 3)

↓↓↑ ↓↑↓ ↑↓↓ two spins $-^1/_2$, one spin $+^1/_2$
(three ways; relative probability 3)

↓↓↓ all spins $-^1/_2$
(one way; relative probability 1)

Figure SG13.3 Spin possibilities for three neighboring protons to accompany the solution to Problem 13.12.

13.12 There are four possibilities for the spin of three neighboring equivalent protons *b*; these are shown in Fig. SG13.3. This diagram shows that the resonance for protons *a* would be split by the three protons *b* into a quartet whose lines are in the intensity ratio 1:3:3:1.

13.13 (a) This is ruled out by several facts. First, there are *eight* nonequivalent sets of protons in this candidate; note that the two CH_3O— groups on the left, as well as the CH_2 protons, are diastereotopic, and hence, nonequivalent. The actual NMR spectrum shows five sets of protons. Also, this candidate should have three methoxy singlets unless two of them accidentally overlap.

(b) This candidate, would also be expected to have *eight* nonequivalent sets of protons. (Can you find them all?) This would also not show two triplets, and there should be four methoxy singlets, and three if the two diastereotopic ones accidentally overlapped, but certainly not two.

13.14 (a)

δ 1.03 δ 1.88 δ 3.40

$H_3C—CH_2—CH_2—Br$

(b)

δ 3.98 δ 5.87

$ClCH_2—CHCl_2$

(c)

$C(CH_2Br)_4$ (drawn as: CH_2Br above, $BrCH_2CCH_2Br$, CH_2Br below)

13.15 (a) The reaction is the peroxide-promoted HBr addition to the alkene (Sec. 5.5 in the text):

$$BrCH_2CH{=}CH_2 + HBr \xrightarrow{\text{peroxides}} BrCH_2CH_2CH_2Br$$

3-bromopropene 1,3-dibromopropane

(b) The chemical shifts and splittings are indeed consistent with this structure. The resonance of the central methylene group is predicted to be a quintet in the δ 0.7–1.7 range (Fig. 13.4); the β-bromines will probably bring this to the high end of this range or beyond; and the resonance for the two methylenes α to the bromines is predicted to be a triplet near δ 3. The splittings agree exactly, and the observed chemical shifts are acceptably close.

13.16 The resonances for protons H^c and H^a will be doublets at their respective chemical shifts, and the resonance of proton H^b will be split into a quartet ($J = 6$ Hz) by the three protons H^c, and each line of this quartet will be split into a doublet ($J = 3.5$ Hz) by the single proton H^a. The resonance for H^b can be analyzed with a splitting diagram as shown in Fig. SG13.4(a). The complete spectrum is sketched in Fig. SG13.4(b).

13.17 (a) Because protons H^b and H^c are diastereotopic, they are chemically nonequivalent; therefore, counting the methyl group, there are four chemically nonequivalent sets of protons. The methyl group is split into a doublet by H^a. The resonance for proton H^a is split into a quartet by the methyl protons, and each line of

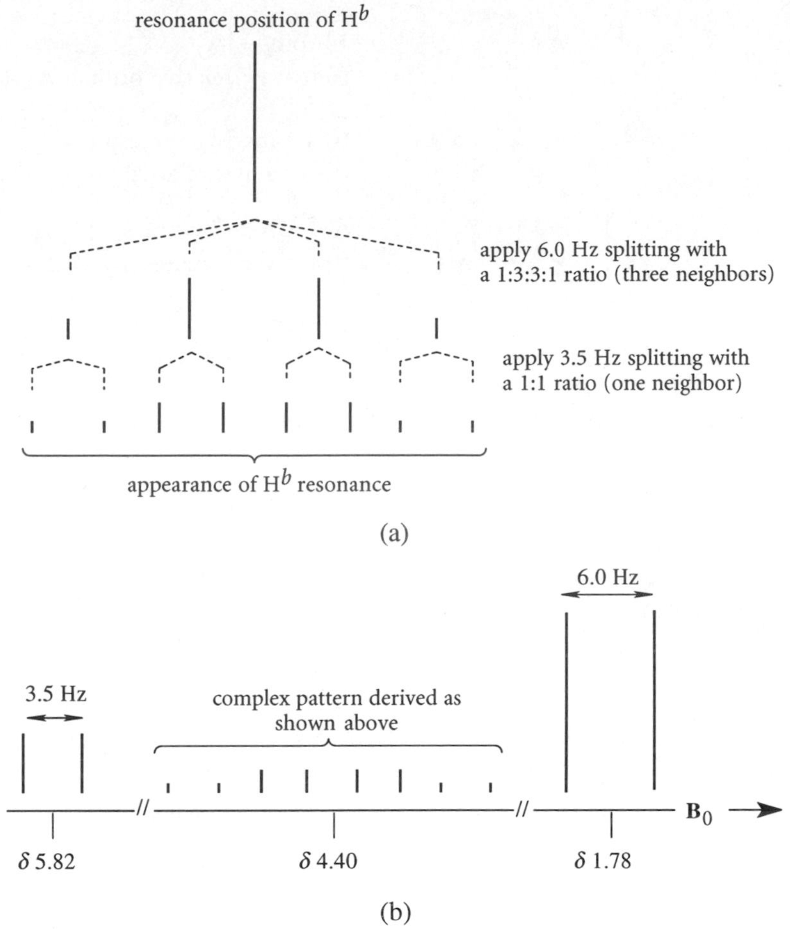

resonance position of H^b

apply 6.0 Hz splitting with a 1:3:3:1 ratio (three neighbors)

apply 3.5 Hz splitting with a 1:1 ratio (one neighbor)

appearance of H^b resonance

(a)

6.0 Hz

3.5 Hz

complex pattern derived as shown above

B_0

δ 5.82 δ 4.40 δ 1.78

(b)

Figure SG13.4 (a) Splitting diagram for proton H^b to accompany the solution to Problem 13.16. (b) The complete NMR spectrum for 1,1,2-trichloropropane. Compared with the diagram in part (a), horizontal and vertical scales have been compressed.

---

this quartet is split into a triplet by H^b and H^c, to give a "quartet of triplets," or twelve lines, for proton H^a. (The triplet is the result of two overlapping doublets with $J_{ab} = J_{ac}$; see Fig. 13.12 on text p. 569 for a related case.) Because H^b and H^c are diastereotopic, they absorb at different chemical shifts. Proton H^b is split by proton H^a into a doublet, and each line of this doublet is split into another doublet by H^c. Therefore, proton H^b is a doublet of doublets, or four equally intense lines. The same is true of H^c—its resonance will be four equally intense lines at a slightly different chemical shift from the resonance of H^b. Of course, several of these patterns may overlap to create a very complex-looking spectrum.

(b) The methyl group, as in part (a), is split into a doublet by H^a. The resonance for proton H^a is split into a quartet by the methyl group; each line of that quartet is split into a doublet by proton H^b, for a subtotal of eight lines; and each line of the eight is split into a doublet by proton H^c, for a grand total of sixteen lines. (Actually, all sixteen lines are discernible in the spectrum of this compound.) The resonances for H^b and H^c would appear as in part (a): each would be four equally intense lines, although the J_{ab} and J_{bc} coupling constants would differ.

The above analysis assumes validity of the $n + 1$ rule. This is a spectrum in which breakdown of the $n + 1$ rule might occur, and this will add further complexity, as discussed in Sec. 13.5B.

13.18 (a) The integration of the δ 3.42 resonance and its splitting show that compound *A* has the partial structure —$CH_2\mathbf{CH_2}Br$. The δ 0.91 resonance indicates the two methyls of an isopropyl group. This means that compound *A* can only be

$(CH_3)_2CHCH_2CH_2Br$ **1-bromo-3-methylbutane (isoamyl bromide)**
(compound *A*)

Compound *B*, from the chemical shifts, has no hydrogens α to the bromine. Hence, this is a tertiary alkyl halide. We see two methyl singlets and an ethyl quartet, which determine the structure:

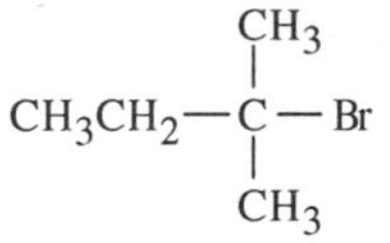

$$CH_3CH_2{-}\overset{\overset{\large CH_3}{|}}{\underset{\underset{\large CH_3}{|}}{C}}{-}Br$$

2-bromo-2-methylbutane
(compound *B*)

(b) Running the spectrum of compound *A* at higher field could simplify the complexity of the δ 1.7-1.8 resonance. This is the combined resonances for the following protons:

$$(CH_3)_2\overbrace{CH-CH_2}^{\delta\ 1.7\text{–}1.8}-CH_2-Br$$

The difference in chemical shift between these protons, in Hz, would increase (double), but the coupling constants would remain unchanged. Thus, the likelihood that Eq. 13.18 would be valid would be increased. In the event that all resonances are first-order, the —CH_2— group would become a quartet and the $\gt$CH— group would remain complex (21 lines even if first-order).

13.19 The δ 1.2-1.5 resonance correspond to the following protons in 1-chlorohexane:

$$H_3C-\overbrace{CH_2-CH_2}^{\delta\ 1.2\text{–}1.5}-CH_2-CH_2-CH_2-Cl$$

1-chlorohexane

The complexity and hence, the inability to assign an exact chemical shift, arises because of complex splitting of the protons of interest. To solve this problem, we would prepare the deuterium-substituted analogs in which this splitting is not present.

$$D_3C-CH_2-CD_2-CH_2-CH_2-CH_2-Cl \qquad H_3C-CD_2-CH_2-CD_2-CH_2-CH_2-Cl$$

compound *A* compound *B*

The NMR spectra of compounds *A* and *B* would show resonances for the protons of interest as singlets.

13.20 The total integral is 2219 + 2451 + 7202 + 21,459 = 33,331. The proton with an integral of 2219 corresponds to $\dfrac{2219}{33{,}331} \times 14 = 0.93$ proton, or 1 proton. By a similar calculation, the proton with the integral of 2451 corresponds to 1.03 proton, or 1 proton. These are vinylic protons, and they are not chemically equivalent. Their very small splitting is consistent only with geminal (that is, $H_2C{=}$) protons. Thus, we have the partial structure

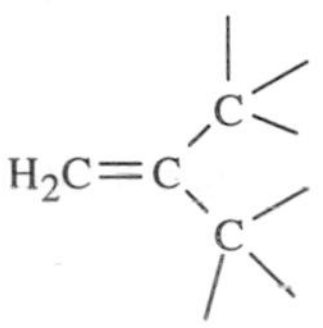

$$H_2C{=}C\begin{matrix}\diagup C\lessdot \\ \diagdown C\lessdot\end{matrix}$$

The resonance at δ 1.8 integrates for 3 protons, and it is in the allylic region. This enriches our partial structure as follows:

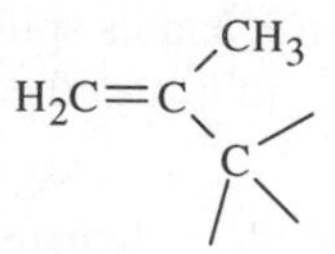

$$H_2C{=}C(CH_3){-}C{<}$$

The resonance at δ 1.03 integrates for nine protons. Because it is a singlet, this can only be a *tert*-butyl group. This completes the structure:

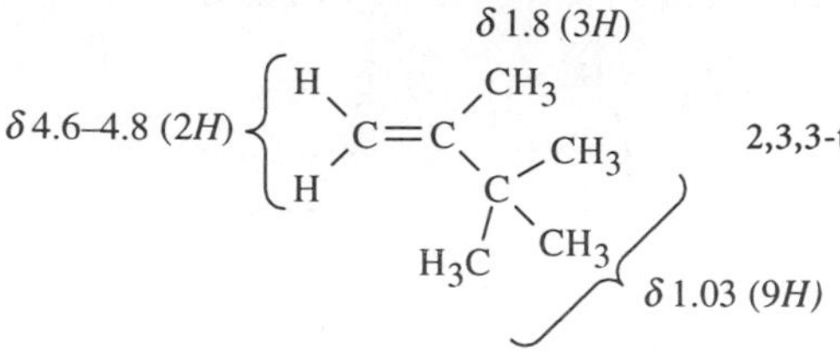

$$\underbrace{(H)_2}_{\delta\,4.6\text{–}4.8\,(2H)}C{=}C(\underbrace{CH_3}_{\delta\,1.8\,(3H)}){-}C(\underbrace{CH_3)_3}_{\delta\,1.03\,(9H)}$$

2,3,3-trimethyl-1-butene

13.21 (a) Nothing but ethyls and one oxygen! This can only be

$$CH_3CH_2{-}O{-}CH_2CH_3 \qquad \text{diethyl ether}$$

(b) The large coupling constants for both sets of protons indicate splitting by fluorine. The fact that they are triplets suggest that they are adjacent to the carbon bearing both fluorines, but not on carbons adjacent to each other; otherwise they would split each other as well. The δ 3.63 protons are clearly α to the chlorine. These considerations demand the following structure:

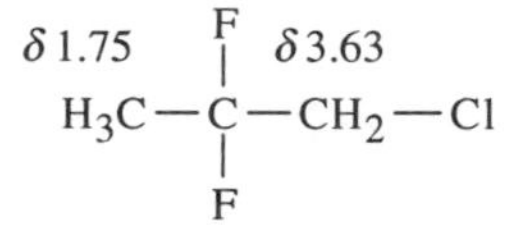

$$\underset{\delta\,1.75}{H_3C}{-}CF_2{-}\underset{\delta\,3.63}{CH_2}{-}Cl \qquad \text{1-chloro-2,2-difluoropropane}$$

Remember that the fluorines are not observed in proton NMR. What would the ^{19}F NMR spectrum look like? *Answer:* a quartet of triplets (or triplet of quartets) with two splittings, coupling constants J = 13 Hz and 17 Hz, respectively.

13.22 The spectrum of ethyl chloride would resemble that of ethyl bromide (Fig. 13.6, text p. 558). It would consist of a typical ethyl pattern—a triplet (for the CH_3) and a quartet (for the CH_2), except that the quartet for the —CH_2— in the ethyl chloride spectrum would be at lower field than it is in ethyl bromide. In ethyl fluoride, the resonance of the methyl group would be a *triplet of doublets.* (Splitting by the —CH_2— group gives three lines; each of these are split into two by the fluorine for a total of six.) The resonance of the —CH_2— group would be a *doublet of quartets.* (Splitting by the fluorine gives a widely spaced doublet; each line of the doublet is split into a quartet by the —CH_3 group.) The chemical shift of the —CH_2— resonance in ethyl fluoride would be expected to be somewhat greater than that of the —CH_2— group in ethyl fluoride.

13.23 (a) The unknown is *tert*-butyl alcohol, $(CH_3)_3C$—OH. Notice the absence of an α-proton.

(b) The exchangeable proton suggests the presence of an alcohol. The resonance at δ 5.40 is a vinylic proton adjacent to two other protons, and the resonance at δ 4.41 arises from two protons adjacent to both the oxygen and the vinyl group. (Their chemical shift is too great to be caused by either the oxygen or the vinyl group alone.) Evidently the sample is wet enough that the splitting between the —OH proton and the α-proton is not observed. 3-Methyl-2-buten-1-ol, $(CH_3)_2C{=}CHCH_2OH$, is consistent with the data.

13.24 At room temperature, the three conformations of 1-bromo-1,1,2-trichloroethane are in rapid equilibrium. Hence, the NMR spectrum of this compound is a singlet whose chemical shift reflects the environment of the proton averaged over time across all conformations. Any changes observed at low temperature would be related to the fact that the interconversion between the different conformations is retarded, and thus certain conformations can be individually observed at the lower temperature. There are three such conformations, two of which are enantiomers:

A ⇌ B ⇌ C

(B and C: enantiomeric conformations)

In conformation *A*, protons H^a and H^b are equivalent and have the same chemical shift. In either one of the enantiomeric conformations, H^a and H^b are chemically nonequivalent because they are diastereotopic. In either of these conformations, the resonance of H^a is split by H^b and is therefore a doublet; likewise, the resonance of H^a is split by H^b and is also a doublet. Hence, the spectrum of conformation *B* should consist of two doublets, that is, four lines of equal intensity, assuming no overlap. Because conformations *B* and *C* are enantiomeric, their spectra will be identical. Hence, the total spectrum should consist of one line for conformation *A*, and two doublets for conformations *B* and *C*, for a total of five lines at low temperature, assuming no overlaps. The relative integrals depend on the amount of form *A* relative to the amounts of forms *B* and *C*; the latter two forms will be present in equal amounts.

13.25 The resonance for the methyl group at room temperature is a singlet. When the temperature is lowered, the resonance of this methyl group should consist of two singlets, one for the conformation of 1-chloro-1-methylcyclohexane in which the methyl group is axial, and one for the conformation in which the methyl group is equatorial. The relative integrals of the two singlets will be proportional to the relative amounts of the two conformations. Because chlorine and methyl are about the same size, there should be about equal amounts of the two conformations.

13.26 Because of its symmetry, 4-heptanol has only four chemically nonequivalent sets of carbons, and hence its spectrum consists of four lines, and is therefore spectrum 1. All of the carbons of 3-heptanol are chemically nonequivalent; hence, the CMR spectrum of this compound consists of seven lines, and is therefore spectrum 2.

13.27 (a) This compound has five nonequivalent sets of carbons and therefore should give five lines, not three as observed.

(b) This compound should have a CMR spectrum with three lines, as observed. Furthermore, the attached hydrogen ratio is also consistent with this structure. What is *not* consistent, however, is the chemical shift information. The δ 112.9 resonance suggests a carbon bound to *more than one* oxygen. In this structure, no carbon is bound to more than one oxygen, whereas in the correct structure, the methine (CH) carbon is bound to three oxygens.

13.28 (a) First draw the structures!

$ClCH_2(H)C{=}C(H)CH_3$ — *trans*-1-chloro-2-butene (*A*)

$H_2C{=}C(Cl)CH_2CH_3$ — 2-chloro-1-butene (*B*)

Two of the most obvious reasons that *trans*-1-chloro-2-butene (*A*) is not consistent with the data are that this alkene would have a spectrum in which the integral indicates a total of *two,* not three, vinylic protons, and *two,* not one, protons on the carbon α to the chlorine.

Among the reasons that 2-chloro-1-butene (*B*) does not fit the data are that it offers no explanation for the δ 4.5 absorption, and it has only two vinylic protons, whereas the observed NMR spectrum indicates three.

When considering possible structures for consistency with NMR data, there is a tendency among beginning students to try to wrestle with splitting data before they have considered other aspects of the data that are more straightforward and at least as powerful. Notice in the foregoing solution that the alternative structures can be clearly ruled out on the basis of relative integral or number of absorptions.

(b) The assignments for the vinylic protons are easily justified by both relative integrals and chemical shifts. Terminal vinylic protons always occur, other things being equal, at a smaller chemical shift than internal vinylic protons, because alkyl substitution results in a downfield shift. Secondly, the integral for the δ 5.0–5.3 resonances is twice that of the δ 5.9–6.0 resonances, and there are twice as many terminal vinylic protons as internal ones. Finally, the δ 5.0–5.3 resonances show splittings consistent with *both* a cis and a trans relationship to other vinylic protons, and the only way this can occur is if there are two such protons—one cis to the remaining vinylic proton and the other trans.

And this brings us to the issue of detailed assignment. Trans splittings are larger than cis splittings. Because the vinylic proton at lower chemical shift (about δ 5.1) in the δ 5.0–5.3 set has the smaller splitting, it is the one that has the cis relationship to the internal vinylic proton. The vinylic proton at δ 5.3 has the trans relationship.

split by both vinylic Hs and by adjacent alkyl H (8 lines in theory; in practice two lines overlap; thus 7 lines)

Cl
CH_3CH C=C H H H
δ 5.9–6.0
larger trans splitting δ 5.3
smaller cis splitting δ 5.1

13.29 It would be easy to believe that the methyl groups of *A* could account for the twelve-proton singlet at δ 1.22, and the two OH groups for the two-proton exchangeable resonance at δ 1.96; but this leaves no hydrogens to account for the δ 1.57 resonance. In other words, structure *A* has two nonequivalent sets of hydrogens, whereas the NMR spectrum indicates at least three. Structure *A* also has two nonequivalent sets of carbons, whereas the CMR spectrum indicates three.

Structure *B* is ruled out by its molecular formula, which is $C_8H_{16}O_2$. But the NMR spectrum is also not consistent with structure *B*, which should have two methyl singlets, each integrating for 6*H*, and one methyl singlet integrating for 3*H*, along with one exchangeable OH proton. Although it would have the required three resonances, the integration ratio (6:3:1) is different from that observed (12:4:2 or 6:2:1). This difference is well within the ability of the spectrometer to differentiate. But the CMR is even more definitive. Structure *B* has five nonequivalent sets of carbons, whereas the CMR spectrum indicates three sets. Even if the methyl resonances overlapped accidentally, which is unlikely, the attached-hydrogen analysis for *B* would predict no carbon with two attached protons.

equivalent
CH_3
O CH_3
H_3C CH_3
H_3C OH
equivalent

structure *B*

Solutions to Additional Problems

13.30 The three pieces of information are

1. the chemical shift, which is used to determine what functional groups are near the proton(s) under observation;
2. the integral, which is used to determine how many protons are being observed relative to the total number in the molecule;
3. splitting, which is used to determine how many protons are on carbons adjacent to the carbon that bears the proton being observed.

13.31 (a) The spectrum of cyclohexane is a singlet. The spectrum of its constitutional isomer *trans*-2-hexene has many resonances, including vinylic resonances in the δ 5–δ 6 region that are not present in cyclohexane.

note vinylic protons not present in cyclohexane

$H_3C(H)C{=}C(H)CH_2CH_2CH_3$

cyclohexane

trans-2-hexene

(b) Only 1-hexene will have a complex vinylic proton absorption that integrates for 25% of the total absorption (that is, 3 protons); the vinylic proton absorption of *trans*-3-hexene will consist of a triplet integrating for 17% of the total absorption (that is, 2 protons).

note three vinylic protons

$H_2C{=}CH{-}CH_2CH_2CH_2CH_3$

1-hexene

note two vinylic protons

$CH_3CH_2(H)C{=}C(H)CH_2CH_3$

trans-3-hexene

(c) The resonance at greatest chemical shift in the spectrum of 1,1-dichlorohexane would be a one-proton triplet; this resonance would be at a chemical shift of at least δ 5.9. 1,6-Dichlorohexane would show a four-proton triplet at greatest chemical shift, but this would be at a chemical shift of about δ 3.3. The resonances at highest chemical shift in the spectrum of 1,2-dichlorohexane would be a complex pattern in the δ 3.3–4.3 region. How complex depends on whether the diastereotopic protons at carbon-1 have different chemical shifts and whether they have different coupling constants with the methine proton at carbon-2.

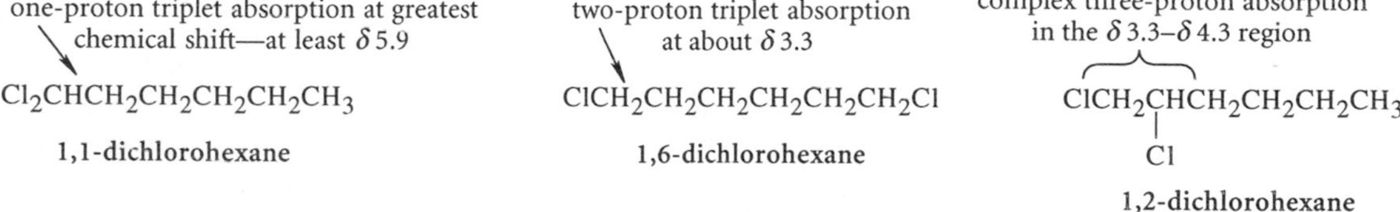

(d) The spectrum of *tert*-butyl methyl ether consists of two singlets; that of isopropyl methyl ether contains a singlet for the methoxy group, but a more complex doublet-septet pattern for the isopropyl group.

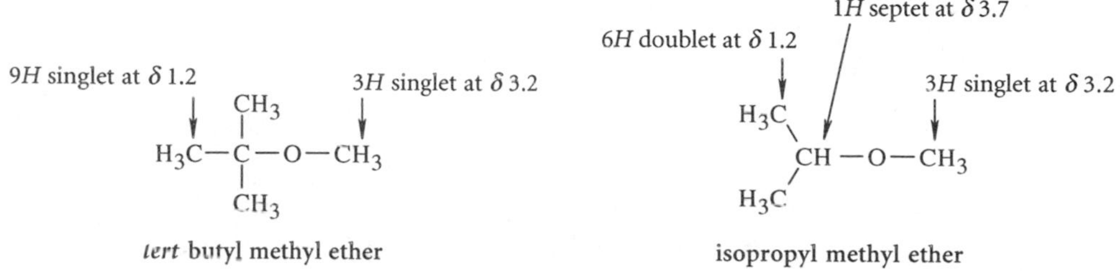

(e) The first compound, $Cl_3CCH_2CH_2CHF_2$, has the absorption at greatest chemical shift and the more complex spectrum; furthermore, this high-shift absorption will show not only H–H splitting, but also the typically large H-F splitting, and will thus appear as a triplet of triplets—that is, a triplet in which each individual line is a more closely spaced triplet. Only the second compound, $CH_3CH_2CCl_2CClF_2$, will show a three-proton triplet at about δ 1 as part of a typical ethyl triplet-quartet pattern.

13.32 (a) NMR spectroscopy requires that the sample be situated in a magnetic field before absorption of electromagnetic radiation can occur. Other forms of absorption spectroscopy do not require the presence of a magnetic field. The field establishes the energy difference between nuclear spins. NMR spectroscopy also differs conceptually in the phenomenon responsible for absorption: the "flipping" of nuclear spins.

(b) Some of the protons change from a lower-energy spin to a higher-energy spin.

(c) The chemical shift in Hz is proportional to the magnitude of the applied field. (See text p. 547; see also the solution to Problem 13.51(a) on p. 427 of this manual.)

(d) The coupling constant J is independent of operating frequency.

(e) The chemical shift in ppm (δ) does not change with operating frequency ν_0 because it is *defined* as the proportionality constant that relates the chemical shift in Hz, $\Delta\nu$, to the operating frequency ν_0. (See Eq. 13.1 on text p. 540 or Eq. 13.12 on text p. 547.) Because $\Delta\nu$ and ν_0 are proportional, δ for a given resonance is the same at any operating frequency.

(f) The chemical shift difference *in frequency units (Hz)* between two protons must be considerably greater than the coupling constant between these protons (in the same units). This relationship is embodied in Eq. 13.18 on text p. 571.

13.33

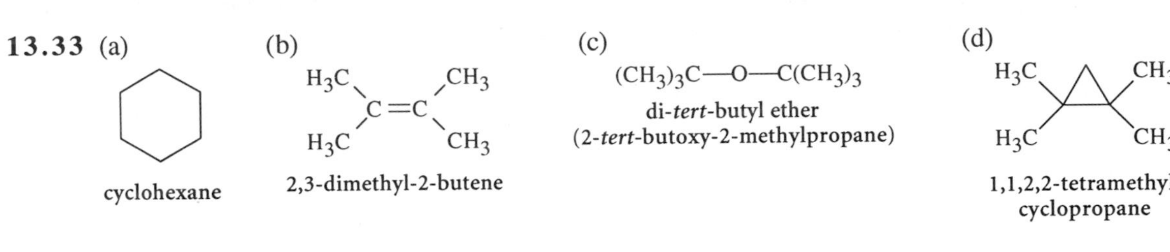

13.34 (a)

CH_3
H_3C—C—CH_3
CH_3

2,2-dimethylpropane (neopentane)

(b)

cyclopentane

(c)

OCH_3
H_3C—CH
OCH_3

1,1-dimethoxyethane

(d) The total integral is 19,986. Given 16 hydrogens, the integral of the δ 0.8 triplet is $\frac{11{,}238}{19{,}986} \times 16H = 9H$; the integral of the δ 1.4 singlet is $1H$; and the integral of the δ 1.5 quartet is $6H$. The D_2O exchange result indicates the presence of an alcohol. Given that the splitting indicates an ethyl group, the integral indicates three equivalent ethyl groups. Because no resonance occurs in the δ 3–4 region, there is no α proton, that is, no CH—O—. Hence, the alcohol is a tertiary alcohol. The compound is

CH_2CH_3
CH_3CH_2—C—OH   3-ethyl-3-pentanol
CH_2CH_3

(e) The spectrum shows indicates vinylic protons with no cis or trans splittings. Hence, a partial structure is

H_2C=C—

The resonances at δ 2.0 and δ 1.8 are nearly singlets and integrate for 2 and 3 protons, respectively. These protons, from their chemical shifts, must be allylic. We now have the partial structure

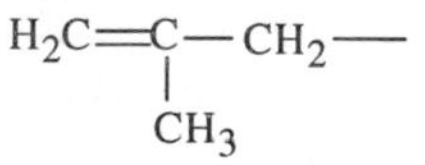

$$H_2C{=}\underset{\displaystyle CH_3}{\underset{|}{C}}{-}CH_2{-}$$

The very small splittings of these groups are evidently allylic splittings (Table 13.3). The remaining resonance is a nine-proton singlet, which can only be a *tert*-butyl group. This completes the structure:

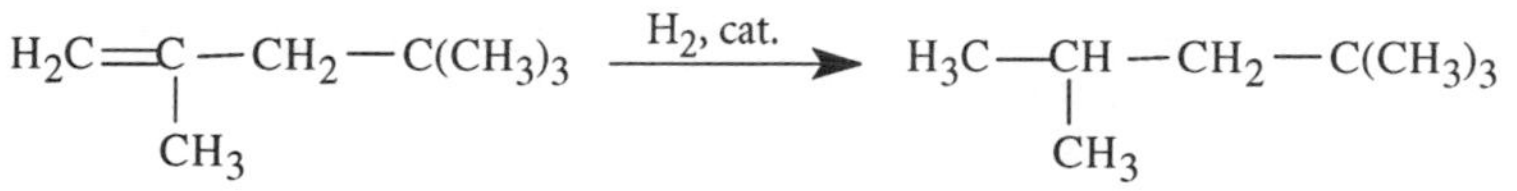

$$\underset{\text{2,4,4-trimethyl-1-pentene}}{H_2C{=}\underset{\displaystyle CH_3}{\underset{|}{C}}{-}CH_2{-}C(CH_3)_3} \xrightarrow{H_2,\ \text{cat.}} \underset{\text{2,4,4-trimethylpentane}}{H_3C{-}\underset{\displaystyle CH_3}{\underset{|}{CH}}{-}CH_2{-}C(CH_3)_3}$$

Clearly, this compound hydrogenates to give 2,4,4-trimethylpentane.

(f) First, convert the integrals into actual numbers of hydrogens:

δ 1.07 (9*H*, s); δ 2.28 (2*H*, d, J = 6 Hz); δ 5.77 (1*H*, t, J = 6 Hz).

The compound has one degree of unsaturation and a *tert*-butyl group. This, plus the requirement for two chlorines and a partial structure —CH_2—CH required by the splitting leaves only the following possibility:

$$(CH_3)_3C{-}CH_2{-}CH{=}CCl_2$$

1,1-dichloro-4,4-dimethyl-1-pentene

(g) The two protons are evidently equivalent (a single chemical shift), and they must be very close to the fluorines (large splitting). The compound is

$BrCH_2CF_2Br$ **1,2-dibromo-1,1-difluoroethane**

Why not the following compound?

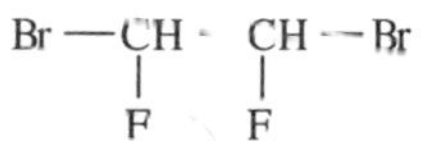

$$Br{-}\underset{\displaystyle F}{\underset{|}{CH}}{-}\underset{\displaystyle F}{\underset{|}{CH}}{-}Br$$ **1,2-dibromo-1,2 difluoroethane**

The proton NMR spectrum of this compound would be a doublet of doublets (why?). In addition, with *two* halogens on the same carbon, the chemical shift would be much greater. Furthermore, one of the H—F coupling constants would be considerably larger because of the location of a fluorine on the same carbon.

(h) Since there are sixteen protons, the integrals account for two, twelve, and two protons, respectively. The compound is

$(CH_3O)_2CHCH_2CH(OCH_3)_2$ **1,1,3,3-tetramethoxypropane**

better known to chemists as malonaldehyde dimethyl acetal. What makes this problem tricky—and the whole point of the problem—is that not all equivalent protons of a given type are on the same carbon. For example, the two CH protons are equivalent and therefore have exactly the same chemical shift; they are split into a triplet by the adjacent CH_2 protons. Likewise, the CH_2 protons are split by the two adjacent protons into a triplet.

(i) The D_2O exchange indicates an alcohol, which must be tertiary. (How do we know this?) The doublet-septet pattern indicates an isopropyl group, and the 6-proton singlet indicates two methyl groups. The structure is

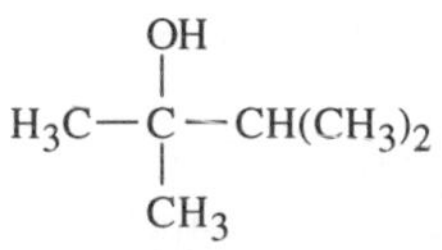

$$H_3C{-}\overset{\displaystyle OH}{\overset{|}{\underset{\displaystyle CH_3}{\underset{|}{C}}}}{-}CH(CH_3)_2$$ **2,3-dimethyl-2-butanol**

13.35 (a) The NMR spectrum of the starting material has two sets of resonances, one of which is in the vinylic-proton region (δ 4.5–5.0). The other would be a singlet in the allylic methyl range, perhaps around δ 1.8. The NMR of the product is a singlet near δ 1.7. To follow the reaction, look for the disappearance of the vinylic proton resonance and one singlet and the appearance of the other singlet.

(b) The NMR spectrum of the starting material is a singlet at δ 1.7; see the solution to Problem 13.33(b). The NMR spectrum of the product should contain an isopropyl doublet-septet pattern at about δ 0.9 and δ 2, respectively, as well as a six-proton methyl singlet near δ 1.4. To follow the reaction, look for the disappearance of the singlet at δ 1.7 and the appearance of a doublet around δ 0.9. (Note that the exact values of the chemical shifts should not concern you.)

13.36 The most obvious difference is that the spectrum of 1-methylcyclohexene should show *one* vinylic proton and a three-proton singlet (neglecting any allylic splitting); whereas the spectrum for 3-methylcyclohexene should show *two* vinylic protons and a three-proton doublet.

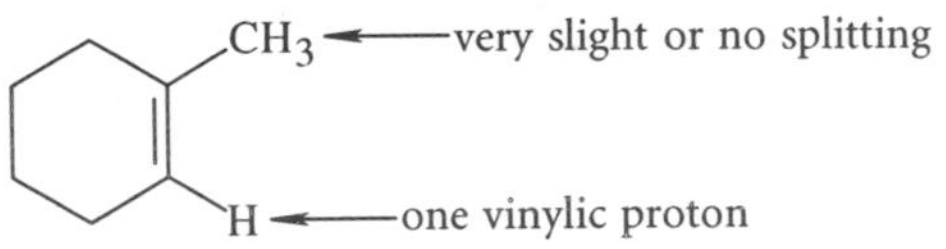

CH_3 ← very slight or no splitting

H ← one vinylic proton

1-methylcyclohexene

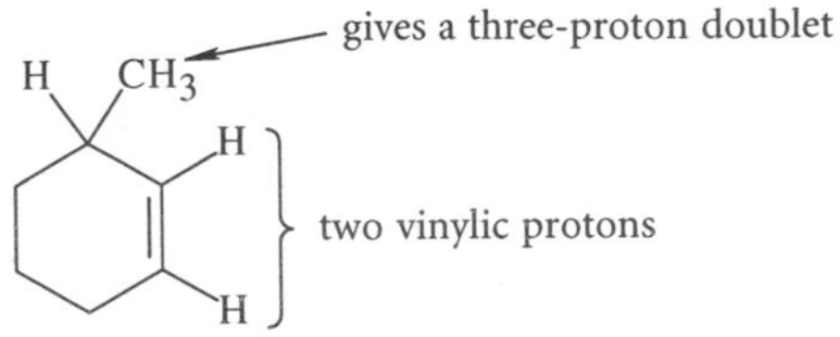

H CH_3 ← gives a three-proton doublet

H, H } two vinylic protons

3-methylcyclohexene

13.37 (a) Because the chair flip is rapid, the relationships of the carbons in these compounds can be decided from the planar projections shown. The compound shown in this part should have six CMR absorptions because there are six chemically nonequivalent sets of carbons (unless there are accidental overlaps between some absorptions). In the structure of this compound shown below, carbons with the same number are enantiotopic, and therefore chemically equivalent. The following pairs of carbons are diastereotopic, and therefore chemically nonequivalent: 3 and 4; 5 and 6; 1 and 2. All other pairs are constitutionally nonequivalent.

3 CH_3; 1, 5, 1; 4 H_3C, 6, 6, 4 CH_3; 2

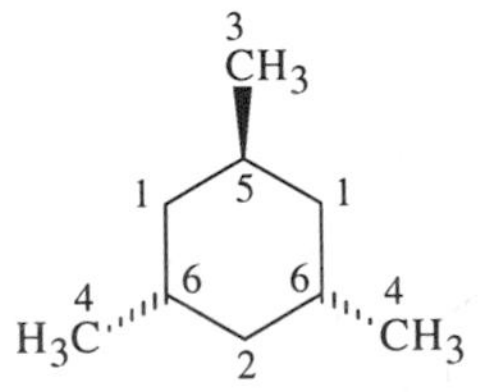

(b) All methyl groups are homotopic; all methine carbons are homotopic; and all methylene carbons are homotopic. (Prove this by a substitution test if this isn't clear!) Each carbon of one type is constitutionally nonequivalent to all carbons of other types. Consequently, the CMR spectrum of this compound should consist of three resonances.

13.38 (a) The NMR spectrum of the first compound, $(CH_3)_2CH—Cl$, should consist of the doublet-septet pattern characteristic of isopropyl groups. The NMR spectrum of the second compound, $(CH_3)_2CD—Cl$, should consist of a singlet at essentially the same chemical shift as the doublet in the first compound. (The splitting between H and D nuclei on adjacent carbons is nearly zero.)

(b) The NMR spectrum of the compound containing deuterium should consist of two triplets of equal intensity. (The coupling constant between H and D nuclei on adjacent carbons is nearly zero.) The NMR spectrum of the other compound, as shown in Fig. 13.8 on text p. 561, should consist of a triplet at low field and a quintet at higher field; the triplet should have twice the integrated intensity of the quintet. Each set of resonances in one spectrum should have the same chemical shift as the corresponding set of resonances in the other.

(c) To decide on the spectra for the first two compounds, first analyze the third compound. The resonance of the methyl group will be a three-proton singlet. The two methylene hydrogens are diastereotopic and are therefore chemically nonequivalent. Hence, they should have different (if only slightly different) chemical shifts, and each will split the other into a doublet. Thus, the resonances for these two protons should consist of a closely spaced leaning doublet of doublets.

(Solution continues on p. 422)

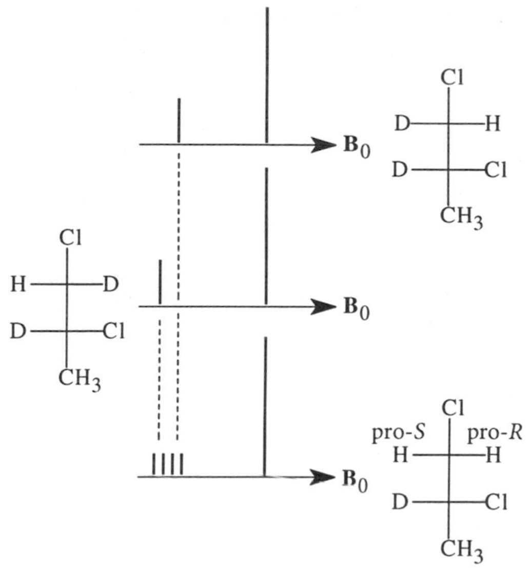

Figure SG13.5. Schematic spectra to illustrate the differences expected in the NMR spectra of the three compounds in Problem 13.38(c). In these spectra, the pro-*S* hydrogen is arbitrarily assigned the greater chemical shift, but the chemical shifts of the two diastereotopic hydrogens might be reversed. Splitting by deuterium is neglected.

Because the first two compounds are diastereomers, they should have different NMR spectra. To begin with, neglect the splitting by deuterium. (Deuterium splitting, which is considered in Problem 13.42, can be eliminated by an instrumental technique.) Each spectrum should consist of a three-proton singlet for the methyl group and a one-proton singlet for the methylene hydrogen. Although the methyl singlets will likely have identical chemical shifts in the two compounds, the chemical shifts of the other hydrogens will differ by essentially the same amount as the chemical shifts of the two methylene hydrogens in the third compound. The resonance of the methylene hydrogen in the first compound will have the same chemical shift as the resonance of the pro-*S* hydrogen in the third compound; and the resonance of the methylene hydrogen in the second compound will have the same chemical shift as the resonance of the pro-*R* hydrogen in the third compound. (This analysis neglects very slight chemical-shift differences that may be caused by the α-deuterium in each case.) A summary of these spectra is shown in Fig. SG13.5.

13.39 Compound *A* is an alkene with twelve equivalent allylic protons and no vinylic protons. Compound *A* is

$(CH_3)_2C{=}C(CH_3)_2$ 2,3-dimethyl-2-butene

(See Problem 13.33(b)).

The IR spectrum of compound *B* indicates a $H_2C{=}C$ group. The NMR spectrum shows a typical isopropyl pattern: a septet/doublet pattern (δ 2.2, δ 1.0, respectively); from its chemical shift it appears that the isopropyl CH group is allylic. The isopropyl group must be attached to a carbon that bears no protons, since the CH of the isopropyl group is split only by the adjacent methyls. The presence of two vinylic hydrogens (δ 4.6–4.7) with very weak splitting is in agreement with the geminal vinylic hydrogens suggested by the IR spectrum, and the three equivalent hydrogens remaining probably correspond to a methyl group. Compound *B* is

$(CH_3)_2CH{-}C(CH_3){=}CH_2$ 2,3-dimethyl-1-butene (compound *B*)

Compound *C* shows a nine-proton singlet characteristic of a *tert*-butyl group. The remaining three protons are vinylic. The IR spectrum shows C═C stretching and C—H bending absorptions that confirm the presence of a $—CH{=}CH_2$ group. Although it is not necessary to interpret the splitting to define the structure, it is very similar to that in Figs. 13.10–13.11 on text pp. 566–567. Compound *C* is

$(CH_3)_3C—CH{=}CH_2$ **3,3-dimethyl-1-butene**

That compound *D* is not an alkene follows from both the chemical data and the NMR absorption. A six-carbon hydrocarbon that is a singlet in the NMR, that has one degree of unsaturation, and that is not an alkene can be only cyclohexane. (You may have seen this one before: Problem 13.33(a).)

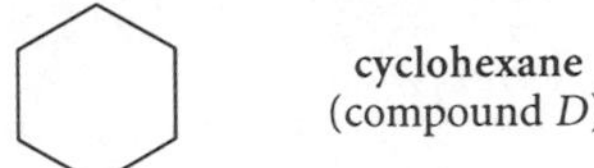

cyclohexane
(compound *D*)

The structures of compound *A* and *D* would be difficult to distinguish on the basis of their proton NMR spectra alone; note that additional reactivity data were provided in this problem. However, distinguishing these two compounds would be trivial with CMR; can you see why?

13.40 The spectrum shows at least four nonequivalent sets of protons in the ratio 1:1:4:6. Let's see how the possible structures stack up against this analysis.

$CH_3CH_2(H)C{=}C(H)CH_2CH_3$
A
cis-3-hexene
3 nonequivalent sets of Hs

$CH_3CH_2O(H)C{=}C(H)CH_2CH_3$
B
(*Z*)-1-ethoxy-1-butene
6 nonequivalent sets of Hs

$H_2C{=}C(CH_2CH_3)_2$
C
2-ethyl-1-butene
3 nonequivalent sets of Hs

$Cl_2CH—CH(OCH_2CH_3)_2$
D
1,1-dichloro-2,2-diethoxyethane
4 constitutionally nonequivalent sets of Hs

$Cl—CH(OCH_2CH_3)—CH(OCH_2CH_3)—Cl$
E
1,2-dichloro-1,2-diethoxyethane
3 constitutionally nonequivalent sets of Hs

Compounds *A–C* are ruled out immediately. The structures of *D* and *E* bear further analysis. The two methine hydrogens of *D* would definitely give a pair of doublets at fairly high chemical shift, as observed. The two methine hydrogens of *E* are completely equivalent and would give a singlet as the absorption at greatest chemical shift. Therefore, compound *E* is ruled out.

Notice that the two CH_2 hydrogens in either ethyl group on both *D* and *E* are *diastereotopic,* and hence, chemically nonequivalent! For example, in the case of *D,*

CH_3 — C(H^a)(H^b) — O — C(H)($CHCl_2$) — O — C(H^c)(H^d) — CH_3

H^a and H^b are diastereotopic; H^c and H^d are also diastereotopic. On the other hand, H^a is enantiotopic to H^c, and H^b is enantiotopic to H^d. (You should verify these statements by a substitution test.) Therefore, H^a and H^c are chemically equivalent, as are H^b and H^d. But H^a and H^b are chemically *nonequivalent,* as are H^c and H^d. Hence, there are *five* nonequivalent sets of hydrogens. However, we should not be surprised to find that the chemical shifts of the diastereotopic protons are not very different.

Let's now consider the splitting patterns of these diastereotopic protons. We've already said that H^a and H^b occur at slightly different chemical shifts. This would account for two absorptions. Each of these is split into a doublet by the other to give four lines. Each of these is split into a quartet by the neighboring methyl group. This gives $4 \times 4 = 16$ lines. Because H^c is chemically equivalent to H^a, and H^d is chemically equivalent to H^b, the analysis is complete. You can actually see all sixteen lines for these diastereotopic hydrogens in the spectrum.

13.41 Compounds *A* and *B* have four sets of nonequivalent carbons and compound *C* has three. Because the given spectrum indicates four nonequivalent sets of carbons we can rule out compound *C*. We now have to decide between compounds *A* and *B*. An important difference in these structures is that only *A* contains a carbon connected to *two* oxygens. From Fig. 13.21, we see that a typical carbon in the environment —C—O— has a chemical shift between δ 40 and δ 80. The maximum chemical shift in the spectrum, however, is nearly δ 100, and this would be consistent with a compound in which a carbon is bound to two oxygens. Hence, the compound is *A*. To further rule out *B*, note that the spectrum contains no methyl (3-proton) carbon in the ether region.

13.42 (a) The nitrogen can take on three values of spin with equal probability: +1, 0, and –1. Hence, the *protons* near a nitrogen of each spin state is shifted accordingly (or not shifted, in the case of spin = 0.) The *proton* NMR spectrum of $^{+}NH_4$ should (and does) consist of three lines of equal intensity.

(b) The reason the CMR spectrum of $CDCl_3$ is a 1:1:1 triplet is the same reason that the proton NMR spectrum of $^{+}NH_4$ is a 1:1:1 triplet [see part (a)]: deuterium, like nitrogen, can have spins of +1, 0, and –1, and splits ^{13}C resonances just as it splits proton signals.

(c) The compound with one deuterium, H_2CD—I, would show a triplet proton resonance with lines that have a 1:1:1 intensity ratio, because the deuterium can have three different values of spin (+1, 0, and –1). Because these are equally probable, they give rise to lines in the proton splitting pattern of equal intensity. The compound D_3C—I has no proton NMR spectrum. For HCD_2—I, the two deuteriums have the following combinations of spin values:

Total spin = +2:	+1, +1 (one way)
Total spin = +1:	0, +1 and +1, 0 (two ways)
Total spin = 0:	0, 0; +1, –1; and –1, +1 (three ways)
Total spin = –1:	0, –1 and –1, 0 (two ways)
Total spin = –2:	–1, –1 (one way)

Each value of the total deuterium spin makes a different contribution to the position of the proton resonance; thus, there are five lines, with relative intensities proportional to relative probabilities of deuterium spin (that is, to the number of different ways that each spin can be achieved): 1:2:3:2:1.

Mass spectrometry could easily determine the identity of each sample, as the samples each differ in molecular mass. Furthermore, only D_3C—I will show a C—D stretching absorption and no C—H stretching absorption in the IR spectrum.

13.43 The IR spectrum and the D_2O shake results indicate that compound *A* is an alcohol, and the absence of a resonance in the δ 3–4 region indicates that the alcohol is tertiary. Treatment of compound *A* with H_2SO_4 yields a compound *B*, which, from its NMR spectrum, is evidently an alkene; note the resonance at δ 5.1. Evidently, the reaction with H_2SO_4 is a dehydration. If so, the molecular mass of 84 for compound *B* means that compound *A* has the molecular mass of compound *B* plus 18 (the molecular mass of H_2O). Hence, the molecular mass of *A* is 84 + 18 = 102.

Now let's turn to the integral for compound *A*. The integrals (from high to low shift) are in the approximate ratio 1:4:6:3. Let's adopt the hypothesis that the O—H proton accounts for 1*H*. If so, then compound *A* contains 14 hydrogens, and its formula is $C_6H_{14}O$, which gives the molecular mass of 102 hypothesized above. Why couldn't the alcohol have a formula $C_7H_{14}O$? Because this would give the wrong molecular mass.

In the NMR of compound *A*, the singlet at δ 1.2 integrates for 6*H*. This can only be two methyl groups. The triplet just below δ 1.0 integrates for 3*H*, and, from its splitting, corresponds to a CH_3CH_2— group. However, the methylene protons in this group must be split by other protons because they are not a simple quartet. The following structure would account for the complex splitting at δ 1.4, as well as the other facts:

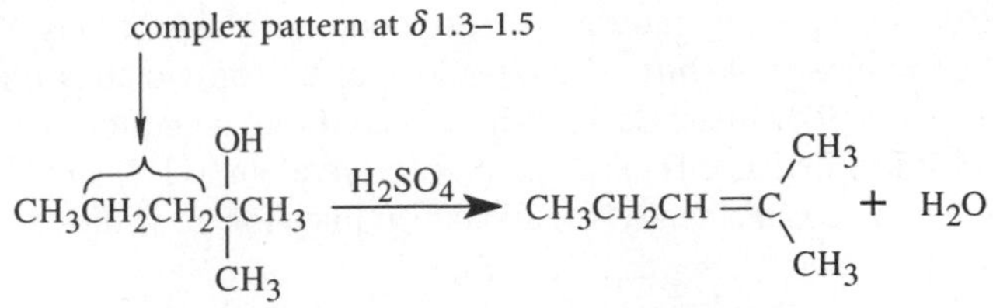

complex pattern at δ 1.3–1.5

$$CH_3CH_2CH_2C(OH)(CH_3)CH_3 \xrightarrow{H_2SO_4} CH_3CH_2CH{=}C(CH_3)_2 + H_2O$$

2-methyl-2-pentanol
compound *A*

2-methyl-2-pentene
compound *B*

The dehydration would then give compound *B* shown above. In the NMR of this compound we expect a single vinylic proton split into a triplet, and this is what we observe at δ 5.1. (Some additional, very small, allylic splitting is also observed.) The NMR of compound *B* is rationalized as follows:

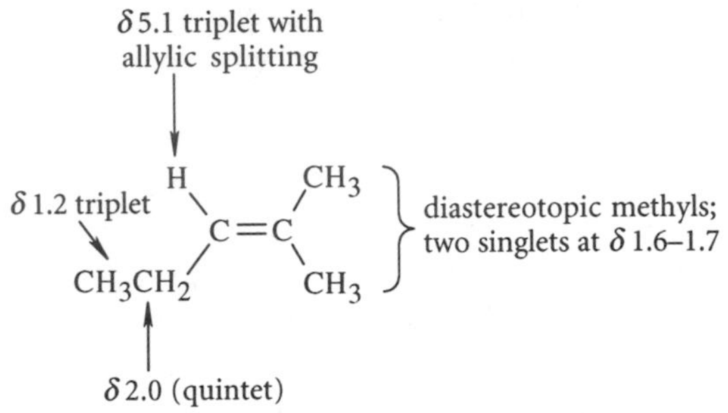

δ 5.1 triplet with allylic splitting (H)

δ 1.2 triplet (CH_3 of CH_3CH_2)

δ 2.0 (quintet) (CH_2)

diastereotopic methyls; two singlets at δ 1.6–1.7 (CH_3, CH_3)

$$CH_3CH_2(H)C{=}C(CH_3)_2$$

13.44 The formula, the IR spectrum, and the D_2O shake indicate that compound *A* is a five-carbon alcohol with one degree of unsaturation. This could be a ring or a double bond, but the NMR spectrum clearly indicates two vinylic protons near δ 5.5. Moreover, the IR shows weak C═C stretch and a strong absorption at 965 cm^{-1} that indicates a *trans*-alkene. The CMR resonances at δ 125.5 and δ 135.5, each with one attached hydrogen, support the diagnosis of an internal alkene. The alcohol is secondary because there is one proton in the α-oxygen region (δ 4.2) and one attached hydrogen for the corresponding carbon at δ 68.8; the latter proton is shifted a little more than normal for this type of proton; hence, this proton is also allylic. Furthermore, this α-proton is adjacent to four other protons. Two methyl groups are adjacent to single hydrogens. One of these methyls is allylic (δ 1.7) and the other is not (δ 1.2). To summarize, we have the following structural elements, which can only be assembled to give *trans*-3-buten-2-ol as the structure. Note that the chirality of the proposed structure is consistent with the fact that compound *A* can be resolved into enantiomers.

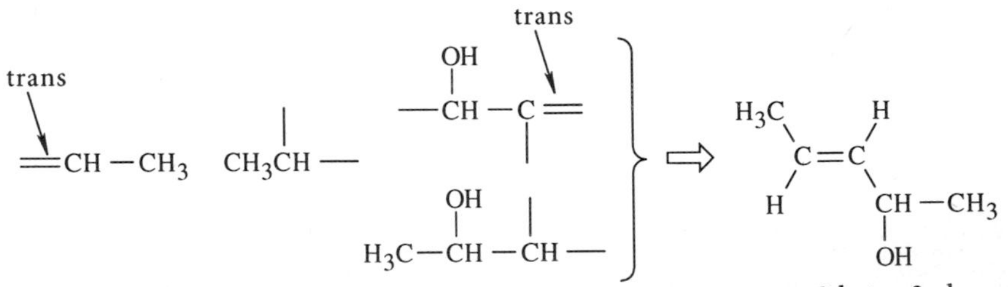

trans: ═CH—CH_3 CH_3CH— —CH(OH)—C═ (trans) H_3C—CH(OH)—CH— ⇒ $H_3C(H)C{=}C(H)CH(OH)CH_3$

trans-3-buten-2-ol
(compound *A*)

13.45 The mass spectrum indicates that the compound contains a single bromine (double molecular ion). The NMR spectrum indicates an ethoxy group (CH_3CH_2O—); this comes from the typical triplet-quartet pattern in which the —CH_2— group is at a typical α-oxygen chemical shift and shows splitting with *only* the CH_3. The spectrum indicates two vinylic hydrogens that split each other with a very small splitting, which, however, is larger than a geminal splitting. The IR spectrum shows no evidence for a trans alkene, nor does it show the characteristic absorption for a ═CH_2 group. Hence, the vinylic hydrogens are cis. The ethoxy group counts for 29 mass units; the ^{79}Br for 79 mass units; two vinylic CH groups would complete the mass of 150. The structure can only be

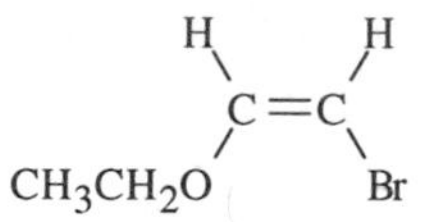

(Z)-1-bromo-2-ethoxyethene

13.46 The sample of 2,5-hexanediol is a mixture of diastereomers, the racemate and the meso compound. Each of these stereoisomers has a CMR spectrum that consists of three resonances, but the two spectra are different, if only slightly so.

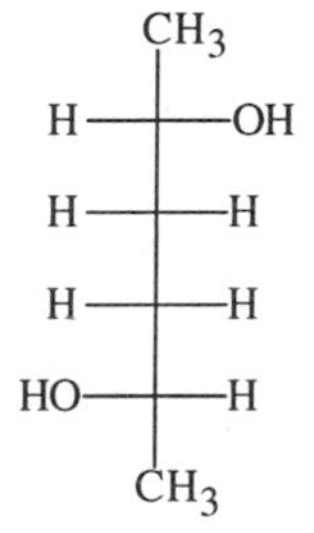

(±)-2,5-hexanediol
(one of the two enantiomers)

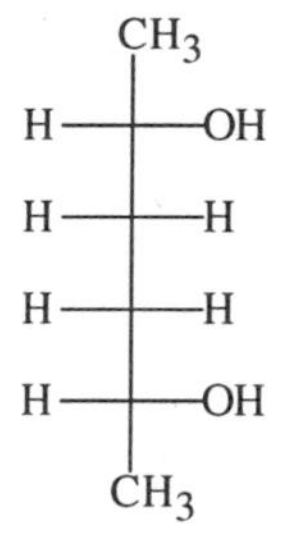

meso-2,5-hexanediol

13.47 (a) As always, we first draw the structure:

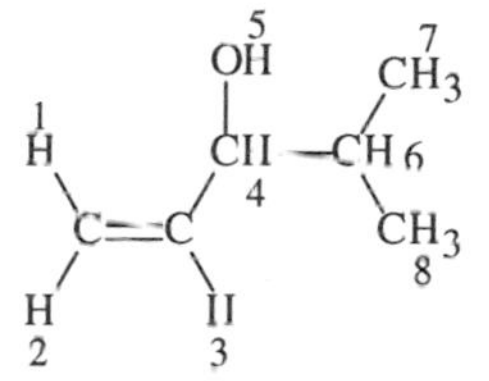

4-methyl-1-penten-3-ol

The proton NMR spectrum of 4-methyl-1-penten-3-ol should contain eight sets of absorptions because the compound contains eight chemically nonequivalent sets of protons (numbered 1–8 in the preceding structure). Notice that protons 1 and 2 are diastereotopic and therefore chemically nonequivalent, as are protons 7 and 8.

(b) Because all carbons are chemically nonequivalent, 4-methyl-1-penten-3-ol should have six carbon resonances in its CMR spectrum. (Notice that the carbons of methyl groups 7 and 8 in the structure shown in part (a), like the protons of these groups, are diastereotopic.)

13.48 First, label the protons for discussion:

$$\overset{3}{\text{Cl—CH}_2}\text{—}\overset{2}{\text{CH}_2}\text{—}\overset{1}{\text{OH}}$$

2-chloroethanol

(a) The NMR spectrum of a very dry sample would show splitting between protons 1 and 2. Hence, the resonance of proton 1 should be a triplet; the resonance of protons 2 should be either a quartet (if $J_{12} = J_{23}$) or a doublet of triplets (if $J_{12} \neq J_{23}$). In a wet sample, the splitting associated with proton 1 is obliterated. In that case, the resonance of proton 1 should be a singlet; the resonance of protons 2, which is split only by protons 3, should be a triplet. The resonance of protons 3 should be a triplet in both the wet and the dry sample.

(b) Following a D_2O shake, the resonance of proton 1 should disappear; and the resonance of protons 2 should be the same as it is in the wet sample, that is, a triplet.

13.49 Vitamin D_3 has six degrees of unsaturation. Each absorption at δ 6.0, δ 6.2, δ 4.8, and δ 5.1 corresponds to 1*H*. We know this by calculating the fraction that each of these absorptions contributes to the total integral and multiplying that fraction by 44, the total number of Hs. The IR shows no evidence of *trans*-alkene absorption. The δ 6.0 and δ 6.2 absorptions could be due to a *cis*-alkene (arrangement *A* below, although the coupling constant of 11.3 Hz, while not impossible for a *cis*-alkene, is on the high side. (See Table 13.3 on text p. 576.)

Another possibility that could explain the same feature of the spectrum is arrangement *B*, which would require two double bonds.

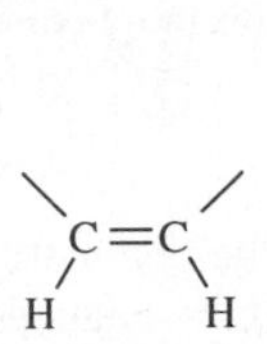

A 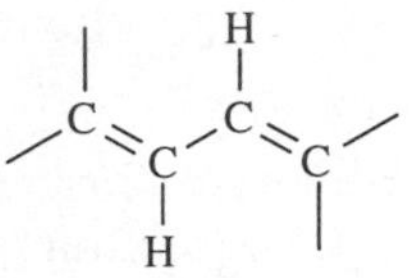 *B*

(It turns out that coupling constants are a function of dihedral angle, and coupling constants for protons in C—H bonds with a dihedral angle of 180° are particularly large.) The chemical shifts and virtual absence of splitting in the δ 4.8 and δ 5.1 absorptions are consistent with a *terminal* $H_2C{=}$. The δ 3.9 absorption corresponds to a single proton α to the alcohol indicated by the IR. The chemical shift of this proton indicates that it might be allylic.

Turning to the low-shift region, the δ 0.6 singlet integrates for 3*H* and is therefore a methyl group on a quaternary carbon—remember, it is unsplit. The apparent doublet of doublets at δ 0.9 is either two overlapping doublets, or two equivalent allylic methyl groups that show splitting to both an adjacent CH and an allylic H. The shift is not great enough for allylic methyls, which typically appear near δ 1.7. Hence these must be two partially overlapping and nonequivalent methyls in the environment H_3C—CH; the other doublet is a similar type of methyl group.

Hence, we conclude that Vitamin D_3 can be classified in one of the two following ways. It has either a cis double bond (arrangement *A*), a $\rangle C{=}CH_2$ group, and four rings; or it has *two* double bonds in arrangement *B*, a $\rangle C{=}CH_2$ group, and three rings. In either case, it has three methyl groups on tertiary carbons, one quaternary methyl group, and a secondary alcohol.

Let's have a look at the structure:

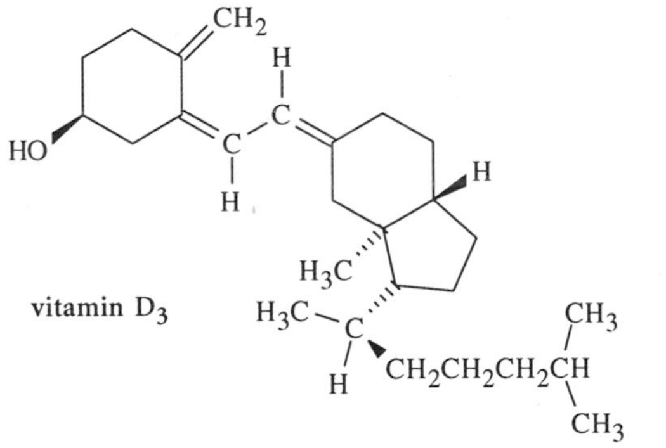

vitamin D_3

We leave it to you to show how the structure is consistent with each feature of the spectrum. In particular, why are there *three* tertiary methyl groups?

13.50 The ^{17}O NMR resonance of ^{17}O-water is split into a triplet by the attached two protons when a small amout of water is dissolved in CCl_4.. The splitting pattern suggests that, in the presence of acid, the ^{17}O resonance is split by *three* protons. Evidently, this splitting is due to the hydronium ion, H_3O^+.

13.51 (a) Start with Eq. 13.4 on text p. 544:

$$\nu_p = \frac{\gamma_H}{2\pi}\mathbf{B_p} \qquad \text{(Text Eq. 13.4, p. 544)}$$

Next, we invoke Eq. 13.5 on text p. 545, which is

$$\mathbf{B_p} = (1 - \sigma_p)\mathbf{B_0} \qquad \text{(Text Eq. 13.5, p. 545)}$$

where σ_p is the shielding constant for proton p. Substituting (13.5) into (13.4), we have

$$\nu_p = \frac{\gamma_H}{2\pi}(1 - \sigma_p)\mathbf{B}_0 \qquad \text{(SG13.2)}$$

Now the chemical shift of any proton p, in Hz, is defined as the frequency difference between its resonance and that of TMS as shown in Eq. 13.11 on text p. 547.

$$\text{chemical shift in Hz} = \nu_p - \nu_{TMS} \qquad \text{(Text Eq. 13.11, p. 547)}$$

The resonance frequency for TMS follows Eq. SG13.2, with p = TMS. That is,

$$\nu_{TMS} = \frac{\gamma_H}{2\pi}(1 - \sigma_{TMS})\mathbf{B}_0 \qquad \text{(SG13.3)}$$

Substituting Eqs. SG13.2 and SG13.3 into text Eq. 13.11 gives

$$\text{chemical shift in Hz} = \Delta\nu_p = \nu_p - \nu_{TMS} = \frac{\gamma_H}{2\pi}[(1 - \sigma_p) - (1 - \sigma_{TMS})]\mathbf{B}_0$$

$$= \frac{\gamma_H}{2\pi}(\sigma_{TMS} - \sigma_p)\mathbf{B}_0 \qquad \text{(SG13.4)}$$

This proves the assertion, because σ_{TMS} and σ_p are both constants; hence, their difference is a constant.

(b) Use Eq. 13.13, text p. 547, which is

$$\nu_0 = \frac{\gamma_H}{2\pi}\mathbf{B}_0 \qquad \text{(Text Eq. 13.13, p. 547)}$$

Substitute this into Eq. SG13.4 from the previous part to give

$$\Delta\nu_p = \nu_0(\sigma_{TMS} - \sigma_p) \qquad \text{(SG13.5)}$$

Using the definition of δ_p (Eq. 13.12, text p. 547), which is

$$\delta_p = \frac{\Delta\nu_p}{\nu_0} \qquad \text{(Text Eq. 13.12, p. 547)}$$

Eq. SG13.5 becomes

$$\delta_p = \sigma_{TMS} - \sigma_p \qquad \text{(SG13.6)}$$

This says that the chemical shift in ppm is nothing more than the difference in shielding constants between the TMS protons and the protons of interest *p*. Of course, we discuss chemical shifts in terms of relative magnetic shielding. TMS protons are *more* shielded than most other protons; hence, $\sigma_{TMS} > \sigma_p$, and δ_p is generally positive, as observed.

13.52 We can use the result of Eq. SG13.6 in the previous problem. Alternatively, we can simply observe that shielding decreases chemical shift. Because a "naked" proton is completely *unshielded* by electrons, its chemical shift should be very large. In terms of Eq. SG13.6, its shielding constant is zero. Hence, its chemical shift is equal to σ_{TMS}, which is a larger number than any chemical shift we are likely to encounter for any proton in a chemical compound, in which electrons are present—that is, for any proton for which $\sigma_p > 0$. Because the range of proton chemical shifts within organic compounds extends to about 11–12 ppm (Fig. 13.4 on text p. 550), we expect the chemical shift of a "naked" proton to be even greater than this. Hence, the first answer, $\delta > 8$, is the correct one.

13.53 Note that the ^{13}C resonances from *unenriched* compounds will be much weaker than those from the isotopically enriched compounds and can be ignored. The proton-decoupled CMR spectrum will be a composite spectrum that consists of the spectra of the individual species present. These species, and their relative abundances, will be as follows ($\overset{*}{C} = {}^{13}C$):

$\overset{*}{C}H_3$—$\overset{*}{C}H_2$—Br Relative probability = (0.5)(0.5) = 0.25 two doublets
CH_3—$\overset{*}{C}H_2$—Br Relative probability = (0.5)(0.5) = 0.25 one singlet
$\overset{*}{C}H_3$—CH_2—Br Relative probability = (0.5)(0.5) = 0.25 one singlet

CH_3—CH_2—Br Relative probability = (0.5)(0.5) = 0.25 no ^{13}C signal

Each resonance will show an apparent "triplet" pattern: the line from the singly-enriched species will be in the center of a doublet for the doubly-enriched species. The singly-enriched species are singlets because there is so little ^{13}C resulting from the natural abundance of ^{13}C at each carbon that it can be ignored. Each line of the doublet will have half the total intensity of the doublet, and therefore half the intensity of the singlet. A diagram of the spectrum is as follows:

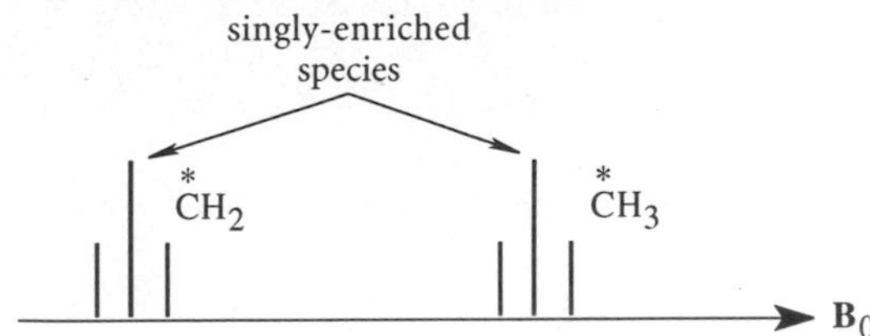

13.54 (a) The resonance of the unpaired electron is split into four lines by the three neighboring protons according to the same $n + 1$ rule that applies to splitting between nuclei.

(b) Use Eq. 13.22 on text p. 584 with $\gamma_n = \gamma_{electron}$ = 17.6 × 10^6 radians·gauss^{-1}·sec^{-1}, and $\mathbf{B_n} = \mathbf{B_0}$ = 3400 gauss. We obtain

$$\nu_{electron} = \frac{\gamma_{electron}}{2\pi}\mathbf{B_0} = \frac{17.6 \times 10^6 \text{ radians·gauss}^{-1}\text{·sec}^{-1}}{2\pi \text{ radians}}(3400 \text{ gauss}) = 9.5 \times 10^9 \text{ sec}^{-1} \approx 10^{10} \text{ Hz}$$

Fig. 12.2 on text p. 499 shows that a frequency of 10^{10} Hz is in the microwave region of the electromagnetic spectrum. Indeed, ESR instruments employ microwave radiation to detect the magnetic resonance of electrons.

13.55 (a) At 25 °C the conformations equilibrate too rapidly to be separately observed in the NMR spectrum; a single resonance is observed that is the "time-average" resonance of all conformations. At low temperature, however, the rates of the internal rotations are considerably reduced, and certain conformations can be separately observed:

enantiomeric conformations have identical NMR spectra

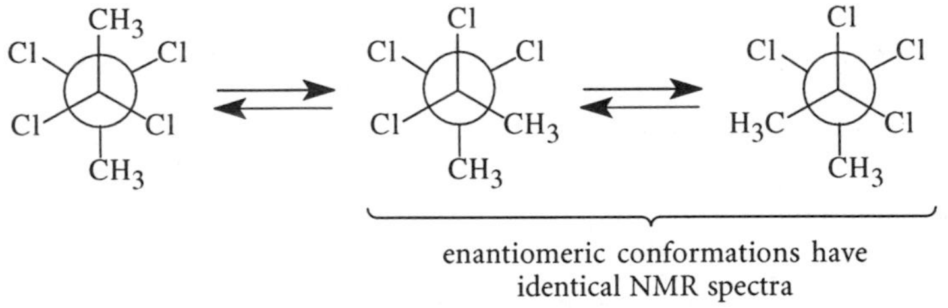

The first of these conformations is the conformational diastereomer of the other two and hence has a different NMR spectrum. The other two conformations are enantiomeric and therefore have identical NMR spectra. Consequently, two resonances are observed: one for the first conformation, and the second for the other two.

(b) The two lines at low temperature have different intensities because the diastereomers are present in different amounts. The intensity of each resonance is proportional to the amount of the conformation of which it is characteristic.

14 Chemistry of Alkynes

TERMS

Term	Section	Term	Section
acetylene	Introduction	kairomone	14.9
acetylenic anion	14.7A	Lindlar catalyst	14.6A
acetylenic carbon	14.3C	pheromone	14.9
acetylenic proton	14.3C	propargyl group	14.1
alkyl anion	14.7A	propargylic carbon	14.3C
alkyne	Introduction	propargylic proton	14.3C
alkynyl group	14.1	radical anion	14.6B
amide anion, $^{-}NH_2$	14.7A	*sp* hybrid orbital	14.2
catalyst poison	14.6A	solvated electron	14.6B
disiamylborane	14.5B	transmetallation	14.7A
enol	14.5A	vinylic anion	14.7A
hydration	14.5A	vinylic radical	14.6B

CONCEPTS

I. INTRODUCTION TO ALKYNES

A. General

1. An alkyne (acetylene) is a hydrocarbon containing a carbon–carbon triple bond; the simplest alkyne is acetylene (ethyne), $H{-}C{\equiv}C{-}H$.
   a. The chemistry of the carbon-carbon triple bond is similar in many respects to that of the carbon-carbon double bond.
   b. The unique chemistry of alkynes is mostly associated with the bond between hydrogen and the triply bonded carbon, the $\equiv C{-}H$ bond.
2. Naturally occurring alkynes are relatively rare and do not occur as constituents of petroleum.
   a. Acetylene can be produced by heating coke (carbon from coal) with calcium oxide in an electric furnace to yield calcium carbide, CaC_2.

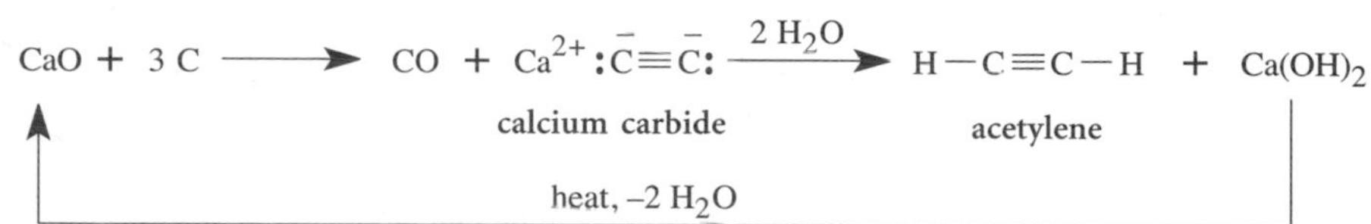

$$CaO + 3\,C \longrightarrow CO + Ca^{2+}\,:\bar{C}{\equiv}\bar{C}: \xrightarrow{2\,H_2O} H{-}C{\equiv}C{-}H + Ca(OH)_2$$

i. Calcium carbide is an organometallic compound that can be conceptually regarded as the calcium salt of the acetylene dianion.

ii. Calcium carbide reacts vigorously with water to yield acetylene and calcium hydroxide, which, on heating, gives calcium oxide.

b. Acetylene can also be made by thermal cracking of ethylene.

B. Common Nomenclature of Alkynes

1. In common nomenclature, simple alkynes are named as derivatives of the parent compound acetylene.

$$CH_3CH_2—C{\equiv}C—CH(CH_3)_2 \quad \textbf{ethyl isopropyl acetylene}$$

2. Certain compounds are named as derivatives of the propargyl group, $H—C{\equiv}C—CH_2—$ (the triple-bond analog of the allyl group).

$$H—C{\equiv}C—CH_2—OH \quad \textbf{propargyl alcohol}$$

C. Substitutive Nomenclature of Alkynes

1. The substitutive nomenclature of alkynes is very much like that of alkenes.
   a. The suffix *ane* in the name of the corresponding alkane is replaced by the suffix *yne*.
   b. The triple bond is given the lowest possible number.
2. Substituent groups that contain a triple bond (called alkynyl groups) are named by replacing the final *e* in the name of the corresponding alkyne with the suffix *yl*; they are numbered from the point of attachment to the main chain.

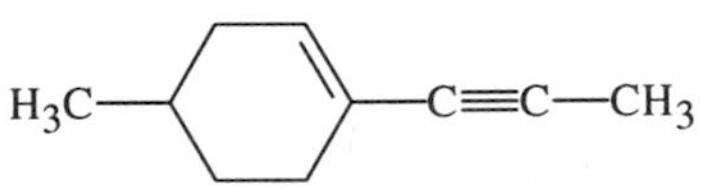

H_3C—(cyclohexene ring)—$C{\equiv}C—CH_3$ **4-methyl-(1-propynyl)cyclohexene**

3. Groups that can be cited as principal groups are given numerical precedence over the triple bond.

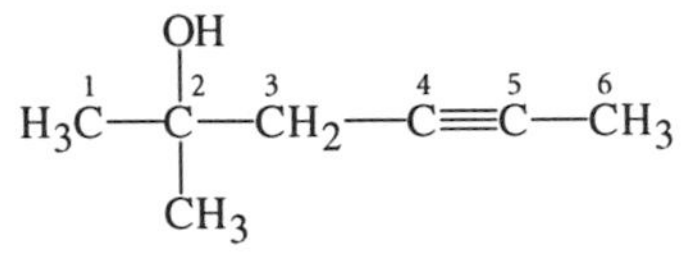

$$\overset{1}{H_3C}—\overset{2}{C}(OH)(CH_3)—\overset{3}{C}H_2—\overset{4}{C}{\equiv}\overset{5}{C}—\overset{6}{C}H_3 \quad \textbf{2-methyl-4-hexyn-2-ol}$$

II. PHYSICAL PROPERTIES OF ALKYNES

A. Structure and Bonding in Alkynes

1. The bonding in alkynes involves the hybridization of the $2s$ orbital with one $2p$ orbital to form two sp hybrid orbitals.
   a. An sp-hybridized carbon atom has two sp orbitals at a relative orientation of 180° (linear).
      i. An sp orbital has a shape that is similar to that of an sp^2 or sp^3 orbital (one small lobe, one large lobe).
      ii. Electrons in an sp orbital are somewhat closer to the carbon nucleus than they are in sp^2 or sp^3 orbitals.
   b. The two remaining unhybridized $2p$ orbitals lie along axes that are at right angles both to each other and to the sp orbitals.

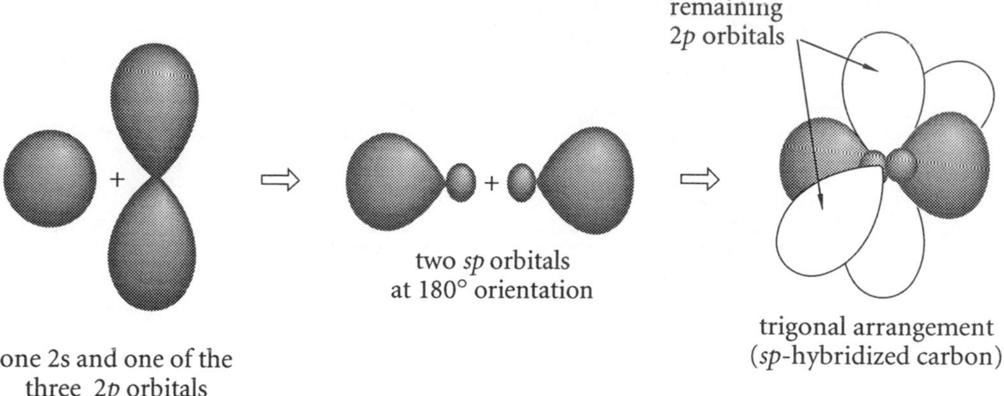

2. The acetylene molecule is linear.
   a. One bond between the carbon atoms results from the overlap of two *sp* hybrid orbitals and is called an *sp-sp* σ bond.
   b. Two π bonds are formed by the side-to-side overlap of the unhybridized 2*p* orbitals.
      i. These bonding π molecular orbitals are mutually perpendicular, and each contains two electrons.
      ii. The total electron density from all the π electrons taken together forms a cylinder, or barrel, about the axis of the molecule.

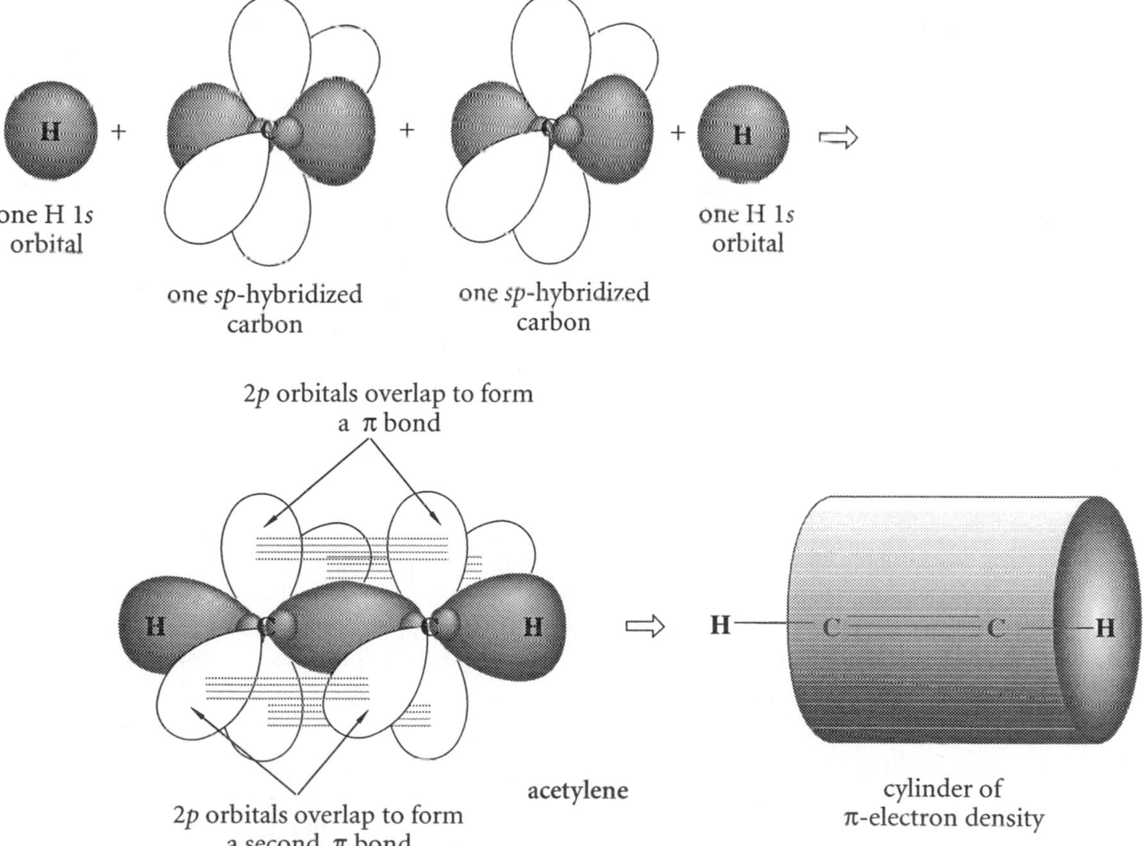

   c. Cis-trans isomerism cannot occur in alkynes because the bond angles between the C≡C bond and the attached atoms are 180°.
   d. Cycloalkynes smaller than cyclooctyne cannot be isolated under ordinary conditions.

3. The C≡C bond is shorter than the C=C and C—C bonds.
4. Alkynes are less stable than isomeric dienes
   a. The *sp* hybridization state is inherently less stable than the sp^2 hybridization state, other things being equal
   b. Alkynes with internal triple bonds are more stable than isomeric alkynes with terminal triple bonds.

B. Boiling Points and Solubilities

1. The boiling points of most alkynes are not very different from those of analogous alkenes and alkanes.
2. Alkynes, like alkanes and alkenes, have much lower densities than water.
3. Alkynes, like alkanes and alkenes, are insoluble in water.

III. SPECTROSCOPY OF ALKYNES

A. IR Spectroscopy of Alkynes

1. Many alkynes have a C≡C stretching absorption in the 2100-2200 cm^{-1} region of the infrared spectrum; this absorption is very weak or absent in the IR spectra of many symmetrical or nearly symmetrical alkynes because of the dipole moment effect.
2. A useful absorption of 1-alkynes is the ≡C—H stretching absorption, which occurs at about 3300 cm^{-1}. (The narrow peak for this absorption distinguishes it from the broad peak for the O—H absorption, which occurs at about the same position.)

B. NMR Spectroscopy of Alkynes

1. The proton NMR chemical shifts of allylic and propargylic protons are very similar (δ 1.8–δ 2.2).
2. The proton NMR chemical shifts of acetylenic protons (δ 1.7–δ 2.5) and the CMR chemical shifts of carbons involved in triple bonds (δ 70–δ 80) are much smaller than those of vinylic protons (δ 4.7–δ 6.2) or vinylic carbons (δ 105–δ 150).
   a. In response to a magnetic field, an induced electron circulation is set up in the cylinder of π electrons that encircles the alkyne molecule.
   b. The resulting induced field $\mathbf{B_i}$ opposes the applied field along the axis of this cylinder.
      i. Acetylenic protons and carbons of triple bonds lie in this region and thus are shielded from the applied field $\mathbf{B_0}$.
      ii. Greater shielding causes acetylenic protons and carbons to have smaller chemical shifts than they would have otherwise.

IV. ACIDITY OF 1-ALKYNES

A. Acetylenic Anions

1. Carbanions are extremely strong bases (that is, hydrocarbons are very weak acids), and 1-alkynes are the most acidic of all the aliphatic hydrocarbons.
   a. An alkyl anion, the conjugate base of an alkane, has an unshared electron pair in an sp^3 orbital.
   b. A vinylic anion, the conjugate base of an alkene, has an unshared electron pair in an sp^2 orbital.
   c. An acetylenic anion, the conjugate base of an alkyne, has an unshared electron pair in an *sp* orbital.

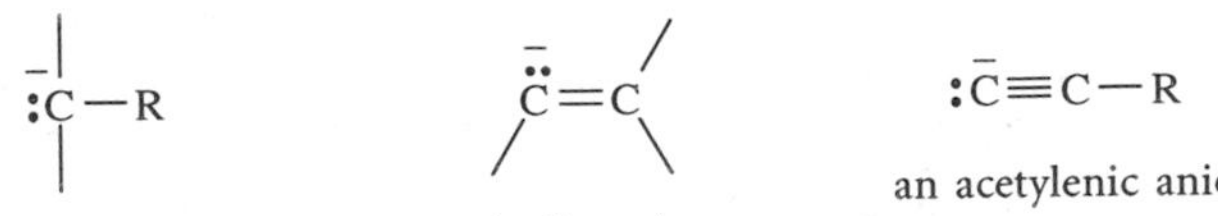

:C̄—R (an alkyl anion) C̄=C (a vinylic anion) :C̄≡C—R (an acetylenic anion)

an alkyl anion   a vinylic anion   an acetylenic anion

2. 1-Alkynes are sufficiently acidic ($pK_a \approx 25$) that their conjugate-base acetylide anions can be formed with strong bases.

a. The sodium salt of an alkyne can be formed from a 1-alkyne quantitatively with sodium amide (sodamide), $Na^+ \ {}^-\!:\ddot{N}H_2$, dissolved in its conjugate acid, liquid ammonia.

$$Na^+ \ :\ddot{\bar{N}}H_2 \ + \ R{-}C{\equiv}C{-}H \longrightarrow R{-}C{\equiv}\bar{C}: \ Na^+ \ + \ \ddot{N}H_3$$

b. The reaction of an alkylmagnesium halide with an alkyne gives an acetylenic Grignard reagent.

$$R{-}C{\equiv}C{-}H \ + \ CH_3CH_2{-}MgBr \longrightarrow \underset{\text{an acetylenic Grignard reagent}}{R{-}C{\equiv}C{-}MgBr} \ + \ \underset{\text{ethane a gas}}{CH_3CH_3}$$

i. This reaction (transmetallation) is another Brønsted acid-base reaction.
ii. The release of ethane gas in the reaction of a 1-alkyne with ethylmagnesium bromide is a useful test for 1-alkynes.

c. Alkynes with internal triple bonds do not react because they lack an acidic alkyne 1-hydrogen.

3. The electronegativity of the *sp*-hybridized carbon is responsible for the enhanced acidity of alkynes.
   a. The electrons in *sp*-hybridized orbitals are closer to the nucleus, on the average, than sp^2 electrons, which in turn are closer than sp^3 electrons; the proximity of *sp* electrons to the nucleus is a stabilizing effect.
   b. Unshared electron pairs have lower energy when they are in orbitals of greater *s* character.

REACTIONS

I. ADDITION REACTIONS OF ALKYNES

A. Addition of Hydrogen Halides and Halogens

1. Additions of halogens or hydrogen halides to the triple bond are somewhat slower in most cases than the same reactions of comparably substituted alkenes.
   a. The first addition to an alkyne gives a substituted alkene.
   b. The second addition of a hydrogen halide or a halogen to an alkyne is usually slower than the first because a halo alkene is less reactive than an alkyne.
   c. It is possible to isolate the substituted alkene resulting from addition of only one equivalent of halogen or hydrogen halide.

$$-C{\equiv}C- \xrightarrow{Br_2} \underset{\text{Br}}{\overset{\text{Br}}{>C{=}C<}} \xrightarrow{Br_2} -\underset{Br}{\overset{Br}{C}}-\underset{Br}{\overset{Br}{C}}-$$

2. HBr can be added to the triple bond in a regioselective addition that is analogous to the addition of HBr to alkenes.
   a. The bromine adds to the carbon of the triple bond that bears the alkyl substituent group.
   b. The regioselectivity is reversed in the presence of peroxides because free-radical intermediates are involved.

$$R{-}C{\equiv}C{-}H \xrightarrow{HBr} \underset{R\quad\ H}{\overset{Br\quad\ H}{C{=}C}} \xrightarrow{HBr} R{-}\underset{Br}{\overset{Br}{C}}{-}\underset{H}{\overset{H}{C}}{-}H$$

$$R{-}C{\equiv}C{-}H \xrightarrow[\text{peroxides}]{\text{HBr}} \underset{R\quad\ \ H}{\overset{H\quad\ Br}{C{=}C}} \xrightarrow[\text{peroxides}]{\text{HBr}} R{-}\underset{H}{\overset{H}{C}}{-}\underset{Br}{\overset{Br}{C}}{-}H$$

B. Hydration of Alkynes

1. The addition of water to the triple bond of an alkyne is catalyzed by strong acid or, more commonly, by a combination of dilute acid and mercuric ion, Hg^{2+}.

$$\underset{\text{an alkyne}}{{-}C{\equiv}C{-}} + H_2O \xrightarrow{H_2SO_4,\ HgSO_4} \underset{\text{a ketone}}{{-}CH_2{-}\overset{O}{\overset{\|}{C}}{-}}$$

a. The addition of water to a triple bond is called hydration.

b. Enols (alcohols containing an —OH group on a carbon of a double bond) are formed in the hydration of alkynes.

i. Most enols are unstable and are rapidly converted into the corresponding ketones or aldehydes.

ii. Most aldehydes and ketones are in equilibrium with small amounts of the corresponding enols.

iii. Because most enols are unstable, any synthesis designed to give an enol gives instead the corresponding aldehyde or ketone.

$$\underset{\text{an enol}}{C{=}C{-}OH} \rightleftharpoons \underset{\text{an aldehyde or a ketone}}{H{-}C{-}C{=}O}$$

2. The mechanism of alkyne hydration is very similar to that of oxymercuration of alkenes.

a. In the first part the mechanism, a mercuric ion and an —OH group from the solvent water add to the triple bond.

i. The —OH group of water adds to the carbon of the triple bond that bears the alkyl substituent.

ii. The bond between the mercury and this carbon in the cyclic ion intermediate is weaker and thus more readily broken.

$$AcO{-}Hg{-}OAc + R{-}C{\equiv}C{-}H \xrightarrow{H_2O} \text{[cyclic Hg ion, } R(C{=}C)H\text{, } AcO{-}Hg^{+}\text{, } {}^{-}{:}OAc\text{, } H_2O{:}\text{]} \longrightarrow \underset{HO^{+}H\ \ \ H}{\overset{R\quad\ \ HgOAc}{C{=}C}} + {}^{-}{:}OAc \longrightarrow \underset{HO\quad\ H}{\overset{R\quad\ HgOAc}{C{=}C}} + HOAc$$

b. The second part of the mechanism involves protonolysis, which occurs under the conditions of hydration.

i. This protonation of the double bond occurs at the carbon bearing the mercury because the resulting carbocation is resonance-stabilized.

ii. Dissociation of mercury from this carbocation liberates the catalyst Hg^{2+} along with the enol.

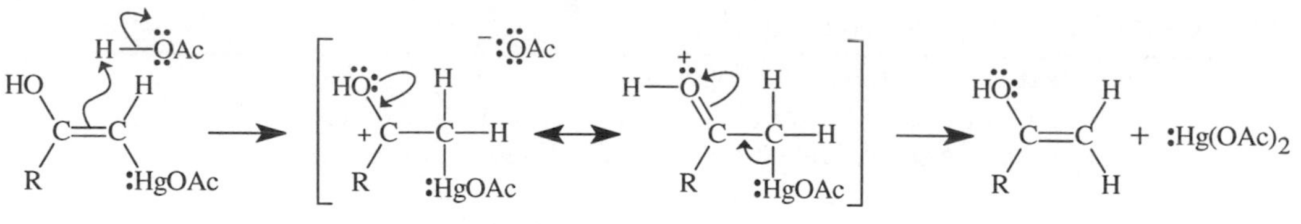

a resonance-stabilized carbocation

c. Conversion of the enol into the ketone is a rapid, acid-catalyzed process.
   i. Protonation of the double bond gives another resonance-stabilized carbocation, which is also the conjugate acid of a ketone.
   ii. Loss of a proton from the carbocation gives the ketone product.

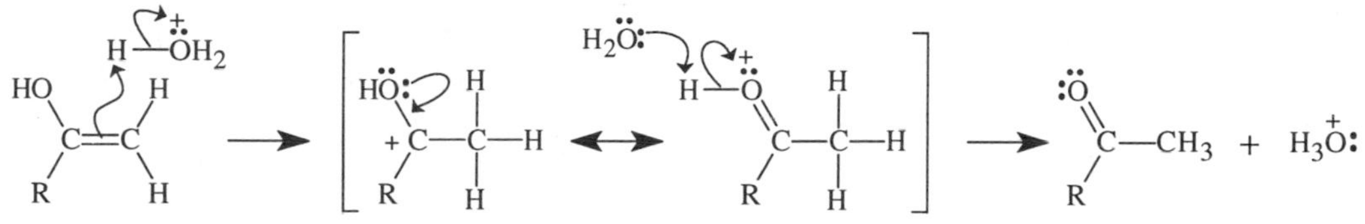

a resonance-stabilized carbocation

3. The hydration of alkynes is a useful way to prepare ketones provided that the starting material is a 1-alkyne or a symmetrical alkyne; most other alkynes give difficult-to-separate mixtures of isomers.

C. Hydroboration of Alkynes

1. The hydroboration of alkynes is analogous to the same reaction of alkenes.
2. When alkynes react with diborane (B_2H_6), the elements of BH_3 are added to the triple bond.
   a. Oxidation of the organoborane with alkaline hydrogen peroxide yields the corresponding "alcohol," an enol.
   b. The enol reacts further to give the corresponding aldehyde or ketone.

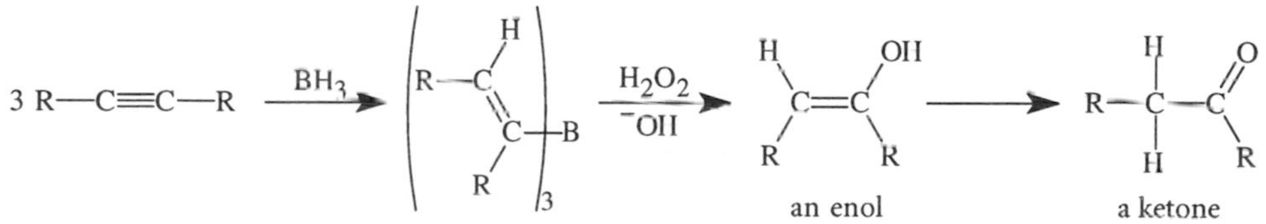

an enol a ketone

3. The reaction conditions can be controlled so that only one addition takes place provided that the alkyne is not a 1-alkyne.
4. The hydroboration of 1-alkynes can be stopped after a single addition provided that organoborane containing highly branched groups is used instead of BH_3.
   a. One reagent developed for this purpose is called disiamylborane.
   b Boron adds to the unbranched carbon atom of the triple bond, and hydrogen adds to the branched carbon.
   c. The disiamylborane molecule is so large and highly branched that only one equivalent can react with a 1-alkyne.

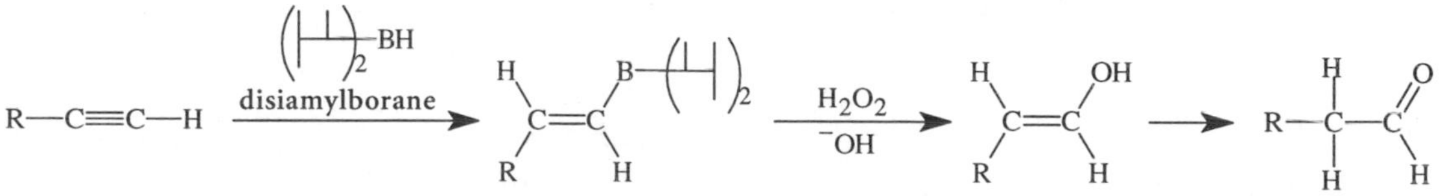

disiamylborane

an aldehyde

D. Catalytic Hydrogenation of Alkynes

1. Alkynes undergo catalytic hydrogenation.
   a. Addition of one molar equivalent of H_2 yields a *cis*-alkene.
   b. Addition of a second molar equivalent of H_2 gives an alkane.

$$\underset{\text{an alkyne}}{R-C\equiv C-R} \xrightarrow{H_2,\ \text{catalyst}} \underset{\text{a } cis\text{-alkene}}{(R)(H)C=C(R)(H)} \xrightarrow{H_2,\ \text{catalyst}} \underset{\text{an alkane}}{R-CH_2-CH_2-R}$$

2. Hydrogenation of an alkyne may be stopped at the alkene stage if the reaction mixture contains a catalyst poison.
   a. A catalyst poison is a compound that disrupts the action of a catalyst. Some poisons can selectively block the hydrogenation of *cis*-alkenes without preventing the hydrogenation of alkynes to *cis*-alkenes.
   b. Useful catalyst poisons are salts of Pb^{2+} and certain nitrogen compounds such as pyridine, quinoline, or other amines.
   c. The Lindlar catalyst ($Pd/CaCO_3$ that has been washed with $Pb(OAc)_2$) is a commonly used poisoned catalyst.

$$R-C\equiv C-R' \xrightarrow[\substack{Pd/CaCO_3 \\ [\text{washed with } Pb(OAc)_2]}]{H_2} (H)(R)C=C(H)(R')$$

3. Hydrogenation of alkynes is a stereoselective *syn*-addition and, when carried out with a poisoned catalyst, is one of the best methods for preparing *cis*-alkenes.

E. Reduction of Alkynes with Alkali Metals in Ammonia

1. Reaction of an alkyne with a solution of an alkali metal (usually sodium) in liquid ammonia gives an alkene. If the alkyne is not a 1-alkyne, the product is a *trans*-alkene.

$$2\,NH_3 + \underset{\text{an alkyne}}{R-C\equiv C-R} \xrightarrow[\text{liquid } NH_3]{\text{Na metal}} \underset{\text{a } trans\text{-alkene}}{(R)(H)C=C(H)(R)} + 2\,Na^+\ {}^-NH_2$$

2. The mechanism of this reaction involves a two-electron reduction:
   a. The deep blue solution of alkali metals dissolved in pure liquid ammonia is a source of electrons complexed to ammonia (solvated electrons).

$$n\,NH_3 + Na \xrightarrow[\text{liquid } NH_3]{} Na^+ + \underset{\text{solvated electron}}{e^-(NH_3)_n}$$

   b. The addition of an electron to the triple bond results in a species (a radical anion) that has both an unpaired electron and a negative charge.

$$R-C\equiv C-R \xrightarrow{e^-} \underset{\text{a radical anion}}{\left[R\dot{C}=\ddot{C}^{-}R \longleftrightarrow R\,{}^{-}\ddot{C}=\dot{C}R \right]}$$

c. The radical anion is such a strong base that it removes a proton from ammonia to give a vinylic radical (a radical in which the unpaired electron is associated with one carbon of a double bond).

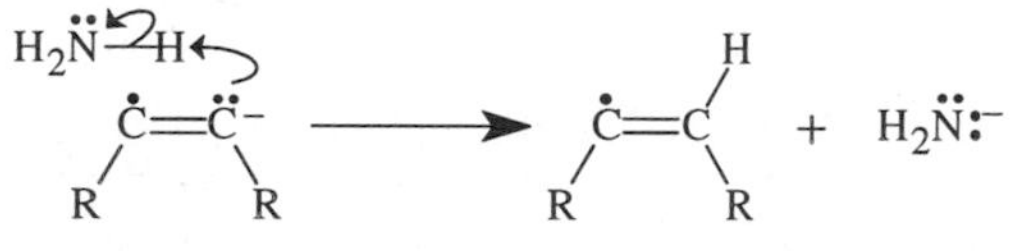

a radical anion a vinylic radical

d. The resulting vinylic radical rapidly undergoes inversion.

i. The equilibrium between the cis and trans radicals favors the trans radical for steric reasons.

ii. Because there is much more of the trans radical, the ultimate product of the reaction is the *trans*-alkene.

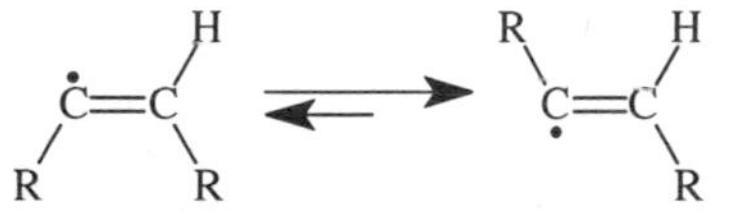

a cis vinylic radical a trans vinylic radical

e. The vinylic radical accepts an electron to form a vinylic anion.

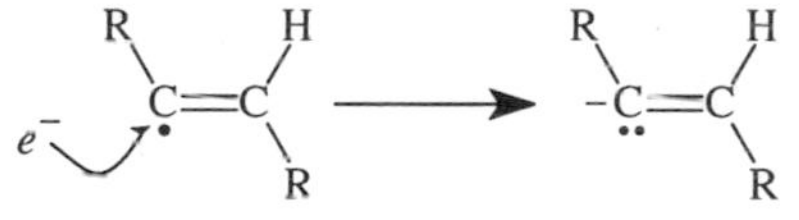

a vinylic radical a vinylic anion

f. This anion is more basic than the solvent and removes a proton from ammonia to complete the addition.

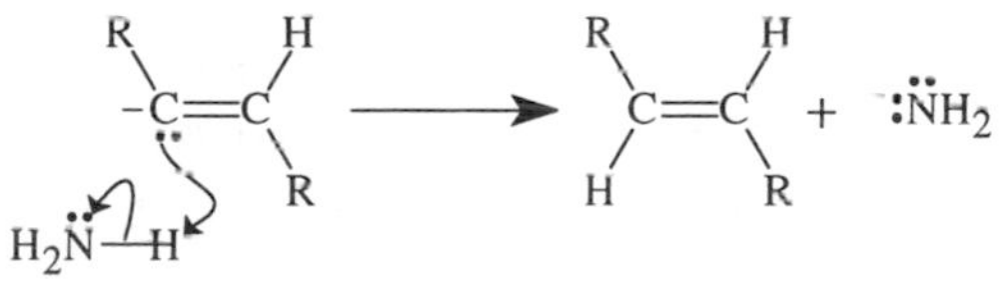

F. Acetylenic Anions as Nucleophiles

1. Acetylenic anions are strong bases; they are much stronger than hydroxide or alkoxides. (The pK_a of a 1-alkyne is about 25.)
2. Acetylenic anions can be used as nucleophiles in S_N2 reactions to prepare other alkynes.
   a. When an alkyl halide or sulfonate is used in this reaction, it must be primary or unbranched secondary.

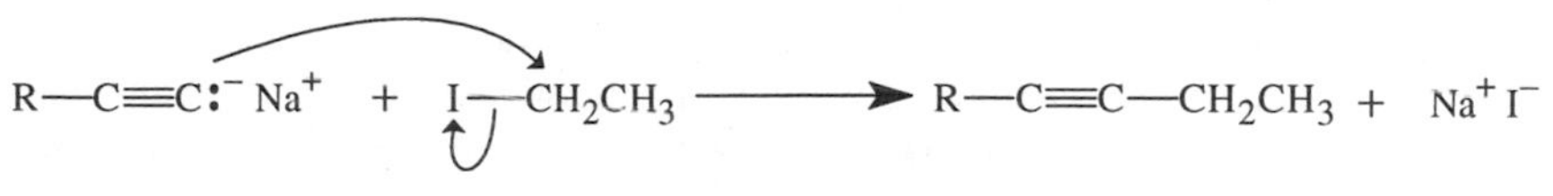

$$R{-}C{\equiv}C{:}^{-}\,Na^{+} + I{-}CH_2CH_3 \longrightarrow R{-}C{\equiv}C{-}CH_2CH_3 + Na^{+}\,I^{-}$$

   b. This is another method of carbon-carbon bond formation.
   c. Acetylenic anions can also react with ethylene oxide.

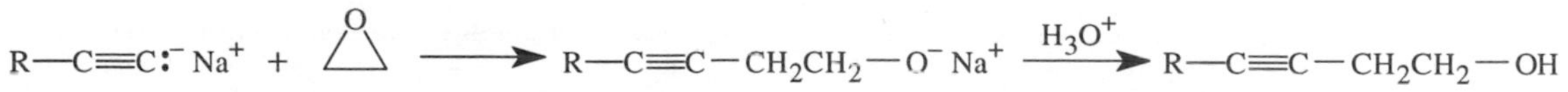

$$R{-}C{\equiv}C{:}^{-}\,Na^{+} + \text{(ethylene oxide)} \longrightarrow R{-}C{\equiv}C{-}CH_2CH_2{-}O^{-}\,Na^{+} \xrightarrow{H_3O^{+}} R{-}C{\equiv}C{-}CH_2CH_2{-}OH$$

STUDY GUIDE LINKS

✓14.1 Functional Group Preparations

Section 14.5 in the text discusses two ways to prepare aldehydes and ketones. At this point, you should be asking yourself, "What *other* ways do I know to make aldehydes and ketones?" You should recall, for example, that oxidation of alcohols is an important method for preparation of aldehydes and ketones. Keep a separate review list that groups reactions by the *type of product formed.* Thus, you'll have a list of reactions for preparing aldehydes and ketones, a list for preparing alcohols, etc. You should write out each reaction in detail, noting the limitations on each. For example, what are the limitations of alkyne hydration as a method of ketone preparation? (Can *any* ketone be prepared with this method? Look at Problems 14.9–14.10 on text pp. 617–18.) You can check how complete your list is by comparing it to the abbreviated list in Appendix V (p. A-8) of the text.

Use your list when working problems in synthesis. When your synthesis requires an aldehyde or ketone, consult your list; review the reactions if they start to look unfamiliar. Remember that putting together such lists keeps your studying *active* rather than passive.

✓14.2 Ammonia, Solvated Electrons, and Amide Ion

Sections 14.6 and 14.7 of the text have discussed several different uses of ammonia. The purpose of this Study Guide Link is to review and contrast these so that there is no confusion.

The first reagent is *liquid ammonia.* Ammonia can be condensed at –33° to a liquid. Once condensed, ammonia is easy to maintain in the liquid state because of its relatively high heat capacity. Liquid ammonia should not be confused with *aqueous* ammonia, which is a solution of ammonia in water. ("Household ammonia" is an aqueous ammonia solution.) If an alkali metal (such as sodium) is dissolved in liquid ammonia, a blue solution is obtained that is an excellent source of electrons—that is, an excellent reducing agent. This solution contains "solvated electrons"—electrons solvated by the hydrogens of the ammonia. The solution of solvated electrons is used in the reduction of alkynes to *trans*-alkenes (Sec. 14.6B).

If a trace of Fe^{3+} ion is added to a solution of sodium in liquid ammonia, a *reaction* occurs between the sodium and the ammonia:

$$2\,Na\cdot + 2\,\ddot{N}H_3 \xrightarrow{FeCl_3} 2\,Na^+ \;{}^{-}\!:\ddot{N}H_2 + H_2$$

sodium amide
(sodamide)

Just as sodium reacts with water to give hydrogen gas and sodium hydroxide, sodium also reacts with ammonia to give hydrogen gas and sodium amide. The difference is that the reaction of sodium with water requires no catalyst, but the reaction with ammonia does. It is this solution—sodium amide in liquid ammonia—that is used in forming acetylenic anions from alkynes. Solid sodium amide can be purchased as a gray powder. Solutions of sodium amide can also be formed by adding this material to liquid ammonia or other solvents.

Do not confuse the basicity of amide ion with that of ammonia. Amide ion is the *conjugate base* of ammonia, which has a pK_a of about 35. Ammonia is also a base, but is the conjugate base of the *ammonium ion,* NH_4^+, which has a pK_a of 9.2. Most people normally think of ammonia as a base, but in the present context, ammonia is the *acid*, and the amide ion is the conjugate base. (See the discussion of amphoteric substances in Sec. 3.4D on text pp. 95–96.)

SOLUTIONS TO PROBLEMS

Solutions to In-Text Problems

14.1 (a) $(CH_3)_2CHC{\equiv}CH$
isopropylacetylene

(b) cyclononyne

(c) $HC{\equiv}C{-}CH_2CH(CH_3)_2$
4-methyl-1-pentyne

(d) 1-ethynylcyclohexanol (cyclohexane ring bearing OH and $C{\equiv}CH$ on the same carbon)

(e) $CH_3CH(OCH_2CH_2CH_2CH_3)C{\equiv}CCH_2CH_2CH_3$
2-butoxy-3-heptyne

(f) $HC{\equiv}C{-}C{\equiv}C{-}CH_2CH_3$
1,3-hexadiyne

14.2 (a) 1-hexyne (common: butylacetylene)
(b) 5-decyne (common: dibutylacetylene)
(c) 2-methyl-3-pentyn-2-ol

14.3 (a) The *sp* hybridization of the carbon atoms of the triple bond requires that these two atoms and the two attached carbon atoms lie on a straight line. With a bond length of 1.2 Å for the triple bond and bond lengths of 1.47 Å for the attached carbon-carbon single bonds, the distance to be bridged by two remaining carbons of cyclohexyne is about 4.1 Å. This distance is too great to be connected by two carbon atoms that lie *on the same line,* let alone at reasonable bond angles.
(b) Cyclodecyne is much more stable than cyclohexyne (and in fact can be isolated), because the distance of 4.1 Å referred to in the previous problem can be bridged by six carbons with reasonable bond lengths and bond angles.

14.4 A nine-proton singlet cries out, "*tert*-butyl group." The other resonance is in the position expected for an acetylenic proton. The 3300 cm^{-1} absorption in the IR spectrum confirms the presence of an acetylenic C—H. *Tert*-butylacetylene (3,3-dimethyl-1-butyne, $(CH_3)_3C{-}C{\equiv}CH$), which has a molecular mass of 82, is consistent with all the data.

14.5 (a) Because of its symmetry, 3-hexyne should have three chemically nonequivalent sets of carbons; in contrast, all six carbons of 2-hexyne are chemically nonequivalent. Thus, spectrum *A* is that of 2-hexyne, and spectrum *B* is that of 3-hexyne.
(b) Use the model compound 2-heptyne on text p. 613 to assist you.

$$\underset{\delta 3.3}{H_3C}{-}\underset{\delta 79.1}{C}{\equiv}\underset{\delta 75.4}{C}{-}\underset{\delta 22.9}{CH_2}{-}\underset{\delta 21.1}{CH_2}{-}\underset{\delta 13.6}{CH_3}$$

$$\underset{\delta 14.6}{H_3C}{-}\underset{\delta 12.7}{CH_2}{-}\underset{\delta 81.0}{C}{\equiv}\underset{\delta 81.0}{C}{-}\underset{\delta 12.7}{CH_2}{-}\underset{\delta 14.6}{CH_3}$$

(Interchanging the δ 14.6 and δ 12.7 assignments in 3-hexyne is also a reasonable answer, although the assignments above are correct. How could one verify the correct assignments? (*Hint:* See text p. 589 and Fig. 13.24 on text p. 590.)

14.6 The proton NMR spectrum of propyne should consist of two resonances. Evidently, the two happen to have the same chemical shift of δ 1.8, which is in fact a reasonable chemical shift for both acetylenic and propargylic protons.

14.7 Bromine addition to 3-hexyne could give two stereoisomers:

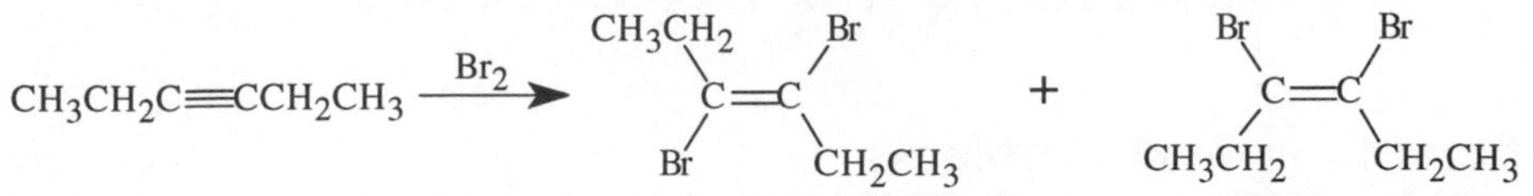

$$CH_3CH_2C{\equiv}CCH_2CH_3 \xrightarrow{Br_2} (CH_3CH_2)(Br)C{=}C(Br)(CH_2CH_3) + (Br)(CH_3CH_2)C{=}C(Br)(CH_2CH_3)$$

(*E*)-3,4-dibromo-3-hexene (from *anti*-addition)

(*Z*)-3,4-dibromo-3-hexene (from *syn*-addition)

14.8 The product is (*Z*)-3-chloro-3-hexene:

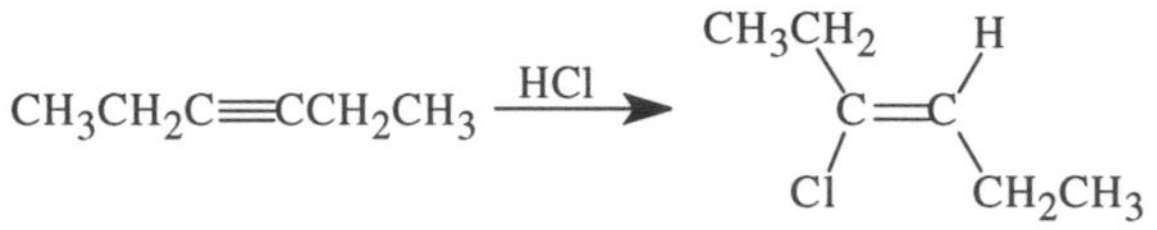

$$CH_3CH_2C{\equiv}CCH_2CH_3 \xrightarrow{HCl} (CH_3CH_2)(Cl)C{=}C(H)(CH_2CH_3)$$

(*Z*)-3-chloro-3-hexene

14.9 (a) Only 1-pentyne would undergo hydration to give 2-pentanone free of constitutional isomers. (As shown in Study Problem 14.1 on text p. 617, 2-pentyne would give a difficult-to-separate mixture of 2-pentanone and 3-pentanone.)

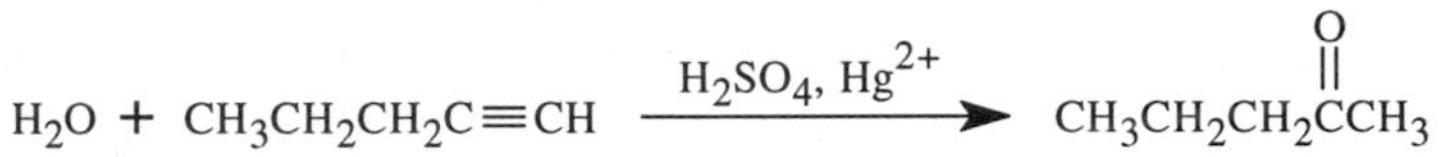

$$H_2O + \underset{\text{1-pentyne}}{CH_3CH_2CH_2C{\equiv}CH} \xrightarrow{H_2SO_4,\ Hg^{2+}} \underset{\text{2-pentanone}}{CH_3CH_2CH_2\overset{O}{\overset{\|}{C}}CH_3}$$

(b) $(CH_3)_3C{-}C{\equiv}CH$ *tert*-butylacetylene (3,3-dimethyl-1-butyne)

(c) Only 5-decyne, $CH_3(CH_2)_3C{\equiv}C(CH_2)_3CH_3$, would give the indicated ketone (5-decanone) free of constitutional isomers. 4-Decanone would not; why?

14.10 (a) The three-carbon alkyne propyne hydrates to give acetone, not propanal (the compound shown in the problem).

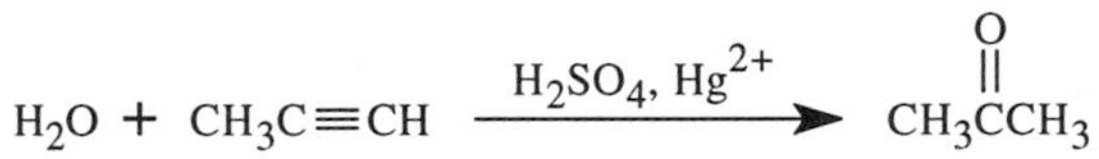

$$H_2O + \underset{\text{propyne}}{CH_3C{\equiv}CH} \xrightarrow{H_2SO_4,\ Hg^{2+}} \underset{\text{acetone}}{CH_3\overset{O}{\overset{\|}{C}}CH_3}$$

(b) Hydration can only be used to prepare ketones that have at least two hydrogens on the carbon α (that is, adjacent) to the carbonyl carbon:

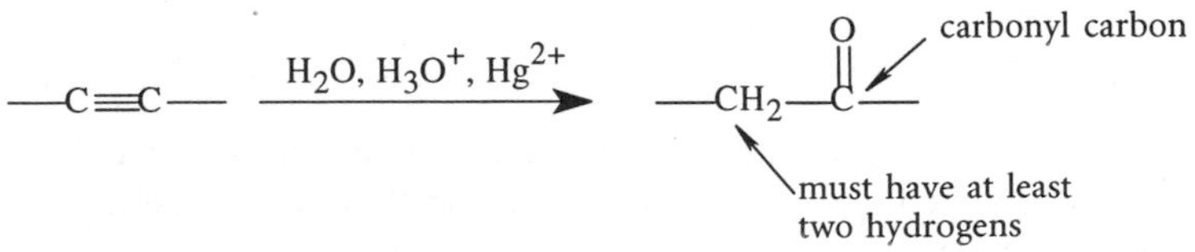

$${-}C{\equiv}C{-} \xrightarrow{H_2O,\ H_3O^+,\ Hg^{2+}} {-}CH_2{-}\overset{O}{\overset{\|}{C}}{-}$$

carbonyl carbon

must have at least two hydrogens

The ketone shown does not fulfill this requirement.

(c) The alkyne that would hydrate to this ketone (cyclohexanone) is cyclohexyne. Since cyclohexyne is too unstable to isolate, such a hydration reaction is impractical. (See the solution to Problem 14.3a for a discussion of the instability of cyclohexyne.)

14.11 (a) Two constitutionally isomeric enols are possible, one of which can exist as *E,Z*-stereoisomers.

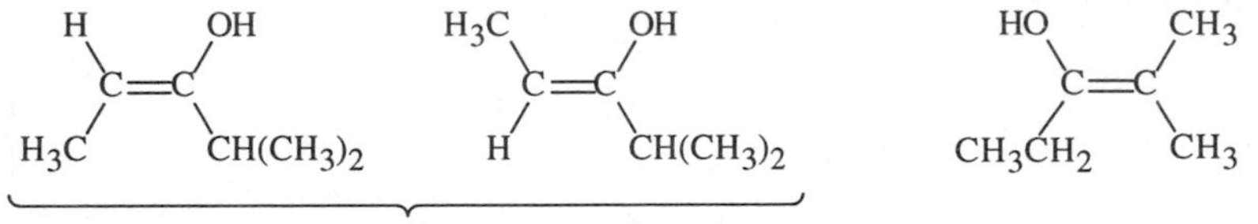

H, OH / H_3C, $CH(CH_3)_2$ (C=C); H_3C, OH / H, $CH(CH_3)_2$ (C=C); HO, CH_3 / CH_3CH_2, CH_3 (C=C)

E,Z-stereoisomers

(b) Hydration would not be a good preparative method for this compound. The only alkyne that might give this compound as a product is 4-methyl-2-pentyne; but, because both carbons of its triple bond have one alkyl substituent, that alkyne would give a mixture of two constitutionally isomeric products of which the desired compound is only one component.

$$CH_3C{\equiv}CCH(CH_3)_2 \xrightarrow{H_2O,\ H_3O^+,\ Hg^{2+}} CH_3CH_2\overset{O}{\overset{\|}{C}}CH(CH_3)_2 + CH_3\overset{O}{\overset{\|}{C}}CH_2CH(CH_3)_2$$

4-methyl-2-pentyne

14.12 (a) Because it is symmetrical, 2-butyne gives the same product in the two reactions:

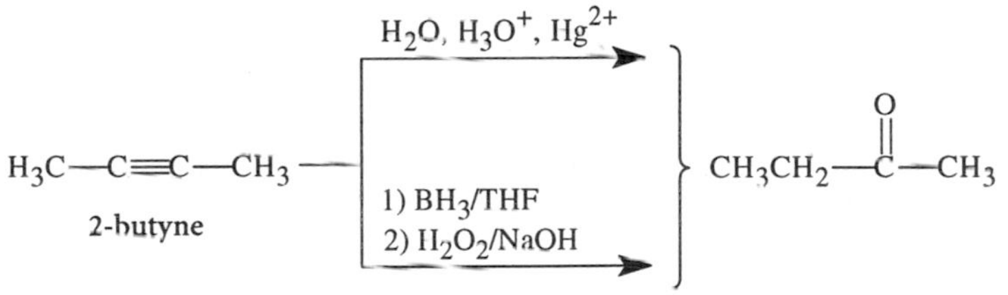

$$H_3C{-}C{\equiv}C{-}CH_3 \xrightarrow[\text{or 1) } BH_3/THF\text{; 2) } H_2O_2/NaOH]{H_2O,\ H_3O^+,\ Hg^{2+}} CH_3CH_2{-}\overset{O}{\overset{\|}{C}}{-}CH_3$$

2-butyne

(b) Cyclohexylacetylene gives different products in the two reactions:

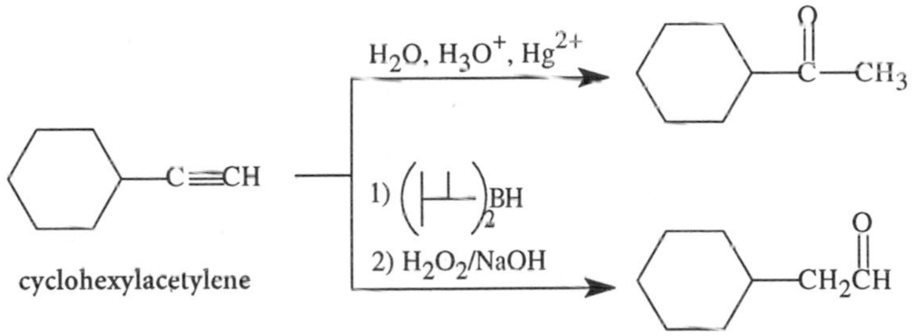

cyclohexyl–C≡CH (cyclohexylacetylene)

$\xrightarrow{H_2O,\ H_3O^+,\ Hg^{2+}}$ cyclohexyl–$\overset{O}{\overset{\|}{C}}$–$CH_3$

1) $(\text{sec-isoamyl})_2BH$ 2) $H_2O_2/NaOH$ → cyclohexyl–$CH_2\overset{O}{\overset{\|}{C}}H$

14.13 (a) The poisoned catalyst causes hydrogenation to stop at the alkene stage:

$$CH_3(CH_2)_5C{\equiv}CH \xrightarrow[\text{Lindlar catalyst}]{H_2} CH_3(CH_2)_5CH{=}CH_2$$

1-octyne 1-octene

(b) With no poison, hydrogenation of the double bond is not inhibited, and the hydrogenation proceeds all the way to the alkane.

$$CH_3(CH_2)_5C{\equiv}CH \xrightarrow{H_2,\ Pd/C} CH_3(CH_2)_5CH_2CH_3$$

1-octyne octane

(c) Because the molecule contains its own catalyst poison, hydrogenation stops at the alkene. The product is

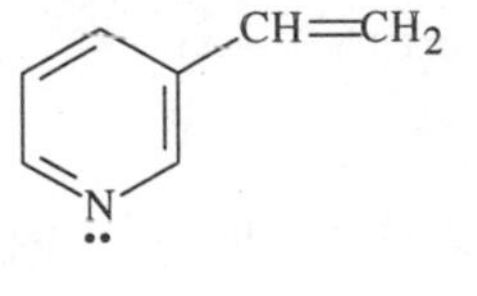

pyridine ring with $CH{=}CH_2$ at the 3-position

3-vinylpyridine

14.14 (a) The first reaction is an *anti*-addition that gives *trans*-3-hexene, and the second reaction, a *syn*-addition, gives (±)-hexane-3,4-d_2.

$$CH_3CH_2C{\equiv}CCH_2CH_3 \xrightarrow[\textit{anti}\text{-addition}]{Na,\ NH_3(liq.)} \underset{\text{(trans)}}{(CH_3CH_2)HC{=}CH(CH_2CH_3)} \xrightarrow[\textit{syn}\text{-addition}]{D_2,\ Pd/C}$$

Fischer projection: top CH_3CH_2; H—C—D; D—C—H; bottom CH_3CH_2 — and its enantiomer

(±)-hexane-3,4-*d2*

(b) The first reaction is a *syn*-addition that gives *cis*-3-hexene; the second is another *syn*-addition that gives *meso*-hexane-3,4-d_2.

$$CH_3CH_2C{\equiv}CCH_2CH_3 \xrightarrow[\textit{syn}\text{-addition}]{H_2,\ Pd/C} \underset{\text{(cis)}}{(CH_3CH_2)HC{=}CH(CH_2CH_3)} \xrightarrow[\textit{syn}\text{-addition}]{D_2,\ Pd/C}$$

Fischer projection: top CH_3CH_2; D—C—H; D—C—H; bottom CH_3CH_2

meso-hexane-3,4-d_2

14.15 The conjugate acids are as follows:

	$CH_3—CH{=}\overset{+}{N}H_2$	$CH_3—C{\equiv}\overset{+}{N}H$	$CH_3—CH_2—\overset{+}{N}H_3$
conjugate acid of	*A*	*B*	*C*

The order of increasing acidity of the conjugate acids is $B > A > C$. The order of increasing basicity of the compounds themselves is then $B < A < C$. This order follows from the principle developed in this section: the hydrogen with the greatest amount of *s* character in its C—H bond is most acidic. This problem demonstrates that it doesn't matter whether the atom to which the hydrogen is attached is charged or neutral, or whether it is a carbon or some other atom.

14.16 (a) *Part (1):* The logarithm of the equilibrium constant K_{eq} for an acid-base reaction equals the pK_a value of the acid on the right minus that of the acid on the left. Assuming that the pK_a of ammonia is about 35, and that the pK_a of an alkyne is about 25, log K_{eq} for the reaction in Eq. 14.22 (text p. 624) is $(35 - 25) = 10$. Consequently, the equilibrium constant for this reaction is 10^{10}. This equilibrium lies far to the right.

Part (2): Use the same procedure employed in the solution to part (a); assume that the pK_a of an alkane is about 55. Therefore, log K_{eq} for the reaction of the amide ion with an alkane C—H is about –20, and the equilibrium constant is therefore 10^{-20}.

(b) As the previous two calculations show, the equilibrium for the reaction of the base sodium amide with an alkyne strongly favors ionization of the alkyne and protonation of the base; the equilibrium for the reaction of sodium amide with an alkane strongly favors the un-ionized alkane and unprotonated sodium amide. In other words, sodium amide is a strong enough base to remove a proton from an alkyne, but is far too weak a base to remove a proton from an alkane.

14.17 (a) This is an S_N2 reaction of the anion with ethyl iodide to give

$$CH_3C{\equiv}CCH_2CH_3 + Na^+\ I^-$$

2-pentyne

(b) The acetylenic anion reacts in an S_N2 reaction with butyl tosylate:

$$\underset{\text{butyl tosylate}}{CH_3CH_2CH_2CH_2—OTs} + {}^{-}\!:C{\equiv}CPh \longrightarrow CH_3CH_2CH_2CH_2C{\equiv}CPh + {}^{-}OTs$$

(c) The acetylenic Grignard reagent, like any other Grignard reagent, opens the epoxide to give an alcohol:

$$CH_3C{\equiv}CMgBr \xrightarrow[\text{2) } H_3O^+]{\text{1) ethylene oxide}} \underset{\text{3-pentyn-1-ol}}{CH_3C{\equiv}CCH_2CH_2OH}$$

(d) Sodium acetylide reacts with both alkyl halide groups to give a diyne:

$$\underset{\text{1,8-nonadiyne}}{HC{\equiv}C(CH_2)_5C{\equiv}CH} + 2\ Na^+\ Br^-$$

14.18 Choke forgot the lessons of Sec. 9.5F: tertiary alkyl halides react with strong bases to give elimination products, not substitution products. Acetylenic anions are strong bases; thus, Choke's products were 2-methylpropene and propyne.

$$(CH_3)_3CBr + :\bar{C}{\equiv}CCH_3 \xrightarrow{\text{(E2 elimination)}} \underset{\text{2-methylpropene}}{(CH_3)_2C{=}CH_2} + \underset{\text{propyne}}{HC{\equiv}CCH_3} + Br^-$$

14.19 Either alkyl group can be introduced in an S_N2 reaction of an acetylenic anion with an alkyl halide. The two preparations of 2-pentyne:

Method 1:

$$\underset{\text{1-butyne}}{HC{\equiv}CCH_2CH_3} \xrightarrow{NaNH_2} Na^+\ :\bar{C}{\equiv}CCH_2CH_3 \xrightarrow[\text{iodomethane}]{H_3C{-}I} \underset{\text{2-pentyne}}{CH_3C{\equiv}CCH_2CH_3}$$

Method 2:

$$\underset{\text{propyne}}{CH_3C{\equiv}CH} \xrightarrow{NaNH_2} CH_3C{\equiv}\bar{C}:\ Na^+ \xrightarrow[\text{iodoethane}]{I{-}CH_2CH_3} \underset{\text{2-pentyne}}{CH_3C{\equiv}CCH_2CH_3}$$

Although either method would work, method 1 would perhaps be better because iodomethane (methyl iodide) is more reactive in an S_N2 reaction than iodoethane (why?).

14.20 A synthesis of undecanone from compounds containing five or fewer carbons:

$$HC{\equiv}\bar{C}:\ Na^+ \xrightarrow{Br(CH_2)_4CH_3} HC{\equiv}C(CH_2)_4CH_3 \xrightarrow{NaNH_2} Na^+\ :\bar{C}{\equiv}C(CH_2)_4CH_3 \xrightarrow[\text{2) } H_3O^+]{\text{1) ethylene oxide}}$$

(prepared in Eq. 14.29c, text p. 628)

$$HOCH_2CH_2C{\equiv}C(CH_2)_4CH_3 \xrightarrow{H_2,\ Pd/C} HO(CH_2)_8CH_3 \xrightarrow{\text{conc. HBr, } H_2SO_4} Br(CH_2)_8CH_3 \xrightarrow{HC{\equiv}\bar{C}:\ Na^+}$$

$$HC{\equiv}C(CH_2)_8CH_3 \xrightarrow{H_2O,\ H_3O^+,\ Hg^{2+}} \underset{\text{2-undecanone}}{CH_3\overset{O}{\overset{\|}{C}}(CH_2)_8CH_3}$$

14.21 (a) When a *cis*-alkene is called for, catalytic hydrogenation of an alkyne is the preferred route.

$$Br(CH_2)_8{-}O{-}\text{THP} \xrightarrow{CH_3CH_2C{\equiv}\bar{C}:\ Na^+} CH_3CH_2C{\equiv}C(CH_2)_8O{-}\text{THP} \xrightarrow{H_2,\ \text{Lindlar catalyst}}$$

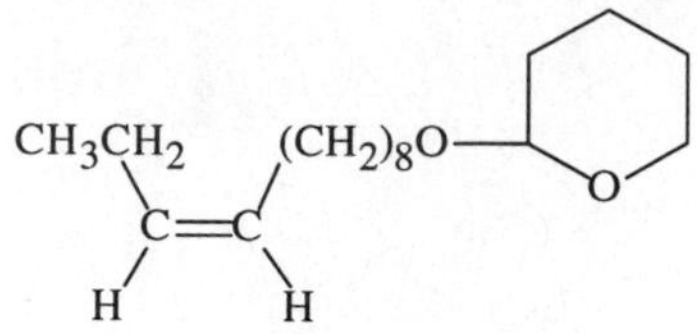

CH_3CH_2 $(CH_2)_8O$– (tetrahydropyranyl)
C=C
H H

(b) Because the required acetylene is already in hand from the part (a), treating it with Na in liquid ammonia will achieve the desired result:

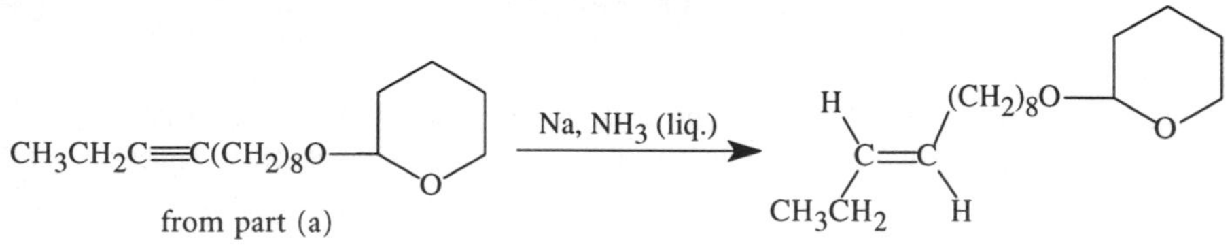

$CH_3CH_2C{\equiv}C(CH_2)_8O$– (tetrahydropyranyl) $\xrightarrow{\text{Na, NH}_3\text{ (liq.)}}$ H, $(CH_2)_8O$– (tetrahydropyranyl); C=C; CH_3CH_2, H

from part (a)

Solutions to Additional Problems

14.22 (a) $CH_3CH_2CH_2CH_2C(Br){=}CH_2$

(b) $CH_3(CH_2)_4CH_3$

(c) $CH_3CH_2CH_2CH_2CH{=}CH_2$

(d) $CH_3CH_2CH_2CH_2CH{=}O$ + $O{=}CH_2$

(e) $CH_3CH_2CH_2CH_2CH_2CH_2OH$

(f) $CH_3CH_2CH_2CH_2CH(Br)CH_2Br$

(g) $CH_3CH_2CH_2CH_2C{\equiv}\bar{C}\text{:}\ Na^+$

(h) $CH_3CH_2CH_2CH_2C{\equiv}CCH_2CH_3$ + $Na^+\ I^-$

(i) $CH_3CH_2CH_2CH_2C({=}O)CH_3$

(j) $CH_3CH_2CH_2CH_2CH(OH)CH_3$

14.23 (a) $CH_3C{\equiv}CCH(OH)CH_2CH_3$
~~2-hexyn-4-ol~~
4-hexyne-3-ol

(b) $HC{\equiv}CCH_2CH_2C{\equiv}COCH_3$
~~6-methoxy-1,5-hexadiyne~~
1-methoxy-1,5-hexadiyne

(c) $CH_3CH_2CH_2CH_2C{\equiv}CH$
~~5-hexyne~~
1-hexyne

14.24 (a) Because the triple bond cannot be contained within the ring, the compound can only be cyclopropylacetylene.

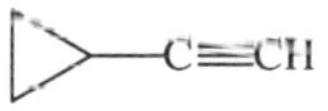

cyclopropyl–$C{\equiv}CH$ cyclopropylacetylene

(b) Several answers are possible:

$CH_3CH_2CH(CH_3)C{\equiv}CH$ or H_3C–cyclopropane–$C{\equiv}CH$ or H_3C–cyclopropane–$C{\equiv}CH$

3-methyl-1-pentyne

(c) For two reactions with different regioselectivities to give the same product, the alkyne must be symmetrical.

$CH_3CH_2C{\equiv}CCH_2CH_3$ 3-hexyne

(d) The cyclopropane derivatives in (b) are good answers. Another answer is

$CH_3CH{=}CHCH_2C{\equiv}CH$ 4-hexen-1-yne

which can exist as cis and trans isomers. (Remember that cis-trans isomers are diastereomers.)

14.25 (a) Bonds with greater *s* character are shorter. (See text p. 609.) Therefore, the order of increasing C—H bond length is

acetylene < ethylene < ethane

(b) Hybridization affects the lengths of C—C bonds in the same way that it affects the lengths of C—H bonds. Therefore, the lengths of C—C bonds increase in the following order:

propyne < propene < propane

14.26 (a) The order of increasing basicity is $F^- < CH_3CH_2O^- < HC{\equiv}C^-$. The reason is essentially the element effect as it operates across a row of the periodic table: the more electronegative the atom, the less basic it is. (Note that the element effect on acidity is more important than the effect of hybridization.)

(b) Because alkynes are more acidic than alkenes, which are more acidic than alkanes, the acetylenic anion is less basic than the vinylic anion, which is less basic than the alkyl anion. Thus, the basicity order is $CH_3(CH_2)_3C{\equiv}C^- < CH_3(CH_2)_3CH{=}\bar{C}H < CH_3(CH_2)_4\bar{C}H_2$.

14.27 (a) The alkyne should give off a gas (methane or ethane, respectively) when treated with CH_3MgBr or C_2H_5MgBr. (See Eq. 14.25 on text p. 625.) The alkene is not acidic enough to react in this way.
(b) Only the 1-alkyne should react with C_2H_5MgBr to release a gas (ethane).
(c) The alkyne should react rapidly with Br_2 in an inert solvent, and should therefore rapidly decolorize such a bromine solution; the alkane will not react.
(d) Forget about chemical tests; propyne is a gas and 1-decyne is a liquid at room temperature.

How would you know this? You know that *propane* is a gas, right? (It is used instead of natural gas in rural areas for heating or in barbecue grills as a fuel for cooking.) You also know that the presence of a double or triple bond has little effect on the physical properties of hydrocarbons. Thus, propyne is a gas also.

14.28 Note that in many syntheses more than one acceptable route may be possible.

(a)

$$\underset{\text{(excess)}}{HC{\equiv}CH} \xrightarrow{NaNH_2} HC{\equiv}\bar{C}{:}\ Na^+ \xrightarrow{CH_3(CH_2)_3Br} \underset{\text{1-hexyne}}{CH_3(CH_2)_3C{\equiv}CH}$$

(b)

$$\underset{\text{prepared in part (a)}}{HC{\equiv}\bar{C}{:}\ Na^+} \xrightarrow{CH_3CH_2Br} \underset{\text{1-butyne}}{HC{\equiv}CCH_2CH_3} \xrightarrow{NaNH_2}$$

$$Na^+\ {:}\bar{C}{\equiv}CCH_2CH_3 \xrightarrow{CH_3CH_2Br} \underset{\text{3-hexyne}}{CH_3CH_2C{\equiv}CCH_2CH_3} \xrightarrow{H_2,\ \text{Lindlar catalyst}}$$

$$\text{(cis) } CH_3CH_2(H)C{=}C(H)CH_2CH_3 \xrightarrow[\text{2) } H_2O_2/NaOH]{\text{1) } BH_3/THF} CH_3CH_2\overset{OH}{\overset{|}{C}}HCH_2CH_2CH_3$$

Oxymercuration/reduction would work equally well in the last step because the alkene is symmetrical and reaction regioselectivity is not an issue.

(c)

$$\underset{\text{1-hexyne prepared in part (a)}}{CH_3CH_2CH_2CH_2C{\equiv}CH} \xrightarrow{H_2,\ \text{Lindlar catalyst}} \underset{\text{1-hexene}}{CH_3(CH_2)_3CH{=}CH_2}$$

(d)

$$\underset{\text{3-hexyne prepared in part (b)}}{CH_3CH_2C{\equiv}CCH_2CH_3} \xrightarrow{D_2,\ Pd/C} CH_3CH_2CD_2CD_2CH_2CH_3$$

(e) One synthesis uses a two-carbon chain extension with 1-heptyne as the alkyne starting material.

$$\underset{\text{prepared in part (a)}}{Na^+\ {:}\bar{C}{\equiv}CH} \xrightarrow[\text{1-bromopentane}]{CH_3CH_2CH_2CH_2CH_2{-}Br} \underset{\text{1-heptyne}}{CH_3CH_2CH_2CH_2CH_2C{\equiv}CH} \xrightarrow{NaNH_2/NH_3}$$

$$CH_3CH_2CH_2CH_2CH_2C{\equiv}\bar{C}{:}Na^+ \xrightarrow[2)\ H_3O^+]{1)\ \text{ethylene oxide}} CH_3CH_2CH_2CH_2CH_2C{\equiv}CCH_2CH_2OH \xrightarrow{H_2/\text{catalyst}}$$

$$CH_3(CH_2)_7CH_2OH \xrightarrow[2)\ H_3O^+]{1)\ KMnO_4,\ ^-OH} CH_3(CH_2)_7\overset{O}{\overset{\|}{C}}OH$$

Of course, the final oxidation in this synthesis would work equally well with aqueous Cr(VI) reagents such as $K_2Cr_2O_7$ in acid. Another synthesis provides all of the carbons necessary in the starting alkyl halide:

$$\underset{\text{prepared in part (a)}}{Na^+\ {:}\bar{C}{\equiv}CH} \xrightarrow[\text{1-bromoheptane}]{CH_3(CH_2)_6Br} CH_3(CH_2)_6C{\equiv}CH \xrightarrow[2)\ H_2O_2/NaOH]{1)\ (\text{disiamyl})_2BH}$$

$$CH_3(CH_2)_7CH{=}O \xrightarrow[2)\ H_3O^+]{1)\ KMnO_4,\ ^-OH} CH_3(CH_2)_7\overset{O}{\overset{\|}{C}}OH$$

(f)

$$\underset{\text{prepared in part (a)}}{Na^+\ {:}\bar{C}{\equiv}CH} \xrightarrow{(CH_3)_2CHCH_2CH_2Br}$$

$$(CH_3)_2CHCH_2CH_2C{\equiv}CH \xrightarrow[2)\ H_2O_2/NaOH]{1)\ (\text{disiamyl})_2BH} (CH_3)_2CHCH_2CH_2CH_2CH{=}O$$

(g)

$$\underset{\text{prepared in part (b) from 1-butyne}}{Na^+\ {:}\bar{C}{\equiv}CCH_2CH_3} \xrightarrow{CH_3I} \underset{\text{2-pentyne}}{CH_3C{\equiv}CCH_2CH_3} \xrightarrow{H_2,\ \text{Lindlar catalyst}} \underset{\textit{cis}\text{-2-pentene}}{(H_3C)(H)C{=}C(H)(CH_2CH_3)}$$

(h)

$$\underset{\text{prepared in part (b) from 1-butyne}}{Na^+\ {:}\bar{C}{\equiv}CCH_2CH_3} \xrightarrow{CH_3(CH_2)_5Br} CH_3(CH_2)_5C{\equiv}CCH_2CH_3 \xrightarrow{Na,\ NH_3\ (liq.)} \underset{\textit{trans}\text{-3-decene}}{(CH_3(CH_2)_5)(H)C{=}C(H)(CH_2CH_3)}$$

(i)

$$\underset{\text{prepared in part (a)}}{Na^+\ {:}\bar{C}{\equiv}CH} \xrightarrow{CH_3(CH_2)_2Br} CH_3(CH_2)_2C{\equiv}CH \xrightarrow[NH_3]{NaNH_2} CH_3(CH_2)_2C{\equiv}\bar{C}{:}\ Na^+ \xrightarrow{Br(CH_2)_2CH_3}$$

$$\underset{\text{4-octyne}}{CH_3(CH_2)_2C{\equiv}C(CH_2)_2CH_3} \xrightarrow[\text{Lindlar catalyst}]{H_2} \underset{\textit{cis}\text{-4-octene}}{(CH_3(CH_2)_2)(H)C{=}C(H)((CH_2)_2CH_3)} \xrightarrow[\text{NMMO},\ (CH_3)_3COH/H_2O]{OsO_4\ (cat.)}$$

$$\underset{\textit{meso}\text{-4,5-octanediol}}{CH_3(CH_2)_2CH(OH){-}CH(OH)(CH_2)_2CH_3}$$

An alternate route involves an *anti*-addition of two —OH groups to *trans*-4-octene by acid-catalyzed epoxide hydrolysis:

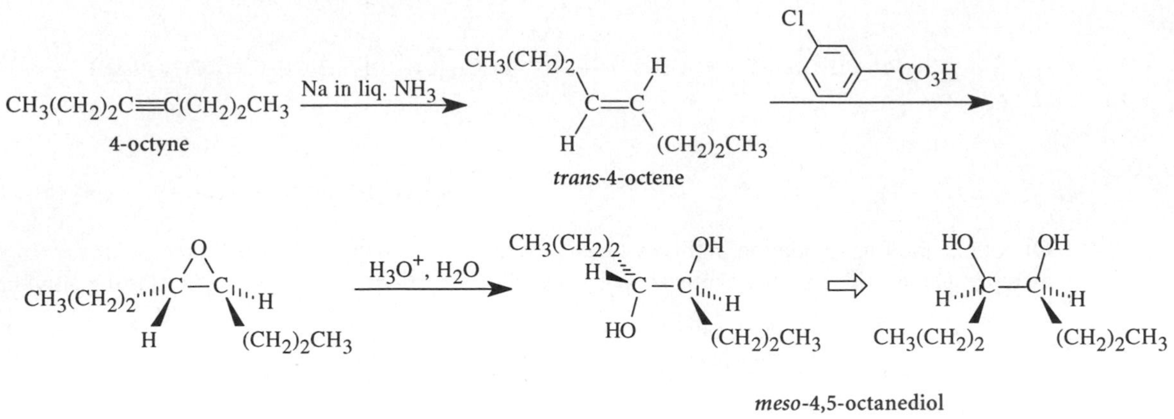

$CH_3(CH_2)_2C{\equiv}C(CH_2)_2CH_3$ (4-octyne) $\xrightarrow{\text{Na in liq. } NH_3}$ *trans*-4-octene [$CH_3(CH_2)_2(H)C{=}C(H)(CH_2)_2CH_3$] $\xrightarrow{m\text{-}ClC_6H_4CO_3H}$ epoxide $\xrightarrow{H_3O^+,\ H_2O}$ *meso*-4,5-octanediol

(j)

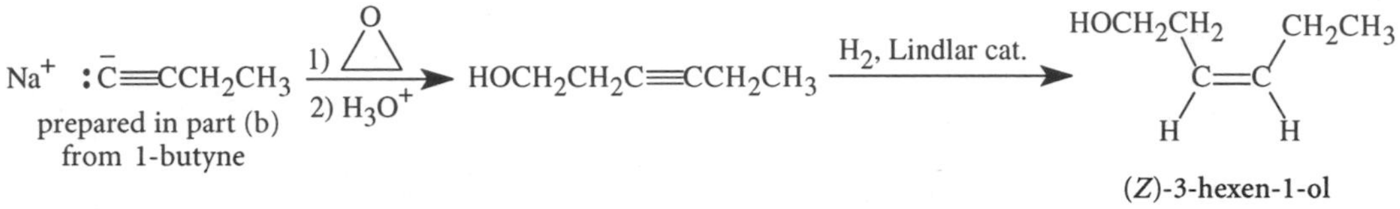

$Na^+ :\bar{C}{\equiv}CCH_2CH_3$ (prepared in part (b) from 1-butyne) $\xrightarrow[\text{2) } H_3O^+]{\text{1) ethylene oxide}}$ $HOCH_2CH_2C{\equiv}CCH_2CH_3$ $\xrightarrow{H_2,\ \text{Lindlar cat.}}$ (*Z*)-3-hexen-1-ol [$HOCH_2CH_2(H)C{=}C(H)CH_2CH_3$]

14.29 (a) Take advantage of the basicity of an acetylenic Grignard reagent towards D_2O:

$$CH_3CH_2C{\equiv}CH \text{ (1-butyne)} \xrightarrow[-CH_3CH_3]{C_2H_5MgI} CH_3CH_2C{\equiv}CMgI \xrightarrow{D_2O} CH_3CH_2C{\equiv}CD$$

(b)

$$CH_3CH_2C{\equiv}CD \text{ (prepared in part (a))} \xrightarrow{D_2,\ Pd/C} CH_3CH_2CD_2CD_3$$

(c)

$$CH_3CH_2C{\equiv}CH \xrightarrow{H_2,\ \text{Lindlar catalyst}} CH_3CH_2CH{=}CH_2 \xrightarrow[\text{2) } H_2O_2/NaOH]{\text{1) } BH_3/THF}$$

$$CH_3CH_2CH_2CH_2OH \text{ (1-butanol)} \xrightarrow[\text{2) } H_3O^+]{\text{1) } KMnO_4,\ ^-OH} CH_3CH_2CH_2CO_2H$$

(d)

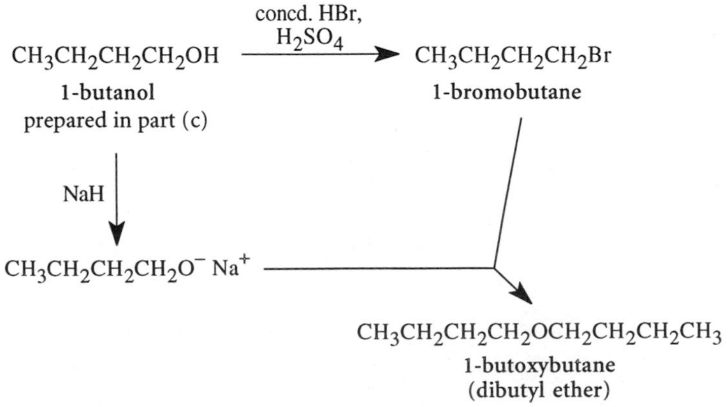

$$CH_3CH_2CH_2CH_2OH \text{ (1-butanol, prepared in part (c))} \xrightarrow{\text{concd. HBr, } H_2SO_4} CH_3CH_2CH_2CH_2Br \text{ (1-bromobutane)}$$

$$CH_3CH_2CH_2CH_2OH \xrightarrow{NaH} CH_3CH_2CH_2CH_2O^-\ Na^+$$

$CH_3CH_2CH_2CH_2O^-\ Na^+$ + 1-bromobutane → $CH_3CH_2CH_2CH_2OCH_2CH_2CH_2CH_3$

1-butoxybutane (dibutyl ether)

(e)

$$CH_3CH_2C{\equiv}CH \xrightarrow{NaNH_2} CH_3CH_2C{\equiv}\bar{C}\!:\ Na^+ \xrightarrow{BrCH_2CH_2CH_2CH_3} CH_3CH_2C{\equiv}CCH_2CH_2CH_2CH_3$$

3-octyne

$\xrightarrow{H_2,\text{ Lindlar cat.}}$ cis-alkene: C=C with CH_3CH_2 and $CH_2CH_2CH_2CH_3$ on top, H and H below $\xrightarrow{D_2,\text{ Pd/C}}$ CH_3CH_2, H, D on one carbon; $CH_2CH_2CH_2CH_3$, H, D on the other

+ an equal amount of its enantiomer

In assigning configurations, notice that a $CH_3CH_2CHD—$ group receives priority over a $—CH_2CH_2CH_2CH_3$ group.

(f)

$$CH_3CH_2C{\equiv}CCH_2CH_2CH_2CH_3 \xrightarrow{H_2,\text{ Pd/C}} CH_3CH_2CH_2CH_2CH_2CH_2CH_2CH_3$$

3-octyne prepared in part (e) — octane

(g)

$$CH_3CH_2C{\equiv}CH \xrightarrow{H_2O,\ H_3O^+,\ Hg^{2+}} CH_3CH_2\overset{O}{\overset{\|}{C}}CH_3$$

14.30 First look for the telltale C—H and C≡C stretching absorptions; these are present in the spectrum of compound *B* near 2100 cm^{-1} and 3300 cm^{-1}. Hence, compound *B* is 1-hexyne. Compound *A* has no trace of C≡C absorption and, more importantly, it has a 1640 cm^{-1} C=C stretching absorption and 910 cm^{-1} and 990 cm^{-1} C—H bending absorptions, all of which are consistent with a $—CH{=}CH_2$ group. Therefore, compound *A* is 3-methyl-1,4-pentadiene. By elimination, compound *C* must be 2-hexyne.

$H_2C{=}CH\overset{CH_3}{\overset{|}{C}}HCH{=}CH_2$ — 3-methyl-1,4-pentadiene, compound *A*

$CH_3CH_2CH_2CH_2C{\equiv}CH$ — 1-hexyne, compound *B*

$CH_3C{\equiv}CCH_2CH_2CH_3$ — 2-hexyne, compound *C*

14.31 Because muscalure is a *cis*-alkene, we can prepare it by hydrogenation of an alkyne. The required alkyne is unbranched and can be prepared by a series of alkylations involving acetylenic anions.

First, prepare 1-bromooctane.

$$CH_3CH_2C{\equiv}CH \xrightarrow[\text{Lindlar catalyst}]{H_2} CH_3CH_2CH{=}CH_2 \xrightarrow{\text{HBr, peroxides}} CH_3CH_2CH_2CH_2Br \xrightarrow[\text{[for preparation see solution to Problem 14.28(a)]}]{Na^+\ :\bar{C}{\equiv}CH}$$

$$CH_3CH_2CH_2CH_2C{\equiv}CH \xrightarrow[\text{Lindlar catalyst}]{H_2} CH_3(CH_2)_3CH{=}CH_2 \xrightarrow{\text{HBr, peroxides}} CH_3(CH_2)_5Br \xrightarrow{Na^+\ :\bar{C}{\equiv}CH}$$

$$CH_3(CH_2)_5C{\equiv}CH \xrightarrow[\text{Lindlar catalyst}]{H_2} CH_3(CH_2)_5CH{=}CH_2 \xrightarrow{\text{HBr, peroxides}} CH_3(CH_2)_7Br$$

1-bromooctane

As you can see from this example, the carbon chain of an alkyne can be extended two carbons at a time using the following sequence (referred to below as the *alkyne n + 2 sequence*):

$$CH_3(CH_2)_nC{\equiv}CH \xrightarrow[\text{Lindlar catalyst}]{H_2} CH_3(CH_2)_nCH{=}CH_2 \xrightarrow{\text{HBr, peroxides}} CH_3(CH_2)_{n+2}Br \xrightarrow{Na^+\ :\bar{C}{\equiv}CH} CH_3(CH_2)_{n+2}C{\equiv}CH$$

Consequently, 1-bromotridecane, $CH_3(CH_2)_{12}Br$, can be prepared by applying four successive "alkyne $n + 2$ sequences" to 1-pentyne ($n = 2$), and finishing with the hydrogenation and HBr-addition steps:

$$\underset{\text{1-pentyne}}{CH_3(CH_2)_2C{\equiv}CH} \xrightarrow{\text{four alkyne } n+2 \text{ sequences}} CH_3(CH_2)_{10}C{\equiv}CH \xrightarrow[\text{Lindlar catalyst}]{H_2} CH_3(CH_2)_{10}CH{=}CH_2 \xrightarrow{\text{HBr, peroxides}} \underset{\text{1-bromotridecane}}{CH_3(CH_2)_{12}Br}$$

To complete the synthesis of muscalure, carry out two successive alkylations of acetylenic anions with the alkyl bromides prepared above and hydrogenate the resulting product.

$$HC{\equiv}\bar{C}{:}\ Na^+ \xrightarrow{Br(CH_2)_{12}CH_3} HC{\equiv}C(CH_2)_{12}CH_3 \xrightarrow[NH_3]{NaNH_2} \xrightarrow{CH_3(CH_2)_7Br} CH_3(CH_2)_7C{\equiv}C(CH_2)_{12}CH_3 \xrightarrow[\text{Lindlar cat.}]{H_2}$$

(cis) $CH_3(CH_2)_7CH{=}CH(CH_2)_{12}CH_3$ (H, H on the same side)

muscalure

14.32 This problem is similar to the previous one. Disparlure is a *cis*-epoxide, which can be prepared by epoxidation of a *cis*-alkene. (Recall that this reaction proceeds with retention of stereochemistry; Sec. 11.2A.) The *cis*-alkene can be prepared by hydrogenation of an alkyne.

First prepare 1-bromodecane from 1-bromooctane, which was prepared in the solution to Problem 14.31.

$$CH_3(CH_2)_7Br \xrightarrow{Na^+\ :\bar{C}{\equiv}CH} CH_3(CH_2)_7C{\equiv}CH \xrightarrow[\text{Lindlar cat.}]{H_2} CH_3(CH_2)_7CH{=}CH_2 \xrightarrow{\text{HBr, peroxides}} \underset{\text{1-bromodecane}}{CH_3(CH_2)_9Br}$$

Next, alkylate sodium acetylide with 1-bromo-3-methylbutane (isoamyl bromide). Apply the alkyne $n + 2$ sequence described in the previous solution, alkylate the conjugate base of the resulting alkyne with 1-bromodecane, hydrogenate, and form the epoxide to complete the synthesis.

$$BrCH_2CH_2CH(CH_3)_2 \xrightarrow{HC{\equiv}\bar{C}{:}\ Na^+} HC{\equiv}CCH_2CH_2CH(CH_3)_2 \xrightarrow{\text{alkyne } n+2 \text{ sequence}}$$

$$HC{\equiv}C(CH_2)_4CH(CH_3)_2 \xrightarrow[\text{2) } CH_3(CH_2)_9Br]{\text{1) } NaNH_2} CH_3(CH_2)_9C{\equiv}C(CH_2)_4CH(CH_3)_2 \xrightarrow{H_2\text{, Lindlar cat.}}$$

(cis) $CH_3(CH_2)_9CH{=}CH(CH_2)_4CH(CH_3)_2$ (H, H on the same side) $\xrightarrow{\text{mCPBA}}$ $CH_3(CH_2)_9$–CH–O–CH–$(CH_2)_4CH(CH_3)_2$ (cis-epoxide)

disparlure

14.33 (a) The "anionic carbon" of ethylmagnesium bromide acts as a base towards the C—H bond of ethynylmagnesium bromide:

$$CH_3CH_2{-}MgBr + H{-}C{\equiv}CMgBr \longrightarrow CH_3CH_3 + BrMgC{\equiv}CMgBr$$

According to Eq. (2) in the problem, an excess of acetylene reacts with the product of Eq. (1) to regenerate ethynylmagnesium bromide. Thus, the formation of BrMgC≡CMgBr, the undesired by-product, is avoided by "recycling" it back to the desired product HC≡CMgBr.

(b) The "anionic carbon" of one Grignard reagent acts as a base toward the C—H bond of the other.

$$H{-}C{\equiv}C{-}MgBr + H{-}C{\equiv}CMgBr \rightleftharpoons H{-}C{\equiv}C{-}H + BrMgC{\equiv}CMgBr$$

An excess of acetylene, by LeChatelier's principle, drives the equilibrium to the left.

(c) Precipitation of BrMgC≡CMgBr would pull both reactions (1) and (2) to the right, thus depleting the desired product HC≡CMgBr and forming more of the undesired one. Using a solvent that avoids this precipitation thus avoids these side-reactions.

14.34 (a) The radical anion forms in this case just as it does with internal alkynes:

$$R{-}C{\equiv}C{-}H + \dot{Na} \longrightarrow R{-}\dot{C}{=}\ddot{C}^{-}{-}H$$

radical anion

This radical anion is a strong base, and, in the case of a 1-alkyne, the alkyne can serve as an acid. (Internal alkynes, of course, lack this acidic hydrogen.) Recall that 1-alkynes have a pK_a of 25, whereas ammonia, the solvent, has a pK_a of about 32. Ammonia is the only acid available in the reaction of internal alkynes, but in the reaction of 1-alkynes, the alkyne itself is the strongest acid present. This is the source of the side reaction.

$$Na^+\ R{-}\dot{C}{=}\ddot{C}^{-}{-}H + H{-}C{\equiv}C{-}R \longrightarrow R{-}\dot{C}{=}CH_2 + Na^+\ {:}\bar{C}{\equiv}C{-}R$$

radical anion; vinylic radical; acetylenic anion

The vinylic radical is reduced as in the case of internal alkynes to give a vinylic anion:

$$R{-}\dot{C}{=}CH_2 + \dot{Na} \longrightarrow Na^+{:}\bar{C}(R){=}CH_2$$

vinylic radical; vinylic anion

Here we have another strong base; the pK_a of a vinylic proton is about 42. This anion therefore consumes yet another molecule of 1-alkyne as the alkene is formed.

$$Na^+{:}\bar{C}(R){=}CH_2 \quad H{-}C{\equiv}C{-}R \longrightarrow RCH{=}CH_2 + Na^+\ {:}\bar{C}{\equiv}C{-}R$$

vinylic anion; alkene; acetylenic anion

Hence, we form *two* acetylenic anions for every one alkene formed; the acetylenic anions do not react further.

(b) Presumably, the side reactions above could be avoided if a proton source were present that is a stronger acid than the acetylide ion. Ammonium sulfate (that is, $^+NH_4$ ion) was used for this purpose; the pK_a of

$^+NH_4$ is 9.2. In the mechanism of part (a), simply substitute this acid everywhere a 1–alkyne is acting as an acid. With this ploy, all of the alkyne is reduced.

14.35 (a) The IR spectrum suggests a 1-alkyne. In the NMR spectrum, the δ 3.31 singlet indicates a methoxy group, and the six-proton δ 2.41 singlet suggests two equivalent methyl groups bound to a quaternary carbon. The compound is 3-methoxy-3-methyl-1-butyne.

$$HC{\equiv}C\overset{\displaystyle OCH_3}{\underset{\displaystyle CH_3}{\overset{|}{\underset{|}{C}}}}CH_3$$ 3-methoxy-3-methyl-1-butyne

(b) The IR data suggest a 1-alkyne, and this diagnosis is confirmed by the formation of a gas (ethane) when the compound is treated with C_2H_5MgBr. The three-proton δ 3.41 singlet suggests a methoxy group. The structure is 3-methoxypropyne.

$HC{\equiv}CCH_2OCH_3$ 3-methoxypropyne

(c) The IR data suggest that the compound is both a 1-alkyne and an alcohol. The δ 4.6 resonance for 1*H* indicates that the alcohol is secondary. Its splitting also shows that this α-hydrogen is adjacent to a methyl group, and its chemical shift shows that other functionality, probably the C≡CH group, is adjacent. The dry/wet contrast also indicates that this is the resonance of an alcohol α-proton. A small splitting of the α-hydrogen by the acetylenic hydrogen would account for the secondary splitting of each quartet line into a doublet. (Recall that vinylic, and by extension, acetylenic, splitting can occur over four bonds.) The spectrum is consistent with only the following structure.

$$HC{\equiv}C\overset{\displaystyle OH}{\overset{|}{C}}HCH_3$$ 3-butyn-2-ol

The methyl doublet at δ 1.5, the slightly split acetylenic-proton resonance at δ 2.5, and the D_2O-exchangeable resonance at δ 6.5 are all consistent with this structure.

(d) The problem indicates an alkyne. The presence of a methoxy group is indicated by the three-proton singlet at δ 3.79. There is one additional degree of unsaturation; because three carbons are already accounted for, the unsaturation must be in the remaining two. Consequently, there is a double bond. Cis stereochemistry is suggested by the small 6-Hz coupling constant between the vinylic protons. (See Table 13.3 on text p. 576). The splitting suggests that one of the vinylic protons (δ 4.52) is also coupled to an acetylenic proton. The chemical shift of the other vinylic proton (δ 6.38) suggests that it is α to the methoxy oxygen. All these data conspire to give the following structure:

$CH_3O(H)C{=}C(H)C{\equiv}CH$ (cis: CH_3O and C≡CH on the same side) (*Z*)-1-methoxy-1-buten-3-yne

14.36 (a) The IR absorptions indicate a carbon-carbon triple bond (2100 cm^{-1}) and a 1-alkyne (3300 cm^{-1}). The CMR data show that there are only four nonequivalent sets of carbons. The compound is

$(CH_3)_3CC{\equiv}CH$ 3,3-dimethyl-1-butyne (*tert*-butylacetylene)

(See also the solution to Problem 14.4.)

(b) Assuming that all resonances for chemically nonequivalent carbons are separately observable, 1-hexyne should have a CMR spectrum consisting of six resonances; 4-methyl-2-pentyne should have a CMR spectrum consisting of five, because two of the methyl groups are chemically equivalent.

chemically equivalent methyl groups → $(CH_3)_2CHC{\equiv}CCH_3$

4-methyl-2-pentyne

14.37 (a) This reaction consists of two successive Brønsted acid-base reactions.

$$PhC{\equiv}C{-}H + {:}\ddot{O}D^{-} \rightleftharpoons PhC{\equiv}C{:}^{-} + H{-}\ddot{O}D \rightleftharpoons PhC{\equiv}C{-}D + {:}\ddot{O}D^{-}$$

(b) A bromohydrin is formed; however, because the bromohydrin is also an enol, it is converted into the α-bromo ketone. (See Eqs. 14.6d–e on text p. 616 for the mechanism of the enol-to-ketone conversion.)

$$PhC{\equiv}CH \xrightarrow{Br_2} PhC^{+}{=}CH(Br) \xrightarrow{H_2\ddot{O}:} PhC(\overset{+}{O}H_2){=}CH(Br) \xrightarrow{H_2\ddot{O}:} PhC(\ddot{O}H){=}CH(Br) \longrightarrow PhC(=O){-}CH_2Br$$

enol form of product

$$+\ H_2\overset{+}{\ddot{O}}{-}H$$

A carbocation rather than a bromonium ion is shown above as the reactive intermediate in bromine addition, although a bromonium ion is a possible alternative. Because of its carbon-carbon double bond, a bromonium ion derived from bromine addition to an alkyne would be more strained than a bromonium ion derived from bromine addition to an alkene, and might be unstable enough to open to a carbocation.

14.38 The ozonolysis results define compound *B* as 1,5-hexadiene. Note that the formula of compound *B* is C_6H_{10}. Because compound *B* is produced by hydrogenation of compound *A*, and because *two equivalents* of H_2 must be added to compound *A* (C_6H_6) to give compound *B* (C_6H_{10}), compound *A* must be 1,5-hexadiyne.

$$\underset{\text{1,5-hexadiyne (compound } A)}{HC{\equiv}CCH_2CH_2C{\equiv}CH} \xrightarrow{H_2,\ \text{Lindlar cat.}} \underset{\text{1,5-hexadiene (compound } B)}{H_2C{=}CHCH_2CH_2CH{=}CH_2} \xrightarrow[2)\ H_2O,\ H_2O_2]{1)\ O_3} \underset{\text{succinic acid}}{HO\overset{O}{\overset{\|}{C}}CH_2CH_2\overset{O}{\overset{\|}{C}}OH} + \underset{\text{formic acid}}{2\ H\overset{O}{\overset{\|}{C}}OH}$$

14.39 The elemental analysis of *A* gives the empirical formula C_5H_7. Because the hydrogenation product of *A* contains ten carbons, the likely molecular formula is $C_{10}H_{14}$. Compound *A* has the carbon skeleton of butylcyclohexane (a six-membered ring bearing an unbranched chain of four carbons). Compound *A* also has four degrees of unsaturation. One degree of unsaturation is accounted for by the ring. Two others are accounted for by the triple bond; remember, it was given that compound *A* is an alkyne. Because there are no other rings, the one remaining degree of unsaturation must be a double bond. The triple bond cannot, of course, be within the ring, nor can it be at the end of the four-carbon chain because of the absence of a reaction of *A* with ethylmagnesium bromide. Hydrogenation of *A* under conditions that hydrogenate triple bonds to double bonds but leave double bonds unaffected gives an alkene which, on ozonolysis, loses *two* carbons. The fact that a *tricarboxylic* acid is formed on ozonolysis shows that the double bond of *A* is within the ring. It is now possible to write several preliminary structures for *A*:

$CH_2C{\equiv}CCH_3$ (on cyclohexene ring) — *A1*

$CH_2C{\equiv}CCH_3$ (on cyclohexene ring) — *A2*

$CH_2C{\equiv}CCH_3$ (on cyclohexene ring) — *A3*

Compound *A1* is ruled out because it is not chiral; recall that compound *A* is optically active. To distinguish between *A2* and *A3*, imagine the results of hydrogenation over a poisoned catalyst followed by ozonolysis:

$$\text{A2 (3-(2-butynyl)cyclohexene, } CH_2C{\equiv}CCH_3\text{)} \xrightarrow[\text{quinoline}]{H_2,\ Pd/C} \text{(cis) cyclohexenyl}-CH_2CH{=}CHCH_3 \xrightarrow[2)\ H_2O_2]{1)\ O_3} HO_2C-CH_2CH_2CH_2-CH(CO_2H)-CH_2CO_2H + CH_3CO_2H$$

chiral; would be optically active if *A2* were optically active

$$\text{A3 (4-(2-butynyl)cyclohexene, } CH_2C{\equiv}CCH_3\text{)} \xrightarrow[\text{quinoline}]{H_2,\ Pd/C} \text{(cis) cyclohexenyl}-CH_2CH{=}CHCH_3 \xrightarrow[2)\ H_2O_2]{1)\ O_3} HO_2C-CH_2-CH(CH_2CO_2H)-CH_2CO_2H + CH_3CO_2H$$

achiral; cannot be optically active

The observation of optical activity in the ozonolysis product rules out *A3* and is consistent only with *A2* as the structure of compound *A*.

compound *A* = *A2* (cyclohexene ring bearing $CH_2C{\equiv}CCH_3$)

3-(2-butynyl)cyclohexene

The data, of course, do not determine which enantiomer compound *A* is.

14.40 (a) The Grignard reagent converts 1-hexyne into the acetylenic Grignard reagent; see Eq. 14.24 on text p. 625. Protonolysis of the Grignard reagent by D_2O gives $CH_3CH_2CH_2CH_2C{\equiv}CD$ as the final product.

(b) Butyllithium is a strong base; like a Grignard reagent, it converts a 1-alkyne into a lithium acetylide. Lithium acetylides, like acetylenic Grignard reagents, react as if they were acetylenic anions. Consequently, the lithium acetylide, as suggested by the hint, reacts as a nucleophile, displacing chloride ion from $(CH_3)_3Si{-}Cl$.

$$CH_3CH_2CH_2CH_2C{\equiv}CH \xrightarrow{CH_3CH_2CH_2CH_2Li} CH_3CH_2CH_2CH_2C{\equiv}C{-}Li + CH_3CH_2CH_2CH_3 \xrightarrow{(CH_3)_3SiCl}$$

$$CH_3CH_2CH_2CH_2C{\equiv}CSi(CH_3)_3 + Li^+\ Cl^-$$

(c) The acetylenic anion, which is formed by the reaction of 1-octyne with $NaNH_2$, is alkylated by diethyl sulfate. The product is 3-undecyne.

$$CH_3(CH_2)_6C{\equiv}\overset{-}{C}{:}\ Na^+ \quad CH_3CH_2{-}\ddot{O}{-}S(=O)_2{-}O{-}CH_2CH_3 \longrightarrow$$

$$CH_3(CH_2)_6C{\equiv}CCH_2CH_3 + {}^{-}{:}\ddot{O}{-}S(=O)_2{-}O{-}CH_2CH_3\ Na^+$$

3-undecyne

(d) The lithium acetylide undergoes an S_N2 reaction with the alkyl halide; the better leaving group (chloride) is expelled. The product is

$$F(CH_2)_5C{\equiv}CH + Li^+ \; Cl^-$$

7-fluoro-1-heptyne

(e) Bromine adds to the double bond; both bromines are subsequently eliminated in consecutive $NaNH_2$-promoted E2 reactions to give phenylacetylene. (Remember, $^-NH_2$ is a strong base and, like many strong bases, can promote the E2 reaction.)

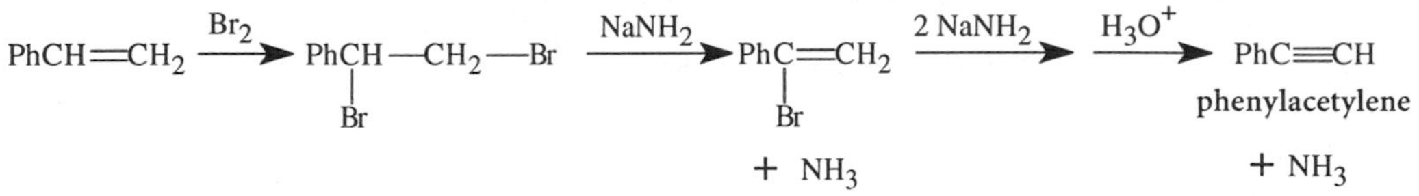

$$PhCH{=}CH_2 \xrightarrow{Br_2} PhCH(Br){-}CH_2{-}Br \xrightarrow{NaNH_2} PhC(Br){=}CH_2 + NH_3 \xrightarrow{2\ NaNH_2} \xrightarrow{H_3O^+} PhC{\equiv}CH + NH_3$$

phenylacetylene

Two equivalents of $NaNH_2$ are required in the last step because one equivalent would be consumed by reaction with the phenylacetylene product to give its conjugate-base acetylenic anion. This anion is converted back into phenylacetylene by addition of acid (or water) at the conclusion of the reaction.

(f) As suggested by the hint, chloroform and a strong base give dichloromethylene, a carbene, which adds to the triple bond to give a cyclopropene.

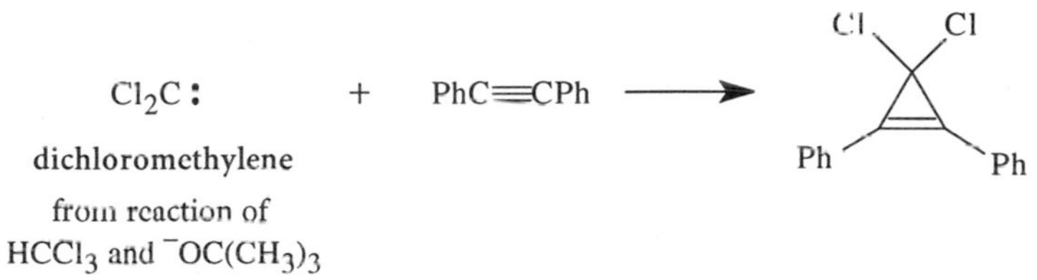

$$Cl_2C\!: \; + \; PhC{\equiv}CPh \longrightarrow \text{3,3-dichloro-1,2-diphenylcyclopropene (Cl, Cl; Ph, Ph)}$$

dichloromethylene

from reaction of $HCCl_3$ and $^-OC(CH_3)_3$

15

Dienes, Resonance, and Aromaticity

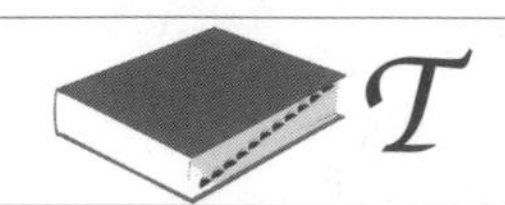

TERMS

absorbance, A 15.2A
1,4-addition 15.3A, 15.4A
1,2-addition 15.4A
allene .... Introduction
allylic 15.4B
allylic carbocation 15.4B
antiaromatic 15.7E
antibonding π^* molecular orbital 15.2B
aromatic 15.7A, 15.7D
aromatic compound 15.7A
aromaticity 15.7D
Beer's law 15.2A
Beer-Lambert law 15.2A
benzene 15.7A
bonding π molecular orbital 15.2B
chromophore 15.2B
cisoid conformation 15.1A
conjugate addition 15.3A, 15.4A
conjugated diene .... Introduction, 15.1A
conjugated double bond .... Introduction, 15.1A
copolymer 15.5
crosslink 15.5
cumulated double bond .... Introduction, 15.1B
cumulene .... Introduction, 15.1B
cycloaddition reaction 15.3A
1,3-cyclobutadiene 15.7E
1,3,5,7-cyclooctatetraene (COT) 15.7A
delocalization of electrons 15.6A
Diels-Alder reaction 15.3A
diene .... Introduction, 15.3A
diene polymer 15.5
dienophile 15.3A
electronic spectrum 15.2B
empirical resonance energy 15.7C
endo 15.3C
excited state 15.2B
exo 15.3C
ferrocene 15.7D
ground state 15.2B
Hückel $4n + 2$ rule 15.7D
kinetically controlled reaction 15.4C
Ladenburg benzene 15.7A
lambda-max (λ_{max}) 15.2A
millimicron, mμ 15.2A
molar absorptivity, ϵ 15.2A
molar extinction coefficient, ϵ 15.2A
$4n + 2$ rule 15.7D
nanometer, nm 15.2A
near ultraviolet 15.2
optical density, OD 15.2A
pericyclic reaction 15.3A
$\pi \rightarrow \pi^*$ transition 15.2B
prismane 15.7A
resonance-stabilized 15.6
s-cis conformation 15.1A
s-trans conformation 15.1A
styrene-butadiene rubber, SBR 15.5
thermodynamically controlled reaction 15.4C
transoid conformation 15.1A
ultraviolet-visible (UV-VIS) spectroscopy 15.2
UV spectrometer 15.2A
UV spectroscopy 15.2
UV spectrum 15.2
vulcanization 15.5

CONCEPTS

I. STRUCTURE AND REACTIVITY OF DIENES

A. Introduction

1. Dienes are compounds with two carbon-carbon double bonds.
   a. Conjugated dienes have two double bonds that are separated by one single bond; the double bonds are called conjugated double bonds.

$$H_3C—CH{=}CH—CH{=}CH—CH_3 \qquad \text{a conjugated diene}$$

   b. Cumulenes are compounds in which one carbon participates in two carbon-carbon double bonds; the double bonds are called cumulated double bonds.

$$H_3C—CH{=}C{=}CH—CH_3 \qquad \text{a cumulated diene}$$

      i. Allene is the simplest cumulene.
      ii. The term allene is also sometimes used as a family name for compounds containing only two cumulated double bonds.
2. Conjugated dienes and allenes have unique structures and chemical properties.
3. Dienes in which the double bonds are separated by two or more single bonds ("ordinary" dienes) have structures and chemical properties more or less like those of simple alkenes.

B. Conjugated Dienes

1. Conjugated dienes are more stable than their unconjugated isomers because:
   a. The carbon-carbon single bond between the two double bonds in a conjugated diene is derived from the overlap of two carbon sp^2 orbitals.
      i. An sp^2-sp^2 single bond is a stronger bond than the sp^2-sp^3 single bond of an ordinary diene and thus contributes to the stability of a conjugated diene.
      ii. The sp^2-sp^2 single bond is shorter than the sp^2-sp^3 or sp^3-sp^3 carbon-carbon single bond.
   b. There is overlap of 2*p* orbitals across the carbon-carbon bond connecting the two alkene units.
      i. There is π bonding between as well as within each of the two alkene units.
      ii. The additional bonding associated with this overlap provides additional stability to the molecule.

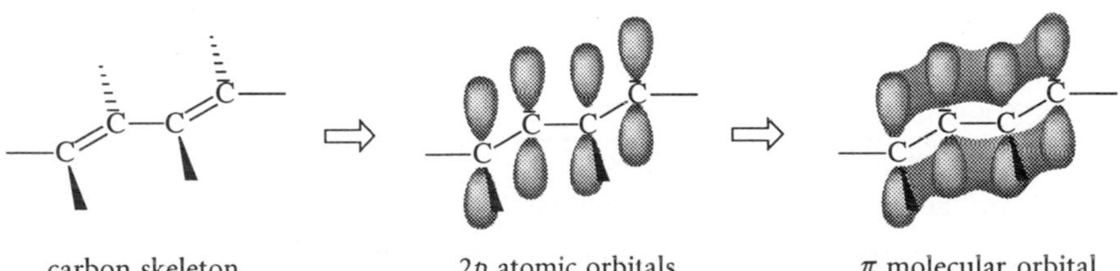

carbon skeleton — 2*p* atomic orbitals — π molecular orbital

2. Internal rotation about the single bond of 1,3-butadiene results in two stable conformations.
   a. The transoid or *s*-trans conformation.
   b. The cisoid or *s*-cis conformation.

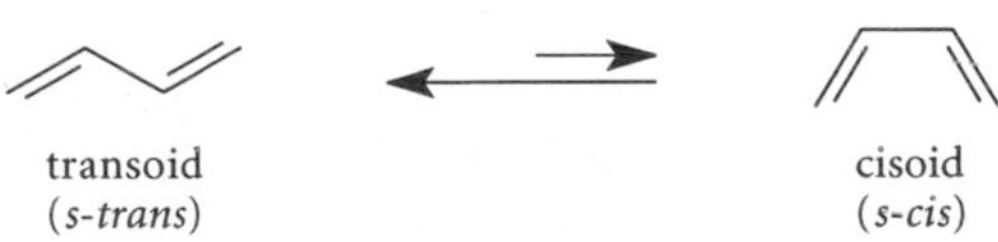

transoid (*s-trans*) — cisoid (*s-cis*)

   c. The *s*- prefix refers to the conformation about the *single* bond.

d. Both forms are planar, or nearly so.
   i. The *s*-trans form of the molecule of an acyclic diene is more stable than the *s*-cis form.
   ii. The internal rotation that interconverts these two forms is very rapid at room temperature.

C. Cumulated Dienes

1. The carbon skeleton of allene is linear.
   a. The central carbon of allene is *sp*-hybridized; the two remaining carbons are sp^2-hybridized and have trigonal geometry.
   b. The two bonding π molecular orbitals in allenes are mutually perpendicular; these two π systems do not overlap.
   c. The H—C—H plane at one end of the allene molecule is perpendicular to the H—C—H plane at the other end.

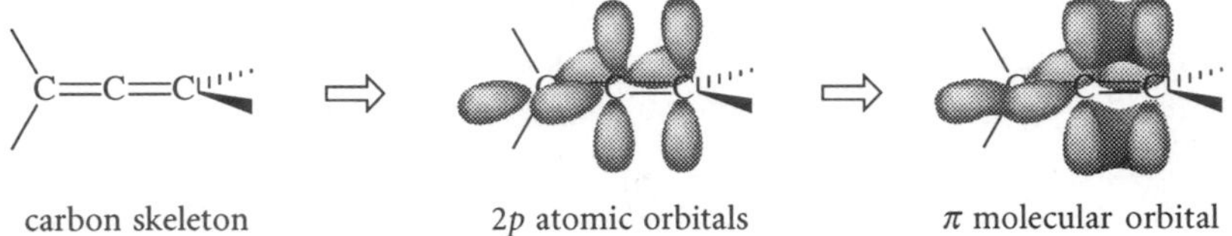

carbon skeleton 2*p* atomic orbitals π molecular orbital

2. Because of their geometries, some allenes are chiral even though they do not contain an asymmetric carbon atom.
   a. The two sp^2-hybridized carbons can be stereocenters.
   b. The perpendicularity of the planes of the two trigonal carbons can give rise to enantiomeric allenes.

H_3C(H)C=C=C(H)(CH_3) ← enantiomers → (H)(H_3C)C=C=C(CH_3)(H)

   c. Enantiomeric allenes can be isolated since internal rotation about a double bond does not occur.
3. The bond length of each cumulated double bond is somewhat less than that of an ordinary alkene double bond.
4. The cumulated arrangement is the least stable arrangement of two double bonds; allenes also have greater heats of formation than isomeric alkynes.
5. The C=C stretching absorption in the infrared spectrum of allenes is around 1950 cm^{-1}.
6. Naturally occurring allenes are relatively rare.

II. RESONANCE

A. Resonance Structures

1. Molecules that can be represented as resonance hybrids are said to be resonance-stabilized; that is, they are more stable than they would be if they had any one of their contributing structures.
   a. The additional stabilization results from additional bonding, which in turn results from the delocalization of electrons and additional orbital overlap.
   b. Resonance structures can be derived by the curved-arrow notation.
   c. Resonance structures show the delocalization of electrons.
2. The derivation and use of resonance structures is important for understanding both molecular structure and molecular stability.
3. Resonance structures can be drawn when bonds, unshared electron pairs, or single electrons can be delocalized (moved) by the curved-arrow notation without moving any atoms.
   a. Resonance structures are usually placed in brackets to emphasize the fact that they are being used to describe a single species.

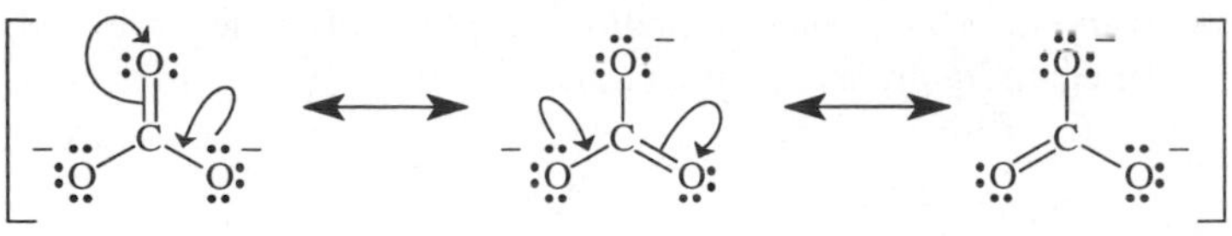

b. Resonance structures are used to describe single molecules; a molecule is the weighted average of its resonance structures.

c. The resonance arrow means that two or more fictitious structures are being used to represent a compound for which a single Lewis structure is inadequate.

B. Relative Importance of Resonance Structures

1. The structure of a molecule is most closely approximated by its most important resonance structures.
2. A molecule is a weighted average of its contributing resonance structures; some resonance structures are more important than others.
   a. To evaluate the relative importance of resonance structures, compare the stabilities of the resonance structures for a given molecule by treating each structure as a separate molecule.
   b. The most stable structures are the most important ones.
3. Guidelines for important resonance structures:
   a. Identical structures are equally important descriptions of a molecule.

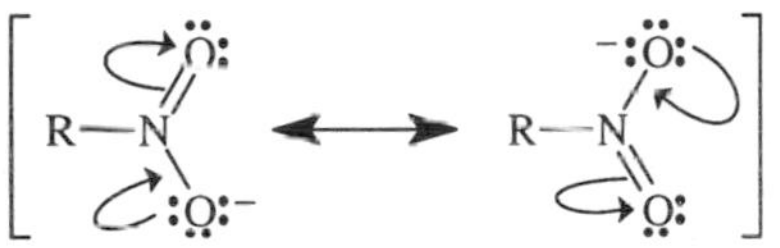

identical resonance structures (equally important)

   b. Structures with the greater number of bonds are more important; bonding is energetically favorable.

$H_2C{=}CH_2 \longleftrightarrow H_2\dot{C}{-}\dot{C}H_2$

not an important structure (fewer bonds)

   c. Structures that require the separation of opposite charges are less important than those that do not.

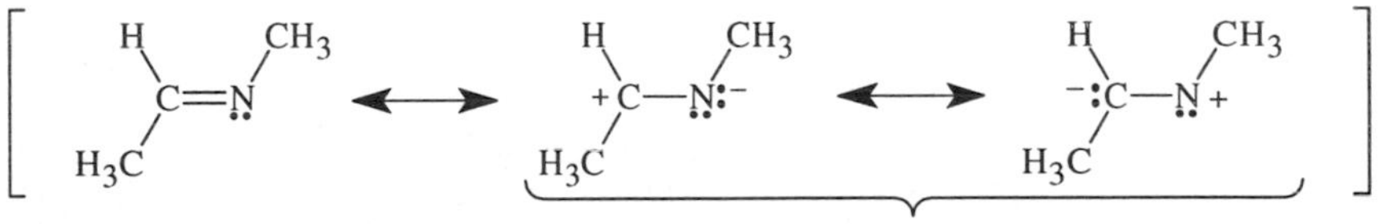

separation of charges (less important)

      i. When a charge is separated, two opposite charges are moved away from each other.

$H_3C{-}C{\equiv}N: \longleftrightarrow H_3C{-}\overset{+}{C}{=}\ddot{N}:^-$ charge separation

      ii. When a charge is delocalized by resonance, charge of a given type is moved to different locations within a molecule.

$H_2\ddot{C}^-{-}C{\equiv}N: \longleftrightarrow H_2C{=}C{=}\ddot{N}:^-$ charge delocalization

d. Structures in which charges and/or electron deficiency are assigned to atoms of appropriate electronegativity are more important.

most important (more bonds) | next most important (charges match electronegativities) | unimportant (charges do not match electronegativites)

e. If the orbital overlap symbolized by a resonance structure is not possible, the resonance structure is not important.

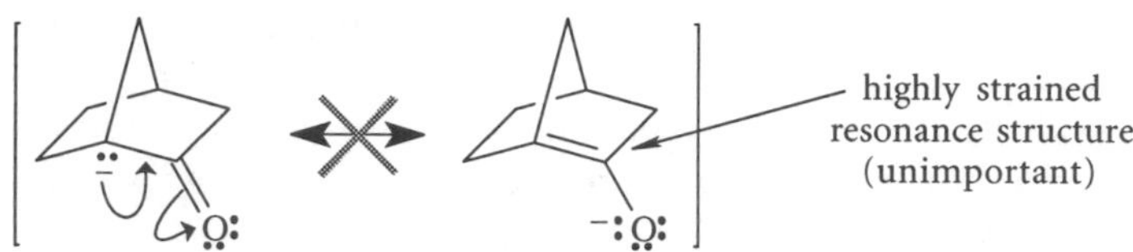

4. All other things being equal, a molecule with a greater number of important resonance structures is more stable than an isomeric molecule with fewer resonance structures.

III. BENZENE

A. Introduction

1. Benzene and its derivatives constitute the class of organic substances called aromatic compounds.
2. Benzene undergoes few of the addition reactions that are associated with either conjugated dienes or ordinary alkenes. When it does undergo addition, harsh conditions are required.

B. Structure of Benzene

1. The structure of benzene has only one type of carbon-carbon bond with a bond length intermediate between the lengths of single and double bonds.
   a. All atoms in benzene lie in one plane.
   b. Each carbon atom is trigonal (sp^2-hybridized).
      i. All six $2p$ orbitals are parallel.
      ii. These $2p$ orbitals overlap to form a continuous bonding π molecular orbital.

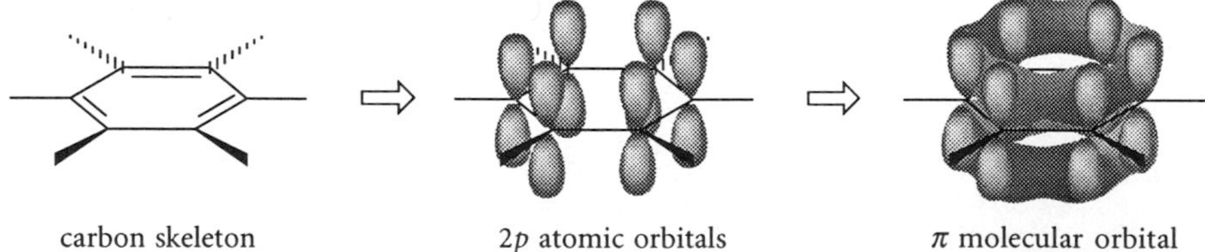

carbon skeleton | $2p$ atomic orbitals | π molecular orbital

2. Benzene can be depicted as the hybrid of two equally contributing resonance structures.

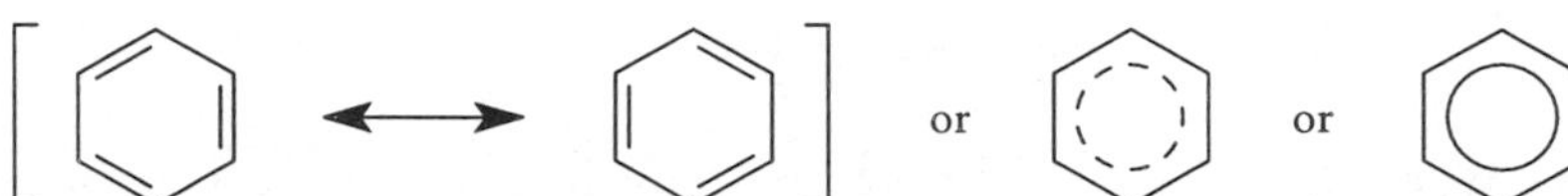

3. Benzene is more stable than a hypothetical six-carbon cyclic triene by 126–172 kJ/mol (30–41 kcal/mol); this energy difference is called the empirical resonance energy.
4. The empirical resonance energy of benzene is an estimate of how much special stability is implied by the resonance structures for benzene.

IV. AROMATICITY

A. Aromaticity and the Hückel $4n + 2$ Rule

1. The unusual stability of benzene is called aromaticity; aromaticity is a property of certain cyclic compounds.
2. To be aromatic, a compound must conform to all of the following criteria (often called collectively the Hückel $4n + 2$ rule or simply the $4n + 2$ rule):
   a. Aromatic compounds contain one or more rings that have a cyclic arrangement of *p* orbitals.
   b. Every atom of an aromatic ring has a *p* orbital; thus, aromatic compounds have a continuous cycle of *p* orbitals.
   c. Aromatic compounds are planar; that is, the *p* orbitals must overlap to form π molecular orbitals.
   d. The cyclic arrangement of *p* orbitals in an aromatic compound must contain $4n + 2$ π electrons, where n is any positive integer (0, 1, 2, ...); thus, an aromatic ring must contain 2, 6, 10, ... π electrons.
3. Electron counting rules:
   a. Each atom that is part of a double bond contributes one π electron.
   b. Vinylic unshared electrons (unshared electrons on doubly bonded atoms) do not contribute to the π-electron count.
   c. Allylic unshared electrons contribute to the π-electron count if they occupy an orbital parallel to the other *p* orbitals in the molecule.
   d. An atom with an empty *p* orbital can be part of a continuous aromatic π-electron system, but contributes no electrons.
4. Aromatic compounds are of many different types:
   a. Aromatic compounds may contain heteroatoms (aromatic heterocycles).
   b. Some ionic species are aromatic.
   c. Fused bicyclic and polycyclic compounds can be aromatic.
   d. Some remarkable organometallic compounds have aromatic character.

pyrrole
(an aromatic heterocycle)

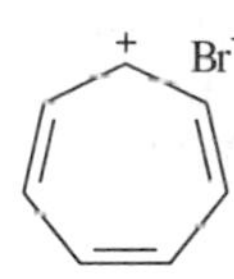

tropylium bromide
(an ionic aromatic compound)

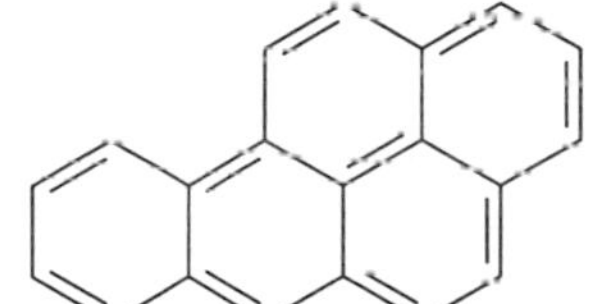
benzo[*a*]pyrene
(a fused polycyclic aromatic compound)

ferrocene
(an aromatic organometallic compund)

5. The source of aromatic stability is the complete filling of the bonding molecular orbitals of cyclic π-electron systems and the particularly low energy of these molecular orbitals.

B. Antiaromatic compounds

1. Compounds that contain planar, continuous rings of $4n$ π electrons are especially unstable and are said to be antiaromatic.

1,3-cyclobutadiene
antiaromatic (4 π electrons)

2. The overlap of *p* orbitals in molecules with cyclic arrays of $4n$ π electrons is a destabilizing effect.
3. The distortion of antiaromatic molecules occurs, when possible, so that the destablizing overlap is minimized.

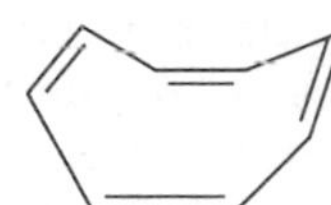
1,3,5,7-cyclooctatetraene
nonplanar to avoid antiaromaticity

V. ULTRAVIOLET SPECTROSCOPY

A. Introduction

1. In ultraviolet-visible (UV-VIS) spectroscopy, the absorption of radiation by a compound of interest in the ultraviolet or visible region of the spectrum is recorded as a function of wavelength.
   a. Both UV and visible spectroscopy are considered together as one type of spectroscopy, often called simply UV spectroscopy.
   b. The part of the ultraviolet spectrum of greatest interest to organic chemists is the near ultraviolet range (200–400 nm).
2. The UV spectrum of a substance is the graph of radiation absorption by a substance versus the wavelength of radiation. The UV spectrum is determined in an instrument called a UV spectrophotometer.
   a. The wavelength λ of the ultraviolet radiation (in nanometers, nm) is plotted on the horizontal axis of the UV spectrum.
   b. The absorbance A (or optical density, OD) is plotted on the vertical axis of a UV spectrum; the absorbance is a measure of the amount of radiation energy absorbed.
      i. The radiation entering a sample has intensity I_0.
      ii. The light emerging from the sample has intensity I.
      iii. The absorbance A is defined as the logarithm of the ratio I_0/I.

$$A = \log\left(\frac{I_0}{I}\right)$$

   c. Absorption peaks in the UV spectra of compounds in solution are generally quite broad.
   d. The wavelength at the maximum of an absorption peak is called the λ_{max}; some compounds have several absorption peaks and a corresponding number of λ_{max} values.
3. Ultraviolet-visible spectroscopy can be used to identify organic compounds containing conjugated π-electron systems.
4. UV spectroscopy is frequently used for quantitative analysis.
   a. The absorbance at a given wavelength depends on the number of molecules in the light path.
   b. The Beer-Lambert law (usually called Beer's law) states that the absorbance A is proportional to the product of the path length (l) and the concentration (c).

$$A = \epsilon lc$$

   where ϵ, the constant of proportionality, is called the molar extinction coefficient or molar absorptivity.
      i. Each absorption in a given spectrum has a unique extinction coefficient.
      ii. Each molar extinction coefficient depends on wavelength, solvent, and temperature.
      iii. The larger the ϵ, the greater is the light absorption at a given concentration c and path length l; thus, strong absorptions can be obtained from very dilute solutions if the absorbing molecule has a high molar extinction coefficient ϵ.
      iv. Extinction coefficients of conjugated alkenes are typically in the range of 10^3–10^5 $M^{-1} \cdot \text{cm}^{-1}$.
5. Some UV spectra are presented in abbreviated form by citing the λ_{max} values of their principal peaks, the solvent used, and the extinction coefficients.
6. UV spectroscopy is especially useful for the diagnosis of conjugated double or triple bonds.

B. Physical Basis of UV Spectroscopy

1. Ultraviolet and/or visible radiation is absorbed by the π electrons and, in some cases, by the unshared electron pairs in organic compounds.
   a. Absorptions by compounds containing only unshared electron pairs are generally quite weak.
   b. UV and visible spectra are sometimes called electronic spectra.
2. The structural feature of a molecule responsible for its UV or visible absorption is called a chromophore.
3. UV absorptions of conjugated alkenes are due to $\pi \rightarrow \pi^*$ transitions.

a. When a chromophore absorbs energy from light, a π electron is elevated from a bonding molecular orbital (ground state) to an antibonding π^* molecular orbital (excited state) in a process called a $\pi \rightarrow \pi^*$ transition.
b. The energy required for $\pi \rightarrow \pi^*$ absorption must match the difference in energy between the π and π^* orbitals. (See Study Guide Link 15.1.)

C. UV Spectroscopy of Conjugated Alkenes

1. When UV spectroscopy is used to determine chemical structure, the most important aspect of a spectrum is the λ_{max} values.
2. The structural features of a compound that are most important in determining the λ_{max} are
   a. the number of consecutive conjugated double or triple bonds (the longer the conjugated π-electron system, the higher the wavelength of the absorption).
      i. The λ_{max} increases with increasing number of conjugated double bonds.
      ii. Molecules that contain many conjugated double bonds generally have several absorption peaks; the λ_{max} usually quoted for such compounds is the one at highest wavelength.
      iii. When one or more of the λ_{max} values of a compound is large enough to fall within the visible region of the electromagnetic spectrum, the compound appears colored. This situation can occur, for example, with alkenes that have many double bonds in conjugation.
   b. the conformation of a diene unit about its central single bond.
      i. Noncyclic dienes generally assume the lower energy *s*-trans, or transoid, conformation.
      ii. Dienes that are locked into *s*-cis, or cisoid, conformations have higher values of λ_{max} and lower extinction coefficients than comparably substituted *s*-trans, or transoid, compounds.
   c. the presence of substituent groups on the double bond.
      i. Each alkyl substituent on a double bond adds +5 nm to the base λ_{max} of the conjugated system.
3. Light absorption by a pigment, rhodopsin, in the rod cells (as well as a related pigment in the cone cells) of the eye triggers the series of physiological events associated with vision.

VI. GENERAL CONCEPTS

A. Allylic Carbocations

1. An allylic carbocation is a carbocation adjacent to a double bond.

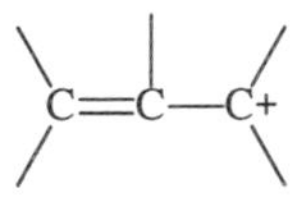

$$\mathrm{C{=}C{-}C^{+}}$$ allylic carbocation

2. Allylic cations are more stable than comparably substituted nonallylic alkyl cations.
   a. The unusual stability of allylic carbocations lies in their electronic structures.
   b. The electron-deficient carbon and the carbons of the double bond are all sp^2-hybridized (each carbon has a $2p$ orbital).
      i. The overlap of these $2p$ orbitals provides additional bonding in allylic carbocations and hence, additional stability.
      ii. Both the positive charge and the double-bond character are delocalized.

$$\left[\mathrm{H_2\overset{+}{C}{-}C{=}CH_2} \longleftrightarrow \mathrm{H_2C{=}C{-}\overset{+}{C}H_2} \right]$$

3. Allylic carbocations are represented as resonance hybrids.
4. Relative stability of carbocations:

primary alkyl < primary allylic ≈ secondary alkyl < secondary allylic ≈ tertiary alkyl < tertiary allylic

B. Kinetic and Thermodynamic Control

1. When the product distribution in a reaction differs substantially from the product distribution that would be observed if the products were at equilibrium, the reaction is said to be kinetically controlled.
2. If the products of a reaction come to equilibrium under the reaction conditions, the product distribution is said to be thermodynamically controlled.
3. When the less stable product of a reaction is the major product, then two things must be true:
   a. The less stable product must be formed more rapidly than the other products.
   b. The products must not come to equilibrium under the reaction conditions.
4. The precise reason for kinetic control varies from reaction to reaction; the relative amounts of products in a kinetically controlled reaction are determined by the relative energies of the transition states for each of the product-determining steps.
5. In a kinetically-controlled reaction, the most stable product is not always the one formed in greatest amount.

$$HCl + H_2C{=}CH{-}CH{=}CH_2 \longrightarrow H_3C{-}\underset{}{\overset{Cl}{CH}}{-}CH{=}CH_2 + (H_3C)(H)C{=}C(H)(CH_2Cl)$$

predominant product (less stable alkene)

minor product (more stable alkene)

REACTIONS

I. ADDITION OF HYDROGEN HALIDES TO CONJUGATED DIENES

A. 1,2- and 1,4-Additions

1. Conjugated dienes react with hydrogen halides to give two types of addition product.
   a. The major product results from 1,2-addition. In a 1,2-addition, the hydrogen and halogen add to adjacent carbons.
   b. The minor product results from 1,4-addition, or conjugate addition. In a 1,4-addition, the hydrogen and halogen add to carbons that have a 1,4-relationship.

[Scheme: 1-methylcyclopentadiene $\xrightarrow{HCl}$ 1,2-addition product (H, Cl, CH_3) + 1,4-addition product (H, Cl, CH_3)]

1,2-addition product 1,4-addition product

2. The 1,2-addition mechanism is analogous to the addition of HX with an ordinary alkene.
3. The 1,4-addition, or conjugate-addition, mechanism involves:
   a. protonation of a double bond to give an allylic carbocation in which the positive charge in this ion is not localized, but is instead shared by two different carbons.
   b. formation of two constitutional isomers by attack of halide ion on either of the electron-deficient carbons.

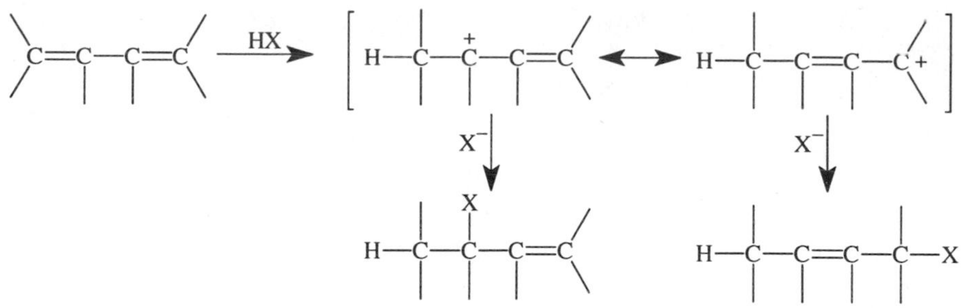

1,2-addition product

1,4-addition product

4. The product distribution in hydrogen-halide addition to a conjugated diene is determined by the relative rates of the product-determining steps.
   a. Formation of the 1,2-addition product is faster but reversible.
   b. Formation of the 1,4-addition product is slower but virtually irreversible.

B. Diene Polymers

1. The polymerization of 1,3-butadiene gives polybutadiene, an important type of synthetic rubber used in the manufacture of tires.

$$n\ H_2C{=}CH{-}CH{=}CH_2 \xrightarrow[\text{initiator}]{\text{free-radical}} \left(CH_2{-}CH{=}CH{-}CH_2\right)_n$$

1,3-butadiene — polybutadiene

   a. Polybutadiene is referred to as a diene polymer because it comes from polymerization of a diene monomer.
   b. Polybutadiene has only one double bond per unit, because one double bond is lost through the addition that takes place in the polymerization process.
2. Polymerization of dienes is a free-radical reaction. Although the polymerization product is largely the result of 1,4-addition, a small amount of 1,2-addition can occur as well.
3. Styrene-butadiene rubber, SBR, is an example of a copolymer, a polymer produced by the simultaneous polymerization of two or more monomers, in this case, styrene ($PhCH{=}CH_2$) and 1,3-butadiene.

II. THE DIELS–ALDER REACTION

A. Introduction

1. Many conjugated dienes undergo addition reactions, called Diels–Alder reactions, with certain alkenes or alkynes.
   a. The conjugated diene component of this reaction is referred to simply as the diene.
   b. The alkene component of this reaction is referred to as the dienophile.

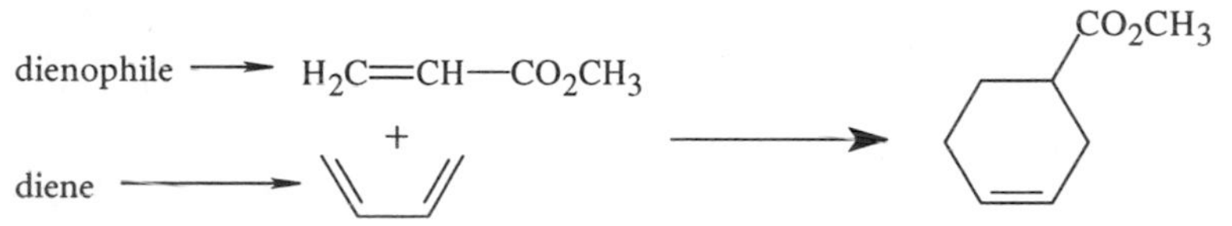

dienophile → $H_2C{=}CH{-}CO_2CH_3$

\+

diene →

CO_2CH_3

2. Some of the dienophiles that react most readily in the Diels–Alder reaction bear substituent groups such as esters ($-CO_2R$), nitriles ($-CN$), or certain other unsaturated, electronegative groups that are conjugated with the alkene double bond. (Some alkynes also serve as dienophiles.)
3. When the diene is cyclic, bicyclic products are obtained in the Diels–Alder reaction.

CO_2CH_3 + → CO_2CH_3 ≡ CO_2CH_3

4. The Diels–Alder reaction is
   a. a cycloaddition reaction (an addition reaction that results in the formation of a ring).
   b. a pericyclic reaction (a concerted reaction that involves a cyclic flow of electrons).

$H_2C{=}CH{-}CO_2CH_3$ → CO_2CH_3

   c. a 1,4-addition, or conjugate addition, reaction (an addition that occurs across the outer carbons of the diene; 1,4-addition is a characteristic type of reaction of conjugated dienes).
   d. a stereoselective *syn*-addition to both diene and dienophile.

B. Effect of Diene Conformation on the Diels–Alder Reaction

1. In the Diels–Alder transition state, the diene component is in the *s*-cis conformation.
   a. Dienes "locked" into *s*-trans, or transoid, conformations are unreactive in Diels–Alder reactions.
   b. Dienes "locked" into *s*-cis, or cisoid, conformations are unusually reactive and in many cases much more reactive than corresponding noncyclic dienes.

1,3-cyclopentadiene
"locked" into *s*-cis diene conformation by its structure
more reactive in Diels-Alder reactions

H_3C H C=C H / H C=C H CH_3

(2*E*,4*E*)-2,4-hexadiene
prefers *s*-trans conformation
less reactive in Diels-Alder reactions

2. The Diels-Alder reactions of a *cis*-diene are much slower than the corresponding reactions of its *trans*-diene stereoisomer, which does not have the destabilizing repulsion in its cisoid conformation.

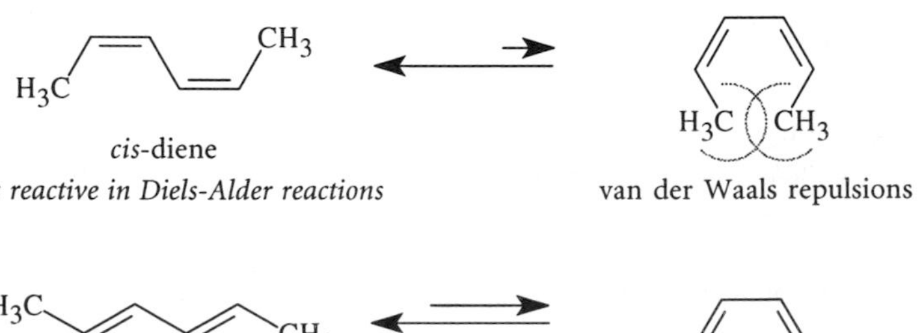

cis-diene
less reactive in Diels-Alder reactions
van der Waals repulsions

H_3C CH_3 ⇌ H_3C CH_3

trans-diene
more reactive in Diels-Alder reactions
no methyl-methyl repulsions

   a. Severe van der Waals repulsion destabilizes the cisoid conformation of the diene.
   b. Severe van der Waals repulsion destabilizes the transition states for the Diels–Alder reactions of the cisoid conformation of the diene.

C. Stereochemistry of the Diels–Alder Reaction

1. When a diene and dienophile react in a Diels–Alder reaction, they approach each other in parallel planes.
   a. This type of approach allows the π-electron clouds of the two components to overlap and form the bonds of the product.
   b. Each component adds to the other at one face.
      i. The diene undergoes a *syn*-addition to the dienophile.
      ii. The dienophile undergoes a 1,4-*syn*-addition to the diene.
   c. Groups that are cis in the alkene starting material are also cis in the product.
   d. *Syn*-addition to the diene is apparent if the terminal carbons of the diene unit are stereocenters.
2. A *syn*-addition requires that in the Diels–Alder product—
   a. the two inner substituents of the diene always have a cis relationship.
   b. the two outer substituents of the diene also always have a cis relationship.
   c. an inner substituent on one carbon of the diene is always trans to an outer substituent on the other.
   d. Cis substituents in the dienophile are cis in the product; trans substituents in the dienophile are trans in the product.

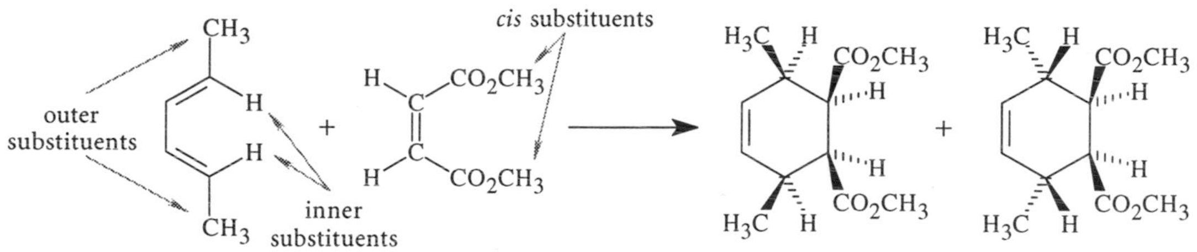

3. For a given diene and dienophile, two diastereomeric *syn*-addition products are possible and are abbreviated with the terms endo and exo.
   a. In the endo product, the dienophile substituents are cis to the outer diene substituents.

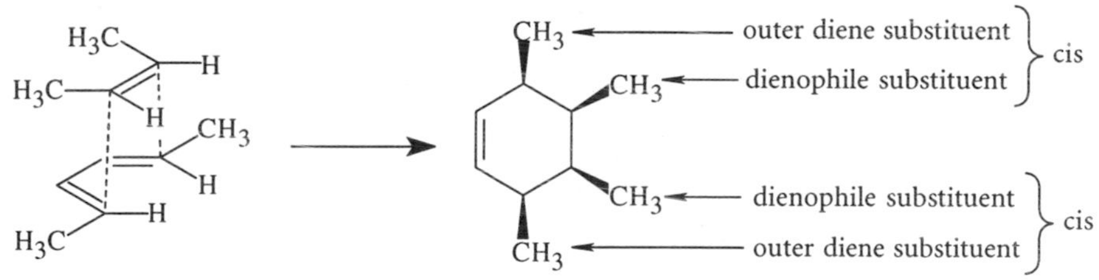

   b. In the exo product, the dienophile substituents are trans to the outer diene substituents.

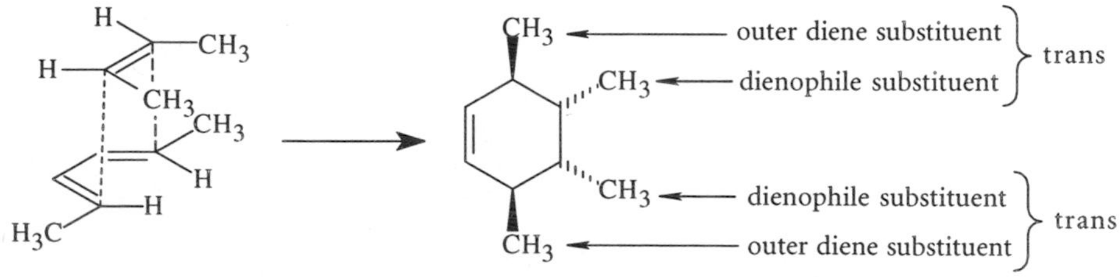

   c. In many cases, the endo mode of addition is favored over the exo mode, particularly when cyclic dienes are used.

SCIENTISTS

Kurt Adler (1902–1958) [Chemistry 1950] 15.3A
Otto Paul Hermann Diels (1876–1954) [Chemistry 1950] 15.3A
Charles Nelson Goodyear (1800–1860) 15.5
Erich Hückel (1896–1980) 15.7D
Thomas J. Katz (b. 1936) 15.7A
Fredrich August Kekulé (1829–1996) 15.7A
Albert Ladenburg (1842–1911) 15.7A

Dates in brackets refer to Nobel Prize awards.

STUDY GUIDE LINKS

15.1 More on UV Spectroscopy

This Study Guide Link considers two questions about UV-visible spectroscopy.

1. Why do the wavelengths of $\pi \rightarrow \pi^*$ transitions increase with the number of conjugated double bonds?
2. Why are the peaks in UV spectra so broad?

Why do wavelengths of $\pi \rightarrow \pi^*$ transitions increase with increasing numbers of conjugated double bonds? The reason lies in molecular orbital theory. Specifically, the answer lies in the *relative energies* of the molecular orbitals. To see this, let's examine the molecular orbitals of 1,3-butadiene in relation to those of ethylene. Remember what you learned in Secs. 1.8A and 4.1A: If we start with a system of j atomic orbitals and allow them to interact within a molecule, j molecular orbitals result. Thus, ethylene has two $2p$ orbitals; overlap of these in both bonding and antibonding combinations gives two molecular orbitals. Similarly, 1,3-butadiene has four $2p$ orbitals. Overlap of these gives four molecular orbitals; two are bonding and two are antibonding. It happens (for reasons we won't examine here) that the energies of these molecular orbitals have a simple relationship to the energies of the ethylene molecular orbitals, as shown in Fig. SG15.1. Notice that the energies of the two butadiene bonding MOs are distributed—one higher, one lower—about the energy of the ethylene bonding MO; the same is true for the antibonding MO. This relationship between the MOs of different alkenes is the key to the wavelength effect. The energy required for an electronic transition is the energy that it would take to promote an electron from the occupied MO of *highest* energy to the antibonding MO of *lowest* energy. These energies, labeled with the appropriate $\Delta\varepsilon$, are shown with "squiggly arrows." These energies are provided by the light source in UV spectroscopy. That is, for absorption to occur, the energy of the light E must equal the appropriate $\Delta\varepsilon$. Because $\Delta\varepsilon = E = h\nu = hc/\lambda$, where λ is the wavelength of light, you can see that the wavelengths for the $\pi \rightarrow \pi^*$ transitions *decrease* as the number of conjugated double bonds increases. The same molecular-orbital arguments apply for compounds with increasing numbers of conjugated double bonds.

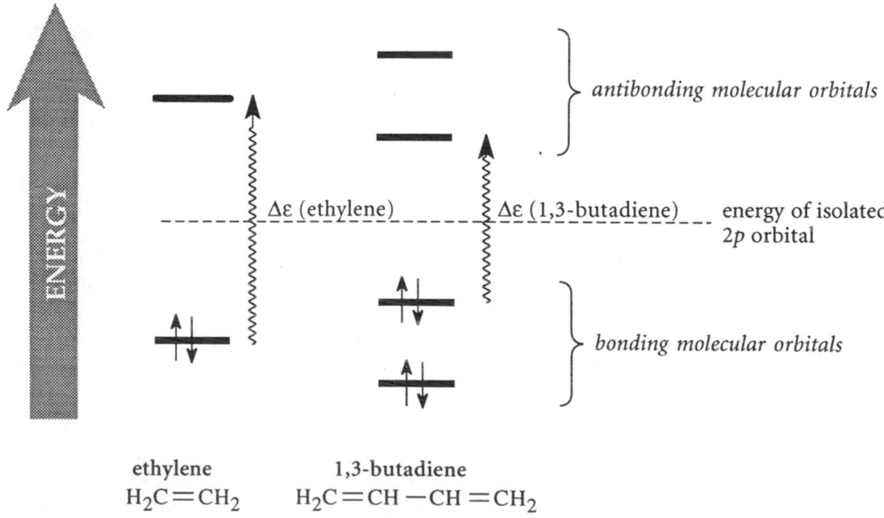

Figure SG15.1. Molecular orbitals of ethylene and 1,3-butadiene. The amount of energy required for promotion of an electron from a bonding molecular orbital of highest energy to the antibonding orbital of lowest energy is shown by the "squiggly arrow." Notice that this energy is smaller for 1,3-butadiene.

Why are peaks in UV spectra so broad? We are led to ask this question because the molecular-orbital theory of UV spectroscopy discussed above implies that a $\pi \rightarrow \pi^*$ transition should occur at a very precise value of the energy, that is, at $E = \Delta\varepsilon$.

The answer lies in the *rate* at which electronic transitions take place. It turns out that the motions of electrons are thousands of times faster than the motions of nuclei. As you learned in Chapter 12, the nuclei in molecules are vibrating constantly. When energy is absorbed from UV light by a molecule, this act of absorption "catches" the molecule in a variety of vibrational states. That is, some bonds are stretched, some are compressed, and all possible combinations of these states for all possible bonds can occur. Because the electronic transition occurs so much faster than bond vibrations, the "molecule" that absorbs UV energy is actually a complex mixture of molecules in various vibrational states. These states differ in energy, as do the excited states into which they are converted. These energy differences are reflected in a broadening of the energy at which the UV transition is observed. In gas-phase UV spectra, one can in many cases see very narrow peaks associated with the different vibrational states. (Spectroscopists call these peaks "vibrational fine structure.") Interaction with solvent further blurs these vibrational energies, so that the absorption observed in solution is something of an "envelope" of all of these states. Thus, UV absorptions in solution are typically very broad.

✓15.2 A Terminology Review

The purpose of this Study Guide Link is to review and to distinguish between the following terms used to describe addition reactions:

1. cycloaddition reactions
2. concerted reactions
3. pericyclic reactions

These terms in some cases may seem synonymous, but they are in fact different, and have very precise meanings. A *cycloaddition* is an addition that results in ring formation. Ozone addition and the Diels-Alder reaction are two examples of cycloadditions. The term *cycloaddition* succinctly describes the *outcome* of a reaction, but says nothing about the *mechanism* of the reaction. The terms *concerted* and *pericyclic*, in contrast, are descriptions of *mechanism.* Reactions that occur by mechanisms involving a single step are referred to as *concerted reactions.* Concerted reactions that occur by a cyclic flow of electrons are called *pericyclic reactions* (pericyclic = "around the circle"). All pericyclic reactions are concerted, but some concerted reactions are not pericyclic. Thus, hydroboration of alkenes (Sec. 5.3B of the text) is a concerted pericyclic reaction (although it is not a cycloaddition because a ring is not formed). The S_N2 reaction (Sec. 9.3 of the text) is concerted, but it is not pericyclic because the flow of electrons in the mechanism is not cyclic. The Diels-Alder reaction is a concerted cycloaddition, and it is also a pericyclic reaction. (Chapter 25 is devoted to a more detailed study of pericyclic reactions.)

✓15.3 Resonance Structures

Remember that resonance structures have *two* major implications:

1. A *structural implication:* resonance structures show that a compound cannot be represented by a single Lewis structure.
2. *An energy implication:* molecules for which there are more important resonance structures are more stable than isomers for which there are fewer important resonance structures.

Another way to state the energy implication is to say that *a molecule represented by resonance structures is more stable than any of its fictitious resonance contributors.* Thus, the allyl cation is more stable than it would be if it were accurately described by *either one* of the following structures in brackets:

$$\left[H_2C{=}CH{-}\overset{+}{C}H_2 \longleftrightarrow H_2\overset{+}{C}{-}CH{=}CH_2 \right]$$

allyl cation

Think about this philosophical point. The term "resonance-stabilized" might sound as if the stability associated with resonance structures is somehow a *consequence of our ability to draw these structures.* The behavior and properties of a chemical substance are, of course, totally independent of how we represent the substance in our minds or on paper. Our pencil-and-paper representations of resonance structures—or any structures, for that matter—are merely symbols for bonding patterns that occur in nature. When you think about it, it is quite remarkable that such simple symbols can be used to predict chemical phenomena or to portray something that requires considerable mathematical sophistication to describe physically. Resonance is a structural representation that has been recruited to say something about bonding. The extra stability associated with resonance is a consequence of this extra bonding. *Resonance structures have energy implications only because the extra bonding they represent has such implications.* We could do without resonance structures if we were willing to sketch out the π-molecular orbitals involved every time we draw a resonance-stabilized species. However, it is easier to use the inherently simpler Lewis structures to represent the same thing.

Because resonance structures are a symbolic representation, they must not be used blindly; rather, you must always have in the back of your mind what resonance structures mean. They convey information about the overlap of orbitals; and, if such overlap is not possible, resonance structures are not appropriate, even if they may seem to be derivable on paper with the curved-arrow notation. In order to use resonance structures really effectively you must *think about the orbital overlap involved* every time you draw resonance structures.

15.4 The π Molecular Orbitals of Benzene

Only the benzene bonding π molecular orbital of lowest energy is shown in Fig. 15.10 on text p. 676. Benzene has five other π molecular orbitals, and they are shown in Fig. SG15.2 (following page).

As you can see, there are three bonding and three antibonding orbitals—six $2p$ orbitals, six MOs. Because there are six π electrons, there are just enough electrons to fill the bonding MOs. All three MOs, because they are bonding, are lower in energy than an isolated $2p$ orbital, whose energy is indicated by the dashed line. Hence, benzene is stabilized by molecular-orbital formation. This π-electron stabilization is a major contributor to the stability of benzene. (Undoubtedly, the fact that the bond angles in benzene are exactly right for sp^2-hybridized carbons contributes to the molecule's stability as well.)

A related relationship between the number of bonding π molecular orbitals and the number of π electrons exists for all cyclic conjugated molecules with $4n + 2$ π electrons—thus the so-called $4n + 2$ rule for aromaticity discussed in Sec. 15.7D.

Notice that MOs ψ_2 and ψ_3 have the same energy (and, as expected, the same number of nodes—one). Molecular orbitals of the same energy are termed *degenerate* molecular orbitals. The antibonding MOs ψ_4^* and ψ_5^* are also degenerate molecular orbitals; these orbitals each have two nodes.

Notice that the π-electron density for all of these MOs lies in the region above and below the ring.

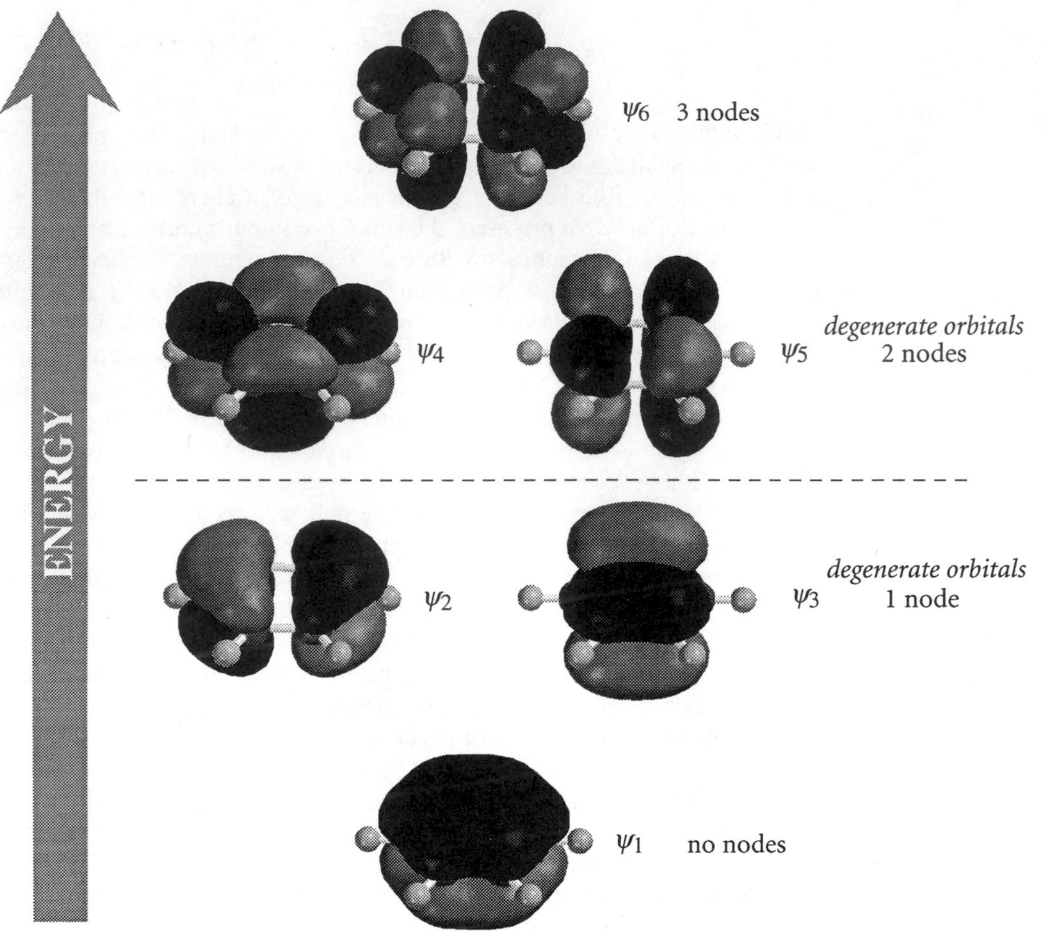

Figure SG15.2. The π molecular orbitals of benzene. The molecular orbitals are superimposed on a ball-and-stick model. The number of nodes given does not include the planar node present in the component 2*p* orbitals.

SOLUTIONS TO PROBLEMS

Solutions to In-Text Problems

15.1 When the 1,3-butadiene molecule undergoes internal rotation about the carbon-carbon single bond such that the π systems of the two double bonds become perpendicular, π-electron overlap is lost at the cost of some stabilization. Notice, however, that the 31 kJ/mol cost of rotation about the central single bond is very modest compared with the cost of rotation about a true double bond, which is about 264 kJ/mol. In other words, the π-orbital overlap across the central single bond, while real, is relatively modest.

15.2 (a) In the *s*-cis conformation of 1,3-butadiene, the inner hydrogens are close enough to have van der Waals repulsions (indicated by crossed arcs in the structure below). These repulsions raise the energy of the *s*-cis form. These van der Waals repulsions are absent in the *s*-trans conformation. Hence, the *s*-cis conformation has a higher energy than the *s*-trans conformation.

van der Waals repulsions

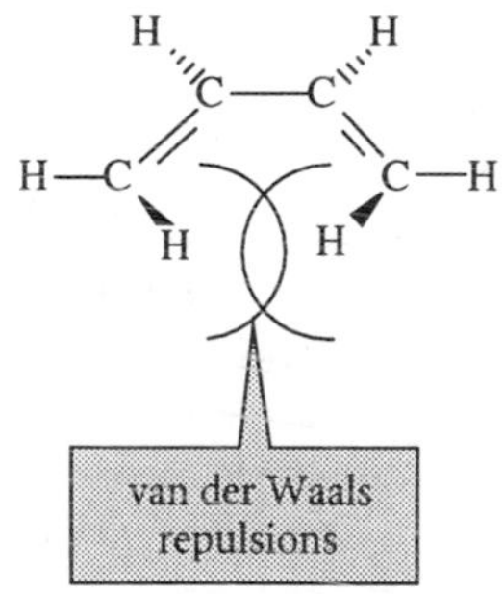

(b) The small internal rotation of 15° alleviates the repulsions shown above at the cost of some overlap of 2*p* orbitals across the single bond. These opposing factors are compromised so as to give the molecule the lowest possible energy.

The overlap between *p* orbitals decreases as $\cos^2 \theta$, where θ is the angle from which the π-electron system is distorted from planarity. A 15° distortion from planarity costs the molecule only 7% of its stabilization from orbital overlap.

15.3 The conformations of (2*E*,4*E*)-2,4-hexadiene:

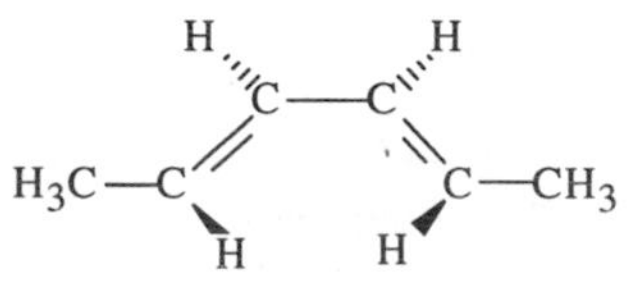

s-cis

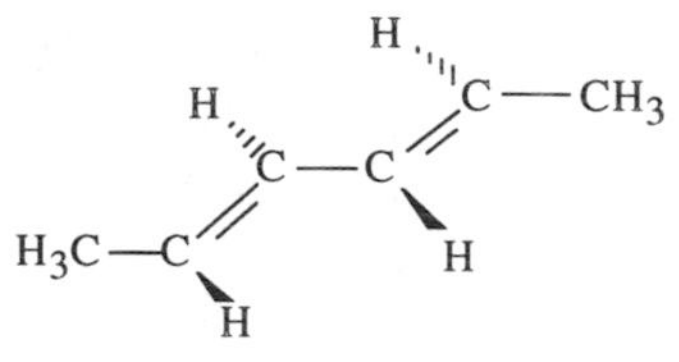

s-trans

The conformations of (2*E*,4*Z*)-2,4-hexadiene:

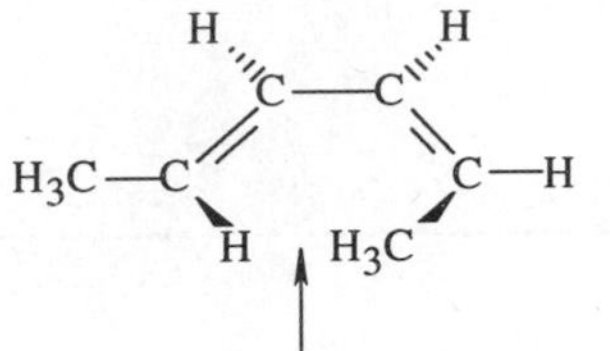

greater van der Waals repulsions

s-cis

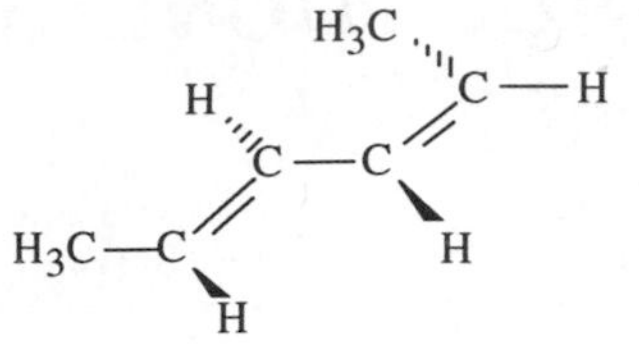

s-trans

(2*E*,4*E*)-2,4-hexadiene has the greater proportion of *s*-cis conformation because the van der Waals repulsions between inner groups (see diagram in the solution to Problem 15.2) are smaller in the (2*E*,4*E*)-stereoisomer. The reason is that the inner groups in the (2*E*,4*E*)-stereoisomer are both hydrogens, whereas, in the (2*E*,4*Z*)-stereoisomer, one of these groups is a methyl group, which is considerably larger than hydrogen.

15.4 First, draw the structures:

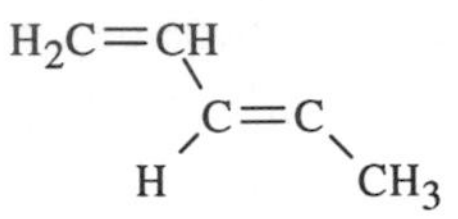

trans-1,3-pentadiene

$H_2C{=}CHCH_2CH{=}CH_2$

1,4-pentadiene

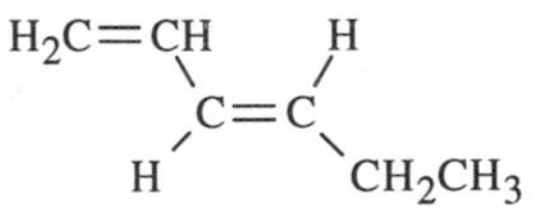

trans-1,3-hexadiene

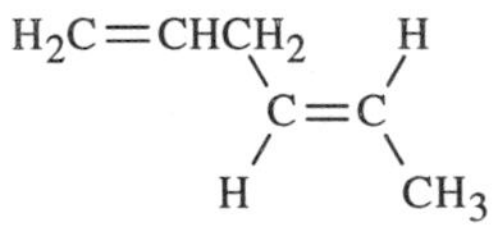

trans-1,4-hexadiene

Trans-1,3-pentadiene has a double bond with one branch and a double bond with two branches, whereas 1,4-pentadiene has two double bonds with a single branch. Hence, the energy difference between these two pentadiene isomers reflects not only the effect of conjugation but also the effect of additional double-bond branching. These two effects *both* stabilize the conjugated isomer. The two hexadiene isomers contain identical types of branching; hence, the energy difference between these two isomers reflects only the effect of conjugation. Because the comparison of pentadiene isomers contains two effects that operate in the same direction, the energy difference between these isomers is expected to be greater, as observed.

15.5 (a) The two enantiomers of an allene:

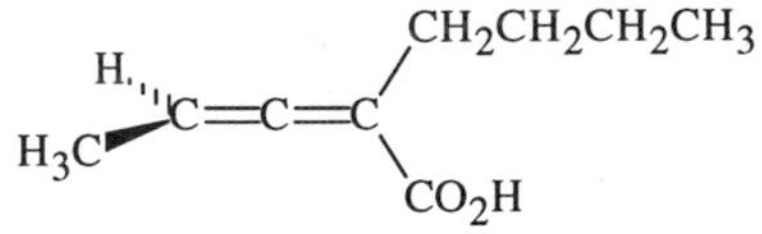

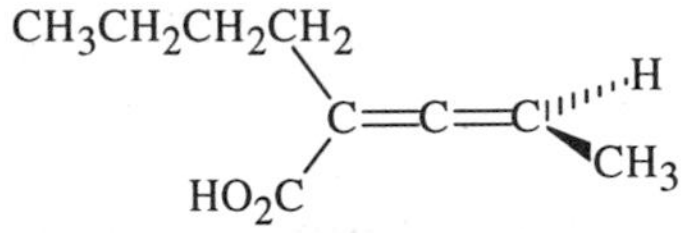

(b) Because enantiomers have specific rotations of equal magnitudes and opposite signs, the other enantiomer has a specific rotation of +30.7°.

15.6 (a) The energy of light is given by Eq. 12.3 on text p. 499 with the value of h, in kJ·sec·mol^{-1}, given in Eq. 12.5.

$$E = \frac{hc}{\lambda} = \frac{(3.99 \times 10^{-13}\ \text{kJ·sec·mol}^{-1})(3.00 \times 10^{8}\ \text{m·sec}^{-1})}{(450 \times 10^{-9}\ \text{m})} = 266\ \text{kJ/mol}$$

(The corresponding value in kcal/mol is 63.6.)

(b) The calculation is identical to that in part (a) with different numbers:

$$E = \frac{hc}{\lambda} = \frac{(3.99 \times 10^{-13}\ \text{kJ·sec·mol}^{-1})(3.00 \times 10^{8}\ \text{m·sec}^{-1})}{(250 \times 10^{-9}\ \text{m})} = 479\ \text{kJ/mol}$$

(The corresponding value in kcal/mol is 114.)

Notice the large effect of wavelength on the energy of light.

15.7 (a) Use Eq. 15.2 in the text. An absorbance $A = 1.0$ means that $(I_0/I) = 10$; that is, one-tenth, or 10%, of the incident radiation is transmitted. When $A = 0$, all of the incident radiation is transmitted.

(b) Use Eq. 15.2 in the text, with $(I_0/I) = 2$. Thus, $A = \log(2) = 0.30$.

15.8 The piece with greater absorbance transmits less of the incident radiation. Therefore, the thick piece of glass has greater absorbance.

15.9 (a) Apply Beer's law (Eq. 15.3 on text p. 644), with $\epsilon = 10{,}750$ absorbance units $\text{cm}^{-1}\cdot\text{L}\cdot\text{mol}^{-1}$, $l = 1.0$ cm, and, from Fig. 15.4 on text p. 643, $A = 0.800$ absorbance units. The concentration of isoprene is therefore

$$\frac{0.800 \text{ absorbance units}}{(10{,}750 \text{ absorbance units}\cdot\text{cm}^{-1}\cdot\text{L}\cdot\text{mol}^{-1})(1 \text{ cm})} = 7.44 \times 10^{-5} \text{ mol/L}.$$

(b) The absorbance of the isoprene sample in Fig. 15.4 at 235 nm is 0.225. With the concentration determined from part (a), Beer's law gives $A = 0.225$ absorbance units $= \epsilon(7.44 \times 10^{-5} \text{ mol/L})(1 \text{ cm})$, or $\epsilon = 3.02 \times 10^3$ absorbance units$\cdot\text{L}\cdot\text{mol}^{-1}\cdot\text{cm}^{-1}$. Another way to determine the extinction coefficient at a different wavelength is based on the fact that the ratio of absorbances at different wavelengths equals the ratio of the extinction coefficients. Hence, the extinction coefficient at 235 nm is

$$\epsilon_{235} = \epsilon_{222.5}\frac{A_{235}}{A_{222.5}}$$

or

$$\epsilon_{235} = (10{,}750 \text{ absorbance units}\cdot\text{L}\cdot\text{mol}^{-1}\cdot\text{cm}^{-1}) \times \frac{0.225 \text{ absorbance units}}{0.800 \text{ absorbance units}}$$

$$= 3.02 \times 10^3 = 3023 \text{ absorbance units}\cdot\text{L}\cdot\text{mol}^{-1}\cdot\text{cm}^{-1}$$

15.10 (a) The two alkyl substituents contribute +10 nm to the base λ_{max} of 217 for a predicted λ_{max} value of 227 nm.

(b) The —CH_2— groups of the ring attached to the double bond are alkyl substituents. (If this point is not clear, see Study Guide Link 9.5 on p. 265 of this manual.) For three alkyl substituents the predicted λ_{max} is (217 + 15) nm = 232 nm. (This calculation assumes that the molecule exists predominantly in the *s*-trans conformation.)

15.11 (a) The Diels-Alder reaction of 1,3-butadiene and ethylene gives cyclohexene.

CH_2 / CH_2 (1,3-butadiene) + $CH_2{=}CH_2$ (ethylene) ⟶ cyclohexene

(b) The stereochemistry of this addition, also shown below, is covered in Sec. 15.3C.

CO_2CH_3, CO_2CH_3, H, H

15.12 (a) An analysis like that employed in Study Problem 15.1 suggests two possibilities.

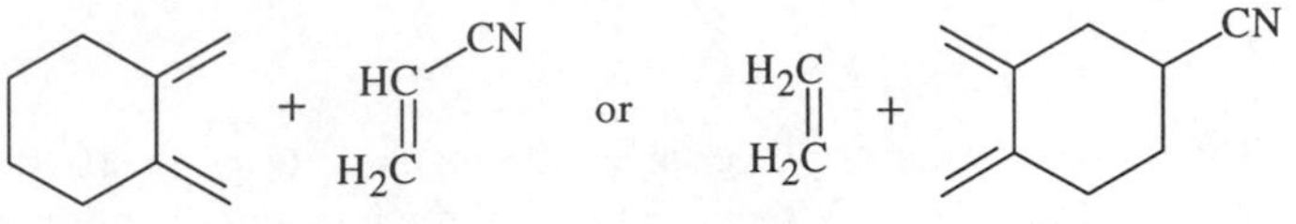

HC—CN, H_2C + ... or H_2C=H_2C + ... CN

A *B*

Pair *A* is preferred because, in many cases, the most reactive dienophiles are those with conjugated electronegative substituents. But if your answer was pair *B*, you have analyzed the problem correctly.

(b) Again, there are two possibilities; the more practical alternative is the following one. Can you find the other?

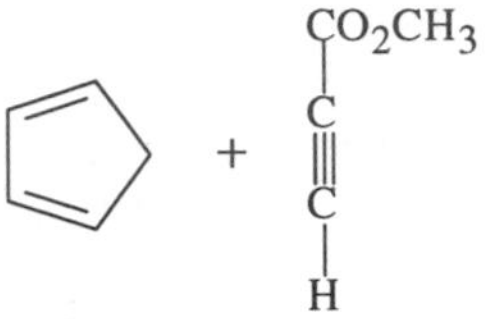

+ CO_2CH_3—C≡C—H

When you want to prepare a bicyclic compound by the Diels-Alder reaction, a cyclic diene is often required.

(c) This product, 4-vinylcyclohexene, results from the Diels-Alder reaction of two molecules of 1,3-butadiene:

4-vinylcyclohexene

15.13 (a) The dienophile can be oriented in two different ways relative to the diene.

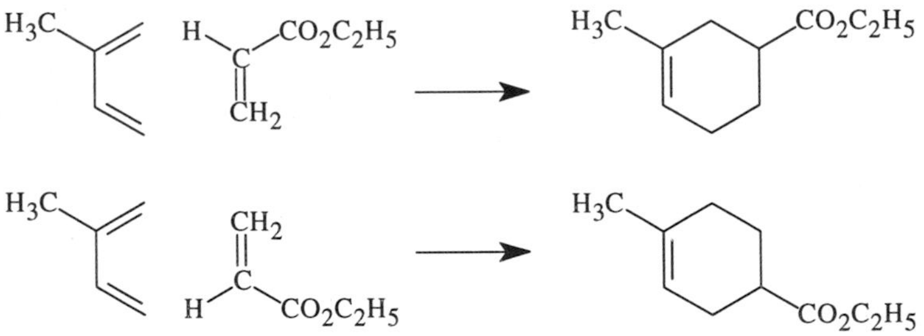

H_3C + H—C($CO_2C_2H_5$)=CH_2 → H_3C ... $CO_2C_2H_5$

H_3C + CH_2=C(H)—$CO_2C_2H_5$ → H_3C ... $CO_2C_2H_5$

(b) As in part (a), two possible orientations of the diene and dienophile lead to the following two possible constitutional isomers:

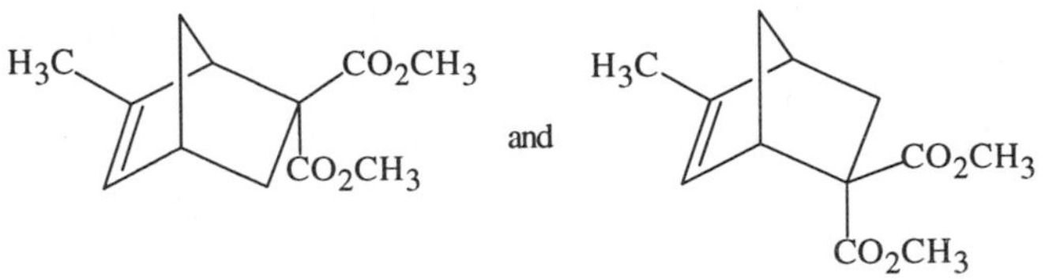

H_3C ... CO_2CH_3, CO_2CH_3 and H_3C ... CO_2CH_3, CO_2CH_3

15.14 (a) The Diels-Alder reaction requires that the diene assume an *s*-cis conformation in the transition state. In this conformation, the (2*E*,4*Z*)-diene has a methyl-hydrogen van der Waals repulsion that is absent in the (2*E*,4*E*)-diene. (This point is covered in the solution to Problem 15.3.) This repulsion raises the energy of the Diels-Alder transition state of this diene, and therefore reduces the rate of its reaction. Since the (2*E*,4*Z*)-diene reacts more slowly than the (2*E*,4*E*)-diene, it follows that the (2*E*,4*E*)-diene is consumed in the reaction, and the (2*E*,4*Z*)-diene remains unreacted.

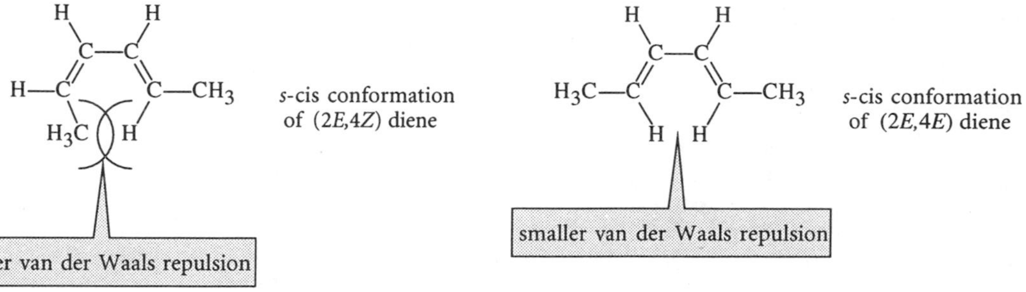

s-cis conformation of (2*E*,4*Z*) diene

greater van der Waals repulsion

s-cis conformation of (2*E*,4*E*) diene

smaller van der Waals repulsion

15.15 The triene contains two diene units with one double bond common to both. The dienophile reacts with the diene unit that is locked in an *s*-cis conformation.

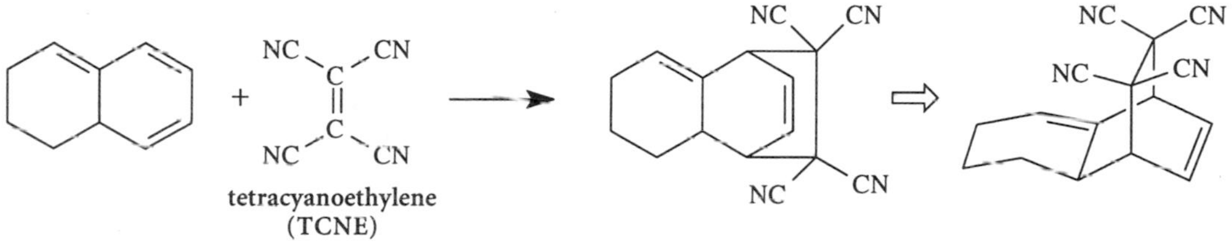

tetracyanoethylene (TCNE)

15.16 (a) The two acetoxy groups are cis in the product because they are both outer substituents in the *s*-cis conformation. (Of course, the racemate is formed; only one enantiomer is shown below.)

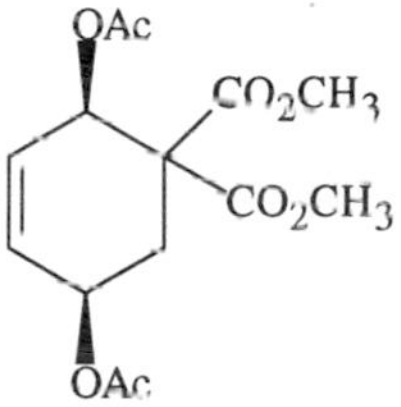

(b) The two products correspond to the two possibilities in Eq. 15.14. They result from addition of the diene at either of the two faces of the alkene (or the alkene at either of the two faces of the diene).

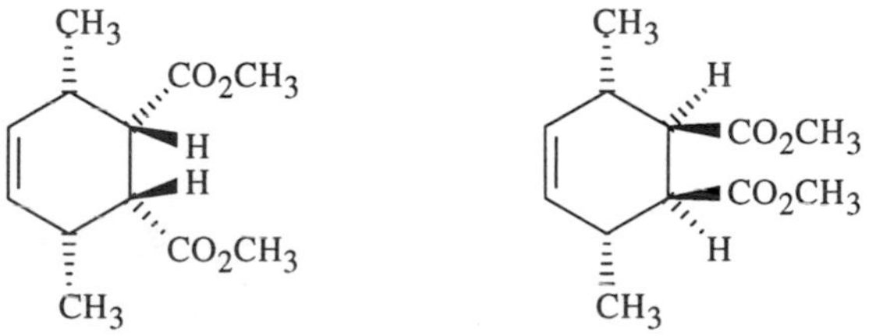

endo stereoisomers

exo stereoisomers

Notice that both stereoisomers result from *syn*-addition. In other words, there are *two* possible modes of *syn*-addition.

(c) The "inner" groups of the diene are "tied together" as the —CH_2— group of the five-membered ring.

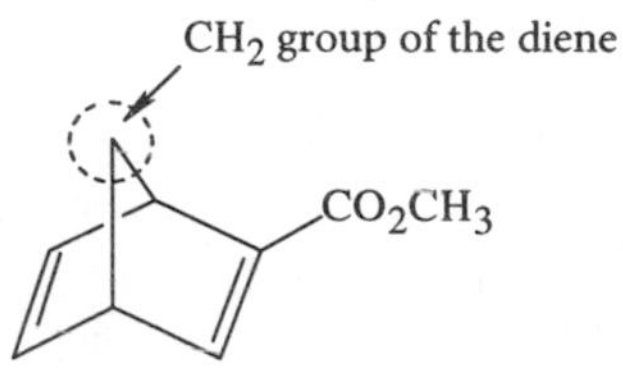

CH_2 group of the diene

15.17 (a)

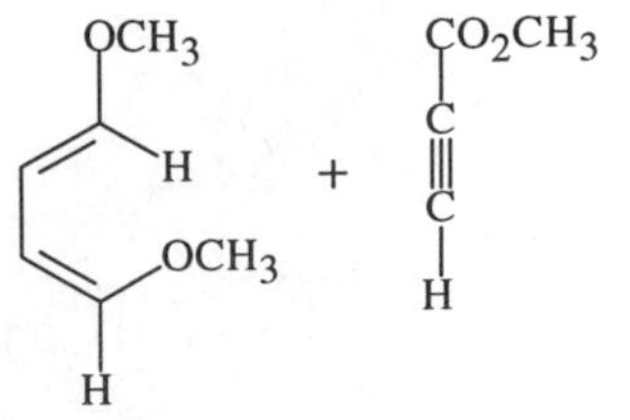

(b)

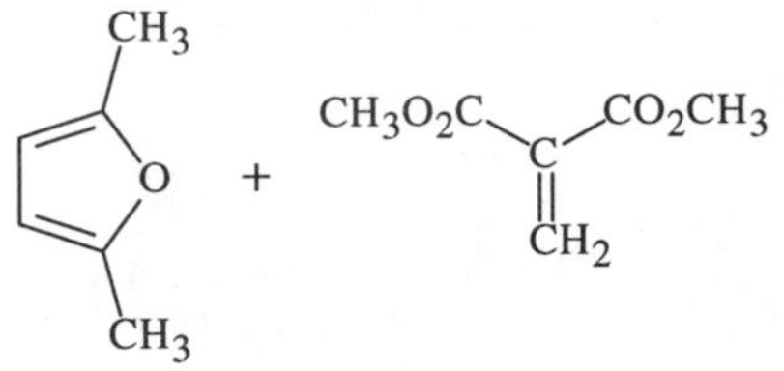

(c)

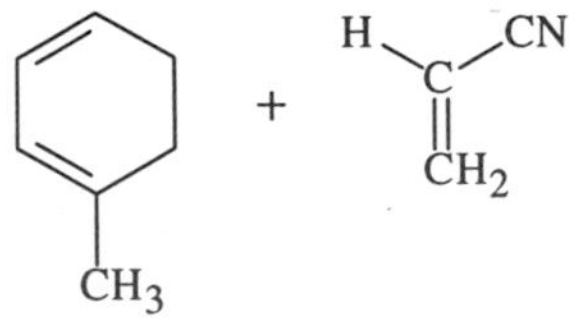

(d)

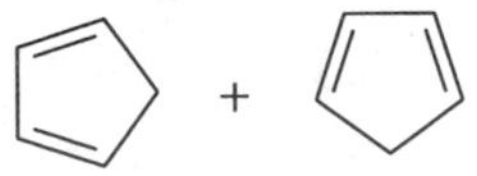

(that is, the reaction of two 1,3-cyclopentadiene molecules)

15.18 (a) As shown by Eq. 15.14 on text p. 657, the *endo* diastereomer is the one in which the dienophile substituents (which, in this case, are the arms of the five-membered ring) are cis to the outer substituents when the diene is drawn in the *s*-cis conformation.

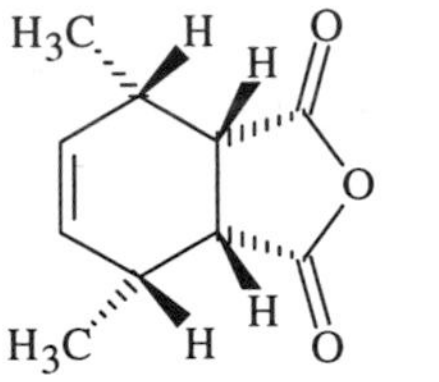

(b) Here are the diagrams. Notice that the product is formed initially in a boatlike conformation that subsequently changes to its more stable conformation. Be sure to make models if you have difficulty seeing these relationships.

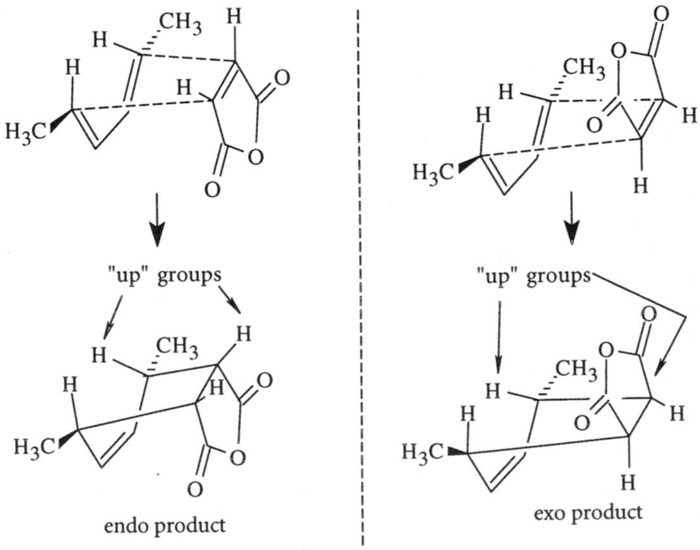

"up" groups

endo product

"up" groups

exo product

15.19 We'll use a diagram like the one in the previous solution:

1,3-cyclohexadiene | maleic anhydride | endo product | or

15.20 (a) The carbocation intermediate in the addition of HCl has two sites of electron deficiency, either of which can be attacked by the chloride ion.

$$H_2C{=}CH{-}CH{=}CH_2 \xrightarrow{H{-}Cl} \left[H_2C{=}CH{-}\overset{+}{C}H{-}CH_3 \longleftrightarrow H_2\overset{+}{C}{-}CH{=}CH{-}CH_3 \right] \xrightarrow{Cl^-}$$

1,3-butadiene

$$\underset{A}{H_2C{=}CH{-}CH(Cl){-}CH_3} + \underset{B}{H_2C(Cl){-}CH{=}CH{-}CH_3}$$

(b) The products are a mixture of constitutionally isomeric ethers because the carbocation intermediate in the S_N1 reaction has two sites of electron deficiency that can be attacked by the nucleophile ethanol. The details for the formation of 3-ethoxy-3-methylcyclohexene are shown below; you should provide the additional details of the curved-arrow mechanism for the formation of 3-ethoxy-1-methylcyclohexene.

3-chloro-1-methylcyclohexene → [carbocation resonance structures] + $:\ddot{Cl}:^-$; with $HO{-}C_2H_5$ → 3-ethoxy-1-methylcyclohexene; with $HO{-}C_2H_5$ → protonated ether, deprotonated by $HO{-}C_2H_5$ → 3-ethoxy-3-methylcyclohexene + $H_2\overset{+}{O}{-}C_2H_5$

It is a common (and useful) to show the derivation of different products by showing the separate "reactions" of different resonance structures. Be sure you remember, however, that the resonance structures are different structures used to depict *one* species. Thus, in the structures above, the carbocation is *one* species that reacts in two different ways, *not* a mixture of different species, each reacting in a separate reaction.

15.21 This is an S_N1 reaction; the two products result respectively from attack of water on the two electron-deficient carbons of the carbocation intermediate. (The acetone is a co-solvent that helps to dissolve the alkyl chloride in the aqueous solution.)

$$(CH_3)_2C{=}CH{-}CH(Cl){-}CH_3 \longrightarrow \left[(CH_3)_2C{=}CH{-}\overset{+}{C}H{-}CH_3 \longleftrightarrow (CH_3)_2\overset{+}{C}{-}CH{=}CH{-}CH_3\right] + :\ddot{Cl}:^-$$

$$\xrightarrow{\ddot{O}H_2} (CH_3)_2C{=}CH{-}CH(\overset{+}{O}H_2){-}CH_3 \quad + \quad :\ddot{Cl}:^- \quad + \quad (CH_3)_2C(\overset{+}{O}H_2){-}CH{=}CH{-}CH_3$$

$$\xrightarrow{\ddot{O}H_2} \underbrace{:\ddot{Cl}:^- + H_3\overset{+}{\ddot{O}}}_{\text{HCl in water}} + (CH_3)_2C{=}CH{-}CH(:\ddot{O}H){-}CH_3 + (CH_3)_2C(:\ddot{O}H){-}CH{=}CH{-}CH_3$$

15.22 The two constitutional isomers result respectively from "normal" addition (1,2-addition, product *A*) and conjugate addition (product *B*). The conjugate-addition product *B* is the major one at equilibrium because it has an internal double bond (that is, the double bond with the greater number of alkyl branches). Although the stereochemistry of product *B* was not requested, it turns out that it is the trans isomer.

$$\underset{\text{1,3-butadiene}}{H_2C{=}CH{-}CH{=}CH_2} \xrightarrow{Br_2} \underset{A}{H_2C(Br){-}CH(Br){-}CH{=}CH_2} + \underset{B}{H_2C(Br){-}CH{=}CH{-}CH_2(Br)}$$

15.23 Start with a radical derived from attack of initiator on 1,3-butadiene, the alkene present in greatest amount. (1,3-butadiene and styrene are present in a 3:1 ratio; see text p. 665.) Incorporation of the first two butadiene units is shown in Eqs. 15.27a–b on text p. 665. The mechanism for incorporation of the third butadiene unit and styrene are as follows:

$$-CH_2CH{=}CHCH_2CH_2CH{=}CH\dot{C}H_2 \quad H_2C{=}CH{-}CH{=}CH_2 \longrightarrow$$

$$\left[-CH_2CH{=}CHCH_2CH_2CH{=}CHCH_2{-}CH_2\dot{C}H{-}CH{=}CH_2 \longleftrightarrow -CH_2CH{=}CHCH_2CH_2CH{=}CHCH_2{-}CH_2CH{=}CH{-}\dot{C}H_2\right] \xrightarrow{H_2C{=}CHPh}$$

$$-CH_2CH{=}CHCH_2CH_2CH{=}CHCH_2{-}CH_2CH{=}CH{-}CH_2CH_2\dot{C}HPh \quad \text{radical } R$$

Radical *R* propagates the chain by adding to another 1,3-butadiene molecule, and so on. Notice that only conjugate addition is shown; 1,2-addition of a chain-propagating radical to 1,3-butadiene is possible at any point. (See the following problem.)

15.24 The structure of polybutadiene if every other unit of the polymer resulted from 1,2-addition:

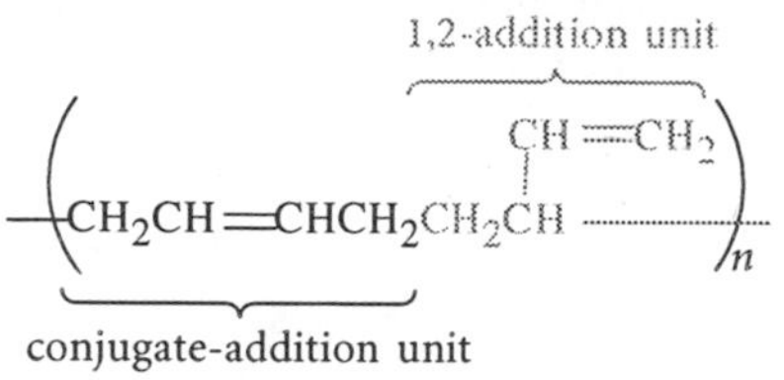

1,2-addition unit

$$\left(CH_2CH{=}CHCH_2CH_2\underset{}{CH}(CH{=}CH_2) \right)_n$$

conjugate-addition unit

15.25 (a) The first structure is most important; the others have fewer formal bonds and involve separation of opposite charges or unpaired electrons.

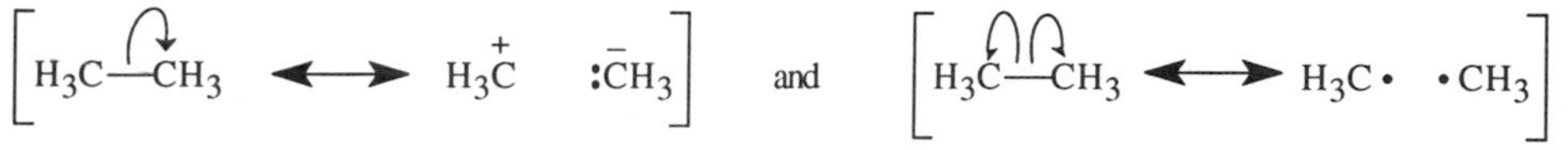

$$\left[H_3C{-}CH_3 \longleftrightarrow H_3\overset{+}{C} \quad :\overset{-}{C}H_3 \right] \quad \text{and} \quad \left[H_3C{-}CH_3 \longleftrightarrow H_3C\cdot \quad \cdot CH_3 \right]$$

(b) The second structure has more formal bonds than the others; the first structure assigns positive charge to the more electropositive atom, carbon. Both structures are important, and both are much more important than the third structure, in which there is separation of charge and fewer formal bonds.

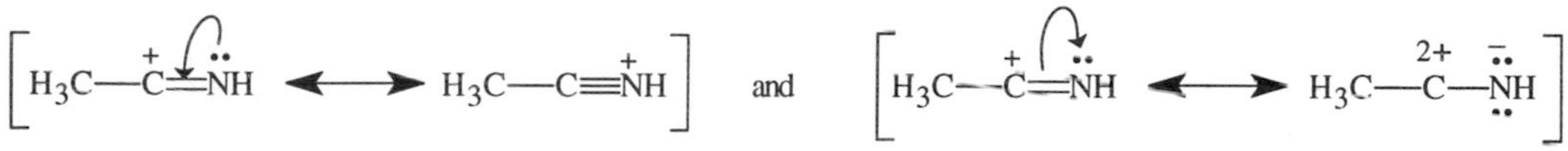

$$\left[H_3C{-}\overset{+}{C}{=}\ddot{N}H \longleftrightarrow H_3C{-}C{\equiv}\overset{+}{N}H \right] \quad \text{and} \quad \left[H_3C{-}\overset{+}{C}{=}\ddot{N}H \longleftrightarrow H_3C{-}\overset{2+}{C}{-}\overset{-}{\ddot{N}}H \right]$$

(c) Both structures are important, but the second structure is more important because it is a tertiary carbocation; the first structure is a primary carbocation.

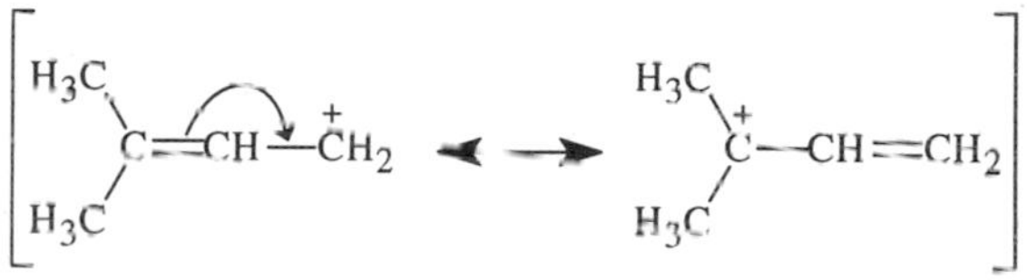

$$\left[(H_3C)_2C{=}CH{-}\overset{+}{C}H_2 \longleftrightarrow (H_3C)_2\overset{+}{C}{-}CH{=}CH_2 \right]$$

15.26 The electron-deficient carbon has an empty $2p$ orbital that overlaps with a filled $2p$ orbital on the oxygen. The other oxygen electron pair in an sp^2 orbital.

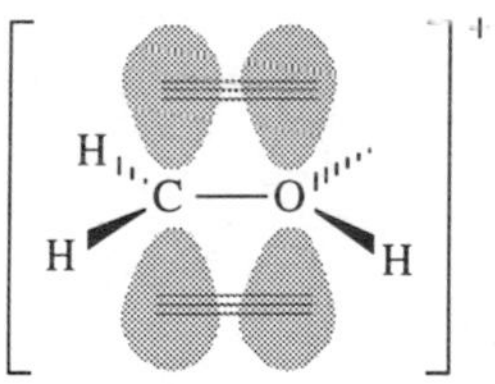

$$\left[H_2C{-}O{-}H \right]^+$$

15.27 (a) The ion on the right is more stable; it has two important resonance structures (shown below); the one on the left has only the one structure shown in the problem.

$$\left[H_3C{-}\overset{:\ddot{O}:^-}{C}{=}CH{-}CH_3 \longleftrightarrow H_3C{-}\overset{\ddot{O}:}{\overset{\|}{C}}{-}\overset{..}{\underset{-}{C}}H{-}CH_3 \right]$$

(b) The radical on the left has three important resonance structures, whereas the radical on the right has two. The resonance structures for the more stable radical are shown here. (You should draw the resonance structure of the less stable radical.) In the more stable radical, the unpaired electron can be delocalized by both double bonds, whereas, in the less stable radical, the unpaired electron is delocalized by only one double bond.

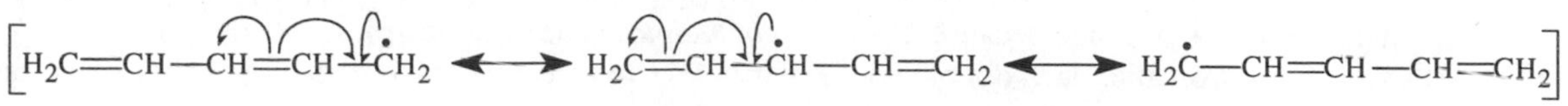

$$\left[H_2C{=}CH{-}CH{=}CH{-}\dot{C}H_2 \longleftrightarrow H_2C{=}CH{-}\dot{C}H{-}CH{=}CH_2 \longleftrightarrow H_2\dot{C}{-}CH{=}CH{-}CH{=}CH_2 \right]$$

(c) Both ions have two resonance structures. However, the ion on the right (ion *B* below) is more stable because one of its resonance contributors is a secondary carbocation, whereas both contributors for the ion on the right (ion *A* below) are primary carbocations.

$$\left[H_2C{=}C(CH_3){-}\overset{+}{C}H_2 \longleftrightarrow H_2\overset{+}{C}{-}C(CH_3){=}CH_2 \right] \qquad \left[H_3C{-}CH{=}CH{-}\overset{+}{C}H_2 \longleftrightarrow H_3C{-}\overset{+}{C}H{-}CH{=}CH_2 \right]$$

A

both resonance contributors are primary carbocations

B

this resonance contributor is a secondary carbocation

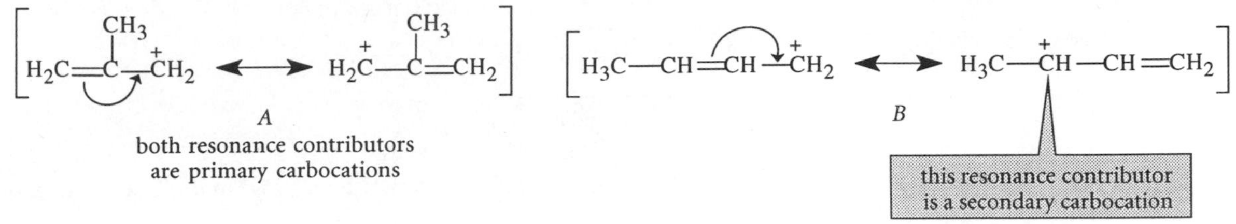

15.28 Hammond's Postulate (Sec. 4.8C) says that the reaction which involves the more stable carbocation intermediate will be faster. The carbocation intermediate involved in the solvolysis of compound *B* is resonance-stabilized; the structures of this ion are shown below. The intermediate involved in the solvolysis of *A* is not resonance-stabilized. Hence, the ion derived from *B* is *much* more stable than the one derived from *A*, and the solvolysis of *B* is therefore *much* faster than the solvolysis of *A*.

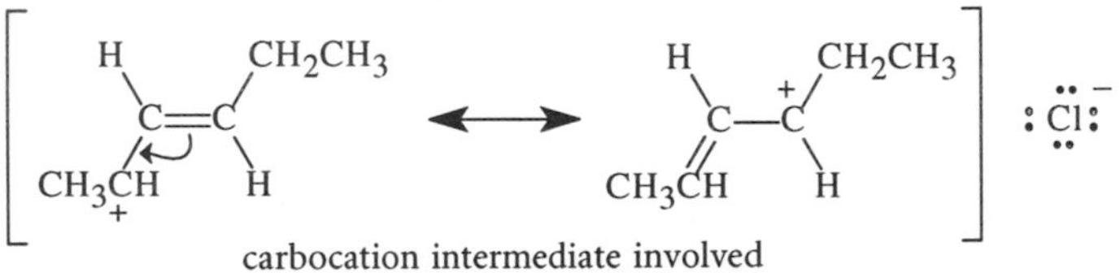

$$\left[CH_3\overset{+}{C}H{-}CH{=}CH{-}CH_2CH_3 \longleftrightarrow CH_3CH{=}CH{-}\overset{+}{C}H{-}CH_2CH_3 \right] \quad :\ddot{\underset{..}{Cl}}:^-$$

carbocation intermediate involved in the solvolysis of compound *B*

15.29 Given that furan is aromatic, it can have $4n + 2$ π-electrons (six electrons) if one of the electron pairs on oxygen is in a $2p$ orbital that is part of the π-electron system of the ring. The other electron pair on oxygen is in an sp^2 orbital whose axis is in the plane of the ring. (The electronic structure of furan is drawn in Fig. 24.5 on text p. 1131.)

15.30 An aromatic compound *cannot* have a single unpaired electron as part of its π-electron system, because the number of π-electrons required for aromaticity, $4n + 2$, must be an even number; a single unpaired electron would result in an odd number of electrons.

However, a free radical could certainly be aromatic if the unpaired electron were not part of the π-electron system. An example is the *phenyl radical,* which could be formed conceptually by abstraction of a hydrogen atom from benzene:

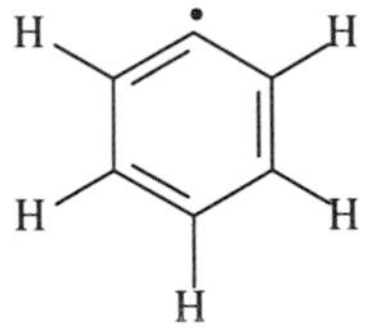

phenyl radical

However, in such a radical, the unpaired electron itself does not contribute to the aromatic stability because it is not part of the $4n + 2$ π-electron system.

15.31 (a) Thiophene is aromatic. One electron pair on sulfur is part of the aromatic π-electron system; the other is not (as in the oxygen of furan; problem 15.29.)

(b) This ion contains $4n$, not $4n + 2$, π-electrons, and is therefore not aromatic.

(c) Isoxazole is aromatic. Each double bond contributes two electrons to the π-electron system. One electron pair on the oxygen is also part of the π-electron system, but the other electron pair on the oxygen is not. (See the solution to Problem 15.29, above.) The electron pair on the nitrogen is vinylic and, like the electron pair in pyridine, is not part of the π-electron system.

(d) This ion is not aromatic because it is not a *continuous* cycle of π-electrons.

(e) This ion is aromatic. Each double bond contributes two electrons to the π-electron system. The cationic carbon bears an empty $2p$ orbital that contributes no electrons but is part of the π-electron system. One electron pair on the oxygen is part of the π-electron system, and the other is not, as in furan (Problem 15.29)

(f) This compound is not aromatic because it has $4n$ rather than $4n + 2$ π-electrons. The empty p orbital on boron, although part of the π-electron system, contributes no electrons.

15.32 Because compounds *A* and *B* are antiaromatic, they distort as much as possible to *avoid* overlap of the *p* orbitals involved in the two double bonds. (See the discussion of cyclobutadiene on text p. 683.) The single bonds are much longer than the double bonds, and the two types of carbon-carbon bonds are not equivalent. Because the deuteriums bridge a double bond in *A* and a single bond in *B*, and because the two types of carbon-carbon bond are different, *A* and *B* are different compounds and are therefore in principle capable of independent existence. In contrast, all carbon-carbon bonds in benzene are equivalent—each is half single bond, half double bond. Consequently, all molecules with deuteriums on adjacent positions are identical and indistinguishable. To summarize:

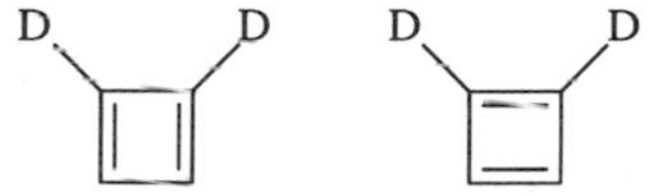

D D D D

different molecules

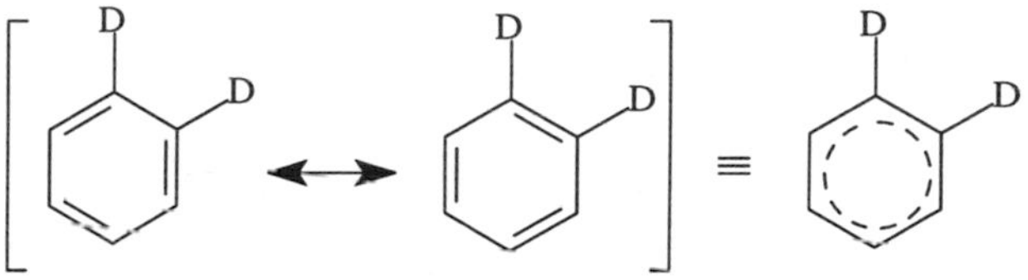

D D D D D D

one molecule

15.33 Compounds (b) and (f) contain $4n$ π-electrons and are in principle antiaromatic.

Solutions to Additional Problems

15.34 (a) The structure on the right is somewhat more important because it is a tertiary free radical, whereas the structure on the left is a primary free radical..

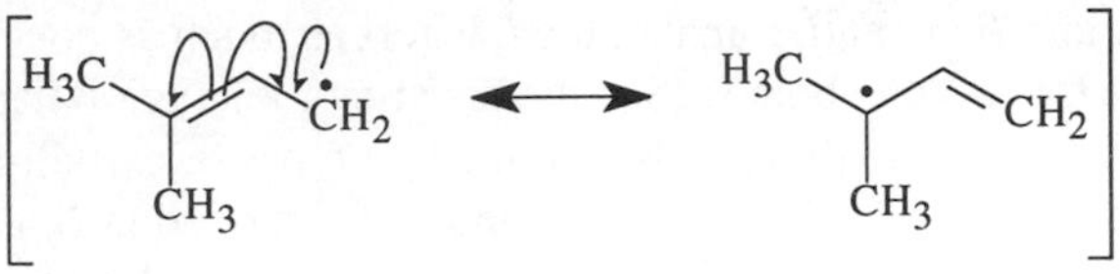

$$\left[H_3C\text{—}C(CH_3)\text{=}CH\text{—}\dot{C}H_2 \longleftrightarrow H_3C\text{—}\dot{C}(CH_3)\text{—}CH\text{=}CH_2 \right]$$

(b) The middle structure is somewhat more important because the electronegative atom (oxygen) bears the negative charge and the double bonds are conjugated.

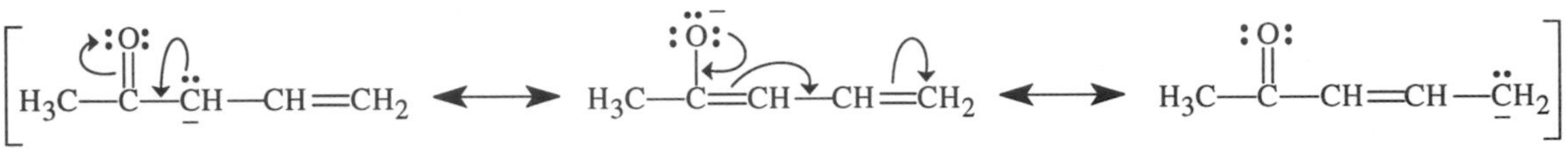

$$\left[H_3C\text{—}\overset{\displaystyle :O:}{\overset{\|}{C}}\text{—}\overset{..}{\underset{-}{C}}H\text{—}CH\text{=}CH_2 \longleftrightarrow H_3C\text{—}\overset{\displaystyle :\ddot{O}:^-}{\overset{|}{C}}\text{=}CH\text{—}CH\text{=}CH_2 \longleftrightarrow H_3C\text{—}\overset{\displaystyle :O:}{\overset{\|}{C}}\text{—}CH\text{=}CH\text{—}\overset{..}{\underset{-}{C}}H_2 \right]$$

(c) The first and second structures are both secondary carbocations, but the first and third structures have conjugated double bonds; in the third structure, the double bonds have the greatest number of alkyl substituents. All structures are important; no structure has clearly greater importance.

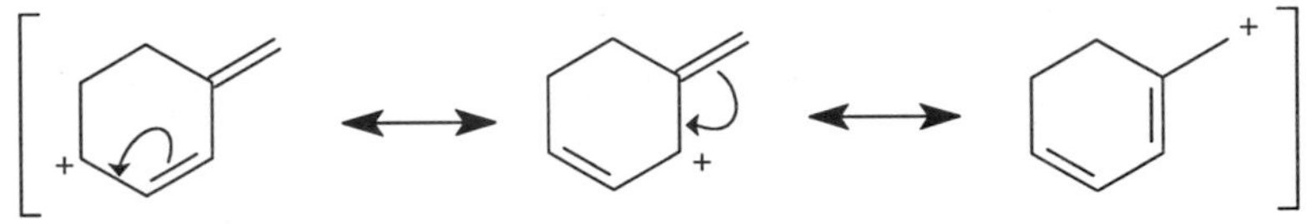

[Resonance structures of the methylenecyclohexenyl cation: $\longleftrightarrow$ $\longleftrightarrow$]

15.35 First draw the structure of the starting material:

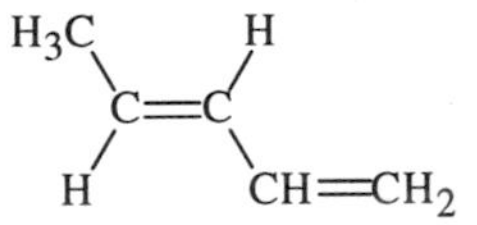

$$\underset{trans\text{-1,3-pentadiene}}{(H_3C)(H)C\text{=}C(H)(CH\text{=}CH_2)}$$

(a)

$$\underbrace{CH_3\underset{}{\overset{Br}{C}}H\text{—}\overset{Br}{C}H\text{—}CH\text{=}CH_2 + (H_3C)(H)C\text{=}C(H)\text{—}\underset{Br}{C}H\text{—}CH_2Br}_{\text{1,2-addition products}} + \underset{\substack{\text{(cis and trans)}\\ \text{1,4-addition products}}}{CH_3\underset{Br}{C}HCH\text{=}CHCH_2Br}$$

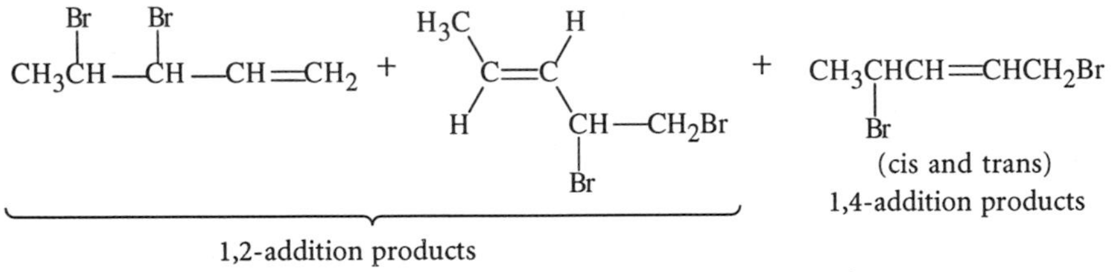

(b) Within each pair, the first product is derived from the more stable carbocation intermediate (why?), and is therefore likely to be formed in greater amount. Notice that one of the 1,2-addition products is identical to one of the 1,4-addition products; that is, this compound can be formed by either process.

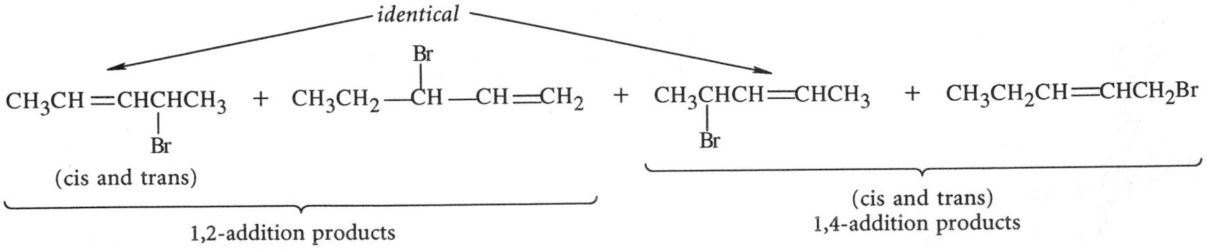

identical

$$\underbrace{\underset{\text{(cis and trans)}}{CH_3CH\text{=}CH\underset{Br}{C}HCH_3} + CH_3CH_2\text{—}\overset{Br}{C}H\text{—}CH\text{=}CH_2}_{\text{1,2-addition products}} + \underbrace{CH_3\underset{Br}{C}HCH\text{=}CHCH_3 + CH_3CH_2CH\text{=}CHCH_2Br}_{\substack{\text{(cis and trans)}\\ \text{1,4-addition products}}}$$

(c) $CH_3CH_2CH_2CH_2CH_3$

(d) Hydration follows the same pattern as HBr addition; notice that, as in part (b), one of the 1,2-addition products is identical to one of the 1,4-addition products.

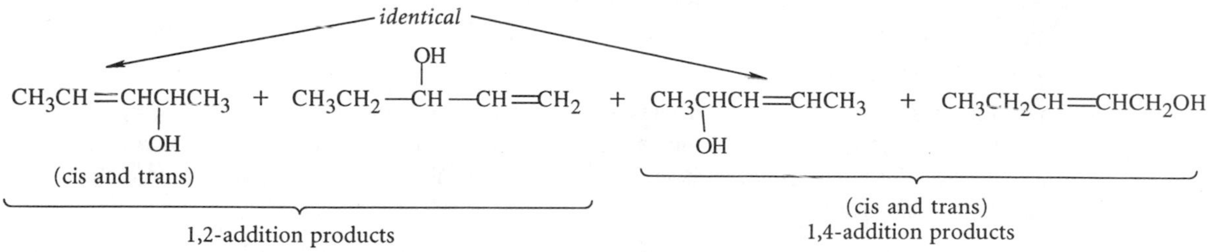

identical

$CH_3CH{=}CHCH(OH)CH_3$ + $CH_3CH_2{-}CH(OH){-}CH{=}CH_2$ + $CH_3CH(OH)CH{=}CHCH_3$ + $CH_3CH_2CH{=}CHCH_2OH$

(cis and trans) 1,2-addition products; (cis and trans) 1,4-addition products

(e) No reaction occurs.

(f) A Diels-Alder reaction takes place, and the two possible adducts are formed. (Typically, the endo adduct predominates.)

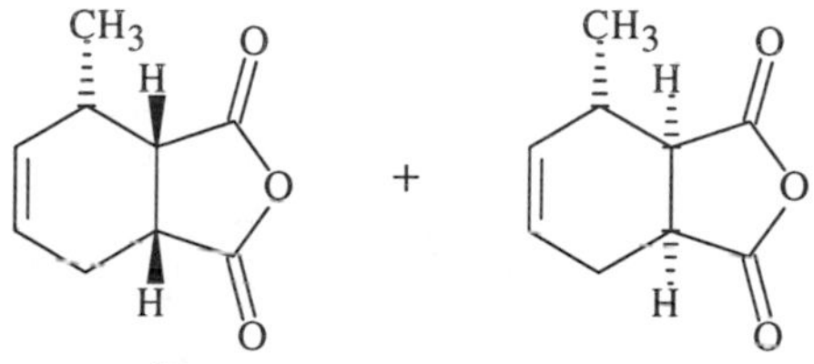

endo + exo

15.36 1,3-Cyclohexadiene gives 3-bromocyclohexene by either 1,2- or 1,4-addition of HBr.

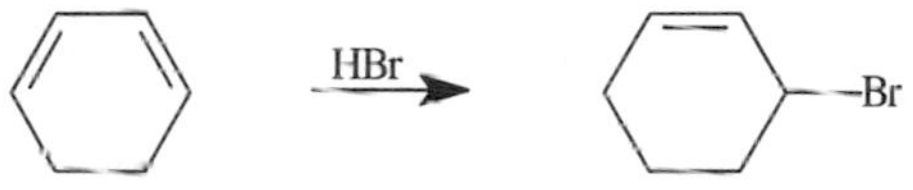

1,3-cyclohexadiene $\xrightarrow{HBr}$ 3-bromocyclohexene

15.37 (a) The answers to both (a) and (b) follow from the geometry of cumulenes. Each additional cumulated double bond results in a 90° twist of one end of a molecule with respect to the other. Thus, in alkenes, the atoms connected to the double bond are in the same plane; in allenes, the atoms connected to the ends of the cumulated double bonds are in *perpendicular* planes; in cumulenes with three contiguous double bonds, the atoms attached to the terminal double bonds are in the same plane, as they are in alkenes. Since the perpendicular relationship is necessary for chirality of a cumulene, allenes, as well as other cumulenes with an even number of double bonds, can be chiral. Indeed, 2,3-heptadiene is a chiral allene, and therefore exists as a pair of enantiomers:

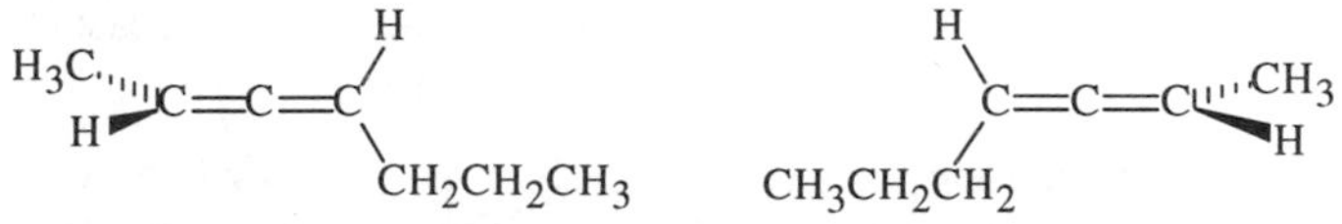

H_3C, H $C{=}C{=}C$ H, $CH_2CH_2CH_3$ and H, $CH_3CH_2CH_2$ $C{=}C{=}C$ CH_3, H

enantiomeric 2,3-heptadienes

(b) Because 2,3,4-heptatriene has an *odd* number of cumulated double bonds, the atoms attached to the terminal double bonds are in the same plane. Hence, the twist necessary for chirality is not present, and 2,3,4-heptatriene therefore cannot exist as enantiomers. On the other hand, 2,3,4-heptatriene can exist as cis-trans isomers, which are diastereomers. (They are stereoisomers; they are not enantiomers; hence, they are diastereomers.)

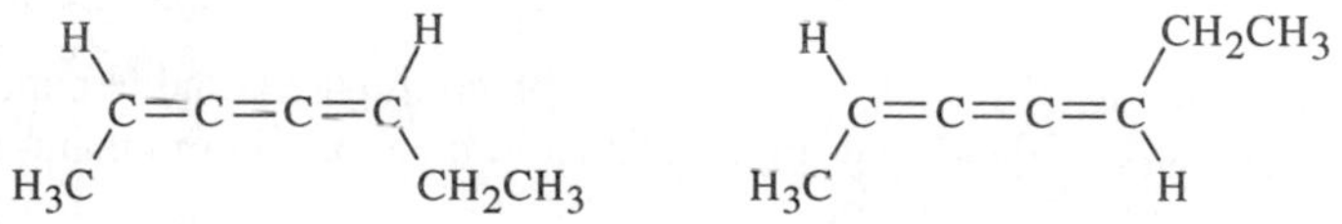

H, H_3C $C{=}C{=}C{=}C$ H, CH_2CH_3 — (*Z*)-isomer

H, H_3C $C{=}C{=}C{=}C$ CH_2CH_3, H — (*E*)-isomer

the diastereomeric 2,3,4-heptatrienes

15.38 (a) This species contains six π electrons: two from the double bond, two from the anion, and two from the oxygen. Because $6 = 4n + 2$ for $n = 1$, this species is aromatic.

(b) This radical has three π electrons: two from the double bond and one from the unpaired electron. It is therefore not aromatic.

(c) Each double and triple bond contributes a pair of π electrons for a total of 18; therefore this compound is likely to be aromatic. Notice that the second pair of π electrons in a triple bond is in a π orbital that is in the plane of the ring; therefore, only the electrons in the π orbital that is perpendicular to the plane of the ring contribute to the aromatic π-electron system.

(d) This anion has ten π electrons: two from each of the four double bonds and two from the anionic carbon. It is likely to be aromatic.

15.39 The species with $4n + 2$ π electrons are likely to be planar because the planar species are aromatic. Thus, tropylium ion and cyclooctatetraenyl dianion are planar species. Oxepin, with eight π electrons, avoids planarity and, like cyclooctatetraene, thus avoids antiaromaticity.

15.40 Severe van der Waals repulsions between the inner hydrogens (shown in the following structure) force the compound out of planarity. These van der Waals repulsions are so great in the planar conformation that not even aromaticity can compensate for the resulting destabilization.

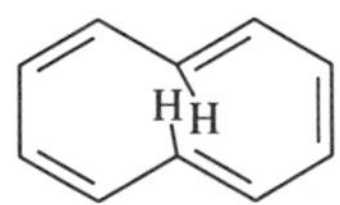

15.41 Remember that increasing heat of formation corresponds to decreasing stability.

(a) *Heats of formation:* 1 < 2 < 4 << 3. *Reasons:* Compound (1) is more stable than compound (2) because a molecule with conjugated double bonds is more stable than an isomer with ordinary double bonds. Compound (2) is more stable than compound (4) because *trans*-alkenes are more stable than *cis*-alkenes. Compound (3) is least stable because an allene is the least stable type of diene.

(b) *Heats of formation:* 3 < 1 < 2. *Reasons:* Compound (3) is most stable because it is aromatic. Compound (1) is more stable than compound (2) because conjugated alkenes are more stable than isomeric alkynes.

(c) *Heats of formation:* 2 < 3 < 1. *Reasons:* Compound (2) is most stable because it is aromatic. Compound (1) is least stable because molecules with cumulated double bonds are less stable than isomeric molecules with conjugated double bonds.

15.42 The principles are that conjugation increases the λ_{max} by 40–50 nm per double bond; each alkyl substituent increases λ_{max} by 5 nm; and an *s*-cis conformation of the diene unit increases the λ_{max} and decreases the absorption intensity relative to that of an analogous *s*-trans diene.

(a) 1,4-Cyclohexadiene is not conjugated and therefore has no UV absorption in the useful region of the spectrum; 1,3-cyclohexadiene, an *s*-cis diene, has absorption at 256 nm (see text page 647). The compound with the UV spectrum is therefore the 1,3-diene.

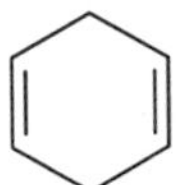

1,4-cyclohexadiene
no UV spectrum

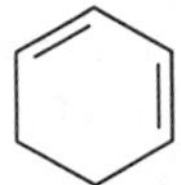

1,3-cyclohexadiene
conjugated; has a UV spectrum

(b) The first compound has three alkyl substituents on conjugated double bonds; the second compound has two. Hence, the λ_{max} of the first compound should be about 5 nm greater than the λ_{max} of the second.

(c) The double bonds of the second compound have two more alkyl branches than the double bonds of the first. (Note that a branch must be *on* a carbon of a double bond to affect the λ_{max}.) Hence, the λ_{max} of the second compound is greater than that of the first by about 10 nm.

(d) The first compound has three conjugated double bonds, whereas the second compound has two; one double bond is not conjugated with the other two. Hence, the first compound should have a considerably greater λ_{max} than the second.

(e) Although the first compound is shown in the *s*-trans conformation, it doesn't exist in that conformation because of severe van der Waals repulsions between the *tert*-butyl groups. It can avoid these repulsions by internally rotating into its *s*-cis conformation. Of course, the *s*-cis conformation has a repulsion between the inner hydrogen and a *tert*-butyl group, but this is much less severe than the repulsions between two *tert*-butyl groups in the *s*-trans conformation. Nevertheless, it is likely that the *s*-cis conformation will be distorted from planarity to relieve this repulsion. This will significantly lower the intensity of its UV absorption.

H H
H C CH₂ H C C(CH₃)₃
C C C C
(CH₃)₃C C(CH₃)₃ (CH₃)₃C CH₂

s-trans conformation *s*-cis conformation

The *s*-trans conformation of the second diene, however, is more stable than the *s*-cis conformation; therefore, the second diene exists mostly in its *s*-trans conformation. Because *s*-cis dienes have considerably greater λ_{max} values than related *s*-trans dienes, the spectrum of the first diene should have a greater λ_{max} value, but a much lower intensity, than the spectrum of the second diene.

15.43 First, use Beer's law to determine the concentration of the diene:

$$A = 0.356 \text{ absorbance units} = \epsilon c l = (10{,}750) \text{ absorbance units} \cdot \text{L} \cdot \text{mol}^{-1} \cdot \text{cm}^{-1} (c)(1.0 \text{ cm})$$

or

$$c = 3.31 \times 10^{-5} \text{ mol/L}$$

Since the entire sample was diluted to one liter, the entire sample contains 3.31×10^{-5} mole of isoprene. (Notice that any hydrogenation product lacks conjugated double bonds and does not have UV absorption.) The amount of isoprene present is

$$(1.00 \text{ L})(3.31 \times 10^{-5} \text{ mol/L})(68.12 \text{ g/mol}) = 2.25 \times 10^{-3} \text{ g} = 2.25 \text{ mg}$$

Because the mass of the entire sample is 75 mg, the mass percent of isoprene is

$$\frac{2.25 \text{ mg}}{75.0 \text{ mg}} \times 100\% = 3.00\%$$

That is, 3% of the isolated material is unreacted isoprene.

15.44 The color of β-carotene is due to its chromophore of extensively conjugated double bonds. Catalytic hydrogenation would result in addition of hydrogen to these double bonds; hence, catalytic hydrogenation of a β-carotene sample would convert it from a red-orange conjugated alkene into a colorless alkane.

15.45 The carbon skeleton of compound *A*, from the hydrogenation data, is that of hexane; that is, it has an unbranched chain of six carbons. The two degrees of unsaturation indicate that the compound is either a diene or an alkyne; the IR absorption and the observation of optical activity indicate that the compound is an allene. The only possible chiral allene is 2,3-hexadiene, $H_3C—CH{=}C{=}CH—CH_2CH_3$. Indeed, partial hydrogenation of this diene would give a mixture of *cis*-2-hexene and *cis*-3-hexene.

15.46 Although mycomycin has no asymmetric carbons, it does contain stereocenters: the outer carbons of the allene unit. Evidently, mycomycin is one enantiomer of this chiral allene. The chirality of certain allenes is discussed in Sec. 15.1B.)

15.47 In order for a diene to react in a Diels-Alder reaction, it must be able to assume an *s*-cis conformation in the transition state.

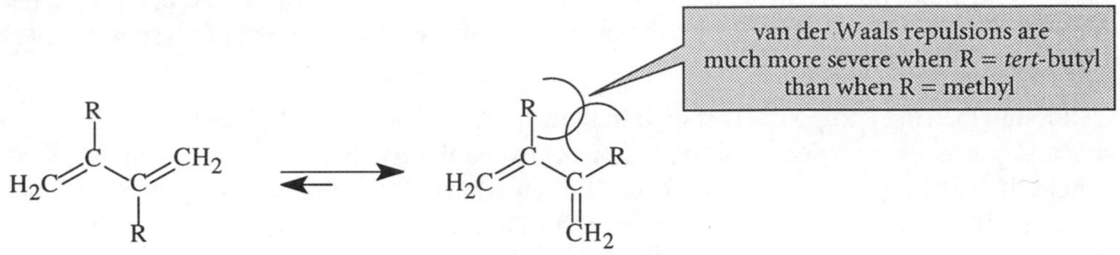

van der Waals repulsions are much more severe when R = *tert*-butyl than when R = methyl

s-trans conformation

s-cis conformation

When R = *tert*-butyl, the van der Waals repulsions in the *s*-cis conformation are much more severe than when R = methyl. Consequently, when R = *tert*-butyl, the transition state of any Diels-Alder reaction is destabilized by these van der Waals repulsions to a greater extent than when R = methyl. Evidently, the destabilization is so great that the Diels-Alder reaction with maleic anhydride does not occur at all when R = *tert*-butyl.

15.48 The analysis in Eqs. 15.9a–b on text p. 654 shows that the *s*-cis conformations of dienes with cis double bonds are destabilized by van der Waals repulsions, whereas the *s*-cis conformations of dienes with trans double bonds do not suffer the same repulsions.

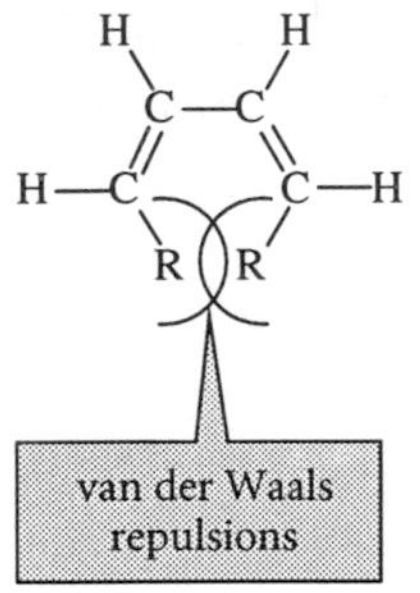

s-cis conformation of all-cis diene

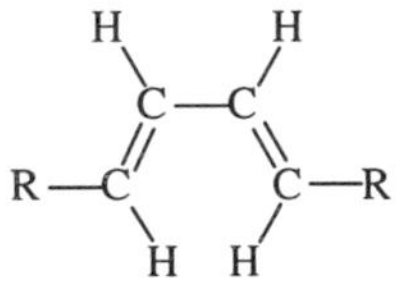

van der Waals repulsions

s-cis conformation of all-trans diene

These van der Waals repulsions also destabilize the transition states of Diels-Alder reactions, which require *s*-cis conformations of the reacting dienes. Because the diene in the problem undergoes the Diels-Alder reaction, it is probably the all-trans diene.

15.49 The product of such a Diels-Alder reaction would be most interesting:

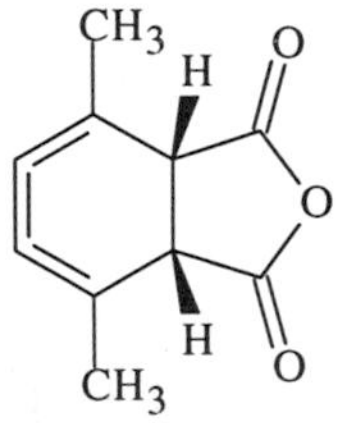

Wow! A cyclic triene containing three cumulated double bonds within the ring. Alas, such an adduct could never form, because the requirement for linear geometry in the cumulated double bonds of the product would result in a compound with immense angle strain. A related explanation is that the 2*p* orbitals of maleic anhydride and those of the alkyne must overlap as shown in Figure 15.6 on text p. 653. Because the alkyne is constrained to linear geometry, the transition state for such a reaction would either be very strained (if the alkyne units were bent) or would contain extremely poor electronic overlap (if the alkyne units remained linear).

15.50 The answer to this problem lies in the *s*-cis conformations of the three dienes, which are as follows:

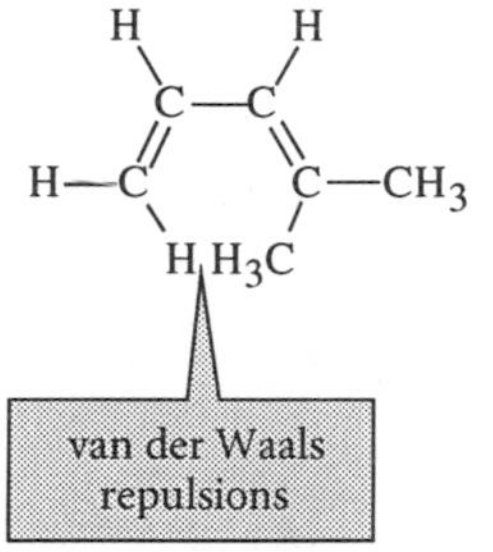

van der Waals repulsions

van der Waals repulsions

4-methyl-1,3-pentadiene
s-cis conformation

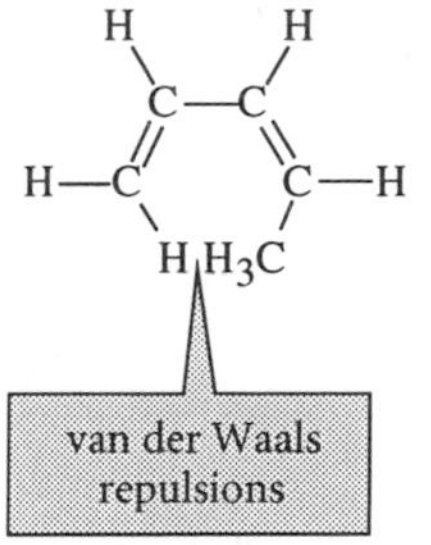

(*Z*)-1,3-pentadiene
s-cis conformation

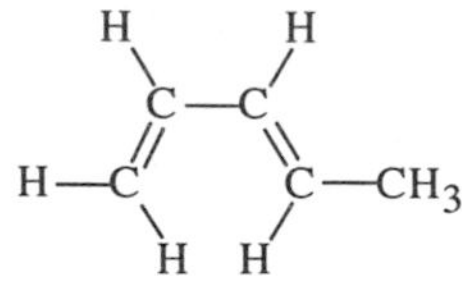

(*E*)-1,3-pentadiene
s-cis conformation

The reactivity of these three dienes in the Diels-Alder reactions correlates nicely with the accessibility of their *s*-cis conformations. (Remember, the transition state of the Diels-Alder reaction involves the *s*-cis conformation of the diene.) The van der Waals repulsions shown above destabilize the *s*-cis conformations of both 4-methyl-1,3-pentadiene and (*Z*)-1,3-pentadiene to the same extent because the interacting groups are identical. Hence, their reactivity is reduced by about the same amount relative to that of 1,3-pentadiene, whose *s*-cis conformation has no such repulsions.

15.51 (a) First, draw the structures of the two carbocations that would be produced by the alternative modes of protonation:

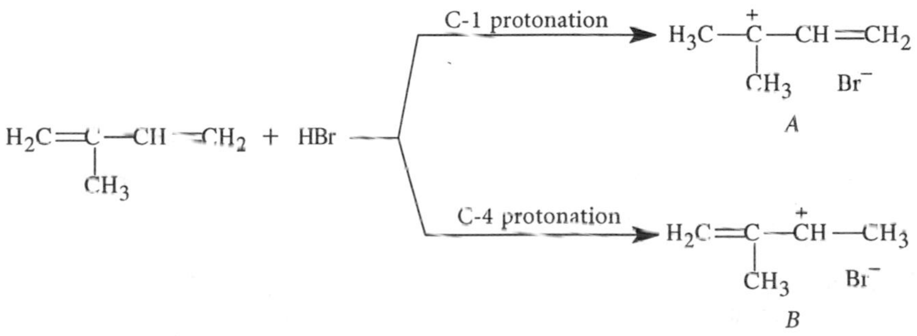

$H_2C{=}C(CH_3){-}CH{=}CH_2 + HBr$

C-1 protonation → $H_3C{-}\overset{+}{C}(CH_3){-}CH{=}CH_2$ Br^- *A*

C-4 protonation → $H_2C{=}C(CH_3){-}\overset{+}{C}H{-}CH_3$ Br^- *B*

Although both carbocations are allylic, carbocation *A* has a *tertiary* carbocation resonance structure, shown above, while neither resonance form of carbocation *B* is tertiary, Hence, carbocation *A*, from protonation at carbon-1, is the more stable carbocation.

(b) Because carbocation *A* is the more stable ion, it is the carbocation intermediate involved in HBr addition, and the products are thus derived from it. The two products come from attack of bromide ion on the carbons that bear positive charge, which we deduce from the resonance structures of the ion.

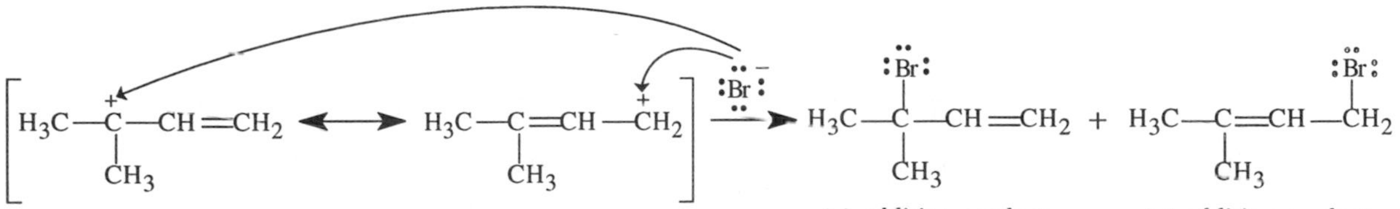

$\left[H_3C{-}\overset{+}{C}(CH_3){-}CH{=}CH_2 \longleftrightarrow H_3C{-}C(CH_3){=}CH{-}\overset{+}{C}H_2\right] \xrightarrow{:\ddot{Br}:^-} H_3C{-}C(Br)(CH_3){-}CH{=}CH_2 + H_3C{-}C(CH_3){=}CH{-}CH_2Br$

resonance structures of carbocation *A*

1,2-addition product

1,4-addition product

(c) Protonation at carbon-1 of *trans*-1,3,5-hexatriene gives the carbocation with the greatest number of resonance structures. (Verify this point by examining the carbocations obtained from protonation at other carbons.)

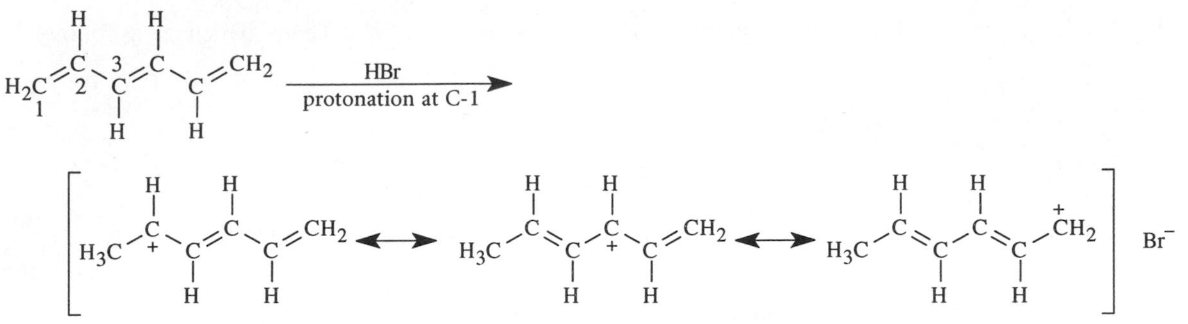

HBr

protonation at C-1

Br^-

The three possible products are derived from attack of Br^- on the positively charged carbons.

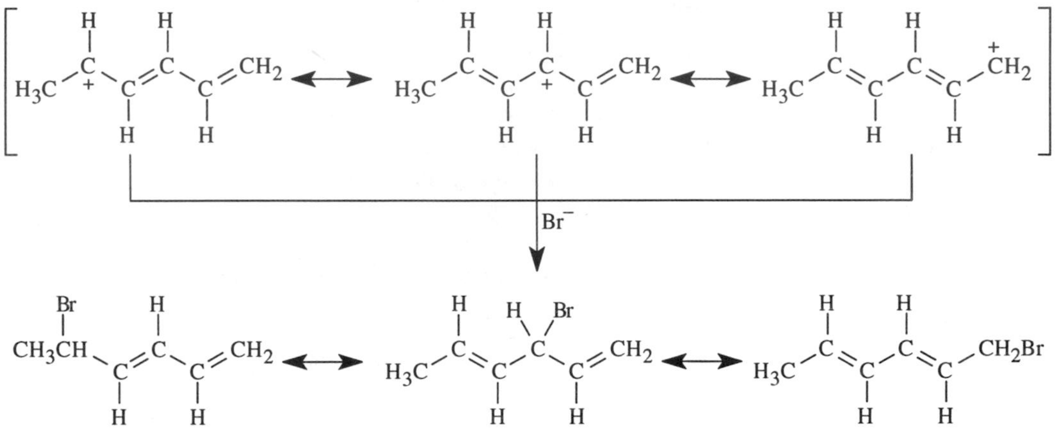

Br^-

A *B* *C*

(d) The kinetically controlled product is the 1,2-addition product in each case; this is labeled explicitly in part (b), and is compound *A* in part (c). The reason is given in the text; the 1,2-product results from collapse of the initially formed ion pair at the nearest site of positive charge. (See Eq. 15.26 on text p. 664.) In part (b), the thermodynamically controlled product is the 1,4-addition product, because it contains the double bond with the greater number of branches. The thermodynamically controlled product in part (c) is compound *C*, a 1,6-addition product, because it derives special stability from the conjugated arrangement of its double bonds.

15.52 (a) The 1,2- and 1,4-addition products formed in the reaction of 1,3-pentadiene and HCl are identical if we assume that all double bonds retain their *E* stereochemistry.

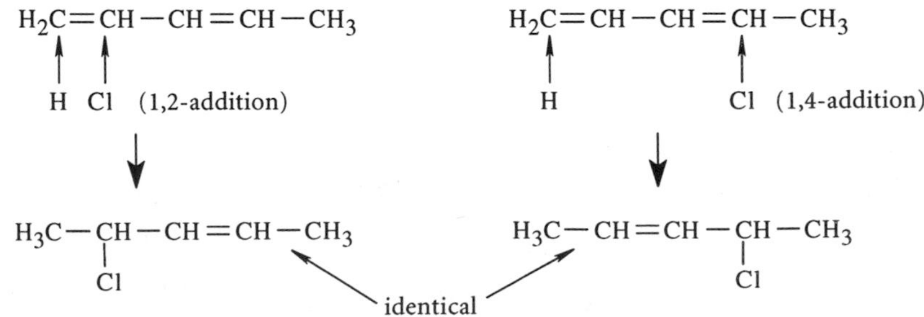

(b) You can see in part (a) that there can be no preference for 1,2- versus 1,4-addition on the basis of the product stability, because the products are the same. The problem, of course, is that we can't tell one mode of addition from the other! The use of D—Cl, however, solves this problem. This allows us to tell the hydrogen that has added from the ones that started off in the diene.

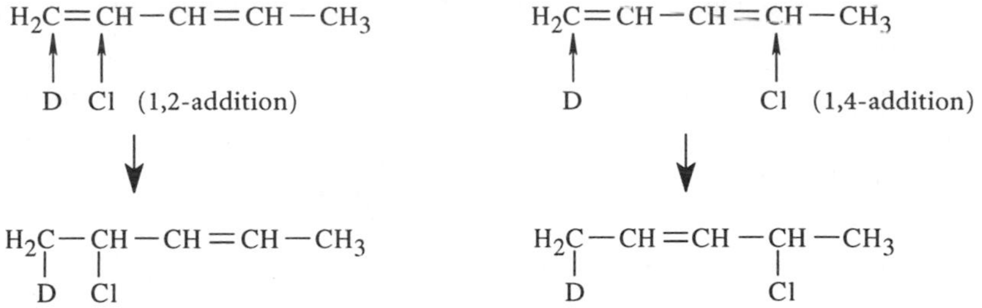

$H_2C{=}CH{-}CH{=}CH{-}CH_3$ + D, Cl (1,2-addition) → $H_2C(D){-}CH(Cl){-}CH{=}CH{-}CH_3$

$H_2C{=}CH{-}CH{=}CH{-}CH_3$ + D, Cl (1,4-addition) → $H_2C(D){-}CH{=}CH{-}CH(Cl){-}CH_3$

The two products are different, if only subtly so. The isotope has a negligible effect on product stability. According to the text, the product of 1,2-addition is preferred, and in fact, this is what was observed experimentally. The actual ratio is about 70:30 in favor of 1,2-addition. (How would we tell one product from the other?) This experiment, reported in 1979 by J. E. Nordlander of Case Western Reserve University, established clearly that product stability has nothing to do with the kinetic preference. The arguments in the text are the only reasonable alternative.

15.53 (a) Unless the reaction is allowed to run for a very long time, this is a kinetically controlled reaction, because the ratio of the two products is different from the ratio at equilibrium.

(b) Compound *C* has a double bond with four branches; compound *B* has a double bond with three branches. Because branching at the double bond enhances alkene stability, it stands to reason that *C* would be favored at equilibrium.

(c) The reaction occurs through the following carbocation intermediate by a typical E1 process (see Sec. 10.1).

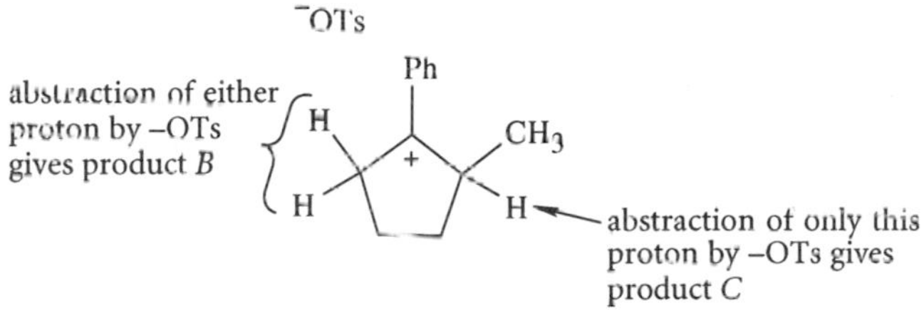

The formation of the product involves loss of a proton from this intermediate. Alkene *B* can form in two ways, namely, by abstraction of either one of two protons. Alkene *C* can form in only one way. Hence, the formation of alkene *B* is favored on statistical grounds. It is interesting that, in Fig. P15.53, the ratio of products at zero time is indeed about 2:1.

15.54 (a) The transformations involved in the reaction between 1,3-cyclopentadiene and maleic anhydride:

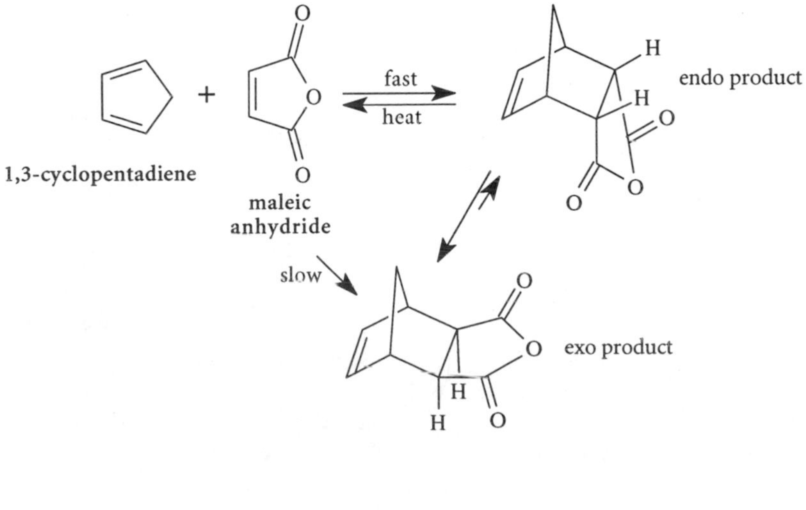

(b) Because the initially formed product distribution is markedly different from the equilibrium distribution, the reaction is kinetically controlled.

(c) At low temperature, the formation of product is kinetically controlled. The problem tells us that the endo stereoisomer is the major one formed. The transition state for this process is shown on the left side of the following diagram. The transition state for the other process is shown at right.

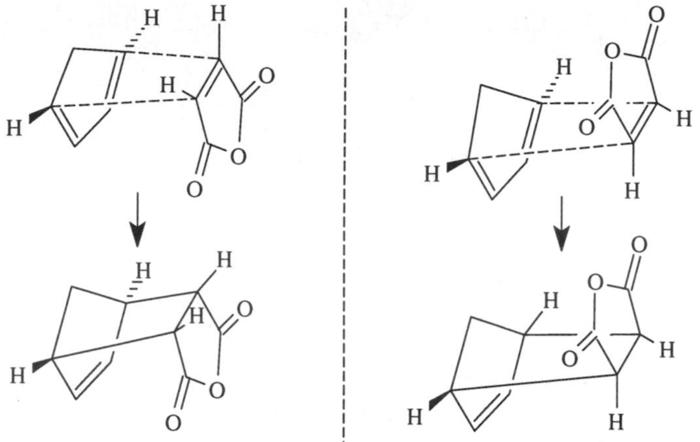

endo product

exo product

It has been suggested that kinetic control in the Diels-Alder reaction is due to "maximum accumulation of unsaturation," or "secondary orbital interactions." You can see that in the endo transition state, the $2p$ orbitals of the anhydride carbonyl groups and some of the $2p$ orbitals of the diene unit are "face-to-face."

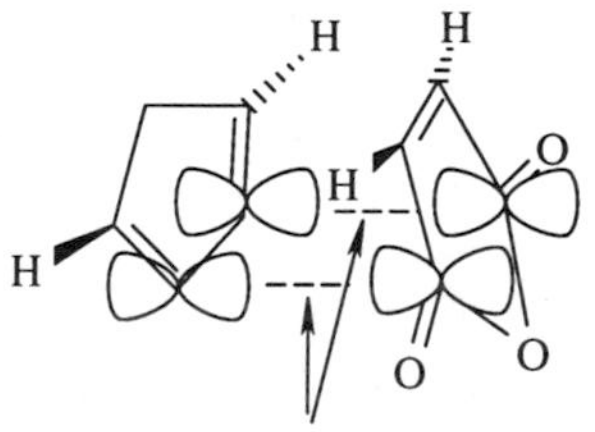

secondary orbital interactions

The interaction between these orbitals is evidently a stabilizing one. The endo product is formed almost exclusively. However, when the two products are allowed to come to equilibrium, there is little difference in their stabilities.

15.55 Given that the addition occurs to the triple bond, the addition should occur regiospecifically to give the product in which the chlorine ends up on the internal carbon of the triple bond.

$$HC{\equiv}C{-}CH{=}CH_2 \xrightarrow{HCl} H_2C{=}C(Cl){-}CH{=}CH_2$$

vinylacetylene (1-buten-3-yne) → chloroprene

To deduce the structure of neoprene, follow the pattern for 1,4-addition polymerization shown for polybutadiene in Eq. 15.27b on text p. 665.

$$\left(CH_2{-}C(Cl){=}CH{-}CH_2\right)_n$$

neoprene

15.56 Assume that *A* and *B* are 1,2- and 1,4-addition products.

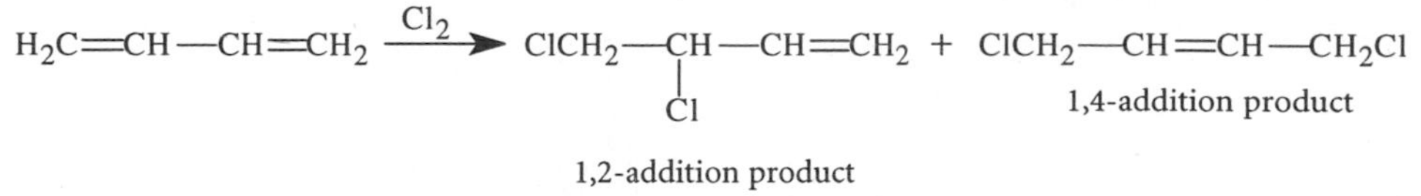

$$H_2C{=}CH{-}CH{=}CH_2 \xrightarrow{Cl_2} ClCH_2{-}\underset{\displaystyle Cl}{CH}{-}CH{=}CH_2 + ClCH_2{-}CH{=}CH{-}CH_2Cl$$

1,2-addition product; 1,4-addition product

Which is *A* and which is *B*? And what is the stereochemistry of the 1,4-addition product? To answer these questions, note that further addition of Cl_2 to each gives a 1,2,3,4-tetrachlorobutane. Assume that addition of Cl_2 is an *anti*-addition. Because the 1,4-addition product has two stereocenters, *anti*-addition of Cl_2 will give the meso tetrachloride if the alkene is trans, and will give the racemic tetrachloride if the alkene is cis. (If this point is not clear, review Sec. 7.9C.) Because compound *B* gives a single tetrachloride *C* with meso stereochemistry, it follows, then, that the 1,4-addition product is compound *B*, and that it has trans stereochemistry. The 1,2-addition product, then, is compound *A*. Addition of Cl_2 to compound *A* gives both diastereomers of the tetrachloride. To summarize:

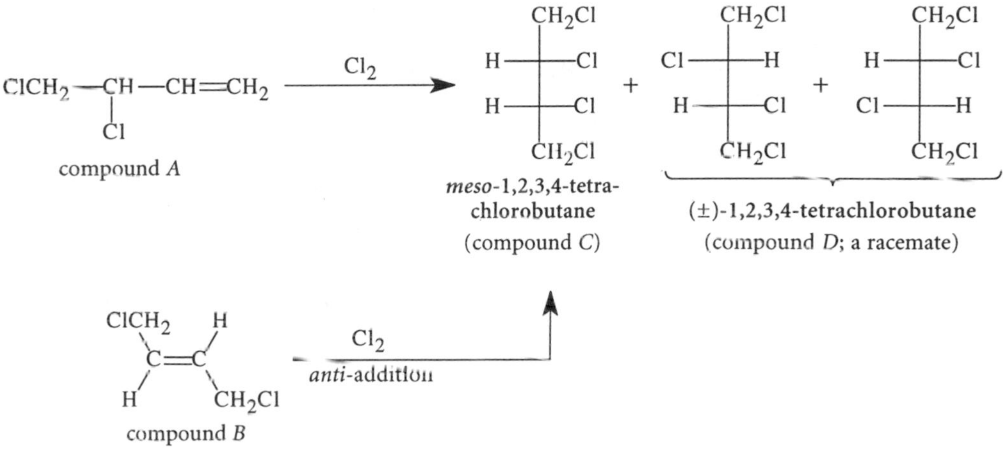

ClCH₂—CH(Cl)—CH=CH₂ (compound *A*) $\xrightarrow{Cl_2}$ meso-1,2,3,4-tetrachlorobutane (compound *C*) + (±)-1,2,3,4-tetrachlorobutane (compound *D*; a racemate)

Fischer projections (top to bottom: CH₂Cl, C, C, CH₂Cl):
- *meso*-1,2,3,4-tetrachlorobutane (compound *C*): H—C—Cl, H—C—Cl
- (±)-1,2,3,4-tetrachlorobutane (compound *D*; a racemate): Cl—C—H, H—C—Cl + H—C—Cl, Cl—C—H

compound *B* (trans ClCH₂CH=CHCH₂Cl) $\xrightarrow[\textit{anti}\text{-addition}]{Cl_2}$ compound *C*

15.57 Reaction of KH with cyclopentadiene forms the potassium salt of the cyclopentadienyl anion (structure *X* below) plus H_2 (the gas). In this anion, all carbons, except for the isotope, are equivalent by resonance. The isotope makes no detectable difference in the relative importance of the resonance contributors; except for the position of the isotope, all resonance contributors are identical. Hence, protonation of this anion by H_2O occurs at each carbon with equal (20%) probability. Protonation at carbons 2 and 5 gives the same product; and protonation at carbons 3 and 4 gives the same product. Hence, each of these products is obtained in (2 × 20%) = 40% yield.

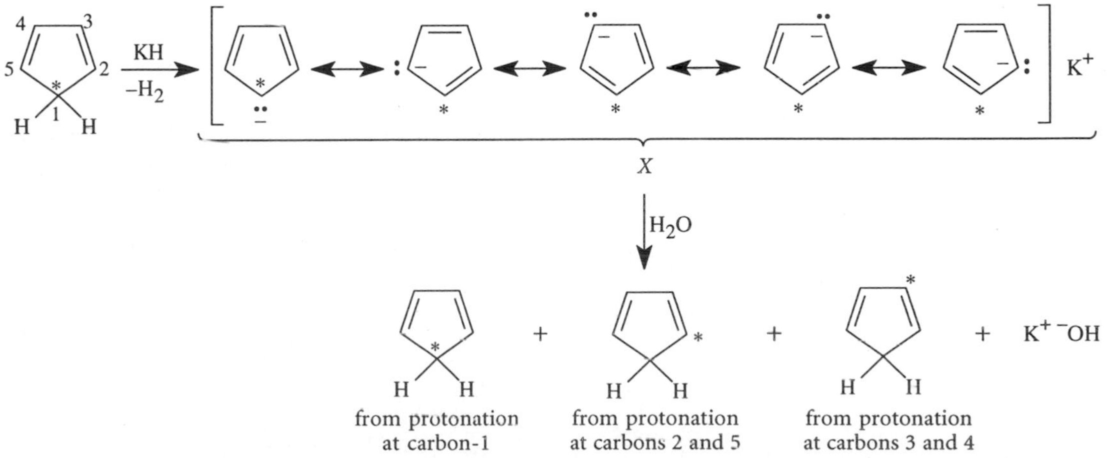

Cyclopentadiene (C-1 labeled *, CH₂ with H, H; carbons numbered 1–5) $\xrightarrow[-H_2]{KH}$ [five resonance contributors of the labeled cyclopentadienyl anion] K⁺ = *X*

X $\xrightarrow{H_2O}$ (products) + $K^{+}\ {}^{-}OH$

from protonation at carbon-1 + from protonation at carbons 2 and 5 + from protonation at carbons 3 and 4

15.58 Borazole is very stable because it is aromatic; each nitrogen contributes two π electrons and each boron contributes zero electrons, for a total of six π electrons in the aromatic system. The resemblance of borazole to benzene is more obvious from its other two resonance structures:

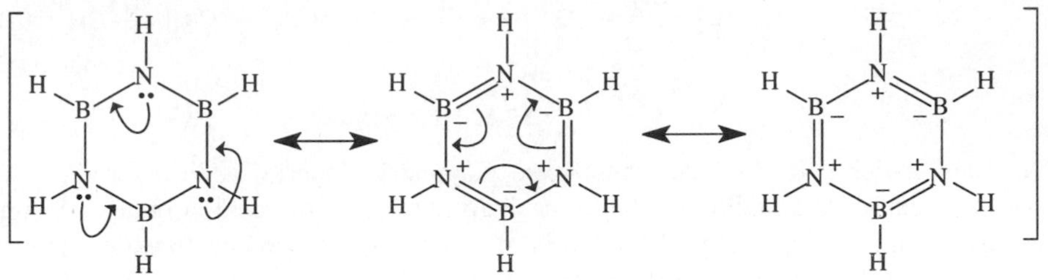

15.59 Compound *A* should react much more rapidly because the carbocation intermediate (shown below) has three important resonance contributors. In particular, an unshared pair of electrons on the oxygen can be delocalized in this cation; in the solvolysis of the other compound, the unshared pairs on oxygen have no resonance interaction with the positive charge in the carbocation.

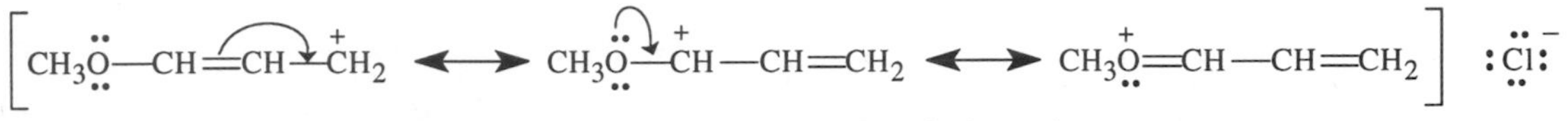

$$\left[CH_3\ddot{O}{-}CH{=}CH{-}\overset{+}{C}H_2 \longleftrightarrow CH_3\ddot{O}{-}\overset{+}{C}H{-}CH{=}CH_2 \longleftrightarrow CH_3\overset{+}{O}{=}CH{-}CH{=}CH_2\right] \; :\ddot{Cl}:^-$$

carbocation intermediate involved in the solvolysis of *A*

By Hammond's postulate, the reaction involving the more stable carbocation intermediate is faster.

15.60 The carbocation intermediate in the S_N1 solvolysis of compound *A* is resonance-stabilized, but the carbocation intermediate in the solvolysis of compound *B* is not. Because the relative energies of transition states reflect the relative energies of the corresponding carbocation intermediates (Hammond's postulate), the solvolysis of *A*, which gives the more stable carbocation, is faster.

Br^- Br^-

carbocation intermediate in the solvolysis of compound *A*

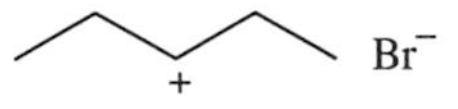

carbocation intermediate in the solvolysis of compound *B*

If compound *C* gives a doubly allylic carbocation, why should it be solvolytically inert? The answer is that the cation has a continuous cycle of π electrons. Because this cation contains $4n$ ($n = 1$) π electrons, it is *antiaromatic* and therefore very unstable. In fact, it is so unstable that it does not form.

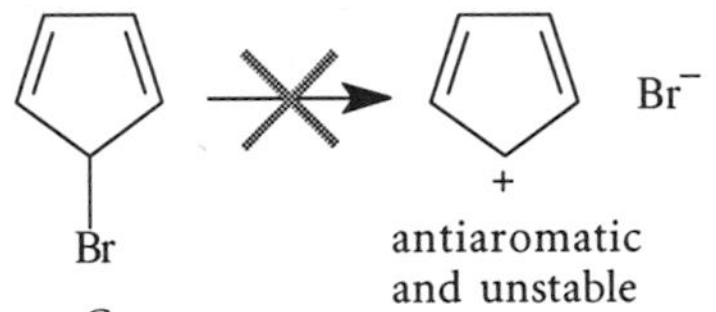

Br

C

Br^-

antiaromatic and unstable

15.61 This compound behaves like a salt because it *is* a salt. Tropylium bromide ionizes readily to bromide ion and the *tropylium cation,* which, because it has a continuous cycle of six ($4n + 2$, $n = 1$) π electrons, is aromatic, very stable, and very easily formed.

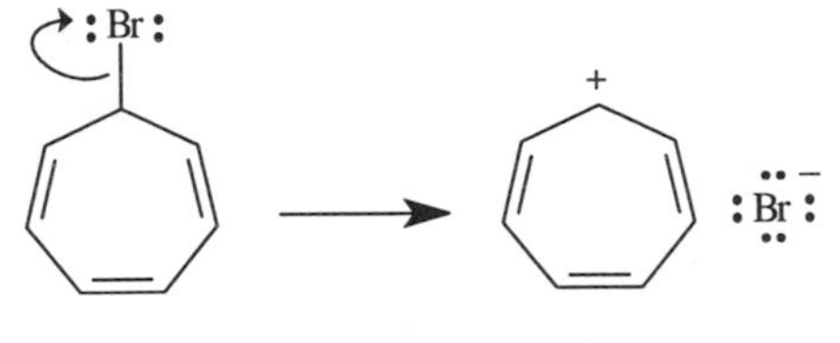

$:\ddot{Br}:$ $:\ddot{Br}:^-$

tropylium bromide

tropylium cation; aromatic and stable

15.62 (a) The triple bond hydrogenates, but the phenyl rings resist hydrogenation because they have aromatic character. The product, of course, is the cis alkene.

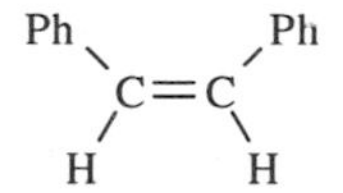

cis-stilbene
(*cis*-1,2-diphenylethylene)

(b) This is an ordinary Diels-Alder reaction; because the diene is cyclic, a bicyclic product is formed. Two diastereomers are possible. (Each, of course, is formed as the racemate, only one enantiomer of which is shown below.)

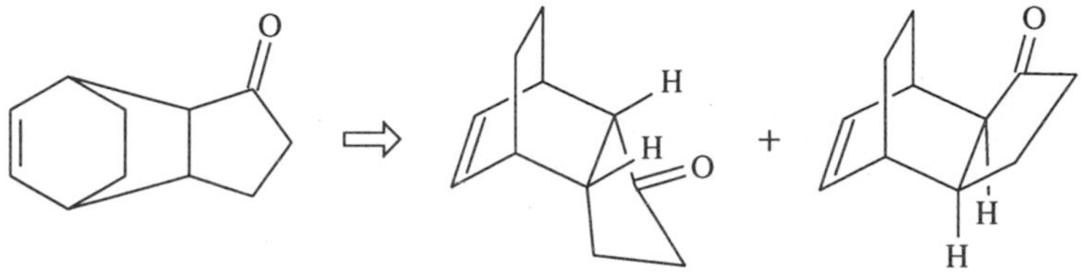

endo exo

(c) This is a Diels-Alder reaction with the alkyne as the dienophile. The phenyl rings are cis because, in the *s*-cis conformation of the diene, they are both "outer" groups. (See text Eq. 15.11b, p. 655.)

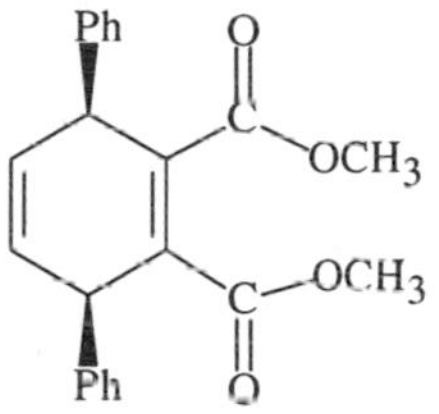

(d) The conjugated-diene part of the triene undergoes a Diels-Alder reaction with a double bond of the benzoquinone. Because there are two double bonds in the quinone, a second Diels-Alder reaction also occurs. Two constitutional isomers result from the second addition because the diene can add in two different orientations.

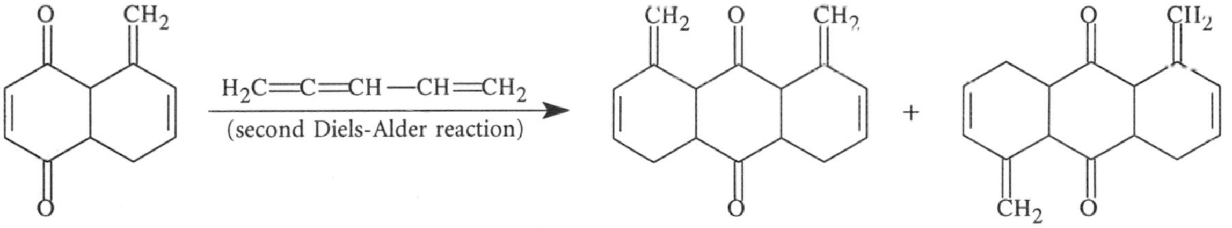

$H_2C{=}C{=}CH{-}CH{=}CH_2$
(second Diels-Alder reaction)

product of first
Diels-Alder reaction

Moreover, each of these constitutional isomers can exist as two diastereomers:

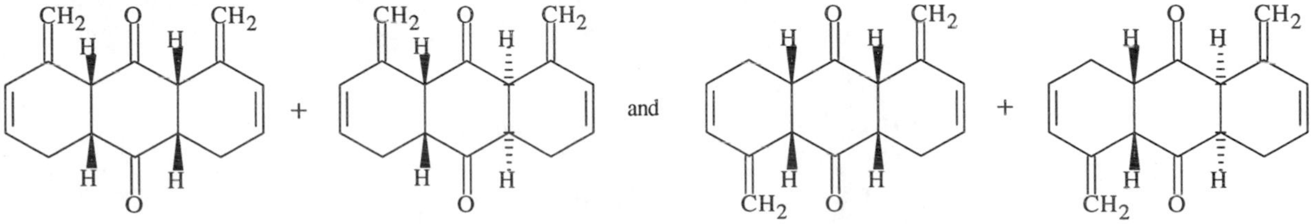

(Which of *these* can exists as enantiomers, and which are meso compounds?)

(e) The maleic anhydride reacts with the diene unit that is "locked" in an *s*-cis conformation. The diastereomer of the product is formed that results from addition to the face of the steroid opposite to the angular methyl groups, which block the β-face of the molecule. (See text p. 274.) In this mode of addition, van der Waals repulsions between maleic anhydride and the angular methyl groups are avoided.

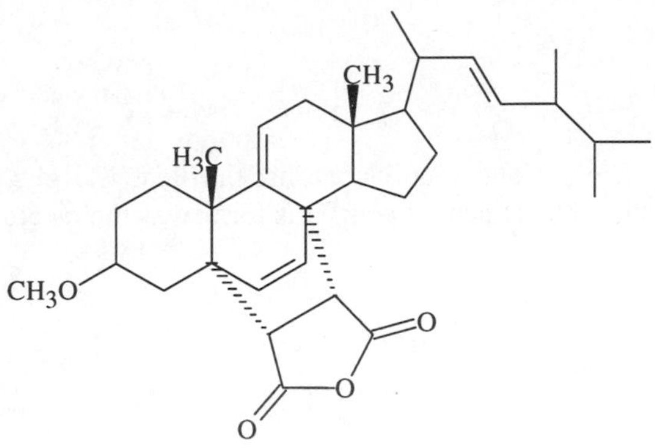

(f) Because the product has the same number of carbons as the starting material, the reaction is intramolecular. In fact, it is an intramolecular Diels-Alder reaction.

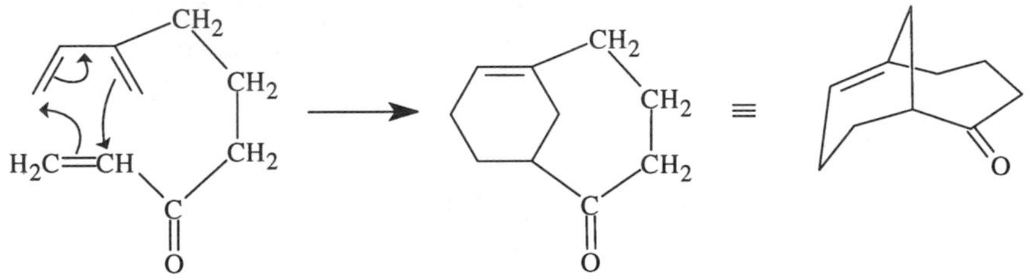

The bridgehead double bond in this product does not violate Bredt's rule because it is part of a large ring.

(g) The product is *nickelocene,* the nickel analog of ferrocene. (See Eq. 15.45, text p. 681.)

nickelocene

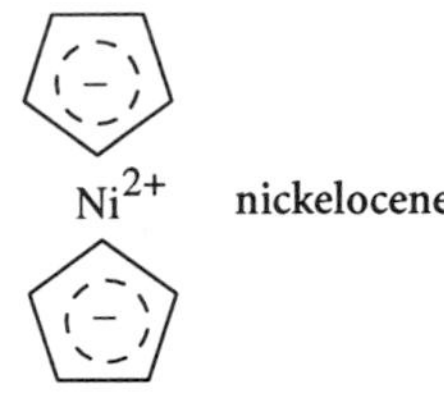

15.63 The structure of the Diels-Alder adduct suggests the transient formation of *cyclopentyne.* This conclusion is obtained by mentally reversing the Diels-Alder reaction in such a way that diphenylisobenzofuran is obtained; cyclopentyne is the other starting material:

cyclopentyne

diphenylisobenzofuran

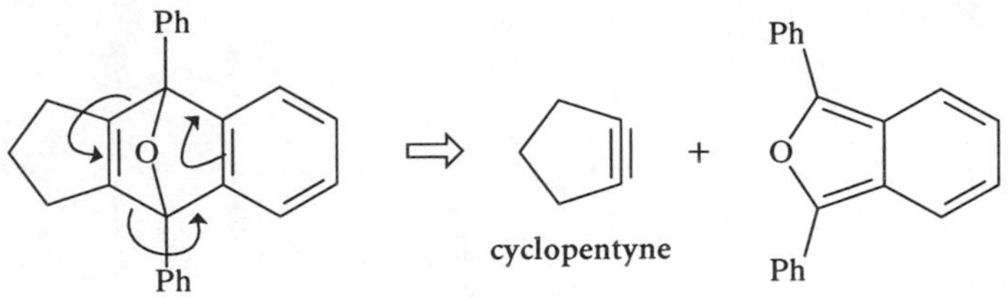

Cyclopentyne is formed by a β-elimination of the Grignard reagent formed from 1,2-dibromocyclopentene. We can more readily understand the elimination by thinking of the Grignard reagent as a carbanion:

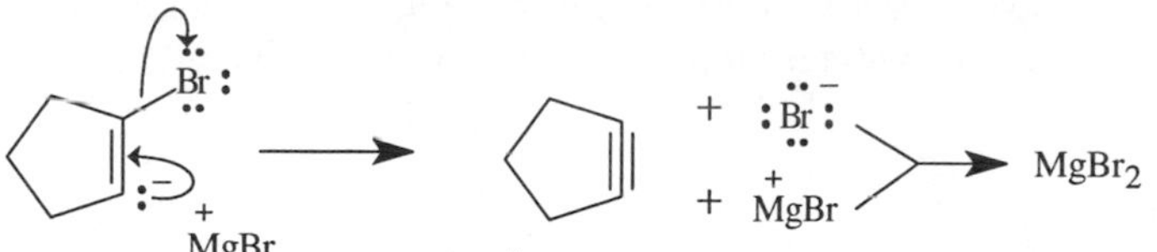

Br : + :Br: MgBr MgBr$_2$

cyclopentyne reacts with diphenylisobenzofuran

15.64 Follow the procedure used in solving Problem 15.63: mentally reverse the Diels-Alder reaction so that maleic anhydride is obtained; the other fragment obtained is compound *X*, which thus has the following structure:

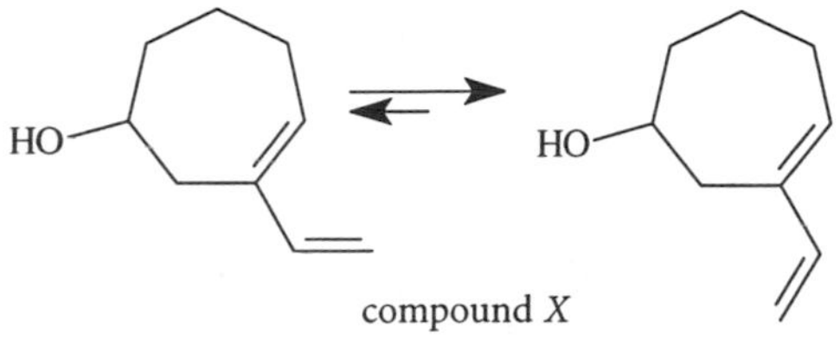

HO HO

compound *X*

The formation of *X* results from ionization of the starting chloride to a tertiary allylic carbocation *A* followed by ring opening of *A* to a secondary carbocation *B*, which is captured by the nucleophile water. (Why should the tertiary allylic carbocation *A* form a secondary carbocation *B*?)

:Cl: :Cl:

A

B :OH$_2$:OH H :OH$_2$:OH + H—OH$_2$

compound *X*

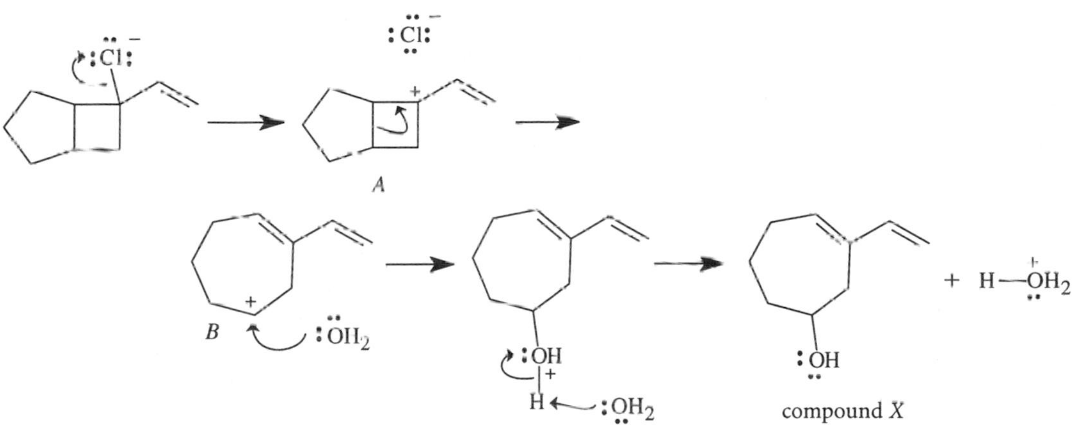

15.65 The structure of spiropentadiene is shown below; it undergoes a Diels-Alder reaction with two molar equivalents of 1,3-cyclopentadiene. Spiropentadiene is unstable because of its great ring strain.

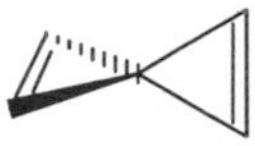

spiropentadiene

15.66 (a) The mechanism below is shown beginning with the protonated alcohol. (See, for example, Eq. 10.9a on text p. 412 for a typical alcohol protonation.)

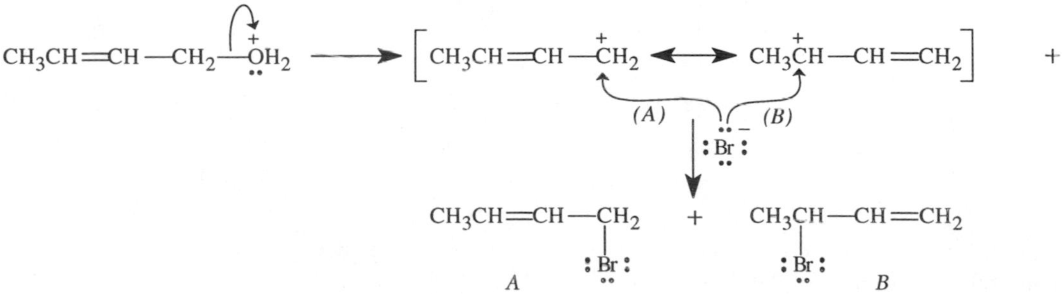

$$CH_3CH{=}CH{-}CH_2{-}\overset{+}{O}H_2 \longrightarrow \left[CH_3CH{=}CH{-}\overset{+}{C}H_2 \longleftrightarrow CH_3\overset{+}{C}H{-}CH{=}CH_2\right] +$$

(*A*) (*B*) :Br:

$$CH_3CH{=}CH{-}CH_2Br \; (A) \quad + \quad CH_3CHBr{-}CH{=}CH_2 \; (B)$$

(b) This reaction involves a rearrangement of carbocation *X1* to a more stable carbocation *X2* followed by loss of a β-proton from the rearranged ion. (Only the part of the molecule involved in the reaction is shown below.)

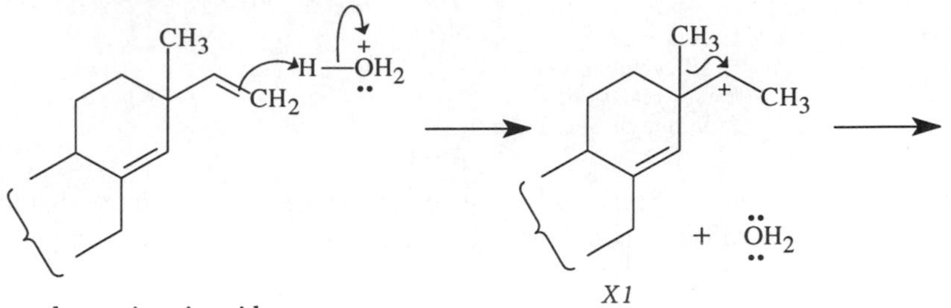

CH_3 H—$\overset{+}{O}H_2$ CH_2 → CH_3 $\overset{+}{}$ CH_3 + $\ddot{O}H_2$ →

dextropimaric acid *X1*

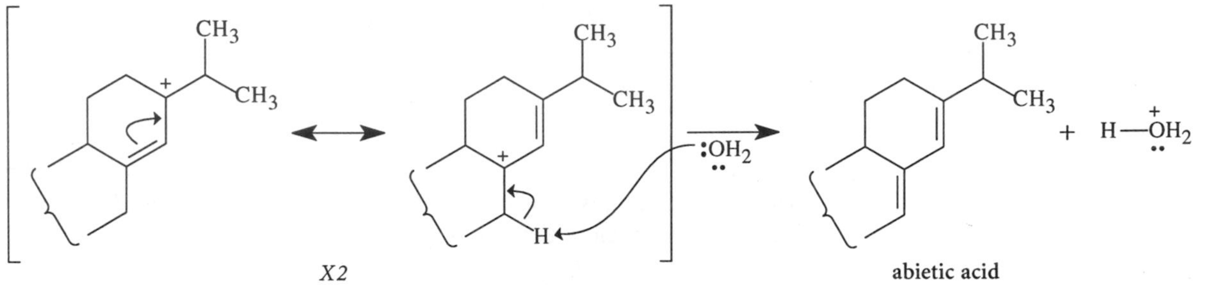

CH_3 + CH_3 ↔ CH_3 CH_3 + H → $:OH_2$ → CH_3 CH_3 + H—$\overset{+}{O}H_2$

X2 abietic acid

(c) This is a reverse Diels-Alder reaction followed by another Diels-Alder reaction. (The solution to Problem 15.54 discusses the reversal of this Diels-Alder reaction.)

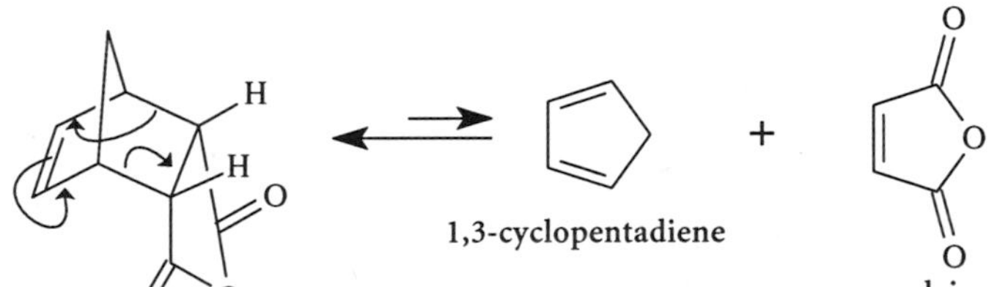

H H O O O ⇌ 1,3-cyclopentadiene + maleic anhydride

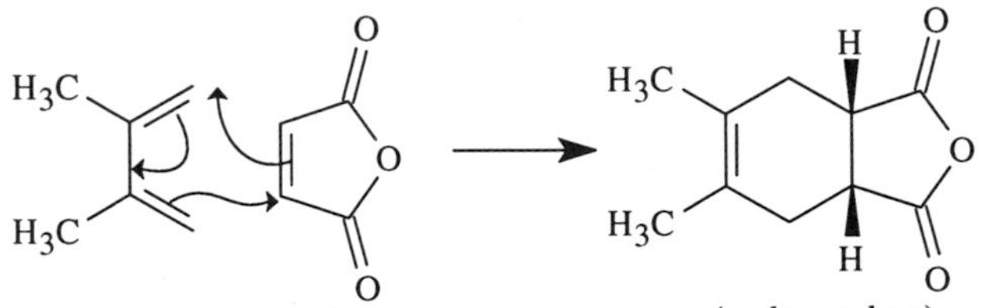

H_3C H_3C + maleic anhydride → H_3C H_3C H H (endo product)

(d) The product *X* is a Diels-Alder adduct between α-phellandrene and the alkyne:

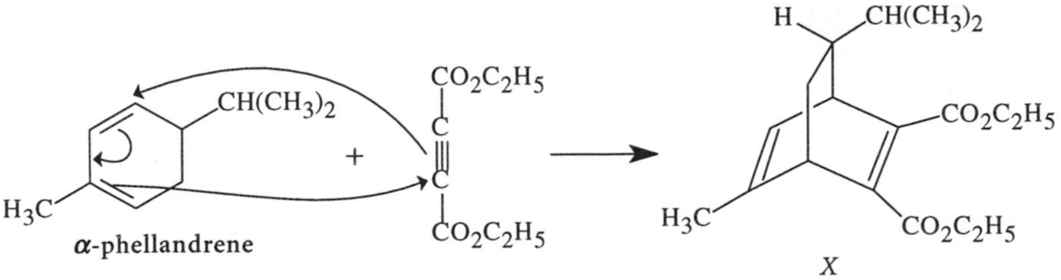

$CH(CH_3)_2$ H_3C α-phellandrene + $CO_2C_2H_5$–C≡C–$CO_2C_2H_5$ → H $CH(CH_3)_2$ $CO_2C_2H_5$ H_3C $CO_2C_2H_5$ *X*

(Because the absolute configuration of α-phellandrene is not specified, it could be either of the two enantiomers, or the racemate, and the same is true of the Diels-Alder adduct *X*.)

Compound *X* undergoes a reverse Diels-Alder in a different manner to lose 3-methyl-1-butene and the aromatic compound diethyl 4-methylphthalate. The driving force for this reaction is, first, the formation of a volatile product (the alkene), and, second, the aromatic stability of the ultimate product.

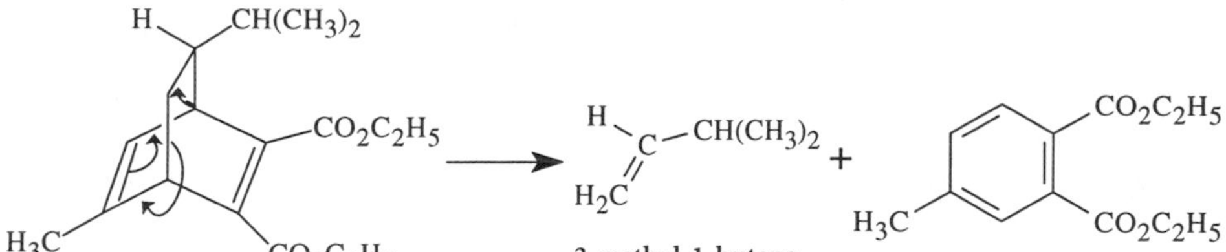

H $CH(CH_3)_2$ $CO_2C_2H_5$ H_3C $CO_2C_2H_5$ *X* → H C $CH(CH_3)_2$ H_2C 3-methyl-1-butene + $CO_2C_2H_5$ H_3C $CO_2C_2H_5$ diethyl 4-methylphthalate

15.67 (a) The exocyclic double bond is protonated in the manner shown below because this process gives the more stable carbocation—the carbocation with the greatest number of resonance structures. (The strong acid is represented by H—X.)

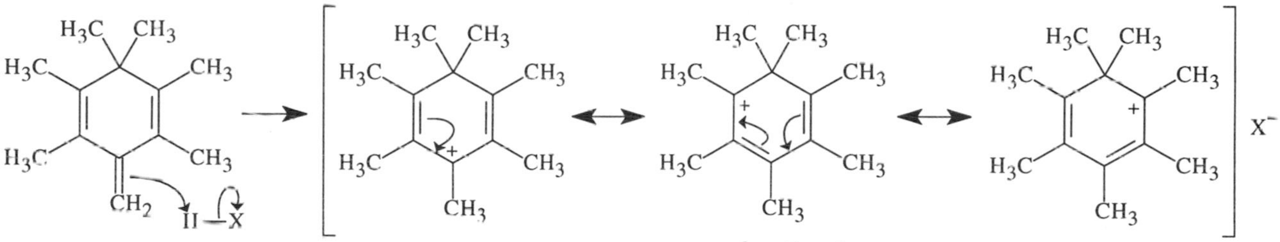

H_3C CH_3 H_3C CH_3 H_3C CH_3 CH_2 H—X → [↔ ↔] X^-

carbocation *A*

(b) The protons nearest the positive charge have the greatest chemical shifts. These are immediately apparent from the resonance structures above. The assignment of the methyl protons at δ 2.82 is clear from their relative integral; this methyl is the only one not equivalent to another methyl by symmetry. The two non-vinylic methyls have the smallest chemical shift. To summarize:

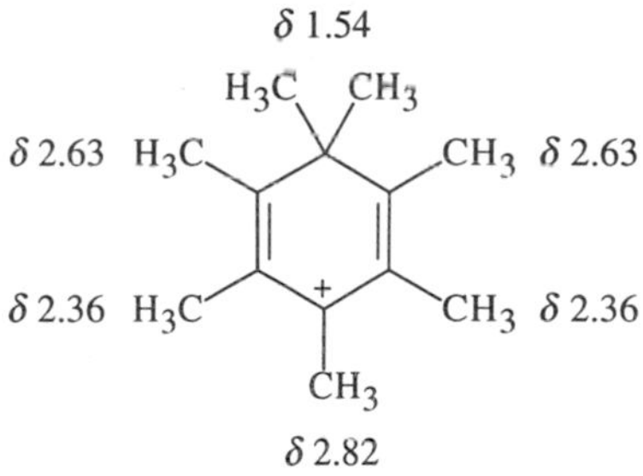

δ 1.54 H_3C CH_3; δ 2.63 H_3C CH_3 δ 2.63; δ 2.36 H_3C CH_3 δ 2.36; CH_3 δ 2.82

(c) The signals coalesce when the temperature is raised because a chemical process takes place that makes the methyl groups equivalent over time. This reaction is nicknamed the "methyl walk."

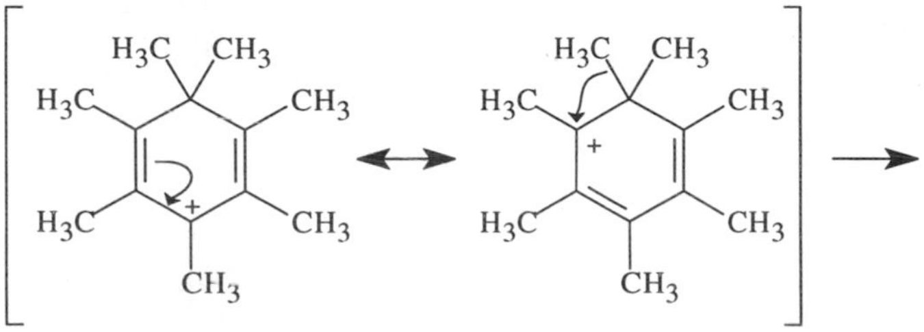

[H_3C CH_3 H_3C CH_3 H_3C CH_3 CH_3 ↔ H_3C CH_3 H_3C CH_3 H_3C CH_3 CH_3] →

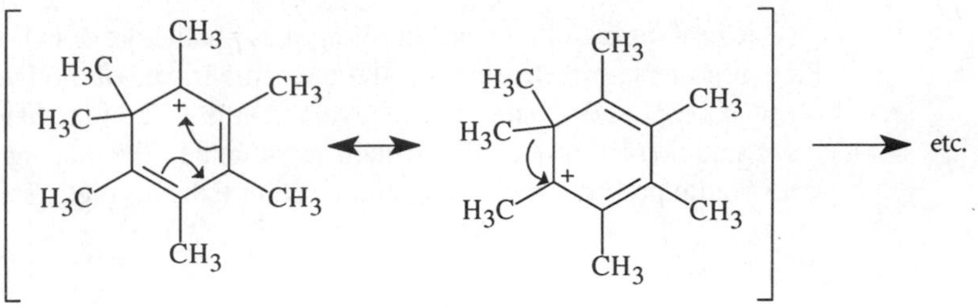

15.68 Remember that benzene rings have resonance structures in which the double bonds are completely delocalized. Let's draw [4]phenylene using the other resonance structures for the benzene rings:

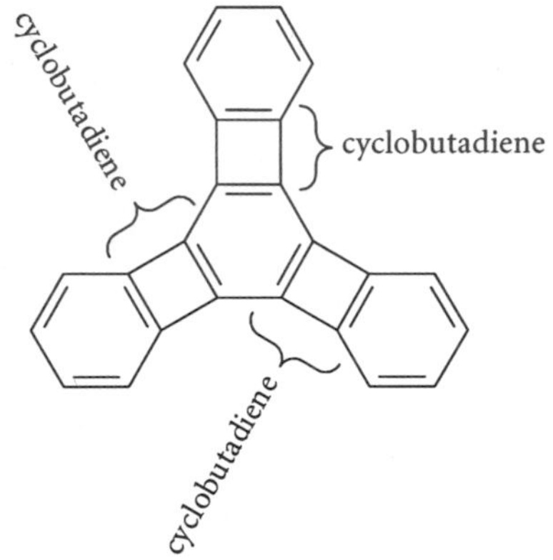

When we do this, we see that [4]phenylene incorporates three cyclobutadiene rings! Because these are antiaromatic, they are *very* unstable. What this means is that all of the benzene rings have no resonance stabilization, because when their π bonds are delocalized, cyclobutadienes result.

Hydrogenation solves this problem. When the central benzene ring is hydrogenated, the π bonds in the remaining benzene rings can be delocalized. (Cyclobutenes, unlike cyclobutadienes, are *not* antiaromatic.) In other words, hydrogenation of the central benzene ring is especially favorable because it is accompanied by restoration of the aromaticity of all three outer benzene rings.

16

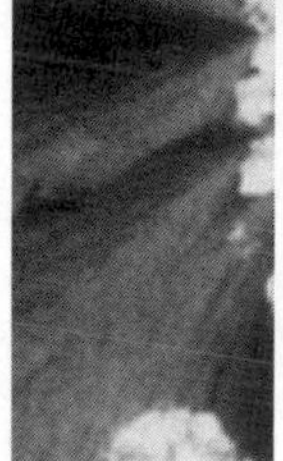

Chemistry of Benzene and Its Derivatives

TERMS

CONCEPTS

I. INTRODUCTION TO BENZENE AND ITS DERIVATIVES

A. Nomenclature of Benzene Derivatives

1. The nomenclature of benzene derivatives follows the same rules used for other substituted hydrocarbons.
2. Some monosubstituted benzene derivatives have well-established common names that must be learned.

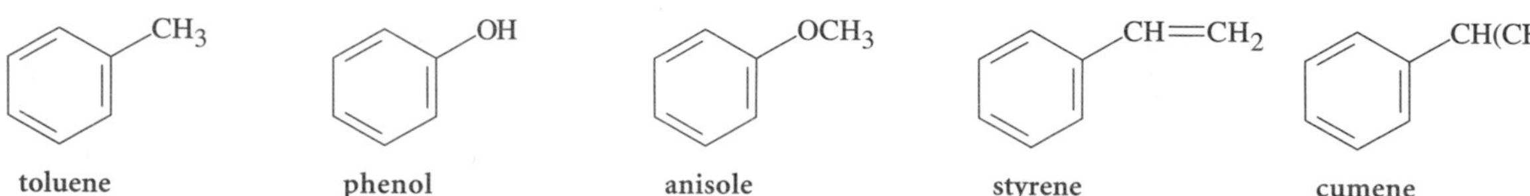

3. The positions of substituent groups in disubstituted benzenes can be designated in two ways:
   a. Substitutive nomenclature utilizes numerical designations in the same manner used for other compound classes.

b. An older system employs special prefixes that can be used *only* for disubstituted benzene derivatives.
   i. *o* (*ortho*) for a 1,2 relationship.
   ii. *m* (*meta*) for a 1,3 relationship.
   iii. *p* (*para*) for a 1,4 relationship.

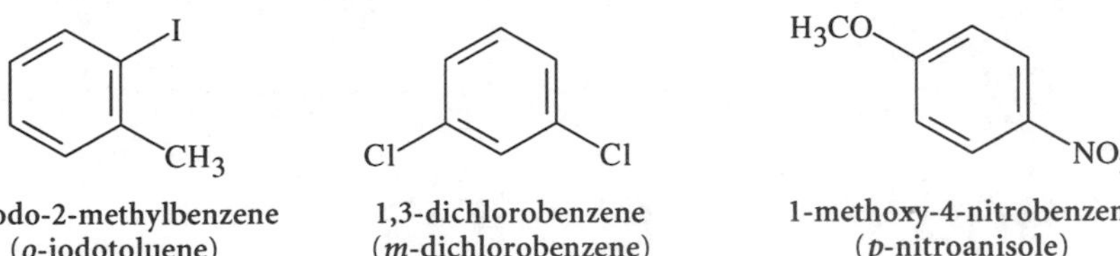

1-iodo-2-methylbenzene (*o*-iodotoluene) | 1,3-dichlorobenzene (*m*-dichlorobenzene) | 1-methoxy-4-nitrobenzene (*p*-nitroanisole)

4. Some disubstituted benzenes derivative have time-honored common names:
   a. The dimethylbenzenes are called xylenes.
   b. The methylphenols are called cresols.

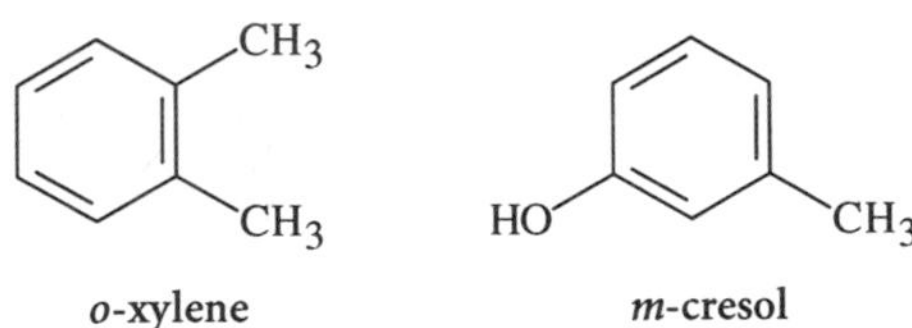

o-xylene | *m*-cresol

   c. The hydroxyphenols are:
      i. catechol (1,2-disubstituted).
      ii. resorcinol (1,3-disubstituted).
      iii. hydroquinone (1,4-disubstituted).

catechol | resorcinol | hydroquinone

5. When a benzene derivative contains more than two substituents on the ring, only numbers may be used to designate the positions of substituents.
   a. When none of the substituents qualify as principal groups, they are cited and numbered in alphabetical order. (See examples under Item 3b.)
   b. If a substituent is eligible for citation as a principal group, it is assumed to be at position-1 of the ring.

4-chloro-2-methylphenol

6. Sometimes it is simpler to name the benzene ring as a substituent group.
   a. When an unsubstituted benzene ring is a substituent, it is called a phenyl group, abbreviated Ph— or C_6H_5—.
   b. A benzene ring or substituted benzene ring cited as a substituent is referred to generally as an aryl group.
7. The Ph—CH_2— group is called the benzyl group.

Ph_3COH | Ph—CH_2—Br
triphenylmethanol | benzyl bromide

B. Physical Properties of Benzene Derivatives

1. Boiling points of benzene derivatives are similar to those of other hydrocarbons with similar shapes and molecular weights.
2. Melting points of para-disubstituted benzene derivatives are typically higher than those of the corresponding ortho or meta isomers.
3. Many para-substituted compounds can be separated from their ortho and meta isomers by recrystallization.
4. Benzene and other aromatic hydrocarbons are less dense than water, but more dense than alkanes and alkenes of about the same molecular mass.
5. Benzene and its hydrocarbon derivatives are insoluble in water; benzene derivatives with substituents that form hydrogen bonds to water are more soluble.

C. Source and Industrial Use of Aromatic Hydrocarbons

1. The most common source of aromatic hydrocarbons is petroleum.
   a. Benzene is obtained by separation from petroleum fractions and from demethylation of toluene.
   b. Toluene is obtained by separation from reformates, which are the products of hydrocarbon interconversion over certain catalysts.
2. Coal tar, a tarry residue obtained when coal is heated in the absence of oxygen, is currently a minor source of aromatic hydrocarbons.

II. SPECTROSCOPY OF BENZENE DERIVATIVES

A. IR Spectroscopy

1. The most useful absorptions in the infrared spectra of benzene derivatives are the carbon-carbon stretching absorptions of the ring, which occur at lower frequency than the C=C absorption of alkenes, one is near 1600 cm^{-1} and another is near 1500 cm^{-1}.
2. Weak absorptions, called overtone and combination bands, are observed in the 1600–2000 cm^{-1} region. (These absorptions were once used to determine the substitution patterns of aromatic compounds.)
3. NMR spectroscopy is more useful than IR spectroscopy for diagnosing the presence of a benzene derivative.

B. NMR Spectroscopy

1. Benzene derivatives containing ring hydrogens typically have characteristic NMR absorptions at large chemical shifts.
   a. Protons on aromatic rings have chemical shifts that are greater than those of typical alkenes by about 1.5–2 ppm.
   b. The NMR spectrum of benzene consists of a singlet at a chemical shift of δ 7.4.
2. When benzene is oriented relative to the applied field $\mathbf{B}_0$, a circulation of π electrons around the ring, called a ring current, is induced. (See Fig. 16.2, text p. 699.)
   a. The ring current induces a magnetic field that forms closed loops through the ring.
   b. The induced field opposes the applied field $\mathbf{B}_i$ along the axis of the ring.
   c. The induced field augments the applied field $\mathbf{B}_i$ outside of the ring and thus deshields the protons from the external field; as a result, these protons have greater chemical shifts than many other protons.
   d. The ring current and the large chemical shift are characteristic of compounds that are aromatic by the Hückel $4n + 2$ rule.
3. When the protons in a substituted benzene derivative are nonequivalent, they split each other.
   a. Their coupling constants depend on their positional relationships.
   b. The NMR spectra of many monosubstituted benzene derivatives have complex absorptions in the aromatic region.
      i. A symmetrical pattern resembling "two leaning doublets" is very typical of benzene derivatives in which two different substituents have a para relationship.

ii. The chemical shifts of ring protons are affected by both the polar and resonance effects of substituents.

4. The chemical shifts of benzylic protons (δ 2–3) are slightly greater than those of allylic protons (δ 1.8–2.2).
5. The O—H absorptions of phenols (δ 5–6) are typically at lower field than that of alcohols (δ 2–3); O—H protons of phenols, like those of alcohols, undergo exchange in D_2O.
6. In CMR spectra, the chemical shifts of aromatic carbons are in the carbon-carbon double bond region (δ 110–160); the exact values depend on the ring substituents that are present.
7. The chemical shifts of benzylic carbons (δ 18–30) are not appreciably different from the chemical shifts of ordinary alkyl carbons.

C. UV Spectroscopy

1. Simple aromatic hydrocarbons have two absorption bands in their UV spectra:
   a. A relatively strong bond near 210 nm.
   b. A much weaker band near 260 nm.
2. Substituent groups on the ring alter both the λ_{max} values and the intensities of both peaks, particularly if—
   a. the substituent has an unshared electron pair.
   b. the substituent has *p* orbitals that can overlap with the π-electron system of the aromatic ring.
3. More extensive conjugation is associated with an increase in both λ_{max} and intensity.

III. INTRODUCTION TO ELECTROPHILIC AROMATIC SUBSTITUTION REACTIONS OF SUBSTITUTED BENZENES

A. Electrophilic Aromatic Substitution

1. Benzene and its derivatives are aromatic compounds and thus do not undergo most of the usual addition reactions of alkenes; instead, they undergo reactions in which a ring hydrogen is substituted by another group.
2. Electrophilic aromatic substitution reactions are typical of benzene and other aromatic compounds.
   a. The reaction is a substitution because hydrogen is replaced by another group.
   b. The reaction is electrophilic because it involves the reaction of an electrophile, or Lewis acid, with the benzene π electrons.
3. All electrophilic aromatic substitution reactions have the following three mechanistic steps:
   a. Generation of an electrophile.
   b. Attack of the π electrons of the aromatic ring on the electrophile and formation of a resonance-stabilized carbocation intermediate.
   c. Loss of a proton from the carbocation intermediate to form the substituted aromatic compound.

(E^+ = electrophile)
(B**:** = base)

+ E^+ ⇌ [H E … ↔ H E … ↔ B: H E …] → E … + H—B^+

resonance-stabilized carbocation intermediate

B. Directing Effects of Substituents

1. Further substitution on a substituted benzene ring occurs in one of two ways.
   a. If a substituted benzene undergoes further substitution at the ortho and para positions, the original substituent is called an ortho, para-directing group.
   b. If a substituted benzene undergoes further substitution mostly at the meta position, the original substituent group is called a meta-directing group.
2. A substituent group is either an ortho, para-directing group or a meta-directing group in all electrophilic aromatic substitution reactions.

3. These effects occur because electrophilic substitution reactions at one position of a benzene derivative are much faster than the same reactions at another position; the substitution reactions at the different ring positions are in competition.

C. Ortho, Para-Directing Groups

1. Substituents containing atoms with unshared electron pairs on atoms directly attached to the benzene ring are ortho, para-directing substituents because the electron pairs can be involved in the resonance stabilization of the carbocation intermediate.
   a. The reaction of an electrophile at either the ortho or para position gives a carbocation with more resonance structures; in one resonance structure, the substituent electrons are delocalized into the ring and charge is delocalized onto the substituent.

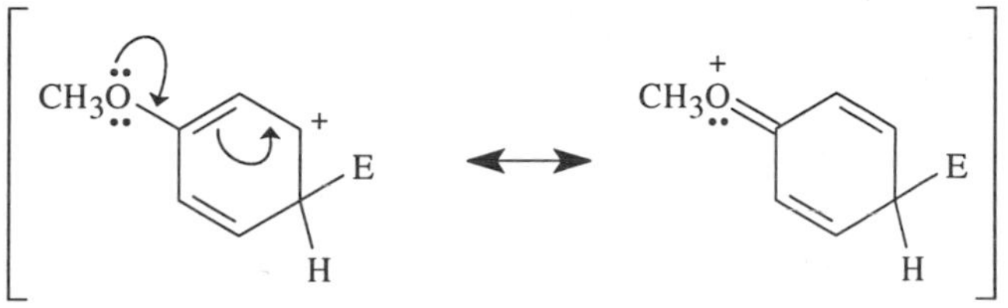

delocalization of charge by an ortho, para-directing substituent

(E = electrophile)

   b. The substituent electrons cannot be used to delocalize charge when reaction occurs at the meta position.
2. All alkyl groups are ortho, para-directing substituents.
   a. Reaction at a position that is ortho or para to an alkyl group gives a carbocation intermediate that has one tertiary carbocation resonance structure.

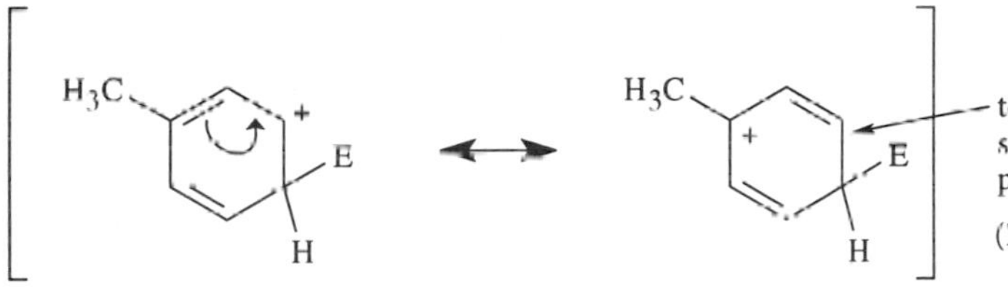

tertiary carbocation resonance structure resulting from para substitution on an alkylbenzene

(E = electrophile)

   b. Reaction of the electrophile meta to the alkyl group gives only secondary carbocation resonance structures.
3. The rate-limiting step in many electrophilic aromatic substitution reactions is formation of the carbocation intermediate.
   a. Hammond's postulate suggests that the more stable carbocation should be formed more rapidly.
   b. The basis of the directing effect is that the products derived from the more rapidly formed carbocation (the more stable carbocation) are the ones observed.

D. The Ortho:Para Ratio

1. If a benzene derivative bears an ortho, para-directing group, most electrophilic aromatic substitution reactions give much more para than ortho product; the reasons for the ortho, para ratio vary from case to case.
   a. In some cases, predominance of the para isomer can be explained by van der Waals repulsions (steric effects) in the ortho isomer.
   b. In a few cases, such as nitration of toluene, the reaction is so fact that it occurs on every encounter of the reagents. As a result, a statistical mixture of products (twice as much ortho as para) is obtained.
2. The ortho and para isomers obtained in many electrophilic aromatic substitution reactions have sufficiently different physical properties that they are readily separated. (See I.B.3. above.)

E. Meta-Directing Groups

1. Meta-directing groups are all electronegative groups that do not have an unshared electron pair on an atom adjacent to the benzene ring.

a. Substituents that have positive charges adjacent to the aromatic ring are meta directors because meta substitution gives the carbocation intermediate in which like charges are farther apart.
b. Para substitution is slower than meta substitution because the carbocation in intermediate in para substitution is especially unstable. (See Eqs. 16.33–16.34 on text p. 721.)

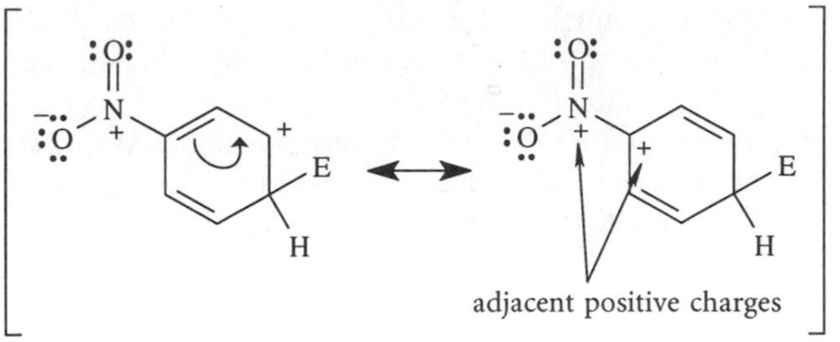

adjacent positive charges

carbocation intermediate from electrophilic substitution at the para position of nitrobenzene

(E = electrophile)

c. All meta-directing groups have bond dipoles that place a substantial amount of positive charge next to the benzene ring.

F. Activating and Deactivating Effects of Substituents

1. Different benzene derivatives have greatly different reactivities in electrophilic aromatic substitution reactions.
   a. If a substituted benzene derivative reacts more rapidly than benzene itself, then the substituent is said to be an activating group.
   b. If a substituted benzene derivative reacts more slowly than benzene itself, then the substituent is said to be a deactivating group.
2. A given substituent group is either activating in all electrophilic aromatic substitution reactions or deactivating in all such reactions. (See Table 16.2, text p. 717.)
   a. All meta-directing groups are deactivating groups.
   b. All ortho, para-directing groups except for the halogens are activating groups.
   c. The halogens are deactivating groups.
3. Except for the halogens, there is a correlation between the activating and directing effects of substituents.
   a. Directing effects are concerned with the relative rates of substitution at different positions of the same compound.
   b. Activating or deactivating effects are concerned with the rates of substitution of different compounds relative to the rate of benzene itself.
4. Two properties of substituents must be considered in order to understand activating and deactivating effects:
   a. The resonance effect of the substituent—the ability of the substituent to stabilize the carbocation intermediate in electrophilic substitution by delocalization of electrons from the substituent into the ring.
      i. The resonance effect is the same effect that is responsible for the ortho, para-directing effects of substituents with unshared electron pairs.
      ii. The electron-donating resonance effect of a substituent group with unshared electron pairs stabilizes the carbocation intermediate and activates further substitution, provided that the substitution occurs in the ortho or para position.
   b. The polar effect of the substituent—the tendency of the substituent group, by virtue of its electronegativity, to pull electrons away from the ring.
      i. This is the same effect discussed in connection with substituent effects on acidity.
      ii. The electron-withdrawing polar effect of an electronegative group destabilizes the carbocation intermediate and deactivates further substitution.
5. Whether a substituted derivative of benzene is activated or deactivated toward further substitution depends on the balance of the resonance and polar effects of the substituent(s).
6. Alkyl groups on a benzene ring—
   a. stabilize carbocation intermediates in electrophilic substitution.
   b. are activating groups.

7. The deactivating effects of halogen substituents reflect a different balance of resonance and polar effects.
   a. The orbitals on Cl, Br, and I and the carbon $2p$ orbitals of the benzene ring have different sizes and different numbers of nodes and thus do not overlap effectively; thus, the resonance effects of these substituents are relatively weak, and the halogens are thus deactivating groups.
   b. Meta substitution in halobenzenes is deactivated even more than ortho or para substitution; this is why the halogens are ortho, para-directing groups.

G. Use of Electrophilic Aromatic Substitution in Organic Chemistry

1. Both activating/deactivating and directing effects of substituents can come into play in planning an organic synthesis that involves electrophilic substitution reactions.
2. Directing effects of the substituents must be considered carefully when designing an organic synthesis that involves electrophilic aromatic substitution reactions.
3. When an electrophilic substitution reaction is carried out on a benzene derivative with more than one substituent:
   a. The activating and directing effects are roughly the sum of the effects of the separate substituents.
   b. If one group is much more strongly activating than the other, the directing effect of the more powerful activating group generally predominates.
   c. In other cases, mixtures of isomers are typically obtained.
4. The activating or deactivating effects of substituents in an aromatic compound determine the conditions that must be used in an electrophilic substitution reaction.
   a. When a deactivating group is being introduced by an electrophilic substitution reaction (for example, Friedel-Crafts acylation), it is easy to introduce one group at a time.
   b. When an activating group is introduced by electrophilic substitution (for example, Friedel-Crafts alkylation), additional substitutions can occur easily under the conditions of the first substitution and mixtures of products substituted to different extents are obtained.
   c. Deactivating substituents retard some reactions to the point that they are not useful.

REACTIONS

I. ELECTROPHILIC AROMATIC SUBSTITUTION REACTIONS OF BENZENE

A. Halogenation of Benzene

1. Benzene reacts with bromine in the presence of a Lewis acid catalyst ($FeBr_3$) to yield a product in which one bromine is substituted for a ring hydrogen.

$$Br_2 + C_6H_6 \xrightarrow{FeBr_3} C_6H_5\text{—}Br + HBr$$

2. An analogous chlorination reaction using Cl_2 and $FeCl_3$ gives chlorobenzene.
3. Halogenation of benzene differs from the halogenation of alkenes by the type of product obtained and the reaction conditions.
   a. Alkenes give addition products and react spontaneously under mild conditions.
   b. Benzene gives substitution products and requires a Lewis acid as well as relatively severe conditions.
4. The mechanism of the bromination of benzene consists of the following steps:
   a. The first step is formation of a complex between Br_2 and the Lewis acid $FeBr_3$.

$$:\ddot{\underset{..}{Br}}\text{—}\ddot{\underset{..}{Br}}: + FeBr_3 \rightleftharpoons :\ddot{\underset{..}{Br}}^{\delta+}\text{-----}\ddot{Br}^{\delta+}\text{-----}FeBr_3^{\delta-}$$

i. This complexation makes one of the bromines a much better leaving group.
ii. This complex of Br_2 and $FeBr_3$ reacts as if it contained the electron-deficient species $:\ddot{Br}^+$.

b. In the second step, this complex is attacked by the π electrons of the benzene ring.

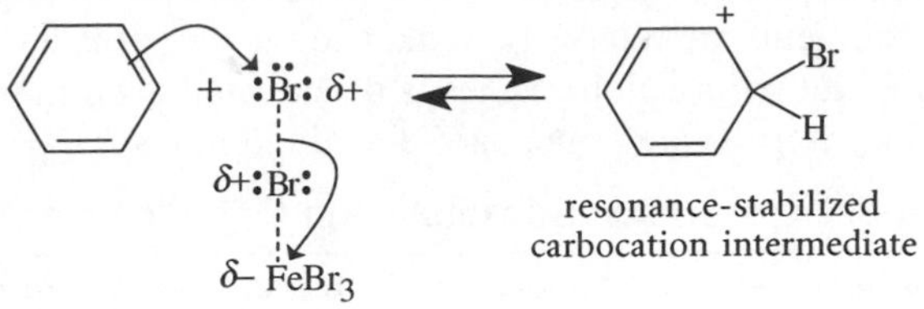

resonance-stabilized carbocation intermediate

i. This step disrupts the aromatic stabilization of the benzene ring (thus the need for the harsh conditions).
ii. The carbocation intermediate forms only because it is resonance-stabilized.

c. The reaction is completed when a bromide ion acts as a base to remove the ring proton and give the products; by losing a β-proton, the carbocation can form a stable aromatic compound.

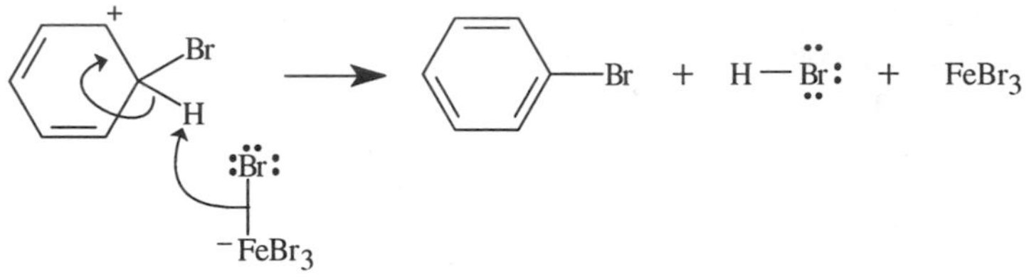

B. Nitration of Benzene

1. Benzene reacts with concentrated nitric acid, usually in the presence of a sulfuric acid catalyst, to form nitrobenzene.
2. In this reaction, called nitration, the nitro group, —NO_2, is introduced into the benzene ring by electrophilic substitution; the electrophile is the nitronium ion, $^+NO_2$.

$$H_2SO_4 + HNO_3 \rightleftharpoons {}^+NO_2\ HSO_4^- + H_2O$$

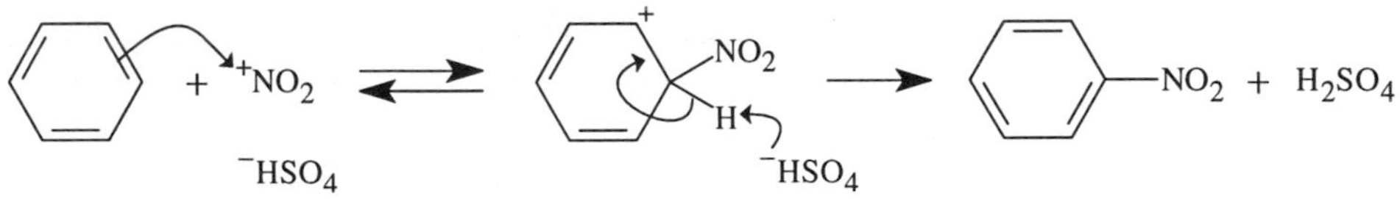

C. Sulfonation of Benzene

1. Benzene reacts with a solution of sulfur trioxide in H_2SO_4 called fuming sulfuric acid to form benzenesulfonic acid.
2. In this reaction, called sulfonation, the sulfonic acid group, —SO_3H, is introduced into the benzene ring by electrophilic substitution; the electrophile is the neutral compound sulfur trioxide, SO_3. 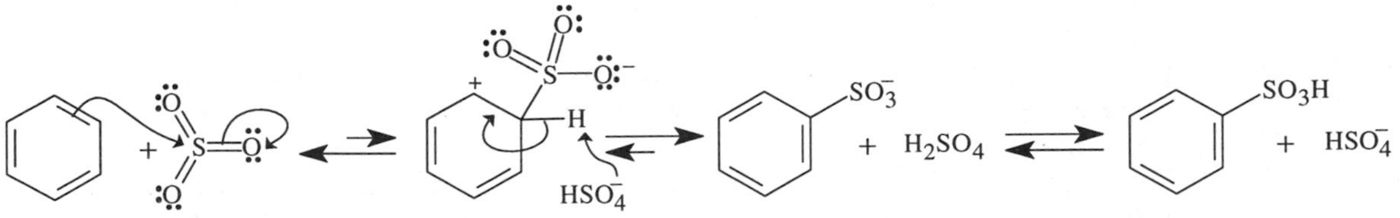
3. Sulfonic acids such as benzenesulfonic acid are rather strong acids.

4. Sulfonation, unlike many electrophilic aromatic substitution reactions, is reversible; the sulfonic acid group is replaced by a hydrogen when sulfonic acids are heated with steam.

$$C_6H_5\text{–}SO_3H + \underset{(steam)}{H_2O} \longrightarrow C_6H_6 + H_2SO_4$$

D. Friedel-Crafts Alkylation of Benzene

1. The reaction of an alkyl halide with benzene in the presence of a Lewis acid gives an alkylbenzene; this type of reaction is called Friedel-Crafts alkylation.

$$C_6H_6 + R\text{—}Cl \xrightarrow[2)\ H_3O^+]{1)\ AlCl_3} C_6H_5\text{—}R$$

2. The electrophile in a Friedel-Crafts reaction is formed by complexation of the Lewis acid $AlCl_3$ with the halogen of an alkyl halide.
   a. Either the alkyl halide-Lewis acid complex, or the carbocation derived from it, can serve as the electrophile in a Friedel-Crafts reaction.
   b. Rearrangements of alkyl groups are observed in some Friedel-Crafts alkylations if the carbocation intermediate is prone to rearrangement.

$$H_3C\text{—}CH(CH_3)\text{—}CH(Cl)\text{—}CH_3 + AlCl_3 \longrightarrow H_3C\text{—}CH(CH_3)\text{—}\overset{+}{C}H\text{—}CH_3\ \ {}^-AlCl_4 \longrightarrow H_3C\text{—}\overset{+}{C}(CH_3)\text{—}CH_2\text{—}CH_3\ \ {}^-AlCl_4$$

$$C_6H_6 + H_3C\text{—}\overset{+}{C}(CH_3)\text{—}CH_2\text{—}CH_3\ \ {}^-AlCl_4 \longrightarrow C_6H_5\text{—}C(CH_3)_2\text{—}CH_2CH_3 + AlCl_3 + HCl$$

3. A catalytic amount (much less than one equivalent) of the $AlCl_3$ catalyst can be used in this reaction.
4. The alkylbenzene products are more reactive than benzene itself.
   a. The product can undergo further alkylation, and mixtures of products alkylated to different extents are observed.
   b. A monoalkylation product can be obtained in good yield if a large excess of the starting material is used; this strategy is practical only if the starting material is cheap, and if it can be readily separated from the product.
5. Alkenes or alcohols with an acid catalyst can also be used as the alkylating agents in Friedel-Crafts alkylation reactions.

$$(H_3C)_2C\text{=}CH_2 + H_2SO_4 \longrightarrow (H_3C)_2\overset{+}{C}\text{—}CH_3\ \ HSO_4^- \xrightarrow{benzene} C_6H_5\text{—}C(CH_3)_2\text{—}CH_3 + H_2SO_4$$

6. Friedel-Crafts alkylation is generally not useful on compounds that are more deactivated than benzene itself.

E. Friedel-Crafts Acylation of Benzene

1. When benzene reacts with an acid chloride in the presence of a Lewis acid catalyst such as aluminum trichloride, $AlCl_3$, a ketone is formed.

$$\text{C}_6\text{H}_6 + \text{Cl}-\text{C(=O)}-\text{R} \xrightarrow[\text{2) } H_3O^+]{\text{1) } AlCl_3} \text{C}_6\text{H}_5-\text{C(=O)}-\text{R} + \text{HCl}$$

(an acyl group)

a. This reaction is an example of a Friedel-Crafts acylation.
b. An acyl group, typically derived from an acid chloride, is introduced into an aromatic ring in the presence of a Lewis acid catalyst.

2. The electrophile in the Friedel-Crafts acylation reaction is a carbocation called an acylium ion.
   a. This ion is formed when the acid chloride reacts with the Lewis acid $AlCl_3$.

$$\text{R}-\text{C(=O)}-\text{Cl} + AlCl_3 \rightleftharpoons \left[\text{R}-\overset{+}{\text{C}}=\ddot{\text{O}}: \longleftrightarrow \text{R}-\text{C}\equiv\overset{+}{\text{O}}:\right]\ {}^{-}AlCl_4$$

an acylium ion

   b. The substitution occurs when the benzene π electrons attack the acylium ion, and a proton is lost from the resulting carbocation intermediate.

$$\text{C}_6\text{H}_6 + \text{R}-\text{C}\equiv\overset{+}{\text{O}}: \ (\text{Cl}-\overset{-}{AlCl_3}) \longrightarrow \text{carbocation intermediate} \longrightarrow \text{C}_6\text{H}_5-\text{C(=O)}-\text{R} + \text{H}-\ddot{\text{Cl}}: + AlCl_3$$

3. The ketone product of the Friedel-Crafts acylation reacts with the Lewis acid catalyst to form a complex that is catalytically inactive; this has two consequences:
   a. Slightly more than one equivalent of the catalyst must be used:
      i. one equivalent to react with the product.
      ii. an additional catalytic amount to ensure the presence of catalyst throughout the reaction.
   b. The complex must be destroyed before the ketone product can be isolated; this is usually accomplished by pouring the reaction mixture into ice water.
4. Because the ketone products of acylation are much less reactive than the benzene starting material, acylation occurs only once.
5. Friedel-Crafts acylation does not occur on a benzene ring substituted solely with one or more meta-directing groups. Thus, for example, nitrobenzene does not undergo Friedel-Crafts acylation.
6. The Friedel-Crafts acylation occurs intramolecularly when acylation results in the formation of a five- or six-membered ring.

$$\text{C}_6\text{H}_5\text{CH}_2\text{CH}_2\text{C(=O)Cl} \xrightarrow[\text{2) } H_3O^+]{\text{1) } AlCl_3} \text{1-indanone} + \text{HCl}$$

   a. The intramolecular process is much faster than attack of the acylium ion on the phenyl ring of another molecule.
   b. This type of reaction can only occur at an adjacent ortho position.
7. The Friedel-Crafts acylation reaction is important for two reasons.
   a. It is an excellent method for the synthesis of aromatic ketones.
   b. It is another method for the formation of carbon-carbon bonds (see a partial list on text page 715 or a complete list in Appendix VI, text page A-13).

II. ADDITION REACTIONS OF BENZENE DERIVATIVES

A. Hydrogenation of Benzene Derivatives

1. Because of its aromatic stability, the benzene ring is resistant to conditions used to hydrogenate ordinary double bonds.
2. Aromatic rings can be hydrogenated under extreme conditions of temperature and/or pressure.
3. Catalytic hydrogenation of benzene derivatives gives the corresponding cyclohexanes, and cannot be stopped at the cyclohexadiene or cyclohexene stages.

$$\text{1,2-}(CH_3)_2C_6H_4 \xrightarrow[\text{high pressure, heat}]{H_2/Ni} \text{1,2-}(CH_3)_2C_6H_{10}$$

4. Hydrogenation of the first double bond of benzene is an endothermic (thermodynamically unfavorable) reaction and requires energy (heat or pressure) to take place.
5. Once hydrogenation of the first double bond has taken place, the aromatic character of the benzene ring is lost; consequently, hydrogenation of the remaining double bonds under the harsh reaction conditions is instantaneous.

TABLE SG16.1 Electrophilic Aromatic Substitution Summary

Electrophile	Reaction Name	Product
$Cl-\overset{+}{Cl}\text{---}\overset{-}{Fe}Cl_3$	chlorination	chlorobenzene
$Br-\overset{+}{Br}\text{---}\overset{-}{Fe}Br_3$	bromination	bromobenzene
SO_3 in H_2SO_4	sulfonation	benzenesulfonic acid
$^{+}NO_2$	nitration	nitrobenzene
$R-C\equiv O:^{+}\ ^{-}AlCl_4$	Friedel-Crafts acylation	an aryl ketone
$R-\overset{+}{Cl}\text{----}\overset{-}{Al}Cl_3$ or $R^{+}\ ^{-}AlCl_4$	Friedel-Crafts alkylation	an alkylbenzene

SCIENTISTS

Sir Henry Trentham Butlin (1845–1912) 16.7
James Mason Crafts (1839–1917) 16.4F
Charles Friedel (1832–1899) 16.4F
Sir John Percivall Pott (1714–1788) 16.7
Charles Adolphe Wurtz (1817–1884) 16.4F

STUDY GUIDE LINKS

✓16.1 NMR of Para-Substituted Benzene Derivatives

Be sure to realize that the NMR spectrum of a para-disubstituted benzene derivative shows the "two leaning doublet" pattern when the two para substituents are different. When the substituents are the same, all ring protons are equivalent, and the resonance for the ring protons is a singlet.

The splitting pattern for para-disubstituted benzene derivatives is a little more subtle than might first appear, and can be rationalized as follows. Each ring proton has an ortho relationship to one nonequivalent proton and a para relationship to the other nonequivalent proton. Consider, for example, *p*-bromoethylbenzene, the structure of which is shown below, and the NMR spectrum of which is shown in Fig. 16.3 on text p. 701. In this structure, H^a is ortho to one H^b and para to the other H^b. Actually, both ortho and para splittings are present, but the para splittings are typically very small and not readily discernible in the spectrum. The major splitting observed is the one between adjacent protons—about 8 Hz in Fig. 16.3, a value within the range for ortho splitting given in Table 16.1 on text p. 700. Thus, to a useful approximation, the resonance of each proton is split into a doublet by its neighboring ortho proton. Since there are two chemically different types of protons—H^a and H^b—the spectrum consists of two such doublets.

two equivalent protons H^a give one resonance split into a doublet by the neighboring ortho proton H^b

Br, H^a, H^a, H^b, H^b, CH_2CH_3

two equivalent protons H^b give one resonance split into a doublet by the neighboring ortho proton H^a

✓16.2 Lewis-Acid Assistance for Leaving Groups

In some cases a Lewis acid can be used to make a molecule more reactive by complexation with the leaving group. In the case described in the text, the leaving bromine does not have to depart as *bromide,* but rather departs as $^-FeBr_4$ (*bromide complexed with $FeBr_3$*):

:Br—Br:

bromide ion as a leaving group (very slow reaction)

:Br—Br^+—$\bar{F}eBr_3$

$^-FeBr_4$ (bromide ion complexed with $FeBr_3$) as a leaving group (much faster reaction)

Why is the complex more reactive? Because when the complexed bromine, an electronegative atom, is forced to assume a positive charge, its tendency to accept a pair of electrons—that is, its tendency to act as a leaving group—is increased. Just as we think of a nucleophile as a species that provides an electronic "push," we can think of a Lewis acid catalyst as something that provides an electronic "pull." *A Lewis acid helps remove a leaving group by making it a better attractor of electrons.*

When you studied S_N1 and S_N2 reactions, you learned that bromide is a good leaving group. The fact that such a good leaving group has to be complexed to a Lewis acid in order for bromination of benzene to take place shows just how unreactive the benzene ring is.

Lewis-acid complexation can be used to accelerate other reactions. For example, alkyl halides that are normally unreactive or not very reactive in S_N1 reactions can be induced to react by addition of Ag^+ (in the form of acidic silver nitrate solution) to the reaction mixture:

$$\underset{\text{not very reactive}}{H_3C{-}\overset{:\ddot{Br}:}{CH}{-}CH_3} \underset{}{\overset{Ag^+}{\rightleftharpoons}} \underset{\text{much more reactive}}{H_3C{-}\overset{\overset{+}{:\ddot{Br}}{-}Ag}{CH}{-}CH_3} \xrightarrow[(S_N1\text{ reaction})]{H_2O/\text{acetone}} H_3C{-}\overset{+}{C}H{-}CH_3 + \underset{\text{(precipitates)}}{Ag{-}\ddot{Br}:}$$

$$H_3C{-}\overset{+}{C}H{-}CH_3 \xrightarrow{H_2O} H_3C{-}\overset{OH}{CH}{-}CH_3 + H_3O^+$$

You also learned that primary alcohols react with concentrated HBr to give alkyl bromides, but that their reactions with HCl are usually very slow. The reaction of primary alcohols with HCl can be accelerated by adding the Lewis acid zinc chloride, because the Zn^{2+} ion forms a complex with the —OH group of the alcohol and thus makes it a better leaving group.

$$\underset{\text{not very reactive}}{H_3C{-}\overset{:\ddot{O}H}{CH}{-}CH_3} \overset{Zn^{2+}}{\rightleftharpoons} \underset{\text{much more reactive}}{H_3C{-}\overset{\overset{Zn^+}{|}}{\overset{+}{:OH}}\!\!\!\!\!\!\!\!\!\!\!\!\!\!CH{-}CH_3} \xrightarrow[(S_N1\text{ reaction})]{\text{conc. HCl}} H_3C{-}\overset{+}{C}H{-}CH_3 + \overset{+}{Zn}{-}\ddot{O}H$$

$$H_3C{-}\overset{+}{C}H{-}CH_3 \xrightarrow{Cl^-} H_3C{-}\overset{Cl}{CH}{-}CH_3$$

In Sec. 10.1 you learned that alcohol dehydration is acid-catalyzed because protonation of the —OH group makes this group a better leaving group. If protons were not present, the leaving group would be hydroxide ($^-$OH); protonation of the —OH group allows it to leave as the much less basic water molecule. In this case, the *proton* is the Lewis acid catalyst (and is also a Brønsted acid). In fact, it is useful to view Lewis acids conceptually as "fat protons!" Lewis acids are used instead of Brønsted acids when the latter would cause undesired side reactions.

✓16.3 Different Sources of the Same Reactive Intermediates

The example of Eq. 16.21 on text p. 713 is a further illustration of an idea that has been discussed in an earlier chapter (see Study Guide Link 11.2): *Different starting materials can serve as a source of the same reactive intermediate.*

In the Friedel-Crafts alkylation, the reactive intermediate is a carbocation. In this section you've learned that carbocation intermediates can be generated from alkyl halides and Lewis acids. But you've also learned that carbocations are formed when alkenes are subjected to strongly acidic conditions:

$$\mathrm{>C{=}C<} + \mathrm{H{-}\ddot{O}SO_3H} \longrightarrow \mathrm{>\overset{+}{C}{-}\overset{H}{\overset{|}{C}}{-}} \quad {:}\ddot{O}SO_3H^{-}$$

carbocation intermediate

The carbocation intermediates formed in such reactions can serve as electrophiles in electrophilic substitution just as carbocations generated in other ways can. Similarly, you've also learned that secondary and tertiary alcohols react with strong acids to generate carbocation intermediates (Sec. 10.1). These carbocations are transformed into alkene dehydration products if the reaction conditions are designed to remove the alkene from the reaction mixture. But if the alkene is not removed, and an aromatic hydrocarbon is present, these carbocation intermediates from alcohol dehydration can also serve as electrophiles in aromatic substitution. Problem 16.14(b) on text p. 713 is an illustration of this idea in practice.

✓16.4 Reaction Conditions and Reaction Rate

What do we mean by "harsh" or "mild" reaction conditions? In electrophilic aromatic substitution, harsh conditions include high temperature, strong Lewis acids, high concentrations of reagents, and/or the use of reagents that generate high concentrations of electrophiles. Mild conditions include lower temperature, weaker Lewis acids (or no Lewis acids), and lower concentrations of reagents.

Why is it that such conditions affect reaction rate? Remember, to say that an aromatic compound is highly activated means that it *reacts relatively rapidly* under a given set of conditions. If it is deactivated, it *reacts much more slowly* under the same conditions. Recall (Sec. 4.8A) that reaction rates increase with increasing temperature. Recall also from your study of rate laws (Sec. 9.3B) that many reaction rates increase with increasing reagent concentration. Thus, raising the temperature or the reagent concentration, or using a reagent that generates a high concentration of an electrophile, increases the reaction rate. Such strategies are necessary in order to get a highly *deactivated* (unreactive) compound to react at a convenient rate. However, use of harsh reaction conditions on a highly *activated* (reactive) compound in many cases leads to a greater degree of substitution than desired (or to other side reactions); hence, for highly activated compounds, milder conditions are used, such as weaker Lewis acids (or no Lewis acids), lower temperatures, or lower concentrations of reagents.

The examples in the text illustrate in a practical sense what is meant by "harsh" or "mild" conditions. It is *not* important to memorize the exact conditions for each reaction. It *is* important for you to understand why certain compounds are more activated toward substitution than others, and to understand conceptually that the success of reaction can depend on the rational choice of reaction conditions.

SOLUTIONS TO PROBLEMS

Solutions to In-Text Problems

16.1 (a) *m*-chloroethylbenzene, or 1-chloro-3-ethylbenzene
(b) *o*-diethylbenzene or 1,2-diethylbenzene
(c) *p*-nitrostyrene or 1-nitro-4-vinylbenzene
(d) 2,4-dichlorophenol
(e) 2-bromo-1-chloro-5-fluoro-4-iodo-3-nitrobenzene
(f) benzylbenzene or (phenylmethyl)benzene (also commonly called diphenylmethane)

16.2 (a) OCH_3, Cl

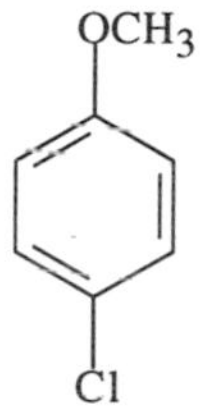

p-chloroanisole (1-chloro-4-methoxybenzene)

(b) CH_3, NO_2

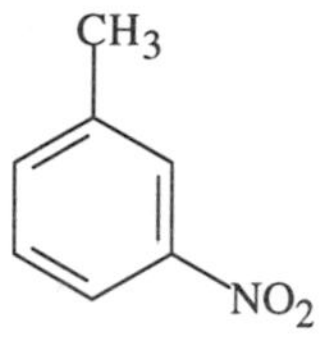

m-nitrotoluene (3-nitrotoluene)

(c) CH_3, Cl, Cl

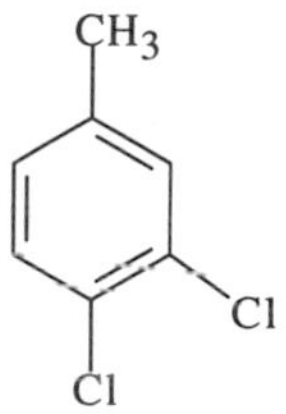

3,4-dichlorotoluene

(d) Br, $CH_2CH_2CH_3$

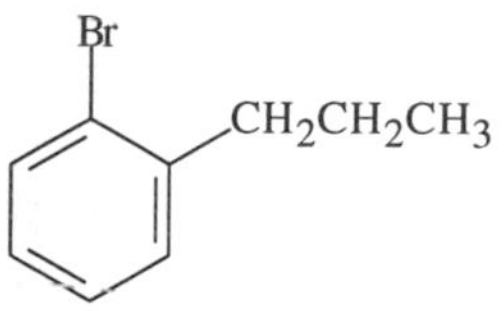

1-bromo-2-propylbenzene

(e) OCH_3

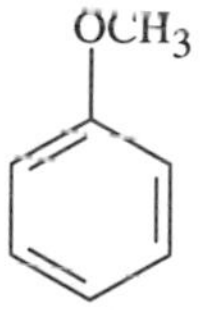

methyl phenyl ether (anisole, or methoxybenzene)

(f) CH_2OCH_3

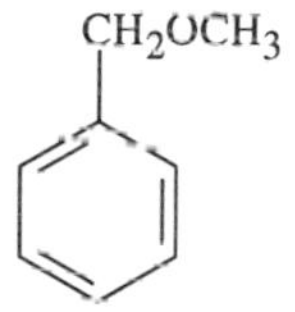

benzyl methyl ether, (methoxymethyl)benzene

(g) CH_3, CH_3

p-xylene (1,4-dimethylbenzene)

(h) OH, CH_3

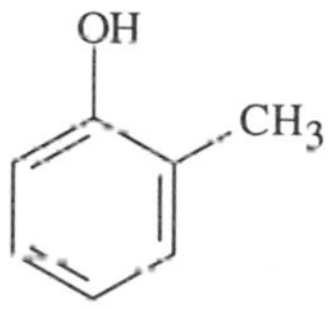

o-cresol (2-methylphenol)

16.3 The aromatic compound has NMR absorptions with greater chemical shift in each case because of the ring current (Fig. 16.2).

(a) Because thiophene is aromatic, its NMR absorptions occur at greater chemical shift than those of divinyl sulfide.
(b) The chemical shift of the benzene protons is at considerably lower field (greater chemical shift) because benzene is aromatic and 1,4-cyclohexadiene is not.

16.4 (a) This hydrocarbon is a continuous cycle of 18 π electrons. Because all atoms are trigonal and all bond angles are 120°, this compound can be planar. It meets the $4n + 2$ criterion for aromaticity for $n = 4$.
(b) The ring current brings about the expected increase in the local field at protons *outside* the ring, but, from Fig. 16.2 on text p. 699, it should have exactly the opposite effect on the protons *inside* the ring: These should experience a smaller local field. The resonance at δ 9.28 is that of the outer protons, and the resonance at δ (–2.99) is that of the inner protons. Because the local field at the outer protons is so large, the resonance frequency, and thus the chemical shift, of these protons is large. Because the local field at the inner protons is so small, the resonance frequency, and thus the chemical shift, of these protons is small. The relative integral is in agreement with this assignment.

16.5 (a) Mesitylene should give an NMR spectrum containing only two singlets in the integral ratio 3:1 for the methyl groups and ring protons, respectively. *p*-Ethyltoluene should contain a methyl triplet at highest chemical shift (around δ 1.0), a methyl singlet at the next highest chemical shift (around δ 2.3), and a methylene quartet at somewhat greater chemical shift. There are two possibilities for its ring protons. The most general one is a typical para-substitution pattern (apparent pair of doublets) for the ring protons in the δ 6.5–7.5 region. However, because the methyl and ethyl groups are so similar in their electronic characteristics, another possibility is that the nonequivalence of the ring protons is so slight as not to be apparent. In fact, this is the case; the ring protons are a singlet.

(b) Among other features, the NMR spectrum of 1-bromo-4-ethylbenzene has a typical ethyl quartet and a typical para-substitution pattern for the ring protons, as shown in Fig. 16.3, whereas the spectrum of (2-bromoethyl)benzene should show a pair of triplets for the methylene protons and a complex pattern for the ring protons. And if this isn't enough to distinguish the two compounds, the integral of the ring protons relative to the integral of the remaining protons is different in the two compounds.

16.6 (a) The signal for an exchangeable proton at δ 3.76 indicates that the compound contains an —OH group. It also contains aromatic absorptions with a splitting pattern that suggests para substitution. The splitting patterns of the δ 1.27 and δ 4.60 resonances indicate the partial structure H_3C—CH, and the chemical shift of the CH is consistent with its being α to both the ring and the OH group. The compound is

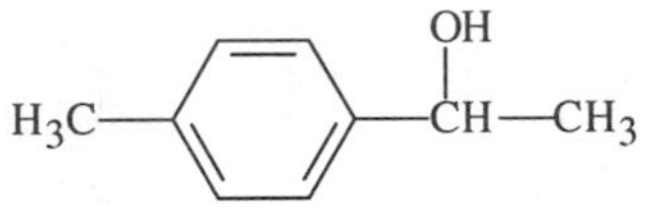

H_3C—C_6H_4—CH(OH)—CH_3 1-(4-methylphenyl)ethanol

(b) The IR spectrum indicates the presence of an —OH group, and the chemical shift of the broad NMR resonance (δ 6.0) suggests that this could be a phenol. The splitting patterns of the δ 1.17 and δ 2.58 resonances show that the compound also contains an ethyl group, and the splitting pattern of the ring protons shows that the compound is a para-disubstituted benzene derivative. The compound is *p*-ethylphenol.

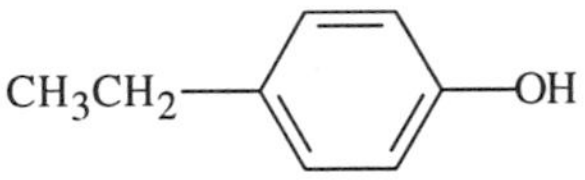

CH_3CH_2—C_6H_4—OH *p*-ethylphenol (4-ethylphenol)

16.7 The following two methyl ethers with the formula $C_7H_6OCl_2$ would show five lines in their proton-decoupled CMR spectra.

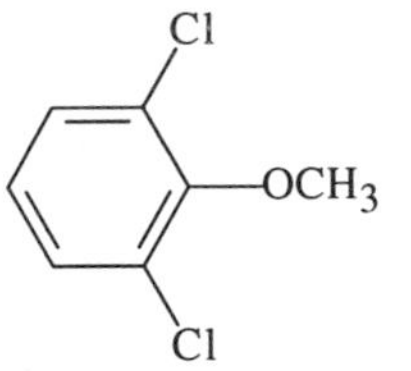

Cl, Cl, OCH_3 — 1,3-dichloro-2-methoxybenzene

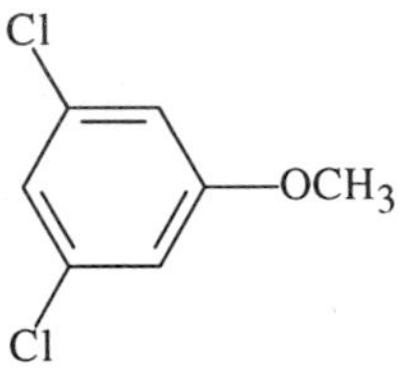

Cl, Cl, OCH_3 — 1,3-dichloro-5-methoxybenzene

16.8 Count the resonances. Mesitylene has three resonances; isopropylbenzene has six.

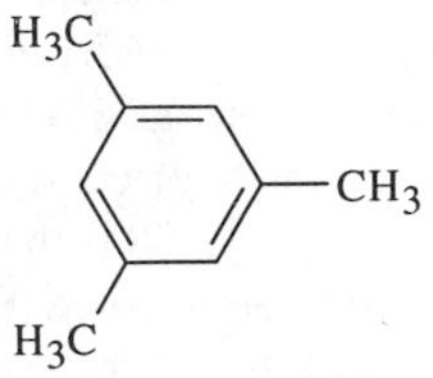

H_3C, CH_3, H_3C — mesitylene

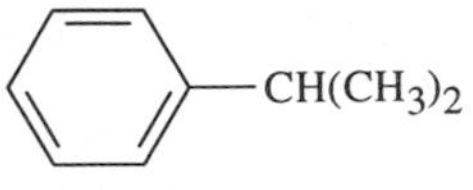

—$CH(CH_3)_2$ — isopropylbenzene

16.9 (a) The two benzene rings are conjugated in compound *A*; because there are more π bonds in conjugation, the UV absorption occurs at longer wavelength than it does in ethylbenzene.

(b) Compound *B* is like two connected molecules of compound *C* in the same sense that compound *A* is like two connected ethylbenzene molecules. It might seem that the absorption of compound *B* should occur at longer wavelength than that of compound *C* for the reason given in part (a). That this is *not* the case means that the two rings in compound *B* are not conjugated. They are not conjugated because the rings are not coplanar; and they are not coplanar because the ortho methyl groups on the two rings have severe van der Waals repulsions that are relieved when the molecule adopts a nonplanar conformation:

van der Waals repulsions between methyl groups

van der Waals repulsions are relieved

planar conformation

nonplanar conformation

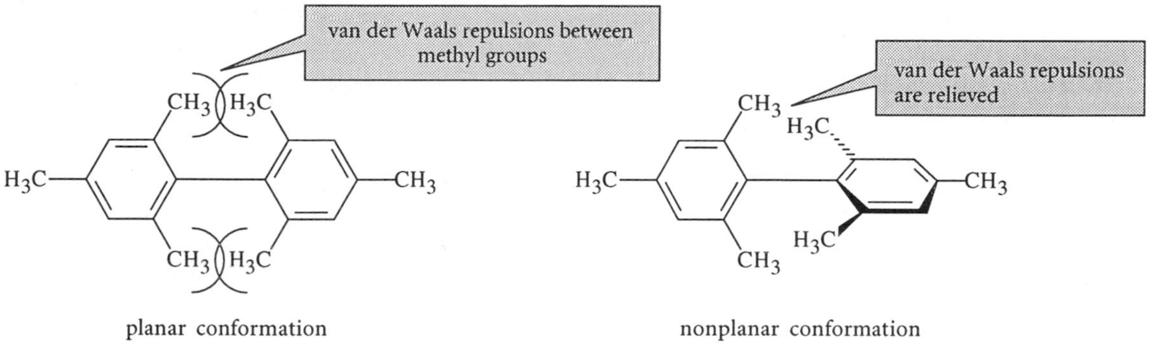

16.10 Because styrene has a double bond in conjugation with the ring and ethylbenzene does not, styrene has a greater λ_{max} in its UV spectrum.

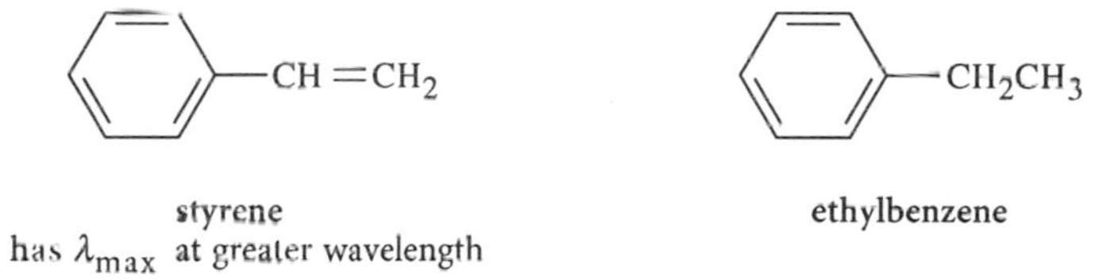

styrene has λ_{max} at greater wavelength

ethylbenzene

16.11 Apply the steps shown in Eqs. 16.6–16.7 on text pp. 705–6 to the para position of bromobenzene.

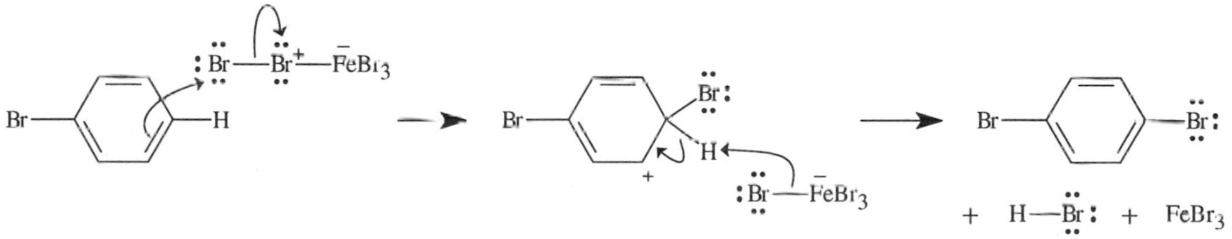

16.12 The mechanism of sulfonation with the conjugate acid of SO_3 as the electrophile:

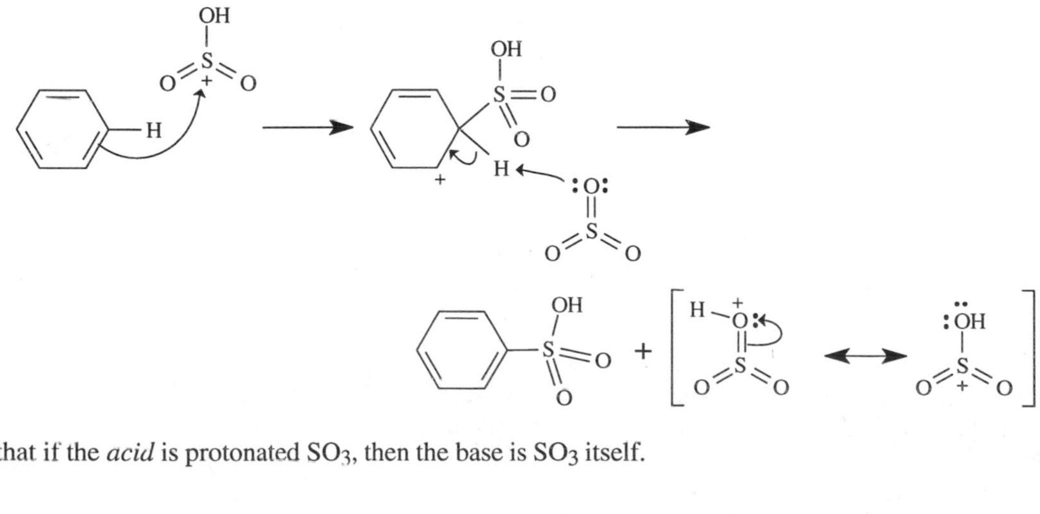

Notice that if the *acid* is protonated SO_3, then the base is SO_3 itself.

16.13 Apply the sulfonation mechanism shown in Eq. 16.13 on text p. 710to the para position of toluene.

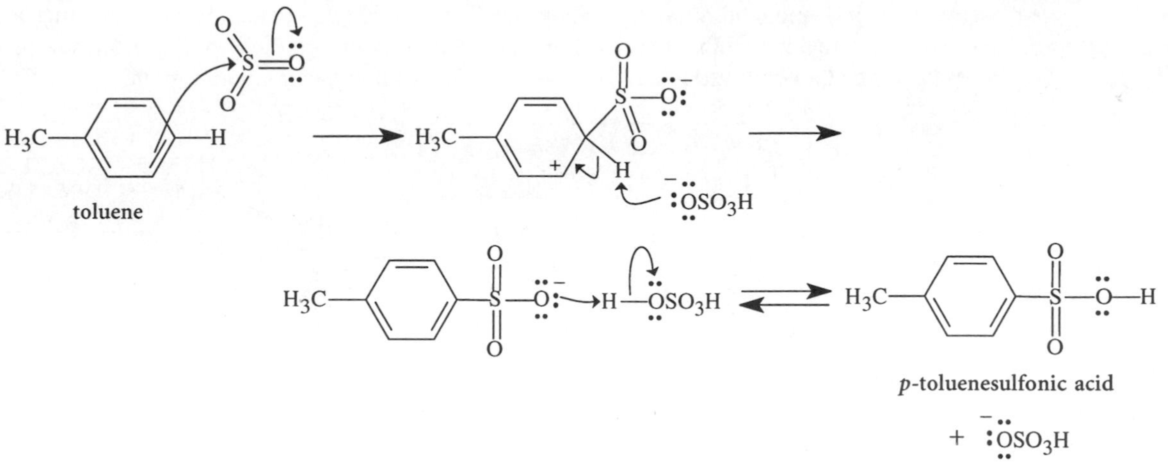

H_3C— toluene —H → H_3C— … $:\ddot{O}SO_3H$ →

H_3C— … —S—$\ddot{O}$:→H—$\ddot{O}SO_3H$ ⇌ H_3C— … —S—$\ddot{O}$—H

p-toluenesulfonic acid

+ $:\ddot{O}SO_3H$

A mechanism involving protonated SO_3 as the electrophile (as in the solution to the previous problem) is equally acceptable.

16.14 (a) As discussed in Study Guide Link 16.3, carbocations generated in a variety of ways can be used as electrophiles in Friedel-Crafts reactions. In this case, the carbocation produced by the protonation of cyclohexene is the electrophile.

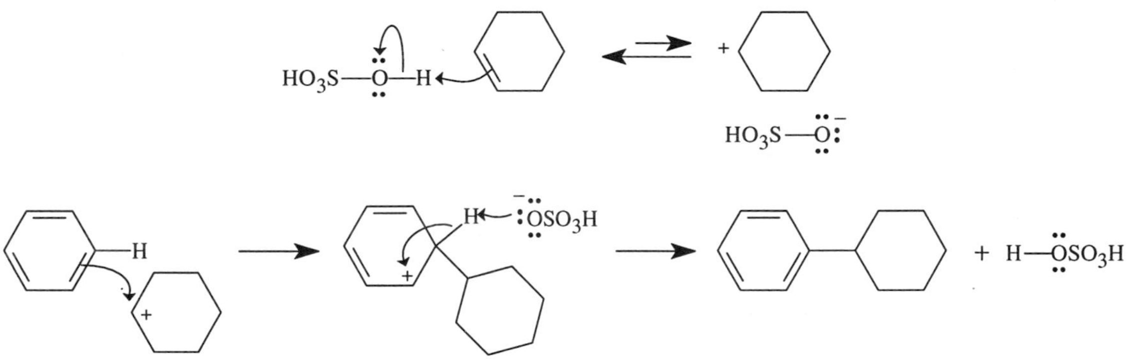

HO_3S—$\ddot{O}$—H ⇌ + HO_3S—$\ddot{O}$:$^-$

—H → H←$:\ddot{O}SO_3H$ → + H—$\ddot{O}SO_3H$

(b) The same product is obtained because the same carbocation electrophile is involved. It is generated from cyclohexanol by protonation and loss of water:

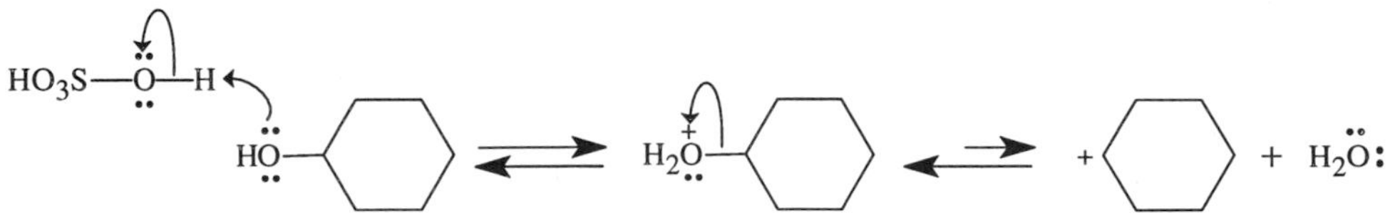

HO_3S—$\ddot{O}$—H HÖ— ⇌ $H_2\overset{+}{O}$— ⇌ + + $H_2\ddot{O}$:

HO_3S—$\ddot{O}$:$^-$

the same carbocation electrophile involved in part (a)

16.15 The product is *tert*-butylbenzene. The role of the Lewis acid BF_3 is to promote the ionization of HF. (See Study Guide Link 16.2.) The mechanism of the reaction is as follows:

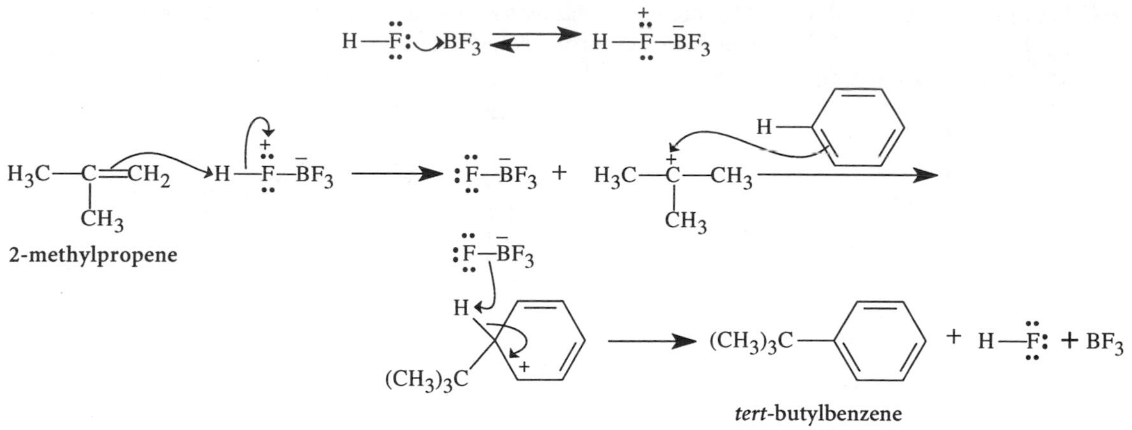

2-methylpropene

tert-butylbenzene

16.16 The hint and the fact that the product has the same number of carbons as the starting material suggest an intramolecular Friedel-Crafts alkylation.

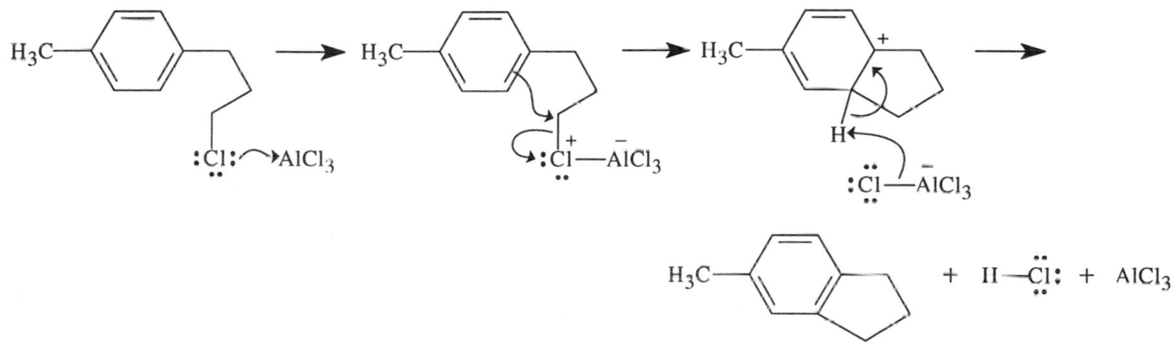

Remember that intramolecular reactions involving the formation of five- and six-membered rings are particularly rapid. This is another example of neighboring-group participation; see Sec. 11.7 of the text.

16.17 (a) (b)

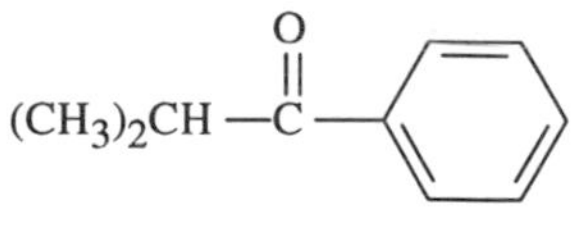

isobutyrophenone

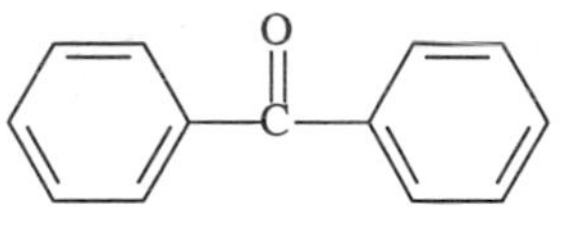

benzophenone

16.18 Either ring can be the one that undergoes substitution.

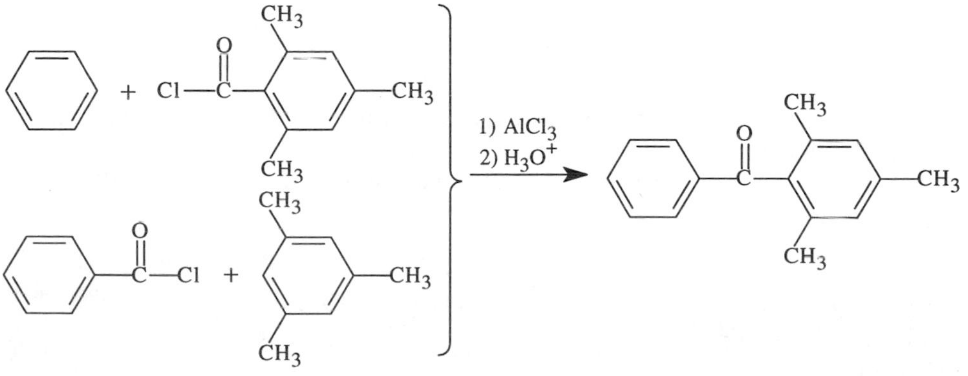

1) $AlCl_3$
2) H_3O^+

16.19 That the product contains the same number of carbons as the starting material suggests an intramolecular Friedel-Crafts reaction. The mechanism below begins with the acylium ion. (For the formation of this ion from the acid chloride, see Eq. 16.23 on text p. 713).

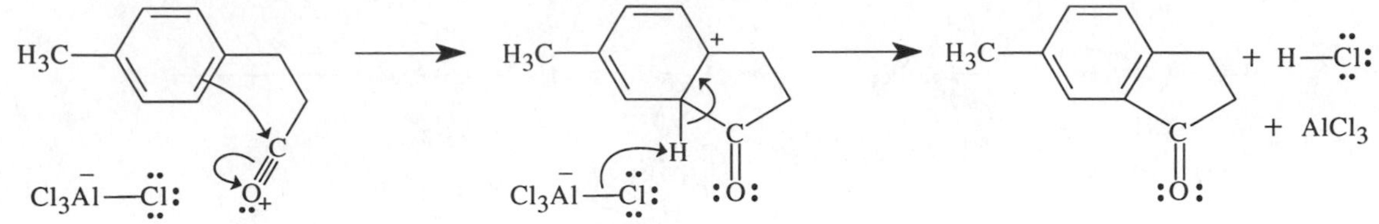

H_3C— ; $Cl_3\bar{Al}$—$\ddot{Cl}$: ; an acylium ion → H_3C— ; $Cl_3\bar{Al}$—$\ddot{Cl}$: ; H ; :O: → H_3C— ; :O: + H—$\ddot{Cl}$: + $AlCl_3$

(This is very similar to the case shown in Problem 16.16 for Friedel-Crafts alkylation.)

16.20 (a) Table 16.2 indicates that the methoxy group is an ortho, para-directing group.

CH_3O—C$_6$H$_4$—C(=O)—CH_3 + CH_3O—C$_6$H$_4$—C(=O)—H_3C

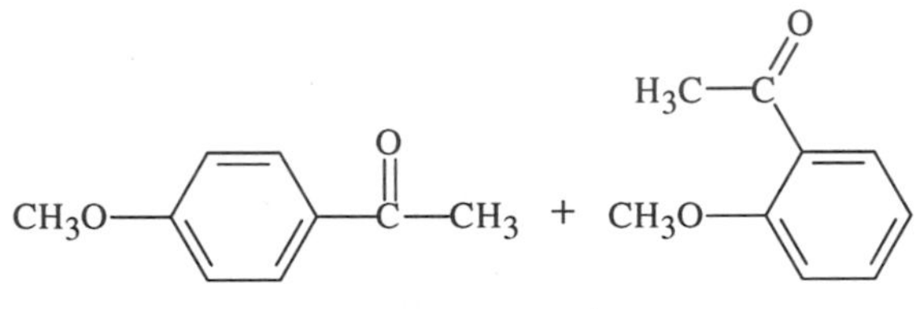

p-methoxyacetophenone *o*-methoxyacetophenone

(b) Table 16.2 indicates that alkyl groups are ortho, para-directing groups, and the ethyl group is a typical alkyl group:

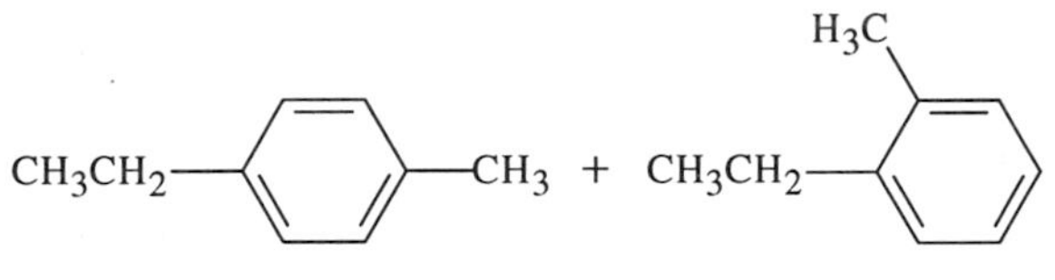

CH_3CH_2—C$_6$H$_4$—CH_3 + CH_3CH_2—C$_6$H$_4$—H_3C

1-ethyl-4-methylbenzene (*p*-ethyltoluene) 1-ethyl-2-methylbenzene (*o*-ethyltoluene)

16.21 Let E^+ be a general electrophile. The four resonance structures of the carbocation intermediate that results from reaction of E^+ at the position ortho to the methoxy group of anisole are as follows:

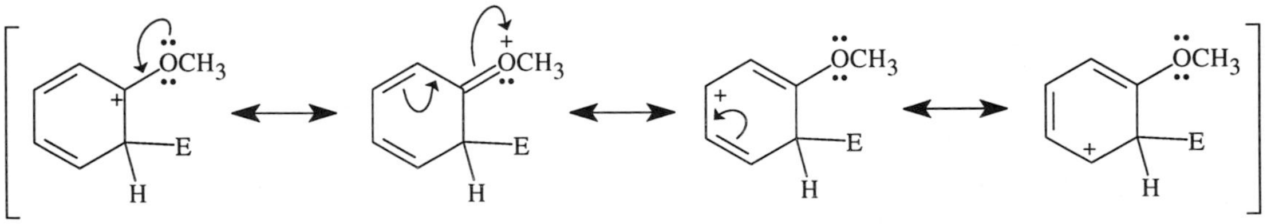

[$\ddot{O}CH_3$, E, H ↔ $\overset{+}{O}CH_3$, E, H ↔ $\ddot{O}CH_3$, E, H ↔ $\ddot{O}CH_3$, E, H]

16.22 (a) Because substitution occurs para to the phenyl group, the phenyl group is evidently an ortho, para-directing substituent.

(b) The following resonance structures show that the electrons of the phenyl substituent can be used to stabilize the carbocation intermediate when substitution occurs at the para position. The electrons of the phenyl substituent cannot be delocalized in this way when substitution occurs at the meta position.

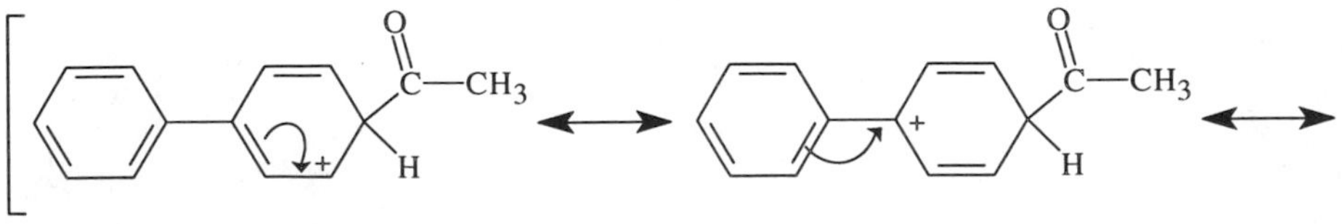

[C(=O)—CH_3, H ↔ C(=O)—CH_3, H ↔

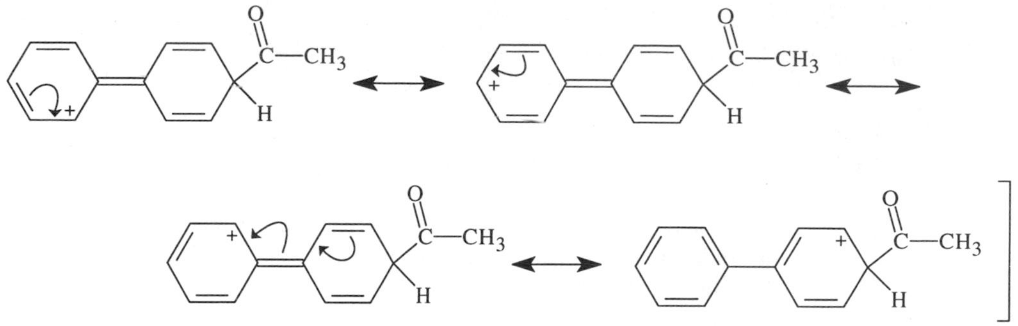

$C-CH_3$ H $\longleftrightarrow$ $C-CH_3$ H $\longleftrightarrow$

$C-CH_3$ H $\longleftrightarrow$ $C-CH_3$ H]

16.23 Notice that, except for alkyl groups, all ortho, para-directing groups have electron pairs on atoms adjacent to the benzene ring.

(a) This substituent, like the methoxy group, is an ortho, para-directing group, a point that is confirmed by entry number 4 of Table 16.2.

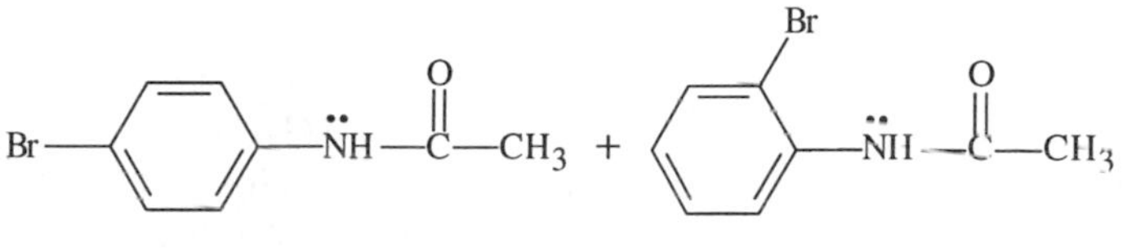

Br—C₆H₄—$\ddot{N}H-C(=O)-CH_3$ + (2-Br)C₆H₄—$\ddot{N}H-C(=O)-CH_3$

4-bromoacetanilide, or *N*-(4-bromophenyl)acetamide

2-bromoacetanilide, or *N*-(2-bromophenyl)acetamide

(b) The three strong carbon-fluorine bond dipoles result in substantial positive charge on the carbon of the $-CF_3$ group; consequently, this is a meta-directing group.

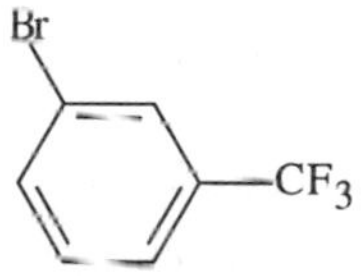

Br, $-CF_3$ 1-bromo-3-trifluoromethylbenzene

(c) The carbocation intermediate involved in meta substitution has a greater separation between the positive charge of the carbocation and the positive charge of the substituent than does the carbocation intermediate involved in ortho, para substitution. (The argument is similar to that used with electrophilic substitution reactions of nitrobenzene on text p. 721.) Therefore, the substituent is a meta-directing group.

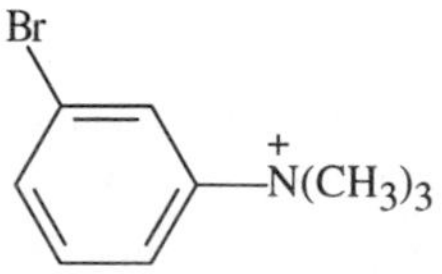

Br, $-\overset{+}{N}(CH_3)_3$ (3-bromophenyl)trimethylammonium ion

(d) The *tert*-butyl group, like all other alkyl groups, is an ortho, para-directing substituent.

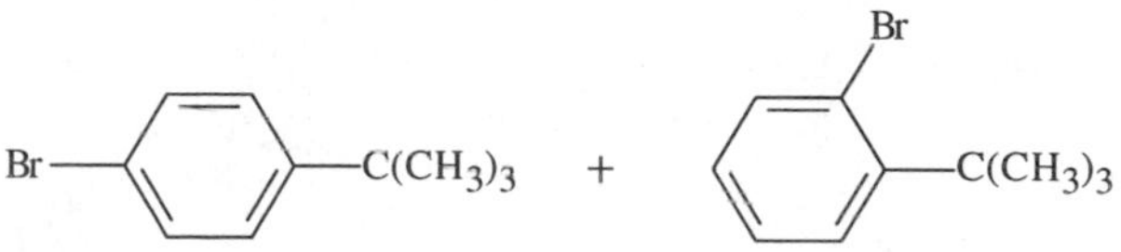

Br—C₆H₄—$C(CH_3)_3$ + (2-Br)C₆H₄—$C(CH_3)_3$

1-bromo-4-*tert*-butylbenzene

1-bromo-2-*tert*-butylbenzene

(e) The $-O^-$ group has both unshared electron pairs and a negative charge with which to stabilize the carbocation intermediate involved in ortho, para substitution. Consequently, this group is an ortho, para-directing group

4-bromophenoxide ion + 2-bromophenoxide ion

Because of the negative charge on the oxygen, the "carbocation intermediate" is not a cation at all, but rather a neutral compound. Moreover, the —O^- group is so strongly activating that substitution does not stop at the monobromo derivatives but continues until all ortho and para positions have been filled. (See Eq. 18.72 on text p. 816.)

16.24 (a) The reaction-free energy profiles for electrophilic substitution of benzene, chlorobenzene at the para position, and chlorobenzene at the meta position are shown in Fig. SG16.1(a) on the top of the following page. Notice that chlorobenzene is less reactive (that is, has a greater standard free energy of activation) than benzene because chlorine is a deactivating substituent. Notice also that para-substitution reactions on chlorobenzene are faster than meta-substitution reactions because chlorine is an ortho, para-directing group.

(b) The reaction-free energy profiles for electrophilic substitution of benzene, nitrobenzene at the meta position, and nitrobenzene at the para position are shown in Fig. SG16.1(b) on the top of the following page. Notice that nitrobenzene is less reactive than benzene because the nitro group is a deactivating substituent. Notice also that meta-substitution reactions on nitrobenzene are faster than para-substitution reactions because the nitro group is a meta-directing group.

16.25 Although both the CH_3O— and the CH_3S— groups are activating, ortho, para-directing groups, the orbitals on sulfur containing the unshared electron pairs, like those on chlorine, are derived from quantum level 3. Consequently, the overlap of sulfur orbitals with the carbon orbitals of the ring is poorer than the overlap of the oxygen orbitals. (See Fig. 16.8 on text page 726; the orbitals on sulfur are analogous to those on chlorine in the figure.) In other words, the electron-donating resonance interaction of the sulfur is weaker than that of the oxygen. Although the electron-withdrawing (and therefore rate-retarding) polar effect of oxygen is much greater than that of sulfur, the resonance interaction of oxygen is so much more powerful that it is the dominant effect.

16.26 Bromination of *N,N*-dimethylaniline is faster because nitrogen has an unshared electron pair that can stabilize the carbocation intermediate by resonance. As in the case of oxygen, the electron-withdrawing polar effect of nitrogen is much less important than its electron-donating resonance effect.

16.27 In both parts (a) and (b), each substituent is an ortho, para-directing group. Two products in each case satisfy the directing effects of both groups. (A third possibility is the product in which the entering group [—SO_3H in (a), —NO_2 in (b)] is ortho to both substituents; little of this product should be formed in either case because of the strong van der Waals repulsions that would result.)

(a)

2-bromo-4-methylbenzenesulfonic acid + 4-bromo-2-methylbenzenesulfonic acid

(b)

2-bromo-1-nitro-4-iodobenzene + 4-bromo-1-nitro-2-iodobenzene

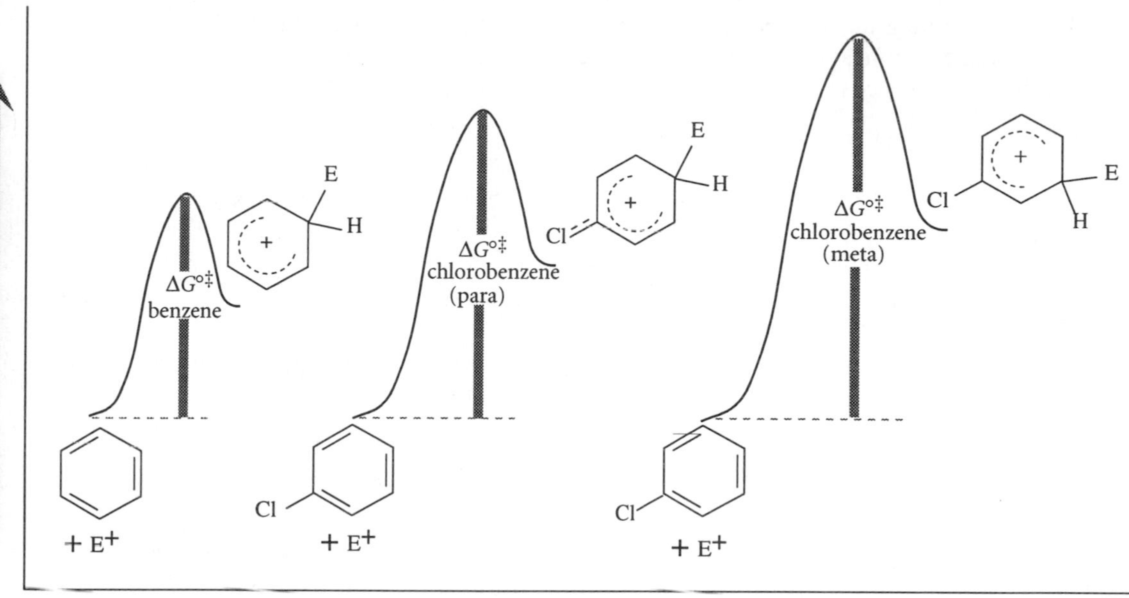

(a)

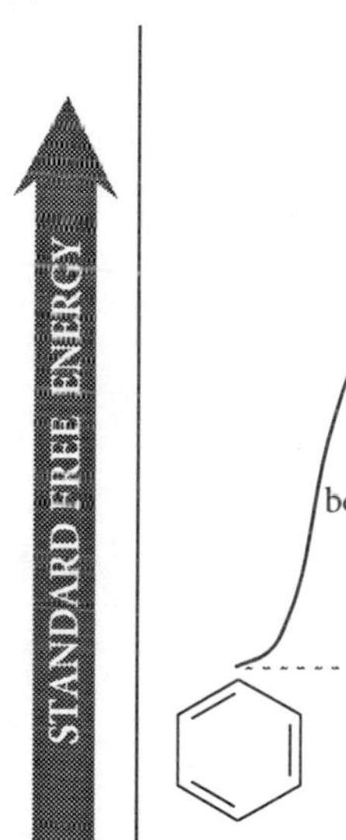

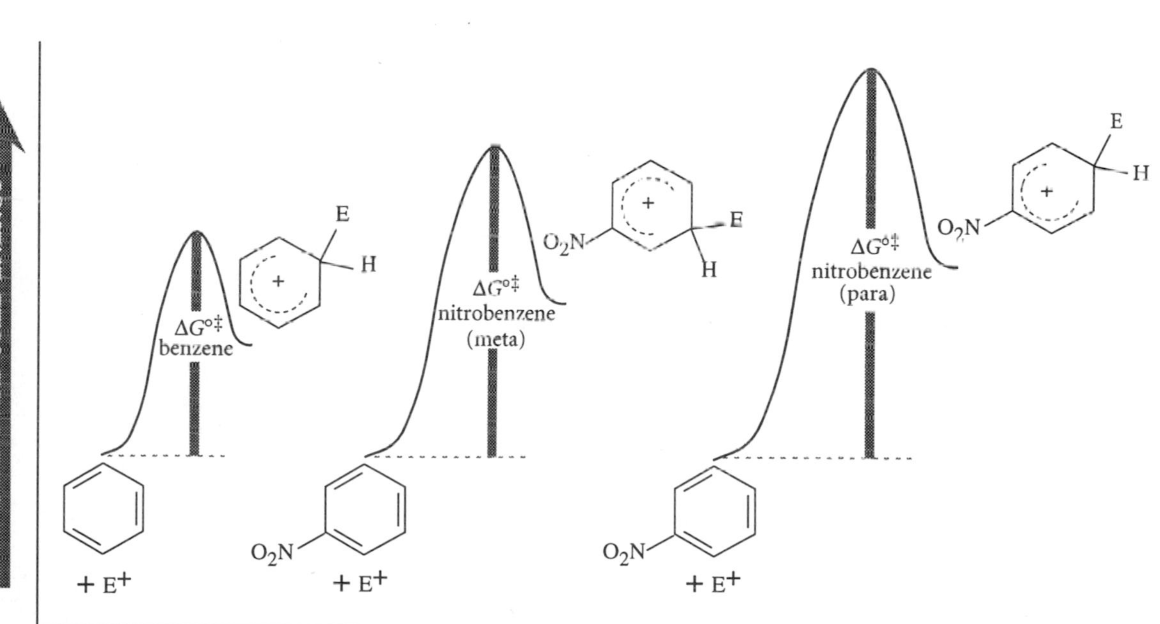

(b)

Figure SG16.1 (a) Reaction-free energy profiles to accompany the solution to Problem 16.24(a).
(b) Reaction-free energy profiles to accompany the solution to Problem 16.24(b).

16.28 (a) The order of increasingly harsh reaction conditions is *m*-xylene < benzene < *p*-dichlorobenzene. The reason is that the methyl groups of *m*-xylene are activating groups and the chlorines of *p*-dichlorobenzene are deactivating groups.

(b) The order is anisole < toluene < chlorobenzene. Chlorobenzene requires the harshest conditions because chlorine is a deactivating group. Anisole requires the mildest conditions because the methoxy group is more activating than the methyl group of toluene. (See Table 16.2 on text p. 717.)

16.29 Because both groups are *meta*-directing groups, it might seem that either could be introduced first. However, the Friedel-Crafts acylation cannot be carried out on nitrobenzene because, as discussed on text p. 730, the nitro group is too deactivating. Consequently, the acetyl group must be introduced first.

CH_3CCl (O above C), $AlCl_3$, then H_3O^+ → ; HNO_3, H_2SO_4, heat →

benzene — acetophenone — *m*-nitroacetophenone

16.30 (a) Hydrogenate *tert*-butylbenzene, which, in turn, is prepared by Friedel-Crafts alkylation as shown in Eq. 16.18 on text p. 712 or by the reaction shown in the solution to Problem 16.15.

$C(CH_3)_3$ — H_2, Ni, heat, pressure → — $C(CH_3)_3$

tert-butylbenzene — *tert*-butylcyclohexane

(b) Hydrogenate cyclohexylbenzene, which, in turn, is prepared as shown in Eq. 16.21 on text p. 713. Note that chlorocyclohexane and $AlCl_3$, or cyclohexanol and either H_2SO_4 or H_3PO_4, may be used as the source of the electrophile instead of cyclohexene; see Problem 16.14(b) and Study Guide Link 16.3.

H_2, Ni, heat, pressure →

cyclohexylbenzene — cyclohexylcyclohexane

Solutions to Additional Problems

16.31 (a) No reaction.

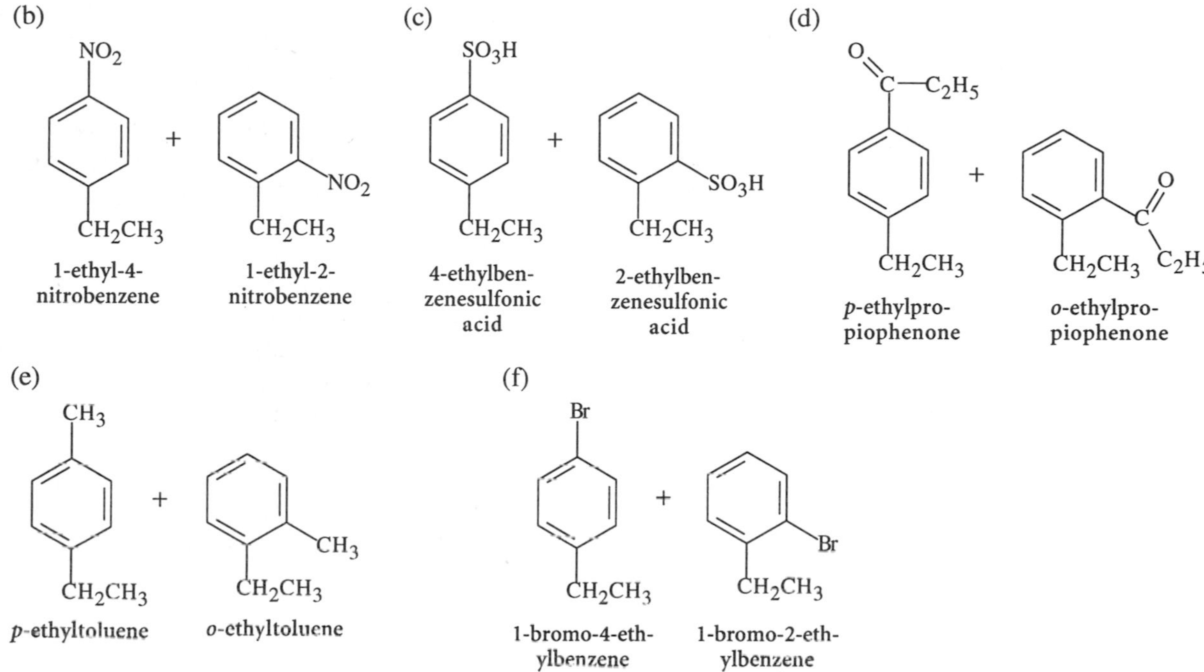

In part (e), ethylbenzene must be present in large excess for the monosubstitution products to be obtained. If ethylbenzene is not present in excess, products of dimethylation, trimethylation, etc., will also be formed; see Eq. 16.22 on text p. 759 and associated discussion.

16.32 (a) (b)

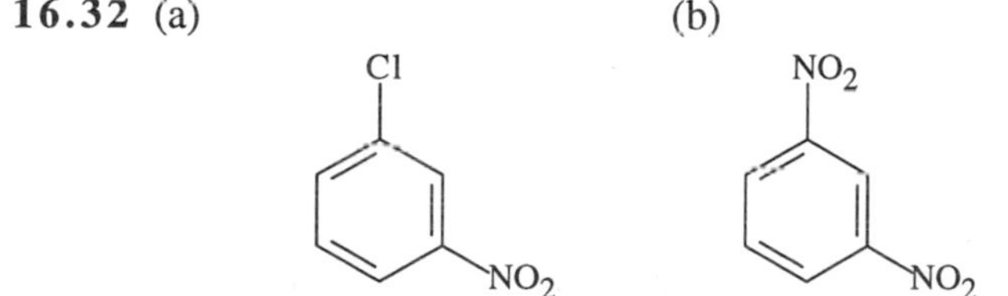

1-chloro-3-nitrobenzene   1,3-dinitrobenzene or *m*-dinitrobenzene

(c) No electrophilic aromatic substitution reaction takes place. Friedel-Crafts acylation does not take place on any benzene derivative less reactive than the halobenzenes. In fact, nitrobenzene can be used as an inert solvent for Friedel-Crafts acylation, as noted near the bottom of text p. 730.

16.33 A compound containing a benzene ring must have at least six carbon atoms and four degrees of unsaturation. Compound *A*, with three degrees of unsaturation, cannot contain a benzene ring; compound *C* has only five carbons and therefore cannot contain a benzene ring. Compounds *B* and *D* have more than six carbons and have five and four degrees of unsaturation, respectively, and therefore could contain a benzene ring.

16.34 (a) Assume that the carbon-chlorine bond dipoles dominate the dipole moment. In the following diagram, the small gray arrows are the C—Cl bond dipoles and the darker arrows are the resultant molecular dipole moments. This diagram shows that the order of increasing dipole moment is *p*-dichlorobenzene ($\mu = 0$) < *m*-dichlorobenzene < *o*-dichlorobenzene.

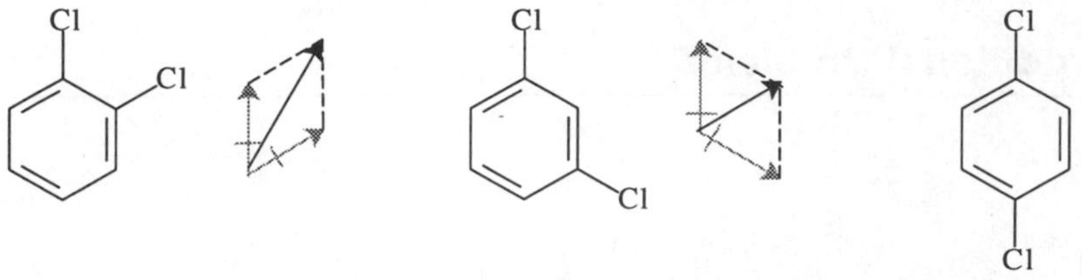

dipole moment = 0

(b) Boiling points are increased by increasing attractions between molecules. To the extent that molecular attractions are dominated by attractions between molecular dipole moments, the order of increasing boiling point is predicted to be the same as the order of increasing dipole moment. (In fact, the boiling point of the ortho isomer (180°) is 6–7° higher than the boiling points of the meta and para isomers, which are within one degree of each other.)

16.35 Only compound *A* should have a proton NMR spectrum containing one six-proton singlet in the alkyl region (about δ 1.5). Only compound *B* should have a proton NMR spectrum that contains two closely spaced singlets in the integral ratio 2:1 in the benzylic proton region of the spectrum (about δ 2.3). Only compound *C* should have a proton NMR spectrum that contains two well-separated singlets in the integral ratio 3:1, the larger in the benzylic proton region and the smaller at about δ 4.5. All three compounds are also readily distinguished by the relative integrals of their aromatic proton absorptions.

16.36 (a) The cyclopentadienyl anion is aromatic; its five resonance structures show that all carbons (and hydrogens) are chemically equivalent. Hence, the proton NMR spectrum of the ion consists of one singlet.

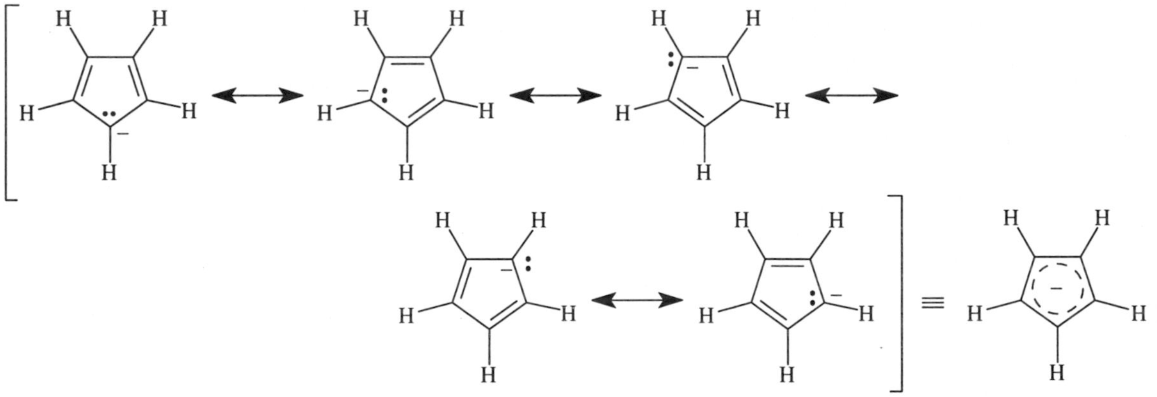

(b) Fig. 16.2 on text p. 699 shows that aromatic protons located in the plane of the ring and outside of the ring experience an augmented local field and thus a greater chemical shift. However, in the region above and below the ring, the induced field has the opposite direction, and consequently protons located in this region experience a reduced local field and thus a smaller chemical shift. Such is the case with the methyl group in the problem; the local field at this group is so small that its resonance occurs at nearly 1.7 ppm *smaller* chemical shift than that of TMS.

16.37 The electron density of the unshared electron pairs on the oxygen of the methoxy group is delocalized to the ortho and para positions on the ring, but not to the meta position:

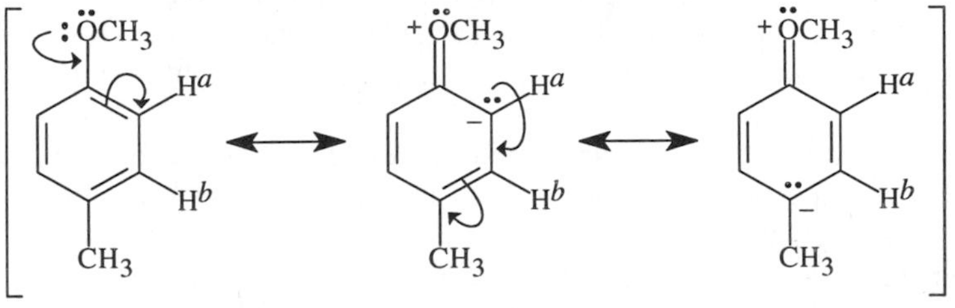

Thus, there is more electron density at the ortho proton H^a than at the meta proton H^b. Since greater electron density results in a reduced local field $\mathbf{B}_0$ (see Sec. 13.3A), the chemical shift of proton H^a is smaller.

16.38 In each synthesis that involves substitution on a benzene derivative that contains an ortho, para-directing group, only the product resulting from para substitution is shown.

(a)

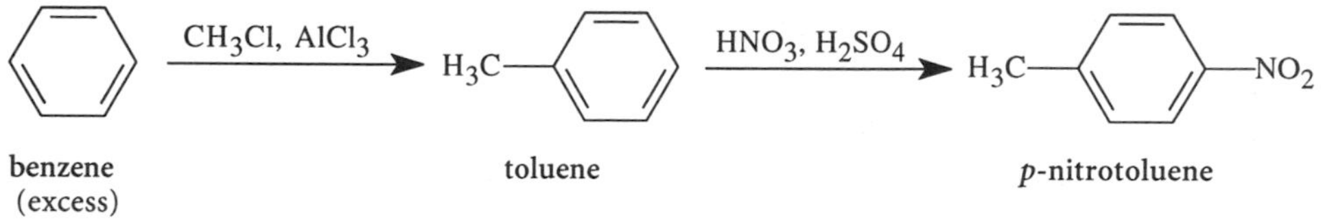

benzene (excess) —[CH_3Cl, $AlCl_3$]→ toluene —[HNO_3, H_2SO_4]→ *p*-nitrotoluene

(b)

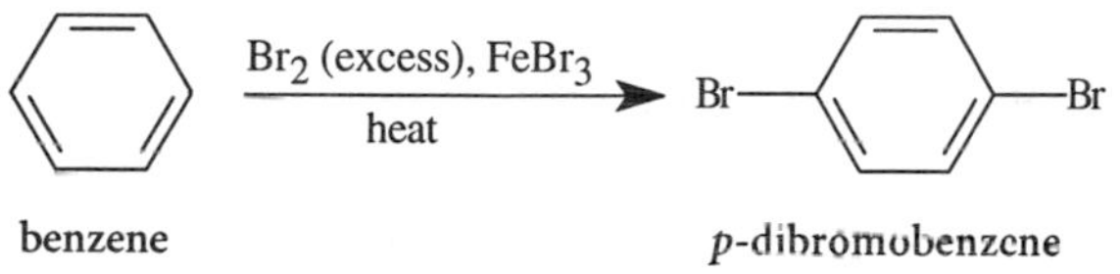

benzene —[Br_2 (excess), $FeBr_3$, heat]→ *p*-dibromobenzene

(c)

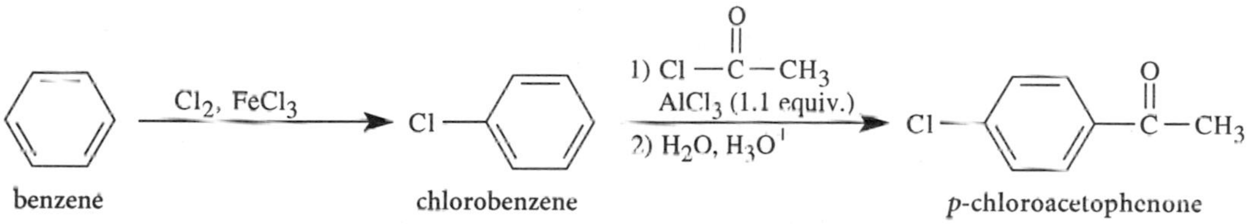

benzene —[Cl_2, $FeCl_3$]→ chlorobenzene —[1) $Cl-\overset{O}{\overset{\|}{C}}-CH_3$, $AlCl_3$ (1.1 equiv.); 2) H_2O, H_3O^+]→ *p*-chloroacetophenone

(d)

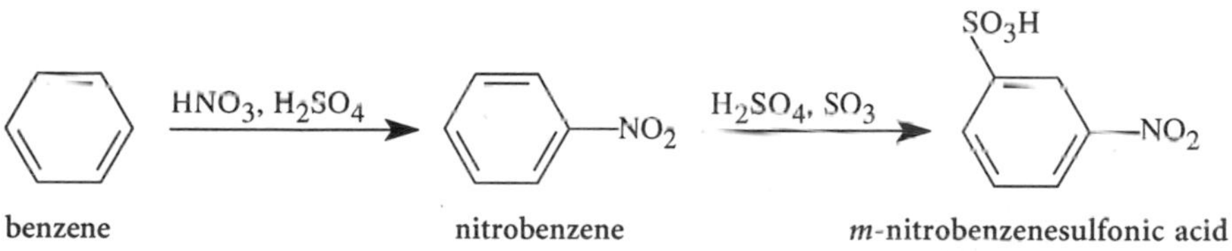

benzene —[HNO_3, H_2SO_4]→ nitrobenzene —[H_2SO_4, SO_3]→ *m*-nitrobenzenesulfonic acid

(e)

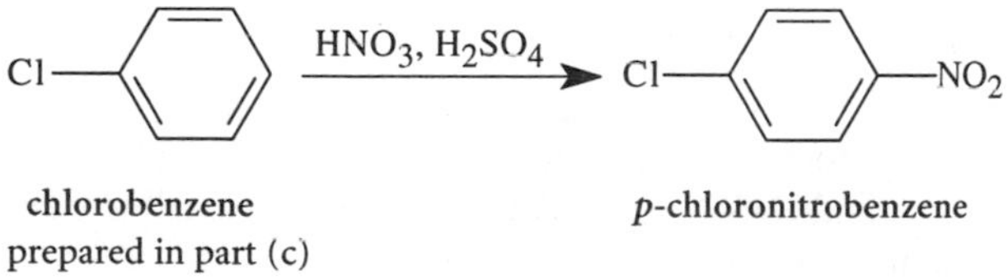

chlorobenzene prepared in part (c) —[HNO_3, H_2SO_4]→ *p*-chloronitrobenzene

(f)

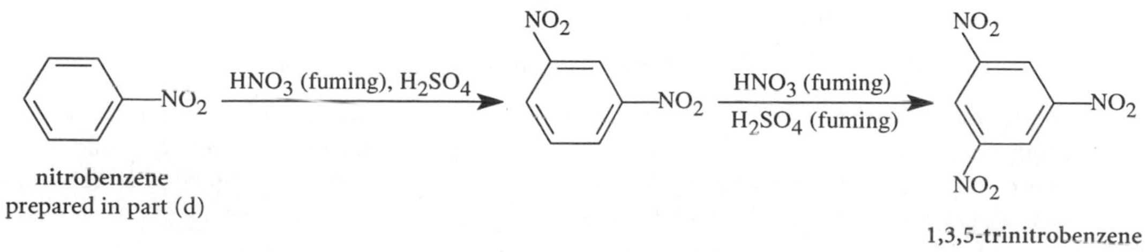

nitrobenzene prepared in part (d) —[HNO_3 (fuming), H_2SO_4]→ *m*-dinitrobenzene —[HNO_3 (fuming), H_2SO_4 (fuming)]→ 1,3,5-trinitrobenzene

(g)

H_3C–(benzene ring)–NO_2 $\xrightarrow{Br_2 \text{ (excess)}, FeBr_3}$ 2,6-dibromo-4-nitrotoluene (Br, Br, H_3C, NO_2)

p-nitrotoluene
prepared in part (a)

2,6-dibromo-4-nitrotoluene

(h)

CH_3, NO_2 $\xrightarrow{Br_2 \text{ (excess)}, FeBr_3, \text{ heat}}$ CH_3, Br, NO_2, Br

o-nitrotoluene
by-product of the
synthesis in part (a)

2,4-dibromo-6-nitrotoluene

Nitration of toluene actually gives more ortho isomer than para isomer of toluene, and the two nitrotoluene isomers are readily separated by fractional distillation. (See text p. 722.) *o*-Nitrotoluene is a good starting material for a number of ortho-substituted benzene derivatives.

(i)

benzene (excess) $\xrightarrow[AlCl_3]{C_2H_5Cl}$ ethylbenzene (C_2H_5) $\xrightarrow[2)\ H_3O^+]{1)\ CH_3\overset{O}{\overset{\|}{C}}Cl,\ AlCl_3}$ *p*-ethylacetophenone ($CH_3\overset{O}{\overset{\|}{C}}$, C_2H_5) $\xrightarrow{HNO_3,\ H_2SO_4}$ 4-ethyl-3-nitroacetophenone ($CH_3\overset{O}{\overset{\|}{C}}$, NO_2, C_2H_5)

benzene
(excess)

ethylbenzene

p-ethylacetophenone

4-ethyl-3-nitroacetophenone

(j)

benzene + cyclopentene $\xrightarrow{H_2SO_4}$ cyclopentylbenzene

benzene
(excess)

cyclopentene

cyclopentylbenzene

Cyclopentanol may be substituted for cyclopentene in this synthesis, or chlorocyclopentane and $AlCl_3$ catalyst may be used instead of cyclopentene and H_2SO_4. Note that each of these possible starting materials serves as a source of the same carbocation, the cyclopentyl cation.

16.39 (a) The reactivity order follows from the relative activating effects of the substituents. (See the last column of Table 16.2 on text p. 717.)

nitrobenzene < chlorobenzene < benzene

(b) In *p*-chloroanisole, the chlorine and methoxy group direct to different positions; that is, the chlorine strongly deactivates the positions that the methoxy group activate, and the methoxy group deactivates the positions that are least deactivated by the chlorine. In the meta isomer, the directing effects of the two substituents do not conflict. In anisole, the deactivating chloro group is absent altogether. The order is

p-chloroanisole < *m*-chloroanisole < anisole

(c) In mesitylene, the three methyl groups all direct nitration to the same position; that is, they activate the same position of the ring. In 1,2,4-trimethylbenzene, only two of the three methyl groups activate the same ring position. Both trimethylbenzenes are more activated than toluene. Hence, the order is

toluene < 1,2,4-trimethylbenzene < mesitylene

(d) The reactivity order follows from the relative activating effects of the substituents. (See the last column of Table 16.2 on text p. 717.)

p-bromoacetophenone < acetophenone < *p*-methoxyacetophenone

16.40 (a) The boron on benzeneboronic acid has an empty $2p$ orbital and is not capable of stabilizing a carbocation by resonance. It would be difficult to introduce electron deficiency into a compound that is already electron deficient; hence, this compound undergoes substitution much more slowly than benzene itself. Moreover, the boron-oxygen bond dipoles place substantial positive charge on boron. The boronic acid group thus has the characteristics of a meta-directing group, and the principal mononitration product is *m*-nitrobenzeneboronic acid (compound *A* in the structures following part (c)).

(b) For the reasons given in the solution to Problem 16.23(c), the $—\overset{+}{N}(CH_3)_3$ group is a meta-directing group. This compound reacts more slowly than benzene, and the product is the *meta*-nitro derivative *B*.

(c) One benzene ring serves as an ortho, para-directing substituent on the other. (The directing effect of the phenyl substituent is explained in the solution to Problem 16.22.) Because the phenyl substituent can stabilize the carbocation intermediate in nitration, it is an activating group. Hence, this compound is more reactive than benzene, and the product is the *para*-nitro derivative *C* (plus perhaps a small amount of the *ortho*-nitro isomer).

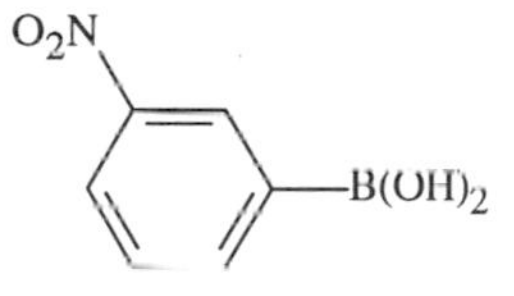

O_2N — $B(OH)_2$ O_2N — $\overset{+}{N}(CH_3)_3$ O_2N—

A 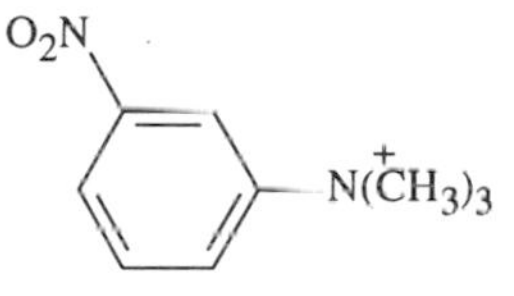   *B* 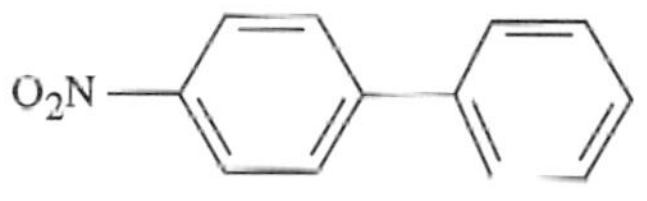   *C*

(d) Only the positions marked with an asterisk (*) are activated by *both* the phenyl and methoxy groups. (You should show that a carbocation intermediate formed in the other ring at *any* position cannot be resonance-stabilized by the unshared pair of the methoxy group.) Because of this activation, nitration at the asterisked positions is more rapid than nitration of benzene itself; the products are as follows:

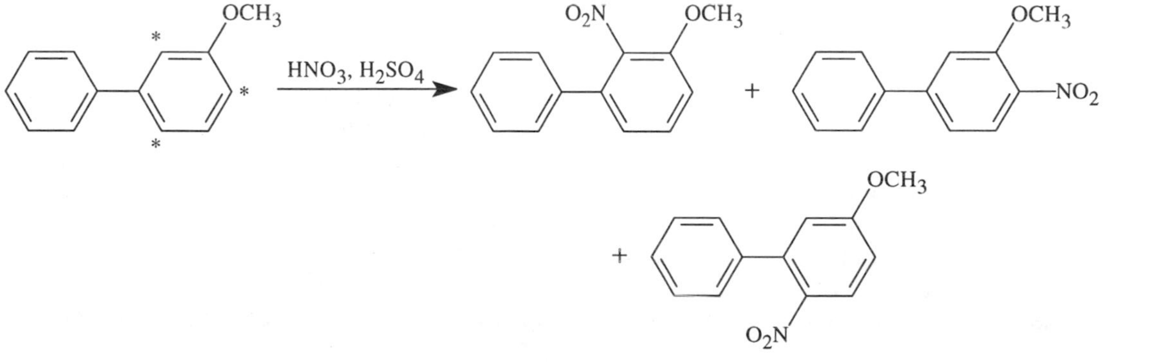

OCH_3 * * * $\xrightarrow{HNO_3,\ H_2SO_4}$ O_2N OCH_3 + OCH_3 $-NO_2$

+ OCH_3 O_2N

16.41 The reactivity order is *A* < *B* < *D* < *C*. Compound *C* is most reactive because the substituent has an unshared electron pair that can be used to stabilize the intermediate carbocation by resonance. Compounds *B* and *D* have alkyl substituents, which stabilize carbocations; however, the alkyl group of compound *B* contains a positively charged group that would interact unfavorably with a carbocation, offsetting the stabilizing effect of the alkyl carbon. Compound *A* has a positively charged, electronegative substituent attached directly to the ring that would interact most unfavorably with the carbocation. (See the solution to Problem 16.40b.) Compounds *C* and *D* undergo bromination at the ortho and para positions; compound *A* undergoes bromination at the meta position; and the position of substitution in compound *B* depends on the balance of the stabilizing effect of the alkyl group and the destabilizing effect of the positive charge. (In fact, this compound brominates in the ortho and para positions.)

16.42 First draw the resonance structure of compound *B* that has an electronic octet around phosphorus:

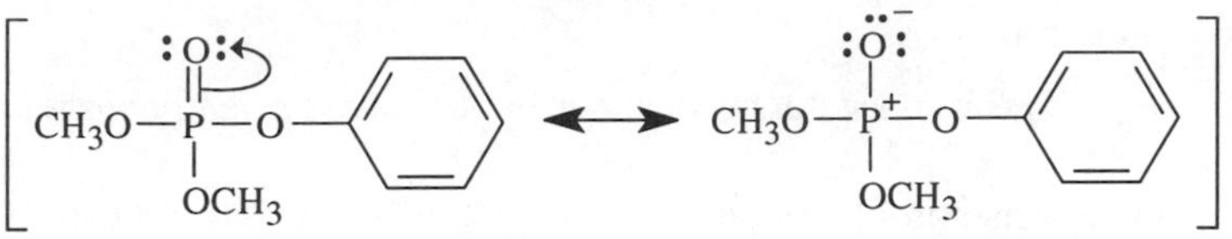

This is an important resonance structure because of its adherence to the octet rule. With this structure in mind, consider then the carbocation intermediate in nitration:

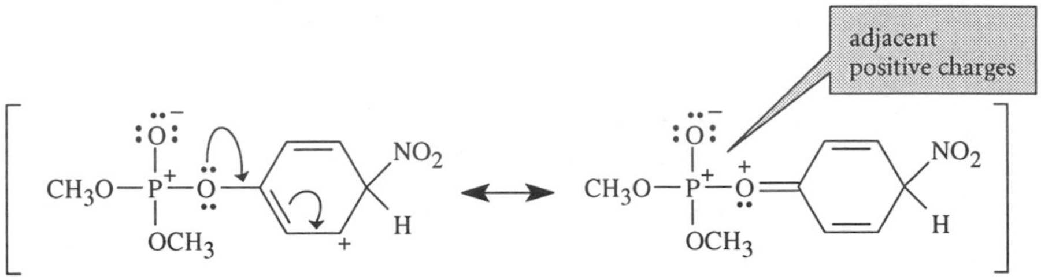

For an oxygen unshared pair to exert a stabilizing resonance effect on the carbocation, a destabilizing interaction must be simultaneously introduced as a result of the adjacent positive charges on oxygen and phosphorus. Hence, this resonance interaction is relatively unimportant—or much less important than in compound *A*. Because the resonance interaction of its electron pairs is weak or nonexistent, the oxygen acts more like an electronegative substituent *without* unshared pairs. Such a substituent (like the nitrogen in a nitro group) should be a deactivating, meta-directing group, as observed.

16.43 The formula $C_{10}H_{10}$ corresponds to an unsaturation number of 6; therefore, any candidate compound could contain, in addition to the benzene ring, either two double bonds or one triple bond. There are only two possibilities:

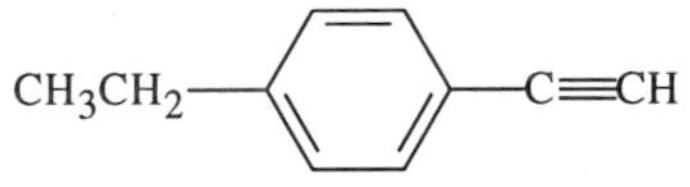

16.44 The unsaturation number of compounds *A* and *B* is 5; hence, both compounds could be aromatic. Compound *C* is a hydrocarbon; because hydrogenation of *C* gives a compound (compound *D*) with the molecular formula C_9H_{10}, then the formula of compound *C* could be C_9H_8. If so, then the reaction of compounds *A* and *B* with acid is a dehydration. If compounds *A* and *B* are alcohols, their —OH groups must be on adjacent carbons in order for them to dehydrate to the same alkene. Compound *D*, the hydrogenation product, has an unsaturation number of 5; nitration confirms that it and, by deduction, the other compounds, are aromatic. Since compound *D* can contain no ordinary double bonds (they would have been hydrogenated), it must contain a ring *in addition to* the aromatic ring. The structures of the compounds that meet all these criteria are as follows:

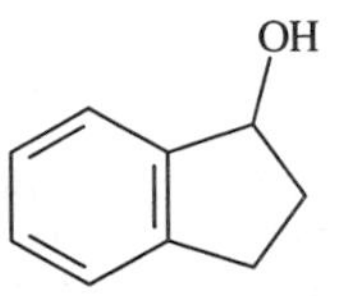

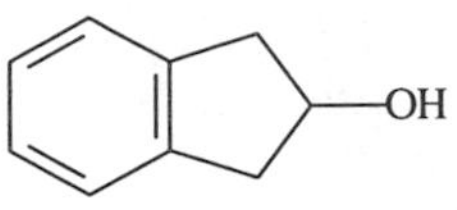

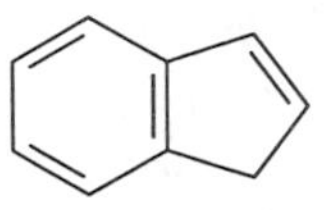

Note that compound *D*, because of its symmetry, would give only two nitration products:

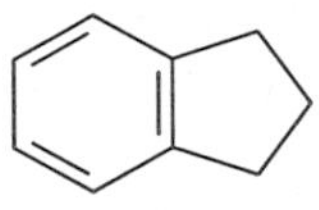

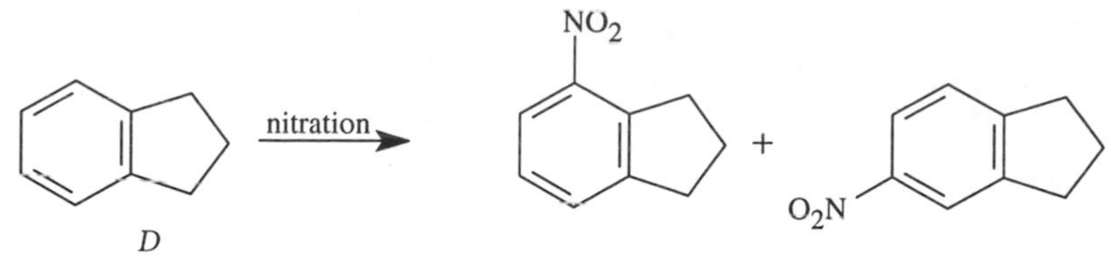

The nitration results rule out isomers such as the following:

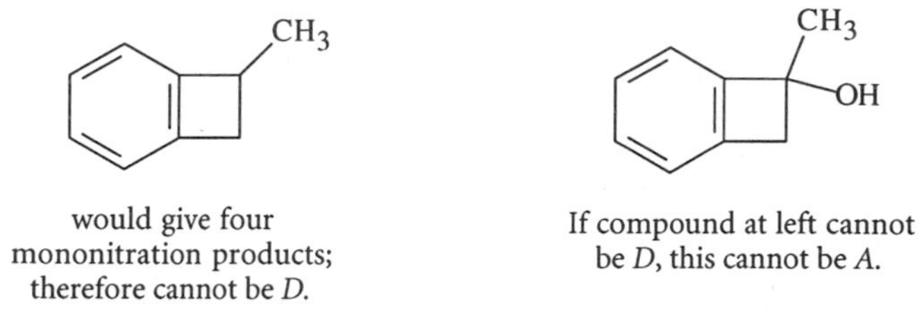

would give four mononitration products; therefore cannot be *D*.

If compound at left cannot be *D*, this cannot be *A*.

16.45 The λ_{max} of styrene is due to the conjugation of a π bond with the benzene ring. Both π bonds of phenylacetylene cannot be simultaneously conjugated with the benzene ring. Hence, the λ_{max} of phenylacetylene is also due to the conjugation of a *single* π bond with the benzene ring. Consequently, the structural features of both compounds responsible for their UV spectra are identical.

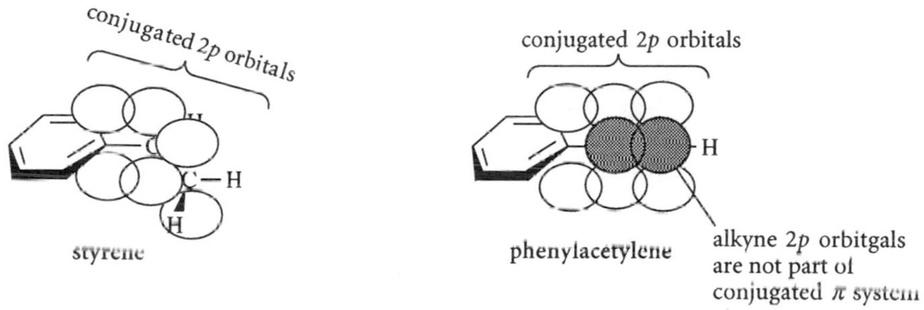

16.46 Essentially we run the electrophilic aromatic substitution reaction in reverse.

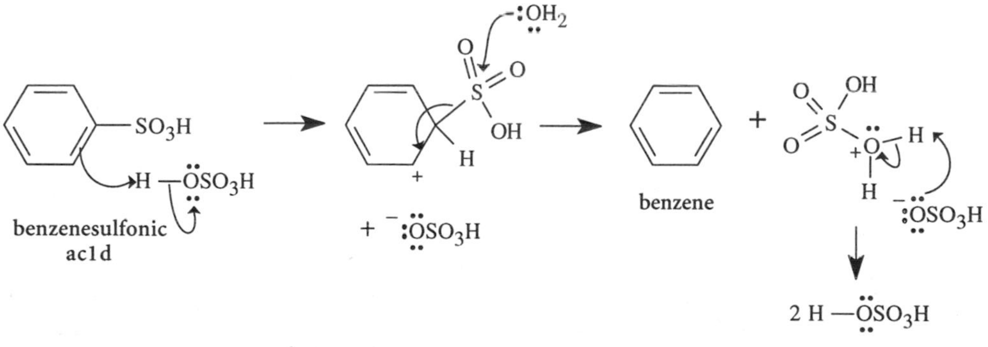

There are several variations on this mechanism. For example, SO_3H could be lost from the carbocation intermediate to give protonated SO_3 (that is, $^+SO_3H$; see Eq. 16.12 on text p. 709) which could react with water to give SO_3 (sulfur trioxide) and H_3O^+. Sulfur trioxide reacts vigorously with water to give H_2SO_4 (sulfuric acid). The important aspect of the mechanism is the protonation of the ring and loss of a species which would serve as an electrophile in the reverse reaction.

16.47 The product has an unsaturation number of 6, which means that, relative to the starting material, the product has an additional ring or double bond. The fact that it does not react with Br_2 suggests that a new ring has been formed. In fact, formation of a carbocation by protonation of the alcohol and loss of water leads to an electrophilic aromatic substitution reaction that forms a new six-membered ring.

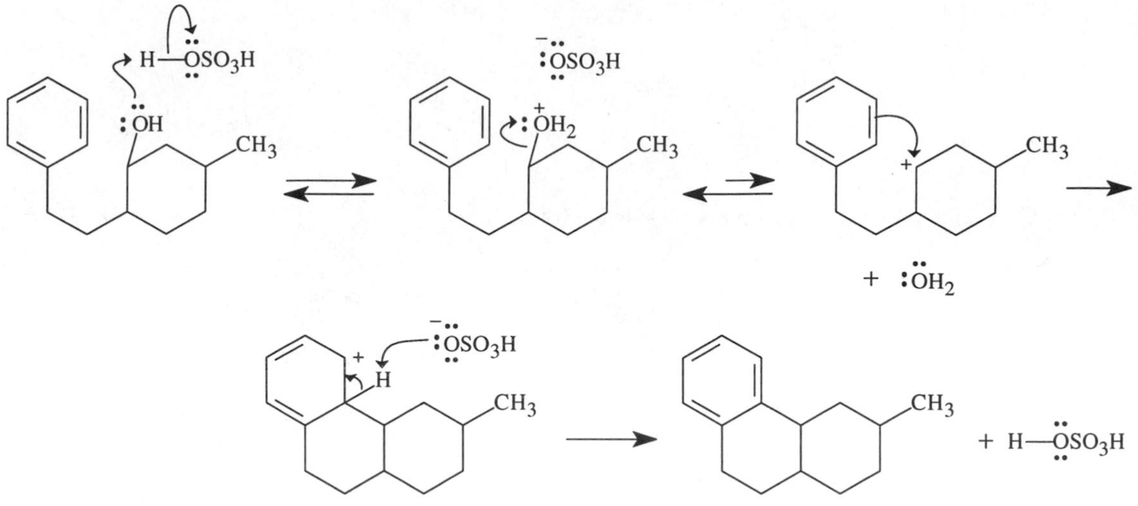

$H—\ddot{O}SO_3H$ $:\ddot{O}H$ CH_3 ⇌ $:\ddot{O}SO_3H^-$ $:\overset{+}{O}H_2$ CH_3 ⇌ CH_3 + $:\ddot{O}H_2$ →

$:\ddot{O}SO_3H^-$ H CH_3 → CH_3 + $H—\ddot{O}SO_3H$

16.48 (a) and (b)

Generation of the electrophile: The electrophile is the carbocation *A* generated by protonation of the alcohol oxygen and loss of water.

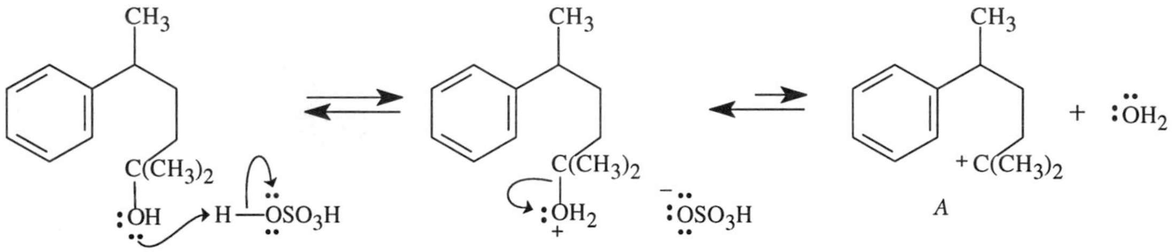

CH_3 $C(CH_3)_2$ $:\ddot{O}H$ $H—\ddot{O}SO_3H$ ⇌ CH_3 $C(CH_3)_2$ $:\overset{+}{O}H_2$ $:\ddot{O}SO_3H^-$ ⇌ CH_3 $+C(CH_3)_2$ *A* + $:\ddot{O}H_2$

Attack of the benzene π electrons on the electrophile to generate another carbocation:

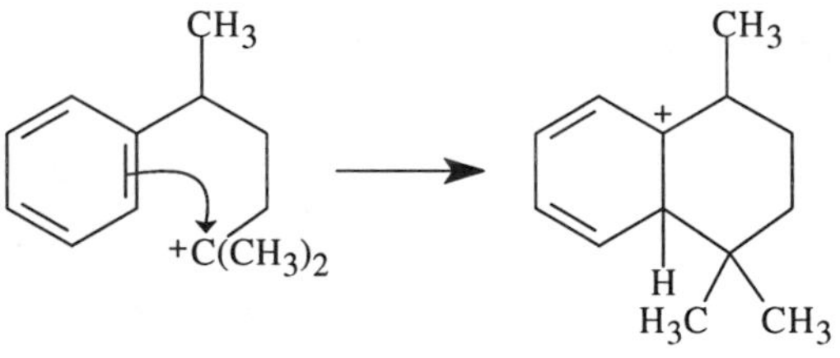

CH_3 $+C(CH_3)_2$ → CH_3 + H H_3C CH_3

Loss of a β-proton to form the new aromatic compound:

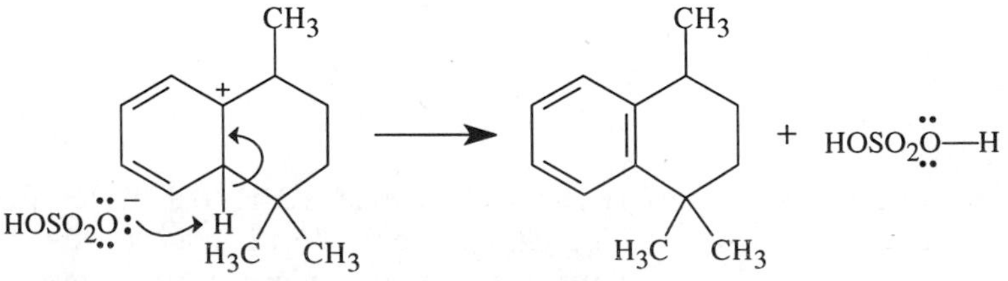

CH_3 + $HOSO_2\ddot{O}:^-$ H H_3C CH_3 → CH_3 H_3C CH_3 + $HOSO_2\ddot{O}—H$

(c) The aromatic ring of compound *A* has three alkyl substituents. Two of them—the *tert*-butyl group and one of the ring bonds—direct substitution by their electronic effects to the positions indicated by the asterisk:

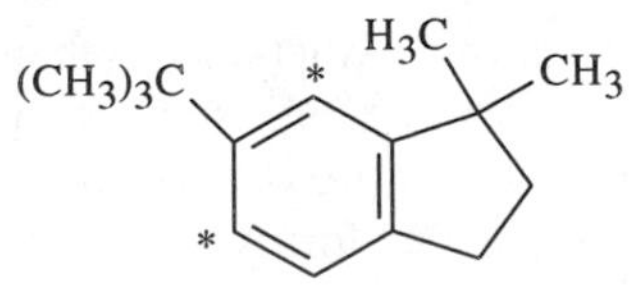

$(CH_3)_3C$ * H_3C CH_3 *

Both of these positions, however, are ortho to the very large *tert*-butyl group, and one of them is ortho to *two* highly branched groups. For steric reasons substitution cannot occur at these very congested positions. Hence, it occurs at the remaining position. Note that ring position meta to alkyl substituents are not deactivated; they are simply less activated than positions that are ortho and para to alkyl substituents. Furthermore, the remaining ring position is activated by one alkyl substituent.

16.49 The formula of compound *B* would be useful in solving this problem. Consider the integral in the NMR spectrum of compound *B*. The entire spectrum integrates for ten hydrogens, and we can hypothesize that no carbons are lost on the reaction with sodium ethoxide. Hence, the formula of compound *B* is thus C_9H_{10}. Thus, compound *A* loses the elements of HBr when it reacts with a base—clearly an elimination reaction—and adds one molar equivalent of H_2 on catalytic hydrogenation. Therefore, compound *B* is an alkene formed by an E2 reaction from compound *A*. The NMR of compound *B* shows that it has five aromatic hydrogens; hence, the remaining unsaturation of compound *B* is due to a monosubstituted benzene ring. The resonances for the vinylic hydrogens in the NMR spectrum centered at δ 5.1 integrate for two protons. These two protons are chemically nonequivalent, and they show no large splittings typical of a *cis*- or *trans*-alkene. Hence, the vinylic protons are in the part structure H_2C═. The chemical shift and integration of the δ 2.1 resonance suggests a vinylic methyl group. The methyl group cannot be benzylic because this would require one fewer aromatic hydrogen and one more vinylic hydrogen. The structure of *B* is determined and the structure of *C*, its hydrogenation product, follows:

C_6H_5–C(CH_3)=CH_2 — *B* isopropenylbenzene (α-methylstyrene)

C_6H_5–$CH(CH_3)_2$ — *C* isopropylbenzene

Two possible alkyl halides can be considered for *A*. The chirality of *A* allows us to decide between them.

C_6H_5–C(CH_3)(Br)–CH_3

C_6H_5–CH(CH_2Br)–CH_3 — *A* (2-bromo-1-methylethyl)benzene

16.50 (a) The IR spectrum indicates the presence of an aromatic ring. The singlet at δ 3.72 suggests a methoxy group, and the apparent pair of doublets indicates a para-disubstituted benzene ring. The M + 2 peak in the mass spectrum with about one-third the intensity of the M = 142 peak suggests the presence of chlorine. A chlorine, a —OCH_3 group, and a para-disubstituted benzene ring fully account for the mass of 142. The compound *p*-chloroanisole (1-chloro-4-methoxybenzene) is consistent with all the data.

Cl–C_6H_4–OCH_3 — *p*-chloroanisole (1-chloro-4-methoxybenzene)

When interpreting mass spectra, don't forget that you must use exact masses for isotopes. Thus, mass M = 142 corresponds to the compound containing ^{35}Cl, and the mass of the M + 2 peak corresponds to the compound containing ^{37}Cl.

(b) The singlet at δ 3.8 suggests a methyl ether. The doublet of doublets centered at δ 7 suggest a *para*-disubstituted benzene ring. The complex pattern centered at δ 6.1 corresponds to two vinylic hydrogens, and the UV spectrum suggests a vinylic group conjugated with the benzene ring. The chemical shift of the δ 1.9 resonance suggests a vinylic methyl group, and its splitting of 6.6 Hz shows that this methyl group is coupled to one of the vinylic hydrogens. The 15.7 Hz splitting within the vinylic proton resonance shows that these protons are trans; the 965 cm^{-1} IR absorption confirms this deduction. All of these data conspire to suggest that compound *B* is (*E*)-4-methoxy-1-(1-propenyl)benzene, known more commonly as *trans-p*-methoxy-β-methylstyrene.

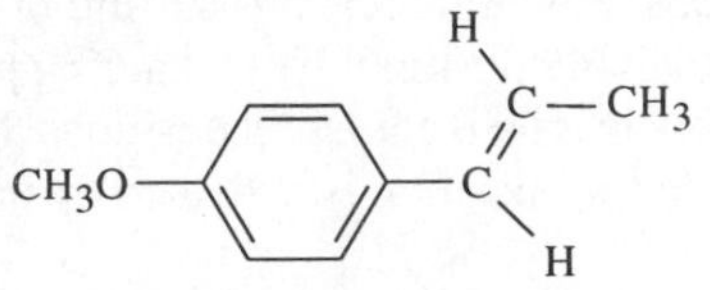

(*E*)-4-methoxy-1-(1-propenyl)benzene
trans-p-methoxy-β-methylstyrene

16.51 (a) *p*-Dibromobenzene can give only one mononitro derivative; hence, it must be compound *A*. *o*-Dibromobenzene can give two mononitro derivatives, and is therefore compound *B*. *m*-Dibromobenzene is compound *C*.

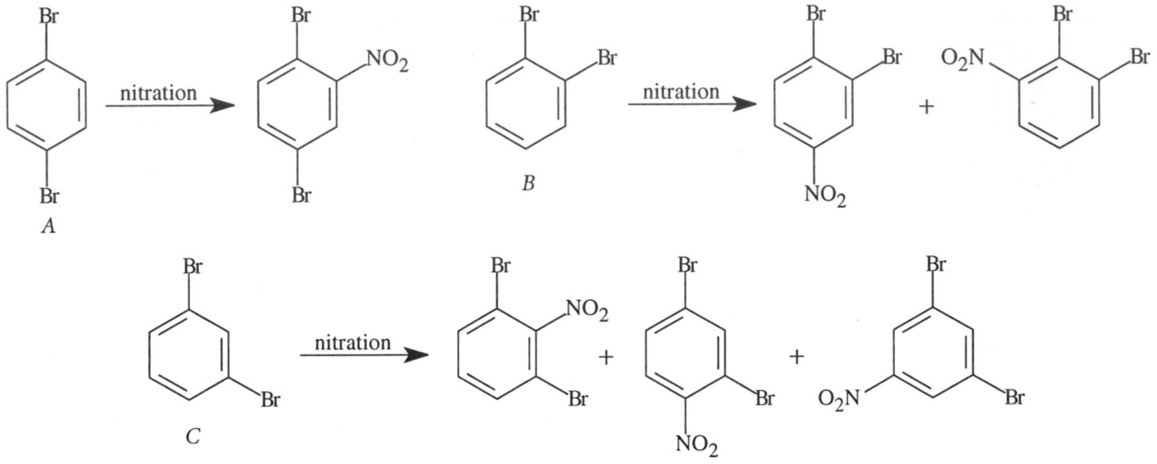

(b) Answer this question by deciding in how many different ways a single nitro group may be substituted for a hydrogen in each isomer. Because isomer separation methods (such as crystallization) based on conventional physical properties were used to differentiate isomers, enantiomeric differences between products were not evident; that is, only constitutional isomers or diastereomers are considered to be different compounds.

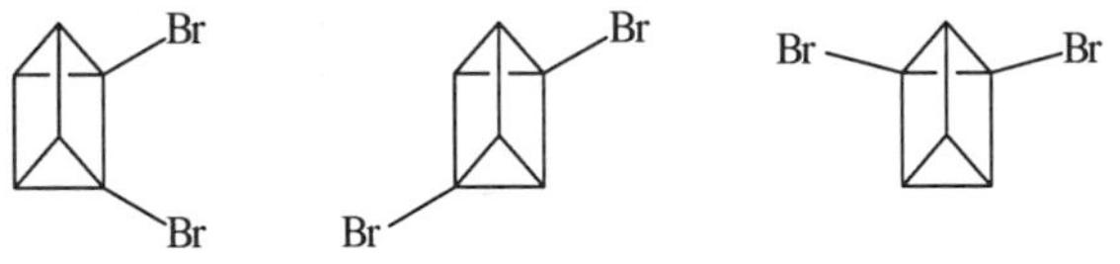

Compound *A* could give only one mononitro derivative; compound *B* could give two separable mononitro derivatives; and compound *C* could give three. (Can you draw the structures of these derivatives?)

(c) The following two products derived from nitration of compound *C* were probably formed in smallest amount. In the formation of compound *C1*, the nitro group and the two bromines are involved in severe van der Waals repulsions; and the formation of compound *C2* satisfies the directing effect of neither bromine substituent.

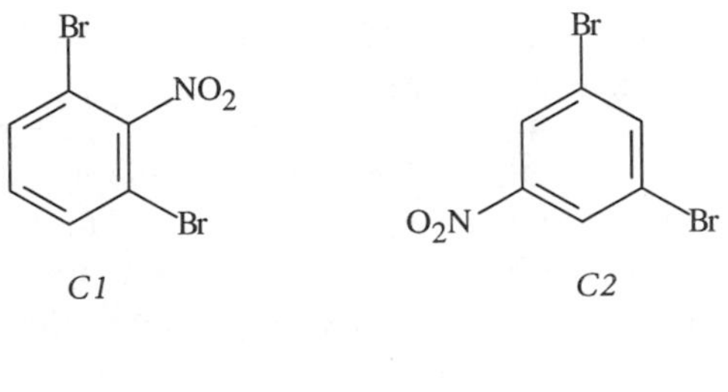

(d) The three dibromobenzene isomers would be differentiated most readily by the number of resonances in their proton-decoupled CMR spectra. Compound *A*, the para isomer, has two resonances; compound *B*, the ortho isomer, has three resonances; and compound *C*, the meta isomer, has four resonances. The number of resonances in each case, of course, corresponds to the number of chemically nonequivalent sets of carbons.

16.52 (a) The required $\Delta H°$ is the sum of the $\Delta H°$ values in Eqs. 16.46a–c: –205 kJ/mol (–49.0 kcal/mol).

(b) The $\Delta H°$ for hydrogenation of three cyclohexenes is $3 \times (-118) = -354$ kJ/mol (–84.6 kcal/mol).

(c) The discrepancy between these two quantities is 149 kJ/mol (35.6 kcal/mol); that is, about 149 kJ/mol (36 kcal/mol) less heat is liberated when benzene is hydrogenated than would be expected from the hydrogenation of three cyclohexenes. This number is remarkably close to the empirical resonance energy of benzene calculated by comparison of its heat of formation with that of cyclooctatetraene (COT) on text page 677.

16.53 (a) Each alkyl halide group reacts with benzene in a separate Friedel-Crafts alkylation reaction.

(b) The reaction is an intramolecular Friedel-Crafts alkylation that forms a ring. Note that the oxygen activates the substitution.

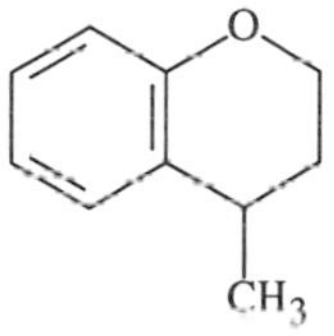

(c) Comparison of the formula of the product to that of naphthalene shows that one molar equivalent of the acylating agent has been introduced. Since there is no chlorine in the product, the reaction must involve a double acylation of the naphthalene ring by both ends of the acid chloride. The only way that this can occur with the formation of rings of reasonable size is for the acylation to occur across ortho positions of one benzene ring, or across the peri positions, the two positions on either side of the ring junction. These three possibilities account for the three products:

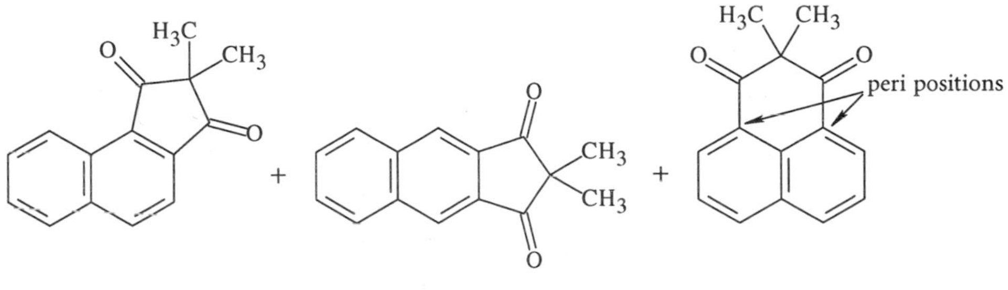

(d) The cyclohexyl group, an alkyl group, directs nitration to the ortho and para positions of the benzene ring.

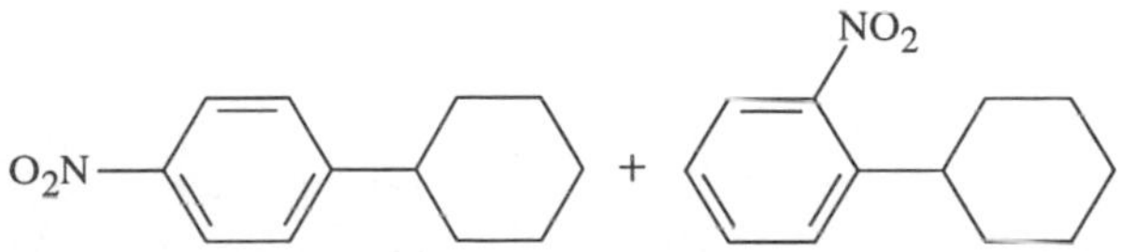

(mostly)

(e) Ferrocene, like other aromatic compounds, undergoes Friedel-Crafts acylation reactions.

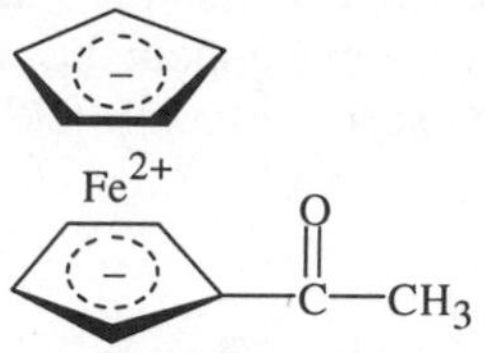

(f) The nitro group is directed by both substituents to the position ortho to the methoxy group, and the bromine in the second reaction is directed to the other position ortho to the methoxy group.

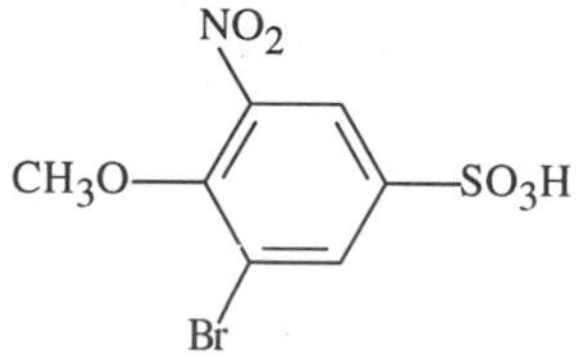

3-bromo-4-methoxy-5-nitrobenzenesulfonic acid

16.54 (a) Because carbon-4 is para to the methoxy group, the carbocation intermediate involved in nitration at this carbon can be stabilized by electron delocalization from the oxygen, as it is in nitration at the para position of anisole. (See the colored structure in Eq. 16.29 on text p. 718) Hence, the methoxy group activates substitution at carbon-4, and nitration at carbon-4 of 1-methoxynaphthalene is faster than nitration of naphthalene.

(b) When nitration occurs at carbon-5, the unshared electrons of the oxygen can be used to stabilize the carbocation intermediate by resonance; consequently, nitration at carbon-5 of 1-methoxynaphthalene is faster than nitration of naphthalene itself.

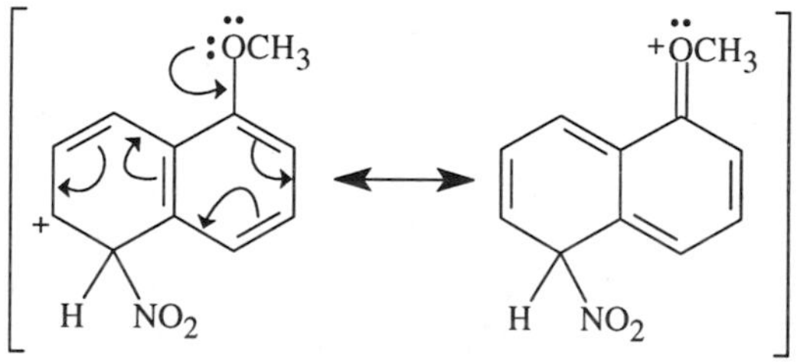

(c) Nitration at carbon-6, in contrast, is analogous to nitration at a meta position of anisole; see Eq. 16.30 on text p. 719. In this case, the oxygen electrons *cannot* be delocalized to stabilize the carbocation. The only effect of the methoxy group in this case is its rate-retarding polar effect. Therefore, nitration at carbon-6 of 1-methoxynaphthalene is slower than nitration of naphthalene.

positive charge must be delocalized to this carbon for oxygen electrons to be involved in resonance stabilization

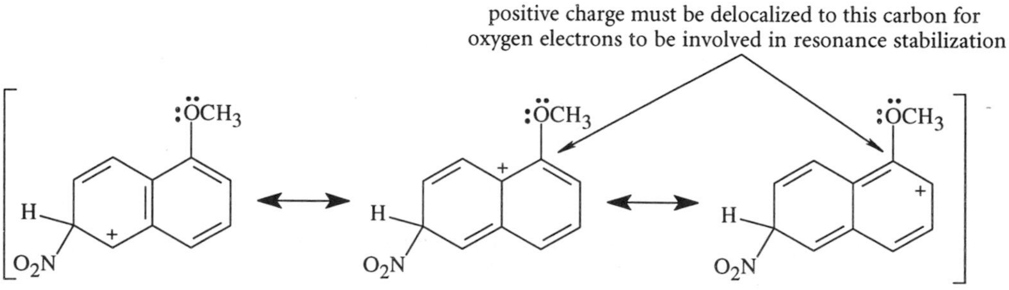

16.55 Friedel-Crafts acylation (or any other electrophilic substitution reaction) at carbon-2 of furan gives a carbocation intermediate with more resonance structures than acylation at carbon-3. Consequently, the carbocation intermediate from acylation at carbon-2 is more stable, and, by Hammond's postulate, acylation at carbon-2 is faster.

Carbocation intermediate from acylation at carbon-2:

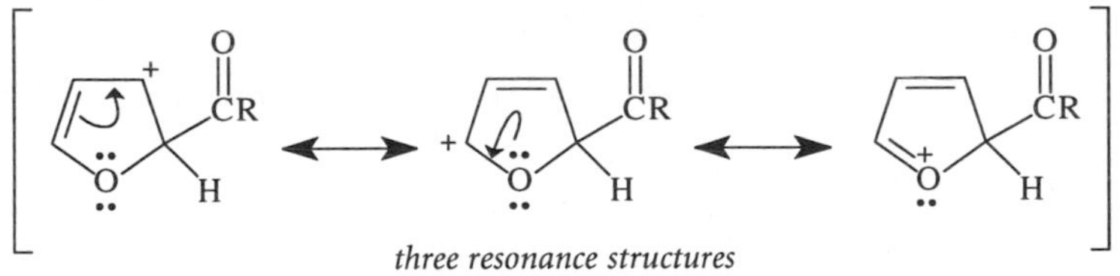

three resonance structures

Carbocation intermediate from acylation at carbon-3:

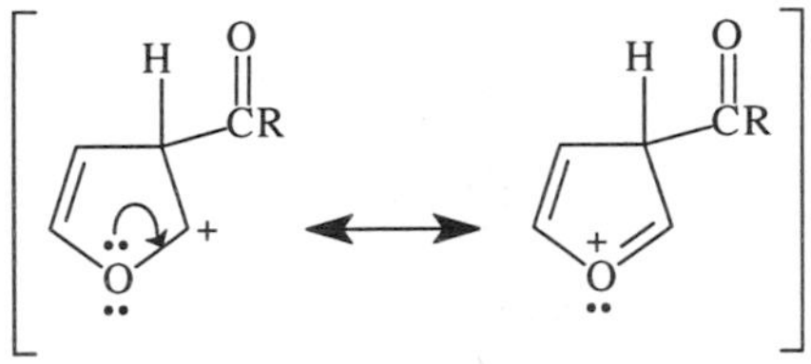

two resonance structures

16.56 The balance between protonation on oxygen and protonation on a carbon of the ring depends on the relative basicities of an aromatic "double bond" and the oxygen of an aromatic ether. The fact that anisole protonates on oxygen shows that an ether oxygen is more basic than an aromatic "double bond."

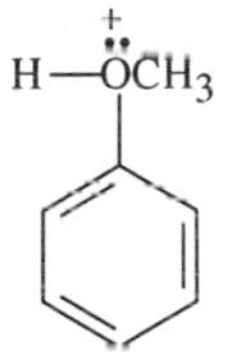

conjugate acid of anisole

The fact that 1,3,5-trimethoxybenzene protonates on carbon, then, means that the resulting conjugate-acid cation is stabilized sufficiently that the "double bonds" of the aromatic ring have enhanced basicity. The stability of this cation is due to resonance stabilization involving all three oxygens:

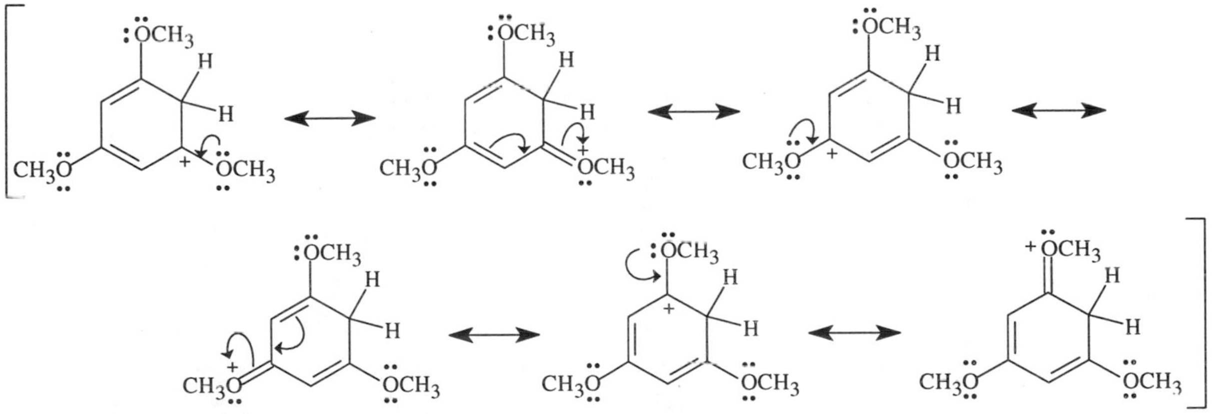

When anisole protonates on a ring carbon, the oxygen can also stabilize the resulting carbocation by analogous resonance interaction; however, in that case, the resonance effect of only one oxygen is available for such stabilization, and evidently that is not sufficient to bring about protonation on carbon.

16.57 (a) Furan derivatives, like 1,3-cyclopentadiene derivatives, are in effect conjugated dienes locked into *s*-cis conformations. Such dienes are reactive in Diels-Alder reactions. The product *A*, shown at the top of the following page, results from such a reaction.

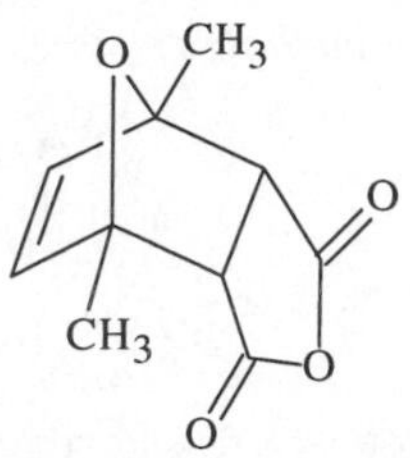

compound *A*

(b) Evidently, the acidic conditions promote loss of the bridging ether oxygen as water. The mechanistic steps involved are protonation, carbocation formation, loss of a hydrogen, and then repetition of a similar sequence. The driving force for this reaction is, of course, the formation of the aromatic ring.

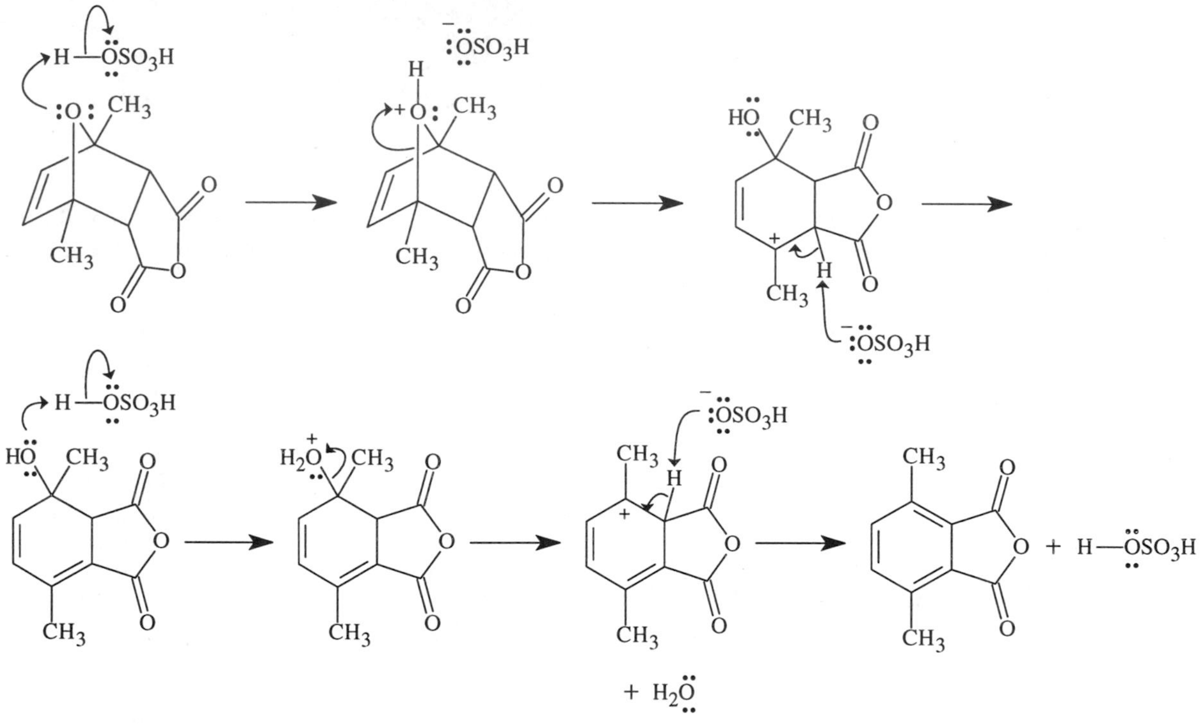

16.58 In this reaction a *tert*-butyl cation is lost rather than a proton from the carbocation intermediate. The electrophile, a nitronium ion $^{+}NO_2$, is generated by the mechanism shown in Eqs. 16.10a–b on text p. 708.

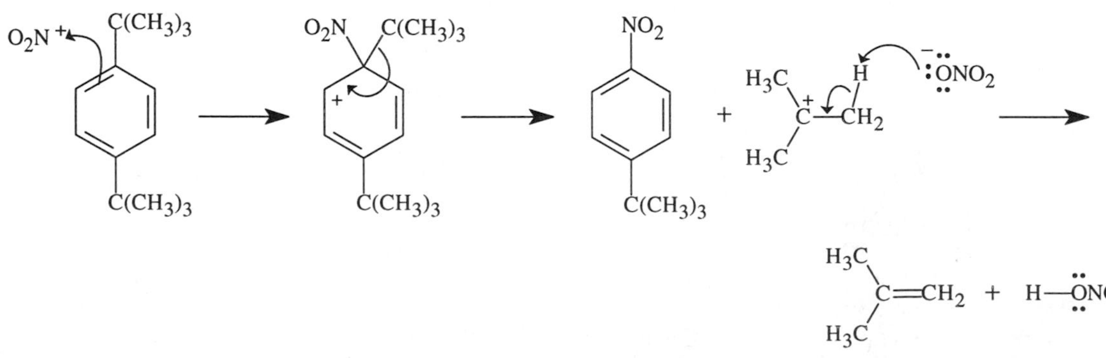

16.59 For this product to form, benzene must react with the electrophile shown in the following equation:

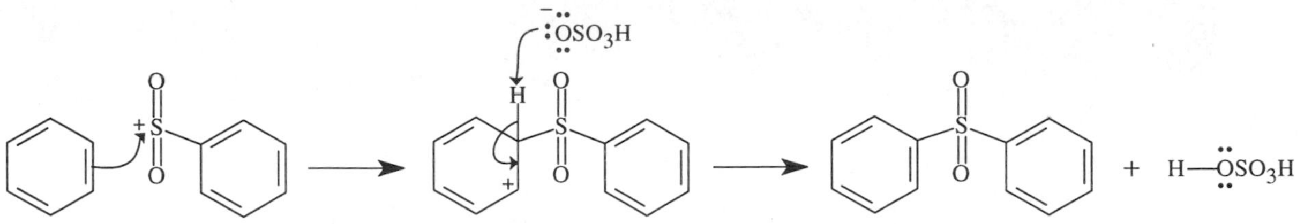

The electrophile, in turn, is formed from benzenesulfonic acid by the following mechanism:

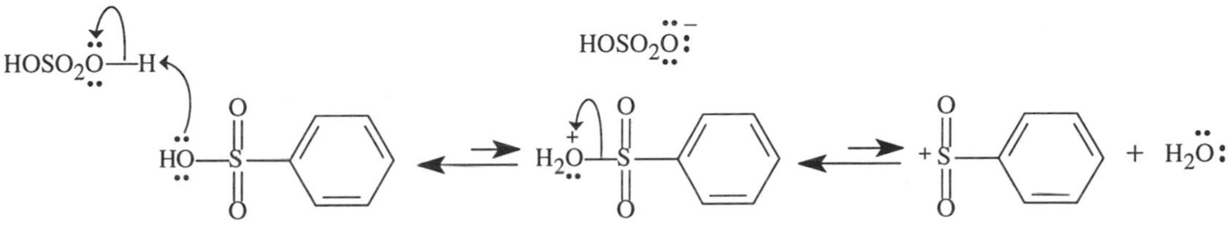

16.60 At the higher temperature, the two ortho methyl groups (labeled *a* in the problem) are chemically equivalent on the NMR time scale because rotation of the isopropyl group is rapid; the resonance of these groups occurs at δ 2.25. However, at the lower temperature, rotation of the isopropyl group is slow, and the two ortho methyl groups are no longer chemically equivalent. Hence, the resonances of the ring methyl groups occur at different chemical shifts at the lower temperature. Evidently, one ortho methyl group is wedged between the two isopropyl methyls, and the other is in the same plane as the isopropyl C—H group, as shown by the structure in the problem.

16.61 (a) The two terminal rings of hexahelicene cannot lie in the same plane because, if they did, their hydrogens would have severe van der Waals repulsions. Consequently, the molecule is somewhat bent out of plane so that one of the terminal rings lies over the other. This causes the molecule to trace one turn of a helix, which is a chiral object. This optical rotation, by the way, is *huge*!

(b) Normally the two rings of a biphenyl derivative lie in the same plane to maximize conjugation, that is, to maximize overlap of their π-electron systems. In this case, however, if the rings were to lie in the same plane, the very large sulfonic acid ($-SO_3H$) group on one ring would have severe van der Waals repulsions with an ortho hydrogen of the other ring. Consequently, the molecule adopts the conformation shown in the problem, in which the planes of the two rings are perpendicular. However, this conformation is chiral and is thus capable of showing optical activity. Evidently, rotation about the central carbon-carbon bond is slow enough that the molecule can be resolved into isolable enantiomeric conformations! When the molecule is heated, the internal rotation "reaction" (like all reactions) is accelerated, and the optically active molecule racemizes. This situation is also discussed in Study Guide Link 6.7 on page 161 of this manual; a similar situation is the subject of Problem 16.9(b) earlier in this chapter.

17

Allylic and Benzylic Reactivity

TERMS

CONCEPTS

I. ALLYLIC AND BENZYLIC SPECIES

A. Introduction

1. An allylic group is a group on a carbon adjacent to a double bond.
2. A benzylic group is a group on a carbon adjacent to a benzene ring or substituted benzene ring.
3. In many situations allylic and benzylic groups are unusually reactive.

B. Allylic and Benzylic Cations, Radicals, and Anions

1. Allylic and benzylic carbocations are resonance-stabilized.
   a. The charge on an allylic carbocation is shared between two carbons.

$$\left[R—CH{=}CH—\overset{+}{C}H—R' \longleftrightarrow R—\overset{+}{C}H—CH{=}CH—R' \right]$$

an allylic cation

   b. The charge on a benzylic carbocation is shared not only by the benzylic carbon, but also by alternate carbons of the ring.

[benzyl cation resonance structures]

benzyl cation

2. Allylic and benzylic radicals are resonance-stabilized; they are more stable, and thus more readily formed as reactive intermediates, than ordinary alkyl radicals.

   a. The unpaired electron of an allylic radical is shared between two carbons.

$$\left[R-\dot{C}H-CH=CH-R' \longleftrightarrow R-CH=CH-\dot{C}H-R' \right]$$

an allylic radical

   b. The unpaired electron of a benzylic radical is shared by the benzylic carbon and alternate carbons of the ring.

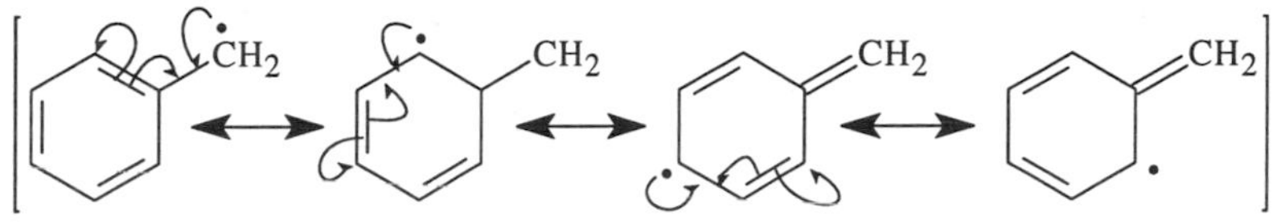

a benzylic radical

3. Allylic and benzylic anions are resonance-stabilized and thus more stable than their nonallylic and nonbenzylic counterparts.

   a. The negative charge in such species is shared by alternate carbons.

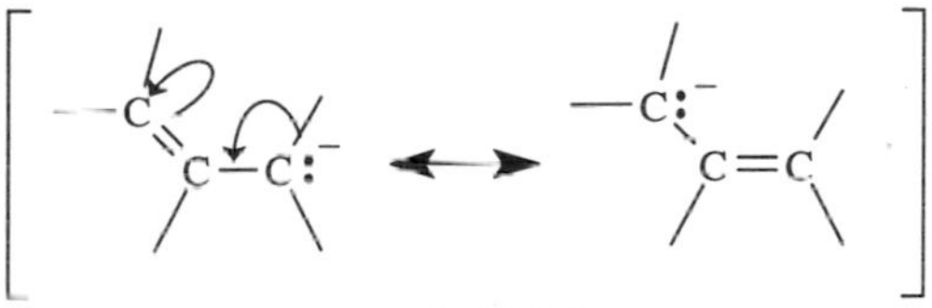

an allylic anion

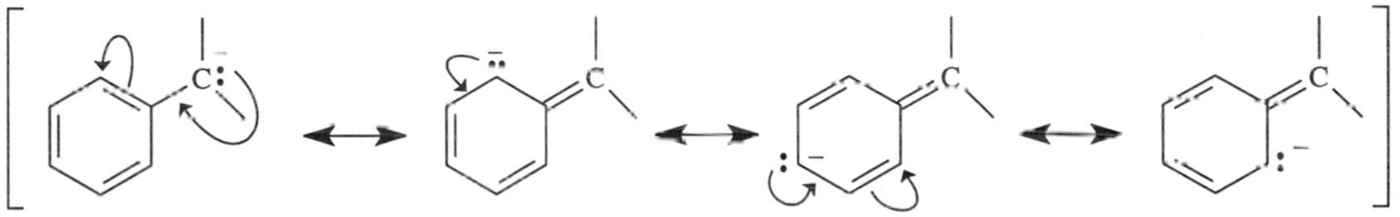

a benzylic anion

   b. The polar effect of the double bond (in the allyl anion) or the phenyl ring (in the benzyl anion) also stabilizes the anion.

   c. Although hydrocarbons with allylic or benzylic hydrogens are very weak acids, their acidities are much greater than the acidities of alkanes that do not contain such hydrogens.

B. Allylic and Benzylic S_N1 Reactions

1. Solvolysis of an allylic or benzylic alkyl halide or sulfonate ester involves an allylic or benzylic carbocation intermediate, respectively.

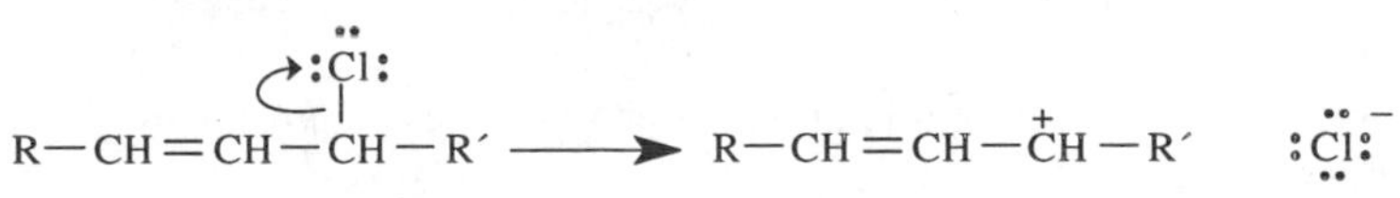

$$R-CH=CH-\underset{}{CH}(Cl)-R' \longrightarrow R-CH=CH-\overset{+}{C}H-R' \quad :\ddot{Cl}:^-$$

2. Two different products are formed when a nucleophile attacks an unsymmetrical allylic cation, because there are two different electron-deficient carbons.

$$\left[R\text{—}CH\text{=}CH\text{—}\overset{+}{C}H\text{—}R' \longleftrightarrow R\text{—}\overset{+}{C}H\text{—}CH\text{=}CH\text{—}R' \right]$$

$$\downarrow \text{Nuc:}^- \qquad\qquad \downarrow \text{Nuc:}^-$$

$$R\text{—}CH\text{=}CH\text{—}\underset{\text{Nuc}}{CH}\text{—}R' \qquad R\text{—}\underset{\text{Nuc}}{CH}\text{—}CH\text{=}CH\text{—}R'$$

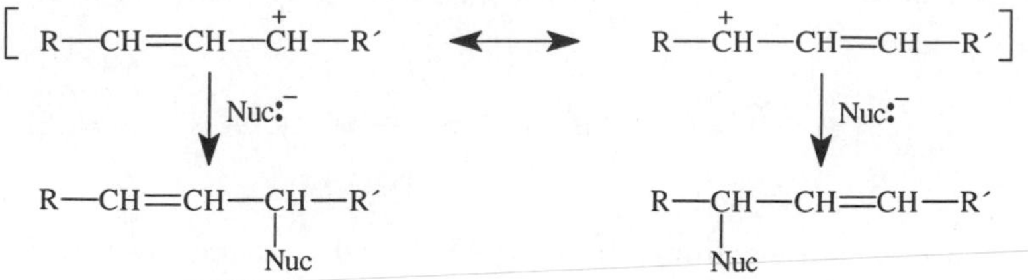

3. One product is formed when a nucleophile attacks an unsymmetrical benzylic cation, because the products formed by attack of the nucleophile at the other electron-deficient carbons are not aromatic and thus lack the stability associated with the aromatic ring.

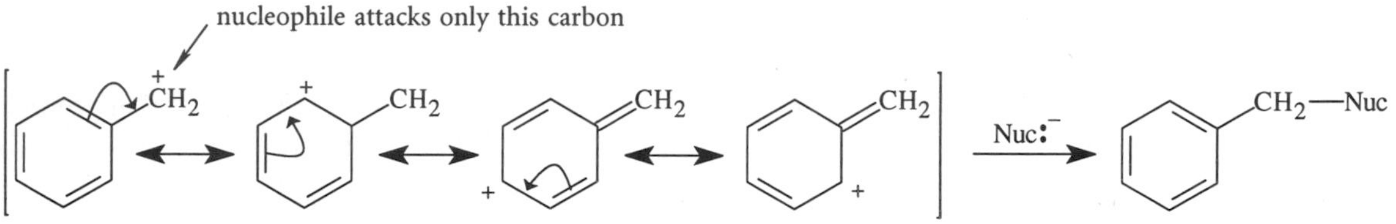

nucleophile attacks only this carbon

$\overset{+}{C}H_2$ CH_2 CH_2 CH_2 Nuc:$^-$ CH_2—Nuc

4. Reactions involving benzylic or allylic carbocations as intermediates are generally considerably faster than analogous reactions involving comparably substituted nonallylic or nonbenzylic carbocations.
   a. The greater reactivities of allylic and benzylic halides in S_N1-E1 reactions are due to stabilities of the carbocation intermediates that are formed when they react; more stable carbocation are formed more rapidly than less stable ones (Hammond's postulate).
   b. Ortho and para substituent groups on the benzene ring that activate electrophilic aromatic substitution accelerate S_N1 reactions at the benzylic position; in the intermediate carbocation, additional resonance structures in which charge can be delocalized onto the substituent group itself are possible.

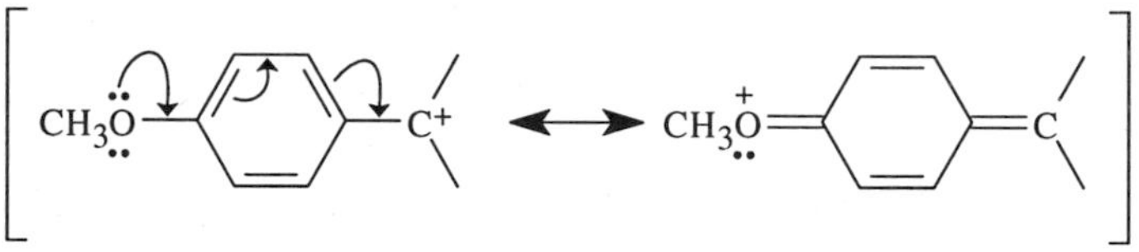

$CH_3\ddot{O}$ C^+ $\longleftrightarrow$ $CH_3\overset{+}{O}$ C

stabilizing interaction of a para substituent with a benzylic cation

C. Allylic and Benzylic S_N2 Reactions

1. S_N2 reactions of allylic and benzylic halides are relatively fast even though they do not involve reactive intermediates.

2. Allylic and benzylic S_N2 reactions are accelerated because the energies of their transition states are reduced by *p*-orbital overlap.
   a. In the transition state of the S_N2 reaction, the carbon at which substitution occurs is sp^2-hybridized.
   b. The incoming nucleophile and the departing leaving group are partially bonded to a 2*p* orbital on this carbon.
   c. Overlap of this 2*p* orbital with the 2*p* orbitals of an adjacent double bond or phenyl ring provides additional bonding that lowers the energy of the transition state and accelerates the reaction.

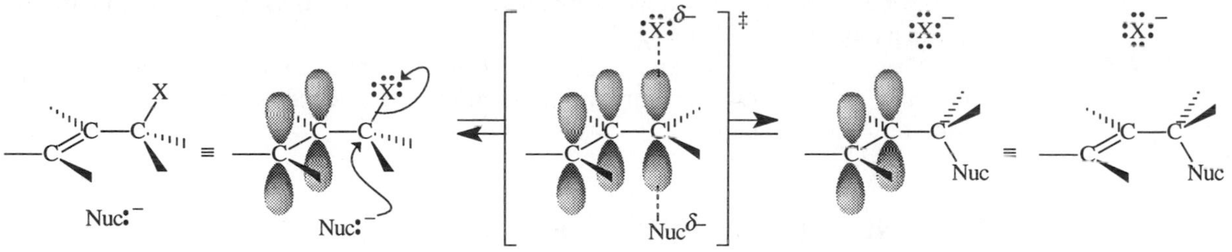

D. Allylic and Benzylic E2 Reactions

1. The acidity of the β-hydrogens is a structural effect in an alkyl halide that tends to promote a greater fraction of elimination.
   a. In the E2 reaction, the base removes a β-proton, and the transition state of the reaction thus has carbanion character at the β-carbon atom.
   b. A greater ratio of elimination to substitution is observed when the β-hydrogens of an alkyl halide have higher than normal acidity.

2. Another factor that promotes elimination is that the alkene double bond, which is partially formed in the transition state, is conjugated with the benzene ring or the double bond of the allylic group.

E. Allylic and Benzylic Radicals

1. Free-radical reactions at allylic and benzylic positions are accelerated because of the relative stability of the allylic or benzylic free-radical intermediates that are involved.

2. Free-radical substitution at the allylic position of an alkene is in competition with bromine addition to the alkene double bond.
   a. Addition of bromine is the predominant reaction if:
      i. free-radical substitution is suppressed by avoiding conditions that promote free-radical reactions (light, heat, or free radical initiators).
      ii. the reaction is carried out in solvents of even slight polarity that promote the ionic mechanism for bromine addition.
   b. When a compound with allylic hydrogens is treated with *N*-bromosuccinimide (NBS) in CCl_4 under free-radical conditions, allylic bromination takes place, and addition to the double bond is not observed. (See Reactions I.A.)

F. Allylic Grignard Reagents

1. Allylic Grignard reagents undergo a rapid equilibrium in which the —MgBr group and the double bond change positions. This is an example of an allylic rearrangement.

$$H_3C{-}CH{=}CH{-}CH_2{-}Br \xrightarrow[\text{ether}]{\text{Mg}} \underbrace{H_3C{-}CH{=}CH{-}\underset{|}{\overset{\text{MgBr}}{C}}H_2 \rightleftharpoons H_3C{-}\overset{\text{MgBr}}{\underset{|}{C}}H{-}CH{=}CH_2}_{\text{allylic rearrangement}}$$

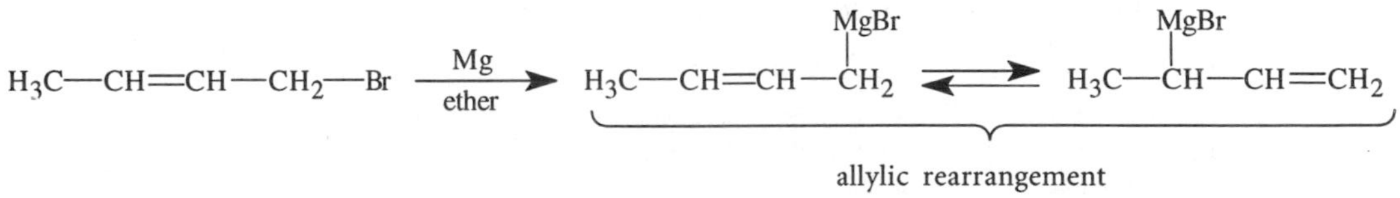

   a. An allylic rearrangement involves the simultaneous movement of an allylic group and a double bond so that one allylic isomer is converted into another.
      i. These two structures are not resonance structures; they are two distinct species in rapid equilibrium.
      ii. Because of this equilibrium, an unsymmetrical allylic Grignard reagent is a mixture of two different reagents.
   b. The same mixture of reagents can be obtained from either of two allylically related alkyl halides.

c. The same mixture of products is obtained when the Grignard reagent is prepared from either allylic halide.
2. Allylic Grignard reagents resemble allylic carbanions; because the allylic carbanion is resonance-stabilized, the transition state for equilibration has relatively low energy, and consequently the equilibration occurs rapidly.

II. THE ISOPRENE RULE; BIOSYNTHESIS OF TERPENES

A. Essential Oils and Terpenes

1. Some pleasant-smelling substances found in nature have come to be called essential oils.
   a. Terpenes (also called isoprenoids) are a class of natural products with similar atomic composition; many of these compounds are familiar natural flavorings or fragrances.
   b. All terpenes consist of repeating units that have the same carbon skeleton as the five-carbon diene isoprene.
2. The basis of the terpene or isoprenoid classification is only the connectivity of the carbon skeleton.
   a. The presence of double bonds and other functional groups, or the configurations of double bonds and asymmetric carbons, have nothing to do with the terpene classification.
   b. All terpenes consist of repeating units that have the same carbon skeleton as the five-carbon diene isoprene (the isoprene rule).
      i. The carbons at the ends of the isoprene skeleton are classified as carbon-1 and carbon-4, with carbon-4 being either carbon of the dimethyl branch.

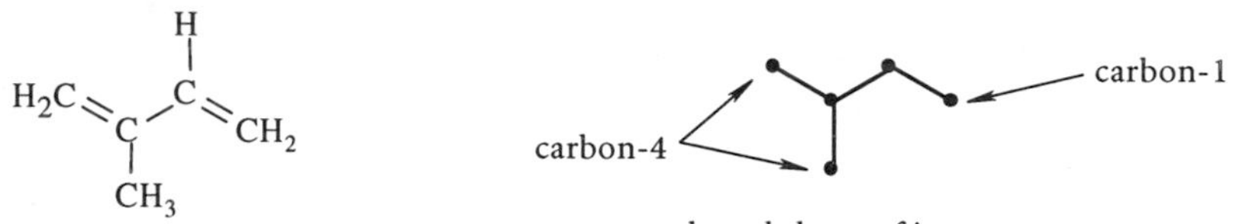

H
H_2C C C CH_2
CH_3
isoprene

carbon-1
carbon-4
carbon skeleton of isoprene

      ii. In many terpenes, the isoprene units are connected in a "head-to-tail" (or carbon-1´ to carbon-4) arrangement. (The prime (´) on one number and its absence on the other mean that the connection is between different isoprene units.)
      iii. Many examples are known in which the isoprene units have a "head-to-head" (or carbon-1´ to carbon-1) connectivity.
   c. Some compounds are derived from the conventional terpene structures by skeletal rearrangements.
   d. Terpene carbon skeletons contain multiples of five carbon atoms (10, 15, 20, ... 5*n*).
      i. Monoterpenes have ten carbon atoms in their carbon chains.
      ii. Sesquiterpenes have fifteen carbon atoms in their carbon chains.
      iii. Diterpenes have twenty carbon atoms in their carbon chains.
3. Criteria for terpene classification are:
   a. A multiple of five carbon atoms in the main carbon chain.
   b. The carbon connectivity of the isoprene carbon skeleton within each five-carbon unit.

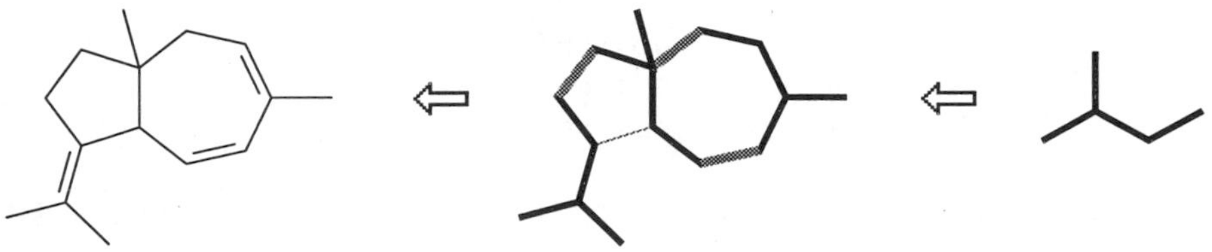

daucene
(a sesquiterpene)

isoprene
carbon skeleton

B. Biosynthesis of Terpenes

1. Biosynthesis is the synthesis of chemical compounds by living organisms.
2. The repetitive isoprene units in all terpenes have a common origin in two simple five-carbon compounds (structures on text page 809, Eq. 17.35):
   a. Isopentenyl pyrophosphate (IPP).
   b. γ,γ-Dimethylallyl pyrophosphate (DMAP).
   c. Alkyl pyrophosphates (R—OPP) are esters of the inorganic acid pyrophosphoric acid; —OPP is an abbreviation of the pyrophosphate group.
      i. Pyrophosphate and phosphate are "nature's leaving groups."
      ii. IPP and DMAP are readily interconverted in living systems by enzyme catalysis.
3. The reactions that connect isoprene units in nature typically involve allylic carbocation intermediates. (B: = a basic group on an enzyme.)

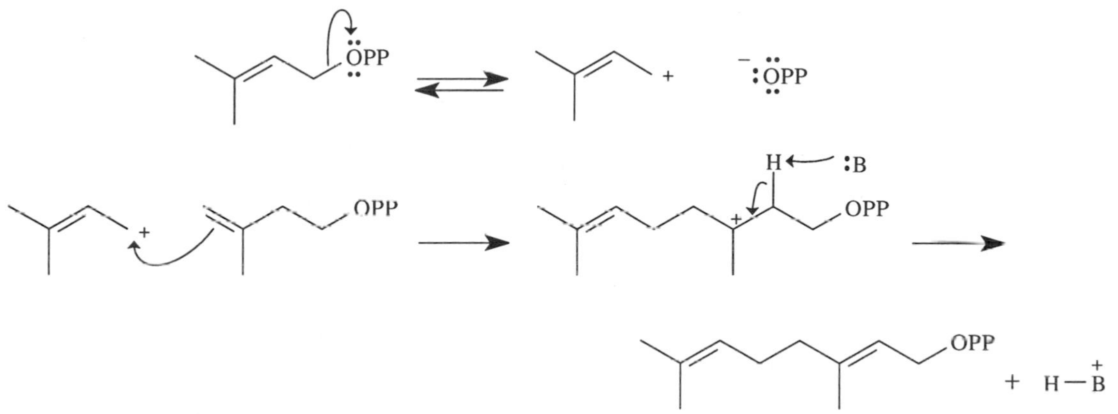

OPP ⇌ + + :OPP⁻

OPP → + OPP, H ← :B →

OPP + $H-\overset{+}{B}$

REACTIONS

I. ALLYLIC AND BENZYLIC SUBSTITUTION REACTIONS

A. Allylic and Benzylic Bromination with NBS

1. *N*-Bromosuccinimide (NBS) can be used to bring about substitution of allylic and benzylic hydrogens with bromine.

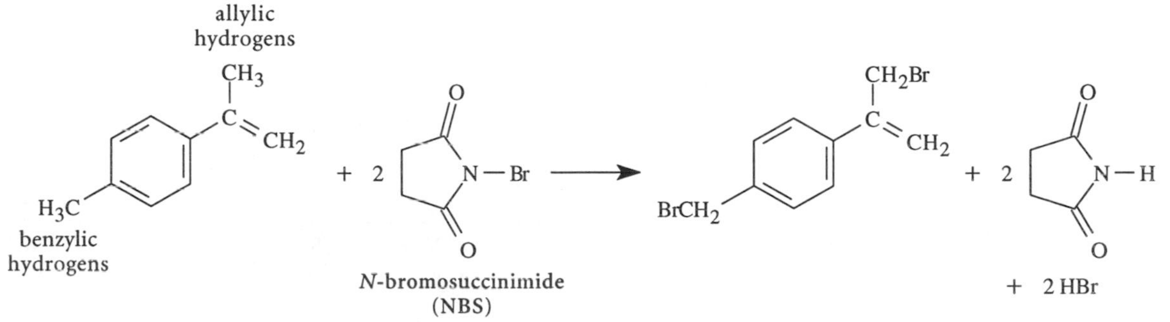

allylic hydrogens; CH_3; $C=CH_2$; H_3C; benzylic hydrogens; + 2 N—Br (*N*-bromosuccinimide (NBS)) → CH_2Br; $C=CH_2$; $BrCH_2$; + 2 N—H; + 2 HBr

2. The mechanism of allylic or benzylic bromination with NBS entails the following steps:
   a. The initiation step is the formation of a bromine atom by homolytic cleavage of the N—Br bond in NBS itself.

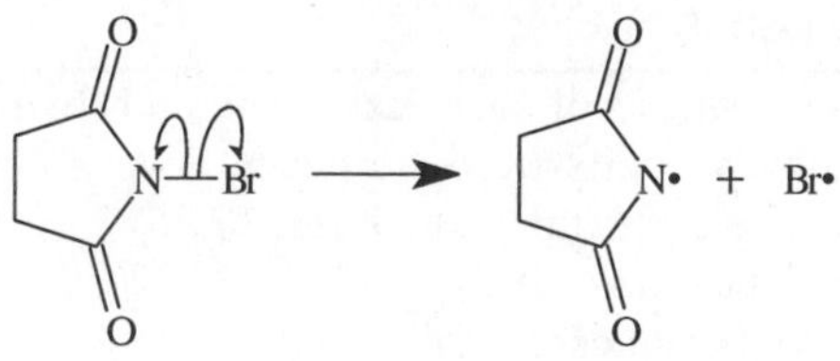

b. The ensuing substitution reaction has three propagation steps:

i. The bromine atom abstracts an allylic or a benzylic hydrogen.

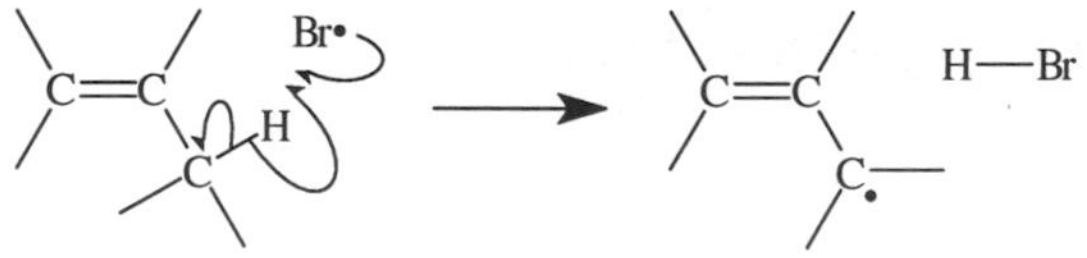

ii. The HBr thus formed reacts with the NBS in the second propagation step (by an ionic mechanism) to produce a Br_2 molecule.

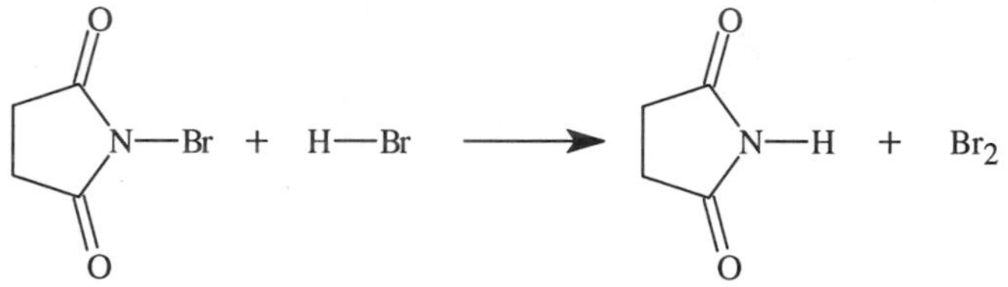

iii. The last propagation step is the reaction of this bromine molecule with the radical formed in the first propagation step.

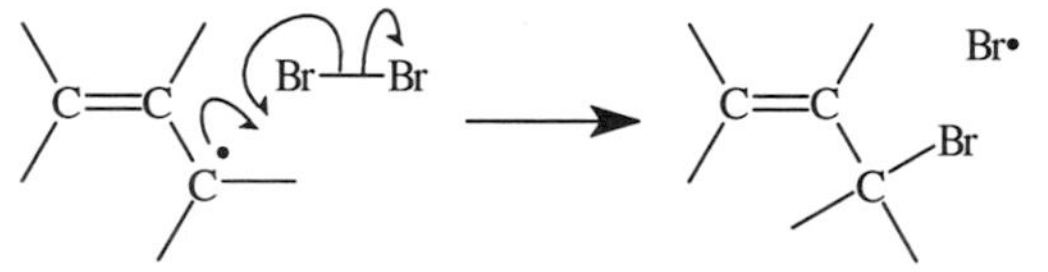

c. The Br_2 concentration remains low because it can be generated no faster than an HBr molecule and an allylic radical are generated.

d. The low solubility of NBS in CCl_4 is crucial to the success of allylic bromination with NBS.

3. Because the unpaired electron of an unsymmetrical allylic radical is shared by two chemically nonequivalent carbons, such a radical can react to give two constitutionally isomeric products.

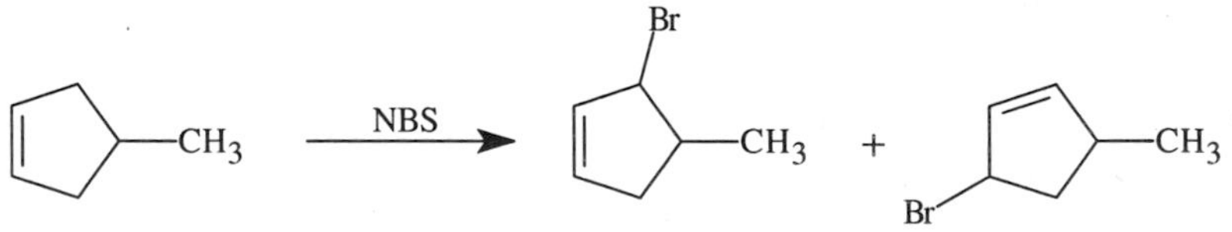

B. Benzylic Bromination

1. Benzylic hydrogens can be substituted at allylic positions with Br_2 in the presence of light.

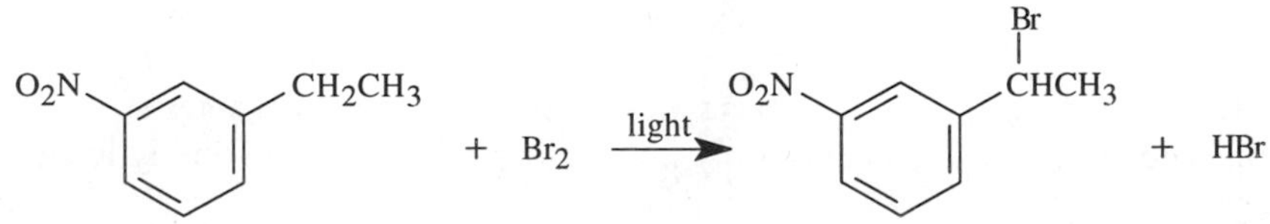

2. This is a free-radical chain reaction that involves the following propagation steps:
   a. abstraction of a benzylic hydrogen by a bromine atom in preference to a nonbenzylic hydrogen.

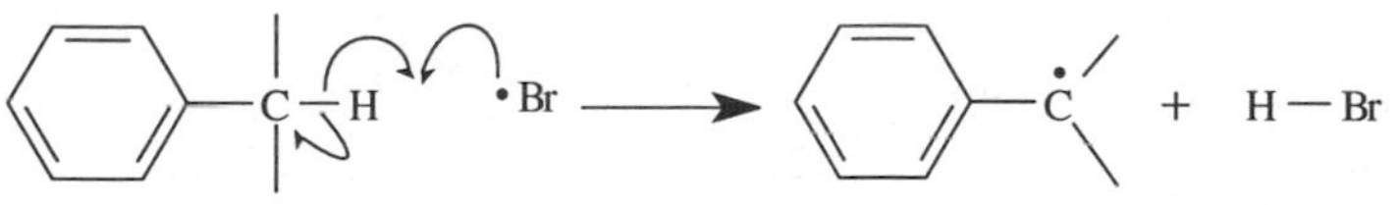

   i. This propagation step gives rise to the selectivity for substitution of the benzylic hydrogen.
   ii. The reason for this selectivity is that the benzylic radical which is formed has greater stability than nonbenzylic radicals.

   b. reaction of the benzylic radical with a molecule of bromine to generate a molecule of product along with another bromine atom, which can then react with another benzylic hydrogen.

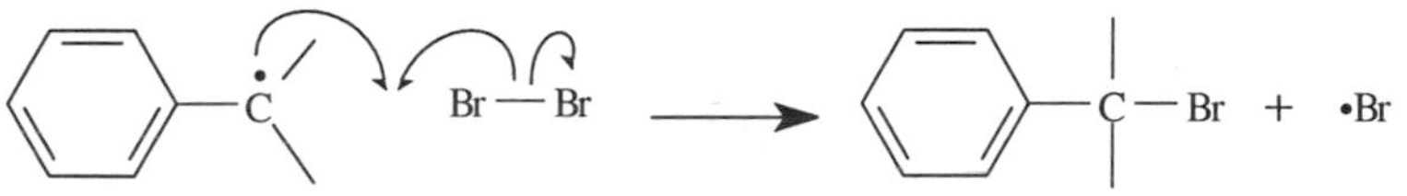

3. Allylic bromination also occurs under these conditions but is generally not used because of the competition with bromine addition to the double bond; NBS bromination is the preferred method for carrying out allylic bromination.

II. BENZYLIC OXIDATION REACTIONS

A. Side-Chain Oxidation of Alkylbenzenes

1. Treatment of alkylbenzene derivatives with strong oxidizing agents under vigorous conditions converts the alkyl side chain into a carboxylic acid group.

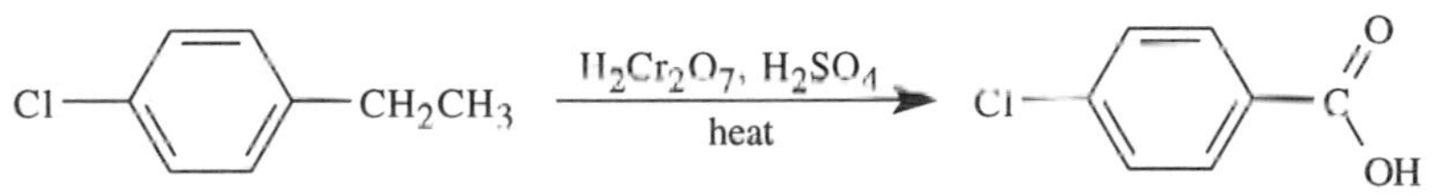

   a. The benzene ring is left intact.
   b. The alkyl side chain, regardless of length, is converted into a carboxylic acid group.
   c. The conditions for this side-chain oxidation are generally vigorous: heat, high concentrations of oxidant, and/or long reaction times.
   d. Common oxidants are Na_2CrO_7 (sodium dichromate), CrO_3 (chromium trioxide), $KMnO_4$ (potassium permanganate), or O_2 and special catalysts.

2. Oxidation of alkyl side chains requires the presence of a benzylic hydrogen; benzene derivatives with no benzylic hydrogens are resistant to side-chain oxidation.

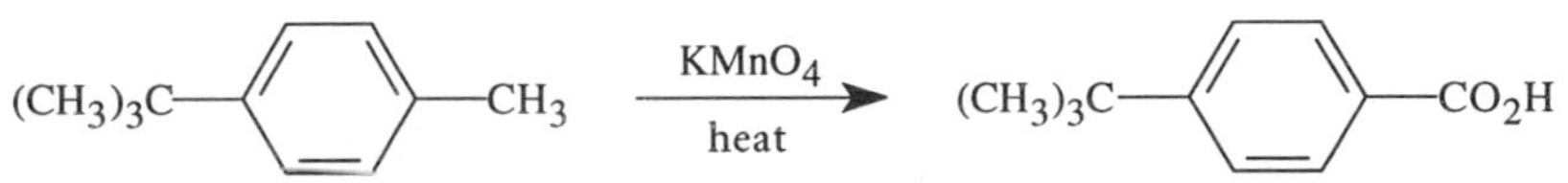

SCIENTISTS

Moses Gomberg (1866–1947) P17.38
Charles Dale Poulter (b. 1942) 17.6A
Hans Christopher Rilling (b. 1933) 17.6B
Otto Wallach (1847–1931) [Chemistry 1910] 17.6A

Dates in brackets refer to Nobel Prize awards.

STUDY GUIDE LINKS

17.1 Addition versus Substitution with Bromine

The purpose of this Study Guide Link is to explain further the effect of low bromine concentration in promoting free-radical substitution. The explanation of this effect lies in the rate laws for addition and substitution. The rate law for *ionic addition* (Sec. 5.1A) at low bromine concentrations is

$$\text{rate of ionic addition} = k_{\text{addition}}[\text{alkene}][\text{Br}_2]$$

Although it is not discussed in the text, addition of Br_2 can also take place by a free-radical chain mechanism. At low Br_2 concentrations, the rate of free-radical addition follows the rate law

$$\text{rate of free-radical addition} = k_{\text{fr-addition}}[\text{alkene}][\text{Br}_2]^{3/2}$$

The rather strange-looking 3/2-order in bromine, roughly speaking, is due to the involvement of a bromine atom ("half of a bromine molecule") produced in the initiation step together with the involvement of a bromine molecule in the rate-limiting propagation step—one and one-half bromine molecules total.

Finally, the rate of free-radical substitution at low Br_2 concentrations has the rate law

$$\text{rate of free-radical substitution} = k_{\text{substitution}}[\text{alkene}][\text{Br}_2]^{1/2}$$

The half-order dependence in bromine is due to the involvement of a bromine atom that is produced in the initiation step.

If you compare these rate laws, you will notice that the substitution reaction has the least dependence on bromine concentration. Hence, the rate of substitution not only *increases less rapidly* with *increasing* bromine concentration than the other reactions, but also *decreases less rapidly* with *decreasing* bromine concentration. At very low bromine concentration, then, free-radical substitution has the greatest rate, and is therefore the observed reaction.

The success of *N* bromosuccinimide (NBS) in CCl_4 as a reagent for substitution of allylic and benzylic hydrogens by bromine is due to the fact that it provides an experimentally convenient way to maintain a low bromine concentration. It wasn't planned that way; NBS in CCl_4 for many years was known as a good reagent for carrying out allylic and benzylic brominations with no competing additions. Careful research later developed the rationale presented here and in the text.

17.2 Polar Effect of Double Bonds

Why should the polar effect of a nearby double bond stabilize negative charge? Recall (Sec. 4.4, text p. 127) that the bond dipole between an sp^3-hybridized carbon atom and an sp^2-hybridized carbon is directed toward the double bond:

bond dipole
$H_2C{=}CH$
sp^3-hybridized carbon
sp^2-hybridized carbon

If a negative charge happens to be on the sp^3-hybridized carbon atom, it is stabilized by interaction with the positive end of the bond dipole.

At a deeper level, we might ask why a bond dipole is oriented from an sp^3-hybridized carbon toward an sp^2-hybridized carbon, as shown above. In an sp^2-hybridized carbon atom, one of the electrons is in a *p* orbital, and is thus relatively far from the nucleus. (Think of this electron as being partially "pulled away" or "peeled back" from the nucleus.) As a result, the electronic shielding of the positive nucleus in an sp^2-hybridized carbon atom is reduced—that is, the nucleus is more exposed. Such a nucleus has a stronger attraction for the electrons in neighboring σ-bonds than the nucleus of an sp^3-hybridized carbon atom, in which there is no *p* electron. The electrons in the σ-bond are thus pulled toward the sp^2-hybridized carbon and away from the sp^3-hybridized carbon, and a bond dipole results.

✓17.3 Synthetic Equivalence

The benzylic oxidation provides an excellent way to introduce a carboxylic acid group into a benzene ring. (None of the electrophilic aromatic substitution reactions in Chapter 16 provide a way to introduce this group directly.) Suppose, for example, we wish to prepare *p*-nitrobenzoic acid from benzene. The desired acid can be formed by benzylic oxidation of any *p*-alkylnitrobenzene with benzylic hydrogens—say, *p*-ethylnitrobenzene:

$$O_2N\text{-}C_6H_4\text{-}CH_2CH_3 \xrightarrow[2)\ H_3O^+]{1)\ KMnO_4,\ ^-OH} O_2N\text{-}C_6H_4\text{-}CO_2H$$

p-ethylnitrobenzene *p*-nitrobenzoic acid

The *p*-ethylnitrobenzene comes, in turn, from nitration of ethylbenzene:

$$C_6H_5\text{-}CH_2CH_3 \xrightarrow{HNO_3,\ H_2SO_4} O_2N\text{-}C_6H_4\text{-}CH_2CH_3$$

ethylbenzene *p*-ethylnitrobenzene

And the ethylbenzene comes from Friedel-Crafts alkylation of benzene (Eq. 16.20, text p. 712).

If we wanted to prepare *m*-nitrobenzoic acid, we would *first* oxidize ethylbenzene to benzoic acid, and then nitrate the benzoic acid, thus taking advantage of the *meta*-directing effect of the carboxylic acid group:

$$C_6H_5\text{-}CH_2CH_3 \xrightarrow[2)\ H_3O^+]{1)\ KMnO_4,\ ^-OH} C_6H_5\text{-}CO_2H \xrightarrow{HNO_3,\ H_2SO_4} m\text{-}O_2N\text{-}C_6H_4\text{-}CO_2H$$

ethylbenzene benzoic acid *m*-nitrobenzoic acid

In these schemes, an alkyl group—specifically, the ethyl group—has been utilized as the *synthetic equivalent* of a carboxylic acid group. That is, the ethyl group is something that is *easily converted into* a carboxylic acid group. (Some people use the word *synthon* to mean the same thing as *synthetic equivalent.*) A person who is skilled in synthesis tends to see various functional groups in terms of their synthetic equivalents. For example, when such a person sees a carboxylic acid group on a benzene ring, an alkyl group pops into mind. For an aldehyde, a primary alcohol comes to mind. (What would you think of as a synthetic equivalent of a *cis*-alkene?) Of course, a group may have more than one synthetic equivalent, and the appropriate synthetic equivalency will generally depend on the exact situation at hand.

By coupling the Friedel-Crafts alkylation with benzylic oxidation, a connection has been made that was not explicitly discussed in the text. As you improve in your mastery of organic chemistry, this is the sort of connection that you should begin to make on your own. A good student will begin to think of compounds "out of context," that is, in ways not directly related to the text material. For example, a student will see a structure and ask, "How could I synthesize this?" even though its synthesis might not be the context in which the structure is encountered. Professional chemists tend to think this way. A structure or a reaction presented in one context will trigger the imagination to think of a wholly new context in which it might be useful. This is one way that new ideas are born. Awareness of this "intellectual triggering" process is one of the best reasons that scientists study the professional literature. And this works not only in chemistry; it works in fields as diverse as physics and medicine.

17.4 Essential Oils

The history of the essential oils is an important part of the early history of both chemistry and medicine. A Swiss alchemist, Theophrastus Bombastus von Hohenheim, better known as Paracelsus (*ca.* 1493–1541), believed that everything had a chemical essence. For example, he could demonstrate that the odor of spearmint could be liberated from the plant as a volatile oil on heating; this was the "essence of spearmint"—its quintessence, or "reason for being." (Paracelsus tried unsuccessfully to find the essences of rocks and other refractory objects.) He believed that a person who was ill was missing part of his or her essence—and Paracelsus sought to restore the essence with chemical cures, among them mercury and sulfur. Paracelsus was reputed to have effected some very dramatic cures. His firm belief in the chemical essence of all things was reflected in public tirades against physicians of the day, whom (he said) ". . . strut about with haughty gait, dressed in silk with rings upon their fingers displayed ostentatiously, or silver poignards fixed upon their loins and sleek gloves upon their hands . . ." while chemists "sweat whole nights and days over fiery furnaces, do not waste time with empty talk, but express delight in the laboratory." Paracelsus was driven out of Basel when he angered some local clergy during a quarrel over his fees; he died in exile.

Many (but certainly not all) of the essential oils turned out to be terpenes. And it is known today that many terpenes are neither volatile nor pleasant-smelling, and many come from animal sources rather than plant sources. Nevertheless, it was the fascination with, and curiosity about, the essential oils that ultimately led chemists to their understanding of how chemical substances are synthesized in the natural world.

✓17.5 Skeletal Structures

Skeletal structures are used in this section and with increasing frequency throughout the rest of the text. Skeletal structures are explained in Sec. 2.5 on text pp. 62–64. Be especially careful to remember that a carbon atom is located not only at *each vertex* of a skeletal structure but also at *each end of a chain.* If you have difficulty interpreting a skeletal structure, do not hesitate to draw out the structure with all the carbons and hydrogens.

SOLUTIONS TO PROBLEMS

Solutions to In-Text Problems

17.1 The allylic carbons are indicated with an asterisk (*).

(a) (b)

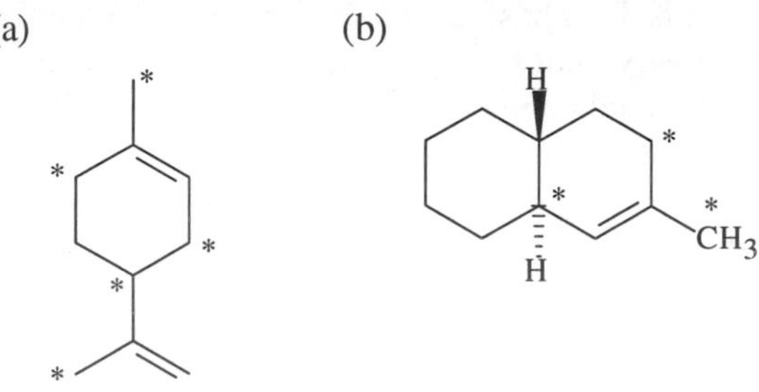

17.2 The benzylic carbons are indicated with an asterisk (*).

(a) (b)

17.3 (a) The order of increasing S_N1 reactivity is (2) < (1) < (3). The rates depend on the stabilities of the respective carbocation intermediates. One resonance structure for the carbocation intermediate in the solvolysis of compound (3) is a tertiary carbocation; the resonance structures of the carbocation intermediates derived from compounds (1) and (2) are secondary carbocations.

carbocation intermediate in the solvolysis of compound (3)

Consequently, the carbocation intermediate derived from compound (3) is more stable, relative to the starting alkyl halide, than the intermediates in the solvolysis reactions of the other compounds. The intermediate in the solvolysis of compound (2) is least stable because of the electron-withdrawing polar effect of the oxygen. Note that an oxygen in the meta position cannot be involved in resonance; consequently, only its rate-retarding polar effect operates.

Note that the rates of S_N1 solvolysis reactions parallel the rates of electrophilic aromatic substitution reactions. Just as a methoxy group activates an aromatic substitution reaction at an ortho or para position, it also activates a reaction involving a benzylic cation at an ortho or para position. Just as a methoxy group *deactivates* aromatic substitution at a meta position, it also deactivates a reaction involving a benzylic cation at a meta position.

(b) The reactivity order is (2) < (3) < (1). The S_N1 reaction of compound (2) is slowest because the polar effect of the *meta*-chloro substituent destabilizes the intermediate carbocation. The reaction of compound (3) is faster because the resonance effect of the *para*-chloro group partially offsets its polar effect.

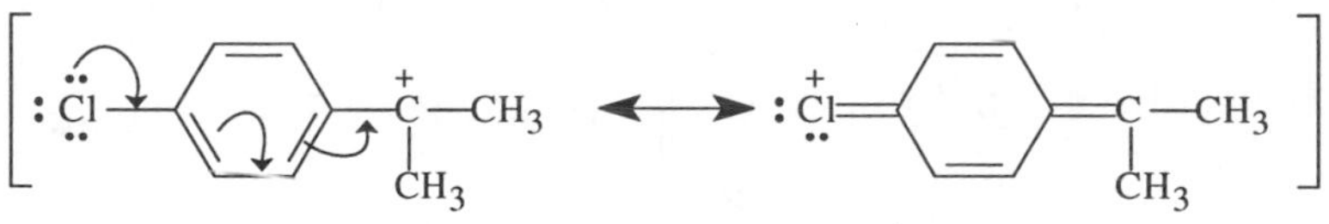

carbocation intermediate in the solvolysis of compound (3)

Compound (1) reacts most rapidly because the carbocation intermediate is not destabilized by the deactivating polar effect of a chloro substituent, which outweighs its resonance effect.

17.4 The alkyl halide is the one that reacts to give the same carbocation intermediate, that is, a carbocation with the same resonance structures:

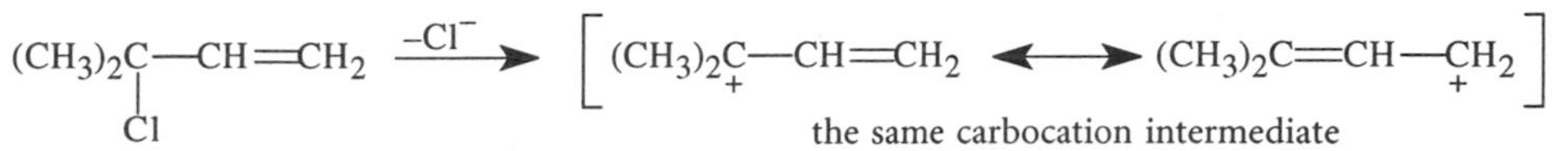

$$(CH_3)_2C(Cl)—CH=CH_2 \xrightarrow{-Cl^-} [(CH_3)_2\overset{+}{C}—CH=CH_2 \longleftrightarrow (CH_3)_2C=CH—\overset{+}{C}H_2]$$

the same carbocation intermediate that is involved in text Eq. 17.7

17.5 The carbocation formed when trityl chloride ionizes, the *trityl cation* (Ph_3C^+), is stabilized by delocalization of electrons from *all three* phenyl rings. This carbocation has more resonance structures than the carbocations formed from the other alkyl halides in the table, and is thus so stable that the transition state leading to its formation also has very low energy; consequently, it is formed very rapidly.

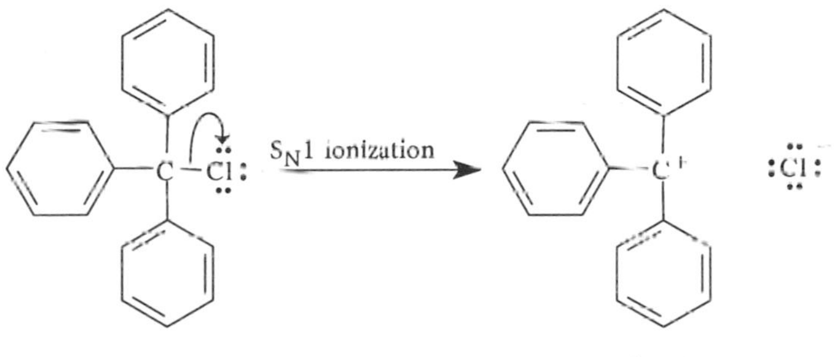

S_N1 ionization

trityl chloride (triphenylmethyl chloride) — trityl cation (triphenylmethyl cation)

17.6 The number of products depends on (1) whether all of benzylic or allylic positions are equivalent, and (2) whether the resonance structures of the free-radical intermediate are identical.

(a) All allylic positions of cyclohexene are chemically equivalent, and the two resonance structures are identical. Hence, only one allylic bromination product is possible.

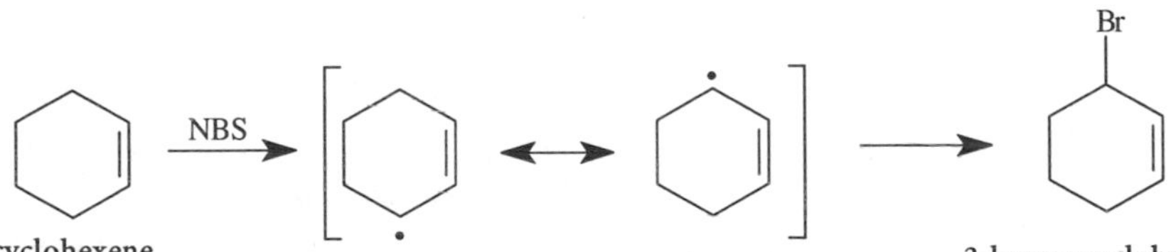

NBS

cyclohexene starting material — free-radical intermediate (identical resonance structures) — 3-bromocyclohexene product

(b) The two allylic positions are nonequivalent, but only one has allylic hydrogens; hence, only one can react. The two resonance structures of the resulting free radical are nonequivalent; hence, two different products can be formed.

3,3-dimethylcyclohexene starting material —NBS→ [free-radical intermediate] → 6-bromo-3,3-dimethylcyclohexene + 3-bromo-4,4-dimethylcyclohexene (products)

(c) *Trans*-2-pentene contains two chemically distinguishable allylic positions. In each of the allylic radical intermediates that result from reactions at the two positions, the unpaired electron is delocalized to two different carbons. (See Study Problem 17.1.) Reaction at H^a leads to two products, and reaction at H^b leads to only one product because the resonance structures of the intermediate radical are identical (if it is assumed that the double bond retains the more stable trans stereochemistry).

trans-2-pentene —NBS→ [free-radical intermediate from abstraction of hydrogen *(a)*] → (*E*)-1-bromo-2-pentene + 3-bromo-1-pentene (products from abstraction of H^a)

trans-2-pentene —NBS→ [free-radical intermediate from abstraction of hydrogen *(b)* (identical resonance structures)] → (*E*)-4-bromo-2-pentene (products from abstraction of H^b)

It is also assumed that the free radicals adopt the more stable trans configuration about their double bonds.

(d) A benzylic hydrogen is abstracted rather than a hydrogen of the *tert*-butyl group because a more stable benzylic free-radical intermediate is obtained.

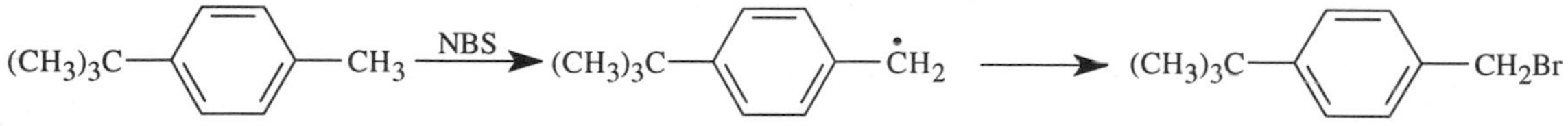

$(CH_3)_3C$—C₆H₄—CH_3 $\xrightarrow{\text{NBS}}$ $(CH_3)_3C$—C₆H₄—$\dot{C}H_2$ ⟶ $(CH_3)_3C$—C₆H₄—CH_2Br

4-*tert*-butyltoluene free-radical intermediate 1-bromomethyl-4-*tert*-butylbenzene

17.7 (a) The initially formed Grignard reagent undergoes a rapid allylic rearrangement; each Grignard reagent in the equilibrium can react with D_2O.

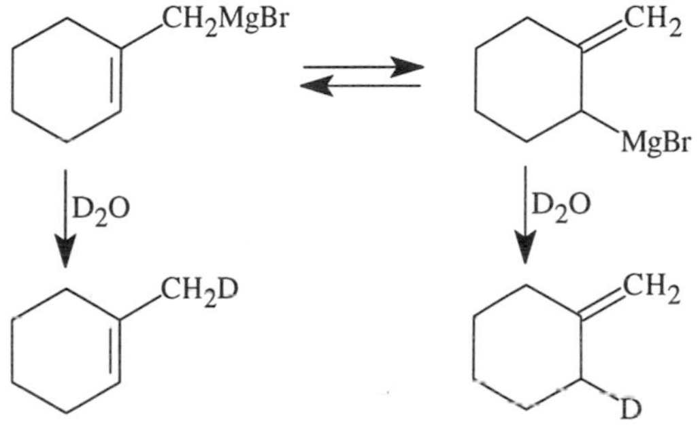

CH_2MgBr ⇌ CH_2 / MgBr

D_2O ↓ D_2O ↓

CH_2D CH_2 / D

(b) Because the two Grignard reagents in rapid equilibrium are identical, only one product is obtained:

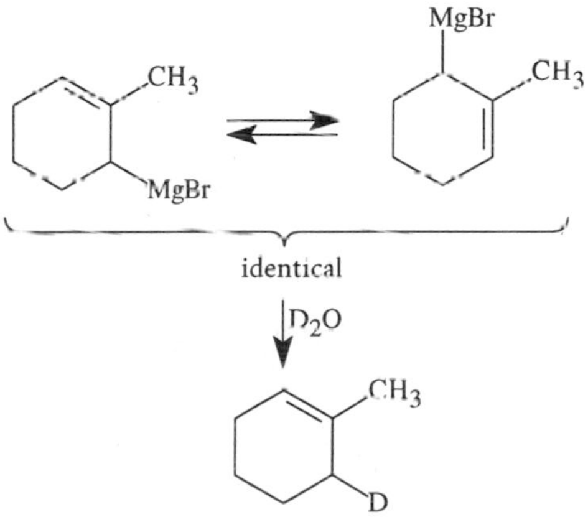

CH_3 / MgBr ⇌ MgBr / CH_3

identical

D_2O ↓

CH_3 / D

17.8 (a) Although this alkyl halide has two chemically nonequivalent types of β-hydrogens, an allylic hydrogen is more acidic than a nonallylic one; hence, the product is the conjugated diene derived from elimination of the allylic hydrogen and the bromine.

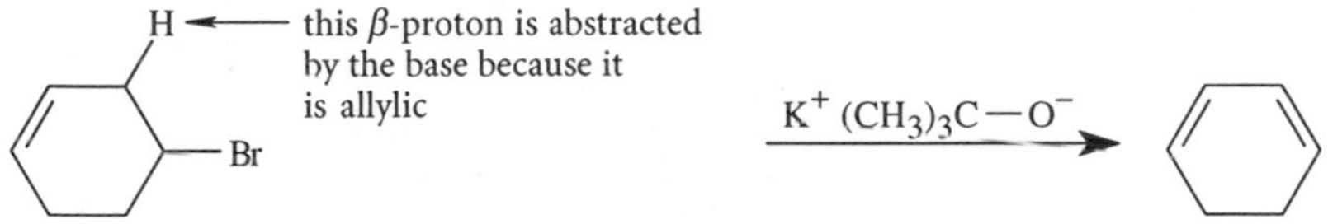

H ⟵ this β-proton is abstracted by the base because it is allylic

Br

$\xrightarrow{K^+\ (CH_3)_3C—O^-}$

1,3-cyclohexadiene

(b) The benzylic proton is abstracted; β-elimination gives a vinylic ether.

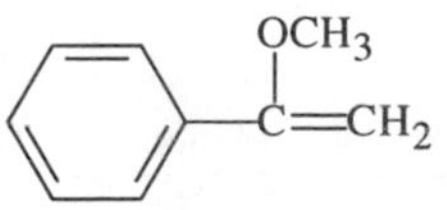

OCH_3 / $C{=}CH_2$

(1-methoxyvinyl)benzene
(α-methoxystyrene)

17.9 Reaction with concentrated HBr can involve an allylic carbocation intermediate; consequently, a mixture of products, *A* and *B*, could be formed:

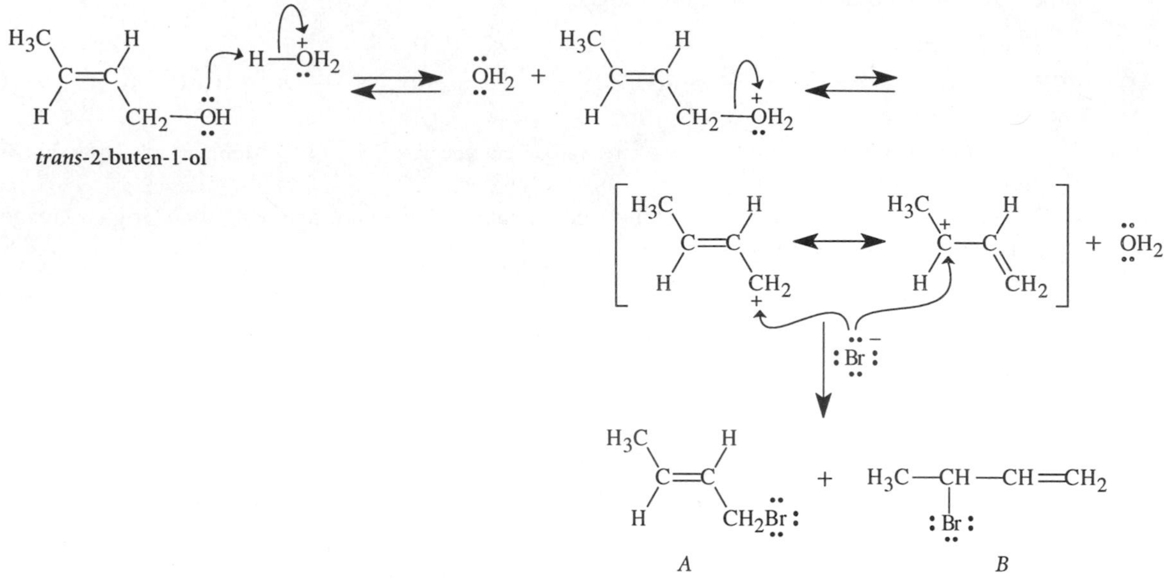

$$\text{H}_3\text{C(H)C=C(H)CH}_2\text{—}\ddot{\text{O}}\text{H} + \text{H—}\overset{+}{\text{O}}\text{H}_2 \rightleftharpoons \ddot{\text{O}}\text{H}_2 + \text{H}_3\text{C(H)C=C(H)CH}_2\text{—}\overset{+}{\text{O}}\text{H}_2 \rightleftharpoons$$

trans-2-buten-1-ol

$$\left[\text{H}_3\text{C(H)C=C(H)}\overset{+}{\text{C}}\text{H}_2 \longleftrightarrow \text{H}_3\text{C(H)}\overset{+}{\text{C}}\text{—C(H)=CH}_2\right] + \ddot{\text{O}}\text{H}_2$$

$$\xrightarrow{:\ddot{\text{Br}}:^-} \text{H}_3\text{C(H)C=C(H)CH}_2\ddot{\text{Br}}: + \text{H}_3\text{C—CH(:}\ddot{\text{Br}}\text{:)—CH=CH}_2$$

A *B*

Reaction of the tosylate with NaBr in acetone, in contrast, involves an S_N2 reaction; consequently, there is no reactive intermediate, and only product *A* is formed.

17.10 (a)

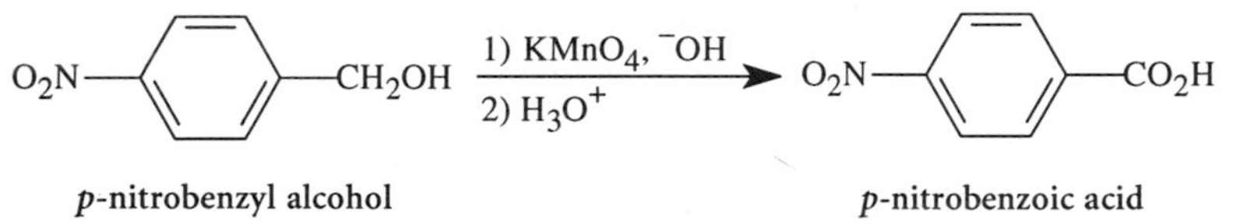

$$O_2N\text{—C}_6\text{H}_4\text{—CH}_2\text{OH} \xrightarrow[\text{2) H}_3\text{O}^+]{\text{1) KMnO}_4,\ ^-\text{OH}} O_2N\text{—C}_6\text{H}_4\text{—CO}_2\text{H}$$

p-nitrobenzyl alcohol *p*-nitrobenzoic acid

(b) The butyl group is oxidized because it has α-hydrogens; the *tert*-butyl group is not affected.

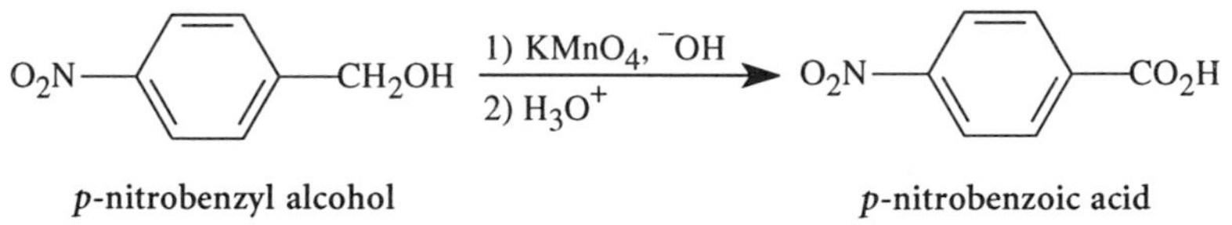

$$O_2N\text{—C}_6\text{H}_4\text{—CH}_2\text{OH} \xrightarrow[\text{2) H}_3\text{O}^+]{\text{1) KMnO}_4,\ ^-\text{OH}} O_2N\text{—C}_6\text{H}_4\text{—CO}_2\text{H}$$

p-nitrobenzyl alcohol *p*-nitrobenzoic acid

17.11 (a) The oxidation product shows that compound *A* is an *ortho*-disubstituted dialkylbenzene. Only compound *A* = *o*-xylene is consistent with this analysis and with the formula.

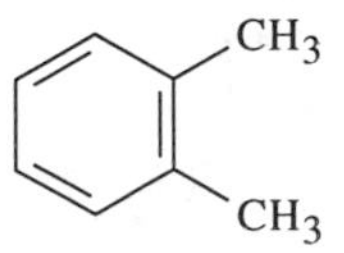

$$\text{C}_6\text{H}_4(\text{CH}_3)_2$$

o-xylene

(b) Because one carbon is lost as a result of the oxidation, and because the benzene ring accounts for all four degrees of unsaturation, compound *B* must be ethylbenzene.

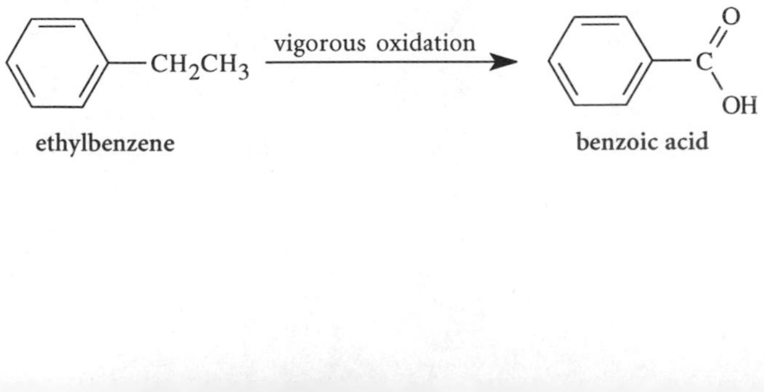

$$\text{C}_6\text{H}_5\text{—CH}_2\text{CH}_3 \xrightarrow{\text{vigorous oxidation}} \text{C}_6\text{H}_5\text{—C(=O)OH}$$

ethylbenzene benzoic acid

17.12 (a) Vitamin A is a diterpene because it contains four isoprene skeletons, which are shown as heavy bonds.

OH

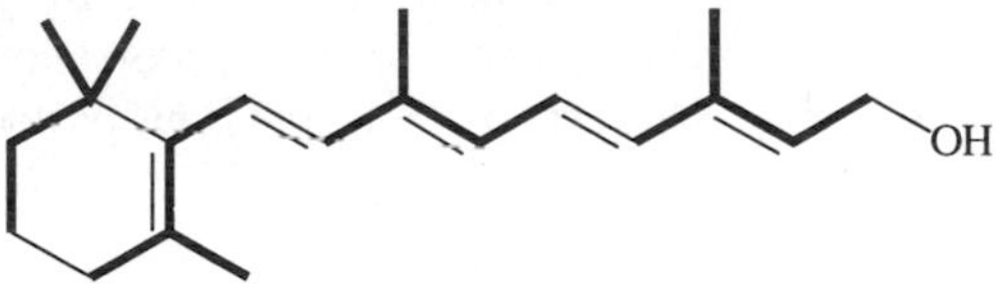

vitamin A

(b) Caryophyllene is a sesquiterpene because it contains three isoprene skeletons, which are shown as heavy bonds.

H

H

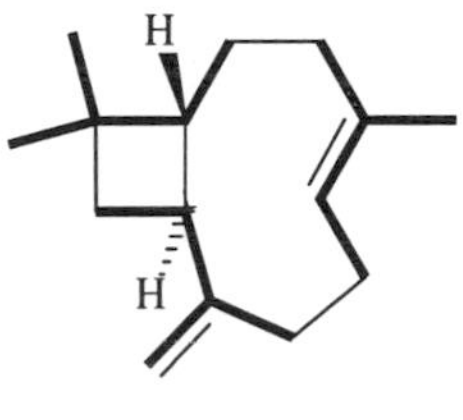

caryophyllene

17.13 (a) A biosynthetic mechanism for limonene is as follows: (B: = a base.)

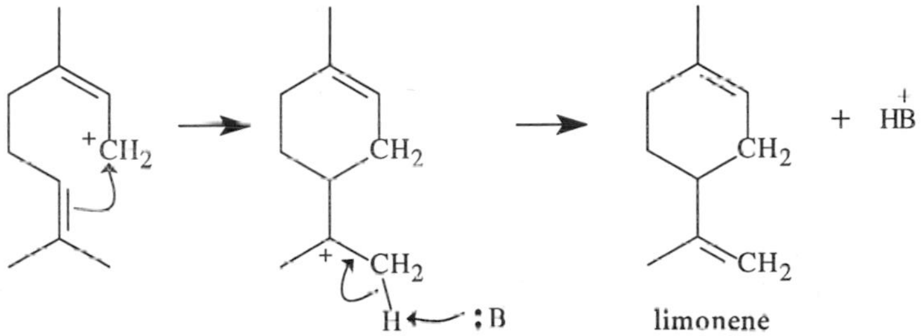

$^+CH_2$ → CH_2 $^+CH_2$ H ← :B → CH_2 CH_2 + $H\overset{+}{B}$

limonene

(b) Ionization of geranyl pyrophosphate is followed by attack of the pyrophosphate anion on the other electron-deficient carbon of the resonance-stabilized carbocation; rotation about a single bond is followed by ionization of pyrophosphate to give the desired carbocation.

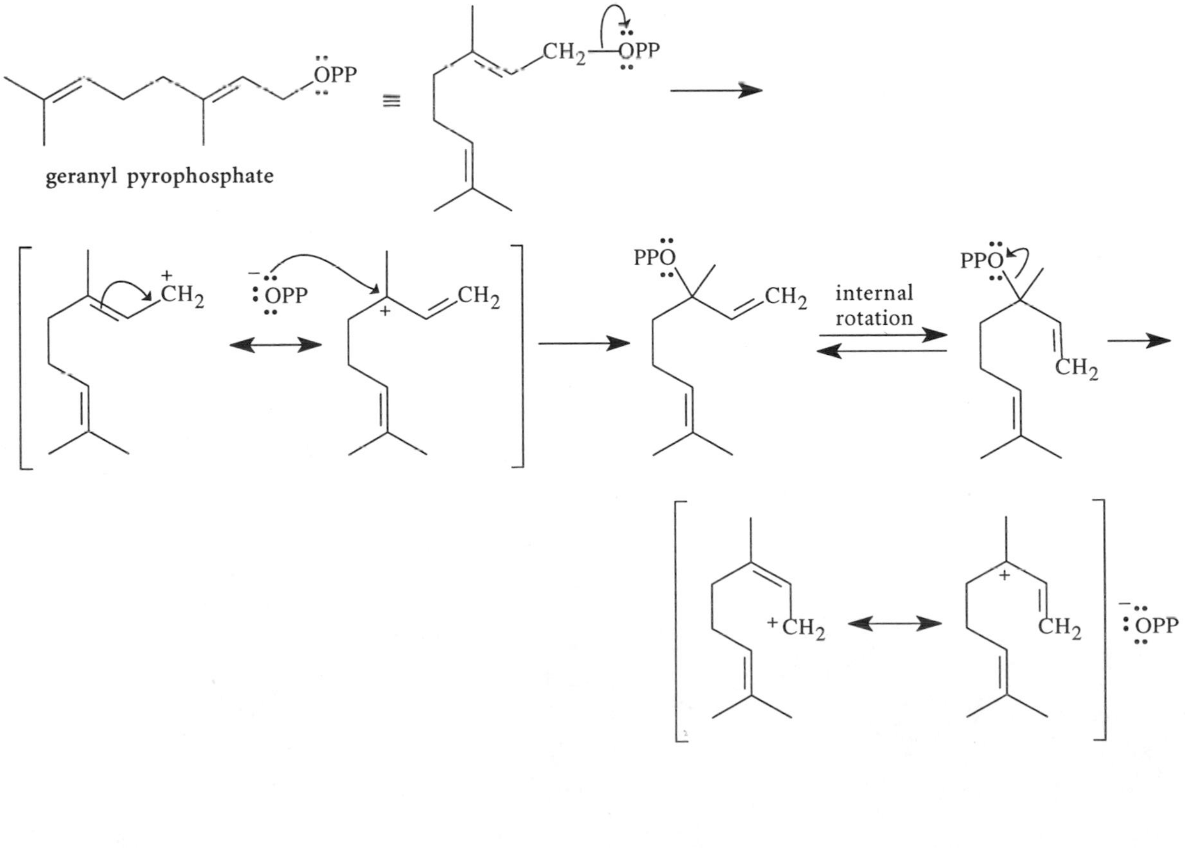

OPP ≡ CH_2—OPP →

geranyl pyrophosphate

[$^+CH_2$:OPP$^-$ ↔ + CH_2] → PPO CH_2 internal rotation ⇌ PPO CH_2 →

[$^+CH_2$ ↔ + CH_2] :OPP$^-$

17.14 (a) To see the new bond connections that have to be made, draw α-pinene in a planar projection. The projection on the left shows the molecule as it is in the text; turning this projection clockwise 120° in the plane of the page yields the second projection; and turning this projection over yields the third projection, which can be related more easily to the carbocation in the previous problem.

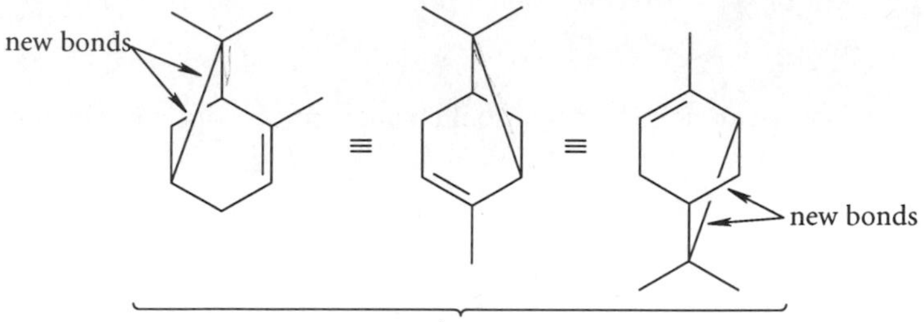

new bonds

new bonds

α-pinene

Here is the mechanism:

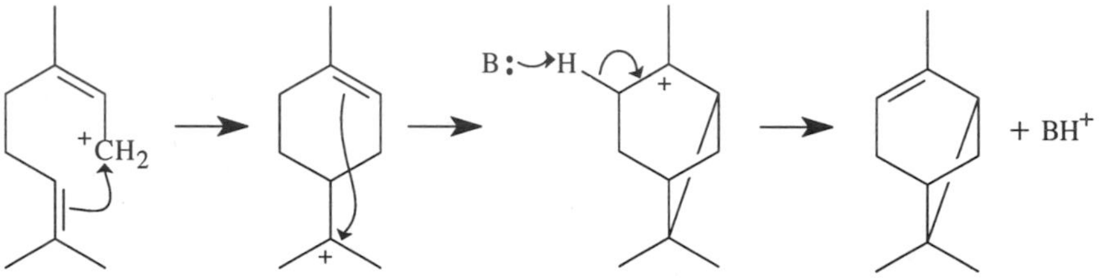

$^{+}CH_2$

B: H

+ BH^+

α-pinene

(b) Geranyl pyrophosphate is converted into farnesyl pyrophosphate by a mechanism exactly analogous to the one shown in Eq. 17.38 on text p. 765. Then farnesyl pyrophosphate hydrolyzes to farnesol; see text Eq. 17.39.

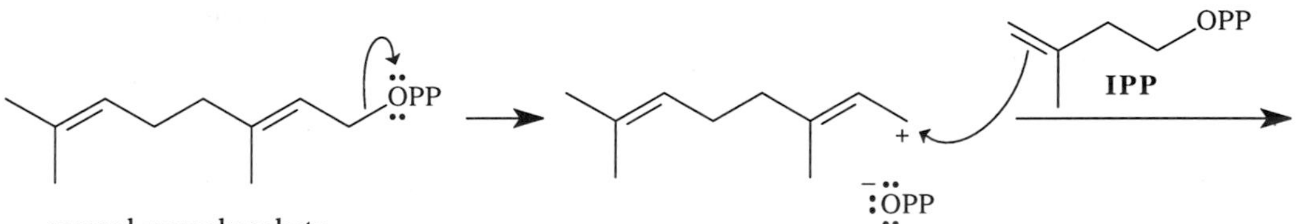

OPP

OPP

IPP

:OPP⁻

geranyl pyrophosphate

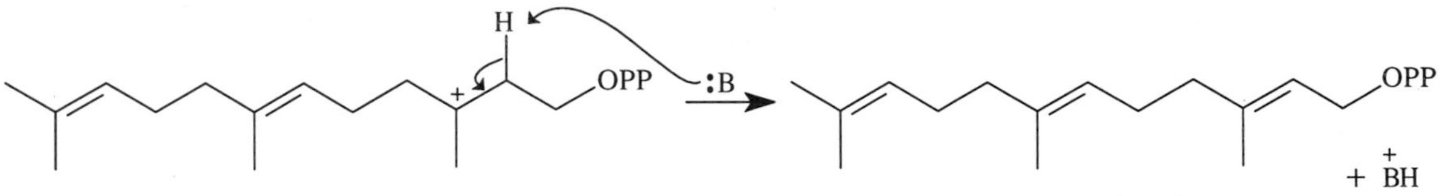

H

OPP

:B

OPP

+ $\overset{+}{B}H$

farnesyl pyrophosphate

H_2O (see text Eq. 17.39)

HOPP +

OH

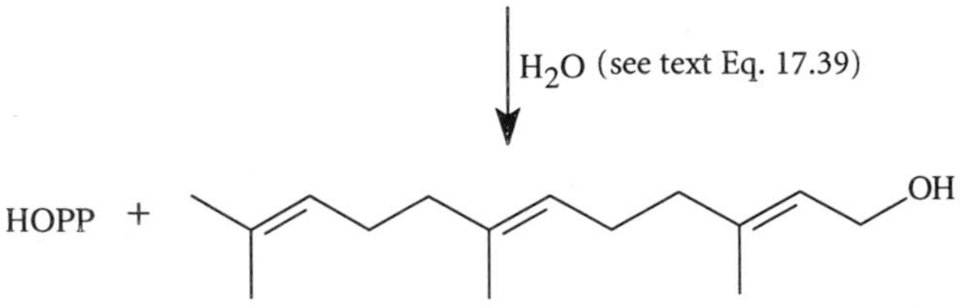

farnesol

Solutions to Additional Problems

17.15 The structure of the starting material is

$H_3C(H)C{=}C(H)CH_3$ *trans*-2-butene

(a) *meso*-$CH_3CHCHCH_3$ (Br Br)
meso-2,3-dibromobutane

(b) $H_3C(H)C{=}C(H)CH_2Br$ + $CH_3CH(Br)CH{=}CH_2$
(*E*)-1-bromo-2-butene   3-bromo-1-butene

(c) $H_3C(H)C{=}C(H)CH_2OH$ + $CH_3CH(OH)CH{=}CH_2$
(*E*)-3-buten-1-ol   1-buten-3-ol

(d) $H_3C(H)C{=}C(H)CH_2MgBr$ + $CH_3CH(MgBr)CH{=}CH_2$

(e) $H_3C(H)C{=}C(H)CH_2D$ + $CH_3CH(D)CH{=}CH_2$

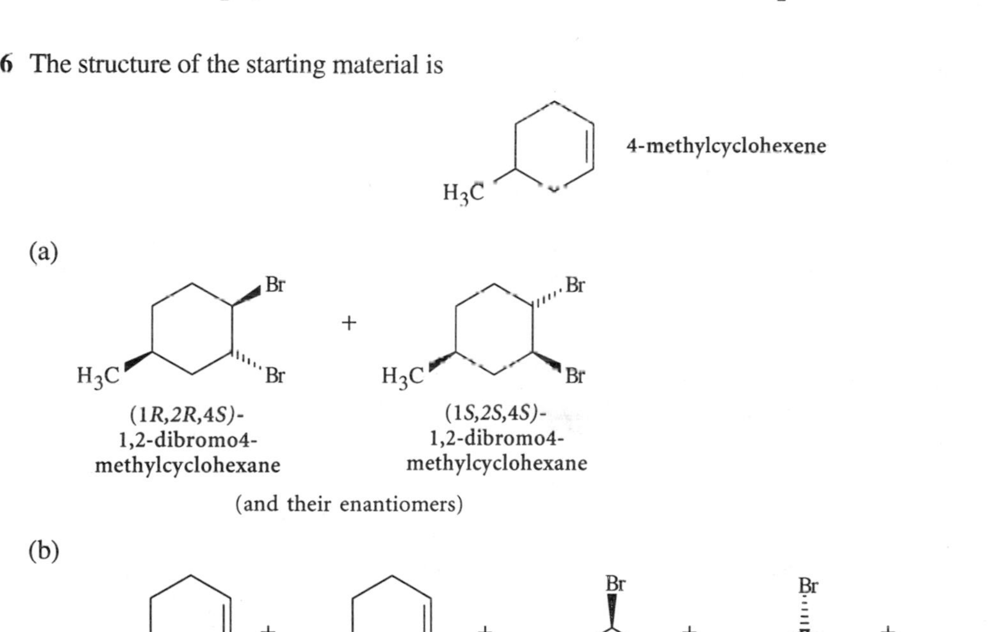

(b)

H3C Br + H3C Br + Br H3C + Br H3C +

3-bromo-4-methylcyclohexene
(two diastereomers and their enantiomers)

3-bromo-5-methylcyclohexene
(two diastereomers and their enantiomers)

H3C Br + H3C Br

3-bromo-6-methylcyclohexene
(two diastereomers and their enantiomers)

(c)

6-methyl-3-cyclohexenol
(two diastereomers and their enantiomers)

4-methyl-2-cyclohexenol
(two diastereomers and their enantiomers)

5-methyl-2-cyclohexenol
(two diastereomers and their enantiomers)

(d) As in the previous parts, all four stereoisomers of each compound are formed.

Don't forget that Grignard reagents undergo a very rapid allylic rearrangement (Eq. 17.24, text p. 754).

(e) As in the previous part, all four stereoisomers of each compound are formed.

17.17 Compounds (2), (4), and (5) are terpenes. (Compound (3), modhephene, is believed to be formed from a terpene by a skeletal rearrangement, but its terpene origin is not obvious from its structure.) The isoprene skeletons are shown with heavy bonds.

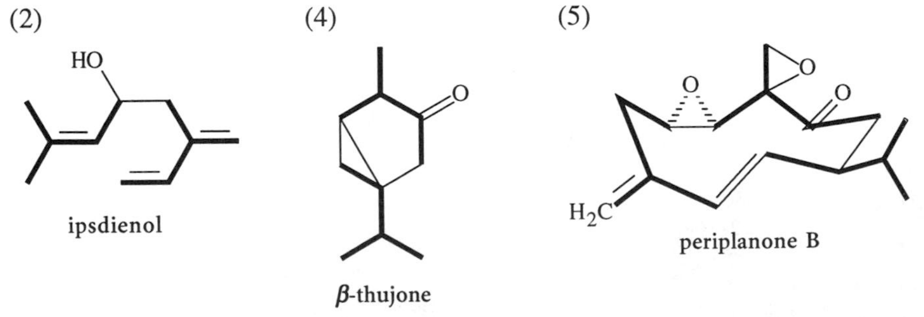

17.18 Zoapatanol is a diterpene; the isoprene units are shown with heavy bonds.

17.19 In reaction (1), the acidic conditions promote protonation of the alcohol and ionization to a carbocation, which is trapped by ethanol at the two nonequivalent electron-deficient carbons. (The mechanism, which is discussed in Sec. 11.1C on text pp. 459–60, is outlined below; fill in the curved-arrow notation.)

OH, H_2SO_4, $\overset{+}{O}H_2$, + OH_2, CH_3CH_2OH, OCH_2CH_3 + OCH_2CH_3 + $CH_3CH_2\overset{+}{O}H_2$

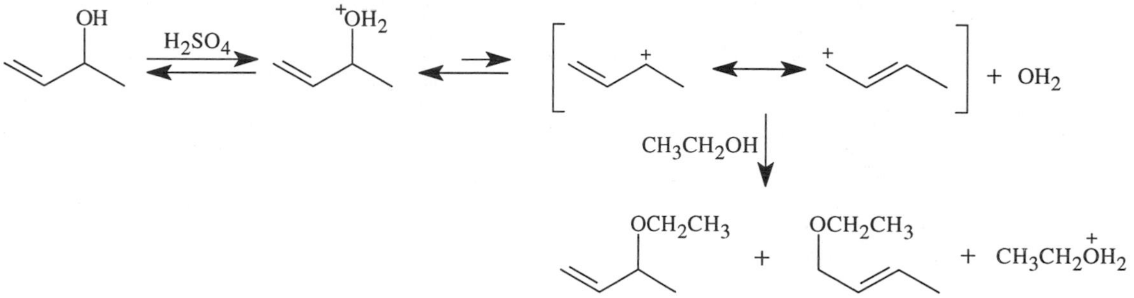

In reaction (2), the carbon-oxygen bond is never broken; consequently, no allylic rearrangement can occur. This is a Williamson ether synthesis that involves formation of the alkoxide anion, which, in turn, is alkylated by ethyl iodide. (See Sec. 11.1A, text pp. 455; Problem 11.46 on text pp. 490 illustrates the same idea.)

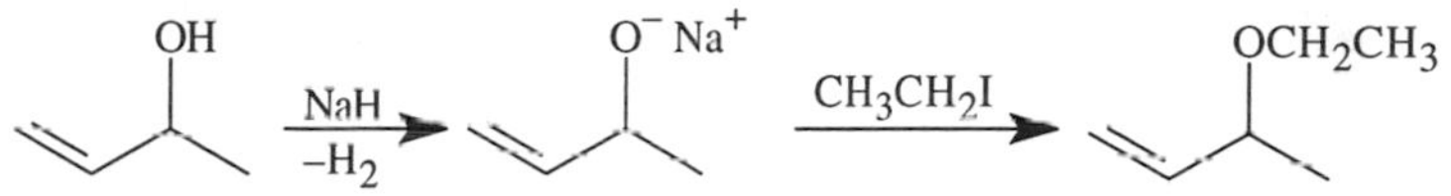

OH $\xrightarrow[-H_2]{NaH}$ $O^-\ Na^+$ $\xrightarrow{CH_3CH_2I}$ OCH_2CH_3

17.20 (a) The allylic-rearrangement product *B* [(*E*) 1-bromo-2-butene] could be formed by ionization to a carbocation and bromide ion followed by re-attack of the bromide ion on the other electron-deficient carbon.

(b) The curved-arrow notation is shown in the following scheme.

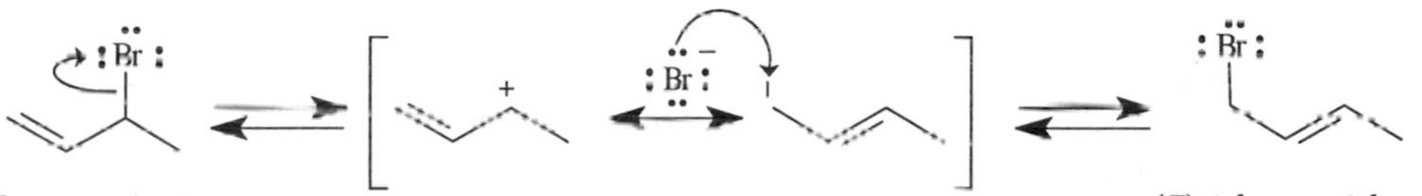

:Br: ⇌ [+ ⟷ :Br:⁻ +] ⇌ :Br:

3-bromo-1-butene (compound *A*)

(*E*)-1-bromo-2-butene (compound *B*)

(c) Compound *B*, the rearrangement product, is favored at equilibrium because it has the double bond with the greater number of alkyl branches.

17.21 (a)

$$\underset{\text{toluene}}{PhCH_3} \xrightarrow[\text{peroxides}]{\text{NBS, }CCl_4} PhCH_2Br \xrightarrow[CH_3OH]{Na^{+}\ {}^{-}OCH_3} \underset{\text{benzyl methyl ether}}{PhCH_2OCH_3}$$

(b)

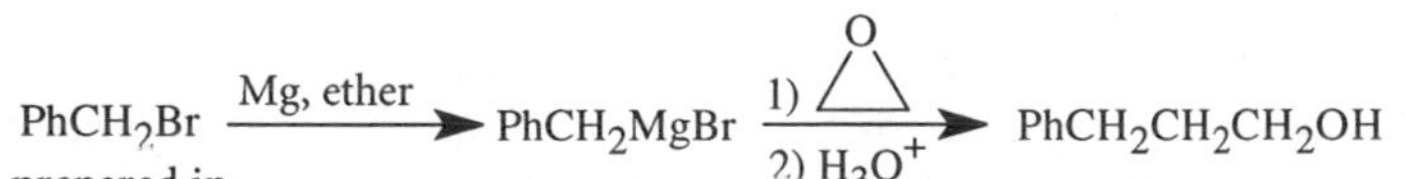

$$\underset{\text{prepared in part (a)}}{PhCH_2Br} \xrightarrow{\text{Mg, ether}} PhCH_2MgBr \xrightarrow[\text{2) } H_3O^+]{\text{1) ethylene oxide}} \underset{\text{3-phenyl-1-propanol}}{PhCH_2CH_2CH_2OH}$$

(c) In the first set of reactions, note that allyl bromide is a particularly reactive alkylating agent.

$$\underset{\text{1-hexyne}}{CH_3(CH_2)_3C{\equiv}CH} \xrightarrow[\text{2) } BrCH_2CH{=}CH_2 \text{ allyl bromide}]{\text{1) } NaNH_2} CH_3(CH_2)_3C{\equiv}CCH_2CH{=}CH_2 \xrightarrow[\text{Lindlar catalyst}]{H_2}$$

$CH_3CH_2CH_2CH_2$ / C=C \ $CH_2CH{=}CH_2$, H, H

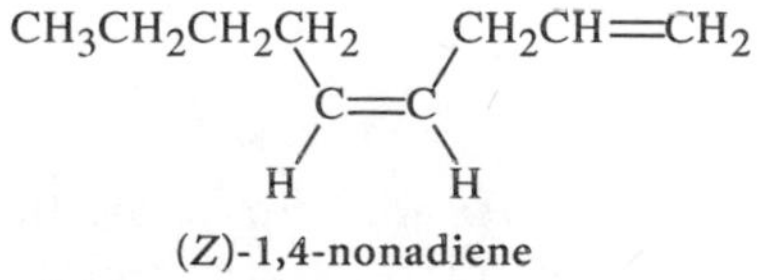

(*Z*)-1,4-nonadiene

(d)

cyclopentene $\xrightarrow[\text{peroxides}]{NBS,\ CCl_4}$ 3-bromocyclopentene $\xrightarrow{\text{Mg, ether}}$ cyclopentenyl–MgBr $\xrightarrow[2)\ H_3O^+]{1)\ \text{ethylene oxide}}$

cyclopentenyl–CH_2CH_2OH $\xrightarrow{CrO_3(pyridine)_2}$ cyclopentenyl–$CH_2CH{=}O$

(e)

cumene $\xrightarrow[2)\ H_3O^+]{1)\ KMnO_4,\ ^-OH}$ C_6H_5–CO_2H $\xrightarrow{HNO_3,\ H_2SO_4}$ O_2N–C_6H_4–CO_2H

m-nitrobenzoic acid

(f)

cumene $\xrightarrow{HNO_3,\ H_2SO_4}$ O_2N–C_6H_4–$CH(CH_3)_2$ $\xrightarrow[2)\ H_3O^+]{1)\ KMnO_4,\ ^-OH}$ O_2N–C_6H_4–CO_2H

p-nitrobenzoic acid

Notice how reversing the sequence of the oxidation and nitration steps brings the directing effects of different substituents into play during nitration.

17.22 The compounds that give the most stable carbocation intermediates are the ones that undergo the most rapid solvolysis. This problem deals with the effect of substituent on the stability of the carbocation intermediate. The key is to analyze the balance of resonance and polar substituent effects just as you would for electrophilic aromatic substitution. The order of increasing reactivity is (4) < (1) < (3) < (2). Thus, compound (2) reacts most rapidly because the carbocation intermediate is stabilized by the *electron-donating resonance effect* of the *p*-methoxy substituent:

$CH_3\ddot{O}$–C_6H_4–$C(CH_3)_2$–$\ddot{Cl}:$ ⟶ [$CH_3\ddot{O}$–C_6H_4–$\overset{+}{C}(CH_3)_2$ ⟷ $CH_3\overset{+}{O}{=}C_6H_4{=}C(CH_3)_2$] $:\ddot{Cl}:^-$

As in electrophilic substitution, the resonance effect of the *p*-methoxy group strongly outweighs its electron-withdrawing polar effect. In compound (3), there is a similar resonance effect; however, the polar effects of halogen substituents outweigh their resonance effects. Consequently, compound (3) reacts more slowly. The nitro group exerts no resonance effect in the carbocation intermediates derived from compounds (1) and (4); the question is then whether its polar effect is stronger from the meta or para position. As in electrophilic aromatic substitution, a *para*-nitro group destabilizes a carbocation intermediate more than a *meta*-nitro group because, in a *para*-nitro carbocation, positive charge is on adjacent atoms:

$O_2\overset{+}{N}$–C_6H_4–$C(CH_3)_2$–$\ddot{Cl}:$ ⟶ [$O_2\overset{+}{N}$–C_6H_4–$\overset{+}{C}(CH_3)_2$ ⟷ $O_2\overset{+}{N}$–$\overset{+}{C_6H_4}{=}C(CH_3)_2$] $:\ddot{Cl}:^-$

positive charge is on adjacent atoms

In the *meta*-nitro carbocation, positive charge does *not* reside on adjacent atoms. Consequently, the *meta*-nitro carbocation is more stable (or perhaps we should say *less unstable)* than the *para*-nitro carbocation, and *m*-nitro-*tert*-cumyl chloride solvolyzes more rapidly than *p*-nitro-*tert*-cumyl chloride.

17.23 Acid-catalyzed dehydration involves a carbocation intermediate; see Sec. 10.1, text p. 409. The problem, then, really asks, "Which compound gives the most stable carbocation intermediate?" Compounds *B* and *C* both involve a *p*-methoxy substituent; the carbocation-stabilizing effect of this substituent in a similar situation is analyzed in the solution to the previous problem. In the dehydration of *B*, the carbocation intermediate, a tertiary carbocation, is more stable than the carbocation in the dehydration of *C*, a secondary carbocation. The *p*-nitro carbocation is the least stable for the reasons given in the solution to Problem 17.22. Therefore, the order of increasing reactivity is $A < C < B$.

17.24 The question is, essentially, "Which —OH group, when protonated and lost as water (Sec. 10.1A), gives the more stable carbocation?" Loss of the OH on the left gives a carbocation that is both tertiary and doubly benzylic. Loss of the OH on the right gives a carbocation that is secondary and singly benzylic. The choice is clear; the —OH group on the left is lost.

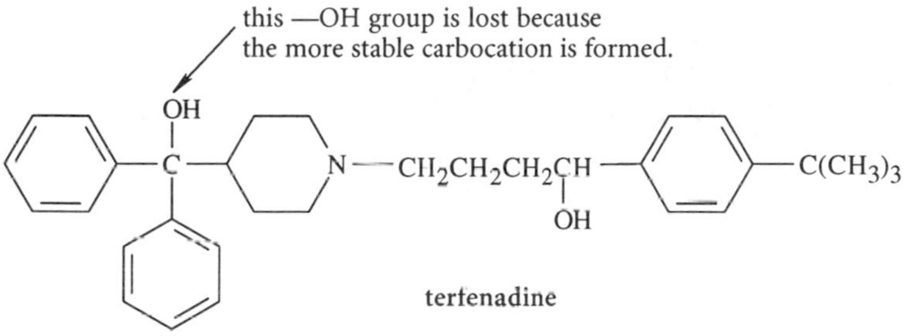

this —OH group is lost because the more stable carbocation is formed.

OH — C — N — $CH_2CH_2CH_2CH$ — C(CH_3)$_3$; OH

terfenadine

17.25 The solution to this problem, like the previous three solutions, hinges on an analysis of the relative stabilities of the carbocation intermediates involved in the S_N1 reactions of the two compounds. The carbocation intermediate in the solvolysis of compound *A* is resonance-stabilized:

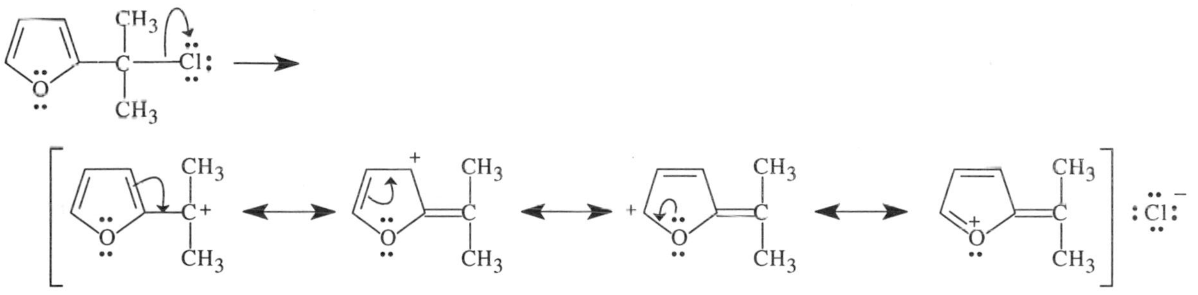

CH_3 C Cl: CH_3 →

[CH_3 C+ CH_3 ⟷ + CH_3 C CH_3 ⟷ + CH_3 C CH_3 ⟷ CH_3 C CH_3] :Cl:$^-$

The carbocation intermediate involved in the solvolysis of compound *B* is not resonance-stabilized, and in fact is somewhat destabilized by the electron-withdrawing polar effect of the oxygen. The greater stability of the carbocation derived from compound *A* results in a greater solvolysis rate.

17.26 The fact that benzoic acid is obtained by chromic-acid oxidation shows that all compounds contain a monosubstituted benzene ring. The NBS reaction is a benzylic bromination, and the alcohol produced by solvolysis of the resulting bromide must be tertiary, since it cannot be oxidized with CrO_3 and pyridine. The structures of compound *A*, *B*, and *C* are therefore as follows:

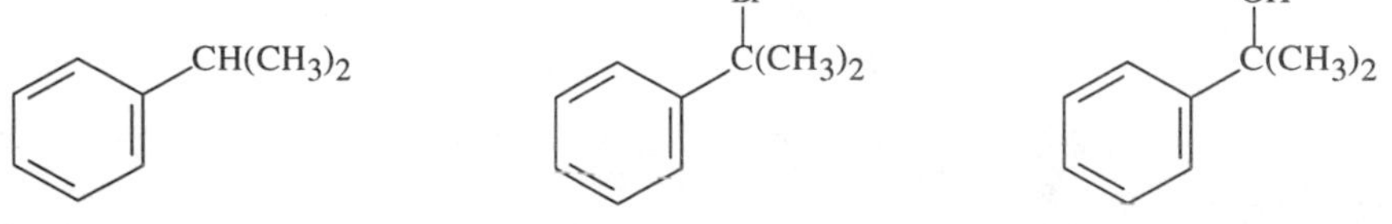

$CH(CH_3)_2$ — *A*

Br, $C(CH_3)_2$ — *B*

OH, $C(CH_3)_2$ — *C*

17.27 Compound *A* has an unsaturation number of 5. Because it ultimately affords phthalic acid, it is an *ortho*-disubstituted benzene derivative. The enantiomeric resolution and oxidation results show that compound *C* is a chiral secondary alcohol, and the α-carbon of this alcohol is the asymmetric carbon, because conversion to a ketone destroys its chirality. Compound *C*, in turn, arises from solvolysis of *B*, a product of either allylic or benzylic bromination. Because *A* contains a benzene ring, the formation of compound *B* is probably a benzylic bromination. A by-product of the formation of compound *C* is compound *D*, presumably an alkene, because it can be hydrogenated back to compound *A*. The only way to accommodate all of these data is for the additional degree of unsaturation in compound *A* to be due to a ring. Structures of all compounds that fit these data are as follows:

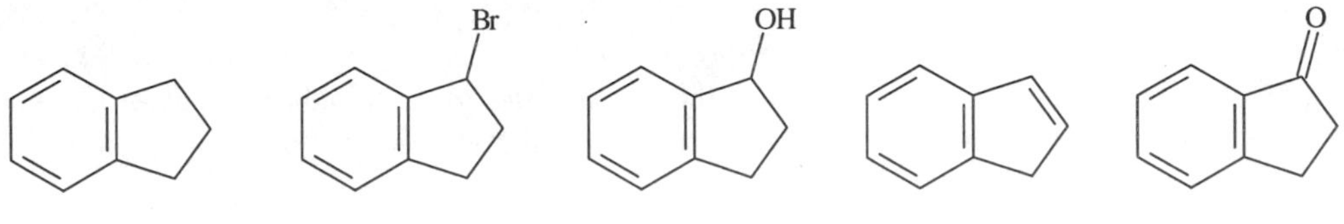

Br OH O

A B C D E

The following structure for compound *A* is ruled out because it would give a *tertiary* alkyl halide, which, in turn, would give a *tertiary* alcohol on solvolysis; a tertiary alcohol could not be oxidized to a ketone.

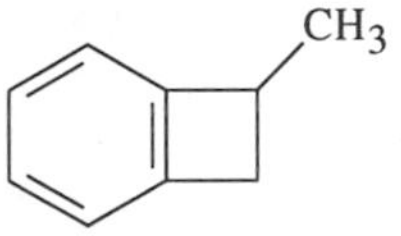

CH_3

17.28 Alkyl halides with allylic or benzylic β-hydrogens undergo more rapid E2 reaction because the allylic or benzylic hydrogens are more acidic than ordinary hydrogens (See Sec. 17.3B on text p. 757.) Because compound *A* has allylic β-hydrogens and compound *B* does not, the E2 reaction of compound *A* is faster, and this reaction proceeds in the following manner:

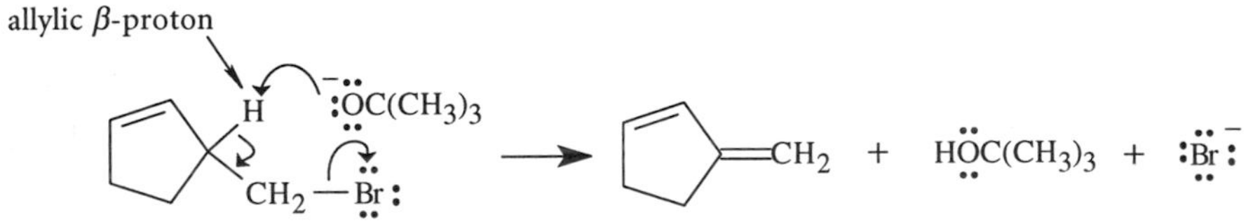

allylic β-proton

H $:\ddot{O}C(CH_3)_3$ CH_2—Br: → $=CH_2$ + $H\ddot{O}C(CH_3)_3$ + $:\ddot{Br}:^-$

In the presence of excess base this reaction does not stop here but would undoubtedly give the following *anion* as a product. Can you see why? Can you give a curved-arrow mechanism for this transformation?

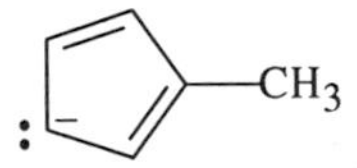

CH_3

17.29 The greater λ_{max} value indicates the formation of a species with more extensive conjugation. Protonation of the hydroxy group and loss of water gives this species, which is the benzyl cation. Addition of ^-OH regenerates benzyl alcohol by a Lewis acid-base association reaction.

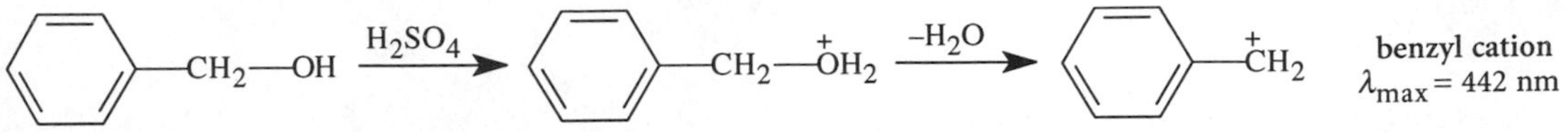

CH_2—OH $\xrightarrow{H_2SO_4}$ CH_2—$\overset{+}{O}H_2$ $\xrightarrow{-H_2O}$ $\overset{+}{C}H_2$ benzyl cation λ_{max} = 442 nm

17.30 First analyze the relationship of the isoprene skeletons. Then use steps like the ones shown in Eqs. 17.37–17.39 on text pp. 764–5 to assemble the parts from IPP and DMAP. Start with farnesyl pyrophosphate, the biosynthesis of which is shown in the solution to Problem 17.14(b). Note that B: = a base.

eudesmol

:OPP

farnesyl pyrophosphate biosynthesis is in the solution to Problem 17.14(b)

:OPP

:OH_2

B—H

B:

$\overset{+}{O}H_2$

OH

B:

OH

OH

H ← :B

OH

+ $H\overset{+}{B}$

eudesmol

A different arrangement of isoprene units in eudesmol can also be envisioned:

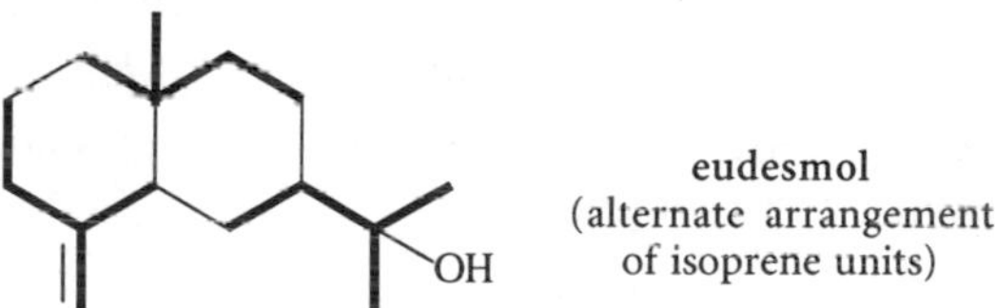

eudesmol (alternate arrangement of isoprene units)

A biosynthetic scheme based on this arrangement would be an equally correct answer. An isotope-labeling experiment would be required to distinguish between the two arrangements.

17.31 (a) Although the conjugate-base anion of 1,4-pentadiene is doubly allylic and resonance-stabilized, the conjugate-base anion of 1,3-cyclopentadiene is in addition aromatic. (See text p. 680 for a discussion of this case.) Consequently, much less energy is required for the ionization of 1,3-cyclopentadiene, and its pK_a is therefore much lower. (The pK_a difference between these two compounds is estimated to be about five units.)

(b) The carbocation intermediate in the solvolysis of 3-bromo-1,4-pentadiene is doubly allylic and therefore resonance-stabilized. Consequently, formation of this ion is relatively fast. Although the carbocation intermediate that results from the ionization of 5-bromo-1,3-cyclopentadiene *appears* to be doubly allylic and resonance-stabilized, it is in fact a cyclic system of $4n$ π electrons, and is therefore *antiaromatic* and particularly *unstable*. Consequently, this ion does not form. (Problem 15.60 on text p. 689 also deals with this case.)

17.32 (a) Two allylic Grignard reagents are in equilibrium, and, each reacts with D_2O.

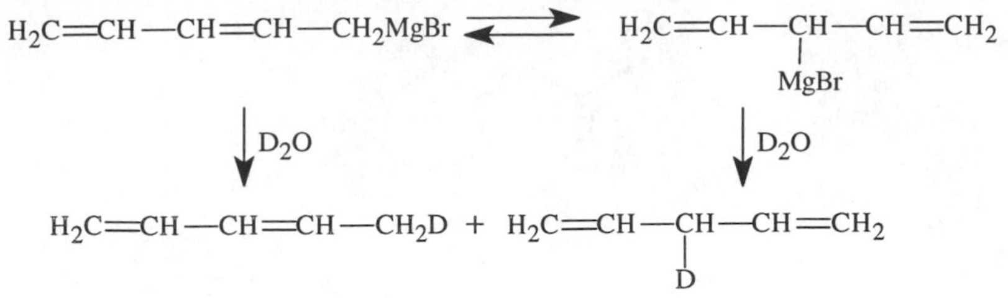

$$H_2C{=}CH{-}CH{=}CH{-}CH_2MgBr \rightleftharpoons H_2C{=}CH{-}\underset{\displaystyle MgBr}{\underset{|}{CH}}{-}CH{=}CH_2$$

$$\downarrow D_2O \qquad\qquad \downarrow D_2O$$

$$H_2C{=}CH{-}CH{=}CH{-}CH_2D \;+\; H_2C{=}CH{-}\underset{\displaystyle D}{\underset{|}{CH}}{-}CH{=}CH_2$$

(Note in this and the following part that a third product resulting from allylic-rearrangement of the Grignard reagent is identical to the first.)

(b) Two allylic Grignard reagents are in equilibrium, and each reacts with ethylene oxide:

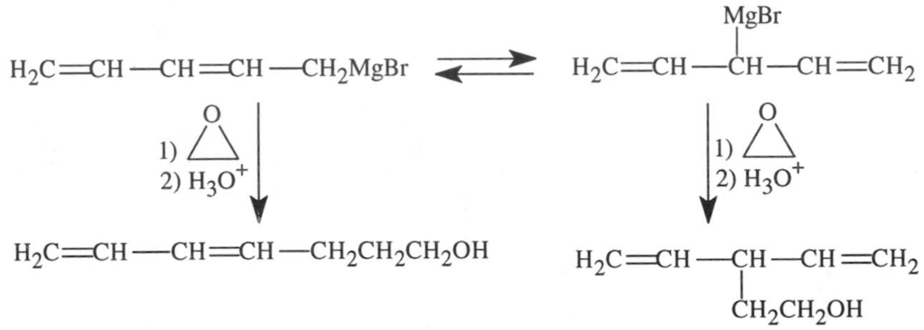

$$H_2C{=}CH{-}CH{=}CH{-}CH_2MgBr \rightleftharpoons H_2C{=}CH{-}\overset{\displaystyle MgBr}{\overset{|}{CH}}{-}CH{=}CH_2$$

1) ethylene oxide; 2) H_3O^+ (each)

$$H_2C{=}CH{-}CH{=}CH{-}CH_2CH_2CH_2OH \qquad H_2C{=}CH{-}\underset{\displaystyle CH_2CH_2OH}{\underset{|}{CH}}{-}CH{=}CH_2$$

(c) This is an S_N2 reaction of the ethanethiolate anion; because there is no reactive intermediate, only one product is formed:

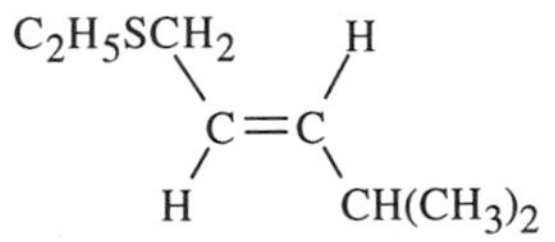

$$\begin{array}{ccc} C_2H_5SCH_2 & & H \\ & C{=}C & \\ H & & CH(CH_3)_2 \end{array}$$

(*E*)-1-ethylthio-4-methyl-2-pentene

(d) In the absence of a good nucleophile, the allylic bromide undergoes a slow S_N1 ionization, and ethanol attacks the two electron-deficient carbons of the resonance-stabilized carbocation to give two constitutionally isomeric ethyl ethers.

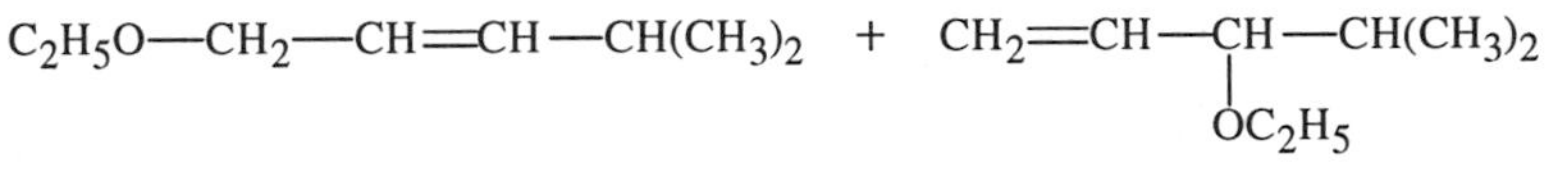

$$C_2H_5O{-}CH_2{-}CH{=}CH{-}CH(CH_3)_2 \;+\; CH_2{=}CH{-}\underset{\displaystyle OC_2H_5}{\underset{|}{CH}}{-}CH(CH_3)_2$$

(e) The unpaired electron in the free-radical intermediate is shared between two chemically nonequivalent carbons; two corresponding products are formed.

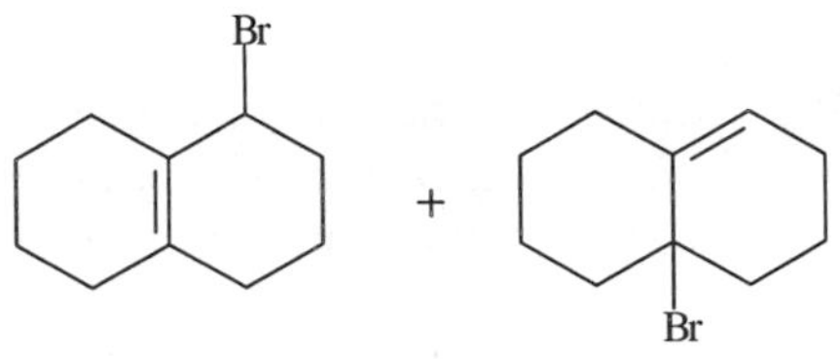

[Bicyclic structures: Br-substituted octahydronaphthalenes] Br + Br

(f) The observed product is derived from substitution of the tertiary benzylic hydrogen. This substitution involves a tertiary benzylic free-radical intermediate, whereas substitution of a *p*-methyl hydrogen would involve a less stable primary benzylic free-radical intermediate.

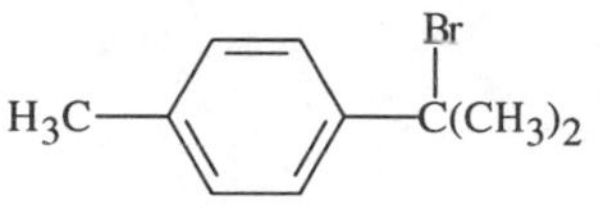

$H_3C-C_6H_4-CBr(CH_3)_2$

(g) The dibromide formed in the bromine-addition reaction undergoes two successive E2 reactions to give naphthalene, an aromatic compound. Both eliminations involve benzylic protons, and the final one is driven by aromaticity of the product.

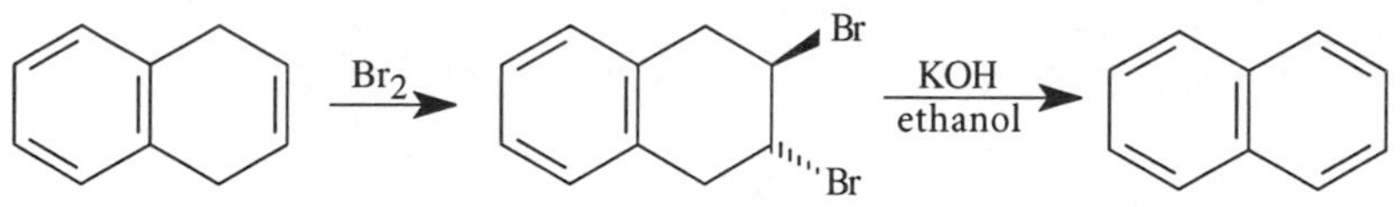

$\xrightarrow{Br_2}$ $\xrightarrow[\text{ethanol}]{KOH}$ naphthalene

(h) This product results from protonation of the alkene at the terminal carbon, because this gives the most stable carbocation intermediate. In this example, the 1,2-addition product not only is the more rapidly formed (kinetic) product but also is the more stable (thermodynamic) product. It is more stable thermodynamically because it is the only product that has a double bond conjugated with the benzene ring.

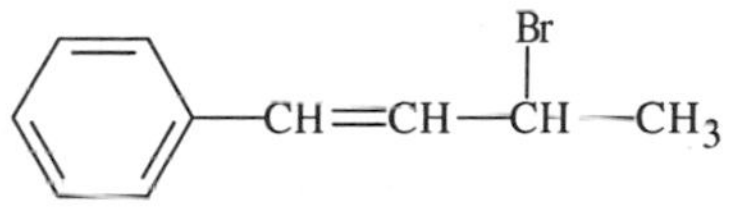

$C_6H_5-CH{=}CH-CHBr-CH_3$

(i) All benzylic carbons—even the ones in the ring—are oxidized to carboxylic acid groups.

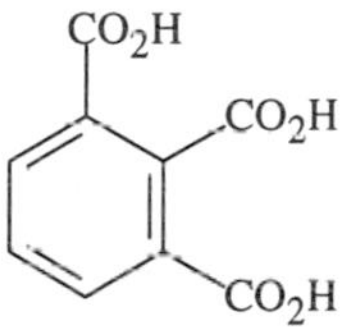

1,2,3 benzenetricarboxylic acid
(hemimellitic acid)

17.33 (a) Propargylic Grignard reagents, like allylic Grignard reagents, are an equilibrium mixture of two constitutional isomers. Each reacts with H_2O. (For the mechanism of Grignard reagent formation, see Eqs. 8.21a–b on text p. 336.)

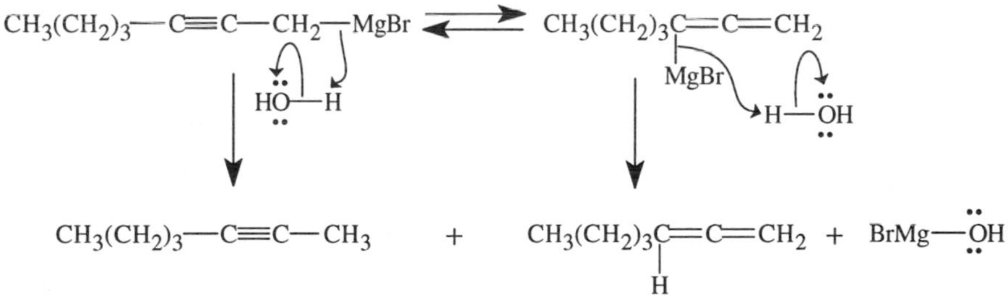

$$CH_3(CH_2)_3-C{\equiv}C-CH_2-MgBr \rightleftharpoons CH_3(CH_2)_3C(MgBr){=}C{=}CH_2$$

$$CH_3(CH_2)_3-C{\equiv}C-CH_3 \quad + \quad CH_3(CH_2)_3C(H){=}C{=}CH_2 + BrMg-OH$$

(b) Protonation of the triple bond gives a resonance-stabilized vinylic cation which is attacked intramolecularly by the oxygen of the —OH group. (H—OTs = *p*-toluenesulfonic acid; see text p. 415 for its structure.)

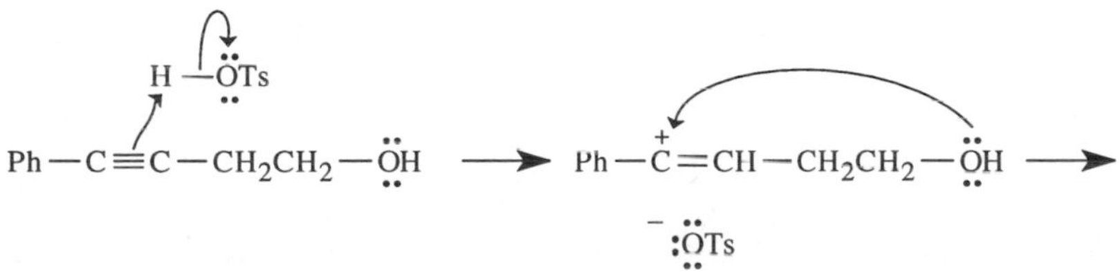

$$Ph-C{\equiv}C-CH_2CH_2-OH \xrightarrow{H-OTs} Ph-\overset{+}{C}{=}CH-CH_2CH_2-OH \quad {}^{-}:OTs \longrightarrow$$

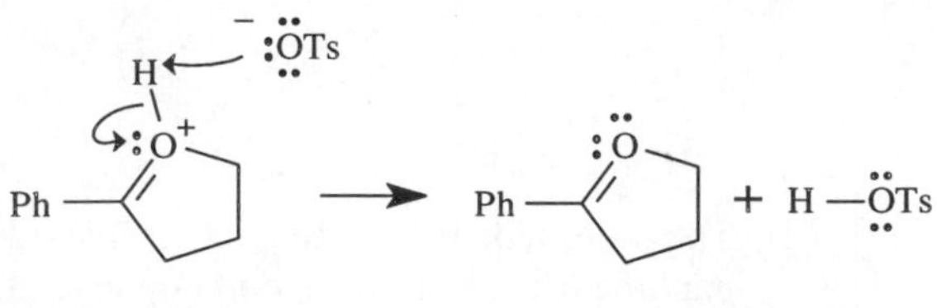

(c) A hydrogen on the central carbon is more acidic than an acetylenic hydrogen because the conjugate-base anion resulting from removal of the central hydrogen is both allylic and propargylic, and is therefore doubly resonance-stabilized. The conjugate-base anion is alkylated by allyl bromide.

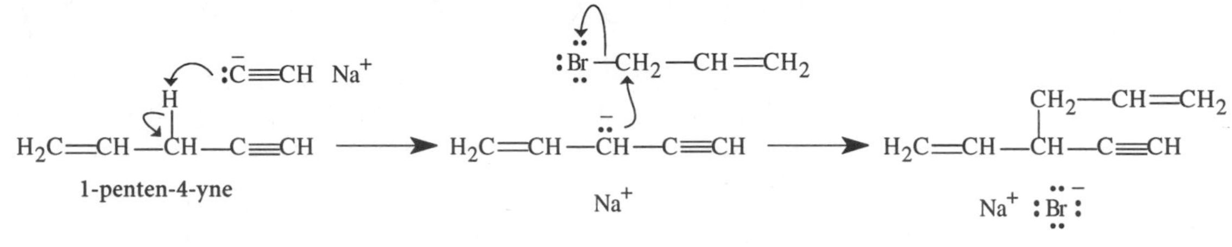

1-penten-4-yne

(d) Protonation of the alcohol and loss of water give an allylic carbocation that can be attacked by ethanol at either of two electron-deficient carbons to give a mixture of two constitutionally isomeric ethyl ethers. The following mechanism begins with the protonated alcohol.

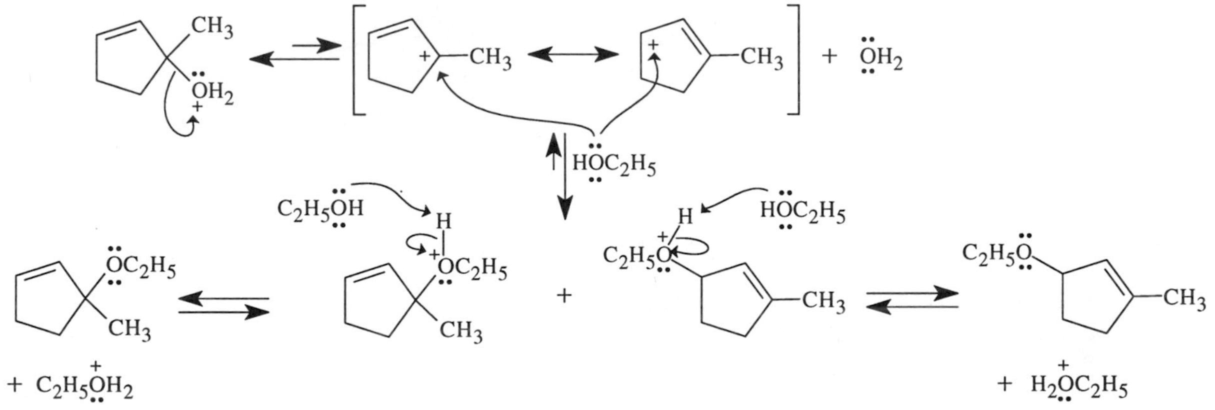

(e) The carbocation formed by protonation of the —OH group and loss of water adds to a neighboring double bond to give a new carbocation; this carbocation adds to a neighboring double bond to give yet another carbocation; and so on, until the product is formed by attack of water on the last carbocation. The mechanism below begins with the protonated alcohol:

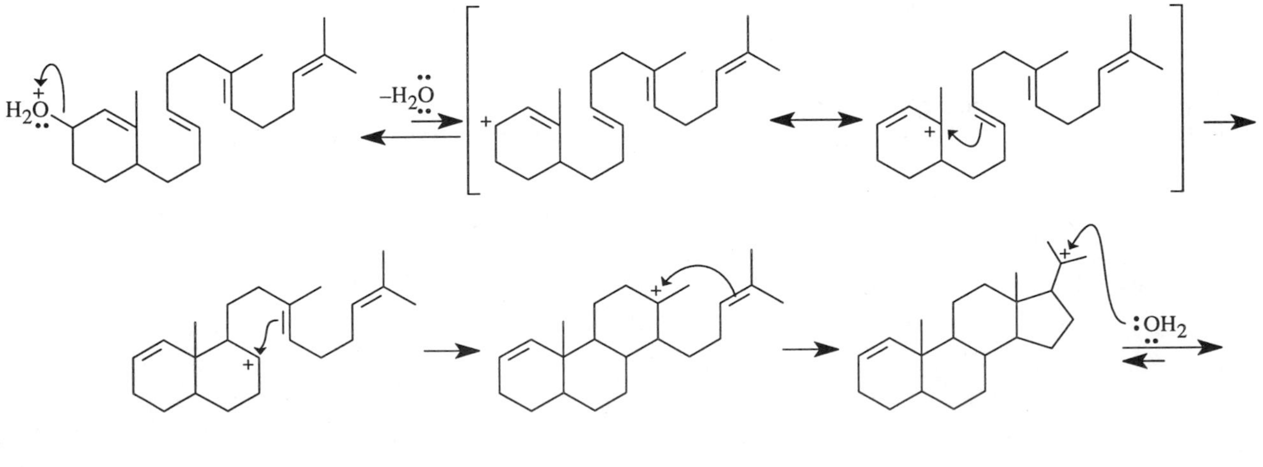

As you can see, the reaction in part (e) forms a steroid ring system. (See Sec. 7.6D.) Conceptually similar cyclization reactions are involved in the biosynthesis of steroids.

(f) The formation of a cyclopropane suggests the intermediacy of the following carbene:

$$(CH_3)_2C{=}C{=}C:$$

This carbene is formed by loss of chloride ion from an acetylenic ion. The carbene then adds to cyclohexene to give the cyclopropane. (See Eq. 9.62, text p. 396.)

$(CH_3)_2C(Cl)—C≡C—H$ + $^-$:$\ddot{O}C(CH_3)_3$ ⟶ $H\ddot{O}C(CH_3)_3$ + $(CH_3)_2C(Cl)—C≡\bar{C}$: ⟶ $(CH_3)_2C{=}C{=}C$: + :$\ddot{Cl}$:$^-$

$(CH_3)_2C{=}C{=}C$: + cyclohexene ⟶ $(CH_3)_2C{=}C{=}C$(cyclopropane fused to cyclohexane)

17.34 (a) Formation of a resonance-stabilized propargylic anion (that is, an anion that is "allylic" to a triple bond) is followed by protonation at the internal carbon. Formation of another resonance-stabilized anion followed by protonation gives the 1-alkyne, which is the most acidic species in the reaction mixture. Consequently, its conjugate-base anion is formed irreversibly, and all equilibria are thus pulled toward this anion. The net result is migration of the triple bond to the end of the carbon chain.

$H_2C(H)—C≡CCH_2CH_2CH_3$ (2-hexyne) + :$\bar{B}$ ⇌ [$H_2\bar{C}—C≡CCH_2CH_2CH_3$ ⟷ $H_2C{=}C{=}\bar{C}CH_2CH_2CH_3$] + B—H ⇌

$HC(H){=}C{=}CHCH_2CH_2CH_3$ + :$\bar{B}$ ⇌ [$H\bar{C}{=}C{=}CHCH_2CH_2CH_3$ ⟷ $HC≡C—\bar{C}HCH_2CH_2CH_3$] + B—H ⇌

$\bar{B}$: + $H—C≡C—CH_2CH_2CH_2CH_3$ ⇌ :$\bar{C}≡C—CH_2CH_2CH_2CH_3$ + B—H

Addition of water to the reaction mixture gives the 1-alkyne. Thus, the reaction shown here followed by addition of water results in the migration of a triple bond from an interior position to the 1-position of a carbon chain provided there are no branches. It has been shown that the triple bond migrates with equal frequency to both ends of the chain.

You might ask why a 1-alkyne is more acidic than a propargylic hydrogen when the conjugate-base anion of the latter is resonance-stabilized, and an acetylenic anion is not. This is a reflection of the effect of hybridization on acidity discussed on text p. 623–5. A *doubly allylic* hydrogen is more acidic than an acetylenic hydrogen, but a singly allylic hydrogen is not.

(b) The question is whether the triple bond migrates to the end of the carbon chain nearer to the methyl branch or to the end of the chain farther from the methyl branch. Once we consider the mechanism shown in part (a), the answer becomes clear. The migration of the triple bond occurs *away* from the methyl branch, because the mechanism of the reaction requires a stepwise migration of the triple bond, and a triple bond cannot form at a carbon that bears a branch because a carbon have no more than four bonds..

triple bond cannot form between these two carbons

$CH_3CH_2CH(CH_3)-C\equiv C-CH_2CH_2CH_3 \xrightarrow[\text{zipper reaction [see part (a)]}]{B:^-} CH_3CH_2CH(CH_3)CH_2CH_2CH_2C\equiv \bar{C}:$

3-methyl-4-octyne

17.35 This reaction, like the one in the previous problem, is a "zipper" reaction brought about by a series of acid-base equilibria. In this case, however, the equilibrium is not driven by ionization of the product, because all C—H bonds are less acidic than *tert*-butyl alcohol. Instead, the equilibrium lies to the right because the product is more stable than the starting material. This, in turn, is so for two reasons: (1) because the double bonds in the products are conjugated with the benzene ring and the double bond in the starting material is not; and (2) because the double bond in the product has more branches than the double bond in the starting material.

$^-:\ddot{O}C(CH_3)_3$ + Ph—CH(H)—CH=CH_2 ⇌ [Ph—$\ddot{C}H^-$—CH=CH_2 ⟷ Ph—CH=CH—$\ddot{C}H_2^-$] + $(CH_3)_3C\ddot{O}$—H ⇌ $(CH_3)_3C\ddot{O}:^-$ + Ph—CH=CH—CH_3

double bond is *not* conjugated and has one branch

double bond *is* conjugated and has two branches

17.36 The equilibrium lies to the right because the double bond has four branches whereas, in the starting material, it has three. Recall that branching at double bonds is a stabilizing effect (Sec. 4.5B) The mechanism involves simply protonation of the double bond to give the benzylic cation and loss of a proton to give the product.

TsÖ—H + 1-phenyl-2-methylenecyclopentane... ⇌ benzylic cation (Ph, H, H, CH_3, +) + $^-:\ddot{O}Ts$ ⇌ 1-phenyl-2-methylcyclopentene (CH_3) + H—ÖTs

17.37 From the structure of the Diels-Alder product *C*, compound *B* must be the conjugated diene 2-chloro-1,3-butadiene. Because compound *B* is an allylic rearrangement product of compound *A*, compound *A* must be the allene 4-chloro-1,2-butadiene. (Note that 1-chloro-1,3-butadiene would not give the same Diels-Alder product.)

Cl-substituted diene ≡ H_2C=CH—C(Cl)=CH_2

2-chloro-1,3-butadiene
compound *B*

$H_2C(Cl)$—CH=C=CH_2

4-chloro-1,2-butadiene
compound *A*

17.38 (a) The stability of the trityl radical is due to delocalization of the unpaired electron into all three benzene rings.

$\longleftrightarrow$ many other structures

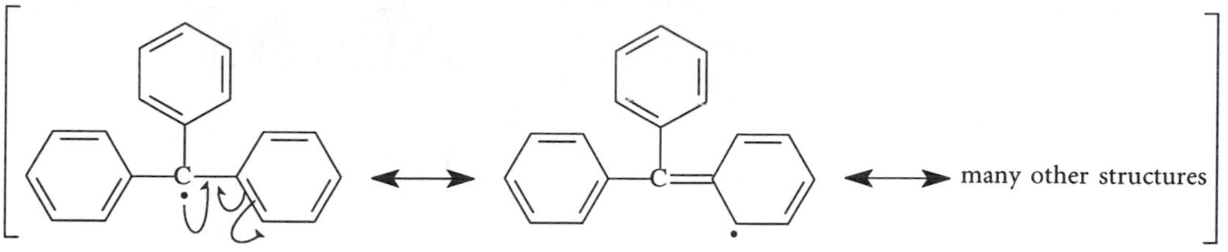

(b) Hexaphenylethane could be formed by the recombination of two trityl radicals:

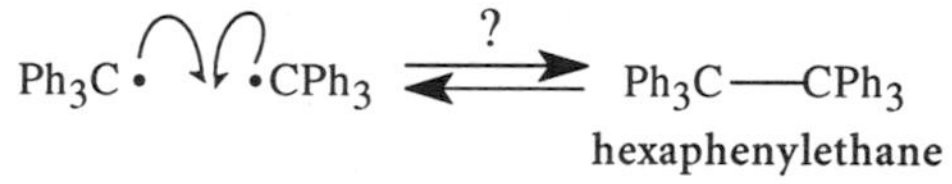

$$Ph_3C\cdot \; \cdot CPh_3 \overset{?}{\rightleftharpoons} Ph_3C{-}CPh_3$$

hexaphenylethane

(c) The dimer of the trityl radical is formed by the following mechanism:

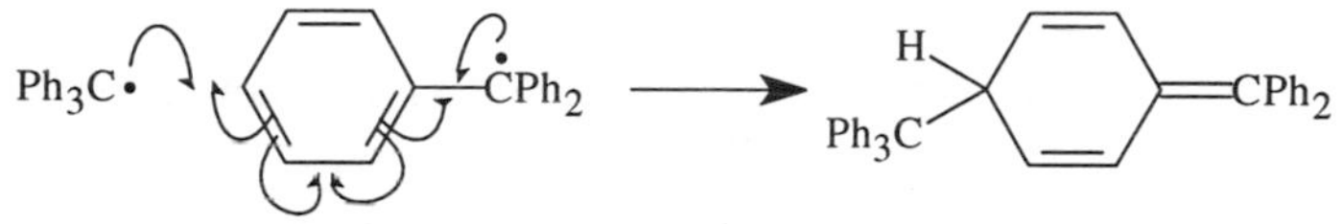

$Ph_3C\cdot$ ... $\dot{C}Ph_2$ $\longrightarrow$ H, Ph_3C ... $=CPh_2$

Hexaphenylethane is destabilized by van der Waals repulsions between gauche pairs of phenyl groups.

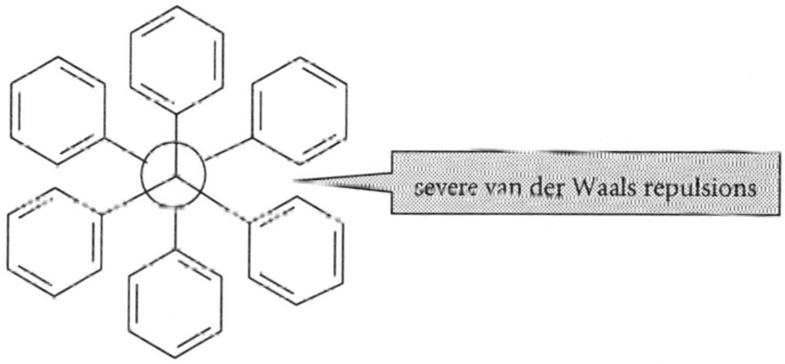

severe van der Waals repulsions

This destabilization is evidently so great, and the drive for recombination so powerful, that aromatic stability in one of the benzene rings can be sacrificed in forming the dimer shown in part (c).

17.39 (a) The triphenylmethyl anion (trityl anion) is stabilized by resonance interaction with all three benzene rings:

$\longleftrightarrow$ many other structures

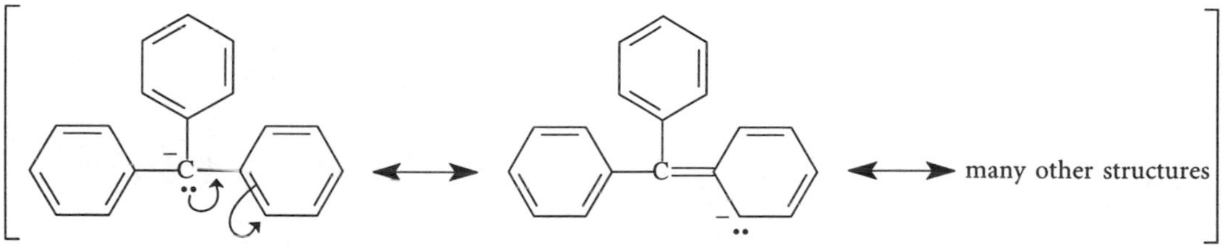

(b) The resonance interaction shown in part (a) is optimized if the three rings are coplanar; when the rings are coplanar, overlap is optimum between the 2*p* orbitals of the central carbon and those of the rings. However, when the rings are coplanar, there are significant van der Waals repulsions between hydrogens on different rings:

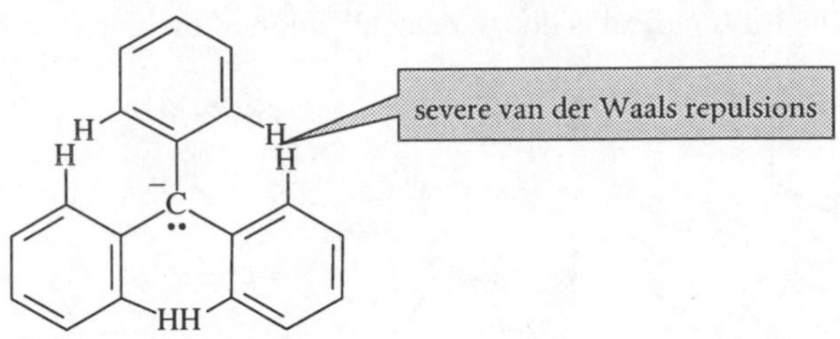

severe van der Waals repulsions

These van der Waals repulsions force the rings in the triphenylmethyl anion to twist significantly away from coplanarity, with a resulting cost in orbital overlap. (The molecule resembles a molecular "propeller.") In other words, in the triphenylmethyl anion, conjugation is not as effective as it would be if the rings were all coplanar. In fluoradene, the hydrogens that are the source of these repulsions are replaced by bonds that constrain the rings to coplanarity. The resulting increase in orbital overlap brings about a concomitant increase in stabilization of the anion. As a result, the pK_a of the conjugate acid is lowered significantly relative to that of triphenylmethane. The conjugate-base anion of fluoradene is really the "ultimate benzylic anion."

17.40 Section 7.9C shows that anti stereochemistry is one of the major pieces of evidence that a bromonium ion is involved in bromine addition to alkenes; similar conclusions hold for chlorine addition. The text also shows that a mechanism involving carbocation intermediates predicts mixed *syn-* and *anti*-addition (Eqs. 7.48a–b, text p. 289).

This problem deals with the *competition* between the two mechanisms: one involving a cyclic *chloronium ion* intermediate and the other involving a carbocation intermediate. The loss of stereoselectivity is evidence for the involvement of a carbocation intermediate. When the carbocation intermediate is resonance-stabilized (last two entries in the table), it is stable enough to compete energetically with the chloronium ion as a reactive intermediate. When R = *p*-methoxyphenyl, the carbocation intermediate is further resonance-stabilized by the para substituent. (See, for example, Eq. 17.6 on text p. 746.) In this case, the carbocation intermediate is stable enough, and the carbocation mechanism thus important enough, that nearly complete loss of stereoselectivity is observed.

17.41 Compound *A* is a benzylic bromination product, and it appears that compounds *B* and *C* are S_N1 products, each derived respectively from reaction of one of the nucleophiles present with a carbocation intermediate derived from *A*. Recall that S_N1 reactions are generally accompanied by E1 reactions; hence, it is reasonable to suppose that MPTP is such a product, namely, an alkene. The molecular formula of MPTP, $C_{12}H_{15}N$, indicates an unsaturation number of 6. Because five unsaturations are accounted for by the two rings in the starting material, the formula is consistent with the hypothesis that MPTP is an alkene.

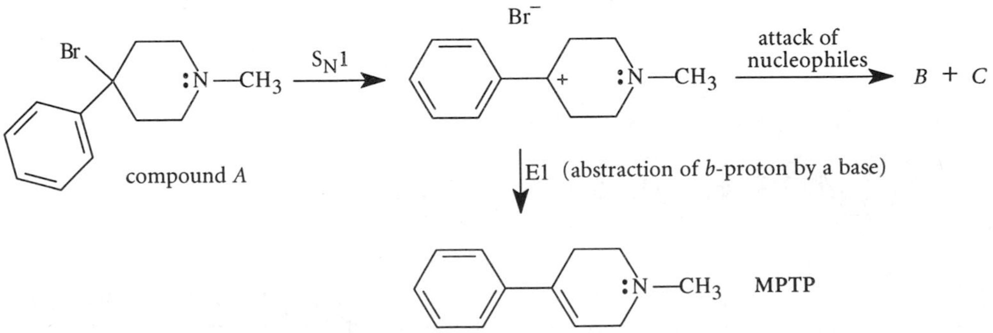

compound *A* —S_N1→ (carbocation, Br^-) —attack of nucleophiles→ *B* + *C*

E1 (abstraction of *b*-proton by a base) → MPTP

17.42 (a) Fact (1) indicates that the transition state involves a molecule of $(C_2H_5)_2NH$ and a molecule of alkyl halide. Fact (2) indicates that the indicated starting materials undergo the reaction shown; had this not been established, one could have postulated that one of the starting alkyl halides is converted into the other by the reaction conditions, and that the observed product could have originated from only one of the two starting compounds. Fact (3) establishes that the observed product is the actual product of the reaction, and is not derived from a subsequent reaction of its allylic isomer.

Now to the mechanisms. In the reaction of the second alkyl chloride an allylic rearrangement has occurred. Such allylic rearrangements generally suggest the involvement of carbocation intermediates, which have two resonance forms, and can be attacked at two different carbons. However, the observation of second-order kinetics rules out an S_N1 reaction. Hence, the reaction is evidently a direct substitution at the *allylic carbon*, thus:

$$(C_2H_5)_2\ddot{N}H \quad H_2C{=}CH{-}\underset{:\ddot{Cl}:}{\underset{|}{CH}}{-}CH_3 \longrightarrow (C_2H_5)_2\overset{+}{N}H{-}CH_2{-}CH{=}CH{-}CH_3 \quad :\ddot{\underset{..}{Cl}}:^-$$

This concerted allylic substitution mechanism has been termed the **S_N2' mechanism**. The other isomer reacts by the conventional S_N2 mechanism, in which the nucleophile attacks at the carbon bearing the halide.

(b) The large size of the nucleophile dictates that its attack occurs at the unbranched carbon to avoid van der Waals repulsions. In one case, this is the allylic carbon; in the other, it is the α-carbon.

17.43 Imagine, in the transition state of the S_N2' reaction, that the three carbons which either start or finish the reaction as part of a double bond, as well as the atoms attached to these carbons, lie in a common plane with the nucleophile and the leaving group either above or below this plane. Placing all groups with their stereochemical configurations corresponding to those of the reactant and the first product shows that approach of the nucleophile and departure of the leaving group occur from the *same face* of the plane. Hence, the major product is formed by a *syn*-substitution, that is, a *frontside* substitution.

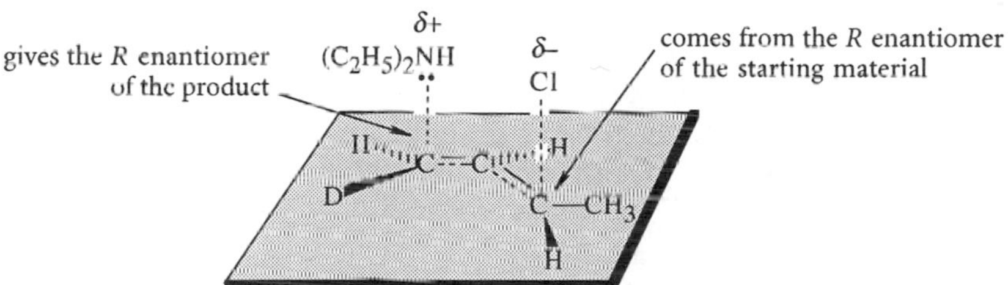

Note that the product is a *trans*-alkene with *R* stereochemistry at the asymmetric carbon. The minor product, a *cis*-alkene with *S* stereochemistry, also comes from a *syn*-substitution in which attack of the nucleophile and loss of the leaving group both occur from the lower face of the same plane. In order for this to happen, the transition state must adopt a conformation that leads to the *cis*-alkene. (You should demonstrate this to yourself with a diagram like the one above.) Because *trans*-alkenes are more stable than *cis*-alkenes, the "transoid" transition state—the one shown above—is more stable and accounts for 95% of the reaction. However, the point is that *both* products result from *syn*-substitution. You should show that if the nucleophile and leaving group were situated on opposite faces, the *R* enantiomer of the product would have a cis double bond and the *S* enantiomer of the product would have a trans double bond, a result not consistent with the observations given in the problem. The syn stereochemistry observed in the S_N2' reaction contrasts with the stereochemistry of the S_N2 reaction, which is an *anti*-substitution; that is, attack of the nucleophile and loss of the leaving group occur from opposite sides of the molecule. (See Fig. 9.2 on text p. 364; note the relationship of the nucleophile and the leaving group with respect to the plane shown in color.)

3. The reactivities of aryl or vinylic halides are quite different from those of ordinary alkyl halides.

B. Lack of Reactivity in S_N1 and E1 Reactions

1. Vinylic and aryl halides are virtually inert to the conditions that promote the S_N1 or E1 reactions of alkyl halides.
2. To undergo S_N1 or E1 reactions, vinylic halides must ionize to form vinylic cations.
   a. A vinylic cation is a carbocation in which the electron-deficient carbon is part of a carbon-carbon double bond.
      i. The geometry at a vinylic cation carbon is linear.
      ii. The electron-deficient carbon is *sp*-hybridized.
      iii. The vacant $2p$ orbital is not conjugated with the π-electron system of the double bond.
   b. Vinylic cations are considerably less stable than alkyl carbocations because of their *sp* hybridization and because of the electron-withdrawing polar effect of the double bond.
   c. Carbon-halogen bonds are stronger in vinylic halides than they are in alkyl halides.
      i. A vinylic carbon-halogen bond has more *s* character.
      ii. It takes more energy to break the carbon-halogen bond of a vinylic halide; this energy is reflected in a smaller rate of ionization.
3. To undergo S_N1 or E1 reactions, aryl halides must ionize to aryl cations.
   a. An aryl cation is a carbocation in which the electron-deficient carbon is part of an aromatic ring.
      i. The aryl cation prefers a linear geometry, which is virtually impossible to achieve because it would introduce too much strain in the six-membered ring.
      ii. The vacant orbital cannot become a $2p$ orbital and must remain an sp^2 orbital.
   b. Because an aryl cation is forced to assume a nonoptimal geometry and hybridization, it has a very high energy.
   c. The electron-withdrawing polar effect of the ring double bonds also destabilizes an aryl cation.

C. Lack of Reactivity in S_N2 Reactions

1. Vinylic and aryl halides are inert under S_N2 conditions.
2. Two factors within the transition state retard the S_N2 reactions of vinylic halides to such an extent that S_N2 reactions do not occur.
   a. Hybridization:
      i. In reaching the transition state, the carbon in the carbon-halogen bond must rehybridize from sp^2 to *sp*.
      ii. The relatively high energy of the transition state caused by *sp* hybridization reduces the rate of S_N2 reactions of vinylic halides.
   b. van der Waals repulsions (steric effects):
      i. The attacking nucleophile must approach the vinylic halide at the backside of the halogen-bearing carbon and in the plane of the alkene.
      ii. The resulting van der Waals repulsions raise the energy of the transition state and lower the reaction rate.
3. S_N2 reactions of aryl halides have the same problems as those of vinylic halides; in addition:
   a. Backside attack on the carbon-halogen bond would place the nucleophile on a path that goes through the plane of the benzene ring.
   b. Since the carbon that is attacked undergoes a stereochemical inversion, the reaction would necessarily yield a twisted and highly strained benzene derivative.

D. Elimination Reactions of Vinylic Halides

1. Base-promoted β-elimination reactions of vinylic halides do occur and can be useful in the synthesis of alkynes.
2. Vinylic eliminations require rather harsh conditions, and some of the more useful examples of this reaction involve elimination of β-hydrogens that have enhanced acidity.

II. NUCLEOPHILIC SUBSTITUTION REACTIONS OF ARYL HALIDES

A. Nucleophilic Aromatic Substitution

1. Aryl halides that have one or more strongly electronegative groups (especially nitro groups) ortho or para to the halogen undergo nucleophilic aromatic substitution reactions under relatively mild conditions.
2. A nucleophilic aromatic substitution reaction is a substitution that occurs at a carbon of an aromatic ring by a nucleophilic mechanism.
   a. It involves a nucleophile and a leaving group.
   b. It obeys a second-order rate law.
   c. It does *not* involve a one-step backside attack.
   d. It is faster when there are electron-withdrawing groups such as nitro groups ortho and/or para to the halogen leaving group that can stabilize the anionic intermediate by resonance.
   e. The effect of the halogen on the rate of this type of reaction is quite different from that in the S_N1 or S_N2 reaction of alkyl halides.
3. The nucleophilic aromatic substitution mechanism involves—
   a. attack of a nucleophile (Lewis base) on the π-electron system above (or below) the plane of the aromatic ring to yield a resonance-stabilized anion called a Meisenheimer complex.

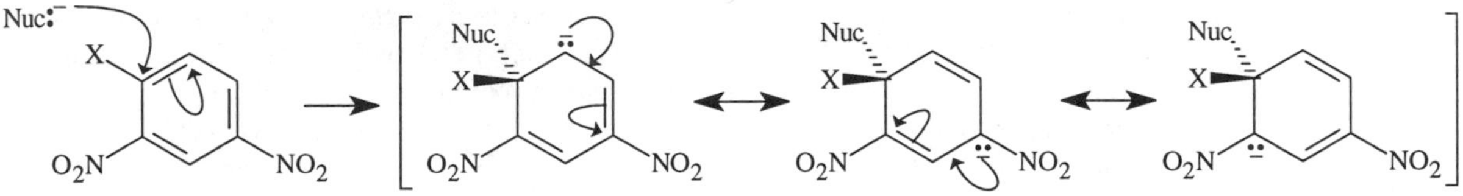

Nuc:⁻ X O₂N NO₂ → [Nuc X O₂N NO₂ ↔ Nuc X O₂N NO₂ ↔ Nuc X O₂N NO₂]

Meisenheimer complex

   i. The negative charge is delocalized throughout the π-electron system of the ring.
   ii. Formation of this anion is the rate-limiting step in many nucleophilic aromatic substitution reactions.
   iii. The negative charge in this complex is also delocalized into an ortho or para nitro group if present.

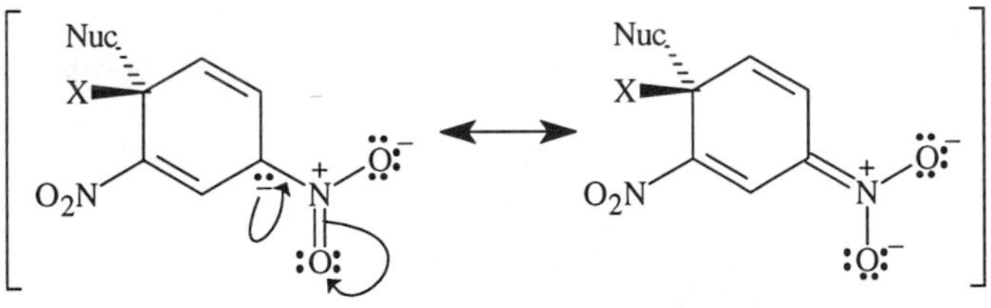

[Nuc X O₂N N⁺ O⁻ O ↔ Nuc X O₂N N⁺ O⁻ O⁻]

   b. breakdown of the Meisenheimer complex to products by loss of the halide ion.

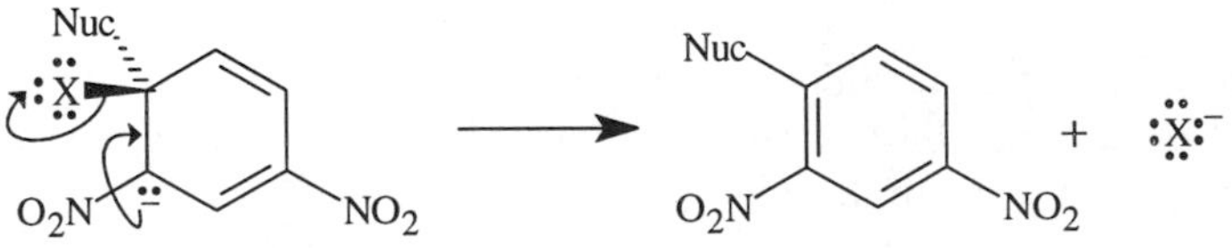

Nuc X O₂N NO₂ → Nuc O₂N NO₂ + :X:⁻

4. Ortho and para nitro groups accelerate the reaction because—
   a. the transition state resembles the Meisenheimer complex
   b. the nitro groups stabilize this complex by resonance.
5. In nucleophilic aromatic substitution reactions, aryl fluorides are most reactive.
   a. The reactivity of aryl halides is

$$\text{Ar—F} >> \text{Ar—Cl} \approx \text{Ar—Br} \approx \text{Ar—I}$$

b. Fluorine stabilizes the negative charge by its electron-withdrawing polar effect.
c. Because the loss of halide is not rate-limiting, the basicity of the halide, or equivalently, the strength of the carbon-halogen bond, is not important in determining the reaction rate.

6. The nucleophilic aromatic substitution reaction is an overall frontside substitution; it requires no inversion of configuration.

B. Substitution by Elimination-Addition: Benzyne

1. β-Elimination of an aryl halide gives a very unstable "alkyne" called benzyne.
   a. One of the two π bonds in the triple bond of benzyne is perpendicular to the π-electron system of the aromatic ring.
   b. The orbitals from which benzyne is formed are more like sp^2 orbitals than p orbitals.
   c. Benzyne is highly strained.
   d. Benzyne is a reactive intermediate in certain reactions of aryl halides.

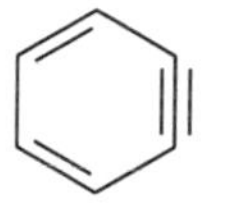

benzyne

2. The mechanism of this reaction involves:
   a. Formation of an anion at the ortho position; this step requires a strong base because benzene derivatives are only weakly acidic.
   b. Expulsion of a halide ion completes the β-elimination reaction and forms benzyne.

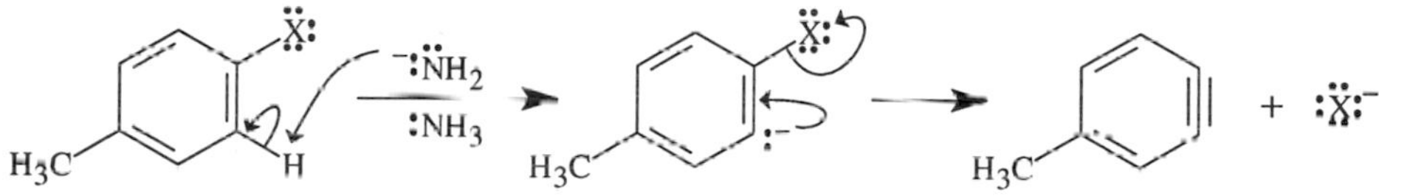

H_3C X: H $^-$:NH$_2$ / :NH$_3$ → H_3C X: :$^-$ → H_3C + :X:$^-$

   c. Attack by amide ion ($^-NH_2$) can occur at either carbon of the triple bond to give a new anion. If the benzyne intermediate is unsymmetrical, two different anions can form.

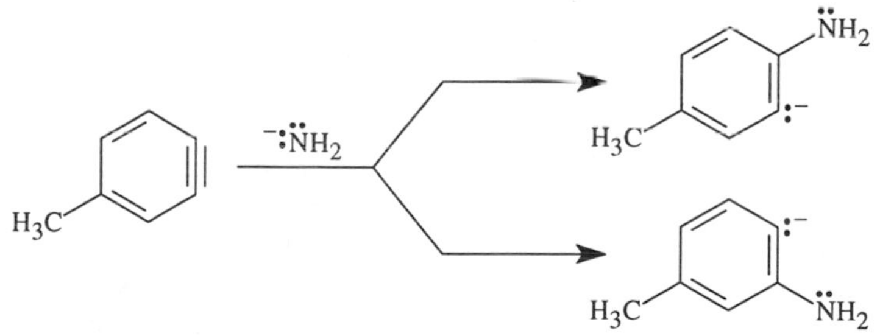

H_3C $^-$:NH$_2$ → H_3C NH$_2$:$^-$; H_3C :$^-$ NH$_2$

   d. Protonation of the two anions gives a mixture of products.

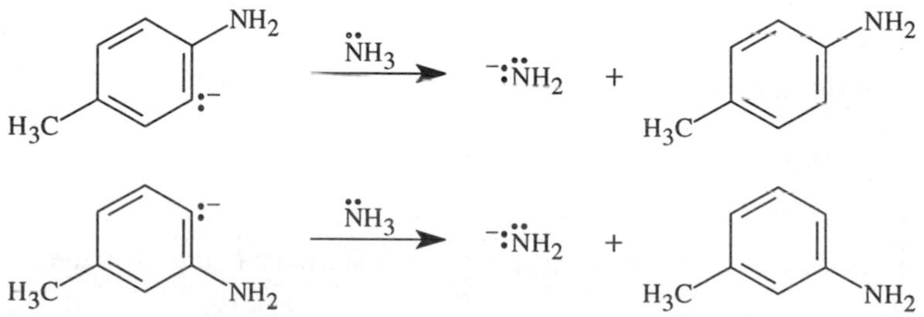

H_3C NH$_2$:$^-$ —NH$_3$→ $^-$:NH$_2$ + H_3C NH$_2$ p-methylaniline

H_3C :$^-$ NH$_2$ —NH$_3$→ $^-$:NH$_2$ + H_3C NH$_2$ m-methylaniline

   e. The overall substitution is really an elimination-addition process.

3. Substitution at a site different from the one occupied by the leaving group is called *cine*-substitution.

III. TRANSITION METALS

A. Transition Metals and Their Complexes

1. Reactions that look very much like nucleophilic substitutions can be carried out using certain transition-metal catalysts.
   a. This field is part of the larger field of organometallic chemistry—the chemistry of carbon–metal bonds.
   b. Transition metals are the elements in the "d block" or "B" group of the periodic table.
   c. In a given period n, elements are characterized by the progressive filling of d orbitals in quantum level $n - 1$ and of the s orbital in quantum level n.
2. Central to transition-metal chemistry are a wide variety of compounds containing transition metals surrounded by several groups, called ligands. Such compounds are called coordination compounds or transition-metal complexes.
3. In transition-metal chemistry, ligands are of two types (see Table 18.1, text page 791):
   a. L-Type ligands are ligands that give neutral molecules when removed from the metal complex.
      i. The electron pair in an L-type ligand is assigned completely to the ligand.
      ii. A dative bond is denoted by leaving the bonding electron pair on the ligand and depicting the ligand-metal bond as an arrow for these electrons to the metal.

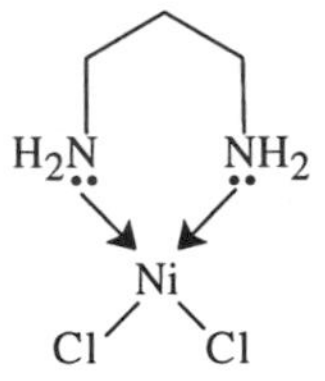

H_2N NH_2 Ni Cl Cl

dichloro(1,3-propanediamine)nickel(II)

      iii Because electrons on an L-type ligand belong to the ligand, removal of the ligand does not change the formal charge on either the ligand or the metal.
   b. X-Type ligands are ligands that give negative ions when removed from the metal complex.
      i. In bonds to X-type ligands, one electron is assigned to the ligand and one to the metal.
      ii. When an X-type ligand is removed, it takes on an additional negative charge and the metal takes on a compensating positive charge.
4. Alkenes can act as ligands by donating their π electrons to a metal.
   a. Allyl and cyclopentadienyl (Cp) are classified as both L-type and X-type ligands.
   b. Cyclopentadienyl (Cp) is an example of an L_2X ligand, a ligand with one X-type bond, which accounts for one negative charge when removed with a bonding pair from the metal, and two L-type bonds, which accounts for the four remaining π electrons.

B. Oxidation State

1. The oxidation state of the metal is an important concept in organometallic chemistry.
   a. For a metal M

Oxidation state of M = number of bonds to X-type ligands + charge on M

   b. L-Type ligands do not contribute to the oxidation state.

C. The d^n Notation

1. The number of unshared valence electrons on the metal in a complex is the number n in a notation called d^n.
2. The number n is calculated by determining the number of electrons remaining on the metal after removing all ligand with their electron pairs.
   a. Start with the number of valence electrons in the neutral transition element.

b. Remove an electron for each positive charge.
c. Add an electron for each negative charge.
d. Remove an electron for each bond to an X-type ligand.
e. L-Type ligands have no effect on d^n.

$$n = \text{valence electrons in neutral M} - \text{oxidation state of M}$$

D. Electron Counting: The 16- and 18-Electron Rules

1. To obtain the electron count in a complex, start with the electron count of the neutral metal, subtract the charge on the metal, add the number of X-type ligands, and add twice the number of L-type ligands.

$$\text{electron count} = \text{valence electrons in neutral M} - \text{charge on M} + \text{number of X-type ligands} + 2(\text{number of L-type ligands})$$

   a. In transition-metal chemistry, the most stable complexes in many cases have electron counts of 18 electrons. This statement is called the 18-electron rule.
   b. Frequently, transition metals in the 8–11 valence-electron group have a tendency to surround themselves with 16 electrons. This tendency is called the 16-electron rule.
   c. Hybridization arguments can be made to understand the basis of the 16- and 18-electron rules.
2. The 18-electron rule is important in octahedral complexes in which one s orbital + 3 p orbitals + 2 d orbitals hybridize to give six equivalent d^2sp^3 hybrid orbitals. For example, for $Ni^{2+}(H_2O)_6$:

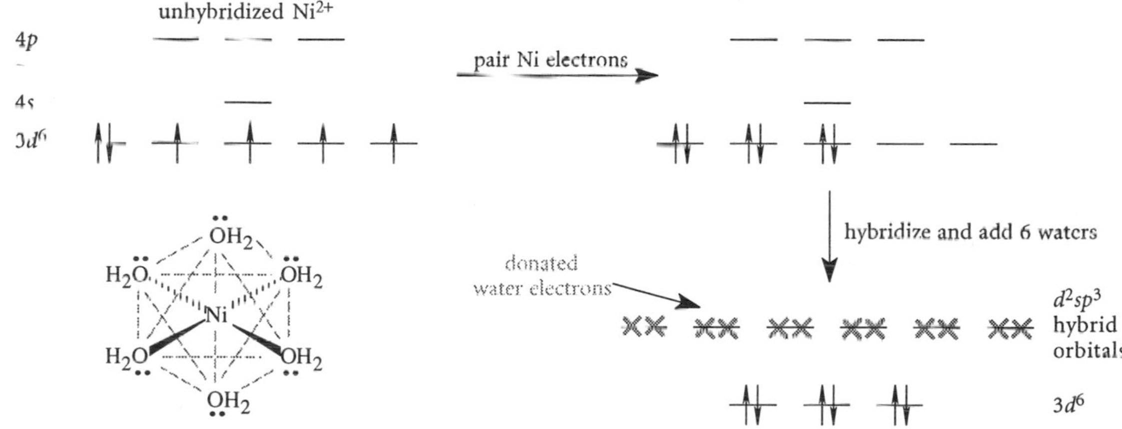

   a. These six hybrid orbitals are directed to the corners of a regular octahedron in the same sense that the four sp^3 carbon orbitals in methane are directed to the corners of a regular tetrahedron.
   b. Because these orbitals are directed in space, they can form stronger bonds than nonhybridized orbitals, and the strength of these bonds more than compensates for the energy cost of electron pairing and hybridization.
3. The 18-electron rule results from the rehybridization and maximal occupancy of all valence orbitals.
4. Just as the octet represents the number of valence electrons (8) in the outermost s and p orbitals of the nearest noble gas, 18 electrons is also the number of total $s + p + d$ valence electrons in the nearest noble gas.
5. The 16-electron rule is important in square-planar complexes of the 10-electron elements Ni, Pd, and Pt.
   a. Hybridization of one s orbital + two p orbitals + one d orbital gives four dsp^2 hybrid orbitals; a third unhybridized, relatively high-energy empty p orbital remains.
   b. The four hybrid orbitals are directed to the corners of a square and accept the electron pairs from four ligands to give a square-planar complex.

IV. FUNDAMENTAL REACTIONS OF TRANSITION-METAL COMPLEXES

Transition-metal complexes undergo a relatively small number of fundamental reaction types, and many reaction are readily understood simply as combination of these fundamental processes.

A. Lewis Base Dissociation-Association; Ligand Substitution

1. One of the most common reactions of transition-metal complexes is ligand dissociation and its reverse, ligand association.
2. In ligand dissociation, a ligand simply departs from the metal with its pair of electrons, leaving a vacant site (orbital) on the metal. This process does not change the oxidation state of the metal, but it does change the electron count.

$$\underset{\text{18-electron complex}}{Ni(CO)_4} \rightleftharpoons CO + \underset{\text{16-electron complex}}{Ni(CO)_3}$$

3. A ligand substitution can occur by—
   a. the dissociation of one ligand and the association of another, which is somewhat analogous to an S_N1 reaction in alkyl halides

$$[NiCl_4]^{2-} \rightleftharpoons Cl^- + [NiCl_3]^{1-} \overset{^-Br}{\rightleftharpoons} [NiBrCl_3]^{2-}$$

   b. or by a direct substitution, in which one ligand "kicks out" another, somewhat analogous to the S_N2 reaction.

$$[NiCl_4]^{2-} \xrightarrow{^-CN} [NiCl_3CN]^{2-}$$

4. In the most common ligand substitution reaction, ligands of the same type are exchanged: X-type ligands for X-type ligands and L-type ligands for L-type ligands.

B. Oxidative Addition

1. In oxidative addition, a metal M reacts with a compound X—Y to form a compound X—M—Y; the metal "inserts" into the X—Y bond.

$$\underset{\text{a 14-electron complex}}{[(C_2H_5)_3P]_2Ni} + I—CH_3 \longrightarrow \underset{\text{a 16-electron complex}}{[(C_2H_5)_3P]_2Ni(CH_3)(I)}$$

   a. Formation of a Grignard reagent from Mg metal and an alkyl halide is an example of the same reaction type in which the Mg is oxidized.
   b. Both the electron count and the oxidation number of the metal increase by two units.
2. Oxidative addition lies at the heart of transition-metal catalysis with aryl and vinylic halides.
   a. The carbon-halogen bond has an associated bonding molecular orbital, which is occupied by the two bonding electrons, and an antibonding molecular orbital, which is unoccupied.
   b. The bonding molecular orbital can serve as a ligand, donating its electrons to one of the empty hybrid orbitals on the metal.
   c. At the same time, one of the filled *d* orbitals of the metal overlaps with the antibonding molecular orbital of the carbon-halogen bond. This is often called back donation.
   d. This additional overlap strengthens the metal-ligand interaction, but weakens the carbon-halogen bond, because addition of electrons to an antibonding molecular orbital removes the energetic advantage of bonding.
   e. The carbon-halogen bond is weakened sufficiently that it actually breaks.
   f. Hence, electrons flow from the aryl halide to the metal and, at the same time, from the metal to the aryl halide.

3. Oxidative addition can occur by a variety of mechanisms, but a concerted process is fairly common.
4. Concerted oxidative addition occurs with retention of stereochemistry.

C. Reductive Elimination

1. Reductive elimination is conceptually the reverse of oxidative addition.
   a. Two ligands bond to each other and their bonds to the metal are broken.
   b. The metal is reduced, and its electron count is decreased.

$$(H)(H_3C)Ni[P(OC_2H_5)_3]_2 \longrightarrow CH_4 + Ni[P(OC_2H_5)_3]_2$$

a 16-electron complex — a 14-electron complex

2. Reductive elimination in general occurs with retention of stereochemistry.

$$(Ph_3P)_2Ni(CH_3)(CH{=}CHPh) \longrightarrow (Ph_3P)_2Ni + H_3C(H)C{=}C(H)Ph$$

D. Ligand Insertion

1. In ligand insertion, a ligand inserts into a metal-ligand bond and the inserting ligand gains a bond.

$$(Et)_3SnNMe_2 + CO_2 \longrightarrow (Et)_3Sn{-}O{-}C({=}O)NMe_2$$

2. Two types of ligand insertion are most frequently observed in transition-metal chemistry.
   a. In a 1,1-insertion, the new bond is formed at the same atom that was bound to the metal.
      i. Insertions of CO ligands are frequently observed examples of this type.

$$(CO)_5MnCH_3 + CO \longrightarrow (CO)_5Mn{-}C({=}O)CH_3$$

      ii. Ligand insertion leaves a vacant site (that is, an empty orbital) on the metal.
      iii. The empty metal orbital is filled by another molecule of the ligand from solution.
   b. In a 1,2-insertion, the migrating group moves to an atom adjacent to the one bound to the metal.

$$Cl_2Ti(CH_2CH_3)(H_2C{=}CH_2) \longrightarrow Cl_2Ti{-}CH_2CH_2CH_2CH_3$$

a 9-electron complex — a 7-electron complex

      i. The electron count is reduced by two.
      ii. This process results in an additional empty orbital on the metal.
      iii. This orbital can then gain an extra electron pair by ligand association.
      iv. A 1,2-insertion of an alkene is essentially a concerted addition of the metal and the migrating ligand to the alkene π bond.
      v. Because 1,2-ligand insertion is a concerted intramolecular addition reaction, the two new bonds must be formed at the same face of the π bond.

3. Ligand insertion is a *syn*-addition.

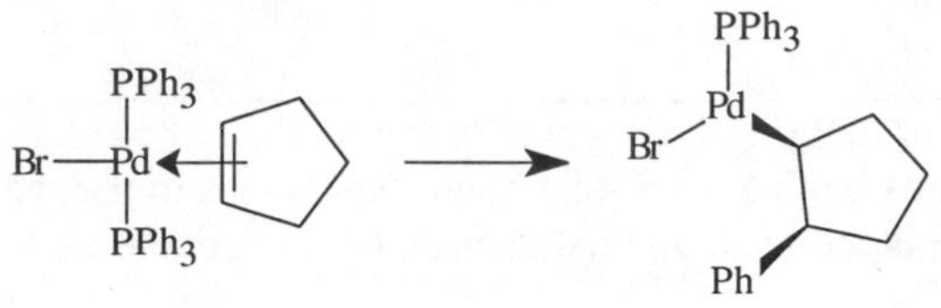

E. β-Elimination

1. In β-elimination, a group β to the metal migrates with its bonding electron pair to the metal.

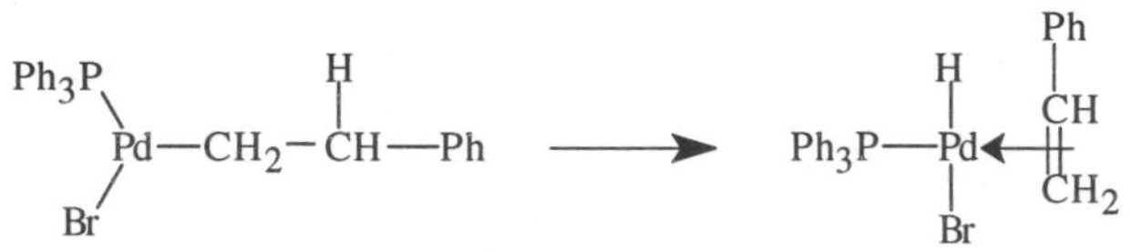

   a. This process is conceptually the reverse of ligand insertion.
   b. It often happens that β-elimination involves a hydride migration. In this reaction the hydride (a hydrogen with its bonding electrons) is eliminated.

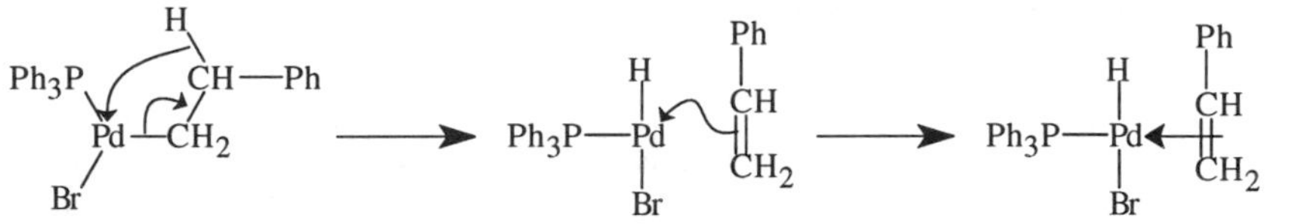

a 14-electron complex

a 16-electron complex

   c. Notice that β-elimination requires an empty orbital on the metal, because, as a result of this process, the electron count is increased by two units.

2. Because β-elimination is intramolecular, it must occur as a *syn*-elimination.

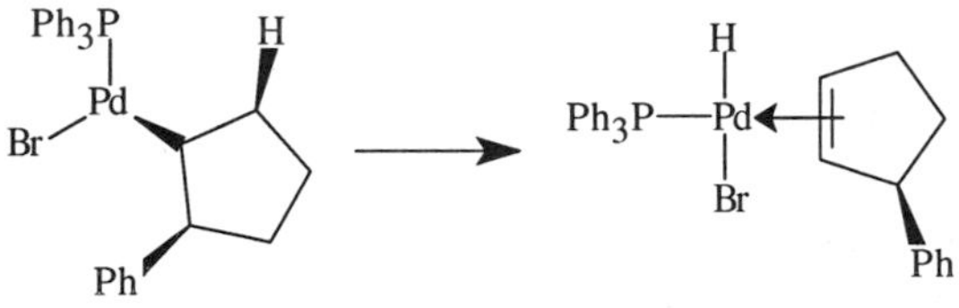

3. β-Elimination is conceptually the reverse of a 1,2 ligand insertion, which is a *syn*-addition.

V. PHENOLS

A. Introduction

1. Phenols are compounds in which a hydroxy (—OH) group is bound to a benzene ring.
2. Phenols, like alcohols, can ionize.
3. Phenols are much more acidic than alcohols.
   a. Typical alcohol pK_a values are 15–17; typical phenol pK_a values are 9–11.
   b. The enhanced acidity of phenols is due to stabilization of their conjugate-base anions.
   c. The conjugate base of a phenol is named as a phenoxide ion (common nomenclature) or as a phenolate ion (substitutive nomenclature).

phenol

sodium phenolate
(sodium phenoxide)

4. The reason for the lower pK_a values of phenols is the stability of their conjugate-base anions.
   a. Phenolate anions are stabilized by resonance.

   b. The polar effect of the benzene ring also stabilizes the negative charge.
5. Phenols are completely converted into their conjugate-base anions by NaOH solution.

$$\text{Ar—OH} \xrightarrow[\text{H}_2\text{O}]{\text{NaOH}} \text{Ar—O}^{-}\ \text{Na}^{+}$$

   a. Alkali metal salts of phenols have considerable solubility in water because they are ionic compounds.
   b. Acidification of an aqueous phenolate solution gives the phenol; if it is water-insoluble, it separates from the solution because, after acidification, it is no longer ionized.
   c. The acidities of water-insoluble phenols can sometimes be used to separate them from mixtures with other organic compounds.

B. Oxidation of Phenols to Quinones

1. Phenols undergo oxidation to quinones, which are compounds having two oxygens doubly bonded to a cyclohexadiene.
   a. If the quinone oxygens have a 1,2 (ortho) relationship, the quinone is called an *ortho*-quinone.
   b. If the quinone oxygens have a 1,4 (para) relationship, the quinone is called a *para*-quinone.

an *ortho*-quinone   a *para*-quinone

   c. *Ortho*-quinones are considerably less stable than the isomeric *para*-quinones because the C═O bond dipoles in *ortho*-quinones are nearly aligned and therefore have a repulsive, destabilizing interaction.
2. Hydroquinone and other phenols are sometimes used to inhibit free-radical reactions that result in the oxidation of other compounds. (See Eqs. 18.66a–b, text page 813.)
   a. Many free radicals abstract a hydrogen from hydroquinone to form a very stable radical called a semiquinone; the semiquinone radical is resonance-stabilized.
   b. A second free radical can react with the semiquinone to complete its oxidation to quinone.
   c. Hydroquinone terminates free-radical chain reactions by intercepting free-radical intermediates R• and reducing them to R—H.
3. Oxidation of phenols is a key element in the operation of some food preservatives and photographic developers.

C. Lack of Reactivity of the Aryl-Oxygen Bond

1. Phenols and phenyl ethers do not react under conditions used for S_N2 reactions of alcohols and their derivatives for the same reasons that aryl halides do not react (see Sec. 18.1).

2. Phenols also do not react under conditions used for S_N1 or E1 reactions of alcohols and their derivatives for the same reasons that aryl halides do not react (see Sec. 18.3).

VI. MECHANISM SUMMARY

A summary of mechanisms for the reactions of various organic halides is shown in Table SG18.1.

TABLE SG18.1 Organohalide Reaction Mechanism Summary

Type of Organohalide	Reaction Mechanism	Comments
methyl	S_N2	usually a very rapid reaction
primary	S_N2	inversion of configuration
	E2	observed with strong, highly branched bases only
secondary	S_N1 and S_N2	S_N1 slower than S_N1 of tertiary halides; S_N2 slower than S_N2 of primary halides
	E1 and E2	highly branched products observed except in E2 reactions with highly branched bases
tertiary	S_N1	predominant racemization
	E1 and E2	Saytzeff products observed except in E2 reaction with hindered bases
allylic	S_N1 and S_N2	products of allylic rearrangement observed in S_N1 reactions
benzylic	S_N1 and S_N2	rearrangement does not occur
vinylic	S_N1 or S_N2	not observed
	β-elimination	requires strong base
	metal-catalyzed substitution	Enabled by oxidative addition; bromides or iodides are generally required; also works with triflates, which can be prepared from phenols.
aryl	S_N1 or S_N2	not observed
	nucleophilic aromatic substitution	requires electron-withdrawing groups, typically nitro groups, in ortho and/or para positions
	elimination-addition (benzyne)	requires strong base; both direct and *cine*-substitution occur
	electrophilic aromatic substitution	proton is substituted by electrophile in *ortho* or *para* positions; deactivated towards substitution relative to benzene
	metal-catalyzed substitution	Enabled by oxidative addition; bromides or iodides are generally required; also works with triflates, which can be prepared from phenols.

REACTIONS

I. REACTIONS OF ARYL AND VINYL HALIDES

A. Elimination Reactions of Vinylic Halides

1. Base-promoted β-elimination reactions of vinylic halides do occur and can be useful in the synthesis of alkynes.

$$\text{Ph—CH═C(Br)—Ph} \xrightarrow[(CH_3)_3COH]{(CH_3)_3CO^- K^+} \text{Ph—C≡C—Ph} + K^+ Br^- + (CH_3)_3COH$$

B. Nucleophilic Aromatic Substitution

1. Aryl halides that have one or more nitro groups ortho or para to the halogen undergo nucleophilic substitution reactions under relatively mild conditions.

$$\text{2,6-}(O_2N)_2\text{-4-}CF_3C_6H_2Cl + 2\ HN[CH(CH_3)_2]_2 \longrightarrow \text{2,6-}(O_2N)_2\text{-4-}CF_3C_6H_2N[CH(CH_3)_2]_2 + H_2\overset{+}{N}[CH(CH_3)_2]_2\ Cl^-$$

trifluralin
(a preemergence herbicide)

2. The mechanism involves a resonance-stabilized anionic intermediate called a Meisenheimer complex. (See discussion of mechanism and reactivity under "Concepts," II.A.3–6.)
3. The reaction is faster when there are more nitro groups ortho and para to the halogen leaving group.
   a. Ortho and para nitro groups accelerate the reaction because the rate-limiting transition state resembles the Meisenheimer complex, and the nitro groups stabilize this complex by resonance.
   b. Other substituents that can provide resonance stabilization to the Meisenheimer complex can also activate nucleophilic aromatic substitution.
4. The effect of the halogen on the rate is quite different from that in the S_N1 or S_N2 reaction of alkyl halides.

$$\text{Ar—F} \gg \text{Ar—Cl} \sim \text{Ar—Br} \sim \text{Ar—I}$$

C. Elimination Reactions of Aryl Halides

1. Strong bases such as sodium amide ($Na^+\ {}^-NH_2$) can promote β-elimination in aryl halides to give benzyne intermediates, which are rapidly consumed by addition reactions with the bases; a mixture of products is usually observed. (B:$^-$ = base; B—H = its conjugate acid, often used as the solvent.)

$$R\text{-}C_6H_4\text{-}X \xrightarrow[-HX]{B:^-\ \text{(strong base)}} [R\text{-benzyne}] \xrightarrow[2)\ B—H]{1)\ B:^-} R\text{-}C_6H_3(H)(B) + R\text{-}C_6H_3(B)(H)$$

(See mechanistic discussion of this reaction under "Concepts," II.B.)

D. The Heck Reaction

1. In the Heck reaction, an alkene is coupled to an aryl bromide, aryl iodide, or aryl triflate under the influence of a Pd(0) catalyst.

$$\text{Ar—Br} + \text{C=C} + N(C_2H_5)_3 \xrightarrow[CH_3CN]{PdL_4} \text{Ar—C=C} + Br^- \; \overset{+}{H}N(C_2H_5)_3 \qquad L = P\text{—}(\text{2-}H_3C\text{-}C_6H_4)_3$$

a. The actual catalytically active species is PdL_2, formed by two ligand dissociations.

$$PdL_4 \rightleftarrows PdL_3 + L \rightleftarrows PdL_2 + L$$

b. The PdL_2 thus generated enters into the catalytic cycle, which consists of the following steps:

i. oxidative addition of the aryl halide

$$L_2Pd + \text{Ar—Br} \longrightarrow L_2Pd(Ar)(Br)$$

ii. ligand substitution of the alkene for one of the catalyst ligands

$$L_2Pd(Ar)(Br) + \text{C=C(H)} \longrightarrow \text{Br—Pd(Ar)(L)} \leftarrow \text{C=C} + L$$

iii. 1,2-insertion of the alkene in which the aryl group migrates to carbon-2 of the alkene

$$\text{Br—Pd(Ar)(L)} \leftarrow \text{C=C(H)} \longrightarrow \text{Br—Pd(L)—C(H)—C—Ar}$$

iv. β-elimination of the newly formed aryl substituted alkene followed by ligand association

$$\text{Br—Pd(L)—C(H)—C—Ar} \rightleftarrows \text{Br—Pd(H)(L)} \leftarrow \text{C=C(Ar)}$$

v. ligand dissociation of the aryl substituted alkene

$$\text{Br—Pd(H)(L)} \leftarrow \text{C=C(Ar)} \rightleftarrows \text{Br—Pd(H)(L)} + \text{C=C(Ar)}$$

vi. ligand association of a catalyst ligand

$$\mathrm{Br{-}PdH(L)} + \mathrm{L} \rightleftharpoons \mathrm{L_2Pd(Br)(H)}$$

vii. reductive elimination of hydrogen halide, which is neutralized by the basic triethylamine present

$$\mathrm{L_2Pd(Br)(H)} \rightleftharpoons \mathrm{L_2Pd} + \mathrm{HBr}$$

$$\mathrm{HBr} + \mathrm{N(C_2H_5)_3} \longrightarrow \mathrm{Br^-}\ \mathrm{H\overset{+}{N}(C_2H_5)_3}$$

c. Pd(II) catalysts can be used with aryl iodides, but it is thought that the Pd(II) is reduced to Pd(0) during the reaction.

2. With cyclic alkenes, β-elimination occurs at the β-carbon to which the aryl group was not added owing to the stereochemical constraints of the reaction mechanism.

$$\mathrm{Ph{-}Br} + \text{cyclopentene} \xrightarrow[\text{N(C}_2\text{H}_5)_3\ \text{dimethylformamide}]{\text{Pd(OAc)}_2} \text{3-phenylcyclopentene} + \mathrm{HBr}$$

3. When the Heck reaction is applied to unsymmetrically substituted alkenes, two products are in principle possible, because insertion might occur at either of the alkene carbons.
   a. When R = phenyl, CO_2R (ester), CN, or other relatively electronegative group, the aryl halide tends to react at the unsubstituted carbon, to usually give the *E* (trans) stereoisomer.

$$\mathrm{Ph{-}Br} + \mathrm{H_2C{=}CH{-}C(=O)OCH_3} \xrightarrow[\text{N(C}_2\text{H}_5)_3\ \text{dimethylformamide}]{\text{Pd(OAc)}_2} \mathrm{Ph{-}CH{=}CH{-}C(=O)OCH_3} + \mathrm{HBr}$$

   b. When R = alkyl, mixtures of products are often observed.
4. The Heck reaction is another example of a reaction that can be used to form carbon-carbon bonds.

II OTHER EXAMPLES OF TRANSITION-METAL CATALYZED REACTIONS

A. Ziegler-Natta Catalyst

1. One of the most important transition-metal catalysts in commerce is a catalyst formed from $TiCl_3$ and $(CH_3CH_2)_2AlCl$, called the Ziegler-Natta catalyst.

$$n\ \mathrm{H_2C{=}CH_2} \xrightarrow[(\mathrm{C_2H_5})_2\mathrm{AlCl}]{\mathrm{TiCl_3}} \left(\mathrm{CH_2CH_2}\right)_n$$

2. This catalyst brings about the polymerization of ethylene and alkenes at 25 °C and 1 atm pressure to give high-density polyethylene.
3. The mechanism involves a two-step iterative process:

a. a ligand association

$$Cl_2Ti(CH_2CH_2R) + H_2C{=}CH_2 \longrightarrow Cl_2Ti(CH_2CH_2R)(\leftarrow CH_2{=}CH_2)$$

b. a 1,2-ligand insertion.

$$Cl_2Ti(CH_2CH_2R)(\leftarrow CH_2{=}CH_2) \longrightarrow Cl_2Ti(CH_2CH_2CH_2CH_2R)$$

B. Hydroformylation

1. Hydroformylation is another commercially important process that involves a transition-metal catalyst, in this case, tetracarbonylhydridocobalt(I) catalyst, sometimes called the oxo process. (See Problem 18.22.)

$$\text{cyclopentene} + H_2 + CO \xrightarrow{HCo(CO)_4} \text{cyclopentane-CHO}$$

C. Catalytic Hydrogenation

1. An important transition-metal catalyzed reaction is the homogeneous catalytic hydrogenation of akenes using a soluble rhodium(I) catalyst called Wilkinson's catalyst, $ClRh(PPH_3)_3$. (See Problem 18.16.)

$$\text{cyclopentene} + H_2 \xrightarrow{ClRh(PPh_3)_3} \text{cyclopentane}$$

2. Catalytic hydrogenation is an extremely important reaction that occurs over carbon-supported transition metals such as Ni, Pd, and Pt. The mechanism of catalytic hydrogenation is not definitively known.

$$\text{cyclopentene} + H_2 \xrightarrow{Pd/C} \text{cyclopentane}$$

III. REACTIONS OF PHENOLS AND PHENOLATES

A. Use of Phenoxides as Nucleophiles

1. Phenoxides can be used as nucleophiles.
2. Aryl ethers can be prepared by the reaction of a phenoxide anion and an alkyl halide (another example of the Williamson ether synthesis).

$$H_3C\text{-}C_6H_4\text{-}OH \xrightarrow[\text{2-butanone, } -KHCO_3]{K_2CO_3} [H_3C\text{-}C_6H_4\text{-}O^- K^+] \xrightarrow[-KI]{CH_3CH_2I} H_3C\text{-}C_6H_4\text{-}OCH_2CH_3$$

B. Substitution at the Aryl-Oxygen Bond: The Stille Reaction

1. Pd(0) catalysts can catalyze substitution at the aryl-oxygen bond if the phenolic —OH group is converted into a trifluoromethanesulfonate ester, which is a very good leaving group.
   a. Sulfonate ester methodology is used for this purpose.

$$\text{Ar—OH} + \text{F}_3\text{C—S(=O)}_2\text{—O—S(=O)}_2\text{—CF}_3 + \text{pyridine} \longrightarrow \text{Ar—O—S(=O)}_2\text{—CF}_3 + \text{pyridinium (N—H)}^+ \; \text{F}_3\text{C—SO}_3^-$$

a phenol; trifluoromethane-sulfonic anhydride; a base; a trifluoromethanesulfonate ester (a triflate); a salt

   b. Trifluoromethanesulfonate esters are nicknamed triflates; the triflate group is often abbreviated —OTf.
   c. The triflate group is one of the best leaving groups because it is an extremely weak base.
2. Aryl triflates react readily with organotin derivatives in the presence of Pd(0) catalysts to give coupling products.

$$\text{Ar—O—S(=O)}_2\text{—CF}_3 + \text{Ph—Sn(CH}_3)_3 + \text{LiCl} \xrightarrow[\text{dioxane}]{\text{Pd(PPh}_3)_4} \text{Ar—Ph} + (\text{CH}_3)_3\text{SnCl} + \text{LiOTf}$$

   a. The organotin compounds used in the Stille reaction are either commercially available or readily prepared from Grignard Reagents and commercially available trialkyltin chlorides.

$$\text{PhBr} \xrightarrow[\text{ether}]{\text{Mg}} \text{PhMgBr} \xrightarrow{(\text{CH}_3)_3\text{SnCl}} \text{PhSn(CH}_3)_3 + \text{ClMgBr}$$

   b. In general, vinylic groups and aryl groups are transferred from the tin more rapidly than alkyl groups.
   c. If a tetraalkylstannane is used, alkyl groups can also be transferred.
3. The mechanism of the Stille reaction probably involves the following steps:
   a. an oxidative addition of the aryl triflate to the 14-electron $Pd(PPh_3)_2$

$$\text{L}_2\text{Pd} \xrightarrow{\text{Ar—OTf}} \text{L}_2\text{Pd(Ar)(OTf)}$$

   b. a ligand substitution by excess chloride ion (as LiCl) to the resulting unstable complex

$$\text{L}_2\text{Pd(Ar)(OTf)} + \text{LiCl} \longrightarrow \text{L}_2\text{Pd(Ar)(Cl)} + \text{LiOTf}$$

   c. a second ligand substitution in which the aryl or alkyl groups on the organotin compound (which resembles a carbanion and is nucleophilic) substitutes for the chloride on the Pd

$$\text{L}_2\text{Pd(Ar)(Cl)} + \text{R—Sn}\!\lt \longrightarrow \text{L}_2\text{Pd(Ar)(R)} + \text{Cl—Sn}\!\lt$$

   d. a reductive elimination to give the product and regenerate the catalyst.

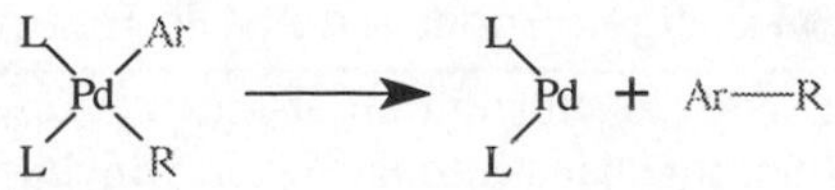

4. The Stille reaction can be carried out with aryl halides instead of aryl triflates.
5. The transferred alkyl group does not rearrange in the Stille reaction.
6. This is another method that can be used to form carbon-carbon bonds.

C. Electrophilic Aromatic Substitution Reactions of Phenols

1. Phenols undergo electrophilic aromatic substitution reactions.
2. Phenols react rapidly with bromine.
   a. Phenol can be halogenated once under mild conditions that do not affect benzene itself.

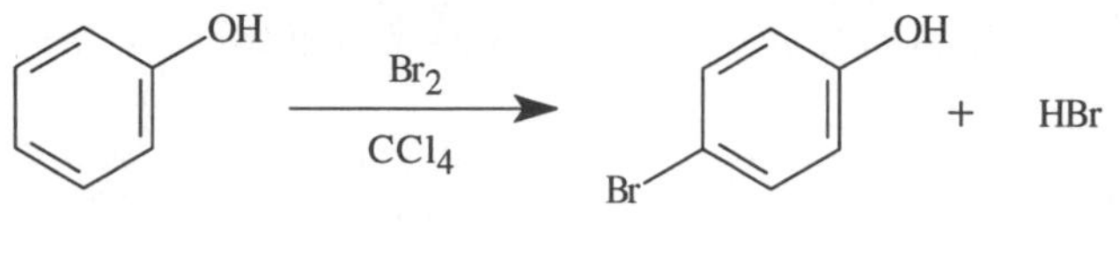

   b. Phenol reacts with Br_2 in H_2O (bromine water) to give 2,4,6-tribromophenol.
      i. Bromine reacts with water to give protonated hypobromous acid, a more potent electrophile than bromine itself.
      ii. In aqueous solutions near neutrality, phenol partially ionizes to its conjugate-base phenoxide anion; this anion is very reactive and brominates instantly, thereby pulling the phenol-phenolate equilibrium to the right.
      iii. Phenoxide ion is much more reactive than phenol because the reactive intermediate is not a carbocation, but is instead a more stable neutral molecule.

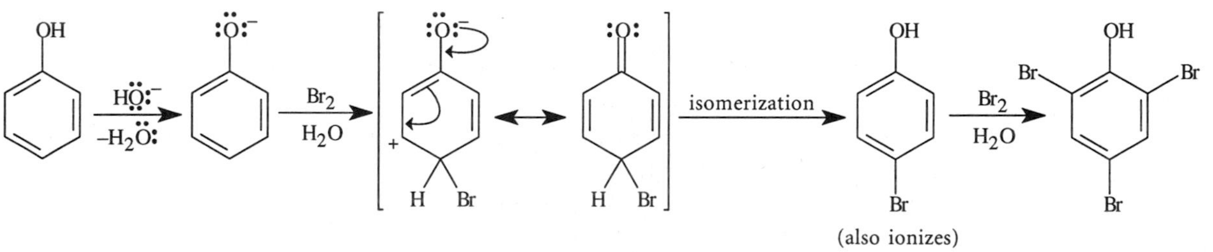

(also ionizes)

   c. In strongly acidic solution, in which formation of the phenolate anion is suppressed, bromination can be stopped at the 2,4-dibromophenol stage.

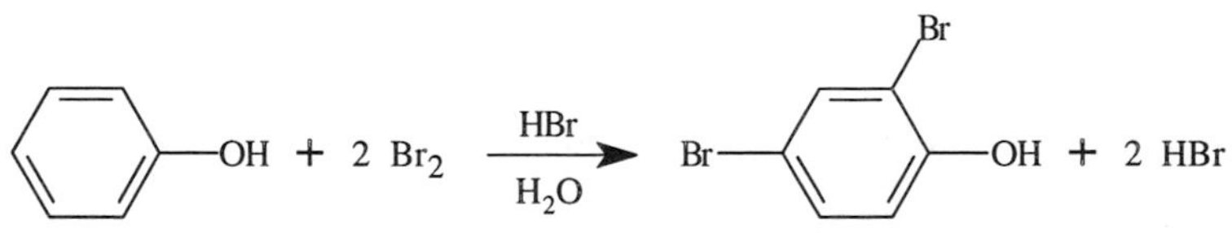

phenol 2,4-dibromophenol

3. Phenol is very reactive in other electrophilic substitution reactions.
   a. Phenol can be nitrated once under mild conditions.
   b. Direct nitration is not the preferred method for synthesis of di- and trinitrophenol, because the concentrated HNO_3 required for multiple nitrations is an oxidizing agent. Instead, nucleophilic aromatic substitution is used. (See "Concepts," II.A for a discussion of mechanism.)

$$O_2N\text{-}C_6H_3(NO_2)\text{-}Cl + Na^+ \,^-OH \longrightarrow \xrightarrow{H_3O^+} O_2N\text{-}C_6H_3(NO_2)\text{-}OH$$

c. The great reactivity of phenol in electrophilic aromatic substitution does not extend to the Friedel-Crafts acylation reaction.

i. Friedel-Crafts acylation of phenol occurs slowly, but may be carried out successfully at elevated temperatures; the ring is acylated only once.

ii. Phenols are less reactive because they react rapidly with the $AlCl_3$ catalyst.

$$C_6H_5\ddot{O}H \xrightarrow{AlCl_3} \left[C_6H_5\ddot{O}(H)\text{-}AlCl_2 \longleftrightarrow C_6H_5\overset{+}{\ddot{O}}{=}\overset{-}{Al}Cl_2 \right] + HCl$$

D. Oxidation of Phenols to Quinones

1. *p*-Hydroxyphenols (hydroquinones), *o*-hydroxyphenols (catechols), and phenols with an unsubstituted position ortho or para to the hydroxy group are oxidized to quinones.

$$HO\text{-}C_6H_4\text{-}OH \xrightarrow[HOAc]{CrO_3} O{=}C_6H_4{=}O$$

hydroquinone $\qquad$ *p*-benzoquinone

2. The oxidation of hydroquinone and its derivatives to the corresponding *p*-benzoquinone can also be carried out reversibly in an electrochemical cell.

E. Industrial Preparation and Use of Phenol

1. Phenol and acetone are prepared from a single starting material, cumene, which in turn comes from benzene and propene, two compounds obtained from petroleum.
2. The production of phenol and acetone is a two-step process.
   a. Cumene undergoes an autoxidation with molecular oxygen to form cumene hydroperoxide.

$$C_6H_5\text{-}C(CH_3)_2\text{-}H \xrightarrow{O_2} C_6H_5\text{-}C(CH_3)_2\text{-}O\text{-}O\text{-}H$$

cumene $\qquad$ cumene hydroperoxide

   b. The cumene hydroperoxide undergoes an acid-catalyzed rearrangement that yields both acetone and phenol.

$$C_6H_5\text{-}C(CH_3)_2\text{-}O\text{-}O\text{-}H \xrightarrow{H_3O^+} C_6H_5\text{-}OH + (H_3C)_2C{=}O$$

cumene hydroperoxide $\qquad$ phenol $\qquad$ acetone

SCIENTISTS

Richard F. Heck (b. 1931)	18.5F
Jakob Meisneheimer (1876–1934)	18.4A
Tsutomu Mizoroki	18.5F
Giulio Natta (1903–1979) [Chemistry 1963]	18.5G
John D. Roberts (b. 1918)	18.4B
John K. Stille (d. 1989)	18.9B
Sir Geoffrey R. Wilkinson (1921–1996) [Chemistry 1973]	18.5G
Karl Waldemar Ziegler (1898–1973) [Chemistry 1963]	18.5G

Dates in brackets refer to Nobel Prize awards.

STUDY GUIDE LINKS

18.1 Vinylic Cations

Despite their instability, vinylic carbocations are reactive intermediates in some reactions, but typically in these situations the vinylic cation is stabilized by some factor. For example, in the acid-catalyzed hydration of phenylacetylene, the carbocation intermediate is not only vinylic, but also benzylic.

$$\text{Ph—C}\equiv\text{CH} + \text{H—}\overset{+}{\text{O}}\text{H}_2 \longrightarrow \text{Ph—}\overset{+}{\text{C}}\text{=CH}_2 \xrightarrow{:\text{OH}_2} \text{Ph—}\underset{}{\overset{\text{HO}^{+}\text{—H}}{\text{C}}}\text{=CH}_2 \xrightarrow{:\text{OH}_2} \text{Ph—}\overset{\text{HO:}}{\text{C}}\text{=CH}_2 + \text{H}_3\text{O}^+$$

phenylacetylene

a vinylic carbocation (also benzylic)

an enol; spontaneously converted into a ketone (Sec. 14.5A)

Notice also that it takes less energy to form a vinylic carbocation from an alkyne than from a vinylic halide, because the alkyne is already *sp* hybridized; an unfavorable hybridization change does not have to occur in order to form the carbocation.

✓18.2 Contrast of Aromatic Substitution Reactions

The name "aromatic substitution" may suggest some connection to the reactions you studied in Chapter 16, such as nitration, halogenation, sulfonation, etc. Other than the fact that a group is replaced on an aromatic ring, *nucleophilic aromatic substitution and electrophilic aromatic substitution have little in common.* In *electrophilic* aromatic substitution, an electrophile, or Lewis acid, is attacked by the benzene π electrons, and a *proton* is lost from the benzene ring. In *nucleophilic* aromatic substitution, a *nucleophile,* or Lewis base, attacks a carbon of the benzene ring and a *halide* is lost from the benzene ring.

Electrophilic aromatic substitution involves introduction of *electron deficiency* (and positive charge) into the benzene ring. Nitro groups and other electron-attracting groups *destabilize* this charge and thus retard the reaction. Nucleophilic aromatic substitution involves introduction of *more electrons* (and negative charge) into the benzene ring. Nitro groups and other electron-attracting groups *stabilize* this charge and thus accelerate this type of reaction.

Notice that *substitution,* like *addition,* is a type of reaction that can occur by a variety of mechanisms. The word *substitution* by itself conveys no information about mechanism. Thus, a substitution can occur by a concerted (S_N2) mechanism, an ionization (S_N1) mechanism, a two-step nucleophilic mechanism (as in nucleophilic aromatic substitution), a two-step electrophilic mechanism (as in electrophilic aromatic substitution), a free-radical mechanism, or (as shown in Section 18.4B of the text) an elimination-addition mechanism. The mechanism that operates depends on the reactants and the reaction conditions. Students sometimes forget this point and assume that the word *substitution* means "S_N2 reaction," because that is the first substitution reaction they learn.

18.3 Resonance Effects on Phenol Acidity

There is an important subtlety in the resonance effect on phenol acidity. *Un-ionized* phenols are also stabilized by resonance. However, the resonance structures for an un-ionized phenol involve *separation of charge,* but the structures of a phenoxide ion involve *dispersal of charge.*

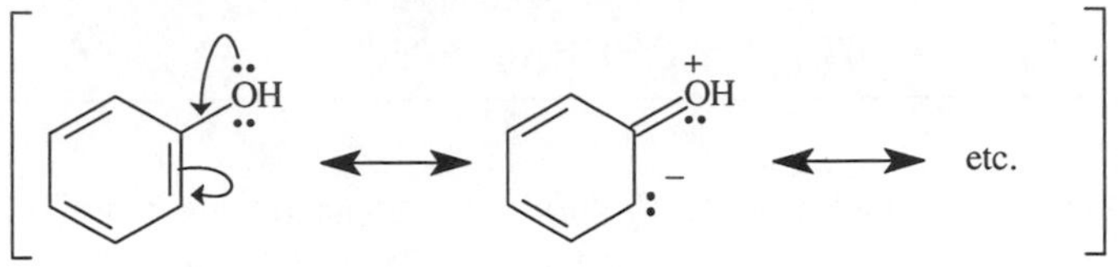

structures for un-ionized phenol involve *separation* of charge

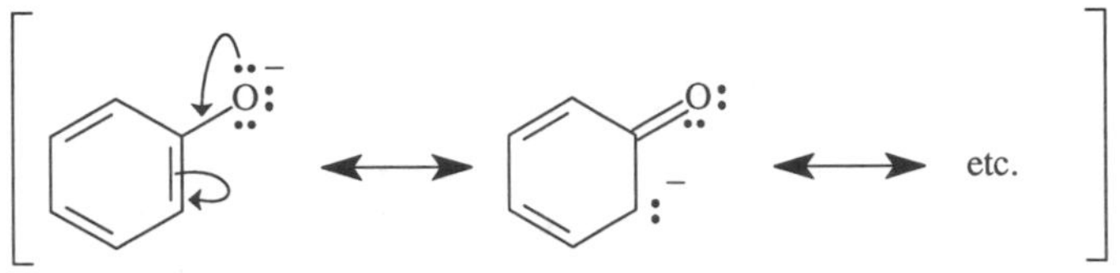

structures for phenolate ion involve *dispersal* of charge

Because structures that disperse charge are more important than those that separate charge (Guideline 3, Sec. 15.6B, text p. 668), a phenoxide ion is stabilized *more* by resonance than its conjugate-acid phenol. If the two were equally stabilized by resonance, the resonance effect would cancel when the free energies of phenol and phenoxide are compared, and it would have no effect on the pK_a.

Notice how this situation differs from the ionization of a benzylic hydrogen in toluene: There is no resonance stabilization in un-ionized toluene involving the benzylic carbon to offset the resonance stabilization of the benzylic anion. *Only the anion* is stabilized by resonance involving the benzylic carbon.

no resonance interaction of this carbon with the ring

stabilized by resonance

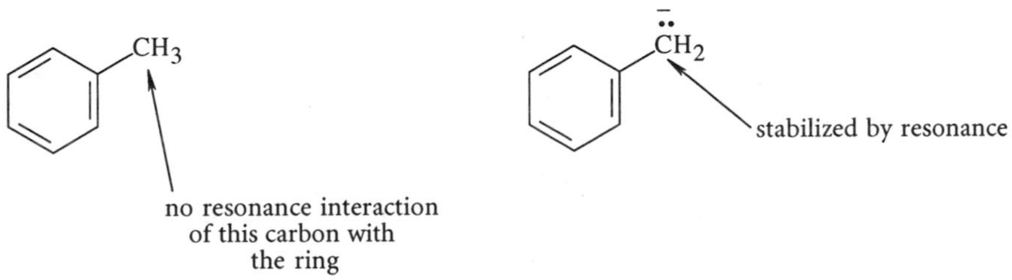

Consequently, the pK_a of toluene should be affected more by resonance than that of phenol. As shown in the text, a phenol is about seven pK_a units more acidic than an alcohol; however, toluene (p$K_a \approx 41$), is about 14–19 pK_a units more acidic than an alkane (pK_a 55–60).

18.4 Fries Rearrangement

There's more to the Friedel-Crafts reactions of phenols than meets the eye. It turns out that phenols react with acid chlorides to give phenyl esters; this is a reaction you'll study in Chapter 21.

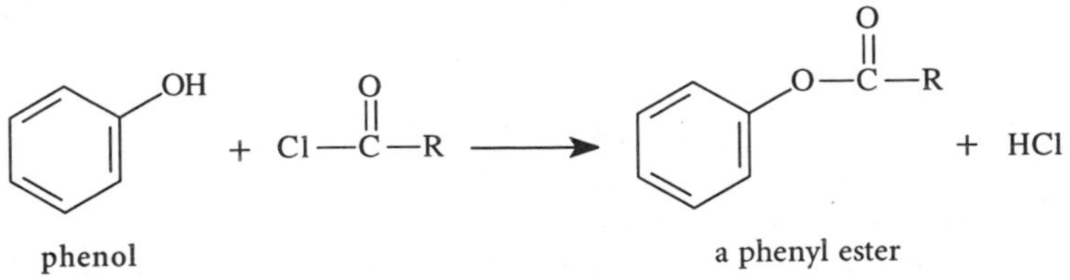

phenol + Cl—C(=O)—R ⟶ a phenyl ester + HCl

(Acid chlorides are *electrophiles* in this reaction just as they are in the Friedel-Crafts reaction. The phenol oxygen, like the aromatic ring, is a source of electrons because of its unshared pairs.) In the presence of $AlCl_3$—the catalyst used in Friedel-Crafts reactions—phenyl esters rearrange to ketones:

$$\text{PhO–C(=O)–R} \xrightarrow[\text{2) } H_3O^+]{\text{1) } AlCl_3} o\text{-(RCO)C}_6H_4OH + p\text{-(RCO)C}_6H_4OH$$

This reaction, known as the *Fries rearrangement,* can be clearly demonstrated because phenyl esters, which can be prepared as pure compounds in the absence of $AlCl_3$, can be shown to rearrange when treated with $AlCl_3$. The products of the Fries rearrangement, as you can see, are exactly what would be expected from a direct Friedel-Crafts reaction. Hence, the "Friedel-Crafts reaction" of a phenol almost surely involves ester formation followed by the Fries rearrangement, or a *combination* of the Fries rearrangement and a direct Friedel-Crafts reaction. In any event, the *end result* is as if a Friedel-Crafts reaction had taken place directly.

It is interesting that the "Friedel-Crafts reactions" of phenols in many cases give significant amounts of *ortho*-substituted products. These may be separated from the *para* isomers by application of the principles of hydrogen bonding. The *ortho* isomers contain *intramolecular* hydrogen bonds:

intramolecular hydrogen bond

Such intramolecular hydrogen bonding is not possible for the isomeric *para*-substituted phenol. Because the hydrogen-bonding capability of the phenolic —OH group is satisfied within the *ortho*-substituted molecule, this —OH group forms hydrogen bonds less effectively to acceptor atoms of *other molecules*—for example, solvents or chromatography supports such as silica gel—than the —OH group of the *para*-substituted isomer. Hence, the *ortho*-substituted isomer is generally much less soluble in solvents that accept hydrogen bonds, and moves through a silica gel chromatography column more rapidly, than the *para*-substituted isomer. Thus, the great difference in hydrogen-bonding properties of the two isomers makes separation a relatively simple matter. Exactly the same principles can be utilized in designing separations of *o*- and *p*-nitrophenol.

✓18.5 The Cumene Hydroperoxide Rearrangement

The formation of phenol and acetone (Eq. 18.88b on text p. 823) is an acid-catalyzed rearrangement of cumene hydroperoxide. This Study Guide Link discusses the mechanism of this reaction.

In the first step of the mechanism, an oxygen of the hydroperoxide is protonated:

$$\text{Ph–C(CH}_3)_2\text{–O–}\ddot{O}H + H\text{–}\overset{+}{O}H_2 \longrightarrow \text{Ph–C(CH}_3)_2\text{–O–}\overset{+}{O}H_2 + \ddot{O}H_2$$

Loss of water gives an *electron-deficient oxygen,* the oxygen analog of a carbocation. Rearrangement occurs to give a much more stable resonance-stabilized carbocation. (The electron-deficient oxygen structure is so unstable that loss of water and the rearrangement undoubtedly take place in a concerted manner.)

$$\mathrm{Ph{-}\overset{CH_3}{\underset{:\ddot{O}-\overset{+}{O}H_2}{C}}{-}CH_3 \longrightarrow Ph{-}\overset{CH_3}{\underset{:\ddot{O}^+}{C}}{-}CH_3 + \ddot{O}H_2 \longrightarrow \left[\ \overset{CH_3}{\underset{Ph-\ddot{O}:}{\overset{+}{C}}}{-}CH_3 \longleftrightarrow \overset{CH_3}{\underset{Ph-\ddot{O}+}{\overset{}{C}}}{-}CH_3\ \right]}$$

resonance-stabilized carbocation

This rearrangement is very similar to the rearrangement that occurs in the oxidation of organoboranes, the mechanism of which was discussed in Study Guide Link 5.3.

Attack of water on the carbocation gives a type of alcohol known as a hemiacetal.

$$\mathrm{H_2\ddot{O}: + \overset{CH_3}{\underset{Ph\ddot{O}:}{\overset{+}{C}}}{-}CH_3 \longrightarrow H_2\ddot{O}: + H\overset{+}{O}{-}\overset{CH_3}{\underset{Ph\ddot{O}:}{C}}{-}CH_3 \longrightarrow H_3\ddot{O}^+ + H\ddot{O}{-}\overset{CH_3}{\underset{Ph\ddot{O}:}{C}}{-}CH_3}$$

a hemiacetal

You should now be able to complete the acid-catalyzed conversion of the hemiacetal into phenol and acetone. Start by protonating the oxygen of the PhO— group. What has to happen next to form phenol? Finish the mechanism to give acetone.

SOLUTIONS TO PROBLEMS

Solutions to In-Text Problems

18.1 (a) *p*-Bromotoluene reacts most slowly (that is, not at all) because it is an aryl halide; aryl halides do not undergo S_N2 reactions. Benzyl bromide is a benzylic halide and therefore reacts most rapidly in an S_N2 reaction (see text Sec. 17.4).

Br—C$_6$H$_4$—CH_3 << C$_6$H$_5$—$CH_2CH_2CH_2Br$ < C$_6$H$_5$—CH_2Br

p-bromotoluene (3-bromopropyl)benzene benzyl bromide

(b) 1-Bromocyclohexene, a vinylic halide, does not react by the S_N2 mechanism; 1-(bromomethyl)cyclohexene, an allylic halide, reacts most rapidly. (See text Sec. 17.4.)

1-bromocyclohexene << bromocyclohexane < 1-(bromomethyl)cyclohexene

18.2 Assume that E2 reactions of *vinylic* halides follow the same trends as E2 reactions of *alkyl* halides. Therefore, because an *anti*-elimination is faster than the corresponding *syn*-elimination (Sec. 9.5D, text pp. 374–6), the reaction of compound *B* is faster than that of compound *A*; and because bromide is a better leaving group than chloride, the reaction of compound *A* is faster than that of compound *C*. Because eliminations involving benzylic β-hydrogens are faster than other eliminations, compounds *A*–*C*, in which the β-hydrogens are benzylic, react more rapidly than compound *D*, in which the β-hydrogens are not benzylic. Furthermore, the elimination reaction of compound *D* gives a triple bond that is conjugated with only one benzene ring (and thus has only one phenyl branch; see structures that follow), whereas all of the others give triple bonds that are conjugated with two benzene rings (and thus have two phenyl branches). Hence, the desired order is $D < C < A < B$.

The product of the elimination reactions of compounds *A*–*C* is diphenylacetylene, Ph—C≡C—Ph; the product of the elimination reaction of compound *D* is phenylacetylene, Ph—C≡C—H.

18.3 (a) The order of increasing S_N1-E1 reactivity is $A << B < C$. The benzylic halide *C* reacts fastest because its ionization gives a resonance-stabilized benzylic carbocation; and the vinylic halide *A* reacts slowest because vinylic halides are nearly inert in S_N1 reactions for the reasons given on pp. 779–81 of the text.

(b) The reactivity order is $B << C < A$. The reaction of compound *B* is slowest because vinylic halides are virtually inert in S_N1 reactions; and the reaction of compound *A* is fastest because its ionization gives a resonance-stabilized allylic carbocation.

18.4 (a) The nitrogen of the amine acts as a nucleophile in a nucleophilic aromatic substitution reaction.

O_2N—C$_6$H$_3$(NO_2)—$\overset{+}{N}H_2C_2H_5$ Cl^- ⇌ O_2N—C$_6$H$_3$(NO_2)—$\ddot{N}HC_2H_5$ + HCl

The nitro groups reduce the basicity of the product to the point that it is deprotonated under the reaction conditions. (Can you give a reason why?) Because you have not yet studied the basicity of aromatic amines, however, either answer should be considered correct.

(b) The product results from nucleophilic aromatic substitution by the thiolate group:

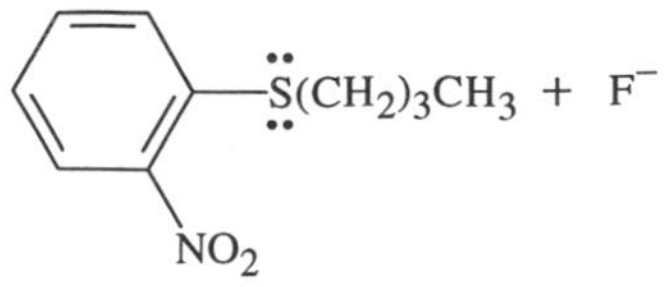

$$o\text{-}NO_2C_6H_4\text{—}\ddot{S}(CH_2)_3CH_3 + F^-$$

(c) No reaction, because the compound contains no substituent that can activate nucleophilic aromatic substitution by stabilizing the anionic intermediate (the Meisenheimer complex).

18.5 (a) The second compound, *p*-fluoronitrobenzene, reacts most rapidly because only in the reaction of this compound is the intermediate Meisenheimer complex stabilized by resonance interaction of an unshared electron pair with the nitro substituent.

(b) The first compound should react more rapidly because the Meisenheimer complex is stabilized by the resonance interaction of both substituents; the complex in the case of the second compound is stabilized by resonance involving only a nitro substituent.

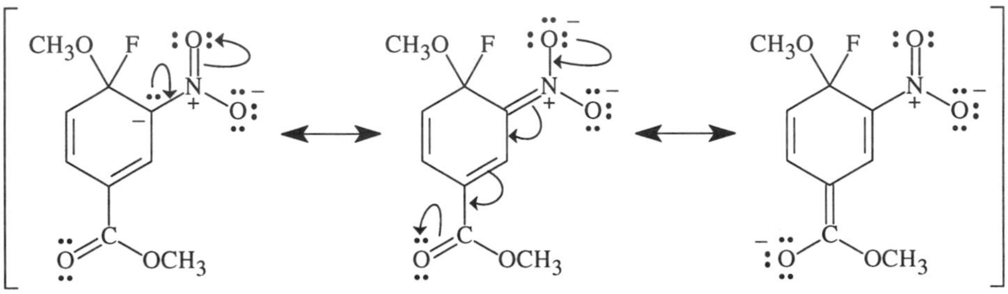

[Resonance structures: three structures of the Meisenheimer complex bearing CH_3O, F, NO_2, and $C(=O)OCH_3$ groups, connected by ↔]

18.6 (a) Because this compound has two chemically nonequivalent *β*-hydrogens, two possible benzyne intermediates, *A* and *B*, can be formed. Each can give two products, one of which is common to both intermediates.

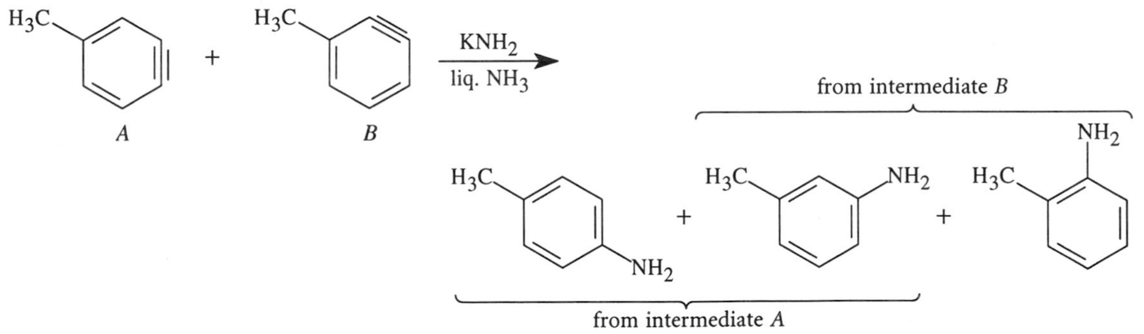

$$A + B \xrightarrow[\text{liq. } NH_3]{KNH_2} \underbrace{p\text{-}H_3CC_6H_4NH_2 + m\text{-}H_3CC_6H_4NH_2}_{\text{from intermediate } A} + o\text{-}H_3CC_6H_4NH_2$$

(*m*-isomer and *o*-isomer: from intermediate *B*)

(b) This compound has one type of *β*-hydrogen; therefore one benzyne intermediate can be formed. This intermediate reacts to give two products, one from attack at each of the two nonequivalent carbons.

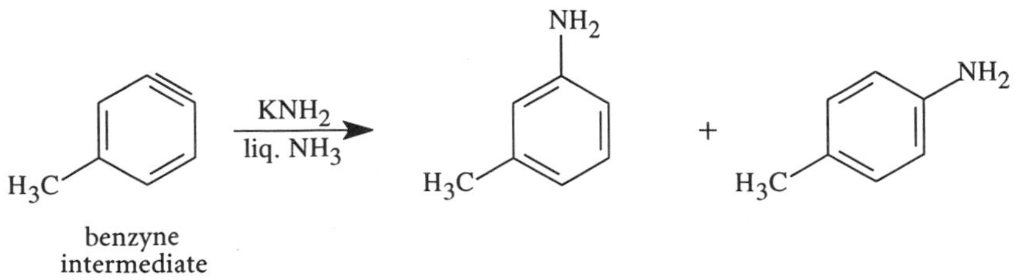

$$\text{benzyne intermediate} \xrightarrow[\text{liq. } NH_3]{KNH_2} m\text{-}H_3CC_6H_4NH_2 + p\text{-}H_3CC_6H_4NH_2$$

(c) Because there is no *β*-hydrogen, no reaction occurs.

18.7 The bond to one allylic carbon is an X-type bond. That is, if we break the metal-carbon bond, a carbon anion is formed. The double bond is an L-type ligand. Breaking its coordination gives a neutral alkene.

L-type interaction → $H_2C{=}CH{-}CH_2{-}M$ ← X-type interaction

As with the case of the cyclopentadienyl anion discussed immediately preceding the problem in the text, this is a bookkeeping fiction; the L- and X-type character is actually spread out over the allyl system:

(allyl anion, H–C(H)–C(H)–C(H)–H, delocalized, bonded to M)

It also happens that allyl can bond to a metal in some cases as an X-type ligand; that is, the π electrons in some cases aren't involved in coordination at all. (See Footnote ** in Table 18.1.) However, we know that this situation is not relevant to the present problem, as the problem tells us to be concerned with LX-type coordination.

18.8 (a) If we were to move an electron pair out of one of the metal-oxygen bonds onto the oxygen the oxygen would take on a negative charge. Hence, by definition, each bond to oxygen counts as one X-type interaction. The number of X-type bonds to Mn is 7, and the charge is on oxygen, so the charge in Eq. 18.29 is 0. Hence, the oxidation state is +7.

The following is an equivalent structure for permanganate in which a charge is on Mn:

$$O{=}\overset{O}{\overset{\|}{Mn^{-}}}{=}O \quad (\text{with a fourth } {=}O \text{ below})$$

In this case the oxidation state is +8 – 1 = +7; obviously, we get the same result from either structure.

(b) The PPh_3 ligands are L-type ligands; hence, there are no X-type ligands, and, because the charge on Pd is 0, the oxidation state of Pd is 0.

You might be wondering about the prefix *tetrakis* in the name of this complex. The prefixes *bis, tris,* and *tetrakis* are used as numerical prefixes instead of *di, tri,* and *tetra* when the group that is enumerated itself contains multiple substituents. Thus, the ligand triphenylphosphine has three phenyl groups on the phosphorus (thus the prefix *tri* in the name of this ligand). There are four triphenylphosphine ligands—thus the prefix *tetrakis.*

(c) The Cp ligand (Table 18.1) can be viewed as an L_2X ligand for bookkeeping purposes. Hence with two X-type bonds and no charge on the iron, the oxidation state of Fe is +2.

18.9 With one X-type ligand (Cl) and zero charge, oxidation state of rhodium (Rh) in the starting material is +1. In the product, two new X-type bonds have been added; hence, the oxidation state is +3. (This is an example of an *oxidative addition,* an important type of reaction that is considered later in this section.)

18.10 (a) Pd has ten valence electrons in the neutral atom. There are no charges and no X-type ligands in the complex; hence, this is a d^{10} complex. Using Eq. 18.31, with an oxidation state of 0, we get the same answer.

(b) Consulting Fig. 18.3, we see that neutral tungsten (W) has 6 valence electrons. Add two electrons for the charge. The ligands are L-type ligands. Hence, this is a d^8 complex. Alternatively, the oxidation state of tungsten is, by Eq. 18.29, $0 - 2 = -2$. We then apply Eq. 18.31: n = valence electron count in neutral W (6) – the oxidation state (–2) = 6 – (–2) = 8.

(c) We've seen this complex before, in Problem 18.9. We calculated in the solution to that problem that the oxidation state of rhodium (Rh) is +3. Fig. 18.3 shows that neutral Rh has 9 valence electrons. Hence, application of Eq. 18.31 shows that this is a d^6 complex. Alternately, we start with 9 electrons and remove 3 for the X-type ligands to get $9 - 3 = 6$ for the value of n.

18.11 For Rh, start with the number of electrons in the neutral atom (9); subtract charges on the metal (0); add the number of X-type ligands (3); and add twice the number of L-type ligands ($2 \times 3 = 6$) to get $9 + 3 + 6 = 18$ electrons. Alternatively, because we know $n = 6$ from the solution to Problem 18.10(c), eq. 18.32 on p. 793 can be applied to obtain the electron count. This is an 18-electron complex.

18.12 Neutral iron (Fe) has 8 electrons. Because CO is an L-type ligand, it is counted twice in the electron count. We simply solve for x in $8 + 2x = 18$ and obtain $x = 5$. $Fe(CO)_5$, or pentacarbonyliron(0), is in fact a stable complex that can be purchased commercially.

18.13 Vanadium (V) in $V(CO)_5$ has an electron count of $5 + (2 \times 6) = 17$. If the tendency of metals is to have a total of 18 electrons, then V should have a significant driving force to obtain an extra electron, that is, to be reduced. Indeed, the total electron count in $[V(CO)_6]^-$ is 18, and this anion is relatively stable.

18.14 First, note that cyanide (^-CN) is an X-type ligand. If we strip the four cyanide ligands from the Zn (zinc), we see that what is left is a Zn^{2+} ion. From Fig. 18.1 on text p. 789 we note that Zn has 12 valence electrons; hence, Zn^{2+} has 10. This is exactly the number needed to fill all of the 3d orbitals with two electrons each. (See Fig. SG18.1, top of following page.) This leaves four valence orbitals—the 4s and the 4p orbitals empty. These are hybridized to form more directed orbitals. (This situation is exactly like carbon hybridization in methane, except that we are using orbitals from period 4.) If we hybridize one 4s and three 4p orbitals, we get four sp^3 hybrid orbitals, which, as we know from methane, are directed to the corners of a regular tetrahedron. Each of these empty orbitals accepts a pair of electrons from a cyanide ligand. Thus, the $Zn(CN)_4]^{2-}$ complex is tetrahedral.

18.15 The student has neglected to understand that when an X-type ligand attacks a metal to displace an L-type ligand, the charge changes by one unit, because electrons from X-type ligands are assumed to be divided between the metal and the ligand. In other words, charge is not balanced in this equation. If we put a negative charge on the Pd, as required by this reaction, the oxidation state of Pd is unchanged.

18.16 (a) *Oxidative addition of H_2 to the catalyst:*

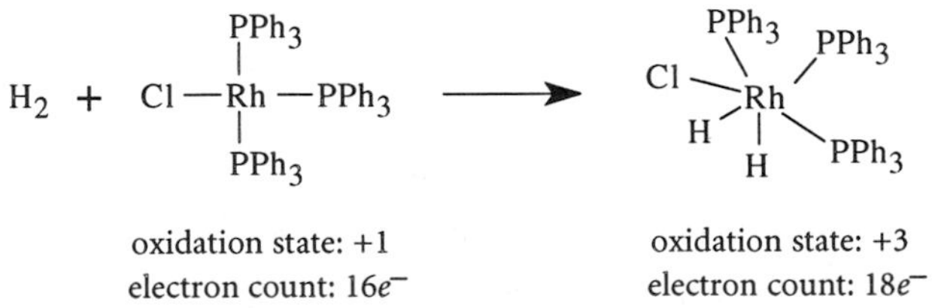

H_2 + $Cl-Rh(PPh_3)_3$ ⟶ $RhCl(H)_2(PPh_3)_3$

oxidation state: +1
electron count: $16e^-$

oxidation state: +3
electron count: $18e^-$

Ligand substitution of one PPh_3 by the alkene:

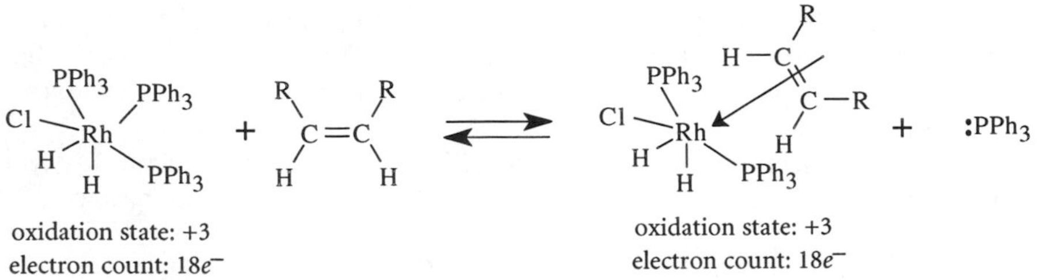

$RhCl(H)_2(PPh_3)_3$ + RHC=CHR ⇌ $RhCl(H)_2(PPh_3)_2$(RHC=CHR) + :PPh_3

oxidation state: +3
electron count: $18e^-$

oxidation state: +3
electron count: $18e^-$

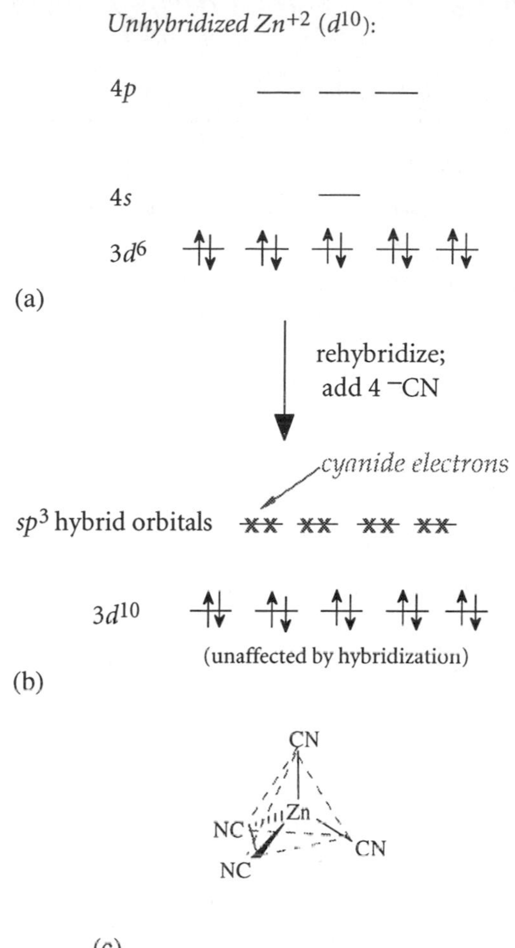

Figure SG18.1 Development of the hybrid-orbital picture for $[Zn(CN)_4]^{2-}$ to accompany the solution to Problem 18.14. (a) The ^{-}CN ligands are removed, leaving Zn^{2+}, a d^{10} ion. The ten electrons are paired in five 3*d* orbitals. (b) The remaining orbitals (one 4*s* and three 4*p*) are hybridized to give four sp^3 hybrid orbitals. (c) Because sp^3 hybridization gives tetrahedral geometry, the $[Zn(CN)_4]^{2-}$ complex ion has tetrahedral geometry. (The gray dashed lines show the tetrahedron; the black lines are bonds.)

1,2-Insertion of the alkene into an Rh—H bond and addition of the previously expelled PPh_3:

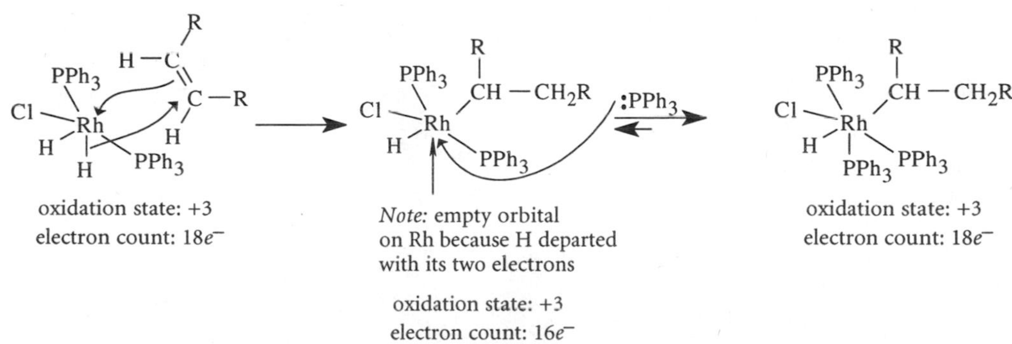

Reductive elimination of the product to regenerate the catalyst:

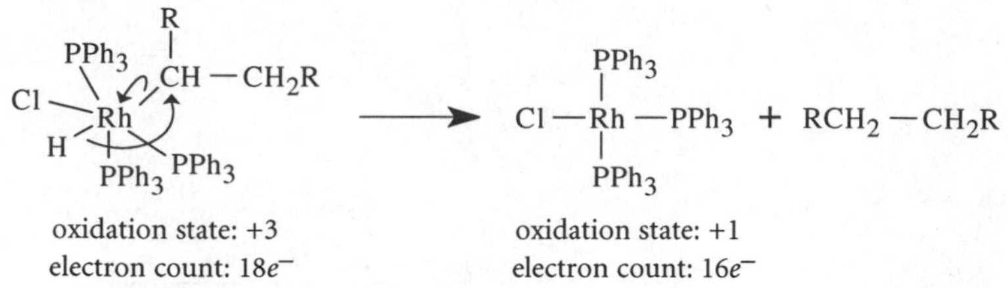

oxidation state: +3
electron count: $18e^-$

oxidation state: +1
electron count: $16e^-$

(b) The 1,2-insertion is a *syn*-addition:

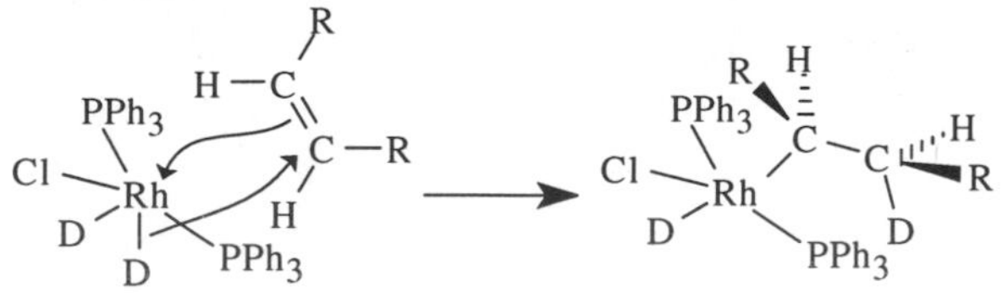

And the reductive elimination is also syn:

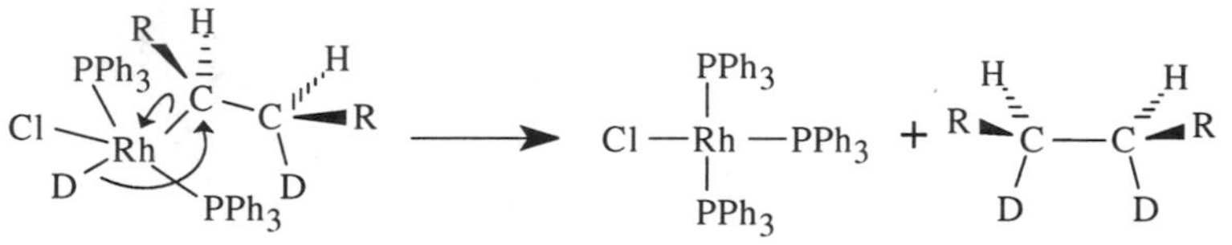

a meso compound

Because we started with a symmetrical *cis*-alkene, the product is the meso stereoisomer.

18.17 The steps in Eq. 18.50b of the text are numbered for reference.

Step 1:

Fundamental process: oxidative addition
Oxidation state of Pd starting catalyst (PdL_2): 0
Electron count of Pd in the starting catalyst: $14e^-$. (Note that neutral Pd is a 10-electron atom.)
Oxidation state of Pd in the product: +2
Electron count of Pd in the product: $16e^-$

Step 2: (From here on, the reactant has the same properties as the product of the previous step.)

Fundamental process: ligand substitution
Oxidation state of Pd in the product: +2
Electron count of Pd in the product: $16e^-$

Step 3:

Fundamental process: 1,2-ligand insertion
Oxidation state of Pd in the product: +2
Electron count of Pd in the product: $14e^-$

Step 4:

Fundamental process: β-elimination
Oxidation state of Pd in the product: +2
Electron count of Pd in the product: $16e^-$

Step 5:

Fundamental process: ligand dissociation
Oxidation state of Pd in the product: +2
Electron count of Pd in the product: $14e^-$

Step 6:

Fundamental process: ligand association
Oxidation state of Pd in the product: +2
Electron count of Pd in the product: $16e^-$

(Note that steps 5 and 6 together result in a ligand substitution.)

Step 7:

Fundamental process: reductive elimination
Oxidation state of Pd in the product: 0
Electron count of Pd in the product: $14e^-$

(Note that the catalyst is regenerated in this step.)

18.18 The two products result from coupling at the two chemically nonequivalent carbons of the alkene:

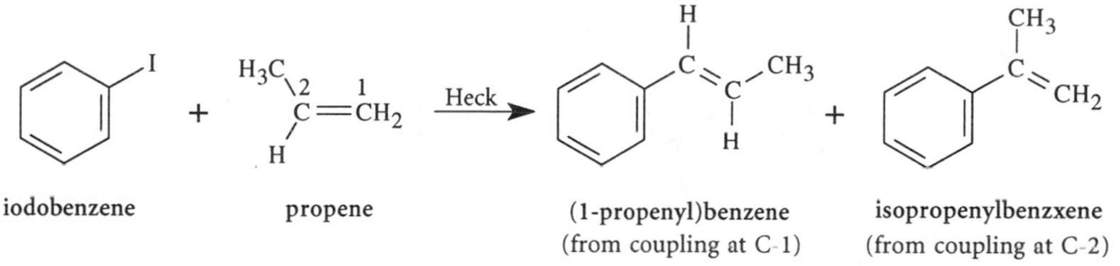

iodobenzene + propene —Heck→ (1-propenyl)benzene (from coupling at C-1) + isopropenylbenzxene (from coupling at C-2)

That is, in step 3 of the mechanism described in Eq. 18.50b and in the previous solution, 1,2-ligand insertion can occur to either carbon of the double bond. Because these two carbons are nonequivalent, two products are obtained.

18.19 The environment around the Pd following the initial oxidative addition is rather congested with the large PPh_3 ligands, the aryl group, and the halide. Ligand replacement by the alkene is more favorable for alkenes containing fewer bulky substituents. As a consequence of the Heck reaction itself, the product is always more branched than the starting material, and hence, the starting alkene will "win" the competition for ligand replacement over the product alkene more often than not. If we assume that the subsequent rate constants for substitution are not very different for the two alkenes, the higher concentration of the catalyst-starting alkene complex leads to a greater reaction rate for this alkene. (Remember, rate = k[complex], where k = the rate constant. Given equal k values, the complex present in higher concentration wins the rate competition.)

18.20 Either aryl substituent could originate from the aryl halide or from the alkene. Remember, if the alkene substituent is aryl (as in these cases), substitution occurs mainly at the less branched carbon.

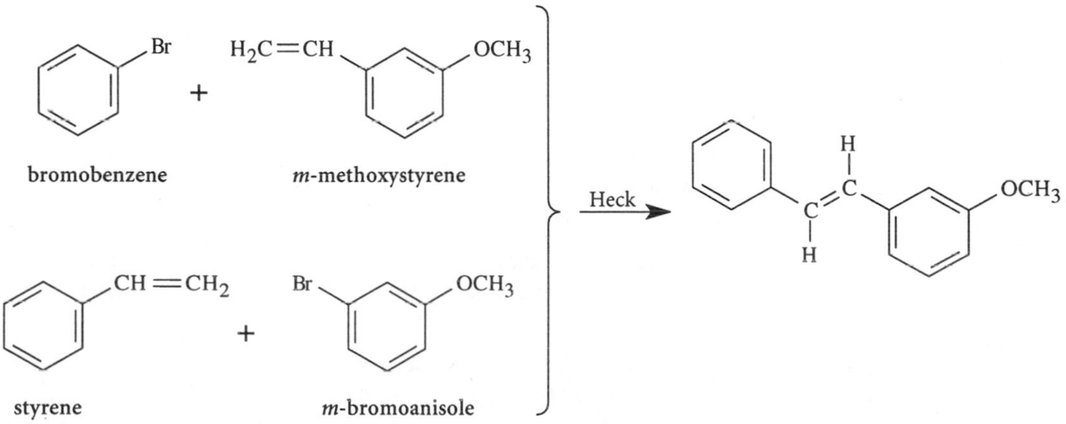

bromobenzene + *m*-methoxystyrene
styrene + *m*-bromoanisole
—Heck→ product

18.21 This is essentially like the cyclohexene case given in Eq. 18.51 (text p. 803) and subsequent discussion.

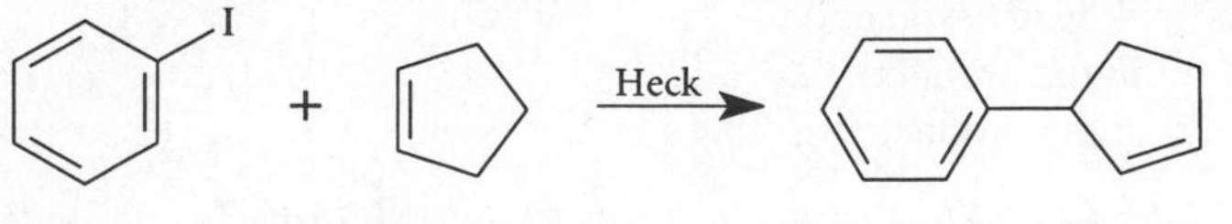

iodobenzene + cyclopentene →(Heck) (2-cyclopentenyl)benzene

18.22 A mechanistic outline is as follows.

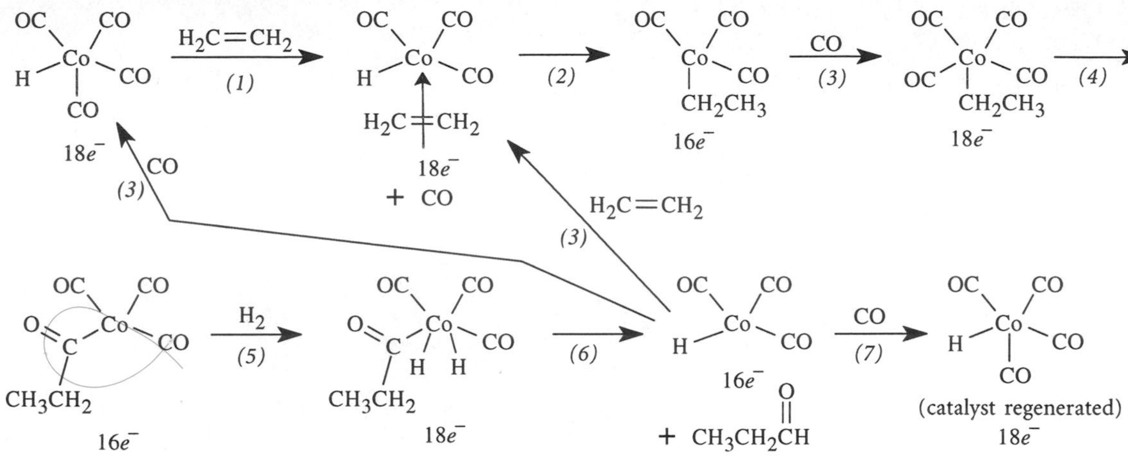

$HCo(CO)_4$ (18e⁻) $\xrightarrow[(1)]{H_2C{=}CH_2}$ $HCo(CO)_3(H_2C{=}CH_2)$ (18e⁻) + CO $\xrightarrow{(2)}$ $(CO)_3Co{-}CH_2CH_3$ (16e⁻) $\xrightarrow[(3)]{CO}$ $(CO)_4Co{-}CH_2CH_3$ (18e⁻) $\xrightarrow{(4)}$ $(CO)_3Co{-}C({=}O)CH_2CH_3$ (16e⁻) $\xrightarrow[(5)]{H_2}$ $(CO)_3Co(H)_2{-}C({=}O)CH_2CH_3$ (18e⁻) $\xrightarrow{(6)}$ $HCo(CO)_3$ (16e⁻) + CH_3CH_2CHO

$HCo(CO)_3$ (16e⁻) $\xrightarrow[(7)]{CO}$ $HCo(CO)_4$ (catalyst regenerated) (18e⁻)

$HCo(CO)_3$ (16e⁻) $\xrightarrow[(3)]{H_2C{=}CH_2}$ $HCo(CO)_3(H_2C{=}CH_2)$ (18e⁻); $HCo(CO)_3$ $\xrightarrow[(3)]{CO}$ $HCo(CO)_4$ (18e⁻)

The classifications of the various steps are:

(*1*) Substitution of one L-type ligand for another
(*2*) 1,2-Ligand insertion
(*3*) Association with L-type ligand
(*4*) 1,1-Ligand insertion
(*5*) Oxidative addition
(*6*) Reductive elimination
(*7*) Association with L-type ligand

18.23 (a) 2,4-Dinitrophenol is more acidic because its conjugate-base anion has more important resonance structures. In the 2,4-dinitrophenoxide anion, both nitro substituents are involved in resonance stabilization; in 2,5-dinitrophenoxide anion, only the 2-nitro group is involved in resonance stabilization. (The 5-nitro group is not involved for the reasons shown in the discussion of Eq. 18.60 on text p. 880.) The actual pK_a values are

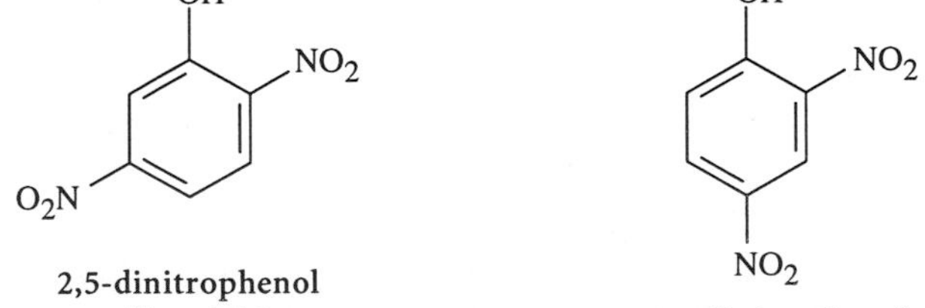

2,5-dinitrophenol pK_a = 5.22

2,4-dinitrophenol pK_a = 4.11

(b) *Meta*-chlorophenol is more acidic because the conjugate-base anion, *m*-chlorophenoxide, is stabilized by the electron-withdrawing polar effect of the chloro substituent. The actual pK_a values are

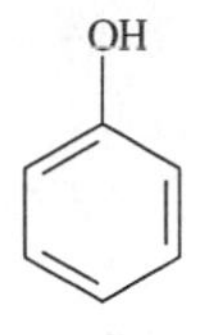

OH OH

Cl

phenol
$pK_a = 9.95$

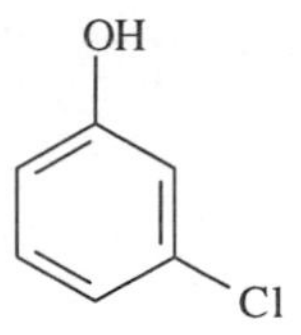

m-chlorophenol
$pK_a = 9.02$

(c) The *para*-isomer of the conjugate-base phenoxide anion has more important resonance structures. The key resonance structure is the one below, in which the substituent group participates in the resonance stabilization. (The actual pK_a values are 7.7 and 8.0 for the *para*- and *meta*-isomers, respectively.)

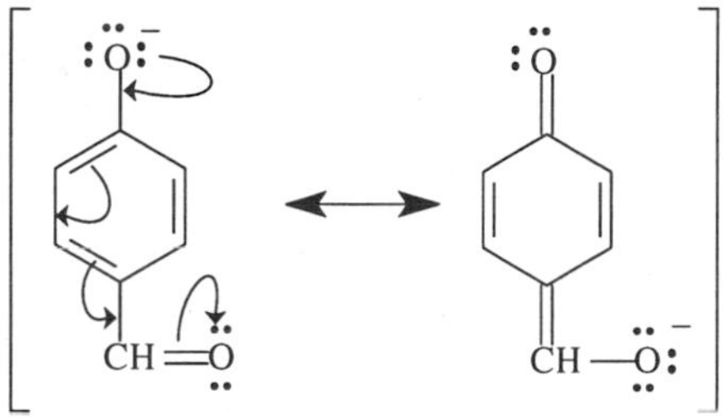

CH=O ⟷ CH—O⁻

18.24 (a)

O_2N—C₆H₄—OH $\xrightarrow{\text{NaOH}}$ O_2N—C₆H₄—$O^- Na^+$ $\xrightarrow{CH_3I}$ O_2N—C₆H₄—OCH_3

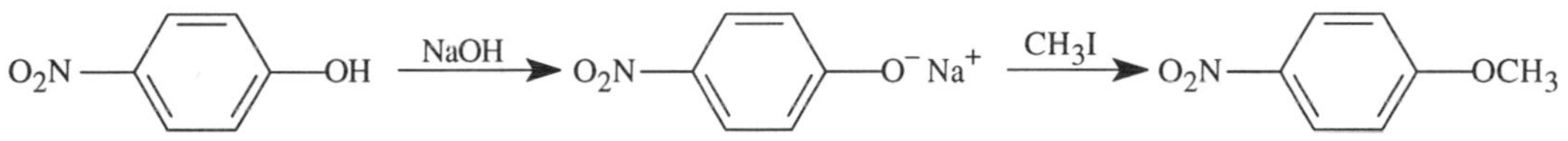

4-nitrophenol
(*p*-nitrophenol)

p-nitroanisole

(b)

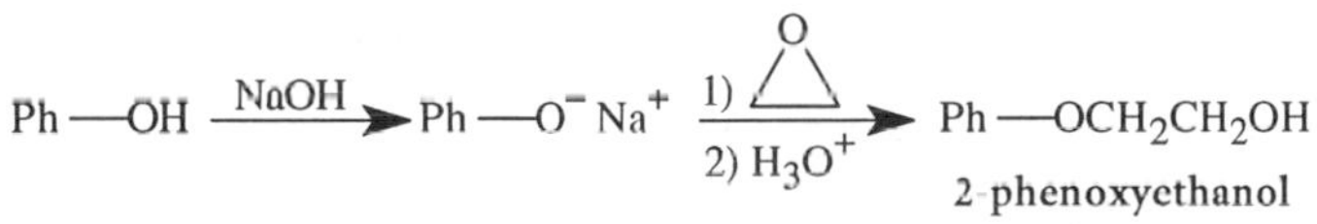

Ph—OH $\xrightarrow{\text{NaOH}}$ Ph—$O^- Na^+$ $\xrightarrow[2)\ H_3O^+]{1)\ \text{ethylene oxide}}$ Ph—OCH_2CH_2OH

2-phenoxyethanol

18.25 The solubility of most phenols in base due to their –1 charge when ionized. Their conjugate bases are ionic compounds, and ionic compounds have greater solubility than nonionic ones. The compound shown in the problem is cationic by virtue of its positively charged nitrogen, Ionization of the OH group converts it into a *neutral* compound. Neutral species are less soluble in water than compounds with a net charge.

Compounds that contain both a positive and a negative charge, such as the conjugate base of the compound shown in the problem, are called *zwitterions* ("double-ions"), or *inner salts.* These types of compounds are discussed in Chapter 26.

18.26 Because quinones are easily reduced to phenols, we postulate that the two-electron reduction of ubiquinone would give the corresponding hydroquinone.

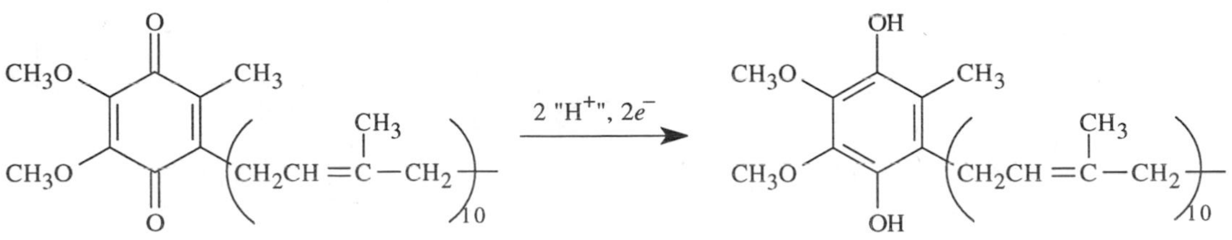

O, CH_3O, CH_3, CH_3O, O, $(CH_2CH=C(CH_3)-CH_2)_{10}$ $\xrightarrow{2\ \text{"H}^+\text{"},\ 2e^-}$ OH, CH_3O, CH_3, CH_3O, OH, $(CH_2CH=C(CH_3)-CH_2)_{10}$

ubiquinone

18.27 (a)

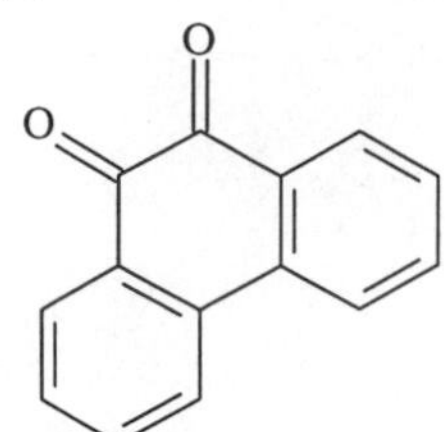

9,10-phenanthraquinone
(an *o*-quinone)

(b)

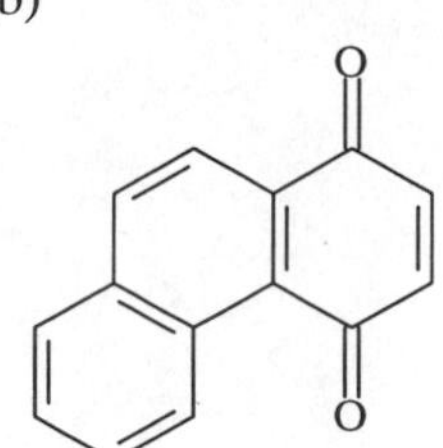

1,4-phenanthraquinone
(a *p*-quinone)

18.28 (a) A *para*-quinone is formed as in Eq. 18.63 (text p. 811); the nitro group is unaffected.

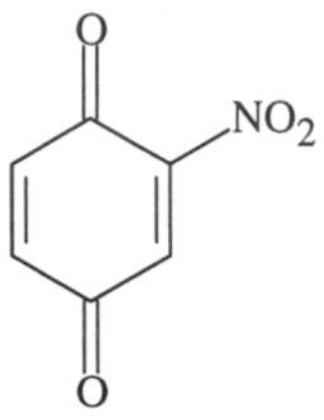

NO_2

(b) The *para*-quinone is formed because it is more stable than the alternative *ortho*-quinone.

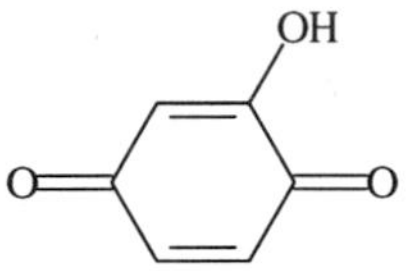

OH

18.29 (a) Let the long alkyl chain in vitamin E (structure on text p. 814) be abbreviated as —alkyl. The most stable radical is formed by abstraction of the phenolic hydrogen atom.

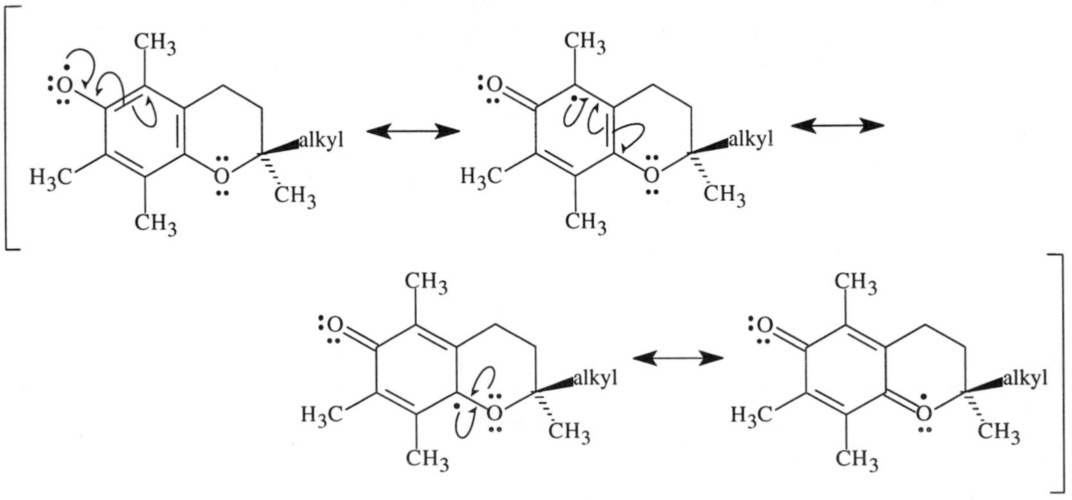

CH_3 H_3C alkyl CH_3 CH_3 ⟷ CH_3 H_3C alkyl CH_3 CH_3 ⟷

CH_3 H_3C alkyl CH_3 CH_3 ⟷ CH_3 H_3C alkyl CH_3 CH_3

(b) As in part (a), the phenolic hydrogen atom is abstracted because a resonance-stabilized radical is formed.

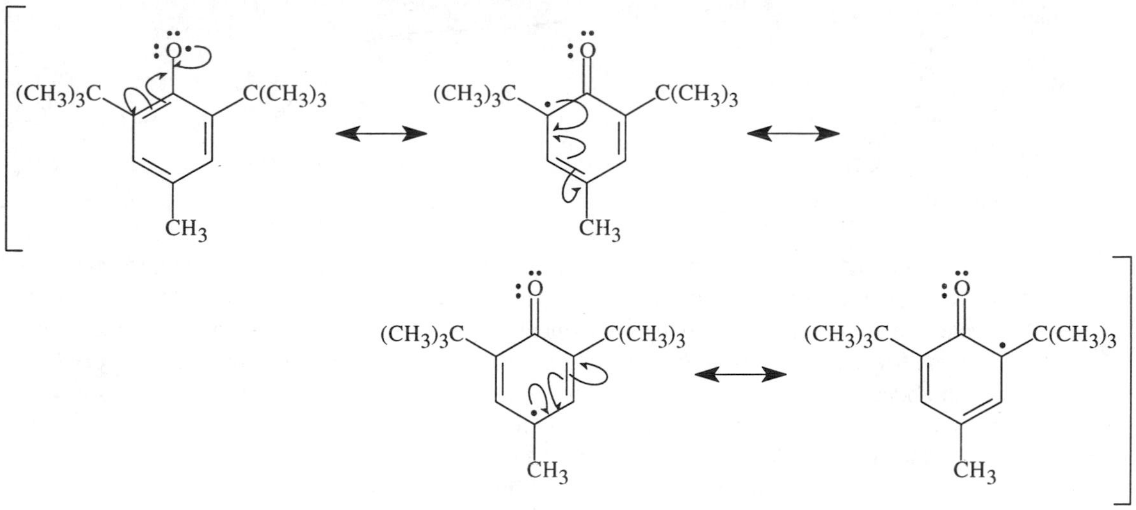

18.30 (a) (b) (c)

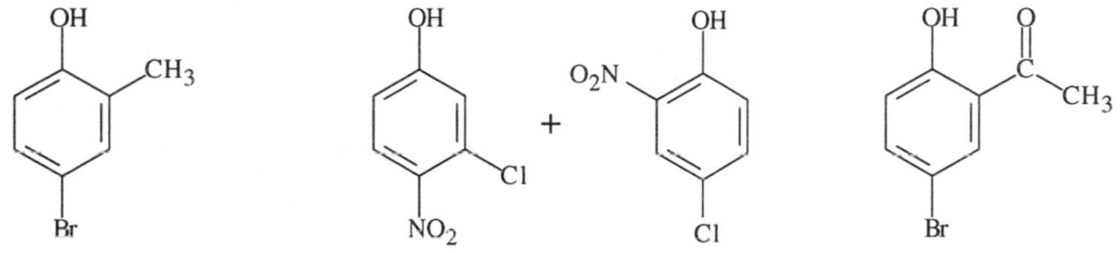

18.31 The electrophile is the *tert*-butyl cation, which is formed by protonation of the alcohol and loss of water.

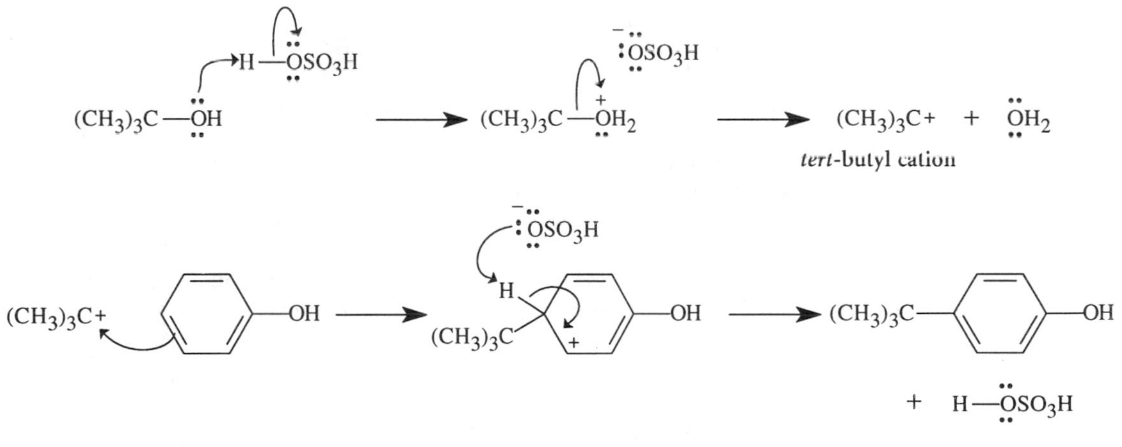

$(CH_3)_3C—\ddot{O}H + H—OSO_3H \longrightarrow (CH_3)_3C—\overset{+}{O}H_2 + {}^-:OSO_3H \longrightarrow (CH_3)_3C^+ + \ddot{O}H_2$

tert-butyl cation

$(CH_3)_3C^+ + C_6H_5—OH \longrightarrow \longrightarrow (CH_3)_3C—C_6H_4—OH + H—\ddot{O}SO_3H$

18.32 (a) Diphenyl ether does not cleave with hot, concentrated HBr, because such a cleavage would require either an S_N1 reaction or an S_N2 reaction at a phenyl-oxygen bond; as this section of the text shows, such reactions do not occur. In contrast, *tert*-butyl phenyl ether cleaves by an S_N1 mechanism involving protonation of the oxygen and loss of phenol to form a *tert*-butyl cation, which reacts with bromide ion to form *tert*-butyl bromide.

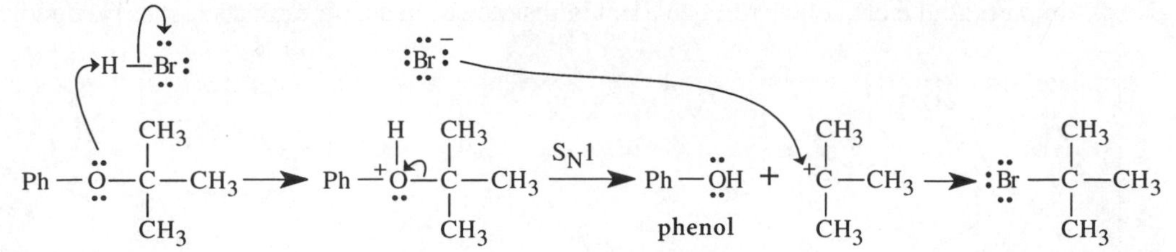

$$\mathrm{Ph{-}O{-}C(CH_3)_3} \xrightarrow{\mathrm{H{-}Br}} \mathrm{Ph{-}\overset{+}{O}H{-}C(CH_3)_3} \xrightarrow{S_N1} \mathrm{Ph{-}OH} + \mathrm{{}^{+}C(CH_3)_3} \xrightarrow{:\ddot{Br}:^-} \mathrm{:\ddot{Br}{-}C(CH_3)_3}$$

tert-butyl phenyl ether — phenol — tert-butyl bromide

(b) Neopentyl phenyl ether does not readily cleave with hot concentrated HBr. Such a reaction would require either an S_N1 or S_N2 reaction of bromide ion at a phenyl-oxygen bond, which does not occur, or an S_N1 or S_N2 reaction at the primary carbon of a neopentyl group. The S_N1 reaction does not occur because it would require the formation of a primary carbocation. (Such a reaction could be forced with very long reaction times and heat, and would occur with rearrangement of the neopentyl group.) The transition state of an S_N2 reaction at the primary carbon of a neopentyl group is so beset by van der Waals repulsions that such a reaction does not take place; see Fig. 9.3(b) on text p. 365.

In contrast, benzyl methyl ether can cleave by both S_N1 and S_N2 mechanisms at the benzylic carbon, or by an S_N2 mechanism at the methyl carbon. Here is the cleavage by an S_N2 mechanism involving attack of Br^- at the methyl carbon of the protonated ether:

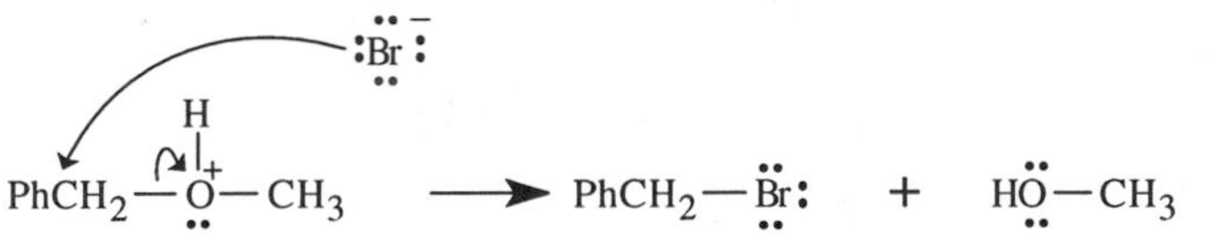

$$\mathrm{PhCH_2{-}\overset{+}{O}H{-}CH_3} + \mathrm{:\ddot{Br}:^-} \longrightarrow \mathrm{PhCH_2{-}\ddot{Br}:} + \mathrm{H\ddot{O}{-}CH_3}$$

benzyl methyl ether (after protonation by HBr) — benzyl bromide — methyl alcohol

And here is the S_N2 mechanism at the benzylic carbon (Sec. 17.4):

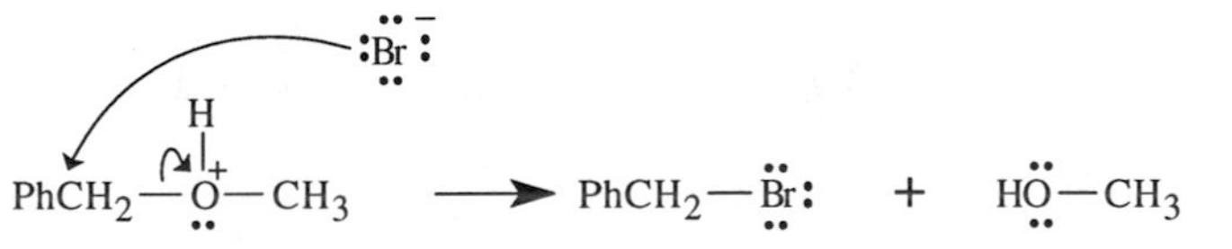

$$\mathrm{PhCH_2{-}\overset{+}{O}H{-}CH_3} + \mathrm{:\ddot{Br}:^-} \longrightarrow \mathrm{PhCH_2{-}\ddot{Br}:} + \mathrm{H\ddot{O}{-}CH_3}$$

benzyl methyl ether (after protonation by HBr) — benzyl bromide — methyl alcohol

An S_N1 mechanism at the benzylic carbon is also possible that parallels the mechanism shown in part (a): protonation of the ether, formation of a benzylic cation, and attack of Br^- on this carbocation. The products are the same as those derived from the S_N2 reaction immediately previous.

p-Methoxytoluene can cleave at the methyl-oxygen bond by an S_N2 mechanism.

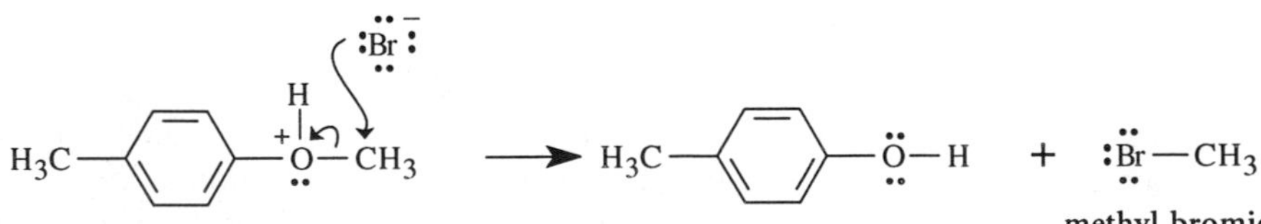

$$\mathrm{H_3C{-}C_6H_4{-}\overset{+}{O}H{-}CH_3} + \mathrm{:\ddot{Br}:^-} \longrightarrow \mathrm{H_3C{-}C_6H_4{-}\ddot{O}{-}H} + \mathrm{:\ddot{Br}{-}CH_3}$$

p-methoxytoluene (after protonation by HBr) — *p*-cresol — methyl bromide

18.33 The product is phenylacetylene, Ph—C≡CH, along with the usual byproducts, Li^{+} $^{-}$OTf and $ClSi(CH_3)_3$.

18.34 The triflate derivative of *p*-nitrophenol and the (*Z*)-stereoisomer of the appropriate trimethylstannyl derivative would be required.

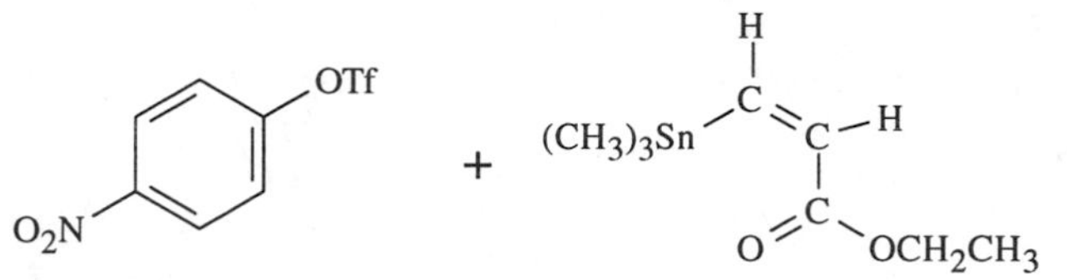

O_2N–C₆H₄–OTf + $(CH_3)_3Sn$–CH=C(H)–C(=O)–OCH_2CH_3

p-nitrophenyl triflate
(prepared from *p*-nitrophenol and triflic anhydride; Eq. 18.82, text p. 820)

18.35 The alcohol *A* can form a relatively stable tertiary benzylic carbocation under the acidic conditions of the reaction. This serves as a Friedel-Crafts alkylating agent for phenol. (Note the similarity to the mechanism shown in the solution to Problem 18.31.)

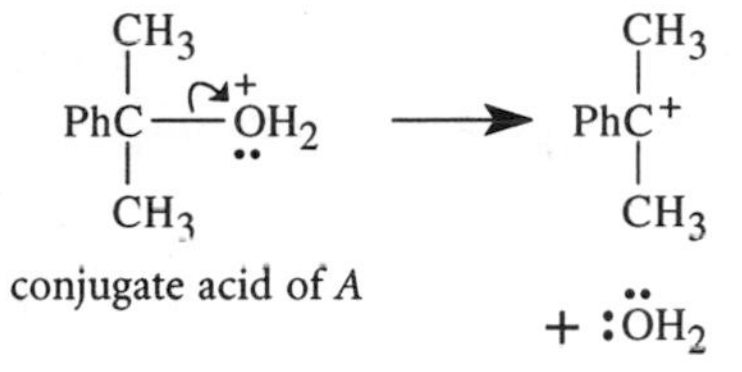

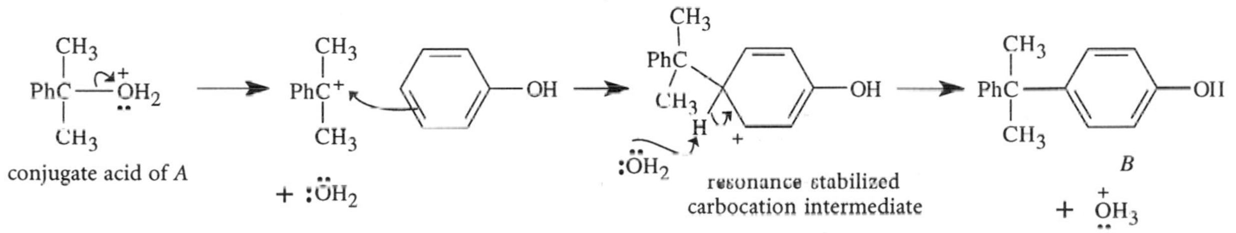

$PhC(CH_3)_2$–$\ddot{O}H_2^{+}$ ⟶ $PhC^{+}(CH_3)_2$ + :$\ddot{O}H_2$

conjugate acid of *A*

$PhC(CH_3)_2$–$\ddot{O}H_2^{+}$ ⟶ $PhC^{+}(CH_3)_2$ + :$\ddot{O}H_2$ + C₆H₅–OH ⟶ resonance stabilized carbocation intermediate (+ :$\ddot{O}H_2$) ⟶ $PhC(CH_3)_2$–C₆H₄–OH (*B*) + $\ddot{O}H_3^{+}$

conjugate acid of *A*

resonance stabilized carbocation intermediate

B

Solutions to Additional Problems

18.36 (a) No reaction (b) No reaction

(c) These are the forcing conditions associated with a benzyne mechanism.

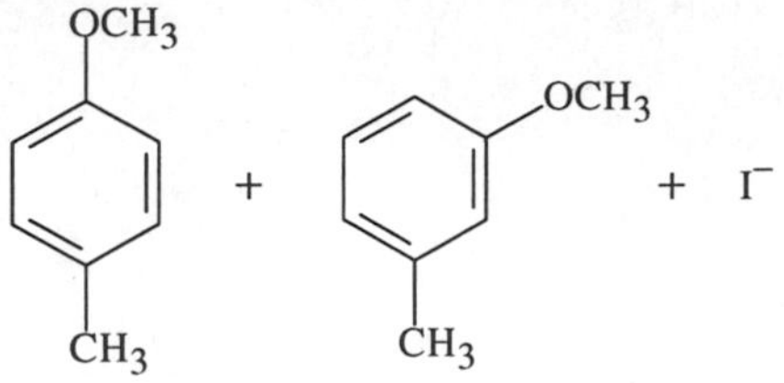

OCH_3, CH_3 + OCH_3, CH_3 + I^-

(d) MgI, CH_3

(e) $Sn(CH_3)_3$, CH_3 + ClMgI

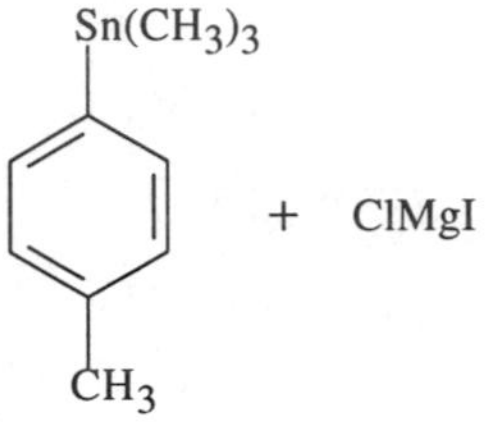

(f) Li, CH_3

(g) $CH{=}CH_2$, CH_3

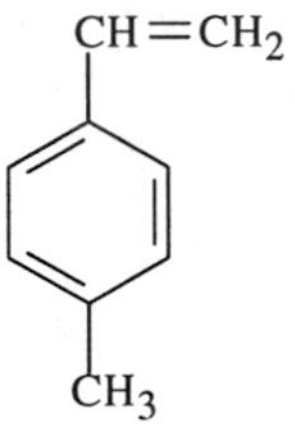

(h) H_3C—

+ $(CH_3)_3SnCl$

+ Li^+ ^-OTf

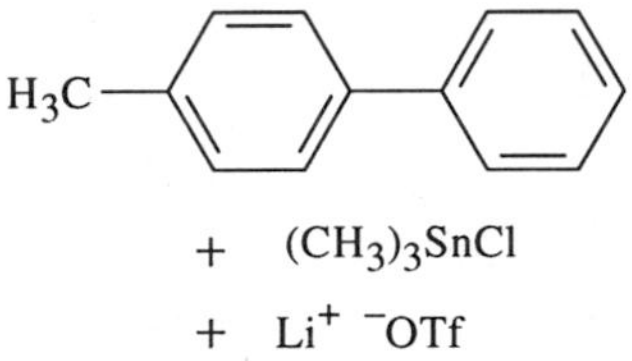

18.37 (a) OH, CH_3, SO_3H + OH, HO_3S, CH_3

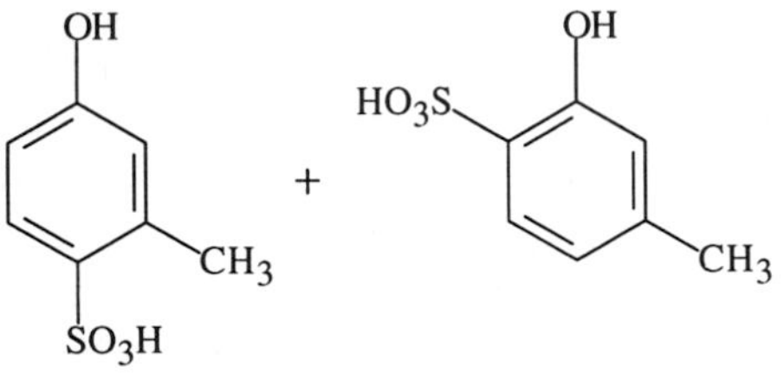

(b) OH, CH_3, Br + OH, Br, CH_3

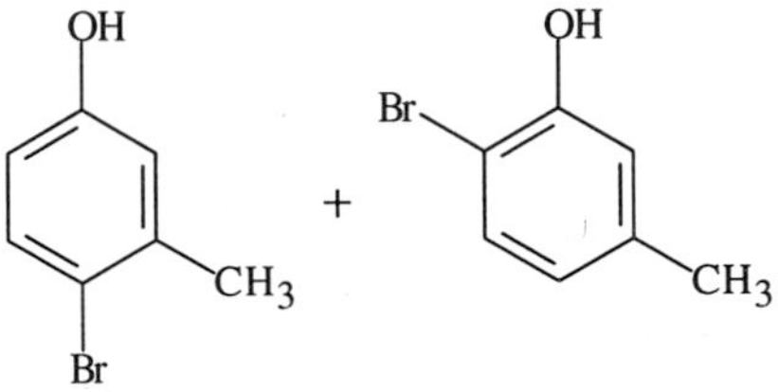

(c) Both benzylic bromination and ring bromination take place; see Eq. 18.69 on text p. 81.

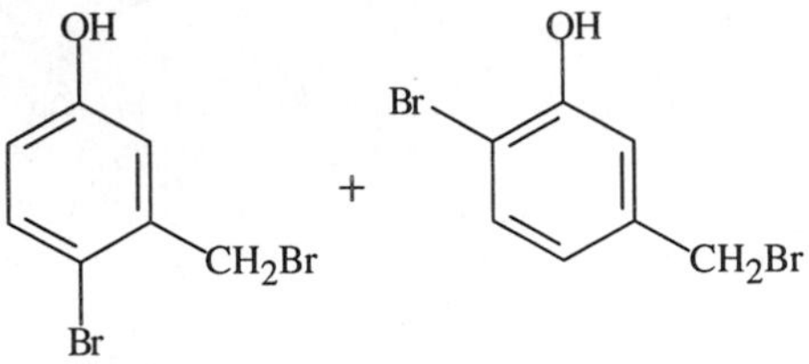

OH, CH_2Br, Br + OH, Br, CH_2Br

(d) no reaction

(e) $O^- Na^+$, CH_3

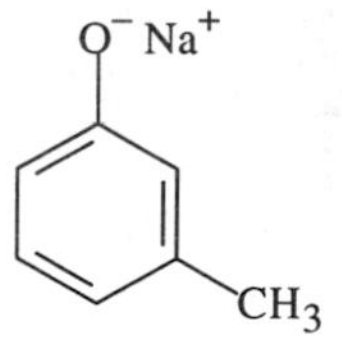

(f) OH, CH_3, NO_2 + OH, O_2N, CH_3

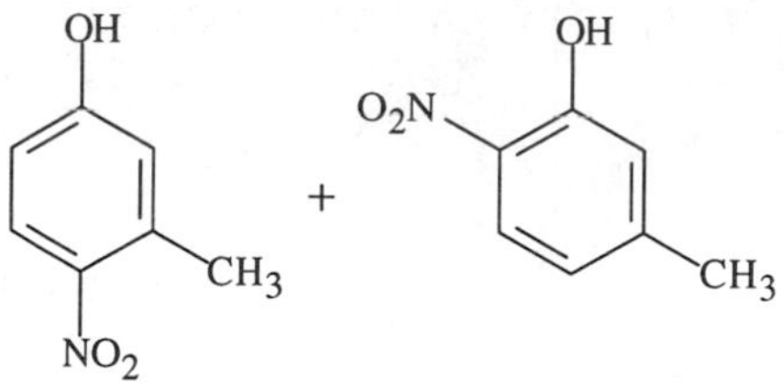

(g) O, OH, C_2H_5C, CH_3 + OH, CH_3, O, C, C_2H_5

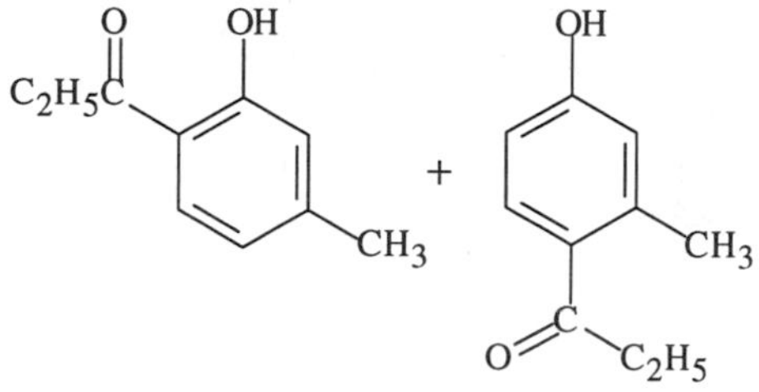

(h) O, CH_3, O

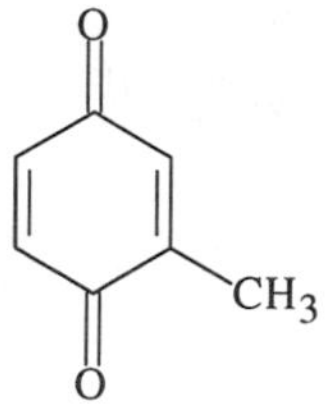

(i) OTf, CH_3

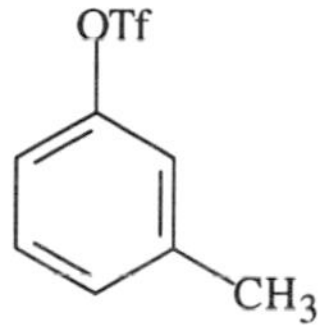

(j) CH_3, CH_3

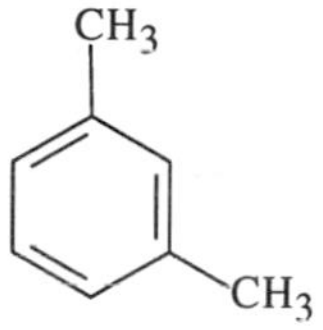

18.38 (a) Mercaptans are more acidic than phenols; and aryl mercaptans (thiophenols) are more acidic than alkyl mercaptans for the same reason that phenols are more acidic than alcohols. The acidity order is

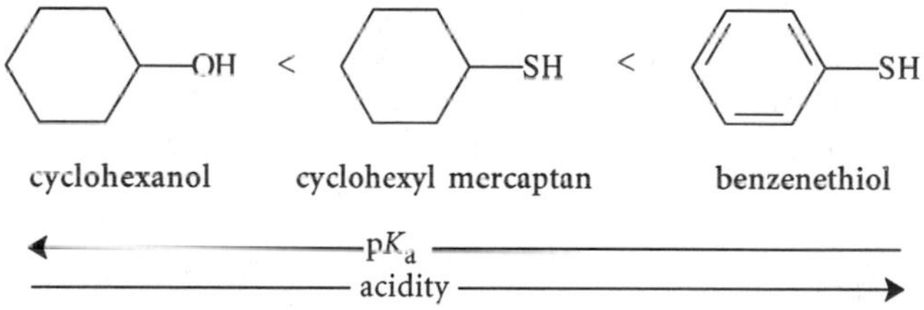

cyclohexanol (—OH) < cyclohexyl mercaptan (—SH) < benzenethiol (—SH)

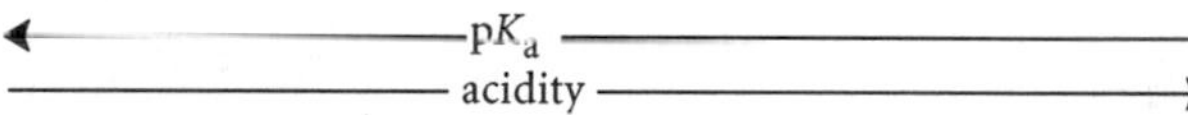

(b) Phenol is most acidic because its conjugate-base anion is stabilized by both the polar and resonance effects of the phenyl group. The conjugate-base anion of benzyl alcohol is stabilized by the polar effect of the phenyl group. The conjugate-base anion of cyclohexanol has none of these stabilizing contributions. The acidity order is

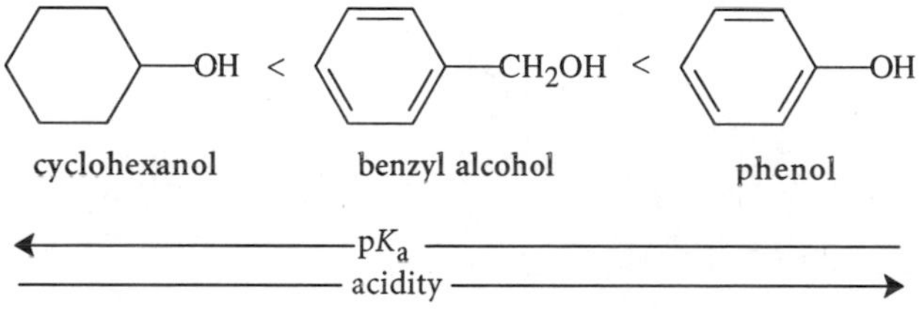

cyclohexanol (—OH) < benzyl alcohol (—CH_2OH) < phenol (—OH)

← pK_a
acidity →

(c) The *p*-nitro substituent stabilizes the *p*-nitrophenolate ion by both its resonance effect (see text Eq. 18.59 on p. 808) and its polar effect; the *p*-chloro substituent stabilizes *p*-chlorophenolate ion only by its polar effect. The third compound is nitric acid, a strong acid. Its conjugate-base anion, the nitrate ion, is stabilized both by the polar effect of the nitro group, which is much closer to the anionic oxygen than it is in the *p*-nitrophenolate ion, and by a resonance effect:

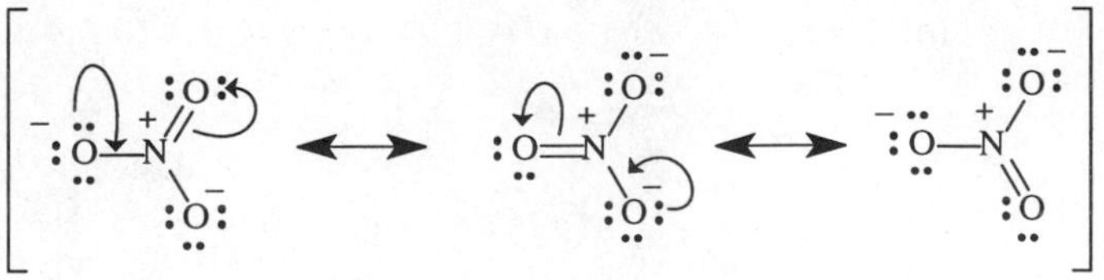

nitrate anion

Here is the acidity order:

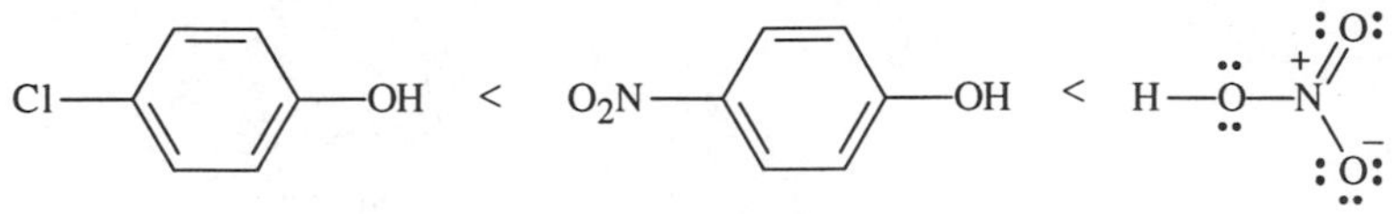

p-chlorophenol < p-nitrophenol < nitric acid

pKa ← ; acidity →

(d) 4-Nitrophenol is more acidic than phenol (see text p. 808), and benzenethiols are more acidic than phenols (element effect.) The acidity order is

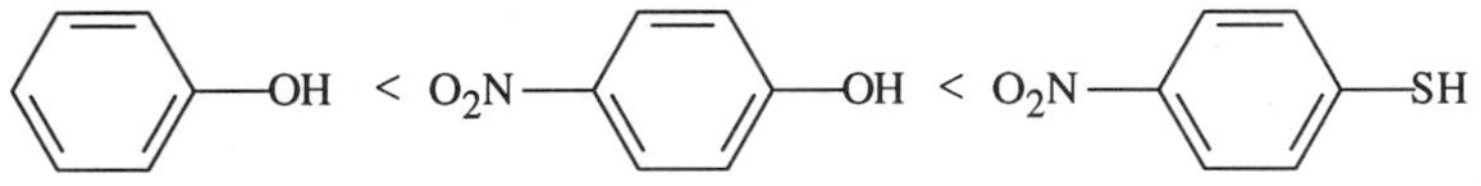

phenol < 4-nitrophenol < 4-nitrobenzenethiol

pKa ← ; acidity →

(e) The *para*-aryl substituents all stabilize the respective conjugate-base alkoxide ions by both a resonance and a polar effect. The nitro group of the *meta*-nitrophenyl substituent exerts an additional stabilizing polar effect; and the nitro group of the *para*-nitrophenyl substituent stabilizes the conjugate-base phenoxide ion by both a polar effect and a resonance effect:

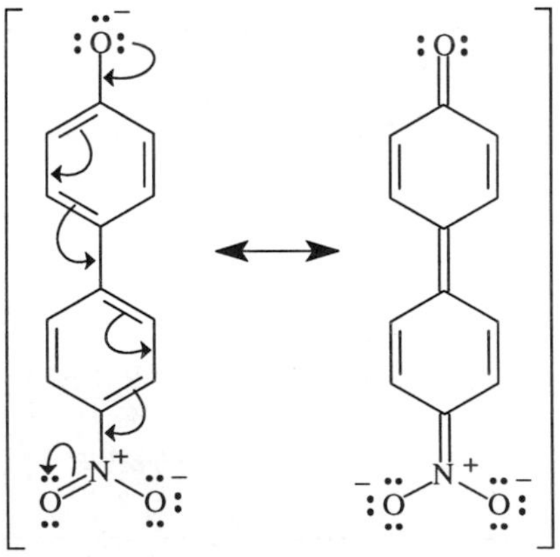

The acidity order is:

$$C < A < D < B$$

pKa ← ; acidity →

18.39 The enol *A* is more acidic (by about five pK_a units). The reason is that its conjugate-base anion is stabilized by resonance as shown below, whereas the conjugate-base anion of compound *B* is not resonance-stabilized.

$$\left[H_3C-\overset{\ddot{O}:^-}{C}=CH_2 \longleftrightarrow H_3C-\overset{\ddot{O}:}{\overset{\|}{C}}-\ddot{C}H_2^- \right]$$

conjugate base of compound *A*

18.40 (a) 3,5-dimethylphenol (compound *A*) (b) *p*-methylbenzyl alcohol (compound *B*) (c) *m*-ethoxytoluene (compound *C*)

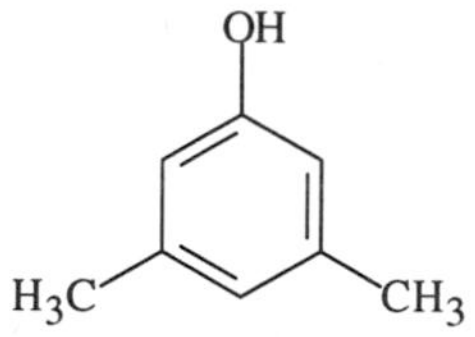

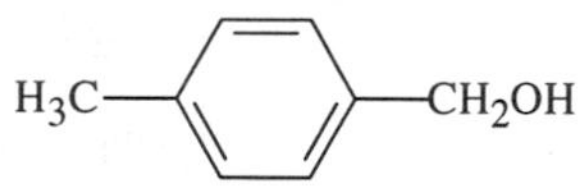

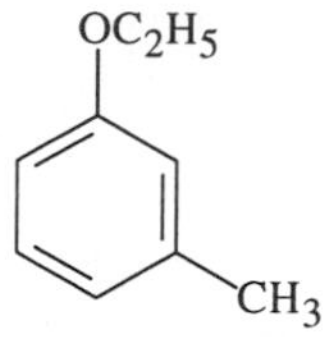

Hydrogenation of phenol *A* yields the corresponding cyclohexanol. Compound *B* reacts readily with HBr to give the benzylic bromide; the corresponding Grignard reagent undergoes protonolysis to give *p*-xylene. The ether group in compound *C* is cleaved by HBr to give C_2H_5Br (which is volatile under the high-temperature reaction conditions) and *m*-cresol.

18.41 (a) Cyclohexanol is ionized only to a very small extent in aqueous NaOH solution; phenol ionizes almost completely.

Cyclohexyl–OH $\xrightarrow{NaOH}$ Cyclohexyl–O^- Na^+ Phenyl–OH $\xrightarrow{NaOH}$ Phenyl–O^- Na^+

(b) Both compounds react rapidly with NaH to give H_2 and their conjugate bases: sodium cyclohexanolate in the case of cyclohexanol, and sodium phenolate (sodium phenoxide) in the case of phenol.

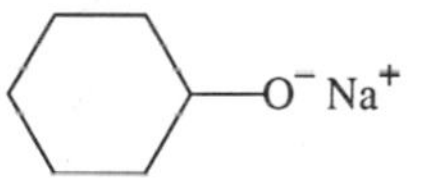

sodium cyclohexanolate 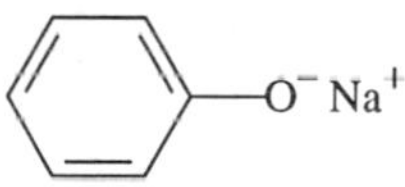 sodium phenolate

(c) Both compounds are converted into their triflate derivatives.

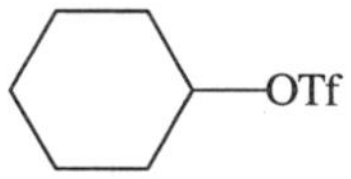

cyclohexyl triflate 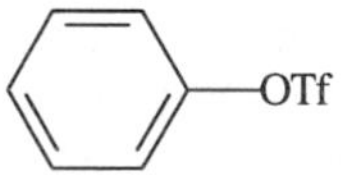 phenyl triflate

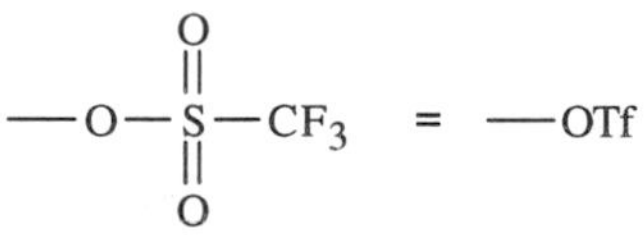

$-O-S(=O)_2-CF_3$ = —OTf

triflate (trifluoromethanesulfonate) group

(d) Only cyclohexanol reacts with concentrated HBr; because the C—O bond in phenol cannot be broken by S_N1 or S_N2 reactions, phenol is inert, except for a small degree of protonation of the phenol oxygen. The product in the case of cyclohexanol is bromocyclohexane.

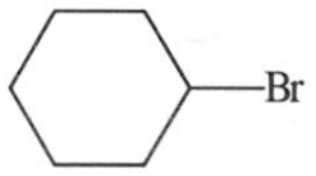

bromocyclohexane

(e) The phenol is brominated on the ring to give the following products; the alcohol is unaffected.

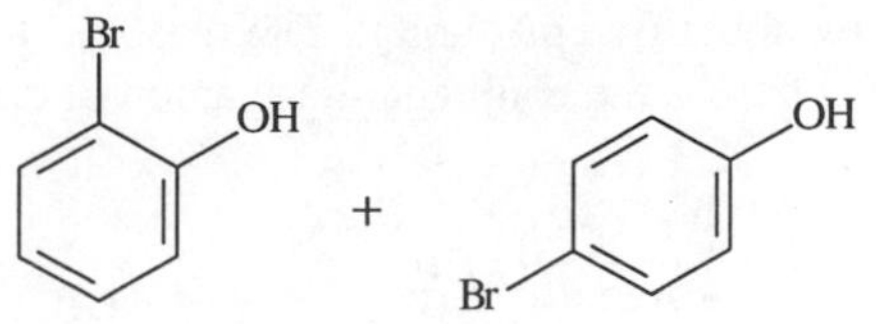

2-bromophenol + 4-bromophenol

(f) Cyclohexanol is oxidized to cyclohexanone, whereas phenol is oxidized to *p*-benzoquinone.

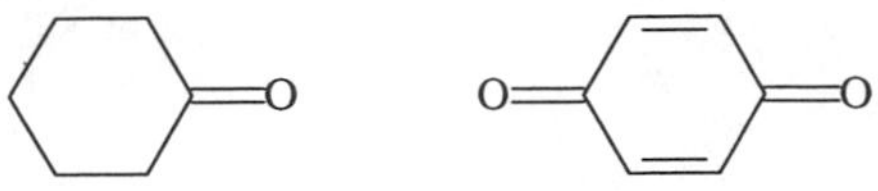

cyclohexanone *p*-benzoquinone

(g) Cyclohexanol undergoes dehydration to cyclohexene; phenol undergoes sulfonation, an electrophilic aromatic substitution reaction.

cyclohexene 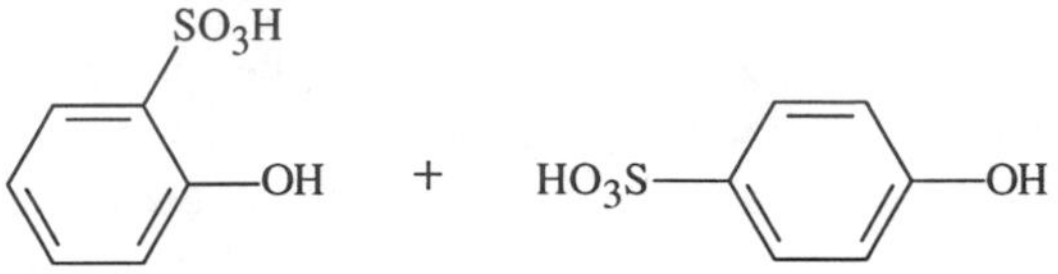 2-hydroxybenzenesulfonic acid + 4-hydroxybenzenesulfonic acid

18.42 (a) The second compound reacts in a nucleophilic aromatic substitution reaction. The first compound does not react in this way because it is not activated toward nucleophilic aromatic substitution; the nitro group is not attached to the ring.

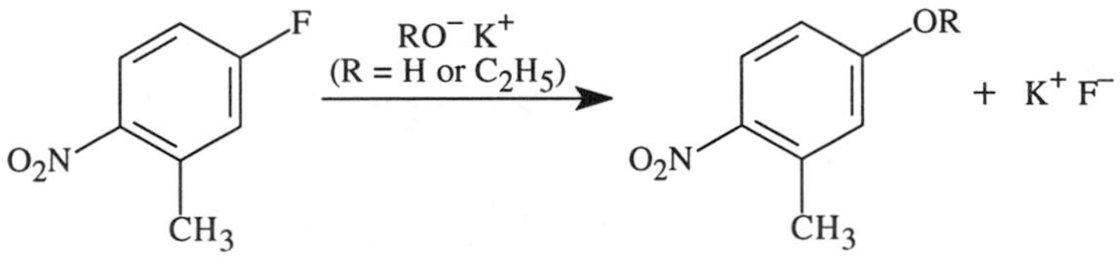

$RO^- K^+$ ($R = H$ or C_2H_5) → $+ K^+ F^-$

In the interest of total accuracy we point out that the first compound actually does react, but in a different way. A hydrogen of the —CH_2— group is acidic enough to ionize in alcoholic KOH; can you see why? But it does not liberate F^-.

(b) The last compound, diphenyl ether, cannot be prepared by a Williamson ether synthesis, because such a synthesis would require an S_N2 reaction to occur at the carbon of a benzene ring. Notice that the first ether, tetrahydrofuran, can be prepared by an *intramolecular* Williamson synthesis from 4-bromo-1-butanol, Br—$(CH_2)_4$—OH. The second ether, ethoxybenzene, can be prepared by a reaction of sodium phenolate, $PhO^- Na^+$, with ethyl iodide.

(c) The first compound undergoes solvolysis by an S_N1 mechanism to give a mixture of an alcohol, an ether, an alkene, and HBr (thus the acidic solution). The second compound is inert, because aryl halides do not undergo solvolysis under neutral conditions.

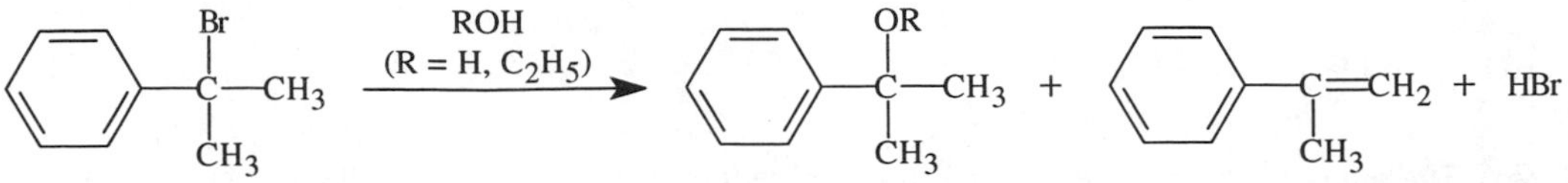

ROH ($R = H, C_2H_5$) → + HBr

(d) Diphenyl ether will not cleave with HI because such a cleavage would require an S_N1 or S_N2 reaction at the carbon of a benzene ring, and such reactions do not occur. Phenyl cyclohexyl ether, however, would cleave to phenol and iodocyclohexane:

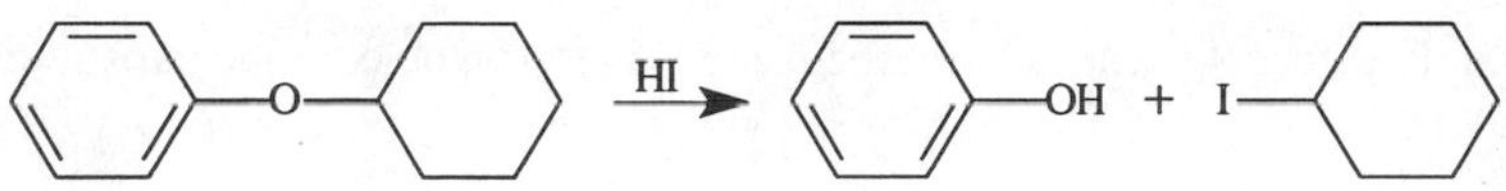
HI
OH + I

cyclohexyl phenyl ether

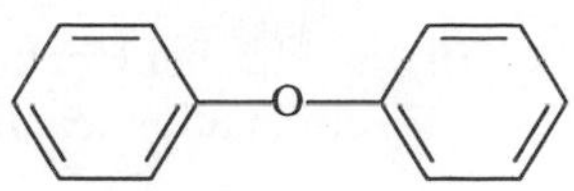

diphenyl ether
(inert to HI)

(e) Because of the symmetry of the benzyne intermediate *A*, the first compound would give only a single product; the last compound does not react because it has no *β*-hydrogen. Only the second compound gives two products.

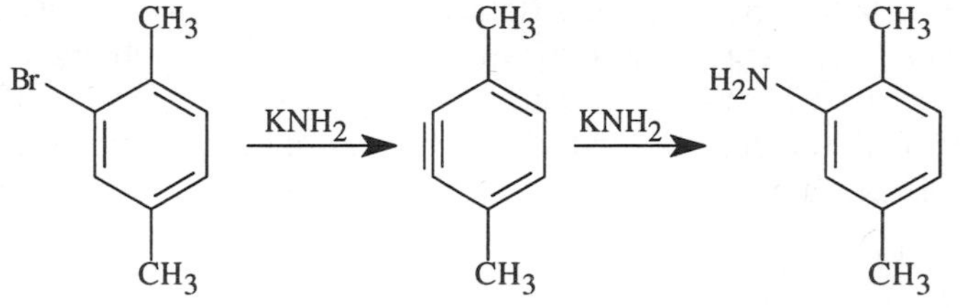
Br
CH3
KNH2
KNH2
H2N

A

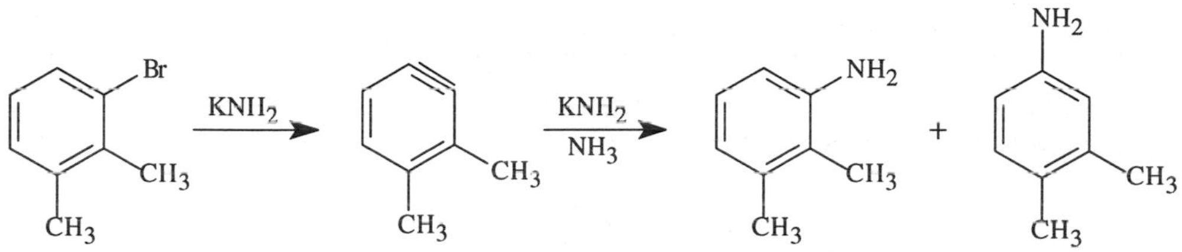
Br
CH3
KNH2
KNH2
NH3
NH2
+

18.43 The phenol group is unaffected in both compounds (except for a small amount of protonation of the phenol oxygen); the alcohol and ether groups react.

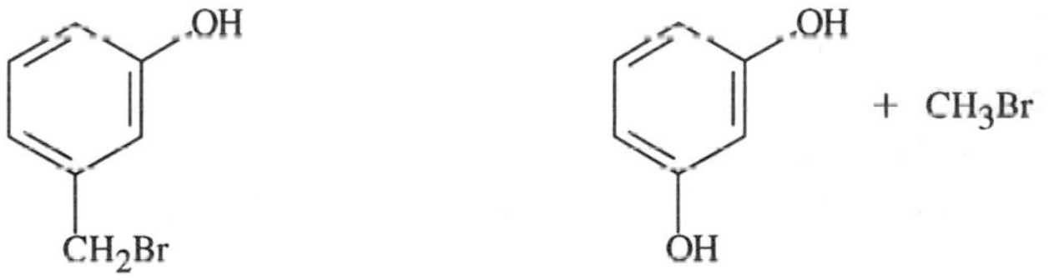
OH
CH2Br
OH
+ CH3Br

18.44 Potassium benzenethiolate, a base, reacts with the tertiary alkyl halide in an E2 reaction:

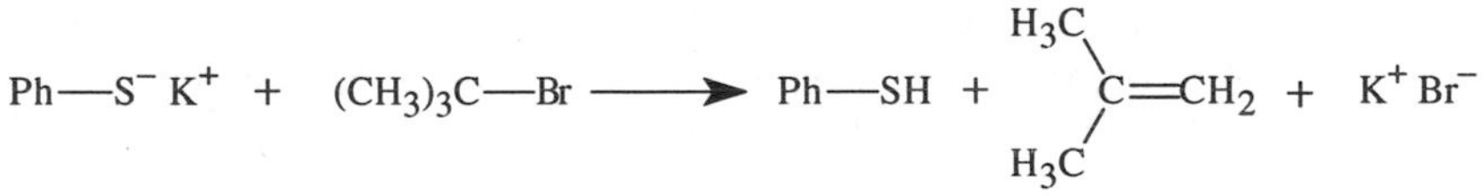
Ph—S⁻ K⁺ + (CH3)3C—Br
Ph—SH +
H3C
C=CH2 + K⁺ Br⁻
H3C

18.45 Sodium ethoxide converts the thiol completely into its conjugate-base thiolate ion.

(a) The thiolate ion is alkylated. (Remember from Sec. 17.4 that allyl bromide is a particularly reactive alkylating agent.)

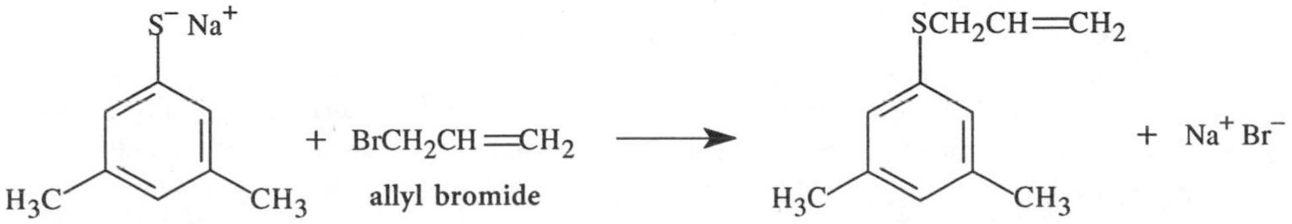
S⁻ Na⁺
H3C
CH3
+ BrCH2CH=CH2
allyl bromide
SCH2CH=CH2
+ Na⁺ Br⁻

product of thiol ionization

(b) No further reaction of the thiolate occurs because bromobenzene is inert to nucleophilic substitution and elimination reactions.

18.46 (a) The reaction of phenol as a base with the acid H_2SO_4:

$$C_6H_5\ddot{O}H + H_2SO_4 \rightleftharpoons C_6H_5\overset{+}{O}H_2 \quad HSO_4^-$$

(b) First, consider polar effects. The phenyl group has an *electron-withdrawing* polar effect; this destabilizes the positively-charged conjugate acid of phenol. Such an effect is absent in the conjugate acid of the alcohol cyclohexanol because its ring has no double bonds. Therefore, the polar effect reduces the basicity of the phenol relative to the alcohol. Second, consider the resonance effect of the benzene ring. The source of this effect is the delocalization of an oxygen unshared electron pair into the ring:

$$\left[C_6H_5\ddot{O}H \longleftrightarrow \overset{-}{C_6H_5}{=}\overset{+}{O}H \longleftrightarrow \text{other structures} \right]$$

resonance structures of phenol

Such an effect is absent in the conjugate acid of the phenol because it would place *two* positive charges on the oxygen, an electronegative atom.

$$\left[C_6H_5\overset{+}{O}H_2 \not\longleftrightarrow \overset{-}{C_6H_5}{=}\overset{2+}{O}H \right]$$

conjugate acid of phenol

Therefore, resonance stabilizes phenol relative to its conjugate acid and therefore reduces its basicity. (Another way to think of this situation is to imagine that delocalization of the oxygen unshared electron pair makes it "less available" for protonation.) Such a resonance effect is not present in the alcohol. The conclusion is that a phenol is much less basic than an alcohol because of both the polar effect and the resonance effect of the ring. Therefore, the alcohol is the stronger base.

18.47 Because NaOH can ionize a phenol, particularly a relatively acidic phenol such as *p*-nitrophenol, which has a pK_a of 7.21, the spectrum *B* (and the corresponding yellow color) is due to the conjugate base of the phenol. The UV absorption reflects conjugation of the extra unshared pair with the ring and its delocalization into the *p*-nitro group, as shown in Eq. 18.59, text p. 808. Addition of acid lowers the pH so that the un-ionized phenol is regenerated, and the UV spectrum of the phenoxide disappears. To summarize:

$$\underset{\lambda_{max} = 315 \text{ nm (spectrum } A)}{O_2N{-}C_6H_4{-}OH} \underset{H_3O^+}{\overset{NaOH}{\rightleftharpoons}} \underset{\lambda_{max} = 392 \text{ nm (spectrum } B)}{O_2N{-}C_6H_4{-}O^- \, Na^+}$$

18.48 It is the *un-ionized* form of vanillin that has the typical odor. In NaOH solution, the phenol group of vanillin ionizes to its conjugate-base phenoxide ion; because vanillin is no longer present, and because ionic compounds such as the conjugate-base phenoxide are not volatile, the odor disappears. (A compound has to be volatile—that is, it must have a significant vapor pressure—to enter the gas phase and thus reach the nostrils.) Acidification of the solution brings about protonation of the phenoxide and regeneration of vanillin and, hence, the characteristic odor.

OCH$_3$ | HC(=O)—C$_6$H$_3$—OH $\xrightarrow{\text{NaOH}}$ OCH$_3$ | HC(=O)—C$_6$H$_3$—O$^-$ Na$^+$

neutral compound: volatile, has vanilla odor

ionic compound: nonvolatile, has no odor.

18.49 The two principles involved in the solution to this problem are, first, that phenoxide ions are more soluble in water than are their conjugate-acid phenols; and second, that an acid is ionized when the pH of the solution is well above the pK_a of the acid. The second point can be demonstrated simply with the concentration expression for the dissociation constant of an acid HA to its conjugate base A$^-$ (Eq. 3.14, text p. 93):

$$K_a = \frac{[H_3O^+][A^-]}{[HA]}$$

Taking logarithms, and letting –log [H_3O^+] = pH,

$$\text{pH} - \text{p}K_a = \log \frac{[A^-]}{[HA]}$$

$$\text{or } 10^{(\text{pH} - \text{p}K_a)} = \frac{[A^-]}{[HA]}$$

This formula shows that in order for the ratio [A$^-$]/[HA] to be large, the pH of the *solution* must be much greater (typically two or more units greater, or *more basic*) than the pK_a of the *acid*.

Condition (1) corresponds to a pH of about 1; neither phenol is ionized, and therefore neither is soluble in the buffer. Under condition (2), some fraction of 2,4-dinitrophenol is ionized, but under condition (3), which is a pH value almost three units higher than the pK_a of the phenol, 2,4-dinitrophenol is >99% ionized. Because this pH value is more than three units *below* the pK_a of *p*-cresol, this phenol is *not* ionized at pH = 7. Therefore, 2,4-dinitrophenol should be relatively soluble at this pH because it is ionized, and *p*-cresol remains insoluble. This is the best condition to separate the two phenols by extraction. Condition (4) corresponds to a pH of about 13; both phenols are ionized at this pH, and both should be relatively soluble (as their conjugate-base phenoxides).

18.50 (a) Because the S_N2 reaction requires approach of the nucleophile from the backside of the C—Br bond, and because this would require approach of the nucleophile along the axis of the triple bond, the S_N2 reaction is impossible.

(b) 1-Haloalkynes cannot undergo an S_N1 reaction. Such a reaction would require that a carbocation be formed at an *sp*-hybridized carbon. Yet carbocations require sp^2 hybridization, which, in turn, requires trigonal-planar geometry. An alkyne carbon cannot achieve this geometry because its three bonds are connected to the same atom. Hence, the sp^2 hybridization required for carbocation formation is impossible.

18.51 (a) The oxidation state is the number of valence electrons in the neutral atom (9) less the charge (+1) less the number of bonds to X-type ligands (2). (Note that the nitrogens that have three bonds *in addition to* the metal-nitrogen bond are L-type ligands; the other nitrogen is an X-type ligand. The second X-type ligand is the cyano (—CN) group.) The oxidation state of Co is +3.

(b) The number *n* is the valence electrons in the neutral atom (9) less the oxidation state (+3). Hence, Co in this complex is a d^6 metal.

(c) The electron count is the number of electrons in the neutral atom (9) less the charge (+1) plus the number of X-type ligands (2) plus twice the number of L-type ligands (4). The electron count is therefore $18e^-$. This complex thus follows the eighteen-electron rule.

18.52 By loss of a proton from the hydroxy group, the carbocation intermediate becomes the neutral compound that precipitates. (Formation of the electrophile is shown in Eq. 18.71 on text p. 816.)

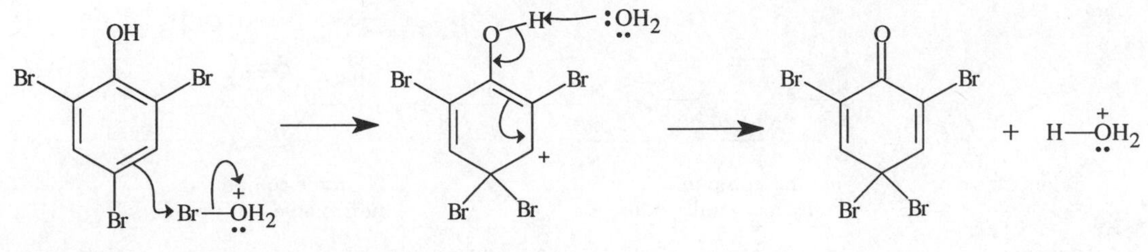

$$\text{2,4,6-tribromophenol} + Br-\overset{+}{O}H_2 \longrightarrow \text{[cation intermediate]} \longrightarrow \text{2,4,4,6-tetrabromocyclohexa-2,5-dienone} + H-\overset{+}{O}H_2$$

18.53 Solvolysis of the benzylic bromide results in a carbocation intermediate that can lose a proton to give the bracketed intermediate, an example of a *quinone methide*, which reacts with hydroxide ion as shown in the following equation. (See also Problem 18.69 for the formation of a similar type of intermediate.)

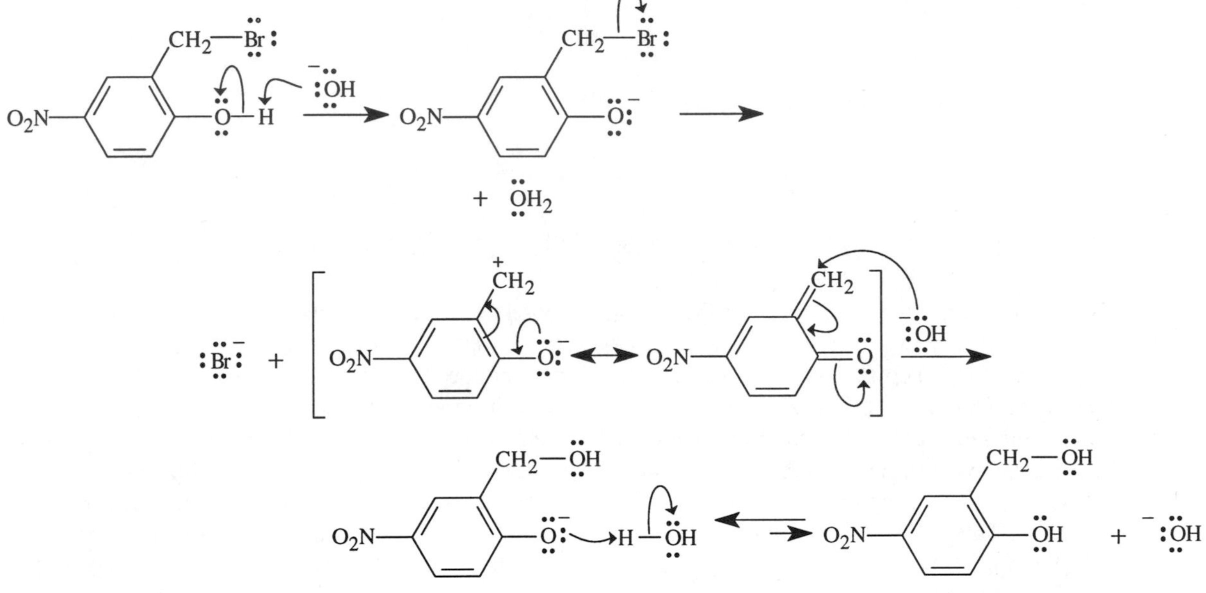

$$O_2N\text{–}C_6H_3(CH_2Br)\text{–}OH \xrightarrow{^-OH} O_2N\text{–}C_6H_3(CH_2Br)\text{–}O^- + OH_2 \longrightarrow$$

$$:Br:^- + \left[O_2N\text{–}C_6H_3(\overset{+}{C}H_2)\text{–}O^- \longleftrightarrow O_2N\text{–}C_6H_3(=CH_2)=O \right] \xrightarrow{^-OH}$$

$$O_2N\text{–}C_6H_3(CH_2OH)\text{–}O^- \; H\text{–}OH \rightleftarrows O_2N\text{–}C_6H_3(CH_2OH)\text{–}OH + {}^-OH$$

18.54 (a)

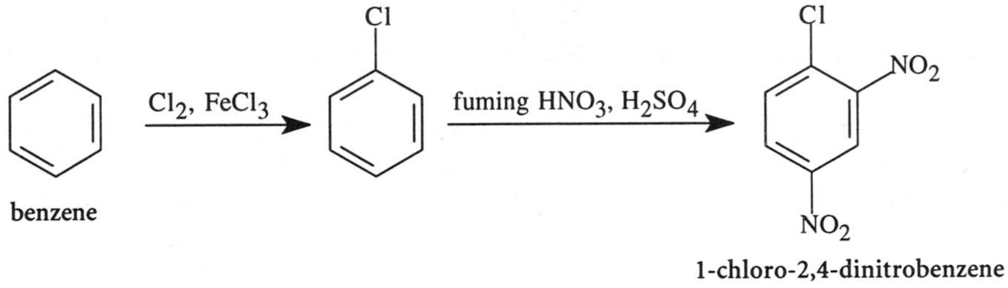

$$\text{benzene} \xrightarrow{Cl_2,\ FeCl_3} \text{chlorobenzene} \xrightarrow{\text{fuming } HNO_3,\ H_2SO_4} \text{1-chloro-2,4-dinitrobenzene}$$

(b)

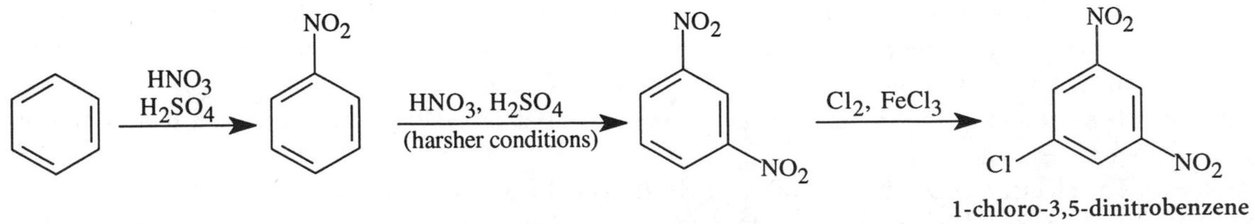

$$\text{benzene} \xrightarrow{HNO_3,\ H_2SO_4} \text{nitrobenzene} \xrightarrow[\text{(harsher conditions)}]{HNO_3,\ H_2SO_4} \text{1,3-dinitrobenzene} \xrightarrow{Cl_2,\ FeCl_3} \text{1-chloro-3,5-dinitrobenzene}$$

(c)

$$\text{(E)-PhCH=CHPh} \xrightarrow{Br_2} PhCHBrCHBrPh \xrightarrow{K^+\ (CH_3)_3CO^-} PhC{\equiv}CPh$$

diphenylacetylene

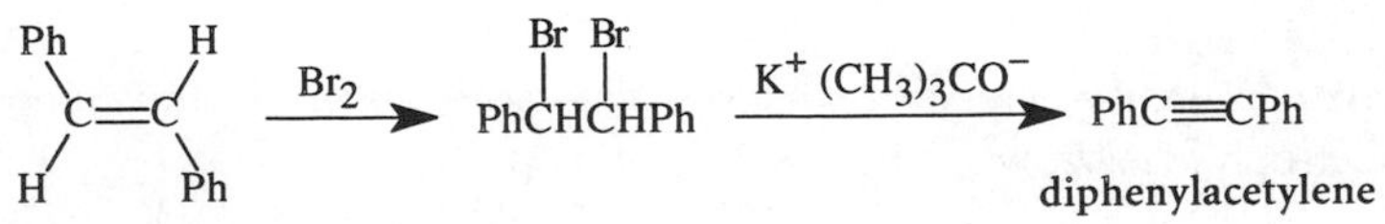

(d)

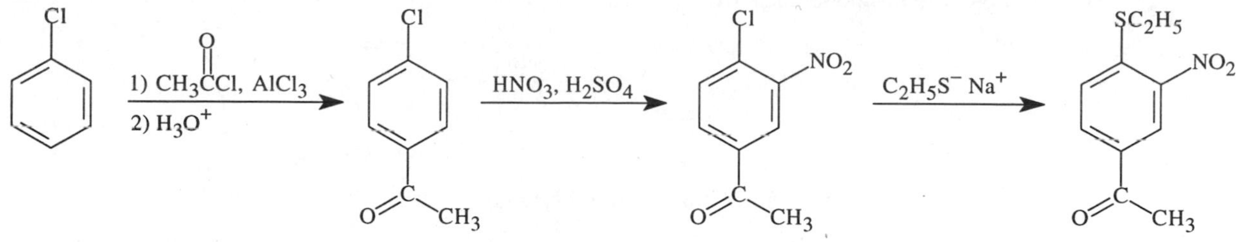

Cl-benzene → 1) CH_3CCl (O), $AlCl_3$; 2) H_3O^+ → p-chloroacetophenone → HNO_3, H_2SO_4 → 4-chloro-3-nitroacetophenone → $C_2H_5S^-\ Na^+$ → 4-(SC_2H_5)-3-nitroacetophenone

(e)

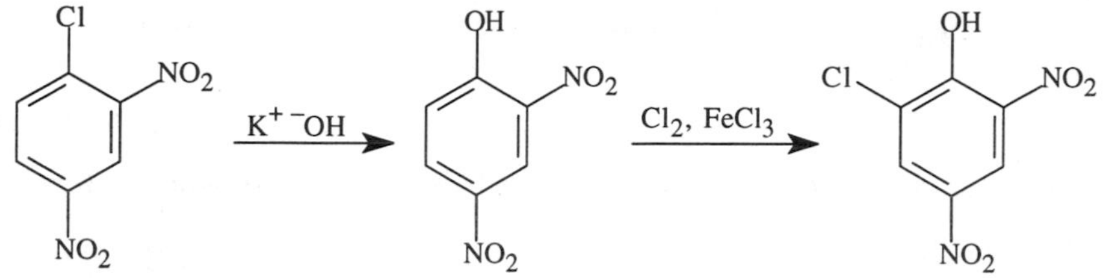

1-chloro-2,4-dinitrobenzene (prepared in part (a)) → $K^+\ ^-OH$ → 2,4-dinitrophenol → Cl_2, $FeCl_3$ → 2-chloro-4,6-dinitrophenol

(f) See Eq. 18.79, text p. 818.) Other sources of the *tert*-butyl cation electrophile could also be used, such as 2-methylpropene and acid. *Tert*-butyl chloride and $AlCl_3$ would probably not work well because of the tendency of phenols to form complexes with $AlCl_3$.

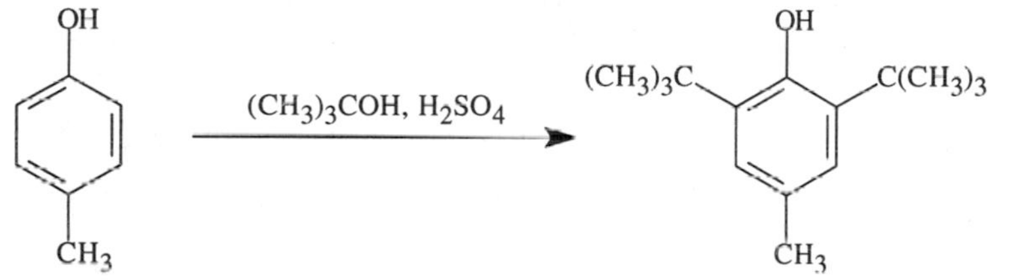

p-cresol → $(CH_3)_3COH$, H_2SO_4 → 2,6-di-*tert*-butyl-4-methylphenol (BHT)

(g)

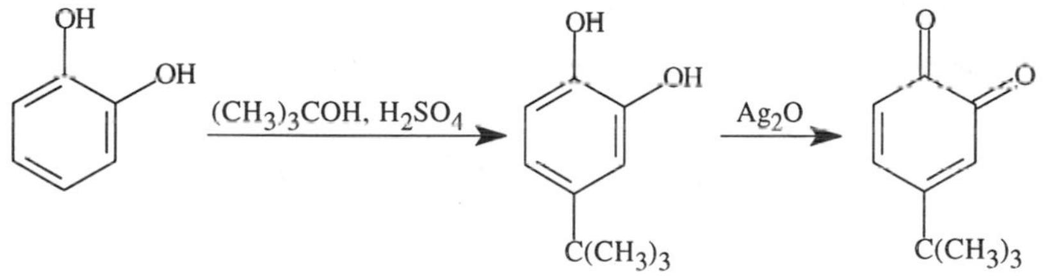

catechol → $(CH_3)_3COH$, H_2SO_4 → 4-$C(CH_3)_3$-catechol → Ag_2O → 4-$C(CH_3)_3$-1,2-benzoquinone

(h)

PhBr (bromobenzene) → Mg, ether → PhMgBr → 1) ethylene oxide; 2) H_3O^+ → $PhCH_2CH_2OH$ (2-phenyl-1-ethanol)

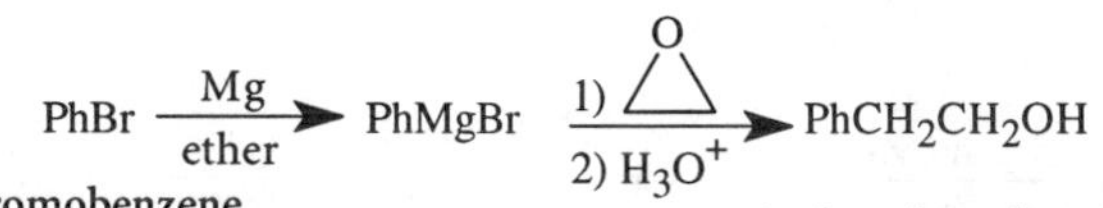

(i)

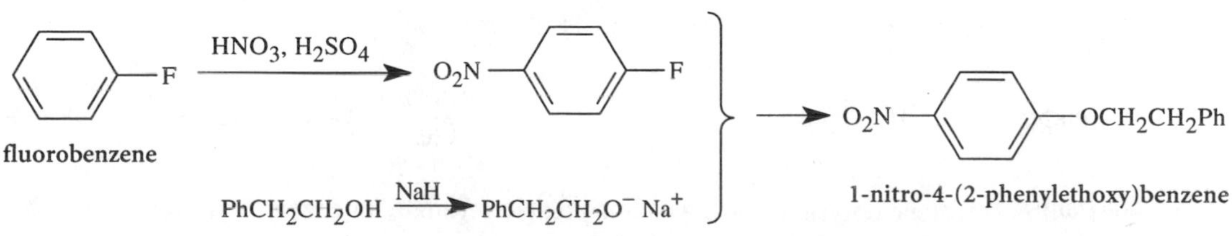

fluorobenzene → HNO_3, H_2SO_4 → O_2N–C_6H_4–F

$PhCH_2CH_2OH$ → NaH → $PhCH_2CH_2O^-\ Na^+$

→ O_2N–C_6H_4–OCH_2CH_2Ph (1-nitro-4-(2-phenylethoxy)benzene)

(j) First, prepare bromobenzene from benzene.

$$C_6H_6 \xrightarrow{Br_2,\ FeBr_3} C_6H_5\text{—}Br$$

bromobenzene

Next, prepare 1-bromo-3-nitrobenzene from benzene.

$$C_6H_6 \xrightarrow{HNO_3,\ H_2SO_4} C_6H_5\text{—}NO_2 \xrightarrow{Br_2,\ FeBr_3} m\text{-}Br\text{—}C_6H_4\text{—}NO_2$$

1-bromo-3-nitrobenzene

Then carry out a Heck reaction of either aryl halide with ethylene. We'll use 1-bromo-3-nitrobenzene.

$$m\text{-}O_2N\text{—}C_6H_4\text{—}Br + H_2C{=}CH_2 \xrightarrow{Pd(PPh_3)_4,\ (C_2H_5)_3N} m\text{-}O_2N\text{—}C_6H_4\text{—}CH{=}CH_2$$

1-bromo-3-nitrobenzene — *m*-nitrostyrene

Finally, carry out another Heck reaction using *m*-nitrostyrene and bromobenzene (using somewhat harsher conditions).

$$m\text{-}O_2N\text{—}C_6H_4\text{—}CH{=}CH_2 \xrightarrow{Pd(PPh_3)_4,\ (C_2H_5)_3N} m\text{-}O_2N\text{—}C_6H_4\text{—}C(H){=}C(H)\text{—}C_6H_5$$

m-nitrostyrene

(k) First, prepare the Grignard reagent from the alkyne, and use it to prepare the corresponding stannane.

$$PhC{\equiv}C\text{—}H + CH_3MgI \longrightarrow PhC{\equiv}C\text{—}MgBr + CH_4$$

$$PhC{\equiv}C\text{—}MgBr + ClSn(CH_3) \longrightarrow PhC{\equiv}C\text{—}Sn(CH_3)_3 + ClMgBr$$

Next, prepare the triflate of the phenol.

$$HO\text{—}C_6H_4\text{—}OCH_3 \xrightarrow[\text{pyridine}]{F_3C\text{—}SO_2\text{—}O\text{—}SO_2\text{—}CF_3\ \text{(triflic anhydride)}} TfO\text{—}C_6H_4\text{—}OCH_3$$

4-methoxyphenol

Finally, couple the triflate and the stannane in a Stille reaction.

$$PhC{\equiv}C\text{—}Sn(CH_3)_3 + TfO\text{—}C_6H_4\text{—}OCH_3 \xrightarrow[LiCl]{Pd(PPh_3)_4} PhC{\equiv}C\text{—}C_6H_4\text{—}OCH_3$$

(l) Start with (2-cyclohexenyl)benzene, which is prepared from iodobenzene and cyclohexene by a Heck reaction as shown in Eq. 18.51, text p. 803.

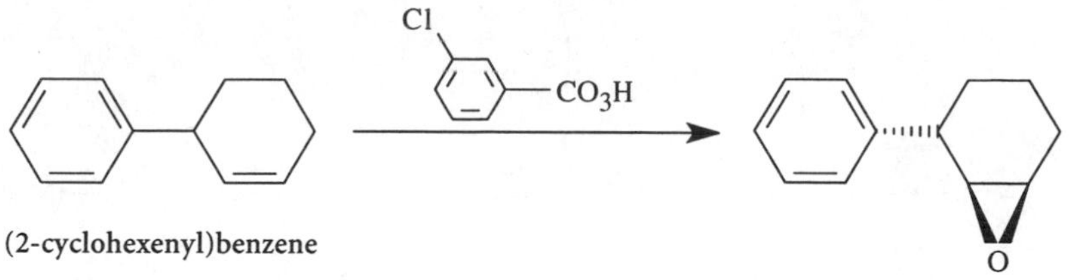

(2-cyclohexenyl)benzene

It is reasonable to suppose that epoxidation will occur at the face of the ring opposite to that occupied by the bulky phenyl group.

18.55 (a) This is an overall substitution reaction that occurs by a benzyne mechanism. (Although the formation of benzyne may be a two-step process, as shown in Eqs. 18.22a–b on text pp. 785–6, it is shown as a concerted elimination here and in subsequent mechanisms to save space.)

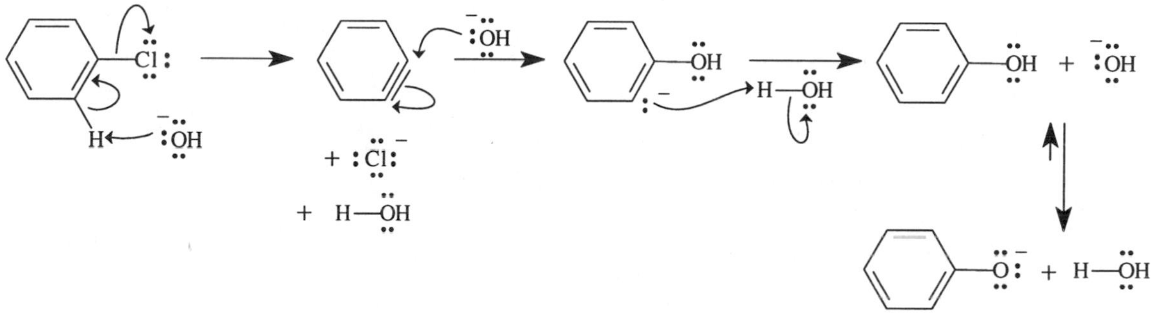

(b) Diphenyl ether is formed by attack of the phenoxide product on the benzyne intermediate.

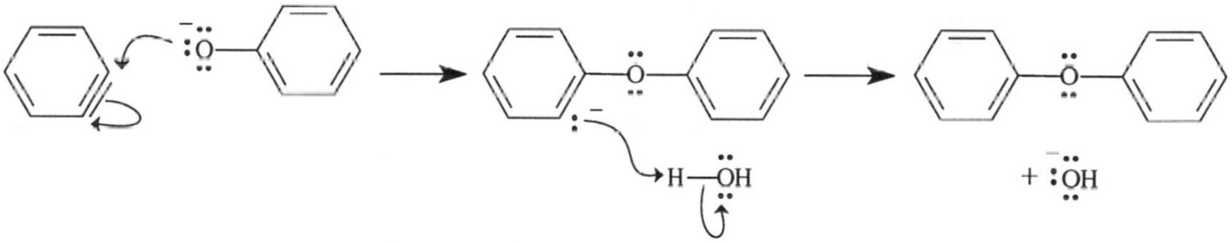

18.56 (a) Fluoride ion is displaced in a nucleophilic aromatic substitution reaction by ethanethiolate ion.

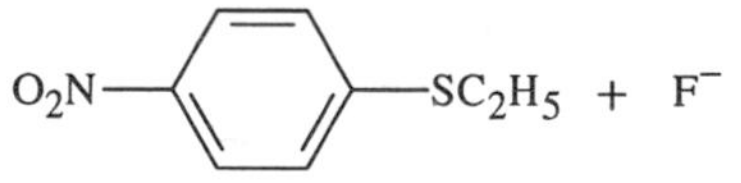

(b) The product is formed by a nucleophilic aromatic substitution reaction in which chloride is displaced by hydroxide ion. The product is ionized by ^{-}OH present in the reaction mixture; addition of acid (H_3O^+) forms the neutral phenol.

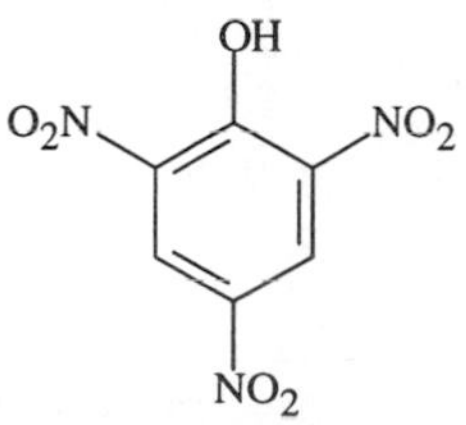

(c) The NaOH ionizes the phenol. The resulting phenolate ion is methylated by dimethyl sulfate to form the methyl ether, which is then hydrogenated to the corresponding cyclohexane.

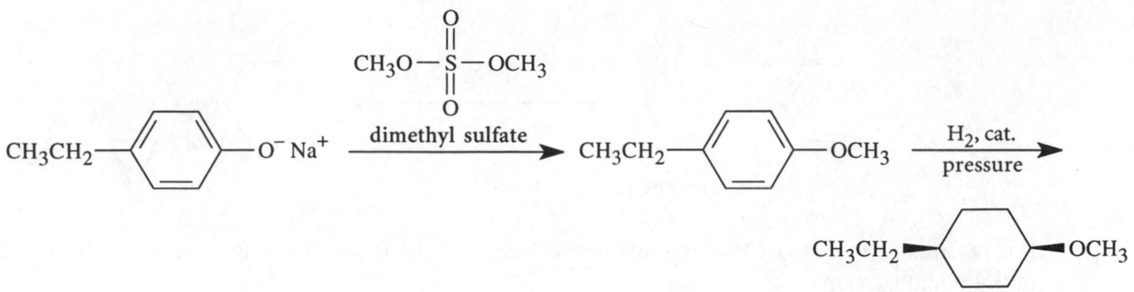

$CH_3O-S(=O)_2-OCH_3$

$CH_3CH_2-C_6H_4-O^-\ Na^+$ $\xrightarrow{\text{dimethyl sulfate}}$ $CH_3CH_2-C_6H_4-OCH_3$ $\xrightarrow[\text{pressure}]{H_2,\ \text{cat.}}$

$CH_3CH_2-C_6H_{10}-OCH_3$

cis-1-ethyl-4-methoxycyclohexane

(d) No reaction occurs under the mild conditions.

(e) The products result from a benzyne mechanism.

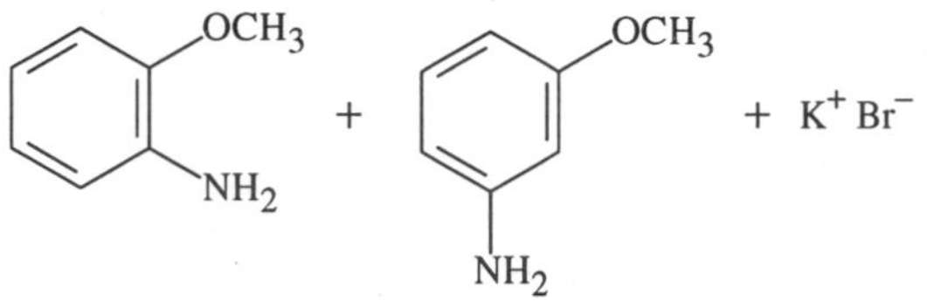

OCH_3 / NH_2 + OCH_3 / NH_2 + $K^+\ Br^-$

(f) If the conditions are severe enough, a reaction will occur by a benzyne mechanism.

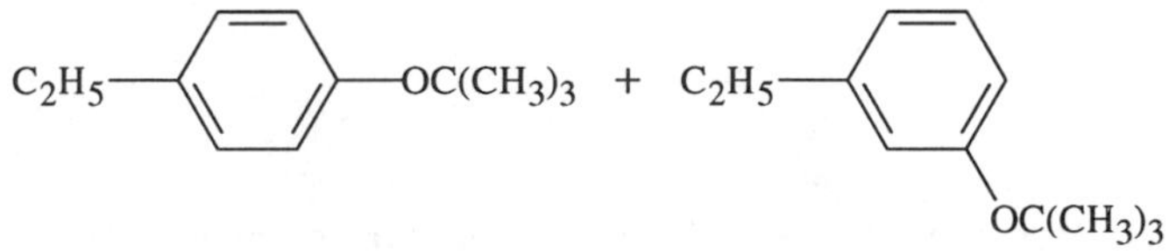

$C_2H_5-C_6H_4-OC(CH_3)_3$ + $C_2H_5-C_6H_4-OC(CH_3)_3$

(g) Bromine is introduced at the ring position that is activated by two hydroxy groups.

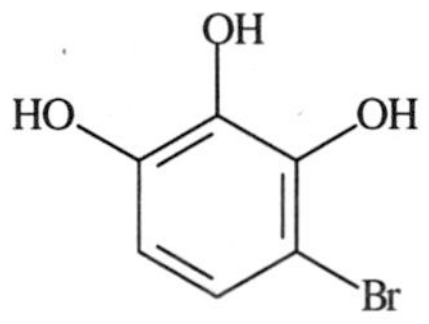

OH, HO, OH, Br

(h) Oxidation occurs to give the quinone.

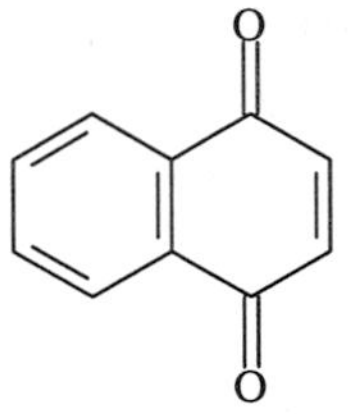

1,4-naphthoquinone

(i) This is a Stille reaction that results in a coupling between the stannane and the naphthol.

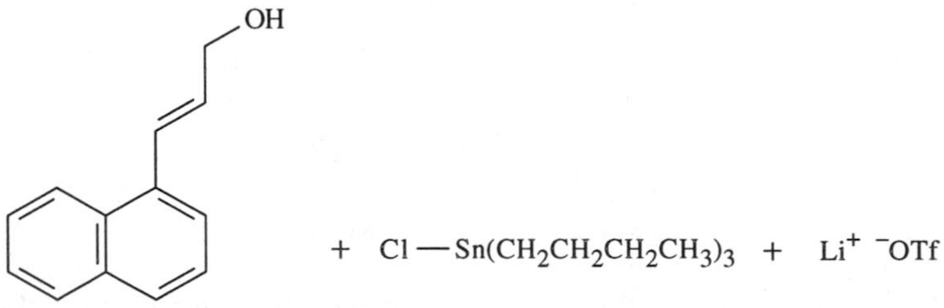

OH

+ $Cl-Sn(CH_2CH_2CH_2CH_3)_3$ + $Li^+\ ^-OTf$

(j) This is a Heck reaction between a *vinylic* halide and an alkene. Vinylic halides react in many cases like aryl halides in Pd-mediated coupling reactions. The coupling with cyclic alkenes result in coupling at an allylic position, as in Eqs. 18.51 and 18.52 on text p. 803-804.

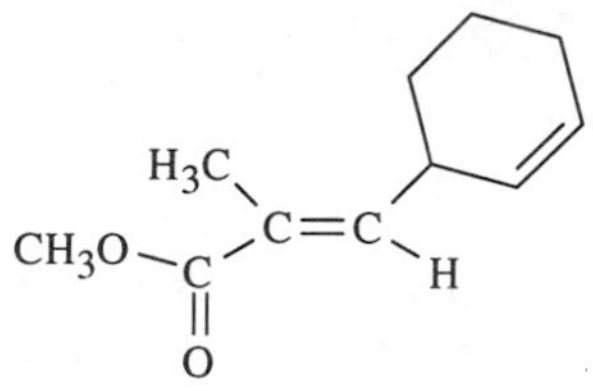

H_3C, C=C, CH_3O—C, O, H

(k) Oxidation occurs to give the quinone.

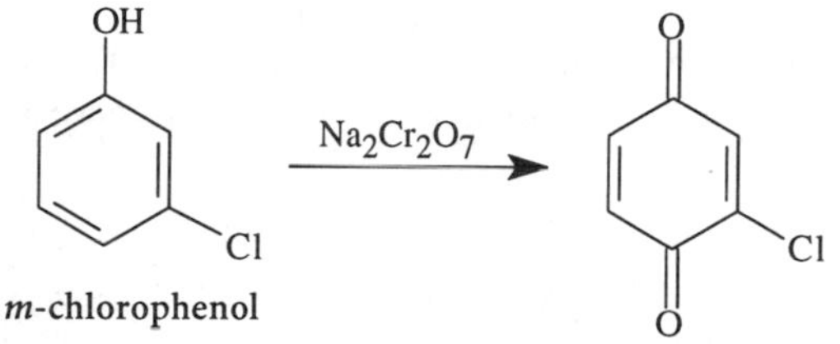

OH, Cl, $Na_2Cr_2O_7$, O, Cl, O

m-chlorophenol

2-chloro-2,5-cyclohexadien-1,4-dione

(l) The methanesulfonate derivative of 2,4-dinitrophenol undergoes a nucleophilic substitution reaction in which the methanesulfonate group is displaced by methoxide.

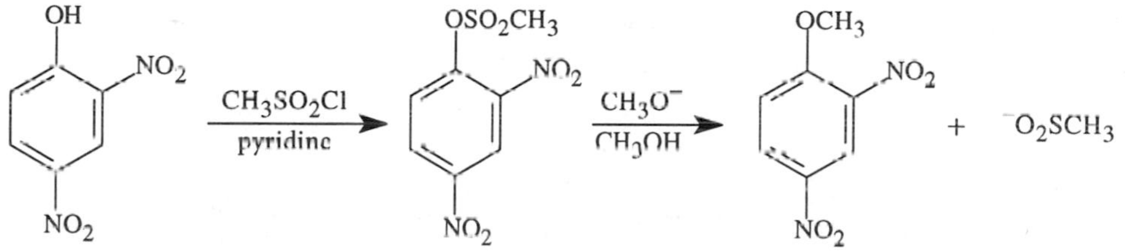

OH, NO_2, NO_2, CH_3SO_2Cl / pyridine, OSO_2CH_3, NO_2, NO_2, CH_3O^- / CH_3OH, OCH_3, NO_2, NO_2, + $^-O_2SCH_3$

2,4-dinitrophenol

1-methoxy-2,4-dinitrobenzene

(m) Nucleophilic aromatic substitution occurs at the position that is *ortho* and *para* to the nitro groups, because this involves the more stable anionic intermediate (Meisenheimer complex).

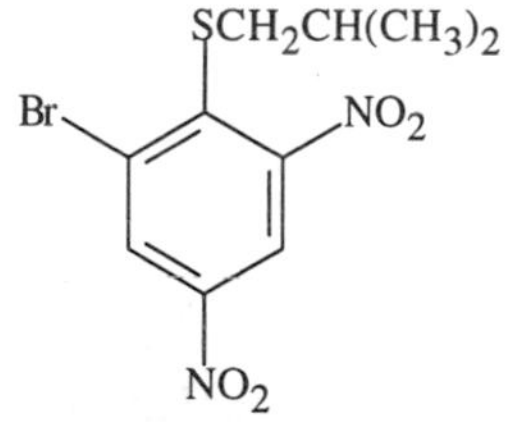

$SCH_2CH(CH_3)_2$, Br, NO_2, NO_2

(n) The methyl ethers are cleaved but the phenyl ethers are not. (See the solution to Problem 18.42(d).)

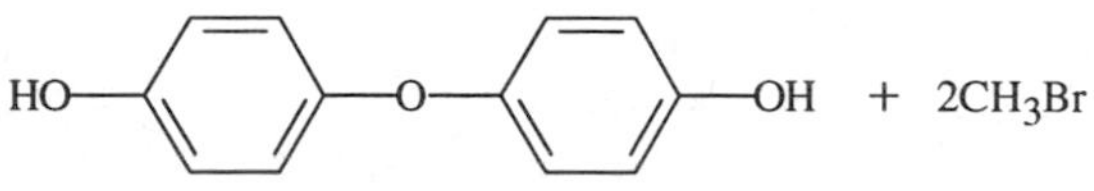

HO—, O, —OH + $2CH_3Br$

(o) Just as vinylic halides react like aryl halides in the Heck reaction (see part (j) of this problem), vinylic triflates react like aryl triflates in the Stille reaction.

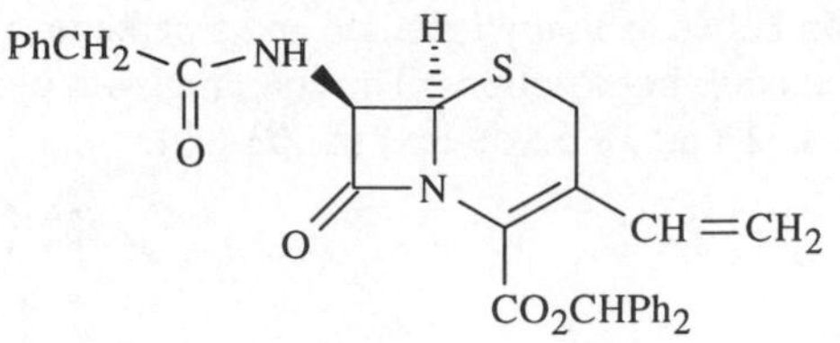

18.57 The basic principle needed to understand the results is that elimination is most rapid when it occurs with anti stereochemistry. In the first reaction, *anti*-elimination leads to the observed product. In the second reaction, formation of the alkyne requires a slower *syn*-elimination; hence, another process can compete, namely, elimination of a methyl hydrogen and the bromine to form the allene. In the first reaction, elimination of the methyl hydrogen to form the allene is a slower process because allenes are not so stable as alkynes. The transition state for elimination is destabilized by its allenelike character. (The methyl hydrogens are approximately as acidic as the vinylic hydrogens because they are allylic.) It is also possible that the allene is the only product formed in the second reaction and that the alkyne is formed by a base-catalyzed isomerization of the allene:

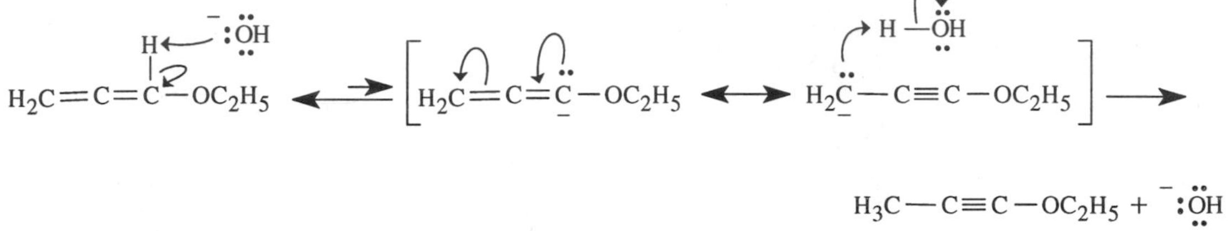

$$H_2C{=}C{=}C(H){-}OC_2H_5 + {}^-:\ddot{O}H \rightleftharpoons \left[H_2C{=}C{=}\bar{C}{-}OC_2H_5 \longleftrightarrow H_2\bar{C}{-}C{\equiv}C{-}OC_2H_5\right] \xrightarrow{H-\ddot{O}H} H_3C{-}C{\equiv}C{-}OC_2H_5 + {}^-:\ddot{O}H$$

18.58 Phenyllithium, a strong base, promotes the formation of benzyne from bromobenzene. The benzyne intermediate reacts with another molecule of phenyllithium.

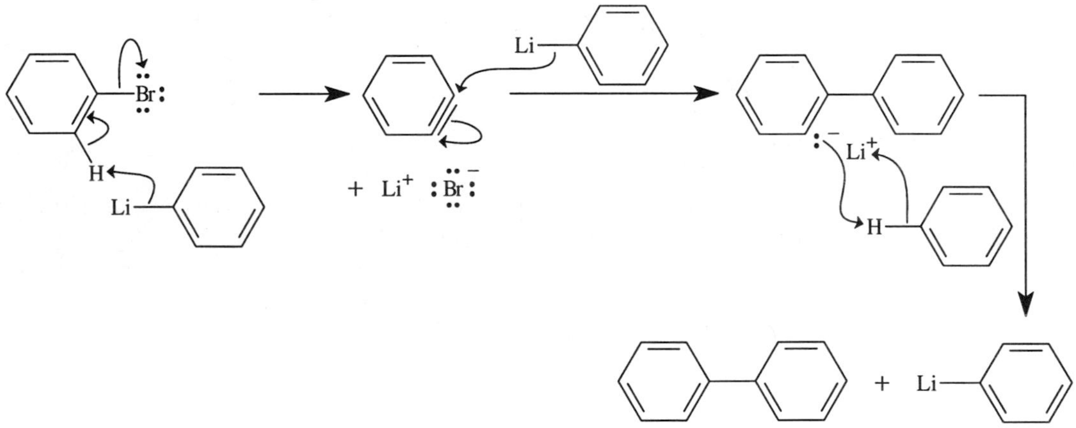

18.59 In base, chloroacetic acid and 2,4,5-trichlorophenol are ionized. Attack of the conjugate base of 2,4,5-trichlorophenol on chloroacetate ion in an S_N2 reaction gives the conjugate base of 2,4,5-T; un-ionized 2,4,5-T is formed on acidification.

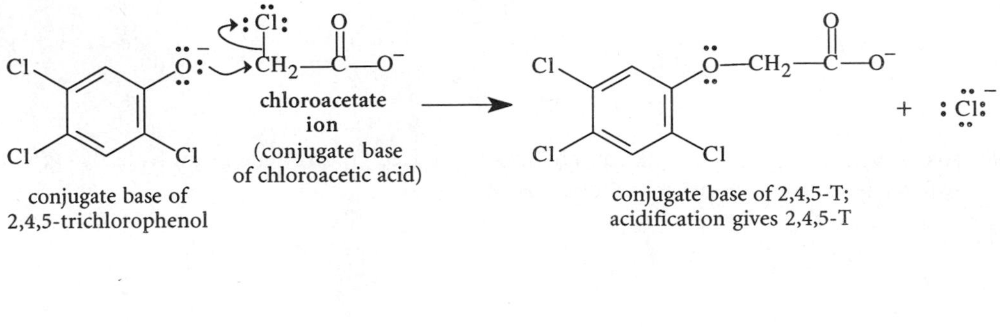

The formation of dioxin is accounted for by a benzyne mechanism similar to that in the solution to Problem 18.55, except that in this case there are two consecutive benzyne reactions, the second of which is intramolecular. The first benzyne intermediate *A* is formed by hydroxide-promoted elimination of HCl from the conjugate base of 2,4,5-trichlorophenol.

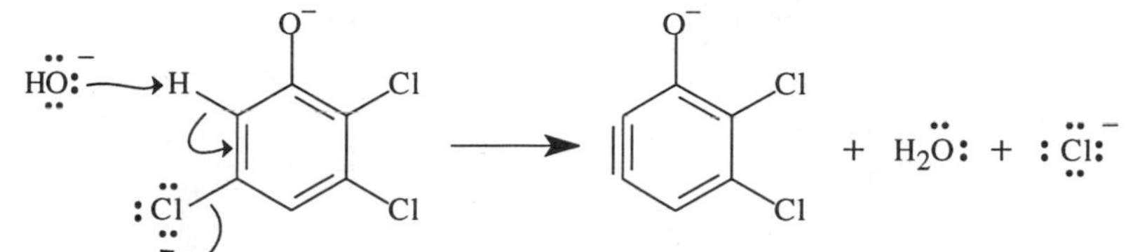

2,4,5-Trichlorophenolate ion attacks this benzyne intermediate to form ether *B*. (Although the reaction of the benzyne intermediate may be a two-step process, as shown in Eqs. 18.22c–d on text p. 786, it is shown below and in subsequent mechanisms as a concerted process to save space.)

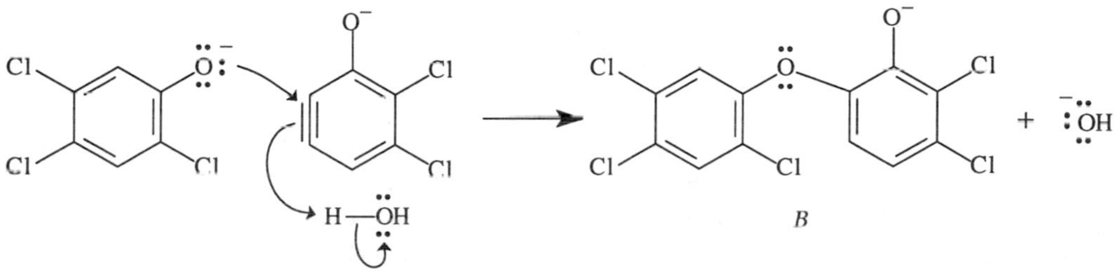

A benzyne intermediate derived from ether *B* then undergoes an intramolecular reaction to form dioxin.

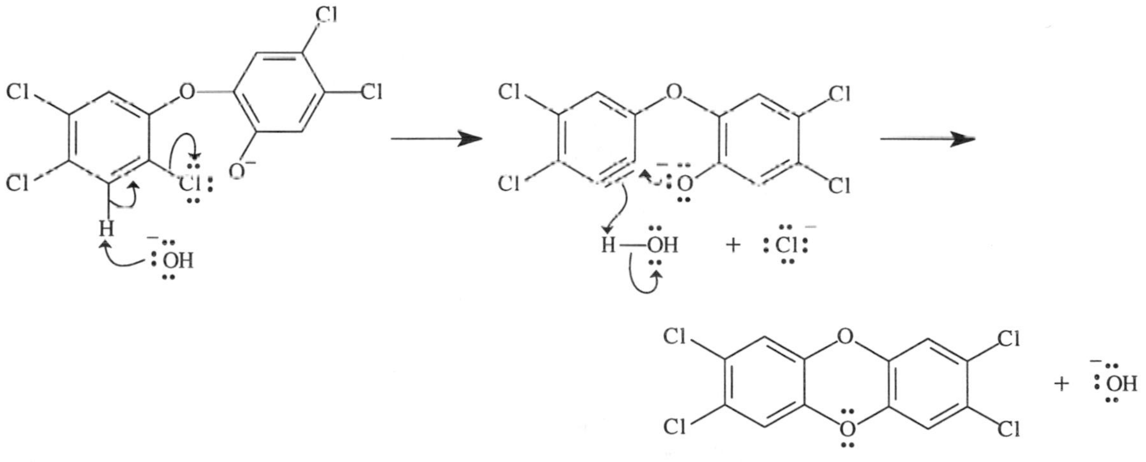

18.60 Because vinylic halides do not undergo S_N2 or S_N1 reactions, this substitution must occur by an unusual mechanism. The key to this solution is to recognize that the leaving-group effect and the effect of a *para*-nitro substituent are much the same as they are in nucleophilic aromatic substitution. This reaction, in fact, occurs by a mechanism that is much like the mechanism of nucleophilic aromatic substitution, except that it takes place at a vinylic carbon. The *para*-nitro group accelerates the reaction by affording additional resonance stabilization of the anionic intermediate *A*.

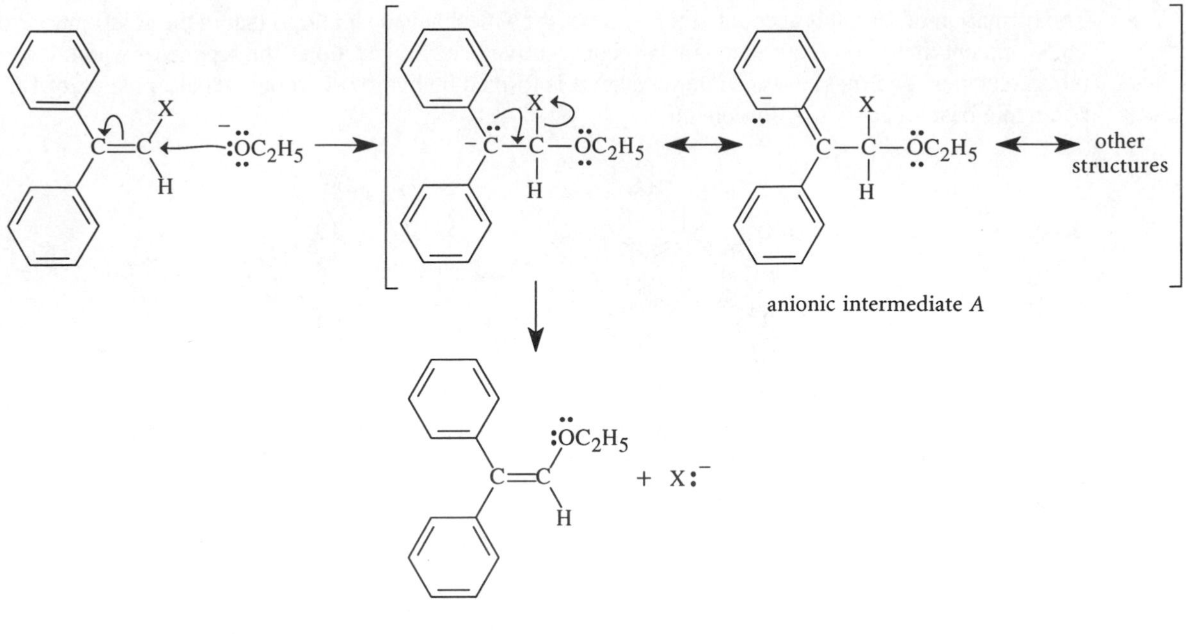

anionic intermediate *A*

18.61 The spectrum is consistent with the formation of an anionic intermediate—a stable Meisenheimer complex. (The dotted lines symbolize resonance delocalization of the negative charge.) If you draw out the resonance structures for this ion, you will see that charge is delocalized to the carbons bearing the nitro groups, and it can also be delocalized into the nitro groups. However, charge is *not* delocalized to the other carbons. Hence, the protons on these carbons do not show the smaller chemical shift that would be expected if there were high electron density on these carbons.

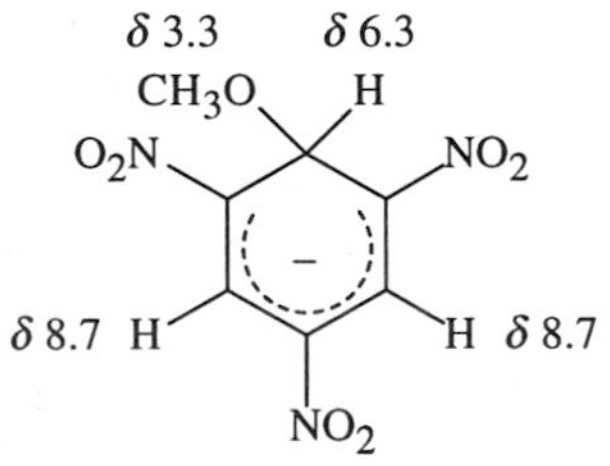

18.62 (a) In the elimination, a proton could be removed from either β-carbon atom:

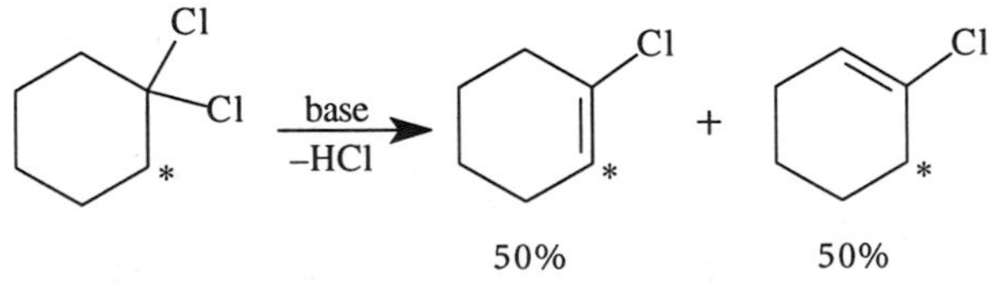

(b) A *cyclohexyne* intermediate—the cyclohexane analog of benzyne—is formed from each of the labeled 1-chlorocyclohexenes, and each cyclohexyne can react with a "phenyl anion" (that is, phenyllithium) at two different carbon atoms with equal probability. This leads to the labeling pattern shown in the problem.

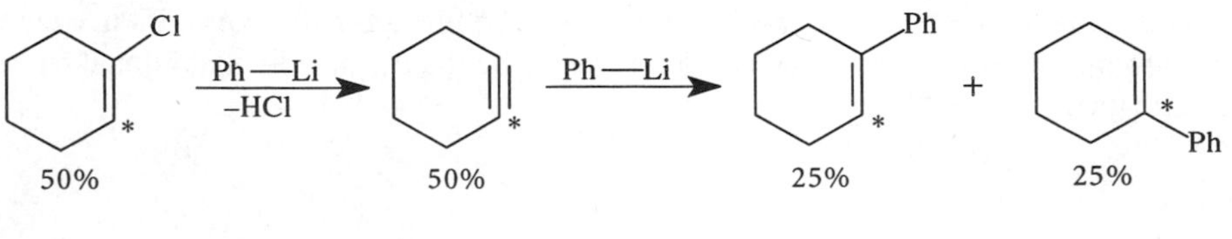

Cl (50%) $\xrightarrow[-HCl]{Ph—Li}$ (50%) $\xrightarrow{Ph—Li}$ Ph (25%) + Ph (25%)

This labeling experiment was used to gather evidence for or against a cyclic allene intermediate, 1,2,-cyclohexadiene:

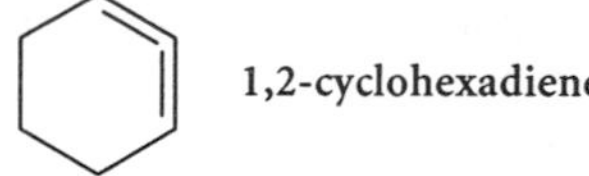

The labeling pattern of the products *clearly excludes* this intermediate. Can you see why? What labeling pattern would be predicted if this were the intermediate?

18.63 The reaction that forms compound *X* is a *hydroalumination,* and it is completely analogous to hydroboration. The structure of *X* is

$C_5H_{11}(H)C{=}C(H)Al[(CH_2CH(CH_3)_2]_2$ compound *X*

The second reaction is very much like a Stille coupling. We expect the carbon-aluminum bond to behave much like other carbon-metal bonds, for example, the carbon-tin bond in a stannane. And the vinylic iodide group is much like a triflate group in the sense that it is a good leaving group. The structure of the coupling product *Y* is

$C_5H_{11}(H)C{=}C(H)–(H)C{=}C(H)C_4H_9$ compound *Y*

The by-product of the reaction is the iododialkylalane, $I—Al[CH_2CH(CH_3)_2]_2$.

18.64 If PPh_3 attacks a chlorine, an additional electron pair can be moved to the Pd.

$PdCl_2$ + $:PPh_3$ ⟶ $\overset{-}{Pd}—Cl:$ $Cl—\overset{+}{P}Ph_3$ ⟶ $\ddot{Pd}$ (Pd(0)) + $:\ddot{Cl}:^-$ + $Cl—\overset{+}{P}Ph_3$ ⟶ Cl_2PPh_3

18.65 (a) This is a nucleophilic aromatic substitution reaction. Notice that the cyano group (:N≡C—) can stabilize the intermediate *A* by resonance.

:N≡C–C₆H₄–F: + H–N(piperidine) ⟶ :N≡C– (intermediate *A*) ⟶ :N≡C–C₆H₄–$\overset{+}{N}$H(piperidine) + $:\ddot{F}:^-$

(b) The observation of *cine*-substitution suggests that this reaction involves a benzynelike mechanism; the intermediate is *2,3-benzofuryne,* a benzofuran analog of benzyne. Presumably a second substitution product is also formed, but it was not mentioned in the publication that reported this work.

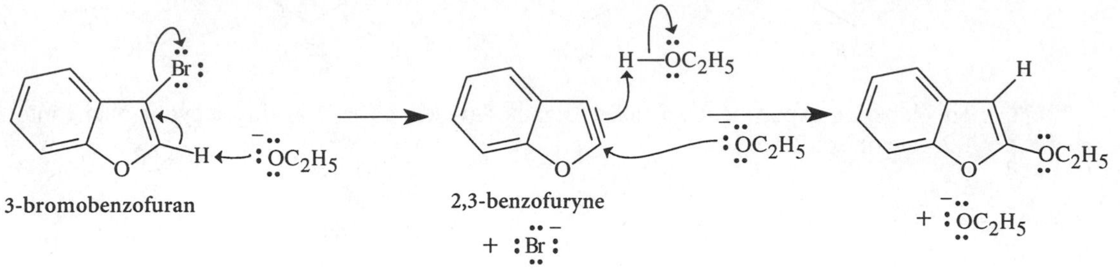

3-bromobenzofuran

2,3-benzofuryne

(c) We've learned that ethers don't cleave in base, but the formation of dimethyl ether in this reaction clearly shows that the O—CH_3 bond of the ether is cleaved. Why should this occur? First, notice the forcing conditions (heat, pressure). Second, the phenolate ion that is expelled is a fairly good leaving group, as the polar effects of the three chlorines considerably reduce its basicity relative to that of an unsubstituted phenol.

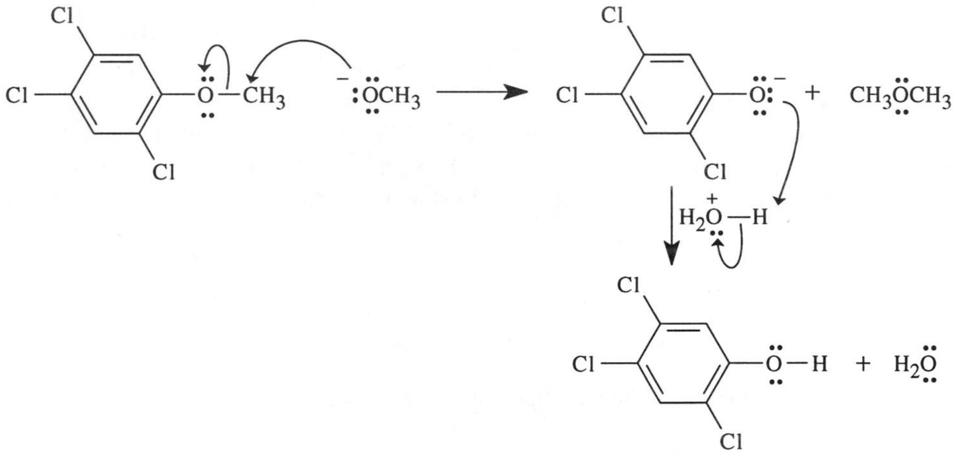

(d) Protonation of 2-methylpropene gives the *tert*-butyl cation, which serves as the electrophile in an electrophilic aromatic substitution reaction.

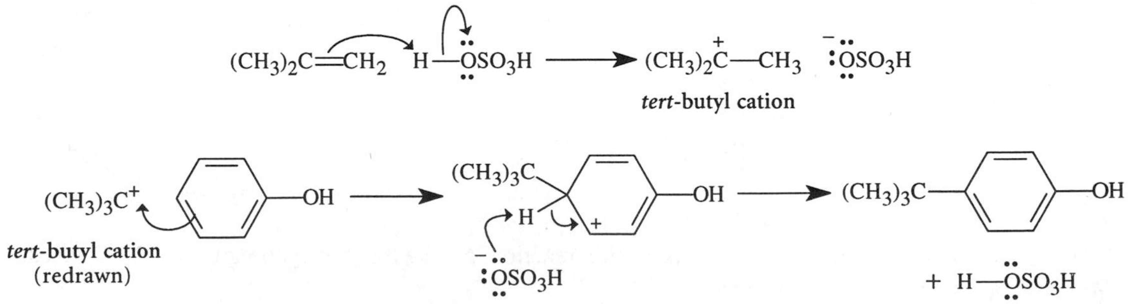

tert-butyl cation

tert-butyl cation (redrawn)

18.66 In 1-chloro-4-nitrobenzene, the dipole contributions of the carbon-chlorine bond and the nitro group are oriented in opposite directions; consequently, the net dipole moment is smaller than it is in nitrobenzene, in which only a dipole contribution of the nitro group is present.

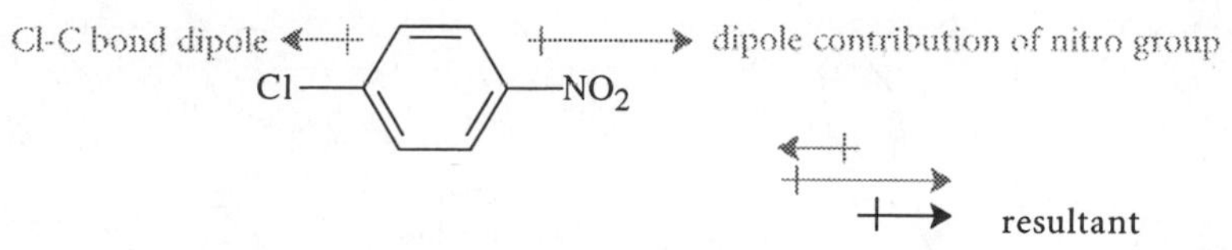

Cl-C bond dipole

dipole contribution of nitro group

resultant

The analysis of *p*-nitroanisole is similar, except that there is an additional effect: the very powerful electron-donating resonance effect of the methoxy group. Delocalization of an unshared electron pair from the oxygen of the methoxy group into the nitro group creates separation of charge that results in a large dipole moment. A similar resonance effect is present in 1-chloro-4-nitrobenzene, but, because the resonance effect of a chloro group is much weaker than that of a methoxy group (Fig. 16.8, text p. 726), it is not sufficient to offset the opposing contribution of the chlorine-carbon bond dipole.

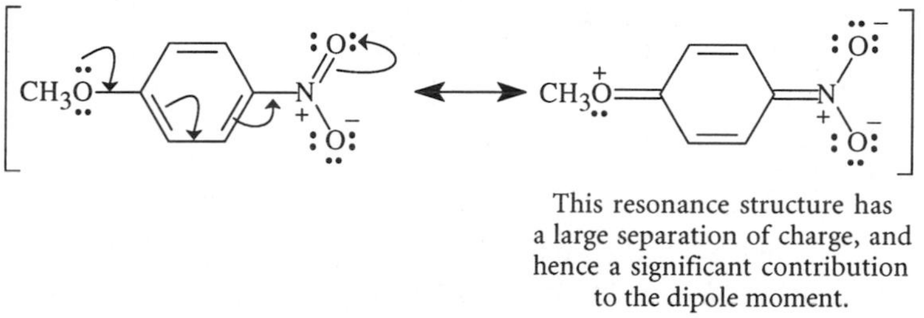

This resonance structure has a large separation of charge, and hence a significant contribution to the dipole moment.

18.67 Because the glycol starting material is also a chlorohydrin, and because the phenoxide salt is a base, epoxide formation takes place rapidly, accompanied by the formation of NaCl. (See Sec. 11.2B, text pp. 463.) It is the epoxide that actually alkylates the phenol more slowly. The hydroxide ion produced in the last step maintains the phenol in its ionized form.

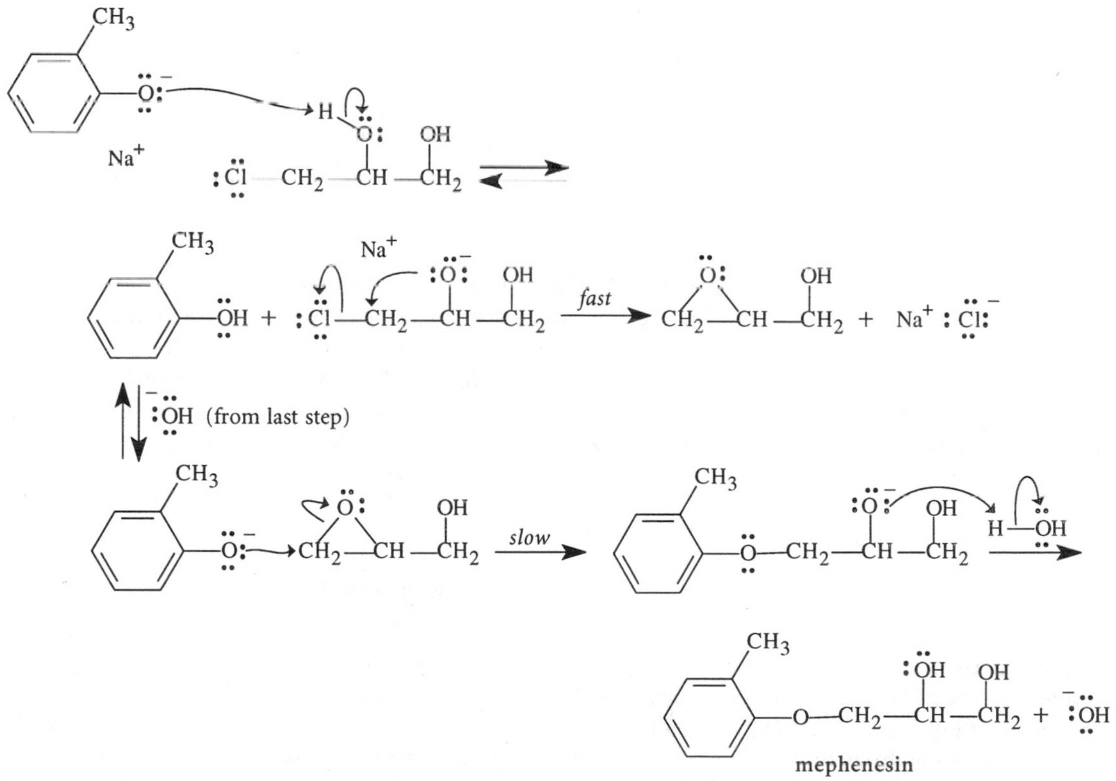

fast

:OH (from last step)

slow

mephenesin

18.68 Deduce the structure of the "very interesting intermediate" by mentally imagining a "reverse Diels-Alder" reaction of triptycene that yields anthracene and the intermediate, which is benzyne:

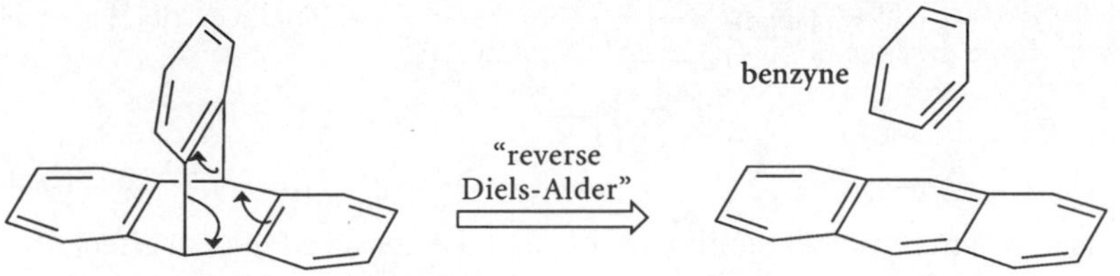

triptycene "reverse Diels-Alder" → benzyne + anthracene

The Grignard reagent has carbanion character, and this "carbanion" is a strong base. Elimination of the weaker base fluoride gives benzyne:

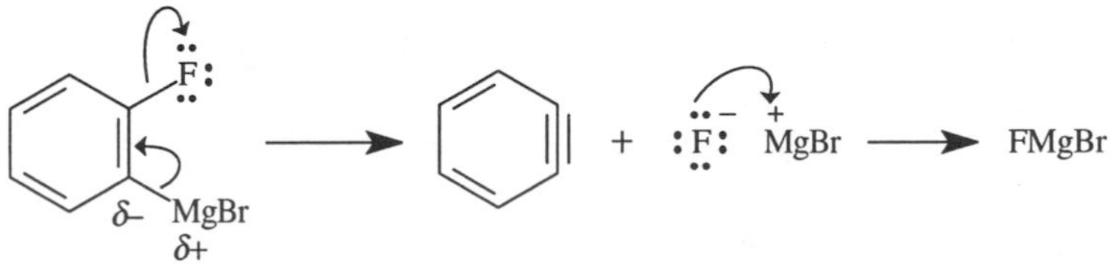

$$\text{(2-fluorophenyl)MgBr } (\delta- \text{ C}, \delta+ \text{ MgBr}) \longrightarrow \text{benzyne} + :\ddot{F}:^- \ \overset{+}{M}gBr \longrightarrow FMgBr$$

18.69 The six-proton singlet in the NMR spectrum indicates that the two methyl groups ortho to the oxygen have been retained; this resonance cannot be due to one ortho methyl and one para methyl, because these would not be chemically equivalent. The two-proton singlet at δ 5.49 indicates two alkene protons, and the two-proton singlet at δ 6.76 is accounted for by the two ring protons, which evidently remain. The total number of protons accounted for by the NMR spectrum are two fewer than are present in the starting material. The conditions are much like those for oxidation of a phenol to a quinone. The structure below for compound *A* fits the data; the singlets are all broad because of very slight splitting over more than three bonds.

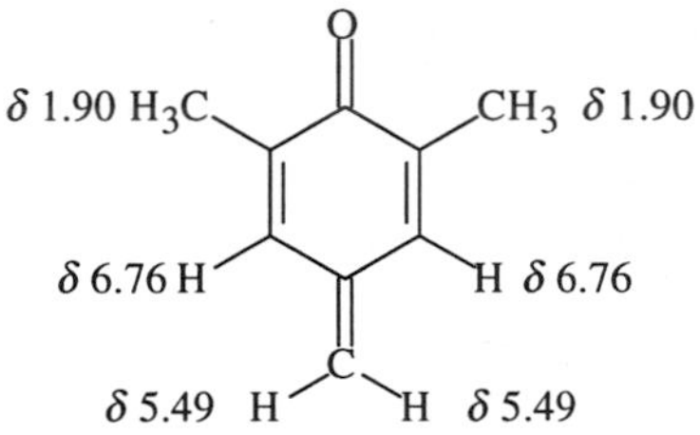

δ 1.90 H_3C, CH_3 δ 1.90; δ 6.76 H, H δ 6.76; =C; δ 5.49 H, H δ 5.49

Compound *A*

18.70 The two reactions are intramolecular Friedel-Crafts acylations. The first equivalent of $AlBr_3$, a strong Lewis acid, forms a complex with the acid chloride group to generate the electrophile, an acylium ion, as discussed in the text on p. 713, and is also consumed by formation of a complex with the product. As the lower reaction in the problem shows, the acylium ion reacts at a position para to a methoxy group, because a methoxy group is both ortho, para-directing and strongly activating. The question, then, is why reaction occurs at the ring that does *not* have methoxy substituents when more Lewis acid is present. The answer is that the second and third equivalents of Lewis acid form complexes with the methoxy groups:

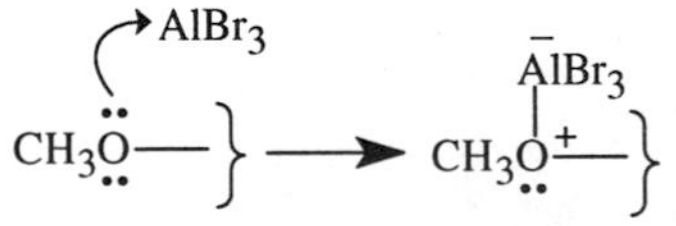

$$CH_3\ddot{O}— + AlBr_3 \longrightarrow CH_3\overset{+}{\ddot{O}}(—\overset{-}{A}lBr_3)—$$

A complexed methoxy group cannot donate an unshared electron pair by resonance, because its unshared pair is already "donated" to the aluminum. Consequently, the methoxy group, through this complexation, is transformed into an electronegative group that is incapable of a resonance effect. An electronegative group that cannot donate electrons by resonance is a *deactivating group*, and a ring containing such a group (or two such groups) undergoes substitution more slowly than a ring without such substituents. In other words, acylation of the ring that does not contain such deactivating substituents is the faster acylation.

18.71 (a) The epoxide slowly and spontaneously opens for three reasons: (1) strain is relieved in the epoxide; (2) a resonance-stabilized carbocation is formed; and (3) the heterolysis of a chemical bond (that is, its dissociation into fragments of opposite charge) is favored by the polar, protic, donor solvent that can solvate both the carbocation and the oxygen anion. The mechanism, then, is as follows:

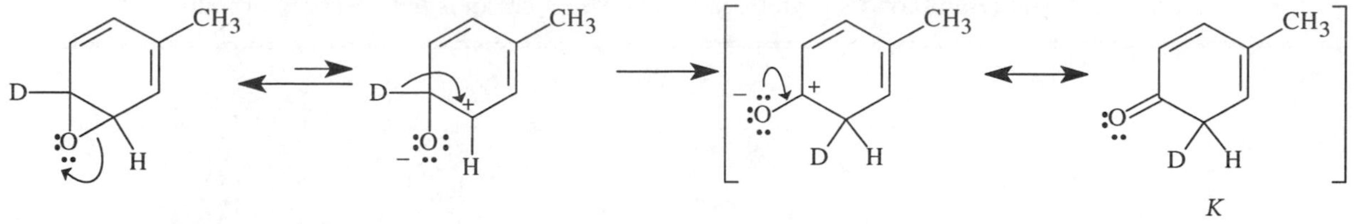

The driving force for this rearrangement is formation of the neutral ketone *K*. The "enol" of ketone *K* is the product phenol; this enol forms because of its aromaticity. In the formation of this product either the H or the D must be lost; loss of the H gives the product that contains deuterium; loss of the deuterium gives the product that contains hydrogen. Because of the primary deuterium isotope effect (see text p. 372), a carbon-deuterium bond is broken more slowly than a carbon-hydrogen bond. (The product distribution suggests that the isotope effect is 75/25, or 3.) A mechanism for conversion of *K* into the deuterium-containing product is as follows:

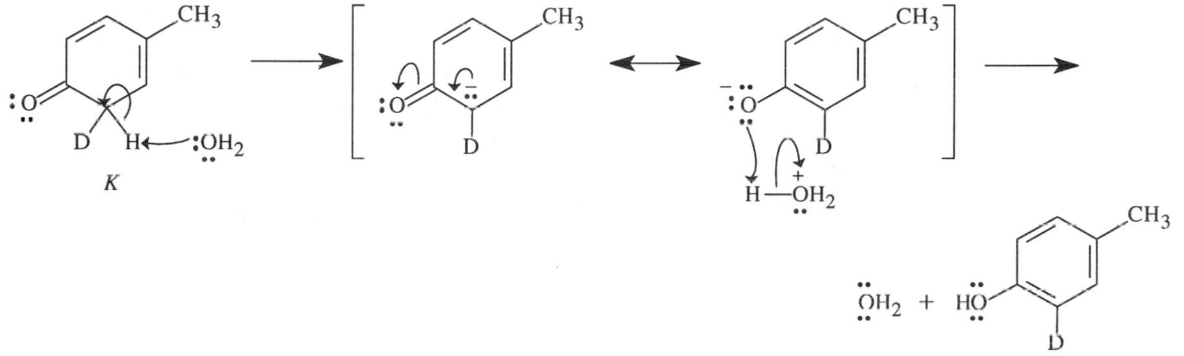

Some of the deuterium is lost from intermediate *K* (by the same mechanism) because the primary isotope effect is not very large. (Only an "infinite" isotope effect would result in the retention of all deuterium.)

(b) The position of the methyl group relative to the oxygen of the product is determined by which bond of the epoxide breaks in the first step of the mechanism. Breaking different bonds leads to carbocations of different stabilities. The carbocation intermediate involved in the formation of *p*-cresol has the following resonance structures:

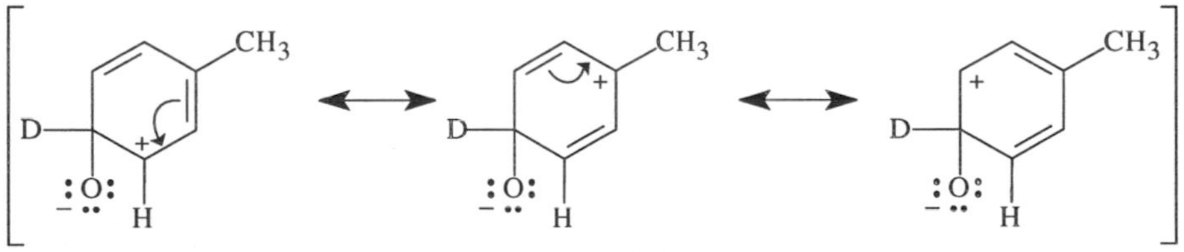

Notice that the middle structure is a *tertiary* carbocation. The carbocation intermediate involved in the formation of *m*-cresol has the following resonance structures:

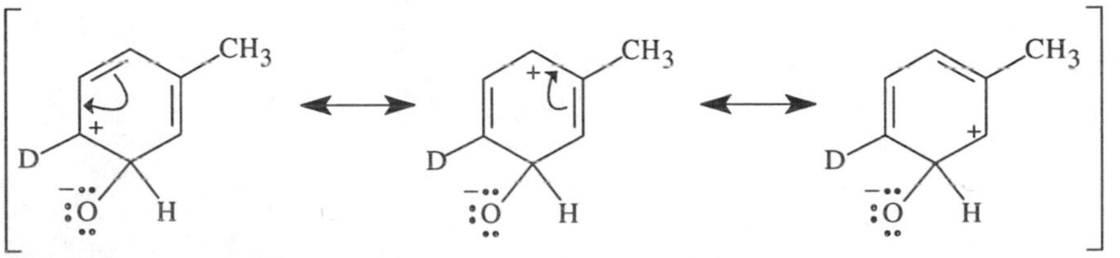

(Convince yourself that application of a mechanism to this carbocation similar to the one shown above gives *m*-cresol.) All of the resonance structures of this carbocation are *secondary* carbocations. Consequently, the carbocation involved in the formation of *p*-cresol is more stable than the one that leads to *m*-cresol. As is usually the case, the pathway involving the more stable intermediate—the pathway that gives *p*-cresol—is faster (Hammond's postulate).This is why *p*-cresol is the observed product.

19

Chemistry of Aldehydes and Ketones. Carbonyl-Addition Reactions

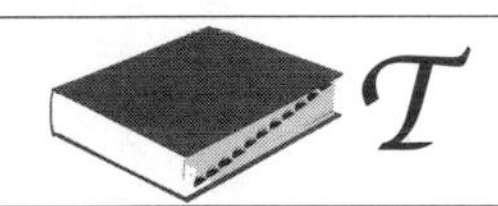

TERMS

acetal 19.10A
acetal hydrolysis 19.10A
acetyl group 19.1A
acyl group 19.1A
aldehyde .... Introduction
α-alkoxy carbocation 19.6
α-carbon 19.1A
α-hydrogen 19.1A
azeotrope 19.10A
Bakelite 19.15
benzoyl group 19.1A
carbinolamine 19.11A
carbonyl compound .... Introduction
carbonyl group .... Introduction
Clemmensen reduction 19.12
cyanohydrin 19.7
cyclic hemiacetal 19.10A
2,4-dinitrophenylhydrazine (2,4-DNP) 19.11A
2,4-dinitrophenylhydrazone (2,4-DNP derivative) 19.11A
enamine 19.11B
"forbidden" absorption 19.3D
formyl group 19.1A
gem-diol 19.7
geminal 19.7
hemiacetal 19.10A
hydrate 19.7
hydrazine 19.11A, 19.12
hydrazone 19.12
hydride ion 19.8
hydride reduction 19.8
hydrolysis 19.10A
hydroxylamine 19.11A
imine 19.11A
ketone .... Introduction
McLafferty rearrangement 19.3E
methylene group (—CH_2—) 19.12
miscible 19.2
n electron 19.3D
$n \rightarrow \pi^*$ absorption 19.3D
nitrile 19.7
nucleophilic addition 19.7A
oxaphosphetane 19.13
oxime 19.11A
phenol-formaldehyde resin 19.15
phenylhydrazine 19.11A
phenylhydrazone 19.11A
phosphonium salt 19.13
pinacol rearrangement 19.6
primary amine 19.11A
propionyl group 19.1A
protecting group 19.10B
resin 19.15
Schiff base 19.11A
secondary amine 19.11B
semicarbazide 19.11A
semicarbazone 19.11A
tertiary amine 19.11B
Tollens test 19.14
Wittig alkene synthesis 19.13
Wolff-Kishner reduction 19.12
ylid 19.13
zinc amalgam, Zn/Hg 19.12

CONCEPTS

I. INTRODUCTION TO ALDEHYDES AND KETONES

A. General

1. Carbonyl compounds contain the carbonyl group (C═O) and include
   a. aldehydes
   b. ketones

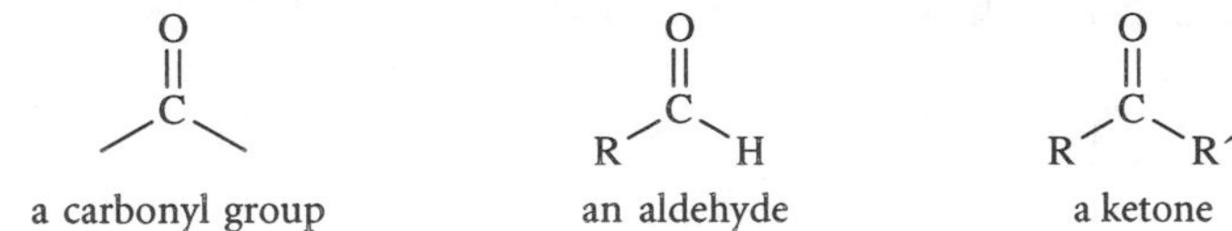

   c. carboxylic acids and most carboxylic acid derivatives:
      i. esters
      ii. amides
      iii. anhydrides
      iv. acid chlorides

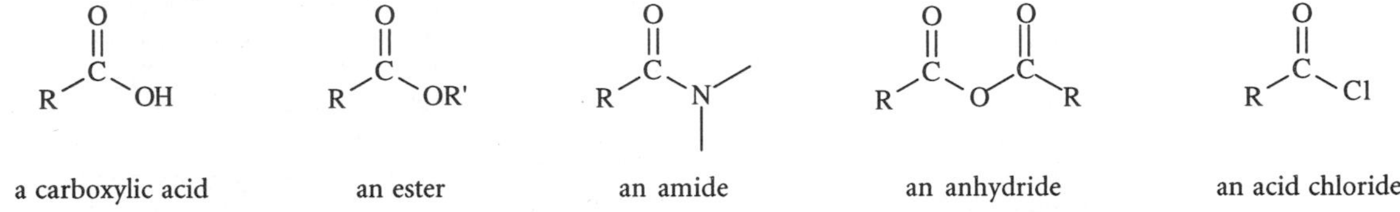

2. In an aldehyde, at least one of the groups at the carbonyl carbon atom is a hydrogen, and the other may be alkyl, aryl, or a second hydrogen.
3. In a ketone, the groups bound to the carbonyl carbon are alkyl or aryl groups.
4. The carbonyl carbon of a typical aldehyde or ketone is sp^2-hybridized with bond angles approximating 120°.
5. A carbon-oxygen double bond:
   a. consists of a σ bond and a π bond.
   b. is shorter than a carbon-carbon double bond.

B. Common Nomenclature

1. In the common nomenclature of aldehydes, the suffix *aldehyde* is added to a prefix that indicates the chain length of the group attached to the carbonyl group.

form	H—	isobutyr	$(CH_3)_2CH_2$—
acet	CH_3—	valer	$CH_3CH_2CH_2CH_2$—
propion	CH_3CH_2—	isovaler	$(CH_3)_2CHCH_2$—
butyr	$CH_3CH_2CH_2$—	benz	Ph—

   a. Common names are almost always used for the simplest aldehydes.
   b. Benzaldehyde is the simplest aromatic aldehyde.

2. The common name of some ketones are constructed by citing the two groups on the carbonyl carbon followed by the word *ketone*.

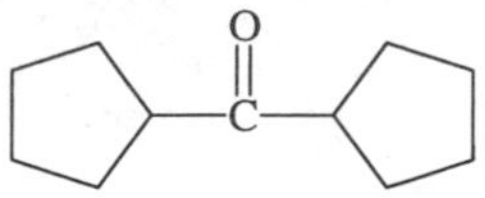

dicyclopentyl ketone

3. Certain aromatic ketones are named by attaching the suffix *ophenone* to the appropriate prefix.

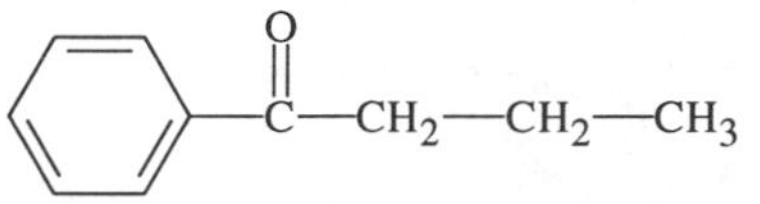

$C_6H_5-C(=O)-CH_2-CH_2-CH_3$ butyrophenone

4. Simple substituted aldehydes and ketones can be named in the common system by designating the position of substituents with Greek letters, beginning at the position adjacent to the carbonyl group.

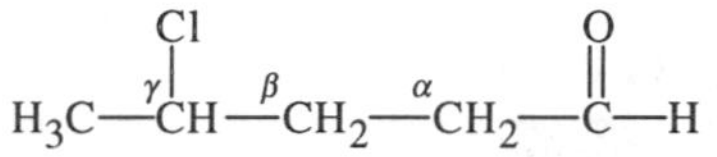

$H_3C-\overset{\gamma}{C}H(Cl)-\overset{\beta}{C}H_2-\overset{\alpha}{C}H_2-C(=O)-H$ γ-chlorovaleraldehyde

a. A carbon adjacent to the carbonyl group is termed the α-carbon.
b. The hydrogens on the α-carbon are termed α-hydrogens.

5. Many common carbonyl-containing substituent groups are named by adding *yl* or *oyl* to the appropriate prefix.

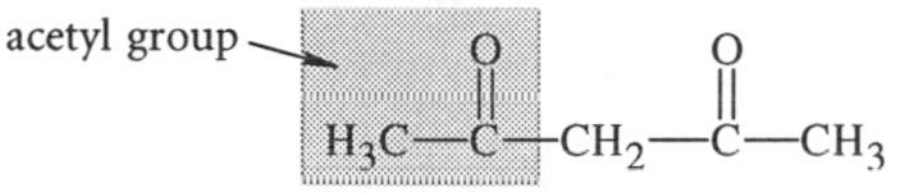

acetyl group → $H_3C-C(=O)-CH_2-C(=O)-CH_3$ acetylacetone

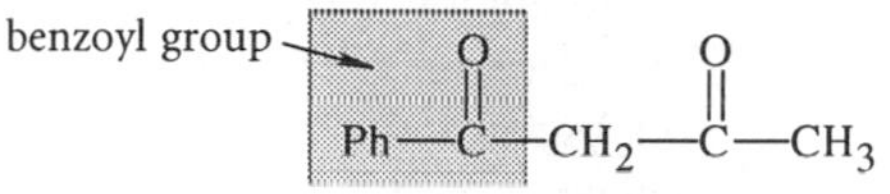

benzoyl group → $Ph-C(=O)-CH_2-C(=O)-CH_3$ benzoylacetone

a. Such groups are called in general acyl groups.
b. To be named as an acyl group, a substituent group must be connected to the remainder of the molecule at its carbonyl carbon.

C. Substitutive Nomenclature

1. The substitutive name of an aldehyde is constructed from a prefix indicating the length of the carbon chain followed by the suffix *al*; the prefix is the name of the corresponding hydrocarbon without the final *e*.
   a. The carbonyl carbon is designated as carbon-1 in numbering the carbon chain of an aldehyde.
   b. The final *e* is not dropped when the carbon chain has more than one aldehyde group.
   c. When an aldehyde group is attached to a ring, the suffix *carbaldehyde* is appended to the name of the ring.
      i. In aldehydes of this type, carbon-1 is not the carbonyl carbon, but rather the ring carbon attached to the carbonyl group.
      ii. In older nomenclature, the suffix carboxaldehyde was used.

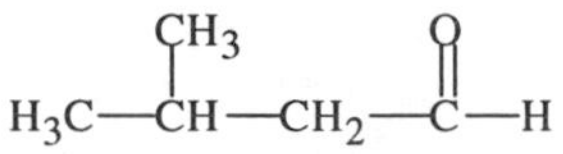

$H_3C-CH(CH_3)-CH_2-C(=O)-H$ 3-methylbutanal

$H-C(=O)-CH_2-C(=O)-H$ propanedial

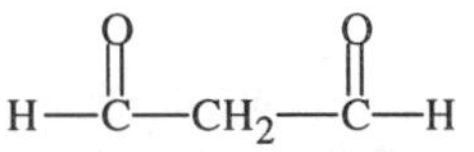

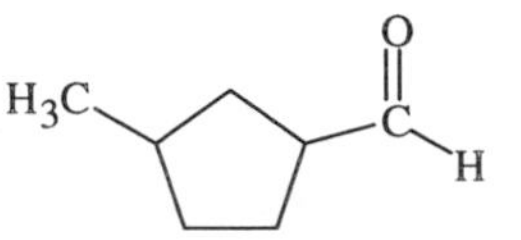

3-methylcyclopentanecarbaldehyde

   d. The name benzaldehyde is used in both common and substitutive nomenclature.
2. A ketone is named by giving the hydrocarbon name of the longest carbon chain containing the carbonyl group, dropping the final *e*, and adding the suffix *one*.
   a. The position of the carbonyl group is given the lowest possible number.
   b. The final *e* of the hydrocarbon name is not dropped in the nomenclature of diones, triones, etc.
   c. When a ketone carbonyl group is treated as a substituent, its position is designated by the term *oxo*.

1,2-cyclobutanedione

$H_3C-C(=O)-C_6H_4-C(=O)-H$

4-(1-oxoethyl)benzaldehyde

3. Aldehyde and ketone carbonyl groups receive higher priority than —OH or —SH groups for citation as principal groups. (A complete list of group priorities is given in Appendix I, text page A-1.)

$$-CH{=}O \text{ (aldehyde)} > \;\rangle C{=}O \text{ (ketone)} > -OH > -SH$$

$HS-CH_2-C(=O)-C(CH_3)_2-C{\equiv}C-C(=O)-H$

6-mercapto-4,4-dimethyl-5-oxo-2-hexynal

D. Physical Properties of Aldehydes and Ketones

1. Most simple aldehydes and ketones are liquids.
   a. Formaldehyde is a gas.
   b. Acetaldehyde has a boiling point very near room temperature.
   c. Aldehydes and ketones have higher boiling points than alkenes or alkanes with similar molecular weights and shapes because of their polarity, which is result of the C=O bond dipole.
   d. Aldehydes and ketones are not hydrogen-bond donors and thus have boiling points that are considerably lower than those of the corresponding alcohols.
2. Aldehydes and ketones with four or fewer carbons have considerable solubilities in water because they can accept hydrogen bonds from water at the carbonyl oxygen.
   a. Acetaldehyde and acetone are miscible with water (soluble in all proportions).
   b. The water solubility of aldehydes and ketones along a series diminishes rapidly with increasing molecular mass.
   c. Acetone and 2-butanone are especially valued as solvents because they dissolve not only water but also a wide variety of organic compounds.

II. SPECTROSCOPY OF ALDEHYDES AND KETONES

A. IR Spectroscopy

1. The principal infrared absorption of aldehydes and ketones is the C=O stretching absorption, a strong absorption that occurs in the vicinity of 1700 cm^{-1}.
2. The position of the C=O stretching absorption varies predictably for different types of carbonyl compounds.
   a. 1710–1715 cm^{-1} for simple ketones
   b. 1720–1725 cm^{-1} for simple aldehydes
3. The stretching absorption of the carbonyl-hydrogen bond of aldehydes near 2710 cm^{-1} is another characteristic absorption.
4. Compounds in which the carbonyl group is conjugated with aromatic rings, double bonds, or triple bonds have lower carbonyl stretching frequencies than unconjugated carbonyl compounds.
5. In cyclic ketones with rings containing fewer than six carbons, the carbonyl absorption frequency increases significantly as the ring size decreases.

B. NMR Spectroscopy

1. The characteristic proton NMR absorption common to both aldehydes and ketones is that of the protons on the carbons adjacent to the carbonyl group (α-protons), which occurs in the δ 2.0–2.5 region of the spectrum.
2. The absorption of the aldehyde proton is quite distinctive (δ 9–10 region of the NMR spectrum).
   a. The position of this absorption results from deshielding that is similar to the deshielding effects of the carbon-carbon double bond.

b. A carbonyl group has a greater deshielding effect than a carbon-carbon double bond because of the electronegativity of the carbonyl oxygen.

3. The most characteristic absorption of aldehydes and ketones in CMR spectroscopy is that of the carbonyl carbon, which occurs typically in the δ 190–220 range.
   a. This large downfield shift is due to the induced electron circulation in the π bond and the additional chemical-shift effect of the electronegative carbonyl oxygen.
   b. Because the carbonyl carbon of a ketone bears no hydrogens, its CMR absorption, like that of other quaternary carbons, is characteristically rather weak.
4. The α-carbon CMR absorptions of aldehydes and ketones show modest downfield shifts, typically in the δ 30–50 range, with, as usual, greater shifts for more branched carbons.

C. UV Spectroscopy

1. The $\pi \rightarrow \pi^*$ absorptions of unconjugated aldehydes and ketones occur at about 150 nm (well below the operating range of common UV spectrometers).
2. A much weaker absorption occurs in the 260–290 nm region in simple aldehydes and ketones.
   a. This absorption is caused by excitation of the unshared electrons (sometimes called the *n* electrons) on oxygen.
   b. This high-wavelength absorption is usually referred to as an $n \rightarrow \pi^*$ absorption.
   c. An $n \rightarrow \pi^*$ absorption arises from promotion of one of the *n* (unshared) electrons on the carbonyl oxygen to a π^* molecular orbital.
      i. $n \rightarrow \pi^*$ absorptions are weak; extinction coefficients (ϵ) are typically 10–100 $M^{-1}cm^{-1}$.
      ii. These absorptions are termed "forbidden."
3. The π electrons of compounds in which carbonyl groups are conjugated with double or triple bonds have strong UV absorptions (large extinction coefficients).
   a. The $\pi \rightarrow \pi^*$ absorptions of conjugated carbonyl compounds arise from the promotion of a π electron from a bonding to an antibonding (π^*) molecular orbital.
   b. The λ_{max} of a conjugated aldehyde or ketone is governed by the same variables that affect the λ_{max} of conjugated dienes:
      i. the number of conjugated double bonds.
      ii. substitution on the double bond.
   c. When an aromatic ring is conjugated with a carbonyl group, the typical aromatic absorptions are more intense and shifted to higher wavelengths than those of benzene.

D. Mass Spectrometry

1. Important fragmentations of aldehydes and ketones arise from—
   a. cleavage of the molecular ion at the carbonyl-α-carbon bond by—
      i. inductive cleavage (the alkyl fragment carries the charge and the carbonyl fragment carries the unpaired electron).
      ii. α-cleavage (the carbonyl fragment carries the charge and the alkyl fragment carries the unpaired electron).
   b. cleavage at the aldehyde carbon-hydrogen bond (which accounts for the fact that many aldehydes show a strong M – 1 peak).
   c. McLafferty rearrangement (abstraction of a hydrogen atom from a carbon five atoms away by the oxygen radical in the molecular ion) and subsequent α-cleavage. McLafferty rearrangement is a common mechanism for the production of odd-electron fragment ions in the mass spectrometry of carbonyl compounds.

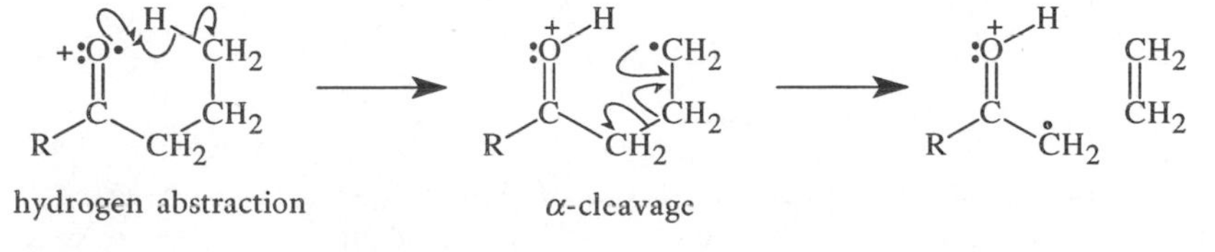

hydrogen abstraction

α-cleavage

III. SYNTHESIS AND REACTIVITY OF ALDEHYDES AND KETONES

A. Review of Ketone and Aldehyde Synthetic Methods

1. The three most important preparations of aldehydes and ketones already presented are
   a. Oxidation of alcohols (Sec. 10.6A, text page 430).
      i. Primary alcohols can be oxidized to aldehydes.

$$\text{a primary alcohol} \xrightarrow[\text{pyridine, } CH_2Cl_2]{CrO_3} \text{an aldehyde}$$

      ii. Secondary alcohols can be oxidized to ketones.

$$\text{a secondary alcohol} \xrightarrow[H_3O^+]{K_2CrO_7} \text{a ketone}$$

   b. Friedel-Crafts acylation (Sec. 16.4F, text page 713).

$$\underset{\text{an aromatic compound}}{Ar{-}H} + \underset{\text{an acid chloride}}{Cl{-}C(=O){-}R} \xrightarrow{AlCl_3} \xrightarrow{H_3O^+} \underset{\text{an aryl ketone}}{Ar{-}C(=O){-}R} + HCl$$

   c. Hydration of alkynes (Sec. 14.5A, text page 615).

$$\underset{\text{an alkyne}}{R{-}C{\equiv}C{-}H} + H_2O \xrightarrow[H_2SO_4]{Hg^{2+}} \underset{\text{a methyl ketone}}{R{-}C(=O){-}CH_3}$$

   d. Hydroboration-oxidation of alkynes (Sec. 14.5B, text page 618).

$$\underset{\text{an symmetric alkyne}}{3\ R{-}C{\equiv}C{-}R} + BH_3 \longrightarrow \underset{\text{organoborane intermediate}}{(R(H)C{=}C(R){-})_3B} \xrightarrow[^{-}OH]{H_2O_2} \underset{\text{a ketone}}{3\ R{-}CH_2{-}C(=O){-}R} + {}^{-}B(OH)_4$$

2. Less important synthetic methods that have already been presented are
   a. Ozonolysis of alkenes (Sec. 5.4, text page 173).

$$C{=}C(H) \xrightarrow[Me_2S]{O_3} C{=}O + O{=}C(H)$$

$$C{=}C(H) \xrightarrow[H_2O_2]{O_3} C{=}O + O{=}C(OH)$$

b. Periodate cleavage of glycols (Sec. 11.5B, text page 477).

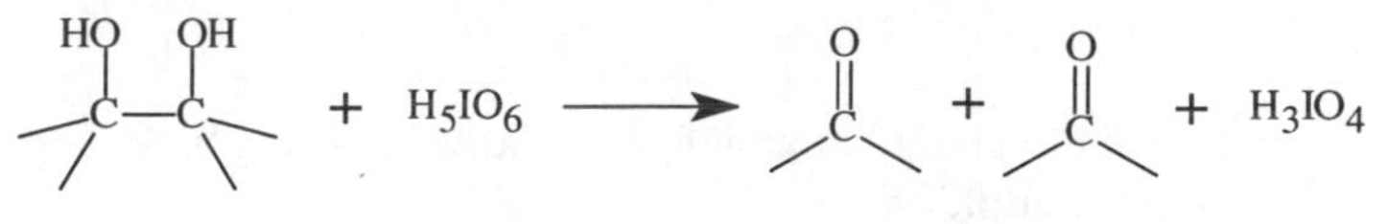

$$\text{a glycol} + H_5IO_6 \longrightarrow \text{aldehydes or ketones} + H_3IO_4$$

B. Basicity of Aldehydes and Ketones

1. Aldehydes and ketones are weakly basic and react at the carbonyl oxygen with protons or Lewis acids.
   a. The conjugate acid of an aldehyde or ketone is resonance-stabilized. (In some cases, the conjugate acids of aldehydes and ketones undergo typical carbocation reactions.)

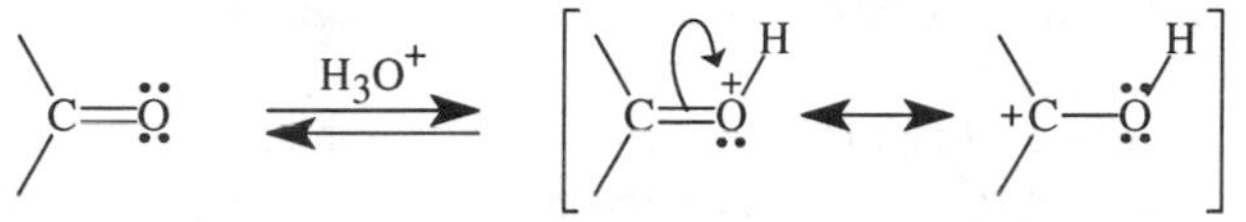

   H_3O^+

   an α-hydroxy carbocation

   b. Closely related to protonated aldehydes and ketones are α-alkoxy carbocations (cations in which the acidic proton of a protonated ketone is replaced by an alkyl group).

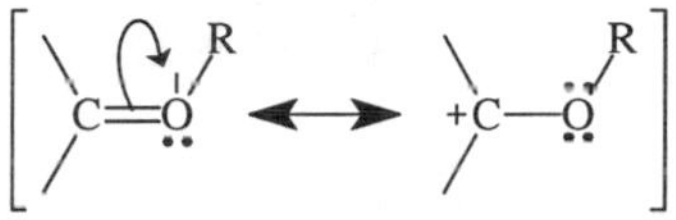

   an α-alkoxy carbocation

   c. α-Hydroxy carbocations and α-alkoxy carbocations are considerably more stable than ordinary alkyl cations.
      i. The greater stability is due to resonance interaction of the electron-deficient carbon with the unshared electrons of the neighboring oxygen.
      ii. This resonance effect far outweighs the electron-attracting polar effect of the oxygen which, by itself, would destabilize these carbocations.
2. Aldehydes and ketones in solution are considerably less basic than alcohols; their conjugate acids are more acidic than those of alcohols.
   a. The relative acidity of protonated alcohols and carbonyl compounds is an example of a solvent effect.
   b. One reason for the greater basicity of alcohols in solution is that protonated alcohols have more O—H hydrogens to participate in hydrogen bonding to solvent than do protonated aldehydes or ketones.

C. Introduction to Carbonyl-Addition Reactions

1. Addition to a carbonyl group is regioselective.
   a. The more electropositive species (for example, the proton of water in the case of hydration) adds to the carbonyl oxygen.
   b. The more electronegative species (for example, —OH in the case of hydration or —CN in the case of HCN addition) adds to the carbonyl carbon.

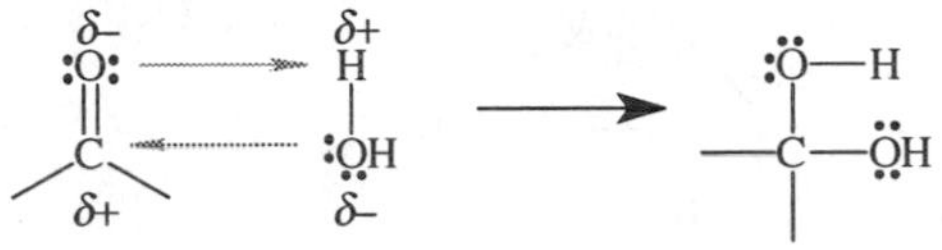

1. Carbonyl-addition reactions occur by two general types of mechanism.
   a. Under basic conditions:
      i. A nucleophile attacks the carbonyl group at the carbonyl carbon, and the carbonyl oxygen becomes negatively charged.

Nuc:⁻ + C=O ⇌ Nuc—C—O⁻

ii. The negatively charged oxygen is a relatively strong base, and is protonated by a weak acid to complete the addition.

Nuc—C—O⁻ + H—OH ⇌ Nuc—C—OH + ⁻OH

iii. This mechanism, nucleophilic addition, has no analogy in the reactions of ordinary alkenes.
iv. This pathway is enabled with aldehydes and ketones because, in the transition state, negative charge is placed on oxygen, an electronegative atom.
v. Attack of the nucleophile occurs on the carbon of the carbonyl group rather than the oxygen for the same reason—the negative charge is "pushed" onto the more electronegative atom—oxygen.
vi. As a result of this attack, the carbonyl carbon changes hybridization from sp^2 to sp^3.
vii. The nucleophile is typically the more electronegative partner of the groups that add, and the nucleophile always attacks the carbonyl carbon.

Nuc⁻ + C=O → [Nucδ– ··· C==O δ–]‡ → Nuc—C—O⁻

In this Study Guide, we are now beginning to write some mechanisms (as the one above) without showing the unshared electron pairs. Be sure to read the boxed section on p. 677 in the "Additional Solutions" section of this chapter, which covers important points that you should be aware of when writing reaction mechanisms in this way.

b. Under acidic conditions:
i the mechanism is closely analogous to the mechanism for the addition of acids to alkenes. Acid-catalyzed hydration of aldehydes and ketones is an example of this mechanism.
ii. The first step in acid-catalyzed addition is usually protonation of the carbonyl oxygen.

C=O + H—$\overset{+}{O}H_2$ ⇌ [C=$\overset{+}{O}$—H ⟷ $\overset{+}{C}$—O—H] + H_2O

iii. The carbon of a protonated carbonyl group is a much stronger Lewis acid than the carbon of an unprotonated carbonyl group. As a result, relatively weak bases such as H_2O can react at the carbonyl carbon.

H_2O + C=$\overset{+}{O}$—H ⇌ $H_2\overset{+}{O}$—C—O—H

iv. Loss of a proton to solvent completes the reaction.

H_2O + H—$\overset{+}{O}$(H)—C—OH ⇌ H_3O^+ + HO—C—OH

D. Equilibria and Rates in Carbonyl-Addition Reactions

1. Some carbonyl-addition reactions, such as hydration and cyanohydrin formation, are reversible. (Not all carbonyl additions are reversible.)
2. Whether the equilibrium for a reversible addition favors the addition product or the carbonyl compound depends strongly on the structure of the carbonyl compound. (See Table 19.2, text p. 858.)
   a. Addition is more favorable for aldehydes than for ketones.
   b. Electronegative groups near the carbonyl carbon of an aldehyde or ketone make carbonyl addition more favorable.
   c. Addition is less favorable when groups are present that donate electrons by resonance to the carbonyl carbon.
3. The rates of carbonyl-addition reactions (the reactivities of carbonyl compounds) follow similar trends; that is, the more a compound favors addition at equilibrium, the more rapidly it reacts in addition reactions.
   a. The stability of the carbonyl compound relative to that of the addition product governs both the $\Delta G°$ and the $\Delta G°^{\ddagger}$ for addition.
   b. Added stability in the carbonyl compound increases the energy change ($\Delta G°$ or $\Delta G°^{\ddagger}$), and hence decreases both the equilibrium constant and the rate, for formation of an addition product.
4. Anything that stabilizes carbocations also tends to stabilize carbonyl compounds.
   a. Electronegative groups such as halogens destabilize carbonyl compounds by their polar effect, just as they destabilize carbocations.
   b. Groups that are conjugated with the carbonyl group stabilize carbonyl compounds by resonance, just as they stabilize carbocations.
      i. Resonance interaction with the carbonyl group cannot occur in an addition product such as a hydrate because the carbonyl group is no longer present.
      ii. Consequently, aryl aldehydes and ketones have relatively unfavorable addition equilibria.
5. A steric effect also operates in carbonyl addition.
   a. The groups at the carbonyl carbon are closer together in the addition compound (bond angles ≈ 109.5°) than they are in the carbonyl compound (bond angles ~ 120°).
   b. These van der Waals repulsions
      i. raise the energy of the addition compound relative to the carbonyl compound
      ii. decrease the $\Delta G°$ and the $\Delta G°^{\ddagger}$, and thus the equilibrium constant and rate, for addition.
6. Aldehydes are generally more reactive than ketones in addition reactions; formaldehyde is more reactive than many other simple aldehydes.

E. Protecting Groups

1. A common tactic of organic synthesis is the use of protecting groups.
2. Carbonyl groups react with a number of reagents used with other functional groups but can be rendered inert to these reagents by the use of protecting groups.
3. Acetals are commonly used to protect the carbonyl groups of aldehydes and ketones from basic, nucleophilic reagents.
   a. Once the protection is no longer needed, the acetal protecting group is easily removed, and the carbonyl group re-exposed, by treatment with dilute aqueous acid.
   b. Because acetals are unstable in acid, they do not protect carbonyl groups under acidic conditions.

RO–C(–OR) (an acetal) —[H_3O^+, H_2O]→ C=O + 2 ROH

an acetal —[^-OH, H_2O]→ no reaction when R ≠ H

F. Manufacture and Use of Aldehydes and Ketones

1. Formaldehyde is manufactured by the oxidation of methanol over a silver catalyst.

$$H_3C\text{—}OH \xrightarrow[O_2,\ 600\text{–}650\ ^\circ C]{\text{Ag catalyst}} H_2C\text{=}O$$

2. Formaldehyde is used in the synthesis of a class of polymers known as phenol-formaldehyde resins.
   a. A resin is a polymer with a rigid three-dimensional network of repeating units.
   b. Phenol-formaldehyde resins are produced by heating phenol and formaldehyde with acidic or basic catalysts.
   c. One phenol-formaldehyde resin, Bakelite, was the first useful synthetic polymer.

3. The simplest ketone, acetone, is co-produced with phenol by the autoxidation-rearrangement of cumene (Sec. 18.10, text page 823).

$$Ph\text{—}C(CH_3)_2\text{—}H \xrightarrow[H_3O^+]{O_2} Ph\text{—}OH + O\text{=}C(CH_3)_2$$

IV. INTRODUCTION TO AMINES AND AMINE DERIVATIVES OF ALDEHYDES AND KETONES

A. General

1. A primary amine (general structure $R\ddot{N}H_2$) is an organic derivative of ammonia in which only one ammonia hydrogen is replaced by an alkyl or aryl group.
2. A secondary amine (general structure $R_2\ddot{N}H$) is an organic derivative of ammonia in which two ammonia hydrogens are replaced by alkyl or aryl groups; the nitrogen may be part of a ring.
3. A tertiary amine (general structure $R_3\ddot{N}$) is an organic derivative of ammonia in which all three ammonia hydrogens are replaced by alkyl or aryl groups; the nitrogen may be part of a ring.

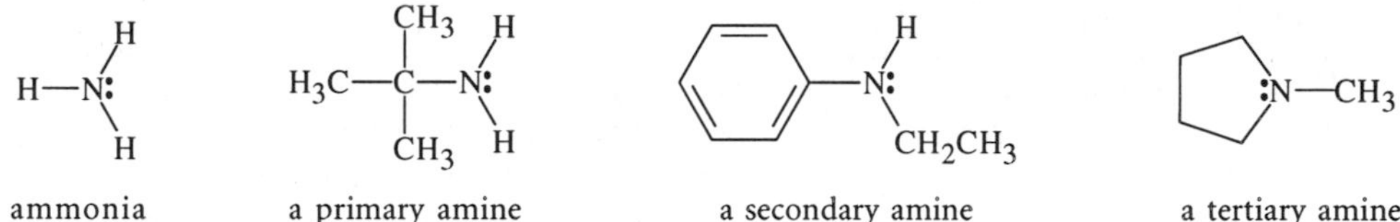

ammonia a primary amine a secondary amine a tertiary amine

B. Amine Derivatives of Aldehydes and Ketones

1. An imine is a nitrogen analog of an aldehyde or ketone; the imine functional group is C=N—R, where R = alkyl, aryl, or H.
   a. Imines are sometimes called Schiff bases.
   b. Imines are prepared by the reactions of aldehydes or ketones with primary amines.

$$\text{>C=O} + H_2NR \xrightleftharpoons{-H_2O} \text{>C=N—R}$$

an imine

   c. Imines revert to the corresponding carbonyl compounds and amines in aqueous acid.

2. A carbinolamine is a compound with an amine group (—NH_2, —NHR, or —NR_2) and a hydroxy group on the same carbon.

$$\mathrm{R_2C{=}O} + \mathrm{H_2NR} \rightleftharpoons \mathrm{R_2C(OH)(HNR)} \xrightleftharpoons{H_3O^+} \mathrm{R_2C{=}NR} + \mathrm{H_2O}$$

a carbinolamine — an imine

a. Carbinolamines are intermediates in imine formation; most carbinolamines are not isolated.
b. Carbinolamines undergo acid-catalyzed dehydration to form imines.

3. An enamine is the nitrogen analog of an enol in which the hydroxy group is replaced by an amine group bearing two R groups, where R can be alkyl, aryl, or part of a ring.
   a. Formation of an enamine occurs when a secondary amine reacts with an aldehyde or ketone, provided that the carbonyl compound has an α-hydrogen.
   b. Enamines, like imines, revert to the corresponding carbonyl compounds and amines in aqueous acid.

$$\mathrm{H{-}C{-}C{=}O} + \mathrm{HNR_2} \rightleftharpoons \mathrm{C{=}C(NR_2)}$$

an enamine

4. A hydrazone is a nitrogen analog of an aldehyde or ketone; the hydrazone functional group is C=N—NR_2, where R = alkyl, aryl, or H.
   a. Hydrazones are prepared by the reactions of aldehydes or ketones with hydrazines (H_2N—NR_2).
   b. 2,4-Dinitrophenylhydrazine (2,4-DNP) reacts with aldehydes and ketones to give 2,4-dinitrophenylhydrazones (2,4-DNP derivatives). These derivatives are generally solids whose characteristic melting points were used, before the advance of NMR spectroscopy, to aid in the identification of new organic compounds.

$$\mathrm{C{=}O} + \mathrm{H_2N{-}NH{-}C_6H_3(NO_2)_2} \longrightarrow \mathrm{C{=}N{-}NH{-}C_6H_3(NO_2)_2} + \mathrm{H_2O}$$

2,4-dinitrophenylhydrazine (2,4-DNP) — 2,4-DNP derivative

REACTIONS

I. CARBONYL-ADDITION REACTIONS

A. Hydration of Aldehydes and Ketones

1. Aldehydes and ketones react reversibly with water; this reaction is called hydration.

$$\mathrm{R_2C{=}O} + \mathrm{H_2O} \xrightarrow{H_3O^+ \text{ or } {}^-OH} \mathrm{R_2C(OH)_2}$$

a carbonyl hydrate

2. Acid-catalyzed hydration of aldehydes and ketones is reminiscent of acid-catalyzed hydration of alkenes, although hydration of aldehydes and ketones is much faster.
   a. The first step is protonation of the carbonyl oxygen; the protonated carbonyl compound is a much stronger Lewis acid (electron acceptor) than an unprotonated carbonyl compound.

b. The electron-deficient carbon is attacked by H_2O, a nucleophile.
c. Loss of a proton to the solvent (water) completes the reaction.

a carbonyl hydrate

3. Hydration of aldehydes and ketones also occurs in neutral and basic solution.

B. Cyanohydrin Formation

1. Hydrogen cyanide, HCN, reacts reversibly with aldehydes and ketones by a nucleophilic addition mechanism to give cyanohydrins.
   a. A cyanide ion, formed by ionization of the weak acid HCN, attacks the carbonyl group at the carbonyl carbon; the carbonyl oxygen becomes negatively charged.
   b. The negatively charged oxygen is a relatively strong base, and is protonated by either water or HCN to complete the addition.

a cyanohydrin

2. Cyanohydrin formation favors the cyanohydrin addition product in the case of aldehydes and methyl ketones, but not in the case of aryl ketones.
3. Cyanohydrin formation is another method of forming carbon-carbon bonds.

C. Reduction of Aldehydes and Ketones to Alcohols

1. Aldehydes and ketones are reduced to alcohols with either lithium aluminum hydride ($LiAlH_4$) or sodium borohydride ($NaBH_4$), which serve as sources of nucleophilic hydride ion ($H:^-$).
   a. These reactions, which are not reversible, result in the net addition of the elements of H_2 across the C=O bond.
   b. Reduction of an aldehyde gives a primary alcohol.
   c. Reduction of a ketone gives a secondary alcohol.

an aldehyde — hydride reduction → a primary alcohol

a ketone — hydride reduction → a secondary alcohol

2. $LiAlH_4$ and $NaBH_4$ reductions are generally referred to as hydride reductions and are examples of nucleophilic addition.
3. The reaction of $LiAlH_4$ with aldehydes and ketones involves the nucleophilic attack of hydride (delivered from $^-AlH_4$) on the carbonyl carbon; all four hydride equivalents of $LiAlH_4$ are active in this reaction.
   a. A lithium ion coordinated to the carbonyl oxygen acts as a Lewis-acid catalyst.
   b. Hydride attacks the carbonyl carbon to give an alkoxide addition compound.
   c. The alkoxide salt is converted by protonation in a separate step into the alcohol product; the proton source is water (or an aqueous solution of a weak acid such as $^+NH_4\,^-Cl$).

$$\text{R}_2\text{C=O} + \text{Li}^+ \ {}^-\text{AlH}_4 \longrightarrow [\text{C=O}\cdots\text{Li}^+,\ \text{H}-\text{AlH}_3^-] \xrightarrow{-\text{AlH}_3} \text{R}_2\text{C(H)}-\text{O}^-\ \text{Li}^+ \xrightarrow{\text{H}_3\text{O}^+} \text{R}_2\text{C(H)}-\text{OH} + \text{Li}^+$$

d. $LiAlH_4$ reacts violently with water and therefore must be used in dry solvents such as anhydrous ether or THF.

4. The reaction of $NaBH_4$ with aldehydes and ketones involves the nucleophilic attack of hydride on the carbonyl carbon; all four hydride equivalents of $NaBH_4$ are active in this reaction.
   a. The sodium ion of $NaBH_4$ does not form as strong a bond to the carbonyl oxygen as the lithium ion.
   b. $NaBH_4$ reductions are carried out in protic solvents, such as alcohols; hydrogen bonding between the alcohol solvent and the carbonyl group serves as a weak acid catalysis that activates the carbonyl group.
   c. $NaBH_4$ reacts only slowly with alcohols, and it can be used in water if the solution is not acidic.

$$\text{R}_2\text{C=O}\cdots\text{H}-\text{OR},\ \text{H}-\text{BH}_3^-\ \text{Na}^+ \longrightarrow \text{R}_2\text{C(OH)}-\text{H} + \text{RO}-\text{BH}_3^-\ \text{Na}^+$$

5. $LiAlH_4$ is a much more reactive reagent than $NaBH_4$.
   a. A number of functional groups react with $LiAlH_4$ but not $NaBH_4$:
      i. alkyl halides
      ii. alkyl tosylates
      iii. esters
      iv. nitro groups
   b. The greater selectivity and safety of $NaBH_4$ make it the preferred reagent in many applications.
6. Aldehydes and ketones can also be reduced to alcohols by catalytic hydrogenation.
   a. This reaction is analogous to the catalytic hydrogenation of an alkene.

$$\text{H}_3\text{C}-\text{C(=O)}-\text{CH(CH}_3)_2 \xrightarrow[\text{pressure}]{\text{H}_2,\ \text{Ni catalyst}} \text{H}_3\text{C}-\text{CH(OH)}-\text{CH(CH}_3)_2$$

   b. It is usually possible to use catalytic hydrogenation for the selective reduction of an alkene double bond in the presence of a carbonyl group; palladium catalysts are particularly effective for this purpose.

$$\text{Ph}-\text{CH=CH}-\text{CH=O} \xrightarrow[\text{5\% Pd/C}]{\text{H}_2} \text{Ph}-\text{CH}_2-\text{CH}_2-\text{CH=O}$$

D. Reactions of Aldehydes and Ketones with Grignard Reagents

1. The reaction of Grignard reagents with carbonyl groups is the most important application of Grignard reagents in organic chemistry; addition to aldehydes and ketones in an ether solvent, followed by protonolysis, gives alcohols. (See Appendix V, text page A-10, for a review of other syntheses of alcohols.)

$$\underset{\text{an aldehyde or ketone}}{\text{R}_2\text{C=O}} + \text{R}-\text{MgX} \longrightarrow \xrightarrow{\text{H}_3\text{O}^+} \underset{\text{an alcohol}}{\text{R}_2\text{C(OH)}-\text{R}} + \text{Mg}^{2+} + \text{X}^-$$

2. The reaction of Grignard reagents with aldehydes and ketones is another example of carbonyl addition; the addition of Grignard reagents to aldehydes and ketones is not reversible.

3. The net effect of the Grignard reaction, followed by protonolysis, is addition of R—H (R = an alkyl or aryl group) across the C═O double bond; this addition is not reversible.
   a. Primary alcohols are synthesized by the addition of Grignard reagents to formaldehyde.

$$\text{formaldehyde}\quad \mathrm{H{-}C({=}O){-}H} \xrightarrow[\text{2) } H_3O^+]{\text{1) R—MgX}} \mathrm{H{-}C(OH)(H){-}R}\quad \text{a primary alcohol}$$

   b. Secondary alcohols are synthesized by the addition of Grignard reagents to aldehydes other than formaldehyde.

$$\text{an aldehyde}\quad \mathrm{H{-}C({=}O){-}R} \xrightarrow[\text{2) } H_3O^+]{\text{1) R'—MgX}} \mathrm{H{-}C(OH)(R){-}R'}\quad \text{a secondary alcohol}$$

   c. Tertiary alcohols are synthesized by the addition of Grignard reagents to ketones.

$$\text{a ketone}\quad \mathrm{R{-}C({=}O){-}R'} \xrightarrow[\text{2) } H_3O^+]{\text{1) R''—MgX}} \mathrm{R{-}C(OH)(R'){-}R''}\quad \text{a tertiary alcohol}$$

4. The Grignard synthesis of a tertiary alcohol, or in some cases a secondary alcohol, can be extended to an alkene synthesis by dehydration of the alcohol with strong acid during the protonation step.

$$\text{cyclopentanone} \xrightarrow[\text{2) } H_2SO_4/H_2O]{\text{1) } CH_3MgI/Et_2O} \text{1-methylcyclopentene } (CH_3)$$

5. The Grignard reaction is an excellent method of carbon-carbon bond formation. (See Appendix VI, text page A-13, for a review of other reactions used to form carbon-carbon bonds.)
6. In the mechanism of the Grignard addition—
   a. the magnesium of the Grignard reagent, a Lewis acid, bonds to the carbonyl oxygen; this bonding makes the carbonyl carbon more electrophilic.
   b. the carbon group of the Grignard reagent attacks the carbonyl carbon to form a halomagnesium alkoxide.
   c. addition of dilute acid in a separate protonolysis step to the reaction mixture gives an alcohol.

$$\mathrm{C{=}O} \xrightarrow{\text{R—MgX}} \mathrm{\overset{\delta+}{C}{=}O\cdots MgX{-}\overset{\delta-}{R}} \longrightarrow \mathrm{R{-}C{-}O^-\ {}^+MgX} \xrightarrow{H_3O^+} \mathrm{R{-}C{-}OH} + Mg^{2+} + X^-$$

7. The reactions of organolithium and sodium acetylide reagents with aldehydes and ketones are fundamentally similar to the Grignard reaction.

E. Preparation and Hydrolysis of Acetals

1. When an aldehyde or ketone reacts with a large excess of an alcohol in the presence of a trace of strong acid, an acetal is formed.
   a. An acetal is a di-ether in which both ether oxygens are bound to the same carbon.
   b. Acetals are ethers of hydrates or *gem*-diols.

$$2\ \mathrm{R{-}OH} + \mathrm{C{=}O} \overset{H_3O^+}{\rightleftharpoons} \mathrm{(RO)(RO)C} + H_2O$$

an acetal

2. Two equivalents of alcohol are consumed in the formation of acetals, but one equivalent of a 1,2- or 1,3-diol can react to form a cyclic acetal, in which the acetal group is part of a five- or six-membered ring, respectively.

HO⁀⁀OH + (C=O) $\xrightleftharpoons{H_3O^+}$ (cyclic acetal, O–C–O ring) + H_2O

a cyclic acetal

3. Acetal formation is reversible and involves an acid-catalyzed carbonyl addition followed by a substitution that occurs by an S_N1 mechanism:
   a. The first step in the mechanism of acetal formation is acid-catalyzed addition of the alcohol to the carbonyl group to give a hemiacetal (a compound with an —OR and —OH group on the same carbon).
   b. The hemiacetal is converted into an acetal by substitution of the —OH group by another —OR group.

C=O $\xrightleftharpoons[\text{addition}]{ROH/H_3O^+}$ RO–C–OH (a hemiacetal) $\xrightleftharpoons[\text{substitution } (S_N1 \text{ mechanism})]{ROH/H_3O^+}$ RO–C–OR (an acetal)

4. The reaction is driven to the right by applying LeChatelier's principle in one or both of the following ways:
   a. Use of excess alcohol as the solvent.
   b. Removal of the water by-product (usually by azeotropic distillation).
5. Acetals in the presence of acid and excess water are transformed rapidly back into the corresponding carbonyl compounds and alcohols.

HO–C–OH $\xrightleftharpoons{H_3O^+}$ C=O + 2 R—OH

   a. This process is called acetal hydrolysis.
   b. By the principle of microscopic reversibility, the mechanism of acetal hydrolysis is the reverse of the mechanism of acetal formation.
   c. Acetal hydrolysis, like hemiacetal formation, is acid-catalyzed.
6. The interconversion of hemiacetals and aldehydes is catalyzed not only by acids, but by bases as well; however, the interconversion of hemiacetals and acetals is catalyzed only by acids. For this reason, the hydrolysis of acetals is catalyzed only by acids; acetals are stable in basic and neutral solutions.
7. Hemiacetals in most cases cannot be isolated because they react further to yield acetals or decompose to aldehydes or ketones plus water.
   a. Simple aldehydes form appreciable amounts of hemiacetals in alcohol solution, just as they form appreciable amounts of hydrates in water.
   b. Five- and six-membered cyclic hemiacetals form spontaneously from the corresponding hydroxy aldehydes, and most are stable, isolable compounds.

(benzene ring with CH=O and CH_2—OH) $\rightleftharpoons$ (cyclic hemiacetal with OH and ring O)

II. OXIDATION OF ALDEHYDES AND KETONES

A. Oxidation of Aldehydes

1. Aldehydes can be oxidized to carboxylic acids.

$$CH_3-C(=O)-H \xrightarrow[^{-}OH]{KMnO_4} \xrightarrow{H_3O^+} CH_3-C(=O)-OH$$

an aldehyde — a carboxylic acid

$$CH_3-C(=O)-H \underset{}{\overset{H_2O,\ H_3O^+}{\rightleftharpoons}} CH_3-C(OH)_2-H \xrightarrow[H_3O^+]{K_2CrO_7} CH_3-C(=O)-OH$$

an aldehyde — an aldehyde hydrate — a carboxylic acid

2. Common oxidants, such as aqueous Cr(VI) reagents, nitric acid, or aqueous $KMnO_4/NaOH$, can be used in the oxidation of aldehydes to carboxylic acids; these oxidizing agents are the same ones used for oxidizing alcohols.
3. Some aldehyde oxidations begin as addition reactions.
   a. In the oxidation of aldehydes by Cr(VI) reagents, the hydrate, not the aldehyde, is actually the species oxidized.
   b. For this reason, some water should be present in solution in order for aldehyde oxidations with Cr(VI) to occur at a reasonable rate.

$$R-C(=O)-H \overset{H_2O,\ H_3O^+}{\rightleftharpoons} -C(OH)-OH \xrightarrow[H_3O^+]{K_2Cr_2O_7} R-C(=O)-OH$$

4. In the laboratory, aldehydes can be conveniently oxidized to carboxylic acids with Ag(I) reagents.
   a. Aldehydes that contain double bonds or alcohol —OH groups (functional groups that react with other oxidizing reagents) are oxidized by Ag_2O without oxidizing these other functional groups.
   b. If the silver ion is solubilized as its ammonia complex, $^{+}Ag(NH_3)_2$, oxidation of the aldehyde is accompanied by the deposition of a metallic silver mirror on the walls of the reaction vessel. (This observation can be used as a convenient test for aldehydes, known as the Tollens test.)

$$R-C(=O)-H \xrightarrow[NH_3/H_2O]{AgNO_3} R-C(=O)-O^{-}\ {}^{+}NH_4 \quad + \quad Ag\ (mirror)$$

ammonium salt of a carboxylic acid

5. Many aldehydes are oxidized by the oxygen in air upon standing for long periods of time; this process is another example of autoxidation.

B. Oxidation of Ketones

1. Ketones cannot be oxidized without breaking carbon-carbon bonds.
2. Ketones are resistant to mild oxidation with Cr(VI) reagents.

III. REACTIONS OF ALDEHYDES AND KETONES WITH AMINES

(See also "Concepts," Section IV,B.)

A. Imine Formation with Primary Amines

1. Imines are prepared by the reaction of aldehydes or ketones with primary amines.

a. Formation of imines is reversible and generally takes place with acid or base catalysis, or with heat.

$$\gt C{=}O + H_2NR \underset{}{\overset{H_3O^+ \text{ or } HO^-}{\rightleftharpoons}} \gt C{=}N{-}R + H_2O$$

an imine

b. Imine formation is typically driven to completion in one or both of the following ways:
   i. Precipitation of the imine.
   ii. Removal of water.

2. Imine formation consists of a carbonyl-addition reaction followed by β-elimination.
   a. The first step of the mechanism is a nucleophilic addition to the carbonyl group in which the nucleophile (an amine) reacts with an aldehyde or ketone to give an unstable addition compound called a carbinolamine.
   b. The carbinolamine undergoes acid-catalyzed dehydration to form an imine.
   c. Dehydration of the carbinolamine is typically the rate-limiting step of imine formation.

$$\gt C{=}O + RNH_2 \overset{\text{addition}}{\rightleftharpoons} \gt C(NHR)(OH) \overset{\text{dehydration}}{\rightleftharpoons} \gt C{=}N{-}R + H_2O$$

a carbinolamine an imine

3. Certain types of imine adducts sometimes find use as derivatives of aldehydes and ketones because they are in most cases solids with well-defined melting points. (See Table 19.3, text page 876, for the corresponding structures.)
   a. Aldehydes or ketones + hydroxylamine → oximes.
   b. Aldehydes or ketones + hydrazine → hydrazones.
   c. Aldehydes or ketones + phenylhydrazine → phenylhydrazones.
   d. Aldehydes or ketones + 2,4-dinitrophenylhydrazine → 2,4-dinitrophenylhydrazones (2,4-DNP derivatives).
   e. Aldehydes or ketones + semicarbazide → semicarbazones.

B. Enamine Formation with Secondary Amines

1. Formation of an enamine occurs when a secondary amine reacts with an aldehyde or ketone that has an α-hydrogen.

$$\text{R}{-}C({=}O){-}C{-}H + R_2N{-}H \overset{H_3O^+}{\rightleftharpoons} R{-}C(NR_2){=}C\lt + H_2O$$

an aldehyde or ketone with an α-hydrogen a secondary amine an enamine

   a. Just as most aldehydes and ketones are more stable than their corresponding enols, most imines are more stable than their corresponding enamines.
   b. Because secondary amines cannot form imines, they form enamines instead.

2. Like imine formation, enamine formation is reversible and must be driven to completion by the removal of one of the reaction products (usually water).
   a. The mechanism of enamine formation begins like the mechanism of imine formation, as a nucleophilic addition to give a carbinolamine intermediate.
   b. Dehydration of a carbinolamine involves proton loss from an adjacent carbon.

H—C—C(=O) + HNR_2 —addition→ H—C—C(NR_2)—OH (a carbinolamine) —$-H_2O$, dehydration→ C=C—NR_2 (an enamine)

3. Enamines, like imines, revert to the corresponding carbonyl compounds and amines in aqueous acid.

1-(cyclopent-1-enyl)pyrrolidine —H_3O^+, H_2O→ cyclopentanone + pyrrolidine (N—H)

C. Reactions of Aldehydes and Ketones with Tertiary Amines

1. Tertiary amines do not react with aldehydes and ketones to form stable derivatives.

IV. CONVERSION OF KETONES AND ALDEHYDES INTO ALKANES AND ALKENES

A. Reduction of Carbonyl Groups to Methylene Groups

1. A carbonyl group of an aldehyde or ketone can be reduced completely to a methylene (—CH_2—) group.
2. One procedure for effecting this transformation involves heating the aldehyde or ketone with hydrazine (H_2N—NH_2) and strong base.

2-methyl-1-tetralone (CH_3) —H_2N—NH_2, KOH, $HOCH_2CH_2OH$→ 2-methyltetralin (CH_3)

   a. This reaction, called the Wolff-Kishner reduction, typically utilizes ethylene glycol or similar compounds as co-solvents.
   b. The high boiling points of these solvents allow the reaction mixtures to reach the high temperatures required for the reduction to take place at a reasonable rate.
3. The Wolff-Kishner reduction is an extension of imine formation; an intermediate in the reduction is a hydrazone (an imine of hydrazine).
4. The Wolff-Kishner reduction takes place under strongly basic conditions.

C=O —H_2N—NH_2, hydrazone formation→ C=N—NH_2 (a hydrazone) —HO^-, loss of nitrogen, $-N_2$→ H—C:$^-$ —H_2O, protonation→ H—C—H

5. The same overall transformation can be achieved under acidic conditions by a reaction called the Clemmensen reduction, which involves reduction of an aldehyde or ketone with zinc amalgam (a solution of zinc metal in mercury) in the presence of HCl.

Ph—C(=O)—CH_3 —Zn/Hg, HCl/H_2O, heat→ Ph—CH_2—CH_3

6. The Wolff-Kishner and Clemmensen reactions are particularly useful for the introduction of alkyl substituents into benzene rings by the following sequence:
   a. Friedel-Crafts acylation to give an aryl ketone.
   b. Wolff-Kishner or Clemmensen reduction to yield the corresponding alkyl group.

$$\text{benzene} + Cl{-}C(=O){-}C_5H_9 \xrightarrow[2)\ H_3O^+]{1)\ AlCl_3} C_6H_5{-}C(=O){-}C_5H_9 \xrightarrow[\text{HCl/H}_2\text{O, heat}]{\text{Zn/Hg}} C_6H_5{-}CH_2{-}C_5H_9$$

B. The Wittig Alkene Synthesis

1. Another addition-elimination reaction, called the Wittig reaction, is an important method for preparing alkenes from aldehydes and ketones.

$$\underset{\text{a Wittig reagent}}{Ph_3\overset{+}{P}{-}\overset{-}{C}\lt} + \underset{\text{an aldehyde or ketone}}{O{=}C\lt} \longrightarrow \underset{\text{triphenylphosphine oxide}}{Ph_3P{=}O} + \underset{\text{an alkene}}{\gt C{=}C\lt}$$

2. The nucleophile in the Wittig reaction is a type of ylid.
   a. An ylid is any compound with opposite charges on adjacent covalently bound atoms, each of which has an electronic octet.
   b. Because phosphorus can accommodate more than eight valence electrons, a phosphorus ylid has an uncharged resonance structure.
   c. Although the structures of phosphorus ylids are sometimes written with phosphorus-carbon double bonds, the charged structures, in which each atom has an octet of electrons, are very important contributors.

$$\left[\underset{\text{an ylid}}{Ph_3\overset{+}{P}{-}\ddot{C}H_2^-} \longleftrightarrow Ph_3P{=}CH_2\right]$$

3. Preparation of phosphorus ylids:
   a. An alkyl halide reacts with triphenylphosphine (Ph_3P) in an S_N2 reaction to give a phosphonium salt. (Because the alkylation of triphenylphosphine is a typical S_N2 reaction, it is limited for the most part to methyl and primary alkyl halides.)
   b. The phosphonium salt is converted into its conjugate base, the ylid, by reaction with a strong base such as an organolithium reagent.

$$Ph_3\ddot{P}\ \ CH_3{-}I \longrightarrow Ph_3\overset{+}{P}{-}CH_3 + I^- \xrightarrow{C_4H_9Li} Ph_3\overset{+}{P}{-}\ddot{C}H_2^- + LiI + C_4H_{10}$$

4. The mechanism of the Wittig reaction:
   a. A nucleophile (the anionic carbon of the ylid) attacks the carbonyl carbon.
      i. The anionic oxygen in the resulting species reacts with phosphorous to form an oxaphosphetane intermediate.
      ii. An oxaphosphetane is a saturated four-membered ring containing both oxygen and phosphorus as ring atoms.
   b. Under the usual reaction conditions, the oxaphosphetane spontaneously decomposes to the alkene and the by-product triphenylphosphine oxide.

$$\text{cyclopentanone} \xrightarrow[\text{cycloaddition}]{H_2\ddot{C}^-{-}\overset{+}{P}Ph_3} \text{(}O^-\text{, }CH_2\overset{+}{P}Ph_3\text{ adduct)} \longrightarrow \underset{\text{an oxaphosphetane}}{\text{spiro O–PPh}_3\text{ ring}} \xrightarrow{\text{elimination}} \text{methylenecyclopentane} + \underset{\text{triphenylphosphine oxide}}{O{=}PPh_3}$$

5. The Wittig reaction is especially important because it gives alkenes in which the position of the alkene double bond is unambiguous.
   a. The reaction is thus completely regioselective.

b. The reaction can be used for the preparation of alkenes that would be difficult to prepare by other reactions.

6. To plan the synthesis of an alkene by the Wittig reaction, consider the origin of each part of the product, and then reason deductively.

a. One carbon of the alkene double bond originates from the alkyl halide used to prepare the ylid; the other is the carbonyl carbon of the aldehyde or ketone.

b. In principle, two Wittig syntheses are possible for any given alkene.

c. Most Wittig syntheses are planned so that the most reactive alkyl halide (a methyl or primary alkyl halide) can be used as one of the starting materials. [See (3a) above.]

$$\text{(cyclopropyl)}\text{—}CH_2\text{—}CH{=}C(CH_3)_2 \Longrightarrow \text{(cyclopropyl)}\text{—}CH_2\text{—}CH{=}O \;+\; Br\text{—}CH(CH_3)_2$$

a secondary alkyl halide

$$\text{(cyclopropyl)}\text{—}CH_2\text{—}CH{=}C(CH_3)_2 \Longrightarrow \text{(cyclopropyl)}\text{—}CH_2\text{—}CH_2\text{—}Br \;+\; O{=}C(CH_3)_2$$

a primary alkyl halide ← better synthesis

d. The Wittig reaction in many cases gives mixtures of E and Z isomers, although certain modifications of the Wittig reaction that avoid this problem have been developed.

SCIENTISTS

Leo Hendrik Baekeland (1863–1944) 19.15
Herbert Charles Brown (b. 1912) [Chemistry 1979] 19.8
Erik Christian Clemmensen (1876–1941 .... 19.12
Nikolai Matveeviè Kishner (1867–1935) 19.12
Fred Warren McLafferty (b. 1923) 19.3E
Hugo Josef Schiff (1834–1915) 19.11A
Bernhard Christian Gottfried Tollens (1841–1918) 19.14
Georg Wittig (1897–1987) [Chemistry 1979] 19.13
Ludwig Wolff (1857–1919) 19.12

Dates in brackets refer to Nobel Prize awards.

STUDY GUIDE LINKS

19.1 IR Absorptions of Cyclic Ketones

The increase of the infrared carbonyl absorption frequency with ring size is a very useful trend that can be used to diagnose the presence of small rings. The main reason for this trend is a coupling phenomenon. In the presentation of IR spectroscopy in Chapter 12, IR absorption is treated as a phenomenon associated with isolated bonds. Even though this is a highly useful approximation, the effects of nearby bonds in some cases cannot be ignored, and the carbonyl stretching frequencies of the cyclic ketones is one of those cases.

Imagine two extreme situations. In the first, the carbonyl group forms an angle of 180° with adjacent C—C bonds (as in ketene, $H_2C{=}C{=}O$). In such a situation, because the vibrations of both bonds occur along the same line, expansion of the carbon-oxygen bond that occurs during a vibration also compresses the carbon-carbon bond; and compression of the carbon-oxygen bond also expands the carbon-carbon bond. The two stretching vibrations are *coupled:*

$H_2C{=}C{=}O$ ⇄ (expansion of C=O bond; compression of C=C bond) $H_2C{=}C{=}O$

$H_2C{=}C{=}O$ ⇅ (compression of C=O bond; expansion of C=C bond) $H_2C{=}C{=}O$

The additional energy required to cause both bonds to vibrate at the same time is reflected in a higher vibrational frequency for both bonds.

Now consider a hypothetical situation in which the C═O bond is perpendicular to adjacent bonds. In such a situation, the C═O bond can stretch and compress without stretching the adjacent C—C bond:

$H_2C{-}C({=}O){-}CH_2$ ⇄ (C=O bond stretch) $H_2C{-}C({=}O){-}CH_2$

(Remember that bending of the C—C bond requires much less energy than stretching.) At this angle, the stretching vibrations of the two bonds are completely uncoupled. Hence, it requires much less energy to stretch the C═O bond in this situation than when the two bonds are collinear.

An intermediate situation occurs at bond angles between 180° and 90°: the stretching vibrations are partially coupled. The more closely the C—C═O angle approaches 180°, the greater the coupling and the higher the vibration frequency. Notice that the bond angle in question is the one between the C═O bond and the ring C—C bond, not the internal angle within the ring. This angle varies from 120° in cyclohexanone to greater values for the ketones with smaller rings. As the data in Eq. 19.4 on p. 845 of the text show, the carbonyl stretching frequencies increase toward smaller ring sizes.

Another factor that contributes to carbonyl stretching frequencies is the relative strengths of the C═O bonds. The C═O bond in ketene involves an *sp*-hybridized carbon atom, whereas the C═O bond in cyclohexanone involves an sp^2-hybridized carbon. The amount of *s* character in the C═O σ bond is therefore greater in ketene, and the amount of *s* character increases from cyclohexanone through the smaller rings. As with C—H bonds (Eq. 14.26, text p. 625), the strengths of C=O bonds, and hence, their absorption frequencies, increase with increasing *s* character.

The coupling effect discussed above is believed to dominate the trend toward higher frequencies, with the bond-strength effect making a secondary contribution.

A similar trend can be seen with carbon–carbon double bond stretching frequencies:

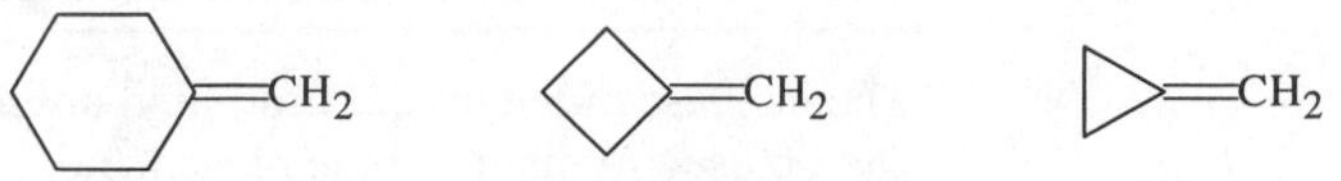

	methylenecyclohexane (=CH$_2$)	methylenecyclobutane (=CH$_2$)	methylenecyclopropane (=CH$_2$)
C=C stretching frequencies:	1650 cm^{-1}	1672 cm^{-1}	1781 cm^{-1}

✓19.2 Why Nucleophiles React at the Carbonyl Carbon

It is tempting to use resonance structures to rationalize the reactivity of a carbonyl group as follows:

$$\left[\ \mathrm{C{=}\ddot{O}{:}} \longleftrightarrow \mathrm{\overset{+}{C}{-}\ddot{O}{:}^{-}}\ \right]$$

It is sometimes erroneously said that carbonyl compounds react with nucleophiles at the carbonyl carbon because there is positive charge at this carbon, and hence, it is a logical point of attack for electron pairs of nucleophiles.

There are two fallacies in this argument. First, when two species collide, they do so randomly, not from any preferred direction. For example, a nucleophile in solution can collide randomly with any of the atoms in a carbonyl compound. In other words, a nucleophile is not selectively directed to a carbonyl carbon. Rather, reaction occurs the way it does because when the nucleophile, in one of many random collisions, happens to collide with the carbonyl carbon from the proper direction, *electrons (and charge) can be shifted onto the electronegative carbonyl oxygen.*

The second fallacy in the charge-separation argument above is its implication that resonance increases reactivity. In fact, *resonance stabilization has the opposite effect.* Carbonyl compounds are actually *less reactive* than they would be if such resonance stabilization did not exist. Many studies have shown that when molecules are prepared in which resonance stabilization of double bonds cannot occur (for example, molecules in which π-electron systems are twisted out of coplanarity), these bonds become more reactive than double bonds in similar molecules in which resonance interaction can occur. The carbon-oxygen π bond is actually about 105 kJ/mol (25 kcal/mol) *stronger* than a carbon-carbon π bond, and this fact, taken alone, means that it should be *less reactive* than a carbon-carbon double bond, exactly as the resonance structures suggest.

If the carbonyl π bond is so stable, why does it react? Remember from Sec. 3.6A of the text that when a Brønsted acid-base reaction occurs, it is not only the strength of the bond to the hydrogen, but also *how well the other atom in the bond accepts electrons,* that governs how easily the reaction takes place. Attack on a carbonyl carbon is no different, except that it is a carbon rather than a hydrogen which is attacked by the base (nucleophile), and a π bond rather than a σ bond which is broken. *Carbonyl compounds react with nucleophiles at the carbonyl carbon because the electronegative oxygen readily accepts negative charge.*

It nevertheless is true that resonance structures *do* suggest ways in which molecules can react. In the case of a carbonyl group, the dipolar resonance structure suggests that the carbon, with its partial positive charge, is the site of electron attack. The reason this works is that the atom which accepts electrons in the dipolar resonance structure (oxygen) is the same one that accepts electrons in the transition state for nucleophilic addition. *Such correspondences between reactivity patterns and resonance structures invariably occur throughout organic chemistry.* For this reason, organic chemists find themselves using resonance structures to predict sites of reactivity in molecules. Resonance structures are undeniably useful for this purpose; but when we use them this way, we must remember that resonance is not the *reason for reactivity.*

19.3 Molecular-Orbital Basis for Nucleophilic Attack on the Carbonyl Carbon

Study Guide Link 9.3 discussed the molecular-orbital basis for backside attack in S_N2 reactions of alkyl halides. Because all bonding molecular orbitals are filled, and the nucleophile is providing an additional pair of electrons, a problem arises: where does the extra electron pair attack the alkyl halide? If you read Study Guide Link 9.3, you learned that the nucleophile attacks an *antibonding* molecular orbital of the alkyl halide.

We have a similar situation with nucleophilic attack on a carbonyl group. With Lewis structures, we show the nucleophile attacking the π bond:

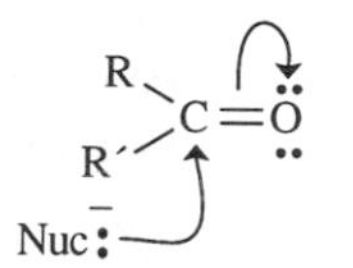

R
C=Ö
R′
Nuc:

However, the π molecular orbital of the carbonyl compound is filled and cannot accommodate additional electrons. Where, then, does attack occur? The answer is that the nucleophile electrons are directed to an *unoccupied* molecular orbital, and, furthermore, they will enter the unoccupied molecular orbital of lowest energy, which is commonly called the **LUMO:** "lowest unoccupied molecular orbital." Because all bonding molecular orbitals are filled, the LUMO for a carbonyl compound, as it is for an alkyl halide, is an *antibonding* molecular orbital. The LUMO for a carbonyl compound is simply the π^* molecular orbital of the carbonyl group, and it is this molecular orbital that is attacked. Figure SG19.1 shows this molecular orbital under attack by a nucleophile. Notice that this molecular orbital has lobes above and below the plane defined by the trigonal carbon and its three attached atoms. This is why the carbonyl group undergoes attack from the direction shown in Fig. 19.8 of the text: this is the direction along which the LUMO is concentrated. Notice also how much larger the lobes are on carbon. This is a reflection of the electronegativity of oxygen.

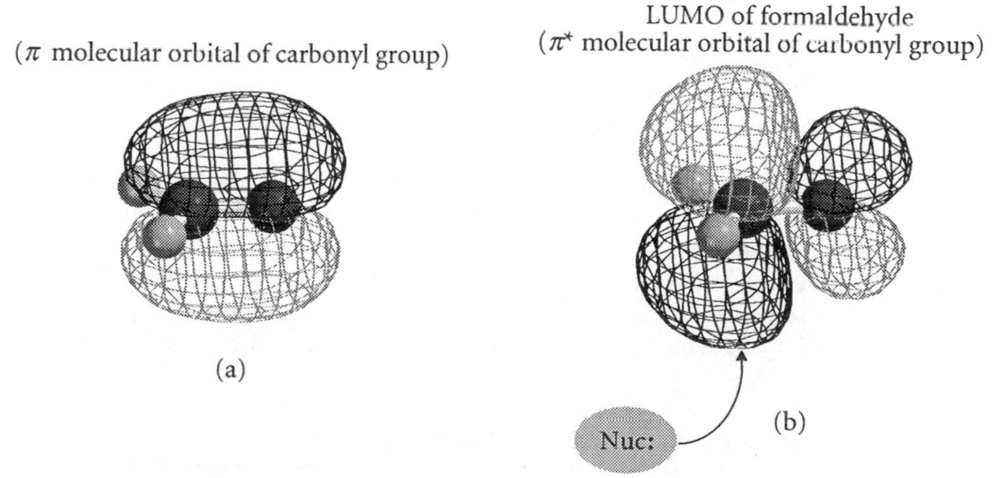

(π molecular orbital of carbonyl group)

(a)

LUMO of formaldehyde
(π^* molecular orbital of carbonyl group)

Nuc:

(b)

Figure SG19.1 (a) The bonding π molecular orbital of formaldehyde, the simplest carbonyl compound. (b) The LUMO of formaldehyde, which is essentially the antibonding (π^*) molecular orbital of the carbonyl group. The preferred direction of attack of the nucleophile is shown. (The nucleophile could also attack from above.) The orbitals are shown in a "wire mesh" representation so that the nuclei are not obscured.

When the LUMO is filled, even with electrons from another molecule, the C═O π bond is weakened. (Remember that when an antibonding molecular orbital is populated, the energetic advantage of bonding disappears.) Thus, attack of the nucleophile on the LUMO weakens the π bond, and that is why the π bond breaks. The energetic "tradeoff" for loss of this bonding is formation of the new bond to the nucleophile.

When the carbonyl group is protonated on oxygen, the LUMO is virtually identical in shape. Hence, attack of nucleophiles on *protonated* carbonyl groups also occurs from the same direction—above or below the plane of the molecule.

✓19.4 Acids and Bases in Reaction Mechanisms

Notice that under acidic conditions, as in the hydration mechanism in the text (Eqs. 19.18a–b, text p. 857), the molecule that acts as a base is *water*, not the hydroxide ion. Although hydroxide ion would be a stronger base, hydroxide is not present in any significant concentration in acidic solution because of the very small ion-product constant of water.

Likewise, in basic solution, in which hydroxide is present in significant concentration, significant amounts of hydronium ions cannot exist; hence, when an acid is required for reactions in basic solution, either the conjugate acid of hydroxide—water—or other weak acids must serve the purpose. (In the case of cyanohydrin formation, HCN serves as the acid, because it is weak enough to exist in mildly alkaline solution.)

To summarize: when we write mechanisms for reactions requiring strong acids we have to use weak bases, because such bases are the only ones that can exist in acidic solution. When we write mechanisms requiring strong bases, we must use weak acids, because these are the only acids that can exist in basic solution. Specifically, if a reaction involves H_3O^+ as an acidic catalyst, the base involved is water, not ^-OH; if a reaction involves ^-OH as a basic catalyst, the acid involved is water, not H_3O^+.

✓19.5 Ground-State Energies and Reactivity

In many cases you have learned to gauge reactivity by applying Hammond's postulate to the relative stability of *reactive intermediates* such as carbocations. The more stable the intermediate, the more reactive a compound is. However, you should remember that reactivity involves the *difference* in two standard free energies: the standard free energy of the transition state and the standard free energy of the reactant (see Fig. 4.11, text p. 143). In some cases, it is the relative free energies of the *reactants* that provide the major effects on reactivity. When one reactant is less stable than another relative to its transition state the compound is more reactive because it is pushed further up the "free-energy hill" towards its transition state (Fig. SG19.2). It often happens that the major effects on the reactivity of carbonyl compounds (not only aldehydes and ketones, but also esters, carboxylic acids, and other carbonyl compounds) come from the relative stabilities of the starting material. Because we'll continue to return to the point in Chapters 21 and 22, it is important for you to understand the relative reactivities of carbonyl compounds in these terms.

✓19.6 Lewis Acid Catalysis

Study Guide Link 16.2 focused on Lewis-acid interactions with leaving groups. What you should notice about both Grignard reagents and lithium aluminum hydride is the important role of Lewis acid catalysis in promoting reactions of these reagents with carbonyl compounds (as well as with epoxides; Sec. 11.4C). Each reagent has its own "built-in" Lewis acid catalyst. As explained in the text, the Lewis acid catalyst in $LiAlH_4$ reductions is the lithium ion, Li^+; and the Lewis acid catalyst in Grignard reactions is the magnesium of the Grignard reagent.

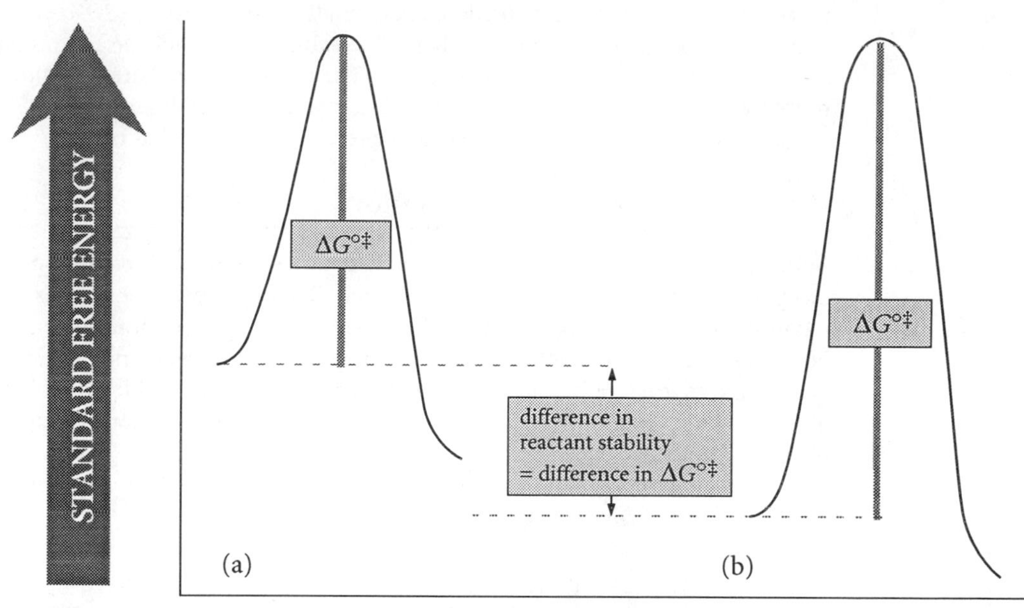

Figure SG19.2 The effect of reactant stability on rates of reactions at a given transition-state stability. (a) The less stable reactant reacts more rapidly. (b) The more stable reactant reacts more slowly. These effects are dominant when reactant stability varies more with structure than transition-state stability.

Why not use proton (Brønsted) acids as catalysts in these reactions? Don't forget that both $LiAlH_4$ and Grignard reagents react instantaneously with protons of even weak acids such as water and alcohols. Consequently, the use of proton acids is not an option, because such acids destroy the reagents. Only *after* the addition of these reagents takes place can a proton source be added tothe reaction mixture to replace the Lewis acid (Li^+ or ^+MgBr) with a hydrogen.

What you should notice here is that the Lewis acid serves the role of a "substitute proton"—or "fat proton," as one of the author's colleagues likes to call it—and fulfills *exactly* the same catalytic role that a proton would fulfill if a proton *could* be used.

✓19.7 Reactions that Form Carbon-Carbon Bonds

It's important to pay special attention to the reactions that can be used to form carbon-carbon bonds, for these are the reactions that are used to build up carbon skeletons. A complete list of these reactions, in the order in which they occur in the text, can be found in Appendix VI. The three reactions for forming carbon-carbon bonds you either have encountered, or will encounter, in this chapter are cyanohydrin formation (Sec. 19.7), reactions of Grignard and related reagents with aldehydes and ketones (Sec. 19.9), and the Wittig alkene synthesis (Sec. 19.13). How many others can you list? Can you give examples of each? Do you understand the limitations of each, if any, and the reasons for these limitations?

✓19.8 Alcohol Syntheses

Study Guide Link 19.7 suggested making special note of reactions that form carbon-carbon bonds. Another approach to reviewing reactions is to classify them by the types of products they give. For example, the reactions of Grignard reagents with aldehydes and ketones, as you've seen, give alcohols as products. The other reaction type you've studied in this chapter that can be used to prepare alcohols is the hydride reduction of aldehydes and ketones (Sec. 19.8). Both reactions are methods for alcohol synthesis of *major* importance. What's more, a great variety of other compounds can be prepared from alcohols. Thus, these methods have importance for the preparation of other types of organic compounds as well.

A complete list of alcohol syntheses, as well as the methods used to prepare every other major functional group, is found in Appendix V. The reactions listed are given in the order that they are presented in the text. How many of these can you name without looking at the list? What are the limitations of each? Do you understand the reasons for the limitations?

✓19.9 Hemiacetal Protonation

Some students ask why, in Eq. 19.47b, text p. 870, the hemiacetal is protonated on the —OH oxygen rather than on the $—OCH_3$ oxygen. This is a reasonable question because ethers and alcohols have similar basicities. The answer is that protonation on the $—OCH_3$ oxygen *does* occur, as does loss of methanol from the resulting protonated species. However, these steps are simply the reverse of hemiacetal formation and lead back to aldehyde or ketone and the starting alcohol. Under conditions of excess alcohol and removal of water, the equilibrium shifts to favor the alcohol.

When writing mechanisms, we usually do not write out reasonable steps that are not important in the formation of the product of interest.

✓19.10 Mechanism of Carbinolamine Formation

You are asked to write the mechanism for acid-catalyzed carbinolamine formation because it is another example of *carbonyl addition.* It is important to understand this reaction mechanistically because you'll see it repeatedly in organic chemistry and biochemistry. If you *really* want to make progress, take these suggestions seriously.

Are you having trouble getting started? In any carbonyl addition, *identify the nucleophile.* The nucleophilic atom is the nitrogen of the amine. *The nucleophile attacks at the carbonyl carbon* because electrons can flow onto the electronegative oxygen. (See Study Guide Link 19.2 on page 656 of this manual.)

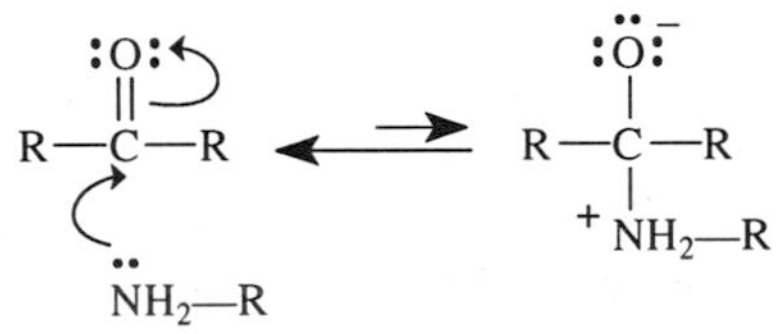

$$R{-}C({=}\ddot{O}{:}){-}R + \ddot{N}H_2{-}R \rightleftharpoons R{-}C({-}\ddot{O}{:}^-)({-}\overset{+}{N}H_2{-}R){-}R \qquad \text{(SG19.1)}$$

Perhaps you added a proton to the carbonyl oxygen first, and *then* let the amine attack. *This is a reasonable step for a beginning student to take,* and follows the mechanism for acid-catalyzed hydration.

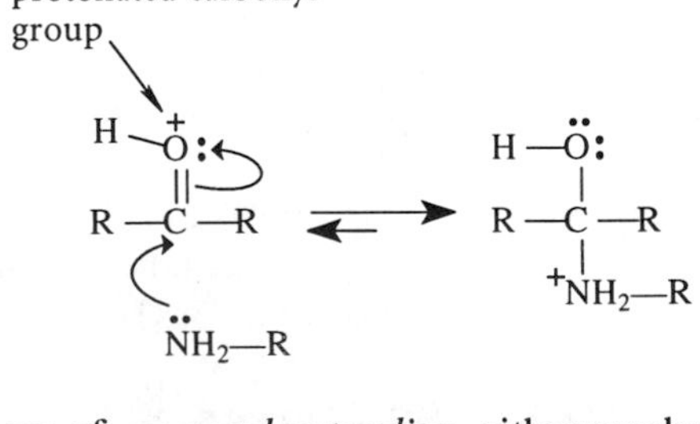

protonated carbonyl group

$$R{-}C({=}\overset{+}{O}{-}H){-}R + \ddot{N}H_2{-}R \rightleftharpoons R{-}C({-}\ddot{O}{-}H)({-}\overset{+}{N}H_2{-}R){-}R$$

Which is correct? For purposes of *your understanding,* either mechanism represents a reasonable first step. However, let's think about this issue in a little more detail. Remember that carbonyl oxygens are *very weak bases* and amine nitrogens are *fairly strong bases*—about like ammonia. If the acid is strong enough to provide a significant concentration of the protonated carbonyl compound, then surely under the same conditions the amine nitrogen would be completely protonated to an ammonium ion (see Eq. 19.57 on text p. 875) and would no longer be nucleophilic. Hence, the first mechanism—attack of the neutral amine on the neutral carbonyl compound (Eq. SG19.1)—is more likely to be correct.

Once attack of the amine has occurred, a proton is transferred *from* H_3O^+ *to* the carbonyl oxygen and *from* the attacking nitrogen to H_2O. Use *two separate steps* for these transfers. It doesn't matter which you write first. Notice that the oxygen, because of its negative charge, is an alkoxide, which is a rather strong base, and is efficiently protonated by rather low concentrations of H_3O^+.

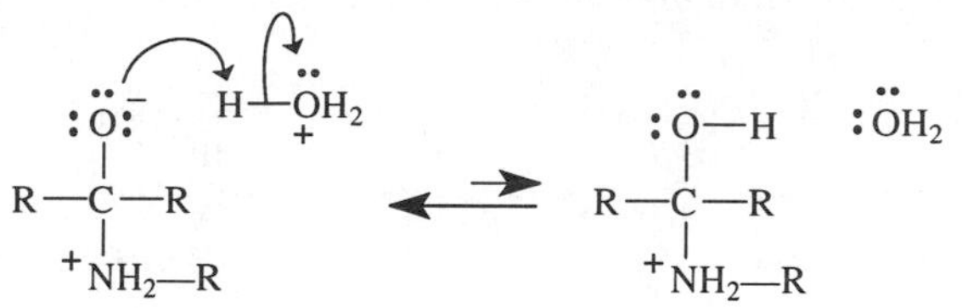

$$R_2C(O^-)(\overset{+}{N}H_2R) + H-\overset{+}{O}H_2 \rightleftharpoons R_2C(OH)(\overset{+}{N}H_2R) + :OH_2 \qquad \text{(SG19.2)}$$

The final step is deprotonation of the nitrogen. Because the nitrogen of the carbinolamine is a base, this final step is an equilibrium. If the acid concentration is high enough, a significant amount of the protonated carbinolamine could be present.

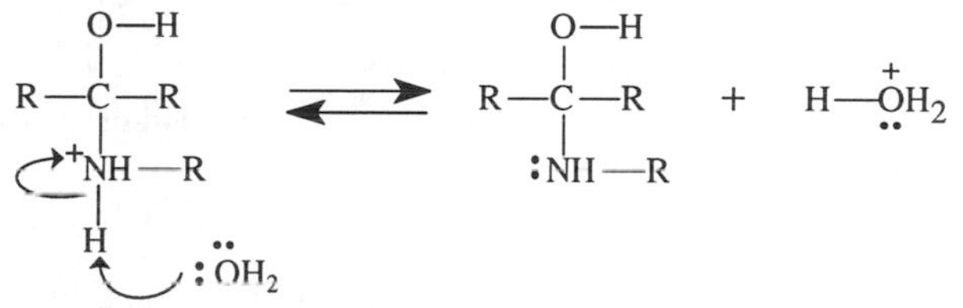

$$R_2C(OH)(\overset{+}{N}H_2R) + :OH_2 \rightleftharpoons R_2C(OH)(:NHR) + H-\overset{+}{O}H_2 \qquad \text{(SG19.3)}$$

Notice that when H_3O^+ is the acid used in the mechanism, then its conjugate base H_2O must be used as the base.

As illustrated here, the mechanism, including the proton-transfer steps, is written *one step at a time.* Students are often tempted to try to show everything in one step. This is not correct because simultaneous collisions of more than two molecules are highly improbable. More important is that consolidating several steps into one can lead to confusion. Remember to write mechanisms *one step at a time.*

Also remember again that all mechanisms involve elementary steps that can only be a Lewis acid-base association, a Lewis acid-base dissociation, or an electron-pair displacement, of which a Brønsted acid-base reaction is a special case. (See Study Guide Link 5.4 on p. 122 of this manual.)

✓19.11 Dehydration of Carbinolamines

You should also try to write the mechanism for dehydration of a carbinolamine in acid, using the dehydration of alcohols as your guide. (Sec. 10.1) When you do this, notice that the carbocation intermediate is resonance-stabilized. This is one reason why the dehydration is so rapid.

Dehydration of carbinolamines, *unlike* dehydration of ordinary alcohols, can also occur in base. In this mechanism, an anion intermediate may be involved:

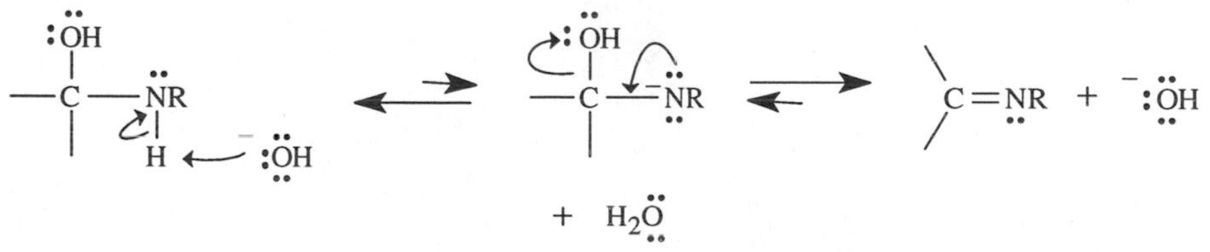

$$-C(OH)-NHR + {}^-:OH \rightleftharpoons -C(OH)-\ddot{N}R^- + H_2O \rightleftharpoons \;>C{=}NR + {}^-:OH$$

Notice that hydroxide ion, ^-OH, acts as a leaving group in the second step of the mechanism. Hydroxide cannot act as a leaving group in the dehydrations of ordinary alcohols because the carbon-oxygen bond is strong, and because hydroxide ion is too basic. In this case, however, the formation of the strong C=N double bond (which is considerably stronger than a C=C double bond) provides an additional driving force for the reaction. In addition, the extra electron pair on the nitrogen anion is very basic (its conjugate-acid pK_a is about 25) and thus provides a strong electronic "push" to expel the leaving group.

✓19.12 Mechanism of the Wolff-Kishner Reaction

The text indicates that the mechanism of the Wolff-Kishner reaction is a series of acid-base reactions. Here they are. Can you supply the curved-arrow notation?

$$R_2C{=}N{-}NH_2 \;\; {}^{-}{:}\ddot{O}H \longrightarrow R_2C{=}N{-}\ddot{N}H^{-} \;\; H{-}\ddot{O}H \longrightarrow R_2C(H){-}N{=}N{-}H \;\; {}^{-}{:}\ddot{O}H \longrightarrow$$

hydrazone → → diazene

$$R_2C(H){-}N{=}\ddot{N}^{-} \;\; H{-}\ddot{O}H \longrightarrow R_2\ddot{C}^{-}{-}H \;\; H{-}\ddot{O}H \;\; + \;\; :N{\equiv}N: \longrightarrow R_2CH_2 \;\; {}^{-}{:}\ddot{O}H$$

a carbanion

Research has suggested that the neutral diazene intermediate may not actually exist—the proton transfers that precede and follow it are concerted (occur together), possibly involving multiple solvent molecules. However, that's a detail.

Notice that the last intermediate is a carbanion. Carbanions, with conjugate-acid pK_a values in the 55–60 range, are generally too unstable to exist. If this one does really exist, it is probably because a very stable molecule (dinitrogen) is produced along with it. That is, the stability of dinitrogen offsets the instability of the carbanion.

SOLUTIONS TO PROBLEMS

Solutions to In-Text Problems

19.1 (a) $(CH_3)_2CHCH{=}O$

(b) O ‖ $CCH_2CH_2CH_2CH_3$ (on a benzene ring)

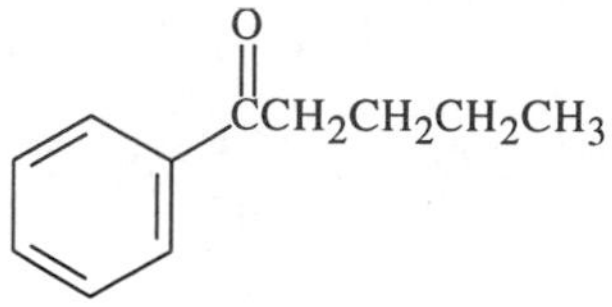

(c) Br, O ‖ C, CH_3 (on a benzene ring)

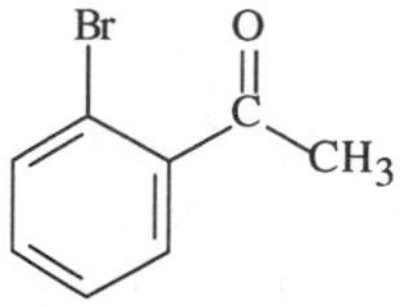

(d) O ‖ $ClCH_2CH_2CH_2CH$

(e) O ‖ CH_3CCHCH_3 with OH

(f) O ‖ CH_3CH_2CHC with Cl —(benzene ring)— O ‖ CH

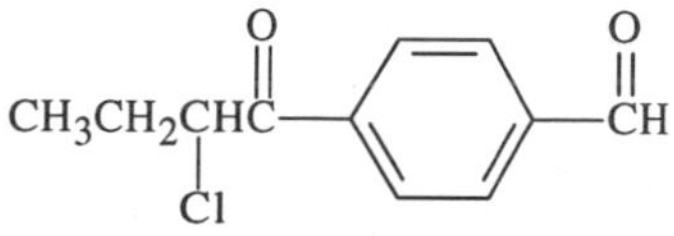

(g) O (cyclohexenone ring)

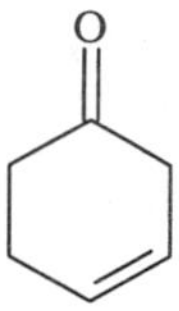

(h) O, O ‖ CH (cyclopentanone ring)

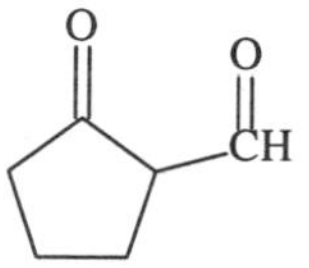

19.2 (a) 2-propanone
(b) 2,4-dimethyl-3-pentanone
(c) 3-allyl-2,4-pentanedione
(d) (*E*)-3-ethoxy-2-propenal
(c) 3-methylcyclobutanecarbaldehyde
(f) 4,4-dimethyl-2,5-cyclohexadienone

19.3 (a) The aldehyde has a somewhat higher carbonyl stretching frequency, and has a C—H stretching absorption near 2700 cm^{-1} that is not present in the IR spectrum of the ketone.

(b) 2-Cyclohexenone has a lower carbonyl stretching frequency because its two double bonds are conjugated.

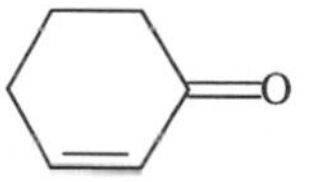

2-cyclohexenone has conjugated double bonds

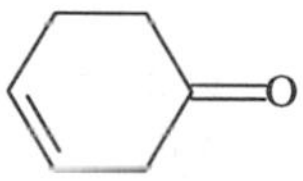

3-cyclohexenone

(c) 3-Buten-2-ol has both O—H and C═C stretching absorptions and 2-butanone does not; 2-butanone has a carbonyl stretching absorption, and 3-buten-2-ol does not.

19.4 (a) The IR spectrum suggests an aldehyde (carbonyl stretch at 1720 cm^{-1} and aldehyde C—H stretch at 2710 cm^{-1} and the low-field absorption at δ 9.65 confirms the presence of the aldehyde hydrogen. The six-proton doublet at δ 1.2 and the septet at δ 2.4 indicate an isopropyl group $(CH_3)_2CH$—. Because the aldehyde carbon is the only remaining carbon, the structure is completely determined:

O ‖ $(CH_3)_2CH—CH$ **isobutyraldehyde (2-methylpropanal)**

Notice that the CH of the isopropyl group and the aldehyde hydrogen split each other slightly.

(b) The compound is 2-butanone:

O ‖ $CH_3CCH_2CH_3$ **2-butanone**

(c) The high frequency of the carbonyl absorption suggests a strained ring. (See Eq. 19.4). In fact, cyclobutanone matches the IR stretching frequency perfectly and the NMR fits as well:

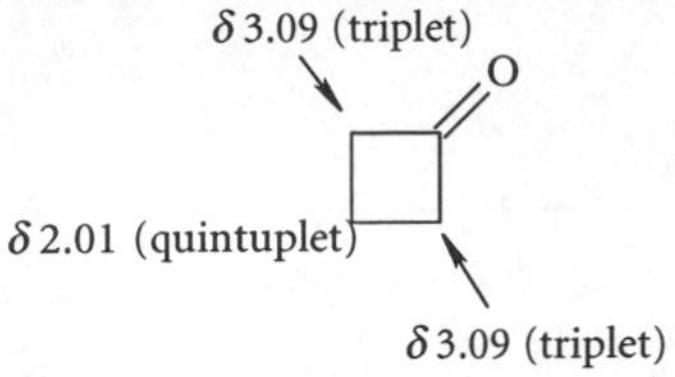

δ 3.09 (triplet)

δ 2.01 (quintuplet)

δ 3.09 (triplet)

cyclobutanone

molecular mass = 70.1

(d) The molecular formula indicates an unsaturation number of 5. The IR spectrum suggests a conjugated carbonyl group and an aromatic ring, and these two features fully account for the unsaturation. The NMR indicates the following partial structures:

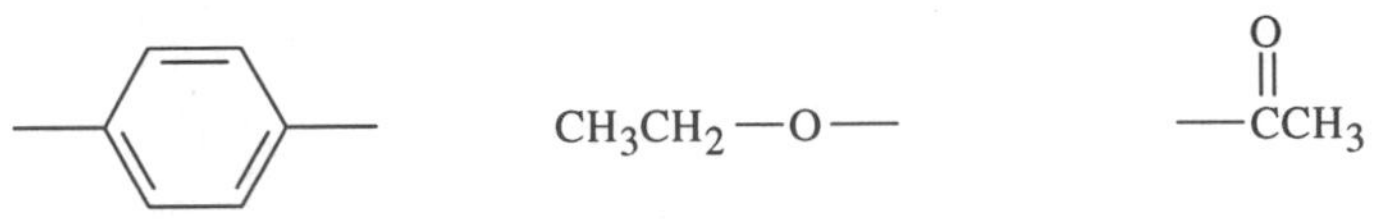

CH_3CH_2—O—   —C(=O)CH_3

and these define the structure as

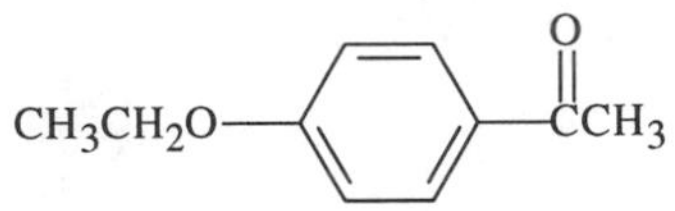

CH_3CH_2O—C₆H₄—C(=O)CH_3 *p*-ethoxyacetophenone

19.5 The low intensities of the carbonyl carbon absorption at δ 212.6 and the α-carbon absorption at δ 44.2 show that these two carbons have no attached hydrogens; therefore this compound must have the following partial structure:

C—C(=O)—C(C)(C)—C

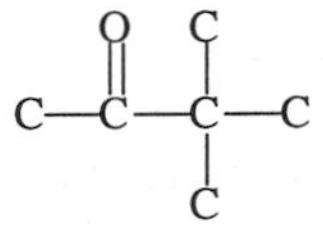

Because all six carbons and their connectivities are accounted for, the only possible choice is to add the appropriate number of hydrogens. The compound is 3,3-dimethyl-2-butanone:

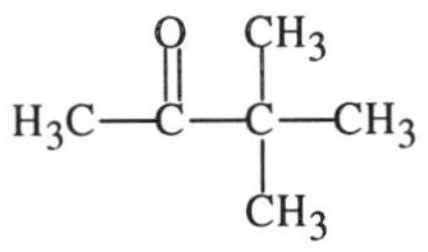

H_3C—C(=O)—C(CH_3)(CH_3)—CH_3 3,3-dimethyl-2-butanone

19.6 The structure and CMR assignments of 2-ethylbutanal are shown below. Note that the two methyl groups are chemically equivalent, and the two methylene groups are chemically equivalent; all carbons with different CMR chemical shifts are chemically nonequivalent.

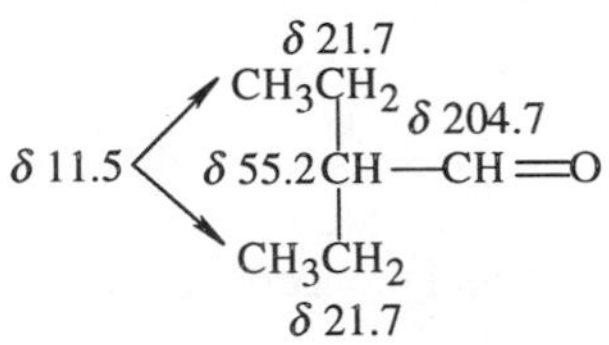

δ 21.7 CH_3CH_2

δ 11.5 (both CH_3) δ 55.2 CH—CH=O δ 204.7 2-ethylbutanal

CH_3CH_2 δ 21.7

19.7 (a) The double bonds in 2-cyclohexenone are conjugated, but the double bonds in 3-cyclohexenone are not. Consequently, 2-cyclohexenone has the UV spectrum with the greater λ_{max}.

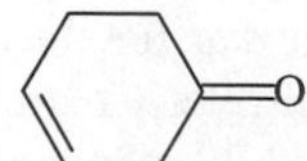

2-cyclohexenone
(conjugated; has UV spectrum with the greater λ_{max})

3-cyclohexenone

(b) The first compound has more conjugated double bonds and therefore will have the UV spectrum with the greater λ_{max} and larger extinction coefficient ϵ.

(c) In 1-phenyl-2-propanone the aromatic ring is not conjugated with the carbonyl group; in *p*-methylacetophenone the aromatic ring and the carbonyl group are conjugated. Because of its additional conjugation, *p*-methylacetophenone has the UV spectrum with the greater λ_{max} and larger extinction coefficient ϵ.

CH_2CCH_3 (O above C)

1-phenyl-2-propanone

H_3C— CCH_3 (O above C)

p-methylacetophenone
(has the UV spectrum with the greater ϵ and λ_{max})

19.8 Addition of NaOH brings about the ionization of the phenolic —OH group of *p*-hydroxyacetophenone. Because the electron pair associated with the anion can be delocalized into the ring and into the carbonyl group, this conjugated species has a unique UV spectrum that is different from that of the un-ionized compound. Because *p*-methoxyacetophenone does not ionize in NaOH solution, its UV spectrum is not affected by addition of base.

HO— —C—CH_3 (:O: above C) $\xrightarrow{^-OH}$ [:O$^-$— —C—CH_3 $\longleftrightarrow$ O= =C—CH_3 (:O:$^-$ above C)]

p-hydroxyacetophenone

electrons associated with negative charge are delocalized;
electron delocalization is reflected in the UV spectrum as a greater λ_{max}.

CH_3O— —C—CH_3 (O above C)

p-methoxyacetophenone
cannot ionize;
base has no effect on the UV spectrum

19.9 (a) In both ketones, inductive cleavage of the molecular ion to give an acyl radical and a carbocation accounts for the $m/z = 57$ peak:

$CH_3\dot{C}$—$CH_2CH_2CH_2CH_3$ (:O:$^+$ above C) ⟶ $CH_3\dot{C}$ (:O: above C) + $^+CH_2CH_2CH_2CH_3$

molecular ion of 2-hexanone

butyl cation
($m/z = 57$)

$CH_3\dot{C}$—$C(CH_3)_3$ (:O:$^+$ above C) ⟶ $CH_3\dot{C}$ (:O: above C) + $^+C(CH_3)_3$

molecular ion of 3,3-dimethyl-2-butanone

tert-butyl cation
($m/z = 57$)

Because the carbocation formed from 3,3-dimethyl-2-butanone is more stable (why?), more of this fragmentation mode is observed for 3,3-dimethyl-2-butanone; that is, the $m/z = 57$ peak is more abundant.

(b) Its even mass suggests that the $m/z = 58$ fragment is an odd-electron ion. Such an ion can be readily produced by a McLafferty rearrangement from the molecular ion of 2-hexanone, but not from the molecular ion of 3,3-dimethyl-2-butanone:

$$CH_3C(=\dot{O}:)CH_2CH_2CHCH_3(H) \longrightarrow CH_3C(=\overset{+}{O}H)-\dot{C}H_2 + H_2C{=}CHCH_3$$

molecular ion of 2-hexanone; $m/z = 58$

19.10 The mass spectrum of 2-heptanone should have major peaks at $m/z = 43$ (from α-cleavage), 71 (from inductive cleavage), and 58 (from McLafferty rearrangement). The mass spectrum of 3-heptanone should have a major peak resulting from both inductive cleavage and α-cleavage at $m/z = 57$ and a major peak resulting from McLafferty rearrangement at $m/z = 72$. Notice that the position of the even-mass, odd-electron ion is a major distinguishing feature. We leave it to you to draw out these fragmentations.

$$H_3C-C(=O)\,\wr\,CH_2CH_2CH_2CH_2CH_3 \quad (43 \mid 71)$$

2-heptanone
McLafferty rearrangement at $m/z = 58$

$$CH_3CH_2-C(=O)\,\wr\,CH_2CH_2CH_2CH_3 \quad (57 \mid 57)$$

3-heptanone
McLafferty rearrangement at $m/z = 72$

19.11 (a) The solvolysis mechanism of (chloromethoxy)methane in ethanol:

$$CH_3\ddot{O}-CH_2-\ddot{Cl}: \longrightarrow \left[CH_3\ddot{O}-\overset{+}{C}H_2 \longleftrightarrow CH_3\overset{+}{O}{=}CH_2\right] :\ddot{Cl}:^- \xrightarrow{H\ddot{O}-C_2H_5}$$

(chloromethoxy)methane

$$CH_3\ddot{O}-CH_2-\overset{+}{O}(H)C_2H_5 \xrightarrow{H\ddot{O}-C_2H_5} CH_3\ddot{O}-CH_2-\ddot{O}C_2H_5 + H_2\overset{+}{O}-C_2H_5$$

(b) (Chloromethoxy)methane undergoes more rapid solvolysis than 1-chlorobutane because the carbocation intermediate is resonance-stabilized [as shown in the mechanism of part (a)] and is therefore more stable relative to starting material than the ordinary primary carbocation intermediate derived from 1-chlorobutane. (In fact, 1-chlorobutane probably solvolyzes by an S_N2 mechanism; the S_N1 reaction would no doubt be even slower than the reaction indicated in the problem.)

19.12 The products follow from a consideration of the mechanism in Study Problem 19.2 on text p. 853. The first steps of this mechanism consist of protonation of an —OH group and loss of water to form a carbocation intermediate. When the two —OH groups are chemically nonequivalent, the —OH group lost is the one that gives the more stable carbocation intermediate.

(a) The reaction is exactly like the one shown in Study Problem 19.2 with phenyl instead of methyl substituents. The product is the following ketone:

$$Ph_3C-C(=O)-Ph$$

(b) The —OH group of the tertiary alcohol is lost because a relatively stable tertiary carbocation is formed.

$$\mathrm{H_3C{-}\overset{OH}{\underset{CH_3}{C}}{-}CH_2OH} \xrightarrow{\text{protonation, }-\mathrm{H_2O}} \mathrm{H_3C{-}\overset{+}{\underset{CH_3}{C}}{-}CH_2OH} \xrightarrow{\text{rearrangement}} \mathrm{H_3C{-}\overset{H}{\underset{CH_3}{C}}{-}CH{=}O}$$

(c) In this case, rearrangement involves a ring expansion. If reasoning through this type of rearrangement gives you difficulty, be sure to consult Study Guide Link 10.2 (p. 307 of this manual) for assistance. The product is the following ketone:

O

19.13 (a) Protonated *p*-methoxybenzaldehyde is essentially a benzylic carbocation; as you learned in Chapter 17, benzylic carbocations are stabilized by substituents that can donate electrons by resonance, and this cation is no exception; the resonance effect of the *p*-methoxy group stabilizes the cationic conjugate acid:

$$\left[\mathrm{CH_3\ddot{O}{-}C_6H_4{-}CH{=}\overset{+}{O}H} \longleftrightarrow \mathrm{CH_3\ddot{O}{-}C_6H_4{-}\overset{+}{C}H{-}\ddot{O}H} \longleftrightarrow \mathrm{CH_3\overset{+}{O}{=}C_6H_4{=}CH{-}\ddot{O}H}\right]$$

conjugate acid of *p*-methoxybenzaldehyde

Stabilization of the conjugate acid of any compound *increases* the compound's basicity. The *p*-nitro group *destabilizes* the corresponding cation that results from protonation of *p*-nitrobenzaldehyde. (See the solution to Problem 17.22 on pp. 564 of this manual, for a very similar case.) Destabilization of the conjugate acid of a compound *decreases* the compound's basicity.

(b) The conjugate acid of 3-buten-2-one has more important resonance structures than the conjugate acid of 2-butanone and is therefore more stable relative to unprotonated ketone than the conjugate acid of 2-butanone. Greater stability of the conjugate acid means that the ketone is more basic.

$$\left[\mathrm{H_2C{=}CH{-}C(\overset{+}{O}H){-}CH_3} \longleftrightarrow \mathrm{H_2C{=}CH{-}\overset{+}{C}(\ddot{O}H){-}CH_3} \longleftrightarrow \mathrm{H_2\overset{+}{C}{-}CH{=}C(\ddot{O}H){-}CH_3}\right]$$

conjugate acid of 3-buten-2-one

19.14 The mechanism of the hydroxide-catalyzed hydration of acetaldehyde:

$$\mathrm{CH_3CH{=}\ddot{O}{:}} + {:}\ddot{O}H^- \rightleftharpoons \mathrm{CH_3CH(\ddot{O}{:}^-)(\ddot{O}H)} \xrightarrow{\mathrm{H{-}\ddot{O}H}} \mathrm{CH_3CH(\ddot{O}H)(\ddot{O}H)} + {:}\ddot{O}H^-$$

19.15 (a) The acid-catalyzed addition of methanol involves mechanistic steps like those involved in acid-catalyzed hydration (Eqs. 19.18a–b on text p. 857). The carbonyl group is protonated, and the carbonyl carbon is then attacked by methanol.

$$\underset{\text{PhCH}}{\overset{\ddot{\text{O}}:}{\|}} \;+\; \text{H}{-}\overset{+}{\ddot{\text{O}}}\text{CH}_3(\text{H}) \rightleftharpoons \text{Ph}\overset{+\ddot{\text{O}}\text{H}}{\text{CH}} + :\ddot{\text{O}}\text{CH}_3(\text{H}) \rightleftharpoons \text{PhCH}(\ddot{\text{O}}\text{H})(\overset{+}{\ddot{\text{O}}}\text{CH}_3\text{H}) \overset{\text{H}\ddot{\text{O}}\text{CH}_3}{\rightleftharpoons} \text{PhCH}(\ddot{\text{O}}\text{H})(\ddot{\text{O}}\text{CH}_3) + \text{H}\overset{+}{\ddot{\text{O}}}\text{CH}_3(\text{H})$$

(b) The methoxide-catalyzed addition of methanol to benzaldehyde is very similar to the hydroxide-catalyzed hydration mechanism (Problem 19.14):

$$\text{PhCH}{=}\ddot{\text{O}}: + {}^{-}:\ddot{\text{O}}\text{CH}_3 \rightleftharpoons \text{PhCH}(:\ddot{\text{O}}:^{-})(:\ddot{\text{O}}\text{CH}_3) + \text{H}{-}\ddot{\text{O}}\text{CH}_3 \rightleftharpoons \text{PhCH}(:\ddot{\text{O}}\text{H})(:\ddot{\text{O}}\text{CH}_3) + {}^{-}:\ddot{\text{O}}\text{CH}_3$$

Be sure to read Study Guide Link 19.4, which deals with the role of acids and bases in mechanisms. We typically invoke as a base the conjugate base of whatever acid is involved in the mechanism. Thus, if H_3O^+ is involved in a mechanism, its conjugate base (H_2O) acts as the base. We would *not* invoke both H_3O^+ and ^{-}OH in the same mechanism because a strong acid and a strong base cannot coexist in solution.

19.16 The data in Table 19.2 of the text show that hydration of benzaldehyde is less than 0.01 times as favorable as hydration of an unconjugated aliphatic aldehyde such as acetaldehyde. The assumption that the same principles apply to cyanohydrin formation leads to the prediction that propanal should have the greater proportion of cyanohydrin at equilibrium. The structure of this cyanohydrin is as follows:

$$CH_3CH_2CH(OH)C{\equiv}N \qquad \text{cyanohydrin derivative of propanal}$$

19.17 The central carbonyl group is hydrated for the following reason. The carbon of any carbonyl group bears a partial positive charge. The central carbonyl group is therefore bonded to two partially positive carbonyl carbons, and the positive charge on each of these carbons has an unfavorable repulsive interaction with the partial positive charge on the central carbon. In addition, the other two carbonyl groups are conjugated with the benzene ring. As explained in the text, adjacent electronegative groups make carbonyl-addition reactions more favorable, and groups that are conjugated with the carbonyl group make carbonyl-addition reactions less favorable. It follows that hydration of the central carbonyl group is the more favorable process. The structure of the hydrate is as follows:

O, OH, OH, O — hydrate of ninhydrin

19.18 We use the same principles to predict reactivity that we use to predict relative equilibrium constants. The more prone a carbonyl compound is to form an addition product, the more reactive it is.

(a) The second compound, bromoacetone, is more reactive because the electronegative atom is closer to the carbonyl group.

(b) The first compound, 2,3-butanedione, is more reactive, because the partial positive charge on one carbonyl destabilizes the molecule by its repulsive interaction with the partial positive charge on the other. (See the solution to Problem 19.17 for a similar situation.)

(c) The first compound, *p*-nitrobenzaldehyde, is more reactive, because the *p*-nitro group raises the energy of the molecule by an unfavorable interaction of the positive charge on the nitrogen with the positive charge on the carbonyl carbon. In contrast, the *p*-methoxy group in the second compound, *p*-methoxybenzaldehyde, stabilizes the molecule by a resonance interaction with the positive charge on the carbonyl carbon.

$CH_3\ddot{O}-C_6H_4-CH{=}O \longleftrightarrow CH_3\ddot{O}-C_6H_4-\overset{+}{C}H-O^- \longleftrightarrow CH_3\overset{+}{O}{=}C_6H_4{=}CH-O^-$

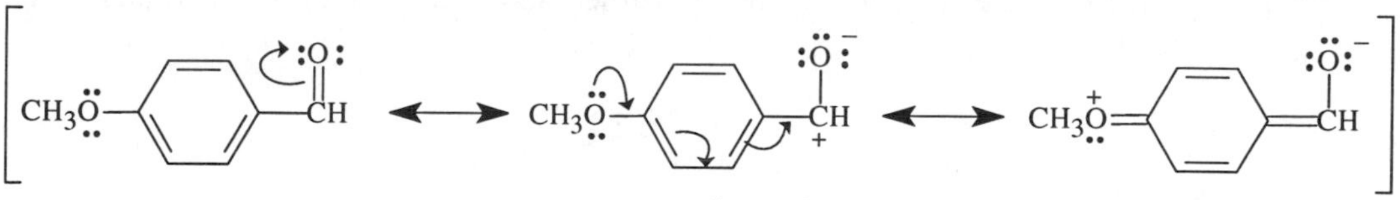

some resonance structures of *p*-methoxybenzaldehyde

(d) The second compound, cyclopropanone, is more reactive for the following reason. A carbonyl compound is most stable when the bond angles at the carbonyl group can be close to 120°; however, in cyclopropanone they are constrained by the three-membered ring to be 60°. As a result, there is considerable strain in the carbonyl form. In the hydrate, the preferred bond angle is approximately tetrahedral (109.5°). Because this angle is closer to the 60° bond angle enforced by the ring than the bond angle in cyclopropanone is, there is less strain in the hydrate than there is in cyclopropanone. Hence, formation of the hydrate relieves some of the strain in cyclopropanone. In other words, the ketone is destabilized relative to the hydrate by strain; there is a driving force to form the hydrate. Because there is little or no strain in cyclopentanone, such effects do not come into play.

19.19 (a) cyclopentyl—CH═O

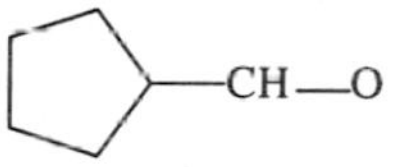

cyclopentanecarbaldehyde

(b) $CH_3\overset{O}{\overset{\|}{C}}CH_2CH_3$

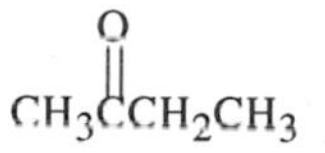

2-butanone

(c) [bicyclic structure with OH at ring-fusion carbon and C═O]

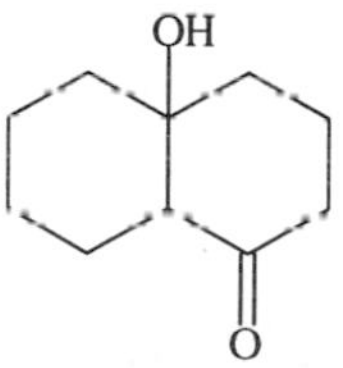

An extra equivalent of $LiAlH_4$ would have to be used with the compound in (c), because the hydride would react vigorously with the tertiary alcohol to form the conjugate-base alkoxide. The tertiary alcohol would be re-formed when a weak acid is added to the reaction mixture. $NaBH_4$, on the other hand, would reduce the compound shown in (c) without this extra complication.

19.20 The tertiary alcohol *C* could not be synthesized by a hydride reduction, because only primary and secondary alcohols can be prepared by this method.

19.21 In each case, ethyl bromide, CH_3CH_2Br, reacts with Mg to give ethylmagnesium bromide, CH_3CH_2MgBr, which is then allowed to react as shown below.

(a)

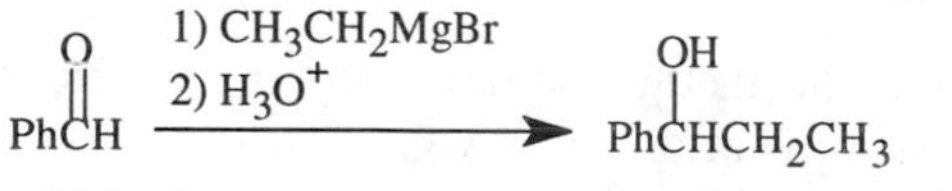

$$Ph\overset{O}{\overset{\|}{C}}H \xrightarrow[2)\ H_3O^+]{1)\ CH_3CH_2MgBr} Ph\overset{OH}{\overset{|}{C}}HCH_2CH_3$$

benzaldehyde → 1-phenyl-1-propanol

(b)

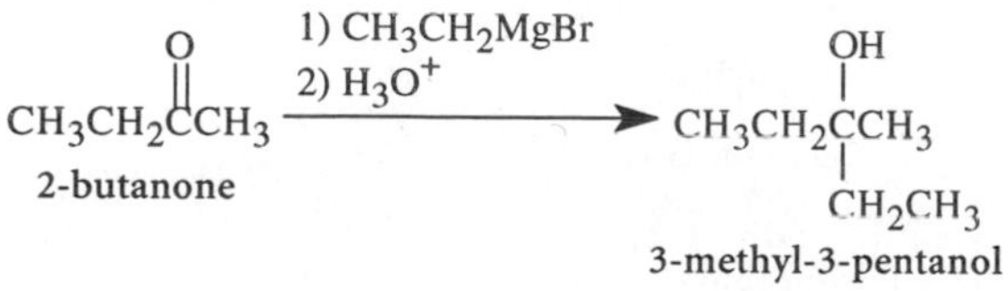

$$CH_3CH_2\overset{O}{\overset{\|}{C}}CH_3 \xrightarrow[2)\ H_3O^+]{1)\ CH_3CH_2MgBr} CH_3CH_2\underset{CH_2CH_3}{\underset{|}{\overset{OH}{\overset{|}{C}}}}CH_3$$

2-butanone → 3-methyl-3-pentanol

(c)

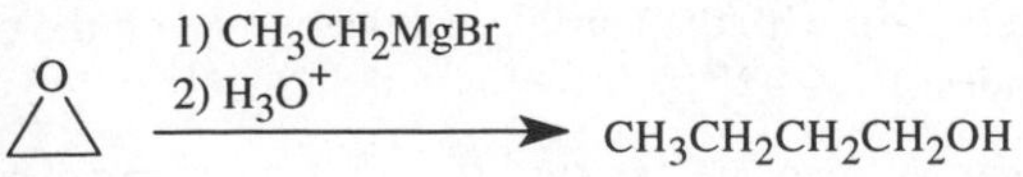

$$\text{ethylene oxide (oxirane)} \xrightarrow[\text{2) } H_3O^+]{\text{1) } CH_3CH_2MgBr} CH_3CH_2CH_2CH_2OH \text{ (1-butanol)}$$

Don't forget what you've already learned about the reactions of Grignard reagents; see Sec. 11.4C.

(d)

$$\underset{\text{formaldehyde}}{H_2C{=}O} \xrightarrow[\text{2) } H_3O^+]{\text{1) } CH_3CH_2MgBr} \underset{\text{1-propanol}}{CH_3CH_2CH_2OH} \xrightarrow{CrO_3(\text{pyridine})_2} \underset{\text{propanal}}{CH_3CH_2CH{=}O}$$

(e)

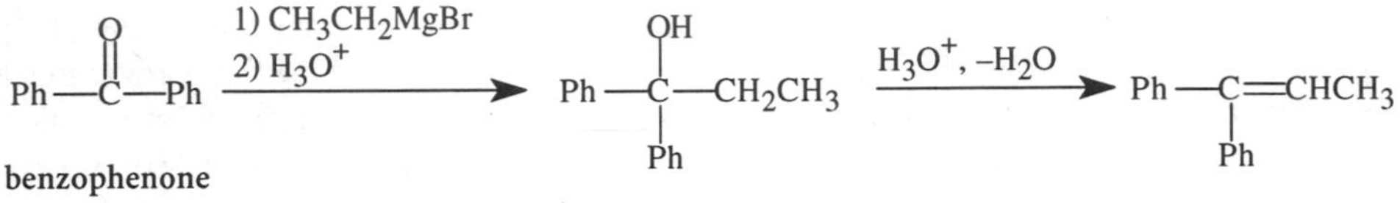

$$\underset{\text{benzophenone}}{Ph{-}C({=}O){-}Ph} \xrightarrow[\text{2) } H_3O^+]{\text{1) } CH_3CH_2MgBr} \underset{\text{1,1-diphenyl-1-propanol}}{Ph{-}C(OH)(Ph){-}CH_2CH_3} \xrightarrow{H_3O^+,\ -H_2O} \underset{\text{(1-phenyl-1-propenyl)benzene}}{Ph{-}C(Ph){=}CHCH_3}$$

(f)

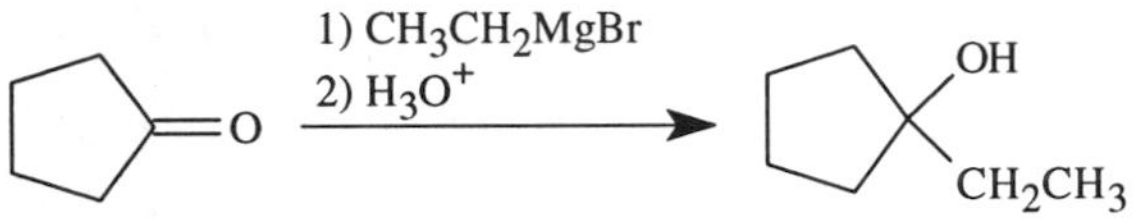

$$\text{cyclopentanone} \xrightarrow[\text{2) } H_3O^+]{\text{1) } CH_3CH_2MgBr} \text{1-ethyl-1-cyclopentanol (cyclopentane ring bearing OH and } CH_2CH_3\text{)}$$

19.22 Either alkyl group bound to the α-carbon of the alcohol can in principle originate from the Grignard reagent.

Synthesis #1:

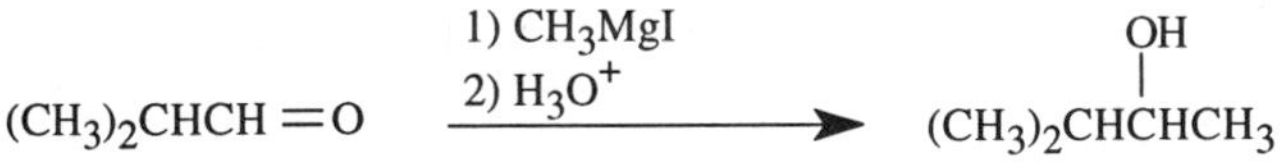

$$\underset{\text{2-methylpropanal}}{(CH_3)_2CHCH{=}O} \xrightarrow[\text{2) } H_3O^+]{\text{1) } CH_3MgI} \underset{\text{3-methyl-2-butanol}}{(CH_3)_2CHCH(OH)CH_3}$$

Synthesis #2:

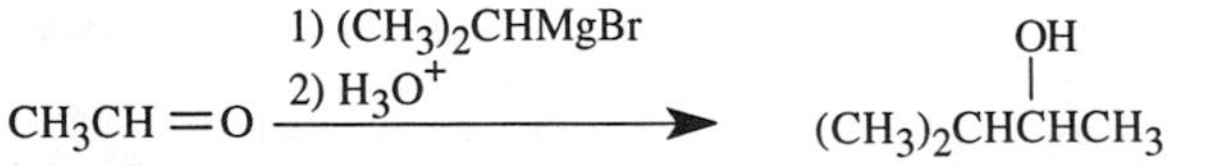

$$\underset{\text{acetaldehyde}}{CH_3CH{=}O} \xrightarrow[\text{2) } H_3O^+]{\text{1) } (CH_3)_2CHMgBr} \underset{\text{3-methyl-2-butanol}}{(CH_3)_2CHCH(OH)CH_3}$$

In synthesis #1, the Grignard reagent is prepared from CH_3I (methyl iodide) + Mg; in synthesis #2, the Grignard reagent is prepared from $(CH_3)_2CH{-}Br$ (isopropyl bromide) + Mg.

19.23 (a)

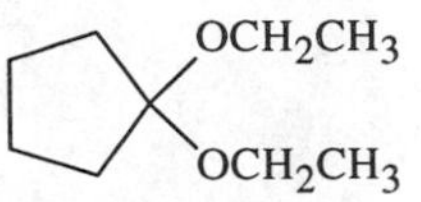

Cyclopentane ring bearing two OCH_2CH_3 groups on one carbon

cyclopentanone diethyl acetal
(1,1-diethoxycyclopentane)

(b)

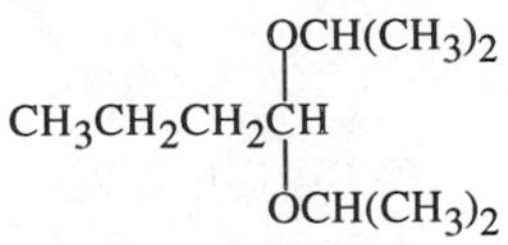

$$CH_3CH_2CH_2CH(OCH(CH_3)_2)_2$$

butyraldehyde diisopropyl acetal
(1,1-diisopropoxybutane)

Note that acetals are named as ethers in the IUPAC system. In the acetals derived from ketones in the common system were once called *ketals.* Thus, in older common nomenclature, the compound in the solution to part (a) would have been called *cyclopentanone diethyl ketal.*

19.24 In each case, a diol reacts with an aldehyde or ketone.

(a)

$$\text{(1,1-bis(hydroxymethyl)cyclohexane)} + H_2C{=}O \xrightarrow{\text{acid}} \text{(spiro 1,3-dioxane)} + H_2O$$

(b)

$$\text{cyclohexanone} + HOCH_2CH_2CH_2OH \xrightarrow{\text{acid}} \text{(spiro 1,3-dioxane)} + H_2O$$

19.25 (a) The formula indicates addition of two carbon atoms and it indicates one degree of unsaturation. If a diethyl acetal were formed and the ring were opened, the formula of the product would be $C_9H_{20}O_3$. The one degree of unsaturation suggests that the ring is intact. The product is a mixed acetal:

$$\text{2-ethoxytetrahydropyran } (C_2H_5O,\ H \text{ on C-2})$$

(b) When five- and six membered rings can be formed, diols generally react with aldehydes and ketones to give cyclic acetals, and this case is no exception:

$$\text{(cyclohexane-fused 1,3-dioxane with } C(CH_3)_2\text{)}$$

19.26 We protect the ketone as an acetal, and then prepare the Grignard reagent, which is allowed to react with acetaldehyde. The protonolysis step produces the alcohol and removes the acetal protecting group. (In the following synthesis, any common alcohol or ethylene glycol can be used instead of methanol.)

$$Br-C_6H_4-\overset{O}{\overset{\|}{C}}CH_3 \xrightarrow{CH_3OH\text{ (solvent), acid}} Br-C_6H_4-C(OCH_3)_2CH_3 \xrightarrow{\text{Mg, ether}}$$

p-bromoacetophenone

$$BrMg-C_6H_4-C(OCH_3)_2CH_3 \xrightarrow[2)\ H_3O^+]{1)\ CH_3CH{=}O} CH_3\overset{OH}{\overset{|}{C}H}-C_6H_4-\overset{O}{\overset{\|}{C}}CH_3 + CH_3OH \xrightarrow{CrO_3(\text{pyridine})_2} CH_3\overset{O}{\overset{\|}{C}}-C_6H_4-\overset{O}{\overset{\|}{C}}CH_3$$

Note that this compound could not be made by a Friedel-Crafts acylation of acetophenone. (Why?) The reaction sequence shown has maneuvered two meta-directing groups into a para relationship!

19.27 (a) (b) 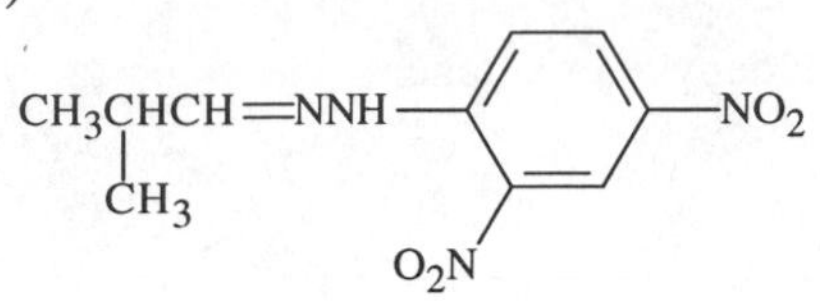 (c)

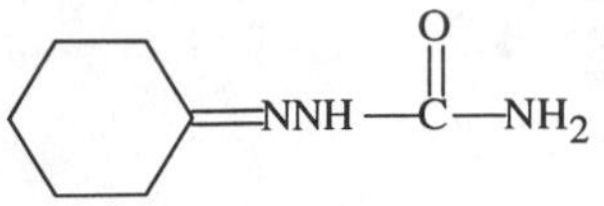

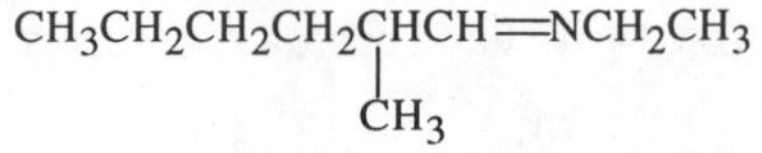

(a) cyclohexylidene $=NNH-\overset{O}{\overset{\|}{C}}-NH_2$

(b) $CH_3\underset{CH_3}{\underset{|}{C}}HCH{=}NNH-$(2,4-dinitrophenyl; O_2N, $-NO_2$)

(c) $CH_3CH_2CH_2CH_2\underset{CH_3}{\underset{|}{C}}HCH{=}NCH_2CH_3$

19.28 First, the carbinolamine intermediate is formed. This intermediate then undergoes acid-catalyzed dehydration to give the hydrazone. (Be sure to consult Study Guide Link 19.10 on p. 660 of this manual.)

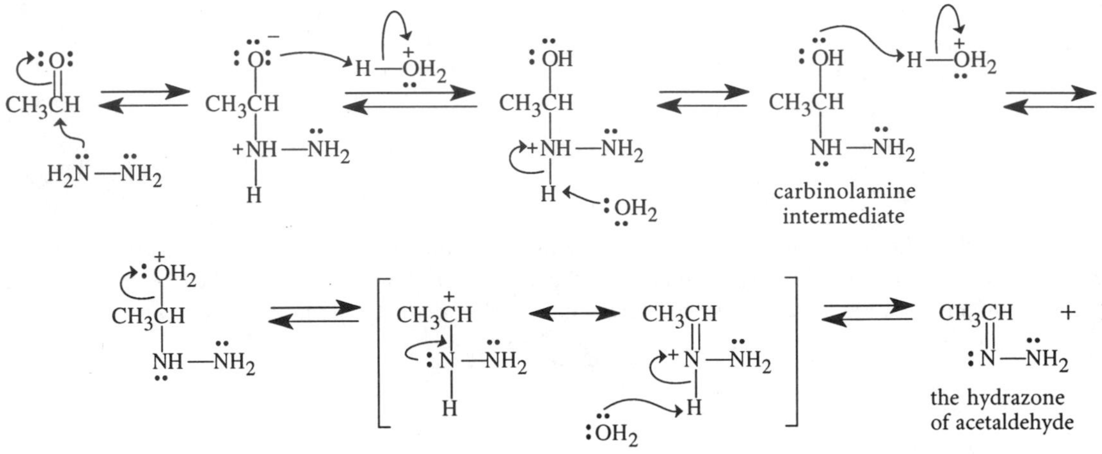

$CH_3CH{=}O$ + H_2N-NH_2 ⇌ $CH_3CH(O^-)-\overset{+}{N}H_2-NH_2$ ⇌ ($H-\overset{+}{O}H_2$) $CH_3CH(OH)-\overset{+}{N}H_2-NH_2$ ⇌ ($:OH_2$) $CH_3CH(OH)-NH-NH_2$ (carbinolamine intermediate) ⇌ ($H-\overset{+}{O}H_2$)

$CH_3CH(\overset{+}{O}H_2)-NH-NH_2$ ⇌ [$CH_3\overset{+}{C}H-NH-NH_2$ ⟷ $CH_3CH{=}\overset{+}{N}H-NH_2$] ($:OH_2$) ⇌ $CH_3CH{=}N-NH_2$ + $H-\overset{+}{O}H_2$

the hydrazone of acetaldehyde

Note that it is equally appropriate to write the loss of water and formation of the carbon-nitrogen double bond as one step, thus avoiding the necessity of drawing resonance structures:

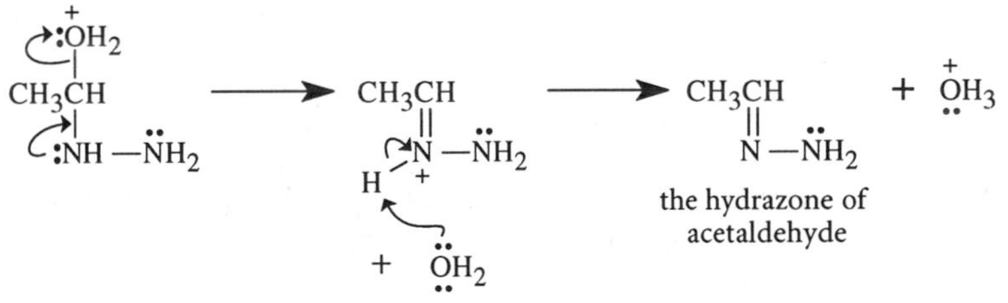

$CH_3CH(\overset{+}{O}H_2)-NH-NH_2$ ⟶ $CH_3CH{=}\overset{+}{N}H-NH_2$ + OH_2 ⟶ $CH_3CH{=}N-NH_2$ + $\overset{+}{O}H_3$

the hydrazone of acetaldehyde

19.29 Imine hydrolysis is the reverse of imine formation. Therefore, retrace the steps of imine formation as illustrated in the solution to Problem 19.28 starting with the imine and working backwards to the aldehyde and the amine.

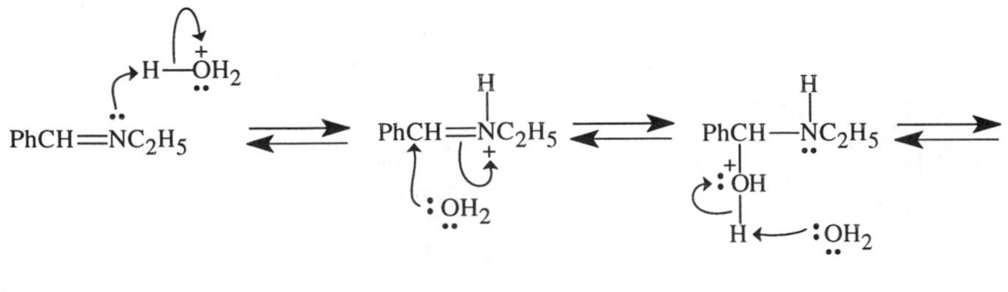

$PhCH{=}NC_2H_5$ ($H-\overset{+}{O}H_2$) ⇌ $PhCH{=}\overset{+}{N}HC_2H_5$ ($:OH_2$) ⇌ $PhCH(\overset{+}{O}H_2)-NHC_2H_5$ ($:OH_2$) ⇌

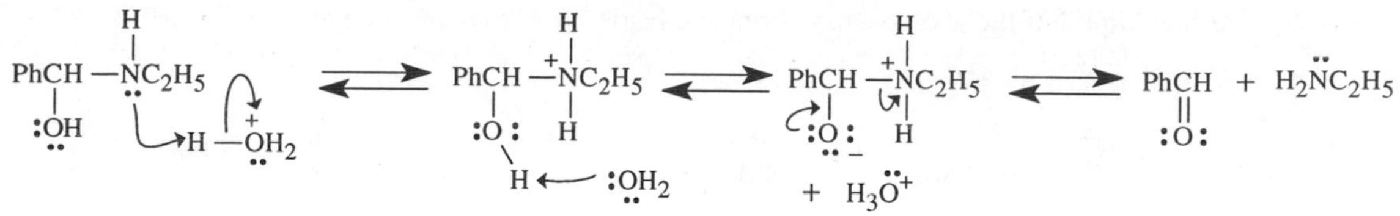

$PhCH(\ddot{O}H)-\ddot{N}HC_2H_5 + H-\overset{+}{O}H_2 \rightleftharpoons PhCH(\ddot{O}:^{-}H)-\overset{+}{N}H_2C_2H_5 + :\ddot{O}H_2 \rightleftharpoons PhCH(:\ddot{O}:^{-})-\overset{+}{N}H_2C_2H_5 + H_3\ddot{O}^+ \rightleftharpoons PhCH{=}\ddot{O}: + H_2\ddot{N}C_2H_5$

19.30 (a) $H_3C-C(=CH_2)-N$ (piperidine ring)

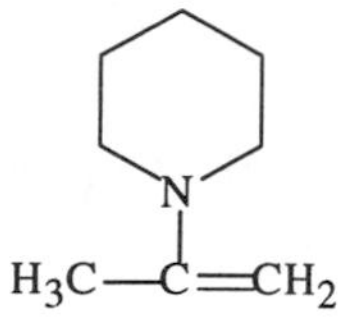

(b) Ph(H)C=C(H)N$(CH_3)_2$

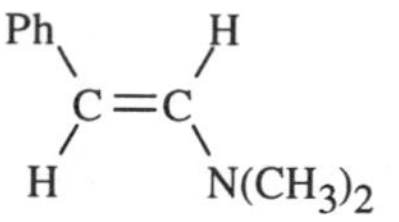

19.31 Any compound with a carbonyl group on any of the prospective alkyl carbons could in principle serve as a starting material. (The answer is restricted to compounds containing only one carbonyl group.)

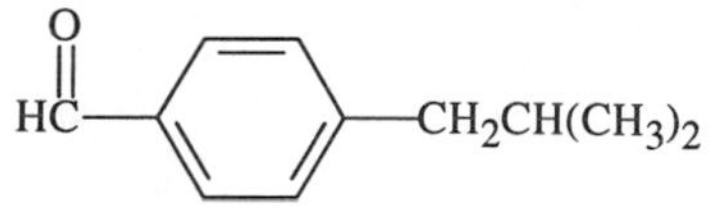

$HC(=O)-C_6H_4-CH_2CH(CH_3)_2$ — 4-isobutylbenzaldehyde

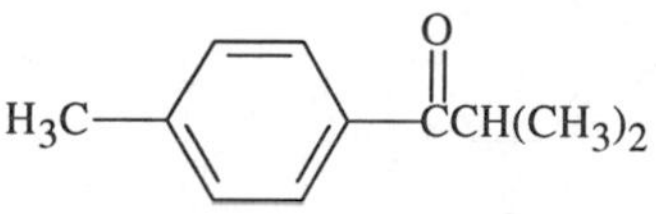

$H_3C-C_6H_4-C(=O)CH(CH_3)_2$ — *p*-methylisobutyrophenone

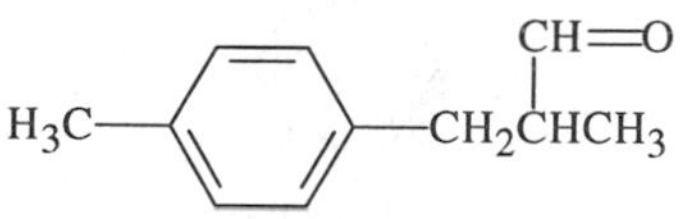

$H_3C-C_6H_4-CH_2CH(CH=O)CH_3$ — 2-methyl-3-(4-methylphenyl)propanal

19.32 Alkylate the phenol, carry out a Friedel-Crafts acylation, and then apply the Wolff-Kishner reaction. Alkylation of the phenol should precede the Friedel-Crafts reaction because of the sluggish reactivity of phenols in the Friedel-Crafts acylation reaction; see text Sec. 18.8. Note that Friedel-Crafts alkylation of the ring with $ClCH_2CH_2CH_3$ would give some rearrangement product in addition to the desired product; see Eq. 19.69a on text p. 880.

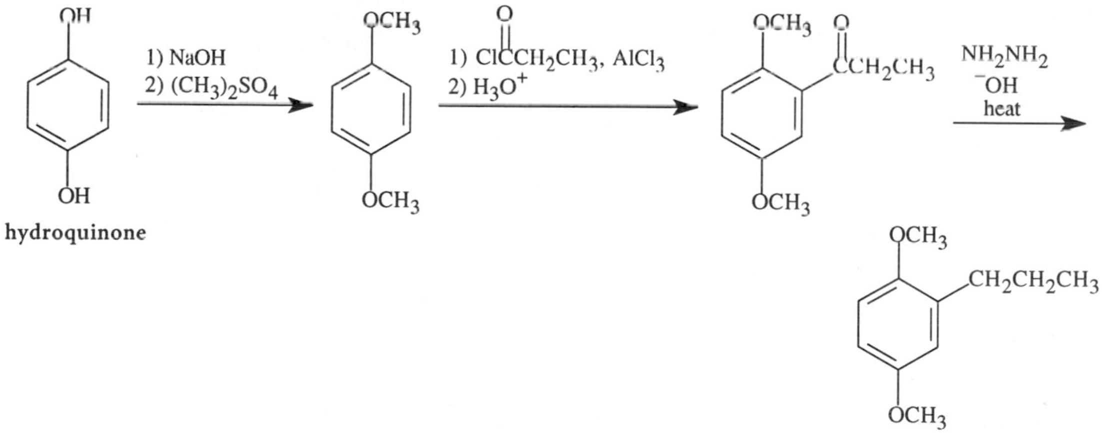

hydroquinone (OH, OH) —[1) NaOH; 2) $(CH_3)_2SO_4$]→ 1,4-dimethoxybenzene (OCH_3, OCH_3) —[1) $ClC(=O)CH_2CH_3$, $AlCl_3$; 2) H_3O^+]→ 2,5-dimethoxyphenyl $C(=O)CH_2CH_3$ —[NH_2NH_2, ^-OH, heat]→ OCH_3, $CH_2CH_2CH_3$, OCH_3

1,4-dimethoxy-2-propylbenzene

19.33 (a) $(H_3C)_2C=C(CH_3)H$

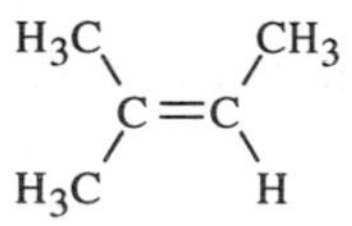

2-methyl-2-butene

(b) $C_6H_5-CH=CH_2$

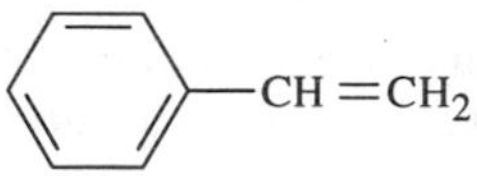

styrene

19.34 (a) Either "half" of the alkene can in principle be derived from an aldehyde.

The first synthesis:

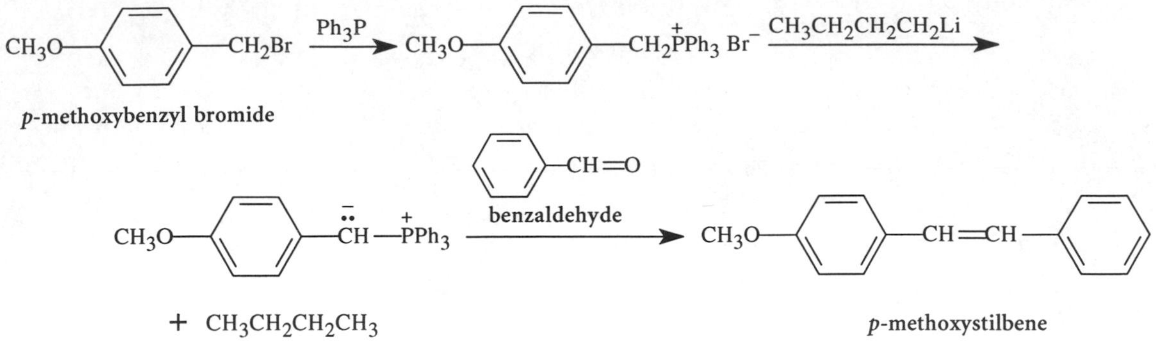

$$CH_3O{-}C_6H_4{-}CH_2Br \xrightarrow{Ph_3P} CH_3O{-}C_6H_4{-}CH_2\overset{+}{P}Ph_3\ Br^- \xrightarrow{CH_3CH_2CH_2CH_2Li}$$

p-methoxybenzyl bromide

$$CH_3O{-}C_6H_4{-}\ddot{\overset{-}{C}}H{-}\overset{+}{P}Ph_3 \xrightarrow[\text{benzaldehyde}]{C_6H_5{-}CH{=}O} CH_3O{-}C_6H_4{-}CH{=}CH{-}C_6H_5$$

$+\ CH_3CH_2CH_2CH_3$ p-methoxystilbene

The second synthesis:

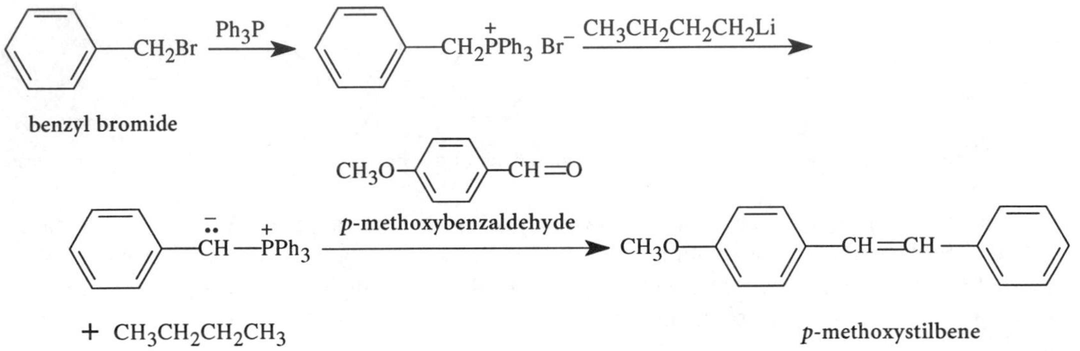

$$C_6H_5{-}CH_2Br \xrightarrow{Ph_3P} C_6H_5{-}CH_2\overset{+}{P}Ph_3\ Br^- \xrightarrow{CH_3CH_2CH_2CH_2Li}$$

benzyl bromide

$$C_6H_5{-}\ddot{\overset{-}{C}}H{-}\overset{+}{P}Ph_3 \xrightarrow[\text{p-methoxybenzaldehyde}]{CH_3O{-}C_6H_4{-}CH{=}O} CH_3O{-}C_6H_4{-}CH{=}CH{-}C_6H_5$$

$+\ CH_3CH_2CH_2CH_3$ p-methoxystilbene

(b)

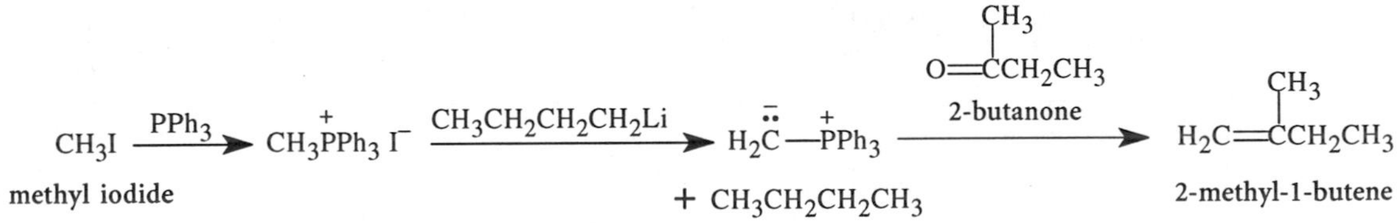

$$CH_3I \xrightarrow{PPh_3} CH_3\overset{+}{P}Ph_3\ I^- \xrightarrow{CH_3CH_2CH_2CH_2Li} H_2\ddot{\overset{-}{C}}{-}\overset{+}{P}Ph_3 \xrightarrow[\text{2-butanone}]{O{=}C(CH_3)CH_2CH_3} H_2C{=}C(CH_3)CH_2CH_3$$

methyl iodide $+\ CH_3CH_2CH_2CH_3$ 2-methyl-1-butene

(c)

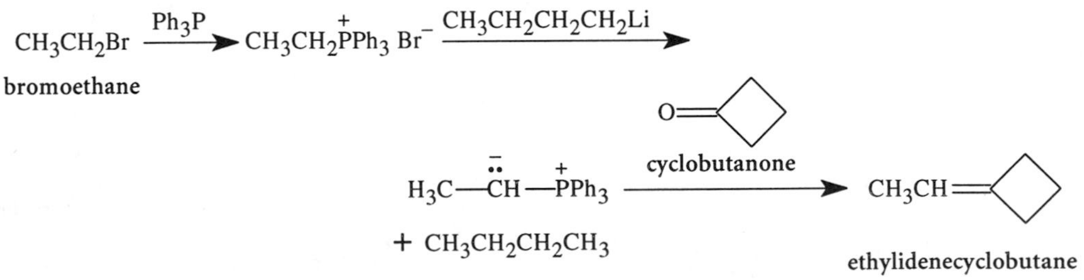

$$CH_3CH_2Br \xrightarrow{Ph_3P} CH_3CH_2\overset{+}{P}Ph_3\ Br^- \xrightarrow{CH_3CH_2CH_2CH_2Li}$$

bromoethane

$$H_3C{-}\ddot{\overset{-}{C}}H{-}\overset{+}{P}Ph_3 \xrightarrow[\text{cyclobutanone}]{O{=}C_4H_6} CH_3CH{=}C_4H_6$$

$+\ CH_3CH_2CH_2CH_3$ ethylidenecyclobutane

19.35 The starting material has five degrees of unsaturation, four of which are accounted for by a benzene ring. Consequently, the compound must be both an aldehyde and an alcohol:

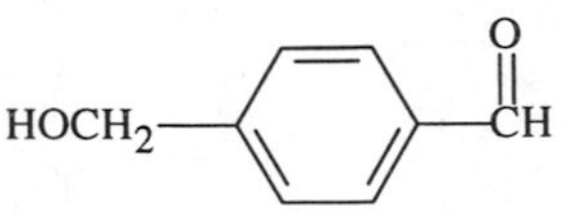

$$HOCH_2{-}C_6H_4{-}\overset{O}{\overset{\|}{C}}H$$ 4-(hydroxymethyl)benzaldehyde

19.36 Silver(I) oxide oxidizes the aldehyde selectively to the following carboxylic acid:

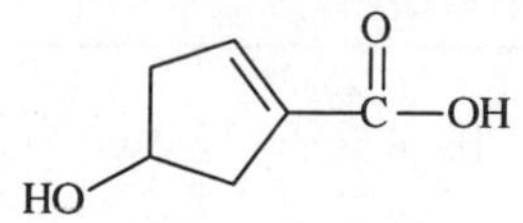

4-hydroxy-1-cyclopentenecarboxylic acid

Solutions to Additional Problems

19.37 Only organic products are shown in the answers below.

(a) Acetone is protonated on oxygen to give the conjugate acid shown below, but unless the acid is very strong, this reaction does not occur to a great extent.

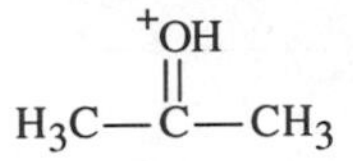

$$H_3C{-}\overset{\overset{+}{O}H}{\overset{\|}{C}}{-}CH_3$$

(b) $CH_3\overset{OH}{\overset{|}{C}}HCH_3$

(c) No reaction

(d) $CH_3\overset{OH}{\overset{|}{C}}CH_3$ with $C{\equiv}N$ on the central carbon

(e) $CH_3\overset{OCH_3}{\overset{|}{C}}CH_3$ with OCH_3 on the central carbon

(f) pyrrolidine ring N attached to $CH_3C{=}CH_2$

(g) $CH_3\overset{NNHCNH_2}{\overset{\|}{C}}CH_3$ (with C=O in the NNHC(=O)NH$_2$ group)

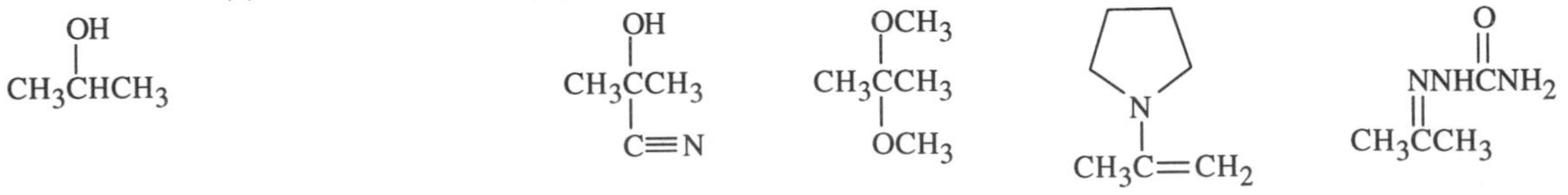

(h) $CH_3\overset{OH}{\overset{|}{C}}CH_3$ with CH_3 on the central carbon

(i) $CH_3\overset{O}{\overset{\|}{C}}CH_3$

(j) $(H_3C)_2C{=}CH_2$

(k) $CH_3\overset{OH}{\overset{|}{C}}HCH_3$

(l) $(CH_3)_2C{=}CH_2$ + $Ph_3P{=}O$

(m) $CH_3CH_2CH_3$

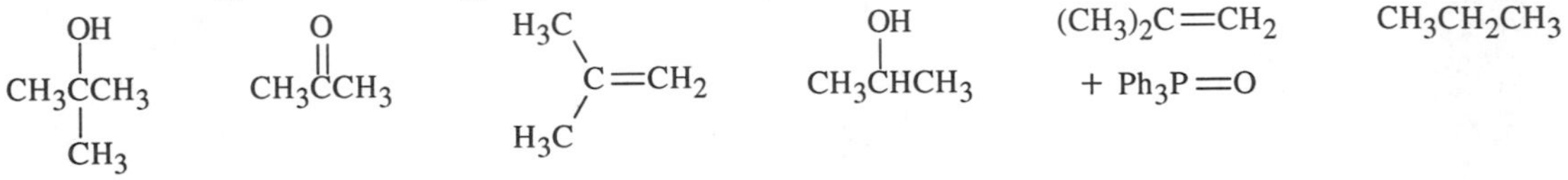

19.38 Only organic products are shown.

(a) $CH_3CH_2CH_2\overset{OH}{\overset{|}{C}}HPh$

(b) $CH_3CH_2CH_2\overset{OH}{\overset{|}{C}}H_2$

(c) $CH_3CH_2CH_2\overset{O}{\overset{\|}{C}}{-}OH$

(d) $CH_3CH_2CH_2\overset{O}{\overset{\|}{C}}{-}OH$

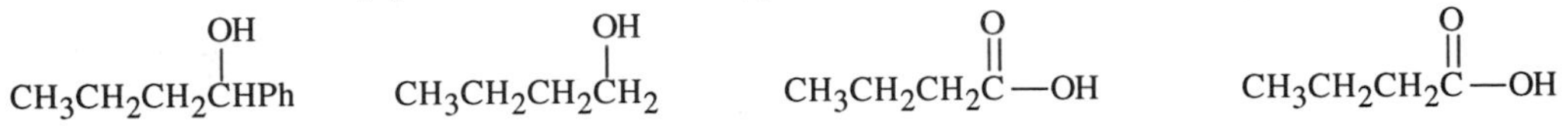

(e) $CH_3CH_2CH_2\overset{NOH}{\overset{\|}{C}}H$

(f) $CH_3CH_2CH_2\overset{O}{\overset{\|}{C}}{-}OH$

(g) $CH_3CH_2CH_2CH_3$

(h) $CH_3CH_2CH_2CH{=}CH_2$

19.39 (a) This is a simple addition reaction in which the sulfur of the bisulfite ion attacks the carbonyl carbon of the aldehyde. In the last step of the mechanism the initially formed addition product *A* ionizes because it is a fairly strong acid.

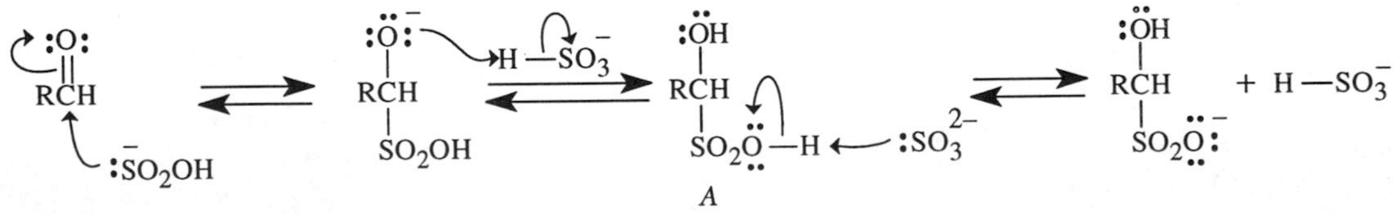

$$RCH{=}O + {}^-{:}SO_2OH \rightleftharpoons R\overset{{:}\ddot{O}{:}^-}{\overset{|}{C}}H{-}SO_2OH \xrightleftharpoons{H-SO_3^-} R\overset{:\ddot{O}H}{\overset{|}{C}}H{-}SO_2\ddot{O}{-}H \;(A) \xrightleftharpoons{:SO_3^{2-}} R\overset{:\ddot{O}H}{\overset{|}{C}}H{-}SO_2\ddot{O}{:}^- + H{-}SO_3^-$$

(b) Acid, that is, H_3O^+, destroys sodium bisulfite by reacting with it and converting it into SO_2, a gas, and H_2O. Destruction of bisulfite in this manner eliminates a reactant in the equilibria shown above and pulls the reaction to the left, that is, toward free aldehyde. Hydroxide ion reacts with bisulfite and converts it into sulfite ion, SO_3^{2-}; because bisulfite is removed as a reactant by this process as well, the equilibrium is also pulled to the left.

In the mechanisms presented thus far in this Study Guide and for the most part in the text, unshared electron pairs have been drawn in explicitly. Omitting these electron pairs can save some time and can eliminate a great deal of tedium in writing mechanisms; most chemists do not draw electron pairs explicitly. This shortcut is useful only if you realize that the unshared electron pairs which are not shown are understood to be present. In many of the mechanisms shown subsequently, electron pairs are omitted in the most common situations. (In unusual or uncommon situations, the electron pairs are retained.) Here are some typical situations in which electron pairs should not be necessary:

$$H_2O \rightarrow C{=}O \longrightarrow H_2\overset{+}{O}{-}C{-}O^- \quad \text{means} \quad H_2\ddot{O}{:} \rightarrow C{=}\ddot{\underset{..}{O}} \longrightarrow H_2\overset{+}{\underset{..}{O}}{-}C{-}\ddot{\underset{..}{O}}{:}^-$$

$$(CH_3)_3C{-}Br \longrightarrow (CH_3)_3C^+ \; Br^- \quad \text{means} \quad (CH_3)_3C{-}\ddot{\underset{..}{Br}}{:} \longrightarrow (CH_3)_3C^+ \; {:}\ddot{\underset{..}{Br}}{:}^-$$

$$R_3N \quad \text{means} \quad R_3N{:}$$

$$R_3C^- \quad \text{means} \quad R_3\bar{C}{:}$$

Thus, an oxygen with three bonds and a positive charge has one unshared electron pair; an oxygen with one bond and a negative charge has three unshared pairs; an uncharged nitrogen with three bonds or a negatively charged carbon with three bonds has one unshared pair; and a halide ion has four unshared pairs.

The reason we do not have to show unshared valence electrons is that if the formal charge on an atom and the number of bonds to an atom are known, the number of unshared valence electrons are automatically known. It follows that the formal charge must be shown; otherwise, the valence-electron count of an atom is undetermined.

Most of the mechanisms shown after this point will be written without showing the unshared pairs. If you ever become confused by the absence of electron pairs, *do not hesitate to draw them in explicitly.*

(c) The bisulfite addition product of 2-methylpentanal:

$$CH_3CH_2CH_2CH(CH_3){-}CH(OH){-}S(=O)_2{-}O^- \; Na^+$$

19.40 (a) Two diastereomeric alcohols are formed by attack of hydride at the upper and lower face, respectively, of the carbonyl group followed by protonation of the resulting alkoxides.

[Structures: H_3C–cyclohexane with axial H and equatorial OH + H_3C–cyclohexane with axial OH and equatorial H]

from attack of hydride at the upper face of the carbonyl carbon

from attack of hydride at the lower face of the carbonyl carbon

(b) Both cis and trans isomers of the alkene PhCH=CHCH$_3$ [(1-propenyl)benzene, also known as β-methylstyrene] are formed.

19.41 (a) Glycerol has three hydroxy groups. Two possible cyclic acetals can be formed; one (*A*) contains a six-membered ring, and the other (*B*) contains a five-membered ring.

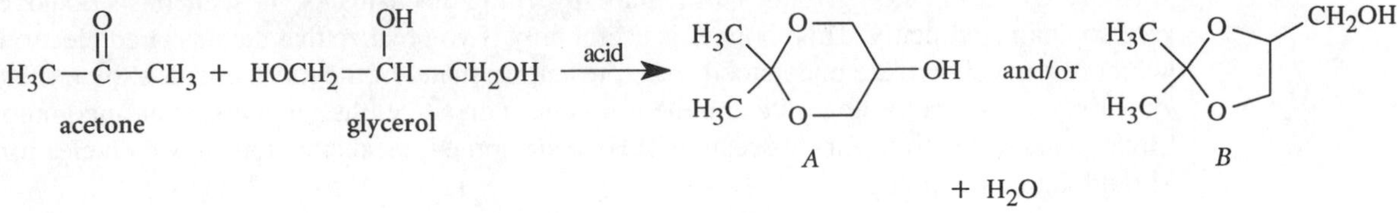

acetone + glycerol $\xrightarrow{\text{acid}}$ *A* and/or *B* + H_2O

(b) Only compound *B* is chiral, and for this reason only compound *B* can be resolved into enantiomers; hence, compound *B* is the observed compound.

19.42 The two separable isomers are the diastereomeric acetals:

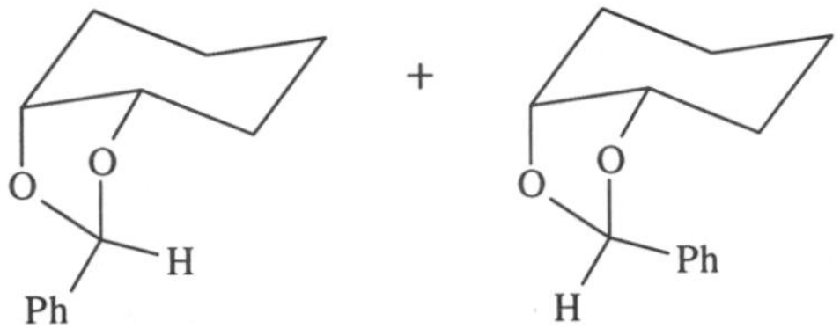

19.43 (a) The reaction is formation of an oxime, a type of imine; see Table 19.3, text p. 876.

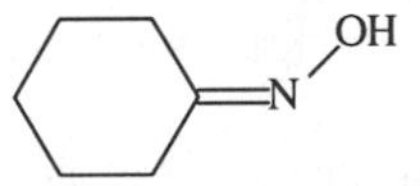

cyclohexanone oxime

(b) The reaction is a straightforward dimethyl acetal formation.

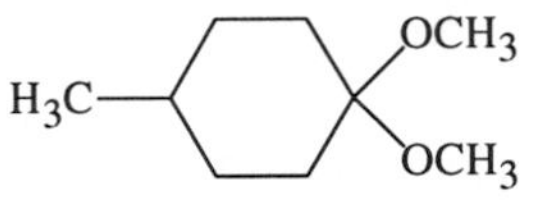

1,1-dimethoxy-4-methylcyclohexane

(c) The formula shows that only one of the carbonyl groups is involved in acetal formation; the aldehyde carbonyl reacts selectively because aldehydes form addition compounds more rapidly than do ketones, and because addition reactions of aldehydes are thermodynamically more favorable than those of ketones.

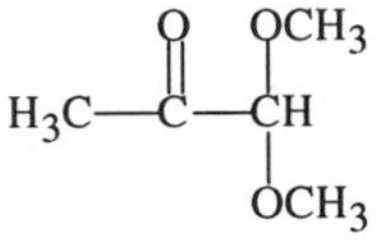

1,1-dimethoxy-2-propanone

(d) This reaction is internal acetal formation, in which the two alcohol groups form an acetal with the carbonyl group within the same molecule.

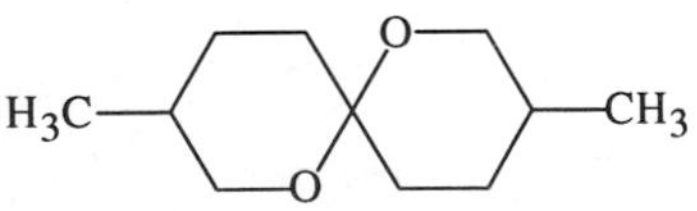

(e) This is a Grignard addition to the ketone to give a tertiary alcohol that subsequently dehydrates under the acidic conditions to an alkene. Whether the dehydration occurs depends on the acid concentration and whether the conditions are designed to remove water.

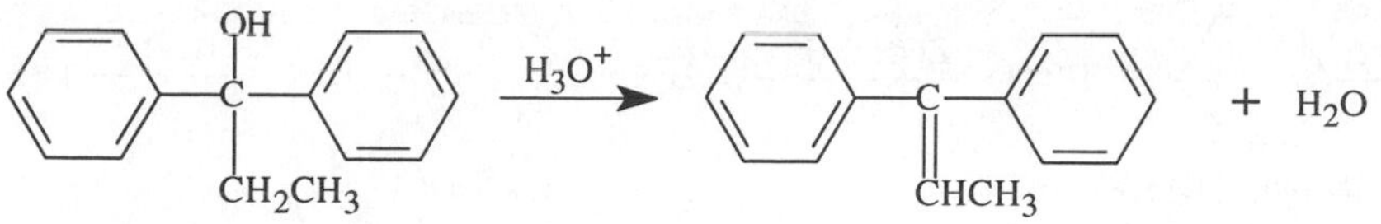

OH, H_3O^+, $+ H_2O$

CH_2CH_3 $CHCH_3$

1,1-diphenyl-1-propanol

(f) This is a Wittig reaction.

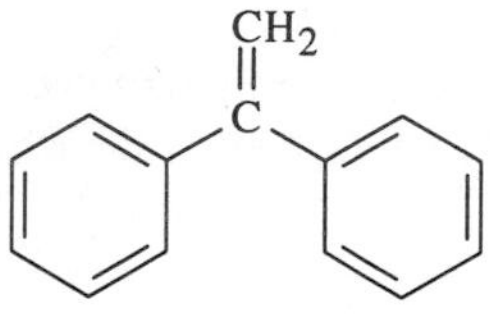

CH_2 C

(g) A "double" Wittig reagent is formed from the two benzylic bromides, and each Wittig reagent reacts at a different carbonyl group of the same aldehyde molecule to form a diene.

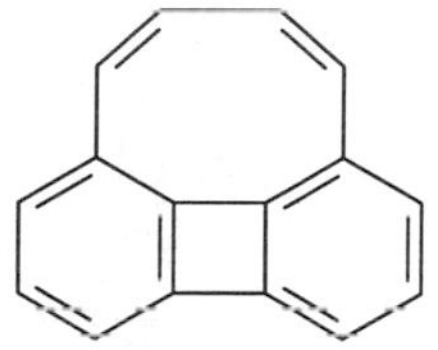

19.44 (a) In this reaction, the carbonyl group is reduced to an alcohol, which is subsequently ionized by NaH. The resulting intermediate is essentially a Wittig oxaphosphetane intermediate, as the hint suggests; this intermediate decomposes to form an alkene.

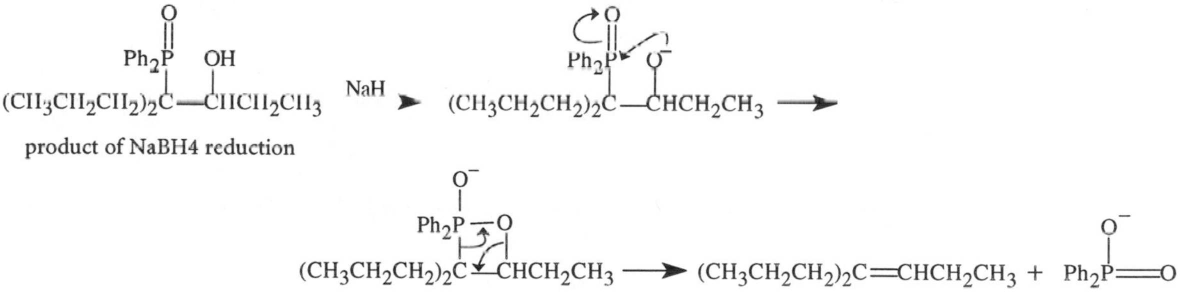

$(CH_3CH_2CH_2)_2C(P(O)Ph_2)—CH(OH)CH_2CH_3$ $\xrightarrow{NaH}$ $(CH_3CH_2CH_2)_2C(P(O)Ph_2)—CH(O^-)CH_2CH_3$ ⟶

product of NaBH4 reduction

$(CH_3CH_2CH_2)_2C—CHCH_2CH_3$ (ring with $Ph_2P(O^-)$—O) ⟶ $(CH_3CH_2CH_2)_2C{=}CHCH_2CH_3 + Ph_2P(O^-){=}O$

an oxaphosphetane
(see Eq. 19.73a)

(b) An initially formed imine reacts with sodium borohydride to form an amine. (Although this reaction is discussed in Sec. 23.7B, the product follows directly from the hint.)

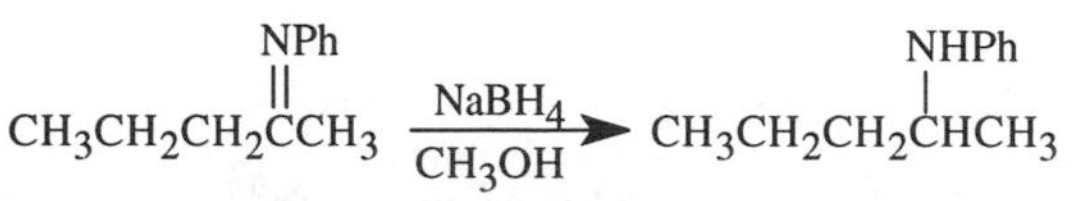

$CH_3CH_2CH_2C({=}NPh)CH_3 \xrightarrow[CH_3OH]{NaBH_4} CH_3CH_2CH_2CH(NHPh)CH_3$

19.45 The data indicate that compound *A* is a benzene derivative with an additional degree of unsaturation. Its reactivity in the Clemmensen reduction and the formation of a xylene suggests that it has two substituents on a benzene ring, one of which is a methyl group and one of which is an aldehyde. Only *p*-methylbenzaldehyde would give, after Clemmensen reduction, a compound (*p*-xylene) that in turn gives, because of its symmetry, one and only one monobromination product, 1-bromo-2,5-dimethylbenzene.

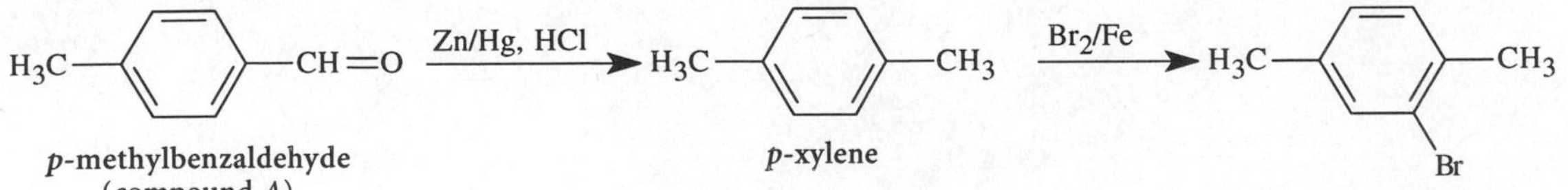

H_3C–C₆H₄–CH=O $\xrightarrow{\text{Zn/Hg, HCl}}$ H_3C–C₆H₄–CH_3 $\xrightarrow{Br_2/Fe}$ H_3C–C₆H₃(Br)–CH_3

p-methylbenzaldehyde (compound A) — p-xylene — 1-bromo-2,5-dimethylbenzene

19.46 (a) Friedel-Crafts acylation gives a ketone, p-methoxyacetophenone, which reacts in a Wolff-Kishner reduction of the carbonyl group to give 1-ethyl-4-methoxybenzene.

H_3CO–C₆H₄–C(=O)–CH_3 $\xrightarrow{\text{Wolff-Kishner}}$ CH_3O–C₆H₄–CH_2CH_3

p-methoxyacetophenone product of Friedel-Crafts acylation — 1-ethyl-4-methoxybenzene (p-ethylanisole)

(b) Prepare the triflate of the phenol and subject it to a Stille reaction (Sec. 18.9B) with tetraethylstannane.

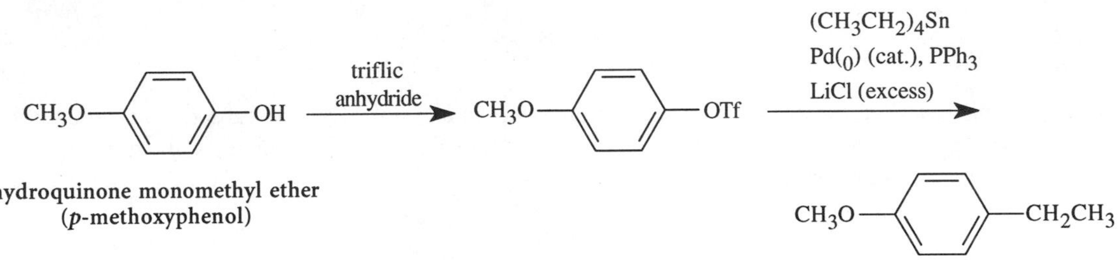

CH_3O–C₆H₄–OH $\xrightarrow{\text{triflic anhydride}}$ CH_3O–C₆H₄–OTf $\xrightarrow{(CH_3CH_2)_4Sn,\ Pd_{(0)}\ (cat.),\ PPh_3,\ LiCl\ (excess)}$ CH_3O–C₆H₄–CH_2CH_3

hydroquinone monomethyl ether (p-methoxyphenol) — 1-ethyl-4-methoxybenzene (p-ethylanisole)

19.47 As always, bear in mind that there is often more than one acceptable synthesis that fits the parameters given.

(a)

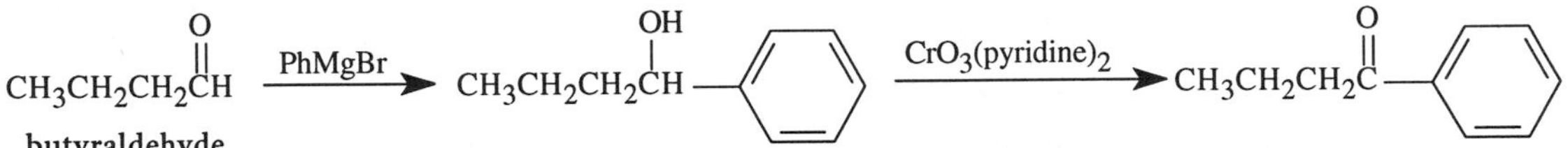

$CH_3CH_2CH_2CH(=O)$ $\xrightarrow{PhMgBr}$ $CH_3CH_2CH_2CH(OH)$–C₆H₅ $\xrightarrow{CrO_3(pyridine)_2}$ $CH_3CH_2CH_2C(=O)$–C₆H₅

butyraldehyde — butyrophenone

(b)

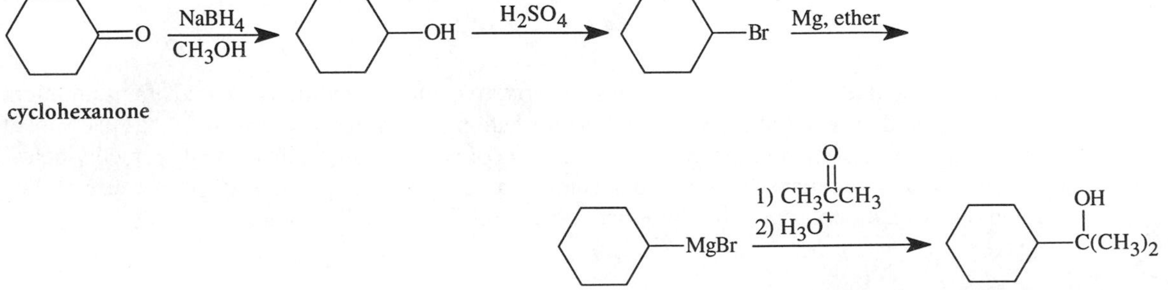

cyclohexanone (C₆H₁₀=O) $\xrightarrow{NaBH_4,\ CH_3OH}$ C₆H₁₁–OH $\xrightarrow{\text{conc. HBr},\ H_2SO_4}$ C₆H₁₁–Br $\xrightarrow{\text{Mg, ether}}$

C₆H₁₁–MgBr $\xrightarrow{1)\ CH_3C(=O)CH_3;\ 2)\ H_3O^+}$ C₆H₁₁–C(OH)$(CH_3)_2$

cyclohexanone — 2-cyclohexyl-2-propanol

(c)

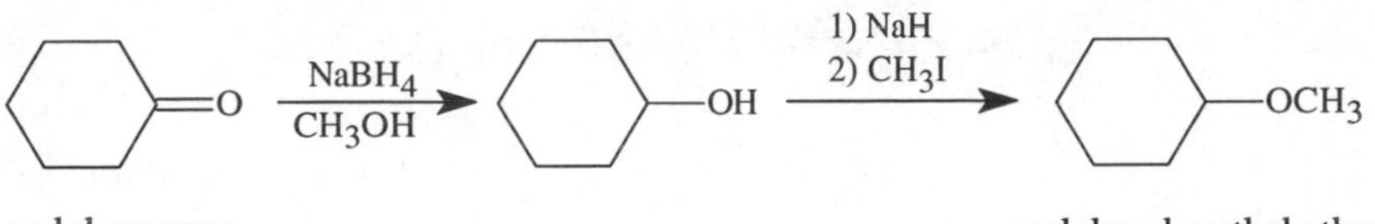

cyclohexanone $\xrightarrow[\text{CH}_3\text{OH}]{\text{NaBH}_4}$ cyclohexanol (–OH) $\xrightarrow[\text{2) CH}_3\text{I}]{\text{1) NaH}}$ –OCH_3

cyclohexyl methyl ether (methoxycyclohexane)

(d) The preparation of the ylid used in this synthesis can be found In Eq. 19.74a–b on text p. 882.

$(CH_3)_2CHCH{=}O \xrightarrow{Ph_3\overset{+}{P}{-}\overset{-}{C}H_2} (CH_3)_2CHCH{=}CH_2$

isobutyraldehyde (2-methylpropanal) — 3-methyl-1-butene

(e)

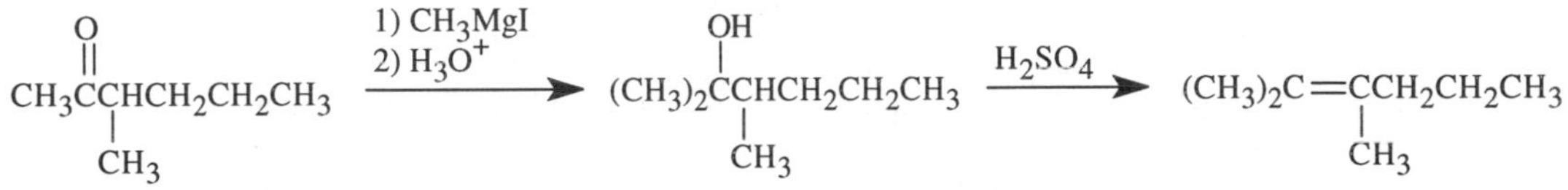

$CH_3\overset{O}{\overset{\|}{C}}\underset{CH_3}{\underset{|}{C}}HCH_2CH_2CH_3 \xrightarrow[\text{2) } H_3O^+]{\text{1) } CH_3MgI} (CH_3)_2\overset{OH}{\overset{|}{C}}\underset{CH_3}{\underset{|}{C}}HCH_2CH_2CH_3 \xrightarrow{H_2SO_4} (CH_3)_2C{=}\underset{CH_3}{\underset{|}{C}}CH_2CH_2CH_3$

3-methyl-2-hexanone — 2,3-dimethyl-2-hexene

(f)

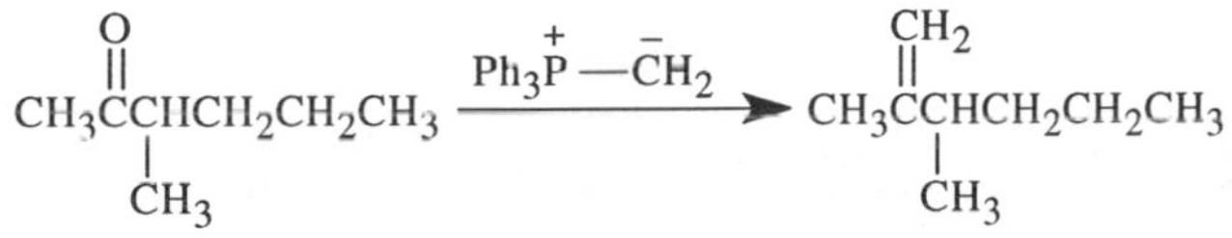

$CH_3\overset{O}{\overset{\|}{C}}\underset{CH_3}{\underset{|}{C}}HCH_2CH_2CH_3 \xrightarrow{Ph_3\overset{+}{P}{-}\overset{-}{C}H_2} CH_3\overset{CH_2}{\overset{\|}{C}}\underset{CH_3}{\underset{|}{C}}HCH_2CH_2CH_3$

3-methyl-2-hexanone — 2,3-dimethyl-1-hexene

(g)

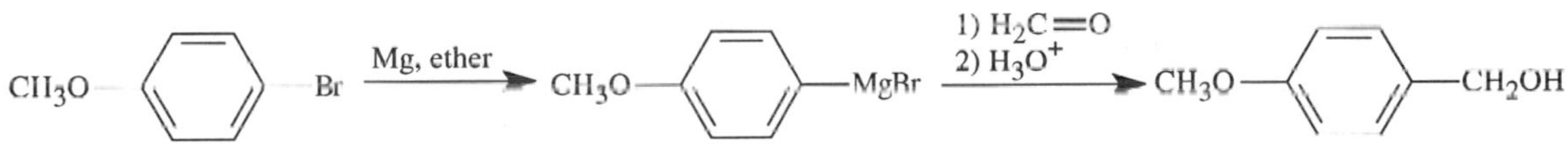

CH_3O–C_6H_4–Br $\xrightarrow{\text{Mg, ether}}$ CH_3O–C_6H_4–MgBr $\xrightarrow[\text{2) } H_3O^+]{\text{1) } H_2C{=}O}$ CH_3O–C_6H_4–CH_2OH

p-bromoanisole (1-bromo-4-methoxybenzene) — p-methoxybenzyl alcohol

(h)

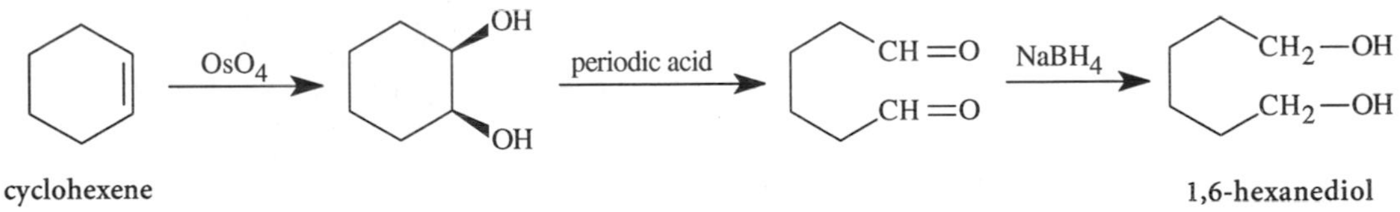

cyclohexene $\xrightarrow{OsO_4}$ cis-1,2-cyclohexanediol (OH, OH) $\xrightarrow{\text{periodic acid}}$ $OHC(CH_2)_4CHO$ (CH=O, CH=O) $\xrightarrow{NaBH_4}$ $HOCH_2(CH_2)_4CH_2OH$ (CH_2—OH, CH_2—OH)

cyclohexene — 1,6-hexanediol

Ozonolysis could also be used to prepare the dialdehyde.

(i) In the following synthesis, the two acyl groups could be introduced in the opposite order.

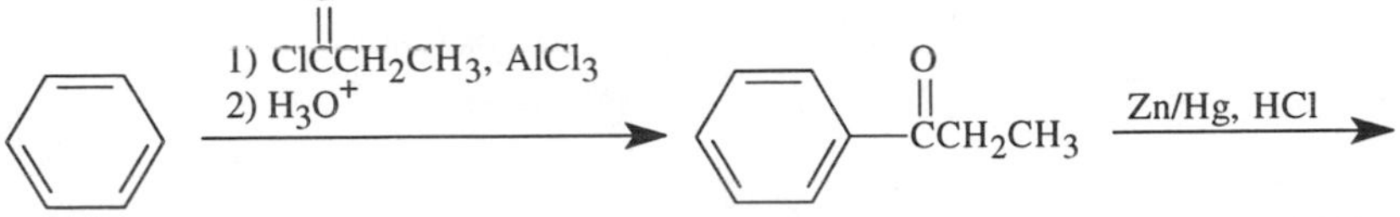

benzene $\xrightarrow[\text{2) } H_3O^+]{\text{1) } Cl\overset{O}{\overset{\|}{C}}CH_2CH_3,\ AlCl_3}$ $C_6H_5\overset{O}{\overset{\|}{C}}CH_2CH_3$ $\xrightarrow{\text{Zn/Hg, HCl}}$

$C_6H_5CH_2CH_2CH_3 \xrightarrow[\text{2) } H_3O^+]{\text{1) } CH_3CH_2CH_2\overset{O}{\overset{\|}{C}}Cl,\ AlCl_3}$

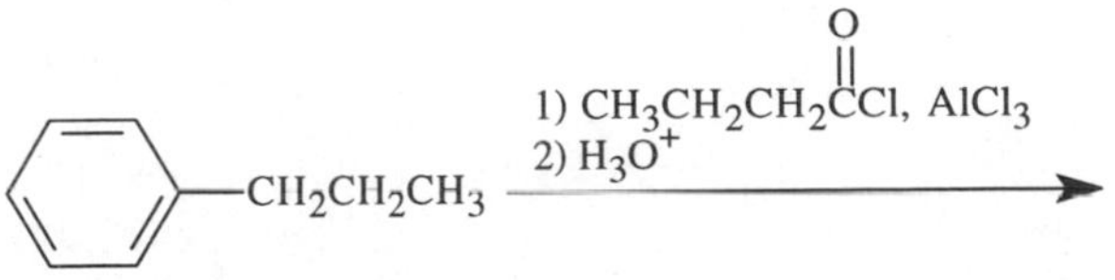

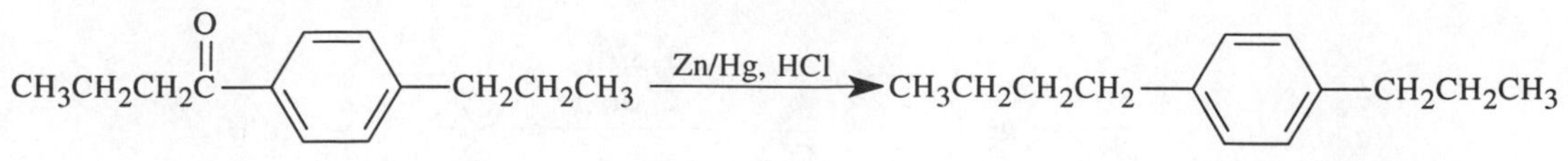

$CH_3CH_2CH_2C(=O)$–C_6H_4–$CH_2CH_2CH_3$ $\xrightarrow{\text{Zn/Hg, HCl}}$ $CH_3CH_2CH_2CH_2$–C_6H_4–$CH_2CH_2CH_3$

1-butyl-4-propylbenzene

(j) First prepare the Grignard reagent benzylmagnesium bromide from benzaldehyde:

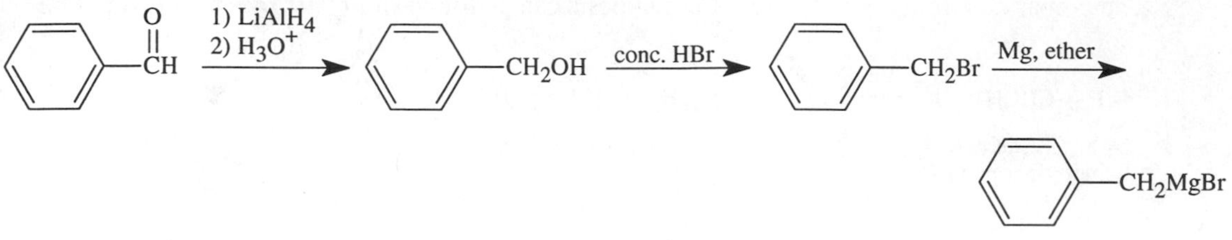

$C_6H_5CH(=O)$ $\xrightarrow{\text{1) LiAlH}_4\text{; 2) H}_3\text{O}^+}$ C_6H_5–CH_2OH $\xrightarrow{\text{conc. HBr}}$ C_6H_5–CH_2Br $\xrightarrow{\text{Mg, ether}}$ C_6H_5–CH_2MgBr

benzylmagnesium bromide

Then let this Grignard reagent react with benzaldehyde and oxidize the resulting alcohol:

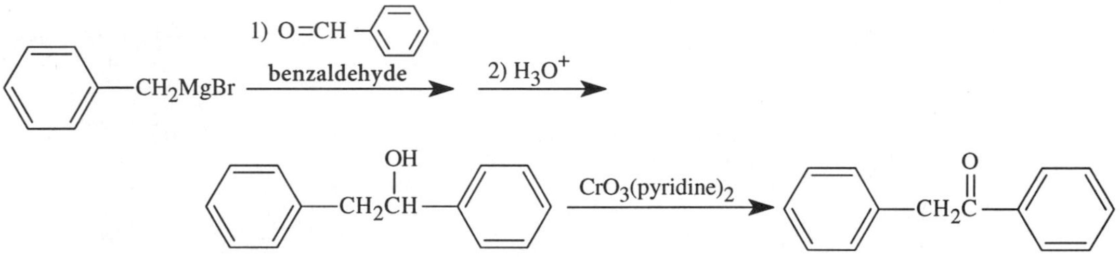

C_6H_5–CH_2MgBr $\xrightarrow{\text{1) O=CH–C}_6\text{H}_5\text{ (benzaldehyde)}}$ $\xrightarrow{\text{2) H}_3\text{O}^+}$

C_6H_5–$CH_2CH(OH)$–C_6H_5 $\xrightarrow{\text{CrO}_3\text{(pyridine)}_2}$ C_6H_5–$CH_2C(=O)$–C_6H_5

(k) The first hint indicates that a protecting group must be used. The second hint is meant to call attention to the hemiacetal linkage which, because it is a five-membered *cyclic* hemiacetal, will form spontaneously from the corresponding aldehyde.

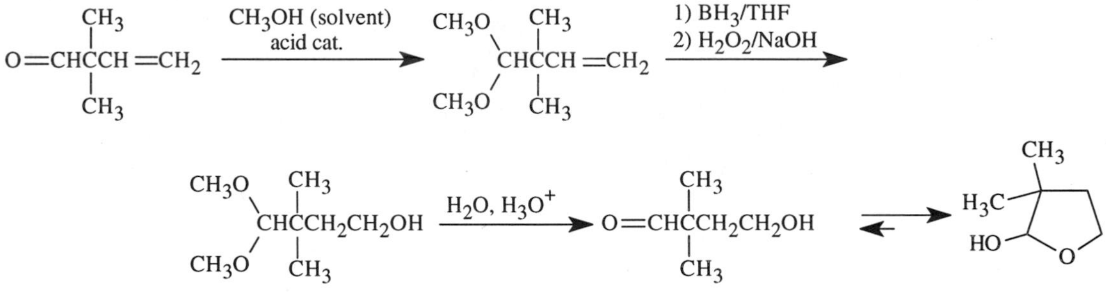

$O{=}CHC(CH_3)_2CH{=}CH_2$ $\xrightarrow{\text{CH}_3\text{OH (solvent), acid cat.}}$ $(CH_3O)_2CHC(CH_3)_2CH{=}CH_2$ $\xrightarrow{\text{1) BH}_3\text{/THF; 2) H}_2\text{O}_2\text{/NaOH}}$

$(CH_3O)_2CHC(CH_3)_2CH_2CH_2OH$ $\xrightarrow{\text{H}_2\text{O, H}_3\text{O}^+}$ $O{=}CHC(CH_3)_2CH_2CH_2OH$ $\rightleftarrows$ (cyclic hemiacetal: five-membered ring with CH_3, H_3C, HO, O)

(l) A protecting group must be used in this synthesis.

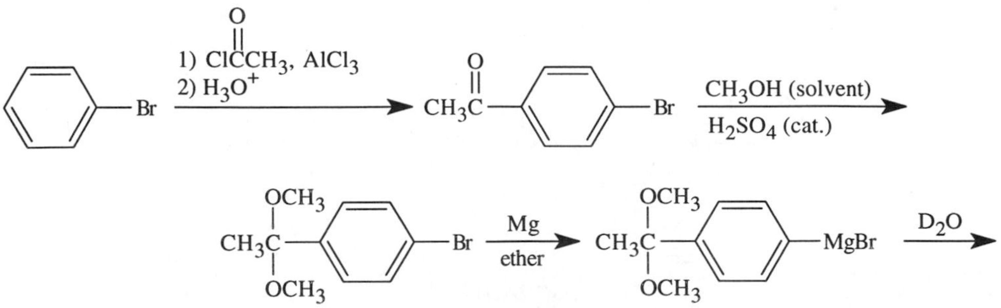

C_6H_5–Br $\xrightarrow{\text{1) ClC(=O)CH}_3\text{, AlCl}_3\text{; 2) H}_3\text{O}^+}$ $CH_3C(=O)$–C_6H_4–Br $\xrightarrow[\text{H}_2\text{SO}_4\text{ (cat.)}]{\text{CH}_3\text{OH (solvent)}}$

$CH_3C(OCH_3)_2$–C_6H_4–Br $\xrightarrow[\text{ether}]{\text{Mg}}$ $CH_3C(OCH_3)_2$–C_6H_4–MgBr $\xrightarrow{\text{D}_2\text{O}}$

$CH_3C(OCH_3)_2$–C_6H_4–D $\xrightarrow{H_3O^+,\ H_2O}$ $CD_3C(=O)$–C_6H_4–D

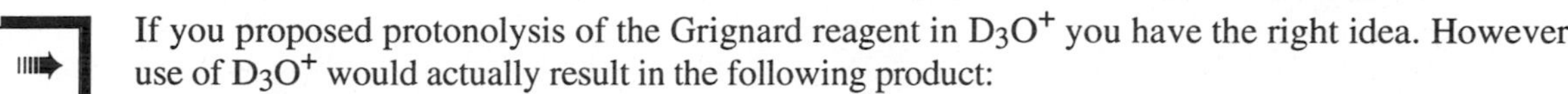

If you proposed protonolysis of the Grignard reagent in D_3O^+ you have the right idea. However, use of D_3O^+ would actually result in the following product:

$CD_3C(=O)$–C_6H_4–D

The reason is that acidic D_2O brings about the exchange of protons that are α to an aldehyde or ketone carbonyl group. This is a reaction you will learn about in Chapter 22. (See Eq. 22.18 on text p. 1006.) To avoid this exchange use D_2O for protonolysis of the Grignard reagent, which gives $Mg(OD)_2$ as a by-product. Because this by-product maintains basic conditions, the acetal is left intact. Hydrolysis of the acetal is then carried out with H_3O^+ in H_2O.

19.48 (a)

Ph—C(OH)(Br)—CH_3 ⟶ [Ph—$\overset{+}{C}$(OH)—CH_3 ⟷ Ph—C(=$\overset{+}{O}$H)—CH_3] Br^- ⟶ Ph—C(=O)—CH_3 + HBr

(b) Osmium tetroxide reacts to form the 1,2-glycol *A*, which also has a bromine on carbon 1 and thus decomposes as shown in the mechanism of part (a):

H_2C=CH—Br $\xrightarrow{OsO_4,\ H_2O}$ $H_2C(OH)$—CH(OH)—Br (*A*) ⟶ $H_2C(OH)$—CH(=O) + H—Br

19.49 The molecular formula of compound *A* shows that two equivalents of methanol are added to the alkyne. The fact that the product hydrolyzes to acetophenone indicates that the two methoxy groups of compound *A* are on the same carbon, that is, that compound *A* is an acetal.

PhC≡CH + H—$\overset{+}{O}$(H)CH_3 $\xrightarrow{(a)}$ Ph$\overset{+}{C}$=CH_2 + $HOCH_3$ ⟶ PhC($\overset{+}{O}$(H)CH_3)=CH_2 ⟶ PhC(OCH_3)=CH_2 + H—$\overset{+}{O}$(H)CH_3 (HBr) $\xrightarrow{(b)}$

Ph$\overset{+}{C}$(OCH_3)—CH_3 + $HOCH_3$ ⟶ PhC(OCH_3)($\overset{+}{O}$(H)CH_3)—CH_3 + $HOCH_3$ ⟶ PhC(OCH_3)$_2$—CH_3 + $H_2\overset{+}{O}CH_3$

compound *A*

acetophenone dimethyl acetal

In the step labeled *(a)*, the acid $CH_3\overset{+}{O}H_2$ is used to protonate the alkyne because it is the major acidic species present when H_2SO_4 is dissolved in methanol (just as H_3O^+ is the major acidic species present when H_2SO_4 is dissolved in water). Protonation occurs on the terminal carbon because it gives a carbocation that is benzylic and therefore resonance-stabilized. In the step labeled *(b)*, protonation again occurs on the terminal carbon because the resulting carbocation is resonance-stabilized by electron donation from both the benzene ring and the neighboring oxygen.

19.50 (a)

$CH_3CH_2CH_2CH_2CH{=}O$ and H_2NNH—C_6H_4—OCH_3

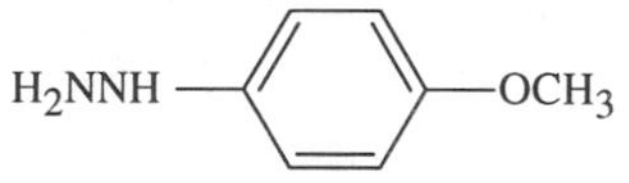

pentanal (valeraldehyde)

p-methoxyphenylhydrazine

(b) Because a cyclic imine is formed, the reaction must be intramolecular; that is, the amine and the carbonyl group that react must be in the same molecule.

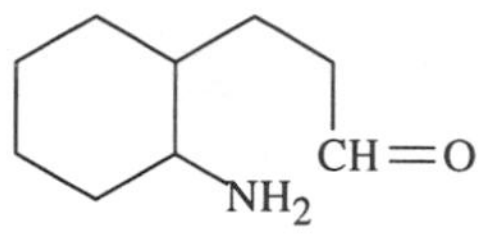

$CH{=}O$

NH_2

19.51 Because compound *A* has one carbon more than the product *D*, a carbon is lost at some point; the most likely place is the benzylic oxidation that gives compound *D* itself. The structure of compound *D* shows that all compounds contain benzene rings with four substituent groups. That compound *A* can be regenerated by oxidation of optically active *B* suggests that *B* is an alcohol and *A* is a ketone. Compound *B* cannot be a primary alcohol because such an alcohol could not be optically active. (The chirality has to be associated with the alcohol functionality because optical activity is lost when the alcohol is oxidized.) It cannot be a tertiary alcohol because it could not be formed by $LiAlH_4$ reduction and could not be oxidized. Compound *A* has one degree of unsaturation in addition to the four accounted for by its benzene ring and the one accounted for by its carbonyl group; the only way to accommodate all the data is for compound *A* to contain a ring. The correct structures are as follows:

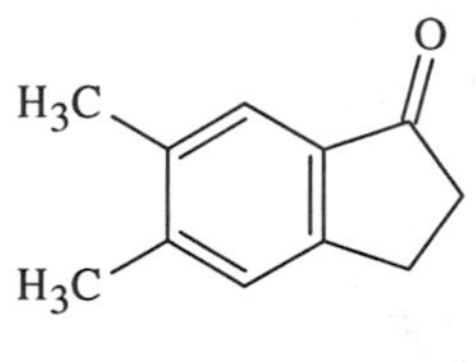

H_3C O H_3C *A*

H_3C H OH H_3C *B*

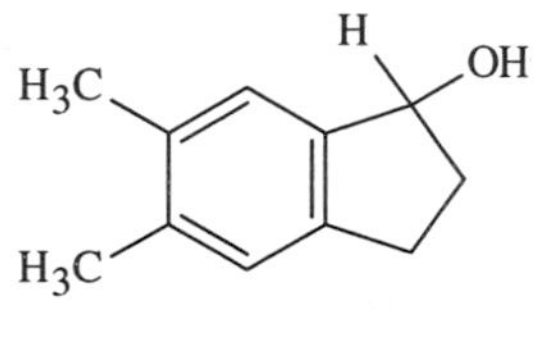

H_3C H_3C *C*

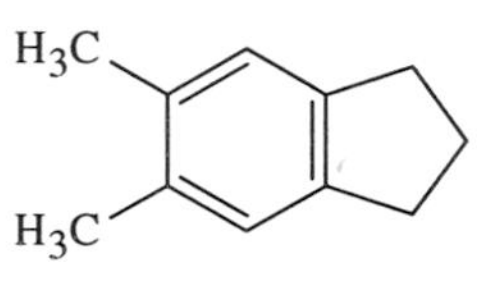

19.52 Compound *A* is an aldehyde because it is oxidized with Ag(I). Because there is no additional unsaturation, the remaining oxygen is accounted for by either an alcohol or an ether functional group. Because Clemmensen reduction of the CrO_3 oxidation product gives a compound without oxygens, the CrO_3 oxidation product of *A* must be a keto aldehyde or a dialdehyde. Because compound *A* is oxidized to a dicarboxylic acid by H_2CrO_4, it must contain a primary alcohol; hence, its CrO_3 oxidation product must be a dialdehyde. Compound *A*, then, is a chiral hydroxy aldehyde that is oxidized by CrO_3 to an achiral dialdehyde. The Clemmensen reduction product shows that all compounds have the carbon skeleton of 3-methylpentane. The compounds with this carbon skeleton that fit all the data are the following.

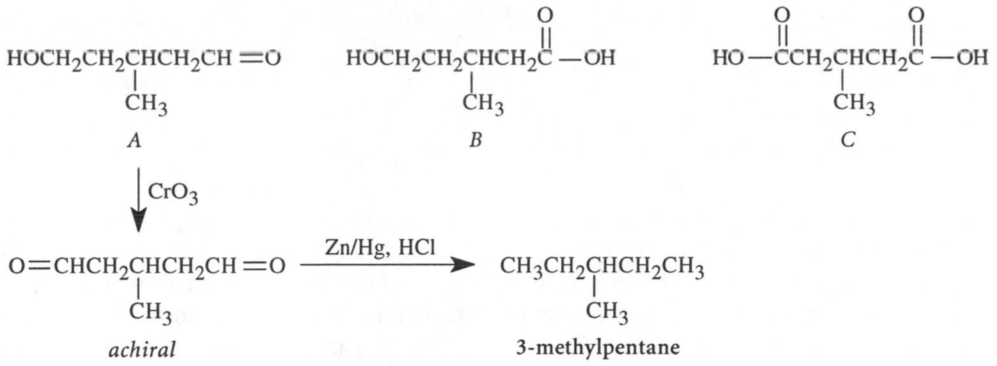

$HOCH_2CH_2CHCH_2CH{=}O$ (with CH_3 on C-3) — *A*

$HOCH_2CH_2CHCH_2C(=O){-}OH$ (with CH_3 on C-3) — *B*

$HO{-}C(=O)CH_2CHCH_2C(=O){-}OH$ (with CH_3 on C-3) — *C*

A $\xrightarrow{CrO_3}$ $O{=}CHCH_2CHCH_2CH{=}O$ (with CH_3; *achiral*) $\xrightarrow{Zn/Hg,\ HCl}$ $CH_3CH_2CHCH_2CH_3$ (with CH_3; 3-methylpentane)

19.53 The mechanism below begins with the protonated aldehyde chloral, which serves as a carbocation electrophile in the ring alkylation of chlorobenzene. The resulting product, an alcohol, dehydrates under the acidic conditions to give another carbocation that alkylates a second chlorobenzene molecule and thus forms the product.

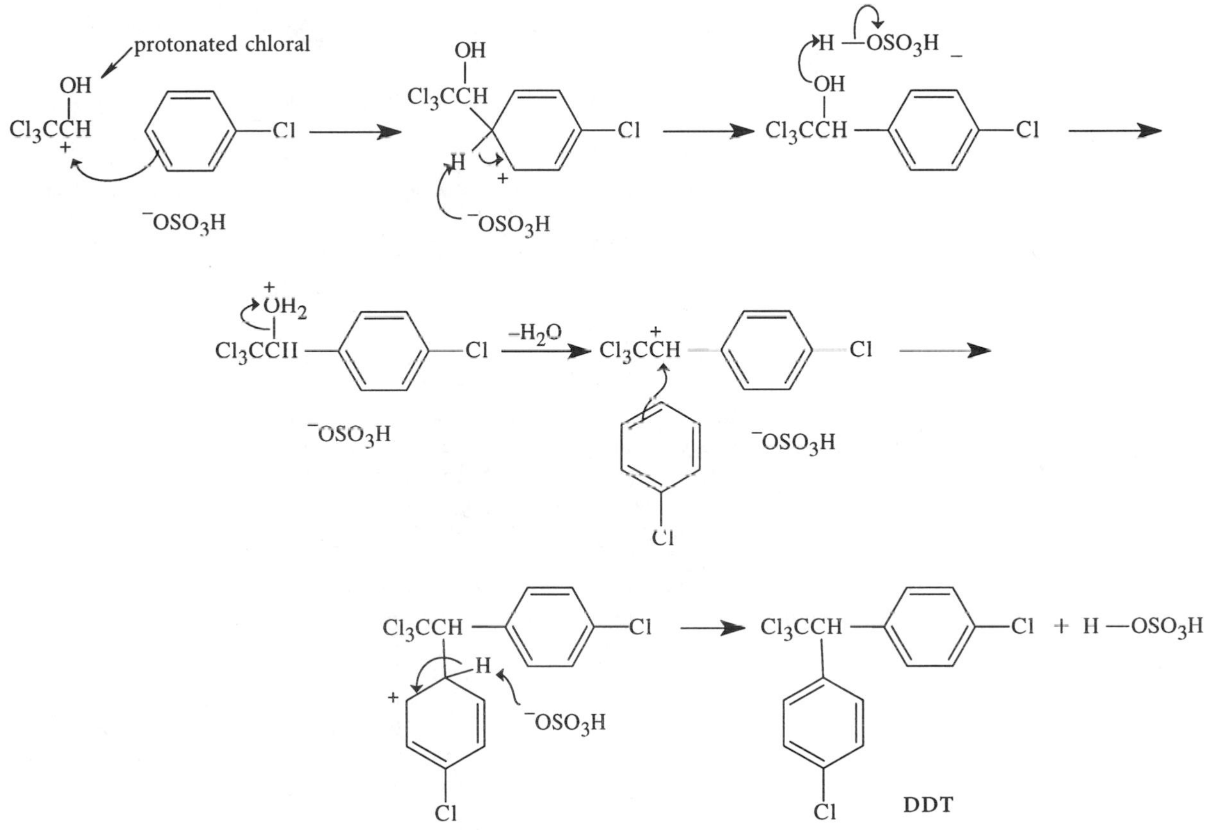

protonated chloral

$Cl_3C\overset{+}{C}H(OH)$ + chlorobenzene ($^{-}OSO_3H$) → $Cl_3CCH(OH)$–cyclohexadienyl cation–Cl (H, $^{-}OSO_3H$) → $Cl_3CCH(OH)$–C_6H_4–Cl ($H{-}OSO_3H$) →

$Cl_3CCH(\overset{+}{O}H_2)$–$C_6H_4$–Cl ($^{-}OSO_3H$) $\xrightarrow{-H_2O}$ $Cl_3C\overset{+}{C}H$–C_6H_4–Cl + chlorobenzene ($^{-}OSO_3H$) →

Cl_3CCH–C_6H_4–Cl (cyclohexadienyl cation, H, Cl; $^{-}OSO_3H$) → $Cl_3CCH(C_6H_4Cl)$–C_6H_4–Cl + $H{-}OSO_3H$

DDT

19.54 (a) In this case, $LiAlD_4$ serves as a source of nucleophilic isotopic hydride (deuteride); deuteride opens the epoxide with inversion of configuration.

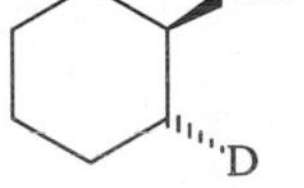

trans-2-deuteriocyclohexanol (OH, D) (racemate)

(b) Hydride ion, delivered from $LiAlH_4$, opens the epoxide with inversion of configuration. Because the epoxide carbons are enantiotopic and therefore chemically equivalent, attack of hydride at each of these carbons occurs at the same rate. As a result, the racemic product is formed. An equivalent view is that the starting epoxide is a meso compound. Hence, the chiral product must be a racemate.

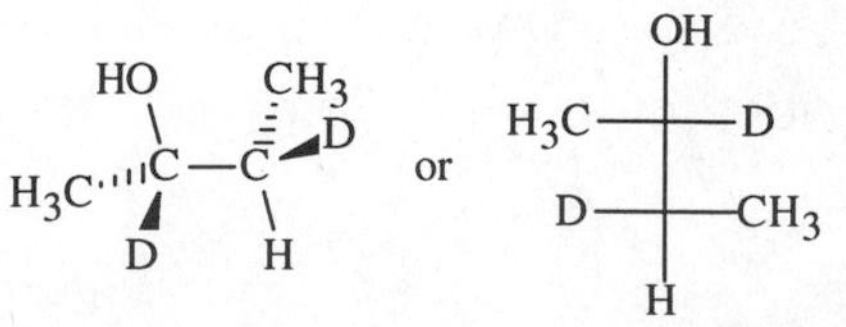

(formed as the racemate)

19.55 (a) Thumbs wants a Grignard reagent to react selectively with a ketone in the presence of an aldehyde. Because aldehydes are more reactive than ketones, the aldehyde, not the ketone, will react most rapidly.

(b) Thumbs is trying to form a Wittig reagent by first carrying out an S_N2 reaction of triphenylphosphine (Ph_3P) with neopentyl bromide. Because neopentyl bromide is virtually unreactive in S_N2 reactions, this reaction will not occur. (See Fig. 9.3B on text p. 365 and associated discussion.)

19.56 (a)

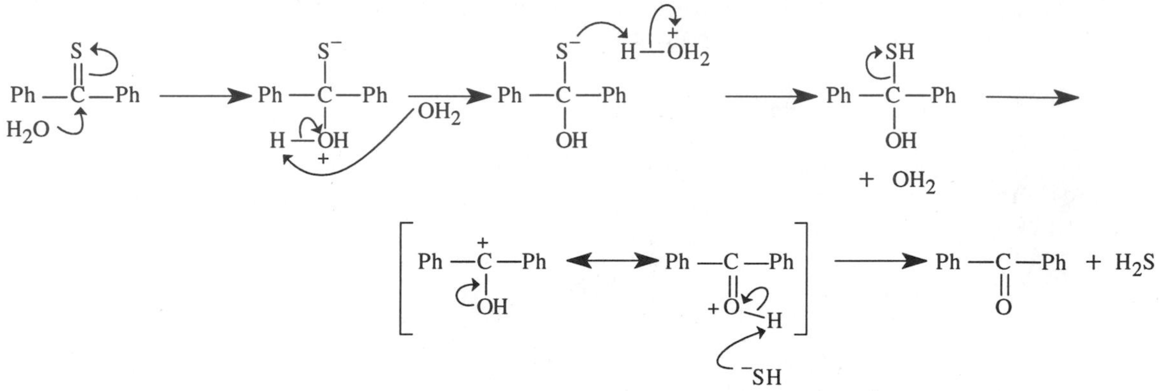

Compare loss of the —SH group, which does not require protonation, to loss of the —OH group in the acid-catalyzed decomposition of a hydrate, which does require protonation. Protonation of the —SH group is not necessary for it to serve as a leaving group for two reasons. First, the C—S bond is substantially weaker than the O—H bond; and second, ^-SH is a much weaker base than ^-OH. Furthermore, the —SH group itself is a much weaker base than the —OH group; consequently, much less —SH than —OH is protonated under neutral or dilute-acid conditions.

(b) S_N1 ionization of the halide gives a resonance-stabilized carbocation that is attacked by water to give a hemiacetal; the hemiacetal hydrolyzes to the aldehyde.

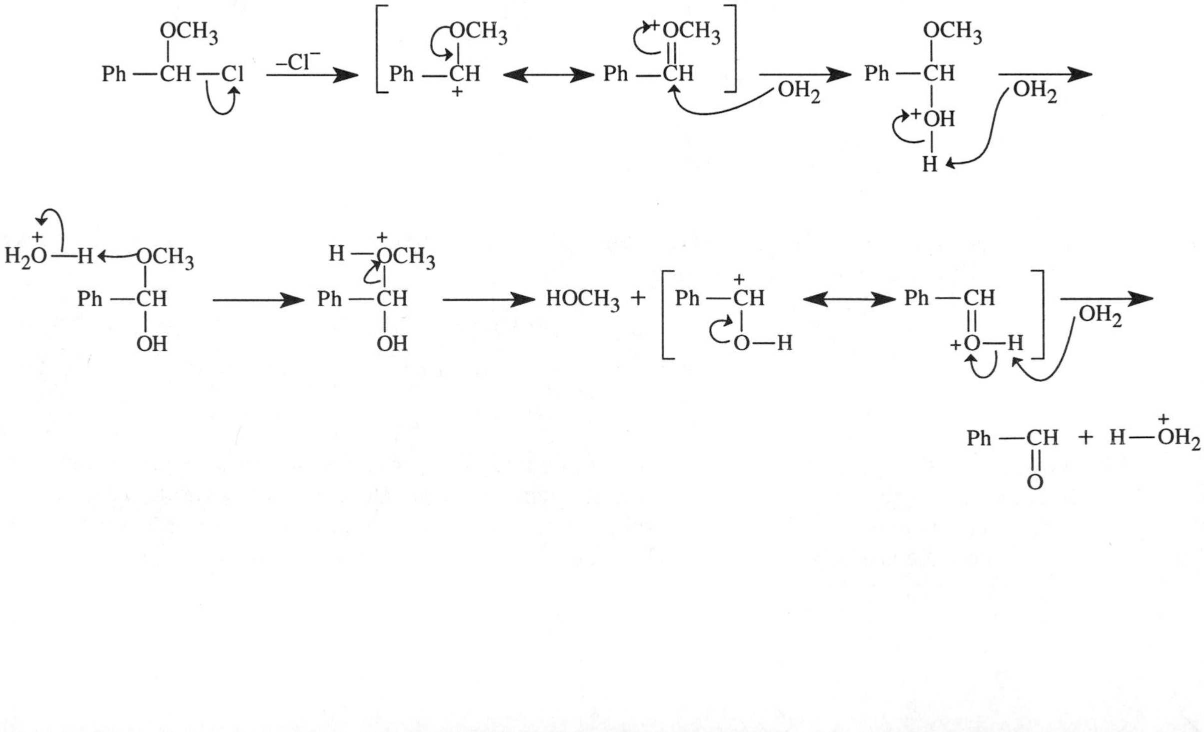

(c) Opening of the epoxide by the phosphorus compound yields an oxaphosphetane, the same type of compound that serves as an intermediate in the Wittig reaction; this compound collapses to an alkene as it does in the Wittig reaction. (See also the solution to Problem 19.44.)

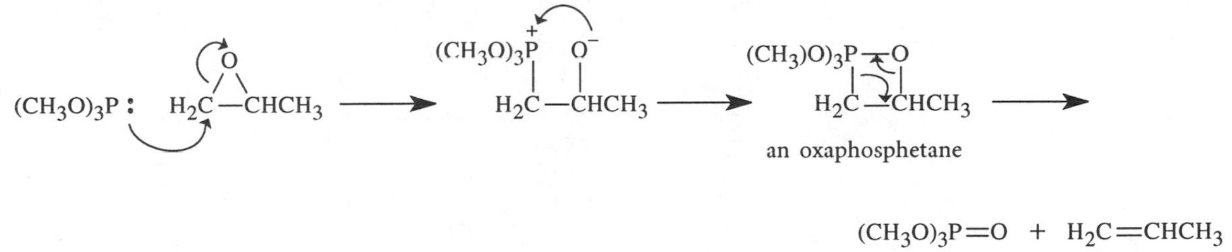

$(CH_3O)_3P$: H_2C—$CHCH_3$ (epoxide) ⟶ $(CH_3O)_3\overset{+}{P}$ O^- / H_2C—$CHCH_3$ ⟶ $(CH_3O)_3P$—O / H_2C—$CHCH_3$ ⟶

an oxaphosphetane

$(CH_3O)_3P{=}O$ + $H_2C{=}CHCH_3$

(d) The imine, formed by the mechanism shown in Eqs. 19.56a–b on text p. 875, is in equilibrium with a small amount of an enamine (just as an aldehyde is in equilibrium with a small amount of enol). The nitrogen of the enamine serves as a nucleophile to attack the second aldehyde carbonyl group intramolecularly to form the ring.

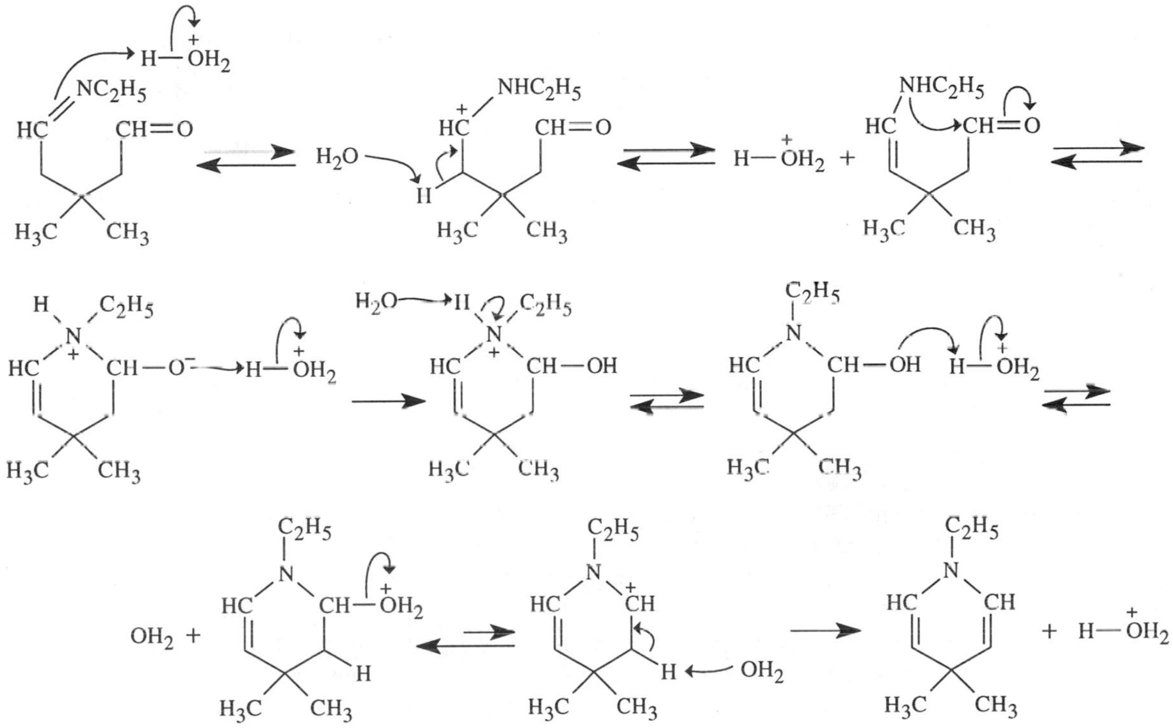

H—$\overset{+}{O}H_2$; NC_2H_5; HC; CH=O; H_3C; CH_3 ⇌ H_2O; NHC_2H_5; $\overset{+}{H}C$; CH=O; H; H_3C; CH_3 ⇌ H—$\overset{+}{O}H_2$ + NHC_2H_5; HC; CH=O; H_3C; CH_3 ⇌

H; C_2H_5; $\overset{+}{N}$; HC; CH—O^- ⟶ H—$\overset{+}{O}H_2$; H_3C; CH_3 ⟶ H_2O ⟶ H; C_2H_5; $\overset{+}{N}$; HC; CH—OH; H_3C; CH_3 ⇌ C_2H_5; N; HC; CH—OH; H—$\overset{+}{O}H_2$; H_3C; CH_3 ⇌

OH_2 + C_2H_5; N; HC; CH—$\overset{+}{O}H_2$; H; H_3C; CH_3 ⇌ C_2H_5; N; HC; $\overset{+}{C}H$; H ⟵ OH_2; H_3C; CH_3 ⟶ C_2H_5; N; HC; CH; H_3C; CH_3 + H—$\overset{+}{O}H_2$

(e) Protonation of the aldehyde gives an α-hydroxy carbocation that alkylates the double bond. The mechanism below starts with the protonated aldehyde.

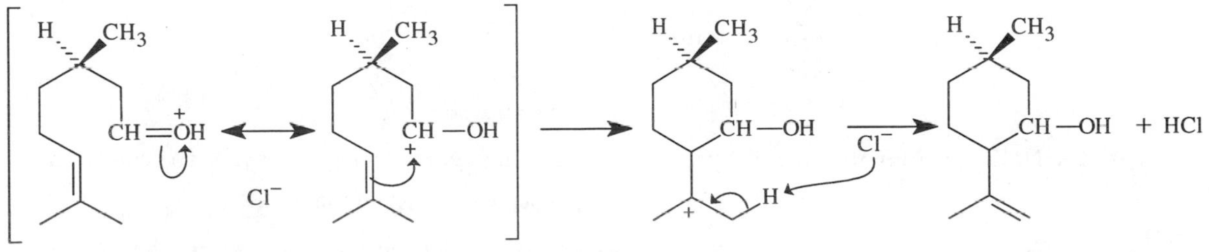

[H; CH_3; CH=$\overset{+}{O}$H ⟷ H; CH_3; $\overset{+}{C}H$ —OH; Cl^-] ⟶ H; CH_3; CH—OH; H; + ⟶ (Cl^-) H; CH_3; CH—OH + HCl

protonated aldehyde

(f) The carbonyl oxygen serves as a nucleophile to attach the carbonyl carbon of a protonated acetaldehyde molecule.

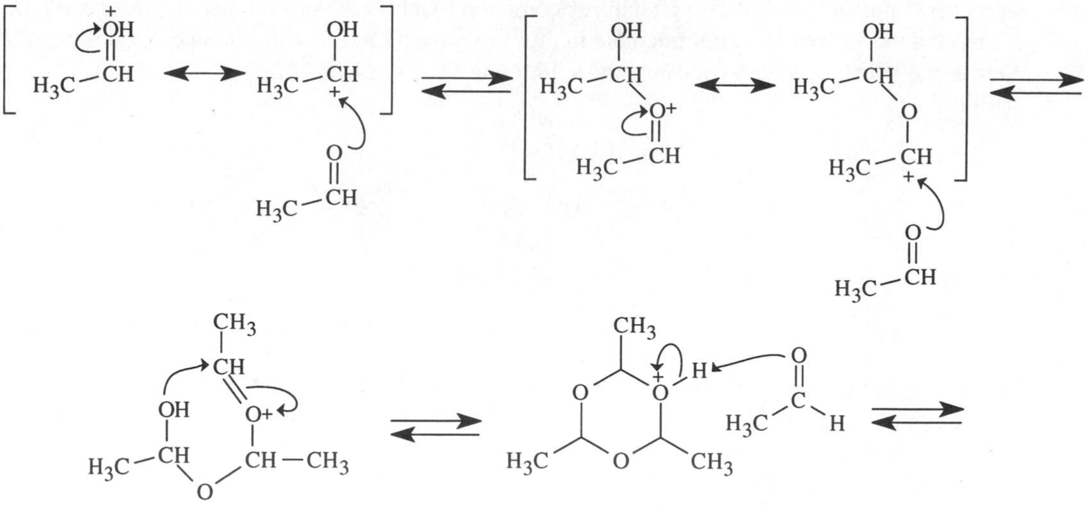

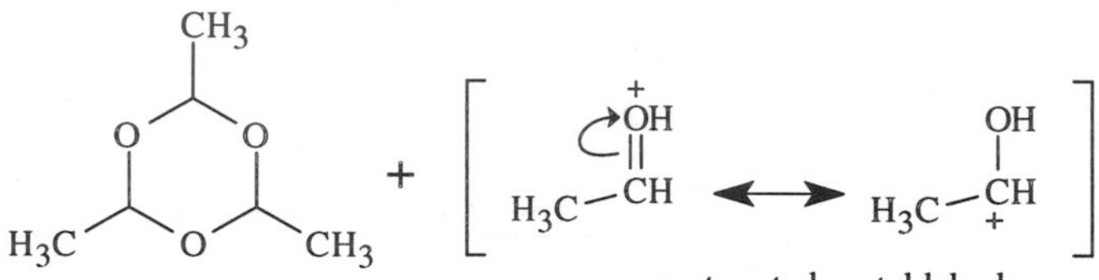

paraldehyde

protonated acetaldehyde; reacts to give more paraldehyde

19.57 (a) The synthesis of 4-methyl-3-heptanol:

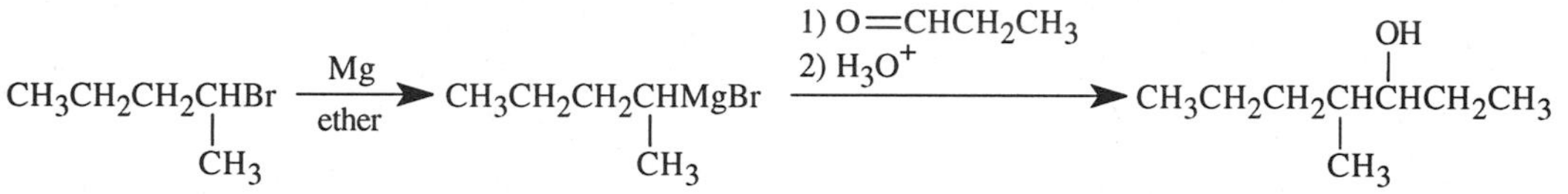

$$CH_3CH_2CH_2CHBr(CH_3) \xrightarrow[\text{ether}]{Mg} CH_3CH_2CH_2CH(CH_3)MgBr \xrightarrow[2)\ H_3O^+]{1)\ O{=}CHCH_2CH_3} CH_3CH_2CH_2CH(CH_3)CH(OH)CH_2CH_3$$

4-methyl-3-heptanol

(b) Because 4-methyl-3-heptanol contains two asymmetric carbons, it can exist as diastereomers. Even if the starting alkyl halide were enantiomerically pure, and even if the Grignard reagent could be prevented from racemizing, the product would likely be a mixture of diastereomers.

19.58 (a) The NMR spectrum suggests a great degree of symmetry: four aromatic protons, all equivalent, and two methyl groups adjacent to a carbonyl group. The IR spectrum suggests the presence of a carbonyl group conjugated with the aromatic ring. The chemical shift of the ring protons is greater than usual because of the electronegative carbonyl groups. Note that all four ring protons are chemically equivalent; thus, they appear in the NMR spectrum as a singlet.

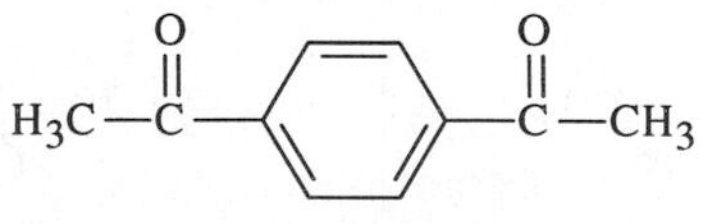

$H_3C{-}C({=}O){-}C_6H_4{-}C({=}O){-}CH_3$

p-diacetylbenzene

(b) The NMR spectrum indicates the presence of a *tert*-butyl group and an aldehyde. The compound is

$(CH_3)_3C{-}CH{=}O$ 2,2-dimethylpropanal (pivalaldehyde)

(c) The NMR indicates two vinylic hydrogens as well as an aldehyde adjacent to a CH. The IR absorption at 970 cm^{-1} shows that the vinylic hydrogens are trans. The double bond and the aldehyde account for all of the unsaturation. The UV absorption at 215 nm indicates that the alkene and aldehyde double bonds are conjugated. The allylic CH_2 absorption at δ 2.3 is essentially a quartet, which suggests three hydrogens on adjacent carbons, that is, a vinylic hydrogen (because the absorption is allylic) and two others. The high-field triplet integrates for three protons, and thus represents a methyl group; its splitting suggests an adjacent CH_2 group. The absorption at δ 5.2 imples a $—CH_2—$ adjacent to five protons. The following structure is thus defined by the data:

$CH_3CH_2CH_2(H)C=C(H)CH=O$ (*E*)-2-hexenal

19.59 The double absorptions of equal intensity in the mass spectrum indicate the presence of one bromine. The IR absorptions indicate a ketone in which the carbonyl group is conjugated to a benzene ring (1678 cm^{-1}) as well as aromatic "double bonds" (1600 cm^{-1}). The NMR spectrum indicates a *para*-disubstituted benzene ring and a methyl group adjacent to a carbonyl group. The compound has structure *A* below (*p*-bromoacetophenone). This compound, of course, has the molecular mass (198 for ^{79}Br) indicated by the mass spectrum.

$Br—C_6H_4—C(=O)—CH_3$

p-bromoacetophenone
A

$H_3C—C_6H_4—C(=O)—Br$

B

If you proposed structure *B*, it is not a bad answer, because you probably don't know about the IR spectra of acid halides. However, the IR carbonyl absorption of such a compound occurs at much higher frequency; furthermore, the chemical shift of the methyl group is more consistent with its being adjacent to a carbonyl group.

19.60 (a) The cyclic trimer can exist as a mixture of diastereomers:

Cl_3C, CCl_3, Cl_3C (trioxane ring) and Cl_3C, CCl_3, Cl_3C (trioxane ring)

β-form *α*-form

two diastereomers of the cyclic trimer of chloral

(You were not expected to specify which form is which.) Notice that the higher-melting form has the more symmetrical structure.

(Note that paraldehyde in Problem 19.55, part (f), can also exist as diastereomers.)

(b) The *α*-form has two chemically nonequivalent sets of protons; hence, its proton NMR spectrum has two resonances in a 2:1 ratio. Because all protons in the *β*-form are chemically equivalent, this form has only one resonance in its NMR spectrum. Related conclusions apply to the CMR spectra: the *α*-form has four lines in its spectrum, whereas the *β*-form has only two.

19.61 (a) Subject the ketone benzyl phenyl ketone to an excess of isotopically enriched water in the presence of an acidic or basic catalyst. Several cycles of formation and decomposition of the hydrate will eventually "wash out" the isotope of lower atomic mass. (Note that $*O = {}^{18}O$)

$$PhC(=O)CH_2Ph + H_2O^* \rightleftharpoons PhC(OH)(O^*H)CH_2Ph \longrightarrow PhC(=O^*)CH_2Ph + H_2O$$

benzyl phenyl ketone; H_2O^* (large excess)

Note that the last step is effectively irreversible only because of the large excess of isotopically enriched water; the unenriched water is present at very low concentration at all times. The ratio of enriched to unenriched ketone will be essentially the same as the ratio of enriched to unenriched water at equilibrium. Once the ketone is formed, reduce it with $LiAlH_4$ or $NaBH_4$ to provide the desired product:

$$\mathrm{PhC(=O^{*})CH_2Ph} \xrightarrow[\text{2) } H_3O^{+}]{\text{1) } LiAlH_4} \mathrm{PhCH(\overset{*}{O}H)CH_2Ph}$$

Another reasonable synthesis is oxymercuration-reduction of *cis*- or *trans*-stilbene ("1,2-diphenylethylene") with isotopically enriched water used as the nucleophile in the oxymercuration step.

$$\underset{\textit{cis- or trans-}\text{stilbene}}{\mathrm{PhCH{=}CHPh}} \xrightarrow[\text{2) } NaBH_4]{\text{1) } Hg(OAc)_2/H_2O^{*}/THF} \mathrm{PhCH(\overset{*}{O}H)CH_2Ph}$$

The S_N1 reaction of the corresponding alkyl bromide with isotopic water would not be a satisfactory synthesis because this particular S_N1 reaction gives mostly *trans*-stilbene (structure above), the elimination (E1) product. This alkene is the major product because it has an internal conjugated double bond and is therefore very stable.

(b) Allow benzyl phenyl ketone to react with $LiAlD_4$.

$$\underset{\text{benzyl phenyl ketone}}{\mathrm{PhC(=O)CH_2Ph}} \xrightarrow[\text{2) } H_3O^{+}]{\text{1) } LiAlD_4} \mathrm{PhCD(OH)CH_2Ph}$$

19.62 (a) In 1,2-cyclopentanedione the C—O bond dipoles are constrained by the geometry of the ring to be aligned at an angle of about 72°. This alignment provides a significant resultant dipole, as shown by diagram *A*:

1,2-cyclopentanedione

A

biacetyl

B

In biacetyl (2,3-butanedione), rotation about the central carbon-carbon bond is possible. Evidently, the two bond dipoles are oriented so that their resultant is less than the resultant in 1,2-cyclopentanedione. Notice that if their dihedral angle were 180°, their resultant would be zero. Hence, the bond dipoles are aligned at an angle somewhere between 72° and 180°, as shown in the Newman projection of diagram *B* above. (The angle turns out to be about 135°.) Aligned bond dipoles cause electrostatic repulsion. By undergoing internal rotation, biacetyl avoids this destabilizing effect.

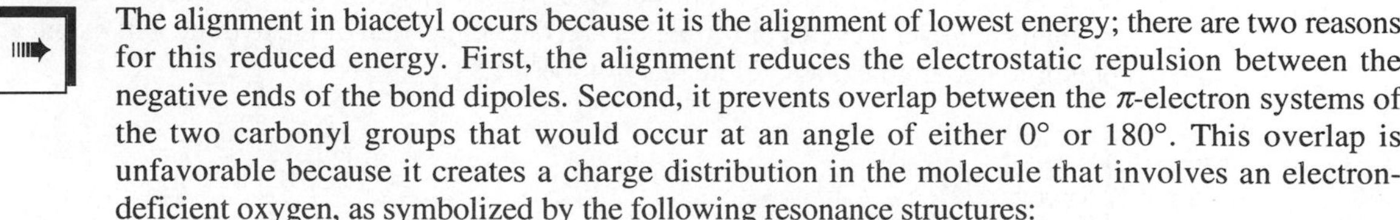

The alignment in biacetyl occurs because it is the alignment of lowest energy; there are two reasons for this reduced energy. First, the alignment reduces the electrostatic repulsion between the negative ends of the bond dipoles. Second, it prevents overlap between the π-electron systems of the two carbonyl groups that would occur at an angle of either 0° or 180°. This overlap is unfavorable because it creates a charge distribution in the molecule that involves an electron-deficient oxygen, as symbolized by the following resonance structures:

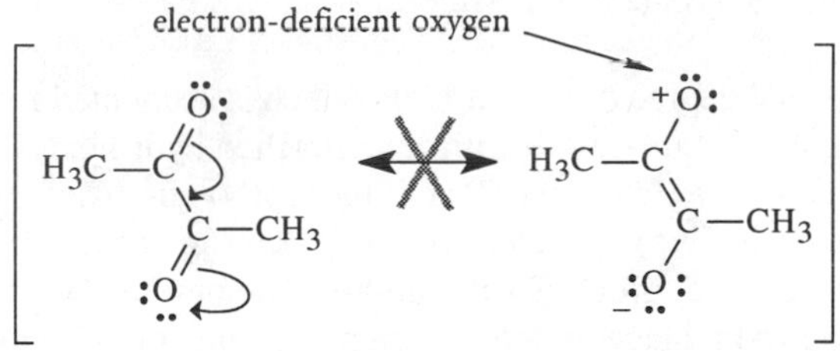

electron-deficient oxygen

The information given in the problem plus a simplifying assumption that the cyclopentane ring has the shape of a regular pentagon allows you to use the law of cosines to calculate the bond dipole of a C═O bond as well as the angle between the two carbonyl bond dipoles in biacetyl (given above as about 135°). Try it!

(b) The $n \rightarrow \pi^*$ absorption is characteristic of the carbonyl group. This absorption disappears because a reaction occurs in which the carbonyl group is converted into another group that does not have this absorption. This reaction is addition of ethanethiol to give the sulfur analog of a hemiacetal:

$$H_3C—CH{=}O + CH_3CH_2SH \rightleftharpoons H_3C—CH(SCH_2CH_3)—OH$$

has a carbonyl group; has $n \rightarrow \pi^*$ absorption

has no carbonyl group; has no $n \rightarrow \pi^*$ absorption

(c) The Tollens test (oxidation with the ammonia complex of Ag^+) is a characteristic reaction of aldehydes. This compound reacts slowly because very little of it is in the aldehyde form; rather, the molecule exists largely as a cyclic hemiacetal.

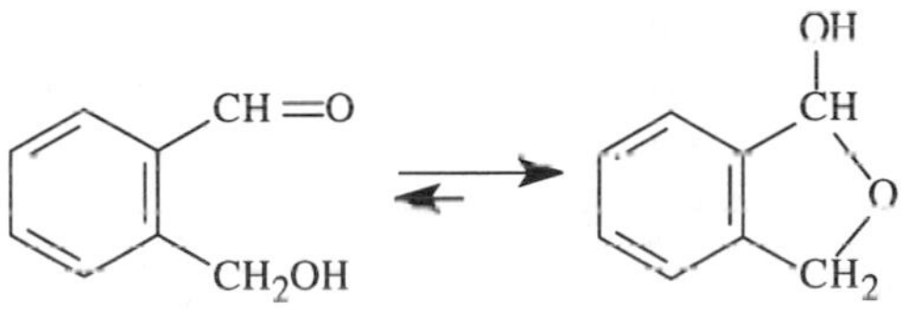

reacts rapidly with the Tollens reagent

The molecule exists mostly in this cyclic hemiacetal form, which is not an aldehyde, and which therefore does not react with the Tollens reagent.

19.63 The molecular formula requires three degrees of unsaturation. The six-proton singlet suggests two chemically equivalent methyl groups. The IR indicates the presence of a carbonyl group. The two doublets at about δ 6 and δ 7.5 are vinylic hydrogens, one of which is significantly shifted by some electronegative group, no doubt the carbonyl group. If the vinylic hydrogens account for one double bond and a carbonyl group accounts for another, we are still missing one degree of unsaturation, which thus must be a ring. The two remaining protons are in the chemical-shift region for protons α to a carbonyl. The data leave only one possibility:

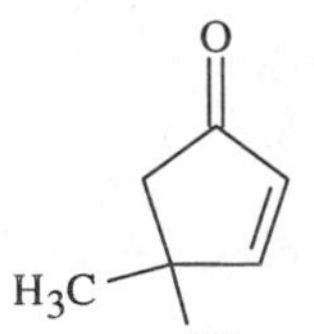

4,4-dimethyl-2-cyclopentenone

If we return to Eq. 19.4 on text p. 645, we see that cyclopentanones should have IR carbonyl absorptions at about 1745 cm^{-1}. However, conjugation reduces this frequency by 40–45 cm^{-1}, as shown by the trends in Eq. 19.2 on text p. 844. This means that the observed carbonyl frequency of 1703 cm^{-1} is exactly what we would expect for this compound. Furthermore, the small vinylic coupling constant is expected for two protons in a cis relationship (see Table 13.3 on text p. 576).

19.64 The formula indicates one degree of unsaturation which is accounted for by a ketone (IR absorption at 1710 cm^{-1}, CMR absorption at δ 204.9 with zero attached hydrogens). The remaining oxygens cannot be alcohols (absence of O—H stretch in the IR) or other carbonyls. The CMR-DEPT data indicate the presence of at least one methyl ether (δ 53.5) as well as a carbon bearing one hydrogen (δ 101.7) that could also be attached to two oxygens; thus, an acetal group is a possibility. The proton NMR data indicate a methyl group attached to a carbonyl. The only carbon unaccounted for is a methylene group with a CMR absorption at δ 47.2, which places it adjacent to the carbonyl and near, but not adjacent to, the acetal. All the data conspire to define the following structure:

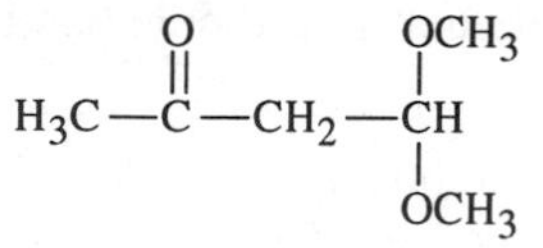

$$H_3C{-}\overset{\overset{\large O}{\|}}{C}{-}CH_2{-}\underset{\underset{\large OCH_3}{|}}{\overset{\overset{\large OCH_3}{|}}{CH}}$$

4,4-dimethoxy-2-butanone

20

Chemistry of Carboxylic Acids

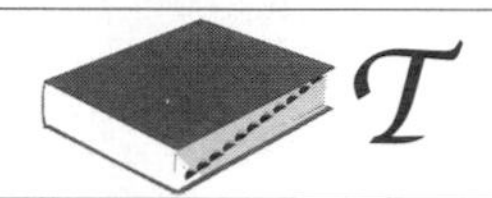

TERMS

acid chloride 20.9A
acid-catalyzed esterification 20.8A
acyloxy group 20.9B
anhydride 20.9B
anionic surfactant 20.5
carbonic acid derivative 20.11
carbonyl oxygen 20.2
carboxy group .... Introduction
carboxylate ion 20.4A
carboxylate oxygen 20.2
carboxylic acid .... Introduction
cationic surfactant 20.5
critical micelle concentration (CMC) 20.5
cyclic anhydride 20.9B
decarboxylation 20.7, 20.11
detergent 20.5
dicarboxylic acid 20.1A
dimer 20.2
ester 20.8A
fatty acid 20.5
Fischer esterification 20.8A
β-keto acid 20.11
malonic acid derivative 20.11
micelle 20.5
soap 20.5
sulfonic acid .... Introduction
surfactant 20.5
tetrahedral addition intermediate 20.8A

CONCEPTS

I. INTRODUCTION TO CARBOXYLIC ACIDS

A. Common Nomenclature of Carboxylic Acids

1. The characteristic functional group in a carboxylic acid is the carboxy group, $-CO_2H$.
2. Common nomenclature is widely used for the simpler carboxylic acids, some names of which owe their origins to natural sources.
3. In common nomenclature, a carboxylic acid is named by adding the suffix *ic* and the word *acid* to the prefix for the appropriate group.

form	$H-$	isobutyr	$(CH_3)_2CH_2-$
acet	CH_3-	valer	$CH_3CH_2CH_2CH_2-$
propion	CH_3CH_2-	isovaler	$(CH_3)_2CHCH_2-$
butyr	$CH_3CH_2CH_2-$	benzo	$Ph-$

4. Substitution in the common system is denoted with Greek letters rather than numbers; the position adjacent to the carboxy group is designated as α; the position of the substituent is omitted if it is unambiguous.

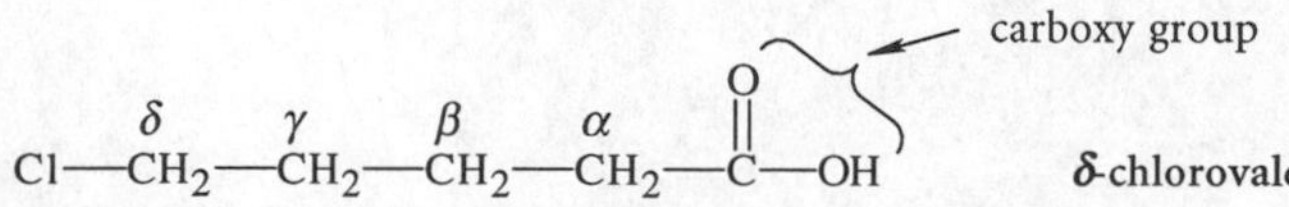

$$\overset{\delta}{Cl-CH_2}-\overset{\gamma}{CH_2}-\overset{\beta}{CH_2}-\overset{\alpha}{CH_2}-\overset{O}{\overset{\|}{C}}-OH$$

carboxy group

δ-chlorovaleric acid

5. Carboxylic acids with two carboxy groups are called dicarboxylic acids.
   a. The unbranched dicarboxylic acids are particularly important, and are invariably known by their common names:

oxalic acid	HO_2CCO_2H	glutaric acid	$HO_2C(CH_2)_3CO_2H$
malonic acid	$HO_2CCH_2CO_2H$	adipic acid	$HO_2C(CH_2)_4CO_2H$
succinic acid	$HO_2C(CH_2)_2CO_2H$	pimelic acid	$HO_2C(CH_2)_5CO_2H$

   b. Phthalic acid is an important aromatic dicarboxylic acid.

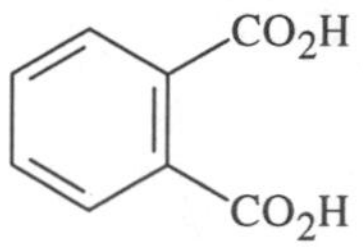

phthalic acid

6. Some carboxylic acids are known by time-honored traditional names.

$Ph-CH=CH-CO_2H$

cinnamic acid

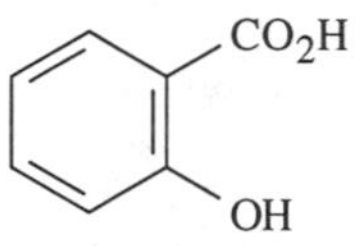

salicylic acid

B. Systematic Nomenclature of Carboxylic Acids

1. A carboxylic acid is named systematically by dropping the final *e* from the name of the hydrocarbon with the same number of carbon atoms and adding the suffix *oic* and the word *acid* (the final *e* is not dropped in the name of dicarboxylic acids).

$CH_3CH_2CH_2CO_2H$ butanoic acid

$HO_2CCH_2CH_2CH_2CO_2H$ pentanedioic acid

2. When the carboxylic acid is derived from a cyclic hydrocarbon, the suffix *carboxylic* and the word *acid* is added to the name of the hydrocarbon. (One exception to this nomenclature is benzoic acid, for which the IUPAC recognizes the common name.)

3. The principal chain in substituted carboxylic acids is numbered by assigning the number 1 to the carbonyl carbon. (In carboxylic acids derived from cyclic hydrocarbons, numbering begins at the ring carbon bearing the carboxy group).

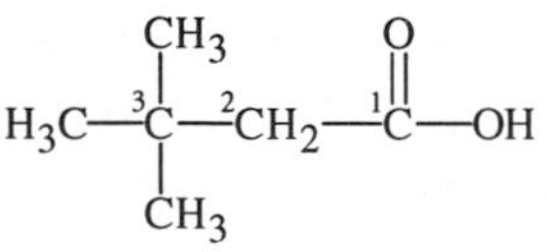

$H_3C-\overset{3}{C}(CH_3)_2-\overset{2}{C}H_2-\overset{1}{C}(=O)-OH$

3,3-dimethylbutanoic acid

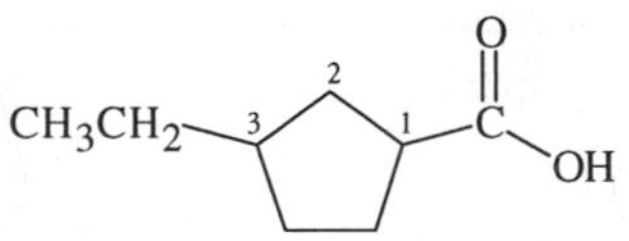

3-ethylcyclopentanecarboxylic acid

4. When carboxylic acids contain other functional groups, the carboxy groups receive priority over aldehyde and ketone carbonyl groups, over hydroxy groups, and over mercapto groups for citation as the principal group. (A complete list of nomenclature priorities for all the functional groups covered in the text is given in Appendix I, text page A-1.)

$$-CO_2H > -CH{=}O > \rangle C{=}O > -OH > -SH$$

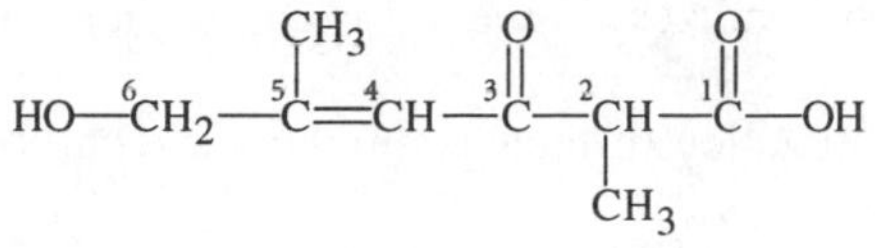

$$HO-\overset{6}{C}H_2-\overset{5}{\underset{}{\overset{CH_3}{\overset{|}{C}}}}=\overset{4}{C}H-\overset{3}{\overset{O}{\overset{||}{C}}}-\overset{2}{\underset{CH_3}{\underset{|}{C}H}}-\overset{1}{\overset{O}{\overset{||}{C}}}-OH$$

6-hydroxy-2,5-dimethyl-3-oxo-4-hexenoic acid

5. The carboxy group is sometimes named as a substituent.

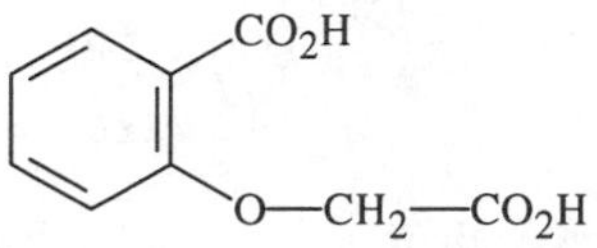

(benzene ring bearing CO_2H and $O-CH_2-CO_2H$ on adjacent carbons)

2-(carboxymethoxy)benzoic acid

C. Structure and Physical Properties of Carboxylic Acids

1. Carboxylic acids have trigonal geometry at their carbonyl carbons.
2. The two oxygens of a carboxylic acid are quite different.
   a. The carbonyl oxygen is part of the C=O double bond, which has about the same bond length as the C=O double bond of aldehydes and ketones.
   b. The carboxylate oxygen is part of the C—O single bond.
   c. Because the carbonyl-oxygen single bond is an sp^2-sp^3 single bond, it is considerably shorter than the C—O bond in an alcohol or ether, which is an sp^3-sp^3 single bond.
3. The carboxylic acids of lower molecular mass have acrid, piercing odors and considerably higher boiling points than many other organic compounds of about the same molecular weight and shape.
   a. The high boiling points of carboxylic acids can be attributed not only to their polarity, but also to the fact that they form very strong hydrogen bonds.
   b. In the solid state, and under some conditions in both the gas phase and solution, carboxylic acids exist as hydrogen-bonded dimers.
4. Many aromatic and dicarboxylic acids are solids.
5. The simpler carboxylic acids have substantial solubilities in water; the unbranched carboxylic acids below pentanoic acid are miscible with water. (Many dicarboxylic acids also have significant water solubilities.)

II. ACIDITY AND BASICITY OF CARBOXYLIC ACIDS

A. Acidity of Carboxylic and Sulfonic Acids

1. The acidity of a carboxylic acid is due to ionization of the O—H group to give the conjugate-base carboxylate ion.

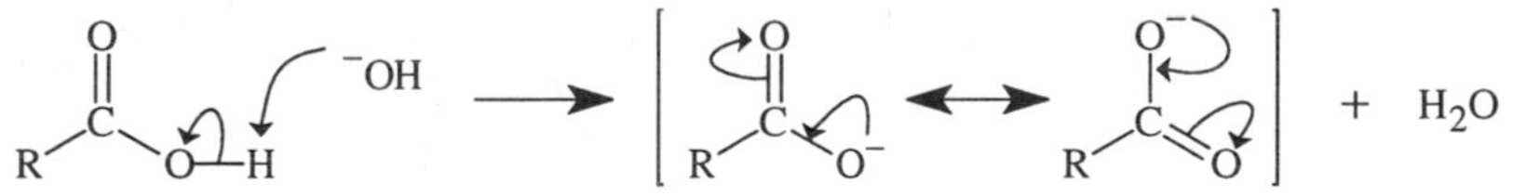

$$R-\overset{O}{\overset{||}{C}}-O-H \;+\; {}^{-}OH \longrightarrow \left[R-\overset{O}{\overset{||}{C}}-O^{-} \longleftrightarrow R-\overset{O^{-}}{\overset{|}{C}}=O \right] + H_2O$$

a carboxylate ion

   a. Carboxylic acids, which have typical pK_a values of 3–5, are among the most acidic organic compounds; they are more acidic than alcohols (pK_a = 15–17) or phenols (pK_a = 9–11).
   b. The conjugate bases of carboxylic acids are called generally carboxylate ions; carboxylate salts are named by replacing the *ic* in the name of the acid with the suffix *ate*.

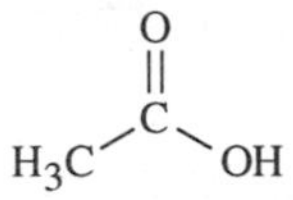

$$H_3C-\overset{O}{\overset{||}{C}}-OH \qquad\qquad H_3C-\overset{O}{\overset{||}{C}}-O^{-}$$

acetic acid 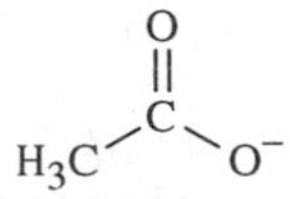 — acetate

2. The acidity of carboxylic acids is due to two factors.
   a. The first factor is the polar effect of the carbonyl group.
      i. The carbonyl group is much more electron-withdrawing than the phenyl ring of a phenol or the alkyl group of an alcohol.

ii. The polar effect of the carbonyl group stabilizes charge in the carboxylate ion (stabilization of a conjugate base enhances acidity).

b. The second factor is the resonance stabilization of their conjugate-base carboxylate ions. (See the resonance structures in item 1 above.)

3. The acidity of carboxylic acids, like that of alcohols, is influenced by the polar effects of substituents.

CH_3CO_2H	FCH_2CO_2H
$pK_a = 4.76$	$pK_a = 2.66$

4. Sulfonic acids ($pK_a \approx -1$) are much stronger than comparably substituted carboxylic acids and are useful as acid catalysts in organic solvents because they are more soluble than most inorganic acids.
   a. The sulfur atom in a sulfonic acid has a high oxidation state.
   b. The octet structure for a sulfonate anion indicates that sulfur has considerable positive charge, which stabilizes the negative charge on the oxygens.

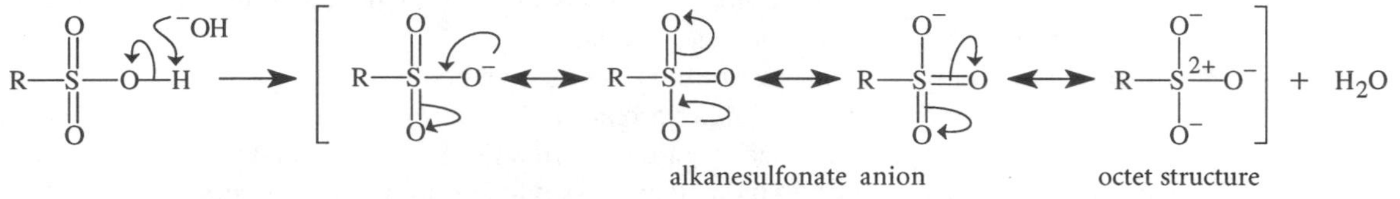

$R{-}SO_2{-}OH + {}^{-}OH \longrightarrow [\text{resonance structures}] + H_2O$

alkanesulfonate anion

octet structure

5. Although many carboxylic acids of moderate molecular weight are not soluble in water, their alkali metal salts are ionic compounds that are much more soluble in water.
   a. Many water-insoluble carboxylic acids dissolve in solutions of alkali metal hydroxides (NaOH, KOH) because the insoluble acids are converted completely into their soluble salts.
      i. A typical carboxylic acid can be separated from mixtures with other water-insoluble, non-acidic substances by extraction with NaOH, Na_2CO_3, or $NaHCO_3$ solution.
      ii. After isolating the basic aqueous solution containing the carboxylate salt, it can be acidified with a strong acid to yield the carboxylic acid, which may be isolated by filtration or extraction with organic solvents.
   b. A carboxylic acid can also be separated from a phenol by extraction of the acid with 5% $NaHCO_3$, if the phenol is not one that is unusually acidic.

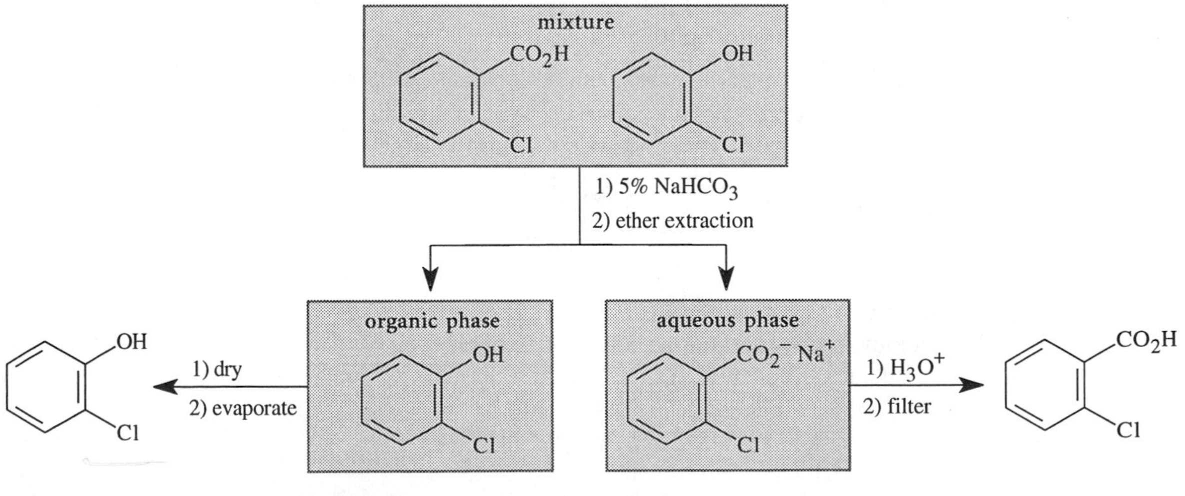

mixture

1) 5% $NaHCO_3$
2) ether extraction

organic phase

aqueous phase

1) dry
2) evaporate

1) H_3O^+
2) filter

B. Basicity of Carboxylic Acids

1. The carbonyl oxygen of a carboxylic acid is weakly basic. A carbonyl-protonated carboxylic acid has a typical pK_a of about –6.
   a. Protonation of a carboxylic acid on the carbonyl oxygen occurs because a resonance-stabilized cation is formed.

$$\mathrm{R{-}C({=}O){-}OH} + \mathrm{H{-}OH_2^+} \rightleftharpoons \left[\mathrm{R{-}C({=}\overset{+}{O}H){-}OH} \leftrightarrow \mathrm{R{-}\overset{+}{C}(OH){-}OH} \leftrightarrow \mathrm{R{-}C(OH){=}\overset{+}{O}H} \right] + H_2O$$

b. Protonation of the carboxylate oxygen is much less favorable because:
   i. it does not give a resonance-stabilized cation
   ii. the positive charge on oxygen is destabilized by the electron-withdrawing polar effect of the carbonyl group.

2. The basicity of carboxylic acids plays a very important role in many of their reactions.

C. Fatty Acids, Soaps, and Detergents

1. Carboxylic acids with long, unbranched carbon chains are called fatty acids.
   a. Fatty acids are liberated from fats and oils by a hydrolytic process called saponification.
   b. Some fatty acids contain carbon-carbon double bonds; fatty acids with cis double bonds occur widely in nature, but those with trans double bonds are rare.
   c. The sodium and potassium salts of fatty acids, called soaps, are the major ingredients of commercial soaps.

$$CH_3CH_2CH_2CH_2CH_2CH_2CH_2CH_2CH_2CH_2CH_2CH_2CH_2CH_2CH_2\overset{O}{\overset{\|}{C}}{-}O^- \; Na^+ \quad \text{or} \quad CH_3(CH_2)_{14}CO_2^- \; Na^+$$

sodium hexadecanoate (sodium palmitate), a soap

   i. Closely related to soaps are synthetic detergents, some of which are the salts of sulfonic acids.
   ii. Many soaps and detergents have not only cleansing properties, but also germicidal characteristics.

   d. Hard-water scum is a precipitate of the calcium or magnesium salts of fatty acids.

2. Soaps and detergents are two examples of a larger class of molecules known as surfactants (molecules with two structural parts that interact with water in opposing ways):
   a. They have a polar head group, which is readily solvated by water. In a soap, the polar head group is the carboxylate anion.
   b. They have a hydrocarbon tail, which is not well solvated by water. In a soap, the hydrocarbon tail is the carbon chain.
3. Soaps and detergents are examples of anionic surfactants, that is, surfactants with an anionic polar head group. (Cationic surfactants are also known.)
   a. When the surfactant concentration in water is raised above the critical micelle concentration (CMC), the surfactant molecules spontaneously form micelles, which are approximately spherical aggregates of 50–150 surfactant molecules.
   b. The micellar structure satisfies the solvation requirements of both the polar head groups, which are close to water (on the outside of the micelle), and the nonpolar tails, which associate with each other (on the inside of the micelle).
4. The antiseptic action of some surfactants owes its success to a phenomenon similar to micelle formation.
   a. A cell membrane is made up of molecules, called phospholipids, that are also surfactants.
   b. When the bacterial cell is exposed to a solution containing a surfactant, phospholipids of the cell membrane tend to associate with the surfactant.
   c. In some cases this disrupts the membrane enough that the cell can no longer function, and it dies.
5. Surfactants are also extremely important as components of fuels and lubricating oils.

III. SPECTROSCOPY OF CARBOXYLIC ACIDS

A. IR Spectroscopy

1. Two important absorptions in the infrared spectrum are hallmarks of a carboxylic acid:

a. A strong C=O stretching absorption occurs near 1710 cm^{-1} for carboxylic acid hydrogen-bonded dimers. (The C=O absorption of carboxylic acid monomers occurs near 1760 cm^{-1} but is rarely observed.)
b. The O—H stretching absorption of a carboxylic acid is much broader than the O—H stretching absorption of an alcohol or phenol, and covers a very wide region of the spectrum—typically 2400–3600 cm^{-1}.

2. A conjugated carbon-carbon double bond affects the position of the carbonyl absorption much less in acids than it does in aldehydes and ketones. (A substantial shift in the carbonyl absorption is observed, however, for acids in which the carboxy group is on an aromatic ring.)

B. NMR Spectroscopy

1. The α-protons of carboxylic acids show NMR absorptions in the δ 2.0–2.5 chemical shift region.
2. The carboxylic acid O—H proton resonance is typically found far downfield, in the δ 9–13 region, and in many cases it is broad.
   a. The O—H proton resonances of carboxylic acids occur at positions that depend on the acidity of the acid and on its concentration.
   b. The O—H proton resonances of carboxylic acids are readily distinguished from the resonances of aldehydic protons because only acid protons rapidly exchange with D_2O.
3. The carbonyl-carbon CMR absorptions of carboxylic acids are observed at somewhat higher field than those of aldehydes or ketones. (This unusual chemical shift is caused by shielding effects of the uncharged electron pairs on the carboxylate oxygen.)

REACTIONS

I. SYNTHESIS AND REACTIVITY OF CARBOXYLIC ACIDS

A. Synthesis of Carboxylic Acids—Review

1. Methods for preparing carboxylic acids introduced in previous chapters:
   a. Oxidation of primary alcohols (Sec. 10.6B, text page 433).

$$\text{—CH}_2\text{—OH} \xrightarrow[H_3O^+]{K_2CrO_7} \text{—C(=O)OH}$$

a primary alcohol → a carboxylic acid

$$\text{—CH}_2\text{—OH} \xrightarrow[^{-}OH]{KMnO_4} \xrightarrow{H_3O^+} \text{—C(=O)OH}$$

a primary alcohol → a carboxylic acid

   b. Side-chain oxidation of alkylbenzenes (Sec. 17.5, text page 758).

$$\text{Ar—CH}_2\text{—R} \xrightarrow[^{-}OH]{KMnO_4} \xrightarrow{H_3O^+} \text{Ar—C(=O)OH}$$

an aromatic hydrocarbon with benzylic hydrogens → an aromatic carboxylic acid

$$\text{Ar—CH}_2\text{—R} \xrightarrow[H_3O^+]{CrO_3} \text{Ar—C(=O)—OH}$$

an aromatic hydrocarbon with benzylic hydrogens → an aromatic carboxylic acid

c. Oxidation of aldehydes (Sec. 19.14, text page 884).

$$\text{—C(=O)—H} \xrightarrow[{}^-OH]{KMnO_4} \xrightarrow{H_3O^+} \text{—C(=O)—OH}$$

an aldehyde → a carboxylic acid

$$\text{—C(=O)—H} \xrightarrow{H_3O^+} \text{—C(OH)}_2\text{—H} \xrightarrow[H_3O^+]{K_2CrO_7} \text{—C(=O)—OH}$$

an aldehyde → an aldehyde hydrate → a carboxylic acid

d. Ozonolysis of alkenes followed by oxidative workup with H_2O_2 (Sec. 5.4, text page 173).

$$\text{>C=C(H)—} \xrightarrow[H_2O_2]{O_3} \text{>C=O} + \text{O=C(OH)—}$$

B. Synthesis of Carboxylic Acids with Grignard Reagents

1. The reactions of Grignard reagents with carbon dioxide, followed by protonolysis, give carboxylic acids.
   a. Addition of a Grignard reagent to carbon dioxide gives the halomagnesium salt of a carboxylic acid.
   b. When aqueous acid is added to the reaction mixture in a separate reaction step, the free carboxylic acid is formed.

$$\text{R—MgX} + \text{O=C=O} \longrightarrow \text{R—C(=O)—O}^- \, {}^+\text{MgX} \xrightarrow{H_3O^+} \text{R—C(=O)—OH} + \text{HOMgX}$$

2. The reaction of Grignard reagents with CO_2 is another method for the formation of carbon-carbon bonds. (See Appendix VI, text page A-13, for a review of other carbon-carbon bond-forming reactions.)

C. Introduction to Carboxylic Acid Reactions

1. Reactions of carboxylic acids can be categorized into four types:
   a. Reactions at the carbonyl group.
      i. The most typical reaction at the carbonyl group is substitution at the carbonyl carbon. In such a reaction, the —OH of the carboxy group is typically substituted by a nucleophilic group.
      ii. Reaction of the carbonyl oxygen with an electrophile (Lewis acid or Brønsted acid), that is, the reaction of the carbonyl oxygen as a base.
   b. Reactions at the carboxylate oxygen, such as reactions of the carboxylate oxygen as a nucleophile.
   c. Loss of the carboxy group as CO_2 (decarboxylation).
   d. Reactions involving the α-carbon.

2. Many substitution reactions at the carbonyl carbon are acid-catalyzed; that is, the reactions of nucleophiles at the carbonyl carbon are catalyzed by the reactions of acids at the carbonyl oxygen.

II. REACTION OF CARBOXYLIC ACIDS AT THE CARBONYL GROUP

A. Acid-catalyzed Esterification of Carboxylic Acids

1. Esters are carboxylic acid derivatives in which the proton on the carboxylate oxygen in effect has been replaced by an alkyl or aryl group.
2. When a carboxylic acid is treated with a large excess of an alcohol in the presence of a strong acid catalyst, an ester is formed; this reaction is called acid-catalyzed esterification (or Fischer esterification).

$$RCOOH + HO{-}R' \underset{}{\overset{H_3O^+}{\rightleftharpoons}} RCOOR' + H_2O$$

an ester

a. Acid-catalyzed esterification is a substitution of —OH at the carbonyl group of the acid by the —OR group of the alcohol; the —OH group leaves as water.
b. The equilibrium constants for esterification with most primary alcohols are near unity.
   i. The reaction is driven toward completion by applying LeChatelier's principle.
   ii. Use of the reactant alcohol as the solvent, which ensures that the alcohol is present in large excess, drives the equilibrium toward the ester product.
c. Acid-catalyzed esterification cannot be applied to the synthesis of esters from phenols or tertiary alcohols.

3. The mechanism of acid-catalyzed esterification serves as a model for the mechanisms of other acid-catalyzed reactions of carboxylic acids and their derivatives.
a. The first step of the mechanism is protonation of the carboxyl oxygen.
   i. The catalyzing acid is the conjugate acid of the solvent, formed by reaction of the solvent with a strong acid such as H_2SO_4.
   ii. Protonation of a carbonyl oxygen makes the carbonyl carbon more electrophilic because the carbonyl oxygen becomes a better electron acceptor.

$$RC(=O)OH + HOR'H^+ \longrightarrow \left[RC(={}^+OH)OH \longleftrightarrow R\overset{+}{C}(OH)OH \longleftrightarrow RC(OH)={}^+OH \right] + HOR'$$

b. Attack of alcohol on the carbonyl carbon of this carbocation, followed by loss of a proton, gives a tetrahedral addition intermediate.

$$RC(={}^+OH)OH + HOR' \longrightarrow R{-}C(OH)_2{-}\overset{+}{O}(H)R' \xrightarrow{HOR'} R{-}C(OH)_2{-}OR' + H_2\overset{+}{O}R'$$

tetrahedral addition intermediate

c. The tetrahedral addition intermediate, after protonation, loses water to give the conjugate acid of the ester.

$$R{-}C(OH)(OH){-}OR' + H\overset{+}{O}R(H) \longrightarrow R{-}C(OH)({}^+OH_2){-}OR' + HOR' \longrightarrow \left[RC(={}^+OH)OR' \longleftrightarrow R\overset{+}{C}(OH)OR' \longleftrightarrow RC(OH)={}^+OR' \right] + H_2O$$

d. Loss of a proton gives the ester product and regenerates the acid catalyst.

$$R\text{-}C(=\overset{+}{O}H)\text{-}OR' + HOR' \longrightarrow R\text{-}C(=O)\text{-}OR' + H_2\overset{+}{O}R'$$

B. Synthesis of Acid Chlorides

1. Acid chlorides are carboxylic acid derivatives in which the —OH group has been replaced by —Cl.
   a. Acid chlorides are often prepared from the corresponding carboxylic acids.
   b. Two reagents used for this purpose are thionyl chloride, $SOCl_2$, and phosphorus pentachloride, PCl_5.

$$R\text{-}C(=O)\text{-}OH \xrightarrow[\text{or } PCl_5]{SOCl_2} R\text{-}C(=O)\text{-}Cl + HCl + SO_2 \text{ (or } POCl_3)$$

2. Acid chloride synthesis fits the general pattern of substitution reactions at a carbonyl group; in this case, —OH is substituted by —Cl.
3. Sulfonyl chlorides are the acid chlorides of sulfonic acids.
   a. Sulfonyl chlorides are prepared by treatment of sulfonic acids or their sodium salts with PCl_5.
   b. Aromatic sulfonyl chlorides can be prepared directly by the reaction of aromatic compounds with chlorosulfonic acid, $ClSO_3H$.
      i. This reaction is a variation of aromatic sulfonation, an electrophilic aromatic substitution reaction.
      ii. Chlorosulfonic acid, the acid chloride of sulfuric acid, acts as an electrophile in this reaction.
      iii. The sulfonic acid produced in the reaction is converted into the sulfonyl chloride by reaction with another equivalent of chlorosulfonic acid.

$$C_6H_6 \xrightarrow{ClSO_3H} \left[C_6H_6(SO_3H)^+ \; Cl^-\right] \xrightarrow{-HCl} C_6H_5\text{-}SO_3H \xrightarrow[-H_2SO_4]{ClSO_3H} C_6H_5\text{-}SO_2Cl$$

C. Synthesis of Anhydrides

1. Carboxylic acid anhydrides are carboxylic acid derivatives in which the hydroxy group has been replaced by an acyloxy group.
   a. Anhydrides are prepared by treatment of carboxylic acids with strong dehydrating agents (usually P_2O_5).

$$R\text{-}C(=O)\text{-}OH \xrightarrow[-H_2O]{P_2O_5} R\text{-}C(=O)\text{-}O\text{-}C(=O)\text{-}R$$

an anhydride (the boxed $\text{-}O\text{-}C(=O)\text{-}R$ portion: acyloxy group)

   b. Most anhydrides may themselves be used as reagents for the preparation of other anhydrides.
2. Some dicarboxylic acids react with acetic anhydride to form cyclic anhydrides—compounds in which the anhydride group is part of a ring.
   a. Cyclic anhydrides containing five- and six-membered rings are readily prepared from the corresponding dicarboxylic acids.
   b. Formation of cyclic anhydrides with five- or six-membered rings is so facile that in some cases it occurs on heating the dicarboxylic acid.

$$HO\text{-}C(=O)\text{-}CH_2CH_2\text{-}C(=O)\text{-}OH \xrightarrow[-H_2O]{\text{heat}} \text{succinic anhydride (five-membered cyclic anhydride)}$$

3. Anhydrides, like acid chlorides, are used in the synthesis of other carboxylic acid derivatives.

D. Reduction of Carboxylic Acids to Alcohols

1. When a carboxylic acid is treated with $LiAlH_4$, then with dilute acid, a primary alcohol is formed.
   a. Before the reduction itself takes place, $LiAlH_4$ reacts with the acidic hydrogen of the carboxylic acid to give the lithium salt of the carboxylic acid and one equivalent of hydrogen gas. (The lithium salt of the carboxylic acid is the species that is actually reduced.)

$$\text{—C(=O)OH} \xrightarrow{LiAlH_4} \xrightarrow{H_3O^+} \text{—CH}_2\text{—OH}$$

2. The reduction occurs in two stages:
   a. AlH_3 (formed from the reaction of $LiAlH_4$ with the acidic proton of the carboxylic acid) reduces the carboxylate ion to an aldehyde.
   b. The aldehyde is rapidly reduced further to give, after protonolysis, the primary alcohol. (Because the aldehyde is more reactive than the carboxylate salt, it cannot be isolated.)

$$R\text{—C(=O)OH} \xrightarrow[-H_2]{LiAlH_4} R\text{—C(=O)O}^-\ Li^+ + AlH_3 \xrightarrow[\text{intermediate}]{\text{via tetrahedral}} [R\text{—C(=O)H}] \xrightarrow[2)\ H_3O^+]{1)\ LiAlH_4} R\text{—CH}_2\text{—OH}$$

3. The $LiAlH_4$ reduction of a carboxylic acid incorporates two different types of carbonyl reaction:
   a. A net substitution reaction at the carbonyl group to give an aldehyde intermediate.
   b. An addition to the aldehyde thus formed.
4. Sodium borohydride, $NaBH_4$, does not reduce carboxylic acids, although it does react with the acidic hydrogens of carboxylic acids.

III. REACTION OF CARBOXYLIC ACIDS AT THE OXYGEN OF THE —OH GROUP

A. Esterification by Alkylation

1. When carboxylic acids are treated with diazomethane in ether solution, they are rapidly converted into their methyl esters.
   a. Protonation of diazomethane by the carboxylic acid gives the methyldiazonium ion; this ion has one of the best leaving groups, dinitrogen.
   b. An S_N2 reaction of the methyldiazonium ion with the carboxylate oxygen results in the displacement of N_2 and formation of the ester.

$$\underset{\text{a carboxylic acid}}{R\text{—C(=O)—O—H}} + \underset{\text{diazomethane}}{H_2\ddot{C}^-\text{—}\overset{+}{N}\equiv N} \longrightarrow R\text{—C(=O)—O}^- + \underset{\text{methydiazonium ion}}{H_3C\text{—}\overset{+}{N}\equiv N} \longrightarrow \underset{\text{a methyl ester}}{R\text{—C(=O)—O—CH}_3} + N_2$$

2. The reaction of certain alkyl halides with carboxylate ions give esters.
   a. This is an S_N2 reaction in which the carboxylate ion, formed by acid-base reaction of the acid and a base (such as K_2CO_3), acts as the nucleophile that attacks the alkyl halide.

$$R\text{—}CO_2H \xrightarrow[-KHCO_3]{K_2CO_3} R\text{—}CO_2^-\ K^+ \xrightarrow[-KBr]{PhCH_2Br} R\text{—}CO_2\text{—}CH_2Ph$$

   b. This reaction works best on alkyl halides that are especially reactive in the S_N2 reaction, such as methyl iodide and benzylic or allylic halides, because carboxylate ions are relatively poor nucleophiles.

IV. DECARBOXYLATION OF CARBOXYLIC ACIDS

1. The loss of carbon dioxide from a carboxylic acid is called decarboxylation.
2. Certain types of carboxylic acids are readily decarboxylated:
   a. β-keto acids (carboxylic acids with a keto group in the β-position).
   b. malonic acid derivatives.
   c. derivatives of carbonic acid.

a β-keto acid: $R-C(=O)-C-C(=O)-OH \longrightarrow R-C(=O)-C-H + CO_2$

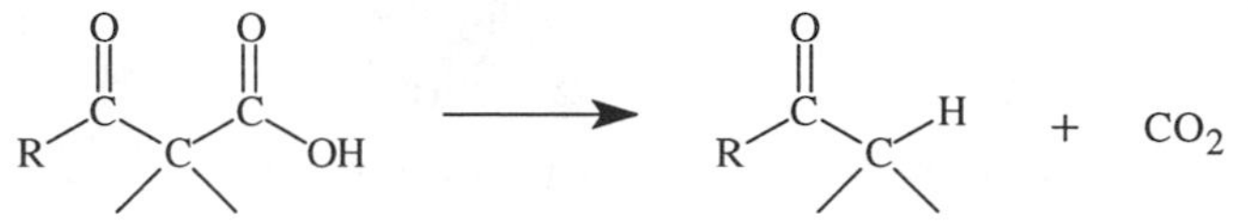

a malonic acid derivative: $HO-C(=O)-C-C(=O)-OH \xrightarrow{\text{heat}} HO-C(=O)-C-H + CO_2$

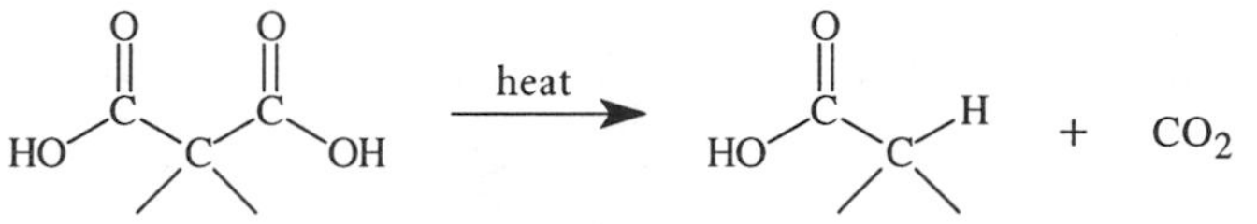

a carbonic acid derivative: $RO-C(=O)-OH \longrightarrow ROH + CO_2$

3. β-Keto acids readily decarboxylate at room temperature in acidic solution.
   a. Decarboxylation of a β-keto acid involves an enol intermediate that is formed by an internal proton transfer from the carboxylic acid group to the carbonyl oxygen atom of the ketone.
   b. The enol is transformed spontaneously into the corresponding ketone.

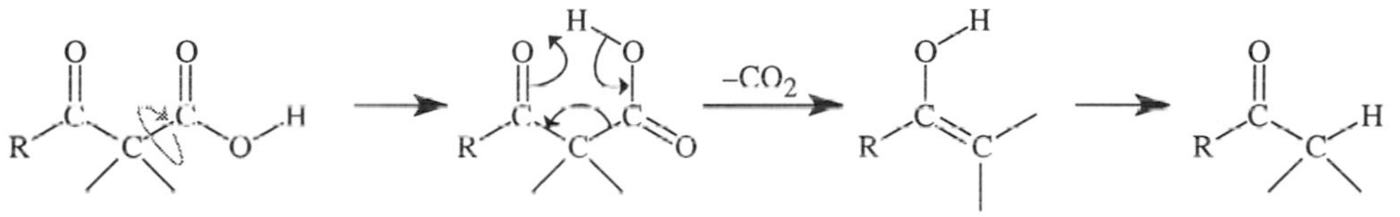

$R-C(=O)-C-C(=O)-O-H \longrightarrow [\text{cyclic transition state}] \xrightarrow{-CO_2} R-C(OH)=C \longrightarrow R-C(=O)-C-H$

enol intermediate

4. The acid form of the β-keto acid decarboxylates more readily than the conjugate-base carboxylate form because the latter has no acidic proton that can be donated to the β-carbonyl oxygen.
5. Malonic acid and its derivatives readily decarboxylate upon heating in acid solution. This reaction, which does not occur in base, bears a close resemblance to the decarboxylation of β-keto acids, because both types of acid have a carbonyl group β to the carboxy group.

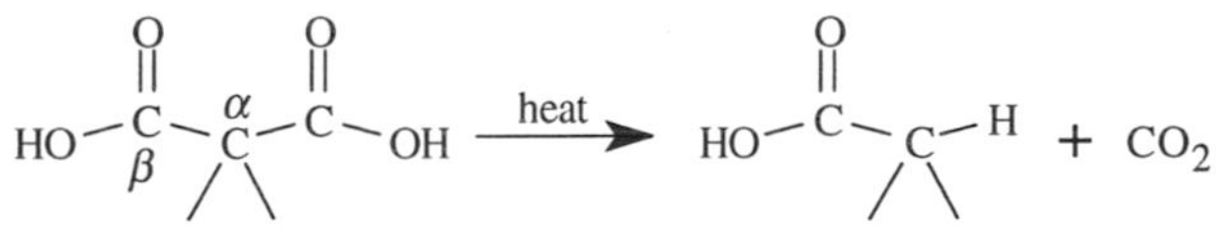

$HO-\underset{\beta}{C}(=O)-\overset{\alpha}{C}-C(=O)-OH \xrightarrow{\text{heat}} HO-C(=O)-C-H + CO_2$

a malonic acid derivative

6. Carbonic acid is unstable and decarboxylates spontaneously in acid solution to carbon dioxide and water.
   a. Carbonic acid derivatives in which only one carboxylate oxygen is involved in ester or amide formation also decarboxylate under acidic conditions.

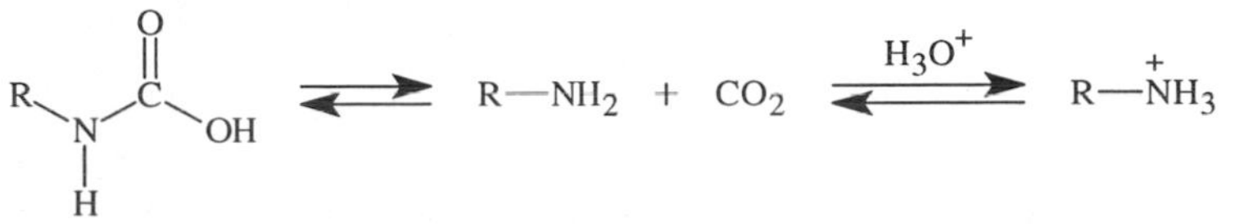

$R-NH-C(=O)-OH \rightleftharpoons R-NH_2 + CO_2 \xrightarrow{H_3O^+} R-\overset{+}{N}H_3$

   b. Under basic conditions, carbonic acid is converted into its salts bicarbonate and carbonate, which do not decarboxylate.

c. Carbonic acid derivatives in which both carboxylic acid groups are involved in ester or amide formation are stable.
   i. Carbonate esters are diesters of carbonic acid.
   ii. Ureas are diamides of carbonic acid.
   iii. Phosgene is the acid chloride of carbonic acid.

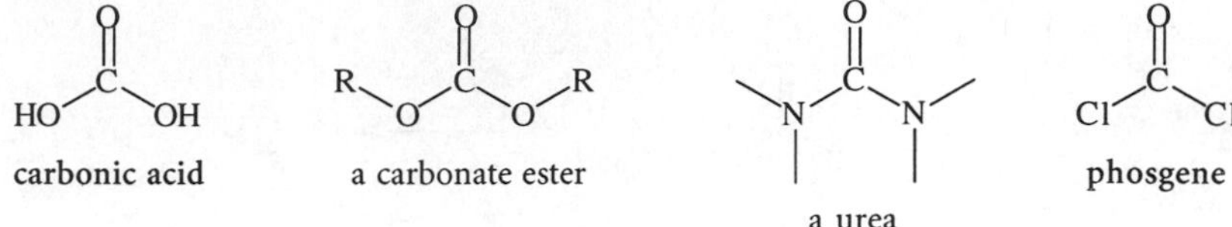

carbonic acid | a carbonate ester | a urea | phosgene

SCIENTISTS

Dates in brackets refer to Nobel Prize awards.

STUDY GUIDE LINKS

20.1 Chemical Shifts of Carbonyl Carbons

As pointed out in the text, the electronegativity of oxygen might lead one to expect the chemical shift for the carbonyl group of acetic acid to be *greater* than that of acetone. An additional puzzle is provided by the carbonyl chemical shift of the acetate ion, the conjugate base of acetic acid:

$$\underset{\text{acetic acid}}{H_3C{-}\overset{\overset{\displaystyle O}{\|}}{C}{-}OH}\ (\delta\,178) \qquad \underset{\text{acetate ion}}{H_3C{-}\overset{\overset{\displaystyle O}{\|}}{C}{-}O^-}\ (\delta\,182)$$

If chemical shifts in CMR spectroscopy correlate with electronegativity, as they do in proton NMR, it seems that the acetate ion, because of its relatively electron-rich —O$^-$, should have a *smaller* carbonyl chemical shift than acetic acid. The data above show that this is not the case.

A related phenomenon is the chemical shift of the O—H proton, which was discussed in Sec. 13.7D. In the gas phase (in which hydrogen bonding and other solvent effects are absent), this proton has a chemical shift of δ 0.8, a much smaller value than would be expected for a proton directly attached to an electronegative atom.

These cases illustrate the point that *chemical shifts are not determined solely by electronegativity of nearby atoms.* You learned, for example, that induced circulation of π electrons in the presence of the applied field accounts for the significant chemical shifts of both aromatic and alkene carbons as well as the shifts of aromatic and alkene protons—shifts far greater than would be predicted on the basis of the electronegativity of an sp^2 hybridized carbon (Fig. 13.15 on text p. 575 and Fig. 16.2 on text p. 699). In the cases discussed here, induced electron circulation also accounts for the results, except that the circulation involved is that of the oxygen unshared electrons, and it occurs in such a way as to oppose, and thus reduce, the applied field in the region of the adjacent atoms; as a result, a lower frequency is required for resonance of these atoms. The result is a smaller chemical shift. The α-carbon atoms of acetone have no such contribution, because they have no unshared electrons. Hence, the carbonyl carbon chemical shifts of ketones are greater than those of carboxylic acids. In acetate ion, the contribution of one unshared pair is partially offset by that of another, because the two orbitals are directed differently in space with respect to the applied field; hence, acetate absorbs at a chemical shift that is closer to the "normal" carbonyl value.

✓20.2 Reactions of Bases with Carboxylic Acids

In Eq. 20.3 (text p. 904), water is the base that reacts with a carboxylic acid, because water is the most abundant base present in the aqueous solution of an acid. However, you should not forget that carboxylic acids can in principle react with *any* base. (For example, write the reaction of a carboxylic acid with one equivalent of a Grignard reagent.)

Another point worth special note is that the reaction of strong bases not only with carboxylic acids, but also with alcohols and water, is very fast—so fast that a reaction occurs at virtually every encounter between an acid molecule and a base molecule. This becomes very important in situations in which proton transfer competes with other reactions for a limiting amount of a base. Suppose, for example, we want the aldehyde group in the following molecule to react with a Grignard reagent. The first equivalent of Grignard reagent reacts not with the aldehyde, but rather with the carboxylic acid, because proton transfer is such a fast reaction.

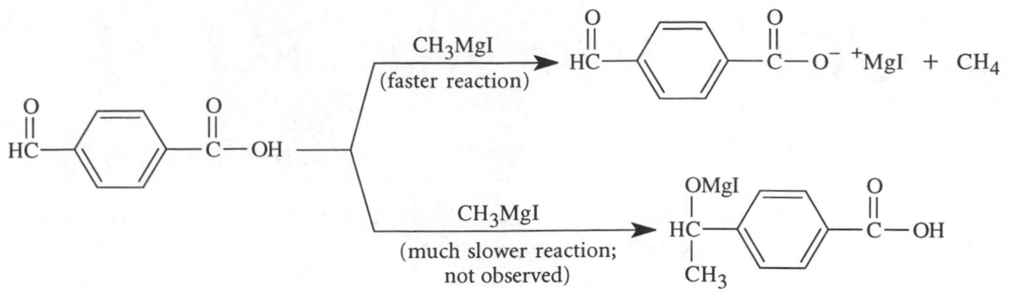

To achieve reaction of the aldehyde, a second equivalent of the Grignard reagent must be added:

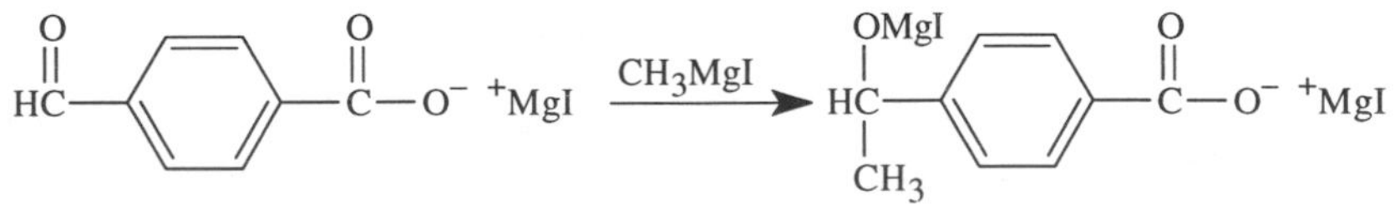

(The aldehyde carbonyl is much more reactive than the carboxylate carbonyl for reasons that are discussed in Chapter 21.)

In summary, proton transfers to strong bases from O—H groups in most cases are much faster than other reactions of strong bases.

✓20.3 Resonance Effect on Carboxylic Acid Acidity

The resonance stabilization of carboxylate ions was once considered the *major* effect responsible for the enhanced acidity of carboxylic acids. It is now believed to be less important than originally thought (although still significant) because *carboxylic acids are also resonance-stabilized.*

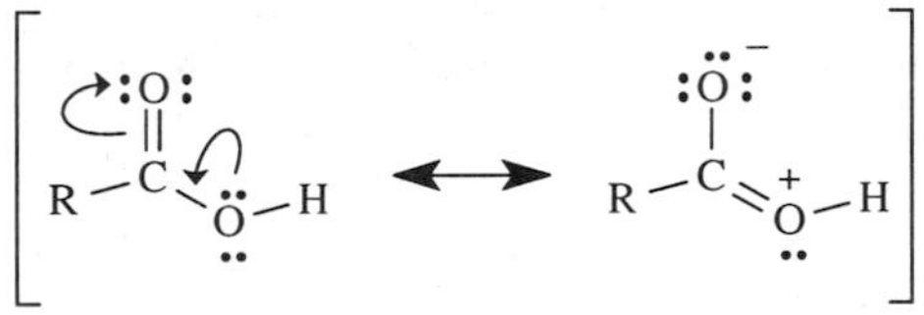

Figure 3.2 on text p. 103 shows that stabilizing a conjugate-base carboxylate decreases the pK_a of the corresponding carboxylic acid. It follows that stabilizing a carboxylic acid, that is, lowering its energy relative to that of its conjugate-base carboxylate, *increases* its pK_a. Hence, the resonance stabilization of an acid and the resonance stabilization of its carboxylate are opposing effects. However, *resonance is more important in the carboxylate ion* because the carboxylate resonance structures *disperse* or *delocalize* charge, whereas the carboxylic acid resonance structures *separate* charge. Guideline 3, text p. 668, discusses the fact that resonance structures which separate charge are less important than those which do not. An exactly analogous situation was discussed for phenols in Study Guide Link 18.3 on page 596 of this manual.

20.4 More on Surfactants

The word "surfactant" means "surface-active compound." To say that something is *surface active* means that is lowers the *surface tension* of a liquid. Without getting into a technical discussion of surface tension, we can say that it is the resistance of a surface to penetration by another object. Water has a particularly high surface tension. (If you've ever seen someone float a coin on the surface of a glass of water, you've witnessed the effects of surface tension.) The high surface tension of water is due to the tight network of strong hydrogen bonds at the water surface. Because water cannot form hydrogen bonds to air, water molecules form particularly strong hydrogen bonds to each other at the air-water interface.

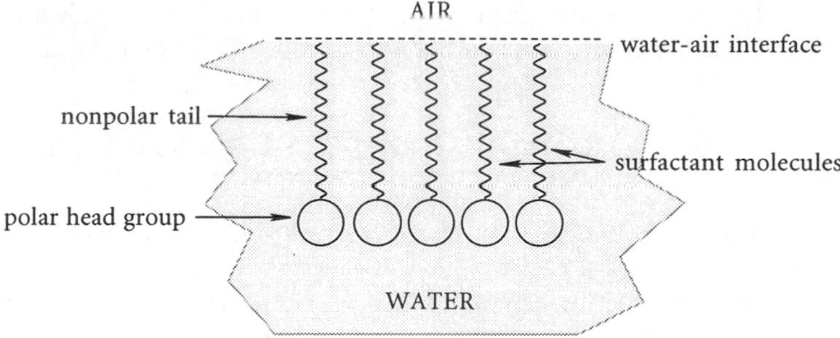

Figure SG20.1 A diagram showing schematically the orientation of surfactant molecules at the interface of an aqueous solution and air. The nonpolar tails are oriented towards the interface.

When a surfactant is dissolved in water, the surfactant molecules tend to collect at the surface of the water with the hydrocarbon tails pointed toward the water-air interface, as shown in Fig. SG20.1. This orientation of surfactant molecules at the surface allows the solvent water molecules to interact with the polar head groups and places the nonpolar tails in contact with air, which is composed mostly of other nonpolar molecules. In this situation the network of tight hydrogen bonds at the surface of pure water is no longer present. As a result, the surface tension of the solution is lowered. One perceptible manifestation of this phenomenon is aqueous solutions of surfactants (for example, soap solutions) are slippery to the touch.

20.5 Orthoesters

Some students ask whether esters react further under the esterification conditions to form compounds analogous to acetals. (See Sec. 19.10A.)

$$R-\overset{O}{\overset{\|}{C}}-OCH_3 + 2\,CH_3OH \;\not\xrightarrow{\text{acid}}\; R-\underset{OCH_3}{\underset{|}{\overset{OCH_3}{\overset{|}{C}}}}-OCH_3 + H_2O \quad \text{(not observed)}$$

an orthoester

Such compounds, called orthoesters, are indeed known. However, they *cannot* be formed in this manner because the equilibrium constants for addition to the carbonyl groups of acid derivatives are only about 10^{-8} times those for formation of aldehyde addition compounds. Such equilibria are simply too unfavorable to be driven to completion in the usual ways.

The reason that such equilibria are unfavorable follows directly from the discussion of the relationship between the stability of a carbonyl compound and its equilibrium constant for addition. (See Sec. 19.7B.) Stabilization of a carbonyl compound reduces its equilibrium constant for addition. Esters are stabilized by resonance interaction of the carboxylate oxygen with the carbonyl group. Ketones, in contrast, have fewer resonance structures and are less stable relative to their addition products.

$$\left[R-\overset{:O:}{\overset{\|}{C}}-\ddot{O}CH_3 \longleftrightarrow R-\underset{+}{\overset{:\ddot{O}:^-}{\overset{|}{C}}}-\ddot{O}CH_3 \longleftrightarrow R-\overset{:\ddot{O}:^-}{\overset{|}{C}}=\overset{+}{\ddot{O}}CH_3 \right]$$

An ester has more resonance structures and is more stable relative to its addition products; addition reactions are less favorable.

$$\left[R-\overset{:O:}{\overset{\|}{C}}-R \longleftrightarrow R-\underset{+}{\overset{:\ddot{O}:^-}{\overset{|}{C}}}-R \right]$$

An aldehyde or ketone has fewer resonance structures and is less stable relative to its addition products; addition reactions are more favorable.

Hence, the equilibrium constant for an addition to a ketone is more favorable than the equilibrium constant for the same addition to an ester. This point is illustrated in Fig. SG20.2.

✓20.6 Mechanism of Acid Chloride Formation

The formation of acid chlorides follows a mechanism that is fundamentally the same as the mechanism for esterification. In esterification, the carbonyl group is activated by protonation. In acid chloride formation, it is activated by reaction with the Lewis acid $SOCl_2$. Notice that chloride ion is displaced from $SOCl_2$ in the process:

As a result of this reaction the carbonyl oxygen is more electrophilic. In other words, the Lewis acid $SOCl_2$ activates the carbonyl carbon toward nucleophilic attack in much the same sense that a proton does. (You've seen that other Lewis acids, such as the lithium ion of $LiAlH_4$ or the magnesium of a Grignard reagent, can also activate attack at a carbonyl carbon.) The nucleophile chloride ion produced in the step above reacts at the carbonyl carbon:

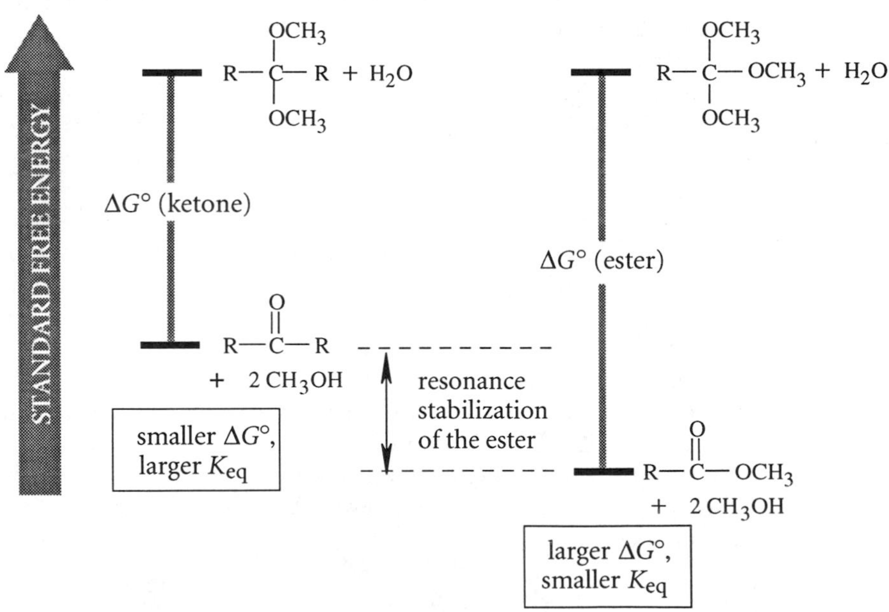

Figure SG20.2 The greater resonance stabilization of esters causes addition reactions of esters to have smaller equilibrium constants than the corresponding addition reactions of ketones.

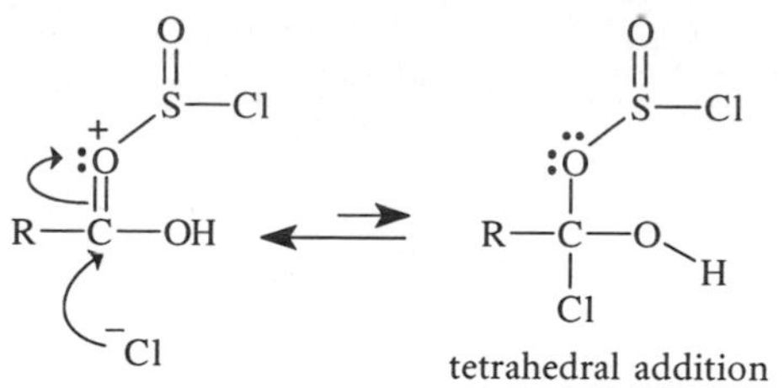

tetrahedral addition intermediate

Notice the formation of a tetrahedral addition intermediate, as in esterification. *This is a common mechanistic thread that runs throughout carbonyl chemistry.* The —OSOCl group is an excellent leaving group and departs to give a protonated acid chloride. Loss of the —OSOCl group is irreversible, because it decomposes to SO_2 (a gas) and chloride ion:

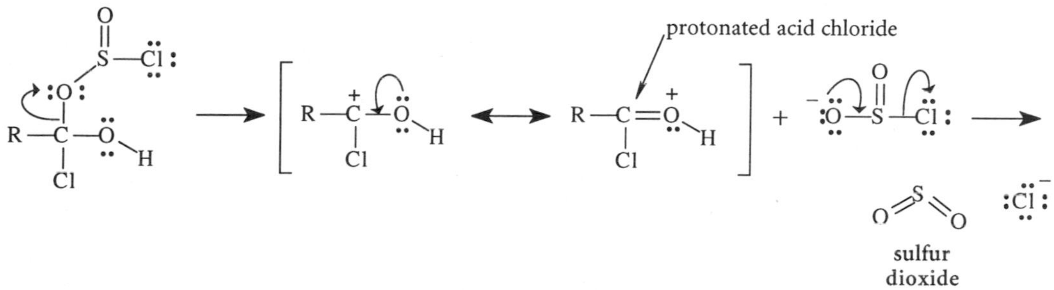

protonated acid chloride

sulfur dioxide

Loss of a proton to chloride ion gives HCl (another gaseous by-product) and the acid chloride:

✓20.7 More on Synthetic Equivalents

Study Guide Link 17.3 (p. 552 of this manual) discussed the notion of a *synthetic equivalent.* Acid chlorides present another opportunity to think about this concept. Recall that acid chlorides are the acylating agents used in Friedel-Crafts acylation (Sec. 16.4F). Because acid chlorides are often prepared from carboxylic acids, it follows, then, that the acyl group of an aromatic ketone can originate from a carboxylic acid. Thus, a carboxylic acid or acid chloride can be viewed as the synthetic equivalent of the acyl group in an aromatic ketone.

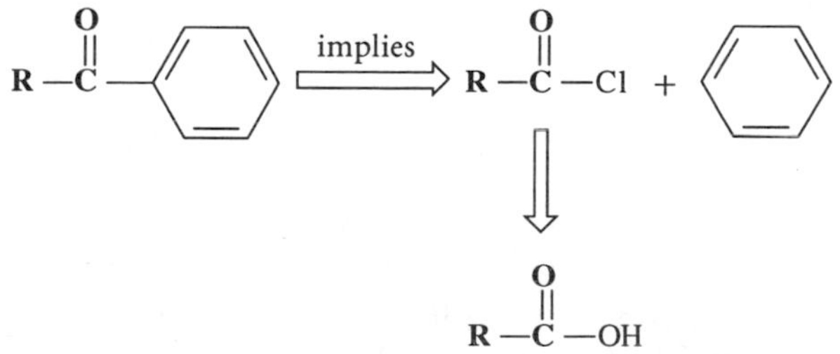

implies

✓20.8 Mechanism of Anhydride Formation

Working through the mechanism of anhydride formation will allow you to see again the patterns involved in carbonyl substitution reactions. Consider the mechanism for the formation of the anhydride of a general carboxylic acid, RCO_2H, by acetic anhydride. The acetic anhydride serves as a Lewis acid to activate the carbonyl group of an R—CO_2H molecule:

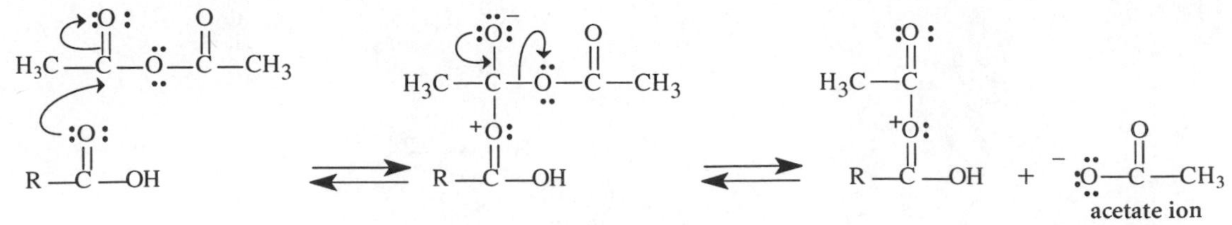

acetate ion

The activated carbonyl group reacts with another molecule of carboxylic acid and loses a proton to the acetate ion formed in the previous step to give a tetrahedral addition intermediate:

tetrahedral addition intermediate

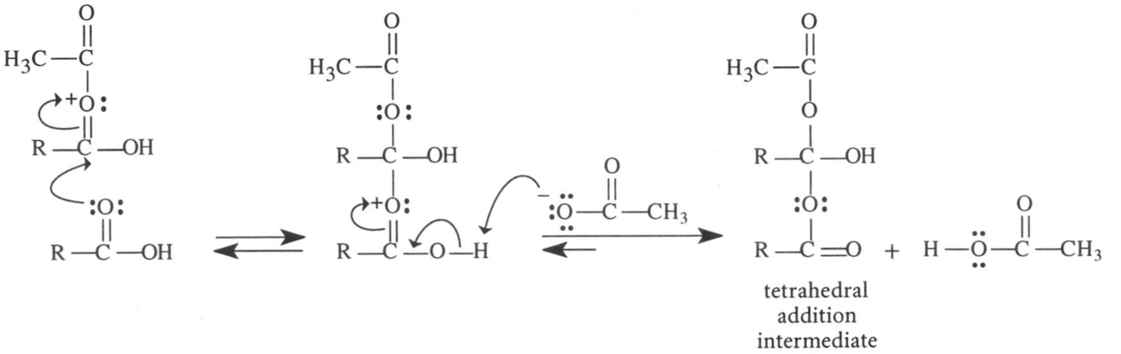

You should now complete the mechanism by showing, first, the loss of another acetate ion as a leaving group, and second, loss of a proton:

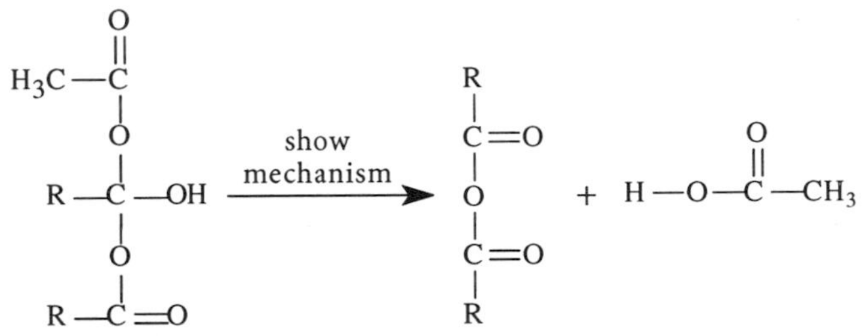

show mechanism

The mechanisms of anhydride formation with P_2O_5 and with $POCl_3$ are similar.

Note that anhydride formation is an equilibrium. When acetic anhydride is used as the reagent, this equilibrium is driven to the right by the use of a large excess of acetic anhydride (LeChatelier's principle in operation). When cyclic anhydrides are being formed, the preference for intramolecular reactions that form five- and six-membered rings (Sec. 11.7) provides not only a kinetic advantage to the reaction—a greater reaction rate—but also an equilibrium advantage. Consequently, reactions that form five- and six-membered cyclic anhydrides in many cases require milder conditions.

20.9 Mechanism of the $LiAlH_4$ Reduction of Carboxylic Acids

The mechanism of the $LiAlH_4$ reduction of carboxylic acids is presented in the text only in outline. This Study Guide Link fills in the details.

Reduction of the carboxylate ion by AlH_3 (step *a* in the following equation) is analogous to the addition of BH_3 to the double bond of an alkene. (BH_3 also reduces carboxylic acids to primary alcohols.)

$$Li^+\ {}^-O{-}C(R){=}O \cdots H{-}AlH_2 \xrightarrow{(a)} R{-}C(O^-\,Li^+)(H){-}OAlH_2 \xrightarrow{(b)} R{-}C({=}O){-}H + Li^+\ H_2AlO^-$$

Notice that a tetrahedral addition intermediate is formed in step *a*. Breakdown of this intermediate (step *b*) expels Li^+ H_2AlO^-, which also can act as a reducing agent. Further reduction of the aldehyde to the alcohol (Sec. 19.8) can occur either by this species or by excess $LiAlH_4$ present in the reaction mixture.

SOLUTIONS TO PROBLEMS

Solutions to In-Text Problems

20.1 (a) $HOCH_2CH_2CH_2CO_2H$
γ-hydroxybutyric acid

(b) $Cl_2CHCH_2CO_2H$
β,β-dichloropropionic acid

(c) $CH_3CH_2(H)C{=}C(H)CH_2CO_2H$
(Z)-3-hexenoic acid

(d) $CH_3CH_2CH(CH_3)CH_2CH_2CO_2H$
4-methylhexanoic acid

(e) HO_2C—(cyclohexane-1,4-diyl)—CO_2H
1,4-cyclohexanedicarboxylic acid

(f) CH_3O—(benzene-1,4-diyl)—CO_2H
p-methoxybenzoic acid

(g) $HO_2CCCl_2CH_2CH_2CH_2CO_2H$
α,α-dichloroadipic acid

(h) $HO_2C—CO_2H$
oxalic acid

20.2 (a) 2,2-dimethylbutanoic acid (common: α,α-dimethylbutyric acid)
(b) 9-methyldecanoic acid (common: ω-methylcapric acid). Note that the term ω (*omega,* the last letter of the Greek alphabet) is used in common nomenclature for a branch at the end of a carbon chain.
(c) 3-hydroxy-4-oxocyclohexanecarboxylic acid
(d) 2,4-dichlorobenzoic acid
(e) 2-methylpropanedioic acid (common: methylmalonic acid)
(f) cyclopropanecarboxylic acid

20.3 The dimer owes its stability to intermolecular hydrogen bonds. The solvent carbon tetrachloride has no donor or acceptor atoms to compete for these hydrogen bonds. However, the solvent water can compete as both a donor and an acceptor for hydrogen bonds. Hence, the dimer has less of an energetic advantage in water and, as a result, there is a lower concentration of dimer (in fact, essentially none) in aqueous solution and a higher concentration in CCl_4.

20.4 The IR spectrum indicates the presence of a carboxy group. A carboxy group accounts for 45 mass units; this leaves 43 mass units unaccounted for. The NMR spectrum indicates seven hydrogens in addition to the carboxy hydrogen at δ 10. A group with seven hydrogens and 43 mass units must be either a propyl or an isopropyl group; the splitting in the NMR spectrum indicates the latter. The compound is 2-methylpropanoic acid (isobutyric acid), $(CH_3)_2CH—CO_2H$.

20.5 The NMR data indicate a *para*-substituted benzoic acid derivative; given this deduction, the para substituent must be a chlorine.

Cl—(benzene-1,4-diyl)—CO_2H
p-chlorobenzoic acid
(4-chlorobenzoic acid)

20.6 First write the structures of the two compounds.

$$HO_2C-\underset{CH_3}{\overset{CH_3}{C}}-CH_2-CO_2H \qquad HO_2CCH_2CH_2CH_2CH_2CO_2H$$

α,α-dimethylsuccinic acid **adipic acid**

(a) α,α-Dimethylsuccinic acid should have five resonances in its CMR spectrum because it has five chemically nonequivalent sets of carbons. Adipic acid, in contrast, should have three resonances because it has only three chemically nonequivalent sets of carbons.

(b) The proton NMR spectrum of only α,α-dimethylsuccinic acid consists of three singlets; the proton NMR spectrum of adipic acid is more complex.

20.7 (a) The first and second ionizations of succinic acid:

$$HO_2CCH_2CH_2CO_2H \underset{H_3O^+}{\overset{H_2O}{\rightleftharpoons}} {}^-O_2CCH_2CH_2CO_2H \underset{H_3O^+}{\overset{H_2O}{\rightleftharpoons}} {}^-O_2CCH_2CH_2CO_2^-$$

$$pK_a = 4.21 \qquad pK_a = 5.64$$

(b) The first pK_a is lower because the electron-withdrawing polar effect of an un-ionized carboxy group stabilizes the anion formed by ionization of the other carboxy group. The second pK_a is higher because the negative charge of an ionized carboxy group destabilizes the anion formed by the second ionization. This destabilization is the result of the electrostatic repulsion of like charges.

20.8 Extract an ether solution of the two compounds with an aqueous solution of $NaHCO_3$, Na_2CO_3, or NaOH. The acid will ionize and its conjugate-base anion will dissolve in the aqueous layer as the sodium salt; *p*-bromotoluene will remain in the ether layer. After isolating the aqueous layer, acidify it with concentrated HCl; neutral *p*-bromobenzoic acid will precipitate.

20.9 (a)

$$\text{C}_5\text{H}_9\text{–OH} \xrightarrow[H_2SO_4]{\text{conc. HBr}} \text{C}_5\text{H}_9\text{–Br} \xrightarrow[\text{ether}]{\text{Mg}} \text{C}_5\text{H}_9\text{–MgBr} \xrightarrow[2)\ H_3O^+]{1)\ CO_2} \text{C}_5\text{H}_9\text{–CO}_2\text{H}$$

cyclopentanol **cyclopentanecarboxylic acid**

(b)

$$CH_3(CH_2)_4CH{=}CH_2 \xrightarrow[2)\ H_2O_2/NaOH]{1)\ BH_3/THF} CH_3(CH_2)_5CH_2OH \xrightarrow[H_2SO_4]{\text{conc. HBr}} CH_3(CH_2)_5CH_2Br \xrightarrow[\text{ether}]{\text{Mg}}$$

1-heptene

$$CH_3(CH_2)_5CH_2MgBr \xrightarrow[2)\ H_3O^+]{1)\ CO_2} CH_3(CH_2)_5CH_2CO_2H$$

octanoic acid

One could also use HBr in the presence of peroxides to form the alkyl bromide directly from the alkene. In general, this alternative scheme will always be applicable in the conversion of a 1-alkene into the corresponding primary alkyl halide.

20.10 (a) The carboxylic acid is esterified to form the ethyl ester:

$$CH_3CH_2CH_2\underset{CH_3}{CH}CH_2\overset{O}{\overset{\|}{C}}OH \xrightarrow[CH_3CH_2OH\ \text{(excess)}]{H_2SO_4\ \text{(cat.)}} CH_3CH_2CH_2\underset{CH_3}{CH}CH_2\overset{O}{\overset{\|}{C}}OCH_2CH_3$$

3-methylhexanoic acid **ethyl 3-methylhexanoate**

(b) Both carboxy groups are esterified; the product is dipropyl adipate:

$$\underset{\text{adipic acid}}{\mathrm{HO\overset{O}{\overset{\|}{C}}(CH_2)_4\overset{O}{\overset{\|}{C}}OH}} \xrightarrow[\mathrm{CH_3CH_2CH_2OH}]{\mathrm{H_2SO_4(cat.)}} \underset{\text{dipropyl adipate}}{\mathrm{CH_3CH_2CH_2O\overset{O}{\overset{\|}{C}}(CH_2)_4\overset{O}{\overset{\|}{C}}OCH_2CH_2CH_3}}$$

20.11 (a) Follow the reverse of the steps shown in Eqs. 20.18a–c of the text with R— = Ph—.

$$\mathrm{Ph\overset{O}{\overset{\|}{C}}{-}OCH_3} \overset{H_2O}{\rightleftharpoons} \mathrm{Ph\overset{\overset{+}{O}H}{\overset{\|}{C}}{-}OCH_3} \overset{H_2O}{\rightleftharpoons} \mathrm{Ph\overset{OH}{\overset{|}{C}}{-}OCH_3}\ (\text{with } \overset{+}{O}H_2) \rightleftharpoons \mathrm{Ph\overset{OH}{\overset{|}{\underset{OH}{\underset{|}{C}}}}{-}OCH_3} \rightleftharpoons$$

$$\mathrm{Ph\overset{OH}{\overset{|}{\underset{OH}{\underset{|}{C}}}}{-}\overset{+}{O}(H)CH_3} + OH_2 \rightleftharpoons \mathrm{Ph\overset{\overset{+}{O}-H}{\overset{\|}{\underset{OH}{\underset{|}{C}}}}} + HOCH_3 \overset{OH_2}{\rightleftharpoons} \mathrm{Ph\overset{O}{\overset{\|}{C}}{-}OH} + H{-}\overset{+}{O}H_2$$

(b) To favor ester hydrolysis rather than ester formation, use a large excess of water as solvent rather than an alcohol. By LeChatelier's principle, this drives the carboxylic acid-ester equilibrium toward the carboxylic acid.

(c) The mechanism of hydroxide-promoted ester hydrolysis:

$$\mathrm{Ph\overset{O}{\overset{\|}{C}}{-}OCH_3} + {}^{-}OH \rightleftharpoons \mathrm{Ph\overset{O^-}{\overset{|}{\underset{OH}{\underset{|}{C}}}}{-}OCH_3} \rightleftharpoons \mathrm{Ph\overset{O}{\overset{\|}{C}}{-}O{-}H}\ \ {}^{-}OCH_3 \rightleftharpoons \mathrm{Ph\overset{O}{\overset{\|}{C}}{-}O^-} + H{-}OCH_3$$

Notice that under the basic conditions the carboxylic acid product is ionized in the last step. This drives base-promoted ester hydrolysis to the right.

The base-promoted hydrolysis of esters is called *saponification,* because soaps are made this way from the base-promoted hydrolysis of fats, which are nothing more than esters of the fatty acids. We'll study saponification in Sec. 21.7A. However, if you were able to construct this mechanism, you're ahead of the game—you understand the reaction already!

(d) As shown in the last step of the mechanism in the solution to part (c), methoxide reacts with the *acidic proton* of benzoic acid to form the benzoate ion much more rapidly than it reacts at the carbonyl carbon. (See Study Guide Link 20.2 on p. 705 of this manual for a discussion of this point.)

20.12 The structures of isobutyric and benzoic acids are given in Table 20.1. The products are as follows:

(a) $\mathrm{(CH_3)_2CH\overset{O}{\overset{\|}{C}}{-}OCH_3}$ — methyl isobutyrate

(b) $\mathrm{(CH_3)_2CH\overset{O}{\overset{\|}{C}}{-}OCH_2Ph}$ — benzyl isobutyrate

(c) $\mathrm{Ph{-}\overset{O}{\overset{\|}{C}}{-}OCH_2CH{=}CH_2}$ — allyl benzoate

20.13 In this reaction the carboxylic acid is alkylated by the *tert*-butyl cation, which is formed by protonation of 2-methylpropene.

$$HO_3SO{-}H + H_2C{=}C(CH_3)_2 \longrightarrow {}^{+}C(CH_3)_3 \;\; HO_3SO^- \xrightarrow{CH_3C(=O){-}OH}$$

$$\left[CH_3C(=\overset{+}{O}{-}C(CH_3)_3){-}O{-}H \longleftrightarrow CH_3C(O{-}C(CH_3)_3){=}\overset{+}{O}{-}H \right] \xrightarrow{HO_3SO^-} CH_3C(O{-}C(CH_3)_3){=}O + HO_3SO{-}H$$

A variation on this mechanism proposed by many students is alkylation of the carboxylate oxygen rather than the carbonyl oxygen by the *tert*-butyl cation:

$$CH_3C(=O){-}OH \;\; {}^{+}C(CH_3)_3 \longrightarrow CH_3C(=O){-}\overset{+}{O}(H){-}C(CH_3)_3 \xrightarrow{HO_3SO^-} CH_3C(=O){-}O{-}C(CH_3)_3 + HO_3SO{-}H$$

Although this appears to be a more direct route to the ester, the mechanism given above is more likely to be correct for the following reason. Just as a carboxylic acid is *protonated* on the carbonyl oxygen rather than the carboxylate oxygen (why? see text p. 907), it should be *alkylated* on the carbonyl oxygen for the same reason.

20.14 (a)

$$CH_3O{-}C_6H_4{-}C(=O){-}Cl$$

p-methoxybenzoyl chloride (anisoyl chloride)

(b)

$$CH_3CH_2CH_2{-}S(=O)_2{-}Cl$$

1-propanesulfonyl chloride

20.15 (a)

$$CH_3CH_2{-}S(=O)_2{-}Cl$$

ethanesulfonyl chloride

(b)

$$C_6H_5{-}C(=O){-}Cl$$

benzoyl chloride

(c)

$$H_3C{-}C_6H_4{-}S(=O)_2{-}Cl$$

p-toluenesulfonyl chloride (tosyl chloride)

20.16 Study Guide Link 20.7 points out that a carboxylic acid is the synthetic equivalent of an acyl group in an aromatic ketone. This concept is realized by applying the Friedel-Crafts acylation reaction to the corresponding acid chloride. This synthesis takes advantage of the fact that the methoxy group is an ortho, para-directing substituent in electrophilic aromatic substitution.

$$HO{-}C(=O){-}C_6H_5 \xrightarrow{SOCl_2} Cl{-}C(=O){-}C_6H_5 \xrightarrow[\text{2) } H_3O^+]{\text{1) } CH_3O{-}C_6H_5,\ AlCl_3} CH_3O{-}C_6H_4{-}C(=O){-}C_6H_5$$

benzoic acid — benzoyl chloride — *p*-methoxybenzophenone

20.17 The corresponding anhydride is formed:

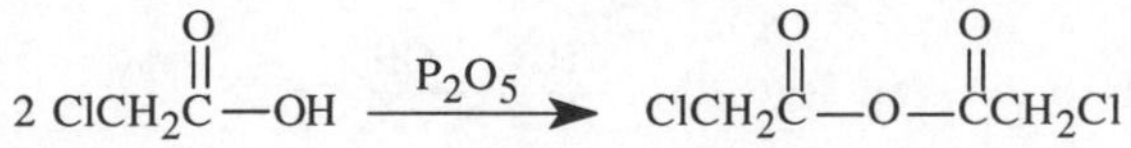

$$2\ ClCH_2\overset{O}{\overset{\|}{C}}—OH \xrightarrow{P_2O_5} ClCH_2\overset{O}{\overset{\|}{C}}—O—\overset{O}{\overset{\|}{C}}CH_2Cl$$

chloroacetic acid — chloroacetic anhydride

20.18 (a) Maleic anhydride forms a cyclic anhydride because its carboxy groups are cis. A cyclic anhydride of fumaric acid would require that a five-membered ring contain a trans double bond.

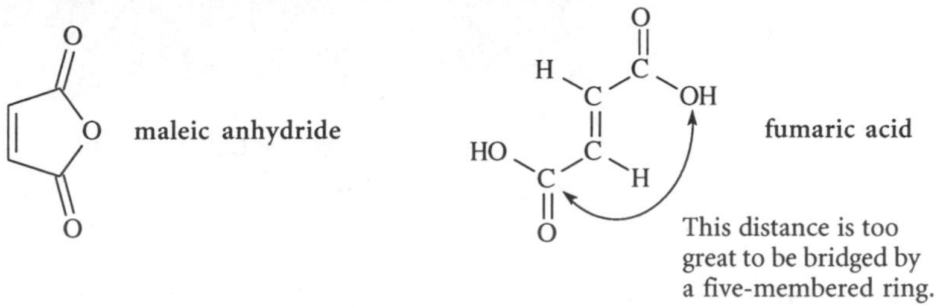

maleic anhydride

fumaric acid

This distance is too great to be bridged by a five-membered ring.

(b) On heating, 2,3-dimethylbutanedioic acid forms a cyclic anhydride containing a five-membered ring. Because a cyclic anhydride of α-methylmalonic acid would contain a very strained four-membered ring, it is not formed on heating.

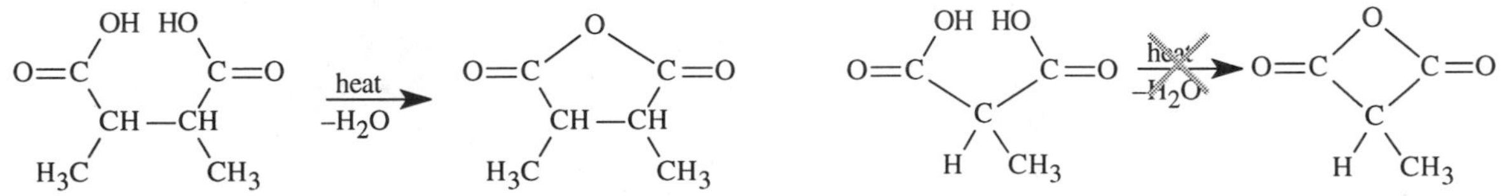

2,3-dimethylbutanedioic acid $\xrightarrow[-H_2O]{heat}$ 2,3-dimethylbutanedioic anhydride

α-methylmalonic acid $\xrightarrow[-H_2O]{heat}$ (crossed out) α-methylmalonic anhydride

20.19 (a) The compound must contain three oxygens and two degrees of unsaturation in addition to the benzene ring. *p*-Formylbenzoic acid meets the indicated criteria.

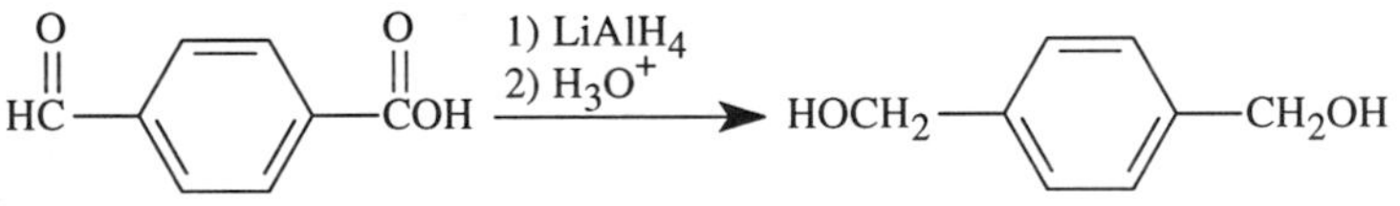

$$HC(=O)—C_6H_4—COH(=O) \xrightarrow[2)\ H_3O^+]{1)\ LiAlH_4} HOCH_2—C_6H_4—CH_2OH$$

p-formylbenzoic acid

(b) 1,4-Benzenedicarboxylic acid (terephthalic acid) is a compound with the formula $C_8H_6O_4$ that gives the indicated diol on treatment with $LiAlH_4$ followed by protonolysis.

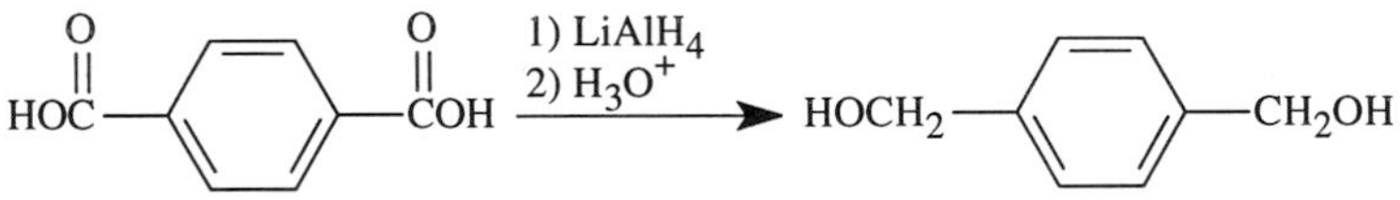

$$HOC(=O)—C_6H_4—COH(=O) \xrightarrow[2)\ H_3O^+]{1)\ LiAlH_4} HOCH_2—C_6H_4—CH_2OH$$

1,4-benzenedicarboxylic acid (terephthalic acid)

20.20 (a) This synthesis requires the addition of one carbon. Follow the general scheme in Study Problem 20.2.

$$PhCO_2H \xrightarrow[2)\ H_3O^+]{1)\ LiAlH_4} PhCH_2OH \xrightarrow{conc.\ HBr} PhCH_2Br \xrightarrow[ether]{Mg} PhCH_2MgBr \xrightarrow[2)\ H_3O^+]{1)\ CO_2} PhCH_2CO_2H$$

benzoic acid — phenylacetic acid

(b) This synthesis requires the addition of two carbons, which can be accomplished by the reaction of a Grignard reagent with oxirane (text Sec. 11.4C).

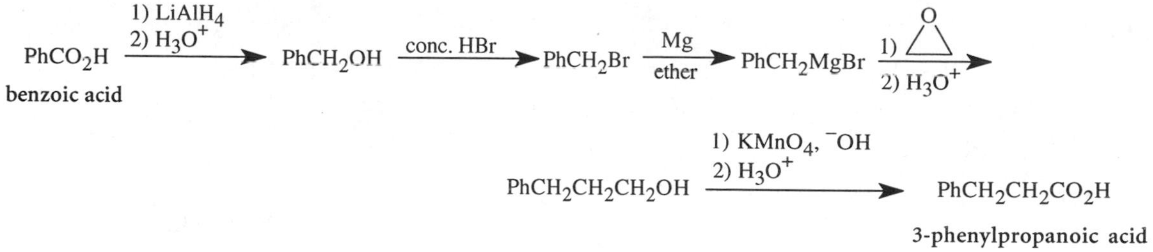

$$\underset{\text{benzoic acid}}{PhCO_2H} \xrightarrow[]{\text{1) } LiAlH_4 \text{ 2) } H_3O^+} PhCH_2OH \xrightarrow{\text{conc. HBr}} PhCH_2Br \xrightarrow[\text{ether}]{Mg} PhCH_2MgBr \xrightarrow[\text{2) } H_3O^+]{\text{1) oxirane}}$$

$$PhCH_2CH_2CH_2OH \xrightarrow{\text{1) } KMnO_4,\ ^-OH \text{ 2) } H_3O^+} \underset{\text{3-phenylpropanoic acid}}{PhCH_2CH_2CO_2H}$$

20.21 Each compound decarboxylates; the net effect is the replacement of the carboxy group by a hydrogen. (In part (c), the amine product is protonated under the acidic conditions.)

(a) $Ph-C(=O)-CH(CH_3)_2$

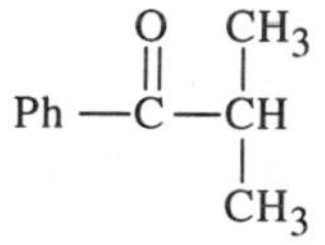

(b) cyclopentane$-CO_2H$

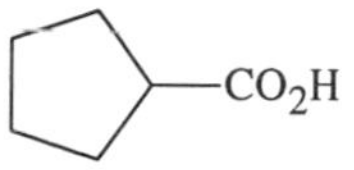

(c) $CH_3CH_2NH_2 \overset{H_3O^+}{\rightleftharpoons} CH_3CH_2\overset{+}{N}H_3$

20.22 The following *β*-keto acids (and their enantiomers) will decarboxylate to give 2-methylcyclohexanone:

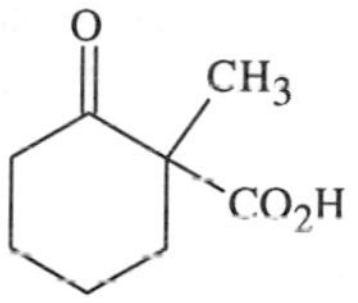

1-methyl-2-oxocyclo-hexanecarboxylic acid

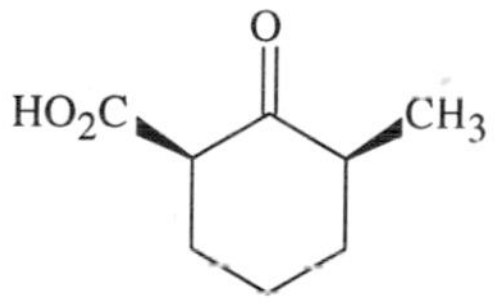

cis-3-methyl-2-oxocyclo-hexanecarboxylic acid

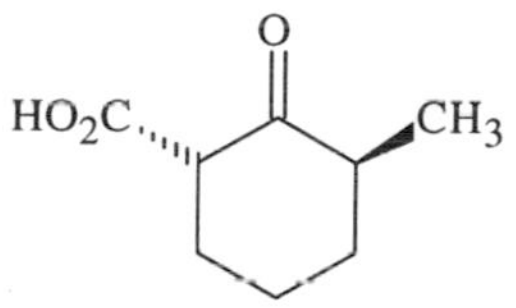

trans-3-methyl-2-oxocyclo-hexanecarboxylic acid

20.23 Were this carboxylic acid to decarboxylate by the mechanism shown in Eq. 20.39 on p. 924 of the text, the following enol would be formed as an intermediate:

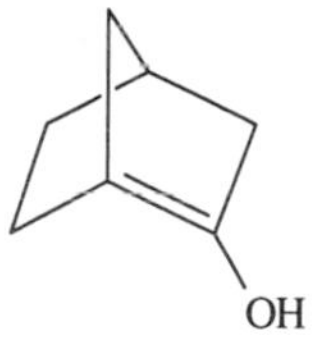

OH

This enol is very unstable because it violates Bredt's rule (Sec. 7.6C); that is, the double bond is twisted to such an extent that the 2*p* orbitals on each carbon cannot overlap.

Solutions to Additional Problems

20.24

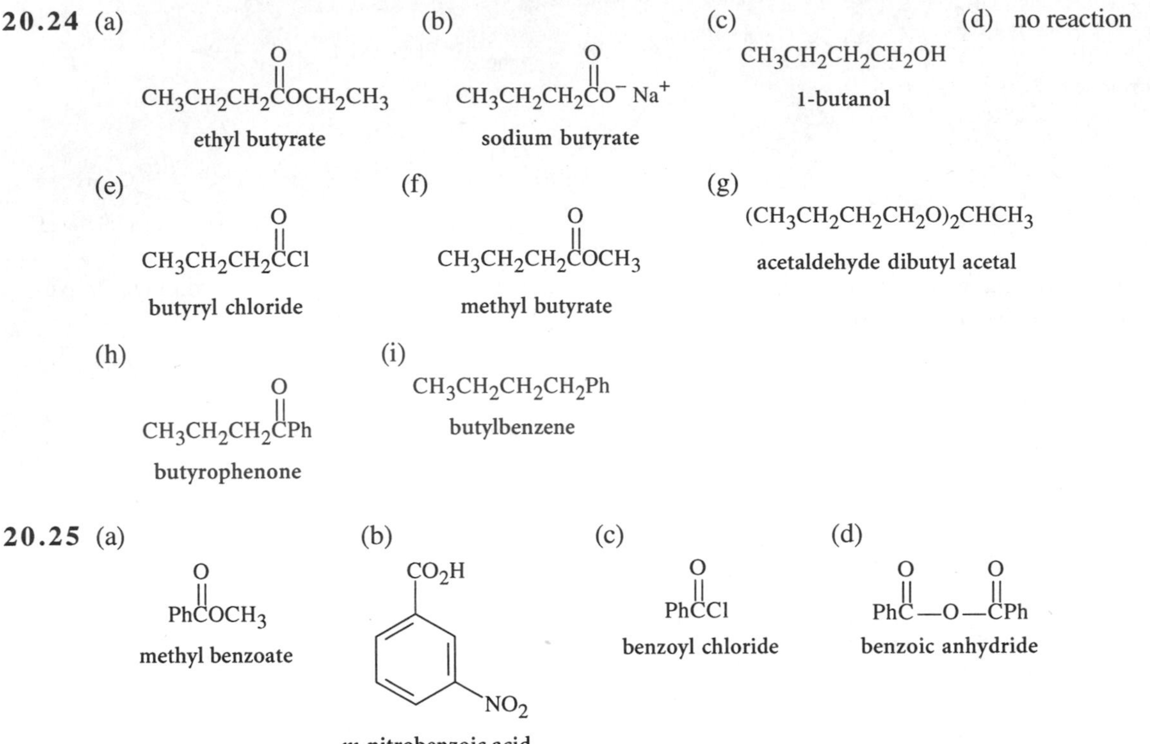

20.26 The dicarboxylic acids with the formula $C_6H_{10}O_4$:

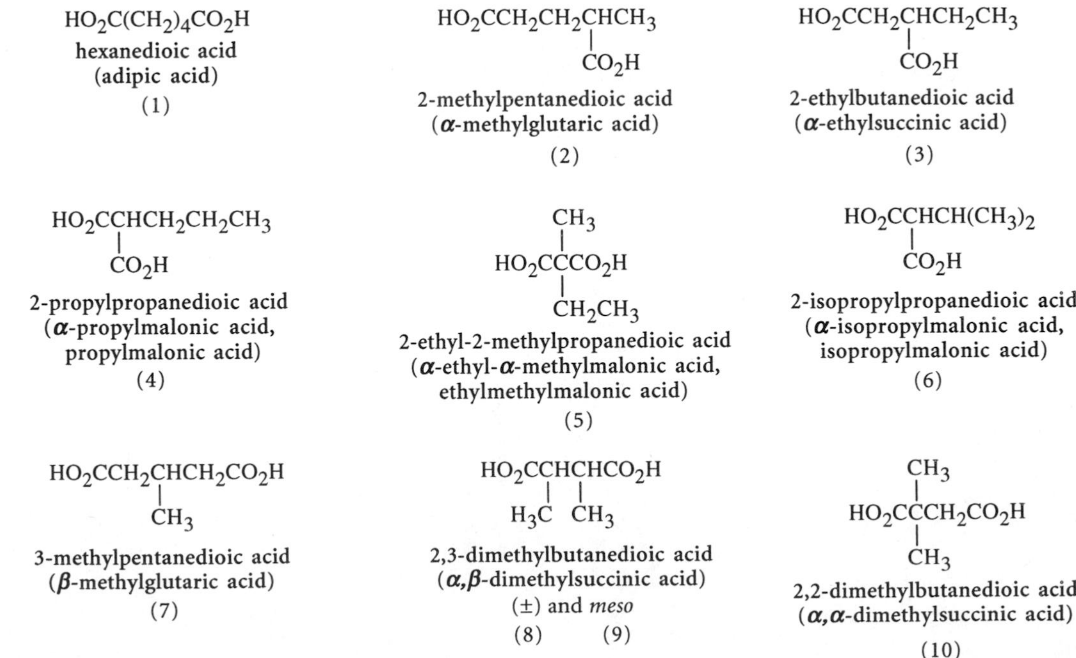

Compounds (2), (3), and (8) are chiral. Compounds (2), (3), (7), (8), (9), and (10) would readily form cyclic anhydrides on heating, because those anhydrides contain five- or six-membered rings. Compounds (4), (5), and (6) would readily decarboxylate on heating because they are malonic acid derivatives.

20.27 First calculate the number of moles of 0.1 *M* NaOH required for the neutralization:

$$\text{mol NaOH} = (0.100 \text{ mol NaOH/L})(8.61 \times 10^{-3} \text{ L}) = 8.61 \times 10^{-4}$$

The same molar amount of the carboxylic acid must be present because one mole of NaOH is required to neutralize one mole of a monocarboxylic acid. The molecular mass is

$$\text{molecular mass of the acid} = (100 \times 10^{-3} \text{ g}) \div (8.61 \times 10^{-4} \text{ mol}) = 116 \text{ g/mol}$$

20.28 In each case, the base reacts with acetic acid in a Brønsted acid-base reaction to form the conjugate base of acetic acid (acetate ion) and a by-product.

(a)

$$CH_3C(=O)-O-H \quad ^{-}OCH_3\ Na^+ \longrightarrow CH_3C(=O)-O^-\ Na^+ + H-OCH_3$$

(b)

$$CH_3C(=O)-O-H \quad ^{-}OH\ Cs^+ \longrightarrow CH_3C(=O)-O^-\ Cs^+ + H_2O$$

(c)

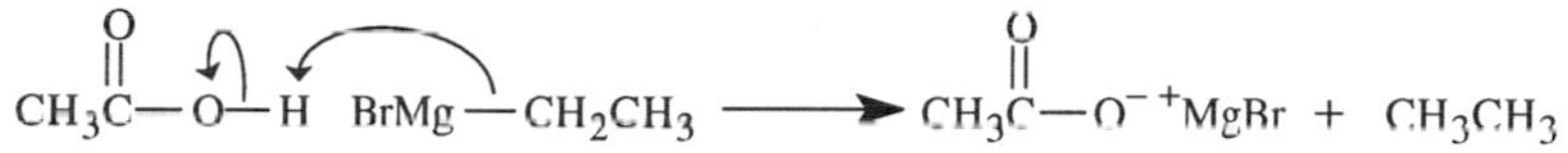

$$CH_3C(=O)-O-H \quad BrMg-CH_2CH_3 \longrightarrow CH_3C(=O)-O^-\ ^{+}MgBr + CH_3CH_3$$

(d)

$$CH_3C(=O)-O-H \quad CH_3-Li \longrightarrow CH_3C(=O)-O^-\ Li^+ + CH_4$$

(e)

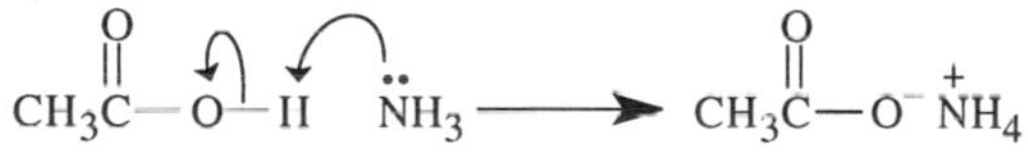

$$CH_3C(=O)-O-H \quad \ddot{N}H_3 \longrightarrow CH_3C(=O)-O^-\ \overset{+}{N}H_4$$

(f)

$$CH_3C(=O)-O-H \quad H-Na \longrightarrow CH_3C(=O)-O^-\ Na^+ + H_2$$

(g)

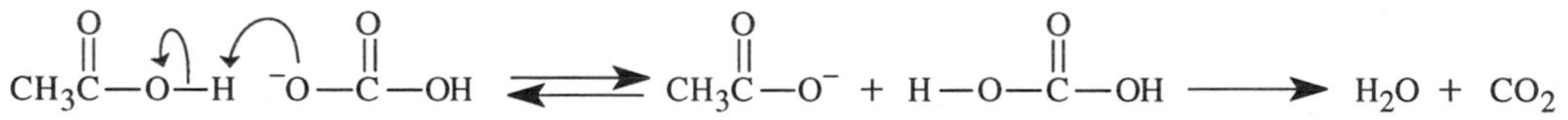

$$CH_3C(=O)-O-H \quad \underset{\text{bicarbonate ion}}{^{-}O-C(=O)-OH} \rightleftharpoons CH_3C(=O)-O^- + \underset{\text{carbonic acid}}{H-O-C(=O)-OH} \longrightarrow H_2O + CO_2$$

20.29 Only enough NaOH is present to ionize one of the carboxy groups. The question, then, is which if any of the carboxy groups is more acidic. Because polar effects of substituents decrease with increasing distance of the substituents from the reaction center, the acid-strengthening polar effect of the chlorines has a greater effect on the carboxy group to which they are closer. Therefore, in the presence of one molar equivalent of NaOH, it is this carboxy group that is principally ionized.

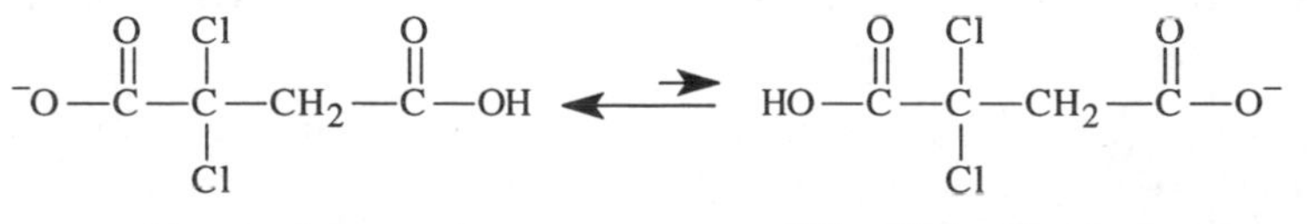

$$\underset{\text{major species present}}{^{-}O-C(=O)-CCl_2-CH_2-C(=O)-OH} \rightleftharpoons \underset{\text{little of this anion is present}}{HO-C(=O)-CCl_2-CH_2-C(=O)-O^-}$$

20.30 As the chain length becomes larger the distance between the two carboxy groups becomes greater. Because polar effects decrease with distance, the polar effect is negligible when the chain length is great, and the two pK_a values are nearly equal. (However, they are not *exactly* equal; for an explanation, see the icon comment following the solution to Problem 3.33 on p. 66 of this manual.)

20.31 The trimethylammonium substituent in *B* has the greatest acid-strengthening polar effect because it carries a full positive charge. The methoxy substituent in *A* has a more modest acid-strengthening polar effect associated with the C—O bond dipole. The polar effect of a methyl group is negligible, and is slightly acid-weakening. Therefore, the order of increasing acidity (and therefore decreasing pK_a is $C < A < B$.

20.32 Carry out a calculation of pH using the dissociation-constant expression for each acid. (See Eq. 20.7a, text p. 906.) Let the concentration of the undissociated acid be $0.1 - x$; the concentrations of both H_3O^+ and $CH_3CO_2^-$ are then x.

$$K_a = 10^{-4.76} = 1.74 \times 10^{-5} = \frac{[H_3O^+][CH_3CO_2^-]}{[CH_3CO_2H]} = \frac{x^2}{0.1 - x}$$

Because the dissociation constant of acetic acid is small, use the approximation that x is small compared with the amount of acetic acid present.

$$\frac{x^2}{0.1} = 1.74 \times 10^{-5}$$
$$x = [H_3O^+] = 1.32 \times 10^{-3}$$
$$-\log x = \text{pH} = 2.88$$

Thus, the pH is less than 3, and acetic acid therefore turns litmus paper red.

An analogous calculation reveals that the pH of a 0.1 *M* solution of phenol ($pK_a = 9.95$) is 5.48. Since this pH is greater than 3, a 0.1 *M* solution of phenol ("carbolic acid") does not turn litmus paper red.

20.33 Use Eq. 20.7c on text p. 906. (This equation is sometimes called the Henderson-Hasselbalch equation.) Let the ratio $[RCO_2H]/[RCO_2^-]$ given in the problem be r. Rearranging Eq. 20.7c, and noting that $\log(1/r) = -\log r$,

$$\text{pH} = pK_a - \log r$$

(a) When $r = ⅓$, pH = 4.76 – log (⅓) = 5.24.
(b) When $r = 3$, pH = 4.76 – log 3 = 4.28.
(c) When $r = 1$, pH 4.76 – log 1 = 4.76.

Many students confuse pK_a, which is a property of a *compound,* with pH, which is a property of the *solution* in which the compound is dissolved. The pK_a is a constant property of an acid that does not vary with concentration; the pH, as text Eq. 20.7c shows, depends on the relative concentrations of the acid and its conjugate base. (See Study Guide Link 3.4.) As the calculation in this problem is designed to illustrate, the pH equals the pK_a when the concentrations of an acid and its conjugate base are identical. When the concentration of the conjugate acid is greater, the pH is less than the pK_a; when the concentration of the acid is less, the pH is greater than the pK_a.

20.34 (a) Consider the following equilibrium, in which AH is an acid, and S is the solvent:

$$AH + S \rightleftharpoons A^- + \overset{+}{S}H$$

Suppose that AH is more acidic than ^+SH. Then this equilibrium lies to the right, and ^+SH, the conjugate acid of the solvent, is the major species present. That is, because AH cannot survive, the strongest acid that can exist is the conjugate acid of the solvent.

(b) Use the result in part (a). When acetic acid is the solvent, the most acidic species that can exist is the *conjugate acid* of acetic acid (that is, protonated acetic acid; see structure in Eq. 20.8, text p. 907 with R— = H_3C—), which has a pK_a of about –6 (text p. 907). Table 3.1 on text p. 94 gives the pK_a of HBr as

–8 to –9.5. Consequently, the HBr-acetic acid system has an effective pK_a of about –6. In aqueous solution, HBr is dissociated to give H_3O^+, the conjugate acid of water, which has a pK_a of –1.7; therefore the effective pK_a of aqueous HBr is –1.7. Hence, HBr in acetic acid is far more acidic than HBr in water because of the greater acidity of the conjugate acid of acetic acid, the solvent.

20.35 The proton and the π bond move back and forth between the two oxygens so rapidly that the two oxygens retain their identities for only a tiny fraction of a second. The mechanism varies with the situation. In a solvent (for example, water) in which the acid readily dissociates, even to a small extent, the two oxygens in the conjugate-base carboxylate ion are equivalent and are protonated with essentially equal frequency; these reactions are very fast. (In the equations that follow, one of the oxygens is marked with an asterisk (*) so that its fate can be traced throughout the reaction.)

$$\mathrm{R{-}C({=}{}^*O){-}O{-}H} \underset{}{\overset{H_2O}{\rightleftharpoons}} \left[\mathrm{R{-}C({=}{}^*O){-}O^-} \longleftrightarrow \mathrm{R{-}C({-}{}^*O^-){=}O}\right] + H_3O^+ \underset{H_2O}{\rightleftharpoons} \mathrm{R{-}C({-}{}^*O{-}H){=}O}$$

In solvents in which the carboxylic acid molecule cannot readily dissociate, protons can jump back and forth within a carboxylic acid hydrogen-bonded dimer (text p. 901):

$$\mathrm{R{-}C({=}{}^*O \cdots H{-}O){-}O{-}H \cdots O^*{=}C{-}R} \rightleftharpoons \mathrm{R{-}C({-}{}^*O{-}H \cdots O{=}C{-}R){=}O \cdots H{-}O^*}$$

20.36 (a)

$$\underset{\substack{\text{isobutyric acid}\\ \text{(2-methylpropanoic acid)}}}{(CH_3)_2CH\overset{O}{\overset{\|}{C}}OH} \xrightarrow[H_2SO_4\ \text{(catalyst)}]{HOCH_2CH_2CH_3\ \text{(solvent)}} \underset{\text{propyl isobutyrate}}{(CH_3)_2CH\overset{O}{\overset{\|}{C}}OCH_2CH_2CH_3}$$

(b)

$$(CH_3)_2CH\overset{O}{\overset{\|}{C}}OH \xrightarrow{CH_3OH\ \text{(solvent)},\ H_2SO_4\ \text{catalyst}} \underset{\text{methyl isobutyrate}}{(CH_3)_2CH\overset{O}{\overset{\|}{C}}OCH_3}$$

(c) Notice in this synthesis that two additional carbons must be added. Incorporate the strategy for a two-carbon chain extension of an alcohol shown in Study Problem 11.5 on text p. 486.

$$\underset{\text{2-methylpropanoic acid}}{(CH_3)_2CH\overset{O}{\overset{\|}{C}}OH} \xrightarrow{\substack{1)\ LiAlH_4\\ 2)\ H_3O^+}} (CH_3)_2CHCH_2OH \xrightarrow[H_2SO_4]{\text{conc. HBr}} (CH_3)_2CHCH_2Br \xrightarrow[\text{ether}]{Mg} \xrightarrow{\substack{1)\ \text{ethylene oxide}\\ 2)\ H_3O^+}}$$

$$(CH_3)_2CHCH_2CH_2CH_2OH \xrightarrow{\substack{1)\ KMnO_4,\ {}^-OH\\ 2)\ H_3O^+}} \underset{\text{4-methylpentanoic acid}}{(CH_3)_2CHCH_2CH_2\overset{O}{\overset{\|}{C}}OH}$$

(d)

$$(CH_3)_2CH\overset{O}{\overset{\|}{C}}OH \xrightarrow{SOCl_2} (CH_3)_2CH\overset{O}{\overset{\|}{C}}Cl \xrightarrow{\substack{1)\ \text{benzene, } AlCl_3\\ 2)\ H_3O^+}} \underset{\substack{\text{isobutyrophenone}\\ \text{(2-methyl-1-phenyl-1-propanone)}}}{(CH_3)_2CH\overset{O}{\overset{\|}{C}}{-}C_6H_5}$$

(e)

$$(CH_3)_2CHCH_2OH \xrightarrow{CrO_3(pyridine)_2} (CH_3)_2CHCH{=}O \xrightarrow[2)\ H_3O^+]{1)\ CH_3MgI} (CH_3)_2CH\underset{}{\overset{OH}{C}}HCH_3 \xrightarrow{CrO_3(pyridine)_2} (CH_3)_2CH\overset{O}{\overset{\|}{C}}CH_3$$

prepared in part (c) — 3-methyl-2-butanone

(f)

$$(CH_3)_2CHCH_2OH \xrightarrow{CrO_3(pyridine)_2} (CH_3)_2CHCH{=}O \xrightarrow[\text{(Wittig reaction)}]{\overset{-}{C}H_2{-}\overset{+}{P}Ph_3} (CH_3)_2CHCH{=}CH_2$$

prepared in part (c) — 3-methyl-1-butene

20.37 Convert each enantiomer of 4-methylhexanoic acid into 3-methylhexane by any process that does not involve breaking a bond to the asymmetric carbon. Because each enantiomer of 4-methylhexanoic acid has known absolute configuration, the configuration of the corresponding enantiomer of 3-methylhexane will be determined. Suppose, for example, that (*S*)-4-methylhexanoic acid gives (–)-3-methylhexane. Then the two molecules should have corresponding configurations, and the configuration of (–)-3-methylhexane would then be *S*:

$$\text{H}\cdots\text{C}(CH_2CH_2CO_2H)(CH_3)(CH_2CH_3) \longrightarrow \text{H}\cdots\text{C}(CH_2CH_2CH_3)(CH_3)(CH_2CH_3)$$

(*S*)-4-methylhexanoic acid — (*S*)-3-methylhexane

The following sequence of reactions would work for such a conversion:

$$CH_3CH_2\underset{CH_3}{\underset{|}{C}}HCH_2CH_2CO_2H \xrightarrow[2)\ H_3O^+]{1)\ LiAlH_4} CH_3CH_2\underset{CH_3}{\underset{|}{C}}HCH_2CH_2CH_2OH \xrightarrow[H_2SO_4]{\text{conc. HBr}}$$

$$CH_3CH_2\underset{CH_3}{\underset{|}{C}}HCH_2CH_2CH_2Br \xrightarrow[\text{ether}]{Mg} CH_3CH_2\underset{CH_3}{\underset{|}{C}}HCH_2CH_2CH_2MgBr \xrightarrow{H_3O^+} CH_3CH_2\underset{CH_3}{\underset{|}{C}}HCH_2CH_2CH_3$$

20.38 (a) The substitutive name of valproic acid is 2-propylpentanoic acid.

(b) The common name of valproic acid is α-propylvaleric acid.

(c) A number of syntheses are possible. Here are two:

Synthesis #1:

$$CH_3CH_2CH_2\overset{O}{\overset{\|}{C}}H \xrightarrow[2)\ H_3O^+]{1)\ CH_3CH_2CH_2MgBr} CH_3CH_2CH_2\overset{OH}{\overset{|}{C}}HCH_2CH_2CH_3 \xrightarrow{CrO_3(pyridine)_2}$$

4-heptanol

$$CH_3CH_2CH_2\overset{O}{\overset{\|}{C}}CH_2CH_2CH_3 \xrightarrow[\text{(Wittig reaction)}]{H_2\overset{-}{C}{-}\overset{+}{P}Ph_3} CH_3CH_2CH_2\overset{CH_2}{\overset{\|}{C}}CH_2CH_2CH_3 \xrightarrow[2)\ H_2O_2/NaOH]{1)\ BH_3/THF}$$

$$CH_3CH_2CH_2\overset{CH_2OH}{\overset{|}{C}}HCH_2CH_2CH_3 \xrightarrow[2)\ H_3O^+]{1)\ KMnO_4,\ ^-OH} CH_3CH_2CH_2\overset{CO_2H}{\overset{|}{C}}HCH_2CH_2CH_3$$

valproic acid

Synthesis #2:

$$CH_3CH_2CH_2CH(OH)CH_2CH_2CH_3 \text{ (4-heptanol)} \xrightarrow[\text{tosyl chloride}]{H_3C\text{—}C_6H_4\text{—}SO_2Cl,\ \text{pyridine}}$$

$$CH_3CH_2CH_2CH(OTs)CH_2CH_2CH_3 \xrightarrow{\text{NaBr in DMSO}} CH_3CH_2CH_2CH(Br)CH_2CH_2CH_3 \xrightarrow{\text{1) Mg, ether; 2) } CO_2\text{; 3) } H_3O^+}$$

$$CH_3CH_2CH_2CH(CO_2H)CH_2CH_2CH_3 \text{ (valproic acid)}$$

20.39 To confirm the labeling pattern, oxidize phenylacetic acid vigorously with $KMnO_4$ or other oxidizing agent that will convert it into benzoic acid. ($\overset{*}{C} = {}^{14}C$.)

$$C_6H_5\text{—}CH_2\overset{*}{C}O_2H \xrightarrow{\text{1) } KMnO_4,\ ^-OH,\ \text{heat; 2) } H_3O^+} C_6H_5\text{—}CO_2H + \overset{*}{C}O_2$$

If all the radiolabel of phenylacetic acid is at the carbonyl carbon, the product benzoic acid should be completely devoid of radioactivity. However, if there is labeling elsewhere, the fraction of radioactivity located in carbons other than the carbonyl carbon is equal to the amount of radioactivity in the product benzoic acid divided by the total radioactivity in the starting phenylacetic acid.

20.40 Use a method that does not involve loss of oxygen from benzoic acid. Thus, acid-catalyzed esterification should *not* be used, because, as Eq. 20.18c on text p. 914 shows, this method results in cleavage of the bond between the carbonyl carbon and the carboxylate oxygen. Esterification with either diazomethane or methyl iodide and K_2CO_3 would be the preferred method because, as the discussion in Sec. 20.8B shows, these methods do not involve loss of oxygen. ($\overset{*}{O} = {}^{18}O$.)

preferred method: $$Ph\text{—}C(=O^*)\text{—}\overset{*}{O}H \xrightarrow{CH_2N_2,\ \text{ether or } CH_3I/\text{acetone}/K_2CO_3} Ph\text{—}C(=O^*)\text{—}\overset{*}{O}CH_3$$ (all isotopic oxygen is retained)

inferior method: $$Ph\text{—}C(=O^*)\text{—}\overset{*}{O}H \xrightarrow[H_2SO_4\ \text{(catalyst)}]{CH_3OH\ \text{(solvent)}} Ph\text{—}C(=O^*)\text{—}OCH_3 + H_2\overset{*}{O}$$ (some isotopic oxygen is lost)

20.41 Dissolve the mixture in a suitable solvent; since the company specializes in chlorinated organic compounds, methylene chloride might be readily available for the purpose. Extract with 5% $NaHCO_3$ solution. *p*-Chlorobenzoic acid is extracted as its conjugate-base anion into the aqueous layer. Acidification of the aqueous layer will give *p*-chlorobenzoic acid itself. Then extract the methylene chloride solution with 5% aqueous NaOH solution. *p*-Chlorophenol will be extracted into the aqueous layer as its conjugate-base , *p*-chlorophenolate anion; acidification will afford *p*-chlorophenol. The basis for this separation is Eq. 20.7c on text p. 906. When the pK_a of an acid is much lower than the pH of the solution, the acid is converted into its conjugate base; that is, the ratio of $[RCO_2^-]$ to $[RCO_2H]$ is large. The pK_a of *p*-chlorobenzoic acid, which is about 4 (text Table 20.2), is much lower than the pH of aqueous sodium bicarbonate, which is about 8.5. Hence, *p*-chlorobenzoic acid is more than 99.99% ionized by the sodium bicarbonate solution. The pK_a of *p*-chlorophenol, 9.4, is higher than the pH of aqueous sodium bicarbonate, and thus this phenol is not appreciably ionized. However, the pK_a of the phenol is much lower than the pH of 0.1 *M* aqueous NaOH solution, which is about 13; hence, the phenol is ionized by, and thus extracted into, this solution.

Once the acid and the phenol are removed, evaporation of the methylene chloride gives a mixture of 4-chlorocyclohexanol and chlorocyclohexane. The boiling point of the alcohol is *much* higher because of hydrogen bonding; consequently, these two remaining substances can be separated by fractional distillation.

20.42 The essence of this solution is to determine the ionization state of penicillin-G at the different pH values. The principles involved are discussed in the solution to Problem 20.41. Because penicillin G is a carboxylic acid, its pK_a should be in the 3–5 range. Because the pH of blood, 7.4, is considerably higher than the pK_a of the drug, penicillin-G is ionized in blood. Because the pH of stomach acid is lower than the pK_a of penicillin-G, the penicillin is largely un-ionized in stomach acid. Because carboxylate ions are generally more soluble in aqueous solution than un-ionized carboxylic acids, penicillin-G is more soluble in blood than it is in stomach acid.

20.43 (a) For this problem, let

Ar— = $(CH_3)_2N$—C_6H_4—

The color of crystal violet is due to the overlap of the empty 2*p* orbital of the positively charged carbon with the π orbitals of the benzene ring. The bleaching is due to the reaction of the carbocation with hydroxide to form the alcohol. In the alcohol, there is no longer an empty 2*p* orbital; hence, the conjugation and the color associated with the carbocation are absent.

Ar_3C^+ (crystal violet (colored)) ^-OH ⟶ Ar_3C—OH (colorless)

(b) SDS, a typical detergent, forms micelles. Despite its charge, crystal violet is incorporated into the interior of the SDS micelles because the "greasy" benzene rings interact favorably with the "greasy" hydrocarbon tails of SDS. (Think of this as a solubility phenomenon; benzene is more soluble in hydrocarbons than it is in water.) In the interior of a micelle, crystal violet is no longer accessible to hydroxide ion. Because hydroxide is strongly solvated by water, hydroxide ion cannot easily enter the interior of the micelles, and therefore it cannot react with the entrapped crystal violet. The very slow reaction that does occur is probably due to hydroxide reacting with the very small amount of crystal violet in solution.

Ar_3C^+ in micelles (colored) ⇄ micelles + Ar_3C^+ in solution $\xrightarrow{^-OH}$ Ar_3C—OH (colorless)

Ar_3C^+ in micelles $\xrightarrow{^-OH}$ no reaction

20.44 (a) (b) (c)

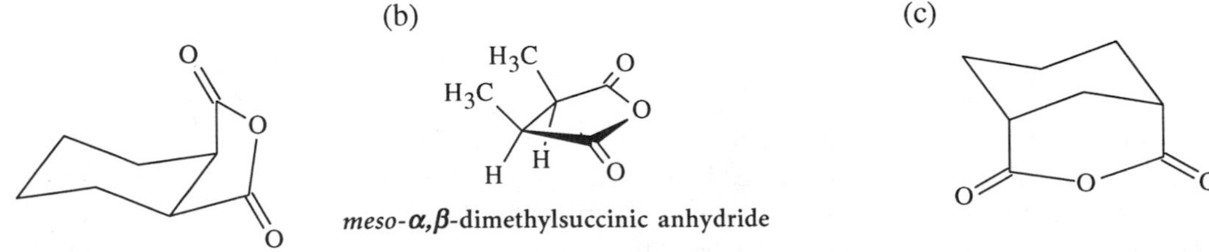

meso-α,β-dimethylsuccinic anhydride

In part (b), Notice the cis relationship of the methyl groups. (The anhydride with trans methyl substituents would be formed from *racemic* α,β-dimethylsuccinic acid).

20.45 Always remember that correct syntheses other than the one given may be possible. Be sure to ask your professor or teaching assistant if your synthesis differs from the one given and you are not sure whether it is correct.

(a)

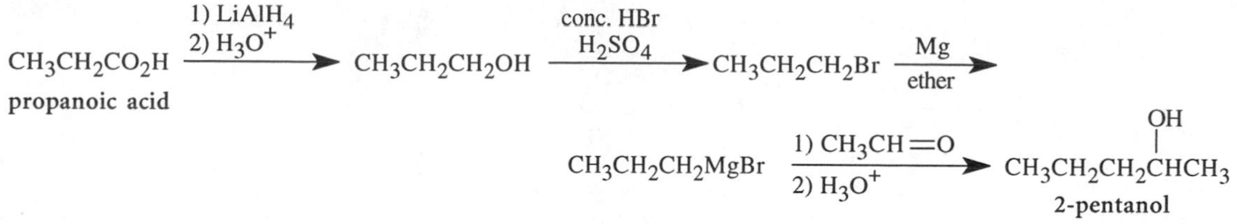

$$\underset{\text{propanoic acid}}{CH_3CH_2CO_2H} \xrightarrow[]{\text{1) LiAlH}_4\text{, 2) H}_3\text{O}^+} CH_3CH_2CH_2OH \xrightarrow[]{\text{conc. HBr, H}_2\text{SO}_4} CH_3CH_2CH_2Br \xrightarrow[\text{ether}]{\text{Mg}}$$

$$CH_3CH_2CH_2MgBr \xrightarrow[\text{2) H}_3\text{O}^+]{\text{1) CH}_3\text{CH=O}} \underset{\text{2-pentanol}}{CH_3CH_2CH_2CH(OH)CH_3}$$

(b)

$$\underset{\text{allyl alcohol}}{H_2C{=}CHCH_2OH} \xrightarrow{\text{H}_2\text{, cat.}} CH_3CH_2CH_2OH \xrightarrow[\text{2) H}_3\text{O}^+]{\text{1) KMnO}_4\text{, }^-\text{OH}} CH_3CH_2CO_2H$$

$$H_2C{=}CHCH_2OH \xrightarrow{\text{HBr}} H_2C{=}CHCH_2Br$$

$$CH_3CH_2CO_2H + H_2C{=}CHCH_2Br \xrightarrow{\text{K}_2\text{CO}_3\text{, acetone}} CH_3CH_3\overset{O}{\overset{\|}{C}}OCH_2CH{=}CH_2$$

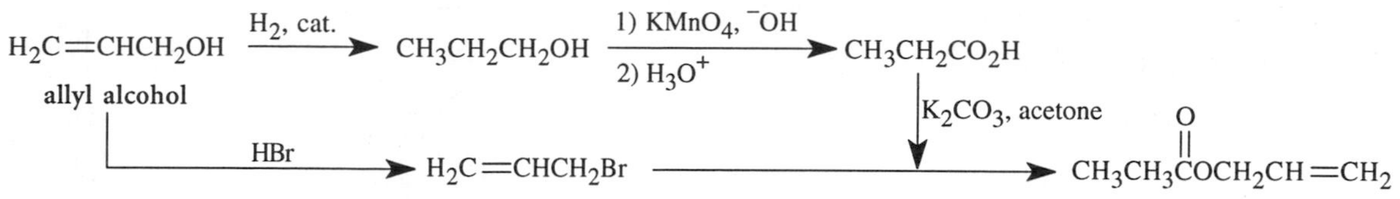

Acid-catalyzed esterification of propionic acid (prepared as shown above) with allyl alcohol could also be used, but this reaction would require an excess of allyl alcohol.

(c)

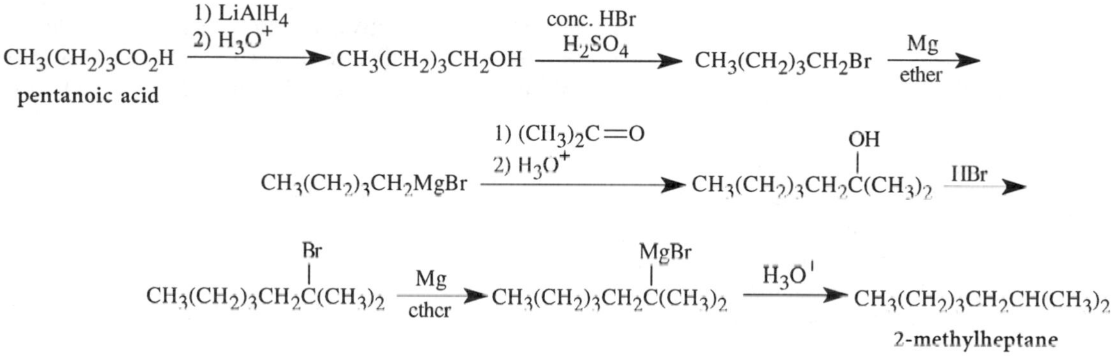

$$\underset{\text{pentanoic acid}}{CH_3(CH_2)_3CO_2H} \xrightarrow{\text{1) LiAlH}_4\text{, 2) H}_3\text{O}^+} CH_3(CH_2)_3CH_2OH \xrightarrow{\text{conc. HBr, H}_2\text{SO}_4} CH_3(CH_2)_3CH_2Br \xrightarrow[\text{ether}]{\text{Mg}}$$

$$CH_3(CH_2)_3CH_2MgBr \xrightarrow{\text{1) (CH}_3)_2\text{C=O, 2) H}_3\text{O}^+} CH_3(CH_2)_3CH_2C(OH)(CH_3)_2 \xrightarrow{\text{HBr}}$$

$$CH_3(CH_2)_3CH_2C(Br)(CH_3)_2 \xrightarrow[\text{ether}]{\text{Mg}} CH_3(CH_2)_3CH_2C(MgBr)(CH_3)_2 \xrightarrow{\text{H}_3\text{O}^+} \underset{\text{2-methylheptane}}{CH_3(CH_2)_3CH_2CH(CH_3)_2}$$

(d)

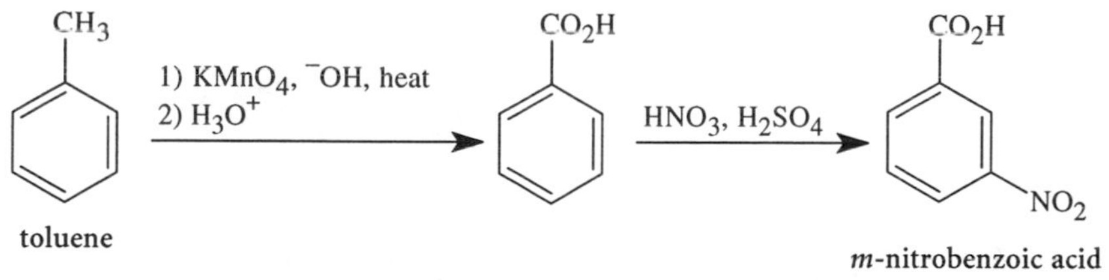

$$\underset{\text{toluene}}{C_6H_5CH_3} \xrightarrow{\text{1) KMnO}_4\text{, }^-\text{OH, heat, 2) H}_3\text{O}^+} C_6H_5CO_2H \xrightarrow{\text{HNO}_3\text{, H}_2\text{SO}_4} \underset{m\text{-nitrobenzoic acid}}{m\text{-}O_2NC_6H_4CO_2H}$$

(e)

$$\underset{\text{benzoic acid}}{PhCO_2H} \xrightarrow{\text{1) LiAlH}_4\text{, 2) H}_3\text{O}^+} PhCH_2OH \xrightarrow{\text{conc. HBr}} PhCH_2Br \xrightarrow[\text{ether}]{\text{Mg}} PhCH_2MgBr \xrightarrow{\text{1) O=CHCH}_2\text{Ph, 2) H}_2\text{SO}_4}$$

$$PhCH{=}CHCHPh \xrightarrow{\text{H}_2\text{, cat.}} \underset{\text{(3-phenylpropyl)benzene}}{PhCH_2CH_2CH_2Ph}$$

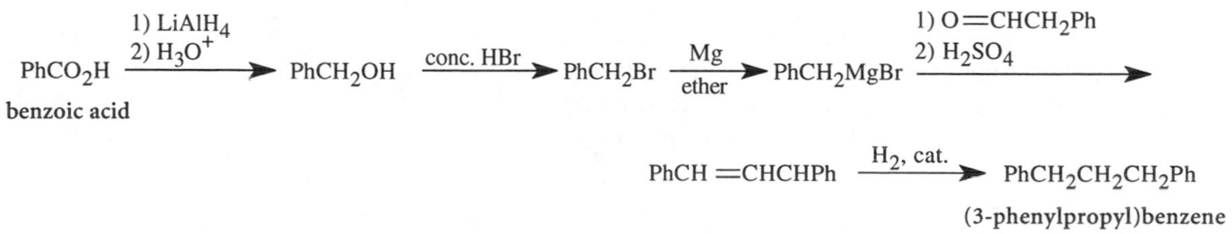

Although the rigorously correct substitutive name for this compound is (3-phenylpropyl)benzene, most chemists would call the product in part (e) 1,3-diphenylpropane.

(f)

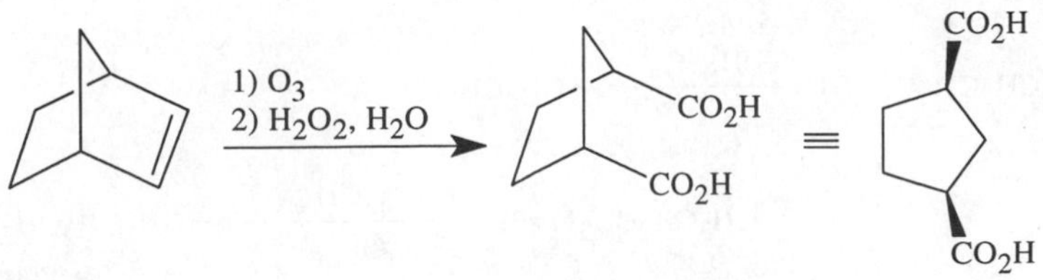

1) O_3
2) H_2O_2, H_2O

CO_2H CO_2H ≡ CO_2H CO_2H

cis-1,3-cyclopentanedicarboxylic acid

(g)

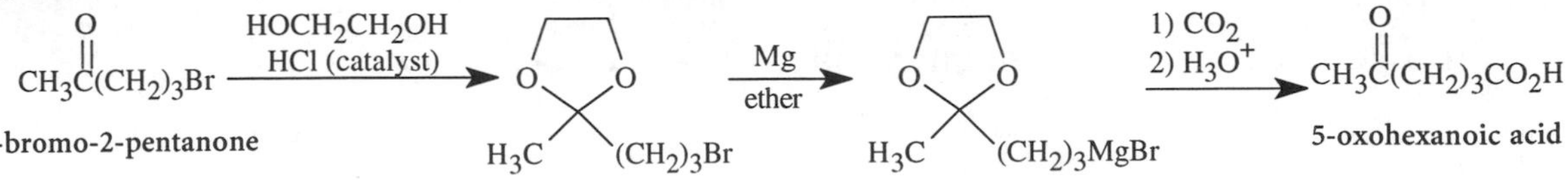

$CH_3\overset{O}{\overset{\|}{C}}(CH_2)_3Br$ — $HOCH_2CH_2OH$, HCl (catalyst) → (ketal) H_3C $(CH_2)_3Br$ — Mg, ether → H_3C $(CH_2)_3MgBr$ — 1) CO_2 2) H_3O^+ → $CH_3\overset{O}{\overset{\|}{C}}(CH_2)_3CO_2H$

5-bromo-2-pentanone 5-oxohexanoic acid

20.46 (a) The two different compounds are the diastereomers that result respectively from loss of each of the two carboxy groups from compound *A*.

Cl, H, CO_2H, H + Cl, H, H, CO_2H

(b) The question is whether loss of the different carboxy groups gives rise to the same compound or to different compounds. Draw the two possible structures and determine whether they are different or identical. In fact, the two are identical; therefore, only one product is formed when compound *B* decarboxylates.

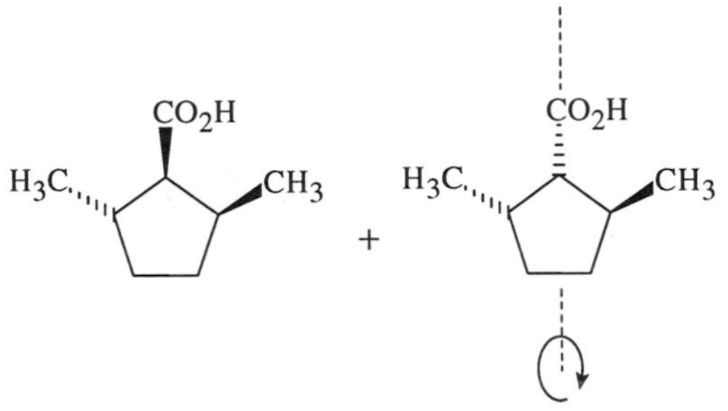

CO_2H, H_3C, CH_3 + CO_2H, H_3C, CH_3

Rotation 180° about the dashed axis shows that these structure are congruent and are therefore identical.

20.47 (a) The first ionization of squaric acid has the lower pK_a, just as the first ionization of a dicarboxylic acid has a lower pK_a than the second. (For the reason, see the solution to Problem 20.7(b) on p. 713 of this manual.)

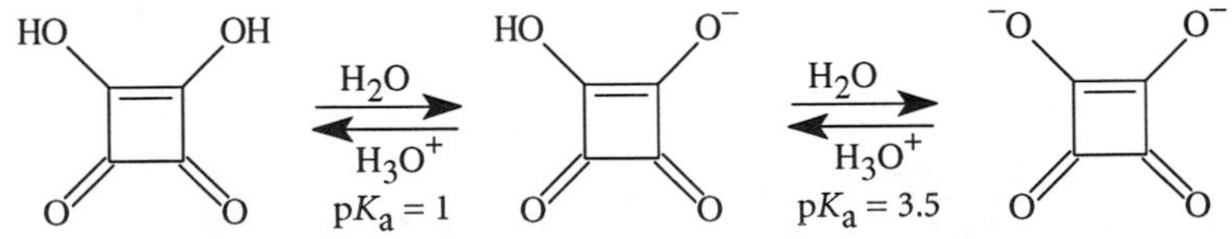

HO, OH, O, O ⇌ (H_2O / H_3O^+, $pK_a = 1$) HO, O^-, O, O ⇌ (H_2O / H_3O^+, $pK_a = 3.5$) ^-O, O^-, O, O

(b) One reason for the greater acidity of squaric acid is the electron-withdrawing polar effect of the two carbonyl groups, which stabilizes the anions. Another reason for the greater acidity of squaric acid is the resonance stabilization of the conjugate-base anions that results from interaction of the unshared pairs on each of the oxygens with a carbonyl group.

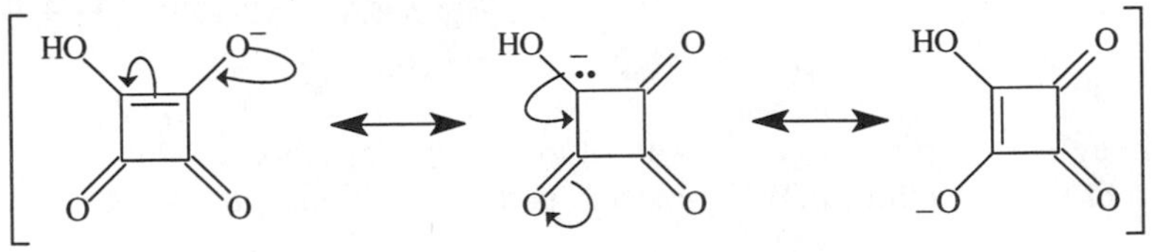

[HO, O^-, O, O ⟷ HO, O, O, O ⟷ HO, O, ^-O, O]

In the conjugate base of an enol, in contrast, the stabilizing polar effect of a carbonyl oxygen is not present, and the only resonance stabilization results from interaction of the electron pair with the double bond.

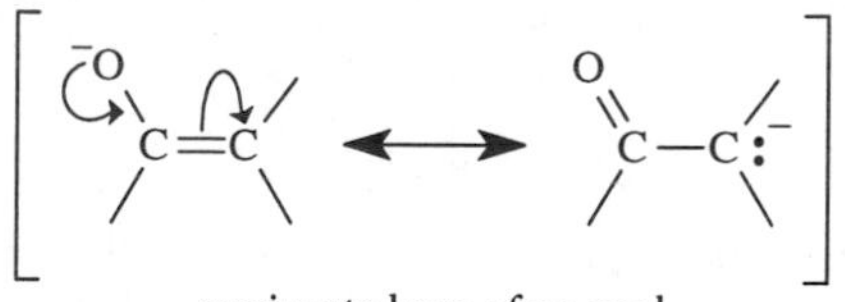

conjugate base of an enol

(c) The reactions of squaric acid with a number of reagents are analogous to the corresponding reactions of carboxylic acids. Thus, reaction with $SOCl_2$ results in the formation of the "di-acid chloride" *A;* and reaction with ethanol and an acid catalyst results in formation of the "di-ester" *B*.

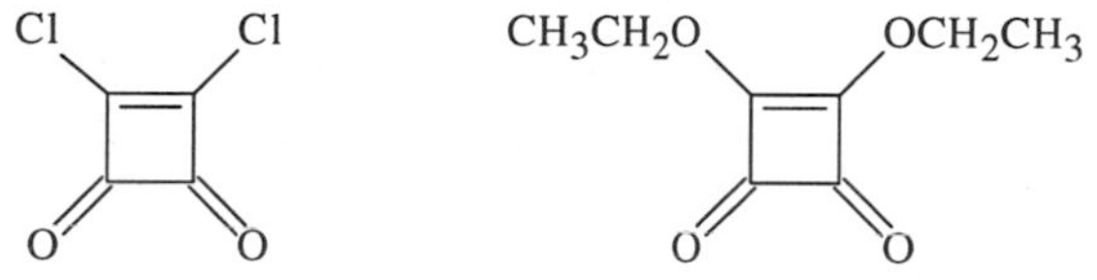

A *B*

20.48 (a) The reagent involved is essentially a diazomethane molecule in which the hydrogens of diazomethane have been replaced by phenyl groups. Just as diazomethane forms methyl esters, this reagent forms a diphenylmethyl ester.

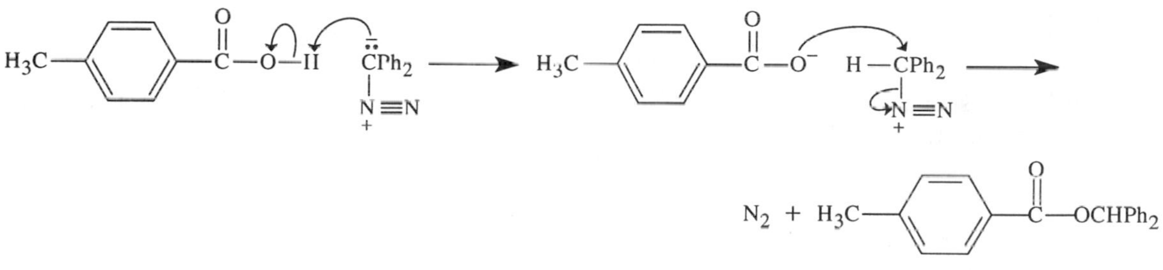

diphenylmethyl 4-methylbenzoate

(b) The KOH converts benzoic acid into its conjugate-base benzoate anion, which is alkylated by benzyl chloride to give benzyl benzoate.

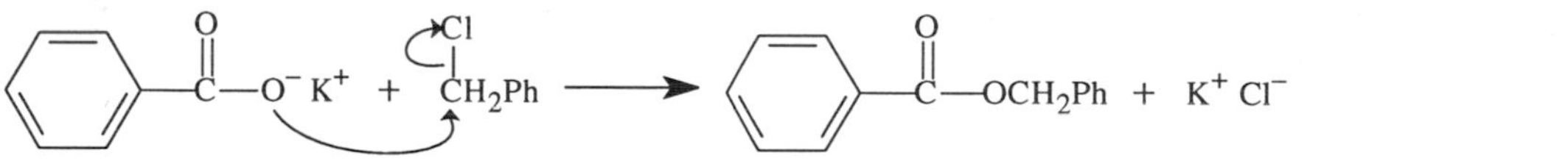

potassium benzoate benzyl benzoate

(c) Ethylene glycol is a diol and the acid, terephthalic acid (1,4-benzenedicarboxylic acid), is a dicarboxylic acid. Each end of both the diacid and the diol can be esterified; the result is a polymeric ester, polyethylene terephthalate. You may be wearing some of this compound, which is known in the industrial world as *polyester.*

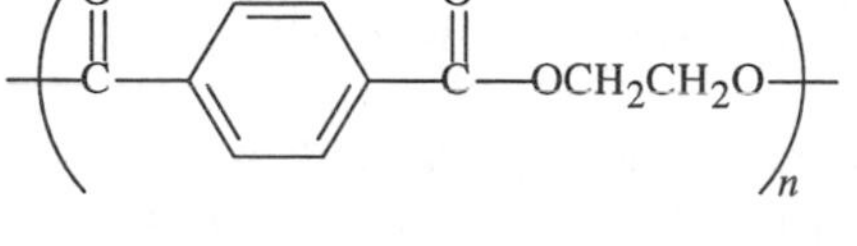

polyethylene terephthalate (polyester)

(d) This is an oxymercuration-reduction reaction in which acetic acid rather than water serves as the nucleophile that opens the mercurinium ion. (See the solution to Problem 5.29(c) on pp. 140 of this manual.)

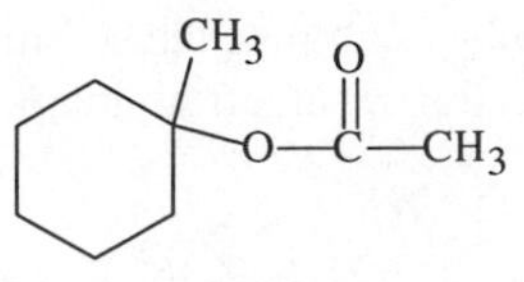

1-methylcyclohexyl acetate

(e) The conjugate-base carboxylate of propionic acid acts as the nucleophile in a ring-opening reaction of the epoxide. Because the conditions are basic, attack occurs at the less branched carbon of the epoxide. Note that KOH reacts *much* more rapidly with the acidic hydrogen of the acid than it does with the epoxide; because only one equivalent of KOH is present, reaction of hydroxide ion with the epoxide is not a competing reaction.

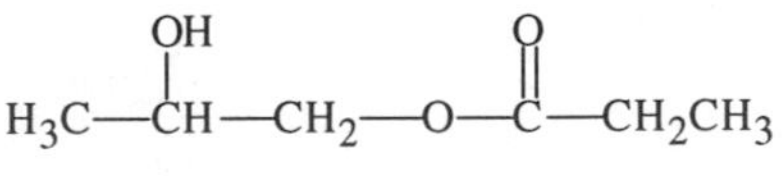

$$H_3C-CH(OH)-CH_2-O-C(=O)-CH_2CH_3$$

2-hydroxy-1-propyl propanoate
(β-hydroxypropyl propionate)

(f) This is an intramolecular variation of the esterification shown in Eq. 20.22, text p. 917. In this case, potassium carbonate converts the carboxylic acid into its conjugate-base potassium carboxylate, which is then intramolecularly alkylated to form the cyclic ester (lactone).

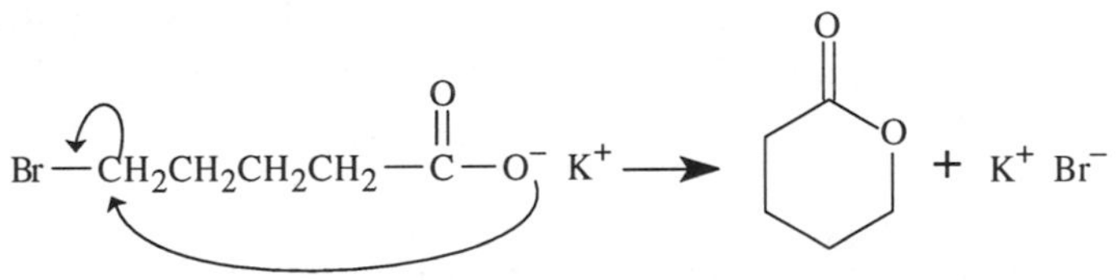

$$Br-CH_2CH_2CH_2CH_2-C(=O)-O^-\ K^+ \longrightarrow \text{(lactone)} + K^+\ Br^-$$

(g) Potassium carbonate converts both the carboxy group and the phenol —OH groups into their respective conjugate-base anions; these are all alkylated by dimethyl sulfate.

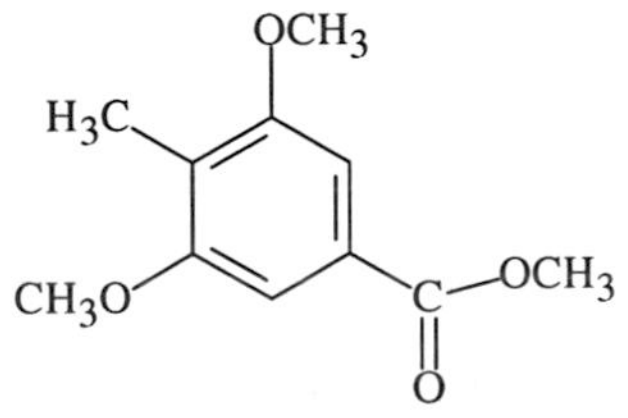

methyl 3,5-dimethoxy-4-methylbenzoate

(h) Chlorosulfonation of chlorobenzene gives electrophilic aromatic substitution at the para position. (See Eq. 20.28a–b on text p. 919.)

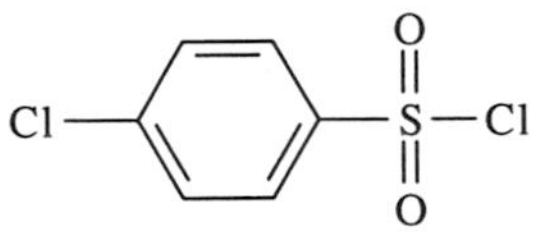

p-chlorobenzenesulfonyl chloride

20.49 (a) The hint refers to the fact that organolithium reagents are strong bases and react instantaneously with carboxylic acids to give their conjugate-base lithium carboxylate salts:

$$R-C(=O)-O-H \quad H_3C-Li \longrightarrow R-C(=O)-O^-\ Li^+ + CH_4$$

A second equivalent of organolithium reagent undergoes a carbonyl addition with the carboxylate salt to give compound *A*, which is the conjugate-base di-anion of a ketone hydrate. Since compound *A* does not contain a good leaving group, it is stable until acid is added. Protonation forms the hydrate, which then decomposes to the ketone by the reverse of the mechanism shown for hydrate formation in Eqs. 19.18a–b on text p. 857.

$$R{-}C(=O){-}O^-Li^+ + CH_3{-}Li \longrightarrow \underset{A}{R{-}C(O^-Li^+)(CH_3){-}O^-Li^+} \xrightarrow{H{-}\overset{+}{O}H_2} R{-}C(O^-Li^+)(CH_3){-}OH + OH_2 \xrightarrow{H{-}\overset{+}{O}H_2}$$

$$OH_2 + R{-}C(OH)(CH_3){-}OH \xrightarrow{\text{(see Eqs. 19.18a–b)}} R{-}C(=O){-}CH_3 + H_2O$$

(b) By analogy to the reaction discussed in part (a), the organic products are the ethyl ketone and ethane.

$$\text{2-}CH_3C_6H_4{-}C(=O){-}OH \xrightarrow{CH_3CH_2Li} \text{2-}CH_3C_6H_4{-}C(=O){-}O^-Li^+ + CH_3CH_3 \xrightarrow[2)\ H_3O^+]{1)\ CH_3CH_2Li} \text{2-}CH_3C_6H_4{-}C(=O){-}CH_2CH_3$$

20.50 (a) Orthoesters have the same relationship to esters that acetals have to ketones, and the hydrolysis mechanisms of both orthoesters and acetals are virtually identical.

$$CH_3C(OC_2H_5)_3 \xrightarrow{H{-}\overset{+}{O}H_2} CH_3C(\overset{+}{O}(H)C_2H_5)(OC_2H_5)_2 + OH_2 \longrightarrow CH_3\overset{+}{C}(OC_2H_5)_2 + HOC_2H_5 \xrightarrow{OH_2} CH_3C(\overset{+}{O}H_2)(OC_2H_5)_2 \xrightarrow{OH_2}$$

$$CH_3C(OH)(OC_2H_5)_2 \xrightarrow{H{-}\overset{+}{O}H_2} CH_3C(OH)(\overset{+}{O}(H)C_2H_5)(OC_2H_5) + OH_2 \longrightarrow CH_3C(=\overset{+}{O}{-}H)(OC_2H_5) + HOC_2H_5 \xrightarrow{OH_2} CH_3C(=O){-}OC_2H_5 + H{-}\overset{+}{O}H_2$$

(b) The mechanism is much like that for acetal formation, except that the carboxy group rather than a second alcohol molecule attacks the α-alkoxy carbocation intermediate. (Recall that many intramolecular reactions that form small rings are faster than related intermolecular reactions; see Sec. 11.7.)

$$\text{2-}(HO{-}C(=O))C_6H_4{-}C(=O){-}CH_3 \xrightarrow{H{-}\overset{+}{O}(H)CH_3} \text{2-}(HO{-}C(=O))C_6H_4{-}C(=\overset{+}{O}H){-}CH_3 \xrightarrow{HOCH_3} \text{2-}(HO{-}C(=O))C_6H_4{-}C(HO)(CH_3){-}\overset{+}{O}(H)CH_3 \xrightarrow{HOCH_3}$$

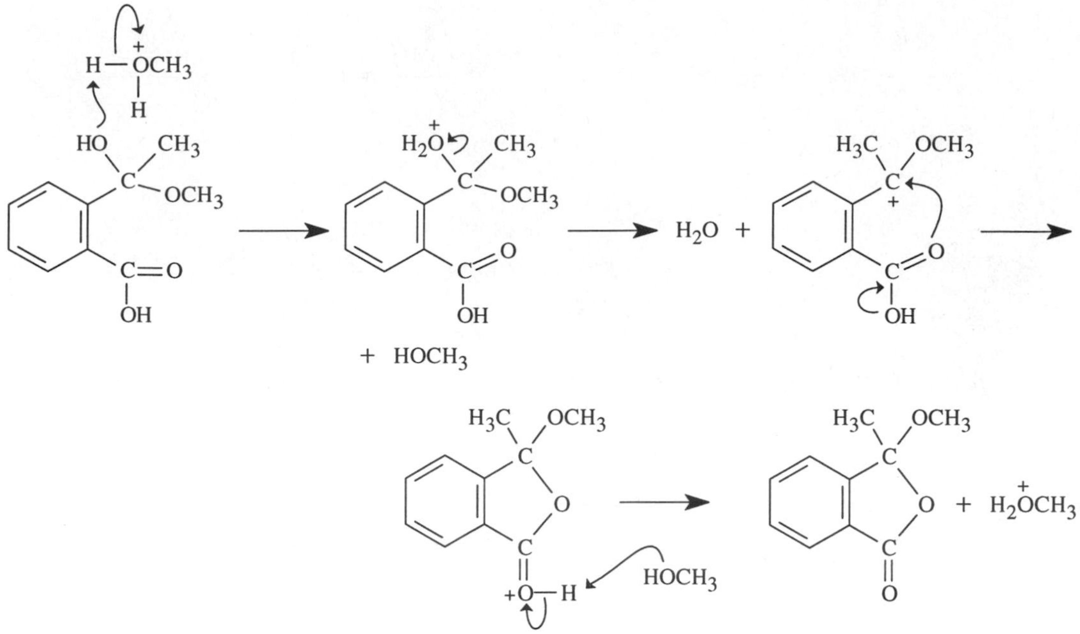

(c) Sulfuric acid protonates the alkene to give a carbocation, which is attacked by carbon monoxide to give an acylium ion (Eq. 16.23, text p. 713). This ion, in turn, is attacked by water to give, after appropriate proton transfers, the carboxylic acid.

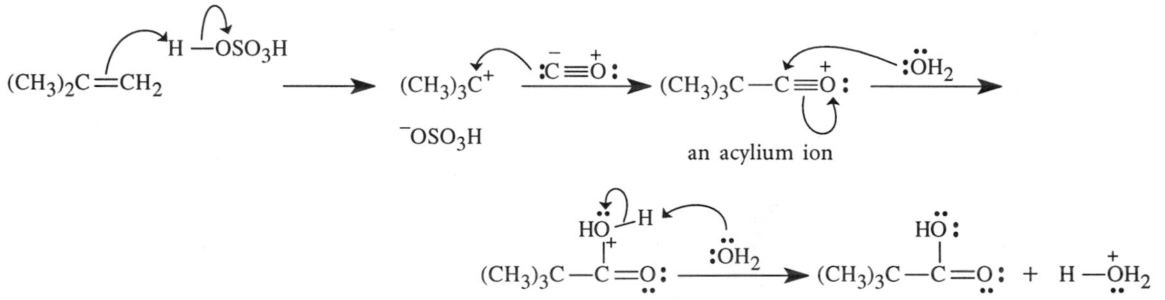

an acylium ion

(d) The pattern is much like that in the solution to part (c), except that acetic acid rather than carbon monoxide attacks the carbocation intermediate.

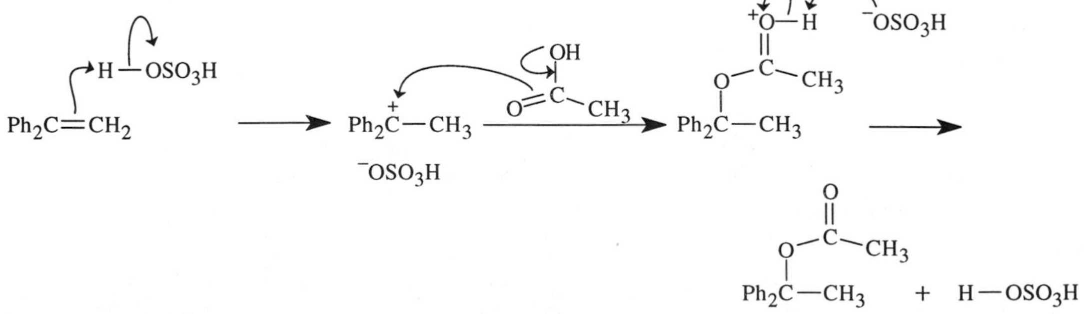

(e) Acid-promoted opening of the epoxide gives a tertiary and benzylic carbocation, which decarboxylates to the enol. The enol then forms the aldehyde by a mechanism like that shown in Eqs. 14.6d–e on text p. 616.

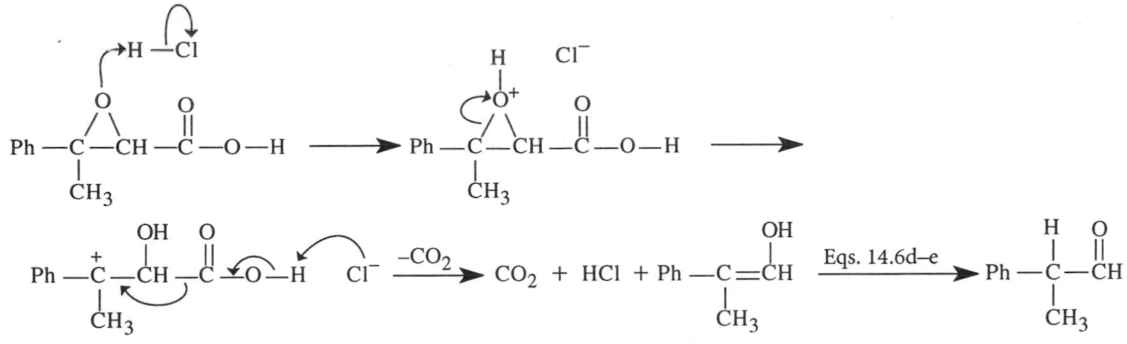

The other C—O bond of the epoxide does not rupture because (1) the carbocation that would result is secondary, and (2) carbocations α to carbonyl groups are very unstable. (The latter point is considered in Chapter 22 on text p. 1013.)

(f) The carbon of diazomethane acts as a nucleophile at the carbonyl carbon to form tetrahedral adduct *T*. This intermediate has two fates. To give the major product, we show loss of N_2 to give a primary carbocation, which then rearranges. In reality, the loss of dinitrogen and the rearrangement occur in a concerted manner, but the transition state undoubtedly has carbocation character.

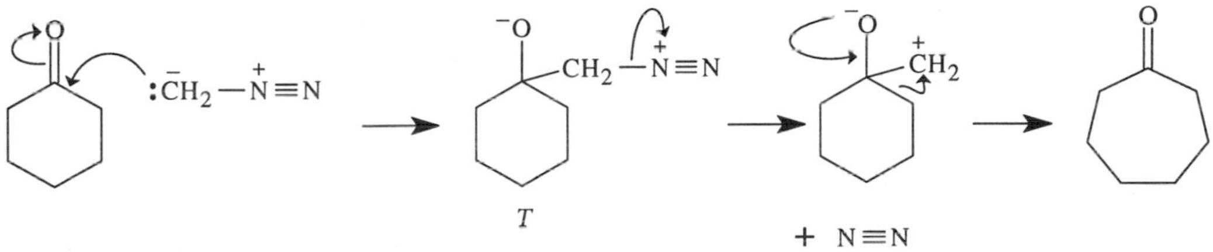

Alternatively, the carbocation can collapse with the anionic oxygen or, more likely, an internal "S_N2" reaction can take place within *T* itself, to give the minor product.

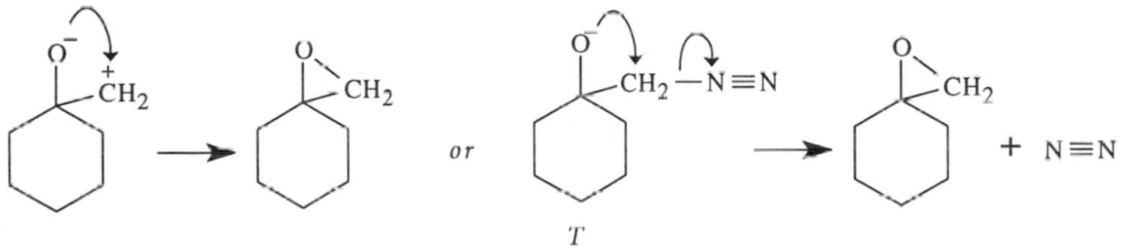

20.51 (a) The even mass of the $m/z = 74$ ion shows that the ion is an odd-electron ion. McLafferty rearrangement is a common mechanism for formation of such ions in carbonyl compounds. (See Sec. 19.3E.)

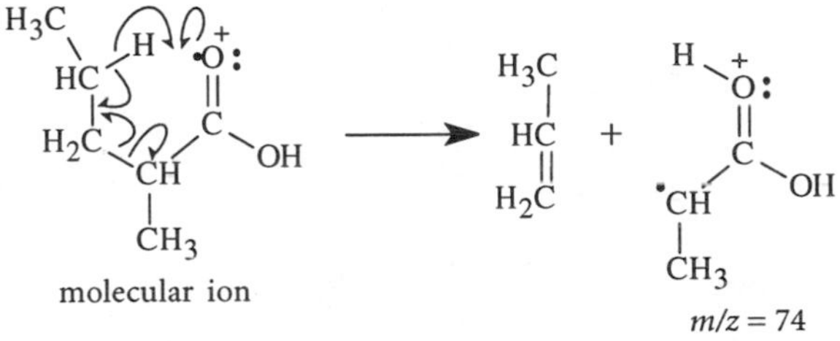

molecular ion

$m/z = 74$

(b) The molecular mass of benzoic acid is 122; therefore, the $m/z = 105$ peak in its mass spectrum represents a mass loss of 17 units, which corresponds to loss of an —OH group. This loss can occur by an α-cleavage mechanism.

Ph—C(=O⁺•)—OH ⟶ Ph—C≡O⁺ + •OH

molecular ion $m/z = 105$

The $m/z = 77$ peak corresponds to a phenyl cation, which can be lost by inductive cleavage at the other side of the carboxy group.

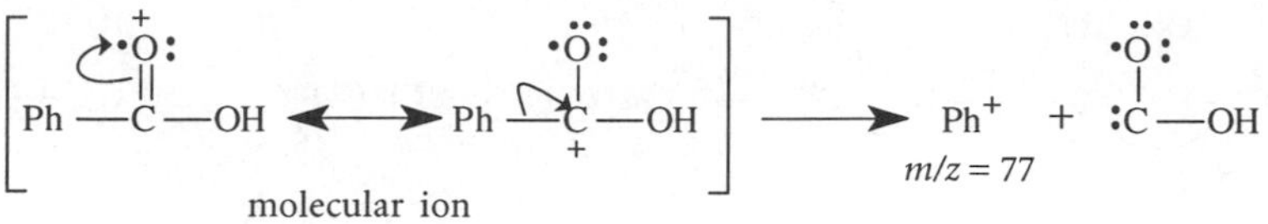

[Ph—C(=O⁺•)—OH ⟷ Ph—C⁺(—O•)—OH] ⟶ Ph⁺ + :C(—O•)—OH

molecular ion $m/z = 77$

20.52 The presence of two exchangeable hydrogens suggests that compound *A* is a dicarboxylic acid. The CMR chemical shifts suggest the presence of a methyl group, a carbonyl group, and a carbon bound to a carbonyl group. Because compound *A* is a dicarboxylic acid, each CMR resonance must correspond to two carbons. Compound *A* is 2,3-dimethylbutanedioic acid; the fact that it can be resolved into enantiomers shows that it is the racemate rather than the *meso* diastereomer.

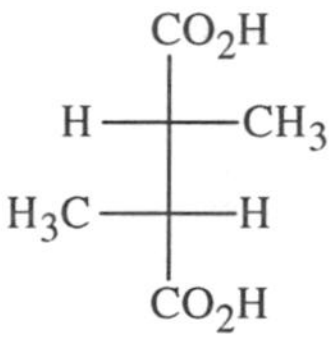

CO_2H; H—C—CH_3; H_3C—C—H; CO_2H

compound *A*
(±)-2,3-dimethylbutanedioic acid
(one of two enantiomers)

20.53 The analytical data dictate an empirical formula of C_3H_5, or a minimum molecular formula of C_6H_{10}. A hydrocarbon cannot have an odd mass; hence, the peak in the mass spectrum at $m/z = 67$ cannot be the molecular ion. Given that the unsaturation number of compound *A* is 2, and that compound *B* is a dicarboxylic acid, compound *A* must contain a double bond within a ring. Calculate the number of moles of carboxy groups titrated as follows:

$$\text{moles of carboxy groups} = (13.7\text{ mL})(0.100\text{ mol/L})(1.00 \times 10^{-3}\text{ L/mL}) = 1.37 \times 10^{-3}$$

Noting that there are two carboxy groups per mole of *B*, we calculate the molecular mass of compound *B* as follows:

$$\text{molecular mass of } B = \frac{(100 \times 10^{-3}\text{ g})(2.00\text{ moles of carboxy groups/mole of } B)}{(1.37 \times 10^{-3}\text{ moles of carboxy groups})} = 146\text{ g/mole}$$

From the following transformations, the atomic masses of the four oxygens introduced, and the molecular mass of compound *B*, the molecular mass of compound *A* can be deduced as follows:

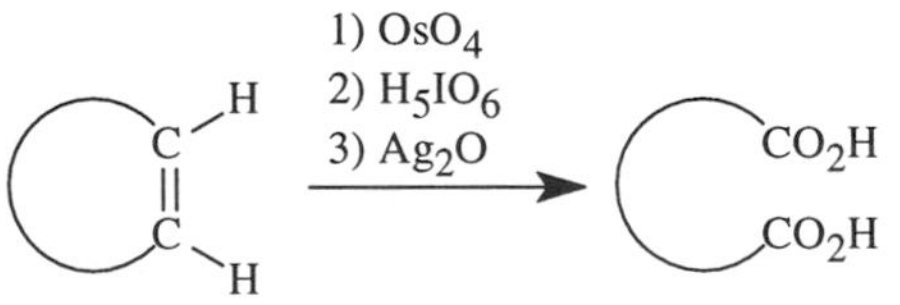

A (ring with HC=CH) —1) OsO_4; 2) H_5IO_6; 3) Ag_2O→ *B* (ring with two CO_2H)

A: molecular mass = x

B: molecular mass = $x + 64 = 146$

$x = 146 - 64 = 82$

If the molecular mass of compound *A* is 82, then the $m/z = 67$ peak in its mass spectrum corresponds to loss of 15 mass units, which corresponds to the loss of a methyl group. Evidently, compounds *A* and *B* contain a ring with a methyl branch. This branch is positioned such that there are three allylic hydrogens (from the NMR integration data) and it is also positioned such that dicarboxylic acid *B* is chiral. Compound *A* is 3-methylcyclopentene.

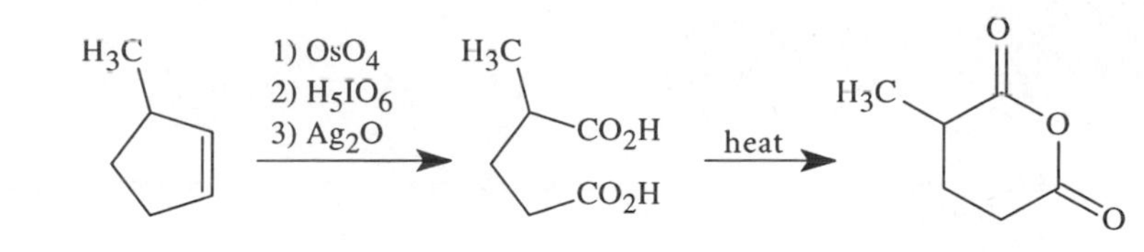

H_3C (cyclopentene) — 1) OsO_4 2) H_5IO_6 3) Ag_2O → H_3C–CH(CO_2H)–CH_2CH_2–CO_2H —heat→ (cyclic anhydride, H_3C)

A
3-methylcyclopentene

B
2-methylpentanedioic acid

20.54 (a) The 2300–3200 cm^{-1} and 1710 cm^{-1} absorptions clearly indicate a carboxylic acid, and this is confirmed by the δ 11.7 absorption in the NMR. The NMR shows five aromatic protons; hence, there is a monosubstituted benzene ring. The two mutually split triplets indicate a partial structure —CH_2CH_2—. We've accounted for all of the carbons and all of the unsaturation. However, what about the third oxygen? The chemical shift of the triplet at δ 4.1 shows that these protons must be adjacent to an oxygen. Hence, our part structure should be modified to —O—CH_2CH_2—. All that it left is to put the benzene ring at one end of the chain and the carboxy group at the other. This can be done in two ways:

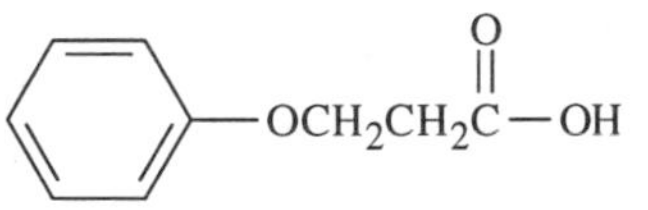

Ph–OCH_2CH_2C(=O)–OH   HO–C(=O)–OCH_2CH_2–Ph

A 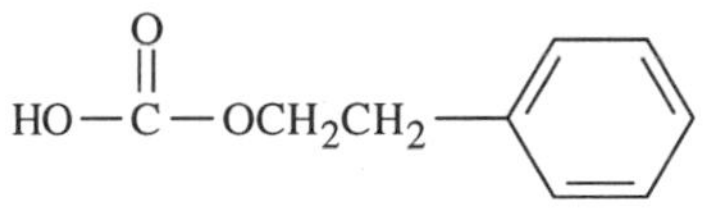   *B*

Only structure *A* is a stable compound. (Why can't *B* exist? See Eq. 20.42 on text p. 925.) Structure *A* is the correct answer.

(b) The IR spectrum indicates the presence of a carboxylic acid, and the NMR indicates the presence of an ethoxy group (—OCH_2CH_3) and a *para*-disubstituted benzene ring. The compound is 4-ethoxybenzoic acid.

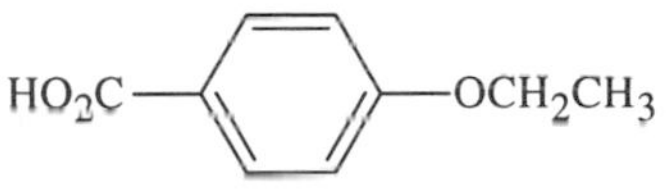

HO_2C–C_6H_4–OCH_2CH_3 4-ethoxybenzoic acid

(c) The IR spectrum indicates a carboxylic acid; the carbonyl absorption indicates that the carbonyl could be conjugated, and this is consistent with the UV absorption. The IR also shows *trans*-alkene absorption at 981 cm^{-1}. Let's now use the mass spectrum to see how much mass remains after subtracting the masses of the structural features we just deduced: 114 – 45 (carboxy group) – 13 (alkene CH) – 13 (alkene CH) = 43. A mass of 43 units is characteristic of a propyl group or an isopropyl group. A propyl group ($CH_3CH_2CH_2$—) would require a three-proton methyl triplet; this is observed at about δ 1. An isopropyl group would require a six-proton methyl doublet; this is not observed. The resonances at δ 1.5 and δ 2.1 could be the remaining protons of a propyl group. All the data are consistent with the following structure:

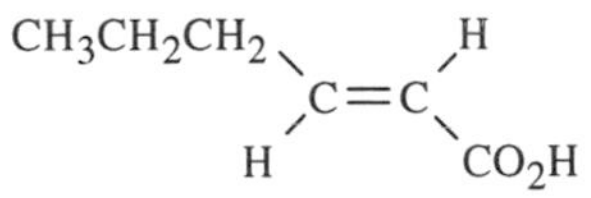

$CH_3CH_2CH_2$(H)C=C(H)CO_2H (*E*)-2-hexenoic acid

(You should account for the remaining splitting that was not discussed above.)

20.55 The titration data suggest that compound *A* is both a carboxylic acid and a phenol, and the IR spectrum shows both —OH and carbonyl absorptions. The change in the UV spectrum at high pH is also consistent with the presence of a group that ionizes at high pH and is conjugated with a benzene ring, namely, the phenolic —OH group. The resonances near δ 7 indicate a *para*-disubstituted benzene ring. The two triplets at δ 2.4 and δ 2.7 integrates for four hydrogens, and they clearly indicate a partial structure —CH_2CH_2—. Add the masses of the groups known to be present: the *para*-disubstituted benzene ring (76 units), the phenol —OH group (17 units), the carboxy group (45 units), and the —CH_2CH_2— group (28 units), and we have accounted for the m/z = 166 peak in the mass spectrum as the molecular ion. Compound *A* is 3-(4-hydroxyphenyl)propanoic acid.

HO—C_6H_4—$CH_2CH_2CO_2H$ **3-(4-hydroxyphenyl)propanoic acid** (compound *A*)

21

Chemistry of Carboxylic Acid Derivatives

TERMS

CONCEPTS

I. NOMENCLATURE AND CLASSIFICATION OF CARBOXYLIC ACID DERIVATIVES

A. Introduction to Carboxylic Acid Derivatives

1. Carboxylic acid derivatives are compounds that can be hydrolyzed under acidic or basic conditions to give the related carboxylic acids.

$R-C(=O)-OR'$	$R-C(=O)-NR'R''$	$R-C(=O)-Cl$	$R-C(=O)-O-C(=O)-R'$	$R-C\equiv N$
ester	amide (R, R′ = alkyl or H)	acid chloride	anhydride	nitrile

2. Carboxylic acids and their derivatives have both structural and chemical similarities.
   a. With the exception of nitriles, all carboxylic acid derivatives contain a carbonyl group.
   b. Many important reactions of these compounds occur at the carbonyl group.
   c. The cyano group ($-C\equiv N$) of nitriles has reactivity that resembles that of a carbonyl group.

B. Esters and Lactones

1. In both common and systematic nomenclature, esters are named as derivatives of their parent carboxylic acids by applying a variation of the system used in naming carboxylate salts.
   a. The group attached to the carboxylate oxygen is named first as a simple alkyl or aryl group.
   b. This name is followed by the name of the parent carboxylate constructed by dropping the final *ic* from the name of the acid and adding the suffix *ate*.
2. In common nomenclature:
   a. Substitution is indicated by numbering the acid portion of the ester as in carboxylic acid nomenclature, beginning with the adjacent carbon as the α-position.
   b. The alkyl or aryl group is numbered (using Greek letters) from the point of attachment to the carboxylate oxygen.
3. In substitutive nomenclature:
   a. Substitution is indicated by numbering the acid portion of the ester as in carboxylic acid nomenclature, beginning with the carbonyl as carbon-1.
   b. The alkyl or aryl group is numbered (using numbers) from the point of attachment to the carboxylate oxygen.

$$\overset{\gamma}{\underset{4}{H_3C}}-\overset{Cl,\ \beta}{\underset{3,\ CH_3}{C}}-\overset{\alpha}{\underset{2}{CH_2}}-\overset{O}{\underset{1}{C}}-OPh$$

common name: **phenyl β-chloroisovalerate**
substitutive name: **phenyl 3-chloro-3-methylbutanoate**

4. Cyclic esters are called lactones. In common nomenclature:
   a. The name of a lactone is derived from the acid with the same number of carbons in its principal chain.
   b. The ring size is denoted by a Greek letter corresponding to the point of attachment of the lactone ring oxygen to the carbon chain.

γ, β, α, δ — δ-valerolactone

C. Acid Halides

1. Acid halides are named in any system of nomenclature by replacing the *ic* ending of the acid with the suffix *yl*, followed by the name of the halide.
2. When the acid halide group is attached to a ring, the compound is named as cycloalkanecarbonyl halide.

$Cl—CH_2—CH_2—C(CH_3)_2—CH_2—C(=O)—Cl$

common name: **δ-chloro-β,β-dimethylvaleryl chloride**

substitutive name: **5-chloro-3,3-dimethylpentanoyl chloride**

CH_3O–(cyclopentane)–$C(=O)$–Br

3-methoxy-1-cyclopentanecarbonyl bromide

D. Anhydrides

1. To name an anhydride, the name of the parent acid is replaced by the word *anhydride*.
2. Mixed anhydrides, which are anhydrides that are derived from two different acids, are named by citing the two parent acids in alphabetical order.

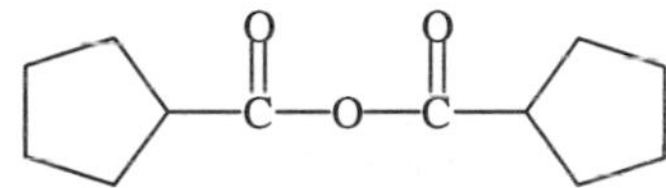

cyclopentanecarboxylic anhydride

$Ph—C(=O)—O—C(=O)—CH_3$

acetic benzoic anhydride
(a mixed anhydride)

E. Nitriles

1. In the common system, nitriles are named by dropping the *ic* or *oic* from the name of the acid with the same number of carbon atoms (counting the nitrile carbon) and adding the suffix *onitrile*. The name of the three-carbon nitrile is shortened in common nomenclature to propionitrile.
2. In substitutive nomenclature, the suffix *nitrile* is added to the name of the hydrocarbon with the same number of carbon atoms.
3. When the nitrile group is attached to a ring, a special *carbonitrile* nomenclature is used.

$I—CH_2—CH_2—C{\equiv}N$

common name: **β-iodopropionitrile**

substitutive name: **3-iodopropanenitrile**

(cyclopropane with CH_3 and $C{\equiv}N$)

1-methylcyclopropanecarbonitrile

F. Amides, Lactams, and Imides

1. Simple amides are named in any system by replacing the *ic* or *oic* suffix of the acid name with the suffix *amide*.
2. When the amide functional group is attached to a ring, the suffix *carboxamide* is used.
3. Amides are classified by the number of hydrogens attached to the amide nitrogen:
   a. Primary—two hydrogens attached.
   b. Secondary—one hydrogen attached.
   c. Tertiary—no hydrogens attached.
4. Substitution on nitrogen in secondary and tertiary amides is designated with the letter *N* (italicized or underlined).

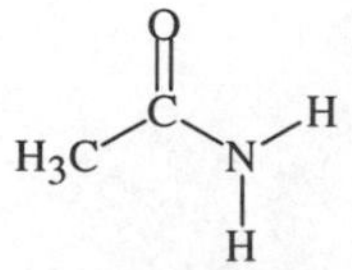

acetamide (a primary amide)

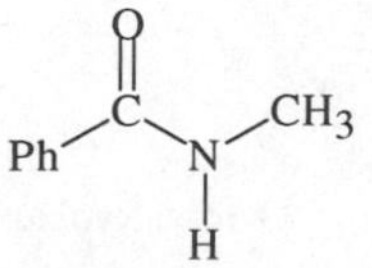

N-methylbenzamide (a secondary amide)

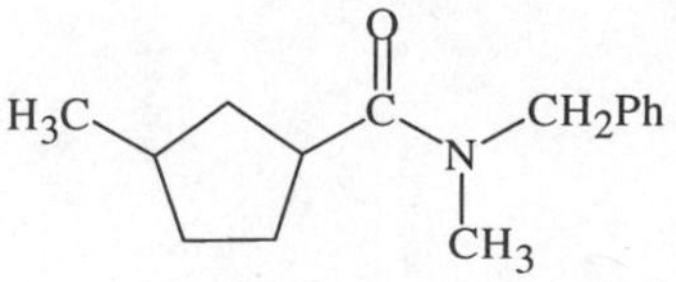

N-benzyl-*N*,3-dimethylcyclopentanecarboxamide (a tertiary amide)

5. Cyclic amides are called lactams. In common nomenclature—
   a. The name of a lactam is derived from the acid with the same number of carbons in its principal chain.
   b. The ring size is denoted by a Greek letter corresponding to the point of attachment of the lactam ring nitrogen to the carbon chain.

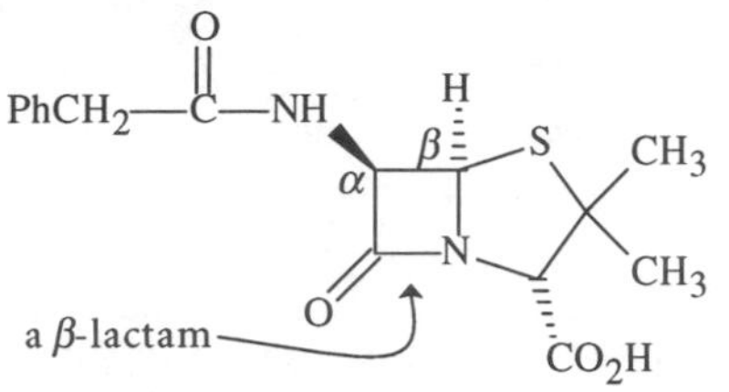

a β-lactam

penicillin G

6. Imides are the nitrogen analogs of anhydrides; cyclic imides are of greater importance than open-chain imides.

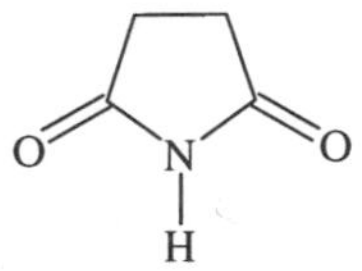

succinimide (an imide)

G. Nomenclature of Substituent Groups

1. The priorities for citing principal groups in a carboxylic acid derivative are as follows. (A complete list of functional group priorities is given in Appendix I, text page A-1.)

acid > anhydride > ester > acid halide > amide > nitrile

2. The names used for citing these groups as substituents are:

—C(=O)—OH	carboxy	—C(=O)—OCH_3	methoxycarbonyl	—C(=O)—OCH_2CH_3	ethoxycarbonyl
—C(=O)—Cl	chloroformyl	—C(=O)—NH_2	carbamoyl	—C≡N	cyano
—CH_2—C(=O)—OH	carboxymethyl	—O—C(=O)—CH_3	acetoxy or acetyloxy[a]	—NH—C(=O)—CH_3	acetamido or acetylamino[a]

[a]Used by Chemical Abstracts.

N≡C—CH_2—CH_2—C(=O)—N(CH_3)—CH_2—CH_2—CH_3 3-cyano-*N*-methyl-*N*-propylpropanamide

H. Carbonic Acid Derivatives

1. Esters of carbonic acid are named like other esters.
2. Other important carbonic acid derivatives have special names that must be learned:

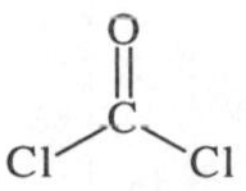

$Cl-C(=O)-Cl$ phosgene

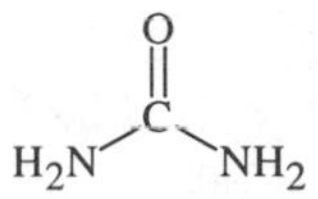

$H_2N-C(=O)-NH_2$ urea

$H_2N-C(=O)-OH$ carbamic acid

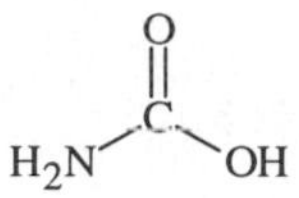

II. STRUCTURE, BASICITY, AND PHYSICAL PROPERTIES OF CARBOXYLIC ACID DERIVATIVES

A. Basicity of Carboxylic Acid Derivatives

1. Carboxylic acid derivatives are weakly basic at the carbonyl oxygen; nitriles are weakly basic at nitrogen.

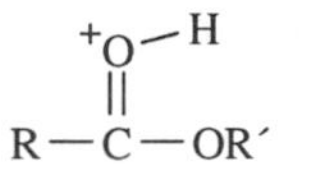

$R-C(=\overset{+}{O}-H)-OR'$ protonated ester $pK_a \approx -6$

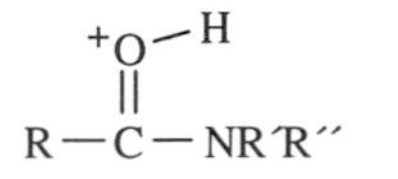

$R-C(=\overset{+}{O}-H)-NR'R''$ protonated amide $pK_a \approx -1$

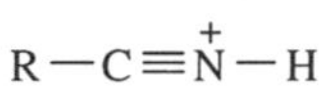

$R-C\equiv\overset{+}{N}-H$ protonated nitrile $pK_a \approx -10$

   a. The basicity of an ester is about the same as the basicity of the corresponding carboxylic acid.
   b. Amides are considerably more basic than other carboxylic acid derivatives (a reflection of the reduced electronegativity of nitrogen relative to oxygen).

B. Physical Properties of Carboxylic Acid Derivatives

1. Esters are polar molecules, but the ester functional group lacks the capability of acids to donate hydrogen bonds.
   a. Most esters are not soluble in water.
   b. Lower esters are typically volatile, fragrant liquids that have lower densities than water.
2. Most of the lower anhydrides and acid chlorides are dense, water-insoluble liquids with acrid, piercing odors.
   a. The boiling points are not very different from those of other polar molecules of about the same molecular mass and shape.
   b. The simplest anhydride, formic anhydride, and the simplest acid chloride, formyl chloride, are unstable and cannot be isolated under ordinary conditions.
3. Nitriles are among the most polar organic compounds.
   a. Their polarities are reflected in their boiling points, which are rather high, despite the absence of hydrogen bonding.
   b. Acetonitrile is miscible with water and propionitrile has a moderate solubility in water, whereas higher nitriles are insoluble in water.
4. Primary and secondary amides, like carboxylic acids, tend to associate into hydrogen-bonded dimers or higher aggregates in the solid state, in the pure liquid state, or in solvents that do not form hydrogen bonds.
   a. This property is of substantial biological importance in the structure of proteins.
   b. With increased substitution at nitrogen—
      i. The capacity for hydrogen bonding is reduced.
      ii. Boiling points decrease in spite of the increase in molecular mass.
   c. The lower amides are water-soluble, polar molecules with high boiling points; a number of amides have high dielectric constants.

C. Structure of Amides

1. In an amide, the bonds to nitrogen have essentially trigonal geometry, which can be understood on the basis of resonance structures in which the bond between the nitrogen and the carbonyl carbon has significant double-bond character.

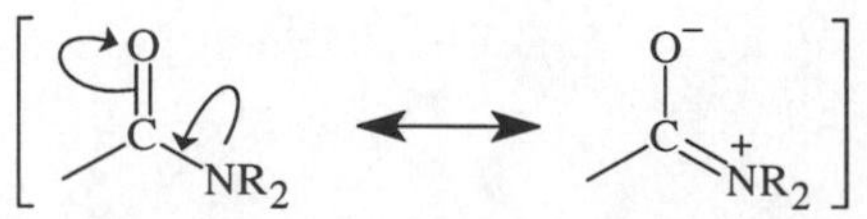

2. Secondary and tertiary amides can exist in both *E* and *Z* conformations about the carbonyl-nitrogen bond.
   a. The *Z* conformation predominates in most secondary amides because, in this form, van der Waals repulsions between the largest groups are avoided.
   b. The interconversion of *E* and *Z* forms of amides is too rapid at room temperature to permit their separate isolation, but is very slow compared to the rotation about ordinary carbon-carbon single bonds.

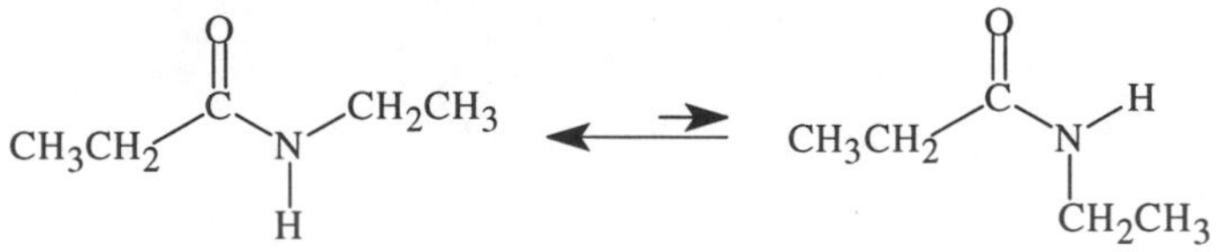

N-ethylpropionamide
Z conformation

N-ethylpropionamide
E conformation

   c. The relatively low rate of internal rotation about the carbon-nitrogen bond is caused by the significant double-bond character in this bond.

III. SPECTROSCOPY OF CARBOXYLIC ACID DERIVATIVES

A. IR Spectroscopy

1. The most important feature in the IR spectra of most carboxylic acid derivatives is the C═O stretching absorption or, for nitriles, the C≡N stretching absorption.
2. Some noteworthy IR absorption trends of carboxylic acid derivatives are as follows (see Table 21.3, text page 944, for a complete listing):
   a. Esters are readily differentiated from carboxylic acids, aldehydes, or ketones by the unique ester carbonyl absorption at 1735 cm^{-1}.
   b. Lactones, lactams, and cyclic anhydrides, like cyclic ketones, have carbonyl absorption frequencies that increase dramatically as the ring size decreases.
   c. Anhydrides and some acid chlorides have two carbonyl absorptions; the two carbonyl absorptions of anhydrides are due to the symmetrical and unsymmetrical stretching vibrations of the carbonyl group.
   d. The carbonyl absorption of amides occurs at much lower frequency than that of other carbonyl compounds.
   e. The C≡N stretching absorption of nitriles generally occurs in the triple-bond region of the spectrum; this absorption is stronger, and occurs at higher frequency, than the C≡C absorption of an alkyne.
   f. Primary and secondary amides show an N—H stretching absorption in the 3200–3400 cm^{-1} and 1640 cm^{-1} regions of the spectrum.
      i. Many primary amides show two N—H absorptions.
      ii. Secondary amides show a single strong N—H absorption.
      iii. A strong N—H bending absorption occurs in the vicinity of 1640 cm^{-1}.
      iv. Tertiary amides lack both of these N—H absorptions.

B. NMR Spectroscopy

1. The α-proton resonances of all carboxylic acid derivatives are observed in the δ 1.9–3 region of the NMR spectrum.
2. In esters, the chemical shifts of protons on the carbon adjacent to the carboxylate oxygen occur at about 0.6 ppm lower field than the analogous protons in alcohols and ethers; this shift is attributed to the electron-attracting character of the carbonyl group.
3. NMR characteristics of amides:

a. The *N*-alkyl protons of amides have chemical shifts in the δ 2.6–3 region; the N—H proton resonances of primary and secondary amides (sometimes broad) are observed in the δ 7.5–8.5 region.
b. The broadening of these proton signals is caused by a slow chemical exchange with the protons of other protic substances and by unresolved splitting with ^{14}N, which has a nuclear spin.
c. Amide N—H resonances can be washed out by exchange with D_2O.
d. The bond between the nitrogen and the carbonyl group of an amide has significant double-bond character. As a result, two identical groups attached to this nitrogen, although constitutionally equivalent, are diastereotopic and thus chemically nonequivalent.

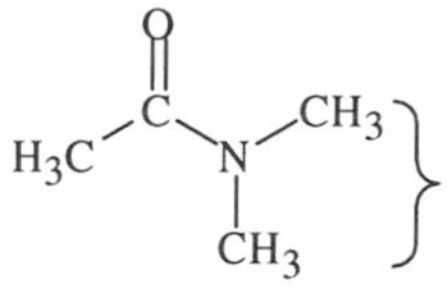

H_3C—C(=O)—N(CH_3)(CH_3) — diastereotopic methyl groups; slightly different chemical shifts

4. In carbon NMR (CMR) spectra, the carbonyl chemical shifts of carboxylic acid derivatives are in the range δ 165–180, very much like those of carboxylic acids.
   a. The chemical shifts of nitrile carbons are considerably smaller, occurring in the δ 115–120 range.
   b. These shifts are much greater, however, than those of acetylenic carbons.

IV. USE AND OCCURRENCE OF CARBOXYLIC ACIDS AND THEIR DERIVATIVES

A. Nylon and Polyesters

1. Two of the most important polymers produced on an industrial scale are nylon and polyesters.
2. Nylon is the general name given to a group of polymeric amides.
   a. Nylon is an example of a condensation polymer, which is a polymer formed in a reaction that liberates a small molecule, in this case H_2O.

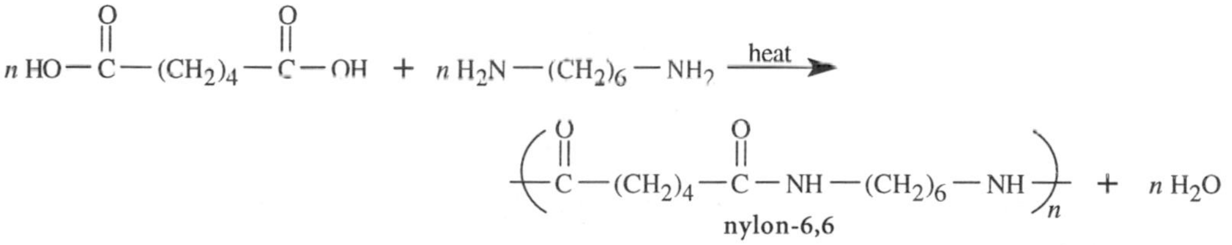

$$n\,\mathrm{HO{-}\overset{O}{\overset{\|}{C}}{-}(CH_2)_4{-}\overset{O}{\overset{\|}{C}}{-}OH} + n\,\mathrm{H_2N{-}(CH_2)_6{-}NH_2} \xrightarrow{\text{heat}} \left(\mathrm{\overset{O}{\overset{\|}{C}}{-}(CH_2)_4{-}\overset{O}{\overset{\|}{C}}{-}NH{-}(CH_2)_6{-}NH}\right)_n + n\,\mathrm{H_2O}$$

nylon-6,6

   b. In contrast, an addition polymer such as polystyrene is formed by the simple addition of one molecule to another.
3. Polyesters are condensation polymers derived from the reactions of diols and dicarboxylic acids.

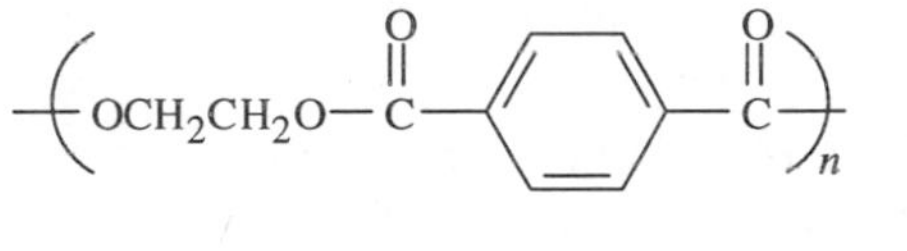

$$\left(\mathrm{OCH_2CH_2O{-}\overset{O}{\overset{\|}{C}}{-}C_6H_4{-}\overset{O}{\overset{\|}{C}}}\right)_n$$

a polyester

B. Waxes, Fats, and Phospholipids

1. Waxes, fats, and phospholipids are important naturally occurring ester derivatives of fatty acids.
2. A wax is an ester of a fatty acid and a fatty alcohol (an alcohol with a long unbranched carbon chain).

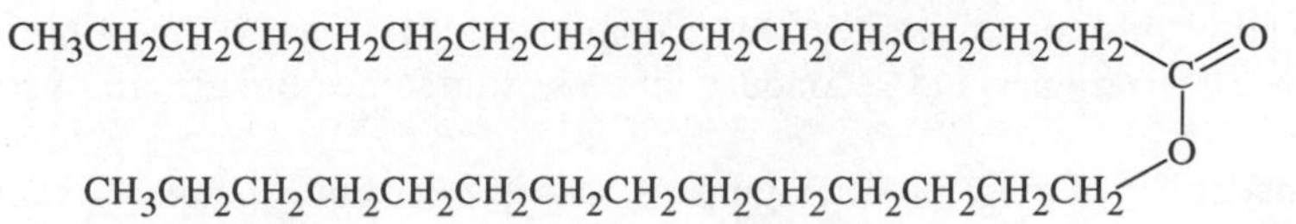

$$CH_3CH_2CH_2CH_2CH_2CH_2CH_2CH_2CH_2CH_2CH_2CH_2CH_2CH_2CH_2{-}C({=}O){-}O{-}CH_2CH_2CH_2CH_2CH_2CH_2CH_2CH_2CH_2CH_2CH_2CH_2CH_2CH_3$$

a wax

3. A fat is an ester derived from a molecule of glycerol and three molecules of fatty acid.
   a. The three acyl groups in a fat may be the same or different, and they may contain unsaturation, which is typically in the form of one or more cis double bonds.

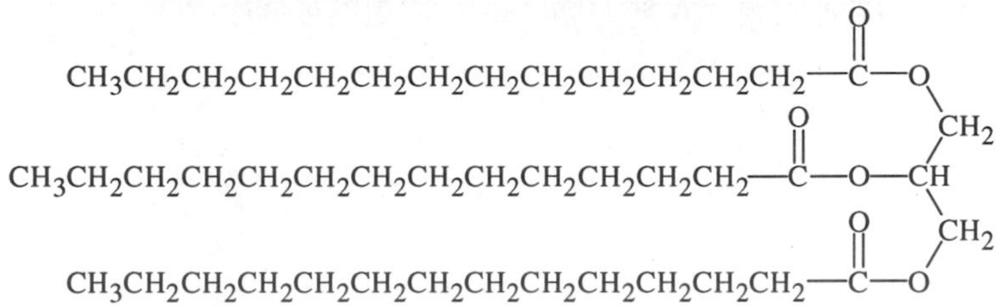

$$\begin{array}{l} CH_3CH_2CH_2CH_2CH_2CH_2CH_2CH_2CH_2CH_2CH_2CH_2CH_2{-}C({=}O){-}O{-}CH_2 \\ CH_3CH_2CH_2CH_2CH_2CH_2CH_2CH_2CH_2CH_2CH_2CH_2CH_2{-}C({=}O){-}O{-}CH \\ CH_3CH_2CH_2CH_2CH_2CH_2CH_2CH_2CH_2CH_2CH_2CH_2CH_2{-}C({=}O){-}O{-}CH_2 \end{array}$$

a fat

   b. Fats with no double bonds, termed saturated fats, are typically solids.
   c. Fats containing double bonds, termed unsaturated fats, are in many cases oily liquids.
4. Phospholipids are closely related to fats, because they too are esters of glycerol.
   a. Phospholipids differ from fats in that one of the terminal oxygens of glycerol in a phospholipid is esterified to a special type of organic phosphate derivative that constitutes the polar head group of the molecule.
   b. Phospholipids closely resemble soaps because both types of molecules are amphipathic; that is, they both have polar and nonpolar ends.
   c. Phospholipids, along with cholesterol (a steroid) and imbedded proteins, make up the phospholipid bilayers of cell membranes, the envelopes that surround living cells.

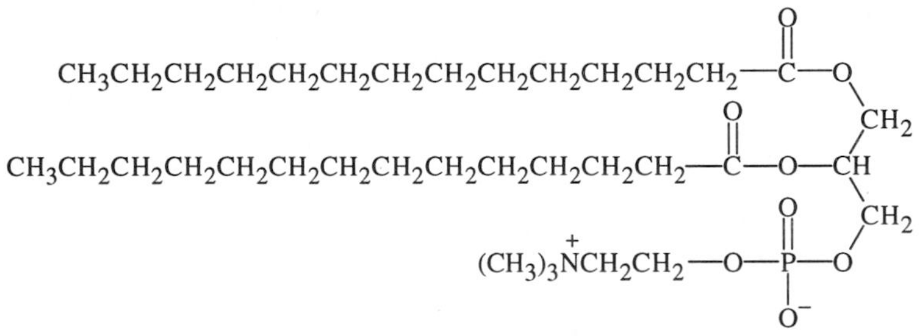

$$\begin{array}{l} CH_3CH_2CH_2CH_2CH_2CH_2CH_2CH_2CH_2CH_2CH_2CH_2CH_2{-}C({=}O){-}O{-}CH_2 \\ CH_3CH_2CH_2CH_2CH_2CH_2CH_2CH_2CH_2CH_2CH_2CH_2CH_2{-}C({=}O){-}O{-}CH \\ (CH_3)_3\overset{+}{N}CH_2CH_2{-}O{-}P({=}O)(O^-){-}O{-}CH_2 \end{array}$$

a phospholipid

V. REACTIVITY OF CARBOXYLIC ACID DERIVATIVES

A. General

1. The reactions of carboxylic acid derivatives can be categorized as follows:
   a. Reactions of the carbonyl group (or cyano group of a nitrile).
      i. Reactions at the carbonyl oxygen or cyano nitrogen.
      ii. Reactions at the carbonyl carbon or cyano carbon, the most common being reactions with nucleophiles, which include:

         water (hydrolysis) or hydroxide ion (saponification)
         alcohols or alkoxides (alcoholysis, or transesterification [in the case of esters])
         amines (aminolysis)
         hydrides (reduction)
         organometallic reagents

   b. Reactions involving the α-carbon.
   c. Reactions at the nitrogen of amides.

2. One carbonyl-group reaction of acid derivatives is the reaction of the carbonyl oxygen (or the nitrile nitrogen) as a base, which often serves as the first step in acid-catalyzed reactions.
3. The major carbonyl-group reaction of carboxylic acid derivatives is substitution at the carbonyl carbon, also called acyl substitution.
4. The C≡N bond behaves chemically much like a carbonyl group.
5. All carboxylic acid derivatives have in common the fact that they undergo hydrolysis to yield carboxylic acids.

B. Mechanisms of Nucleophilic Acyl Substitution Reactions

1. Nucleophilic acyl substitution reactions occur by two types of mechanisms:
   a. Under basic conditions:
      i. A nucleophile attacks the carbonyl carbon to give a tetrahedral addition intermediate in which the carbonyl oxygen assumes a negative charge.
      ii. The leaving group —X is then expelled from the tetrahedral intermediate.

$$\mathrm{R{-}C({=}O){-}X} \xrightarrow{\mathrm{Nuc:^-}} \mathrm{R{-}C(O^-)(Nuc){-}X} \longrightarrow \mathrm{R{-}C({=}O){-}Nuc} + \mathrm{X^-}$$

tetrahedral addition intermediate

   b. Under acidic conditions:
      i. The carbonyl is first protonated.
      ii. This carbon is then attacked by a nucleophile to form a tetrahedral addition intermediate.
      iii. After proton transfer to the leaving group —X, it is expelled as H—X.

$$\mathrm{R{-}C({=}O){-}X} \underset{}{\overset{\mathrm{HA}}{\rightleftharpoons}} \mathrm{R{-}C({=}\overset{+}{O}H){-}X}\ \ \mathrm{A^-} \xrightleftharpoons{\mathrm{:Nuc{-}H}} \mathrm{R{-}C(OH)(\overset{+}{Nuc}H){-}X}\ \ \mathrm{A^-} \rightleftharpoons \mathrm{R{-}C(OH)(Nuc){-}X} + \mathrm{H{-}A} \rightleftharpoons$$

tetrahedral addition intermediate

$$\mathrm{A^-} + \mathrm{R{-}C(OH)(Nuc){-}\overset{+}{X}{-}H} \rightleftharpoons \mathrm{R{-}C({=}\overset{+}{O}H){-}Nuc}\ \ \mathrm{A^-} + \mathrm{HX} \rightleftharpoons \mathrm{R{-}C({=}O){-}Nuc} + \mathrm{HA}$$

2. Stabilization of the carbonyl compound decreases reactivity; destabilization of the carbonyl compound increases reactivity.
3. Higher basicity of the leaving group decreases reactivity; lower basicity increases reactivity.
   a. The reactivity of carboxylic acid derivatives in nucleophilic acyl substitution reactions is:

acid chlorides › anhydrides » esters, acids › amides › nitriles

C. Relative Reactivities in Nucleophilic Acyl Substitution Reactions

1. The conditions under which the different carboxylic acid derivatives are hydrolyzed differ considerably.
   a. Hydrolysis reactions of amides and nitriles require heat as well as acid or base.
   b. Hydrolysis reactions of esters require acid or base but require shorter periods of heating.
   c. Hydrolysis reactions of acid chlorides and anhydrides occur rapidly at room temperature even in the absence of acid or base.
2. The trend in carbonyl reactivity of carboxylic acid derivatives parallels the trend in leaving group effectiveness; the practical significance of this reactivity order is that selective reactions are possible:

carboxylic acid derivative reactivity:	acid chlorides	>	anhydrides	>>	esters, acids	> amides >	nitriles
leaving group effectiveness (basic conditions):	Cl^-	>	RCO_2^-	>>	RO^-, HO^-	>	R_2N^-
leaving group effectiveness (neutral conditions):	Cl^-	>	RCO_2H	>>	ROH, HOH	>	R_2NH

3. Relative reactivity is determined by the stability of each type of compound relative to its transition state for addition or substitution:
   a. The more a compound is stabilized, the less reactive it is.
      i. The major factor affecting the stability of the carbonyl compound is the resonance interaction of the unshared electron pairs of the group X with the carbonyl π electrons.
      ii. Because oxygen is more electronegative than nitrogen, this resonance interaction is less important in an ester than in an amide; hence, amides are stabilized by resonance more than esters are, and are less reactive than esters.
   b. Reactivity is governed by the standard free energy of activation $\Delta G^{\circ\ddagger}$, which is the difference in the standard free energies of the transition state and the reactants.
      i. The major factor accounting for the differences in transition-state stability is the relative base strength of the different leaving groups —X; the best leaving groups are the weakest bases.
      ii. The more a transition state for nucleophilic addition or substitution is stabilized, the more reactive is the compound.
4. Reactions of nitriles in base are slower than those of other acid derivatives; the nitrogen is less electronegative than oxygen and accepts negative charge less readily.
5. Reactions of nitriles in acid are slower because of their extreme low basicities; the protonated form of a nitrile reacts with nucleophiles in acid solution, but there is little of this form present.

REACTIONS

I. HYDROLYSIS OF CARBOXYLIC ACID DERIVATIVES

A. Hydrolysis of Acid Chlorides and Anhydrides

1. Acid chlorides and anhydrides react rapidly with water, even in the absence of acids or bases.
2. The hydrolysis reactions of acid chlorides and anhydrides are almost never used for the preparation of carboxylic acids because these derivatives are themselves usually prepared from acids.

$$\mathrm{R{-}\overset{\overset{\displaystyle O}{\|}}{C}{-}Cl} \xrightarrow{H_2O} \mathrm{R{-}\overset{\overset{\displaystyle O}{\|}}{C}{-}OH} + \mathrm{HCl}$$

$$\mathrm{R{-}\overset{\overset{\displaystyle O}{\|}}{C}{-}O{-}\overset{\overset{\displaystyle O}{\|}}{C}{-}R} \xrightarrow{H_2O} 2\ \mathrm{R{-}\overset{\overset{\displaystyle O}{\|}}{C}{-}OH}$$

B. Basic Hydrolysis (Saponification) of Esters and Lactones

1. Esters hydrolyze in aqueous hydroxide.

$$\underset{\text{an ester}}{\mathrm{R{-}\overset{\overset{\displaystyle O}{\|}}{C}{-}OR'}} + {}^{-}\mathrm{OH} \longrightarrow \underset{\text{a carboxylate anion}}{\mathrm{R{-}\overset{\overset{\displaystyle O}{\|}}{C}{-}O^-}} + \mathrm{HOR'}$$

2. This process is called saponification; the term saponification is sometimes used to refer to hydrolysis in base of any acid derivative.
3. The mechanism of ester saponification involves:
   a. Attack by the nucleophilic hydroxide anion to give a tetrahedral addition intermediate from which an alkoxide ion is expelled.

b. The alkoxide ion thus formed reacts with the carboxylic acid to give the carboxylate salt and the alcohol.

R–C(=O)–OR′ ⇌ ($^{-}$OH; addition) HO–C(O^{-})(R)–OR′ (tetrahedral addition intermediate) → (elimination) R–C(=O)–OH + $^{-}$OR′ ⇌ R–C(=O)–O^{-} + HOR′

4. The equilibrium in this reaction lies far to the right because the carboxylic acid is a much stronger acid than the liberated alcohol; saponification is effectively irreversible.
5. Many esters can be saponified with just one equivalent of $^{-}$OH, although an excess of $^{-}$OH is often used as a matter of convenience.
6. Saponification converts a lactone completely into the salt of the corresponding hydroxy acid; upon acidification, the hydroxy acid forms.

(lactone)–CH_3 ⇌ (HO^{-}) $^{-}$O–C(=O)–CH_2–CH_2–CH(OH)–CH_3 → (H_3O^+) HO–C(=O)–CH_2–CH_2–CH(OH)–CH_3

   a. If a hydroxy acid is allowed to stand in acidic solution, it comes to equilibrium with the corresponding lactone. This reaction is an acid-catalyzed, intramolecular esterification.
   b. Lactones containing five- and six-membered rings are favored at equilibrium over their corresponding hydroxy acids; those with ring sizes smaller than five or larger than six are less stable than their corresponding hydroxy acids.
7. Ester and lactone saponifications are examples of acyl substitution.
   a. The mechanisms of these reactions are classified as nucleophilic acyl substitution mechanisms.
   b. In a nucleophilic acyl substitution reaction, the substituting group attacks the carbonyl carbon as a nucleophile, and a leaving group (another base) departs, usually in a separate step.

C. Acid-Catalyzed Hydrolysis of Esters

1. Esters can be hydrolyzed to carboxylic acids in aqueous solutions of strong acids; in most cases this reaction is slower than base-promoted hydrolysis and must be carried out with an excess of water (in which esters are insoluble).

R–C(=O)–OR′ + H_2O →(acid) R–C(=O)–OH + HOR′

2. By the principle of microscopic reversibility, the mechanism of acid-catalyzed hydrolysis is the exact reverse of acid-catalyzed esterification.
   a. The ester is first protonated by the acid catalyst; protonation makes the carbonyl carbon more electrophilic by making the carbonyl oxygen a better acceptor of electrons.
   b. Water, as the nucleophile, attacks the carbonyl carbon and then loses a proton to give the tetrahedral intermediate.
   c. Protonation of the leaving oxygen converts it into a better leaving group.
   d. Loss of the protonated leaving group gives a protonated carboxylic acid, from which a proton is removed to give the carboxylic acid itself.

R–C(=O)–OR′ + H–$\overset{+}{O}H_2$ ⇌ R–C(=$\overset{+}{O}$H)–OR′ + OH_2 ⇌ R–C(OH)(OR′)–$\overset{+}{O}$(H)–H + OH_2 ⇌ R–C(OH)(OR′)–OH (tetrahedral addition intermediate) + H–$\overset{+}{O}H_2$ ⇌

$$R-C(OH)(OH)-\overset{+}{O}(H)R' \;\; OH_2 \rightleftharpoons R-\overset{+}{C}(=O-H)-OH \;\; OH_2 \;\; HOR' \rightleftharpoons R-C(=O)-OH + H_3O^+$$

3. Saponification, followed by acidification, is a much more convenient method for hydrolysis of most esters than acid-catalyzed hydrolysis because:
   a. It is faster.
   b. It is irreversible.
   c. It can be carried out not only in water but also in a variety of solvents, even in alcohols.
4. Ester hydrolysis is another example of nucleophilic acyl substitution.

D. Hydrolysis of Amides

1. Amides can be hydrolyzed to carboxylic acids and ammonia or amines by heating them in acidic or basic solution.

$$H_2O + R-\overset{O}{\overset{\|}{C}}-NR'R'' \begin{cases} \xrightarrow{H_3O^+} R-\overset{O}{\overset{\|}{C}}-OH + H_2\overset{+}{N}R'R'' \\ \xrightarrow{^-OH} R-\overset{O}{\overset{\|}{C}}-O^- + HNR'R'' \end{cases}$$

2. In acid, protonation of the ammonia or amine by-product drives the hydrolysis equilibrium to completion; the amine can be isolated, if desired, by addition of base to the reaction mixture following hydrolysis.

$$R-C(=O)-NH_2 + H-\overset{+}{O}H_2 \rightleftharpoons R-C(=\overset{+}{O}H)-NH_2 \;\; OH_2 \rightleftharpoons R-C(NH_2)(OH)-\overset{+}{O}H-H \;\; OH_2 \rightleftharpoons R-C(OH)(OH)-NH_2 \;\; H-\overset{+}{O}H_2 \rightleftharpoons$$

tetrahedral addition intermediate

$$R-C(OH)(OH)-\overset{+}{N}H_2(H) \;\; OH_2 \rightleftharpoons R-\overset{+}{C}(=O-H)-OH \;\; NH_3 \rightleftharpoons R-C(=O)-OH + \overset{+}{N}H_4$$

3. Hydrolysis of amides in base is analogous to saponification of esters; the reaction is driven to completion by formation of the carboxylic acid salt.
4. The conditions of both acid- and base-promoted amide hydrolysis are considerably more severe than the corresponding reactions of esters.
5. The mechanisms of amide hydrolysis are typical nucleophilic acyl substitution mechanisms.

E. Hydrolysis of Nitriles

1. Nitriles are hydrolyzed to carboxylic acids and ammonia by heating them in acidic or basic solution.

$$H_2O + R-C\equiv N \begin{cases} \xrightarrow{H_3O^+} R-\overset{O}{\overset{\|}{C}}-OH + \overset{+}{N}H_4 \\ \xrightarrow{^-OH} R-\overset{O}{\overset{\|}{C}}-O^- + NH_3 \end{cases}$$

2. The conditions of nitrile hydrolysis are considerably more severe than the corresponding reactions of esters and amides.
3. The mechanism of nitrile hydrolysis in acidic solution involves:
   a. Protonation of the nitrogen, which makes the nitrile carbon more electrophilic.
   b. Attack of the nucleophile water on the nitrile carbon and loss of a proton, which give an intermediate called an imidic acid, the nitrogen analog of an enol.
   c. Conversion of the unstable imidic acid under the reaction conditions into an amide.
   d. Hydrolysis of the amide to a carboxylic acid and ammonium ion. (See Eqs. 21.20a–c on text p. 956 for the detailed mechanism.)

R—C≡N —(H_3O^+/H_2O; addition)→ HO(R)C=$\overset{+}{N}H_2$ (a protonated imidic acid) → RC(=O)NH_2 —(H_3O^+/H_2O; amide hydrolysis)→ RC(=O)OH + $\overset{+}{N}H_4$

4. In base, the nitrile group is attacked by basic nucleophiles and, as a result, the electronegative nitrogen assumes a negative charge.
   a. Proton transfer gives an imidic acid, which ionizes in base.
   b. The imidic acid reacts further to give the corresponding amide, which in turn, hydrolyzes under the reaction conditions to the carboxylate salt of the corresponding carboxylic acid. (See Eqs. 21.21a–b on text p. 995 for the detailed mechanism.)

R—C≡N —($^-OH/H_2O$; addition)→ ^-O(R)C=NH_2 (conjugate base of an imidic acid) → RC(=O)NH_2 —($^-OH/H_2O$; amide hydrolysis)→ RC(=O)O^- + NH_3

5. The hydrolysis of nitriles is a useful way to prepare carboxylic acids because nitriles, unlike many other carboxylic acid derivatives, are generally synthesized from compounds other than the acids themselves.

R—Br —($^-C≡N$)→ R—C≡N —(hydrolysis)→ R—CO_2H

II. REACTIONS OF CARBOXYLIC ACID DERIVATIVES WITH ALCOHOLS

A. Reaction of Acid Chlorides and Anhydrides with Alcohols and Phenols

1. Esters are formed rapidly when acid chlorides react with alcohols or phenols, usually in the presence of a tertiary amine such as pyridine, or a related base.

RC(=O)Cl —(HOR′, pyridine; addition)→ HO(OR′)C(R)Cl (tetrahedral addition intermediate) —(proton transfer and elimination)→ RC(=O)OR′ + pyridinium chloride ($C_5H_5\overset{+}{N}$—H Cl^-)

2. Esters of tertiary alcohols and phenols, which cannot be prepared by acid-catalyzed esterification, can be prepared by this method.
3. Anhydrides react with alcohols and phenols in much the same way as acid chlorides. Cyclic anhydrides react with alcohols and phenols to give half-esters.

R—C(=O)—O—C(=O)—R + HOR′ → R—C(=O)—OR′ + HO—C(=O)—R

a cyclic anhydride + HOR′ → a half-ester

4. Sulfonate esters are prepared by the analogous reactions of sulfonyl chlorides with alcohols.

B. Reaction of Esters with Alcohols

1. When an ester reacts with an alcohol under acidic conditions, or with an alkoxide under basic conditions, a new ester is formed.

$$R{-}C({=}O){-}OCH_3 \underset{}{\overset{HOCH_2CH_3,\ H_3O^+\ (cat.)}{\rightleftharpoons}} R{-}C({=}O){-}OCH_2CH_3 + HOCH_3$$

2. This type of reaction, called transesterification, typically has an equilibrium constant near unity.
3. The reaction is driven to completion by the use of an excess of the displacing alcohol or by removal of a relatively volatile alcohol by-product as it is formed.

III. REACTIONS OF CARBOXYLIC ACID DERIVATIVES WITH AMINES

A. Reaction of Acid Chlorides and Anhydrides with Ammonia and Amines

1. Acid chlorides react rapidly and irreversibly with ammonia or amines by a nucleophilic acyl substitution reaction mechanism to give amides.
   a. Reaction with ammonia yields a primary amide.
   b. Reaction with a primary amine yields a secondary amide.
   c. Reaction with a secondary amine yields a tertiary amide.

$$R'{-}C({=}O){-}Cl \xrightarrow{2\ NH_3} R'{-}C({=}O){-}NH_2 + \overset{+}{N}H_4\ Cl^-$$

$$R'{-}C({=}O){-}Cl \xrightarrow{2\ RNH_2} R'{-}C({=}O){-}NHR + H_2\overset{+}{N}R\ Cl^-$$

$$R'{-}C({=}O){-}Cl \xrightarrow{2\ R_2NH} R'{-}C({=}O){-}NR_2 + H\overset{+}{N}R_2\ Cl^-$$

2. The amine attacks the carbonyl group to form a tetrahedral intermediate, which expels chloride ion; a proton-transfer step yields the amide.
   a. An important aspect of amide formation is the proton transfer in the last step of the mechanism.
   b. Unless another base is added to the reaction mixture, the starting amine acts as the base in this step; if the only base present is the amine nucleophile, then at least two equivalents must be used:
      i. one equivalent as the nucleophile.
      ii. one equivalent as the base in the final proton-transfer step.

$$R'{-}C({=}O){-}Cl \xrightarrow{R_2NH} R'C(O^-)(\overset{+}{N}HR_2)Cl \xrightarrow{-Cl^-} R'{-}C({=}O){-}\overset{+}{N}HR_2 \xrightarrow{R_2NH} R'{-}C({=}O){-}NR_2 + R_2\overset{+}{N}H_2\ Cl^-$$

addition — tetrahedral addition intermediate — elimination — proton transfer

3. In the Schotten-Baumann technique for amide formation, the reaction is run with a water-insoluble acid chloride and an amine in a separate layer over an aqueous solution of NaOH. The NaOH neutralizes HCl produced as a by-product of the reaction.
4. A tertiary amine such as pyridine can be used as the second amine equivalent. The presence of a tertiary amine does not interfere with amide formation because a tertiary amine itself cannot form an amide.
5. Anhydrides react with amines in much the same way as acid chlorides.

$$R{-}\overset{O}{\overset{\|}{C}}{-}O{-}\overset{O}{\overset{\|}{C}}{-}R \ + \ HNR'R'' \longrightarrow R{-}\overset{O}{\overset{\|}{C}}{-}NR'R'' \ + \ HO{-}\overset{O}{\overset{\|}{C}}{-}R$$

   a. Half-amides of dicarboxylic acids are produced in analogous reactions of amines and cyclic anhydrides.
   b. These compounds can be cyclized to imides by treatment with dehydrating agents, or in some cases, simply by heating.

a cyclic anhydride + H_2NR' → a half-amide (NHR′, OH) —heat→ a cyclic imide (NR′) + H_2O

B. Reaction of Esters with Amines

1. The reaction of esters with ammonia or amines yields amides.

R–C(=O)–OR′ —HNR_2 (addition)→ tetrahedral addition intermediate [R–C(HO)(NR_2)–OR′] —(elimination)→ R–C(=O)–NR_2 + HOR′

2. The reaction of esters with hydroxylamine (NH_2OH) gives *N*-hydroxyamides; these compounds are known as hydroxamic acids.
   a. This chemistry forms the basis for the hydroxamate test, used mostly for esters.
   b. The hydroxamic acid products are easily recognized because they form highly colored complexes with ferric ion.

R–C(=O)–OR′ —$H_2N{-}OH$, $-HOR'$→ R–C(=O)–NH–OH (a hydroxamic acid) —5% $FeCl_3$→ bright color

IV. REACTIONS OF CARBOXYLIC ACID DERIVATIVES WITH HYDRIDES

A. Reduction of Acid Chlorides to Aldehydes and Alcohols

1. Acid chlorides can be reduced to aldehydes by either of two procedures:
   a. Hydrogenation over a catalyst that has been deactivated, or poisoned, with an amine, such as quinoline, that has been heated with sulfur.

R–C(=O)–Cl —H_2, Pd/C, quinoline→ R–C(=O)–H

      i. This reaction is called the Rosenmund reduction.

ii. The poisoning of the catalyst prevents further reduction of the aldehyde product.

b. Reaction at low temperature with lithium tri(*tert*-butoxy)aluminum hydride.

$$\mathrm{RC(=O)Cl} \xrightarrow[\text{addition}]{[(CH_3)_3CO]_3\bar{A}lH\ Li^+ \text{ lithium tri(tert-butoxy)-aluminum hydride}} \underset{\text{tetrahedral addition intermediate}}{[(CH_3)_3CO]_3\bar{A}l{-}O{-}C(H)(R)(Cl)} \xrightarrow[\text{elimination}]{H_3O^+/H_2O} \mathrm{RC(=O)H}$$

i. The hydride reagent used in this reduction is derived by the replacement of three hydrogens of $LiAlH_4$ by *tert*-butoxy groups; this reagent is much less reactive, and is more selective, than $LiAlH_4$.

ii. Because acid chlorides are more reactive than aldehydes toward nucleophiles, the reagent reacts preferentially with the acid chloride reactant rather than with the product aldehyde.

2. Acid chlorides (and anhydrides) react with $LiAlH_4$ to give primary alcohols.

B. Reduction of Esters to Primary Alcohols

1. Esters can be reduced to alcohols with $LiAlH_4$. This reaction involves a nucleophilic acyl substitution reaction followed by a carbonyl addition reaction.

$$\mathrm{RC(=O)OR'} \xrightarrow[\text{addition}]{LiAlH_4} \underset{\text{tetrahedral addition intermediate}}{Li^+\ {}^-O{-}C(H)(R)(OR')} \xrightarrow[\text{elimination}]{} \underset{\text{aldehyde intermediate (not isolated)}}{\mathrm{RC(=O)H}} \xrightarrow{1)\ LiAlH_4 \quad 2)\ H_3O^+/H_2O} R{-}CH_2{-}OH$$

$$+\ Li^+\ {}^-OR' \xrightarrow{H_3O^+/H_2O} R'{-}OH$$

two alcohol products formed

a. The active nucleophile in $LiAlH_4$ reductions is the hydride ion, $H{:}^-$, which replaces alkoxide at the carbonyl group of the ester to give an aldehyde.

b. The aldehyde cannot be isolated because it reacts rapidly with $LiAlH_4$ to give, after protonolysis, the alcohol.

c. Two alcohols are formed:

i. One is derived from the acyl group of the ester

ii. One is derived from the alkoxy group (usually methanol or ethanol, which is discarded).

2. Sodium borohydride reacts very sluggishly or not at all with most esters.

C. Reduction of Amides to Amines

1. Amides are reduced to amines with $LiAlH_4$.

$$R{-}\overset{O}{\overset{\|}{C}}{-}NR'R'' \xrightarrow{LiAlH_4} \xrightarrow{H_3O^+\text{, then neutralize}} RCH_2NR'R''$$

a. Primary amines are obtained from primary amides.

b. Secondary amines are obtained from secondary amides.

c. Tertiary amines are obtained from tertiary amides.

2. The mechanism of the reaction of a secondary amide with lithium aluminum hydride involves—

a. formation of the lithium salt of the amide by reaction of the acidic amide proton with an equivalent of hydride, a strong base.

b. reaction of the lithium salt of the amide, a Lewis base, with AlH_3, a Lewis acid.

c. delivery of hydride to the C=N double bond.

d. loss of $^-OAlH_2$, a fairly good leaving group (better than the nitrogen group), to form an imine.

$$\underset{\text{an amide}}{R\text{-}C(=O)\text{-}NHR'} \xrightarrow[-H_2]{LiAlH_4} R\text{-}C(=O)\text{-}\bar{N}R'\,Li^+ \;(AlH_3) \xrightarrow[\text{addition}]{} \underset{\substack{\text{tetrahedral}\\\text{addition}\\\text{intermediate}}}{H\text{-}C(R)(O\bar{A}lH_2)\text{-}\bar{N}R'Li^+} \xrightarrow[\text{elimination}]{-Li^+\,{}^-OAlH_2} \underset{\text{an imine}}{H(R)C{=}NR'}$$

e. Reduction of the imine and protonolysis in a separate step to yield the product amine. (See Eqs 21.52a–e on text pp. 1009–1010 for the detailed mechanism.)

$$\underset{\text{an imine}}{H(R)C{=}NR'} \xrightarrow[\substack{2)\ H_3O^+/H_2O\\3)\ HO^-}]{1)\ LiAlH_4} \underset{\text{an amine}}{RCH_2NH_2R'}$$

i. An excess of acid will convert the amine, which is a base, into its conjugate-acid ammonium ion.

ii. Hydroxide is required to neutralize this ammonium salt and thus give the neutral amine.

3. The reductions of primary and tertiary amides involve somewhat different mechanisms, but they too involve loss of oxygen rather than nitrogen as a leaving group.

D. Reduction of Nitriles to Amines

1. Nitriles are reduced to primary amines by reaction with $LiAlH_4$, followed by protonolysis in a separate step.

$$R{-}C{\equiv}N \xrightarrow[\substack{2)\ H_3O^+/H_2O\\3)\ HO^-}]{1)\ LiAlH_4} R{-}CH_2{-}NH_2$$

a. An excess of acid will convert the amine, which is a base, into its conjugate-acid ammonium ion.

b. Hydroxide is required to neutralize this ammonium salt and thus give the neutral amine.

2. Nitriles are also reduced to primary amines by catalytic hydrogenation. An intermediate in the reaction is the imine, which is not isolated but is hydrogenated to the amine product.

$$R{-}C{\equiv}N \xrightarrow{H_2,\ (cat.)} R{-}CH{=}NH \xrightarrow{H_2,\ (cat.)} R{-}CH_2{-}NH_2$$

V. REACTIONS OF CARBOXYLIC ACID DERIVATIVES WITH ORGANOMETALLIC REAGENTS

A. Reaction of Acid Chlorides with Lithium Dialkylcuprates

1. The reaction of lithium dialkylcuprates with acid chlorides gives ketones in excellent yield; lithium dialkylcuprates typically react with acid chlorides and aldehydes, very slowly with ketones, and not at all with esters.

$$R{-}C(=O){-}Cl + Li^+\,{}^-CuR'_2 \longrightarrow R{-}C(=O){-}R' + LiCl + CuR'$$

2. Lithium dialkylcuprates react much like Grignard or lithium reagents, but are less reactive; a lithium dialkylcuprate can be considered conceptually as an alkyl anion complexed with copper.

a. Lithium dialkylcuprate reagents are prepared by the reaction of two equivalents of an organolithium reagent with one equivalent of a cuprous halide such as CuBr.

$$RLi + CuI \longrightarrow LiI + CuR \xrightarrow{RLi} LiCuR_2$$

b. The first equivalent forms an alkylcopper compound; the driving force for this reaction is the preference of lithium, the more electronegative metal, to exist as an ion (Li^+).

c. The copper of an alkylcopper reagent is a Lewis acid and reacts accordingly with "alkyl anion" from a second equivalent of the organolithium reagent; the product of this reaction is a lithium dialkylcuprate.

B. Reaction of Esters with Grignard Reagents

1. In the reaction of esters with Grignard reagents, tertiary alcohols are formed after protonolysis. (Secondary alcohols are formed from esters of formic acid after protonolysis.)
   a. Two equivalents of organometallic reagent react per mole of ester.
   b. A second alcohol (derived from the alkoxy group, usually methanol or ethanol) is produced in the reaction and discarded.

$$R'C(=O)OR'' \xrightarrow{RMgX} R'C(R)(O^-\,{}^+MgX)(OR'') \xrightarrow{-R''OMgX} R'C(=O)R \xrightarrow[2)\ H_3O^+]{1)\ RMgX} R'C(OH)R_2$$

addition; tetrahedral addition intermediate; elimination; ketone intermediate (not isolated)

$$+\ R''O^-\ {}^+MgX \xrightarrow{H_3O^+/H_2O} R''\text{—}OH$$

two alcohol products formed

2. This reaction is a nucleophilic acyl substitution followed by an addition.
   a. A ketone is formed in the substitution step along with an alkoxide leaving group.
   b. The ketone intermediate is not isolated because ketones are more reactive than esters toward nucleophilic reagents.
   c. The ketone reacts with a second equivalent of the Grignard reagent to form a magnesium alkoxide, which, after protonolysis, gives the alcohol.
3. This reaction is a very important method for the synthesis of alcohols in which at least two of the groups on the α-carbon of the alcohol product are identical.

VI. REACTIONS OF CARBOXYLIC ACID DERIVATIVES WITH OTHER NUCLEOPHILES

A. Reaction of Acid Chlorides with Carboxylate Salts

1. Even though carboxylate salts are weak nucleophiles, acid chlorides are reactive enough to be attacked by carboxylate salts to give anhydrides.
2. The reaction of acid chlorides with carboxylate salts can be used to prepare mixed anhydrides.

$$RC(=O)Cl + Na^+\ {}^-OC(=O)R' \xrightarrow{heat} RC(=O)OC(=O)R' + NaCl$$

VII. SYNTHESIS OF CARBOXYLIC ACID DERIVATIVES—REVIEW

A. Synthesis of Acid Chlorides

1. Reaction of carboxylic acids with $SOCl_2$ or PCl_5 (Sec. 20.9A, text page 918).

$$R\text{—}C(=O)\text{—}OH \xrightarrow[\text{or } PCl_5]{SOCl_2} R\text{—}C(=O)\text{—}Cl$$

B. Synthesis of Anhydrides

1. Reaction of carboxylic acids with dehydrating agents (Sec. 20.9B, text page 920).

$$2\,R-\overset{O}{\overset{\|}{C}}-OH \xrightarrow[-H_2O]{P_2O_5} R-\overset{O}{\overset{\|}{C}}-O-\overset{O}{\overset{\|}{C}}-R$$

2. Reaction of acid chlorides with carboxylate salts (Sec. 21.8A, text page 963).

$$R-\overset{O}{\overset{\|}{C}}-O^- + Cl-\overset{O}{\overset{\|}{C}}-R' \longrightarrow R-\overset{O}{\overset{\|}{C}}-O-\overset{O}{\overset{\|}{C}}-R'$$

C. Synthesis of Esters

1. Acid-catalyzed esterification of carboxylic acids (Sec. 20.8A, text page 913).

$$R-CO_2H + HOR' \xrightleftharpoons{H_3O^+} R-CO_2R' + H_2O$$

2. Alkylation of carboxylic acids or carboxylates (Sec. 20.8B, text page 916).

$$\left.\begin{array}{l} R-CO_2H + CH_2N_2 \longrightarrow \\ R-CO_2H + H_3C-I \xrightarrow{K_2CO_3} \end{array}\right\} R-\overset{O}{\overset{\|}{C}}-OCH_3$$

3. Reaction of acid chlorides and anhydrides with alcohols or phenols (Sec. 21.8A, text page 963).

$$R-\overset{O}{\overset{\|}{C}}-Cl + HOR' \xrightarrow{\text{base}} R-\overset{O}{\overset{\|}{C}}-OR'$$

4. Transesterification of other esters (Sec. 21.8C, text page 967).

$$R-CO_2R' + HOR'' \rightleftharpoons R-CO_2R'' + HOR'$$

D. Synthesis of Amides

1. Reaction of acid chlorides, anhydrides, or esters with amines (Sec. 21.8, text pp. 963–968).

$$R-\overset{O}{\overset{\|}{C}}-X + HNR'R'' \longrightarrow R-\overset{O}{\overset{\|}{C}}-NR'R'' \qquad -X = -Cl, -O-\overset{O}{\overset{\|}{C}}-R, \text{ or } -OR$$

When X = halide or carboxylate, a base or a second equivalent of an amine is required to neutralize the HX generated as a byproduct. In many cases, a tertiary amine such as pyridine or triethylamine is used.

E. Synthesis of Nitriles

1. S_N2 reaction of cyanide ion with alkyl halides or sulfonate esters. In this reaction, primary or unbranched secondary alkyl halides or sulfonate esters are required (Sec. 9.2, text page 356).

$$R-CH_2-X + {}^-C{\equiv}N \longrightarrow R-CH_2-C{\equiv}N + X^- \qquad X = \text{halide ion}, {}^-OTs$$

2. Cyanohydrin formation (Sec. 19.7A, text page 855).

$$R-\overset{O}{\overset{\|}{C}}-R' + H-C{\equiv}N \longrightarrow R-\underset{C\equiv N}{\underset{|}{\overset{OH}{\overset{|}{C}}}}-R'$$

SCIENTISTS

Eugen Baumann (1846–1996) 21.8A
Karl Wilhelm Rosenmund (1884–1965) 21.9D
Carl Ludwig Johannes Schotten (1853–1910) 21.8A

STUDY GUIDE LINKS

21.1 NMR Evidence for Internal Rotation in Amides

Internal rotation about the carbonyl-nitrogen bond of amides can be conveniently studied by NMR using the principles discussed in Sec. 13.8 on p. 581 of the text. Consider, for example, internal rotation in *N*,*N*-dimethylacetamide:

$$H_3C-\overset{\overset{\displaystyle O}{\|}}{C}-\underset{\underset{\displaystyle CH_3}{|}}{N}-CH_3$$

these methyl groups are diastereotopic and hence chemically nonequivalent

The *N*-methyl groups in this compound are diastereotopic; one is cis to the carbonyl oxygen, and the other is trans. At low temperature, rotation about the C—N bond is slow enough that these methyl groups are observed as separate singlets in the NMR spectrum. However, when the temperature is raised, these signals broaden and, at about 60 °C, they coalesce into one singlet at a chemical shift that is the average of the two individual methyl chemical shifts. This is illustrated in Fig. SG21.1 on the following page.

✓21.2 Solving Structure Problems Involving Nitrogen-Containing Compounds

Amides and nitriles are the first nitrogen-containing functional groups that are considered in the text. When you solve structure problems you should remember certain things about compounds that contain nitrogen. First, compounds containing an odd number of nitrogen atoms have *odd molecular mass*. (If you studied mass spectrometry (Chapter 12), you will appreciate that this means that the parent ion in the mass spectrum of such a compound will occur at an odd *m/z* value.) Second, the special formula given below (see also Eq. 4.7 on text p. 126) is used to calculate the unsaturation number *U* of a compound containing nitrogen. If *C* is the number of carbons, *H* is the number of hydrogens, *X* is the number of halogens, and *N* is the number of nitrogens, then *U* is given by

$$U = \frac{(2C + 2) + N - (H + X)}{2}$$

This formula means that every nitrogen increases the number of hydrogens in a fully saturated molecule by one relative to the number of hydrogens in the corresponding hydrocarbon. Finally, when a compound contains nitrogen, you can't tell whether an ion in its mass spectrum is an odd-electron ion or an even-electron ion simply by its mass, because the number of nitrogens in the ion determines whether its mass is odd or even. When a compound contains zero or an even number of nitrogens, odd-electron ions have odd masses; when a compound contains an odd number of nitrogens, odd-electron ions have even masses.

(text continues on p. 757)

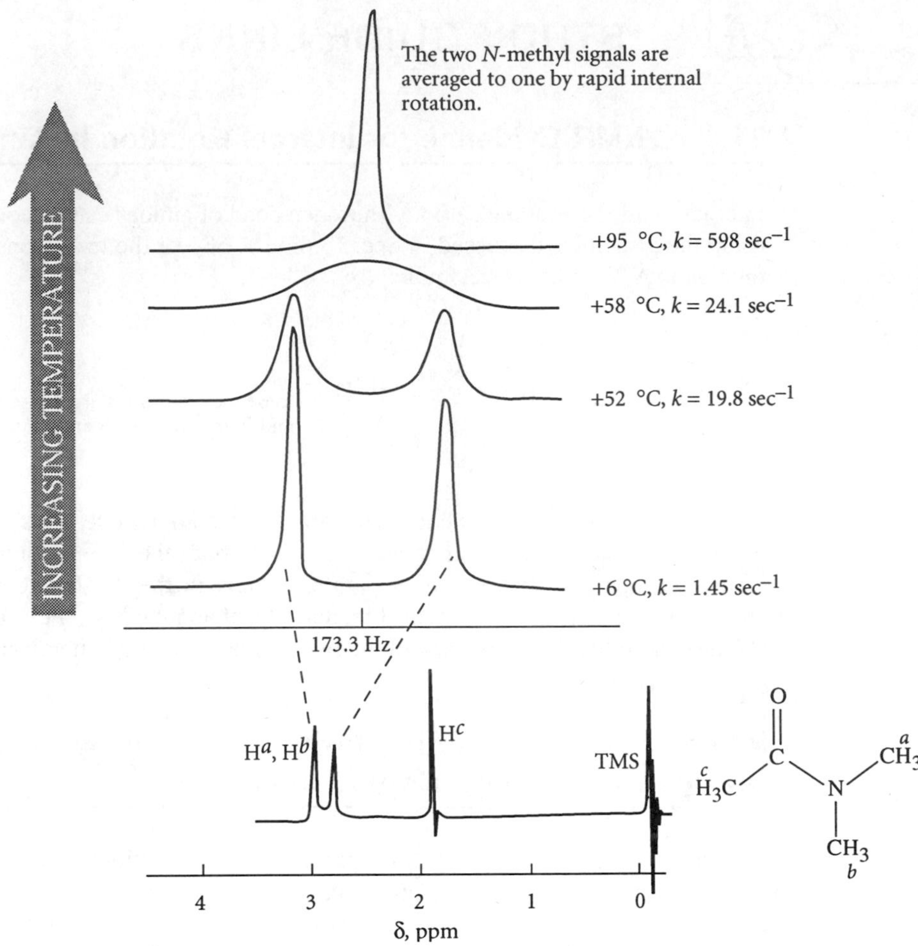

Figure SG21.1 The 60 MHz NMR spectrum of *N,N*-dimethylacetamide changes with increasing temperature as the rate of rotation about the carbonyl-nitrogen bond increases. The full spectrum of this compound is shown in the lower part of the figure, and the pair of singlets near δ 2.9 (173.3 Hz) is expanded in the upper part of the figure and shown as a function of temperature. The two *N*-methyl groups are individually observable at temperatures at or below 52° (lower two spectra) because internal rotation about the carbonyl-nitrogen bond is relatively slow at these temperatures. As the temperature is raised, the two resonances broaden and coalesce into one because the internal rotation becomes too rapid for the NMR experiment to resolve the individual lines. (See Fig. 13.19 on text p. 582 for a related phenomenon.) The *k* values are the first-order rate constants for internal rotation; a *k* of 1.45 s^{-1} corresponds to an approximate lifetime of 0.5 s for the individual conformations.

✓21.3 Basicity of Nitriles

Let's place the basicity of nitriles within the context of other basicities that you've studied. Consider the following series:

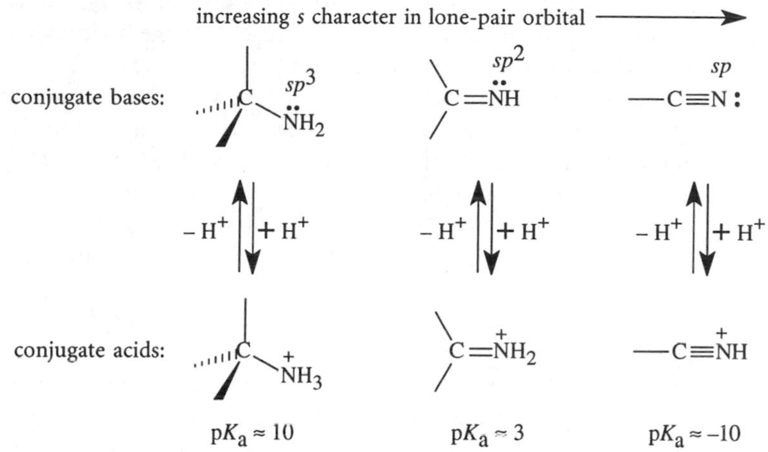

increasing *s* character in lone-pair orbital ⟶

conjugate bases: sp^3 $\ddot{N}H_2$; sp^2 $C{=}\ddot{N}H$; sp $—C{\equiv}N{:}$

$-H^+$ ⇅ $+H^+$

conjugate acids: $C—\overset{+}{N}H_3$; $C{=}\overset{+}{N}H_2$; $—C{\equiv}\overset{+}{N}H$

$pK_a \approx 10$ $pK_a \approx 3$ $pK_a \approx -10$

Notice that as the amount of *s* character in the lone-pair orbital of the conjugate base increases, the pK_a of the conjugate acid decreases. Although the pK_a values are much different, this is exactly the same trend observed in the acidities of hydrocarbons (Sec. 14.7A, text pp. 623):

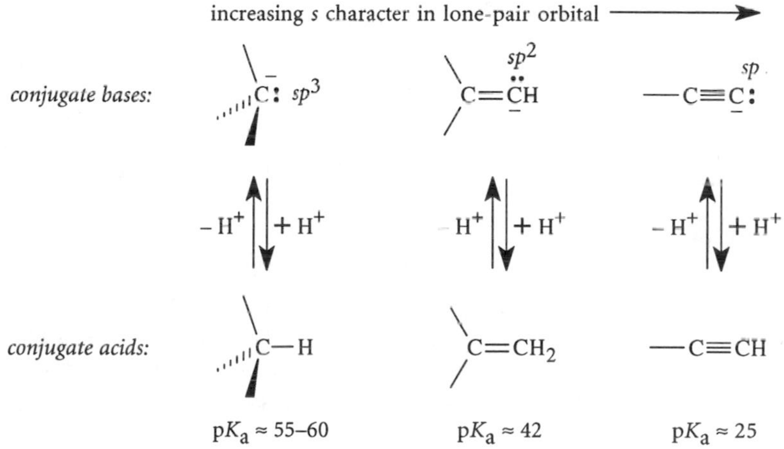

increasing *s* character in lone-pair orbital ⟶

conjugate bases: $\overset{-}{C}{:}$ sp^3; sp^2 $C{=}\ddot{C}H^-$; sp $—C{\equiv}C{:}^-$

$-H^+$ ⇅ $+H^+$

conjugate acids: $C—H$; $C{=}CH_2$; $—C{\equiv}CH$

$pK_a \approx 55–60$ $pK_a \approx 42$ $pK_a \approx 25$

In other words, just as an acetylenic anion, with its lone pair in an *sp* orbital, is the least basic hydrocarbon anion, a nitrile, also with a lone pair in an *sp* orbital, is the least basic among the analogous series of nitrogen compounds.

✓21.4 Mechanism of Ester Hydrolysis

Some students try to combine the features of the acid-catalyzed hydrolysis and saponification by using strong acid to protonate the carbonyl oxygen and hydroxide to act as the nucleophile.

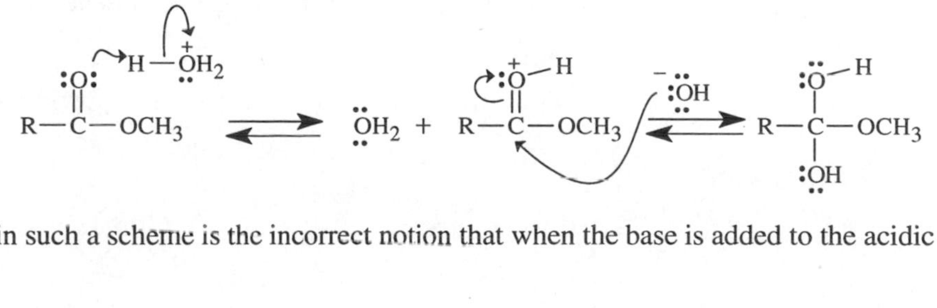

$R—C({=}\ddot{O}{:})—OCH_3 + H—\overset{+}{O}H_2 \rightleftharpoons \ddot{O}H_2 + R—C({=}\overset{+}{O}—H)—OCH_3 \xrightleftharpoons{^-{:}\ddot{O}H} R—C(\ddot{O}—H)(OCH_3)—{:}\ddot{O}H$

Implicit in such a scheme is the incorrect notion that when the base is added to the acidic solution, the carbonyl oxygen of the ester would remain protonated and that all hydronium ions would not dissociate. However, both protonated esters and hydronium ions are strong acids. Remember: *the deprotonation of strong acids by strong bases is instantaneous.* (See Study Guide Link 20.2 on p. 705 of this manual.) The very first event that would occur on addition of base to the solution would be deprotonation of the protonated carbonyl oxygen and rapid neutralization of the acid (H_3O^+) in solution. Attack of ^-OH on the ester carbonyl carbon is a much slower reaction. All of the hydronium ions in solution would react with the hydroxide ion before hydroxide could react to a significant extent at the carbonyl carbon.

In summary: *a solution cannot at the same time be acidic enough to protonate a carbonyl oxygen and basic enough to contain substantial hydroxide ion.* Acid-catalyzed reactions generally involve relatively weak bases such as water as nucleophiles; reactions involving hydroxide as a base generally involve weak acids such as water. The bottom line: *Don't try to use H_3O^+ as an acid and ^-OH as a base simultaneously in the same solution.* This same point was addressed in Study Guide Link 19.4 (p. 658 of this manual).

21.5 Cleavage of Tertiary Esters and Carbonless Carbon Paper

The hydrolysis of tertiary esters has some noteworthy differences from the hydrolysis of other esters discussed in Sec. 21.7A. First, the saponification of tertiary esters is considerably slower than that of primary and secondary esters. For example, the saponification of *tert*-butyl acetate occurs at about 0.01 times the rate of the saponification of methyl acetate. The reason is that in the transition state for saponification, the methyl branches in the *tert*-butyl group are involved in van der Waals repulsions with the attacking base. A practical consequence of this rate difference is that a methyl or ethyl ester can generally be saponified without affecting a *tert*-butyl ester in the same molecule.

$$(CH_3)_3CO{-}\overset{O}{\overset{\|}{C}}{-}C_6H_4{-}\overset{O}{\overset{\|}{C}}{-}OCH_3 \xrightarrow[\text{(1 equiv.)}]{\text{NaOH}} (CH_3)_3CO{-}\overset{O}{\overset{\|}{C}}{-}C_6H_4{-}\overset{O}{\overset{\|}{C}}{-}O^-\,Na^+ + HOCH_3$$

tertiary ester is not hydrolyzed; methyl ester is hydrolyzed

A second difference is that the acid-catalyzed hydrolysis of tertiary esters occurs by a mechanism that is completely different from that of primary and secondary esters. The first step in the hydrolysis mechanism is the same—protonation of the carbonyl oxygen. But there the similarity ends. The protonated ester dissociates by an S_N1 mechanism into a tertiary carbocation and the carboxylic acid.

alkyl-oxygen bond cleavage

$$R{-}\overset{\overset{+}{O}{-}H}{\overset{\|}{C}}{-}O{-}\underset{CH_3}{\overset{CH_3}{\overset{|}{\underset{|}{C}}}}{-}CH_3 \xrightarrow{S_N1} R{-}\overset{\ddot{O}H}{\overset{|}{C}}{=}O \quad {}^{+}\underset{CH_3}{\overset{CH_3}{\overset{|}{\underset{|}{C}}}}{-}CH_3$$

a protonated ester; a tertiary carbocation

Attack of water (or any other nucleophile that might be present) occurs on the tertiary carbocation.

$$H_2\ddot{O}\!: \; {}^{+}\underset{CH_3}{\overset{CH_3}{\overset{|}{\underset{|}{C}}}}{-}CH_3 \rightleftarrows H{-}\overset{+}{\ddot{O}}{-}\underset{CH_3}{\overset{CH_3}{\overset{|}{\underset{|}{C}}}}{-}CH_3 \;(\text{with } H,\ H_2\ddot{O}\!:) \rightleftarrows H\ddot{O}{-}\underset{CH_3}{\overset{CH_3}{\overset{|}{\underset{|}{C}}}}{-}CH_3 + H_3\ddot{O}^+$$

Notice that hydrolysis of tertiary esters by this mechanism involves breaking the *alkyl-oxygen bond.* In contrast, the mechanism of hydrolysis of primary and secondary esters involves breaking the *carbonyl-oxygen bond.* Consequently, the hydrolysis of tertiary esters is *not* a nucleophilic acyl substitution reaction. Rather, it is more like the S_N1 mechanism for substitution of a tertiary alcohol (see Eqs. 10.9b–c on text p. 412). Esters of primary and secondary alcohols in most cases do not cleave by an S_N1 mechanism because the carbocation intermediates that would be involved are much less stable.

"Carbonless carbon paper" (sometimes called NCR paper, for "No Carbon Required") is a very ingenious commercial application of tertiary ester cleavage as applied to lactones. The active component in "carbonless carbon paper" is the colorless lactone *A*, which is encapsulated within microscopic particles of a phenol–formaldehyde resin (Sec. 19.15 on text p. 886) and deposited on a paper base that also contains an acidic layer. When the particles are broken by the pressure of a pen or typewriter, the lactone is brought into contact with the acidic layer. This causes an S_N1-like opening of the lactone ring to give a relatively stable carbocation *B*. This carbocation has a blue-violet color.

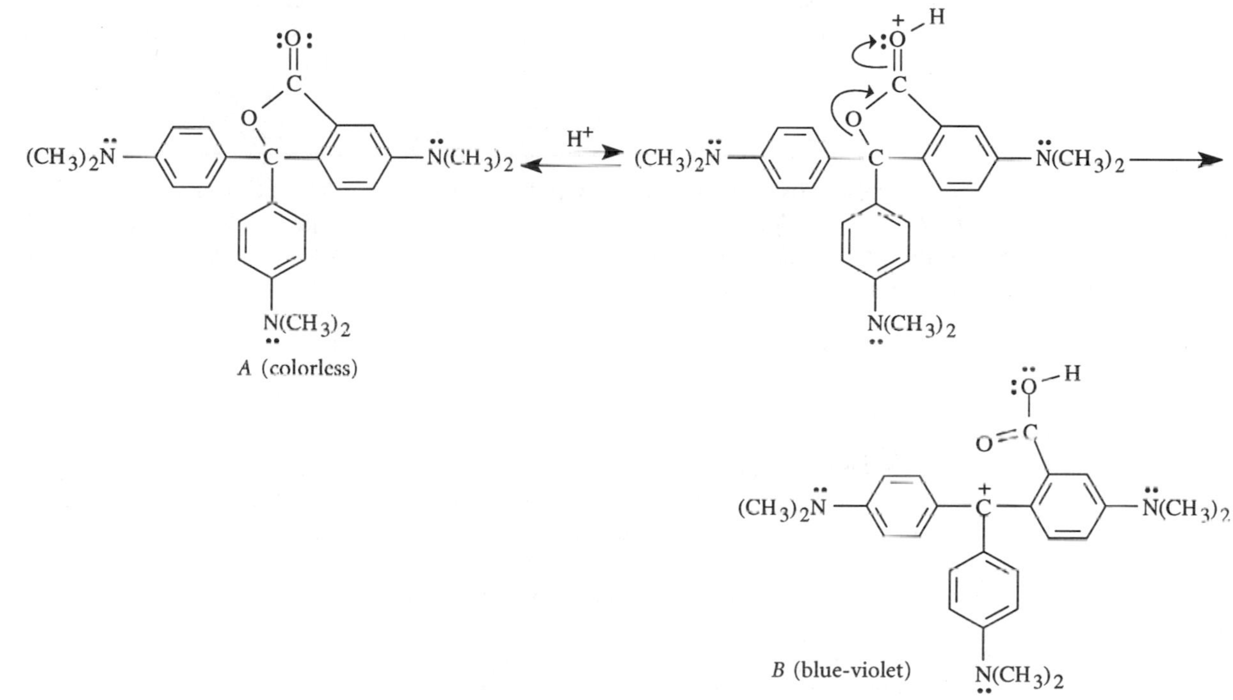

A (colorless)

B (blue-violet)

It takes some time for color to develop because the reaction is not instantaneous. As you may know from experience, the color in carbonless carbon paper intensifies with time.

21.6 Reaction of Tertiary Amines with Acid Chlorides

Tertiary amines react with acid chlorides to form *acylammonium salts.*

pyridine (a tertiary amine)

an acylammonium salt

Acylammonium salts, because of their positive charge adjacent to the carbonyl carbon, are very reactive and are converted into amides by reaction with primary or secondary amines.

$$R{-}\overset{\overset{\Large O}{\|}}{C}{-}\overset{+}{N}C_5H_5\ Cl^- + H_2NR \longrightarrow R{-}\overset{\overset{\Large O}{\|}}{C}{-}NHR + H\overset{+}{N}C_5H_5\ Cl^-$$

pyridinium chloride
(the HCl salt of pyridine)

If the acid chloride has acidic α-hydrogens, another possible reaction is the formation of a *ketene* by a β-elimination:

$$R{-}\underset{\underset{\large H\ \leftarrow :NR_3}{|}}{CH}{-}\overset{\overset{\Large O}{\|}}{C}{-}Cl \longrightarrow R{-}CH{=}C{=}O + H{-}\overset{+}{N}R_3\ Cl^-$$

a ketene

Ketenes also react rapidly with primary and secondary amines to yield amides.

Despite these side reactions, then, tertiary amines can be used as catalysts for amide formation with primary and secondary amines because the products of these side reactions—acylammonium salts and ketenes—themselves react with amines to form amides.

✓21.7 Another Look at Friedel-Crafts Acylation

The Friedel-Crafts acylation reaction is first classified in the text as an *electrophilic aromatic substitution* reaction (Sec. 16.4F on text p. 713), because the reaction is first considered as a reaction of benzene derivatives. However, in terms of what happens to the acid chloride, the reaction can also be viewed as a nucleophilic acyl substitution reaction.

$$C_6H_5{-}H + Cl{-}\overset{\overset{\Large O}{\|}}{C}{-}CH_3 \xrightarrow[2)\ H_3O^+]{1)\ AlCl_3} C_6H_5{-}\overset{\overset{\Large O}{\|}}{C}{-}CH_3 + HCl$$

If the Friedel-Crafts acylation is considered to be a nucleophilic acyl substitution reaction, the nucleophilic electrons are the π electrons of the aromatic ring. Because benzene and its derivatives are very weak nucleophiles, a strong Lewis acid such as $AlCl_3$ is required to activate the acid chloride. Review again the mechanism of Friedel-Crafts acylation (Eq. 16.23 and 16.24 on text pp. 713–14), paying particular attention to what happens to the acid chloride.

✓21.8 Esters and Nucleophiles

Notice that most esters of *carboxylic acids* react with nucleophiles at the carbonyl carbon—that is, they undergo nucleophilic acyl substitution reactions. The result is that the bond between the carbonyl group and the ester oxygen—the *acyl-oxygen bond*—is cleaved. (Study Guide Link 21.5 discusses one exception to this generalization.)

cleavage occurs at the
acyl-oxygen bond

$$R{-}\overset{\overset{\Large O}{\|}}{C}{\wr}OCH_3 + {}^-OH \longrightarrow R{-}\overset{\overset{\Large O}{\|}}{C}{-}O^- + CH_3OH$$

Esters of *sulfonic acids* react in a fundamentally different manner. These reactions are presented in Sec. 10.3A (text p. 418), where the focus is on the sulfonate ester groups essentially as equivalents of halide leaving groups. As the discussion in the text shows, most sulfonate esters react with nucleophiles not by attack of the nucleophile on the sulfur, but by S_N2 attack of the nucleophile on the *alkyl group* (if the alkyl group is primary or secondary); in such a reaction, the bond between the alkyl group and the ester oxygen is cleaved. (See Eq. 10.15 on text p. 417.)

cleavage occurs at the alkyl-oxygen bond

$$R{-}\overset{\overset{\displaystyle O}{\|}}{\underset{\underset{\displaystyle O}{\|}}{S}}{-}O{\wr}CH_3 + {}^{-}OH \xrightarrow{S_N2} R{-}\overset{\overset{\displaystyle O}{\|}}{\underset{\underset{\displaystyle O}{\|}}{S}}{-}O^{-} + CH_3OH$$

The reason sulfonate esters react differently is that they are better leaving groups than carboxylates because they are considerably weaker bases; hence, they are more easily displaced.

You may have noticed that we are returning with increasing frequency to earlier parts of the text and looking at reactions you've studied earlier in somewhat different contexts. If you willingly participate in this type of exercise, you'll gain a lot of insight about organic chemistry. In particular, you'll see how the parts begin to "fit together," that is, you'll begin to see the "forest" and not just the individual "trees."

SOLUTIONS TO PROBLEMS

Solutions to In-Text Problems

21.1 (a) $N{\equiv}CCH_2CH_2CH_2CH_2CO_2H$
5-cyanopentanoic acid

(b) $CH_3CH_2CH_2CH_2\overset{O}{\overset{\|}{C}}OCH(CH_3)_2$
isopropyl valerate

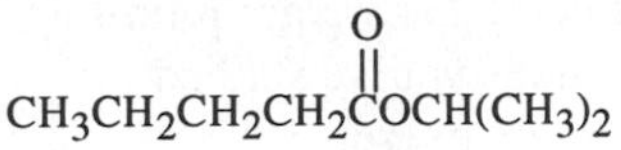

(c) $CH_3CH_2O{-}\overset{O}{\overset{\|}{C}}{-}CH_2{-}\overset{O}{\overset{\|}{C}}{-}OCH_3$
ethyl methyl malonate

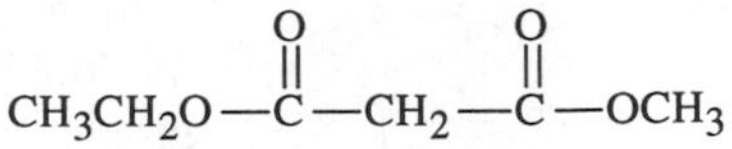

(d) $CH_3\overset{O}{\overset{\|}{C}}O{-}$(cyclohexyl)
cyclohexyl acetate

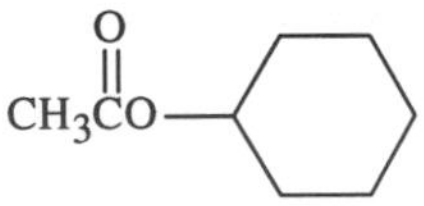

(e) $H\overset{O}{\overset{\|}{C}}N(CH_3)_2$
N,N-dimethylformamide

(f) γ-valerolactone

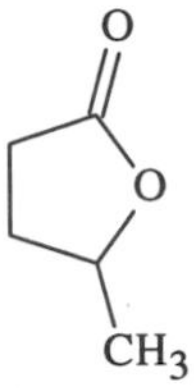

(g)

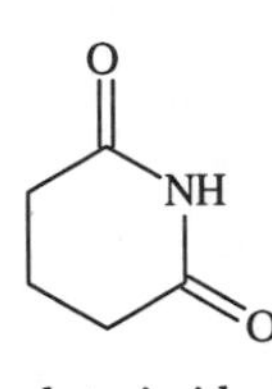

glutarimide

(h) $(CH_3)_2\underset{Cl}{C}\overset{O}{\overset{\|}{C}}Cl$
α-chloroisobutyryl chloride

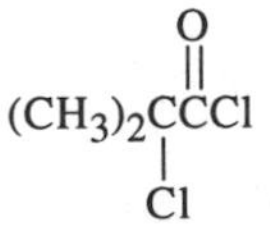

(i) $HO_2CCH_2CHCH_2CH_2CO_2H$ with $C(=O)OCH_2CH_3$ on C-3
3-ethoxycarbonylhexanedioic acid

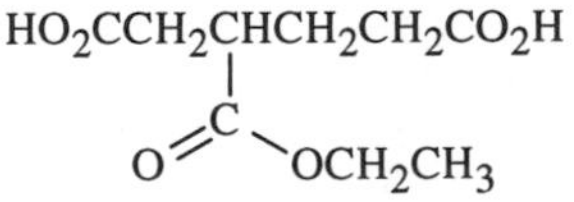

21.2 (a) butanenitrile (*common:* butyronitrile)
(b) *N,N*-dimethylbenzamide
(c) isopentyl 3-methylbutanoate (*common:* isoamyl isovalerate)

The *isoamyl* group is the same as an isopentyl or 3-methylbutyl group:

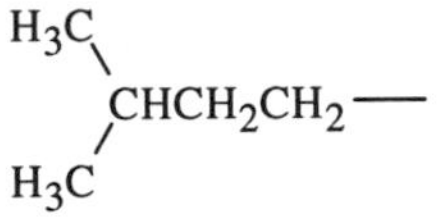

$(H_3C)_2CHCH_2CH_2{-}$ isoamyl group

(d) cyclopropanecarbonyl chloride
(e) (*Z*)-2-butenedioic anhydride (*common name more often used:* maleic anhydride)
(f) ethyl 3-oxopentanoate (*common:* ethyl β-ketovalerate
(g) 1-methyl-3-butenyl propanoate. Notice the numbering *from the point of attachment to the carboxylate oxygen.*)

21.3 The *E* and *Z* conformations of *N*-acetylproline:

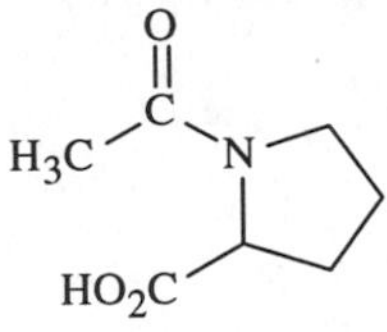

E conformation

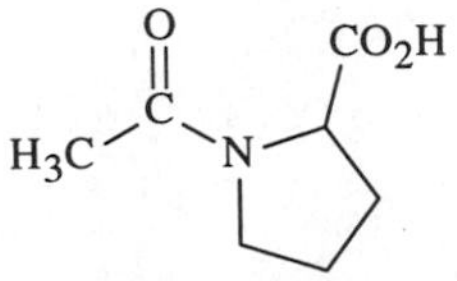

Z conformation

21.4 Any lactam containing a relatively small ring (fewer than eight members) must exist in an *E* conformation. An example is δ-valerolactam. A small-ring lactam cannot exist in a *Z* conformation, because such a conformation would result in a transoid bond within the ring, a very strained and therefore unstable arrangement. (This is the same reason that small-ring (*E*)-cycloalkenes are unstable.) If you try to construct a model of a small-ring lactam with its amide bond in a *Z* conformation, the strain of such a conformation should be obvious.

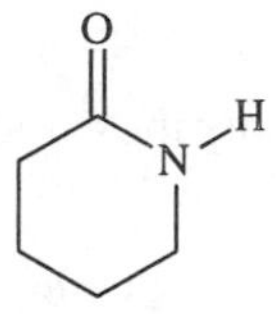

δ-valerolactam

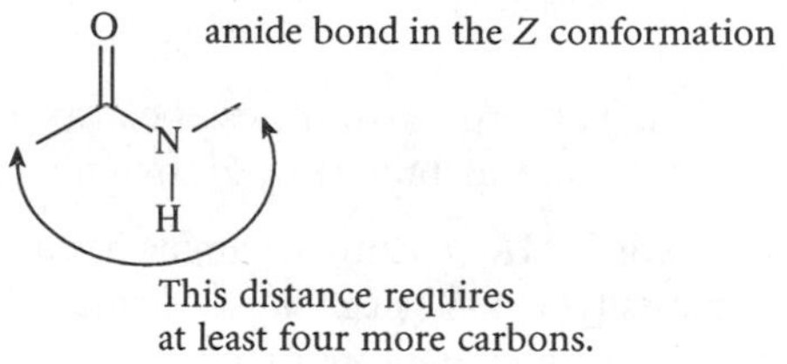

amide bond in the *Z* conformation

This distance requires at least four more carbons.

21.5 As shown by the data above the problem, a carboxylic acid has a higher boiling point than an ester because it can both donate and accept hydrogen bonds within its liquid state; hydrogen bonding does not occur in the ester. Consequently, pentanoic acid (valeric acid) has a higher boiling point than methyl butanoate. Here are the actual data:

$CH_3CH_2CH_2CH_2CO_2H$
pentanoic acid
(valeric acid)
bp 186 °C

$CH_3CH_2CH_2CO_2CH_3$
methyl butanoate
(methyl butyrate)
bp 98 °C

21.6 (a) 2-Butanone should have the higher dipole moment, because, other things being equal, molecules with higher dipole moments have higher boiling points.

(b) The dipole-moment data require that esters, like amides, exist predominantly in the *Z* conformation. In this conformation of an ester, the resultant of the bond dipoles of the two carbon-oxygen single bonds opposes the bond dipole of the C=O bond; the overall dipole moment of the ester is thus reduced. In the ketone, the C=O bond dipole is not opposed by any other significant bond dipole.

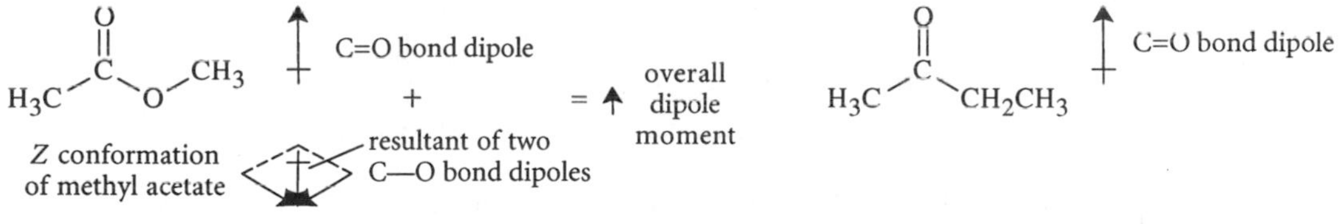

Z conformation of methyl acetate: C=O bond dipole + resultant of two C—O bond dipoles = overall dipole moment

H_3C—CO—CH_2CH_3: C=O bond dipole

If the ester were in an *E* conformation, the resultant bond dipoles of the two carbon-oxygen single bonds would *augment* the bond dipole of the C=O bond. In this case, the dipole moment of the ester would be larger than that of the ketone.

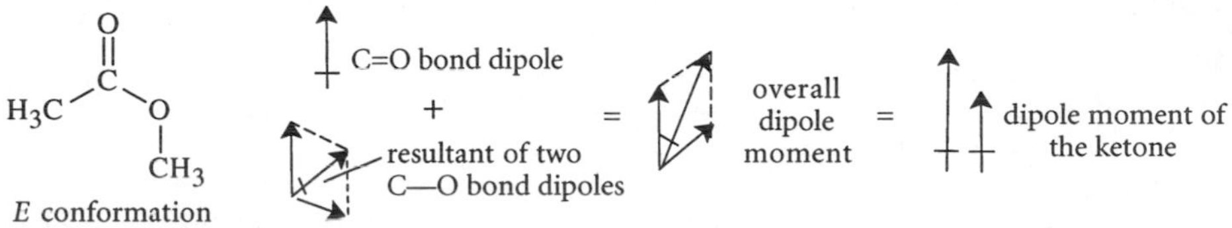

E conformation of methyl acetate: C=O bond dipole + resultant of two C—O bond dipoles = overall dipole moment = dipole moment of the ketone

The experimental facts bear out these predictions! γ-Butyrolactone, an "ester" that is locked into an *E* conformation, has a greater dipole moment than 2-butanone, which has a greater dipole moment than methyl acetate, an ester which exists mostly in the *Z* conformation.

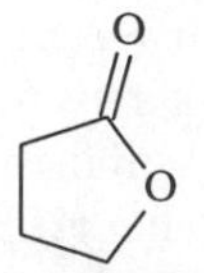

	γ-butyrolactone	$CH_3\overset{O}{\overset{\|}{C}}CH_2CH_3$ 2-butanone	$CH_3\overset{O}{\overset{\|}{C}}OCH_3$ methyl acetate
dipole moments:	4.1D	2.8 D.	1.5 D

21.7 (a) The carbonyl absorption of the ester occurs at higher frequency, and only the carboxylic acid has the characteristic strong, broad O—H stretching absorption in 2400–3600 cm^{-1} region.

(b) In the proton NMR spectra, the amide has a single methyl resonance that is a doublet, whereas the nitrile has two methyl resonances that are singlets. In addition, the integral of the aromatic proton absorptions in the nitrile is smaller in proportion to the integral of the other absorptions because there are fewer aromatic protons.

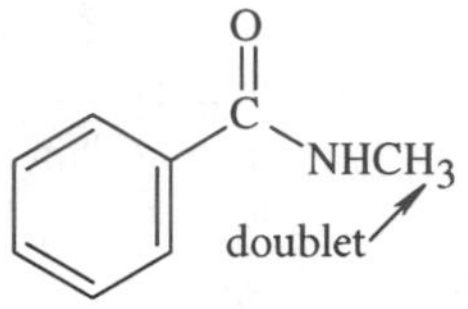

C_6H_5—C(=O)—$NHCH_3$ (doublet)
N-methylbenzamide

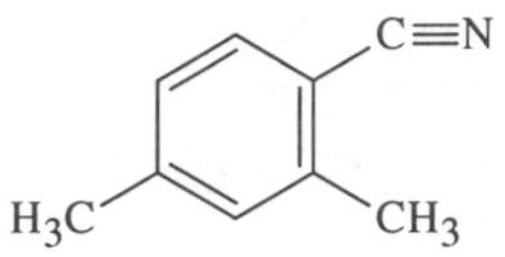

2,4-dimethylbenzonitrile: C≡N; H_3C singlet; CH_3 singlet

(c) In methyl propionate, the singlet for the methyl of the CH_3O— group is farther downfield (near δ 4) than the methyl singlet in ethyl acetate (near δ 2). In ethyl acetate, the quartet for the —CH_2— group is further downfield (near δ 4) than the —CH_2— quartet in methyl propionate (near δ 2).

$H_3C-\overset{O}{\overset{\|}{C}}-OCH_2CH_3$ ethyl acetate

$CH_3CH_2-\overset{O}{\overset{\|}{C}}-OCH_3$ methyl propionate

(d) In *N*-methylpropanamide, the *N*-methyl group is a doublet at about δ 3. *N*-Ethylacetamide has no doublet resonances. In *N*-methylpropanamide, the α-protons are a quartet near δ 2.5. In *N*-ethylacetamide, the α-protons are a singlet at δ 2. The NMR spectrum of *N*-methylpropanamide has no singlets.

$CH_3CH_2-\overset{O}{\overset{\|}{C}}-NHCH_3$ *N*-methylpropanamide

$H_3C-\overset{O}{\overset{\|}{C}}-NHCH_2CH_3$ *N*-ethylacetamide

(e) Ethyl isobutyrate, $(CH_3)_2CHCO_2CH_2CH_3$, has five CMR resonances, whereas ethyl butyrate, $CH_3CH_2CH_2CO_2CH_2CH_3$, has six.

21.8 That the compound is an amide is confirmed by the N—H absorption at 3300 cm^{-1} and by the carbonyl absorption at 1650 cm^{-1}. The compound is *N*-ethylacetamide; the —CH_2— resonance of the ethyl group is split by both the adjacent methyl protons and the N—H proton.

$H_3C-\overset{O}{\overset{\|}{C}}-NHCH_2CH_3$ *N*-ethylacetamide

21.9 (a) The first ester is more basic because its conjugate acid is stabilized not only by resonance interaction with the ester oxygen, but also by resonance interaction with the double bond; that is, the conjugate acid of the first ester has one more important resonance structure than the conjugate acid of the second. (See Problem 19.13(b) on text p. 854 and its solution on p. 667 of this manual for a similar situation.)

$$\left[\mathrm{CH_3CH{=}CH{-}\overset{\overset{+}{O}H}{\overset{\|}{C}}{-}OCH_3} \longleftrightarrow \mathrm{CH_3\overset{+}{C}H{-}CH{=}\overset{OH}{\overset{|}{C}}{-}OCH_3}\right]$$

resonance interaction of the protonated carbonyl group with the double bond

(b) The first ester is more basic because its conjugate acid is stabilized by resonance interaction of the *p*-methoxy oxygen. The meta position of the methoxy group in the second ester does not allow a similar resonance interaction; consequently, only its electron-withdrawing polar effect operates. This polar effect destabilizes the conjugate acid of the second ester. (See Problem 19.13(a) on text p. 854 and its solution on p. 667 of this manual for a similar situation.)

21.10 (a) The mechanism of hydrolysis in acidic solution is much like that for the hydrolysis of an ester. (See the solution to Problem 20.11(a) on p. 714 of this manual.)

$$\mathrm{Ph{-}\overset{O}{\overset{\|}{C}}{-}NHCH_3} + \mathrm{H{-}\overset{+}{O}H_2} \rightleftharpoons \mathrm{Ph{-}\overset{\overset{+}{O}H}{\overset{\|}{C}}{-}NHCH_3} + \mathrm{OH_2} \rightleftharpoons \mathrm{Ph{-}\overset{OH}{\overset{|}{C}}({-}\overset{+}{O}H_2){-}NHCH_3} \overset{OH_2}{\rightleftharpoons}$$

$$\underset{\text{tetrahedral addition intermediate}}{\mathrm{Ph{-}C(OH)_2{-}NHCH_3}} + \mathrm{H{-}\overset{+}{O}H_2} \rightleftharpoons \mathrm{Ph{-}C(OH)_2{-}\overset{+}{N}H_2CH_3} + \mathrm{OH_2} \rightleftharpoons \mathrm{Ph{-}\overset{\overset{+}{O}{-}H}{\overset{\|}{C}}{-}OH} + \underset{\text{(protonated by H3O+)}}{\mathrm{H_2NCH_3}} \overset{OH_2}{\rightleftharpoons} \mathrm{Ph{-}\overset{O}{\overset{\|}{C}}{-}OH} + \mathrm{H{-}\overset{+}{O}H_2}$$

(b) The mechanism of the base-promoted hydrolysis of *N*-methylbenzamide is essentially a saponification mechanism analogous to that shown for esters in Eq. 21.9a–b on p. 952 of the text.

$$\mathrm{Ph{-}\overset{O}{\overset{\|}{C}}{-}NHCH_3} + {}^{-}\mathrm{OH} \rightleftharpoons \underset{\text{tetrahedral addition intermediate}}{\mathrm{Ph{-}\overset{O^-}{\overset{|}{C}}(OH){-}NHCH_3}} \rightleftharpoons \mathrm{Ph{-}\overset{O}{\overset{\|}{C}}{-}O{-}H} + \overset{-}{\mathrm{N}}\mathrm{HCH_3} \longrightarrow \mathrm{Ph{-}\overset{O}{\overset{\|}{C}}{-}O^-} + \mathrm{NH_2CH_3}$$

21.11 The hydrolysis products consist of the conjugate base of a carboxylic acid and an amine:

$$\underset{\text{isobutyrate ion}}{\mathrm{(CH_3)_2CH{-}\overset{O}{\overset{\|}{C}}{-}O^-}} + \underset{\text{pyrrolidine}}{\mathrm{HN(CH_2)_4}}$$

21.12 *Resonance effects:* The application of the resonance effect for predicting the relative rates of acid-catalyzed ester and amide hydrolyses is exactly the same as it is for the corresponding base-promoted hydrolyses. (See Eqs. 21.26 and 21.28 and the discussion on text pp. 959–60.)

Leaving-group basicities: In the acid-catalyzed hydrolysis of an ester, the leaving group is an alcohol; in the acid-catalyzed hydrolysis of an amide, the leaving group is ammonia or an amine. Alcohols are much weaker bases (conjugate-acid $pK_a = -2$ to -3) than amines or ammonia (conjugate-acid $pK_a = 9$–10). Because weaker bases are better leaving groups, esters should hydrolyze more rapidly, and they do.

21.13 First, consider leaving-group ability. The acid chloride should (and does) hydrolyze more rapidly, because chloride is a weaker base (better leaving group) than fluoride. Next, consider the relative stabilities of the starting materials. An acid fluoride is stabilized by resonance interaction of a fluorine unshared electron pair with the π bond of the carbonyl group:

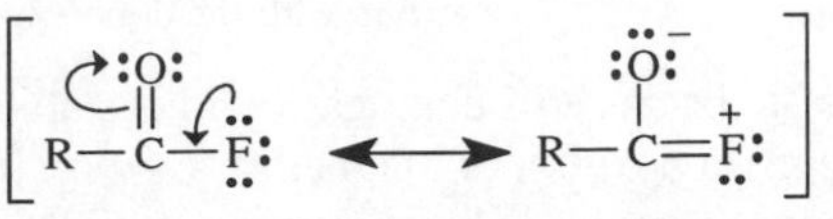

$$\left[R-\overset{\overset{\displaystyle :\ddot{O}:}{\|}}{C}-\ddot{\underset{\cdot\cdot}{F}}: \longleftrightarrow R-\overset{\overset{\displaystyle :\ddot{O}:^{-}}{|}}{C}=\overset{+}{\underset{\cdot\cdot}{F}}: \right]$$

The corresponding resonance structure for an acid chloride is much less important because of the poor overlap of the chlorine orbitals with the carbonyl orbitals. (See Fig. 16.8 on text p. 726 and the accompanying discussion.) Thus, for hydrolysis reactions, acid fluorides are stabilized relative to their tetrahedral addition intermediates more than acid chlorides are, and acid fluorides are thus predicted to (and do) hydrolyze more slowly.

21.14 (a) This is an ordinary ester saponification followed by acidification.

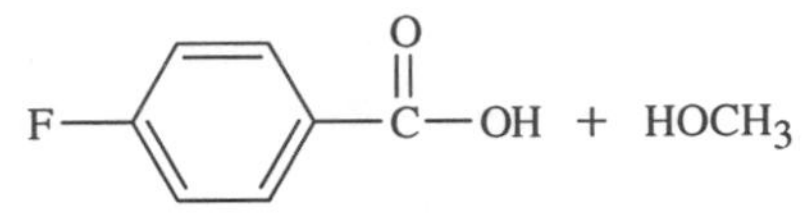

$$F-C_6H_4-\overset{\overset{\displaystyle O}{\|}}{C}-OH + HOCH_3$$

(b) Because esters hydrolyze much more rapidly than nitriles, one equivalent of base brings about selective hydrolysis of the ester.

$$N{\equiv}C-CH_2-\overset{\overset{\displaystyle O}{\|}}{C}-O^- + HOCH_3$$

(c) Hydrolysis of one amide gives carbamic acid, which spontaneously decarboxylates; see Eq. 20.43 on text p. 925.

$$H_2N-\overset{\overset{\displaystyle O}{\|}}{C}-NH_2 + H_2O \xrightarrow{H_3O^+,\ \text{heat}} \overset{+}{N}H_4 + \underset{\text{carbamic acid}}{H_2N-\overset{\overset{\displaystyle O}{\|}}{C}-OH} \xrightarrow{H_3O^+} \overset{+}{N}H_4 + CO_2$$

(d) Because the amide is cyclic—it is a lactam—the amine and carboxylic "products" are part of the same molecule.

$$H_3\overset{+}{N}-CH_2CH_2CH_2-\overset{\overset{\displaystyle O}{\|}}{C}-OH$$

g-aminobutyric acid
(4-aminobutanoic acid)

21.15 First, prepare the acid chloride from hexanoic acid:

$$\underset{\text{hexanoic acid}}{CH_3(CH_2)_4\overset{\overset{\displaystyle O}{\|}}{C}-OH} \xrightarrow{SOCl_2} \underset{\text{hexanoyl chloride}}{CH_3(CH_2)_4\overset{\overset{\displaystyle O}{\|}}{C}-Cl}$$

(a) To prepare the ester, allow the acid chloride to react with ethanol:

$$\underset{\text{hexanoyl chloride}}{CH_3(CH_2)_4\overset{\overset{\displaystyle O}{\|}}{C}-Cl} \xrightarrow{HOCH_2CH_3} \underset{\text{ethyl hexanoate}}{CH_3(CH_2)_4\overset{\overset{\displaystyle O}{\|}}{C}-OCH_2CH_3}$$

(b) To prepare the secondary amide, allow the acid chloride to react with the appropriate amine:

$$CH_3(CH_2)_4C(=O)-Cl \xrightarrow{H_2NCH_3\text{ (excess)}} CH_3(CH_2)_4C(=O)-NHCH_3$$

hexanoyl chloride → *N*-methylhexanamide

21.16 (a) $CH_3CH_2C(=O)-N(CH_3)_2$ + $(CH_3)_2\overset{+}{N}H_2\ Cl^-$

(b) cyclopropyl cyclopropanecarboxylate: cyclopropyl$-C(=O)-O-$cyclopropyl

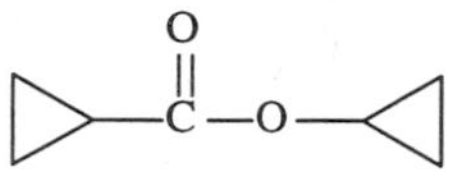

(c) $PhCH_2C(=O)-SCH_2CH_3$

(d) $CH_3CH_2CH_2C(=O)-O-C(=O)CH_3$

(e) $Cl-C(=O)-OCH_3$

(f) $CH_3O-C(=O)-OCH_3$

(g) cyclic carbonate (five-membered ring: O–C(=O)–O with CH_2CH_2) + $2\ C_2H_5OH$

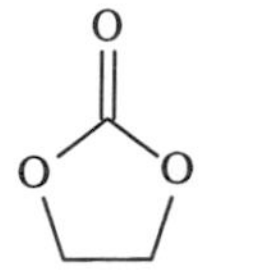

(h) benzene ring bearing $-C(=O)-OCH_3$ and, ortho, $-C(=O)-OH$

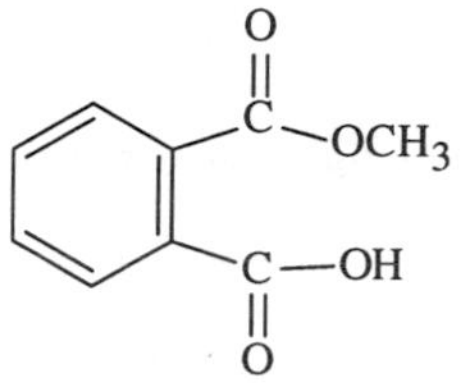

21.17 Because sulfonate esters behave like alkyl halides, cleavage in (a) occurs at the alkyl-oxygen bond. (See Study Guide Link 21.8 on p. 760 of this manual for a discussion.) Consequently, the oxygen isotope ends up in the product alcohol. Cleavage in (b) occurs at the acyl-oxygen bond; therefore, the isotope ends up in the product acetate ion. The products are therefore as follows: (*O = ^{18}O):

(a) $PhCH_2\overset{*}{O}H\ +\ ^{-}OS(=O)_2CH_3$

(b) $PhCH_2OH\ +\ ^{-}\overset{*}{O}C(=O)CH_3$

21.18 (a)

$$CH_3CH(Ph)OH + Cl-SO_2-C_6H_4-CH_3 \xrightarrow{\text{pyridine}} CH_3CH(Ph)O-SO_2-C_6H_4-CH_3$$

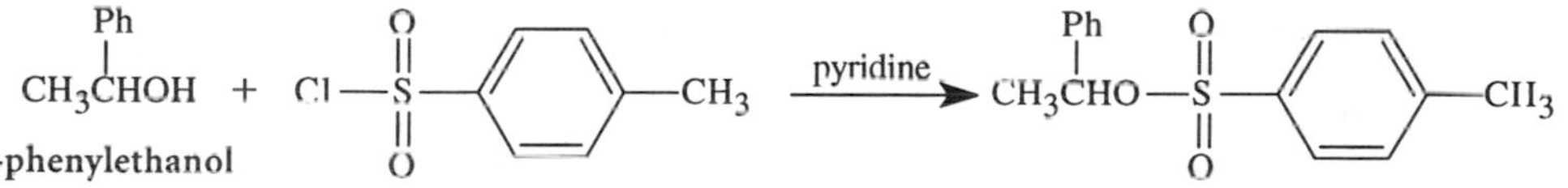

1-phenylethanol + *p*-toluenesulfonyl chloride → 1-phenylethyl tosylate

(b)

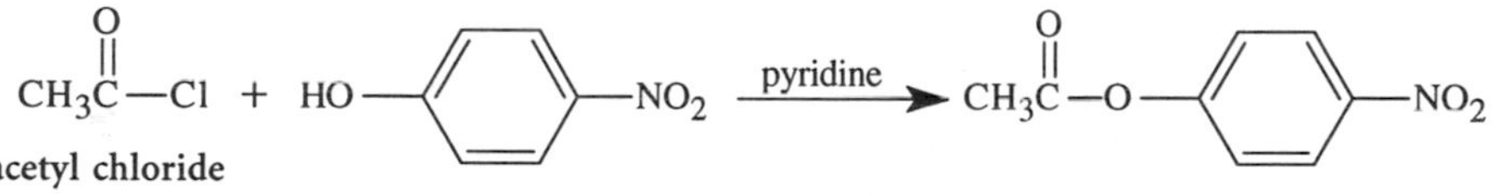

$$CH_3C(=O)-Cl + HO-C_6H_4-NO_2 \xrightarrow{\text{pyridine}} CH_3C(=O)-O-C_6H_4-NO_2$$

acetyl chloride + *p*-nitrophenol → *p*-nitrophenyl acetate

(c)

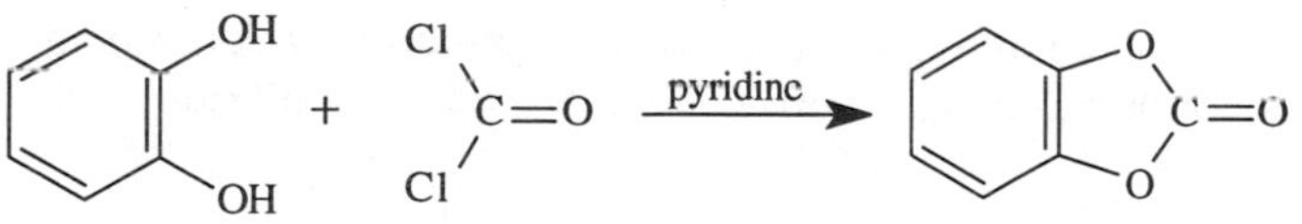

catechol (1,2-dihydroxybenzene, OH, OH) + phosgene ($Cl_2C=O$) $\xrightarrow{\text{pyridine}}$ benzene ring fused to O–C(=O)–O (cyclic carbonate)

(d)

$$2\ (CH_3)_3COH + Cl-C(=O)-CH_2-C(=O)-Cl \xrightarrow{\text{pyridine}} (CH_3)_3CO-C(=O)-CH_2-C(=O)-OC(CH_3)_3$$

tert-butyl alcohol + malonyl dichloride → di-*tert*-butyl malonate

21.19 (a) Although one might be tempted to convert the acid chloride into a carboxylic acid or ester, whose reactions with $LiAlH_4$ we have studied, surely it is reasonable that an acid chloride should react in the same manner as these other derivatives. Hence, we allow an acid chloride to react directly with $LiAlH_4$:

$$\underset{\text{benzoyl chloride}}{\text{PhC}(=\text{O})\text{—Cl}} \xrightarrow[\text{2) } H_3O^+]{\text{1) } LiAlH_4} \underset{\text{benzyl alcohol}}{PhCH_2OH}$$

(b) Carry out one of the reductions in Sec. 21.9D:

$$\underset{\text{benzoyl chloride}}{\text{PhC}(=\text{O})\text{—Cl}} \xrightarrow[\text{2) } H_3O^+]{\text{1) } Li^+\ \overline{H\text{Al}}[OC(CH_3)_3]_3} \underset{\text{benzaldehyde}}{\text{PhC}(=\text{O})\text{—H}}$$

(c) Reduce the tertiary amide with $LiAlH_4$.

$$\underset{\text{benzoyl chloride}}{\text{PhC}(=\text{O})\text{—Cl}} + \underset{\substack{\text{pyrrolidine}\\ \text{(excess)}}}{\text{HN}(CH_2)_4} \longrightarrow \text{PhC}(=\text{O})\text{—N}(CH_2)_4 \xrightarrow[\text{2) } H_3O^+]{\text{1) } LiAlH_4} \underset{N\text{-benzylpyrrolidine}}{PhCH_2N(CH_2)_4}$$

21.20 (a) $PhCH_2CH_2NH_2$

(b) Although the ester reacts most rapidly, the nitrile also reacts because an excess of $LiAlH_4$ is present. The product is the amino alcohol $HOCH_2CH_2CH_2NH_2$ plus the by-product ethanol, C_2H_5OH.

(c) Both esters are reduced.

$$\underset{\substack{\text{ethanol}\\ \text{(from reduction}\\ \text{of both the acetoxy}\\ \text{group and the ethyl ester)}}}{2\ CH_3CH_2OH} + \underset{\text{1-phenyl-1,2-ethanediol}}{PhCH(OH)CH_2OH}$$

21.21 Among the compounds that would give $(CH_3)_2CHCH_2CH_2CH_2NH_2$ as the result of $LiAlH_4$ reduction are the following:

$$\underset{\text{4-methylpentanenitrile}}{(CH_3)_2CHCH_2CH_2C{\equiv}N} \qquad \underset{\text{4-methylpentanamide}}{(CH_3)_2CHCH_2CH_2C(=O)\text{—}NH_2}$$

21.22 (a) As shown in Eq. 21.56, text p. 973, the reduction involves an imine intermediate. This intermediate *A* can be attacked by a molecule of amine product to form the nitrogen analog of an acetal, which then can lose ammonia to form a different imine *B*, which in turn is reduced to the by-product. (You should be able to provide the curved-arrow notation for the reactions involved in the $A \rightleftharpoons B$ equilibrium in the overall process outlined below.)

$$RC{\equiv}N \xrightarrow{H_2,\ cat.} \underset{A}{RCH{=}NH} \underset{}{\overset{H_2NCH_2R}{\rightleftharpoons}} \underset{NHCH_2R}{RCH{-}NH_2} \rightleftharpoons NH_3 + \underset{\underset{B}{NCH_2R}}{RCH{=}} \xrightarrow{H_2,\ cat.} \underset{NHCH_2R}{RCH_2}$$

$$RCH{=}NH\ (A) \xrightarrow{H_2,\ cat.} RCH_2NH_2$$

(b) By LeChatelier's principle, ammonia drives the equilibria in the above equation away from imine *B* and back toward imine *A*, which is eventually reduced.

21.23 (a)

$$\underset{\text{benzoic acid}}{PhCO_2H} \xrightarrow{C_2H_5OH,\ H_2SO_4} PhCO_2CH_2CH_3 \xrightarrow[2)\ H_3O^+]{1)\ PhMgBr\ (excess)} \underset{\text{triphenylmethanol}}{Ph_3C{-}OH}$$

(b)

$$\underset{\text{butyric acid}}{CH_3CH_2CH_2\overset{O}{\overset{\|}{C}}OH} \xrightarrow{PCl_5} CH_3CH_2CH_2\overset{O}{\overset{\|}{C}}Cl \xrightarrow[2)\ H_3O^+]{1)\ Li^+\ {}^-Cu(CH_3)_2} CH_3CH_2CH_2\overset{O}{\overset{\|}{C}}CH_3 \xrightarrow[2)\ H_3O^+]{1)\ CH_3CH_2MgI}$$

$$\underset{\text{3-methyl-3-hexanol}}{CH_3CH_2CH_2\overset{OH}{\overset{|}{\underset{CH_3}{\underset{|}{C}}}}CH_2CH_3}$$

(c) First, the nitrile is hydrolyzed to isobutyric acid, which is converted into its acid chloride:

$$\underset{\text{isobutyronitrile}}{(CH_3)_2CHC{\equiv}N} \xrightarrow{H_2O,\ H_3O^+,\ heat} \underset{\text{isobutyric acid}}{(CH_3)_2CH\overset{O}{\overset{\|}{C}}{-}OH} \xrightarrow{SOCl_2} \underset{\text{isobutyryl chloride}}{(CH_3)_2CH\overset{O}{\overset{\|}{C}}{-}Cl}$$

One synthesis involves reaction of the acid chloride with lithium dimethylcuprate :

$$(CH_3)_2CH\overset{O}{\overset{\|}{C}}{-}Cl \xrightarrow[2)\ H_3O^+]{1)\ Li^+\ {}^-Cu(CH_3)_2} (CH_3)_2CH\overset{O}{\overset{\|}{C}}{-}CH_3 \xrightarrow[2)\ H_3O^+]{1)\ CH_3MgI} \underset{\text{2,3-dimethyl-2-butanol}}{(CH_3)_2CH\overset{OH}{\overset{|}{\underset{CH_3}{\underset{|}{C}}}}{-}CH_3}$$

The second synthesis involves the Grignard reaction of an ester (see Eq. 21.61, text p. 977):

$$(CH_3)_2CH\overset{O}{\overset{\|}{C}}{-}OH \xrightarrow[H_2SO_4]{C_2H_5OH\ (solvent)} (CH_3)_2CH\overset{O}{\overset{\|}{C}}{-}OC_2H_5 \xrightarrow[2)\ H_3O^+]{1)\ CH_3MgI\ (excess)} \underset{\text{2,3-dimethyl-2-butanol}}{(CH_3)_2CH\overset{OH}{\overset{|}{\underset{CH_3}{\underset{|}{C}}}}{-}CH_3}$$

Although you have not studied this reaction explicitly, a more direct route to the alcohol is the reaction of the acid chloride with excess CH_3MgI followed by protonolysis.

$$(CH_3)_2CH\overset{\overset{O}{\|}}{C}-Cl \xrightarrow[2)\ H_3O^+]{1)\ CH_3MgI\ (excess)} (CH_3)_2CH\underset{\underset{CH_3}{|}}{\overset{\overset{OH}{|}}{C}}-CH_3$$

2,3-dimethyl-2-butanol

Can you see why acid chlorides should react the same way as esters in this reaction?

(d)

$$CH_3CH_2\overset{\overset{O}{\|}}{C}OH \xrightarrow{PCl_5} CH_3CH_2\overset{\overset{O}{\|}}{C}Cl \xrightarrow[2)\ H_3O^+]{1)\ Li^+\ {}^-Cu(C_2H_5)_2} CH_3CH_2\overset{\overset{O}{\|}}{C}CH_2CH_3$$

propionic acid → 3-pentanone

21.24 (a) The reaction of a Grignard reagent with ethyl formate gives a secondary alcohol in which the two alkyl groups at the α-carbon are identical.

$$CH_3CH_2O\overset{\overset{O}{\|}}{C}H + 2\ RMgBr \longrightarrow \xrightarrow{H_3O^+} R-\overset{\overset{OH}{|}}{CH}-R + CH_3CH_2OH$$

ethyl formate

(b) Follow the pattern derived in part (a) with R— = CH_3CH_2—.

$$CH_3CH_2O\overset{\overset{O}{\|}}{C}H + 2\ CH_3CH_2MgBr \longrightarrow \xrightarrow{H_3O^+} CH_3CH_2\overset{\overset{OH}{|}}{C}HCH_2CH_3 + CH_3CH_2OH$$

ethyl formate → 3-pentanol

21.25 (a) Because only the acid chloride reacts, the product is the following keto nitrile.

$$N\equiv C(CH_2)_{10}\overset{\overset{O}{\|}}{C}CH_3$$ 12-oxotridecanenitrile

(b) Because only the acid chloride reacts, the product is the following keto ester:

$$CH_3(CH_2)_3O-\overset{\overset{O}{\|}}{C}(CH_2)_4\overset{\overset{O}{\|}}{C}-CH_3$$ butyl 6-oxoheptanoate

21.26 The first method involves carbonation of the corresponding Grignard reagent; the second involves hydrolysis of the nitrile.

$$(CH_3)_2CH(CH_2)_3Br \xrightarrow[ether]{Mg} \xrightarrow[2)\ H_3O^+]{1)\ CO_2} (CH_3)_2CH(CH_2)_3CO_2H$$

1-bromo-4-methylpentane → 5-methylhexanoic acid

$$(CH_3)_2CH(CH_2)_3Br \xrightarrow[DMSO]{Na^+\ {}^-CN} (CH_3)_2CH(CH_2)_3C\equiv N \xrightarrow{H_2O,\ H_3O^+,\ heat} (CH_3)_2CH(CH_2)_3CO_2H$$

21.27 The required nitrile is 2-hydroxypropanenitrile. This is a cyanohydrin, and this cyanohydrin can be prepared by addition of cyanide to acetaldehyde.

$$\underset{\text{acetaldehyde}}{CH_3\overset{O}{\overset{\|}{C}}H} \xrightarrow{NaCN,\ H_2O} \underset{\text{2-hydroxypropanenitrile}}{CH_3\overset{OH}{\overset{|}{C}}H-C\equiv N} \xrightarrow{H_2O,\ H_3O^+,\ heat} \underset{\text{2-hydroxypropanoic acid}}{CH_3\overset{OH}{\overset{|}{C}}H-CO_2H}$$

Notice that the alcohol does *not* dehydrate under the rather severe conditions of nitrile hydrolysis. The reason is that the carbocation intermediate that would be involved is destabilized by both the cyano group and the carboxy group of the product. Can you think of a reason why?

21.28 Amides are more resistant to base-promoted hydrolysis than esters; the same is true of *polymeric* amides and esters. Nylon is more resistant to hydrolysis than polyester.

21.29 Process (*a*) is catalytic hydrogenation. Process (*b*) is ether cleavage with HCl, $ZnCl_2$ (or other acidic catalyst) and heat. Process (*c*) consists of the S_N2 reactions of cyanide ion with the dichloride. Process (*d*) is catalytic hydrogenation or $LiAlH_4$ reduction followed by protonolysis. Process (*e*) is nitrile hydrolysis with aqueous acid and heat. Finally, process (*f*) is to mix the amine and the carboxylic acid and heat (Eq. 21.70), or to form the di-acid chloride and allow it to react with the amine.

21.30 First, water opens the lactam to an amino acid *A*. The amino group of this compound serves as a nucleophile to open another lactam molecule; and the resulting amino group thus liberated repeats the process, thus growing the polymer chain:

ε-caprolactam + OH_2 → tetrahedral intermediate (O^-, $\overset{+}{O}H_2$, NH) —proton transfers→ (O^-, OH, $\overset{+}{N}H_2$) → $HO\overset{O}{\overset{\|}{C}}(CH_2)_5NH_2$ (*A*)

ε-caprolactam + $NH_2(CH_2)_5CO_2H$ (*A*) → (O^-, $\overset{+}{N}H_2(CH_2)_5CO_2H$, NH) —proton transfers→ (O^-, $NH(CH_2)_5CO_2H$, $\overset{+}{N}H_2$) →

$$HO\overset{O}{\overset{\|}{C}}(CH_2)_5NH-\overset{O}{\overset{\|}{C}}(CH_2)_5NH_2 \xrightarrow{\text{similar reactions}} \text{nylon-6}$$

this amino group attacks another ε-caprolactam molecule

Solutions to Additional Problems

21.31 (a) $PhCO_2H$ + CH_3CH_2OH

(b) $PhCO_2^- Na^+$ + CH_3CH_2OH

(c) $PhC(=O)—NH_2$ + CH_3CH_2OH

(d) $PhCH_2OH$ + CH_3CH_2OH

(e) $PhC(OH)(CH_2CH_2CH_3)CH_2CH_2CH_3$ + CH_3CH_2OH

(f) $PhC(OC(=O)CH_3)(CH_2CH_2CH_3)CH_2CH_2CH_3$ + pyridinium ($\overset{+}{N}H$) Cl^-

(g) $PhC(OS(=O)_2Ph)(CH_2CH_2CH_3)CH_2CH_2CH_3$

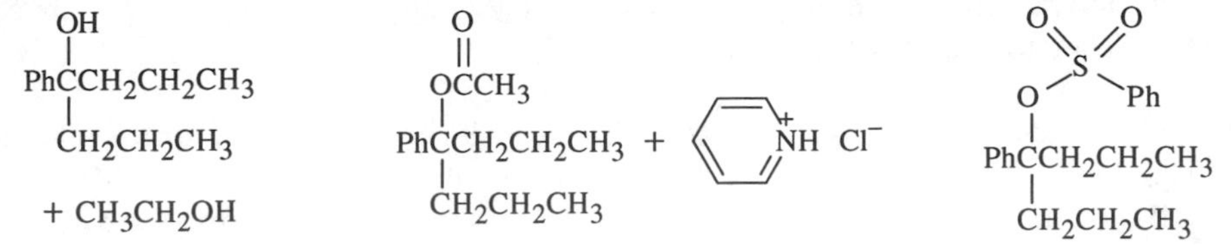

(h) In this reaction, ethoxide displaces ethoxide, but the reaction is "invisible" because the product is the same as the starting material. Thus, there is no *net* reaction.

21.32 (a) $CH_3CH_2CO_2H$

(b) $CH_3CH_2C(=O)SCH_2CH_3$ + pyridinium ($\overset{+}{N}H$) Cl^-

(c) $CH_3CH_2C(=O)OC(CH_3)_3$ + pyridinium ($\overset{+}{N}H$) Cl^-

(d) $CH_3CH_2C(=O)CH_3$

(e) $CH_3CH_2C(=O)H$

(f) $CH_3CH_2C(=O)—C_6H_4—CH_3$ (para)

(g) $CH_3CH_2C(=O)NHCH(CH_3)_2$ + $Cl^- H_2\overset{+}{N}CH(CH_3)_2$

(h) $CH_3CH_2C(=O)OC(=O)Ph$

(i) $CH_3CH_2C(=O)O—C_6H_4—CH_3$ (para) + pyridinium ($\overset{+}{N}H$) Cl^-

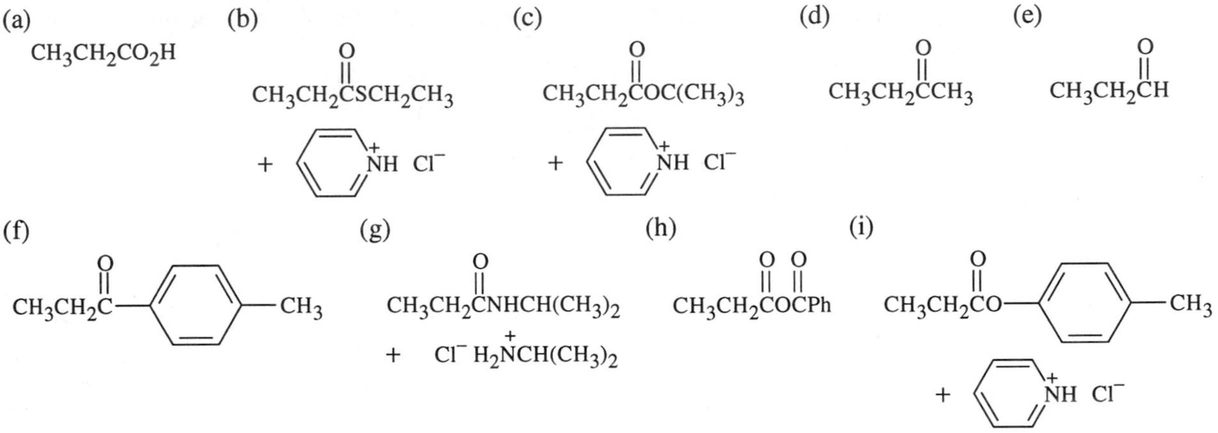

21.33 (a) $CH_3CH_2C{\equiv}N$
propionitrile

(b) $CH_3CH_2C(=O)NH_2$
propionamide

(c) $CH_3C(=O)OCH_2CH_2CH_2CH_3$
butyl acetate

(d) $CH_3(CH_2)_4C(=O)OCH(CH_3)(CH_2)_3CH_3$
1-methylpentyl hexanoate

(e) $N{\equiv}CCH_2C{\equiv}N$
malononitrile

Malononitrile (propanedinitrile) in part (e) hydrolyzes in aqueous acid to malonic acid and ammonium ion, and the malonic acid decarboxylates to acetic acid and carbon dioxide.

21.34 The easiest way to work this problem is to start with the structure and work backwards. Because carbonation of a Grignard reagent gives a carboxylic acid, compound *D* must be the corresponding alkyl bromide. Compound *E* must be the hydrocarbon that gives compound *D* on benzylic bromination. The Wolff-Kishner reaction produces compound *E* from compound *B*, which therefore must be a ketone. Reduction of ketone *B* with sodium borohydride gives alcohol *C*, which is converted into alkyl bromide *D*. Finally, the structure of compound *B* shows that reaction *A* must be a Friedel-Crafts acylation. In summary:

$$\underset{A}{\overset{O}{\overset{\|}{ClCCH_2CH_3}},\ AlCl_3,\ \text{then } H_3O^+} \qquad \underset{B}{PhCCH_2CH_3 \ (C{=}O)} \qquad \underset{C}{PhCH(OH)CH_2CH_3} \qquad \underset{D}{PhCH(Br)CH_2CH_3} \qquad \underset{E}{PhCH_2CH_2CH_3}$$

21.35 The product is (*R*)-(–)-1-phenyl-1,2-ethanediol. Notice that the asymmetric carbon is unaffected by these transformations.

$$\text{(R)-(–)-mandelic acid: } PhCH(OH)CO_2H \xrightarrow{CH_3OH,\ H_2SO_4} PhCH(OH)CO_2CH_3 \xrightarrow{1)\ LiAlH_4\quad 2)\ H_3O^+} PhCH(OH)CH_2OH$$

(*R*)-(–)-mandelic acid

(*R*)-(–)-1-phenyl-1,2-ethanediol (a 1,2-glycol; reacts with periodic acid)

21.36 Both the ketone and the carboxylic acid are reduced with $LiAlH_4$; although the carboxylic acid is ionized by $NaBH_4$ (but regenerated on addition of acid), only the ketone is reduced.

$$CH_3CCH_2CH_2COH \ (\text{4-oxopentanoic acid (levulinic acid)}) \xrightarrow{LiAlH_4} \xrightarrow{H_3O^+} CH_3CH(OH)CH_2CH_2CH_2OH \ (\text{1,4-pentanediol})$$

$$CH_3CCH_2CH_2COH \xrightarrow{NaBH_4,\ \text{methanol}} \xrightarrow{H_3O^+} CH_3CH(OH)CH_2CH_2CO_2H \ (\text{4-hydroxypentanoic acid})$$

4-oxopentanoic acid (levulinic acid)

1,4-pentanediol

4-hydroxypentanoic acid

21.37 Because the two carbonyl groups of the anhydride are not equivalent, two different products are possible; each corresponds to attack of the ethanol at a different carbonyl group. Because ethyl propionate has a higher molecular mass than ethyl acetate, ethyl propionate should have the higher boiling point. Since it is given that the ester with the lower boiling point is formed in greater amount, then ethyl acetate is the major product.

$$CH_3C(=O){-}O{-}C(=O)CH_2CH_3 + CH_3CH_2OH \longrightarrow \underbrace{CH_3C(=O){-}OCH_2CH_3 + HO{-}C(=O)CH_2CH_3}_{\text{major products}} +$$

acetic propionic anhydride

ethyl acetate

major products

$$CH_3CH_2O{-}C(=O)CH_2CH_3 + CH_3C(=O){-}OH$$

ethyl propionate

21.38 (a) Compound *A* has the formula $C_{11}H_{13}NO$. The formula of compound *B* indicates that it differs from *A* by the addition of the elements of water. The fact that compound *C* is a lactone with an oxygen at carbon-4 and an additional hydrogen at carbon-3 suggests that addition of water to the double bond of compound *A*, that is, alkene hydration, occurs metabolically to give an alcohol *B*. The lactone *C* is formed by cyclization of compound *B*, that is, an intramolecular esterification. The structure of compound *B* is given in the mechanism of part (b).

(b) A curved-arrow mechanism for the formation of compound *C* from compound *B*:

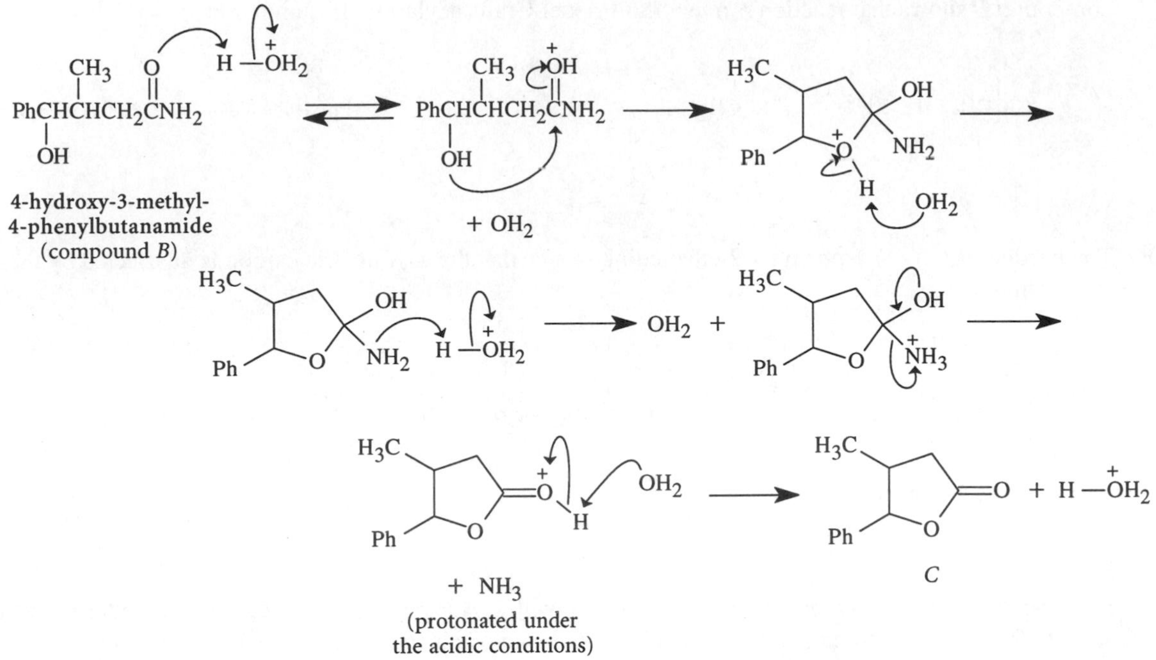

CH3 O H—$\overset{+}{O}H_2$
PhCHCHCH2CNH2
OH
4-hydroxy-3-methyl-4-phenylbutanamide (compound *B*)

CH3 $\overset{+}{O}$H
PhCHCHCH2CNH2
OH
+ OH2

H3C OH Ph O NH2 H OH2

H3C OH Ph O NH2 H—$\overset{+}{O}H_2$ → OH2 + H3C OH Ph O $\overset{+}{N}H_3$

H3C Ph O $\overset{+}{O}$ H OH2 → H3C Ph O =O + H—$\overset{+}{O}H_2$
+ NH3 (protonated under the acidic conditions)
C

21.39 (a) In aqueous base the ester groups are saponified to give glycerol (1,2,3-propanetriol) and three equivalents of sodium oleate, the sodium salt of oleic acid.

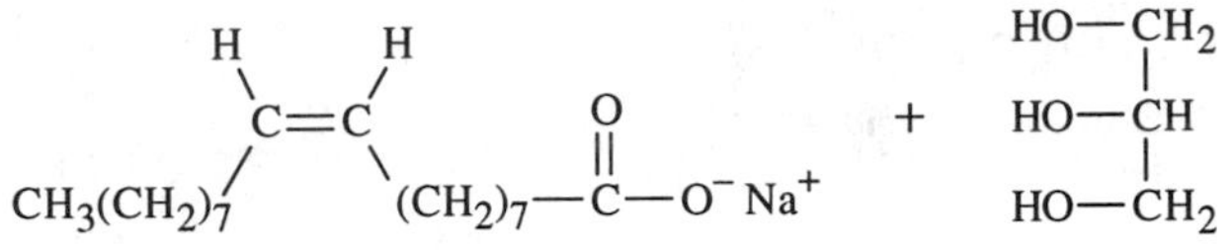

H H C=C CH3(CH2)7 (CH2)7—C(=O)—O⁻ Na⁺ + HO—CH2 HO—CH HO—CH2
sodium oleate glycerol

(b) The structure of glyceryl tristearate:

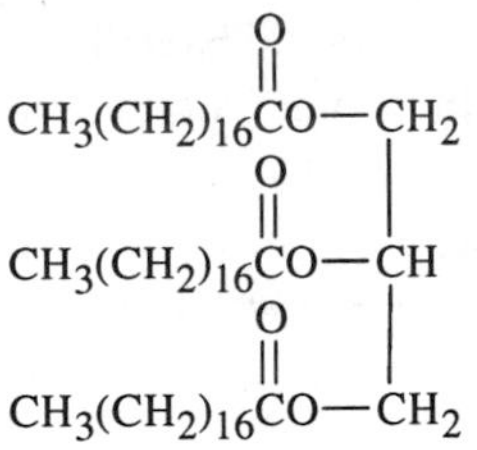

$CH_3(CH_2)_{16}\overset{O}{\overset{\|}{C}}O—CH_2$
$CH_3(CH_2)_{16}\overset{O}{\overset{\|}{C}}O—CH$ glyceryl tristearate
$CH_3(CH_2)_{16}\overset{O}{\overset{\|}{C}}O—CH_2$

Because unsaturated fats are oils and saturated fats are solids, glyceryl trioleate is a liquid, whereas the product of hydrogenation would be a solid. In fact, glyceryl trioleate is a major component of olive oil, whereas glyceryl tristearate is a major component of lard.

21.40 (a)

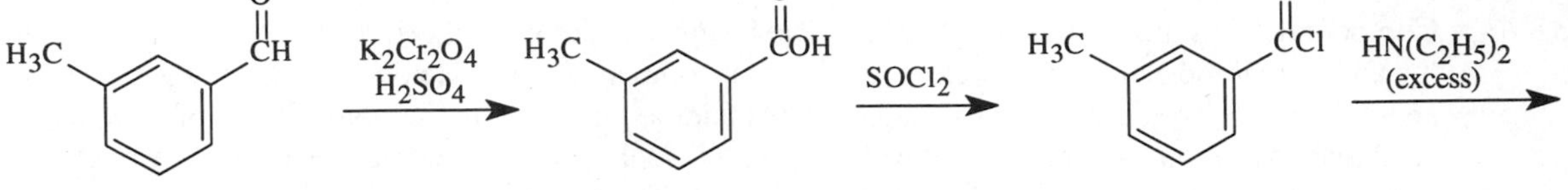

H3C CH(=O) —$K_2Cr_2O_4$, H_2SO_4→ H3C COH(=O) —$SOCl_2$→ H3C CCl(=O) —$HN(C_2H_5)_2$ (excess)→
3-methylbenzaldehyde

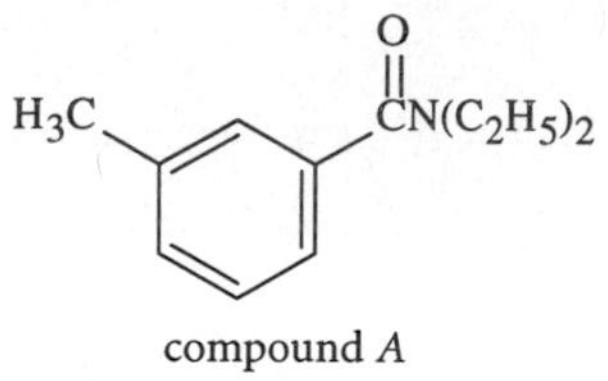

m-$CH_3C_6H_4CON(C_2H_5)_2$

compound *A*

(b)

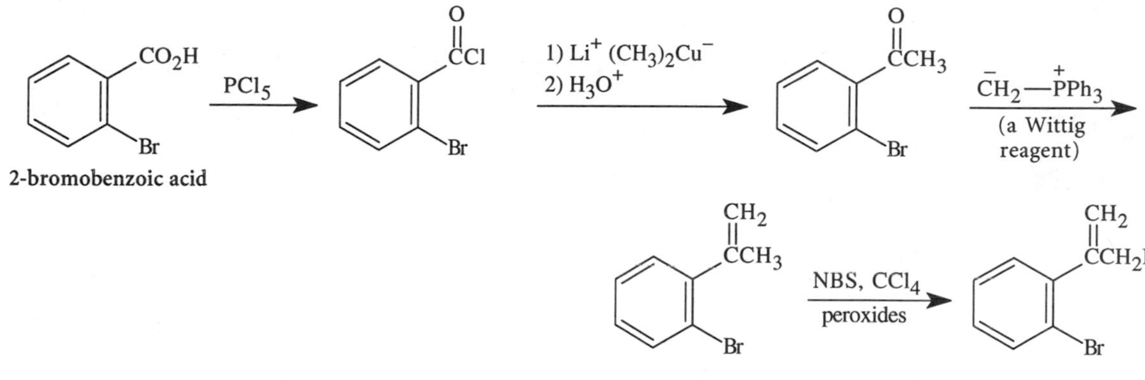

2-bromobenzoic acid (2-$BrC_6H_4CO_2H$) $\xrightarrow{PCl_5}$ 2-BrC_6H_4COCl $\xrightarrow[\text{2) } H_3O^+]{\text{1) } Li^+ (CH_3)_2Cu^-}$ 2-$BrC_6H_4COCH_3$ $\xrightarrow[\text{(a Wittig reagent)}]{\overset{-}{C}H_2{-}\overset{+}{P}Ph_3}$

2-$BrC_6H_4C(=CH_2)CH_3$ $\xrightarrow[\text{peroxides}]{\text{NBS, } CCl_4}$ 2-$BrC_6H_4C(=CH_2)CH_2Br$

compound *B*

21.41 (a)

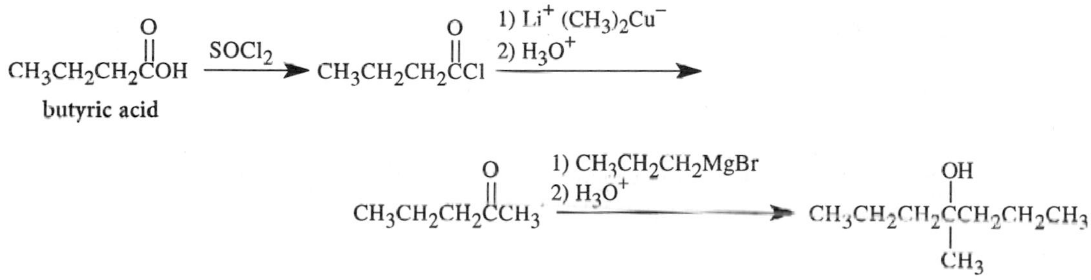

$CH_3CH_2CH_2COOH$ (butyric acid) $\xrightarrow{SOCl_2}$ $CH_3CH_2CH_2COCl$ $\xrightarrow[\text{2) } H_3O^+]{\text{1) } Li^+ (CH_3)_2Cu^-}$

$CH_3CH_2CH_2COCH_3$ $\xrightarrow[\text{2) } H_3O^+]{\text{1) } CH_3CH_2CH_2MgBr}$ $CH_3CH_2CH_2C(OH)(CH_3)CH_2CH_2CH_3$

4-methyl-4-heptanol

(b)

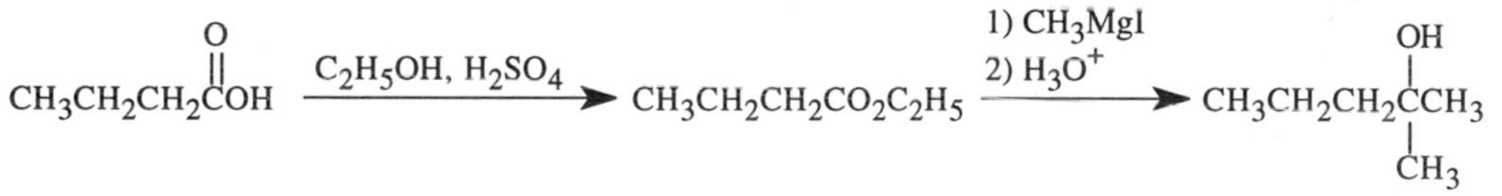

$CH_3CH_2CH_2COOH$ $\xrightarrow{C_2H_5OH,\ H_2SO_4}$ $CH_3CH_2CH_2CO_2C_2H_5$ $\xrightarrow[\text{2) } H_3O^+]{\text{1) } CH_3MgI}$ $CH_3CH_2CH_2C(OH)(CH_3)CH_3$

2-methyl-2-pentanol

(c)

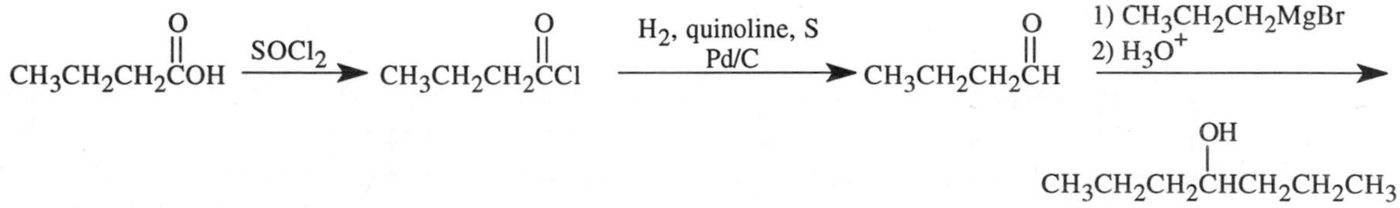

$CH_3CH_2CH_2COOH$ $\xrightarrow{SOCl_2}$ $CH_3CH_2CH_2COCl$ $\xrightarrow[\text{Pd/C}]{H_2\text{, quinoline, S}}$ $CH_3CH_2CH_2CHO$ $\xrightarrow[\text{2) } H_3O^+]{\text{1) } CH_3CH_2CH_2MgBr}$

$CH_3CH_2CH_2CH(OH)CH_2CH_2CH_3$

4-heptanol

(d)

$CH_3CH_2CH_2COOH$ $\xrightarrow[\text{2) } H_3O^+]{\text{1) } LiAlH_4}$ $CH_3CH_2CH_2CH_2OH$ $\xrightarrow[H_2SO_4]{\text{conc. HBr}}$ $CH_3CH_2CH_2CH_2Br$ $\xrightarrow[\text{ether}]{Mg}$

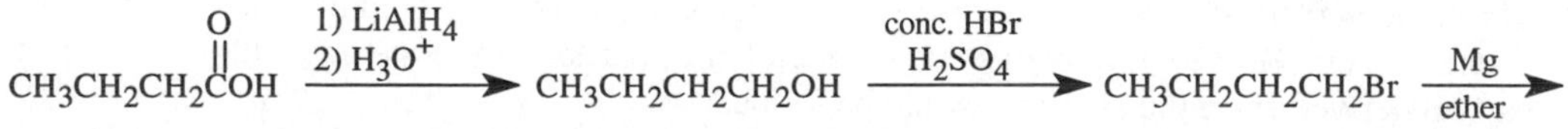

$$CH_3CH_2CH_2CH_2MgBr \xrightarrow[2)\ H_3O^+]{1)\ \text{ethylene oxide}} CH_3CH_2CH_2CH_2CH_2CH_2OH \xrightarrow[2)\ H_3O^+]{1)\ KMnO_4,\ ^-OH}$$

$$CH_3CH_2CH_2CH_2CH_2CO_2H \xrightarrow{SOCl_2} CH_3CH_2CH_2CH_2CH_2\overset{O}{\overset{\|}{C}}Cl \xrightarrow{NH_3\ (excess)}$$

$$CH_3CH_2CH_2CH_2CH_2\overset{O}{\overset{\|}{C}}NH_2 \xrightarrow[2)\ H_2O]{1)\ LiAlH_4} CH_3CH_2CH_2CH_2CH_2CH_2NH_2$$

1-hexanamine (hexylamine)

(e)

$$CH_3CH_2CH_2\overset{O}{\overset{\|}{C}}OH \xrightarrow[2)\ H_3O^+]{1)\ LiAlH_4} CH_3CH_2CH_2CH_2OH \xrightarrow[H_2SO_4]{conc.\ HBr} CH_3CH_2CH_2CH_2Br \xrightarrow[ether]{Mg}$$

$$CH_3CH_2CH_2CH_2MgBr \xrightarrow[2)\ H_3O^+]{1)\ CO_2} CH_3CH_2CH_2CH_2CO_2H \xrightarrow{SOCl_2} CH_3CH_2CH_2CH_2\overset{O}{\overset{\|}{C}}Cl \xrightarrow{NH_3\ (excess)}$$

$$CH_3CH_2CH_2CH_2\overset{O}{\overset{\|}{C}}NH_2 \xrightarrow[2)\ H_2O]{1)\ LiAlH_4} CH_3CH_2CH_2CH_2CH_2NH_2$$

1-pentanamine (pentylamine)

(f)

$$CH_3CH_2CH_2\overset{O}{\overset{\|}{C}}OH \xrightarrow{SOCl_2} CH_3CH_2CH_2\overset{O}{\overset{\|}{C}}Cl \xrightarrow[(excess)]{NH_3} CH_3CH_2CH_2\overset{O}{\overset{\|}{C}}NH_2 \xrightarrow[2)\ H_2O]{1)\ LiAlH_4} CH_3CH_2CH_2CH_2NH_2$$

1-butanamine (butylamine)

21.43 (a) First of all, two sets of constitutional isomers can be obtained as a result of the attack of 1-phenylethanol at either of the nonequivalent carbonyl groups of the anhydride:

α-phenylglutaric anhydride + $CH_3\underset{|}{\overset{OH}{C}}HPh$ (1-phenylethanol) ⟶ A + B

A: $PhCH(\overset{O}{\overset{\|}{C}}OCH(CH_3)Ph)CH_2CH_2CO_2H$

B: $PhCH(CO_2H)CH_2CH_2\overset{O}{\overset{\|}{C}}OCH(CH_3)Ph$

Each of these products has two asymmetric carbon stereocenters and can thus exist as four stereoisomers. All stereoisomers of *A* can in principle be separated from all stereoisomers of *B* because constitutional isomers have different properties. The four stereoisomers of *A* can be separated into two pairs of enantiomers; the same is true of *B*. Consequently, four compounds can in principle be obtained from this reaction mixture without enantiomeric resolution.

(b) The same two constitutional isomers would be obtained, and each would be obtained as a pair of diastereomers. However, only one enantiomer of each diastereomer (the one with the *S* configuration at the carbon derived from the alcohol) would be formed.

If the diastereomeric esters from part (b) were separated from each other and saponified, the α-phenylglutaric acid formed would be enantiomerically pure. The process described here would therefore be useful as an enantiomeric resolution of α-phenylglutaric acid (or any other chiral dicarboxylic acid from which a cyclic anhydride can be readily formed).

21.43 Carboxylate salts are less reactive than esters in nucleophilic acyl substitution reactions for two reasons. First, resonance stabilization of carboxylate salts is particularly important because carboxylates have two identical, important resonance structures; see Eq. 20.4 on text p. 904. Resonance stabilization of esters is less important because one resonance structure separates charge; see the structures in Eq. 21.28 on text p. 960. Because resonance stabilization reduces reactivity, carboxylates are less reactive than esters.

The second and major reason that carboxylate salts are less reactive is that they contain no leaving group. The oxygen anion of the carboxylate, in order to serve as a leaving group, would have to depart as an oxide ion, O^{2-}. This very basic leaving group is unprecedented in the usual reactions of organic chemistry.

Reactions such as $LiAlH_4$ reduction that occur on carboxylate salts might at first glance appear to involve oxide ions as leaving groups. In such reactions, however, the "oxygen anion" is bonded to another species that allows the oxygen to serve as a leaving group. (See the details in Study Guide Link 20.9 on p. 710 of this manual.)

21.44 Thiol esters are less stabilized by resonance than esters because the resonance interaction of sulfur unshared electron pairs with a carbonyl group is less effective than the resonance interaction of oxygen unshared electron pairs. The reason for this difference is that the orbitals on sulfur are derived from quantum level 3, and such orbitals overlap poorly with the π orbitals of a carbonyl group, which are derived from quantum level 2. (See Fig. 16.8 on text p. 726 and the accompanying discussion of this point.) Thus, resonance arguments lead to the conclusion that a thiol ester should hydrolyze more rapidly than an ester. In addition, a thiolate ion is a better leaving group than an alkoxide because thiolates are less basic than alkoxides. Whether resonance stabilization or leaving-group effectiveness is considered, thiol esters are predicted to be more reactive than esters.

Because both sulfur and chlorine are in the third period of the periodic table, the resonance stabilizations of acid chlorides and thiol esters are comparable. However, a chloride ion is a much weaker base, and therefore a much better leaving group, than a thiolate ion. These points argue for increased reactivity of acid chlorides relative to thiol esters.

Hence, the answer is (2): thiol esters are less reactive than acid chlorides, but more reactive than esters.

21.45 (a) Synthesis of nylon-4,6 requires mixing 1,4-butanediamine and adipoyl dichloride.

$$\underset{\text{1,4-butanediamine}}{H_2N(CH_2)_4NH_2} + \underset{\text{adipoyl dichloride}}{ClC(=O)(CH_2)_4C(=O)Cl} \longrightarrow \text{nylon-4,6}$$

(Alternatively, 1,4-butanediamine can be heated with adipic acid; see Eq. 21.70 on text p. 983.) The preparations of these materials are as follows:

$$\underset{\text{succinic acid}}{HOC(=O)CH_2CH_2C(=O)OH} \xrightarrow{SOCl_2} ClC(=O)CH_2CH_2C(=O)Cl \xrightarrow{NH_3\text{ (excess)}} H_2NC(=O)CH_2CH_2C(=O)NH_2 \xrightarrow[\text{2) } H_2O]{\text{1) } LiAlH_4} \underset{\text{1,4-butanediamine}}{H_2N(CH_2)_4NH_2}$$

$$\underset{\text{adipic acid}}{HOC(=O)(CH_2)_4C(=O)OH} \xrightarrow{SOCl_2} \underset{\text{adipoyl dichloride}}{ClC(=O)(CH_2)_4C(=O)Cl}$$

Why do you think heating the dicarboxylic acid with the diamine is preferred industrially to reaction of the diamine with the dicarboxylic acid dichloride? (There are at least two reasons.)

(b) The preparation of this polyester requires heating 1,4-butanediol with glutaric acid.

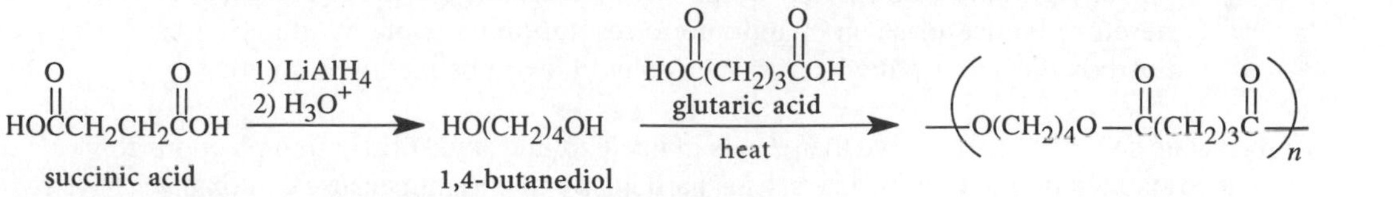

$$\underset{\text{succinic acid}}{HO\overset{O}{\overset{\|}{C}}CH_2CH_2\overset{O}{\overset{\|}{C}}OH} \xrightarrow[\text{2) } H_3O^+]{\text{1) } LiAlH_4} \underset{\text{1,4-butanediol}}{HO(CH_2)_4OH} \xrightarrow[\text{heat}]{HO\overset{O}{\overset{\|}{C}}(CH_2)_3\overset{O}{\overset{\|}{C}}OH \text{ glutaric acid}} \left(O(CH_2)_4O-\overset{O}{\overset{\|}{C}}(CH_2)_3\overset{O}{\overset{\|}{C}} \right)_n$$

(Alternatively, the di-acid chloride of glutaric acid could be mixed with 1,4-butanediol.)

21.46 The IR spectrum of compound *A* indicates that it is an anhydride, and probably a cyclic anhydride containing a five-membered ring. (Compare the carbonyl absorptions of compound *A* with those in Table 21.3 on text p. 944; the 1050 cm^{-1} absorption is a C—O stretching absorption). Addition of methanol to a cyclic anhydride should give a methyl half-ester, compound *B*. Indeed, the IR spectrum of compound *B* indicates the presence of both a carboxylic acid and an ester group. Subtracting the elements of methanol (CH_4O) from the formula of compound *B* gives the formula of compound *A*, $C_4H_4O_3$. Compound *A* is succinic anhydride.

Two points about the NMR spectra are worth noting. First, the methylene protons of compound *B* give a complex resonance at δ 2.7 because they are chemically nonequivalent but have very similar chemical shifts. This is the condition under which complex, non-first-order splitting can occur. (See text Sec. 13.5B, pp. 570–5.) That these protons should have similar chemical shifts is a reasonable observation because both sets of protons are in very similar electronic environments: both are on carbons that are α to carbonyl groups. The second point is that the proton of the carboxy group of compound *B* is not observed because the spectrum is taken in D_2O. Recall (Sec. 13.7D, text pp. 579–80) that O—H protons are rapidly exchanged for deuterium in when D_2O is present, and that deuterium nuclei are "silent" in proton NMR.

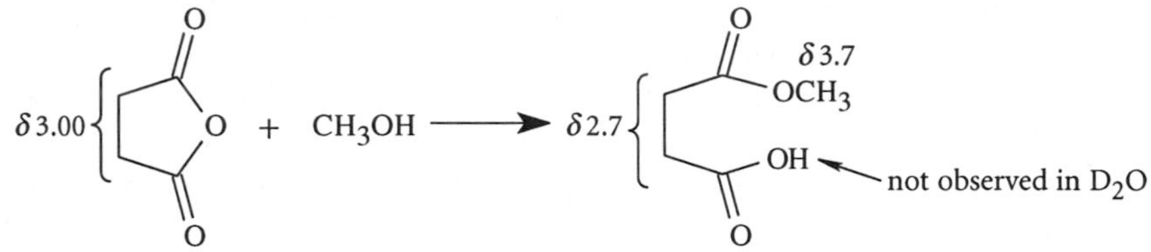

δ 3.00 (succinic anhydride ring CH_2) + CH_3OH → δ 2.7 (CH_2); δ 3.7 (OCH_3); OH ← not observed in D_2O

succinic anhydride
(compound *A*)

methyl hydrogen succinate
(compound *B*)

21.47 The high-frequency IR absorption and the reaction with water to give a dicarboxylic acid suggest that this compound is the anhydride of α,α-dimethylmalonic acid. The formation of a half ester in the reaction with methanol also supports the hypothesis that this compound is an anhydride.

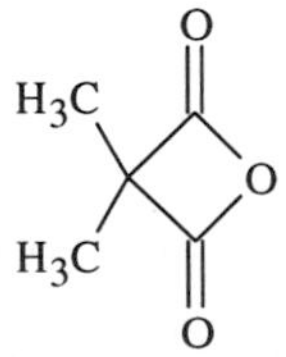

(H_3C, H_3C on four-membered ring with two C=O and O)
α,α-dimethylmalonic anhydride

In most cases cyclic anhydrides show two carbonyl stretching absorptions. Can you think of a reason why this particular anhydride shows only one? (*Hint:* See Study Problem 12.2 on text p. 511.) Is the missing carbonyl absorption at lower or higher frequency than the one that is observed at 1820 cm^{-1}? (*Hint:* See Table 21.3 on text p. 944, and extrapolate from the absorptions of six- and five-membered cyclic anhydrides.)

21.48 (a) Because phenols are very reactive in electrophilic substitution reactions, bromine in CCl_4 will effect bromination of the phenol ring as well as addition to the alkene double bond. (See Eq. 18.69 on text p. 816.)

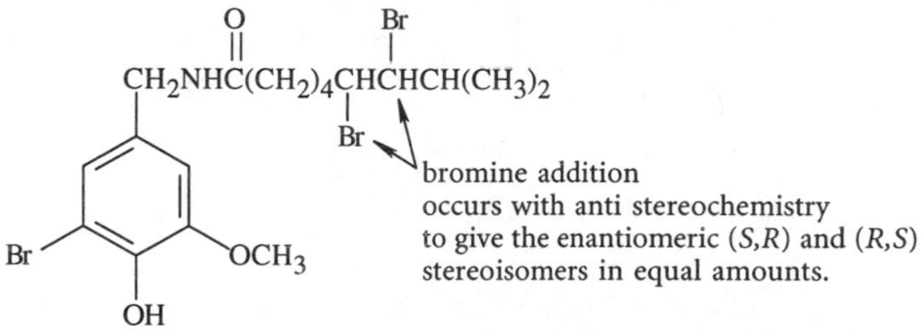

Structure: ring bearing $CH_2NHC(=O)(CH_2)_4CH(Br)CH(Br)CH(CH_3)_2$, Br, OCH_3, OH

bromine addition occurs with anti stereochemistry to give the enantiomeric (*S*,*R*) and (*R*,*S*) stereoisomers in equal amounts.

(b) 5% aqueous NaOH will bring about ionization of the phenol to its conjugate-base phenolate ion. The amide is unaffected, because amide hydrolysis requires heat and strong base.

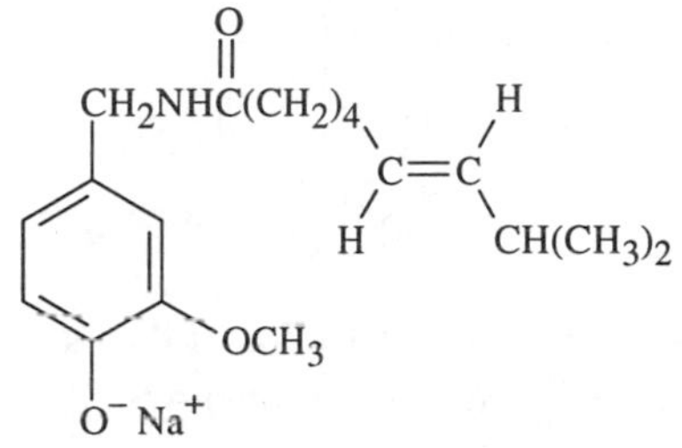

Structure: ring bearing $CH_2NHC(=O)(CH_2)_4CH{=}CHCH(CH_3)_2$ (trans), OCH_3, $O^- Na^+$

(c) No significant reaction occurs with 5% aqueous HCl. (The carbonyl oxygen of the amide is protonated to a small extent.)

(d) Catalytic hydrogenation brings about addition of hydrogen to the alkene double bond.

Structure: ring bearing $CH_2NHC(=O)(CH_2)_4CH_2CH_2CH(CH_3)_2$, OCH_3, OH

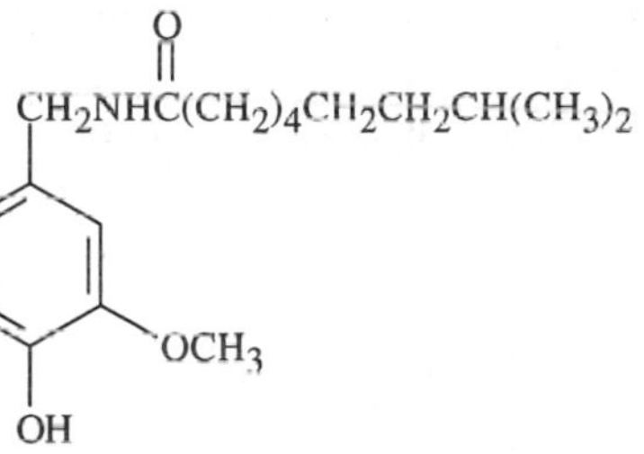

Depending on the conditions, hydrogenation of the *single bond* between the benzylic CH_2 and the nitrogen can also occur; such a reaction is called *hydrogenolysis*. Because you have not studied this reaction, it is ignored.

(e) Heating the product of (d) in 6 *M* HCl brings about amide hydrolysis.

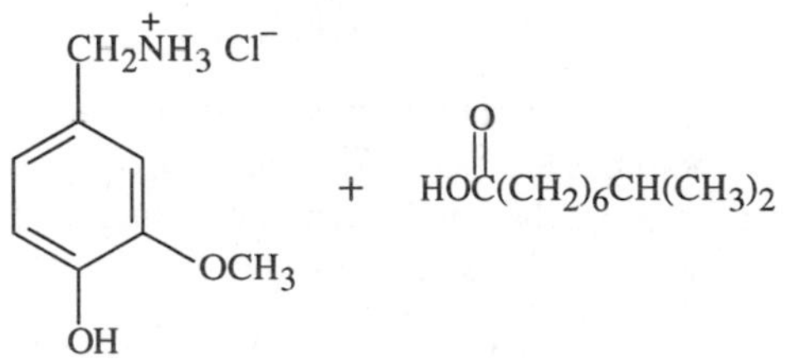

Structure: ring bearing $CH_2\overset{+}{N}H_3\ Cl^-$, OCH_3, OH + $HOC(=O)(CH_2)_6CH(CH_3)_2$

(It is also conceivable that some cleavage of the methyl ether will occur to give methyl chloride and the diphenol.)

(f) The phenolate ion formed in part (b) is alkylated to give a second methyl ether group.

$$\text{3,4-}(OCH_3)_2C_6H_3\text{—}CH_2NH\overset{O}{\overset{\|}{C}}(CH_2)_4\text{—}CH\text{=}CH\text{—}CH(CH_3)_2$$

(g) On heating, the same amide hydrolysis observed in part (e) will take place, and the methyl ether will cleave.

$$CH_3Br \; + \; \text{3,4-}(OH)_2C_6H_3\text{—}CH_2\overset{+}{N}H_3\;Br^- \; + \; HO\overset{O}{\overset{\|}{C}}(CH_2)_6CH(CH_3)_2$$

21.49 Use the titration data to calculate the molecular mass of the ester. The amount of 1 *M* NaOH solution consumed in the saponification reaction is (15.00 mL – 5.30 mL) = 9.70 mL. Make the provisional assumption that there is one ester group per molecule of compound *A*. Then one mole of NaOH is consumed per mole of ester saponified, and the molecular mass of compound *A* is then

$$\text{molecular mass of } A = \frac{2.00 \text{ g}}{(9.70 \times 10^{-3} \text{ L})(1.00 \text{ mol/L})} = 206 \text{ g/mol}$$

Next, try to deduce a structure from the data in the problem. The structure of compound *C*, the alcohol that results from the saponification of compound *A*, can be determined; it is the alcohol that can be oxidized to acetophenone. Therefore, compound *C* is 1-phenylethanol.

$$\underset{\substack{\textbf{1-phenylethanol} \\ (\text{compound } C)}}{Ph\text{—}\overset{OH}{\overset{|}{C}H}\text{—}CH_3} \xrightarrow{K_2Cr_2O_7} \underset{\textbf{acetophenone}}{Ph\text{—}\overset{O}{\overset{\|}{C}}\text{—}CH_3}$$

The molecular mass of compound *C* is 122 g/mol. Now, from the overall transformation of ester hydrolysis, the molecular mass of ester *A* must equal the sum of the molecular masses of compounds *B* and *C* minus the molecular mass of H_2O, that is,

$$\text{molecular mass of } A = 206 \text{ g/mol} = \text{molecular mass of } B + 122 \text{ g/mol} - 18 \text{ g/mol}$$

or

$$\text{molecular mass of } B = 206 \text{ g/mol} + 18 \text{ g/mol} - 122 \text{ g/mol} = 102 \text{ g/mol}$$

Compound *B* is a carboxylic acid; its carboxy group accounts for 45 molecular mass units. Therefore, the remainder of the carboxylic acid has a mass of 57 units, which is the mass of a butyl group. Because acid *B* is optically active, it is chiral; therefore the butyl group must contain an asymmetric carbon stereocenter. Only a *sec*-butyl group meets this criterion. The structures of compounds *B* and *A* are therefore

$$\underset{\substack{\textbf{2-methylbutanoic acid} \\ (\text{compound } B)}}{CH_3CH_2\overset{CH_3}{\overset{|}{C}H}CO_2H} \qquad \underset{\substack{\textbf{1-phenylethyl 2-methylbutanoate} \\ (\text{compound } A)}}{CH_3CH_2\overset{CH_3}{\overset{|}{C}H}\text{—}\overset{O}{\overset{\|}{C}}\text{—}O\underset{CH_3}{\underset{|}{C}H}Ph}$$

As noted in the problem, the absolute configurations of the asymmetric carbons cannot be determined from the data. Hence, compounds *B* and *C* are each one of two possible enantiomers, and compound *A* is the particular stereoisomer with the corresponding absolute configurations of its asymmetric carbons. Compound *D* is the racemate of compound *C*, and compound *E* is the diastereomer of compound *A* with the following configurations:

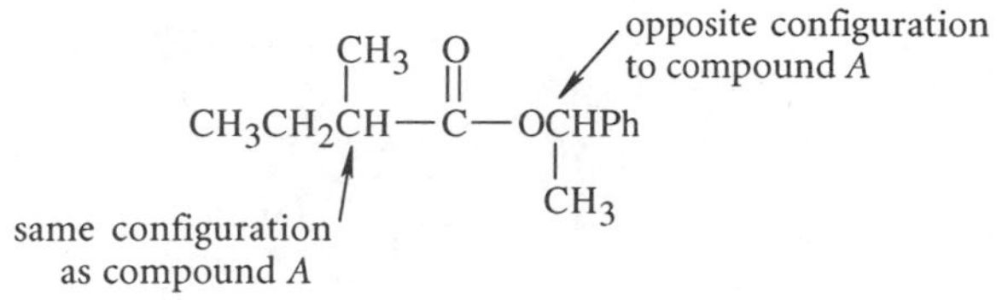

$$\mathrm{CH_3CH_2\underset{\uparrow}{C}H(CH_3)-\overset{O}{\overset{\|}{C}}-O\underset{CH_3}{\underset{|}{C}}HPh}$$

same configuration as compound *A* (at the CH bearing CH_3); opposite configuration to compound *A* (at the OCHPh carbon)

compound *E*

21.50 (a) This is a reprise of Problem 10.53 on text p. 454, which is answered on p. 329 of this manual. Obviously, Klutz Muckfingers has been talking to Buster Bluelip. The very weak acid HCN is not significantly dissociated. Consequently, the solution is insufficiently acidic to effect protonation of the —OH group of the alcohol. This protonation is necessary to convert this group into a good leaving group. Furthermore, there is virtually no cyanide ion ($^{-}C{\equiv}N$) present, and hence virtually no nucleophile to displace the —OH group.

(b) First of all, an excess of an alcohol is generally required to drive acid-catalyzed esterification to completion. However, even if some of the adipic acid is converted into its ester, there is no reason why this monoester would not be essentially as reactive as adipic acid itself, and the reaction mixture would ultimately contain a mixture of adipic acid, its monomethyl ester, and its dimethyl ester. The yield of the desired monoester-ester would be poor, and it would have to be separated from both the di-ester and the unreacted adipic acid.

(c) There is no reason why acetic acid should form an anhydride only with benzoic acid, and vice-versa; substantial amounts of acetic anhydride and benzoic anhydride should also be obtained. If all the reactions were totally random, the desired unsymmetrical anhydride would be formed in a maximum of 50% yield.

(d) The hydroxide ion reacts *much* more rapidly with the O—H proton of the phenol than it does with the ester. (See Study Guide Link 20.2 on p. 705 of this manual for a discussion of this point.) Consequently, the one equivalent of hydroxide is consumed by this reaction, and no base is left to saponify the ester.

(e) Several functional groups in the β-lactam molecule are more reactive toward acid hydrolysis than the amide indicated. Because esters are more reactive than amides, the acetate ester will undoubtedly also hydrolyze under the reaction conditions. The β-lactam ring itself is an amide, and because hydrolysis of this amide relieves substantial ring strain, this amide should also hydrolyze more rapidly. Once the β-lactam hydrolyzes, the resulting enamine also should hydrolyze readily in acid, and the N—C—S linkage is also unstable toward hydrolysis in the same sense that an acetal is unstable. With so many faster competing processes, Klutz has no hope for a selective reaction.

21.51 Determine the structure of compound *B* and reason backwards. Compound *B* is a diol because it gives a diacetate derivative *C*. Because compound *B* is the diol that can be oxidized to β-methylglutaric acid, compound *B* is 3-methyl-1,5-pentanediol.

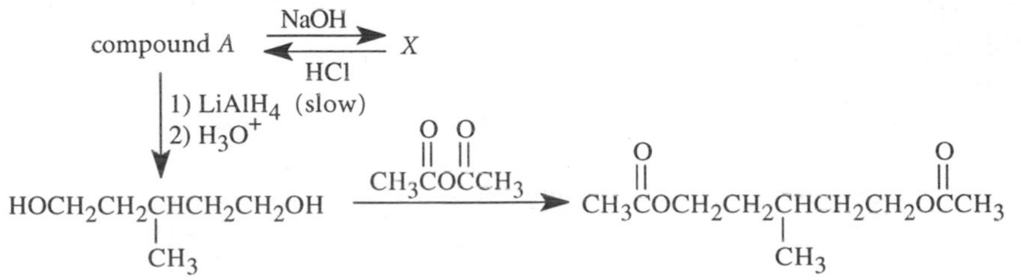

$$\text{compound } A \underset{\text{HCl}}{\overset{\text{NaOH}}{\rightleftharpoons}} X$$

compound *A* —[1) $LiAlH_4$ (slow); 2) H_3O^+]→

$$\mathrm{HOCH_2CH_2\underset{CH_3}{\underset{|}{C}}HCH_2CH_2OH} \xrightarrow{\mathrm{CH_3\overset{O}{\overset{\|}{C}}O\overset{O}{\overset{\|}{C}}CH_3}} \mathrm{CH_3\overset{O}{\overset{\|}{C}}OCH_2CH_2\underset{CH_3}{\underset{|}{C}}HCH_2CH_2O\overset{O}{\overset{\|}{C}}CH_3}$$

3-methyl-1,5-pentanediol (compound *B*) — compound *C*

Compound *A* can be neither a hydroxy acid, a dicarboxylic acid, an oxo acid, nor an ester of these, because it contains only two oxygens. If it has only one carboxy group (either as a carboxylic acid or an ester) then it must contain either a ring or a carbon-carbon double bond. The structure of the $LiAlH_4$ reduction product *B* rules out a carbon-carbon double bond; hence, compound *A* contains a ring. Therefore, compound *A* is a lactone, and *X* is the sodium salt of the corresponding hydroxy acid, which is soluble in aqueous base. Compound *A* appears to dissolve when it is converted into the water-soluble compound *X*.

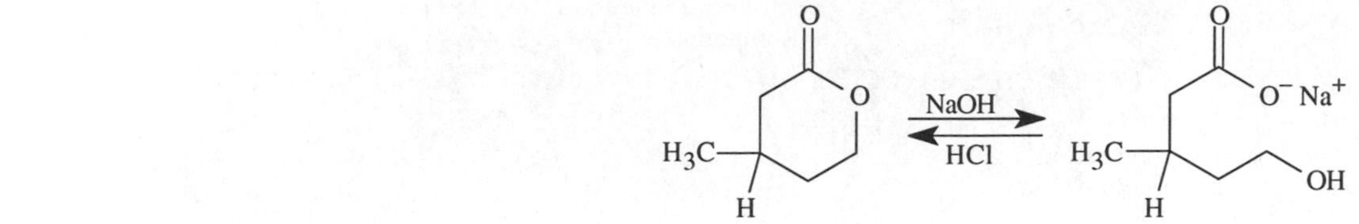

NaOH / HCl

compound *A* ⇌ *X*

21.52 (a) The "alcohol" formed as one of the saponification products is an enol, which spontaneously reverts to the corresponding ketone. (See Eq. 14.5c on text p. 615.)

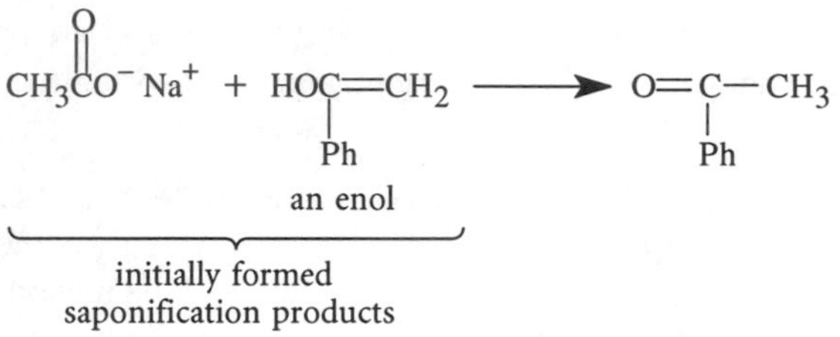

$$CH_3\overset{O}{\overset{\|}{C}}O^- Na^+ + HO\underset{Ph}{\underset{|}{C}}{=}CH_2 \longrightarrow O{=}\underset{Ph}{\underset{|}{C}}{-}CH_3$$

an enol

initially formed saponification products

(b) Transesterification of the formate ester occurs.

$$CH_3\overset{O}{\overset{\|}{C}}CH_2OH + CH_3O\overset{O}{\overset{\|}{C}}H$$

(c) The formula shows that the product includes the carbonyl group of phosgene (the acid chloride) minus two HCl molecules. Because the product has an unsaturation number of 7, a ring must be formed. Because both amines and carboxylic acids can react with acid chlorides, the product, called *isatoic anhydride,* is simultaneously a cyclic anhydride and a cyclic amide (lactam).

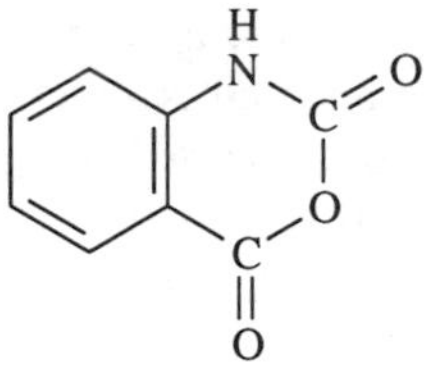

isatoic anhydride

(d) Both amino groups are acylated by the excess of the acid chloride.

$$Ph{-}\overset{O}{\overset{\|}{C}}{-}NH{-}NH{-}\overset{O}{\overset{\|}{C}}{-}Ph$$

(e) Both the amide and the nitrile are hydrolyzed.

$$PhCO_2H + H_3\overset{+}{N}(CH_2)_5CO_2H + \overset{+}{N}H_4$$

(f) The cyclic amide is reduced to a cyclic amine.

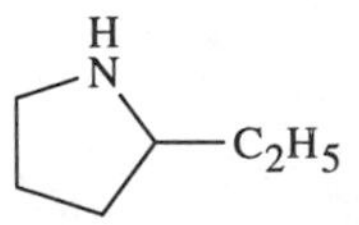

C_2H_5

(g) The Grignard reactions of lactones are much like those of esters, except that the alcohol displaced by the nucleophilic acyl substitution reaction remains as part of the same molecule.

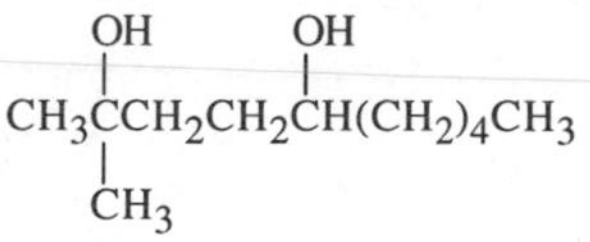

$$CH_3C(OH)(CH_3)CH_2CH_2CH(OH)(CH_2)_4CH_3$$

(h) Diethyl carbonate reacts first to give an ester, then a ketone, and then a tertiary alcohol (after protonolysis) in which all three groups attached to the α-carbon are the same.

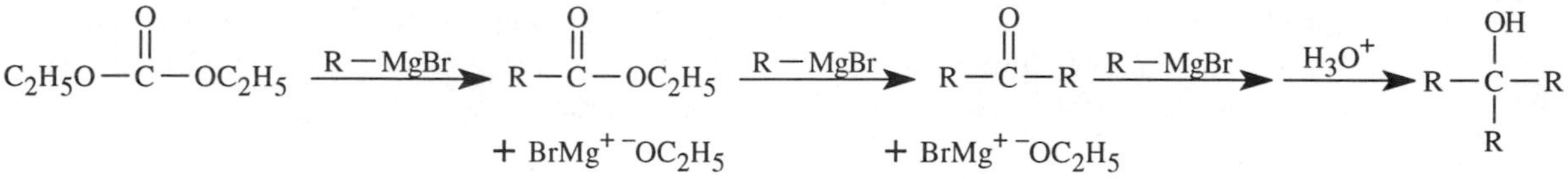

$$C_2H_5O-\overset{O}{\overset{\|}{C}}-OC_2H_5 \xrightarrow{R-MgBr} R-\overset{O}{\overset{\|}{C}}-OC_2H_5 + BrMg^{+}\,{}^{-}OC_2H_5 \xrightarrow{R-MgBr} R-\overset{O}{\overset{\|}{C}}-R + BrMg^{+}\,{}^{-}OC_2H_5 \xrightarrow{R-MgBr} \xrightarrow{H_3O^+} R-\underset{R}{\underset{|}{\overset{OH}{\overset{|}{C}}}}-R$$

From this pattern, the product of the reaction given is 3-ethyl-3-pentanol (R = ethyl in the preceding tertiary alcohol structure).

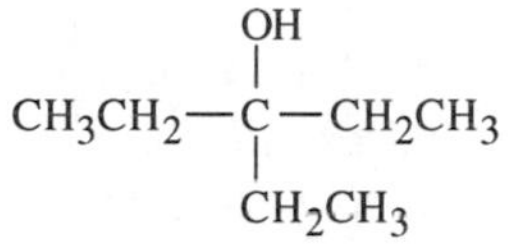

$$CH_3CH_2-\underset{CH_2CH_3}{\underset{|}{\overset{OH}{\overset{|}{C}}}}-CH_2CH_3 \quad \text{3-ethyl-3-pentanol}$$

(i) Lithium aluminum hydride reacts with acid chlorides in the same way that it reacts with esters, only faster. The product is $(CH_3)_3CCH_2OH$ (2,2-dimethyl-1-propanol).

(j) Because aldehydes are much more reactive than esters, only the aldehyde reacts when only one equivalent of the Grignard reagent is used.

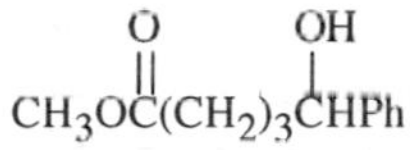

$$CH_3O\overset{O}{\overset{\|}{C}}(CH_2)_3\overset{OH}{\overset{|}{C}}HPh$$

methyl 5-hydroxy-5-phenylpentanoate

(k) Transesterification occurs to give three equivalents of methyl stearate and one of glycerol.

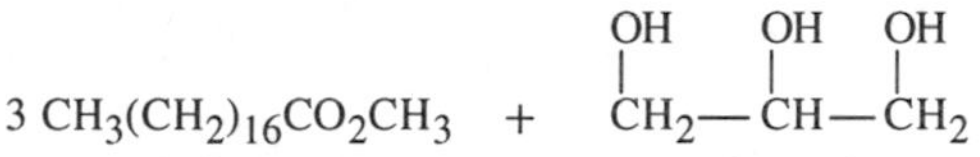

$$3\ CH_3(CH_2)_{16}CO_2CH_3 + \overset{OH}{\overset{|}{C}H_2}-\overset{OH}{\overset{|}{C}H}-\overset{OH}{\overset{|}{C}H_2}$$

methyl stearate glycerol

(l) The ester group undergoes intramolecular aminolysis to form a γ-lactam. Such intramolecular nucleophilic substitution reactions are very common when five- and six-membered rings can be formed. This is another manifestation of the greater rate of intramolecular reactions (Sec. 11.7).

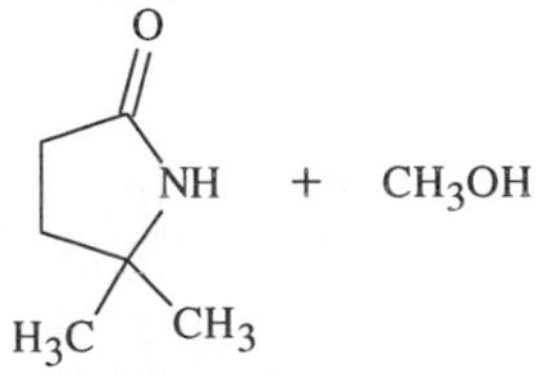

[5,5-dimethyl-2-pyrrolidinone: five-membered ring with C=O, NH, and $C(CH_3)_2$] + CH_3OH

21.53 (a) Isotopically labeled methyl benzenesulfonate (product *A*) is formed in the first reaction, and unlabeled methanol along with benzenesulfonate anion (product *B*) is formed in the second by an S_N2 reaction of hydroxide at the methyl carbon. (Notice that, because of resonance in the by-product benzenesulfonate anion, the three oxygens are equivalent and indistinguishable.)

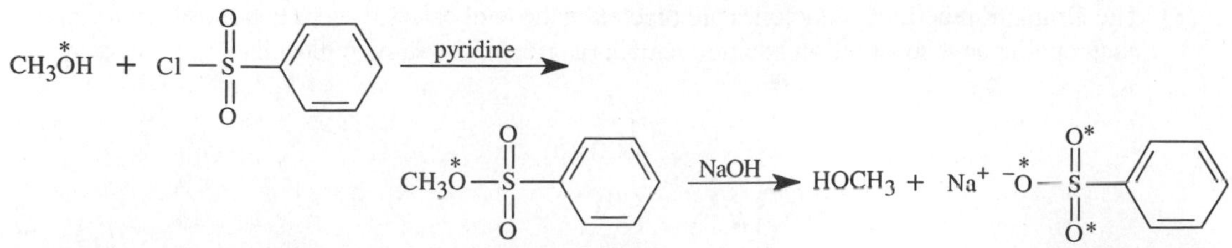

$$CH_3\overset{*}{O}H + Cl-\overset{\overset{O}{\|}}{\underset{\underset{O}{\|}}{S}}-C_6H_5 \xrightarrow{\text{pyridine}} CH_3\overset{*}{O}-\overset{\overset{O}{\|}}{\underset{\underset{O}{\|}}{S}}-C_6H_5 \xrightarrow{\text{NaOH}} HOCH_3 + Na^+\ {}^-\overset{*}{O}-\overset{\overset{O^*}{\|}}{\underset{\underset{O^*}{\|}}{S}}-C_6H_5$$

methyl benzenesulfonate
product *A*

sodium benzenesulfonate
product *B*

(b) Isotopically labeled methyl acetate (compound C), which is formed in the first reaction, is converted into unlabeled sodium acetate (product D) and labeled methanol by saponification, a nucleophilic acyl substitution reaction that breaks the acyl-oxygen bond of the ester.

$$H_3C-\overset{\overset{O}{\|}}{C}-Cl + H\overset{*}{O}CH_3 \xrightarrow{\text{pyridine}} H_3C-\overset{\overset{O}{\|}}{C}-\overset{*}{O}CH_3 \xrightarrow{\text{NaOH}} H_3C-\overset{\overset{O}{\|}}{C}-O^-\ Na^+ + H\overset{*}{O}CH_3$$

methyl acetate
(product *C*)

sodium acetate
(product *D*)

(c) Cleavage of sulfonate esters takes place by an S_N2 mechanism (if the alkyl group is methyl, primary, or unbranched secondary). This mechanism breaks the *alkyl-oxygen* bond. Recall (Sec. 10.3A) that sulfonate esters behave like "fat halides.") Cleavage of carboxylate esters takes place by nucleophilic attack at the carbonyl group and subsequent cleavage of the *acyl-oxygen bond,* that is, the bond between the carbonyl group and the carboxylate oxygen. The reasons for this difference are discussed in Study Guide Link 21.8 on p. 760 of this manual. (See Problem 21.17 on text p. 969 for a similar case.)

21.54 (a) The odd molecular mass indicates the presence of an odd number of nitrogens, and the hydroxamate test as well as the IR carbonyl absorption at 1733 cm^{-1} indicates the presence of an ester. The IR absorption at 2237 cm^{-1} indicates a nitrile. The triplet-quartet pattern in the NMR spectrum clearly indicates that compound *B* is an ethyl ester. Subtracting the masses of all the atoms accounted for leaves 14 mass units, the mass of a CH_2 group. Compound *A* is ethyl cyanoacetate.

$$N\equiv C-CH_2-\overset{\overset{O}{\|}}{C}-OCH_2CH_3$$

ethyl cyanoacetate (compound *A*)

(b) From its IR absorption, compound *B* is an ester. The formula indicates an unsaturation number of 1, which is completely accounted for by the ester carbonyl group. The chemical shift and splitting of the δ 4.0 resonance indicates the part-structure —$CH_2\mathbf{CH_2}O$—. The hint in the figure caption indicates that the δ 1.6 resonance is the other CH_2 of this part-structure; because this consists of six lines, then a more complete part-structure for this group is $CH_3CH_2CH_2O$—. The triplet at δ 0.95 is consistent with the methyl group of this part structure. This leaves an ethyl group unaccounted for; from its chemical shift, the CH_2 of the ethyl group is α to a carbonyl. Compound *B* is propyl propionate.

$$CH_3CH_2-\overset{\overset{O}{\|}}{C}-OCH_2CH_2CH_3$$

propyl propionate
(compound *B*)

(c) The odd molecular mass indicates the presence of nitrogen. The IR spectrum indicates the presence of both a nitrile and an alcohol. The broad, D_2O-exchangeable resonance at δ 3.4 in the NMR spectrum confirms the presence of an O—H group, and the pair of triplets indicates a —CH_2CH_2— group. Compound *C* is 3-hydroxypropanenitrile (β-hydroxypropionitrile).

$$HO-CH_2CH_2C\equiv N$$

3-hydroxypropanenitrile
(compound *C*)

(d) The IR absorption indicates the presence of an ester. The doubled peaks in the mass spectrum indicate the presence of a single bromine. Consequently, compound *D* contains the elements CO_2 (from the ester) and Br. The remaining mass could correspond to C_4H_9. Therefore, adopt $C_5H_9O_2Br$ as a provisional formula. In the NMR spectrum, the triplet and the doublet at low chemical shift cannot be splitting each other, because their integrations are not consistent with such a situation. That is, a mutually split triplet (which indicates two adjacent hydrogens) and doublet (which indicates one adjacent hydrogen) must have an integral ratio of 1:2, not 1:1 as observed. Evidently, the two absorptions at δ 4.23 and δ 4.37 are the resonances that are split by the two triplets at lower chemical shift. From their chemical shifts, some of these protons are probably α to the Br and some are probably α to the oxygen. If the ester is an ethyl ester, then the high-field triplet is the resonance of the methyl group, and, from its integration, the doublet is also due to a methyl group on a carbon adjacent to a CH. These deductions conspire to identify compound *D* as ethyl 2-bromopropanoate.

$$CH_3CH(Br)—C(=O)—OCH_2CH_3$$

ethyl 2-bromopropanoate
(compound *D*)

(e) The mass spectrum shows an odd mass, which indicates the presence of nitrogen. The IR spectrum could indicate the presence of a nitrile, although, if so, the C≡N absorption is about 50 cm^{-1} lower in frequency than the C≡N absorption of an ordinary nitrile. The IR spectrum also indicates a *trans*-alkene. The UV spectrum indicates extensive conjugation. This shift of the nitrile IR absorption is similar to that observed with conjugated carbonyl groups, and it suggests that the nitrile group is conjugated with the double bond. The NMR shows resonances for 7 protons. The large splitting in the resonances at δ 5.85 and δ 7.35 confirms the presence of the *trans*-alkene. Evidently, the apparent singlet at δ 7.4 corresponds to five protons, that is, to a monosubstituted benzene ring. Compound *E* is (*E*)-3-phenylpropenenitrile.

$$Ph(H)C{=}C(H)C{\equiv}N \quad (trans)$$

(*E*)-3-phenylpropenenitrile
(compound *E*)

(f) The unsaturation number is 5, and the NMR resonances in the δ 6.8–7.4 region indicate a para-substituted benzene ring. The broad resonance at δ 7.6 could be an amide N—H, and the IR absorption at 1659 cm^{-1} is confirmatory. The triplet-quartet pattern at δ 1.4 and δ 4.0 could indicate an ethyl ester, but we reject this idea because there is no ester carbonyl absorption. Hence, this quartet is evidently due to a *para*-CH_3CH_2O— substituent on the ring. The singlet at δ 2.1 is at a chemical shift consistent possibly with an *N*-methyl group or with protons α to a carbonyl group. Here are two possible structures:

$$CH_3CH_2O—C_6H_4—C(=O)—NHCH_3 \qquad CH_3CH_2O—C_6H_4—NH—C(=O)—CH_3$$

4-ethoxy-*N*-methylbenzamide

***N*-4-ethoxyphenylacetamide**
(*p*-ethoxyacetanilide)
(compound *F*)

The absence of splitting in the δ 2.1 resonance rules out the first structure, for which we would expect a doublet for the *N*-methyl group. Hence, the second structure is the correct one.

(g) The IR absorptions indicate the presence of a primary amide; the odd mass is consistent with the presence of nitrogen. The weak CMR resonances are probably due to carbons that do not bear hydrogens. The δ 180.5 resonance is the amide carbonyl group; the resonance at δ 38.0 is a quaternary carbon, which, from its chemical shift, could be α to the carbonyl carbon. Subtracting the mass of the carboxamide group from the mass of the molecule gives the remaining mass as 57 units, which is the mass of a C_4H_9 group.

Only a *tert*-butyl group has the symmetry required to fit the CMR data. Compound *G* is 2,2-dimethylpropanamide (pivalamide).

$$(CH_3)_3C—\overset{O}{\overset{\|}{C}}—NH_2$$

2,2-dimethylpropanamide (pivalamide)
(compound *G*)

21.55 (a)

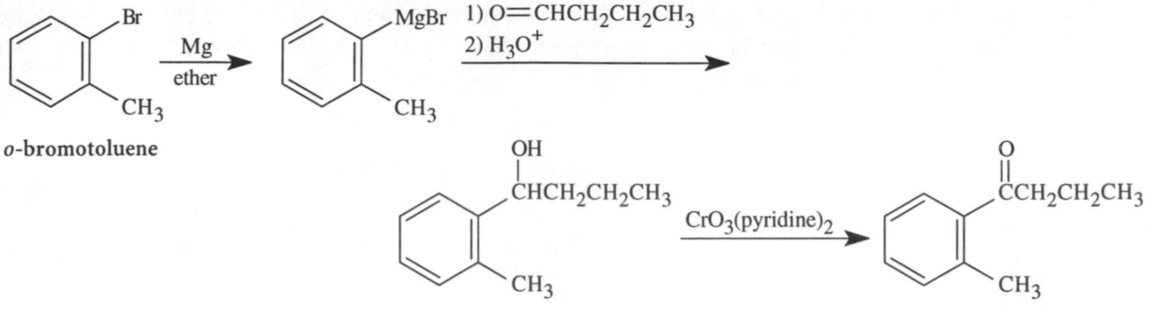

o-bromotoluene $\xrightarrow[\text{ether}]{\text{Mg}}$ *o*-methylphenylmagnesium bromide (MgBr, CH_3) $\xrightarrow{\text{1) } O{=}CHCH_2CH_2CH_3 \text{ 2) } H_3O^+}$ 1-(*o*-methylphenyl)-1-butanol ($\overset{OH}{|}CHCH_2CH_2CH_3$, CH_3) $\xrightarrow{CrO_3(pyridine)_2}$ ($\overset{O}{\|}CCH_2CH_2CH_3$, CH_3)

o-methylbutyrophenone

(b)

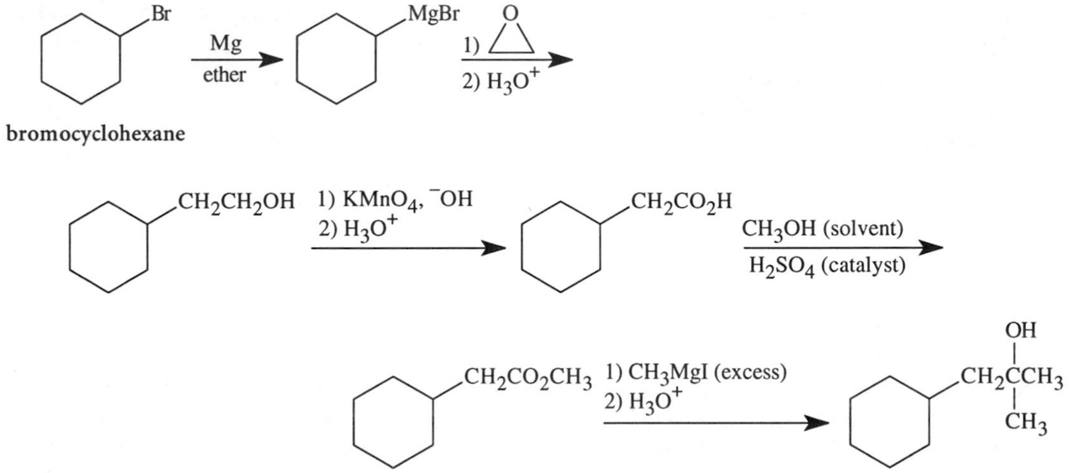

bromocyclohexane $\xrightarrow[\text{ether}]{\text{Mg}}$ cyclohexyl–MgBr $\xrightarrow{\text{1) ethylene oxide 2) } H_3O^+}$ cyclohexyl–CH_2CH_2OH $\xrightarrow{\text{1) } KMnO_4,\ ^-OH \text{ 2) } H_3O^+}$ cyclohexyl–CH_2CO_2H $\xrightarrow[H_2SO_4 \text{ (catalyst)}]{CH_3OH \text{ (solvent)}}$ cyclohexyl–$CH_2CO_2CH_3$ $\xrightarrow{\text{1) } CH_3MgI \text{ (excess) 2) } H_3O^+}$ cyclohexyl–$CH_2\overset{OH}{\overset{|}{C}}CH_3$ (CH_3)

1-cyclohexyl-2-methyl-2-propanol

(c)

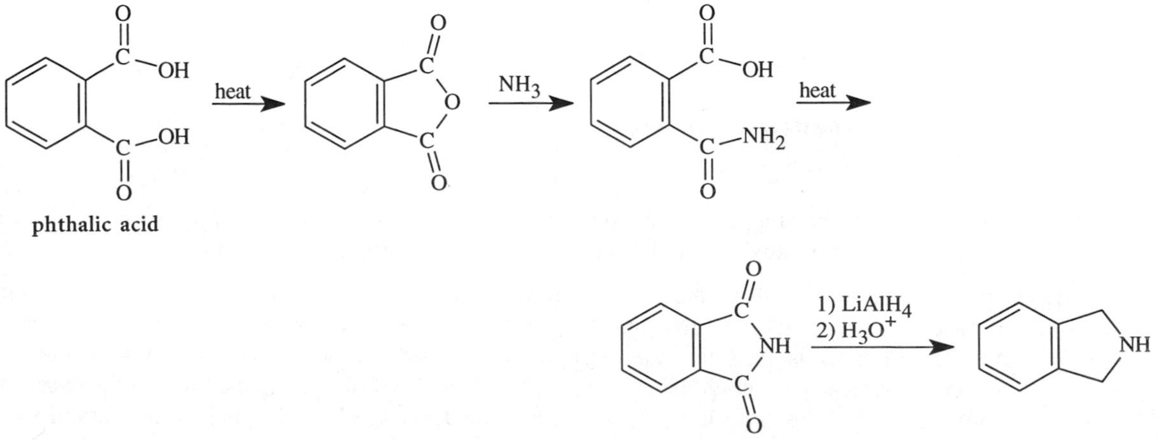

phthalic acid (C(=O)–OH, C(=O)–OH) $\xrightarrow{\text{heat}}$ phthalic anhydride $\xrightarrow{NH_3}$ (C(=O)–OH, C(=O)–NH_2) $\xrightarrow{\text{heat}}$ phthalimide (C=O, NH, C=O) $\xrightarrow{\text{1) } LiAlH_4 \text{ 2) } H_3O^+}$ isoindoline (NH)

(d)

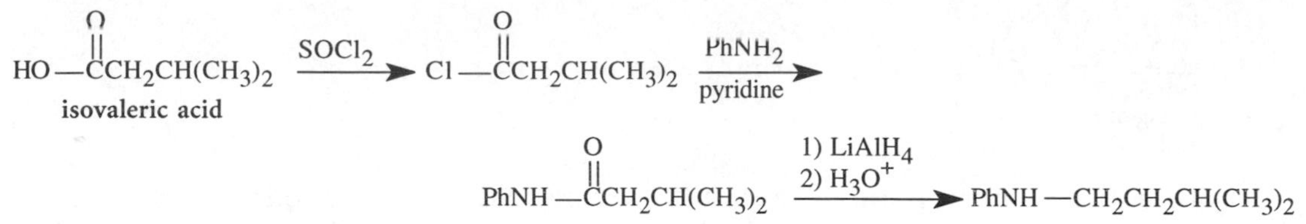

$$\mathrm{HO{-}\overset{O}{\overset{\|}{C}}CH_2CH(CH_3)_2} \xrightarrow{SOCl_2} \mathrm{Cl{-}\overset{O}{\overset{\|}{C}}CH_2CH(CH_3)_2} \xrightarrow[\text{pyridine}]{PhNH_2}$$

isovaleric acid

$$\mathrm{PhNH{-}\overset{O}{\overset{\|}{C}}CH_2CH(CH_3)_2} \xrightarrow{\text{1) LiAlH}_4 \;\; \text{2) } H_3O^+} \mathrm{PhNH{-}CH_2CH_2CH(CH_3)_2}$$

(e)

1-cyclohexenyl–CH=O $\xrightarrow[\text{methanol}]{NaBH_4}$ 1-cyclohexenyl–CH_2OH $\xrightarrow[\text{pyridine}]{\text{tosyl chloride,}}$ 1-cyclohexenyl–CH_2OTs $\xrightarrow[\text{DMSO}]{\text{NaCN,}}$

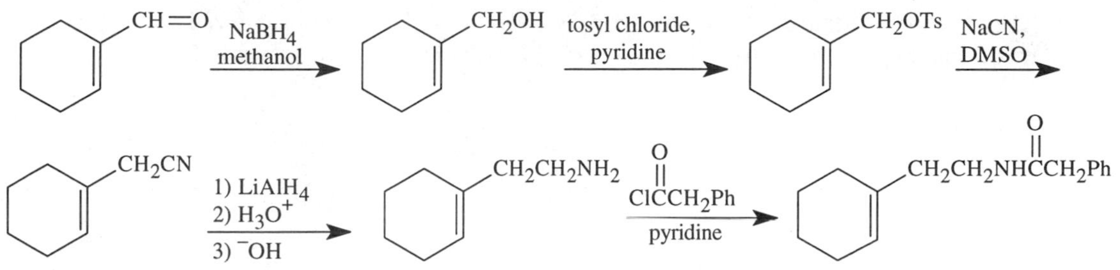

1-cyclohexenyl–CH_2CN $\xrightarrow{\text{1) LiAlH}_4\text{, 2) }H_3O^+\text{, 3) }^-OH}$ 1-cyclohexenyl–$CH_2CH_2NH_2$ $\xrightarrow[\text{pyridine}]{ClCCH_2Ph\ (C{=}O)}$ 1-cyclohexenyl–$CH_2CH_2NH\overset{O}{\overset{\|}{C}}CH_2Ph$

(f)

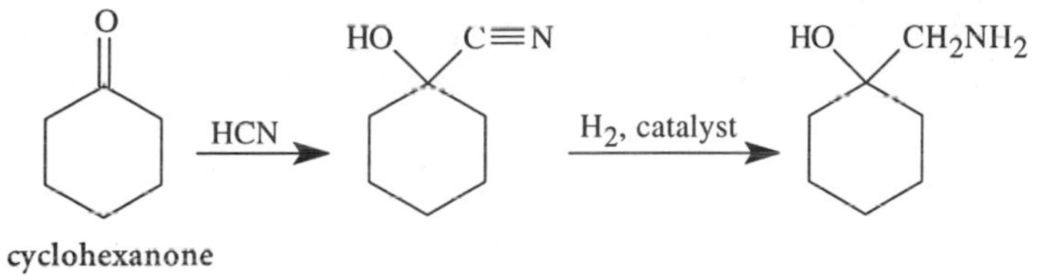

cyclohexanone $\xrightarrow{HCN}$ 1-(HO)(C≡N)cyclohexane $\xrightarrow{H_2\text{, catalyst}}$ 1-(HO)(CH_2NH_2)cyclohexane

(g)

benzoic acid ($C_6H_5CO_2H$) $\xrightarrow{HNO_3,\ H_2SO_4}$ m-$NO_2C_6H_4CO_2H$ $\xrightarrow{PCl_5}$ m-$NO_2C_6H_4\overset{O}{\overset{\|}{C}}Cl$

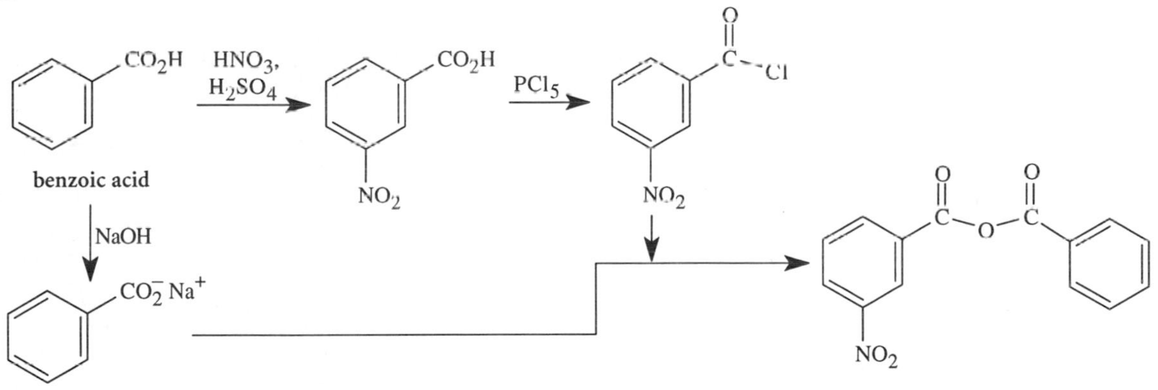

benzoic acid $\xrightarrow{NaOH}$ $C_6H_5CO_2^-\ Na^+$

m-$NO_2C_6H_4COCl$ + $C_6H_5CO_2^-\ Na^+$ ⟶ m-$NO_2C_6H_4\overset{O}{\overset{\|}{C}}{-}O{-}\overset{O}{\overset{\|}{C}}C_6H_5$

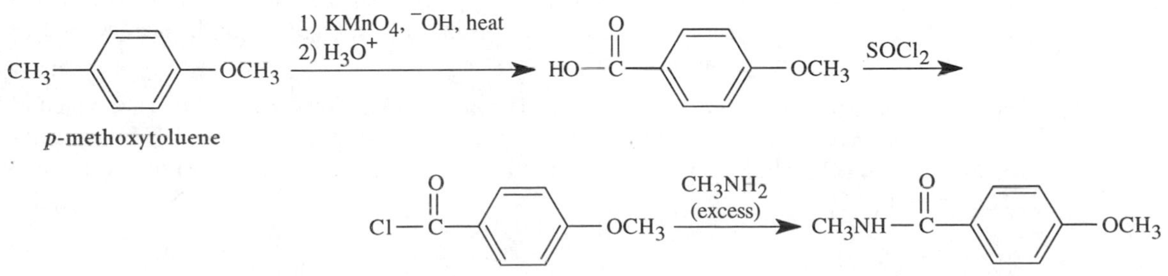

(h)

CH_3–C_6H_4–OCH_3 (*p*-methoxytoluene) $\xrightarrow{\text{1) KMnO}_4,\ ^-OH\text{, heat; 2) }H_3O^+}$ HO–$\overset{O}{\overset{\|}{C}}$–$C_6H_4$–$OCH_3$ $\xrightarrow{SOCl_2}$

Cl–$\overset{O}{\overset{\|}{C}}$–$C_6H_4$–$OCH_3$ $\xrightarrow[\text{(excess)}]{CH_3NH_2}$ CH_3NH–$\overset{O}{\overset{\|}{C}}$–$C_6H_4$–$OCH_3$

(i)

$$\underset{\substack{\textbf{phosgene}\\ \text{(excess)}}}{Cl{-}\overset{O}{\overset{\|}{C}}{-}Cl} \xrightarrow{CH_3OH} Cl{-}\overset{O}{\overset{\|}{C}}{-}OCH_3 \xrightarrow{PhOH} PhO{-}\overset{O}{\overset{\|}{C}}{-}OCH_3$$

An excess of phosgene is used in the first step to ensure that only one chlorine is displaced. It is probably best to use methanol as the first nucleophile because phenyl esters are fairly reactive and could be transesterified by methanol if the order were reversed.

21.56 (a) The reversible hydration of carbon dioxide to carbonic acid is an acid-catalyzed addition of water that is mechanistically identical to the hydration of aldehydes and ketones shown in Eq. 19.18a–b on text p. 857.

$$O{=}C{=}O \quad H{-}\overset{+}{O}H_2 \rightleftharpoons O{=}C{=}\overset{+}{O}{-}H \;(OH_2) \rightleftharpoons O{=}C(\overset{+}{O}H{-}H){-}O{-}H \;\; OH_2 \rightleftharpoons O{=}C(OH){-}OH \; + \; H{-}\overset{+}{O}H_2$$

(b) The first step in this transformation is formation of a Grignard reagent. (See Eqs. 8.21a–b on text p. 336.) Because of its anionic character, the Grignard reagent undergoes an elimination to give the conjugate-base carboxylate ion of the product, which is protonated when acid is added. The Grignard reagent is written as a carbanion in the following mechanism to emphasize its anionic character.

(c) Protonation of the carbonyl oxygen produces a good leaving group which is displaced by bromide ion. The carboxy group is then esterified by the usual mechanism (Eqs. 20.18a–c, text pp. 914–5).

An alternative reasonable mechanism involves transesterification of the lactone by ethanol to give a hydroxy ester, followed by conversion of the alcohol to a bromide by an acid-catalyzed S_N2 reaction. (See Eq. 10.10a–b, text p. 412.) However, it is likely that the mechanism shown above is the correct one, because the equilibrium between lactone and hydroxy ester is likely to strongly favor the lactone, and because a protonated carboxy group is an excellent leaving group.

(d) This reaction is a variation of oxymercuration. The mercurinium ion formed at one double bond is attacked by the π electrons of the second double bond. The resulting carbocation is attacked by the carboxy group. The mechanism below begins with the mercurinium ion. (For the mechanism of its formation, see Eq. 5.12a on text p. 166.)

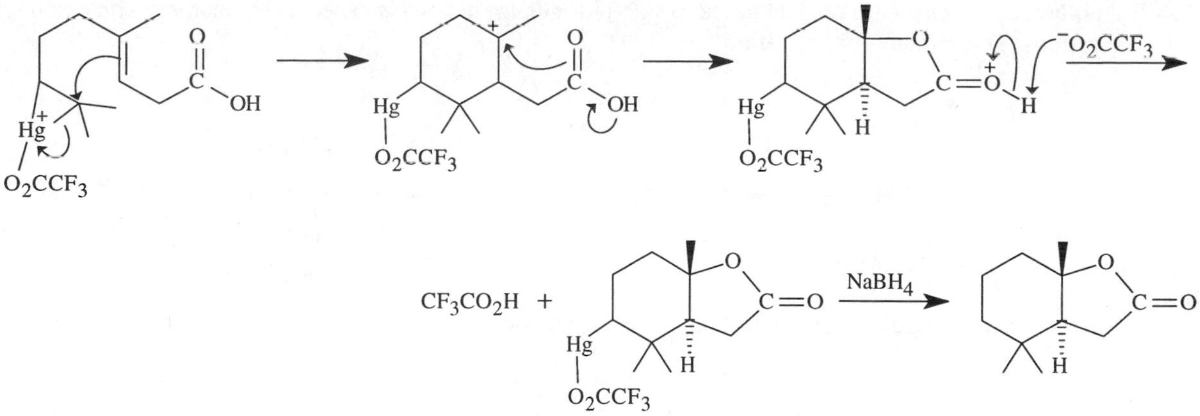

CF_3CO_2H +

Hg

O_2CCF_3

compound *A*

$NaBH_4$

(e) Protonation of the ring double bond gives a tertiary carbocation that is attacked by the carbonyl oxygen of the ester to give the lactone.

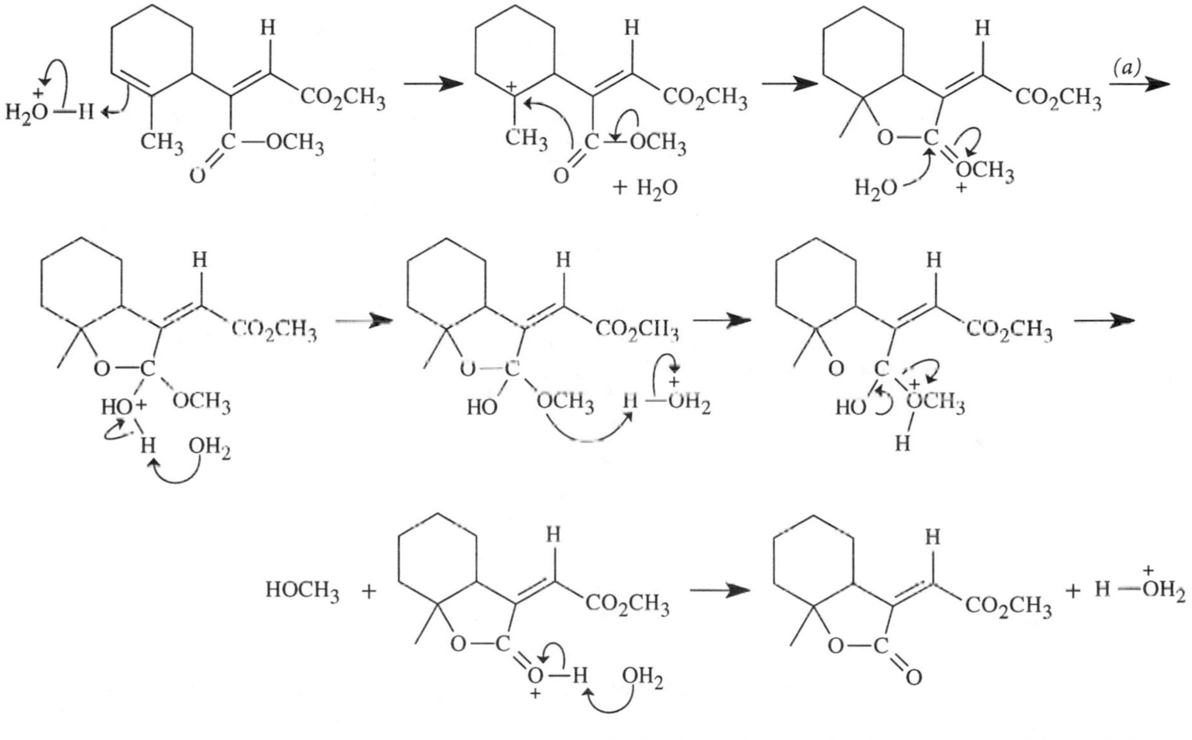

$HOCH_3$ +

+ $H—\overset{+}{O}H_2$

As an alternative to step *(a)* and the subsequent steps in the foregoing mechanism, water could attack the methyl group:

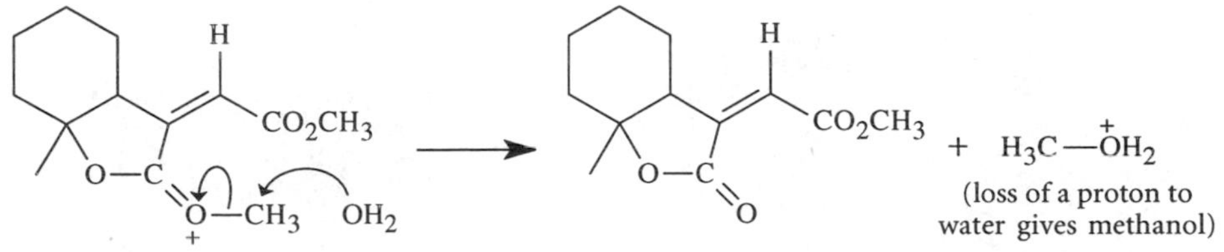

+ $H_3C—\overset{+}{O}H_2$

(loss of a proton to water gives methanol)

How might carrying out the reaction in ^{18}O-water enable one to distinguish between these two mechanisms?

21.57 Addition of the Grignard reagent to the carbon-nitrogen triple bond is expected by analogy with carbonyl-group reactions. The product *A* is an imine.

$$\text{Ph—C}\equiv\text{N:} + \text{Ph:}^{-}\ {}^{+}\text{MgBr} \longrightarrow \text{Ph—C(Ph)=}\ddot{\text{N}}\text{:}^{-}\ {}^{+}\text{MgBr} \xrightarrow{\text{H—OH}} \underset{\text{compound } A}{\text{Ph—C(Ph)=}\ddot{\text{N}}\text{H}} + \text{Br}^{-}\ \text{Mg}^{2+}\ {}^{-}\text{OH}$$

This imine hydrolyzes in acidic solution to benzophenone:

$$\underset{\text{compound } A}{\text{Ph—C(Ph)=}\ddot{\text{N}}\text{H}} \xrightarrow{\text{H—}\overset{+}{\text{O}}\text{H}_2} \text{Ph—C(Ph)=}\overset{+}{\text{N}}\text{H}_2 + \text{H}_2\ddot{\text{O}} \longrightarrow \text{Ph—C(Ph)(}\overset{+}{\text{O}}\text{H—H)—NH}_2 \xrightarrow{\text{proton transfers}}$$

$$\text{Ph—C(Ph)(H}\ddot{\text{O}}\text{:)—}\overset{+}{\text{N}}\text{H}_3 \longrightarrow \left[\text{Ph—C}^{+}\text{(Ph)—}\ddot{\text{O}}\text{—H} \longleftrightarrow \text{Ph—C(Ph)=}\overset{+}{\text{O}}\text{—H}\right] + \text{:NH}_3 \longrightarrow \text{Ph—C(=O)—Ph} + {}^{+}\text{NH}_4$$

21.58 Compound *A* is an *O*-substituted oxime, which is formed by the mechanism summarized in Eqs. 19.56a–b on text p. 875. Compound *A* undergoes a base-promoted *β*-elimination in base to give the nitrile. (In the following mechanism, the 2,4-dinitrophenyl group is abbreviated Ar.)

$$\underset{\text{compound } A}{\text{Ph—C(H)=N—OAr}} + {}^{-}\text{OH} \longrightarrow \text{Ph—C}\equiv\text{N} + \text{H—OH} + {}^{-}\text{OAr}$$

21.59 Compound *A* is an oxime (see Table 19.3 on text p. 876). It is formed by the mechanism for imine formation summarized in Eqs. 19.56a–b on text p. 875. Protonation of the —OH group and loss of water (*a*) gives an electron-deficient nitrogen, a very unstable species. Subsequent rearrangement (*b*) gives a carbocation that is somewhat resonance stabilized. (We say "somewhat" because incorporation of a triple bond within a seven-membered ring introduces considerable strain.) This cation is attacked by water, and, after several proton-transfer steps, *e*-caprolactam is formed.

$$\underset{\substack{\text{cyclohexanone}\\ \text{oxime}\\ (\text{compound } A)}}{\text{:N—OH}} + \text{H—OSO}_3\text{H} \longrightarrow \text{:N—}\overset{+}{\text{O}}\text{H}_2 + {}^{-}\text{OSO}_3\text{H} \xrightarrow{(a)} \text{:N}^{+} + \text{OH}_2 \xrightarrow{(b)}$$

(solution continues)

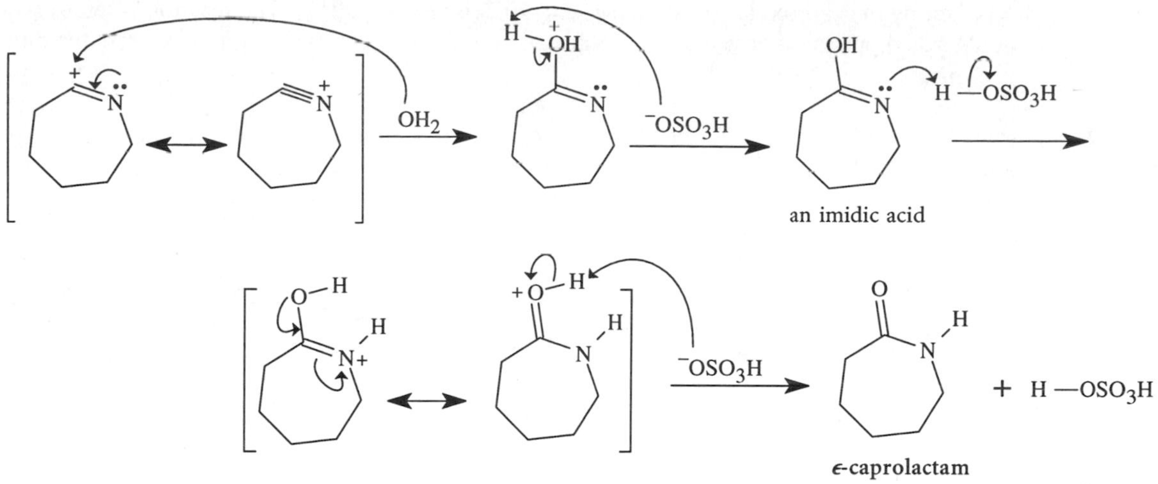

OH_2 $^{-}OSO_3H$ OH H—OSO_3H

an imidic acid

$^{-}OSO_3H$ + H—OSO_3H

ε-caprolactam

Steps (*a*) and (*b*) occur in a single concerted step because a species containing an electron-deficient nitrogen is very unstable. However, two steps are shown above to demonstrate the driving force for the rearrangement.

21.60 This is another example of *neighboring-group participation;* see Sec. 11.7. In this case, the carboxylate ion displaces bromide intramolecularly to give an α-lactone (a three-membered lactone). Because of its strain, this compound reacts rapidly with hydroxide ion in a second displacement that re-inverts the configuration at the asymmetric carbon.

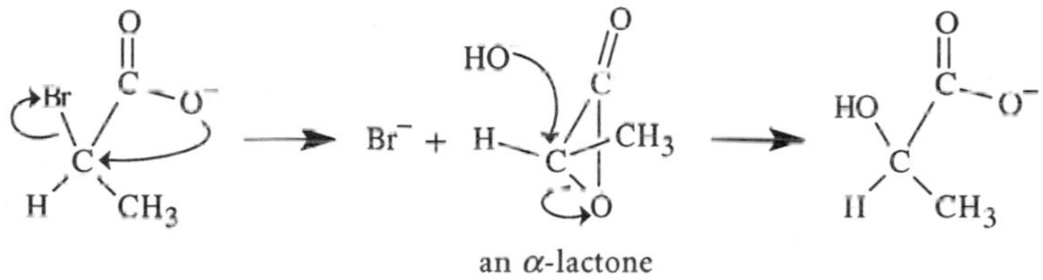

Br^- +

an α-lactone

21.61 (a) Because the carbonyl group has trigonal-planar geometry, it must be turned so that its plane is perpendicular to the plane of the benzene ring in order to avoid significant van der Waals repulsions with the ortho methyl groups. This occurs even though conjugation is lost.

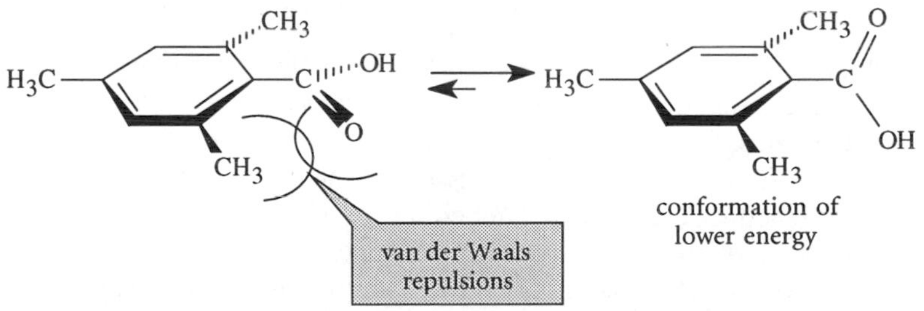

van der Waals repulsions

conformation of lower energy

(b) Remember that nucleophiles attack a carbonyl carbon above or below the plane of the carbonyl group. (See Fig. 19.8 on text p. 856.) In the "perpendicular" conformation shown on the right, the path of a nucleophile to the carbonyl carbon is blocked on both sides by the ortho methyl groups. Because ester hydrolysis occurs by attack of a water molecule at the carbonyl carbon, this mechanism, and thus the hydrolysis reaction, cannot occur.

(c) For the same reason that ester *hydrolysis* cannot occur, ester *formation* also cannot occur by any mechanism that involves attack of a nucleophile at the carbonyl carbon. Because acid-catalyzed esterification involves attack of an alcohol molecule at the carbonyl carbon, this reaction does not take place. However, esterification with diazomethane can occur because the protonated diazomethane molecule is attacked by the carboxylate oxygen (see Eq. 20.21b on text p. 917). The reaction is less susceptible to van der Waals repulsions because it takes place at a site that is much farther removed from the ortho methyl groups.

21.62 (a) Phthalic anhydride can be formed by attack of a carboxy group at the carbonyl carbon of the protonated amide. (Note that the amide carbonyl oxygen is about 10^5 times as basic as the carbonyl oxygen of the carboxylic acid; see Eqs. 21.4a–b on text p. 949.)

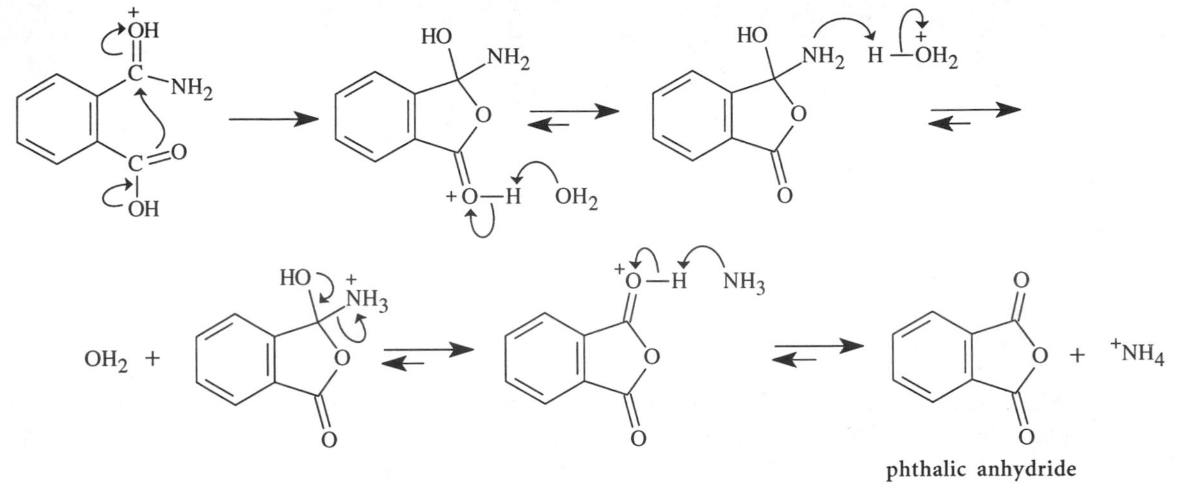

OH_2 + ... + $^{+}NH_4$

phthalic anhydride

(b) The phthalic anhydride is labeled with ^{13}C on only one of its two carbonyl carbons. However, the two carbonyl groups, except for isotopic differences, are chemically indistinguishable and are attacked with essentially equal rates by isotopic water:

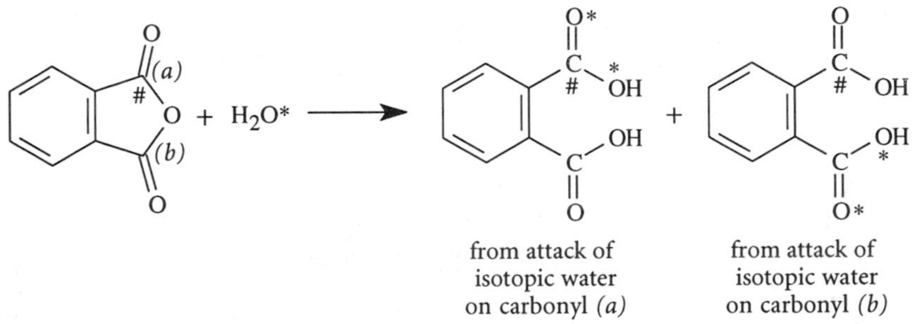

+ H_2O^* ⟶ ... + ...

from attack of isotopic water on carbonyl *(a)*

from attack of isotopic water on carbonyl *(b)*

(c) Formation of the anhydride intermediate must be faster than hydrolysis by the usual mechanism; otherwise it would not occur. Anhydride formation is an *intramolecular* reaction, whereas the ordinary hydrolysis mechanism is *intermolecular.* Intramolecular reactions that involve the formation of five-membered rings are usually much faster than their intermolecular counterparts. (The reasons for the advantage of intramolecular reactions are discussed in Sec. 11.7.) Although the anhydride is formed very rapidly, it is also hydrolyzed rapidly because anhydrides are generally very reactive toward hydrolysis. Thus, the overall hydrolysis is accelerated because an anhydride intermediate is formed rapidly and then destroyed rapidly.

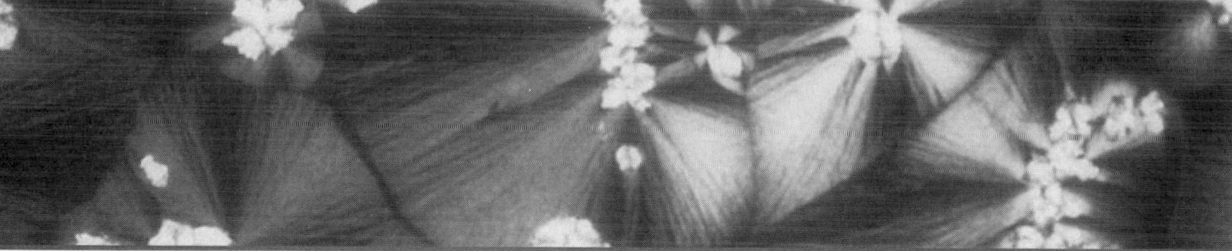

22

Chemistry of Enolate Ions, Enols, and α,β-Unsaturated Carbonyl Compounds

TERMS

CONCEPTS

I. ACIDITY OF CARBONYL COMPOUNDS; ENOLATE IONS

A. Formation of Enolate Anions

1. The α-hydrogens of many carbonyl compounds, as well as those of nitriles, are weakly acidic—much more acidic than other types of hydrogens bound to carbon.
2. When an α-hydrogen is removed, a conjugate-base anion is formed at the α-carbon; the conjugate-base anion of a carbonyl compound formed by removal of an α-hydrogen is called an enolate ion. (B: = base.)

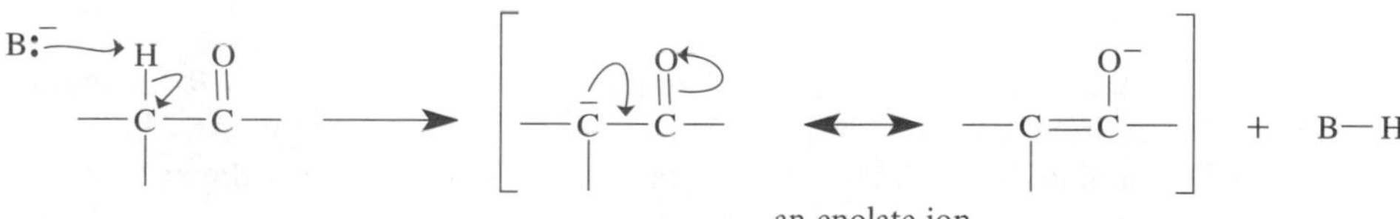

an enolate ion

a. Enolate ions are resonance-stabilized.
b. The anionic carbon of an enolate ion is sp^2-hybridized.
   i. This hybridization allows the electron pair of an enolate anion to occupy a $2p$ orbital, which overlaps with the π orbital of a carbonyl group.
   ii. This additional overlap provides additional bonding and hence additional stabilization.
   iii. The negative charge in an enolate ion is delocalized onto oxygen, an electronegative atom.
   iv. The polar effect of the carbonyl group also stabilizes the enolate anion.

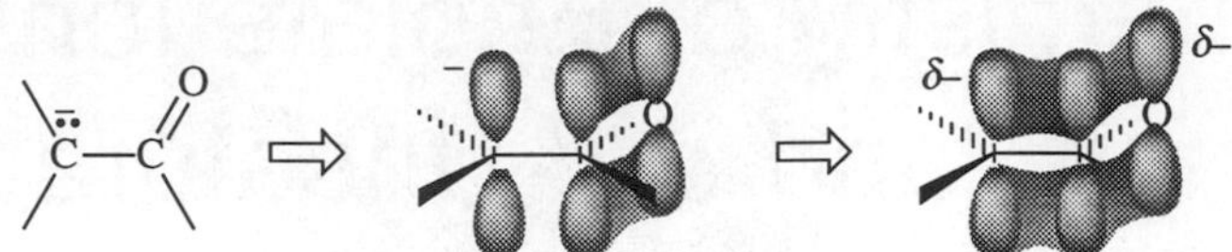

   v. The α-carbon and the carbonyl oxygen are the major sites of electron density.
c. The α-hydrogen pK_a of an aldehyde or ketone is typically in the 16–20 range.
d. Esters are less acidic than aldehydes or ketones; the α-hydrogen pK_a of an ester is about 25.
   i. Esters are stabilized by overlap of the unshared electrons of the carboxylate oxygen with the carbonyl group; ketones lack this type of stabilization.
   ii. This stabilization of esters means that more energy is required to form their conjugate-base enolate ions. (See Fig. 22.2, text p. 1000.)
   iii. The additional resonance stabilization of esters is more important than the polar effect of the carboxylate oxygen, which, in the absence of resonance, would increase the acidity of esters relative to that of aldehydes and ketones.
e. Enolate ions are bases and they can act as nucleophiles; hence, the α-carbon of a carbonyl compound is a site of nucleophilic reactivity.

3. Amide N—H hydrogens (p$K_a \approx 16$) are conceptually α-hydrogens and are the most acidic hydrogens in primary and secondary amides; carboxylic acid O—H hydrogens (p$K_a \approx$ 4–5) are also α-hydrogens.
4. The acidity order of carbonyl compounds corresponds to the reactive electronegativities of the atoms to which the acidic hydrogens are bound:

acidities: carboxylic acids > amides >aldehydes, ketones, esters
electronegativities: oxygen > nitrogen > carbon

B. Introduction to Reactions of Enolate Ions

1. Enolate ions are key reactive intermediates in many important reactions of carbonyl compounds:
   a. They are Brønsted bases and react with Brønsted acids.
   b. They are Lewis bases also; consequently, they react as nucleophiles.
      i. Enolate ions attack the carbon of carbonyl groups. This attack is the first step of a variety of carbonyl addition reactions and nucleophilic acyl substitution reactions of enolate ions.
      ii. Enolate ions, like other nucleophiles, also react with alkyl halides and sulfonate esters.
2. The formation of enolate ions and their reactions with Brønsted acids have two simple but important consequences:
   a. The α-hydrogens, and only the α-hydrogens, of aldehydes and ketones can be exchanged for deuterium by treating the carbonyl compound with acid or base in D_2O.

$$-CH_2-\overset{O}{\overset{\|}{C}}-CH< \xrightarrow[D_2O]{D_3O^+ \text{ or } ^-OD} -CD_2-\overset{O}{\overset{\|}{C}}-CD<$$

   b. If an optically active aldehyde or ketone owes its chirality solely to an asymmetric α-carbon, and if this carbon bears a hydrogen, the compound will be racemized by base.
      i. The enolate ion, which forms in base, is achiral because of the sp^2 hybridization at its anionic carbon.

ii. The ionic α-carbon and its attached groups lie in one plane.
iii. The anion can be reprotonated at either face to give either enantiomer with equal probability.

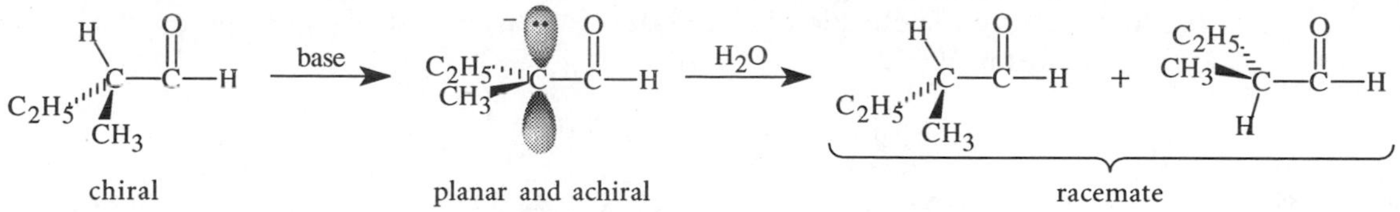

chiral — base → planar and achiral — H_2O → racemate

3. α-Hydrogen exchange and racemization reactions of aldehydes and ketones occur much more readily than those of esters, because aldehydes and ketones are more acidic, and therefore form enolate ions more rapidly and under milder conditions.

C. Enolization of Carbonyl Compounds

1. Carbonyl compounds with α-hydrogens are in equilibrium with vinylic alcohol isomers called enols.
   a. Enols and their parent carbonyl compounds are constitutional isomers that are related by the movement of a proton and a π bond.

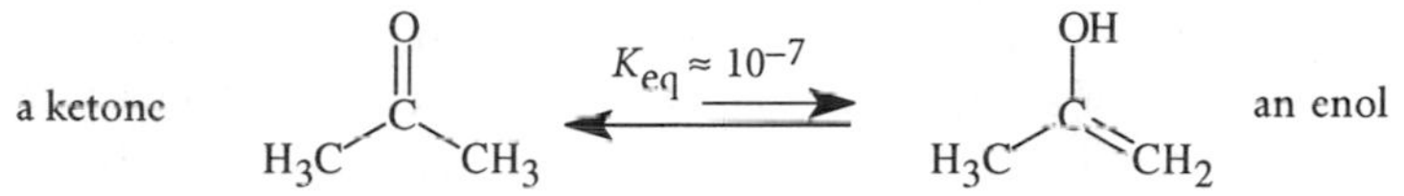

a ketone $\quad K_{eq} \approx 10^{-7} \quad$ an enol

   b. Most carbonyl compounds are considerably more stable than their corresponding enols, primarily because the C=O bond of a carbonyl group is a stronger bond than the C=C bond of an enol.
   c. Phenol is conceptually an enol, but the enol form of phenol is more stable than its keto tautomers because phenol is aromatic.
   d. The enols of β-dicarbonyl compounds (compounds in which two carbonyl groups are separated by one carbon) are also relatively stable.

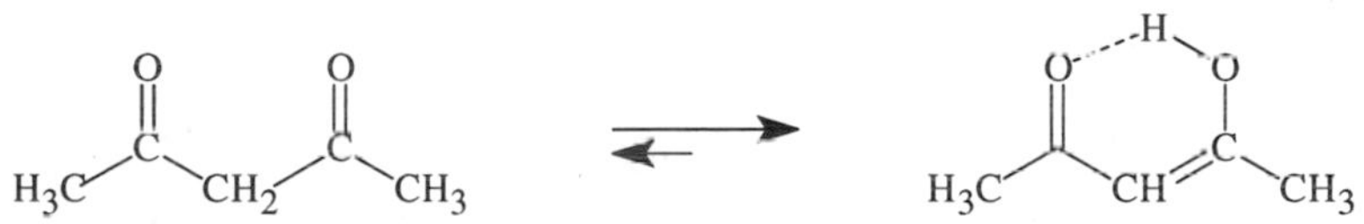

      i. These enol forms are conjugated, but their parent carbonyl compounds are not; the resonance stabilization (π electron overlap) associated with conjugation provides additional bonding that stabilizes the enol.
      ii. The intramolecular hydrogen bond present in such enols provides increased bonding and increased stabilization.
2. The formation of enols and the reverse reaction, conversion of enols into carbonyl compounds, are catalyzed by both acids and bases.
   a. Base-catalyzed enolization involves the intermediacy of the enolate ion, and is thus a consequence of the acidity of the α-hydrogen. (See Eq. 22.17a, text p. 1006.)

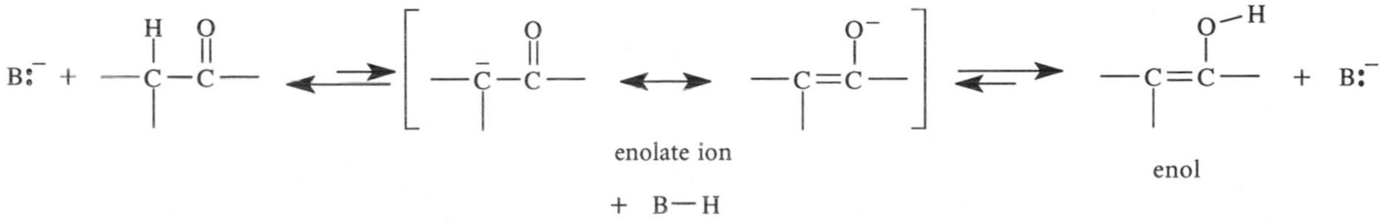

enolate ion + B—H $\quad$ enol

i. Protonation of the enolate anion on the α-carbon gives back the carbonyl compound; protonation on oxygen gives the enol.
ii. The enolate ion is the conjugate base of not only the carbonyl compound, but also the enol.

b. Acid-catalyzed enolization involves the conjugate acid of the carbonyl compound. (See Eq. 22.17b, text p. 1007.)

$$H_3O^+ + -\overset{H}{\underset{|}{C}}-\overset{O}{\overset{\|}{C}}- \rightleftarrows \left[-\overset{H}{\underset{|}{C}}-\overset{\overset{+}{O}H}{\overset{\|}{C}}- \longleftrightarrow -\overset{H}{\underset{|}{C}}-\overset{OH}{\underset{+}{C}}- \right] \rightleftarrows -\underset{|}{C}=\overset{O-H}{C}- + H_3O^+$$

conjugate acid of the carbonyl compound + H_2O — enol

i. Loss of the proton from the α-carbon of a protonated ketone gives the enol.
ii. An enol and its carbonyl isomer have the same conjugate acid.

3. Exchange of α-hydrogens for deuterium as well as racemization at the α-carbon are catalyzed not only by bases but also by acids. Acid-catalyzed exchange can be explained by the intermediacy of enols.
   a. Formation of a carbonyl compound from an enol introduces hydrogen from solvent at the α-carbon; this fact accounts for the observed isotope exchange.
   b. The α-carbon of an enol, like that of an enolate ion, is not an asymmetric carbon.
   c. The absence of chirality in the enol accounts for the racemization observed when an aldehyde or ketone with a hydrogen at an asymmetric α-carbon is treated with acid.

II. APPROACHES TO ORGANIC SYNTHESIS

(See also discussion of reactions in Section R.)

A. Synthetic Short-Hand Notation

1. Chemists often use a compact abbreviation for several commonly occurring organic groups.
2. These abbreviations not only save space, but also make the structures of large molecules less cluttered and more easily read.

methyl	H_3C-	Me	ethyl	CH_3CH_2-	Et			
propyl	$CH_3CH_2CH_2-$	Pr	isopropyl	$(CH_3)_2CH-$	*i*-Pr			
butyl	$CH_3(CH_2)_3-$	Bu	isobutyl	$(CH_3)_2CHCH_2-$	*i*-Bu	*tert*-butyl	$(CH_3)_3C-$	*t*-Bu
acetyl	$H_3C-\overset{O}{\overset{\|}{C}}-$	Ac	acetate or acetoxy	$H_3C-\overset{O}{\overset{\|}{C}}-O-$	AcO			

B. Synthesis with the Aldol Condensation

1. The aldol condensation:
   a. is an addition reaction involving two aldehyde or ketone molecules (aldol addition) followed by a dehydration of the resulting β-hydroxy aldehyde
   b. can be carried out in acidic or basic solution
   c. can be applied to the synthesis of a wide variety of α,β-unsaturated aldehydes and ketones
   d. is a method for the formation of carbon-carbon bonds.

$$R-CH_2-\overset{O}{\overset{\|}{C}}-R' \xrightarrow[\text{aldol addition}]{H_3O^+ \text{ or } ^-OH} R-CH_2-\underset{R'}{\overset{OH}{C}}-\underset{R}{CH}-\overset{O}{\overset{\|}{C}}-R' \xrightarrow[\text{dehydration}]{H_3O^+ \text{ or } ^-OH} R-CH_2-\underset{R'}{C}=\underset{R}{C}H-\overset{O}{\overset{\|}{C}}-R'$$

2. To prepare a particular α,β-unsaturated aldehyde or ketone by the aldol condensation, ask two questions:
   a. What starting materials are required in the aldol condensation?
      i. The starting materials for an aldol condensation can be determined by mentally "splitting" the α,β-unsaturated carbonyl compound at the double bond.
      ii. Replace the double bond on the carbonyl side by two hydrogens and on the other side by a carbonyl oxygen (=O).

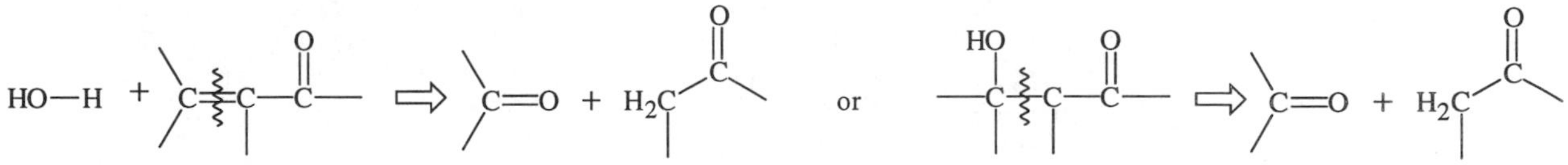

HO—H + C=C—C(=O)— ⇨ C=O + H_2C—C(=O)— or —C(OH)—C—C(=O)— ⇨ C=O + H_2C—C(=O)—

   b. With these starting materials, is the aldol condensation of these compounds a feasible one?

C. Synthesis with the Claisen Condensation

1. The Claisen condensation is a nucleophilic acyl substitution reaction.
   a. The nucleophile is an enolate ion derived from an ester.
   b. One full equivalent of base is required.
   c. The ester starting material must have at least two α-hydrogens.

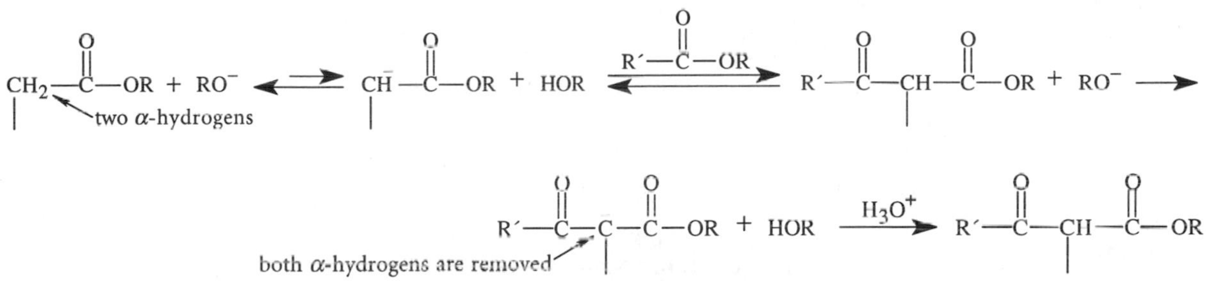

CH_2—C(=O)—OR + RO^- ⇌ $\bar{C}H$—C(=O)—OR + HOR ⇌ (R′—C(=O)—OR) R′—C(=O)—CH—C(=O)—OR + RO^- ⟶
(two α-hydrogens)

R′—C(=O)—$\bar{C}$—C(=O)—OR + HOR $\xrightarrow{H_3O^+}$ R′—C(=O)—CH—C(=O)—OR
(both α-hydrogens are removed)

2. The Claisen condensation can be used for the synthesis of β- dicarbonyl compounds: β-keto esters, β-diketones, and the like.
   a. To determine the starting materials for a Claisen condensation, mentally reverse the Claisen condensation in two different ways by adding the elements of ethanol (or another alcohol) across either of the carbon-carbon bonds between the carbonyl groups.

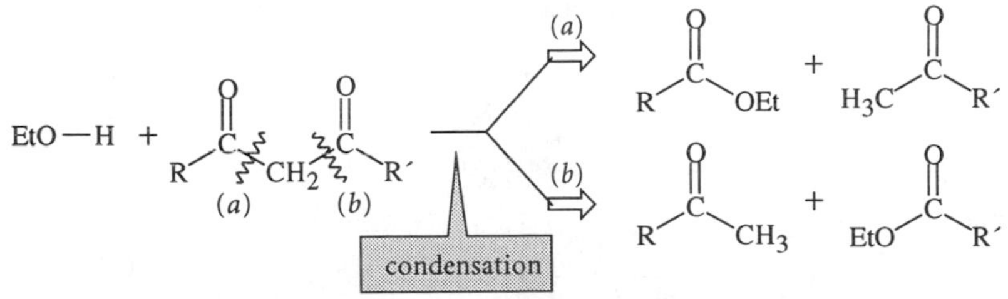

EtO—H + R—C(=O)—CH_2—C(=O)—R′ (bonds (a) and (b)) [condensation]
(a) ⇨ R—C(=O)—OEt + H_3C—C(=O)—R′
(b) ⇨ R—C(=O)—CH_3 + EtO—C(=O)—R′

Remember that the meaning of the open arrow (⇨) in understanding chemical syntheses is, "Implies as starting materials." In the notation used in the foregoing equation, and in similar equations to follow, the process in the box converts the starting materials, shown to the *right* of the arrow, to the products shown to the *left*.

   b. Determine whether the Claisen condensation of the required starting materials will give mostly the desired product and not a complex mixture.

D. Synthesis with the Acetoacetic Ester Reaction

1. The acetoacetic ester synthesis involves alkylation of a β-keto ester. In many cases the alkylated derivative is saponified and decarboxylated to give a branched ketone.
2. Given a target ketone, mentally reverse the acetoacetic ester synthesis.
   a. Replace an α-hydrogen of the target ketone with a carboethoxy group to unveil the β-keto ester required for the synthesis.
   b. The β-keto ester itself is then analyzed in terms of the alkyl halides required for the alkylation.

$$\mathrm{R'(R)CH{-}\overset{O}{\overset{\|}{C}}{-}R''} \;\Longrightarrow\; \mathrm{R'(R)C(\overset{O}{\overset{\|}{C}}{-}R'')(\underset{\underset{O}{\|}}{C}{-}OEt)} \;\Longrightarrow\; \mathrm{H(R)C(\overset{O}{\overset{\|}{C}}{-}R'')(\underset{\underset{O}{\|}}{C}{-}OEt)} \;\Longrightarrow\; \mathrm{H_2C(\overset{O}{\overset{\|}{C}}{-}R'')(\underset{\underset{O}{\|}}{C}{-}OEt)}$$

saponification decarboxylation; alkylation; alkylation

3. Determine whether the β-keto ester is one that can be made by the Claisen condensation.

$$\mathrm{H{-}OEt + R{-}CH_2{-}\overset{O}{\overset{\|}{C}}{-}\underset{\underset{R}{|}}{CH}{-}\overset{O}{\overset{\|}{C}}{-}OEt} \;\Longrightarrow\; \mathrm{2\,R{-}CH_2{-}\overset{O}{\overset{\|}{C}}{-}OEt}$$

or

condensation

$$\mathrm{H{-}OEt + R{-}\overset{O}{\overset{\|}{C}}{-}CH_2{-}\overset{O}{\overset{\|}{C}}{-}OEt} \;\Longrightarrow\; \mathrm{2\,R{-}\overset{O}{\overset{\|}{C}}{-}OEt + H_3C{-}\overset{O}{\overset{\|}{C}}{-}OEt}$$

no α-hydrogens

E. Synthesis with the Malonic Ester Reaction

1. The malonic ester synthesis involves alkylation of a malonic ester derivative. In many cases the alkylated derivative is saponified and decarboxylated to give a substituted acetic acid derivative.
2. To determine whether the malonic ester synthesis can be used for the synthesis of a carboxylic acid:
   a. Determine whether the desired carboxylic acid is an acetic acid with one or two alkyl substituents on its α-carbon.
   b. Mentally reverse the decarboxylation, hydrolysis, and alkylation steps to arrive at the structures of the alkyl halides (or sulfonate esters) that must be used.

an acetic acid with R′ and R α-substituents

$$\mathrm{R'(R)CH{-}\overset{O}{\overset{\|}{C}}{-}OH} \;\Longrightarrow\; \mathrm{R'(R)C(\overset{O}{\overset{\|}{C}}{-}OEt)(\underset{\underset{O}{\|}}{C}{-}OEt)} \;\Longrightarrow\; \mathrm{H(R)C(\overset{O}{\overset{\|}{C}}{-}OEt)(\underset{\underset{O}{\|}}{C}{-}OEt)} \;\Longrightarrow\; \mathrm{H_2C(\overset{O}{\overset{\|}{C}}{-}OEt)(\underset{\underset{O}{\|}}{C}{-}OEt)}$$

saponification, decarboxylation; alkylation; alkylation

   c. If the alkyl halides used in the alkylation step can undergo the S_N2 reaction, then the target carboxylic acid can in principle be prepared by the malonic ester synthesis.

F. Synthesis with Conjugate-Addition Reactions

1. Any group at the β-position of a carbonyl compound (or nitrile) can in principle be delivered as a nucleophile in a conjugate addition.

2. A conjugate addition can be mentally reversed by subtracting a nucleophilic group from the β-position of the target molecule, and a positive fragment (usually a proton) from the α-position.

$$\text{—C(Nuc)—CH—C(=O)—} \Longrightarrow \text{—C=C—C(=O)—} + \text{Nuc—H} \Longrightarrow \text{Nuc}^-$$

conjugate addition

G. Conjugate-Addition versus Carbonyl-Group Reactions

1. A conjugate-addition reaction usually competes with a carbonyl group reaction.
   a. Conjugate addition retains a carbonyl group at the expense of a carbon-carbon double bond.
   b. Carbonyl addition retains a carbon-carbon double bond at the expense of a carbonyl group.
2. In the case of aldehydes and ketones, conjugate addition competes with addition to the carbonyl group.

$$\text{C=C—C(=\ddot{O}:)—} \xrightarrow[\text{conjugate addition}]{\text{Nuc—H or Nuc:}^-} \text{—C(Nuc)—C(H)—C(=\ddot{O}:)—} \text{ or } \text{—C(Nuc)—}\ddot{\text{C}}^-\text{—C(=\ddot{O}:)—}$$

$$\text{C=C—C(=\ddot{O}:)—} \xrightarrow[\text{carbonyl addition}]{\text{Nuc—H or Nuc:}^-} \text{C=C—C(:\ddot{O}—H)(Nuc)—} \text{ or } \text{C=C—C(:\ddot{O}:}^-\text{)(Nuc)—}$$

   a. Relatively weak bases (cyanide ion, amines, thiolate ions, and enolate ions derived from β-dicarbonyl compounds) that give reversible carbonyl-addition reactions with ordinary aldehydes and ketones tend to give conjugate addition with α,β-unsaturated aldehydes and ketones.
      i. Conjugate addition is observed with these nucleophiles primarily because, in most cases, it is irreversible.
      ii. Conjugate-addition products are more stable than carbonyl-addition products primarily because a C=O bond is considerably stronger than a C=C bond. Thus, the conjugate-addition product is the thermodynamic (more stable) product of the reaction.
      iii. This is another case of kinetic versus thermodynamic control of a reaction.
   b. When nucleophiles are used that undergo irreversible carbonyl additions, then the carbonyl-addition product is observed rather than the conjugate-addition product.
      i. Powerful nucleophiles such as $LiAlH_4$ and organolithium reagents add irreversibly to carbonyl groups and form carbonyl-addition products whether the reactant carbonyl compound is α,β-unsaturated or not.
3. Many of the same nucleophiles that undergo conjugate addition with aldehydes and ketones also undergo conjugate addition with esters; in contrast, stronger bases that react irreversibly at the carbonyl carbon react with esters to give nucleophilic acyl substitution products.

III. BIOSYNTHESIS OF COMPOUNDS DERIVED FROM ACETATE

A. Biosynthesis of Fatty Acids

1. The starting material for the biosynthesis of fatty acids is a thiol ester of acetic acid called acetyl-coenzyme-A (abbreviated acetyl-CoA). (For the structure of acetyl-CoA, see Fig. 22.3, text p. 1033.)
   a. Acetyl-CoA is first converted into malonyl-CoA by carboxylation of the α-carbon.

$$\underset{\text{acetyl-CoA}}{H_3C-\overset{O}{\overset{\|}{C}}-SCoA} \xrightarrow{\text{carboxybiotin}} {}^{-}O-\overset{O}{\overset{\|}{C}}-CH_2-\overset{O}{\overset{\|}{C}}-SCoA \; + \text{biotin} \xrightarrow[\text{acyl carrier protein}]{RSH} \underset{\text{malonyl-CoA}}{{}^{-}O-\overset{O}{\overset{\|}{C}}-CH_2-\overset{O}{\overset{\|}{C}}-SR} + HS-CoA$$

i. The —SCoA group in both acetyl- and malonyl-CoA is then replaced by a different group (abbreviated —SR), called the acyl carrier protein.

ii. The resulting compounds react in an enzyme-catalyzed reaction to give an acetoacetyl thiol ester.

b. The nucleophilic electron pair is made available not by proton abstraction, but by loss of CO_2 from malonyl-CoA; the loss of CO_2 as a gaseous by-product also serves to drive the Claisen condensation to completion.

c. The product, an acetoacetyl thiol ester, then undergoes successively a carbonyl reduction, a dehydration, and a double-bond reduction, each catalyzed by an enzyme; the net result is that the acetyl thiol ester is converted into a thiol ester with two additional carbons.

$$H_3C-\overset{O}{\overset{\|}{C}}-SR + {}^{-}O-\overset{O}{\overset{\|}{C}}-CH_2-\overset{O}{\overset{\|}{C}}-SR \xrightarrow[-CO_2]{\text{condensation}} H_3C-\overset{O}{\overset{\|}{C}}-CH_2-\overset{O}{\overset{\|}{C}}-SR \xrightarrow{\text{reduction}}$$

$$H_3C-\overset{OH}{\overset{|}{C}H}-CH_2-\overset{O}{\overset{\|}{C}}-SR \xrightarrow{\text{dehydration}} H_3C-CH=CH-\overset{O}{\overset{\|}{C}}-SR \xrightarrow{\text{reduction}} H_3C-CH_2-CH_2-\overset{O}{\overset{\|}{C}}-SR$$

or

$$H_3C-\overset{O}{\overset{\|}{C}}-SR + {}^{-}O-\overset{O}{\overset{\|}{C}}-CH_2-\overset{O}{\overset{\|}{C}}-SR \xrightarrow[\text{3) dehydration; 4) reduction}]{\text{1) condensation; 2) reduction}} H_3C-CH_2-CH_2-\overset{O}{\overset{\|}{C}}-SR$$

2. This sequence of reactions is then repeated until the proper chain length is obtained.
   a. At each cycle another two carbons are added to the chain.
   b. The fatty-acid thiol ester is then transesterified by glycerol to form fats and phospholipids.
3. Because carbons are added two at a time, the common fatty acids have an even number of carbon atoms; those with an odd number of carbon atoms are relatively rare.

B. Biosynthesis of Isoprenoids, Steroids, and Some Aromatic Compounds

1. Isopentenyl pyrophosphate (the basic building block of isopreneoids and steroids) as well as a number of aromatic compounds found in nature are ultimately derived from acetyl-CoA.

REACTIONS

I. α-HALOGENATION OF CARBONYL COMPOUNDS

A. Acid-Catalyzed α-Halogenation of Aldehydes and Ketones

1. Halogenation of an aldehyde or ketone in acidic solution usually results in the replacement of one α-hydrogen by halogen.
   a. Enols are reactive intermediates in these reactions.

b. Enols add only one halogen atom.

$$-\overset{O}{\overset{\|}{C}}-\underset{|}{\overset{|}{C}H}- \xrightarrow[\text{(rate-limiting step)}]{H_3O^+} -\overset{OH}{\overset{|}{C}}=\underset{|}{C}- \text{ (enol)} \xrightarrow{Br_2} -\overset{O}{\overset{\|}{C}}-\underset{|}{\overset{Br}{\overset{|}{C}}}- + HBr$$

rate-limiting step   enol

2. The rate law implies that even though the reaction is a halogenation, the rate is independent of the halogen concentration.

$$\text{rate} = k[\text{ketone}][H_3O^+]$$

   a. Halogens are not involved in the transition state for the rate-limiting step of the reaction.
   b. Enol formation is the rate-limiting process in acid-catalyzed halogenation of aldehydes and ketones.
3. Introduction of a second halogen is much slower than introduction of the first halogen.

B. Halogenation of Aldehydes and Ketones in Base; the Haloform Reaction

1. Halogenation of aldehydes and ketones with α-hydrogens also occurs in base; all α-hydrogens are substituted by halogen.

$$-CH_2-\overset{O}{\overset{\|}{C}}-\underset{|}{\overset{|}{C}H}- \xrightarrow[^-OH]{Br_2} -\underset{Br}{\underset{|}{\overset{Br}{\overset{|}{C}}}}-\overset{O}{\overset{\|}{C}}-\underset{|}{\overset{Br}{\overset{|}{C}}}-$$

2. When the aldehyde or ketone starting material is either acetaldehyde or a methyl ketone, the trihalomethyl carbonyl compound is not stable under the reaction conditions, and it reacts further to give, after acidification of the reaction mixture, a carboxylic acid and a haloform.

$$-\overset{O}{\overset{\|}{C}}-CH_3 \xrightarrow{Br_2,\ NaOH} -\overset{O}{\overset{\|}{C}}-CBr_3 \ (+\ 3\ NaBr + 3\ H_2O) \xrightarrow{^-OH} -\overset{O}{\overset{\|}{C}}-O^- + HCBr_3$$

bromoform (a haloform)

   a. This reaction is called the haloform reaction.
   b. A carbon-carbon bond is broken.
3. The mechanism of the haloform reaction involves the formation of an enolate ion as a reactive intermediate.
   a. The enolate ion reacts as a nucleophile with halogen to give an α-halo carbonyl compound.
   b. A dihalo and trihalo carbonyl compound are formed more rapidly in successive halogenations.
   c. A carbon-carbon bond is broken when the trihalo carbonyl compound undergoes a nucleophilic acyl substitution reaction.
      i. The leaving group in this reaction is a trihalomethyl anion, which is much less basic than ordinary carbanions.
      ii. The trihalomethyl anion reacts irreversibly with the carboxylic acid by-product to drive the overall haloform reaction to completion.
      iii. The carboxylic acid itself can be isolated by acidifying the reaction mixture.

$$R-\overset{O}{\overset{\|}{C}}-CH_3 \xrightarrow{HO^-} R-\overset{O}{\overset{\|}{C}}-\overset{-}{C}H_2 \text{ (enolate ion)} \xrightarrow{Br_2} R-\overset{O}{\overset{\|}{C}}-CH_2-Br + Br^- \xrightarrow[\text{twice more}]{Br_2/OH^-} R-\overset{O}{\overset{\|}{C}}-CBr_3 + 2\,Br^- \xrightarrow[\text{C—C bond cleavage}]{OH^-}$$

$$\mathrm{RC(=O)OH} + {}^{-}\mathrm{CBr_3} \rightleftharpoons \mathrm{RC(=O)O^-} + \mathrm{HCBr_3}$$

4. The haloform reaction can be used to prepare carboxylic acids from readily available methyl ketones.
5. The haloform reaction is used as a qualitative test for methyl ketones, called the iodoform test.
   a. A compound of unknown structure is mixed with alkaline I_2.
   b. A yellow precipitate of iodoform (HCI_3) is taken as evidence for a methyl ketone (or acetaldehyde).
   c. Alcohols of the structure $\mathrm{R{-}CH(OH){-}CH_3}$ also give a positive iodoform test because they are oxidized to methyl ketones (or to acetaldehyde, in the case of ethanol) by the basic iodine solution.

C. α-Bromination of Carboxylic Acids: the Hell-Volhard-Zelinsky Reaction

1. A bromine is substituted for an α-hydrogen when a carboxylic acid is treated with Br_2 and a catalytic amount of red phosphorus or PBr_3; this reaction is called the Hell-Volhard-Zelinsky (HVZ) reaction.
2. The first stage in the mechanism of this reaction is the conversion of the carboxylic acid into a small amount of acid bromide by the catalyst PBr_3.
   a. The acid bromide enolizes in the presence of acid; this reaction is similar to the acid-catalyzed bromination of ketones.
   b. The enol of the acid bromide is the species that actually brominates.

$$\mathrm{{-}CH{-}C(=O){-}OH} \xrightarrow[\text{catalyst}]{PBr_3} \mathrm{{-}CH{-}C(=O){-}Br} \rightleftharpoons \underset{\text{enol}}{\mathrm{{>}C{=}C(OH)Br}} \xrightarrow{Br_2}$$

$$\mathrm{{-}C(Br){-}C(=O){-}Br} \xrightarrow{RCO_2H} \mathrm{{-}C(Br){-}C(=O){-}OH} + \mathrm{R{-}C(=O){-}Br}$$

   i. When a small amount of PBr_3 catalyst is used, the α-bromo acid bromide reacts with the carboxylic acid to form more acid bromide, which is then brominated.
   ii. If one full equivalent of PBr_3 catalyst is used, the α-bromo acid bromide is the reaction product.
   iii. The reaction mixture can be treated with an alcohol to give an α-bromo ester.

D. Reactions of α-Halo Carbonyl Compounds

1. Most α-halo carbonyl compounds are very reactive in S_N2 reactions, and can be used to prepare other α-substituted carbonyl compounds.
2. In the case of α-halo ketones, nucleophiles used in these reactions must not be too basic.
   a. Stronger bases promote enolate-ion formation; and the enolate ions of α-halo ketones undergo other reactions.
   b. More basic nucleophiles can be used with α-halo acids because, under basic conditions, α-halo acids are ionized to form their carboxylate conjugate-base anions, and these do not undergo the side-reactions that α-halo ketones undergo.
3. α-Halo carbonyl compounds react so slowly by the S_N1 mechanism that this reaction is not useful; reactions that require the formation of carbocations alpha to carbonyl groups generally do not occur.

II. CONDENSATIONS INVOLVING ENOLATE IONS AND ENOLS OF ALDEHYDES AND KETONES

A. Base-Catalyzed Aldol Addition and Condensation Reactions

1. In aqueous base, aldehydes and ketones with α-hydrogens undergo a reaction called the aldol addition.

$$2\ \text{—}\overset{|}{\underset{\text{H}}{\text{C}}}\text{—}\overset{\text{O}}{\overset{\|}{\text{C}}}\text{—R} \xrightleftharpoons{H_2O,\ ^-OH} \text{—}\overset{|}{\underset{\text{H}}{\text{C}}}\text{—}\overset{\text{OH}}{\underset{\text{R}}{\text{C}}}\text{—}\overset{|}{\underset{|}{\text{C}}}\text{—}\overset{\text{O}}{\overset{\|}{\text{C}}}\text{—R}$$

an aldehyde or ketone — a β-hydroxy aldehyde or ketone

a. The term aldol is both a trivial name for 3-hydroxybutanal and a generic name for β-hydroxy aldehydes.
b. The aldol addition is a very important and general reaction of aldehydes and ketones that have α-hydrogens.
c. The aldol addition is another nucleophilic addition to a carbonyl group.

2. The base-catalyzed aldol addition involves an enolate ion as an intermediate; the enolate ion adds to a second molecule of aldehyde or ketone.

$$^-OH + H\text{—}CH_2\text{—}\overset{O}{\overset{\|}{C}}\text{—H} \rightleftharpoons H_2O + \left[H_2\bar{C}\text{—}\overset{O}{\overset{\|}{C}}\text{—H} \longleftrightarrow H_2C\text{=}\overset{O^-}{\overset{|}{C}}\text{—H}\right]$$

an enolate ion

$$H_3C\text{—}\overset{O}{\overset{\|}{C}}\text{—H} + H_2\bar{C}\text{—}\overset{O}{\overset{\|}{C}}\text{—H} \rightleftharpoons H_3C\text{—}\underset{H}{\overset{O^-}{\overset{|}{C}}}\text{—}CH_2\text{—}\overset{O}{\overset{\|}{C}}\text{—H} \ (+\ HO\text{—H}) \rightleftharpoons H_3C\text{—}\underset{H}{\overset{OH}{\overset{|}{C}}}\text{—}CH_2\text{—}\overset{O}{\overset{\|}{C}}\text{—H} + HO^-$$

aldol

3. The aldol addition is reversible.
   a. The equilibrium for the aldol addition is much more favorable for aldehydes than for ketones.
   b. For example, the aldol-addition product of acetone is obtained only if an apparatus is used that allows the product to be removed from the base catalyst as it is formed.
4. Under severe conditions (higher base concentration and/or heat), the product of aldol addition undergoes a dehydration reaction. (The sequence of reactions consisting of the aldol addition followed by dehydration is called the aldol condensation; a condensation is a reaction in which two molecules combine to form a larger molecule with the elimination of a small molecule, in many cases water.)

$$2\ R\text{—}CH_2\text{—}\overset{O}{\overset{\|}{C}}\text{—H} \underset{\text{addition } via \text{ enolate}}{\xrightleftharpoons{^-OH}} R\text{—}CH_2\text{—}\overset{OH}{\overset{|}{CH}}\text{—}\underset{R}{\underset{|}{CH}}\text{—}\overset{O}{\overset{\|}{C}}\text{—H} \xrightarrow[\text{dehydration}]{^-OH} R\text{—}CH_2\text{—}CH\text{=}\underset{R}{\underset{|}{C}}\text{—}\overset{O}{\overset{\|}{C}}\text{—H}$$

   a. The dehydration part of the aldol condensation is catalyzed by base, and occurs in two distinct steps through a carbanion intermediate.
   b. Base-catalyzed dehydration reactions of alcohols are unusual; β-hydroxy aldehydes and β-hydroxy ketones undergo base-catalyzed dehydration because—
      i. the α-hydrogen is relatively acidic.
      ii. the product is conjugated and therefore particularly stable.
5. The product of an aldol addition-dehydration sequence is an α,β-unsaturated carbonyl compound.

6. The aldol condensation is an important method for the preparation of certain α,β-unsaturated carbonyl compounds. Whether the aldol addition product or the condensation product is formed depends on the reaction conditions, which must be determined on a case-by-case basis. (See "Concepts," Section II.B.2.)

B. Acid-Catalyzed Aldol Condensation

1. Aldol condensations are also catalyzed by acid.
   a. Acid-catalyzed aldol condensations generally give α,β-unsaturated carbonyl compounds as products.
   b. The addition products usually cannot be isolated.
2. Both the conjugate acid and the enol isomer of the aldehyde or ketone are key reactive intermediates.
   (See Eqs. 22.46a–b on text page 1017 for a detailed mechanism.)
3. The α-hydroxy aldehyde or ketone formed in the reaction spontaneously undergoes acid-catalyzed dehydration to give an α,β-unsaturated carbonyl compound; this dehydration drives the aldol condensation to completion.

$$2\,RCH_2-\overset{O}{\overset{\|}{C}}-CH_2R \underset{\substack{\text{addition} \\ \textit{via}\text{ enol}}}{\overset{H_3O^+}{\rightleftharpoons}} RCH_2-\underset{RCH_2}{\overset{OH}{\overset{|}{\underset{|}{C}}}}-\underset{R}{\underset{|}{CH}}-\overset{O}{\overset{\|}{C}}-CH_2R \xrightarrow[\text{dehydration}]{H_3O^+} RCH_2-\underset{RCH_2}{\underset{|}{C}}=\underset{R}{\underset{|}{C}}-\overset{O}{\overset{\|}{C}}-CH_2R$$

C. Special Types of Aldol Condensation

1. When two different carbonyl compounds are used, a crossed aldol condensation occurs; in many cases, the result is a difficult-to-separate mixture.
2. Under the usual conditions (aqueous or alcoholic acid or base), useful crossed aldol condensations as a practical matter are limited to situations in which a ketone with α-hydrogens is condensed with an aldehyde without α-hydrogens.
3. An important example of this type of crossed aldol condensation is the Claisen-Schmidt condensation.
   a. A ketone with α-hydrogens is condensed with an aromatic aldehyde that has no α-hydrogens.

$$Ph-\overset{O}{\overset{\|}{C}}-H + H_3C-\overset{O}{\overset{\|}{C}}-CH_3 \underset{\textit{via}\text{ enol}}{\overset{\substack{H_3O^+ \\ \text{condensation}}}{\rightleftharpoons}} \underset{\text{not isolated}}{Ph-\overset{OH}{\overset{|}{CH}}-CH_2-\overset{O}{\overset{\|}{C}}-CH_3} \xrightarrow[]{\substack{H_3O^+ \\ \text{dehydration}}} Ph-CH=CH-\overset{O}{\overset{\|}{C}}-CH_3$$

   b. The Claisen-Schmidt condensation occurs instead of other competing processes for three reasons:
      i. Because the aldehyde has no α-hydrogens, it cannot act as the enolate or enol component of the aldol condensation.
      ii. Addition of the ketone enolate or enol to a ketone occurs more slowly than addition to an aldehyde; aldehydes are more reactive than ketones.
      iii. The aldol addition reaction is reversible; thus, to the extent that this side reaction occurs, it is reversed under the reaction conditions.
   c. The Claisen-Schmidt condensation, like other aldol condensations, can be catalyzed by either acid or base.
4. When a molecule contains more than one aldehyde or ketone group, an intramolecular reaction (a reaction within the same molecule) is possible; the resulting formation of a ring is particularly favorable when five- and six-membered rings can be formed.

$$\text{(2-oxocyclohexyl)}CH_2-\overset{O}{\overset{\|}{C}}-CH_3 \xrightarrow{\text{condensation}} \text{bicyclic enone}$$

III. CONDENSATION REACTIONS INVOLVING ESTER ENOLATE IONS

A. Claisen Condensation

1. The Claisen condensation is used for the formation of β-dicarbonyl compounds (compounds with two carbonyl groups in a β-relationship), particularly—
   a. β-keto esters (compounds in which a ketone carbonyl group is β to an ester carbonyl group)
   b. β-diketones (See "Concepts," Section II.C.2.a.)

$$EtO^- + 2\,\underset{\text{an ester}}{RCH_2\overset{O}{\overset{\|}{C}}-OEt} \longrightarrow RCH_2\overset{O}{\overset{\|}{C}}-\underset{R}{\underset{|}{\bar{C}}}-\overset{O}{\overset{\|}{C}}-OEt + EtOH \xrightarrow{H_3O^+} \underset{\text{a }\beta\text{ keto ester}}{RCH_2\overset{O}{\overset{\|}{C}}-\underset{R}{\underset{|}{CH}}-\overset{O}{\overset{\|}{C}}-OEt}$$

$$EtO^- + \underset{\text{a ketone}}{RCH_2\overset{O}{\overset{\|}{C}}R'} + \underset{\text{an ester}}{-\overset{O}{\overset{\|}{C}}-OEt} \longrightarrow -\overset{O}{\overset{\|}{C}}-\underset{R}{\underset{|}{\bar{C}}}-\overset{O}{\overset{\|}{C}}R' + EtOH \xrightarrow{H_3O^+} \underset{\text{a }\beta\text{-diketone}}{-\overset{O}{\overset{\|}{C}}-\underset{R}{\underset{|}{CH}}-\overset{O}{\overset{\|}{C}}R'}$$

2. The mechanism of the Claisen condensation:
   a. An enolate ion is formed by the reaction of an ester with ethoxide as the base. (Although the ester enolate ion is formed in low concentration, it is a strong base and good nucleophile.)

$$EtO^- + R-\overset{H}{\overset{|}{C}H}-\overset{O}{\overset{\|}{C}}-OEt \rightleftharpoons \left[R-\bar{C}H-\overset{O}{\overset{\|}{C}}-OEt \longleftrightarrow R-CH=\overset{O^-}{\overset{|}{C}}-OEt\right]$$

   b. The ester enolate undergoes a nucleophilic acyl substitution reaction with a second molecule of ester.

$$R-CH_2-\overset{O}{\overset{\|}{C}}-OEt + R-\bar{C}H-\overset{O}{\overset{\|}{C}}-OEt \rightleftharpoons R-CH_2-\overset{O^-}{\overset{|}{C}}(OEt)-\underset{R}{\underset{|}{CH}}-\overset{O}{\overset{\|}{C}}-OEt \rightleftharpoons EtO^- + R-CH_2-\overset{O}{\overset{\|}{C}}-\underset{R}{\underset{|}{CH}}-\overset{O}{\overset{\|}{C}}-OEt$$

   c. The overall equilibrium favors the reactants; that is, all β-keto esters are less stable than the esters from which they are derived.
   d. The Claisen condensation has to be driven to completion by applying LeChatelier's principle.
      i. The most common technique is to use one equivalent of alkoxide catalyst.
      ii. The hydrogens adjacent to both carbonyl groups in the β-keto ester product are especially acidic ($pK_a = 10.7$); the alkoxide (conjugate acid $pK_a = 16$) removes one of these hydrogens to form quantitatively the conjugate base of the product.

$$\mathrm{EtO^-} + \mathrm{R{-}CH_2{-}\overset{O}{\overset{\|}{C}}{-}\underset{R}{\underset{|}{\overset{H}{\overset{|}{C}}}}{-}\overset{O}{\overset{\|}{C}}{-}OEt} \rightleftharpoons \mathrm{R{-}CH_2{-}\overset{O}{\overset{\|}{C}}{-}\underset{R}{\underset{|}{\bar{C}}}{-}\overset{O}{\overset{\|}{C}}{-}OEt} + \mathrm{EtOH}$$

$pK_a = 10$–11 $\qquad pK_a = 16$

e. The un-ionized β-keto ester product is formed when acid is added subsequently to the reaction mixture.

$$\mathrm{R{-}CH_2{-}\overset{O}{\overset{\|}{C}}{-}\underset{R}{\underset{|}{\bar{C}}}{-}\overset{O}{\overset{\|}{C}}{-}OEt} \xrightarrow{H_3O^+} \mathrm{R{-}CH_2{-}\overset{O}{\overset{\|}{C}}{-}\underset{R}{\underset{|}{CH}}{-}\overset{O}{\overset{\|}{C}}{-}OEt}$$

3. Attempts to condense an ester that has only one α-hydrogen result in little or no product, because the desired condensation product has no α-hydrogens acidic enough to react completely with the alkoxide.

B. Dieckmann Condensation

1. Intramolecular Claisen condensations take place readily when five- or six-membered rings can be formed.

$$\text{diester (C(=O)OEt at both chain ends)} \xrightarrow[2)\ H_3O^+]{1)\ EtO^-} \text{2-oxocyclohexane-C(=O)OEt}$$

2. The intramolecular Claisen condensation reaction is called the Dieckmann condensation.
3. As with the Claisen condensation, one equivalent of base is necessary; the α-proton must be removed from the initially formed product in order for the reaction to be driven to completion.

C. Crossed Claisen Condensation

1. The Claisen condensation of two different esters is called a crossed Claisen condensation; the crossed Claisen condensation of two esters that both have α-hydrogens gives a mixture of compounds that are typically difficult to separate.
2. Crossed Claisen condensations are useful if one ester is especially reactive or has no α-hydrogens. Formate esters and diethyl carbonate (a less reactive ester that is used in excess) are frequently used in this type of reaction.

$$\underset{\text{no }\alpha\text{-hydrogens}}{\mathrm{G{-}\overset{O}{\overset{\|}{C}}{-}OEt}} + \mathrm{R{-}CH_2{-}\overset{O}{\overset{\|}{C}}{-}OEt} \xrightarrow[2)\ H_3O^+]{1)\ EtO^-} \mathrm{G{-}\overset{O}{\overset{\|}{C}}{-}\underset{R}{\underset{|}{CH}}{-}\overset{O}{\overset{\|}{C}}{-}OEt}$$

ethyl formate (G = H)
diethyl carbonate (G = EtO)

3. Another type of crossed Claisen condensation is the reaction of ketones with esters.

$$\text{cyclopentanone} + \mathrm{EtO{-}\overset{O}{\overset{\|}{C}}{-}R} \xrightarrow[2)\ H_3O^+]{1)\ EtO^-} \text{2-(}\mathrm{\overset{O}{\overset{\|}{C}}{-}R}\text{)cyclopentanone}$$

a. In this type of reaction the enolate ion of a ketone attacks the carbonyl group of an ester.
   i. The equilibrium for the aldol addition of two ketones favors the reactants, whereas the Claisen condensation is irreversible because one equivalent of base is used to form the enolate ion of the product.
   ii. Ketones are far more acidic than esters; the enolate ion of the ketone is formed in much greater concentration than the enolate ion of the ester.
b. The product is a β-dicarbonyl compound, which is especially acidic and is ionized completely by the one equivalent of base; ionization makes this reaction irreversible.

IV. ALKYLATION OF ESTER ENOLATE IONS

A. Malonic Ester Synthesis

1. Diethyl malonate (malonic ester), like many other β-dicarbonyl compounds, has unusually acidic α-hydrogens (pK_a = 12.9).
   a. Its conjugate-base enolate ion can be formed completely with alkoxide bases such as sodium ethoxide, whose conjugate acid, ethanol, has a pK_a of about 16.
   b. The conjugate-base anion of diethyl malonate is nucleophilic, and it reacts with alkyl halides and sulfonate esters (RX, R´X in the following equation) in typical S_N2 reactions.
   c. This reaction can be used to introduce alkyl groups at the α-position of malonic ester.

$$\underset{\substack{\text{diethyl malonate}\\ pK_a = 12.9}}{\text{EtO—C(=O)—CH}_2\text{—C(=O)—OEt}} \xrightarrow[\text{2) RX}]{\text{1) EtO}^-} \text{EtO—C(=O)—CH(R)—C(=O)—OEt} + \text{EtOH} + \text{X}^- \xrightarrow[\text{2) R'X}]{\text{1) EtO}^-} \text{EtO—C(=O)—C(R)(R')—C(=O)—OEt} + \text{EtOH} + \text{X}^-$$

2. The malonic ester synthesis can be extended to the preparation of carboxylic acids. (See "Concepts," Section II.E.)
   a. Saponification of the diester and acidification of the resulting solution give a substituted malonic acid derivative.
   b. Heating any malonic acid derivative causes it to decarboxylate.

$$\text{EtO—C(=O)—C(R)(R')—C(=O)—OEt} \xrightarrow[\text{2) H}_3\text{O}^+]{\text{1) HO}^-} \text{HO—C(=O)—C(R)(R')—C(=O)—OH} + 2\,\text{EtOH} \longrightarrow \text{H—C(R')(R)—C(=O)—OH} + \text{CO}_2$$

saponification

decarboxylation

   c. The result of alkylation, saponification, and decarboxylation is a carboxylic acid that conceptually is a disubstituted acetic acid—an acetic acid molecule with two alkyl groups on its α-carbon.
3. The overall sequence of ionization, alkylation, saponification, and decarboxylation starting from diethyl malonate is called the malonic ester synthesis.
   a. The alkylation step of the malonic ester synthesis results in the formation of one or two new carbon-carbon bonds.
   b. The anion of malonic acid can be alkylated twice in two successive reactions with different alkyl halides (if desired) to give, after hydrolysis and decarboxylation, a disubstituted acetic acid.

B. Acetoacetic Ester Synthesis

1. β-Keto esters (pK_a = 10–11), which are substantially more acidic than ordinary esters (pK_a = 25), are completely ionized by alkoxide bases, which have conjugate-acid pK_a values near 16.
   a. The resulting enolate ions can be alkylated by alkyl halides or sulfonate esters.

$$\underset{\substack{\text{a }\beta\text{-keto ester}\\ pK_a = 10\text{–}11}}{\text{—}\overset{O}{\overset{\|}{C}}\text{—}CH_2\text{—}\overset{O}{\overset{\|}{C}}\text{—OEt}} \xrightarrow[\text{2) RX}]{\text{1) EtO}^-} \text{—}\overset{O}{\overset{\|}{C}}\text{—}\underset{R}{\underset{|}{CH}}\text{—}\overset{O}{\overset{\|}{C}}\text{—OEt} + \text{EtOH} + X^- \xrightarrow[\text{2) R}'\text{X}]{\text{1) EtO}^-} \text{—}\overset{O}{\overset{\|}{C}}\text{—}\underset{R\ \ R'}{C}\text{—}\overset{O}{\overset{\|}{C}}\text{—OEt} + \text{EtOH} + X^-$$

b. Dialkylation of β-keto esters is also possible.
c. The alkylated derivatives of ethyl acetoacetate can be hydrolyzed and decarboxylated to give ketones.

$$\text{—}\overset{O}{\overset{\|}{C}}\text{—}\underset{R\ \ R'}{C}\text{—}\overset{O}{\overset{\|}{C}}\text{—OEt} \xrightarrow[\text{2) H}_3\text{O}^+]{\text{1) HO}^-} \text{—}\overset{O}{\overset{\|}{C}}\text{—}\underset{R\ \ R'}{C}\text{—}\overset{O}{\overset{\|}{C}}\text{—OH} + \text{EtOH} \xrightarrow{H_3O^+} \text{—}\overset{O}{\overset{\|}{C}}\text{—}\underset{R\ \ R'}{C}\text{—H} + CO_2$$

2. The alkylation of ethyl acetoacetate followed by saponification, protonation, and decarboxylation to give a ketone is called the acetoacetic ester synthesis and involves the construction of one or two new carbon-carbon bonds.

3. Whether a target ketone can be prepared by the acetoacetic ester synthesis can be determined by mentally reversing the synthesis: (See "Concepts," Section II.D.)
   a. Replace an α-hydrogen on the target ketone with a —CO_2Et group.
   b. Determine whether the resulting β-keto ester can be prepared by a Claisen condensation or from other β-keto esters by alkylation or di-alkylation with appropriate alkyl halides.

C. Direct Alkylation of Enolate Ions Derived from Monoesters

1. A family of very strong, highly branched nitrogen bases can be used to form stable enolate ions rapidly at –78° from esters.
   a. The nitrogen bases themselves are generated from the corresponding amines and butyllithium at –78° in tetrahydrofuran (THF).

$$CH_3CH_2CH_2CH_2\text{—Li} + \underset{pK_a = 32\text{–}35}{i\text{-}Pr_2NH} \xrightarrow[-78^\circ]{} i\text{-}Pr_2N^-\ Li^+ + \underset{pK_a = 55\text{–}60}{CH_3CH_2CH_2CH_3}$$

   b. These amide bases are strong enough to convert esters completely into their conjugate-base enolate ions.

$$\text{—}\underset{|}{CH}\text{—}\overset{O}{\overset{\|}{C}}\text{—OEt} \xrightarrow[\text{2) RX}]{\text{1) } i\text{-Pr}_2\text{N}^-\ \text{Li}^+/-78^\circ} \text{—}\underset{|}{\overset{R}{\overset{|}{C}}}\text{—}\overset{O}{\overset{\|}{C}}\text{—OEt} + i\text{-}Pr_2NH + Li^+\ X^-$$

      i. The ester is added to the base; the reaction of esters with strong amide bases is much faster, even at –78°, than the Claisen condensation; the enolate is formed rapidly.
      ii. Attack of the amide base on the carbonyl carbon of the ester, a reaction that competes with proton removal, is retarded by van der Waals repulsions between groups on the carbonyl compound and the large branched groups on the amide base.
      iii. These van der Waals repulsions have been aptly termed F-strain, or "front-strain."
   c. The ester enolate anions formed with these bases can be alkylated with alkyl halides.
   d. Esters with quaternary α-carbon atoms can be prepared with this method.

2. This method of ester alkylation is considerably more expensive than the malonic ester synthesis and requires special inert-atmosphere techniques.

V. CONJUGATE-ADDITION REACTIONS

A. Conjugate Addition to α,β-Unsaturated Carbonyl Compounds

1. Addition of a nucleophile (Nuc^- in the following equation) to the double bond of an α,β-unsaturated carbonyl compound is an example of conjugate addition (1,4-addition); nucleophilic conjugate addition has no parallel in the reactions of simple conjugated dienes.

$$Nuc^- + \text{C}{=}\text{C}{-}\text{C}{=}\text{O} \xrightarrow{\text{conjugate addition}} Nuc{-}\text{C}{-}\text{C}{=}\text{C}{-}\text{O}^- \xrightarrow[\text{protonation}]{Nuc{-}H} Nuc{-}\text{C}{-}\text{C}{=}\text{C}{-}\text{OH} + Nuc^- \xrightarrow{\text{enol-ketone isomerization}} Nuc{-}\text{C}{-}\text{CH}{-}\text{C}{=}\text{O}$$

2. Nucleophilic addition to the carbon-carbon double bonds of α,β-unsaturated aldehydes, ketones, esters, and nitriles is a rather general reaction that can be observed with a variety of nucleophiles.
   a. The addition of a nucleophile to acrylonitrile ($H_2C{=}CH{-}C{\equiv}N$) is a useful reaction called cyanoethylation.

$$\underset{\text{nucleophile}}{Nuc{-}H} + \underset{\text{acrylonitrile}}{H_2C{=}CH{-}CN} \xrightarrow{\text{base}} \underset{\text{cyanoethylated nucleophile}}{Nuc{-}CH_2{-}CH_2{-}CN}$$

   b. Quinones are α,β-unsaturated carbonyl compounds, and they also undergo similar addition reactions.
   c. When cyanide is the nucleophile, a new carbon-carbon bond is formed; the nitrile group can then be converted into a carboxylic acid group by acid hydrolysis.
3. Nucleophilic addition to the double bond in an α,β-unsaturated carbonyl compound occurs because it gives a resonance-stabilized enolate ion intermediate.
4. Acid-catalyzed conjugate additions to α,β-unsaturated carbonyl compounds are also known.

B. Conjugate Addition of Enolate Ions: Michael Addition

1. Enolate ions derived from malonic ester derivatives, β-keto esters, and the like undergo conjugate-addition reactions with α,β-unsaturated carbonyl compounds.
   a. Conjugate additions of carbanions to α,β-unsaturated carbonyl compounds are called Michael additions.

$$H_2C(CO_2Et)_2 \xrightarrow[EtOH]{EtO^-} \bar{H}C(CO_2Et)_2 \xrightleftharpoons{H_3C{-}C({=}O){-}CH{=}CH_2}$$

$$H_3C{-}C({=}O){-}\bar{C}H{-}CH_2{-}CH(CO_2Et)_2 \xrightarrow{H_2C(CO_2Et)_2} H_3C{-}C({=}O){-}CH_2{-}CH_2{-}CH(CO_2Et)_2 + \bar{H}C(CO_2Et)_2$$

   b. Many Michael additions can originate from either of two pairs of reactants.
   c. To maximize conjugate addition, choose the pair of reactants with the less basic enolate ion.
2. In one useful variation of the Michael addition, called the Robinson annulation, the immediate product of the reaction can be subjected to an aldol condensation that closes a ring. (An annulation is a ring-forming reaction.)

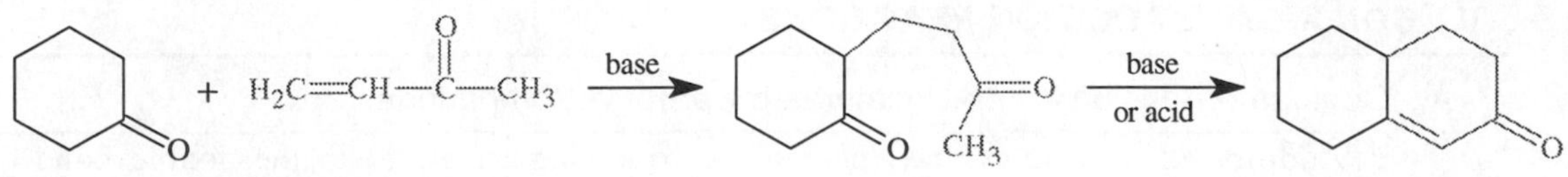

+ $H_2C{=}CH{-}\overset{O}{\overset{\|}{C}}{-}CH_3$ $\xrightarrow{\text{base}}$ $\xrightarrow[\text{or acid}]{\text{base}}$

VI. REACTIONS OF α,β-UNSATURATED CARBONYL COMPOUNDS WITH OTHER NUCLEOPHILES

(See also "Concepts," Section II.F and II.G.)

A. Reduction of α,β-Unsaturated Carbonyl Compounds

1. The carbonyl group of an α,β-unsaturated aldehyde or ketone is reduced to an alcohol with lithium aluminum hydride.

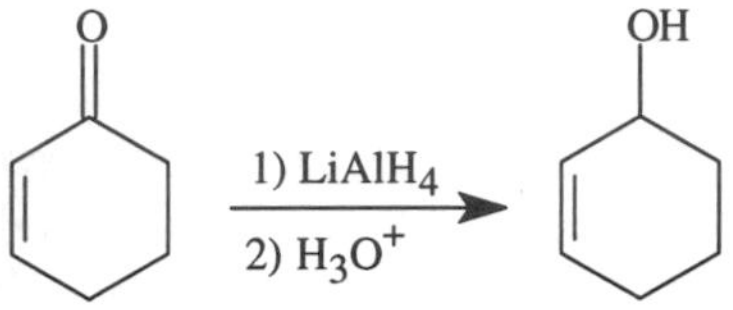

$\xrightarrow[\text{2) } H_3O^+]{\text{1) } LiAlH_4}$

   a. This reaction involves the attack of hydride at the carbonyl carbon and is therefore a carbonyl addition.
   b. Carbonyl addition is not only faster than conjugate addition but, in this case, also irreversible; reduction of the carbonyl group with $LiAlH_4$ is a kinetically controlled reaction.
2. Many α,β-unsaturated carbonyl compounds are reduced by $NaBH_4$ to give mixtures of both carbonyl-addition products and conjugate-addition products.
3. The carbon-carbon double bond of an α,β-unsaturated carbonyl compound can in most cases be reduced selectively by catalytic hydrogenation.

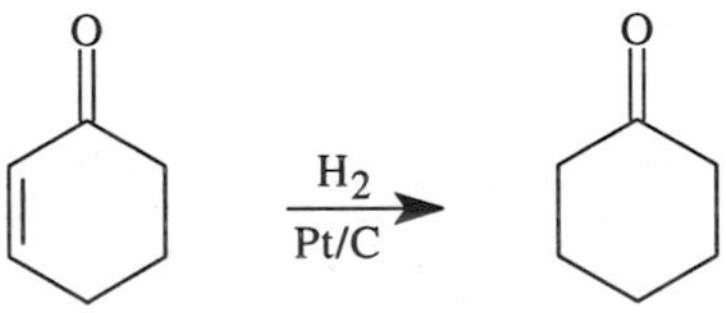

$\xrightarrow[\text{Pt/C}]{H_2}$

B. Addition of Organolithium Reagents to the Carbonyl Group

1. Organolithium reagents react with α,β-unsaturated carbonyl compounds to yield products of carbonyl addition.
2. The product is the result of kinetic control, since carbonyl addition is more rapid than conjugate addition and it is also irreversible.

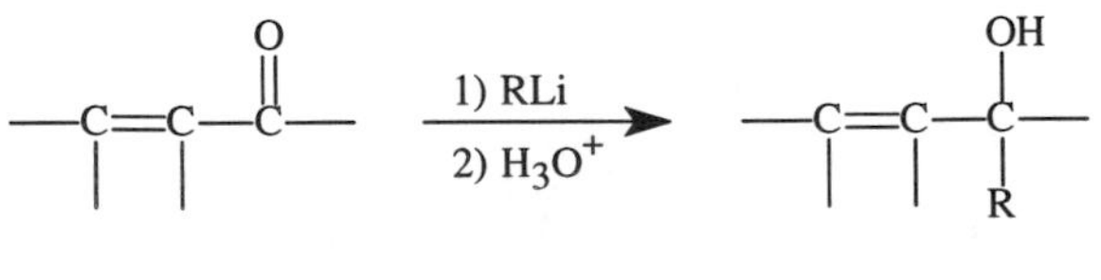

$$-\overset{|}{C}{=}\overset{|}{C}{-}\overset{O}{\overset{\|}{C}}{-} \xrightarrow[\text{2) } H_3O^+]{\text{1) RLi}} -\overset{|}{C}{=}\overset{|}{C}{-}\overset{OH}{\underset{R}{C}}{-}$$

C. Conjugate Addition of Lithium Dialkylcuprate Reagents

1. Lithium dialkylcuprate reagents give exclusively products of conjugate addition when they react with α,β-unsaturated esters and ketones.

$$—\overset{|}{C}=\overset{|}{C}—\overset{O}{\overset{||}{C}}— \xrightarrow[\text{2) } H_3O^+]{\text{1) } R_2CuLi} —\overset{R}{\overset{|}{\underset{|}{C}}}—CH_2—\overset{O}{\overset{||}{C}}—$$

2. Although conjugate addition of lithium dialkylcuprates actually involves a radical mechanism (see Study Guide Link 22.10), the reaction can be conceptualized mechanistically to be similar to other conjugate additions.
   a. Attack of an "alkyl anion" of the dialkylcuprate on the double bond gives a resonance-stabilized enolate ion.
   b. Protonation of the enolate ion by the addition of water gives the conjugate-addition product.
3. Conjugate addition also occurs with Grignard reagents in the presence of added CuCl.

SCIENTISTS

Aleksandr Porfirevich Borodin (1833–1887) 22.4A
Rainer Ludwig Claisen (1851–1930) 22.4C, 22.5A
George Darzens (1867–1954) P22.84
Walter Dieckmann (1869–1925) 22.5B
Carl Magnus von Hell (1849–1926) 22.3C
Arthur Michael (1853–1942) 22.8C
Sir Robert Robinson (1886–1975) [Chemistry 1947] 22.8C
J. Gustav Schmidt 22.4C
Jacob Volhard (1834–1910) 22.3C
Friedrich Wöhler (1800–1882) 22.6
Charles Adolphe Wurtz (1817–1884) 22.4A
Nikolai Dimitrievič Zelinsky (1861–1952) 22.3C

Dates in brackets refer to Nobel Prize awards.

SGL

STUDY GUIDE LINKS

✓22.1 Ionization versus Nucleophilic Attack on the Carbonyl Carbon

One of the things you learned about Brønsted bases is that they can usually act as nucleophiles, and that nucleophiles attack the carbon atoms of carbonyl groups. Yet in Eqs. 22.1 and 22.2 of the text (p. 997), bases are shown removing protons. You might be wondering, "What is the major reaction: nucleophilic attack on the carbonyl carbon, or removal of an α-hydrogen to give an enolate ion?" The answer is that both can occur. This section of the text focuses on the proton removal, but later sections of this chapter show the interplay between both types of reaction. Nevertheless, it might help you to see just one example here.

Suppose an ester is treated with aqueous NaOH. Recall (Sec. 21.7A) that NaOH saponifies esters. Saponification results from nucleophilic attack of hydroxide ion at the carbonyl carbon of an ester. However, while saponification is taking place, hydroxide ion removes the α-protons of a few ester molecules to form some enolate ions. This reaction has no consequences for saponification because saponification occurs much more rapidly than the reactions of enolate ions (many of which are discussed in this chapter). For this reason, it was not important to consider enolate-ion formation in the discussion of saponification.

The point of this chapter is to focus on situations in which enolate ion formation is significant, and ultimately on what happens when *both* proton removal and nucleophilic attack on a carbonyl carbon can occur.

✓22.2 Kinetic versus Thermodynamic Stability of Enols

When we think of unstable organic compounds we often think of molecules containing strained rings or molecules containing van der Waals repulsions. The structures of enols have no such destabilizing features. Yet enols in most cases cannot exist for significant lengths of time, while other less stable molecules can be put in bottles and stored for years. Why this seeming contradiction?

The essence of this problem is the meaning of the word *stability* as applied to chemical substances. *Thermodynamic stability* is a term used to describe the relative energy contents of molecules. In general, thermodynamic stability has very little to do with whether a compound can be isolated. *Kinetic stability* is a term used to describe how rapidly molecules revert to more stable molecules. In order for a molecule to be *kinetically unstable*, a mechanism must exist for its conversion into a more stable molecule. For a compound to be kinetically unstable, it must be thermodynamically unstable, but the converse need not be true. Some *thermodynamically unstable compounds* are easy to isolate because they are *kinetically stable*. For example, almost all hydrocarbons in the presence of oxygen are thermodynamically unstable relative to CO_2 and H_2O by hundreds of kJ/mol; yet hydrocarbons do not rapidly revert to CO_2 and H_2O because there is no readily available mechanism by which this conversion can be accomplished. If we supply a flame, however, such a conversion mechanism becomes available—we call it *combustion*—and we see very graphically how unstable hydrocarbons really are!

Examples of kinetic stability and instability in carbonyl chemistry are provided by acetals (Sec. 19.10, text pp. 868–71) and carbonyl hydrates (Sec. 19.7B, text pp. 857–9). The hydrate of cyclohexanone is both thermodynamically and kinetically unstable with respect to the ketone and water:

$$\text{the hydrate of cyclohexanone} \underset{}{\overset{\text{fast}}{\rightleftharpoons}} \text{cyclohexanone} + H_2O$$

Yet replacing the O—H hydrogens with alkyl groups gives compounds called *acetals* that are thermodynamically unstable, but *kinetically stable*, in neutral and basic solution.

$$H_2O + \text{cyclohexanone dimethyl acetal (1,1-dimethoxycyclohexane)} \underset{}{\overset{\text{too slow to measure}}{\rightleftharpoons}} \text{cyclohexanone} + 2\,CH_3OH$$

The base-catalyzed mechanism of hydrate decomposition requires the presence of an O—H hydrogen, which acetals lack. Hence, the base-catalyzed pathway of hydrate decomposition cannot operate with acetals. Therefore, acetals, although thermodynamically unstable relative to their corresponding carbonyl compounds, are kinetically stable in basic solution. This is why acetals are good protecting groups for carbonyl groups under basic or neutral conditions. Under acidic conditions, however, acetals are kinetically unstable because an acid-catalyzed mechanism exists for their conversion into ketones.

The thermodynamic stabilities of most enols are not much less than the stabilities of their corresponding carbonyl isomers—typically, about 30–40 kJ/mol (7–10 kcal/mol). What makes enols difficult to isolate is their *kinetic* instabilities. Simple paths, or mechanisms, exist for their rapid transformation into carbonyl compounds in the presence of acids or bases; these mechanisms are shown in Eqs. 22.17a–b in the text (p. 1006). Enols are kinetically unstable molecules because the conditions under which they are converted into carbonyl compounds—the presence of acids and/or bases—are so prevalent. It's hard to generate an enol without an acid or base around. In fact, enols themselves are weak acids, and can catalyze their own slow conversion into ketones. Enols simply don't stand much of a chance for survival.

Not many years ago chemists figured out ways to generate simple enols in solution in the absence of other acids and bases and, indeed, found that they could be isolated and observed for significant periods of time.

✓22.3 Dehydration of β-Hydroxy Carbonyl Compounds

At first sight, the dehydration of β-hydroxy carbonyl compounds would appear to be an ordinary acid-catalyzed dehydration as discussed in Sec. 10.1 of the text. Recall that in the mechanism of such a dehydration, protonation of the hydroxy group is followed by carbocation formation:

$$H_3C-\underset{CH_3}{\overset{OH}{C}}-CH_2-\overset{O}{\overset{\|}{C}}-CH_3 \;(+\; H-\overset{+}{O}H_2) \rightleftharpoons H_3C-\underset{CH_3}{\overset{\overset{+}{O}H_2}{C}}-CH_2-\overset{O}{\overset{\|}{C}}-CH_3 + H_2O \rightleftharpoons \underset{A}{H_3C-\underset{CH_3}{\overset{+}{C}}-CH_2-\overset{O}{\overset{\|}{C}}-CH_3} + H_2O$$

Loss of a C—H proton from carbocation A gives the α,β-unsaturated ketone. However, a different mechanism—and, as it turns out, the correct one—is possible. This mechanism involves dehydration not of the ketone, but instead of the *enol.*

$$\mathrm{H_3C{-}C(OH)(CH_3){-}CH_2{-}C({=}O){-}CH_3} \rightleftharpoons \mathrm{H_3C{-}C(OH)(CH_3){-}CH{=}C(OH){-}CH_3}$$

an enol form

The dehydration involves protonation of the —OH group, formation of carbocation *B*, and loss of a proton from this carbocation:

$$\mathrm{H_3C{-}C(\overset{+}{O}H_2)(CH_3){-}CH{=}C(OH){-}CH_3} \rightleftharpoons \underset{B}{\mathrm{H_3C{-}\overset{+}{C}(CH_3){-}CH{=}C(OH){-}CH_3}} + \mathrm{H_2O} \longrightarrow \mathrm{H_3C{-}C(CH_3){=}CH{-}C({=}O){-}CH_3} + \mathrm{H_3\overset{+}{O}}$$

The reason the enol mechanism is favored is that *carbocation B is much more stable than carbocation A.* Can you see why? Draw resonance structure(s) for carbocation *B*. Is carbocation *A* resonance-stabilized? What is the *polar effect* of the carbonyl group on the stability of carbocation *A*?

✓22.4 Understanding Condensation Reactions

Condensations such as the aldol condensation and its variants, as well as the other reactions you'll encounter in Chapter 22, present a particular challenge to the beginning student, because the products are somewhat more complex than the products of many typical organic reactions, and because more than one functional-group transformation is involved (an addition followed by a dehydration). The way to develop an understanding of these reactions is to write the mechanism of each new reaction you see. First, identify the sources of the atoms in the product. (See Study Guide Link 4.9 on p. 89 of this manual; the discussion in Sec. 22.4D of the text should also be useful.) Use the base-catalyzed reaction of acetaldehyde (Eqs. 22.40a–b, text p. 1014, and 22.44, text p. 1016) and the acid-catalyzed reaction of acetone (Eqs. 22.46a–c, text p. 1017) as models. If you still have difficulty, *seek assistance!* You should have taken this approach with the crossed-aldol products in Study Problem 22.1. Be sure you understand the reactions shown in Eqs. 22.47 and 22.48 on text p. 1019. If you work through these examples, the reaction in Eq. 22.49 on text p. 1020 should be clear. The only thing different in this case is that the condensation occurs internally. Which carbons of the starting material are joined in the product? What is the structure of the enol intermediate involved? Use Eq. 22.46c on text p. 1017 to help you understand the formation of the carbon-carbon bond.

Just as in a sport, or in music, or in any other worthwhile endeavor, improving your skill requires practice. If you are concerned or discouraged that these transformations are not immediately obvious, don't be! They are more likely to become easier to understand, however, if you'll take the approach described here.

✓22.5 Variants of the Aldol and Claisen Condensations

Many variations of the aldol and Claisen condensations exist. The text has considered two: the Claisen-Schmidt variation of the aldol condensation, and the Dieckmann variation of the Claisen condensation. Possibly you'll hear of the Perkin condensation, the Stobbe condensation, the Knoevenagel condensation, or other variations that have no name. To cover all these in a text of reasonable (or even unreasonable) size is neither possible nor even necessary. For if you master the principles of reactivity contained in Chapters 21 and 22, understanding variations of these condensation reactions will only be a matter of re-applying the principles. In fact, mastering these principles might even allow you to invent your own condensation reaction!

22.6 Malonic Ester Alkylation

Notice in Eq. 22.64b of the text (p. 1036) that a *secondary* alkyl halide is used to alkylate the enolate ion of malonic ester. When you studied the S_N2 reaction, you learned that a side reaction is E2 elimination, and that this side reaction is particularly prevalent with secondary and tertiary alkyl halides. Because the product of Eq. 22.64b is formed in high yield, elimination is evidently not a problem in this case, and one might ask why.

Recall (Sec. 9.5F) that, although most strong bases are good nucleophiles, a number of relatively weak bases are also good nucleophiles, and that branched or secondary alkyl halides undergo S_N2 reactions with such nucleophiles with little interference from E2 reactions. (Note particularly the example in Eq. 9.45 on text p. 380.) The enolate ion of diethyl malonate is in this category. Chemists have found that delocalized (that is, resonance-stabilized) carbanions such as enolate ions are particularly good nucleophiles (for reasons that are not completely understood). Consequently, the S_N2 reaction is particularly rapid for these anions, and the E2 reaction does not interfere unless highly branched secondary or tertiary alkyl halides are used.

22.7 Alkylation of Enolate Ions

The acetoacetic ester synthesis requires the presence of the ester group of a β-keto ester; this group is then "thrown away" by hydrolysis and decarboxylation at the end of the synthesis and replaced by a hydrogen. For the same reasons that carboxylic acids with quaternary α-carbons cannot be prepared by the malonic ester synthesis, ketones with quaternary α-carbons cannot be prepared by the acetoacetic ester synthesis. For example, suppose we want to prepare 2,2-dimethylcyclohexanone:

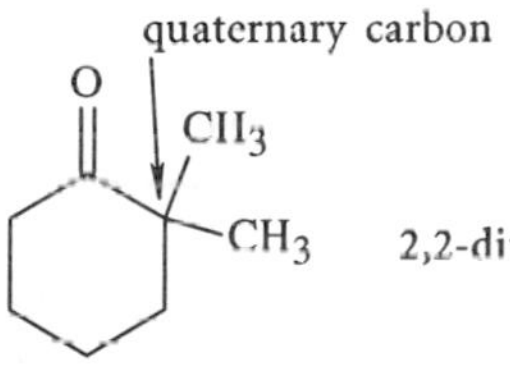

quaternary carbon

O, CH_3, CH_3 2,2-dimethylcyclohexanone

Although 2-methylcyclohexanone is easily prepared by a Dieckman condensation-alkylation sequence, introduction of the second methyl group is not possible in such a reaction scheme.

Why not, then, resort to the strategy used in Sec. 22.7B of the text? Methylcyclohexanone could be alkylated *directly* by forming its enolate ion with a strong base such as LCHIA and treatment of that enolate with methyl iodide:

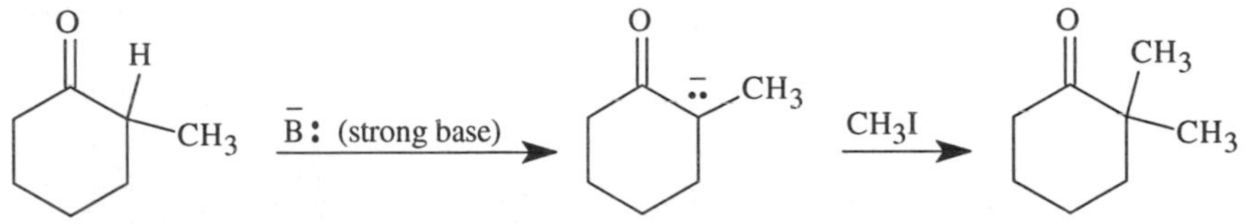

O H CH_3 $\xrightarrow{\bar{B}\text{: (strong base)}}$ O CH_3 $\xrightarrow{CH_3I}$ O CH_3 CH_3

Can you see the problem with this approach? With an ester, only one enolate ion is possible. But with a ketone, in general *two* enolate ions are possible:

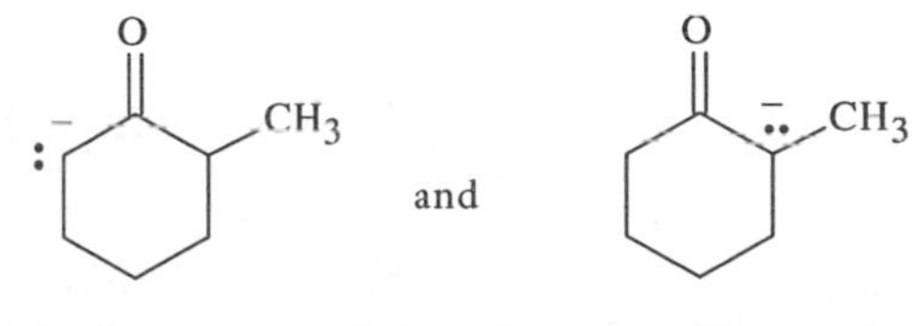

O CH_3 and O CH_3

The problem, then, is how to form one enolate and not the other, and to keep the two enolate ions from equilibrating once the desired one is formed.

But the problems don't stop here. Suppose a way could be found to form the desired enolate ion. An additional problem is that enolate ions can react in more than one way. The enolate ions of ketones are not only nucleophilic at carbon; they are also nucleophilic at *oxygen*:

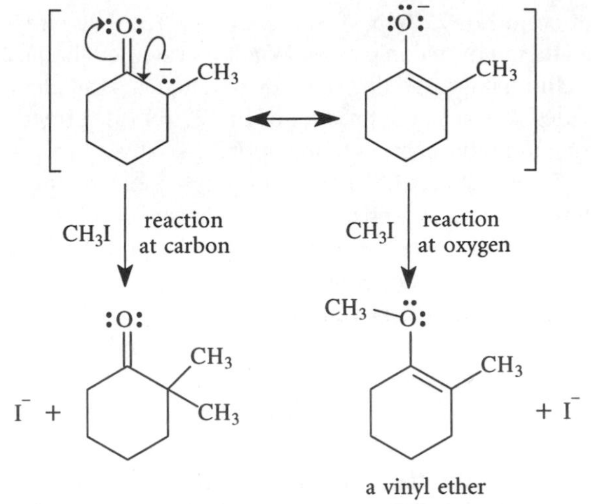

CH₃I | reaction at carbon

CH₃I | reaction at oxygen

I^- +

+ I^-

a vinyl ether

Alkylation at oxygen gives a vinyl ether—sometimes called an *enol ether,* because conceptually it is derived by replacement of the —OH hydrogen of an enol with an alkyl group. In general, both products are formed. (Alkylation of oxygen is not usually observed with esters because enols of esters—and enol ethers of esters—are *much* less stable relative to their parent esters than enols of ketones are; see Eq. 22.13 on text p. 1004.)

Direct alkylation of ketones, then, is a much more complex problem than alkylation of esters. Nevertheless, chemists have discovered ways to control, in many cases, which enolate ion is formed, and to control which alkylation product is produced. These solutions, although beyond the scope of an introductory text, are conceptually not difficult to understand if you understand the principles of enolate-ion chemistry.

✓22.8 Further Analysis of the Claisen Condensation

If you imagine the Claisen condensation on an ester of general structure *A*, you'll see that saponification and decarboxylation *must* yield a *symmetrical ketone* of general structure *C*. (A symmetrical ketone is a ketone in which the two groups attached to the carbonyl carbon are the same.)

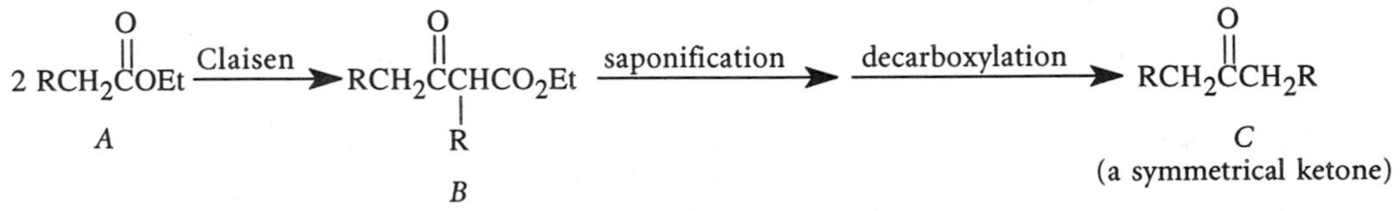

$$2\ \mathrm{RCH_2\overset{O}{\overset{\|}{C}}OEt}\ (A) \xrightarrow{\text{Claisen}} \mathrm{RCH_2\overset{O}{\overset{\|}{C}}\underset{R}{\underset{|}{C}}HCO_2Et}\ (B) \xrightarrow{\text{saponification}} \xrightarrow{\text{decarboxylation}} \mathrm{RCH_2\overset{O}{\overset{\|}{C}}CH_2R}\ (C)$$

(a symmetrical ketone)

(Be sure to go through this case in detail to verify the foregoing statement.) Alkylation of the Claisen product *B* can lead to a variety of *substituted derivatives* of *C*. For example, treatment of *B* with NaOEt, then a general alkyl halide R′—I, would give, after saponification, acidification, and decarboxylation, a general derivative *D*.

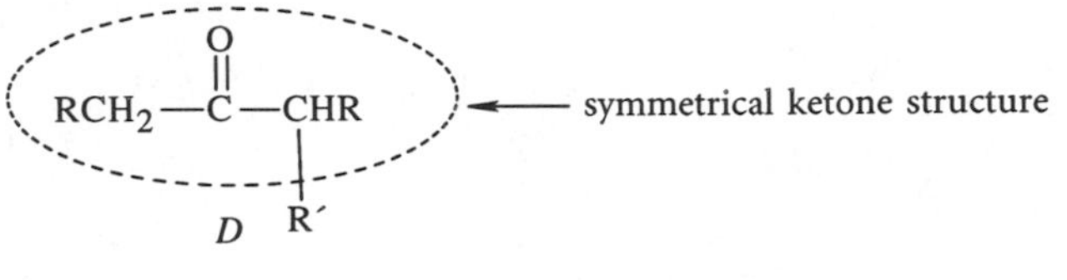

$$\mathrm{RCH_2{-}\overset{O}{\overset{\|}{C}}{-}\underset{R'}{\underset{|}{C}}HR}\quad (D)$$

← symmetrical ketone structure

Notice the symmetrical ketone structure "lurking" as the circled part of *D*.

When you are asked to prepare a ketone by the Claisen condensation, look for such symmetrical-ketone "substructures" within the target molecule. Problem 22.38 of the text (p. 1043) gives you practice in applying this strategy.

✓22.9 Synthetic Equivalents in Conjugate Addition

Study Guide Links 17.3 (p. 552 of this manual) and 20.7 (p. 709 of this manual) discussed the notion of *synthetic equivalents* (and the corresponding property, *synthetic equivalence*), that is, thinking of one group in terms of other groups from which it might be formed. This notion is particularly useful in planning conjugate additions. In Study Problem 22.6, for example, the di(ethoxycarbonyl)methyl anion (the conjugate base of diethyl malonate, $^{-}$:CH(CO_2Et)$_2$) is used as the *synthetic equivalent* of a carboxymethyl anion ($^{-}$:CH_2CO_2H), which itself cannot exist. (Why can't this anion exist?).

What are some other synthetic equivalents that are useful in conjugate-addition reactions?

$$:\bar{C}N \text{ is equivalent to } :\underline{C}(=O)\text{—OH} \quad \text{(see Eq. 22.79 on text p. 1045)}$$

$$:\bar{C}N \text{ is also equivalent to } :\bar{C}H_2\text{—}NH_2 \quad \text{(see Sec. 21.9C on text pp. 972–3)}$$

$$:\bar{C}H(CO_2Et)\text{—}C(=O)\text{—}R \text{ is equivalent to } :\bar{C}H_2\text{—}C(=O)\text{—}R \quad \text{(see Eq. 22.75 on text p. 1042)}$$

(You should show how each group can be converted into the other, for example, how a —C≡N group can be converted into a —CH_2NH_2 group.)

When you can start thinking in terms of synthetic equivalents you really have started to master the art of organic synthesis.

22.10 Conjugate Addition of Organocuprate Reagents

Although the mechanism presented in the text is conceptually useful, it does not explain the effect of copper in promoting conjugate addition. Because carbonyl addition rather than conjugate addition is observed in the *absence* of copper, it is evident that copper has a special effect. A substantial body of evidence has shown that conjugate-addition reactions of lithium organocuprate reagents involve free-radical intermediates, and that conjugate addition is a consequence of a free-radical mechanism.

SOLUTIONS TO PROBLEMS

Solutions to In-Text Problems

22.1 The acidic hydrogens are the α-hydrogens on the carbons between the two carbonyl groups. (—OEt = ethoxy group = $—OCH_2CH_3$.)

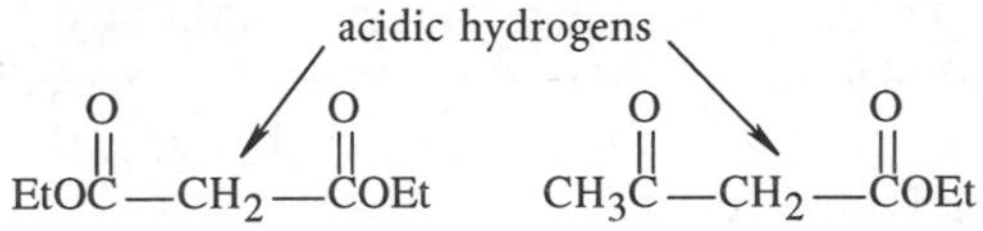

acidic hydrogens

$EtO\overset{O}{\overset{\|}{C}}—CH_2—\overset{O}{\overset{\|}{C}}OEt$ — diethyl malonate

$CH_3\overset{O}{\overset{\|}{C}}—CH_2—\overset{O}{\overset{\|}{C}}OEt$ — ethyl acetoacetate

The reason that these hydrogens are particularly acidic is that their conjugate-base enolate ions are stabilized by the polar effects and resonance effects of two carbonyl groups, whereas the conjugate-base enolate ion of an ordinary ester is stabilized by the corresponding effects of only one carbonyl group. The resonance structures of the conjugate-base enolate ion of diethyl malonate are as follows; the resonance structures of the conjugate-base enolate ion of ethyl acetoacetate are similar.

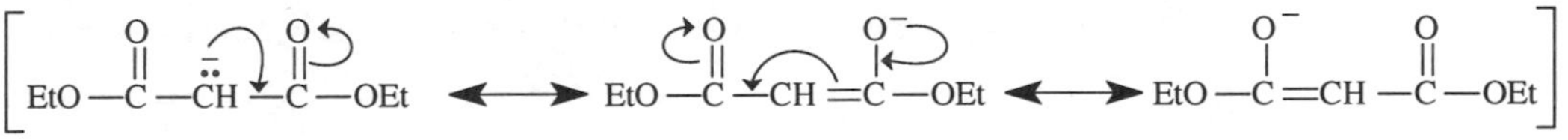

$$\left[EtO—\overset{O}{\overset{\|}{C}}—\ddot{C}H^{-}—\overset{O}{\overset{\|}{C}}—OEt \longleftrightarrow EtO—\overset{O}{\overset{\|}{C}}—CH=\overset{O^-}{\overset{|}{C}}—OEt \longleftrightarrow EtO—\overset{O^-}{\overset{|}{C}}=CH—\overset{O}{\overset{\|}{C}}—OEt\right]$$

resonance structures for the conjugate-base enolate ion of diethyl malonate

22.2 The imide of succinic acid (succinimide) is more acidic, because its conjugate-base anion is stabilized by the polar effects and resonance effects of two adjacent carbonyl groups. The conjugate-base anion of succinic acid diamide is stabilized by resonance interaction with only one carbonyl group.

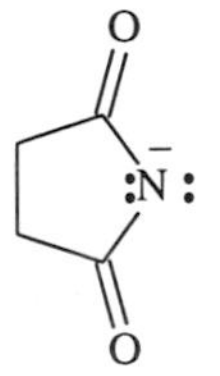

conjugate-base anion of succinimide

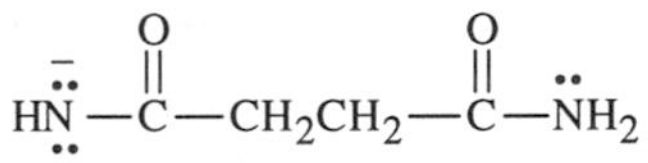

$$H\ddot{N}^{-}—\overset{O}{\overset{\|}{C}}—CH_2CH_2—\overset{O}{\overset{\|}{C}}—\ddot{N}H_2$$

conjugate-base anion of succinic acid diamide

Note that the resonance effect in the conjugate base of succinimide far outweighs the statistical effect of two amide groups in succinic acid diamide.

22.3 A mechanism for replacement of one hydrogen in the reaction of Eq. 22.6 is shown in the following equation. (The mechanisms for replacement of the others are identical.) Only the α-hydrogens are replaced because the enolate ion is the only carbanion stable enough to be formed. The carbanion intermediates required in a similar mechanism for the replacement of the hydrogens other than the α-hydrogens are not resonance-stabilized.

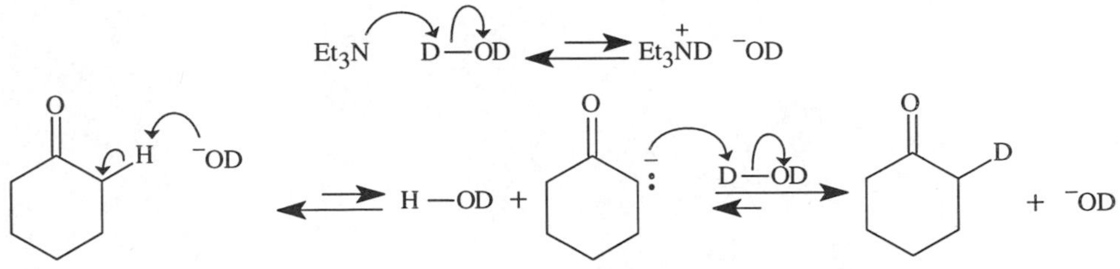

$$Et_3N \quad D—OD \rightleftharpoons Et_3\overset{+}{N}D \quad {}^{-}OD$$

$$\text{(cyclohexanone)}—H \quad {}^{-}OD \rightleftharpoons H—OD + \text{(enolate)} \quad D—OD \rightleftharpoons \text{(2-deuteriocyclohexanone)} + {}^{-}OD$$

resonance-stabilized enolate ion

22.4 The proton NMR spectrum would change as follows:

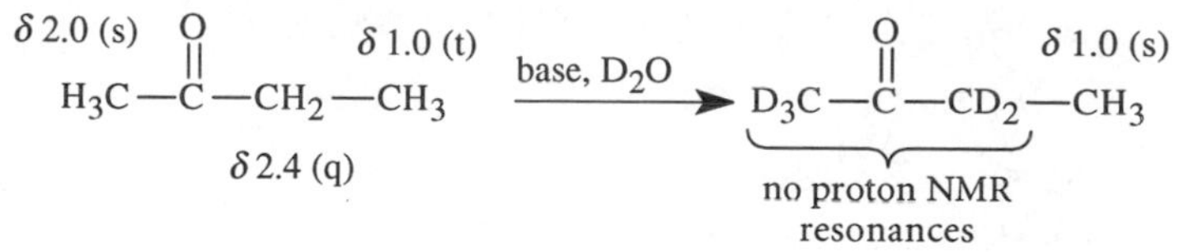

$$\underset{\delta\,2.0\,(s)}{H_3C}-\overset{O}{\overset{\|}{C}}-\underset{\delta\,2.4\,(q)}{CH_2}-\underset{\delta\,1.0\,(t)}{CH_3} \xrightarrow{\text{base, }D_2O} \underbrace{D_3C-\overset{O}{\overset{\|}{C}}-CD_2}_{\text{no proton NMR resonances}}-\underset{\delta\,1.0\,(s)}{CH_3}$$

That is, in basic D_2O, the resonances for the α-protons disappear and the triplet at δ 1.0 becomes a singlet.

22.5 Exchange does not occur because the orbital containing the unshared electron pair on the α-carbon of the enolate ion is perpendicular to the π-electron system of the carbonyl group; therefore, the enolate ion is not stabilized by resonance. Another way to view this situation is suggested by the hint: the resonance structure of the enolate ion violates Bredt's rule. This situation is discussed in rule 5, text pp. 669.

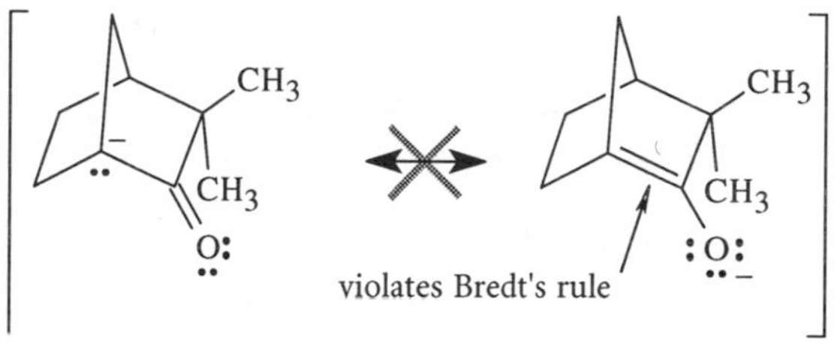

CH3 CH3 O: ⟷ (crossed out) CH3 CH3 :O:⁻

violates Bredt's rule

22.6 (a) All α-hydrogens as well as the O—H hydrogen are exchanged for deuterium:

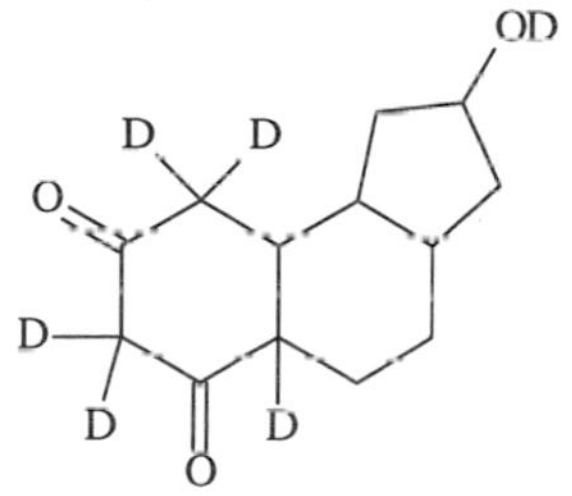

OD D D O D D D O

(b) The only possible enolate ion is not resonance-stabilized for the same reason that the enolate ion in Problem 22.5 is not resonance-stabilized; consequently, it is too unstable to form. Because deuterium exchange of the α-hydrogens would involve an enolate ion intermediate, deuterium exchange cannot occur.

22.7 (a)

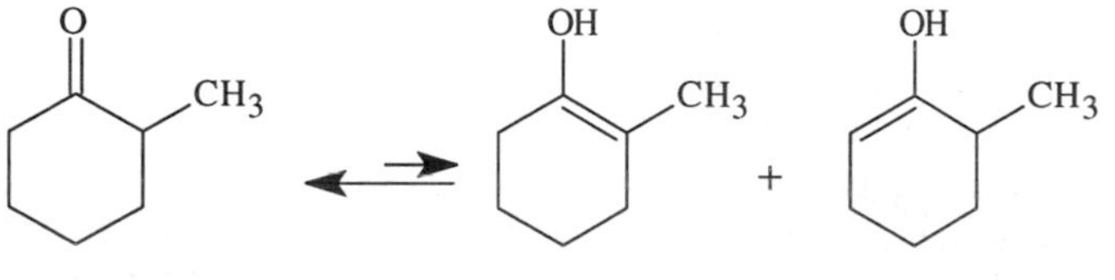

O CH3 ⇌ OH CH3 + OH CH3

2-methylcyclohexanone — enol forms of 2-methylcyclohexanone

(b)

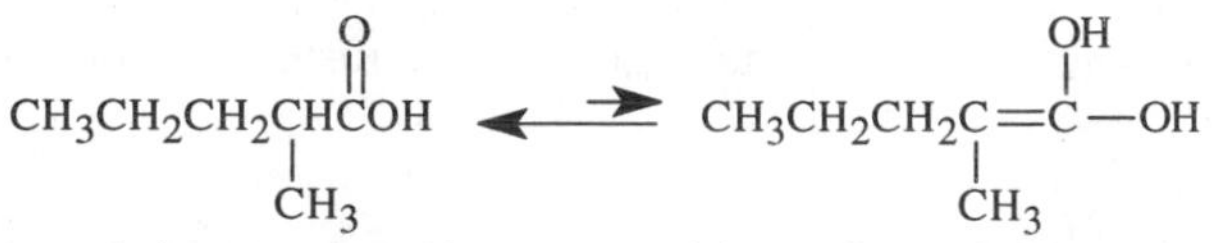

$$CH_3CH_2CH_2\underset{\underset{CH_3}{|}}{CH}\overset{O}{\overset{\|}{C}}OH \rightleftharpoons CH_3CH_2CH_2\underset{\underset{CH_3}{|}}{C}=\overset{OH}{\overset{|}{C}}-OH$$

2-methylpentanoic acid — enol form of 2-methylpentanoic acid

(c) Benzaldehyde, PhCH=O, has no enol forms because it has no α-hydrogens.

(d)

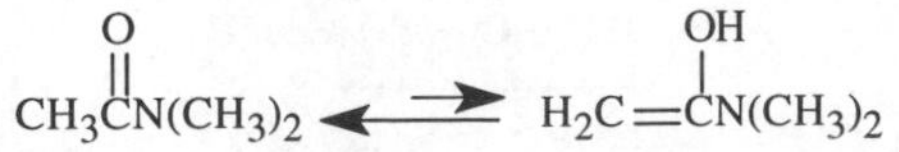

$$\underset{N,N\text{-dimethylacetamide}}{CH_3\overset{O}{\overset{\|}{C}}N(CH_3)_2} \rightleftharpoons \underset{\text{enol form of } N,N\text{-dimethylacetamide}}{H_2C{=}\overset{OH}{\overset{|}{C}}N(CH_3)_2}$$

22.8 The enol forms of 2-butanone.

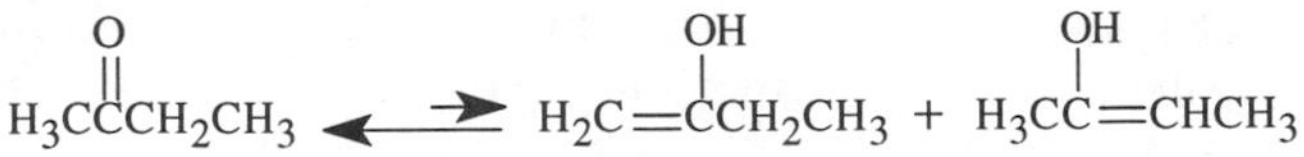

$$\underset{\text{2-butanone}}{H_3C\overset{O}{\overset{\|}{C}}CH_2CH_3} \rightleftharpoons \underset{\textit{least stable}}{H_2C{=}\overset{OH}{\overset{|}{C}}CH_2CH_3} + \underset{(E)\text{ and }(Z)\ \textit{most stable}}{H_3C\overset{OH}{\overset{|}{C}}{=}CHCH_3}$$

enol forms of 2-butanone

We know that double bonds are stabilized by alkyl substitution (text pp. 130–3). Hence the enol on the left in the previous equation is the least stable.

22.9 (a) The "enol" form of a nitro compound:

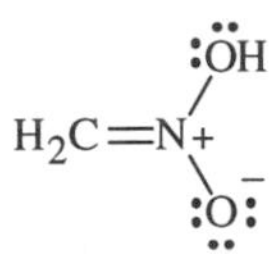

$$H_2C{=}\overset{+}{N}(\ddot{O}H)(\ddot{O}:^-)$$

"enol" form of nitromethane
(an *aci*-nitro compound)

(b) The "enol" form of an amide that results from tautomerization involving the nitrogen is called an *imidic acid;* see Eq. 21.20b on text p. 956. Notice that this particular amide cannot enolize toward the α-carbon because it has no α-hydrogens on that carbon.

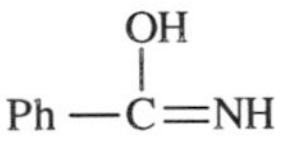

$$Ph{-}\overset{OH}{\overset{|}{C}}{=}NH$$

"enol" form of benzamide
(an imidic acid)

22.10 (a) One source of stabilization of the enols of β-diketones is their intramolecular hydrogen bonds; see the structure preceding Problem 22.7 on text p. 1005. In aqueous solution, the solvent can serve as a hydrogen-bond donor to the carbonyl oxygen; hence the energetic advantage of the enol that results from intramolecular hydrogen bonding is reduced in aqueous solution. In a solvent that cannot donate hydrogen bonds, such as hexane, there are no external hydrogen bonds, and the internal hydrogen bond thus provides an energetic advantage to the enol.

(b) As the structure above Problem 22.7 on text p. 1005 shows, the enol form of 2,4-pentanedione has a conjugated π-electron system involving a carbonyl double bond and a carbon-carbon double bond. This conjugated system in the enol is the source of the UV spectrum. The absorption is stronger in hexane because there is a greater proportion of the enol form in hexane for the reasons discussed in the solution to part (a) of this problem.

22.11 (a) The enol is formed by the mechanism shown in Eq. 22.17b on text p. 1006. The enol is achiral, and it can be protonated at either face of the double bond at the same rate to give the racemate of the protonated ketone.

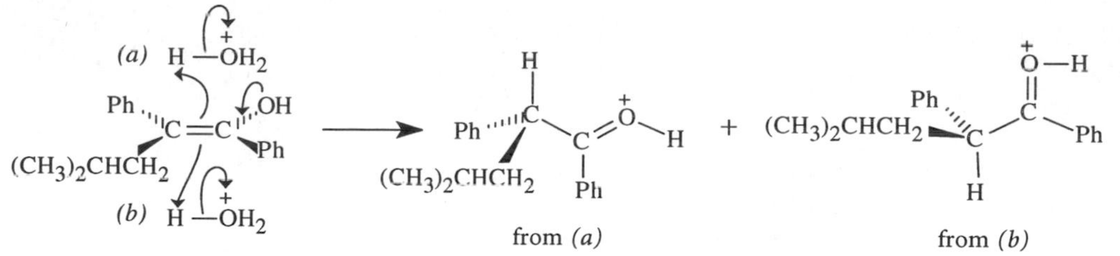

Loss of a proton from each of these conjugate acids gives the racemate.

(b) The enol is formed by the mechanism shown in Eq. 22.17b on text p. 1006, except that D_3O^+ is the acid. This results in the "washout" of one α-hydrogen into the large excess of deuterated solvent.

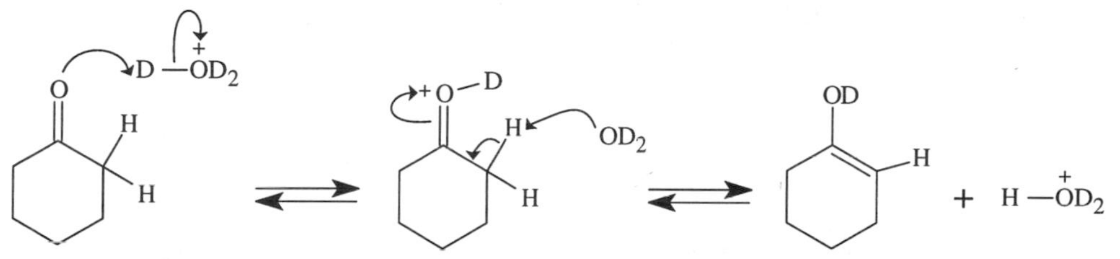

Protonation of the double bond by D_3O^+ gives a deuterium at the α-position. Replacement of one α-hydrogen by deuterium is shown; the mechanism for replacement of the other α-hydrogens is identical.

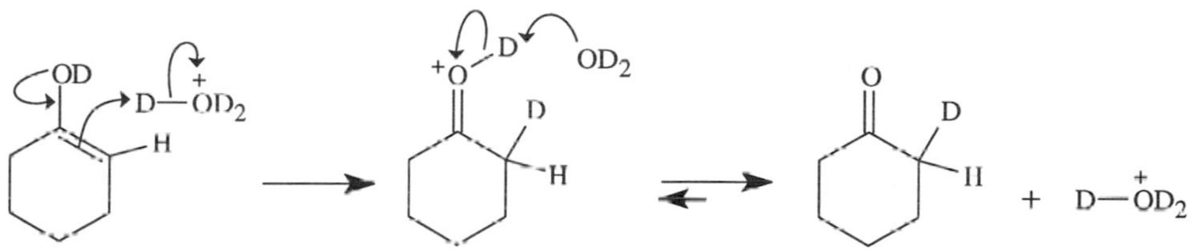

22.12 (a) The reaction free-energy diagram is given in Fig. SG22.1.

(b) The reaction free-energy diagram is given in Fig. SG22.2. Notice that the first two steps are identical to those shown in Fig. 22.1, because the process is identical.

22.13 (a) Because enolization is the process responsible for acid-catalyzed racemization, and because the rate-limiting process in acid-catalyzed ketone halogenation is enolization, the rates of the two processes are the same, and corresponds to the rate of enolization.

(b) Because the rate of ketone halogenation is independent of the halogen concentration, the rates of halogenation of the same ketone with two different halogens are also independent of halogen concentration and therefore independent of the identity of the halogen itself. In fact, the rate in both cases is the rate of enolization, as in part (a). The two processes are compared at the same acid concentration because the enolization process is acid-catalyzed (Eq. 22.2, text p. 1008).

22.14 In part (b), benzophenone does not halogenate because it cannot form an enolate ion (it has no α-hydrogens), and the aromatic rings are deactivated by the carbonyl group toward electrophilic halogenation. The answers to parts (a) and (c) are as follows:

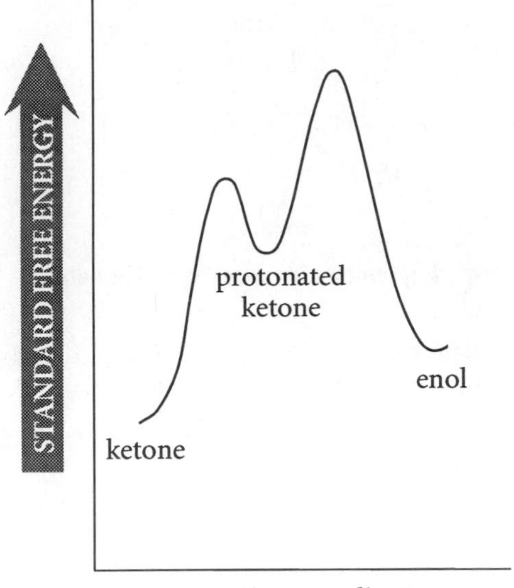

Figure SG22.1 Reaction free-energy diagram for enolization of a ketone to accompany the solution to Problem 22.12(a).

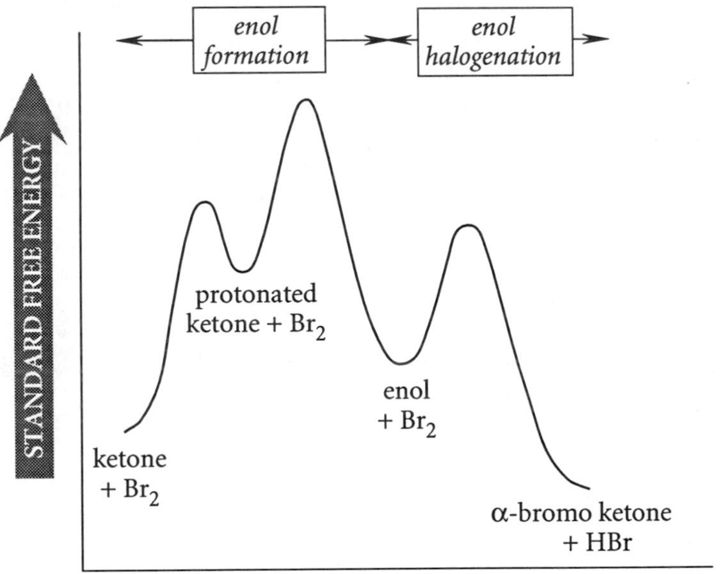

Figure SG22.2 Reaction free-energy diagram to accompany the solution to Problem 22.12(b). Notice that the rate-limiting steps for enol formation and overall halogenation are the same.

(a)

$$\text{(decalin-4a-yl)}-C(=O)-O^- \, Na^+ \quad + \quad HCBr_3$$

(c)

$$Ph-\overset{\overset{\displaystyle O}{\|}}{C}-O^-\,Na^+ \;+\; HCBr_3$$

In part (c), the alcohol is first oxidized to the corresponding ketone, acetophenone, which then undergoes the haloform reaction.

22.15 2,4-Hexanedione gives succinic acid and iodoform when subjected to the iodoform reaction.

$$\underset{\text{2,4-hexanedione}}{CH_3\overset{O}{\overset{\|}{C}}CH_2CH_2\overset{O}{\overset{\|}{C}}CH_3} \xrightarrow[2)\ H_3O^+]{1)\ I_2,\ \text{base}} \underset{\text{succinic acid}}{HO\overset{O}{\overset{\|}{C}}CH_2CH_2\overset{O}{\overset{\|}{C}}OH} + \underset{\text{iodoform}}{2HCI_3}$$

It is possible (and even likely) that this ketone could also undergo iodination at the methylene carbons, followed by iodination at the methyl carbons. Hence, succinic acid may not be the only carboxylic acid product in addition to iodoform. More than likely some iodinated succinic acids (as well as by-products from other reactions not discussed in this chapter) would be produced as well.

22.16 (a) The α-bromo acid bromide reacts with ethanol to give the α-bromo ester.

$$\underset{\text{phenylacetic acid}}{PhCH_2\overset{O}{\overset{\|}{C}}OH} \xrightarrow{Br_2,\ PBr_3} PhCH(Br)\overset{O}{\overset{\|}{C}}Br \xrightarrow{HOCH_2CH_3} \underset{\text{ethyl }\alpha\text{-bromo-}\alpha\text{-phenylacetate}}{PhCH(Br)\overset{O}{\overset{\|}{C}}OCH_2CH_3} + HBr$$

(b) The α-bromo acid bromide is formed, and it reacts with ammonia in an S_N2 reaction to give the amide of the α-amino acid alanine.

$$\underset{\text{propionic acid}}{CH_3CH_2\overset{O}{\overset{\|}{C}}OH} \xrightarrow{Br_2,\ PBr_3} CH_3CH(Br)\overset{O}{\overset{\|}{C}}Br \xrightarrow{NH_3} \underset{\substack{\text{alaninamide}\\ \text{(2-aminopropanamide)}}}{CH_3CH(NH_2)\overset{O}{\overset{\|}{C}}NH_2} + 2\ \overset{+}{N}H_4\ Br^-$$

In part (a), *neutral* ethanol is not basic enough to act as a nucleophile and displace the α-bromine.

22.17 (a) The base pyridine displaces the α-bromine is an S_N2 reaction.

$$CH_3CH_2\overset{O}{\overset{\|}{C}}CH_2-\overset{+}{N}C_5H_5 \quad Br^-$$

(b) The base acetate ion displaces the α-bromine to give a compound that is an example of a *phenacyl ester.*

$$\underset{\substack{\text{benzoylmethyl acetate}\\ \text{(a phenacyl ester)}}}{CH_3\overset{O}{\overset{\|}{C}}-O-CH_2\overset{O}{\overset{\|}{C}}Ph} + Na^+\ Br^-$$

22.18 One equivalent of NaOH is consumed by its reaction with the carboxylic acid. The second equivalent forms the 2,4-dichlorophenolate ion, which is the nucleophile in the ensuing S_N2 reaction.

$$Cl-CH_2\overset{O}{\overset{\|}{C}}-O-H \quad {}^-OH \longrightarrow Cl-CH_2\overset{O}{\overset{\|}{C}}-O^- + H-OH$$

$$\text{2,4-}Cl_2C_6H_3-O-H \quad {}^-OH \longrightarrow \text{2,4-}Cl_2C_6H_3-O^- + H-OH$$

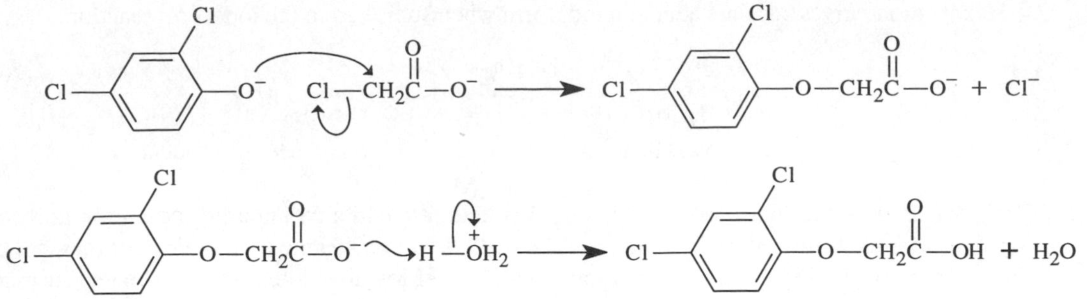

$$\text{Cl–}C_6H_3(\text{Cl})\text{–O}^- + \text{Cl–CH}_2\text{C(=O)–O}^- \longrightarrow \text{Cl–}C_6H_3(\text{Cl})\text{–O–CH}_2\text{C(=O)–O}^- + \text{Cl}^-$$

$$\text{Cl–}C_6H_3(\text{Cl})\text{–O–CH}_2\text{C(=O)–O}^- + \text{H–}\overset{+}{\text{O}}\text{H}_2 \longrightarrow \text{Cl–}C_6H_3(\text{Cl})\text{–O–CH}_2\text{C(=O)–OH} + H_2O$$

Note that the final step does not take place until acid is added to the reaction mixture; under the basic conditions of the substitution reaction the product is ionized.

22.19 (a) The aldol addition reaction of phenylacetaldehyde:

$$\text{PhCH}_2\text{CH=O} + \text{Ph}\bar{\text{C}}\text{HCH=O} \longrightarrow \text{PhCH}_2\text{CH(O}^-\text{)–PhCHCH=O} \xrightarrow{\text{H–OH}} \text{PhCH}_2\text{CH(OH)–PhCHCH=O} + {}^-\text{OH}$$

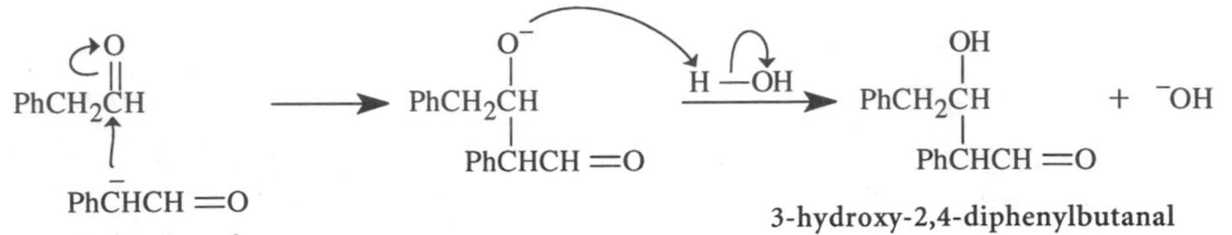

enolate ion of phenylacetaldehyde

3-hydroxy-2,4-diphenylbutanal

(b) The aldol addition reaction of propionaldehyde:

$$\text{CH}_3\text{CH}_2\text{CH=O} + \text{CH}_3\bar{\text{C}}\text{HCH=O} \longrightarrow \text{CH}_3\text{CH}_2\text{CH(O}^-\text{)–CH}_3\text{CHCH=O} \xrightarrow{\text{H–OH}} \text{CH}_3\text{CH}_2\text{CH(OH)–CH}_3\text{CHCH=O} + {}^-\text{OH}$$

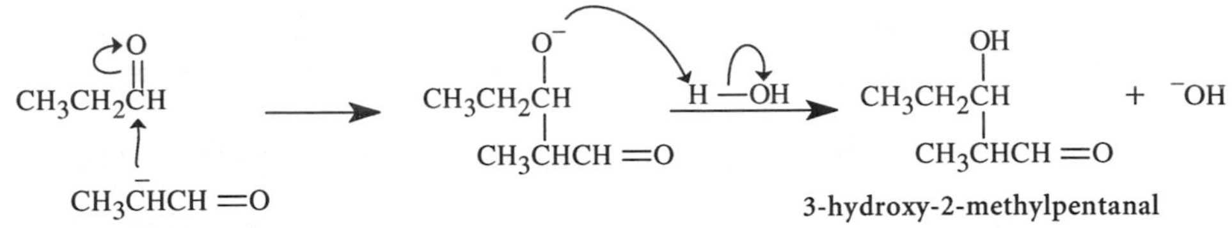

enolate ion of propionaldehyde

3-hydroxy-2-methylpentanal

22.20 (a) The problem states that equal molar amounts of each component. (Otherwise two moles of the aldehyde might react—one mole at each set of ketone α-hydrogens.) The enolate ion of the ketone reacts with the aldehyde because aldehyde carbonyl groups are more reactive in additions than ketone carbonyl groups.

(2-furyl)CH=CH–C(=O)CH$_3$ (structure with H, H, O, CH$_3$ labels)

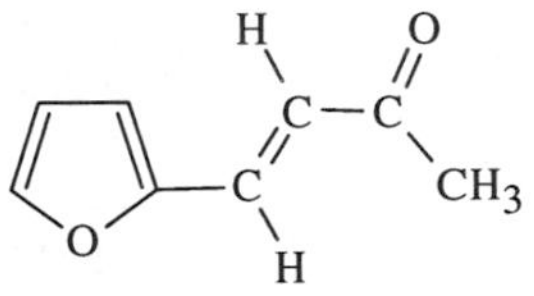

(b) As in part (a), the enolate ion of the ketone adds to the aldehyde carbonyl group to give product *A* because aldehyde carbonyl groups are more reactive than ketone carbonyl groups. It is also likely that aldol condensation of two molecules of hexanal will occur to give product *B*.

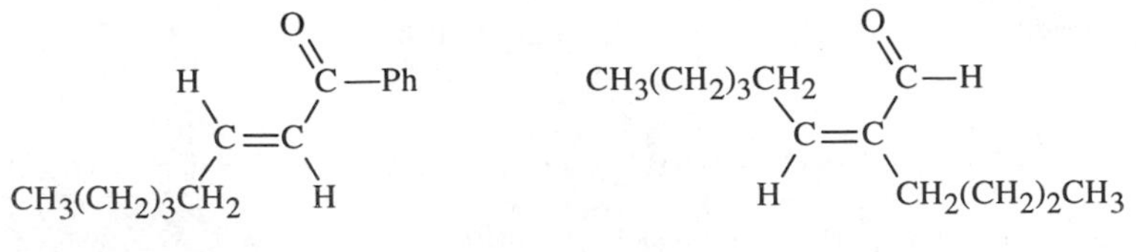

A: $CH_3(CH_2)_3CH_2$(H)C=C(H)–C(=O)–Ph

B: $CH_3(CH_2)_3CH_2$(H)C=C($CH_2(CH_2)_2CH_3$)–C(=O)–H

(c) This is an intramolecular aldol condensation.

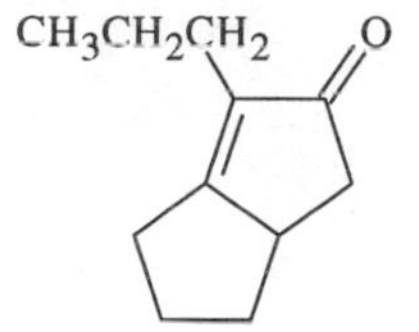

CH3CH2CH2 O

22.21 The fact that each product contains three carbons indicates the point of cleavage. The reverse aldol addition "undoes" the β-hydroxy ketone:

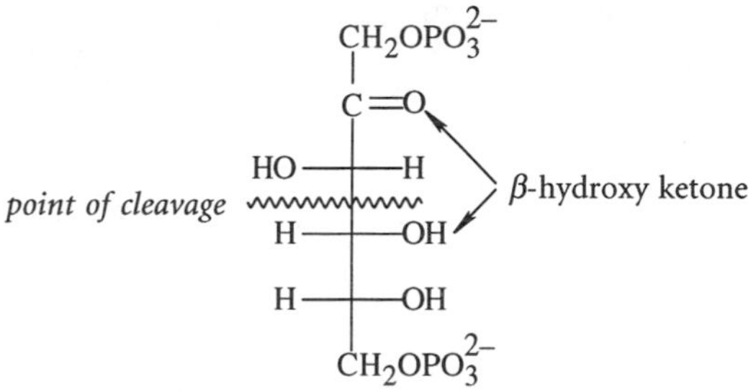

point of cleavage
β-hydroxy ketone

The curved-arrow mechanism is generated by running the aldol addition (Eqs. 22.40a–b, text p. 1014) in reverse.

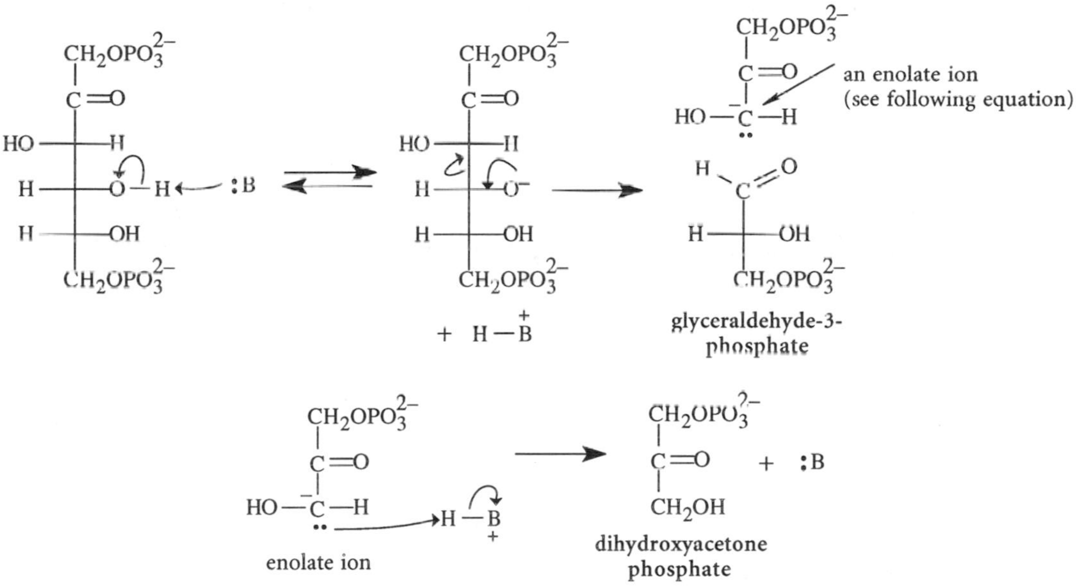

an enolate ion (see following equation)

glyceraldehyde-3-phosphate

enolate ion

dihydroxyacetone phosphate

22.22 (a) In this case, the enolate ion of a symmetrical ketone, 3-pentanone, would have to react with *p*-methoxybenzaldehyde. There is only one possible enolate ion, and the aldehyde carbonyl group is much more reactive than the ketone carbonyl group. Hence, the reaction is reasonable.

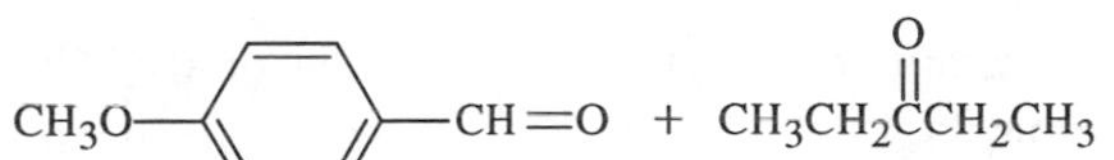

$$CH_3O-C_6H_4-CH{=}O \; + \; CH_3CH_2\overset{O}{\overset{\|}{C}}CH_2CH_3$$

p-methoxybenzaldehyde 3-pentanone

(b) The product shown would require the following starting materials. However, since the enolate ion required for the synthesis is only one of two that could form, at least two isomeric products are possible of which the desired product is only one. The desired product would therefore be formed as one component of a mixture of isomers.

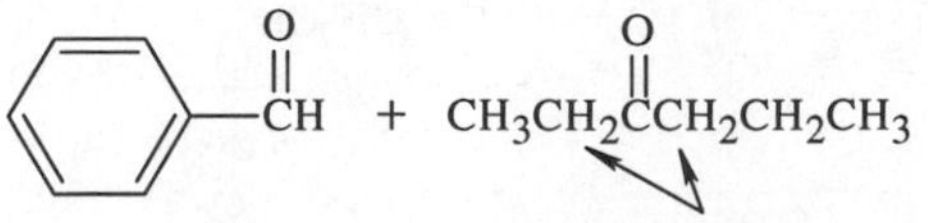

$$\text{C}_6\text{H}_5\text{CH}{=}\text{O} + \text{CH}_3\text{CH}_2\overset{\text{O}}{\overset{\|}{\text{C}}}\text{CH}_2\text{CH}_2\text{CH}_3$$

an enolate ion could form at either carbon

(c) This product would require the intramolecular aldol condensation of the following keto aldehyde. In this case, the enolate ion formed at the α-carbon of the aldehyde would have to react with the ketone carbonyl group. Because the aldehyde carbonyl group is more reactive, and because its intramolecular reaction with an enolate ion of the ketone would also give a six-membered ring, synthesis of the compound shown in the problem is not reasonable. The following reaction would likely occur instead:

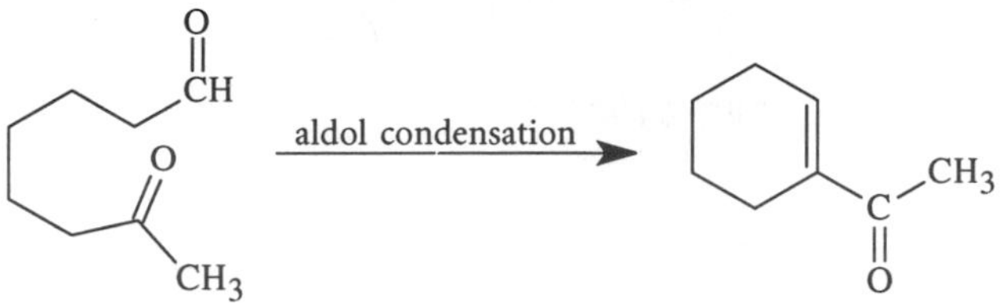

CH, O, CH_3 —aldol condensation→ CH_3, C, O

This compound is required for the product given.

This compound would be formed instead.

(d) This product is the result of an intramolecular reaction of a ketone enolate and an aldehyde carbonyl group; hence, its synthesis from the following starting material is reasonable.

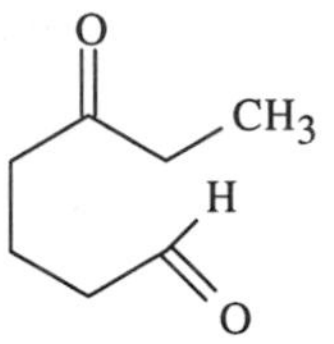

O, CH_3, H, O

(e) A double aldol condensation of the following starting materials would give the desired product. The reaction is particularly favorable because the product is highly conjugated.

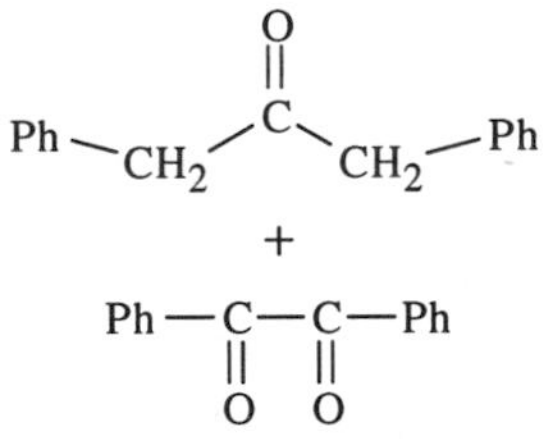

$$\text{Ph}{-}\text{CH}_2{-}\overset{\text{O}}{\overset{\|}{\text{C}}}{-}\text{CH}_2{-}\text{Ph}$$

$$+$$

$$\text{Ph}{-}\underset{\text{O}}{\underset{\|}{\text{C}}}{-}\underset{\text{O}}{\underset{\|}{\text{C}}}{-}\text{Ph}$$

(f) This is a reaction of two equivalents of benzaldehyde with one of acetone. Because an aldehyde without α-hydrogens is involved, the reaction is a reasonable example of the Claisen-Schmidt condensation.

$$2\ \text{PhCH}{=}\text{O} + \text{CH}_3\overset{\text{O}}{\overset{\|}{\text{C}}}\text{CH}_3$$

(g) Because the diketone starting material is symmetrical, the product shown is the only possible one, and thus the synthesis is reasonable.

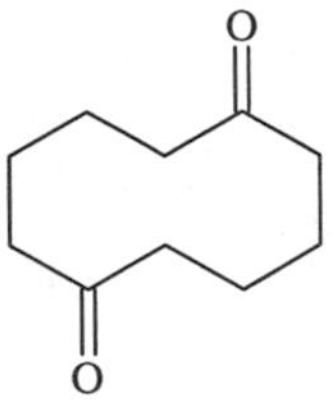

O, O

(h) This product would require that the carbonyl group of acetone, $(CH_3)_2C{=}O$, react with the conjugate-base enolate ion of acetaldehyde, $CH_3CH{=}O$. Because the aldehyde carbonyl group is more reactive than the ketone carbonyl group, the more likely products would be the aldol condensation product of acetaldehyde along with the aldol condensation product derived from the reaction of the conjugate-base enolate ion of acetone with the carbonyl group of acetaldehyde.

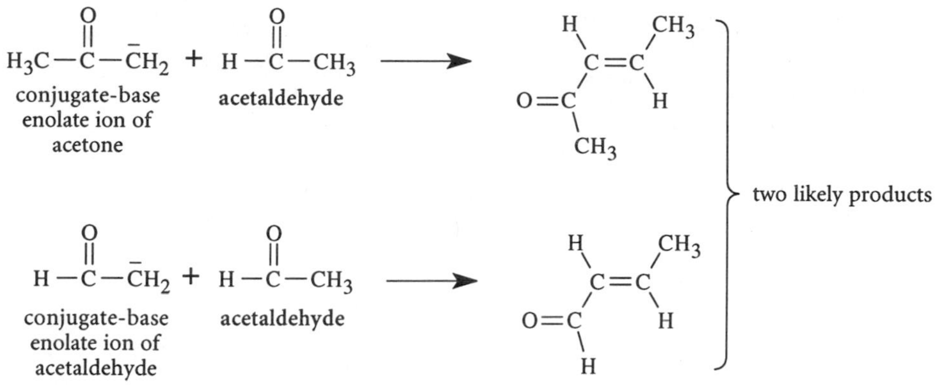

$$H_3C{-}\overset{O}{\overset{\|}{C}}{-}\overset{-}{C}H_2 + H{-}\overset{O}{\overset{\|}{C}}{-}CH_3 \longrightarrow$$

conjugate-base enolate ion of acetone; acetaldehyde → product: (H)(O=C(CH₃))C=C(CH₃)(H)

$$H{-}\overset{O}{\overset{\|}{C}}{-}\overset{-}{C}H_2 + H{-}\overset{O}{\overset{\|}{C}}{-}CH_3 \longrightarrow$$

conjugate-base enolate ion of acetaldehyde; acetaldehyde → product: (H)(O=C(H))C=C(CH₃)(H)

two likely products

22.23 First, break the double bond as shown in Eq. 22.50 on text p. 1021 to reveal the possible starting material.

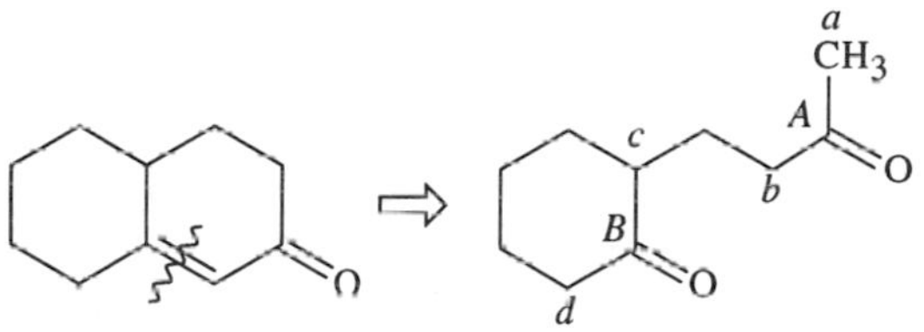

a, CH_3, A, c, b, B, O, O, d

The product shown in the problem results from the reaction of an enol formed at carbon *a* with carbonyl group *B*. Reaction of the enol formed at carbon *b* with carbonyl *B* would require formation of a strained four-membered ring (product *aB* below) and is thus not likely. Reaction of an enol formed at carbon *c* with carbonyl *A* also would require the formation of a strained four-membered ring (product *cA* below). Reaction of an enol formed at carbon *d* with carbonyl *A* would give bicyclic compound *dA*:

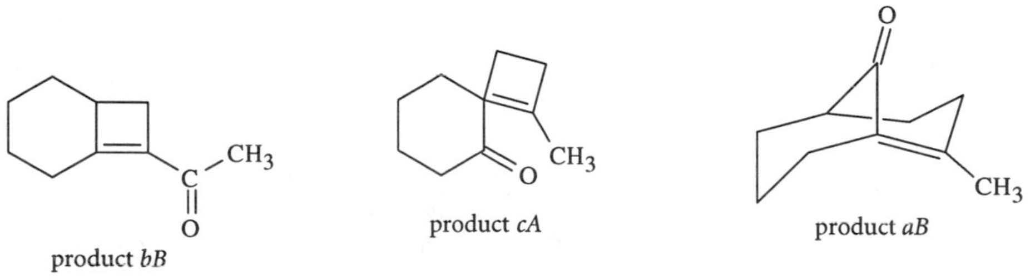

CH_3, C, O — product *bB*

O, CH_3 — product *cA*

O, CH_3 — product *aB*

This compound is a modest violation of Bredt's rule (Sec. 7.6C of the text; build a model to verify that it is strained). The aldol *addition* product could form, thus avoiding dehydration to the strained alkene, but aldol additions of ketones are highly reversible. The product shown in the problem is thus the only one that can readily be driven to completion by dehydration.

22.24 (a)

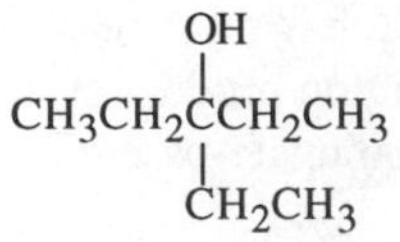

(b)

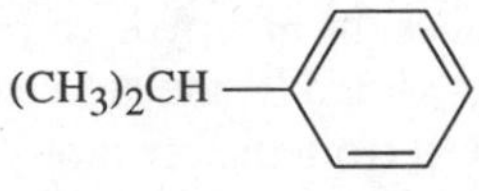

(c)

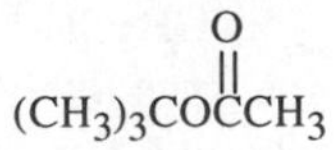

(d) $CH_3CH_2CH_2OH$

(e)

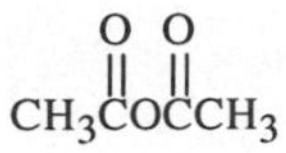

(f)

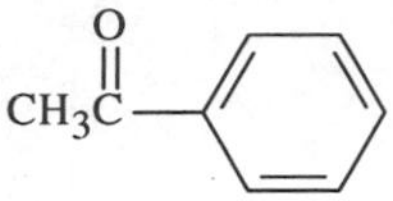

22.25 (a)

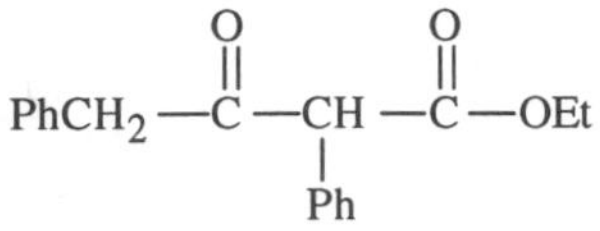

ethyl 2,4-diphenyl-3-oxobutyrate

(b)

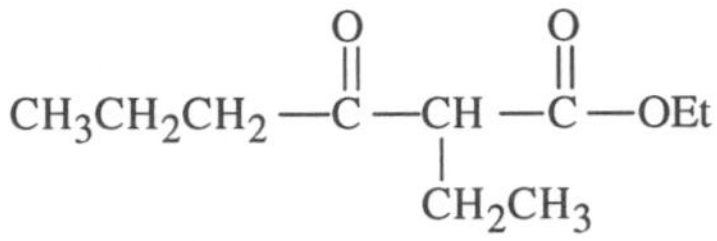

ethyl 2-ethyl-3-oxohexanoate

22.26 Hydroxide ion would *not* be a good base for the Claisen condensation because it would saponify ethyl acetate and convert it into acetate ion. The role of the ethoxy group in ethyl acetate is to serve as a leaving group; acetate ion lacks this leaving group and therefore will not react in the Claisen condensation.

22.27 (a) An ordinary Claisen condensation is driven to completion by ionization of an α-hydrogen more acidic than ethanol, the conjugate acid of the base ethoxide. Compound *A* lacks such an acidic α-hydrogen. Consequently, ethoxide attacks the ketone carbonyl group and converts it into *B*. This reaction is driven to completion by ionization of the acidic α-hydrogen of compound *B*. Compound *B* itself is formed on acidification of the reaction mixture.

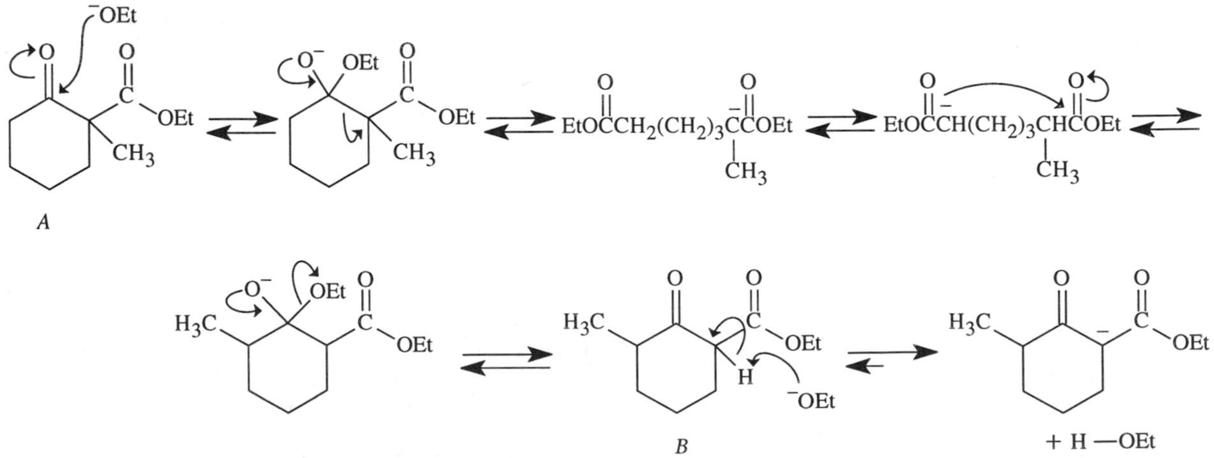

(b) Use reasoning similar to that used in solving part (a). Two possible enolate ions, *X* and *Y*, can form; the product derived from ion *X* is ultimately observed because only this product is formed irreversibly by ionization.

$$EtO_2C\bar{C}H(CH_2)_2CH(CH_3)CO_2Et \rightleftharpoons EtO_2CCH_2(CH_2)_2\bar{C}(CH_3)CO_2Et$$

X ⇅ (cyclopentanone enolate with EtO_2C and CH_3)

(observed product; acidification gives the keto ester)

Y ⇅ (2-carbethoxy-2-methylcyclopentanone)

22.28 (a) $EtOC(=O)CH_2C(=O)CMe_3$ (b) $PhC(=O)CH_2C(=O)Ph$ (c) (cyclohexane-1,3-dione bearing CO_2Et, and Me, Me on the adjacent carbon)

In all four parts, ethanol is a by-product. In part (c), the carbanion formed at carbon *a* leads to the product. Hydrogens *b* are more acidic than hydrogens *a*, but a carbanion formed at carbon *b* cannot lead to a product containing a ring of reasonable size. Hence, the two anions are in an equilibrium that is driven by ionization of the product (LeChatelier's principle).

$$\overset{a}{C}H_3C(=O)CH_2C(Me)_2\overset{b}{C}H(CO_2Et)_2$$

22.29 (a) Apply the analysis shown in Eq. 22.62 on text p. 1030. The two possible sets of starting materials are as follows:

$$\underset{A}{CH_3C(=O)OEt + CH_3CH_2C(=O)CH_2CH_3} \quad \text{and} \quad \underset{B}{CH_3C(=O)CH_2CH_3 + EtOC(=O)CH_2CH_3}$$

Set *A* could work provided that an excess of the ester is used. The analysis is similar to the analysis of Eq. 22.60 on text p. 1028. Because the ketone is symmetrical, only one enolate ion is possible. Set *B* would give a mixture of products because the ketone could form two different conjugate-base enolate ions, and there is no reason for one to be strongly preferred over the other.

(b) The two possible sets of starting materials are as follows:

$$\underset{A}{CH_3CH_2C(=O)CH_2CH_3 + EtOC(=O)OEt} \quad \text{and} \quad \underset{B}{2\ CH_3CH_2C(=O)OEt}$$

Either set would work. In set *A*, a large excess of diethyl carbonate must be used. (See Eq. 22.58 and the discussion that follows on text p. 1028.) However, set *B* would be more convenient because only one ester is required as a starting material and because we would not have to separate the product from an excess of starting material, as in set *A*.

(c) Apply the analysis shown in Eq. 22.61 on text p. 1030. The two possible sets of starting materials are as follows:

O, OEt, CO_2Et, CH_3 — *A* — and — O, CH_3 + $EtO\overset{O}{\overset{\|}{C}}OEt$ — *B*

Neither set of starting materials is a good choice. Because the desired product has no α-hydrogen more acidic than ethanol, it cannot be formed in a Claisen condensation because the equilibrium would be unfavorable. In fact, *A could* react in a Claisen condensation, but a different product would be formed. (Can you give its structure?) Reactants *B* are also not a good choice for the same reason: the desired product is not acidic enough to be ionized by ethoxide. Furthermore, in this case a different product would be formed.

(d) The two possible sets of starting materials are as follows:

most acidic hydrogens → $PhCH_2\overset{O}{\overset{\|}{C}}OEt + CH_3CH_2CH_2CH_2\overset{O}{\overset{\|}{C}}OEt$ (*A*) and most acidic hydrogens → $PhCH_2\overset{O}{\overset{\|}{C}}CH_2CH_2CH_2CH_3 + EtO\overset{O}{\overset{\|}{C}}OEt$ (*B*)

Set *A* consists of two esters with α-hydrogens; four different products are possible. Furthermore, the α-hydrogens of ethyl phenylacetate are considerably more acidic than those of ethyl valerate (why?); therefore, the major product is likely to be derived from the enolate ion formed at this position, but this is not the desired product. In set *B*, there are two sets of α-hydrogens, but the hydrogens that are α to the phenyl ring are considerably more acidic (why?). Unfortunately, the desired product is *not* derived from the enolate ion formed at this position. Hence, both sets of starting materials are unsatisfactory. In summary, the desired product cannot be made by a Claisen condensation.

22.30 (a)

O, EtO, CH_3, CH_3, O — *A* — or — O, H_3C, CH_3, OEt, O — *B*

Either answer is satisfactory, but perhaps compound *B* would be a better choice because, in compound *A*, attack of the ketone enolate on the ester carbonyl carbon might be somewhat retarded by van der Waals repulsions with the α-methyl group.

(b)

O, OEt, CH_3, O — ethyl 2-acetylbenzoate

22.31 Follow the pattern in Eqs. 22.63c–d on text p. 1034.

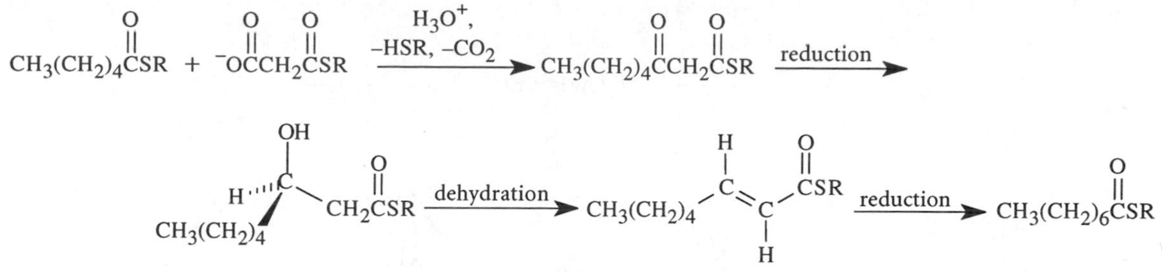

$$CH_3(CH_2)_4\overset{O}{\overset{\|}{C}}SR + {}^{-}O\overset{O}{\overset{\|}{C}}CH_2\overset{O}{\overset{\|}{C}}SR \xrightarrow[-HSR,\ -CO_2]{H_3O^+,} CH_3(CH_2)_4\overset{O}{\overset{\|}{C}}CH_2\overset{O}{\overset{\|}{C}}SR \xrightarrow{\text{reduction}}$$

$$CH_3(CH_2)_4CH(OH)CH_2\overset{O}{\overset{\|}{C}}SR \xrightarrow{\text{dehydration}} CH_3(CH_2)_4CH{=}CH\overset{O}{\overset{\|}{C}}SR \xrightarrow{\text{reduction}} CH_3(CH_2)_6\overset{O}{\overset{\|}{C}}SR$$

22.32 The carboxylation of acetyl-CoA by carboxybiotin is a nucleophilic acyl substitution reaction. A curved-arrow mechanism is as follows:

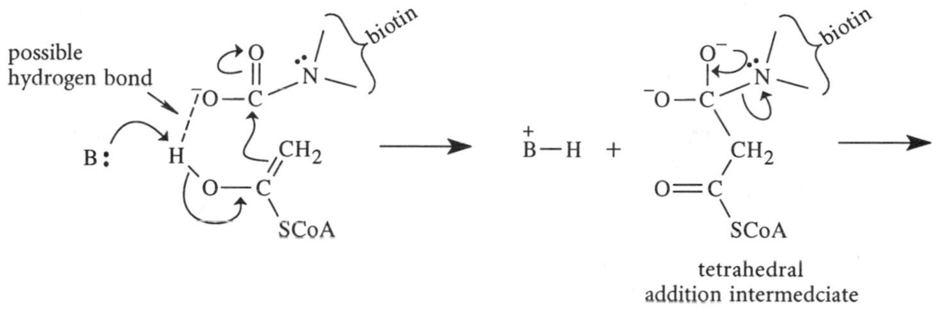

possible hydrogen bond

B: H biotin CH_2 SCoA $\longrightarrow$ $\overset{+}{B}$—H + biotin ^{-}O—C CH_2 O=C SCoA $\longrightarrow$

tetrahedral addition intermedciate

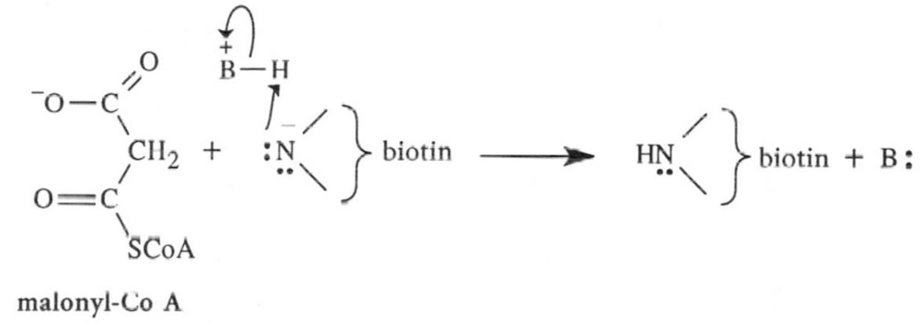

$\overset{+}{B}$—H biotin

^{-}O—C(=O)—CH_2—C(=O)—SCoA + :N biotin $\longrightarrow$ HN biotin + B:

malonyl-Co A

Several variations on the mechanism are possible. The proton transfers are probably concerted. The first transfer in this mechanism is shown as concerted, but the second one is shown as stepwise (by analogy to the other nucleophilic acyl substitution reactions we have studied). Note that carboxybiotin is an *imide*—the nitrogen analog of an anhydride—and thus, the conjugate-base anion of biotin is stabilized by resonance, and is therefore a much better leaving group than an ordinary amine would be. Nevertheless, this leaving group is still rather basic—it is the conjugate base of an amide, which has a pK_a of about 15. A concerted proton transfer avoids the formation of such a basic anion. The concertedness of proton transfers is one way that enzyme chemistry differs from laboratory chemistry. This can happen with enzymes because the necessary acidic and basic groups are actually part of the enzyme structure and are thus present in the correct position in the active site. Hence, they do not have to find their co-reactants by random diffusion.

The negatively charged oxygen on the carboxy group of carboxybiotin is undoubtedly involved in hydrogen-bonding, either to the enol of acetyl-CoA (as shown above), to the enzyme, or both. Otherwise, the resonance stabilization of the carboxylate ion would strongly reduce the reactivity of the carbonyl group towards nucleophilic attack.

Despite these subtleties, you should be able to see that this reaction bears a close analogy to the other nucleophilic acyl substitutions we have studied.

22.33 (a) 3-Phenylpropanoic acid can be prepared by a malonic ester synthesis:

$$CH_2(CO_2Et)_2 \xrightarrow[\text{benzyl chloride}]{\text{1) NaOEt, EtOH; 2) } PhCH_2Cl} PhCH_2CH(CO_2Et)_2 \xrightarrow[-EtOH,\ -CO_2]{H_3O^+,\ \text{heat}} PhCH_2CH_2CO_2H$$

3-phenylpropanoic acid

Recall that benzyl chloride, $PhCH_2Cl$, is exceptionally reactive in S_N2 reactions. (See Eq. 17.31a–b, text p. 757.)

(b) 2-Ethylbutanoic acid can be prepared by a malonic ester synthesis:

$$CH_2(CO_2Et)_2 \xrightarrow{\text{1) NaOEt, EtOH; 2) EtBr}} EtCH(CO_2Et)_2 \xrightarrow{\text{1) NaOEt, EtOH; 2) EtBr}} Et_2C(CO_2Et)_2 \xrightarrow[-EtOH,\ -CO_2]{H_3O^+,\ \text{heat}} Et_2CHCO_2H$$

2-ethylbutanoic acid

(c) 3,3-Dimethylbutanoic acid cannot be prepared by a malonic ester synthesis because the requisite alkyl halide, *tert*-butyl bromide, is virtually unreactive in S_N2 reactions. (See Sec. 9.4C in the text.)

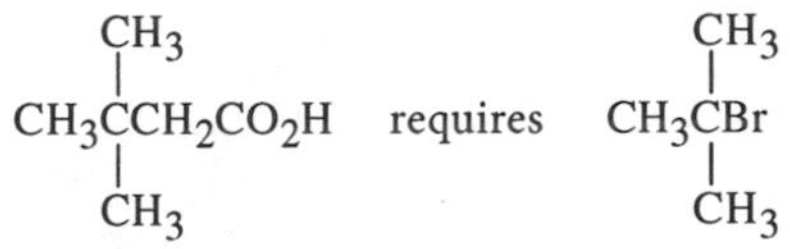

$$CH_3C(CH_3)_2CH_2CO_2H \quad \text{requires} \quad CH_3C(CH_3)_2Br$$

3,3-dimethylbutanoic acid — *tert*-butyl bromide

22.34 On the assumption that hydrolysis of the ester groups and decarboxylation of one carboxy group have occurred, the unsaturation number (2) of the product calculated from the molecular formula indicates that a ring has been formed. The formation of cyclobutanecarboxylic acid by the following reaction sequence accounts for the results:

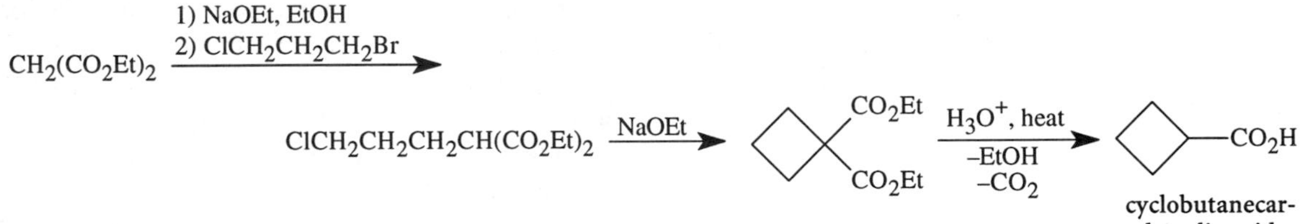

$$CH_2(CO_2Et)_2 \xrightarrow{\text{1) NaOEt, EtOH; 2) } ClCH_2CH_2CH_2Br} ClCH_2CH_2CH_2CH(CO_2Et)_2 \xrightarrow{NaOEt} \text{diethyl cyclobutane-1,1-dicarboxylate} \xrightarrow[-EtOH,\ -CO_2]{H_3O^+,\ \text{heat}} \text{cyclobutane}-CO_2H$$

cyclobutanecarboxylic acid

The ring closure occurs by an internal "S_N2" reaction in which the enolate ion is the nucleophile:

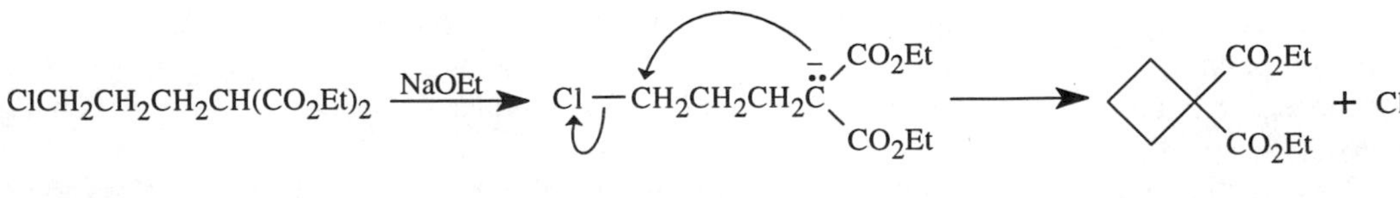

$$ClCH_2CH_2CH_2CH(CO_2Et)_2 \xrightarrow{NaOEt} Cl-CH_2CH_2CH_2\ddot{C}^-(CO_2Et)_2 \longrightarrow \text{diethyl cyclobutane-1,1-dicarboxylate} + Cl^-$$

22.35 Because part (c) involves formation of a carboxylic acid with a quaternary α-carbon, and because such compounds cannot be prepared by a malonic ester synthesis, we'll use our one allowed reaction of amide bases to form this compound. All of the other compounds can be prepared by the malonic ester synthesis, as follows:

(a) Note that allylic halides such as allyl bromide ($H_2C{=}CHCH_2Br$) are very reactive in S_N2 reactions. (See Eq. 17.30a-b, text p. 757.)

$$CH_2(CO_2Et)_2 \xrightarrow[]{\text{1) NaOEt, EtOH; 2) } CH_3I} CH_3CH(CO_2Et)_2 \xrightarrow[]{\text{1) NaOEt, EtOH; 2) } H_2C{=}CHCH_2Br}$$

$$CH_3C(CO_2Et)_2(CH_2CH{=}CH_2) \xrightarrow[-EtOH,\ -CO_2]{\text{1) NaOH, } H_2O\text{; 2) } H_3O^+\text{, heat}} CH_3CH(CO_2H)CH_2CH{=}CH_2$$

2-methyl-4-pentenoic acid

(b)

$$CH_2(CO_2Et)_2 \xrightarrow[]{\text{1) NaOEt, EtOH; 2) PrBr}} PrCH(CO_2Et)_2 \xrightarrow[]{\text{1) NaOEt, EtOH; 2) PrBr}} Pr_2C(CO_2Et)_2 \xrightarrow[-EtOH,\ -CO_2]{H_3O^+\text{, heat}} Pr_2CHCO_2H$$

valproic acid (2-propylpentanoic acid)

(See the solution to Problem 20.38, text p. 928, on pp. 722–3 of this manual.)

(c)

$$CH_2(CO_2Et)_2 \xrightarrow[]{\text{1) NaOEt, EtOH; 2) EtBr}} EtCH(CO_2Et)_2 \xrightarrow[]{\text{1) NaOEt, EtOH; 2) EtBr}} Et_2C(CO_2Et)_2 \xrightarrow[-EtOH,\ -CO_2]{H_3O^+\text{, heat}}$$

$$Et_2CHCO_2H \xrightarrow[H_2SO_4\text{ (catalyst)}]{\text{EtOH (solvent)}} Et_2CHCO_2Et \xrightarrow[]{\text{1) LDA; 2) } H_2C{=}CHCH_2Br} Et_2C(CO_2Et)CH_2CH{=}CH_2$$

2,2-diethyl-4-pentenoic acid

22.36 (a) The enolate ion of *tert*-butyl acetate, formed by reaction of the ester with LCHIA, adds to the carbonyl group of acetone to give the lithium alkoxide, which is protonated when acid is added to give the β-hydroxy ester.

$$CH_3C(=O)CH_3 + Li^+\ {}^-CH_2C(=O)OC(CH_3)_3 \longrightarrow CH_3C(O^-\,Li^+)(CH_3)CH_2C(=O)OC(CH_3)_3 \xrightarrow{\text{dil. } H_3O^+} CH_3C(OH)(CH_3)CH_2C(=O)OC(CH_3)_3$$

acetone; enolate ion of *tert*-butyl acetate; *tert*-butyl 3-hydroxybutanoate

The tertiary alcohol addition product can be isolated if very dilute acid is used in the workup. ("Dilute" is abbreviated "dil." in chemical equations.) The dehydration product can be obtained if stronger acid is used. (The same considerations hold for part (b) that follows.)

(b) By the same logic used in part (a), the β-hydroxy ester is formed.

$$PhCH{=}O + Li^+\ {}^-CH_2C(=O)OC(CH_3)_3 \longrightarrow PhCH(O^-\,Li^+)CH_2C(=O)OC(CH_3)_3 \xrightarrow{\text{dil. } H_3O^+} PhCH(OH)CH_2C(=O)OC(CH_3)_3$$

benzaldehyde; enolate ion of *tert*-butyl acetate; *tert*-butyl 3-phenyl-3-hydroxypropanoate

22.37 (a)

$$\underset{\text{ethyl acetoacetate}}{CH_3\overset{O}{\overset{\|}{C}}CH_2CO_2Et} \xrightarrow[\text{isobutyl bromide}]{\text{1) NaOEt, EtOH; 2) } BrCH_2CH(CH_3)_2} CH_3\overset{O}{\overset{\|}{C}}\underset{CH_2CH(CH_3)_2}{\underset{|}{C}H}CO_2Et \xrightarrow{\text{1) NaOH; 2) } H_3O^+, -CO_2} \underset{\text{5-methyl-2-hexanone}}{CH_3\overset{O}{\overset{\|}{C}}CH_2CH_2CH(CH_3)_2}$$

(b)

$$\underset{\text{ethyl acetoacetate}}{CH_3\overset{O}{\overset{\|}{C}}CH_2CO_2Et} \xrightarrow[\text{benzyl bromide}]{\text{1) NaOEt, EtOH; 2) } BrCH_2Ph} CH_3\overset{O}{\overset{\|}{C}}\underset{CH_2Ph}{\underset{|}{C}H}CO_2Et \xrightarrow{\text{1) NaOH; 2) } H_3O^+, -CO_2} \underset{\text{4-phenyl-2-butanone}}{CH_3\overset{O}{\overset{\|}{C}}CH_2CH_2Ph}$$

22.38 (a) Follow the hint in Study Guide Link 22.8.

$$CH_3\overset{CO_2Et}{\overset{|}{C}H}-\overset{O}{\overset{\|}{C}}-CH_2CH_3 \xrightarrow{\text{1) NaOEt, EtOH; 2) } PhCH_2Br} CH_3\underset{CH_2Ph}{\underset{|}{\overset{CO_2Et}{\overset{|}{C}}}}-\overset{O}{\overset{\|}{C}}-CH_2CH_3 \xrightarrow{\text{1) NaOH; 2) } H_3O^+, -CO_2} CH_3\underset{CH_2Ph}{\underset{|}{C}H}\overset{O}{\overset{\|}{C}}CH_2CH_3$$

The starting β-keto ester is prepared by a Claisen condensation of ethyl propionate:

$$\underset{\text{ethyl propionate}}{2\ EtO\overset{O}{\overset{\|}{C}}CH_2CH_3} \xrightarrow{\text{1) NaOEt, EtOH; 2) } H_3O^+} CH_3\overset{CO_2Et}{\overset{|}{C}H}-\overset{O}{\overset{\|}{C}}CH_2CH_3$$

(b)

$$Ph\overset{CO_2Et}{\overset{|}{C}H}-\overset{O}{\overset{\|}{C}}CH_2Ph \xrightarrow{\text{1) NaOEt, EtOH; 2) } CH_3I} Ph\underset{CH_3}{\underset{|}{\overset{CO_2Et}{\overset{|}{C}}}}-\overset{O}{\overset{\|}{C}}CH_2Ph \xrightarrow{\text{1) NaOH; 2) } H_3O^+, -CO_2} Ph\underset{CH_3}{\underset{|}{C}H}\overset{O}{\overset{\|}{C}}CH_2Ph$$

The starting β-keto ester is prepared by a Claisen condensation of ethyl phenylacetate:

$$\underset{\text{ethyl phenylacetate}}{2\ EtO\overset{O}{\overset{\|}{C}}CH_2Ph} \xrightarrow{\text{1) NaOEt, EtOH; 2) } H_3O^+} Ph\overset{CO_2Et}{\overset{|}{C}H}-\overset{O}{\overset{\|}{C}}CH_2Ph$$

22.39 Compound *A* is the enolate ion of diethyl malonate. Attack of the enolate ion on the methylene carbon of the epoxide gives the alkoxide *B*, which, in turn, undergoes an intramolecular transesterification to give a lactone.

$$(CH_3)_2\overset{O}{C-CH_2}\ \ \underset{A}{^-CH(CO_2Et)_2} \longrightarrow \underset{B}{(CH_3)_2\overset{O^-}{\overset{|}{C}}-CH_2\overset{EtO-C=O}{\overset{|}{C}H}CO_2Et} \xrightarrow{-EtO^-} \xrightarrow{H_3O^+} \text{lactone (}O\text{, }CO_2Et\text{, }H_3C\text{, }CH_3\text{)}$$

22.40 (a)

$$\underset{\text{methyl methacrylate}}{H_2C=\underset{CH_3}{\underset{|}{C}}-CO_2CH_3} \xrightarrow{^-CN/HCN,\ MeOH} N{\equiv}CCH_2-\underset{CH_3}{\underset{|}{C}H}-CO_2CH_3$$

(b)

$$C_2H_5SCH_2-\underset{CH_3}{\underset{|}{C}H}-\overset{O}{\overset{\|}{C}}-OCH_3$$

(c)

$$\mathrm{BrCH_2{-}\underset{\displaystyle CH_3}{\underset{|}{C}H}{-}CO_2CH_3}$$

(d)

$$\mathrm{H_2C{=}\underset{\displaystyle CH_3}{\underset{|}{C}}{-}\overset{\displaystyle O}{\overset{\|}{C}}{-}O^- + CH_3OH}$$

In part (d), saponification occurs instead of conjugate addition because saponification is an *irreversible* nucleophilic acyl substitution reaction.

When considering acid-catalyzed additions to α,β-unsaturated carbonyl compounds, pay careful attention to the mechanism in Eq. 22.86 on text p. 1046. In part (c), it is tempting to add HBr to the double bond to give the tertiary carbocation *A*; however, such a mechanism would lead to a product *B* that is different from the product observed:

$$\mathrm{Br{-}H \quad H_2C{=}\underset{\displaystyle CH_3}{\underset{|}{C}}{-}\overset{\displaystyle O}{\overset{\|}{C}}{-}OCH_3 \;\not\longrightarrow\; Br^- \quad H_3C{-}\underset{\displaystyle CH_3}{\underset{|}{\overset{+}{C}}}{-}\overset{\displaystyle O}{\overset{\|}{C}}{-}OCH_3 \longrightarrow H_3C{-}\underset{\displaystyle CH_3}{\underset{|}{\overset{\displaystyle Br}{\overset{|}{C}}}}{-}\overset{\displaystyle O}{\overset{\|}{C}}{-}OCH_3}$$

A *B*

Remember that carbocations α to carbonyl groups are quite unstable and are almost never formed; see Eq. 22.38 on text p. 1013. In contrast, the mechanism in Eq. 22.86 on text p. 1046 involves a ketone that is protonated on the carbonyl oxygen; this is a resonance-stabilized species in which positive charge is shared at the terminal carbon. Attack of bromide at this carbon leads to the observed product.

22.41 (a) The product results from two successive conjugate additions. (Proton transfers are shown for simplicity as intramolecular processes, but they could be mediated by other acids and bases in solution.)

$$\mathrm{MeNH_2 \quad H_2C{=}CHCO_2Me \rightleftharpoons Me\overset{H}{\underset{+}{N}}H{-}CH_2{-}\bar{C}HCO_2Me \rightleftharpoons MeNHCH_2CH_2CO_2Me \xrightleftharpoons{H_2C=CHCO_2Me}}$$

$$\mathrm{CH_3O_2C\bar{C}HCH_2\overset{+|H}{N}CH_2CH_2CO_2CH_3 \ (N{-}CH_3) \rightleftharpoons CH_3O_2CCH_2CH_2\underset{\displaystyle CH_3}{\underset{|}{N}}CH_2CH_2CO_2CH_3}$$

(b) This is a conjugate addition of the benzenethiolate anion to the carbon-carbon double bond. The benzenethiolate nucleophile, PhS^-, is formed by the reaction of benzenethiol, PhSH, with ^-OEt. A mixture of diastereomers is formed because the final protonation step can occur from the same face as the methyl group or from the face opposite the methyl group.

PhS⁻ + H₂C= (lactone, CH₃) $\xrightleftharpoons{\text{HOEt}}$ PhSCH₂ (carbanion, CH₃) + H—OEt $\rightleftharpoons$ PhSCH₂ (H, lactone, CH₃) + ^-OEt

(mixture of cis and trans)

(c) The conjugate-addition product has no carbon-carbon double bond. Internal rotation within this adduct, followed by reversal of the addition, gives the isomerized product.

EtO₂C, CO₂Et, H, H, Et₂NH ⇌ EtO₂C, H, CO₂Et, H, Et₂NH₂⁺ ⇌ EtO₂C, H, H, CO₂Et, Et₂NH₂⁺ ⇌ EtO₂C, H, H, CO₂Et + Et₂NH

22.42 (a) The key is to recognize that the desired product results from the addition of *two equivalents* of acrylonitrile, which end up as the two —$CH_2CH_2CO_2H$ groups of the product. The conjugate-base anion of ethyl acetoacetate would be a suitable nucleophile *Y*. Reaction of *Y* with one equivalent of acrylonitrile would give compound *A*. Ionization of *A* gives anion *B*, which reacts with a second equivalent of acrylonitrile to give compound *C*. The nitrile groups are hydrolyzed to carboxy groups and the ester group is hydrolyzed and decarboxylated to give the final product when compound *C* is heated in acid.

$CH_3C(=O)\bar{C}HCO_2Et$ (*Y*, enolate ion from ethyl acetoacetate) $\xrightarrow[\text{EtOH}]{H_2C{=}CHCN \text{ (acrylonitrile)}}$ (*Michael addition*) $CH_3C(=O)CH(CH_2CH_2CN)CO_2Et$ (*A*) $\xrightarrow{EtO^-}$ $CH_3C(=O)\bar{C}(CH_2CH_2CN)CO_2Et$ (*B*) $\xrightarrow[\text{EtOH}]{H_2C{=}CHCN \text{ (acrylonitrile)}}$ $CH_3C(=O)C(CH_2CH_2CN)_2CO_2Et$ (*C*)

(b) The nucleophile must be a species that can be converted into a carboxymethyl (HO_2CCH_2—) group. Because the nucleophile ends up as a "substituted acetic acid," the conjugate-base enolate ion of diethyl malonate can serve as species *X*. Addition of *X* to the ester gives compound *D*. All ester groups of *D* are hydrolyzed and one of the resulting carboxy groups is decarboxylated when *D* is heated in aqueous acid.

$(EtO_2C)_2\bar{C}H$ (*X*, enolate ion from diethyl malonate) + $H_2C{=}C(CH_3){-}CO_2Et$ $\xrightarrow{\text{EtOH}}$ $(EtO_2C)_2CHCH_2CH(CH_3)CO_2Et$ (*D*)

one of these groups decarboxylates following hydrolysis in acid

22.43 (a) The first part of the mechanism is a Michael addition followed by loss of formic acid to give, following protonation, intermediate *A*.

O=CH, O, $H_2C{=}CHCCH_3$ (O) $\xrightarrow{H_2O}$ HO—H, O=CH, $CH_2\bar{C}HCCH_3$ (O), O ⟶ ⁻OH, O=CH, $CH_2CH_2CCH_3$ (O), O ⟶

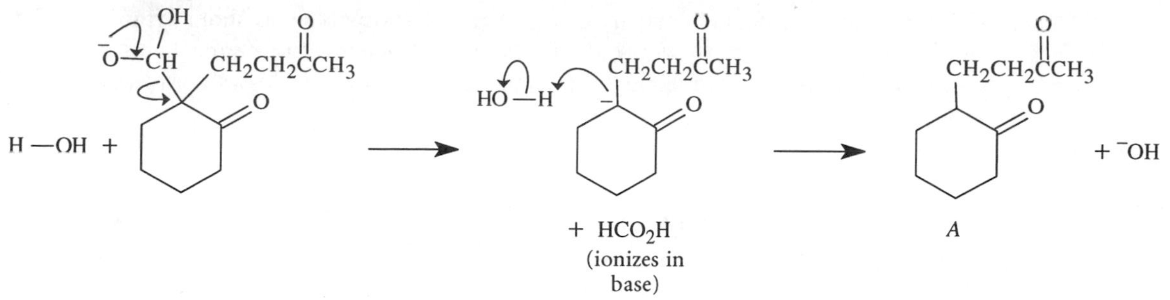

H—OH + ... ⟶ HO—H ... + HCO_2H (ionizes in base) ⟶ A + ^{-}OH

Compound *A* then undergoes an intramolecular aldol condensation reaction.

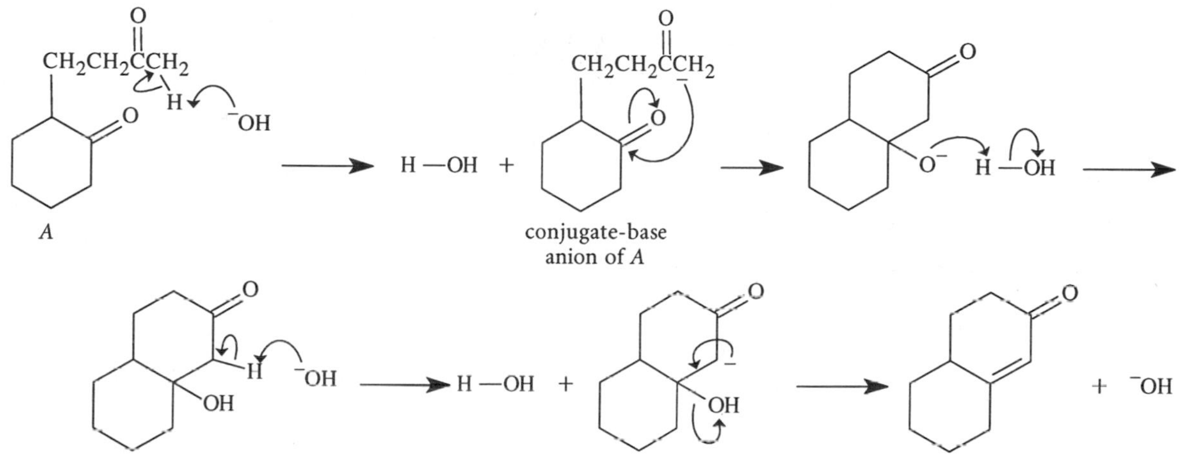

A ⟶ H—OH + conjugate-base anion of *A* ⟶ ... H—OH ⟶

... ^{-}OH ⟶ H—OH + ... ⟶ ... + ^{-}OH

(b) A Michael addition of the diethyl malonate conjugate-base anion gives compound *B*, which undergoes an intramolecular Claisen (Dieckmann) condensation. This Dieckmann condensation, like all Claisen-type condensations, is driven to completion by ionization of the product; the un-ionized product is formed when acid is added. These final steps are not shown in the mechanism below. Note that the tertiary hydrogen of compound *B* is the most acidic hydrogen, but although the anion resulting from removal of this proton is formed in the reaction mixture, it is depleted by the reaction that gives the product.

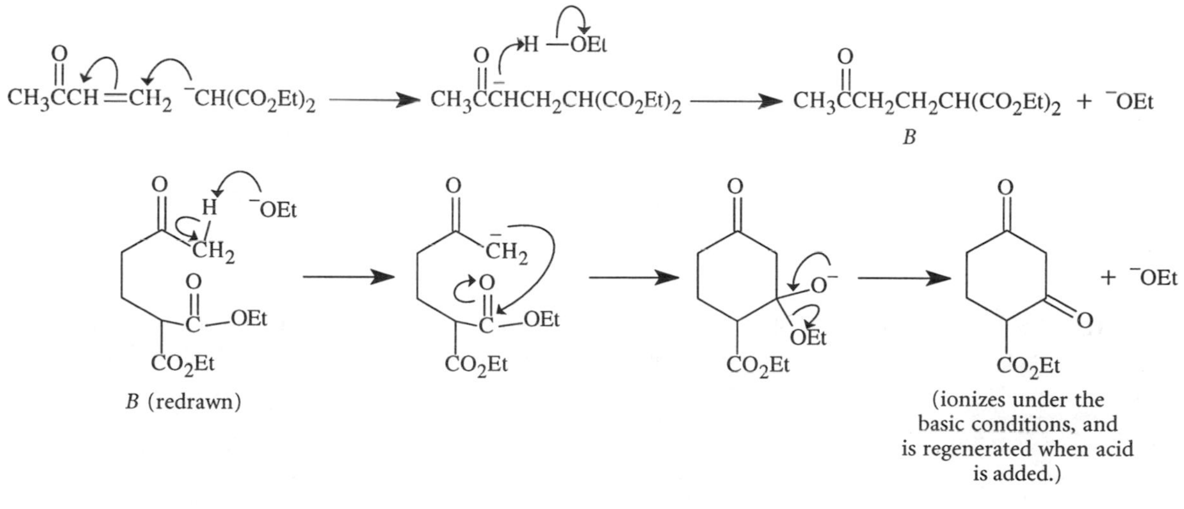

$CH_3\overset{O}{\overset{\|}{C}}CH{=}CH_2 \;\; {}^{-}CH(CO_2Et)_2 \longrightarrow CH_3\overset{O}{\overset{\|}{C}}\overset{-}{C}HCH_2CH(CO_2Et)_2 \longrightarrow CH_3\overset{O}{\overset{\|}{C}}CH_2CH_2CH(CO_2Et)_2 + {}^{-}OEt$

B

B (redrawn) ⟶ ⟶ ⟶ + ^{-}OEt

(ionizes under the basic conditions, and is regenerated when acid is added.)

In the foregoing mechanisms many of the steps are reversible and should rigorously be shown with equilibrium arrows. However, the focus of the problem is on the mechanism of the forward reaction and not on the reversibility of the reaction. In such cases, forward arrows only are shown.

22.44 (a)

$$CH_3CH{=}CH\overset{O}{\overset{\|}{C}}OEt \xrightarrow{H_2,\ \text{catalyst}} CH_3CH_2CH_2\overset{O}{\overset{\|}{C}}OEt$$

ethyl 2-butenoate → ethyl butanoate

(b)

$$CH_3CH{=}CH\overset{O}{\overset{\|}{C}}OEt \xrightarrow[2)\ H_3O^+]{1)\ LiAlH_4} CH_3CH{=}CHCH_2OH + EtOH$$

ethyl 2-butenoate → 2-buten-1-ol

22.45 (a)

$$(CH_3)_2C{=}CH\overset{O}{\overset{\|}{C}}CH_3 \xrightarrow[2)\ H_3O^+]{1)\ Li^+\ (CH_3)_2Cu^-} (CH_3)_3CCH_2\overset{O}{\overset{\|}{C}}CH_3$$

mesityl oxide (4-methyl-3-penten-2-one) → 4,4-dimethyl-2-pentanone

(b)

$$(CH_3)_2C{=}CH\overset{O}{\overset{\|}{C}}CH_3 \xrightarrow[2)\ \text{dil. } H_3O^+]{1)\ CH_3Li} (CH_3)_2C{=}CH\underset{CH_3}{\overset{OH}{C}}CH_3$$

2,4-dimethyl-2-penten-2-ol

(c)

$$(CH_3)_2C{=}CH\overset{O}{\overset{\|}{C}}CH_3 \xrightarrow[2)\ H_3O^+]{1)\ Li^+\ (C_2H_5)_2Cu^-} C_2H_5\underset{CH_3}{\overset{CH_3}{C}}CH_2\overset{O}{\overset{\|}{C}}CH_3 \xrightarrow[2)\ H_2SO_4]{1)\ CH_3MgI} C_2H_5\underset{CH_3}{\overset{CH_3}{C}}CH{=}C(CH_3)_2$$

2,4,4-trimethyl-2-hexene

Notice that the conjugate addition must *precede* the carbonyl addition. (If the carbonyl addition were carried out first, the product would not be an α,β-unsaturated carbonyl compound, and conjugate addition could not occur.)

22.46 (a) The question in this problem is what type of conjugate addition occurs (1,4 or 1,6). The 1,6-addition occurs because it gives the enolate ion intermediate with the greater number of resonance structures and hence, the more stable enolate ion.

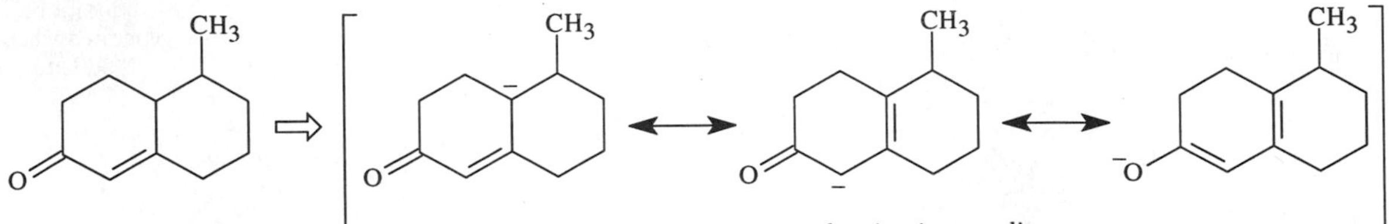

1,6-addition product (observed)

enolate ion intermediate

1,4-addition product ⇨ [enolate ion intermediate]

(b) Conjugate addition also occurs with α,β-unsaturated compounds in which a carbon-carbon triple bond is part of the conjugated π-electron system. The product is $(CH_3)_2C{=}CHCO_2Me$ (methyl 3-methyl-2-butenoate).

22.47 (a) An analysis similar to that used in Study Problem 22.7 on text p. 1056 reveals that either a methyl group or an ethyl group can be added in the conjugate addition. Addition of an ethyl group is illustrated first.)

$$CH_3\overset{O}{\overset{\|}{C}}\underset{CH_3}{\underset{|}{C}}{=}CHCH_3 \xrightarrow[2)\ H_3O^+]{1)\ Li^+\ (C_2H_5)_2Cu^-} CH_3\overset{O}{\overset{\|}{C}}\underset{CH_3}{\underset{|}{C}}H\overset{CH_3}{\overset{|}{C}}HC_2H_5$$

3-methyl-3-penten-2-one → 3,4-dimethyl-2-hexanone

(b) Addition of the methyl group is illustrated here.

$$CH_3\overset{O}{\overset{\|}{C}}\underset{CH_3}{\underset{|}{C}}{=}CHC_2H_5 \xrightarrow[2)\ H_3O^+]{1)\ Li^+\ (CH_3)_2Cu^-} CH_3\overset{O}{\overset{\|}{C}}\underset{CH_3}{\underset{|}{C}}H\overset{CH_3}{\overset{|}{C}}HC_2H_5$$

3-methyl-3-hexen-2-one → 3,4-dimethyl-2-hexanone

(c) An analysis along the lines of Study Problem 22.7 reveals the following two possibilities:

$$\underbrace{\text{“}CH_3\overset{O}{\overset{\|}{C}}\bar{C}H_2\text{”} + H_2C{=}CHCO_2R}_{(a)} \quad \text{or} \quad \underbrace{CH_3\overset{O}{\overset{\|}{C}}CH{=}CH_2 + \text{“}\bar{C}H_2CO_2H\text{”}}_{(b)}$$

A practical synthetic equivalent to the anion in (*a*) is the conjugate base of ethyl acetoacetate:

$$CH_3\overset{O}{\overset{\|}{C}}\bar{C}HCO_2Et + H_2C{=}CHCO_2Et \xrightarrow{EtOH} CH_3\overset{O}{\overset{\|}{C}}\underset{CH_2CH_2CO_2Et}{\underset{|}{C}H}CO_2Et \xrightarrow[-EtOH,\ -CO_2]{H_3O^+,\ heat} CH_3\overset{O}{\overset{\|}{C}}CH_2CH_2CH_2CO_2H$$

conjugate-base anion of ethyl acetoacetate; ethyl acrylate (ethyl 2-propenoate); 5-oxohexanoic acid

The practical synthetic equivalent to the anion in (*b*) is the conjugate base of diethyl malonate:

$$CH_3\overset{O}{\overset{\|}{C}}CH{=}CH_2 + {}^-CH(CO_2Et)_2 \xrightarrow{EtOH} CH_3\overset{O}{\overset{\|}{C}}CH_2CH_2CH(CO_2Et)_2 \xrightarrow[-EtOH,\ -CO_2]{H_3O^+,\ heat} CH_3\overset{O}{\overset{\|}{C}}CH_2CH_2CH_2CO_2H$$

methyl vinyl ketone (3-buten-2-one); conjugate-base anion of diethyl malonate; 5-oxohexanoic acid

(d) An analysis similar to that in Study Problem 22.7 suggests the following possibilities:

$$\underbrace{\text{“}CH_3\overset{O}{\overset{\|}{C}}{}^-\text{”} + H_2C{=}CHCO_2R}_{(a)} \quad \text{or} \quad \underbrace{CH_3\overset{O}{\overset{\|}{C}}CH{=}CH_2 + \text{“}\overset{O}{\overset{\|}{\underline{C}}}OH\text{”}}_{(b)}$$

A practical equivalent for the anion in (*a*) might be an acetylide anion, $^-C{\equiv}CH$. The resulting acetylene could then be hydrated to the desired ketone. Unfortunately, acetylenic cuprate reagents do not work in conjugate additions because the acetylenic ligands do not transfer from the copper. Nevertheless, if you came up with this possibility, you are analyzing the problem with considerable sophistication.

Anion (*b*) does have a simple practical equivalent: the cyanide ion:

$$\underset{\substack{\textbf{methyl vinyl ketone}\\ \textbf{(3-buten-2-one)}}}{CH_3\overset{O}{\overset{\|}{C}}CH{=}CH_2} + {}^{-}C{\equiv}N \xrightarrow[\text{trace of acid}]{\text{EtOH}} CH_3\overset{O}{\overset{\|}{C}}CH_2CH_2C{\equiv}N \xrightarrow{H_3O^+,\ \text{heat}} \underset{\substack{\textbf{levulinic acid}\\ \textbf{(4-oxopentanoic acid)}}}{CH_3\overset{O}{\overset{\|}{C}}CH_2CH_2\overset{O}{\overset{\|}{C}}OH}$$

Solutions to Additional Problems

22.48 The structure of the starting material is

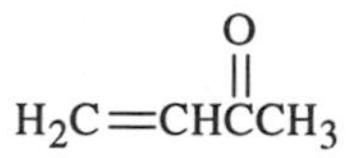

$H_2C{=}CHCCH_3$ (C=O on the carbonyl carbon)

3-buten-2-one

(a) $BrCH_2CH_2CCH_3$ (C=O)

(b) $CH_3CH_2CCH_3$ (C=O)

(c) $H_2C{=}CHCHCH_3$ (OH on CH)

(d) $N{\equiv}CCH_2CH_2CCH_3$ (C=O)

(e) $CH_3CH_2CH_2CH_2CCH_3$ (C=O)

(f) $(EtO_2C)_2CHCH_2CH_2CCH_3$ (C=O)

(g) cyclic ketal (1,3-dioxolane) bearing $H_2C{=}CH$ and CH_3

(h) cyclohexene bearing $C(=O)CH_3$

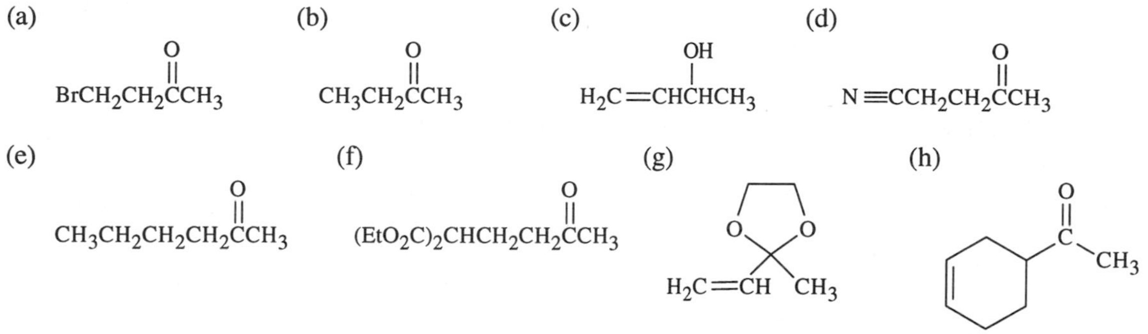

In part (h), the α,β-unsaturated ketone serves as a Diels-Alder dienophile.

22.49 The structure of the starting material is

H, CO_2Et / C=C / H_3C, H

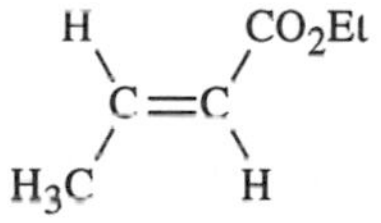

ethyl *trans*-2-butenoate

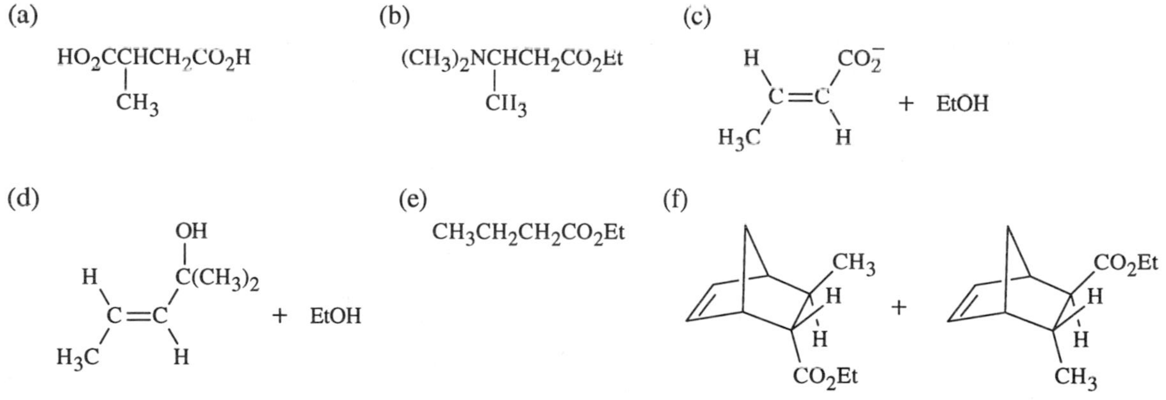

(a) $HO_2CCHCH_2CO_2H$ with CH_3 on CH

(b) $(CH_3)_2NCHCH_2CO_2Et$ with CH_3 on CH

(c) H, CO_2^- / C=C / H_3C, H + EtOH

(d) H, $C(CH_3)_2$OH / C=C / H_3C, H + EtOH

(e) $CH_3CH_2CH_2CO_2Et$

(f) bicyclic Diels-Alder adducts: CH_3, H, H, CO_2Et + CO_2Et, H, H, CH_3

(both compounds are racemates)

As in Problem 22.48(h), the α,β-unsaturated ester in part (f) serves as a Diels-Alder dienophile.

22.50 (a) $CH_3CH_2CHCCH_3$ (C=O; CH_3 on CH)

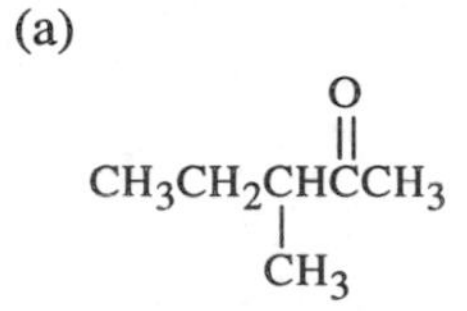

(either enantiomer)

(b) cyclobutane bearing CH_3 and OH on one carbon and H_3C on another; $CH_3CH_2CCH_2CH_2CH_3$ (C=O)

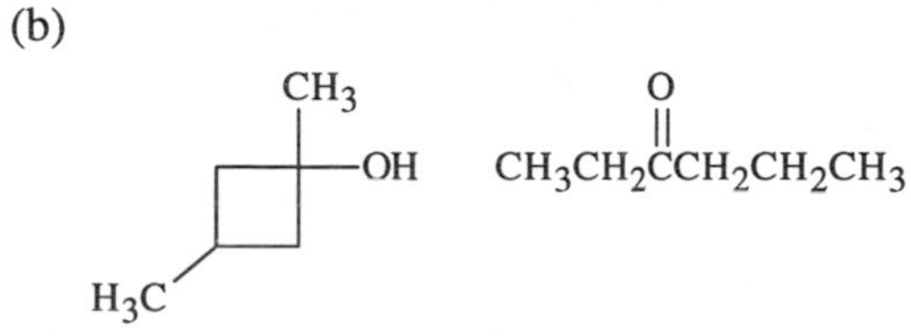

(two of several possibilities)

(c)

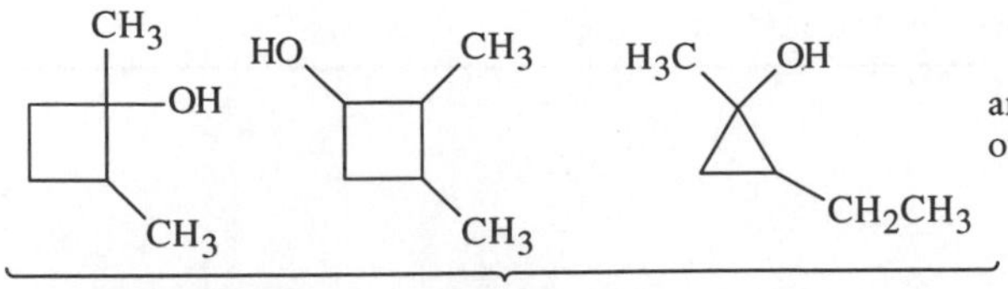

and many others

(Each could be any one of several stereoisomers—six possible stereoisomers for the middle structure and four possible stereoisomers for the other two.)

(d) $CH_3CH_2CH(CH_3)CH_2CH{=}O$

22.51 (a) $CH_3C(=O)CH_2CH_3$

(b) $(CH_3)_2CHCH{=}O$ or $CH_3CH_2CH_2CH{=}O$

(c) $H_2C{=}CHCH(OH)CH_3$

(d) (one of several possibilities)

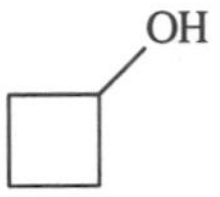

For part (c), recall (Eq. 22.28, text p. 1010) that alcohols of the form

$$—CH(OH)—CH_3$$

give a positive haloform text, because the basic iodine solution used in the iodoform reaction oxidizes secondary alcohols. A secondary alcohol of this form is thus oxidized to a methyl ketone, which then gives the iodoform test.

22.52 (a) This compound is an "ynol," the acetylenic analog of an enol. It is spontaneously converted into its isomeric ketene derivative, $CH_3CH{=}C{=}O$.

(b) This compound, 1,3,5-cyclohexanetrione, exists as its "triene-triol" tautomer phloroglucinol because the latter is aromatic.

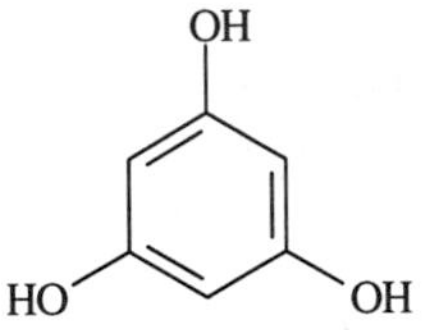

1,3,5-benzenetriol (phloroglucinol)

(c) This compound is a hemiacetal, and it spontaneously decomposes to acetaldehyde and an enol *A*, which, in turn, spontaneously forms propionaldehyde.

$$CH_3CH(OH)OCH{=}CHCH_3 \longrightarrow \underset{\text{acetaldehyde}}{CH_3CH{=}O} + \underset{A}{HOCH{=}CHCH_3} \longrightarrow \underset{\text{propionaldehyde}}{O{=}CHCH_2CH_3}$$

22.53 (a) The conjugate bases of the two compounds, because they are resonance structures, are identical.

conjugate base of *A* — conjugate base of *B*

resonance structures, and therefore identical ions

(b) Because compound *B* (phenol) is aromatic, it is more stable, and thus more energy is required to convert it into the conjugate base shown in part (a). From Eq. 3.27 on text p. 103, $pK_a = \Delta G_a^\circ/2.3RT$. Thus, compound *B* has the greater pK_a (it is less acidic). This is demonstrated in Fig. SG22.3.

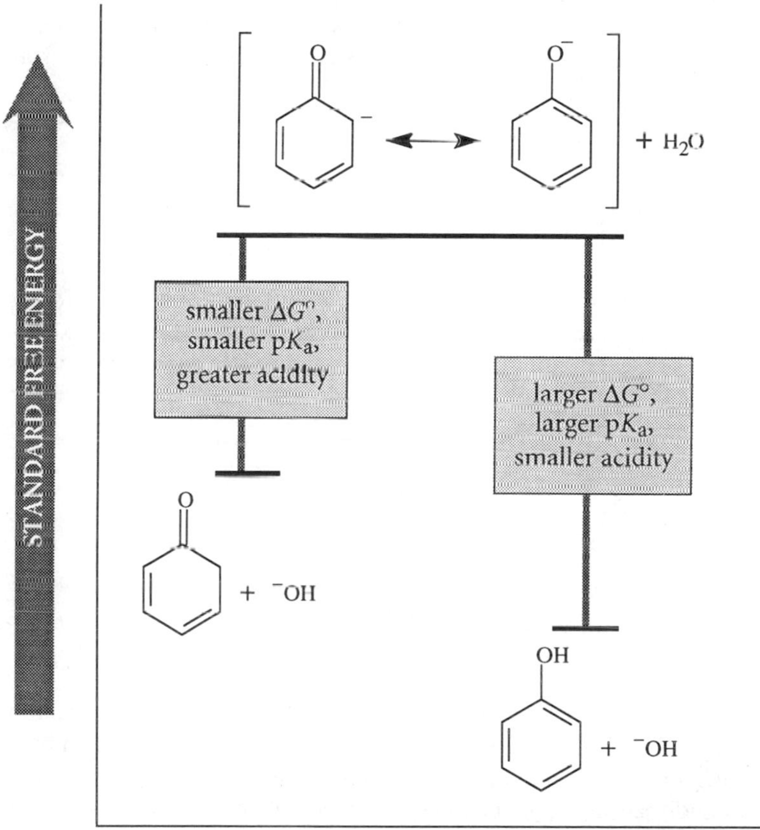

Figure SG22.3 Diagram for the solution to Problem 22.53, which shows the effect of relative stability on the acidity of two isomers. Because the conjugate acids of the two isomers are identical, the more stable isomer is less acidic. The phenol isomer is more stable because it is aromatic.

22.54 (a) Removal of the O—H proton from *A* and a C—H proton from *B* gives conjugate-base anions that are resonance structures and therefore identical. (B:$^-$ = a general base.)

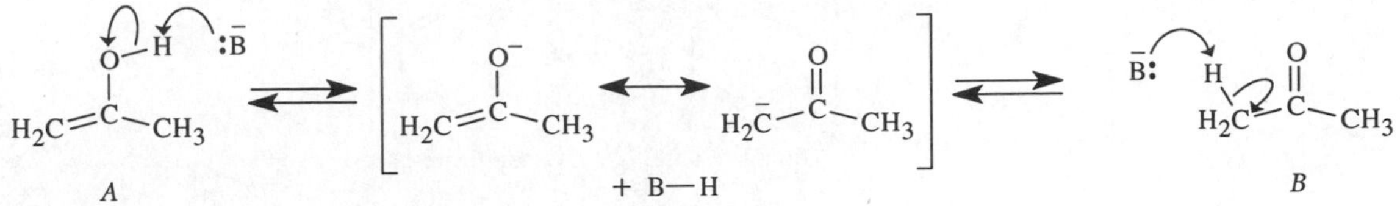

$$H_2C{=}C(OH){-}CH_3 \;(A) \rightleftharpoons \left[H_2C{=}C(O^-){-}CH_3 \longleftrightarrow H_2\bar{C}{-}C({=}O){-}CH_3 \right] + B{-}H \rightleftharpoons H_2C(H){-}C({=}O){-}CH_3 \;(B)$$

(b) Because the ketone is more stable than its enol, more energy must be expended for it to ionize. Consequently, the ketone is less acidic. The logic is similar to that in the solution to the previous problem.

22.55 (a) The second compound is most acidic because the conjugate-base anion (structure below) has the greatest number of resonance structures and is therefore most stable. (Draw these structures!) This anion is stabilized by resonance interaction with two carbonyl groups and a benzene ring. In the first compound, the conjugate-base anion lacks the resonance interaction with a benzene ring; and in the third compound, the conjugate-base anion at the central carbon also lacks the resonance interaction with a benzene ring, and the conjugate-base anion at the benzylic carbon lacks the resonance interaction with a second carbonyl group.

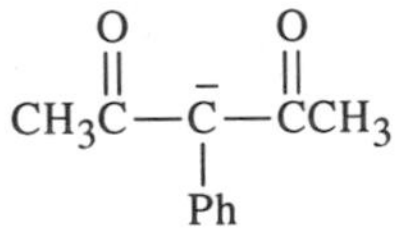

$$CH_3C({=}O){-}\bar{C}(Ph){-}C({=}O)CH_3$$

conjugate-base anion
of the most acidic compound

(b) The first compound is more acidic because the conjugate-base anion (see following structure) has the greater number of resonance structures. It is stabilized by resonance interaction with both the carbon-carbon double bond and the carbonyl group. The two possible enolate ions of the second compound lack the resonance interaction with a carbon-carbon double bond.

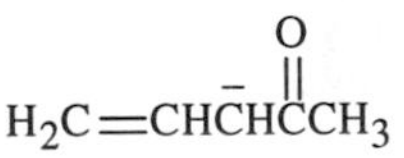

$$H_2C{=}CH\bar{C}HC({=}O)CH_3$$

conjugate-base anion
of the most acidic compound

22.56 The order of increasing acidity (decreasing pK_a) is as follows. [Approximate pK_a values are in brackets.]

(3) [42] < (4) ≈ (5) [25] < (1)[15] < (6) [10] < (2) [4.5]

Toluene is much less acidic than phenol, and amides are much less acidic than carboxylic acids (element effect). The pK_a values of 1-alkynes and esters are very similar (about 25). The remaining rankings require knowledge of approximate pK_a values. The point of this problem is that it is very important to know the typical pK_a values of organic compounds because so much of organic reactivity is based on acid-base principles.

22.57 (a) Bromination can occur at either of the two α-carbons; compounds *B* and *C* are diastereomers. (Each of the three compounds can also exist as enantiomers.)

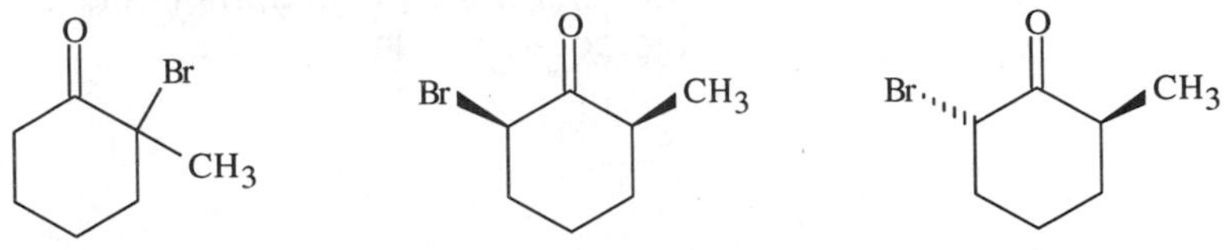

A: 2-bromo-2-methylcyclohexanone (O, Br, CH_3)

B: Br and CH_3 (both wedged) on the two α-carbons of cyclohexanone

C: Br (hashed) and CH_3 (wedged) on the two α-carbons of cyclohexanone

(b) The enol leading to *A* is the more stable one because it has the greater number of alkyl branches on the double bond. Hence, compound *A* is the α-bromo derivative that is formed.

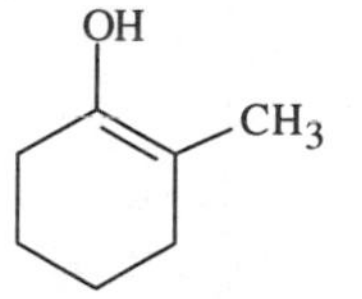

enol intermediate involved in the formation of *A* (more stable enol)

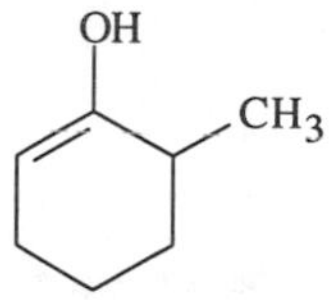

enol intermediate involved in the formation of compounds *B* and *C*

22.58 The dibromo derivative is a mixture of diastereomers: the (±)-diastereomer (that is, the racemate) and the meso diastereomer. (The structures below are Fischer projections.)

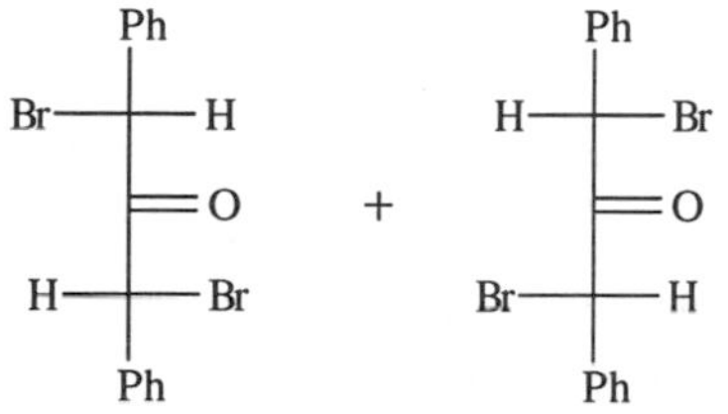

(±)-1,3-dibromo-1,3-diphenyl-2-propanone

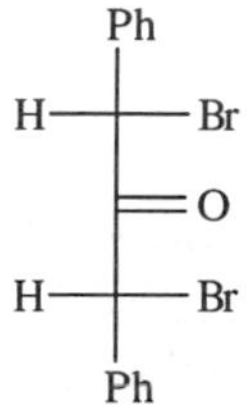

meso-1,3-dibromo-1,3-diphenyl-2-propanone

22.59 The discussion on text p. 1008 explains that enols are intermediates in the acid-catalyzed bromination of aldehydes and ketones, and that enol formation is the rate-limiting process. As explained on text p. 924, an enol intermediate is also involved in the decarboxylation of a β-keto acid. Thus, the bromination reaction described in the problem evidently involves an enol intermediate, and formation of the enol is the rate-limiting process. Hence, the rate of the bromination reaction is zero-order in bromine.

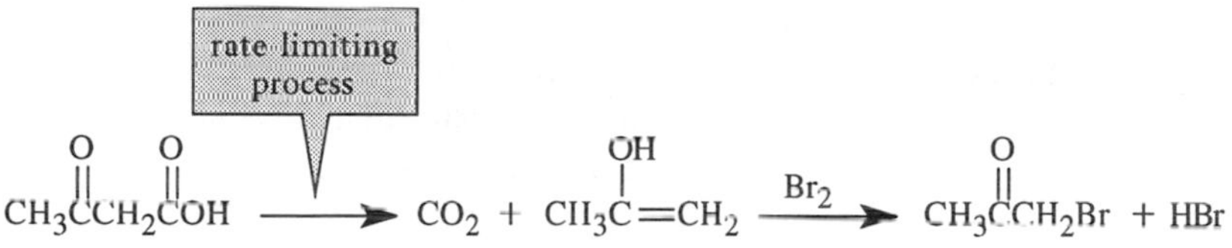

rate limiting process

$$CH_3\overset{O}{\overset{\|}{C}}CH_2\overset{O}{\overset{\|}{C}}OH \longrightarrow CO_2 + CH_3\overset{OH}{\overset{|}{C}}{=}CH_2 \xrightarrow{Br_2} CH_3\overset{O}{\overset{\|}{C}}CH_2Br + HBr$$

22.60 It is reasonable to hypothesize that the carbon atom of the iodoform must originate from the central carbon of the β-diketone:

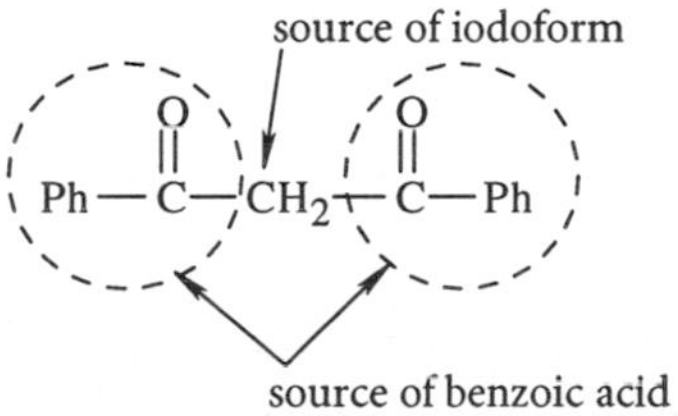

source of iodoform

$$Ph-\overset{O}{\overset{\|}{C}}-CH_2-\overset{O}{\overset{\|}{C}}-Ph$$

source of benzoic acid

Base-promoted iodination occurs at the central carbon by the usual mechanism involving enolate ions to give the diiodo ketone *A*. Hydroxide then displaces an α-diiodo enolate anion *B* in a nucleophilic acyl substitution reaction; this resonance-stabilized anion is a fairly good leaving group. Anion *B* is then iodinated, and the resulting triiodo ketone *C* undergoes the final steps of the haloform reaction illustrated in Eq. 22.27e on text p. 1010 to give iodoform and benzoate anion. The two equivalents of benzoate anion produced in this sequence give benzoic acid when the reaction mixture is acidified.

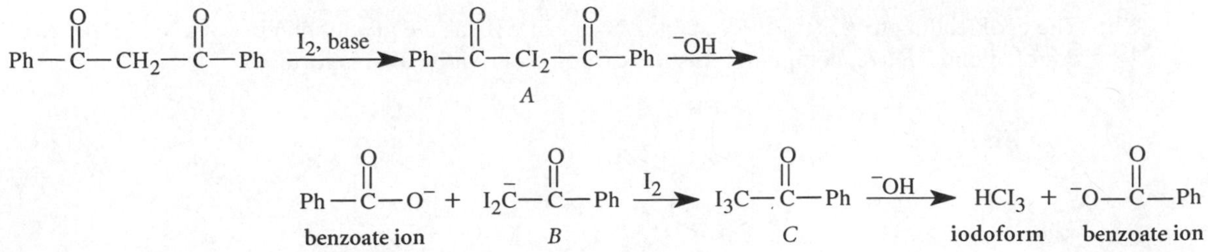

22.61 (a) The α-hydrogens and the N—H hydrogen are acidic enough to be replaced. These hydrogens are shown as deuteriums in the following structure.

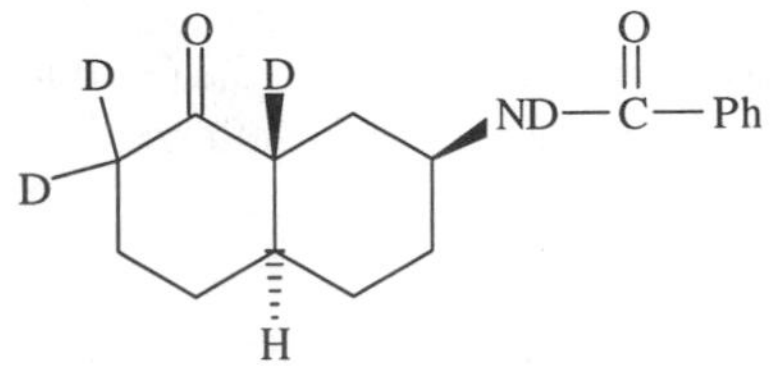

The stereochemistry of the deuterium at the ring junction could be a mixture of "up" and "down" (that is, trans and cis), because the planar carbanion intermediate could be protonated from either face. (See the solution to Problem 22.11(a) on p. 820 of this manual.) However, because the *trans*-decalin derivatives are more stable than *cis*-decalin derivatives, the stereoisomer shown above should be the predominant one.

(b) The two α-hydrogens are acidic enough to be replaced. These hydrogens are shown as deuteriums in the following structure.

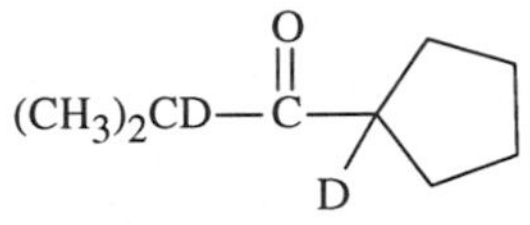

22.62 (a) The isomerization favors compound *B* because *trans*-decalin derivatives are more stable than *cis*-decalin derivatives. (See Sec. 7.6B.) The isomerization mechanism involves formation of an enolate ion intermediate, which can be protonated on either face to give *A* or *B*, respectively.

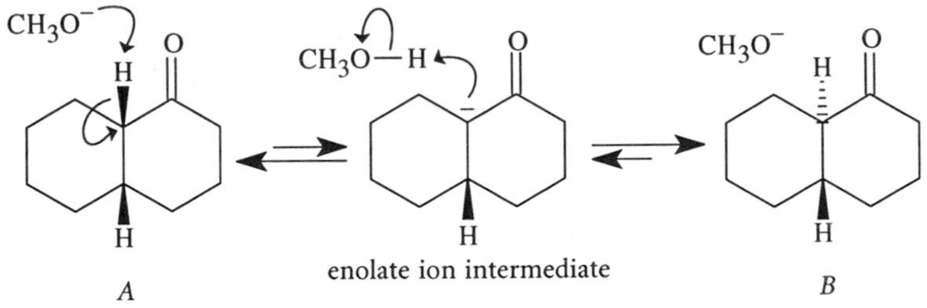

(b) The analogous isomerization of compound *C* cannot occur because there is no α-hydrogen at the ring junction. Only removal of an acidic hydrogen at a ring junction could lead to isomerization, and the hydrogen at the other carbon of the ring junction is not acidic enough to ionize under the conditions given.

22.63 (a) The equilibrium favors the α,β-unsaturated isomer because it is conjugated. Conjugation is a stabilizing effect because of the additional π-electron overlap that is possible in conjugated compounds. (See Sec. 15.1A.)

(b) The mechanism in aqueous base involves a resonance-stabilized enolate ion intermediate, which can be protonated on either the α-carbon or the γ-carbon to give the respective unsaturated ketones.

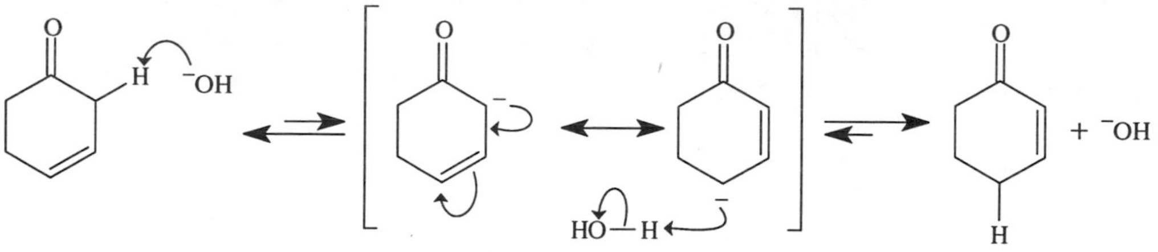

$$\text{H} \quad {}^{-}\text{OH} \rightleftharpoons \left[\longleftrightarrow \right] \rightleftharpoons \quad + {}^{-}\text{OH}$$

HO—H

H

(c) The mechanism below commences with the dienol 1,3-cyclohexadienol, which is formed by the usual acid-catalyzed enolization mechanism (Eq. 22.17b on text p. 1006). Protonation of this dienol gives a resonance-stabilized carbocation, which is the same as the protonated α,β-unsaturated ketone. Loss of the O—H proton gives the α,β-unsaturated ketone itself.

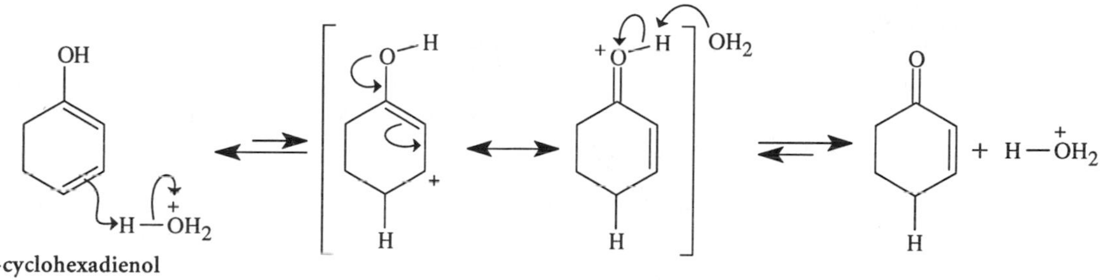

OH

$\text{H}-\overset{+}{\text{O}}\text{H}_2$

1,3-cyclohexadienol

$$\rightleftharpoons \left[\longleftrightarrow \right] \rightleftharpoons \quad + \text{H}-\overset{+}{\text{O}}\text{H}_2$$

(d) The equilibrium constant for conversion of 4-methyl-3-cyclohexenone into 4-methyl-2-cyclohexenone should be smaller, because the additional alkyl branch on the β,γ-double bond tends to offset the stabilizing effect of conjugation in the α,β-unsaturated derivative.

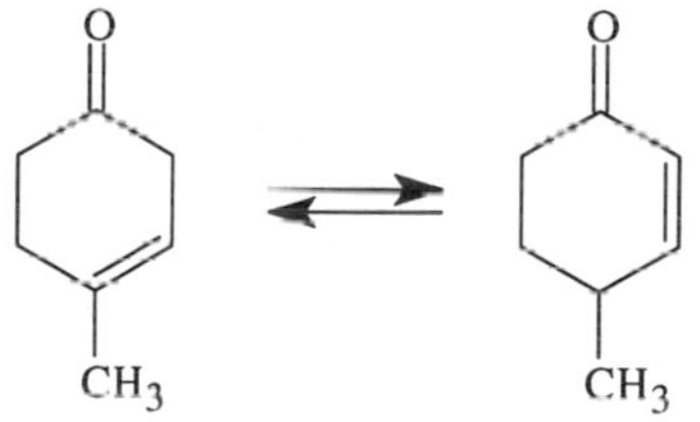

CH_3 $\rightleftharpoons$ CH_3

4-methyl-3-cyclohexenone 4-methyl-2-cyclohexenone

22.64 (a) The exchange of H^a occurs by the usual enolate-ion mechanism.

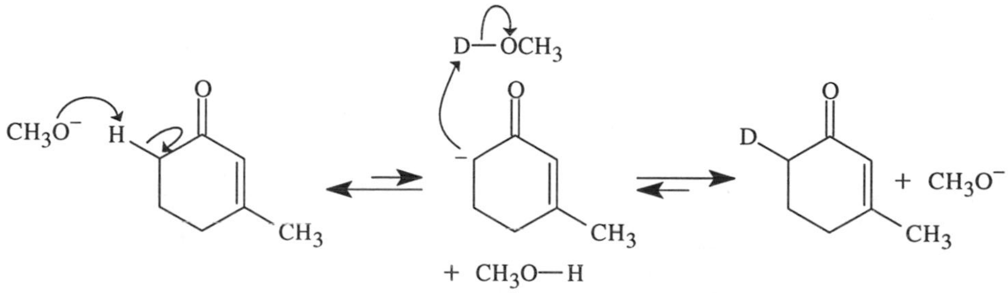

$$CH_3O^{-} \quad H \rightleftharpoons \quad (D-OCH_3) \rightleftharpoons D \quad + CH_3O^{-}$$

CH_3 CH_3 CH_3

$+ \ CH_3O-H$

Removal of either hydrogen H^c or H^d gives an anion that has resonance structures which show that it is an enolate ion. Hence, these hydrogens are acidic enough to undergo exchange. The exchange of H^c is shown explicitly below; you should show the exchange of H^d, which occurs by essentially the same mechanism.

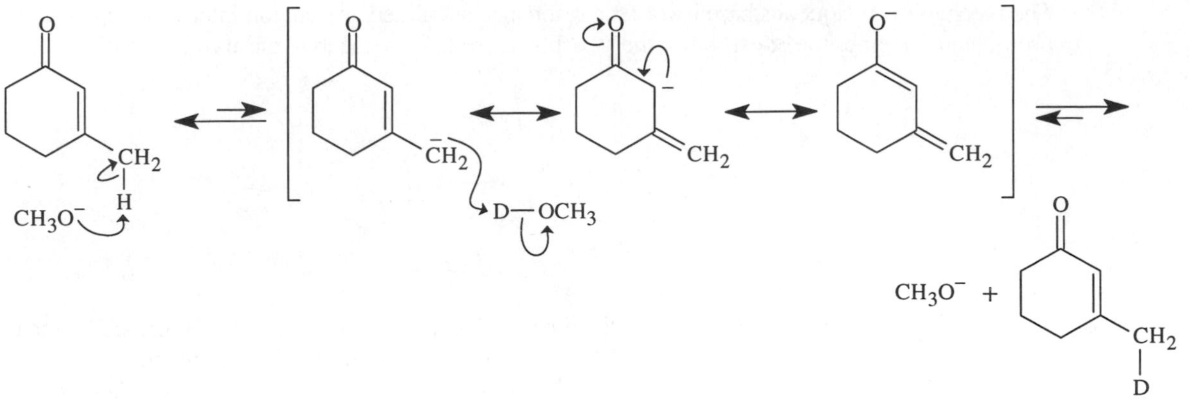

CH$_3$O$^-$ + CH$_2$—D

(b) Although H^b is an α-proton, it is not acidic because the anion that would result from its removal is not resonance-stabilized. The reason is that the orbital containing the unshared electron pair in this anion is perpendicular to the π-electron system of the carbonyl group and cannot overlap with it. An equivalent resonance argument is that if the overlap which lies at the basis of resonance were to occur, the nuclei would have to move and a cumulated double bond and the attendant large amount of strain would be introduced into the six-membered ring.

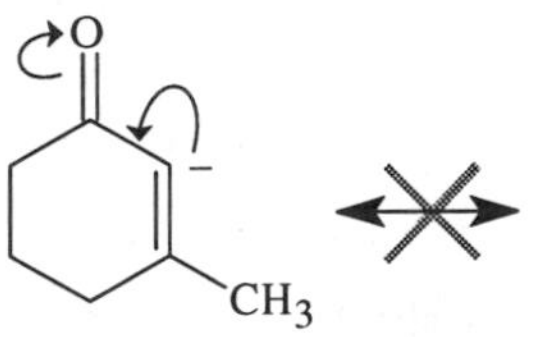

overlap introduces strained cumulated double bond

(c) Hydrogen H^b is not acidic in the α,β-unsaturated ketone. However, as shown in Problem 22.63, the α,β-unsaturated ketone readily isomerizes to a β,γ-unsaturated ketone. Ionization of H^b in this compound gives a resonance-stabilized anion, as shown in the solution to Problem 22.63(b). Hence, H^b exchanges, and the β,γ-unsaturated ketone isomerizes back to the α,β-unsaturated ketone.

$^-$OCH$_3$ (Problem 22.63) CH$_3$OD + H—OCH$_3$

CH$_3$O$^-$ +

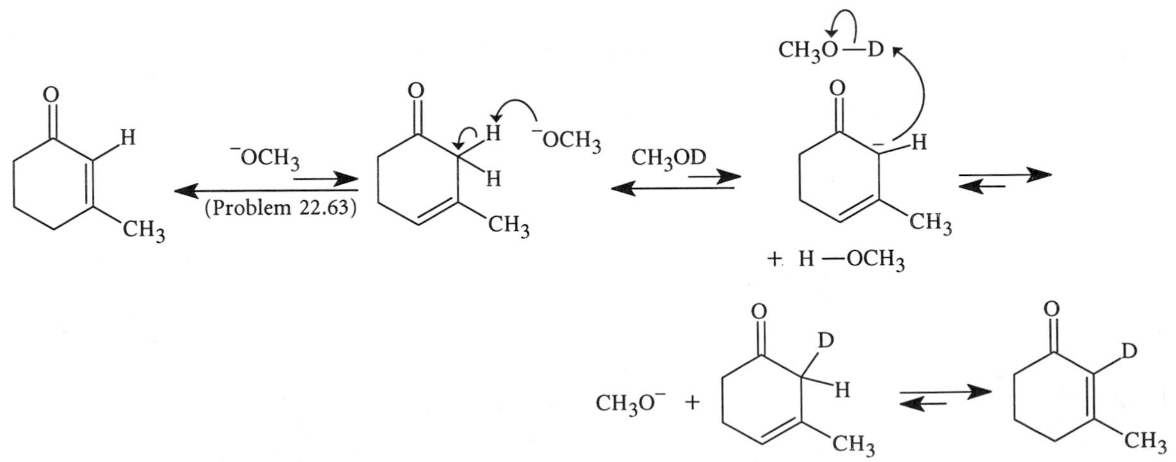

22.65 (a) Although ethers are usually more basic than carbonyl oxygens, in compound *A* the carbonyl oxygen is protonated because the resulting carbocation has $4n + 2$ π electrons; that is, it is aromatic. Remember that increasing the stability of the conjugate acid in an acid-base equilibrium decreases its acidity (or increases the basicity of the conjugate base). The aromaticity of its conjugate acid thus explains why compound *A* is more basic than compound *B*. Compound *B* is protonated on the ether oxygen and has the basicity typical of an ether.

conjugate acid of compound *A*
(an aromatic species)

conjugate acid
of compound *B*

The relative basicities of these two compounds are deceptive because they involve protonation on different oxygens. The conjugate-acid pK_a of a ketone is typically –6. Hence, aromaticity actually raises the pK_a of the conjugate acid of *A* by about 6 units.

(b) Tropone is unusually basic because its conjugate acid is an aromatic carbocation.

conjugate acid of tropone
(an aromatic α-hydroxy carbocation)

The consequences of aromaticity on the pK_a of a conjugate acid were explored in part (a), and these apply equally here.

22.66 (a) As shown in Fig. SG19.2 on page 659 of this manual, the more stable a species is relative to its conjugate base the less acidic it is. Consequently, resonance stabilization of an α,β-unsaturated carboxylic acid decreases its acidity.

(b) The pK_a of the last compound, butanoic acid, is the reference value—the pK_a to be expected for a four-carbon carboxylic acid containing no carbon-carbon double bonds. In the first compound, 3-butenoic acid, the carbon-carbon double bond is not conjugated with the carbonyl group. Hence, the difference between its pK_a value and that of butanoic acid is due to the electron-withdrawing polar effect of the carbon-carbon double bond. If the polar effect were the only effect operating in the middle compound, *trans*-2-butenoic acid (crotonic acid), it should be the most acidic compound in the series, because the polar effect of any group on acidity increases when the group, in this case a carbon-carbon double bond, is closer to the site of ionization. Because *trans*-2-butenoic acid is in fact *less* acidic than 3-butenoic acid, some acid-weakening effect must also be operating, and this is the resonance effect discussed in part (a).

22.67 (a) The α-hydrogen of any nitro compound is particularly acidic because the conjugate-base anion is stabilized both by resonance and by the polar effect of the nitro group (note the positive charge next to the negative charge of the carbanion).

$(CH_3)_2\ddot{C}^{-}-\overset{+}{N}(=O)O^{-} \longleftrightarrow (CH_3)_2C=\overset{+}{N}(O^{-})O^{-}$

the conjugate-base anion of 2-nitropropane

(b) To answer this question, ask where besides the anionic carbon there is negative charge and a pair of electrons that can be protonated. As the resonance structures above show, the negative charge of the carbanion is shared by the oxygens of the nitro group. Protonation of either oxygen gives the nitro analog of an enol, which is called an *aci*-nitro compound. This is the isomer that is requested in the problem. (See also

Problem 22.9(a), text p. 1005, and its solution.

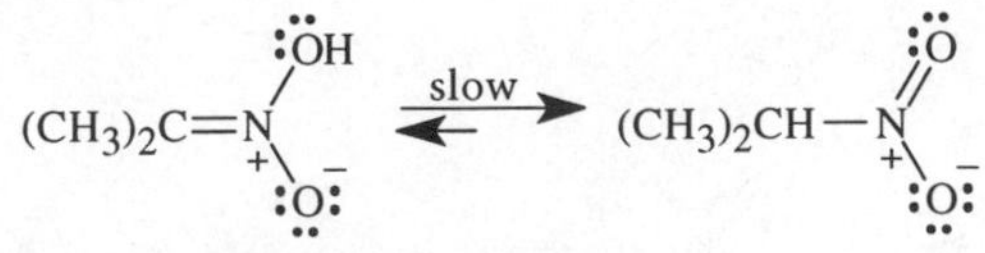

$$(CH_3)_2C{=}\overset{+}{N}(\ddot{O}H)(\ddot{O}:^-) \xleftarrow[]{\text{slow}}\!\!\rightarrow (CH_3)_2CH{-}\overset{+}{N}(=\ddot{O})(\ddot{O}:^-)$$

an *aci*-nitro compound

(c) Because 2-nitropropane is much more acidic than ethanol, it is completely converted into its anion by sodium ethoxide. This anion, like many other "enolate ions," undergoes a Michael addition to ethyl acrylate to give ethyl 4-methyl-4-nitropentanoate.

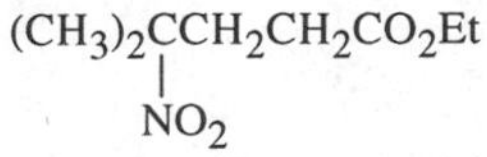

$$(CH_3)_2C(NO_2)CH_2CH_2CO_2Et$$

ethyl 4-methyl-4-nitropentanoate

22.68 (a) The conjugate-base enolate ion of diethyl malonate reacts at the carbonyl group of acetone. The addition product dehydrates to give the following product.

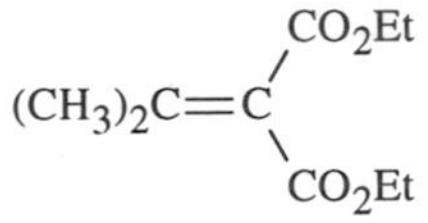

$$(CH_3)_2C{=}C(CO_2Et)_2$$

(Be sure to go through the mechanism if the derivation of this product isn't clear.)

(b) The most acidic hydrogen in either of the two carbonyl compounds is the α-hydrogen of diethyl malonate; thus, formation of the conjugate-base anion of diethyl malonate consumes the sodium ethoxide. Formation of the conjugate-base enolate ion of acetone therefore will not be a competing reaction. Ketone carbonyl groups are more reactive than ester carbonyl groups; hence, the conjugate-base anion of diethyl malonate reacts with acetone. The condensation is pulled to the right by dehydration.

(c) The synthesis of 3,3-dimethylbutanoic acid:

$$(CH_3)_2C{=}C(CO_2Et)_2 \xrightarrow[\text{2) } H_3O^+]{\text{1) } Li^+\,(CH_3)_2Cu^-} (CH_3)_3CCH(CO_2Et){-}CO_2Et \xrightarrow{\text{NaOH}} \xrightarrow[-\text{EtOH},\ -CO_2]{H_3O^+,\ \text{heat}} (CH_3)_3CCH_2CO_2H$$

3,3-dimethylbutanoic acid

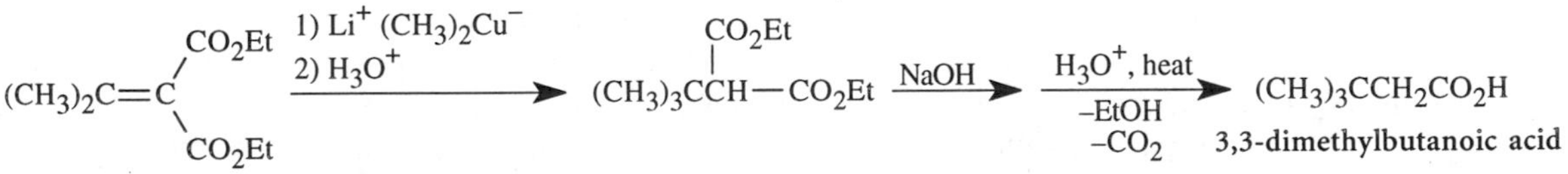

Note that this product cannot be prepared by a conventional malonic ester synthesis involving alkylation with *tert*-butyl bromide. (Why? See Sec. 9.4C.)

22.69 Intermediate *A* is the anionic product of a crossed Claisen condensation. This anion undergoes a Michael addition to the α,β-unsaturated ketone, and the product of that reaction undergoes an intramolecular aldol condensation to give compound *B*. Hydroxide ion effects a reverse Claisen condensation that removes the formyl group. The steps are outlined below; you should provide the mechanistic details using the curved-arrow notation.

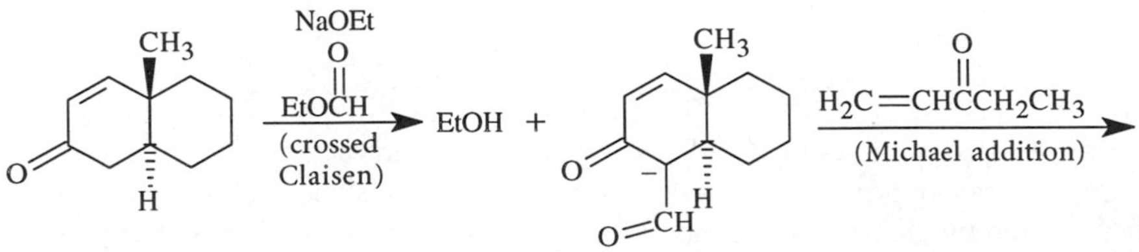

[bicyclic enone with CH_3 and H] $\xrightarrow[\text{(crossed Claisen)}]{\text{NaOEt},\ EtOCH{=}O}$ EtOH + [enolate anion bearing $O{=}CH$ group] *A* $\xrightarrow[\text{(Michael addition)}]{H_2C{=}CHCOCH_2CH_3}$

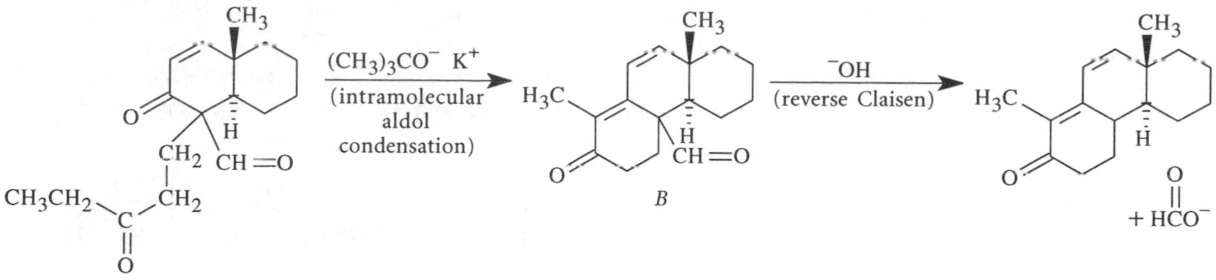

$$\xrightarrow[\text{(intramolecular aldol condensation)}]{(CH_3)_3CO^-\ K^+} B \xrightarrow[\text{(reverse Claisen)}]{^-OH} \text{product} + HCO^-$$

22.70 (a) One full equivalent of base is required in the Claisen or Dieckmann condensation to ionize the product. Without this ionization, the condensation equilibrium is unfavorable. (See text pp. 1024–6.)

(b) Sodium phenoxide, the conjugate base of phenol, is about 10^{-5} as basic as sodium ethoxide. Thus, sodium phenoxide is not basic enough to ionize completely the Claisen condensation product. Without such an ionization, the condensation equilibrium is unfavorable.

(c) The equilibrium for the Claisen condensation is rather unfavorable and requires the final ionization of the product to proceed to completion. Under acidic conditions, no base is available to cause this ionization. In addition, an acid-catalyzed Claisen condensation would require the enol of an ester, because enols are the α-carbon nucleophiles under acidic conditions. As shown in Eq. 22.13, p. 1004, the formation of enols from esters is *much* less favorable than the formation of enols from aldehydes or ketones. The much lower concentration of reactive intermediate would translate into a much reduced rate. In summary, both the rate of the reaction is very low, and its equilibrium constant is highly unfavorable; hence, no reaction is observed.

22.71 (a) An alkylation of this type requires a base that will rapidly and completely convert the ester into its conjugate-base enolate anion so that the enolate ion is not present simultaneously with the ester. In the presence of sodium ethoxide, however, small amounts of the enolate ion are present together with large amounts of the un-ionized ester; consequently, the Claisen condensation will occur as the major reaction. In all likelihood methyl iodide will alkylate the anionic product of the Claisen condensation to give the following compound:

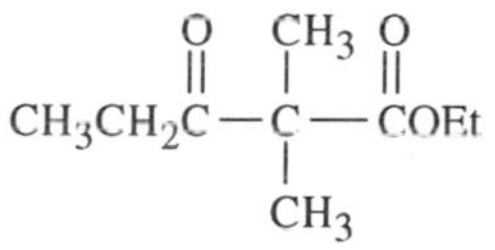

$$CH_3CH_2\overset{O}{\overset{\|}{C}}-\underset{CH_3}{\overset{CH_3}{\overset{|}{\underset{|}{C}}}}-\overset{O}{\overset{\|}{C}}OEt$$

(b) The conjugate-base anion of diethyl malonate would be formed with sodium ethoxide, but it would not react with bromobenzene because the latter compound, an aryl halide, does not undergo S_N2 reactions. (See Sec. 18.1.)

(c) Alcohol dehydration involves a carbocation intermediate, and carbocations α to carbonyl groups are particularly unstable. The desired dehydration will not occur. The more likely reaction would be polymerization of the ester by reaction of the hydroxy group of one molecule with the carbonyl of the other.

(d) The first step will effect α-halogenation of the carboxylic acid. However, an attempt to form the Grignard reagent will fail, because the Grignard reagent will be destroyed instantly by a protonolysis reaction with the α-bromo acid. Because the Grignard reagent does not survive, its reaction with acetaldehyde cannot occur.

(e) This is a crossed aldol condensation in which the conjugate-base enolate ion of the ketone reacts with the aldehyde. A mixture of isomeric products will result because there are two possible conjugate-base enolate ions of the ketone that can react. Moreover, a very significant, if not predominant, competing reaction will be the aldol condensation of acetaldehyde itself.

(f) In addition to bromination of the benzene ring, the acidic conditions will also promote α-bromination at the methyl group. Lewis acids can catalyze this reaction just as Brønsted acids can. In addition, HBr is a by-product of ring halogenation, and this can also catalyze α-halogenation.

22.72 (a) The gas is H_2 and the species *A* is the sodium salt of the conjugate-base enolate ion of 2,4-pentanedione. As the resonance structures indicate, the negative charge, and therefore the nucleophilic character, of this anion is shared by both the anionic carbon and the oxygens.

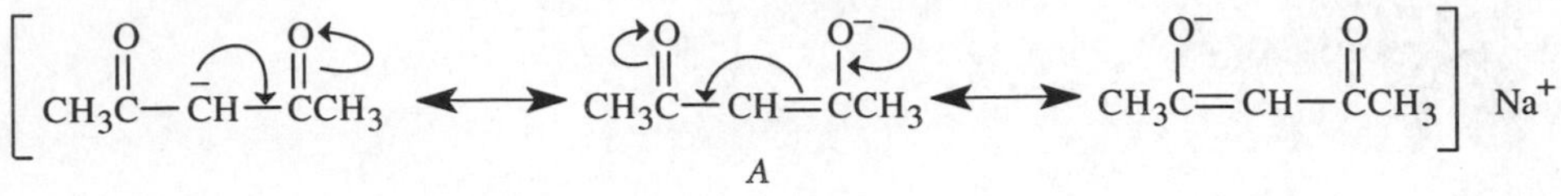

$$\left[CH_3\overset{O}{\overset{\|}{C}}-\overset{-}{C}H-\overset{O}{\overset{\|}{C}}CH_3 \longleftrightarrow CH_3\overset{O}{\overset{\|}{C}}-CH=\overset{O^-}{\overset{|}{C}}CH_3 \longleftrightarrow CH_3\overset{O^-}{\overset{|}{C}}=CH-\overset{O}{\overset{\|}{C}}CH_3 \right] Na^+$$

A

(b) The three species that are formed all result from reaction with a nucleophilic atom with methyl iodide in an S_N2 reaction. The products are the two stereoisomeric ethers *B* and *C* and the alkylated β-diketone *D*.

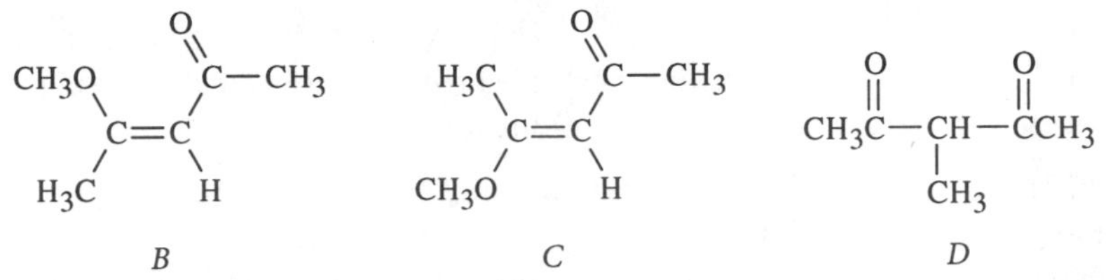

B: $(CH_3O)(H_3C)C=C(H)(C(=O)CH_3)$

C: $(H_3C)(CH_3O)C=C(H)(C(=O)CH_3)$

D: $CH_3\overset{O}{\overset{\|}{C}}-\underset{CH_3}{\underset{|}{C}H}-\overset{O}{\overset{\|}{C}}CH_3$

B *C* *D*

22.73 (a)

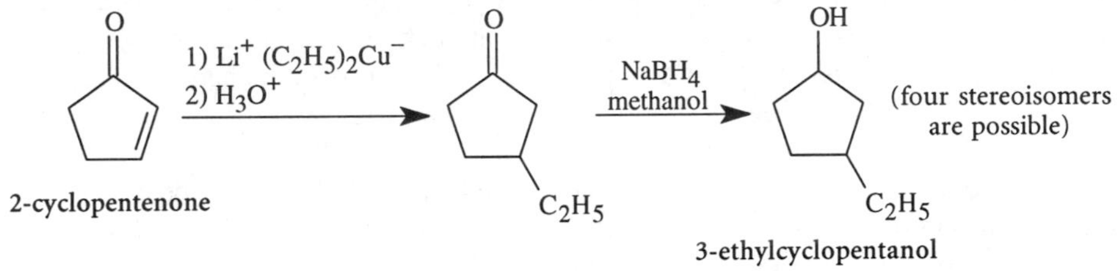

2-cyclopentenone $\xrightarrow[\text{2) } H_3O^+]{\text{1) } Li^+ (C_2H_5)_2Cu^-}$ 3-ethylcyclopentanone $\xrightarrow[\text{methanol}]{NaBH_4}$ 3-ethylcyclopentanol (four stereoisomers are possible)

(b)

3-methyl-2-cyclohexenone $\xrightarrow[\text{2) } H_3O^+]{\text{1) } Li^+ (CH_3)_2Cu^-}$ 3,3-dimethylcyclohexanone $\xrightarrow[\text{2) } H_3O^+]{\text{1) } CH_3MgI}$ 1,3,3-trimethylcyclohexanol

(c)

$EtO_2C(CH_2)_5CO_2Et$ (diethyl pimelate) $\xrightarrow[EtOH]{NaOEt}$ 2-carbethoxycyclohexanone enolate ($^-$, CO_2Et) $\xrightarrow{ClCH_2Ph}$ 2-(CO_2Et)-2-(CH_2Ph)cyclohexanone $\xrightarrow[\text{2) } H_3O^+,\ -CO_2]{\text{1) NaOH}}$ 2-benzylcyclohexanone (CH_2Ph)

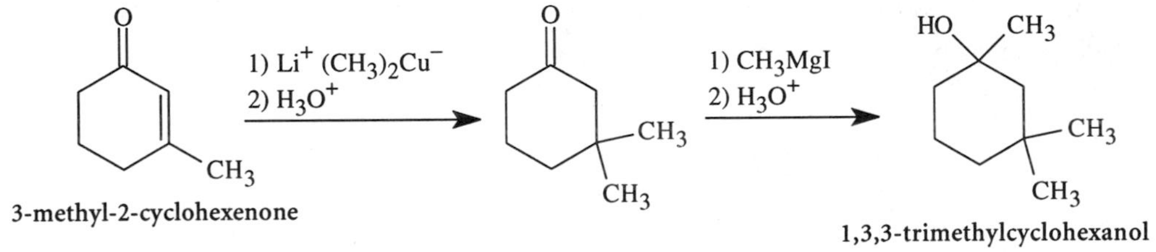

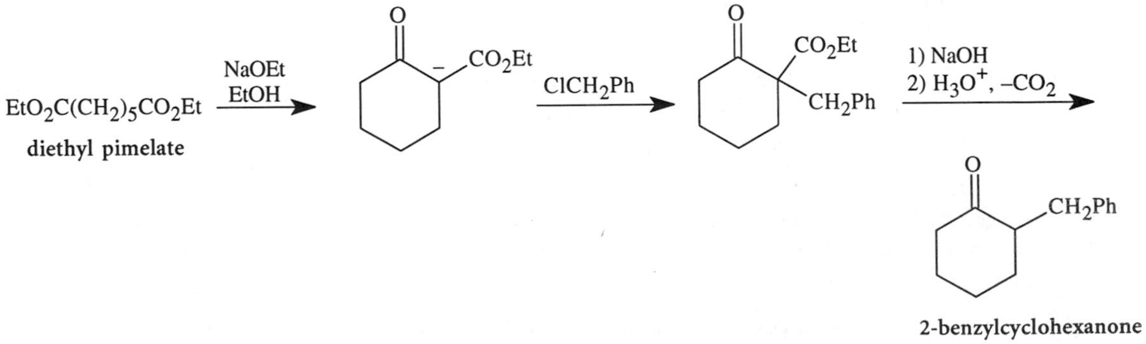

(d)

$$\underset{\text{diethyl malonate}}{EtO_2CCH_2CO_2Et} \xrightarrow[\text{2) } CH_3I]{\text{1) NaOEt, EtOH}} EtO_2C\underset{CH_3}{\underset{|}{C}}HCO_2Et \xrightarrow[\text{2) } CH_3I]{\text{1) NaOEt, EtOH}}$$

$$EtO_2C\overset{CH_3}{\overset{|}{\underset{CH_3}{\underset{|}{C}}}}CO_2Et \xrightarrow[\text{2) } H_3O^+]{\text{1) } LiAlH_4} \underset{\text{2,2-dimethyl-1,3-propanediol}}{HOCH_2\overset{CH_3}{\overset{|}{\underset{CH_3}{\underset{|}{C}}}}CH_2OH}$$

(e)

$$\underset{\text{ethyl acrylate}}{H_2C{=}CHCO_2Et} \xrightarrow{HCN} N{\equiv}CCH_2CH_2CO_2Et \xrightarrow[\text{3) } {}^-OH]{\text{1) } LiAlH_4 \text{ 2) } H_3O^+} H_2NCH_2CH_2CH_2CH_2OH$$

(f)

$$\underset{\text{acrylonitrile}}{H_2C{=}CH{-}CN} \xrightarrow{(C_2H_5)_2NH} (C_2H_5)_2NCH_2CH_2CN \xrightarrow[\text{3) } {}^-OH]{\text{1) } LiAlH_4 \text{ 2) } H_3O^+} (C_2H_5)_2NCH_2CH_2CH_2NH_2$$

(g)

$$\underset{\alpha\text{-phenylacetic acid}}{PhCH_2\overset{O}{\overset{\|}{C}}OH} \xrightarrow[H_2SO_4 \text{ (catalyst)}]{\text{EtOH (solvent)}} PhCH_2\overset{O}{\overset{\|}{C}}OEt \xrightarrow[\text{NaOEt}]{EtO\overset{O}{\overset{\|}{C}}OEt \text{ (excess)}} PhCH\overset{O}{\overset{\|}{C}}OEt \text{ (with } CO_2Et \text{ on CH)} \xrightarrow[\text{2) } CH_3CH_2Br]{\text{1) NaOEt, EtOH}}$$

$$Ph\overset{CH_3CH_2}{\overset{|}{\underset{CO_2Et}{\underset{|}{C}}}}{-}\overset{O}{\overset{\|}{C}}OEt \xrightarrow[-EtOH,\ -CO_2]{H_3O^+\text{, heat}} \underset{\text{2-phenylbutanoic acid}}{CH_3CH_2\underset{Ph}{\underset{|}{C}}HCO_2H}$$

(h)

$$\underset{\text{acetophenone}}{2\ Ph{-}\overset{O}{\overset{\|}{C}}{-}CH_3} \xrightarrow{H_2SO_4} Ph{-}\overset{O}{\overset{\|}{C}}{-}CH{=}\underset{Ph}{\underset{|}{C}}{-}CH_3 \xrightarrow{H_2\text{, cat.}} \underset{\text{1,3-diphenyl-1-butanone}}{Ph{-}\overset{O}{\overset{\|}{C}}{-}CH_2\underset{Ph}{\underset{|}{C}}HCH_3}$$

(i)

$$\underset{\text{phenylacetic acid}}{HO\overset{O}{\overset{\|}{C}}CH_2Ph} \xrightarrow{SOCl_2} Cl\overset{O}{\overset{\|}{C}}CH_2Ph \xrightarrow[\text{2) } H_3O^+]{\text{1) benzene, } AlCl_3} Ph\overset{O}{\overset{\|}{C}}CH_2Ph \xrightarrow{D_2O\text{, base}}$$

$$Ph\overset{O}{\overset{\|}{C}}CD_2Ph \xrightarrow[\text{2) dil. } H_3O^+]{\text{1) } LiAlH_4} Ph\overset{OH}{\overset{|}{C}}HCD_2Ph$$

(j)

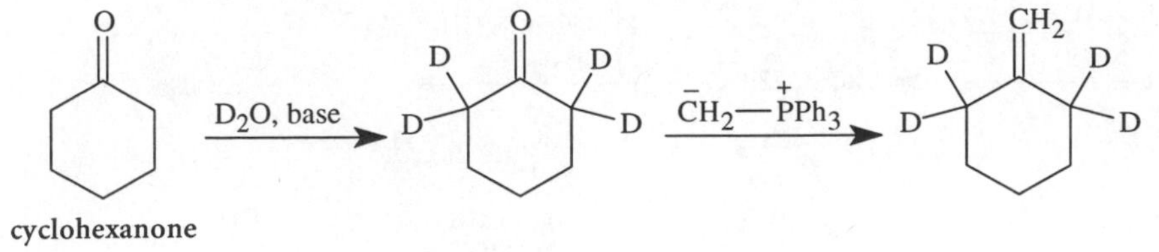

cyclohexanone $\xrightarrow{D_2O,\ base}$ 2,2,6,6-tetradeuteriocyclohexanone $\xrightarrow{\bar{C}H_2-\overset{+}{P}Ph_3}$ 2,2,6,6-tetradeuterio-1-methylenecyclohexane

(k)

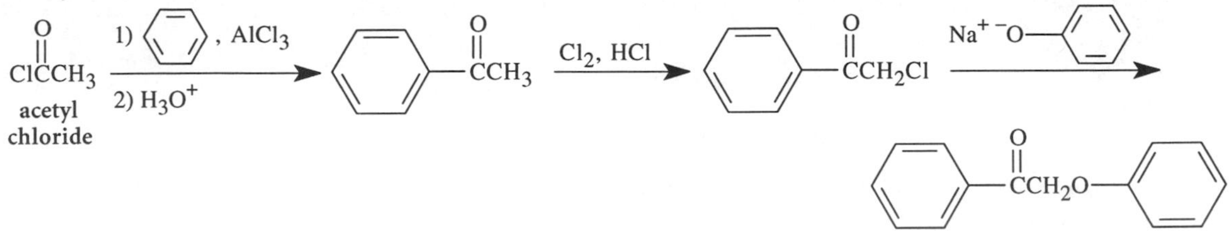

$ClCCH_3$ (O above C) acetyl chloride $\xrightarrow[2)\ H_3O^+]{1)\ C_6H_6,\ AlCl_3}$ $C_6H_5CCH_3$ (O) $\xrightarrow{Cl_2,\ HCl}$ $C_6H_5CCH_2Cl$ (O) $\xrightarrow{Na^+\ {}^-O-C_6H_5}$ $C_6H_5CCH_2O-C_6H_5$ (O)

(l)

$(EtO_2C)_2CH_2$ diethyl malonate $\xrightarrow[2)\ CH_3CH_2CH_2Br]{1)\ NaOEt,\ EtOH}$ $(EtO_2C)_2CHCH_2CH_2CH_3$ $\xrightarrow[2)\ CH_3I]{1)\ NaOEt,\ EtOH}$

$(EtO_2C)_2C(CH_3)CH_2CH_2CH_3$ $\xrightarrow[-EtOH,\ -CO_2]{H_3O^+,\ heat}$ $HO_2CCH(CH_3)CH_2CH_2CH_3$ $\xrightarrow{SOCl_2}$

$ClCCH(CH_3)CH_2CH_2CH_3$ (O) $\xrightarrow[2)\ H_3O^+]{1)\ Li^+\ (CH_3CH_2CH_2CH_2)_2Cu^-}$ $CH_3CH_2CH_2CH_2CCH(CH_3)CH_2CH_2CH_3$ (O)
4-methyl-5-decanone

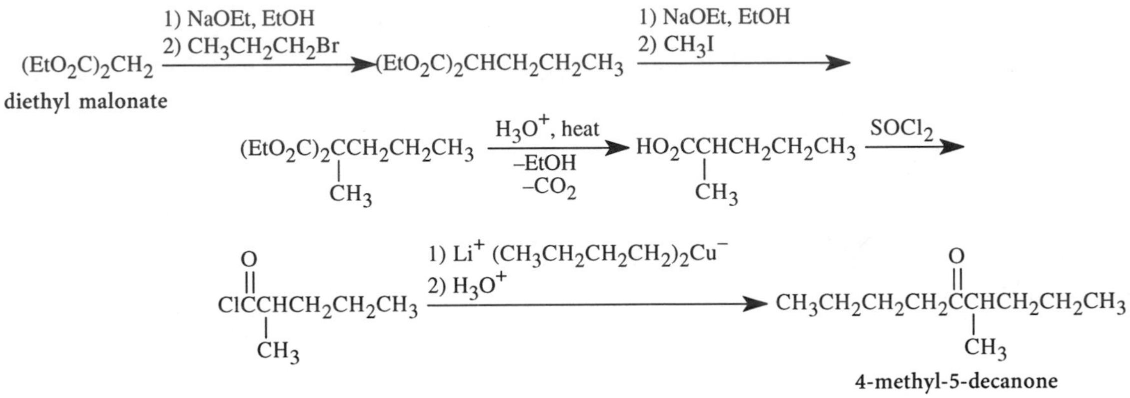

22.74 In the first step, the conjugate-base anion of diethyl malonate reacts with mesityl oxide, an α,β-unsaturated ketone, in a Michael addition.

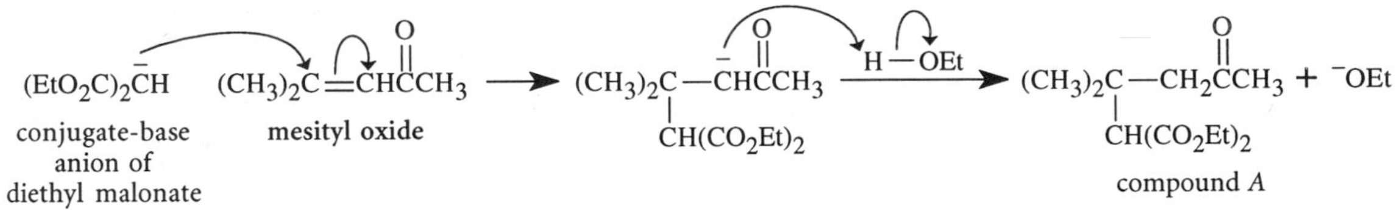

$(EtO_2C)_2\bar{C}H$ (conjugate-base anion of diethyl malonate) + $(CH_3)_2C{=}CHCCH_3$ (O) (mesityl oxide) $\longrightarrow$ $(CH_3)_2C(CH(CO_2Et)_2)-\bar{C}HCCH_3$ (O) $\xrightarrow{H-OEt}$ $(CH_3)_2C(CH(CO_2Et)_2)-CH_2CCH_3$ (O) + ^-OEt
compound *A*

A crossed Claisen between the enolate ion of the methyl ketone and one of the ester groups then ensues.

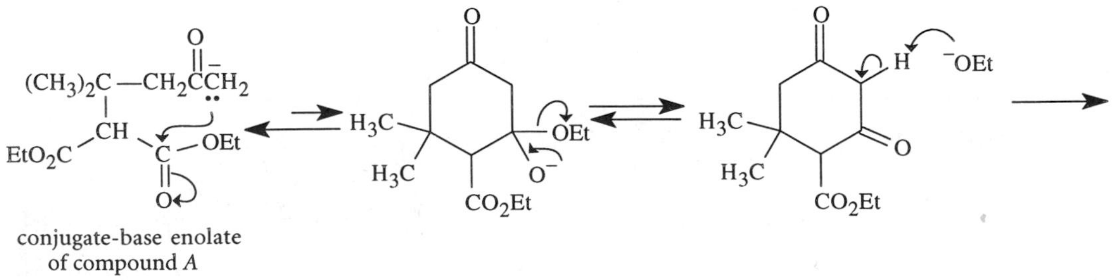

$(CH_3)_2C(-CH_2C(O)\bar{C}H_2)-CH(CO_2Et)-C(O)OEt$ (conjugate-base enolate of compound *A*) $\rightleftharpoons$ cyclic tetrahedral intermediate (H_3C, H_3C, CO_2Et, O^-, OEt) $\rightleftharpoons$ cyclic β-diketone (H_3C, H_3C, CO_2Et, =O) + ^-OEt $\longrightarrow$

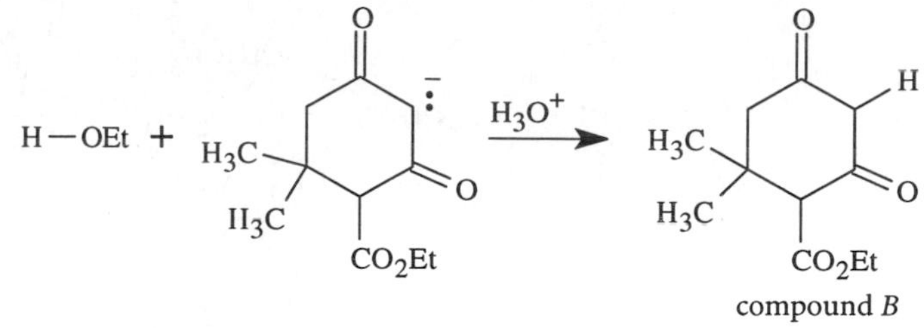

H—OEt + (enolate of compound) $\xrightarrow{H_3O^+}$ compound *B*

Dimedone results from the saponification and decarboxylation of the remaining ester group (a β-keto ester) in compound *B*.

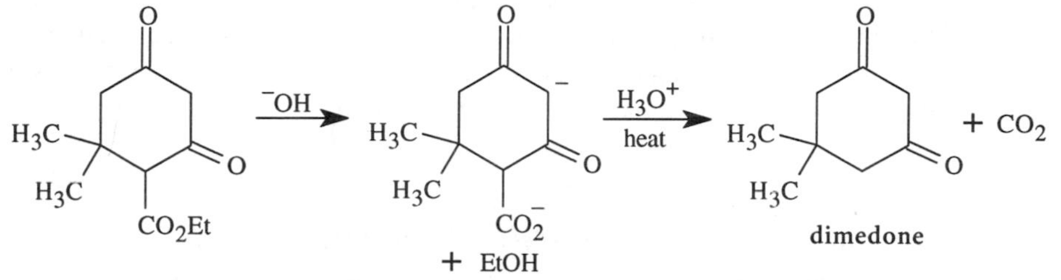

compound *B* $\xrightarrow{^{-}OH}$ (anion, CO_2^-) + EtOH $\xrightarrow[\text{heat}]{H_3O^+}$ dimedone + CO_2

Note that the acidic C—H ionizes again during the saponification. This ionization is what prevents a reverse Claisen condensation from occurring in which ^{-}OH attacks one of the carbonyls of the β-diketone.

22.75 This reaction is essentially a type of crossed Claisen condensation in which the "enolate ion" is the conjugate-base anion of urea. A second reaction of the same type closes the ring. The following mechanism starts with the attack of the ionized urea on one of the ester groups.

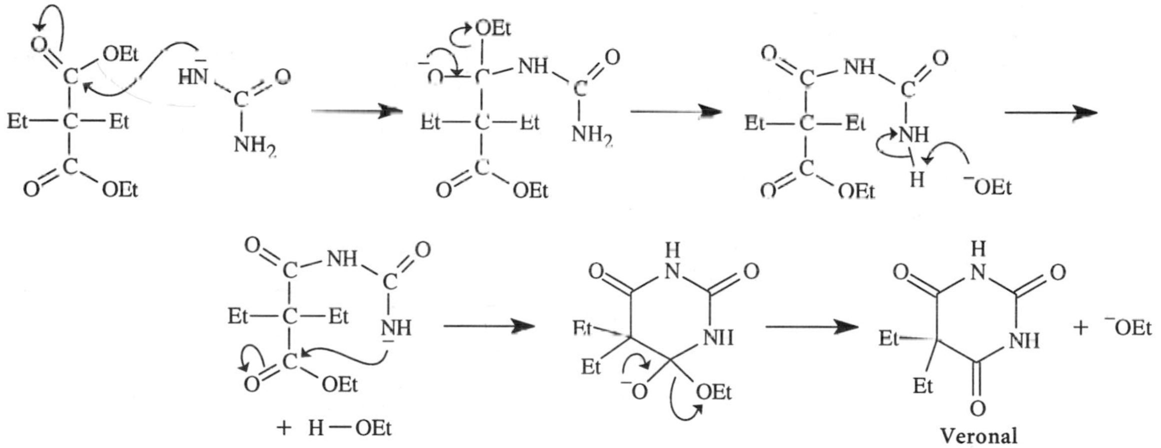

+ H—OEt

Veronal

+ ^{-}OEt

As in other Claisen-type condensation reactions, the product Veronal is ionized by the ethoxide base, and Veronal is regenerated when acid is added. (These final steps are not shown in the foregoing mechanism.)

22.76 The synthesis of pentothal involves, first, preparation of the substituted malonic ester derivative, and then condensing it with thiourea, the sulfur analog of urea.

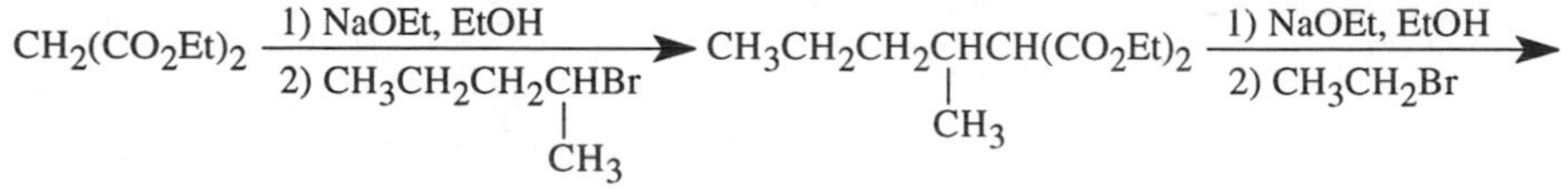

$$CH_2(CO_2Et)_2 \xrightarrow[\text{2) } CH_3CH_2CH_2CHBr(CH_3)]{\text{1) NaOEt, EtOH}} CH_3CH_2CH_2CH(CH_3)CH(CO_2Et)_2 \xrightarrow[\text{2) } CH_3CH_2Br]{\text{1) NaOEt, EtOH}}$$

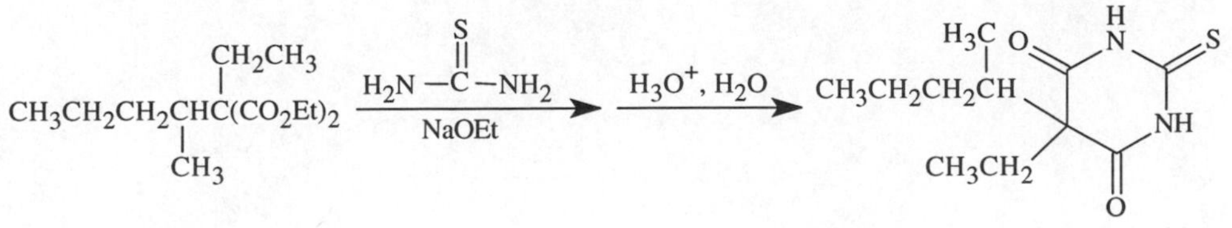

$$CH_3CH_2CH_2CH(CH_3)C(CH_2CH_3)(CO_2Et)_2 \xrightarrow[\text{NaOEt}]{H_2N-C(=S)-NH_2} \xrightarrow{H_3O^+,\ H_2O}$$ pentothal

pentothal

22.77 Compound *B* is an aldehyde (it gives a positive Tollens test); therefore, compound *A* is a primary alcohol. Compound *B* is conjugated (it has a strong UV spectrum), and it has the same carbon skeleton as octanoic acid (an unbranched chain of eight carbons). Therefore, compound *B* is 2-octenal, and compound *A* is 2-octenol. (The data do not differentiate between the *E* and *Z* stereoisomers.) In order to form 2-octenol, lithium dibutylcuprate must undergo a *conjugate addition* to the epoxide:

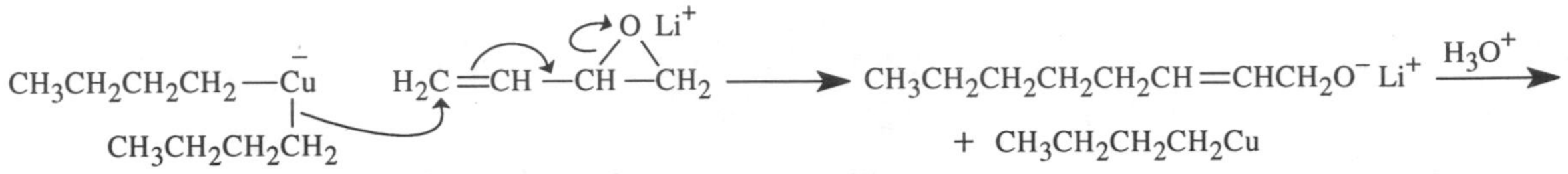

$$(CH_3CH_2CH_2CH_2)_2\bar{C}u \quad H_2C{=}CH{-}CH{-}CH_2(O)\ Li^+ \longrightarrow CH_3CH_2CH_2CH_2CH_2CH{=}CHCH_2O^-\ Li^+ + CH_3CH_2CH_2CH_2Cu \xrightarrow{H_3O^+}$$

$$CH_3CH_2CH_2CH_2CH_2CH{=}CHCH_2OH \xrightarrow{CrO_3(\text{pyridine})_2} CH_3CH_2CH_2CH_2CH_2CH{=}CHCH{=}O$$

2-octenol (compound *A*) 2-octenal (compound *B*)

22.78 (a) This is an aldol addition; dehydration cannot occur because the resulting alkene would violate Bredt's rule. The mechanism below begins with the enolate ion of the methyl ketone, which is formed when a proton is removed by potassium carbonate.

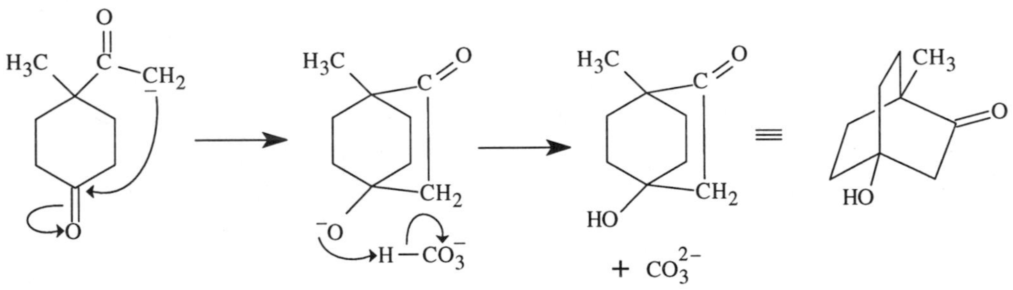

H_3C, $\bar{C}H_2$, ^-O, $H-CO_3^-$, HO, $+\ CO_3^{2-}$, CH_3

(b) This reaction is crossed aldol condensation followed by an intramolecular crossed aldol condensation followed by the same sequence with another molecule of dialdehyde. The following mechanism begins with the conjugate-base enolate ion of the diketone. Although the dehydration step is shown for brevity as a concerted process, it probably involves an enolate-ion intermediate. (See Eq. 22.44 on text p. 1016.)

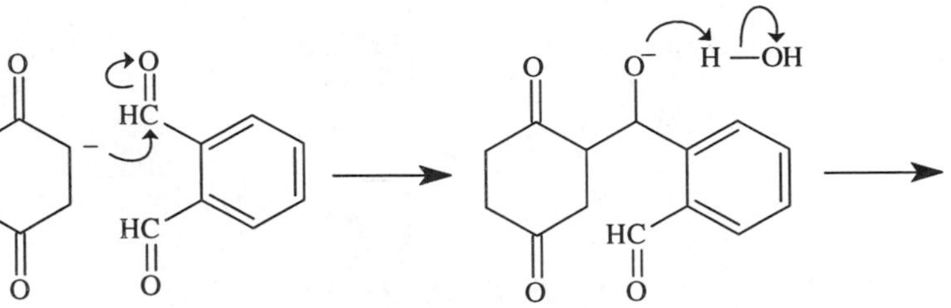

HC, HC, O^-, H—OH, HC

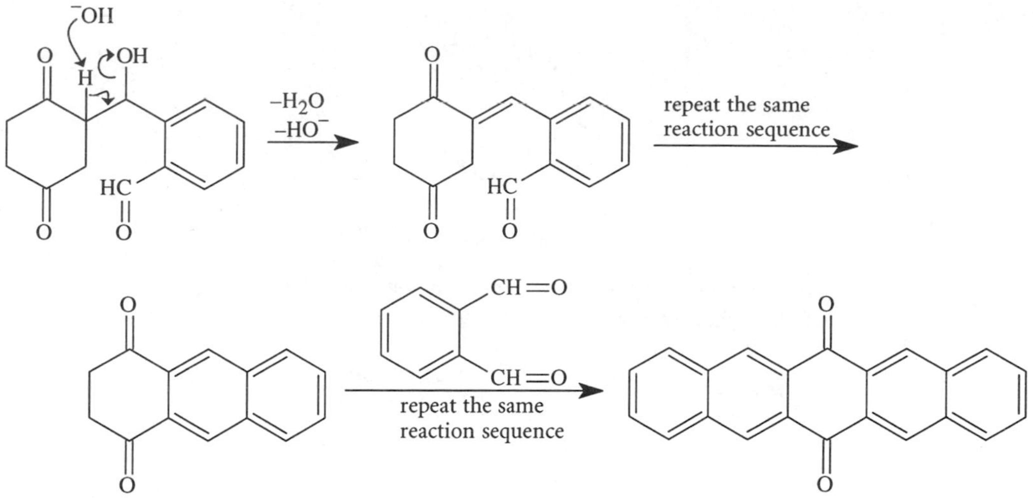

(c) The mechanism shown below begins with an α-bromo enolate ion, formed by the removal of an α-proton by NaH. This ion displaces a bromine in an internal nucleophilic substitution reaction to form the ring. An E2 reaction of the resulting compound with NaH as the base gives the product. (Recall that E2 reactions are particularly rapid when the proton undergoing elimination is particularly acidic; see Sec. 17.3B)

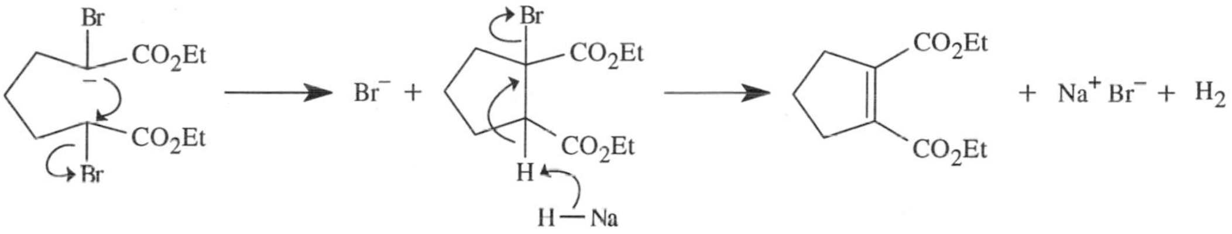

(d) This is a double Claisen condensation followed by hydrolysis and decarboxylation of the ester groups. The mechanism below begins with the conjugate-base enolate ion of the diester.

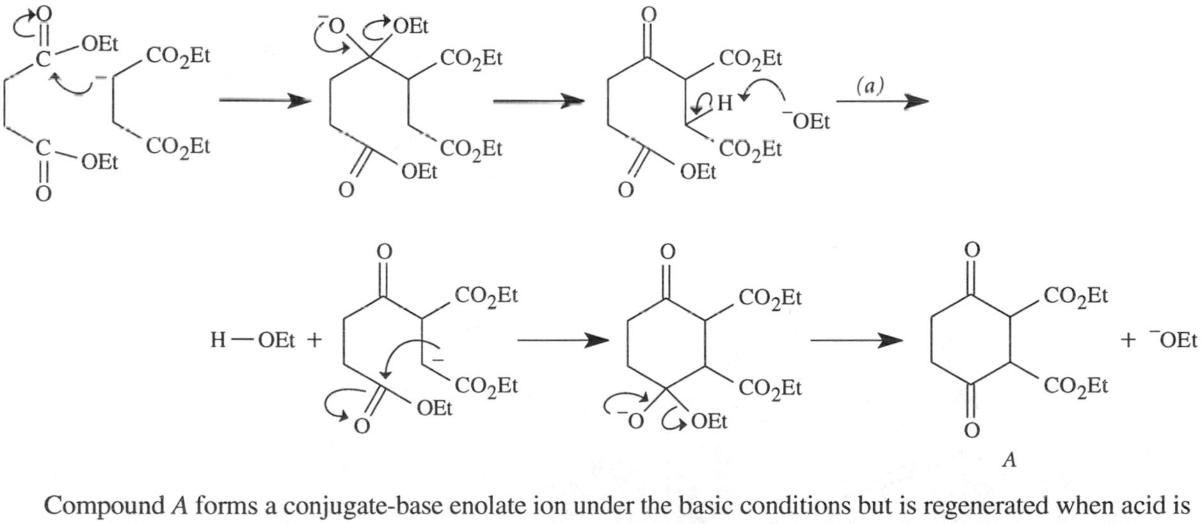

Compound *A* forms a conjugate-base enolate ion under the basic conditions but is regenerated when acid is added. Hydrolysis of the two ester groups and decarboxylation of the resulting β-keto acid groups give the product. In the step labeled *(a)*, the hydrogen that is removed is not the most acidic hydrogen in the molecule. (Which hydrogen is most acidic?) However, removal of the most acidic hydrogen does not lead to the observed product. Removal of the hydrogen shown, although occurring less frequently, leads rapidly to product. Note also that the second condensation reaction could involve a different enolate ion and a different ester group so that the β-keto ester shown on the right would be formed:

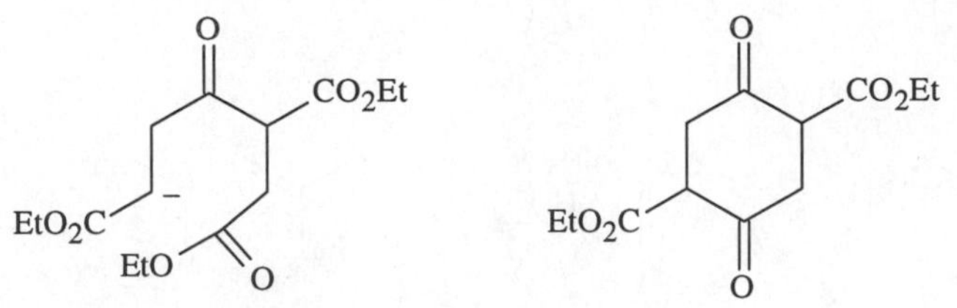

alternate enolate for second condensation

(e) A carbon-carbon double bond of a dienol on one ring attacks the protonated carbonyl group on the other. The mechanism below begins with a protonated ketone.

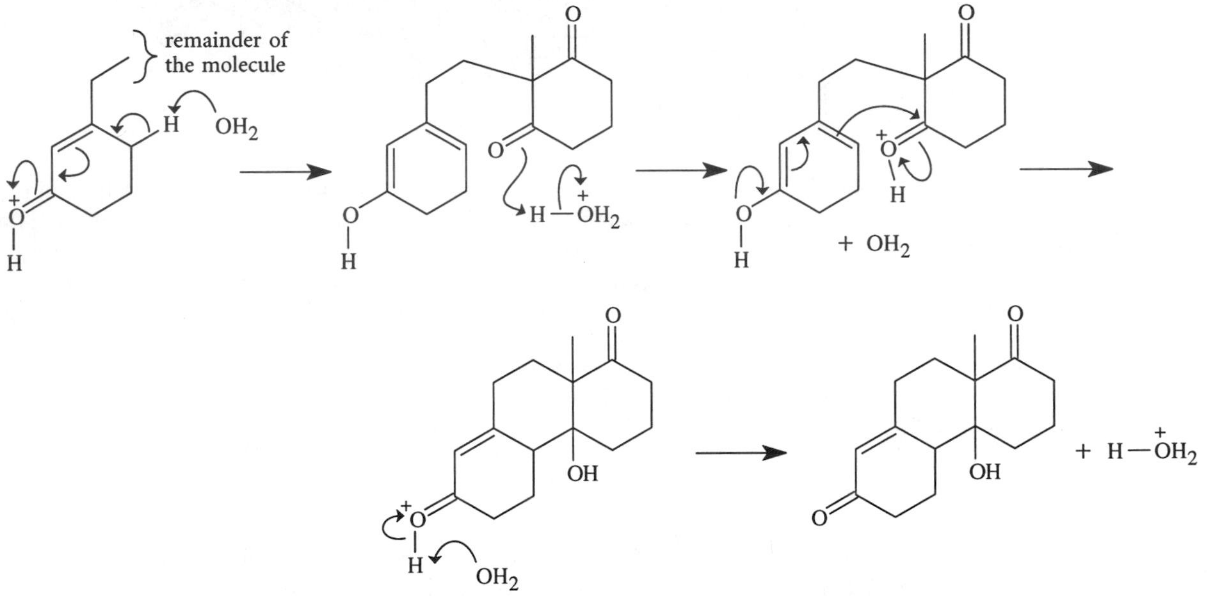

remainder of the molecule

(f) This reaction is a double crossed-Claisen condensation followed by an intramolecular transesterification reaction in which the anionic oxygen of an enolate ion serves as the nucleophile. (The oxygen rather than the anionic carbon acts as the nucleophile because a six-membered ring is formed; attack of the anionic carbon would give a strained four-membered ring.) The mechanism that follows begins with the conjugate-base enolate ion of acetone. Acid-catalyzed hydrolysis of the diester by the usual mechanism gives chelidonic acid.

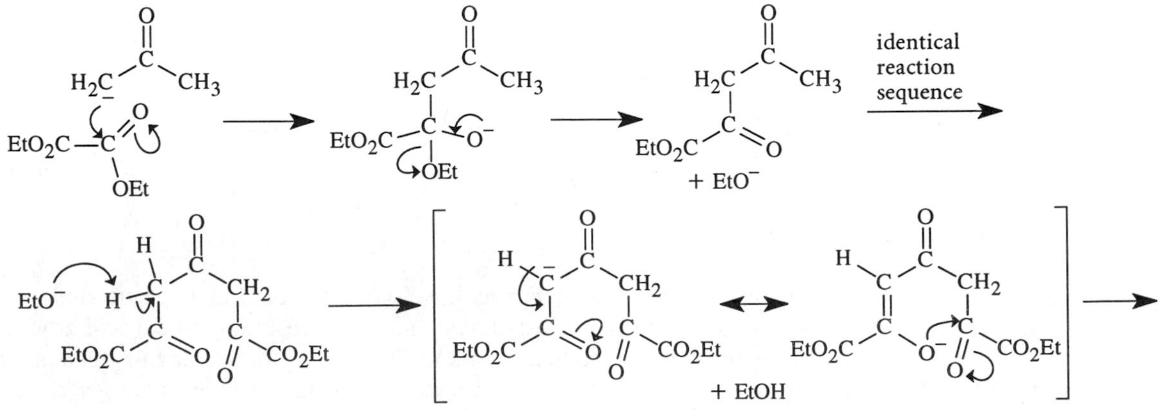

identical reaction sequence

$+ EtO^-$

$+ EtOH$

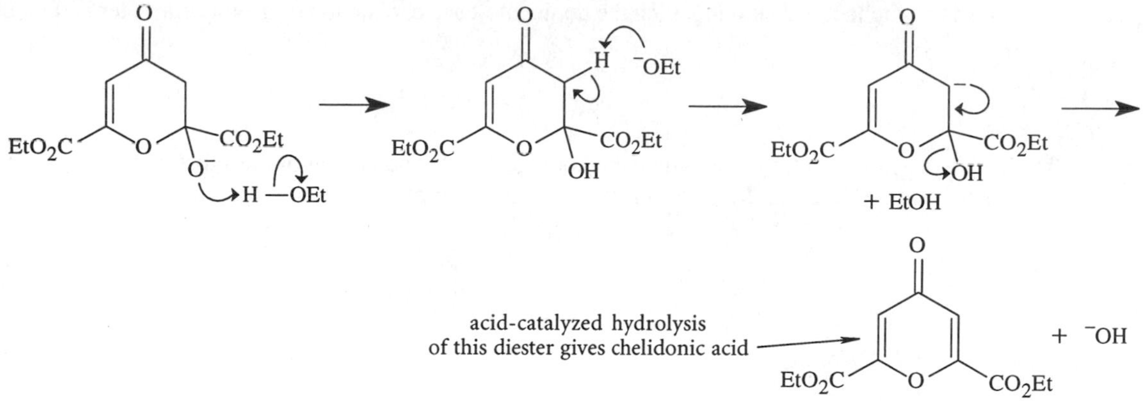
EtO2C
CO2Et
H —OEt
OH
H
-OEt
+ EtOH
acid-catalyzed hydrolysis
of this diester gives chelidonic acid
+ -OH

diethyl ester of chelidonic acid

22.79 (a) The conjugate-base enolate ion is alkylated by the alkyl halide. Then the conjugate-base enolate ion of the alkylation product is formed and reacts intramolecularly with the second alkyl halide group to form a ring. Hydrolysis and decarboxylation of the ester group give the following product.

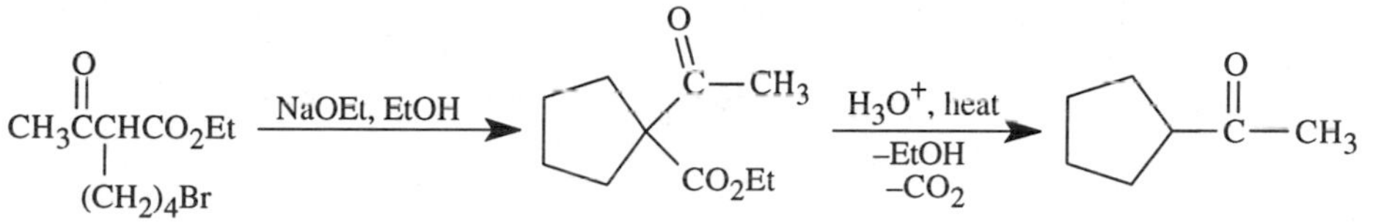
CH3CCHCO2Et
(CH2)4Br
NaOEt, EtOH
C—CH3
CO2Et
H3O+, heat
-EtOH
-CO2
C—CH3

product of the first alkylation

product of the second alkylation

(b) Lithium aluminum hydride reduces the carbonyl group to an alcohol.

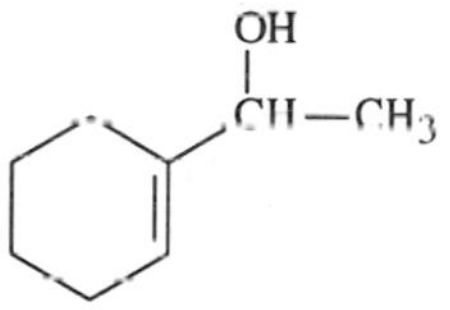
OH
CH—CH3

(c) The lactone, a cyclic ester, can form an enolate ion, and this ion is alkylated by methyl iodide. (The structure of γ-butyrolactone is given on text p. 936.)

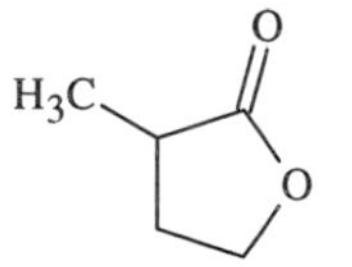
H3C

(d) This is a nucleophilic aromatic substitution reaction in which the nucleophile is an enolate ion. The ester groups are hydrolyzed, and one of the resulting carboxylic acids is decarboxylated, by the hot acidic conditions.

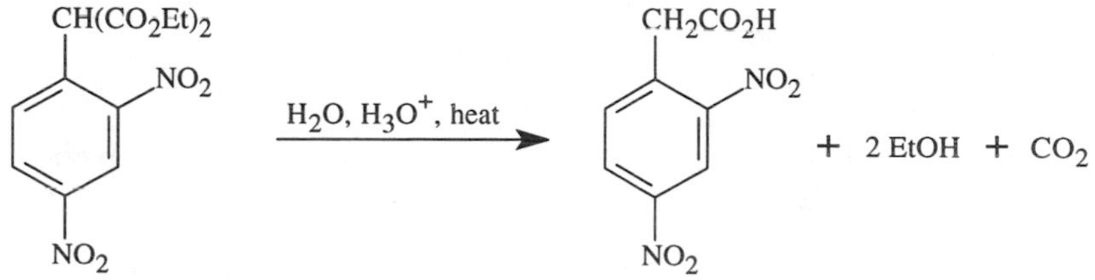
CH(CO2Et)2
NO2
NO2
H2O, H3O+, heat
CH2CO2H
NO2
NO2
+ 2 EtOH + CO2

initial substitution product

(e) This is a Michael addition in which the conjugate-base enolate ion of acetoacetic ester is the nucleophile.

$$H_3C-\overset{O}{\overset{\|}{C}}-\overset{CO_2Et}{\overset{|}{C}}HCH_2CH_2CO_2Et$$

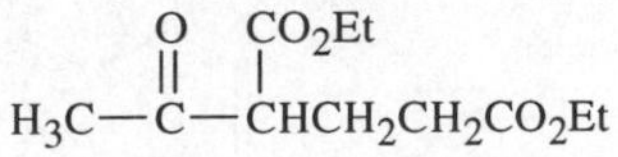

(f) This is a Michael addition in which the conjugate-base enolate ion of the substituted malonic ester is the nucleophile. (Note that four diastereomers of this compound are possible.)

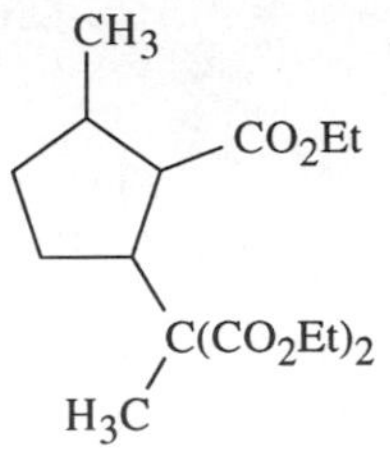

CH_3, CO_2Et, $C(CO_2Et)_2$, H_3C

(g) The magnesium organocuprate reagent undergoes conjugate addition; the acetal group hydrolyzes; and an acid-catalyzed intramolecular aldol condensation ensues.

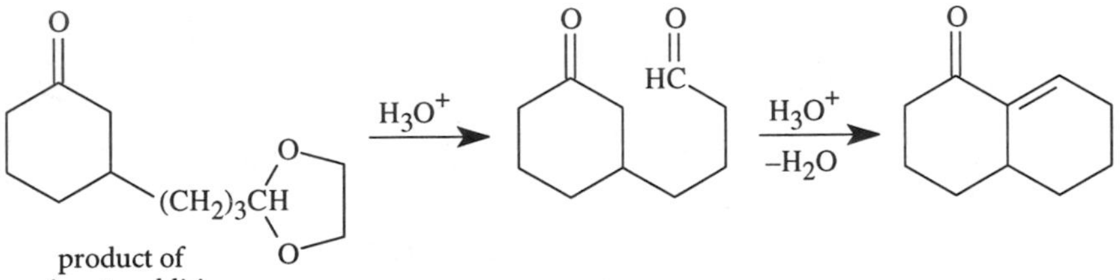

$(CH_2)_3CH$ — H_3O^+ → HC — H_3O^+, $-H_2O$ →

product of conjugate addition

(h) This is a conjugate addition of the magnesium organocuprate reagent. Assume that the isopropenyl group (the group delivered from the organocuprate reagent) enters trans to the angular methyl group to avoid van der Waals repulsions.

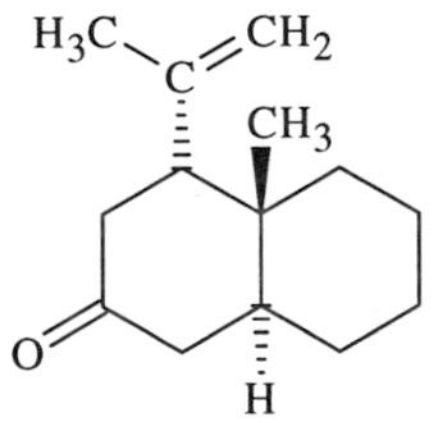

H_3C, CH_2, CH_3, O, H

(i) The benzene π electrons serve as the nucleophile in an acid-catalyzed conjugate addition, which can also be viewed as a variation of Friedel-Crafts alkylation.

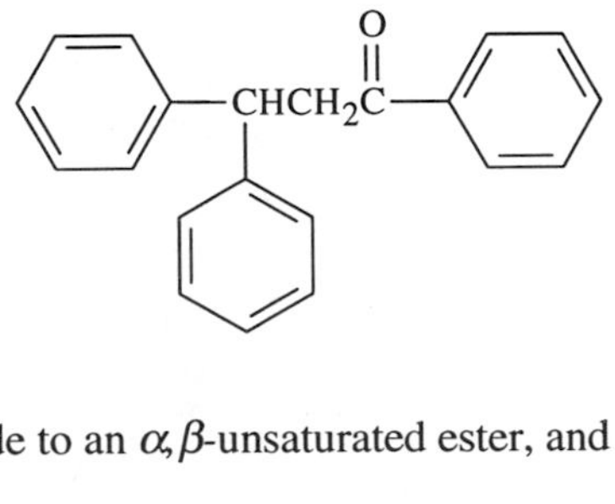

$$\text{Ph}-\overset{\text{Ph}}{\overset{|}{C}}HCH_2\overset{O}{\overset{\|}{C}}-\text{Ph}$$

22.80 (a) Use a conjugate addition of cyanide to an α,β-unsaturated ester, and then hydrolyze. (Either stereoisomer of the starting ester can be used.)

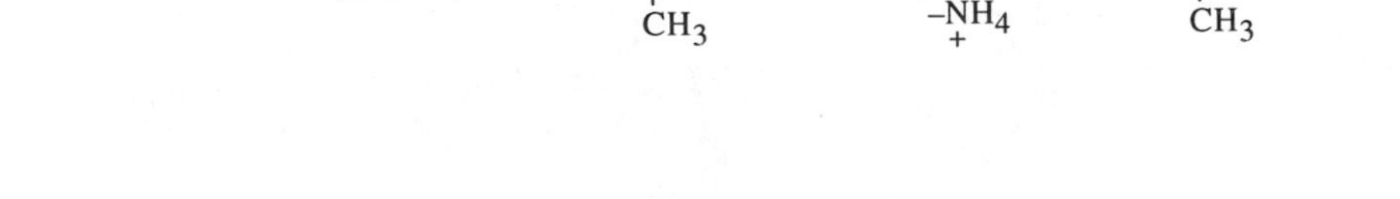

$$CH_3CH{=}CHCO_2Et \xrightarrow[\text{dil. HOAc}]{\text{Na}\overset{*}{\text{C}}\text{N}} N{\equiv}\overset{*}{C}\underset{CH_3}{\underset{|}{C}}HCH_2CO_2Et \xrightarrow[-\text{EtOH},\ -\overset{+}{\text{N}}\text{H}_4]{H_3O^+,\ \text{heat}} HO_2\overset{*}{C}\underset{CH_3}{\underset{|}{C}}HCH_2CO_2H$$

(b) Take the same approach as in part (a), but with a different ester.

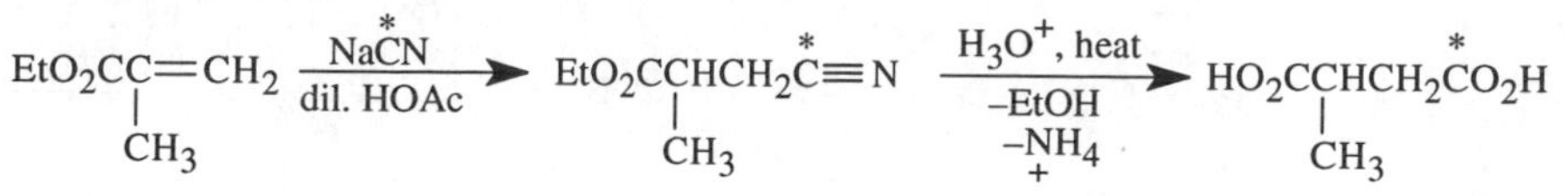

22.81 (a) A conjugate addition of hydroxide to pulegone gives a β-hydroxy ketone, which then undergoes the reverse of an aldol addition to give acetone and 3-methylcyclohexanone.

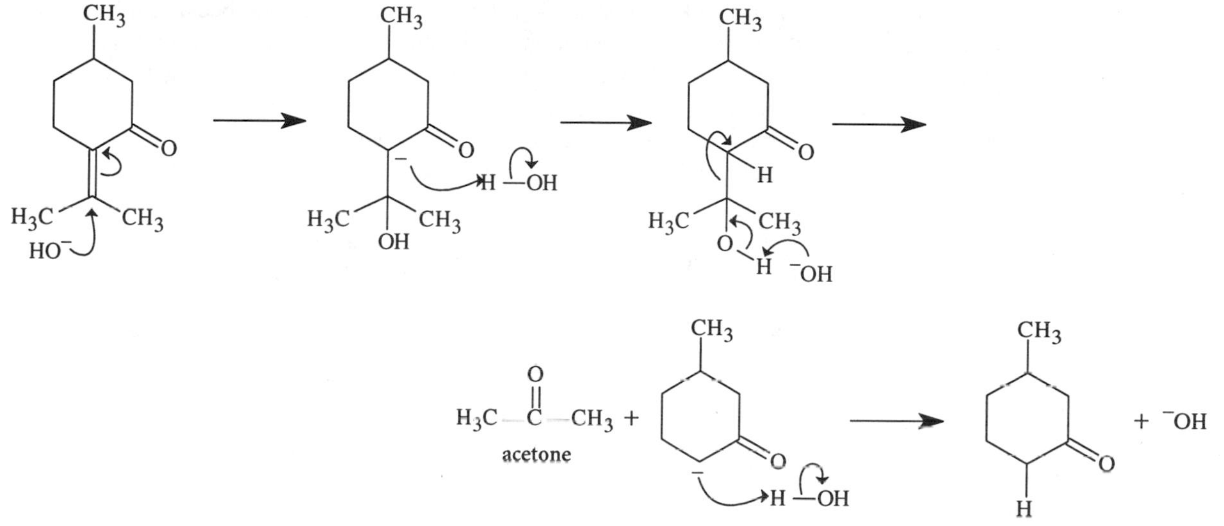

3-methylcyclohexanone

(b) The starting material is a β-hydroxy ketone. Ionization of the hydroxy group followed by the reverse of an aldol addition gives the product.

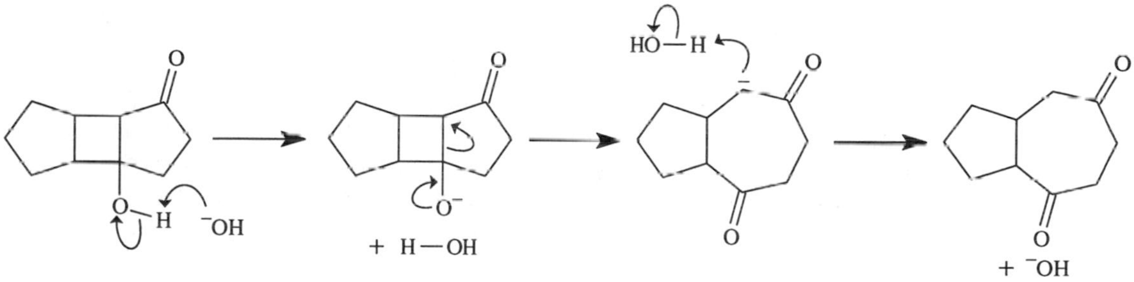

22.82 (a) A crossed Claisen condensation is followed by a reverse Claisen condensation, driven by expulsion of the more volatile ester, ethyl acetate, from the reaction mixture. The tertiary proton in compound *A* is very acidic and is removed to give a large amount of a conjugate-base anion (not shown) in equilibrium with a small amount of compound *A*; however, by LeChatelier's principle, the reaction of *A* shown in the following mechanism eventually depletes this anion.

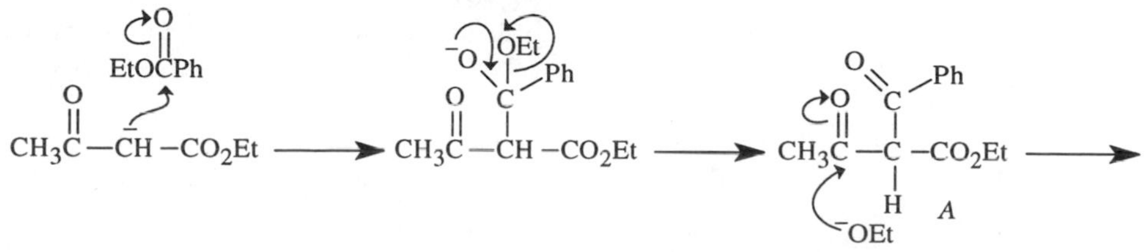

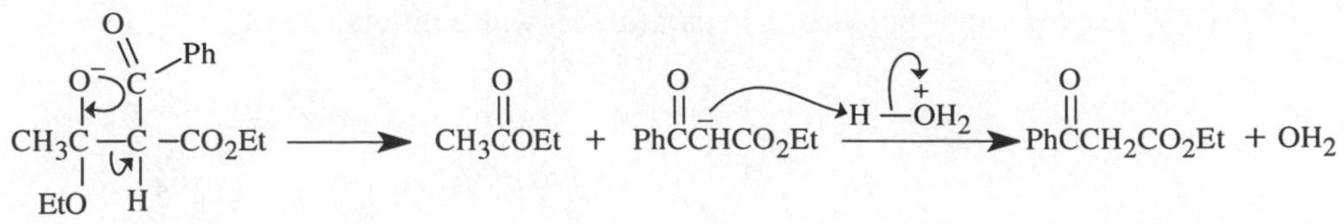

$$CH_3C(O^-)(OEt)-C(H)(CO_2Et)-C(=O)Ph \longrightarrow CH_3\overset{O}{\overset{\|}{C}}OEt + Ph\overset{O}{\overset{\|}{C}}\overset{-}{C}HCO_2Et \xrightarrow{H-\overset{+}{O}H_2} Ph\overset{O}{\overset{\|}{C}}CH_2CO_2Et + OH_2$$

(b) The conjugate-base anion of diethyl malonate serves as the nucleophile in an epoxide-opening reaction. Normally, epoxide opening under basic conditions involves attack of the nucleophile at the less branched carbon. In this case, however, attack also occurs at the more branched carbon because it is benzylic and therefore activated toward nucleophilic substitution. (Only attack at the benzylic carbon is shown below; attack at the other carbon occurs by a similar mechanism.) The resulting alkoxide undergoes intramolecular transesterification to give a lactone; hydrolysis and decarboxylation of the remaining ester give the product. The lactone does not hydrolyze because the equilibrium favors lactone formation when five- and six-membered rings are involved; see Eqs. 21.12–21.13 on text p. 954.

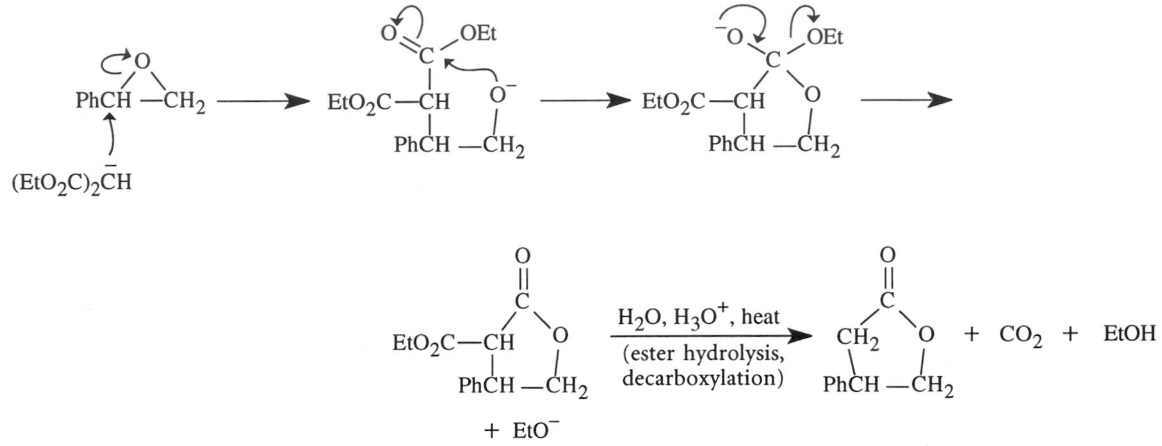

$$PhCH\overset{O}{—}CH_2 + (EtO_2C)_2\overset{-}{C}H \longrightarrow EtO_2C-CH(C(=O)OEt)-CH(Ph)-CH_2-O^- \longrightarrow \text{tetrahedral intermediate} \longrightarrow$$

$$\text{lactone (EtO}_2\text{C—CH, C=O, O, PhCH—CH}_2) + EtO^- \xrightarrow[\text{(ester hydrolysis, decarboxylation)}]{H_2O,\ H_3O^+,\ \text{heat}} \text{lactone (CH}_2\text{, C=O, O, PhCH—CH}_2) + CO_2 + EtOH$$

(c) The triple bond hydrates first by the usual mechanism (see Eqs. 14.6a–e on text p. 616). The resulting ketone enolizes by the mechanism shown in Eq. 22.17b on text p. 1006. The enol then, after protonation of the —OH group, loses water to give a resonance-stabilized carbocation *A*. This carbon cyclizes, and, after formation of the ketone by the reverse of the enolization mechanism and loss of a proton, the product is formed. Notice that cyclization must occur *after* the alkyne is hydrated; why?

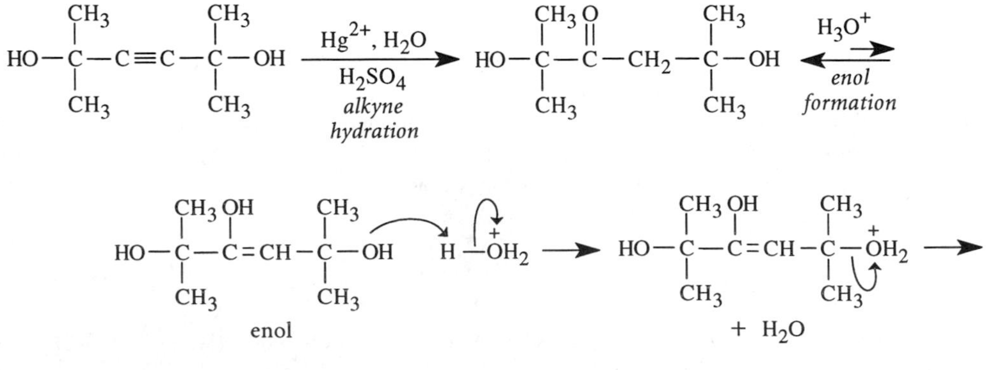

$$HO-C(CH_3)_2-C\equiv C-C(CH_3)_2-OH \xrightarrow[\substack{\text{alkyne}\\ \text{hydration}}]{Hg^{2+},\ H_2O,\ H_2SO_4} HO-C(CH_3)_2-\overset{O}{\overset{\|}{C}}-CH_2-C(CH_3)_2-OH \underset{\text{enol formation}}{\overset{H_3O^+}{\rightleftharpoons}}$$

$$\underset{\text{enol}}{HO-C(CH_3)_2-\overset{OH}{\overset{|}{C}}=CH-C(CH_3)_2-OH} + H-\overset{+}{O}H_2 \longrightarrow HO-C(CH_3)_2-\overset{OH}{\overset{|}{C}}=CH-C(CH_3)_2-\overset{+}{O}H_2 + H_2O \longrightarrow$$

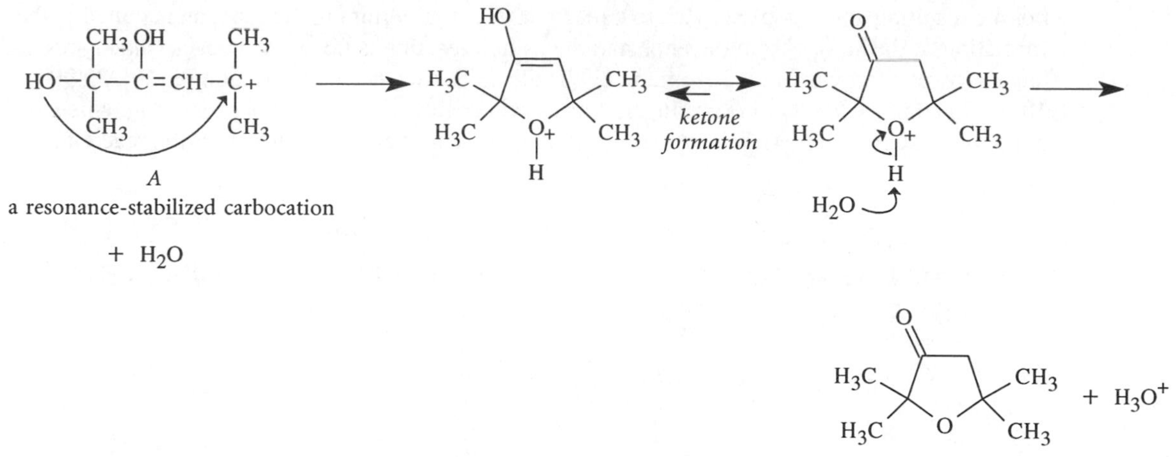

$$HO-C(CH_3)_2-C(OH)=CH-C^{+}(CH_3)_2 \longrightarrow \text{(cyclic oxonium ion, enol)} \rightleftarrows \text{(ketone formation)} \longrightarrow$$

A

a resonance-stabilized carbocation

+ H_2O

$$\text{2,2,5,5-tetramethyl-3-oxotetrahydrofuran} + H_3O^+$$

(d) This is an acid-catalyzed conjugate addition of water followed by elimination of phenol. The resulting enol is then spontaneously converted into the corresponding aldehyde.

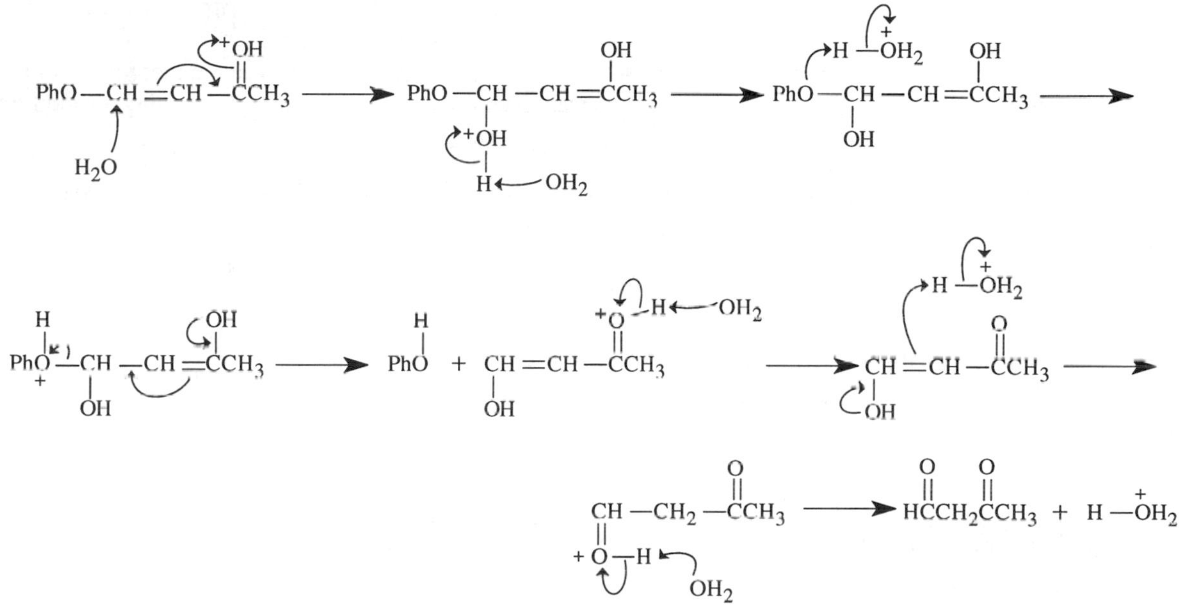

$$PhO-CH=CH-\overset{+OH}{\overset{\|}{C}}CH_3 \xrightarrow{H_2O} PhO-CH(\overset{+}{O}H_2)-CH=C(OH)CH_3 \longrightarrow PhO-CH(OH)-CH=C(OH)CH_3 \longrightarrow$$

$$Ph\overset{+}{O}H-CH(OH)-CH=C(OH)CH_3 \longrightarrow PhOH + CH(OH)=CH-\overset{+OH}{\overset{\|}{C}}CH_3 \longrightarrow CH(OH)=CH-\overset{O}{\overset{\|}{C}}CH_3 \longrightarrow$$

$$\overset{+OH}{\overset{\|}{C}}H-CH_2-\overset{O}{\overset{\|}{C}}CH_3 \longrightarrow H\overset{O}{\overset{\|}{C}}CH_2\overset{O}{\overset{\|}{C}}CH_3 + H-\overset{+}{O}H_2$$

(e) This mechanism is much like that in part (d), except that basic conditions are involved in all steps but the last.

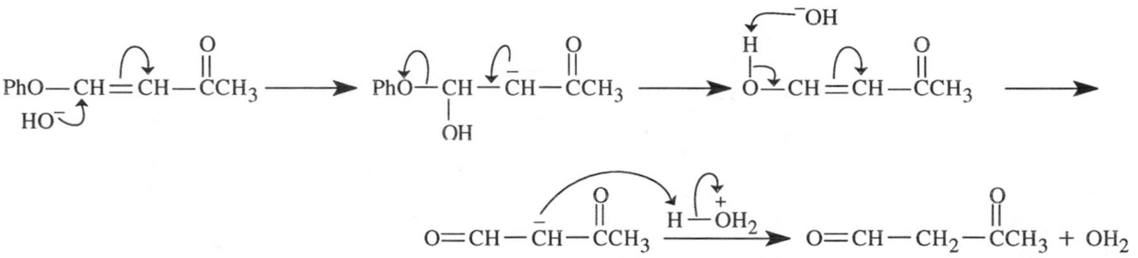

$$PhO-CH=CH-\overset{O}{\overset{\|}{C}}CH_3 \xrightarrow{HO^-} PhO-CH(OH)-\overset{-}{C}H-\overset{O}{\overset{\|}{C}}CH_3 \longrightarrow \overset{H}{O}-CH=CH-\overset{O}{\overset{\|}{C}}CH_3 \xrightarrow{^-OH}$$

$$O=CH-\overset{-}{C}H-\overset{O}{\overset{\|}{C}}CH_3 \xrightarrow{H-\overset{+}{O}H_2} O=CH-CH_2-\overset{O}{\overset{\|}{C}}CH_3 + OH_2$$

(f) Ionization of the benzylic hydrogen gives a carbanion that not only is allylic and benzylic but also is stabilized by the polar effect of the α-oxygen. The O—H hydrogen is also acidic, and removal of it occurs, but the resulting anion does not lead to a reaction. The equilibrium in the reaction is pulled to the right by formation of the ketone from the enol, and the overall reaction is favorable because aldehydes and ketones have a lower energy than their isomeric allylic alcohols. (See heats of formation in Eq. 22.90 on text p. 1048.) In addition, the ketone product in this case is conjugated with the benzene ring, whereas the double bond in the starting material is not. This provides an additional driving force for the reaction.

$$\mathrm{PhC(OH)(H)—CH{=}CHCH_3} \;\rightleftharpoons\; \left[\mathrm{Ph\bar{C}(OH)—CH{=}CHCH_3} \longleftrightarrow \mathrm{PhC(OH){=}CH—\bar{C}HCH_3}\right] \;\rightleftharpoons$$

$$\underset{\text{enol}}{\mathrm{PhC(OH){=}CH—CH_2CH_3}} \;\rightleftharpoons\; \left[\mathrm{PhC(O^-){=}CH—CH_2CH_3} \longleftrightarrow \mathrm{PhC(=O)—\bar{C}H—CH_2CH_3}\right] \longrightarrow {}^-\mathrm{OH} + \mathrm{PhC(=O)—CH_2CH_2CH_3}$$

(g) Conjugate addition with the nitrogen as a nucleophile gives compound *A*, which undergoes intramolecular transesterification. Alternatively, transesterification could occur first followed by conjugate addition.

$$\mathrm{CH_3CH{=}CH—C(=O)—OMe} + \mathrm{CH_3NH—OH} \longrightarrow \mathrm{CH_3CH(\overset{+}{N}H(CH_3)OH)—\bar{C}H—C(=O)—OMe} \longrightarrow \underset{A}{\mathrm{CH_3CH(N(CH_3)OH)—CH_2—C(=O)—OMe}} \longrightarrow$$

[cyclic tetrahedral intermediate, CH_3, CH_3–N, $\overset{+}{O}$–H, OMe, O^-] $\longrightarrow$ [cyclic ketone with $\overset{+}{O}$–H, $^-$OMe] $\longrightarrow$ [isoxazolidinone, CH_3, CH_3–N–O, =O] + HOMe

(h) Formation of the α,β-unsaturated ester releases HCl, which undergoes conjugate addition.

$$\mathrm{CH_3CH{=}CH—C(=O)—Cl} + \mathrm{H—OEt} \longrightarrow \mathrm{CH_3CH{=}CH—C(O^-)(Cl)—\overset{+}{O}(H)Et} \longrightarrow \mathrm{CH_3CH{=}CH—C(=O)—\overset{+}{O}(H)Et}\;\;\mathrm{Cl^-} + \mathrm{EtOH} \longrightarrow$$

$$\mathrm{CH_3CH{=}CH—C(=O)—OEt}\;\;\mathrm{Cl^-} + \mathrm{H—\overset{+}{O}(H)Et} \longrightarrow \mathrm{CH_3CH{=}CH—C(=\overset{+}{O}—H)—OEt}\;\;\mathrm{Cl^-} \longrightarrow \mathrm{CH_3CH(Cl)—CH{=}C(OH)—OEt} \xrightarrow{\mathrm{H_3O^+}} \mathrm{CH_3CH(Cl)CH_2C(=O)OEt}$$

(i) α-Cyano esters have acidities comparable to those of β-keto esters. (Why?) The conjugate-base enolate ion of the α-cyano ester adds to the aldehyde carbonyl group. Elimination of water gives an α,β-unsaturated nitrile, which undergoes conjugate addition with cyanide ion. Hydrolysis of the ester and decarboxylation of the resulting acid give the product.

$$EtO_2CCHC{\equiv}N \rightleftharpoons EtO_2C\bar{C}HC{\equiv}N + H{-}OH$$

$$Ph(CH_2)_3CH{=}O + EtO_2C\bar{C}HC{\equiv}N \longrightarrow Ph(CH_2)_3CH(O^-){-}CH(CO_2Et)C{\equiv}N \xrightarrow{H-OH} Ph(CH_2)_3CH(OH){-}CH(CO_2Et)C{\equiv}N \longrightarrow Ph(CH_2)_3CH(OH){-}\bar{C}(CO_2Et)C{\equiv}N + H{-}OH \longrightarrow$$

$$^-OH + Ph(CH_2)_3CH{=}C(CO_2Et){-}C{\equiv}N \xrightarrow[EtOH]{^-CN} Ph(CH_2)_3CH(C{\equiv}N){-}\bar{C}(CO_2Et){-}C{\equiv}N \xrightarrow{H-OEt} Ph(CH_2)_3CH(C{\equiv}N){-}CH(CO_2Et){-}C{\equiv}N + EtO^- \xrightarrow[\text{ester hydrolysis}]{\text{heat, } H_2O,\ H_3O^+}$$

$$Ph(CH_2)_3CH(C{\equiv}N){-}CH(CO_2H){-}C{\equiv}N + H{-}\overset{+}{O}H_2 \longrightarrow Ph(CH_2)_3CH(C{\equiv}N){-}CH(CO_2H){-}\overset{+}{C}{=}NH \longrightarrow$$

$$CO_2 + H_2\overset{+}{O}{-}H + Ph(CH_2)_3CH(C{\equiv}N){-}CH{=}C{=}N{-}H \xleftarrow{} OH_2 \longrightarrow H_2O + H{-}\overset{+}{O}H_2 + Ph(CH_2)_3CH(C{\equiv}N)CH_2C{\equiv}N$$

(j) This reaction is a "conjugate addition" of cyanide ion on a cyclopropane. Notice that attack on the least branched carbon of the cyclopropane occurs and that the immediate product of ring opening is a very stable anion. Heating in acid protonates this anion, hydrolyzes the ester and nitrile groups, and decarboxylates the resulting β-keto acid.

[bicyclic cyclopropyl ketone with CO_2Me + ^-CN] $\longrightarrow$ [cyclopentanone enolate anion with CO_2Me and CH_2CN] $\xrightarrow[\text{(ester and nitrile hydrolysis followed by decarboxylation)}]{H_2O,\ H_3O^+,\ \text{heat}}$ [3-oxocyclopentyl CH_2CO_2H] $+\ CO_2 + {}^+NH_4 + MeOH$

(k) Conjugate addition of triphenylphosphine to the anhydride starting material followed by proton transfer to the initially formed anion gives an ylid *A*, which undergoes a Wittig reaction with the aldehyde butanal. Hydrolysis of the resulting anhydride gives the dicarboxylic acid.

$Ph_3P{:}$ + maleic anhydride $\longrightarrow$ [$Ph_3\overset{+}{P}$ adduct anion] $\longrightarrow$ [$Ph_3\overset{+}{P}$ ylid] *A* $\xrightarrow{CH_3CH_2CH_2CH{=}O}$

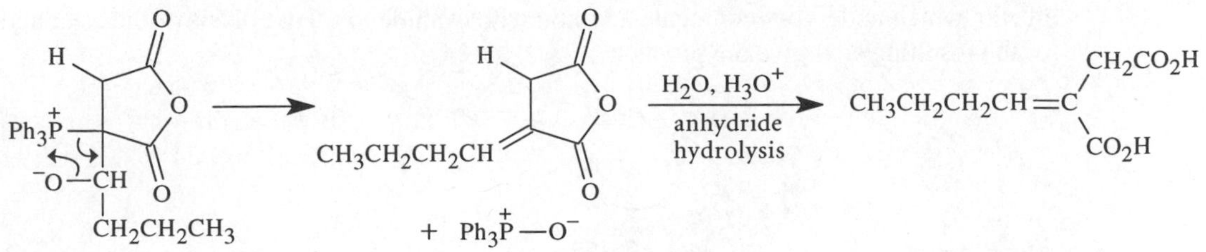

H_2O, H_3O^+ anhydride hydrolysis

$CH_3CH_2CH_2CH$ + $Ph_3\overset{+}{P}—O^-$

$CH_3CH_2CH_2CH{=}C(CH_2CO_2H)(CO_2H)$

22.83 (a) The conjugate-base enolate ion of the α-bromo ester, formed by reaction of the α-bromo ester with potassium *tert*-butoxide, attacks the carbonyl carbon of the aldehyde to give β-bromo alkoxide *A*, the conjugate base of a bromohydrin. As shown in Eq. 11.20–11.21 on text p. 463, an alkoxide of this type readily cyclizes to an epoxide by an internal nucleophilic substitution mechanism.

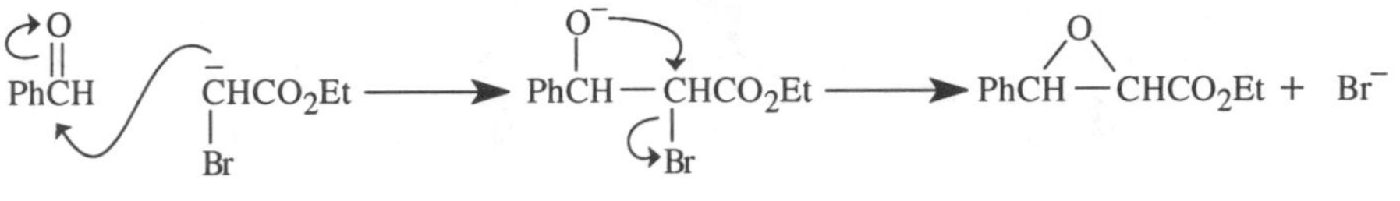

PhCH=O + $\overset{-}{C}HCO_2Et$(Br) ⟶ PhCH(O⁻)—CH(Br)CO₂Et (*A*) ⟶ PhCH—CHCO₂Et (epoxide) + Br⁻

(b) One of the two possible conjugate-base enolate ions undergoes an intramolecular nucleophilic substitution reaction to give the cyclopropane. Although the other enolate is undoubtedly also formed, its cyclization would lead to a different product. In addition, the enolate that leads to product is the more stable enolate (why?). In any case, the enolate ion shown evidently reacts more rapidly.

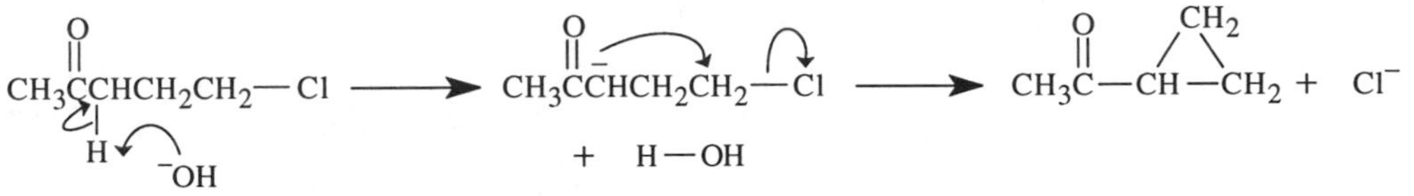

$CH_3\overset{O}{\overset{\|}{C}}CHCH_2CH_2—Cl$ ⟶ $CH_3\overset{O}{\overset{\|}{C}}\overset{-}{C}HCH_2CH_2—Cl$ + H—OH ⟶ $CH_3\overset{O}{\overset{\|}{C}}—CH—CH_2$ (cyclopropane, CH_2) + Cl^-

22.84 Conjugate addition of lithium dimethylcuprate to the α,β-unsaturated ketone can in principle give two diastereomers. Only product *A* and its Clemmensen reduction product *C* could be optically active. Compound *B* and its reduction product are meso compounds and are therefore achiral. Therefore, compound *A*, which has the (3*S*,5*S*) configuration, is the product of conjugate addition, and product *C*, which has the (1*R*,3*R*) configuration, is the Clemmensen reduction product.

(*S*)-(+)-5-methyl-2-cyclohexenone —1) $Li^+ (CH_3)_2Cu^-$ 2) H_3O^+⟶ (3*S*,5*S*)-(+)-3,5-dimethylcyclohexanone *A* + *meso*-3,5-dimethylcyclohexanone *B*

A —Zn/Hg, HCl⟶ (1*R*,3*R*)-(+)-1,3-dimethylcyclohexane *C*

B —Zn/Hg, HCl⟶ *cis*-1,3-dimethylcyclohexane

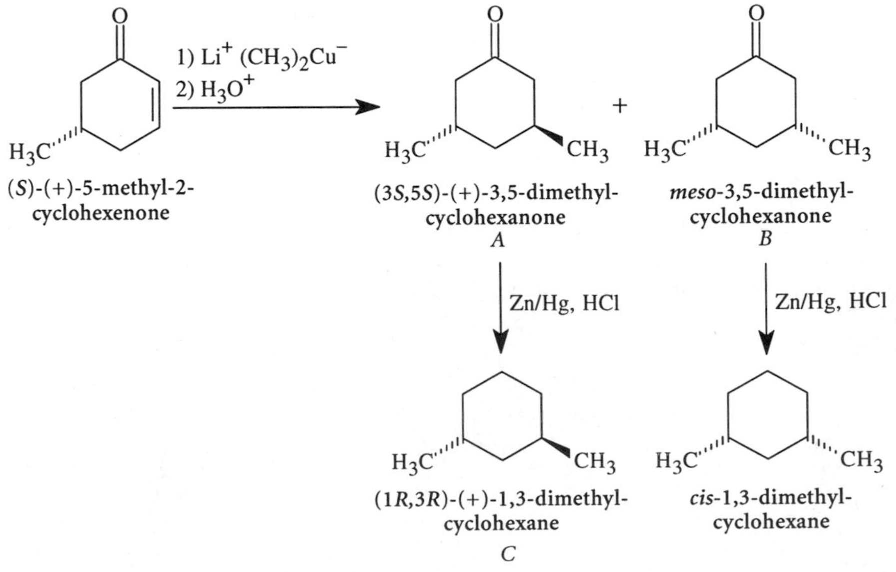

22.85 The deuterium incorporation results show that the carbon-carbon double bond is protonated. This is exactly the way that enols are protonated (see the reverse of Eq. 22.17b on text p. 1006). As a result, a relatively stable α-alkoxy carbocation is formed, and this is attacked by water to give a hemiacetal. The hemiacetal breaks down by the usual mechanism (see text p. 870) to give acetaldehyde and the alcohol. Hydrolysis of vinyl ethers is much faster than hydrolysis of ordinary ethers because vinyl ether hydrolysis involves a relatively stable carbocation intermediate, an α-alkoxy carbocation.

In the following mechanism, the asterisk indicates the position of the deuterium that is incorporated into the product if the reaction were run in D_2O. (Of course, all the other "hydrogens" of H_2O and H_3O^+ would be deuteriums as well, but these are not shown in the mechanism.)

$$C_2H_5O{-}CH{=}CH_2 \xrightarrow{\overset{*}{H}{-}\overset{+}{O}H_2} C_2H_5O{-}\overset{+}{C}H{-}CH_3 \xrightarrow{OH_2} C_2H_5O{-}\underset{}{CH}(\overset{+}{O}H{-}H){-}\overset{*}{C}H_3 \xrightarrow{H_2O}$$

$$\underset{\text{a hemiacetal}}{C_2H_5O{-}CH(OH){-}\overset{*}{C}H_3} \xrightarrow{H{-}\overset{+}{O}H_2} C_2H_5\overset{+}{O}(H){-}CH(OH){-}\overset{*}{C}H_3 + OH_2 \longrightarrow C_2H_5OH + \overset{+}{C}H(O{-}H){-}\overset{*}{C}H_3 \xrightarrow{OH_2} O{=}CH{-}\overset{*}{C}H_3 + H{-}\overset{+}{O}H_2$$

22.86 The Wittig reaction gives a vinylic ether, which, as shown in the previous problem, hydrolyzes rapidly to the corresponding aldehyde.

$$CH_3OCH_2{-}Cl \xrightarrow{:PPh_3} CH_3OCH_2{-}\overset{+}{P}Ph_3\ Cl^- \xrightarrow{Li{-}CH_2CH_2CH_2CH_3} \underset{\text{the ylid}}{CH_3O{-}\overset{-}{C}H{-}\overset{+}{P}Ph_3} + Li^+\,Cl^- + CH_3CH_2CH_2CH_3 \xrightarrow{\text{cyclohexanone}}$$

$$\underset{\text{product of the Wittig reaction (a vinylic ether)}}{(\text{cyclohexylidene}){=}CHOCH_3} \xrightarrow{H_3O^+} (\text{cyclohexyl}){-}CH{=}O + CH_3OH$$

22.87 First, relate the connectivities of the reactant and product. The $-CO_2Me$ group at the end of the side chain and the ring carbonyl group are keys to this process. The symbols and numbers in the following structures show this relationship. (Note that the numbers have nothing to do with nomenclature.)

[Structures: reactant — cyclopentenone with ring carbons #, 6, 7, 8, 9; side chain at C6: $CH_2CH_2CH_2$ (5 4 3), then H–C=C–H (C2=C1) bearing CO_2Me (*). Product — tricyclic ketone with carbons #, 6, 5, 4, 3, 2, 7, 9, 8, 1 and CO_2Me (*) at C1.]

First, we see that the product involves a new connection between C2 and C6. This could be made with a Michael addition that starts with formation of the enolate ion at C6:

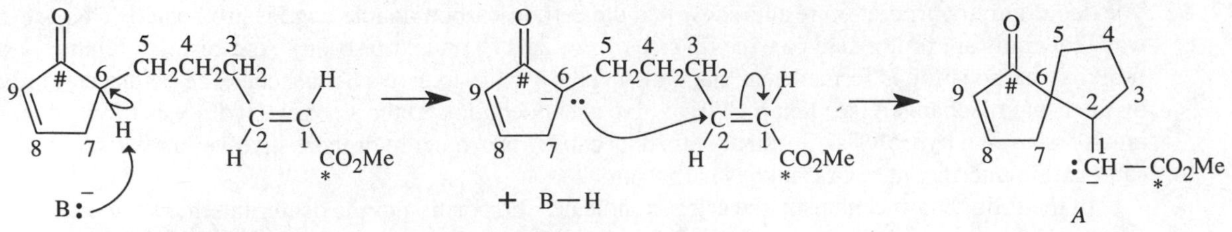

The remaining carbon-carbon bond that must form is the C1-C8 bond. Anion *A* can simply do another Michael addition by attacking C8. Protonation gives the product. Voila!

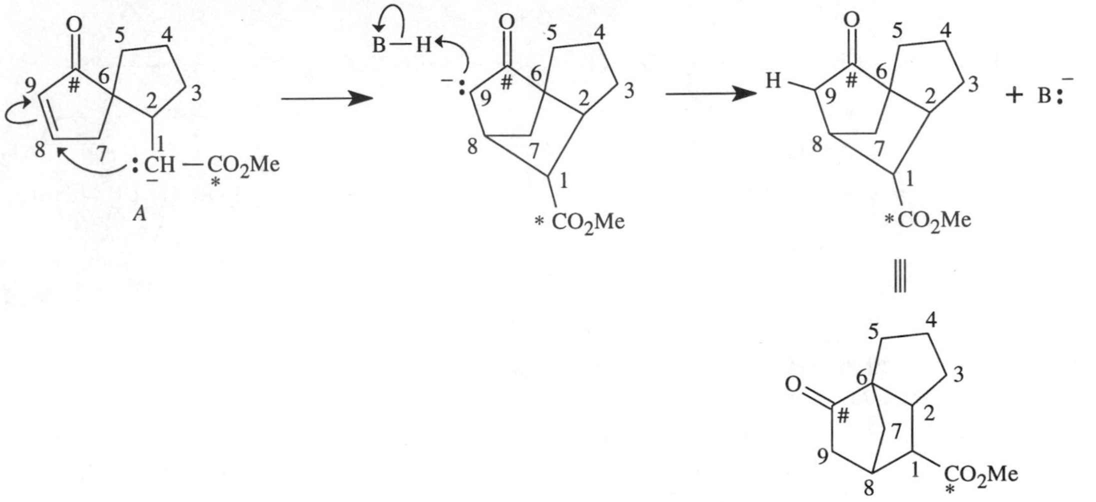

23 Chemistry of Amines

TERMS

Term	Section
acyl azide	23.11C
acyl hydrazide	23.11C
acylation	23.7C
alkaloid	23.12B
amide base	23.5D
amine	Introduction
amino group	23.1B
ammonium salt	23.5A
azo dye	23.10B
azobenzene	23.10B
Borch reaction	23.7B
α-cleavage	23.4C
Curtius rearrangement	23.11C
diazonium salt	23.10A
diazotization	23.10A
exhaustive methylation	23.7A
Gabriel synthesis	23.11A
Hofmann elimination	23.8
Hofmann hypobromite reaction	23.11C
Hofmann rearrangement	23.11C
hormone	23.12B
imminium ion	23.7B
inversion (of amines)	23.2
isocyanate	23.11C
neurochemistry	23.12B
neurotransmitter	23.12B
nitrosamine	23.10C
primary amine	Introduction
quaternary ammonium salt	23.6
quaternization	23.7A
polarization effect on basicity	23.5B
reductive amination	23.7B
Sandmeyer reaction	23.10A
secondary amine	Introduction
synapse	23.11B
tertiary amine	Introduction

CONCEPTS

I. INTRODUCTION TO AMINES

A. General

1. Amines are organic derivatives of ammonia in which the ammonia hydrogens are replaced by alkyl or aryl groups.
2. Amines are classified by the number of alkyl or aryl substituents (R groups) on the amine nitrogen.
   a. A primary amine has one substituent.
   b. A secondary amine has two substituents.
   c. A tertiary amine has three substituents.

$H_3C—NH_2$ — a primary amine

(pyrrolidine) N—H — a secondary amine

(benzene ring)—NMe_2 — a tertiary amine

B. Common Nomenclature

1. In common nomenclature an amine is named by appending the suffix *amine* to the name of the alkyl group; the name of the amine is written as one word.
2. When two or more alkyl groups in a secondary or tertiary amine are different, the compound is named as an *N*-substituted derivative of the larger group. (An *N* or N designates that the substituent is on the amine nitrogen.)
3. Aromatic amines are named as derivatives of aniline.

$(CH_3)_2CHNH_2$

isopropylamine (isopropyl = name of the alkyl group)

$H{-}N(CH_2CH_3)(CH_2CH_2CH_2CH_3)$

***N*-ethylbutylamine**
(a secondary amine with two different alkyl groups)

$Cl{-}C_6H_4{-}NEt_2$

***p*-chloro-*N*,*N*-diethylaniline**
(an aromatic amine)

C. Systematic Nomenclature

1. The most widely used system of substitutive amine nomenclature is that of *Chemical Abstracts*.
   a. In this system, an amine is named in the same way as the analogous alcohol, except that the suffix *amine* is used.
   b. In diamine nomenclature, the final *e* of the hydrocarbon name is retained.

$C_6H_{11}{-}NH_2$

cyclohexylamine

$H_2N{-}CH_2CH_2CH_2CH(CH_3){-}NH_2$

1,4-pentanediamine

   c. When the amine nitrogens of a diamine bear substituents, these are treated by designating the nitrogens by the number of the carbon on which they are located or, when the positions of the nitrogens on the principal chain are indistinguishable, by an *N*,*N*′ designation.

$CH_3NH-\overset{2}{C}H(\overset{3}{C}H_3)-\overset{1}{C}H_2-NHCH_2CH_3$

***N*¹-ethyl-*N*²-methyl-1,2-propanediamine**

$CH_3NH-CH_2CH_2CH_2-NHCH_2CH_3$

***N*-ethyl-*N*′-methyl-1,3-propanediamine**

2. The priority for citation of amine groups as principal groups is just below that of alcohols. (A complete list of group priorities is given in Appendix I, text page A-1.)

$$—CO_2H \text{ and derivatives} > —CH{=}O,\ {>}C{=}O > —OH > —NR_2$$

3. When cited as a substituent, the $—NH_2$ group is called the amino group.

$H_2NCH_2CH_2OH$

2-aminoethanol

$CH_3CH_2NH{-}C_6H_4{-}CO_2H$

4-(ethylamino)benzoic acid

4. Although *Chemical Abstracts* calls aniline *benzenamine*, the more common practice is to use the common name *aniline* in substitutive nomenclature.
5. Many important nitrogen-containing heterocyclic compounds are known by specific names that should be learned (numbering generally begins with the heteroatom).

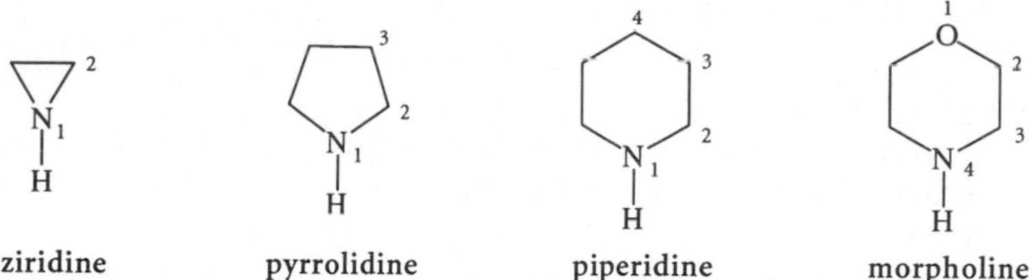

D. Structure of Amines

1. The C—N bonds of aliphatic amines are longer than the C—O bonds of alcohols, but shorter than the C—C bonds of alkanes.
2. Aliphatic amines have a pyramidal shape (or approximately tetrahedral shape, if the electron pair is considered to be a "group").
   a. Most amines undergo rapid inversion at nitrogen; this occurs through a planar transition state and converts an amine into its mirror image.
   b. Because of this inversion, amines in which the only asymmetric atom is the amine nitrogen cannot be resolved into enantiomers.

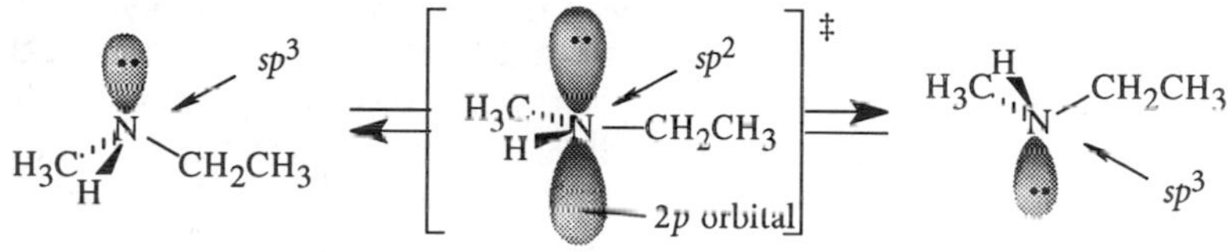

3. The C—N bond in aniline is shorter than that in aliphatic amines because of—
   a. sp^2 hybridization of the adjacent carbon.
   b. overlap of the unshared electrons on nitrogen with the π-electron system of the ring; this overlap gives double-bond character to the C—N bond.

E. Physical Properties of Amines

1. Most amines are somewhat polar liquids with unpleasant odors.
2. The boiling points of amines depend on whether they are primary, secondary, or tertiary.
   a. Primary and secondary amines, which can both donate and accept hydrogen bonds, have higher boiling points than isomeric tertiary amines, which cannot donate hydrogen bonds.
   b. Primary and secondary amines have higher boiling points than ethers but lower boiling points than alcohols. (Alcohols are better hydrogen-bond donors.)
3. Most primary and secondary amines with four or fewer carbons, as well as trimethylamine, are miscible with water; amines with large carbon groups have little or no water solubility.

F. Uses and Occurrence of Amines

1. Among the relatively few industrially important amines is hexamethylenediamine (1,6-hexanediamine), which is used in the synthesis of nylon-6,6.
2. Among the many types of naturally occurring amines are the alkaloids (nitrogen-containing bases that occur naturally in plants).
   a. Most alkaloids are amines, and many are heterocyclic amines. (See Fig. 23.3 on text p. 1108 for examples.)
   b. Many alkaloids have biological activity; others have no known activity, and their functions within the plants from which they come are, in many cases, obscure.
3. Some hormones, such as neurotransmitters, are amines.
   a. Hormones are compounds that regulate the biochemistry of multicellular organisms, particularly vertebrates.
   b. Neurotransmitters are molecules that are important in the communication between nerve cells, or between nerve cells and their target organs.

II. SPECTROSCOPY OF AMINES

A. IR Spectroscopy

1. The most important absorptions in the infrared spectra of primary amines are the N—H stretching absorptions, which usually occur as two or more peaks at 3200–3375 cm^{-1}.
2. Primary amines show an NH_2 scissoring absorption near 1600 cm^{-1}.
3. Most secondary amines show a single N—H stretching absorption rather than the multiple N—H stretching peaks observed for primary amines.
4. Tertiary amines show no N—H absorptions.

B. NMR Spectroscopy

1. The characteristic resonances in the NMR spectra of amines are those of the protons adjacent to the nitrogen (the α-protons) and the N—H protons.
   a. In alkylamines, the α-protons are observed in the δ 2.5–3.0 region of the spectrum.
   b. In aromatic amines, the chemical shifts of the α-protons of *N*-alkyl groups are somewhat greater, near δ 3.
2. The chemical shift of the N—H proton depends on the concentration of the amine and on the conditions of the NMR experiment.
   a. In alkylamines, this resonance typically occurs at rather low chemical shift, typically around δ 1.
   b. In aromatic amines, this resonance is at considerably higher chemical shift.
3. The N—H protons of amines under most conditions undergo rapid exchange.
   a. The N—H absorption can be obliterated from the spectrum by exchange with D_2O (the "D_2O shake").
   b. In some amine samples the N—H resonance is broadened.
   c. Splitting between the amine N—H and adjacent C—H groups is usually not observed.
4. The characteristic CMR absorptions of amines are those of the α-carbon; these absorptions occur in the δ 30–50 chemical shift range.

C. Mass Spectrometry

1. α-Cleavage is a particularly important fragmentation mode of aliphatic amines. (See Eq. 23.4, text p. 1075, for an example.)
2. The molecular ion occurs at an odd mass if the amine contains an odd number of nitrogens because compounds containing an odd number of nitrogens have odd molecular weights.
   a. Odd-electron ions containing an odd number of nitrogens are observed at odd mass.
   b. Even-electron ions containing an odd number of nitrogens are observed at even mass.

III. ACIDITY AND BASICITY OF AMINES

A. Acidity of Amines

1. Primary and secondary amines are very weakly acidic.
2. The conjugate base of an amine is called an amide base. Amide bases are very strong.
   a. The amide conjugate base of ammonia is usually prepared by dissolving an alkali metal such as sodium in liquid ammonia in the presence of a trace of ferric ion.
   b. The conjugate bases of alkylamines are prepared by treating the amine with butyllithium in an ether solvent such as THF.

$$\underset{}{CH_3CH_2CH_2CH_2{-}Li} + \underset{pK_a = 32\text{–}35}{R_2NH} \xrightarrow[-78^\circ]{} \underset{\text{an amide base}}{R_2N^- \, Li^+} + \underset{pK_a = 55\text{–}60}{CH_3CH_2CH_2CH_3}$$

3. The pK_a of a typical amine is about 35. This pK_a refers to the process:

$$R_2\ddot{N}{-}H + B^- \rightarrow R_2\ddot{N}{:}^- + BH$$

where B is a base, and not to $R_3\overset{+}{N}$—H + B$^-$ → R_3N: + BH.

B. Basicity of Amines

1. Amines are strong enough bases that they are completely protonated in dilute acid solutions.
2. The salts of protonated amines are called ammonium salts.
   a. The ammonium salts of simple alkylamines are named as substituted derivatives of the ammonium ion (for example, methylammonium chloride).
   b. Other ammonium salts are named by replacing the final *e* in the name of the amine with the suffix *ium* (for example, pyridinium chloride).
3. Ammonium salts are fully ionic compounds; the N—H bonds are covalent, but there is no covalent bond between the nitrogen and the counter-ion.
4. The basicity of an amine is expressed in terms of the pK_a of its conjugate-acid ammonium salt; the higher the pK_a of an ammonium ion, the more basic is its conjugate-base amine.
5. The conjugate acids of typical alkylamines have pK_a values in the 9–11 range. The conjugate acids of typical arylamines have pK_a values in the 4–5 range.

C. Factors Affecting Amine Basicity

1. Three effects influence the basicity of amines (the same effects that govern the acidity and basicity of other compounds):
   a. the effect of alkyl substitution
   b. the polar effect
   c. the resonance effect
2. In alkyl substitution two opposing factors are involved:
   a. The first is the tendency of alkyl groups to stabilize charge through a polarization effect.
      i. The polarization of alkyl groups can act to stabilize either positive or negative charge.
      ii. This effect governs the acidity of ammonium ions in the gas phase, in which the acidity of ammonium ions decreases regularly with increasing alkyl substitution.
   b. The second factor is a solvent effect.
      i. Ammonium ions in solution are stabilized by hydrogen-bond donation to the solvent.
      ii. Primary ammonium ions are stabilized by hydrogen bonding more than tertiary ones; primary ammonium salts have three hydrogens that can be donated to form hydrogen bonds, but a tertiary ammonium salt has only one.
3. The polar effect operates largely on the conjugate acid of the amine because this cation is the charged species in the acid-base equilibrium.
   a. An electronegative (electron-withdrawing) group destabilizes an ammonium ion because of a repulsive electrostatic interaction between the positive charge on the ammonium ion and the positive end of the substituent bond dipole.
   b. The base-weakening effect of electron-withdrawing substituents decreases rapidly with distance between the substituent and the charged nitrogen.
   c. The electron-withdrawing polar effect of the aromatic ring also contributes significantly to the reduced basicity of aromatic amines.
4. The basicity of aniline derivatives is affected by resonance:
   a. Aniline is stabilized by resonance interaction of the unshared electron pair on nitrogen with the aromatic ring.
      i. When aniline is protonated, this resonance stabilization is no longer present, because the unshared electron pair is bound to a proton.
      ii. The resonance stabilization of aniline relative to its conjugate acid reduces its basicity.
   b. The resonance stabilization of aniline lowers the energy required for its formation from its conjugate acid, and thus lowers its basicity relative to that of cyclohexylamine.
5. The pK_a of an ammonium ion is directly related to the standard free-energy difference $\Delta G°$ between it and its conjugate base.

$$\Delta G° = 2.3RT\,pK_a$$

a. If a substituent stabilizes an amine ion more than it stabilizes the conjugate-acid ammonium ion, the standard free energy of the amine is lowered, $\Delta G°$ is decreased, the pK_a of the ammonium ion is reduced, and the basicity of the amine is decreased.
b. If a substituent stabilizes the ammonium ion more than its conjugate-base amine, the opposite effect is observed: the pK_a is increased, and the amine basicity is also increased. (See Fig. 23.2 on text p. 1078 for an illustration of these ideas.)

D. Quaternary Ammonium Salts

1. Ammonium salts in which all four hydrogens of $^{+}NH_4$ are replaced by alkyl or aryl groups are called quaternary ammonium salts.

$$CH_3CH_2CH_2CH_2CH_2CH_2CH_2CH_2CH_2CH_2{-}\overset{+}{N}(CH_3)_2{-}CH_2Ph \quad Cl^-$$

benzyldecyldimethylammonium chloride
(a quaternary ammonium salt)

2. Quaternary ammonium salts are ionic compounds.
3. Many quaternary ammonium salts containing large organic groups are soluble in nonaqueous solvents.

E. Separations Using Amine Basicity

1. Because ammonium salts are ionic compounds, many have appreciable water solubilities.
   a. When a water-insoluble amine is treated with dilute aqueous acid, the amine dissolves as its ammonium salt.
   b. Upon treatment with base, an ammonium salt is converted back into the corresponding amine.
2. This property of ammonium salts can be used to design separations of amines from other compounds.

$PhNH_2 + PhNO_2$ — 1) dilute HCl 2) ether extraction →
- organic layer → $PhNO_2$ — remove solvent → $PhNO_2$
- aqueous layer → $Ph\overset{+}{N}H_3$ — dilute NaOH → $PhNH_2$

3. Amine basicities can play a key role in the design of enantiomeric resolutions.

REACTIONS

I. SYNTHESIS OF AMINES; ALKYLATION REACTIONS

A. General

1. Amines are good nucleophiles (Lewis bases).
2. Three reactions of nucleophiles have been studied in previous sections:
   a. S_N2 reaction with alkyl halides, sulfonate esters, or epoxides (Sec. 9.1A, text page 351; Sec. 9.4, text page 360; Secs. 10.3A–C, text pages 415–419; and Sec. 11.4, text page 467).
   b. Addition to aldehydes, ketones, and α,β-unsaturated carbonyl compounds (Sec. 19.7, text page 855; Sec. 19.11, text page 874; and Sec. 22.8A, text page 1043).
   c. Nucleophilic acyl substitution at the carbonyl groups of carboxylic acid derivatives (Sec. 21.8, text page 963).

B. Direct Alkylation of Amines

1. Treatment of ammonia or an amine with an alkyl halide or other alkylating agent (RX in the following equation) results in alkylation of the nitrogen. This is an example of an S_N2 reaction in which the amine acts as the nucleophile.

$$NH_3 + RX \longrightarrow RNH_2 + R_2NH + R_3N + R_4N^+ X^-$$

   a. The immediate product of the reaction is an alkylammonium ion.
   b. If this ammonium ion has N—H bonds, further alkylations can take place to give a complex product mixture.
   c. A mixture of products is formed because the alkylammonium ion produced initially can be deprotonated by the amine starting material.

2. Epoxides, as well as α,β-unsaturated carbonyl compounds and α,β-unsaturated nitriles, also react with amines or ammonia; multiple alkylation can occur with these alkylating agents as well.

$$NH_3 \xrightarrow{\text{ethylene oxide}} HOCH_2CH_2NH_2 \xrightarrow{\text{ethylene oxide}} (HOCH_2CH_2)_2NH \xrightarrow{\text{ethylene oxide}} (HOCH_2CH_2)_3N$$

$$RNH_2 \xrightarrow{H_2C{=}CH{-}CN} RNH{-}CH_2CH_2CN \xrightarrow[\text{excess}]{H_2C{=}CH{-}CN} RN(CH_2CH_2CN)_2$$

3. In an alkylation reaction, the exact amount of each product obtained depends on the precise reaction conditions and on relative amounts of starting amine and alkyl halide; the utility of alkylation as a preparative method for amines is limited.

C. Quaternization of Amines

1. Amines can be converted into quaternary ammonium salts with excess alkyl halide or sulfonate ester (R—X) under forcing conditions.

$$R_3N + R{-}X \longrightarrow R{-}\overset{+}{N}R_3\ X^-$$

   a. This process, called quaternization, is one of the most important synthetic applications of amine alkylation.
   b. The reaction is particularly useful when especially reactive alkyl halides are used.

2. Conversion of an amine into a quaternary ammonium salt with excess methyl iodide is called exhaustive methylation.

$$\text{pyrrolidine (N—H)} \xrightarrow{\text{excess MeI}} \text{N,N-dimethylpyrrolidinium } (N^+\text{Me}_2)\ I^-$$

D. Reductive Amination

1. Primary and secondary amines form imines and enamines, respectively, when they react with aldehydes and ketones; in the presence of a reducing agent, imines and enamines are reduced to amines.

$$\text{>C=O} \xrightarrow[-H_2O]{RNH_2} \text{>C=N-R} \xrightarrow{NaBH_4,\ NaCNBH_3,\ \text{or } H_2/\text{catalyst}} \text{-CH-NH-R}$$

imine formation — an imine — reduction — an amine

a. Reduction of the C═N double bond is analogous to reduction of the C═O double bond.
b. The imine or enamine does not have to be isolated; because imines and enamines are reduced much more rapidly than carbonyl compounds, reduction of the carbonyl compound is not a competing reaction.

2. The process in which an amine reacts with a carbonyl compound and the resulting imine is reduced to an amine is called reductive amination. (It has also been referred to as reductive alkylation.)
   a. When the reduction step is carried out by hydride reducing agents such as $NaBH_4$ and $NaBH_3CN$, a proton source such as a protic solvent or dilute acid is required.
   b. An imminium ion is formed in solution by protonation and dehydration of the carbinolamine intermediate in the reductive amination reaction of a secondary amine and formaldehyde; this ion is rapidly and irreversibly reduced.

$$H_2C{=}O \underset{}{\overset{R_2NH}{\rightleftharpoons}} \underset{\text{a carbinolamine}}{R_2N{-}CH_2{-}OH} \overset{H_3O^+}{\rightleftharpoons} \underset{\text{an iminium ion} + H_2O}{R_2\overset{+}{N}{=}CH_2} \xrightarrow[\text{reduction}]{NaBH_3CN} R_2N{-}CH_3$$

   c. The reaction of an amine with an excess of formaldehyde is a useful way to introduce methyl groups to the level of a tertiary amine; quaternization does not occur in this reaction.

3. When the reducing agent is sodium cyanoborohydride ($NaBH_3CN$), the reaction is known as the Borch reaction.

$$\text{>C=O} \xrightarrow{RNH_2} \text{>C=N-R} \xrightarrow[\text{MeOH, HCl}]{NaBH_3CN} \text{-CH-NH-R}$$

4. To determine the required starting materials for reductive amination:
   a. Starting with the desired compound, mentally reverse the reductive amination process by breaking one of the C—N bonds:
      i. On the nitrogen side, replace it with an N—H bond
      ii. On the carbon side, remove a hydrogen from the carbon and add a carbonyl oxygen.
   b. Repeat this process for each C—N bond and evaluate the best pathway.

H_2O + $(CH_3)_2CHCH_2CH_2{-}NH{-}CH(CH_3)_2$ ⇨ $(CH_3)_2CHCH_2CHO$ + $H_2N{-}CH(CH_3)_2$

H_2O + $(CH_3)_2CHCH_2CH_2{-}NH{-}CH(CH_3)_2$ ⇨ $(CH_3)_2CHCH_2CH_2NH_2$ + $O{=}C(CH_3)_2$

II. OTHER SYNTHESES OF AMINES

A. Gabriel Synthesis of Primary Amines

1. The alkylation of phthalimide anion followed by hydrolysis of the alkylated derivative to the primary amine is called the Gabriel synthesis.

$$\text{phthalimide (N—H)} \xrightarrow[\text{2) RX}]{\text{1) KOH}} \text{N—R} \xrightarrow[\text{heat}]{\text{HO}^-} H_2N\text{—}R + \text{phthalate } (O^-, O^-)$$

phthalimide

a. The nitrogen in phthalimide has only one acidic hydrogen, and thus it can be alkylated only once.
   i. The conjugate base of phthalimide ($pK_a \approx 9$) is easily formed with KOH or NaOH.
   ii. The phthalimide anion can be alkylated by alkyl halides or sulfonate esters (primary or unbranched secondary; shown as RX in the preceding equation) in S_N2 reactions.
   iii. The *N*-alkylated phthalimide formed is converted into the free amine by amide hydrolysis in either strong acid or base.
b. Although *N*-alkylphthalimides also have a pair of unshared electrons on nitrogen, they do not alkylate further, because neutral imides are much less basic, and therefore less nucleophilic, than the phthalimide anion.

2. Multiple alkylation, which occurs in the direct alkylation of ammonia, does not occur in the Gabriel synthesis.

B. Reduction of Nitro Compounds

1. Nitro compounds can be reduced to amines under a variety of conditions.
   a. The nitro group is usually reduced very easily by catalytic hydrogenation.
   b. The reduction of aromatic nitro compounds to primary amines can also be accomplished with finely divided metal powders and HCl.

$$C_6H_5\text{—}NO_2 \xrightarrow[\text{Fe}]{\text{HCl}} C_6H_5\text{—}\overset{+}{N}H_3\ Cl^-$$

   i. Iron or tin powder is frequently used.
   ii. In this reaction the nitro compound is reduced at nitrogen, and the metal, which is oxidized to a metal salt, is the reducing agent.

2. Aromatic nitro compounds react with $LiAlH_4$, but the reduction products are azobenzenes, not amines.
3. Nitro groups do not react at all with $NaBH_4$ under the usual conditions.

C. Hofmann Rearrangement

1. Treatment of a primary amide with bromine in base gives rise to a rearrangement called the Hofmann rearrangement or Hofmann hypobromite reaction.

$$Br_2 + 2\,^{-}OH + R\text{—}\overset{O}{\overset{\|}{C}}\text{—}NH_2 \longrightarrow R\text{—}NH_2 + CO_2 + 2\,Br^- + H_2O$$

2. The first step in the mechanism is ionization of the amide N—H; the resulting anion is then brominated.
   a. The *N*-bromoamide product is more acidic than the amide starting material and it ionizes.

i. The *N*-bromo anion rearranges rapidly to an isocyanate. (See Eq. 23.76a–d, text p. 1105, for details.)

ii. Because the Hofmann rearrangement is carried out in aqueous base, the isocyanate cannot be isolated.

b. The isocyanate formed spontaneously hydrates to a carbamate ion, which then decarboxylates to the amine product under the strongly basic reaction conditions.

c. When the reaction mixture is acidified, the carbamate decarboxylates to give the amine.

$$\underset{\text{an amide}}{\text{Me(Et)(H)C—C(=O)NH}_2} \xrightarrow[\text{HO}^-]{\text{Br}_2} \underset{\text{an } N\text{-bromoamide}}{\text{Me(Et)(H)C—C(=O)NHBr}} \xrightarrow[\text{deprotonation and rearrangement}]{^-\text{OH}} \underset{\text{an isocyanate}}{\text{Me(Et)(H)C—N=C=O}}$$

$$\xrightarrow{\text{HO}^-} \underset{\text{a carbamate}}{\text{Me(Et)(H)C—NH—C(=O)—O}^-} \xrightarrow{\text{H}_2\text{O}} \underset{\text{an amine}}{\text{Me(Et)(H)C—NH}_2} + \text{CO}_2 + \text{HO}^-$$

3. The Hofmann rearrangement takes place with retention of stereochemical configuration in the migrating alkyl group. (See foregoing equation for an example.)

D. Curtius Rearrangement

1. When an acyl azide is heated in an inert solvent such as benzene or toluene, it is transformed with loss of nitrogen into an isocyanate; this concerted reaction is called the Curtius rearrangement.

$$\underset{\text{an acyl azide}}{\text{R—C(=O)—N}_3} \xrightarrow{\text{heat}} \underset{\text{an isocyanate}}{\text{R—N=C=O}} + \text{N}_2\uparrow$$

2. The overall transformation that occurs as a result of the Curtius rearrangement followed by hydration is the removal of the carbonyl carbon of the acyl azide as CO_2.

   a. The isocyanate product of the Curtius rearrangement can be transformed into a carbamic acid by hydration in either acid or base.

   b. Spontaneous decarboxylation of the carbamic acid gives the amine, which is protonated under the acidic conditions of the reaction.

   c. The free amine is obtained by neutralization.

$$\underset{\text{an acyl azide}}{\text{Me(Et)(H)C—C(=O)N}_3} \xrightarrow[\text{heat}]{\text{rearrangement}} \underset{\text{an isocyanate}}{\text{Me(Et)(H)C—N=C=O}}$$

$$\xrightarrow{\text{H}_3\text{O}^+} \underset{\text{a carbamic acid}}{\text{Me(Et)(H)C—NH—C(=O)—OH}} \xrightarrow{-\text{CO}_2} \xrightarrow{^-\text{OH}} \underset{\text{an amine}}{\text{Me(Et)(H)C—NH}_2}$$

3. The Curtius rearrangement takes place with retention of stereochemical configuration in the migrating alkyl group. (See preceding equation.)

4. The Curtius reaction can be run under mild, neutral conditions.
   a. The isocyanate can be isolated if desired.
   b. Some acyl azides in the pure state can detonate without warning, and extreme caution is required in handling them.
5. An important use of the Curtius rearrangement is for the preparation of carbamic acid derivatives. (See summary equation 23.71, text p. 1104.)
   a. Reaction of isocyanates with alcohols or phenols yields carbamate esters.
   b. Reaction of isocyanates with amines yields ureas.
6. Acyl azides can be prepared in two ways:

a. The most straightforward method is the reaction of an acid chloride with sodium azide.

$$\underset{\text{an acid chloride}}{R{-}\overset{O}{\overset{\|}{C}}{-}Cl} + NaN_3 \longrightarrow \underset{\text{an acyl azide}}{R{-}\overset{O}{\overset{\|}{C}}{-}\overset{-}{N}{-}\overset{+}{N}{\equiv}N} + NaCl$$

b. An ester can be converted into an acyl derivative of hydrazine by aminolysis; the resulting amide, an acyl hydrazide, is then diazotized with nitrous acid to give the acyl azide.

$$\underset{\text{an ester}}{R{-}\overset{O}{\overset{\|}{C}}{-}OR} \xrightarrow[\text{hydrazine}]{H_2N{-}NH_2} \underset{\text{an acyl hydrazide}}{R{-}\overset{O}{\overset{\|}{C}}{-}NH{-}NH_2} \xrightarrow[HCl]{NaNO_2} \underset{\text{an acyl azide}}{R{-}\overset{O}{\overset{\|}{C}}{-}\overset{-}{N}{-}\overset{+}{N}{\equiv}N}$$

E. Aromatic Substitution Reactions of Aniline Derivatives

1. Aromatic amines can undergo electrophilic aromatic substitution reactions on the aromatic ring.

$$C_6H_5{-}NH_2 + 3\ Br_2 \longrightarrow \text{2,4,6-}Br_3C_6H_2{-}NH_2 + 3\ HBe$$

   a. The amino group is one of the most powerful ortho, para-directing groups in electrophilic aromatic substitution provided that it is not protonated under the reaction conditions.
   b. A protonated amino group does not have the unshared electron pair on nitrogen that gives rise through resonance to the activating, ortho, para-directing effect of a free amino group.
   c. Ammonium groups are meta-directing groups.
2. Aniline can be nitrated regioselectively at the para position if the amino group is first protected from protonation by conversion into an acetamide derivative by acetylation.

$$C_6H_5{-}NH_2 \xrightarrow[\text{pyridine}]{H_3C{-}\overset{O}{\overset{\|}{C}}{-}Cl} C_6H_5{-}NH{-}\overset{O}{\overset{\|}{C}}{-}CH_3 \xrightarrow[H_2SO_4,\ 20\ ^\circ C]{HNO_3} p\text{-}O_2N{-}C_6H_4{-}NH{-}\overset{O}{\overset{\|}{C}}{-}CH_3 \xrightarrow[H_2O]{^-OH} p\text{-}O_2N{-}C_6H_4{-}NH_2$$

protection of nitrogen

removal of the protecting group

F. Synthesis of Amines: Summary

1. Methods of amine synthesis:
   a. Reduction of amides with $LiAlH_4$ (Sec. 21.9B, text page 970).
   b. Reduction of nitriles with $LiAlH_4$ (Sec. 21.9C, text page 972).
   c. Direct alkylation of amines (Sec. 23.7A, text page 1084). This reaction is of limited utility, but it is useful for preparing quaternary ammonium salts.
   d. Reductive amination (Sec. 23.7B, text page 1085).
   e. Aromatic substitution reactions of anilines (Sec. 23.9, text page 1092).
   f. Gabriel synthesis of primary amines (Sec. 23.11A, text page 1100).
   g. Reduction of nitro compounds (Sec. 23.11B, text page 1102).
   h. Hofmann and Curtius rearrangements (Sec. 23.11C, text page 1103).
2. Methods c, d, and e represent methods of preparing amines from other amines.
3. Methods b, f, g, and h are limited to the preparation of primary amines.
4. Methods a, b, d, g, and h can be used to obtain amines from other functional groups.

III. SYNTHESES OF ORGANIC COMPOUNDS WITH AMINES

A. Acylation of Amines

1. Amines can be converted into amides by acylation with acid chlorides, anhydrides or esters.

$$CH_3C(=O)OC(=O)CH_3 + HN(CH_3)_2 \xrightarrow{\text{base}} CH_3C(=O)N(CH_3)_2 + CH_3C(=O)O^-$$

2. The reaction of an amine with an acid chloride or anhydride requires either two equivalents of the amine or one equivalent of the amine and an additional equivalent of another base such as a tertiary amine or hydroxide ion.

B. Hofmann Elimination of Quaternary Ammonium Hydroxides

1. Alkenes can be formed from amines by a three-step process:
   a. exhaustive methylation
   b. conversion of the ammonium salt to the hydroxide
   c. Hofmann elimination.

$$H-C-C-NH_2 \xrightarrow{\text{MeI}} H-C-C-\overset{+}{N}Me_3\ I^- \xrightarrow{\text{AgOH}} H-C-C-\overset{+}{N}Me_3\ HO^- \xrightarrow[-NMe_3]{\text{heat}} C{=}C + H_2O$$

2. The Hofmann elimination involves a quaternary ammonium hydroxide as the starting material; an amine acts as the leaving group.
   a. When a quaternary ammonium hydroxide is heated, a β-elimination reaction takes place to give an alkene, which distills from the reaction mixture.
   b. A quaternary ammonium hydroxide is formed by treating a quaternary ammonium salt with silver hydroxide (AgOH, formed from water and silver oxide, AgO_2).
   c. The Hofmann elimination is conceptually analogous to the E2 reaction of alkyl halides, in which a proton and a halide ion are eliminated; in the Hofmann elimination, a proton and a tertiary amine are eliminated.
   d. Hofmann elimination generally occurs as an *anti*-elimination.
   e. The conditions of the Hofmann elimination are typically harsh.
3. The elimination reactions of alkyl halides and those of quaternary ammonium salts show distinct differences in regiochemistry.

a. E2 elimination of most alkyl halides gives a predominance of the alkene with the greatest amount of branching at the double bond.
b. In contrast, Hofmann elimination of the corresponding trialkylammonium salt generally occurs so that the base abstracts a proton from the β-carbon with the least branching.

$$\text{2-methylcyclohexylamine} \xrightarrow[\text{2) AgOH; 3) heat}]{\text{1) MeI}} \text{3-methylcyclohexene} \quad \text{(major alkene product)}$$

c. This behavior is accounted for by the preference for *anti*-elimination and the minimization of van der Waals repulsions in the transition state of the reaction.

4. Especially acidic β-hydrogens tend to be eliminated even if they are on a more highly branched carbon.

more acidic β-hydrogen

$$Ph-CH(CH_3)-CH(NH_2)-CH_3 \xrightarrow[\text{2) AgOH; 3) heat}]{\text{1) MeI}} Ph-C(CH_3)=CH-CH_3$$

C. Diazotization; Reactions of Diazonium Ions

1. Oxidation of amines generally occurs at the amino nitrogen; an important oxidation reaction of amines is called diazotization.

$$\underset{\text{a primary amine}}{R-NH_2} + \underset{\text{nitrous acid}}{HNO_2} + HCl \longrightarrow \underset{\text{a diazonium salt}}{R-\overset{+}{N}\equiv N\ Cl^-} + 2\,H_2O$$

a. Diazotization is the reaction of a primary amine with nitrous acid (HNO_2) to form a diazonium salt.
b. A diazonium salt is a compound of the form $R-\overset{+}{N}\equiv N{:}\ X^-$, in which X^- is a typical anion.
c. Both aliphatic and aromatic primary amines are readily diazotized.

2. Diazonium ions incorporate one of the best leaving groups, molecular nitrogen.
3. Aliphatic diazonium ions react immediately as they are formed by S_N1, E1, and/or S_N2 mechanisms to give substitution and elimination products along with nitrogen gas. (The rapid liberation of nitrogen gas on treatment with nitrous acid is a qualitative test for primary alkylamines.)
4. Aromatic diazonium salts may be isolated and used in a variety of reactions; however, they are usually prepared in solution at 0–5 °C and used without isolation, because they lose nitrogen on heating and they are explosive in the dry state.

D. Aromatic Substitution with Diazonium Ions

1. Aryldiazonium ions react with aromatic compounds containing strongly activating substituent groups, such as amines and phenols, to give substituted azobenzenes.

$$C_6H_5-\overset{+}{N}_2 + C_6H_5-OH \longrightarrow C_6H_5-N=N-C_6H_4-OH \quad \text{an azobenzene}$$

2. This is an electrophilic aromatic substitution reaction in which the terminal nitrogen of the diazonium ion is the electrophile. (See Eqs. 23.54a–b, text p. 1098, for a detailed mechanism.)
3. The azobenzene derivatives formed in these reactions have extensive conjugated π-electron systems, and most of them are colored.

4. Some of these compounds are used as dyes and indicators; as a class they are known as azo dyes. (An azo dye is a colored derivative of azobenzene.)

E. Substitution Reactions of Aryldiazonium Salts

1. Among the most important reactions of aryldiazonium salts are substitution reactions with cuprous halides; in these reactions the diazonium group is replaced by a halogen.

H_3C–C_6H_4–NH_2 → ($NaNO_2$, HBr) → H_3C–C_6H_4–$\overset{+}{N_2}$ Br^- → (CuBr) → H_3C–C_6H_4–Br; → (CuCN) → H_3C–C_6H_4–CN → (H_2O, H_3O^+, heat) → H_3C–C_6H_4–CO_2H

a. An analogous reaction occurs with CuCN.
   i. This reaction is another way of forming a carbon-carbon bond, in this case to an aromatic ring.
   ii. The resulting nitrile can be converted by hydrolysis into a carboxylic acid.
b. The reaction of an aryldiazonium ion with a cuprous salt is called the Sandmeyer reaction.
c. This reaction is an important method for the synthesis of aryl halides and nitriles.

2. Aryl iodides can also be made by the reaction of diazonium salts with potassium iodide; a cuprous salt is not required.

$C_6H_4(CO_2H)(NH_2)$ → (1) $NaNO_2$/HCl; 2) KI) → $C_6H_4(CO_2H)(I)$

3. Diazonium salts can be hydrolyzed to phenols by heating them with water or by treating them with cuprous oxide (Cu_2O) and an excess of aqueous cupric nitrate [$Cu(NO_3)_2$] at room temperature.

O_2N–C_6H_4–NH_2 → (1) $NaNO_2$/HCl; 2) H_2O, heat; or CuO, H_2O, excess $Cu(NO_3)_2$) → O_2N–C_6H_4–OH

4. The diazonium group is replaced by hydrogen when the diazonium salt is treated with hypophosphorous acid, H_3PO_2.

$Br_3C_6H_2$–NH_2 → (1) $NaNO_2$/HCl; 2) H_3PO_2) → $Br_3C_6H_3$

5. All of the diazonium salt reactions listed above are substitution reactions but none are S_N2 or S_N1 reactions because aromatic rings do not undergo substitution by these mechanisms.
   a. The Sandmeyer and related reactions occur by radical-like mechanisms mediated by the copper.
   b. The reaction of diazonium salts with KI probably occurs by a similar mechanism.

c. The reaction of diazonium salts with H_3PO_2 has been shown definitely to be a free-radical chain reaction.

6. The substitution reactions of diazonium salts achieve ring-substitution patterns that cannot be obtained in other ways.

F. Reactions of Secondary and Tertiary Amines with Nitrous Acid

1. Secondary amines react with nitrous acid to yield *N*-nitrosoamines, usually called simply nitrosamines. Nitrosamines are known to be potent carcinogens.

$$R_2NH + HNO_2 \longrightarrow R_2N{-}N{=}O \quad \text{a nitrosamine}$$

2. The nitrogen of tertiary amines does not react under the strongly acidic conditions used in diazotization reactions.
3. *N,N*-Disubstituted aromatic amines undergo electrophilic aromatic substitution on the benzene ring; the electrophile is the nitrosyl cation, which is generated from nitrous acid under acidic conditions.

$$C_6H_5{-}NMe_2 \xrightarrow[HCl]{HNO_2} O{=}N{-}C_6H_4{-}NMe_2$$

SCIENTISTS

STUDY GUIDE LINKS

23.1 Alkyl-Group Polarization in Ionization Reactions

The text points out that alkyl groups can stabilize *either* positive charge or negative charge, as evidenced by studies of gas-phase acidity of alcohols (Sec. 8.6C) and gas-phase basicity of amines (this chapter). The purpose of this section is to give you a more concrete feeling for the nature of these alkyl-group effects.

Consider first the stabilization of negative charge by alkyl groups. Let's consider the ionization reaction of *tert*-butyl alcohol:

$$(CH_3)_3C{-}OH \rightleftharpoons (CH_3)_3C{-}O^- + H^+$$

tert-butyl alcohol — conjugate base of *tert*-butyl alcohol

Using a quantum-mechanics program, we can examine the charge on each atom of the molecule before and after the reaction. These charges aren't *formal charges,* which, as you recall, are calculated by a bookkeeping procedure, but are the *actual* charges in the molecule that are present as a result of bond dipoles.

tert-butyl alcohol: H +0.11; C −0.50; **C +0.93**; O −0.85; O–H +0.43

conjugate base of *tert*-butyl alcohol: H +0.02; C −0.40; **C +1.05**; O −1.06 (overall charge −)

(You can add the charges on the left to verify that they add to zero, which is the overall charge on the molecule; and you can add the charges on the right to verify that they add to –1. Remember that two of the three methyl groups are not shown). Notice that, in response to the greater negative charge on oxygen following ionization, the α-carbon (bold) takes on greater positive charge, and the positive charge on the hydrogens is reduced. This is another way of saying that *electrons flow away from the oxygen onto the hydrogens of the methyl groups.*

Now let's look at the protonation of trimethylamine in a similar way:

$$(CH_3)_3N\colon + H^+ \rightleftharpoons (CH_3)_3\overset{+}{N}{-}H$$

trimethylamine — conjugate acid of trimethylamine

Here is the result of the charge calculation:

trimethylamine: H +0.10; **C −0.15**; N −0.39

conjugate acid of trimethylamine: H +0.18; **C −0.39**; N +0.14; N–H +0.31 (overall charge +)

Notice in this case that introduction of positive charge on the electronegative nitrogen causes the α-carbons (bold numbers) to become more *negative*; the additional negative charge is clearly coming from the hydrogens, which have become more positive.

This shows graphically how the electron distribution in methyl groups can be polarized by nearby charges. Polarization can occur either way, and can therefore stabilize both positive and negative charges.

23.2 Structures of Amide Bases

Although the structures of lithium amide bases are written as ionic compounds for conceptual simplicity, recent research has shown that the structures of these species are considerably more complex. For example, lithium diisopropylamide in tetrahydrofuran (THF) has a symmetrical dimer structure (that is, it is a symmetrical aggregate of two units):

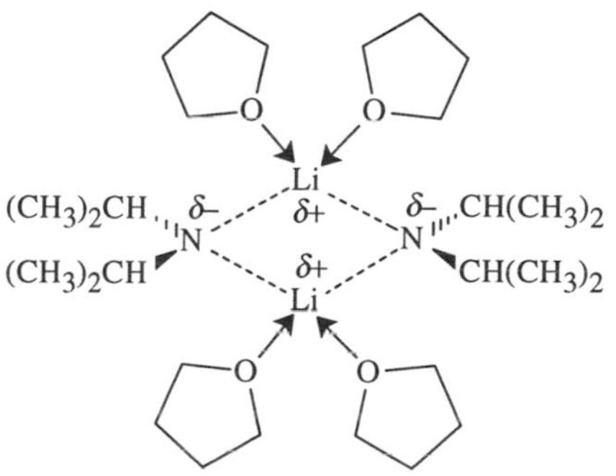

$(CH_3)_2CH$, $(CH_3)_2CH$ — $N^{\delta-}$ ⋯ $Li^{\delta+}$ ⋯ $N^{\delta-}$ — $CH(CH_3)_2$, $CH(CH_3)_2$; each Li coordinated by two THF (O)

Each lithium is partially bonded to two nitrogens with "half bonds" (dashed lines in the structure above). The nitrogen-lithium bonds are covalent, although they have significant ionic character. Each lithium is strongly solvated by two solvent molecules (THF in this case). The degree to which these amide bases are aggregated into larger structures varies with their concentration and with the nature of the solvent.

✓23.3 Nitration of Aniline

Remember that rates of reactions are affected by both intrinsic reactivity (reflected in the magnitude of the rate constant) and concentrations of the species involved (Sec. 9.3B). Under the acidic conditions of nitration, the highly reactive free aniline is present in very small concentration; the much less reactive protonated aniline is present in much greater concentration.

$$\text{rate of formation of } \textit{para}\text{-nitroaniline} = \underbrace{k_{para}}_{\text{very large}}\underbrace{[PhNH_2]}_{\text{very small}}$$

$$\text{rate of formation of } \textit{meta}\text{-nitroaniline} = \underbrace{k_{meta}}_{\text{very small}}\underbrace{[Ph\overset{+}{N}H_3]}_{\text{very large}}$$

These two competing rates must be about the same, because comparable amounts of *meta*- and *para*-substituted products are formed in text Eq. 23.42 (text p. 1093).

23.4 Mechanism of Diazotization

Diazotization at first sight may seem to be a very strange reaction, but once you understand the reactivity of nitrous acid, diazotization should make more sense.

Nitrous acid in solution is in equilibrium with its anhydride, dinitrogen trioxide.

$$2\,HNO_2 \rightleftharpoons O{=}N{-}O{-}N{=}O + H_2O$$

dinitrogen trioxide

Dinitrogen trioxide is the actual diazotizing agent under conditions of dilute acid; under conditions of very strong acid, the diazotizing agent is the nitrosyl cation, $^+N{=}\ddot{O}:$, formed by dehydration of nitrous acid. The dinitrogen trioxide mechanism is explored here.

The N═O double bonds of dinitrogen trioxide have a reactivity somewhat analogous to that of a carbonyl group. They react at nitrogen with nucleophiles such as amines with loss of nitrite.

$$R\ddot{N}H_2 + O{=}N{-}O{-}N{=}O \longrightarrow R\overset{+}{N}H_2{-}N(O^-){-}O{-}N{=}O \longrightarrow$$

$$R\overset{+}{N}H(H){-}N{=}O \;\; {}^-:\ddot{O}{-}N{=}O \longrightarrow R\ddot{N}H{-}N{=}O + H{-}\ddot{O}{-}N{=}O$$

nitrite anion — a nitrosamine — nitrous acid

(The pK_a of nitrous acid is 3.2; hence, nitrite is a weak base, and a good leaving group.) The product of this sequence is a *nitrosamine*. Because the amine nitrogen loses only one hydrogen in forming a nitrosamine, secondary amines as well as primary amines form nitrosamines when they react with nitrous acid (Sec. 23.10C).

In the case of primary amines, nitrosamines react further under the acidic conditions. First, they undergo isomerization to give another intermediate called a diazoic acid. (You should draw the curved-arrow notation for this step, which is analogous to enol formation from a ketone.)

$$\underset{\text{a nitrosamine}}{RNH{-}N{=}O} \underset{}{\overset{H_2O,\ H_3O^+}{\rightleftharpoons}} \underset{\text{a diazoic acid}}{R{-}N{=}N{-}OH}$$

Under the acidic conditions of diazotization, diazoic acids dehydrate to give diazonium ions:

$$R{-}\ddot{N}{=}\ddot{N}{-}\ddot{O}H + H{-}\overset{+}{O}H_2 \rightleftharpoons \ddot{O}H_2 + R{-}\ddot{N}{=}\ddot{N}{-}\overset{+}{O}H(H) \longrightarrow \underset{\text{a diazonium ion}}{R{-}\overset{+}{N}{\equiv}N:} + \ddot{O}H_2$$

✓23.5 Mechanism of the Curtius Rearrangement

Although the Curtius rearrangement is concerted, as shown in the text, it is helpful to *think* of the reaction in a stepwise fashion to see why the rearrangement occurs. There is a strong driving force for loss of dinitrogen, a very stable molecule:

$$R{-}\overset{O}{\overset{\|}{C}}{-}\ddot{N}^-{-}\overset{+}{N}{\equiv}N \longrightarrow R{-}\overset{O}{\overset{\|}{C}}{-}\ddot{N} + :N{\equiv}N:$$

electron-deficient nitrogen

Loss of dinitrogen leaves behind an *electron-deficient* nitrogen; it is electron-deficient because it is short one electron pair of an octet.

Just as rearrangements of alkyl groups occur to the electron-deficient *carbons* of carbocations, so they also occur to electron-deficient nitrogens.

$$R-\overset{O}{\overset{\|}{C}}-\ddot{N}: \longrightarrow \left[O{=}\overset{+}{C}-\overset{-}{\ddot{N}}-R \longleftrightarrow O{=}C{=}\ddot{N}-R \right]$$

an isocyanate

In other words, the Curtius rearrangement is driven by the very strong tendency of nitrogen to have a complete octet.

✓23.6 Formation and Decarboxylation of Carbamic Acids

The formation of carbamic acids by hydration of isocyanates is an acid-catalyzed addition to the C=N double bond:

$$R-\ddot{N}{=}C{=}O \xrightleftharpoons{H-\overset{+}{O}H_2} R-\overset{+}{\underset{|}{\overset{H}{N}}}{=}C{=}O \xrightarrow{:\ddot{O}H_2} R-\ddot{N}H-\overset{H\overset{+}{O}-H}{C}{=}O \xrightarrow{:\ddot{O}H_2} R-\ddot{N}H-\overset{H\ddot{O}:}{C}{=}O + H_3\ddot{O}^+$$

Hydration can also occur in base by attack of ^-OH on the carbonyl carbon. (Write the mechanism of this reaction.) The hydration in basic solution gives the carbamate ion, the conjugate base of a carbamic acid. The hydration of isocyanates under basic conditions is important in the Hofmann rearrangement, which is also discussed in Sec. 23.11C of the text.

Decarboxylation is a concerted reaction in which the same water molecule acts as both an acid and base catalyst at the same time.

$$RNH-\overset{O}{\overset{\|}{C}}-O-H \;\; (H-O-H) \rightleftharpoons RNH_2 + O{=}C{=}O + H_2O$$

The decarboxylation reaction is reversible; that is, amines react readily with CO_2 to give carbamic acids. However, the decarboxylation reaction in acidic solution is driven to completion by two factors: (1) protonation of the product amine, and (2) loss of CO_2 from solution as a gas.

Decarboxylation is fastest in acidic solution, but also occurs in base:

$$RNH-\overset{O}{\overset{\|}{C}}-O^- \;\; (H-O-H) \rightleftharpoons RNH_2 + O{=}C{=}O + {}^-OH$$

This reaction is driven to completion by reaction of the hydroxide ion with CO_2 to form bicarbonate ion:

$$O{=}C{=}O + {}^{-}OH \longrightarrow O{=}C(O^{-})OH$$

bicarbonate ion

If the solution is strongly basic, bicarbonate ion itself ionizes further to carbonate ion.

Decarboxylation of carbamates in basic solution is important in the Hofmann rearrangement.

SOLUTIONS TO PROBLEMS

Solutions to In-Text Problems

23.1 (a) C_6H_5—NHCH$(CH_3)_2$

N-isopropylaniline

(b) $(CH_3)_3C—NH_2$

tert-butylamine

(c) Piperidine ring (N—H) with OCH_3 at C-3

3-methoxypiperidine

(d) $CH_3C(CH_3)_2—CH(NH_2)CH_2CH_2CH_3$

2,2-dimethyl-3-hexanamine

(e) $CH_3CH_2CH_2CH(N(CH_2CH_3)_2)C(=O)OCH_2CH_3$

ethyl 2-(diethylamino)pentanoate

(f) $CH_3CH_2CH(N(CH_2CH_3)_2)CH_2CH_2CH_2CH_3$

N,N-diethyl-3-heptanamine

23.2 (a) *N*-Ethyl-*N*-methylisobutylamine (common), or *N*-ethyl-*N*,2-dimethyl-1-propanamine
(b) *N,N*-Dimethyl-*p*-nitroaniline (common), or 1-dimethylamino-4-nitrobenzene
(c) Dicyclohexylamine (common), or *N*-cyclohexylcyclohexanamine
(d) 3-Methylamino-1-pentanol. Note that the —OH group is the principal group.
(e) 1-(2-Chloroethyl)-3-propylpyrrolidine

23.3 (a) The principle to apply is that the more double-bond character the carbon-nitrogen bond has, the shorter it is. The order of decreasing double-bond character, and hence increasing carbon-nitrogen bond length, is

p-nitroaniline < aniline < cyclohexylamine

Because of resonance interaction of their nitrogen lone pairs with the benzene ring, aniline and *p*-nitroaniline have the most double-bond character; cyclohexylamine lacks this interaction and therefore has the least carbon-nitrogen double-bond character. In *p*-nitroaniline, the unshared electron pair on the amine nitrogen can be delocalized into the *p*-nitro group:

[$O=\overset{+}{N}(O^-)$—C_6H_4—$\ddot{N}H_2$ ⟷ $^-O—\overset{+}{N}(O^-)$=C_6H_4=$\overset{+}{N}H_2$]

This interaction is in addition to the resonance interaction of the unshared pair with the ring of the type shown in Eq. 23.3 on p. 1073 of the text. This additional delocalization increases the carbon-nitrogen double-bond character and decreases the carbon-nitrogen bond length.

(b) Apply the principle used in part (a). The order of increasing carbon-nitrogen bond length is

$$A < C < B$$

Compound *B* has a true carbon-nitrogen single bond, and hence has the longest carbon-nitrogen bond. Compound *A* has a true carbon-nitrogen double bond, and hence has the shortest carbon-nitrogen bond. The carbon-nitrogen bond of *C* has some double-bond character because of resonance interaction of the nitrogen unshared pair with the double bond. (Draw the appropriate resonance structure.) Another way to compare the C—N bond lengths of compounds *B* and *C* is to note that the C—N bond in *B* is an sp^3-sp^3 single bond, whereas the C—N bond in *C* is an sp^2-sp^3 single bond, and to remember that σ bonds with greater *s* character are shorter. (See discussion on text pp. 115 and 639.)

23.4 The NMR spectrum indicates the presence of only two phenyl hydrogens, and the singlet absorptions at δ 2.07 and δ 2.16 indicate the presence of chemically nonequivalent methyl groups in the ratio 2:1. Only compound (2), 2,4,6-trimethylaniline, fits the data. The two resonances at δ 2.07 and δ 2.16 correspond to the ortho and para methyl groups, respectively, and the δ 3.19 resonance to the —NH_2 protons of the amine.

23.5 The peaks arise from α-cleavage of the molecular ions derived from the respective amines. 2-Methyl-2-heptanamine can form a cation with $m/z = 58$ by this mechanism, and is therefore compound *B*. *N*-Ethyl-4-methyl-2-pentanamine can form a cation of $m/z = 72$ by this mechanism, and is therefore compound *A*. The α-cleavage mechanism is shown below for *B*; you should show it for compound *A*.

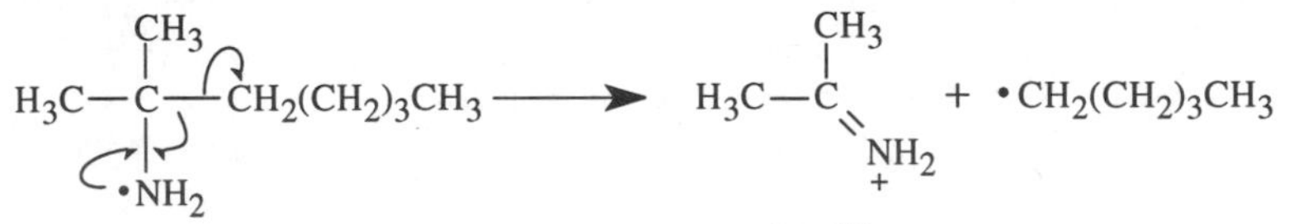

$H_3C—C(CH_3)(\cdot NH_2^+)—CH_2(CH_2)_3CH_3$ ⟶ $H_3C—C(CH_3)=\overset{+}{N}H_2$ + $\cdot CH_2(CH_2)_3CH_3$

molecular ion from 2-methyl-2-heptanamine (compound *B*)

$m/z = 58$

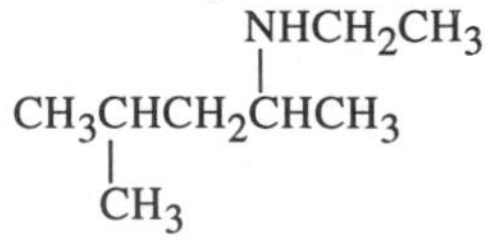

$CH_3CH(CH_3)CH_2CH(NHCH_2CH_3)CH_3$

N-ethyl-4-methyl-2-pentanamine (compound *A*)

23.6 The five-proton resonance at δ 7.18 in the NMR spectrum indicates the presence of a monosubstituted benzene ring (which accounts for 77 mass units), and the mutually split triplet-quartet pattern at δ 1.07 and δ 2.60 indicates an ethyl group (which accounts for 29 mass units). The IR spectrum indicates the presence of an amine; if it is a secondary amine, as suggested by the integral of the δ 0.91 resonance in the NMR spectrum, the NH accounts for 15 mass units, leaving 14 mass units unaccounted for. A CH_2 group could account for this remaining mass, and the two-proton singlet at δ 3.70 could correspond to such a resonance. On the assumption that the N—H proton is undergoing rapid exchange and does not show splitting with neighboring protons, *N*-ethylbenzylamine fits the data.

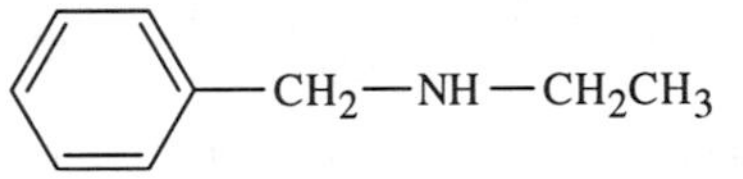

$C_6H_5—CH_2—NH—CH_2CH_3$ *N*-ethylbenzylamine

The predicted chemical shift (Fig. 13.4) for the benzylic CH_2 group is greater than that of ordinary benzylic protons because these protons are α to a nitrogen; the chemical shift should also be greater than protons that are only α to a nitrogen. The chemical shift of δ 3.6 is quite reasonable. The $m/z = 120$ peak in the mass spectrum corresponds to loss of a methyl radical by α-cleavage; and the $m/z = 91$ peak corresponds to a benzyl cation, which is formed by inductive cleavage. (See text pp. 525–6 for a discussion of these cleavage mechanisms.)

23.7 (a) 2,2-Dimethyl-1-propanamine (neopentylamine, $(CH_3)_3CCH_2NH_2$) has a maximum of three resonances in its CMR spectrum. 2-Methyl-2-butanamine, $CH_3CH_2C(CH_3)_2NH_2$, should have four resonances in its CMR spectrum.

(b) *Trans*-1,4-cyclohexanediamine has only two nonequivalent sets of carbons and thus should have only two resonances in its CMR spectrum. *Trans*-1,2-cyclohexanediamine has three nonequivalent sets of carbons and therefore should have three resonances in its CMR spectrum. (Chemically equivalent (and in these cases, homotopic) carbons in each compound are indicated by the same symbol.)

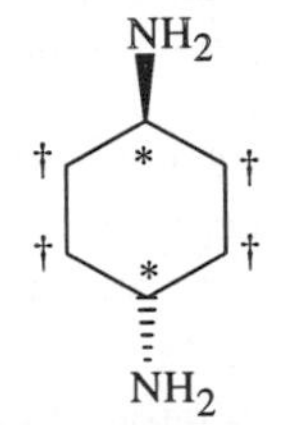

trans-1,4-cyclohexanediamine
two nonequivalent sets of carbons

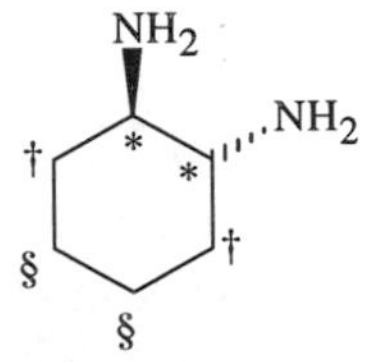

trans-1,2-cyclohexanediamine
three nonequivalent sets of carbons

23.8 (a) The order of increasing basicity (and increasing conjugate-acid pK_a) is NH_3(ammonia) < $PrNH_2$ (propylamine) < Pr_2NH_2 (dipropylamine). As the text discussion on p. 1077 shows, within a series of amines of increasing alkyl substitution basicity increases from ammonia to the secondary amine.

(b) The order of increasing basicity is

$$H_3\overset{+}{N}CH_2CH_2NH_2 < H_2NCH_2CH_2CO_2Me < CH_3CH(NH_2)CH_2CH_3$$

methyl 3-aminopropanoate; *sec*-butylamine

(Note that *basicity* refers to the basicity of the uncharged amino group in the first compound, not the basicity of the *conjugate base* of the ammonium ion.) The electron-withdrawing polar effect of the partial positive charge on the carbonyl carbon of the ester reduces the basicity of its β-amino group, and the polar effect of the full positive charge in the ammonium ion reduces the basicity of its β-amino group even more. Note that the polar effect of a substituent containing a fully charged atom is generally greater than the polar effect of a substituent containing a dipolar group.

(c) The order of increasing basicity is

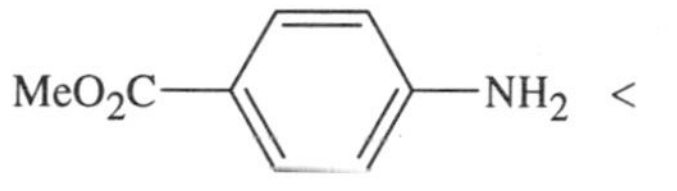

methyl *p*-aminobenzoate 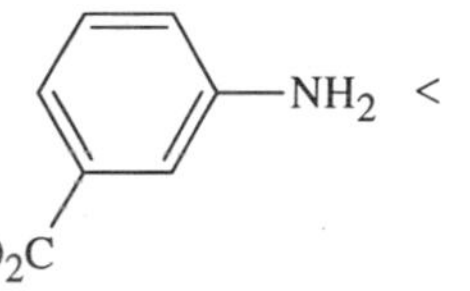 < methyl *m*-aminobenzoate < aniline

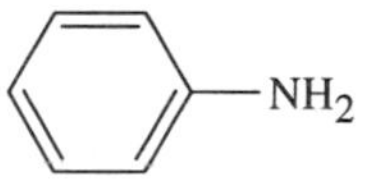

The methoxycarbonyl group (that is, the methyl ester group) decreases basicity in the first two derivatives by a polar effect. Furthermore, methyl *p*-aminobenzoate is stabilized by resonance interaction of the nitrogen unshared pair with the carbonyl group:

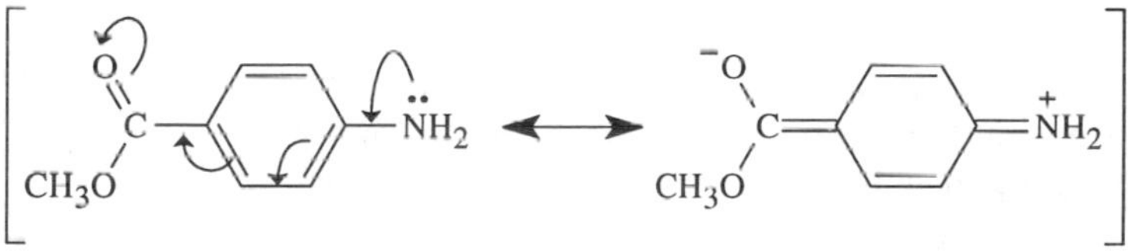

[$CH_3O(O=)C-C_6H_4-\ddot{N}H_2 \longleftrightarrow CH_3O(^-O)C=C_6H_4=\overset{+}{N}H_2$]

Because there is no unshared pair on nitrogen in the conjugate-acid ammonium ion, such resonance stabilization of the conjugate acid is not possible. As Fig. 23.2 on text p. 1078 shows, anything that stabilizes an amine relative to its conjugate-acid ammonium ion lowers its basicity. Note that resonance interaction between the carbonyl group and a *meta*-amino group is not possible.

(d) The order of increasing basicity is

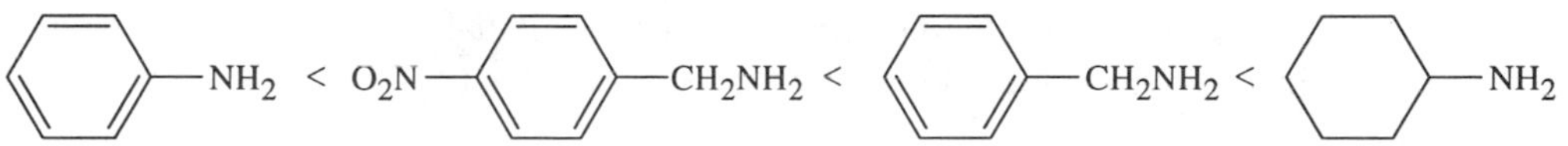

aniline < *p*-nitrobenzylamine < benzylamine < cyclohexylamine

The resonance and electron-withdrawing polar effect of the benzene ring severely reduces the nitrogen basicity of aniline relative to that of an alkylamine. The electron-withdrawing polar effect of the nitro group of *p*-nitrobenzylamine reduces its basicity relative to that of benzylamine. And the electron-withdrawing polar effect of the benzene rings reduces the basicity of both benzylamines relative to cyclohexylamine, an ordinary alkylamine.

23.9 From Table 23.1, we see that the basicity order is as follows:

$$A < B < C$$

The basicities of all these amines are reduced by resonance interaction of the nitrogen unshared electron pair with the ring. The electron-withdrawing polar effect of the nitro group lowers basicity further, and resonance interaction of the nitrogen unshared pair with the *p*-nitro group stabilizes *p*-nitroaniline and lowers its basicity even more. (Resonance structures depicting this interaction are shown in the solution to Problem 23.2(a) on p. 889 of this manual.)

23.10 Dissolve the mixture in a low-boiling, water-insoluble organic solvent such as methylene chloride. Extract this solution with 5% sodium bicarbonate or 5% sodium hydroxide solution; *p*-chlorobenzoic acid will dissolve in the aqueous layer as its sodium salt. It can be recovered as the free acid by acidifying the aqueous extracts.

The methylene chloride solution now contains only *p*-chloroaniline and *p*-chlorotoluene. Extract the methylene chloride solution with 5% HCl solution. *p*-Chloroaniline will dissolve in the aqueous layer as its hydrochloride salt. It can be recovered as the free amine by neutralizing the aqueous layer with dilute NaOH.

p-Chlorotoluene remains in the methylene chloride layer; it can be isolated by drying the solution and evaporating the methylene chloride. (Note that the order of the acid and base extractions could be interchanged.)

23.11 Mix the racemic acid with one equivalent of the enantiomerically pure amine. (The *S* enantiomer of the amine is utilized in the equation below, but either *pure* enantiomer of the amine could be used.) The carboxylic acids will react with the amine to give a mixture of *diastereomeric* salts.

$$\underbrace{(S)\text{-PhCH(CH}_3\text{)CO}_2\text{H} + (R)\text{-PhCH(CH}_3\text{)CO}_2\text{H}}_{(\pm)\text{-2-phenylpropanoic acid}} + 2\,\underbrace{(S)\text{-PhCH(CH}_3\text{)NH}_2}_{(S)\text{-1-phenylethanamine}} \longrightarrow$$

$$\underbrace{(S)\text{-PhCH(CH}_3\text{)CO}_2^-\ (S)\text{-PhCH(CH}_3\text{)}\overset{+}{\text{N}}\text{H}_3 + (R)\text{-PhCH(CH}_3\text{)CO}_2^-\ (S)\text{-PhCH(CH}_3\text{)}\overset{+}{\text{N}}\text{H}_3}_{\text{diastereomeric salts}}$$

Because these salts have different properties, they can be separated by differential solubility, for example, by fractional crystallization from some solvent, probably an alcohol. After separating the salts, aqueous acid is added to each. The amine resolving agent dissolves in the aqueous acid solution (from which it can be recovered), leaving the enantiomerically pure free carboxylic acid as a solid that can be isolated by filtration and recrystallized.

$$\underbrace{(S)\text{-PhCH(CH}_3\text{)CO}_2^-\ (S)\text{-PhCH(CH}_3\text{)}\overset{+}{\text{N}}\text{H}_3}_{\text{salt}} + \text{HCl} \xrightarrow{H_2O} \underbrace{(S)\text{-PhCH(CH}_3\text{)CO}_2\text{H}}_{\text{crystallizes}} + \underbrace{(S)\text{-PhCH(CH}_3\text{)}\overset{+}{\text{N}}\text{H}_3\ \text{Cl}^-}_{\text{an ionic compound; remains in solution}}$$

Notice that not only is amine basicity important in forming the initial salt, but also it is important in separating the amine resolving agent from the resolved carboxylic acid.

23.12 (a) $(CH_3CH_2)_4N^+\ F^-$

tetraethylammonium fluoride

(b) $(PhCH_2)_2\overset{+}{N}(CH_3)_2\ Br^-$

dibenzyldimethylammonium bromide

23.13 Compound *A* has an asymmetric nitrogen stereocenter; therefore, it is chiral and can exist as two enantiomers. Compound *B* also contains an asymmetric nitrogen; however, compound *B* undergoes an acid-base equilibrium by which it is in equilibrium with a very small amount of the free amine:

$$\underbrace{\text{PhCH}_2\overset{+}{\text{N}}(\text{CH}_3)(\text{C}_2\text{H}_5)\text{—H}\ \ \text{Cl}^-}_{\text{compound } B} \rightleftharpoons \text{PhCH}_2\ddot{\text{N}}(\text{CH}_3)(\text{C}_2\text{H}_5) + \text{H—Cl}$$

This equilibrium takes place so rapidly, and the minuscule amount of free amine undergoes nitrogen inversion so quickly, that the ammonium salt is rapidly racemized. Because compound *A* does not have a hydrogen that can be involved in an equilibrium of this type, it is not racemized.

23.14 Either the ethyl group can originate from acetaldehyde, or the cyclohexyl group can originate from cyclohexanone.

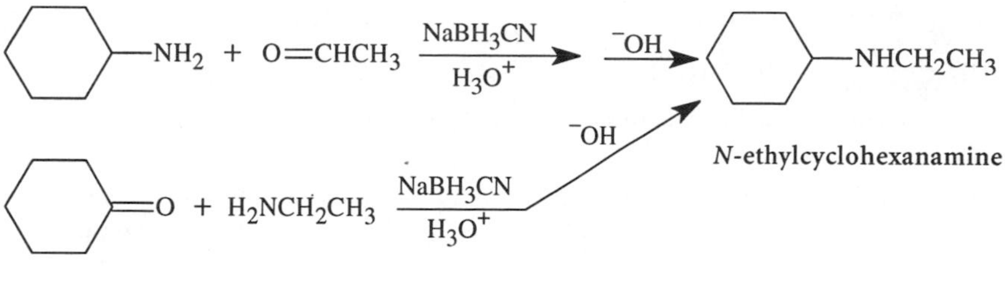

cyclohexyl—NH_2 + $O{=}CHCH_3$ $\xrightarrow[H_3O^+]{NaBH_3CN}$ $\xrightarrow{^-OH}$ cyclohexyl—$NHCH_2CH_3$

cyclohexanone ($=O$) + $H_2NCH_2CH_3$ $\xrightarrow[H_3O^+]{NaBH_3CN}$ $\xrightarrow{^-OH}$ *N*-ethylcyclohexanamine

23.15 The ethyl group could originate from acetaldehyde:

C_6H_5—$NH(CH_3)$ + $O{=}CHCH_3$ $\xrightarrow[MeOH]{NaBH_3CN}$ C_6H_5—$N(CH_3)CH_2CH_3$

N-ethyl-*N*-methylaniline

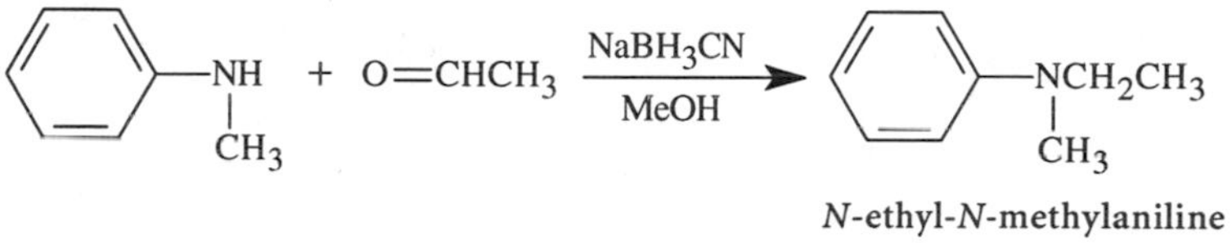

23.16 (a) Prepare the tertiary amine by reductive amination and alkylate it to give the ammonium salt.

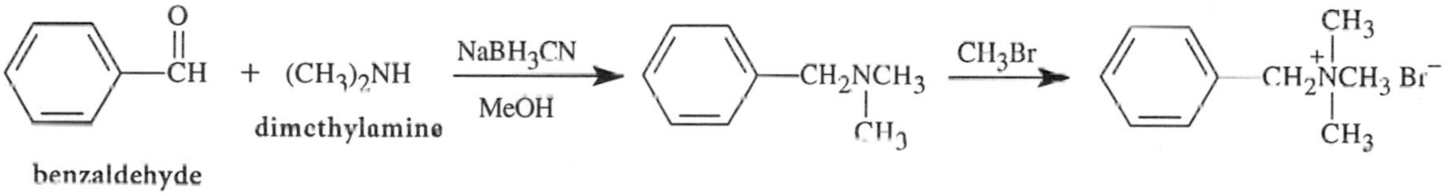

C_6H_5—$CH{=}O$ (benzaldehyde) + $(CH_3)_2NH$ (dimethylamine) $\xrightarrow[MeOH]{NaBH_3CN}$ C_6H_5—$CH_2N(CH_3)_2$ $\xrightarrow{CH_3Br}$ C_6H_5—$CH_2\overset{+}{N}(CH_3)_3\ Br^-$

An alternative synthesis would be to reductively aminate the same amine with formaldehyde to give trimethylamine and then alkylate this tertiary amine with benzyl bromide ($PhCH_2Br$).

(b) Exhaustively methylate benzylamine with a large excess of methyl bromide.

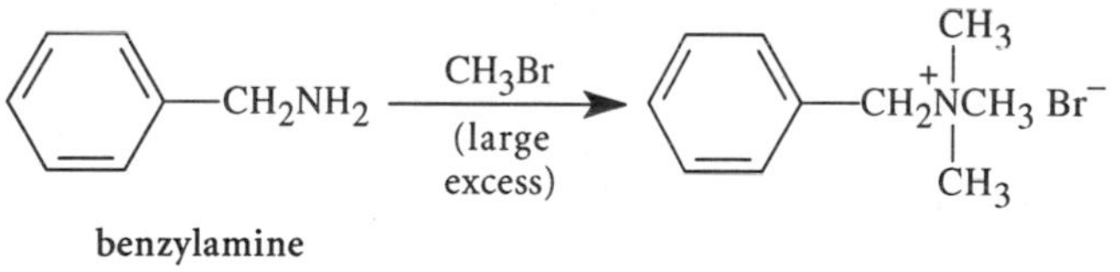

C_6H_5—CH_2NH_2 (benzylamine) $\xrightarrow[\text{(large excess)}]{CH_3Br}$ C_6H_5—$CH_2\overset{+}{N}(CH_3)_3\ Br^-$

23.17 Caleb has forgotten, if he ever knew, that aryl halides such as bromobenzene do not undergo S_N2 reactions (Sec. 18.1).

23.18 The dimethylammonium ion can lose a proton to give dimethylamine, which can then be alkylated by methyl iodide.

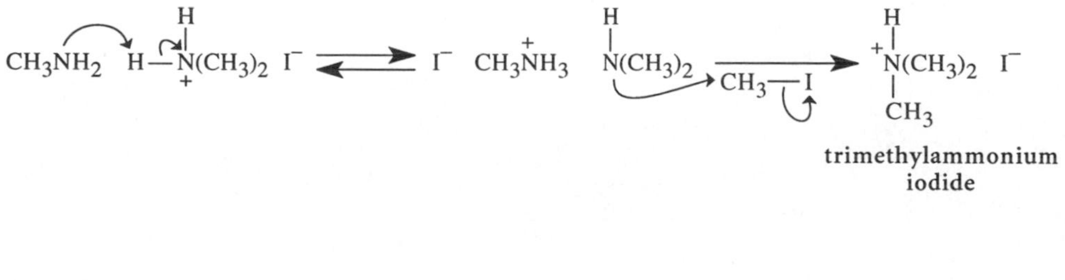

CH_3NH_2 $H—\overset{+}{N}H(CH_3)_2\ I^-$ $\rightleftharpoons$ I^- $CH_3\overset{+}{N}H_3$ $HN(CH_3)_2$ $\xrightarrow{CH_3—I}$ $H\overset{+}{N}(CH_3)_3\ I^-$

trimethylammonium iodide

23.19 (a)

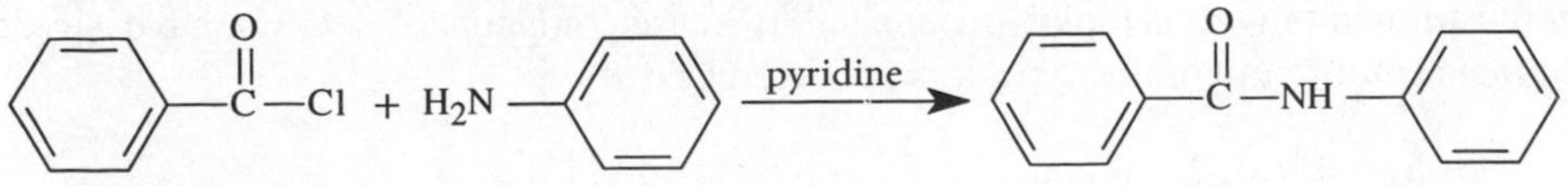

$$C_6H_5COCl + H_2NC_6H_5 \xrightarrow{\text{pyridine}} C_6H_5CONHC_6H_5$$

benzoyl chloride   aniline   N-phenylbenzamide

(b)

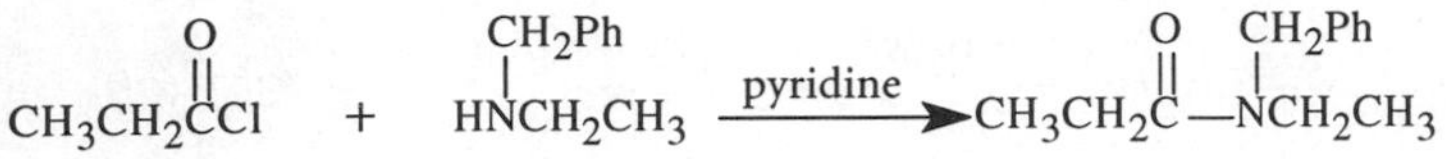

$$CH_3CH_2CCl(=O) + HN(CH_2Ph)CH_2CH_3 \xrightarrow{\text{pyridine}} CH_3CH_2C(=O)N(CH_2Ph)CH_2CH_3$$

propanoyl chloride   benzylethylamine   N-benzyl-N-ethylpropanamide

23.20 (a) Draw the molecule in a conformation in which the trimethylammonium group is anti to the β-hydrogen, and examine the relative positions of the phenyl groups. In this part, the alkene product must have the Z configuration.

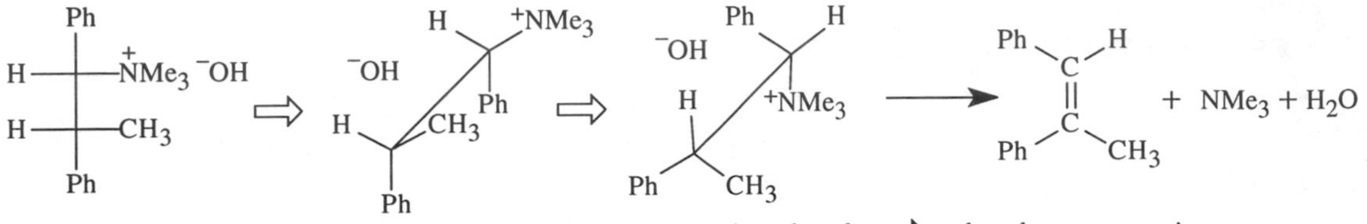

Ph; H, $\overset{+}{N}Me_3$ ^{-}OH; H, CH_3; Ph ⇨ H, $^{+}NMe_3$, ^{-}OH, Ph, H, CH_3, Ph ⇨ Ph, H, ^{-}OH, H, $^{+}NMe_3$, Ph, CH_3 ⟶ Ph(H)C=C(Ph)CH_3 + NMe_3 + H_2O

Note that phenyl groups are gauche ⇨ phenyl groups are cis; therefore, alkene has Z configuration

(b) Follow the procedure used in part (a).

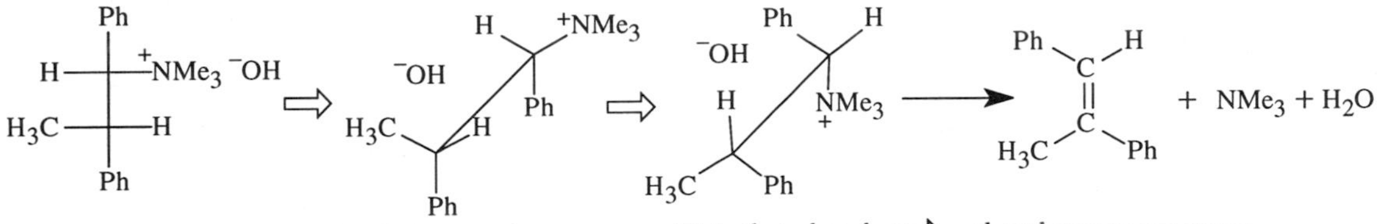

Ph; H, $\overset{+}{N}Me_3$ ^{-}OH; H_3C, H; Ph ⇨ H, $^{+}NMe_3$, ^{-}OH, Ph, H_3C, H, Ph ⇨ Ph, H, ^{-}OH, H, $\overset{+}{N}Me_3$, H_3C, Ph ⟶ Ph(H)C=C(H_3C)Ph + NMe_3 + H_2O

Note that phenyl groups are anti ⇨ phenyl groups are trans; therefore, alkene has E configuration

23.21 (a) The hydroxide ion removes the proton from the β-carbon with the least amount of branching, that is, from the methyl carbon.

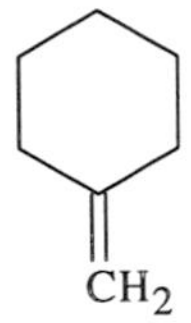

(methylenecyclohexane: cyclohexane ring =CH_2)

(b) The hydrogen α to the carbonyl group is removed because it is considerably more acidic than the other β-hydrogens.

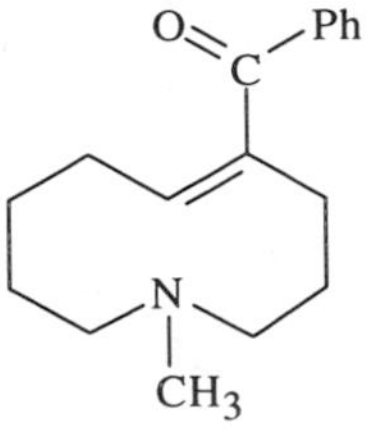

(ring containing N–CH_3 and C=C bearing C(=O)Ph; labels: O, C, Ph, N, CH_3)

23.22 Because there are three chemically nonequivalent β-hydrogens, three different alkenes are possible, shown below in order of the decreasing amount expected.

A (methylenecyclopentane with CH_2CH_3; exocyclic $=CH_2$) > B (1-methylcyclopentene isomer with CH_2CH_3, CH_3) > C (cyclopentene isomer with CH_2CH_3, CH_3)

Compound *A* is expected to predominate because it arises from loss of a proton at the least branched β-carbon. Compound *C* should be obtained in smallest amount because it arises from loss of the hydrogen from the most branched β-carbon.

23.23 Coniine has one degree of unsaturation. Because exhaustive methylation-Hofmann elimination introduces a double bond, and because each of the products *A–C* contains only one double bond, it follows that coniine is a cyclic amine. Product *C* shows that the nitrogen in coniine is attached to the fourth carbon from the end of the alkyl side-chain, and products *A* and *B* show that the nitrogen in coniine is attached to the end carbon of the alkyl side-chain. The simultaneous attachment of this nitrogen to two different points in the carbon chain is consistent with the ring structure for coniine deduced earlier, and that the nitrogen is part of the ring. Because five carbons intervene between points of attachment of the nitrogen, coniine must be a piperidine derivative (see text p. 1071):

(piperidine ring, N–H, with $CH_2CH_2CH_3$ at C-2)

2-propylpiperidine
(coniine)

Exhaustive methylation, Ag_2O treatment, and Hofmann elimination give three alkenes that result from the three possible points of β-elimination:

(N,N-dimethyl-2-propylpiperidinium ion, H_3C, CH_3 on $\overset{+}{N}$, $CH_2CH_2CH_3$, ^{-}OH)

elimination of this β hydrogen gives *C*

elimination of this β-hydrogen gives *B*

elimination of this β-hydrogen gives *A*

product resulting from exhaustive methylation and Ag_2O treatment of coniine

In a number of problems, the text uses Ag_2O (silver(I) oxide) for the conversion of quaternary ammonium halides into quaternary ammonium hydroxides. Attempts to balance these reactions will show that a hydrogen is unaccounted for. The resolution of this problem is that silver(I) oxide is typically used in a *hydrated* form; hydrated Ag_2O is equivalent to AgOH. Consequently, you should assume in all problems involving Hofmann elimination that Ag_2O is operationally the same thing as AgOH.

23.24 Use an acetyl protecting group as outlined in Study Problem 23.4.

H_2N–C_6H_5 $\xrightarrow[\text{pyridine}]{CH_3\overset{O}{\overset{\|}{C}}Cl}$ $CH_3\overset{O}{\overset{\|}{C}}NH$–$C_6H_5$ $\xrightarrow{ClSO_3H}$ $CH_3\overset{O}{\overset{\|}{C}}NH$–$C_6H_4$–$SO_2$–Cl $\xrightarrow{NH_3\text{ (excess)}}$

p-acetamidobenzenesulfonyl chloride

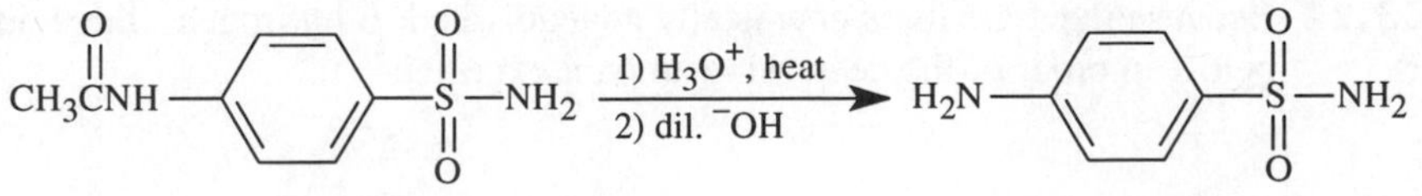

CH_3CNH—C₆H₄—SO_2NH_2 → (1) H_3O^+, heat; 2) dil. ^-OH) → H_2N—C₆H₄—SO_2NH_2

sulfanilamide

Note that if the acetyl protecting group were not used, the amine group on one molecule would react with a chlorosulfonyl group on the other and a polymer would be formed.

23.25 (a) Begin with *p*-nitroacetanilide, prepared as shown in Study Problem 23.4. Then convert it into 2,4-dinitroaniline. The acetyl protecting group is used to avoid protonation of the nitrogen, which would result in a significant amount of unwanted meta substitution. (See Eq. 23.42 on text p. 1093.)

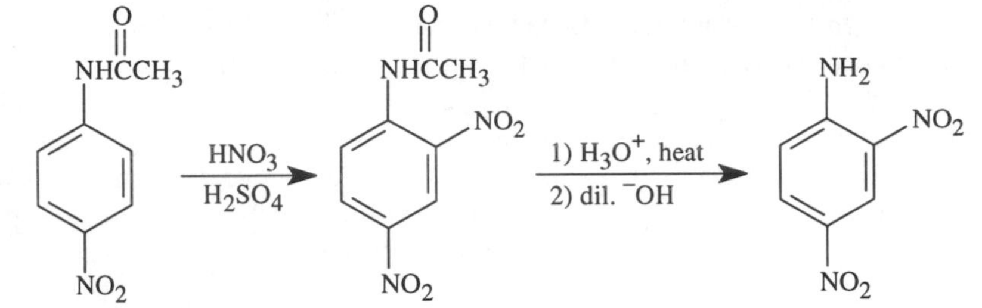

p-nitroacetanilide → (HNO_3, H_2SO_4) → 2,4-dinitroacetanilide → (1) H_3O^+, heat; 2) dil. ^-OH) → 2,4-dinitroaniline

(b) Begin with *p*-acetamidobenzenesulfonyl chloride, prepared as shown in the solution to Problem 23.24.

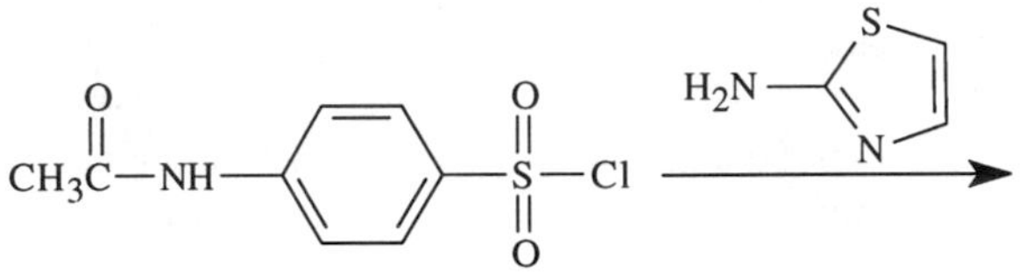

p-acetamidobenzenesulfonyl chloride → (2-aminothiazole, H_2N–thiazole) → $CH_3C(=O)$—NH—C₆H₄—SO_2—NH—(2-thiazolyl) → (1) H_3O^+, heat; 2) dil. ^-OH) → H_2N—C₆H₄—SO_2—NH—(2-thiazolyl)

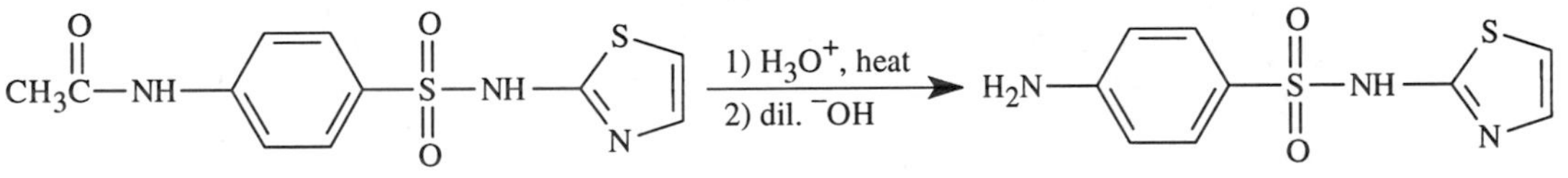

sulfathiazole

23.26 (a)

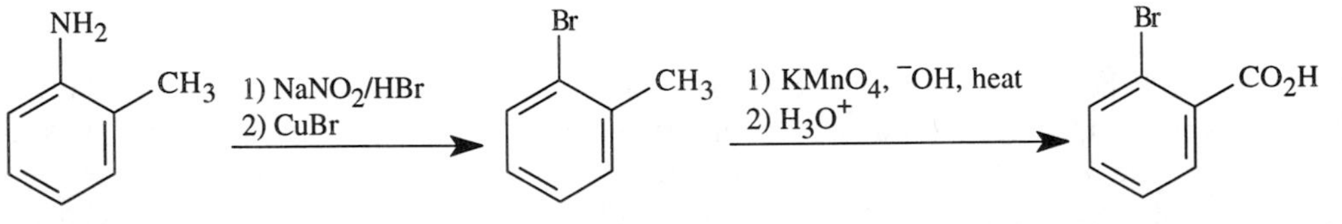

o-toluidine → (1) $NaNO_2$/HBr; 2) CuBr) → 2-bromotoluene → (1) $KMnO_4$, ^-OH, heat; 2) H_3O^+) → 2-bromobenzoic acid

(b)

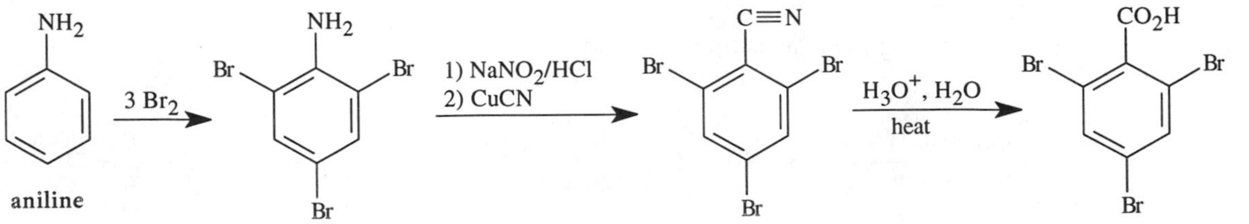

aniline → (3 Br_2) → 2,4,6-tribromoaniline → (1) $NaNO_2$/HCl; 2) CuCN) → 2,4,6-tribromobenzonitrile (C≡N) → (H_3O^+, H_2O, heat) → 2,4,6-tribromobenzoic acid

23.27 (a) Because diazotization does not break the carbon-nitrogen bond, the diazonium ion intermediate, like the starting amine, has the *R* configuration. Hence, inversion of stereochemical configuration occurs in the reaction with water.

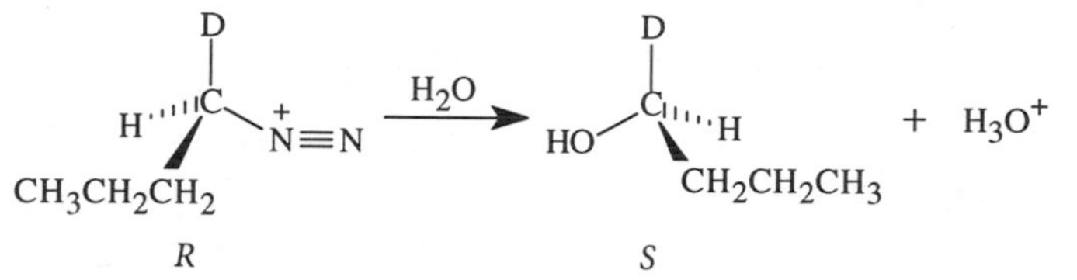

D, H, $CH_3CH_2CH_2$, C, $\overset{+}{N}{\equiv}N$ (*R*) $\xrightarrow{H_2O}$ D, HO, C, H, $CH_2CH_2CH_3$ (*S*) + H_3O^+

(b) Because inversion of stereochemical configuration is observed, an S_N2 reaction in which water is the nucleophile and N_2 is the leaving group is likely to be involved. An S_N1 mechanism involving a carbocation intermediate would be expected to result in some racemization. (Why?)

23.28 *p*-Aminobenzenesulfonic acid is prepared and diazotized to give the diazonium ion *A*. This synthesis begins with *p*-acetamidobenzenesulfonyl chloride, prepared as shown in the solution to Problem 23.24 on p. 895 of this manual. Formaldehyde is reductively aminated with aniline to give *N,N*-dimethylaniline, *B*. Compounds *A* and *B* are coupled to give methyl orange.

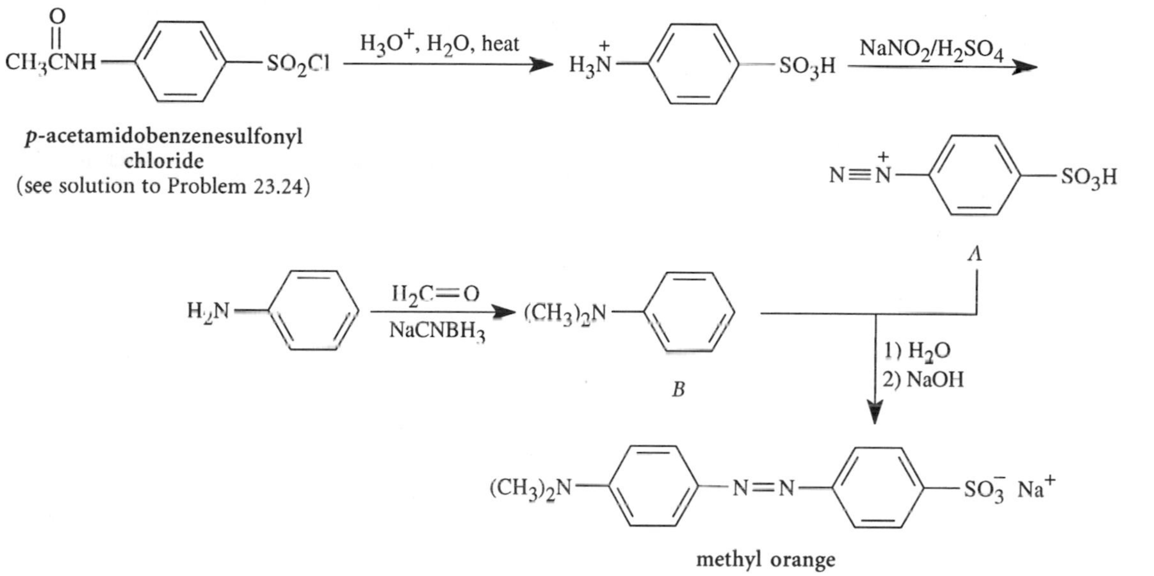

$CH_3\overset{O}{\overset{\|}{C}}NH-C_6H_4-SO_2Cl$ $\xrightarrow{H_3O^+,\ H_2O,\ heat}$ $H_3\overset{+}{N}-C_6H_4-SO_3H$ $\xrightarrow{NaNO_2/H_2SO_4}$ $N{\equiv}\overset{+}{N}-C_6H_4-SO_3H$ (*A*)

p-acetamidobenzenesulfonyl chloride
(see solution to Problem 23.24)

$H_2N-C_6H_5$ $\xrightarrow[NaCNBH_3]{H_2C{=}O}$ $(CH_3)_2N-C_6H_5$ (*B*)

A + *B* $\xrightarrow{1)\ H_2O;\ 2)\ NaOH}$ $(CH_3)_2N-C_6H_4-N{=}N-C_6H_4-SO_3^-\ Na^+$

methyl orange

23.29 The following compounds would react to give FD & C yellow #6:

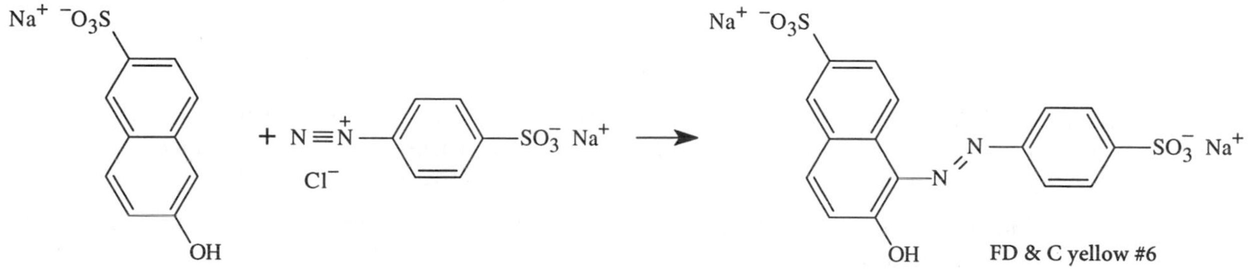

$Na^+\ {}^-O_3S$ (naphthalene) OH + $N{\equiv}\overset{+}{N}-C_6H_4-SO_3^-\ Na^+$ Cl^- ⟶ $Na^+\ {}^-O_3S$ (naphthalene) OH, $-N{=}N-C_6H_4-SO_3^-\ Na^+$

FD & C yellow #6

As you will learn in Chapter 24, the position next to the ring junction (called the α-position) in a naphthalene ring is more reactive in electrophilic substitution than the ring position one carbon removed. Hence, of the two positions "ortho" to the phenolic —OH group, the α-position is the more reactive and gives the observed product.

23.30 (a) Protonation of the —OH oxygen and loss of water give the nitrosyl cation. (The chloride counter-ion is not shown.)

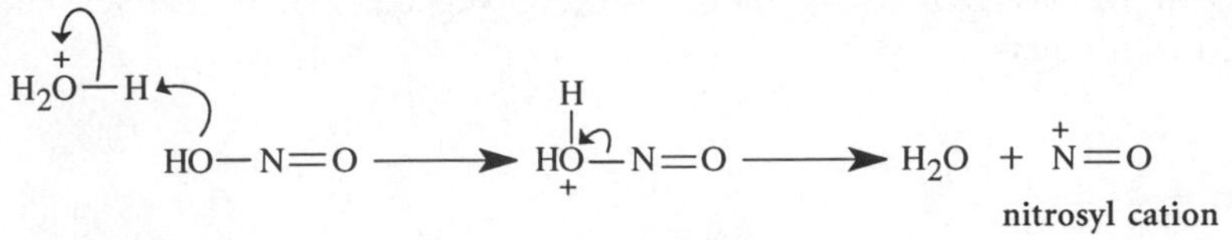

$$H_2\overset{+}{O}-H \quad HO-N=O \longrightarrow H\overset{+}{O}(H)-N=O \longrightarrow H_2O + \overset{+}{N}=O$$

nitrosyl cation

(b) The curved-arrow mechanism:

$$O=\overset{+}{N} + C_6H_5-N(CH_3)_2 \longrightarrow \text{resonance-stabilized carbocation intermediate} \xrightarrow{OH_2}$$

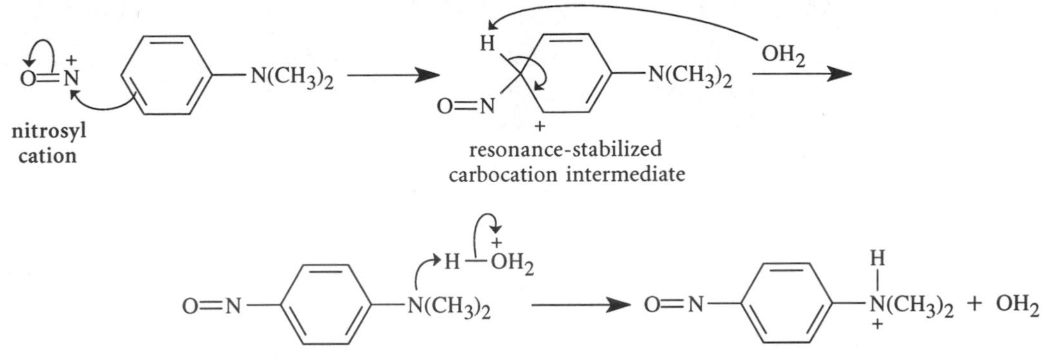

nitrosyl cation

resonance-stabilized carbocation intermediate

$$O=N-C_6H_4-N(CH_3)_2 + H-\overset{+}{O}H_2 \longrightarrow O=N-C_6H_4-\overset{+}{N}H(CH_3)_2 + OH_2$$

23.31 (a)

$$\text{phenol} \xrightarrow[\text{(Eq. 18.75, text p. 817)}]{HNO_3} O_2N-C_6H_4-OH \xrightarrow[\text{2) } CH_3I]{\text{1) } K_2CO_3} O_2N-C_6H_4-OCH_3 \xrightarrow{H_2,\ Pd/C}$$

$$H_2N-C_6H_4-OCH_3 \xrightarrow[\text{2) KI}]{\text{1) } NaNO_2/H_2SO_4} I-C_6H_4-OCH_3$$

p-iodoanisole

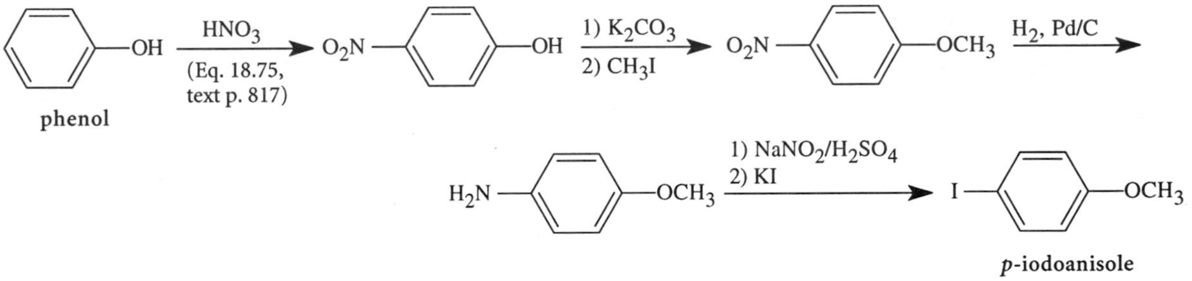

Note that ether formation is carried out before reduction of the nitro group because otherwise the amino group could also be alkylated with methyl iodide.

(b)

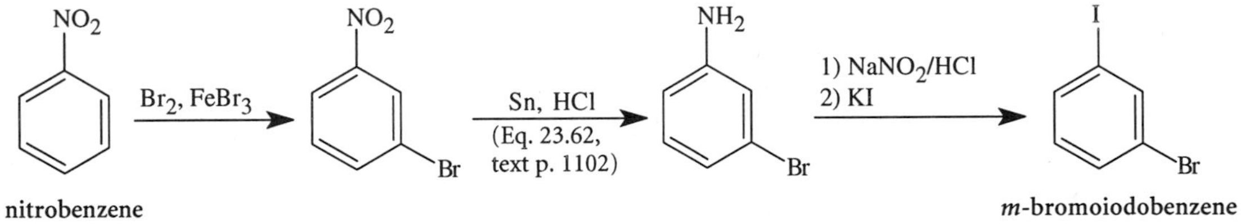

$$\text{nitrobenzene} \xrightarrow{Br_2,\ FeBr_3} m\text{-}BrC_6H_4NO_2 \xrightarrow[\text{(Eq. 23.62, text p. 1102)}]{Sn,\ HCl} m\text{-}BrC_6H_4NH_2 \xrightarrow[\text{2) KI}]{\text{1) } NaNO_2/HCl} m\text{-}BrC_6H_4I$$

nitrobenzene

m-bromoiodobenzene

23.32 (a) *Tert*-butylamine cannot be prepared by the Gabriel synthesis because it would require that the nitrogen of phthalimide be alkylated with *tert*-butyl bromide in an S_N2 reaction. Tertiary alkyl halides generally do not undergo S_N2 reactions but, under basic conditions, undergo E2 reactions instead.

(b) The Hofmann or Curtius rearrangement would be an ideal way to prepare *tert*-butylamine. (The Hofmann rearrangement is illustrated below.)

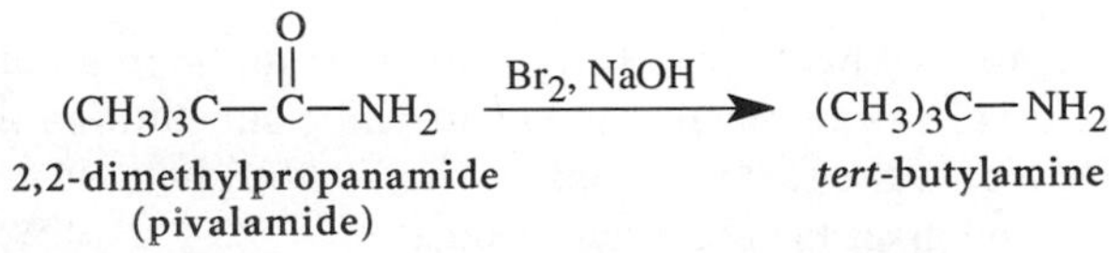

$$(CH_3)_3C-\overset{O}{\overset{\|}{C}}-NH_2 \xrightarrow{Br_2,\ NaOH} (CH_3)_3C-NH_2$$

2,2-dimethylpropanamide (pivalamide)

tert-butylamine

23.33 (a) The curved-arrow mechanism for the conversion of ethyl isocyanate into ethyl *N*-ethylcarbamate:

$$\text{Et—N=C=O} \longrightarrow \left[\text{Et—N=C—O}^- \longleftrightarrow \text{Et—}\overset{-}{\text{N}}\text{—C=O}\right] \longrightarrow \text{Et—NH—C(=O)—OEt}$$

ethyl isocyanate + HO—Et; intermediates bearing H—$\overset{+}{\text{O}}$Et on carbon; product: ethyl *N*-ethylcarbamate

(b) The curved-arrow mechanism for the reaction of ethylamine with ethyl isocyanate to give *N,N*′-diethylurea, is as follows.

$$\text{Et—N=C=O} + \text{H}_2\text{N—Et} \longrightarrow \left[\text{Et—N=C(—}\overset{+}{\text{N}}\text{H}_2\text{Et)—O}^- \longleftrightarrow \text{Et—}\overset{-}{\text{N}}\text{—C(=O)—}\overset{+}{\text{N}}\text{H}_2\text{Et}\right] \longrightarrow \text{Et—NH—C(=O)—HNEt}$$

N,N-diethylurea

Note that the proton-transfer steps, shown in the forgoing reactions for brevity as intramolecular processes, probably are intermolecular. For example, in part (b), an ethylamine molecule takes the proton and an ethylammonium ion delivers it. (What would the base and acid be in part (a)?) Of course, intramolecular proton transfers are possible, but they tend to take place fastest through 6, 7, and 8-membered rings. Nevertheless, we can abbreviate the more complex intermolecular processes as intramolecular processes, and we'll sometimes follow this practice for convenience, and you can, too.

23.34 (a) In ethanol solvent, ethanol (or its conjugate base ethoxide) is the most abundant nucleophile present, and thus an ethyl carbamate is obtained as the product.

$$\underset{\text{2-methylpropanamide}}{\text{CH}_3\text{CH(CH}_3\text{)—C(=O)—NH}_2} \xrightarrow{\text{Br}_2,\ \text{NaOEt}} \underset{\text{ethyl } N\text{-isopropylcarbamate}}{\text{CH}_3\text{CH(CH}_3\text{)—NH—C(=O)—OEt}}$$

(b) In aqueous NaOH, hydroxide is the nucleophile that attacks the isocyanate, and the resulting carbamic acid (or its salt) decarboxylates to give the corresponding amine.

$$\underset{\text{2-methylpropanamide}}{\text{CH}_3\text{CH(CH}_3\text{)—C(=O)—NH}_2} \xrightarrow{\text{Br}_2,\ \text{NaOH}} \underset{\text{isopropylamine}}{\text{CH}_3\text{CH(CH}_3\text{)—NH}_2}$$

23.35 Pentyl isocyanate is an intermediate in the reaction. Because it is largely insoluble in water, it forms a separate phase. A small amount of pentyl isocyanate reacts with the aqueous NaOH to form pentylamine, which is then extracted into the isocyanate, with which it reacts to give the urea.

$$\underset{\text{hexanamide}}{\text{CH}_3(\text{CH}_2)_4\text{C(=O)NH}_2} \xrightarrow[\text{H}_2\text{O}]{\text{Br}_2,\ \text{NaOH}} \text{CH}_3(\text{CH}_2)_4\text{N=C=O} \xrightarrow{\text{H}_2\text{O},\ \text{NaOH}} \underset{A}{\text{CH}_3(\text{CH}_2)_4\text{NH}_2} + \text{CO}_2$$

$$\downarrow$$

$$\underset{\substack{N,N'\text{-dipentylurea} \\ B}}{\text{CH}_3(\text{CH}_2)_4\text{NHC(=O)NH(CH}_2)_4\text{CH}_3}$$

23.36 (a)

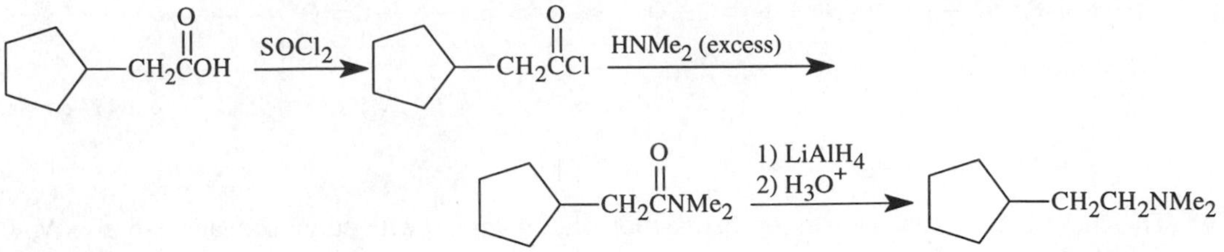

$$\text{C}_5\text{H}_9\text{-CH}_2\overset{\text{O}}{\overset{\|}{\text{C}}}\text{OH} \xrightarrow{\text{SOCl}_2} \text{C}_5\text{H}_9\text{-CH}_2\overset{\text{O}}{\overset{\|}{\text{C}}}\text{Cl} \xrightarrow{\text{HNMe}_2\text{ (excess)}}$$

$$\text{C}_5\text{H}_9\text{-CH}_2\overset{\text{O}}{\overset{\|}{\text{C}}}\text{NMe}_2 \xrightarrow[\text{2) H}_3\text{O}^+]{\text{1) LiAlH}_4} \text{C}_5\text{H}_9\text{-CH}_2\text{CH}_2\text{NMe}_2$$

2-cyclopentyl-*N,N*-dimethylethanamine

(b)

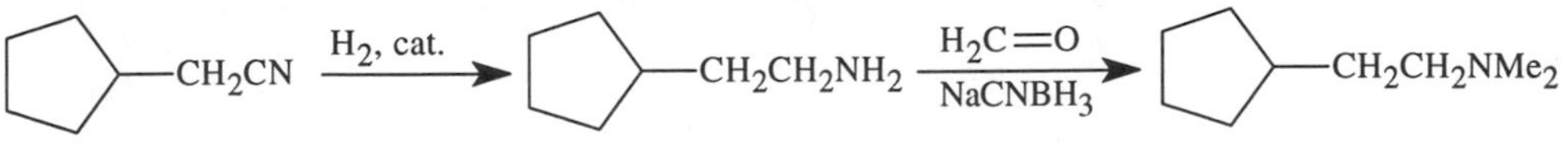

$$\text{C}_5\text{H}_9\text{-CH}_2\text{CN} \xrightarrow{\text{H}_2\text{, cat.}} \text{C}_5\text{H}_9\text{-CH}_2\text{CH}_2\text{NH}_2 \xrightarrow[\text{NaCNBH}_3]{\text{H}_2\text{C=O}} \text{C}_5\text{H}_9\text{-CH}_2\text{CH}_2\text{NMe}_2$$

(c)

$$\text{C}_5\text{H}_9\text{-CH}_2\text{CH}_2\overset{\text{O}}{\overset{\|}{\text{C}}}\text{OH} \xrightarrow{\text{SOCl}_2} \text{C}_5\text{H}_9\text{-CH}_2\text{CH}_2\overset{\text{O}}{\overset{\|}{\text{C}}}\text{Cl} \xrightarrow{\text{NaN}_3} \text{C}_5\text{H}_9\text{-CH}_2\text{CH}_2\overset{\text{O}}{\overset{\|}{\text{C}}}\text{N}_3 \xrightarrow{\text{heat}}$$

$$\text{C}_5\text{H}_9\text{-CH}_2\text{CH}_2\text{N=C=O} \xrightarrow{\text{H}_2\text{O, H}_3\text{O}^+} \text{C}_5\text{H}_9\text{-CH}_2\text{CH}_2\text{NH}_2 \xrightarrow[\text{NaCNBH}_3]{\text{H}_2\text{C=O}} \text{C}_5\text{H}_9\text{-CH}_2\text{CH}_2\text{NMe}_2$$

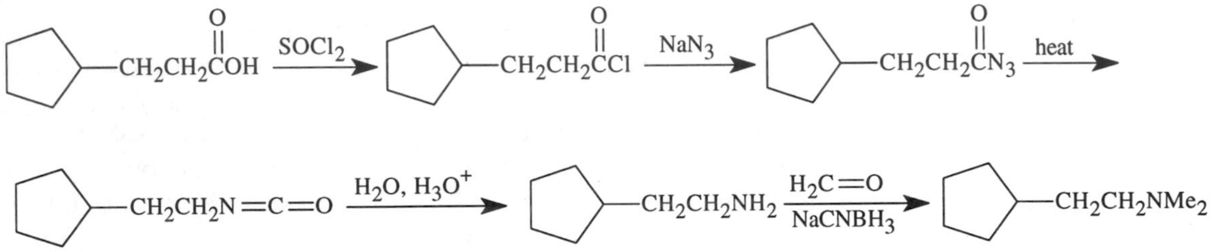

(d) One synthesis involves a reductive amination of the aldehyde:

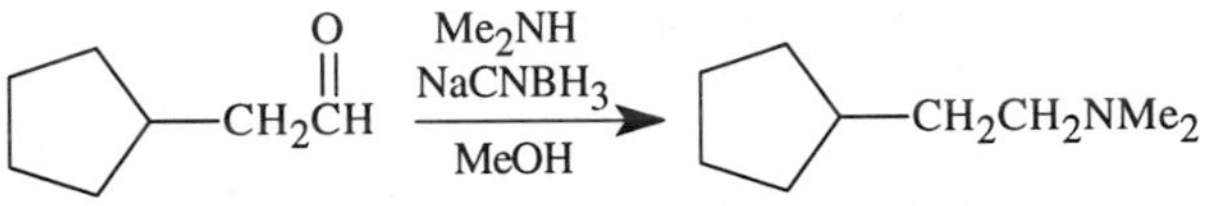

$$\text{C}_5\text{H}_9\text{-CH}_2\overset{\text{O}}{\overset{\|}{\text{C}}}\text{H} \xrightarrow[\text{MeOH}]{\text{Me}_2\text{NH, NaCNBH}_3} \text{C}_5\text{H}_9\text{-CH}_2\text{CH}_2\text{NMe}_2$$

A second synthesis involves oxidation of the aldehyde to the carboxylic acid starting material of part (a) with aqueous H_2CrO_4, $KMnO_4$, or other common oxidant, and then proceeding as in part (a).

23.37 (a) In morphine, the nitrogen of the tertiary amine is the most basic atom. Therefore the conjugate acid of morphine has the following structure:

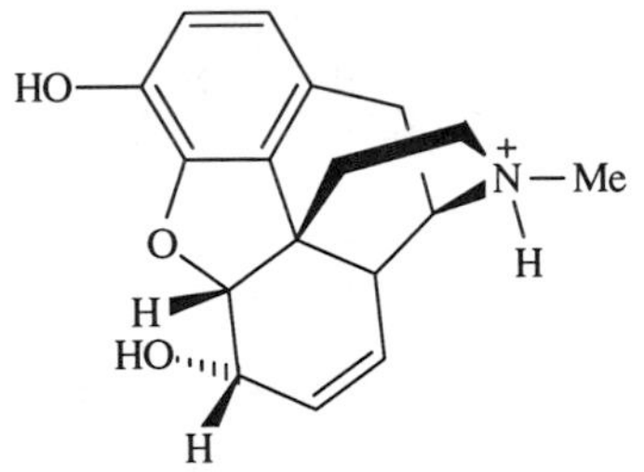

HO, O, H, HO, H, $\overset{+}{\text{N}}$—Me, H

conjugate acid of morphine

(b) The most basic atom in mescaline is the amine nitrogen. Thus, the conjugate acid of mescaline has the following structure:

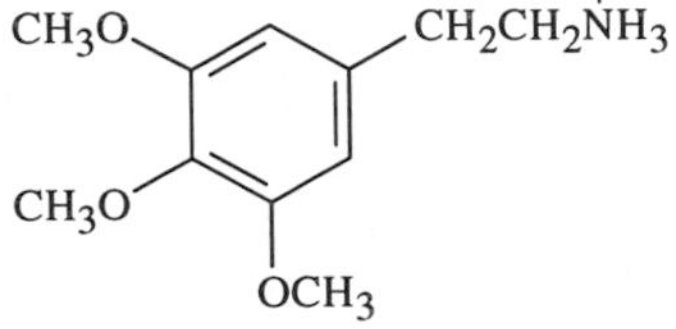

CH_3O, CH_3O, OCH_3, $CH_2CH_2\overset{+}{N}H_3$

conjugate acid of mescaline

Solutions to Additional Problems

23.38 The structure of *p*-chloroaniline is

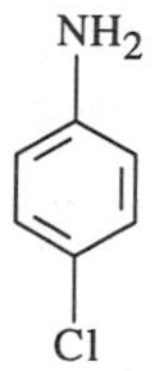

NH_2–C_6H_4–Cl

p-chloroaniline

(a) $\overset{+}{N}H_3\ Br^-$ on p-chlorophenyl
(b) $\overset{-}{N}H\ {}^{+}MgBr$ on p-chlorophenyl + CH_3CH_3
(c) $\overset{+}{N}{\equiv}N\ Cl^-$ on p-chlorophenyl
(d) p-Cl–C_6H_4–NH–SO_2–C_6H_4–CH_3
(e) OH on p-chlorophenyl
(f) Br on p-chlorophenyl
(g) chlorobenzene
(h) CN on p-chlorophenyl
(i) Na^+ p-Cl–C_6H_4–$\overset{-}{N}$–SO_2–C_6H_4–CH_3

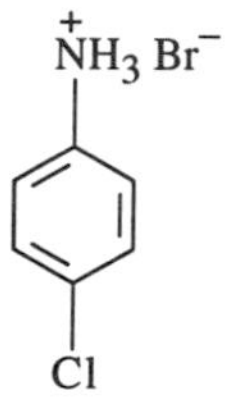

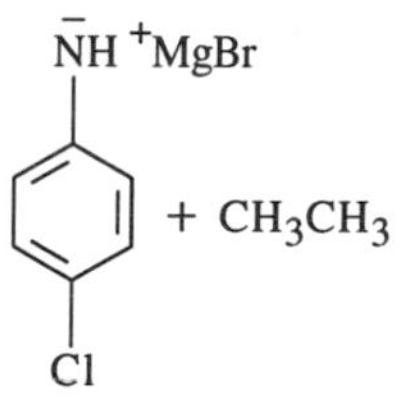

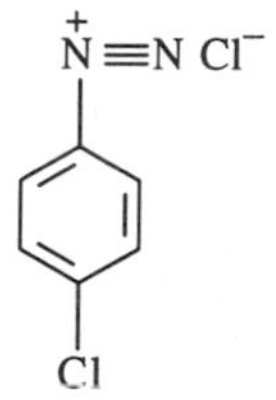

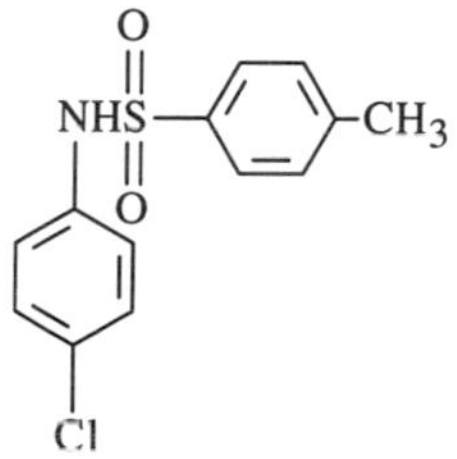

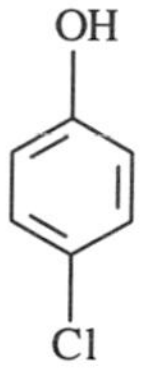

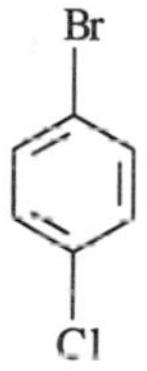

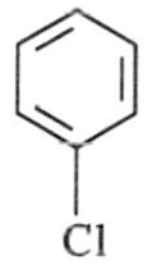

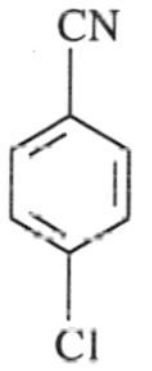

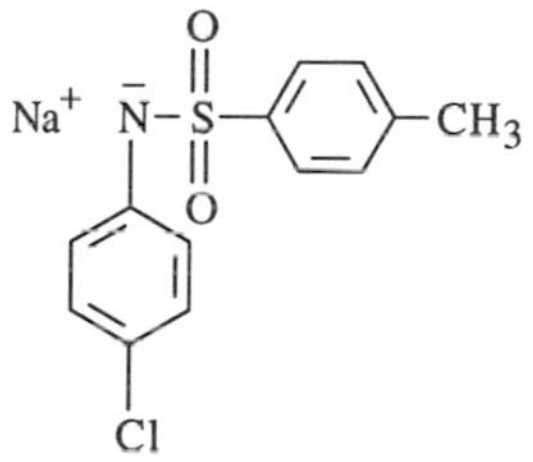

Sulfonamides are acidic enough to be ionized by NaOH. How would you know this? An amide has a pK_a of about 16. We know that sulfonic acids are stronger than carboxylic acids by about 6 pK_a units; assuming the same difference hold for sulfonamides versus carboxamides, then a sulfonamide should have an N—H pK_a of about 16–6 = 10. This is about as acidic as a phenol, and acidic enough to be fully ionized by NaOH (which has a conjugate-acid pK_a = 15.7). Furthermore, the conjugate-base anion of this particular sulfonamide is resonance-stabilized by the phenyl group. We know that phenols are more acidic than alcohols by about 5 pK_a units; if the phenyl group affects the acidity of the sulfonamide in the same way, the pK_a of this compound could be as low as 5.

23.39 The structure of *N*-methylaniline is

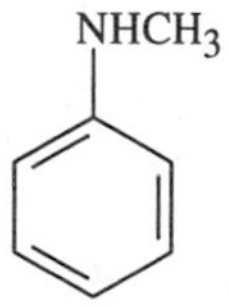

$NHCH_3$–C_6H_5

N-methylaniline

(a) 2,4,6-tribromo ring bearing $NHCH_3$ (Br, Br, Br substituents)

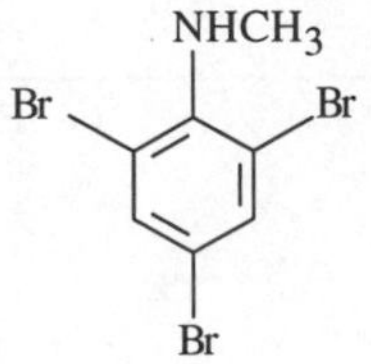

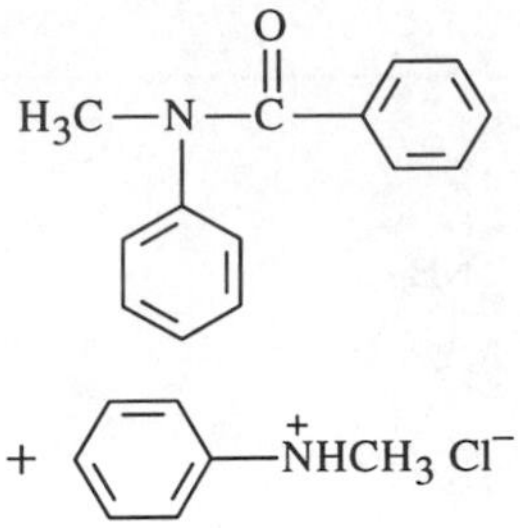

(b) $H_3C—N(C_6H_5)—\overset{O}{\overset{\|}{C}}—C_6H_5$ + $C_6H_5—\overset{+}{N}HCH_3$ Cl^-

(c) $H_3C—\overset{+}{N}(CH_2Ph)(C_6H_5)—CH_2Ph$ ^-OH

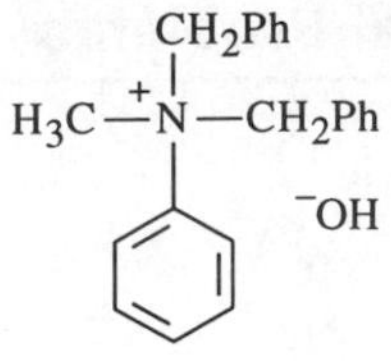

(d) $H_3C—\overset{+}{N}H_2(C_6H_5)$ $^-OS(=O)_2—C_6H_4—CH_3$

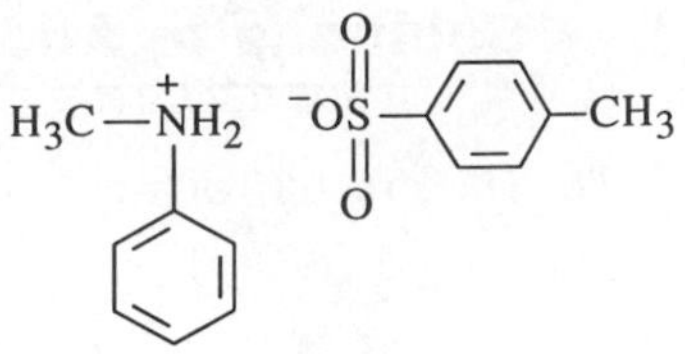

(e) $H_3C—N(C_6H_5)—N{=}O$

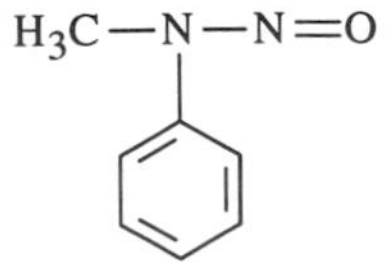

(f) $C_6H_5—\overset{+}{N}(CH_3)_3$ ^-OH

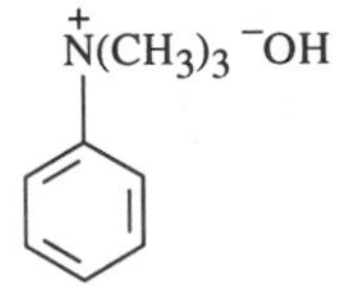

(g) $H_3C—N(C_6H_5)—CH_2CH_3$

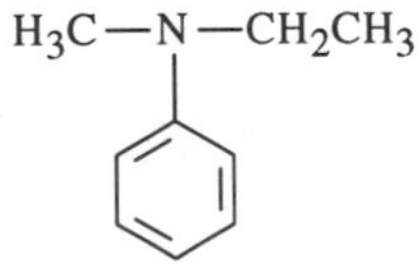

23.40 The structure of isopropylamine is $(CH_3)_2CH—NH_2$

(a) $(CH_3)_2CH\overset{+}{N}H_3$ $^-OSO_3H$

(b) no reaction

(c) $(CH_3)_2CH\overset{-}{N}H$ Li^+ + $CH_3CH_2CH_2CH_3$

(d) $(CH_3)_2CHNH\overset{O}{\overset{\|}{C}}CH_3$ + pyridinium ($C_5H_5\overset{+}{N}H$) Cl^-

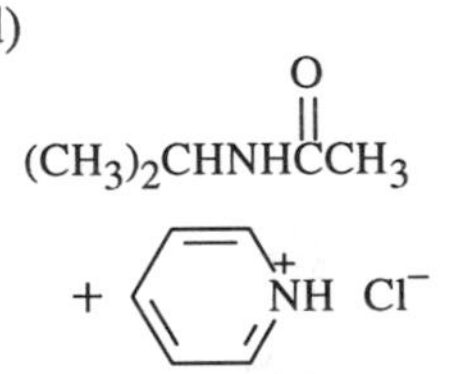

(e) $(CH_3)_2CHBr$ + $(CH_3)_2CHOH$ + $H_2C{=}CHCH_3$

(f) $(CH_3)_2CHNHCH(CH_3)_2$

(g) $(CH_3)_2CH\overset{+}{N}(CH_3)_3$ I^-

(h) $(CH_3)_2CH\overset{+}{N}H_3$ $^-O\overset{O}{\overset{\|}{C}}Ph$

(i) $(CH_3)_2CHN(CH_3)_2$

(j) $H_2C{=}CHCH_3$ + $N(CH_3)_3$

(k) $(CH_3)_2CHNHCH_2CH_3$

23.41 (a) NH_2 on cyclobutene ring — **2-cyclobutenamine** — or NH_2 on cyclopropene ring with CH_3 — **2-methyl-2-cyclopropenamine**

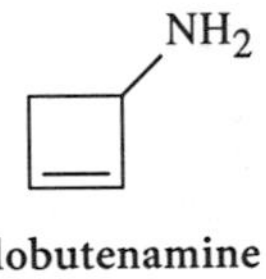

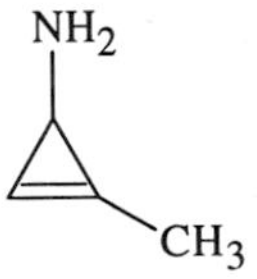

(b) $CH_3\underset{NH_2}{\underset{|}{C}}HCH_2CH_3$ — ***sec*-butylamine (2-butanamine)**

(c) $C_6H_5—NHCH(CH_3)_2$ (***N*-isopropylaniline**) + $C_6H_5—NHCH_2CH_2CH_3$ (***N*-propylaniline**)

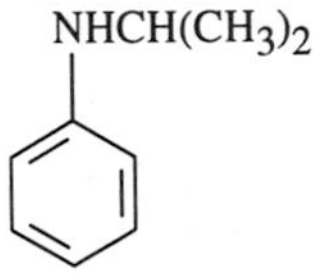

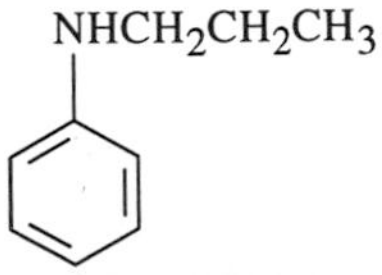

(d) The required compound could be either $(CH_3)_2C{=}NCH_3$, the imine that results from the reaction of acetone and methylamine, or $(CH_3)_2CHN{=}CH_2$, the imine that results from the reaction of isopropylamine and formaldehyde.

23.42 (a)

$CH_3CH_2CH_2CH_2CH_2\overset{O}{\overset{\|}{C}}NCH_3$ — ***N*-methylhexanamide**

$CH_3CH_2CH_2CH_2CH_2CH_2CH_2CH_2NH_2$ — **1-octanamine**

$CH_3CH_2CH_2CH_2CH_2CH_2N(CH_3)_2$ — ***N.N*-dimethyl-1-hexanamine**

N-Methylhexanamide is an amide, and it is therefore not significantly soluble in dilute HCl; the two amines are soluble as their conjugate-acid ammonium salts in dilute HCl solution. 1-Octanamine, like all primary alkylamines, gives off a gas (N_2) when diazotized with $NaNO_2$ in aqueous acid. *N,N*-Dimethyl-1-hexanamine does not give off a gas when treated with nitrous acid.

(b)

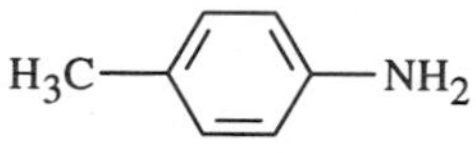

H_3C–C_6H_4–NH_2 — *p*-methylaniline (*p*-toluidine)

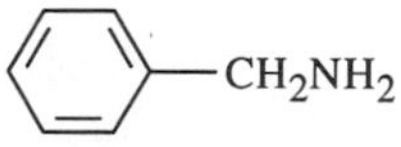

C_6H_5–CH_2NH_2 — benzylamine

H_3C–C_6H_4–OH — *p*-cresol

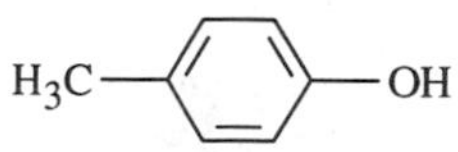

C_6H_5–OCH_3 — anisole

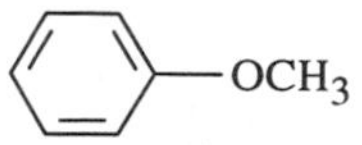

Only the two amines, *p*-methylaniline and benzylamine, are soluble (as their conjugate-acid ammonium ions) in dilute HCl. Only the phenol *p*-cresol is soluble (as its conjugate-base phenolate) in dilute NaOH solution. Anisole is soluble in neither dilute acid nor dilute base. Only benzylamine gives off a gas (N_2) when diazotized at 0° with HNO_2. *p*-Methylaniline gives off N_2 when the corresponding diazonium salt is heated.

23.43 (a) The amine nitrogen (but *not* the amide nitrogen) is the most basic atom in the molecule. Hence, labetalol hydrochloride is the ammonium ion that results from protonation of this nitrogen.

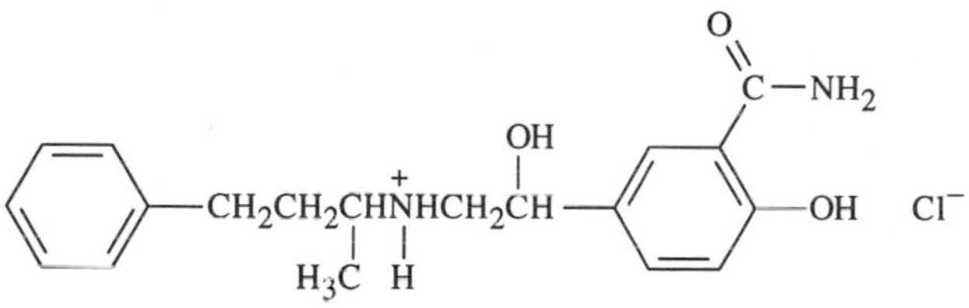

C_6H_5–$CH_2CH_2CH(CH_3)\overset{+}{N}H_2CH_2CH(OH)$–$C_6H_3$(C(=O)–$NH_2$)–OH Cl^- — labetalol hydrochloride

(b) When labetalol hydrochloride is treated with one molar equivalent of NaOH, a proton on the positively charged nitrogen is removed and neutral labetalol is formed.

(c) Hot NaOH ionizes the phenol group and also brings about hydrolysis of the amide group to a carboxylate.

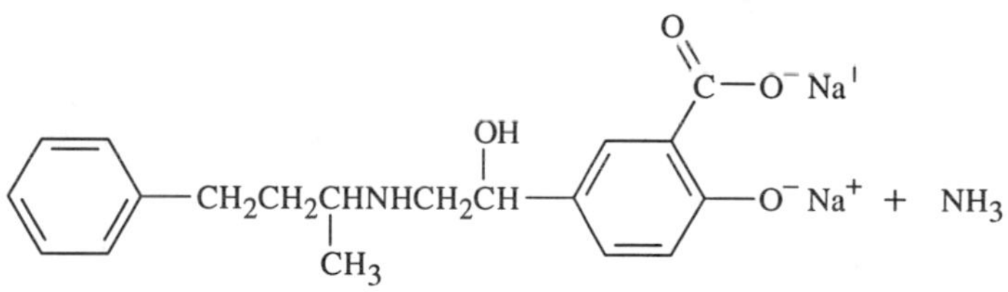

C_6H_5–$CH_2CH_2CH(CH_3)NHCH_2CH(OH)$–$C_6H_3$(C(=O)–$O^-$ Na^+)–O^- Na^+ + NH_3

(d) Hot 6 *M* aqueous HCl solution results in protonation of the amine nitrogen (as in part (a) and hydrolysis of the amide.

C_6H_5–$CH_2CH_2CH(CH_3)\overset{+}{N}H_2CH_2CH(OH)$–$C_6H_3$(C(=O)–OH)–OH + $\overset{+}{N}H_4$

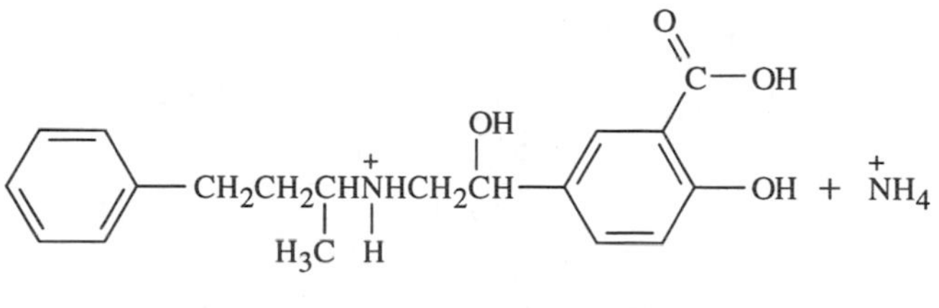

Depending on the severity of the conditions, the alcohol could dehydrate to give compound *A*, which is the conjugate acid of an enamine. This would hydrolyze to an amine and an aldehyde under the aqueous conditions.

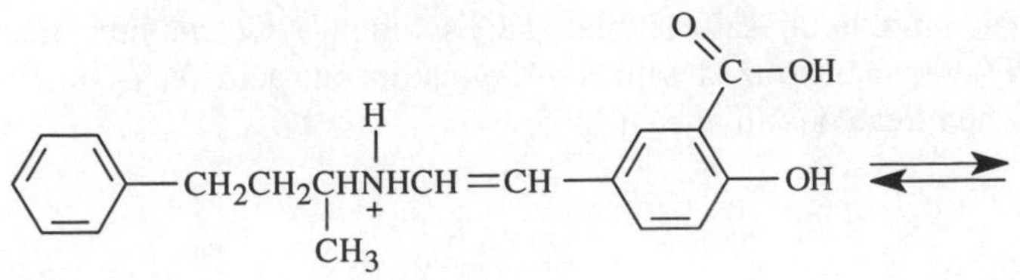

compound *A*

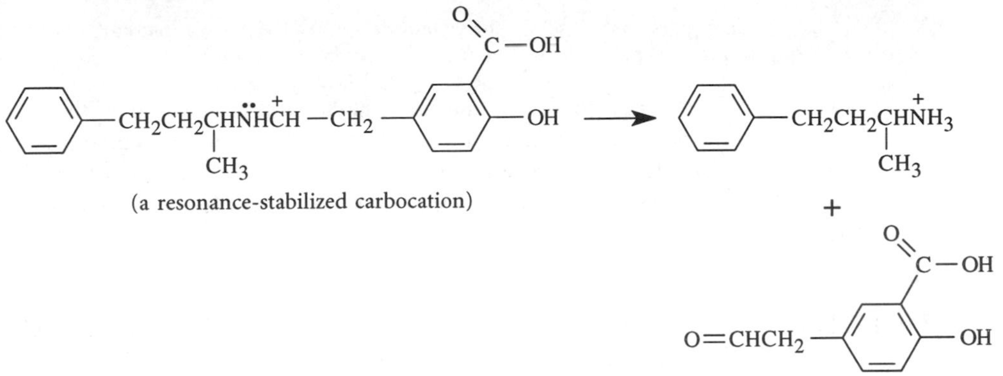

(a resonance-stabilized carbocation)

23.44 (a) In 1 *M* HCl solution, cocaine is protonated on its amino nitrogen.

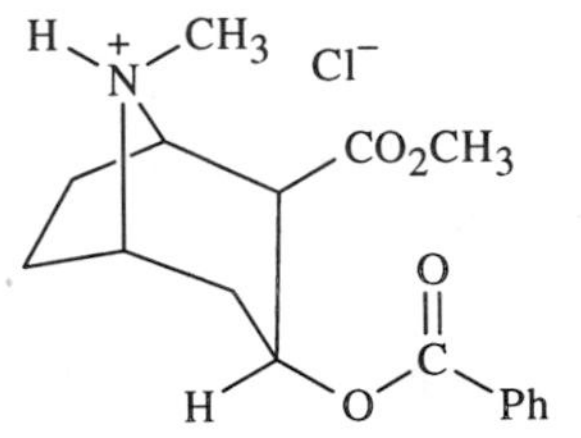

(b) Hot aqueous NaOH would bring about saponification of the ester groups:

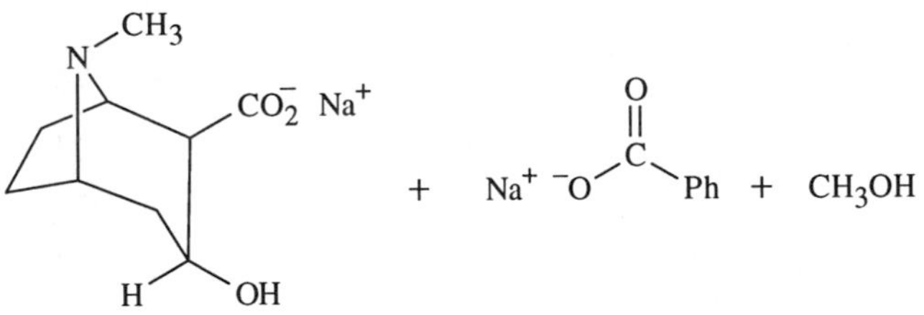

$$+ \quad Na^{+}\ {}^{-}O-C(=O)-Ph \quad + \quad CH_3OH$$

(c) Concentrated, hot aqueous HCl would protonate the amine nitrogen (as in part (a)) and would bring about hydrolysis of the two ester groups.

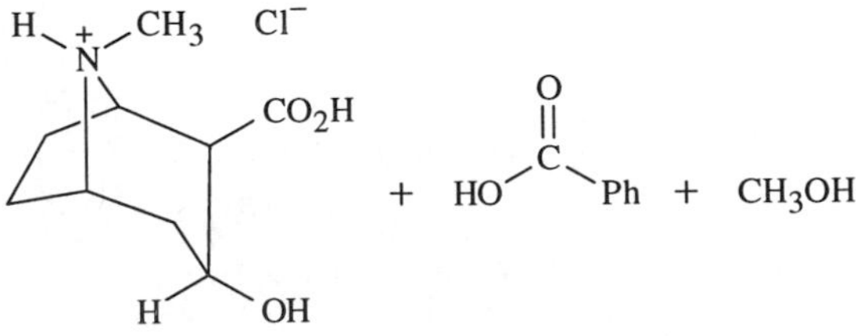

$$+ \quad HO-C(=O)-Ph \quad + \quad CH_3OH$$

23.45 Dissolve the mixture in a suitable low-boiling solvent such as ether or methylene chloride. Extract with 5% aqueous sodium bicarbonate solution. The carboxylic acid *p*-nitrobenzoic acid will dissolve in the aqueous solution as its sodium salt. Isolate this aqueous solution and acidify it with HCl to precipitate the carboxylic acid. Then extract the remaining organic solution with 5% aqueous NaOH solution. *p*-Chlorophenol will dissolve in the aqueous solution as its conjugate-base phenolate ion. Isolate this aqueous layer and acidify it to obtain the neutral phenol. Then extract the organic solution with 5% aqueous HCl solution. The amine aniline will dissolve in the aqueous layer as its conjugate-acid ammonium ion. Separate the aqueous layer and neutralize it with NaOH. Aniline will form a separate layer that can be separated, dried, and distilled. Finally, dry the remaining organic layer and remove the solvent to obtain nitrobenzene. (The acidic extraction could be carried out as the first step.)

23.46 The basicity of trifluralin is much less than that of *N,N*-diethylaniline. The electron-withdrawing polar effects of both the nitro substituents and the trifluoromethyl substituent destabilize the conjugate acid of trifluralin, and resonance interaction of the nitrogen unshared electron pair with the two nitro groups stabilizes trifluralin itself; this interaction is absent in the ammonium ion. (In fact, trifluralin is not basic enough to dissolve in dilute HCl.)

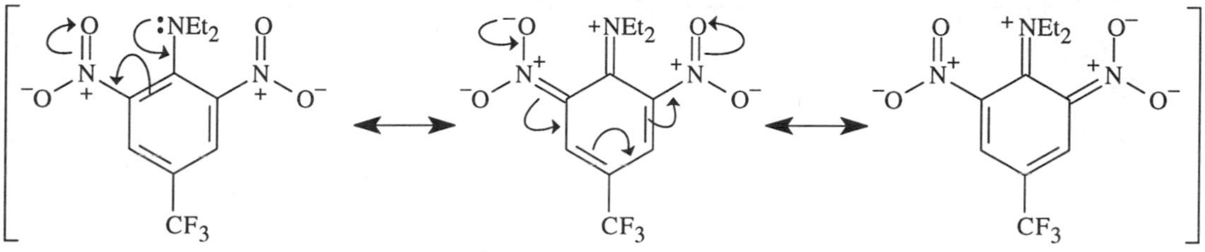

some resonance structures of trifluralin

23.47 (a) Nitrous acid forms a diazonium ion from anthranilic acid. This diazonium ion undergoes coupling with *N,N*-dimethylaniline to give methyl red.

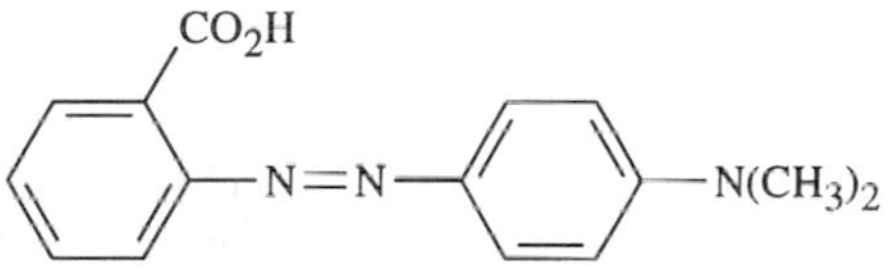

methyl red

(b) In acidic solution, methyl red is protonated to give its conjugate acid. Protonation occurs on one of the diazo nitrogens because the resulting conjugate acid is resonance-stabilized.

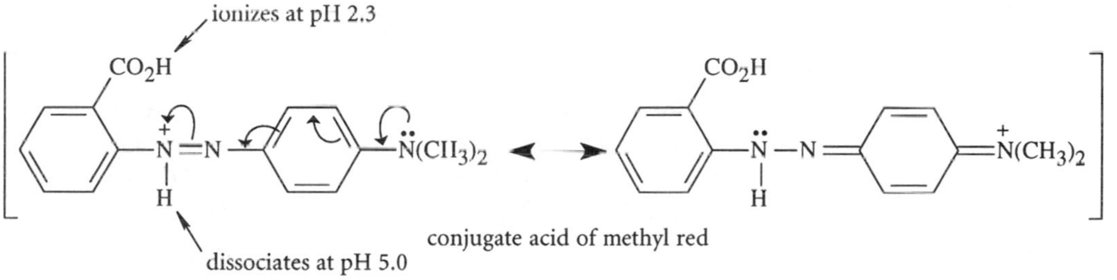

ionizes at pH 2.3

dissociates at pH 5.0

conjugate acid of methyl red

You may have noticed that the pK_a of the carboxy group is unusually low for the pK_a of a carboxylic acid. Can you explain why?

23.48 (a) First prepare *p*-nitroaniline from aniline as shown in Study Problem 23.4 on text p. 1093. Then diazotize this amine to give the diazonium salt. Couple the diazonium salt to salicylic acid to obtain alizarin yellow R.

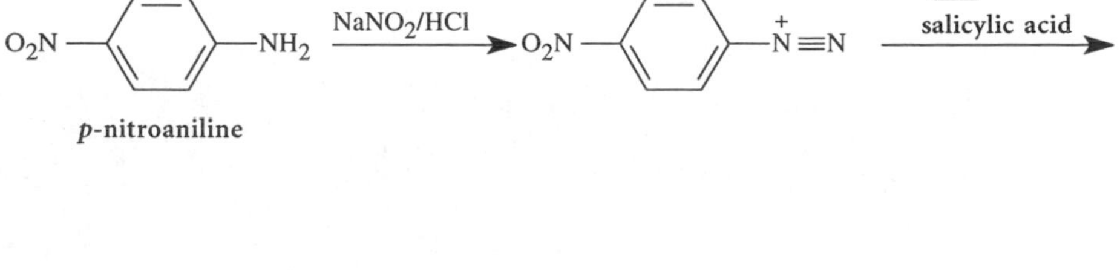

$NaNO_2/HCl$

salicylic acid

p-nitroaniline

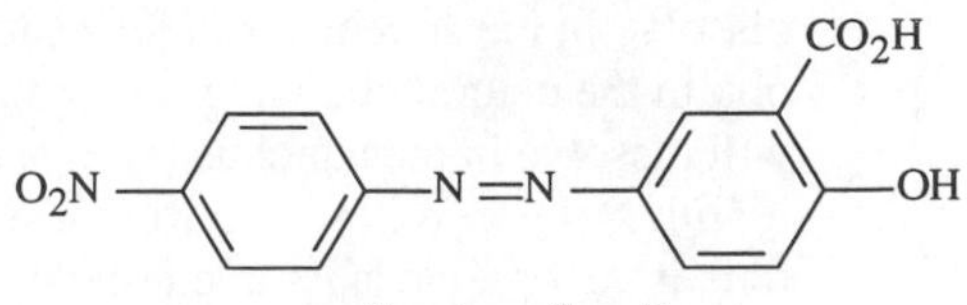

alizarin yellow R

(b) From the solution to the previous problem (23.47), or from the information in the problem, the pK_a of a protonated diazo group should be around 5. Consequently, at a pH value well above 5, the diazo nitrogens are unprotonated. The pK_a of the carboxy group should be around 4. Therefore, at pH = 9, the carboxy group should be ionized. The group that ionizes between 10 and 12 should affect the conjugated π-electron system because it affects the color. Since this group is not the diazo group, it must be the phenol. If the phenol has a pK_a of about 11, then at pH = 9 it is un-ionized.

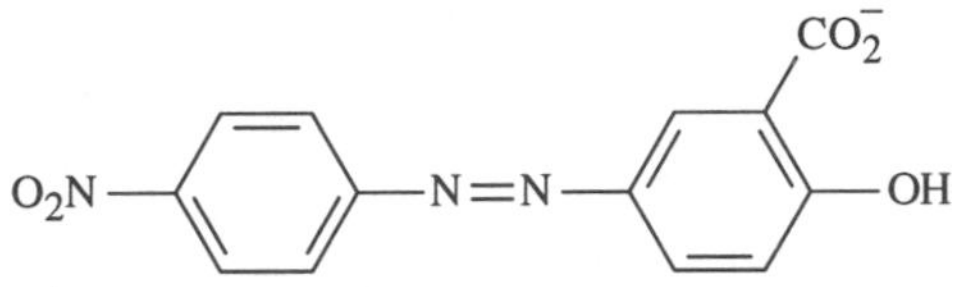

structure of alizarin yellow R at pH = 9

(c) At pH > 12, both the carboxy group and the phenol are ionized.

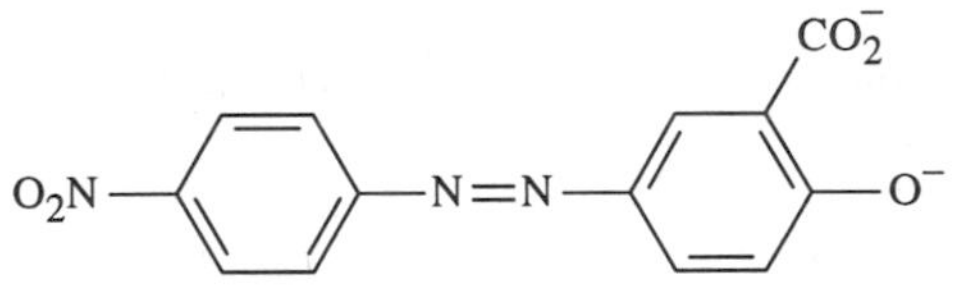

structure of alizarin yellow R at pH > 12

23.49 (a) Amanda is evidently trying to run a Friedel-Crafts acylation on aniline. This cannot work as shown for two reasons. First, the aluminum trichloride, a Lewis acid, will react with the electron pair on the amino group, thus deactivating the ring. (This is much like its effect on phenol acylation; see text p. 818.) Second, a competing reaction is acylation of the amino group to form acetanilide, a reaction that will undoubtedly be catalyzed by the aluminum trichloride.

(b) Excess methyl iodide will form the quaternary salt $(CH_3)_3C\overset{+}{N}(CH_3)_3\ I^-$; the reaction will not stop at the secondary amine shown in the problem.

(c) Amanda is attempting to nitrate the ring and is relying on the activating, ortho, para-directing effect of the dimethylamino group to direct the substitution to the para position. The fallacy here is that under the very acidic conditions the dimethylamino group is protonated, and the protonated dimethylamino group is expected to be a meta director. (See Eq. 23.42 and related discussion on text p. 1093.)

(d) Although the aldehyde will indeed be reduced, so will the nitro group. (See Eq. 23.63 on text p. 1102.)

(e) Amanda is attempting to displace water from the protonated alcohol with dimethylamine, which is a good nucleophile. Unfortunately, the acidic conditions required to convert the —OH group of the alcohol into a good leaving group also protonate the amine, which is much more basic than the alcohol. The protonated amine no longer has the unshared electron pair that is responsible for its nucleophilicity.

(f) Amanda is attempting to carry out an S_N2 reaction on a tertiary alkyl halide. Unfortunately, tertiary alkyl halides do not undergo S_N2 reactions. If a reaction occurs at all, it will be the competing E2 process.

(g) Amanda is attempting an exhaustive methylation followed by Hofmann elimination. An elimination would occur, but unfortunately, the alkene shown would not be the major product of the reaction. (Why?)

23.50 (a)

$$PhNH_2 \xrightarrow[\text{2) CuCN}]{\text{1) NaNO}_2\text{/HCl}} PhC{\equiv}N \xrightarrow{H_2,\ \text{cat.}} PhCH_2NH_2$$

benzylamine

(b)

$$\text{PhC}{\equiv}\text{N} \xrightarrow{\text{H}_3\text{O}^+,\ \text{H}_2\text{O, heat}} \text{Ph}\overset{\text{O}}{\overset{\|}{\text{C}}}\text{OH} \xrightarrow[\text{2) H}_3\text{O}^+]{\text{1) LiAlH}_4} \text{PhCH}_2\text{OH}$$

PhC≡N prepared in part (a); $PhCH_2OH$ benzyl alcohol

(c)

$$\text{PhC}{\equiv}\text{N} \xrightarrow{\text{H}_3\text{O}^+,\ \text{H}_2\text{O, heat}} \text{Ph}\overset{\text{O}}{\overset{\|}{\text{C}}}\text{OH} \xrightarrow[\text{2) H}_3\text{O}^+]{\text{1) LiAlH}_4} \text{PhCH}_2\text{OH} \xrightarrow{\text{conc. HBr}} \text{PhCH}_2\text{Br} \xrightarrow[\text{ether}]{\text{Mg}} \xrightarrow[\text{2) H}_3\text{O}^+]{\text{1) CO}_2}$$

PhC≡N prepared in part (a)

$$\text{PhCH}_2\overset{\text{O}}{\overset{\|}{\text{C}}}\text{OH} \xrightarrow{\text{SOCl}_2} \text{PhCH}_2\overset{\text{O}}{\overset{\|}{\text{C}}}\text{Cl} \xrightarrow{\text{NH}_3\text{ (excess)}} \text{PhCH}_2\overset{\text{O}}{\overset{\|}{\text{C}}}\text{NH}_2 \xrightarrow[\text{2) H}_2\text{O}]{\text{1) LiAlH}_4} \text{PhCH}_2\text{CH}_2\text{NH}_2$$

2-phenylethanamine

(d)

$$\text{PhNH}_2 + \text{CH}_3\overset{\text{O}}{\overset{\|}{\text{C}}}\text{CH}_2\text{CH}_3 \xrightarrow[\text{CH}_3\text{OH}]{\text{NaCNBH}_3} \text{PhNH}\underset{\text{CH}_3}{\underset{|}{\text{C}}}\text{HCH}_2\text{CH}_3$$

N-phenyl-2-butanamine

(e)

$$\text{C}_6\text{H}_5\text{–NH}_2 \xrightarrow{\text{Ac}_2\text{O}} \text{C}_6\text{H}_5\text{–NHAc} \xrightarrow{\text{Cl}_2,\ \text{FeCl}_3} \text{Cl–C}_6\text{H}_4\text{–NHAc} \xrightarrow[\text{2) NaOH (dilute)}]{\text{1) H}_3\text{O}^+,\ \text{H}_2\text{O, heat}}$$

$$\text{Cl–C}_6\text{H}_4\text{–NH}_2 \xrightarrow[\text{2) CuCN}]{\text{1) NaNO}_2\text{/HCl}} \text{Cl–C}_6\text{H}_4\text{–C}{\equiv}\text{N} \xrightarrow[\text{H}_2\text{O}]{\text{H}_3\text{O}^+,\ \text{heat}} \text{Cl–C}_6\text{H}_4\text{–CO}_2\text{H}$$

p-chlorobenzoic acid

23.51 The acetic anhydride treatment acetylates either the amino group or the hydroxy group. Because the initially formed product *A* can form an ethyl ether, its —OH group is not affected in the first reaction. Consequently, compound *A* is *p*-acetamidophenol and compound *B* is its ethyl ether. This is reasonable because, so long as the hydroxy group is not ionized, the amino group is the most basic group in the molecule, and is thus the more nucleophilic group. Compound *A* dissolves in base because the hydroxy proton can ionize.

$$\text{H}_2\text{N–C}_6\text{H}_4\text{–OH} \xrightarrow{\text{Ac}_2\text{O}} \text{AcHN–C}_6\text{H}_4\text{–OH} \underset{}{\overset{\text{NaOH}}{\rightleftharpoons}} \text{AcHN–C}_6\text{H}_4\text{–O}^- \xrightarrow{\text{EtI}} \text{AcHN–C}_6\text{H}_4\text{–OEt}$$

p-aminophenol

p-acetamidophenol (4-acetylaminophenol, or acetaminophen) (compound *A*)

N-(4-ethoxyphenyl)acetamide (compound *B*)

In compound *A* the acidity of the amide N—H is likely to be comparable to that of the phenol (why?). Hence, the conjugate-base anion of the phenoxide is in equilibrium with the phenoxide ion. If both ions are indeed present, evidently the conjugate base phenolate is alkylated more rapidly. Alkylation of the oxygen then pulls the equilibrium between the anions towards the phenoxide:

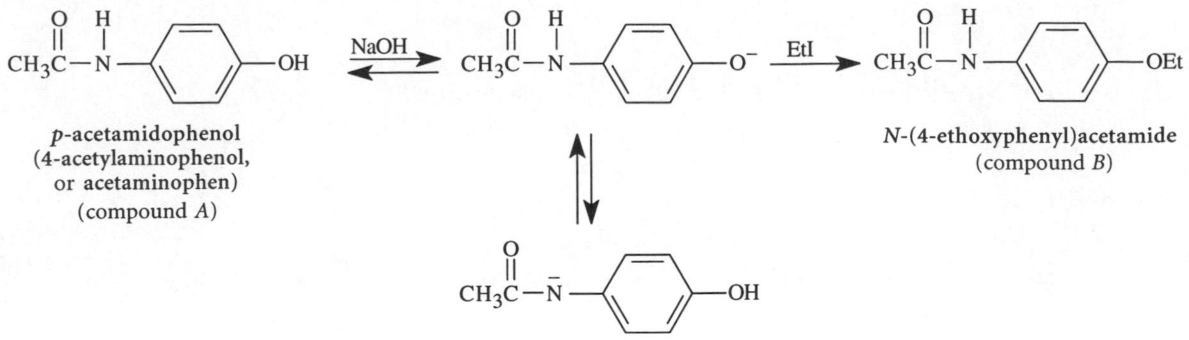

$CH_3C(=O)-N(H)-C_6H_4-OH$ $\underset{}{\overset{NaOH}{\rightleftharpoons}}$ $CH_3C(=O)-N(H)-C_6H_4-O^-$ $\xrightarrow{EtI}$ $CH_3C(=O)-N(H)-C_6H_4-OEt$

p-acetamidophenol (4-acetylaminophenol, or acetaminophen) (compound *A*)

N-(4-ethoxyphenyl)acetamide (compound *B*)

$\updownarrow$

$CH_3C(=O)-\bar{N}-C_6H_4-OH$

23.52 First identify compound *B* by completing the reaction sequence in Fig. P23.52:

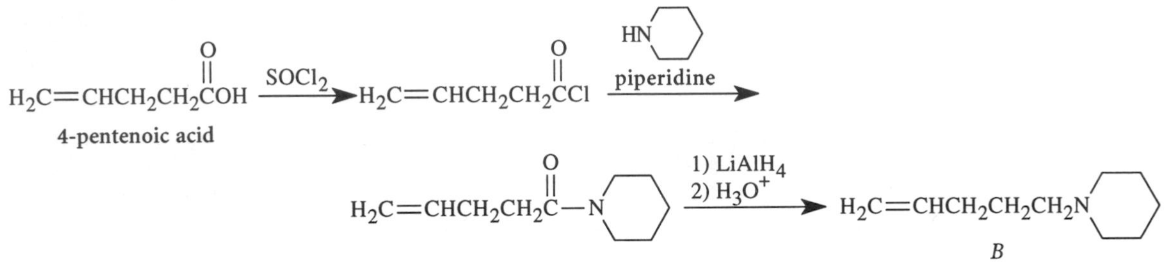

$H_2C{=}CHCH_2CH_2COH$ (4-pentenoic acid, C=O) $\xrightarrow{SOCl_2}$ $H_2C{=}CHCH_2CH_2CCl$ (C=O) $\xrightarrow{\text{HN (piperidine)}}$

$H_2C{=}CHCH_2CH_2C(=O)-N(\text{piperidine ring})$ $\xrightarrow{1)\ LiAlH_4;\ 2)\ H_3O^+}$ $H_2C{=}CHCH_2CH_2CH_2N(\text{piperidine ring})$ *B*

Compound *A*, because of its reaction with Ag^+ to give AgBr, appears to be an ionic compound. The formula of compound *B* shows that it must be derived from two molecules of 1,5-dibromopentane. Because the alkene *B* is obtained by an elimination reaction, it is reasonable to suppose that compound *A* is a quaternary ammonium salt. A reasonable structure of such a salt that could result from the reaction of two equivalents of 1,5-dibromopentane and one of ammonia is as follows. (This equation also shows the reaction of compound *A* with hydroxide ion to give compound *B*.)

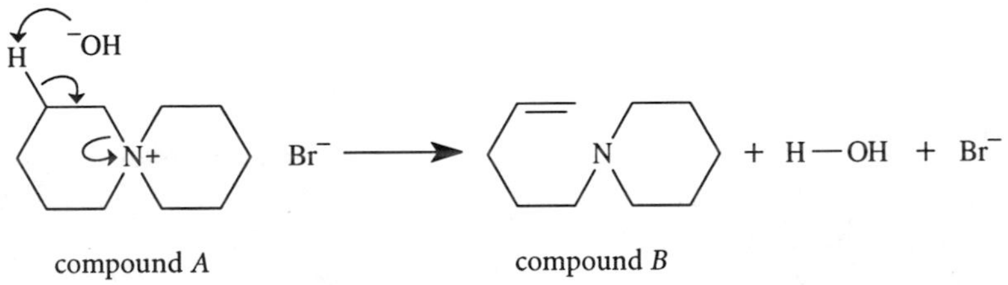

^-OH removes H; spiro N^+ (compound *A*) Br^- $\longrightarrow$ $H_2C{=}CHCH_2CH_2CH_2N$(piperidine) (compound *B*) $+$ H—OH $+$ Br^-

compound *A* compound *B*

The formation of compound *A* from 1,5-dibromopentane and ammonia thus must involve the following steps.

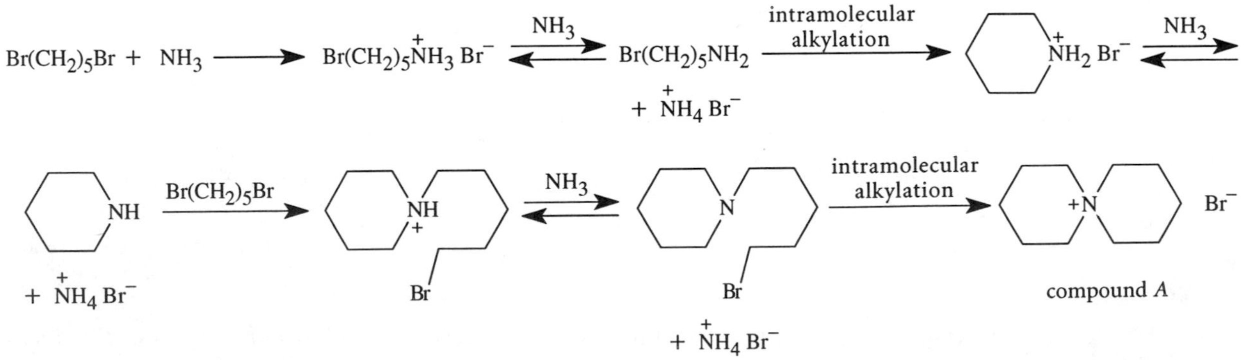

$Br(CH_2)_5Br + NH_3 \longrightarrow Br(CH_2)_5\overset{+}{N}H_3\ Br^- \underset{}{\overset{NH_3}{\rightleftharpoons}} Br(CH_2)_5NH_2 + \overset{+}{N}H_4\ Br^- \xrightarrow{\text{intramolecular alkylation}}$ (piperidinium) $\overset{+}{N}H_2\ Br^- \overset{NH_3}{\rightleftharpoons}$

(piperidine) NH $+ \overset{+}{N}H_4\ Br^-$ $\xrightarrow{Br(CH_2)_5Br}$ $\overset{+}{N}H(CH_2)_5Br$ (piperidinium) $\overset{NH_3}{\rightleftharpoons}$ N$(CH_2)_5Br$ (piperidine) $+ \overset{+}{N}H_4\ Br^-$ $\xrightarrow{\text{intramolecular alkylation}}$ spiro $+N$ Br^- (compound *A*)

23.53 (a) The barrier to internal rotation about the N-phenyl bond in *N*-methyl-*p*-nitroaniline is greater because there is more double bond character in this bond than there is in *N*-methylaniline. There is more double bond character because of the resonance interaction between the unshared electron pair on the amine nitrogen and the *p*-nitro group.

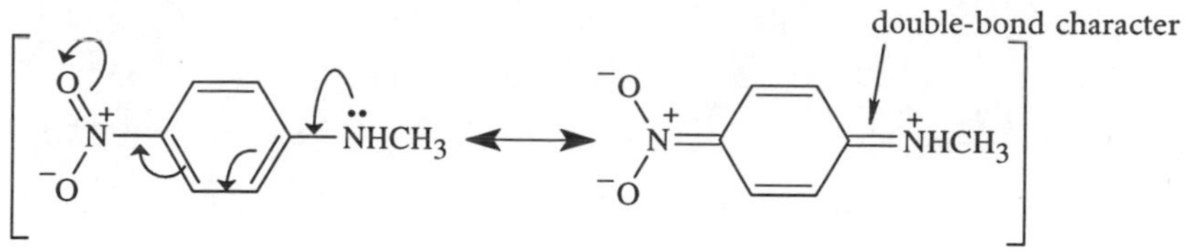

double-bond character

(b) *Cis*- and *trans*-1,3-dimethylpyrrolidine rapidly interconvert because all that is required for this process is amine inversion, which is very rapid.

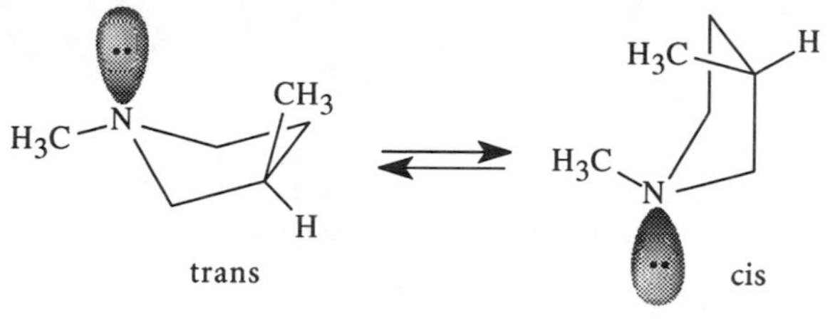

trans cis

(c) This compound is the nitrogen analog of an acetal and, by a mechanism similar to that for the hydrolysis of an acetal, decomposes to formaldehyde and two equivalents of methylamine in aqueous solution.

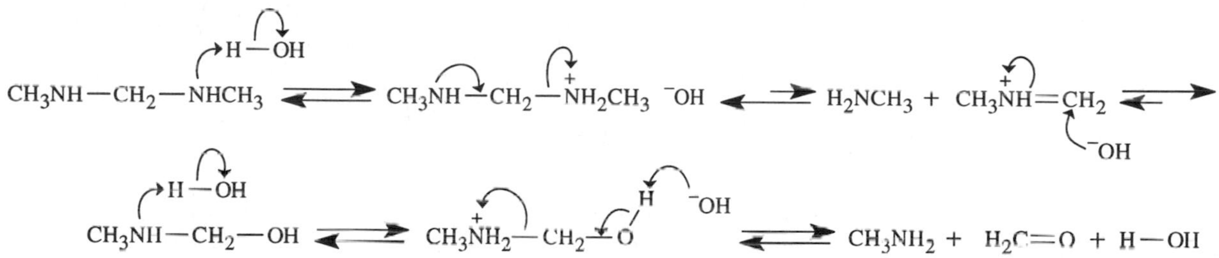

$CH_3NH—CH_2—NHCH_3 \rightleftharpoons CH_3NH—CH_2—\overset{+}{N}H_2CH_3\ \ ^-OH \rightleftharpoons H_2NCH_3 + CH_3\overset{+}{N}H{=}CH_2 \rightleftharpoons$

$CH_3NH—CH_2—OH \rightleftharpoons CH_3\overset{+}{N}H_2—CH_2—O^-\ \ ^-OH \rightleftharpoons CH_3NH_2 + H_2C{=}O + H—OH$

(d) This compound exists as an enamine for the same reason that β-diketones exist as enols: internal hydrogen bonding and conjugation stabilize the enamine form.

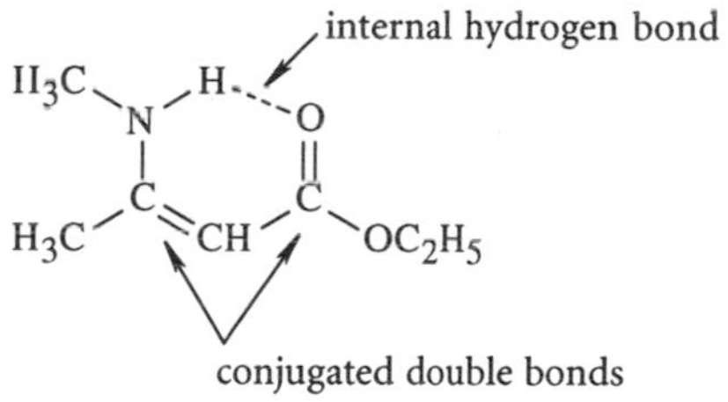

internal hydrogen bond

conjugated double bonds

(e) The diazonium salt does not decompose to a carbocation because the carbocation has a $4n$ π-electron system and is therefore antiaromatic and very unstable.

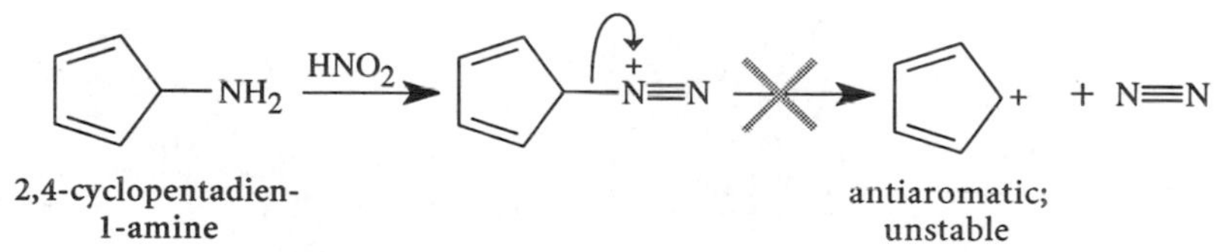

2,4-cyclopentadien-1-amine $\xrightarrow{HNO_2}$ diazonium salt $\not\rightarrow$ carbocation + $N{\equiv}N$

antiaromatic; unstable

23.54 First, resolve 2-phenylbutanoic acid into enantiomers using an enantiomerically pure amine as the resolving agent. Then allow the appropriate enantiomer of the carboxylic acid to undergo the reactions shown in each part below. The key element in all these syntheses is that the Hofmann and Curtius rearrangements take place with retention of stereochemical configuration.

(a)

$$\underset{\text{Et}}{\overset{R}{\text{PhCH}}}\overset{O}{\overset{\|}{\text{C}}}\text{OH} \xrightarrow{SOCl_2} \underset{\text{Et}}{\overset{R}{\text{PhCH}}}\overset{O}{\overset{\|}{\text{C}}}\text{Cl} \xrightarrow{NH_3} \underset{\text{Et}}{\overset{R}{\text{PhCH}}}\overset{O}{\overset{\|}{\text{C}}}\text{NH}_2 \xrightarrow[\text{MeOH}]{Br_2,\ \text{NaOMe}} \underset{\text{Et}}{\overset{R}{\text{PhCH}}}\text{NH}\overset{O}{\overset{\|}{\text{C}}}\text{OCH}_3$$

(b) In this case, esterify the *S* enantiomer of the carboxylic acid that was obtained from the enantiomeric resolution.

$$\underset{\text{Et}}{\overset{S}{\text{PhCH}}}\overset{O}{\overset{\|}{\text{C}}}\text{OH} \xrightarrow{SOCl_2} \underset{\text{Et}}{\overset{S}{\text{PhCH}}}\overset{O}{\overset{\|}{\text{C}}}\text{Cl} \xrightarrow{\text{EtOH}} \underset{\text{Et}}{\overset{S}{\text{PhCH}}}\overset{O}{\overset{\|}{\text{C}}}\text{OEt}$$

(c) In this part, the isocyanate product from the Curtius rearrangement is divided in half, and part of it is converted into an amine. This amine is then allowed to react with the remaining isocyanate.

$$\underset{\text{Et}}{\overset{R}{\text{PhCH}}}\overset{O}{\overset{\|}{\text{C}}}\text{OH} \xrightarrow{SOCl_2} \underset{\text{Et}}{\overset{R}{\text{PhCH}}}\overset{O}{\overset{\|}{\text{C}}}\text{Cl} \xrightarrow{NaN_3} \underset{\text{Et}}{\overset{R}{\text{PhCH}}}\overset{O}{\overset{\|}{\text{C}}}\text{N}_3 \xrightarrow{\text{heat}} \underset{\text{Et}}{\overset{R}{\text{PhCH}}}\text{N}{=}\text{C}{=}\text{O}$$

$$\underset{\text{Et}}{\overset{R}{\text{PhCH}}}\text{N}{=}\text{C}{=}\text{O} \xrightarrow[\text{2) NaOH (dilute)}]{\text{1) } H_3O^+,\ H_2O,\ \text{heat}} \underset{\text{Et}}{\overset{R}{\text{PhCH}}}\text{NH}_2$$

$$\underset{\text{Et}}{\overset{R}{\text{PhCH}}}\text{N}{=}\text{C}{=}\text{O} + \underset{\text{Et}}{\overset{R}{\text{PhCH}}}\text{NH}_2 \longrightarrow \underset{\text{Et}}{\overset{R}{\text{PhCH}}}\text{NH}\overset{O}{\overset{\|}{\text{C}}}\text{NH}\underset{\text{Et}}{\overset{R}{\text{CH}}}\text{Ph}$$

(d) Use the same reactions in part (c), except prepare the (*S*)-amine from the (*S*)-carboxylic acid. Then let the (*R*)-isocyanate react with the (*S*)-amine to give the meso product. (The same product would be obtained from the reaction of the (*S*)-isocyanate with the (*R*)-amine.)

23.55 In all three parts of this problem the Curtius rearrangement is used. Reasonable routes involving the Hofmann rearrangement can also be devised.

(a)

$$m\text{-}H_3C\text{-}C_6H_4\text{-}\overset{O}{\overset{\|}{\text{C}}}\text{OH} \xrightarrow{SOCl_2} m\text{-}H_3C\text{-}C_6H_4\text{-}\overset{O}{\overset{\|}{\text{C}}}\text{Cl} \xrightarrow{NaN_3} m\text{-}H_3C\text{-}C_6H_4\text{-}\overset{O}{\overset{\|}{\text{C}}}\text{N}_3 \xrightarrow{\text{heat}}$$

$$m\text{-}H_3C\text{-}C_6H_4\text{-}\text{N}{=}\text{C}{=}\text{O} \xrightarrow{CH_3OH} m\text{-}H_3C\text{-}C_6H_4\text{-}\text{NH}\overset{O}{\overset{\|}{\text{C}}}\text{OCH}_3$$

(b)

$$m\text{-}H_3C\text{-}C_6H_4\text{-}\text{N}{=}\text{C}{=}\text{O} \xrightarrow[\text{2) NaOH (dilute)}]{\text{1) } H_3O^+,\ H_2O,\ \text{heat}} m\text{-}H_3C\text{-}C_6H_4\text{-}\text{NH}_2 \xrightarrow{Ac_2O} m\text{-}H_3C\text{-}C_6H_4\text{-}\text{NHAc}$$

prepared in part (a)

Treatment of the isocyanate with acetic acid also gives the same product plus CO_2 (the *Dakin-West* reaction). Try to write a mechanism for this process.

(c)

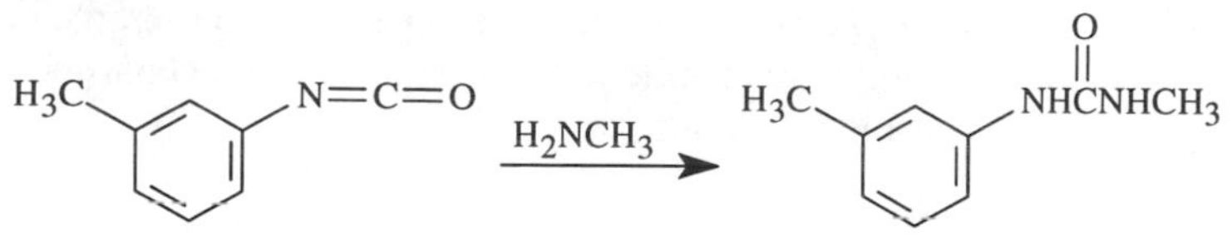

H_3C–C₆H₄–N=C=O $\xrightarrow{H_2NCH_3}$ H_3C–C₆H₄–NHCNHCH$_3$ (C=O)

prepared in part (a)

23.56 Carbaryl can be prepared by the reaction of 1-naphthol with methyl isocyanate.

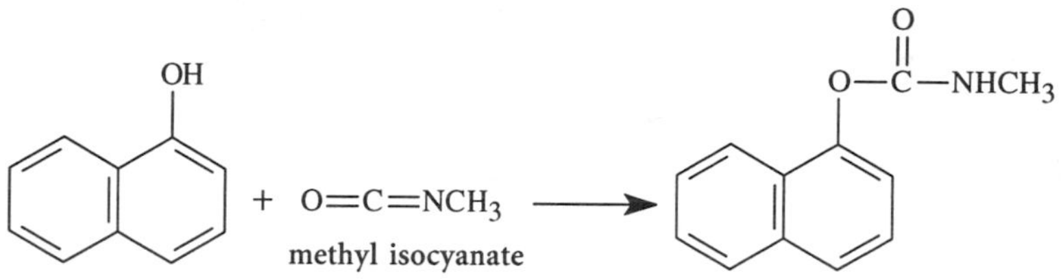

OH (1-naptho) + O=C=NCH$_3$ ⟶ O–C(=O)–NHCH$_3$

methyl isocyanate

1-napthol

carbaryl

23.57 Compound *A* appears to be an *N*-benzoyl amide of compound *B* because it is hydrolyzed to benzoic acid, and because it is re-formed when *B* is treated with benzoyl chloride. Compound *A* cannot be an ester because it contains only one oxygen. Compound *B* is therefore an achiral primary or secondary amine. It cannot be a primary amine because it does not liberate nitrogen gas after treatment with nitrous acid; thus, it must be secondary because a tertiary amine cannot form an amide. The formula of *B* is obtained by subtracting the formula of a benzoyl group (Ph—CO—, C_7H_5O) from the formula of *A* and adding a hydrogen. Thus, the formula of *B* is $C_{15}H_{23}N$. Compound *B* therefore contains five degrees of unsaturation. One degree of unsaturation is accounted for by a cyclohexane ring, reflected in the degradation product *D*, and the remaining four by a phenyl ring, reflected in the degradation product styrene, which is formed along with product *C*. There are only two structures for compound *B*, shown below as *B1* and *B2*, that would give the indicated degradation products:

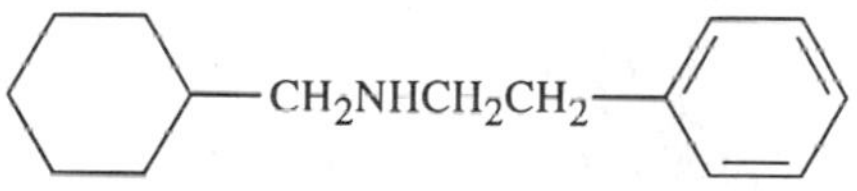

CH$_2$NHCH$_2$CH$_2$–

B1

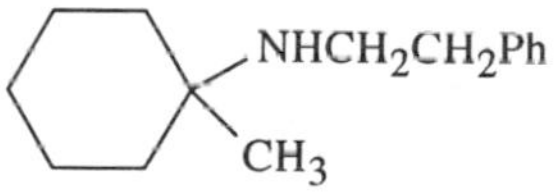

NHCH$_2$CH$_2$Ph

CH$_3$

B2

Although both are reasonable answers, structure *B1* is better because it would lead to *one* alkene *D*, whereas *B2* would lead to *two* different alkenes of which *D* is the major one. Taking *B1* as the correct structure for *B*, the identities of the unknown compounds are as follows:

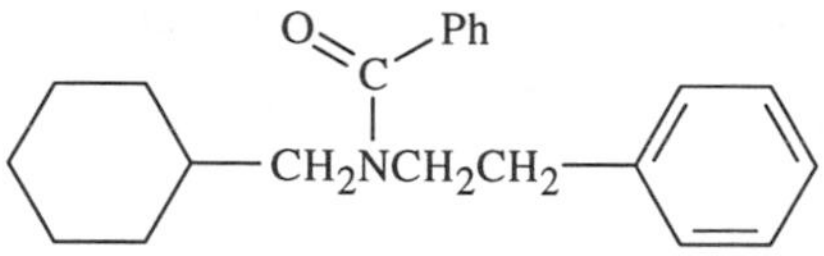

O=C–Ph

CH$_2$NCH$_2$CH$_2$–

A

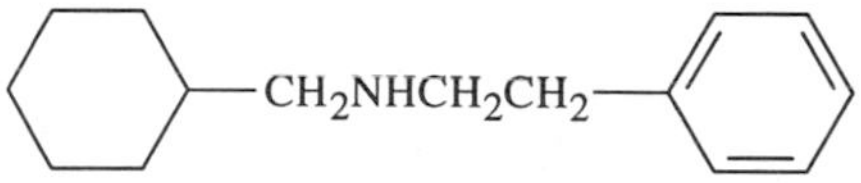

CH$_2$NHCH$_2$CH$_2$–

B

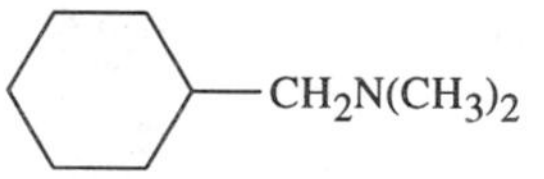

CH$_2$N(CH$_3$)$_2$

C

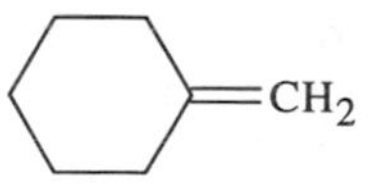

=CH$_2$

D

23.58 The oxidation data reveal that compound *A* contains a monosubstituted benzene ring, compounds *B* and *C* contain *para*-disubstituted benzene rings, and compound *C* additionally has an amino group directly attached to the ring. The pK_a of compound *C* additionally confirms that it is an aniline derivative, and the fact that compound *C*, following diazotization, gives off a gas (N_2) on warming also indicates that it is an aniline derivative. Compound *A* is chiral, and furthermore, it must be the racemate, because it evidently forms diastereomeric salts with (+)-tartaric acid. Because six carbons are involved in a benzene ring, the problem is to determine how the remaining two carbons are arranged so that the resulting structures are consistent with the data. The following structures are the only reasonable possibilities:

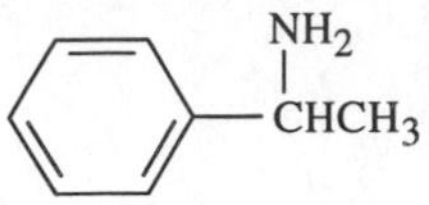

A: C_6H_5–CH(NH_2)CH_3 *B*: H_3C–C_6H_4–CH_2NH_2 *C*: CH_3CH_2–C_6H_4–NH_2

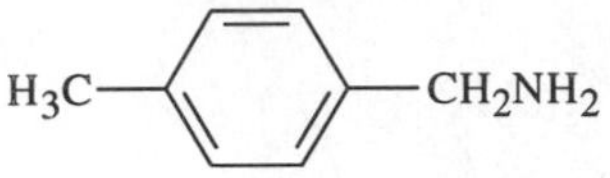

23.59 (a) Since diazotization of R—NH_2 gives N_2 and products derived from R^+, then diazotization of H—NH_2 should give H^+, that is, H_3O^+, and N_2.

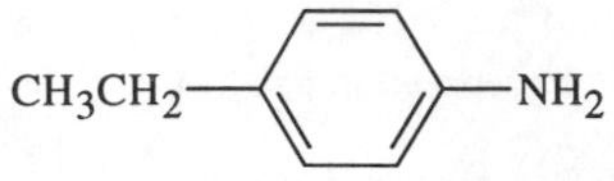

(b) Exhaustive methylation is followed by Hofmann elimination in which a proton is removed from the β-carbon with the fewest branches.

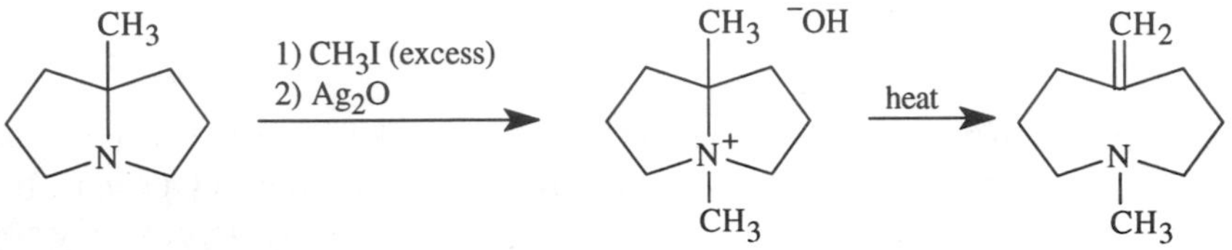

1) CH_3I (excess) 2) Ag_2O; heat

CH_3 ^-OH N^+ CH_3 → CH_2 N CH_3

(c) The triamide resulting from treatment of the acid chloride is reduced with $LiAlH_4$ to give triamine *A*, which is exhaustively methylated and then subjected to Hofmann elimination to give the product, which is an isomer of benzene.

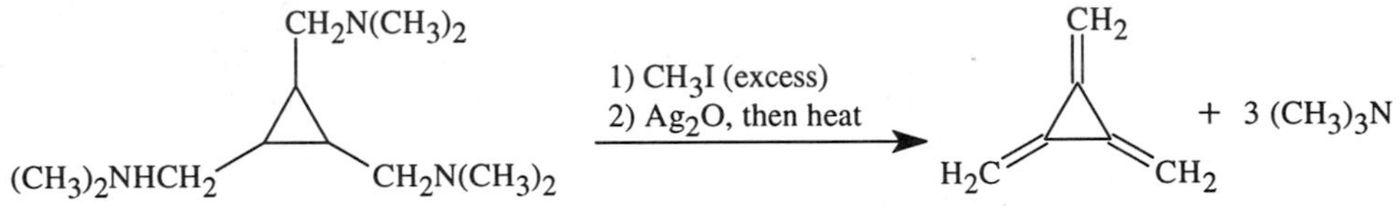

$CH_2N(CH_3)_2$ $(CH_3)_2NHCH_2$ $CH_2N(CH_3)_2$ *A*

1) CH_3I (excess) 2) Ag_2O, then heat

CH_2 H_2C CH_2 + 3 $(CH_3)_3N$

(d) The nitrosyl cation acts as the electrophile in an electrophilic aromatic substitution reaction of phenol.

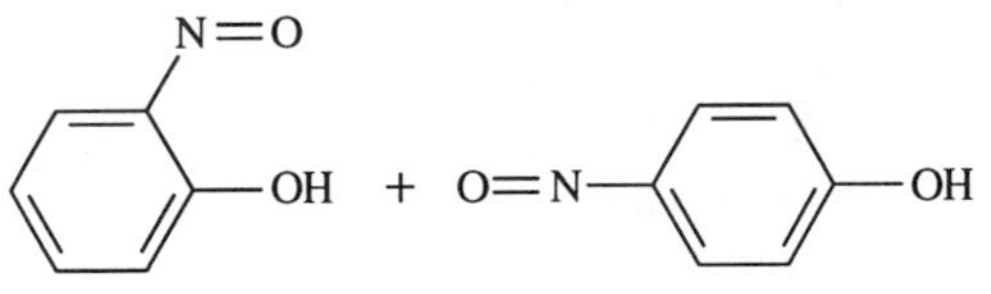

N=O OH + O=N– –OH

(e) In this variation of the Sandmeyer reaction, copper(I) nitrite introduces a nitro group into the ring.

O_2N– –NO_2

p-Dinitrobenzene would be very difficult to prepare by electrophilic aromatic substitution; can you see why?

(f) The amino group is alkylated twice by ethylene oxide to give $CH_3CH_2CH_2CH_2N(CH_2CH_2OH)_2$.

(g) The amine attacks the epoxide at the less branched carbon.

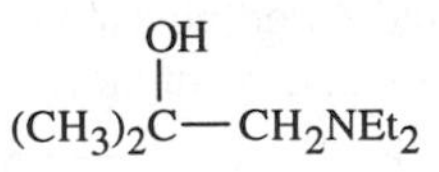

$(CH_3)_2C(OH)$—CH_2NEt_2

(h) The conjugate-base anion of phthalimide reacts at the less branched carbon of the epoxide to give compound *A*; hydrolysis of the amide groups gives the amine product.

$$A \text{ (N-(2-hydroxypropyl)phthalimide, } NCH_2CH(OH)CH_3) \xrightarrow{NaOH,\ H_2O,\ heat} C_6H_4(CO_2^-)_2 + H_2NCH_2CH(OH)CH_3$$

(i) The amine attacks the epoxide with inversion of configuration. Because both starting materials are achiral (the epoxide is the meso stereoisomer), the product is the racemate; one enantiomer results from attack at one carbon of the epoxide, and the other enantiomer results from attack at the other carbon.

$$\text{HO(H)(Ph)C—C(H)(Ph)(N-piperidinyl)} + \text{(piperidinyl-N)(Ph)(H)C—C(OH)(H)(Ph)}$$

(j) Vinylic halides are inert in S_N2 reactions, and allylic halides are very reactive. Because an excess of the amine is used, only one halide molecule reacts with each amine, and the allylic bromine is displaced.

$$H_2C{=}C(Br){-}CH_2\overset{+}{N}H_2Et\ Br^-$$

(k) Catalytic hydrogenation reduces nitrobenzene to aniline, which then reductively aminates the aldehyde; because excess aldehyde is present, the resulting secondary amine then reductively aminates a second molecule of aldehyde. The final product is the tertiary amine *N*,*N*-dibutylaniline, $PhN(CH_2CH_2CH_2CH_3)_2$.

23.60 (a)

$$\underset{\text{pentanoic acid}}{CH_3(CH_2)_3\overset{O}{\overset{\|}{C}}OH} \xrightarrow{1)\ LiAlH_4\ \ 2)\ H_3O^+} CH_3(CH_2)_3CH_2OH \xrightarrow[H_2SO_4]{conc.\ HBr} CH_3(CH_2)_3CH_2Br \xrightarrow[ether]{Mg} \xrightarrow{1)\ CO_2\ \ 2)\ H_3O^+}$$

$$CH_3(CH_2)_3CH_2\overset{O}{\overset{\|}{C}}OH \xrightarrow{SOCl_2} CH_3(CH_2)_3CH_2\overset{O}{\overset{\|}{C}}Cl \xrightarrow{H_2NCH_3\ (excess)} CH_3(CH_2)_3CH_2\overset{O}{\overset{\|}{C}}NHCH_3 \xrightarrow{1)\ LiAlH_4\ \ 2)\ H_3O^+\ \ 3)\ NaOH}$$

$$\underset{N\text{-methyl-1-hexanamine}}{CH_3(CH_2)_3CH_2CH_2NHCH_3}$$

(b)

$$\underset{\text{pentanoic acid}}{CH_3(CH_2)_3\overset{O}{\overset{\|}{C}}OH} \xrightarrow{SOCl_2} CH_3(CH_2)_3\overset{O}{\overset{\|}{C}}Cl \xrightarrow{NH_3\ (excess)} CH_3(CH_2)_3\overset{O}{\overset{\|}{C}}NH_2 \xrightarrow{1)\ LiAlH_4\ \ 2)\ H_3O^+\ \ 3)\ NaOH} \underset{\text{pentylamine}}{CH_3(CH_2)_3CH_2NH_2}$$

(c)

$$CH_3(CH_2)_3\overset{O}{\overset{\|}{C}}OH \xrightarrow{SOCl_2} CH_3(CH_2)_3\overset{O}{\overset{\|}{C}}Cl \xrightarrow{HN(CH_3)_2\ (excess)}$$

$$CH_3(CH_2)_3\overset{O}{\overset{\|}{C}}N(CH_3)_2 \xrightarrow[\text{3) NaOH}]{\text{1) LiAlH}_4\text{, 2) H}_3\text{O}^+} CH_3(CH_2)_3CH_2N(CH_3)_2$$

N,N-dimethyl-1-pentanamine

(d)

$$CH_3(CH_2)_3\overset{O}{\overset{\|}{C}}NH_2 \xrightarrow[\text{H}_2\text{O}]{\text{Br}_2\text{, NaOH}} CH_3(CH_2)_3NH_2$$

prepared in part (b) — butylamine

(e)

$$CH_3(CH_2)_3CH_2\overset{O}{\overset{\|}{C}}Cl \xrightarrow{\text{NH}_3\text{ (excess)}} CH_3(CH_2)_3CH_2\overset{O}{\overset{\|}{C}}NH_2 \xrightarrow[\text{3) NaOH}]{\text{1) LiAlH}_4\text{, 2) H}_3\text{O}^+} CH_3(CH_2)_3CH_2CH_2NH_2$$

prepared in part (a) — hexylamine (1-hexanamine)

(f)

$$O{=}CHCH_2CH_2CH_3 \xrightarrow[\text{CH}_3\text{OH}]{(\text{CH}_3)_2\text{NH, NaCNBH}_3} (CH_3)_2NCH_2CH_2CH_2CH_3 \xrightarrow{\text{PhCH}_2\text{Br}} PhCH_2-\overset{+}{N}(CH_3)_2-CH_2CH_2CH_2CH_3\ Br^-$$

butyraldehyde

(g)

$$PhCH_3 \xrightarrow[\text{peroxides}]{\text{NBS, CCl}_4} PhCH_2Br \xrightarrow[\text{ether}]{\text{Mg}} \xrightarrow[\text{2) H}_3\text{O}^+]{\text{1) ethylene oxide}} PhCH_2CH_2CH_2OH \xrightarrow{\text{1) KMnO}_4\text{, }^-\text{OH; 2) H}_3\text{O}^+}$$

toluene

$$PhCH_2CH_2\overset{O}{\overset{\|}{C}}OH \xrightarrow{\text{SOCl}_2} PhCH_2CH_2\overset{O}{\overset{\|}{C}}Cl \xrightarrow[\text{(excess)}]{\text{H}_2\text{NEt}} PhCH_2CH_2\overset{O}{\overset{\|}{C}}NHEt \xrightarrow[\text{3) NaOH}]{\text{1) LiAlH}_4\text{, 2) H}_3\text{O}^+} PhCH_2CH_2CH_2NHEt$$

N-ethyl-3-phenyl-1-propanamine

(h)

$$CH_2(CO_2Et)_2 \xrightarrow{\text{1) NaOEt, EtOH; 2) CH}_3\text{CH}_2\text{CH}_2\text{Br}} CH_3CH_2CH_2CH(CO_2Et)_2 \xrightarrow{\text{1) NaOEt, EtOH; 2) CH}_3\text{I}}$$

diethyl malonate

$$CH_3CH_2CH_2\underset{CH_3}{\underset{|}{C}}(CO_2Et)_2 \xrightarrow{\text{H}_3\text{O}^+\text{, H}_2\text{O, heat}} CH_3CH_2CH_2\underset{CH_3}{\underset{|}{C}H}\overset{O}{\overset{\|}{C}}OH \xrightarrow[\text{2) NH}_3]{\text{1) SOCl}_2}$$

$$CH_3CH_2CH_2\underset{CH_3}{\underset{|}{C}H}\overset{O}{\overset{\|}{C}}NH_2 \xrightarrow[\text{H}_2\text{O}]{\text{Br}_2\text{, NaOH}} CH_3CH_2CH_2\underset{CH_3}{\underset{|}{C}H}NH_2$$

2-pentanamine

(i)

$$(CH_3)_2C{=}O \xrightarrow{\text{1) LiAlH}_4\text{; 2) H}_3\text{O}^+} (CH_3)_2CHOH \xrightarrow[\text{H}_2\text{SO}_4]{\text{conc. HBr}} (CH_3)_2CHBr \xrightarrow[\text{ether}]{\text{Mg}} \xrightarrow[\text{2) H}_3\text{O}^+]{\text{1) ethylene oxide}}$$

acetone

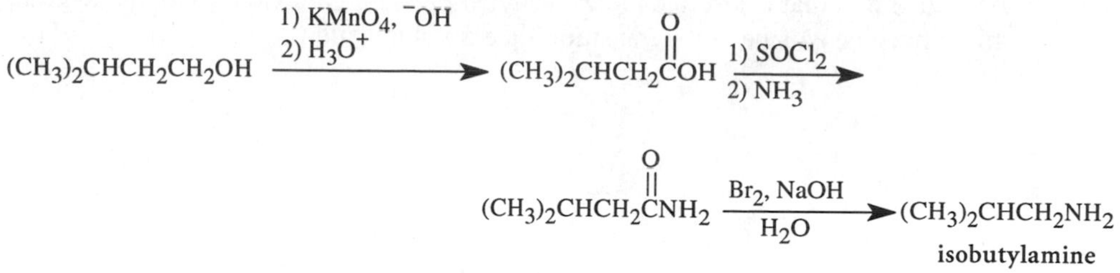

$(CH_3)_2CHCH_2CH_2OH \xrightarrow[2)\ H_3O^+]{1)\ KMnO_4,\ ^-OH} (CH_3)_2CHCH_2\overset{O}{\overset{\|}{C}}OH \xrightarrow[2)\ NH_3]{1)\ SOCl_2}$

$(CH_3)_2CHCH_2\overset{O}{\overset{\|}{C}}NH_2 \xrightarrow[H_2O]{Br_2,\ NaOH} (CH_3)_2CHCH_2NH_2$

isobutylamine

(j)

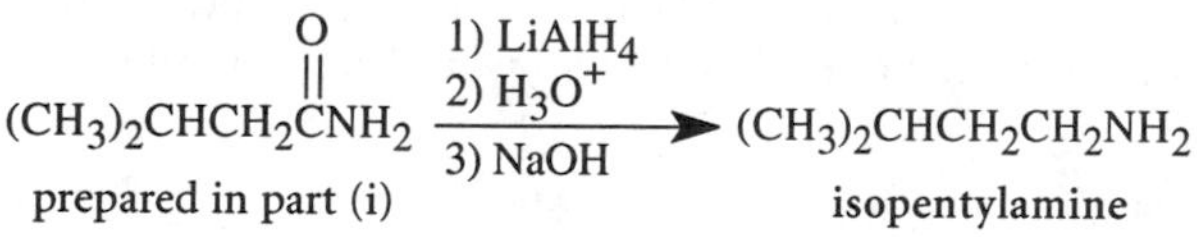

$(CH_3)_2CHCH_2\overset{O}{\overset{\|}{C}}NH_2 \xrightarrow[3)\ NaOH]{1)\ LiAlH_4;\ 2)\ H_3O^+} (CH_3)_2CHCH_2CH_2NH_2$

prepared in part (i) — isopentylamine

(k)

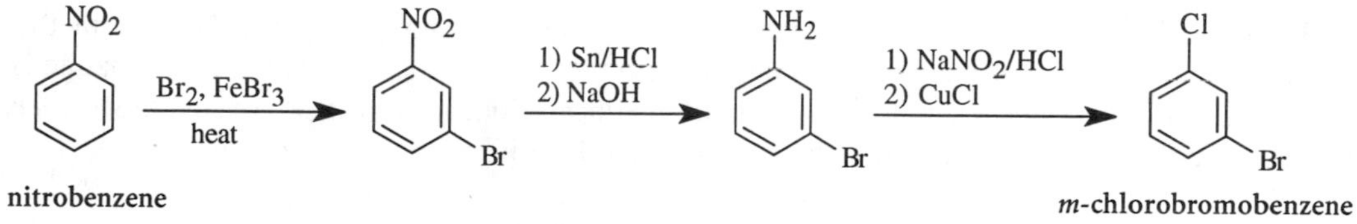

nitrobenzene $\xrightarrow[heat]{Br_2,\ FeBr_3}$ m-bromonitrobenzene $\xrightarrow[2)\ NaOH]{1)\ Sn/HCl}$ m-bromoaniline $\xrightarrow[2)\ CuCl]{1)\ NaNO_2/HCl}$ *m*-chlorobromobenzene

(l)

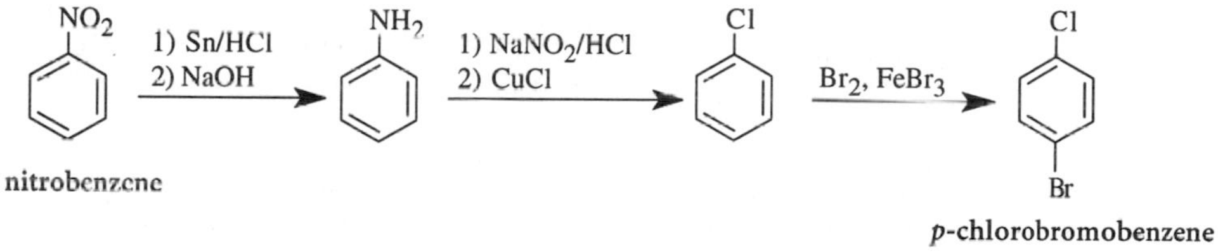

nitrobenzene $\xrightarrow[2)\ NaOH]{1)\ Sn/HCl}$ aniline $\xrightarrow[2)\ CuCl]{1)\ NaNO_2/HCl}$ chlorobenzene $\xrightarrow{Br_2,\ FeBr_3}$ *p*-chlorobromobenzene

(m)

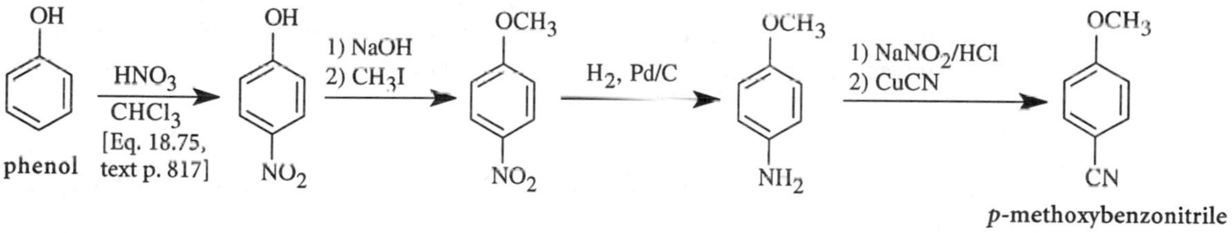

phenol $\xrightarrow[CHCl_3]{HNO_3}$ [Eq. 18.75, text p. 817] *p*-nitrophenol $\xrightarrow[2)\ CH_3I]{1)\ NaOH}$ *p*-nitroanisole (OCH_3, NO_2) $\xrightarrow{H_2,\ Pd/C}$ *p*-methoxyaniline (OCH_3, NH_2) $\xrightarrow[2)\ CuCN]{1)\ NaNO_2/HCl}$ *p*-methoxybenzonitrile (OCH_3, CN)

(n)

(D)(H)(CH_3)C–OH $\xrightarrow[2)\ NaCN]{1)\ TsCl,\ pyridine}$ N≡C–C(D)(H)(CH_3) $\xrightarrow[2)\ H_2O]{1)\ LiAlH_4}$ H_2NCH_2–C(D)(H)(CH_3)

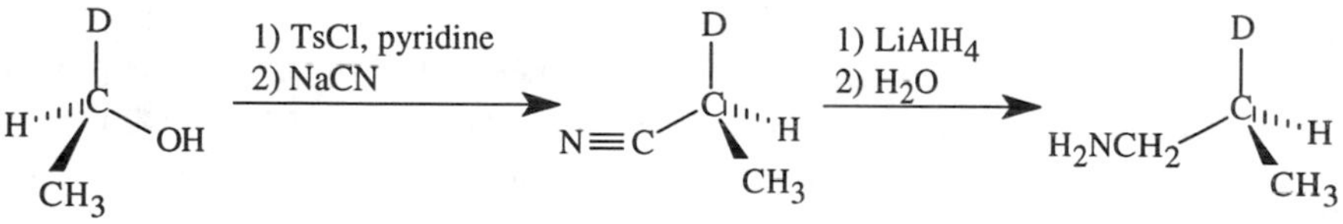

Note that the S_N2 reaction of cyanide ion proceeds with inversion of configuration.

(o)

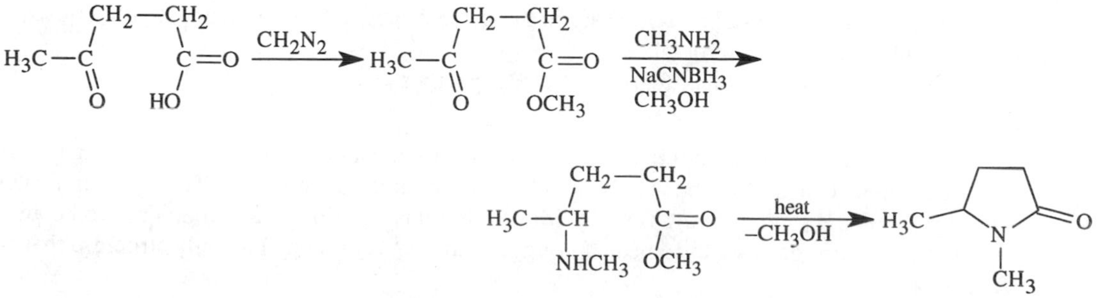

$H_3C-\overset{O}{\overset{\|}{C}}-CH_2-CH_2-\overset{O}{\overset{\|}{C}}-OH \xrightarrow{CH_2N_2} H_3C-\overset{O}{\overset{\|}{C}}-CH_2-CH_2-\overset{O}{\overset{\|}{C}}-OCH_3 \xrightarrow[NaCNBH_3,\ CH_3OH]{CH_3NH_2}$

$H_3C-CH(NHCH_3)-CH_2-CH_2-\overset{O}{\overset{\|}{C}}-OCH_3 \xrightarrow[-CH_3OH]{heat}$ 5-methyl-1-methyl-2-pyrrolidinone (H_3C, N–CH_3, =O)

23.61 For the proton that is lost and the trimethylammonium group to be anti, the R-group and the trimethylammonium group must be gauche in the transition state for elimination:

H, H, H, H, R, $\overset{+}{N}(CH_3)_3$ — gauche

As the size of the R-group is increased, the van der Waals repulsions in this transition state are more severe, the transition state is less stable, and the rate of the Hofmann elimination is slower. Therefore, the rates of Hofmann elimination increase in the order

$$R = (CH_3)_3C \;<\; R = H_3C \;<\; R = H$$

23.62 At pH 1, methylamine exists almost completely as the methylammonium ion, $CH_3\overset{+}{N}H_3$. The methyl quartet is expected for the presence of three neighboring nitrogen protons provided that chemical exchange is slow. As the pH is raised, more free methylamine is present and chemical exchange of the protons on the nitrogen can occur rapidly by the following mechanism. (The asterisk is used to differentiate the two nitrogens.)

$$CH_3\underset{*}{\overset{+}{N}}H_3 + CH_3NH_2 \rightleftharpoons CH_3\underset{*}{N}H_2 + CH_3\overset{+}{N}H_3$$

Evidently, this exchange is so fast on the NMR time scale that splitting is obliterated. (See Secs. 13.6D and 13.7.)

23.63 The λ_{max} of aniline occurs at longer wavelength than that of benzene because of the conjugation of the aniline unshared electron pair with the π-electron system of the benzene ring. Recall that the more atoms are involved in a conjugated π-electron system, the greater is the intensity of the UV spectrum and the greater is the λ_{max} (Sec. 15.2C). When aniline is protonated, the nitrogen no longer has an unshared pair, and the conjugated π-electron system is restricted to the phenyl ring, in which case the UV spectrum looks much as it does in benzene. In other words, protonated aniline, as far as the π-electron system and the UV spectrum are concerned, is essentially the same as benzene.

23.64 Assume that the —NH_2 or —NH protons of the amine rapidly exchange, so that they are not coupled to adjacent protons on carbon.

Compound *C* should have the resonance at greatest chemical shift, because the proton α to the nitrogen is also benzylic and is a methine. This proton should be a triplet.

Compound *B* should have the resonance at the next greatest chemical shift, because the protons α to the nitrogen are also benzylic, but are methylenes, which have smaller chemical shifts than methine protons by about 0.5–0.7 ppm. These α-protons should be a singlet. Of course, the relative integrations of the α-protons of *B* and *C* are different, as well.

Between compounds *A* and *D*, compound *A* has the resonance at smallest chemical shift—the methyl group, which should be a doublet. These methyl protons are neither benzylic nor α to a nitrogen; thus, they should have a resonance not far from δ 1.

All of the resonances of compound *D* are in the δ 2–3.5 range. This spectrum should consist of two triplets and a singlet (plus the NH proton).

Of course, one might cite other features of these spectra as well.

23.65 (a) With an unsaturation number of 4 and clear evidence for a para-substituted benzene ring (the four-proton pair of doublets centered at about δ 7), no other unsaturations are possible. The NMR indicates partial structures —OCH_3 and —CH_2CH_2—. With no other unsaturation, the nitrogen has to be an amine, and the broad, two-proton resonance near δ 1 suggests an —NH_2 group. The only structure that fits all the data is

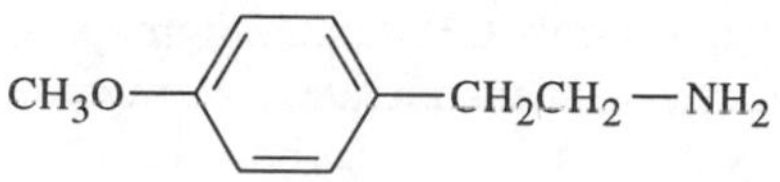

$CH_3O-C_6H_4-CH_2CH_2-NH_2$

2-(4-methoxyphenyl)ethylamine
(compound *A*)

(b) The compound has no unsaturation and therefore must be an amine. The mutually split triplet-quartet pattern suggests an ethyl group, and the integration suggests two identical ethyl groups. If so, the δ 1.1 resonance has buried under it (as suggested by the hint) a broad resonance for two additional protons. Indeed, the chemical shift is reasonable for an $-NH_2$ group. (See the shift for the $-NH_2$ protons in part (a)). Because there are two amine nitrogens, these have to be two chemically identical secondary amines. Two resonances have four identical protons; this suggests two chemical identical $-CH_2-$ groups for each resonance. One of these resonances is a quartet and is part of the ethyl group discussed above; hence, there are two chemically equivalent ethyl groups. The other $-CH_2-$ resonance is a singlet; hence this corresponds to two $-CH_2-$ groups in identical environments that do not give rise to splitting. Clearly, this is a molecule with some internal symmetry. The structure is

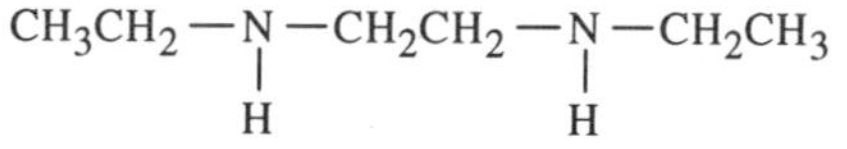

$CH_3CH_2-NH-CH_2CH_2-NH-CH_2CH_3$

N,N'-diethyl-1,2-ethanediamine
(compound *B*)

Notice that, although the two $-CH_2-$ groups in the center of the molecule are adjacent, they do not show any splitting because they are chemically equivalent.

(c) This compound has only one degree of unsaturation, and both the NMR (δ 4.9) and the IR (898 cm^{-1}) suggest a vinylic methylene group ($H_2C=$) in which the protons are coupled to no other protons. The chemical shift of the three-proton singlet (δ 1.8) suggests an allylic methyl group. We have a two-proton singlet at δ 3.2 that is clearly α to the nitrogen, and an ethyl absorption in which the $-CH_2-$ group is also α to the nitrogen. These data conspire to define the following structure:

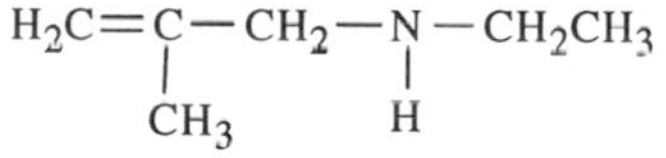

$H_2C=C(CH_3)-CH_2-NH-CH_2CH_3$

N-ethyl-2-methyl-2-propenamine
(compound *C*)

23.66 The unsaturation number of this compound is 0. The instability of this compound in aqueous acid and the presence of two oxygens is consistent with the presence of an acetal. The six-proton singlet at δ 3.27 could be due to the methyl protons of a dimethyl acetal; the δ 53.2 resonance in the CMR spectrum could be due to the carbons of these methyl groups, and the δ 102.4 resonance could be due to the carbon of the acetal group, that is, $-O-CH-O-$. If so, the resonance in the proton NMR at δ 4.50 is due to the proton of this CH group, and its splitting indicates two adjacent protons. The resonance for these two protons is found at δ 2.45. This leaves a six-proton singlet at δ 2.30 unaccounted for; this could be due to the six methyl protons of a dimethylamino group. The compound is

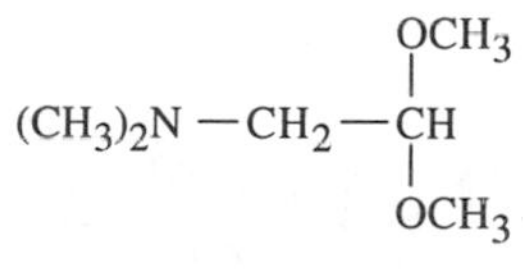

$(CH_3)_2N-CH_2-CH(OCH_3)_2$

dimethylaminoacetaldehyde dimethyl acetal
(2,2-dimethoxy-*N*,*N*-dimethylethanamine)

23.67 (a) The chemical data indicate the presence of a primary amine. The compound has one degree of unsaturation, and the chemical shifts in the CMR are too small for any carbon to be involved in a double bond. Hence, compound *A* contains a ring, and only CH and CH_2 groups are present. The presence of only three resonances means that two carbons are equivalent. Compound *A* is cyclobutylamine. Cyclopropylmethanamine is ruled out by the fact that the carbon bound to one hydrogen has the largest chemical shift and therefore must be bound to the nitrogen.

cyclobutylamine (compound *A*)   cyclopropylmethanamine (ruled out)

(b) The IR data for this compound indicate a —CH=CH_2 group. The molecule is neither a primary amine (chemical data) nor a tertiary amine (it has an N—H). Hence it is a secondary amine. Two structures worth considering are *A* and *B*:

$CH_3CH_2NHCH{=}CH_2$ (*A*) $CH_3NHCH_2CH{=}CH_2$ (*B*)

Because the CMR spectrum contains no resonances near δ 15, which would be expected for the methyl carbon of compound *A*, the structure must be *B*, *N*-methyl-2-propenamine. The chemical shifts of the carbon resonances are consistent with this structure.

23.68 All of these reactions are related to the Hofmann and Curtius rearrangements.

(a) Potassium hydride removes the NH proton to form an anion that rearranges to an isocyanate. This is exactly like the Hofmann rearrangement shown in Eq. 23.76d on text p. 1105, except that the leaving group is benzoate. (This reaction is an example of the *Lossen rearrangement.* It is like the Hofmann rearrangement, except that the leaving group in the rearrangement step is a carboxylate ion rather than a bromide ion.) The amine is formed by hydration of the isocyanate and decarboxylation of the resulting carbamate. (For brevity, the proton transfer in step *(a)* is shown as an intramolecular concerted process, but it could instead be a two-step process involving ^-OH and H_2O.)

$PhCH_2C(=O)NH{-}O{-}C(=O)Ph \xrightarrow{H{-}K} H_2 + PhCH_2C(=O)\bar{N}{-}O{-}C(=O)Ph \longrightarrow {}^-O{-}C(=O)Ph + PhCH_2N{=}C{=}O \xrightarrow{^-OH}$

$PhCH_2\bar{N}{-}C(OH){=}O \xrightarrow[(a)]{H_2O} PhCH_2NH{-}C(O^-){=}O \longrightarrow PhCH_2NH_2 + O{=}C{=}O + {}^-OH$

(b) Attack of hydroxide on the carbonyl carbon of *N*-bromosuccinimide gives an *N*-bromo anion, which undergoes a Hofmann rearrangement to the corresponding isocyanate. Hydrolysis of the isocyanate to a carbamate and decarboxylation of the carbamate to the amine occur by the mechanism shown in the solution to part (a).

N-bromosuccinimide + ^-OH ⟶ tetrahedral intermediate ⟶ $Br{-}\bar{N}{-}C(=O){-}CH_2CH_2C(=O)OH$ (ionizes in base) ⟶ $O{=}C{=}N{-}CH_2CH_2C(=O){-}O^-$

hydrolysis followed by decarboxylation gives the amine product

(c) In this case, the amide nitrogen attacks the isocyanate following a Curtius rearrangement. As in part (a), the proton transfer is shown for brevity as an intramolecular process.

$$H_2NCCH_2CH_2-\overset{O}{\overset{\|}{C}}-\overset{-}{N}-\overset{+}{N}\equiv N \longrightarrow N_2 + H_2N\overset{O}{\overset{\|}{C}}CH_2CH_2-N=C=O \longrightarrow$$

23.69 Evidently, an anion is generated at the α-carbon of the ester, and this then adds to the aldehyde. Although one might be tempted to use the amine as a base to remove an α-hydrogen, vinylic hydrogens that are α to carbonyl groups are not particularly acidic, because the electron pair of the enolate ion that would form and the double bond cannot simultaneously be conjugated with the carbonyl. Rather, the amine acts as a nucleophile in a conjugate addition, which generates the required anion. Once the addition occurs, the amine is expelled again to regenerate the α,β-unsaturated product. (As in previous problems, proton transfers are shown as intramolecular processes for convenience.)

$$Et_3N \quad H_2C=CH-CO_2Et \rightleftharpoons H_2\underset{^{+}NEt_3}{C}-\ddot{\overset{-}{C}}H-CO_2Et \;(+\; H_3C-CH=O) \rightleftharpoons$$

$$H_3C-\overset{O^-}{CH}-\underset{H_2C(^{+}NEt_3)}{C}(H)CO_2Et \rightleftharpoons H_3C-\overset{OH}{CH}-\overset{-}{C}(CO_2Et)-CH_2{}^{+}NEt_3 \longrightarrow H_2C=C(CO_2Et)-\overset{OH}{CH}-CH_3 + Et_3N$$

23.70 An intermediate in the formation of 1-pentanamine is the imine *A*. The other products arise from the fact that 1-pentanamine, *B*, can react with either the starting aldehyde, pentanal, or with imine *A* to give a new imine *C*. (Give the mechanism for this reaction.) Imine *C* can also undergo hydrogenation to the second product, dipentylamine, *D*. Dipentylamine can also react with other imine intermediates to form immonium ion *E*, which can undergo hydrogenation to form tripentylamine *F*, the tertiary amine by-product.

$$\underset{\text{pentanal}}{CH_3(CH_2)_3\overset{O}{\overset{\|}{C}}H} + NH_3 \underset{}{\overset{-H_2O}{\rightleftharpoons}} \underset{A}{CH_3(CH_2)_3\overset{NH}{\overset{\|}{C}}H} \xrightarrow{H_2,\ cat.} \underset{B}{CH_3(CH_2)_3CH_2NH_2}$$

$$\underset{A}{CH_3(CH_2)_3\overset{NH}{\overset{\|}{C}}H} + \underset{B}{H_2NCH_2(CH_2)_3CH_3} \rightleftharpoons NH_3 + \underset{C}{CH_3(CH_2)_3\overset{NCH_2(CH_2)_3CH_3}{\overset{\|}{C}}H} \xrightarrow{H_2,\ cat.} \underset{D}{[CH_3(CH_2)_3CH_2]_2NH}$$

$$\underset{C}{CH_3(CH_2)_3CH{=}NCH_2(CH_2)_3CH_3} + \underset{D}{[CH_3(CH_2)_3CH_2]_2NH} \rightleftharpoons \underset{B}{H_2NCH_2(CH_2)_3CH_3} + \underset{E}{CH_3(CH_2)_3CH{=}\overset{+}{N}[CH_2(CH_2)_3CH_3]_2}$$

$$\xrightarrow{H_2,\ \text{cat.}} \underset{F}{[CH_3(CH_2)_3CH_2]_3N}$$

The yield of pentylamine (compound *B*) can be increased, and the amounts of the other by-products reduced, by including ammonia in the reaction mixture. Can you see why? (*Hint:* Apply LeChatelier's principle.)

23.71 (a) Amides, like amines, can be diazotized. The result is an acyldiazonium ion *A*, which rapidly hydrolyzes; remember that N_2 is a superb leaving group. This produces carbamic acid, which decomposes to ammonia and CO_2. Then ammonia is itself diazotized to give N_2 as shown in the solution to Problem 23.59(a) on p. 912 of this manual.

$$H_2N\overset{O}{\overset{\|}{C}}NH_2 \xrightarrow{HNO_2} \underset{A}{H_2N\overset{O}{\overset{\|}{C}}\overset{+}{N}{\equiv}N} \xrightarrow{H_2O} \underset{\text{carbamic acid}}{H_2N\overset{O}{\overset{\|}{C}}OH} \longrightarrow CO_2 + NH_3 \xrightarrow{HNO_2} N_2 + 2\,H_2O$$

(b) The amine reacts intramolecularly with the carbonyl group to form a cyclic imine *A*, which is reduced by the sodium borohydride.

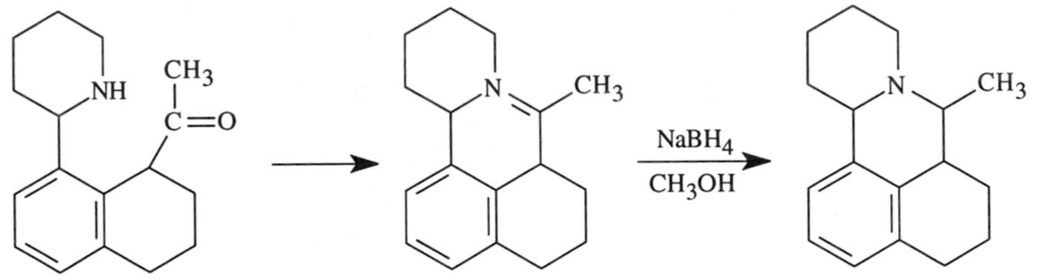

NH, CH_3, C=O → N, CH_3 (*A*) $\xrightarrow[CH_3OH]{NaBH_4}$ N, CH_3

The $NaBH_4$ reduction could also take place on the imminium cation *A1* or the enamine *A2*:

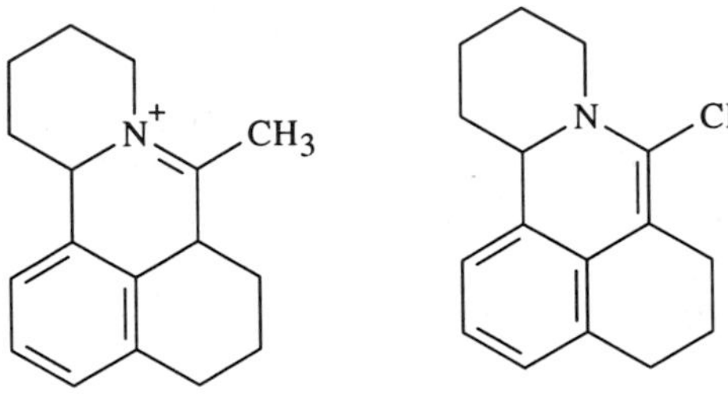

$\overset{+}{N}$, CH_3 (*A1*)   N, CH_3 (*A2*)

(c) Cyanide ion serves as a base to promote a β-elimination of trimethylamine and form an α,β-unsaturated ketone *A*. This undergoes conjugate addition of cyanide ion to give the product.

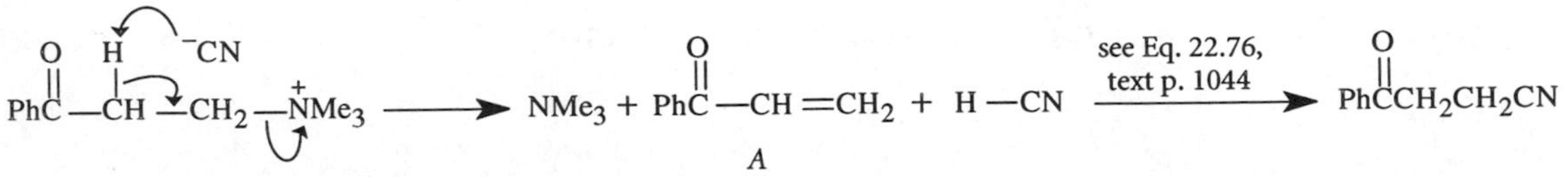

$$Ph\overset{O}{\overset{\|}{C}}{-}\overset{H}{\overset{|}{C}H}{-}CH_2{-}\overset{+}{N}Me_3 \;\; {}^{-}CN \longrightarrow NMe_3 + \underset{A}{Ph\overset{O}{\overset{\|}{C}}{-}CH{=}CH_2} + H{-}CN \xrightarrow[\text{text p. 1044}]{\text{see Eq. 22.76,}} Ph\overset{O}{\overset{\|}{C}}CH_2CH_2CN$$

(d) The acyl azide *A* undergoes Curtius rearrangement to an enamine *B*, which hydrolyzes under the aqueous reaction conditions to the product ketone.

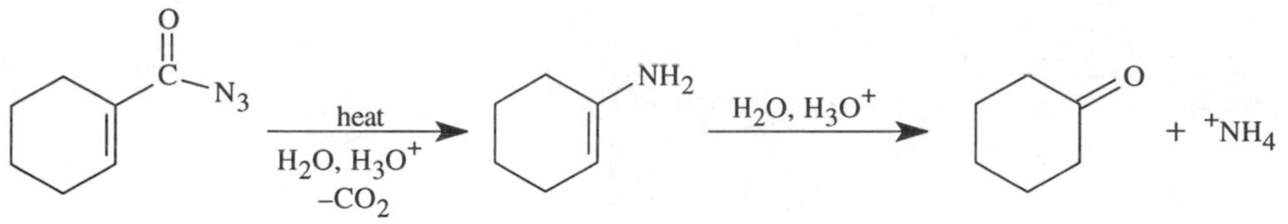

(e) The product shows that a 1,4-addition of benzyne to furan (that is, a Diels-Alder reaction) has occurred. (See the solution to Problem 15.63 on p. 498 of this manual for the approach to this type of problem.) The benzyne is formed by elimination of CO_2 and N_2 from the diazotization product *A* of the starting material, anthranilic acid.

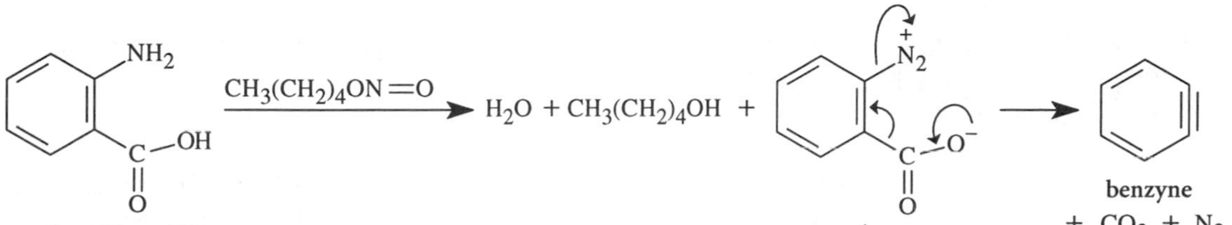

anthranilic acid

$CH_3(CH_2)_4ON{=}O$ → $H_2O + CH_3(CH_2)_4OH$ + *A* → benzyne + CO_2 + N_2

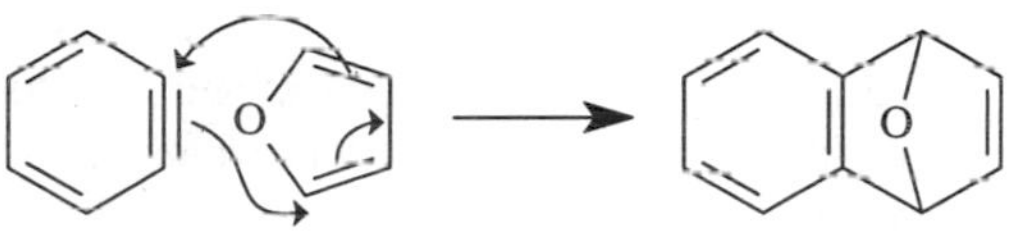

benzyne furan

(f) The diazonium ion *A* formed from the starting material loses nitrogen to give a carbocation *B*. This carbocation undergoes a rearrangement to the product. (See Study Problem 19.2 on text p. 853 for a discussion of a closely related reaction.)

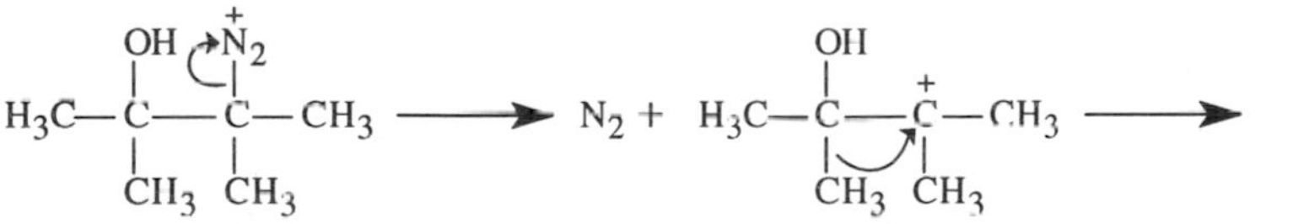

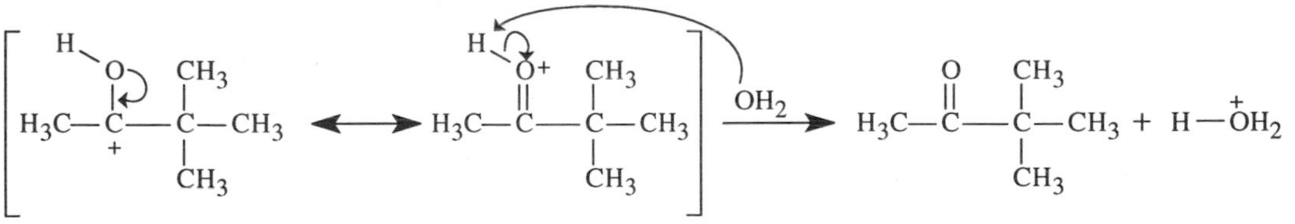

23.72 The conjugate-acid pK_a of amine *C* is a normal conjugate-acid pK_a for a tertiary amine. (See, for example, the pK_a of the conjugate acid of triethylamine in Table 23.1 on text p. 1077.) This observation shows that incorporation of an amine into a bicyclic structure has no significant effect on its conjugate-acid pK_a. The conjugate-acid pK_a of amine *A* is not unusual for an *N,N*-dialkylaniline. (See, for example, the pK_a of $PhN(CH_3)_2$ in Table 23.1.) Recall (text p. 1080) that the conjugate-acid pK_a values of aniline derivatives are considerably lower than those of tertiary alkylamines because of resonance interaction of the unshared electron pair on the nitrogen with the benzene ring. The conjugate-acid pK_a of compound *B*, then, has the unusual value that requires explanation. In this compound, the unshared pair on nitrogen *cannot* overlap with the π-electron system of the benzene ring because the π orbitals of the benzene ring and the orbital on the nitrogen are forced by the bicyclic structure out of coplanarity.

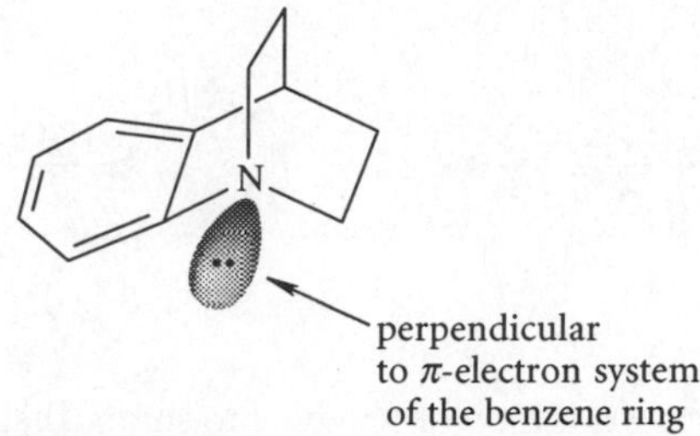

Consequently, the basicity-lowering resonance effect is absent. For this reason, the conjugate-acid pK_a of compound *B* is higher than that of compound *A*. However, the basicity-lowering polar effect of the benzene ring remains; for this reason, the conjugate-acid pK_a of compound *B* is lower than that of compound *C*.

The pK_a data in the problem provide good estimates of the polar and resonance effects of a benzene ring on amine basicity. These data show that polar and resonance effects are about equally important.

23.73 The data show that the conjugate-acid pK_a values of a bicyclic tertiary amine *C* (which is the same as amine *C* in the previous problem) and a monocyclic tertiary amine *B* are not very different. Hence, the additional ring has no unusual effect on pK_a. The conjugate-acid pK_a of amide *A* is also fairly normal for an amide that is protonated on the carbonyl oxygen. (See Eq. 21.4b on text p. 949.) The pK_a of amide *D*, then, is unusual. Normally, amides protonate on the carbonyl oxygen because the resulting carbocation is resonance-stabilized. In this case, however, the bicyclic structure of the ring prevents the overlap of the nitrogen unshared electron pair with the π-electron system of the carbonyl group:

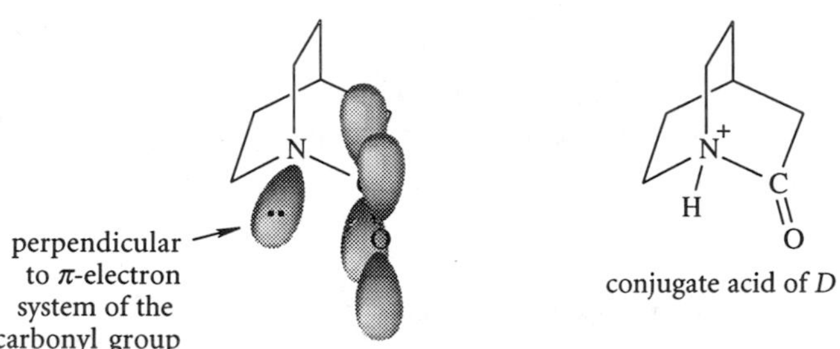

Hence, the electronic character of this nitrogen is more like that of an amine with a neighboring electron-withdrawing substituent (the carbonyl group) than it is like that of an amide. Thus, this amide behaves like an amine: it protonates on *nitrogen.* What about the hydrolysis rate? Recall that ordinary amides hydrolyze slowly because they are resonance-stabilized. (See Sec. 21.7E.) When resonance stabilization is not present, as in the case of compound *D*, an amide hydrolyzes rapidly. In acidic solution, when the nitrogen is protonated, its hydrolysis rate is more like that of an acid chloride!

24

Chemistry of Naphthalene and the Aromatic Heterocycles

TERMS

CONCEPTS

I. INTRODUCTION TO POLYCYCLIC AROMATIC HYDROCARBONS

A. General

1. Polycyclic aromatic hydrocarbons are aromatic hydrocarbons that contain two or more fused rings.
   a. The aromaticity of such compounds is not covered by the Hückel $4n + 2$ rule.
   b. The Lewis structures of these compounds consist of alternating single and double bonds.
   c. Naphthalene is the simplest example; others are shown in Figure 24.1, text page 1121.
2. Naphthalene can be represented by three resonance structures.

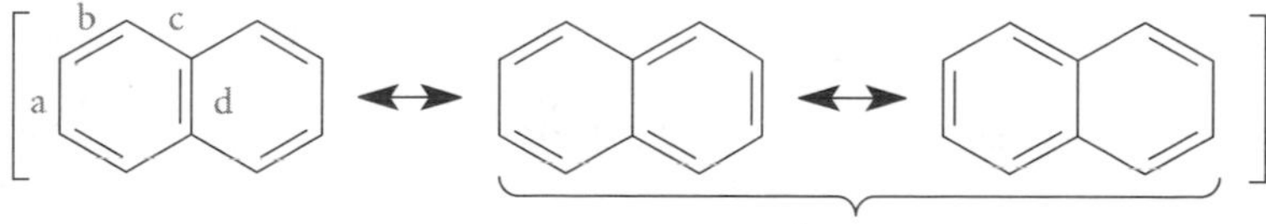

   a. Two are equivalent and one is unique.
   b. Naphthalene has four nonequivalent sets of carbon-carbon bonds (labeled a–d in the structure above).
3. Graphite, a carbon polymer that consists of layers of fused benzene rings, is an excellent electrical conductor because of the ease with which π electrons can be delocalized across its structure.

4. Polycyclic aromatic compounds having ball-shaped structures of various sizes have been nicknamed fullerenes after buckminsterfullerene (the name of the first isolated compound of this type).

B. Nomenclature of Naphthalene Derivatives

1. In substitutive nomenclature, carbon-1 of naphthalene is the carbon adjacent to the bridgehead carbon (a vertex at which the rings are fused); substituents are given the lowest numbers consistent with this scheme and their relative priorities.
2. Naphthalene also has a common nomenclature that uses Greek letters: the 1-position is designated as α and the 2-position as β.
3. Common and substitutive nomenclature should never be mixed.
4. As a substituent, the naphthalene ring is called the naphthyl group.

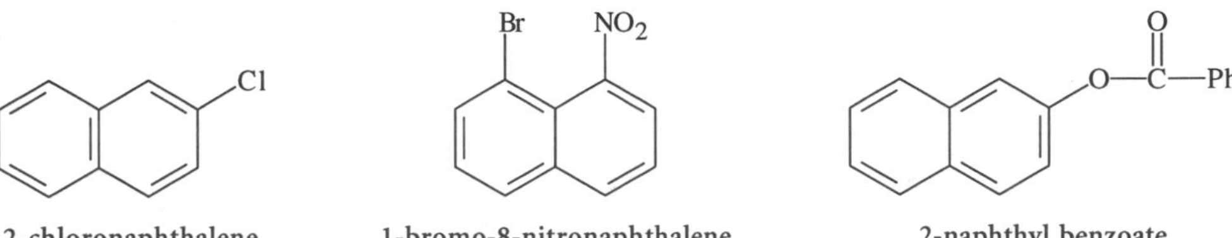

2-chloronaphthalene (β-chloronaphthalene) | 1-bromo-8-nitronaphthalene | 2-naphthyl benzoate

II. INTRODUCTION TO THE AROMATIC HETEROCYCLES

A. General

1. Heterocyclic compounds are compounds with rings that contain more than one type of atom; the heterocyclic compounds of greatest interest to organic chemists have carbon rings containing one or two heteroatoms (atoms other than carbon).
2. The chemistry of many saturated heterocyclic compounds is analogous to that of their noncyclic counterparts.
3. A significant number of unsaturated heterocyclic compounds exhibit aromatic behavior.

B. Nomenclature of the Heterocycles

1. The names and structures of some common aromatic heterocyclic compounds are given in Fig. 24.4, text page 1129.
2. The same rules used in numbering and naming saturated heterocyclic compounds are used for numbering and naming aromatic heterocyclic compounds.
   a. In all but a few cases, a heteroatom is given the number 1 (isoquinoline is an exception).
   b. Oxygen and sulfur are given a lower number than nitrogen when a choice exists.
   c. Substituent groups are given the lowest numbers consistent with this scheme.

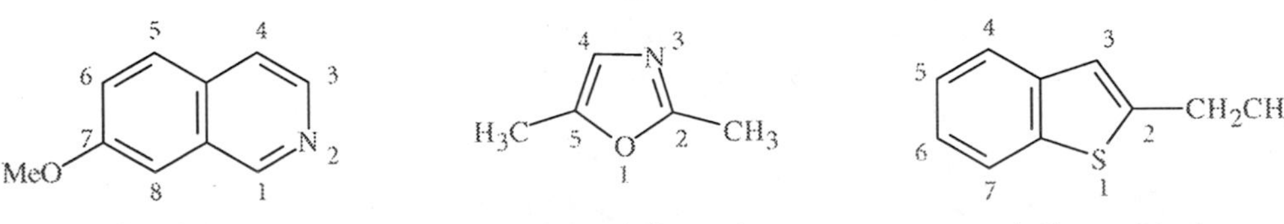

7-methoxyisoquinoline | 2,5-dimethyloxazole | 2-ethylbenzothiophene

C. Structure and Aromaticity of the Heterocycles

1. The aromatic heterocyclic compounds furan, thiophene, and pyrrole can be written as resonance hybrids.

resonance structures of pyrrole

a. The importance of the charge-separated structures is evident in a comparison of the dipole moments of furan and tetrahydrofuran.
   i. Electrons in the σ bonds are pulled toward the oxygen because of its electronegativity.
   ii. The resonance delocalization of the oxygen unshared electrons into the ring tends to push electrons away from the oxygen into the π-electron system of the ring.
b. These two effects nearly cancel in furan; thus furan has a very small dipole moment.
c. That tetrahydrofuran has a considerably higher boiling point than furan reflects the greater dipole moment of tetrahydrofuran.

2. Pyridine can be represented by two equivalent neutral resonance structures; three additional structures of less importance reflect the relative electronegativity of nitrogen.

minor contributors

3. Heteroatoms involved in formal double bonds of the Lewis structure (such as the nitrogen of pyridine) contribute one π electron to the six π-electron aromatic system The orbital containing the unshared electron pair of the pyridine nitrogen is perpendicular to the $2p$ orbitals of the ring and is therefore not involved in π bonding.

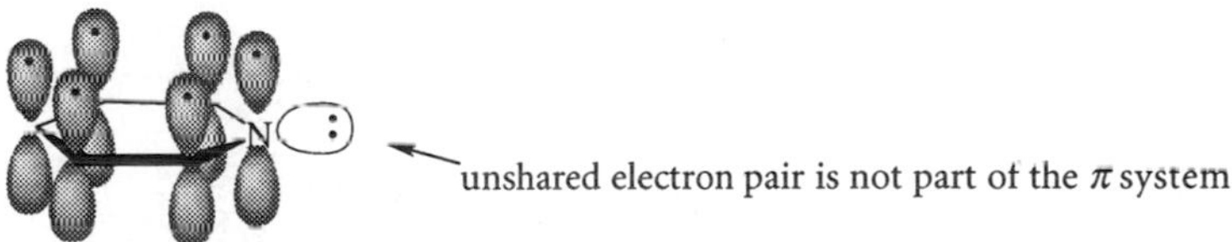

4. An unshared electron pair on a heteroatom in an allylic position (such as the unshared pair on the nitrogen of pyrrole) is part of the aromatic π system; the hydrogen of pyrrole lies in the plane of the ring.

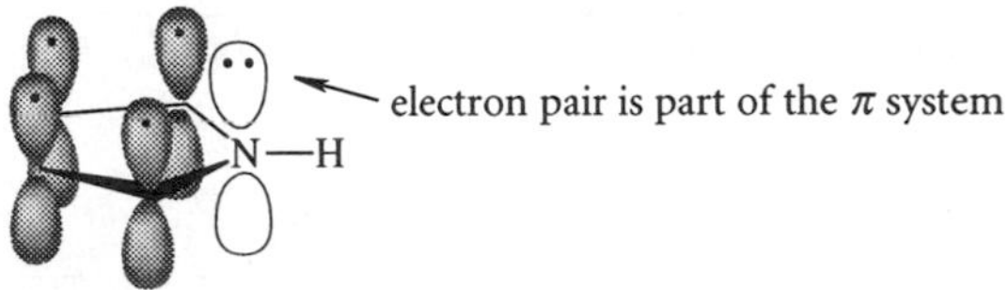

5. The oxygen of furan contributes one unshared electron pair to the aromatic π-electron system; the other unshared electron pair occupies a position analogous to the hydrogen of pyrrole (in the ring plane, perpendicular to the $2p$ orbitals of the ring).

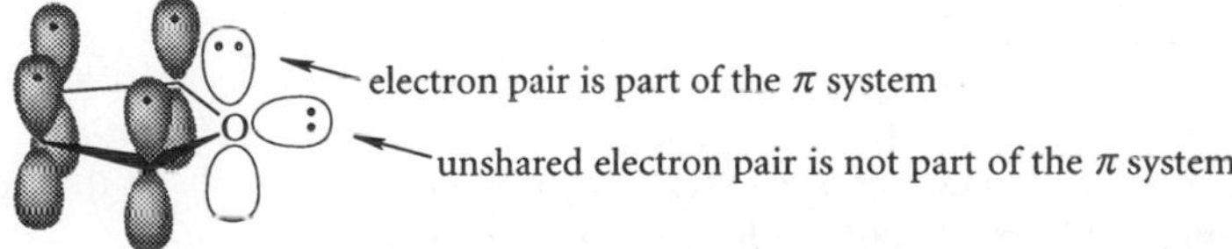

6. Aromatic stabilization (empirical resonance energy) of the heterocycles and benzene is in the order

benzene > pyridine ≈ thiophene > pyrrole > furan

(See Table 24.1, text p. 1132.)

III. BASICITY AND ACIDITY OF THE NITROGEN HETEROCYCLES

A. Basicity

1. Pyridine and quinoline act as ordinary aromatic amine bases; they are less basic than aliphatic tertiary amines because of the sp^2 hybridization of their nitrogen unshared electron pairs.
   a. Protonation of the pyridine unshared electron pair occurs because this electron pair is not part of the π-electron system.
   b. Protonation of this electron pair does not destroy aromaticity.
2. Pyrrole and indole are not very basic.
   a. These compounds are protonated only in strong acid, and protonation occurs on carbon, not nitrogen. Although protonation of the carbon of pyrrole disrupts the aromatic π-electron system, the resulting cation is resonance-stabilized.
   b. Protonation of the pyrrole nitrogen would disrupt the aromatic six π-electron system by taking the nitrogen's unshared pair "out of circulation" and would give a conjugate acid that is not resonance-stabilized.

B. Acidity

1. Pyrrole and indole are weak acids.
   a. The N—H protons of pyrrole and indole are about as acidic as alcohol O—H protons.
   b. Pyrrole and indole are acidic enough to behave as acids toward basic organometallic compounds.

N—H (pyrrole) —CH_3MgI→ N$^-$ $^+$MgI + CH_4

2. The greater acidity of these compounds, relative to the acidity of amines, is a consequence of the resonance stabilization and aromaticity of the conjugate-base anions.

IV. SOURCES OF AROMATIC NITROGEN HETEROCYCLIC COMPOUNDS

A. Indoles

1. The Fischer indole synthesis is an important method for the preparation of indoles that involves the reaction of a phenylhydrazine derivative with an aldehyde or ketone that has α-hydrogens.

phenylhydrazine (Ph–NH–NH_2) + R–CH_2–C(=O)–R′ —acid→ a protonated phenylhydrazone → an indole

   a. The mechanism of the Fischer indole synthesis begins with the conversion of the carbonyl compound into a phenylhydrazone, a type of imine.

i. The phenylhydrazone, which is protonated under the reaction conditions, is in equilibrium with a small amount of a protonated enamine tautomer.
ii. The latter species undergoes a pericyclic reaction involving three electron pairs (six electrons) to give a new intermediate in which the N—H bond of the phenylhydrazone has been broken (the intermediate formed is an imine).
iii. The imine, after protonation on the imine nitrogen, undergoes nucleophilic addition with the amine group in the same molecule with loss of ammonia.
iv. The resulting "enamine" derivative is the product indole.

phenylhydrazine + ketone —(acid; phenylhydrazone formation)→ a protonated phenylhydrazone —(isomerization)→

a protonated enamine —(pericyclic reaction)→ a protonated imine —(acid; "enamine" formation)→ an indole

b. The Fisher synthesis is most often used with phenylhydrazine itself to obtain indoles that are substituted at the 2- or 3-position.

2. The Reissert synthesis occurs under basic conditions; the key starting materials for this synthesis are diethyl oxalate and *o*-nitrotoluene or a substituted derivative.

substituted *o*-nitrotoluene + EtO—C(=O)—CO_2Et —(1) NaOEt; 2) acid; 3) H_2, cat.)→ ethyl indole-2-carboxylate —(1) NaOH; 2) H_3O^+; 3) heat, $-CO_2$)→ indole

a. The mechanism of the Reissert synthesis begins with the condensation of the *o*-nitrotoluene with diethyl oxalate in a variation of the Claisen condensation. (The *o*-nitro group is an essential element in the success of this reaction because its presence makes the benzylic methyl hydrogens acidic enough to be removed by ethoxide.)
i. The nitro group is converted into an amino group in a separate reduction step.
ii. The amino group thus formed reacts with neighboring ketone to yield, after acid-base equilibria, an "enamine," which is the product indole.

condensation

reduction

acid

"enamine" formation

1) saponification
2) decarboxylation

b. 2-Indolecarboxylic acid can be decarboxylated to prepare indole itself.

3. The Reissert synthesis is complementary to the Fischer indole synthesis.
   a. If substituted *o*-nitrotoluene derivatives are used in the Reissert reaction, this reaction, in conjunction with the final decarboxylation step, can be used to prepare indoles that are substituted in the benzene ring and unsubstituted at the 2- or 3-positions.
   b. Although many substituted phenylhydrazines work in the Fischer synthesis, some are difficult to prepare; thus, the Fischer synthesis is most often used to prepare indoles that are substituted at the 2- or 3-positions rather than in the phenyl ring.

B. Pyridines

1. The Chichibabin reaction can be used to prepare 2-aminopyridines.

+ $NaNH_2$ 1) heat 2) H_2O → + NaOH + H_2

   a. In the first step of the mechanism, the amide ion (nucleophile) attacks the 2-position of the ring to form a tetrahedral addition intermediate.
      i. The C=N linkage of the pyridine ring is somewhat analogous to the C=O of a carbonyl group.
      ii. The carbon at the 2-position of pyridine has some of the character of a carbonyl carbon, and can thus be attacked by nucleophiles.
      iii. However, the C=N group of pyridine is much less reactive than a carbonyl group because it is part of an aromatic system.
   b. In the second step of the mechanism, the leaving group, a hydride ion, is lost; hydride ion is a very poor leaving group because it is very basic.
      i. The aromatic pyridine ring is re-formed; aromaticity lost in the formation of the tetrahedral addition intermediate is regained when the leaving group departs.
      ii. The basic sodium hydride produced in the reaction reacts with the $—NH_2$ group irreversibly to form hydrogen gas and the resonance-stabilized conjugate-base anion of 2-aminopyridine.
   c. The neutral 2-aminopyridine is formed when water is added in a separate step.

$NaNH_2$ → tetrahedral addition intermediate → + NaH → $-H_2$ → H_2O →

d. The 2-aminopyridines formed in the Chichibabin reaction serve as starting materials for a variety of other 2-substituted pyridines.

2. Nicotinic acid, prepared by oxidation of nicotine, is a good source of 3-substituted pyridine derivatives.

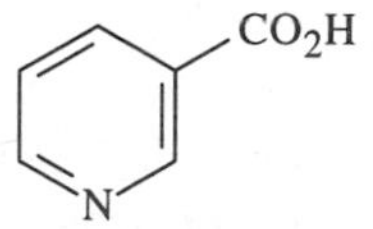

nicotinic acid
(pyridine-3-carboxylic acid)

3. Nitration of pyridine-*N*-oxide followed by reduction provides an entry into 4-substituted pyridines. (See "Reactions," Section II.B.4.)

C. Quinolines

1. One of the best known quinoline syntheses is the Skraup synthesis:

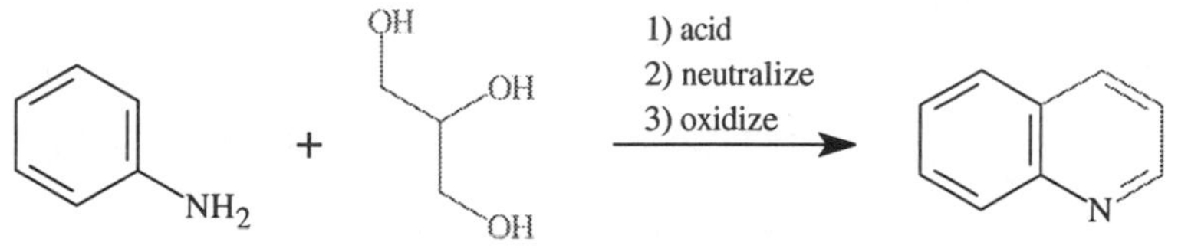

aniline + glycerol —(1) acid; 2) neutralize; 3) oxidize)→ quinoline

a. In the synthesis of quinoline itself, glycerol undergoes an acid-catalyzed dehydration to provide a small but continuously replenished amount of acrolein, an α,β-unsaturated aldehyde. (If acrolein itself were used as a reactant at high concentrations, it would polymerize.)
b. Aniline undergoes a conjugate addition with the acrolein.
c. The resulting aldehyde, after protonation, acts as an electrophile in an intramolecular electrophilic aromatic substitution reaction.
d. Dehydration of the resulting alcohol yields a 1,2-dihydroquinoline.
e. The 1,2-dihydroquinoline product differs from quinoline by only one degree of unsaturation, and is readily oxidized to the aromatic quinoline by mild oxidants such as As_2O_5, nitrobenzene, or ferric ion.

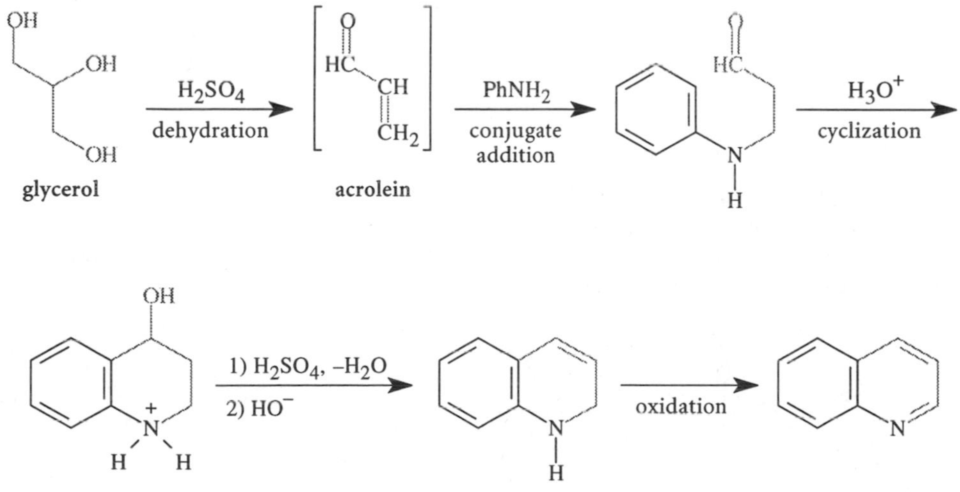

glycerol —(H_2SO_4, dehydration)→ [acrolein] —($PhNH_2$, conjugate addition)→ —(H_3O^+, cyclization)→

—(1) H_2SO_4, $-H_2O$; 2) HO^-)→ 1,2-dihydroquinoline —(oxidation)→ quinoline

D. Natural Occurrence of Heterocyclic Compounds

1. A number of monosubstituted pyridines are available from natural sources.
   a. Methylpyridines (or picolines) are obtained from coal tar.

b. Nicotinic acid (pyridine-3-carboxylic acid) can be prepared by the side-chain oxidation of nicotine, an alkaloid present in tobacco.

2. Nitrogen heterocycles occur widely in nature, for example, the alkaloids (many of which contain heterocyclic ring systems).
3. The naturally occurring amino acids proline, histidine, and tryptophan contain respectively a pyrrolidine, imidazole, and indole ring.
4. A number of vitamins are heterocyclic compounds.
5. The nucleic acids contain purine and pyrimidine rings in combined form.
6. The color of blood is due to an iron complex of heme, a heterocycle of the porphyrin type that is composed of pyrrole units.
7. The green color of plants is caused by chlorophyll, a class of compounds closely related to the porphyrins.

REACTIONS

I. ELECTROPHILIC AROMATIC SUBSTITUTION REACTIONS OF NAPHTHALENE AND ITS DERIVATIVES

A. Electrophilic Aromatic Substitution Reactions of Naphthalene

1. Naphthalene is an aromatic compound that undergoes electrophilic aromatic substitutions much like those of benzene, generally at the 1-position.
   a. Seven resonance structures can be drawn for the carbocation intermediate in electrophilic aromatic substitution at the 1-position, and four of these contain intact benzene rings (black).

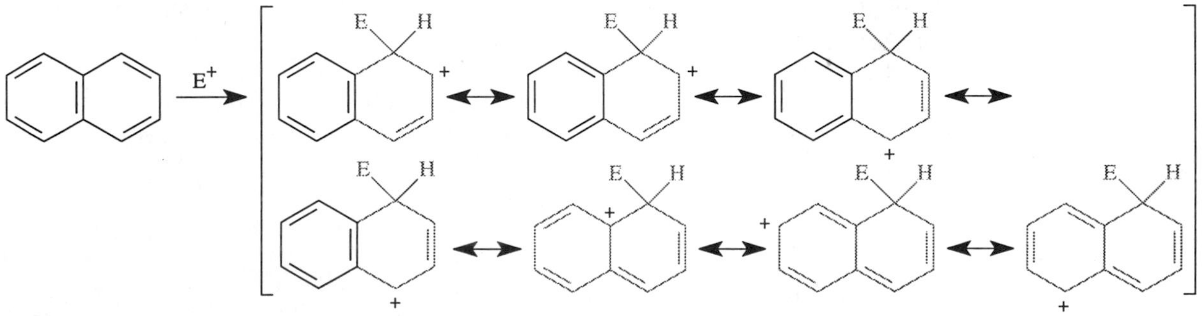

   i. Structures in which benzene rings are left intact are more important than those in which the formal double bonds are moved out of the ring.

   ii. Structures lacking the intact benzene rings are not aromatic, and thus they are less stable.

   b. Six resonance structures can be drawn for the carbocation intermediate in electrophilic aromatic substitution at the 2-position, and only two contain intact benzene rings (black).

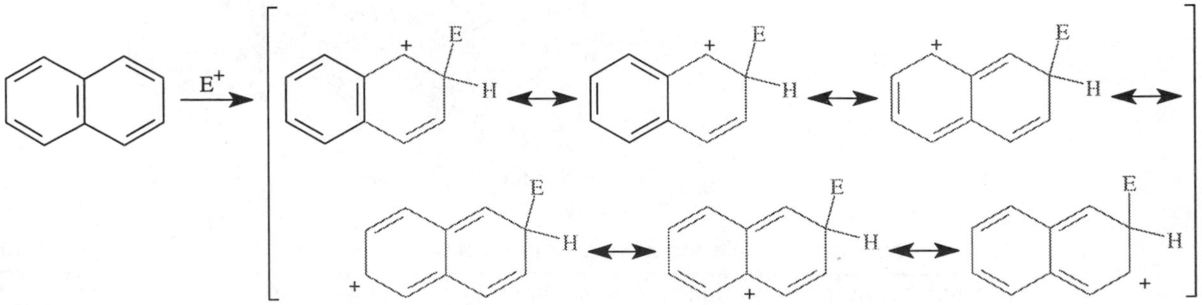

c. Substitution at the 1-position gives the more stable carbocation intermediate and, by Hammond's postulate, occurs more rapidly than substitution at the 2-position.

2. As with benzene, sulfonation of naphthalene is a reversible reaction.

   a. Sulfonation of naphthalene under mild conditions gives mostly 1-naphthalenesulfonic acid; under more vigorous conditions, sulfonation yields mostly 2-naphthalenesulfonic acid.

H_2SO_4; 0–40° reversible → SO_3H (1-position); 160° → SO_3H (2-position)

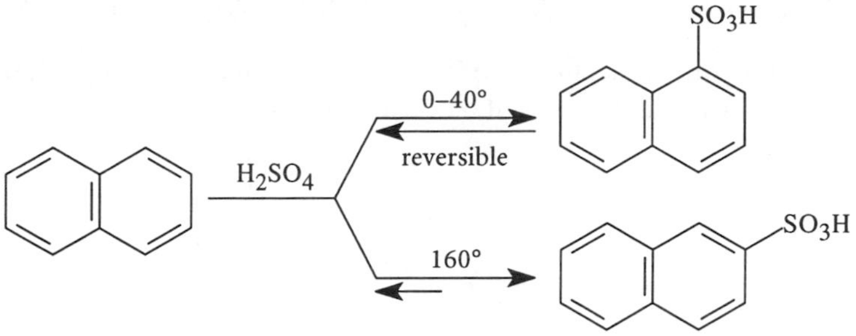

   b. This is a case of kinetic versus thermodynamic control of a reaction.
      i. At low temperature, substitution at the 1-position is observed because it is faster.
      ii. At higher temperature, formation of 1-naphthalenesulfonic acid is reversible, and the more stable but more slowly formed 2-naphthalenesulfonic acid is observed.

   c. In the 1-position, van der Waals repulsions occur between the large sulfonic acid group and the adjacent hydrogen in the 8-position.

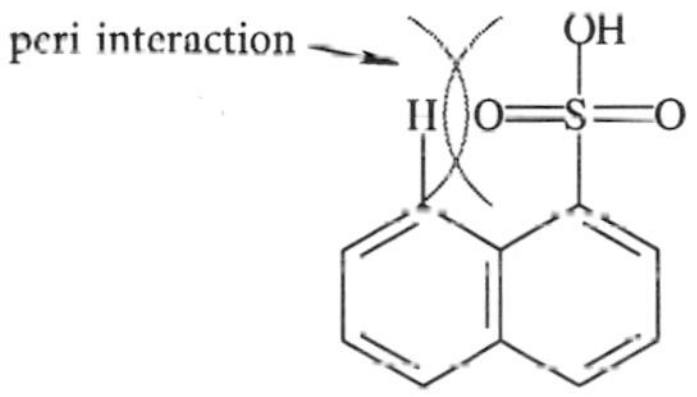

peri interaction; OH; H O=S=O

      i. This interaction, called a peri interaction, destabilizes the 1-isomer.
      ii. A peri interaction is much more severe than the interaction of the same two groups in ortho positions.
      iii. 1-Naphthalenesulfonic acid, if allowed to equilibrate, is converted into the 2-isomer to avoid this unfavorable steric interaction.

3. Naphthalene is considerably more reactive than benzene in electrophilic aromatic substitution.

   a. Naphthalene is readily brominated in CCl_4 without a catalyst.

   b. The greater reactivity of naphthalene in electrophilic aromatic substitution reactions reflects the considerable resonance stabilization of the carbocation intermediate.

B. Electrophilic Aromatic Substitution Reactions of Substituted Naphthalenes

1. The position of a second substitution on a monosubstituted naphthalene depends on the substituent present.

2. The following trends are observed in most cases:

   a. When one ring of naphthalene is substituted with deactivating groups, further substitution occurs on the unsubstituted ring at an open α-position (if available).

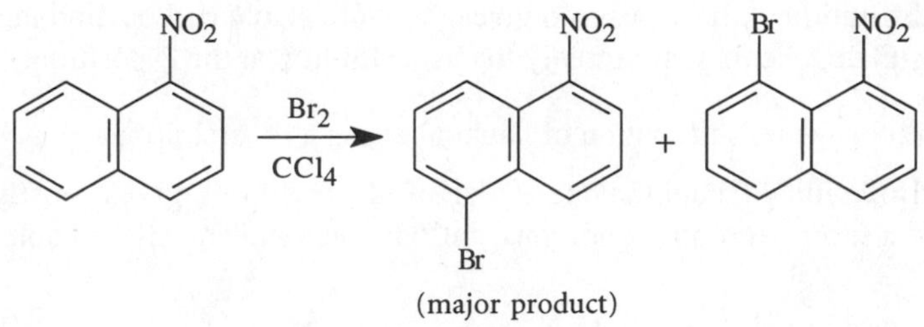

b. When one ring of naphthalene is substituted with activating groups, further substitution occurs in the substituted ring at the ortho or para positions.

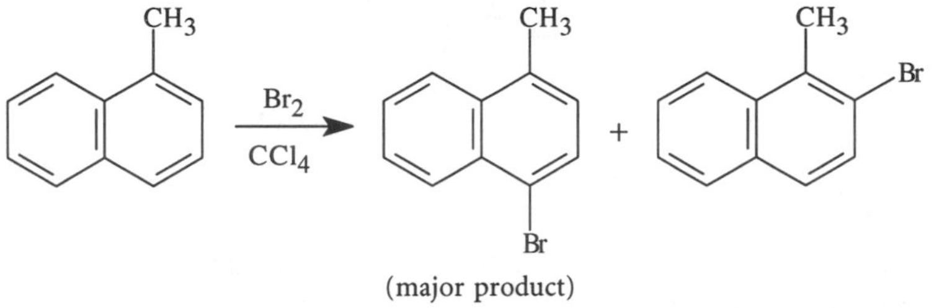

II. ELECTROPHILIC AROMATIC SUBSTITUTION REACTIONS OF AROMATIC HETEROCYCLIC COMPOUNDS

A. Furan, Pyrrole, and Thiophene

1. Furan, thiophene, and pyrrole undergo electrophilic substitution predominantly at the 2-position of the ring.
   a. The carbocation resulting from substitution at carbon-2 has more important resonance structures, and is therefore more stable, than the carbocation resulting from attack at carbon-3.

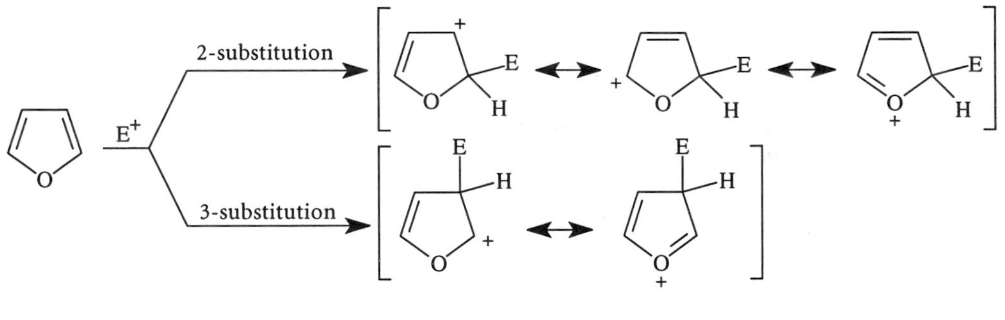

   b. Some carbon-3 substitution product accompanies the major carbon-2 substitution product in many cases.
2. Furan, pyrrole, and thiophene are all much more reactive than benzene in electrophilic aromatic substitution reactions.
   a. The reactivity order of the heterocycles is a consequence of the relative abilities of the heteroatoms to stabilize positive charge in the intermediate carbocations.

pyrrole > furan > thiophene >> benzene

      i. Both pyrrole and furan have heteroatoms from the first row of the periodic table.
      ii. Because nitrogen is better than oxygen at delocalizing positive charge (it is less electronegative), pyrrole is more reactive than furan.

iii. The sulfur of thiophene is a second row element and, although it is less electronegative than oxygen, its $3p$ orbitals overlap less efficiently with the $2p$ orbitals of the aromatic π-electron system.

b. The reactivity order of the heterocycles in aromatic substitution parallels the reactivity order of the corresponding substituted benzene derivatives:

$$(CH_3)_2N\text{—}Ph > CH_3O\text{—}Ph > CH_3S\text{—}Ph$$

3. Milder reaction conditions must be used with more reactive compounds.

furan (more reactive) is acetylated with $H_3C-\overset{O}{\overset{\|}{C}}-O-\overset{O}{\overset{\|}{C}}-CH_3/BF_3$ (milder conditions) to give 2-furyl$-\overset{O}{\overset{\|}{C}}-CH_3$

benzene (less reactive) is acetylated with $H_3C-\overset{O}{\overset{\|}{C}}-Cl/AlCl_3$ (harsher conditions) to give phenyl$-\overset{O}{\overset{\|}{C}}-CH_3$

4. The usual activating and directing effects of substituents in aromatic substitution apply; superimposed on these effects is the normal effect of the heterocyclic atom in directing substitution to the 2-position.
   a. Count around the carbon framework of the heterocycle, not through the heteroatom, when using the ortho, meta, para analogy.

2-fluorofuran (F: ortho, para-directing group) $\xrightarrow[HOAc]{HNO_3}$ O_2N-furan-F (*para* to fluorine)

   b. When the directing effects of substituents and the ring compete, it is not unusual to observe mixtures of products.
   c. If both 2-positions are occupied, 3-substitution takes place.

2,5-dimethylthiophene (H_3C, CH_3, S) $\xrightarrow[HOAc]{HNO_3}$ 3-nitro-2,5-dimethylthiophene (NO_2, H_3C, CH_3, S)

B. Pyridine

1. In general, pyridine has very low reactivity in electrophilic aromatic substitutions; it is much less reactive than benzene.
   a. An important reason for this low reactivity is that pyridine is protonated under the very acidic conditions of most electrophilic aromatic substitution reactions.
   b. The resulting positive charge on nitrogen makes it difficult to form a carbocation intermediate, which would place a second positive charge within the same ring.
2. Pyridine rings substituted with activating groups such as methyl groups do undergo electrophilic aromatic substitution reactions.
   a. When substitution in pyridine does occur, it generally takes place at the 3-position.

b. The preference for 3-substitution in pyridines can be understood by considering the resonance structures for the possible carbocation intermediates.

i. Substitution in the 3-position gives a carbocation with three resonance structures.

E^+

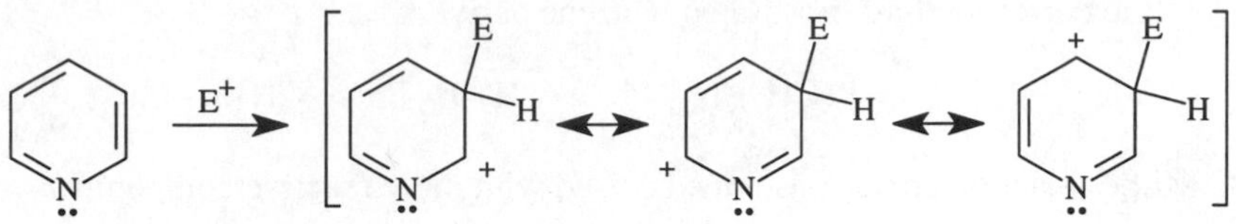

ii. Substitution at the 4-position also involves an intermediate with three resonance structures, but the one shown in gray is particularly unfavorable because the nitrogen, an electronegative atom, is electron deficient.

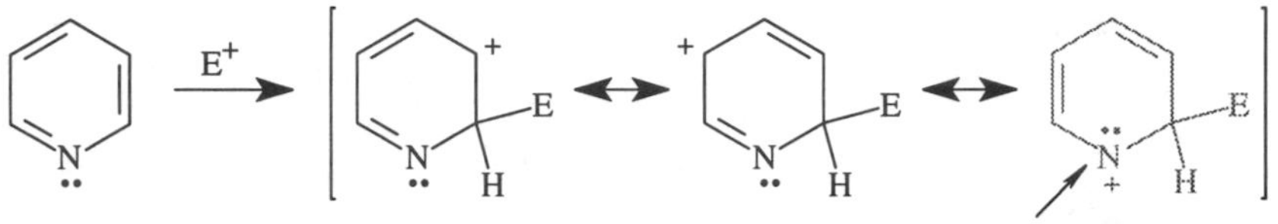

E^+

no octet of electrons

3. Pyridine can be oxidized to give pyridine-*N*-oxide.

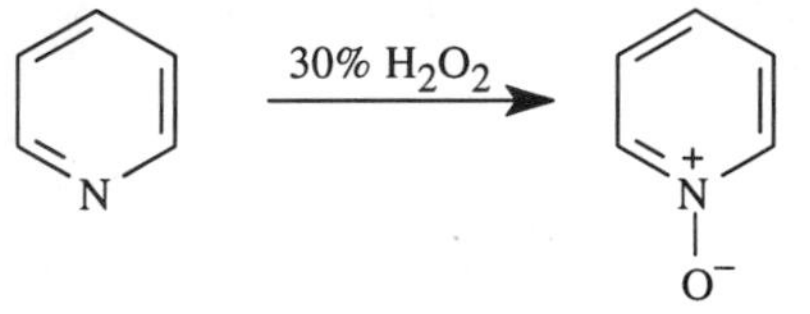

30% H_2O_2

pyridine-*N*-oxide

4. Pyridine-*N*-oxide is more reactive than pyridine and undergoes nitration in the 4-position.

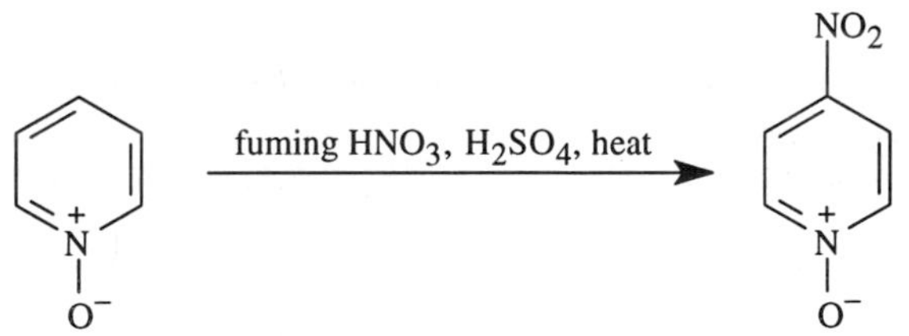

fuming HNO_3, H_2SO_4, heat

a. The *N*-oxide function can be removed by catalytic hydrogenation, which will also reduce any nitro groups present.

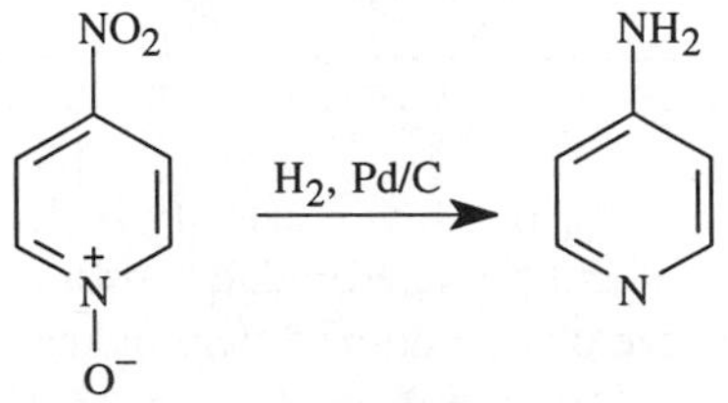

H_2, Pd/C

b. Reaction with trivalent phosphorous compounds, such as PCl_3, removes the *N*-oxide function without reducing nitro groups.

4-nitropyridine-N-oxide $\xrightarrow{PCl_3}$ 4-nitropyridine + $POCl_3$

5. Quinoline-*N*-oxide undergoes reactions similar to pyridine-*N*-oxide.

III. NUCLEOPHILIC AROMATIC SUBSTITUTION REACTIONS OF AROMATIC HETEROCYCLIC COMPOUNDS

A. Pyridine

1. Treatment of a pyridine derivative with the strong base sodium amide ($Na^+ {}^-NH_2$) brings about the direct substitution of an amino group for a ring hydrogen; this reaction is called the Chichibabin reaction.

pyridine $\xrightarrow[2)\ H_2O]{1)\ NaNH_2,\ heat}$ 2-aminopyridine (NH_2)

2. A reaction similar to the Chichibabin reaction occurs with organolithium reagents.

pyridine $\xrightarrow[2)\ H_2O]{1)\ PhLi,\ heat}$ 2-phenylpyridine (Ph)

3. When pyridine is substituted with a better leaving group than hydride at the 2-position, it reacts more rapidly with nucleophiles; thus, the 2-halopyridines readily undergo substitution of the halogen by other nucleophiles under conditions milder than those used in the Chichibabin reaction.

2-chloropyridine $\xrightarrow{Na^+ {}^-SMe}$ 2-(methylthio)pyridine (SMe) + $Na^+ {}^-Cl$

4. Nucleophilic substitution reactions on pyridine rings can be classified as nucleophilic aromatic substitution reactions.
   a. The "electron-withdrawing group" in the reaction of pyridines is the pyridine nitrogen itself; consider the ring nitrogen of pyridine as if it were a nitro group attached to a benzene ring.
   b. The tetrahedral addition intermediate is analogous to the Meisenheimer complex of nucleophilic aromatic substitution.

2-X-pyridine + $:Nuc^-$ → tetrahedral addition intermediate → 2-Nuc-pyridine + $X:^-$

tetrahedral
addition intermediate

5. 3-Substituted pyridines are not reactive in nucleophilic substitution because negative charge in the addition intermediate cannot be delocalized onto the electronegative nitrogen.

6. 2-Pyridone contains a significant amount of both carbonyl and hydroxy isomers, and it often undergoes some reactions reminiscent of hydroxy compounds.

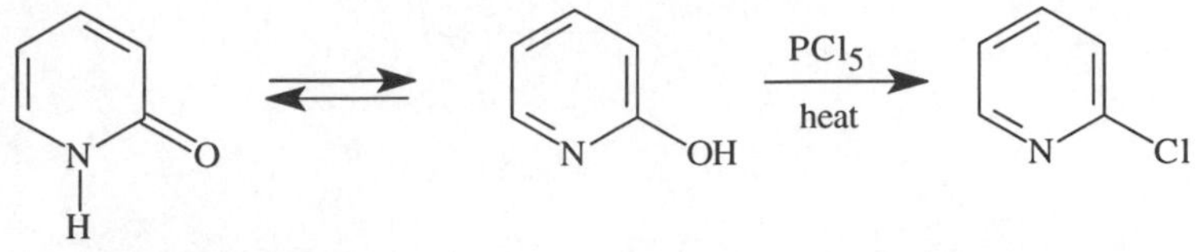

PCl_5 / heat

2-pyridone

B. Pyridinium Salts and Their Reactions

1. Pyridine is a nucleophile and reacts in S_N2 reactions with alkyl halides or sulfonate esters to form quaternary ammonium salts, called pyridinium salts.

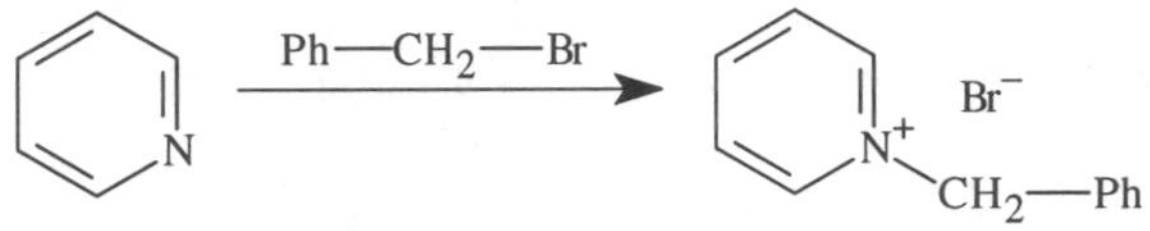

Ph—CH_2—Br → Br^-, N^+—CH_2—Ph

2. Pyridinium salts are activated toward nucleophilic displacement of groups at the 2- and 4-positions of the ring much more than are pyridines themselves, because the positively charged nitrogen is much more electronegative than the neutral nitrogen of a pyridine.

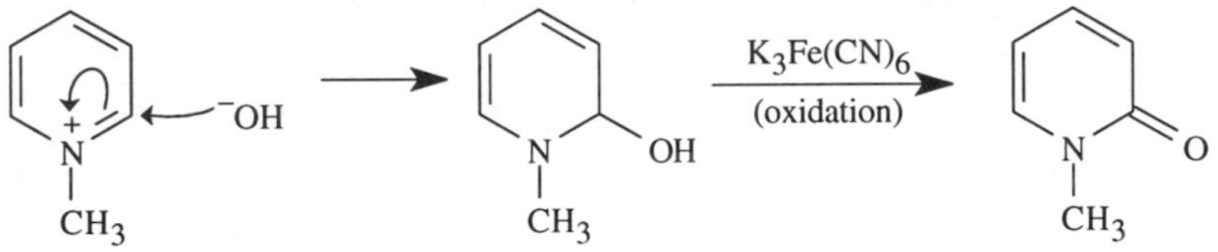

$K_3Fe(CN)_6$ (oxidation)

a. When the nucleophiles in such displacement reactions are anions, charge is neutralized.
b. Pyridine-*N*-oxides are in one sense pyridinium ions, and they react with nucleophiles in much the same way as quaternary pyridinium salts.

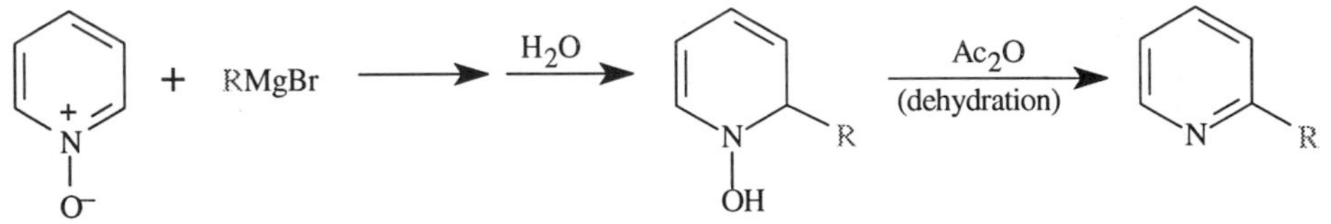

\+ RMgBr → H_2O → ; Ac_2O (dehydration)

IV. OTHER REACTIONS OF HETEROCYCLIC AROMATIC COMPOUNDS

A. Addition Reactions to Furan

1. Furan, pyrrole, or thiophene can be viewed as a 1,3-butadiene with its terminal carbons "tied down" by a heteroatom bridge.

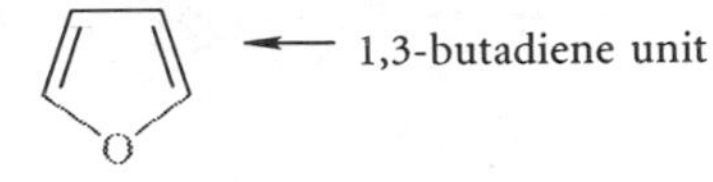

← 1,3-butadiene unit

2. Of the three heterocyclic compounds furan, pyrrole, and thiophene, furan has the least resonance energy.

a. Furan has the greatest tendency to behave like a conjugated diene.
b. Furan undergoes some conjugate-addition reactions.
c. Furan undergoes Diels-Alder reactions with reactive dienophiles.

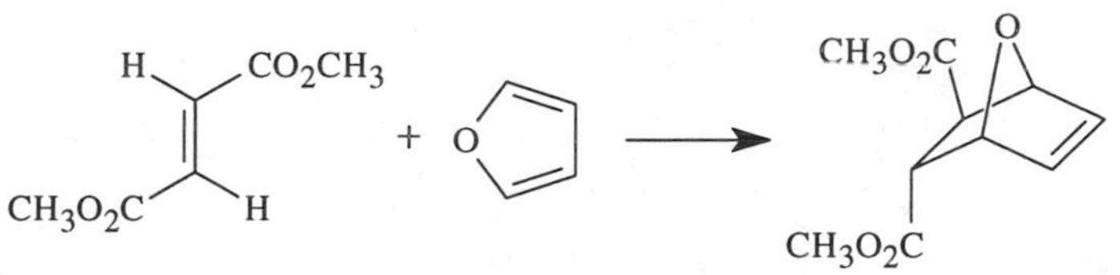

H, CO_2CH_3, CH_3O_2C, H + O → CH_3O_2C, O, CH_3O_2C

B. Side-Chain Reactions of Furan, Pyrrole, and Thiophene

1. Many reactions occur at the side chains of heterocyclic compounds without affecting the rings.
2. Decarboxylation of carboxylic acid groups directly attached to a heterocyclic ring is important in the synthesis of some unsubstituted heterocyclic compounds.

S, CO_2H —heat→ S + CO_2 ↑

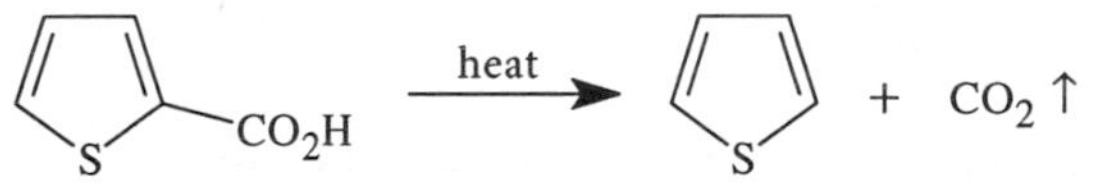

C. Side-Chain Reactions of Pyridine Derivatives

1. The "benzylic" hydrogens of an alkyl group at the 2- or 4-position of a pyridine ring are more acidic than ordinary benzylic hydrogens because the electron pair (and charge) in the conjugate-base anion is delocalized onto the electronegative pyridine nitrogen.
   a. Strongly basic reagents such as organolithium reagents or $NaNH_2$ abstract the benzylic proton from 2- or 4-alkylpyridines.

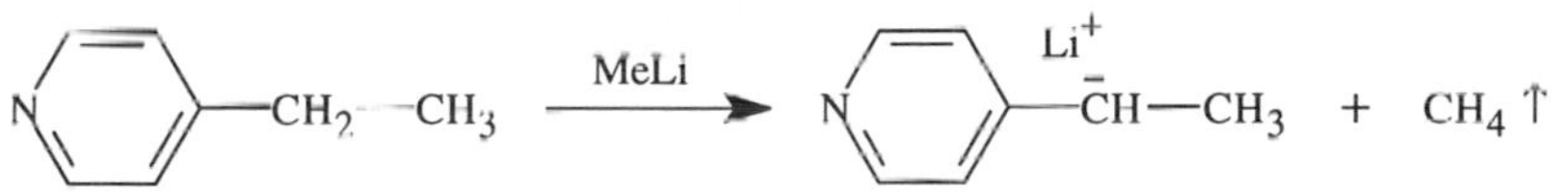

N, CH_2—CH_3 —MeLi→ N, Li^+, $\bar{C}H$—CH_3 + CH_4 ↑

   b. The anion formed in this way has a reactivity much like that of other organolithium reagents.
   c. These anions are somewhat analogous to enolate ions, and thus they undergo some of the reactions of enolate anions, such as aldol condensations.
2. The "benzylic" hydrogens of 2- or 4-alkylpyridinium salts are much more acidic than those of the analogous pyridines.
   a. One resonance form of the conjugate-base "anion" is an uncharged species.

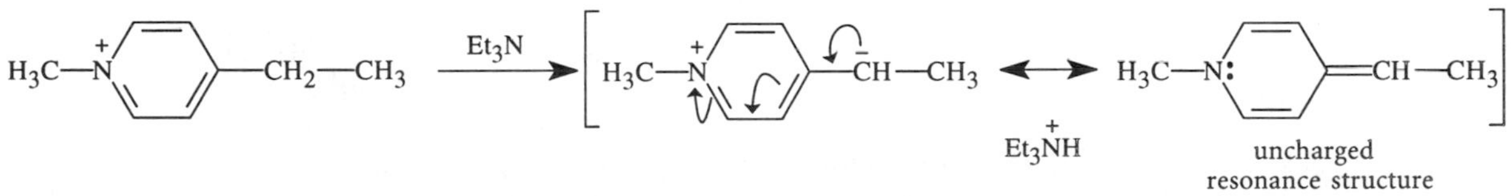

H_3C—$\overset{+}{N}$, CH_2—CH_3 —Et_3N→ [H_3C—$\overset{+}{N}$, $\bar{C}H$—CH_3 ←→ H_3C—N:, =CH—CH_3]

$Et_3\overset{+}{N}H$

uncharged resonance structure

   b. The conjugate base "anion" can be formed in useful concentrations by aqueous NaOH or amines.
3. Many side-chain reactions of pyridines, for example, side-chain oxidation, are analogous to those of the corresponding benzene derivatives.

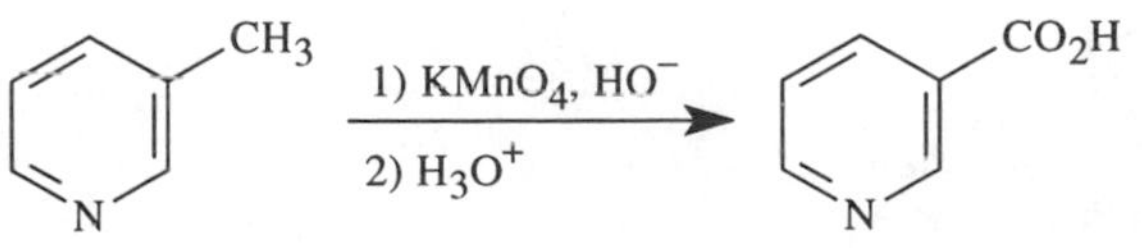

CH_3, N —1) $KMnO_4$, HO^- 2) H_3O^+→ CO_2H, N

D. Pyridinium Ions in Biology: Pyridoxal Phosphate

1. Pyridoxal phosphate is one of the several forms of vitamin B_6.

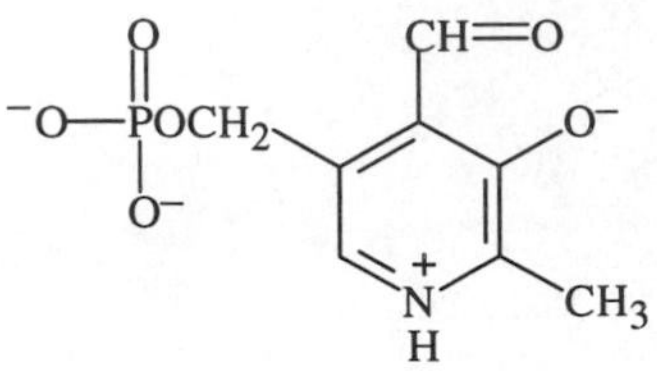

pyridoxal phosphate

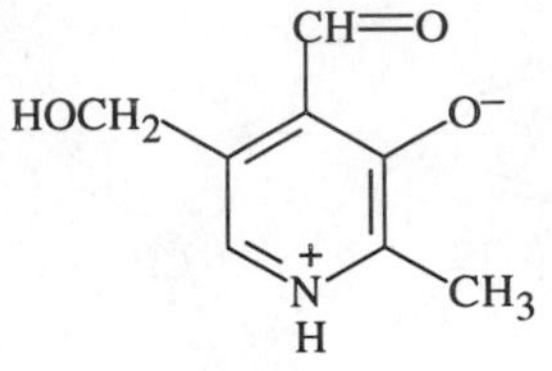

pyridoxal

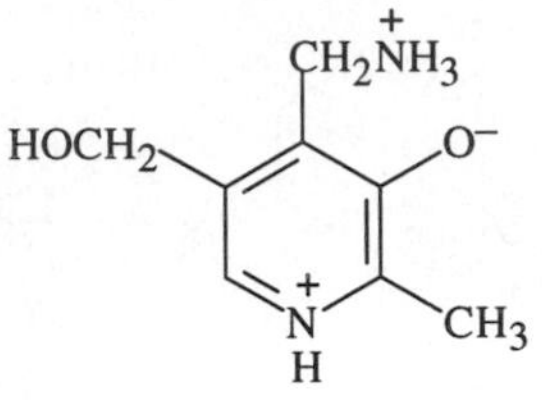

pyridoxamine

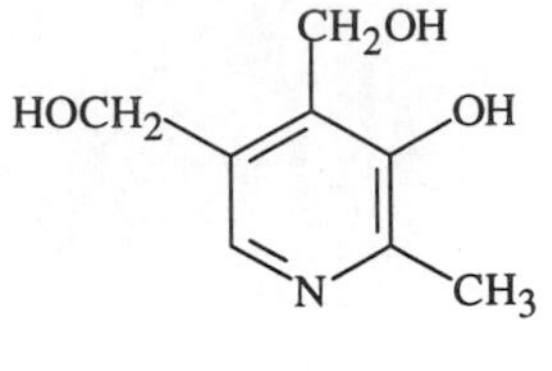

pyridoxol (pyridoxine)

2. Pyridoxal phosphate is an essential reactant in several important biochemical transformations.
   a. interconversion of α-amino acids and α-keto acids

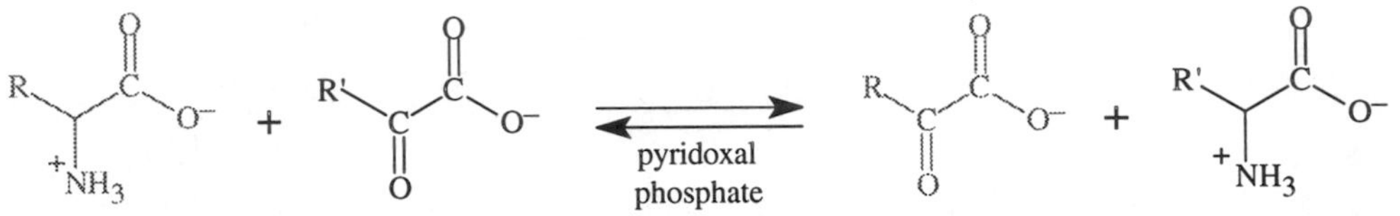

an α-amino acid + an α-keto acid ⇌ (pyridoxal phosphate) an α-keto acid + an α-amino acid

   b. decarboxylation of amino acids

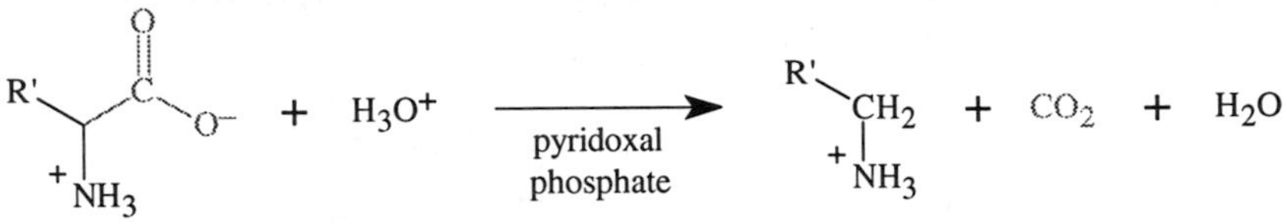

an α-amino acid + H_3O^+ → (pyridoxal phosphate) $R'CH_2\overset{+}{N}H_3$ + CO_2 + H_2O

   c. loss of formaldehyde from serine

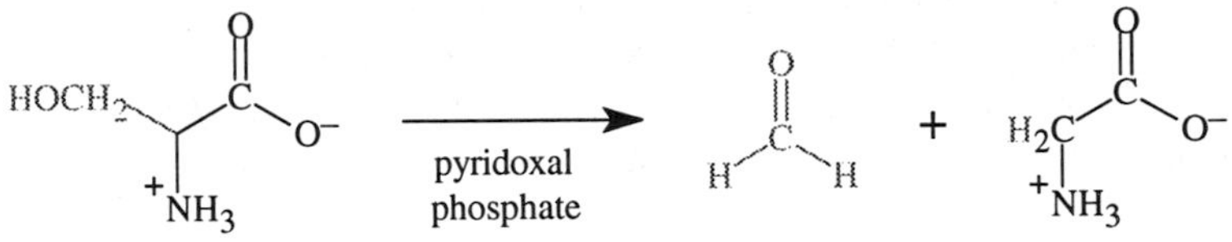

serine (an α-amino acid) → (pyridoxal phosphate) formaldehyde + glycine (an α-amino acid)

3. In the biological world, pyridoxal phosphate exists as imine derivatives of various amines, including enzymes.

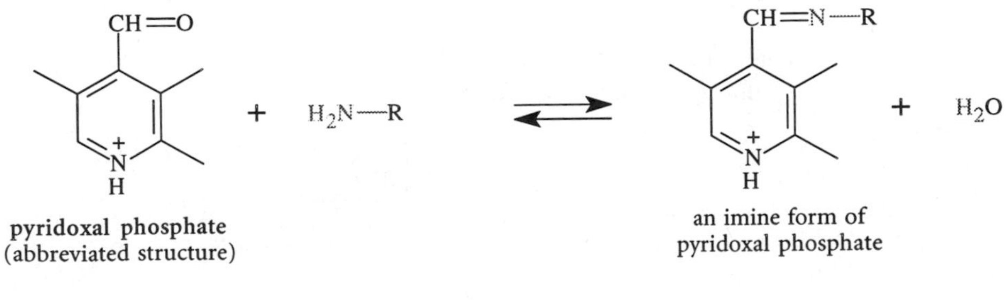

pyridoxal phosphate (abbreviated structure) + $H_2N—R$ ⇌ an imine form of pyridoxal phosphate + H_2O

4. Biological reactions involving pyridoxal phosphate hinge on the ability of the pyridinium ion to stabilize carbon anions.

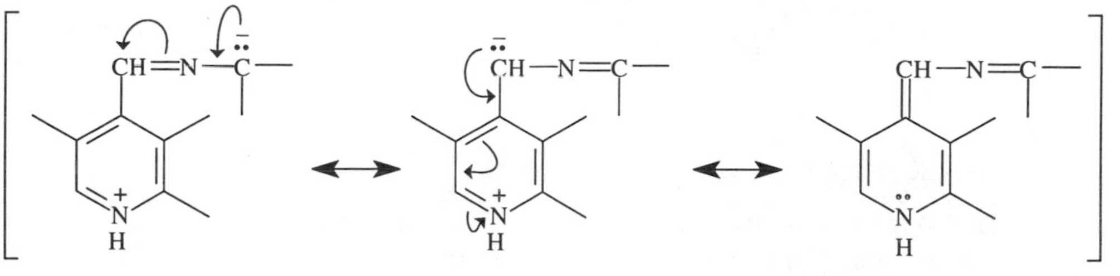

uncharged resonance structure

5. The *N*-protonated form of pyridoxal phosphate is thus very important to its mechanism of action.
   a. The predominant form of pyridoxal phosphate at physiological pH is the form in which the phenol is ionized and the pyridine is protonated.
   b. The —OH group in the 3-position and its ortho relationship to the aldehyde makes the phenolic —OH group of pyridoxal phosphate unusually acidic.
   c. Ionization of the phenolic —OH group raises the pK_a of the pyridinium ion.
   d. The pyridinium form is stabilized when pyridoxal phosphate is bound to the enzymes that catalyze its reactions.

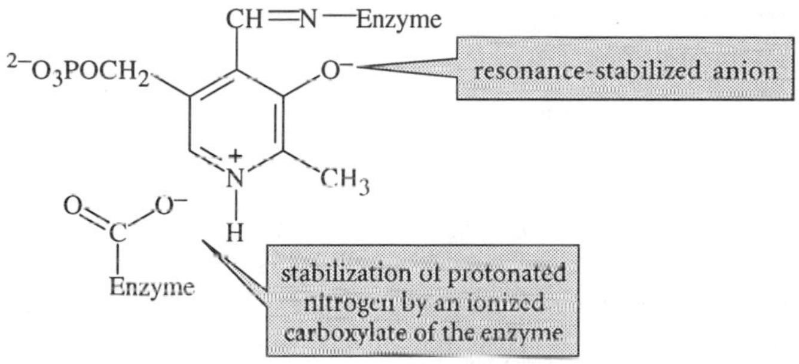

resonance-stabilized anion

stabilization of protonated nitrogen by an ionized carboxylate of the enzyme

V. SYNTHESIS OF INDOLE AND QUINOLINE FROM ACYCLIC STARTING MATERIALS

A. Synthesis of Indoles—the Fischer and Reissert Syntheses

1. In the Fischer indole synthesis, an aldehyde or ketone with at least two α-hydrogens is reacted with a phenylhydrazine derivative in the presence of an acid catalyst and/or heat. (See "Concepts," Section IV.A.1., for a summary of the mechanism.)

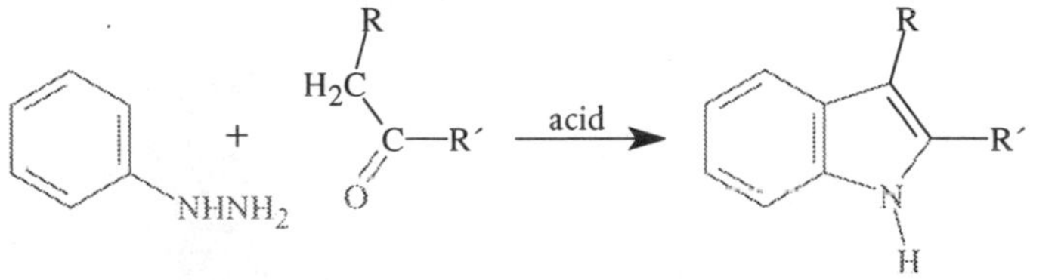

   a. A variety of Brønsted or Lewis acid catalysts can be used.
   b. The reaction works with many different substituted phenylhydrazines and carbonyl compounds.
   c. Acetaldehyde, however, does not work in this reaction, probably because it polymerizes under the reaction conditions.
2. The Reissert synthesis gives substituted 2-indolecarboxylic acids from diethyl oxalate and substituted *o*-nitrotoluenes; decarboxylation yields a substituted indole. (See "Concepts," Section IV.A.2., for a summary of the mechanism.)

CH_3, NO_2 + EtO–C(=O)–CO_2Et —[1) base, 2) reduction, 3) acid]→ (indole)–C–CO_2Et (N–H)

B. Synthesis of Quinolines

1. The Skraup synthesis of quinolines occurs under acid catalysis; the key starting materials for this reaction are glycerol and aniline or an aniline derivative. (See "Concepts," Sec. IV.C.2., for a summary of the mechanism.)

(aniline, NH_2) + (glycerol, OH, OH, OH) —[H_2SO_4, $PhNO_2$]→ (quinoline, N)

2. α,β-Unsaturated aldehydes and ketones that are less prone to polymerize than acrolein can be used instead of glycerol in the Skraup synthesis to give substituted quinolines.

SCIENTISTS

Alexei Yevgenievich Chichibabin (1871–1945)	24.5B
Robert Floyd Curl, Jr. (b. 1933) [Chemistry 1996]	Introduction
Hermann Emil Fischer (1852–1919) [Chemistry 1902]	24.4A
Harold Walter Kroto (b. 1939) [Chemistry 1996]	Introduction
Arnold Reissert	24.4B
Zdenko Hans Skraup (1850–1910)	24.5F
Richard Errett Smalley (b. 1943) [Chemistry 1996]	Introduction
Esmond Emerson Snell (b. 1914)	24.5E

Dates in brackets refer to Nobel Prize awards.

STUDY GUIDE LINKS

24.1 Relative Acidities of 1,3-Cyclopentadiene and Pyrrole

It is interesting that pyrrole, a nitrogen acid, is somewhat *less* acidic than 1,3-cyclopentadiene, a carbon acid.

1,3-cyclopentadiene
$pK_a \approx 15$

pyrrole
$pK_a \approx 17$

This contrasts with the general expectation from the *element effect* (text p. 99) that hydrogens on more electronegative atoms are more acidic than those on less electronegative atoms. (For example, the pK_a of NH_3 is about 35; the pK_a of CH_4 is about 55; the N—H pK_a of amides is about 16; the α-hydrogen pK_a of a ketone is about 19.) The greater acidity of 1,3-cyclopentadiene shows the importance of aromaticity. 1,3-Cyclopentadiene itself is not aromatic, but its conjugate-base anion is. Consequently, 1,3-cyclopentadiene "gains aromaticity" when it ionizes. This effect lowers the pK_a of 1,3-cyclopentadiene significantly from what it otherwise would be. (The effect of aromaticity might be worth as much as 20 pK_a units.) In contrast, both pyrrole and its conjugate-base anion are aromatic; ionization of pyrrole does not gain any aromatic stability for the molecule. In summary, then, the increase in stability resulting from the formation of an aromatic species causes the pK_a of the carbon acid to be lower than that of the nitrogen acid.

✓24.2 Fischer Indole Synthesis

Although the reaction in Eq. 24.38c on text p. 1141 is a concerted reaction, thinking of it as a stepwise process helps to understand why it takes place.

The weakest bond in the protonated enamine is the bond between the two nitrogens; bonds between two electronegative atoms are typically rather weak. Moreover, the protonated nitrogen is especially electronegative, and electrons from the N—N bond are drawn to that nitrogen:

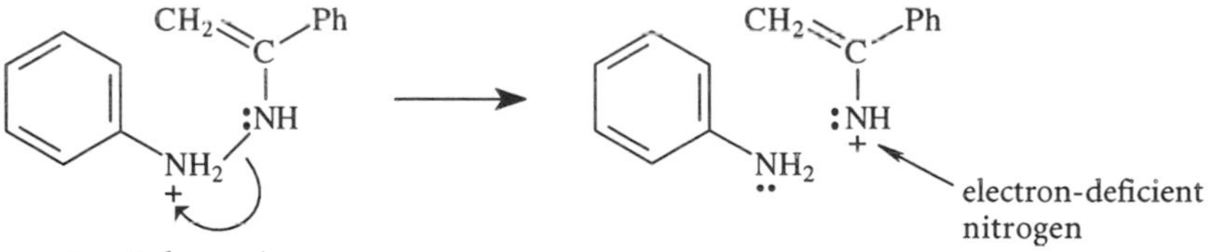

protonated enamine

electron-deficient nitrogen

The species with the electron-deficient nitrogen is really a carbocation:

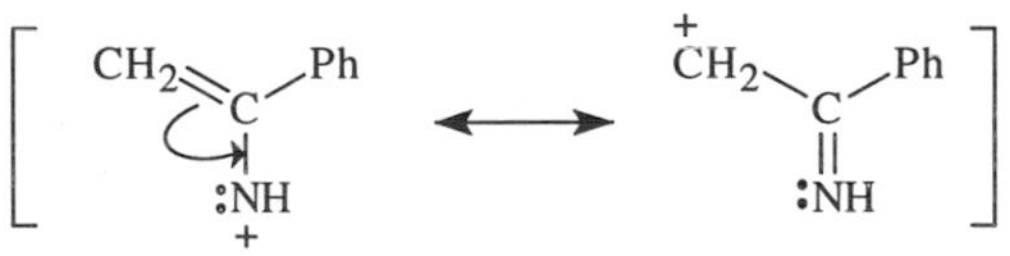

The carbocation carries out an electrophilic substitution on the neighboring aniline, which is very reactive in such reactions (Sec. 23.9):

H $\overset{+}{C}H_2$ Ph C :NH $\ddot{N}H_2$ $\longrightarrow$ H CH_2 Ph C :NH $\overset{+}{N}H_2$

This is the product of the pericyclic process, which goes on to form the imine intermediate as shown in the last part of text Eq. 24.38c.

24.3 Dehydration of Glycerol

The dehydration of glycerol that is part of the Skraup quinoline synthesis is actually an example of what is called the *pinacol rearrangement.* This rearrangement was introduced in Study Problem 19.2 on text p. 853. The steps in this rearrangement can be outlined as follows; you draw the curved-arrow notation.

$$\underset{\text{OH}}{CH_2}-\underset{\text{OH}}{CH}-\underset{\text{OH}}{CH_2} \xrightleftharpoons{H_2SO_4} \underset{\text{OH}}{CH_2}-\underset{\overset{+}{O}H_2}{CH}-\underset{\text{OH}}{CH_2} + HSO_4^- \xrightleftharpoons{-H_2O} \underset{\text{OH}}{CH_2}-\overset{+}{CH}-\underset{\text{OH}}{CH_2} \longrightarrow$$

$$\overset{+OH}{\overset{\|}{CH}}-CH_2-\underset{\text{OH}}{CH_2} \xrightleftharpoons{HSO_4^-} \overset{O}{\overset{\|}{CH}}-CH_2-\underset{\text{OH}}{CH_2} + H_2SO_4$$

3-hydroxypropanal

Dehydration of 3-hydroxypropanal to acrolein probably proceeds by way of the enol; this type of reaction is discussed in Study Guide Link 22.3 on p. 813 of this manual.

SOLUTIONS TO PROBLEMS

Solutions to In-Text Problems

24.1 Naphthalene has four nonequivalent sets of carbon-carbon bonds, shown as *a–d* in the structure below. Focus on *one* bond of each set and determine the number of resonance structures in which it appears as a double bond. Only bond *b* is a double bond in two of the three resonance structures (Eq. 24.1 on text p. 1122). Because no other bond has so much double-bond character, and because double bonds are shorter than single bonds, bond *b* is the shortest bond. (This hypothesis is confirmed by the data: the bond lengths are *a* and *d*, 1.42 Å; *b,* 1.36 Å, and *c,* 1.41 Å.)

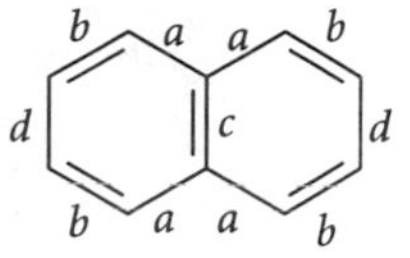

24.2 (a) 5,8-Dinitro-2-naphthalenecarboxylic acid
(b) 2-Bromonaphthalene (common: *β*-bromonaphthalene)
(c) 2-Naphthyl ethanoate (common: *β*-naphthyl acetate)
(d) 1,8-Naphthalenediol

24.3 (a) Because the substituted ring is deactivated, and because sulfonation is reversible, the sulfonic acid group is introduced into the *β* positions of the unsubstituted ring.

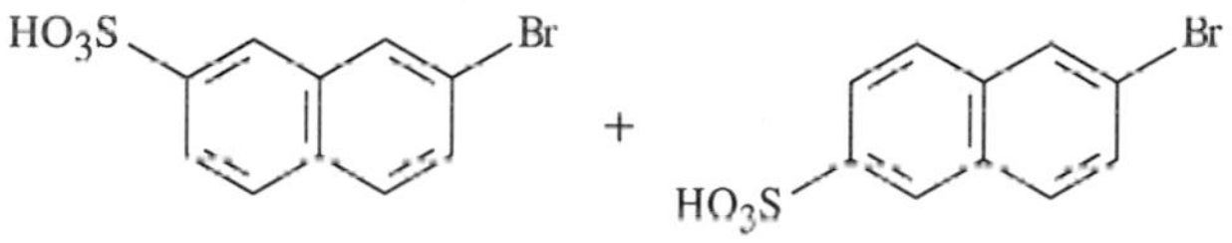

(b) Because the substituted ring is activated, and because the methyl group is an ortho, para-directing substituent, substitution in part (b) occurs predominantly at the 4-position.

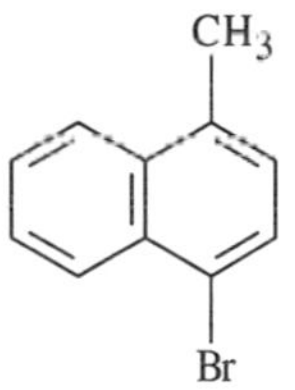

(c) Because the substituted ring in (c) is activated, and because the hydroxy group is an ortho, para-directing substituent, diazo coupling occurs at the 4-position.

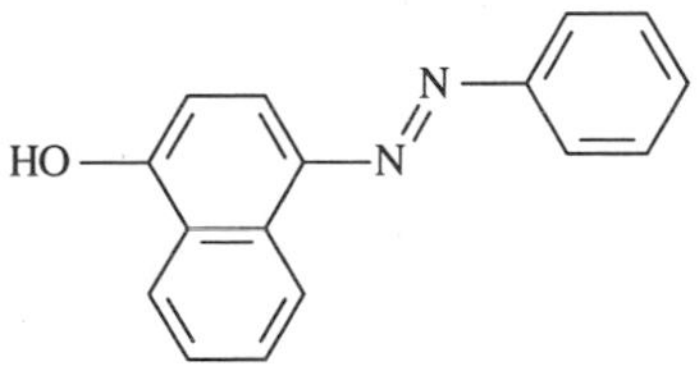

(d) Because the substituted ring is activated, because the methoxy group is an ortho, para-directing substituent, and because the α-position is more reactive than the β-position, substitution in part (d) occurs primarily at the 1-position.

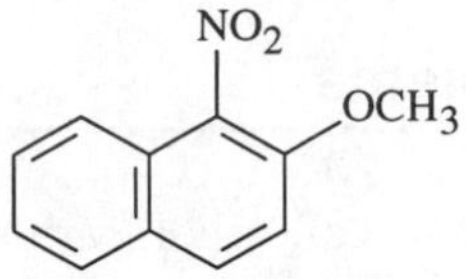

NO_2 OCH_3

24.4 Use a Grignard reaction to form the carbon-carbon bond.

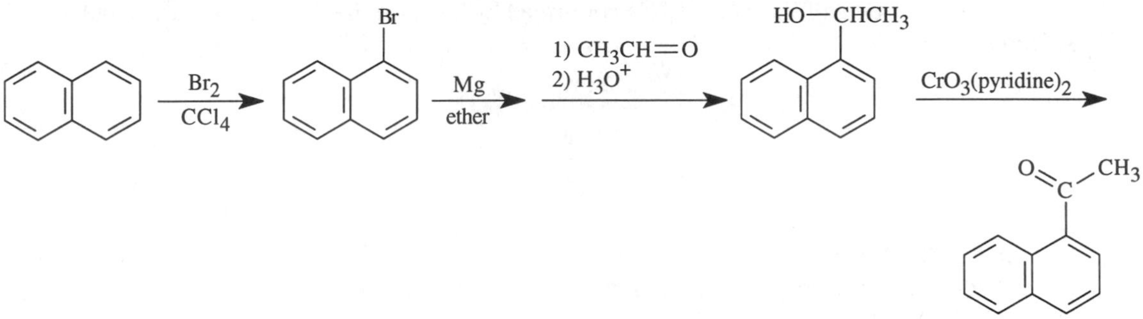

$\xrightarrow[CCl_4]{Br_2}$ (Br) $\xrightarrow[ether]{Mg}$ $\xrightarrow{1)\ CH_3CH{=}O \quad 2)\ H_3O^+}$ ($HO{-}CHCH_3$) $\xrightarrow{CrO_3(pyridine)_2}$ ($O{=}C{-}CH_3$)

24.5 (a)

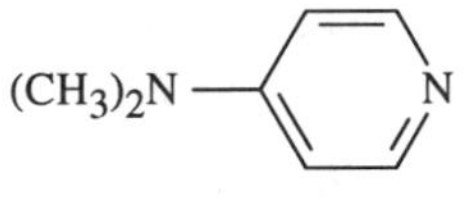

$(CH_3)_2N$— N

4-(dimethylamino)pyridine

(b)

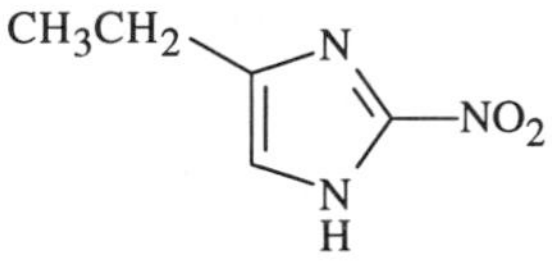

CH_3CH_2 N NO_2 N H

4-ethyl-2-nitroimidazole

24.6 (a) 5-Bromo-2-methylthiazole
(b) 2-Bromo-4-nitropyrrole
(c) 8-Methoxyquinoline

24.7 The resonance structures of pyrrole:

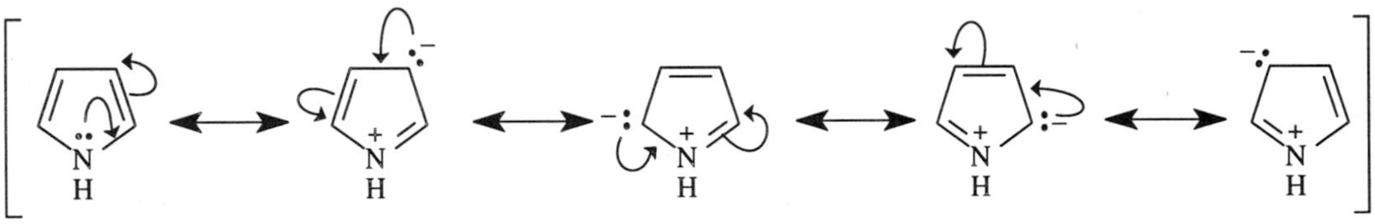

24.8 (a) Because nitrogen is an electronegative atom, the C—N bond dipoles in pyrrolidine are directed towards the nitrogen, and their resultant is also directed towards the nitrogen:

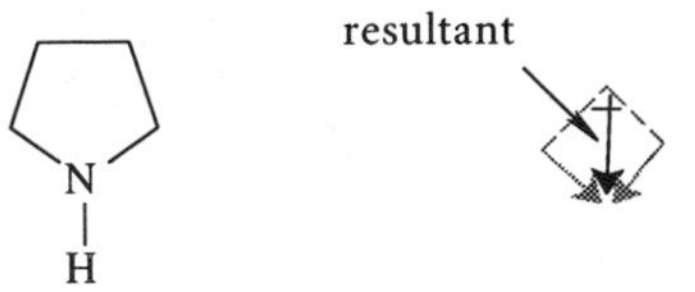

It is given that the dipole moment of pyrrole is in the opposite direction, that is, directed away from the nitrogen. Although there is a dipole component directed towards the nitrogen as shown for pyrrolidine, the major component is due to the charge separation shown in the resonance structures given in the solution to Problem 24.7. In all resonance structures but the first, positive charge resides on the nitrogen and negative charge is on one of the carbons. Evidently, these contributions significantly outweigh those from the carbon-nitrogen bond dipoles.

(b) Oxygen is more electronegative than nitrogen. Consequently, the contribution of the carbon-oxygen bond dipoles in furan is greater than the contribution of the carbon-nitrogen bond dipoles in pyrrole, and the contribution of the charge-separated resonance structures for furan is smaller than the contribution of the analogous structures for pyrrole. Notice that the resonance energy of furan is less than that of pyrrole (Table 24.1 on text p. 1132); this is consistent with the idea that the resonance structures of furan are less important than those of pyrrole.

In summary: In furan, the sum of the C—O bond dipoles outweighs the dipole-moment contribution of π-electron delocalization (see Eq. 24.13 on text p. 1130). In pyrrole, the dipole-moment contribution of π-electron delocalization outweighs the sum of its C—N bond dipoles.

(c) Because the dipole vector of pyrrole is directed away from the nitrogen, and because the resultant of the two carbon-chlorine bond dipoles is also directed away from the nitrogen, the dipole moment of 3,4-dichloropyrrole should be greater than that of pyrrole:

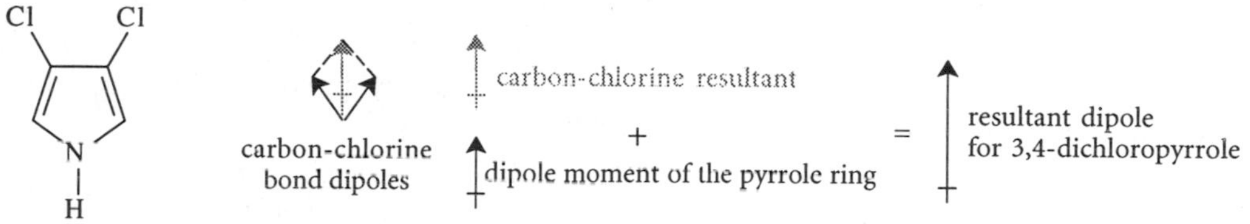

Cl Cl
N
H

carbon-chlorine bond dipoles

carbon-chlorine resultant
+
dipole moment of the pyrrole ring
=
resultant dipole for 3,4-dichloropyrrole

24.9 Both pyrrole and pyridine are aromatic, and therefore both should have ring current-induced downfield chemical shifts. Consequently, the two larger shifts are associated with the two aromatic compounds, and the δ 2.82 resonance is that of pyrrolidine. This chemical shift is not far from the value predicted for the α-proton resonance of an alkylamine. (See Fig. 13.4 on text p. 550.) The resonance structures for pyrrole (see the solution to Problem 24.7, above) as well as its dipole moment (see the solution to Problem 24.8(a), above) show that pyrrole has partial negative charge on carbon-2; in contrast, the resonance structures of pyridine (Eq. 24.14, text p. 1131) show that carbon-2 has partial positive charge. Since electron deficiency is associated with large chemical shifts, it follows that the δ 8.51 resonance is that of pyridine and the δ 6.41 resonance is that of pyrrole.

24.10 (a) The pyridine nitrogen is basic. Since bases can accept hydrogen bonds, pyridine can accept hydrogen bonds from water. This hydrogen-bonding capability increases the solubility of pyridine in water. Pyrrole, in contrast, is not basic, and the N—H hydrogen is not particularly acidic. Hence, pyrrole does not form strong hydrogen bonds with water, and it is therefore not very soluble.

(b) Imidazole has one nitrogen that has the electronic character of the nitrogen in pyrrole, and one that is like the nitrogen in pyridine, except that it is somewhat more basic than the nitrogen of pyridine (Study Problem 24.1 on text p. 1134). Because of its basicity, imidazole can accept hydrogen bonds from water and, like pyridine, it is very soluble in water.

24.11 (a) Protonation on the pyridine nitrogen occurs because the conjugate acid is resonance-stabilized.

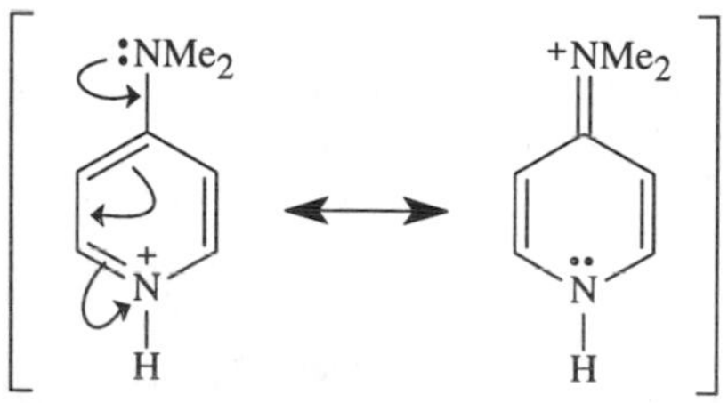

:NMe$_2$ ⟷ +NMe$_2$
N+ H, N: H

conjugate acid of 4-(dimethylamino)pyridine

Protonation on the dimethylamino nitrogen, in contrast, gives a conjugate acid that is not resonance-stabilized. (Be sure to demonstrate this for yourself.) 4-(Dimethylamino)pyridine is more basic than pyridine because of the electron-donating resonance interaction of the dimethylamino substituent in the conjugate-base cation, shown above. The conjugate-base cation of pyridine itself, of course, lacks this stabilizing interaction.

(b) Because the pyridine nitrogen is the more basic nitrogen, it is also the more nucleophilic nitrogen. Consequently, alkylation occurs on this nitrogen.

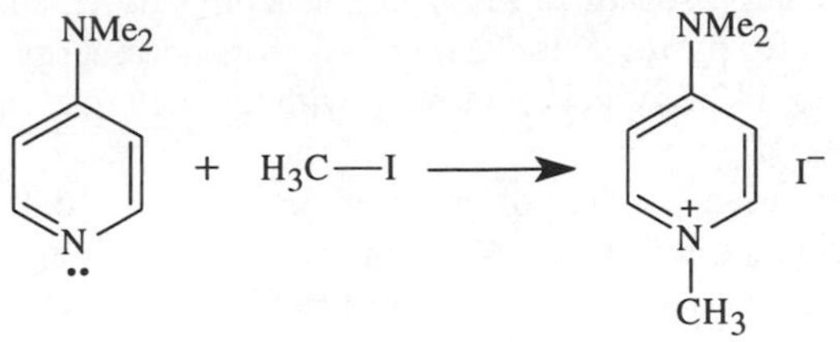

NMe_2 (pyridine) + H_3C—I ⟶ NMe_2 (N-methylpyridinium, $\overset{+}{N}$—CH_3) I^-

24.12 The unshared electron pair on the nitrogen of aniline is conjugated with the π-electron system of the ring. When aniline is protonated, this electron pair is no longer available for conjugation with the ring. Consequently, the reduction in conjugation causes a change in the UV spectrum. (See the solution to Problem 23.63 on pp. 916 of this manual.) The pyridine unshared electron pair, in contrast, is not conjugated with the π-electron system of the ring because its orbital is in the plane of the ring. Consequently, protonation does not affect the conjugated system and therefore does not affect the UV spectrum.

24.13 (a) The bromine directs the incoming nitro group to the "ortho" position, and the thiophene ring is more activated at the same position.

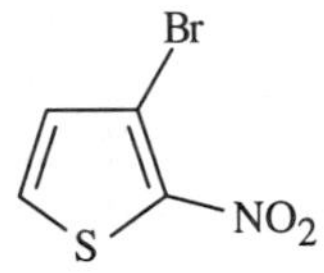

Br, NO_2, S

3-bromo-2-nitrothiophene

(b) This is a Friedel-Crafts acylation; both the methyl group and the ring direct substitution to the open "para" position (5-position).

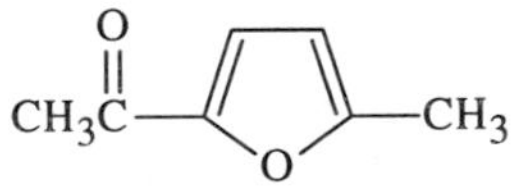

$CH_3\overset{O}{\overset{\|}{C}}$—(furan, O)—$CH_3$

(c) This is an aldol condensation.

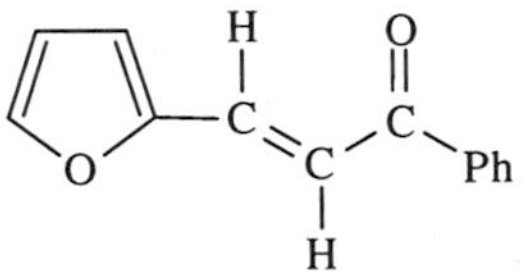

(furan, O)—C(H)=C(H)—C(=O)—Ph

(d) This is essentially a "benzylic" bromination: a free-radical bromination at a carbon adjacent to an aromatic ring (Sec. 17.2).

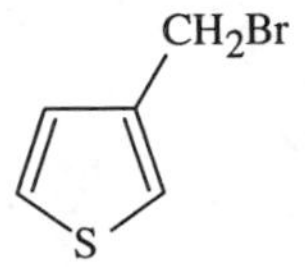

CH_2Br, S

3-(bromomethyl)thiophene

24.14 This is an electrophilic aromatic substitution reaction in which the electrophile is the protonated aldehyde. (The carbonyl-protonated aldehyde shown in the mechanism below is undoubtedly in equilibrium with its nitrogen-protonated isomer, but the latter is not shown since it is not involved directly in the mechanism.)

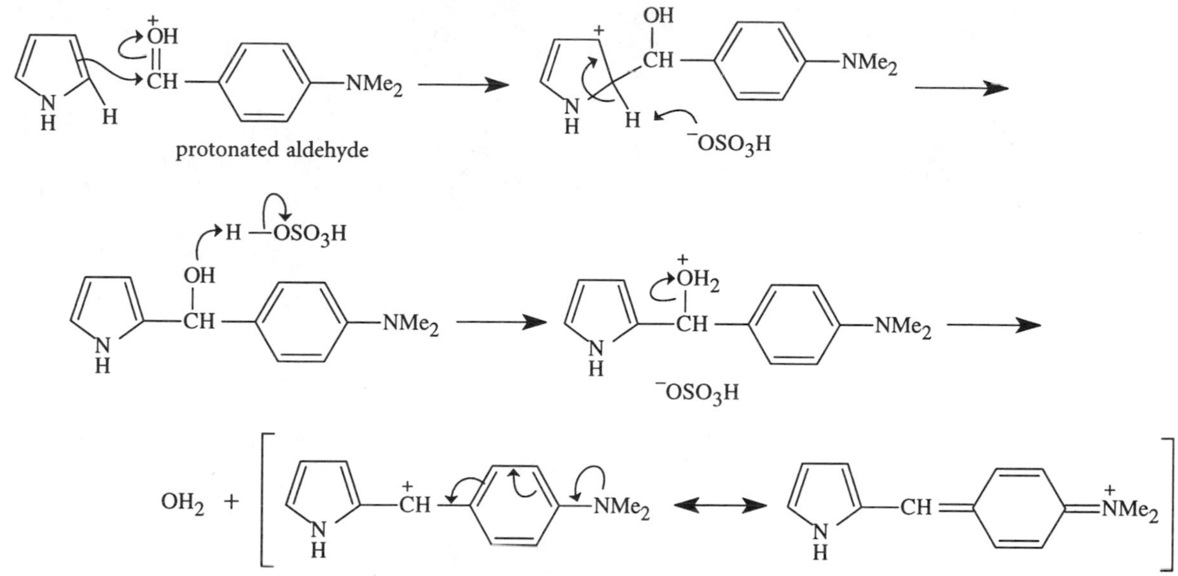

protonated aldehyde

OH_2 +

24.15 (a) $PhNHNH_2$ + $HCCH_2CH(CH_3)_2$ (with C=O)

(b) $PhNNH_2$ (N–CH_3) + CH_3CPh (with C=O)

(c) $PhNHNH_2$ + cyclohexanone

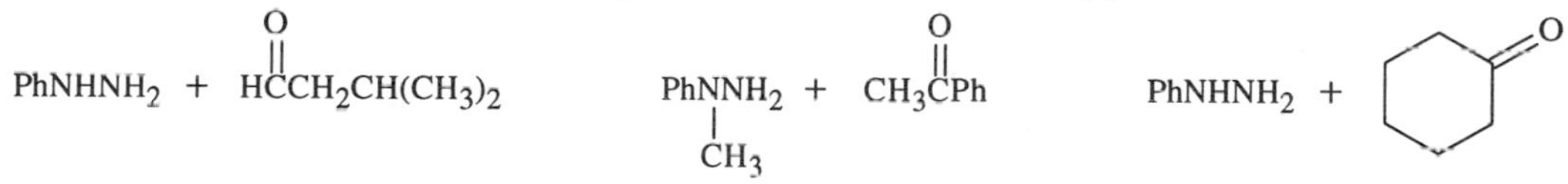

In addition to these starting materials an acid catalyst is required in each case.

24.16 In the mechanistic step of Eq. 24.38c, a protonated enamine is involved. 2-Butanone can give two possible enamines and, hence, two products.

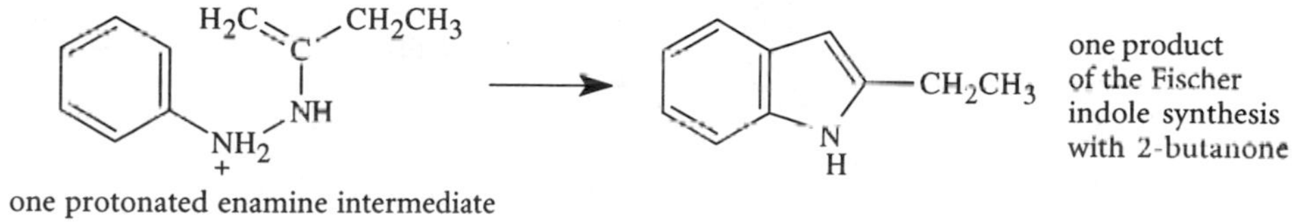

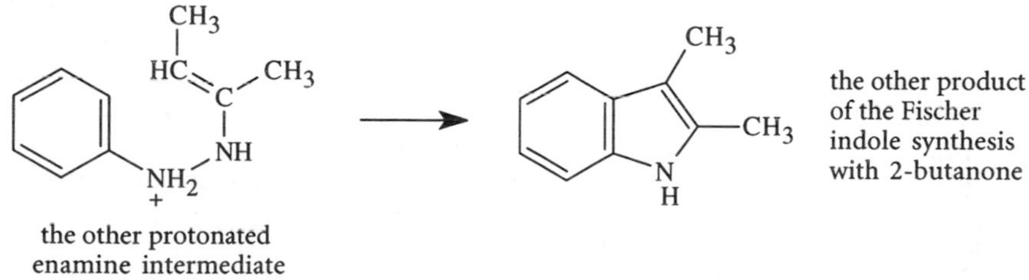

one protonated enamine intermediate

one product of the Fischer indole synthesis with 2-butanone

the other protonated enamine intermediate

the other product of the Fischer indole synthesis with 2-butanone

24.17 (a)

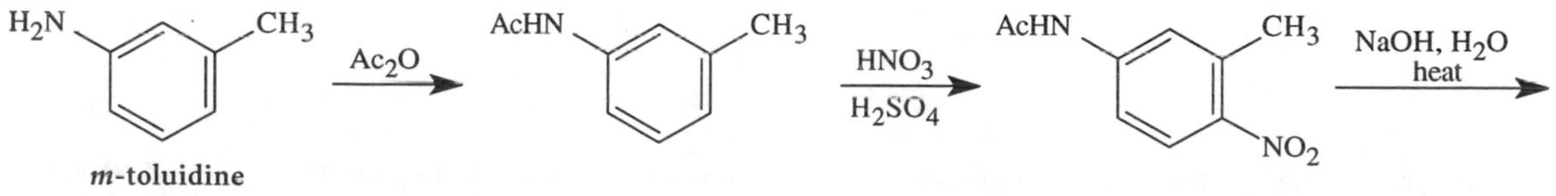

m-toluidine $\xrightarrow{Ac_2O}$ $\xrightarrow[H_2SO_4]{HNO_3}$ $\xrightarrow[\text{heat}]{NaOH,\ H_2O}$

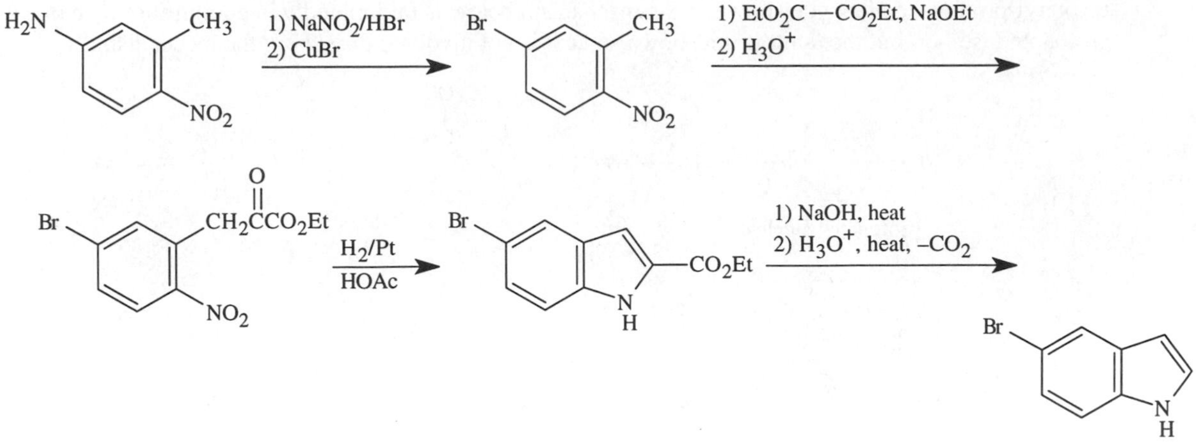

H_2N, CH_3, NO_2 — 1) $NaNO_2$/HBr 2) CuBr → Br, CH_3, NO_2 — 1) $EtO_2C—CO_2Et$, NaOEt 2) H_3O^+ →

Br, $CH_2\overset{O}{\overset{\|}{C}}CO_2Et$, NO_2 — H_2/Pt, HOAc → Br, CO_2Et, N H — 1) NaOH, heat 2) H_3O^+, heat, $-CO_2$ → Br, N H

5-bromoindole

Notice that the amino group of *m*-toluidine must be protected for the nitro group to be directed to the proper position; see Study Problem 23.4 on text p. 1093.

(b)

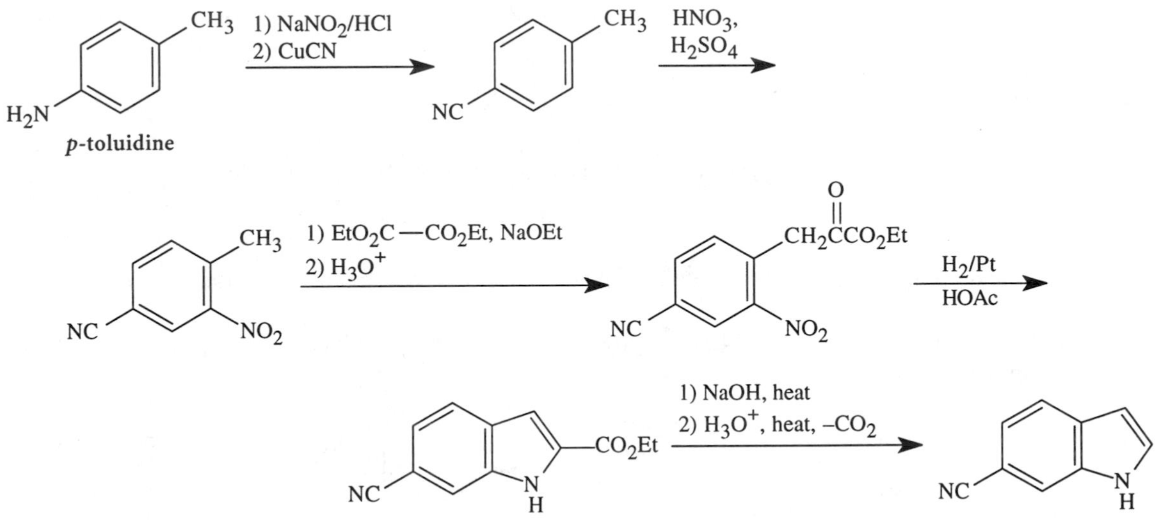

H_2N, CH_3 (*p*-toluidine) — 1) $NaNO_2$/HCl 2) CuCN → NC, CH_3 — HNO_3, H_2SO_4 →

NC, CH_3, NO_2 — 1) $EtO_2C—CO_2Et$, NaOEt 2) H_3O^+ → NC, $CH_2\overset{O}{\overset{\|}{C}}CO_2Et$, NO_2 — H_2/Pt, HOAc →

NC, CO_2Et, N H — 1) NaOH, heat 2) H_3O^+, heat, $-CO_2$ → NC, N H

6-indolecarbonitrile

Notice that a nitro group is much more reactive to catalytic hydrogenation than a cyano group (nitrile).

24.18 The structures of the picolines are given on text p. 1144. α-Picoline is more reactive because both the methyl substituent group and the ring direct substitution to the 3-position. The carbocation intermediate for substitution at the 3-position has a resonance structure in which the positive charge is on the carbon bearing the methyl group:

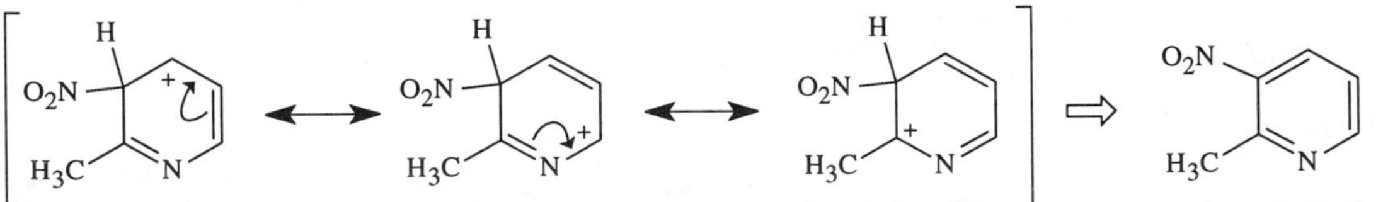

[H, O_2N, +, H_3C, N ⟷ H, O_2N, H_3C, N, + ⟷ H, O_2N, +, H_3C, N] ⇨ O_2N, H_3C, N

tertiary carbocation resonance structure

product of nitration at carbon-3

In contrast, nitration of β-picoline gives a carbocation intermediate for which all the resonance structures are secondary carbocations.

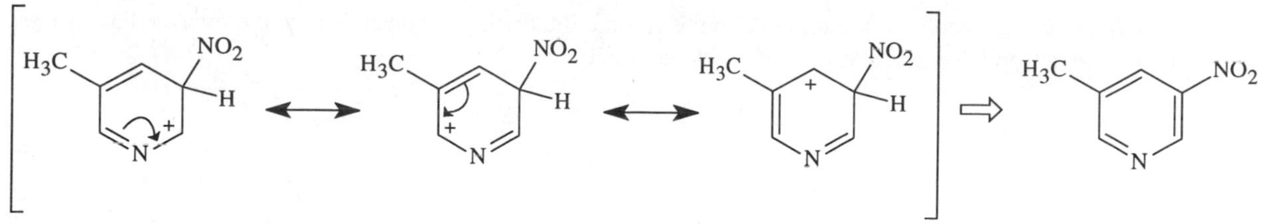

(Nitration at the 2- or 4-position gives a carbocation that has a tertiary resonance structure; however, one of the resonance structures of this carbocation is unimportant because it has an electron-deficient nitrogen.)

24.19 Substitution at the 4-position gives a carbocation that is stabilized by resonance interaction with an unshared electron pair on the oxygen. Substitution at the 3-position gives a carbocation that is not stabilized by such an interaction.

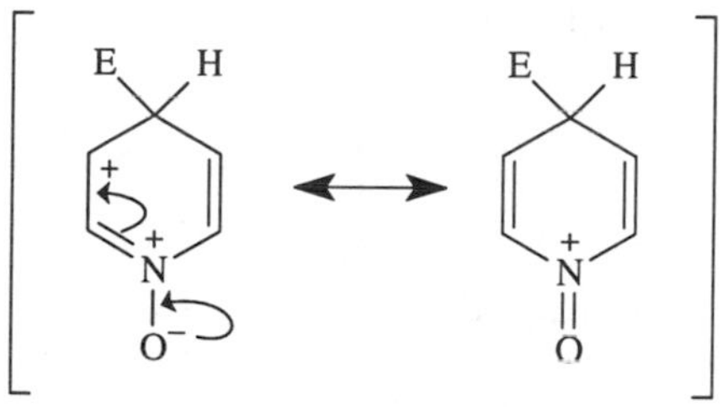

two resonance structures of the carbocation intermediate that results from substitution at the 4-position

24.20 Electrophilic aromatic substitution on a substituted naphthalene derivative occurs on the more activated ring. Because a pyridine ring is deactivated toward electrophilic aromatic substitution, the pyridine ring of quinoline is less reactive than the phenyl ring; hence, electrophilic substitution in quinoline occurs in the phenyl ring. In naphthalene, the α-positions are more activated towards electrophilic aromatic substitution than the β-positions. Thus, electrophilic aromatic substitution of quinoline occurs on the two α-positions of the phenyl ring. Consequently, nitration of quinoline yields a mixture of nitroquinolines in which the two possible α-positions of the phenyl ring are nitrated, that is, 5-nitroquinoline and 8-nitroquinoline.

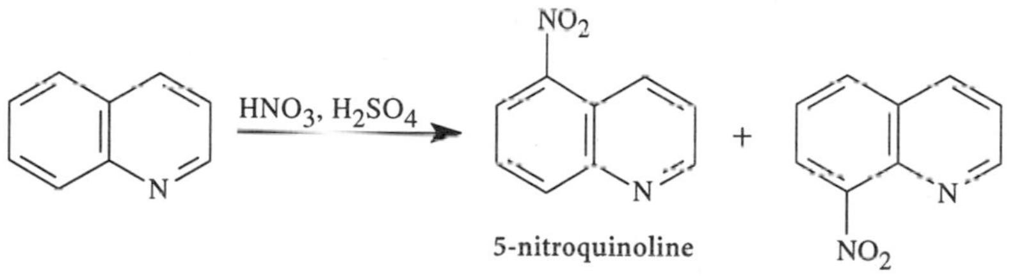

HNO_3, H_2SO_4

5-nitroquinoline

8-nitroquinoline

24.21 The curved-arrow mechanism for the nucleophilic aromatic substitution reaction of 4-chloropyridine with methoxide ion is as follows. Notice that negative charge in the anionic intermediate is delocalized onto the nitrogen, an electronegative atom.

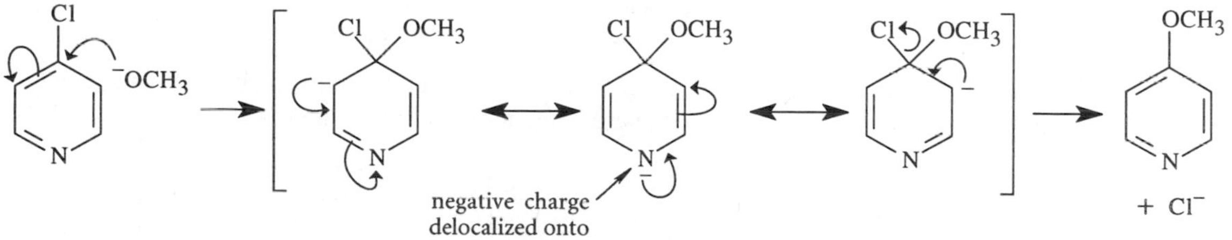

negative charge delocalized onto nitrogen

24.22 4-Bromopyridine undergoes nucleophilic aromatic substitution by phenolate ion. The anionic intermediate is stabilized because negative charge is delocalized onto the nitrogen, as shown in Eq. 24.58b on text p. 1150 (with Y = Br, $^{-}$:Nuc = $^{-}$OPh). In the analogous substitution reaction of 3-bromopyridine, negative charge cannot be delocalized onto the nitrogen in the anionic intermediate; consequently, the intermediate is less stable and the reaction is so much slower that it doesn't occur.

4-bromopyridine + ^{-}OPh ⟶ 4-phenoxypyridine + ^{-}Br

24.23 (a) (b) (c)

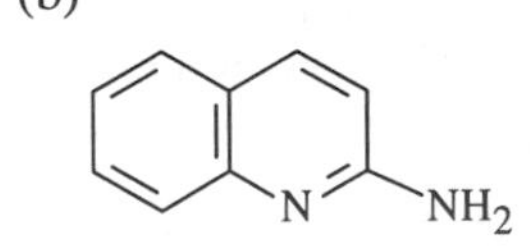

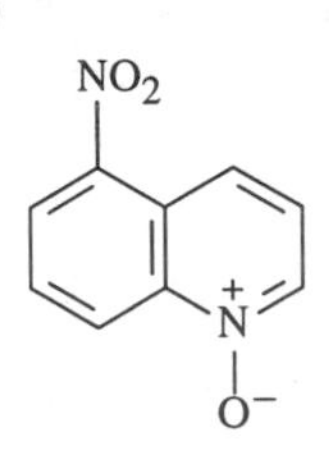

If you thought about the relationship of pyridine to quinoline, you may have proposed that quinoline should nitrate in the 4-position (that is, para to the *N*-oxide) because pyridine nitrates this way. However, the more apt relationship is that of quinoline and naphthalene; remember that substitution occurs in the more reactive ring of naphthalene, and so it is with quinoline: the "benzene" ring is more reactive than the "pyridine" ring.

24.24 (a) Because nitration of a pyridine ring generally occurs in a 3- or 5-position, the pyridine must first be oxidized to a pyridine-*N*-oxide so that nitration is directed to the 4-position. Once nitrated, the *N*-oxide is reduced back to a substituted pyridine. (See Eq. 24.47 on text p. 1146.)

3-picoline (β-picoline) $\xrightarrow{30\%\ H_2O_2}$ N-oxide $\xrightarrow{HNO_3,\ H_2SO_4}$ 4-nitro N-oxide $\xrightarrow{PCl_3}$ 3-methyl-4-nitropyridine

(b) In this case, the pyridine ring is nitrated on the 3-position by direct nitration.

4-picoline (γ-picoline) $\xrightarrow{HNO_3,\ H_2SO_4}$ 4-methyl-3-nitropyridine

(c) The "anion" formed by abstraction of a methyl proton by butyllithium can be carbonated with CO_2 much as any other organolithium reagent or Grignard reagent can be carbonated.

2-picoline (α-picoline) $\xrightarrow{LiCH_2CH_2CH_2CH_3}$ 2-pyridyl-$\bar{C}H_2\ Li^+$ + $CH_3CH_2CH_2CH_3$ $\xrightarrow{1)\ CO_2;\ 2)\ H_3O^+}$ 2-pyridyl-CH_2CO_2H

(d) A Hofmann or Curtius rearrangement can be used to introduce an amino group with loss of a carbon atom.

3-picoline (β-picoline) $\xrightarrow{HNO_3,\ heat}$ (3-pyridinecarboxylic acid, CO_2H) $\xrightarrow{1)\ SOCl_2;\ 2)\ NH_3}$ (3-pyridinecarboxamide, CNH_2 with C=O) $\xrightarrow{Br_2,\ NaOH}$ 3-aminopyridine

24.25 (a) The formula of the product shows that one methyl has been introduced. The conjugate-base anion of 3,4-dimethylpyridine is formed by proton removal from the 4-methyl group, and this anion is alkylated by methyl iodide. A 4-methyl hydrogen is considerably more acidic than a 3-methyl hydrogen because the anion *A* that results from removal of a 4-methyl hydrogen is more stable. It is more stable because the negative charge can be delocalized to the pyridine nitrogen. (Draw the appropriate resonance structures.) The negative charge in the anion derived from ionization of the 3-methyl group cannot be delocalized to the nitrogen.

3,4-dimethylpyridine $\xrightarrow{CH_3CH_2CH_2CH_2Li}$ *A* ($\bar{C}H_2\ Li^+$ at 4-position, CH_3 at 3-position) $\xrightarrow{CH_3I}$ 4-ethyl-3-methylpyridine (CH_2CH_3, CH_3)

(b) The bromine at the 4-position is displaced by ammonia for the reasons discussed in the solution to Problem 24.22 on p. 949 of this manual.

3,4-dibromopyridine (Br, Br) $\xrightarrow{NH_3,\ heat}$ 4-amino-3-bromopyridine (NH_2, Br)

24.26 Begin with the imine product of Eq. 24.71 formed from pyridoxal phosphate and the amino group of serine. (Note that we abbreviate the structure of pyridoxal phosphate as in the text.)

$CH{=}N{-}CH(CO_2^-){-}CH_2{-}O{-}H \leftarrow :B$ (on pyridinium ring, $\overset{+}{N}{-}H$) $\longrightarrow$ $CH{=}N{-}\ddot{C}H(CO_2^-)$ (on pyridinium ring, $\overset{+}{N}{-}H$) $\quad H{-}\overset{+}{B}$ $\longrightarrow$

resonance-stabilized "anion"
(Draw the structures!)

+ $H_2C{=}O$ formaldehyde

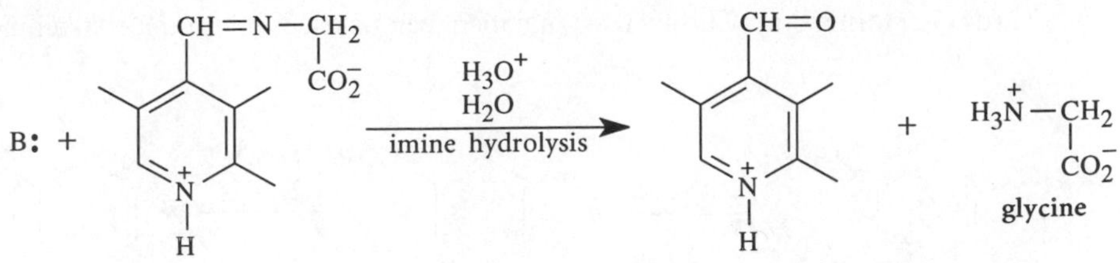

B: + [pyridinium imine: $CH{=}N{-}CH_2CO_2^-$] $\xrightarrow[\text{imine hydrolysis}]{H_3O^+,\ H_2O}$ [$CH{=}O$ pyridinium] + $H_3\overset{+}{N}{-}CH_2CO_2^-$

pyridoxal phosphate

glycine

The first step is a β-elimination that produces formaldehyde and a resonance-stabilized "carbanion." in which the charge is essentially neutralized by delocalization of electrons onto the nitrogen of the pyridinium ring. (Be sure to show this!) This "carbanion" is then protonated to give an imine, which, in turn, hydrolyzes to glycine and pyridoxal phosphate.

24.27 Isoniazid is a type of amide called a *hydrazide*. Semicarbazide (Table 19.3, text p. 876) is the same type of compound. As Table 19.3 shows, such compounds form imine derivatives called *hydrazones*. In this case the structure of the hydrazone is as follows:

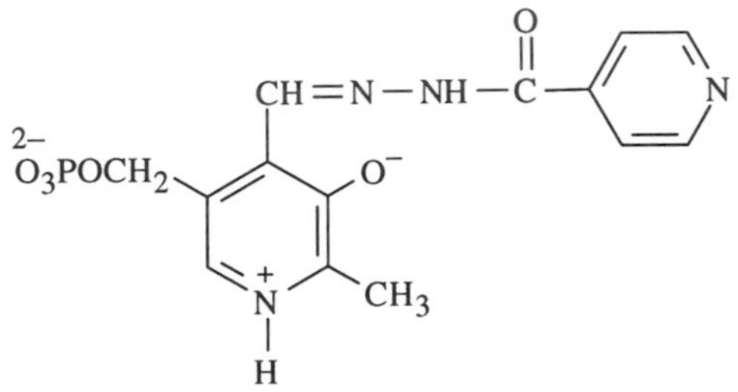

[Structure: $^{2-}O_3POCH_2$-substituted pyridinium (N–H, O^-, CH_3) bearing $CH{=}N{-}NH{-}C(=O)$–4-pyridyl]

It is reasonable to ask why isoniazid does not harm the host organism (*i.e.*, humans) by destroying pyridoxal phosphate in the host. The reason is that the bacterium has a very low level of the enzyme that converts pyridoxal into pyridoxal phosphate, whereas the host has a much higher level. Hence, replenishing the pool of pyridoxal phosphate takes a relatively long time in the Mycobacterium. Isoniazid is typically co-dosed with large amounts of Vitamin B_6 (that is, unphosphorylated pyridoxal and its congeners), which the human body recruits to form pyridoxal phosphate much faster than the Mycobacterium does. Hence, the Mycobacterium's reserve of available pyridoxal phosphate is destroyed by the drug and, unlike the human host, it cannot replenish pyridoxal phosphate fast enough to survive.

24.28 (a) Glycerol undergoes dehydration to acrolein as shown in Eq. 24.76a on text p. 1159, and acrolein reacts with *p*-anisidine to give the following product:

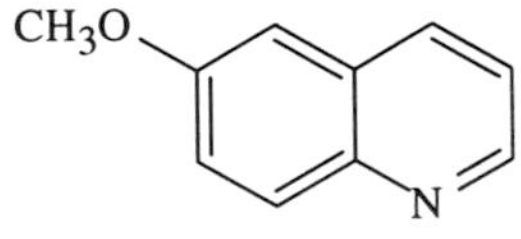

[Structure: CH_3O-substituted quinoline]

6-methoxyquinoline

(b)

[Reaction: CH_3O–C$_6$H$_4$–NH_2 + $O{=}C(Ph){-}CH{=}CH{-}CH_3$ $\xrightarrow{\text{Skraup synthesis}}$ 6-methoxy-4-phenyl-2-methylquinoline]

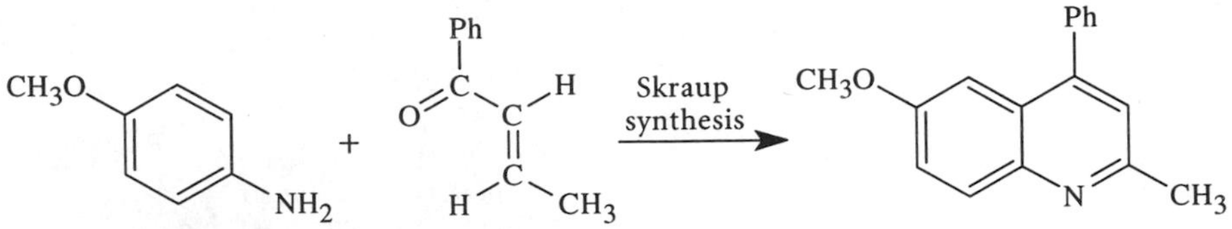

p-methoxyaniline (*p*-anisidine)

1-phenyl-2-buten-1-one

24.29 The reactants required for a Skraup synthesis of 6-chloro-3,4-dimethylquinoline are *p*-chloroaniline and 3-methyl-3-buten-2-one.

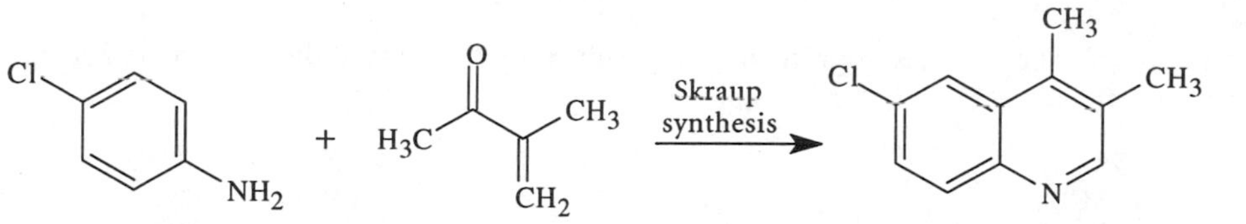

p-chloroaniline + 3-methyl-3-buten-2-one $\xrightarrow{\text{Skraup synthesis}}$ 6-chloro-3,4-dimethylquinoline

24.30 In the electrophilic aromatic substitution step in this Skraup synthesis (Eq. 24.76b on text p. 1160), the conjugate-addition product can cyclize in two distinguishable ways. One product arises from cyclization at the ring position para to the methyl group; the other arises from cyclization at the ring position ortho to the methyl group.

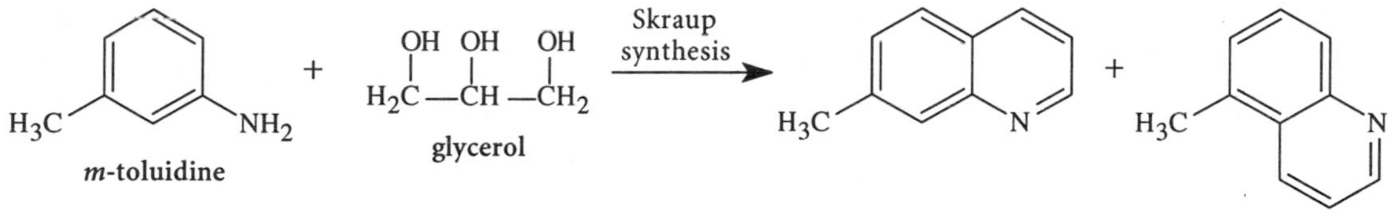

m-toluidine + glycerol $\xrightarrow{\text{Skraup synthesis}}$ 7-methylquinoline + 5-methylquinoline

Solutions to Additional Problems

24.31 In parts (a), (b), and (d), electrophilic aromatic substitution occurs at the more activated ring. In part (c), the reaction is a free-radical benzylic bromination.

(a) (b) (c) (d)]

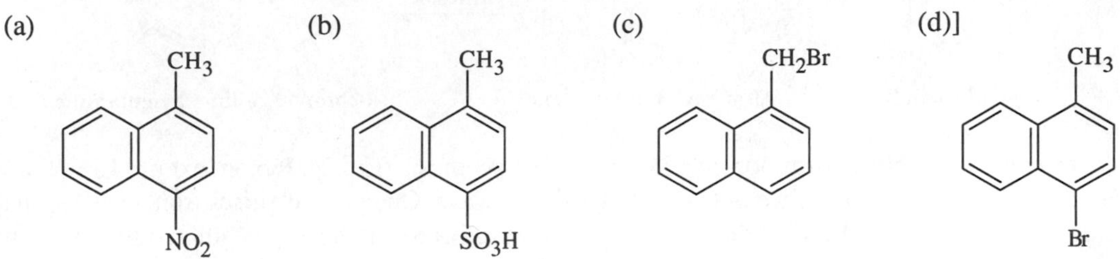

24.32 (a) H_3C— (S) —$\overset{O}{\|}$$CCH_3$ (b) H_3C— (S) —NO_2 (c) $BrCH_2$— (S) (d) no reaction

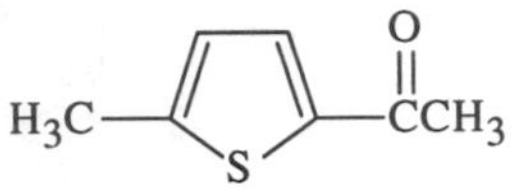

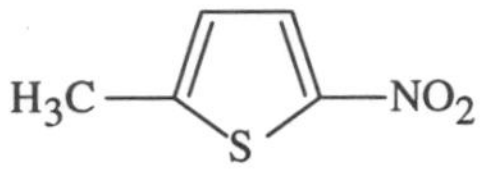

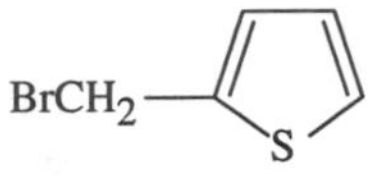

(e) no reaction (f) HO_2CCH_2— (S) (g) H_3C— (S) —$\overset{O}{\|}$C—CH=CH—Ph (H, H, Ph)

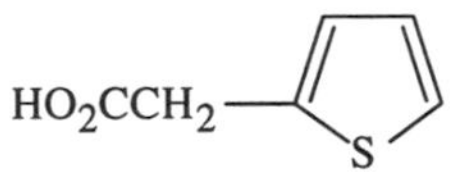

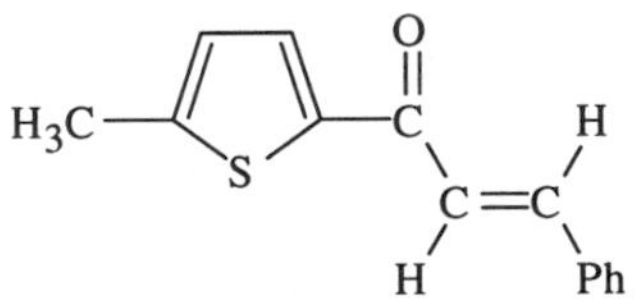

24.33 (a) $\overset{+}{N}$(H) CH_3 Cl^- (b) no reaction (c) N $\bar{C}H_2$ Li^+ + $CH_3CH_2CH_2CH_3$ (d) O_2N N CH_3

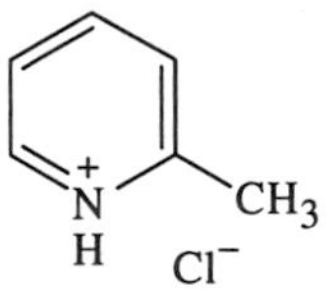

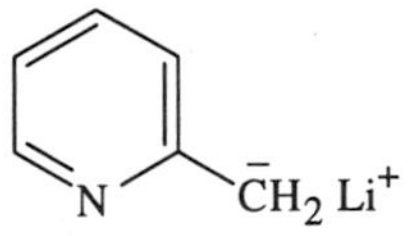

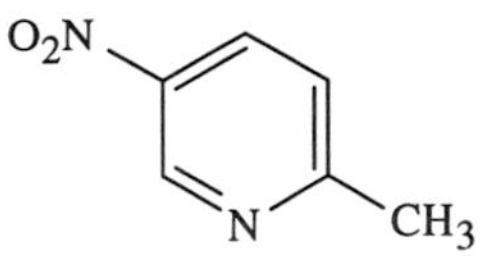

(e) $\overset{+}{N}$ CH_3 O^- (f) $\overset{+}{N}$ CH_3 CH_3 I^- (g) N $CH_2\overset{OH}{|}CHPh$ (h) N CH_3

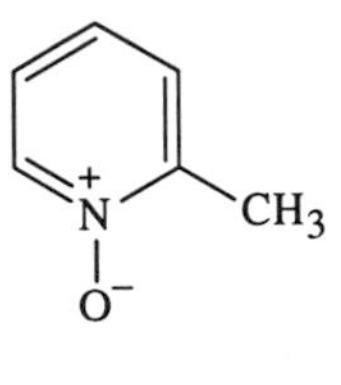

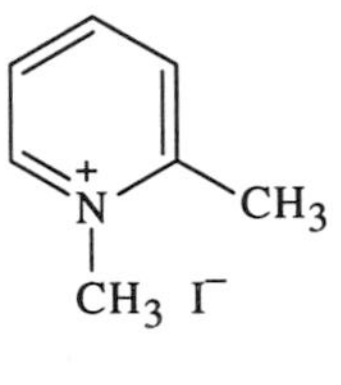

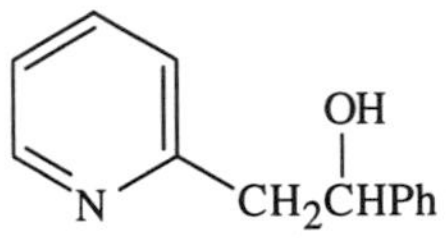

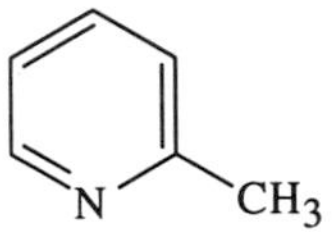

24.34 (a) The order of increasing reactivity is

pyridine < quinoline < naphthalene

The reason for the low reactivity of pyridine is discussed on text p. 1145, and the reason for the high reactivity of naphthalene is discussed on text p. 1126. Quinoline is more reactive than pyridine for the same reasons that naphthalene is more reactive than benzene. However, quinoline is also less reactive than naphthalene because the carbocation intermediates in electrophilic aromatic substitution reactions are destabilized by the polar effect of nitrogen, an electronegative atom.

(b) The order of increasing reactivity is

benzene << thiophene < 3-methylthiophene

The relative reactivity of thiophene and benzene is discussed on text pp. 1137. Another reason that thiophene is more reactive is that it has a smaller aromatic resonance energy than benzene, and therefore it "loses less aromatic stabilization" when attacked by an electrophile. 3-Methylthiophene is more reactive than thiophene because it contains an activating substituent, the methyl group.

24.35 The order of increasing S_N1 reactivity is as follows:

$$B < A < D < C$$

These reactivities parallel the corresponding reactivities of the respective ring systems themselves in electrophilic aromatic substitution reactions. For example, furan is more reactive than benzene in electrophilic aromatic substitution, and thus compound *C* is more reactive than compound *A* in S_N1 reactions, because of the involvement of an oxygen unshared electron pair in the resonance stabilization of the carbocation intermediate:

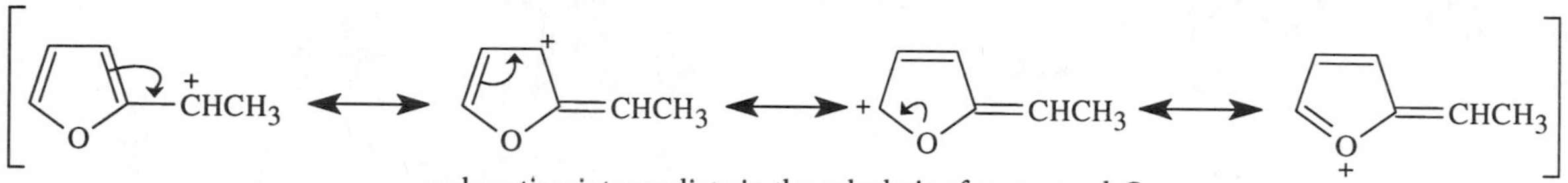

carbocation intermediate in the solvolysis of compound *C*

Just as a *para*-methoxy group accelerates solvolysis reactions (see Eq. 17.6 on text p. 746), atoms that donate electrons by resonance within the ring also accelerate solvolysis reactions. In contrast, a pyridine ring retards solvolysis reactions because the nitrogen unshared electron pair *cannot* be used to stabilize the carbocation by resonance; this electron pair is in the plane of the ring and cannot overlap with the rest of the π-electron system. Furthermore, one resonance structure of the carbocation intermediate places positive charge and electron deficiency on the nitrogen, an electronegative atom. Such a structure is not important. Thus, the nitrogen exerts only its rate-retarding polar effect. The reactivity of the pyridine derivative is further reduced by the formation of HCl as a by-product of the solvolysis reaction. The HCl protonates the pyridine nitrogen and further enhances its electronegativity, thus retarding the solvolysis rate even more.

The carbocation intermediate in the solvolysis of the thiophene derivative has resonance structures analogous to those shown above for the furan derivative. However, because sulfur uses orbitals from period 3, its overlap is less effective in stabilizing the carbocation.

24.36 (a) The carbocation intermediate involved in the Friedel-Crafts acylation of benzofuran at carbon-2:

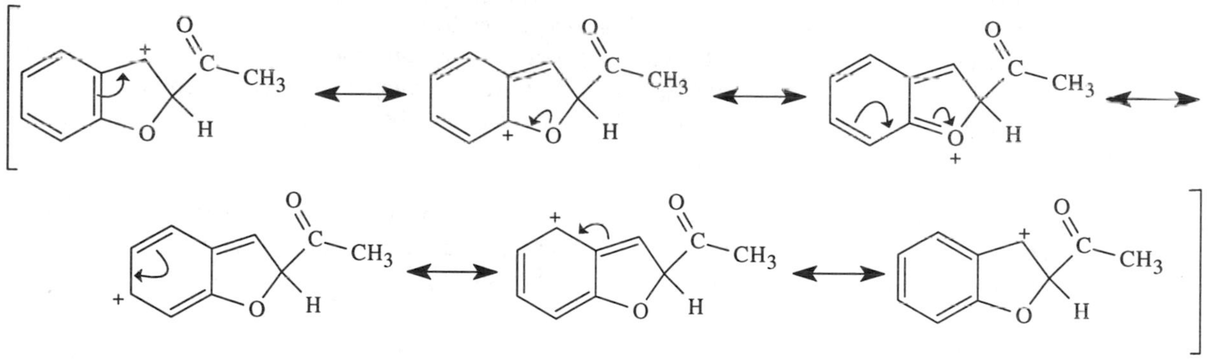

(b) The carbocation intermediate involved in the nitration of benzothiophene at carbon-3:

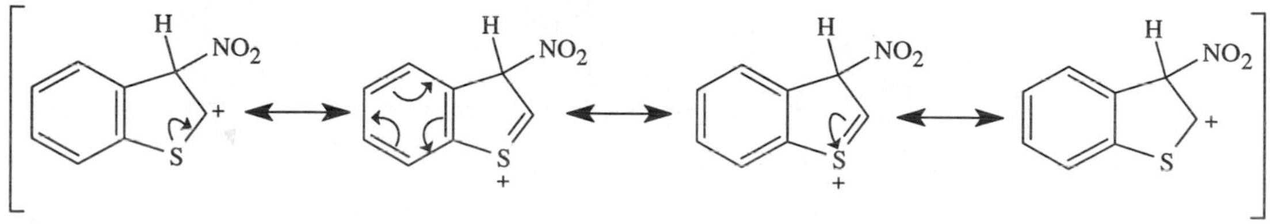

24.37 The hydroxy isomer in each case is aromatic, and the carbonyl isomer is not. The principle to apply is that the compound with the greater resonance energy (Table 24.1 on text p. 1132) has the greater amount of hydroxy isomer.

(a) Because pyrrole has more resonance energy (that is, more stabilization due to aromaticity) than furan, 2-hydroxypyrrole contains a greater percentage of hydroxy isomer than 2-hydroxyfuran.

2-hydroxyfuran

2-hydroxypyrrole

(b) Because benzene has greater aromatic stabilization (greater resonance energy) than pyridine, phenol contains the greater percentage of hydroxy isomer.

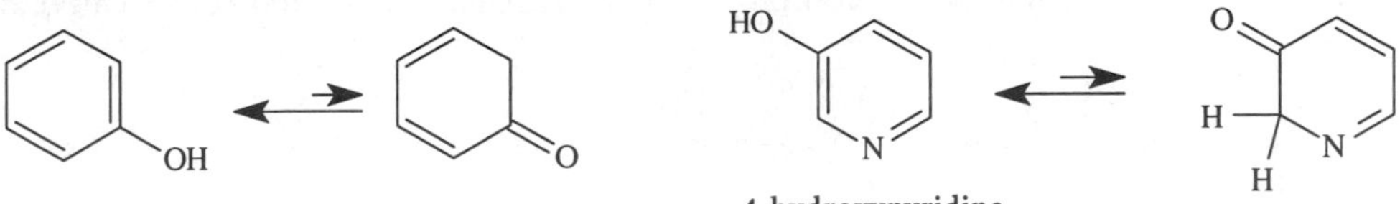

phenol

4-hydroxypyridine

(Note that a second carbonyl isomer of phenol, shown in Eq. 22.14 on text p. 1004, is not shown above; an analogous second carbonyl isomer of 4-hydroxypyridine is also not shown. These are less important; why?)

24.38 The following three isomers are formed in the bromination of 1,6-dimethylnaphthalene.

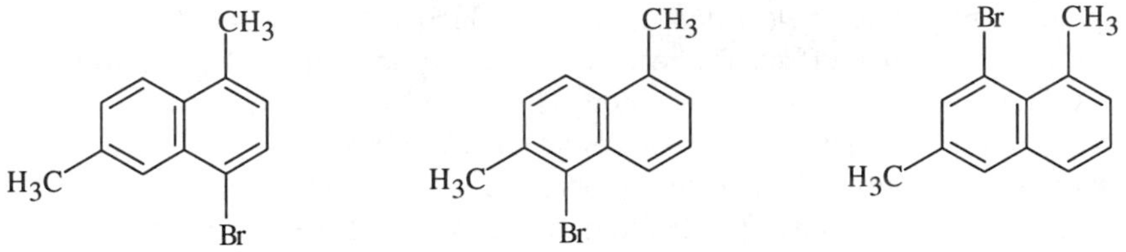

A *B* *C*

All products result from bromination at an α-position; recall that α-positions of naphthalene rings are more activated than are β-positions towards electrophilic aromatic substitution (text pp. 1124–5. Compounds *A* and *B* result from bromination at an α-position that is also activated by the methyl substituent. The substituent effect of the methyl group and the tendency of the ring to brominate in an α-position are in opposition in product *C*; furthermore, there is a severe methyl-bromine peri interaction in compound *C*. Thus, compound *C* is formed in least amount. Compound *A* is probably formed in greatest amount because there is no methyl-bromine ortho van der Waals repulsion as there is in compound *B*.

24.39 At 40 °C sulfonation is kinetically controlled, and the product *A* results from α-substitution in the more activated (that is, the unsubstituted) ring. At the higher temperature, the thermodynamically favored products are compounds *B* and *C*, in which the sulfonic acid groups occupy β-positions.

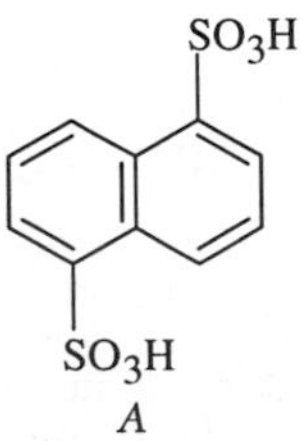

A
product of sulfonation at 40 °C

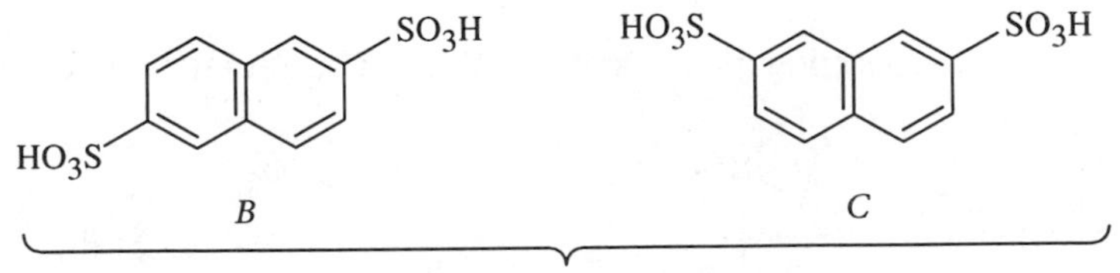

B *C*
products of sulfonation at 180 °C

24.40 The order of increasing basicity (increasing conjugate-acid pK_a) is as follows:

5-methoxyindole << 3-methoxypyridine < pyridine < 4-methoxypyridine

All pyridines are much more basic than the indole, which is not basic at nitrogen. 4-Methoxypyridine is more basic than either of the other pyridines, because an unshared electron pair of the methoxy substituent is available to stabilize the conjugate acid by resonance:

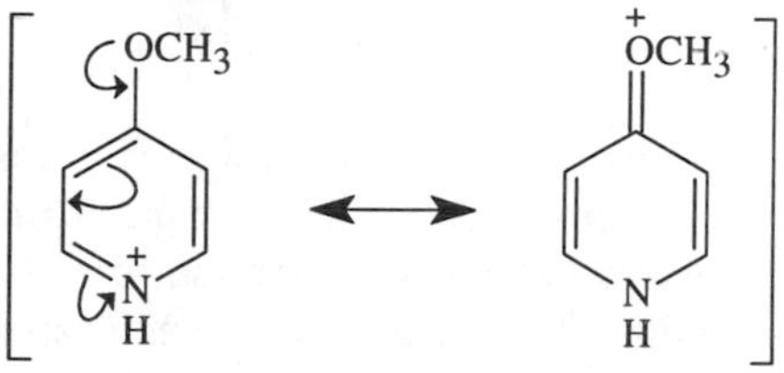

conjugate acid of 4-methoxypyridine

Such resonance is not possible when the methoxy group is in the 3-position; hence, the effect of a 3-methoxy substituent is limited to its acid-strengthening (base weakening) polar effect. Pyridine itself has neither the base-strengthening resonance effect nor the base-weakening polar effect of a methoxy substituent. Hence, its conjugate-acid pK_a is between those of the methoxypyridines.

(b) The order of increasing basicity is

3-nitropyridine < 3-chloropyridine < pyridine

The order is a consequence of the relative polar effects of the nitro and chloro substituents, both of which are electron-withdrawing and base-weakening. As shown by the pK_a values in Table 23.1 on text p. 1077, the electron-withdrawing polar effect of the nitro group on the pK_a of an ammonium ion is greater than that of the chloro group.

(c) The first "anion" in the problem is actually a neutral compound.

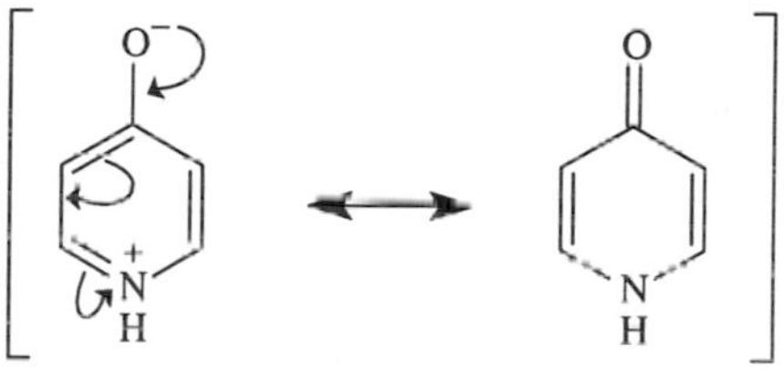

A B

That is, it is a carbonyl compound. It has much of the characteristics of an amide, because, as structures *A* and *B* above show, a nitrogen unshared electron pair is delocalized into the carbonyl group, as in an amide. Thus, the pK_a of its conjugate acid is near –1. The other anion in the problem is a phenolate ion; its conjugate acid is a phenol, with a pK_a of about 10. Consequently, the phenolate ion is considerably more basic.

(d) The resonance structures of protonated imidazole are shown in Eq. 24.19 on text p. 1134. Compare the structure of protonated imidazole with the structure of protonated oxazole:

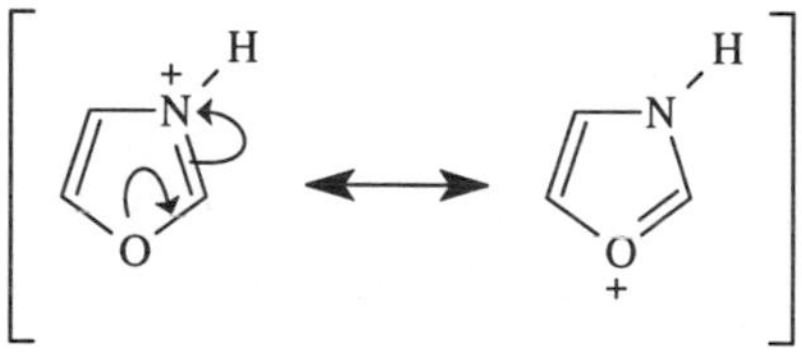

conjugate acid of oxazole

In protonated oxazole, the positive charge is shared between an oxygen and a nitrogen; in protonated imidazole, the positive charge is shared between two nitrogens. Because oxygen is more electronegative than nitrogen, it supports positive charge less effectively; consequently, oxazole is less basic, and imidazole is more basic. (The conjugate-acid pK_a of imidazole is 6.95; that of oxazole is 0.38.)

(e) In the case of thiazole, an electron pair of a sulfur is involved in the resonance stabilization of the conjugate acid:

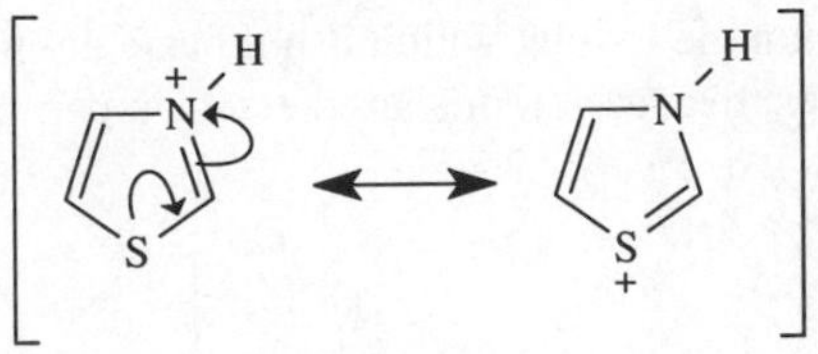

conjugate acid of thiazole

The orbitals of sulfur overlap more poorly than those of nitrogen with carbon 2*p* orbitals because they are derived from a different quantum level; consequently, the resonance structure on the right is less important than the corresponding structure of protonated imidazole. The sulfur in thiazole is actually less electronegative than the nitrogen in imidazole, but, as in electrophilic aromatic substitution, the electron-donating resonance effect of nitrogen far outweighs its electron-withdrawing polar effect. Consequently, imidazole is considerably more basic than thiazole. (The actual conjugate-acid pK_a of imidazole is 6.95, and that of thiazole is 2.44.)

24.41 The conjugate acid is formed by protonation on the nitrogen of the double bond. Protonation on either of the other nitrogens does not yield a resonance-stabilized cation.

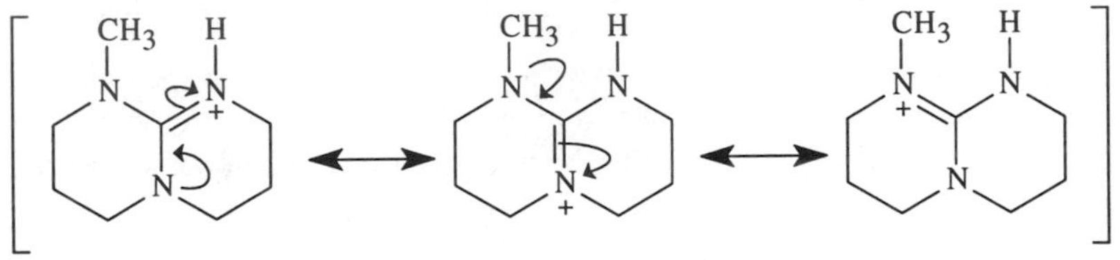

24.42 (a) Because alkylamines are more basic than pyridines (or quinolines), the conjugate acid of quinine is protonated on the nitrogen of the bridged bicyclic ring:

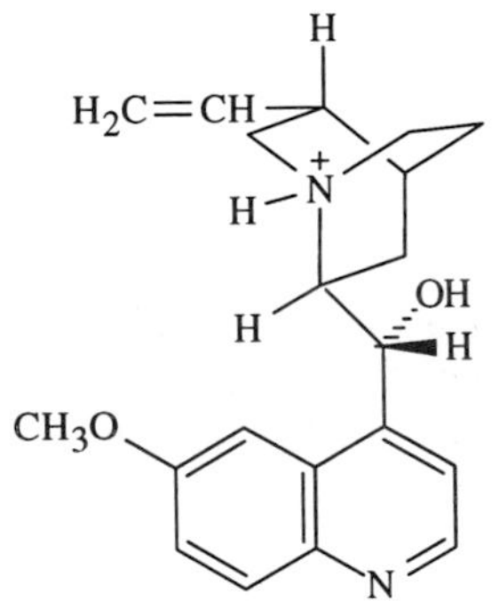

conjugate acid of quinine

(b) The alkylamine group is more basic than the pyridine:

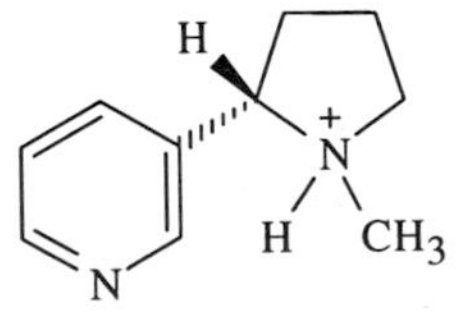

conjugate acid of nicotine

(c) Indoles are not appreciably basic; tryptamine protonates on the nitrogen of the primary amino group.

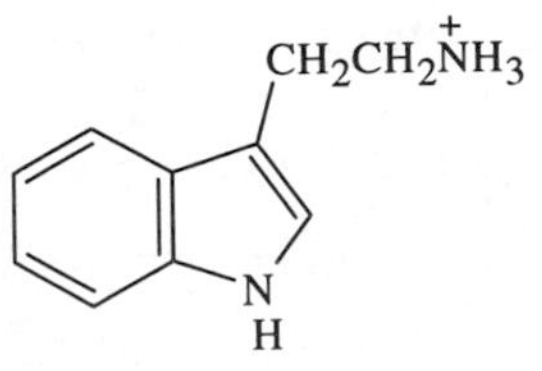

$CH_2CH_2\overset{+}{N}H_3$

conjugate acid of tryptamine

(d) Although it might seem that the nitrogen of an amino group might be more basic than the nitrogen of a pyridine ring, the pyridine nitrogen protonates because the conjugate acid is resonance-stabilized. (See the solution to Problem 24.11(a) on page 945 of this manual for a similar situation.)

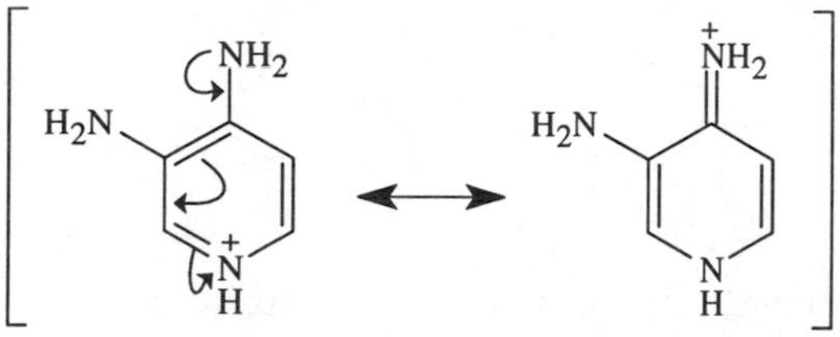

conjugate acid of 3,4-diaminopyridine

(e) One nitrogen resembles the nitrogen of indole, which is not basic; the other resembles the nitrogen of pyridine, which is basic. Furthermore, protonation of the pyridine nitrogen gives a conjugate-acid cation that is stabilized by resonance interaction with the unshared pair on the "indole" nitrogen. (Draw the appropriate resonance structures.) In fact, the conjugate-acid pK_a of 1,4-diazaindene is 6.92, whereas the conjugate-acid pK_a of pyridine is 5.2. The greater basicity of 1,4-diazaindene reflects the resonance stabilization of its conjugate acid.

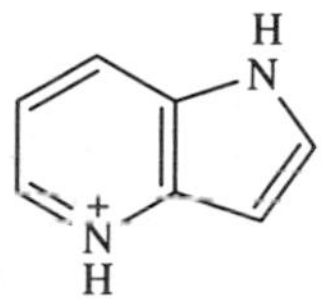

conjugate acid of 1,4-diazaindene

(f) Think of a 1,2,3-triazole as an imidazole with an extra nitrogen. Protonation occurs on the terminal nitrogen of the double bond as it does in imidazole because a resonance-stabilized cation is formed:

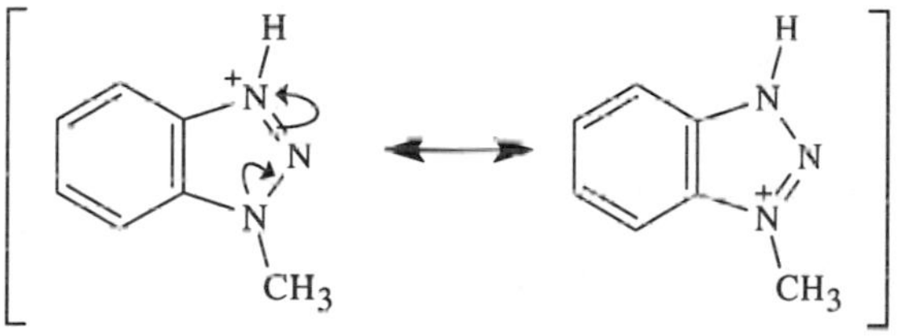

conjugate acid of 1-methyl-1,2,3-benzotriazole

This resonance stabilization notwithstanding, the conjugate-acid pK_a of benztriazole is 1.6; the 1-methyl group will raise that somewhat, but not a great deal. Compare this with the conjugate-acid pK_a of benzimidazole, which is 5.52. Why is 1-methyl-1,2,3-benztriazole so much less basic?

24.43 (a) The text has stressed that many reactions of pyridines can be rationalized by thinking of the N=C bond of the pyridine as if it were a carbonyl group. Thus, the hydrogens of the methyl group, like the α-hydrogens of a methyl ketone, are acidic enough to exchange for deuterium in basic D_2O.

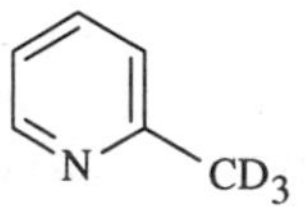

(b) The indole N—H is acidic enough to be removed by strong bases such as phenyllithium:

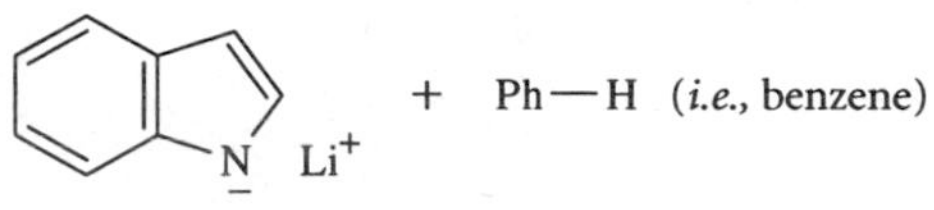

+ Ph—H (*i.e.*, benzene)

(c) Naphthalene rings are more activated towards electrophilic aromatic substitution than are phenyl rings; furthermore, a phenyl group is an activating, ortho, para-directing substituent. (The rationale for this conclusion is explored in detail in the solution to Problem 16.22 on p. 522 of this manual.)

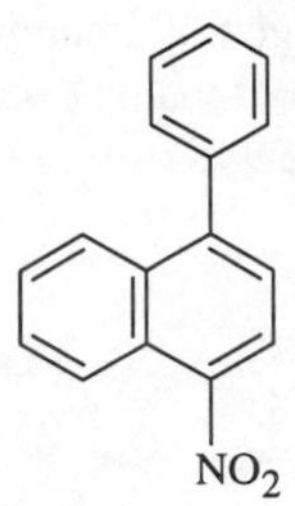

NO_2

(d) The methoxy group is an activating, ortho, para-directing group; hence substitution occurs in the ring that is already substituted.

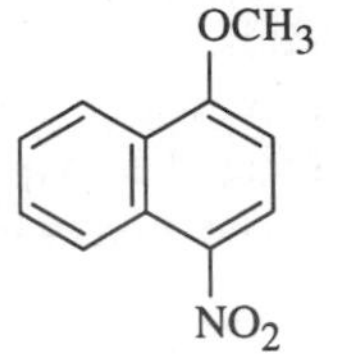

OCH_3

NO_2

(e) The fact that the amino group is part of a five-membered ring does not alter the fact that it is an activating, ortho, para-directing substituent. Under the conditions shown, the aromatic ring brominates once. (See Eq. 18.69, text p. 816, for bromination of phenol under similar conditions.)

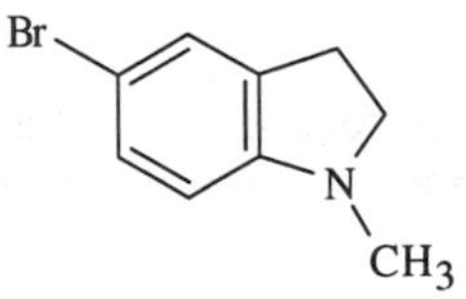

Br

N

CH_3

(f) The product results from hydrogenation of the double bond that is not part of an aromatic system:

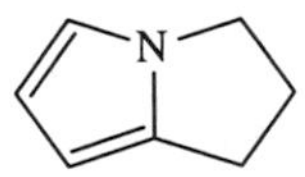

N

(g) The ethoxycarbonylamino substituent and the pyridine ring itself direct aromatic nitration to carbon-5. The stringent conditions are necessary because the pyridine ring is protonated, and thus is highly deactivated, under the acidic conditions of the reaction.

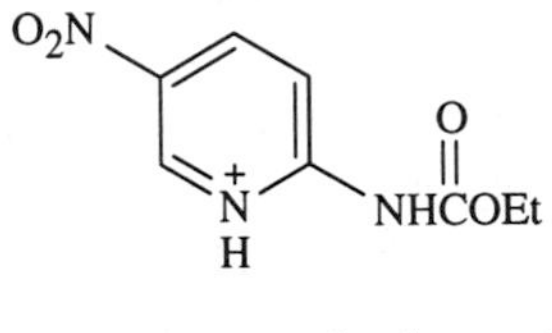

O_2N

$\overset{+}{N}$ H

O ‖ NHCOEt

(h) Because furan is more activated than thiophene towards electrophilic aromatic substitution (Eq. 24.26 on text p. 1136), substitution occurs predominantly in the furan ring. Notice that the directing effect of the electron-releasing ring oxygen overrides the directing effect of the carbonyl group, which, by itself, would result in substitution at the 4-position of the furan ring. (See the discussion of the relative importance of directing effects in the last paragraph of text p. 728.)

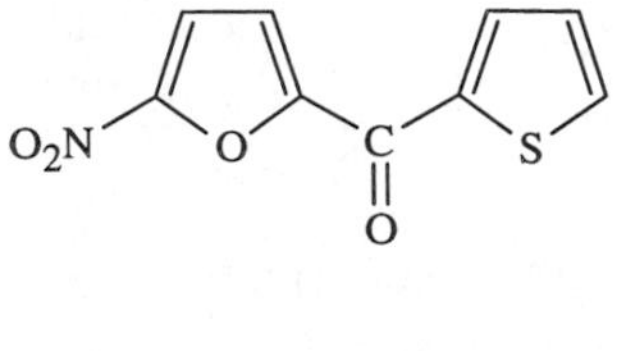

O_2N O C ‖ O S

(i) Pyridine-*N*-oxides are nitrated at carbon-4. (See Eq. 24.46 on text p. 1146.) The directing effect of methyl is not strong enough to override the effect of an OH group. (Note that the *N*-oxide oxygen is protonated under the conditions of nitration; this point was not mentioned in the text.)

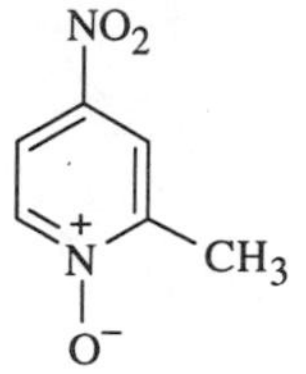

NO_2 CH_3 $\overset{+}{N}$ O^-

(j) Wolff-Kishner reductions occur on acylated pyridines just as they do on acylated benzenes.

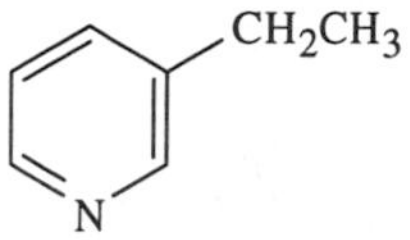

CH_2CH_3 N

(k) This is a Fischer indole synthesis in which the product is an *N*-methyl indole:

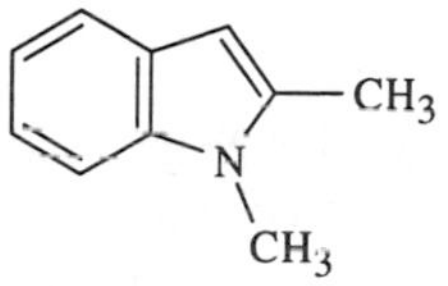

CH_3 N CH_3 1,2-dimethylindole

(l) This compound, 8-aminoquinoline, is to the conditions of the Skraup synthesis nothing more than a substituted aniline. The As_2O_5 is a mild oxidizing agent required for the final step in the Skraup synthesis (see text p. 1160).

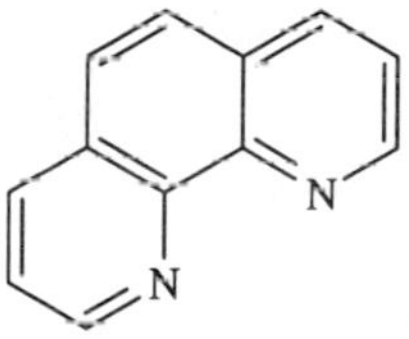

N N phenanthroline

The product, phenanthroline, is a potent chelator of metal ions; can you see why?

24.44 The *p*-nitrobenzenediazonium ion, which is formed in the diazotization reaction, acts as the electrophile in an electrophilic aromatic substitution at carbon-3 of indole:

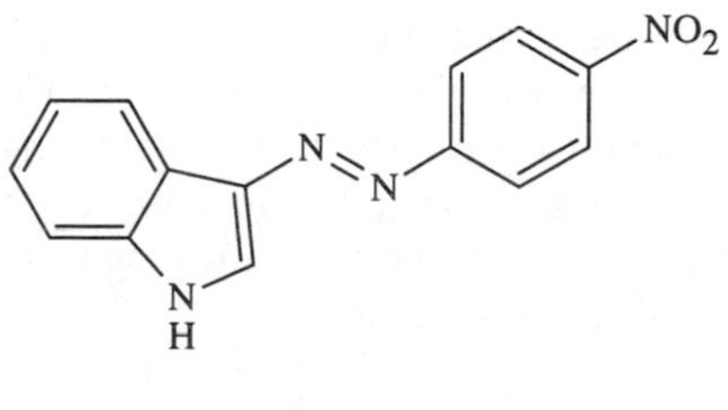

NO_2 N N N H

24.45 (a) Given that anthracene adds benzyne across carbons 9 and 10, the structure of triptycene can be deduced by the curved-arrow notation:

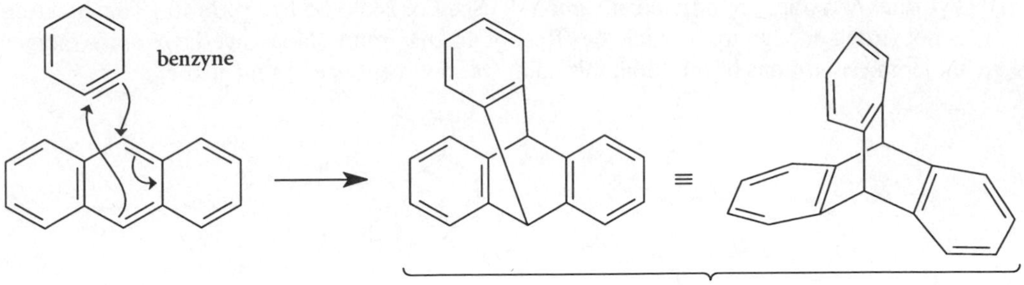

benzyne

triptycene

(b) Use the same approach as in part (a).

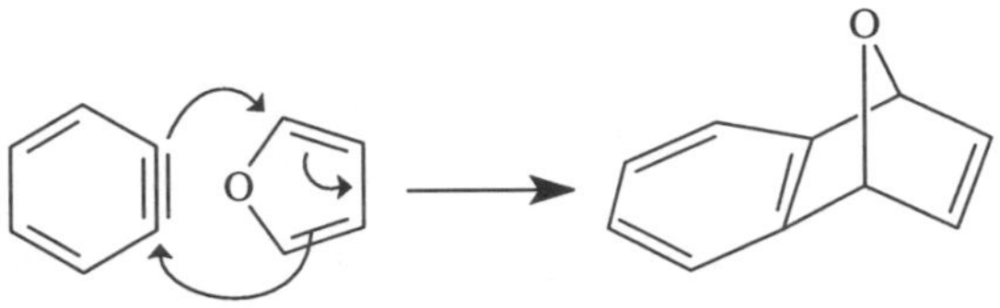

benzyne

24.46 Use the principle that the carbocation intermediate with the greatest number of resonance structures is most stable, and that the reaction which involves the most stable carbocation intermediate is fastest. First, consider the carbocation intermediate in bromination at an α-position of a terminal ring:

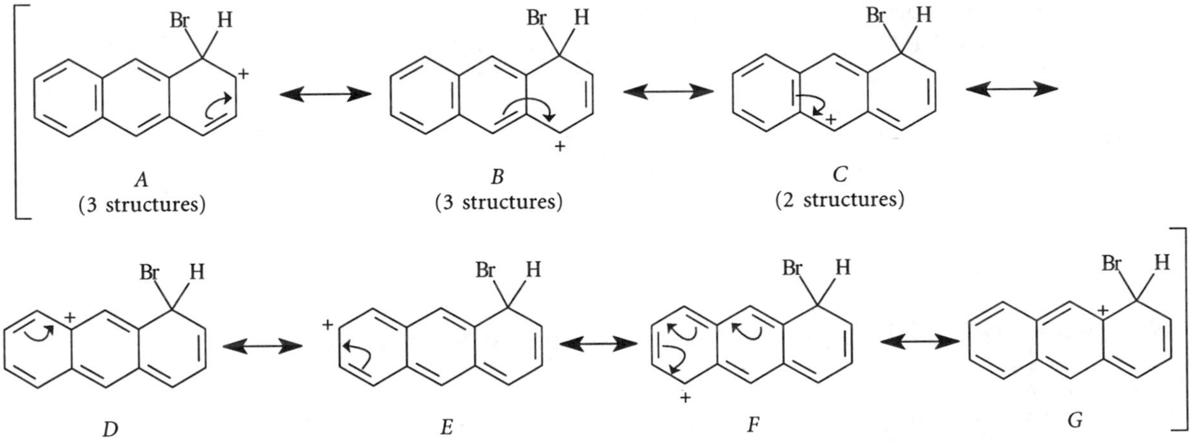

A (3 structures) *B* (3 structures) *C* (2 structures)

D *E* *F* *G*

For each structure in which there is an intact naphthalene ring there are two additional structures (as in Eq. 24.1 on text p. 1122). Thus, structures *A* and *B* actually represent three resonance structures each. For the structure *C*, in which there is an intact benzene ring, there is one additional structure (as in Eq. 15.40 on text p. 675). Thus, structure *C* actually represents two resonance structures. Hence, the carbocation shown above has twelve resonance structures. It is shown on text p. 1125 that substitution at a β-position of a naphthalene ring results in a carbocation with fewer resonance structures; the same is true for substitution at the β-position of an anthracene ring. (If you don't believe it, draw them!) The one remaining possibility is bromination at a central carbon. This results in the following carbocation intermediate:

A (2 structures) *B* (2 structures) *C* (2 structures)

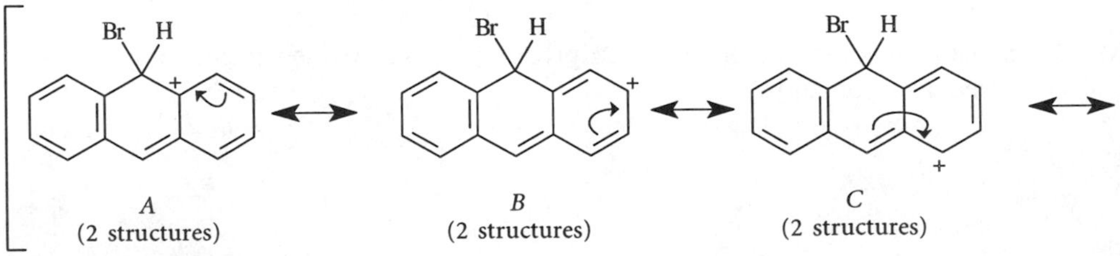

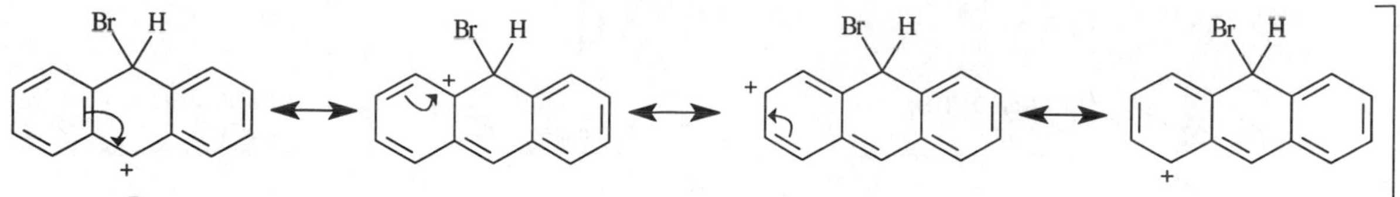

D (4 structures) ⟷ E (2 structures) ⟷ F (2 structures) ⟷ G (2 structures)

Again, each intact benzene ring represents two structures. This carbocation has a total of sixteen resonance structures! Because this carbocation is the more stable intermediate, bromination therefore occurs at carbon-9 of anthracene.

24.47 (a)

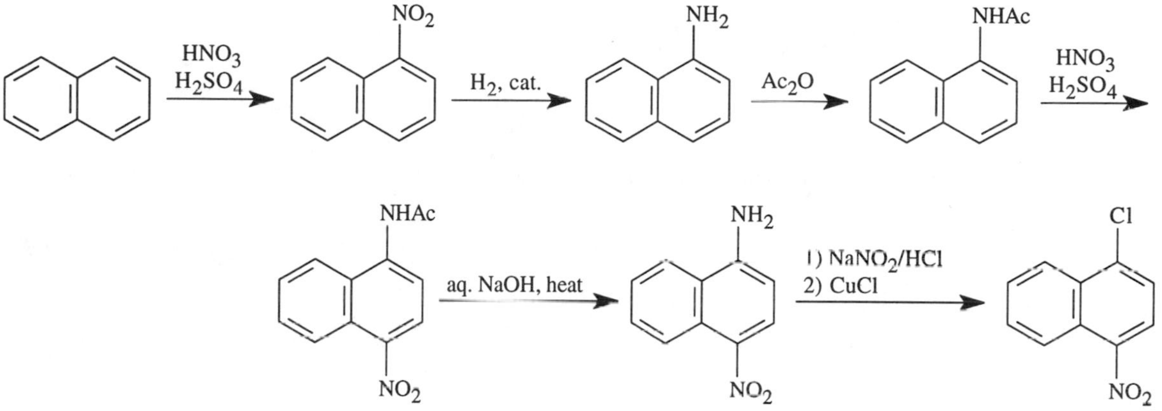

HNO_3, H_2SO_4 → H_2, cat. → Ac_2O → HNO_3, H_2SO_4 → aq. NaOH, heat → 1) $NaNO_2$/HCl 2) CuCl →

1-chloro-4-nitronaphthalene

Note that chlorination of naphthalene followed by nitration would not work. (Why?)

(b)

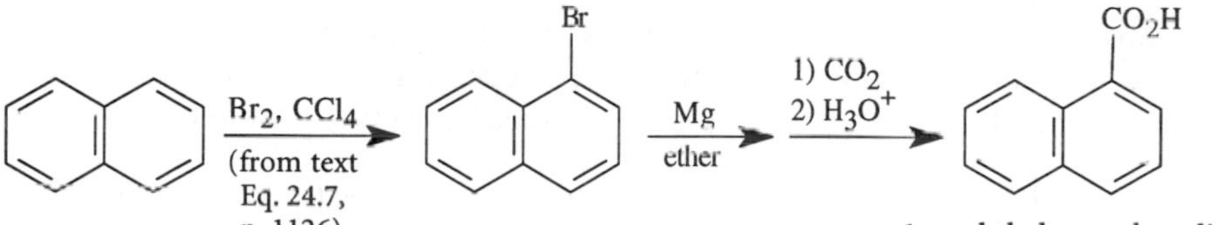

Br_2, CCl_4 (from text Eq. 24.7, p. 1126) → Mg ether → 1) CO_2 2) H_3O^+ →

1-naphthalenecarboxylic acid

(c)

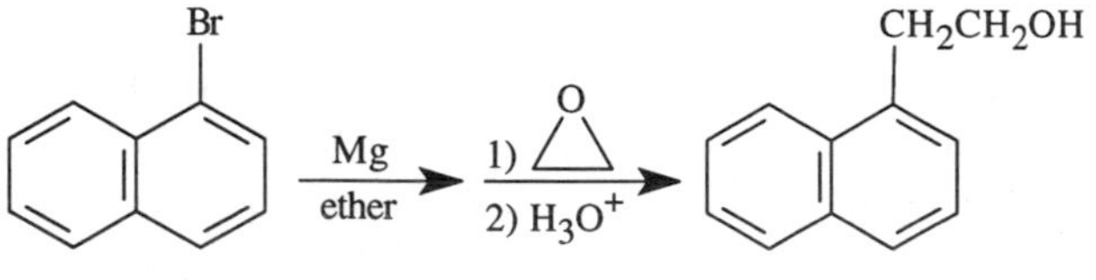

prepared in part (b) → Mg ether → 1) ethylene oxide 2) H_3O^+ →

2-(1-naphthyl)ethanol

(d)

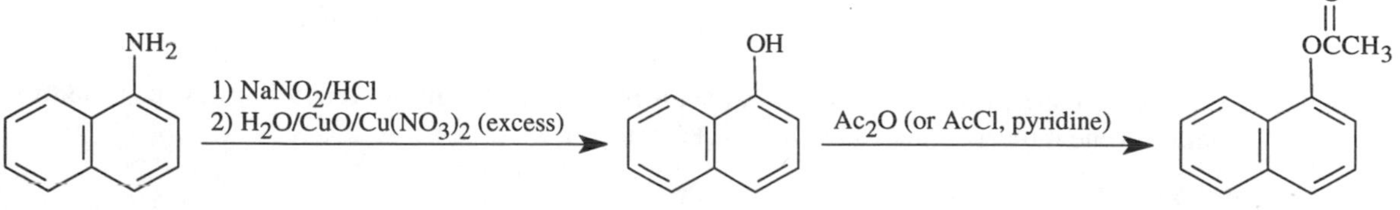

prepared in part (a) → 1) $NaNO_2$/HCl 2) $H_2O/CuO/Cu(NO_3)_2$ (excess) → Ac₂O (or AcCl, pyridine) →

1-naphthyl acetate

(e)

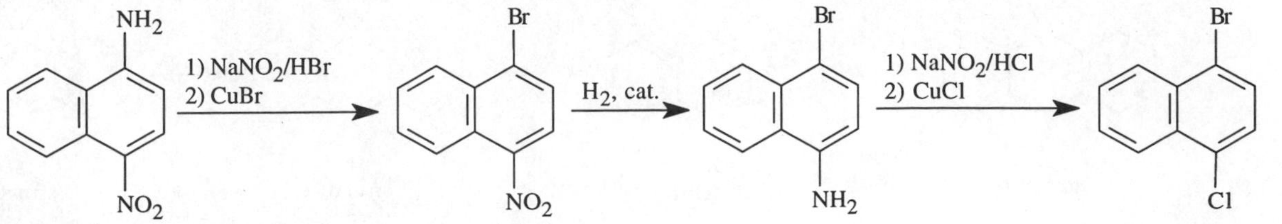

NH_2 / NO_2 — 1) $NaNO_2$/HBr 2) CuBr → Br / NO_2 — H_2, cat. → Br / NH_2 — 1) $NaNO_2$/HCl 2) CuCl → Br / Cl

prepared in part (a)

1-bromo-4-chloro-naphthalene

(f)

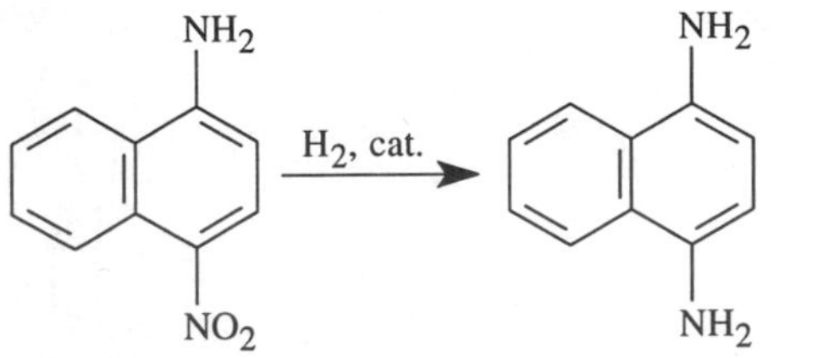

NH_2 / NO_2 — H_2, cat. → NH_2 / NH_2

prepared in part (a)

1,4-naphthalenediamine

(g)

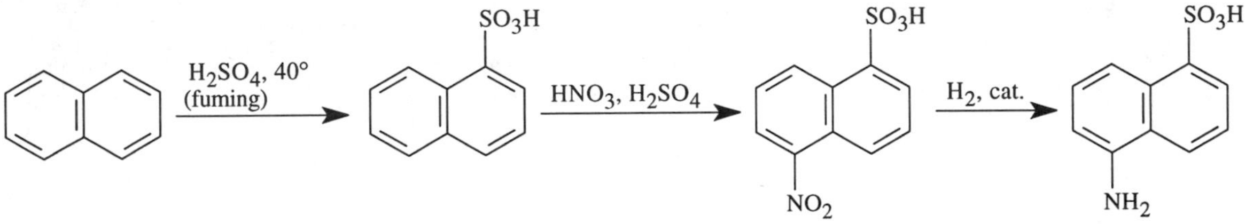

naphthalene — H_2SO_4, 40° (fuming) → SO_3H — HNO_3, H_2SO_4 → SO_3H / NO_2 — H_2, cat. → SO_3H / NH_2

5-amino-1-naph-thalenesulfonic acid

24.48 (a) Doreen is attempting to apply a reaction of pyridines and quinolines to indole. Because attack on a double bond of indole by the amide ion cannot result in charge delocalization to the nitrogen, the reaction does not take place. The reaction that occurs instead is removal of the N—H proton of indole.

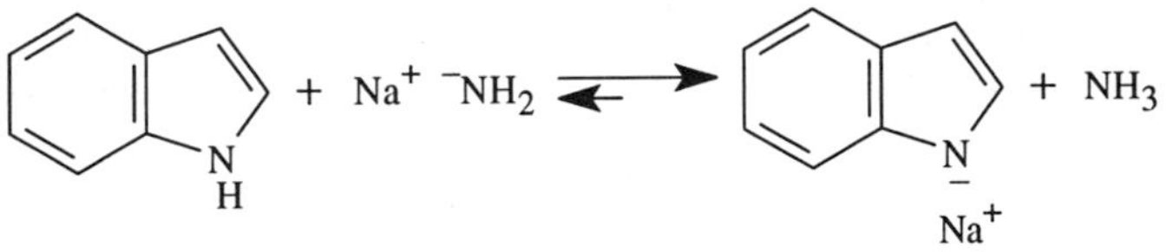

$$\text{indole (N–H)} + Na^+ \ {}^-NH_2 \rightleftharpoons \text{indole anion } (N^- \ Na^+) + NH_3$$

(b) Doreen is attempting the Chichibabin reaction on a pyridine derivative, but this derivative contains a chlorine in the 2-position, which, as chloride ion, is a vastly superior leaving group to hydride. The chlorine is lost rather than hydride to give 2-aminopyridine.

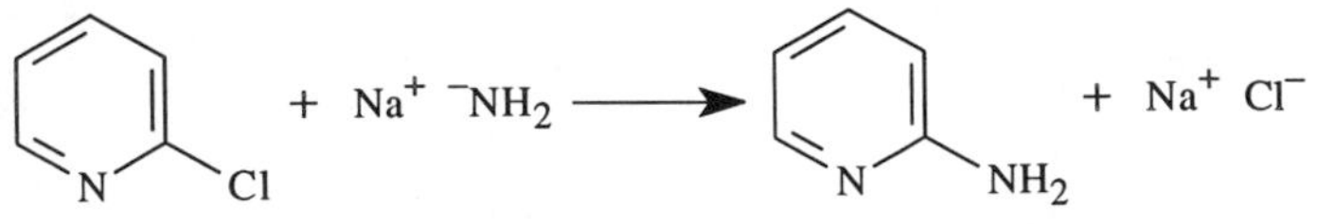

$$\text{2-chloropyridine} + Na^+ \ {}^-NH_2 \longrightarrow \text{2-aminopyridine} + Na^+ \ Cl^-$$

2-aminopyridine

24.49 The conjugate-base anion of 2-aminopyridine, formed as shown in Eq. 24.49d on text p. 1148, reacts with another pyridine molecule in a Chichibabin-like reaction:

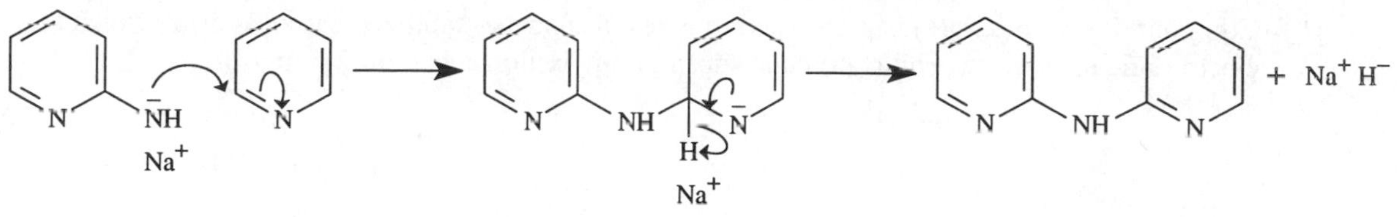

N NH Na$^+$ N → N NH N H Na$^+$ → N NH N + $Na^+ H^-$

24.50 This is a variation of the Fischer indole synthesis. The mechanism that follows begins with the protonated enamine, formed in an acid-catalyzed reaction from the ketone and phenylhydrazine starting materials as shown in Eqs. 24.38a–b on text pp. 1141. All proton transfers are shown as intramolecular processes, although some of them could be intermolecular. Notice that a protonated indolenine is formed as an intermediate in the normal Fischer indole synthesis (Eq. 24.38d), and that isomerization of this intermediate leads to the indole itself. In the mechanism below, the final isomerization is not possible because carbon-3 has no hydrogen.

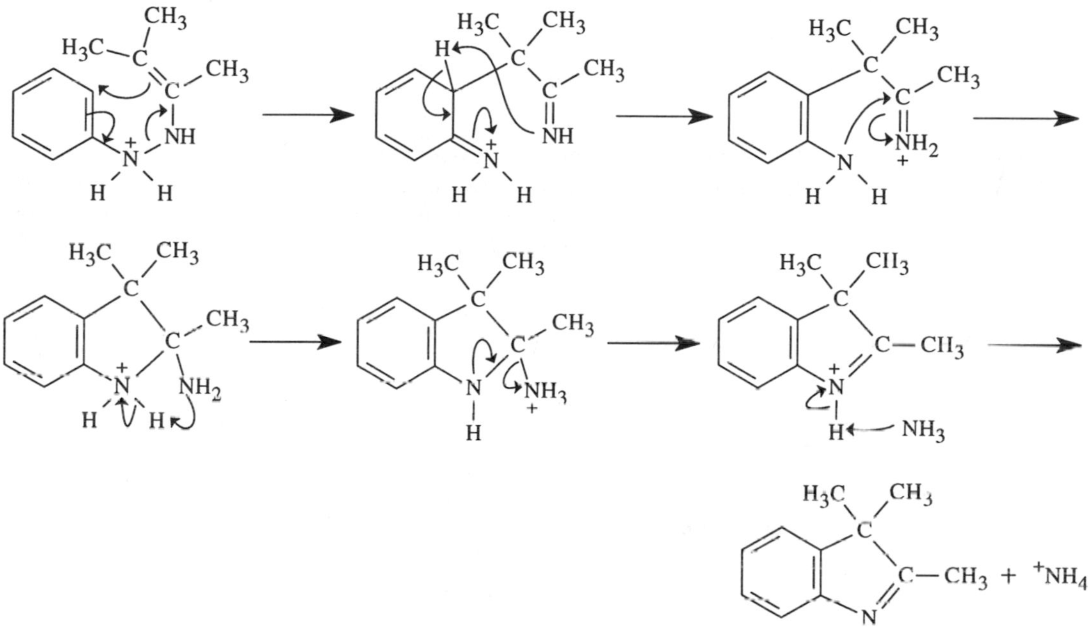

24.51 (a) Protonation of one pyrrole molecule at carbon-2 gives a carbocation that reacts with a second indole. The resulting product *X* undergoes an acid-catalyzed double-bond shift to give *A*. This rearrangement is favorable because, in compound *A*, the double bond is conjugated with a nitrogen unshared electron pair whereas, in compound *X*, such conjugation is absent.

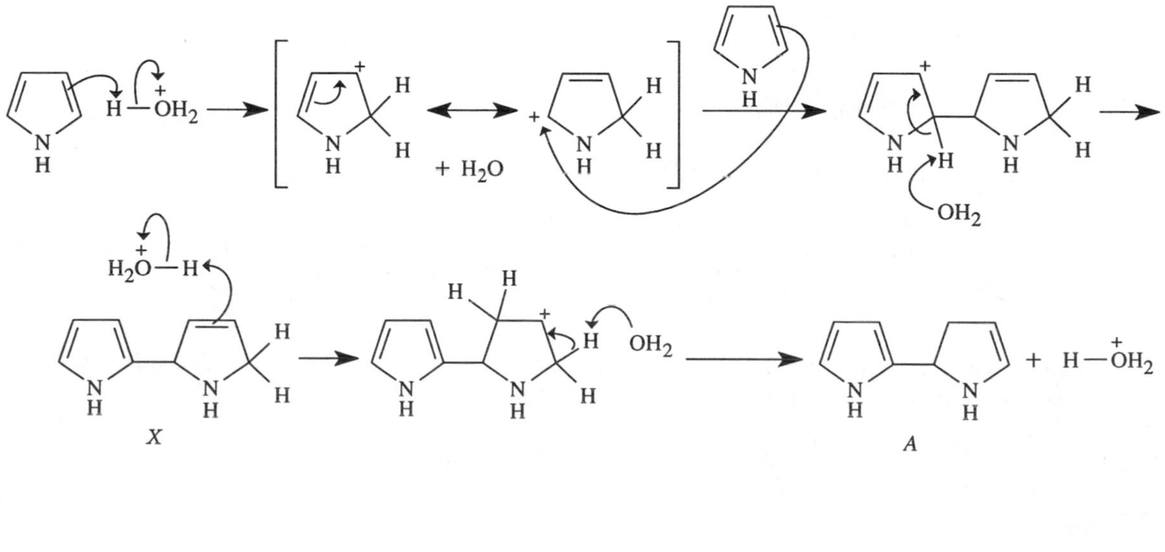

X → *A* + $H-\overset{+}{O}H_2$

(b) Compound *A* from part (a) is protonated to give a resonance-stabilized carbocation *Y*, which serves as the electrophile in an electrophilic aromatic substitution reaction at carbon-2 of pyrrole.

A — $H—\overset{+}{O}H_2$ → OH_2 + Y → B + $H—\overset{+}{O}H_2$

If you solved this problem by protonating carbon-3 of pyrrole in the initial step, thus avoiding the necessity for the final tautomerization, you have also written a reasonable and more direct mechanism. However, because pyrrole adds electrophiles (including the proton) predominantly at carbon-2, the mechanism above is probably preferable.

24.52 (a) First prepare 4-bromo-2-nitrotoluene and then use it in a Reissert synthesis.

o-nitrotoluene —(Br_2, Fe)→ 4-bromo-2-nitrotoluene —(NaOEt, EtOC(=O)—C(=O)OEt)→ —(AcOH)→ $CH_2C(=O)—C(=O)OEt$ (Br, NO_2) —(H_2/Pt)→ 6-bromoindole-2-CO_2Et —(1) NaOH 2) H_3O^+ 3) heat, $-CO_2$)→ 6-bromoindole

(b) This is a Fischer indole synthesis involving 3-pentanone and 4-methylphenylhydrazine.

$H_3C—C_6H_4—NHNH_2$ + $CH_3CH_2CCH_2CH_3$ (=O) —(H_2SO_4)→ 2-ethyl-3,5-dimethylindole

4-methylphenylhydrazine; 3-pentanone; 2-ethyl-3,5-dimethylindole

(c) Prepare pyridine-2-carbonitrile and reduce it to the amine.

pyridine —(1) $NaNH_2$ 2) H_2O)→ 2-aminopyridine (NH_2) —(1) $NaNO_2/H_2SO_4$ 2) CuCN)→ pyridine-2-C≡N —(1) $LiAlH_4$ 2) H_3O^+ 3) NaOH)→ pyridine-2-CH_2NH_2

(d) Use the *N*-oxide to direct nitration to the 4-position. Then convert the nitrated *N*-oxide into 4-chloropyridine, which is then displaced with ethanethiolate anion.

30% H_2O_2 → HNO_3, H_2SO_4 → H_2, Pd/C → 1) $NaNO_2$/HCl 2) CuCl → CH_3CH_2SH, NaOEt →

(e) Prepare 1-(2-furyl)-1-ethanone ("2-acetylfuran") and carry out an aldol condensation with furfural.

furfural → 1) CH_3MgI 2) H_3O^+ → $CrO_3(pyridine)_2$ → "2-acetylfuran" → furfural, NaOH →

(f) Oxidize furfural to the carboxylic acid and convert the acid into the ester. Notice that the acid chloride method of esterification is preferable to acid-catalyzed esterification with propanol because furans are somewhat sensitive to strongly acidic conditions. (See, for example, Problem 24.54.)

Ag_2O → $SOCl_2$ → $HOCH_2CH_2CH_3$, pyridine →

(g) Oxidize the methyl group to carboxylic acid, convert it into an acyl azide, and carry out a Curtius rearrangement to the isocyanate. Divide the isocyanate in half; convert one part into the amine, and then let the amine react with the remaining isocyanate to give the urea.

1) $KMnO_4$, ^-OH, heat 2) H_3O^+ → 1) $SOCl_2$ 2) NaN_3 → heat $-N_2$ →

NaOH, H_2O $-CO_2$ → (+ A) →

(h) Prepare 1-aminonaphthalene and use it as the "aniline" derivative in the Skraup synthesis.

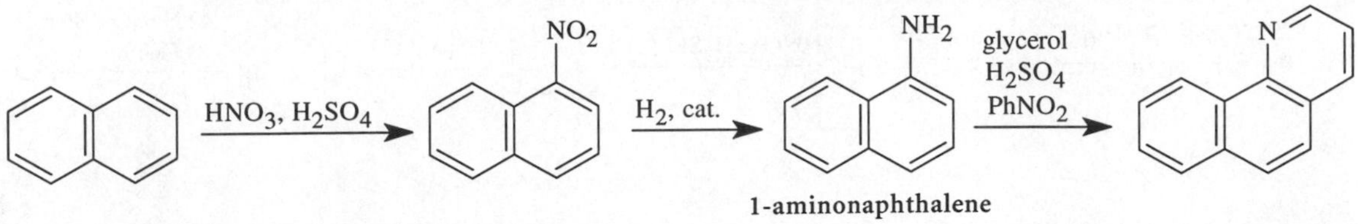

NO_2 NH_2 N

HNO_3, H_2SO_4 → ; H_2, cat. → ; glycerol, H_2SO_4, $PhNO_2$ →

1-aminonaphthalene

(i) Use the acidity of *N*-methylpyridinium salts to form a nucleophilic anion at the benzylic carbon of the ethyl group, which is then cyanoethylated twice. (Compare to the reactions in Eqs. 24.65 and 24.66 on text p. 1153. Cyanoethylation is discussed at the bottom of text p. 1045.

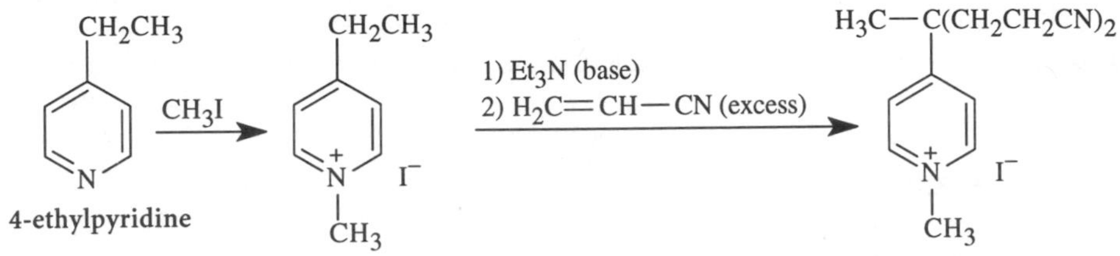

CH_2CH_3 CH_2CH_3 $H_3C—C(CH_2CH_2CN)_2$

CH_3I → ; 1) Et_3N (base) 2) $H_2C=CH—CN$ (excess) →

N $\overset{+}{N}$ I^- $\overset{+}{N}$ I^-

4-ethylpyridine CH_3 CH_3

(j) Alkylate the conjugate-base anion of 2-methylpyridine with propyl iodide; then convert the resulting 2-butylpyridine into its conjugate-base anion, which is carboxylated with CO_2 like any other organolithium or Grignard reagent.

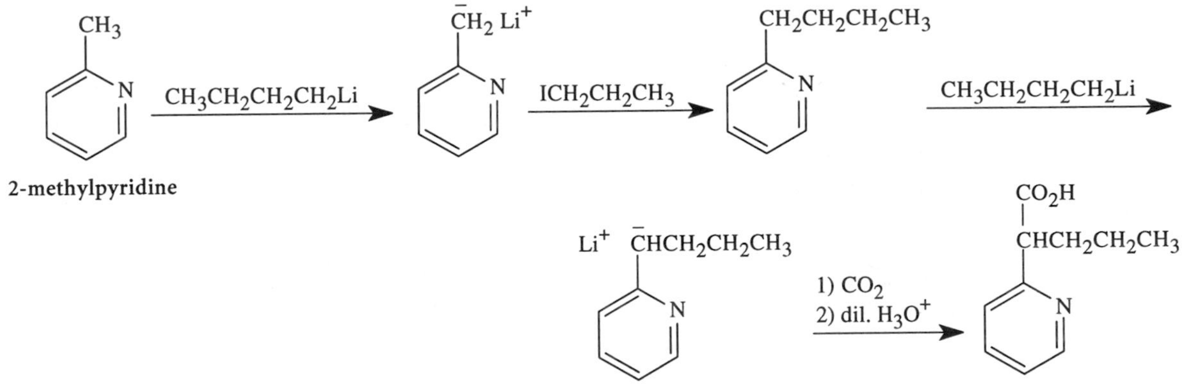

CH_3 $\bar{C}H_2$ Li^+ $CH_2CH_2CH_2CH_3$

N $CH_3CH_2CH_2CH_2Li$ → N $ICH_2CH_2CH_3$ → N $CH_3CH_2CH_2CH_2Li$ →

2-methylpyridine

CO_2H

Li^+ $\bar{C}HCH_2CH_2CH_3$ $CHCH_2CH_2CH_3$

N 1) CO_2 2) dil. H_3O^+ → N

Note that if a β-halo nitrile such as $ICH_2CH_2C{\equiv}N$ were used as the alkylating agent, $H_2C{=}CHC{\equiv}N$ would be formed and would be the effective alkylating agent. (Why? See Sec. 17.3B.)

24.53 The unsaturation number of compound *A* is 4; its oxidation to nicotinic acid shows that it contains a pyridine ring with a 3-substituent. The pyridine ring accounts for all four degrees of unsaturation. Because pyridine contains five carbons, the side-chain has three carbons. Compound *A* cannot be a primary alcohol, because the oxidation product of such an alcohol would be an aldehyde, which could not have five exchangeable α-hydrogens. (An aldehyde of the form $RCH_2CH{=}O$ has only two α-hydrogens.) Therefore, compound *A* is a secondary alcohol, and ketone *B* is its oxidation product.

OH O

CH_2CHCH_3 CH_2CCH_3

N N

A *B*

24.54 A compound with two degrees of unsaturation that has an NMR spectrum consisting entirely of singlets which is completely obliterated in base, and that gives hexane as a Clemmensen reduction product, must be a six-carbon unbranched diketone in which all hydrogens are α-hydrogens, that is, 2,5-hexanedione. A mechanism for the formation of this compound from 2,5-dimethylfuran is as follows.

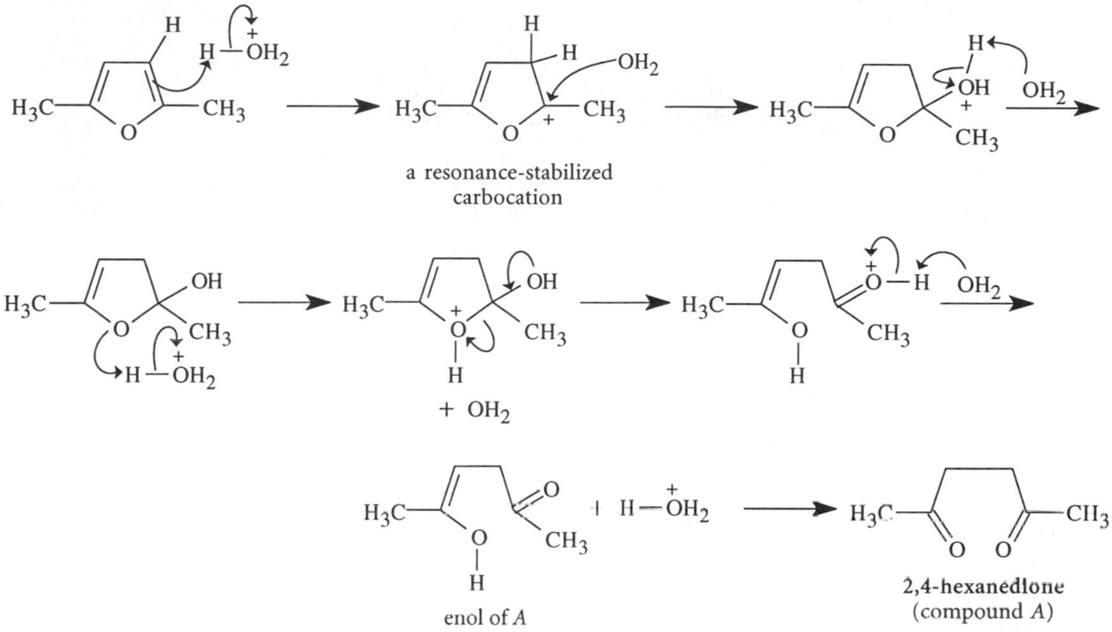

a resonance-stabilized carbocation

$+\ OH_2$

enol of *A*

2,4-hexanedione (compound *A*)

The final step, formation of the ketone from the enol, occurs by the reverse of the mechanism shown in Eq. 22.17b on text p. 1006.

24.55 (a) The reaction of thiophene with chlorosulfonic acid gives compound *A*, 2-thiophenesulfonyl chloride (see Eq. 20.27 on text p. 919). Because the chlorosulfonyl group exerts a strong *meta*-directing effect, nitration of *A* gives the 4-nitro derivative *B*. Heating *B* in water converts the sulfonyl chloride into the sulfonic acid, which then loses its sulfonyl group to give 3-nitrothiophene, *C*. (Recall that sulfonation is reversible; see Eq. 24.4, text page 1136.)

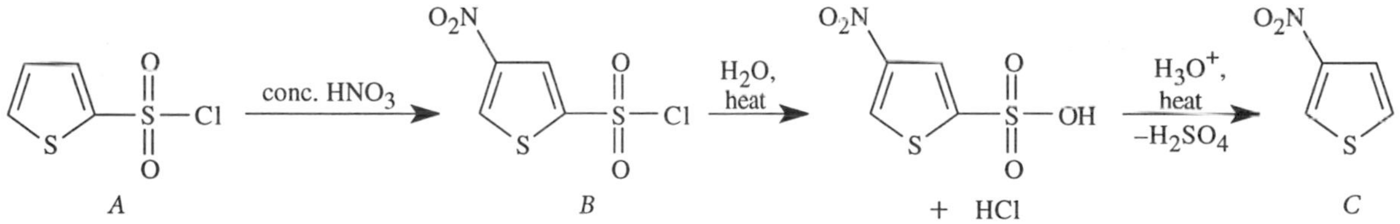

A —conc. HNO_3→ *B* —H_2O, heat→ (+ HCl) —H_3O^+, heat, $-H_2SO_4$→ *C*

(b) 3-Nitrothiophene cannot be made directly from thiophene because thiophene nitrates mostly in the 2-position to give 2-nitrothiophene. (See Eq. 24.25 on text page 1136.)

24.56 Compound *A* is nicotinic acid, the side-chain oxidation product; compound *B* is nicotinamide, the amide of *A*; compound *C* is 3-aminopyridine, the Hofmann rearrangement product of *C*; and compound *D* is the product of the Skraup synthesis. (As_2O_5 serves as the oxidant in the last step of the Skraup synthesis; see text p. 1160.)

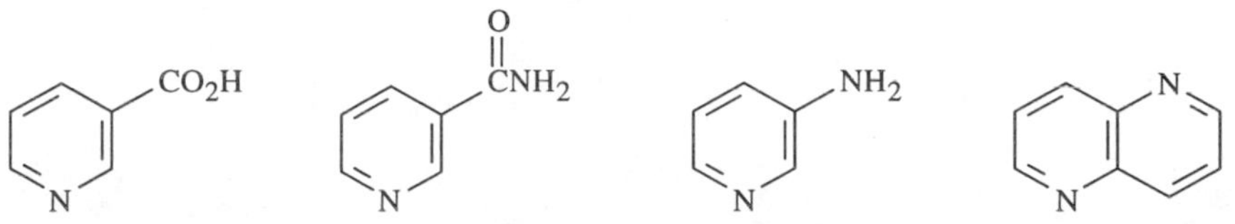

nicotinic acid (compound *A*)

nicotinamide (compound *B*)

3-aminopyridine (compound *C*)

5-azaquinoline (compound *D*)

The alternate possibility for compound *D* alluded to in the problem is

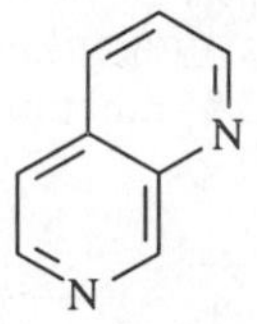

This compound lacks the symmetry necessary to ensure a zero dipole moment.

24.57 (a) Diethylamine and formaldehyde react to form an imminium ion by a mechanism completely analogous to the one shown in Eq. 23.28 on text p. 1087. This ion then serves as the electrophile in an electrophilic aromatic substitution reaction to the 3-position of indole.

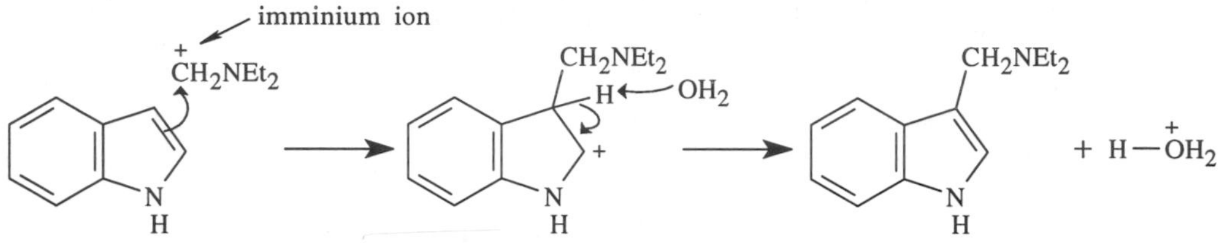

imminium ion

$\overset{+}{C}H_2NEt_2$ CH_2NEt_2 OH_2 CH_2NEt_2 $+ H—\overset{+}{O}H_2$

Note that indole, benzofuran, and benzothiophene tend to give a much greater percentage of 3-substitution than pyrrole, furan, and thiophene.

(b) Exchange is initiated with protonation by the deuterated acid at carbon-3 of the indole. In any exchange reaction of this sort, the experiment is set up so that there is a large excess of deuterium in the solvent; thus, the hydrogen that washes out into the solvent is swamped by the large excess of deuterium. In the last step, hydrogen is lost much more rapidly than deuterium because of the primary isotope effect (Sec. 9.5C). However, the relatively few molecules from which deuterium is lost are continually recycled through the mechanism until hydrogen is eventually washed out into the solvent.

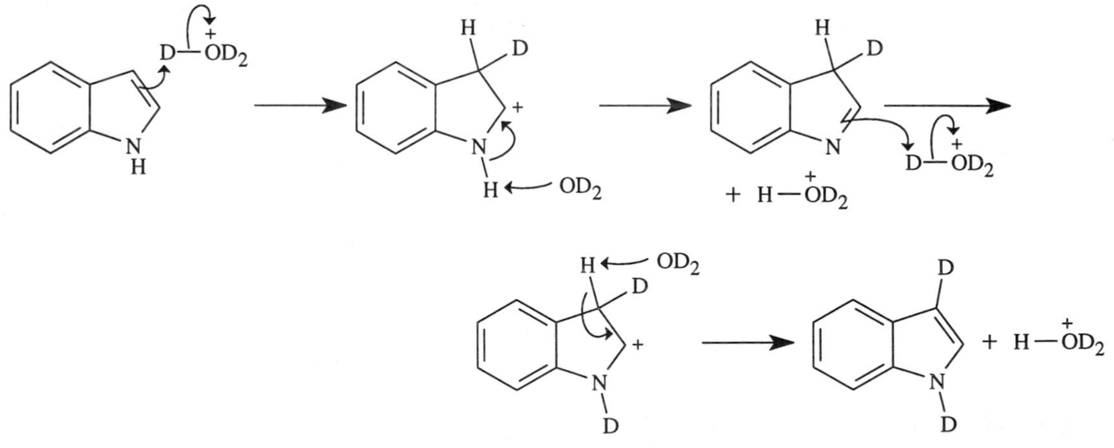

$D—\overset{+}{O}D_2$ OD_2 $+ H—\overset{+}{O}D_2$ $D—\overset{+}{O}D_2$

OD_2 $+ H—\overset{+}{O}D_2$

(c) This reaction is a carbocation rearrangement known as the *indolenine rearrangement.* The driving force for this reaction is the aromaticity of the product.

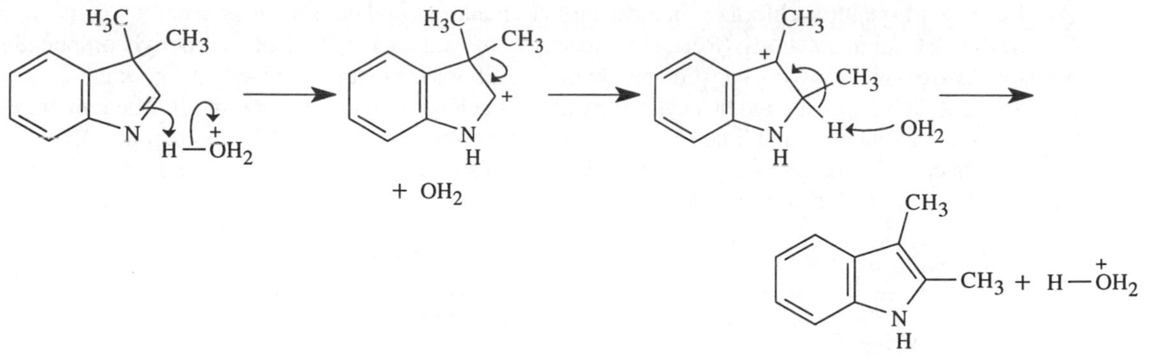

H_3C CH_3 N H—$\overset{+}{O}H_2$ → H_3C CH_3 + N H + OH_2 → CH_3 + CH_3 N H H← OH_2 → CH_3 CH_3 + H—$\overset{+}{O}H_2$ N H

(d) Alkylation of the nitrogen to give compound *A* (a quaternary ammonium salt) activates the carbon-nitrogen double bond toward nucleophilic attack by hydroxide ion, which, in turn, leads to ring opening.

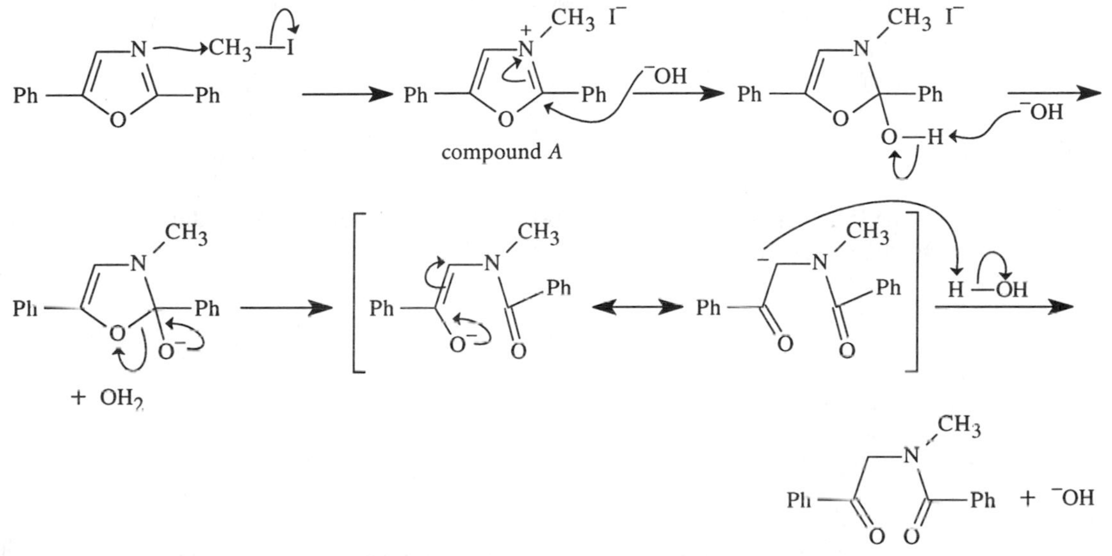

Ph N→CH_3—I O Ph → Ph $\overset{+}{N}$ CH_3 I^- O Ph compound *A* ^-OH → Ph N CH_3 I^- O Ph O—H← ^-OH →

Ph N CH_3 O Ph O^- + OH_2 → [Ph N CH_3 Ph O^- O ↔ Ph N CH_3 Ph O O] H—OH →

Ph N CH_3 Ph O O + ^-OH

(e) Sulfonylation of the *N*-oxide by the sulfonyl chloride activates the carbon-nitrogen double bond toward attack by chloride ion, which is produced in the sulfonation reaction. Hydroxide-promoted elimination forms a double bond and liberates chloride ion, which then attacks the terminal carbon of the double bond and displaces tosylate ion in an allylic variation of an S_N2 reaction (sometimes called an S_N2' reaction). This last step is undoubtedly driven by the aromatic stability of the product. The mechanism below begins with the sulfonate ester and chloride ion, produced in the formation of the sulfonate ester. (For the mechanism of sulfonate ester formation, see Study Guide Link 10.3 on p. 308 of this manual.) The toluenesulfonate group is abbreviated —OTs.

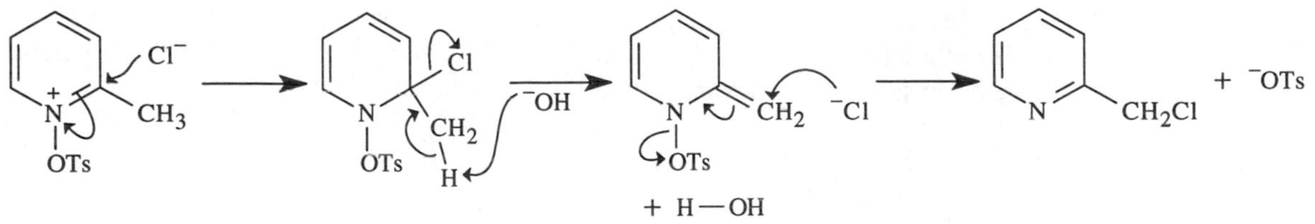

$\overset{+}{N}$ OTs Cl^- CH_3 → N OTs Cl CH_2 H ^-OH → N OTs CH_2 ^-Cl + H—OH → N CH_2Cl + ^-OTs

(f) Because of the three chlorines on its α-carbon, trichloroacetyl chloride is unusually reactive in nucleophilic acyl substitution reactions. (Recall that reactivity toward nucleophiles of a carbonyl compound is enhanced by electron-withdrawing substituents on the α-carbon; see text pp. 857–860.) The π-electrons of pyrrole are nucleophilic enough to be acylated by this acid chloride to give compound *B*. The remaining mechanistic steps are the same as the final steps of a haloform reaction (see Eq. 22.27e on text p. 1010), except that the base is ethoxide rather than hydroxide. Ethoxide displaces the trichloromethyl anion, which reacts with ethanol to give chloroform and ethoxide.

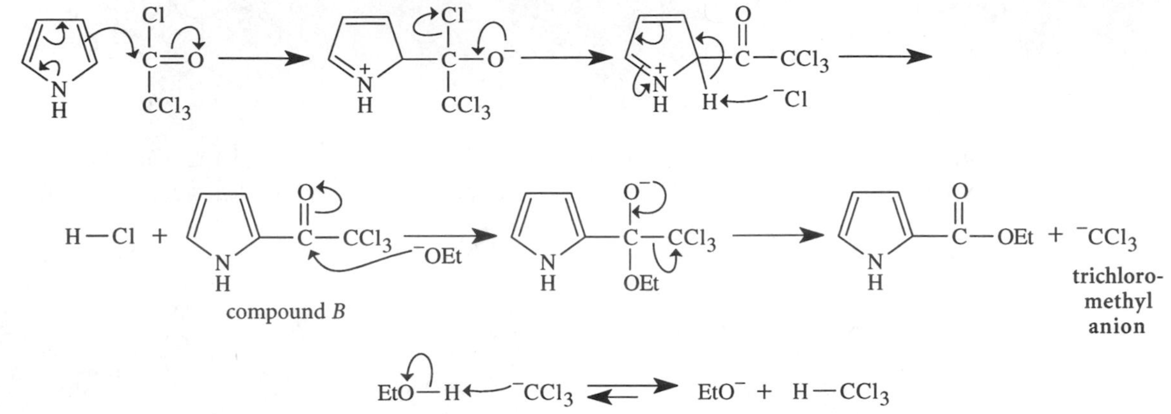

compound *B*

trichloromethyl anion

$$\text{EtO—H} + {}^{-}\text{CCl}_3 \rightleftharpoons \text{EtO}^{-} + \text{H—CCl}_3$$

(g) This reaction is a nucleophilic aromatic substitution reaction by the benzenethiolate anion (PhS^-), which is formed by the reaction of benzenethiol with triethylamine.

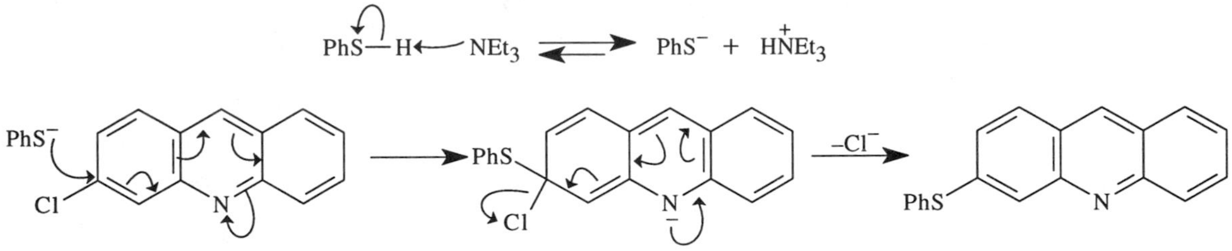

$$\text{PhS—H} + \text{NEt}_3 \rightleftharpoons \text{PhS}^{-} + \text{H}\overset{+}{\text{N}}\text{Et}_3$$

Notice that this is like a nucleophilic aromatic substitution reaction on pyridine in the sense that a pyridine nitrogen stabilizes the anionic intermediate.

24.58 For simplicity, represent the histidine side-chain as R; that is,

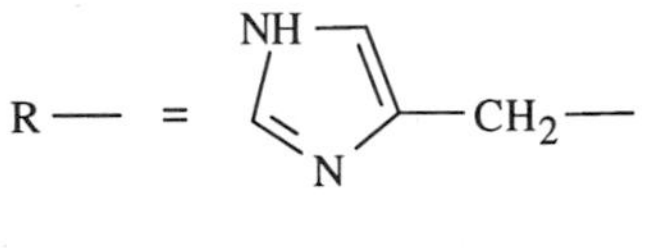

In this mechanism, the pyruvyl group is acting as a "surrogate" of pyridoxal phosphate. The first step is formation of the imine *A* of the pyruvyl group. This forms at the ketone carbonyl (why?). The mechanism of imine formation is outlined in Eq. 19.56a–b on text p. 875. This imine then loses CO_2 to give carbanion *B*, which is stabilized because its electron pair is delocalized into the other carbonyl group.

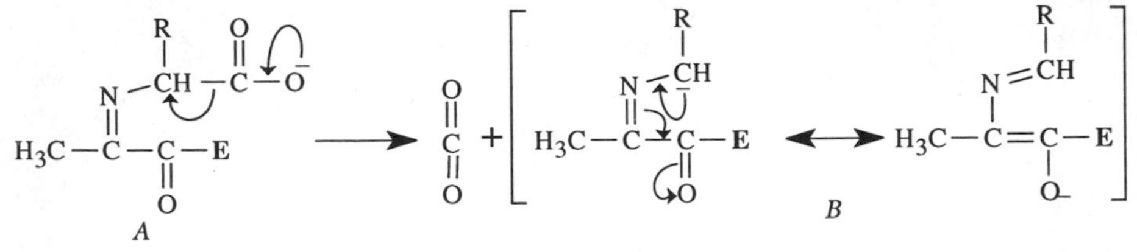

Protonation of this carbanion at carbon yields a new imine, which hydrolyses by a mechanism that is the reverse of imine formation to give histamine:

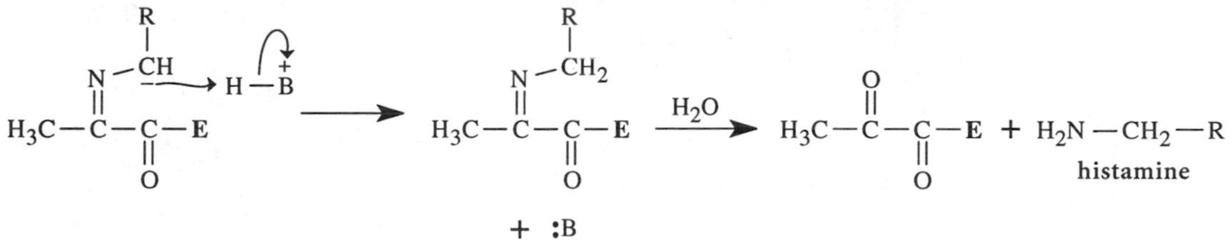

$$H_3C-C(=N-CH(R)^{-})-C(=O)-E + H-\overset{+}{B} \longrightarrow H_3C-C(=N-CH_2R)-C(=O)-E + :B \xrightarrow{H_2O} H_3C-C(=O)-C(=O)-E + H_2N-CH_2-R$$

histamine

24.59 (a) First, the imine of serine forms with pyridoxal phosphate by the mechanism shown in Eq. 24.71 on text p. 1156. This imine undergoes a β-elimination of the elements of water to give the imine of dehydroalanine (compound *A* in the problem). The role of imine formation is to make the proton that is eliminated quite acidic; recall (Sec. 17.3B) that E2 reactions involving acidic β-protons are particularly rapid. (Why is this β proton acidic?) A base on the enzyme B**:** abstracts the proton, and an acidic group on the enzyme $\overset{+}{B}$—H protonates the —OH group as it leaves to form water.

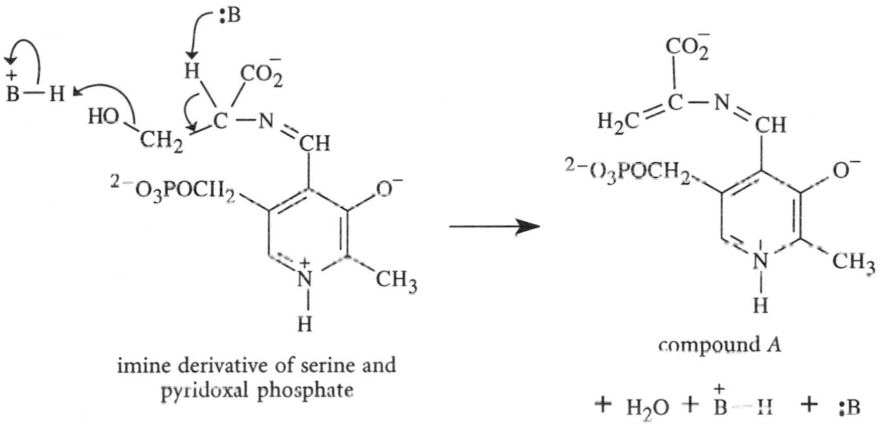

imine derivative of serine and pyridoxal phosphate

compound *A*

$+ H_2O + \overset{+}{B}-H + :B$

(b) Delocalization of electrons from the double bond on the colored carbon to the pyridinium nitrogen reveals that this carbon has carbocation character, and should thus be electrophilic:

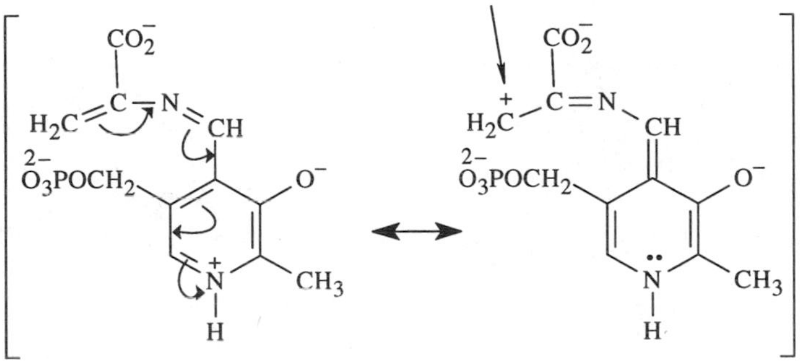

carbocation character

(c) The electrophilic carbon of the imine of β-alanine is attacked by the electrons of the indole ring:

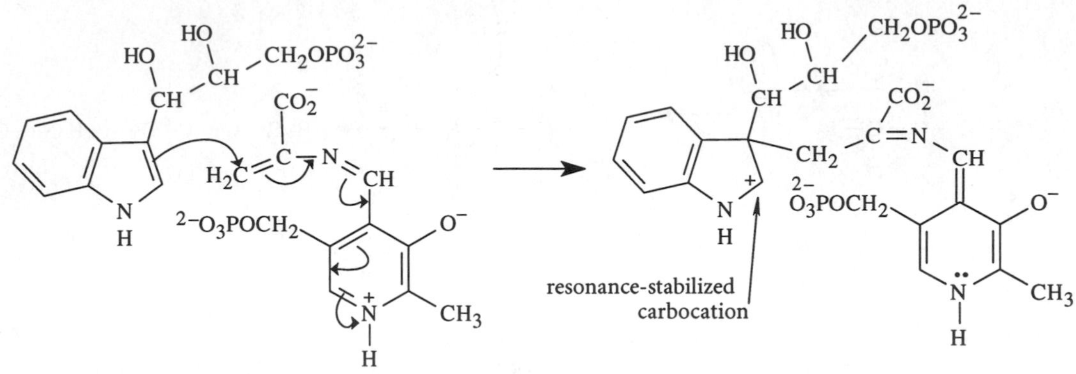

resonance-stabilized carbocation

The resulting carbocation intermediate is resonance-stabilized by the electrons of the indole nitrogen. A basic group removes an O—H proton to start an electronic cascade that generates glyceraldehyde-3-phosphate and an imine *C*:

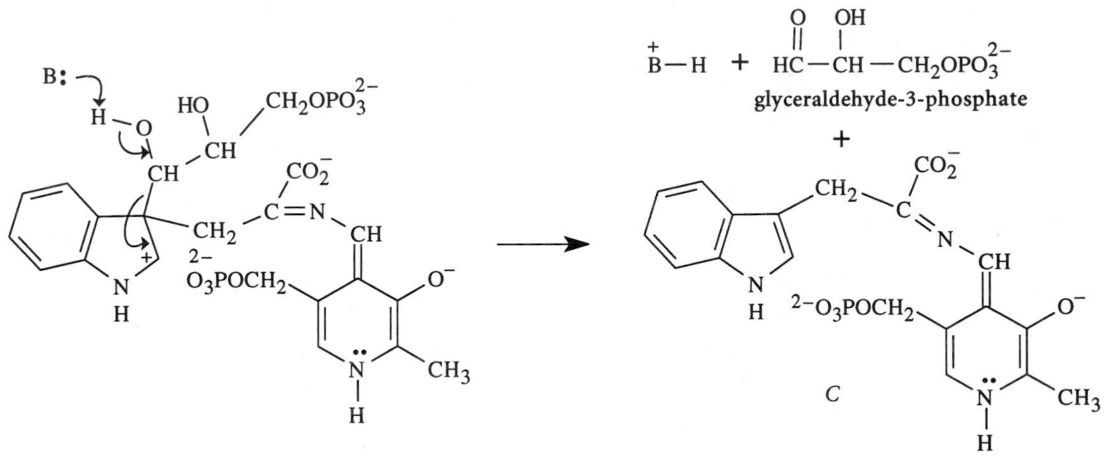

glyceraldehyde-3-phosphate

C

Finally, compound *C* is protonated on carbon to form an imine *D* between pyridoxal phosphate and trypotophan. Hydrolysis of this imine gives pyridoxal phosphate and the amino acid.

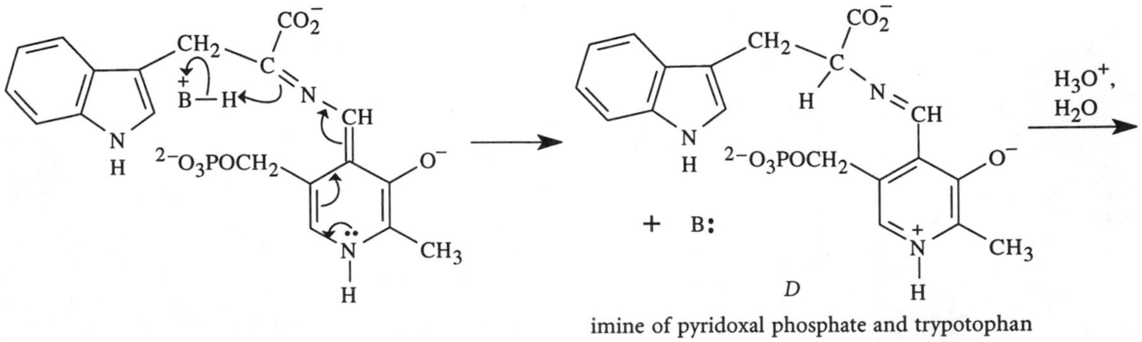

D

imine of pyridoxal phosphate and trypotophan

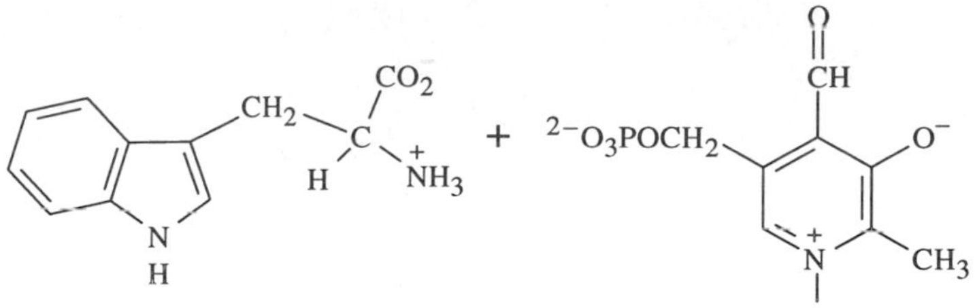

troptophan

pyridoxal phosphate

The stereochemical issues have been ignored in this problem. Tryptophan produced in this way is solely the *S* enantiomer. At what step is the configuration of tryptophan established?

24.60 Form the carbanion intermediate as shown in Eqs. 24.71 and 24.72a on p. 1156 of the text. This carbanion can be protonated from either face to give enantiomers. The enantiomeric amino acids are freed by imine hydrolysis.

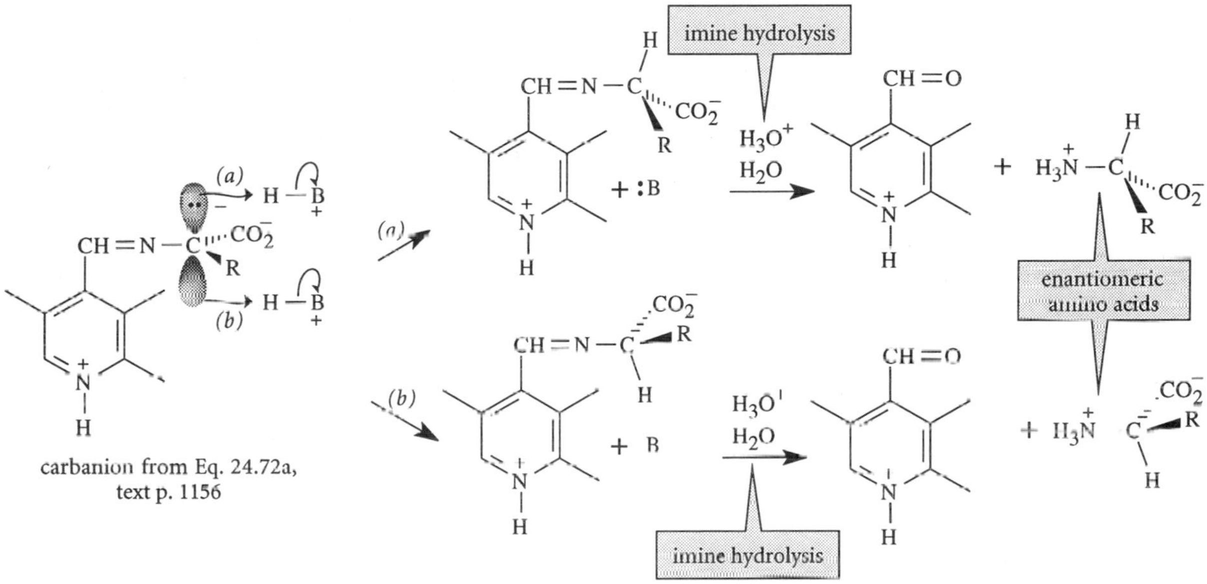

imine hydrolysis

enantiomeric amino acids

carbanion from Eq. 24.72a, text p. 1156

imine hydrolysis

24.61 (a) Exchange occurs when the conjugate-base anion is formed by abstraction of one of the colored protons, and this anion removes a deuteron (a deuterium nucleus) from the solvent. The anionic intermediate involved in this exchange is resonance-stabilized in such a way that negative charge is delocalized onto two electronegative atoms—one of the nitrogens and the carbonyl oxygen. In contrast, removal of a proton from the other methyl group gives an anion in which the negative charge is not delocalized onto electronegative atoms; consequently, this anion is less stable. Thus, exchange of a colored hydrogen occurs because this exchange involves the more stable anionic intermediate.

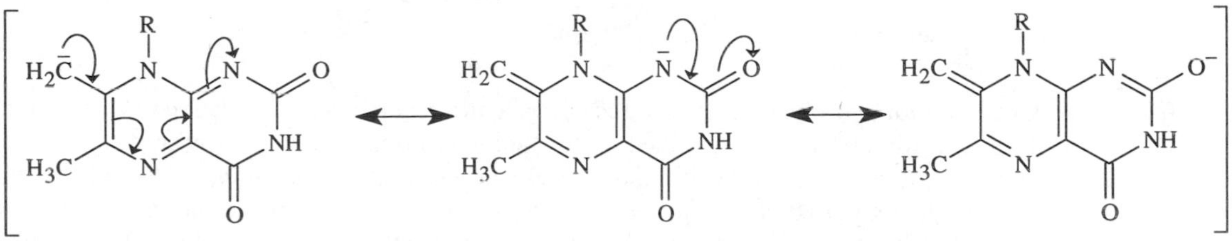

anionic intermediate involved in the exchange of the colored protons

The imide hydrogen is also exchanged by a base-catalyzed mechanism that involves the conjugate-base anion of the imide. (Imides, like other β-dicarbonyl compounds, are acidic; see Problem 22.2 on text

p. 1001.)

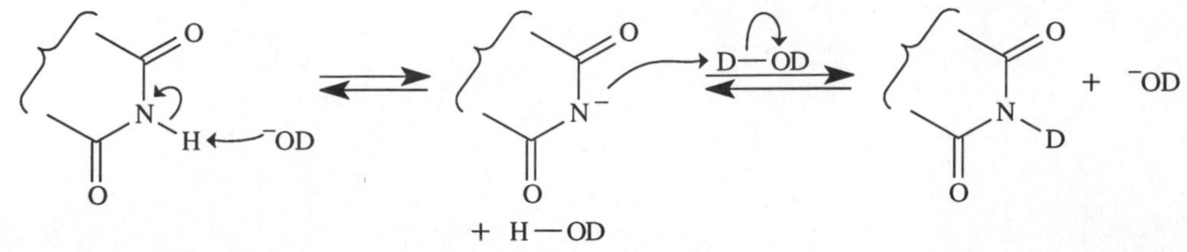

(b) Removal of a hydrogen from the black methyl group gives an anion that is stabilized by the polar effect of the attached positively charged nitrogen and by resonance interaction with the adjacent double bond. However, the resonance structure of this anion (draw it) delocalizes charge to a carbon atom. In contrast, removal of a hydrogen from the colored methyl group gives an "anion" which is actually a neutral compound; that is, the negative charge is delocalized to the positively charged nitrogen. Such charge neutralization is a significant stabilizing effect, and is the reason for the greater acidity of the colored hydrogen.

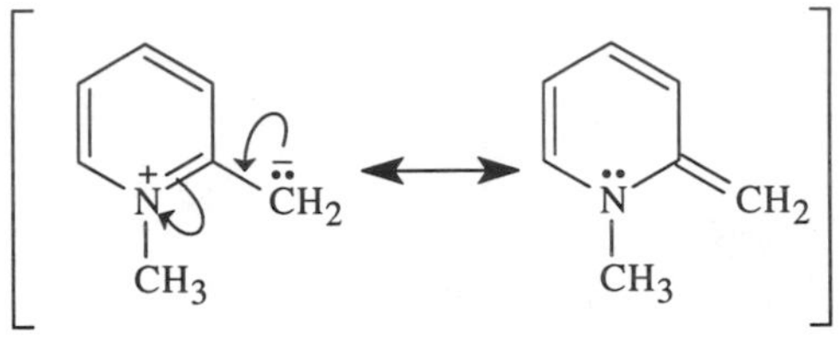

resonance structures of the "anion" formed by removal of a hydrogen from the colored methyl group

(c) The positive charge on the nitrogen activates the carbon-nitrogen double bond to nucleophilic attack by hydroxide ion. This forms a tetrahedral addition intermediate that breaks down to a substituted formamide *A*, which then undergoes amide hydrolysis to the products shown in the problem.

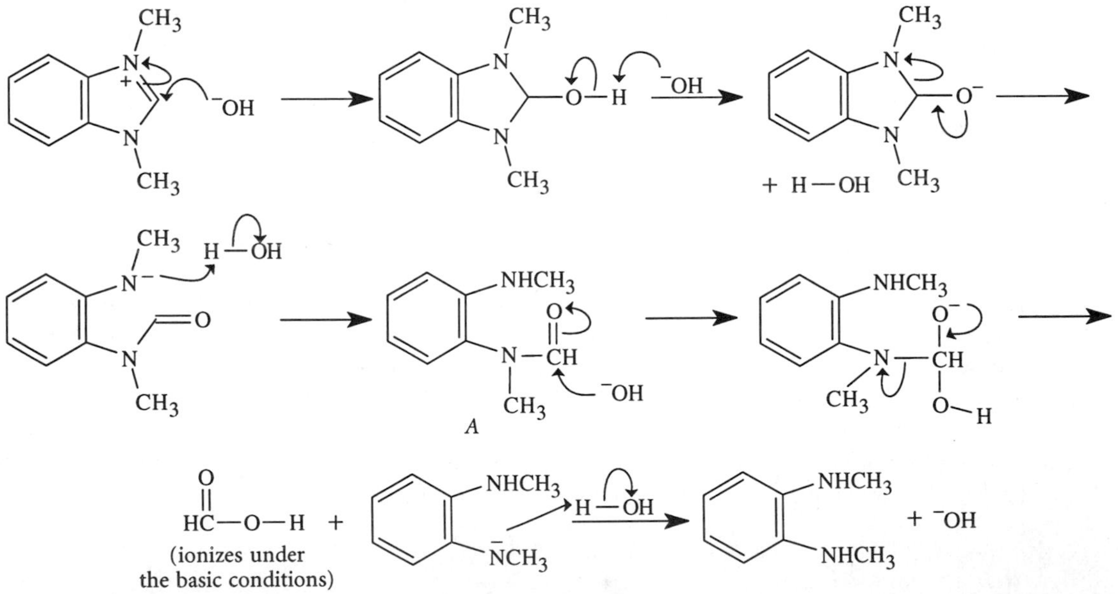

(ionizes under the basic conditions)

(d) 2-Pyridone does not hydrolyze because it undergoes a different reaction: ionization to its conjugate-base anion. This ionization occurs because the resulting anion is aromatic and is therefore very stable. The ion doesn't hydrolyze because the ionized amide is resonance stabilized *and aromatic*. Remember that the more stable a carbonyl compound is, the slower it hydrolyzes (Sec. 21.7E). Furthermore, hydrolysis would require that a second negative charge be introduced into the molecule when $^{-}$OH attacks the carbonyl. Ionization of γ-butyrolactam is much less favorable; consequently, ionization does not compete with the hydrolysis reaction.

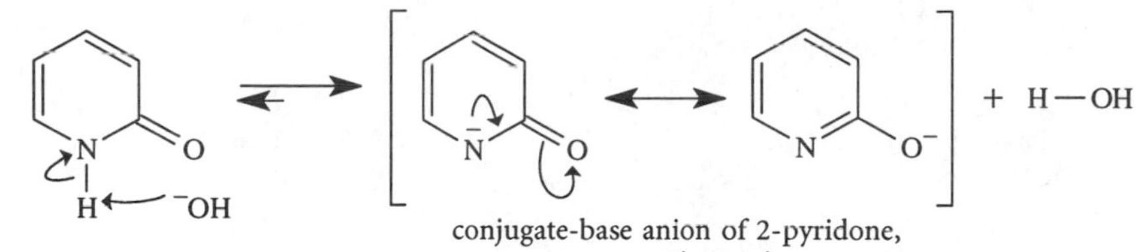

$+ \text{H—OH}$

conjugate-base anion of 2-pyridone, an aromatic species

(e) The reaction of 4-chloropyridine with ammonia is a nucleophilic aromatic substitution reaction that occurs readily because the negative charge in the addition intermediate is delocalized to the pyridine nitrogen, an electronegative atom, as shown in Eq. 24.58b on text p. 1150 (with :Nuc = $:NH_3$ and Y = Cl). In contrast, the negative charge in the corresponding intermediate for nucleophilic aromatic substitution of 3-chloropyridine can only be delocalized to carbon atoms. (See Eq. 24.58c on text page 1151.) As a result, 3-chloropyridine is much less reactive. [When it is forced to react, it does so as described in part (f).]

(f) The observation of *cine*-substitution suggests that a pyridine analog of benzyne (3,4-pyridyne) is involved as an intermediate. (You should supply the curved-arrow formalism.)

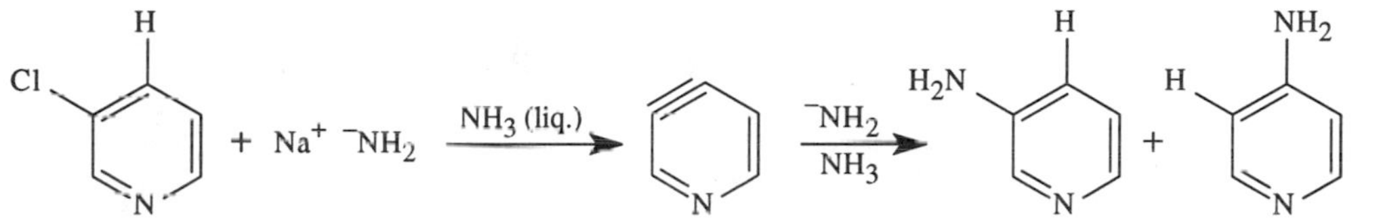

3-chloropyridine $+ \ Na^+ \ ^-NH_2 \xrightarrow{NH_3 \text{ (liq.)}}$ 3,4-pyridyne $\xrightarrow[NH_3]{^-NH_2}$ 3-aminopyridine + 4-aminopyridine

24.62 Step (1) is a nitration reaction. Because 2-pyridone derivatives such as this starting material contain significant amounts of their aromatic hydroxypyridine isomers (Eq. 24.54 on text p. 1149), the nitro group can be introduced by aromatic nitration. The conditions of step (1) are HNO_3 and H_2SO_4. Step (2) can be effected with PCl_5 and heat. (See Eq. 24.55 on text p. 1150.) Step (3) is a reduction of both the nitro group and the nitrile to amine groups; catalytic hydrogenation is the method of choice. Evidently, the chlorine is also removed under the conditions of this hydrogenation. To see why this is reasonable, recall that 2-chloropyridines have some of the reactivity characteristics of acid chlorides, and that acid chlorides are reduced to aldehydes (that is, the chlorine is replaced by a hydrogen) in a type of catalytic hydrogenation (Rosenmund reduction; text p. 974). Step (4) is an ether cleavage, which can be effected with aqueous H_2SO_4; water is the nucleophile in this reaction. This ether is activated toward cleavage by either an S_N1 or an S_N2 mechanism because it is benzylic. Neutralization with aqueous hydroxide gives the free amine shown. Finally, step (5) can be realized by diazotization with $NaNO_2$ in aqueous H_2SO_4, which converts the benzylic amine into a benzylic diazonium ion, and the arylamine into an aryldiazonium ion; nitrogen is rapidly displaced from the benzylic diazonium ion by water. Elimination, usually a competing reaction in the decomposition of alkyldiazonium ions, is not observed here because there are no β-hydrogens. The aryldiazonium ion is also converted into the phenolic hydroxy group by heating the aqueous solution. (See the discussion above Eq. 23.51 on text p. 1096.)

24.63 (a) Bromine adds to the double bond, and KOH saponifies the resulting dibromo lactone. An internal nucleophilic substitution reaction then closes the five-membered ring with loss of one bromine as bromide ion. (Recall that α-halo carbonyl compounds are particularly reactive in nucleophilic substitution reactions; see Sec. 22.3D.) An E2 reaction forms the double bond with loss of the second bromine as bromide ion. (This is a very favorable E2 reaction; why? See Sec. 17.3B.) The mechanism of bromine addition, not shown below, is exactly analogous to the mechanism shown in Eqs. 5.3–5.4 on text p. 161; and the mechanism of saponification, also not shown below, is exactly analogous to the mechanism shown in Eqs 21.9a–b on text p. 952.)

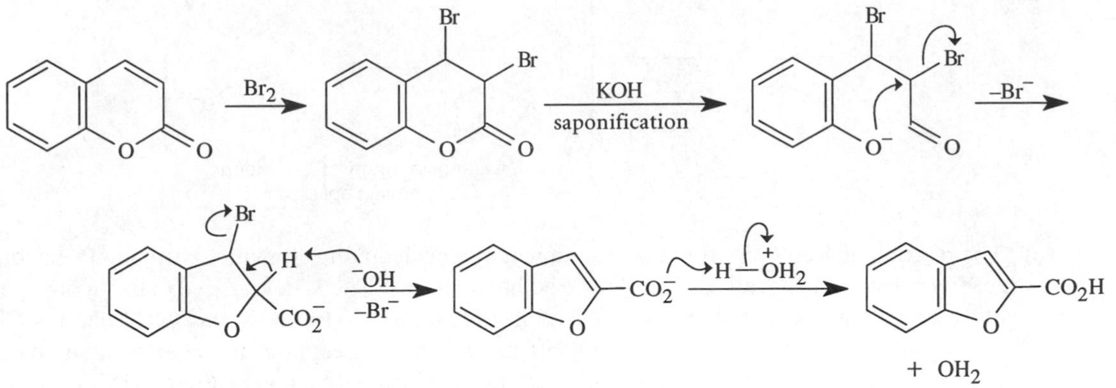

Br$_2$ KOH saponification $-Br^-$ ^-OH $-Br^-$ CO_2^- $H-\overset{+}{O}H_2$ CO_2H + OH_2

(b) This reaction begins as an aldol condensation between the conjugate-base enolate ion *A* of the diester and the diketone. Normally esters undergo Claisen condensations, but, because 1,2-diketones are particularly reactive (why?), the aldol condensation is observed instead. Once the two molecules are joined by the first aldol condensation, a second aldol condensation by an analogous mechanism closes the ring.

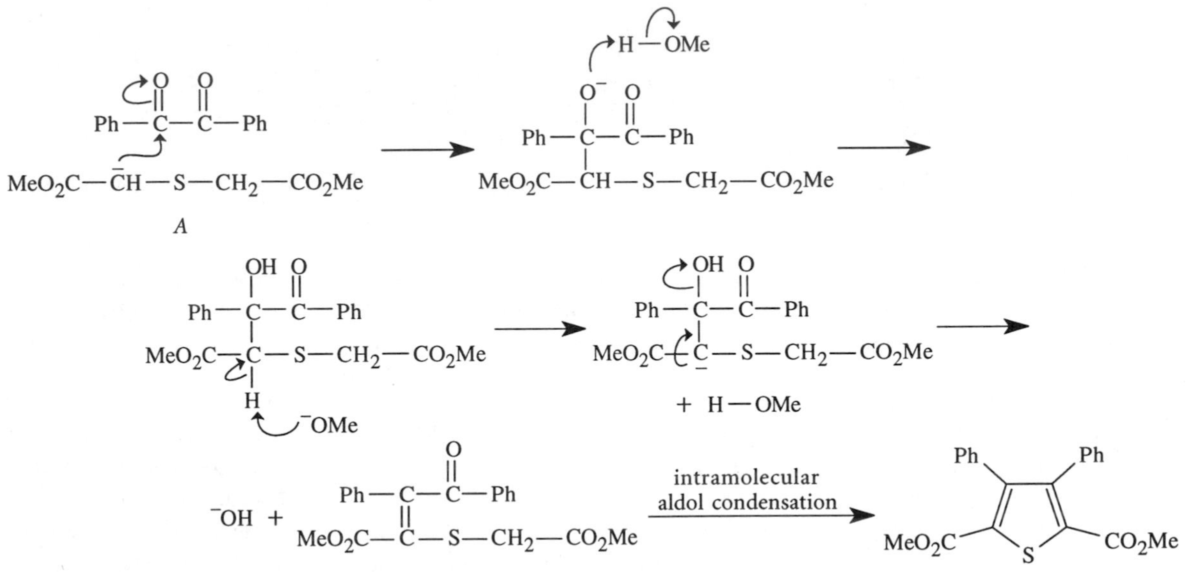

$Ph-CO-CO-Ph$ + $MeO_2C-\bar{C}H-S-CH_2-CO_2Me$ (*A*) ⟶ $Ph-C(O^-)(-CH(CO_2Me)-S-CH_2-CO_2Me)-CO-Ph$ + $H-OMe$ ⟶

$Ph-C(OH)(-CH(CO_2Me)-S-CH_2-CO_2Me)-CO-Ph$ + ^-OMe ⟶ $Ph-C(OH)(-\bar{C}(CO_2Me)-S-CH_2-CO_2Me)-CO-Ph$ + H—OMe ⟶

^-OH + $Ph-C(=C(CO_2Me)-S-CH_2-CO_2Me)-CO-Ph$ —intramolecular aldol condensation→ dimethyl 3,4-diphenylthiophene-2,5-dicarboxylate (Ph, Ph, MeO_2C, S, CO_2Me)

(c) The mechanism shown below starts with the enamine formed between the amino group and the carbonyl group of 2-butanone. (The mechanism of enamine formation is discussed on text p. 878.) Note that, although primary amines form imines with ketones, imines are in equilibrium with enamines just as ketones are in equilibrium with enols. (See Eq. 19.62a–b on text p. 878.) Water is produced as a by-product of enamine formation. The conjugate acid of water, H_3O^+, is shown as the catalyzing acid, but any of the other acids present could also catalyze the reaction.

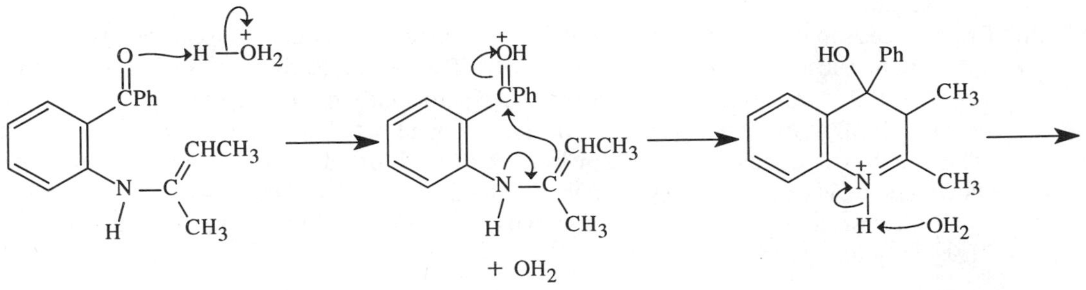

O=CPh, $H-\overset{+}{O}H_2$, N—C(=CHCH$_3$)CH$_3$, H ⟶ $\overset{+}{O}H$=CPh, CHCH$_3$, N—C, H, CH$_3$, + OH_2 ⟶ HO Ph, CH$_3$, $\overset{+}{N}$, CH$_3$, H, OH_2 ⟶

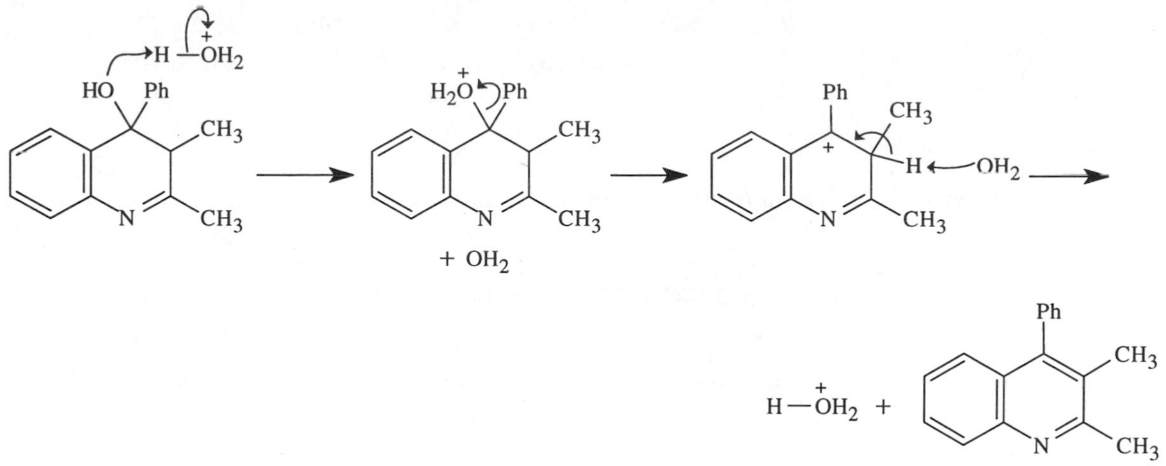

(d) Imine formation between the amine and one of the ketone groups is followed by an intramolecular electrophilic aromatic substitution reaction in which the protonated ketone serves as the electrophile. The mechanism below begins with the imine; the mechanism of imine formation is discussed on text p. 875.

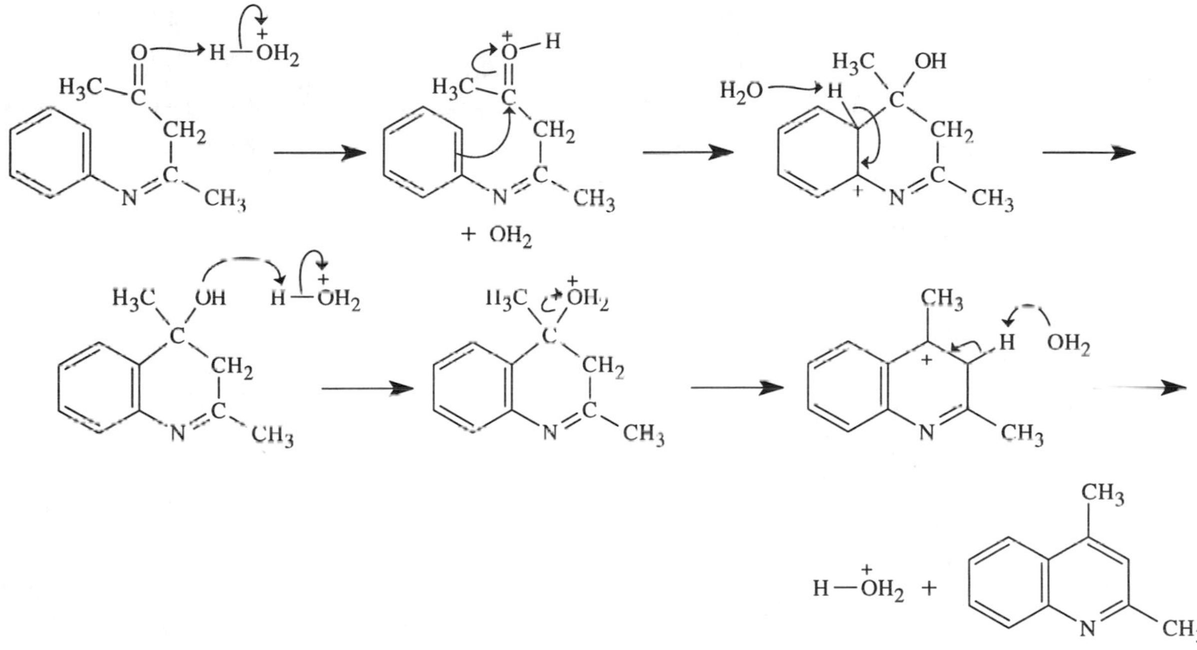

(e) Diethylamine serves as a base catalyst; its first function is to form the conjugate-base enolate ion *A* of ethyl acetoacetate:

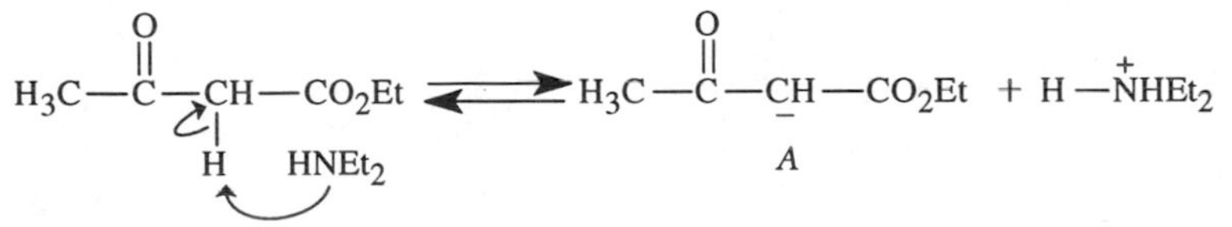

$$H_3C—\overset{O}{\overset{\|}{C}}—\underset{H}{\underset{|}{CH}}—CO_2Et \;\; HNEt_2 \rightleftharpoons H_3C—\overset{O}{\overset{\|}{C}}—\underset{-}{CH}—CO_2Et + H—\overset{+}{N}HEt_2$$

A

The fact that a secondary amine rather than a tertiary amine is used suggests that the role of diethylamine is also to form the enamine, which, like an enol, can serve as a nucleophile:

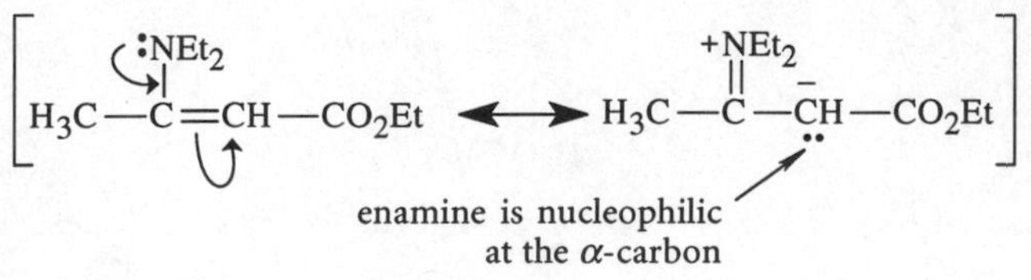

$$\left[H_3C{-}\overset{:NEt_2}{\overset{|}{C}}{=}CH{-}CO_2Et \longleftrightarrow H_3C{-}\overset{+NEt_2}{\overset{\|}{C}}{-}\overset{-}{\underset{\cdot\cdot}{C}H}{-}CO_2Et \right]$$

enamine is nucleophilic at the α-carbon

Diethylamine can also form an imminium ion by reaction with formaldehyde (Eq. 23.28, text p. 1087):

$$Et_2NH + H_2C{=}O \rightleftharpoons Et_2N{-}CH_2{-}OH \xrightleftharpoons{Et_2\overset{+}{N}H_2} Et_2\overset{+}{N}{=}CH_2 + H_2O$$

imminium ion

The first stage of the mechanism is a crossed-aldol condensation in which enolate *A* (or the enamine shown above) condenses with formaldehyde, (or with the imminium ion shown above). The mechanism of the aldol condensation is given in Eqs. 22.40b and 22.44 on text pp. 1014 and 1016, respectively. The product of this aldol condensation then undergoes a conjugate-addition reaction (that is, a Michael addition) with a second enolate ion *A* (or a second enamine). Ammonia reacts with one of the ketone functional groups to form an enamine, and the ring is closed when the amino group of that enamine forms an intramolecular enamine with the second ketone group. (The mechanism of enamine formation is discussed on text p. 878; see the note about the equilibration of imines and enamines in the solution to part (c) of this problem.)

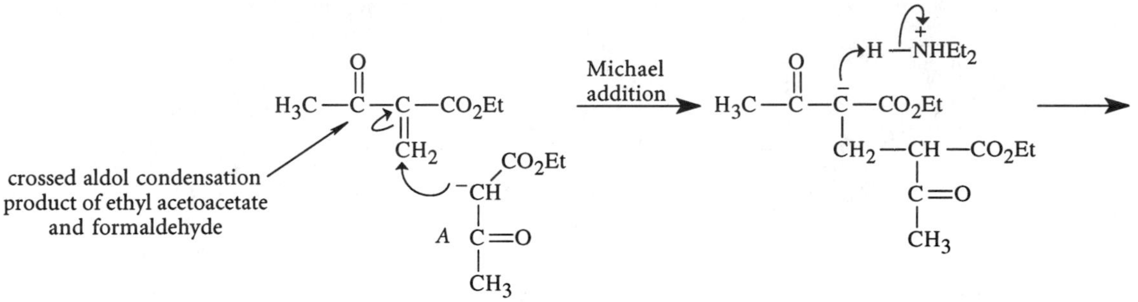

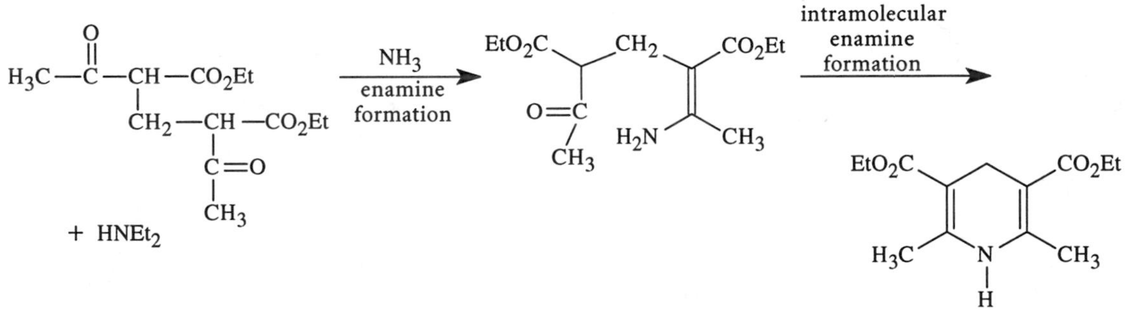

$H_3C{-}C(=O){-}C(=CH_2){-}CO_2Et$ + $^{-}CH(CO_2Et){-}C(=O){-}CH_3$ (*A*) $\xrightarrow{\text{Michael addition}}$ $H_3C{-}C(=O){-}\overset{-}{C}(CO_2Et){-}CH_2{-}CH(CO_2Et){-}C(=O){-}CH_3$ (H—$\overset{+}{N}HEt_2$) $\longrightarrow$

crossed aldol condensation product of ethyl acetoacetate and formaldehyde

$H_3C{-}C(=O){-}CH(CO_2Et){-}CH_2{-}CH(CO_2Et){-}C(=O){-}CH_3$ + $HNEt_2$ $\xrightarrow[\text{enamine formation}]{NH_3}$ $EtO_2C{-}CH(C(=O)CH_3){-}CH_2{-}C(CO_2Et){=}C(NH_2){-}CH_3$ $\xrightarrow{\text{intramolecular enamine formation}}$ (1,4-dihydropyridine: EtO_2C, CO_2Et, H_3C, CH_3, N—H)

We leave it to you to complete the similar mechanism with an enamine as the nucleophile.

25

Pericyclic Reactions

TERMS

allowed reaction 25.2A
antarafacial 25.3
antibonding molecular orbital (ψ^*) 25.1A
antisymmetric (A) molecular orbital 25.1A
bonding molecular orbital (ψ) 25.1A
Claisen rearrangement 25.4B
concerted reaction .... Introduction
conrotatory 25.2A
Cope rearrangement 25.4B
cycloaddition reaction .... Introduction
disrotatory 25.2A
electrocyclic reaction .... Introduction, 25.3
excited state 25.1C
fluxional molecule 25.6
forbidden reaction 25.2A
frontier-orbital theory .... Introduction
frontier orbitals 25.1A
ground state 25.1C
highest occupied molecular orbital (HOMO) 25.1A
lowest unoccupied molecular orbital (LUMO) 25.1A
nonbonding molecular orbital 25.1B
oxyCope reaction 25.4B
pericyclic reaction .... Introduction
pericyclic selection rules 25.2C
phase 25.1A
photochemical electrocyclic reaction 25.2B
photochemical pericyclic reaction 25.2B
principle of microscopic reversibility 25.2C
reference plane (for molecular orbital symmetry classification) 25.1A
sigmatropic reaction .... Introduction, 25.4
singularly occupied molecular orbital (SOMO) 25.1B
suprafacial 25.3
symmetric (S) molecular orbital 25.1A
thermal electrocyclic reaction 25.2A
thermal pericyclic reaction 25.2B

CONCEPTS

I. MOLECULAR ORBITALS OF CONJUGATED π-ELECTRON SYSTEMS

A. Molecular Orbitals of Conjugated Alkenes

1. Atomic $2p$ orbitals on adjacent atoms within molecules overlap to give π molecular orbitals.
2. The π molecular orbitals for ethylene and conjugated alkenes can be constructed according to the following generalizations:
   a. When a number (say m) of p orbitals interact, the resulting π-electron system contains the same number m of molecular orbitals (MOs), all with different energies.
   b. Ethylene and conjugated alkenes have two types of π molecular orbitals:
      i. Bonding π molecular orbitals (half of the π molecular orbitals) have lower energy than the isolated $2p$ orbitals.
      ii. Antibonding π molecular orbitals (the other half of the π molecular orbitals) have higher energy than the isolated $2p$ orbitals. (Antibonding MOs are indicated with asterisks.)

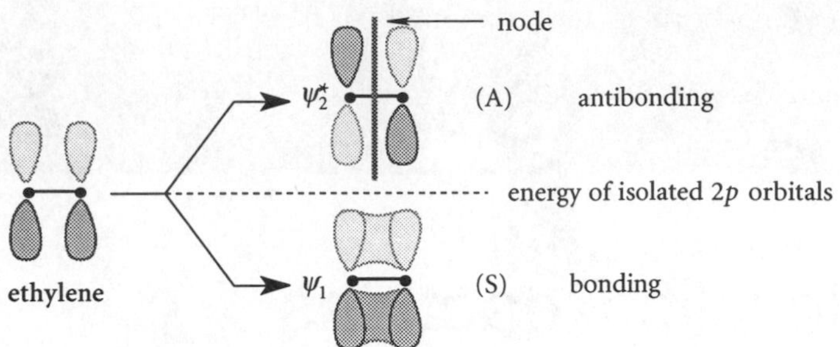

(See Figs. 25.1 and 25.2, text pp. 1176–1177)

c. The bonding molecular orbital of lowest energy, ψ_1, has no nodes between the atoms; each molecular orbital of increasingly higher energy has one additional node.
   i. A node is a plane at which any wave is zero; that is, when an electron is in a given MO there is zero probability of finding the electron, and thus zero electron density, at the node.
   ii. The electron wave has a peak on one side of the node and a trough on the other side.
   iii. The orbital changes phase at the node.

d. The nodes occur between atoms and are arranged symmetrically with respect to the center of the π-electron system.

e. With respect to an imaginary reference plane at the center of the π-electron system and perpendicular to the plane of the molecule—
   i. odd-numbered MOs are symmetric (S) (molecular orbitals in which peaks reflect across the reference plane into peaks, and troughs into troughs)
   ii. even-numbered MOs are antisymmetric (A) (molecular orbitals in which peaks reflect across the reference plane into troughs, and troughs into peaks)
   iii. Within any symmetric MO, the phase at the two terminal carbons is the same; within any antisymmetric MO the phase at the two terminal carbons is different.

f. Electrons are placed pairwise into each molecular orbital, beginning with the orbital of lowest energy (Aufbau principle).

3. The presence of unconjugated substituents, to a useful approximation, does not alter the π molecular orbital structure of a conjugated alkene.
4. The π-electron contribution to the energy of a molecule is determined by the energies of its occupied MOs. Bonding MOs have lower energy than isolated 2*p* orbitals; hence, there is an energetic advantage to π molecular orbital formation.
5. Two MOs are of particular importance in understanding pericyclic reactions:
   a. The highest occupied π molecular orbital (HOMO).
   b. The lowest unoccupied π molecular orbital (LUMO).
   c. The HOMO and LUMO of a conjugated alkene have opposite symmetry.
   d. The HOMO and LUMO are sometimes collectively termed frontier orbitals because they are the molecular orbitals at the high- and low-energy extremes, respectively, of the occupied and unoccupied molecular orbitals. The HOMO has a lower energy than the LUMO.

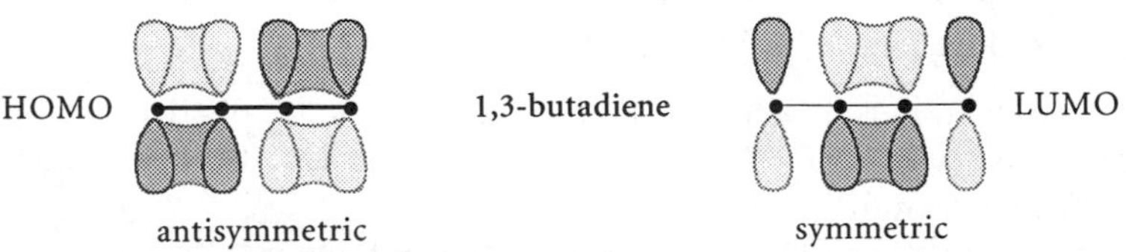

B. Molecular Orbitals of Conjugated Ions and Radicals

1. Conjugated unbranched ions and radicals have an odd number of carbon atoms.
2. The MOs of such species follow many of the same patterns as those of conjugated alkenes, with two important differences:
   a. One MO is neither bonding nor antibonding but has the same energy as isolated 2*p* orbitals.

i. This MO, a nonbonding molecular orbital, is the LUMO of the cation and the HOMO of the anion.
ii. The nonbonding MO of a radical is occupied by a single unpaired electron and is called a SOMO (singly occupied molecular orbital).
iii. The remaining orbitals are either bonding or antibonding, and there is an equal number of each type.

b. In some of the MOs, nodes pass through carbon atoms.

3. The charge in a conjugated carbanion can be associated with the electrons in its HOMO.
4. Cations, radicals, and anions involving the same π system have the same MOs; they differ only in the number of π electrons.

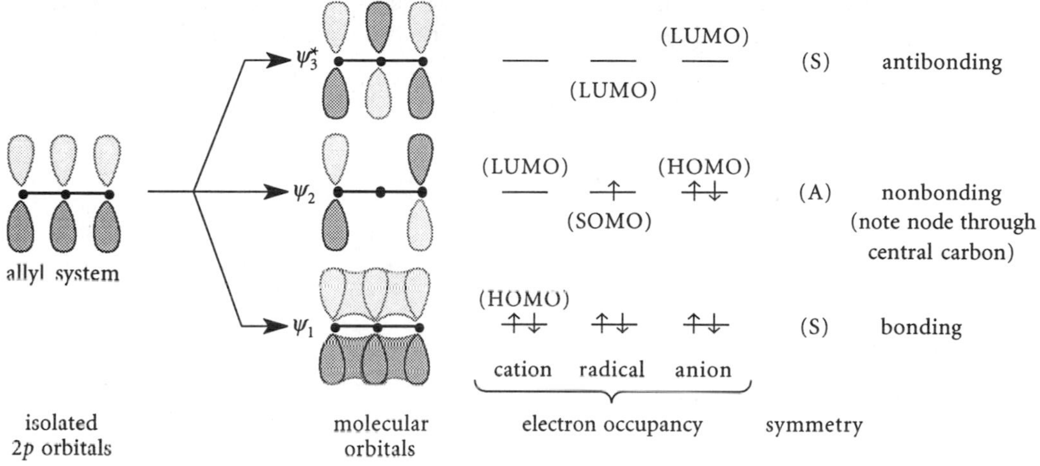

C. Excited States

1. The normal electronic configuration of any molecule is called the ground state.
2. Energy from absorbed light is used to promote an electron from the HOMO of the ground state into the LUMO.
   a. The species with the promoted electron is called an excited state.
   b. The HOMOs of the ground state and the excited state have opposite symmetries.

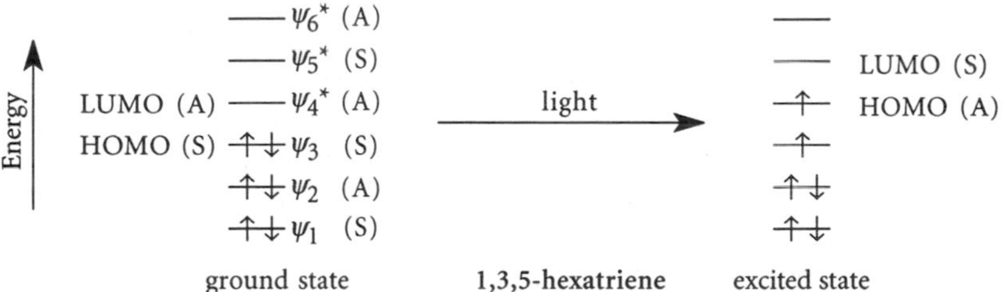

II. PERICYCLIC REACTIONS

A. Introduction

1. Pericyclic reactions occur by a concerted cyclic shift of electrons; that is, reactant bonds are broken and product bonds are formed at the same time, without intermediates.
2. There are three major types of pericyclic reactions:

a. Electrocyclic reaction: an intramolecular reaction of an acyclic π-electron system in which a ring is formed with a new σ bond, and the product has one fewer π bond than the starting material.

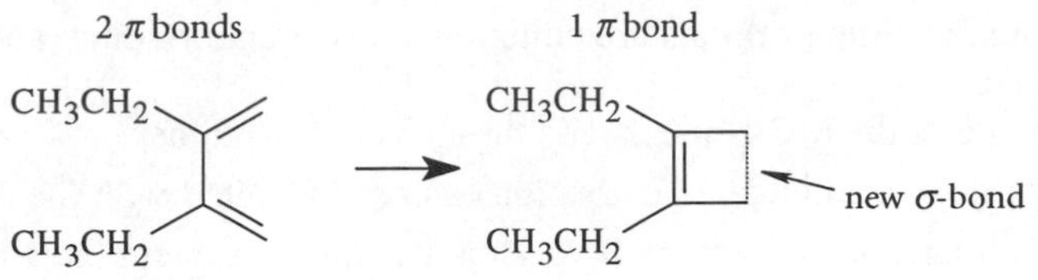

2 π bonds — 1 π bond

CH_3CH_2, CH_3CH_2 → CH_3CH_2, CH_3CH_2 ← new σ-bond

b. Cycloaddition reaction: a reaction of two separate π-electron systems in which a ring is formed with two new σ bonds, and the product has two fewer π bonds than the reactants.

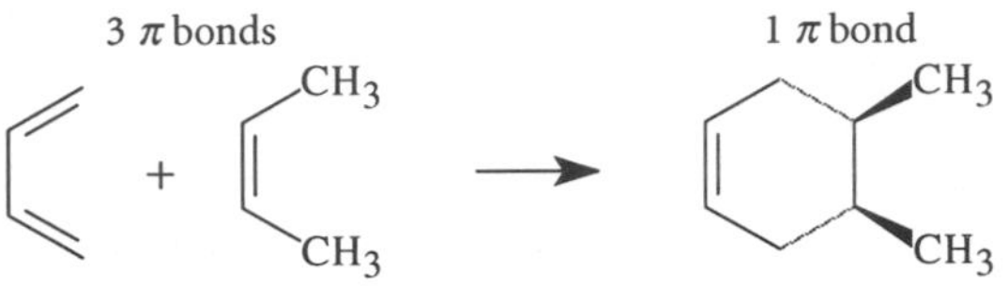

3 π bonds — 1 π bond

+ CH_3, CH_3 → CH_3, CH_3

c. Sigmatropic reaction: a reaction in which a σ bond at one end of a π-electron system appears to migrate to the other end of the π-electron system; the π bonds change position in the process, and their total number is unchanged.

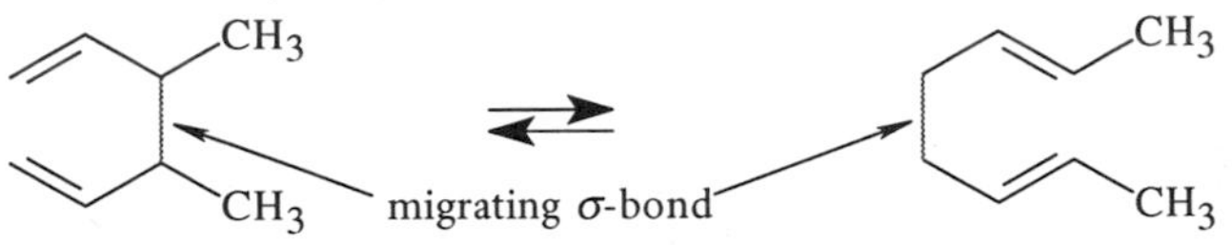

CH_3, CH_3 ⇄ CH_3, CH_3 migrating σ-bond

3. Three features of any type of pericyclic reaction are intimately related; specifying any two of the features specifies the third:
   a. The way the reaction is activated (heat or light)
      i. Many pericyclic reactions require no catalysts or reagents other than the reacting partners.
      ii. Such reactions take place either on heating or on irradiation with ultraviolet light.
      iii. Many reactions activated by heat are not activated by light, and vice-versa.
   b. The number of electrons involved in the reaction
      i. The number of electrons involved in a pericyclic reaction is twice the number of curved arrows required to write the reaction mechanism in the curved-arrow notation.
      ii. The direction of "electron flow" in pericyclic reactions indicated by the curved arrows is arbitrary.
   c. The stereochemistry of the reaction

B. Excited-State Pericyclic Reactions

1. When a molecule absorbs light, it reacts through its excited state.
2. The mode of ring closure in photochemical electrocyclic and cycloaddition reactions differs from that of thermal electrocyclic reactions because the HOMO of the excited state is different from the HOMO of the ground state and has different symmetry.
   a. Photochemical reactions are reactions that occur through electronically excited states.
   b. Thermal reactions occur through the ground state.

C. Classification of Sigmatropic Reactions

1. Sigmatropic reactions are classified by using bracketed numbers to indicate the number of atoms over which a σ bond appears to migrate. (Count the point of original attachment as atom #1.)

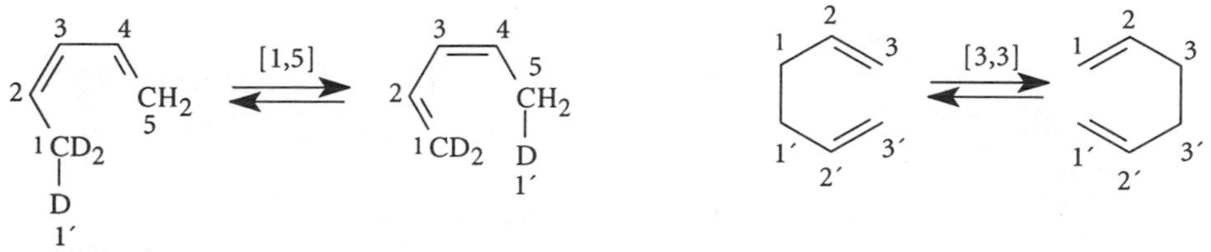

   a. In some reactions, each end of a σ bond migrates.
   b. In other reactions, one end of a σ bond remains fixed to the same group and the other end migrates.

D. Fluxional Molecules

1. A number of compounds continually undergo rapid sigmatropic rearrangements at room temperature.
2. Molecules that undergo rapid bond shifts are called fluxional molecules; their atoms are in a continual state of motion associated with the rapid changes in bonding.

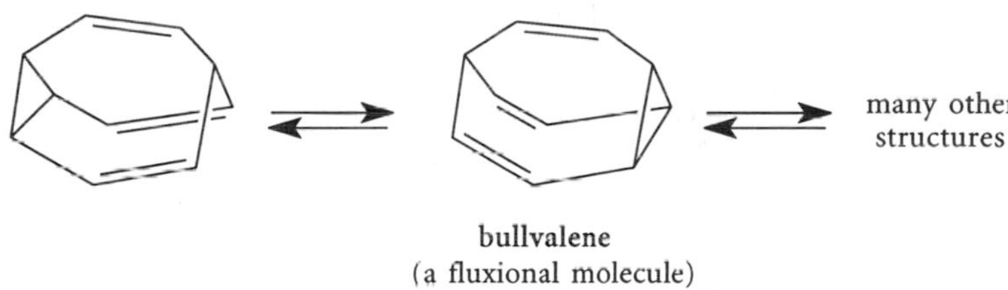

bullvalene
(a fluxional molecule)

E. Pericyclic Selection Rules; Summary

1. It is important to understand that the selection rules refer to the rates of pericyclic reactions, but have nothing to say about the positions of the equilibria involved.
   a. It is common for a photochemical reaction to favor the less stable isomer of an equilibrium because the energy of light is harnessed to drive the equilibrium energetically "uphill."
   b. The selection rules do not indicate which component of an equilibrium will be favored, only whether the equilibrium will be established at a reasonable rate.
   c. According to the principle of microscopic reversibility, the selection rules apply equally well to the forward and reverse of any pericyclic reaction, because the reaction in both directions must proceed through the same transition state.
   d. "Allowed" reactions are sometimes prevented from occurring for reasons having nothing to do with the selection rules, such as van der Waals repulsions in the product.
   e. The allowed stereochemistry of a pericyclic reaction follows from the phase relationships within the molecular orbitals involved.
2. The selection rules for the three types of pericyclic reactions are summarized in:
   a. Table 25.1 (text page 1186) for electrocyclic reactions
   b. Table 25.2 (text page 1191) for cycloaddition reactions
   c. Table 25.3 (text page 1202) for sigmatropic reactions
3. A convenient way to remember the selection rules is to assign either +1 or –1 to each of the following aspects of the reaction:
   a. For the mode of activation assign
      i. +1 for a thermal reaction
      ii. –1 for a photochemical reaction

b. For the number of reacting electrons (twice the number of arrows needed to describe the reaction in conventional curved-arrow notation) assign
   i. +1 for $4n + 2$ electrons
   ii. –1 for $4n$ electrons

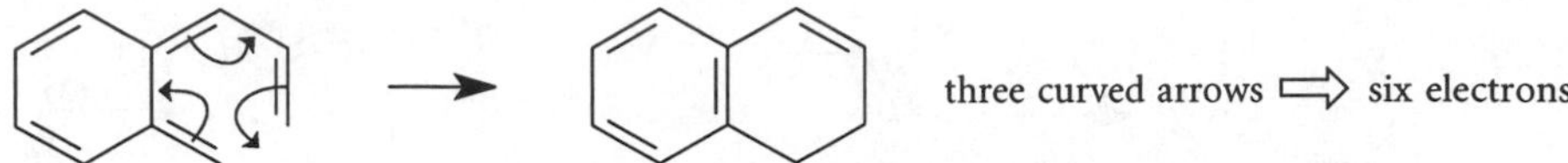

c. For the stereochemistry of each component assign
   i. +1 for a suprafacial or disrotatory reaction or for retention of stereochemistry
   ii. –1 for an antarafacial or conrotatory reaction or for inversion of stereochemistry
d. Multiply together the resulting numbers:
   i. if the product is +1, the reaction is allowed.
   ii. if the product is –1, the reaction is forbidden.

F. Summary of Sigmatropic Reaction Selection Rules

1. The stereochemistry of sigmatropic reactions is a function of the number of electrons involved which, in turn, is determined from the curved-arrow notation by counting the curved arrows and multiplying by two. (See Table 25.4, text page 1203.)
   a. All-suprafacial sigmatropic reactions occur when $4n + 2$ electrons are involved in the reaction (an odd number of electron pairs, or curved arrows).

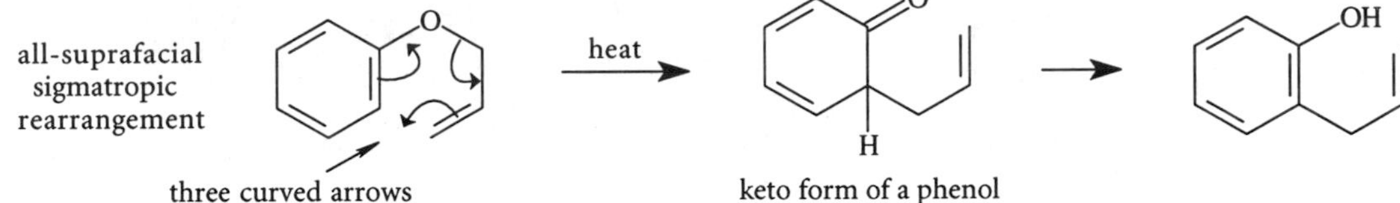

   b. A sigmatropic reaction must be antarafacial on one component and suprafacial on the other when $4n$ electrons are involved (an even number of electron pairs, or curved arrows).

R^1 R^2 R^3 D, H C=CH—C D H → [1,3] → R^2 R^1 R^3 C—CH=C D H D H

inversion occurs at migrating carbon

reaction is suprafacial with respect to the alkene

   c. For a single migrating carbon, the term "suprafacial" is taken to mean "retention of configuration," and the term "antarafacial" is taken to mean "inversion of configuration."

REACTIONS

I. ELECTROCYCLIC REACTIONS

1. When an electrocyclic reaction takes place, the carbons at each end of the conjugated π system turn in a concerted fashion so that the $2p$ orbitals can overlap (and rehybridize) to form the σ bond that closes the ring.
2. This turning can occur in two stereochemically distinct ways.

a. In a conrotatory closure the two carbon atoms turn in the same direction, either both clockwise or both counterclockwise.

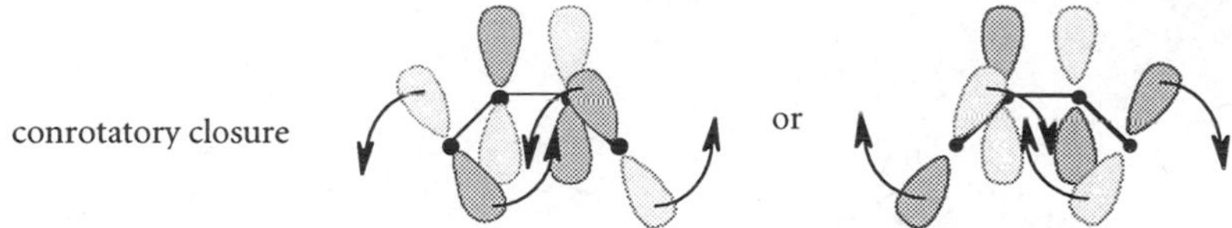

b. In a disrotatory closure the two carbon atoms turn in opposite directions so that either the upper lobes of the $2p$ orbitals overlap or the lower lobes of the $2p$ orbitals overlap.

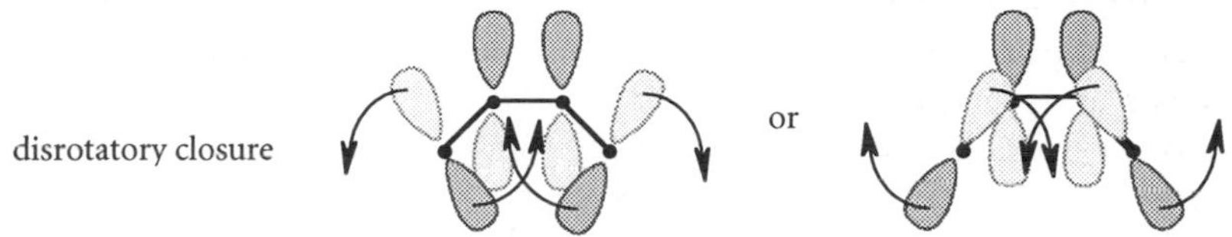

3. The HOMO of the conjugated alkene contains the π electrons of highest energy and governs the course of pericyclic reactions.
   a. When the ring closure takes place, the two $2p$ orbitals on the ends of the π system must overlap in phase.
   b. The wave peak on one carbon must overlap with the wave peak on the other, or a wave trough must overlap with a wave trough.
4. The relative orbital phase at the terminal carbon atoms of the HOMO (the orbital symmetry) determines whether the reaction is conrotatory or disrotatory.
   a. Conjugated alkenes with $4n$ π electrons (n = any integer) have antisymmetric HOMOs and undergo conrotatory ring closure. Conrotatory ring closure is allowed for systems with $4n$ π electrons; it is forbidden for systems with $4n + 2$ π electrons.
   b. Conjugated alkenes with $4n + 2$ π electrons have symmetric HOMOs and undergo disrotatory ring closure. Disrotatory ring closure is allowed for systems with $4n + 2$ π electrons; it is forbidden for systems with $4n$ π electrons.
5. For photochemical (excited-state) electrocyclic reactions, the selection rules are reversed; that is, conrotatory closure is allowed for a $4n + 2$ π-electron reaction and disrotatory closure is allowed for a $4n$ π-electron reaction.

II. CYCLOADDITION REACTIONS

1. Cycloadditions are classified, first, by the number of electrons involved in the reaction with respect to each component.
   a. The number of electrons involved is determined by writing the reaction mechanism in the curved-arrow notation.
   b. The number of electrons contributed by each given reactant is equal to twice the number of curved arrows originating from that component (two electrons per arrow).

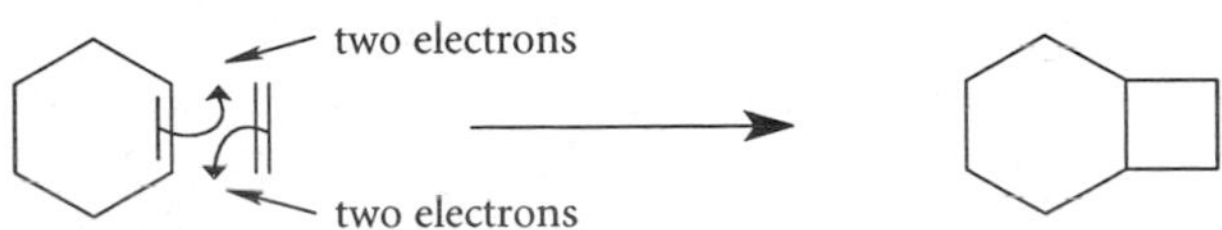

2. Cycloaddition reactions are also classified by their stereochemistry with respect to the plane of each reacting molecule and may in principle occur either across the same face, or across opposite faces, of the planes in each reacting component.

a. If the reaction occurs across the same face of a π system, the reaction is said to be suprafacial with respect to that π system; that is, a *syn*-addition that occurs in a single mechanistic step.

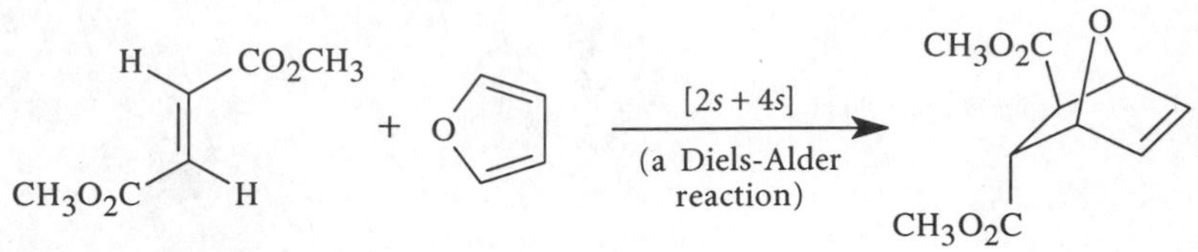

b. If the reaction bridges opposite faces of a π system, it is said to be antarafacial; that is, it is an *anti*-addition that occurs in one mechanistic step.

3. The reaction classification is written as a combination of numbers and letters. Thus, a $[4s + 4a]$ cycloaddition involves 4 π electrons on each component and is suprafacial (s) on one component and antarafacial (a) on the other.
4. In order for a cycloaddition to occur, bonding overlap must take place between the $2p$ orbitals at the terminal carbons of each π electron system.
   a. The bonding overlap begins when the HOMO of one component interacts with the LUMO of the other.
   b. The LUMO of the other component is the empty orbital of lowest energy into which the electrons from the HOMO must flow.
   c. These two frontier MOs involved in the interaction must have matching phases if bonding overlap is to be achieved.
   d. It does not matter which component provides the HOMO and which provides the LUMO.
5. All-suprafacial cycloadditions are allowed thermally for systems in which the total number of reacting electrons is $4n + 2$, and they are allowed photochemically for systems in which the number is $4n$.

III. SIGMATROPIC REACTIONS

A. Stereochemistry of Sigmatropic Reactions

1. Sigmatropic reactions can be classified by their stereochemistry according to whether the migrating bond moves over the same face, or between opposite faces, of the π-electron system.
   a. If the migrating bond moves across one face of the π system the reaction is said to be suprafacial.
   b. If the migrating bond moves from one face of the π system to the other the reaction is said to be antarafacial.
   c. When both ends of a σ bond migrate, then two π-electron systems are involved, and the reaction can be suprafacial or antarafacial with respect to each π system.
2. The stereochemistry of a sigmatropic reaction is revealed experimentally only if the molecules involved have stereocenters at the appropriate carbons.
3. Molecular orbital theory provides the connection between the type of sigmatropic reaction and its stereochemistry.

B. [1,3] and [1,5] Sigmatropic Rearrangements

1. The interaction of the LUMO of the migrating group with the HOMO of the π system, or vice-versa, controls the stereochemistry of the reaction.
   a. In the migration of a hydrogen, the orbital involved is a $1s$ orbital, which has no nodes.
   b. In the migration of a carbon, the orbital involved is a $2p$ orbital, which has one node.
   c. The shift of a carbon can occur in two stereochemically distinct ways:
      i. Migration with retention of configuration.
      ii. Migration with inversion of configuration.
   d. In the allyl anion, the HOMO is antisymmetric:

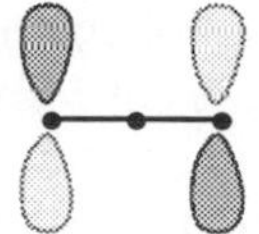

i. The suprafacial [1,3] shift of a hydrogen is forbidden by orbital symmetry because the phases of the π-system orbitals and the migrating groups do not match.

ii. The antarafacial [1,3] migration of hydrogen is allowed by orbital symmetry but is virtually nonexistent in organic chemistry because unreasonably long bonds would be required in the transition state.

iii. The suprafacial [1,3] shift of a carbon with inversion of configuration is allowed by orbital symmetry.

iv. The antarafacial [1,3] shift of a carbon with retention of configuration is allowed by orbital symmetry but is rarely observed.

e. In the 2,4-pentadienyl anion, the HOMO is symmetric:

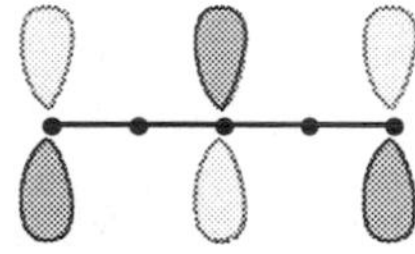

i. The suprafacial [1,5] migration of a hydrogen is allowed by orbital symmetry.

ii. The antarafacial [1,5] migration of a hydrogen is forbidden by orbital symmetry.

iii. The suprafacial [1,5] migration of a carbon with retention of configuration is allowed by orbital symmetry.

iv. The antarafacial [1,5] migration of a carbon with inversion of configuration is allowed by orbital symmetry.

C. Sigmatropic Rearrangements; Cope and Claisen Rearrangements

1. A [3,3] sigmatropic rearrangement is a reaction in which both ends of a σ bond change positions (migrate).
   a. The transition state of a [3,3] sigmatropic rearrangement can be visualized as the interaction of two allylic systems, one a cation and one an anion.

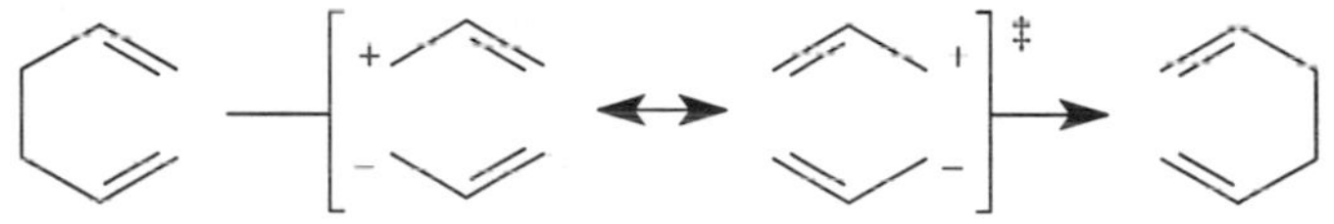

   b. The two MOs involved achieve bonding overlap when the [3,3] sigmatropic rearrangement occurs suprafacially on both components.
2. The Cope rearrangement is a [3,3] sigmatropic rearrangement in which a 1,5-diene isomerizes.

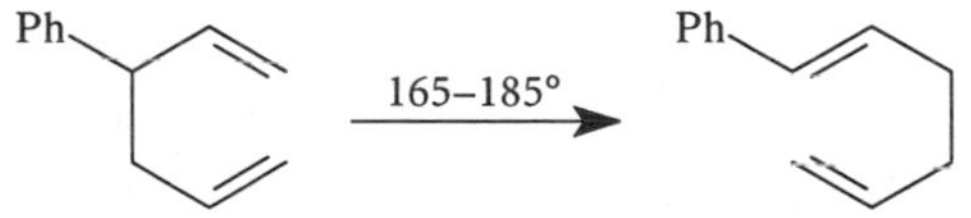

   Ph — 165–185° → Ph

3. The oxyCope reaction involves the initial formation of an enol; isomerization of the enol into the corresponding carbonyl compound is a very favorable equilibrium that drives the reaction to completion.

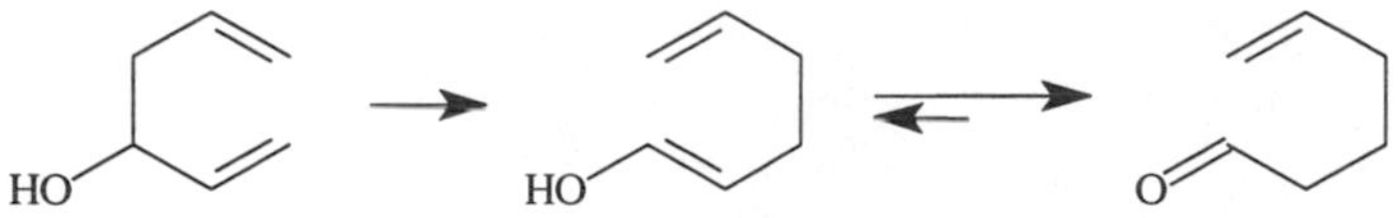

   HO — HO — O

4. In the Claisen rearrangement, an ether that is both allylic and vinylic (or an allylic aryl ether) undergoes a [3,3] sigmatropic rearrangement. If both ortho positions of an aryl allylic ether are blocked by substituent groups, the *para*-substituted derivative is obtained by a sequence of two Claisen rearrangements followed by tautomerization of the product to the phenol.

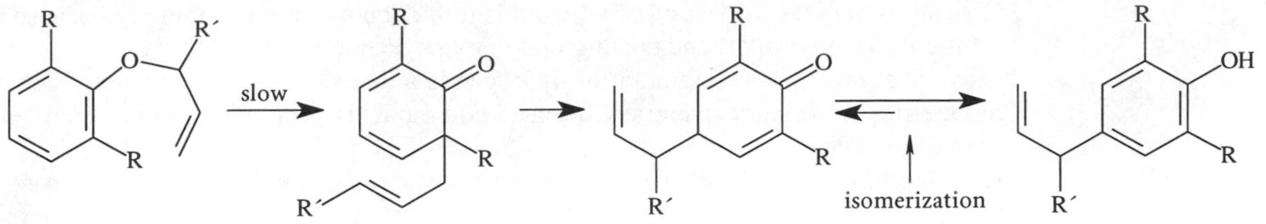

SCIENTISTS

Rainer Ludwig Claisen (1851–1930) 25.4B
Arthur Clay Cope (1909–1966) 25.4B
Sir James Dewar (1842–1923) 25.2C
William von E. Doering (b. 1917) 25.6
Kenichi Fukui (1918–1998) [Chemistry 1981] .... Introduction
Roald Hoffman (b. 1937) [Chemistry 1981] .... Introduction
Wolfgang R. Roth 25.4A
Gerhard F. Schröder 25.6
Robert Burns Woodward (1917–1979) [Chemistry 1965] .... Introduction

Dates in brackets refer to Nobel Prize awards.

STUDY GUIDE LINKS

✓25.1 Frontier Orbitals

You've seen that the analysis of pericyclic reactions in the text has focused on frontier orbitals. The discussion of electrocyclic reactions focused on one of the frontier orbitals, the HOMO, and the analysis of cycloadditions focused on the HOMO of one component and the LUMO of the other. Yet each of the frontier orbitals houses only two of the many reacting π electrons. Because *all* π electrons in the reacting molecules are involved in pericyclic reactions, it can perhaps be appreciated that all of the π molecular orbitals are involved as well. Why focus, then, on frontier orbitals? This focus is justified because it has been shown that frontier-orbital interactions play the *major* role in determining the energies of transition states for concerted pericyclic reactions. However, these are not the only orbital interactions that occur; they are simply the *most important* interactions.

Some students try to associate frontier molecular orbital interactions with formation of *a particular bond* in the reactant or product. This temptation is understandable because of the association of a π molecular orbital with the double bond in a simple alkene, or the association of a σ molecular orbital with a particular single bond in saturated molecules. These early encounters with molecular orbitals fail to take account of the fact that all molecular orbitals within a molecule must have certain symmetry properties, namely, they must be symmetric or antisymmetric with respect to the symmetry elements of the molecule itself. (For example, if a molecule contains an internal mirror plane, molecular orbitals of the molecule must be either antisymmetric or symmetric with respect to that plane.) For this reason, a one-to-one correspondence in general does not exist between molecular orbitals and the localized bonds used to write Lewis structures. Molecular orbitals are orbitals of *molecules,* and changes in a particular molecular orbital may involve more than one bond, and vice versa.

✓25.2 Orbital Analysis of Sigmatropic Reactions

Sigmatropic reactions are concerted and do not involve free ions, but the orbital relationships are nevertheless understood by thinking of the two components of the transition state as ions of opposite charge. However, for the reaction on text p. 1195 we could just as well think of the reaction as the migration of a hydrogen atom across a 2,4-pentadien-1-yl radical, or as the migration of a hydrogen anion (hydride ion) across a 2,4-pentadien-1-yl cation; these different approaches give the same predictions. Let's verify this point by viewing the transition state of the [1,5] hydrogen migration as the migration of a hydrogen atom across a 2,4-pentadien-1-yl radical:

$$\left[\overset{\cdot}{C}H_2 \cdots H\cdot \quad CH_2 \longleftrightarrow CH_2 \quad \overset{\cdot}{C}H_2 \cdots \cdot H \right]^{\ddagger}$$

In this case, analyze the symmetries of the orbitals containing the unpaired electrons, that is, the SOMO of each radical. For a hydrogen atom, this orbital is just the 1*s* orbital. For the 2,4-pentadien-1-yl radical, this is ψ_3 (see Fig. 25.5 on text p. 1181). Because this molecular orbital is symmetric, the migration is predicted to occur suprafacially.

Now let's analyze the reaction as if it were the migration of a hydrogen anion (hydride ion, $H{:}^-$) across the termini of a 2,4-pentadien-1-yl cation and show that such an analysis makes the same prediction. In this case, the 1*s* orbital of the migrating hydride contains two electrons; thus, this orbital is classified as the HOMO of the hydride ion. It interacts with the LUMO of the 2,4-pentadien-1-yl cation, which, from Fig. 25.5 on text p. 1181, is ψ_3. Because this molecular orbital is symmetric, the migration hydride is predicted to occur suprafacially.

Notice that whether we deal with the migration of a proton across an anion, a radical across another radical, or an anion across a cation, the relevant orbitals of the species involved are the same: the 1*s* orbital of the hydrogen and ψ_3 of the pentadienyl system. Hence, the predictions of orbital symmetry must also be the same.

SOLUTIONS TO PROBLEMS

Solutions to In-Text Problems

25.1 (a) This is an electrocyclic reaction; six electrons are involved.

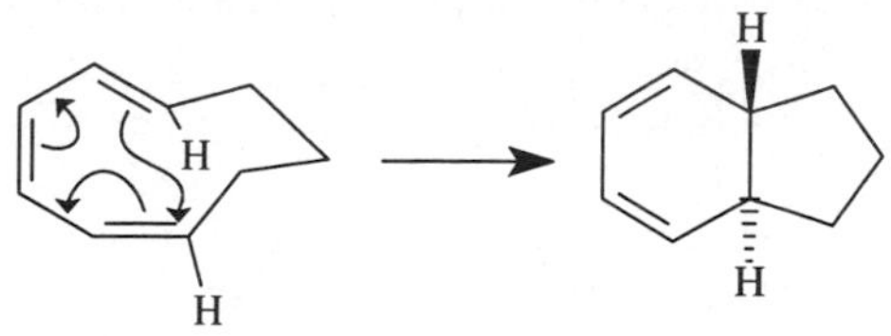

(b) This is a sigmatropic reaction; two electrons are involved; two electrons are involved.

$$H_3C-\overset{+}{C}H-C(CH_3)(H)-CH_3 \longrightarrow H_3C-CH_2-\overset{+}{C}(CH_3)-CH_3$$

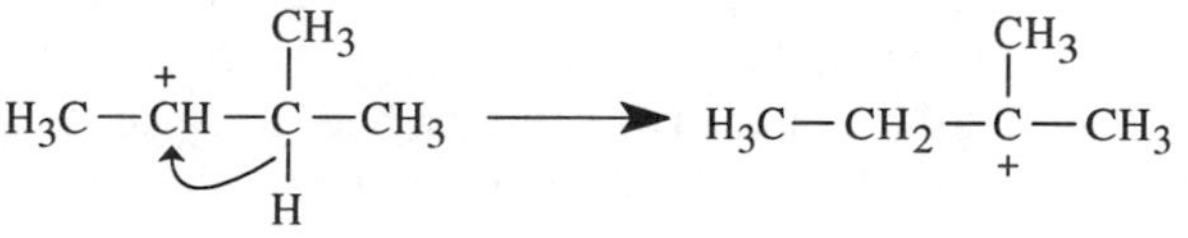

(c) This is an intramolecular cycloaddition reaction; four electrons are involved.

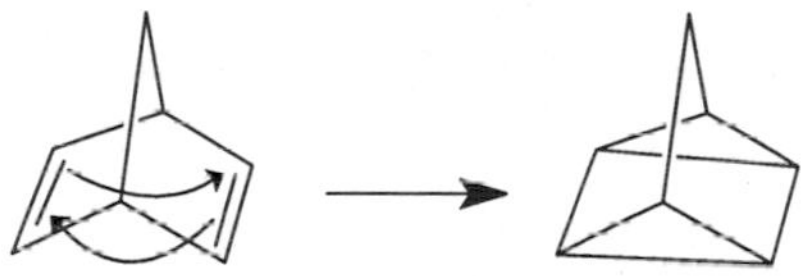

(d) This is a sigmatropic reaction; six electrons are involved.

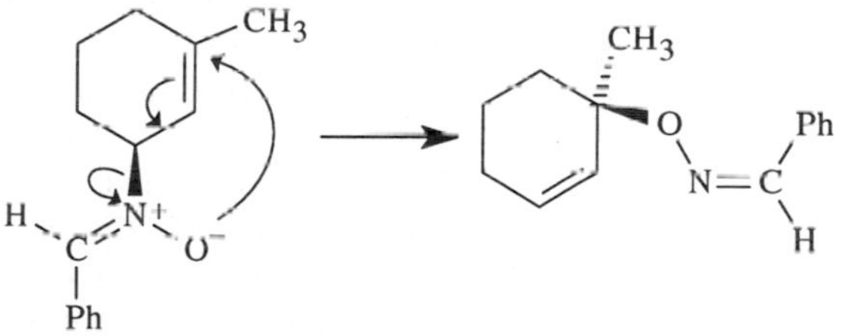

(e) This is an electrocyclic reaction; two electrons are involved.

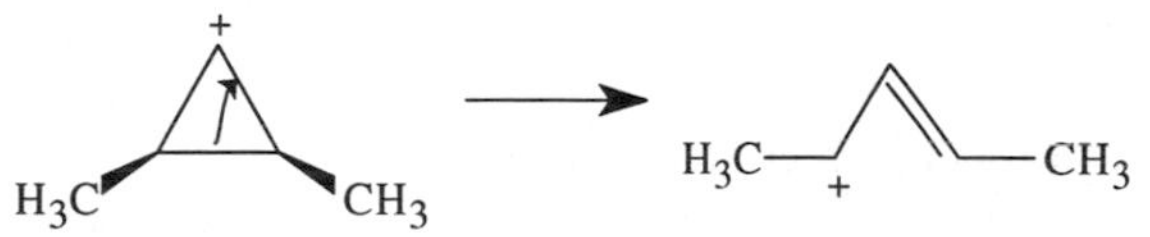

25.2 (a) 1,3,5-Hexatriene has six π molecular orbitals.

(b) The symmetry of the 1,3,5-hexatriene MOs alternates; that is, ψ_1, ψ_3, and ψ_5^* are symmetric, and ψ_2, ψ_4^*, and ψ_6^* are antisymmetric.

(c) ψ_1, ψ_2, and ψ_3 are bonding and ψ_4^*, ψ_5^*, and ψ_6^* are antibonding.

(d) Because 1,3,5-hexatriene has six π electrons, the highest occupied molecular orbital (HOMO) is ψ_3; the lowest unoccupied molecular orbital (LUMO) is ψ_4^*.

(e) As indicated in the solution to part (b), the HOMO (ψ_3) is symmetric; therefore, the phase of the HOMO at the terminal carbons is the same.

(f) The phase of the LUMO (ψ_4^*) at the terminal carbons is different.

25.3 Because there are ten π electrons, there are five occupied MOs in this alkene, each containing two electrons. The

MOs alternate in symmetry, the odd-numbered ones are symmetric, and the even-numbered ones are antisymmetric. Hence, ψ_6^* is antisymmetric (A); it is antibonding; and it is the LUMO and is thus one of the frontier orbitals.

25.4 (a) The MO ψ_4 is nonbonding. (If there is an odd number n of molecular orbitals in an unbranched acyclic system, the nonbonding MO is always the one with the number $(n + 1)/2$.)

(b) The symmetry of the MOs alternates; that is, ψ_1, ψ_3, ψ_5^*, and ψ_7^* are symmetric, and ψ_2, ψ_4, and ψ_6^* are antisymmetric.

(c) According to its resonance structures, this cation has sites of positive charge at alternating carbons:

$$\left[H_2C{=}CH{-}CH{=}CH{-}CH{=}CH{-}\overset{+}{C}H_2 \longleftrightarrow H_2C{=}CH{-}CH{=}CH{-}\overset{+}{C}H{-}CH{=}CH_2 \longleftrightarrow\right.$$

$$\left.H_2C{=}CH{-}\overset{+}{C}H{-}CH{=}CH{-}CH{=}CH_2 \longleftrightarrow H_2\overset{+}{C}{-}CH{=}CH{-}CH{=}CH{-}CH{=}CH_2\right]$$

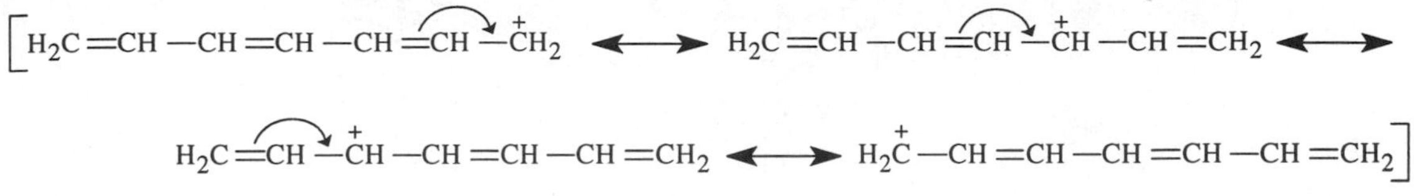

In molecular orbital terms, the positive charge in the cation results from the absence of an electron in the LUMO; the LUMO *is* the nonbonding MO. Positive charge is shared at alternating carbons because the LUMO has nodes at the other carbons. That is, positive charge can only exist on carbons at which there is no node in the LUMO. If this is so, the LUMO must have three nodes: one at each carbon that does not share the positive charge. As shown in the solution to part (a), ψ_4 is the nonbonding MO and is therefore the LUMO, and it has three nodes, one at each carbon that does not share positive charge in the above structures.

25.5 (a) Resonance structures of the allyl radical show that the unpaired electron is shared between the terminal carbons:

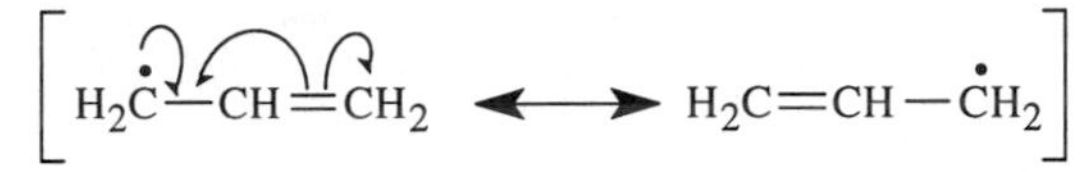

$$\left[H_2\dot{C}{-}CH{=}CH_2 \longleftrightarrow H_2C{=}CH{-}\dot{C}H_2\right]$$

(b) Figure 25.4 on text p. 1180 shows that the unpaired electron of the allyl radical resides in the nonbonding MO, which has a node at the central carbon. The unpaired electron density therefore exists only at the carbons on either side of a node, that is, at the terminal carbons, as the resonance structures show.

25.6 *Reaction 1* is a disrotatory reaction involving $4n + 2$ electrons, and is allowed. Therefore it can occur readily by a concerted mechanism. The reaction is disrotatory because the hydrogens shown in the product are both on the outside of the starting material. (Draw these hydrogens in the starting material, if necessary.) In order for these hydrogens to end up on the same face of the product, the ends of the π-electron system must turn in opposite directions.

Reaction 2 is a conrotatory reaction involving $4n + 2$ electrons, and is not allowed; it does not occur readily.

25.7 The two different conrotatory processes are as follows:

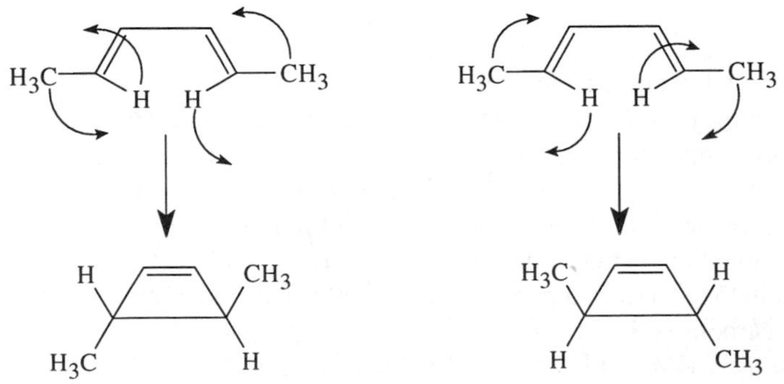

H₃C CH₃ H H → H CH₃ H₃C H

H₃C CH₃ H H → H₃C H H CH₃

They are equally likely because they are enantiomeric; enantiomers have equal energies, and enantiomeric pathways have identical energies at all points and therefore identical rates (Sec. 7.8A).

25.8 The thermal ring opening of *trans*-3,4-dimethylcyclobutene must be a conrotatory process. The two possible conrotatory processes are as follows:

severe van der Waals repulsions

A *B*

Product *A* has severe van der Waals repulsions between the methyl groups; product *B* does not. The repulsions raise the energy of product *A* and the transition state for its formation; consequently, product *B* is observed.

25.9 The opening of the cyclobutene is conrotatory and therefore gives (1*E*,3*Z*)-cyclodecadiene. The trans double bond does not introduce as much strain into the large ring as was present in the starting cyclobutene.

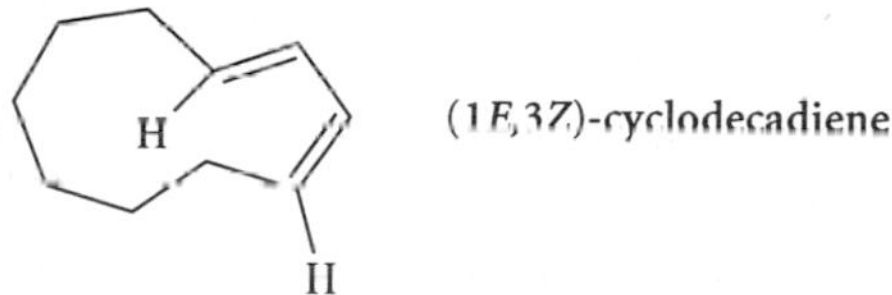

Here is one way to use models to visualize this transformation. To avoid having to build two models, construct a cyclodecane (the cyclic *alkane* with a ten-membered ring); don't worry about the missing double bond. Now connect two carbons to make a cis-fused four-membered ring. Identify the "up" hydrogens on these two carbons. Using a pencil, draw a single line across the bond that is a double bond in the starting material (even though it is a single bond in your model). Now draw two lines across each of the bonds that will become double bonds as a result of the reaction. Now break the cyclobutane ring and turn the carbons bearing the "up" hydrogens in a conrotatory manner. Examine the stereochemistry at the bonds marked with two lines; one should have an *E* configuration and the other a *Z* configuration. Voila!

25.10 The HOMO of the 2π-electron component is symmetric, and the LUMO of the 4π-electron component is symmetric. As the diagram in Fig. SG25.1 on the following page shows, the ends of these two MOs have matching phases in a $[4s + 2s]$ cycloaddition.

25.11 Use the HOMO of the 4π-electron component and the LUMO of the 2π-electron component. The $[4a + 2s]$ process would involve the overlap shown in Fig. SG25.2(a) (following page). The $[4s + 2a]$ process would involve the overlap shown in Fig. SG25.2(b). In both cases an out-of-phase overlap between the π-electron systems is required at one end. Consequently, these process are not allowed.

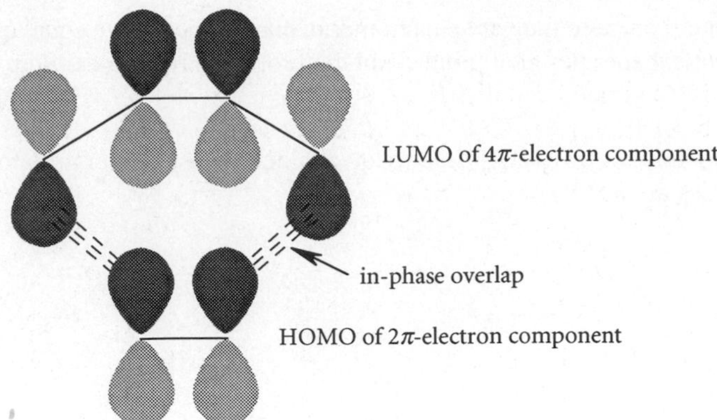

Figure SG25.1 Orbital diagram to accompany the solution to Problem 25.10. Interaction of the LUMO of the 4π-electron component and the HOMO of the 2π-electron component in a [4*s* + 2*s*] cycloaddition gives in-phase overlap at both ends of the two π-electron systems.

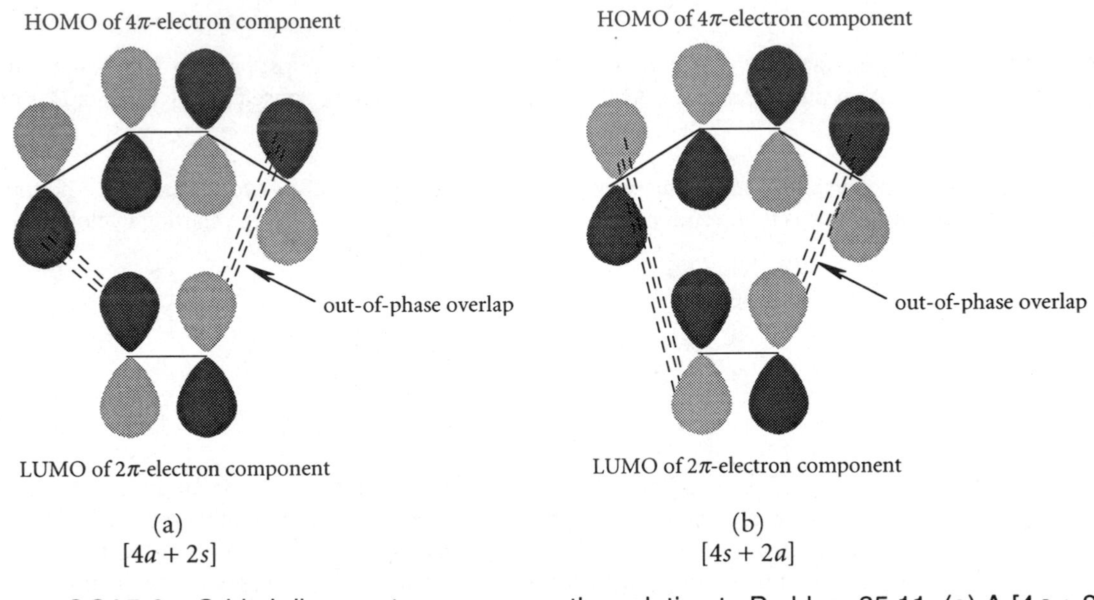

Figure SG25.2 Orbital diagram to accompany the solution to Problem 25.11. (a) A [4*a* + 2*s*] cycloaddition; (b) a [4*s* + 2*a*] cycloaddition. Both processes require nonbonding overlap at one end of the two π-electron systems; consequently, neither is allowed.

25.12 The [8*s* + 2*s*] cycloaddition and its product:

CO_2Et C ≡ C CO_2Et → CO_2Et CO_2Et

25.13 The four products are the diastereomeric 1,2,3,4-tetramethylcyclobutanes. Each of the pure alkene stereoisomers can undergo an allowed photochemical [2*s* + 2*s*] cycloaddition in two distinguishable ways. *Cis*-2-butene reacts to give compounds *A* and *B*; *trans*-2-butene reacts to give compounds *B* and *C*; and the mixture of *cis*- and *trans*-2-butene can give these three compounds plus a fourth, compound *D*, that results from the [2*s* + 2*s*] cycloaddition of *cis*-2-butene to *trans*-2-butene.

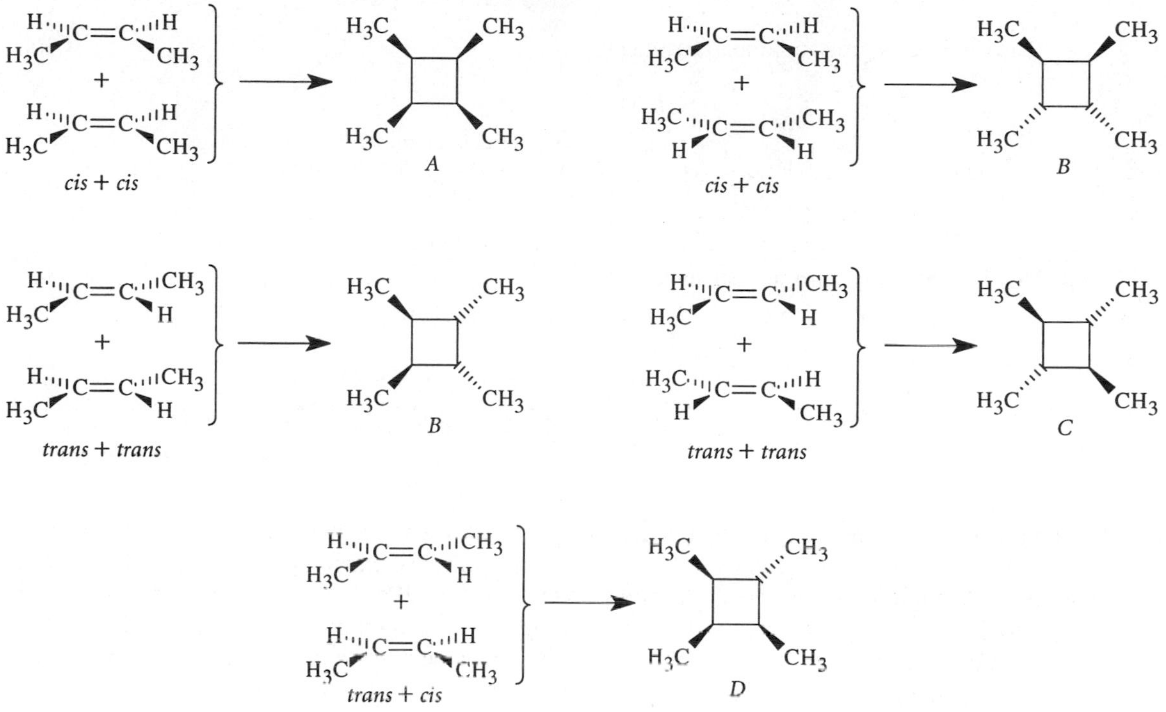

cis + cis → *A*

cis + cis → *B*

trans + trans → *B*

trans + trans → *C*

trans + cis → *D*

25.14 (a) Since exchanging the positions of two groups at a stereocenter changes the configuration of the stereocenter, exchange the positions of the T and the D in the starting material; this will lead to product with the opposite, that is, the *R*, configuration at the indicated carbon.

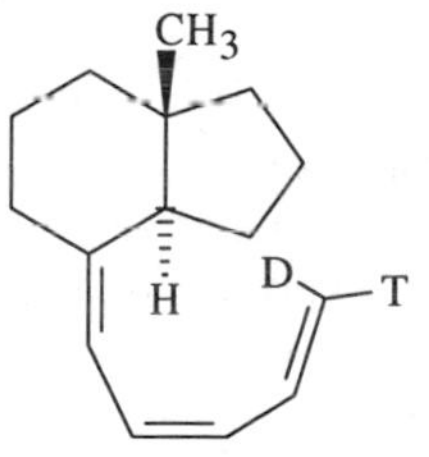

(b) Two other starting materials that would give the same product as in part (a):

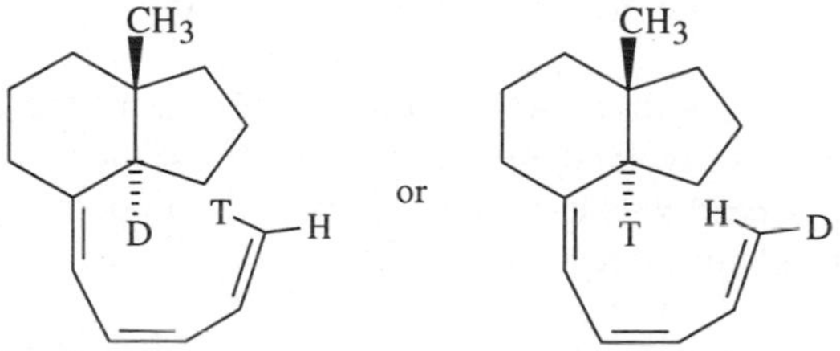

or

25.15 (a) This is a [1,4] sigmatropic rearrangement.

(Solution continues)

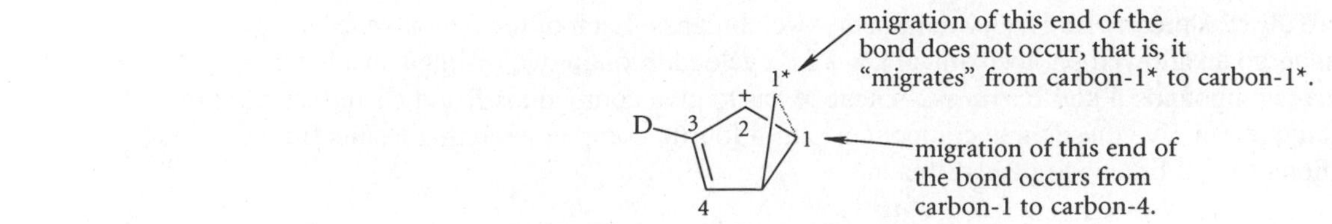

migration of this end of the bond does not occur, that is, it "migrates" from carbon-1* to carbon-1*.

migration of this end of the bond occurs from carbon-1 to carbon-4.

(b) This is a [2,3] sigmatropic rearrangement.

One end of this bond moves from carbon-1 to carbon-3; the other end moves from the sulfur (1*) to the oxygen (2*).

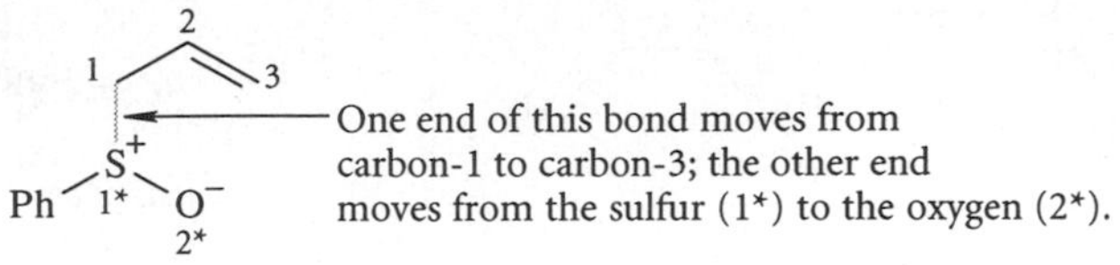

(c) This is a [5,5] sigmatropic rearrangement.

This bond migrates from carbons 1 and 1* to carbons 5 and 5*, respectively.

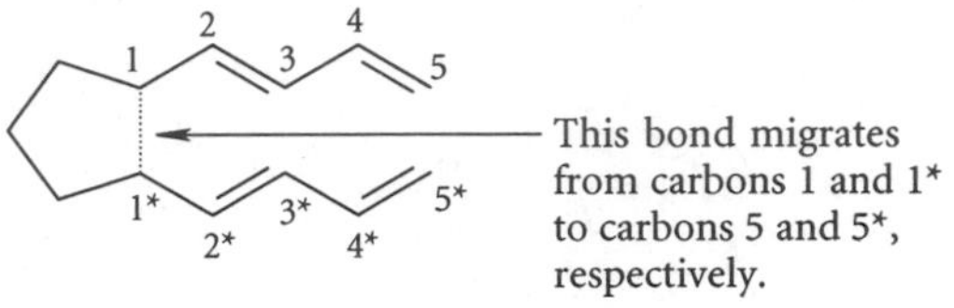

25.16 Reaction (1) occurs readily because it is a [1,5] sigmatropic rearrangement, which is allowed when the hydrogen migrates suprafacially over the π-electron system. This reaction pathway does not introduce significant strain or twist into the π-electron system. Reaction (2) does not occur because it is a [1,3] sigmatropic rearrangement, which is allowed only if the hydrogen migrates antarafacially. Such an antarafacial migration requires that the hydrogen simultaneously bridge the upper and lower faces of the π-electron system. This requirement cannot be met while at the same time maintaining normal bond lengths and angles.

25.17 If the migration were antarafacial, the opposite stereochemical result would have been observed; that is, the *3E* stereoisomer of the product would have the *S* configuration at the asymmetric carbon stereocenter, and the 3*Z* stereoisomer of the product would have the *R* configuration at the asymmetric carbon stereocenter. Of course, this is the result that is *not* observed.

25.18 (a) Think of the five-carbon π-electron system as a 2,4-pentadienyl cation, whose MOs are shown in Fig. 25.5 on text p. 1181, and think of the migrating group as a carbon anion in which the unshared electron pair resides in a 2*p* orbital. The orbitals involved in the rearrangement are the LUMO of the pentadienyl system and the HOMO of the carbon anion. The LUMO of the pentadienyl system, as shown by Fig. 25.5, is symmetric; that is, at each end of the system, the LUMO has the same phase on a given face. The HOMO of the carbon anion is simply the filled 2*p* orbital. The migration is allowed only if the carbon anion migrates suprafacially such that each end of the pentadienyl LUMO interacts with the same lobe of the carbon anion 2*p* orbital. (See Fig. SG25.3 on the following page.)

Alternatively, you can think of this problem in terms of the 2,4-pentadienyl *anion* and a methyl *cation*. The orbitals involved are now the HOMO of the anion and the LUMO of the cation. But these are the same orbitals used in the foregoing solution (only their occupancies are different); hence, the stereochemical result is the same.

(b) The first step (conversion of the starting material into compound *A*) is a [1,5] sigmatropic rearrangement. Because the stereochemistry of the methyl group on the migrating carbon does not change, the rearrangement occurs with retention of configuration. This is the result predicted for an allowed suprafacial process. The remaining two steps are [1,5] sigmatropic hydrogen migrations, which presumably occur in a suprafacial manner, although the stereochemical course of the reaction cannot be determined from the structures of the products.

HOMO of migrating carbon anion

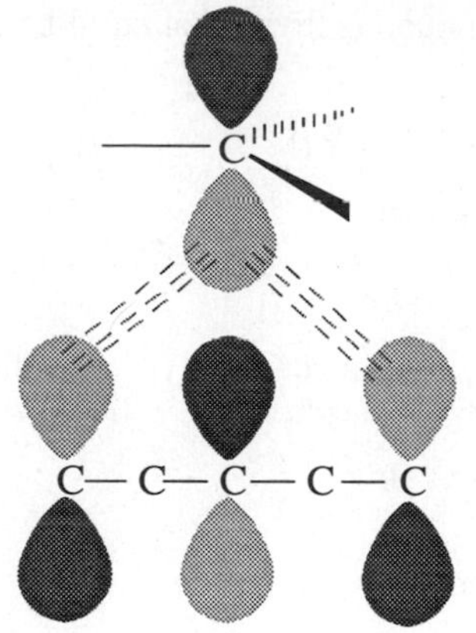

LUMO of 2,4-pentadienyl cation

Migration occurs suprafacially on both systems (and thus with retention of configuration at the migrating carbon).

Figure SG25.3 Orbital diagram to accompany the solution to Problem 25.18(a). For continuous orbital overlap to occur, migration of the carbon must occur suprafacially on the pentadienyl system and at the same lobe of the 2*p* orbital of the migrating carbon (that is, suprafacially and with retention of configuration).

25.19 (a) This transformation involves two successive [1,5] sigmatropic hydrogen migrations—sort of a "hydrogen walk," as shown by the following curved arrows. These are allowed suprafacial processes. (See Problem 25.18b.)

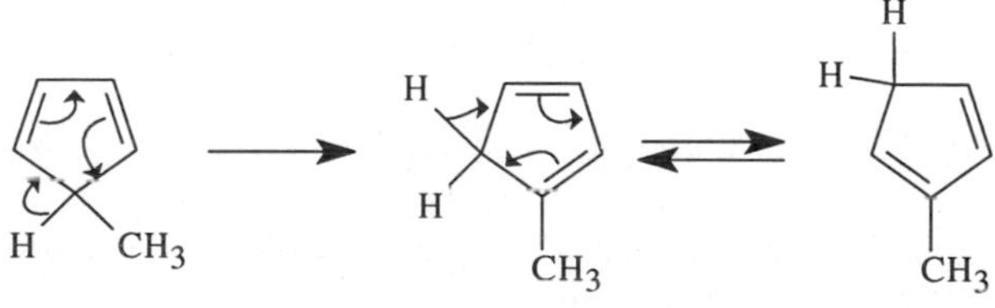

Why does the equilibrium favor the two products over the starting material? See Sec. 4.5B.

(b) The "hydrogen walk" on 2,3-dimethyl-1,3-cyclopentadiene should proceed as shown in the following equation. Although a hydrogen that starts out on carbon-1 could migrate to four other carbons, only two of the possible products are unique. It can be argued that compounds *A* and *B* should be the predominant alkenes at equilibrium, and that compound *C* should be present in least amount. (Why? See Sec. 4.5B.) Remember that the selection rules have nothing to say about which compound is favored at equilibrium, only whether the equilibrium can be attained at a reasonable rate.)

H_3C CH_3 ⇌ H_3C CH_3 ⇌ H_3C CH_3

2,3-dimethyl-1,3-cyclopentadiene *A*; 1,2-dimethyl-1,3-cyclopentadiene *B*; 1,5-dimethyl-1,3-cyclopentadiene *C*

25.20 (a) Deduce the product by using the curved-arrow notation:

CH_3, CH_3, CH_2, O, H ⟶ H_3C CH_3, CH_2, O, H 3,3-dimethyl-4-pentenal

(b) Deduce the starting material by drawing the curved arrows for the reverse of a Claisen rearrangement:

O, CH_2—CH ⟶ O, CH_2=CH ≡ O—CH, CH_2

25.21 See Fig. 25.4 on text p. 1180 for the relevant orbital diagram. The SOMOs of the two allylic radicals (ψ_2) interact because it is these molecular orbitals that contain the unpaired electrons. Since the two orbitals are the same, they have the same symmetry, and therefore interact suprafacially on both components with positive overlap.

25.22 An orbital symmetry analysis involves the interaction of ψ_2 of one allylic system with ψ_2 of the other as shown in Fig. SG25.4. (See following page.) Certainly the all-antarafacial rearrangement gives positive overlap at both ends of the allylic systems, but it requires either the initial formation of π bonds rather than σ bonds or introduction of significant strain in the two systems akin to having two trans double bonds in a six-membered ring. Neither alternative is as energetically favorable as the direct formation of two σ bonds that results from the all-suprafacial process.

25.23 (a) Thermal reaction: (+1); stereochemistry: (+1)(–1); number of electrons ($8 = 4n$): (–1). Result = +1; the reaction is allowed.

(b) Thermal reaction: (+1); stereochemistry: (+1); number of electrons ($8 = 4n$): (–1). Result = –1; the reaction is forbidden.

(c) Photochemical (excited state) reaction: (–1); stereochemistry: (+1); number of electrons ($8 = 4n$): (–1). Result = +1; the reaction is allowed.

25.24 (a) The σ bond that "moves" is indicated with an asterisk (*). It moves suprafacially on both π systems; [3,3] sigmatropic rearrangements are thermally allowed processes.

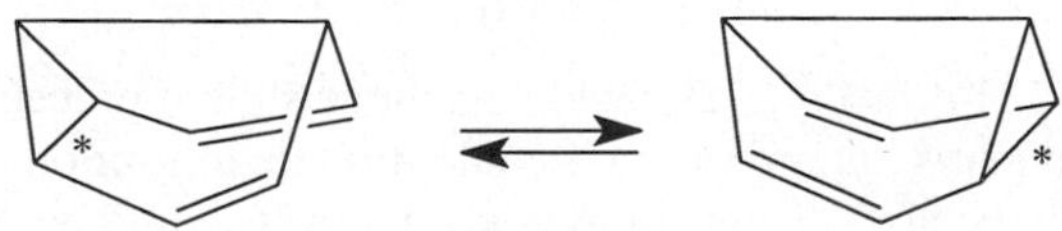

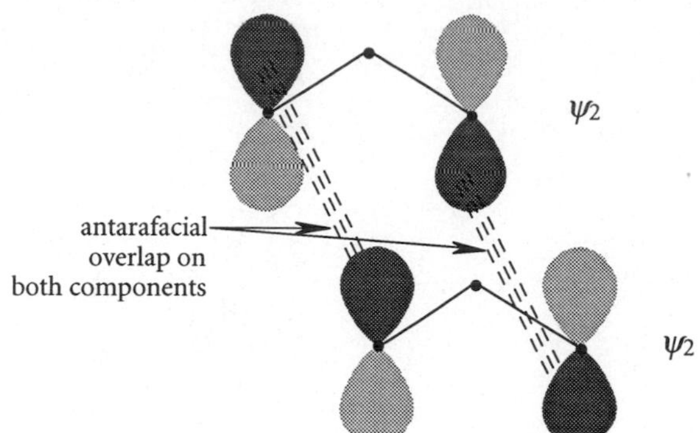

Figure SG25.4 Orbital diagram to accompany the solution to Problem 25.22. Antarafacial overlap of both allylic components gives bonding overlap but requires side-to-side (π) rather than head-to-head (σ) overlap. The formation of σ bonds would require severe distortions of both systems.

(b) This is a carbocation rearrangement in which either methyl group migrates. (This is the "methyl walk" discussed in the solution to Problem 15.67(c) on p. XR of this manual.)

This can also be thought of as an all-suprafacial [1,6] sigmatropic rearrangement, which is an allowed process.

25.25 Note that previtamin D_2 is the same as previtamin D_3 except for the R-group. (See the last sentence on text p. 1206.) Comparing the structures of previtamin D_3 (or D_2) in Eq. 25.38 on text p. 1205, ergosterol at the bottom of text p. 1206, and lumisterol in the problem reveals that the latter two compounds result from conrotatory electrocyclic ring closures, ergosterol being formed by a clockwise conrotatory motion (as viewed edge-on by the "eye" below) and the lumisterol from a counterclockwise conrotatory motion. Photochemical conrotatory electrocyclic reactions involving $4n + 2$ electrons are allowed.

previtamin D_2 = light → ergosterol; light → lumisterol

25.26 Heating should promote a disrotatory reaction; that is, the reaction should have a stereochemical course which is opposite to that of the photochemical reaction.

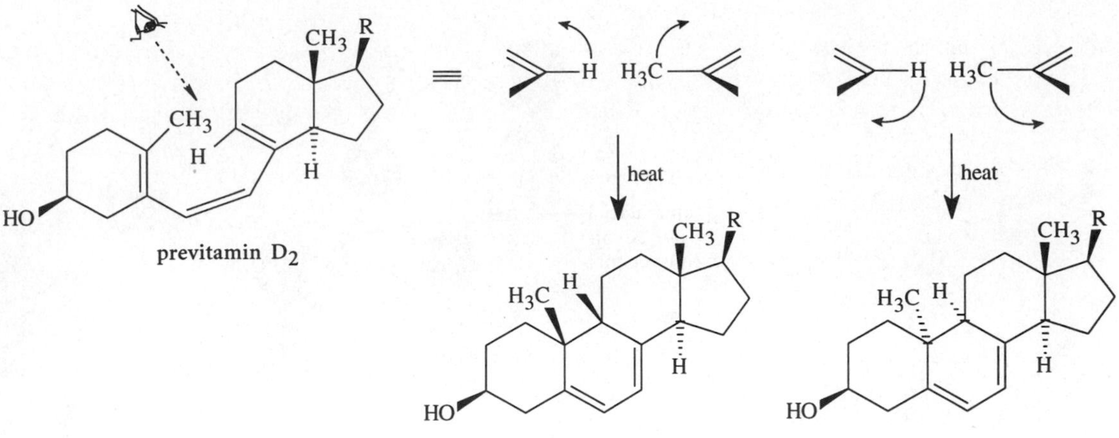

CH_3 R CH_3 H H HO previtamin D_2 ≡ H H_3C H H_3C heat heat CH_3 R H_3C H H HO *A* CH_3 R H_3C H H HO *B*

25.27 First determine why irradiation of *A* or *B* does not give back previtamin D_2. One reason is mechanistic. Since *A* and *B* originate from *heating* previtamin D_2, then previtamin D_2 can only result from the exact reverse of the same reaction, which must occur as a disrotatory process. However, ring opening by irradiation, if it were to occur, would have to take place with opposite stereochemistry, that is, as a conrotatory process. Either of the two possible conrotatory processes would result in a product containing a trans double bond within a six-membered ring. Even irradiation does not provide enough energy for this to occur.

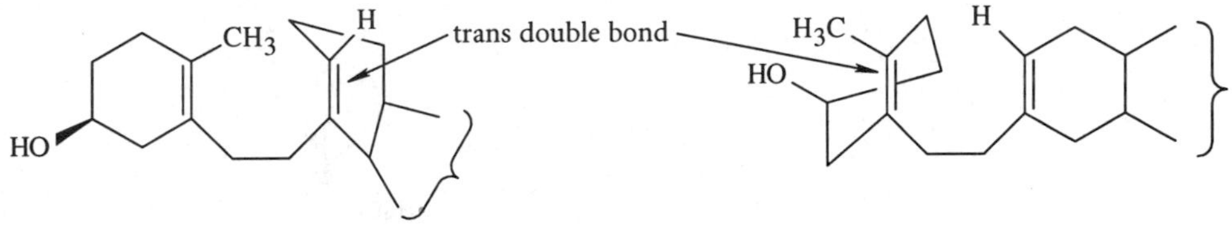

CH_3 H trans double bond H_3C H HO HO

the two possible products of conrotatory ring opening

The only alternative to this process is a four-electron disrotatory process that gives a cyclobutene:

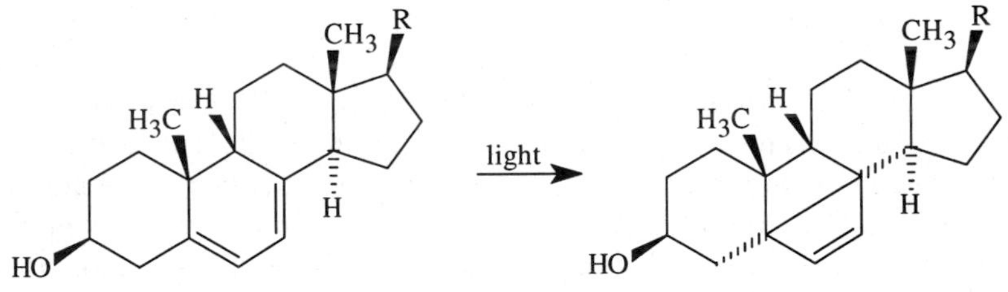

CH_3 R H_3C H H HO *A* light CH_3 R H_3C H H HO *C*

CH_3 R H_3C H H HO *B* light CH_3 R H_3C H H HO *D*

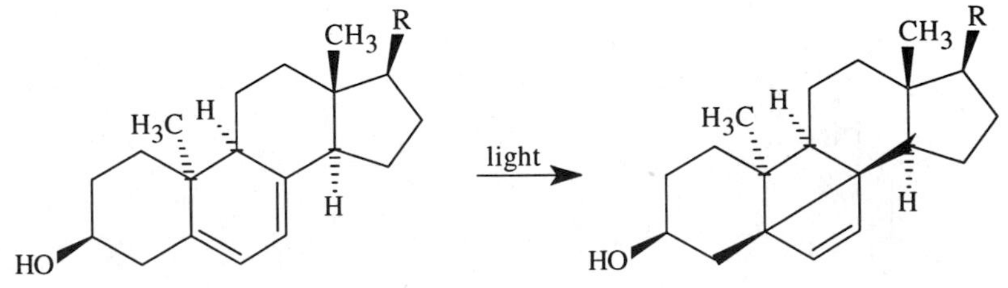

25.28 A [1,7] thermal sigmatropic hydrogen shift should be antarafacial. That is, a hydrogen from the methyl group migrates from a conformation in which it is "up" to the lower face of the ring, or a hydrogen from the methyl group migrates from a conformation in which it is "down" to the upper face of the ring, or both.

Problem 25.51 (text p. 1213) describes an effort to elucidate the stereochemistry of the previtamin D rearrangement.

Solutions to Additional Problems

25.29 (a) Follow the procedure used in Study Problem 25.3 on text p. 1203. This reaction is a thermal disrotatory process that involves $4n$ electrons. Result: $(+1)(+1)(-1) = (-1)$; the reaction is forbidden.

(b) The reaction is a photochemical, $[8s + 4s]$ process involving $4n$ electrons. Result: $(-1)(-1)(+1)(+1) = (+1)$; the reaction is allowed.

(c) The highest occupied molecular orbital (HOMO) of a conjugated triene is ψ_3. Since orbitals of conjugated dienes alternate in symmetry, this MO is symmetric. (See the solution to Problem 25.2 on p. 993 of this manual.) The methyl groups, to a useful approximation, have no effect on the nodal properties of the molecular orbitals.

25.30 The pericyclic selection rules say *absolutely nothing* about the position of equilibrium in each case. These rules refer to *rates* of reactions, not to equilibrium constants. Other considerations must be used to decide on the position of equilibrium. The point of the problem is for you to use what you have learned about the relative energies of molecules; the molecule of lower energy is favored in each equilibrium.

(a) The right side of the equation is favored at equilibrium because the double bonds are conjugated with the phenyl rings.

(b) The left side of the equation is favored at equilibrium because the double bonds have more alkyl branches. (Alkyl substitution stabilizes a double bond.)

(c) The right side of the equation is favored at equilibrium because a C═O double bond is formed at the expense of a C═C double bond. (C═O bonds are stronger than C═C bonds; see text page 1048.)

(d) The right side of the equation is favored at equilibrium because one product is aromatic and therefore particularly stable, and the other product is volatile (ethylene is a gas).

(e) The left side of the equation is favored at equilibrium because the double bonds are conjugated not only with each other but also with the carbonyl group.

25.31 (a) Because it involves $4n$ π electrons and thermal conditions, the electrocyclic reaction that converts compound *A* into compound *B* is conrotatory. (A "thermal" process can occur at low temperature if the process is particularly rapid.) Hence, the methyl groups in compound *B* are trans. The thermal process that converts compound *B* into compound *C* involves six ($4n + 2$) electrons and is therefore disrotatory. (Both compounds *B* and *C* are, of course, formed as racemates.)

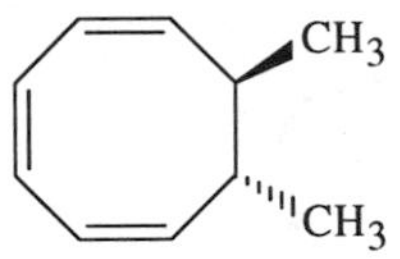

(b) The stereoisomer of compound *A* that also gives compound *C* on heating is the one in which both of the terminal double bonds have the *Z* configuration.

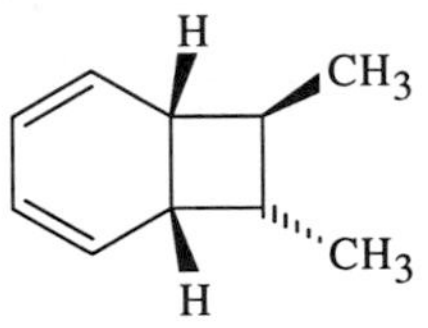

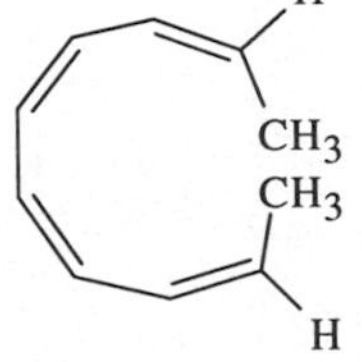

25.32 (a) This reaction can be viewed as either a [1,9] or a [1,13] sigmatropic rearrangement. In either of these classifications the reaction is an allowed process that is suprafacial on the π-electron system and occurs with retention at the migrating carbon. The [1,9] migration is as follows:

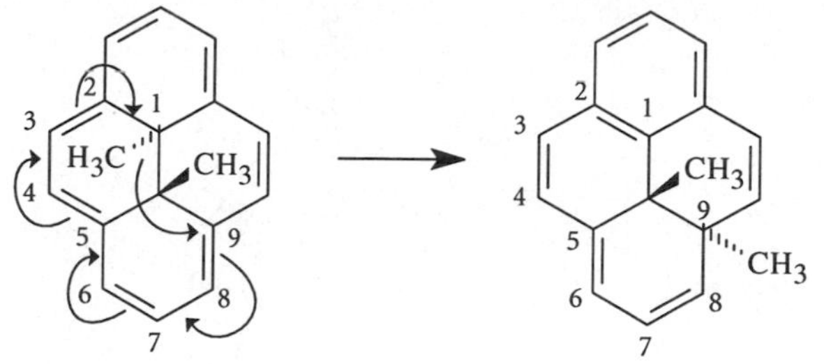

And the [1,13] migration is as follows:

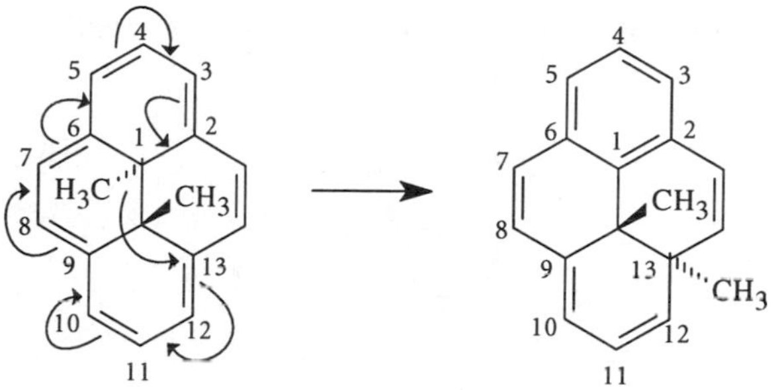

The two final structures shown above differ only in the positions of double bonds in the benzene ring; that is, they are resonance structures and hence are the same compound.

(b) Because a thermal suprafacial [1,9] or [1,13] methyl migration must occur with retention of configuration at the migrating carbon, the methyl group that migrates, when isotopically substituted, must have the *S* configuration in both starting material and product.

25.33 (a) The formula indicates that an addition has taken place. This is an allowed photochemical [2*s* + 2*s*] cycloaddition:

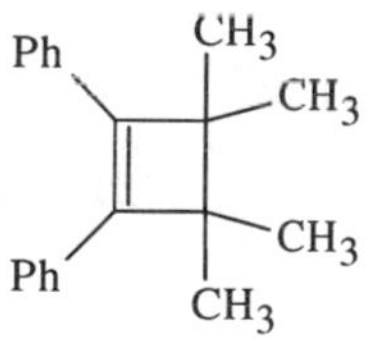

(b) This allylic vinylic ether undergoes a [3,3] sigmatropic rearrangement (that is, a Claisen rearrangement) on heating to give ketone *A*, which isomerizes to its aromatic "enol" form *B*.

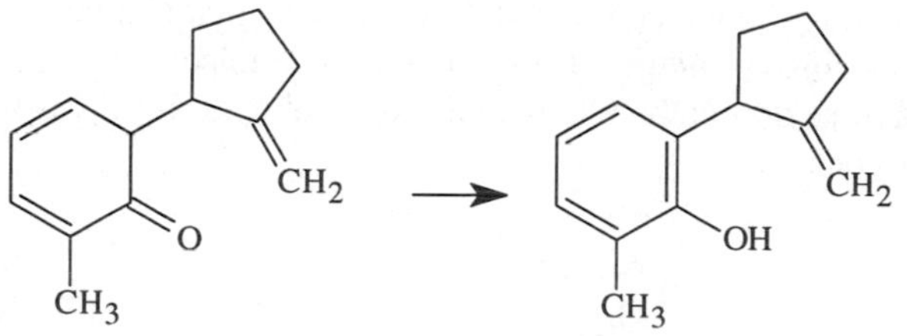

A *B*

(c) This is a [4*s* + 2*s*] cycloaddition, that is, a Diels-Alder reaction. The exocyclic double bond is not involved in the reaction, and *endo* stereochemistry is assumed:

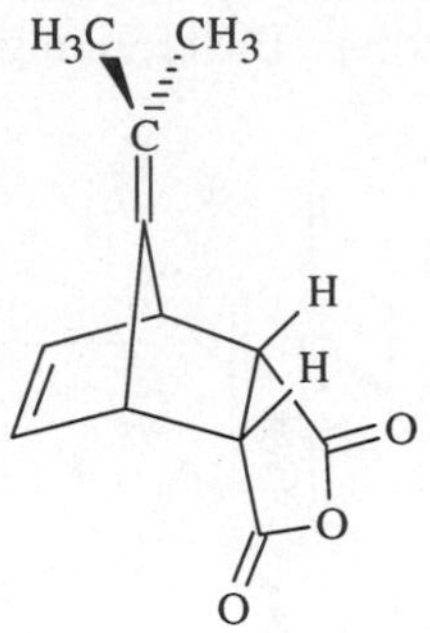

Two other reactions are allowed in principle that *do* involve the exocyclic double bond. One is a [2*s* + 6*a*] cycloaddition to give product *A*, and the other is a [2*a* + 6*s*] cycloaddition to give product *B*:

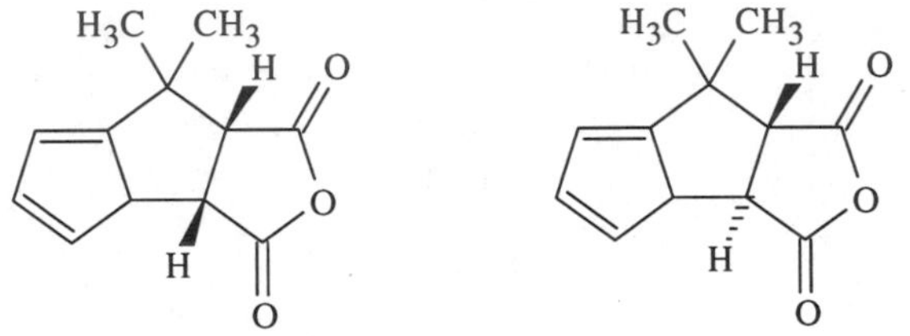

A *B*

If you construct models of the starting materials, and if you bring them together so that the appropriate orbital interactions can occur, you will see that the transition states required to form these products (as well as the products themselves) are considerably strained. Hence, the reactions that form these products are much slower than the Diels-Alder process.

25.34 (a) The structure of the ozonolysis product *C* shows that compound *B* is a cyclobutene, which must be formed in a disrotatory photochemical electrocyclic reaction. This defines the stereochemistry of *B*, which, in turn, defines the stereochemistry of *C*:

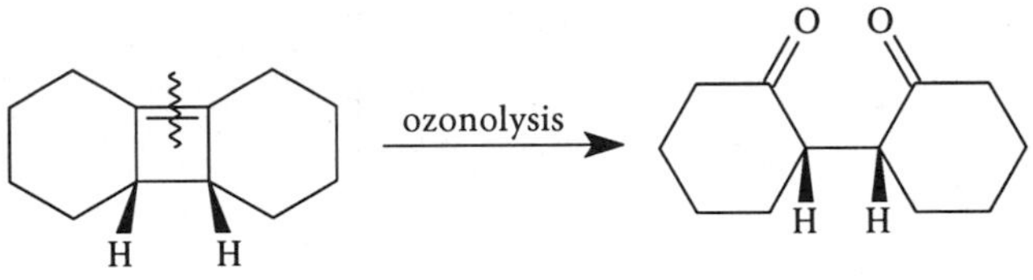

B —ozonolysis→ *C*

(b) Compound *D* is the trans-fused stereoisomer of the cyclobutene derivative, which undergoes conrotatory opening upon heating to give compound *A*. The reason that compound *B* is inert under the same conditions is that conrotatory ring opening of *B* would give a stereoisomer of *A* in which one of the cyclohexene rings would contain a trans double bond within a six-membered ring. This introduces so much strain that the reaction does not occur.

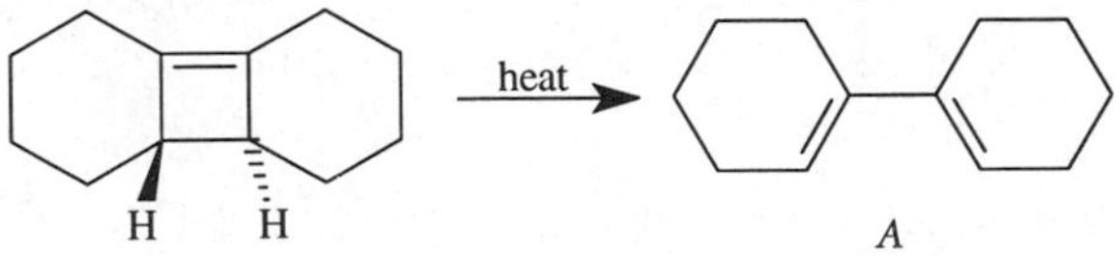

D —heat→ *A*

25.35 As usual, it helps to analyze first the origins of the carbons. The carbonyl groups are a good place to start, as they can have originated only from the quinone. The carbons marked with an asterisk are derived from the quinone. The remaining carbons, marked with a dagger (†) are derived from 1,3-cyclopentadiene. The new bonds are shown with heavy shading.

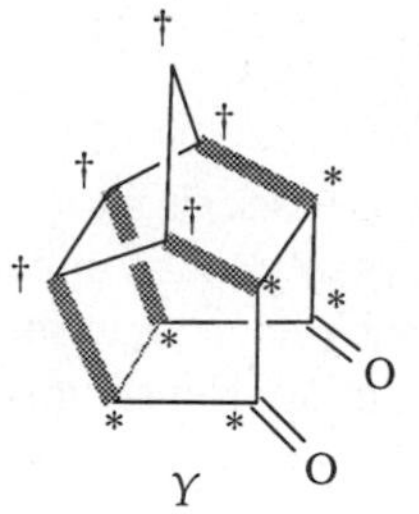

Y

The first reaction is a Diels-Alder reaction because we have a very reactive Diels-Alder diene and dienophile:

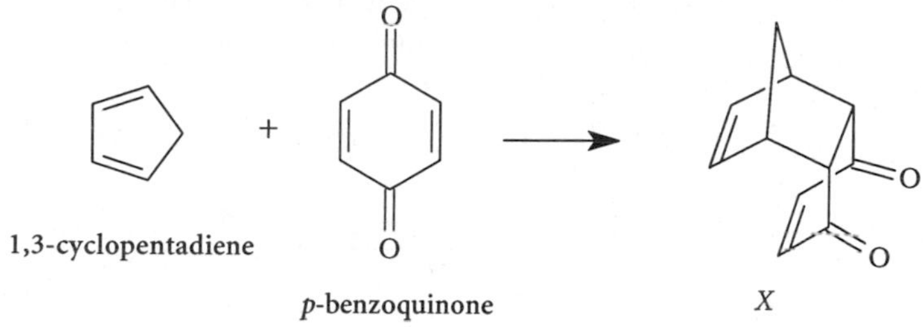

1,3-cyclopentadiene + *p*-benzoquinone → *X*

Compound *X* is all set up for a [2*s* + 2*s*] allowed photochemical cycloaddition to complete the formation of compound *Y*:

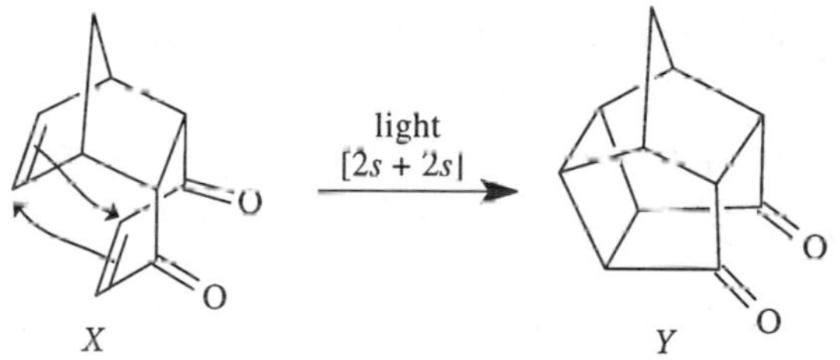

X —light [2*s* + 2*s*]→ *Y*

25.36 This cycloaddition reaction involves sixteen electrons; hence, it must be suprafacial on one component and antarafacial on the other. Thus, it must be a [14*s* + 2*a*] or a [14*a* + 2*s*] cycloaddition. The heptafulvene molecule is large enough that its π-electron system can twist without introducing too much strain or without losing too much π-electron overlap; hence, the cycloaddition is a [14*a* + 2*s*] process. In either case, the product has the following stereochemistry:

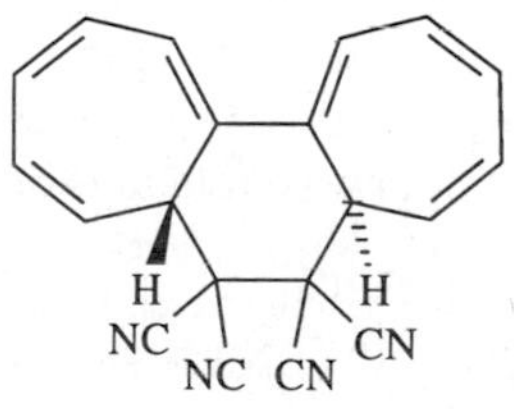

H H
NC CN
NC CN

25.37 (a) Toluene has added stability associated with aromaticity; compound *A*, although conjugated, is not aromatic. Because any equilibrium favors the more stable compound, the added stability of toluene causes it to be strongly favored in the equilibrium.

(b) If the reaction were concerted, it would have to be a [1,3] thermal sigmatropic rearrangement, which must occur in an antarafacial manner:

In this diagram, the "up" hydrogen (asterisk) is shown migrating to the lower face of the $=CH_2$ group. (Of course, migration of the "down" hydrogen to the upper face of the $=CH_2$ group is equally likely.) As discussed on text p. 1196, such an antarafacial migration requires that the migrating hydrogen bridge too great a distance for proper orbital overlap. Hence, this process is so slow that compound *A* has significant thermal stability.

(c) We would expect the conversion of compound *C* to compound *D* under thermal conditions to be extremely rapid, because this is a [1,5] sigmatropic rearrangement that could (and must) occur suprafacially. (Likewise, we would expect compound *A* to be rapidly transformed into *B* under *photochemical* conditions; why?)

25.38 (a) This is a [3,3] sigmatropic rearrangement, that is, a Claisen rearrangement.

The solution to the foregoing part and the solution to part (c) that follows illustrate a useful two-step process for drawing the products of complicated rearrangements. It usually makes sense to draw the bond connections first and thus obtain a highly distorted structure of the product. Then convert the distorted structure into a more conventional structure, using models if necessary.

(b) This reaction superficially appears to involve internal rotation about both double bonds, a process that is virtually impossible. However, a more reasonable pericyclic mechanism is available: a sequence of two conrotatory electrocyclic reactions, the first one a ring closing, and the second one a ring opening. (Note that the curved arrows refer to the clockwise rotations of groups and not to electron flow.)

(c) This is a [3,3] sigmatropic rearrangement in which the product requires a little redrawing!

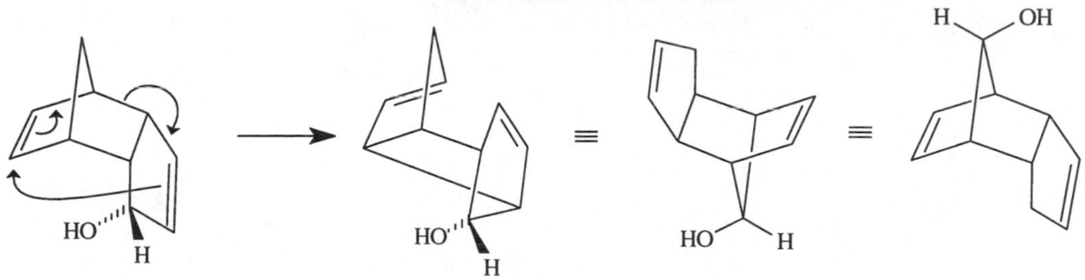

HO H → HO H ≡ HO H ≡ H OH

25.39 This is a [1,3] sigmatropic rearrangement with inversion in the migrating carbon, as required for the thermal [1,3] rearrangement.

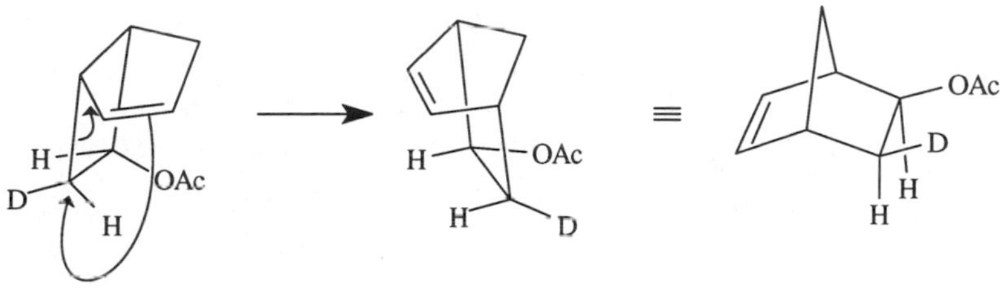

H D H OAc → H OAc H D ≡ OAc D H H

25.40 Take a cue from the solution to Problem 25.31(a) on p. 1004 of this manual, in which the last step involves a thermal reaction of 1,3,5-cyclooctatriene with two methyl groups. Replace the methyl groups with hydrogens and the result is the same. The resulting compound *B* undergoes a Diels-Alder reaction (that is, a [4*s* + 2*s*] cycloaddition) with the alkyne, and the product of that reaction, *C*, undergoes a reverse Diels-Alder to generate the final products.

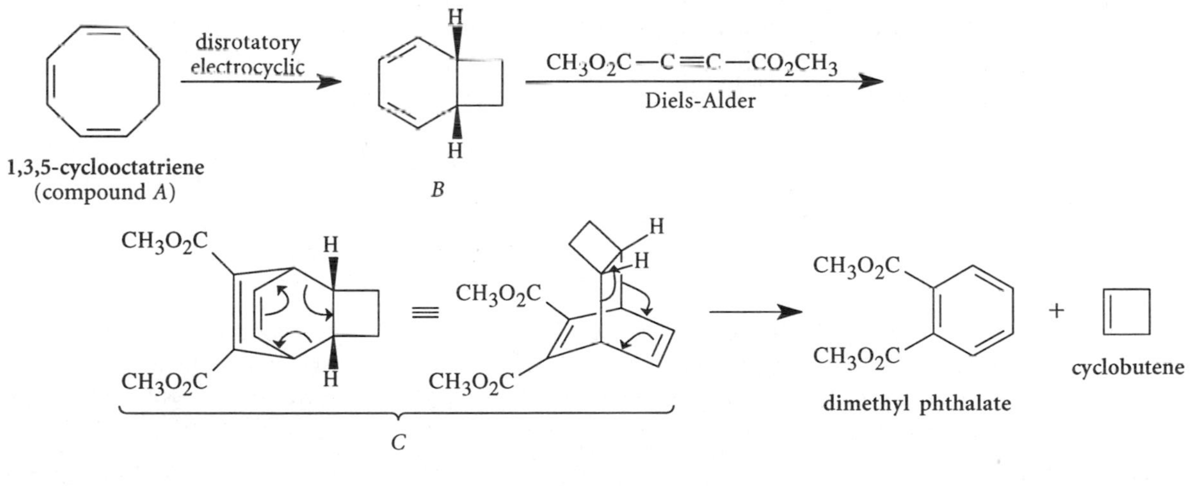

1,3,5-cyclooctatriene (compound *A*) —disrotatory electrocyclic→ *B* —$CH_3O_2C—C{\equiv}C—CO_2CH_3$, Diels-Alder→

C: CH_3O_2C, CH_3O_2C, H, H ≡ CH_3O_2C, CH_3O_2C, H, H → dimethyl phthalate (CH_3O_2C, CH_3O_2C) + cyclobutene

25.41 These transformations are very much like the last two in the solution to Problem 25.40. A Diels-Alder reaction is followed by a reverse Diels-Alder reaction. The driving force for the reverse Diels-Alder, of course, is formation of a product that is stabilized by aromaticity.

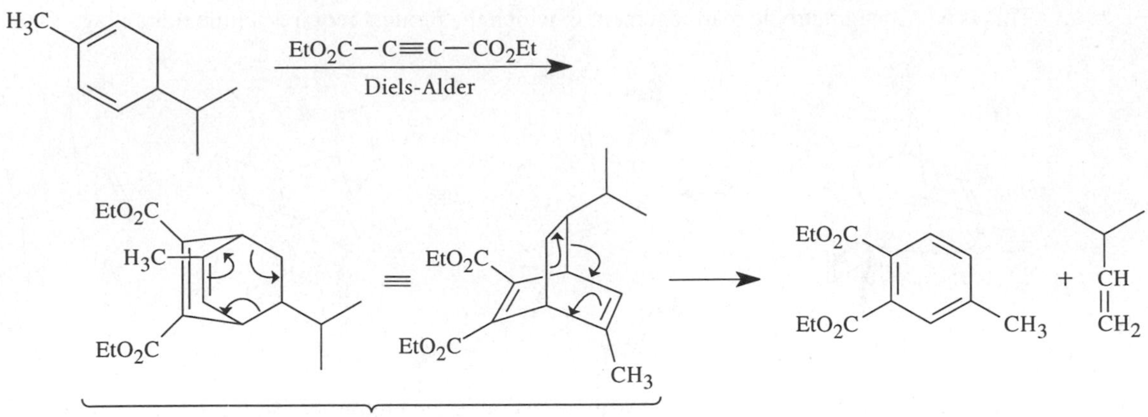

H_3C — $EtO_2C—C{\equiv}C—CO_2Et$, Diels-Alder →

EtO_2C, H_3C, EtO_2C ≡ EtO_2C, EtO_2C, CH_3 → EtO_2C, EtO_2C, CH_3 + CH=CH_2

compound *A*

25.42 Heating an allylic vinylic ether generally results in a [3,3] sigmatropic (Claisen) rearrangement.

Et, O, OEt → Et, O, OEt (substance *A*)

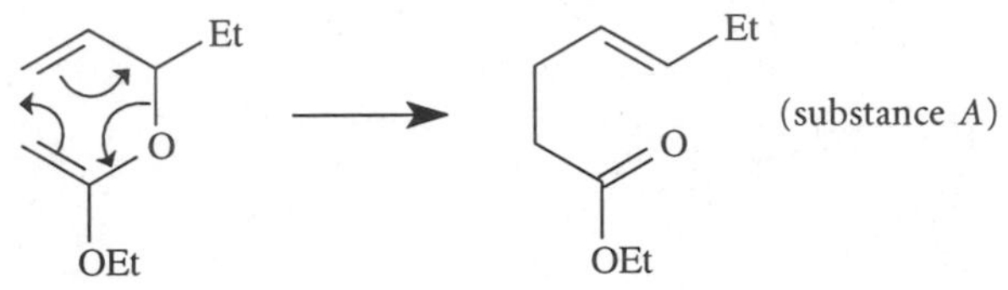

Why the product should have *E* stereochemistry is not obvious from the mechanism shown above, and an answer that gives either the *E* or the *Z* stereoisomer (or a mixture of both) is satisfactory. If you work Problem 25.48(a) you will find that the Claisen rearrangement proceeds through a chairlike conformation. If the ethyl group is placed in the more favorable equatorial position of the chairlike conformation, the product is the *E* stereoisomer of substance *A* shown above.

25.43 (a) The approach to solving this type of problem is described in the solution to Problem 15.63 and 15.64 on pp. 498–9 of this manual. The intermediate *X* is formed from the starting material by an allowed thermal disrotatory ring closure. Because *X* is a conjugated diene, it reacts in a Diels-Alder reaction with maleic anhydride, a good dienophile, to give the product shown in the problem.

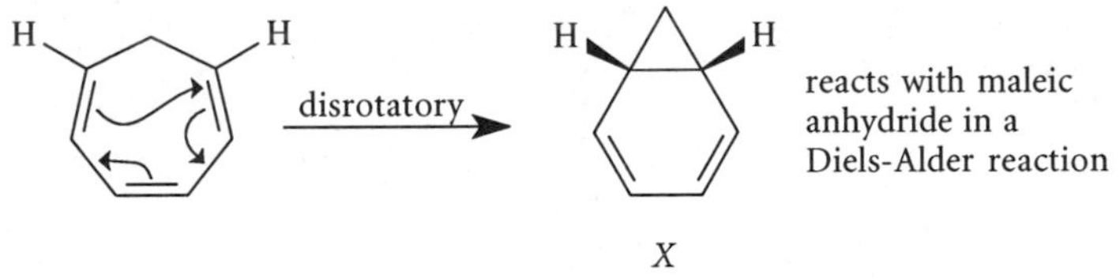

H, H —disrotatory→ H, H (*X*) reacts with maleic anhydride in a Diels-Alder reaction

(b) In this case maleic anhydride is used to trap the trienone intermediate *Y* formed in the first of what would, in the absence of maleic anhydride, be two successive [3,3] sigmatropic rearrangements. (See Eqs. 25.36a–b on text p. 1201.) Maleic anhydride reacts with the conjugated-diene unit of intermediate *Y*.

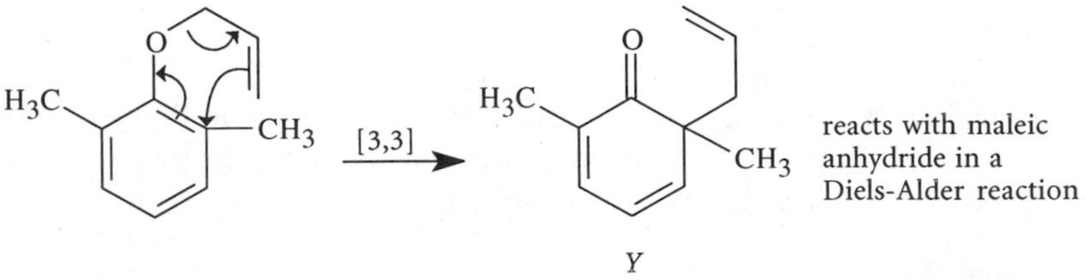

O, H_3C, CH_3 —[3,3]→ O, H_3C, CH_3 (*Y*) reacts with maleic anhydride in a Diels-Alder reaction

(c) The intermediate trapped by maleic anhydride is the tetraene *Z* formed by a [1,9] (presumably suprafacial) sigmatropic hydrogen migration.

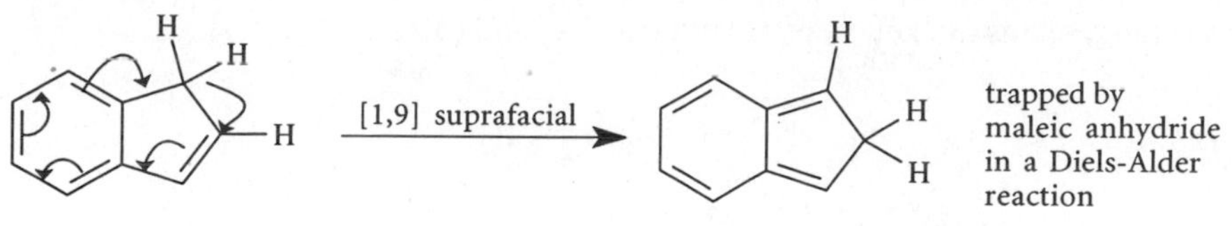

H H H [1,9] suprafacial → H H H trapped by maleic anhydride in a Diels-Alder reaction

Z

25.44 Compound *A* is a secondary alcohol, and the aldehyde *B* is formed from an enol intermediate (shown in parentheses) that results from a [3,3] sigmatropic (oxyCope) rearrangement.

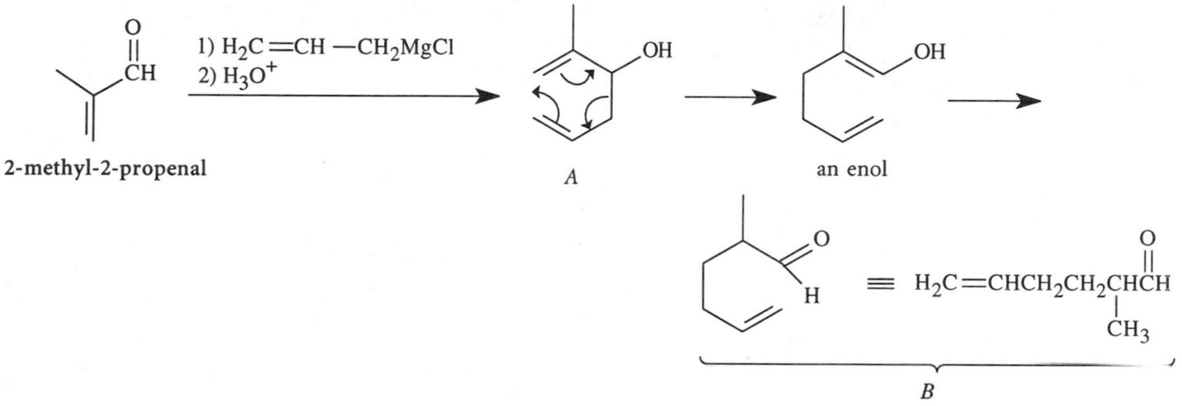

2-methyl-2-propenal —1) $H_2C{=}CH{-}CH_2MgCl$ 2) H_3O^+→ *A* (OH) → an enol (OH) →

$\equiv H_2C{=}CHCH_2CH_2CH(CH_3)CH{=}O$

B

Evidently in this case the Grignard reagent undergoes mostly 1,2-addition to the α,β-unsaturated aldehyde; none of the 1,4-addition product is formed. (See Sec. 22.10A.)

25.45 The structure of the ozonolysis product *D* and the formula of elemicin show that elemicin has the structure *C*. Compound *B* and elemicin differ by one methyl group, and compound *B* ionizes in base. Consequently, a reasonable hypothesis is that one of the three oxygens of compound *B* is part of a phenol —OH group, and that this is methylated by dimethyl sulfate treatment to give elemicin. Compound *A* is *not* a phenol because it is insoluble in base, but forms a phenol *B* when heated. This series of observations suggests that compound *A* is an allyl ether that undergoes a Claisen rearrangement when heated. Which oxygen is part of the allyl ether? If one of the two outer oxygens of *A* is an allyl ether, its rearrangement would result in transfer of the allyl group to the adjacent ortho position. Since the allyl group in *C* is not ortho to one of the methoxy groups, the rearrangement must have occurred from the central oxygen. Since the ortho positions are blocked, rearrangement occurs to the para position. The structures of *A* and *B* are thus defined.

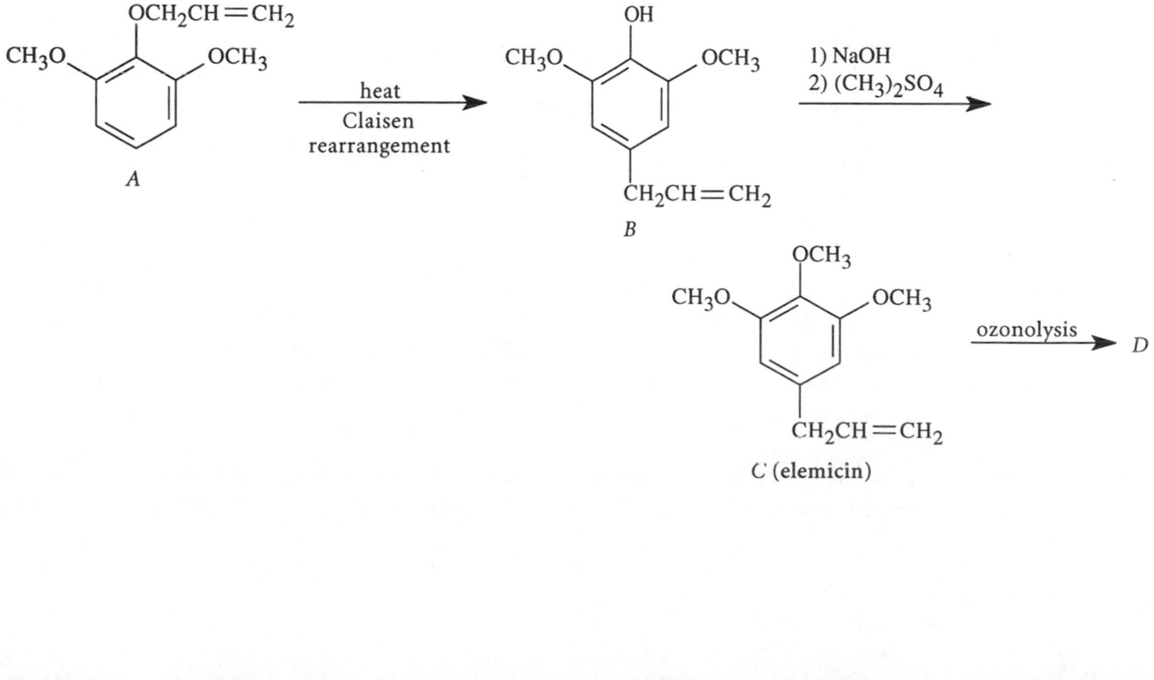

$OCH_2CH{=}CH_2$, CH_3O, OCH_3 (*A*) —heat, Claisen rearrangement→ OH, CH_3O, OCH_3, $CH_2CH{=}CH_2$ (*B*) —1) NaOH 2) $(CH_3)_2SO_4$→

OCH_3, CH_3O, OCH_3, $CH_2CH{=}CH_2$ — *C* (elemicin) —ozonolysis→ *D*

25.46 Claisen rearrangements are key steps in both parts (a) and (b).

(a)

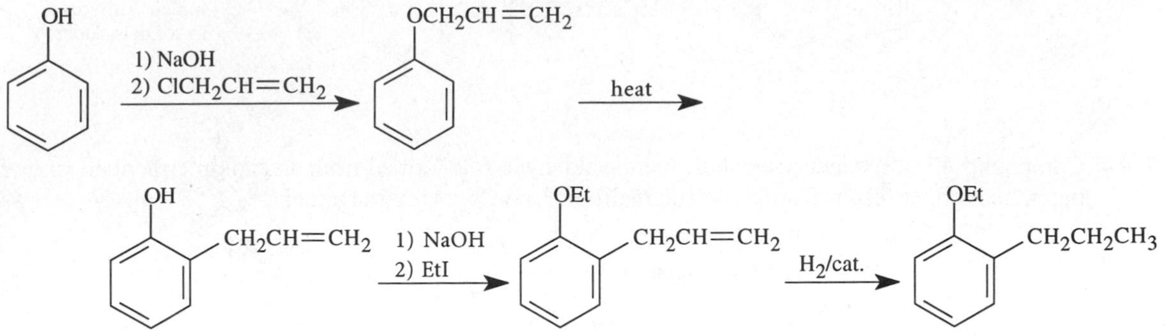

Phenol (OH) → 1) NaOH, 2) $ClCH_2CH{=}CH_2$ → allyl phenyl ether ($OCH_2CH{=}CH_2$) → heat →

2-allylphenol (OH, $CH_2CH{=}CH_2$) → 1) NaOH, 2) EtI → (OEt, $CH_2CH{=}CH_2$) → H_2/cat. → (OEt, $CH_2CH_2CH_3$)

1-ethoxy-2-propylbenzene

(b)

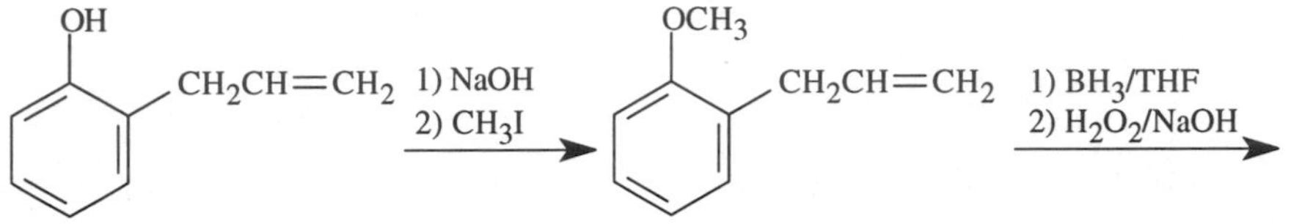

(OH, $CH_2CH{=}CH_2$) → 1) NaOH, 2) CH_3I → (OCH_3, $CH_2CH{=}CH_2$) → 1) BH_3/THF, 2) H_2O_2/NaOH →

prepared in part (a)

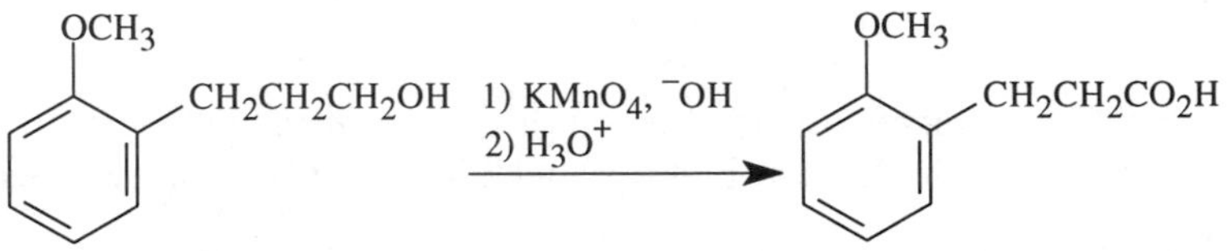

(OCH_3, $CH_2CH_2CH_2OH$) → 1) $KMnO_4$, ^-OH, 2) H_3O^+ → (OCH_3, $CH_2CH_2CO_2H$)

25.47 (a) The first transformation is a thermal conrotatory electrocyclic ring opening, and the second transformation is a thermal disrotatory electrocyclic ring closure.

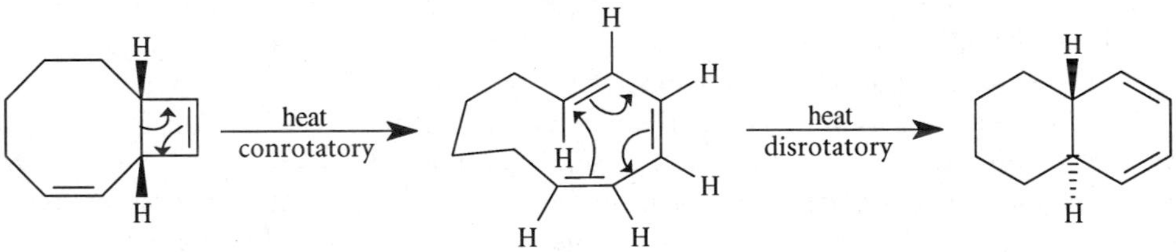

heat, conrotatory → *X* → heat, disrotatory →

(b) Two successive photochemical disrotatory electrocyclic reactions account for the products.

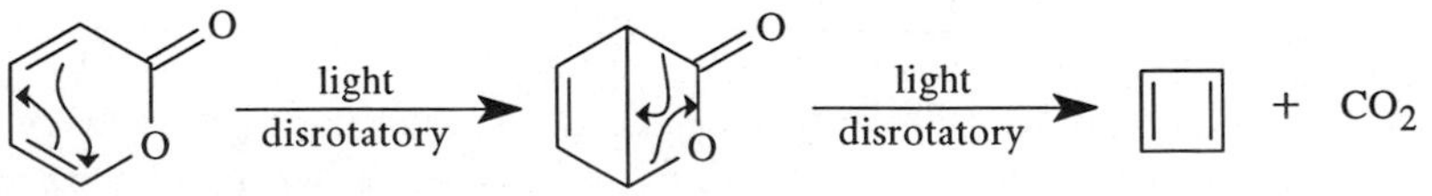

light, disrotatory → *Y* → light, disrotatory → □ + CO_2

The product, the very unstable antiaromatic diene 1,3-cyclobutadiene (see text p. 683), was generated in 1972 by this sequence of reactions carried out at a temperature of 8 K in the laboratory of Professor Orville L. Chapman (then at Iowa State University, now at the University of California, Los Angeles). This technique allowed the observation and characterization of 1,3-cyclobutadiene by infrared spectroscopy.

25.48 (a) Place the starting material in a chairlike conformation, carry out the [3,3] sigmatropic rearrangement, and compare the stereochemistry of the product with that obtained experimentally. (Models are useful to ensure that the asymmetric carbons have the proper configuration.)

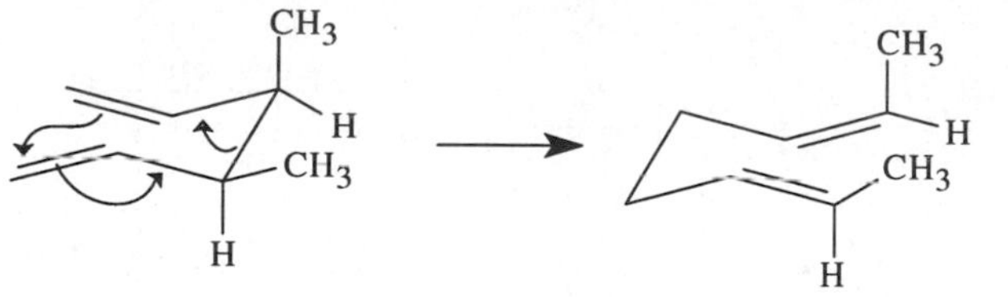

One double bond is *E* and one is *Z*; this is identical to the product obtained experimentally.

This verifies that the product could be formed from a chairlike conformation. However, the boatlike conformation must also be examined:

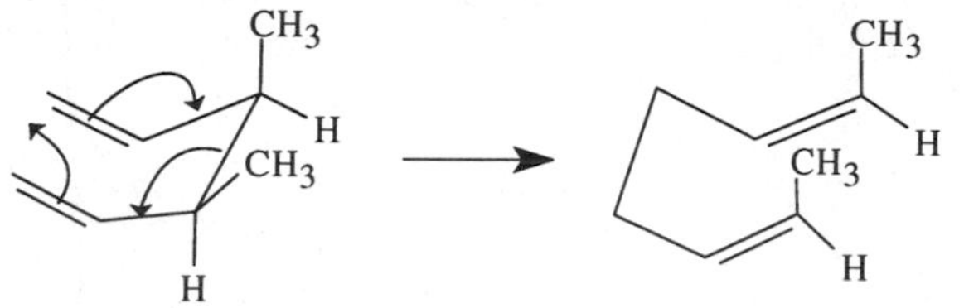

Both double bonds are *Z*; this is different from the product obtained experimentally.

(A second boatlike conformation is possible, and it would lead to formation of the all-*E* product, which is also different from that observed experimentally.) Thus, of the two possible conformations, only a chairlike transition state fits the observations. You should use similar reasoning to verify that a chairlike conformation of the second reactant in the problem gives the all-*E* product observed and that a boatlike conformation does not.

Notice that experiments with *both* alkene stereoisomers are required to establish that a chairlike transition state is a general phenomenon.

(b) Given that germacrone assumes a chairlike conformation in its transition state when it reacts to give β-elemenone, deduce the structure of germacrone by mentally reversing the Cope rearrangement.

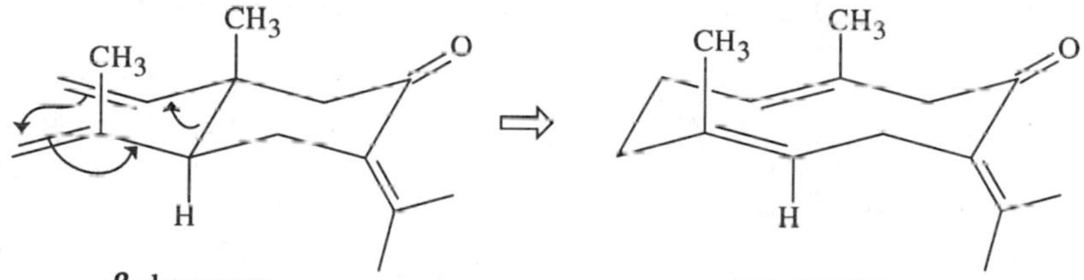

β-elemenone

germacrone

Now re-examine (or work) Problem 25.42 to see why the product in that problem has *E* stereochemistry.

25.49 A thermal electrocyclic reaction of the carbocation, which is formed by protonation of the ketone starting material, is followed by loss of a proton to give an enol, which spontaneously isomerizes to the corresponding ketone. Since four electrons are involved, the reaction is predicted to be conrotatory, and the methyl groups in the product are therefore trans:

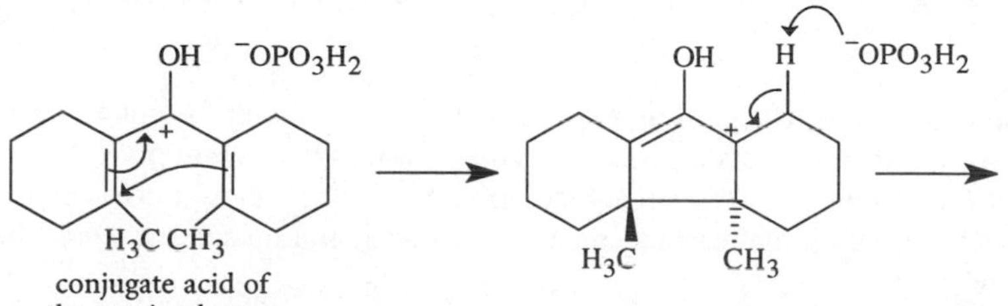

conjugate acid of the starting ketone

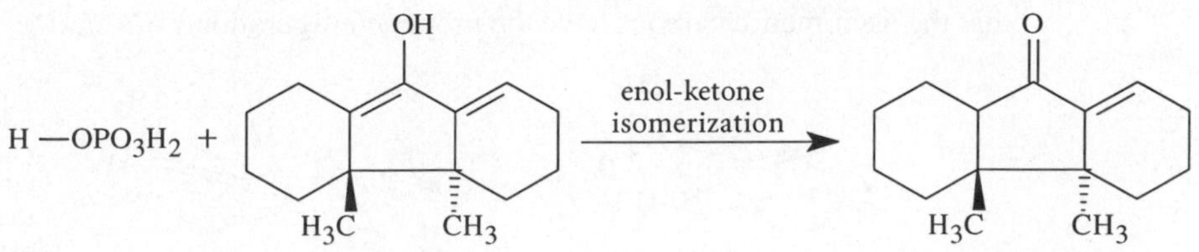

H —OPO_3H_2 + [structure: an enol] —enol-ketone isomerization→ [structure]

an enol

25.50 (a) A (presumably) conrotatory ring opening gives intermediate *A*. Although this compound is no longer aromatic, its formation is driven by the relief of strain in the four-membered ring. Compound *A* undergoes an *intramolecular* Diels-Alder reaction that restores aromaticity and gives the product with precisely the desired stereochemistry.

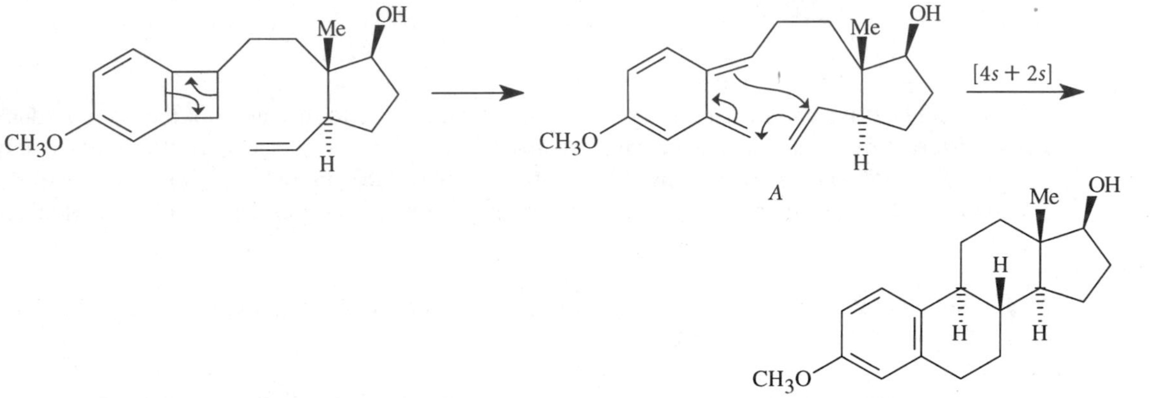

A

[4*s* + 2*s*]

(b) Oxidation of the secondary alcohol and cleavage of the ether gives estrone:

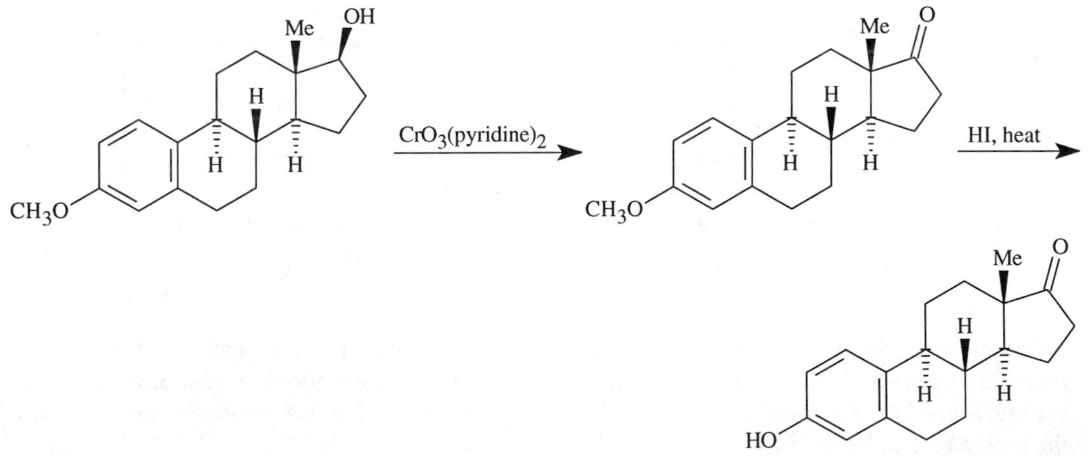

$CrO_3(pyridine)_2$

HI, heat

estrone

25.51 This is a [1,7] sigmatropic rearrangement, which, according to the selection rules, should be antarafacial. (See Problem 25.28 and its solution.) Migration of hydrogen gives *B* and migration of deuterium gives *C*. (Presumably, the predominance of *B* is due in part to a primary isotope effect, which should favor more rapid migration of hydrogen, all things being equal. If the rearrangements are indeed antarafacial, the stereochemistry of the products should be as follows:

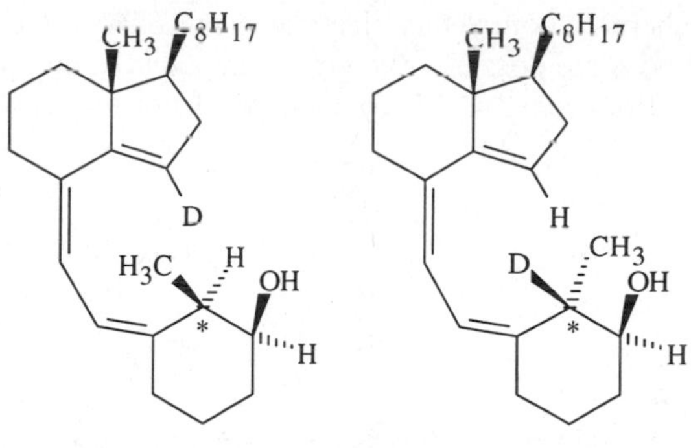

B *C*

This was indeed the stereochemistry observed.

This work was carried out to adduce evidence for the stereochemistry of the previtamin D rearrangement to vitamin D (Eq. 25.39 on text p. 1206). Compound *A* was constructed to be a "model" for previtamin D. As noted in Problem 25.28, the structure of the previtamin itself does not reveal the stereochemistry of the rearrangement.

25.52 First deduce the structure of compound *C* from the structure of the Diels-Alder adduct. Then deduce the structures of compound *A* and *B* from the reactions. Compound *A* is the product of bromine addition; compound *B* is a conjugated diene that results from two successive E2 eliminations of compound *A*. Compound *C* is formed from compound *B* by a disrotatory electrocyclic ring opening.

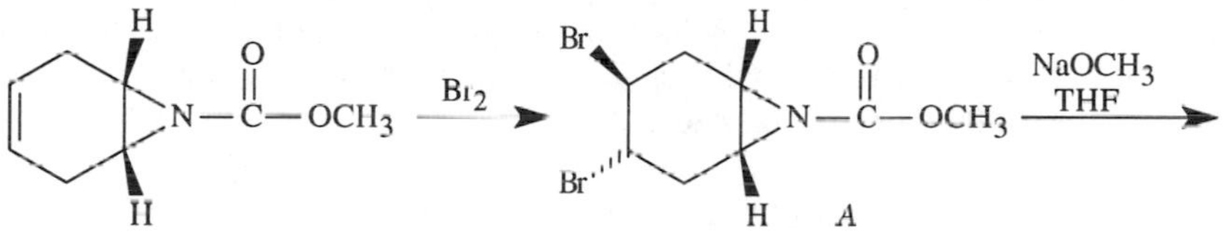

Br_2 → *A* $\xrightarrow[\text{THF}]{NaOCH_3}$

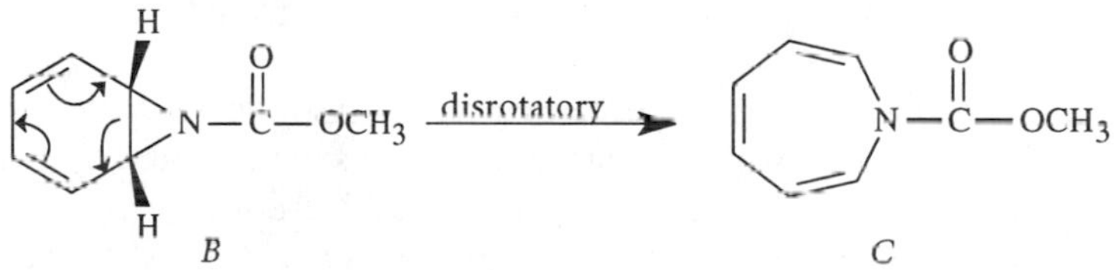

B $\xrightarrow{\text{disrotatory}}$ *C*

Compound *C* is interesting because if it were planar, it would have antiaromatic character. Like 1,3,5,7-cyclooctatetraene (COT; text pp. 674–7), it avoids antiaromaticity by avoiding planarity. However, this means that it reacts like an alkene.

Notice that compound *C* reacts with tetracyanoethylene as a *diene* in a Diels-Alder reaction rather than as a triene in a $[6a + 2s]$ or $[6s + 2a]$cycloaddition. Why?

25.53 (a) If compound *B* is aromatic, then it must be planar. If it is planar, the "inner" hydrogens (the ones shown explicitly in the problem) fall essentially on top of each other. (A model can be used to demonstrate this point.) The resulting van der Waals repulsions are so severe that compound *B* does not exist in spite of its aromaticity. (Compound *B* was also the subject of Problem 15.40 on text p. 686.)

(b) The formation of compound *B* from compound *A* involves $4n + 2$ (6) electrons and is one of two possible allowed conrotatory photochemical electrocyclic ring-opening reactions. (The other is discussed in the solution to part (c).)

(c) Compound *C* could be formed in two ways. First, compound *B* could be formed as an unstable intermediate, and then compound *C* could be formed from *B* by an allowed *thermal* disrotatory electrocyclic reaction. Such a reaction would be very rapid even at low temperature because of the instability of compound *B*. The second pathway is that the all-cis pentaene *D* could be formed from compound *A* in an allowed photochemical conrotatory electrocyclic reaction. (Note that two different allowed conrotatory reactions of compound *A* are possible: one gives compound *B*, and one gives compound *D*.) Compound *C* could then be formed from compound *D* by a thermal disrotatory electrocyclic reaction.

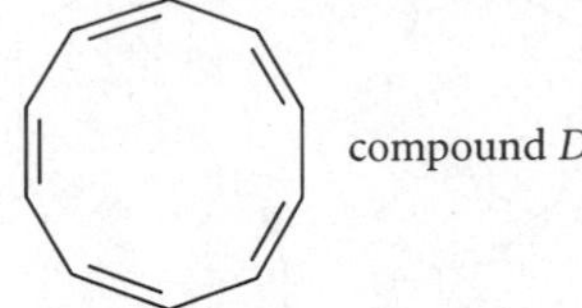

One could potentially distinguish between the two pathways for the formation of compound *C* by preparing compound *D* by another route and determining whether it spontaneously closes to compound *C*. If it does not, then the first pathway must be operating. In fact, compound *D* was prepared in 1971 by chemists at the University of Alberta. They found that, indeed, compound *D* is spontaneously transformed into compound *C* in about 3.5 hr at 6 °C. The rate of this process was found to be at least ten times grater than the rate of the photochemical conversion of *A* to *C* described in the problem. Hence, compound *D* is a competent intermediate. Furthermore, when they carried out the photochemical reaction of *A* described in the problem at temperatures so low that compound *D* would not react further, compound *D* accumulated in the reaction mixture. The simplest interpretation of these observations is that compound *B* is not an intermediate, and the pathway for the observations recorded in the problem is the one involving compound *D*.

25.54 (a) If the formation of benzene were concerted, it would have to be a thermal disrotatory electrocyclic ring-opening reaction involving four electrons.

(b) The reaction that gives benzene is not allowed by the selection rules; therefore, it does not occur. In spite of the high degree of strain of Dewar benzene, and in spite of the great stability of benzene, Dewar benzene is effectively constrained to exist by the selection rules!

Prismane, or Ladenburg benzene, which is discussed on text p. 1187, is another very unstable constitutional isomer of benzene that is effectively trapped into existence because its concerted conversion into benzene would violate the selection rules for pericyclic reactions.

25.55 (a) Bromine addition to the starting material is followed by four successive base-promoted E2 reactions to give compound *A*, which spontaneously forms compound *B* by an allowed disrotatory thermal electrocyclic reaction. Notice that compound *B* is a "tied-down" variation on hydrocarbon *B* in Problem 25.53. Because the offending inner hydrogens are absent in *B*, its continuous π-electron system can nearly achieve planarity.

$2Br_2$ → Br Br Br Br → KOH, EtOH

(probably a mixture of stereoisomers)

A —disrotatory→ *B*

(b) As discussed in part (a), compound *B* contains a planar (or nearly planar) $4n + 2$ π-electron system. Consequently, it shows the typical ring-current effects expected of aromatic compounds in its NMR spectra. (See Fig. 16.2 on text p. 699.) The protons on the double bonds, like those in benzene, show typical aromatic chemical shifts near δ 7.1. The two methylene protons, however, occupy the region that is strongly shielded by the ring current. Thus, it is these protons that have the negative chemical shift.

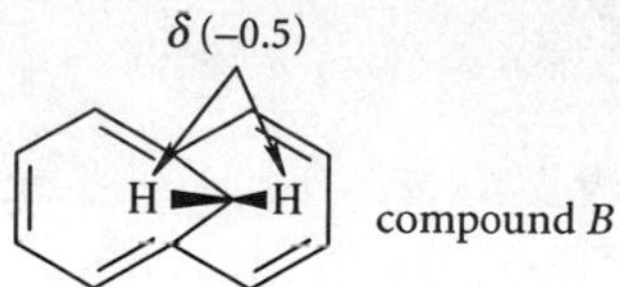

26

Amino Acids, Peptides, and Proteins

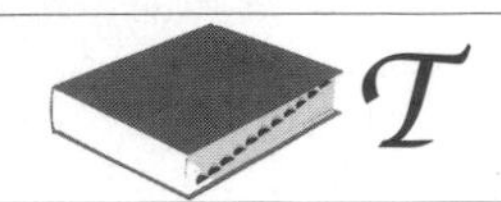

TERMS

acidic amino acid 26.3B
active site 26.10
acyl-enzyme 26.10
O-acylisourea 26.7
amino acid .... Introduction
α-amino acid .... Introduction
amino acid analysis 26.6A
amino acid sequence 26.6B
amino end 26.1B
amino terminus (Pep^N) 26.1B, 26.6B
anion-exchange resin 26.3C
antiparallel pleated sheet 26.9B
AQS-NHS (1-[[(6-quinolylamino)carbonyl]-oxy]-2,5-pyrrolidinedione 26.6A
β barrel 26.9C
basic amino acid 26.3B
Boc (*tert*-butyloxycarbonyl) group 26.7
carboxy end 26.1B
carboxy terminus (Pep^C) 26.1B, 26.6B
cation-exchange resin 26.3C
chemical library 26.8C
chromatography 26.3C
chymotrypsin 26.6C
Cleland's reagent (DTT) 26.9A
coenzyme 26.10
combinatorial chemistry 26.8A
competitive inhibitor 26.10B
cyanogen bromide 26.6C
DCC 26.7
denaturation 26.9C
dipeptide 26.1B
disulfide bond (in a protein) 26.9A
diversity element 26.8C
D,L system 26.2
domain 26.9C
Edman degradation 26.6B
Edman reagent 26.6B
electrostatic interactions 26.9C
endopeptidase 26.6C
exopeptidase 26.6C
enzyme 26.10A
enzyme-substrate complex 26.10A
Fmoc [(9-fluorenyl-methyloxy)carbonyl] group 26.7
fluorescent 26.6A
four-helix bundle 26.9C
guanidino group 26.3B
α-helix (right-handed) 26. 9B
C18 high-performance liquid chromatography (C18 HPLC) 26.6A
high-throughput screening 26.8B
homoserine lactone 26.6C
hydrophilic residue 26.9C
hydrophobic bond 26.9C
hydrophobic residue 26.9C
induced fit 26.10B
inhibitor 26.10B
instance (of diversity elements) 26.8C
ion-exchange chromatography 26.3C
ion-exchange resin 26.3C
isoelectric pH 26.3B
isoelectric point, p*I* 26.3B
isothiocyanate 26.6B
liquid chromatography 26.3D
neutral amino acid 26.3B
NHS 26.7
ninhydrin 26.6A
oligopeptide 26.1B
parallel pleated sheet 26.9B
peptidase 26.6C
peptide backbone 26.1B
peptide bond .... Introduction
peptide .... Introduction
phenylthiohydantoin (PTH) 26.6B
pleated sheet 26.9B
primary sequence 26.6B
primary structure 26.9A
protease 26.6C
protein .... Introduction

(Glossary continues)

CONCEPTS

I. INTRODUCTION TO AMINO ACIDS AND PEPTIDES

A. General

1. Compounds that contain both an amino group and a carboxylic acid group are called amino acids.
   a. An amino acid with an overall charge of zero (neutral) can contain within the same molecule two groups of opposite charge.
   b. Molecules containing oppositely charged groups are known as zwitterions.
2. An α amino acid has an amino group on the α-carbon, the carbon adjacent to the carboxylic acid group.

$$H_3\overset{+}{N}-\underset{\displaystyle R}{\underset{|}{CH}}-\overset{\displaystyle O}{\overset{\|}{C}}-O^-$$

an α-amino acid in zwitterionic form

3. Peptides are polymers in which α-amino acids are joined into chains through amide bonds (peptide bonds)

$$\sim NH-\underset{\displaystyle R}{\underset{|}{CH}}-\overset{\displaystyle O}{\overset{\|}{C}}-NH-\underset{\displaystyle R}{\underset{|}{CH}}-\overset{\displaystyle O}{\overset{\|}{C}}-NH-\underset{\displaystyle R}{\underset{|}{CH}}-\overset{\displaystyle O}{\overset{\|}{C}}-NH-\underset{\displaystyle R}{\underset{|}{CH}}-\overset{\displaystyle O}{\overset{\|}{C}}\sim$$

(R groups may be the same or different)

   a. A peptide bond is derived from the amino group of one amino acid and the carboxylic acid group of another.
   b. Proteins are large peptides, and some proteins are aggregates of more than one peptide.

B. Nomenclature of Amino Acids

1. Some amino acids are named substitutively as uncharged compounds (carboxylic acids with amino substituents).
2. Twenty α-amino acids occur commonly as constituents of most proteins and are known by widely accepted traditional names which, with their structures, are given in Table 26.1, text pages 1218–1219.
   a. With the exception of proline, all α-amino acids have the same general structure, differing only in the identity of their side chains R.
      i. Proline is the only naturally occurring amino acid with a secondary amino group.
      ii. In proline the —NH— and the side chain are "tied together" in a ring.

proline
(in zwitterionic form)

b. The amino acids can be organized into six groups according to the nature of their side chains:
   i. hydrogen or aliphatic hydrocarbons
   ii. aromatic groups
   iii. thiol, sulfide, or alcohol groups
   iv. carboxylic acid or amide groups
   v. basic side chains
   vi. proline

3. The α-amino acids are often designated by either three-letter or single-letter abbreviations, which are given in Table 26.1, text pages 1218–1219.

C. Nomenclature of Peptides

1. The peptide backbone is the repeating sequence of nitrogen, α-carbon, and carbonyl groups.
   a. The characteristic amino acid side chains are attached to the peptide backbone at the respective α-carbon atoms.
   b. Each amino acid unit in a peptide is called a residue.
   c. The ends of a peptide are labeled as the amino end or amino terminus (Pep^N) and the carboxy end or carboxy terminus (Pep^C).
   d. A peptide can be characterized by the number of residues it contains.
      i. A prefix *di*, *tri*, *tetra*, etc., is attached to the word *peptide* to indicate the number of amino acids that are contained within the peptide.
      ii. A relatively short peptide of unspecified length containing a few amino acids is sometimes referred to as an oligopeptide.
2. A peptide is conventionally named by giving successively the names of the amino acid residues, starting at the amino end.
   a. The names of all but the carboxy-terminal residue are formed by dropping the final ending of the amino acid and replacing it with *yl*.
   b. This type of nomenclature is used only for the smallest peptides.
3. Peptides can be represented by connecting with hyphens the three-letter (or one-letter) abbreviations of the component amino acid residues beginning with the amino-terminal residue.

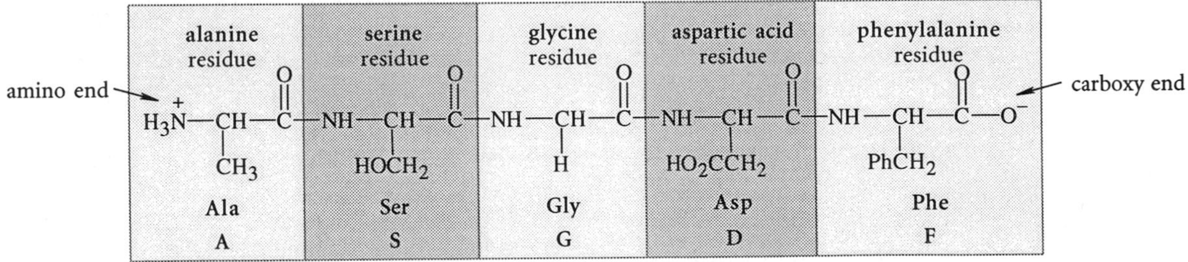

alanylserylglycylaspartylphenylalanine (abbreviated **Ala-Ser-Gly-Asp-Phe** or **A-S-G-D-F**)

4. Large peptides and proteins of biological importance are known by common names.

D. Stereochemistry of the α-Amino Acids

1. With the exception of glycine, all common naturally occurring α-amino acids have an asymmetric carbon atom; all have the *S* configuration at the α-carbon except for cysteine.
2. The stereochemistry of α-amino acids is often specified with an older system, the D,L system.
   a. An L amino acid by definition has the amino group on the left and the hydrogen on the right when the carboxylic acid group is up and the side chain is down in a Fischer projection of the α-carbon.

b. The naturally occurring amino acids have the L configuration (the correspondence between *S* and L is not general).

mirror plane

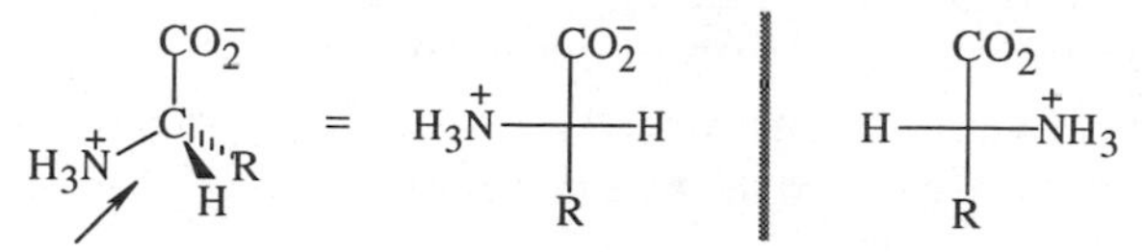

CO_2^- ... $H_3\overset{+}{N}$—C—R, H   = $H_3\overset{+}{N}$—C(—CO_2^-)(—H)—R | H—C(—CO_2^-)(—$\overset{+}{N}H_3$)—R

S configuration   L-α-amino acid   D-α-amino acid

c. The D or L designation for an α-amino acid refers to the configuration of the α-carbon regardless of the number of asymmetric carbons in the molecule.

d. In the D,L system, diastereomers are given different names. For example, L-threonine and L-allothreonine are diastereomers that differ in configuration at carbon-3.

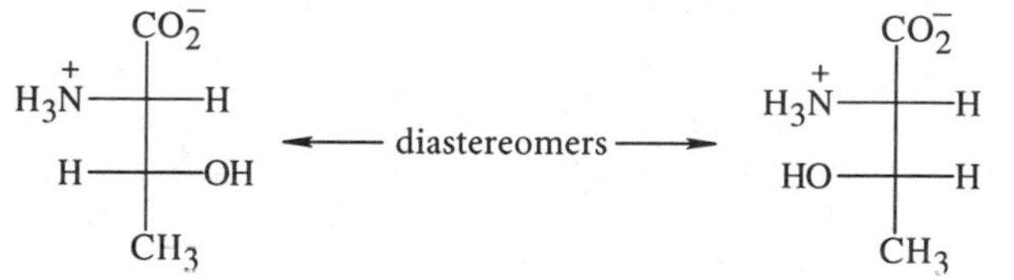

CO_2^-; $H_3\overset{+}{N}$—H; H—OH; CH_3 ⟵ diastereomers ⟶ CO_2^-; $H_3\overset{+}{N}$—H; HO—H; CH_3

L-threonine   L-allothreonine

II. ACID-BASE PROPERTIES OF AMINO ACIDS AND PEPTIDES

A. Zwitterionic Structures of Amino Acids and Peptides

1. The major neutral form of any α-amino acid is the zwitterion.
   a. The high melting points of amino acids and their greater solubilities in water than in ether are characteristics of salts (charged species). Water is the best solvent for most amino acids because it solvates ionic groups.
   b. The dipole moments of the amino acids are much larger than those of similar-size molecules with only one amine or carboxylic acid group. This observation suggests a great deal of separated charge.
   c. The pK_a values for amino acids are what one would expect for the zwitterionic forms of the neutral molecules.
2. Peptides also exist as zwitterions; at neutral pH, the amino and guanidino groups are protonated and the carboxylic acid groups are ionized.

B. Isoelectric Points of Amino Acids and Peptides

1. Amino acids and peptides are amphoteric substances; they contain both acidic and basic groups.
2. An important measure of the acidity or basicity of an amino acid or peptide is its isoelectric point, or isoelectric pH; this is the pH of a dilute aqueous solution of the amino acid or peptide at which the total charge on all molecules is exactly zero. (The isoelectric points of the α-amino acids are given in Table 26.1, text pages 1218–1219.)
3. At the isoelectric point, two conditions are met:
   a. The concentration of negatively charged species equals the concentration of positively charged species.
   b. The relative concentration of the neutral form is greater than at any other pH.
4. The isoelectric point, pI, of an amino acid is the average of the two pK_a values of the amino acid:

$$\text{isoelectric point} = \text{p}I = \frac{\text{p}K_{a1} + \text{p}K_{a2}}{2}$$

5. The isoelectric point indicates not only the pH at which a solution of the amino acid or peptide contains the greatest amount of neutral form but also the sign of the net charge on the amino acid or peptide at any pH.
   a. At pH values less than the isoelectric point, more molecules of the amino acid or peptide are positively charged than are negatively charged.
   b. At pH values greater than the isoelectric point, more molecules of an amino acid or peptide are negatively charged than are positively charged.
   c. At pH = p*I*, the amount of neutral form is greater than at any other pH, and any positively charged molecules are balanced by an equal number of negatively charged ones.

6. When an amino acid or peptide has a side chain containing an acidic or basic group, the isoelectric point is markedly changed.
   a. When peptide side chains contain more amino and guanidino groups than carboxy groups, the isoelectric point of the peptide is the average of the two highest pK_a values.
   b. When peptide side chains contain more carboxy groups than amino and guanidino groups, the isoelectric point of the peptide is the average of the two lowest pK_a values.
      i. Amino acids with high isoelectric points are classified as basic amino acids.
      ii. Amino acids with low isoelectric points are classified as acidic amino acids.
      iii. Amino acids with isoelectric points near 6–8 are classified as neutral amino acids.

7. A peptide can be classified as acidic, basic, or neutral by examining the number of acidic and basic groups that it contains.
   a. A peptide with more amino and guanidino groups than carboxy groups will have a high isoelectric point.
   b. A peptide with more carboxy groups than amino and guanidino groups will have a low isoelectric point.

C. Separation of Amino Acids and Peptides Using Acid-Base Properties

1. Separations of amino acids and peptides can be based on their isoelectric points.
   a. Most amino acids and peptides are most soluble when they carry a net charge and least soluble in their neutral forms.
   b. Most amino acids and peptides are least soluble at pH values equal to their isoelectric points because they are neutral.
   c. In some cases, peptides precipitate when the pH is adjusted to their isoelectric points.

2. Separations of amino acids and peptides by ion-exchange chromatography depend on their isoelectric points.
   a. Chromatography is a separation technique based on the relative adsorptions of compounds to a material called a stationary phase.
   b. In ion-exchange chromatography, a hollow tube or column is filled with a buffer solution in which is suspended a finely powdered, insoluble polymer called an ion-exchange resin; the resin bears charged groups.
      i. Resins that bear negatively charged pendant groups adsorb cations, and are called cation-exchange resins.
      ii. Resins that bear positively charged pendant groups adsorb anions, and are called anion-exchange resins.
   c. Whether an amino acid or peptide is adsorbed by the column depends on its charge, which, in turn, depends on the relationship of its isoelectric point to the pH of the buffer.
   d. Compounds that are not adsorbed to the column emerge from the column first; compounds that are adsorbed most strongly to the column emerge from the column last.

III. STRUCTURES OF PEPTIDES AND PROTEINS

A. Primary Structure

1. One description of a peptide or protein structure is its covalent structure or primary structure.
2. The most important aspect of any primary structure is the amino acid sequence.
3. Disulfide bonds link the cysteine residues in different parts of a sequence and serve as cross-links between different parts of a peptide chain.

```
~Val—Ala—Cys—Arg~
          |
          S
          |  <———————— disulfide linkage
          S
          |
~Asn—Ser—Cys—His—Lys~
```

a. The disulfide bonds of a protein are readily reduced to free cysteine thiols by other thiols.
b. Two commonly used thiol reagents are 2-mercaptoethanol ($HSCH_2CH_2OH$), and dithiothreitol (DTT, or Cleland's reagent), $HSCH_2CH(OH)CH(OH)CH_2SH$.

B. Secondary Structure

1. The description of a peptide or protein in terms of the conformations of its peptide chains is called secondary structure.
2. Three conformations occur commonly:
   a. In a right-handed α-helix, the side-chain groups are positioned on the outside of the helix, and the helix is stabilized by hydrogen bonds between the amide N—H of one residue and the carbonyl oxygen four residues further along the helix. (See Fig. 26.9 on text p. 1264 for a more detailed view.)

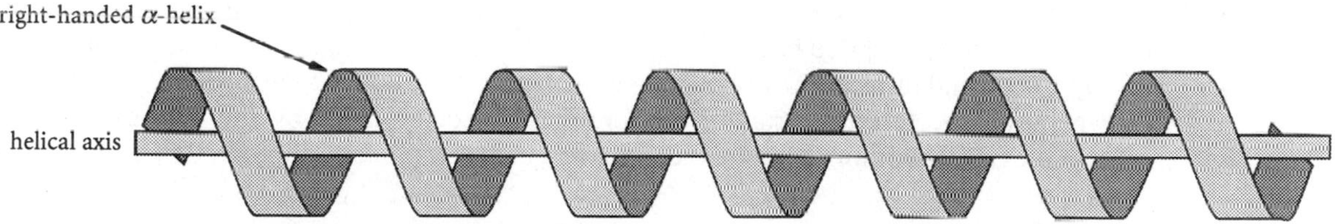

   b. In a β-structure (pleated sheet), the peptide chain adopts an open, zigzag conformation, and it is engaged in hydrogen bonding with another peptide chain (or a different part of the same chain) in a similar conformation.
      i. Successive hydrogen-bonded chains can run (in the amino terminal to carboxy terminal sense) in the same direction (parallel pleated sheet) or in opposite directions (antiparallel pleated sheet).
      ii. The positions of the side-chain R-groups alternate above and below the sheet.

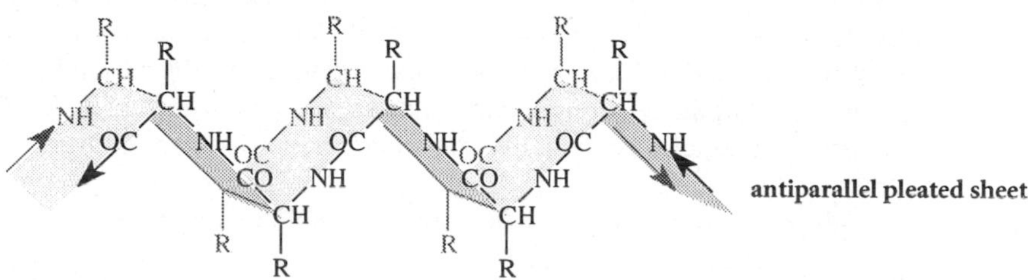

   c. Peptides that adopt a random-coil conformation show no discernible pattern in their conformation.

3. Some peptides and proteins exist entirely as α-helix or pleated sheet; however, most proteins contain different types of secondary structure in different parts of their peptide chains.

C. Tertiary and Quaternary Structures

1. The complete three-dimensional description of protein structure at the atomic level is called tertiary structure and is determined by X-ray crystallography.
2. The tertiary structure of any given protein is an aggregate of α-helix, β-sheet, random-coil, and other structural elements.
   a. In many proteins certain higher-order structural motifs are common.
      i. A common motif is a bundle of four helices (called a four-helix bundle), each running approximately antiparallel to the next and separated by short turns in the peptide chain.
      ii. Another common structural motif is the β-barrel, a "bag" consisting of β-sheets connected by short turns.
   b. Several motifs can occur within a given protein, so that a protein might consist of several smaller, relatively ordered structures connected by short random loops; these ordered substructures are sometimes called domains.
      i. A useful way to portray β-structure as part of an overall protein structure is a ribbon structure in which the peptide backbone is portrayed as a ribbon (see Fig. 26.11, text page 1266).
      ii. Twists and turns in the ribbon are defined by angles shown by arrows.
3. In general, the tertiary structures of proteins are determined by the noncovalent interactions between groups within protein molecules, and between groups of the protein with the surrounding solvent:
   a. Hydrogen bonds
      i. Hydrogen bonds stabilize both α-helices and β-sheets as well as other conformations.
      ii. Protein conformations are also stabilized in part by hydrogen bonding of certain groups to solvent water.
   b. Van der Waals interactions
      i. Van der Waals attractions, or dispersion forces, are the same interactions that provide the cohesive force in a liquid hydrocarbon, and can be regarded as examples of the "like-dissolves-like" phenomenon.
      ii. The van der Waals attractions between hydrocarbonlike residues are sometimes called hydrophobic bonds, because the hydrocarbon groups energetically prefer each other to water.
      iii Residues such as the side chain of phenylalanine or isolucine are sometimes called hydrophobic residues. (In contrast, polar residues are often found to interact with other polar residues or with the aqueous solvent and are sometimes termed hydrophilic residues.)
   c. Electrostatic interactions: noncovalent interactions between charged groups governed by the electrostatic law.
4. A protein adopts a tertiary structure in which favorable interactions are maximized and unfavorable interactions are minimized.
   a. Although there are exceptions, most soluble proteins are globular and compact rather than extended, because a near-spherical shape minimizes the amount of protein surface exposed to water.

b. The reason for minimizing the exposed surface is that the majority of the residues in most proteins are hydrophobic, and the interactions of hydrophobic side chains with solvent water is unfavorable.
   i. The hydrocarbonlike amino acid residues tend to be found on the interior of a protein, away from solvent water.
   ii. The polar residues tend to be on the exterior of a protein, where they can form hydrogen bonds with water.
5. When a protein is denatured, it is converted entirely into a random-coil structure; chemical denaturation of a protein is typically brought about by breaking its disulfide bonds with thiols, such as DTT or 2-mercaptoethanol, and treating it with 8 *M* urea, detergents, or heat.
6. The amino acid sequence of a protein specifies its conformation; that is, the native structure is the most stable structure. In many cases it appears that primary structure dictates tertiary structure.
7. Some proteins are aggregates of other proteins, called subunits; the description of subunit arrangement in a protein is called quaternary structure.

IV. ENZYMES: BIOLOGICAL CATALYSTS

1. Many naturally occurring proteins are enzymes, which are catalysts for biological reactions.
   a. Enzymes are true catalysts; their concentrations are typically much lower than the concentrations of the compounds in the reactions they catalyze.
   b. They do not affect the equilibrium constants of the reactions they catalyze.
   c. They catalyze equally both the forward and reverse reactions of an equilibrium.
2. Two characteristics of all enzymes are
   a. catalytic efficiency—the rate acceleration caused by enzyme catalysis of a particular reaction.
   b. specificity—the selectivity of an enzyme for the compounds on which it acts and for the reaction it catalyzes.
3. When an enzyme catalyzes the reaction of a certain compound, called a substrate of the enzyme, it acts in at least three stages:
   a. The enzyme binds the substrate in a noncovalent enzyme-substrate complex.
      i. The binding occurs at a part of the enzyme called the active site.
      ii. Within the active site are groups that attract the substrate by interacting favorably with it; the interactions involved are electrostatic interactions, hydrogen bonding, and van der Waals attractions (hydrophobic bonds).
   b. The enzyme promotes the appropriate chemical reaction(s) on the bound substrate to give an enzyme-product complex. The necessary chemical transformations are brought about by groups in the active site of the enzyme.
   c. The product(s) depart from the active site, the enzyme is thus ready to repeat the process on a new substrate molecule.

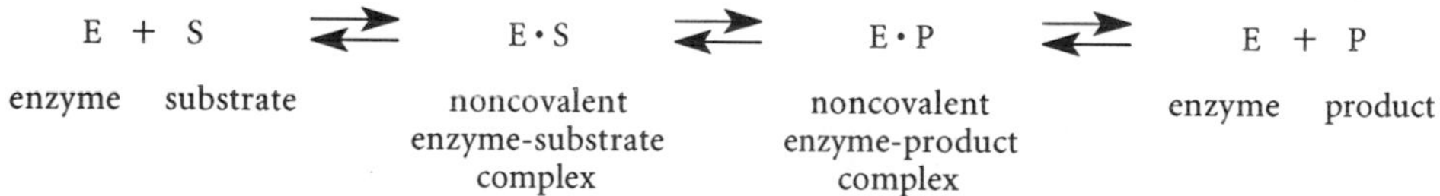

4. Enzymes interact differently with enantiomers because enzymes are chiral reagents; the enzyme-substrate complex derived from the enantiomer of the substrate is a diastereomer of the complex derived from the substrate itself.
   a. Because the two complexes are diastereomers, they have different energies, and one is more stable than the other.
   b. The rates at which these complexes are converted into products also differ.
   c. In most cases enzymes catalyze the reaction of only one enantiomer of a pair.
5. One way to "knock out" an enzyme is to subject it to an inhibitor, a compound that prevents an enzyme from fulfilling its catalytic role.
   a. The most common inhibitors are competitive inhibitors, a compound that binds to the enzyme's active site so tightly that the enzyme can no longer bind its usual substrate.

i. Inhibitors of trypsin have positively charged groups that interact with a carboxylate ion deep in the enzyme's active site, and hydrocarbon groups that interact with similar groups that line the active-site cavity.
ii. Inhibitors of HIV protease rely on van der Waals attractions and strategically placed hydrogen bonds to bind to the active site.

b. Molecular modeling, a computer-based technique that calculates the interaction energies between molecules, has been used to simulate the active sites of enzymes and to design inhibitors for those sites.

6. Most proteins are not static when they interact with small molecules; this interaction in many cases induces a conformational change in the protein. This phenomenon is termed induced fit.

V. COMBINATORIAL CHEMISTRY

1. Combinatorial chemistry is the simultaneous synthesis and assay of many compounds in such a way that compounds with a desired property can be quickly identified and their structures determined.
   a. The variable elements in a combinatorial synthesis are termed diversity elements.
      i. Each diversity element has a certain number of instances at each step; an instance is a unique way in which the diversity element is expressed.
      ii. If the same number of instances I is used at each step, and if a combinatorial synthesis had n steps, then the number of possible compounds that can be prepared in a combinatorial synthesis is I^n.
   b. A strategy used to obtain all possible peptides is called split-pool synthesis.
      i. The sample must be split into I pools at each step.
      ii. The same reaction is carried out on each pool with a different set of reagents.
      iii. The pools are combined and mixed thoroughly, and the process is repeated.
2. The simultaneous array of arrays of compounds is termed high-throughput screening.
3. The set of compounds produces in a systematic combinatorial synthesis is called a chemical library.

REACTIONS

I. SYNTHESIS AND OPTICAL RESOLUTION OF α-AMINO ACIDS

A. Alkylation of Ammonia

1. Some α-amino acids can be prepared by alkylation of ammonia with α-bromo carboxylic acids.

$$\text{—C(Br)—}CO_2H \xrightarrow{NH_3\text{ (large excess)}} \text{—C(}\overset{+}{N}H_3\text{)—}CO_2^- + \overset{+}{N}H_4 + Br^-$$

   a. This is an S_N2 reaction in which ammonia acts as the nucleophile.
   b. The use of a large excess of ammonia in the synthesis favors monoalkylation.
2. Amino acids are less reactive toward alkylating agents than simple alkylamines because the amino groups of amino acids are less basic (nucleophilic) than ammonia and simple alkylamines, and because branching in amino acids retards further alkylation.

B. Alkylation of Aminomalonate Derivatives

1. Another method for preparing α-amino acids is a variation of the malonic ester synthesis.

$$\mathrm{AcNH-\underset{H}{\overset{CO_2Et}{C}}-CO_2Et \xrightarrow[\text{2) R—X}]{\text{1) NaOEt/HOEt}} AcNH-\underset{R}{\overset{CO_2Et}{C}}-CO_2Et \xrightarrow[\text{heat}]{H_3O^+} \xrightarrow{\text{neutralize}}}$$

diethyl acetamidomalonate

$$\mathrm{H_3\overset{+}{N}-\underset{R}{CH}-CO_2^- + 2\,EtOH + CO_2\uparrow + HOAc}$$

2. The malonic ester derivative used is one in which a protected amino group is already in place: diethyl α-acetamidomalonate.
   a. Treatment of diethyl α-acetamidomalonate with sodium ethoxide in ethanol forms the conjugate base enolate ion, which is then alkylated with an alkyl halide.
   b. The resulting compound is then treated with hot aqueous HCl or HBr which accomplishes three things:
      i. The ester groups are hydrolyzed to give a substituted malonic acid.
      ii. The malonic acid derivative is decarboxylated under the reaction conditions.
      iii. The acetamido group, an amide, is also hydrolyzed.
   c. Neutralization affords the α-amino acid.

C. Strecker Synthesis

1. Hydrolysis of α-amino nitriles gives α-amino acids.

$$\mathrm{H_2N-\underset{R}{CH}-C{\equiv}N \xrightarrow[\text{heat}]{H_3O^+} H_3\overset{+}{N}-\underset{R}{CH}-CO_2H + \overset{+}{N}H_4}$$

α-amino nitrile

2. α-Amino nitriles are prepared by treatment of aldehydes with ammonia in the presence of cyanide ion, a reaction that probably involves an imine intermediate.

$$\mathrm{O{=}\underset{R}{CH} \xrightarrow[\text{NaCN}]{NH_4Cl} HN{=}\underset{R}{CH} \longrightarrow H_2N-\underset{R}{CH}-C{\equiv}N}$$

an imine (not isolated) — α-amino nitrile

   a. The conjugate acid of the imine reacts with cyanide under the conditions of the reaction to give the α-amino nitrile. (See Eq. 26.14a–b on text p. 1231.)
   b. The addition of cyanide to an imine is analogous to the formation of a cyanohydrin from an aldehyde or ketone.

3. The preparation of α-amino acids by α-amino nitrile formation followed by hydrolysis is called the Strecker synthesis.

D. Enantiomeric Resolution of α-Amino Acids

1. α-Amino acids synthesized by common laboratory methods are racemic; since many applications require the pure enantiomers, the racemic compounds must be resolved.

2. An alternative approach to the preparation of enantiomerically pure amino acids is the synthesis of amino acids by microbiological fermentation.

3. Certain enzymes can be used to resolve racemic amino acid derivatives by selectively catalyzing a reaction of one enantiomer.

II. REACTIONS OF AMINO ACIDS AND SYNTHESIS OF PEPTIDES

A. Acylation and Esterification

1. Amino acids undergo many of the characteristic reactions of both amines and carboxylic acids.
   a. They can be acylated by acid chlorides or anhydrides.
   b. They are easily esterified by heating with an alcohol and a strong acid catalyst.

B. Solid-Phase Peptide Synthesis—General

1. In a method called solid-phase peptide synthesis, the carboxy-terminal amino acid is covalently anchored to an insoluble polymer, called a resin, and the peptide is "grown" by adding one residue at a time to this polymer.
   a. Solutions containing the appropriate reagents are allowed to contact the resin with shaking.
   b. At the conclusion of each step, the resin containing the peptide is simply filtered away from the solution, which contains soluble by-products and impurities.
   c. The completed peptide is removed from the resin by a reaction that cleaves the peptide-resin bond.
2. The advantage of the solid-phase method is the ease with which dissolved impurities and by-products are removed from the resin-bound peptide by simple filtration.
   a. The same reagents used in solid-phase peptide synthesis can also be used for peptide synthesis in solution, but removal of by-products from the product peptide is sometimes difficult.
   b. In order to avoid impurities, each step in the solid-phase synthesis must occur with virtually 100% yield, an ideal that is often approached in practice.

B. Solid-Phase Peptide Synthesis—Fmoc Protecting Group

1. In a solid-phase peptide synthesis, the amino group of the amino acids is protected with a special acyl group, the (9-fluorenylmethyloxy)carbonyl (Fmoc) group. (For complete details, see Sec. 26.7, text pages 1246–1254.)
   a. Fmoc-Amino acids are prepared by allowing an *N*-hydroxysuccinimide ester derivative, Fmoc-NHS, to react with the amino group of the amino acid.

$$\text{Fmoc-NHS} + H_2N\text{—CH(R)—C(=O)—O}^- \xrightarrow[\text{MeOCH}_2\text{CH}_2\text{OMe/H}_2\text{O (solvent)}]{\text{Na}_2\text{CO}_3} \text{Fmoc-amino acid} + N\text{-hydroxysuccinimide}$$

Fmoc-NHS (fluorenyl—CH_2O—C(=O)—O—N(succinimide)); an α-amino acid; Fmoc-amino acid (fluorenyl—CH_2O—C(=O)—NH—CH(R)—C(=O)—O^-); *N*-hydroxysuccinimide (HO—N(succinimide))

   b. An S_N2 reaction between the cesium salt of an Fmoc-amino acid and the chloromethyl group of the resin results in the formation of an ester linkage to the resin. (See Eq. 26.24, text p. 1247, for a more detailed structure of the resin.)

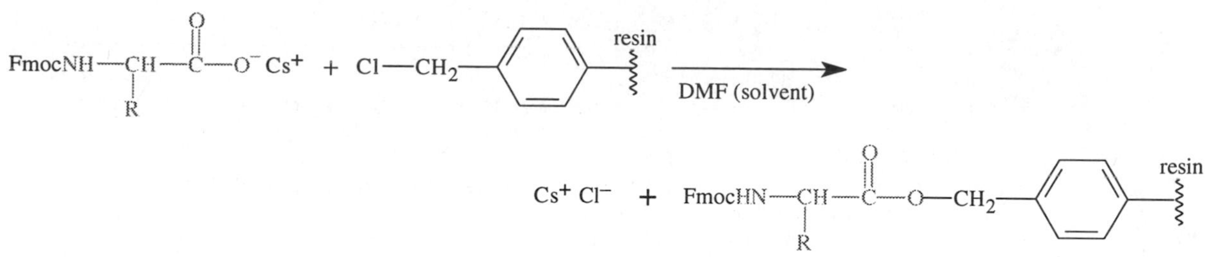

FmocNH—CH(R)—C(=O)—O⁻ Cs^+ + Cl—CH_2—C₆H₄—resin $\xrightarrow{\text{DMF (solvent)}}$ Cs^+ Cl^- + FmocHN—CH(R)—C(=O)—O—CH_2—C₆H₄—resin

2. Once the Fmoc-amino acid is anchored to the resin, the Fmoc-protecting group is removed by treatment with an amine base, piperidine, in an E2 reaction.

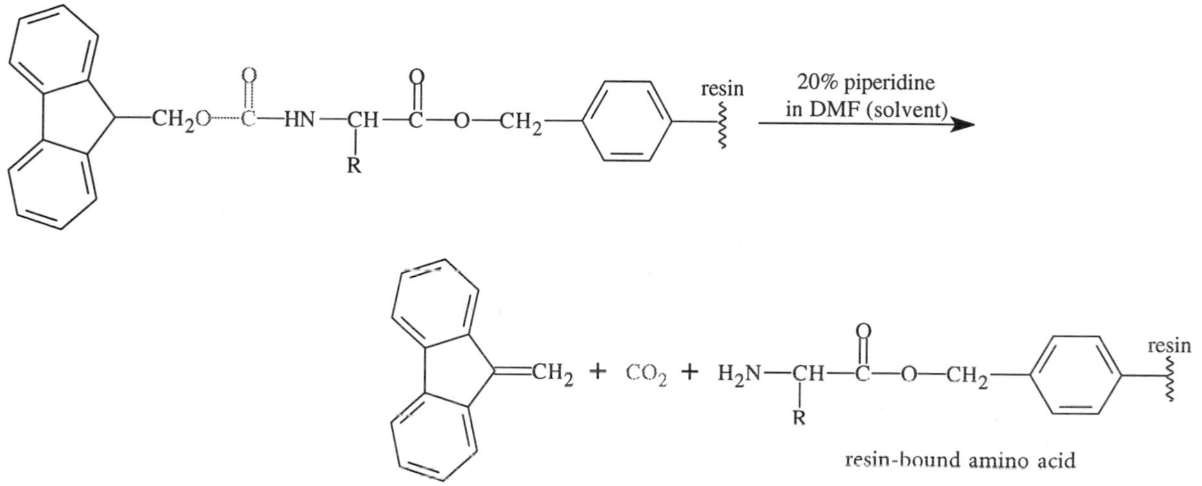

Fluorenyl—CH_2O—C(=O)—HN—CH(R)—C(=O)—O—CH_2—C₆H₄—resin $\xrightarrow{\text{20\% piperidine in DMF (solvent)}}$ Fluorenylidene=CH_2 + CO_2 + H_2N—CH(R)—C(=O)—O—CH_2—C₆H₄—resin

resin-bound amino acid

a. The product of the β-elimination is a carbamate anion.
b. The carbamate anion decarboxylates under the reaction conditions.

3. Coupling of the next Fmoc-amino acid to the free amino group of the resin-bound amino acid is effected by the reagent 1,3-dicyclohexylcarbodiimide (DCC) in the presence of *N*-hydroxysuccinimide (NHS).

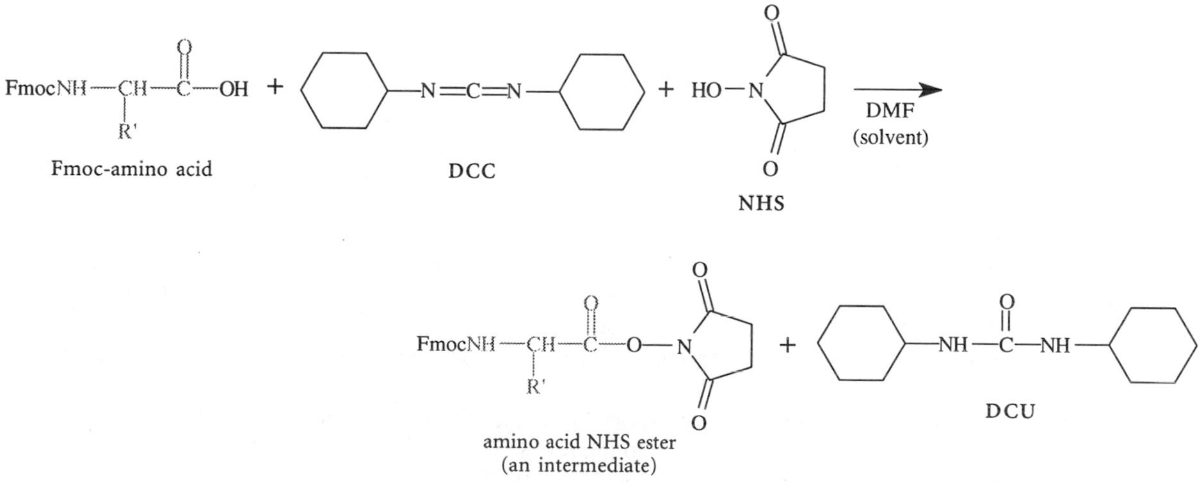

FmocNH—CH(R')—C(=O)—OH + C₆H₁₁—N=C=N—C₆H₁₁ + HO—N(succinimide) $\xrightarrow{\text{DMF (solvent)}}$

Fmoc-amino acid; DCC; NHS

FmocNH—CH(R')—C(=O)—O—N(succinimide) + C₆H₁₁—NH—C(=O)—NH—C₆H₁₁

amino acid NHS ester (an intermediate); DCU

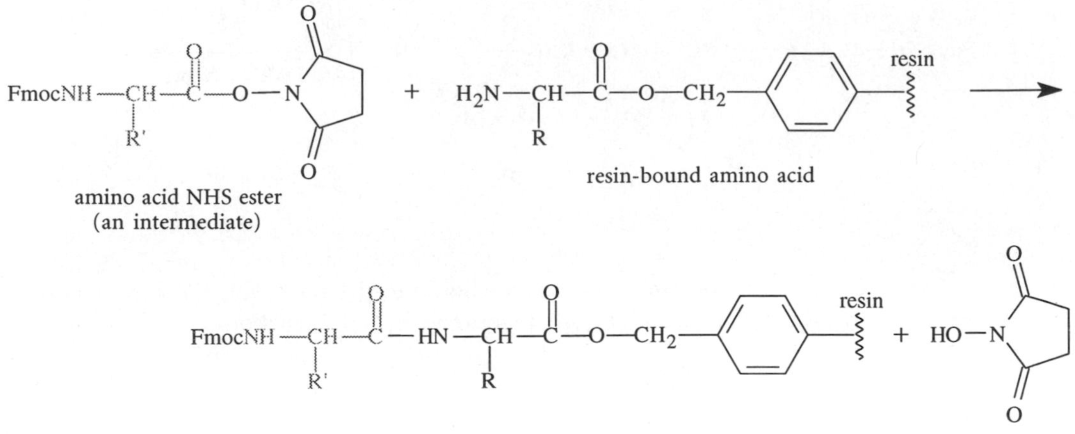

$$\text{FmocNH—CH(R')—C(=O)—O—N(succinimide)} + \text{H}_2\text{N—CH(R)—C(=O)—O—CH}_2\text{—C}_6\text{H}_4\text{—resin} \longrightarrow$$

amino acid NHS ester (an intermediate)

resin-bound amino acid

$$\text{FmocNH—CH(R')—C(=O)—HN—CH(R)—C(=O)—O—CH}_2\text{—C}_6\text{H}_4\text{—resin} + \text{HO—N(succinimide)}$$

a. DCC brings about acylation of NHS by the Fmoc-amino acid. (See Eq. 26.34a–b, text p. 1250, for more detail.)
b. The NHS ester of the Fmoc-amino acid is an intermediate that is not iisolated.
c. The NHS ester of the Fmoc-amino acid reacts with the amino group of the resin-bound amino acid to form the peptide bond (an amide linkage).

4. The ester linkage used to bind the peptide to the resin is broken by a carbocation mechanism using 50–60% trifluoroacetic acid (TFA) in dichloromethane.

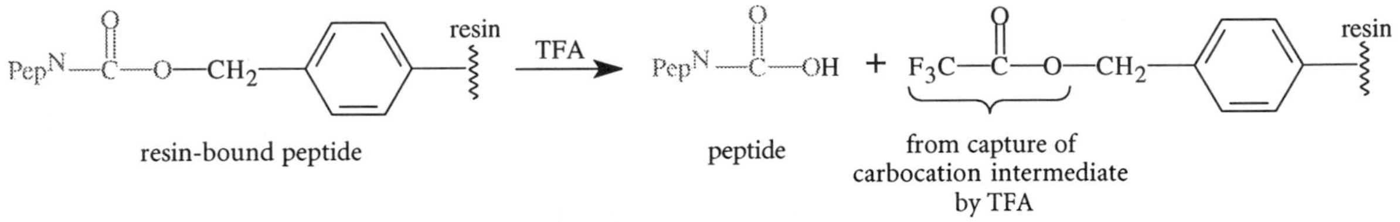

$$\text{PepN—C(=O)—O—CH}_2\text{—C}_6\text{H}_4\text{—resin} \xrightarrow{\text{TFA}} \text{PepN—C(=O)—OH} + \text{F}_3\text{C—C(=O)—O—CH}_2\text{—C}_6\text{H}_4\text{—resin}$$

resin-bound peptide; peptide; from capture of carbocation intermediate by TFA

C. Solid-Phase Peptide Synthesis—Boc Protecting Group

1. In another method of solid-phase peptide synthesis, the amino group of the amino acids is protected with another special acyl group, the *tert*-butyloxycarbonyl (Boc) group.
   a. The Boc group is introduced by allowing an amino acid to react with the anhydride di-*tert*-butyl dicarbonate.

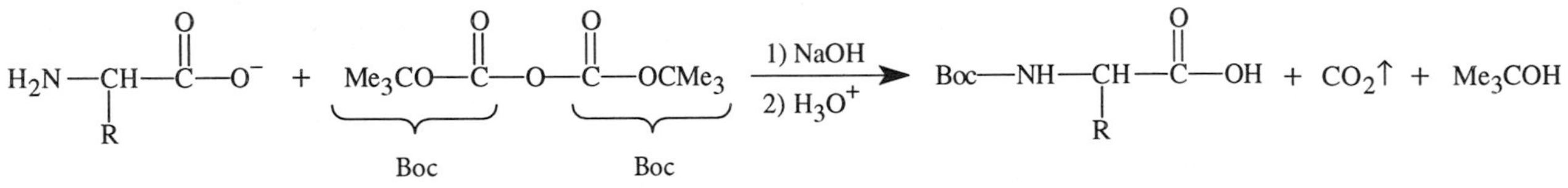

$$\text{H}_2\text{N—CH(R)—C(=O)—O}^- + \underbrace{\text{Me}_3\text{CO—C(=O)}}_{\text{Boc}}\text{—O—}\underbrace{\text{C(=O)—OCMe}_3}_{\text{Boc}} \xrightarrow[\text{2) H}_3\text{O}^+]{\text{1) NaOH}} \text{Boc—NH—CH(R)—C(=O)—OH} + \text{CO}_2\uparrow + \text{Me}_3\text{COH}$$

   b. The amino group of the amino acid rather than the carboxylate group reacts with the anhydride because the amino group is the more basic, and therefore the more nucleophilic, group.

2. At the start of the solid-phase peptide synthesis, a Boc-protected amino acid is anchored to the insoluble solid support (called a Merrifield resin) using the reactivity of its free carboxylic acid group.
   a. An S_N2 reaction between the cesium salt of the Boc-amino acid and the resin results in the formation of an ester linkage to the resin.
   b. Once the Boc-amino acid is anchored to the resin, the Boc protecting group is removed with anhydrous trifluoroacetic acid. This deprotection step, followed by neutralization of the resulting ammonium salt, exposes the free amino group of the resin-bound amino acid, which is used as a nucleophile in the next reaction.

$$\text{BocNH—CH(R)—C(=O)—O}^- \text{Cs}^+ + \text{Cl—CH}_2\text{—C}_6\text{H}_4\text{—resin} \xrightarrow[\text{3) Et}_3\text{N}]{\text{1) DMF; 2) CF}_3\text{CO}_2\text{H}} \text{H}_2\text{N—CH(R)—C(=O)—O—CH}_2\text{—C}_6\text{H}_4\text{—resin}$$

$$+ \text{(CH}_3)_3\text{C—O—C(=O)CF}_3 + \text{CO}_2$$

3. Coupling of another Boc-protected amino acid to the free amino group of the resin-bound amino acid is similar to that of an Fmoc-protected amino acid.

$$\text{BocNH—CH(R}'\text{)—C(=O)—OH} + \text{H}_2\text{N—CH(R)—C(=O)—O—CH}_2\text{—C}_6\text{H}_4\text{—resin} \xrightarrow[\text{3) Et}_3\text{N}]{\text{1) DCC/CH}_2\text{Cl}_2\text{; 2) CF}_3\text{CO}_2\text{H}}$$

$$\text{H}_2\text{N—CH(R}'\text{)—C(=O)—NH—CH(R)—C(=O)—O—CH}_2\text{—C}_6\text{H}_4\text{—resin}$$

4. Completion of the peptide synthesis requires deprotection of the resin-bound peptide in the usual way, a final coupling step with the amino-terminal Boc-amino acid and DCC, and removal of the peptide from the resin.
   a. The ester linkage that connects the peptide to the resin, like most esters, is more easily cleaved than the peptide (amide) bonds, and is typically broken by liquid HF.
   b. This acidic reagent also removes the Boc group from the product peptide.

benzylic ester linkage

$$\text{BocNH—CH(R)—C(=O)—OH} + \text{H}_2\text{N—Pep}^{\text{C}}\text{—resin} \xrightarrow[\text{3) HF}]{\text{1) DCC/CH}_2\text{Cl}_2\text{; 2) CF}_3\text{CO}_2\text{H}} \text{H}_3\overset{+}{\text{N}}\text{—CH(R)—C(=O)—NH—Pep}^{\text{C}} + \text{resin—F}$$

III. ANALYTICALLY IMPORTANT REACTIONS OF PEPTIDES

A. Hydrolysis of Peptides; Amino Acid Analysis

1. An important reaction used to determine the structures of unknown peptides is hydrolysis of the peptide (amide) bonds of a peptide to give its constituent amino acids.
2. When a peptide or protein is hydrolyzed, the product amino acids can be separated, identified, and quantitated by a technique called amino acid analysis.
   a. In the method most commonly used today, the amino acids in the hydrolyzed mixture are allowed to reacted with AQC-NHS to form AQC-amino acids.

$$\text{AQC-NHS} + \text{H}_2\text{N—CH(R)—C(=O)—O}^- \xrightarrow[\text{H}_2\text{O}]{\text{pH 9 buffer}}$$

1-[[(6-quinolylamino)carbonyl]-oxy]-2,5-pyrrolidinedione (AQC-NHS)

α-amino acid

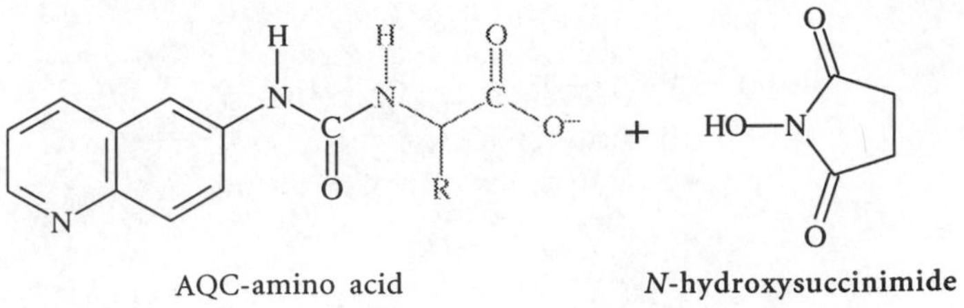

AQC-amino acid N-hydroxysuccinimide

i. The mixture of AQC-amino acids are separated by passing them through a C18 HPLC column under very carefully defined conditions; the time at which each derivatized amino acid emerges from the column is accurately known.

ii. The AQC-amino acids with the greatest hydrocarbon character are eluted last.

iii. The AQC group is fluorescent; the absorption of light at 254 nm (UV detector) or emission of light at 395 nm (fluorescence detector) is used to determine the relative amounts of each amino acid present.

b. In a second and older method, the amino acids in the hydrolyzed mixture are separated by passing them through a cation-exchange column under very carefully defined conditions; the time at which each amino acid emerges from the column is accurately known.

i. As each amino acid emerges, it is mixed with ninhydrin to give an intense blue-violet dye called Ruhemann's purple; the intensity of the resulting color is proportional to the amount of the amino acid present.

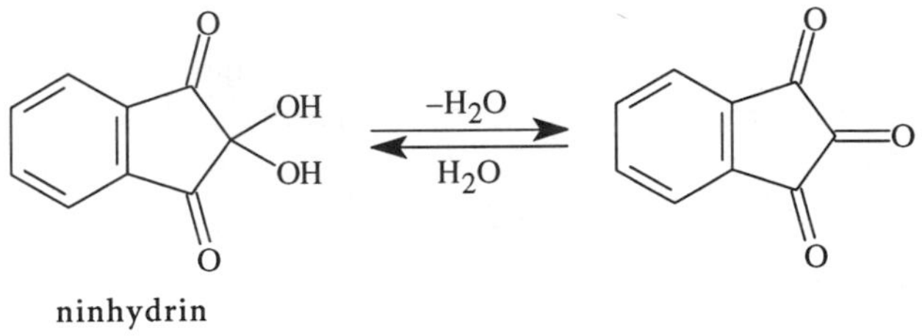

$-H_2O$ / H_2O

ninhydrin

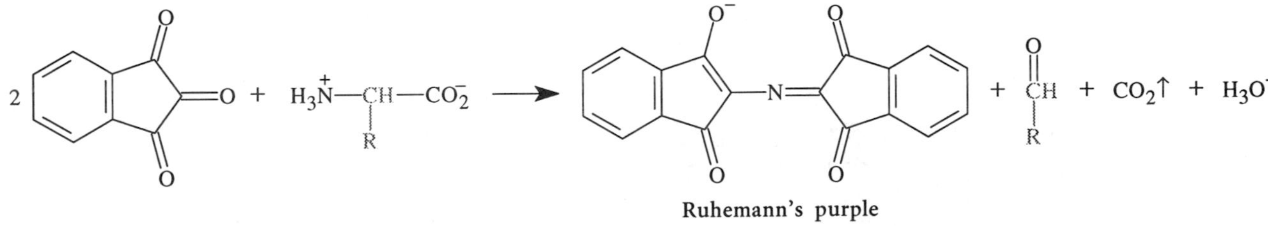

$2 \text{ (indanetrione)} + H_3\overset{+}{N}\text{—CH(R)—}CO_2^- \longrightarrow \text{Ruhemann's purple} + RCHO + CO_2\uparrow + H_3O^+$

Ruhemann's purple

ii. The color intensity is recorded as a function of time; the area of the peak is proportional to the amount of the amino acid.

c. Mass spectrometry may be used to determine the sequence of short peptides since peptides undergo residue-by-residue fragmentation, in many cases from both ends simultaneously, in a mass spectrometer.

3. Amino acid analysis can determine the identity and relative amounts of amino acid residues, but not their relative orders within a peptide.

B. Sequential Degradation of Peptides

1. The actual arrangement, or sequential order, of amino acid residues in a peptide is called the amino acid sequence, or primary sequence, of the peptide.
2. It is possible to remove one residue at a time from the amino end of the peptide, identify it, and then repeat the process sequentially on the remaining peptide; the standard method for implementing this strategy is called the Edman degradation.

a. The peptide is treated with phenyl isothiocyanate (Edman reagent), with which it reacts at its amino groups to give a thiourea derivative. (Only the reaction at the terminal amino group is relevant to the degradation.)
b. Any remaining phenyl isothiocyanate is removed, and the modified peptide is then treated with anhydrous trifluoroacetic acid.

Ph—N=C=S + H_2N—CH(R)—C(=O)—NH—Pep^C →[1) pyridine/H_2O/Me_2NPh (gives the thiourea derivative); 2) CF_3CO_2H]

phenyl isothiocyanate (Edman reagent)

$H_3\overset{+}{N}$—Pep^C + (thiazolinone: Ph—NH, S, N, O, R) →[H_3O^+] (S, Ph, N, H—N, O, R)

peptide shorter by one residue

a thiazolinone

a phenylthiohydantoin (PTH) derivative of the amino terminal residue

c. The sulfur of the thiourea, which is nucleophilic, displaces the amino group of the adjacent residue to yield a five-membered heterocycle called a thiazolinone; the other product of the reaction is a peptide that is one residue shorter.
d. When treated subsequently with aqueous acid, the thiazolinone derivative forms an isomer called a phenylthiohydantoin (PTH).
e. Because the PTH derivative carries the characteristic side chain of the amino-terminal residue, the structure of the PTH derivative identifies the amino acid residue that was removed.

3. In practice, this type of analysis is limited to 20–60 consecutive residues because the yields at each step are not perfectly quantitative.

C. Specific Cleavage of Peptides: Introduction

1. The amino acid sequence of most large proteins is determined by breaking the protein into smaller peptides and sequencing these peptides individually. (The sequence of the protein is reconstructed from the sequences of the peptides.)

2. When breaking a larger protein into smaller peptides, it is desirable to use reactions that cleave the protein in high yield at well-defined points so that a relatively small number of peptides are obtained.

3. One method (cleavage with cyanogen bromide) uses ordinary chemical reagents; another method involves the use of enzymes to catalyze peptide-bond hydrolysis.

D. Peptide Cleavage at Methionine with Cyanogen Bromide

1. When a peptide reacts with cyanogen bromide (Br—C≡N) in aqueous HCl, a peptide bond is cleaved specifically at the carboxy side of each methionine residue.

Pep^N—NH—CH(CH_2CH_2SMe)—C(=O)—NH—Pep^C →[BrCN] Pep^N—NH—CH($CH_2CH_2\overset{+}{S}Me$—C≡N)—C(=O)—NH—Pep^C →

a sulfonium ion

$$MeSCN \; + \; Pep^N{-}NH{-}(\text{iminolactone ring}){=}\overset{+}{N}H{-}Pep^C \xrightarrow{H_3O^+} Pep^N{-}NH{-}(\text{lactone ring}){=}O \; + \; H_3\overset{+}{N}{-}Pep^C$$

homoserine lactone

2. Although cyanogen bromide has the character of an acid chloride, under acidic conditions only its reaction at methionine leads to a peptide cleavage.
   a. The sulfur in the methionine side chain acts as a nucleophile, displacing bromide from cyanogen bromide to give a type of sulfonium ion.
   b. The sulfonium ion, with its electron-withdrawing cyanide, is an excellent leaving group, and is displaced by the oxygen of the neighboring amide bond to form an imminium ion containing a five-membered ring; only methionine has a side chain that can form a five-membered ring by such a mechanism.
   c. Hydrolysis of the imminium ion formed cleaves the peptide bond.
3. The amino-terminal fragment formed in the cleavage has a carboxy-terminal homoserine lactone residue instead of the starting methionine.
4. Methionine is a relatively rare amino acid; hence, when a typical protein is cleaved with BrCN, relatively few cleavage peptides are obtained, and all of them are derived from cleavage at methionine residues.

E. Peptide Cleavage with Proteolytic Enzymes

1. A number of enzymes (called proteases, peptidases, or proteolytic enzymes) catalyze the hydrolysis of peptide bonds at specific points in an amino acid sequence.
2. One of the most widely used proteases is the enzyme trypsin, which catalyzes the hydrolysis of peptides or proteins at the carbonyl group of arginine or lysine residues provided that—
   a. these residues are not at the amino end of the protein.
   b. these residues are not followed by a proline residue.
3. Because trypsin catalyzes the hydrolysis of peptides at internal rather than terminal residues, it is called an endopeptidase; enzymes that cleave peptides only at terminal residues are termed exopeptidases.
4. Chymotrypsin, a protein related to trypsin, is used to cleave peptides at amino acid residues with aromatic side chains and, to a lesser extent, residues with large hydrocarbon side chains; thus, chymotrypsin cleaves peptides at Phe, Trp, Tyr, and in some cases Leu and Ile residues.

SCIENTISTS

Christian Boehmer Anfinsen (1916–1995) [Chemistry 1972] 26.9C
Louis Albert Carpino (b. 1927) 26.7
William Wallace Cleland (b. 1930) 26.9A
Pehr Victor Edman (1916–1977) 26.6B
Hendrik Mario Geysen (b. 1944) 26.8A
George Simms Hammond (b. 1921) 26.8A
Robert Bruce Merrifield (b. 1921) [Chemistry 1984] 26.7
Linus Carl Pauling (1901–1994) [Chemistry 1954, Peace 1962] 26.9B
Seigfried Ruhemann (1895–1945) 26.6A
Adolf Friedrich Ludwig Strecker (1822–1871) 26.4C

Dates in brackets refer to Nobel Prize awards.

STUDY GUIDE LINKS

✓26.1 Neutral Amino Acids

It is very important in this chapter to make a clear distinction between the terms *neutral* and *uncharged* when they are used with amino acids and their derivatives. A *neutral* amino acid can have charged atoms; if so, then it is a *zwitterion*, that is, it bears two equal charges of opposite sign (+ and –). An uncharged amino acid has no atoms bearing formal charge.

✓26.2 Names of the Amino Acids

A student in the author's organic chemistry course proposed the following mnemonics for remembering the twenty common naturally occurring amino acids. (The groups are those listed on text p. 1217.)

Group 1 and proline:	LIAGVP (Loudon *Is A G*hastly *V*olleyball *P*layer)
Group 2:	HTTP (*H*ail *T*o *T*errific *P*urdue)
Group 3:	CTMS (*C*hemistry *T*akes *M*uch *S*tudy)
Groups 4 and 5:	AGAGLA (*A*ll *G*irls *A*nd *G*uys *L*ove *A*rtichokes)

The substitution of other adjectives in the group 1 and group 2 mnemonics are, of course, acceptable, provided that they begin with the same letter.

26.3 Reaction of α-Amino Acids with Ninhydrin

To understand the reaction of ninhydrin with α-amino acids, first notice that *two* molecules of ninhydrin react per molecule of amino acid. Second, notice that ninhydrin is the hydrate of a tricarbonyl compound.

[indane-1,2,3-trione] + H_2O ⇌ [2,2-dihydroxyindane-1,3-dione]

ninhydrin

The ninhydrin reaction with α-amino acids is a combination of simpler reactions that you have already studied. The first step is the formation of an imine (Sec. 19.11A). In this reaction, the amino group of the amino acid reacts with the central carbonyl group of the tricarbonyl form of ninhydrin. (Why is the central carbonyl group more reactive than the other carbonyl groups? See Sec. 19.7B in the text.)

[indane-1,2,3-trione] + $H_2N{-}CH(R){-}CO_2^-$ ⟶ [indane-1,3-dione]$={N}{-}CH(R){-}CO_2^-$ + H_2O

an imine

This imine is converted into a different imine by loss of CO_2 (decarboxylation), and the new imine hydrolyzes to an aldehyde and an amine. (Recall that imine formation is reversible.) Note that the aldehyde bears the side-chain R of the original amino acid.

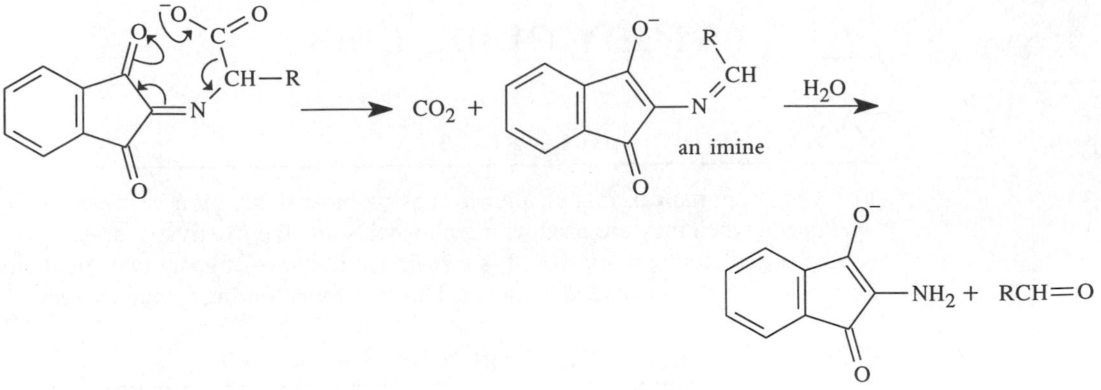

CO_2 + an imine $\xrightarrow{H_2O}$ NH_2 + RCH=O

The resulting amine then forms Ruhemann's purple, the final product, by reacting with a second molecule of ninhydrin in another imine-forming reaction.

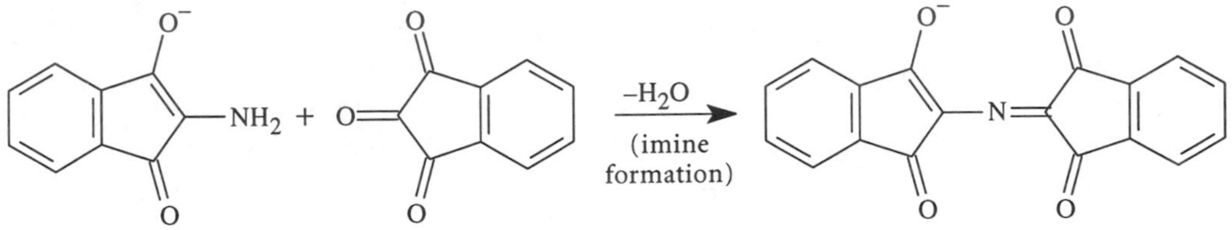

$\xrightarrow{-H_2O}$ (imine formation)

Ruhemann's purple

Simple primary amines also react with ninhydrin to give Ruhemann's purple. The only alteration in the mechanism is the loss of a proton instead of decarboxylation in the first equation at the top of this page.

26.4 Solid-Phase Peptide Synthesis

The reactions of solid-phase peptide synthesis must be carried out in solvents that *solvate* the Merrifield resin, that is, solvents that interact well with the groups of the resin. If these groups are not solvated, reagents in solution cannot penetrate the resin and no reaction takes place. Because the polystyrene-based Merrifield resin is essentially a polymeric ethylbenzene, it is mostly a hydrocarbon. Yet many of the reactions of solid-phase peptide synthesis involve ionic compounds and intermediates. Water and alcohols do not work as solvents in solid-phase peptide synthesis because they do not solvate the resin. Dimethylformamide, *N*-methylpyrrolidone, and methylene chloride have been found empirically to solvate the Merrifield resin adequately and to dissolve the reagents used in solid-phase peptide synthesis. In fact, when such solvents are added to a Merrifield resin, a visually perceptible swelling of the resin bed occurs; this is caused by incorporation of the solvent into the resin.

The use of a cesium salt in Eq. 26.30 on text p. 1248 is related to the requirement for aprotic solvents. In DMF, ionic compounds tend to form ion pairs and other aggregates, and carboxylate salts have metal-oxygen bonds with a significant amount of covalent character. Yet the alkylation shown in this equation is fastest when the carboxylate oxygen has as much *ionic character* as possible, because a carboxylate oxygen with a full negative charge is more nucleophilic and therefore reacts much more rapidly in S_N2 reactions than an ion-paired (or covalently bonded) carboxylate oxygen. Because cesium is the most electropositive of the readily available alkali metals, its carboxylate salts have more ionic character than lithium, sodium, or potassium salts and hence are most reactive in S_N2 reactions.

✓26.5 An All-D Enzyme

Although all naturally occurring enzymes are proteins that consist entirely of L-amino acids, a protein consisting entirely of D-amino acids is conceivable. What would be predicted for the catalytic activity of such a protein? It should catalyze only the reactions of the enantiomers of the substrates of its all-L counterpart, but on those enantiomeric substrates, its catalytic activity should be identical to that of the all-L enzyme.

In 1993 this idea was tested for the very first time by the synthesis of HIV protease, a small protein from human immunodeficiency virus (HIV), the virus that causes AIDS. The viral protein as it occurs in nature consists of only L-amino acids, and a research group at the Scripps Research Institute in La Jolla, California, had chemically synthesized this protein using solid-phase methods. The same group prepared the enantiomeric protein from D-amino acids and found that this enzyme has no catalytic activity on substrates of the L-enzyme, but has identical activity on *enantiomeric* substrates. This finding, of course, was completely expected. However, what made the work worth doing is that the expectation stems from the fundamental principles of stereochemistry, heretofore untested with enzymes. Had the expectation *not* been realized, these principles would have been called into question.

SOLUTIONS TO PROBLEMS

Solutions to In-Text Problems

26.1 Each peptide is drawn in the form that exists at neutral pH.

(a) The structure of tryptophylglycylisoleucylaspartic acid:

tryptophylglycylisoleucylaspartic acid
(Trp-Gly-Ile-Asp, or W-G-I-D)

(b) The structure of Glu-Gln-Phe-Arg:

Glu-Gln-Phe-Arg
(glutamylglutaminylphenylalanylarginine, or E-Q-F-R)

26.2 This peptide is His-Ile-Tyr-Met-Ser (histidylisoleucyltyrosylmethionylserine, or H-I-Y-M-S).

26.3 A Fischer projection of L-isoleucine:

(b) L-Alloisoleucine has the 2*S* configuration, that is, the same configuration at the α-carbon as L-leucine, but has a different configuration at the β-carbon. D-Alloisoleucine is the enantiomer of L-alloisoleucine.

L-alloisoleucine D-alloisoleucine

26.4 (a) The α-carbon of L-cysteine has the *R* configuration.

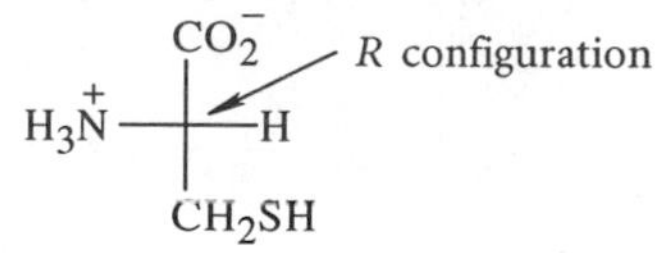

CO_2^- ← *R* configuration

$H_3\overset{+}{N}$—C—H

CH_2SH

L-cysteine

(b) In terms of the sequence rules for determining configuration, the two carbons attached to the α-carbons of cysteine or serine can be represented as C(O,O,O) and C(X,H,H), where X = S or O, respectively. When X = S (cysteine), the latter carbon receives priority. When X = O (serine), the former receives priority. Sulfur receives priority over oxygen because it has higher atomic number, and the priority is decided *at the first point of difference.* (See Sec. 4.2B.)

26.5 (a) In the major neutral form of A-K-V-E-M (Ala-Lys-Val-Glu-Met), all amino groups are protonated and all carboxy groups are ionized.

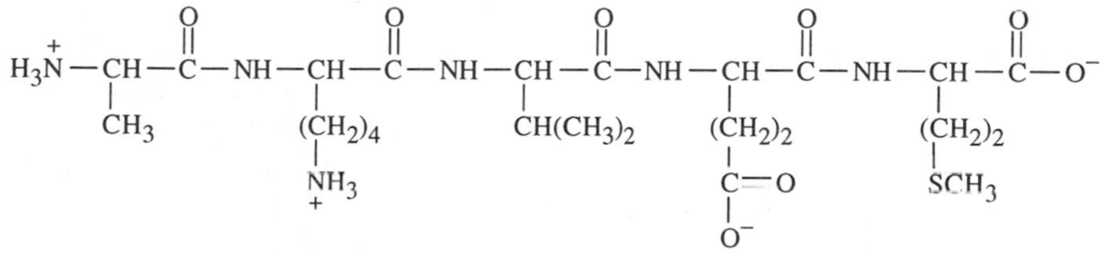

$H_3\overset{+}{N}$—CH(CH_3)—C(=O)—NH—CH($(CH_2)_4\overset{+}{N}H_3$)—C(=O)—NH—CH($CH(CH_3)_2$)—C(=O)—NH—CH($(CH_2)_2$C(=O)$O^-$)—C(=O)—NH—CH($(CH_2)_2SCH_3$)—C(=O)—$O^-$

(b) In the major neutral form of G-D-G-L-F (Gly-Asp-Gly-Leu-Phe), the side-chain carboxy group of Asp is un-ionized and the carboxy-terminal carboxy group is ionized because it has the lower pK_a of the two carboxy groups (Table 26.1). This form would exist at a pH between the pK_a values of the two carboxy groups. That is, the pH at which this neutral form predominates is relatively acidic. (See Sec. 26.3B.) Notice that the major neutral form of a peptide (such as this one) may not be the form that predominates at *neutral pH*.

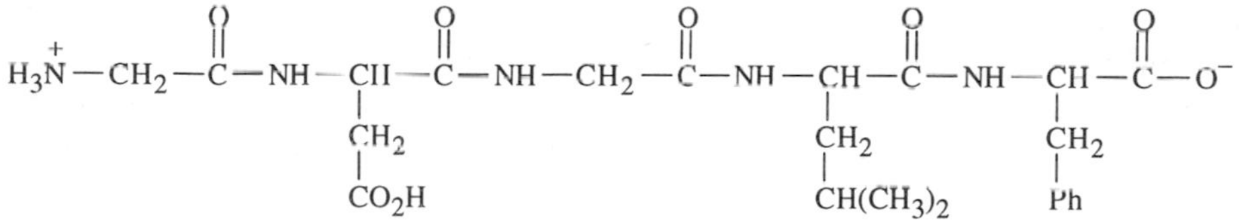

$H_3\overset{+}{N}$—CH_2—C(=O)—NH—CH(CH_2CO_2H)—C(=O)—NH—CH_2—C(=O)—NH—CH($CH_2CH(CH_3)_2$)—C(=O)—NH—CH(CH_2Ph)—C(=O)—O^-

26.6 (a) The amino group of tyrosine can be protonated, and both the carboxy group and the phenolic O—H group can be ionized.

(b) At pH 6, the net charge on tyrosine is zero. A pH value of 6 is below the pK_a of the conjugate acid of the amino group, and the amino group is therefore protonated; a pH value of 6 is above the pK_a of the carboxy group, which is therefore ionized; and a pH value of 6 is below the pK_a of the phenolic O—H group, which is therefore un-ionized.

(c) The structure of tyrosine in aqueous solution at pH 6:

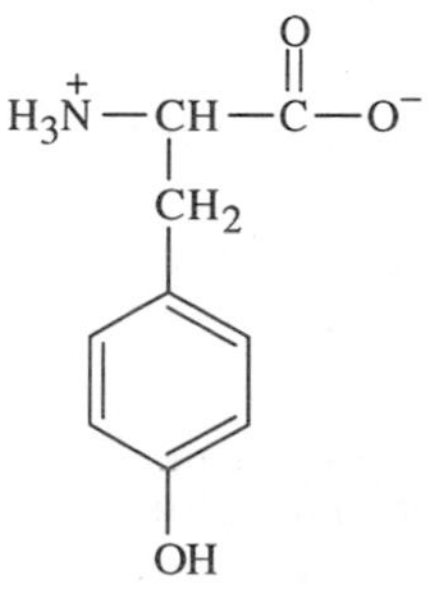

$H_3\overset{+}{N}$—CH(CH_2—C_6H_4—OH)—C(=O)—O^-

26.7 (a) and (b) The acid-base equilibria of histidine are as follows. The ionizable groups are the ones that are charged in any of the species shown in these equilbria and are indicated with asterisks (*).

$$H_3\overset{+}{N}{*}-CH(CH_2\text{-imidazolium})-C(=O)-OH \rightleftharpoons H_3\overset{+}{N}{*}-CH(CH_2\text{-imidazolium})-C(=O)-O^{-}{*} \rightleftharpoons H_3\overset{+}{N}{*}-CH(CH_2\text{-imidazole})-C(=O)-O^{-}{*} \rightleftharpoons H_2N-CH(CH_2\text{-imidazole})-C(=O)-O^{-}{*}$$

major forms at pH 6 (equal amounts)

increasing pH

Don't forget that a protonated imidazole has two resonance structures, which are shown in Eq. 24.19 on text p. 1134. If you drew the protonated histidines in the peptide above with the other resonance structure, it is equally correct.

(c) and (d) The nitrogen of the α-amino group is protonated at pH 6 because a pH value of 6 is below the pK_a value of this ammonium ion. The conjugate-acid pK_a of the imidazole group is 6 (Table 26.1 on text p. 1218). Consequently, at pH 6, this group is "half-protonated," that is, half of the histidine side-chain imidazoles in solution are protonated. The carboxy group is ionized at pH 6 because its pK_a value is below 6. The major forms of histidine at pH 6 are labeled in the equilibria shown above in the solution to parts (a) and (b). Because there are equal amounts of the two forms with net charges of +1 and 0, respectively, the overall net charge on histidine in solution at pH 6 is +0.5.

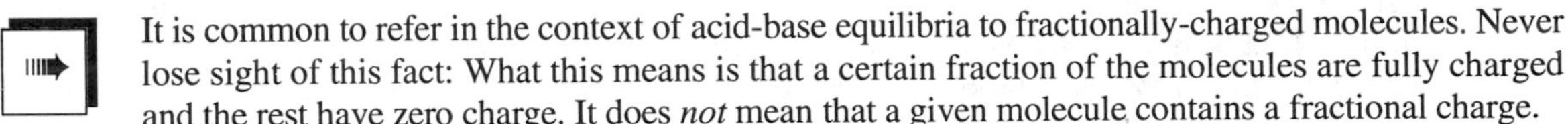

It is common to refer in the context of acid-base equilibria to fractionally-charged molecules. Never lose sight of this fact: What this means is that a certain fraction of the molecules are fully charged and the rest have zero charge. It does *not* mean that a given molecule contains a fractional charge.

26.8 The general rule of thumb is that if the peptide contains more acidic than basic groups it is an acidic peptide; if it contains more basic than acidic groups it is a basic peptide; and if the number of acidic and basic groups are equal, the peptide is neutral. (An acidic group is a group that is in its conjugate-base form at neutral pH, such as a carboxy group; a basic group is a group that is in its conjugate-acid form at neutral pH, such as an amino group.)

(a) The peptide is neutral because it contains one basic group (the terminal amino group, which is protonated at pH 6) and one acidic group (the terminal carboxy group, which is ionized at pH 6). Its net charge at pH 6 is 0.

(b) This peptide contains three basic groups: the α-amino group of the amino-terminal residue Leu and the side-chain amino groups of the Lys residues. The peptide contains one acidic group: the carboxy group of the carboxy-terminal Lys residue. Because the peptide contains more basic residues than acidic residues; it is a basic peptide. Its net charge at pH 6 is +2.

(c) Acetylation eliminates the basicity of the terminal amino group. The peptide contains two basic groups (the Arg residues) and two acidic groups (the Asp residue and the terminal carboxy group). Thus, the peptide is neutral and its net charge is 0.

(d) This peptide contains three acidic groups: the side-chain carboxy groups of the Asp and Glu residues and the carboxy group of the carboxy-terminal Ile residue. The peptide contains two basic groups: the α-amino group of the amino-terminal residue Glu and the side-chain amino group of the Lys residue. Because the peptide contains more acidic residues than basic residues, it is an acidic peptide. Its net charge at pH 6 is −1.

26.9 Use Eqs. 26.5 on text p. 1224. Multiplication of the two equations gives

$$\frac{K_{a1}K_{a2}}{[H_3O^+]^2} = \frac{[N][B]}{[A][N]}$$

When the pH is equal to the isoelectric point, by definition $[A] = [B]$. Substituting this equality, and letting $[H_3O^+]_i$ be the hydronium-ion concentration at this pH, the equation above becomes

$$\frac{K_{a1}K_{a2}}{[H_3O^+]_i^2} = 1$$

or

$$K_{a1}K_{a2} = [H_3O^+]_i^2$$

Taking negative logarithms of both sides and letting $-\log\,[H_3O^+]_i = pI$,

$$pK_{a1} + pK_{a2} = 2pI$$

Division by 2 gives Eq. 26.7a on p. 1225:

$$\frac{pK_{a1} + pK_{a2}}{2} = pI$$

26.10 (a) A resin containing a cationic group will serve as an anion exchanger. A quaternary ammonium salt is a suitable cationic group:

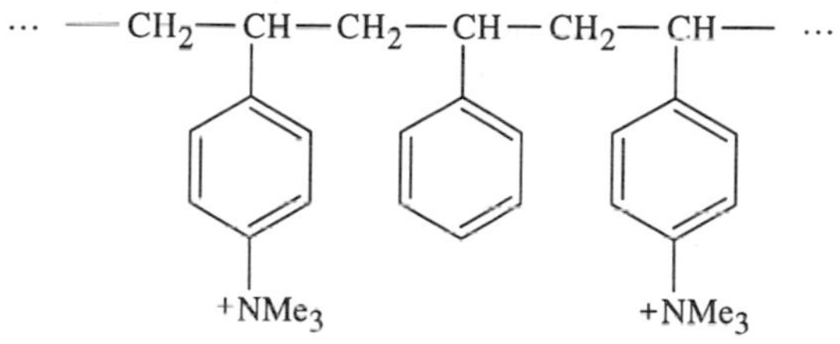

$$\cdots\ \text{—}CH_2\text{—}CH\text{—}CH_2\text{—}CH\text{—}CH_2\text{—}CH\text{—}\ \cdots$$

(each CH bears a benzene ring; the first and third rings carry $^+NMe_3$ para substituents)

(b) The ion-exchange column is positively charged, as shown by the solution to part (a). At pH 6, the peptide A-V-G has a net charge of zero, is not retained by the column, and will emerge first, the peptide D-N-N G has a net charge of –1, is retained by the column and will emerge next; and the peptide D-E-E-G has a charge of –3, is most strongly retained by the column, and will emerge last.

26.11 At pH 6 the ion-exchange column is negatively charged because it bears strongly acidic sulfonic acid pendant groups. At this pH, the peptide Ac-Leu Gly bears a negative charge, is repelled by the column, and will emerge first; Lys-Gly-Leu has a net charge of +1, is attracted to the column, and will emerge next; and Lys-Gly-Arg has a net charge of +2, is most strongly attracted to the column, and will emerge last.

26.12 (a) Alkylation of ammonia by α-bromophenylacetic acid will work, particularly since the alkyl halide is benzylic.

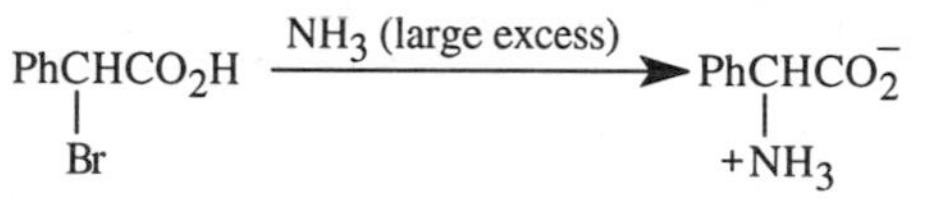

$$\underset{\alpha\text{-bromophenylacetic acid}}{PhCH(Br)CO_2H} \xrightarrow{NH_3\ \text{(large excess)}} \underset{\alpha\text{-phenylglycine}}{PhCH(\overset{+}{N}H_3)CO_2^-}$$

The acetamidomalonate method would not work because it would require alkylation of a malonate anion by bromobenzene. Bromobenzene does not undergo S_N2 reactions. The Strecker synthesis would work.

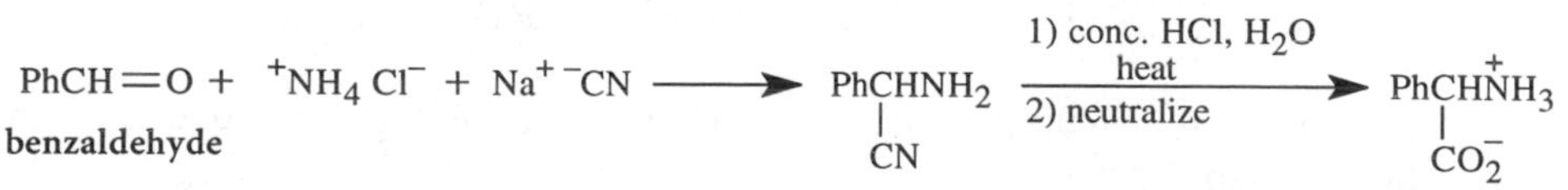

$$\underset{\text{benzaldehyde}}{PhCH{=}O} + {}^+NH_4\,Cl^- + Na^+\,{}^-CN \longrightarrow PhCH(CN)NH_2 \xrightarrow[\text{2) neutralize}]{\text{1) conc. HCl, }H_2O\text{, heat}} \underset{\alpha\text{-phenylglycine}}{PhCH(CO_2^-)\overset{+}{N}H_3}$$

(b) All three methods would work in principle for leucine.

Alkylation of ammonia:

$$\underset{\text{2-bromo-4-methylpentanoic acid}}{(CH_3)_2CHCH_2\underset{\text{Br}}{\underset{|}{C}}HCO_2H} \xrightarrow{NH_3\text{ (excess)}} \underset{\text{leucine}}{(CH_3)_2CHCH_2\underset{+NH_3}{\underset{|}{C}}HCO_2^-}$$

Acetamidomalonate method:

$$CH_3\overset{O}{\overset{||}{C}}NHCH(CO_2Et)_2 \xrightarrow[2)\ (CH_3)_2CHCH_2Br]{1)\ NaOEt,\ EtOH} CH_3\overset{O}{\overset{||}{C}}NH\underset{CH_2CH(CH_3)_2}{\underset{|}{C}}(CO_2Et)_2 \xrightarrow[2)\ NaOH\ (dilute)]{1)\ H_3O^+,\ H_2O,\ heat} H_3\overset{+}{N}\underset{CH_2CH(CH_3)_2}{\underset{|}{C}}HCO_2^-$$

Strecker synthesis:

$$\underset{\text{3-methylbutanal}}{(CH_3)_2CHCH_2CH{=}O} + {}^+NH_4\ Cl^- + Na^+\ {}^-CN \longrightarrow (CH_3)_2CHCH_2\underset{CN}{\underset{|}{C}}HNH_2 \xrightarrow[2)\ neutralize]{1)\ conc.\ HCl,\ H_2O,\ heat} (CH_3)_2CHCH_2\underset{CO_2^-}{\underset{|}{C}}H\overset{+}{N}H_3$$

26.13 (a) Tosyl chloride, the acid chloride of a typical sulfonic acid, reacts in the same way as other acid chlorides.

$$\underset{\text{leucine}}{(CH_3)_2CHCH_2\underset{CO_2^-}{\underset{|}{C}}H\overset{+}{N}H_3} + \underset{\text{tosyl chloride}}{Cl{-}SO_2{-}C_6H_4{-}CH_3} \xrightarrow{base} \underset{N\text{-tosylleucine}}{(CH_3)_2CHCH_2\underset{CO_2H}{\underset{|}{C}}HNH{-}SO_2{-}C_6H_4{-}CH_3}$$

(b) Alanine is esterified to give its methyl ester:

$$\underset{\text{alanine methyl ester}}{CH_3\underset{CO_2CH_3}{\underset{|}{C}}H\overset{+}{N}H_3}$$

26.14 Polyglycine results from aminolysis of the ester group on one molecule by the amine on another:

$$H_2NCH_2\overset{O}{\overset{||}{C}}OCH_3 \quad H_2NCH_2\overset{O}{\overset{||}{C}}OCH_3 \xrightarrow{-CH_3OH} H_2NCH_2\overset{O}{\overset{||}{C}}{-}NHCH_2\overset{O}{\overset{||}{C}}OCH_3 \xrightarrow[H_2NCH_2CO_2CH_3]{-CH_3OH}$$

$$H_2NCH_2\overset{O}{\overset{||}{C}}{-}NHCH_2\overset{O}{\overset{||}{C}}{-}NHCH_2\overset{O}{\overset{||}{C}}OCH_3 \longrightarrow \text{etc.}$$

In this reaction the amino group of glycine acts as a nucleophile. In acidic solution the amino group of glycine is protonated and cannot act as a nucleophile; hence, the reaction does not occur.

26.15 (a) The amide group of Asn hydrolyzes to give ammonium ion and Asp. Consequently, Asn appears in amino acid analysis as Asp.

(b) Glu and Gln are not differentiated by amino acid analysis because the side-chain carboxamide group of Gln, like that of Asn in part (a), is hydrolyzed; that is, Gln is converted into Glu.

26.16 The indole side-chain of tryptophan is very hydrocarbonlike, more so than the side chain of any other amino acid. (Recall (Sec. 24.2C) that pyrrole and indole are neither basic nor acidic.) Hence, we would expect to see AQC-Trp emerge on the far right of the amino acid analysis chromatogram, beyond AQC-Tyr.

26.17 The side chains of these amino acids as well as the α-amino groups are nucleophilic. Hence, they are acylated as well by AQC-NHS. The structures of these derivatives are as follows, with the carboxylic acid group in its ionized form:

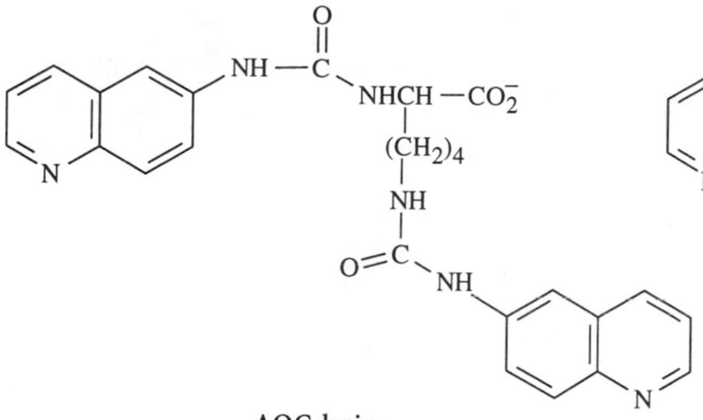

AQC-lysine

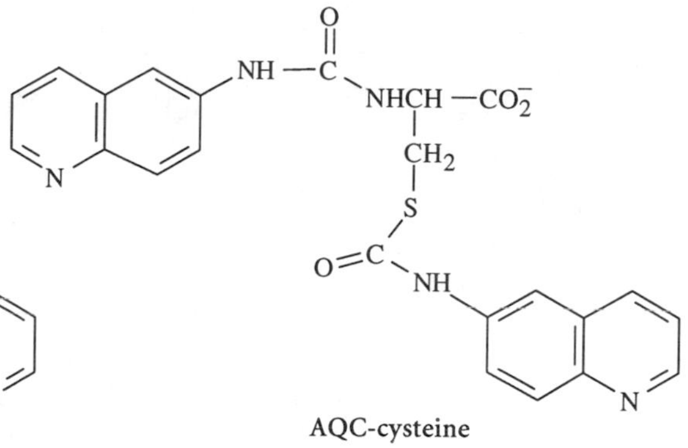

AQC-cysteine

26.18 (a) The amino group serves as a nucleophile in an addition to the isothiocyanate.

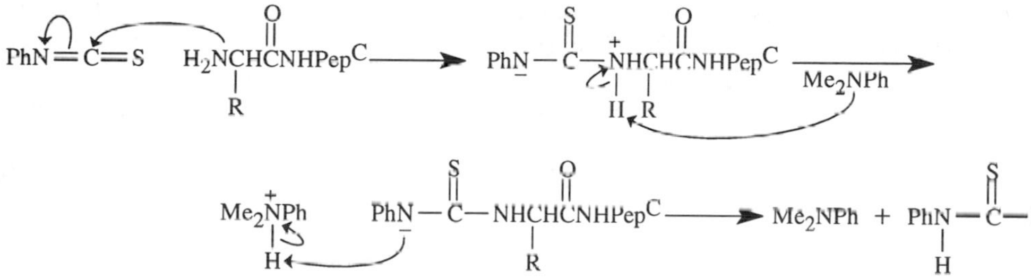

(b) The sulfur serves as the nucleophile in an intramolecular nucleophilic acyl substitution reaction. The following mechanism begins with the carbonyl-protonated thiourea derivative.

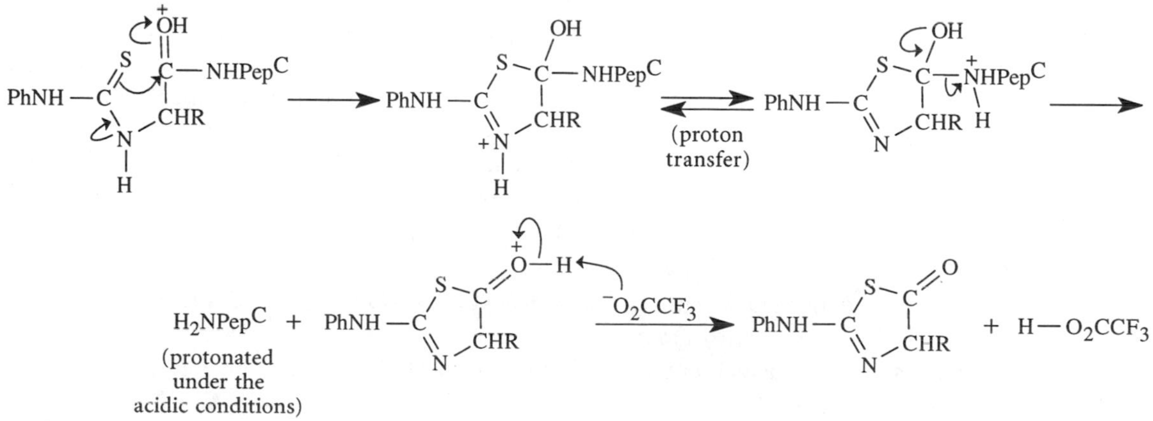

(proton transfer)

H_2NPep^C (protonated under the acidic conditions)

(c) Water serves as a nucleophile in opening the thiazolinone, and nitrogen of the resulting thiourea serves as the nucleophile in closing the ring to the PTH. The mechanism below begins with the carbonyl-protonated thiazolinone.

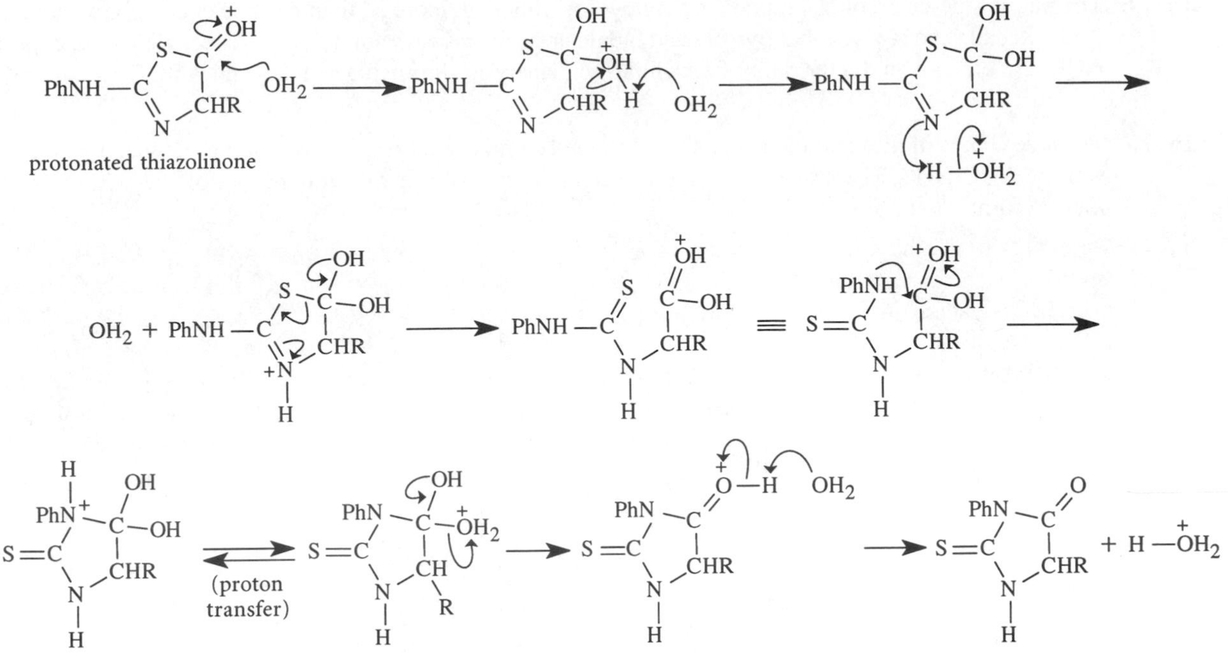

protonated thiazolinone

(proton transfer)

26.19 As Eq. 26.24a on text p. 1238 as well as its mechanism in the solution to Problem 26.18(a) show, the Edman degradation depends on the presence of a free terminal amino group in the peptide. Because this group is blocked as an amide in acetylated peptides, such peptides cannot undergo the Edman degradation. (This is one reason that the development of a carboxy-terminal degradation is an important topic of current research.)

26.20 (a) Because cyanogen bromide is an acid halide of HC≡N, it reacts with amines in the same way that other acid halides react with amines:

$$—(CH_2)_4NH_2 + Br—C{\equiv}N \longrightarrow —(CH_2)_4NH—C{\equiv}N + HBr$$

lysine side chain

(b) The acidic conditions of the reaction prevent the reaction. Under such conditions the amino group of lysine is protonated and therefore cannot act as a nucleophile.

26.21 The peptide *C* is derived from the amino end of *Q* because it contains Leu, the amino-terminal residue of *Q* itself. Peptide *D* must be at the carboxy end of *Q* because it contains no Lys or Arg, and therefore must have resulted from cleavage at its amino-terminal residue only. The order of *A* and *B*, however, must be established by other data; the required order is provided by cleavage with chymotrypsin. Peptide *E* shows that the Lys residue in *Q* is followed by a Gly residue; since the other Gly residue follows a Pro (peptide *B*), this establishes that peptide *A* follows peptide *C* in the sequence of *Q*. The Pro-Gly-Arg-Ser sequence in peptide *F* confirms that peptide *D* follows peptide *B* in the sequence of *Q*. The final sequence of *Q*, then, is *C-A-B-D*, or

Peptide *Q* = Leu-Lys-Gly-Arg-Ile-Trp-Phe-Pro-Gly-Arg-Ser-Glu-Ile

26.22 First write what is known as a diagram. (Hse = homoserine lactone)

$$C\ (\text{Arg,Gly}) + D\ (\text{Ala}_2,\text{Leu,Trp}) \xleftarrow{\text{trypsin}} A + B \xrightarrow{\text{trypsin}} E(\text{Gly,Lys}) + F(\text{Asp,Hse})$$

(where $A + B$ arise from R by BrCN cleavage: $R \xrightarrow{\text{BrCN}} A + B$)

$$R \xrightarrow{\text{chymotrypsin}} G(\text{Ala}_2,\text{Leu}) + H(\text{Arg,Asp,Gly}_2,\text{Lys,Met,Trp}) + I(=H + \text{Leu}) + J(\text{Ala})$$

Peptide *B* is the amino-terminal peptide of *R* because it contains homoserine lactone; and peptide *F* is the carboxy-terminal peptide of *B* for the same reason. The sequence of peptide *E* is Gly-Lys because it arises from trypsin cleavage, which occurs at the carboxy-terminal side of Lys and Arg residues. The sequence of peptide *B* is thus established as Gly-Lys-Asp-Met. Peptide *C* is the amino-terminal peptide of *A* because it contains an Arg, which must be at the carboxy terminus of *C*. Thus, a partial sequence of *A* is Gly-Arg-*D*, and a partial sequence of *R* is *B*-Gly-Arg-*D*. Comparing the compositions of peptides *H* and *I* with the sequences of peptides *A* and *B* gives the sequence of peptide *I* as Gly-Lys-Asp-Met-Gly-Arg-Trp-Leu. Only the two Ala residues are unaccounted for; peptides *G* and *J* show that they occur together. The final sequence of *R* is therefore

Peptide *R*: Gly-Lys-Asp-Met-Gly-Arg-Trp-Leu-Ala-Ala

26.23 If the average yield of each step is *Y*, then the yield of the first step (assuming it is average) is *Y*; the yield of the second is $Y{\cdot}Y$; and that of the *n*th step is Y^n. Hence,

$$Y^{369} = 0.17$$
$$369 \log Y = \log 0.17 = -0.7696$$
$$\log Y = -0.0021, \text{ or } Y = 0.995$$

Therefore, the average yield of each step is 99.5%!

The high yield of each step demonstrates two points about peptide synthesis. First, it has been developed into a remarkably efficient process. The second point, however, is that even with yields in excess of 99%, the overall yields of large proteins prepared by this method will be extremely small. Furthermore, these will be contaminated by large numbers of impurities that will be difficult to separate from the desired material. Fortunately, genetic-engineering methods (which you will study if you take biochemistry) allow scientists to prepare pure naturally occurring proteins in large amounts. At the present time, the chemical synthesis of peptides is most useful for the preparation of peptides containing about 2–50 residues, although a few noteworthy successes have been achieved in the synthesis of larger peptides and some proteins. (See, for example, Study Guide Link 26.5 on p. 1037 of this manual.) Furthermore, chemical synthesis can provide peptides that contain modified or unnatural amino acids that cannot be produced by the biological methods.

26.24 The box on text p. 1246 shows that solid-phase peptide synthesis requires about three synthetic steps per residue. Hence, about 300 steps are required for the synthesis of a protein containing 100 amino acid residues. If the average yield of each step is *Y*, then

$$Y^{300} = 0.50$$
$$300 \log Y = \log 0.50 = -0.3010$$
$$\log Y = -0.001003, \text{ or } Y = 0.998$$

Thus, an average yield of 99.8% per step would be required.

If you used 100 steps for your calculation, the average yield is 99.3%. Either way, the yield must be nearly perfect at each step to achieve even a 50% overall yield. Notice that a drop of a small fraction of a percent in average yield per step has a large effect on the overall yield because there are so many steps.

26.25 (a) Because lysine contains two amino groups, both must be protected in order to prevent nucleophilic side reactions, such as the reaction of the amino group of one lysine molecule with the DCC-activated carboxy group of another.

(b) The 20% piperidine step removes both Fmoc protecting groups. Hence, both amino groups react in subsequent acylation reactions. (The Lys residue is drawn in more structural detail for clarity.)

Gly—NHCH(C=O)—Ala with side chain $(CH_2)_4$—NH_2 H_2NCH(C=O)—Ala with side chain $(CH_2)_4$—NH—Gly   Gly—NHCH(C=O)—Ala with side chain $(CH_2)_4$—NH—Gly

26.26 (a) Compound *A* results from the S_N2 coupling reaction to the resin (analogous to Eq. 26.30 on text p. 1248), and compound *B* results from deprotection of *A*:

A: $FmocNHCH_2C(=O)—O—CH_2$—resin   *B*: $H_2NCH_2C(=O)—O—CH_2$—resin

Compound *C* is the coupling product that results from coupling of the lysine derivative reacting at its carboxylic acid group with the free amino group of *B*, and compound *D* results from removal of the Fmoc protecting group, *but not the Boc group,* from compound *C*.

C: FmocNH—CH[$(CH_2)_4$NHBoc]—C(=O)—$NHCH_2C(=O)—O—CH_2$—resin

D: H_2N—CH[$(CH_2)_4$NHBoc]—C(=O)—$NHCH_2C(=O)—O—CH_2$—resin

Compound *E* results from coupling of the carboxy terminus of Boc-Val with the free amino group of *D*. Peptide *P* results from both removal of *E* from the resin and removal of the Boc group.

E: BocVal—NH—CH[$(CH_2)_4$NHBoc]—C(=O)—$NHCH_2C(=O)—O—CH_2$—resin

Val—Lys—Gly (*P*) + $F_3C—C(=O)—O—CH_2$—resin + CO_2 + $F_3C—C(=O)—O—C(CH_3)_3$ (from reaction of *tert*-butyl cation of the Boc group with trifluoroacetic acid)

Note that the tripeptide *P* is in the dicationic, amino-protonated form as a result of the acidic deprotection conditions. The neutral (zwitterionic) form would be obtained by neutralization.

(b) The Boc group prevents the side-chain amino group of lysine from reacting with the carboxylic acid group of another molecule of itself under the coupling conditions.

(c) Boc-Val introduces a protecting group that can be removed at the same time as the peptide is released from the resin. Use of Fmoc-Val would require an additional deprotection step with 20% piperidine.

26.27 We must remember that each bead contains multiple sites for peptide attachment. (If each bead contained only one site, it wouldn't matter; but then, the process would produce so little material that it would be useless.) The splitting into thirds ensures that all sites *on a given bead* contain the same peptide. We want diversity in this synthesis, but we want the diversity to be among the different beads, *not* within a single bead.

26.28 (a) We would get 27 different peptides, but we are eliminating all possibilities in the original synthesis that contain more than one identical amino acid. Let's enumerate them to be sure.

From Round 1: We get F, G, and V (attached to the resin, of course).

From Round 2: We get W-F, T-F, and I-F; W-G, T-G, and I-G; and W-V, T-V, and I-V

From Round 3: We get R-W-F, K-W-F, H-W-F, R-T-F, K-T-F, H-T-F, R-I-F, K-I-F, and H-I-F; we get also R-I-F, K-I-F, H-I-F, R-W-G, K-W-G, H-W-G, R-I-G, K-I-G, and H-I-G; and we get also R-W-V, K-W-V, H-W-V, R-T-V, K-T-V, H-T-V, R-I-V, K-I-V, and H-I-V.

(b) All peptides must have V, G, or F in the C-terminal position; this eliminated peptide (1). All peptides must have W, T, or I in the second position; this eliminates peptides (2) and (4). All peptides must have R, K, or H in the amino-terminal position, Peptide (2) (K-W-F) is consistent with all of these constraints, and, indeed, it is the second peptide listed under "Round 3" in the previous part.

26.29 (a) In this case, the number of instances in each step is 6, and there are 6 steps. Hence, we expect 6^6 peptides, that is, 46,656 peptides.

Note that the log of 6^6 is 6 log 6, which is $6 \times 0.7781 = 4.6689$. This is the logarithm of the number of peptides. The antilog of 4.6689 is the number of peptides.

(b) We have in this case six instances at each of 5 steps plus one instance at one step. The number of peptides is therefore $6^5 + 6^1 = 7776 + 6 = 7782$ peptides.

26.30 One gram of beads corresponds to 1 g/$(0.2 \times 10^{-6}$ g/bead$) = 5 \times 10^6$ beads. The number of compounds cannot exceed the number of beads. However, the problem specifies that there should be at least ten beads per compound. Thus, the maximum number of compounds is 5×10^5. With ten instances of diversity and six steps, the number of compounds in the synthesis will be 10^6. Hence, there are too many compounds for the number of beads used to meet the desired goal.

26.31 Because the four chains of hemoglobin are held together by the same noncovalent forces that account for the tertiary structures of the individual chains, the subunits of hemoglobin would dissociate and would unfold into two random-coil α chains and two random-coil β chains.

26.32 One end of aeruginosin-B resembles an arginine side chain; one can hypothesize the following correspondences:

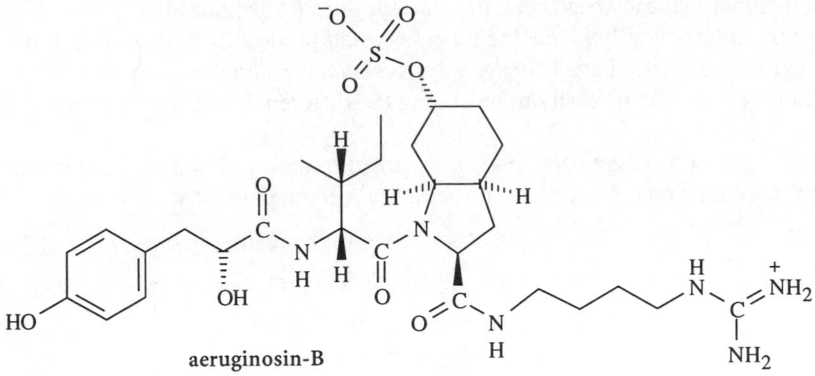

aeruginosin-B

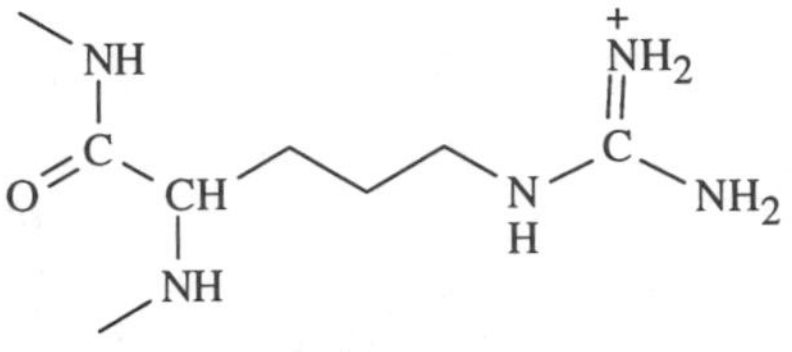

arginine

Aeruginosin-B has a positively charged guanidino group, like arginine, that can interact with the ionized Asp-189 in the binding site; and this is attached to a hydrocarbon chain that can interact favorably with the "walls" of the hydrophobic pocket.

26.33 In the following mechanism, the enzyme groups are shown in shaded type.

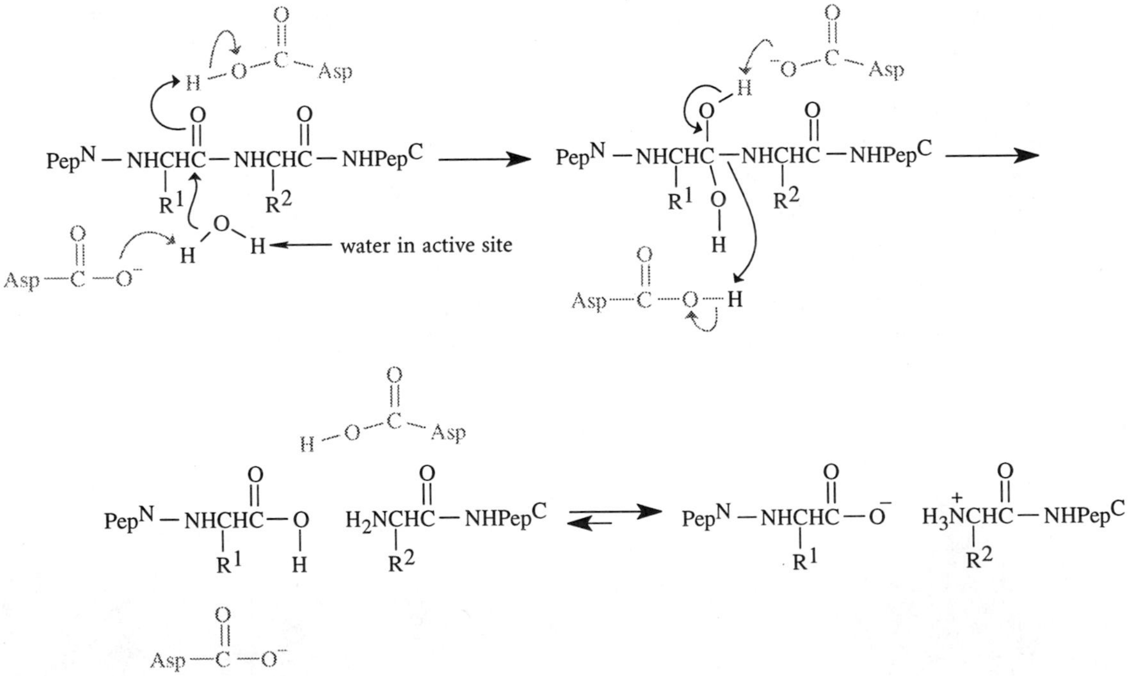

water in active site

26.34 Although the sulfur has two unshared pairs, the nitrogen of the thiazole is like the nitrogen of pyridine or imidazole and is the most basic site on the ring; hence, it is likely to be the hydrogen-bond acceptor. An amide backbone N—H is the only hydrogen bond donor site on a peptide backbone.

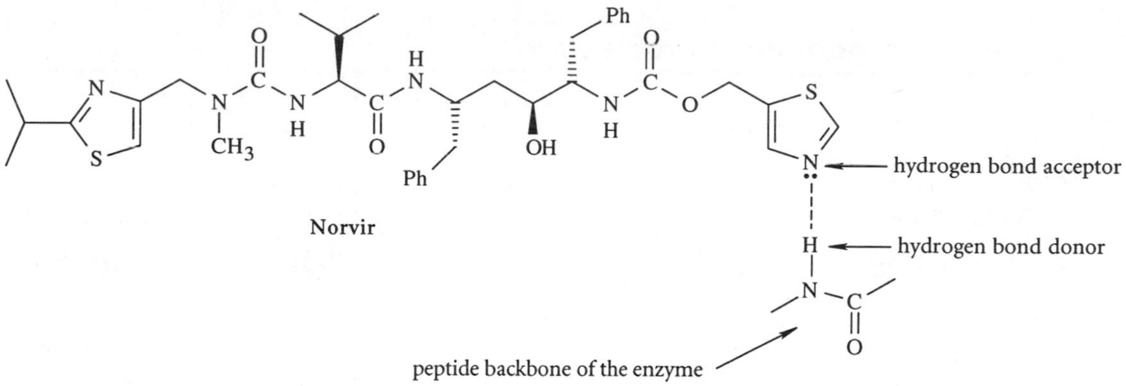

26.35 We use the criteria for hydrogen-bond donors and acceptors discussed in Sec. 8.3C.

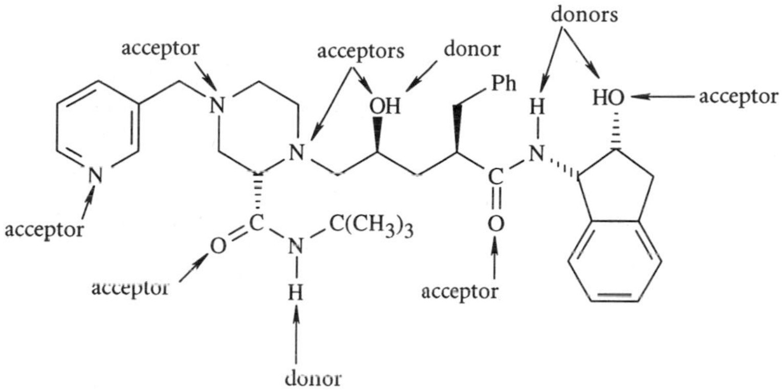

Solutions to Additional Problems

26.36 Products from the reactions of valine:

(a) $H_3\overset{+}{N}CHCO_2Et\ HSO_4^-$ with $CH(CH_3)_2$ on the α-carbon

(b) $PhC(=O)NHCHC(=O)O^-\ Et_3\overset{+}{N}H$ with $CH(CH_3)_2$ on the α-carbon

(c) $H_3\overset{+}{N}CHCO_2H\ Cl^-$ with $CH(CH_3)_2$ on the α-carbon

(d) $H_2NCHCO_2^-\ Na^+$ with $CH(CH_3)_2$ on the α-carbon

The reaction in (e) is a variation of the first reaction of the Strecker synthesis in which an amine—in this case, the amino acid valine—rather than ammonia is the source of nitrogen. (See Eq. 26.13 on text p. 1231.)

(e) $PhCH(C{\equiv}N)NHCH(CH(CH_3)_2)CO_2^-\ Na^+$

(f) Fluorenyl–$CH_2O-C(=O)-NHCH(CH(CH_3)_2)C(=O)OH$

Fmoc-Val

(g) Fluorenyl–$CH_2O-C(=O)-NHCH(CH(CH_3)_2)C(=O)-NHCH_2C(=O)OC(CH_3)_3$

Fmoc-Val-Gly *tert*-butyl ester

(h) Fluorenyl–$CH_2O-C(=O)-NHCH(CH(CH_3)_2)C(=O)-NHCH_2C(=O)OH$

Fmoc-Val-Gly

(i) Dibenzofulvene ($=CH_2$) + $H_2NCH(CH(CH_3)_2)C(=O)-NHCH_2C(=O)O^-\ H_2\overset{+}{N}$(piperidine ring)

(reacts further with piperidine)

Val-Gly (piperidinium salt)

(j) $H_3\overset{+}{N}CH(CH(CH_3)_2)C(=O)OH\ Cl^-$ + $H_3\overset{+}{N}CH_2C(=O)OH\ Cl^-$

valine hydrochloride; glycine hydrochloride

26.37 (a) Aspartic acid. It has the lowest isoelectric point.
(b) Arginine. It has the highest isoelectric point.
(c) Isoleucine and threonine. Each has two asymmetric carbon stereocenters.
(d) Glycine. It is not chiral and therefore it cannot be optically active.
(e) Asparagine and glutamine. Their side-chain amides are hydrolyzed to give aspartic acid and glutamic acid, respectively.

26.38 (a) Because the amino group of leucine is not protected, a certain amount (maybe most) of attachment to the resin will occur by attack of the α-amino group on the chloromethyl groups of the resin. Subsequent peptide synthesis would then not proceed as planned.
(b) Fmoc-Leu, once attached to the resin, must be deprotected before the subsequent coupling reaction can be expected to succeed. Polly forgot the deprotection step with 20% piperidine, which is essential for

removing the Fmoc group.

26.39 Since lysozyme has many more basic residues (Lys and Arg) than acidic residues (Asp and Glu), lysozyme is expected to be (and is) a basic protein. Its isoelectric point should be >>6. (In fact, its isoelectric point is about 12.)

26.40 Cysteic acid is essentially an aspartic acid in which the side-chain carboxy group has been substituted with a sulfonic acid group. Because sulfonic acids are more acidic than carboxylic acids, cysteic acid is expected to be (and is) more acidic than aspartic acid. Thus, the correct answer is (1): cysteic acid has a lower isoelectric point than aspartic acid.

26.41 The amino-terminal residue is lysine. Both the side-chain amino group and the α-amino group react with phenyl isothiocyanate (the Edman reagent). Thus, the side-chain amino group in the resulting PTH derivative (shown in the problem) is present as a thiourea derivative.

26.42 The amino-terminal residue of the peptide *P* is valine. Because dansyl-valine is obtained, the α-amino group of valine must have been free in the peptide, and hence, valine must have been the amino-terminal residue. The sequence of the other residues cannot be determined from the data given.

"Dansylation" is a method for determining the amino-terminal residue of a peptide.

26.43 Number the residues of *Q* from the amino terminus as 1,2,3, . . . First, the Edman degradation of *Q* shows that leucine is the amino-terminal residue. The formation of only dipeptides, including the dipeptide Leu-Val, shows that DPAP catalyzes the hydrolysis of peptide *Q* at every even-numbered residue from the amino terminus. Thus, this enzyme also catalyzes the hydrolysis of peptide *R* at every other residue — the *odd* residues in the numbering of *Q*:

Q: 1—2 ≀ 3—4 ≀ 5—6 ≀ 7—8 ≀ 9 ≀ = DPAP cleavage site

↓ Edman

R: 2—3 ≀ 4—5 ≀ 6—7 ≀ 8—9

This analysis shows that Gly is residue 9, the carboxy-terminal residue of *Q*. To get the order of the dipeptides, work back and forth between the dipeptides derived from *Q* and those derived from *R*. Since Gly has to be at the carboxy terminus of one of the dipeptides from *R*, and since the only dipeptide that meets this criterion is Ala-Gly, then Ala is the next residue in from the carboxy terminus. If Ala is in position 8, then position 7, from the *Q* dipeptides, must be Gln; position 6, from the *R* dipeptides, must be Asp; and so on. Alternatively, work from the amino terminus of *Q*: Leu-Val of *Q* and Val-Arg of *R* establish the sequence Leu-Val-Arg; Arg-Gly of *Q* shows that the next residue is Gly; and so on. The final sequence of peptide *R*, then, is:

Leu-Val-Arg-Gly-Val-Asp-Gln-Ala-Gly

(Remember that the amino terminus by convention is on the left when peptides are written this way.) The ammonia arises from the hydrolysis of Gln to glutamic acid.

26.44 (a) Trypsin should catalyze the cleavage of glucagon at Lys and Arg residues. The following five peptides are expected. (Remember that, by convention, peptides written in this way have their amino ends on the left.)

(1) His-Ser-Gln-Gly-Thr-Phe-Thr-Ser-Asp-Tyr-Ser-Lys
(2) Tyr-Leu-Asp-Ser-Arg

(Solution continues)

(3) Arg
(4) Ala-Gln-Asp-Phe-Val-Gln-Trp-Leu-Met-Asn-Thr
(5) Arg-Ala-Gln-Asp-Phe-Val-Gln-Trp-Leu-Met-Asn-Thr

Note that if cleavage occurs after the first Arg in the Arg-Arg sequence, then the second Arg becomes the amino terminus of the resulting peptide (5). Cleavage does not occur after this Arg because trypsin is an endopeptidase. (Eq. 26.27b on text p. 1243.)

(b) Cyanogen bromide cleavage would yield two peptides:

(1) His-Ser-Gln-Gly-Thr-Phe-Thr-Ser-Asp-Tyr-Ser-Lys-Tyr-Leu-Asp-Ser-Arg-Arg-Ala-Gln-Asp-Phe-Val-Gln-Trp-Leu-homoserine lactone
(2) Asn-Thr

(c) The products of this Edman cleavage would be the PTH derivative of histidine and glucagon minus its amino-terminal residue.

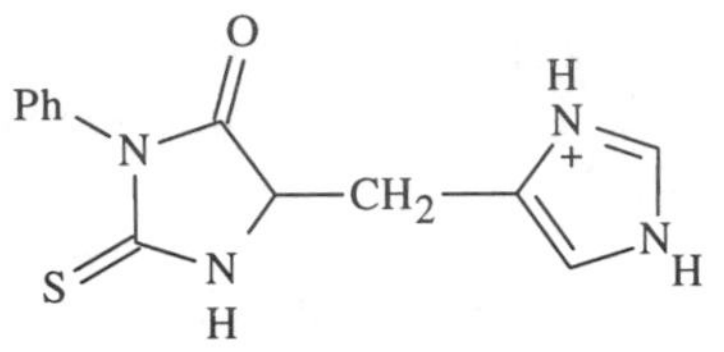

+ Ser-Gln-Gly- ··· (remainder of glucagon sequence)

PTH derivative of histidine

26.45 A molecular mass of about 1100 is consistent with the actual composition (Ala_4,Arg_2,Gly_2,Ser_2); amino acid analysis gives only the *relative* amounts of each amino acid. The absence of a reaction with the Edman reagent suggests that there is no terminal amino group. A *cyclic* peptide would give these results. Cleavage of a cyclic peptide at two arginine residues would give two peptides; but if the two peptides were identical, two equivalents of a single peptide would be obtained. The results are in accord with the following structure:

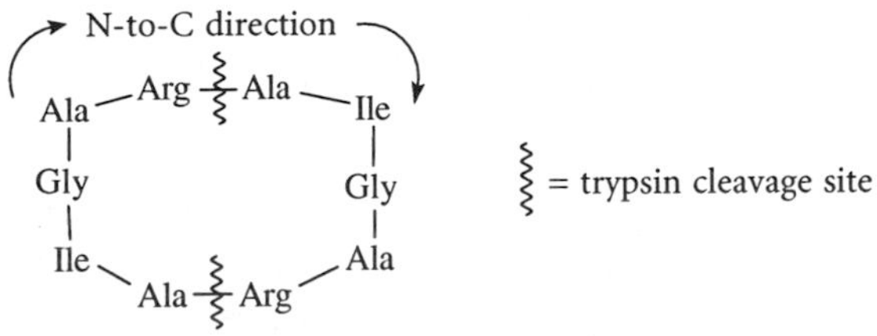

N-to-C direction

Ala—Arg ⌇ Ala—Ile
Gly Gly
Ile—Ala ⌇ Arg—Ala

⌇ = trypsin cleavage site

26.46 The presence of two amino-terminal residues suggests that either *(a)* insulin is an approximately equimolar mixture of two proteins, or *(b)* it is a single protein containing two peptide chains. In fact, the latter is correct; insulin is two polypeptide chains connected by disulfide bonds. Glycine is the amino-terminal residue of one chain, and phenylalanine is the amino-terminal residue of the other.

26.47 Amino acid *A* is 2,4-diaminobutanoic acid. The key piece of structural data is the Hofmann rearrangement, which converts the side-chain amide of glutamine into an amino group. Compound *A* is expected to be basic (and thus have a high isoelectric point) because it has two amino groups and only one carboxy group.

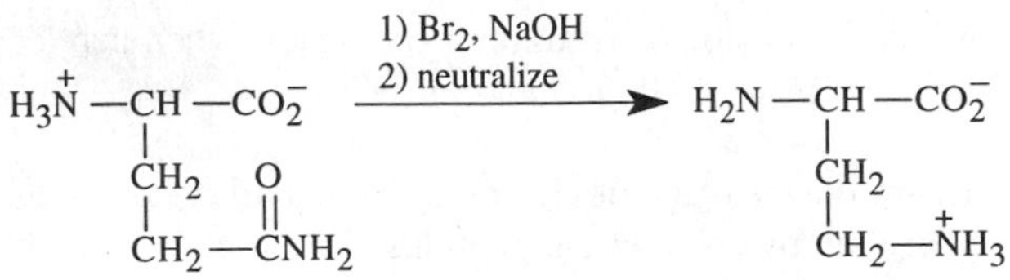

$$H_3\overset{+}{N}-CH(CH_2CH_2CONH_2)-CO_2^- \xrightarrow[\text{2) neutralize}]{\text{1) } Br_2,\ NaOH} H_2N-CH(CH_2CH_2\overset{+}{N}H_3)-CO_2^-$$

2,4-diaminobutanoic acid
(compound *A*)

26.48 This amino acid is really a substituted malonic acid. When heated in acid, it, like most malonic acid derivatives, decarboxylates. In this case, the product is glutamic acid. (See Eq. 20.40 on text p. 924.)

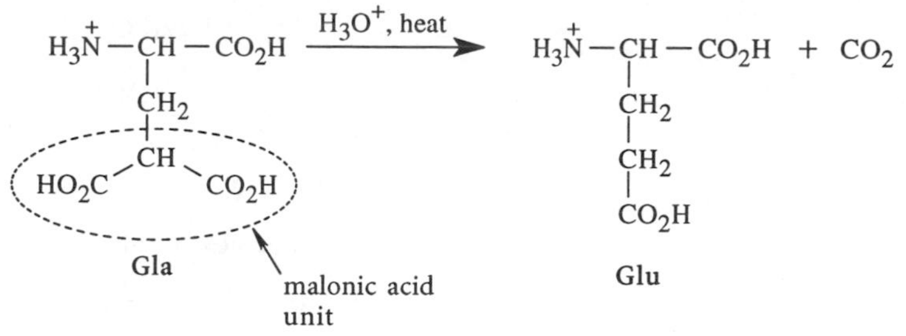

$$\underset{\text{Gla}}{\overset{+}{H_3N}-\underset{\substack{|\\ CH_2\\ |\\ HO_2C-CH-CO_2H}}{CH}-CO_2H} \xrightarrow{H_3O^+,\ \text{heat}} \underset{\text{Glu}}{\overset{+}{H_3N}-\underset{\substack{|\\ CH_2\\ |\\ CH_2\\ |\\ CO_2H}}{CH}-CO_2H} + CO_2$$

(The $HO_2C-CH-CO_2H$ portion of Gla is circled and labelled "malonic acid unit".)

26.49 (a) Reaction of chloromethylated polystyrene with trimethylamine would give the resin shown:

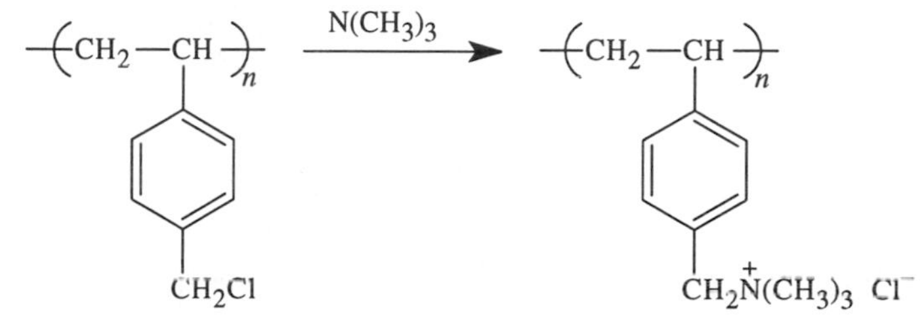

$$\left(CH_2-\underset{\substack{|\\ C_6H_4\\ |\\ CH_2Cl}}{CH}\right)_n \xrightarrow{N(CH_3)_3} \left(CH_2-\underset{\substack{|\\ C_6H_4\\ |\\ CH_2\overset{+}{N}(CH_3)_3\ Cl^-}}{CH}\right)_n$$

Notice that the alkyl halide is benzylic and is thus particularly reactive in S_N2 reactions.

(b) This resin is an *anion-exchange resin.* (See the solution to Problem 26.10 on p. 1041 of this manual.) At pH 6, Arg is positively charged and is therefore repelled by the column; it emerges earliest from the column. Leu has zero charge at pH 6; it emerges next. Glu is negatively charged at pH 6 and is therefore attracted to the positively-charged groups on the column; it emerges last from the column.

26.50 The principle is that a peptide with a given charge migrates to the electrode of opposite charge; a peptide with zero charge does not migrate. Because Gly-Lys is a basic peptide, it has a net positive charge at pH 6, and it therefore migrates to the cathode, the negatively charged electrode. Because Gly-Asp is an acidic peptide, it has a net negative charge at pH 6, and it migrates to the anode, the positively charged electrode. Gly-Ala is a neutral peptide; it has net zero charge at pH 6, and therefore it does not migrate.

26.51 The resin contains a large number of pendant phenyl groups. Phenylalanine also contains a phenyl group. The resin has dual characteristics. Not only is it an anion-exchange resin, but also it is like an HPLC resin in which the side-chains are phenyl instead of C-18 hydrocarbons. The "like-dissolves-like" principle suggests that Phe, the amino acid containing the phenyl side chain will be partitioned into the resin phase more extensively than Gly. The selective adsorption of phenylalanine is due to this effect.

26.52 Such a resin should be less effective in partitioning the amino acids with hydrocarbon side chains, because they will be "less soluble" in the resin phase. In effect, there are fewer $-CH_2-$ groups on the resin to provide favorable interactions with the amino acid side chains. Furthermore, the amino acid side-chains would be forced to be closer to the more polar and hydrophilic resin surface, with which they might not interact as favorably.

26.53 (a) Alanine (like the other amino acids) is a different compound in HCl, NaOH, and neutral H_2O because its ionization state is different, and different compounds have different optical rotations. In acidic solution, the optical rotation is that of the acidic form; in base, the optical rotation is that of the basic form; and in water, the optical rotation is that of the neutral (zwitterion) form.

(b) Lysine has two amino groups, and acetylation of each is possible.

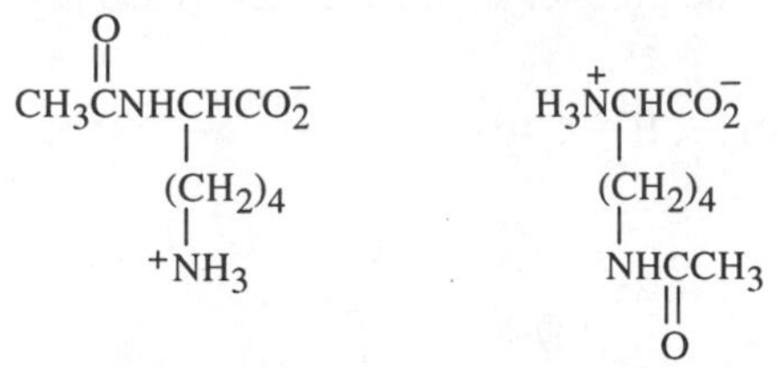

$$CH_3\overset{O}{\overset{\|}{C}}NH\underset{\underset{^{+}NH_3}{|}}{\underset{(CH_2)_4}{|}}{CH}CO_2^- \qquad H_3\overset{+}{N}\underset{\underset{NH\overset{}{C}CH_3\,(C{=}O)}{|}}{\underset{(CH_2)_4}{|}}{CH}CO_2^-$$

the two mono-*N*-acetylated lysine derivatives

(c) The peptide is cleaved at the Arg-Ala bond at pH 8. However, in the presence of 8*M* urea, trypsin, like most enzymes, is denatured. Denatured enzymes are devoid of catalytic activity because their tertiary structures are disrupted. Thus, trypsin, when denatured, cannot catalyze cleavage of the peptide.

(d) The 2-mercaptoethanol treatment ensures that disulfide bonds are reduced. Aziridine reacts much like an epoxide when the nitrogen is protonated. Aziridine is basic enough that a significant amount of it is protonated in aqueous solution. Reaction of protonated aziridine with the thiol group of a cysteine residue gives an amine:

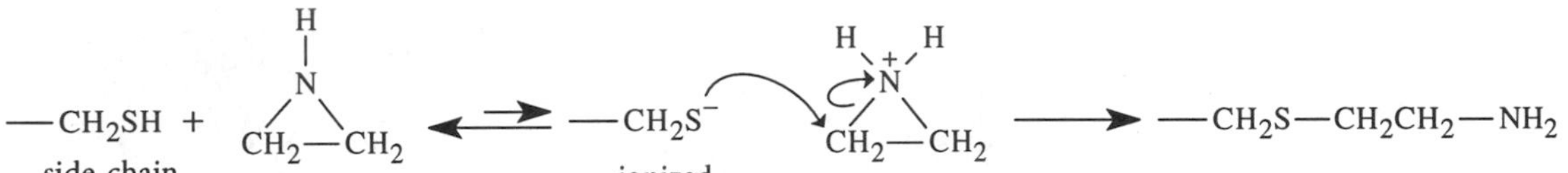

$$—CH_2SH + \text{aziridine } (CH_2—CH_2 \text{ ring with N—H}) \rightleftharpoons —CH_2S^- + \text{conjugate acid of aziridine } (H_2N^+ \text{ ring}) \longrightarrow —CH_2S—CH_2CH_2—NH_2$$

side chain of cysteine; aziridine; ionized side chain of cysteine; conjugate acid of aziridine

$$—CH_2S—CH_2CH_2—NH_2 \overset{\text{pH 8}}{\rightleftharpoons} —CH_2S—CH_2CH_2—\overset{+}{N}H_3$$

The amino group of the modified residue is protonated at pH 8, the pH at which trypsin digestion is carried out. The side chain of the modified residue resembles the side chain of lysine in both length and charge, and trypsin hydrolyzes peptides at the modified residue much as it hydrolyzes peptides at lysine residues.

(e) The two sulfoxides are diastereomers. They both have the same configuration at the α-carbon, but differ in configuration at the sulfur of the sulfoxide group. This sulfur is an asymmetric atom and stereocenter because it has four different "groups" attached: the CH_2, the CH_3, the O, and the electron pair. Evidently, inversion at sulfur in a sulfoxide, unlike amine inversion, is very slow, because the individual diastereomers can be isolated.

26.54 (a) The diversity elements are ring substituents R^A and R^B; the Fmoc-amino acid side-chain R^C; and the alkyl group R^D of the alkylating agent. The stereochemical configuration of the Fmoc-amino acid derivative could also be varied (*R* or *S*).

(b) Forty-eight compounds are possible. We start with three pure samples corresponding to each of the different R^A structures (all having the same R^B), mix them, and separate the resulting mix into four pools. Each pool is subjected to reactions *a*, *b*, and *c* with a different one of the four instances of the Fmoc-amino acid. This gives rise to twelve different compounds in four pools. The four pools are mixed and separated again into four pools, which are subjected to reactions *d* and *e*, with each of the four pools subjected to a different instance of R^D—I. $(3 \times 4 \times 4) = 48$.

(c) The amino group of the starting compound is particularly unreactive because it is not very basic. First, it is an arylamine, which (from the conjugate-acid pK_a values in Table 23.1) are considerably less basic than alkylamines. However, this amine is stabilized by an additional resonance effect in which the unshared electron pair of the nitrogen is delocalized into the *ortho*-carbonyl group:

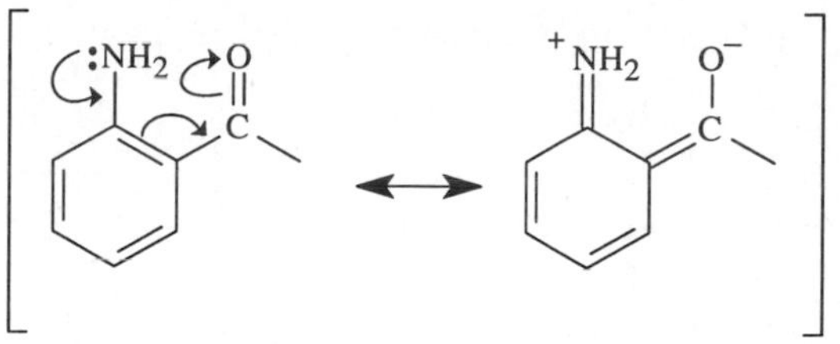

Because of this effect, the basicity of this amino group is further reduced. Remember that nucleophilicity is correlated with basicity; that is, the weaker the base, the weaker the nucleophile. Hence, this amino group is a rather poor nucleophile; hence, a very reactive acylating reagent must be used.

(d) The acylation reaction liberates HF. (Write the balanced reaction to see this.) HF can protonate the amino group and shut down the acylation reaction. Hence, a base is needed to "scavenge" the HF. (This point was discussed thoroughly in Sec. 21.8A.)

Notice that the pyridine derivative used to scavenge the HF has two ortho *tert*-butyl groups. These groups provide steric hindrance so that this amine will not compete as a nucleophile *at carbon*; however, they do not impede the reaction of the basic nitrogen with the tiny proton. A similar strategy was used in Sec. 22.7B in the formation of ester enolate ions with hindered amide bases.

(e) The role of the piperidine is to deprotect the nitrogen of the amino acid derivative, that is, to remove the Fmoc group (Eq. 26.31, text p. 1248).

(f) Compound *W* is the acylated and deprotected amino ketone:

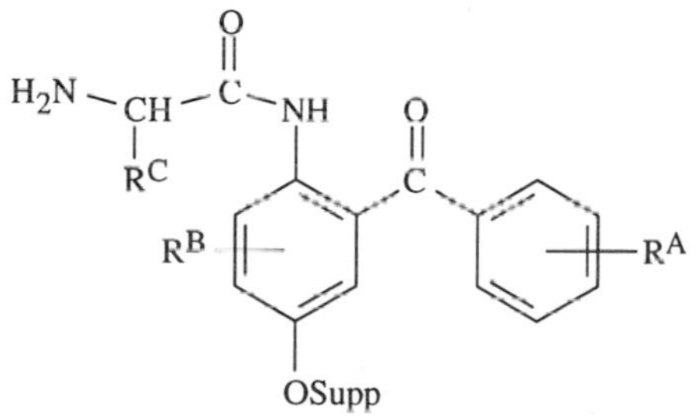

structure of compound *W*

(g) We have a ketone and we have an amine; conditions are ripe for acid-catalyzed imine formation. The major protonated species of compound *W* is the conjugate acid of the amine, *WH1*. However, this is in equilibrium with a small amount (how small??) of the compound *WH2* in which the nitrogen is unprotonated and the carbonyl is protonated:

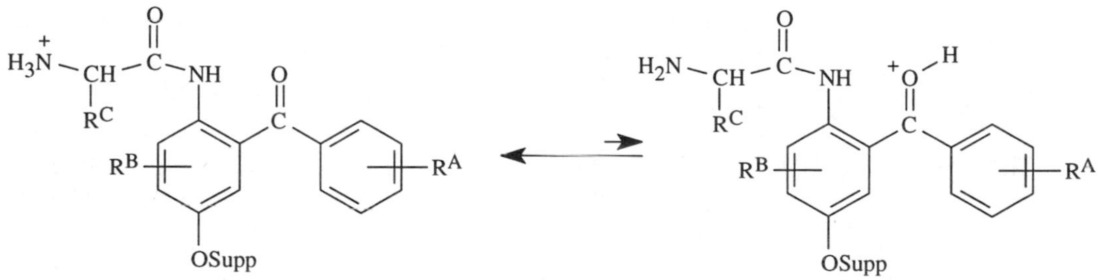

WH1 $\rightleftharpoons$ *WH2*

Compound *WH2* is very reactive, in part because the resulting imine formation is *intramolecular* (Sec. 11.7).

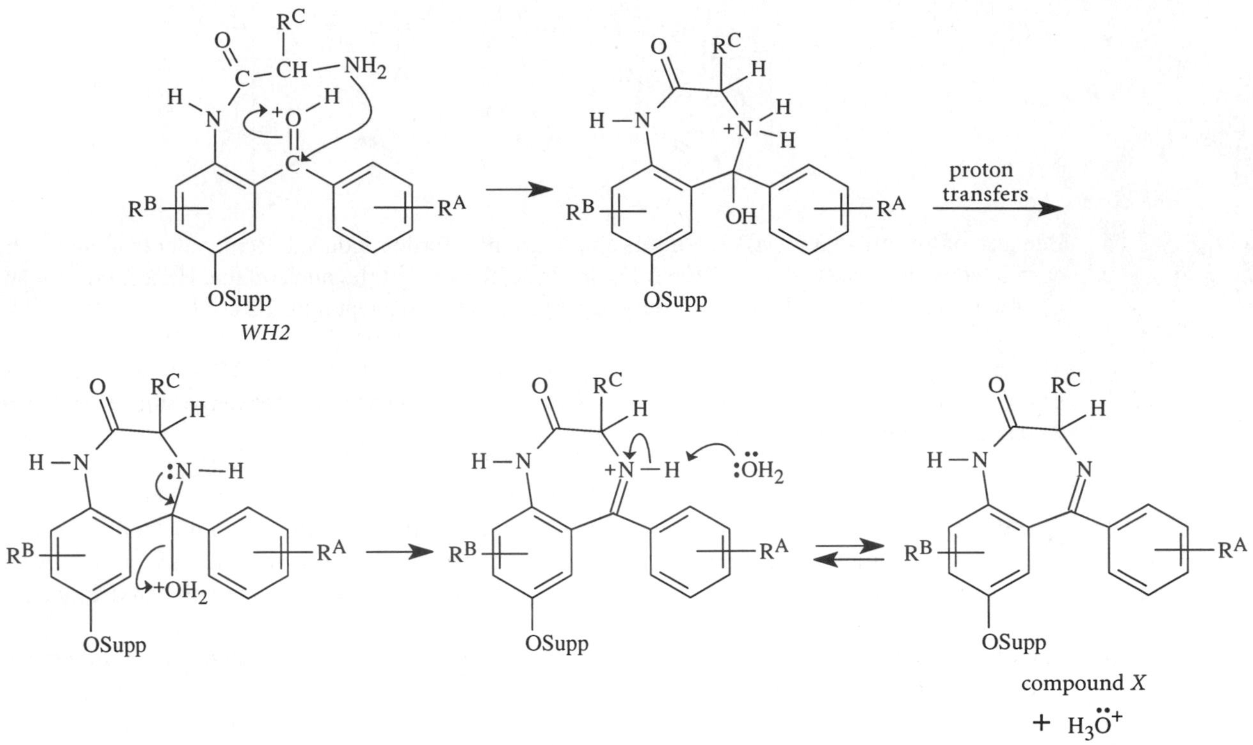

WH2

proton transfers

compound X

$+ \ H_3\ddot{O}^+$

(h) The proton α to the carbonyl is acidic because removal of it would form a resonance-stabilized enolate ion:

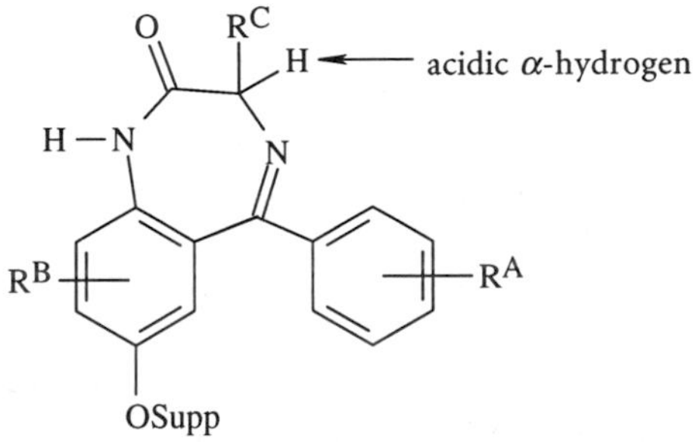

acidic α-hydrogen

(i) The base *Y* was used in excess to ensure complete deprotonation of the amide hydrogen. Hence, it has to be strong enough for this purpose, yet weak enough that the other acidic hydrogens are not affected. Notice that *Y* is the conjugate base of a cyclic carbamate, which is *more* acidic than an ordinary amide (why?). Hence, *Y* is *less* basic than an ordinary amide anion. However, the amide N—H in the compound above is *much* more acidic than an ordinary amide; why? Hence, it is completely deprotonated by the excess of *Y*.

(j) The restriction on R^D is that it must be an alkyl iodide that is sufficiently reactive in an S_N2 reaction. This means that R^D—I must be a primary alkyl bromide without too much β-branching, an unbranched secondary alkyl iodide, an allylic alkyl halide, or a benzylic alkyl halide.

26.55 The mechanism of cleavage in trifluoroacetic acid involves a benzylic carbocation intermediate. This carbocation is resonance-stabilized not only by the *para*-oxygen, but also by the *ortho*-oxygen as well. (Draw the resonance structures!). Because the carbocation intermediate is more stable, the cleavage reaction is faster. Hence, milder conditions (lower acid concentration) can be used for the cleavage.

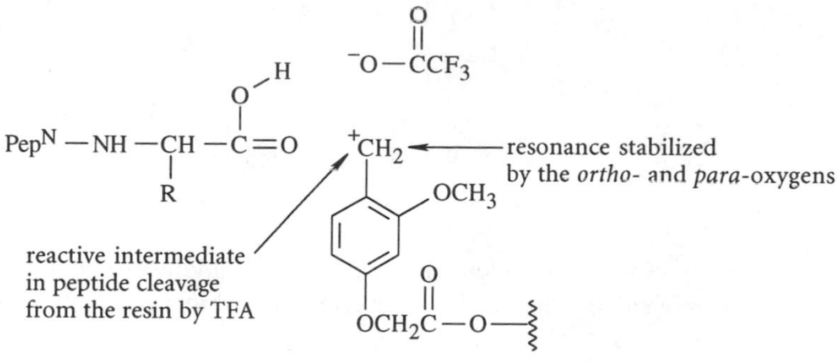

$$\text{Pep}^{N}-\text{NH}-\underset{\text{R}}{\text{CH}}-\overset{\text{O-H}}{\text{C}}=\text{O} \qquad {}^{-}\text{O}-\overset{\text{O}}{\overset{\|}{\text{C}}}\text{CF}_3 \qquad {}^{+}\text{CH}_2$$

resonance stabilized by the *ortho*- and *para*-oxygens

reactive intermediate in peptide cleavage from the resin by TFA

$$\text{OCH}_3 \qquad \text{OCH}_2\overset{\text{O}}{\overset{\|}{\text{C}}}-\text{O}-$$

26.56 The two Asp residues would be expect to interact with this hydroxy group by hydrogen bonding. Perhaps one Asp in protonated form could serve as a donor and the other, in ionized form, could serve as an acceptor:

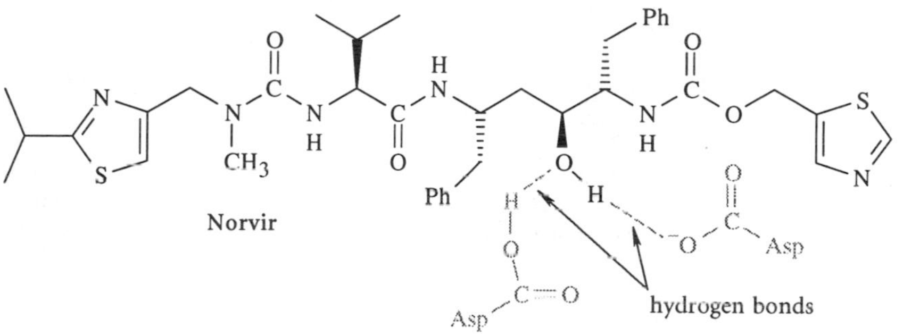

Norvir

hydrogen bonds

(The interaction with Crixivan is similar.)

26.57 Figure 26.9 on text p. 1264 shows that the amino acid side chains extend outward from the periphery of the helix and are actually rather close to each other in space. The pK_a of a protonated lysine side-chain amino group is about 10.5. At pH values below 10, the side-chain amino groups in polylysine are protonated; thus, these side chains are positively charged. The peptide avoids the α-helical conformation at pH values below 10 to avoid the repulsive electrostatic interactions between the adjacent positively charged groups. At pH values above 11, the side-chain amino groups are unprotonated and hence uncharged. As a result, there are no charge-charge repulsions to destabilize the α-helix.

The same principles operate in polyglutamic acid, but at the other end of the pH scale. In polyglutamic acid, the side chains are negatively charged at pH values above the pK_a of the carboxy groups. Hence, the repulsions between negative charges cause the helix to be destabilized, and it unfolds at high pH. At low pH, the carboxy groups are un-ionized and hence uncharged. Consequently, the helix can form at low pH because there are no charge-charge repulsions to destabilize it.

26.58 (a) This is a reductive amination of formaldehyde by the amino group of the lysine residue. Because excess formaldehyde is present, the amino group of lysine is methylated twice.

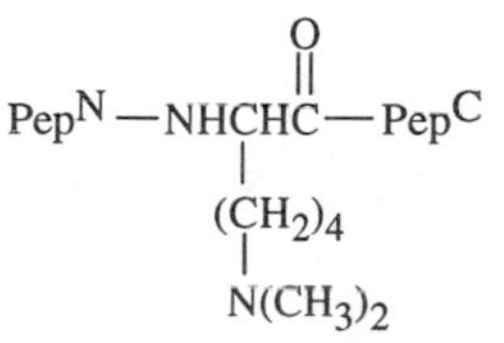

$$\text{Pep}^{N}-\text{NH}\underset{\underset{\text{N(CH}_3)_2}{|}}{\underset{(\text{CH}_2)_4}{\text{CH}}}\overset{\text{O}}{\overset{\|}{\text{C}}}-\text{Pep}^{C}$$

(b) The side-chain amino group of lysine serves as a nucleophile to open the anhydride. The resulting carboxy group is ionized at the pH of the reaction.

$$\text{Pep}^{N}\text{—NHCH(C=O)—Pep}^{C};\ \text{side chain: }(CH_2)_4\text{—NH—C(=O)—}CH_2CH_2\text{—C(=O)—O}^-$$

(c) The thiol of cysteine serves as a nucleophile in a conjugate-addition reaction to maleimide. The pH must be high enough to form a small amount of the conjugate-base thiolate anion of cysteine, which is the actual nucleophilic species.

$$\text{Pep}^{N}\text{—NHCH(C=O)—Pep}^{C};\ \text{side chain: }CH_2\text{—S—(succinimide ring, NH, two C=O)}$$

(d) The side-chain sulfhydryl group of the cysteine residue is alkylated by iodoacetate. Recall (Sec. 22.3D) that α-halo carbonyl compounds are very reactive in S_N2 reactions. The pH of the reaction ensures a reasonable concentration of the conjugate-base thiolate anion, which is the nucleophilic species.

$$\text{Pep}^{N}\text{—NHCH(C=O)—Pep}^{C};\ \text{side chain: }CH_2\text{—}SCH_2C(=O)O^-$$

➡ Reduction of disulfide bonds to cysteine thiols (Eq. 26.40, text p. 1261) followed by treatment with iodoacetic acid at pH 8 is a coon technique used for breaking disulfide bonds irreversibly in proteins.

(e) The carbodiimide promotes a condensation between the side-chain carboxy group of aspartic acid and the amino group of the glycine ester.

$$\text{Pep}^{N}\text{—NHCH(C=O)—Pep}^{C};\ \text{side chain: }CH_2\text{—C(=O)—}NHCH_2CO_2CH_3 \quad + \quad C_6H_{11}\text{—NH—C(=O)—}NHCH_2CH_2\text{—}\overset{+}{N}(H_3C)(\text{morpholine O})\ Cl^-$$

(f) The phenol ring of the tyrosine residue undergoes electrophilic substitution by the diazonium ion (see Sec. 23.4B). Because the para position is blocked, substitution occurs at the position ortho to the hydroxy group.

$$\text{Pep}^{N}\text{—NHCH(C=O)—Pep}^{C};\ \text{side chain: }CH_2\text{—}C_6H_3(OH)\text{—N=N—Ph}$$

26.59 (a)

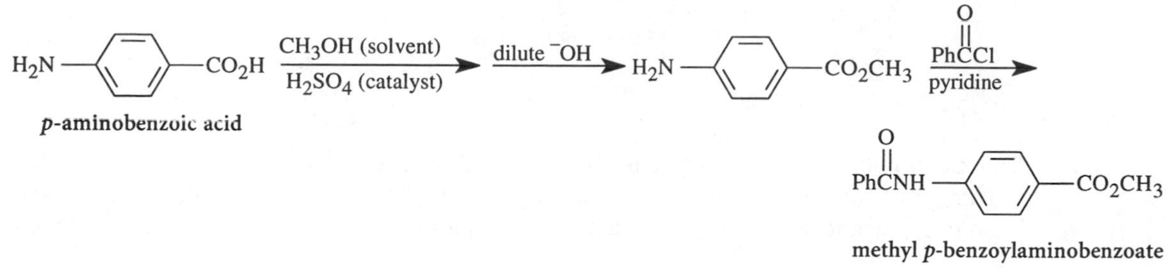

$$H_2N\text{—}C_6H_4\text{—}CO_2H \xrightarrow[H_2SO_4\text{ (catalyst)}]{CH_3OH\text{ (solvent)}} \xrightarrow{\text{dilute }^-OH} H_2N\text{—}C_6H_4\text{—}CO_2CH_3 \xrightarrow[\text{pyridine}]{PhCCl\ (C{=}O)} PhCNH\text{—}C_6H_4\text{—}CO_2CH_3$$

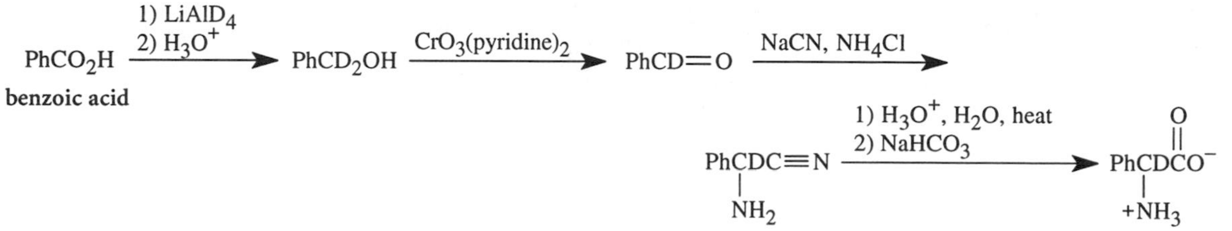

p-aminobenzoic acid

methyl *p*-benzoylaminobenzoate

(b)

$$PhCO_2H \xrightarrow{1)\ LiAlD_4;\ 2)\ H_3O^+} PhCD_2OH \xrightarrow{CrO_3(\text{pyridine})_2} PhCD{=}O \xrightarrow{NaCN,\ NH_4Cl} PhCD(NH_2)C{\equiv}N \xrightarrow{1)\ H_3O^+,\ H_2O,\ \text{heat};\ 2)\ NaHCO_3} PhCD(\overset{+}{N}H_3)CO^-\ (C{=}O)$$

benzoic acid

(c)

$$CD_3CH{=}O \xrightarrow[D_2O]{NaCN,\ \overset{+}{N}D_4\ Cl^-} CD_3CH(NH_2)C{\equiv}N \xrightarrow{1)\ H_3O^+,\ H_2O,\ \text{heat};\ 2)\ NaHCO_3} CD_3CH(\overset{+}{N}H_3)CO^-\ (C{=}O)$$

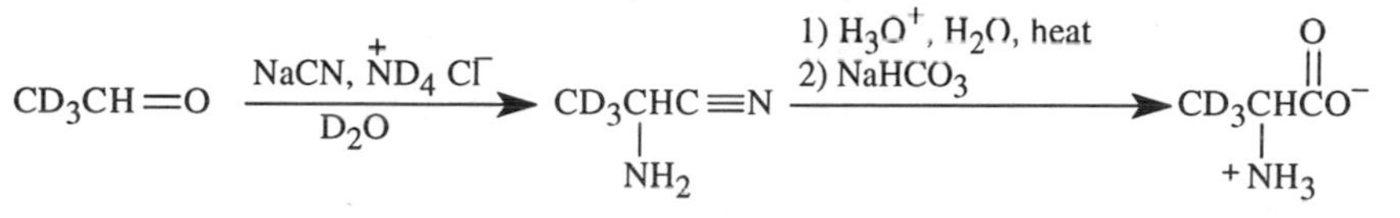

The isotopically substituted reagents ($^+ND_4\ Cl^-$, D_2O) are necessary in the first step because the pH of the reaction is sufficiently basic that the α-deuteriums of the starting trideuteroacetaldehyde would exchange for hydrogens if the analogous protic reagents $^+NH_4\ Cl^-$ and H_2O were used. (See Eq. 22.6 on text p. 1001.)

(d)

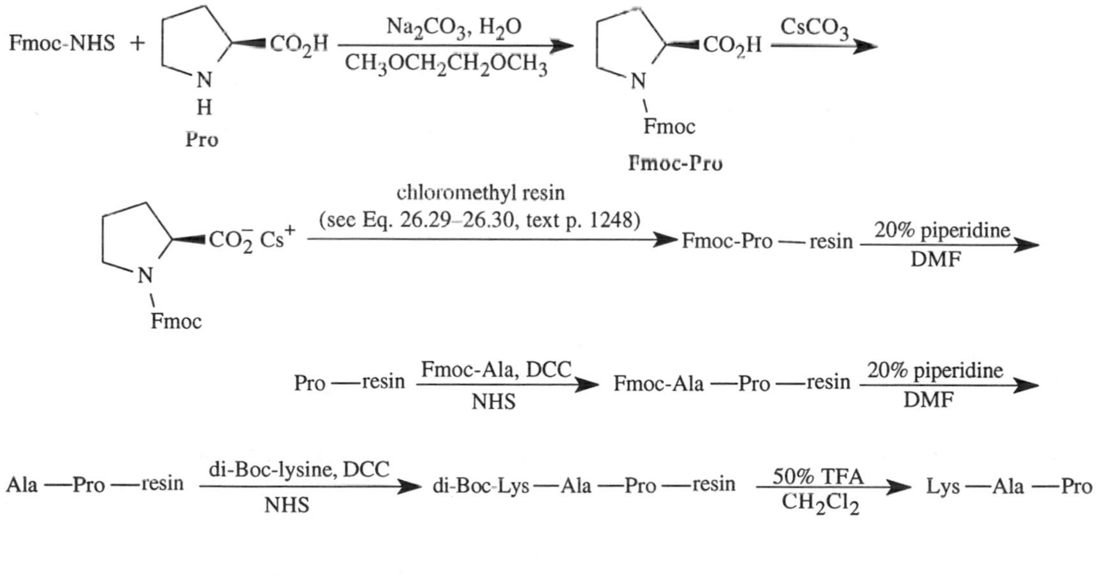

$$\text{Fmoc-NHS} + \text{Pro (pyrrolidine–}CO_2H\text{, N–H)} \xrightarrow[CH_3OCH_2CH_2OCH_3]{Na_2CO_3,\ H_2O} \text{Fmoc-Pro (N–Fmoc, }CO_2H) \xrightarrow{CsCO_3}$$

Pro — Fmoc-Pro

$$\text{(N–Fmoc pyrrolidine–}CO_2^-\ Cs^+) \xrightarrow{\text{chloromethyl resin (see Eq. 26.29–26.30, text p. 1248)}} \text{Fmoc-Pro—resin} \xrightarrow[DMF]{20\%\ \text{piperidine}}$$

$$\text{Pro—resin} \xrightarrow[NHS]{\text{Fmoc-Ala, DCC}} \text{Fmoc-Ala—Pro—resin} \xrightarrow[DMF]{20\%\ \text{piperidine}}$$

$$\text{Ala—Pro—resin} \xrightarrow[NHS]{\text{di-Boc-lysine, DCC}} \text{di-Boc-Lys—Ala—Pro—resin} \xrightarrow[CH_2Cl_2]{50\%\ TFA} \text{Lys—Ala—Pro}$$

(e)

$$EtO_2CCH{=}O \xrightarrow{NaCN,\ NH_4Cl} EtO_2CCH(NH_2)C{\equiv}N \xrightarrow{Ac_2O} EtO_2CCH(NHCCH_3)C{\equiv}N\ (C{=}O)$$

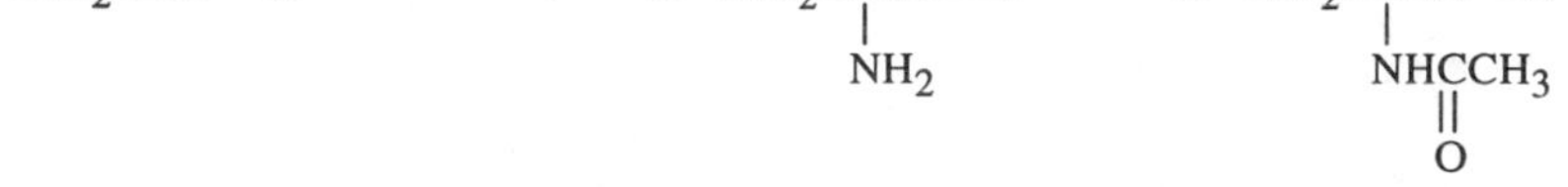

(f) First prepare 3-bromocyclopentene from cyclopentene:

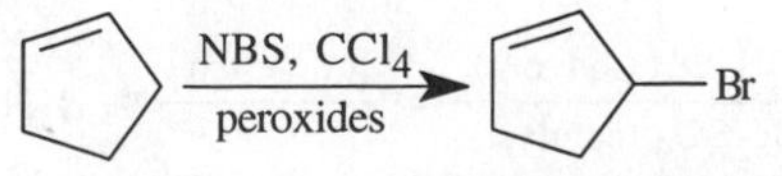

$$\text{cyclopentene} \xrightarrow[\text{peroxides}]{\text{NBS, }CCl_4} \text{3-bromocyclopentene}$$

Then use it in the following synthesis. Because the α-hydrogen of the product of (e) is about as acidic as an α-hydrogen of diethyl malonate, the techniques of the acetamidomalonate method can be used. Notice that 3-bromocyclopentene is an allylic halide and is therefore very reactive in S_N2 reactions. (See Sec. 17.4.)

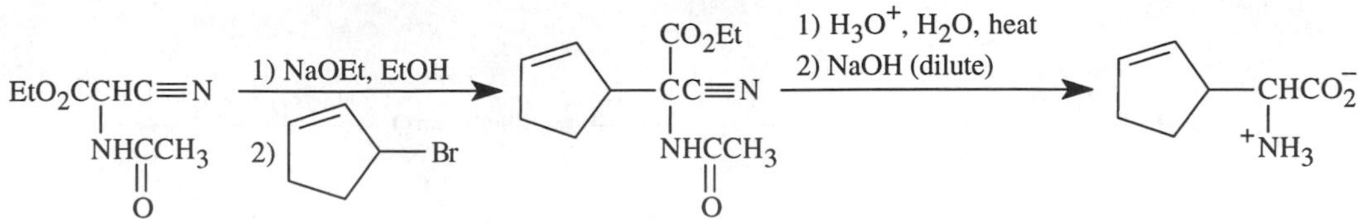

$$EtO_2CCH(NHCCH_3)C{\equiv}N \;(\text{NHCCH}_3 \text{ with C=O}) \xrightarrow[\text{2) 3-bromocyclopentene}]{\text{1) NaOEt, EtOH}} \text{(2-cyclopentenyl)}C(CO_2Et)(NHCCH_3)C{\equiv}N \xrightarrow[\text{2) NaOH (dilute)}]{\text{1) }H_3O^+,\ H_2O,\text{ heat}} \text{(2-cyclopentenyl)}CH(\overset{+}{N}H_3)CO_2^-$$

(g)

$$HO_2C{-}C_6H_4{-}CO_2H \xrightarrow{SOCl_2} ClC(=O){-}C_6H_4{-}C(=O)Cl \xrightarrow{NH_3\text{ (excess)}} H_2NC(=O){-}C_6H_4{-}C(=O)NH_2$$

$$H_2NC(=O){-}C_6H_4{-}C(=O)NH_2 \xrightarrow[\text{(Hofmann rearrangement; Sec. 23.11C)}]{Br_2,\ H_2O} H_2N{-}C_6H_4{-}NH_2 \xrightarrow{\;+\ ClC(=O){-}C_6H_4{-}C(=O)Cl\;} p\text{-aramid}$$

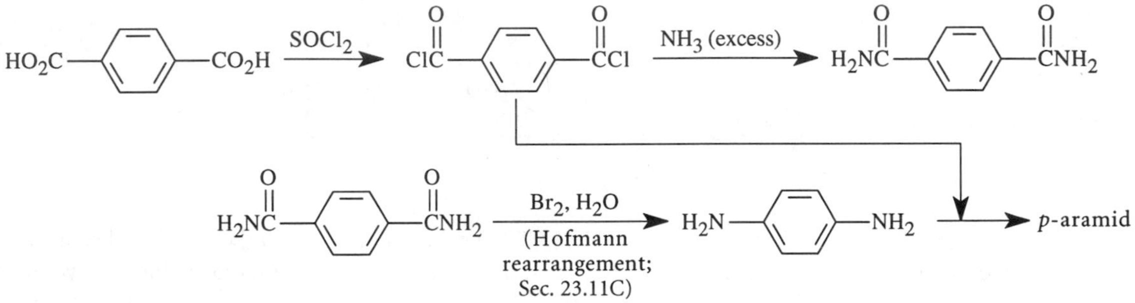

26.60 Every synthesis has the same final steps:

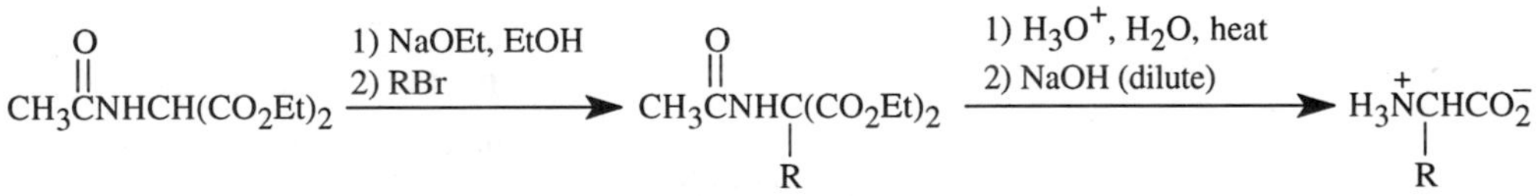

$$CH_3CNHCH(CO_2Et)_2 \;(\text{C=O}) \xrightarrow[\text{2) RBr}]{\text{1) NaOEt, EtOH}} CH_3CNHC(CO_2Et)_2 \;(\text{C=O};\ \text{R on C}) \xrightarrow[\text{2) NaOH (dilute)}]{\text{1) }H_3O^+,\ H_2O,\text{ heat}} H_3\overset{+}{N}CH(R)CO_2^-$$

The challenge in each case is to prepare the alkyl halide R—Br from the given starting material.

(a)

$$(CH_3)_2C{=}CH_2 \xrightarrow{\text{DBr, peroxides}} (CH_3)_2CD{-}CH_2Br$$

2-methylpropene (isobutylene)

Alternatively, the same alkene could be hydroborated with BD_3, the resulting organoborane oxidized to the primary alcohol by alkaline H_2O_2, and the alcohol converted into the alkyl bromide with concentrated HBr and H_2SO_4 catalyst.

(b)

$$PhCH{=}O \xrightarrow[\text{2) }H_3O^+]{\text{1) }LiAlD_4} PhCHD{-}OH \xrightarrow{HBr} PhCHD{-}Br$$

benzaldehyde

(c)

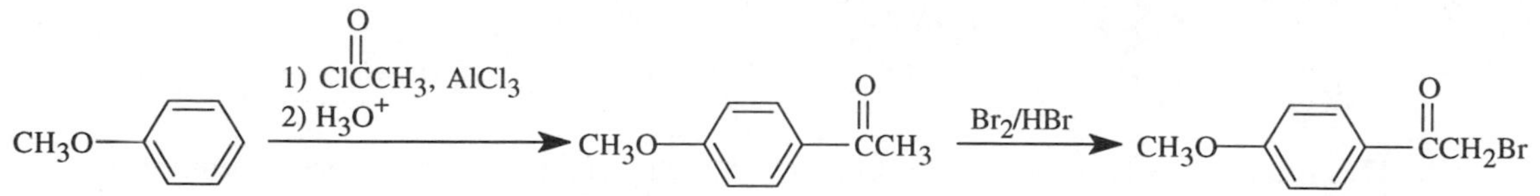

$CH_3O-C_6H_5$ (anisole) $\xrightarrow[\text{2) } H_3O^+]{\text{1) } ClC(=O)CH_3,\ AlCl_3}$ $CH_3O-C_6H_4-C(=O)CH_3$ $\xrightarrow{Br_2/HBr}$ $CH_3O-C_6H_4-C(=O)CH_2Br$

anisole

Alkylate the acetamidomalonate ion with this halide. Then, in the final hydrolysis step, use concentrated HBr, which will also cleave the methyl ether.

26.61 In the first step, the amine adds to the Edman reagent; for the mechanism of this reaction, see the solution to Problem 26.18(a) on page 1043 of this manual. The sulfur of the thiourea then serves as a nucleophile to close a six-membered ring and cleave the peptide. The mechanism below begins with the product of the reaction between the Edman reagent and the peptide, which is protonated on the carbonyl oxygen by the catalyzing acid.

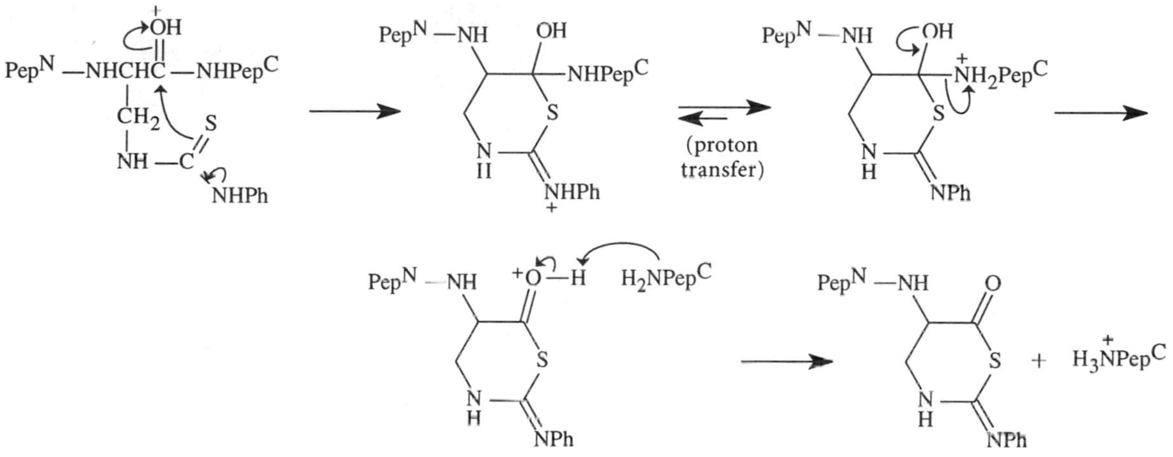

$Pep^N-NHCH(CH_2NH-C(=S)NHPh)C(=\overset{+}{O}H)-NHPep^C$ ⟶ six-membered ring (Pep^N—NH, OH, NHPep^C, S, N—H, $=\overset{+}{N}HPh$) ⇌ (proton transfer) ring with OH, $\overset{+}{N}H_2Pep^C$, S, N—H, =NPh ⟶

ring with Pep^N—NH, $+\ddot{O}-H$, H_2NPep^C, S, N—H, =NPh ⟶ Pep^N—NH, C=O, S, N—H, =NPh + $H_3\overset{+}{N}Pep^C$

compound *X*

26.62 This reaction is an intramolecular ester aminolysis. The mechanism below begins with the conjugate base of the peptide formed by loss of a proton from the terminal amino group.

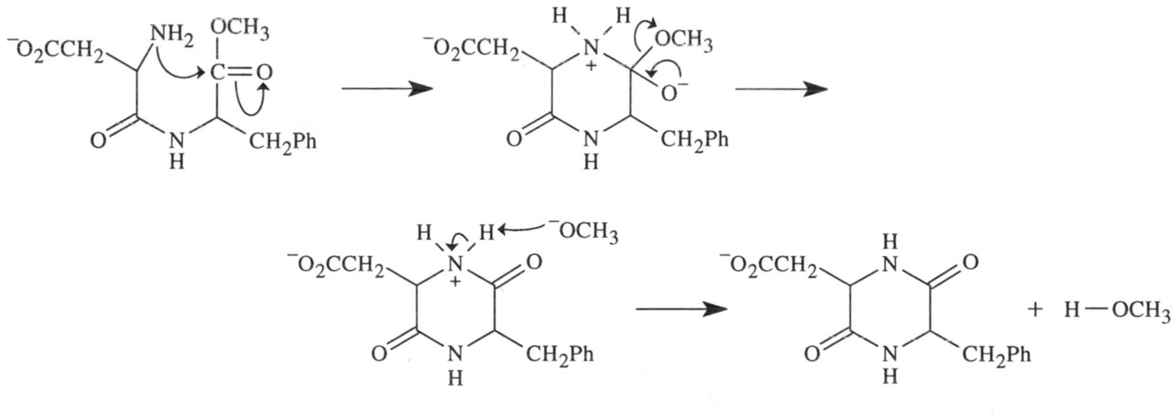

$^-O_2CCH_2$—, NH_2, OCH_3, C=O, O, N—H, CH_2Ph ⟶ $^-O_2CCH_2$—, H, H, $\overset{+}{N}$, OCH_3, O^-, O, N—H, CH_2Ph ⟶

$^-O_2CCH_2$—, H, H, $\overset{+}{N}$, $^-OCH_3$, O, O, N—H, CH_2Ph ⟶ $^-O_2CCH_2$—, H—N, O, O, N—H, CH_2Ph + $H-OCH_3$

You may have noticed that diet soft drinks that go on sale in the supermarket are not quite as sweet as really fresh ones. This is because some of the sweetener has degraded by the mechanism just depicted.

26.63 (a) This is like the reaction of an amine with an isocyanate; see Eq. 23.71 on text p. 1104.

$$\mathrm{EtNH{-}\overset{\overset{\displaystyle S}{\|}}{C}{-}NHPh}$$

(b) This is a modified Strecker synthesis in which methylamine is used in place of ammonia.

$$\mathrm{Ph{-}\underset{\underset{\displaystyle H_2\overset{+}{N}CH_3}{|}}{CH}{-}CO_2^-}$$

(c) This is a reductive amination in which the amine is the α-amino group of the amino acid.

$$\mathrm{(CH_3)_2CHCH_2NH\underset{\underset{\displaystyle CH_3}{|}}{CH}{-}CO_2^-}$$

(d) The methyl ester is saponified much more rapidly that the *tert*-butyl ester for two reasons. First, the methyl branches of the *tert*-butyl group impede the approach of hydroxide to the carbonyl group. Second, the *tert*-butyl "ester" is also an amide, and amides hydrolyze much more slowly than esters.

$$\mathrm{(CH_3)_3CO\overset{\overset{\displaystyle O}{\|}}{C}NH\underset{\underset{\displaystyle CH_3}{|}}{CH}\overset{\overset{\displaystyle O}{\|}}{C}O^-\ Na^+ \;+\; CH_3OH}$$

(e) This is a diazotization to form the *N*-diazo compound, that is, the acyl azide. (See Eq. 23.73 on text p. 1104.) [*Note:* The products of parts (e), (f), and (g) are summarized in the solution to part (g).]

(f) Heating the acyl azide gives a Curtius rearrangement to the isocyanate.

(g) The amino group of the amino acid ester adds to the isocyanate to give a urea. To summarize parts (e)–(g):

$$\underset{\substack{\text{benzoyl azide}\\\text{product of part (e)}}}{\mathrm{Ph\overset{\overset{\displaystyle O}{\|}}{C}{-}\overset{-}{N}{-}\overset{+}{N}{\equiv}N}} \xrightarrow[-N_2]{\text{heat}} \underset{\substack{\text{phenyl isocyanate}\\\text{product of part (f)}}}{\mathrm{PhN{=}C{=}O}} \xrightarrow[\text{valine methyl ester}]{\mathrm{H_2N\underset{\underset{\displaystyle CH(CH_3)_2}{|}}{C}HCO_2CH_3}} \underset{\substack{N\text{-phenylaminocarylvaline}\\\text{methyl ester}\\\text{product of part (g)}}}{\mathrm{PhNH\overset{\overset{\displaystyle O}{\|}}{C}NH\underset{\underset{\displaystyle CH(CH_3)_2}{|}}{C}HCO_2CH_3}}$$

(h) This is a formylamidomalonate reaction, a variation of the acetamidomalonate synthesis.

$$\mathrm{H_3\overset{+}{N}\underset{\underset{\displaystyle \underset{\underset{\displaystyle H_2C{=}C{-}CH_3}{|}}{CH_2}}{|}}{C}HCO_2H \;+\; HCO_2H \;+\; EtOH}$$

(i) Hydrazine ($H_2N{-}NH_2$) displaces the ethoxy group of the ester to form a hydrazide *A*; diazotization forms the acyl azide *B*; heating this in ethanol gives a Curtius rearrangement in which the isocyanate is trapped by reaction with the solvent to give the ethyl carbamate *C*; and acidic hydrolysis generates the amino-protonated amino acid *D*.

$$\underset{A}{\mathrm{N{\equiv}C\underset{\underset{\displaystyle CH(CH_3)_2}{|}}{C}H\overset{\overset{\displaystyle O}{\|}}{C}NHNH_2}} \xrightarrow{\mathrm{NaNO_2/HCl}} \underset{B}{\mathrm{N{\equiv}C\underset{\underset{\displaystyle CH(CH_3)_2}{|}}{C}H\overset{\overset{\displaystyle O}{\|}}{C}N_3}} \xrightarrow{\text{EtOH, heat}} \underset{C}{\mathrm{N{\equiv}C\underset{\underset{\displaystyle CH(CH_3)_2}{|}}{C}HNH\overset{\overset{\displaystyle O}{\|}}{C}OEt}} \xrightarrow[\text{heat}]{\mathrm{HCl/H_2O}} \underset{D}{\mathrm{HO_2C\underset{\underset{\displaystyle CH(CH_3)_2}{|}}{C}H\overset{+}{N}H_3\ Cl^-}}$$

26.64 The chemistry is very similar to that of Problem 26.63(i). Compound *A* is the hydrazide; compound *B* is the acyl azide; compound *C* is the conjugate acid of an *N*-(1-aminoalkyl)amide, which is a nitrogen analog of a hemiacetal. This hydrolyzes in base to the aldehyde shown, ammonia, and acetamide *D*. (If the conditions were harsh enough, acetamide *D* would be converted into acetate ion and ammonia. If you postulated that compound *D* is acetate ion, your answer is equally satisfactory.)

$$\underset{A}{CH_3\overset{O}{\overset{\|}{C}}NH\underset{CH(CH_3)_2}{CH}\overset{O}{\overset{\|}{C}}NHNH_2} \xrightarrow{NaNO_2/HCl} \underset{B}{CH_3\overset{O}{\overset{\|}{C}}NH\underset{CH(CH_3)_2}{CH}\overset{O}{\overset{\|}{C}}N_3} \xrightarrow[\text{heat}]{HCl/H_2O} \underset{C}{CH_3\overset{O}{\overset{\|}{C}}NH\underset{CH(CH_3)_2}{CH}\overset{+}{N}H_3} \xrightarrow[\text{heat}]{H_2O,\ {}^-OH}$$

$$\underset{D}{CH_3\overset{O}{\overset{\|}{C}}NH_2} + NH_3 + (CH_3)_2CHCH{=}O$$

26.65 (a) The first step of the mechanism is formation of an imine. (This is discussed in Sec. 19.11A.) Reaction of this imine with the conjugate base of the thiol gives an addition product which is eventually transformed into the product. The following mechanism begins with the imine.

CH=NCH(CH₃)CO₂⁻ / CH=O (benzene ring) + H—SCH₂CH₂OH ⇌ CH=N⁺(H)CH(CH₃)CO₂⁻ / CH=O + ⁻SCH₂CH₂OH → CH(SCH₂CH₂OH)—NHCH(CH₃)CO₂⁻ / CH=O →

cyclic: S–CH₂CH₂OH, H, +N(H)–CHCO₂⁻(CH₃), O⁻ $\xrightleftharpoons[\text{transfer})]{\text{(proton}}$ S–CH₂CH₂OH, H, N—CHCO₂⁻(CH₃), OH $\xrightarrow{^-OH}$ isoindole with S—CH₂CH₂OH, N—CHCO₂⁻(CH₃)

+ ⁻OH + H—OH

(b) In the first part of the mechanism, ⁻SH displaces chloride in an S_N2 reaction. Recall that α-halo carbonyl compounds are particularly reactive in S_N2 reactions (Sec. 22.3D). Ammonia then reacts to forms an enamine. (Enamine formation is discussed in Sec. 19.11B.) Although imine formation is favored, imines and enamines are in equilibrium just as aldehydes and enols are in equilibrium. Addition of both the —SH group and the —NH_2 group of this enamine to acetone (in a reaction much like acetal formation) and proton transfers give the first product *A*.

HS⁻ + CH=O / CH₂—Cl → CH=O / CH₂—SH + Cl⁻ $\xrightarrow[\text{formation})]{NH_3 \ \text{(enamine}}$ CH—NH₂ ‖ CH—SH + (H₃C)₂C=O →

CH—N⁺H₂—C(CH₃)₂(O⁻) ‖ CH—SH $\xrightleftharpoons[\text{transfer})]{\text{(proton}}$ CH—NH—C(CH₃)₂(OH) ‖ CH—SH → CH—N⁺(H)=C(CH₃)₂ ‖ CH—SH + ⁻OH →

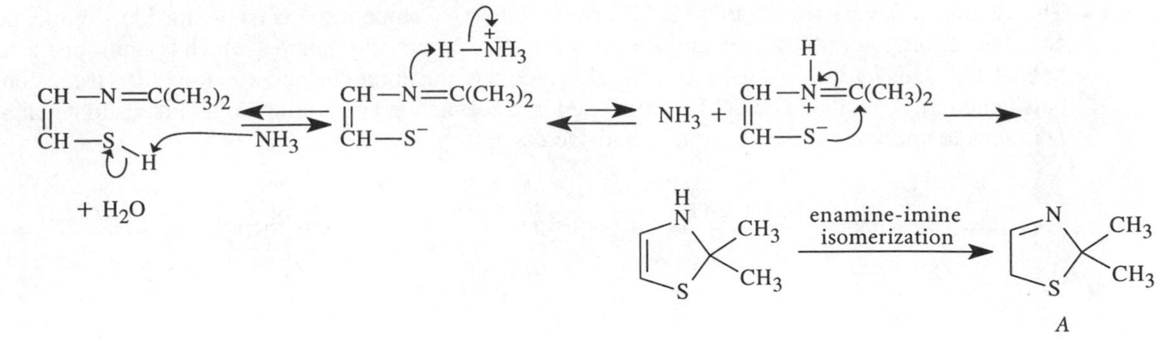

The isomerization leading to compound *A* is analogous to the conversion of an enol into a carbonyl compound. Compound *A* then reacts with cyanide ion in a variation of the Strecker synthesis. Hydrolysis in acid liberates acetone and the conjugate acid *B* of the α-cyano amine. The cyano group hydrolyzes in acid to give the conjugate acid of cysteine. (The mechanism of nitrile hydrolysis is given in Eqs. 21.20a–c on text p. 956.)

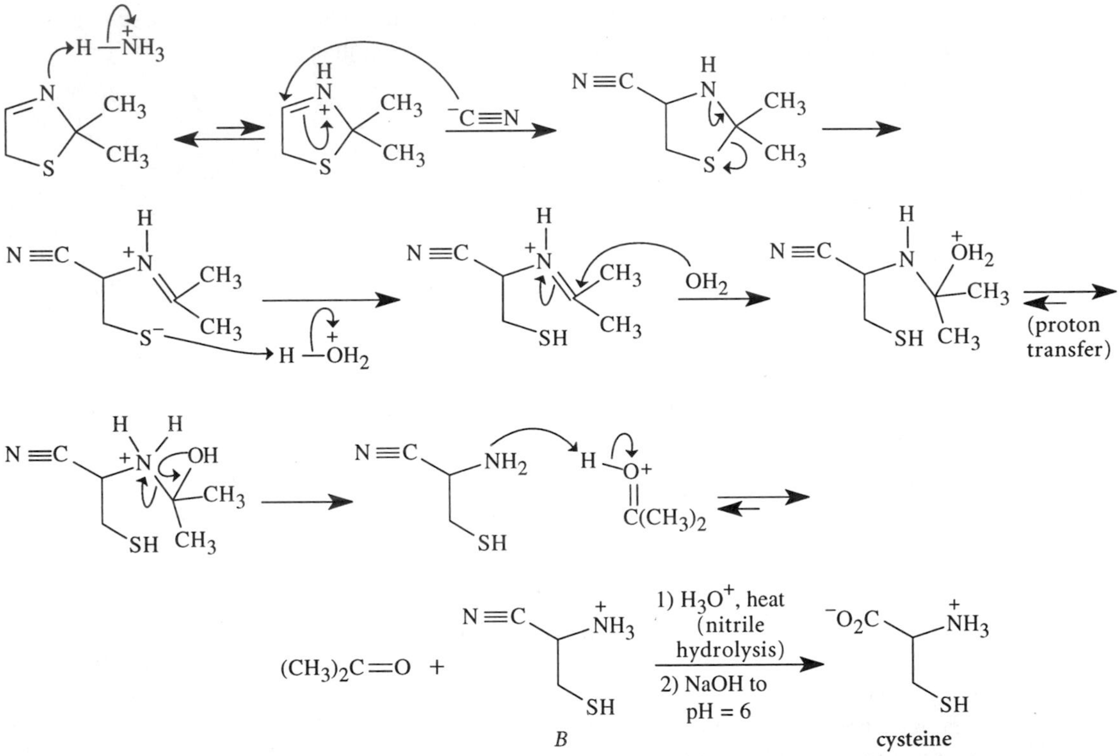

(c) Formaldehyde and the amino group of tryptophan react to give the conjugate acid of an imine, which acts as an electrophile in an intramolecular electrophilic aromatic substitution reaction to give the product. The mechanism below begins with the conjugate acid of the imine.

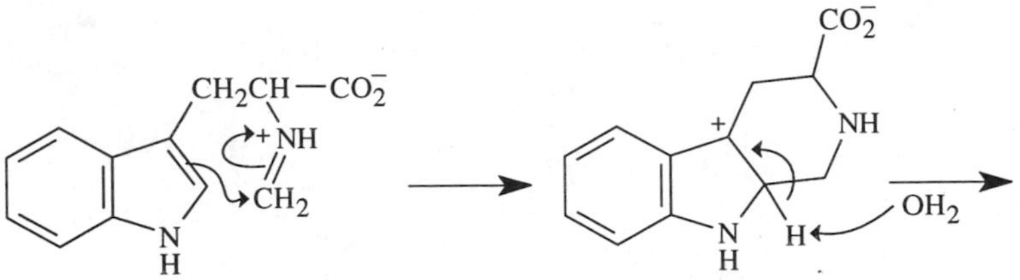

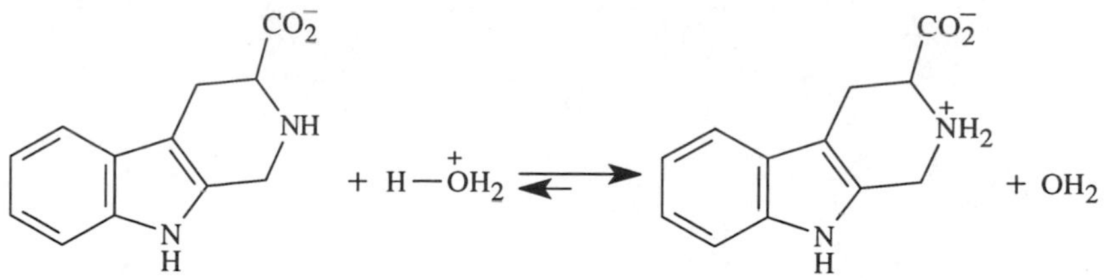

(d) This reaction is a Curtius rearrangement (with concomitant loss of N_2) to give an isocyanate, which reacts intramolecularly with the side-chain hydroxy group. The mechanism of the Curtius rearrangement is given in Eq. 23.67 on text p. 1103. The mechanism below begins with the isocyanate.

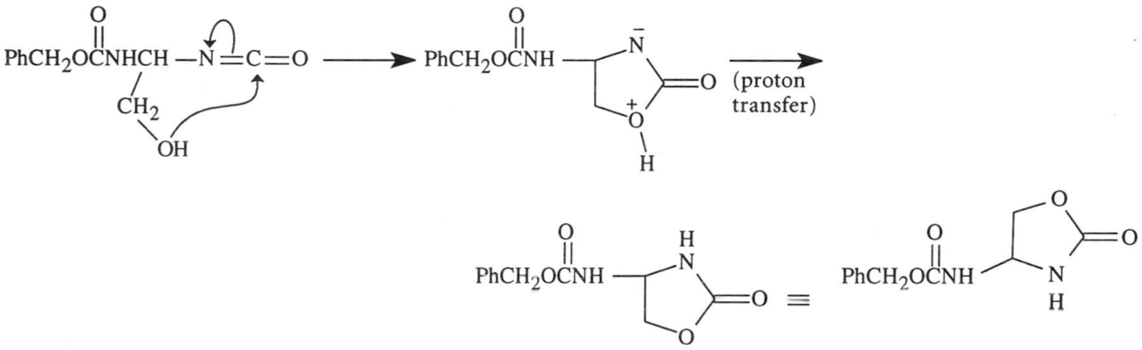

(proton transfer)

26.66 Peptide *I* results from intramolecular nucleophilic attack of the conjugate-base anion of the neighboring amide bond. This occurs mostly at Asn-Gly because glycine has no carbon side chain; in residues other than glycine, the carbon side chains can cause van der Waals repulsions in the ring-closure step.

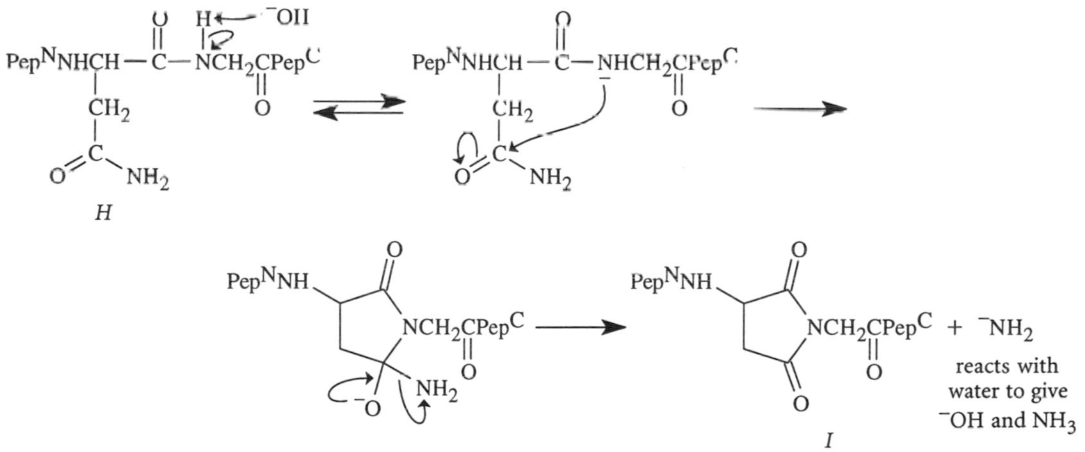

H

reacts with water to give ^-OH and NH_3

I

Derivative *I* is an imide, which is the nitrogen analog of an anhydride. Hydroxide can attack either carbonyl carbon of the imide ring to open the ring, thus generating either peptide *J* or peptide *K*. The mechanism for the formation of peptide *K* is as follows:

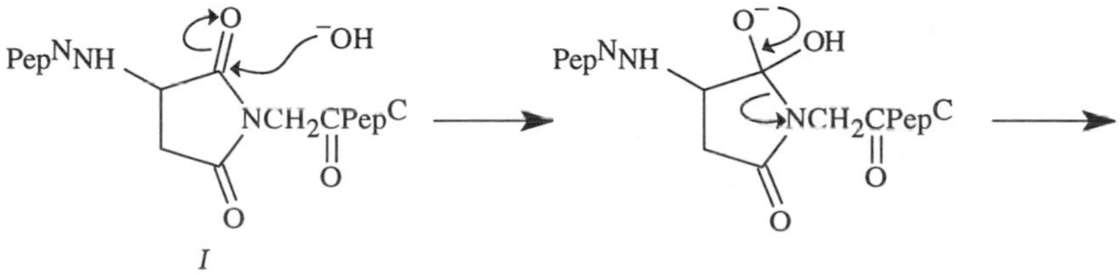

I

(Solution continues)

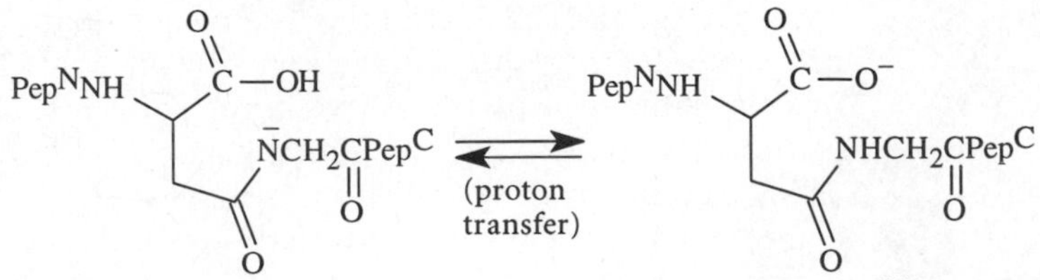

The structure of peptide *J* is as follows; you should show the mechanism for its formation.

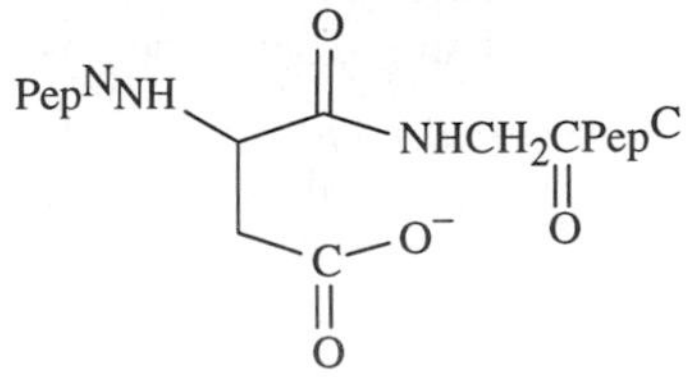

26.67 (a) In the *Z* conformation, the two large groups Pep^N and the *N*-alkyl group Pep^C are trans, and thus van der Waals repulsions between these groups cannot occur. In the *E* conformation. these two groups are close enough that van der Waals repulsions can result. These raise the energy of the peptide; hence, the *Z* conformation is energetically preferred.

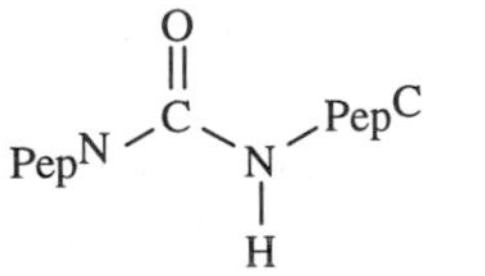

Z conformation
large groups are trans

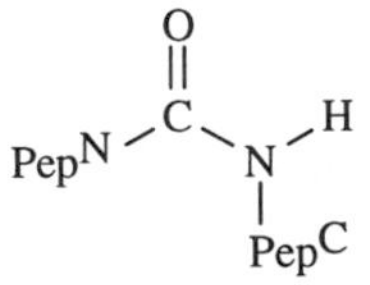

E conformation
large groups are cis

(b) The proline nitrogen bears two *N*-alkyl groups, whereas the peptide-bond nitrogens of other residues have only one. Thus, in either the *E* or the *Z* conformation, an alkyl group is cis to the Pep^N group. Although the group that is cis to Pep^N in the *E* configuration is larger, there is a much smaller difference in energy between *E* and *Z* conformations for proline residues than there is for other residues. Hence, *E* conformations in peptide bonds are most often found at proline.

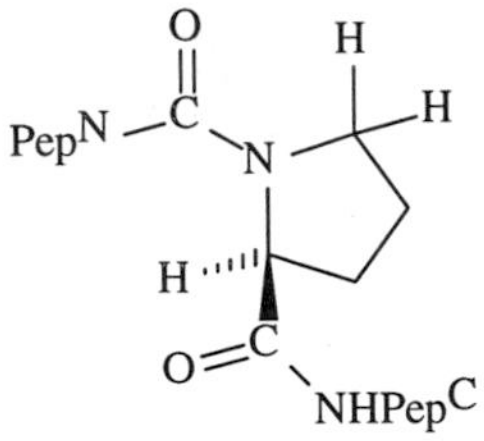

proline residue
in an *E* conformation

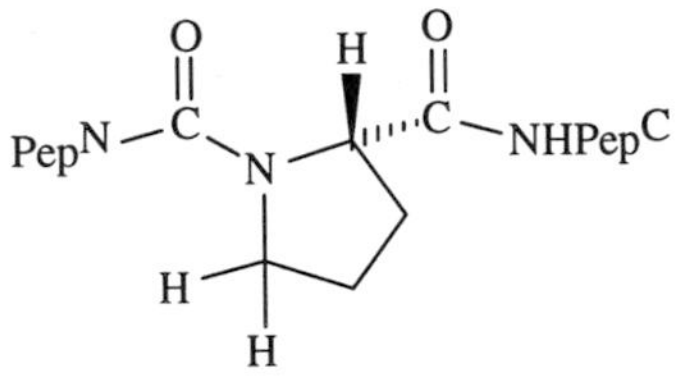

proline residue
in a *Z* conformation

26.68 (a) The two monomethyl esters result from ester formation at each of the two carboxy groups.

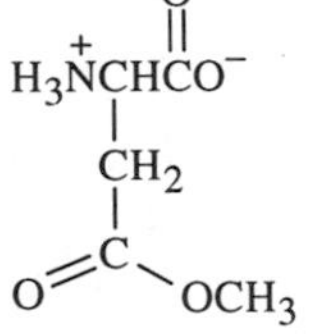

A

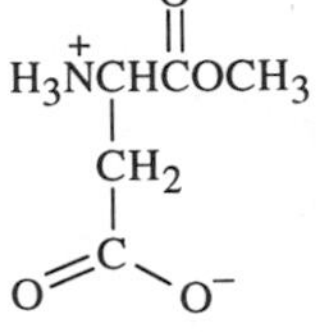

B

(b) According to Table 26.1, the pK_a values of the two carboxy groups in aspartic acid are about 1.9 (for the α-carboxy group) and 3.6 (for the β-carboxy group). (How do we know which is which?) Assume that the pK_a values for the respective esters are similar. At pH 3.0, the carboxy group of compound *A* is largely ionized; at the same pH, the carboxy group of *B* is mostly un-ionized. Hence, at pH 3.0, compound *B* carries a positive charge, whereas compound *A* is neutral. Therefore, compound *B* will be retained by a cation-exchange column (which contains negatively-charged groups) and will be eluted more slowly than compound *A*. At pH 7, however, both compounds are negatively charged. Since they have the same charge, neither is retained by a cation-exchange column; consequently, these compounds are therefore not separated.

26.69 The formula of compound *A* is the same as that of *N*-acetylaspartic acid minus the elements of water. Since treatment of a dicarboxylic acid with acetic anhydride is a good way to prepare a cyclic anhydride (see Eq. 20.31 on text p. 921), it is reasonable to suppose that compound *A* is the cyclic anhydride of aspartic acid. When this anhydride is treated with L-alanine, the amino group of alanine can attack either of the two carbonyl groups to give either of the two isomeric peptides *B* and *C*. Both would hydrolyze to give alanine and aspartic acid (as well as acetic acid).

$CH_3C(=O)NH$–(cyclic anhydride) + $H_3\overset{+}{N}—CH(CH_3)—CO_2^-$ ⟶

N-acetylaspartic anhydride (compound *A*); alanine

$CH_3C(=O)NH—CH(CH_2COOH)—C(=O)—NH—CH(CH_3)—CO_2H$ + $CH_3C(=O)NH—CH(COOH)—CH_2—C(=O)—NH—CH(CH_3)—CO_2H$

B; *C*

Carbohydrates and Nucleic Acids

TERMS

aldaric acid 27.7, 27.7B
alditol 27.7, 27.7D
aldohexose 27.1
aldonic acid 27.7, 27.7A
aldonolactone 27.7A
aldose 27.1
alkylating agent 27.12B
amino sugar 27.10B
amylopectin 27.10B
amylose 27.10B
anomalous dispersion 27.9B
anomer 27.2B
α-anomer 27.2B
β-anomer 27.2B
anomeric carbon 27.2B
ATP 27.11B
base (in nucleic acids) 27.11B
carbohydrate .... Introduction
cellulose 27.10B
Chargaff's rules 27.11B
chitin 27.10B
cyclic hemiacetal 27.2B
deoxyribonucleic acid (DNA) 27.11B
deoxyribonucleoside 27.11A
deoxyribonucleotide 27.11A
dextrose 27.10A
disaccharide 27.1, 27.10A
D,L system 27.2A
double helix 27.11B
enediol 27.4
epimer 27.2A
furanose 27.2B
furanoside 27.5
genetic code 27.12A
D-glyceraldehyde 27.2A
glycoside 27.5
Haworth projection 27.2B
hexose 27.1
ketopentose 27.1
ketose 27.1
Kiliani-Fischer synthesis 27.8
lignin 27.10B
Lobry de Bruyn-
  Alberda van Ekenstein reaction 27.4
messenger RNA (mRNA) 27.12A
monosaccharide 27.1
mutarotation 27.3
nonreducing end 27.10A
nonreducing sugar 27.10A
nucleic acid 27.11B
nucleotide 27.11A
oligosaccharide 27.1
pentose 27.1
pentulose 27.1
polysaccharide 27.1, 2710B
purine 27.11A
pyranose 27.2B
pyranoside 27.5
pyrimidine 27.11A
reducing end 27.10A
reducing sugar 27.10A
ribonucleic acid (RNA) 27.11B
ribonucleoside 27.11A
ribonucleotide 27.11A
Ruff degradation 27.9B
squiggly bond 27.2B
starch 27.10B
sugar .... Introduction
transcription (of DNA) 27.12A
translation (of mRNA) 27.12A
uronic acid 27.7
Watson-Crick base pair 27.11B

CONCEPTS

I. INTRODUCTION TO CARBOHYDRATES

A. Classification and Properties of Carbohydrates

1. Carbohydrates are defined as aldehydes and ketones containing a number of hydroxy groups on an unbranched carbon chain (sugars), as well as their chemical derivatives.
   a. Most of the common sugars have formulas that fit a "hydrate of carbon" pattern, that is, a formula of the form $C_n(H_2O)_m$.
2. Carbohydrates can be classified using a variety of systems:
   a. by the type of carbonyl group in the carbohydrate:
      i. A carbohydrate with an aldehyde carbonyl group is called an aldose.
      ii. A carbohydrate with a ketone carbonyl group is called a ketose.
   b. by the number of carbon atoms:
      i. A six-carbon carbohydrate is called a hexose.
      ii. A five-carbon carbohydrate is called a pentose.
   c. by a combination of the two systems cited in (a) and (b) above:
      i. A six-carbon carbohydrate containing an aldehyde carbonyl group is called an aldohexose.
      ii. A five-carbon carbohydrate containing a ketone carbonyl group is called a ketopentose or pentulose.
      iii. A ketose can be indicated with the suffix *ulose*; thus, a six-carbon ketose is also termed a hexulose.

$CH_2(OH)—CH(OH)—CH(OH)—CH(OH)—CH(OH)—C(=O)—H$

an aldose
a hexose
an aldohexose

$CH_2(OH)—CH(OH)—CH(OH)—C(=O)—CH_2(OH)$

a ketose
a pentose
a ketopentose or pentulose

   d. by their hydrolysis to simpler carbohydrates:
      i. Monosaccharides cannot be converted into simpler carbohydrates.
      ii. Disaccharides are converted into two monosaccharides.
      iii. Trisaccharides are converted into three monosaccharides.
      iv. Oligosaccharides are converted into several monosaccharides.
      v. Polysaccharides are converted into a large number of monosaccharides.
3. Carbohydrates are very soluble in water because of their many hydroxy groups but virtually insoluble in nonpolar solvents.

II. STRUCTURE OF MONOSACCHARIDES, DISACCHARIDES, AND POLYSACCHARIDES

A. Stereochemistry and Configuration

1. The aldopentoses have three asymmetric carbons and 2^3 or eight possible stereoisomers and can be divided into two enantiomeric sets of four diastereomers.
2. The aldohexoses have four asymmetric carbons and 2^4 or sixteen possible stereoisomers and can be divided into two enantiomeric sets of eight diastereomers.
3. Each diastereomer is a different carbohydrate with different properties, known by a different name.
4. Although the *R*,*S* system could be used to describe the configuration of the asymmetric carbon atoms of a carbohydrate, it is more convenient to use the D,L-system, in which the configuration of a carbohydrate enantiomer is specified by applying the following conventions:

a. The 2*R* enantiomer of the naturally occurring aldotriose glyceraldehyde is arbitrarily said to have the D configuration; its enantiomer, the 2*S* enantiomer, is then said to have the L configuration.

R configuration — CHO, H, HO, C, CH_2OH = CHO; H—C—OH; CH_2OH (a D carbohydrate)

CHO; HO—C—H; CH_2OH (an L carbohydrate) = CHO, $HOCH_2$, C, H, OH — S configuration

b. Other aldoses or ketoses are written in a Fischer projection with their carbon atoms in a straight vertical line, and the carbons are numbered consecutively as they would be in systematic nomenclature, so that the carbonyl carbon receives the lowest number.

$CH_2(OH)$—CH(OH)—CH(OH)—C(=O)—$CH_2(OH)$ = ${}^{1}CH_2OH$; 2 =O; H—3—OH; H—4—OH; ${}^{5}CH_2OH$

c. The asymmetric carbon of highest number is designated as a reference carbon.
   i. If this carbon has the H, OH, and CH_2OH groups in the same relative configuration as the same three groups of D-glyceraldehyde, the carbohydrate is said to have the D configuration.
   ii. If this carbon has the same configuration as L-glyceraldehyde, the carbohydrate is said to have the L configuration.

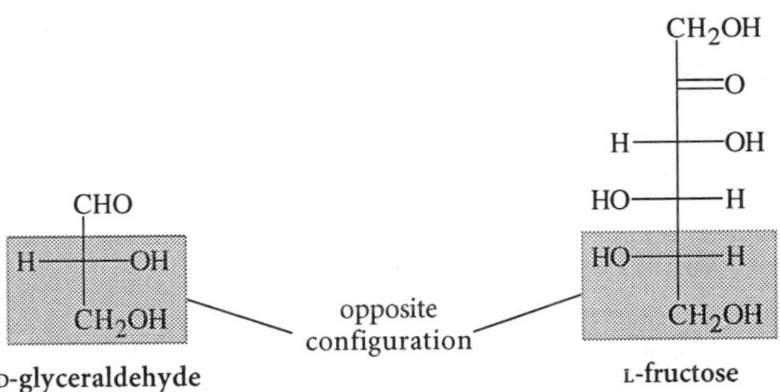

5. There is no general correspondence between configuration and the sign of the optical rotation; there is also no simple relationship between the D,L system and the *R*,*S* system.
   a. The *R*,*S* system is used to specify the configuration of each asymmetric carbon atom in a molecule.
   b. The D,L system specifies a particular enantiomer of a molecule that can contain any number of asymmetric carbons.
6. A few of the aldoses and ketoses are particularly important, and their structures should be learned.

D-glucose: CHO; H—OH; HO—H; H—OH; H—OH; CH_2OH

D-mannose: CHO; HO—H; HO—H; H—OH; H—OH; CH_2OH

D-galactose: CHO; H—OH; HO—H; HO—H; H—OH; CH_2OH

D-fructose: CH_2OH; =O; HO—H; H—OH; H—OH; CH_2OH

D-ribose: CHO; H—OH; H—OH; H—OH; CH_2OH

7. Compounds that differ in configuration at only one of several asymmetric carbons are called epimers. For example, mannose and galactose are epimers of glucose.

B. Cyclic Structures of the Monosaccharides

1. Aldoses and ketoses exist predominantly as cyclic hemiacetals.
2. In many carbohydrates cyclic hemiacetals that involve either five-membered or six-membered rings are possible, depending on which hydroxy group undergoes cyclization.
   a. A five-membered cyclic acetal form of a carbohydrate is called a furanose.
   b. A six-membered cyclic acetal form of a carbohydrate is called a pyranose.

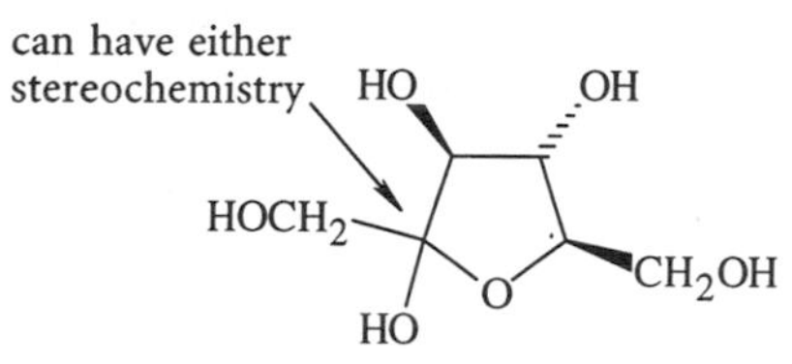

can have either stereochemistry

D-fructofuranose

can have either stereochemistry

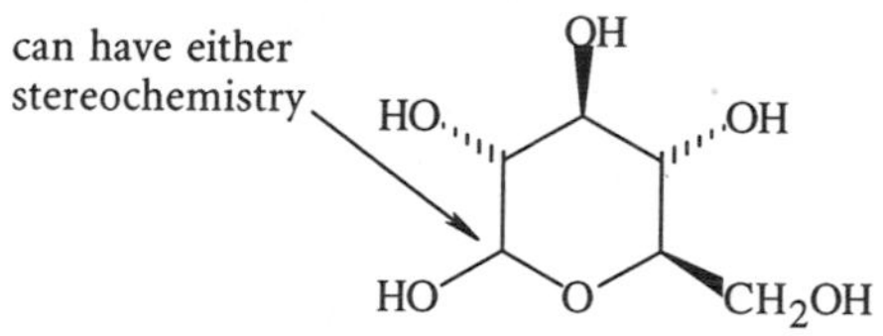

D-glucopyranose

   c. The aldohexoses and aldopentoses exist predominantly as pyranoses, but the furanose forms of some carbohydrates are important.
3. A name such as glucose is used when referring to any or all forms of the carbohydrate.
4. To name a cyclic hemiacetal form of a carbohydrate, start with a prefix derived from the name of the carbohydrate followed by a suffix that indicates the type of hemiacetal ring.
   i. furanose for a five-membered ring
   ii. pyranose for a six-membered ring
5. A widely used convention for representing the cyclic forms of sugars is the Haworth projection.
   a. In this convention, the cyclic form of a sugar is represented as a planar ring perpendicular to the page.
   b. The shaded ring bonds are in front of the page, and the other ring bonds are in back.
   c. The substituents are indicated with up or down bonds.
   d. "Squiggly bonds" are used to indicate mixed or uncertain stereochemistry, usually at the hemiacetal carbon.

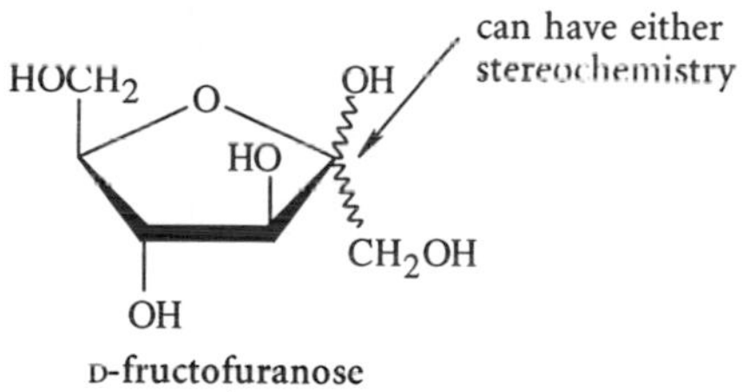

can have either stereochemistry

D-fructofuranose

can have either stereochemistry

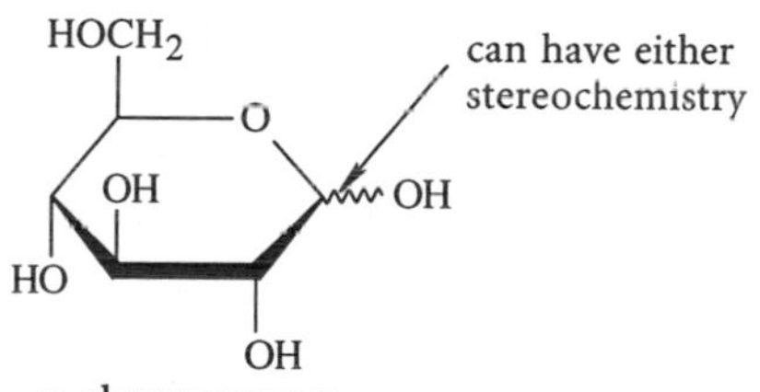

D-glucopyranose

   e. Haworth projections show the stereochemical configurations and the cyclic structures of carbohydrates without showing conformational details. Although still used, they offer little advantage over wedge/dashed wedge formulas.

C. Anomers and Mutarotation

1. The furanose or pyranose form of a carbohydrate has one more asymmetric carbon (carbon-1) than the open-chain form.
   a. When two cyclic forms of a carbohydrate differ in configuration only at the hemiacetal carbon (carbon-1 of an aldose), they are said to be anomers.
   b. The hemiacetal or acetal carbon of a carbohydrate is called the anomeric carbon.
2. The ring in a Fischer projection of these cyclic compounds is represented as a long bond.
3. Anomers are named with the Greek letters α and β.
   a. In the α-anomer, the hemiacetal —OH group is on the same side of the Fisher projection as the oxygen at the configurational carbon.

b. In the β-anomer, the hemiacetal —OH group is on the side of the Fisher projection opposite the oxygen at the configurational carbon.

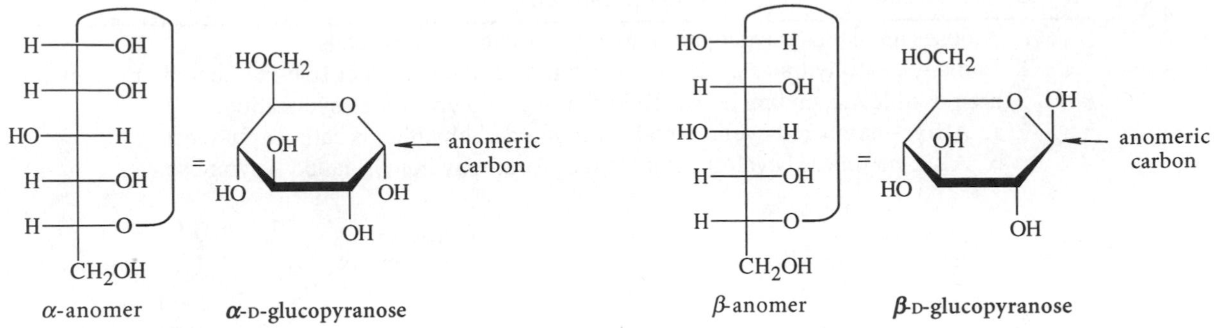

anomeric carbon

α-anomer α-D-glucopyranose

anomeric carbon

β-anomer β-D-glucopyranose

4. When a pure anomer of a carbohydrate is dissolved in aqueous solution, its optical rotation changes with time.
   a. This change, called mutarotation, is catalyzed by both acid or base, and it results from the formation of an equilibrium mixture of both the α- and β-anomers.
   b. Mutarotation is characteristic of the cyclic hemiacetal forms of carbohydrates.
5. The mechanism of mutarotation begins as the reverse of hemiacetal formation.
   a. A 180° rotation about the bond to the carbonyl group permits attack of the hydroxy group on the opposite face of the carbonyl carbon.
   b. Hemiacetal formation then gives the other anomer.

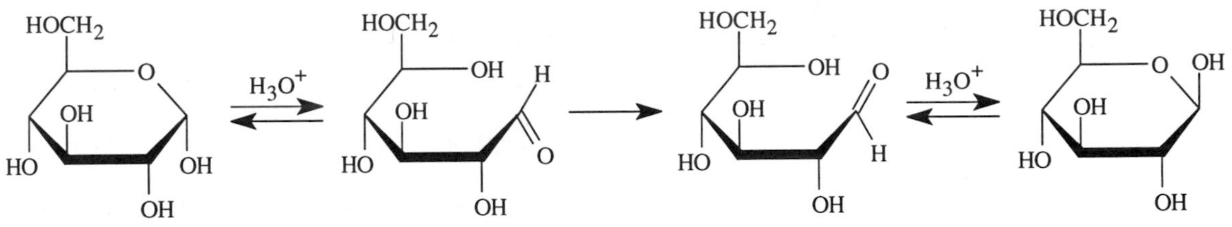

H_3O^+ H_3O^+

6. Some general conclusions:
   a. Most aldohexoses exist primarily as pyranoses, although a few have substantial amounts of furanose forms.
   b. There are relatively small amounts of noncyclic carbonyl forms of most monosaccharides.
   c. Mixtures of α- and β-anomers are usually found, although the exact amounts of each vary from case to case.

D. Conformational Representations of Pyranoses

1. The six-membered ring of a pyranose exists in two chair conformations related by the chair-chair interconversion.
2. To go from a Fisher projection to a chair conformation:
   a. Using an allowed manipulation of Fisher projections, first redraw the molecule in an equivalent Fisher projection in which the ring oxygen is in a down position.
   b. Draw a Haworth projection by turning the plane of the ring 90° so that the anomeric carbon is on the right and the ring oxygen is in the rear.
      i. The groups in the up positions are those that are on the left in the Fischer projection.
      ii. The groups in the down positions are those that are on the right in the Fischer projection.
   c. Draw either one of the two chair conformations in which the anomeric carbon and the ring oxygen are in the same relative positions as they are in the Haworth projection above.
   d. Place the up and down groups in axial or equatorial positions, as appropriate.

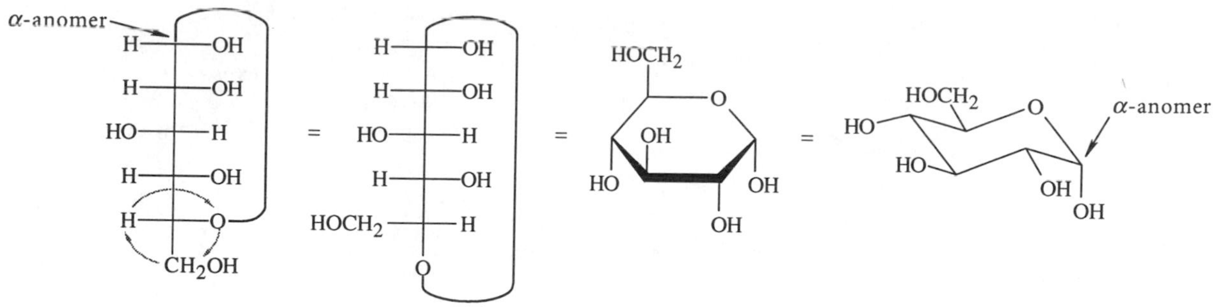

e. If the configuration of the anomeric carbon is uncertain, or if there is a mixture of anomers, the bond is represented by a squiggly line.

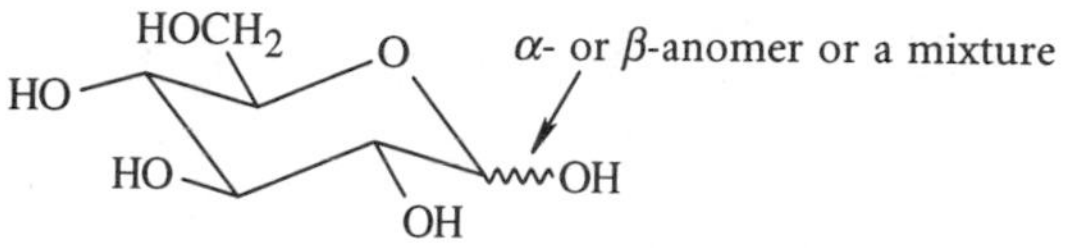

3. The five-membered rings of furanoses are nonplanar, but they are close enough to planarity that Haworth formulas are good approximations to their actual structures and are frequently used to portray them.

4. In some cases it is simpler to derive a cyclic structure of one carbohydrate from its relationship to another carbohydrate.

E. Glycosides

1. Most monosaccharides react with alcohols under acidic conditions to yield cyclic acetals called glycosides, which are named as derivatives of the parent carbohydrate.
   a. The term pyranoside indicates that the glycoside ring is a six-membered ring.
   b. The term furanoside indicates that the glycoside ring is a five-membered ring.

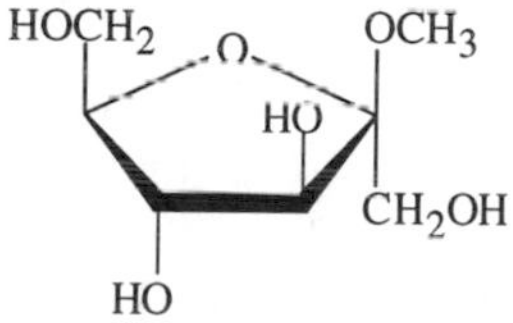

2. Glycoside formation is catalyzed by acid.

3. Because glycosides are acetals, they are stable to base, but are hydrolyzed in dilute aqueous acid back to their parent carbohydrates and alcohols.

4. Many compounds occur naturally as glycosides; the glycoside of a natural product can be hydrolyzed to its component alcohol or phenol and carbohydrate.

F. Structures of Disaccharides and Polysaccharides

1. Disaccharides consist of two monosaccharides connected by a glycosidic linkage.
   a. (+)-Lactose (milk sugar) is a disaccharide in which galactose is linked by a β-glycosidic bond to the oxygen at carbon-4 of glucose.
   b. (+)-Sucrose (table sugar), a disaccharide in which glucose is linked by an α-glycosidic bond to the oxygen at carbon-2 of fructose, is a nonreducing sugar.

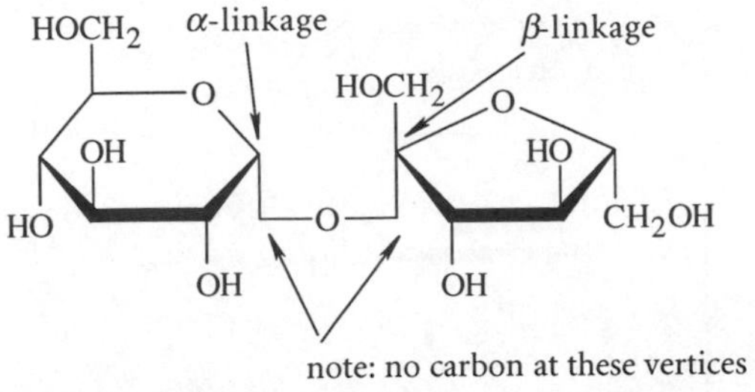

HOCH$_2$ α-linkage β-linkage HOCH$_2$ O O OH HO HO CH$_2$OH O OH OH

note: no carbon at these vertices

sucrose
β-D-fructofuranosyl-α-D-glucopyranoside
α-D-glucopyranosyl-β-D-fructofuranoside

Note that the bonds to the oxygens connecting the rings, in order to maintain the linear representation of the disaccharide structure, have vertices. In ordinary skeletal structures, each vertex would indicate an additional —CH_2— group. In Haworth projections, there are *no* carbons at these vertices. The cellulose structure in item 2 following is drawn with similarly stylized bonds. Note the conformational interpretation of the cellulose structure just below the Haworth structure, which shows that there is only a single oxygen joining each pair of rings.

2. Any number of monosaccharide residues can be linked together with glycosidic bonds to form chains; long chains of connected monosaccharide residues are called polysaccharides.
   a. Cellulose is a polymer of D-glucopyranose residues connected by β-1,4-glycosidic linkages.

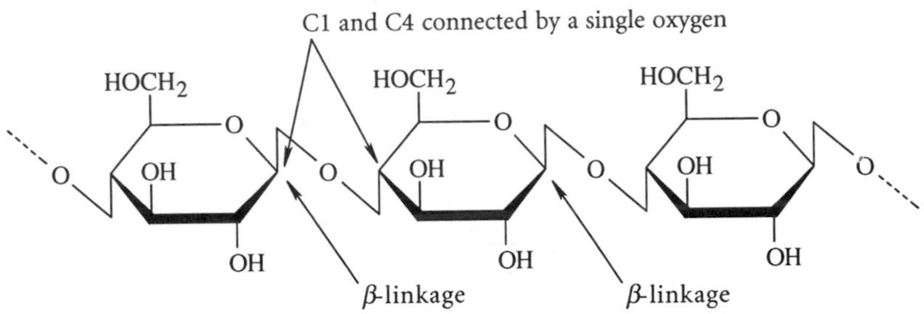

C1 and C4 connected by a single oxygen

HOCH$_2$ HOCH$_2$ HOCH$_2$ O O O O OH O OH O OH O OH OH OH β-linkage β-linkage

cellulose

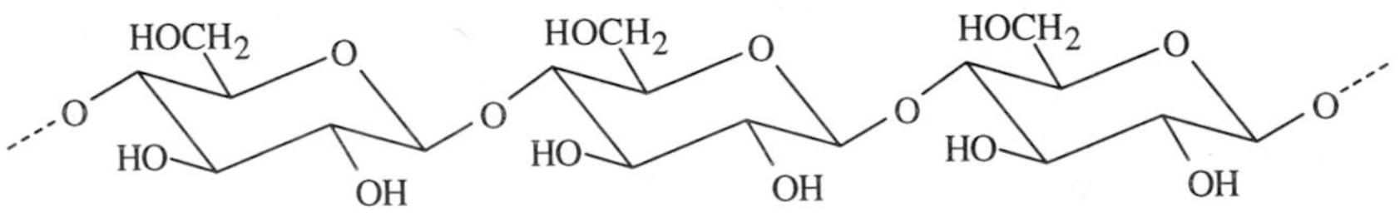

HOCH$_2$ O HOCH$_2$ O HOCH$_2$ O O O O O HO OH HO OH HO OH

cellulose

   b. Starch is a polymer of D-glucopyranose. Starch consists of two components:
      i. Amylose has glucose residues connected by α-1,4-glycosidic linkages.
      ii. Amylopectin has glucose residues connected by α-1,4-glycosidic and α-1,6-glycosidic linkages.
   c. Chitin is a polysaccharide of *N*-acetyl-D-glucosamine connected by β-1,4-glycosidic linkages.

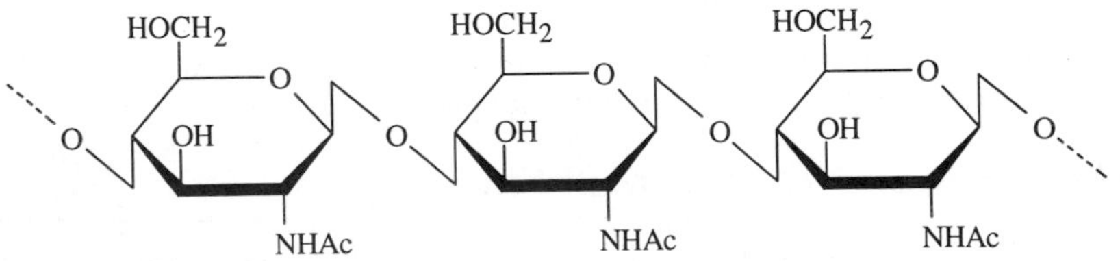

HOCH$_2$ HOCH$_2$ HOCH$_2$ O O O O OH O OH O OH O NHAc NHAc NHAc

chitin

HOCH$_2$ O HOCH$_2$ O HOCH$_2$ O O O O O HO NHAc HO NHAc HO NHAc

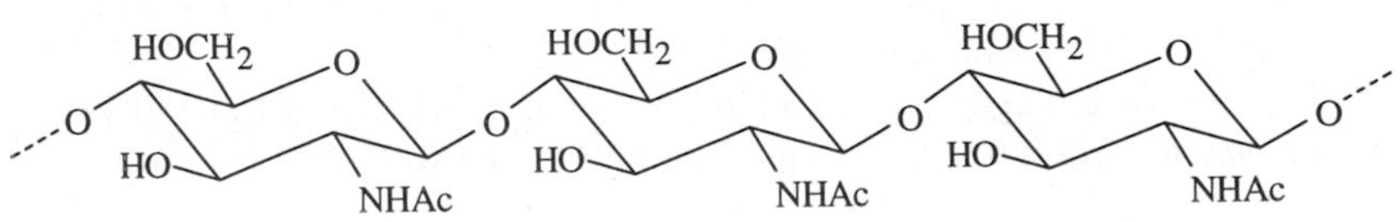

chitin

3. Polysaccharides are mostly long chains with some branches.
   a. There are no highly cross-linked, three-dimensional networks.
   b. Cyclic oligosaccharides are known.

4. The linkages between monosaccharide units are in every case glycosidic linkages; thus, all polysaccharides can be converted into their component monosaccharides by acid-catalyzed hydrolysis.

5. A given polysaccharide incorporates only one stereochemical type of glycoside linkage. Thus, the glycoside linkages in cellulose are all β; those in starch are all α.

III. PROOF OF GLUCOSE STEREOCHEMISTRY

A. The Fischer Proof

1. Fischer arbitrarily assumed that carbon-5 (the configurational carbon in the D,L system) of (+)-glucose has the —OH on the right in the standard Fischer projection.

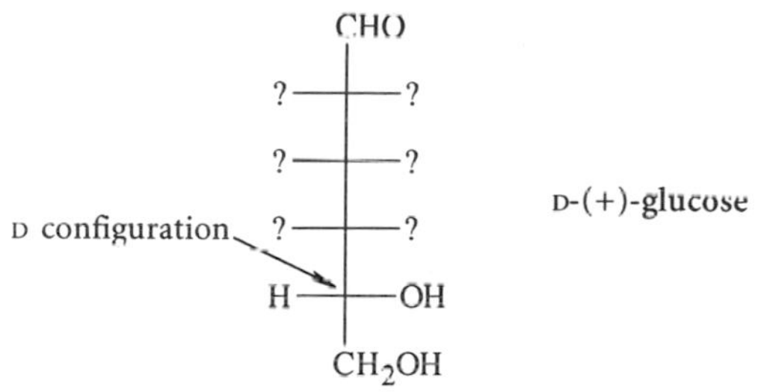

CHO
? — ?
? — ?
? — ?
H — OH
CH_2OH

D configuration → (C-5)

D-(+)-glucose

2. The subsequent logic involved can be summarized in four steps:
   a. (–) Arabinose is converted into both (+)-glucose and (+)-mannose by a Kiliani-Fischer synthesis. Therefore:
      i. (+)-Glucose and (+)-mannose are epimeric at carbon-2.
      ii. (–)-Arabinose has the same configuration at carbons-2, 3, and 4 as that of (+)-glucose and (+)-mannose at carbons-3, 4, and 5, respectively.

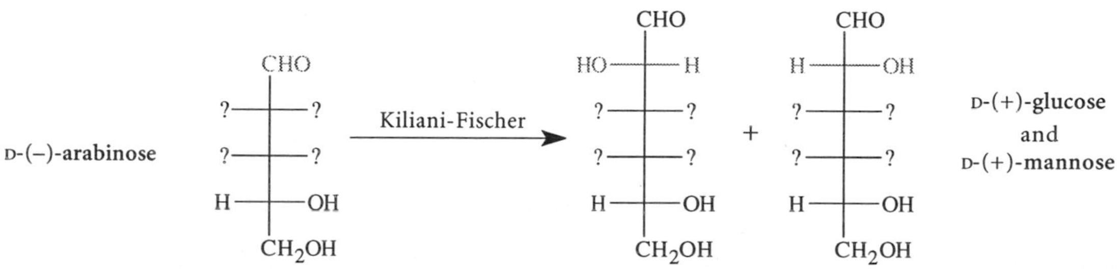

D-(–)-arabinose: CHO; ? — ?; ? — ?; H — OH; CH_2OH

Kiliani-Fischer →

CHO; HO — H; ? — ?; ? — ?; H — OH; CH_2OH + CHO; H — OH; ? — ?; ? — ?; H — OH; CH_2OH

D-(+)-glucose and D-(+)-mannose

   b. (–)-Arabinose can be oxidized by dilute HNO_3 to an optically active aldaric acid. Therefore:
      i. The —OH group at carbon-2 of arabinose must be on the left; otherwise, arabinose would be meso, regardless of the configuration of the —OH group at carbon-3.
      ii. The —OH group at carbon-3 of (+)-glucose must be on the left.

```
      CHO                         CO2H
  ?---+---?                 HO ---+--- H
  ?---+---?     HNO3 -->     ?---+---?     HO— group must
  H---+---OH                 H---+---OH    be on the left side
     CH2OH                       CO2H
D-(−)-arabinose          optically active aldaric acid
```

```
            CO2H
 (H,OH)-----+-----(OH,H)
     HO-----+-----H
      ?-----+-----?
      H-----+-----OH
            CO2H
D-(+)-glucose and D-(+)-mannose
```

c. Oxidation of both (+)-glucose and (+)-mannose with HNO_3 give optically active aldaric acids. Therefore:

i. The —OH group at carbon-4 is on the right in both (+)-glucose and (+)-mannose.

ii. The configuration at carbon-4 of (+)-glucose and (+)-mannose is the same as that at carbon-3 of (–)-arabinose.

```
      CHO                     CO2H
  H---+---OH              H---+---OH     —OH group
 HO---+---H    HNO3 -->  HO---+---H      must be on
  ?---+---?               H---+---OH     the right side
  H---+---OH              H---+---OH
     CH2OH                   CO2H
D-(+)-glucose or         optically active
D-(+)-mannose            aldaric acid
```

```
      CHO                                         CHO
 HO---+---H                                    H---+---OH
 HO---+---H       —OH group is on the same    HO---+---H
  H---+---OH  <--  side in both  -->           H---+---OH
  H---+---OH                                   H---+---OH
     CH2OH                                       CH2OH
D-(+)-glucose or D-(+)-mannose
```

iii. The structure of D-(–)-arabinose is thus determined.

```
      CHO
 HO---+---H
  H---+---OH      D-(−)-arabinose
  H---+---OH
     CO2H
```

d. A completely different aldose, (+)-gulose, can be oxidized with HNO_3 to the same aldaric acid as (+)-glucose.

i. The structure with the —OH group at carbon-2 on the right forms an aldaric acid upon oxidation with HNO_3 that is identical to the aldaric acid formed by the oxidation of (+)-gulose with HNO_3; this must be (+)-glucose.

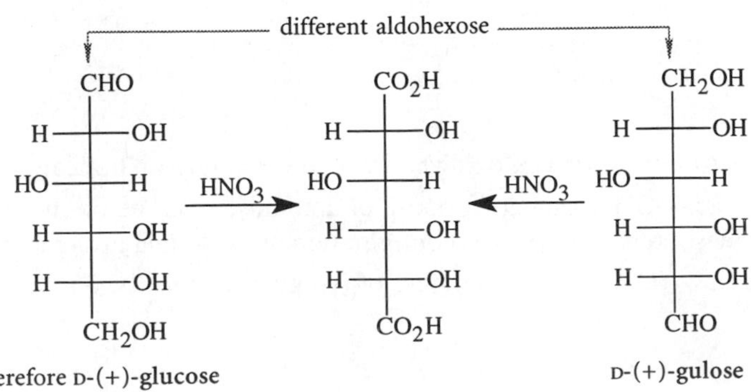

therefore D-(+)-glucose

D-(+)-gulose

ii. Had the —OH group at carbon-2 been on the left, (+)-glucose would have been the only aldohexose that could give the corresponding aldaric acid upon oxidation with HNO_3. Hence, the aldohexose with the —OH group on the left must be the remaining possibility, that is, it must be (+)-mannose.

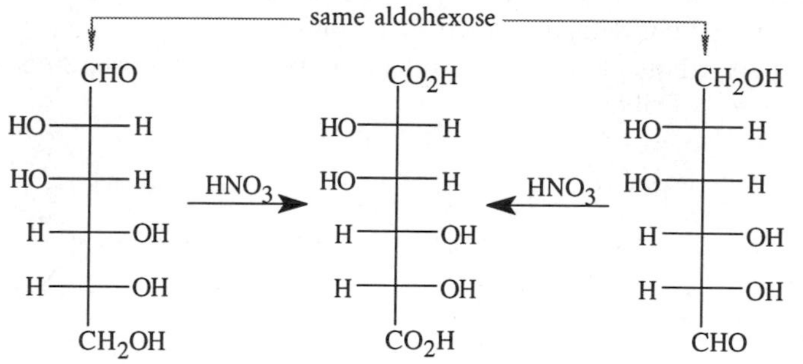

same aldohexose

HNO_3 → ← HNO_3

∴D-(+)-mannose

B. Absolute Configuration of Glucose

1. Two cycles of the Ruff degradation convert (+)-glucose into (–)-erythrose.
   a. D-Glyceraldehyde, in turn, is related to (–)-erythrose by a Kiliani-Fischer synthesis.

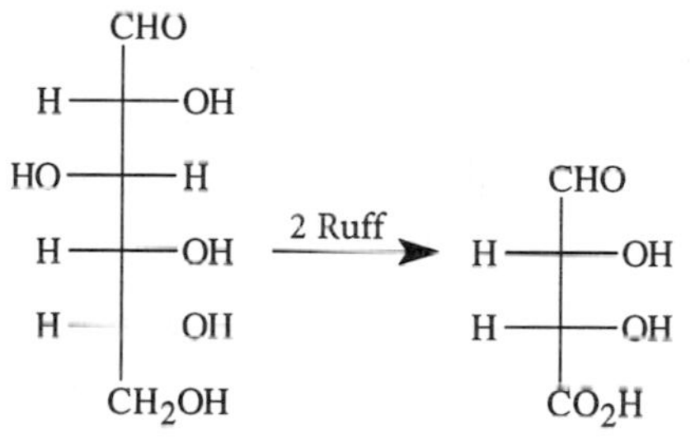

2 Ruff →

D-(+)-glucose D-(–)-erythrose

   b. (+)-Glucose, (–)-erythrose, (–)-threose, and (+)-glyceraldehyde are all of the same stereochemical series—the D series.

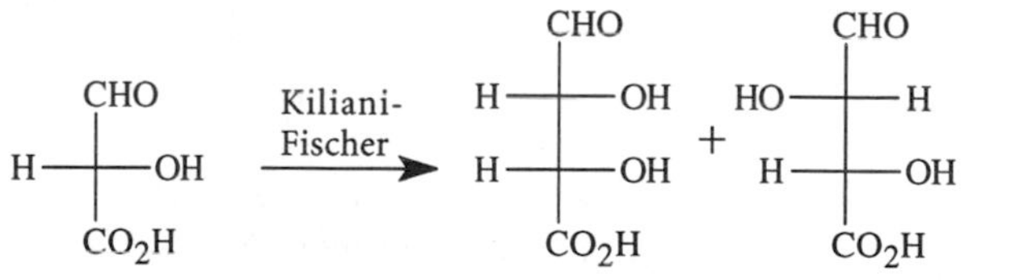

Kiliani-Fischer →

D-(+)-glyceraldehyde D-(–)-erythrose + D-(–)-threose

2. Oxidation of D-(–)-threose with dilute HNO_3 gives (–)-tartaric acid. (+)-Tartaric acid was shown by X-ray crystallography to possess the L configuration; hence, (–)-tartaric acid, and therefore (+)-glucose, have the D configuration.

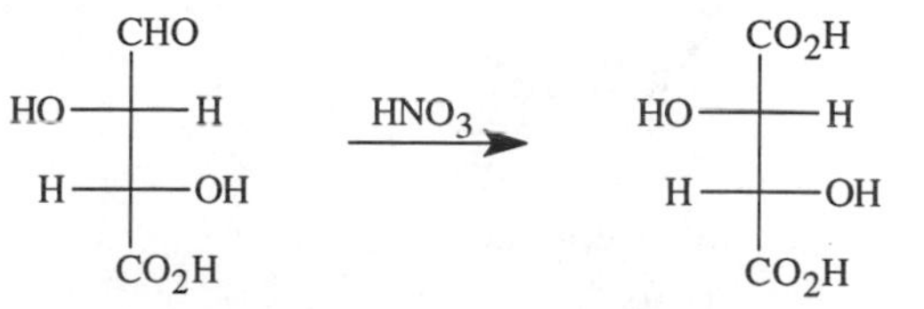

HNO_3 →

D-(–)-threose D-(–)-tartaric acid

IV. NUCLEOSIDES, NUCLEOTIDES, AND NUCLEIC ACIDS

A. Nomenclature of Nucleosides, Nucleotides, and Nucleic Acids

1. A β-glycoside of a heterocyclic nitrogen base is called a nucleoside; the base and the sugar ring systems are numbered separately; primes (´) are used to refer to the sugar carbon atoms.
   a. A ribonucleoside is derived from D-ribose.
   b. A deoxyribonucleoside is derived from D-2-deoxyribose; 2-deoxyribose lacks the —OH group at carbon-2 of ribose.

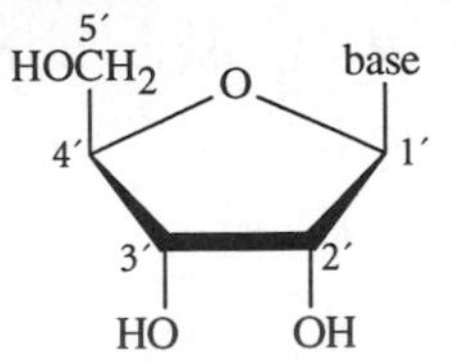

5´ HOCH₂ O base 4´ 1´ 3´ 2´ HO OH

a ribonucleoside

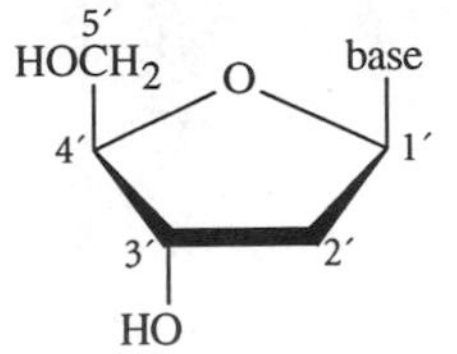

5´ HOCH₂ O base 4´ 1´ 3´ 2´ HO

a deoxyribonucleoside

2. The bases that occur most frequently in nucleosides are derived from two heterocyclic ring systems:
   a. Three pyrimidines occur most commonly and are attached to the sugar at the *N*-1 position.
   b. Two purines occur most commonly and are attached to the sugar at the *N*-9 position.

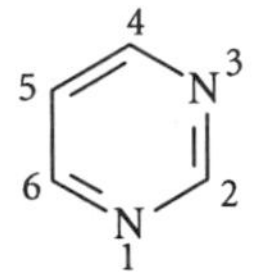

pyrimidine

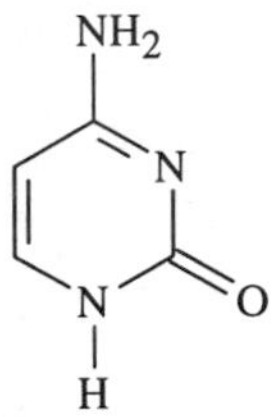

cytosine (C)

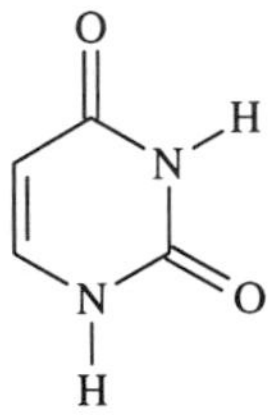

uracil (U)

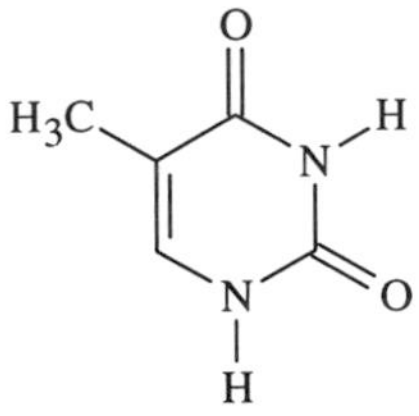

thymine (T)

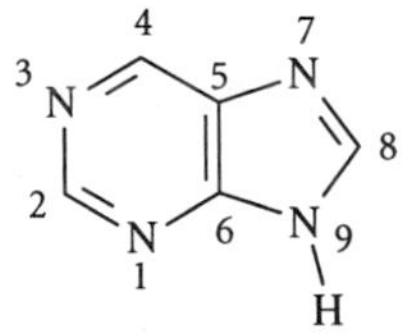

purine

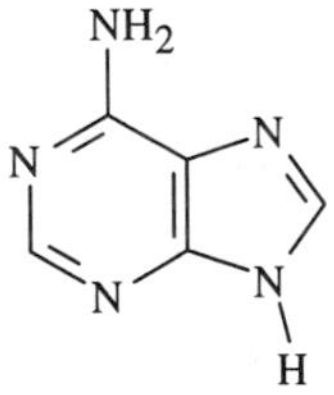

adenine (A)

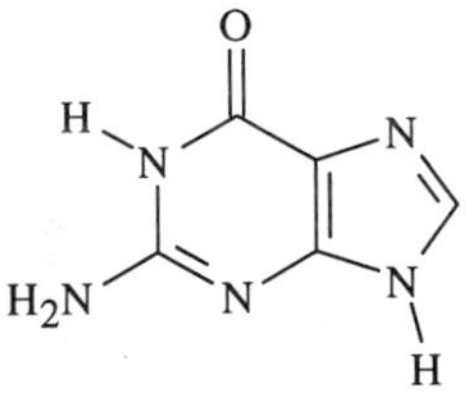

guanine (G)

3. The 5´—OH group of the ribose in a nucleoside is often found esterified to a phosphate group; a 5´-phosphorylated nucleoside is called a nucleotide.
   a. A ribonucleotide is derived from D-ribose.
   b. A deoxyribonucleotide is derived from 2´-deoxyribose.

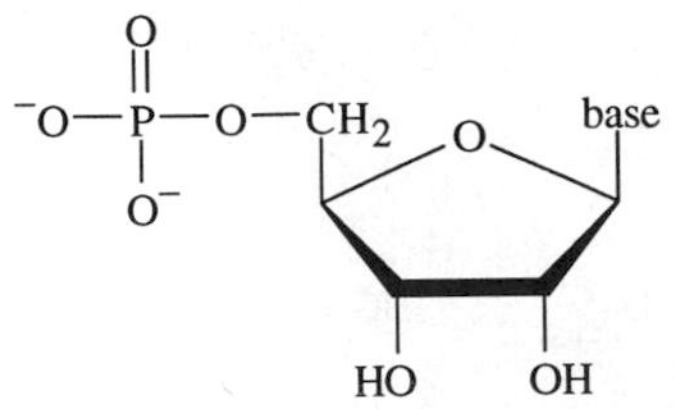

a ribonucleotide

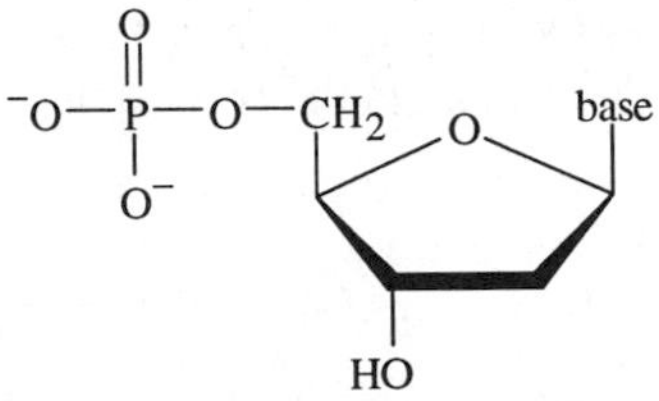

a deoxyribonucleotide

4. Some nucleotides contain a single phosphate group; others contain two or three phosphate groups condensed in phosphoric anhydride linkages.

5. The nomenclature and abbreviations of the five common bases and their corresponding nucleosides and nucleotides are summarized in Table 27.3, text page 1334. The corresponding 2′-deoxy derivatives are named by appending the prefix *2′-deoxy* (or *deoxy*) to the names of the corresponding ribose derivatives, or by appending a *d* prefix to the abbreviation.

thymidylic acid
or thymidine monophosphate (TMP)

deoxyguanosine diphosphate (*d*GDP)

6. One of the most ubiquitous nucleotides is ATP (adenosine triphosphate), which serves as the fundamental energy source for the living cell. (See Eq. 27.48 on text p. 1335 for the structure.)
7. Nucleic acids are of two general types:
   a. Deoxyribonucleic acid (DNA) is a principal component of the cell nucleus and is the storehouse of genetic information in the cell.
   b. Ribonucleic acid (RNA) serves various roles in translating and processing the information encoded in the structure of DNA.

B. Structures of DNA and RNA

1. Deoxyribonucleic acid (DNA) is a polymer of deoxyribonucleotides.
   a. The individual nucleotide residues are connected by a phosphate group that is esterified both to the 3′—OH group of one ribose and the 5′—OH of another.

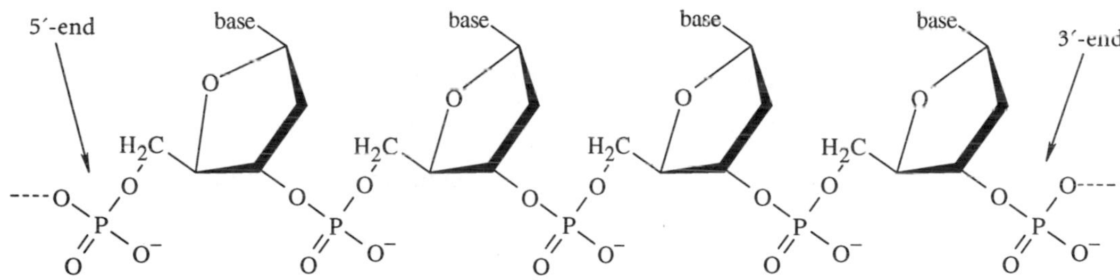

a deoxyribonucleic acid

   b. DNA incorporates adenine, thymine, guanine, and cytosine as the bases; each residue in a polynucleotide is distinguished by the identity of its base.
   c. The ratios of adenine to thymine, and guanosine to cytosine, in DNA are both 1.0; these observations are called Chargaff's rules.
   d. A typical strand of DNA, which carries genetic information, might be thousands of nucleotides long and is replicated, or copied, during cellular reproduction.
2. Ribonucleic acid (RNA) polymers are much like DNA polymers, except that ribose is the sugar; RNA incorporates essentially the same bases as DNA, except that uracil occurs in RNA instead of thymine, and some rare bases are found in certain types of RNA.
3. The Watson-Crick structure of DNA has the following important features:
   a. The structure contains two right-handed helical polynucleotide chains that run in opposite directions, coiled around a common axis; the structure is therefore that of a double helix. (See Fig. 27.5 on text p. 1338 for the structure.)

b. The sugars and phosphates, which are rich in —OH groups and charges, are on the outside of the helix.
c. The chains are held together by hydrogen bonds between bases, which are on the inside of the double helix.
    i. Adenine (A) in one chain always hydrogen-bonds to thymine (T) in the other.
    ii. Guanine (G) in one chain always hydrogen-bonds to cytosine (C) in the other.
    iii. This hydrogen-bonding complementarity accounts for Chargaff's rules.
d. The planes of the complementary base pairs are stacked, one on top of the other, and are perpendicular to the axis of the helix.
e. There is no restriction on the sequence of bases in a polynucleotide; however, the sequence of one polynucleotide strand in the double helix is complementary to that of the other strand.

4. The proper sequence of each new DNA strand during cellular reproduction is assured by hydrogen-bonding complementary.

C. DNA, RNA, and the Genetic Code

1. A strand of DNA is the code for the synthesis of a complementary strand of RNA; this RNA is called messenger RNA (mRNA), and the process by which it is assembled is called transcription.
    a. The sequence of the mRNA transcript is complementary to one DNA strand of the gene and runs in the opposite direction to that of its parent DNA.
    b. The mRNA sequence is used by the cell as the code for the synthesis of a specific protein from its component amino acids; this process is called translation.
        i. Each successive three-residue sequence of mRNA is translated as a specific amino acid in the sequence of a protein according to the genetic code given in Table 27.4, text page 1342; some amino acids have multiple codes.
        ii. The precise sequence of bases in DNA (by way of its complementary mRNA transcription product) codes for the successive amino acids of a protein.
    c. There is a specific start signal (either of the nucleotide sequences AUG or GUG) at the appropriate point in the mRNA; specific stop signals (UAA, UGA, or UAG) cause protein synthesis to be terminated.
    d. It is possible for the change of only one base in the DNA of an organism to cause the change of an amino acid in the corresponding protein.
2. There are many different types of RNA besides messenger RNA, each with a specific function in the cell.
3. The new sciences of genomics and proteomics are concerned with—
    a. identifying the products of each gene in an organism;
    b. identifying the functions of each protein gene product.;
    c. determining the effect of external stimuli on the DNA transcription of each gene.

D. DNA Modification and Chemical Carcinogenesis

1. There is strong circumstantial evidence that chemical damage to DNA can interfere with its hydrogen-bonding complementary and can trigger the state of uncontrolled cell division known as cancer.
    a. One type of chemical damage is caused by alkylating agents.
    b. Another type of chemical damage is caused by ultraviolet light. In adjacent positions on a strand of DNA, ultraviolet light promotes the [2 + 2] cycloaddition of the two pyrimidines. People who lack the enzymes to restore the original DNA structure die at an early age.

REACTIONS

I. REACTIONS OF CARBOHYDRATES

A. Base-Catalyzed Isomerization of Carbohydrates

1. In base, aldoses and ketoses rapidly equilibrate to mixtures of other aldose and ketoses; this transformation is an example of the Lobry de Bruyn-Alberda van Ekenstein reaction.
   a. An aldose can ionize to give a small amount of its enolate ion in base.
   b. Protonation of this enolate ion at one face gives back the aldose; protonation at the other face gives an epimer.

D-(+)-glucose ⇌ (HO^- / H_2O) enolate ion ⇌ (H_2O / HO^-) D-(+)-mannose

   c. The enolate ion can also protonate on the oxygen to give a new enol, called an enediol, which contains a hydroxy group at both ends of the double bond.
   d. The enediol is simultaneously the enol of two epimeric aldoses and an isomeric ketose, and can be converted into any one of these.

enolate ion of D-(+)-glucose or D-(+)-mannose ⇌ (H_2O / HO^-) an enediol ⇌ (HO^- / H_2O) D-(+)-fructose

2. Several transformations of this type are important in metabolism.

B. Ether and Ester Derivatives of Carbohydrates

1. In the presence of concentrated base, carbohydrates are converted into ethers by reactive alkylating agents (Williamson ether synthesis).

→ (Me_2SO_4, HO^-) methyl 2,3,4,6-tetra-*O*-methyl-D-glucopyranoside

2. The alkoxy group at the anomeric carbon is different from other alkoxy groups in the alkylated carbohydrate.
   a. The alkoxy group at the anomeric carbon is part of a glycosidic linkage.

b. Since it is an acetal, it can be hydrolyzed in aqueous acid under mild conditions; the other alkoxy groups are ordinary ethers and do not hydrolyze under these conditions.

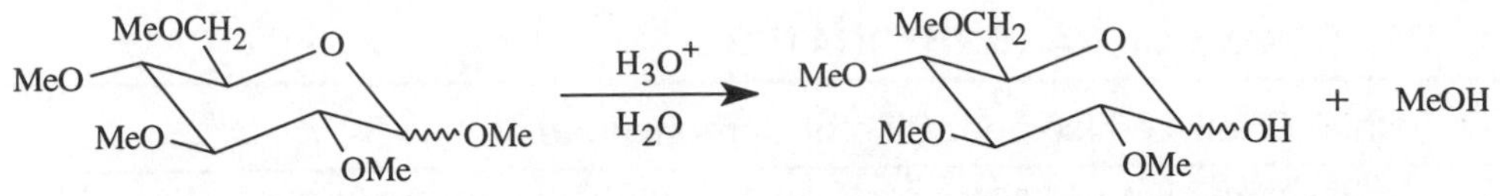

MeOCH$_2$ ... OMe → (H_3O^+, H_2O) → 2,3,4,6-tetra-*O*-methyl-D-glucopyranose + MeOH

3. The hydroxy groups of carbohydrates can be esterified; the resulting esters can be saponified in base or can be removed by transesterification with an alkoxide.

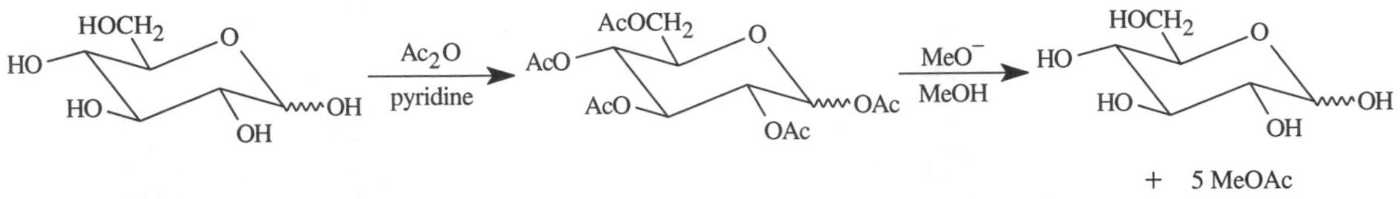

HOCH$_2$ ... OH → (Ac_2O, pyridine) → AcOCH$_2$ ... OAc → (MeO^-, MeOH) → HOCH$_2$ ... OH + 5 MeOAc

4. Ethers and esters are used as protecting groups in reactions involving carbohydrates; furthermore, they have broader solubility characteristics and greater volatility than the carbohydrates themselves.

C. Oxidation and Reduction Reactions of Carbohydrates

1. Treatment of an aldose with bromine water oxidizes aldoses to aldonic acids; this reaction is a useful test for aldoses. (Carbohydrates that can be oxidized by bromine water are called reducing sugars.)

   a. Aldonic acids exist in acidic solution as five-membered lactones called aldonolactones.

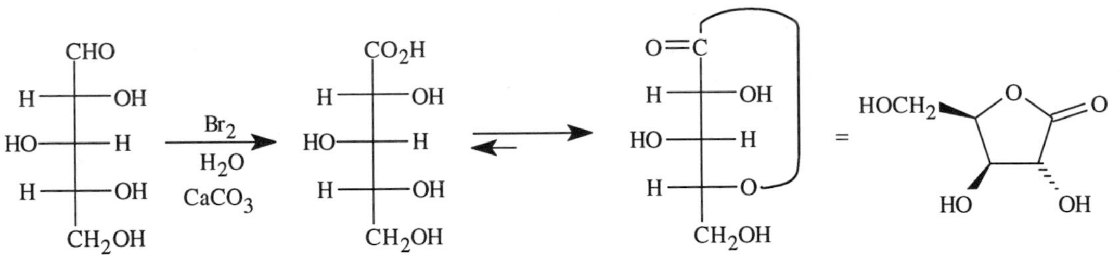

CHO / H–OH / HO–H / H–OH / CH$_2$OH → (Br_2, H_2O, $CaCO_3$) → CO_2H / H–OH / HO–H / H–OH / CH$_2$OH ⇌ O=C / H–OH / HO–H / H–O / CH$_2$OH = HOCH$_2$ (lactone) HO, OH

an aldonic acid — an aldolactone

   b. Glycosides are not oxidized by bromine water because the aldehyde carbonyl group is protected as an acetal. (Carbohydrates that cannot be oxidized by bromine water are called nonreducing sugars.)

2. Aldoses can also be oxidized with other reagents; the alkaline conditions of Tollens test, however, also promote the equilibration of ketoses and aldoses; thus, ketoses also give a positive Tollens test.

3. Both ends of an aldose are oxidized to carboxylic acid groups by dilute HNO_3, but the secondary alcohol groups are not affected; the oxidation product is an aldaric acid.

   a. Aldaric acids in acidic solution form lactones.

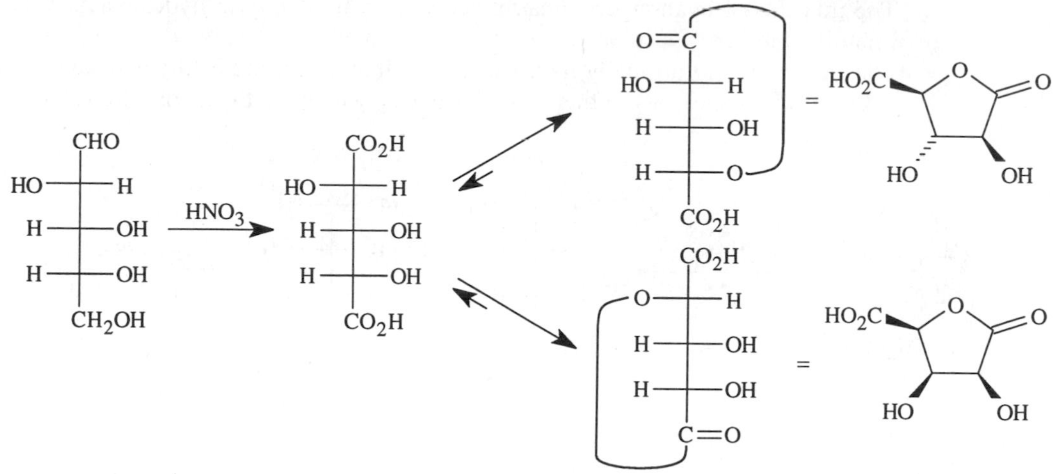

b. Two different five-membered lactones are possible from a hexose or pentose, depending on which carboxylic acid group undergoes lactonization.

c. Under certain conditions, some aldaric acids derived from aldohexoses can be isolated as dilactones, in which both carboxylic acid groups are lactonized. (An example can be found in Eq. 27.27 on text p. 1316.)

4. Many carbohydrates contain vicinal glycol units and are oxidized by periodic acid.

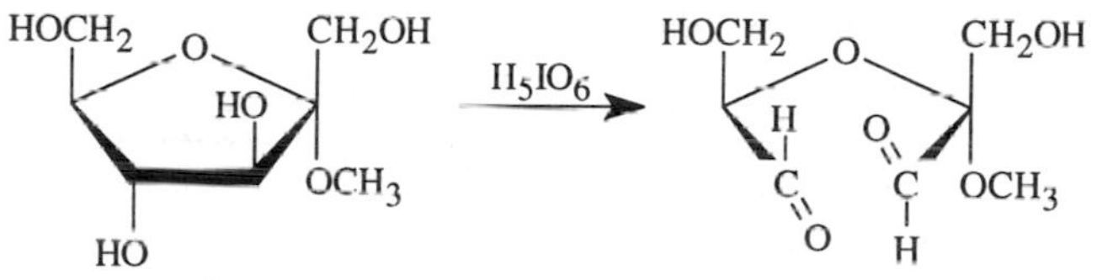

a. α-Hydroxy aldehydes are oxidized to formic acid and another aldehyde with one fewer carbon. (See Eq. 27.28 on text p. 1317.)

b. α-Hydroxymethyl ketones are oxidized to formaldehyde and a carboxylic acid.

c. Because it is possible to determine accurately both the amount of periodic acid consumed and the amount of formic acid produced, periodic acid oxidation can be used to differentiate between pyranose and furanose structures for saccharide derivatives.

5. Aldoses and ketoses undergo many of the usual carbonyl reductions; an aldose is reduced to a primary alcohol known as an alditol.

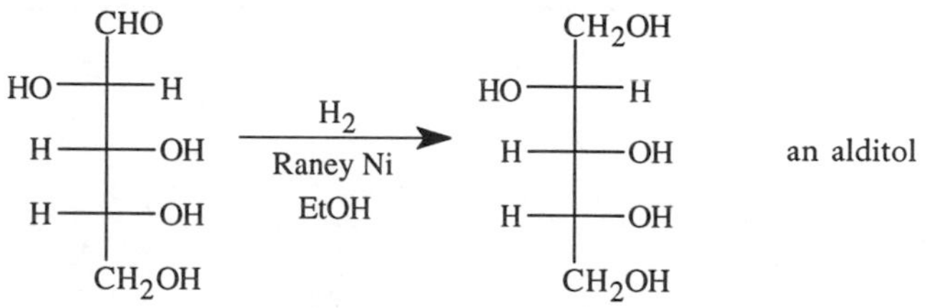

an alditol

II. SYNTHESIS OF SUGARS FROM OTHER SUGARS

A. Kiliani-Fischer Synthesis; Increasing the Length of the Aldose Chain

1. Addition of hydrogen cyanide to aldoses gives cyanohydrins; because the cyanohydrin product of such a reaction has an additional asymmetric carbon, it is formed as a mixture of two epimers.

a. These epimers are diastereomers and are typically formed in different amounts.

b. The mixture of cyanohydrins can be converted by catalytic hydrogenation into a mixture of imines, which are not isolated.

c. The imines would normally hydrogenate to amines, but the acidic aqueous reaction conditions intercept them and convert them into aldehydes, which are the aldose products.

CHO; H—OH; H—OH; CH_2OH —($NaCN/H_2O$)→ CN; H—OH; H—OH; H—OH; CH_2OH + CN; HO—H; H—OH; H—OH; CH_2OH —($H_2/Pd/BaSO_4$, H_3O^+, H_2O)→

CH=NH; HO—H; H—OH; H—OH; CH_2OH + CH=NH; H—OH; H—OH; H—OH; CH_2OH —(H_3O^+, H_2O)→ CH=O; HO—H; H—OH; H—OH; CH_2OH + CH=O; H—OH; H—OH; H—OH; CH_2OH

imines (not isolated)

2. The cyanohydrin formation-reduction sequence converts an aldose into two epimeric aldoses with one additional carbon; this process is known as the Kiliani-Fischer synthesis.

B. Ruff Degradation; Decreasing the Length of the Aldose Chain

1. In the Ruff degradation, the calcium salt of an aldonic acid is oxidized with hydrogen peroxide in the presence of Fe^{3+}.
   a. An aldose is degraded to another aldose with one fewer carbon atom.
   b. The stereochemistry of the remaining groups remains the same.

CHO; HO—H; H—OH; H—OH; CH_2OH —(1) Br_2/H_2O; 2) $Ca(OH)_2$; 3) $Fe(OAc)_3/30\%\ H_2O_2$)→ CHO; H—OH; H—OH; CH_2OH

SCIENTISTS

Dates in brackets refer to Nobel Prize awards.

STUDY GUIDE LINKS

27.1 Nomenclature of Anomers

Long before the actual structures of the anomeric forms of many carbohydrates were known, the α-anomer of a D-carbohydrate was simply defined as "the more dextrorotatory of the two anomers." This convention, proposed by C. S. Hudson (1881–1952), an American carbohydrate chemist, is cumbersome because, in order to apply it, one has to know the optical rotations of the two anomers. When the structures of several anomers became known, it was found that in most cases the more dextrorotatory anomer of each D-carbohydrate has the configuration shown in the text for the α-anomer. Hudson's definition in terms of a physical property—optical rotation—was subsequently replaced by the structural definition used in the text.

✓27.2 Acid Catalysis of Carbohydrate Reactions

Acid catalysis of glycoside formation involves protonation of the hydroxy group at the anomeric carbon; acid catalysis of mutarotation (Eqs. 27.8 on text p. 1304 and Problem 27.5 on text p. 1306) involves protonation of the ring oxygen. You may be asking why these specific oxygens are protonated—why not others? The answer is that *any one* of the oxygens can indeed be protonated to small extents in acidic solution. However, *only* protonation at the ring oxygen leads to mutarotation, and *only* protonation of the —OH group at the anomeric carbon leads to glycoside formation. In other words, the various protonated forms are in rapid equilibrium and are present whether they are shown explicitly or not. Only certain of the protonated forms can react further, and these are the ones shown in the equations.

✓27.3 Configurations of Aldaric Acids

Consideration of the stereochemistry of aldaric acids reveals a subtle yet important aspect of the D,L configurational system. Certain aldaric acids can be derived from either a D or an L carbohydrate.

	CO_2H				CO_2H	
H	—	OH		HO	—	H
HO	—	H	—(180°)→	HO	—	H
H	—	OH		H	—	OH
H	—	OH		HO	—	H
	CO_2H				CO_2H	

D-glucaric acid (from D-glucose) or **L-gularic acid** (from L-gulose)
(the same compound)

In this case, whether the aldaric acid is classified as D or L depends on which carbon is specified as carbon-1. Because the two ends of the molecule are *constitutionally equivalent* (see text p. 437), the choice is completely arbitrary! The names D-glucaric acid and L-gularic acid are both correct names for this aldaric acid. The two names reflect the fact that it can be formed by nitric acid oxidation of either D-glucose or L-gulose.

This situation arises because the —OH groups on the endmost asymmetric carbons are on the same side of the Fischer projection. However, when these —OH groups are on opposite sides, the configuration is unambiguous. For example, D-tartaric acid also has constitutionally equivalent ends, but it has the D configuration no matter how it is turned.

CO_2H / HO—H / H—OH / CO_2H → (180°) → CO_2H / HO—H / H—OH / CO_2H

D-tartaric acid

As you might expect, a similar situation arises in other derivatives with constitutionally equivalent ends, such as alditols.

This issue is explored in Problem 27.17 on text p. 1317.

27.4 More on the Fischer Proof

The Fischer proof of glucose stereochemistry has a number of interesting facets that are not discussed in the text. One of the most important experimental aspects of carbohydrate chemistry is that many carbohydrates are notoriously difficult to crystallize; they are frequently isolated as "syrups" and "gums." Because chemists in Fischer's day did not have available the sophisticated purification methods available today, they had to rely almost exclusively on crystallization as a purification method. Thus, because carbohydrates were difficult to crystallize, they were very difficult to purify. Few chemists were interested in working on problems fraught with so many experimental difficulties; consequently, little progress had been made in the field of carbohydrate chemistry when Fischer began his work. In fact, Fischer himself, in some of his correspondence, deplored the dreadfully slow pace of carbohydrate research in his own laboratory brought about by the experimental difficulties of handling carbohydrates.

Two solutions to the problem of carbohydrate crystallization emerged from Fischer's long-standing interest in phenylhydrazine, $PhNHNH_2$. He had already shown that this compound could be used to prepare indoles (Fischer indole synthesis; Sec. 24.4A). When he allowed phenylhydrazine to react with aldohexoses, he found that very interesting compounds called *osazones* were obtained.

CH=O / H—OH / HO—H / H—OH / H—OH / CH_2OH + 3 $PhNHNH_2$ → CH=N—NHPh / C=N—NHPh / HO—H / H—OH / H—OH / CH_2OH + $PhNH_2$ + NH_3 + H_2O

(+)-glucose ... glucosazone

This reaction is very reminiscent of phenylhydrazone formation (Table 19.3 on text p. 876). However, an unusual aspect of this reaction is that one equivalent of phenylhydrazine serves as an oxidizing agent, with the net result that a second substituted imine group is introduced. (For the mechanism of this reaction, see Loudon, *Organic Chemistry,* First edition, Addison-Wesley Publishing Co., 1984, pp. 1417–1418.) Osazones proved to be much more easily crystallized, and thus more easily characterized, than aldoses themselves. Furthermore, they served to confirm some of Fischer's stereochemical deductions. Notice that *neither carbon-1 nor carbon-2* are asymmetric in the osazone. Consequently, isomeric aldoses that are epimeric only at carbon-2 give the same osazone. Fischer found that (+)-glucose and (+)-mannose give *different* phenylhydrazones (formed under milder conditions with phenylhydrazine), but give the same osazone. This fact, along with their simultaneous preparation in the Kiliani-Fischer synthesis from (–)-arabinose, showed that they are epimers at carbon-2.

Phenylhydrazine figured prominently in Fischer's research in a second way when Fischer found that the aminolysis reactions of aldonic acids (undoubtedly reacting as their lactones) with

compounds.

D-gluconic acid $\xrightarrow{-H_2O}$ D-γ-gluconolactone $\xrightarrow{PhNHNH_2 \text{ (ester aminolysis)}}$ D-gluconic acid phenylhydrazide

One other interesting aspect of the Fischer proof has to do with the last step, which was made possible by a remarkable bit of serendipity. It is noted in the text that Fischer had prepared L-(+)-gulose in the course of his research; recall that it was the oxidation of D-glucose and L-gulose to the same aldaric acid that completed the Fischer proof. Equation 27.27 on text p. 1316 shows that D-glucaric acid can exist as a mixture of two different lactones. Fortuitously, Fischer's procedure for isolating the lactones gave mainly the 1,4-lactone. His synthesis of L-gulose began with this lactone.

D-glucaric acid 1,4-lactone $\xrightarrow{Na/Hg}$ D-glucuronic acid $\xrightarrow{Na/Hg}$

L-gulonic acid $\xrightarrow{-H_2O}$ L-gulonolactone $\xrightarrow{Na/Hg}$ L-gulose

"Na/Hg" is sodium amalgam, a solution of sodium in mercury, which was widely used as a reducing agent until lithium aluminum hydride and sodium borohydride were developed. The sodium-amalgam reductions shown above can be carried out in stages by controlling the conditions of each step. Recall that D-glucose is readily oxidized to D-glucaric acid-1,4 lactone with dilute nitric acid. Consequently, beginning with D-glucose, the overall sequence of glucose oxidation followed by the sodium amalgam reductions effects the *net interchange* of the CH=O and CH_2OH groups, and thus gives L-gulose from D-glucose. The serendipitous element was this: had Fischer isolated the 3,6-lactone of D-glucaric acid, it would have given back D-glucose when taken through this sequence of reactions, and the last step of the proof would not have been possible.

An intriguing account of Fischer's life and work, including details of the Fischer proof, was written by F. W. Lichtenthaler and published in *Angewandte Chemie International Edition in English,* vol. 31, No. 12, pp. 1541–1596 (1992). This journal is available in most chemistry libraries.

✓27.5 DNA Transcription

The process of DNA transcription presented in the text is simplified. The DNA transcript in higher organisms contains long sections of "intervening RNA," or *introns,* that are not part of the final mRNA product; these are excised out of the transcript, and the resulting shortened strand of RNA undergoes certain chemical modifications to produce mRNA. (You can find a description of this process in any modern biology or biochemistry text.) However, the presence of introns does not alter the fundamental idea, namely, that information flows from DNA to RNA to protein.

SOLUTIONS TO PROBLEMS

Solutions to In-Text Problems

27.1 (a) Use a cyclic permutation to transform the given Fischer projection into a standard form in which all of the backbone carbons are in a vertical line. This will show that the —OH group on carbon-5 is on the right. Thus, this aldose has the D-configuration.

H_3C—┼—H (OH below; carbon-5) ⇨ H—┼—OH (CH_3 below)

carbon-5

OH on the right; therefore D-configuration

(b) An analysis of the carbon-3 configuration of D-glucose shows that carbon-3 has the *S* configuration. Consequently, the L-enantiomer of glucose has the *R* configuration at carbon-3.

Remember that the configuration of an asymmetric carbon in a Fischer projection can be analyzed directly. To review, see Study Guide Link 6.8 on p. 161 of this manual.

27.2 Put all the aldoses in the standard Fischer projection to see the relative location of corresponding groups:

A	*B*	*C*
CH=O	CH=O	CH=O
H—OH	H—OH	HO—H
HO—H	HO—H	H—OH
H—OH	H—OH	HO—H
HO—H	H—OH	HO—H
CH_2OH	CH_2OH	CH_2OH

Compounds *B* and *C* are enantiomers. Compound *A* is an epimer of *B* and a diastereomer of both *B* and *C*. (Remember that an epimer is a particular type of diastereomer.)

27.3 (a) Recognize that α-D-glucopyranose is an anomer of β-D-glucopyranose, and convert the β-D-glucopyranose structures in Study Problem 27.2 on text p. 1300 to α-D-glucopyranose structures by inverting the configuration of carbon-1.

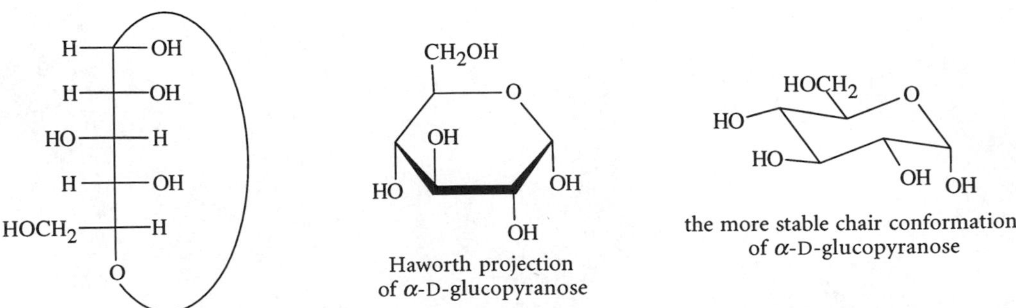

Fischer projection of α-D-glucopyranose

Haworth projection of α-D-glucopyranose

the more stable chair conformation of α-D-glucopyranose

(b) Recognize that D-mannose is epimeric to D-glucose at carbon-2, and thus convert the β-D-glucopyranose structures on text p. 1300 to β-D-mannopyranose structures by inverting the configuration of carbon-2.

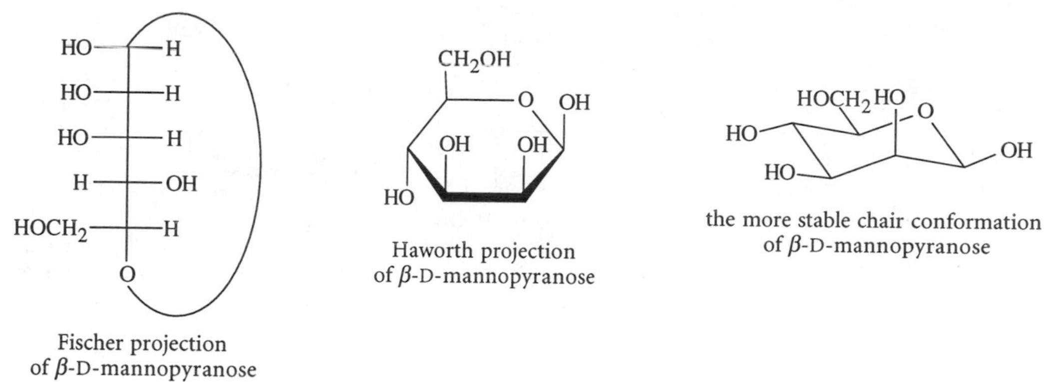

Fischer projection of β-D-mannopyranose

Haworth projection of β-D-mannopyranose

the more stable chair conformation of β-D-mannopyranose

(c) Start with the structure of D-xylose in Figure 27.1 (text p. 1301) and perform the usual manipulations. An easier solution is to recognize that D-xylose is epimeric to D-ribose at carbon-3. Thus, the structures for β-D-ribofuranose in Eq. 27.5 on text p. 1301 can be used if the configuration of carbon-3 is inverted.

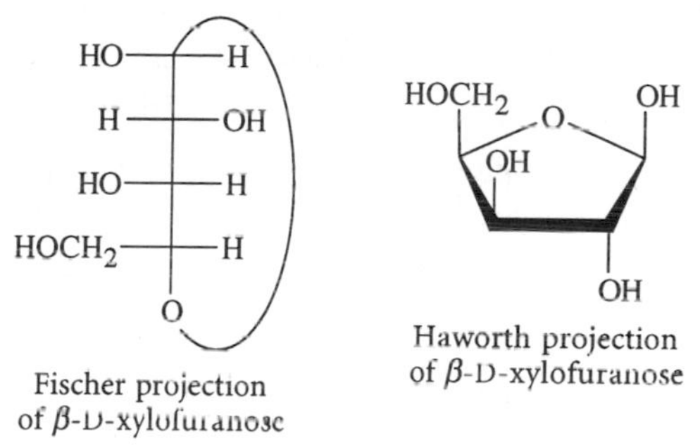

Fischer projection of β-D-xylofuranose

Haworth projection of β-D-xylofuranose

(d) Start with the Fischer projection for D-fructose from text p. 1296.

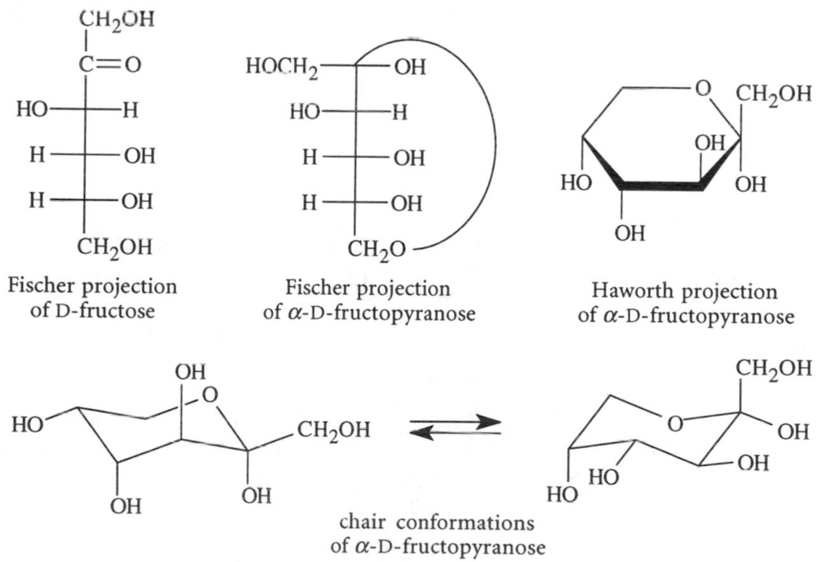

Fischer projection of D-fructose

Fischer projection of α-D-fructopyranose

Haworth projection of α-D-fructopyranose

chair conformations of α-D-fructopyranose

(e) Draw the enantiomers of the corresponding structures for α-D-glucopyranose, which are shown on text pp. 1300–1 or in the answers to part (a) of this problem.

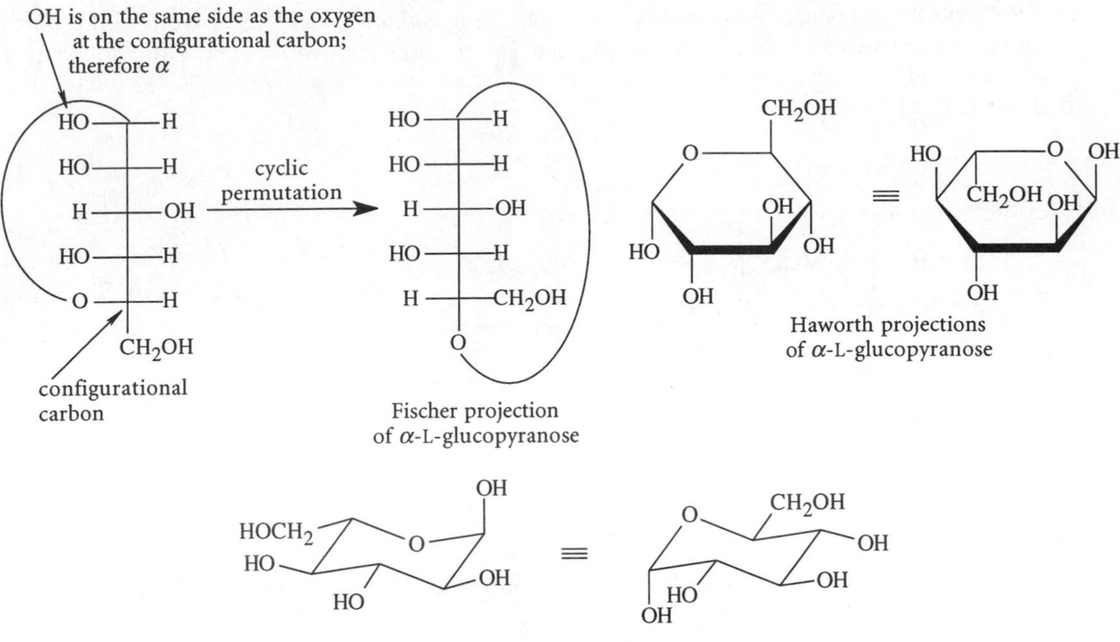

Fischer projection of α-L-glucopyranose

(f) Start with the structures of α-D-glucopyranose in the previous part and make the bond to the anomeric carbon a "squiggly bond." For the Fischer projection simply do not show the configuration. ("Squiggly bonds" are not used in Fischer projections.)

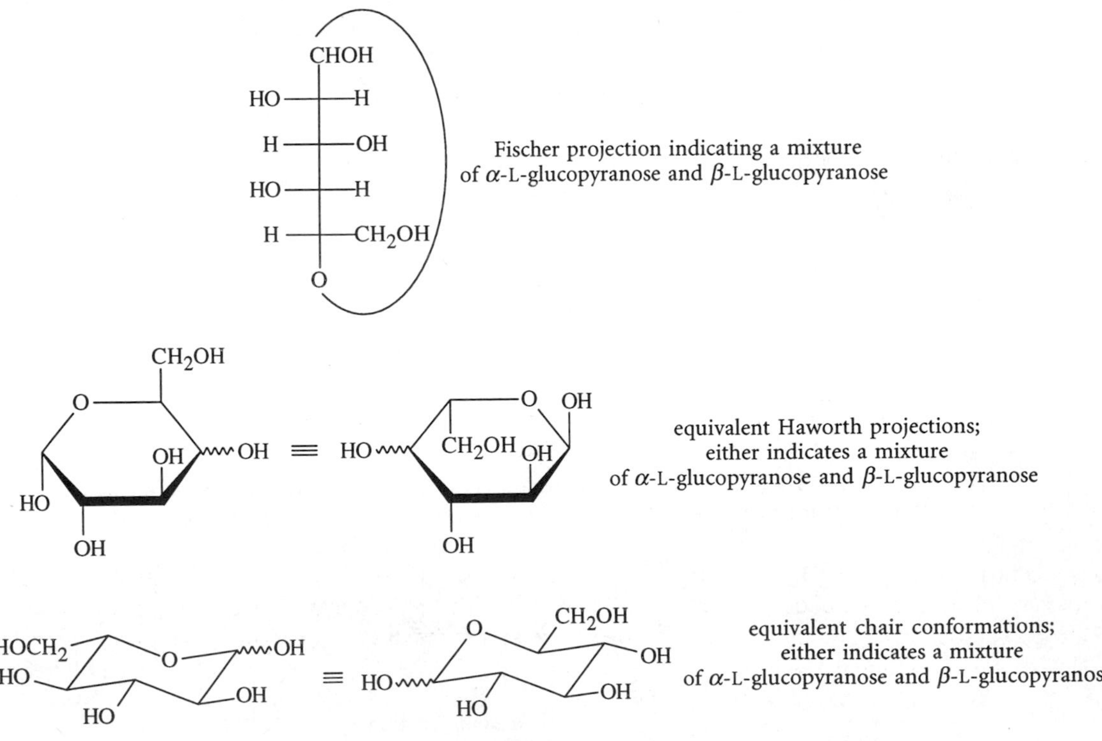

27.4 (a) Perform the reverse of the analysis in Study Problem 27.2 on text p. 1300 to reveal that this is α-D-ribose. But you can also look at Eq. 27.5 on text p. 1301 and see that the compound given is the α-anomer of β-D-ribose, which means that it is α-D-ribose.

(b) This is the β-anomer of a D-hexopyranose that differs from D-glucose in its configuration at carbons 3 and

4; therefore, it is β-D-gulopyranose.

27.5 The following mechanism begins with the α-anomer protonated on the ring oxygen and ends with the β-anomer, also protonated on the ring oxygen.

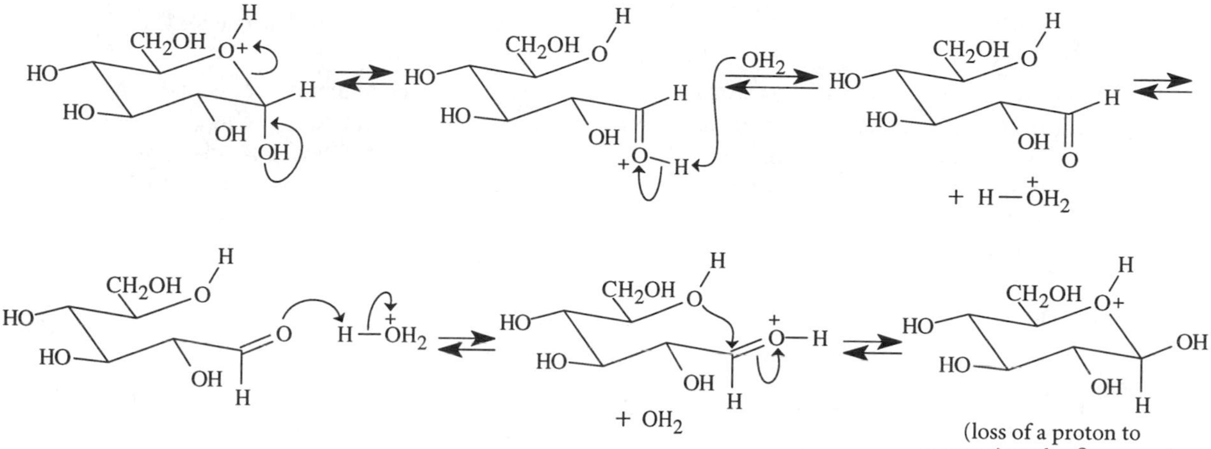

(loss of a proton to water gives the β-anomer)

27.6 The mechanism for base-catalyzed mutarotation of glucose:

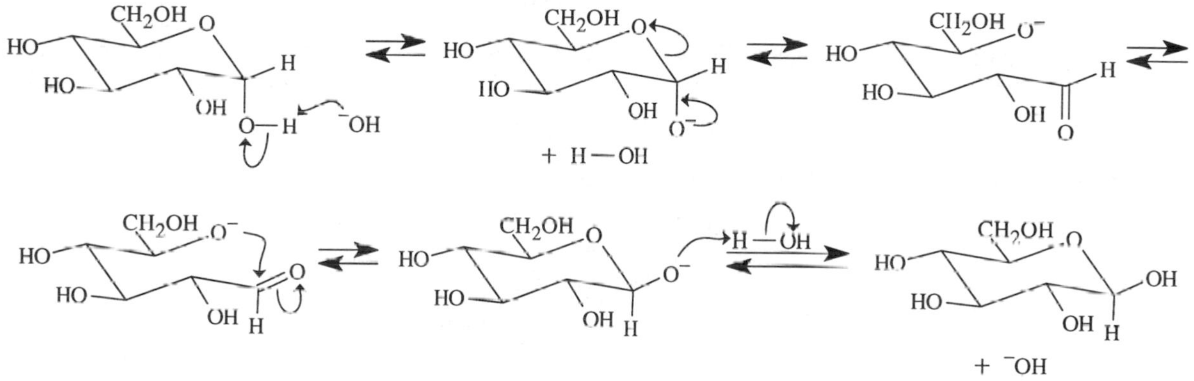

27.7 The β-pyranose forms of D-glucose and D-talose are as follows:

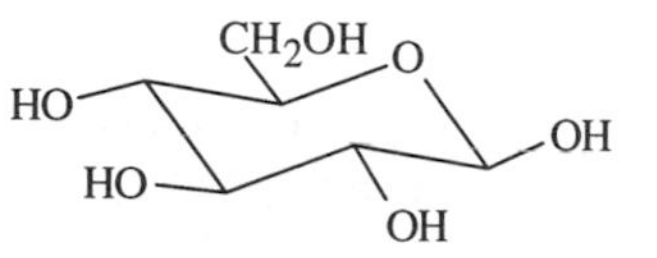

β-D-glucopyranose

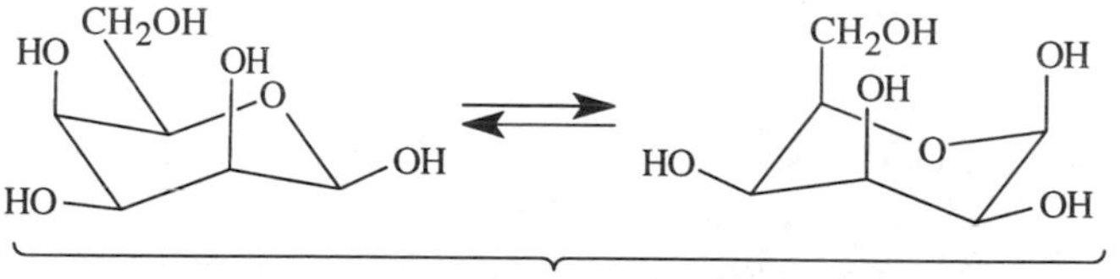

β-D-talopyranose

Both chair forms of the β-D-talopyranose molecule have a significantly greater number of 1,3-diaxial interactions than β-D-glucopyranose. These unfavorable interactions in β-D-talopyranose raise its energy and thus lower the relative amounts of these forms at equilibrium. Hence, talose contains a lower percentage of the β-pyranose form (and a lower percentage of the α-pyranose form for the same reason) and a higher percentage of the furanose and aldehyde forms at equilibrium.

27.8 (a) The easiest way to do this is to consider the relationship of D-allose to D-glucose and modify the conformational representation of β-D-glucopyranose (in which all of the ring substituents are equatorial) accordingly. Thus, Fig. 27.1 on text p. 1294 shows that D-allose is epimeric to D-glucose at carbon-3. Consequently, configurational inversion of the β-D-glucopyranose structure at carbon-3 gives the desired conformational representation of β-D-allopyranose.

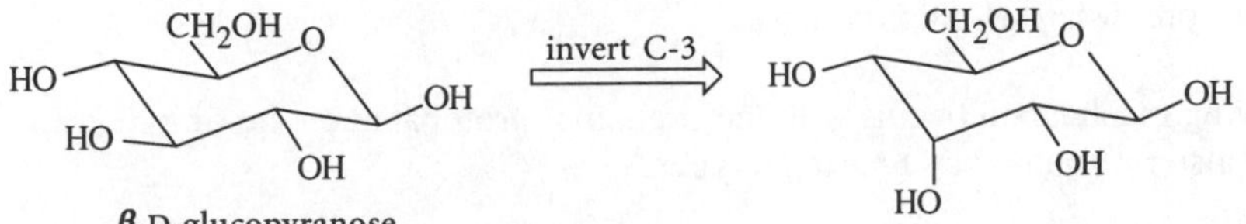

β-D-glucopyranose $\xrightarrow{\text{invert C-3}}$ β-D-allopyranose

(b) Follow the procedure in Study Problem 27.2 on text p. 1300. Note that in the second and third Fischer projections the configuration of carbon-5 cannot be represented without violating the Fischer-projection conventions. We must remember that this carbon has the *R* configuration, and this configuration must be shown explicitly in the conformational representation.

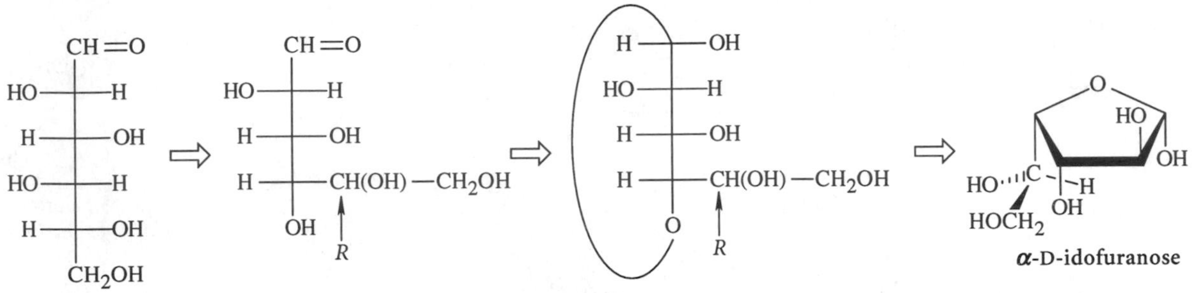

α-D-idofuranose

27.9 Each form of the sugar contributes its own optical rotation in proportion to the amount that is present. Let N_i = the fraction of form *i*. (The dimensions of the specific rotations are abbreviated simply as degrees.)

$$N_{\text{total}} = 52.7° = N_\alpha(112.0°) + N_\beta(18.7°)$$

Since $N_\alpha + N_\beta = 1$, substitute $N_\beta = (1 - N_\alpha)$ and obtain

$$52.7° = N_\alpha(112.0°) + (1 - N_\alpha)(18.7°)$$

$$N_\alpha(112.0° - 18.7°) = 52.7° - 18.7° = 34.0°$$

$$N_\alpha = \frac{34.0°}{93.3°} = 0.36$$

$$N_\beta = 1 - N_\alpha = 0.64$$

Table 27.1 on text p. 1305 confirms that D-glucopyranose at equilibrium contains 36% of the α-form and 64% of the β-form.

27.10 (a) D-Galactose would be transformed into the aldohexose that is epimeric at carbon-2, namely, D-talose. (See the structures in Fig. 27.1 on text p. 1294.) Also formed would be the ketose D-tagatose, which has the following structure:

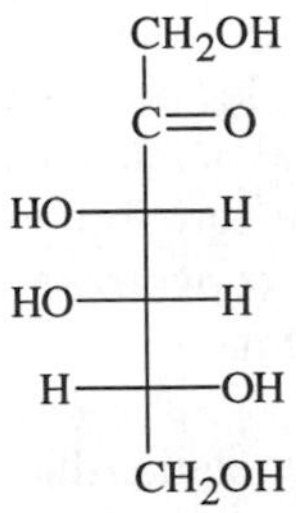

D-tagatose

(b) D-Allose would be transformed by base into a mixture of D-altrose (structure in Fig. 27.1 on text p. 1294) and the ketose D-psicose.

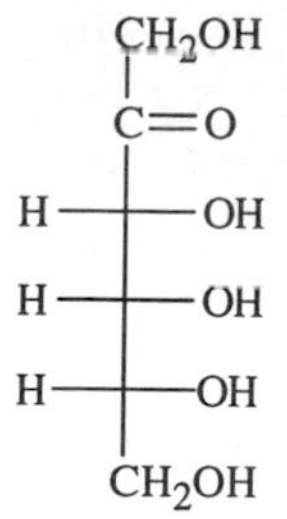

CH_2OH, C=O, H—OH, H—OH, H—OH, CH_2OH

D-psicose

27.11 (a) The compound is the *β-p*-nitrophenyl glycoside of D-galactopyranose; its name is *p*-nitrophenyl *β*-D-galactopyranoside.

(b) This compound is hydrolyzed in aqueous acid to *p*-nitrophenol and a (mutarotated) mixture of *α*- and *β*-D-galactopyranose.

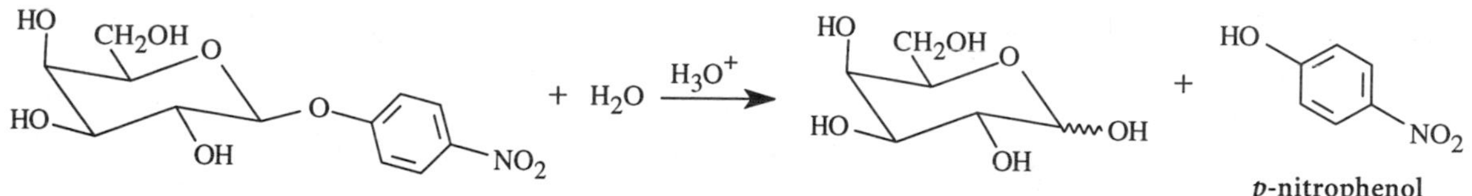

HO, CH_2OH, O, HO, O, OH, NO_2 + H_2O $\xrightarrow{H_3O^+}$ HO, CH_2OH, O, HO, OH, OH + HO, NO_2

α- and *β*-D-galactopyranose

p-nitrophenol

27.12 Because naturally-occurring glycosides generally have bonds between an alcohol or phenol oxygen and the anomeric carbon of a carbohydrate, it is reasonable to propose that the *β*-D-glycoside of vanillin and glucose has the following structure:

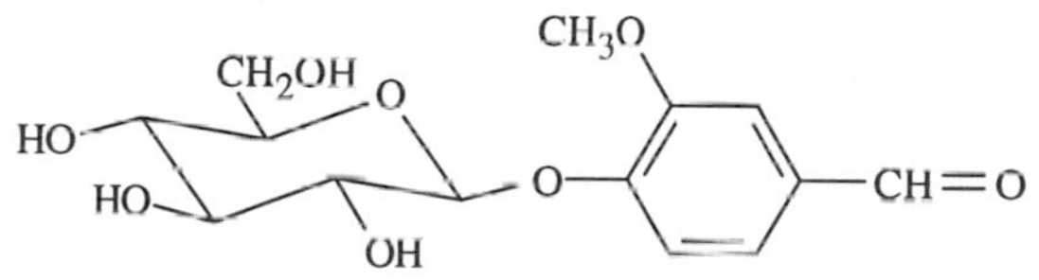

CH_3O, CH_2OH, O, HO, HO, O, OH, CH=O

27.13 (a) Start with the structure of *β*-D-fructofuranose on text p. 1305 and use it to draw the *β*-glycoside of methanol.

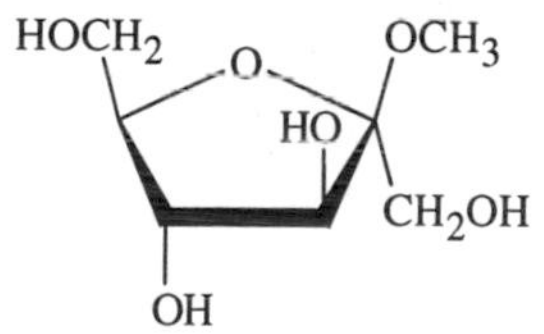

$HOCH_2$, O, OCH_3, HO, CH_2OH, OH

methyl *β*-D-fructofuranoside

(b) Start with the structure of *α*-D-galactopyranose, which is the same as *α*-D-glucopyranose except that carbon-4 is inverted. Then substitute an isopropoxy group for the hydroxy group at the anomeric carbon.

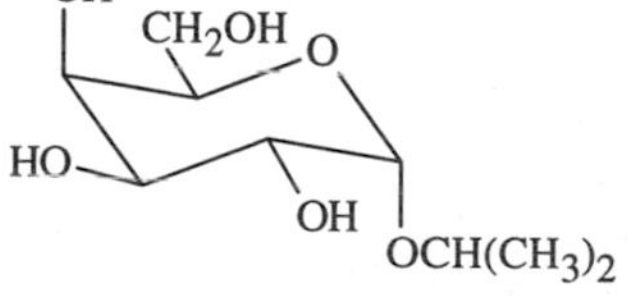

OH, CH_2OH, O, HO, OH, $OCH(CH_3)_2$

isopropyl *α*-D-galactopyranoside

27.14 The protonation of an oxygen of the acetal group (the glycosidic oxygen in the example below) and loss of an alcohol (methanol in the example below) gives a *resonance-stabilized* carbocation intermediate. A similar mechanism applied to ordinary ethers gives carbocations that are not resonance-stabilized. (In fact, the cleavage of such ethers in carbohydrates by an S_N1 mechanism is retarded by the polar effects of the surrounding oxygens, which destabilize carbocations.) Hammond's postulate implies that the hydrolysis reaction which involves the more stable carbocation intermediate should occur more rapidly.

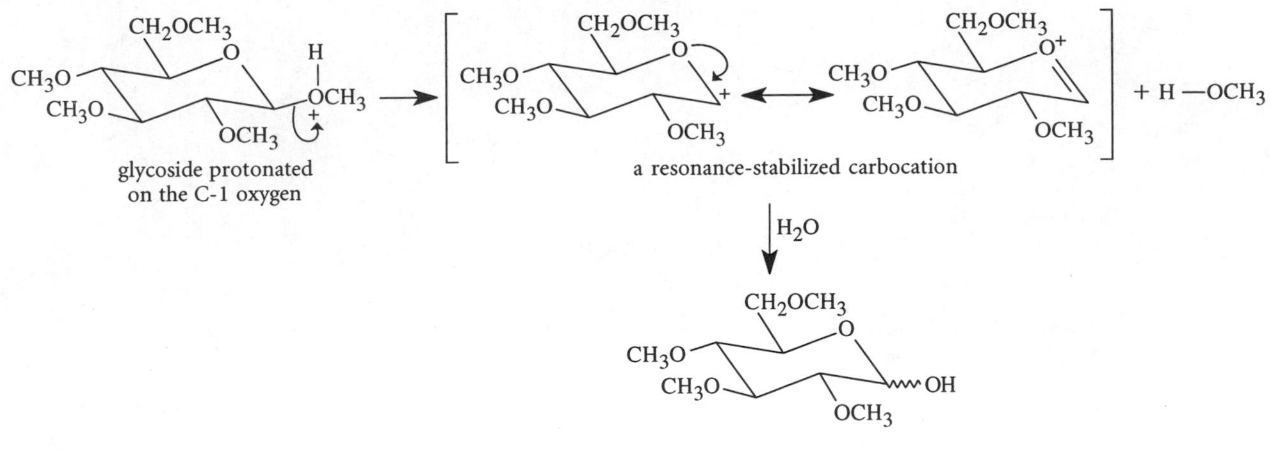

glycoside protonated on the C-1 oxygen → [a resonance-stabilized carbocation] + $H—OCH_3$

H_2O ↓

27.15 (a) Prepare the 1-*O*-ethyl derivative and use the Williamson synthesis to introduce the methyl groups.

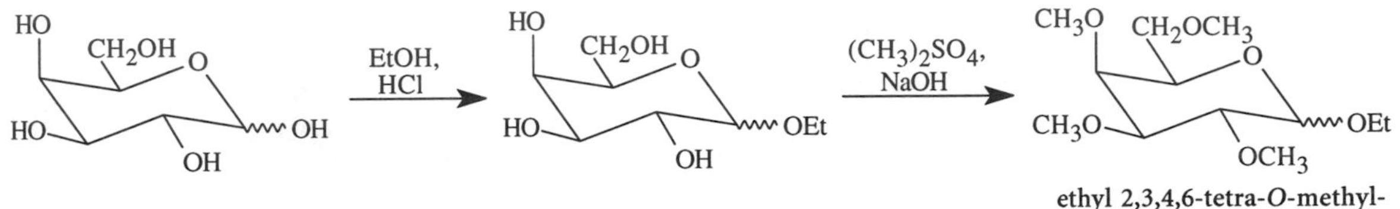

EtOH, HCl → ; $(CH_3)_2SO_4$, NaOH →

ethyl 2,3,4,6-tetra-*O*-methyl-D-galactopyranoside

(b) Alkylate all of the hydroxy groups with benzyl chloride ($PhCH_2Cl$), a very reactive alkyl halide; then hydrolyze the benzyl glycoside. (Why is benzyl chloride very reactive? See Sec. 17.4.)

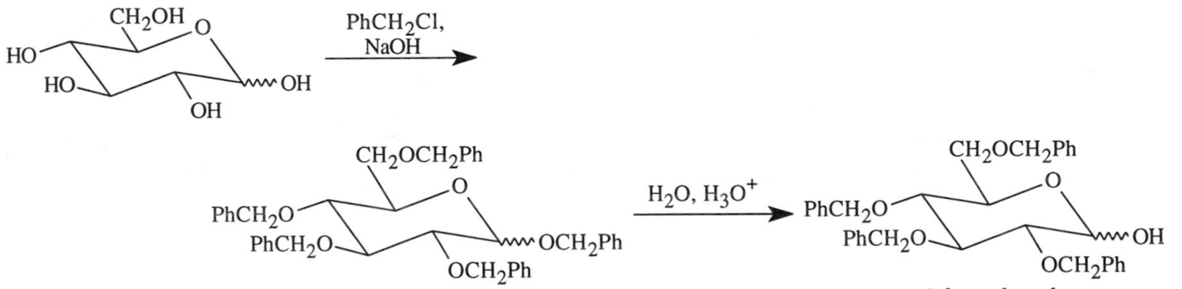

$PhCH_2Cl$, NaOH →

H_2O, H_3O^+ →

2,3,4,6-tetra-*O*-benzyl-D-glucopyranose

27.16 (a) (b)

	CH=O	
H		OH
HO		H
HO		H
H		OH
	CO_2H	

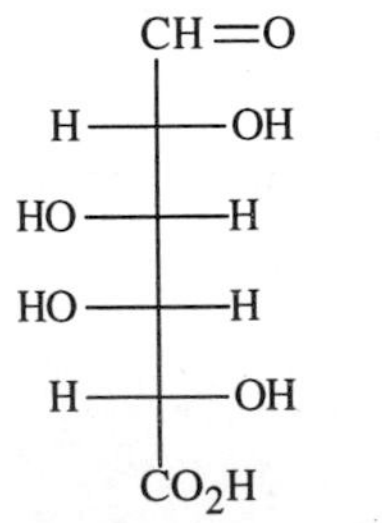

D-galacturonic acid

	CH_2OH	
H		OH
H		OH
H		OH
	CH_2OH	

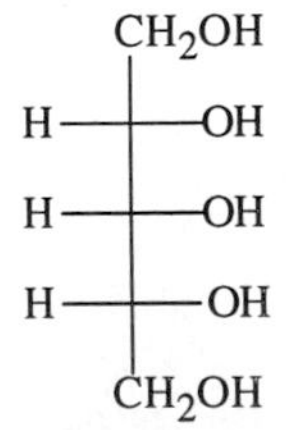

ribitol

Notice that ribitol is a meso compound.

27.17 The structures of D-glucaric and L-gularic acids are as follows. Turning either structure 180° in the plane of the page shows that the two structures are identical. (This identity was a key element in the Fischer proof of glucose stereochemistry, which is discussed in text Sec. 27.9A.)

D-glucaric acid (derived from D-glucose)

L-gularic acid (derived from L-gulose)

27.18 (a) The structure of the aldaric acid derived from oxidation of D-galactose, and its 1,4-lactone:

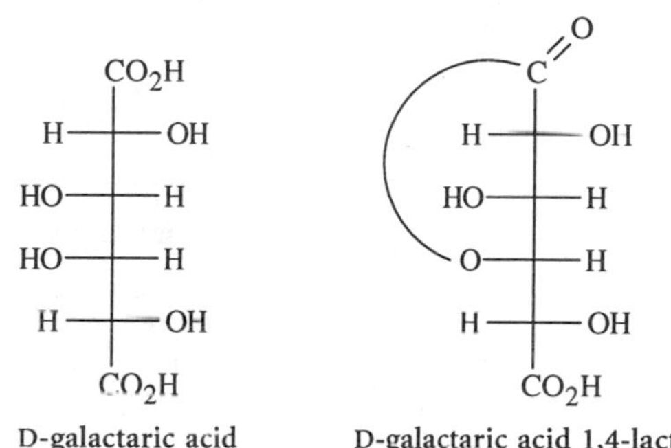

D-galactaric acid D-galactaric acid 1,4-lactone

(b) The structure of the aldaric acid derived from oxidation of D-mannose, and the structure of its 1,4-lactone:

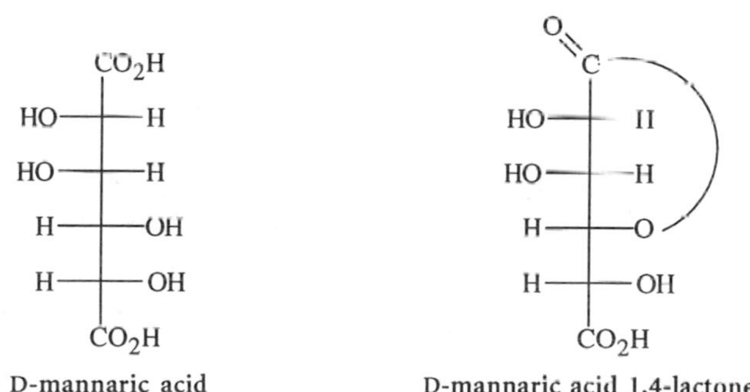

D-mannaric acid D-mannaric acid 1,4-lactone

27.19 As is the case with carbohydrates, the primary alcohol groups are selectively oxidized by HNO_3

(a) (b)

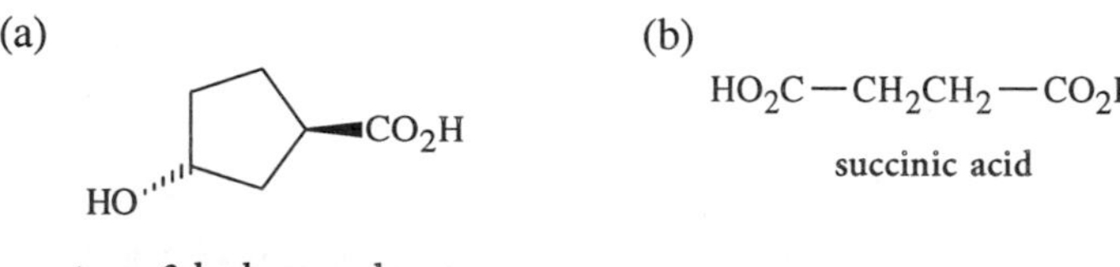

trans-3-hydroxycyclopentane-1-carboxylic acid

succinic acid

27.20 As Eq. 27.29a on text p. 1317 shows, periodate oxidation of a methyl pyranoside in the D-series gives a product containing carbons 1, 2, 4, 5, and 6 of the pyranoside. The secondary alcohols at carbons 2 and 4 are converted into aldehyde groups in all cases. In the oxidation products, only the carbons corresponding to carbons 1 and 5 in the pyranoside starting material are asymmetric. Since the configuration of carbon-5 determines whether the pyranoside has the D or the L configuration, and the configuration of carbon-1 determines whether the pyranoside is α or β, it follows that all methyl α-D-pyranosides give the same oxidation product with periodate.

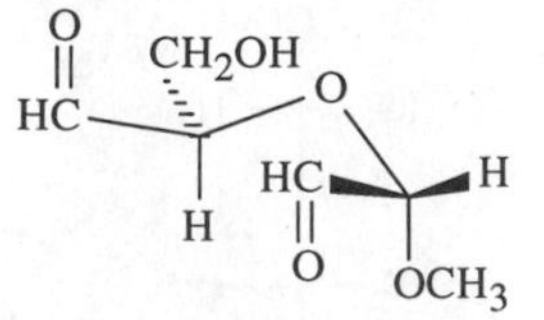

fragment obtained from periodate oxidation of any methyl α-D-pyranoside

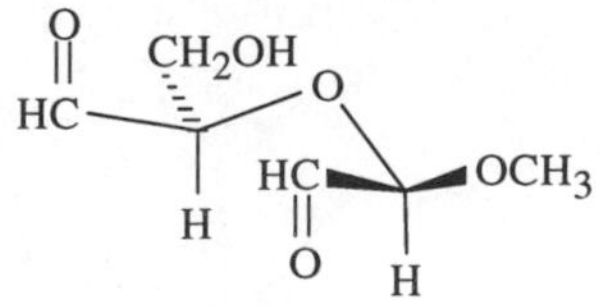

fragment obtained from periodate oxidation of any methyl β-D-pyranoside

(See the solution to Problem 27.51 on p. 1111 of this manual for a more detailed consideration of the structures of the periodate oxidation products of carbohydrates.)

27.21 The discussion in the solution to Problem 27.20 shows that the fragment obtained from oxidation of a β-D-pyranoside is the *diastereomer* of the fragment obtained from oxidation of an α-D-pyranoside. Diastereomers have different properties. Hence, these fragments, which can in principle be identified by their properties, including their optical rotations, can be used to determine whether the original pyranoside was α or β. This method works not only for the methyl galactopyranosides, but for *all* pyranosides. (Periodate oxidation was used to determine the anomeric configurations of many pyranosides before modern spectroscopic tools were available, and these results have withstood the scrutiny of modern analytical methods.)

27.22 The aldopentose *A* is D-ribose, and the aldohexose *B* is D-allose. The other hexose formed in the Kiliani-Fischer synthesis is D-altrose. (See Fig. 27.1 on text page 1294 for the structures of these aldoses. Note also in this figure the "Kiliani-Fischer relationships" among the aldoses.) Although D-xylose would also be oxidized to an optically inactive aldaric acid, its Kiliani-Fischer products, D-gulose and D-idose, would both be oxidized to optically active aldaric acids.

27.23 The aldopentose *A* is D-lyxose, and the aldohexose *B* is D-galactose. The other hexose *C* formed in the Kiliani-Fischer synthesis is D-talose. (See Fig. 27.1 on text page 1294 for the structures of these aldoses.) Although D-arabinose would also be oxidized to an optically active aldaric acid, its Kiliani-Fischer products, D-glucose and D-mannose, would both be oxidized to optically active aldaric acids.

27.24 Compound *A* is D-gulose, because it gives the same aldaric acid as L-glucose. The fact that some aldaric acids can be derived from two different aldoses is also addressed in Problem 27.17, the solution to which is found on pp. 1096–7 of this manual. The same point is discussed in Study Guide Link 27.3 on p. 1086 of this manual. In contrast, D-idose is the only aldohexose that can be oxidized to its aldaric acid.

27.25 Oxidize both (with dilute nitric acid) to their respective aldaric acids (which in this case are stereoisomeric tartaric acids). The compound that gives the optically active tartaric acid is D-threose; the compound that gives *meso*-tartaric acid (which is achiral and therefore optically inactive) is D-erythrose.

27.26 In such a case, the structure of D-(+)-glucose, indeed, the structures of *all* D-carbohydrates, would be the enantiomers of the structures shown in Fig. 27.1 on text p. 1294.

27.27 (a) Aqueous HCl (1 *M*) would bring about the hydrolysis of lactobionic acid into one equivalent each of D-galactose and D-gluconic acid, which, under the acidic conditions, exists primarily as D-γ-gluconolactone (structure in Eq. 27.25 on text p. 1315).

(b) Dimethyl sulfate in the presence of NaOH methylates all hydroxy groups.

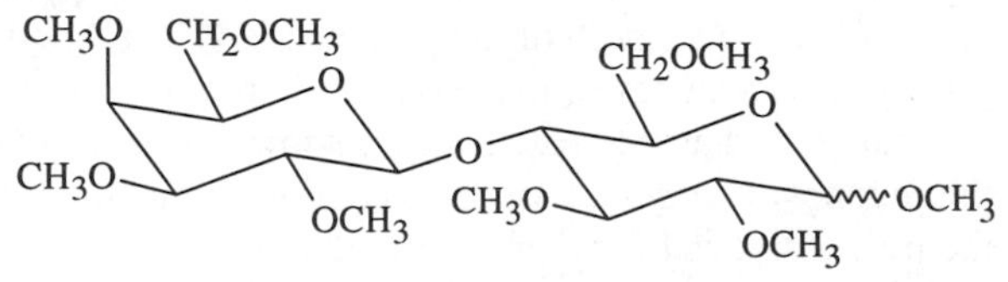

(c) Hydrolysis of all acetal and hemiacetal groups (but not the other ethers) takes place.

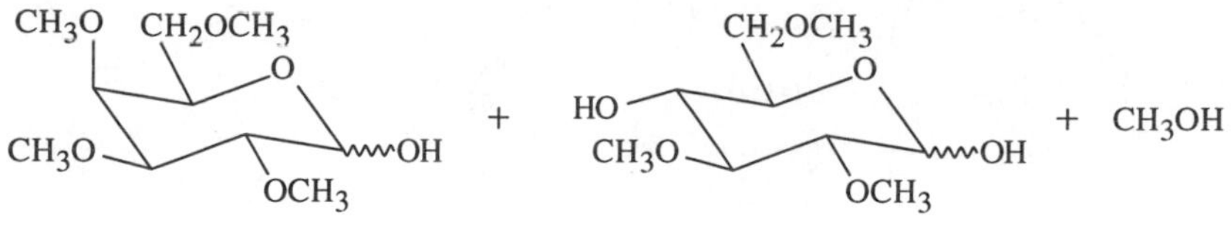

CH_3O CH_2OCH_3 ... CH_3O ... OH ... OCH_3 + HO ... CH_2OCH_3 ... CH_3O ... OH ... OCH_3 + CH_3OH

27.28 Because cellobiose is a dimer of D-glucose, it is hydrolyzed by aqueous acid into two molar equivalents of D-glucose.

27.29 The products would be a small amount of 2,3,4,6-tetra-*O*-methyl-D-glucopyranose (*A*, from the residue at the nonreducing end of the polymer) and mostly 2,3,6-tri-*O*-methyl-D-glucopyranose (*B*):

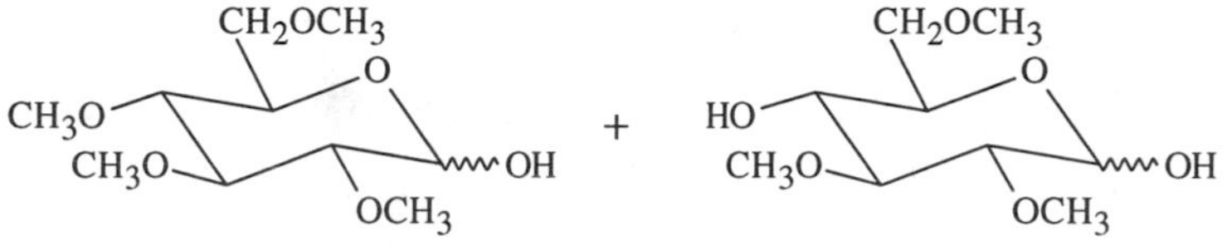

A + *B*

27.30 (a) deoxythymidine monophosphate (dTMP) (b) guanosine diphosphate (GDP)

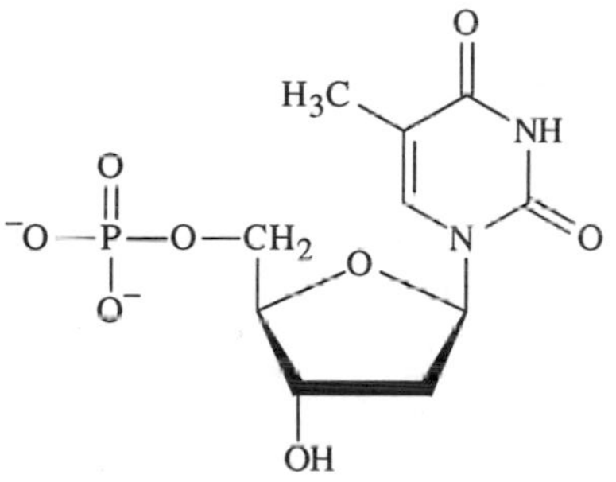

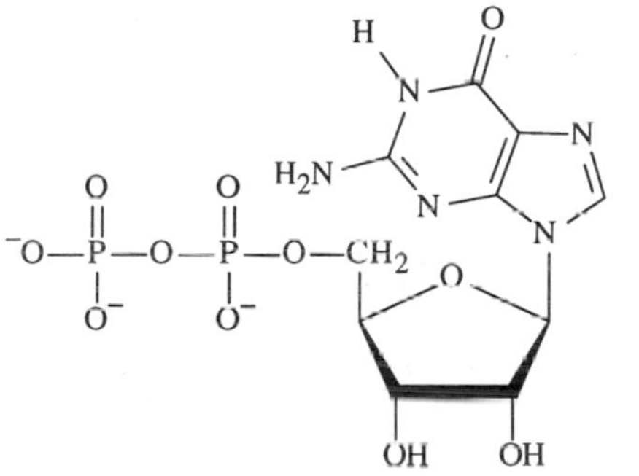

27.31 The structure of a tetranucleotide (a four-residue segment of RNA) with the sequence A-U-C-G is shown in Fig. SG27.1 on the following page.

27.32 Chargaff's rules do not apply to individual DNA strands because the rules are a consequence of the base-pair complementarity that occurs *between* strands.

27.33 (a) The structure of an O-4-methylated thymine residue in DNA is as follows. (For the numbering of purine and pyrimidine rings see text p. 1333.)

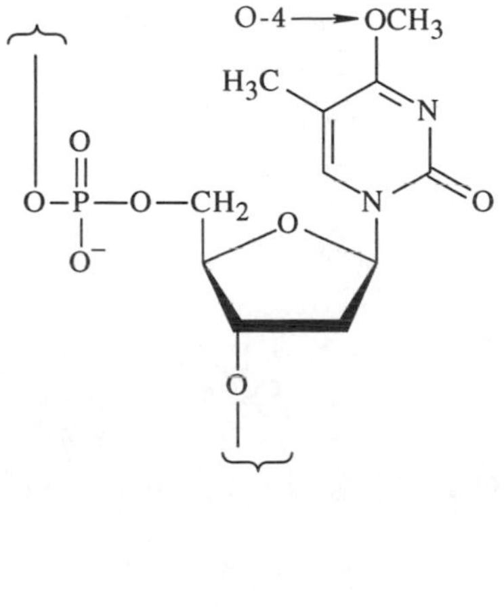

O-4 → OCH_3

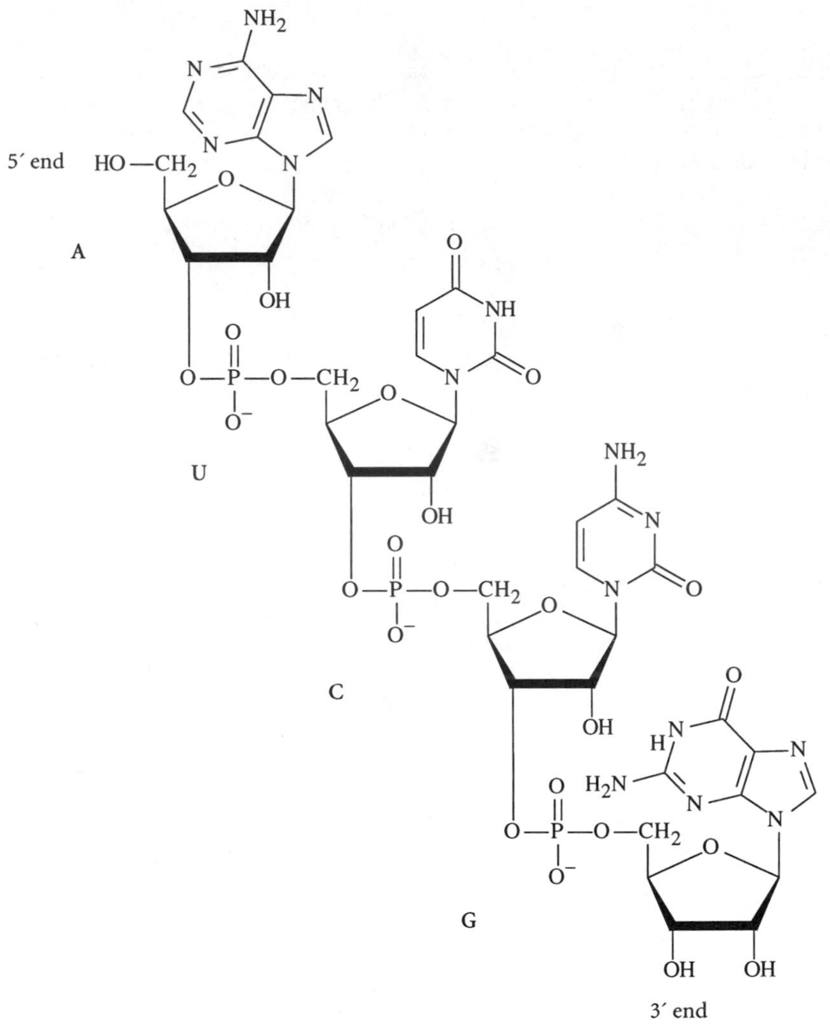

Figure SG27.1 The structure of a tetranucleotide A-U-C-G to accompany the solution to Problem 27.31.

(b) Refer to Fig. 27.6(b) on text p. 1339. The answer to part (a) shows that *O*-methylation eliminates the hydrogen on nitrogen-3 of thymine that is involved in hydrogen bonding with nitrogen-1 of adenine. Because this hydrogen bonding is an essential part of base pairing, disruption of this hydrogen bond also disrupts base pairing.

Solutions to Additional Problems

27.34 Note that carboxylic acids derived from carbohydrates usually exist as lactones, and aldoses exist primarily as cyclic hemiacetals. For simplicity they are represented below in their noncyclic forms.

(a)

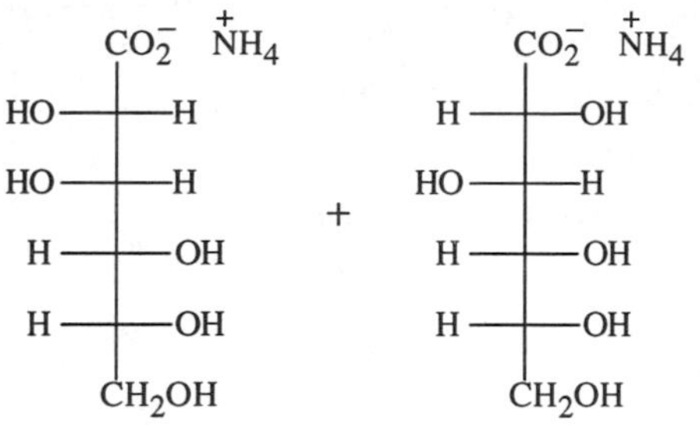

(b) No reaction.

Note that in part (a), the basic conditions under which the Tollens reagent is used (aqueous ammonium hydroxide) promote the de Bruyn-van Ekenstein rearrangement (see text pp. 1306–7 and p. 1315). Thus, carbon-2 of D-mannose is epimerized.

(c)

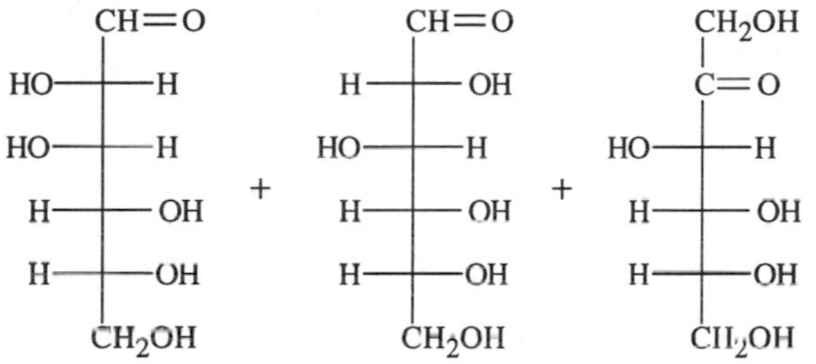

(d)

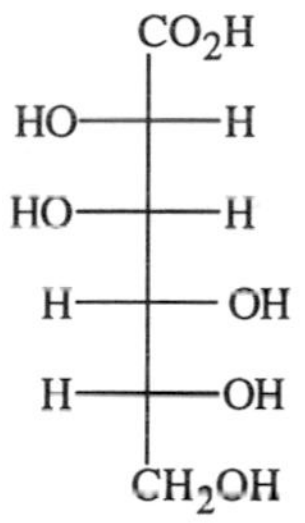

(e)

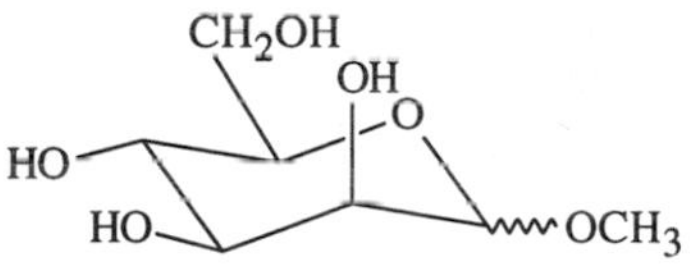

(f)

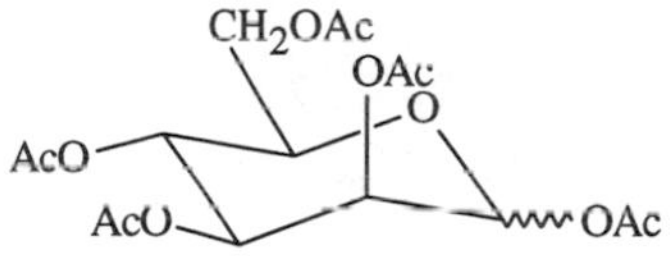

(g)

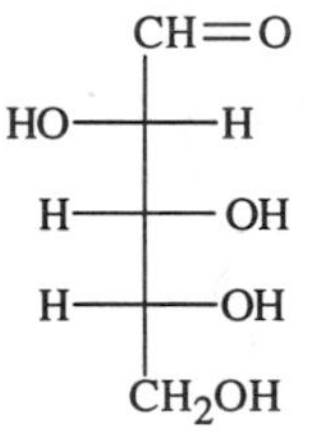

(h)

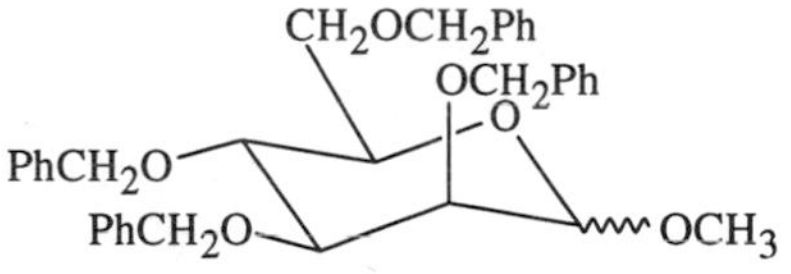

Note that reaction (g) is a Ruff degradation.

27.35 Note that carboxylic acids derived from carbohydrates usually exist as lactones, and aldoses exist primarily as cyclic hemiacetals. For simplicity they are represented below in their noncyclic forms.

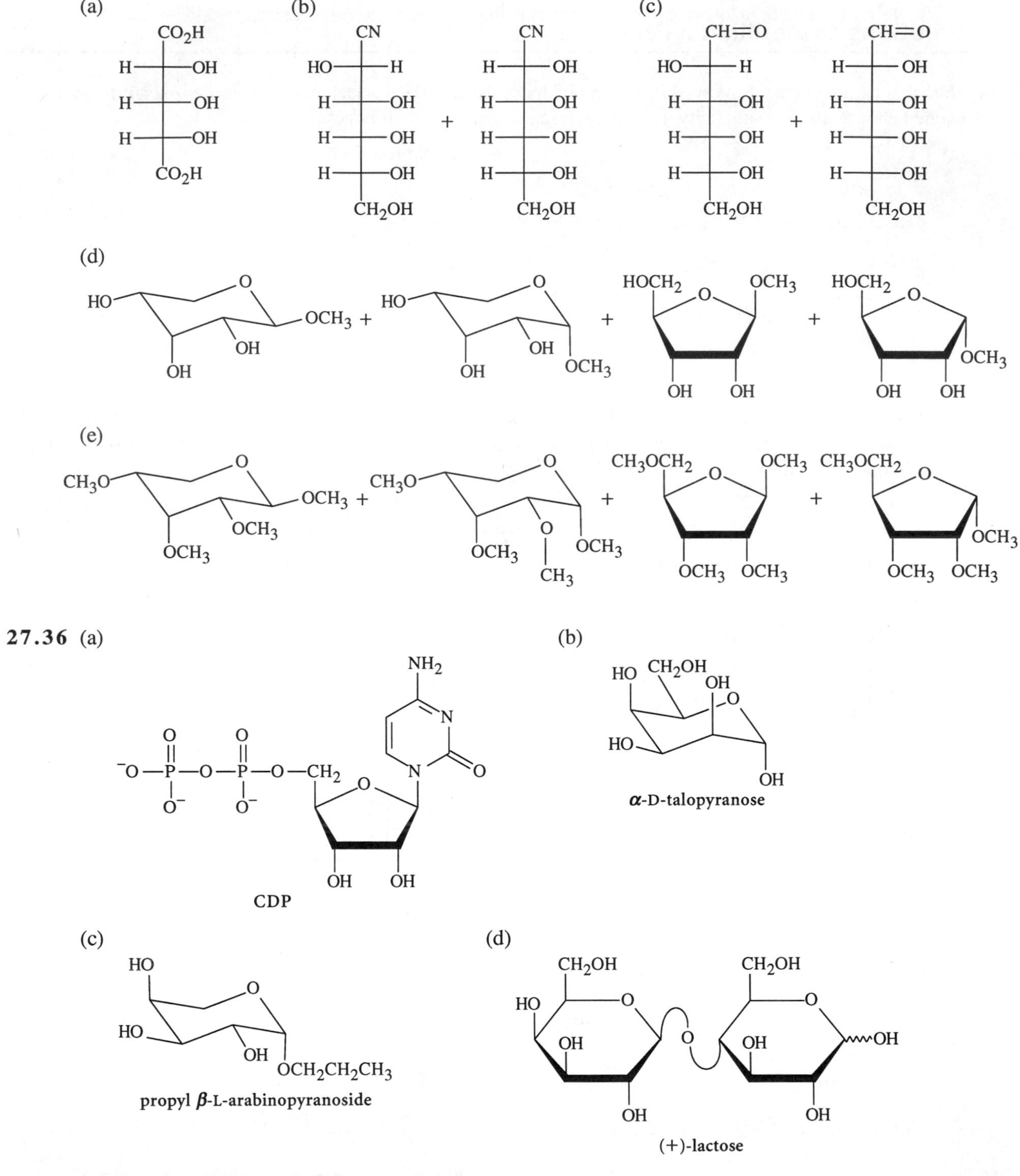

Remember that there are several valid ways to draw the chair conformations of pyranoses. Because the text probably has made you accustomed to seeing β-anomers with equatorial groups at carbon-1, you may be thinking that the chair conformation in the solution to part (c) shows the α-anomer. However, this *is* a correct representation of the β-anomer. Whether the β-anomer has an equatorialor axial group at the anomeric carbon depends, of course, on which chair conformation you choose to draw. The following equally valid representations of a different chair conformation of the same compound do have the group at the anomeric carbon in an equatorial position.

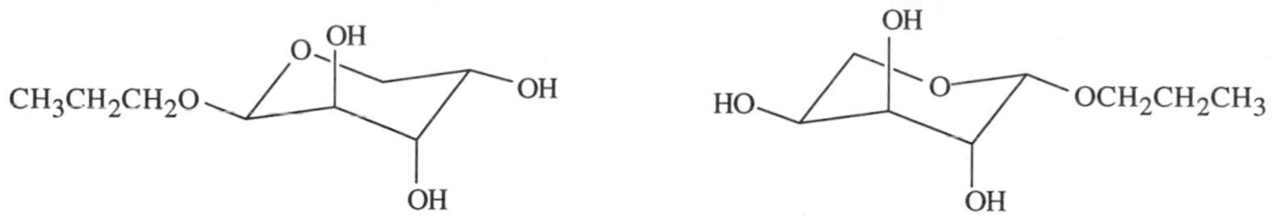

$CH_3CH_2CH_2O$ OH OH OH — HO OH OH $OCH_2CH_2CH_3$

You probably noticed that the bonds that connect the rings look very strange in Haworth projections. Another well-accepted (but equally strange) way to draw these bonds is illustrated in the Haworth structures of the disaccharides shown on p. 1074 of this manual

27.37 (a) Turning this structure over 180° shows that it is the noncongruent mirror image of β-D-galactopyranose; therefore it is β-L-galactopyranose.

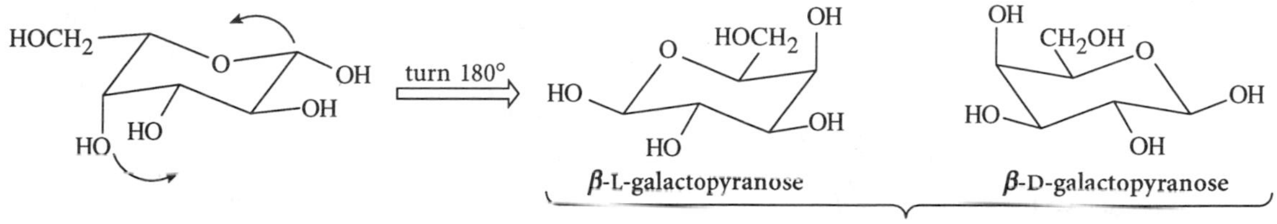

$HOCH_2$ OH OH HO HO — turn 180° — HO $HOCH_2$ OH OH HO — OH CH_2OH OH HO OH

β-L-galactopyranose β-D-galactopyranose

enantiomers

(b) This pyranose is epimeric to β-D-glucopyranose at carbon-3. Consult Fig. 27.1 on text p. 1294 to discover that the aldohexose epimeric to glucose at carbon-3 is allose. Therefore, this compound is β-D-allopyranose.

(c) First, notice that carbon-5 of this furanose has the *R* configuration. Since the D-hexoses have the *R* configuration at carbon-5 (the configurational carbon), and since the hemiacetal carbon bears a hydrogen, this must be a D-aldohexose. This can be converted into a Fischer projection as follows:

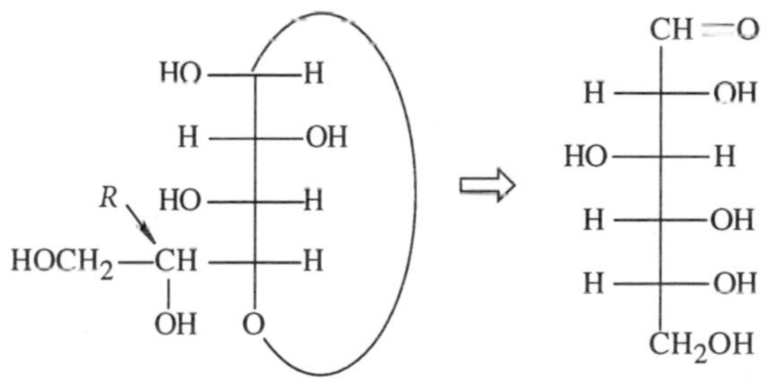

HO—H; H—OH; *R* HO—H; $HOCH_2$—CH(OH)—H; O ⟹ CH=O; H—OH; HO—H; H—OH; H—OH; CH_2OH

This shows that this compound is a form of D-glucose; it is β-D-glucofuranose.

27.38 (a) The D-series of the stereoisomeric 2-ketohexoses are shown below. (There is, of course, an enantiomeric L-series.) The problem didn't ask for the names, but they are given as part of our policy of extra value at no added cost.

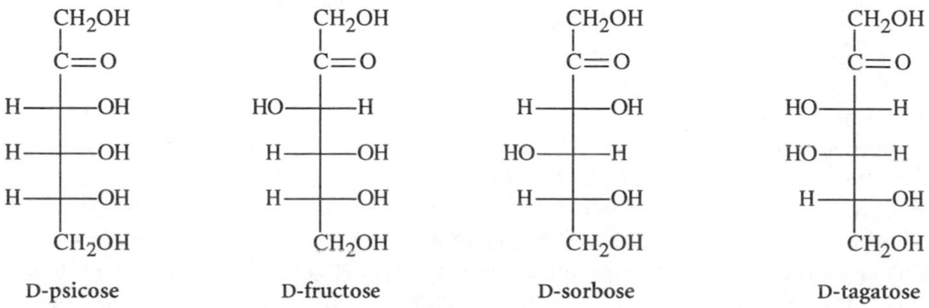

D-psicose	D-fructose	D-sorbose	D-tagatose
CH_2OH	CH_2OH	CH_2OH	CH_2OH
C=O	C=O	C=O	C=O
H—OH	HO—H	H—OH	HO—H
H—OH	H—OH	HO—H	HO—H
H—OH	H—OH	H—OH	H—OH
CH_2OH	CH_2OH	CH_2OH	CH_2OH

(b) An achiral ketopentose must be a meso compound:

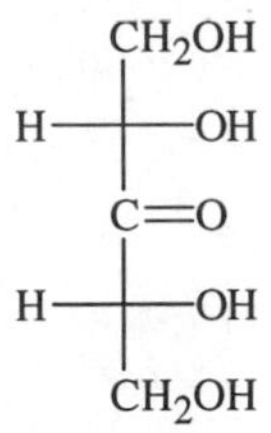

CH_2OH
H—C—OH
C=O
H—C—OH
CH_2OH

(c) The structure of α-D-galactofuranose is derived from the Fischer projection as follows:

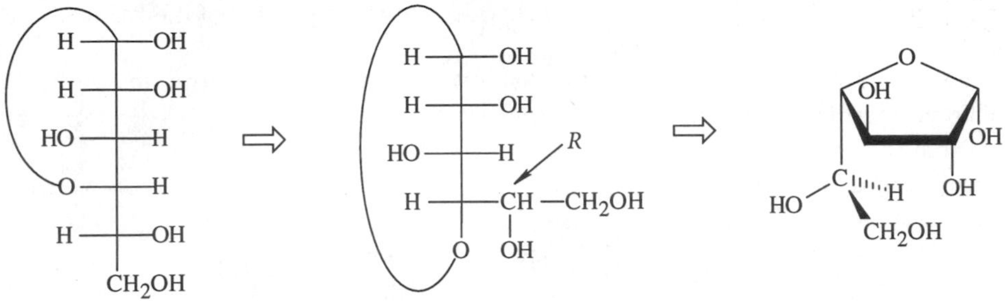

H—OH, H—OH, HO—H, O—H, H—OH, CH_2OH ⇨ H—OH, H—OH, HO—H, H—CH—CH_2OH (R), O, OH ⇨ O, OH, OH, HO, C····H, OH, CH_2OH

(d) To derive the structure of β-D-idofuranose, you can work from the Fischer projection as shown for α-D-galactofuranose in part (c); or you can recognize that D-idose differs in configuration from D-galactofuranose at every carbon except carbons 4 and 5. Working from the solution to part (c),

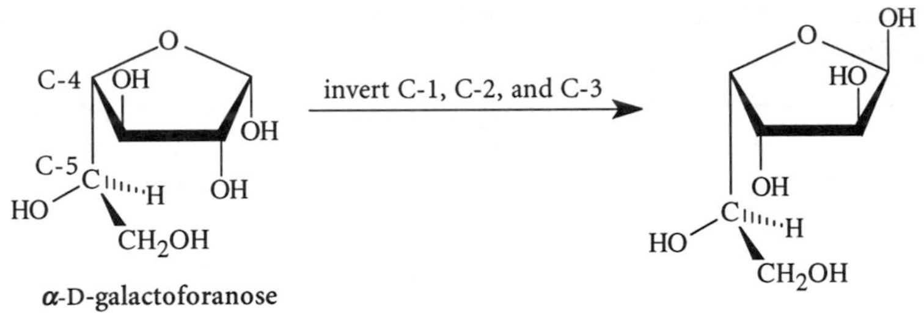

C-4, O, OH, OH, C-5, HO, C····H, OH, CH_2OH — invert C-1, C-2, and C-3 → O, OH, HO, OH, HO, C····H, CH_2OH

α-D-galactoforanose

β-D-idoforanose

27.39 (a) Diastereomers, epimers, and anomers. Remember that epimers and anomers are special types of diastereomers.
(b) Diastereomers and epimers
(c) Enantiomers
(d) Constitutional isomers
(e) Constitutional isomers
(f) The two names describe the same structure.

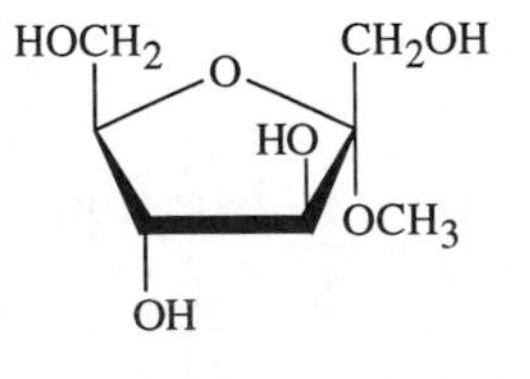

$HOCH_2$, O, CH_2OH, HO, OCH_3, OH

methyl α-D-fructofuranoside
(2-O-methyl-α-D-fructofuranose)

27.40 (a) L-Sorbose is a hexose.
(b) L-Sorbose is a ketohexose, or hexulose.
(c) L-Sorbose is not a glycoside because it is not involved in acetal formation with an alcohol or a phenol.
(d) L-Sorbose is not an aldohexose; it's a ketohexose, as noted in part (b).
(e) This is another valid Fischer projection of L-sorbose obtained by a cyclic permutation at carbon-5.
(f) This compound is epimeric to L-sorbose at carbon-5, and thus it is not a form of L-sorbose. In fact, it is

D-fructose.

(g) This is the α-furanose form of L-sorbose (α-L-sorbofuranose). The simplest way to see this is to work from the Haworth projection given in the problem to the Fischer projection:

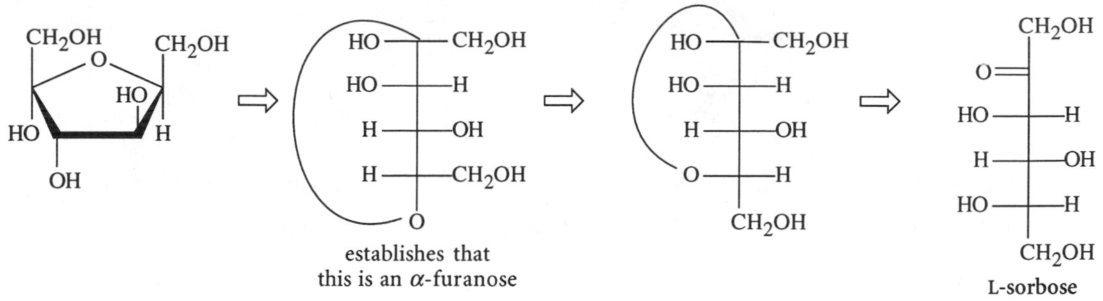

establishes that this is an α-furanose

L-sorbose

(h) This structure is the enantiomer of the structure in part (g), and thus it is not a form of L-sorbose. (It is the α-furanose form of D-sorbose.)

(i) The relationship of this furanose to the one shown in part (g) is best discerned by turning either of the structures over 180°. This operation reveals that this structure differs in configuration from the one in part (g) at carbons 3 and 4. Hence, this is not a form of L-sorbose.

(j) This is an aldopyranose and therefore cannot be a form of L-sorbose. (It is β-D-glucopyranose.)

(k) This is the α-pyranose form of L-sorbose, α-L-sorbopyranose. As with part (g), the easiest way to see this is by working from the chair to the Haworth to the Fischer.

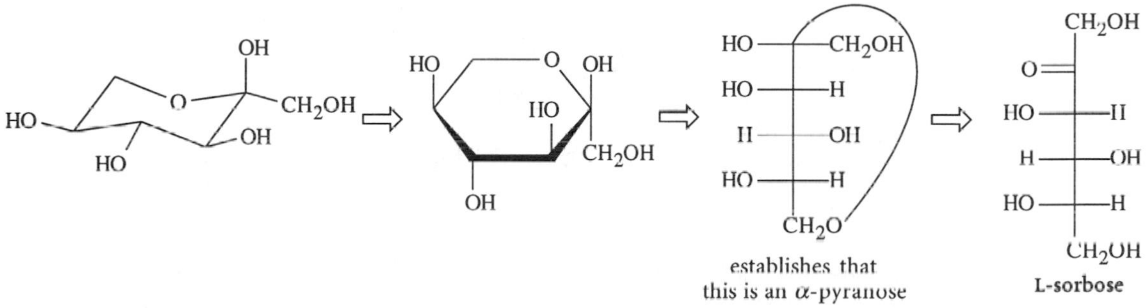

establishes that this is an α-pyranose

L-sorbose

27.41 (a) Raffinose is a nonreducing sugar because it has no hemiacetal groups; the saccharide residues at both ends are involved in acetal linkages.

(b) The glycoside linkages in raffinose:

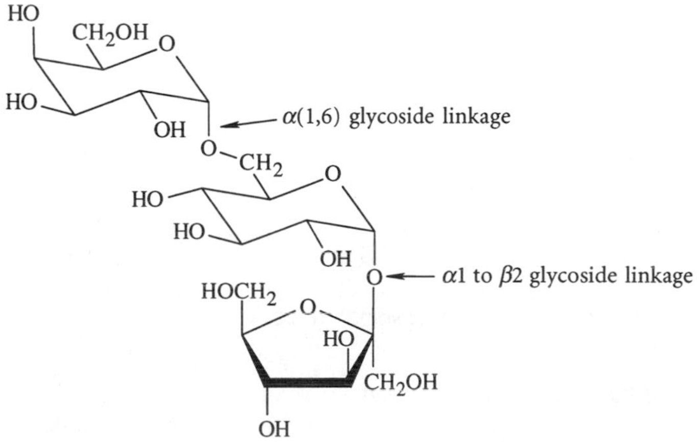

α(1,6) glycoside linkage

α1 to β2 glycoside linkage

The type of linkage in the furanose residue can be determined systematically by using the Haworth projection to form the standard Fischer projection. (The glycosidic linkage is indicated as RO—.)

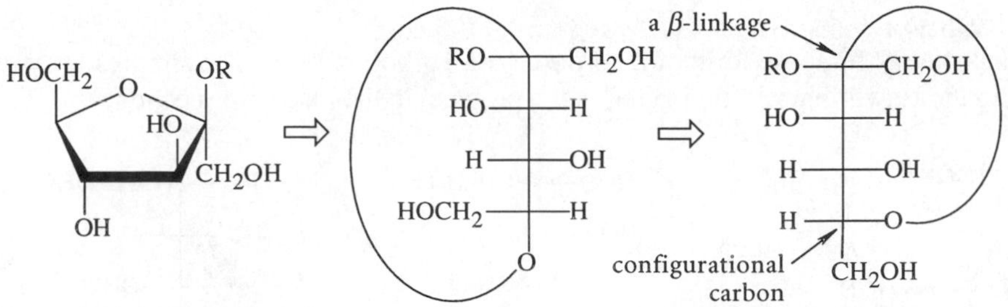

a β-linkage

configurational carbon

(c) Raffinose is hydrolyzed into equal amounts of D-galactose, D-glucose, and D-fructose.

(d) The products of methylation followed by hydrolysis are the following; each is a mixture of anomers.

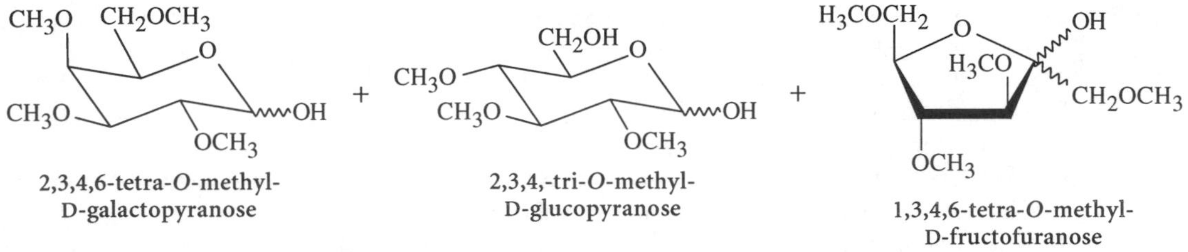

2,3,4,6-tetra-*O*-methyl-D-galactopyranose + 2,3,4,-tri-*O*-methyl-D-glucopyranose + 1,3,4,6-tetra-*O*-methyl-D-fructofuranose

27.42 The structure of 3-*O*-β-D-glucopyranosyl-α-D-arabinofuranose:

3-*O*-β-D-glucopyranosyl-α-D-arabinofuranose

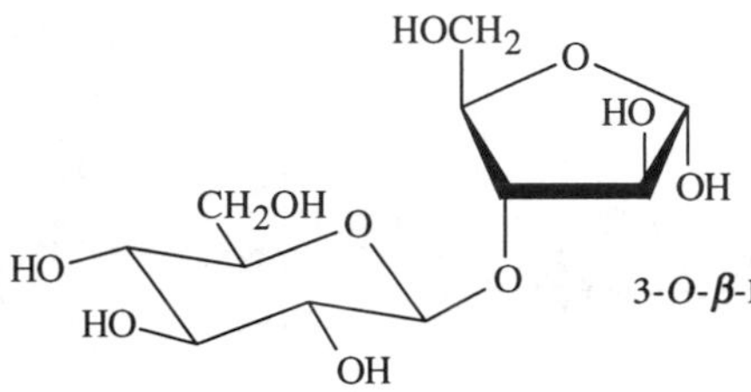

27.43 (a) Turn the chair conformation 180° about a vertical axis in the page to get the equivalent chair structure below. Then convert this to a Fischer projection as shown. These manipulations reveal that, when we draw the Fischer projection in standard form (carbon-6 vertical), the oxygen at carbon-5 is on the left. Hence, this enantiomer of fucose has the L configuration.

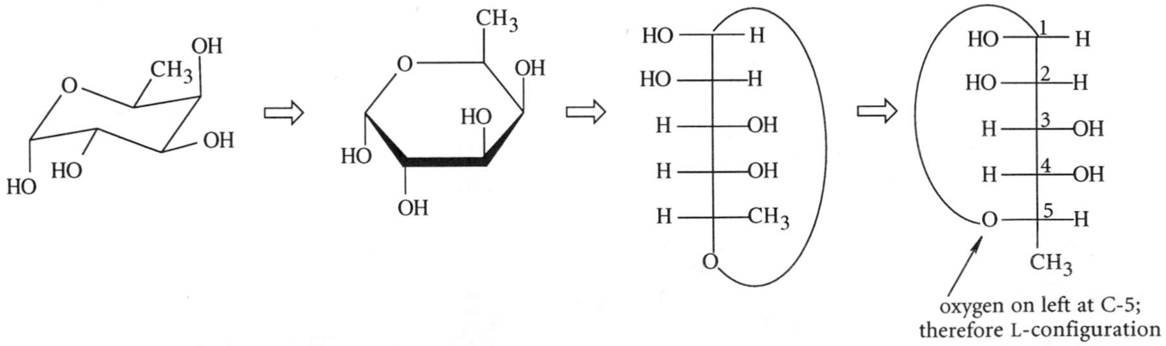

oxygen on left at C-5; therefore L-configuration

(b) Because the —OH at carbon-1 in this standard representation is on the same side as the oxygen at carbon-5, this is by definition the α-anomer.

(c) Fucose is an aldose, because the hemiacetal carbon bears a hydrogen; thus, in the carbonyl form, there is also a hydrogen at carbon-1. Hence, the carbonyl form must be an aldehyde.

(d) A Fischer projection of the aldehyde form of L-fucose:

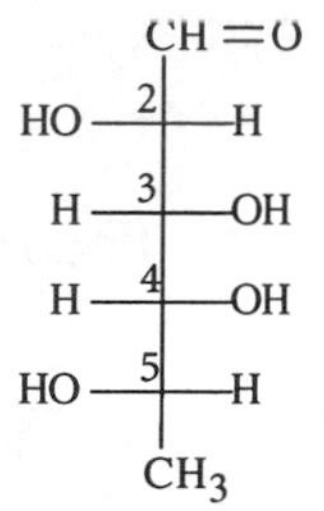

CH=O
HO —2— H
H —3— OH
H —4— OH
HO —5— H
CH_3
L-fucose

Fucose is classified as a *deoxy* carbohydrate; that is, it has one fewer —OH group than an ordinary aldohexose. (An aldohexose would have an —OH group at carbon-6.)

27.44 (a) Because the osazones from the two carbohydrates are the same, and given that both are aldohexoses, the only stereochemical difference between them—a difference that is obliterated by osazone formation—is at carbon-2. Consequently, the two carbohydrates must be epimeric at carbon-2.

(b) D-Ribose gives the same osazone as D-arabinose because D-ribose is epimeric to D-arabinose at carbon-2.

27.45 (a) Just as the acetal units in carbohydrates undergo *hydrolysis* in water, they also undergo *methanolysis* in methanol. The result is a mixture of the methyl glycosides along with the by-product phenol.

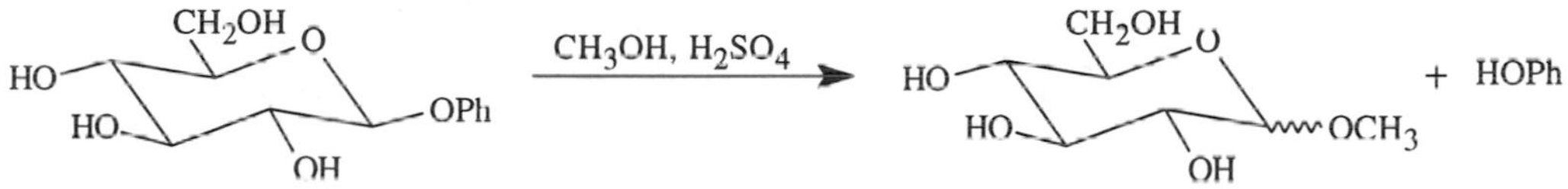

phenyl β-D-glucopyranoside —(CH_3OH, H_2SO_4)→ methyl D-glucopyranoside (mixture of α and β) + HOPh

(b) Any 3- or 4-hydroxy aldehyde, like a carbohydrate, will form a cyclic acetal in acidic methanol.

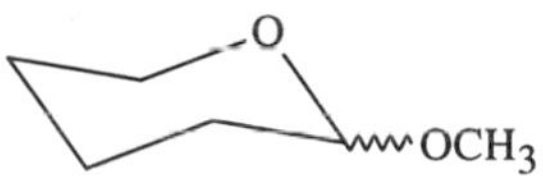

(c) Osmium tetroxide forms *cis*-1,2-cyclohexanediol; periodic acid cleaves this diol to form the dialdehyde 1,6-hexanedial, $O{=}CHCH_2CH_2CH_2CH_2CH{=}O$; and this dialdehyde is reduced by sodium borohydride to 1,6-hexanediol, $HOCH_2CH_2CH_2CH_2CH_2CH_2OH$.

(d) Cyclohexene reacts with OsO_4 to give *cis*-1,2-cyclohexanediol, which forms a cyclic acetal with acetone.

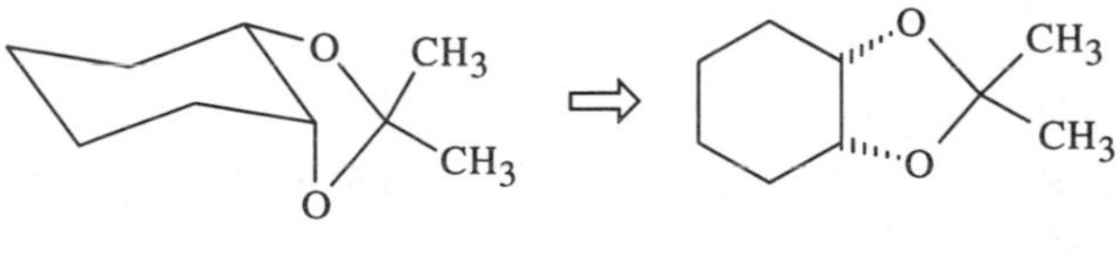

(e) All of the free hydroxy groups in sucrose are methylated. (For the structure of (+)-sucrose see text p. 1328.)

(Solution continues)

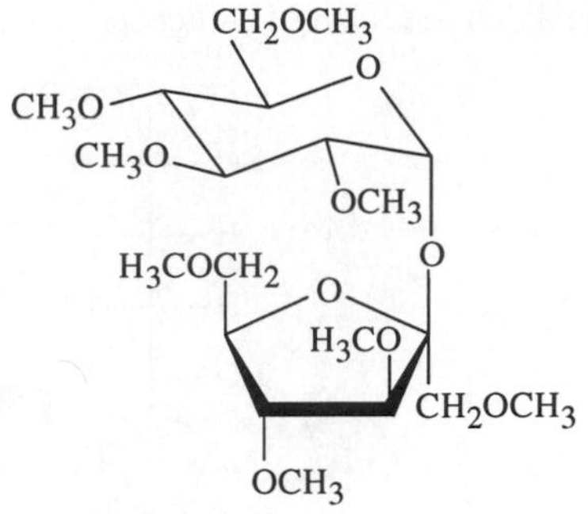

CH2OCH3 O CH3O CH3O OCH3 O H3COCH2 O H3CO CH2OCH3 OCH3

(f) Ethanolysis of the acetal linkage occurs in ethanol just as hydrolysis occurs in water. An ethyl pyranoside is also formed from the glucose residue at the reducing end of lactose.

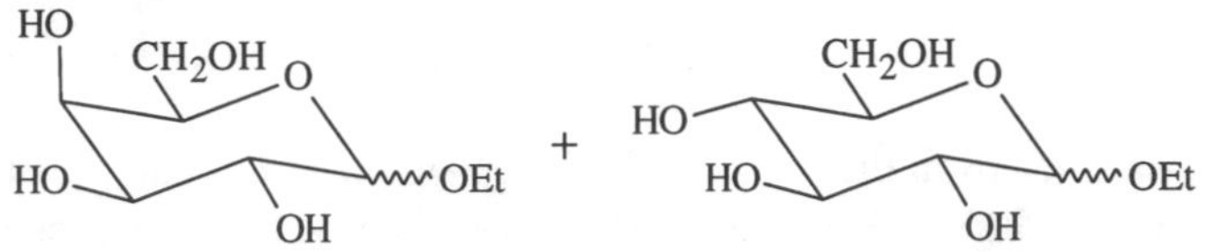

HO CH2OH O HO OEt OH + HO CH2OH O HO OEt OH

(g) This reaction is not unlike the one in part (a), in which one acetal is converted into another in methanol. Acid-catalyzed loss of methanol is followed by cyclization to give a mixture of the two methyl pyranosides. (Note that $CH_3\overset{+}{O}H_2\ Cl^-$, the conjugate acid that results from protonation of the solvent by HCl, is the effective acid catalyst.)

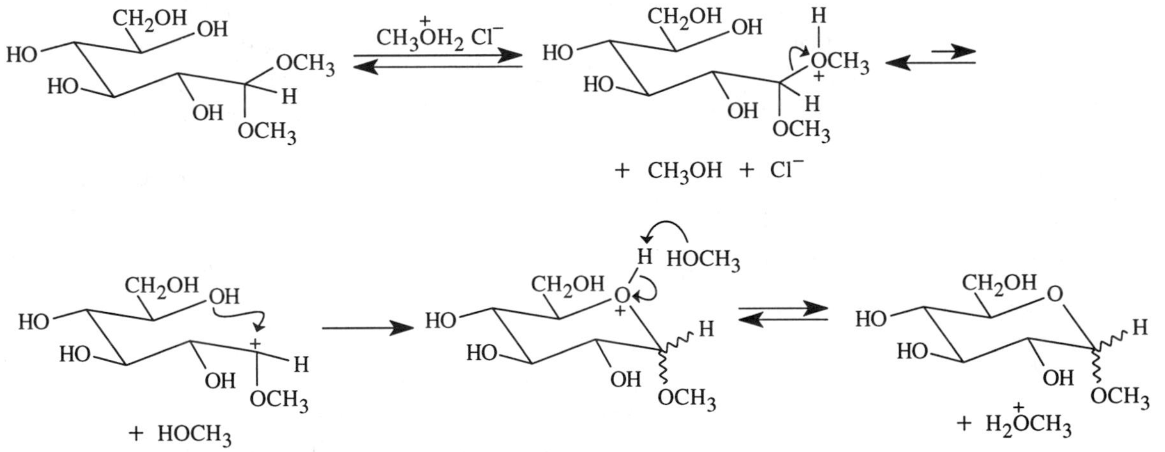

HO CH2OH OH HO OCH3 H OH OCH3 $CH_3\overset{+}{O}H_2\ Cl^-$ HO CH2OH OH H OCH3 HO H OH OCH3 + CH_3OH + Cl^-

HO CH2OH OH HO H OH OCH3 + $HOCH_3$ → HO CH2OH H HOCH3 O H HO OH OCH3 ⇌ HO CH2OH O HO H OH OCH3 + $H_2\overset{+}{O}CH_3$

27.46 (a) This is an isotopically labeled analog of glucose, and is prepared by the Kiliani-Fischer synthesis shown in Eqs. 27.32 and 27.33 on text p. 1319, except that radioactive sodium cyanide ($Na^{14}CN$) is used instead of ordinary NaCN. In this synthesis, ^{14}C-D-mannose will be a by-product.

(b) This compound is prepared in the same way as the compound in (a), except that ordinary (unlabeled) NaCN is used, and the reduction is carried out with tritium-enriched hydrogen ($^3H_2$). In this synthesis, ^{3}H-D-mannose will be isolated as a by-product.

(c) The synthesis of compound (c) requires removal of an *unlabeled* carbon from D-arabinose, replacing it with a labeled carbon, and then completion of an ordinary Kiliani-Fischer synthesis to give labeled mannose. (* = ^{14}C)

Ruff degradation (Eq. 27.41, text p. 1325); Kiliani-Fischer synthesis with $\overset{*}{NaCN}$; Kiliani-Fischer synthesis

Each of the last two reactions also gives epimeric sugars—D-ribose and D-glucose, respectively—as by-products.

27.47 The results of the oxidation of *A* indicate that compound *A* is not a methyl pyranoside or furanoside, that is, the methyl ether is not at carbon-1. The Ruff degradation yields another reducing sugar. This means that the methyl ether cannot be at carbon-2. Only if the ether is at carbon-3 can one of the Kiliani-Fischer products be oxidized to an optically inactive (that is, a meso) compound:

B ← (Br_2, H_2O) *A* → Kiliani-Fischer synthesis

HNO_3 oxidation gives a meso ompound

27.48 The extract of mouse spleen evidently contains enzymes that catalyze the removal of a phosphate as well as the conversion of ribose into an optically inactive isomer. Ribitol and xylitol are the alditols derived from ribose and xylose, respectively, and therefore are epimeric at carbon-3. Compound *X* has one degree of unsaturation, and, because reduction gives alcohols that are epimeric at carbon-3, compound *X* is a 3-ketopentose. The structure for *X* shown in the following transformation fits the data:

H_2O + ribose-5-phosphate → (extract of mouse spleen) compound *X* + $HOPO_3^{2-}$ → (periodic acid) $OHC{-}CO{-}CHO$ + 2 $H_2C{=}O$

27.49 Compound *A* is the corresponding oxime (see Table 19.3 on text p. 876), and compound *B* results from acetylation of the hydroxy groups of the oxime. Compound *C* results from a sodium acetate-promoted *β*-elimination reaction. The conversion of compound *C* to D-arabinose is the result of a transesterification reaction, which removes all acetyl groups as methyl acetate, followed by the reverse of cyanohydrin formation.

Identity of compounds A and B:

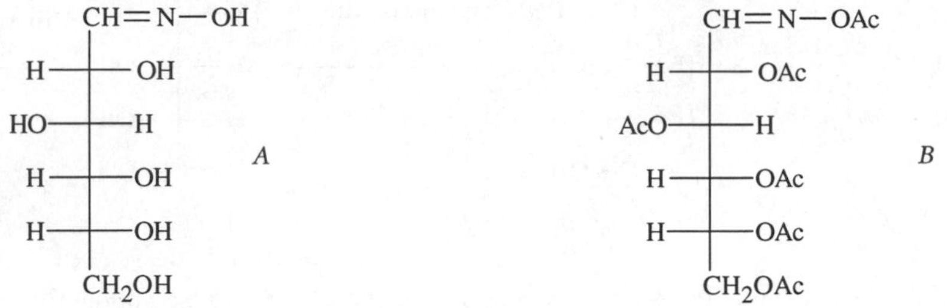

The mechanism for the conversion of compound B into compound C:

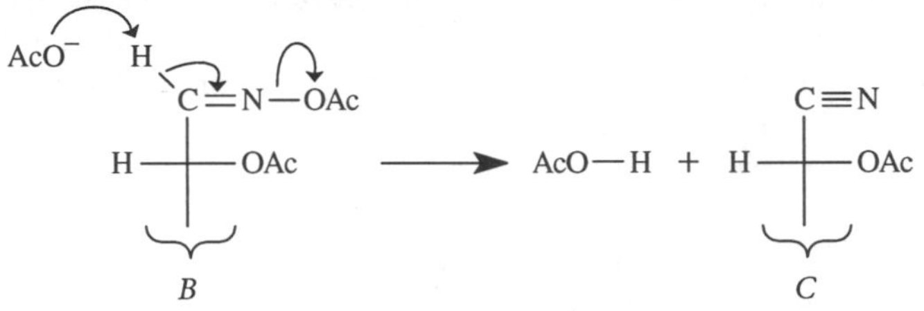

$$AcO-H + H-\overset{C\equiv N}{C}-OAc$$

The mechanism for the conversion of compound C into D*-arabinose:*

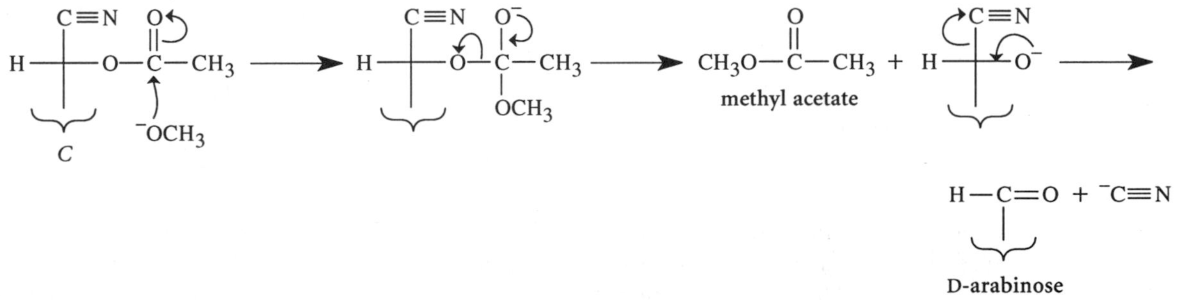

methyl acetate

D-arabinose

Note that the other acetate groups are removed in transesterification reactions, the mechanisms of which are identical to the first two steps of the foregoing mechanism.

27.50 Compound *A* is the lactone of the corresponding aldonic acid, and compound *B* is the amide that results from ester (lactone) ammonolysis. Chlorine and NaOH bring about a Hofmann rearrangement of amide *B* to give *C*, a carbinolamine, which, under the aqueous reaction conditions, breaks down spontaneously to ammonia and an aldehyde. (See Eq. 19.56a on text p. 875.) This aldehyde is the aldose that has one fewer carbon atom than the starting aldose. (This is illustrated below with D-glucose, but any aldose whose aldonic acid can form a lactone would undergo the same chemistry.)

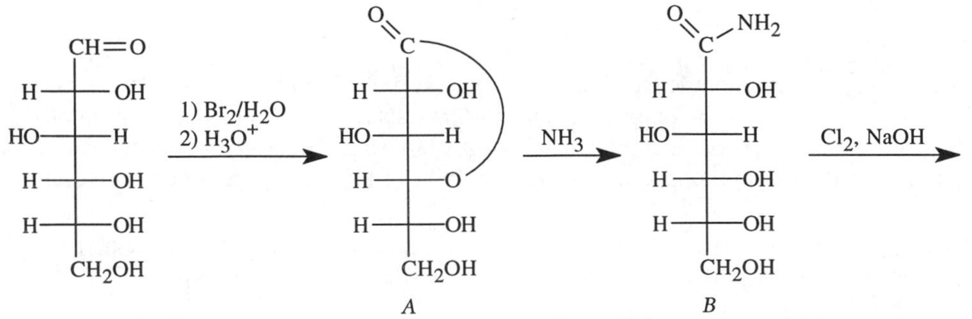

1) Br_2/H_2O
2) H_3O^+

NH_3

Cl_2, NaOH

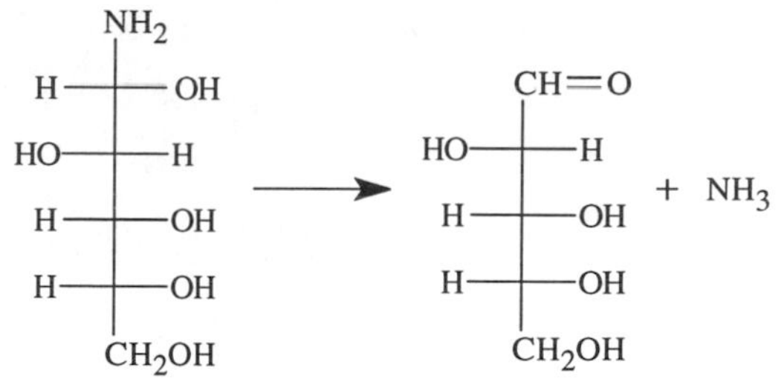

C
(a carbinolamine)

$+ NH_3$

27.51 The product *X* might be reasonably expected to result from treatment of methyl α-L-rhamnopyranoside with periodic acid. (See the solution to Problem 27.20 on p. 1097 of this manual.)

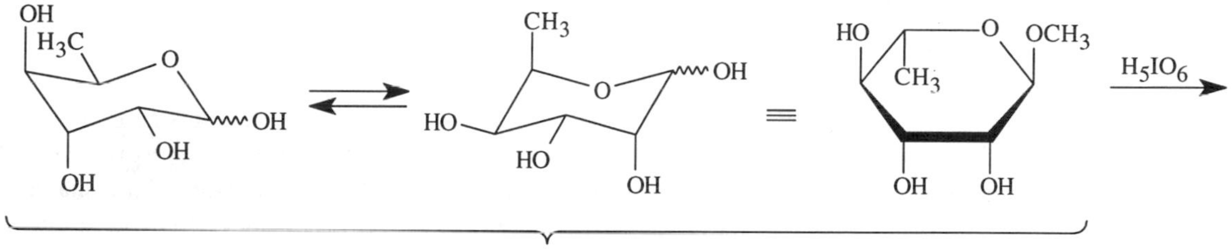

H_5IO_6

methyl α-L-rhamnopyranoside

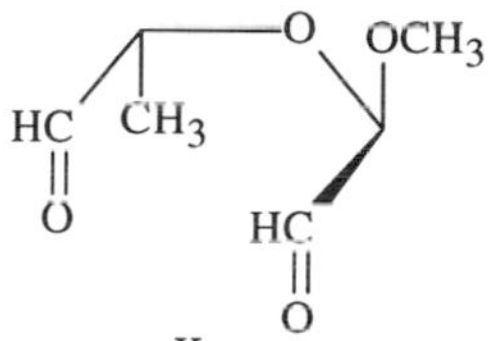

X

The formula of *X*, $C_6H_{10}O_4$, is short of the formula of actual product *A* by the elements of H_2O. Nevertheless, reduction of this product by $NaBH_4$ would give compound *C*. Evidently, compound *A* has two —OH groups, because methylation introduces two carbon atoms. Compound *A* results, first, from hydration of one of the carbonyl groups of *X*; an —OH group of the hydrate then attacks the other carbonyl carbon to give a "double hemiacetal" *A*. The structure of *A* and its methylation product *B* are as follows:

$H_2O + X \rightleftharpoons$

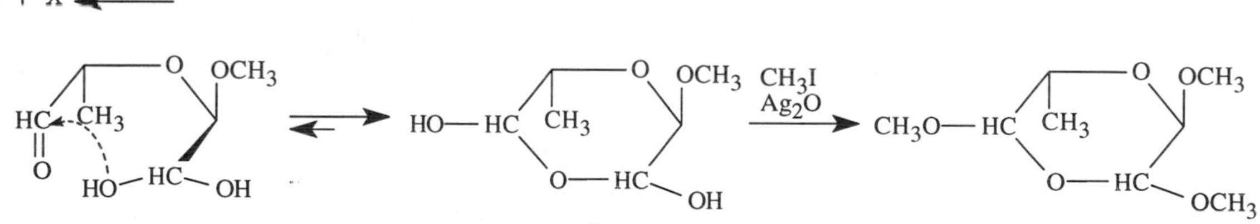

a hydrate of *X*   *A*   (CH_3I, Ag_2O)   *B*

(Compound *B* is undoubtedly a mixture of stereoisomers.) Because compound *A* is a hemiacetal, it shows no carbonyl absorption. However, it is in equilibrium with a small amount of the dialdehyde *X*, which is reduced by $NaBH_4$ to give compound *C*. The reduction pulls the equilibrium between *A* and *X* toward *X* until all of this compound has been reduced.

27.52 The residue at the nonreducing ("left") end undergoes the following transformation:

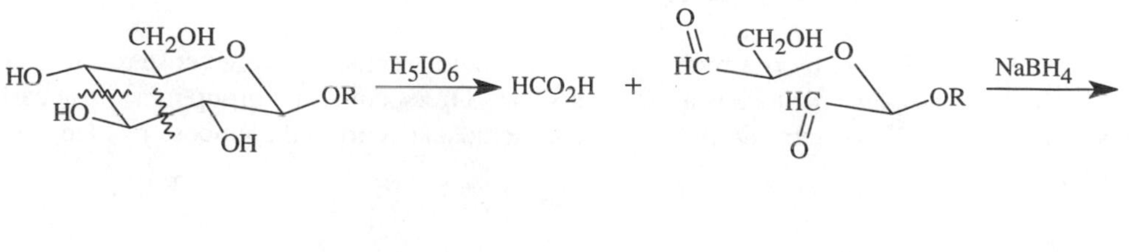

H_5IO_6 → HCO_2H + ... $NaBH_4$ →

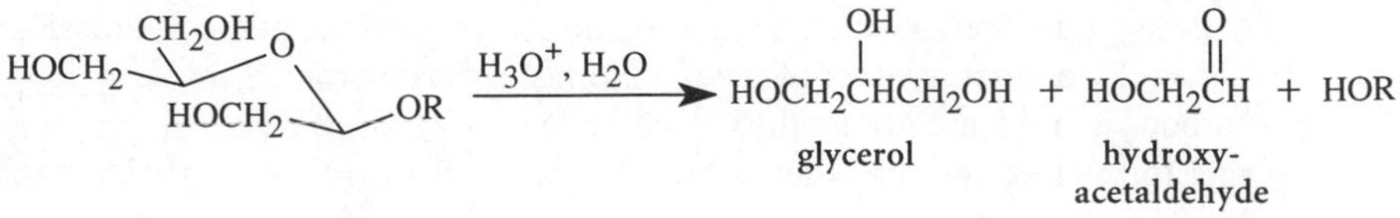

$$\text{HOCH}_2\text{–}(\text{CH}_2\text{OH})\text{–O–HOCH}_2\text{–OR} \xrightarrow{H_3O^+,\ H_2O} \underset{\text{glycerol}}{\text{HOCH}_2\text{CH(OH)CH}_2\text{OH}} + \underset{\text{hydroxy-acetaldehyde}}{\text{HOCH}_2\text{CH=O}} + \text{HOR}$$

The internal residues undergo the following transformation:

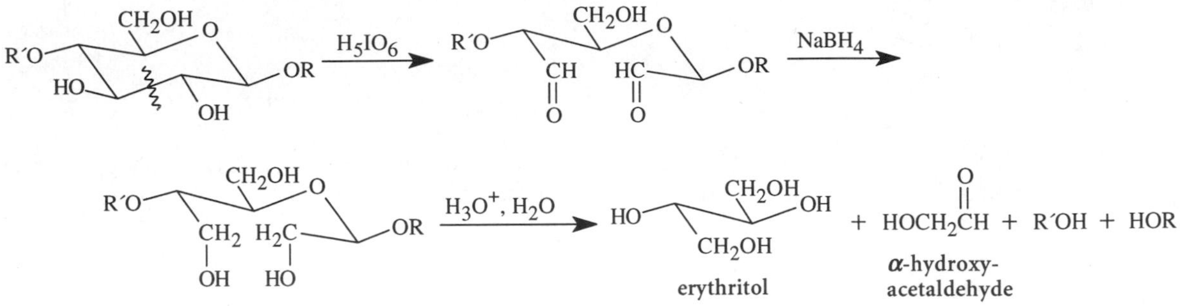

$$\text{R'O–(ring with CH}_2\text{OH, HO, OH)–OR} \xrightarrow{H_5IO_6} \text{R'O–(CH}_2\text{OH)(CH=O)(HC=O)–OR} \xrightarrow{NaBH_4}$$

$$\text{R'O–(CH}_2\text{OH)(CH}_2\text{OH)(H}_2\text{C–HO)–OR} \xrightarrow{H_3O^+,\ H_2O} \underset{\text{erythritol}}{\text{HO–CH(CH}_2\text{OH)–CH(CH}_2\text{OH)–OH}} + \underset{\alpha\text{-hydroxy-acetaldehyde}}{\text{HOCH}_2\text{CH=O}} + \text{R'OH} + \text{HOR}$$

The residue at the reducing ("right") end gives erythritol, R´OH, and two equivalents of formic acid (HCO_2H). (Demonstrate this for yourself; see Eq. 27.28 on text p. 1317.) Erythritol therefore results from the cleavage reactions of all residues except for the one at the nonreducing end; and glycerol results from cleavage of that residue. Thus, glycerol is produced from one residue, and erythritol is produced from $n - 2 + 1 = n - 1$ residues. Since $n = 12$, eleven moles of erythritol per mole of glycerol are formed.

27.53 (a) The pairing of an imine isomer of C with A:

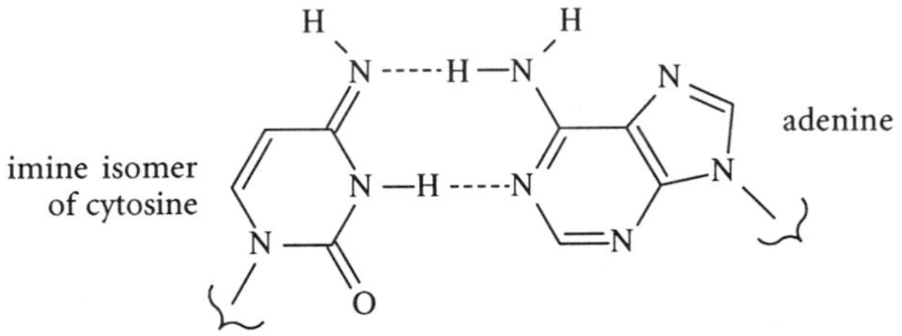

imine isomer of cytosine — adenine

(b) The pairing of an enol isomer of T with G:

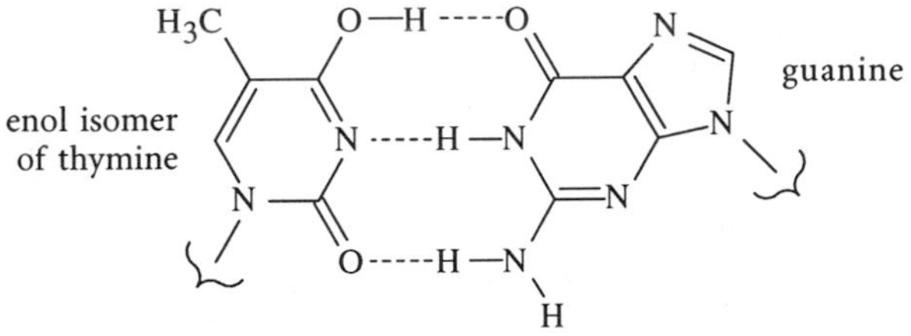

enol isomer of thymine — guanine

27.54 (a) Figure 27.6 on text p. 1339 shows that a G-C pair has three hydrogen bonds, but an A-T pair has only two. The hydrogen bonds hold the strands of the double helix together. Because the melting temperature is a measure of the forces holding the strands of the double helix together, the higher melting temperature of the polyG-polyC double helix is accounted for by the greater number of hydrogen bonds per residue.

(b) The solution to part (a) leads to the conclusion that, because of the greater amount of hydrogen bonding in G-C base pairs, G-C rich DNA should have a higher melting temperature than A-T rich DNA. Hence, the human adenovirus I DNA contains the greater ratio of G + C.

27.55 The hydrolysis of maltose by α-amylase shows that maltose contains two glucose residues connected by an α-glycosidic bond. Because maltose is a reducing sugar, a free hemiacetal group must be at carbon-1 of one of the glucose units. The 2,3,4,6-tetra-*O*-methyl-D-glucose must arise from the glucose residue at the nonreducing end. The question is which oxygen in the glucose residue at the reducing end is involved in the glycosidic linkage; the other product of methylation-hydrolysis provides evidence on this point. Because the oxygens at carbons 4 and 5 are not methylated in this product, one of these oxygens is involved in pyranoside or furanoside ring formation, and the other is involved in the glycosidic linkage. Two structures satisfy these requirements:

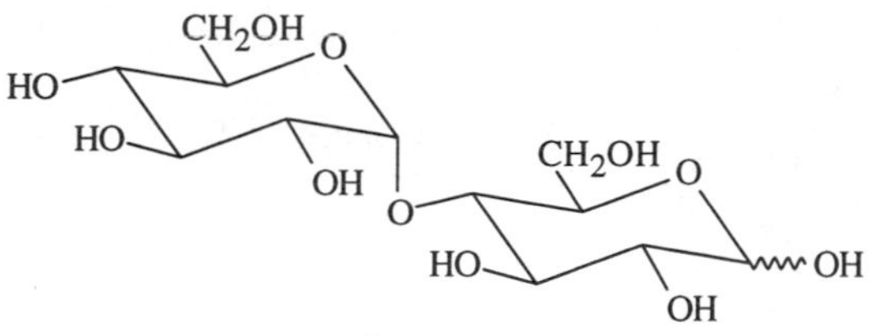

A and *B*

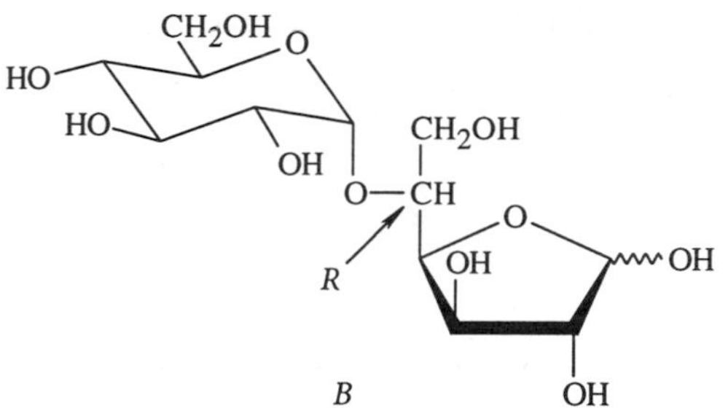

Methylation of maltobionic acid occurs under basic conditions; under these conditions any lactone present would be saponified. Hence, the additional oxygen that is methylated under these conditions must be the oxygen that was *within* the glycoside ring at the reducing end of maltose. In other words, oxidation and saponification expose this oxygen as an —OH group and thus make it available for methylation. Since the oxygen at carbon-5 is methylated, it must have been the oxygen within the saccharide ring. Since the oxygen at carbon-4 is *not* methylated, it must have been the oxygen involved in the glycosidic bond. Hence, structure *A* is the correct one for maltose. To summarize the chemistry using the correct structure:

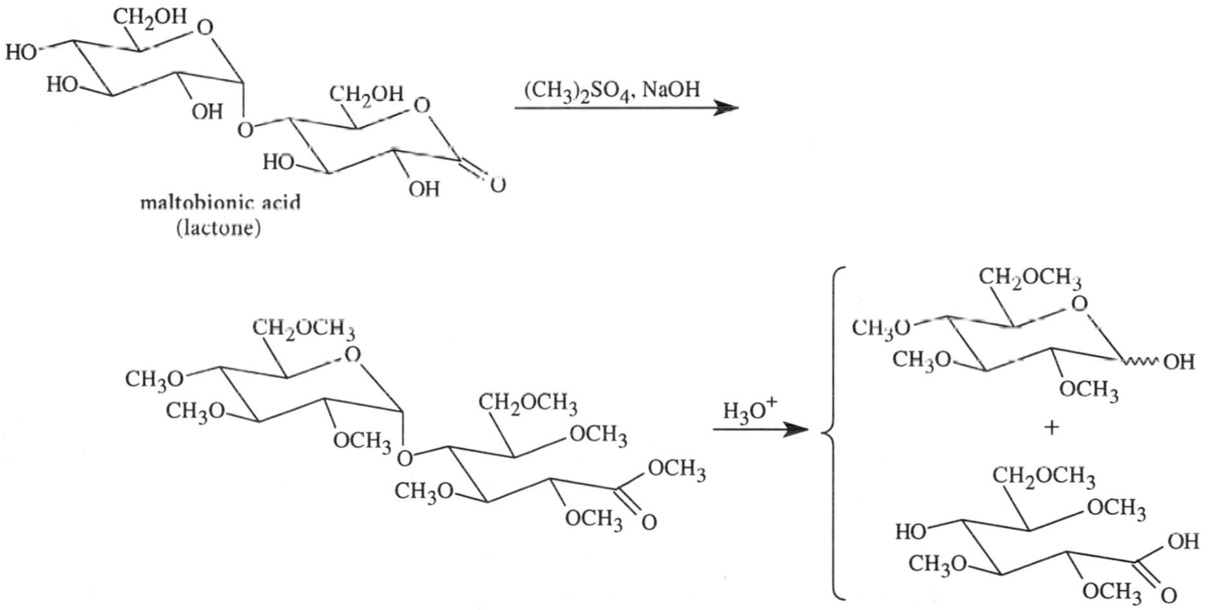

maltobionic acid (lactone) $\xrightarrow{(CH_3)_2SO_4,\ NaOH}$

$\xrightarrow{H_3O^+}$ +

27.56 The hydrolysis of planteose by almond emulsin establishes that an intact sucrose unit is connected to a galactose residue by an α-galactosyl glycosidic bond. The question, then, is which oxygen of the sucrose residue is involved in the glycosidic linkage to the galactose. Methylation followed by hydrolysis provides the answer. (It will be helpful to refer to the structure of sucrose on text p. 1328.) If the fructose residue of sucrose were *not* connected to the galactose, it would have four oxygens available for methylation: the oxygens at carbons 1, 3, 4, and 6. Because the oxygen at carbon-6 is not methylated, it must be the point of attachment of the galactose residue. Hence, the structure of planteose is as follows:

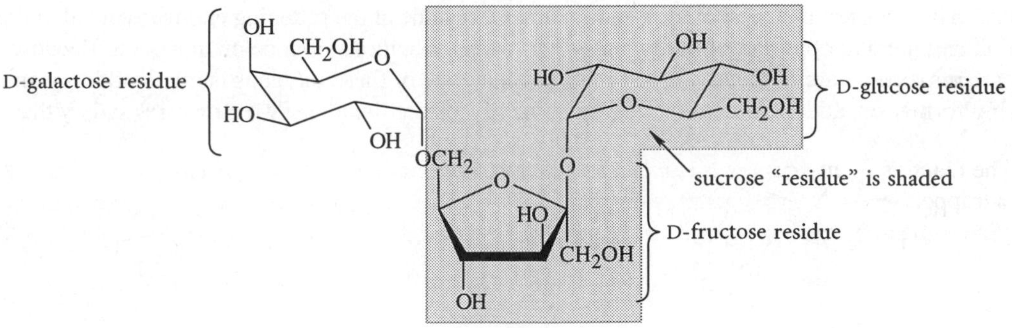

D-galactose residue

D-glucose residue

sucrose "residue" is shaded

D-fructose residue

planteose

27.57 (a) The ribonucleotide sequence is translated from the 5'-end according to the genetic code in Table 27.4 on text p. 1342:

5′ end A-U-G-A-A-A-C-A-A-G-A-U-U-U-U-A-U-U-G-G-G-G-G 3′ end

Met—Lys—Gln—Asp—Phe—Tyr—Trp—Gly

(b) The DNA sequence is

3′ end T-A-C-T-T-T-G-T-T-C-T-A-A-A-A-A-T-A-A-C-C-C-C-C 5′ end

(c) The codon at positions 16, 17, and 18 is U-A-U; the indicated mutation would change this to U-A-A. Because this is a "stop" codon, the resulting peptide would be Met-Lys-Gln-Asp-Phe.

27.58 (a) The hydroxy groups of cellulose react with the anhydride to introduce succinate half-esters into the cellulose structure, as follows. (In these equations. HO—cellulose means any of the hydroxy groups of cellulose.)

+ HO—cellulose → ; pH 7 ; O—cellulose ; OH ; $pK_a \approx 4–5$

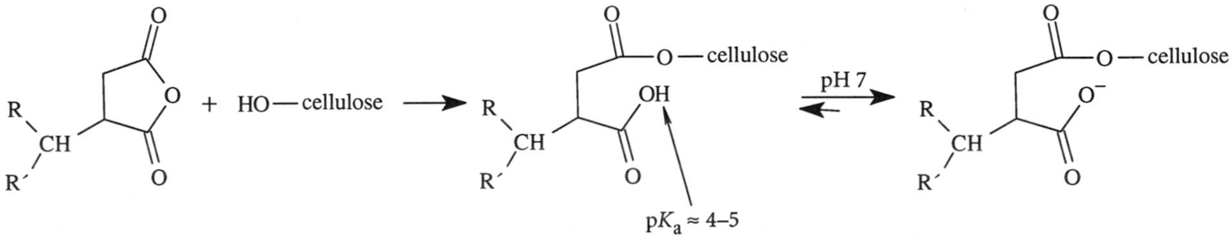

(b) Sizing introduces a large number of carbon groups into the cellulose structure. These resist solvation by water for the same reasons that hydrocarbons are insoluble in water; they are "hydrophobic" groups. Hence, these groups tend to resist the incorporation of solvent water into the cellulose structure. Furthermore, many of the —OH groups in the cellulose, which are hydrogen-bond donors before the reaction, are no longer hydrogen-bond donors after the reaction.

(c) As the reaction in the solution to part (a) shows, the carboxylic acid groups introduced by the reaction are ionized by the pH 7 treatment because their pK_a values are well below the pH of the solution. When the paper is dried, these groups are left in their ionized state (with an appropriate counterion). The carboxylate groups are the slightly basic groups.

For example, when they come into contact with water, they react with water to produce a small amount of hydroxide ion:

$$R-\overset{O}{\overset{\|}{C}}-O^- + H_2O \rightleftharpoons R-\overset{O}{\overset{\|}{C}}-OH + {}^-OH$$

You can use the pK_a value for the carboxylic acid and the ion-product constant of water to calculate that the pH of an aqueous solution of a carboxylic acid salt should have a pH of around 9. Hence, when the "sized" paper comes into contact with acid, the acid is neutralized by the small amount of hydroxide ion present. So many carboxylate groups are present that the hydroxide ion is immediately replenished, by LeChatelier's principle.

27.59 The reaction of an amine with an aldose gives a carbinolamine, which dehydrates to an iminium ion *A*. This ion is trapped by the hydroxy group at carbon-5 to give an aminoglycoside, which is shown as its conjugate acid *B*. (The *p*-methylphenyl group of the amine is abbreviated R.)

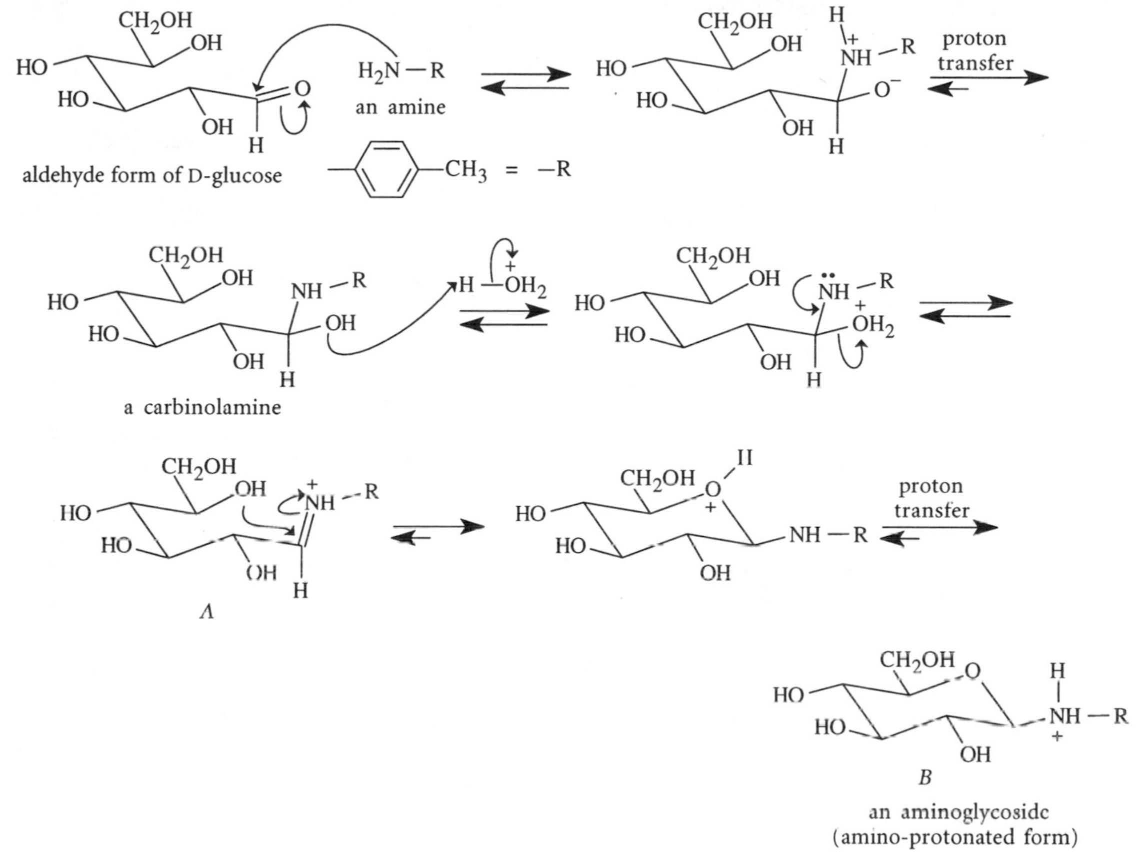

aldehyde form of D-glucose

$H_2N—R$ an amine

$-C_6H_4-CH_3 = -R$

proton transfer

a carbinolamine

A

proton transfer

B
an aminoglycoside
(amino-protonated form)

This reaction is called the *Maillard reaction.* From here on, only the part of D-glucose that is involved in the reaction is shown explicitly. The iminium ion (which, by resonance, is also a carbocation) isomerizes to an enamine *C* which is also an enol. Acid-catalyzed formation of the isomeric ketone completes the reaction. This isomerization is called the *Amadori rearrangement.*

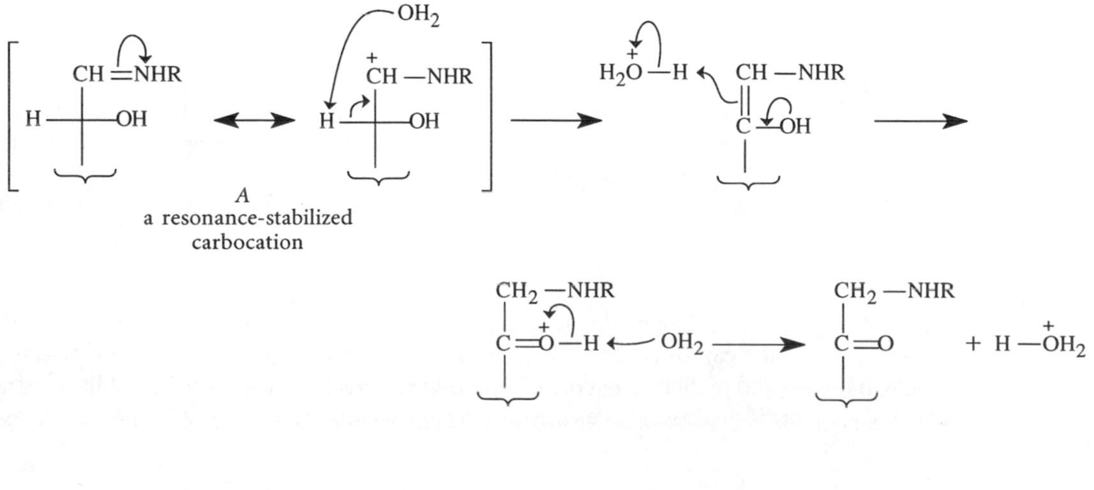

A
a resonance-stabilized carbocation

This reaction is similar to the Lobry de Bruyn-Alberda van Ekenstein rearrangement (text p. 1307), except that imine and enamine intermediates are involved rather than aldehyde and enol intermediates. The driving force for the rearrangement is that the C=N double bond is lost and a stronger C=O double bond is formed.

27.60 (a) Periodic acid cleaves the *cis*-2′,3′-diol group at the 3′-end of RNA to form a dialdehyde. (Note that only the 3′ residue of RNA contains a 2′,3′-diol.) The presence of an aldehyde carbonyl group makes the α-hydrogen at carbon-4′ more acidic. Removal of this hydrogen sparks an E2 reaction in which the internucleotide bond is cleaved.

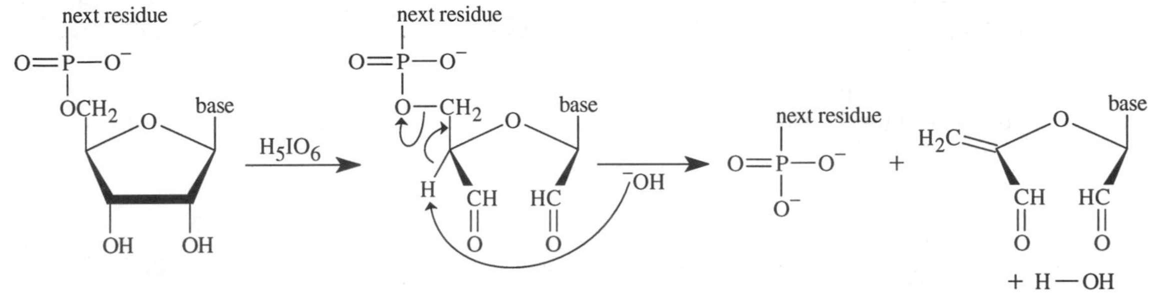

(b) DNA cannot undergo this reaction because a cis 2′,3′-diol group is needed for the periodate cleavage reaction. DNA, of course, has no 2′-hydroxy group at the 3′ end, and thus, no 2′,3′-diol.

27.61 This transformation is essentially a Lobry de Bruyn-Alberda van Ekenstein reaction, the detailed mechanism of which is shown in Eqs. 27.11–27.12 on text p. 1307. The aldehyde form of D-glucosamine is in equilibrium with a small amount of its enol *A*. This enol is also an enamine, which is in equilibrium with an imine *B*. Because imine formation is reversible in aqueous acid and base, the imine hydrolyzes to give the corresponding ketose, D-fructose, along with the by-product ammonia. Of course, under the basic conditions, D-fructose is also in equilibrium with D-glucose and D-mannose, as shown on text p. 1307.

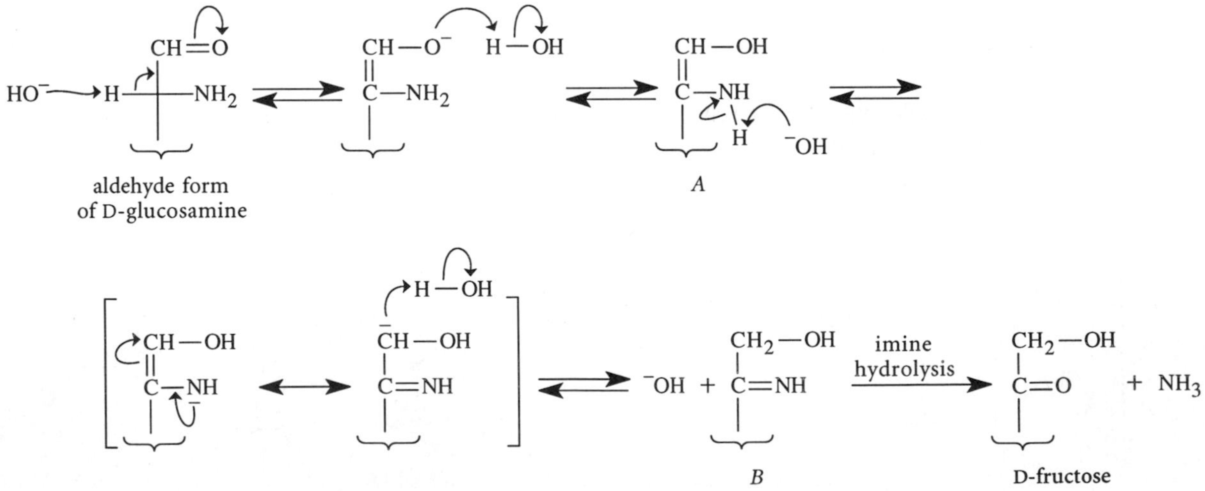

27.62 (a) The acidic hydrogen is the one that ionizes to give a conjugate-base anion in which the negative charge is delocalized into the carbonyl group. This anion is stabilized by the electron-withdrawing polar effects of nearby oxygens and by the same type of resonance interactions that are present in a carboxylate ion. Consequently, ascorbic acid is about as acidic as a carboxylic acid.

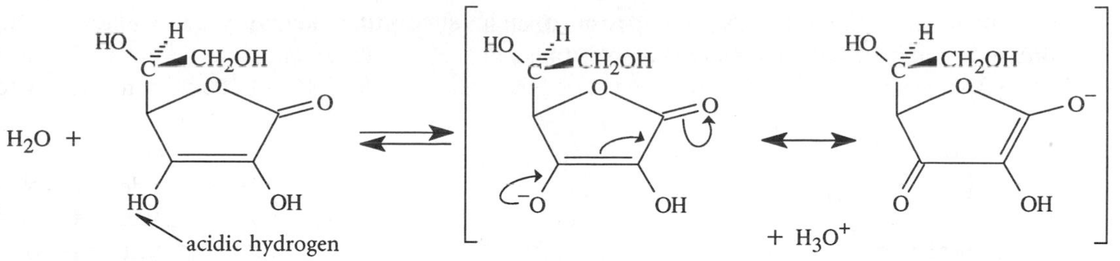

H_2O + [structure] ⇌ [resonance structures] + H_3O^+

acidic hydrogen

(b) The reaction sequence with the missing structures added begins with compound *A*. The fact that carbon-2 of L-sorbitol (or carbon-5 of D-glucitol) becomes the acetal carbon as a result of the reaction labeled "*C*" in the problem shows that the —OH group on this carbon is the one that is oxidized; this carbon is therefore at the ketone level of oxidation. $KMnO_4$ oxidizes the —CH_2OH group to a carboxylate; because the other —OH groups are protected as acetals, they are stable under the basic conditions of permanganate oxidation. Hydrolysis of the oxidation product in acid removes the acetal protecting groups, protonates the carboxylate group, and promotes lactone formation.

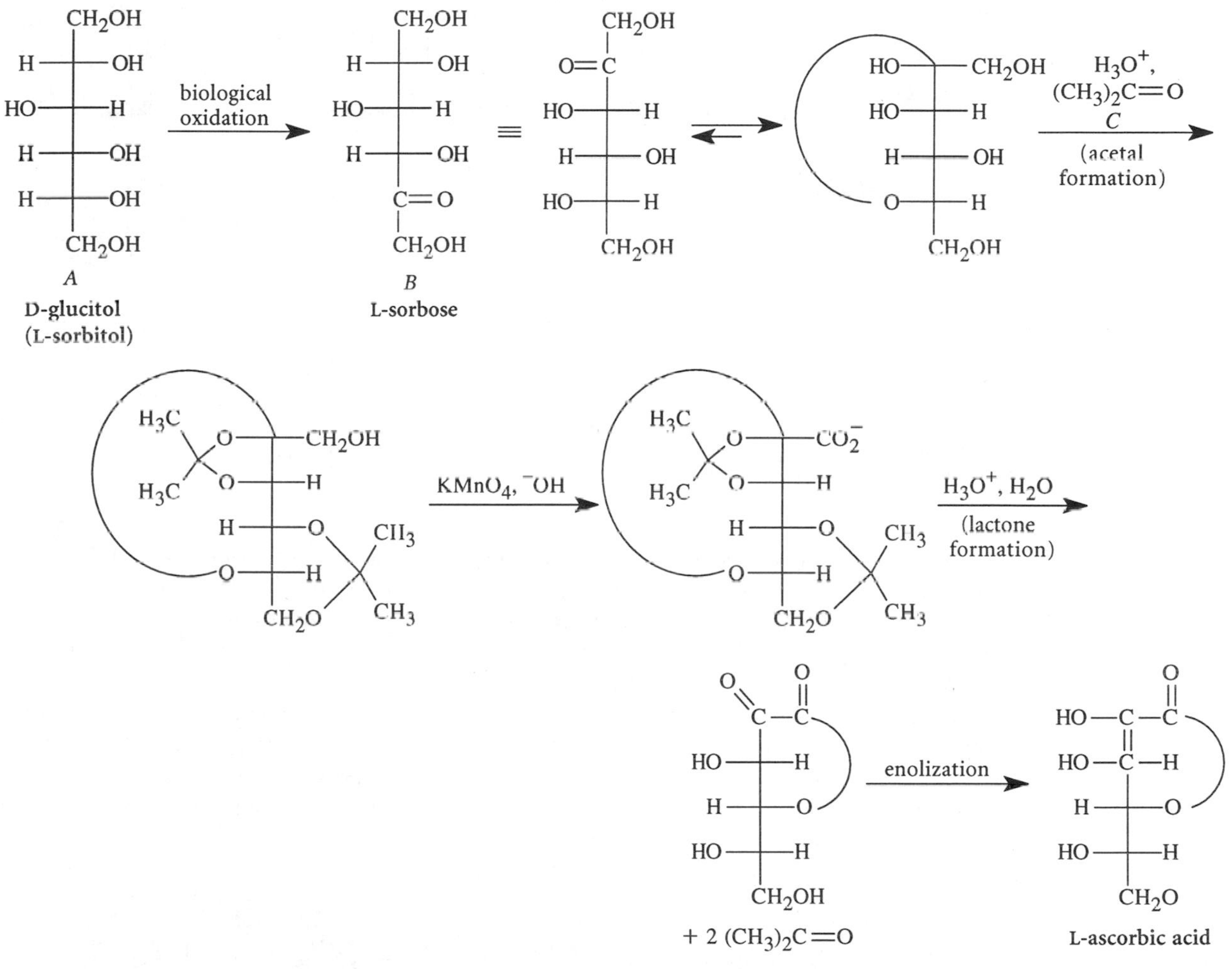

A D-glucitol (L-sorbitol) —biological oxidation→ *B* L-sorbose ⇌ [cyclic form] —H_3O^+, $(CH_3)_2C{=}O$, *C* (acetal formation)→

—$KMnO_4$, ^-OH→ —H_3O^+, H_2O (lactone formation)→

+ 2 $(CH_3)_2C{=}O$ —enolization→ L-ascorbic acid

Can you think of one or more reasons why ascorbic acid should exist as an enediol rather than as an α-keto lactone? (See Sec. 22.2 and the discussion of carbonyl dipoles on text p. 812.)

27.63 The reaction of RNA involves ionization of the 2′-hydroxy group within each residue and reaction of the resulting alkoxide ion as a nucleophile at the neighboring phosphorus. This reaction is particularly rapid because it is *intramolecular*; it is an example of neighboring-group participation. (See Sec. 11.7 in the text.) Loss of the leaving group splits the internucleotide bond. DNA cannot undergo this reaction because the reaction depends on the presence of the 2′-hydroxy group to act as an intramolecular nucleophile; DNA lacks this hydroxy group. (That is why it is called *deoxy*ribonucleic acid.)

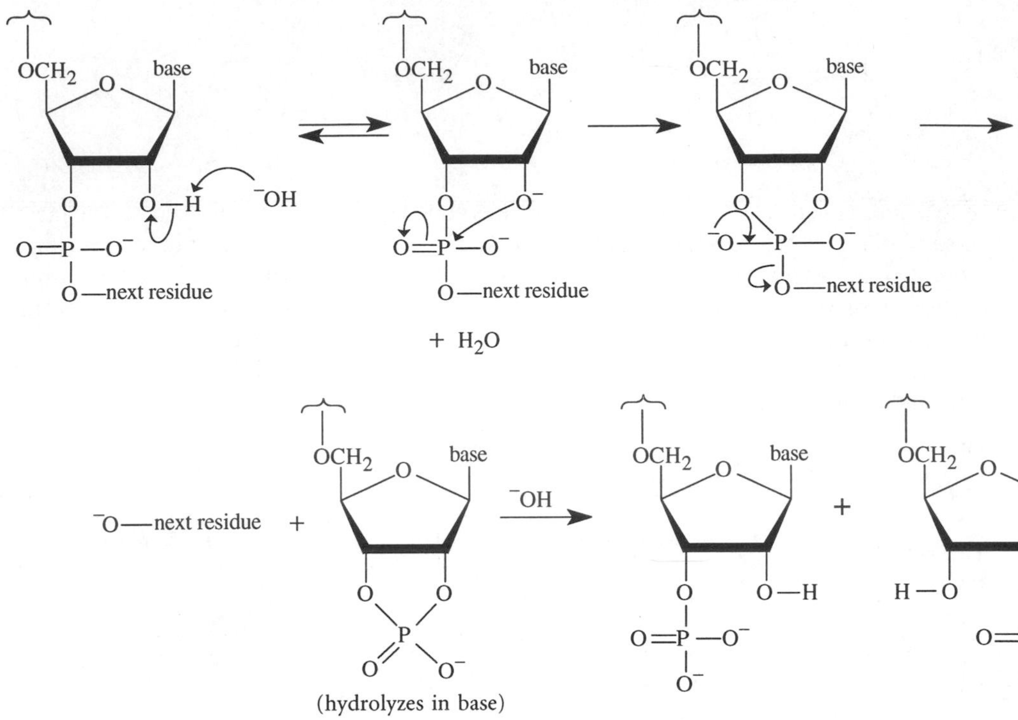

OCH$_2$ O base; O—H; $^-$OH; O=P—O$^-$; O—next residue ⇄ OCH$_2$ O base; O$^-$; O=P—O$^-$; O—next residue + H_2O → OCH$_2$ O base; $^-$O—P—O$^-$; O—next residue →

$^-$O—next residue + OCH$_2$ O base; P; O; O$^-$ (hydrolyzes in base) $\xrightarrow{^-OH}$ OCH$_2$ O base; O=P—O$^-$; O$^-$; O—H + OCH$_2$ O base; H—O; O=P—O$^-$; O$^-$

27.64 (a) The anhydro form of D-idose has the following chair conformation:

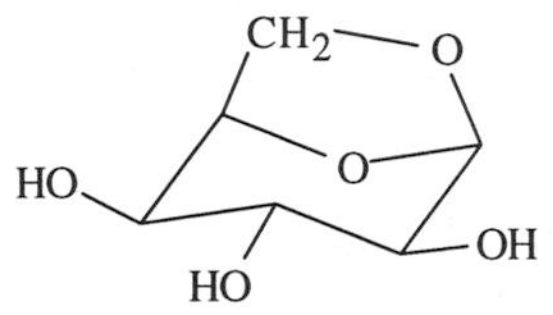

CH$_2$—O; HO; O; OH; HO — 1,6-anhydro-D-idopyranose

(b) The issue is the relative stabilities of the anhydro and ordinary pyranose forms of D-idose and D-glucose. In the chair conformation of 1,6-anhydro-D-glucopyranose, all of the hydroxy groups are axial. The many resulting 1,3-diaxial interactions destabilize this form of glucose. In the more stable chair form of D-glucopyranose, in contrast, all of the hydroxy groups are equatorial. Thus, the pyranose form of D-glucose is much more stable than the 1,6-anhydro form, and D-glucose exists primarily as an ordinary pyranose. In 1,6-anhydro-D-idopyranose, all of the hydroxy groups are equatorial. In the more stable of the chair conformations of the ordinary pyranose form of D-idose, the —CH_2OH group is forced into an axial position. Formation of the ether bond between carbons 1 and 6 relieves one set of 1,3-diaxial interactions in this form of D-idose. Thus, the 1,6-anhydro form of D-idose is more stable than the pyranose form, and therefore D-idose exists mostly in the 1,6-anhydro form.

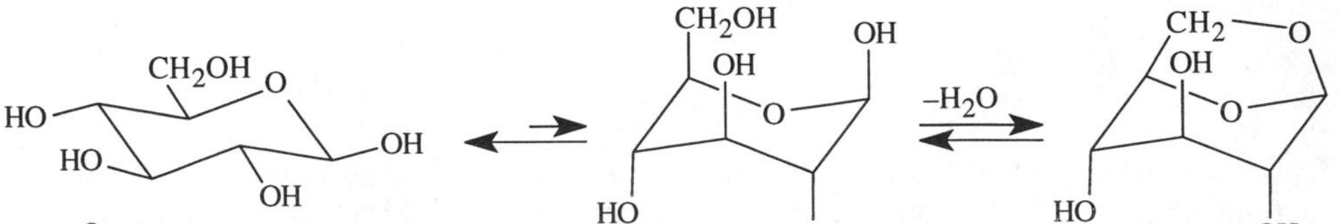

HO; CH$_2$OH O; HO; OH; OH — *β*-D-glucopyranose ⇄ CH$_2$OH; OH; OH; O; HO; OH $\xrightarrow{-H_2O}$ CH$_2$—O; OH; O; HO; OH — 1,6-anhydro-D-glucopyranose

β-D-idopyranose

1,6-anhydro-D-idopyranose